Congratulations...

on your purchase of the most complete interstate services guide
ever printed!

How many times have you:

1) _passed_ an exit and _looked back_ to see the gas station/restaurant/ motel you wanted?

2) _taken_ an exit and found _no suitable_ services?

3) _wondered_ if there were services located ahead and _how far_ ?

4) filled up/eaten/slept somewhere simply because you _didn't know_ any thing else was available?

After you have used *the Next EXIT* you will wonder how and why you ever traveled without it!

Developed over many years and updated annually, *the Next EXIT* will save time, money and frustration. This guide will serve you as an invaluable aid in finding services along the interstate highway system like nothing you have ever used.

ALL MAJOR ROUTES and MORE, ALL EXITS!

PO Box 888
Garden City, UT 84028
theNextEXIT.com

How to use *the Next EXIT*:

① ... state locator tab
② ... city locator strip
③ ... map with interstate highways
④ ... directional arrow, inside margin
⑤ ... services accessed from the exit

⑤ **Arkansas Interstate 40**

Exit #	Services
285mm	Mississippi River, Arkansas/Tennessee state line
281	AR 131, S to Mound City
280	Club Rd, Southland Dr, **N**...Fina/diesel/mart/24hr, Williams/Blimpie/Perkins/diesel/mart, Goodyear, **S**...Flying J/Conoco/LP/diesel/mart/rest., Petro/diesel/rest./24hr, Pilot/Subway/DQ/diesel/mart, KFC/Taco Bell, McDonald's, Waffle House, Best Western, Deluxe Inn, Express Inn, Super 8, Blue Beacon, SpeedCo Lube
279b	I-55 S(from eb)
a	Ingram Blvd, **N**...Classic Inn, Comfort Inn, Day's Inn, Greyhound Park, U-Haul, **S**...Citgo/diesel/mart, Exxon, Shell/mart, Waffle House, Western Sizzlin, Best Western, Econolodge, Hampton Inn, Holiday Inn, Howard Johnson, Motel 6, Relax Inn, Chevrolet, Chrysler/Plymouth/Dodge/Jeep, same as 280
278	AR 77, 7th St, Missouri St, **N**...Citgo/Blimpie/mart/24hr, **S**...HOSPITAL, DENTIST, Citgo/diesel/mart, Coastal/mart, Express/mart, Exxon/mart, Love's/diesel/24hr, Phillips 66/diesel/mart, RaceTrac, Shell/diesel/rest./24hr, Texaco, Voss/diesel/mart/24hr, Backyard Burger, Baskin-Robbins, BBQ, Bonanza, Burger King, Cracker Barrel, Domino's, KFC, Krystal, Little Caesar's, Mrs Winner's, McDonald's, Pizza Hut, Pizza Inn, Shoney's, Subway, Taco Bell, TCBY, Wendy's, Quality Inn, Ramada Ltd, Cato's, Chief Parts, Goodyear/auto, Hancock Fabrics, Kroger, Mr FastLube, Radio Shack, Sawyers RV Park, Sears, Walgreen, Wal-Mart SuperCtr/24hr
277	I-55 N, to Jonesboro, no facilities

④ E ⬆⬇ W

① ② ③

W Memphis

Exit

Most states number exits by the nearest mile marker(mm), but some use consecutive numbers, in which case mile markers are given. Mile markers are the little green vertical signs beside the interstate at one mile intervals, except in California. They tell how far you are from the southern or western border of the state you are in. Odd numbered interstates run north/south, even numbers run east/west.

Services

Services are listed alphabetically by name in the order...gas, food, lodging, other services(including camping). "HOSPITAL" indicates an exit from which a hospital may be accessed, but it may not be close to the exit. Services located away from the exit may be referred to by "access to," or "to...," and a distance may be given. A directional notation is also given, such as "N...".

Read services from the page according to your direction of travel; i.e., if traveling from North to South or East to West, read services DOWN the page. If traveling S to N or W to E, then read services UP the page.

the Next EXIT, inc. makes no warranty regarding the accuracy of the information contained in this publication. This information is subject to change without notice.

Table of Contents

Abbreviations used in *the Next EXIT:*

AFB	Air Force Base	NF	National Forest
Bfd	Battlefield	NP	National Park
bldg	building	NRA	Nat Rec Area
Ctr	Center	pk	park
Coll	College	pkwy	parkway
Gen	General	rest.	restaurant
$	Dollar	nb	northbound
Mem	Memorial	sb	southbound
Mkt	Market	eb	eastbound
Mtn	Mountain	wb	westbound
mm	mile marker	SP	state park
N...	north side of exit	SF	state forest
S...	south side of exit	Sprs	springs
E...	east side of exit	st	street, state
W...	west side of exit	sta	station
NM	National Monument	TPK	Turnpike
NHS	Nat Hist Site	USPO	Post Office
NWR	Nat Wildlife Reserve	whse	warehouse

Four Wheels, Please

It was afternoon in Virginia Beach and the traffic on Interstate 264 had begun to pick up. I looked in my rearview mirror to see a helmeted set of handlebars approaching at about the speed of light. Just when I thought my car would be bisected by the laser-like single beam of his headlight, a motorcycle flashed around me and through traffic gaps ahead and was gone. I imagined what I would soon see: cars pulled over, red and blue lights, torn clothes and parts for hundreds of feet. Either the horrific rainstorm that hit a few minutes later washed the debris away, or the rider got lucky because I never did come upon his wreckage. I crossed under the mouth of the James River and continued up Interstate 64 into the evening, but by now the memories were passing me almost as quickly as that biker.

If you grew up on the Darlington Highway, almost any adventure started at Becky's Store. It was here on a Saturday morning in the spring of 1967 that Life imparted one of its great lessons to me. This one came cheaply, but it was by no means free. When I got to the store that morning, out front on the gravel sat one beautiful motorcycle. As I stood admiringly wondering who was the lucky owner, the screen door to Becky's opened and out walked Donnie. His proud countenance told all. "Get on and let's ride down to Flinns Crossroads," he said. Whereas speed, freedom and machismo combine in the minds of most teenage boys into an exhilarating substance, fear makes an impression on people like few other emotions. In the imagination, one can do strange and thrilling things on a motorbike.

We stood there for a minute taking in all its features. Then came the repeated offer: "Get on the back and let's go to the crossroads." You know those little feelings we get that try to warn us? But Flinns Crossroads was only one mile away, and what could possibly happen in just one mile? "Put your arms around my waist," said Donnie. I had this thing about putting my arms around other boys. Against all reason I climbed onto the back of the oblong seat, placing just the tips of my fingers around the front of his hips. With a smooth motion, he kick started, blasting my ears with a thunder-warning. "Hear that?" he beamed proudly. I could barely hear anything.

One rule quickly became obvious: when riding double, the person on the back is responsible for his own safety. Hold on if you want to, or don't if you don't. This particular bike, like most machines of that era, had no back support, a fact that came to me vividly a few moments later. With a casual glance over his left shoulder toward town, we pulled onto the Darlington Highway headed toward Flinns.

From the time this road leaves the railroad tracks in town until it reaches the crossroads three miles later, it warps slightly only once. It was constructed in sections of poured concrete, and later highway crews came along and poured tar into the joints and any cracks. In more innocent times I had ridden my bicycle upon every piece of that road in both directions, counting the cracks both by sight and by the bumps they made on my tires.

This day, however, innocence was gone, having been replaced by something else. There was no counting the cracks on this bike trip. I did feel a few bumps of the road momentarily as we slid through the lower gears, but that was before the G-forces set in. With no helmets and no windshield, there was only the raw, gut-leaving feeling of speed, freedom and machismo, combined with lots of little bugs that smacked us constantly. Somewhere near the Culpepper Sisters' place, I began to see, I mean really *see* things for the first time. By the time we reached the Lewis' I had begun to repent. The road beneath us had changed colors by now, taking on the grayish hue of a runway. Pushing away thoughts of a blowout, I realized that I had tightened my grip on Donnie's waist, thing or no thing. I could see the crossroads coming up pretty quickly, so I began to wonder two things. Were the brakes as good as the motor? And had Donnie ridden enough to know that he should use both the front *and* rear brakes at the same time?

Thankfully, the answer to both questions was yes. After slowing enough to make a wide turn at Flinns Crossroads, we headed back to the store in the same fashion, that being to see if we could get there before we started. Back at Becky's, I made the customary admiring comments of one truly impressed by someone's new possession, said I'd see him later and went home. I felt like I'd been born again, you know, given another chance. It was all over in less than three minutes, if you include the time it took to slow down and turn around, but it seemed *much* longer.

About 20 years passed before I saw Donnie again. After talking a few minutes about the intervening years, I asked him about his limp and the cane in his right hand. "Motorsickle wreck," he grinned. "Hit a train track going about a hundred."

Seeing that our interests had diverged I made a few more polite inquiries, said goodbye and left. Wisdom road home with me that day, whispering "Two roads converged in a yellow wood," and "but for the Grace of God, there go I," and so on. I began to see again.

Sometimes it's hard to tell exactly how Life will choose to teach us. We seldom go into anything wondering what great lesson is going to be learned. Normally these things are seen better looking back, and often this takes years. Whenever I think about events that taught me something of value, I place a cost on the experience and then weigh whether the price was well invested. When I occasionally give myself a retest it shouldn't surprise me that I come up with the same answer time after time.

Once in a while I see a motorcycle streaking down the interstate with its rider basking in the ecstasy of speed, freedom and machismo, and I think to myself, "yeah, there I go." After a brief moment of fantasy, someone inside me does a reality check. "Think about it," he says, holding up the scales. That's when I remember what I learned on the back of a motorcycle on the way to Flinns Crossroads.

Mark Watson
spring, 2001

Alabama Interstate 10

Exit #	Services
66.5mm	Alabama/Florida state line
66mm	**Welcome Ctr/weight sta wb, full(handicapped)facilities, phone, vending, picnic tables, litter barrels, petwalk**
53	AL 64, Wilcox Rd, **N**...BP/Oasis/diesel/mart/atm/24hr, Mirage Café, Hilltop RV Park, Wilderness RV Park, Styx Water World, **S**...Chevron/diesel/mart
44	AL 59, Loxley, **N**...Texaco/diesel/mart, **S**...Chevron/mart/24hr, Exxon/diesel/mart/24hr, Hardee's, McDonald's, Waffle House, Day's Inn(2mi), Loxley Motel(3mi), WindChase Inn, to Gulf SP
38	AL 181, Malbis, **N**...Amoco/mart, **S**...Chevron/diesel/mart, LA Subs, Malbis Motor Inn(1mi)
35	US 90, US 98, **N**...MEDICAL CARE, DENTIST, BP/mart, Shell/mart, Rite Aid, USPO, bank, cleaners, **S**...CHIROPRACTOR, FOOTCARE, BP, Chevron/mart, Conoco/mart/wash, Exxon/diesel/mart, Shell/mart, Circle K, Angelo's Seafood, Arby's, Burger King, Checker's, Domino's, Godfather's, Hardee's, IHOP, Krystal, McDonald's, Nautilis Rest., Pizza Hut, Quincy's, Shoney's, South China Rest., Subway, Taco Bell, Waffle House, Wendy's, Comfort Suites, Eastern Shore Motel(1mi), Hampton Inn, Legacy Inn, Dillard's, Home Depot, MailBag, Office Depot, bank, cleaners
30	US 90/98, Battleship Pkwy, **S**...Amoco, Citgo/mart, Texaco/diesel/mart, Pier 4 Seafood, Ramada
29mm	tunnel begins wb
28mm	tunnel begins eb
27	US 90/98, Battleship Pkwy, Gov't St, **S**...Capt's Table Seafood, Best Western, to USS Alabama
26b	Water St, Mobile, downtown, **N**...Riverview Café, Rousso's Seafood Rest., Adam's Mark Hotel, Holiday Inn, Holiday Inn Express, Radisson
a	Canal St(from eb), same as 26b
25b	Virginia St, Mobile, **N**...Texaco/diesel, **S**...tires
a	Texas St(from wb, no return), no facilities
24	Duval St, Broad St, Mobile, **N**...Chevron/mart/wash, Citgo/mart
23	Michigan Ave, **N**...Exxon/mart, Travel Inn, **S**...$General
22b a	AL 163, Dauphin Island Pkwy, **N**...Amoco, Villager Lodge, **S**...Citgo, Exxon/Subway/mart/24hr, Texaco/café/mart, Checker's, Gone Fishin Catfish, Waffle House, Gone With the Wind Motel, $General
20	I-65 N, to Montgomery, **N**...to Bates Airport
17	AL 193, Tillmans Corner, to Dauphin Island, **N**...HOSPITAL, Chevron/Blimpie/mart, Citgo/mart, BBQ, Burger King, Golden Corral, McDonald's, Ruby Tuesday, Wal-Mart SuperCtr/24hr, Webb's RV Ctr, bank
15b a	US 90, Tillmans Corner, to Mobile, **N**...CHIROPRACTOR, BP/mart/wash, Chevron/mart/24hr, Conoco/diesel, RaceTrac/mart, Shell/mart/wash, Arby's, Burger King, Checker's, Domino's, Godfather's, KFC, McDonald's, Neighbor's Seafood, Papa John's, Pizza Hut, Pizza Inn, Popeye's, Quincy's, Subway, Taco Bell, Waffle House, Best Western, Comfort Inn, Day's Inn, Hampton Inn, Holiday Inn, Microtel, Motel 6, Red Roof Inn, Shoney's Inn/rest., Suite 1 Motel, Super 8, Travelodge, AutoZone, CarQuest, Family$, Firestone/auto, FoodWorld/24hr, Goodyear, K-Mart, Radio Shack, Rite Aid, SuperLube, Winn-Dixie/deli, bank, cleaners, **S**...Chevron/mart/wash/24hr, Conoco/diesel, Texaco/diesel/mart, Hardee's, Waffle House, RV Ctr, auto repair, tires, transmissions
13	to Theodore, **N**...Amoco/McDonald's/diesel/mart/24hr, Conoco, Pilot/Citgo/Wendy's/KrispyKreme/diesel/atm/24hr, Shell/Subway/mart/wash, Waffle House, Greyhound Park, airport, **S**...Bellingraf Gardens, I-10 Kamping
4	AL 188 E, to Grand Bay, **N**...Citgo/mart, Shell/Dairy Queen/Stuckey's, TA/BP/diesel/rest./24hr, Waffle House, **S**...Chevron/Krispy Kreme/mart/atm/24hr, Hardee's, Trav-L-Kamp
1mm	**Welcome Ctr eb, full(handicapped)facilities, info, phone, picnic tables, litter barrels, petwalk, RV dump**
0mm	Alabama/Mississippi state line

Alabama Interstate 20

Exit #	Services
215mm	Alabama/Georgia state line,Central/Eastern time zone
213mm	**Welcome Ctr wb, full(handicapped)facilities, info, phone, vending, picnic tables, litter barrels, petwalk, RV dump, 24hr security**
210	AL 49, Abernathy, **N**...fireworks, **S**...fireworks
209mm	Tallapoosa River

Alabama Interstate 20

208mm weigh sta wb

205 AL 46, to Heflin, **N**...BP/diesel, Texaco/diesel

199 AL 9, Heflin, **N**...Texaco/Subway/Taco Bell/diesel/mart/24hr, Hardee's, Howard Johnson, **S**...BP/diesel, Chevron/mart, Huddle House, Pop's Charburgers

198mm Talladega Nat Forest eastern boundary

191 US 431, to US 78, **S**...Talladega Scenic Drive, Cheaha SP

188 to US 78, to Anniston, **N**...Exxon/mart, Texaco/mart/wash, Cracker Barrel, Dairy Queen, IHOP, KFC, LoneStar Steaks, Subway, Waffle House, Wendy's, Holiday Inn Express, Jameson Inn, Sleep Inn, Wingate Inn, Lowe's Bldg Supply

185 AL 21, to Anniston, **N**...MEDICAL CARE, Amoco/mart, BP/diesel/mart/wash, Chevron/mart/wash/24hr, Shell/mart, Texaco, Applebee's, Arby's, Burger King, Capt D's, Domino's, Hardee's, Krystal, McDonald's, O'Charley's, Piccadilly's, Pizza Hut, Quincy's, Red Lobster, Shoney's, Taco Bell, Waffle House, Western Sizzlin, Day's Inn, Holiday Inn, Howard Johnson, Liberty Inn, Red Carpet Inn, Books-A-Million, $General, Firestone/auto, FoodMax, JC Penney, Sears, XpressLube, bank, mall, to Ft McClellan, **S**...HOSPITAL, Exxon/mart, RaceTrac, Shell/mart, Texaco/Baskin-Robbins/Subway/diesel/mart, Williams/diesel/mart, Chick-fil-A, Huddle House, Outback Steaks, Waffle House, Wendy's, Comfort Inn, Econolodge, Hampton Inn, Motel 6, Travelodge, Goodyear/auto, Wal-Mart SuperCtr/24hr, antiques, fireworks

179 to Munford, Coldwater, **N**...Anniston Army Depot

173 AL 5, Eastaboga, **S**...Shell/mart, Texaco/DQ/Stuckey's, to Speedway/Hall of Fame

168 AL 77, to Talladega, **N**...Citgo/mart, Phillips 66/diesel, Jack's Rest., KFC/Taco Bell, **S**...Chevron/diesel, PaceCar/diesel/24hr, Texaco/Burger King/mart, McDonald's, Day's Inn, Holiday Inn Express, McCaig Motel, camping, to Speedway, Hall of Fame

165 Embry Cross Roads, **S**...Exxon/diesel, Express/diesel/mart, Texaco/diesel/mart, Huddle House, McCaig Motel/rest.

164mm Coosa River

162 US 78, Riverside, **S**...BP/diesel, Chevron/mart, Hal's Rest., Mary's Pancakes/24hr, Best Western, Safe Harbor Camping

158 US 231, Pell City, **S**...HOSPITAL, BP/mart, Chevron, Citgo/diesel, Burger King, Hardee's, KFC, McDonald's, Pizza Hut, Waffle House, Ramada Ltd, Lakeside Camping

156 US 78 E, to Pell City, **S**...Amoco/mart, Chevron/diesel/mart/24hr, Exxon/diesel/mart/24hr

153 US 78, Chula Vista, no facilities

152 Cook Springs, no facilities

147 Brompton, **N**...Citgo, **S**...Ala Outdoors RV Ctr/LP

144 US 411, Leeds, **N**...Amoco, BP/mart/wash, RaceTrac/mart, Shell/Subway/mart, Arby's, Cracker Barrel, Milo's Café, Pizza Hut, Ruby Tuesday, Waffle House, Wendy's, Comfort Inn, Super 8, Winn-Dixie, XpressLube, RV camping, cleaners, **S**...Exxon/mart/wash/24hr, RaceWay/mart, Speedway/diesel/mart, Burger King, Capt D's, Guadalajara Jalisco Mexican, Hardee's, KFC, McDonald's, Quincy's, Taco Bell, Waffle House, Day's Inn, Advance Parts, AutoZone, ExpressLube, Wal-Mart SuperCtr/24hr

140 US 78, Leeds, **S**...Chevron/mart/wash, Exxon/mart/wash, Shell(1mi)

139mm Cahaba River

136 I-459 S, to Montgomery, Tuscaloosa, no facilities

135 US 78, Old Leeds Rd, **N**...Citgo/mart, B'ham Racetrack

133 US 78, to Kilgore Memorial Dr, **N**...MEDICAL CARE, Chevron, Exxon/diesel/mart/24hr, El Palacio Mexican, Hamburger Heaven, Italian Villa Rest., Jack's Rest., Krystal, Waffle House, Eastwood Inn, Siesta Motel, Super 8, JiffyLube, SuperPetz, carwash, **S**...Amoco/24hr, Arby's, BBQ, Burger King, Chinese Rest., McDonald's, Omelet Shop, Subway, B'ham Inn, Econolodge, Hampton Inn, Hawthorn Suites, Holiday Inn Express, Ford, HomePlace, K-Mart, Sam's Club, Waccamaw, Wal-Mart SuperCtr/24hr, tires, same as 132

132b a US 78, Crestwood Blvd, **N**...Exxon, Krystal, Waffle House, Chevrolet, Midas Muffler, Pennzoil, **S**...HOSPITAL, BP, Crown/mart/24hr, Shell/mart/wash/24hr, Capt D's, Chinese Rest., Denny's, Hooters, KFC, Lee's Chicken, Logan's Roadhouse, McDonald's, O'Charley's, Omelet Shop, Olive Garden, Pizza Hut, Red Lobster, Ruby Tuesday, Shoney's, Steak&Ale, Taco Bell, Wendy's, Holiday Inn Express, KeyWest Inn, Motel B'ham, Circuit City, JC Penney, K-Mart, McRae's, OfficeMax, Sears/auto, Service Merchandise, TJ Maxx, ToysRUs, Waldenbooks, mall

130b US 11, 1st Ave, **N**...Amoco, Chevron/mart/24hr, Huddle House, Food Fair, Advance Parts, AutoZone, Western Auto, RV Ctr, **S**...Capt D's, McDonald's, Mrs Winners, Relax Inn

a I-59 N, to Gadsden

I-59 S and I-20 W run together from B'ham to Meridian, MS

129 Airport Blvd, **N**...Clarion, airport, **S**...Exxon, Shell/mart/wash, Texaco/Blimpie/diesel/mart, Hardee's, Huddle House, Day's Inn, Holiday Inn, transmissions

E

W

2

Alabama Interstate 20

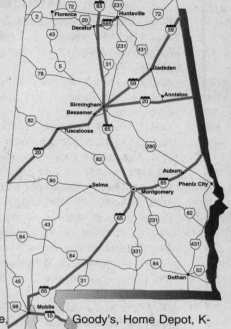

E

W

128   AL 79, Tallapoosa St, **N**...Shell/mart

126b   31st St, downtown, to Sloss Furnaces, **N**...Circle K/gas, Texaco, McDonald's, Best Western, Sheraton

a   US 31, US 280, 26th St, Carraway Blvd, downtown, to UAB, **N**...KFC, Rally's

125b   22nd St, to downtown

a   17th St, to downtown, Civic Ctr

124b a   I-65, S to Montgomery, N to Nashville

123   US 78, Arkadelphia Rd, **N**...Chevron/mart/24hr, JetPep/gas, Pilot/Wendy's/diesel/mart/24hr, Shell/mart, Denny's, Popeye's, La Quinta, Aamco, **S**...HOSPITAL, to Legion Field

121   Bush Blvd(from wb, no return), Ensley, **N**...CHIROPRACTOR, BP, Exxon, FatBurger, **S**...Pure/diesel

120   AL 269, 20th St, Ensley Ave, **N**...Crown Gas, KFC, GMC/Pontiac, SummerBreeze RV, **S**...HOSPITAL, BP/mart, JiffyMart/gas, Chevrolet, Chrysler/Plymouth/Jeep, Hyundai, Toyota

119b   Ave I(from wb), no facilities

a   Lloyd Noland Pkwy, Gary Ave, **N**...Amoco, Chevron/mart/24hr, Burger King, McDonald's, Mrs Winner's, Subway, Taco Bell, Food Fair, Family$, auto repair, **S**...HOSPITAL, Exxon, Texaco, Omelet Shoppe

118   AL 56, Valley Rd, Fairfield, **S**...Papa John's, Fairfield Inn, Villager Lodge, Goody's, Home Depot, K-Mart, NTB, Sears, diesel repair

115   Allison-Bonnett Memorial Dr, **N**...Amoco, Citgo/mart, Philips 66/mart, RaceTrac/mart, Church's Chicken, Golden Corral

113   18th Ave, to Hueytown, **S**...Speedway/diesel/mart, McDonald's

112   18th St, 19th St, Bessemer, **N**...Amoco/24hr, BP/mart, Citgo, Crown/diesel/mart, Church's Chicken, Jack's Rest., auto repair/tires, hardware, **2 mi S**...Chevron, Arby's, Burger King, Krystal, McDonald's, Shoney's

108   US 11, AL 5 N, Academy Dr, **N**...Exxon/mart, Applebee's, Cracker Barrel, Waffle House, Best Western, Jameson Inn, Motel 6, **S**...HOSPITAL, BP/KrispyKreme/mart/wash, Shell, Texaco/Church'sChicken/diesel/mart/wash, Burger King, Hardee's, Omelet Shoppe/24hr, Day's Inn, Hampton Inn, Masters Inn, Ramada Inn, Travelodge, $Tree, Ford, Wal-Mart SuperCtr/24hr, Winn-Dixie/deli/24hr, to civic ctr

106   I-459 N, to Montgomery

104   Rock Mt Lake, **S**...Flying J/Conoco/diesel/LP/rest./24hr

100   to Abernant, **N**...Citgo, KOA, **S**...Exxon/mart, Petro/Chevron/diesel/rest./24hr, Phillips 66, Tannehill SP(3mi)

97   US 11 S, AL 5 S, to W Blocton, **S**...Amoco/KFC/ice cream/mart, Shell/mart, Texaco/diesel/mart, Dot's Farmhouse Rest., Jack's Rest.

89   Mercedes Dr, **S**...Mercedes Auto Plant

86   Vance, to Brookwood, **N**...Shell/diesel/rest./mart/24hr

85mm   rest area both lanes, full(handicapped)facilities, phone, vending, picnic tables, litter barrels, petwalk, RV dump

79   US 11, University Blvd, Coaling, no facilities

77   Cottondale, **N**...Amoco/McDonald's/mart, TA/BP/Subway/Taco Bell/diesel/rest./atm, Pizza Hut, Hampton Inn, Microtel, SpeedCo Lube, USPO, truckwash, **S**...Honda Parts

76   US 11, E Tuscaloosa, Cottondale, **N**...Citgo/mart, Exxon/mart, Shell/diesel/mart, Cracker Barrel, Waffle House, KeyWest Inn, Knight's Inn, RV Ctr, **S**...Speedway/CountryKitchen/diesel/mart/24hr, Texaco/diesel, Sleep Inn, **1-2 mi S**...Super 8, Buick, Toyota

73   US 82, McFarland Blvd, Tuscaloosa, **N**...BP/diesel/mart, Chevron/mart/wash/24hr, Exxon, Parade/mart, Phillips 66/mart, RaceTrac/mart, Shell/diesel/mart/wash, Arby's, Burger King, Cancun Mexican, Capt D's, Harvey's Rest., Krystal, LJ Silver, O'Charley's, Red Lobster, Schlotsky's, Waffle House, Best Western, Holiday Inn, Shoney's Inn/rest., Travelodge, Advance Parts, AllTransmissions, Big 10 Tire, CarPort Parts, Chrysler/Plymouth/Jeep, Circuit City, Firestone/auto, Goodyear/auto, Kinko's, Midas Muffler, OfficeMax, Pontiac/Cadillac/GMC, PrecisionTune, Rite Aid, SteinMart, ToysRUs, bank, cinema, cleaners/laundry, **S**...MEDICAL CARE, Amoco, Exxon/atm, Texaco/diesel/mart/wash, Chili's, Dairy Queen, Guthrie's Café, Hardee's, Huddle House, KFC, Logan's Roadhouse, McDonald's, Piccadilly's, Quincy's, Sonic, Subway, Taco Bell, Taco Casa, Wendy's, Comfort Inn, Country Inn Suites, Day's Inn, La Quinta, Motel 6, Quality Inn, Ramada Inn, Super 8, Books-A-Million, CVS Drug, Dillard's, FoodWorld, Ford, Gayfer's, Goody's, Isuzu, MailBoxes Etc, Michael's, NAPA, Office Depot, Pier 1, Rite Aid, Sam's Club, Service Merchandise, Spiffy Lube/wash, TJ Maxx, U-Haul, Wal-Mart SuperCtr/24hr, Winn-Dixie, bank, cinema, mall

Alabama Interstate 20

71b   I-359, Al 69 N, to Tuscaloosa, **N**...HOSPITAL, U of AL, to Stillman Coll

   a   AL 69 S, to Moundville, **S**...Exxon/diesel/mart, Texaco/diesel/mart, Arby's, LoneStar Steaks, OutBack Steaks, Pizza Hut, Waffle House, Wendy's, Courtyard, Fairfield Inn, Jameson Inn, K-Mart, Lowe's Bldg Supply, to Mound SM

68   Northfort-Tuscaloosa Western Bypass, no facilities

64mm   Black Warrior River

62   Fosters, no facilities

52   US 11, US 43, Knoxville, **N**...Exxon/diesel/mart

45   AL 37, Union, **N**...Cotton Patch Rest., **S**...Amoco/diesel/rest./24hr, Hardee's, Southfork Rest./24hr, Western Inn/rest., Greene Co Greyhound Park, Auto Value Parts

40   AL 14, Eutaw, **N**...to Tom Bevill Lock/Dam, **S**...HOSPITAL

39mm   **rest area sb, full(handicapped)facilities, phone, vending, picnic tables, litter barrels, petwalk, RV dump**

38mm   **rest area nb, full(handicapped)facilities, phone, vending, picnic tables, litter barrels, petwalk, RV dump**

32   Boligee, **N**...BP/diesel/mart/rest./24hr, **S**...Chevron/Subway/mart/24hr/atm

27mm   Tombigbee River, Tenn-Tom Waterway

23   Epes, to Gainesville, no facilities

17   AL 28, Livingston, **S**...BP/diesel/mart/24hr, Chevron/Subway/mart/24hr, Noble/Citgo/Janet's/Waffle King/diesel/mart/24hr, Texaco/diesel/mart, Burger King, Pizza Hut, Comfort Inn, Livingston Motel(3mi), Western Inn(1mi), repair/24hr

8   AL 17, York, **S**...HOSPITAL, BP/diesel/mart/rest., Day's Inn/Briar Patch Rest.

1   to US 80 E, Cuba, **N**...Phillips 66/diesel/rest., **S**...Chevron/mart, Dixie Gas

.5mm   **Welcome Ctr eb, full(handicapped)facilities, phone, vending, picnic tables, litter barrels, petwalk, RV dump**

I-20 E and I-59 N run together from Meridian, MS to B'ham

0mm   Alabama/Mississippi state line

Alabama Interstate 59

Exit #   Services

241.5mm   Alabama/Georgia state line, Central/Eastern time zone

241mm   **Welcome Ctr sb, full(handicapped)facilities, phone, vending, picnic tables, litter barrels, petwalk, RV dump**

239   to US 11, Sulphur Springs Rd, **E**...Freeway Motel, Sequoyah Caverns Camping(4mi)

231   AL 40, AL 117, Hammondville, Valley Head, **E**...Sequoyah Caverns Camping(5mi), antiques, **W**...Texaco, flea mkt

222   US 11, to Fort Payne, **E**...Shell, **1-2 mi E**...Bojangles, Krystal, Subway, Best Western/rest., **W**...Citgo/mart, Exxon/diesel, Texaco/diesel/mart, Comfort Inn, auto/tire repair

218   AL 35, Fort Payne, **E**...Citgo(1mi), Conoco/diesel/mart/24hr, Exxon(1mi), Capt D's, Central Park, McDonald's, Pizza Hut, Shoney's, Taco Bell, Wendy's, Mountain View Motel(2mi), Advance Parts, AutoZone, Buick/Olds/Pontiac, Chrysler/Dodge, Goodyear/auto, Radio Shack, Winn-Dixie Foods, cleaners, **W**...HOSPITAL, Exxon/mart/atm, Phillips 66/mart/Harco Drug, Pure/diesel, Shell/mart, Burger King, Hardee's, Huddle House, Olde Times Café, Quincy's, Ryan's Steaks, Waffle House, Day's Inn, Quality Inn, FoodWorld/24hr, Ford/Lincoln/Mercury, Goody's, K-Mart/Little Caesar's, Wal-Mart SuperCtr/24hr

205   AL 68, Collinsville, **E**...Chevron/mart, Pure Gas, Big Valley Rest., Jack's Rest., HoJo Inn/rest., to Little River Canyon, Weiss Lake, **W**...MEDICAL CARE, Shell/mart, Texaco/ice cream/diesel/mart

188   AL 211, to US 11, Reece City, to Gadsden, **E**...Chevron/mart/24hr, Noccalula Rest., **W**...Amoco/diesel

183   US 431, US 278, Gadsden, **E**...Amoco/diesel/mart, BP/diesel, Conoco, Shell/mart, Texaco/ice cream, Hardee's, Magic Burger, Waffle House, Wendy's, Columbia Inn, Holiday Inn Express, Rodeway Inn, auto parts, bank, st police, **W**...Amoco/mart, Chevron/mart/24hr, Exxon/wash, KFC, McDonald's, Pizza Hut, Quincy's, Shoney's, Subway, Taco Bell, Econolodge, auto repair, tires

182   I-759, to Gadsden, no facilities

181   AL 77, Rainbow City, to Gadsden, **E**...Hospitality Inn, Shell(2mi), **W**...BP/diesel/mart, Hardee's, Subway

174   to Steele, **E**...truck tire service, **W**...Chevron/diesel/mart, Spur/diesel/mart/atm, Steele City/diesel/rest.

168mm   **rest area sb, full(handicapped)facilities, phone, vending, picnic tables, litter barrels, petwalk, RV dump**

166   US 231, Whitney, to Ashville, **E**...Chevron/diesel/mart, Exxon/diesel/mart, **W**...Texaco/Taco Bell/ice cream/diesel/mart/24hr

165mm   **rest area nb, full(handicapped)facilities, phone, vending, picnic tables, litter barrels, petwalk, RV dump**

156   to US 11, AL 23, Springville, to St Clair Springs, no facilities

154   AL 174, Springville, to Odenville, **W**...Chevron/mart/wash/24hr, Exxon, Texaco/diesel/mart, Jack's Rest./playplace, McDonald's

E

W

N

S

Alabama Interstate 59

148	to US 11, Argo, **E**...BP/mart
143	Mt Olive Church Rd, Deerfoot Pkwy, no facilities
141	to Trussville, Pinson, **E**...DENTIST, CHIROPRACTOR, Amoco/diesel/mart/wash, BP/diesel/mart/wash, Shell/mart/wash, Chinese Rest., McDonald's, Pizza Hut, Taco Bell, Waffle House, Wendy's, cleaners, **W**...MEDICAL CARE, VETERINARIAN, Chevron/mart/wash/24hr, Burger King, Ruby Tuesday, CVS Drug, K-Mart/Little Caesar's, Radio Shack, Western Foods, XpressLube, bank
137	I-459 S, to Montgomery, Tuscaloosa
134	to AL 75, Roebuck Pkwy, **W**...HOSPITAL, DENTIST, CHIROPRACTOR, Amoco/mart, BP/diesel/mart, Chevron/mart/wash, Citgo, Crown Gas, Exxon, Phillips 66, Shell/mart/24hr, Texaco/diesel/mart/wash, Arby's, Barnhill's Country Buffet, Baskin-Robbins, Burger King, Capt D's, Cedar Pita, Chick-fil-A, Chinese Rest., ChuckeCheese, Church's Chicken, Costa's Café, Denny's, El Gringo Mexican, Guadalajara Mexican, Krystal, Lee's Chicken, McDonald's, Milo's Burgers, Mrs Winner's, O'Charley's, Pasquale's Pizza, Pioneer Cafeteria, Rally's, Shoney's, Steak&Ale, Subway, SubZone, Taco Bell, Waffle House, Wendy's, Anchor Inn Motel, Econolodge, Parkway Inn, Bruno's Foods, Chevrolet, Chrysler/Jeep, CVS Drug, Dodge, Drugs4Less, Firestone/auto, Food Giant, Ford, Goodyear/auto, Goody's, Harco Drugs, Honda, Jo-Ann Fabrics, Kinko's/24hr, K-Mart, LensCrafters, Lincoln/Mercury, MailCtr, Mr Transmissions, NTB, 1Hr Photo, Pearle Vision, PetSupplies+, SpeedyLube, Winn-Dixie, bank, cinema, cleaners, Hyundai, JiffyLube, Mazda/Kia, Midas Muffler, USPO, VW, Wal-Mart,
133	4th St, to US 11(from nb), **W**...CHIROPRACTOR, Amoco, Phillips 66/wash, Shell, Arby's, Catfish Cabin Rest., Checker's, Krystal, Shoney's, Waffle House, Chrysler/Plymouth/Eagle, Mazda, mall, accesses same as 134
132	US 11 N, 1st Ave, **W**...PODIATRIST, Chevron/mart/24hr, Exxon, El Palacio Mexican, Krispy Kreme, Pizza Hut, SE Meat/Veg Mkt, auto parts, drugs, hardware, accesses same as 131
131	77th Ave(from nb), **E**...Exxon/mart, Burger King, Checker's, Church's Chicken, Rally's, Subway, Taco Bell, NAPA, U-Haul, accesses same as 132
130	I-20, E to Atlanta, W to Tuscaloosa

I-59 S and I-20 W run together from B'ham to Mississippi. **See Alabama Interstate 20.**

Alabama Interstate 65

Exit #	Services
367mm	Alabama/Tennessee state line
365	AL 53, to Ardmore, **E**...Texaco/diesel/mart, Granny's Rest., Budget Inn
364mm	**Welcome Ctr sb, full(handicapped)facilities, info, phone, vending, picnic tables, litter barrels, petwalk, RV dump**
361	Elkmont, **W**...Amoco/Charlie's/diesel/mart/rest./repair, Exxon/diesel/mart, antiques
354	US 31S, to Athens, **W**...Chevron/Subway/mart, Athens Rest., Dairy Queen, Budget Inn, Mark Hotel, **3-5 mi W**...HOSPITAL, facilities in Athens
351	US 72, to Athens, to Huntsville, **E**...VETERINARIAN, BP/diesel/mart/wash, Chevron/mart/wash/24hr, Exxon/mart, RaceTrac/mart, Shell/Subway/mart, Texaco/diesel/mart, Cracker Barrel, Hungry Fisherman, McDonald's, Shoney's, Waffle House, Wendy's, Comfort Inn, Day's Inn, Hampton Inn, Holiday Inn Express, Travelodge, Acme Boots, Ford/Mercury/Lincoln, XpressLube, bank, **W**...HOSPITAL, Chevron, Texaco/diesel/mart, Arby's(2mi), Hardee's, Krystal, Best Western, Wal-Mart(2mi), to Joe Wheeler SP
340b	I-565, to Huntsville, to Ala Space & Rocket Ctr
a	AL 20, to Decatur, **W**...RaceTrac/mart, Conoco/diesel/mart/wash, Jameson Inn(2mi)
337mm	Tennessee River
334	AL 67, Priceville, to Decatur, **E**...Amoco/diesel/mart, RaceTrac/diesel/mart/24hr, Day's Inn, Super 8, diesel repair, **W**...HOSPITAL, BP/diesel/mart, Chevron/mart, Express/diesel/mart, Texaco/LP/mart, Dairy Queen, Hardee's, Libby's Diner, McDonald's, Waffle House, Key West Inn(6mi), Ramada Ltd(4mi)
328	AL 36, Hartselle, **E**...Reeves Farm Produce, **W**...Chevron/diesel/mart/24hr, Shell/mart, Jet-Pep Gas, Huddle House
325	Thompson Rd, to Hartselle, **W**...Express Inn
322	AL 55, to US 31, to Falkville, Eva, **E**...BP/diesel/rest., antiques
318	US 31, to Lacon, **E**...Texaco/Dairy Queen/Stuckey's

Birmingham

Athens

Alabama Interstate 65

Cullman

310   AL 157, Cullman, West Point, **E**...HOSPITAL, Amoco/mart/24hr, Conoco/diesel, Shell/diesel/mart/24hr, Burger King, Cracker Barrel, Dairy Queen, Denny's, McDonald's, Morrison's Café, Shoney's, Taco Bell, Waffle House, Best Western, Comfort Inn, Hampton Inn, Acme Boots, Ford/Lincoln/Mercury, Pontiac/Olds/GMC, WDG Foods, **W**...BP/diesel/mart, Exxon/diesel, Super 8, Cullman Camping(2mi)

308   US 278, Cullman, **E**...Omelet Shoppe/24hr, Day's Inn/rest., **W**...Chevron/mart/24hr, Howard Johnson, flea mkt

304   AL 69, Good Hope, to Cullman, **E**...HOSPITAL, BP/mart, Exxon/diesel/mart, Shell/diesel/mart, Texaco/diesel/mart, Hardee's, Jack's Rest., Miguel's Mexican, Waffle House, Holiday Inn Express, Ramada Inn, Good Hope Camping, Good Hope Truck/wrecker, diesel repair, laundry, **W**...to Smith Lake

302mm   **E...rest area both lanes, full(handicapped)facilities, phone, vending, picnic tables, litter barrels, petwalk, RV dump**

299   AL 69, to Jasper, **W**...Amoco/diesel/mart, Chevron/diesel/mart, 76/Conoco/diesel/mart, Texaco/diesel/mart, antiques

292   AL 91, to Arkadelphia, **E**...Country RV Park(1mi), **W**...Shell/diesel/mart/motel

291mm   Warrior River

289   to Blount Springs, **W**...Texaco/Dairy Queen/Stuckey's, to Rickwood Caverns SP

287   US 31 N, to Blount Springs, no facilities

284   US 31 S, AL 160 E, Hayden, **E**...Conoco/diesel/mart, Phillips 66/mart, Shell/diesel/mart, bank

282   AL 140, Warrior, **E**...BP/mart, Chevron/mart/24hr, Hardee's, McDonald's, Pizza Hut, Taco Bell, **W**...Amoco/mart/24hr

281   US 31, to Warrior

280   to US 31, to Warrior, **E**...Chevrolet

279.5mm   Warrior River

275   to US 31, Morris, no facilities

272   Mt Olive Rd, **E**...Gardendale Kamp(1mi), **W**...Chevron/mart/24hr, Shell/mart, Texaco/mart(1mi), bank

271   Fieldstown Rd, **E**...MEDICAL CARE, DENTIST, BP/mart/wash, Chevron/mart/wash/24hr, RaceTrac/mart, Shell, Texaco/diesel/mart/wash, Arby's, Baskin-Robbins, BBQ, Capt D's, Dairy Queen, Guthrie's Diner, Hardee's, KFC, McDonald's, Milo's Burgers, Shoney's, Subway, Taco Bell, Waffle House, Wendy's, AutoZone, Chevrolet, Delchamps Foods, Eckerd's, K-Mart/auto, NAPA, Pennzoil, Wal-Mart, XpressLube, bank, cleaners, tires, **W**...Cracker Barrel

267   Walkers Chapel Rd, to Fultondale, **1-2 mi E**...Amoco/mart, Texaco/TCBY/Wall St Deli/mart/wash/24hr, Hardee's, Jack's Rest., Jalisco Mexican, Day's Inn, Super 8, CVS Drug, USPO, Winn-Dixie, cleaners

266   US 31, Fultondale, **E**...Chevron/mart/24hr, Day's Inn, Super 8

264   41st Ave, no facilities

263   33rd Ave, **E**...Chevron/24hr, Shell/mart/wash, Hardee's, Apex Lodge, **W**...Amoco/mart, Exxon/mart, BP/mart

262   Finley Ave, **E**...Amoco/mart, Phillips 66/mart, Texaco/diesel/mart, **W**...Chevron/mart/24hr, Express/diesel/mart/24hr, Capt D's, Checker's, McDonald's/grits

261b a   I-20/59, E to Gadsden, W to Tuscaloosa, no facilities

260b a   6th Ave, **E**...Amoco, Shell/mart/atm, Jack's Rest., Hardee's, Mrs Winner's, Popeye's, Tourway Inn, Buick, Nissan, **W**...Chevron/wash, Church's Chicken, Adams Inn, to Legion Field

259b a   University Blvd, 4th Ave, 5th Ave, **E**...HOSPITAL, Waffle House, Best Western, Food Fair, **W**...Chevron/mart/wash, Mazda

Birmingham

258   Green Springs Ave, **E**...Citgo/mart, Exotic Wings, **W**...Quality Inn

256b a   Oxmoor Rd, **E**...Amoco, Exxon/diesel/mart, Shell/mart, Texaco, Burger King, Cuco's Mexican, KFC, Krystal, Lovoy's Italian, McDonald's, Howard Johnson, Big Lots, Firestone/auto, Food World, Goodyear/auto, JiffyLube, Jo-Ann Fabrics, K-Mart/drugs, Office Depot, Reynolds Drug, cleaners/laundry, **W**...BP/mart/wash, Chevron/mart/atm/24hr, BBQ, Hardee's, Shoney's, Comfort Inn, Fairfield Inn, Holiday Inn/rest., Microtel, Red Roof Inn, Shoney's Inn, Super 8, Batteries+

255   Lakeshore Dr, **E**...HOSPITAL, BP/mart/atm, to Samford U, **W**...CHIROPRACTOR, Capt D's, Chili's, Chick-fil-A, IHOP, Landry's Seafood, LoneStar Steaks, Mr Wang's Chinese, O'Charley's, Outback Steaks, Schlotsky's, Taco Bell, Tony Roma, Wendy's, Best Suites, La Quinta, Residence Inn, TownePlace Suites, Academy Sports, BabysRUs, Books-A-Million, Bruno's Foods, $Tree, Goody's, HomePlace, Kinko's, Lowe's Bldg Supply, OfficeMax, Old Navy, Parisian, Sam's Club, SuperPetz, Waccamaw, Wal-Mart SuperCtr/24hr, XpressLube, bank, cinema, mall

254   Alford Ave, Shades Crest Rd, **E**...VETERINARIAN, Chevron, **W**...Amoco/mart, BP/mart, Citgo/mart/wash, cleaners

252   US 31, Montgomery Hwy, **E**...HOSPITAL, VETERINARIAN, Amoco, BP/mart/wash, Chevron, Shell/mart, Arby's, Capt D's, ChuckeCheese, Hardee's, Johnny Ray's Diner, Milo's Burgers, Piccadilly's, Pizza Hut, Taco Bell, Waffle House, Comfort Inn, Hampton Inn, The Motor Lodge/rest., Lincoln/Mercury, Olds/GMC/Saturn/Isuzu, Aamco, cleaners, **W**...MEDICAL CARE, Amoco/mart, Chevron/mart, Exxon/mart, Shell, Texaco, BBQ, Burger King, Chick-fil-A, Damon's, FishMkt Rest., Kenny Rogers, Krystal, Longhorn Steaks, Maestro's Pizza, McDonald's, Outback Steaks, Salvatore's Pizza, Schlotsky's, Shoney's, Subway, Waffle House, Day's Inn, Holiday Inn, Books-A-Million, Bruno's Food, Buick, Chevrolet, Chrysler/Plymouth/Jeep, Circuit City, Eckerd, Express Change/brakes, Goodyear/auto, Hancock Fabrics, Honda, Kia, Midas Muffler, Nissan, Pontiac, Rite Aid, TJ Maxx, Toyota, bank, transmissions

250   I-459, US 280, no facilities

N

S

Alabama Interstate 65

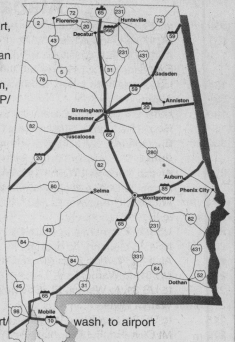

247   AL 17, Valleydale Rd, **E**...BP/mart, Hardee's, **W**...BP/mart, RaceTrac/mart, Shell/mart, Texaco/mart, Backyard Burger, IHOP, Milo's Burgers, Papa John's, RagTime Café, Waffle House, InTown Motel, La Quinta, Suburban Lodge, $General, Rite Aid, Walgreen

246   AL 119, Cahaba Valley Rd, **E**...Texaco/mart, Southern Heritage Museum, to Oak Mtn SP, RV camping, **W**...HOSPITAL, Amoco/diesel/mart/24hr, BP/mart/wash, Chevron, Shell/mart/wash, Speedway/diesel/mart, Applebee's, Arby's, BBQ, Blimpie, Buffalo's Café, Capt D's, Chick-fil-A, Cracker Barrel, Dairy Queen, Hardee's, KFC, Krystal, Le's Dragon Chinese, Logan's Honeyhams, McDonald's, O'Charley's, Pizza Hut, Pizza Planet, San Marcos Mexican, Schlotsky's, Shoney's, Sonic, Taco Bell, Waffle House, Wendy's, Best Western, Comfort Inn, Holiday Inn Express, Ramada Ltd, Sleep Inn, Travelodge, CVS Drug, DriversWay Auto Dealers, ExpressLube/wash, Firestone/auto, Harley-Davidson, Winn-Dixie, bank

242   Pelham, **E**...Chevron/diesel/mart/24hr, Exxon/diesel/mart/24hr, **W**...Shelby Motor Lodge(2mi), KOA(1mi)

238   US 31, Alabaster, Saginaw, **E**...BP/diesel/mart, **W**...Chevron/diesel/mart, Cannon/gas, Shell, Texaco/mart/wash, Waffle House, XpressLube, **2 mi W**...HOSPITAL, Arby's, Shelby Motor Inn

234   Shelby Co Airport, **W**...Chevron/diesel/mart/atm/24hr, Texaco/diesel/mart/wash, to airport

231   US 31, Saginaw, **E**...Holiday Inn Express, Burton Campers Ctr/LP gas

228   AL 25, to Calera, **E**...Citgo/diesel/mart, Texaco/diesel/mart, Best Western, Parkside Motel, to Brierfield Works SP, Day's Inn, **W**...Hardee's(1mi)

227.5mm Buxahatchie Creek

219   Union Grove, Thorsby, **E**...Chevron/diesel/mart/24hr, Citgo/diesel/rest., Headco/diesel/mart, Mims Marine, Peach Queen Camping, **W**...Shell/ice cream/mart, Smokey Hollow Rest.

213.5mm rest area both lanes, full(handicapped)facilities, phone, vending, picnic tables, litter barrels, petwalk, RV dump

212   AL 145, Clanton, **E**...Chevron/diesel/mart, **W**...HOSPITAL, BP/mart, Headco/diesel/mart/rest., Peach Tower Rest., One Big Peach, repair/tires

208   Clanton, to Lake Mitchell, **W**...Exxon/diesel/mart/24hr, Shoney's Inn/rest., Heaton Pecans

205   US 31, AL 22, to Clanton, **E**...Amoco/mart, Shell/diesel/mart, McDonald's, Waffle House, Best Western, Day's Inn, Scottish Inn, Peach Park, to Confed Mem Park, antiques, **W**...BP/diesel/The Store, Chevron/diesel/mart/24hr, Burger King, Capt D's, Hardee's, Heaton Pecans, KFC, Subway, Taco Bell, Key West Inn

200   to Verbena, **E**...BP/diesel/mart, **W**...Texaco/DQ/Stuckey's/diesel, RV camping

186   US 31, Pine Level, **E**...Confederate Mem Park(13mi), **W**...HOSPITAL, BP/diesel/mart, Chevron/mart/24hr, Citgo/diesel, Conoco/diesel/mart, motel

181   AL 14, to Prattville, **E**...Chevron/diesel/mart/24hr, **W**...HOSPITAL, Amoco/mart, Citgo, Conoco/diesel/mart, Exxon/mart, Cracker Barrel, Ria Pizzaria, Best Western, Comfort Inn, Super 8, bank

179   Millbrook, **E**...Chevron/mart/wash, K&K RV Ctr, **W**...Amoco/diesel/mart, BP/diesel/mart, Citgo/mart, Exxon/Subway/mart/24hr, Shell/mart, Burger King, Hardee's, McDonald's, Shoney's, Steak'n Shake, Waffle House, Econolodge, Hampton Inn, Holiday Inn/rest., Jameson Inn

176   AL 143 N(from nb, no return), Millbrook, Coosada, no facilities

173   AL 152, North Blvd, to US 231, no facilities

172.5mm Alabama River

172   Clay St, Herron St, downtown, **E**...Amoco, Petro/mart, Embassy Suites

171   I-85 N, Day St, no facilities

170   Fairview Ave, **E**...Citgo/mart, Checker's, China King, Church's Chicken, KFC, Krystal, McDonald's, Advance Parts, AutoZone, CVS Drug, Food Fair, Rite Aid, cleaners, **W**...Exxon/mart, Hardee's, Family$

169   Edgemont Ave(from sb), **E**...Amoco, carwash

168   US 80 E, US 82, South Blvd, **E**...HOSPITAL, BP/diesel/mart/wash, Entec/diesel/mart, TA/Citgo/diesel/rest./24hr, Arby's, Burger King, Capt D's, KFC, McDonald's, Pizza Hut, Shoney's, Taco Bell, Waffle House, Best Western, Knight's Inn, Travel Inn, Calhoun Foods, **W**...Amoco/mart, Chevron/diesel/24hr, RaceTrac/mart, Shell/Subway/diesel/mart, Speedway/diesel/mart, Dairy Queen, Hardee's, Omelet House, Quincy's, Waffle House, Wendy's, Comfort Inn, Day's Inn, Econolodge, Holiday Inn, Inn South, Super 8, airport

167   US 80 W, to Selma, **1 mi W**...Citgo, PaceCar/mart, Church's Chicken, Subway, CVS Drug

164   US 31, Hope Hull, **E**...Amoco/mart/24hr, Saveway/diesel/mart/24hr, Texaco/mart, Huddle House, Rodeway Inn, KOA, airport, **W**...BP/Burger King/mart, Chevron/mart/wash/24hr, Exxon/FoodCourt/mart, Best Western

Alabama Interstate 65

Gr**e**e**n**v**i**l**l**e	158	to US 31, **E**...Texaco/Dairy Queen/Stuckey's
	151	AL 97, to Letohatchee, **W**...Amoco/diesel/mart, BP/mart
	142	AL 185, to Ft Deposit, **E**...BP/diesel/mart, Shell/Subway/diesel/mart/24hr, Priester's Pecans, Wishing Well Rest., auto parts, **W**...Chevron/mart/atm/24hr, Pure/mart, Village Café
	134mm	**rest areas both lanes, full(handicapped)facilities, phone, vending, picnic tables, litter barrels, petwalk, RV dump**
	130	AL10 E, AL 185, to Greenville, **E**...BP/diesel/mart, Chevron/mart/wash/24hr, Phillips 66/diesel/24hr, Shell/mart/24hr, Arby's, Beaugez's Steaks, Capt D's, Hardee's, KFC, McDonald's, Pizza Hut, Waffle House, Wendy's, Econolodge, Ramada Inn, CVS Drug, $General, Goody's, antiques, auto parts, bank, to Sherling Lake Park, **W**...Exxon/mart, Phillips 66/Subway/mart, Texaco/Baskin-Robbins/diesel/mart, Bates Turkey Rest., Burger King, Cracker Barrel, Krystal, Taco Bell/TCBY, Best Western, Comfort Inn, Hampton Inn, Jameson Inn, Cato's, Chevrolet, Chrysler/Plymouth/Dodge/Jeep, Wal-Mart/auto, Winn-Dixie
	128	AL 10, to Greenville, **E**...HOSPITAL, Exxon/diesel/ice cream, Shell/diesel/mart, Smokehouse/Store, **W**...Amoco/mart
	114	AL 106, to Georgiana, **E**...HOSPITAL, **W**...Amoco/mart, Chevron/mart/24hr, auto repair
	107	AL 8, to Garland, no facilities
	101	AL 29, to Owassa, **E**...BP/diesel/mart, **W**...Exxon/diesel/mart
	96	AL 83, to Evergreen, **E**...HOSPITAL, Chevron/mart/wash/24hr, Shell/mart, Burger King, Church's Chicken(2mi), Floyd's Café, Hardee's, KFC/Taco Bell, McDonald's, **W**...BP/mart/wash, El VaQuero Mexican, Pizza Hut, Waffle House, Comfort Inn, Day's Inn, Evergreen Inn
	93	US 84, to Evergreen, **E**...BP, Exxon, Texaco/diesel/ice cream, PineCrest RV Park, auto repair
	89mm	**rest area sb, full(handicapped)facilities, phone, vending, picnic tables, litter barrels, petwalk, RV dump**
	85mm	**rest area nb, full(handicapped)facilities, phone, vending, picnic tables, litter barrels, petwalk, RV dump**
	83	AL 6, to Lenox, **E**...Exxon/diesel/LP/mart, Texaco/mart/atm, Louise's Rest., Sunshine RV Park(4mi)
	77	AL 41, to Range, **E**...BP/mart, **W**...Amoco/mart, Shell/Stuckey's/diesel, Texaco/diesel, Ranch House Rest.
	69	AL 113, to Flomaton, **E**...Texaco/diesel/mart/24hr, Angel Ridge Rest., **W**...Conoco/diesel/mart/24hr, Huddle House
	57	AL 21, to Atmore, **E**...Exxon/diesel/mart, Texaco/diesel/mart, Creek Family Rest., Best Western, Indian Bingo, **W**...BP/diesel, to Kelley SP
	54	Escambia Co Rd 1, **E**...Citgo/diesel, to Creek Indian Res
	45	to Perdido, **W**...Amoco/diesel/mart, Harville's Café
	37	AL 287, to Bay Minette, **E**...Chevron/mart, Jones Trkstp/diesel/rest.
	34	to AL 59, to Bay Minette, Stockton, **E**...HOSPITAL
	31	AL 225, to Stockton, **E**...to Blakeley SP, Confederate Mem Bfd, **W**...Conoco/diesel/mart
	29mm	Tensaw River
	28mm	Middle River
	25mm	Mobile River
	22	Creola, **E**...antiques, marine ctr, repair, **W**...KOA(4mi)
	19	US 43, to Satsuma, **E**...Chevron, Speedway/Hardee's/diesel/mart/24hr, McDonald's, I-65 RV Park(2mi), **W**...BP/mart, Texaco/diesel/mart, KOA(5mi)
	15	AL 41, **E**...Chevron, Plantation Motel, **W**...Citgo/diesel/mart, Circle K/gas
	13	AL 158, AL 213, to Saraland, **E**...EYECARE, BP/diesel/mart/wash, Starvin Marvin/gas, Texaco/Krystal/diesel/mart/24hr, Shoney's, Waffle House, Bamboo Motel(1mi), Comfort Inn, Day's Inn, Holiday Inn Express, Plantation Motel(1mi), Wal-Mart SuperCtr/24hr, **W**...Exxon/mart, Best Western, Chickasabogue RV Park(4mi)
	10	W Lee St, **E**...Citgo, Conoco/mart, Shell/Subway/mart, Spur/mart, Howard Johnson Express
	9	I-165 S, to Mobile
	8b a	US 45, to Prichard, **E**...Chevron/mart/24hr, Texaco/mart/wash, Church's Chicken, My Seafood Café, Star Motel, Tiger Foods, Chevrolet, Peterbilt, **W**...VETERINARIAN, Amoco/mart/24hr, Conoco/mart, Exxon/mart, RaceWay/mart, Burger King, Domino's, Golden Egg Café, McDonald's, Pride Trucklube/diesel, auto parts
	5b	US 98, Moffett Rd, **E**...Burger King, Church's Chicken, Quincy's, Advance Parts, **W**...Speedway/diesel/mart
	a	Spring Hill Ave, **E**...HOSPITAL, Amoco, Chevron, Shell/diesel/mart, Chinese Kitchen, Dairy Queen, KFC, Love's Seafood, McDonald's, AutoZone, Big O Tire, CarQuest, Family$, Food Tiger Foods, L&M Automotive, SuperLube, cleaners, **W**...DENTIST, Chevron/diesel/24hr, Exxon/mart, Texaco/diesel/mart/atm/wash, Waffle House, Extended Stay America
Mo**b**i**l**e	4	Dauphin St, **E**...HOSPITAL, Amoco/diesel/mart, BP/mart/wash, Checker's, Cracker Barrel, Godfather's, Hong Kong Chinese, Krystal, Lil Jubilee Café, McDonald's, Quizno's, Taco Bell, TCBY, Waffle House, Wendy's, Comfort Suites, Econolodge, Red Roof Inn, Cadillac/Pontiac/GMC, DelChamps Foods, Mercedes, Mr Transmission, Rite Aid, SuperLube, Tanner's Pecans, bank, same as 3 & 5a

Alabama Interstate 65

3   Airport Blvd, to Airport, **E...**HOSPITAL, BP/mart/lube, Burger King, Hooters, Piccadilly's, Schlotsky's, Wendy's, Clarion Hotel, Acura/Jaguar/Infiniti, Audi, Baby Depot, Barnes&Noble, Books-A-Million, Burlington Coats, Dillard's, Firestone/auto, Ford, Goodyear/auto, Harley-Davidson, Hyundai, Kia, Lowe's Bldg Supply, Luxury Linens, McRae's, MensWearhouse, Mitsubishi, Olds/Daewoo, Nissan, Parisian, RV Ctr, Saab, Saturn, Sears/auto, Staples, Target, ToysRUs, VW, bank, cinema, mall, **W...**DENTIST, BP/mart/wash, Exxon/Subway/mart, Shell/mart, Spur/diesel/mart, Texaco/diesel/mart/ wash, American Café, Arby's, Burger King, Chili's, Darryl's, Denny's, Don Pablo's, El Chico, IHOP, Indian Cuisine, Japanese Steaks, Joe's Crabshack, JR's Smokehouse, Krispy Kreme, Little Saigon, LoneStar Steaks, Los Rancheros Mexican, O'Charley's, Olive Garden, Outback Steaks, Pizza Hut, Popeye's, Quincy's, Quizno's, Red Lobster, Scallion's Deli, Shoney's, South China Rest., Tony Roma, Waffle House, Wanfu Mongolian, Best Inn, Best Suites, CountryHearth Inn, Courtyard, Day's Inn, Drury Inn, Fairfield Inn, Family Inn, Hampton Inn, Holiday Inn, Motel 6, Ramada Inn, Residence Inn, Shoney's Inn, Suburban Lodge, Books-A-Million, Bruno's Food/drug, Circuit City, DelChamps Foods, $Tree, Gateway Computers, Home Depot, Jo-Ann Fabrics, Kinko's, K-Mart, Lenscrafters, MailBoxes Etc, Michael's, Midas Muffler, OfficeMax, Office Depot, PepBoys, PetsMart, Phar-Mor Drug, Pier 1, Radio Shack, Sam's Club, 24hr, bank, cleaners, to USAL, SteinMart, U-Haul, Wal-Mart/

1b a   US 90, Government Blvd, **E...**Chevron/mart, Conoco/diesel/mart, Burger King, McAlister's Deli, Steak'n Shake, Guesthouse Inn, Aamco, BMW, Buick/Isuzu/Volvo, Chevrolet, Chrysler/Jeep, Dodge, Family$, Honda, Lexus, Lincoln/Mercury, Mazda, Precision Tune, Toyota, **W...**Citgo/diesel/mart/wash, Waffle House, Rest Inn, All Tune/lube, West Marine

0mm   I-10, E to Pensacola, W to New Orleans. I-65 begins/ends on I-10, exit 20.

Alabama Interstate 85

Exit #	Services
80mm	Chattahoochee River, Alabama/Georgia state line, Eastern/Central time zone

79   US 29, to Lanett, **E...**Amoco/KrispyKreme/diesel/mart, RaceTrac/mart/24hr, Arby's, Burger King, BBQ, Capt D's, Hardee's, KFC, Krystal, Magic Wok Chinese, McDonald's, San Marcos Mexican, Subway, Taco Bell, Waffle House/24hr, Wendy's, Western Sizzlin, Advance Parts, Cato's, $General, Radio Shack, Wal-Mart SuperCtr/gas/ 24hr, bank, transmissions, to West Point Lake, RV camping, **W...**DENTIST, VETERINARIAN, Conoco/mart, Exxon/mart/24hr, Phillips 66/diesel/mart, Domino's, Shoney's, Sonic, Day's Inn, Econolodge, Super 8, AutoZone, CVS Drug, Kroger/drug, Parts+, laundry

78.5mm   **Welcome Ctr sb, full(handicapped)facilities, phone, vending, picnic tables, litter barrels, petwalk**

77   AL 208, to Huguley, **E...**Amoco/diesel/mart/24hr, Texaco/WaffleKing/diesel/24hr, Waffle House, Holiday Inn Express, Chevrolet, Chrysler, Ford/Lincoln/Mercury, West Point Pepperell Outlet

70   AL 388, to Cusseta, **E...**Texaco/Subway/KrispyKreme/diesel/rest./24hr, Bridges Boots

64   US 29, to Opelika, no facilities

62   US 280/431, to Opelika, **E...**Amoco/mart, BP/diesel/mart, Chevron, Texaco/Church's/KrispyKreme/diesel/mart/ 24hr, Burger King, Denny's, McDonald's, Shoney's, Subway, Budget Inn, Day's Inn, Holiday Inn, Knight's Inn, Motel 6, Opelika Inn, Shoney's Inn, **W...**Conoco, Exxon/mart, Shell/mart, Cracker Barrel, Waffle House, Western Sizzlin, Comfort Inn, Mariner Inn, Chevrolet, Chrysler/Plymouth/Dodge, Ford, GNC, Hyundai, Jeep, Toyota, USA Stores/famous brands

60   AL 51, AL 169, to Opelika, **E...**Amoco/diesel/mart/wash, Hardee's, RV camping, **W...**HOSPITAL, Chevron/Krystal/ diesel/mart/24hr, Texaco/diesel/mart, Wendy's

58   US 280 W, to Opelika, **E...**golf, museum, **W...**HOSPITAL, BP, Chevron/Subway/mart, Dairy Queen, Golden Corral, Taco Bell, Guesthouse Inn(4mi), Lowe's Bldg Supply

57   Glenn Ave, **W...**Phillips 66/mart/24hr, Jameson Inn, Plaza Motel, Super 8, to airport

51   US 29, to Auburn, **E...**Amoco/diesel/pizza/mart/24hr, Hampton Inn, to Chewacla SP, camping, **W...**Exxon/diesel/ mart, Raceway/mart, McDonald's, Waffle House, Auburn U Hotel/Conf Ctr, Comfort Inn, Econolodge, Heart of Auburn Motel, Ford/Lincoln/Mercury/Mitsubishi, to Auburn U

45mm   **rest area both lanes, full(handicapped)facilities, phone, vending, picnic tables, litter barrels, petwalk, 24hr security**

42   US 80, AL 186 E, Wire Rd, **E...**to Tuskegee NF, diesel repair/tires, **W...**Amoco/diesel/mart

38   AL 81, to Tuskegee, **E...**Western Inn/rest., NHS

Alabama Interstate 85

32 AL 49 N, to Tuskegee, **E**...Amoco/diesel/mart

26 AL 229 N, to Tallassee, **W**...HOSPITAL

22 US 80, to Shorter, **E**...Amoco/diesel/mart, Chevron/Petro/diesel/rest., Exxon/diesel/mart, Day's Inn, to Macon Co Greyhound Pk, Windrift RV Park

16 Waugh, to Cecil, **E**...BP/diesel/mart, BBQ, auto repair

11 US 80, to Mt Meigs, **W**...Chevron/mart/wash/24hr

9 AL 271, to AL 110, to Auburn U/Montgomery, **E**...Applebee's, **W**...HOSPITAL, Citgo/diesel/mart

6 US 80, US 231, AL 21, East Blvd, **E**...BP/wash, Chevron/mart/24hr, Exxon/diesel/mart/24hr, RaceTrac/mart, Burger King, Cracker Barrel, Darryl's, Don Pablo's, KFC, Luigi's Italian, Piccadilly's, Roadhouse Grill, Schlotsky's, Semolina Pasta, Shogun Japanese, SteakOut, Subway, Taco Bell, Up the Creek Grill, Waffle House, Wings Grill, Baymont Inn, Best Inn, Courtyard, Extended Stay America, Fairfield Inn, Hampton Inn, La Quinta, Radisson, Ramada Inn, Residence Inn, SpringHill Suites, Studio+, Wingate Inn, Goody's, Home Depot, HomePlace, Kinko's/24hr, Lowe's Bldg Supply, PetSupplies+, Pontiac/Cadillac, Radio Shack, Sam's Club, SportsAuthority, USPO, Waccamaw Outlet, XpertTune/lube, antiques, bank, cinema, **W**...Amoco, Citgo/mart, Shell/mart, Church's Chicken, KFC, LoneStar Steaks, McDonald's, OutBack Steaks, Shoney's, Waffle House, Comfort Suites, Motel 6, Holiday Inn, Winfield Inn, BMW, Buick, Chevrolet, Chrysler/Jeep, Ford, Hyundai, Isuzu, Kia, Lexus, Mazda, Midas Muffler, Mitsubishi, Nissan, Toyota, VW, ExpressLube, to Gunter AFB, **1 mi W**...IHOP, Circuit City, Sears/auto, Service Merchandise, mall

4 Perry Hill Rd, **E**...DENTIST, EYECARE, Chappy's Deli, Bruno's Foods, Rite Aid, cleaners, **W**...Chevron/mart/24hr, bank

3 Ann St, **E**...BP/diesel/mart, Arby's, BBQ, Brewster's Grill, Capt D's, Domino's, Great Wall Chinese, Hardee's, Krystal, McDonald's, Pizza Hut, Quincy's, Taco Bell, Waffle House, Wendy's, Day's Inn, Econolodge, Knight's Inn, Big O Tires, bank, **W**...Amoco/mart, Entec/mart, Exxon/Burger King/mart, Shoney's, Villager Lodge

2 Forest Ave, **E**...MEDICAL CARE, VETERINARIAN, CVS Drug, **W**...HOSPITAL, bank

1 Court St, Union St, downtown, **E**...CHIROPRACTOR, MEDICAL CARE, Amoco/diesel/mart, Exxon/mart, **W**...Mr Transmission, bank, to Ala St U

0mm I-85 begins/ends on I-65, exit 171 in Montgomery.

Alabama Interstate 459(Birmingham)

Exit # Services

33b a I-59, N to Gadsden, S to Birmingham. I-459 begins/ends on I-59, exit 137.

32 US 11, Trussville, **E**...Chevron/diesel/mart/wash/24hr, Exxon/diesel/mart/24hr, RaceTrac/mart/atm/24hr, Shell/mart, Texaco/Wendy's/diesel/mart/wash, Jack's Rest., Waffle House, Hampton Inn, Academy Sports, Bed Bath&Beyond, Books-A-Million, Harley-Davidson, Home Depot, Lowe's Bldg Supply, Mazda, Michael's, Pontiac/GMC/Buick, Target, TJ Maxx, XpressLube, cleaners, **W**...VETERINARIAN, Amoco/mart/24hr, Parkway Inn(4mi), auto repair

31 Derby Parkway, **W**...B'ham Race Course

29 I-20, E to Atlanta, W to Birmingham

27 Grants Mill Rd, **S**...Exxon/diesel

23 Liberty Parkway, no facilities

19 US 280, Mt Brook, Childersburg, **1-2 mi S**...Exxon, Arby's, Burger King, Chick-fil-A, McDonald's, Piccadilly's, Ruby Tuesday, Baymont Inn, Fairfield Inn, Hampton Inn, Holiday Inn Express, Homestead Guest Studios, KeyWest Suites, Marriott, Sheraton, Studio Inn, bank, cinema

17 Acton Rd, **N**...Texaco/diesel/mart/24hr, BBQ, Krystal, McDonald's

15b a I-65, N to Birmingham, S to Montgomery

13 US 31, Hoover, Pelham, **N**...DENTIST, MEDICAL CARE, VETERINARIAN, Amoco/mart, Chevron, Citgo, Shell, Texaco, BBQ, Chick-fil-A, Dairy Queen, Kenny Rogers, LongHorn Steaks, McDonald's, Salvatore's Pizza, Schlotsky's, Shoney's, SteakOut, Subway, Comfort Inn, Day's Inn, Hampton Inn, Holiday Inn, RiverChase Inn, Acura, Books-A-Million, Bruno's Foods, Buick, Burlington Coats, Cadillac, Chevrolet, Chrysler/Plymouth/Jeep, Circuit City, Eckerd, Firestone, Goodyear, Hancock Fabrics, Honda, Kia, Luxury Linens, Midas Muffler, Mitsubishi, Mr Transmissions, Spiffy's Carwash, TJ Maxx, Toyota, US Auto Repair, XpressLube, auto parts, bank, **S**...Crown Gas, Shell/diesel/mart/24hr, Circle K, BBQ, Burger King, CiCi's, McDonald's, Olive Garden, Omelet Shoppe, Piccadilly's, Sun's Chinese, Taco Bell, Wendy's, Courtyard, Winfrey Hotel, Barnes&Noble, Bruno's Foods, CVS Drug, Gateway Computers, GNC, Goody's, Infiniti, KidsRUs, K-Mart, Marshall's, MensWearhouse, Mercedes, Michael's, Parisian, Pier 1, ToysRUs, Wal-Mart(1mi), cleaners, mall

10 AL 150, Waverly, **S**...VETERINARIAN, BP/diesel/mart/wash/24hr, Exxon, Mei China, AmeriSuites(1mi), GNC, Walgreen, Winn-Dixie/deli

6 AL 52, to Bessemer, **N**...Amoco, **S**...MEDICAL CARE, BP/mart/wash/24hr, Exxon/mart/wash/24hr, Shell/mart, Texaco/mart, Arby's, McDonald's, Pizza Hut, Taco Bell, Waffle House, Wendy's, Sleep Inn, CVS Drug, Winn-Dixie/deli/24hr, airport, bank

1 AL 18, Bessemer, **N**...Citgo/diesel/mart, **S**...Amoco/diesel/mart/24hr, to Tannehill SP

0mm I-459 begins/ends on I-20/59, exit 106.

10

Arizona Interstate 8

Exit #	Services
178b a	I-10, E to Tucson, W to Phoenix. I-8 begins/ends on I-10, exit 199.
174	Trekell Rd, to Casa Grande, **2-4 mi N**...HOSPITAL, gas, food, lodging
172	Thornton Rd, to Casa Grande, **2-4 mi N**...Sizzler, Best Western, Holiday Inn, Super 8, **S**...Francisco Grande Resort
171mm	Santa Cruz River
169	Bianco Rd, no facilities
167	Montgomery Rd, no facilities
163mm	Santa Rosa Wash
161	Stanfield Rd, no facilities
151	AZ 84 E, Maricopa Rd, to Stanfield, **S**...Pullman Motel/diesel/rest., RV camping
150mm	picnic area wb, picnic tables, litter barrels
149mm	picnic area eb, picnic tables, litter barrels
144	Vekol Rd, no facilities
140	Freeman Rd, no facilities
119	Butterfield Trail, to AZ 85, I-10, Gila Bend, **N**...Chevron/diesel, Texaco/diesel/mart/RV dump, American/Mexican Café, Exit West Café, Subway(3mi), Best Western/rest.(3mi), Super 8, RV camping, laundry
117mm	Sand Tank Wash
115	AZ 85, to Gila Bend, **1-2 mi N**...HOSPITAL, Chevron/diesel/mart, Gas4Less, Exxon/mart, 76/Circle K, Texaco/diesel/mart, A&W, Dairy Queen, McDonald's, Best Western, Desert Rest Motel, El Coronado Motel, Pay Less Inn, Wheel Inn, Yucca Motel, Big A Parts, Goodyear/auto, NAPA, SavMart Foods, bank, laundry, RV park, tires
111	Citrus Valley Rd, no facilities
106	Paloma Rd, no facilities
102	Painted Rock Rd, no facilities
87	Sentinel, **N**...Sentinel Gen Store/gas, RV camping, towing
85mm	**rest area wb, full(handicapped)facilities, phone, picnic tables, litter barrels, vending, petwalk**
84mm	**rest area eb, full(handicapped)facilities, phone, picnic tables, litter barrels, vending, petwalk**
78	Spot Rd, no facilities
73	Aztec, no facilities
67	Dateland, **N**...RV camping, dates, **S**...Exxon/mart/24hr, Oasis RV Park/dump
56mm	**rest area both lanes, full(handicapped)facilities, phone, picnic tables, litter barrels, vending, petwalk**
54	Ave 52 E, Mohawk Valley, no facilities
42	Ave 40 E, to Tacna, **N**...Chevron/diesel/mart, Chaparral Motel
37	Ave 36 E, to Roll, no facilities
30	Ave 29 E, Wellton, **N**...MEDICAL CARE, 76/Circle K, Tier Drop RV Park, auto parts/repair
24mm	Ligurta Wash
23mm	Red Top Wash
22mm	parking area both lanes
21	Dome Valley, **N**...Ligurta Sta Café, RV park
17mm	insp sta eb
15mm	Fortuna Wash
14	Foothills Blvd, **N**...RV Park, **S**...Foothills/diesel/food, Domino's, bank, RV camping, tires
12	Fortuna Rd, to US 95 N, **N**...Chevron/mart/24hr, Shamrock/Barney's/diesel/rest./24hr, Texaco/diesel/mart, Copper Miner's Rest./24hr, Jack-in-the-Box, Pizza Hut, Caravan Oasis Motel/RV camping/LP, **S**...OPTICAL CARE, Texaco/Hardee's/diesel/mart, Dairy Queen, Don Quijote Rest., Subway, IGA Food/76, cleaners, tires
9	32nd St, to Yuma, **N**...Sun Vista RV Park, **7 mi S**...Carrow's Rest., Furr's Cafeteria, Holiday Inn Express, Ramada Inn, Travelodge/rest., MailBoxes Etc, PepBoys, Walgreen, **access to services on 4th St**
7	Araby Rd, **S**...76/Circle K, diesel repair, RV Ctr, RV camping, to AZWU
3	AZ 280 S, Ave 3E, **N**...Texaco/Dairy Queen/diesel/mart, **S**...Big A Parts, **3 mi S**...KFC, Pizza Hut, Wendy's(6mi), to Marine Corp Air Sta, airport
2	US 95, 16th St, Yuma, **N**...76/mart, Texaco/diesel/café/mart, Cracker Barrel, Denny's, Best Western/rest., Day's Inn, La Fuente Inn, Motel 6, Shilo Inn/rest., CarQuest, auto/tire repair, **S**...HOSPITAL, VETERINARIAN, Chevron, Ultramar/gas, Texaco/diesel/mart, Baskin-Robbins, Burger King, China Boy, Jack-in-the-Box, Johnny's Grill, McDonald's, Red Lobster, Shoney's, Taco Bell, Village Inn Pizza, Comfort Inn, Interstate 8 Inn, Motel 6, Super 8, Chief Auto Parts, Radio Shack, Staples, SW Supermkts, U-Haul, bank, **services on 4th Ave S**...Chevron/mart, 76/Circle K, Shell, Applebee's, Arby's, Bella Italian, Bobby's Rest., Carl's Jr, Church's, Dairy Queen, Del Taco, Golden Corral, HomeTown Buffet, JB's, KFC, Little Caesar's, LJ Silver, McDonald's, Mi Rancho Mexican, Pizza Hut, Rally's, Rocky's Pizza, Schlotsky's, Wendy's, Wienerschnitzel, El Rancho Motel, Holiday Inn Express, Radisson, Royal Motel, Travelodge, Tropicana Motel, Yuma Cabana Motel, Albertson's, AutoZone, Barnes&Noble, Checker Parts, Dillard's, Econo Lube'n Tune, Hastings Books, JC Penney, Mervyn's, Michael's, Midas, Pets+, Rite Aid, Sears/auto, Smith's/24hr, Target, XpressLube, hardware, mall

E
W

Yuma

Arizona Interstate 8

1.5mm	weigh sta both lanes
1	Giss Pkwy, Yuma, **N...**to Yuma Terr Prison SP, **S...**Best Western
0mm	Colorado River, Arizona/California state line, Mountain/Pacific time zone

Arizona Interstate 10

Exit #	Services
391mm	Arizona/New Mexico state line
390	Cavot Rd, no facilities
389mm	**rest area both lanes, full(handicapped)facilities, phone, picnic tables, litter barrels, vending, petwalk**
383mm	weigh sta eb, wiegh/insp sta wb
382	Portal Rd, San Simon, **2 mi N...**gas, food, lodging
381.5mm	San Simon River
378	Lp 10, San Simon, **N...**4K/Chevron/Chester Fried Chicken/diesel/24hr, Texaco/Kactus Kafe/diesel/mart, auto/diesel/RV repair, lodging
366	Lp 10, Bowie Rd, **N...**Texaco/Subway/Baskin-Robbins/mart/24hr, **S...**Alaskan RV park
362	Lp 10, Bowie, **N...**gas, lodging, camping, **S...**to Ft Bowie NHS
355	US 191 N, to Safford, **N...**to Roper Lake SP
352	US 191 N, to Safford, same as 355
344	Lp 10, to Willcox, **S...**Willcox Diesel Service/truck repair
340	AZ 186, to Rex Allen Dr, Ft Grant Field, **N...**Rip Griffin/Texaco/Subway/Taco Bell/diesel/24hr, Super 8, Magic Circle RV Park, Stout's CiderMill, **S...**HOSPITAL, Chevron/diesel/mart/24hr, Mobil/diesel/mart, 76/Circle K, Burger King, KFC, La Kachina Mexican, McDonald's, Pizza Express, Pizza Hut, Plaza Rest., Best Western, Day's Inn, Motel 6, IGA Foods, Radio Shack, Safeway, Grande Vista RV Park, Dick's Repair, tires, to Chiricahua NM
336	AZ 186, Willcox, **S...**Chevron/diesel/LP/mart, **1-3 mi S...**Cactus Kitchen Rest., Regal Rest., Desert Inn Motel/RV, Royal Western Lodge, Ft Willcox RV Park
331	US 191 S, to Sunsites, Douglas, **S...**to Cochise Stronghold, no facilities
322	Johnson Rd, **S...**Citgo/Dairy Queen/diesel/mart/gifts
320mm	**rest area both lanes, full(handicapped)facilities, phone, picnic tables, litter barrels, vending, petwalk**
318	Triangle T Rd, to Dragoon, lodging, camping
312	Sibyl Rd, **S...**Stuckey's/gas/RV park
309.5mm	Adams Peak Wash
306	Pomerene Rd, Benson, **N...**Sun Country RV Park, **1 mi S...**Texaco, Pato Blanco RV Park
305.5mm	San Pedro River
304	Ocotillo St, Benson, **N...**Denny's, Day's Inn, Super 8, Benson RV Park, KOA, Red Barn RV Park, **S...**HOSPITAL, Chevron/mart, Berryhill Rest., Burger King, Chinese Rest., Family Inn Rest., Plaza Rest., Wendy's, Best Western, QuarterHorse Inn, NAPA, Safeway, bank, hardware, laundry
303	US 80, to Tombstone, Bisbee, **1 mi S...**Texaco/diesel/mart, Beijing Chinese, Los Amigo's Café, Reb's Rest., Shoot Out Steaks, Wendy's, NAPA, Safeway, auto/diesel/repair, RV camping, to Douglas NHL
302	AZ 90 S, to Ft Huachuca, Benson, **S...**Gas City/A&W/Pizza Hut/TCBY/diesel/mart, Shell/Subway/diesel/mart, KFC/Taco Bell, McDonald's, Holiday Inn Express, Motel 6, Cochise Terrace RV Park, Ft Huachuca NHS
301	ranch exit eb
299	Skyline Rd, no facilities
297	Mescal Rd, J-6 Ranch Rd, **N...**QuickPic/diesel/deli/mart/24hr
292	Empirita Rd, no facilities
289	Marsh Station Rd, no facilities
288mm	Cienega Creek
281	AZ 83 S, to Patagonia, no facilities
279	Vail/Wentworth Rd, **N...**Vail Steakhouse, winery, to Colossal Caves
275	Houghton Rd, **N...**to Saguaro NM, camping, **S...**to fairgrounds
273	Rita Rd, **S...**fairgrounds
270	Kolb Rd, **N...**Chevron/diesel/mart, **S...**Voyager RV Resort
269	Wilmot Rd, **N...**Chevron, Texaco/diesel, Comfort Inn, Travel Inn, RV camping, **S...**RV park
268	Craycroft Rd, **N...**TTT/diesel/rest./atm/24hr, 76/Circle K, **S...**diesel repair, tires
267	Valencia Rd, **N...**Pima Air Museum, **S...**airport
265	Alvernon Way, Davis-Monthan AFB, no facilities
264b a	Palo Verde Rd, **N...**Carl's Jr, Crossland Suites, Day's Inn, Fairfield Inn, Holiday Inn, Red Roof Inn, **S...**Arby's, McDonald's, Motel 6, Ramada Inn, Studio 6 Suites, Beaudry RV Ctr, Factory Outlet/famous brands, RV park
263b	Kino Pkwy N, no facilities

W i l l c o x   B e n s o n

E

W

Arizona Interstate 10

a	Kino Pkwy S, **S**...Shamrock/mart, Fry's Food/drug, to Tucson Intn'l Airport
262	Benson Hwy, Park Ave, **S**...Arco/mart/24hr, Chevron, Texaco/mart, JB's, McDonald's, Navaronne Rest., Silver Saddle Steaks, Waffle House, Howard Johnson, Lazy 8 Motel, Motel 6, Rodeway Inn
261	6th/4th Ave, **N**...Chevron, Budget Inn, Econolodge, **S**...Burger King, Carrow's Rest., Dunkin Donuts, Silver Saddle Steaks, Rodeway Inn, Sandman Inn, Lazy 8 Motel, Spanish Trail
260	I-19 S, to Nogales
259	22nd St, Starr Pass Blvd, **N**...Circle K/gas, EZ 8 Motel, tires, **S**...76, Texaco, Kettle, Waffle House, Comfort Inn, Holiday Inn Express, Howard Johnson, Motel 6, Super 8, Travel Inn
258	Congress St, Broadway St, **N**...Texaco/diesel, Circle K/76, Sizzler, Vittles Rest., Holiday Inn, Motel 6, Quality Hotel Suites, Ramada Inn, Travelodge, **S**...Bennigan's, Carl's Jr, Days Inn, Desert Inn, Sheraton
257a	St Mary's Rd, **N**...HOSPITAL, 76/diesel, **S**...Texaco/24hr, Burger King, Denny's, Furr's Cafeteria, Jack-in-the-Box, La Quinta, Pim Comm Coll
257	Speedway Blvd, **N**...HOSPITAL, 7-11/24hr, Best Western, Gin's Oil Express, Desert Museum, Old Town Tucson, U of AZ, **S**...Arco/diesel/mart/24hr
256	Grant Rd, **N**...Sonic, diesel/transmission repair, **S**...Exxon, 76/Circle K, Texaco, Del Taco, IHOP, Subway, Waffle House, Baymont Inn, Hampton Inn, Motel 6, Rodeway Inn, Shoney's Inn, **1 mi S**...VETERINARIAN, Circle K, McDonald's, Walgreen, groceries, hardware, tires
255	AZ 77 N, to Miracle Mile
254	Prince Rd, **N**...Circle K, U-Haul, tires, **S**...Texaco, Prince of Tucson RV Park, golf
251	Sunset Rd, **N**...auto parts
250	Orange Grove Rd, **N**...DENTIST, Arco/mart/24hr, Circle K/76, Wendy's, carwash, **N on Thornydale**...CHIROPRACTOR, VETERINARIAN, Costco/gas, Casa Molina Mexican, Little Caesar's, AutoNation USA, Big O Tire, Econo Lube'n Tune, Home Depot, PetsMart
248	Ina Rd, **N**...CHIROPRACTOR, OPTOMETRIST, Chevron/diesel/mart, Conoco/diesel/mart, Exxon/Circle K, Arby's, Burger King, Carl's Jr, Chuy's Mesquite Grill, Dairy Queen, Donut Wheel, Keukendutch Rest., LJ Silver, McDonald's, Perkins, Pizza Hut, Taco Bell, Waffle House, Motel 6, CarQuest, Discount Tire, Fry's Food/drug, Hancock Fabrics, JiffyLube, K-Mart/auto, Michael's, Office Depot, OfficeMax, Osco Drug, PepBoys, Rubio's Grill, Target, **S**...Exxon/Circle K, Texaco/diesel/mart, Denny's, Comfort Inn, Holiday Inn Express, Red Roof Inn
246	Cortaro Rd, **S**...Shell/diesel/mart, Burger King, Cracker Barrel, McDonald's, Day's Inn, Quality Inn, Super 8, USPO, golf, access to RV camping
243	Avra Valley Rd, **S**...Saguaro NM(13mi), RV camping, airport
240	Tangerine Rd, to Rillito, **N**...A_A RV Park, **S**...Miller's Mkt, USPO
236	Marana, **S**...Chevron/diesel/LP/mart, 76/Circle K, Mexican Rest., Sun RV Park, auto repair
232	Pinal Air Park Rd, **S**...Pinal Air Park
228mm	wb pulloff, to frontage rd
226	Red Rock, **S**...USPO
219	Picacho Peak Rd, **N**...Citgo/Dairy Queen/diesel, **S**...AZ Nuthouse, Peak Trading Post/gas, Pichaco Peak RV Park, to Picacho Peak SP
212	Picacho(from wb), **N**...Exxon/LP, Picacho Motel/rest., RV camping
211b	AZ 87 N, AZ 84 W, to Coolidge
a	Picacho(from eb), **S**...KOA, state prison
208	Sunshine Blvd, to Eloy, **S**...Flying J/Conoco/diesel/LP/rest./24hr, camping, funpark
203	Toltec Rd, to Eloy, **N**...Chevron, Circle K/diesel/mart, Carl's Jr, McDonald's, Mexican Rest., Waffle House, Super 8, Toltec Inn, Blue Beacon, diesel/tire repair, **S**...Exxon/diesel/mart, TA/A&W/Taco Bell/diesel/24hr, Pizza Hut
200	Sunland Gin Rd, Arizona City, **N**...Petro/Mobil/diesel/mart/rest./24hr, Pilot/Subway/diesel/mart/24hr, Burger King, Day's Inn, Sunland Inn/rest., Blue Beacon, **S**...Love's/Arby's/Baskin-Robbins/diesel/mart/24hr, Texaco/repair, Golden 9 Rest., Motel 6
199	I-8 W, to Yuma, San Diego
198	AZ 84, to Eloy, Casa Grande, **N**...Sunland Inn, **S**...Wendy's, Tanger Outlet Ctr/famous brands, Buena Tierra RV Pk, **4 mi S**...Sizzler, Best Western, Holiday Inn, Sunland Inn

E

W

Arizona Interstate 10

Casa Grande

194   AZ 287, to Casa Grande, **S**...HOSPITAL, Arco/diesel/mart/24hr, Chevron/Dairy Queen/mart, Burger King, Cracker Barrel, Del Taco, Denny's, Comfort Inn, Super 8, Factory Stores/famous brands, **2 mi S**...EYECARE, Chevron, 76/Circle K, Arby's, China King, Church's Chicken, Filberto's Mexican, Golden Corral, Italian Pizza, Jack-in-the-Box, JB's, Little Caesar's, McDonald's, Peter Piper Pizza, Sizzler, Subway, Wendy's, Best Western, Holiday Inn, Albertson's, Basha's Foods, BrakeMasters, Firestone/auto, Fry's Food/drug, Goodyear/auto, Grease'n Go, JC Penney, K-Mart, Radio Shack, SW SuperMkts, Walgreen, Wal-Mart, bank, hardware, to Casa Grande Ruins NM

190   McCartney Rd, **N**...to Central AZ Coll

185   AZ 387, to Casa Grande, Casa Grande Ruins NM, **2 mi S**...**Val Vista Winter Village RV Park(E on Val Vista Blvd, 1/2 mi)**, **9 mi S**...gas, Sizzler, Holiday Inn, Francisco Grande Golf Resort(14mi)

183mm   **rest area wb, full(handicapped)facilities, phone, picnic tables, litter barrels, vending, petwalk**

181mm   **rest area eb, full(handicapped)facilities, phone, picnic tables, litter barrels, vending, petwalk**

175   AZ 587 N, Casa Blanca Rd, **S**...Casa Blanca Gas/mart, Casa Blanca RV Camping

173mm   Gila River

167   Riggs Rd, to Sun Lake, **1 mi N**...Dairy Queen

164   AZ 347 S, Queen Creek Rd, to Maricopa, **N**...to Chandler Airport

162b a   Maricopa Rd N, **S**...Firebird Sports Park, Gila River Casino

Chandler

160   Chandler Blvd, to Chandler, **N**...CHIROPRACTOR, Exxon/Blimpie/Circle K, Mobil/mart, Damon's, Denny's, Marie Callender's, Perkins, Sizzler, Whataburger/24hr, Fairfield Inn, Hampton Inn, Homewood Suites, Red Roof Inn, Southgate Motel, Super 8, Wyndham Garden, Firestone/auto, Grease'n Go, to Compadre Stadium, to Williams AFB, **S**...HOSPITAL, Chevron/mart/wash, Citgo/7-11, 76/Circle K, Cracker Barrel, Holiday Inn Express, La Quinta, Wellesley Inn

159   Ray Rd, **N**...76/Circle K, Texaco/diesel/mart, Carrabba's, Charleston's Rest., McDonald's/playplace, Outback Steaks, AutoNation USA, Ford, Home Depot, **S**...CHIROPRACTOR, DENTIST, EYEMASTERS, URGENT CARE, 76/Circle K/diesel, BBQ, Boston Mkt, Chicago Deli, IHOP/24hr, Jack-in-the-Box, Keegan's Grill, Macaroni Grill, Mimi's Café, On-the-Border, Peter Piper Pizza, Pizza Hut, Ruby Tuesday, Subway, Sweet Tomato Buffet, ValleLuna Cantina, Wendy's, Extended Stay America, Albertson's/24hr, BabysRUs, Barnes&Noble, Best Buy, BrakeMasters, FootHills Carwash, HomePlace, Jo-Ann Fabrics, K-Mart/drugs, MailBoxes Etc, Mervyn's, Michael's, OfficeMax, PetCo, Ross, Smith's Food/drug/24hr, SteinMart, SunDevil AutoCare, Target, bank, cinema, cleaners

158   Warner Rd, **N**...76/Circle K, **S**...VETERINARIAN, CHIROPRACTOR, DENTIST, OPTICAL CARE, Arco/mart/24hr, 76/Circle K, Burger King, Coffee Plantation, Dairy Queen, Malaya Mexican, McDonald's/playplace, Ruffino's Italian, Taco Bell, Basha's Foods, Goodyear/auto, JiffyLube, MailBoxes Etc, Osco Drug, Wild Oats Foods, cleaners, hardware, tires

157   Elliot Rd, **N**...Circle K/gas, Applebee's, Arby's, BlackEyed Pea, Burger King, Chili's, Chinese Rest., Coco's, Country Harvest Buffet, Eastside Mario's, Fuddrucker's, Kenny Rogers, Kyoto Japanese, Mi Amigos Mexican, Olive Garden, Red Robin Rest., Schlotsky's, Sooper Salad, Shoney's, Subway, Taco Bell, Village Inn Rest., Wendy's, Country Suites, Cadillac/GMC/Olds, Circuit City, Costco, Discount Tire, Dodge, Fiddlesticks Funpark, Ford/Lincoln/Mercury, GNC, HomeBase, Honda, Midas Muffler, Nissan, OfficeMax, PetsMart, Saturn, Sports Authority, Staples, Toyota, U-Haul, Wal-Mart, cleaners, **S**...MEDICAL CARE, DENTIST, CHIROPRACTOR, VETERINARIAN, Mobil/diesel/mart, 76/Circle K, Baskin-Robbins, Burrito Co, China Dragon, KFC, McDonald's, Peter Piper Pizza, Pizza Hut, Sub Factory, Taco John's, Best Western, Quality Inn, Checker Parts, 1-Hr Photo, Radio Shack, Walgreen, bank, cinema, cleaners

155   Baseline Rd, Guadalupe, **N**...Mobil, 76/Circle K, Texaco/diesel/mart, Carl's Jr, Jack-in-the-Box, Rusty Pelican Seafood, Shoney's, Waffle House, AmeriSuites, Candlewood Studios, Holiday Inn Express, InnSuites, Red Roof Inn, Residence Inn, Travelers Inn, AZ Mills Factory Shops, Burlington Coats, IMAX Cinema, JC Penney Outlet, K-Mart/Little Caesar's, Marshall's, Pro Auto Parts, Savco Drugs, cleaners/laundry, **S**...Aunt Chilada's Mexican, Denny's, Homestead Village, Fry's Electronics

Phoenix

154   US 60 E, AZ 360, Superstition Frwy, to Mesa, **N**...to Camping World(off Mesa Dr)

153b a   AZ 143 N, Broadway Rd E, to Sky Harbor Airport, **N**...Denny's, Courtyard, Fairfield Inn, Hilton, La Quinta, MainStay Suites, Ramada Inn, Red Roof Inn, Sleep Inn, to Diablo Stadium, **S**...Texaco/A&W/Taco Bell/mart/24hr, Del Taco, Hampton Inn

152   40th St, **N**...Cardlock/diesel/mart, Texaco/diesel/mart, SubMachine, U of Phoenix, **S**...76/Taco Bell/Circle K, Burger King, bank

151mm   Salt River

151b a   28th St, 32nd St, University Ave, **N**...Radisson, Extended Stay America, La Quinta, bank, AZSU, U of Phoenix, **S**...76/Circle K

150b   24th St E(from wb), **N**...Air Nat Guard, Exxon, Durado's Rest., Golden 9 Motel/rest., Knight's Inn, Motel 6, Rodeway Inn, **S**...Best Western/rest.

E

↑

↓

W

Arizona Interstate 10

a	I-17 N, to Flagstaff
149	Buckeye Rd, **N…**Sky Harbor Airport
148	Washington St, Jefferson St, **N…**Motel 6, to Sky Harbor Airport, **S…**HOSPITAL, 76/Circle K, Carl's Jr, McDonald's
147b a	AZ 51 N, AZ 202 E, to Squaw Peak Pkwy, no facilities
146	16th St, **S…**76/Circle K, Shell, KFC, Whataburger
145	7th St, **S…**HOSPITAL, 76/Circle K, Texaco/diesel/mart, Olds, America West Arena
144	7th Ave, Amtrak, central bus dist
143c	19th Ave, US 60, no facilities
b a	I-17, N to Flagstaff, S to Phoenix
142	27th Ave(from eb, no return), **N…**Circle K, 7-11, Comfort Inn, **S…**Pacific Pride/diesel
141	35th Ave, **N…**Circle K, Jack-in-the-Box, Rita's Mexican, **S…**Texaco/mart
140	43rd Ave, **N…**Citgo/7-11, Exxon/CircleK/diesel, Mobil/mart/wash, KFC, Little Caesar's, Mexican Seafood, Smitty's Foods, Subway, Wendy's, Food City, Midas Muffler, Radio Shack, Walgreen
139	51st Ave, **N…**Chevron/diesel/mart/wash, 76/Circle K, Citgo/7-11, Domino's, Little China, McDonald's, Sonic, Waffle House, Holiday Inn, Motel 6, Red Roof Inn, Discount Tire, Goodyear/auto, JiffyLube, **S…**Carl's Jr, IHOP, Taco Bell, Fairfield Inn, Hampton Inn, Travelers Inn
138	59th Ave, **N…**76/Circle K, Shamrock/mart, 7-11, Armando's Mexican/24hr, AutoZone, **S…**Waffle House, Whataburger/24hr
137	67th Ave, **N…**76/Circle K, Texaco/mart/wash, **S…**Flying J/diesel/LP/rest./24hr
136	75th Ave, **N…**Chevron/mart/wash, 76/Circle K, Denny's, Taco Bell, Whataburger, Fashion Bug, Ford, Home Depot, PetsMart, Wal-Mart/drugs, **S…**Arco/mart/24hr
135	83rd Ave, **N…**76/Circle K, Arby's, Burger King, Jack-in-the-Box, Waffle House, Econolodge, K-Mart, Sam's Club
134	91st Ave, Tolleson, no facilities
133a	99th Ave, no facilities
132	107th Ave(from eb), no facilities
131	115th Ave, to Cashion, **S…**to Phoenix International Raceway
129	Dysart Rd, to Avondale, **N…**Wal-Mart, **S…**Baskin-Robbins, Jerry's Rest., KFC, McDonald's/playplace, Subway, Waffle House, Whataburger, Comfort Inn, Super 8, Walgreen, Chevrolet, Chrysler/Plymouth/Dodge/Jeep, tires
128	Litchfield Rd, **N…**Mobil/Blimpie/diesel/mart, Cracker Barrel, Denny's, Wendy's, Holiday Inn Express, Wigwam Outlet Stores/famous brands, Wigwam Resort(3mi), to Luke AFB, **S…**DENTIST, MEDICAL CARE, Chevron/mart, Mobil/mart, Arby's, Burger King, JB's, Schlotsky's, Taco Bell, Best Western, Albertson's, Checker Parts, Fry's Food/drug, Goodyear/auto, Pontiac/GMC, Radio Shack
126	Pebble Creek Pkwy, to Estrella Park, no facilities
125mm	Roosevelt Canal
124	Cotton Lane, to AZ 303, **N…**st prison, **S…**KOA
123	Perryville, no facilities
121	Jackrabbit Trail, **S…**Circle K/gas
114	Miller Rd, to Buckeye, **S…**Love's/Baskin-Robbins/Taco Bell/diesel/mart/24hr, A&W, Burger King, Leaf Verde RV Park
112	AZ 85, to I-8, Gila Bend, no facilities
109	Sun Valley Pkwy, Palo Verde Rd, no facilities
104mm	Hassayampa River
103	339th Ave, **S…**Rip Griffin/Texaco/Subway/Pizza Hut/diesel/mart/LP/24hr, truckwash
98	Wintersburg Rd, **S…**to Palo Verde Nuclear Sta
97mm	Coyote Wash
95.5mm	Old Camp Wash
94	411th Ave, Tonopah, **S…**Chevron/diesel/mart, Texaco/Subway/diesel/mart/LP/24hr, Joe&Alice's Café/24hr, Mineral Wells Motel, USPO, tires, Saddle Mtn RV Park
81	Salome Rd, Harquahala Valley Rd, no facilities
69	Ave 75E, no facilities
53	Hovatter Rd, no facilities
52mm	**rest area both lanes, full(handicapped)facilities, phone, vending, picnic tables, litter barrels, petwalk**
45	Vicksburg Rd, **S…**Tomahawk/diesel/rest./24hr, Jobski's Diesel Repair/towing
31	US 60 E, to Wickenburg, no facilities

E ↕ W

P h o e n i x

Arizona Interstate 10

26	Gold Nugget Rd, no facilities
19	Quartzsite, to US 95, Yuma, **N**...Chevron, Texaco/mart(1mi), RV camping
18mm	Tyson Wash
17	to US 95, Quartzsite, **N**...Mobil/Burger King/LP/mart, Pilot/Subway/Dairy Queen/diesel/mart/24hr, Shell/diesel/mart/24hr, Chinese Rest., McDonald's, Wendy's, RV camping, **S**...Love's/A&W/Baskin-Robbins/Taco Bell/diesel/mart/24hr, Desert Gardens RV Park
11	Dome Rock Rd, no facilities
5	Tom Wells Rd, **N**...Beacon/diesel/mart/24hr
4.5mm	**rest area both lanes, full(handicapped)facilities, phone, vending, picnic tables, litter barrels, petwalk**
3.5mm	**eb**...AZ Port of Entry, **wb**...weigh sta
1	Ehrenberg, to Parker, **S**...Flying J/Wendy's/Cookery/diesel/mart/LP/24hr, Best Western
0mm	Colorado River, Arizona/California state line, Mountain/Pacific time zone

E ↕ W

Arizona Interstate 15

Exit #	Services
29.5mm	Arizona/Utah state line
27	Black Rock Rd, no facilities
21mm	turnout sb
18	Cedar Pocket, Virgin River Canyon RA, **S**...**rest area both lanes, full(handicapped)facilities, phone, picnic tables, litter barrel, camping, petwalk**
16mm	truck parking both lanes
15mm	truck parking nb
14mm	truck parking nb
10mm	truck parking nb
9	Farm Rd, **W**...phone, tire repair
8.5mm	Virgin River
8	Littlefield, Beaver Dam, **1 mi W**...gas, food, lodging, camping
0mm	Arizona/Nevada state line, Pacific/Mountain time zone

N ↕ S

Arizona Interstate 17

Exit #	Services
341	McConnell Dr, I-17 begins/ends. **See Arizona Interstate 40, exit 195b.**
340b a	I-40, E to Gallup, W to Kingman
339	Lake Mary Rd(from nb), Mormon Lake, access to same as 341. **E**...Texaco/diesel/mart, KOA
337	US 89A S, Oak Creek Canyon, to Sedona, Ft Tuthill RA, airport, camping
333	Kachina Blvd, Mountainaire Rd, **W**...76/Subway/diesel/mart
331	Kelly Canyon Rd, no facilities
328	Newman Park Rd, no facilities
326	Willard Springs Rd, no facilities
324mm	**rest area both lanes, full(handicapped)facilities, phone, picnic tables, litter barrels, vending, petwalk**
322	Pinewood Rd, to Munds Park, **E**...Chevron/diesel/mart, Shell/diesel/mart, Buckley's Rest., Lone Pine Rest., Motel in the Pines/RV camp, USPO, **W**...Exxon/mart, Munds RV Park, RV repair
322mm	Munds Canyon
320	Schnebly Hill Rd, no facilities
317	Fox Ranch Rd, no facilities
316mm	Woods Canyon
315	Rocky Park Rd, no facilities
313mm	scenic view sb, litter barrels
306	Stoneman Lake Rd, no facilities
300mm	runaway truck ramp sb
298	AZ 179, to Sedona, Oak Creek Canyon, **15 mi W**...Burger King, Poco Café, Best Western, Sky Ranch Lodge, Wildflower Inn
297mm	**rest area both lanes, full(handicapped)facilities, phone, picnic tables, litter barrels, vending, petwalk**
293mm	Dry Beaver Creek
293	Cornville Rd, McGuireville Rd, to Rimrock, **E**...XpressFuel/mart/atm, McGuireville Café, Beaver Creek Golf Resort(3mi), antiques, auto repair, **W**...76/diesel/mart, Beaver Hollow Café
289	Middle Verde Rd, Camp Verde, **E**...Texaco/diesel/mart, Bo's Rest.(3mi), Best Western(4mi), Ft Verde Motel(4mi), Cliff Castle Casino, to Montezuma Castle NM, RV Park
288mm	Verde River

Flagstaff

N ↕ S

Arizona Interstate 17

287	AZ 260, to AZ 89A, Cottonwood, Payson, E...Shell/mart, Texaco/Subway/ Baskin-Robbins/diesel/mart/RVdump/LP/24hr, Bo's Rest., Burger King, Dairy Queen, Denny's, McDonald's, Pizza Hut/Taco Bell, Comfort Inn, Microtel, Super 8, RV Park, W...Chevron/diesel/mart/24hr, to Jerome SP, RV camping, 12-14 mi W...HOSPITAL, Jack-in-the-Box, Best Western, Quality Inn, to DeadHorse SP
285	Camp Verde, Gen Crook Tr, E...to Ft Verde SP
281mm	brake check area nb
278	AZ 169, Cherry Rd, to Prescott, no facilities
268	Dugas Rd, Orme Rd, no facilities
269mm	Ash Creek
265.5mm	Agua Fria River
262b a	AZ 69 N, Cordes Jct Rd, to Prescott, E...Chevron/mart/24hr, Texaco/ Subway/diesel/mart/24hr, McDonald's, Lights On Motel/RV Park, W...Papa's Place Steaks
262mm	Big Bug Creek
259	Bloody Basin Rd, to Crown King, Horse Thief Basin RA, no facilities
256	Badger Springs Rd, no facilities
252	**Sunset Point, W...scenic view/rest area both lanes, full(handicapped)facilities, phone, picnic tables, litter barrels, vending**
248	Bumble Bee, no facilities
244	Squaw Valley Rd, Black Canyon City, E...Chevron/diesel/24hr, Texaco/diesel, Squaw Peak Steaks, Greyhound Track, KOA
243.5mm	Agua Fria River
242	Rock Springs, Black Canyon City, E...KOA, W...Chevron/diesel/mart/24hr, Rock Springs Café, auto repair
239.5mm	Little Squaw Creek
239mm	Moore's Gulch
236	Table Mesa Rd, no facilities
232	New River, E...gas/mart
231.5mm	New River
229	Anthem Way, Desert Hills Rd, W...Chevron/Burger King/mart, Prime Outlets/famous brands/food court
227mm	Dead Man Wash
225	Pioneer Rd, W...Pioneer RV Park, museum
223	AZ 74, Carefree Hwy, to Wickenburg, W...gas/mart, Lake Pleasant Park, camping
219mm	Skunk Creek
218	Happy Valley Rd, no facilities
217	Pinnacle Peak Rd, E...Phoenix RV Park
217b	Deer Valley Rd, Rose Garden Ln, E...Exxon, 76/Circle K(1/2mi), Texaco/diesel/mart, Burger King, McDonald's, Sonic, Taco Bell, Wendy's, W...76/Circle K, Texaco/diesel/mart, Cracker Barrel, Denny's, Schlotsky's, Waffle House, Wendy's, Country Inn Suites, Day's Inn, Extended Stay America, N Phoenix RV Park
215	AZ 101 W
214b a	Yorkshire Dr, Union Hills Dr, E...76/Circle K, W...HOSPITAL, Circle K, ClaimJumper Rest., 5&Diner, Jack-in-the-Box, Macaroni Grill, Souper Salad, Subway, Tony Roma, Homestead Village, Sleep Inn, Wyndham Gardens, FashionQ, Kinko's, Michael's, OfficeMax, PetsMart, Radio Shack, Ross, Target, cinema
213	Union Hills Dr(from nb), same as 214
212b a	Bell Rd, Scottsdale, to Sun City, E...Chevron/diesel/mart/atm, Exxon, Mobil/mart, Texaco/diesel/LP/mart, Blimpie, Coco's, IHOP/24hr, Jack-in-the-Box, McDonald's, Sonic, Waffle House, Wendy's, Best Western, Comfort Inn, Fairfield Inn, Motel 6, Checker Parts, Chevrolet, Dodge, Drug Emporium, Ford/Lexus/Isuzu, Fry's Food/drug, Lincoln/Mercury, Meineke Muffler, Nissan/Infiniti, Pontiac/Buick/GMC, Smith's Food/drug/24hr, Toyota, U-Haul, W...CHIROPRACTOR, DENTIST, Chevron/wash, Mobil/mart/wash, 76/wash, Applebee's, Denny's, Good Egg Rest., HomeTown Buffet, Hooters, Kyoto Bowl Japanese, Que Pasa Mexican, Sizzler, Village Inn Rest., Fairfield Inn, Red Roof Inn, Albertson's Food/drugs, Fred Meyer, Sam's Club, bank, cinema
211	Greenway Rd, E...MEDICAL CARE, 7-11, Embassy Suites, La Quinta
210	Thunderbird Rd, E...Arco/mart, Exxon, 76/Circle K, Texaco, Barro's Pizza, Jack-in-the-Box, Magic Bowl Chinese, McDonald's, Schlotsky's, Wendy's, Home Depot, Osco Drug, Safeway, Walgreen, cleaners, W...Exxon/mart, Motel 6, Cowtown Boots
209	Cactus Rd, W...MEDICAL CARE, Chevron/mart/24hr, Citgo/7-11, Cousins Subs, Denny's, Ramada, Basha's Foods, Firestone, Mailboxes/24hr, antiques, cleaners

N

S

Scottsdale Phoenix

Arizona Interstate 17

208	Peoria Ave, **E...**Fajita's, LoneStar Steaks, Comfort Suites, Crowne Plaza Hotel, Homewood Suites, **W...**76/Circle K, Bennigan's, Black Angus, Burger King, China Gate, Coco's, El Torito, FlashBacks Café, Italian Oven, Mamma Mia's, Olive Garden, Red Lobster, Sizzler, Souper Salad, TGIFriday, Wendy's, Premier Inn, Wyndham MetroCtr, Barnes&Noble, Burlington Coats, Circuit City, Computer City, Firestone/auto, JiffyLube, Luxury Linens, Macy's, Michael's, Nevada Bob's, PetCo, Pier 1, Sears, Staples, bank, cinema, funpark, mall
208.5mm	Arizona Canal
207	Dunlap Ave, **E...**CHIROPRACTOR, 76/mart, Blimpie, Fuddrucker's, Peter Piper Pizza, Courtyard, Parkway Inn, Sheraton, Sierra Suites, SpringHill Suites, TownPlace Suites, Aamco, Firestone, bank, mall, **W...**Exxon/mart/wash/LP, Circle K, Bobby McGee's, Denny's, Schlotsky's, Subway, Circuit City, ClutchPro, Midas Muffler, OfficeMax, U-Haul, funpark
206	Northern Ave, **E...**CHIROPRACTOR, EYECARE, Mobil/mart, 76/Circle K, Texaco/diesel/mart, Boston Mkt, Burger King, Denny's/24hr, Marie Callender's, McDonald's, Papa John's, Pizza Hut, South China Buffet, Starbucks, Subway, Hampton Inn, Albertson's, Checker Parts, USPO, Walgreen, bank, cleaners, **W...**CHIROPRACTOR, VETERINARIAN, Arco/mart, Dairy Queen, Furr's Dining, Village Inn Rest., Winchell's, Motel 6, Residence Inn, Super 8, K-Mart, NAPA AutoCare, carwash
205	Glendale Ave, **E...**VETERINARIAN, Circle K/24hr, BBQ, El Maya Mexican, carwash, drugs, **W...**Exxon/diesel, Circle K, 7-11, Jack-in-the-Box, RV Park, to Luke AFB
204	Bethany Home Rd, **E...**CHIROPRACTOR, DENTIST, VETERINARIAN, Arco/mart/24hr, 76/Circle K, Brad's Fish'n Chips, Isberto's Mexican/24hr, McDonald's, Subway, Teriyaki Grill, Whataburger, Desert Mkt Foods, cleaners/laundry, drugs, **W...**VETERINARIAN, Chevron/diesel/mart, Exxon, JiffyLube, Savers Foods, transmissions/24hr, tune-up
203	Camelback Rd, **E...**Arco/mart/24hr, Circle K, Blimpie, Burger King, Church's Chicken, Country Boy's Rest./24hr, Denny's/24hr, Pizza Hut, Taco Villa, Buick, Chrysler/Plymouth/Jeep/Dodge, Volvo, Discount Tire, Firestone/auto, Checker Parts, NAPA, Walgreen, **W...**Mobil, QT/mart, Texaco/diesel/mart, Circle K, Dairy Queen, Denny's, Jack-in-the-Box, McDonald's, Sizzler, Taco Bell, Tacos Mexico, Treulich's Rest., Comfort Inn, AutoZone, Chevrolet, Firestone/auto, carwash, to Grand Canyon U
202	Indian School Rd, **E...**Arco/mart/24hr, Alberto's Mexican, Chinese Rest., Filiberto's Mexican/24hr, Jimmy's Rest., Pizza Hut, Pizza Mia, Subway, Albertson's Foods, Ace Hardware, RV Sales, **W...**MEDICAL CARE, Exxon, Texaco/mart, Circle K, 7-11, Hunter Steaks, JB's, Wendy's, Motel 6, Super 8, auto repair, bank, cinema
201	Thomas Rd, **E...**Circle K, Arby's, Denny's, Jack-in-the-Box, McDonald's, Roman's Pizza, Taco Bell/24hr, Day's Inn, EZ Inn, La Quinta, Just Brakes, PetCo, **W...**Burger King, Ford, NAPA
200b	McDowell Rd, Van Buren, **E...**Wendy's, Goodyear, **W...**Travelodge
a	I-10, W to LA, E to Phoenix
199b	Jefferson St(from sb), Adams St(from nb), Van Buren St(from nb), **E...**Circle K/76/24hr, Jack-in-the-Box, McDonald's, Mexican Rest., Econo Lube'n Tune, Sandman Motel, bank, laundry, to st capitol, **W...**Circle K/76, KFC, La Canasta Mexican, Pete's Fish'n Chips, Penny Pincher Parts, PepBoys, SW Foods
a	Grant St, no facilities
198	Buckeye Rd(from nb), **W...**RV Ctr
197	US 60, 19th Ave, Durango St, to st capitol, **E...**Whataburger/24hr
196	7th St, Central Ave, **E...**HOSPITAL, Amtrak Sta, Burger King
195b	7th St, Central Ave, **E...**HOSPITAL, Exxon/24hr, Circle K, Trailside Gas/diesel/mart/24hr, McDonald's/playplace, Taco Bell, EZ 8 Motel/rest., RV Rental
a	16th St(no EZ return from sb), **E...**Checker Parts, Food City, Smitty's Foods, Walgreen, to Sky Harbor Airport
194	I-10 W, to AZ 51, Squaw Peak Pkwy, to Sky Harbor Airport

I-17 begins/ends on I-10, exit 150a.

Phoenix

N ↕ S

Arizona Interstate 19

Exit #	Services
	I-19 uses kilometers (km)
101b a	I-10, E to El Paso, W to Phoenix. I-19 begins/ends on I-10, exit 260.
99	AZ 86, Ajo Way, **E...**MEDICAL CARE, Circle K/gas, Conoco, Dairy Queen, Eegee's Café, Hamburger Stand, Peter Piper Pizza, Pizza Hut, Subway, Taco Bell, El Campo Tires, Fry's Food/drug, GNC, Jo-Ann Fabrics, Mervyn's, Osco Drug, Radio Shack, U-Haul, Walgreen, bank, cleaners, repair, **W...**HOSPITAL, Chevron/diesel/mart/wash/24hr, Conoco, 76/Circle K, Burger King, Abco Foods, RV camping, to Old Tucson, Desert Museum, carwash
98	Irvington Rd, **E...**Arco/mart/24hr, Little Mexico Rest., Fry's Food/drug, **W...**McDonald's, Home Depot
95b a	Valencia Rd, **E...**DENTIST, Conoco, Donut Wheel, Eegee's Café, Jack-in-the-Box, KFC, Little Caesar's, McDonald's, Pizza Hut, Sonic, Taco Loco, Yokohama RiceBowl, Abco Foods, AutoZone, BrakeMasters, Checker Parts, JiffyLube, Radio Shack, USPO, airport, **W...**CHIROPRACTOR, MEDICAL CARE, DENTIST, Chevron/mart, 76/Circle K, Applebee's, Arby's, Burger King, Carl's Jr, Chuy's Mesquite Grill, Denny's, Golden Corral, IHOP, Jalapeno's Mexican, Papa John's, Pizza Hut, Taco Bell, Wendy's, Abco Foods/24hr, Big O Tires, CarQuest, K-Mart/auto, Walgreen, Wal-Mart SuperCtr/24hr, bank, repair, transmissions, **2 mi W...**Arco/mart/24hr, Texaco/mart, KFC, McDonald's, AutoZone, Checker's Parts, Osco Drug, Radio Shack, Safeway
92	San Xavier Rd, **W...**to San Xavier Mission

Tucson

N ↕ S

Arizona Interstate 19

91.5km	Santa Cruz River
87	Papago Rd, no facilities, emergency phones sb
80	Pima Mine Rd(no sb return), no facilities
75	Helmut Peak Rd, to Sahuarita, no facilities
69	US 89 N, Green Valley, **E**...Denny's, Pizza Hut, Schlotsky's, Basha's Food/deli/bakery, Radio Shack, Wal-Mart, RV camping, bank, **W**...DENTIST, VETERINARIAN, 76/mart/wash, Burger King, Dairy Queen, Taco Bell/TCBY, Holiday Inn Express, Big O Tire, Ford/Lincoln/Mercury, Titan Missile Museum, carwash, cinema, RV resort
65	Esperanza Blvd, to Green Valley, **E**...Texaco/repair, **W**...Exxon/diesel, Arizona Family Rest., Armando's Italian, Palms Rest., Stuffed Olive Café, Best Western, Abco Food/deli/bakery, Beall's, bank, hardware
63	Continental Rd, Green Valley, **E**...USPO, **W**...CHIROPRACTOR, DENTIST, MEDICAL CARE, OPTICAL CARE, Exxon, KFC/Taco Bell, McDonald's, Ribs&More, Cobre Tires, Goodyear/auto, HealthFoods, MailBoxes Etc, NAPA, Osco Drug, Safeway, TrueValue, bank, golf, laundry, to Madera Cyn RA, **2 mi E**...San Ignacio Golf Club/rest.
56	Canoa Rd, **W**...Fairfield Green Valley Lodge
54km	**rest area both lanes, full(handicapped)facilities, phone, picnic tables, litter barrels, vending, petwalk**
48	Arivaca Rd, Amado, **2 mi E**...Mtn View RV Park, **W**...Amado Plaza/gas/mart
42	Agua Linda Rd, to Amado, **E**...Mtn View RV Park, food
40	Chavez Siding Rd, Tubac, **3-5 mi E**...Tubac Mkt/diesel/deli, Tubac Golf Resort, **W**...Burrito Inn
34	Tubac, **E**...Tubac Mkt/diesel/deli, Montura Rest., Tubac Golf Resort, USPO, to Tubac Presidio SP, info
29	Carmen, Tumacacori, **E**...gas, food, lodging, to Tumacacori Nat Hist Park
25	Palo Parado Rd, no facilities
22	Peck Canyon Rd, insp sta nb
17	Rio Rico Dr, Calabasas Rd, **W**...Chevron, American Bar/grill, IGA Foods, Rio Rico Resort, USPO, RV camping, laundry
12	AZ 289, to Ruby Rd, **E**...Pilot/Wendy's/diesel/mart/24hr, to Pena Blanca Lake RA
8	US 89, AZ 82(exits left from sb), Nogales, **1-3 mi E**...HOSPITAL, 76/Circle K, Texaco/diesel/mart, Denny's, Americana Motel, Best Western, Goodyear/auto, U-Haul, Mi Casa RV Park, info
4	AZ 189 S, Mariposa Rd, Nogales, **E**...Chevron/DQ/mart, FasTrip/gas, 76/mart, Arby's, Baskin-Robbins, ChinaStar, Denny's, Exquisito Mexican, KFC, McDonald's, Ragazzi's Italian, Shakey's, Taco Bell, Yokohama Rest., Americana Motel, Best Western, Motel 6, Super 8/rest., Chevrolet/Olds, FashionBug, Ford/Lincoln/Mercury, JC Penney, K-Mart, Mexican Insurance, Pontiac/Buick/GMC, Radio Shack, Safeway, Veterans Foods, Walgreen, Wal-Mart/drugs, bank, cleaners/laundry, **W**...Shell/diesel/mart, Carl's Jr, Famous Sam's Rest., Chrysler/Plymouth/Dodge, auto repair
1	Western Ave, Nogales, **E**...HOSPITAL
0km	I-19 begins/ends in Nogales, Arizona/Mexico Border, **1/2 mi**...76/Circle K, Shell/repair, Burger King, Domino's, Jack-in-the-Box, McDonald's, Peter Piper Pizza, AutoZone, Basha's Mercado, CarQuest, Checker Parts, Family$, NAPA, PepBoys, Walgreen, bank, museum

Arizona Interstate 40

Exit #	Services
359.5mm	Arizona/New Mexico state line
359	Grants Rd, to Lupton, **N**...**rest area both lanes, full(handicapped)facilities, phone, picnic tables, litter barrels, petwalk,** Speedy's/diesel/mart/rest./atm/24hr, Tee Pee Trading Post/rest., YellowHorse Indian Gifts, **S**...Indian Galleria/gas
357	AZ 12 N, Lupton, to Window Rock, no facilities
354	Hawthorne Rd, no facilities
351	Allentown Rd, **N**...Indian City Gifts/gas
348	St Anselm Rd, Houck, **N**...Chevron, Pancake House, Taco Bell, Ft Courage Food/gifts
347.5mm	Black Creek
346	Pine Springs Rd, no facilities
345mm	Box Canyon
344mm	Querino Wash
343	Querino Rd, no facilities
341	Ortega Rd, Cedar Point, **N**...Shell/mart/gifts
340.5mm	insp/weigh sta both lanes
339	US 191 S, to St Johns, **S**...Conoco/Taco Bell/diesel/mart/24hr, lube/tires, USPO
333	US 191 N, Chambers, **N**...Chevron/diesel/mart, to Hubbell Trading Post NHS, **S**...Mobil/mart, Best Western/rest., USPO
330	McCarrell Rd, no facilities

Arizona Interstate 40

325	Navajo, **S**...Texaco/Navajo Trading Post/Subway/diesel/24hr, Painted Desert Inn/café
323mm	Crazy Creek
320	Pinta Rd, no facilities
316mm	Dead River, picnic area both lanes, litter barrels
311	Painted Desert, **N**...Chevron/diesel/mart, Petrified Forest NP, Painted Desert
303	Adamana Rd, **S**...Painted Desert Indian Ctr/gas
302.5mm	Big Lithodendron Wash
301mm	Little Lithodendron Wash
300	Goodwater, no facilities
299mm	Twin Wash
294	Sun Valley Rd, **N**...RV camping, **S**...AZ StageStop/diesel/mart, Desert Garden Motel
292	AZ 77 N, to Keams Canyon, **N**...Shell/Burger King/TCBY/diesel/mart/24hr, AZ Country Café/24hr, Red Baron Truckwash, diesel service, tires
289	Lp 40, Holbrook, **N**...Chevron/diesel/mart, Burger King, Denny's, Jerry's Rest., Best Western, Comfort Inn, Day's Inn, Econolodge, Motel 6, Ramada Ltd, Motel 6, Travelodge, KOA
286	Lp 40, Holbrook, **N**...76/Circle K, Shamrock/diesel/mart, Shell, Burger King, KFC, Mandarin Chinese, McDonald's, Pizza Hut, Roadrunner Café, Taco Bell, Best Western, Holiday Inn Express, Rainbow Inn, 66 Motel, Sahara Inn, Super 8, Travelodge, Basha's Foods, Checker Parts, KOA, OK RV Park, Osco Drug, **S**...HOSPITAL, Chevron/diesel/repair, Exxon/diesel/mart, Hensley's Mart/gas, 76/repair, Super Fuels, Dairy Queen, Budget Inn, El Rancho Motel, Moenkopi Motel, Western Holiday Motel/rest., Ford/Lincoln/Mercury, Plymouth/Jeep/Dodge, bank, rockshops, museum, transmissions
285	US 180 E, AZ 77 S, Holbrook, **1 mi S**...Pacific Pride/diesel, Butterfield Steaks, Cholla Rest., Pizza Hut, Best Western, Budget Host, Golden Inn Motel, Holbrook Inn, Sandman Motel, Sun'n Sand Motel, Wigwam Motel, Safeway Foods, RV repair, to Petrified Forest NP
284mm	Leroux Wash
283	Perkins Valley Rd, Golf Course Rd, **S**...Texaco/Country Host Rest./diesel/24hr
280	Hunt Rd, Geronimo Rd, **N**...Geronimo RV Park
277	Lp 40, Joseph City, **N**...Love's/A&W/Subway/diesel/mart/24hr, **S**...to Cholla Lake CP, RV camping
274	Lp 40, Joseph City, **N**...gas, food, lodging, RV camping
269	Jackrabbit Rd, **S**...Jackrabbit Trading Post/gas/diesel
264	Hibbard Rd, no facilities
257	AZ 87 N, to Second Mesa, **N**...rockshop, to Homolovi Ruins SP, camping, **S**...gifts
256.5mm	Little Colorado River

255	Lp 40, Winslow, **N**...Shell/rest./RV park, Holiday Inn Express, **S**...Flying J/CountryMkt/diesel/LP/24hr, Best Western(2mi)
253	N Park Dr, Winslow, **N**...Chevron/mart, Pilot/SenorD's/diesel/mart/RV dump/24hr, Arby's, Capt Tony's Pizza, Denny's, Pizza Hut, Big O Tires, Wal-Mart, info, laundry, truckwash, **S**...HOSPITAL, Texaco/mart, Giant/mart/24hr, KFC, McDonald's, Subway, Taco Bell, Best Western/rest., Comfort Inn, Econolodge/gas, Basha's Foods, NAPA, Safeway/drugs
252	AZ 87 S, Winslow, **S**...Shell/mart, Texaco/pizza/mart, Burger King, Chinese Rest., Dos Amigos Grill, Joe's Café, Best Inn, Best Western, Day's Inn, Delta Motel, Super 8, NAPACare, RV camping
245	AZ 99, Leupp Corner, no facilities
239	Meteor City Rd, Red Gap Ranch Rd, **S**...Meteor City Trading Post, to Meteor Crater
235mm	**rest area both lanes, full(handicapped)facilities, info, phone, picnic tables, litter barrels, petwalk**
233	Meteor Crater Rd, **S**...Mobil/mart/Meteor Crater RV Park, to Meteor Crater NL
230	Two Guns, no facilities
229.5mm	Canyon Diablo
225	Buffalo Range Rd, no facilities
219	Twin Arrows, **S**...Twin Arrows Trading Post/diesel/gifts
218.5mm	Padre Canyon
211	Winona, **N**...Texaco/diesel/mart/repair, Winona Trading Post
207	Cosnino Rd, no facilities

204	to Walnut Canyon NM, no facilities
201	US 89, to Flagstaff, **N**...HOSPITAL, Chevron/mart, Circle K/gas, Exxon/mart, 76/repair, Shell/mart, Texaco/diesel/mart, XpressStop/gas, Arby's, Burger King, Jack-in-the-Box, McDonald's, Pizza Hut, Ruby Tuesday, Sizzler, Taco Bell, Wendy's, Village Inn Rest., Best Western, Day's Inn, Hampton Inn, Howard Johnson, MasterHost Inn, Super 8, Checker's Parts, Dillard's, Goodyear/auto, Honda/Kawasaki, JC Penney, KOA, OfficeMax, Osco Drug, PitStop Lube, Safeway, Savers Foods, Sears/auto, mall, RV Ctr/LP, **S**...Mobil/mart, Fairfield Inn, Residence Inn

E

W

Arizona Interstate 40

198	Butler Ave, Flagstaff, **N**...VETERINARIAN, Exxon, Giant/diesel/mart, Shell, Texaco/mart, Burger King, Cracker Barrel, Denny's, Kettle Rest., McDonald's, Sonic, Taco Bell, Econolodge, Holiday Inn, Howard Johnson, Motel 6, Ramada Ltd, Super 8, Travelodge, NAPA, Sam's Club, **1 mi N on US 89**...76/diesel, Baskin-Robbins, China House, Denny's, Kachina Mexican, KFC, Yippee-ei-o Steaks, Best Western, Frontier Motel, Red Roof Motel, Relax Inn, Rodeway Inn, Royal Inn, 66 Motel, Travelodge, Timberline Motel, TwiLight Motel, Western Hills Motel/rest., Whispering Winds Inn, Wonderland Motel, Albertson's, AutoZone, Big A Parts, Buick/Pontiac/GMC, Econo Lube'n Tune, Firestone, Fry's Foods, High Country Tire/auto, Inn Suites, JiffyLube, Kia, Muffler Magic, Nissan/Subaru, Plymouth/Dodge, Smith's Foods, U-Haul, Wienerschnitzel, **S**...Mobil/mart, Sinclair/Little America/diesel/motel, Black Bart's Steaks
197.5mm	Rio de Flag
195b	US 89A N, McConnell Dr, Flagstaff, **N**...CHIROPRACTOR, Chevron, Conoco/diesel/mart, Exxon/Wendy's/diesel/mart, Gasser/diesel/mart, Giant/diesel/mart, Mobil, 76/diesel/mart, Shell, Texaco/diesel/mart, Circle K, Arby's, August Moon Chinese, Baskin-Robbins, Blimpie, Burger King, Buster's Rest., Carl's Jr, Chili's, Coco's, Dairy Queen, Del Taco, Denny's, Domino's, El Chilito's, Fazoli's, IHOP, India Cuisine, Jack-in-the-Box, Fiddler's Rest., Fuddrucker's, Furr's Café, KFC, McDonald's, Olive Garden, Perkins, Peter Piper Pizza, Pizza Hut, Red Lobster, Roma Pizza, Sizzler, Souper Salad, Strombolli's, Subway, Taco Bell, TCBY, Village Inn Rest., AZ Motel, AmeriSuites, AutoLodge, Comfort Inn, Crystal Inn, Day's Inn, Econolodge, Embassy Suites, Fairfield Inn, Hampton Inn, Hilton, La Quinta, Motel 6, Quality Inn, Ramada Ltd, Rodeway Inn, Sleep Inn, Barnes&Noble, Basha's Foods, Big 5 Sports, CarQuest, Checker Parts, Discount Tire, Hastings Books, JiffyLube, Jo-Ann Crafts, K-Mart, Michael's, New Frontiers Natural Foods, Osco Drug, Pier 1, Safeway, Staples, Target, Walgreen, Wal-Mart, cinema, laundry
a	I-17 S, AZ 89A S, to Phoenix
192	Flagstaff Ranch Rd, no facilities
191	Lp 40, Flagstaff, to Grand Canyon, **5 mi N**...Dairy Queen, Best Western, Budget Host, Comfort Inn, Day's Inn, Embassy Suites, Travelodge, to Woody Mtn Camping
190	A-1 Mountain Rd, no facilities
189.5mm	Arizona Divide, elevation 7335
185	Transwestern Rd, Bellemont, **N**...Texaco/Subway/Pizza Hut/diesel/mart/24hr, Country Host Rest., **S**...Roadhouse Grill, Harley-Davidson
183mm	**rest area wb, full(handicapped)facilities, phone, picnic tables, litter barrels, vending, weather info, petwalk**
182mm	**rest area eb, full(handicapped)facilities, phone, picnic tables, litter barrels, vending, weather info, petwalk**
178	Parks Rd, **N**...Ponderosa Forest RV Park, gas
171	Pittman Valley Rd, Deer Farm Rd, **S**...Quality Inn/rest.
167	Garland Prairie Rd, Circle Pines Rd, **N**...KOA
165	AZ 64, to Williams, to Grand Canyon, **4 mi N**...Texaco/diesel/mart, KOA, Red Lake Camping(9mi), **2-4 mi S**...Red's Steaks, Comfort Inn, Super 8, Travelodge, RV Sales/service
163	Williams, **N**...Chevron/Subway/diesel/mart, Fairfield Inn, Cyn Gateway RV Park, to Grand Canyon, **S**...Mobil/diesel/mart, Texaco/diesel/mart, Circle K, Buckle's Rest., Jack-in-the-Box, McDonald's, Ariz Suites, Downtowner Motel, Econolodge, Gateway Motel, Holiday Inn, Howard Johnson Express, Ramada Inn, Rodeway Inn, Royal American Inn, Rte 66 Inn, USPO, same as 161
161	Lp 40, Golf Course Dr, Williams, **N**...RV camping, **1-3 mi S**...HOSPITAL, Chevron/diesel, Mobil, PicQuik/mart, 76 Gas, Shell/mart/atm, Dairy Queen, Denny's, Old Smoky Rest., Parker House Rest., Red's Steaks, Rosa's Cantina, Best Western, Budget Host, Canyon Country Inn, Comfort Inn, Day's Inn, El Rancho Inn, Highlander Motel, Holiday Inn Express, Motel 6, Norris Motel, Westerner Motel, NAPA, Safeway, auto/truck repair, Railside RV Ranch
157	Devil Dog Rd, no facilities
155.5mm	safety pullout wb, litter barrels
151	Welch Rd, no facilities
149	Monte Carlo Rd, **N**...Monte Carlo Truckstop/diesel/café/repair/24hr
148	County Line Rd, no facilities
146	AZ 89, Ash Fork, to Prescott, **N**...Mobil, Ted's/diesel/rest./repair, Alma Motel, Ash Fork Inn/rest., KOA
144	Ash Fork, **N**...Ash Fork Inn/rest., KOA, **S**...Chevron/Piccadilly's/mart, Exxon/diesel/mart/RV park
139	Crookton Rd, to Rte 66, no facilities
123	Lp 40, to Rte 66, Seligman, **N**...Shell/diesel/mart, Mr J's Coffehouse, Bil-Mar-Den Motel, KOA, to Grand Canyon Caverns, **S**...Mobil/Subway/diesel/mart/wash
121	Lp 40, to Rte 66, Seligman, **N**...Mr J's Coffeehouse, Subway, to gas, lodging, camping

ARIZONA

Arizona Interstate 40

109	Anvil Rock Rd, no facilities
108mm	Markham Wash
103	Jolly Rd, no facilities
96	Cross Mountain Rd, no facilities
91	Fort Rock Rd, no facilities
87	Willows Ranch Rd, no facilities
83mm	Willow Creek
79	Silver Springs Rd, no facilities
75.5mm	Big Sandy Wash
73.5mm	Peacock Wash
71	US 93 S, to Wickenburg, Phoenix, no facilities
66	Blake Ranch Rd, **N**...Petro/Mobil/diesel/rest./24hr, Texaco/diesel/mart, Blue Beacon, repair, **S**...Beacon/diesel/mart/ café, Blake Ranch RV Park
60mm	Frees Wash
59	DW Ranch Rd, no facilities
57mm	Rattlesnake Wash
53	AZ 66, Andy Devine Ave, to Kingman, **N**...Chevron, Flying J/Conoco/diesel/LP/mart/24hr, Mobil/diesel, Herbst/ diesel/24hr, Shell/TCBY/diesel/mart, Arby's, Burger King, Denny's, Jack-in-the-Box, McDonald's, Taco Bell, Day's Inn, 1st Value Inn, Motel 6, Silver Queen Motel, Super 8, Travel Inn, Travelodge, Basha's Foods, Goodyear/auto, K-Mart, KOA(1mi), TireWorld, diesel repair, **S**...MEDICAL CARE, Exxon/repair, Shamrock/mart, Sinclair/diesel/repair, Texaco/repair, ABC Rest., Cheryl Ann's Bakery, JB's, Lo's Chinese, Sonic, Best Western, Comfort Inn, Day's Inn, High Desert Inn, Holiday Inn, Lido Motel, Mojave Inn, Rte 66 Motel, Chrysler/Jeep/Dodge, Kingman Parts, NAPA, Uptown Drug, RV camping
51	Stockton Hill Rd, Kingman, **N**...HOSPITAL, Arco/mart/24hr, Chevron/mart, 76/Circle K, KFC, Mardarin Chinese, Sonic, Taco Bell/24hr, Albertson's/drugs, AutoZone, Checker Parts, Chevrolet/Olds/Buick/Pontiac/Cadillac, Chief Parts, Ford/Lincoln/Mercury, Honda/Yamaha, Nissan, OilCan Henry's Lube, Smith's/drugs/24hr, Stockton Tire, Toyota, TrueValue, Walgreen, Wal-Mart/auto, Winston Tire, KOA(2mi), **S**...MEDICAL CARE, CHIROPRACTOR, EYECARE, 76/Circle K, Texaco/diesel/mart, Angel's Rest., Baskin-Robbins, Dairy Queen, Golden Corral, Little Caesar's, Pizza Hut, Big A Parts, BreadBasket Bakery, Hastings Books, JC Penney, JiffyLube, Radio Shack, Safeway, SavOn Drug, Sears Hardware, bank
48	US 93 N, Beale St, Kingman, **N**...Chevron/mart, Exxon/diesel/mart/RV dump, Pilot/Subway/diesel/mart/24hr, Shell/ diesel/mart/atm/24hr, TA/76/diesel/rest./24hr, Texaco/mart/repair, Woody's/gas, Chan's Chinese, Budget Inn, Frontier Motel, Holiday House Motel/rest., **S**...Chevron/diesel/mart/24hr, Circle K, Calico's Rest., Carl's Jr, Arizona Inn, Motel 6, Quality Inn, CarQuest, Ft Beale RV Park, museum, tires/repair
46.5mm	Holy Moses Wash
44	AZ 66, Oatman Rd, McConnico, to Rte 66, **S**...Crazy Fred's Fuel/diesel/mart/café, truckwash, camping
40.5mm	Griffith Wash
37	Griffith Rd, no facilities
35mm	Black Rock Wash
32mm	Walnut Creek
28	Old Trails Rd, no facilities
26	Proving Ground Rd, **S**...Ford Proving Grounds
25	Alamo Rd, to Yucca, **N**...Micromart/diesel
23mm	**rest area both lanes, full(handicapped)facilities, phone, picnic tables, litter barrels, vending, petwalk**
21mm	Flat Top Wash
20	Gem Acres Rd, no facilities
18.5mm	Illavar Wash
15mm	Buck Mtn Wash
13.5mm	Franconia Wash
13	Franconia Rd, no facilities
9	AZ 95 S, to Lake Havasu City, Parker, London Br, **S**...Pilot/Wendy's/diesel/mart/24hr
4mm	weigh sta both lanes
2	Needle Mtn Rd, no facilities
1	Topock Rd, to Bullhead City, Oatman, Havasu NWR, **N**...gas, food, camping
0mm	Colorado River, Arizona/California state line, Mountain/Pacific time zone

Kingman E / W

Arkansas Interstate 30

Exit #	Services
143b a	I-40, E to Memphis, W to Ft Smith. I-30 begins/ends on I-40, exit 153b.
142	15th St, **S**...Phillips 66/mart
141b	US 70, Broadway St, downtown, **N**...Exxon, US Fuel, Burger King, Taco Bell, AllTel Arena, Kroger, U-Haul, **S**...Total/mart, Arby's, KFC, McDonald's, Popeye's, Wendy's, bank
a	AR 10, Cantrell Rd, Markham St, downtown
141mm	Arkansas River
140	9th St, 6th St, downtown, **N**...Exxon, Shell/mart/wash, Texaco, Pizza Hut, Best Western, bank, **S**...Shell/mart, Waffle House, Master's Inn
139b	I-630, downtown
a	AR 365, Roosevelt Rd, **N**...Exxon/mart, **S**...Texaco/mart/wash, Family$, Kroger, NAPA, bank
138b	US 167 S, US 65 S, to Pine Bluff, no facilities
a	I-440 E, to Memphis, airport
135	W 65th St, **N**...Mapco/mart, Exxon/mart, Texaco/diesel/mart, Day's Inn, **S**...Denny's, La Quinta
134	Scott Hamilton Dr, **S**...Exxon/diesel/mart, Waffle House, Motel 6, Red Roof Inn
133	Geyer Springs Rd, **N**...Exxon, Church's Chicken, Sim's BBQ, Subway, Whataburger, transmissions, **S**...Conoco, Phillips 66, Shell/mart/wash, Total/mart, Arby's, Backyard Burgers, Blimpie, Burger King, Dixie Café, El Chico, KFC, Mazzio's Pizza, McDonald's, Pizza Inn, Shoney's, Taco Bell, TCBY, Waffle House, Wendy's, Western Sizzlin, Comfort Inn, Hampton Inn, Super 8, Goodyear/auto, Harvest Foods, JiffyLube, auto repair, bank, cleaners
132	US 70, University Ave, **N**...Fina/mart, Pizza Hut, Chevrolet, Ford/Lincoln/Mercury
131	Chicot Rd, **S**...Shell/Popeye's/mart, Dixie Café, McDonald's, Ryan's, Santa Fe Café, Waffle House, Knight's Inn, Motel 6, Plantation Inn, Ramada Ltd, Super 7 Inn, Firestone, RV Ctr, Wal-Mart SuperCtr/24hr
130	AR 338, Baseline Rd, Mabelvale, **N**...Mapco/diesel/mart, King Motel, Harley-Davidson, RV Ctr, **S**...Phillips 66/mart, McDonald's, Sonic, same as 131
129	I-430 N, no facilities
128	Otter Creek Rd, Mabelvale West, **S**...HOSPITAL, Michael's Rest., La Quinta/rest., Goodyear
126	AR 111, County Line Rd, to Alexander, **N**...Shell/diesel/mart, US Fuel/diesel/mart, KFC/Taco Bell, Z Motel, Aamco, RV Ctr, **S**...Mapco/diesel/mart, RV Ctr
123	AR 183, Reynolds Rd, to Bryant, Bauxite, **N**...TJ's/mart, Backyard Burgers, Cracker Barrel, Waffle House, Holiday Inn Express, RV Ctr(1mi), **S**...CHIROPRACTOR, DENTIST, Exxon/diesel/mart/wash, Total/diesel/mart, Little Caesar's, McDonald's, Sonic, Subway, Taco Bell, Wendy's, Super 8, Foster's Foods, Isuzu
121	Alcoa Rd, **N**...Buick/Pontiac/GMC, Chrysler/Plymouth/Dodge/Jeep, **S**...Texaco/diesel/mart, Branch Hollow RV Park, Chevrolet, Ford/Lincoln/Mercury, Salem Pines RV
118	Congo Rd, **N**...Chief/diesel/mart, Texco/diesel/mart, Dixie Café, Santa Fe Grill, Sergio's Café, Scottish Inn, **S**...Citgo, Exxon, Shell, Arby's, Backyard Burger, Brown's Country Store/rest., Burger King, Capt D's, Dairy Queen, Hardee's, McDonald's, Sonic, Taco Bell, TCBY, Waffle House, Western Sizzlin, Best Western, Day's Inn, Econolodge, Ramada Inn, Sleep Cheap, AutoZone, Big Lots, Family$, GNC, Goody's, Hastings Books, JC Penney, Kroger, MailBoxes Etc, Olds/GMC, QuikLube, RV City, USPO, Wal-Mart Super Ctr/Wendy's/24hr, antique cars
117	US 64, AR 5, AR 35, **N**...DENTIST, Conoco/diesel/mart, Shell/mart, Bo's BBQ, Denny's, Pizza Hut, Waffle House, Best Western, Econolodge, Ramada Inn, Sleep Cheap, Cedarwood Inn, Benton Inn, cinema, **S**...HOSPITAL, Citgo, Arby's, Dairy Queen
116	Sevier St, **N**...Shell/diesel/mart, Sinclair/diesel, Texaco, Ed&Kay's Rest., Hunan Chinese, Tastee Freeze, Troutt Motel, **S**...BP/mart, Texaco/diesel, US Fuel/diesel, Capri Motel, laundry
114	US 67 S, Benton, **S**...Shell/mart(1mi)
113mm	insp sta eb, st police
111	US 70 W, Hot Springs, no facilities
106	Old Military Rd, **N**...Fina/JJ's Rest./diesel/mart, Motel 106, JBS RV Park
98b a	US 270, Malvern Springs, **N**...Super 8, **S**...HOSPITAL, Shell/Blimpie/diesel/mart, Phillips 66/mart, Total/diesel/mart, Andy's Rest., Burger King, Mazzio's Pizza, McDonald's(3mi), Pizza Hut, Subway, Taco Bell, TCBY, Waffle House, Budget Inn, Economy Inn, AutoZone, Ford/Lincoln/Mercury, GNC, Huckaby's Foods, Radio Shack, Wal-Mart/auto

E

W

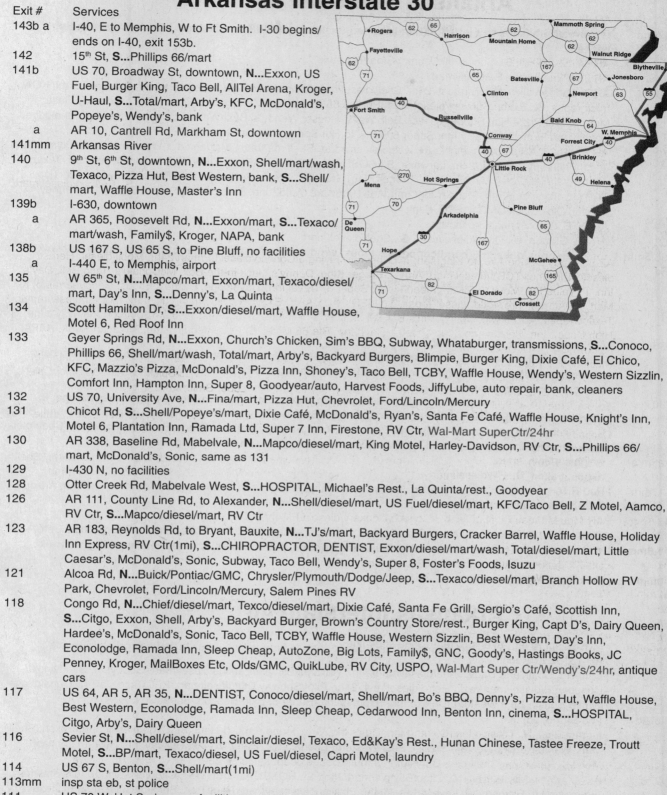

Arkansas Interstate 30

97   AR 84, AR 171, **N**...Lake Catherine SP, RV camping

93mm   **rest area(both lanes exit left), full(handicapped)facilities, phone, picnic tables, litter barrels, vending, petwalk**

91   AR 84, Social Hill, **S**...Social Hill Country Store/RV park

83   AR 283, Friendship, **S**...Shell/mart

78   AR 7, Caddo Valley, **N**...Fina/rest./24hr, Shell/diesel/mart/24hr, Total/A&W/diesel/mart/24hr, Cracker Barrel, KOA, truck repair, **S**...Exxon/Subway/diesel/mart, Phillips 66/diesel/mart, Texaco/diesel/mart, BBQ, Bowen's Rest., McDonald's, Pizza Hut, Shoney's, Taco Bell, Waffle House, Wendy's, Best Western, Day's Inn/rest., Econolodge, Holiday Inn Express, Quality Inn, Super 8, bank, to Hot Springs NP, De Gray SP

73   AR 8, AR 26, AR 51, Arkadelphia, **N**...Citgo/diesel/mart, Shell/mart/atm, McDonald's, Western Sizzlin, Plymouth/Jeep, Wal-Mart SuperCtr/24hr, carwash, **S**...HOSPITAL, VETERINARIAN, DENTIST, Exxon/diesel/mart, Texaco/mart, Total/diesel/mart, Andy's Rest., Burger King, Grandy's, Kreg's Rest., Mazzio's Pizza, Subway, Taco Tico, TCBY, College Inn(2mi), Pioneer Inn(2mi), Ace Hardware, AutoZone, Brookshire's Foods, Cato's, JC Penney, Pontiac/Olds/Buick/GMC, USPO, bank

69   AR 26 E, Gum Springs, no facilities

63   AR 53, Gurdon, **N**...Citgo/diesel/rest., **S**...Shell/diesel/mart

56.5mm   **rest area both lanes, full(handicapped)facilities, vending, picnic tables, litter barrels, vending, petwalk**

54   AR 51, Gurdon, Okolona, no facilities

46   AR 19, Prescott, **N**...Citgo/diesel/mart/24hr, Phillips 66/diesel/mart, **S**...Love's/Hardee's/diesel/mart/24hr, Texaco/mart, Pizza Hut(1mi), Crater of Diamonds SP

44   AR 24, Prescott, **S**...HOSPITAL, Norman's 44 Trkstp/diesel/rest., Pitt Grill, Sonic, Split Rail Café, Econolodge, S Ark U

36   AR 299, to Emmet, **S**...Citgo/mart

31   AR 29, Hope, **N**...Shell/mart, Texaco/diesel, Pitt Grill, Best Western, Quality Inn/rest., Village Inn Motel/rv park, carwash, st police, **S**...Exxon/mart, Texaco/mart, Total/diesel/mart, Andy's Rest., BBQ, Doc's Hot Skillet, KFC, Economy Inn, King's Inn/rest.

30   AR 4, Hope, **N**...Phillips 66, Little B's Mexican, Western Sizzlin, Best Western, Holiday Inn Express, Super 8, Millwood SP, **S**...HOSPITAL, BP/mart, Exxon/Wendy's/Baskin-Robbins/mart, Shell/mart, Texaco/diesel/mart, Burger King, Catfish King, McDonald's, Pitt Grill, Pizza Hut, Taco Bell, Brookshire's Foods, Buick/Pontiac/GMC/Chevrolet/Olds, Chrysler/Jeep/Dodge, Ford/Lincoln/Mercury, Wal-Mart/auto, Old Washington Hist SP

26mm   weigh sta both lanes

18   rd 355, Fulton, **N**...BP/diesel/mart

17mm   Red River

12   US 67, Fulton, no facilities

7   AR 108, Mandeville, **N**...Flying J/Conoco/Cookery/diesel/LP/24hr, KOA

2   US 67, AR 245, Texarkana, **N**...Phillips 66/mart, Total/diesel/café/24hr, **S**...BP/diesel/mart, Peggy Sue's Café,

1.5mm   **Welcome Ctr eb, full(handicapped)facilities, info, phone, picnic tables, litter barrels, vending, petwalk**

1   US 71, Jefferson Ave, Texarkana, **S**...antiques

0mm   Arkansas/Texas state line

(side margin, top to bottom: Arkadelphia · Texarkana)

(right margin: E ↑ ↓ W)

Arkansas Interstate 40

Exit #	Services
285mm	Mississippi River, Arkansas/Tennessee state line
281	AR 131, S to Mound City
280	Club Rd, Southland Dr, **N**...Fina/diesel/mart/24hr, Williams/Blimpie/Perkins/diesel/mart, Goodyear, **S**...Flying J/Conoco/LP/diesel/mart/rest., Petro/diesel/rest./24hr, Pilot/Subway/DQ/diesel/mart, KFC/Taco Bell, McDonald's, Waffle House, Best Western, Deluxe Inn, Express Inn, Super 8, Blue Beacon, SpeedCo Lube
279b	I-55 S(from eb)
a	Ingram Blvd, **N**...Classic Inn, Comfort Inn, Day's Inn, Greyhound Park, U-Haul, **S**...Citgo/diesel/mart, Exxon, Shell/mart, Waffle House, Western Sizzlin, Best Western, Econolodge, Hampton Inn, Holiday Inn, Howard Johnson, Motel 6, Relax Inn, Chevrolet, Chrysler/Plymouth/Dodge/Jeep, same as 280
278	AR 77, 7th St, Missouri St, **N**...Citgo/Blimpie/mart/24hr, **S**...HOSPITAL, DENTIST, Citgo/diesel/mart, Coastal/mart, Express/mart, Exxon/mart, Love's/diesel/24hr, Phillips 66/diesel/mart, RaceTrac, Shell/diesel/rest./24hr, Texaco, Voss/diesel/mart/24hr, Backyard Burger, Baskin-Robbins, BBQ, Bonanza, Burger King, Cracker Barrel, Domino's, KFC, Krystal, Little Caesar's, Mrs Winner's, McDonald's, Pizza Hut, Pizza Inn, Shoney's, Subway, Taco Bell, TCBY, Wendy's, Quality Inn, Ramada Ltd, Cato's, Chief Parts, Goodyear/auto, Hancock Fabrics, Kroger, Mr FastLube, Radio Shack, Sawyers RV Park, Sears, Walgreen, Wal-Mart SuperCtr/24hr

(side margin, top to bottom: W Memphis)

Arkansas Interstate 40

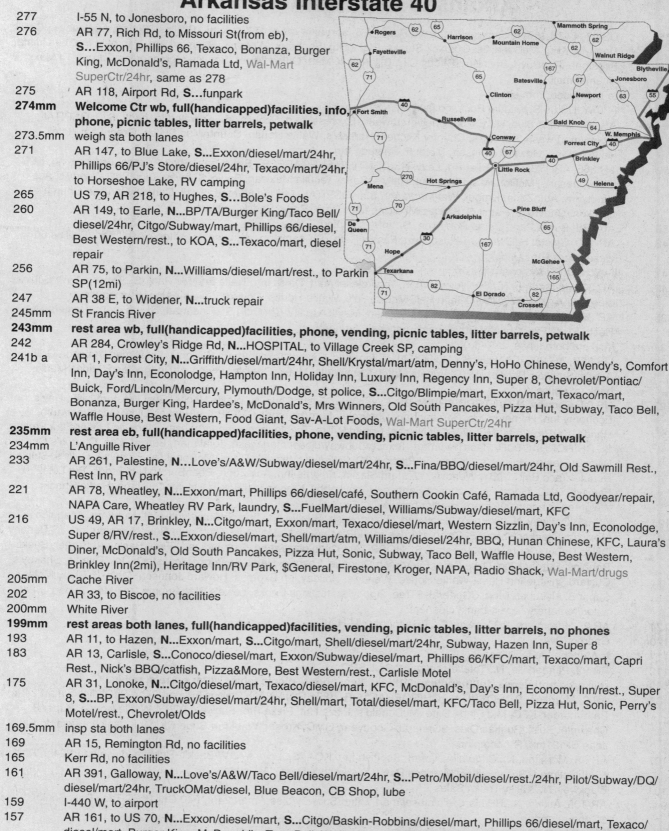

277	I-55 N, to Jonesboro, no facilities
276	AR 77, Rich Rd, to Missouri St(from eb), **S**...Exxon, Phillips 66, Texaco, Bonanza, Burger King, McDonald's, Ramada Ltd, Wal-Mart SuperCtr/24hr, same as 278
275	AR 118, Airport Rd, **S**...funpark
274mm	**Welcome Ctr wb, full(handicapped)facilities, info, phone, picnic tables, litter barrels, petwalk**
273.5mm	weigh sta both lanes
271	AR 147, to Blue Lake, **S**...Exxon/diesel/mart/24hr, Phillips 66/PJ's Store/diesel/24hr, Texaco/mart/24hr, to Horseshoe Lake, RV camping
265	US 79, AR 218, to Hughes, **S**...Bole's Foods
260	AR 149, to Earle, **N**...BP/TA/Burger King/Taco Bell/ diesel/24hr, Citgo/Subway/mart, Phillips 66/diesel, Best Western/rest., to KOA, **S**...Texaco/mart, diesel repair
256	AR 75, to Parkin, **N**...Williams/diesel/mart/rest., to Parkin SP(12mi)
247	AR 38 E, to Widener, **N**...truck repair
245mm	St Francis River
243mm	**rest area wb, full(handicapped)facilities, phone, vending, picnic tables, litter barrels, petwalk**
242	AR 284, Crowley's Ridge Rd, **N**...HOSPITAL, to Village Creek SP, camping
241b a	AR 1, Forrest City, **N**...Griffith/diesel/mart/24hr, Shell/Krystal/mart/atm, Denny's, HoHo Chinese, Wendy's, Comfort Inn, Day's Inn, Econolodge, Hampton Inn, Holiday Inn, Luxury Inn, Regency Inn, Super 8, Chevrolet/Pontiac/ Buick, Ford/Lincoln/Mercury, Plymouth/Dodge, st police, **S**...Citgo/Blimpie/mart, Exxon/mart, Texaco/mart, Bonanza, Burger King, Hardee's, McDonald's, Mrs Winners, Old South Pancakes, Pizza Hut, Subway, Taco Bell, Waffle House, Best Western, Food Giant, Sav-A-Lot Foods, Wal-Mart SuperCtr/24hr
235mm	**rest area eb, full(handicapped)facilities, phone, vending, picnic tables, litter barrels, petwalk**
234mm	L'Anguille River
233	AR 261, Palestine, **N**...Love's/A&W/Subway/diesel/mart/24hr, **S**...Fina/BBQ/diesel/mart/24hr, Old Sawmill Rest., Rest Inn, RV park
221	AR 78, Wheatley, **N**...Exxon/mart, Phillips 66/diesel/café, Southern Cookin Café, Ramada Ltd, Goodyear/repair, NAPA Care, Wheatley RV Park, laundry, **S**...FuelMart/diesel, Williams/Subway/diesel/mart, KFC
216	US 49, AR 17, Brinkley, **N**...Citgo/mart, Exxon/mart, Texaco/diesel/mart, Western Sizzlin, Day's Inn, Econolodge, Super 8/RV/rest., **S**...Exxon/diesel/mart, Shell/mart/atm, Williams/diesel/24hr, BBQ, Hunan Chinese, KFC, Laura's Diner, McDonald's, Old South Pancakes, Pizza Hut, Sonic, Subway, Taco Bell, Waffle House, Best Western, Brinkley Inn(2mi), Heritage Inn/RV Park, $General, Firestone, Kroger, NAPA, Radio Shack, Wal-Mart/drugs
205mm	Cache River
202	AR 33, to Biscoe, no facilities
200mm	White River
199mm	**rest areas both lanes, full(handicapped)facilities, vending, picnic tables, litter barrels, no phones**
193	AR 11, to Hazen, **N**...Exxon/mart, **S**...Citgo/mart, Shell/diesel/mart/24hr, Subway, Hazen Inn, Super 8
183	AR 13, Carlisle, **S**...Conoco/diesel/mart, Exxon/Subway/diesel/mart, Phillips 66/KFC/mart, Texaco/mart, Capri Rest., Nick's BBQ/catfish, Pizza&More, Best Western/rest., Carlisle Motel
175	AR 31, Lonoke, **N**...Citgo/diesel/mart, Texaco/diesel/mart, KFC, McDonald's, Day's Inn, Economy Inn/rest., Super 8, **S**...BP, Exxon/Subway/diesel/mart/24hr, Shell/mart, Total/diesel/mart, KFC/Taco Bell, Pizza Hut, Sonic, Perry's Motel/rest., Chevrolet/Olds
169.5mm	insp sta both lanes
169	AR 15, Remington Rd, no facilities
165	Kerr Rd, no facilities
161	AR 391, Galloway, **N**...Love's/A&W/Taco Bell/diesel/mart/24hr, **S**...Petro/Mobil/diesel/rest./24hr, Pilot/Subway/DQ/ diesel/mart/24hr, TruckOMat/diesel, Blue Beacon, CB Shop, lube
159	I-440 W, to airport
157	AR 161, to US 70, **N**...Exxon/diesel/mart, **S**...Citgo/Baskin-Robbins/diesel/mart, Phillips 66/diesel/mart, Texaco/ diesel/mart, Burger King, McDonald's, Taco Bell, Waffle House, Best Western, Day's Inn, Masters Inn, Red Roof Inn, Super 8
156	Springhill Dr, **N**...Fairfield Inn, Holiday Inn Express, Residence Inn

Arkansas Interstate 40

155	US 67 N, US 167, to Jacksonville, Little Rock AFB, **services off of 167 N 1 mi**...Phillips 66/mart, Applebee's, Chili's, Denny's, KFC, McDonald's, Outback Steaks, Pizza Hut, Shoney's, Wendy's, Baymont Inn, Hampton Inn, La Quinta, Firestone, PepBoys, Best Buy, Circuit City, Dillard's, K-Mart, Office Depot, Service Merchandise, Target, TJ Maxx, cinema, mall
154	to Lakewood(from eb)
153b	I-30 W, US 65 S, to Little Rock, no facilities
a	AR 107 N, JFK Blvd, **N**...Exxon, Shell/mart/wash, Express/mart, Schlotsky's, Travelodge, bank, **S**...HOSPITAL, Exxon/mart/wash, Bonanza, Country Kitchen, Hardee's, Waffle House, Country Inn Suites, GreenLeaf Hotel, Hampton Inn, Howard Johnson, Motel 6, Ramada Inn
152	AR 365, AR 176, Camp Pike Rd, Levy, **N**...DENTIST, EYECARE, EZ Serve/gas, Shell, Texaco, Burger King, KFC, Little Caesar's, McDonald's, Pizza Hut, Rally's, Senor Tequila Mexican, Sonic, Subway, Taco Bell, Wendy's, Ace Hardware, AutoZone, Kroger, Wal-Mart/24hr, **S**...HOSPITAL, Phillips 66/mart, Shell, Cancun Cantina, Church's Chicken, Chevrolet, Family$, Kroger, Medicine Man Drug, Radio Shack, antiques
150	AR 176, Burns Park, Camp Robinson, **S**...info, camping
148	AR 100, Crystal Hill Rd, **N**...Texaco/mart, Family Rest., RV camping, antiques, **S**...Citgo/diesel/mart, Phillips 66/diesel/mart, RV camping
147	I-430 S, to Texarkana
142	AR 365, to Morgan, **N**...Shell/diesel/mart, Total/diesel/mart, Day's Inn, Trails End RV Park, **S**...Texaco/A&W/Subway/diesel/mart, BJ's Rest., KFC/Taco Bell, McDonald's, Waffle House, Comfort Suites, Super 8, antiques
135	AR 365, AR 89, Mayflower, **N**...Phillips 66/mart, Mayflower RV Ctr(1mi), **S**...Fina/mart, Exxon/mart, Farmers Mkt Rest., Pizza Pro, Country Store, bank, carwash, laundromat
134mm	insp sta both lanes
129	US 65B, AR 286, Conway, **S**...HOSPITAL, Citgo/mart, Exxon, Texaco/mart, Williams/diesel/mart, Amazon Grill, Arby's, Subway, Budget Inn, Clements Motel, Continental Motel, Chrysler/Plymouth/Dodge/Jeep, Midas Muffler, bank, st police
127	US 64, Conway, **N**...VETERINARIAN, BP, Exxon/mart/wash, Denny's, Waffle House, Best Western/rest., Day's Inn, Economy Inn, Hampton Inn, Ramada Inn, Chevrolet, Ford/Mercury, GMC/Buick/Pontiac/Olds, Goodyear/auto, Honda, KarPro Parts, Moix RV Ctr, Nissan, Superior Tire, Toyota, bank, repair/transmissions, to Lester Flatt Park, **S**...CHIROPRACTOR, RaceTrac/mart, Texaco/Subway/diesel/mart/wash, Total/mart, Burger King, Church's Chicken, Dillon's Steaks, Hardee's, Hollywood Pizza, KFC, LJ Silver, Mazzio's Pizza, McDonald's, Pizza Inn, Rally's, Shipley Donuts, Taco Bell, TCBY, Wendy's, Western Sizzlin, Knights Inn, AutoZone, BeanSprout Nat Foods, Big Lots, Family$, GNC, Goodyear, Hancock Fabrics, HobbyLobby, Kroger, MailBoxes Etc, Mr Quiklube, Radio Shack, Walgreen, cinema, cleaners, muffler/brakes
125	US 65, Conway, **N**...Conoco/diesel/mart, Exxon/wash, Shell/mart, Texaco/Subway/diesel/mart/24hr, Acapulco Mexican, Cracker Barrel, El Chico, Hardee's, MktPlace Grill, McDonald's, Ole South Rest., Comfort Inn, JC Penney, Office Depot, Sears, bank, cinema, **S**...HOSPITAL, Citgo/pizza/mart, Exxon/diesel, Fina, Backyard Burger, Baskin-Robbins, Burger King, Dixie Café, Evergreen Chinese, Fazoli's, Hart's Seafood, McAlister's Deli, Ryan's, Shakey's Custard, Village Inn Rest., Waffle House, Wendy's, Holiday Inn Express, Howard Johnson, Motel 6, Stacy Motel, Super 8, All Lube&Tune, $General, $Tree, Goody's, Hastings Books, Lowe's Bldg Supply, Wal-Mart SuperCtr/24hr, auto/tire/battery repair, bank, cleaners
124	AR 25 N(from eb), to Conway, **S**...Texaco/mart, U-Haul
120mm	Cadron River
117	to Menifee, no facilities
112	AR 92, Plumerville, **N**...Total/diesel/mart, **S**...USPO
109mm	**rest area both lanes, full(handicapped)facilities, phone, picnic tables, litter barrels, petwalk**
108	AR 9, Morrilton, **S**...HOSPITAL, CHIROPRACTOR, VETERINARIAN, Phillips 66/mart/24hr, Shell/diesel/mart/24hr, Betterburger, KFC, Mkt Place Café, McDonald's, Pizza Hut, Pizza Pro, Subway, TCBY, Waffle House, Super 8, Chevrolet/Buick/Pontiac/Olds, $General, Goodyear, GMC, Kroger, Wal-Mart SuperCtr/24hr, bank, cinema, to Petit Jean SP(21mi), RV camping
107	AR 95, Morrilton, **N**...Citgo/diesel/mart, Scottish Inn, KOA, **S**...Love's/A&W/Baskin-Robbins/diesel/mart/24hr, Phillips 66/diesel/mart, Shell/waffles, Morrilton Rest., Yesterday's Rest., Day's Inn
101	Blackwell, **N**...Utility Trailer Sales
94	AR 105, Atkins, **N**...BP/Hardee's/diesel/mart, Citgo/Subway/diesel/mart/24hr, I-40 Grill, KFC/Taco Bell, Sonic, $General, repair
88	Pottsville, **S**...Texaco/truck repair/wash

Little Rock

Conway

E

W

Arkansas Interstate 40

84	US 64, AR 331, Russellville, **N**...FlyingJ/Conoco/Country Mkt/diesel/LP/24hr, Texaco/diesel/mart, **S**...Phillips 66/mart/atm, Williams/Perkins/Baskin-Robbins/diesel/mart/24hr, CiCi's, Hardee's, Hunan Chinese, McDonald's, Subway, Western Sizzlin, Comfort Inn, Ramada Inn, AutoZone, Chrysler/Jeep, Firestone, JC Penney, Lowe's Bldg Supply, Nissan, PriceCutter Foods/24hr, USPO, Wal-Mart/auto, bank, **2 mi S**...Arby's, Burger King, KFC, Ryan's, Classic Inn, Johnson Motel, Merrick Motel
81	AR 7, Russellville, **N**...Conoco/diesel/mart, Waffle House, Knight's Inn, Motel 6, Sunrise Inn, Outdoor RV Ctr, **S**...Exxon/diesel/mart/24hr, Phillips 66/diesel/mart/24hr, Texaco/mart/24hr, Arby's, BBQ, Burger King, Cracker Barrel, Dixie Café, New China, Ole South Pancakes, Pizza Inn, Ruby Tuesday, Subway, Waffle House, Best Western, Hampton Inn, Southern Inn, Super 8, Travelodge, antiques, **2 mi S**...HOSPITAL, CHIROPRACTOR, Fina, Total/mart, Bailey's Grill, Lee's Chicken, LJ Silver, Mazzio's, McDonald's, Pizza Hut, Taco Bell, Taco John's, Watta-Burger, Wendy's, Goodyear/auto, Goody's, to Lake Dardanelle SP, RV camping
80mm	Dardanelle Reservoir
78	US 64, Russellville, **S**...to Lake Dardanelle SP, RV camping
74	AR 333, London, **S**...gas
72mm	**rest area wb, full(handicapped)facilities, phone, picnic tables, litter barrels, vending, petwalk**
70mm	overlook wb lane
68mm	**rest area eb, full(handicapped)facilities, phone, picnic tables, litter barrels, vending, petwalk**
67	AR 315, Knoxville, **S**...gas/diesel, food, phone
64	US 64, Clarksville, Lamar, **S**...Texaco/diesel/mart/24hr, Red Oak Pizza/subs, Dad's Dream RV Park
58	AR 21, AR 103, Clarksville, **N**...HOSPITAL, CHIROPRACTOR, DENTIST, Phillips 66/mart/24hr, Texaco/mart, BBQ, KFC, Mazzio's, McDonald's, Old South Pancakes, Pizza Hut, Pizza Pro, Sonic, Taco Bell, Waffle House, Wendy's, Woodard's Rest., Best Western, Budget Inn, Comfort Inn, Economy Inn, Holiday Inn Express, Super 8, Big A Parts, Chevrolet/Olds, Cooper Tires, Wal-Mart/auto, bank, cleaners, repair, RV parking, **S**...Texaco/diesel/mart/rest., Chrysler/Dodge/Jeep, Ford/Lincoln/Mercury
57	AR 109, Clarksville, **N**...Citgo, Conoco, Subway, to Subiaco Academy, **S**...Texaco/A&W/diesel/mart
55	US 64, AR 109, Clarksville, **N**...Citgo/diesel/mart, Phillips 66, Hardee's, Lotus Chinese, Pizza Hut, Waffle Inn, Day's Inn, Hampton Inn, cinema, st police, **S**...Exxon/mart, Western Sizzlin
47	AR 164, Coal Hill, no facilities
41	AR 186, Altus, **S**...Pine Ridge RV Park
37	AR 219, Ozark, **S**...HOSPITAL, Exxon, Love's/A&W/Subway/diesel/mart/24hr, KFC/Taco Bell, Day's Inn, McNutt RV Park
36mm	**rest area both lanes, full(handicapped)facilities, phone, picnic tables, litter barrels, petwalk**
35	AR 23, Ozark, **3 mi S**...HOSPITAL, Total/diesel/mart, Hardee's, Oxford Inn
24	AR 215, Mulberry, **2 mi S**...Total, camping
20	Dyer, **N**...Conoco/diesel, **S**...Texaco/diesel/mart, Mill Creek Inn
13	US 71 N, to Fayetteville, **N**...Exxon, Phillips 66, Shell/mart, Burger King, Cracker Barrel, Dairy Queen, KFC, Mazzio's, Taco Bell, Alma Inn, Meadors Inn, Alma RV Park, $General, KOA(2mi), antiques, to U of AR, Lake Ft Smith SP, **S**...Citgo/Piccadilly's/diesel/mart/atm, Conoco/mart, Phillips 66/mart, Texaco/diesel/mart, Braum's, McDonald's, Sonic, Subway, Day's Inn, Coleman Drug, CV's Foods, NAPA/lube/repair, O'Reilly's Parts, Wal-Mart, bank
12	I-540 N, to Fayetteville, **N**...to Lake Ft Smith SP
9mm	weigh sta both lanes

E

W

Russellville

Ft Smith

Arkansas Interstate 40

Ft Smith

7	I-540 S, US 71 S, to Ft Smith, Van Buren, **S**...HOSPITAL, Baymont Inn(3mi), Quality Inn(3mi)
5	AR 59, Van Buren, **N**...Exxon/McDonald's/diesel/mart, Fina/diesel, Phillips 66/mart, Total/diesel/mart, Burger King, Santa Fe Café, Schlotsky's, Holiday Inn Express, FashionBug, Radio Shack, Wal-Mart SuperCtr/gas/24hr, **S**...Citgo/mart, Texaco/diesel/mart/rest./24hr, Braum's, Big Jake's Steaks, Geno's Pizza, Gringo's TexMex, Hardee's, KFC/Taco Bell, Little Caesar's, Mazzio's, Sonic, Subway, Waffle House, Wendy's, Best Western, Motel 6, Super 8, CV's Foods, $General, Ford, Overland RV Park, Pharmacy Express, Pennzoil, bank, diesel repair
3	Lee Creek Rd, no facilities
2.5mm	**Welcome Ctr eb, full(handicapped)facilities, info, phone, picnic tables, litter barrels, vending, petwalk**
1	Dora, to Ft Smith(from wb), no facilities
0mm	Arkansas/Oklahoma state line

Arkansas Interstate 55

Exit #	Services
72mm	Arkansas/Missouri state line
72	State Line Rd, weigh sta sb
71	AR 150, Yarbro, **E**...antiques, **W**...Citgo/mart
68mm	**Welcome Ctr sb, full(handicapped)facilities, phone, picnic tables, litter barrels, petwalk**
67	AR 18, Blytheville, **E**...Phillips 66/diesel/mart, Burger King, Day's Inn, Cato's, $Tree, Wal-Mart SuperCtr/24hr, **W**...HOSPITAL, CHIROPRACTOR, Amoco/Blimpie/diesel/mart/atm, Exxon/diesel/mart, Texaco/mart, Bonanza, El Acapulco Mexican, Grecian Steaks, Hardee's, KFC, Mazzio's Pizza, McDonald's, Olympia Steaks, Perkins, Pizza Inn, Shoney's, Sonic, Subway, Taco Bell, Wendy's, Comfort Inn, Drury Inn, Hampton Inn, Holiday Inn, Advance Parts, CrossRoads Books, K-Mart, Pennzoil, PriceChopper Foods, bank, cinema
63	US 61, to Blytheville, **E**...Phillips 66/diesel/mart, Budget Inn(2mi), **W**...Citgo/diesel/mart/24hr, Dodge's Store/diesel, Shell/Krystal/Popeye's/diesel/mart/atm/24hr, Texaco/diesel/mart, Subway, Best Western, Delta Motel, Royal Inn(2mi), Chevrolet, Nissan, RV Park
57	AR 148, Burdette, **E**...Cotton Bowl Tech
53	AR 158, Victoria, Luxora, no facilities
48	AR 140, to Osceola, **E**...Amoco/diesel/mart, Texaco/Baskin-Robbins/diesel/mart, Best Western/rest., Holiday Inn Express, **3 mi E**...HOSPITAL, Hardee's, McDonald's, Pizza Inn, Sonic, Subway
45.5mm	**rest area nb, full(handicapped)facilities, picnic tables, litter barrels, petwalk**
44	AR 181, Keiser, no facilities
41	AR 14, Marie, **E**... to Hampton SP/museum
36	AR 181, to Wilson, to Bassett, no facilities
35mm	**rest area sb, full(handicapped)facilities, picnic tables, litter barrels, phones, petwalk**
34	AR 118, Joiner, no facilities
23b a	US 63, AR 77, to Marked Tree, Jonesboro, ASU, **W**...Citgo/Baskin-Robbins/pizza/mart
21	AR 42, Turrell, **W**...Fuel Mart/Subway/diesel/24hr
17	AR 50, to Jericho, no facilities
14	rd 4, to Jericho, **E**...Citgo/diesel/mart/24hr, **W**...Citgo/Stuckey's/mart, Trav-L Park Camping
10	US 64 W, Marion, **E**...BP, Citgo/mart, Texaco/McDonald's/diesel/mart, Williams/diesel/mart, Hardee's, KFC/Taco Bell, Sonic, Hallmarc Inn, BigStar Foods, Family$, bank, **W**...Griffin/mart, Odie's Gas/mart, Shell/mart, Best Western, Journey Inn, to Parkin SP(23mi)
9mm	weigh sta both lanes
8	I-40 W, to Little Rock, no facilities
278	AR 77, Missouri St, 7th St, **E**...Citgo/Blimpie/mart/24hr, **W**...Citgo/diesel, Exxon/wash, Love's/diesel/24hr, Phillips 66/mart, Shell/diesel/rest./24hr, Texaco/mart, Williams/mart, Backyard Burger, Bojangles, Bonanza, Burger King, Cracker Barrel, Domino's, Krystal, Mrs Winner's, McDonald's, Pizza Inn, Pizza Hut, Shoney's, Subway, Taco Bell, Wendy's, Ramada Ltd, Blue Beacon, Cato's, Chief Parts, Goodyear/auto, Hancock Fabrics, Kroger, Mr FasLube, Sears, USPO, Walgreen, Wal-Mart SuperCtr/24hr
279b	I-40 E, to Memphis
a	Ingram Blvd, **E**...Comfort Inn, Red Roof Inn, Ford, Greyhound Track, U-Haul, **W**...Citgo/diesel/mart, Shell/mart, Earl's Rest., Waffle House, Western Sizzlin, American Inn, Best Western, Econolodge, Hampton Inn, Holiday Inn, Howard Johnson, Motel 6, Relax Inn, Chevrolet, Chrysler/Plymouth/Dodge/Jeep
4	King Dr, Southland Dr, **E**...Flying J/Conoco/diesel/LP/rest./24hr, Petro/diesel/rest./24hr, Pilot/DQ/Subway/diesel/mart, KFC/Taco Bell, McDonald's, Waffle House, Best Western, Deluxe Inn, Express Inn, Super 8, Blue Beacon, SpeedCo Lube, **1/2 mi E**...Fina/diesel/mart/24hr, Williams/Blimpie/Perkins/diesel/mart/24hr, Goodyear, **W**...Pancho's Mexican
3b a	US 70, Broadway Blvd, AR 131, Mound City Rd
2mm	weigh sta nb
1	Bridgeport Rd, no facilities
0mm	Mississippi River, Arkansas/Tennessee state line

Blytheville

W Memphis

California Interstate 5

California uses road names/numbers. Exit number is approximate mile marker.

Exit #	Services
813mm	California/Oregon state line
812	Hilt, **W**...Texaco/mart/café, B&D Organic Farm
808	Bailey Hill Rd, no facilities
806mm	inspection sta
805	Hornbrook Hwy, Ditch Creek Rd, no facilities
804	A28, Henley, to Hornbrook, **E**...Chevron/diesel/LP, phone, to Iron Gate RA
801	CA 96, Klamath River Hwy, **W...rest area both lanes, full(handicapped)facilities, info, phone, picnic tables, litter barrels, petwalk, to Klamath River RA**
797mm	Anderson Summit, elev 3067
795mm	vista point sb
794mm	Shasta River
791	Yreka, Montague, **W**...DENTIST, CHIROPRACTOR, USA/diesel/mart, Casa Ramos Mexican, Ma&Pa's Rest., Phoenix Nest Rest., Best Western, GoldRush Inn, Heritage Inn, KFC, Super 8, Thunderbird Lodge, Ray's Foods, Tandy's Repair, airport
790	Miner St, Central Yreka, **W**...HOSPITAL, CHIROPRACTOR, Chevron/24hr, Exxon/diesel, 76/repair, Texaco/A&W/diesel/mart, China Dragon, Denny's, Grandma's House, Miner St Sta, RoundTable Pizza, Best Western, Heritage Inn, Rodeway Inn, Ace Hardware, AllPro Parts, CarQuest, PriceLess Foods, Radio Shack, Rite Aid, USPO, Weldon Tire, bank
788	CA 3, to Ft Jones, Etna, **E**...CFN/diesel, RV Ctr, RV park, Schwab Tire, **W**...HOSPITAL, Exxon/diesel/mart, Shell/mart, BlackBear Diner, Boston Shaft Rest., Burger King, Carl's Jr, Food2Go Expresso, KFC, McDonald's, Papa Murphy's, Pizza Hut, Subway, Taco Bell, AmeriHost, Day's Inn, Motel 6, $Tree, Ford/Lincoln/Mercury, JC Penney, NAPA, Nissan, Raley's Food/drug, Schuck's Parts, Wal-Mart, bank, st patrol, **1 mi W**...El Rancho Café, KFC, Bender Motel, Klamath Motel, Chevrolet/Olds/Pontiac/Honda, OilChange, PriceLess Foods
785	Shamrock Rd, Easy St, **W**...Beacon/diesel/mart, RV camping
780	A12, Grenada, to Gazelle, **E**...CFN/diesel, RV camping
770	Louie Rd, no facilities
768	Weed Airport Rd, **W**...Porky Bob's Café, **rest area both lanes, full(handicapped)facilities, picnic tables, litter barrels, phone, petwalk**
767	Stewart Springs Rd, Edgewood, **E**...Lake Shasta RA/RV camp(2mi), **W**...RV camp(7mi)
761	to US 97, Weed, to Klamath Falls, **E**...Chevron/mart/24hr, 76/repair, Shell/diesel/mart, Texaco/mart, Expresso Bakery, Natural Foods Café, Pizza Factory, Hi-Lo Motel/rest., Motel 6, Townhouse Motel, The Y Motel/rest., NAPA, golf, RV camping, Weed Repair
760	Central Weed, **E**...same as 761, Pizza Factory, Ray's Foods, Shastina Golf Motel, auto repair, laundry, Coll of Siskiyou, **W**...bank
759	S Weed Blvd, **E**...Chevron/diesel/mart/24hr, 76/CFN/diesel/mart, Tesoro/diesel/mart/24hr, Burger King, McDonald's, Silva's Rest., Taco Bell, Best Inn, Holiday Inn Express, RV camping
758	Summit Dr, Truck Village Dr, **E**...CFN/diesel
757mm	Black Butte Summit, elev 3912
756	Abrams Lake Rd, **W**...Abrams Lake RV Park
754	Mt Shasta City(from sb), **E**...Pacific Pride/diesel/LP, Shasta Lodge Motel, KOA
752	Central Mt Shasta, **E**...CHIROPRACTOR, HOSPITAL, 76/diesel/mart, Chevron/diesel/mart, Shell/diesel/LP/mart, Texaco/diesel/mart/wash/24hr, BlackBear Diner, Burger King, Ray's FoodPlace, RoundTable Pizza, Say Cheese Pizza, Shasta Family Rest., Subs'n Such, Subway, Taco Shop, Best Western, Choice Inn, Econolodge, Travel Inn, KOA, Ray's Foods, Rite Aid, bank, **W**...Mt Shasta Resort/rest., Lake Siskiyou RV Park
750	CA 89, to McCloud, to Reno, **1 mi E**...Piemont Italian Diner, Econolodge, Evergreen Lodge, Finlandia Motel, Mt Aire Lodge, Pine Needles Motel, Strawberry Valley Inn
749	weigh sta sb
748	Mott Rd, to Dunsmuir, no facilities
745	Dunsmuir Ave, Siskiyou Ave, **E**...Best Choice Inn, GlassHouse Rest., Shasta Alpine Inn/rest., **W**...76/diesel/LP, Shell, Texaco/diesel/mart, Hitching Post Café, Acorn Inn, Cave Springs Motel, Cedar Lodge, Garden Motel, Hedge Creek RV Park, Wiley's Mkt
743	Central Dunsmuir, **E**...Pizza Factory, Ford/Mercury, **W**...Texaco/diesel/LP, Hitching Post Café, Micki's Burgers, Shelby's Dining, Cave Springs Motel, Travelodge, Wiley's Mkt
741	Crag View Dr, Dunsmuir, Railroad Park Rd, **E**...Tesoro/diesel/mart, Burger Barn, Best Choice Inn, Oak Tree Inn, Railroad Park Resort/motel, **W**...Railroad Park RV camping
739	Soda Creek Rd, to Pacific Crest Trail, no facilities
738	Castella, **W**...Chevron/diesel/mart, RV camping, Castle Crags SP

N

S

California Interstate 5

736mm	vista point nb
735	Sweetbrier Ave, no facilities
733	Conant Rd, no facilities
729	Flume Creek Rd, no facilities
726	Sims Rd, **W**...RV camping, phone
724	Gibson Rd, no facilities
722	Pollard Flat, **E**...Exxon/diesel/mart/LP/24hr, Pollard Flat USA Rest.
720	Slate Creek Rd, La Moine, no facilities
718	Delta Rd, Dog Creek Rd, to Vollmers, no facilities
716mm	**rest area both lanes, full(handicapped)facilities, phone, litter barrels, picnic tables**
715	Lakeshore Dr, Riverview Dr, Antlers Rd, to Lakehead, **E**...Shell/diesel/mart/24hr, Top Hat Café, Neu Lodge Motel/café, Antlers RV Park, Lakehead Campground, USPO, auto repair, towing/24hr, **W**...76/LP/mart, Shasta Lake Motel/RV, Lakeshore Villa RV Park, RV service/dump
710	Salt Creek Rd, Gilman Rd, **W**...Salt Creek RV Park, Trail End RV Park
707	Shasta Caverns Rd, O'Brien, camping
706	Packers Bay Rd, boating
705mm	**rest area nb, full(handicapped)facilities, phone, picnic tables, litter barrels, petwalk**
704	Turntable Bay Rd, no facilities
703	Bridge Bay Rd, **W**...Tail of a Whale Rest., Bridge Bay Motel
701	Fawndale Rd, Wonderland Blvd, **E**...Fawndale Lodge, Fawndale Oaks RV Park, **W**...Wonderland RV Park
699	Wonderland Blvd, Mountain Gate, **E**...Bear Mountain RV Resort, Mountain Gate RV Park, Marine/engine repair, Ranger Sta, **W**...76/LP/diesel/mart
697	CA 151, Shasta Dam Blvd, Project City, Central Valley, **W**...Chevron/diesel/LP/mart, Texaco/Burger King/diesel/LP/mart, Bakery&Grill, McDonald's, StageStop Café, Taco Den, Shasta Dam Motel
696	Pine Grove Ave, **E**...Cousin Gary's RV Ctr, **W**...Exxon/diesel/mart, antiques
694	Oasis Rd, **E**...CA RV Ctr/camping, **W**...Arco/mart/atm/24hr, LP, st patrol
693	CA 273, Market St, Johnson Rd, to Central Redding(no sb re-entry), **W**...HOSPITAL, Exxon
692	Twin View Blvd, **E**...76/mart/24hr, Motel 6, Ramada Ltd, Harley-Davidson, **W**...Pacific Pride/diesel, Holiday Inn Express, RV camping
691	CA 299E, **1/2 mi W**...Arco/mart/24hr, Chevron, Exxon/wash, Texaco/mart, BBQ Pit, Arby's, Carl's Jr, Figaro's Pizza, KFC, McDonald's, Murphy's Pizza, Phil's Pizza, Subway, TCBY, Winchell's, River Inn Motel/rest., $Tree, Food Connection, J&K Muffler, Jo-Ann Fabrics, KOA, MailBoxes Etc, Raley's Foods, RV Ctr, ShopKO, transmissions
690	CA 299 W, CA 44, Redding, to Eureka, Burney, **E**...DENTIST, CHIROPRACTOR, Chevron/mart, Texaco/mart, Applebee's, Carl's Jr, Chevy's Mexican, Chinese Rest., Hometown Buffet, Italian Cottage, In-n-Out, Jack-in-the-Box, McDonald's, Olive Garden, Outback Steaks, Pizza Hut, Quizno's, Red Robin, Holiday Inn, Motel 6, Red Lion Inn, Albertson's, Barnes&Noble, Circuit City, Costco, Econo Lube'n Tune, Food4Less, HomeBase, Home Depot, JC Penney, Kragen's Parts, MensWearhouse, Office Depot, OfficeMax, PetCo, Schwab Tire, Sears/auto, Target, ToysRUs, Wal-Mart, WinCo Foods
689	Hilltop Dr, Cypress Ave, Redding, **E**...CHIROPRACTOR, VETERINARIAN, DENTIST, Chevron/mart/24hr, 76, Shell, Applebee's, Burger King, Carl's Jr, KFC, La Palomar Mexican, Little Caesar's, McDonald's, RoundTable Pizza, Taco Shop, Wendy's, Park Terrace Inn, Big 5 Sports, Buick/Pontiac/GMC, K-Mart, Longs Drug, Mervyn's, Rite Aid, Ross, West Marine, bank, mall, **1/4 mi E on Hilltop**...DENTIST, Chevron, Exxon/mart, Black Bear Diner, Chinese Rest., ChuckeCheese, Denny's, IHOP, Marie Callender's, Pizza Hut, Taco Bell, Best Western, Cattlemens Motel, Comfort Inn, Holiday Inn Express, La Quinta, Motel Orleans, Oxford Suites, Batteries4Everything, laundry, **W**...GasAmart, 76/mart, Shell/mart, USA/diesel/mart, BBQ, California Cattle Co, Denny's, Giant Burger, Lyon's Rest., Perko's Rest., Taco Den, Howard Johnson, Motel 6, Vagabond Inn, Big A Parts, Big O Tire, Chevrolet, FabricLand, Hyundai, Lincoln/Mercury, Lube'n Oil, Mazda/Toyota/Subaru, Midas Muffler, Nissan, Office Depot, Radio Shack, Raley's Food/drug, U-Haul, bank, transmissions
686	Bechelli Lane, Churn Creek Rd, **E**...Arco/mart/24hr, Chevron/diesel/mart, 76/mart, Taco Bell, Super 8, diesel repair, **W**...Texaco/Burger King/diesel/mart
684	Knighton Rd, **E**...TA/76/Pizza Hut/Popeye's/diesel/LP/24hr, **W**...JGW RV Park(3mi), airport
682	Riverside Ave, **1 mi W**...JGW RV Park
680	Balls Ferry Rd, Anderson, **E**...CHIROPRACTOR, EYECARE, MEDICAL CARE, VETERINARIAN, Beacon/mart, 76/diesel/mart, Burger King, Denny's, El Mariachi Mexican, McDonald's, Papa Murphy's, Perko's Rest., RoundTable Pizza, Subway, Taco Bell, Walkabout Creek Deli/gifts, Best Western/rest., Valley Inn, CastroLube, $Tree, NAPA, Rite Aid, Safeway/drugs, Schwab Tires, bank, cinema, hardware, **W**...CHIROPRACTOR, Beacon/diesel/mart, Chevron/mart, Exxon/diesel/mart, Texaco, City Grill, Giant Burger, Goodtimes Pizza, KFC, Koffee Korner Rest., Holiday Food/drugs, Kragen Parts, Owen's Drugs, RV Ctr, bank, cleaners/laundry
678	CA 273, Factory Outlet Blvd, **W**...Shell/diesel/mart/wash, TowneMart/diesel, Arby's, LJ Silver, AmeriHost, GNC, Prime Outlets/famous brands, farmers mkt
677	Cottonwood, **E**...Branding Iron Rest., Travelers Motel
676	Gas Point Rd, to Balls Ferry, **E**...CFN/diesel, Chevron/MeanGene's/diesel/LP/mart, Exxon/diesel/LP, Travelers Motel, Jim's Auto Repair, **W**...CHIROPRACTOR, Beacon/diesel/LP/mart, 76/diesel/mart, Sentry Foods, hardware, laundry

N

S

Redding

California Interstate 5

675	Bowman Rd, to Cottonwood, **E**...Pacific Pride/diesel, Texaco/A&W/diesel/mart
668mm	weigh sta both lanes
667	Snively Rd, Auction Yard Rd, no facilities
665	Hooker Creek Rd, Auction Yard Rd, **E**...trailer sales
664mm	**rest area sb, full(handicapped)facilities, phone, picnic tables, litter barrels, petwalk**
663mm	**rest area nb, full(handicapped)facilities, phone, picnic tables, litter barrels, petwalk**
663	Jellys Ferry Rd, **E**...Bend RV Park
662	Wilcox Golf Rd, no facilities
661	CA 36W(from sb), Red Bluff, **3/4 mi W**...Hwy Patrol
657	CA 36, CA 99, Red Bluff, **E**...Exxon/diesel/mart, 76, Shell/diesel/mart, Burger King, Perko's Rest., KFC, McDonald's, RoundTable Pizza, Subway, Best Inn, Motel 6, Super 8, bank, **W**...VETERINARIAN, Gas4Less, USA/diesel/mart, Carl's Jr, Denny's, Egg Roll King, Marie's Rest., Riverside Dining, Shari's/24hr, Subway, Wild Bill's Steaks, Winchell's, Cinderella Motel, Red Bluff Inn, Travelodge, Value Lodge, AutoZone, Food4Less, Idle Wheels RV Park, Sears
656	Diamond Ave, Red Bluff, **E**...HOSPITAL, Exxon/mart, Red Rock Café, Day's Inn, Motel Orleans/rest., **W**...VETERINARIAN, Arco/diesel/mart/24hr, Beacon/diesel/ mart, Chevron/mart, Arby's, Country Waffles, Crystal Oven Rest., Feedbag Rest., Italian Cottage, Jack-in-the-Box, Pepe's Pizza, Taco Bell/24hr, Yogurt Alley, Crystal Motel, Flamingo Motel, Sky Terrace Motel, Triangle Motel, Chevrolet/Olds/Cadillac, $Tree, Ford/Mercury, K-Mart, Kragen Parts, Radio Shack, Raley's Food/drug, Staples, Wal-Mart/auto, XpressLube, tires
655	Flores Ave, to Proberta, Gerber, **2 mi E**...Wal-Mart Dist Ctr
646	CA 811, Gyle Rd, to Tehama, **E**...RV camping(7mi)
644	Finnel Rd, to Richfield, no facilities
643mm	**rest area both lanes, full(handicapped)facilities, phone, picnic tables, litter barrels, petwalk**
641	A9, Corning Ave, Corning, **E**...VETERINARIAN, EYECARE, CHIROPRACTOR, Chevron/mart/ 24hr, Citgo/7-11, Exxon/mart, Shell/diesel/LP, 76/diesel/mart, Burger King, Casa Ramos Mexican, Marcos Pizza, Olive Pit Rest., Papa Murphy's, Rancho Grande Mexican, RoundTable Pizza, Taco Bell, AmeriHost, Best Western, Budget Inn, Economy Inn, Olive City Inn, 7 Inn Motel/ RV Park, Ace Hardware, Clark Drug, Ford/Mercury/Kia, Heritage RV Park, Holiday Mkt/deli, Kragen Parts, NAPA, Pontiac/Chevrolet/Buick/Olds, Rite Aid, Safeway/24hr, Tires+, laundry, tire/brake/lube/muffler, **W**...Giant Burger, Corning RV Park
640	South Ave, Corning, **E**... Petro/Exxon/diesel/rest., TA/Arco/Arby's/Subway/diesel/24hr, Jack-in-the-Box, McDonald's, Day's Inn, Shilo Inn/rest./24hr, Blue Beacon, Goodyear, RV Ctr, RV camping(1-6mi), Woodson Br SRA, hardware, **W**...Ron's Muffler
636	CA 99W, Liberal Ave, no facilities
634	CA 7, no facilities
629	CA 32, Orland, **E**...Gas4Less, 76, Burger King, Berry Patch Rest., Subway, Amberlight Motel, Orlanda Inn, **W**...Sportsmans/diesel/mart, USA/mart, Taco Bell, Green Acres RV Park, Old Orchard RV Park
626	CA 16, **E**...Beacon/diesel/mart/24hr, CFN/diesel, Pizza Factory, Orland Inn, Longs Drug, Schwab Tires, laundry
622	CA 27, no facilities
620	Artois, no facilities
618mm	**rest area both lanes, full(handicapped)facilities, phone, picnic tables, litter barrels, petwalk, RV dump**
616	CA 39, Blue Gum Rd, Bayliss, **2-3 mi E**...Blue Gum Motel/rest.
613	CA 162, Willows, to Oroville, **E**...HOSPITAL, Arco/mart/24hr, Chevron/mart/24hr, Shell/diesel/mart, BlackBear Diner, Burger King, Denny's/24hr, Java Jim's Café, KFC, McDonald's, RoundTable Pizza, Subway/TCBY, Taco Bell/24hr, Best Western/rest., CrossRoads West Inn, Day's Inn, Super 8/RV Park, JiffyLube, antiques, CHP, **W**...Nancy's Café/ 24hr, Wal-Mart, RV Park(8mi), airport
608	CA 57, **E**...CFN/diesel
603	Norman Rd, to Princeton, **E**...to Sacramento NWR
601	Delevan Rd, no facilities
595	to Maxwell(from sb), access to camping
594	Maxwell Rd, **E**...Delavan NWR, **W**...CFN/diesel, Chevron/mart, Maxwell Inn/rest., Maxwell Parts, NAPA
588	CA 20 W, Colusa, **W**...HOSPITAL, Beacon/diesel/mart, Orv's Rest., hwy patrol
587	CA 20 E, Williams, **E**...Texaco/Togo's/diesel/mart/24hr, Carl's Jr, Taco Bell, Holiday Inn Express, **W**...HOSPITAL, Arco/mart/24hr, CFN/diesel, Chevron/diesel/mart/atm/24hr, 76/mart, Shell/diesel/mart, A&W, Burger King, Cairo's Italian, Caliente Café, Dairy Queen, Denny's, Granzella's Deli, McDonald's/RV parking, Wendy's, Williams Chinese Rest., Capri Motel, Comfort Inn, El Rancho Motel, Granzella's Inn, Motel 6, StageStop Motel, Travelers Motel, NAPA, U-Haul, USPO, ValuRite Drug, bank, camping, hwy patrol, tires

California Interstate 5

586	Husted Rd, to Williams, no facilities
581	Hahn Rd, to Grimes, no facilities
577	frontage rd(from nb), to Arbuckle, **E**...Beacon/diesel/mart, El Jaliscience Mexican, **W**...CFN/diesel
576	Arbuckle, to College City, **E**...Pacific Pride/diesel, 76, Boyd's GasHouse Grill, **W**...Beacon/diesel/mart
568	Yolo/Colusa County Line Rd, no facilities
566mm	**rest area both lanes, full(handicapped)facilities, phone, picnic tables, litter barrels, petwalk**
566	E4, Dunnigan, **E**...Chevron/mart/24hr, Shell/diesel/mart, Bill&Kathy's Rest., Jack-in-the-Box, Best Value Inn, Best Western, USPO, **W**...76/mart, Camper's RV Park
563	rd 8, **E**...Pilot/Wendy's/diesel/mart/24hr, Budget 8 Motel, HappyTime RV Park, **W**...Beacon/diesel
562	I-505, to San Francisco, callboxes begin sb
557	Zamora, **E**...Pacific Pride/diesel, Texaco/mart, Zamora Minimart, camping, **W**...USPO
551	Yolo, **1 mi E**...gas
550	CA 16, Woodland, **3 mi W**...HOSPITAL
548	West St, **W**...Denny's
547	CA 113 N, E St, Woodland, **E**...Valley Oaks Inn, **W**...CFN/diesel, Shell, Denny's, Best Western, Woodland Opera House(1mi)
546	CA 113 S, Main St, Woodland, to Davis, **E**...Chrysler/Jeep, Nissan, Olds/Cadillac/Pontiac, tires, **W**...Exxon/diesel/mart, 76/diesel/mart, Burger King, Denny's, McDonald's, Primo's Rest., Subway, Taco Bell, Teriyaki Rest., Wendy's, Comfort Inn, Motel 6, Food4Less
544	CA 102, **E**...Exxon, Shell, Jack-in-the-Box, Target, museum, **W**...Chevron/mart, CHP
539	CA 22, W Sacramento, no facilities
538mm	Sacramento River
537mm	**rest area sb, full(handicapped)facilities, phone, picnic tables, litter barrels, petwalk**
535	Airport Rd, **E**...airport
533	CA 99, 70, to Marysville, Yuba City, no facilities
532	Del Paso Rd, **E**...Arco Arena, no facilities
531	I-80, E to Reno, W to San Francisco
530	Garden Hwy, **W**...Courtyard, Homestead Village, Residence Inn,
529	Richards Blvd, **E**...Chevron/mart/atm/24hr, Buttercup Pantry, Hungry Hunter Rest., Lyon's Rest./24hr, McDonald's, Monterey Rest., Rusty Duck Rest., Day's Inn, Governor's Inn, Hawthorn Suites, Super 8, **W**...Arco/mart/24hr, Shell, Perko's Café, Best Western/rest., Capitol Inn, Crossroads Inn, La Quinta, Motel 6
528	J St, Old Sacramento, Hist SP, **E**...Holiday Inn, Vagabond Inn, **W**...Railroad Museum
527	Q St, Sacramento, downtown
526	US 50, CA 99, Broadway, **E**...multiple facilities downtown
525	Sutterville Rd, **E**...76/diesel/repair, Shell/repair, China Wok, Land Park, zoo
524	Fruitridge Rd, Seamas Rd, no facilities
523	43rd Ave, Riverside Blvd, **E**...76/repair, 7-11, **W**...VETERINARIAN
522	Florin Rd, **E**...CHIROPRACTOR, Arco/mart/24hr, Chevron/mart/24hr, Shell/repair/24hr, Alicia's Mexican, Bill's Yogurt, RoundTable Pizza, Bel Air Foods, Kragen's Parts, Longs Drug, Subway, cleaners, **W**...Burger King, RoundTable Pizza, Shari's, Jo-Ann Crafts, Marshall's, MensWearhouse, Rite Aid, SuperSave Foods, bank
520	CA 160, Pocket Rd, Meadowview Rd, to Freeport, **E**...Shell/mart/wash/24hr, McDonald's, **W**...DENTIST, Chinese Rest., Pizza Hut, postal service, cleaners
516	La Guna Blvd, **E**...Chevron/mart, Shell/mart/wash, Texaco/mart
515	Elk Grove Blvd, no facilities
512	Hood-Franklin Rd, no facilities
504	Twin Cities Rd, to Walnut Grove, no facilities
501	Walnut Grove Rd, Thornton, **E**...76/diesel/mart, CGN/diesel, C&N Tire Repair
498	Peltier Rd, no facilities
495	Turner Rd, no facilities
493	CA 12, Lodi, **E**...Arco/diesel/mart/24hr, Chevron/diesel/LP, Exxon/Wendy's/mart, 76/Pacific Pride/diesel, Baskin-Robbins, McDonald's, Rocky's Rest./24hr, Subway, Taco Bell, Tower Park Marina Camping(5mi)
489	Eight Mile Rd, **3 mi E**...KOA
487	Hammer Lane, Stockton, **E**...CHIROPRACTOR, Arco/mart/24hr, 76/Circle K, Alberto's Mexican, KFC, Little Caesar's, S-Mart Foods, Xpress Parts, bank, cleaners, drugs, laundry, **W**...Exxon/mart/24hr, QuikStop/gas, Arby's, Burger King, Jack-in-the-Box, RoundTable Pizza, Subway, Taco Bell, Yogurt, InnCal, MailBoxes Etc
486	Benjamin Holt Dr, Stockton, **E**...Arco/mart, Chevron/mart/wash/24hr, QuikStop/gas, Pizza Guys, Motel 6, **W**...VETERINARIAN, Shell, 7-11, Lyon's/24hr, McDonald's, RoundTable Pizza, Subway, Ace Hardware, Sundance Sports, bank, cleaners
485	March Lane, Stockton, **E**...MEDICAL CARE, Citgo/7-11, Applebee's, Black Angus, Boston Mkt, Carl's Jr, Denny's, El Torito, Jack-in-the-Box, Marie Callender's, Taco Bell, Tony Roma, Wendy's, Ramada(2mi), Red Roof Inn, Traveler's Inn, Longs Drug, Marshall's, Nevada Bob's, S-Mart Foods, bank, **W**...76/diesel/mart/24hr, 7-11, Carrow's Rest., In-n-Out, Old Spaghetti Factory, La Quinta, Super 8, Travelodge

Sacramento

Stockton

N

S

California Interstate 5

484	Alpine Ave, Country Club Blvd, **E**...Shell/repair, **W**...Citgo/7-11, USA/Baskin-Robbins/Sub- way/ diesel/mart, Safeway
483	Monte Diablo Ave, **W**...Aamco
482	Oak St, Fremont St, **W**...Arco/mart/24hr, Catfish Café, Fremont River Inn
481	CA 4 E, to CA 99, Fresno Ave, downtown
479	CA 4 W, Charter Way, **E**...Chevron/24hr, Shell/mart/wash/24hr, Texaco/diesel/mart/ wash, Burger King, Denny's/24hr, McDonald's, Best Western, Motel 6, Charter Way Drug, Kragen Parts, transmissions, **W**...Vanco/76/diesel/rest./24hr, Exxon/Country MktPlace/TCBY/diesel, Carrow's Rest., Taco Bell, Motel 6
478	8th St, Stockton, **W**...DENTIST, CAStop/diesel/mart, SunRise Foods, Econolodge
477	Downing Ave, **W**...Food4Less
476	French Camp, Stockton Airport, **E**...Exxon/Togo's Eatery/diesel/mart, Pan Pacific RV, **W**...HOSPITAL
474	Mathews Rd, **E**...Exxon/diesel, CFN/diesel, RV Ctr, tires/repair, **W**...HOSPITAL
473	Sharpe Depot, Roth Rd, no facilities
471	Lathrop Rd, **E**...Chevron/mart, Exxon/mart, Joe's/Subway/TCBY/diesel/mart, Papa's Pizza, Day's Inn, **W**...Dos Reis CP, RV camping
470	Louise Ave, **E**...Arco/diesel/mart/24hr, 76/mart, Texaco/diesel, Carl's Jr, Denny's, Jack-in-the-Box, McDonald's, Taco Bell, Holiday Inn Express
468	CA 120, Manteca, to Sonora, **E**...Oakwood Lake Resort Camping
467	Manthey Rd, **E**...Texaco/diesel/mart
466	I-205, to Oakland(from sb, no return), no facilities
465	Defense Depot, to Tracy, **2 mi W**...gas/diesel/food
464	Kasson Rd, to Tracy, **W**...CFN/diesel
459	CA 33, Vernalis, no facilities
456	CA 132, to Modesto, **W**...The Orchard Campground
453	I-580(from nb, no return), no facilities
452mm	**Westley Rest Area both lanes, full(handicapped)facilities, picnic tables, litter barrels, phone, RV dumping, petwalk**
447	Westley, **E**...Beacon/diesel/mart, Chevron/diesel/mart/24hr, 76/diesel/mart/24hr, Shell/diesel, Westley/diesel/rest./24hr, Bobby Ray's Rest., Carl's Jr, McDonald's, PepperTree Rest., Budget Inn, Day's Inn, Holiday Inn Express, **W**...Shell/diesel/mart/Ingram Creek Coffee
439	Patterson, **E**...Arco/mart/24hr, Jack-in-the-Box
435mm	vista point nb
434	Crow's Landing, no facilities
429	Newman, **5 mi E**...HOSPITAL, food, lodging, RV camping
428	vista point sb
422	CA 140, Gustine, **E**...Shell/diesel/mart, Texaco/diesel/mart
413	weigh sta both lanes
412	CA 33, Santa Nella, **E**...76/diesel/rest./24hr, Texaco, Burger King, Carl's Jr, Del Taco, Andersen's Rest./motel, Best Western, Holiday Inn Express, **W**...Beacon/diesel/mart, Chevron/mart, Shell, Denny's, McDonald's, Taco Bell, Motel 6, Ramada Inn
408	CA 152, Los Banos, **6 mi E**...HOSPITAL, **W**...Super 8
396	CA 165, Mercy Springs Rd, **W**...Shell/mart
392	**rest area both lanes, full(handicapped)facilities, phone, picnic tables, litter barrels, petwalk**
391	Nees Ave, to Firebaugh, **W**...Texaco/CFN/diesel/mart
385	Shields Ave, to Mendota, no facilities
378	Russell Ave, no facilities
374	Panoche Rd, **W**...Chevron, Shell/mart, Texaco/Foster's Freeze/diesel/mart/burgers, Apricot Tree Motel/rest.
371	Manning Ave, to San Joaquin, no facilities
363	Kamm Ave, no facilities
354	CA 33 N, Derrick Ave, no facilities
342	CA 33 S, CA 145 N, to Coalinga
339	CA 198, Huron, to Lemoore, **E**...Shell/mart, Texaco/diesel, Harris Ranch Inn/rest., **W**...HOSPITAL, Chevron/diesel/ mart, CFN/diesel, Mobil/diesel, 76/mart, Burger King, Carl's Jr, Denny's, McDonald's, Oriental Express Chinese, Red Robin Rest., Taco Bell, Windmill Mkt/deli, Big Country Inn, Motel 6
329	Jayne Ave, to Coalinga, **E**...Arco/mart/24hr, Almond Tree RV Park, **W**...HOSPITAL
325mm	**rest area both lanes, full(handicapped)facilities, phone, picnic tables, litter barrels, petwalk**
323	CA 269, Lassen Ave, to Avenal, **W**...gas/diesel, food, lodging, st prison
313	CA 41, Kettleman City, **E**...Arco, Beacon/diesel/rest., Chevron/diesel/mart, Exxon/mart/24hr, 76, Shell/mart, Burger King, Carl's Jr, In-n-Out, Jack-in-the-Box, Major Bros Rest./24hr, McDonald's, Subway, Taco Bell, TCBY, Best Western, Super 8, tire/radiator/diesel repair, **W**...Arco/mart/24hr, Kettleman RV Park/dump/LP
309	Utica Ave, no facilities

California Interstate 5

B u t t o n w i l l o w	292	Twisselman Rd, no facilities
	282	CA 46, Lost Hills, **E**...Texaco/diesel/mart, to Kern NWR, **W**...Arco/mart/24hr, Beacon/diesel/mart, 76/mart/24hr, Chevron/diesel/mart/24hr, Carl's Jr, Denny's, Jack-in-the-Box, Day's Inn, Motel 6, KOA, tires
	272	Lerdo Hwy, to Shafter, no facilities
	267	Rowlee Rd, to Buttonwillow, no facilities
	262mm	**Buttonwillow Rest Area both lanes, full(handicapped)facilities, phone, picnic tables, litter barrels, petwalk**
	261	CA 58, Buttonwillow, to Bakersfield, **E**...Arco/mart/24hr, Bruce's/A&W/diesel/mart/24hr, Chevron/diesel/mart/24hr, Mobil/Subway/TCBY/diesel/mart, Shell, TA/76/Taco Bell/diesel/24hr, Texaco/diesel/mart, Burger King, Carl's Jr, Denny's, McDonald's, Willow Ranch Rest., Zippy Freeze, FirstValue Inn, GoodNite Inn, Motel 6, Super 8, **W**...Exxon/diesel/mart
	256	Stockdale Hwy, **E**...76/mart/24hr, Exxon/diesel/mart/24hr, Shell/mart, Jack-in-the-Box, Palm Rest., Perko's Rest., Best Western, **W**...Tule Elk St Reserve
	250	CA 43, to Taft, Maricopa, to Buena Vista RA
	247	CA 119, to Pumpkin Center, no facilities
	242	CA 223, Bear Mtn Blvd, to Arvin, **W**...RV camping
	237	Old River Rd, no facilities
	231	Copus Rd, no facilities
	227	CA 166, to Mettler, **2-3 mi E**...gas/diesel, food
	223	I-5 and CA 99(from nb, no return), no facilities
	220	Laval Rd, Wheeler Ridge, **E**...Chevron, TA/diesel/rest./mart, Burger King, Pizza Hut, Subway, Taco Bell, TCBY, truckwash/repair, **W**...Petro/Mobil/diesel/rest./24hr
	219	truck weigh sta, sb
	215	Grapevine, **E**...Denny's, Jack-in-the-Box, RanchHouse Rest., **W**...76/diesel, Texaco/Taco Bell/diesel/mart, Farmer's Table Rest., Countryside Inn
	211	Ft Tejon, garage/towing, to Ft Tejon Hist SP
	210	brake check area nb
	208	Lebec Rd, **W**...USPO, CHP, antiques, towing
	207mm	**rest areas both lanes, full(handicapped)facilities, phone, vending, picnic tables litter barrels, petwalk**
	206	Frazier Park, **W**...Exxon/Subway/diesel/mart, Flying J/diesel/LP/rest./motel/24hr, Texaco, Jack-in-the-Box, Los Pinos Mexican, Best Rest Inn, Auto Parts+, towing/repair/radiators/transmissions, to Mt Pinos RA
	205	Tejon Pass, elev 4144
	202	Gorman, to Hungry Valley, **E**...Chevron/diesel/LP, Texaco/diesel, Brian's Diner, Carl's Jr, Gorman Plaza Café, Sizzler/motel, **W**...Mobil/mart, McDonald's
	198	CA 138, Lancaster Rd, to Palmdale, no facilities
	196	Quail Lake Rd, no facilities
	193	Smokey Bear Rd, Pyramid Lake, **W**...Pyramid Lake RV Park
	189	Vista del Lago Rd, **W**...visitors ctr
	184mm	brake inspection area sb, motorist callboxes begin sb
	183	Templin Hwy, **W**...Ranger Sta
	176	Lake Hughes Rd, Castaic, **E**...VETERINARIAN, DENTIST, Arco, Citgo/7-11, Castaic Trkstp/diesel/rest./24hr, Giant/diesel/rest./24hr, Shell, Village Fuelstop/diesel, Burger King, Café Mike, Carl's Jr, Del Taco, Domino's, Foster Freeze, McDonald's, Subway, Zorba's Rest., Castaic Inn, Comfort Inn, Day's Inn, Ralph's Food, Rite Aid, to Castaic Lake, **W**...Mobil/mart, 76/Taco Bell/repair, Jack-in-the-Box, auto repair
S a n t a	174	Hasley Canyon Rd, **W**...CHIROPRACTOR, Ameci Pizza/pasta, cleaners
	173	CA 126 W, to Ventura, no facilities
	172mm	weigh sta sb
	171	Rye Canyon Rd, **W**...Shell/mart, Texaco/mart, amusement park
	170	CA 126 E, Saugus, Magic Mtn Pkwy, **E**...Dillon's Rest., Ranch House Inn, **W**...Chevron/mart, El Torito, Hamburger Hamlet, Marie Callender's, Red Lobster, Wendy's, Hilton, Magic Mtn Rec Park
	169	Valencia Blvd, no facilities
C l a r i t a	168	McBean Pkwy, **W**...HOSPITAL, Baskin-Robbins, Chili's, ChuckeCheese, ClaimJumper Rest., JambaJuice, Macaroni Grill, Starbucks, Subway, Bed Bath&Beyond, Circuit City, MailBoxes Etc, Michael's, Old Navy, Sports Chalet, Staples, Vons Foods, cleaners
	167	Lyons Ave, **E**...Chevron, 76, Texaco, Burger King, Chevrolet, **W**...Arco/mart/24hr, Mobil/Blimpie/mart/atm, Shell/repair, Carl's Jr, Foster's Freeze, In-n-Out, Del Taco, Denny's, El Pollo Loco, Fortune Express Chinese, Floridinos Italian, IHOP, Jack-in-the-Box, McDonald's, Outback Steaks, Taco Bell, Comfort Inn, Fairfield Inn, Hampton Inn, Residence Inn, Camping World RV Service, GNC, GoodGuys, JiffyLube, Linens'n Things, PetsMart, Ralph's Foods, Wal-Mart
	165	Calgrove Blvd, **E**...Big Boy, Carrows Rest., Canyon Auto Ctr, Nissan, Subaru
	162	CA 14 N, to Palmdale, no facilities
	161	Balboa Blvd, no facilities
	160	I-210, to San Fernando, Pasadena

N

S

California Interstate 5

159	Roxford St, Sylmar, **E...**Chevron/diesel, Mobil/diesel/atm, Denny's/24hr, McDonald's, Good Nite Inn, Motel 6
158	I-405 S(from sb, no return), no facilities
157	Mission Hill Blvd, **E...**HOSPITAL, Chevron, Mobil/mart/atm, 76/mart, Shell, Carl's Jr, In-n-Out, Pollo Gordo, Popeye's, Chief Parts, JiffyLube, tires, **W...**auto repair
156	CA 118, no facilities
155.5	Paxton St, **E...**Chevron, Shell
155	Van Nuys Blvd, **E...**DENTIST, Mobil, 76, Jack-in-the-Box, KFC, Mexican Rest., McDonald's, Pizza Hut, Popeye's, Chinese Foods, Chief Parts, bank, **W...**Domino's
154.5	Terra Bella St(from nb)
154	Osborne St, to Arleta, **E...**Arco, 76, Shell, Food4Less, Target, laundry **W...**Mobil/ Burger King/mart, 7-11, Mi Taco, Yummy Donuts, drugs, laundry
153	CA 170(from sb), to Hollywood, no facilities
152	Sheldon St, **E...**HOSPITAL, Big Jim's Rest., **W...**donuts/subs
151	Lankershim Blvd, Tucksford, **E...**Texaco/diesel, **W...**Tune-Up Masters, radiators
150.5	Penrose St, no facilities
150	Sunland Blvd, Sun Valley, **E...**Mobil, 76, 7-11, Acapulco Rest., China King, El Pollo Loco, Scottish Inn, **W...**CHIROPRACTOR, Shell/mart, Dimion's Rest., El Mexicano's, McDonald's
149.5	GlenOaks Blvd(from nb), **E...**Arco, motel
149	Hollywood Way, **W...**Texaco/mart, U-Haul, airport
148.5	Buena Vista St, **W...**World Gas, Jack-in-the-Box, Buena Vista Motel, Hilton, Ramada
148	Scott Rd, to Burbank, same as 147
147	Burbank Blvd, **E...**76, Baskin-Robbins, Carl's Jr, El Pollo Loco, Harry's Rest./24hr, IHOP, In-n-Out, Marie Callender's, Kenny Rogers Roasters, McDonald's, Popeye's, Shakey's Pizza, Spoon's Grill, Taco Bell, Tommy's Burgers, Holiday Inn, Barnes&Noble, Circuit City, CompUSA, Ikea Furnishings, K-Mart, Macy's, Mervyn's, Ralph's Foods, Ross, SavOn Foods, Sears, Sports Chalet, Von's Foods, bank, cinema, cleaners, **W...**VETERINARIAN, Chevron, Chinese Rest., Mexican Rest., Subway
146	Olive Ave, Verdugo, **E...**Black Angus, Bobby McGee's, Bombay Bicycle Club, Fuddrucker's, Holiday Inn Suites, cinemas, **W...**HOSPITAL
145	Alameda Ave, **E...**Chevron, **W...**Arco/mart, Shell, Burbank Inn Suites, U-Haul
144.5	Western Ave, **W...**Gene Autrey Museum
144	CA 134 E, Glendale, Pasadena, no facilities
143	Colorado St, **E...**bank
142	Los Feliz Blvd, **E...**HOSPITAL, **W...**Griffith Park, zoo
141	Glendale Blvd, **E...**Arco, Texaco, River Glen Motel, auto repair
140	Fletcher Dr, **E...**U-Haul, **W...**Arco/mart/24hr, Chevron, 76, Charburger, Rick's Drive-In, Ralph's Foods
139.5	CA 2, Glendale Fwy, no facilities
139	Stadium Way, **W...**to Dodger Stadium
138.5	Figueroa St, no facilities
138	CA 110, Pasadena Fwy, no facilities
137.5	Broadway St(from sb), industrial
137	Main St, **E...**HOSPITAL, Chevron/mart/24hr, 76, Burgers+, Chinatown Express, McDonald's, Mr Pizza, Parts+, Tune-up Masters
136	Boyle St, Mission Rd, **E...**HOSPITAL, Chevron/mart, 76, Jack-in-the-Box, Marengo Grill, McDonald's, Howard Johnson
135	I-10, E to San Bernadino, San Bernadino Fwy
134	4th St, Soto St, no facilities
133	I-10, W to Santa Monica, Santa Monica Fwy
132	CA 60 E, Pomona Fwy, no facilities
131	Ditman Ave, Indiana St, **E...**Arco, Texaco/diesel, **W...**HOSPITAL, Mobil, Shell/wash
130	I-710, to Long Beach, no facilities
129.5	Olympic Blvd, Kingface Rest.
129	Triggs St, **E...**Winchell's, outlet mall, **W...**76/diesel, Denny's/24hr, Steven's Steaks, Tru-Value Inn, Ford
128.5	Atlantic Blvd, Eastern Ave
128	Washington Blvd, Commerce, **E...**Chevron/diesel/24hr, Card Club Rest./casino, McDonald's, Wyndham Garden Hotel, Firestone, mall
127.5	Garfield Blvd, industrial area
127	Slauson Ave, Montebello, **E...**Shell, Holiday Inn, **W...**Arco/mart, Denny's, Ramada Inn, Travelodge
126	Paramount Blvd, Downey, no facilities
124	CA 19 S, Lakewood Blvd, HOSPITAL, **E...**Mobil, Arthur's Rest., Ford, bank

N

S

California Interstate 5

123   I-605, no facilities

122   Imperial Hwy, Pioneer Blvd, **E**...CHIROPRACTOR, DENTIST, Chevron, 76, Shell/mart, Bakers Square, Jack-in-the-Box, Mexican Rest., McDonald's, Red Lobster, Subway, Wendy's, Firestone/auto, Thrifty Drug, bank, **W**...EYECARE, Arco, Shell/wash, Denny's, Pizza Hut, Rally's, Sizzler, Whataburger, Comfort Inn, Keystone Motel, Westland Motel, Ford

121.5   San Antonio Dr, to Norwalk Blvd, **E**...Shell, Sheraton, **W**...Bronco Rest., auto parts

121   Rosecrans Ave, HOSPITAL, **E**...Mobil, **W**....Arco/mart/24hr, El Pollo Loco, Chevrolet, Nissan, Subaru/Infiniti, Tune-Up Masters, RV Ctr

120   Carmenita Rd, Buena Park, **E**...76, Carrows Rest., Big Boy, Jack-in-the-Box, Neil's Taco, Paul's Charbroiled Burger, Best Western, Motel 6, Ford Trucks, RV Ctr, **W**...Arco/mart/24hr, Mobil, Dynasty Suites, Super 8

119   Valley View Blvd, **E**...Holiday Inn, **W**...Chevron/mart, Texaco, Denny's, Taco Tio, Residence Inn, RV Sales/service, to Camping World, bank, cinema

118   Knott Ave, Artesia Blvd, **E**...Texaco/diesel, **W**...Chrysler/Plymouth, Knott's Berry Farm, to Camping World RV Service/supplies

117   CA 39, Beach Blvd, HOSPITAL, **E**...Johnny's Auto Parts/repair, BMW, Hyundai, Nissan, Pontiac, Toyota, VW, **W**...VETERINARIAN, Chevron, Arby's, KFC, Korean BBQ, Pizza Hut, Subway, Franklin Motel, Hampton Inn, to Knott's Berry Farm, hardware

116   CA 91 E, Riverside Fwy, airport, multiple services on adjacent rds

115.5   Magnolia Ave, frontage rds have multiple services

115   Brookhurst St, LaPalma, **E**...Chevron, **W**...Arco/mart/24hr, Shell, La Estrella Mexican, MinitLube

114   Euclid St, **W**...Mobil, Texaco/wash, Arby's, Denny's, Marie Callender's, Chevrolet, Ford, Anaheim Plaza

113   Lincoln Ave, to Anaheim, Lincoln Exit Motel

112   Ball Rd, HOSPITAL, **E**...Arco, Chevron/diesel, Citgo/7-11, Burger King, El Pollo Loco, McDonald's, Shakey's Pizza, Subway, Anaheim Motel, Courtesy Lodge, Day's Inn, Traveler's World RV Park, laundry, **W**...Arco/mart/24hr, Spaghetti Sta, Best Western, Rodeway Inn, Sheraton, Travelodge, Camping World RV Service/supplies, bank

111.5   Harbor Blvd, **W**...to Disneyland, same as 111

111   Katella Ave, Anaheim Blvd, **E**...diesel, Econolodge, **W**...Chevron, Mobil, Texaco, 7-11/24hr, Acapulco Mexican, Carl's Jr, Chinese Rest., Denny's, Flakey Jake's, IHOP, McDonald's, Subway, Tony Roma, Best Western, Candy Cane Inn, Caravan Inn, Desert Inn Suites, Econolodge, Grand Hotel, Hampton Inn, Hilton, Holiday Inn, Holiday Inn Express, Howard Johnson, Quality Inn, Residence Inn, Rip van Winkle, Saga Inn, Super 8, Travelodge, Village Inn, 1-Hr Photo, Disneyland, Jaguar, RV Ctr

110.5   St Coll Blvd, City Drive, **E**...Hilton Suites, **W**...HOSPITAL, Doubletree Hotel, bank

110   CA 57 N, Chapman Ave, **E**...Mobil, BeefRigger Rest., Burger King, Denny's, Motel 6, **W**...HOSPITAL, Ramada Inn

109   CA 22, Garden Grove Fwy, Bristol St, bank

107   N Broadway, Main St, **E**...76, Quick King Foods, BMW, Nordstrom's, Suzuki, Dowers Museum, cinema, mall, **W**...bank

106   17th St, Chevron, 76, Circle K, McDonald's, Parks Court Hotel, Top Tune, bank

105   Santa Ana Blvd, Grand Ave, Shell/autocare, Jack-in-the-Box, Winchell's, The Register Hotel, Buick, carwash, groceries

104   4th St, to CA 55 N, no facilities

103   CA 55 S, to Newport Beach, no facilities

102   Newport Ave, Arco, Chevron, Carl's Jr, Pioneer Chicken, Tustin Transmissions

101   Red Hill Ave, **E**...Mobil, Shell, Del Taco, Fat Freddie's, Key Inn, Drug Emporium, **W**...Chevron, Exxon/mart, 76/mart, Circle K, 50's Diner, U-Haul, carwash, bank

100.5   Tustin Ranch Rd, Tustin Auto Ctr/multiple auto dealers

100   Jamboree Rd, **E**...El Pollo Loco, Tustin MarketPlace, TJ Maxx, El Pollo Loco

99   Culver Dr, **E**...Shell/24hr, Bullwinkle's Rest., Denny's

98.5   Jeffrey Rd, no facilities

98   Sand Canyon Ave, Old Towne, **W**...76, La Quinta, Orange Inn/café/mkt, Tiajuana Mexican Rest., RV Ctr

97   CA 133, Laguna Fwy, Laguna Beach, no facilities

96   Alton Pkwy, **E**...Price Club

95.5   I-405 N(from nb)

95   Lake Forest Dr, Laguna Hills, **E**...CHIROPRACTOR, DENTIST, MEDICAL CARE, EYECARE, PODIATRIST, Chevron/24hr, Texaco/diesel/mart, Black Angus, Burger King, Chinese Rest., Country Rock Café, Del Taco, Fresca's Mtn Grill, Hungry Hunter, Jack-in-the-Box, McDonald's, Peppino's Café, Subway, Taco Bell, Teriyaki Japanese, Best Western, Irvine Suites, Travelodge, Audi/Jeep, Chevrolet, BMW, Ford/Lincoln/Mercury, GMC/Kia, Honda, Isuzu, Jaguar, Mazda, Mercedes/Suzuki, Subaru, Toyota, MailWorld, EconoLube/Tune, PepBoys, bank, **W**...CHIROPRACTOR, DENTIST, Chevron/mart/wash/24hr, Shell/mart/wash, Circle K, Carl's Jr, Coco's, Del Taco, McDonald's, NYC Café, Subway, Suzuki Seafood, Taco Villa, Comfort Inn, Courtyard, Quality Suites, Travelodge, AutoNation USA, AZ Leather, Books Etc, JC Penney Homestore, Kinko's/24hr, MailBiz+, cleaners

93   El Toro Rd, **E**...Arco/mart/24hr, Chevron/diesel/mart, Mobil, Shell/autocare/24hr, Texaco/diesel/mart, USA/gas, Arby's, Bakers Square, Baskin-Robbins, Chinese Rest., Denny's, Fuddrucker's, Jack-in-the-Box, KFC, McDonald's,

N

S

California Interstate 5

MegaBurger, Red Lobster, Scarantino's Rest., Souper Salad, Spoon's Grill, Wendy's, House of Fabrics, K-Mart, Office Depot, PetCo, Pier 1, SavOn Drug, bank, cleaners/laundry, cinema, **W...**HOSPITAL, DENTIST, Chevron/diesel/24hr, Shell/24hr, 76, Bennigan's, California Pizza, Carrows, Coco's, El Torito's, KooKooRoo Kitchen, Pizza Hut, RoundTable Pizza, Ruby's Aerodiner, Laguna Hills Lodge, B Dalton's, Circuit City, Firestone/auto, JC Penney, Longs Drugs, Macy's, Marshall's, Mens Wearhouse, Sears/auto, Tire Station, Walgreen, cinema, mall

92 Alicia Pkwy, Mission Viejo, **E...**DENTIST, Chevron/wash/24hr, 76/diesel, Carl's Jr, Denny's, Little Caesar's, Ristorante Italiano, Subway, Wendy's, Winchell's, Albertson's, Buick/Pontiac/Mazda, Firestone, Mervyn's, Kragen's Parts, Target, bank

90 La Paz Rd, Mission Viejo, **E...**VETERINARIAN, Arco/mart/24hr, Chevron, Mobil/mart, UltrMar/gas, Diedrich's Coffee, KFC, Taco Bell, Lucky SavOn, bank, cleaners, **W...**CHIROPRACTOR, OPTOMETRIST, Chevron, 76, Chinese Rest., Claim Jumper Rest., Dairy Queen, Elephant Barn Rest., Flamingo Mexican, Jack-in-the-Box, McDonald's, Outback Steaks, Roma D Italia, Shakey's Pizza, Holiday Inn, Best Buy, Borders Books&Café, Cloth World, CompUSA, Econo Lube'n Tune, Goodyear/auto, Linens'n Things, PetCo, Rei Outdoor Gear, SportMart, Winston Tires

89 Pacific Park Dr, Oso Pkwy, **E...**Chevron/service/mart, 76/service/mart, Fairfield Inn

87 Crown Valley Pkwy, **E...**HOSPITAL, CHIROPRACTOR, VETERINARIAN, Arco, Chevron, 76, Coco's, TJ's Mexican, Aamco, Macy's, bank, cinema, hardware, mall, **W...**Chevron

85 Avery Pkwy, **E...**CHIROPRACTOR, OPTOMETRIST, Texaco/Subway/mart, Blazing Saddle Steaks, Carrow's, Del Taco, Gecko Rest., Jack-in-the-Box, Mandarin Dynasty, McDonald's, Acura/Mercedes/Volvo, Goodyear, Hobby Shack, Kinko's, Land Rover, Lexus, Pier 1, Staples, bank, **W...**VETERINARIAN, Shell/diesel/mart/24hr, A's Burgers, Buffy's Rest., In-n-Out, Laguna Inn, Firestone/auto, Olds/Cadillac/GMC

84 CA 73 N toll

83 Junipero Serra Rd, to San Juan Capistrano, **W...**Shell/mart, UlraMar/diesel

82 CA 74, Ortego Hwy, **E...**DENTIST, MEDICAL CARE, CHIROPRACTOR, Chevron, Shell/mart/service, 76, Denny's, Santa Fe Grill, Best Western, XpressLube, **W...**Arco, Chevron/24hr, Carl's Jr, Del Taco, Jack-in-the-Box, McDonald's, Walnut Grove Rest., Mission Inn

81 San Juan Creek Rd(from sb), **E...**VETERINARIAN, Peugeot/VW, **W...**Chevron, Boston Mkt, Harry's Rest., KFC, Starbucks, Goodyear, Kinko's, Pick'n Save, Rite Aid, Ross, Von's Food, tires

80 CA 1, Pacific Coast Hwy, Capistrano Bch, **1 mi W...**Arco/mart/24hr, Carl's Jr, Del Taco, Jack-in-the-Box, McDonald's, Villa Mexican, Hilton, Ramada Inn, Sheraton, Chrysler/Jeep, Honda, Nissan, Saturn, USPO, hardware

79 Camino de Estrella, San Clemente, **E...**HOSPITAL,OPTOMETRIST,CHIROPRACTOR, PODIATRIST, 76/diesel/mart/24hr, Baker's Square, Big City Bagel, Boston Mkt, Café Expresso, Carl's Jr, China Wall, JuiceStop, RoundTable Pizza, Rubio's Grill, Subway, Wahoo's Fish Taco, Lucky Food, MailBoxes Etc, 1-Hr Photo, Ralph's Foods, SavOn Drug, bank, cinema, **W...**Arco/diesel/mart, K-Mart, bank, cleaners

78 Ave Pico, **E...**DENTIST, Mobil/mart, Carrow's, McDonald's, NY Pizza, Lucky Foods, bank, cleaners, **W...**VETERINARIAN, CHIROPRACTOR, Chevron, Texaco/diesel/mart, Del Taco, Denny's/24hr, Pizza Hut, Stuft Pizza, Surfin Chicken, Subway, Country Side Inn, GNC, MailBoxes Etc, Midas Muffler, Pier 1, Ralph's Foods, SavOn Drug, antiques, bank, cameras

77 Ave Palizada, Ave Presidio, **W...**DENTIST, Arco/mart, UltraMar/gas, 7-11, Antoine's Café, Baskin-Robbins, KFC, Holiday Inn, Albertson's, Ford, auto parts/repair, bank

76 El Camino Real, **E...**Chevron/24hr, Pedro's Tacos, BudgetLodge, same as 75, **W...**Mobil, 76, 7-11, FatBurger, HotDog Heaven/Chinese Rest., LoveBurger, Taco Bell, Tommy's Rest./24hr, Kragen Parts, Ralph's Foods, Top Tune

75 Ave Calafia, Ave Magdalena, **E...**76, Shell/mart, 7-11, Beef Cutter Rest., Burrito Basket Mexican, China Beach Rest., Coco's, El Marianchi Rest., Jack-in-the-Box, Taco Shop, The Shack Rest., C-Vu Inn, El Rancho Motel, LaVista Inn, Quality Suites, San Clemente Motel, Trade Winds Motel, Travelodge, Winston Tires, cleaners, **W...**San Clemente Inn

74 Cristianitios Ave, **W...**to San Clemente SP, RV camping/dumping

73 Basilone Rd, **W...**San Onofre St Beach

71mm weigh sta both lanes

70mm viewpoint sb

64 Las Pulgas Rd, no facilities

62 Aliso Creek rest area both lanes, full(handicapped)facilities, phone, vending, picnic tables, litter barrels, petwalk, RV dump

N

S

Mission Viejo

California Interstate 5

Oceanside

58	Oceanside Harbor Dr, **W**...Chevron, Mobil, Shell, Burger King(1mi), Del Taco, Denny's/24hr, Jolly Roger Rest., Sandman Hotel, Travelodge, The Bridge Motel, to Camp Pendleton
57	Hill St(from sb), to Oceanside, **W**...Shell, Carrow's Rest., Bridge Motor Inn
54	CA 76, Mission Ave, Oceanside, **E**...Arco/mart/24hr, Mobil/diesel, Shell/diesel, 76/LP, Arby's, Burger King, El Charrito Mexican, Jack-in-the-Box, KFC, McDonald's, Mission Donuts, Pizza Hut, Rally's, Taco Shop, Grandee Inn/rest., Welton Inn, Econo Lube'n Tune, NAPA, cleaners, **W**...Burger Joint, Carrow's Rest., El Pollo Loco, Giant Pizza, Long John Silver, Grocery Outlet, Office Depot, Radio Shack, Rite Aid, carwash
53	Oceanside Blvd, **E**...CHIROPRACTOR, DENTIST, Mobil, Boney's MktPlace, Domino's, IHOP, McDonald's, Red Dragon Chinese, RoundTable Pizza, Pizza Hut, Subway, Taco Bell, Ralph's Food, Von's Food, Longs Drug, CHP, **W**...Texaco/24hr, Best Western
52	Cassidy St(from sb), **W**...HOSPITAL
51	CA 78, Vista Way, Escondido, **E**...CHIROPRACTOR, Chevron/mart, Texaco, Applebee's, Boston Mkt, Carl's Jr, Coco's, Denny's, Einstein Bagels, McDonald's, Mimi's Café, Olive Garden, Rubio's Grill, Souplantation, StarBucks, Tony Roma, Bed Bath&Beyond, JC Penney, Kids R Us, LensCrafters, Lucky-SavOn, Macy's, MailBoxes Etc, Marshall's, Michael's, Robinson-May, Saturn, Sears/auto, SportsAuthority, Staples, Super Crown Books, Wal-Mart, mall, **W**...Texaco, Hunter Steaks
50.5	Las Flores Dr, no facilites
50	Elm Ave, Carlsbad Village Dr, **E**...Shell/auto/24hr, Mexican Rest., **W**...Chevron/24hr, 76, Texaco/diesel, Carl's Jr, Denny's/24hr, McDonald's, Albertson's
49.8	Tamarack Ave, **E**...Chevron, Texaco/diesel, Tamarack Rest., Carlsbad Lodge, Super 8, Travel Inn, Thrifty Food/drug, **W**...76, Texaco/24hr, Gerico's Rest.
49.5	Cannon Rd, Car Country Carlsbad, **E**...Acura, Chevrolet/Cadillac, Ford, Isuzu, Lexus, Mazda, Mercedes, Toyota, VW, bank
49	Carlsbad Blvd, Palomar Airport Rd, **E**...Mobil/diesel/mart, Citgo/7-11, PeaSoup Anderson's Rest., Denny's, Hadley Orchard's Foods, Mexican Rest., Subway, Best Western, Motel 6, Travelodge, Costco, Honda, Volvo, cleaners, **W**...Texaco/diesel, ClaimJumper Rest., In-n-Out, Marie Callender's, McDonald's, S Carlsbad St Bch
48	Poinsettia Lane, **W**...CHIROPRACTOR, DENTIST, VETERINARIAN, Jack-in-the-Box, Japanese Rest., Raintree Grill, Subway, Inn of America, Motel 6, Ramada, Travelodge, Ralph's Food, Payless Drug, Honda, Volvo, 1-Hr Photo, MailBoxes Etc, bank
46	La Costa Ave, **E**...vista point, **W**...Chevron/diesel/mart
44	Leucadia Blvd, **E**...Holiday Inn Express, **W**...Shell/auto, Texaco/diesel/mart
42	Encinitas Blvd, **E**...Chevron, Texaco, Baskin-Robbins, Coco's, Del Taco, HoneyBaked Ham, Hungry Hunter Rest., Lucky Foods, SavOn Drug, Nissan, bank, to Quail Botanical Gardens, **W**...Shell/mart/wash, Cinnamon's Rest., Italian Rest., Taco Bar, Wendy's, Budget Motel, cleaners/laundry
40	Santa Fe Dr, to Encinitas, **E**...Shell, 7-11/24hr, Carl's Jr, Hide Away Café, Papa Tony's Pizza, laundry, **W**...HOSPITAL, 76, Straw Hat Pizza, Thrifty Foods, bank, cleaners
39	Birmingham Dr, **E**...Chevron, Mobil/diesel, Texaco, IHOP, Taco Bell, Countryside Inn, **W**...Arco/mart/24hr
38mm	viewpoint sb
37	Manchester Ave, **E**...76, to MiraCosta College
36	Lomas Santa Fe Dr, Solana Bch, **E**...Baskin-Robbins, Roadhouse Grill, **W**...DENTIST, OPTOMETRIST, Mobil/diesel/mart, Texaco/diesel/mart, Carl's Jr, Chinese Rest., RoundTable Pizza, Discount Tires, Ross, SavOn Drug, Von's Foods, bank
35	Via de La Valle, Del Mar, **E**...Arco/mart, Chevron, Mobil, Burger King, Chevy's Mexican, McDonald's, Papa Chino's Italian, Tony Roma's, Albertson's, Pier 1, **W**...Texaco/diesel, Denny's, Hilton, racetrack
34	Del Mar Heights Rd, **E**...Texaco/diesel, **W**...Citgo/7-11, Jack-in-the-Box, Mexican Grill, Von's Foods, Longs Drug, bank
33	Carmel Valley Rd, **E**...Chevron, Shell/mart, Olive Garden, Taco Bell, Tio Leo's Mexican, Doubletree Hotel
32	I-805(from sb), no facilities
31	Genesee Ave, **E**...HOSPITAL, Embassy Suites, **1 mi W**...gas, food, lodging
29	La Jolla Village Dr, **E**...HOSPITAL, Italian Bistro, Embassy Suites, Hyatt, Marriott, to LDS Temple, **W**...Mobil/diesel/mart, BJ's Grill, Calif Pizza, Domino's, El Torito, Island's Burgers, Samson's Deli, TGIFriday, Trader's Joe's, Radisson, Linens'n Things, Marshall's, Ralph's Foods, Sav-On Drug, Whole Foods, cinema, cleaners
28	Nobel Dr(from nb), **E**...Hyatt, LDS Temple, **W**...same as 29
27	Gilman Dr, no facilities
26	CA 52, San Clemente Canyon, Ardath Rd(from nb), no facilities
25	CA 274, Balboa Ave, **E**...Mobil, Travelodge, RV camping, **W**...HOSPITAL, Citgo/7-11, Mobil, Shell/diesel/24hr, 76, Arby's, In-n-Out, Comfort Inn, SleepyTime Motel, Super 8, Ford, Nissan, Plymouth, Toyota, Winston Tires, Mission Bay Pk
24	Grand Ave, Garnet Ave(from nb), **W**...Cadillac
23	Clairemont Dr, Mission Bay Dr, **E**...76, Shell, Jack-in-the-Box, Day's Inn, Chevrolet/VW, DaNino's Pizza, **W**...Chinese Rest., KFC, Best Western, access to Sea World Dr
22	Sea World Dr, Tecolote Dr, **E**...Arco, Shell, Circle K, Hilton, Firestone, Old Town SP

N

S

California Interstate 5

20	I-8, W to Nimitz Blvd, E to El Centro, CA 209 S(from sb), to Rosecrans St
19	Old Town Ave, **E**...Arco/mart/24hr, Shell, Ramada Inn, Travelodge
18	Washington St, **E**...Comfort Inn, laundry
17	Sassafras St, Front St, **E**...Mobil, 76, **W**...Shell, Super 8, civic center, airport
16.5	Front St, **W**...HOSPITAL, Shell, bank
16	CA 163 N, 10th St, **W**...HOSPITAL, Shell, Budget Motel, Flamingo Lodge, Holiday Inn, Radisson, Marriott, bank
15.7	Pershing Dr, B St, downtown
15.5	CA 94 E, no facilities
15	Imperial Ave, **E**...Shell, **W**...downtown
14	Crosby St, **W**...Econolodge
13.5	CA 75, to Coronado, **W**...toll rd to Coronado
13	National Ave SD, 28th St, **E**...Chief Parts, **W**...Texaco/mart
12.8	CA 15 N, to Riverside, no facilities
12.4	Main St, National City, no facilities
12	8th St, National City, **E**...Mobil, Shell/24hr, Budget Inn, EZ 8 Motel, Holiday Inn, Nite Lite Inn, Ramada Inn, Sun Coast Inn, Value Inn, **W**...MEDICAL CARE, Nat City Cars
11	Civic Center Dr, no facilities
10	24th St, **1/2 mi E**...Denny's, In-n-Out
9.5	CA 54 E, no facilities
9	E St, Chula Vista, **E**...Mobil, **W**...Anthony's Fish Grotto, GoodNite Inn
8	H St, **E**...HOSPITAL, Arco/diesel, Chevron, 7-11, Dodge, Goodyear, motel
7	J St(from sb), **W**...Marina Pkwy
6	L St, gas/diesel, food
5	Palomar St, **E**...Arco, Palomar Inn
4	Main St, to Imperial Beach, **E**...Chinese Rest., Mexican Rest.
3.5	CA 75(from sb), Palm Ave, to Imperial Beach, **E**...Arco, Chevron, 7-11, Papa John's, Gator's Parts, Midas Muffler, **W**...CHIROPRACTOR, DENTIST, Arco, Shell/repair/24hr, 7-11, Boll Weevil Diner, Carl's Jr, Carrow's Rest., El Pollo Loco, Lydia's Mexican, McDonald's, Rally's, Si Alberto's Mexican, Subway, Taco Bell, Wienerschnitzel, Silverado Motel, Super 8, Chief Parts, Home Depot, JiffyLube, Mervyn's, SavOn Drug, Taylor Auto Parts, TuneUp Masters, VitaLube, Von's Foods
3.2	Coronado Ave(from sb), **E**...Chevron/service, Shell/service, 7-11, Denny's, EZ 8 Motel, San Diego Motel, **W**...Arco/mart/24hr, Texaco/diesel, South Bay Lodge, to Border Field SP, wildlife viewing
3	CA 905, Tocayo Ave, no facilities
2	Dairy Mart Rd, **E**...Arco/mart/24hr, Circle K, Al's Donuts, Antonio's Pizza, Burger King, Carl's Jr, Coco's, McDonald's, Americana Inn, Super 8, Valli-Hi Motel, E-Z MailBoxes, Parts+, Radio Shack, cleaners, RV camping
1	Via de San Ysidro, **E**...Chevron, Mobil/mart, 76, Shell/mart/24hr, Max's Foods, NAPA, bank, **W**...Chevron/mart, Denny's, KFC, Econo Motel, Frontier Motel, International Inn/RV park, Motel 6
.5	Willow St(from sb), **E**...Burger King, El Pollo Loco, Jack-in-the-Box, KFC, McDonald's, Subway, Gateway Inn, Holiday Motel, Travelodge, Mercado Sonoro, **W**...Taco Bell, factory outlet, border parking
0	US/Mexico Border, California state line, customs, I-5 begins/ends.

California Interstate 8

Exit #	Services
172	Colorado River, California/Arizona state line, Pacific/Mountain time zone
171	4th Ave, Winterhaven Dr, Yuma, **N**...Ft Yuma Casino, **S**...Chevron/mart, 76/Circle K, Shell/mart/24hr, Texaco, Domino's, Jack-in-the-Box, Little Caesar's, Mi Rancho Mexican, Yuma Landing Rest., Best Western, Interstate 8 Inn, Regalodge Motel, Yuma Inn, Rivers Edge RV Park, to Yuma SP
167	Seminole Rd, to Winterhaven, **S**...Rivers Edge RV Park
164	CA 186, Algodones Rd, Andrade, **S**...to Mexico
163mm	CA Insp Sta
162	Sidewinder Rd, **N**...st patrol, **S**...Shell/LP, Pilot Knob RV Park
157	CA 34, Ogilby Rd, to Blythe, no facilities
156	Grays Well Rd, no facilities
155mm	**rest area both lanes, full(handicapped)facilities, picnic tables, litter barrels, petwalk**
154	Gordons Well, no facilities

N

S

E

W

San Diego Area

Yuma

39

California Interstate 8

150	Brock Research Ctr Rd, no facilities
141	CA 98, Midway Well, to Calexico, no facilities
130	CA 115, Vanderlinden Rd, to Holtville, **5 mi N**...gas, food, lodging, **S**...RV camping
127	Bonds Corner Rd, no facilities
124	Orchard Rd, Holtville, **4 mi N**...gas/diesel, food
118	Bowker Rd, no facilities
117	CA 111, to Calexico, **1/2 mi N**...Texaco/diesel/mart/café, RV park
115	Dogwood Rd, **N**...RV camping, funpark
114	4th St, El Centro, **N**...CHIROPRACTOR, Arco/mart/24hr, Chevron, Citgo/7-11/diesel, Texaco/diesel/mart, USA/mart, Carl's Jr, Foster's Freeze, Jack-in-the-Box, Lucky Chinese, McDonald's, Mexican Rest., Rally's, Motel 6, Firestone/auto, Goodyear/auto, Ford/Lincoln/Mercury, MailBox, Midas, SW SuperMercado, U-Haul, radiators, **S**...Mobil/A&W/diesel/mart, Millie's Kitchen, Taco Bell, Best Western, EZ 8 Motel, Motel 31, Buick/Cadillac/Pontiac, Chevrolet, Honda, Max Foods
113	Imperial Ave, El Centro, **N**...MEDICAL CARE, DENTIST, Arco/mart/24hr, Chevron/service, Citgo/7-11/diesel, Texaco, USA/diesel/mart, Burger King, Chinese Rest., Church's Chicken, Del Taco, Denny's/24hr, Domino's, Jack-in-the-Box, KFC, LJ Silver, McDonald's, Pizza Hut, Sizzler, Wendy's, Brunner's Motel, Crown Motel, Day's Inn, Laguna Inn, Ramada, Sands Motel/rest., Travelodge, Vacation Inn/RV Park, Aamco, Chief Parts, Goodyear/auto, JC Penney, Kragen Parts, Longs Drug, Thrifty Foods, XpressLube, auto/brake repair, st patrol
108	Forrester Rd, to Westmorland, no facilities
107mm	**Sunbeam Rest Area both lanes, full(handicapped)facilities, phone, picnic tables, litter barrels, petwalk, RV dump**
106	Drew Rd, Seeley, **N**...RV camping, **S**...RV camping
100	Dunaway Rd, Imperial Valley, elev 0 ft, **N**...st prison
92	Imperial Hwy, CA 98, Ocotillo, **N**...Shell/diesel/mart, Lazy Lizard Saloon, **S**...Desert FuelStop/gas/diesel/café/repair, RV camping
91	CA 98(from eb), to Calexico, no facilities
88mm	runaway truck ramp, eb
87	Mountain Springs Rd, no facilities
81	In-Ko-Pah Park Rd, phone, towing
80mm	brake insp area, phone
78	Jacumba, **S**...Shell/diesel/mart, Texaco/diesel/towing/mart/24hr, NAPA, RV camping
70	CA 94, Boulevard, to Campo, **S**...Buena Vista Motel, gas
67mm	Tecate Divide, elev 4140 ft
65	Crestwood Rd, Live Oak Springs, **S**...Country Broiler Rest., FoodSource/diesel/24hr, Live Oak Sprs Country Inn, RV camping, info
64mm	Crestwood Summit, elev 4190 ft
56	Kitchen Creek Rd, Cameron Station, **S**...gas/diesel/LP, food, RV camping
53	rd 1, Buckman Spgs Rd, to Lake Morena, **S**...gas/diesel/LP, food, lodging, RV camping, Potrero CP(19mi), **rest area both lanes, full(handicapped)facilities, phone, picnic tables, litter barrels, petwalk, RV dump**
49	rd 1, Sunrise Hwy, Laguna Summit, elev 4055 ft, **N**...to Laguna Mtn RA
47	Pine Valley, Julian, **N**...Major's Diner, to Cuyamaca Rancho SP, gas/diesel, lodging
46mm	Pine Valley Creek
44mm	elev 4000 ft
42	CA 79, Japatul Rd, Descanso, **N**...to Cuyamaca Rancho SP, gas, food
39mm	vista point eb, elev 3000 ft
38	E Willows, **N**...Alpine Sprs RV Park, Viejas Indian Res, casino
36	Willows Rd, to Alpine, **N**...Alpine Sprs RV Park, Viejas Res, casino, **S**...ranger sta
34mm	elev 2000 ft
33	Tavern Rd, to Alpine, **N**...Alpine/diesel/mart/24hr, Texaco/diesel/mart, **S**...MEDICAL CARE, PODIATRIST, Circle K/gas, Shell/mart, Carl's Jr, La Carreta Mexican, LJ Silver, Mandarin Chinese, Mediterraneo Italian, Countryside Inn, Radio Shack, bank, cleaners, drugs, groceries, hardware, city park
30	Dunbar Lane, Harbison Canyon, **N**...Flinn Sprs Inn(1mi), RV camping
29mm	elev 1000 ft
28mm	map stop wb, phone
27	Lake Jennings Pk Rd, Lakeside, **N**...to Lake Jennings CP, RV camping, **S**...Citgo/7-11, Marechiaro's Italian, Flinn Sprs CP

E

W

California Interstate 8

26   Los Coches Rd, Lakeside, **N**...Circle K/gas, Citgo/7-11, Mobil/service, Laposta Mexican, Mike's Pizza, PillCo Drug, hardware, RV camping, **S**...DENTIST, Shell/Blimpie/diesel/mart, McDonald's, NY Pizza, Subway, Taco Bell, MailBoxes Etc, Radio Shack, Von's Foods, Wal-Mart, cleaners

25   Greenfield Dr, to Crest, **N**...HOSPITAL, Shell, Texaco/mart, 7-11, Janet's Café, McDonald's, YumYum Donuts, RV Ctr, RV camping, auto repair, st patrol, **S**...Mobil/LP/mart

24   E Main St(from wb, no EZ return), **N**...Main St Grill, Pernicano's Italian, Budget Inn, Embasadora Motel, HP Motel, Ford, RV Ctr, auto repair, **S**...Coco's, Cadillac, Kinko's, RV Ctr, carwash

23   2nd St, CA 54, El Cajon, **N**...DENTIST, Arco/mart/24hr, Chevron/mart, Mobil/diesel, Marechiaro's Italian, Taco Shop, Midas/tires, Parts+, Von's Foods, auto repair, **S**...DENTIST, 76, Shell, Texaco/A&W/diesel/mart, Arby's, Baskin-Robbins, Boll Weevil Diner, Burger King, Carl's Jr, Chinese Cuisine, Dairy Queen, Golden Corral, IHOP, Jack-in-the-Box, KFC, McDonald's, Pizza Hut, Taco Bell/24hr, Wings'n Things, Firestone/auto, Henry's MktPlace, InstaLube, PetCo, Ralph's Foods, Rite Aid, auto/parts/repair, bank

22   Mollison Ave, El Cajon, **N**...Chevron, Citgo/7-11, Denny's, Best Western, **S**...Arco/mart/24hr, Taco Bell/24hr, Super 8, Valley Motel

21   Magnolia Ave, CA 67(from wb), to Santee, **N**...LJ Silver, JC Penney, mall, same as 19, **S**...Texaco/service/mart, Mexican Rest., Perry's Café, Wienerschnitzel, MidTown Motel, Motel 6, Travelodge, Nudo's Drug, cinema, tires

20   CA 67(from eb), **N**...JC Penney, Sears/auto, mall, same as 19 & 21

19   Johnson Ave(from eb), **N**...CHIROPRACTOR, DENTIST, Applebee's, Boston Mkt, Burger King, Chinese Rest., LJ Silver, Subway, Albertson's, Chevrolet, Home Depot, Honda, JC Penney, K-Mart, Marshall's, Mervyn's, Nevada Bob's, PetsMart, Rite Aid, Robinson-May, Sears/auto, bank, mall, **S**...Isuzu, Saturn, Aamco

18   Main St, **N**...Arco/mart/24hr, 7-11, Denny's, Sombrero Mexican, Lexus, **S**...76, Chevron, Nissan, brakes/transmissions

17   El Cajon Blvd(from eb), **N**...Day's Inn, **S**...Mobil/diesel, Shell, BBQ, Chrysler/Plymouth

16.5   Severin Dr, Fuerte Dr(from wb), **N**...Arco/mart/24hr, Mobil, 7-11, Holiday Inn Express, Charcoal House Rest., **S**...Brigantine Seafood Rest.

16   CA 125, to CA 94, no facilities

15   Jackson Dr, Grossmont Blvd, **N**...VETERINARIAN, EYECARE, Arco/mart/24hr, Chevron/wash, Mobil, Shell, 7-11/24hr, Arby's, Burger King, Chili's, Chinese Rest., ChuckeCheese, Fuddrucker's, KFC, Olive Garden, Red Lobster, Taco Bell, Barnes&Noble, Computer City, KidsRUs, Kinko's, Kragen Parts, Longs Drugs, Macy's, Staples, Target, USA SuperSports, USPO, Ward's/auto, bank, cinema, mall, **S**...DENTIST, Chile Bandido, HoneyBaked Ham, Jack-in-the-Box, Circuit City, Discount Tire, Firestone/auto, Ralph's Foods, Sports Authority, ToysRUs, VW

14   Spring St(from eb), El Cajon Blvd(from wb), **N**...Dodge/Kia, Jeep/Eagle, **S**...DENTIST, Travelodge, BEST, Travelodge

13   Fletcher Pkwy, to La Mesa, **N**...MEDICAL CARE, DENTIST, Texaco, 7-11, Baker's Square, Boston Mkt, Chili's, McDonald's, EZ 8 Motel, Babys R Us, Costco Whse, Havoline Lube, Lucky Foods, cleaners, **S**...La Salsa Mexican, Motel 6, Chevrolet, InstaLube, RV Park

12   70th St, Lake Murray Blvd, **N**...MEDICAL CARE, Shell/mart/wash, Pepper's Mexican, Subway, **S**...HOSPITAL, CHIROPRACTOR, VETERINARIAN, Mobil, Shell/24hr, Texaco/diesel/repair, 7-11, Aiken's Deli, Denny's, La Casa Maria Mexican, Marie Callender's, **1/4 mi S**...multiple services on El Cajon Blvd

11   College Ave, **N**...Chevron, **S**...MEDICAL CARE, to San Diego St U

10   Waring Rd, **N**...Nicolo's Italian, GoodNite Inn, Madrid Suites, Motel 7

9   Fairmont Ave, to Mission Gorge Rd, **N**...HOSPITAL, CHIROPRACTOR, Arco/mart/24hr, Citgo/7-11, Mobil/diesel/wash, Shell/mart, UltraMar/diesel, Arby's, Baskin-Robbins, Boll Weevil Diner, Boston Mkt, Burger King, Carl's Jr, Chili's, Chinese Rest., Coco's, Happy Chef, Jack-in-the-Box, El Pollo Loco, KFC, Krazy Pete's Diner, McDonald's, Rally's, Subway, Taco Bell, Tio Leo's Mexican, Super 8, Brake Depot, Discount Tire, Home Depot, Longs Drugs, Midas, NAPA Autocare, PetCo, Radio Shack, Rite Aid, Toyota, TuneUp Masters, Valvoline, Von's Foods, **S**...truck rental

8   I-15 N, CA 15 S, to 40th St

7   I-805, N to LA, S to Chula Vista

6   Texas St, Stadium Way, **N**...Chevron, same as 5

E / **W**

California Interstate 8

San Diego Area

5   Mission Ctr Rd, **N**...Chevron, Bennigan's, Canyon Café, Hogi Yogi, Hooters, In-n-Out, Mandarin Cuisine, Quizno's, Rubio's Grill, Rusty Pelican Rest., Bed Bath&Beyond, Borders Books&Café, Chevrolet, Ford, Macy's, Michael's, Nordstrom Rack, Old Navy, Robinson-May, Sak's Off 5th, Staples, bank, cinema, mall, **S**...DENTIST, Arco/mart/24hr, Denny's, Wendy's, Comfort Inn, Hilton, Radisson, Red Lion Inn, Chrysler/Plymouth, GMC/Pontiac, Mazda, Subaru/Olds

4   CA 163, Cabrillo Frwy, **S**...to downtown, zoo

3a   Hotel Circle, Taylor St, **N**...Chevron, DW Ranch Rest., Villanti's Rest., Best Western Hanalei, Comfort Suites, Handlery Hotel, Motel 6, Premier Inn, cinema, **S**...VETERINARIAN, Chevron/mart, Albie's Rest., Valley Kitchen, Best Western, Day's Inn, Holiday Inn, Hotel Circle Inn, Howard Johnson, King's Inn/rest., Quality Resort, Ramada Inn, Regency Plaza, Travelodge, Vagabond Inn

3   I-5, N to LA, S to San Diego

2   Rosecrans St(from wb), CA 209, **S**...Chevron, Burger King, Del Taco, Denny's, In-n-Out, Jack-in-the-Box, McDonald's, Rally's, Totio Carlos Mexican, Arena Inn, Best Western, Day's Inn, Holiday Inn, Quality Inn, Rio Motel, Super 8, Big 5 Sports, Chrysler/Jeep, Circuit City, Goodyear/auto, House of Fabrics, JiffyLube, MensWhse, Midas Muffler, PetsMart, Staples, SaveOn Drug, TJ Maxx, cleaners

1   W Mission Bay Blvd, Sports Arena Blvd(from wb), **S**...Arby's, Coco's, Embers Café, McDonald's, EZ 8 Motel, Holiday Inn Express, auto supply

0mm   I-8 begins/ends on Sunset Cliffs Blvd, **N**...Mission Bay Park, **1/4 mi W**...Mobil/service, Shell, Anthony's Rest., Chris' Deli, Jack-in-the-Box, Kaiserhof Deli, Valvoline, auto repair

California Interstate 10

California uses road names/numbers. Exit number is approx mile marker.

Exit #	Services
244mm	Colorado River, California/Arizona state line, Pacific/Mountain time zone
243.5mm	inspection sta wb
243	Riviera Dr, **S**...Riviera RV Camp
242	US 95, Intake Blvd, Blythe, **N**...Shell/mart, Texaco/mart, UltraMar/gas, Judy's Rest., Best Western, Desert Winds Motel, diesel repair/24hr, to Needles, **S**...McIntyre Park
241	7th St, **N**...MEDICAL CARE, Chevron/service/mart, EZ Mart/gas, Blimpie, Foster's Freeze, Blythe Inn, Budget Inn, Comfort Suites, Dunes Motel, Albertson's, CarQuest, Chief Parts, Chrysler/Dodge, Ford, Rite Aid, auto repair/tires, RV repair/LP
240	Lovekin Blvd, Blythe, **N**...MEDICAL CARE, Mobil/Subway/diesel/mart, Shell/service, Carl's Jr, Del Taco, Jack-in-the-Box, La Casita Dos Mexican, McDonald's/playplace, Pizza Hut, Popeye's, Sizzler, Best Western, Comfort Inn, Econolodge, EZ 8 Motel, Hampton Inn, Travelodge, Tourest Motel, Checker Auto Parts, K-Mart, Winston Tire/auto, hardware, **S**...Arco/mart/24hr, Chevron/diesel/mart/24hr, 76/diesel, Shell, Texaco/Dairy Queen/mart, UltraMar/gas, Burger King, Denny's, KFC, Phoenix Pizza, Townes Square Café/24hr, Taco Bell, Holiday Inn Express, Motel 6, Super 8, city park/RV dump
237	CA 78, Neighbours Blvd, to Ripley, **N**...Texaco/service, **S**...to Cibola NWR
234	Mesa Dr, **N**...Chevron/mart, 76/diesel/rest./24hr, airport, **S**...Mesa Verde/diesel/mart
233	weigh sta wb
224	Wileys Well Rd, **N**...**rest area both lanes, full(handicapped)facilities, phone, picnic tables, litter barrels, petwalk, S**...to st prison
217	Ford Dry Lake Rd, no facilities
208	Corn Springs Rd, no facilities
197	CA 177, Rice Rd, Desert Center, to Lake Tamarisk, **N**...Stanco/diesel/repair/24hr, Family Café, USPO, camping
194	Eagle Mtn Rd, no facilities
187	Red Cloud Rd, no facilities
180	Hayfield Rd, no facilities
175	Chiriaco Summit, **N**...Chevron/diesel/mart/café/24hr, Patton Museum, truck/tire repair
159	to Twentynine Palms, to Mecca, Joshua Tree NM, **N**...wildlife viewing
155	frontage rd, no facilities
151mm	**Cactus City Rest Area both lanes, full(handicapped)facilities, picnic tables, litter barrels, petwalk**
148mm	0 ft elevation
147	Dillon Rd, to CA 86, to CA 111 S, Coachella, **N**...Chevron/mart/24hr, Shell/mart/24hr, **S**...Burns Bros/Taco Bell/TCBY/diesel/mart/24hr, casino
145	CA 111 N, CA 86 S, Indio, **N**...Holiday Inn Express, Fantasy Sprgs Casino, RV camping, **S**...casino, **1 mi S**...Auto Ctr Dr, Audi/VW, Chevrolet/Olds, Ford/Lincoln/Mercury, Isuzu, Mazda, Nissan

E

W

E

W

Blythe

California Interstate 10

144	Jackson St, Indio, **S**...Circle K/gas, Big A Parts, NAPA, Circle K/gas, auto repair
143	Monroe St, Central Indio, **N**...RV camping, **S**...HOSPITAL, 76, Shell/diesel/mart/ LP, Circle K/gas, Alicia's Mexican, Carrow's, Denny's, Jerry's Rest., Best Western, Comfort Inn, Motel 6, Super 8, Target, auto/transmission repair
141	Jefferson St, Indio Blvd, **N**...RV camping, hwy patrol, **2 mi S**...Best Western, Motel 6
139	Washington St, Country Club Dr, to Indian Wells, **N**...Arco/mart/24hr, Del Taco, Motel 6, Ford, Honda, auto mall, RV camping, **S**...CHIROPRACTOR, DENTIST, MEDICAL CARE, Mobil/diesel/mart, 76/mart, Circle K, Carl's Jr, Goody's Café, Italian Café, Lili's Chinese, Mexican Rest., Subway, Goodyear/auto, cleaners
137	Cook St, to Indian Wells, no facilities
136	Monterey Ave, Thousand Palms, **N**...Arco/mart/24hr, **S**...Bubba Bears Pizza, IHOP, Taco Bell, CostCo/gas, Home Depot, PetsMart, carwash
133	Ramon Rd, Bob Hope Dr, **N**...Chevron/mart/24hr, Flying J/Conoco/diesel/LP/ rest./24hr, Mobil/diesel/mart, UltraMar/gas, Circle K, Burger King, Carl's Jr, Del Taco, Denny's, In-n-Out, McDonald's/playplace, Traveler's Inn, tires, truckwash, **S**...HOSPITAL
128	Date Palm Dr, Rancho Mirage, **S**...Arco/mart/24hr, Mobil/mart/wash, UltraMar/ gas, Domino's
125	Palm Dr, to Desert Hot Sprgs, **3 mi S**...gas, food, lodging, to Gene Autry Trail
122	Indian Ave, to N Palm Spgs, **N**...Arco/mart/24hr, 76/mart, Shell/mart/LP, Denny's, Motel 6, NAPA, **S**...HOSPITAL, Pilot/Dairy Queen/Wendy's/diesel/24hr
119	CA 62, to Yucca Valley, Twentynine Palms, to Joshua Tree NM
117	White Water, many windmills
116mm	**rest area both lanes, full(handicapped)facilities, phone, picnic tables, litter barrels**
114	CA 111(from eb), to Palm Springs, no facilities
112	Verbenia Ave, no facilities
110	Main St, to Cabazon, **N**...Texaco/diesel/mart, Burger King
109	Cabazon, **N**...Desert Hills Factory Stores/famous brands, Shell/diesel/mart/wash/24hr, Hadley Orchard Fruits, Morongo Casino/rest.
106	Fields Rd, **N**...McDonald's, Desert Hills Factory Stores/famous brands, Morongo Reservation/ casino
105mm	Banning weigh sta
103	Ramsey St, no facilities
102	Hargrave St, Banning, **N**...Shell/mart/wash, Texaco/diesel/mart/LP, Desert Star Motel, Parts+, transmissions
101	CA 243, 8ᵗʰ St, Banning, **N**...MEDICAL CARE, DENTIST, UltaMar/diesel, Banning Burger, Chinese Rest., Paradise Pizza, AutoValue, laundry, **S**...RV camping
100	22ⁿᵈ St, to Ramsey St, **N**...Arco/mart/24hr, Mobil/mart, Carl's Jr, Carrow's, Del Taco, KFC, McDonald's, Pizza Hut, Raliberto's Mexican, Sizzler, Taco Bell, Wendy's, Day's Inn, Hacienda Inn, Super 8, Travelodge, Chrysler/ Dodge, Ford, Winston Tire, repair/tuneup
99	Sunset Ave, Banning, **N**...DENTIST, Texaco/diesel/24hr, UltraMar/gas/24hr, Billy T's Rest. Domino's/Donut Factory, Gus Jr #7 Burger, Subway, AutoZone, Chief Parts, Chevrolet/Olds, Radio Shack, Save-U Foods, Thrifty Drug, bank, laundry, hardware, RV Ctr
98	Highland Springs Ave, **N**...Chevron/mart, UltraMar/diesel, Burger King, Denny's, FarmHouse Rest., Guy's Italian, Jack-in-the-Box, Little Caesar's, Subway, Food4Less, Kragen Parts, Radio Shack, Stater Bros Foods, hardware, **S**...DENTIST, Arco/mart/24hr, Mobil/mart, Baskin-Robbins, Carl's Jr, Albertson's, K-Mart, Payless Drug, bank, cleaners, hwy patrol
97	Pennsylvania Ave, Beaumont, **N**...Circle K/gas, ABC Rest., carwash, RV Sales
96	CA 79, Beaumont, **N**...CHIROPRACTOR, Arco, 76, Baker's DriveThru, Donald's Tacos, McDonald's, El Rancho Steaks, YumYum Donuts/24hr, Best Western, Budget Host, auto parts, bank, **S**...Denny's, RV camping
95	CA 60, to Riverside, no facilities
94	San Timoteo Canyon Rd, no facilities
93mm	**rest area wb, full(handicapped)facilities, phone, picnic tables, litter barrels, petwalk**
91	Cherry Valley Blvd, no facilities
90	Singleton Rd(from wb), no facilities
89	Calimesa Blvd, **N**...Arco/mart/24hr, Shell/mart/wash, Burger King, McDonald's, Subway, Taco Bell, Calimesa Inn, Stater Bros Foods, bank, **S**...Big Boy
88	County Line Rd, to Yucaipa, **N**...Texaco, Del Taco, Calimesa Inn, auto repair/tires
87mm	**Wildwood Rest Area eb, full(handicapped)facilities, phone, picnic tables, litter barrels, petwalk**
86	Live Oak Canyon Rd, **N**...Cedar Mill Rest.

E
↑
↓
W

California Interstate 10

San Bernardino

85	Yucaipa Blvd, **N**...Arco/mart/24hr, Baker's DriveThru, **S**...antiques
84	Wabash Ave(from wb), no facilities
83	Redlands Blvd, Ford St, **S**...76, Dillon's Steak/seafood, Griswold's Smorgasbord
82	Cypress Ave, University St, **N**...HOSPITAL, to U of Redlands
81	CA 38, 6th St, Orange St, Redlands, **N**...Arco/mart, Chevron, Taco Ray, Budget Inn, Stater Bros Foods, Winston Tire, **S**...CHIROPRACTOR, EYECARE, Boston Mkt, Albertson's, Kragen Auto Parts, NAPA
80	Tennessee St, **S**...Texaco/mart, Arby's, Bakers DriveThru, Burger King, Carl's Jr, Coco's, El Pollo Loco, Foster's Donuts, LJ Silver, Subway, Taco Bell, Best Western, Ford, Tri-City Mall, USPO, cleaners
79	CA 30, to Highlands, access to Tri-City Mall
78	Alabama St, **N**...Chevron, Denny's, Motel 6, 7 West Motel, Super 8, Ford, VW, U-Haul, **S**...DENTIST, Texaco/A&W/diesel/mart, Chinese Rest., Del Taco, IHOP, McDonald's, Nick's Burgers, Zabella's Mexican, GoodNite Inn, Chief Auto Parts, Big 5 Sports, Chevrolet, Econo Lube'n Tune, Goodyear/auto, JiffyLube, K-Mart, Mervyn's, Midas Muffler, Nissan, PepBoys, Ross, Tri-City Mall, autoplex services, hardware
77	California St, **N**...DENTIST, funpark, museum, **S**...DENTIST, Arco/mart/24hr, 76/wash, Shell/mart/24hr, Applebee's, Jack-in-the-Box, Jose's Mexican, Food4Less, MailBoxes Etc, Wal-Mart, RV camping
76	Mountain View Ave, Loma Linda, **N**...Mobil/diesel/mart, **S**...CHIROPRACTOR, Arco/mart/24hr, FarmerBoy's Burgers, Luby's Mexican, Subway
75	Tippecanoe Ave, Anderson St, **N**...Arco, Shell, In-n-Out, **S**...DENTIST, 76, Baker's DriveThru, Del Taco, HomeTown Buffet, KFC, Kool Kactus Café, Taco Bell, Audi, Honda, Jaguar, Saab, Saturn, antiques, transmissions, to Loma Linda U
74	Waterman Ave, **N**...76, Shell/diesel/mart/24hr, Black Angus, Bobby McGee's, Beef Bowl Chinese, Chili's, Coco's, El Torito, Guadalaharry's, IHOP/24hr, Lotus Garden Chinese, Olive Garden, Outback Steaks, Red Lobster, Sizzler, Souplantation, Spoon's Grill, TGIFriday, Thai Garden, Tony Roma, Yamazato Japanese, Comfort Inn, EZ 8 Motel, Hilton, La Quinta, Super 8, Travelodge, Best Buy, Circuit City, CompUSA, Home Depot, Office Depot, PetsMart, Sam's Club, SportsMart, cinema, **S**...Arco/mart/24hr, Beacon/diesel/rest., Carl's Jr, Gus Jr Burger #8, McDonald's, Popeye's, Taco Bell, Motel 6, CostCo, Econo Lube'n Tune, HomeBase, Nevada Bob's Golf, Staples, Camping World RV Service/supplies, laundry
73	I-215, CA 91, no facilities
72.5	Mt Vernon Ave, Sperry Ave, **N**...Arco/mart/24hr, Valley/diesel/LP/rest., Peppersteak Rest., Pepito's Mexican, Colony Inn, Colton Motel, Rio Inn, brake/muffler
72	9th St, **N**...Mobil/mart/wash, Baskin-Robbins, Burger King, Carrow's Rest./24hr, Denny's, Foster's Donuts, Jeremiah's Steaks, KFC, LaVilla Rest., McDonald's, Taco Bell, ThriftLodge, Colony Inn, Big A Parts, Stater Bros Foods, hardware
70	Rancho Ave, **N**...Arco/diesel/mart, Del Taco, Wienerschnitzel, Winner's Pizza
69.5	Pepper Dr, **N**...Shell, Baker's DriveThru, Ford, frontier town, tires
69	Riverside Ave, to Rialto, **N**...Arco/mart/24hr, Chevron, 10/diesel/rest., China Palace, HomeTown Buffet, Izumi Japanese, Jack-in-the-Box, McDonald's, Taco Joe's, Best Western, Rialto Motel, Ross, Valley View Motel, Wal-Mart/McDonald's, diesel repair
67	Cedar Ave, to Bloomington, **N**...Arco/mart/24hr, Chevron/mart, Mobil/mart, Baker's DriveThru, Burger King, USPO, **S**...Citgo/7-11
65	Sierra Ave, to Fontana, **N**...HOSPITAL, CHIROPRACTOR, DENTIST, Arco/mart/24hr, Mobil/mart, Shell, Texaco/mart, Applebee's, Arby's, Burger King, China Panda, ChuckeCheese, Coco's, Dairy Queen, Denny's, Del Taco, In-n-Out, KFC, McDonald's, Millie's Kitchen, Sizzler, Spires Rest./24hr, Taco Bell, Yoshinoya Japanese, Wienerschnitzel, Comfort Inn, Travelodge, Motel 6, Aamco, Chief Parts, Chevrolet, Food4Less, Goodyear/auto, Honda/GMC, Kids R Us, K-Mart, Lucky SavOn, Mazda, Nissan, PepBoys, Radio Shack, Stater Bros Foods, Von's Foods, bank, cinema, **S**...Circle K, Burger Basket, Fosters Freeze, Ikea Furnishings, Mervyn's, Old Navy, Ross, Target
64	Citrus Ave, **N**...UltarMar/gas, Baker's DriveThru, **S**...Arco/mart/24hr
62	Cherry Ave, **N**...Arco/mart/24hr, Mobil/diesel/24hr, Jack-in-the-Box, Trucktown/diesel/café, Circle Inn Motel, Ford Trucks, **S**...Circle K/gas, FarmerBoys Burgers, Peterbilt
60	Etiwanda Ave, no facilities
59	I-15, N to Barstow, S to San Diego
58	Milliken Ave, **N**...Chevron/mart, Mobil/Subway/diesel/mart, 76/Del Taco/diesel/mart, Texaco/diesel/mart, Burger King, Carl's Jr, Coco's, Cucina Italian, Dave&Buster's, FoodCourt, McDonald's, NY Grill, RainForest Café, Rubio's Rest., Tokyo Japanese, Wienerschnitzel, Wolfgang Puck Café, AmeriSuites, Country Suites, America's Tire, Bed Bath&Beyond, Burlington Coats, JC Penney, Marshall's, SportsAuthority, TJ Maxx, cinema, mall, **S**...TA/76/diesel/rest./24hr, RV Ctr
57	Haven Ave, Rancho Cucamonga, **N**...Mobil/mart, Black Angus, El Torito, Spoon's Grill, Tony Roma, Extended Stay America, Hilton, Holiday Inn, La Quinta, **S**...Panda Chinese, TGIFriday, Fairfield Inn, Econo Lube'n Tune, bank
56	Holt Blvd, to Archibald Ave, **N**...MEDICAL CARE, Mobil/diesel/mart, Circle K, Burger Town, Joey's Pizza, Subway, cleaners, bank, **S**...to airport

E

W

California Interstate 10

E

W

Ontario

Los Angeles Area

55   Vineyard Ave, **N**...DENTIST, Chinese Rest., Del Taco, Popeye's, Rocky's Foods, Sizzler, Taco Bell, Chief Parts, Rite Aid, Stater Bros Foods, **S**...Arco/mart/24hr, Mobil, Texaco, Circle K, Cuisine of India, Denny's, In-n-Out, Marie Callender's, Michael J's Rest., Rosa's Italian, Spires Rest., Yoshinoya Japanese, Best Western, Country Suites, DoubleTree Inn, Express Inn, GoodNite Inn, Red Roof Inn, Residence Inn, Super 8, Chevrolet/Olds, to airport

54   San Bernardino Ave, 4th St, to Ontario, **N**...DENTIST, Chevron/service, Circle K/ gas, Shell/wash, Baskin-Robbins, Burger King, Carl's Jr, Chinese Rest., Del Taco, Jack-in-the-Box, Popeye's, Sizzler, Taco Bell/24hr, Motel 6, Quality Inn, Chief Parts, K-Mart, Radio Shack, Ralph's Foods, Rite Aid, **S**...Arco/mart/24hr, Texaco/A&W/diesel/24hr, Denny's, Gordo's Mexican, KFC, McDonald's, Pizza Hut, Yumyum Donuts, CA Inn, Travelodge, West Coast Inn

53   CA 83, Euclid Ave, to Ontario, Upland, **N**...Coco's Rest.

51   Mountain Ave, to Mt Baldy, **N**...DENTIST, Arco, Chevron/mart, Mobil/mart, Texaco/A&W/diesel/mart, BBQ, Carrow's, Denny's, El Burrito, El Torito, Green Burrito, Happy Wok Chinese, HoneyBaked Ham, Mimi's Café, Mi Taco, Subway, Trader Joe's, Home Depot, Longs Drug, Mervyn's of CA, Staples, Super 8, cinema, cleaners, **S**...VETERINARIAN, 76/diesel, Baskin-Robbins, Carl's Jr, Tacos Mexico, Albertson's, Food4Less, Rite Aid, Target, USPO, cinema, funpark

49   Central Ave, to Montclair, **N**...MEDICAL CARE, Chevron, Mobil, Texaco/24hr, 7-11, Burger King, El Pollo Loco, KFC, McDonald's, Subway, Tom's Burgers, Circuit City, Firestone/auto, Goodyear/auto, Hi-Lo Auto Supply, JC Penney, Linens'n Things, Macy's, Mens Wearhouse, Montgomery Ward/auto, Office Depot, PetCo, Ross, Sears/auto, SportMart, bank, mall, same as 48, **S**...DENTIST, 76, Jack-in-the-Box, LJ Silver, Wienerschnitzel, K-Mart

48   Monte Vista, **N**...HOSPITAL, Texaco, Black Angus, Olive Garden, Red Lobster, Tony Roma, LensCrafters, Nordstrom's, Robinson-May, cinema, mall, same as 49

47   Indian Hill Blvd, to Claremont, **N**...DENTIST, Mobil, Bakers Square, Tony Roma, Howard Johnson, Travelodge, **S**...MEDICAL CARE, DENTIST, CHIROPRACTOR, Chevron/ McDonald's/mart, 76/diesel, Shell, 7-11, Burger King, Carl's Jr, Chili's, In-n-Out, RoundTable Pizza, Wienerschnitzel, Ramada Inn, America's Tire, Chief Parts, Ford, Lucky SavOn, Pick'n Save, Radio Shack, Toyota, laundry

46b   Towne Ave, **N**...Jack-in-the-Box

46a   Garey Ave(from eb), to Pomona, **N**...HOSPITAL, DENTIST, Arco/mart, **S**...Arco/mart/24hr, Chevron, Shell, cleaners

45   White Ave, Garey Ave, to Pomona, same as 46a

44b a   Dudley St, Fairplex Dr, **N**...76, Denny's, Sheraton, **S**...Chevron/mart/24hr, Mobil/mart, Texaco/diesel, 7-11, McDonald's, Lemontree Motel

43   I-210 N, CA 57 S, no facilities

42   Kellogg Dr, to Cal Poly Inst

40   Via Verde, no facilities

39   Holt Ave, to Covina, **N**...Blake's Steaks/seafood, Embassy Suites

38   Barranca St, Grand Ave, **N**...DENTIST, Chevron/service, Shell/autocare/mart, BBQ, Charley Brown Steaks/ lobster, Coco's, El Torito, Magic Recipe Rest., Marie Callender, Mariposa Mexican, Monterrey Rest., Best Western, Hampton Inn, Holiday Inn, cleaners, **S**...In-n-Out Burger, McDonald's, Comfort Inn

37   Citrus Ave, to Covina, **N**...Chevron, Mobil, Shell, Burger King, Del Taco, IHOP, Jack-in-the-Box, TGIFriday, Winchell's, Acura, Babys R Us, Burlington Coats, GMC, Honda, Lincoln/Mercury, Lucky Foods, Marshall's, Mervyn's, Office Depot, Old Navy, Ross, Ralph's Foods, Target, Volvo/VW, bank, **S**...HOSPITAL, 76/autocare, Trader Joe's, Comfort Inn, 5 Star Inn, Buick/Cadillac

36   CA 39, Azusa Ave, to Covina, **N**...DENTIST, Arco/24hr, Chevron, 76, Black Angus, McDonald's, Papa John's Pizza, Red Lobster, Steak Corral, Subway, El Dorado Motel, Big 5 Sports, Chrysler/Plymouth, Circuit City, Econo Lube'n Tune, Kinko's, Midas Muffler, **S**...Mobil, Shell, Carrow's, Jaguar, Mazda, Mercedes, Mitsubishi, Nissan, Saab, Saturn, Toyota

35   Vincent Ave, Glendora Ave, **N**...Arco/mart/24hr, Chevron/mart/24hr, Mobil, 76, KFC, Pizza Hut, Wienerschnitzel, **S**...Applebee's, Chevy's Mexican, JC Penney, LensCrafters, Linens'n Things, Macy's, Mens Wearhouse, Nevada Bob's, Robinson-May, Sears/auto, bank, cinema, laundry, mall, tires

34   Pacific Ave, **N**...Covina Motel, Hi-Lo Auto Supply, **S**...HOSPITAL, Mobil, K-Mart, Goodyear/auto, House of Fabrics, JC Penney, Sears, bank, mall, same as 35

33a   La Puente Ave, **N**...Chevron/mart, China House, McDonald's, Radisson, Staples, **S**...Arco, Texaco/diesel, Howard Johnson, Mercedes, U-Haul

California Interstate 10

32	Francisquito Ave, to La Puente, **N**...Tuneup Masters, 1-Hr Photo, hwy patrol, **S**...Chevron, Carl's Jr, Wienerschnitzel, Travelodge
32a	Frazier St, Baldwin Pk Blvd, **N**...HOSPITAL, Arco, Chevron, Shell/mart, 7-11, Jack-in-the-Box, McDonald's, Taxcos Mexican, Aristocrat Motel, Royal Knight Motel, Food4Less, OfficeMax, **S**...In-n-Out Burger, RV Ctr
31	I-605 S, to Long Beach
30	Valley Blvd, Peck Rd, **N**...MEDICAL CARE, DENTIST, Chevron, Motel 6, Ford, Honda, Nissan, Toyota/Lexus, **S**...Mobil, Texaco, Del Taco, McDonald's, PepBoys, Pontiac/GMC Trucks
29.5	S Peck Rd(from eb), no facilities
29	Santa Anita Ave, to El Monte, **N**...Texaco/diesel, Chevrolet, **S**...76, 7-11
28	Baldwin Ave(from eb), **S**...Arco/mart/24hr, Denny's, Edward's Steaks, same as 27b a
27b a	CA 19, Rosemead Blvd, Pasadena, **N**...MEDICAL CARE, Rodeway Inn, Goodyear/auto, KidsRUs, ToysRUs, **S**...bank
27	Walnut Grove Ave, no facilities
26	San Gabriel Blvd, **N**...MEDICAL CARE, DENTIST, CHIROPRACTOR, Arco, Mobil/mart, Shell/autocare, Carl's Jr, Popeye's, Taco Bell, Teriyaki Bowl, Wienerschnitzel, Best Motel, Chief Parts, San Gabriel Foods, **S**...Arco, Burger King, SavOn Drug
25b	Del Mar Ave, to San Gabriel, **N**...76, auto repair
25	New Ave, to Monterey Park, **N**...Mobil, to Mission San Gabriel
24	Garfield Ave, to Alhambra, **N**...HOSPITAL, **S**...Shell/mart
23	Atlantic Blvd, Monterey Park, **N**...HOSPITAL, DENTIST, Mobil/mart, 76, Del Taco, Pizza Hut, Popeye's, **S**...Firestone/auto, carwash
22b	Fremont Ave, **N**...HOSPITAL, tuneup, **S**...7-11, donuts
22	I-710, Long Beach Fwy, Eastern Ave(from wb)
21	Eastern Ave(from eb), City Terrace Dr, **S**...Chevron/service/mart, McDonald's
20	Soto St, **N**...HOSPITAL, Shell, **S**...Mobil, Shell/mart, Burger King
19	I-5(from wb), N to Burbank, S to San Diego
18	State St, **N**...HOSPITAL
17	US 101, to LA
16	4th St, downtown
15.5	I-5 S(from eb)
15	Santa Fe Ave, San Mateo St, **S**...Shell, Hertz Trucks, industrial area
14	Alameda St, **N**...downtown, **S**...industrial area
13	Central Ave, **N**...Shell/repair
12.5	San Pedro Blvd, **S**...industrial
12.3	LA Blvd, **N**...conv ctr, **S**...Kragen Parts, Radio Shack, Rite Aid, services on Washington Blvd
12	I-110, Harbor Fwy, no facilities
11a	Hoover St, Vermont Ave, **N**...Mobil, Texaco, Burger King, King Donuts, PepBoys, Thrifty Drug, Toyota, **S**...Chevron, Shell, Jack-in-the-Box, Office Depot, Staples
11	Western Ave, Normandie Ave, **N**...Chevron, Mobil/mart, McDonald's, Winchell's, Chief Parts, Food4Less, Radio Shack, SavOn Drug, **S**...Chevron, Texaco, Goodyear
10	Arlington Ave, **N**...Chevron, Mobil/mart
9	Crenshaw Blvd, **S**...Chevron, Thrifty Gas, U-Haul, tires
8	La Brea Ave, **N**...Chevron, Shell/diesel/repair, Walgreen, transmissions, **S**...Chevron, Chief Parts, Ralph's Foods
7	Washington Blvd, Fairfax Ave, **S**...Mobil, same as 8
6	La Cienega Blvd, Venice Ave(from wb), **N**...Chevron/mart/24h, Firestone/auto, **S**...Arco/mart/24hr, Mobil/mart, Carl's Jr, McDonald's, Pizza Hut, Subway, Staples
5	Robertson Blvd, Culver City, **N**...Arco/mart, Chevron/mart, Mobil, EZLube, museum, tires, **S**...Del Taco, OfficeMax, Lucky Sav-On Drug, Ross, cleaners
4	National Blvd, **N**...76, 7-11, Rite Aid, Von's Foods, 1-Hr Photo, laundry
3	Overland Ave, **S**...Arco/mart/24hr, Mobil/mart, 76, Shell, Winchell's, JiffyLube
2	I-405, N to Sacramento, S to Long Beach, no facilities
1d	Bundy Dr, **N**...76, Eddy's Café, Taco Bell, Camera City
1c	Centinela Ave, to Santa Monica, **N**...Don Antonio's Mexican, Taco Bell, **1/4 mi S**...McDonald's, Trader Joe's, Santa Monica Hotel, bank
1b	20th St(from eb), Cloverfield Blvd, 26th St(from wb), **N**...HOSPITAL, Arco, Shell/repair/mart
1a	Lincoln Blvd, CA 1 S, **N**...Denny's, Holiday Inn, House of Fabrics, Macy's, Midas, Sears, auto repair, mall, **S**...DENTIST, Chevron/mart/24hr, Jack-in-the-Box, Firestone/auto, TuneUp Masters, U-Haul
.5	4th, 5th,(from wb) **N**...Macy's, Sears
0	Santa Monica Blvd, to beaches, I-10 begins/ends on CA 1.

Los Angeles Area

E

W

California Interstate 15

California uses road names/numbers. Exit number is approximate mile marker.

Exit #	Services
292	California/Nevada state line, facilities located at state line, Nevada side.
285	Yates Well Rd, no facilities
280	Nipton Rd, **E**...E Mojave Nat Scenic Area, to Searchlight
275	Bailey Rd, no facilities
274mm	brake check area for trucks, nb
268	Cima Rd, **E**...gas/diesel/mart/LP
266mm	**Valley Wells Rest Area both lanes, full(handicapped)facilities, phone, picnic tables, litter barrels, petwalk**
253	Halloran Summit Rd, **E**...Hilltop Gas/diesel, towing/tires/repair
244	Halloran Springs Rd, **E**...gas/diesel/café
241	to Baker(from sb), access to same as 239
239	CA 127, Kel-Baker Rd, Baker, to Death Valley, **W**...Arco/mart/24hr, Chevron/Taco Bell/mart/atm/24hr, Mobil/Denny's/diesel/mart, 76/diesel/mart, Shell/diesel/mart, Texaco/mart, Arby's, BunBoy Rest., Burger King, Del Taco, Los Dos Toritos Mexican, Mad Greek Café, Royal Hawain Motel, Arnold's Mkt/repair, Baker Auto Parts, Baker Mkt Foods, Gen Store/NAPA/atm, World's Largest Thermometer, hardware, repair
238	to Baker(from nb), access to same as 239
236	Zzyzx Rd, no facilities
233	Rasor Rd, **E**...gas/diesel/mart
223	Basin Rd, no facilities
215	Afton Rd, to Dunn, no facilities
211mm	**rest area both lanes, full(handicapped)facilities, phone, picnic tables, litter barrels, petwalk**
208	Field Rd, no facilities
201	Harvard Rd, to Newberry Springs, **W**...Jeremy's/diesel/mart, Lake Dolores Resort/WaterPark(1mi), Twin Lakes RV Park
195	Minneola Rd, **W**...Mobil/diesel/mart/café, to Lake Dolores Resort(5mi)
194mm	agricultural insp sta sb
190	to Yermo, no facilities
189	Calico Rd, **E**...International Café
187	Ghost Town Rd, **E**...Arco/mart/24hr, Vegas/diesel/rest./atm/24hr, Peggy Sue's 50s Diner, Calico Motel, **W**...Clink's #2/Shell/diesel/mart, Texaco/mart, Jenny Rose' Rest., Calico GhostTown(3mi), KOA, to USMC Logistics
185	Ft Irwin Rd, **W**...76/mart
181	CA 58 W, to Bakersfield, **W**...RV camping
180	E Main, Barstow, to I-40, **E**...DENTIST, Herbst/mart, Mobil/diesel/mart, Shell/autocare, Burger King, McDonald's, MegaTom 's Burgers, Straw Hat Pizza, Taco Bell, Best Western, Gateway Motel, mall, **S of I-40**...Arco/mart/24hr, Texaco/diesel/mart, Wal-Mart/McDonald's/auto, **W**...DENTIST, EYECARE, VETERINARIAN, Arco/mart/24hr, Chevron/mart, 76/Circle K/Subway/TCBY/atm, Shell/mart, Arby's, Burger King, Carl's Jr, Carrow's Rest., China Gourmet, Coco's, Del Taco, Denny's, Golden Dragon, IHOP, Jack-in-the-Box, KFC, Sizzler, Taco Bell, AstroBudget Motel, Best Motel, Brant's Motel, Cactus Motel, Comfort Inn, Desert Inn, Econolodge, Executive Inn, Holiday Inn/rest., Quality Inn, Stardust Inn, Kragen Parts, Radio Shack, Thrifty Discount, U-Haul/LP, Von's Food/drug, bank, laundry
179	I-40 E(from nb), I-40 begins/ends
178	CA 247, Barstow Rd, **E**...UltraMar/Pizza Hut/gas, **W**...HOSPITAL, DENTIST, Chief Auto Parts, Food4Less, K-Mart/Little Caesar's, Sears, Mojave River Valley Museum, st patrol
177	L St, W Main, Barstow, **W**...Arco/mart/24hr, Burns Bros/diesel/rest., Heartland/diesel/rest./24hr, Thrifty Gas, BunBoy Rest., Rte 66 Pizza, Desert Lodge, Holiday Inn Express, tires/towing
176	CA 58, to Bakersfield
175	Lenwood, to Barstow, **E**...Chevron/diesel/mart, Exxon/mart, Mobil/mart, Shell/mart/24hr, 76/diesel, Arby's, Baskin-Robbins, Carl's Jr, Del Taco, Denny's, El Pollo Loco, Hana Grill, Harvey House Rest., In-n-Out, Jack-in-the-Box, KFC, Panda Express, Quigley's Rest., Starbucks, Subway, Wendy's, Tanger Outlet/famous brands, **W**...Arco/mart/24hr, Pilot/Dairy Queen/diesel/mart/24hr, Rip Griffin/Texaco/Subway/diesel/24hr, McDonald's, GoodNite Inn, truck repair/wash
165	Sidewinder Rd, Outlet Ctr Dr, **4 mi E**...factory outlets
161	Hodge Rd, no facilities
157	Wild Wash Rd, no facilities
153	Boulder Rd, **W**...airport
150	Stoddard Wells Rd, to Bell Mtn, **E**...Peggy Sue's 50s Diner, KOA, **W**...Mobil/mart, 76, Texaco, Denny's, Motel 6, Travelodge, Queens Motel

N

S

Baker

Barstow

California Interstate 15

Victorville	149.5mm	Mojave River
	149	E St, no facilities
	148	CA 18 E, D St, to Apple Valley, **E**...HOSPITAL, 76/mart, **W**...Arco/mart/24hr
	146	Mojave Dr, Victorville, **E**...76/diesel/mart/24hr, Budget Inn, **W**...Chevron/mart, Economy Inn, Sunset Inn
	145	Roy Rogers Dr, **E**...Texaco/A&W/TCBY/mart, USA/mart, Carl's Jr, China Palace, Dairy Queen, HomeTown Buffet, IHOP, Jack-in-the-Box, New Corral Motel, Asian Mkt, Costco, Food4Less/24hr, Goodyear/auto, Harley-Davidson, HomeBase, House of Fabrics, Midas Muffler, Pic'n Save, Rite Aid, Winston Tires, bank, same as 144, **W**...Arco/mart/24hr
	144	CA 18 W, Palmdale Rd, Victorville, **E**...DENTIST, CHIROPRACTOR, Arco/mart/24hr, Chevron, Shell, Texaco/diesel/mart/24hr, Antoni's Café, Baker's Drive-Thru, Burger King, Carl's Jr, Casa Delicias Mexican, Denny's, Don's Rest., Jack-in-the-Box, KFC, Richie's Diner, YumYum Donuts, Best Western, Chief Parts, K-Mart, Dodge/Plymouth, Olds/Cadillac/GMC, PepBoys, **W**...Arco/mart/24hr, Chevron/mart, Mobil/mart/wash, Shell/mart/wash, Thrifty Gas/mart, Andrew's Rest., Chinese Rest., Coco's, Del Taco, La Casita Mexican, LJ Silver, McDonald's, Pizza Hut, Subway, Taco Bell, Tom's #21 Rest., Budget Inn, EZ 8 Motel, Holiday Inn, Aamco, AutoZone, Chevrolet/Hyundai, Chief Parts, Chrysler/Jeep, Ford/Lincoln/Mercury, Honda/Toyota, Kamper's Korner RV, Nissan, Ralph's Foods, Town&Country Tire, Target, cinema, mall
	140	Bear Valley Rd, to Lucerne Valley, **E**...CHIROPRACTOR, DENTIST, VETERINARIAN, Arco/mart/24hr, Citgo/7-11, Mobil/mart/wash, Shell/mart/wash/atm, Baker's Drive-Thru, Boston Mkt, Burger King, Carl's Jr, Del Taco, Golden Gate Chinese, Hogi Yogi, Kenny Rogers, KFC, LJ Silver, Los Toritos Mexican, Marie Callender's, McDonald's, Panda Express, Red Robin Rest., Pizza Palace, Shakey's Pizza, Steer In Rest., Steer&Stein, Straw Hat Pizza, TNT Café, Wienerschnitzel, Winchell's/Taco Shop, Best Value Motel, Day's Inn, Econolodge, Super 8, America's Tires, AutoZone, Circuit City, FashionBug, Firestone/auto, HealthFoods, Home Depot, Kwik Lube'n Tune, Michael's, Pier 1, Range RV, Staples, Wal-Mart/auto, cinema, cleaners, **W**...El Tio Pepe Mexican, Greenhouse Café, Jack-in-the-Box, Red Lobster, Tony Roma's, Best Buy, Big 5 Sports, JC Penney, Mervyn's, OfficeMax, PetsMart, Sears/auto, cinema, mall
	138	Main St, to Hesperia, Phelan, **E**...Shell/mart, Texaco/A&W/Popeyes/diesel/mart/atm, UltraMar/Dairy Queen/pizza, Burger King, Denny's, In-n-Out, Jack-in-the-Box, **W**...Arco/mart/atm/24hr, Chevron/Taco Maker/mart, Big Boy, RV camping
	136	US 395, to Adelanto, **W**...Newt's Outpost/diesel/mart
	133	Oak Hill Rd, **E**...Arco/mart/24hr, Texaco/mart, Summit Inn/café, **W**...RV camping, LP
	127	Cajon Summit, elevation 4260, brake check sb
	126	CA 138, to Palmdale, Silverwood Lake, **E**...Chevron/McDonald's/24hr, repair, **W**...76/Del Taco/mart/LP, Texaco/diesel/mart, Economy Inn
	125mm	weigh sta both lanes, elevation 3000
	123	Cleghorn Rd, no facilities
	119	Kenwood Ave, no facilities
	118	I-215 S, to San Bernardino, **S**...Arco/mart/24hr, to Glen Helen Park
	116	Glen Helen Parkway, no facilities
	114	Sierra Ave, **W**...Arco/Jack-in-the-Box/mart/24hr, Shell/Taco Bell/Pizza Hut/TCBY/diesel/mart, to Lytle Creek RA
	113	CA 30, Highland Ave, **E**...to Lake Arrowhead
Ontario	112	Base Line Rd, no facilities
	111	CA 66, Foothill Blvd, **E**...DENTIST, EYECARE, Chevron, ClaimJumper Rest., Coco's, Golden Spoon, In-n-Out, Stuft Bagel, Subway, Taco Bell, Wienerschnitzel's, Circuit City, Costco, Food4Less, MailBoxes Etc, Michael's, Office Depot, OfficeMax, OilMax, PetsMart, Radio Shack, Sport Chalet, Wal-Mart/auto, cleaners, **W**...Home Depot, stadium
	108	4th St, **W**...Arco/mart/24hr, Mobil/Subway/mart/wash, Texaco/Burger King/diesel/mart, Carl's Jr, Chevy's FreshMex, Coco's, CucinaCucina, FoodCourt, Jack-in-the-Box, KFC, McDonald's, NY Grill, Rubio's, Starbucks, TokyoTokyo, Wienerschnitzel's, AmeriSuites, America's Tire, AutoNation USA, Bed Bath&Beyond, JC Penney, JiffyLube, Lube'n Tune, Ontario Mills Mall, SportsAuthority, TJ Maxx, cinema
	105	I-10, E to San Bernardino, W to LA
	103	Jurupa St, **E**...Ontario Auto Ctr/BMW, Buick, Chrysler/Jeep/Dodge, GMC/Pontiac, Honda, Lexus, Nissan, Saturn, Toyota, Volvo, Sun Country Marine, **W**...Arco/mart/24hr, Carl's Jr, Ford, Lincoln/Mercury, funpark
	101	CA 60, E to Riverside, W to LA
	97	Limonite Ave, no facilities
	95	6th St, Norco Dr, **E**...Arco/mart/24hr, Chevron/mart/24hr, Jack-in-the-Box, McDonald's, Old Town Norco, **W**...Arco/mart/24hr, UltraMar/gas, Country Cookin', Brakemasters
	92	2nd St, **W**...MEDICAL CARE, Shell/diesel/atm, Texaco, 7-11, Burger King, Domino's, In-n-Out, Del Taco, Howard Johnson Express, Little Caesar's, Sizzler, Steer Grill, Texas Loosey's, Wienerschnitzel, Norco Auto Mall/Chrysler/Plymouth, Dodge, Ford, Jeep/Eagle, Mitsubishi, Pontiac, Norco RV, Target, antiques, carwash, cleaners
	91	Yuma Dr, **W**...Texaco/A&W/TCBYdiesel/mart, Carl's Jr, Dairy Queen, McDonald's, Wendy's, Lucky Sav-On
	90	CA 91, to Riverside, beaches, no facilities

N

S

California Interstate 15

88	Magnolia Ave, **E**...76/diesel, tire/truck parts, **W**...DENTIST, Mobil/wash, Shell/mart/wash, Baskin-Robbins, Burger King/playland, Carl's Jr, Donut Star, Little Caesar's, Lotus Garden Chinese, McDonald's, Pizza Palace, Sizzler, Subway, Zendejas Mexican, Ralph's Foods, Rite Aid, Sav-On Drug, Stater Bros Foods, Kragen Parts, House of Fabrics, MailBoxes Etc, 1-Hr Photo, bank
85	Ontario Ave, to El Cerrito, **W**...VETERINARIAN, Arco/mart/24hr, Chevron/mart/24hr, Jack-in-the-Box, Wienerschnitzel, Albertson's, USPO, cleaners
84	El Cerrito Rd, **E**...Circle K/gas
83	Cajalco Rd, no facilities
81	Weirick Rd, **E**...diesel repair, hardware
75	Temescal Cyn Rd, **W**...Arco/diesel/mart/24hr, Carl's Jr, Tom's Farms, RV camping
73	Indian Truck Trail, no facilities
72	Lake St, no facilities
71	Nichols Rd, **W**...Arco/mart/24hr, VF Outlet/85 famous brands, auto repair
70	CA 74, Central Ave, Lake Elsinore, **E**...Arco/mart/24hr, Chevron/mart/wash/atm, Mobil/mart, Douglas Burgers, Italian Deli, Off Ramp Café, **W**...craft mall
69	Main St, Lake Elsinore, **E**...76, **W**...Circle K, motel
67	Railroad Cyn Rd, to Lake Elsinore, **E**...DENTIST, 76/Circle K, Denny's, Donut Depot, In-n-Out, KFC, Von's Foods, Wal-Mart/auto, cleaners, copyshop, **W**...MEDICAL CARE, CHIROPRACTOR, VETERINARIAN, Arco/mart, Chevron, Mobil/diesel/mart, 7-11, Carl's Jr, Chinese Rest., Del Taco, Don Jose's Mexican, Green Burrito, McDonald's, Pizza Hut, Sizzler, Subway, Taco Bell, Vista Donuts, Lake Elsinore Hotel/casino, Lake View Inn, Travelodge, Albertson's, Big O Tires, Chief Parts, Chevrolet, Do-It Hardware, Econo Lube'n Tune, Firestone, Goodyear/auto, NAPA, 1-Hr Photo, Postal Annex, Radio Shack, SavOn Drug, Stater Bros Foods, bank, cinema, cleaners, crafts
65	Bundy Cyn Rd, **W**...Arco/mart/24hr, Jack-in-the-Box
63	Baxter Rd, no facilities
60	Clinton Keith Rd, **E**...HOSPITAL
59	California Oaks Rd, Kalmia St, **E**...DENTIST, CHIROPRACTOR, VETERINARIAN, Chevron, Mobil/diesel/mart, Shell/Burger King/diesel/mart, Texaco/Subway/Taco Bell/diesel/mart, Carl's Jr, Dairy Queen, KFC, McDonald's, Numero Uno Pizza, Albertson's, Chief Parts, Goodyear/auto, Kragen Parts, Rite Aid, cinema, cleaners, **W**...Arco/mart/24hr, Jack-in-the-Box, funpark
58	Murrieta Hot Springs Rd, to I-215, **E**...CHIROPRACTOR, Sizzler, HomeBase, Ralph's Foods, Rite Aid, **W**...Texaco/Subway/Popeyes/diesel/mart, IHOP, McDonald's, Best Buy, Big 5 Sports, Home Depot, PetsMart, Pic'n Save, Staples
56	I-215 N, to Riverside
55	CA 79 N, Winchester Rd, **E**...MEDICAL CARE, CHIROPRACTOR, DENTIST, EYECARE, Chevron/mart/wash, Carl's Jr, Coco's, McDonald's, Mimi's Café, Taco Bell, YellowBasket Burgers, America's Tires, Big O Tires, Chief Parts, Costco, Food 4Less, Goodyear, Honda/Acura, K-Mart, Kragen Parts, Mervyn's, Michael's, Midas, PepBoys, Pier 1, Ralph's Food, TJ Maxx, Toyota, VW, XpressLube, cinema, **W**...VETERINARIAN, CHIROPRACTOR, MEDICAL CARE, Arco/mart/24hr, Chevron/diesel/mart, Mobil/mart/wash, Arby's, Chinese Rest., Dairy Queen, Del Taco, El Pollo Loco, Ernie's Hogy's, Golden Corral, Guadalajara Mexican, Hungry Hunter, In-n-Out, Jack-in-the-Box, Margarita Grill, NY Pizza, Richie's Diner, Sizzler, Tecate Grill, Tony Roma's, Tony's Steer, Wendy's, Best Western, Comfort Inn, Temecula Valley Inn, Ace Hardware, Big A Parts, Econo Lube'n Tune, Kinko's/24hr, MailMart, Nevada Bob's, Richardson's RV, Sir Speedy, Stater Bros Foods, Winston Tire, st patrol
54	Rancho California Rd, **E**...MEDICAL CARE, CHIROPRACTOR, VETERINARIAN, DENTIST, Mobil, 76/mart/wash, Texaco/A&W/diesel/mart, Bagel Bakery, Black Angus, Buddie's Pizza, Chili's, ClaimJumper, Great Grains Bread, Marie Callender's, Oscar's Rest., RoundTable Pizza, Starbucks, Embassy Suites, Albertson's, SavOn Drug, Target, Von's Foods, bank, cleaners, **W**...Chevron/mart/repair, 76/diesel/mart, Circle K, Denny's, Domino's, KFC, McDonald's, Mexico Chiquito, OldTown Donuts, Penfold's Mexican, SW Grill, Taco Bell, Taco Grill, Best Western, Motel 6, USPO
52	CA 79 S, Temecula, to Indio, **E**...Mobil/mart, Carl's Jr, Temecula Pizza, Rancho Motor Inn, **W**...DENTIST, Texaco/diesel/mart/24hr, Alberto's Drive-Thru Mexican, Baskin-Robbins, Hungry Howie's, Wienerschnitzel, Butterfield Inn, Ramada Inn, Firestone, Goodyear/auto, RV service
50mm	check sta nb
49	Rainbow Valley Blvd, **2 mi E**...gas, food, **W**...CA Insp Sta
46	Mission Rd, to Fallbrook, **2 mi E**...Rainbow Oaks Rest., gas, **W**...HOSPITAL
41	CA 76, Pala, to Oceanside, **W**...Mobil/mart, La Estancia Inn, Pala Mesa Resort, RV camp
39mm	La Estancia River

N

S

California Interstate 15

E s c o n d i d o

38	Old Hwy 395, no facilities
35	Gopher Canyon Rd, Old Castle Rd, **E**...All Seasons Camping, Texaco, food, lodging
31	Deer Springs Rd, Mountain Meadow Rd, **W**...Arco/mart/24hr
30	Centre City Pkwy(from sb)
28	El Norte Pkwy, **E**...Arco/mart/24hr, Mobil/mart, Shell/diesel/mart/wash, Best Western, Precision Tune, RV camping, **W**...CHIROPRACTOR, 76/Circle K, Wendy's, Von's Food/drugs, MailBox
26	CA 78, to Oceanside, **1 mi E on Mission**...Mobil, 7-11, Carl's Jr, ChuckeCheese, Denny's, El Mexicano Tacos, Jack-in-the-Box, LJ Silver, Wienerschnitzel, Motel 6, Lucky Food, Speedee Lube, Valvoline, PepBoys, K-Mart, U-Haul, bank
25	Valley Pkwy, **E**...HOSPITAL, Arco/mart/24hr, Chili's, Coco's, McDonald's, Olive Garden, Yoshinoya Beef Bowl, Circuit City, Marshall's, PetWay, cleaners, mall, **W**...Boston Mkt, Burger King, Carl's Jr, Del Taco, Subway, Comfort Inn, Holiday Inn Express, Home Depot, Lenscrafters, Sports Authority, Target
24	9th Ave, AutoPark Way, **W**...Mervyn's, same as 25
23	Felicita Rd, **W**...gas, food, lodging
22.5	Center City Pkwy(from nb, no return), **E**...Center City Café, Palms Inn
22	Via Rancho Pkwy, to Escondido, **E**...Chevron/mart/24hr, Shell/mart/wash/atm, FoodCourt, Red Lobster, Red Robin, JC Penney, Macy's, Nordstrom's, Robinsons-May, Sears/auto, Animal Zoo, mall, **W**...DENTIST, Texaco/Subway/Taco Bell/diesel/mart, CA Coffee, McDonald's, Olive House Café, Tony's Steer, MailBox Annex, cleaners
21	W Bernardo Dr, to Highland Valley Rd, Palmerado Rd, no facilities
20	Rancho Bernardo Rd, to Lake Poway, **E**...Arco/mart/24hr, Mobil/mart, Pier 1, cinema, cleaners, **W**...76/Circle K, Shell/mart/repair, Texaco/diesel/repair, Elephant Bar Rest., Holiday Inn, Travelodge
19	Bernardo Ctr Dr, **E**...VETERINARIAN, Chevron/mart/wash, Burger King, Carl's Jr, Coco's, Denny's, El Torito, Hunan Chinese, Jack-in-the-Box, Quizno's, Roberto's Tacos, Rubio's Grill, Sesame Donuts, Taco Bell, Firestone/auto, Goodyear/auto, SavOn Drug, Valvoline, bank, cleaners
18	Camino del Norte, **E**...HOSPITAL, to same as 17, **W**...Bernardo Pizza, Togo Eatery
17	Carmel Mtn Rd, **E**...OPTOMETRIST, Chevron/wash, Shell/mart/atm, Texaco, Baskin-Robbins, Bruebagger's Bagels, Calif Pizza, Carl's Jr, Chevy's Mexican, ClaimJumper, In-n-Out, Islands Burgers, Marie Callender's, McDonald's, Olive Garden, Oscar's Rest., Rubio's Grill, Schlotsky's, Subway, TGIFriday, Wendy's, Residence Inn, Barnes&Noble, Borders Books, Circuit City, Home Depot, K-Mart, Linens'n Things, Marshall's, Mervyn's, Michael's, OilMax, Pearle Vision, PetsMart, Ralph's Foods, Rite Aid, Ross, SportMart, Staples, Trader Joes, USPO, bank, cleaners
16	CA 56 W, Ted Williams Pkwy, **E**...DENTIST, Chinese Take-Out, cleaners
15	Rancho Penasquitos Blvd, Poway Rd, **E**...Arco/mart/24hr, Manhattan Deli, **W**...CHIROPRACTOR, DENTIST, Mobil/LP, 76/mart, Shell/mart/24hr, 7-11, Burger King, IHOP, Little Caesar's, McDonald's, Subway, Taco Bell, La Quinta, cleaners

S a n D i e g o A r e a

14	Mercy Rd, Scripps Pkwy, no facilities
12	Mira Mesa Blvd, to Lake Miramar, **E**...EYECARE, DENTIST, Denny's, Filippi's Pizza, Golden Crown Chinese, Taco Bell, Quality Suites, Medico Drug, USPO, cleaners, **W**...DENTIST, CHIROPRACTOR, MEDICAL CARE, Arco/mart/24hr, Mobil, 76, Shell/mart/wash, Applebee's, Arby's, Burger King, Caprezio Italian, Carrow's, In-n-Out, Jack-in-the-Box, Little Caesar's, McDonald's, Schlotsky's, Subway, Szechuan Chinese, Wendy's, Albertson's, Longs Drug, Pic'n Save Foods, Ralph's Foods, Rite Aid, USPO, Valvoline, bank
11	Carroll Canyon Rd, to Miramar College
10	Palmerado Rd, Miramar Rd, **W**...CHIROPRACTOR, DENTIST, EYECARE, Arco/mart/24hr, Chevron/mart, Mobil/mart/wash, Shell/mart/24hr, Texaco/diesel/mart, Acapulco Mexican, Keith's Rest., Marie Callender's, Maxwell's Rest., Pizza Hut, Rice King Japanese, SubMarina, Subway, Best Western, Budget Inn, Holiday Inn, Aamco, Kragen Parts, Land Rover, MailBoxes Etc, NAPA, Porsche, VW, bank, cleaners/laundry
9	Miramar Way, US Naval Air Station
8	CA 163 S(from sb), to San Diego, no facilities
6	CA 52, no facilities
5	Clairemont Mesa Blvd, **W**...DENTIST, Chinese Rest., Kathy's Kitchen, Mkt Foods
4	CA 274, Balboa Ave, access to gas, food
3	Aero Dr, **W**...Arco/mart/24hr, Shell/Taco Bell/wash, Chinese Café, D'Amato's Pizza, Einstein Bros Bagels, Jack-in-the-Box, McDonald's, Sizzler, Starbucks, SubMariner, Best Western, Holiday Inn, Fry's Electronics, Kinko's, PetsMart, Von's Foods, Wal-Mart/auto, cleaners, funpark
2	Friars Rd, **W**...San Diego Stadium
1	Friars Rd E, no facilities
.5	I-8, E to El Centro, W to beaches
0mm	Adams Ave, I-15 begins/ends.

N

↑
↓

S

California Interstate 40

California uses road names/numbers. Exit number is approximate mile marker.

Exit #	Services
151	Colorado River, California/Arizona state line, Pacific/Mountain time zone
149	Park Moabi Rd, to Rte 66, **N**...boating, camping
146mm	weigh sta eb, insp wb
145	5 Mile Rd, to Topock, to Rte 66, no facilities
140	US 95 S, E Broadway, Needles, **N**...MEDICAL CARE, Arco/mart/24hr, Chevron/diesel/mart, Shell/mart/repair, Texaco/diesel/mart, Burger King, Vito's Pizza, Basha's/bakery/deli, Rite Aid, U-Haul, laundry, **S**...Super 8, auto/RV/transmission repair
138	J St, Needles, **N**...Shell/mart, 76/24hr, Jack-in-the-Box, McDonald's, Travelers Inn, Big A Parts, NAPA, bank **S**...HOSPITAL, Denny's, Day's Inn, Motel 6, st patrol
137	W Broadway, River Rd, Needles, **N**...Chevron/diesel/mart/24hr, Denny's, KFC, Best Motel, El Rancho Motel, LeBrun Motel, Kiva Motel, River Valley Lodge, Stardust Motel, Travelodge, Buick/Olds/Cadillac, Goodyear/auto, diesel repair, **S**...Arco/mart/24hr, Chevron/diesel/mart/24hr, Mobil/diesel/mart, Texaco/Dairy Queen/mart/24hr, Calif Pantry Rest./24hr, Carl's Jr/Green Burrito, Hui's Rest., Taco Bell, WagonWheel Rest., Best Western, Chalet Lodge, Day's Inn, Relax Inn, Chevrolet, RV/tire/repair
135	River Rd Cutoff(from eb), **N**...rec area, KOA, Hist Rte 66
132	US 95 N, to Searchlight, Las Vegas, to Rte 66, no facilities
122	Water Rd, no facilities
117	Mountain Springs Rd, High Springs Summit, elev 2770
109	Goffs Rd, **N**...gas/diesel/food, Hist Rte 66
105mm	**rest area both lanes, full(handicapped)facilities, phone, picnic tables, litter barrels, petwalk**
101	Essex Rd, Essex, **N**...to Providence Mtn SP, Mitchell Caverns
80	Kelbaker Rd, to Amboy, Kelso, to E Mojave NSA, **S**...Hist Rte 66, RV camping
51	Ludlow, **N**...76/Dairy Queen/mart/24hr, **S**...Chevron/mart/24hr, Ludlow Truckstop/diesel/rest., coffeeshop, Ludlow Motel
33	Hector Rd, to Hist Rte 66, no facilities
28mm	**rest area both lanes, full(handicapped)facilities, phone, picnic tables, litter barrels, petwalk**
24	Ft Cady Rd, **N**...Wesco/diesel/mart/24hr, camping
19	Newberry Springs, **N**...UltraMar/diesel/mart, **S**...Shell/mart, auto/diesel repair, RV camp
10	Barstow-Daggett Airport, **N**...airport
7	Daggett, **N**...Daggett Truckstop/diesel/mart, RV camping(2mi), to Calico Ghost Town
5	Nebo St, to Hist Rte 66, no facilities
3	USMC Logistics Base, **N**...Barstow Lodge
1	Montara Rd, Barstow, **N**...DENTIST, Herbst/mart, Mobil/diesel/mart, Shell/autocare, Burger King, McDonald's, MegaTom 's Burgers, Straw Hat Pizza, Taco Bell, Best Western, Gateway Motel, **1 mi N**...DENTIST, EYECARE, VETERINARIAN, Arco/mart/24hr, Chevron/mart, 76/Circle K/Subway/TCBY/atm, Shell/mart, Arby's, Burger King, Carl's Jr, Carrow's Rest., China Gourmet, Coco's, Del Taco, Denny's, Golden Dragon, IHOP, Jack-in-the-Box, KFC, Sizzler, Taco Bell, AstroBudget Motel, Best Motel, Brant's Motel, Cactus Motel, Comfort Inn, Desert Inn, Econolodge, Executive Inn, Holiday Inn/rest., Quality Inn, Stardust Inn, Kragen Auto Parts, Radio Shack, Thrifty Discount, U-Haul/LP, Von's Food/drug, bank, laundry, **S**...Arco/mart/24hr, Texaco, Wal-Mart/auto/24hr
0mm	I-40 begins/ends on I-15 in Barstow.

E ↕ W (margin direction indicator)

California Interstate 80

California uses road names/numbers. Exit number is approximate mile marker.

Exit #	Services
208	California/Nevada state line
201	Farad, no facilities
197	Floristan, no facilities
193	Hirschdale Rd, **N**...to Boca Dam, Stampede Dam, **S**...United Trails Gen Store/gas, RV camping, boating, camping
191	weigh sta wb
187	Prosser Village Rd, no facilities
185	CA 89 N, CA 267, Truckee, to N Shore Lake Tahoe, **N**...RV camping, USFS, **S**...76, Best Western, Major Muffler, Truckee Hotel, camping, info, airport, same as 184
184	Central Truckee(no eb return), **S**...CHIROPRACTOR, DENTIST, 76, CB Whitehouse Rest., El Toro Bravo Mexican, Ponderosa Café, Wagontrain Café, Star Hotel, Swedish House Bed/breakfast, Major Muffler, bank

E ↕ W (margin direction indicator)

California Interstate 80

T r u c k e e

183   CA 89 S, to N Lake Tahoe, **N**...MEDICAL CARE, EYECARE, DENTIST, Sierra Superstop/gas, 7-11, Burger King, Dairy Queen, La Bamba Mexican, Little Caesar's, Pizza Jct, Pizza Shack, Port of Subs, Sizzler, Sunset Inn, Ace Hardware, Allied Parts, Book Shelf, Gateway Pets, GNC, MailCtr, NAPA, New Moon Natural Foods, Rite Aid, Safeway, bank, laundry, **S**...Shell/mart/wash, Burger King, KFC, McDonald's, Pizzaria, Subway, Super 8, Truckee Inn, Albertson's, Longs Drugs, auto repair, to Squaw Valley, RV camping

182   Donner Pass Rd, Truckee, **N**...Shell/diesel/mart/atm/wash, factory outlet/famous brands, **S**...Chevron/diesel/mart/24hr, 76/mart, Beginning Rest., Donner House Rest., Donner Lake Pizza, Alpine Village Motel, Holiday Inn Express, chain service, to Donner SP, RV camping

176mm vista point both lanes

175   Donner Lake(from wb), **S**...Donner Lake Village Resort

174mm Donner Summit, elev 7239, rest area both lanes, full(handicapped)facilities, view area, phone, picnic tables, litter barrels, petwalk

173   Castle Park, Boreal Ridge Rd, **S**...Boreal Inn/rest., Pacific Crest Trailhead, skiing

172   Soda Springs, **S**...76/LP/diesel/mart, Donner Summit Lodge, chain services

170   Kingvale, **S**...Shell/mart

168   Rainbow Rd, to Big Bend, **S**...Rainbow Lodge/rest., RV camping

165   Big Bend(from eb), ranger sta, RV camping, same as 168

162   Cisco Grove, **N**...RV camping, skiing, snowmobiling, **S**...Chevron/mart/24hr, chain serv

159   Eagle Lakes Rd, **N**...RV camping

158   CA 20 W, to Nevada City, Grass Valley, no facilities

157   Yuba Gap, **S**...snowpark, phone, picnic tables, skiing, boating, camping

155   Laing Rd, **S**...Rancho Sierra Inn/café

154mm vista point wb

153   Emigrant Gap(from eb), **S**...Shell/Burger King/diesel/mart/24hr, Rancho Sierra Inn/café, phone

151   Nyack Rd, Emigrant Gap, **S**...Shell/Burger King/diesel/mart, Nyack Café, chainup services

149   Blue Canyon, no facilities

147   Drum Forebay, no facilities

144   Baxter, **N**...RV camping, chainup services, food, phone

143   Crystal Springs, no facilities

142   Alta, no facilities

141   Dutch Flat, **N**...Monte Vista Inn, RV camping, **S**...Tesoro/gas, CHP, chainup services

140   Gold Run(from wb), **N**...gas/diesel, food, phone, chainup

139mm rest area both lanes, full(handicapped)facilities, phone, picnic tables, litter barrels, petwalk

138   Gold Run, **N**...gas/diesel, food, phone, chainup services

135   Magra Rd, Rollins Lake Rd, Secret Town Rd, RV camping

134   Rollins Lake Road(from wb), RV camping

C o l f a x

131   CA 174, Colfax, to Grass Valley, **N**...Chevron/mart, 76/mart, A&W, Classic Kitchen, McDonald's, Pizza Factory, Rosy's Café, Taco Bell, Sierra Mkt Foods, Colfax Motel, NAPA, cleaners, **S**...Chevron/mart, Sierra/diesel/mart, Tesoro/mart, Chinese Rest., Subway, Sierra RV Ctr

130   Canyon Way, to Colfax, **N**...Dingus McGee's, **S**...California Cantina, Mexican Villa Rest., Chevrolet, Sierra Tire/repair

127   Cross Rd, to Weimar, park&ride, RV camping

126   W Paoli Lane, to Weimar, **S**...Weimar Store/gas/diesel

125   Heather Glen, elev 2000 ft, no facilities

124   Applegate, **N**...Beacon/diesel/mart, Firehouse Motel, chainup services

122   Clipper Gap, Meadow Vista, **S**...RV camping, park&ride

121   Dry Creek Rd, **N**...CA Conservation Corps, **S**...park&ride

120   Bell Rd, **N**...HOSPITAL, VETERINARIAN, KOA(3mi), airport, **S**...HQ House Rest.

118   Foresthill Rd, Ravine Rd, Auburn, **N**...RV camping/dump, **S**...76, Burger King, Ikeba's Burgers, Jack-in-the-Box, Sizzler, Best Western, Country Squire Inn/rest., same as 117

117   Lincolnway, Auburn, **N**...Flyers Gas, Arby's, Denny's, Pizza Hut, Sam's Rest., Taco Bell/24hr, Wienerschnitzel, Wimpy's Burgers, Best Inn, Foothills Motel, Motel 6, Sleep Inn, Super 8, carwash, **S**...Arco/diesel/mart/24hr, Chevron/diesel/mart/24hr, 76/mart, Shell/diesel, Bakers Square, Baskin-Robbins, Burger King, Burrito Shop, Carl's Jr, Chinese Rest., Country Waffle, Dairy Queen, Izzy's Burgers, Jack-in-the-Box, KFC, LaBonte's Rest., Lyon's Rest., McDonald's, Sizzler, Thai Cuisine, Raley's Food/drug, Best Western, Country Squire Inn, Travelodge, 1-Hr Photo, carwash, cleaners

A u b u r n

116.5   Russell Ave(from wb), **S**...Texaco/diesel/mart, Ryder Trucks, U-Haul, same as 117

116   Elm Ave, Auburn, **N**...76, Shell/mart, Blimpie, Foster's Freeze, Taco Bell, Holiday Inn, Albertson's, Gottchaulk's, Kinko's, Longs Drug, Staples, Thrifty Foods, U-Haul, bank, cleaners, mufflers, **S**...Sierra/gas, bank, tires

115   CA 49, Auburn, to Grass Valley, Placerville, **N**...MEDICAL CARE, Shell, In-n-Out, Marie Callender's, Holiday Inn, Staples, Thrifty Foods, **S**...Chevy's FreshMex, bank

114   Maple St, Nevada St, Old Town Auburn, **N**...CHIROPRACTOR, **S**...Budget Gas/mart, Tiopete Mexican

E ↑ ↓ W

California Interstate 80

113.5 Ophir Rd, **N**...Pizza Lunch, Pop's Place Foods

113 CA 193, to Lincoln, **S**...Gamel RV Ctr

112 Newcastle, **N**...transmissions, **S**...Arco/diesel/mart/24hr, Exxon/diesel/mart, Denny's, CHP, golf

110 Penryn, **N**...Beacon/diesel/LP/mart, CattleBaron's Café

107 Horseshoe Bar Rd, to Loomis, **N**...Burger King, Taco Bell, Raley's Food/drug, RoundTable Pizza, bank, cleaners

105 Sierra College Blvd, **N**...Chevron/McDonald's/diesel/mart, Citgo/7-11, 76/mart, Day's Inn, Camping World RV Service/supplies, KOA

104 Rocklin Rd, **N**...MEDICAL CARE, DENTIST, CHIROPRACTOR, EYECARE, VETERINARIAN, Beacon/mart, Exxon/mart, Arby's, Baskin-Robbins, Blimpie, Burger King, Carl's Jr, Denny's, Hacienda del Roble Mexican, Jack-in-the-Box, Jasper's Giant Burgers, KFC, Papa Murphy's, Red Pepper Café, RoundTable Pizza, Subway, Swank's Dinner Theatre, Taco Bell, 1st Choice Inn, Microtel, Ramada, Camping World RV Service, CarQuest, Fabric Shop, Gamel RV Ctr, GNC, Harley-Davidson, Longs Drug, Pottery World, Radio Shack, Safeway, bank, cleaners, **S**...VETERINARIAN, Arco/mart/24hr, Rocklin Park Hotel, Susanne Bakery

103 CA 65, to Lincoln, Marysville, **1 mi N on Stanford Ranch Rd**...Exxon, Shell/mart/wash, Applebee's, Carl's Jr, Jack-in-the-Box, FoodSource, Costco, Linens'n Things, SportsAuthority, Wal-mart, 5Star Lube/wash

102 Taylor Rd, to Rocklin(from eb), **N**...Cattlemen's Rest., Albertson's(1mi), Holiday RV Ctr, K-Mart(1mi), **S**...76/Burger King/mart, Hilton Garden, Residence Inn

101.5 Atlantic St, Eureka Rd, **N**...VETERINARIAN, **S**...Pacific Pride/diesel, 76/mart, Shell, Brookfield's Rest., Black Angus, In-n-Out, Taco Bell, Tarver's Steaks, Wendy's, Marriott, Buick/GMC, Chevrolet, Ford, CompUSA, Home Depot, OfficeMax, PetsMart, Sam's Club, mall

101 Douglas Blvd, **N**...HOSPITAL, CHIROPRACTOR, DENTIST, Arco/mart/24hr, Exxon/mart, 76/mart, Circle K, BBQ, Burger King, ClaimJumper, Delicia's Mexican, Jack-in-the-Box, KFC, McDonald's, Mtn Mike's Pizza, Taco Bell, Taquiera Mexican, Best Western, Extended Stay America, Heritage Inn, Ace Hardware, Big O Tire, Chevrolet, Firestone/auto, Goodyear, Kragen Parts, Michael's, Midas Muffler, 98¢ Store, Old Navy, Radio Shack, Ross, SportMart, Trader Joe's, bank, **S**...Arco/mart/24hr, Chevron/mart/24hr, Shell, Carrow's/24hr, Carl's Jr, Del Taco, Denny's, Outback Steaks, Yamasushi, Quality Inn, Oxford Suites, Albertson's, Lincoln/Mercury, Mervyn's, MotoPhoto, Office Depot, Rite Aid

100 Riverside Ave, to Roseville, **N**...MEDICAL CARE, VETERINARIAN, Arco/mart/24hr, Meineke Muffler, **S**...DENTIST, CHIROPRACTOR, Exxon/diesel/mart, Shell/mart/wash, BBQ, California Burgers, Golden Donuts, Jack-in-the-Box, Super K-Mart, Village RV Ctr

98 Antelope Rd, to Citrus Heights, **N**...VETERINARIAN, OPTOMETRIST, DENTIST, 76/mart, 7-11, Burger King, Carl's Jr, Chinese Rest., Chubby's Diner, Giant Pizza, KFC, Little Caesar's, LJ Silver, McDonald's, RoundTable Pizza, Subway, Taco Bell, Wendy's, Albertson's, Mail4U, Raley's Food/drug, Rite Aid, USPO, hardware, **S**...CHIROPRACTOR

97mm weigh sta both lanes

96 Greenback Lane, Elkhorn Blvd, Orangedale, Citrus Heights, **N**...MEDICAL CARE, DENTIST, CHIROPRACTOR, 76/service, Circle K, Carl's Jr, Chinese Rest., Leyva's Mexican, McDonald's, Pizza Hut, Subway, Taco Bell, Longs Drug, Safeway, **S**...cinema

94 Madison Ave, **N**...VETERINARIAN, Beacon/mart/24hr, Shell/mart, Brookfield's Rest., Denny's, Foster's Freeze, Motel 6, Super 8, Scandia Funpark, to McClellan AFB, **S**...CHIROPRACTOR, Arco/mart/24hr, 76, Shell/repair, 7-11, A&W, Boston Mkt, Burger King, Carl's Jr, El Pollo Loco, Eppie's Rest., Humberto's Mexican, IHOP, Jack-in-the-Box, LJ Silver, McDonald's, Subway, Taco Bell, T-Bonz Steaks, Wienerschnitzel, Holiday Inn, La Quinta, Acura/Porsche, America's Tire, Bedroom Superstore, Chevrolet, CompuSet, Ford/Isuzu, Goodyear, HomeBase, Kinko's, Marco Muffler, Midas Muffler, Nevada Bob's, Office Depot, PepBoys, PetZone, Target, U-Haul, bank, carwash

93 Auburn Blvd, **S**...Burger King, Travelodge, to same as 94

92 Watt Ave, **N**...Day's Inn, McClellan AFB, **S**...Arco/mart/24hr, 76/diesel, Shell/mart, 7-11, Burger King, Carl's Jr, Carrow's Rest., Church's Chicken, Dairy Queen, Denny's, Golden Egg Café, KFC, Pizza Hut, Taco Bell, Wendy's, Best Inn, Motel 6, Travelodge, Firestone, Lube

91 Longview Dr, no facilities

90.5 Winters St, no facilities

90 Raley Blvd, Marysville Blvd, to Rio Linda, **N**...Arco/mart/24hr, Chevron/diesel/mart/24hr, Angelina's Pizza, **S**...Mkt Basket Foods, tires

88 Norwood Ave, **N**...Arco/Jack-in-the-Box/mart/24hr, Texaco/diesel/mart, McDonald's, Subway, Chief Parts, Rite Aid, SavMax Food/bakery/deli

E (arrow up/down) W

California Interstate 80

Sacramento

87 Northgate Blvd, Sacramento, **N**...Fry's Electronics, **S**...Circle K/gas, Shell/mart/wash, Texaco/diesel/mart, Valero/diesel, Burger King, Carl's Jr, IHOP, KFC, Lamp Post Pizza, LJ Silver, McDonald's, Taco Bell, Extended Stay America, Red Roof Inn, Goodyear/auto, JiffyLube, K-Mart, PepBoys, tires, transmissions

85 Truxel Rd, **N**...DENTIST, Shell/diesel/mart, Applebee's, Del Taco, In-n-Out, Jamba Juice, On the Border, Quizno's, Starbucks, Steve's Place Pizza, Arco Arena, Home Depot, Michael's, PetsMart, Raley's Depot, Ross, Staples, Wal-Mart, cinema, mall

82 I-5, N to Redding, S to Sacramento, to CA 99 N, to airport

81 W El Camino, **N**...Chevron/diesel/mart/24hr, Shell/diesel/rest., Burger King, Subway, Microtel

80 Reed Ave, **N**...Jack-in-the-Box, Tony's Rest., Ford Trucks, diesel repair, **S**...Arco/mart/24hr

79 US 50 E, W Sacramento, no facilities

78 Enterprise Blvd, W Sacramento, **N**...Chevron/mart/24hr, Shell/mart/24hr, Valero/diesel, Eppie's Rest/24hr, Granada Inn, **S**...76/diesel/mart, Burger King, Denny's, KOA

76 Chiles Rd, no facilities

75 Mace Blvd, **S**...VETERINARIAN, Chevron/mart/24hr, DieselFuel, 76/mart, Shell/mart/wash, Burger King, Cindy's Rest., Denny's, Lamp Post Pizza, McDonald's, Subway, Taco Bell/24hr, Wendy's, Howard Johnson, Motel 6, Chevrolet/Toyota, Chrysler/Jeep, Ford/Mercury/Nissan, Honda, La Mesa RV Ctr, Mazda/VW, Nugget Mkt Foods, Olds/Pontiac/Buick/GMC, to Mace Ranch

74 Olive St(from wb, no EZ return), **N**...Fair Deal RV Ctr, Lube/tune/repair

73 Richards Blvd, Davis, **N**...Shell/mart/24hr, In-n-Out, Davis Motel/Café Italia, University Park Inn, NAPA, **S**...Chevron/mart, Del Taco, KFC, RoundTable Pizza, Wendy's, Hawthorn Suites, Holiday Inn Express, AggieLube, Kragen Parts

72 CA 113 N, to Woodland, **N**...HOSPITAL

69 Kidwell Rd, no facilities

68 Pedrick Rd, **N**...76/LP/mart

67 Milk Farm Rd(from wb), no facilities

66 CA 113 S, Currey Rd, to Dixon, **S**...Arco/mart, CFN/diesel, 76/Popeye's/mart, Shell/diesel, Cattlemen's Rest., Jack-in-the-Box, **1 mi S**...Beacon/gas, Citgo/7-11, Ford

64 Pitt School Rd, to Dixon, **S**...MEDICAL CARE, CHIROPRACTOR, Chevron/mart/24hr, Valero/mart, Arby's, Burger King, Chevy's Mexican, Chinese Rest., Denny's, Domino's, IHOP/24hr, LaBella's Pizza, Mary's Pizza, McDonald's, Pizza Hut, Solano Bakery, Subway, Taco Bell, Valley Grill, Best Western, Microtel, Kragen Parts, MailBoxes Etc, PetSupply, Radio Shack, Safeway, cleaners

63 Dixon Rd, Midway Rd, **N**...Chevron/diesel/mart, **S**...Arco/mart/24hr, Shell/mart/wash/tune, Carl's Jr, KFC, Los Altos Mexican, Super 8, Dixon Fruit Mkt

62 Midway Rd, Lewis Rd, **N**...RV camping

60 Meridian Rd, Weber Rd, no facilities

58 Leisure Town Rd, **S**...Beacon/gas, Black Oak Rest., Jack-in-the-Box, Joe's Rest., Hick'ry Pit BBQ, SplitFire Rest., Fairfield Inn, Motel 6, Residence Inn, Vaca Valley Inn, Chevrolet/Olds, Chrysler/Plymouth/Dodge/Jeep, HomeBase, Honda, Mazda, Nissan, Toyota, VW

57 I-505 N, to Winters, no facilities

56 Orange Dr, same as 55 and 58

Vacaville

55 Monte Vista Ave, **N**...HOSPITAL, Beacon/diesel, Citgo/7-11, 76/mart, Shell/mart/wash, Arby's, Burger King, City Sports Grill, Denny's, Hisui Japanese, IHOP, McDonald's, Murillo's Mexican, Nations Burger, NutTree Rest., Pelayo's Mexican, RoundTable Pizza, Taco Bell, Wendy's, Best Western, Royal Motel, Super 8, America Tire, Firestone/auto, Goodyear/auto, Midas Muffler, U-Haul, Winston Tire, transmissions, **S**...Arco/mart/24hr, Chevron/mart/wash/atm/24hr, Applebee's, Carl's Jr, Chevy's FreshMex, Chili's, Chubby's Diner, CoffeeTree Rest., HomeTown Buffet, In-n-Out, Italian Café, Jack-in-the-Box, KFC, Starbucks, Tahoe Joe's Rest., Togo's, Courtyard, Big 5 Sports, CompUSA, Goodyear, Kinko's, Mervyn's, Michael's, Old Navy, PetCo, Ritz Camera, Ross, Safeway, Sam's Club, SportMart, Staples, Target, Vacaville Stores/famous brands, Wal-Mart/auto, bank, cinema

54 Mason St, Peabody Rd, **N**...Beacon/Taco Time/diesel/mart, Petro/mart/wash, Hawthorn Suites, AutoZone, NAPA, Schwab Tire, **S**...Arco/mart, Shell/repair, Carl's Jr, Solano's Bakery, Wok'n Roll Chinese, Aegean Tires, Costco, Ford/Mercury, Goodyear/auto, PepBoys, bank

53.5 Davis St, **N**...Chevron/McDonald's/mart, Outback Steaks, CarQuest, bank, cinema, **S**...QuikStop/gas, Cecil's Repair

53 Merchant St, Alamo Dr, **N**...Chevron/mart/wash, Shell/diesel/mart/wash, Adalberto's Mexican, Bakers Square, Lyon's/24hr, Alamo Inn, Monte Vista Motel, JiffyLube, **S**...CHIROPRACTOR, Jack-in-the-Box, KFC, McDonald's, Pizza Hut, Port Subs, MailBoxes Etc, Radio Shack, SavMax Foods/24hr, cleaners

52 Cherry Glen Rd(from wb), no facilities

50 Pena Adobe Rd, **S**...Ranch Hotel

49 Lagoon Valley Rd, Cherry Glen, no facilities

48 N Texas St, Fairfield, **S**...Arco/mart/24hr, Chevron/mart, Shell, Burger King, Lou's Jct Rest., RoundTable Pizza, EZ 8 Motel, GNC, K-Mart/auto, Raley's Foods, cinema

47 Waterman Blvd, **N**...Dynasty Chinese, Hungry Hunter Rest., RoundTable Pizza, Strings Italian, TCBY, Buick/Pontiac/GMC, Chevrolet/Cadillac/Olds, MailBoxes Etc, Safeway, Village RV Ctr, bank, cleaners, **S**...to Travis AFB, museum

E

W

California Interstate 80

E

W

46   Travis Blvd, Fairfield, **N**...CHIROPRACTOR, Arco/mart/24hr, Chevron/mart/24hr, Shell/diemart, Burger King, Denny's, In-n-Out, McDonald's, NY Pizza, Taco Bell, Holiday Inn, Motel 6, Raley's Foods, Harley-Davidson, Ford, Hyundai/Daewoo, Nissan, Pier 1, CHP, **S**...HOSPITAL, DENTIST, Shell, Blue Frog Brewery/rest., Chevy's Mexican, FreshChoice Rest., Great Wall Chinese, Marie Callender's, Mimi's Café, Red Lobster, Subway, Barnes&Noble, Best Buy, Circuit City, Edwards Cinema, Firestone/auto, Gateway Computers, JC Penney, Macy's, Mazda, Mervyn's, Michael's, OfficeMax, Old Navy, Ross, Sears/auto, ToysRUs, Trader Joe's, bank, mall

45   W Texas St, Fairfield, same as 46, **N**...VETERINARIAN, Shell/diesel/mart, ChuckeCheese, Gordito's Mexican, Sleepy Hollow Motel, Mazda/Subaru, Suzuki, Target, **S**...DENTIST, CHIROPRACTOR, Exxon, Dairy Queen, Jack-in-the-Box, Johanne's Diner, McDonald's, Nations Burgers, Pelayo's Mexican, Travelodge, Acura/Honda, Home Depot, Isuzu, Nissan, PepBoys, PostMasters, SavMax Foods, Target, Toyota, Walgreen

44   CA 12 E, Abernathy Rd, Suisun City, **S**...Wal-Mart, Budweiser Plant

42mm   weigh sta both lanes, phone

41   Suisan Valley Rd, **S**...Arco/mart/24hr, Chevron/mart, 76/diesel/mart/24hr, Shell/diesel/mart/wash, Arby's, Bravo's Pizza, Burger King, Carl's Jr, Denny's, Green Bamboo Chinese, McDonald's, Old SF Expresso, Subway, Taco Bell, Wendy's, Best Western, Hampton Inn, Holiday Inn Express, Inns of America, Overnighter Lodge, Camping World RV Service, Scandia FunCtr

40   I-680(from wb)

39   CA 12 W, to Napa, Green Valley Rd, **N**...Longs Drug, Safeway, **S**...Saturn

38   Red Top Rd, **N**...76/Circle K/24hr, Jack-in-the-Box

36   American Canyon Rd, no facilities

35mm   **rest area wb, full(handicapped)facilities, info, phone, picnic tables, litter barrels, petwalk, vista parking**

34   CA 37, to San Rafael, Columbus Pkwy, **N**...Best Western

33   Redwood St, to Vallejo, **N**...HOSPITAL, 76, Denny's, Panda Garden, Day's Inn, Holiday Inn, Motel 6, **S**...DENTIST, VETERINARIAN, Arco/mart, Shell/mart/wash, Applebee's, Black Angus, Cham Thai Rest., Chevy's Mexican, IHOP, Little Caesar's, Lyon's Rest., McDonald's, Mtn Mike's Pizza, Olive Garden, Red Lobster, Subway, Susie's Café, Taco Bell, Wendy's, Comfort Inn, Ramada Inn, AutoZone, Buick/Olds/GMC, Costco, Hancock Fabrics, HomeBase, Home Depot, Honda, Linens'n Things, Longs Drug, OfficeMax, OilChangers, PepBoys, PetCo, Radio Shack, Ross, Safeway, SaveMart Foods, Service Merchandise, Target, Toyota, ToysRUs, ZoomLube, bank, cinema, cleaners

31   Tennessee St, to Vallejo, **N**...CHIROPRACTOR, Baskin-Robbins, Lucky Garden Chinese, Scotty's Rest., Grocery Outlet, Medicine Shoppe, bank, cleaners, **S**...DENTIST, 76, Valero/mart, Jack-in-the-Box, Pacifica Pizza, Great Western Inn, Quality Inn, USPO, carwash

30   Solano Ave, Springs Rd, **N**...Chevron/diesel/mart/24hr, Burger King, Church's Chicken, Nitti Gritti Rest., Szechuan Chinese, Taco Bell, Best Value Inn, Deluxe Inn, Ford, Albertson's, Rite Aid, U-Haul, laundry, **S**...CHIROPRACTOR, Beacon/mart, Chemco/mart, Chevron/mart, QuikStop, Bud's Burgers, Dairy Queen, McDonald's, Panino Italian, Pizza Hut, RoundTable Pizza, SmorgaBob's, Subway, Islander Motel, Kragen Parts, MailBoxes Etc, Walgreen, bank

29.5   Georgia St, Central Vallejo, **N**...Safeway, Ford, **S**...Shell/Starbucks/diesel/mart, Mandarin Chinese, Crest Motel

29   Benicia Rd(from wb), **S**...Shell/diesel/mart

28.5   I-780, to Martinez, no facilities

28   Lincoln Rd(from wb), no facilities

27.5   Magazine St, Vallejo, **N**...Shell, Rod's Hickory Pit/24hr, Economy Inn, El Curtola Motel, 7 Motel, TradeWinds RV Park, **S**...7-11, McDonald's, Knight's Inn

27   CA 29, Maritime Academy Dr, Vallejo, **N**...Arco/diesel, Chevron/mart/24hr, Subway, Motel 6, Vallejo Inn

25mm   toll plaza, pay toll from eb

26   Crockett Rd, Crockett, to Rodeo, **N**...seafood rest.

24   Cummings Skyway, to CA 4(from wb), to Martinez, no facilities

23   Willow Ave, to Rodeo, **N**...DENTIST, CHIROPRACTOR, EYECARE, MEDICAL CARE, 76/diesel/mart, Chinese Rest., Straw Hat Pizza, NAPA, Safeway/24hr, Rodeo Drugs, TrueValue, USPO, bank, laundry, **S**...76/Circle K/diesel/mart, Burger King, Jamalo's Pizza, cleaners

22   CA 4, Hercules, to Stockton, **N**...Shell, Jack-in-the-Box, **S on Sycamore**...MEDICAL CARE, CHIROPRACTOR, DENTIST, EYECARE, Burgerama, Chinese Rest., McDonald's, RoundTable Pizza, Subway, Taco Bell, Valley Ice Cream/deli, Albertson's, Postal Annex, Rite Aid, USPO, bank, cleaners

21   Pinole Valley Rd, **S**...CHIROPRACTOR, DENTIST, Arco/mart/atm/24hr, Beacon Gas, Chevron/diesel, Shell, 76, 7-11, Chinese Rest., Jack-in-the-Box, Pizza Plenty, Red Onion Rest., Rico's Mexican, Ristorante Italiano, Subway, Waffle Shop, Zip's Rest., Albertson's, cleaners

California Interstate 80

Richmond	20	Appian Way, **N**...MEDICAL CARE, Beacon/mart, McDonald's, Kragen Parts, Longs Drug, Safeway, **S**...Valero/diesel/mart, Burger King, Carl's Jr, HomeTown Buffet, HotDog Sta, KFC, LJ Silver, RoundTable Pizza, Sizzler, Starbucks, Taco Bell, Wendy's, Day's Inn, Motel 6, Albertson's, Best Buy, Big 5 Sports, Goodyear/auto, K-Mart, MailBoxes Etc, Radio Shack, cinema, cleaners
	19.5	Richmond Pkwy, to I-580 W, **N**...Chevron/mart, McDonald's, IHOP, Barnes&Noble, Chrysler/Plymouth/Jeep, Circuit City, Ford, OfficeMax, Party City, Pearle Vision, Ross, **S**...Chevron/mart, Shell/diesel/mart, Applebee's, Chuck Steak, In-n-Out, Outback Steaks, RoundTable Pizza, Food4Less, GoodGuys, Kragen Parts, Old Navy, Mervyn's, Staples, Target
	19	Hilltop Dr, to Richmond, **N**...CHIROPRACTOR, DENTIST, Chevron/mart/wash, Chevy's Mexican, Olive Garden, Red Lobster, Subway, Tokyo Rest., Courtyard, Extended Stay America, Albertson's, Buick/Pontiac, Chevrolet/Olds, Firestone, JC Penney, Jo-Ann Fabrics, Macy's, Nissan, OilChangers, Sears/auto, bank, cinema, cleaners, mall, **S**...CHIROPRACTOR, 7-11
	18	El Portal Dr, to San Pablo, **S**...Shell/mart/wash, 76, KFC, McDonald's, Raley's Foods, SpeeDee Lube
	17	San Pablo Dam Rd, **N**...HOSPITAL, DENTIST, Arco/mart/24hr/atm, 76/mart, 7-11, Burger King, Chinese Rest., Denny's, KFC, McDonald's, Nations Burgers, Taco Bell, Albertson's, K-Mart/Little Caesar's/auto, Longs Drug, **S**...CamperLand RV Ctr
	16.8	Macdonald Ave(from eb), McBryde Ave(from wb), Richmond, **N**...DENTIST, Arco/mart/24hr, Citgo/7-11, Burger King, Church's Chicken, KFC, Taco Bell, OilChanger, TuneUp Masters, bank, **S**...VETERINARIAN, Chevron/mart/24hr, Bakers Square, Wendy's, Albertson's, Safeway, auto repair/wash, cleaners
	16.5	San Pablo Ave, San Pablo, to Richmond, **S**...76, Valero/mart, Wendy's, Aamco, Dodge/Jeep, Toyota
	16	Cutting Blvd, Potrero St, to I-580 Br(from wb), to El Cerrito, **N**...Arco/mart, **S**...Chevron, Shell, Carrow's Rest., Church's Chicken, IHOP, Jack-in-the-Box, McDonald's, Best Inn, Travelodge, FoodsCo Foods, Home Depot, Honda, PepBoys, Staples, Target, Walgreen
	15	Carlson Blvd, El Cerrito, **N**...76, Carlson Foods, 40 Flags Motel, **S**...Super 8
Berkeley	14	Central Ave, El Cerrito, **S**...DENTIST, CHIROPRACTOR, Shell/24hr, 76, Valero/mart, Burger King, Daimo Japanese, KFC, Nations Burgers, Branch Mkt Foods, GreaseMonkey, mall
	13.5	Albany, to I-580(from eb), no facilities
	13	Gilman St, to Berkeley, **N**...Golden Gate Fields Racetrack
	12	University Ave, to Berkeley, **S**...VETERINARIAN, 76, Beacon/gas/mufflers, to UC Berkeley
	11.5	CA 13, to Ashby Ave, no facilities
	11	Powell St, Emeryville, **N**...Shell/mart, Chevy's Mexican, Holiday Inn, **S**...Beacon/mart, Burger King, Denny's, Lyon's Rest., Starbucks, Trader Joe's, Courtyard, Day's Inn, Sheraton, Woodfin Suites, Borders Books, Circuit City, Emery Bay Mkt, Good Guys, Jo-Ann Fabrics, Old Navy, Pier 1, Ross
	8	Oakland, to I-880, I-580, no facilities
	7	W Grand Ave, Maritime St, no facilities
	6mm	toll plaza wb
	4mm	SF Bay
	2	Fremont St, Harrison St, Embarcadero
	1	5ᵗʰ St, downtown SF

I-80 begins/ends at 7ᵗʰ St in San Francisco.

California Interstate 110(LA)

Exit #	Services
25	I-110 begins/ends on I-10.
24	Adams Blvd, **E**...Nissan
23	37ᵗʰ St, Exposition Blvd, **E**...repair, **W**...Chevron/McDonald's/mart, Radisson
22	MLK Blvd, Expo Park
21	Vernon Ave, **E**...Mobil, Shell, **W**...76/24hr, Shell/mart/wash, Jack-in-the-Box, Ralph's Foods, Rite Aid, laundry
20	Slauson Ave, **E**...Mobil/mart, Shell/mart, TuneUp Masters, **W**...76
19	Gage Blvd, **E**...Arco, Church's Chicken
18	Florence Ave, **E**...Mobil, Jack-in-the-Box, hardware, **W**...Arco/mart, Chevron/mart, Shell/24hr, Burger King, Pizza Hut
17	Manchester Ave, **E**...Arco, McDonald's, **W**...76/Circle K, Church's Chicken, Jack-in-the-Box, Pam's Burgers, Popeye's
16	Century Blvd, **E**...Texaco/Subway/mart, Burger King, **W**...76
15	108ᵗʰ St, no facilties
14	Imperial Hwy, **W**...Shell, McDonald's
13	El Segundo Blvd, **E**...Shell/diesel/24hr, TuneUp Masters, auto repair, **W**...Arco/mart/24hr, Shell/mart/wash
12	Rosecrans Ave, **E**...Arco/mart/24hr, Cardlock Fuel, Chevron/service, transmissions/lube/tires, **W**...Chevron/McDonald's/mart, Mobil, 7-11, Jack-in-the-Box, KFC, Subway, Chief Parts, HomeBase

California Interstate 110

E

↕

W

11	Redondo Beach Blvd, **E**...McDonald's, Peter Pan Pizza/burgers, **W**...HOSPITAL, Mobil/mart
10	CA 91, no facilities
9	Artesia Ave, **E**...Home Depot
8	I-405, San Diego Fwy
7	Torrance Blvd, Del Amo, **E**...Burger King, K-mart, **W**...Mobil/mart, Texaco/diesel/pizza/mart
6	Carson St, **E**...VETERINARIAN, KFC, Cali Inn, cleaners, **W**...HOSPITAL, DENTIST, Mobil, Shell/mart, Bakers Square, Hong Kong Garden, In-n-Out, Jack-in-the-Box, Harbor Drug, Kragen Parts, PetShop, Radio Shack, cleaners
5	Sepulveda Blvd, **E**...Target, **W**...Arco/mart/24hr, Chevron/mart, Mobil/mart/atm, Shell, Carl's Jr, Golden Ox Burger, McDonald's, Popeye's, Taco Bell, Motel 6, Food4Less, K-Mart, Rite Aid, Von's Foods, bank, cleaners
4	CA 1, Pacific Coast Hwy, **E**...Texaco/mart/wash, Jack-in-the-Box, **W**...HOSPITAL, Chevron, Mobil/diesel/mart, Denny's, El Pollo Loco, EZ TakeOut Burger, Best Western, Discount Parts, Honda, Midas Muffler, PepBoys
3	Anaheim St, **E**...Texaco/mart
2	C St, **E**...Texaco/diesel, radiators, **W**...76 Refinery
1	Channel St, no facilities
.5	Gaffey Ave, no facilities
0mm	I-110 begins/ends

California Interstate 205(Tracy)

Exit #	Services
14	I-205 begins wb, ends eb, accesses I-5 nb.
11	MacArthur Dr, Tracy, **S**...Prime Outlet Ctr/famous brands
7	Tracy Blvd, Tracy, **N**...Chevron, Shell/diesel/mart, Texaco/diesel/mart, Denny's/24hr, Holiday inn Express, Motel 6, **S**...HOSPITAL, CHIROPRACTOR, Arco/atm/24hr, Arby's, Burger King, ChuckeCheese, In-n-Out, LJ Silver, Lyon's Rest., McDonald's, Nations Burgers, TCBY, Wendy's, Wok King, Best Western, Phoenix Lodge, Albertson's, Food4Less, Jo-Ann Fabrics, Kragen Parts, Longs Drugs, Walgreen, CHP
4	Grant Line Rd, Antioch, **N**...Burger King, IHOP, Taco Bell, Fairfield Inn, Hampton Inn, Chevroelt, Gottchalk's, JC Penney, Ross, Sears/auto, Staples, Target, Toyota, Wal-Mart/McDonald's, carwash, cinema, mall, **S**...Arco/mart/24hr, Citgo/7-11, Dik Tracy/gas, Shell/diesel/mart/wash, Carl's Jr, Orchard Rest., Cadillac/Pontiac/GMC, Tracy Marine
3	11TH St(from eb), to Tracy, Defense Depot
2	Mtn House Rd, to I-580 E
1	Patterson Pass Rd, **N**...fruitstand
0mm	I-205 begins eb/ends wb, accesses I-580 wb.

California Interstate 210(Pasadena)

Exit #	Services
49	I-10, San Bernardino Fwy. CA 57, Orange Fwy. I-210 begins/ends on I-10.
48	CA 71, Corona Fwy, no facilities
47	Raging Waters Dr, Via Verde, **S**...Bonelli Park, gas/LP
46	Covina Blvd, no facilities
45	Arrow Hwy, San Dimas, **N**...Shell, Del Taco, Denny's, McDonald's, **S**...Mobil/mart, OfficeMax, PetCo, Ralph's Foods, Rite Aid, Ross, Target, TJ Maxx, Trader Joe's
44	CA 30 E, Lone Hill Ave, **S**...Wal-Mart/auto, auto ctr
43	Sunflower Ave, no facilities
42	Grand Ave, to Glendora, **N**...HOSPITAL, Mobil, Denny's, Sports Chalet
41	Citrus Ave, to Covina, no facilities
40	CA 39, Azusa Ave, **N**...Arco/mart/24hr, Chevron, Texaco/Subway/Del Taco/mart, Super 8, Western Inn, **S**...76, In-n-Out
39	Vernon Ave, **N**...Carl's Jr, Taco Bell, Costco
38	Irwindale, **N**...MEDICAL CARE, Carl's Jr, Don Ramon's Grill, McDonald's, Taco Bell, Costco
37	Mt Olive Dr
36	I-605 S, no facilities
35	Mountain Ave, **N**...Old Spaghetti Factory, BMW, Buick, Chevrolet, Ford/Lincoln/Mercury, Honda, Infiniti, Isuzu, Mazda, Mitsubishi, Nissan, Saturn, Staples, Subaru, Target, **S**...Home Depot, Wal-Mart
34	Myrtle Ave, **S**...76

California Interstate 210

Pasadena

33	Huntington Dr, Monrovia, **N**...Winchell's, Wyndham Garden Hotel, CompUSA, Marshall's, Office Depot, cinema **S**...ClaimJumper, Macaroni Grill, Embassy Suites, Extended Stay America, Hampton Inn, Holiday Inn, OakTree Inn
32	Santa Anita Ave, Arcadia, **S**...In-n-Out, carwash
31	Baldwin Ave, to Sierra Madre, no facilities
30.5	Rosemead Blvd, **N**...Shell, **S**...76
30	San Gabriel Blvd, **N**...76, Panda Inn, Ralph's Foods, **S**...76, Best Western, Ramada Inn, Buick/Chevrolet/Pontiac/GMC, Cadillac, Circuit City, Toyota
29	Altadena Dr, **S**...Chevron
28	Hill Ave, no facilities
27	Lake Ave, **N**...Mobil/mart
26	CA 134, Ventura
25	Seco St, no facilities
24.5	Washington Blvd, no facilities
24	Lincoln Ave, **S**...Lincoln Motel
23.5	Arroyo Blvd, **N**...Jack-in-the-Box, Rose Bowl Motel, repair, **S**...to Rose Bowl
23	Berkshire Ave, no facilities
22	Gould Ave, **S**...McDonald's, Ralph's Foods, tires
21	Angeles Crest Hwy, no facilities
20	CA 2, Glendale Fwy, **S**...HOSPITAL, cinema
19	Ocean View Blvd, to Montrose
18	Pennsylvania Ave, La Crescenta, **N**...DENTIST, Mobil, Shell, Wienerschnitzel, Play-it-Again Sports, Von's Foods, cleaners, **S**...Freeway Mkt/deli
16	Lowell Ave, no facilities
15	LaTuna Cyn Rd, no facilities
13	Sunland Blvd, Tujunga, **N**...Sizzler, Rite Aid, Ralph's Foods
11	Wheatland Ave, **N**...food mkt
8	Osborne St, Lakeview Terrace, **N**...7-11
7	Paxton St, no facilities
6	CA 118, no facilities
5	Maclay St, to San Fernando, **S**...Chevron, 76, El Pollo Loco, McDonald's, Taco Bell, Home Depot, Office Depot, Sam's Club, XpressLube
4	Hubbard St, **N**...Chevron, Chief Parts, Radio Shack, Rite Aid, **S**...Mobil/diesel/mart, Shell, El Caparal Mexican, Jack-in-the-Box, Subway
3	Polk St, **S**...Arco, Chevron/mart, 7-11
2	Roxford St, **S**...Super 8
1	Yarnell St, no facilities
0mm	I-210 begins/ends on I-5, exit 160.

San Fernando

E ↕ W

California Interstate 215(Riverside)

San Bernardino

Exit#	Services
55	I-215 begins/ends on I-15.
54	Devore, **E**...Arco/mart/24hr, KOA, **W**...Coffeeshop
53	Palm Ave, Kendall Dr, **E**...Citgo/7-11, Mobil, Burger King, Dairy Queen, **W**...Arco/mart/24hr, Denny's
52	University Pkwy, **E**...Chevron, 76, Carl's Jr, Boston Mkt, Del Taco, Green Burrito, McDonald's, Hilltop Hotel, Ralph's Foods, golf, **W**...Shell/mart/wash, IHOP, Jack-in-the-Box, Taco Bell, Motel 6, golf
51	27th St, no facilities
50	CA 30 W, Highland Ave, no facilities
49	Muscuplabe Dr, **E**...Mobil, Shell, Chevrolet, Home Depot, SavOn Foods, Stater Bros
48	CA 30 E, Highlands, no facilities
47	Baseline St, no facilities
46	CA 66 W, 5th St, **E**...Arco/mart/24hr, Sands Motel
45	2nd St, Civic Ctr, **E**...China Hut, Del Taco, Denny's, In-n-Out, Taco Bell, Radisson, Best, Ford, Food4Less, JC Penney, Kelly Tire, Marshall's, **W**...Big C AutoWorks
44	Mill St, SBD Airport, **E**...McDonald's, Suzuki/Honda, **W**...Shell/mart, YumYum Donuts
43	Inland Ctr Dr, **E**...Robinsons-May, mall
42	Orange Show Rd, **E**...Chevron, Shell, Denny's, Pancho Villa's Mexican, Budget Inn, Continental Motel, Chevrolet, Kinko's/24hr, Mazda, Olds, Target, **W**...Villager Lodge, auto plaza/multiple dealers
41	I-10, E to Palm Springs, W to LA
40	Washington St, **E**...Arco/mart/24hr, Chevron/diesel/mart, Arby's, Bluff's Rest., Siquios Mexican, Taco Joe's, Goodyear, Payless Drug, funpark, **W**...Carl's Jr, Del Taco, Denny's, McDonald's, Red Tile Inn, RV Expo, Wal-Mart

N ↕ S

California Interstate 215

39	Barton Rd, **E**...Arco/mart/24hr, Texaco/mart/wash, **W**...Dorothy's Burgers
38	La Cadena Dr, **E**...Shell/mart/24hr, Day's Inn, tires
37	Center St, to Highgrove, **W**...76/diesel/LP/mart
36	Columbia Ave, **E**...Arco, **W**...Circle K
35	CA 91, CA 60, Riverside, to beach cities
34	Blaine St, 3rd St, **E**...76, Shell, **W**...Arco
33	University Ave, Riverside, **W**...MEDICAL CARE, Arco, Mobil/mart, Shell, Baker's Drive-Thru, Baskin-Robbins, Carl's Jr, Chicago Grill, Chinese Food, Del Taco, Domino's, Green Burrito, IHOP, Jilberto's Tacos, Manhattan Bagels, Mike's Seafood, Quizno's, Starbucks, Taco Bell, Tempo Del Sol Mexican, Wienerschnitzel, Winchell's, Courtyard, FarmHouse Motel, Hampton Inn, Highway Host Motel, Motel 6, Super 8, Kragen Parts, Rite Aid, bank, cleaners, cinema, tuneups
32	MLK Blvd, El Cerrito, no facilities
31	Central Ave, Watkins Dr, no facilities
30	Fair Isle Dr, Box Springs, **E**...Shell/mart/atm, Margon's RV, **W**...Ford
29	CA 60 E, to Indio, **E on Day St**...MEDICAL CARE, Arco/mart/24hr, Texaco/Del Taco/Subway/diesel/mart, McDonald's, Costco, Kinko's, PetsMart, Pier 1, Robinsons-May, Sam's Club, Sears, Wal-Mart, cinema, mall
28	Eucalyptus Ave, Eastridge Ave, **E**...PetsMart, Sam's Club, Wal-Mart, same as 29
27	Alessandro Blvd, **E**...Arco/mart/24hr, Mobil, Big O Tires, JiffyLube, auto repair
26	Cactus Air Blvd, **E**...Texaco/Burger King/mart, Subway
24	Van Buren Blvd, **E**...March Field Museum, **W**...Riverside Nat Cem
22	Oleander Ave, no facilities
21	Ramona Expswy, **1 mi E**...Texaco/Subway/mart
18	Nuevo Rd, **E**...Arco/mart/24hr, 76, Carl's Jr, IHOP, Sizzler, Albertson's, Food4Less, Wal-Mart, cinema
16	CA 74 W, 4th St, to Perris, Lake Elsinore, **E**...Shell, **W**...DENTIST, Chevron, Del Taco, Denny's, Jack-in-the-Box, Popeye's, Best Western, Perris Inn, Chrysler/Dodge/Plymouth/Jeep
14	CA 74 E, Hemet, **E**...motel, Ford/Lincoln/Mercury
12	Ethanac Rd, **E**...KFC, Richardson's RV/marine Sales/service, **W**...Mobil/mart
11	McCall Blvd, Sun City, **E**...HOSPITAL, 76, UltraMar, Wendy's, Sun City Motel, Travelodge, D&M Auto/marine, NAPA, RV camp, RV Sales, **W**...MEDICAL CARE, EYECARE, VETERINARIAN, Chevron/diesel/mart, Mobil/mart, 76, Shell/mart, Soco Gas/repair, Burger King, Coco's, McDonald's, Rite Aid, Stater Bros Foods, Von's Foods, bank
9	Newport Rd, Quail Valley, **E**...Texaco/diesel/mart, Cathay Chinese, Jack-in-the-Box, Mexican Rest., Chief Parts, GNC, Ralph's Foods, Target
5	Scott Rd, no facilities
3	Clinton Keith Rd, no facilities
2	Los Alamos, **E**...DENTIST, Shell/mart/atm, Mexican Food, Peonys Chinese, Taco Bell, Wendy's, bank, **W**...Mobil/McDonald's/mart, ChuckeCheese, Green Burrito, Jack-in-the-Box, ShowCase Café, Lucky SavOn
1	Murrieta Hot Springs, **E**...CHIROPRACTOR, Del Taco, Domino's, Sizzler, HomeBase, PostOffice, Ralph's Foods, Rite Aid, Ross, bank, cleaners
0mm	I-215 begins/ends on I-15.

N
S

California Interstate 280(Bay Area)

California uses road names/numbers. Exit # is approximate mile marker.

E
W

Exit #	Services
60	4th St, downtown, I-280 begins/ends.
59	6th St, to I-80, Bay Bridge, downtown
58	Mariposa St, downtown
57	Army St, Port of SF
56	US 101 S, Alemany Blvd, Mission St, **E**...Shell
55	San Jose Ave, Bosworth St(from nb, no return)
54	Geneva Ave, **E**...donuts
53	CA 1, 19th Ave, **W**...Chevron, to Bay Bridge, SFSU
52	Daly City, Westlake Dist, **E**...76/diesel/LP
51	Serramonte Blvd, Daly City(from sb), **E**...Chevrolet, **W**...Beacon Gas, 76, McDonald's, Macy's, Mervyn's, Circuit City, Good Guys, Office Depot, PetsMart, Ross, SportMart, bank

California Interstate 280

50	CA 1, Mission St, Pacifica, **E**...HOSPITAL, Chevron/repair, RoundTable Pizza, Sizzler, Dodge, Drug Barn, Ford, Fresh Choice Foods, Home Depot, Isuzu, Jo-Ann Fabrics, Mitsubishi, Nissan, Nordstrom's, Target, cinema, mall
49	Hickey Blvd, Colma, **E**...Chevron/diesel/mart/wash/24hr, Shell, **W**...DENTIST, VETERINARIAN, Shell/mart/wash/24hr, 7-11/atm, Chinese Rest., El Torito's, Peppermill Rest., RoundTable Pizza, Sizzler, Pak'nSave Foods/24hr, bank, hardware, laundry
48	Avalon Dr, Westborough, **W**...Arco/mart/24hr, Exxon/diesel/mart, Denny's, McDonald's, Safeway, Walgreen/24hr, Skyline Coll
47	I-380 E, to US 101, airport
46	San Bruno Ave, Sneath Lane, **E**...Baskin-Robbins, Petrini's Rest., Longs Drugs, bank, cleaners, **W**...Beacon/gas, 76/mart, Chevron/repair/24hr, 7-11/atm, Bakers Square
45	Crystal Springs(from eb), no facilities
44	CA 35 N, Skyline Blvd(from wb, no EZ return), to Pacifica, **1 mi W**...DENTIST, VETERINARIAN, Beacon Gas, Lunardi's Foods, cleaners, drugs, hardware, pizza
42	Hillcrest Blvd, Larkspur Dr, Millbrae, **E**...Beacon/gas, Chevron
40	Trousdale Dr, to Burlingame, **E**...HOSPITAL
39	Black Mtn Rd, Hayne Rd, **W**...golf, vista point
38	**Crystal Springs rest area wb, full(handicapped)facilties, phone, picnic table, litter barrels, petwalk**
37	CA 35, CA 92W(from eb), to Half Moon Bay, no facilities
36	Bunker Hill Dr, no facilities
35	CA 92, San Mateo, to Half Moon Bay, no facilties
33mm	vista point both lanes
31	Edgewood Rd, Canada Rd, to San Carlos, **E**...HOSPITAL, phone
29	Farm Hill Blvd, **E**...Cañada Coll, phone
26	CA 84, Woodside Rd, Redwood City, **1 mi W**...MEDICAL CARE, DENTIST, Chevron/diesel/repair, USPO, bank
24	Sand Hill Rd, Menlo Park, **E**...Shell, Capriccio Rest., Longs Drug, MailBoxes Etc, Safeway
23	Alpine Rd, Portola Valley, **E**...HOSPITAL, produce, **W**...Chevron/wash, Shell/autocare, Bianchini's Mkt, RoundTable Pizza, bank, cleaners
21	Page Mill Rd, to Palo Alto, **E**...HOSPITAL
17	El Monte Rd, Moody Rd, **W**...phone
14	Magdalena Ave, no facilities
12	Foothill Expswy, Grant Rd, **E**...Chevron/mart/24hr, Chinese Rest., Hick'ry Pit Rest., Pacific Steamer Pizza, Rite Aid, cleaners, photo shop, **W**...Arco/mart/24hr, to Rancho San Antonio CP
11	CA 85, N to Mtn View, S to Gilroy
10	Saratoga-Sunnyvale Rd, Cupertino, Sunnyvale, **E**...Chevron/mart/wash, Carl's Jr, Cupertino Inn, Goodyear, Rite Aid, TJ Maxx, 1-Hr Photo, cleaners/laundry, repair, **W**...Arco, Chevron, Shell/diesel, Chinese Rest., Croutons Rest., Outback Steaks, Togo's Sandwiches, Target, bank
9	Wolfe Rd, **E**...DENTIST, EYECARE, Arco/mart/24hr, Carlos Murphy's Rest., Duke of Edinburgh Rest., Momo House Rest., Sankee Rest., Silver Wing Rest., Courtyard, Hilton Garden, Ranch Mkt, bank, cleaners, **W**...76, El Torito, McDonald's, Pizza, TGIFriday, FabricLand, JC Penney, Sears/auto, Vallco Fashion Park, cleaners
8	Lawrence Expswy, Stevens Creek Blvd(from eb), **N**...Rite Aid, Safeway/drug/deli/24hr, Linens'nThings, Marshall's, **S**...DENTIST, 76, 7-11, Howard Johnson, House of Bagels, IHOP, Rock'n Tacos, Woodcrest Hotel
7	Saratoga Ave, **N**...Arco/mart/24hr, Chevron/mart/24hr, 7-11, Black Angus, Burger King, Happi House, Harry's Hosbrau, McDonald's, Thai Cuisine, Albertson's, JiffyLube, auto dealers, bank, cleaners, **S**...76, Exxon, Shell, Denny's, RoundTable Pizza, Tony Roma's
6.5	Winchester Blvd, Campbell Ave(from eb), **S**...76/mart
6	CA 17 S, to Santa Cruz, I-880 N, to San Jose, Oakland
5.5	Leigh Ave, Bascom Ave, **N**...Exxon, KFC
5	Meridian St(from eb), **S**...Chevron/Fastpay, 76, 7-11, KFC, Taco Bell, Wienerschnitzel, Food4Less, K-Mart
4	Bird Ave, Race St, no facilities
3	CA 87, **N**...Hilton, Holiday Inn, Hotel Sainte Claire, airport, arena, bank
2	7th St, to CA 82, **N**...conv ctr
1	10th St, 11th St, **N**...7-11, to San Jose St U
0mm	I-280 begins/ends on US 101.

San Jose

E
↑
↓
W

California Interstate 405(LA)

CA uses road names/numbers. Exit # is approx mile marker.

Exit #	Services
56	I-5, N to Sacramento
55	Rinaldi St, Sepulveda, **E**...Arco, USA/gas/mart, Presidente Rest., Toyota, **W**...Shell/mart/atm, Grenada Motel
54	CA 118 W, Simi Valley, no facilities
53	Devonshire St, Grenada Hills, **E**...Arco, Mobil, Shell, Holiday Drug
52	Nordhoff St, **E**...MEDICAL CARE, DENTIST, Arco, Mobil/diesel/atm, 7-11/24hr, Mexican Rest., **W**...Arco, Mobil/mart, Baker's Square, Charlie's Burgers, Donut Co
51	Roscoe Blvd, to Panorama City, **E**...Chevron, Exxon/mart, 76, Burger King, Carl's Jr, Denny's, Jack-in-the-Box, McDonald's, Tacos Mexico, Holiday Inn Express, Ford/Lincoln/Mercury/Jaguar, Midas Muffler, Saturn, **W**...Shell/mart, Texaco, Coco's, Tommy's Burgers, Motel 6
50	Sherman Blvd, Reseda, **E**...Chevron, Mobil, 76, **W**...HOSPITAL, CHIROPRACTOR, 76/diesel/mart/24hr, Taco Bell
49	Victory Blvd, Van Nuys, **E**...76, Mobil, **W on VanOwen**...Quickmart, Chinese Rest., cleaners, **W on Victory**...Arco/mart/24hr
48	Burbank Blvd, **E**...Chevron, Shell/mart, 76, Little Caesar's, Zankou Chicken, Carriage Inn
47	US 101, Ventura Fwy, no facilities
46.5	Ventura Blvd, **E**...Mobil, Denny's, Robinson's-May, Thrifty Food, bank, mall, **W**...Chevron, 76, McDonald's, Radisson
46	Valley Vista Blvd, no facilities
45	Mulholland Dr, no facilities
44	Getty Ctr Dr, **W**...to Getty Ctr
43	Sunset Blvd, **E**...Chevron/24hr, Shell/repair, to UCLA, **W**...Holiday Inn, Radisson
42.5	Montana Ave(from nb), **E**...bank
42	Waterford St, no facilities
41	Wilshire Blvd, **E**...downtown, **W**...HOSPITAL
40	CA 2, Santa Monica Blvd, **E**...Exxon, **W**...Chevron, 76, Best Western
39	Olympic Blvd, Peco Blvd(from sb), **W**...Linens'n Things, Marshall's
38	I-10, Santa Monica Fwy
37	Venice Blvd, **E**...Chevron/service/mart, Citgo/7-11, Mobil, 76, Shell/diesel, Baskin-Robbins, Carl's Jr, Winston Tire, **W**...SP/gas, FatBurger, Spudnuts, auto repair, cleaners, multiple services on Sepulveda
36	Culver Blvd, Washington Blvd, **E**...Saab, XpertTune, repair, **W**...76, Thrifty Gas, repair
35	CA 90, Slauson Ave, to Marina del Rey, **E**...CHIROPRACTOR, Arco/mart/24hr, Del Taco, Shakey's Pizza, Circuit City, CompUSA, Firestone/auto, Goodyear/auto, Kinko's, Midas Muffler, Office Depot, Pick'n Save Foods, transmissions, **W**...76, Denny's, Red Lion Hotel, Lucky SavOn
34.5	Jefferson Blvd(from sb), **W**...to LA Airport
34	Sepulveda Blvd, to Centinela Ave, **E**...Mobil/diesel/mart, Sizzler, Ramada Inn, Wyndham Garden, JC Penney, Macy's, Robinson-May, mall, **W**...Chevron, Dinah's Rest., Extended Stay America, Radisson, Ford, Saturn, SavOn Drug, Howard Hughes Ctr, bank, mall
33.5	La Tijera Blvd, **E**...Baskin-Robbins, El Pollo Loco, McDonald's, Von's Foods, **W**...Arco/mart/24hr, Chevron/diesel/24hr, 76, auto repair
33	CA 42, Manchester Ave, to Inglewood, **E**...76/diesel/mart/24hr, Econolodge, Mexican Rest., Best Western, auto repair, cleaners, **W**...Shell/mart, Day's Inn, K-Mart
32	Century Blvd, **E**...76, Burger King, Casa Gamino Mexican, Flower Drum Chinese, Rally's, Subway, Best Western Suites, Motel 6/rest., Tiboli Hotel, Chief Parts, Econo Lube'n Tune, auto repair, **W**...Arco/mart/24hr, Chevron/diesel/mart, 76, Texaco, Carl's Jr, Denny's, McDonald's, Taco Bell, Hampton Inn, Hilton, Holiday Inn, Marriott, Quality Hotel, Travelodge, Westin Hotel, airport, bank
31.5	I-105, Imperial Hwy, **E**...American/diesel, Shell, BBQ, El Pollo Loco, Jack-in-the-Box, McDonald's, auto repair, **W**...Proud Bird Rest.(1mi), airport
31	El Segundo Blvd, to El Segundo, **E**...Arco/mart, Chevron/service/24hr, Christy's Donuts, Jack-in-the-Box, Pizza Hut, Taco Bell, El Segundo Inn, Aloha Drug, TuneUp Masters, cleaners, **W**...Arco/mart, Carl's Jr, Denny's, Luigi's Rest., Ramada Inn, CostCo, Kinko's, Office Depot
30	Rosecrans Ave, to Manhattan Beach, **E**...76/mart, Mobil/diesel, Shell/autocare, Del Taco, Denny's, Subway, Albertson's/SavOn, Best Buy, Food4Less, Home Depot, KidsRUs, MailBoxes Etc, Michael's, PartyCity, Ross, SportMart, Staples, ToysRUs, auto repair

N

S

California Interstate 405

29.5 Inglewood Ave, **E**...VETERINARIAN, DENTIST, Arco, Del Taco, In-n-Out, Albertson SavOn, **W**...76, Mobil, Shell, Texaco/diesel/24hr, Drug Emporium, Goodyear/auto, cleaners

29 CA 107, Hawthorne Blvd, **E**...Golden China, Jack-in-the-Box, McDonald's, Saigon Rest., Spires Rest., Day's Inn, **W**...Arco/mart/24hr, Mobil/diesel, Blimpie, Boston Mkt, Subway, Taco Bell, Woshinoya Japanese, Goodyear, Hobby Shack, Robinson-May

28.5 Redondo Beach Blvd, Hermosa Beach, **E**...Arco/mart/24hr, Thrifty Gas/mart, Amigo's Tacos, Chuck-e-Cheese, golf, **W**...DENTIST, EYECARE, Boston Mkt, RoundTable Pizza, Ralph's Foods, SavOn Drug, U-Haul, cleaners

28 CA 91 E, Artesia Blvd, to Torrance, **W**...76/24hr, Winchell's, CarQuest, XpressLube/Burger King/Taco Bell

27.5 Crenshaw Blvd, to Torrance, **E**...MEDICAL CARE, DENTIST, OPTOMETRIST, Arco/mart/24hr, Chevron/mart/24hr, Mobil/repair, Shell/mart, 7-11, Ralph's Foods, **W**...Mobil/diesel/mart, Texaco/Subway/diesel/mart, Denny's, TuneUp Masters

27 Western Ave, to Torrance, **E**...76, United/diesel, Del Taco, Denny's, Wendy's, Albertson's, Toyota, auto repair, **W**...Mobil/mart, Millie's Kitchen, Courtyard

26.5 Normandie Ave, to Gardena, **E**...Comfort Suites, **W**...Texaco/diesel/mart/wash, Carl's Jr, StarBucks, Subway, Extended Stay America, AutoNation USA, Goodyear, Office Depot

26 Vermont Ave(from sb), **W**...Holiday Inn, hwy patrol

25.8 I-110, Harbor Fwy, no facilities

25.7 Main St(from nb), no facilities

25.6 mm weigh sta both lanes

25.5 Avalon Blvd, to Carson, **E**...MEDICAL CARE, DENTIST, Arco/mart/24hr, Chevron, Mobil, Shell, ChuckeCheese, Denny's, Jack-in-the-Box, Pizza Hut, Shakey's Pizza, Sizzler, Subway, Tony Roma, Ramada Inn, Goodyear/auto, Ikea Furnishing, JC Penney, PepBoys, Sears/auto, Tire Sta, cleaners, mall, **W**...EYECARE, Arco/mart/24hr, Mobil, Shell, Carl's Jr, IHOP, Dodge/Plymouth, Ford/Lincoln/Mercury, Isuzu, Kia, Ralph's Foods, TuneUp, laundry

25.3 Carson St, to Carson, **E**...Comfort Inn, **W**...76/service/24hr, Carl's Jr, Subway, Hilton

25 Wilmington Ave, **E**...CHIROPRACTOR, Arco/service, Chevron/service/24hr, Carson Burgers, TuneUp/lube, **W**...Texaco/Subway/Taco Bell/diesel/mart, Chevrolet

24.9 Alameda St, no facilities

24.7 Santa Fe Ave(from sb), **E**...Arco/mart/24hr, **W**...Chevron/mart/24hr, Shell, donuts

24.5 I-710, Long Beach Fwy, no facilities

24.3 Pacific Ave(from sb), no facilities

24 Long Beach Blvd, **W**...HOSPITAL, Mobil, Shell/autocare/24hr, Toyota

23.7 Atlantic Blvd, **E**...Chevron, Texaco/Subway/diesel/mart, Black Angus, Denny's, El Patio, El Torito, Jack-in-the-Box, Mercedes, Ralph's Foods, Staples, **W**...HOSPITAL, Arco/mart/24hr, Shell, BMW, Dodge, Honda, Mazda, Nissan, Target

23.5 Orange Ave(from sb), no facilities

23.3 Cherry Ave, to Signal Hill, **E**...Mobil/diesel, Fantastic Burgers, Ford, auto repair, **W**...John's Burgers, Charley Brown's Steaks/Lobster, Rib Café, BMW, Dodge, Firestone, Nissan

23 CA 19, Lakewood Blvd, **E**...Marriott, airport, **W**...HOSPITAL, Chevron/mart, Shell/mart/24hr, Spires Rest., Taco Bell, Holiday Inn, Residence Inn, Ford, Goodyear/auto, Olds, Plymouth

22.5 Bellflower Blvd, **E**...Chevron/service, **W**...HOSPITAL, Mobil, 76, Shell/mart/24hr, Hof's Rest., Wendy's, Borders Books, Circuit City, CompUSA, Goodyear/auto, PetCo, Rite Aid/24hr, SavOn Drug, Sears, Target, bank

22 Woodruff Ave(from nb), no facilities

21.5 Palo Verde Ave, **W**...76, Del Taco, Dr Wi Donuts, Subway, Taco Bell, cleaners

21 I-605 N, no facilities

20 CA 22 W, 7th St, to Long Beach, no facilities

19.5 Seal Beach Blvd, Los Alamitos Blvd, **E**...Chevron/repair/24hr, Mobil, 76, Baskin-Robbins, KFC, Panda Panda Chinese, Spaghetti Italian Grill, Tortilla Beach Grill, Winchell's, Goodyear/auto, Lucky Food, Mail&More, Rite Aid, Winston Tire, bank, cinema, hardware

19 CA 22 E, Garden Grove Fwy, Valley View St, **E**...DENTIST, CHIROPRACTOR, Mobil, Texaco/diesel/mart, Coco's, Dairy Queen, Maxwell's Seafood Rest., Sizzler, Chevrolet, Ford, Rite Aid, Von's Foods, cleaners

18 Westminster Ave, to Springdale St, **E**...Arco/mart/24hr, Chevron/diesel/mart, 76/diesel/mart, 7-11/24hr, Baskin-Robbins, Café Westminster, Carl's Jr, Chinese Rest., In-n-Out, KFC, McDonald's, Taco Bell, Motel 6, Travelodge, America's Tires, Chief Parts, Home Depot, Kragen Parts, Lucky Foods, SavOn Drug, auto repair, cinema, **W**...Chevron/diesel/24hr, Shell/diesel/24hr, Pizza Hut, Subway, Best Western, Day's Inn, JiffyLube, cleaners, laundry

17 Bolsa Ave, Golden West St, **W**...Chevron, Mobil, 76, Shell, Bennigan's, Coco's, El Torito, Jack-in-the-Box, Best Buy, Big 5 Sports, House of Fabrics, JC Penney, Las Vegas Golf, Max Foods, PetCo, Robinson-May, Sears/auto, cinema, mall

16 CA 39, Beach Blvd, to Huntington Bch, **E**...HOSPITAL, Shell, Hof's Hut Rest., Jack-in-the-Box, Mei's Chinese, BeachWest Inn, Princess Inn, Super 8, Westminster Inn, Buick/Pontiac/GMC, K-Mart, Midas Muffler, PepBoys, Toyota, **W**...DENTIST, Arco, Mobil/service, 76, Arby's, Burger King, Diedrich's Coffee, El Torito, Jack-in-the-Box, Marie Callender's, Norm's Rest., Popeye's, Starbuck's, Holiday Inn, Barnes&Noble, Burlington Coats, Chevrolet, Chrysler/Jeep, Circuit City, Dodge, Ford/Lincoln/Mercury, Marshall's, Mitsubishi, OfficeMax, PartyCity, Plymouth, Subaru, Target, Tire Sta, VW

California Interstate 405

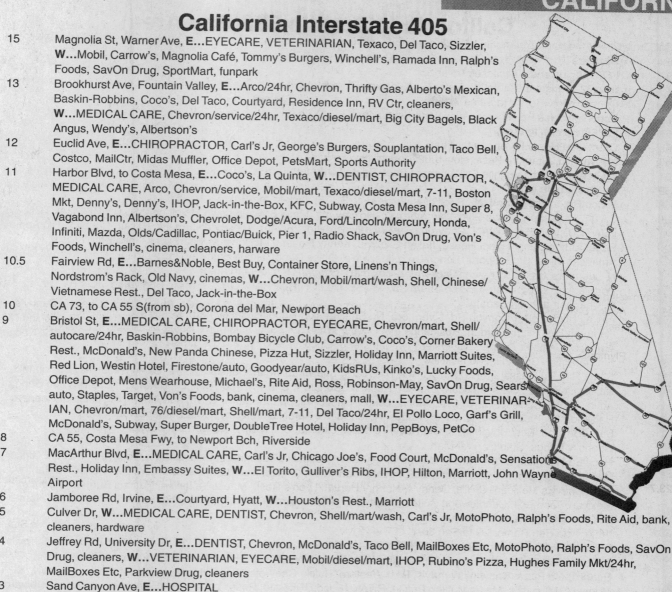

15   Magnolia St, Warner Ave, **E**...EYECARE, VETERINARIAN, Texaco, Del Taco, Sizzler, **W**...Mobil, Carrow's, Magnolia Café, Tommy's Burgers, Winchell's, Ramada Inn, Ralph's Foods, SavOn Drug, SportMart, funpark

13   Brookhurst Ave, Fountain Valley, **E**...Arco/24hr, Chevron, Thrifty Gas, Alberto's Mexican, Baskin-Robbins, Coco's, Del Taco, Courtyard, Residence Inn, RV Ctr, cleaners, **W**...MEDICAL CARE, Chevron/service/24hr, Texaco/diesel/mart, Big City Bagels, Black Angus, Wendy's, Albertson's

12   Euclid Ave, **E**...CHIROPRACTOR, Carl's Jr, George's Burgers, Souplantation, Taco Bell, Costco, MailCtr, Midas Muffler, Office Depot, PetsMart, Sports Authority

11   Harbor Blvd, to Costa Mesa, **E**...Coco's, La Quinta, **W**...DENTIST, CHIROPRACTOR, MEDICAL CARE, Arco, Chevron/service, Mobil/mart, Texaco/diesel/mart, 7-11, Boston Mkt, Denny's, Denny's, IHOP, Jack-in-the-Box, KFC, Subway, Costa Mesa Inn, Super 8, Vagabond Inn, Albertson's, Chevrolet, Dodge/Acura, Ford/Lincoln/Mercury, Honda, Infiniti, Mazda, Olds/Cadillac, Pontiac/Buick, Pier 1, Radio Shack, SavOn Drug, Von's Foods, Winchell's, cinema, cleaners, harware

10.5   Fairview Rd, **E**...Barnes&Noble, Best Buy, Container Store, Linens'n Things, Nordstrom's Rack, Old Navy, cinemas, **W**...Chevron, Mobil/mart/wash, Shell, Chinese/Vietnamese Rest., Del Taco, Jack-in-the-Box

10   CA 73, to CA 55 S(from sb), Corona del Mar, Newport Beach

9   Bristol St, **E**...MEDICAL CARE, CHIROPRACTOR, EYECARE, Chevron/mart, Shell/autocare/24hr, Baskin-Robbins, Bombay Bicycle Club, Carrow's, Coco's, Corner Bakery Rest., McDonald's, New Panda Chinese, Pizza Hut, Sizzler, Holiday Inn, Marriott Suites, Red Lion, Westin Hotel, Firestone/auto, Goodyear/auto, KidsRUs, Kinko's, Lucky Foods, Office Depot, Mens Wearhouse, Michael's, Rite Aid, Ross, Robinson-May, SavOn Drug, Sears/auto, Staples, Target, Von's Foods, bank, cinema, cleaners, mall, **W**...EYECARE, VETERINARIAN, Chevron/mart, 76/diesel/mart, Shell/mart, 7-11, Del Taco/24hr, El Pollo Loco, Garf's Grill, McDonald's, Subway, Super Burger, DoubleTree Hotel, Holiday Inn, PepBoys, PetCo

8   CA 55, Costa Mesa Fwy, to Newport Bch, Riverside

7   MacArthur Blvd, **E**...MEDICAL CARE, Carl's Jr, Chicago Joe's, Food Court, McDonald's, Sensations Rest., Holiday Inn, Embassy Suites, **W**...El Torito, Gulliver's Ribs, IHOP, Hilton, Marriott, John Wayne Airport

6   Jamboree Rd, Irvine, **E**...Courtyard, Hyatt, **W**...Houston's Rest., Marriott

5   Culver Dr, **W**...MEDICAL CARE, DENTIST, Chevron, Shell/mart/wash, Carl's Jr, MotoPhoto, Ralph's Foods, Rite Aid, bank, cleaners, hardware

4   Jeffrey Rd, University Dr, **E**...DENTIST, Chevron, McDonald's, Taco Bell, MailBoxes Etc, MotoPhoto, Ralph's Foods, SavOn Drug, cleaners, **W**...VETERINARIAN, EYECARE, Mobil/diesel/mart, IHOP, Rubino's Pizza, Hughes Family Mkt/24hr, MailBoxes Etc, Parkview Drug, cleaners

3   Sand Canyon Ave, **E**...HOSPITAL

2   CA 133, to Laguna Beach, no facilities

1   Irvine Center Dr, **E**...Chang's Chinese Bistro, Barnes&Noble, cinemas, **W**...AutoNation USA, Ford, JiffyLube, Multi-Dealer Service Ctr, water funpark

0mm   I-405 begins/ends on I-5, exit 132.

California Interstate 505(Winters)

Exit #   Services

33   I-5. I-505 begins/ends on I-5.

31   CA 12A, no facilities

29   CA 14, Zamora, no facilities

25   CA 19, no facilities

22   CA 16, Woodland, to Esparto, **W**...Guy's Food/fuel

17   CA 27, no facilities

15   CA 29A, no facilities

11   CA 128, Putah Creek Rd, **W**...MEDICAL CARE, Chevron/mart/24hr, Mexican Rest., RoundTable Pizza, Eagle Drug, IGA Foods

7   Allendale Rd, no facilities

4   Midway Rd, **E**...RV camping

2   Vaca Valley Pkwy, no facilities

I-505 begins/ends on I-80.

California Interstate 580(Bay Area)

California uses road names/numbers. Exit # is approximate mile marker.

Exit #	Services
82	I-580 begins/ends, accesses I-5 sb.
79	CA 132, Chrisman Rd, to Modesto, **E**...76/diesel, RV camping(5mi)
75	Corral Hollow Rd, no facilities
71	Patterson Pass Rd, **W**...Arco/diesel/mart/atm/24hr
67	I-205(from eb), to Tracy
65	Grant Line Rd, to Byron, no facilities
63	N Flynn Rd, Altamont Pass, elev 1009, no facilities
61	N Greenville Rd, Laughlin Rd, Altamont Pass Rd, to Livermore Lab
59	weigh sta both lanes
58	Vasco Rd, to Brentwood, **N**...Shell/diesel/mart/deli, **S**...76, Jack-in-the-Box
56	CA 84, 1st St, Springtown Blvd, Livermore, **N**...Holiday Inn, Motel 6, Springtown Motel, golf, **S**...CHIROPRACTOR, DENTIST, VETERINARIAN, Chevron, Shell/mart/wash/atm, 76/24hr, Applebee's, Arby's, Baskin-Robbins, Big Apple Bagel, Burger King, Happi House, Italian Express, Kenny Rogers, McDonald's, StarBucks, Taco Bell, Togo's, American Tires, Big 5 Sports, Longs Drug, Mervyn's, Office Depot, Ross, Safeway, Target, bank, tires
55	N Livermore Ave, **S**...Chevron/Jack-in-the-Box/mart, Citgo/7-11, Wal-Mart/auto
54	Portola Ave, Livermore(no EZ eb return), **S**...HOSPITAL, CHIROPRACTOR, Chevron, Shell/mart, JC's RV Ctr, OilChanger, auto/tire repair
53	Airway Blvd, Collier Canyon Rd, Livermore, **N**...Shell/Baskin-Robbins/diesel/mart, Wendy's, Comfort Inn, Courtyard, Hampton Inn, Hilton Garden, Residence Inn, Costco Whse/gas, **S**...Cattlemen's Rest., McDonald's, Extended Stay America, Camelot Funpark, Chrysler/Plymouth/Jeep, Lincoln/Mercury, Mazda, airport
52	El Charro Rd, Croak Rd, no facilities
51	Santa Rita Rd, Tassajara Rd, **S**...CHIROPRACTOR, Shell/mart/wash, Bakers Square, Baskin Robbins, Boston Mkt, California Burger, Ole's Mexican, Chinese Rest., McDonald's, Subway, Taco Bell, Longs Drug, Rose Pavilion Shpg, Acura, BMW, Cadillac, GMC, Olds, Infiniti, Lexus, Mitsubishi, Saab, Saturn, Volvo, auto/tire/repair, cinema, cleaners
49	Hacienda Dr, Pleasanton, **N**...Black Angus, AmeriSuites, AutoNation, BabysRUs, Barnes&Noble, Bed Bath&Beyond, Best Buy, Mimi's, Old Navy, TJ Maxx, cinema, **S**...HOSPITAL, Borders Books, Computer City, Linens'n Things, Staples, Wal-Mart
48	Hopyard Rd, Pleasanton, **N**...76, Minimart, PaknSav Foods, Holiday Inn Express, Dodge, Goodyear, Nissan, Office Depot, 1-Hr Photo, RV Ctr, Toyota, **S**...Chevron/mart/wash, Shell/diesel/mart/wash, Burger King, Buttercup Pantry/24hr, Chevy's Mexican, Chili's, Denny's, Hungry Hunter, Lyon's Rest., Maestro's Italian, Nations Burgers, Pedro's Mexican, Taco Bell, Candlewood Suites, Courtyard, Hilton, Marriott, Motel 6, Sheraton, Super 8, CompUSA, Home Depot, Kinko's, Mercedes, bank
47	I-680, N to San Ramon, S to San Jose
46	Foothills Rd, San Ramon Rd, **N**...VETERINARIAN, Chevron, Exxon, Shell/mart, 76, Burger King, Carl's Jr, Carrow's, Coco's, Foster's Freeze, Frankie&Johnny&Luigi's, India Prince Rest., KFC, McDonald's, Outback Steaks, RoundTable Pizza, Subway, Wendy's, Best Western, Buick, Chevrolet, Ford, Honda, Isuzu, Grand Auto Supply, Mervyn's, Michael's, Midas Muffler, PetCo, Ralph's Foods, Rite Aid, Ross, Target, ToysRUs, bank, cinema, hardware, **S**...CHIROPRACTOR, EYECARE, Black Angus, Chinese Rest., Crown Plaza, Residence Inn, Wyndham Garden, Chrysler/Plymouth, JC Penney, Macy's, Nordstrom's, mall
41	Eden Canyon Rd, Palomares Rd, **S**...rodeo park
39	Center St, Crow Canyon Rd, **N**...mall, same as 38, **S**...Arco/mart/24hr, Chevron/diesel/, 76/mart
38	Redwood Rd, Castro Valley, (no EZ eb return), **N**...CHIROPRACTOR, Chevron, Shell, KFC, McDonald's, Sizzler, Taco Bell, Holiday Inn Express, Longs Drug, NAPA, PetCo, Rite Aid, Safeway, SpeeDee Lube, bank, cleaners, **S**...VETERINARIAN, EYECARE, 7-11, Chinese Rest., bank, laundry
37	I-238 W, to I-880, CA 238, **W off I-238**...Jack-in-the Box, Volvo
36	164th Ave, Miramar Ave, **E**...Chevron/diesel, motel
35	150th Ave, Fairmont, **E**...HOSPITAL, **W**...Arco, Shell/autocare, 76, Denny's, Longs Drug, Macy's
34	Grand Ave(from sb), Dutton Ave, W...Coast Gas/mart, Rite Aid, cleaners
33	106th Ave, Foothill Blvd, MacArthur Blvd, **W**...Arco/mart/24hr, Pennzoil
32.5	98th Ave, Golf Links Rd, **E**...Shell/mart/wash, **W**...76/mart
32	Keller Ave, Mtn Blvd, **E**...repair
31.5	Edwards Ave(from sb, no EZ return), **E**...US Naval Hospital
31	CA 13, Warren Fwy, to Berkeley
30	Seminary Rd, **E**...Observatory/Planetarium, **W**...Arco/mart/24hr
29	High St, to MacArthur Blvd, **E**...Shell, 76, 7-11, Subway, Kragen Parts, **W**...76, Walgreen, cleaners/laundry
28	35th Ave(no EZ sb return), **E**...Exxon, 76, Chinese Rest., Taco Bell, Albertson's, **W**...Chevron/mart, QuikStop
27	Fruitvale, Coolidge Ave, **E**...Shell/24hr, 7-11, McDonald's, Albertson's, Longs Drug, Safeway, bank, **W**...DENTIST, 76
26	Park Blvd, **E**...Shell/mart, **W**...76, OilChangers

E

W

Livermore

Pleasanton

California Interstate 580

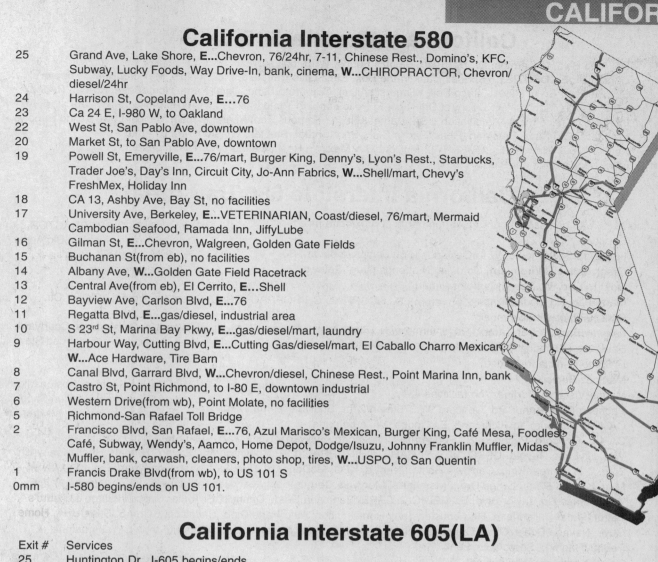

25   Grand Ave, Lake Shore, **E**...Chevron, 76/24hr, 7-11, Chinese Rest., Domino's, KFC, Subway, Lucky Foods, Way Drive-In, bank, cinema, **W**...CHIROPRACTOR, Chevron/diesel/24hr

24   Harrison St, Copeland Ave, **E**...76

23   Ca 24 E, I-980 W, to Oakland

22   West St, San Pablo Ave, downtown

20   Market St, to San Pablo Ave, downtown

19   Powell St, Emeryville, **E**...76/mart, Burger King, Denny's, Lyon's Rest., Starbucks, Trader Joe's, Day's Inn, Circuit City, Jo-Ann Fabrics, **W**...Shell/mart, Chevy's FreshMex, Holiday Inn

18   CA 13, Ashby Ave, Bay St, no facilities

17   University Ave, Berkeley, **E**...VETERINARIAN, Coast/diesel, 76/mart, Mermaid Cambodian Seafood, Ramada Inn, JiffyLube

16   Gilman St, **E**...Chevron, Walgreen, Golden Gate Fields

15   Buchanan St(from eb), no facilities

14   Albany Ave, **W**...Golden Gate Field Racetrack

13   Central Ave(from eb), El Cerrito, **E**...Shell

12   Bayview Ave, Carlson Blvd, **E**...76

11   Regatta Blvd, **E**...gas/diesel, industrial area

10   S 23rd St, Marina Bay Pkwy, **E**...gas/diesel/mart, laundry

9   Harbour Way, Cutting Blvd, **E**...Cutting Gas/diesel/mart, El Caballo Charro Mexican **W**...Ace Hardware, Tire Barn

8   Canal Blvd, Garrard Blvd, **W**...Chevron/diesel, Chinese Rest., Point Marina Inn, bank

7   Castro St, Point Richmond, to I-80 E, downtown industrial

6   Western Drive(from wb), Point Molate, no facilities

5   Richmond-San Rafael Toll Bridge

2   Francisco Blvd, San Rafael, **E**...76, Azul Marisco's Mexican, Burger King, Café Mesa, Foodles Café, Subway, Wendy's, Aamco, Home Depot, Dodge/Isuzu, Johnny Franklin Muffler, Midas Muffler, bank, carwash, cleaners, photo shop, tires, **W**...USPO, to San Quentin

1   Francis Drake Blvd(from wb), to US 101 S

0mm   I-580 begins/ends on US 101.

California Interstate 605(LA)

Exit #	Services
25	Huntington Dr. I-605 begins/ends.
24	I-210, no facilities
23	Live Oak Ave, Arrow Hwy, **E**...Santa Fe Dam, **W**...Irwindale Speedway
22	Lower Azusa Rd, LA St, no facilities
21	Ramona Blvd, **E**...Mobil, Del Taco/24hr
20	I-10, E to San Bernardino, W to LA
19	Valley Blvd, to Industry, **E**...Chevron, McDonald's, Valley Inn
18	CA 60, Pamona Fwy, no facilities
17	Peck Rd, **E**...Shell/mart/atm/wash, **W**...Ford Trucks
16	Beverly Blvd, RoseHill Rd, no facilities
15	Whittier Blvd, **E**...76, Shell, 7-11, Carl's Jr, GoodNite Inn, **W**...Chevron, Buick, Chrysler/Dodge, Ford, GMC, Honda, Isuzu, Jeep, Kia, Olds, Plymouth, Pontiac, Saturn, Toyota, Volvo
14	Washington Blvd, to Pico Rivera, no facilities
13	Slauson Ave, **E**...Mobil, Denny's, Motel 6, **W**...HOSPITAL
12	Telegraph Rd, to Santa Fe Springs, **E**...Chevron, Shell, st patrol
11	I-5, no facilities
10	Florence Ave, to Downey, **E**...Chevrolet, Honda, Sam's Club
9	Firestone Blvd, **E**...76, Shell, ChuckeCheese, KFC, McDonald's, Taco Bell, Best Western, BMW, Costco, Food4Less, Staples, VW/Audi, **W**...Arco, Chevron, 76, Dodge
8	I-105, Imperial Hwy, no facilities
7	Rosecrans Ave, to Norwalk, **E**...Chevron, **W**...Carrow's Rest, Motel 6, auto parts
6	Alondra Blvd, **E**...EYECARE, Chevron, Spires Rest., **W**...Shell/24hr, Texaco/Subway/mart
5	CA 91, no facilities
4	South St, **E**...Burlington Coats, Kinko's, Macy's, Mervyn's, Nordstrom's, Sears/auto, cinema, mall, **W**...76/service, Shell/service, Texaco/diesel/service, UltraMar/service, Daewoo, Dodge, Ford, Honda, Infiniti, Isuzu, Olds

Oakland Area

Los Angeles Area

California Interstate 605

3   Del Amo Blvd, to Cerritos, **E**...Del Taco, Ralph's Foods, **W**...Mobil/service

2   Carson St, **E**...DENTIST, 76, Shell, Jack-in-the-Box, KFC, Little Caesar's, McDonald's, Mexican Rest., Popeye's, Sky Burgers, Spike's Rest., Taco Bell, Wienerschnitzel, CarburetorLand, Chief Parts, HomeBase, Kragen Parts, bank, casino, laundry, **W**...Denny's, Del Taco, El Pollo Loco, El Torito Mexican, HiTop's Mexican, In-n-Out, Jack-in-the-Box, Crazy 8 Motel, AutoNation USA, Barnes&Noble, Famous FootWear, Linens'n Things, Lowe's Bldg Supply, Michael's, Old Navy, PetsMart, Ross, Sam's Club, SportChalet, Staples, Winston Tire, cinema

1   Katella Ave, Willow St, **E**...HOSPITAL, McDonald's, Mexican Rest., cleaners, donuts, racetrack

0mm   I-605 begins/ends on I-405.

California Interstate 680(Bay Area)

California uses road names/numbers. Exit number is approximate mile marker.

Exit #	Services
71	I-80 E, to Sacramento, W to Oakland, I-680 begins/ends on I-80.
70	Green Valley Rd(from eb), Cordelia, **N**...Longs Drug, Safeway
69	Gold Hill Rd, **W**...TowerMart/diesel/mart
66	Marshview Rd, no facilities
64	Parish Rd, no facilities
60	Lake Herman Rd, **E**...Arco/Jack-in-the-Box/diesel/mart, **W**...Shell/Carl's Jr/diesel/mart/wash/24hr, vista point
58	Bayshore Rd, industrial park
57	I-780, to Benicia, toll plaza
55mm	Martinez-Benicia Toll Br
54	Marina Vista, to Martinez, no facilities
53	Pacheco Blvd, Arthur Rd, Concord, **W**...76/mart/24hr, Shell/diesel/mart/24hr
52.5	CA 4 W, Pittsburg, to Richmond, **E**...Dodge, **W**...auto parts
52	CA 4 E, Concord, Pacheco, **E**...Peppermill Coffeeshop, Sullivan's Rest., Taco Bell, Sheraton, Ford, Hyundai, Infiniti/VW, JiffyLube, Toyota, USPO, **W**...EYECARE, CHIROPRACTOR, Chevron/24hr, Exxon, 76/mart, Shell/mart/wash/24hr, 7-11, Burger King, Carrow's Rest., Denny's, KFC, McDonald's, Barnes&Noble, Goodyear/auto, Kinko's, K-Mart, Kragen Parts, Longs Drug, Marshall's, Mervyn's, Target, bank, carwash, repair/tires
51	Willow Pass Rd, Taylor Blvd, **E**...MEDICAL CARE, Benihana Rest., Denny's, El Torito, Grissini Italian, JJ North's Buffet, Marie Callender's, Red Lobster, Tony Roma, Hilton, Auto Parts Club, Circuit City, CompUSA, Cost+, Good Guys, Nevada Bob's, Office Depot, Old Navy, Trader Joe's, Willows Shopping Ctr, auto repair, **W**...Firestone, JC Penney, Macy's, Sears/auto, bank
50	CA 242(from nb), to Concord
49	Gregory Lane, to Pleasant Hill, **E**...Chevron/mart, Jo-Ann Fabrics, **W**...Confetti Rest., Lyon's Rest., Grand Auto Supply
48	Oak Park Blvd, Geary Rd, **E**...CHIROPRACTOR, VETERINARIAN, Chevron, 7-11, Subway, Embassy Suites, Extended Stay America, Best Buy, bank, **W**...Chevron, Shell/diesel/mart, Black Angus, Burger King, China Rest., Sweet Tomatos, Wendy's, Courtyard, Holiday Inn, Bed, Bath&Beyond, Mazda/Subaru, Midas Muffler, Nissan, Staples, Volvo, Walgreen, carwash, cleaners/laundry
47	N Main St, to Walnut Creek, **E**...Chevron/Subway/mart, Shell, Black Diamond Brewery/rest., Fuddrucker's, Jack-in-the-Box, Taco Bell, Vino Ristorante, Marriott, Motel 6, Walnut Cr Motel, Cadillac, Chrysler/Jeep, Harley-Davidson, Mercedes, Target, **W**...76/diesel/24hr, 7-11, Domino's, Honda, NAPA, OfficeMax
46	CA 24, to Lafayette, Oakland
45	Rudgear Rd, no facilities
44	S Main St, Walnut Creek, **E**...HOSPITAL
43	Livorna Rd, no facilities
42	Stone Valley Rd, Alamo, **W**...Chevron, Rotten Robbie/diesel, Shell/diesel/mart/atm, 7-11/gas, Cioni's Italian, Little Caesar's, Mexican Rest., MailBoxes Etc, Safeway/deli, bank, cleaners
41	El Pintado Rd, Danville, no facilities
40	El Cerro Blvd, no facilities
39	Diablo Rd, Danville, **E**...76/mart/24hr, Taco Bell, Albertson's, MailCtr+, cleaners, Mt Diablo SP(12mi), **W**...MEDICAL CARE, EYECARE, Chevron, 76, Chinese Rest., Coco's, Country Waffle Rest., Foster's Freeze, Pizza Hut, Primo's Pizza/pasta, auto repair, bank, cleaners
38	Sycamore Valley Rd, **E**...Shell, Denny's, Econolodge, **W**...CHIROPRACTOR, Arco/diesel, Exxon, 76, Pizza Machine, Tony Roma's, Albertson's, Longs Drug, bank

Concord

California Interstate 680

N

S

37   Crow Canyon Rd, San Ramon, **E**...HOSPITAL, CHIROPRACTOR, Shell, Burger King, Carl's Jr, Chili's, Chinese Rest., Albertson's, JiffyLube, Kinko's, Rite Aid, USPO, bank, cleaners, **W**...CHIROPRACTOR, DENTIST, EYECARE, Chevron/repair/24hr, Exxon/atm, 76, Shell/autocare, Boston Mkt, In-n-Out, Maestro's Rest., McDonald's, Taco Bell, TGIFriday, Harley-Davidson, Longs Drug, Safeway, bank

36   Bollinger Canyon Rd, **E**...Exxon, Subway, Marriott, Residence Inn, Borders Books, Target, Whole Foods, **W**...Chevron/Foodini's/mart, Applebee's, Chevy's Mexican, Marie Callender's, Courtyard, Homestead Village

33   Alcosta Blvd, to Dublin, **E**...76, 7-11, Pizza Hut, TCBY, Albertson's/drugs, Longs Drugs, bank, **W**...CHIROPRACTOR, Chevron/atm, Shell/wash, Chatillon Rest., Chinese Rest., Chubby's Rest., House of Bagels, Pizza Palace, Taco Bell, Albertson's/deli, Walgreen, bank

31   I-580, W to Oakland, E to Tracy

30   Stoneridge, Dublin, **E**...Hilton, **W**...CHIROPRACTOR, EYECARE, Black Angus, Chinese Rest., Doubletree Hotel, Holiday Inn, Big 5 Sports, Chrysler/Plymouth/ Jeep, Good Guys, Macy's, Nordstrom's, Pier 1, Sears, mall

28   Bernal Ave, Pleasanton, **E**...DENTIST, Lindo's Mexican, Ring's Burgers, Vic's Bakery, cleaners

25   Sunol Blvd, Castlewood Dr, Pleasanton, **1 mi E**...Chinese Rest., Jim's Rest., Raley's Food/drugs, PakMail, SpeeDee Lube, bank

22   CA 84, Sunol, W to Dumbarton Bridge

21.5   Calaveras Rd, no facilities

21   Andrade Rd, Sheridan Rd(from sb), **E**...gas/diesel/mart/24hr, **W**...golf

20mm   weigh sta nb

19   Sheridan Rd(from nb), no facilities

18   Vargas Rd, no facilities

17   CA 238, Mission Blvd, to Hayward, **E**...Shell/mart, McDonald's, **W**...HOSPITAL

16   Washington Blvd, Irvington Dist, **E**...VETERINARIAN, QuikStop/gas, Chinese Rest., Mexican Rest., cleaners

15   Durham Rd, to Auto Mall Pkwy, **W**...76/Subway/mart/24hr, Shell/Jack-in-the-Box/mart, Fry's Electronics

13   Mission Blvd, Warm Springs Dist, to I-880, **W**...DENTIST, Exxon/diesel/mart, Burger King, Carl's Jr, Chinese Rest., Denny's, Donut House, Jack-in-the-Box, KFC, Little Caesar's, RoundTable Pizza, Taco Bell, Econolodge, Quality Inn, GNC, Albertson's, Lion Foods, Longs Drug, bank

11   Scott Creek Rd, no facilities

9   Jacklin Rd, **E**...Bonfare Rest., Chinese Rest., pizza, cleaners, **W**...Shell/wash

8   CA 237, Calaveras Blvd, Milpitas, **E**...CHIROPRACTOR, Shell/repair, 76, Flames CoffeeShop, Hungry Hunter, Pizza Hut, Sizzler, Subway, Day's Inn, SuperMkt, bank, **W**...MEDICAL CARE, CHIROPRACTOR, DENTIST, Shell/ wash, Chinese Rest., Dave&Buster's, El Torito, Lyon's Rest., McDonald's, Red Lobster, TCBY, Embassy Suites, Extended Stay America, Albertson's, Longs Drug, MailBoxes Etc, Mervyn's, VisionCare, bank, cinema

7   Landess Ave, Montague Expswy, **E**...CHIROPRACTOR, DENTIST, VETERINARIAN, Arco, 76, Burger King, Jack-in-the-Box, McDonald's, Royal Taco, StrawHat Pizza, Taco Bell, Togo's, Wienerschnitzel, Albertson's, Home Depot, Rite Aid, Target, Walgreen, bank, laundry

6   Capitol Ave, Hostetter Ave, **E**...DENTIST, Shell, Carl's Jr, Popeye's, SaveMart Foods, cleaners, pizza, **W**...MEDICAL CARE, DENTIST, Exxon, 7-11, Chinese Rest., KFC, MailBag, laundry

5   Berryessa Rd, **E**...CHIROPRACTOR, Arco/mart/atm/24hr, Shell, Baskin-Robbins, Chinese Rest., Denny's, McDonald's, Ristorante Italiano, Taco Bell, Albertson's, Longs Drug, MailBoxes Etc, Safeway, cinema, cleaners

4   McKee Rd, **E**...DENTIST, 76/wash, Chevron, Shell/mart/wash, Burger King, Chinese Rest., Country Harvest Buffet, Donut Express, Pizza Hut, Sizzler, Togo's, Wienerschnitzel, Albertson's, PaknSave Foods, Walgreen, cleaners, **W**...HOSPITAL, DENTIST, EYECARE, Baskin-Robbins, Chinese Rest., Foster's Freeze, McDonald's, RoundTable, Winchell's, Fabric Whse, K-Mart

3   Alum Rock Ave, **E**...Shell/diesel/mart/wash/24hr, bank, tires, **W**...Exxon, 76/24hr, Carl's Jr

2   Capitol Expswy, no facilities

1   King Rd, Jackson Ave(from nb), **E**...76, Shell/wash, 7-11, Jack-in-the-Box, Mi Pueblo Mexican, Taco Bell, King's SuperMkt, Quality Tuneup, Walgreen, foreign auto parts, **W**...Mexican Rest., auto repair/lube

0   US 101, I-680 begins/ends.

California Interstate 710(LA)

Los Angeles Area

Exit #	Services
29	I-710 begins/ends on Valley Blvd
28	I-10, no facilities
27	Chavez Ave, no facilities
26	CA 60, Pamona Fwy, **E**...cleaners, **W**...Shell
25	3rd St, no facilities
24	Whittier Blvd, Olympic Blvd, **W**...Shell/mart, McDonald's
23	Washington Blvd, Commerce, **W**...Commerce Trkstp/diesel/rest.
22	Bandini Blvd, Atlantic Blvd, industrial
21	Florence Ave, **E**...Bicycle Club, KFC, Ralph's Foods, Toys R Us
20	CA 42, Firestone Blvd, **E**...Ford, Target
19	Imperial Hwy, **W**...76, Shell
18	I-105, no facilities
17	Rosecrans Ave, no facilities
16	Alondra Ave, no facilities
15	Redondo Beach Blvd
13	CA 91, no facilities
12	Long Beach Blvd, **E**...Mobil/repair, 76, Taco Bell, **W**...Arco/mart/24hr, Day's Inn
10	Del Amo Blvd, no facilities
8	I-405, San Diego Freeway
7	Willow St, **E**...Chevron, Chee Chinese, **W**...MEDICAL CARE, DENTIST, Mobil, Popeye's, Ralph's Foods
5	CA 1, Pacific Coast Hwy, **E**...Arco/mart, Shell, UltraMar/gas, Burger Express, Via Italia Pizza, LaMirage Inn, auto repair, **W**...76/service, Texaco/Carl's Jr/diesel/mart, Xpress Minimart/deli, Alberta's Mexican/24hr, Golden Star Rest., Jack-in-the-Box, McDonald's, Tom's Burgers, Wienerschnitzel, Winchell's, SeaBreeze Motel, SeaBrite Motel, TrueValue, auto repair, tires/brakes
4	Anaheim St, **W**...Shell/LP/24hr, diesel repair, tire/lube
3	Piers B, C, D, E, Pico Ave, Long Beach, no facilities
2	Piers F-J, Queen Mary
1	Piers S, T, Terminal Island, **E**...Hilton, bank
0mm	I-710 begins/ends in Long Beach

California Interstate 780(Vallejo)

Benicia

Exit #	Services
7mm	I-780 begins/ends on I-80.
6	Laurel St, no facilities
5	Glen Cove Pkwy, **S**...DENTIST, Baskin-Robbins, Taco Bell, MailBoxes Etc, Safeway, cleaners/laundry
4.5	Columbus Pkwy, **N**...DENTIST, Shell/Burger King/mart, Napoli Pizza/pasta, Subway, Lube'n Oil, cleaners, **S**...to Benicia RA
4	Military West, **S**...MEDICAL CARE
3	Southampton Rd, Benicia, **N**...VETERINARIAN, DENTIST, EYECARE, Asian Bistro, Burger King, Country Waffle, Papa Murphy's, Rickshaw Express, RoundTable Pizza, Starbucks, Subway, Ace Hardware, AutoZone, MailBoxes Etc, Radio Shack, Raley's Foods, cleaners
2	E 2nd St, Central Benicia, **N**...Exxon/mart/wash, Best Western, **S**...McDonald's
1	E 5th St, Benicia, **N**...Fast&Easy/mart, **S**...Citgo/7-11, GasCity/TCBY/mart, auto repair, carwash
0mm	I-780 begins/ends on I-680.

California Interstate 805(San Diego)

Exit #	Services
31	I-5(from nb). I-805 begins/ends on I-5.
30	Sorrento Valley Rd, Mira Mesa Blvd, **E**...Texaco, McDonald's, Courtyard, Staples
29	Vista Sorrento Pkwy, no facilities
28	La Jolla Village Dr, Miramar Rd, **W**...HOSPITAL
27	Governor Dr, no facilities
26	CA 52, no facilities
25	Clairmont Mesa Blvd, **E**...HomeBase, funpark, **W**...Motel 6, Godfather's
24	CA 274, Balboa Ave, **E**...Arco/mart, Shell/wash, Albertson's
23	CA 163 N, to Escondido, no facilities
22	Mesa College Dr, Kearney Villa Rd, **W**...HOSPITAL

E

W

E

W

N

S

California Interstate 805

21	Murray Ridge Rd, to Phyllis Place, no facilities
20	I-8, E to El Centro, W to beaches
19	El Cajon Blvd, **E...**Shell, **W...**76
18	University Ave, **E...**Chevron, **W...**Arco/mart/24hr, Shell
17	CA 15 N, 40th St, to I-15, no facilities
16	Home Ave, no facilities
15	CA 94, no facilities
14	Market St, no facilities
13	Imperial Ave, no facilities
12	47th St, no facilities
11	43rd St, **W...**Lucky Foods, PepBoys, auto dealers
10	Plaza Blvd, National City, **E...**Dairy Queen, McDonald's, Popeye's, Ralph's Foods, **W...**HOSPITAL, Jimmy's Rest., civic ctr
9	Sweetwater Rd, **W...**Circuit City, Staples
8	CA 54, no facilities
7	E St, Bonita Rd, **E...**Shell, Applebee's, Jack-in-the-Box, JC Penney, Marshall's, Mervyn's, Robinson-May, mall, RV camping, **W...**Chevron, Love's Rest., Ramada Inn
6	H St, same as 7
5	L St, Telegraph Canyon Rd, **E...**HOSPITAL, Shell/mart/wash, McDonald's, Rite Aid, Von's Foods, Olympic Training Ctr, RV camping
4	Orange Ave, no facilities
3	Main St, Otay Valley Rd, **E...**Shell/mart/wash, **W...**Holiday Inn Express
2	Palm Ave, **E...**Arco/mart, Carl's Jr, Subway, Von's Foods, Wal-Mart, **W...**76/diesel, McDonald's
1	CA 905, **E...**Brown Field Airport, Otay Mesa Border Crossing
.5	San Ysidro Blvd, **E...**Arco, Texaco, **W...**Chevron, 76

I-805 begins/ends on I-5.

N ↕ S

San Diego Area

California Interstate 880(Bay Area)

Exit #(name)(mm)Services

I-880 begins/ends in Oakland on I-980.

46	7th St, Grand Ave(from sb), downtown
45	Broadway St, downtown, **E...**Civic Ctr Lodge, **W...**to Jack London Square
44	Oak St, Lakeside Dr, downtown, **E...**Howard Johnson, **W...**Shell/diesel
43	5th Ave, Embarcadero, **W...**Ark Rest., Hungry Hunter Rest., Reef Rest., Executive Inn, Motel 6
42	29th Ave, 23rd Ave, to Fruitvale, **E...**EYECARE, Shell/mart/wash, Boston Mkt, DonutStar, Starbucks, Albertson's, AutoZone, GNC, Office Depot, Radio Shack, **W...**7-11, Buttercup Grill
41	High St, to Alameda, **E...**$Inn, El Monte RV Ctr, **W...**Shell/diesel/mart/wash, McDonald's, K-Mart/Little Caesar's/auto/24hr, bank
40	66th Ave, Zhoney Way, **E...**coliseum
39	Hegenberger Rd, **E...**Arco/mart/atm/24hr, Shell/diesel/mart/wash, Burger King, Denny's, Hungry Hunter Rest., Jack-in-the-Box/24hr, McDonald's, Sam's Hofbrau, Taco Bell, Comfort Inn, Day's Inn/rest., Hampton Inn, Holiday Inn, Motel 6, Pak'n Save Food/deli, Midas Muffler, GMC/Volvo, Freightliner, HomeBase, AutoParts Club, Pennzoil/tuneup, cleaners, **W...**Chevron, Shell/mart, Carrows Rest., Chinese Rest., Francesco's Rest., Hegen Burgers, Hilton, Park Plaza Motel, Ramada Inn, Goodyear, bank, to Oakland Airport
38	98th Ave, **W...**airport, auto rentals, brakes/tuneup
37	Davis St, **E...**Arco, 7-11, Lee's Donuts, Sergio's Pizza, cleaners, **W...**Shell, Burger King, Starbucks, Togo's, Costco, Home Depot, Office Depot, SportMart, Wal-Mart
36	Marina Blvd, **E...**HOSPITAL, Beacon/diesel/mart, Marina/gas, Jack-in-the-Box, La Salsa Mexican, Starbucks, Firestone/auto, Ford, Hyundai, Kia, Kinko's, Marshall's, Nissan, Nordstrom's, OfficeMax, Old Navy, auto/brake repair, **W...**Flyers/diesel/mart, DairyBelle, Denny's, Giant Burger, KFC, Mtn Mike's Pizza, USPO, bank
35	Washington Ave(from nb), Lewellin Blvd(from sb), **W...**DENTIST, Arco, 76/mart, Burrito Shop, Hometown Buffet, Jack-in-the-Box, McDonald's, Subway, Motel Nimitz, MensWearhouse, Longs Drug, Radio Shack, Safeway/24hr, SavMax Foods, SP Parts, Walgreen/24hr, bank, cleaners
34	I-238(from sb), to I-580, Castro Valley
33	Hesperian Blvd, **E...**DENTIST, 76, Bakers Square, KFC, Mr Pizza, Western Superburger, Big 5 Sports, House of Fabrics, Kragen Parts, Target, Wheelworks Repair, **W...**VETERINARIAN, Arco, Chevron, 76, Black Angus, Carrow's Rest., KFC, McDonald's, Taco Bell, Vagabond Inn, Albertson's, bank, cinema, cleaners

N ↕ S

Oakland Area

California Interstate 880

32	A St, San Lorenzo, **E**...76, McDonald's, Best Western, AllPro Parts, Costco, OilChanger, Quality Tuneup, tires, **W**...76, Burger King, StrawHat Pizza, Heritage Inn, cleaners/laundry
31	Winton Ave, **W**...Chevron/mart, Valero/diesel, Applebee's, Sizzler, Circuit City, Firestone/auto, HomeLife, JC Penney, OilChanger, Macy's, Mervyn's, Old Navy, Ross, Sears/auto, bank, cinema, mall
30	CA 92, Jackson St, **E**...Valero/mart, 76, 7-11, Bakers Square, Rickshaw Chinese, Subway, Taco Bell, Albertson's/deli, JiffyLube, Longs Drug, MailBoxes Etc, Rite Aid, Safeway, cleaners/laundry, **W**...San Mateo Br
27	Tennyson Rd, **E**...DENTIST, All American/diesel, 76, Shell, Chinese Rest., Jack-in-the-Box, KFC, Pizza Hut, RoundTable Pizza, Chavez Foods, Grand Parts, Walgreen, laundry, **W**...HOSPITAL, Chevron, 76, Valero/diesel, 7-11, Carl's Jr, McDonald's, PepBoys
25	Whipple Rd, Dyer St, **E**...Chevron/24hr, 76, Denny's, McDonald's, Taco Bell, Motel 6, Super 8, CHP, Food4Less, Home Depot, PepBoys, **W**...DENTIST, QuikStop/24hr, 76, Shell, Chili's, IHOP, Jollibee's Café, Jamba Juice, Krispy Kreme, La Salsa Mexican, Starbucks, Texas Roadhouse, TGIFriday, Togo's, Extended Stay America, Albertson's/SavOn, Borders Books, Kinko's, Michael's, OfficeMax, PetCo, RadioShack, Wal-Mart/auto, cinema, cleaners
24	Alvarado-Niles Rd, **E**...Shell/wash, 7-11, Radisson, **W**...76, Shell, Wal-Mart/drugs, same as 25
23	Alvarado Blvd, Fremont Blvd, **E**...CHIROPRACTOR, DENTIST, Luciano European Café, Phoenix Garden Chinese, Subway, Motel 6, Albertson's/SavOn, Ranch Mkt Foods
22	CA 84 W, Dumbarton Br, Decoto Rd, **E**...Citgo/7-11, McDonald's, Rite Aid, Walgreen, cleaners
21	CA 84 E, Thornton Ave, Newark, **E**...Chevrolet, **W**...Chevron/mart/24hr, Exxon, Shell/mart/wash, 7-11, Bakers Square, Carl's Jr, KFC, RoundTable Pizza, Taco Bell, K-Mart, cleaners/laundry
20	Mowry Ave, Fremont, **E**...HOSPITAL, Chevron/diesel, Exxon, Applebee's, Burger King, Denny's, Hungry Hunter, KFC, Mkt Broiler, Olive Garden, EZ 8 Motel, Studio+ Suites, Albertson's, Cost+, Jo-Ann Fabrics, cleaners, **W**...Bombay Garden, El Burro Mexican, FreshChoice Rest., HomeTown Buffet, Jack-in-the-Box, Lyon's Rest., McDonald's, Red Robin, Subway, Taco Bell, Hawthorn Suites, Holiday Inn Express, Motel 6, Woodfin Suites, Circuit City, Firestone, Ford, Goodyear/auto, Hancock Fabrics, JC Penney, JiffyLube, KidsRUs, Linen Experts, Macy's, Marshall's, MensWearhouse, Mervyn's, PetsMart, Sears/auto, Staples, Stroud's, Target, TJ Maxx, ToysRUs, bank, cinema, mall
19	Stevenson Blvd, **W**...Carl's Jr, Chevy's Mexican, ChuckeCheese, Nijo Castle Rest., Sizzler, Hilton Garden, AAA, Costco, Econo Lube'n Tune, Food4Less, Ford, Harley-Davidson, Home Depot, JC Penney, Macy's, Mervyn's, Nissan, Olds/Pontiac/GMC, PepBoys, Saturn, Sears/auto, TJ Maxx, mall
18	Auto Mall Pkwy, **E**...Chevron, **W**...Crawford Suites, BMW, Dodge/Isuzu, Honda, Hyundai, Kia, Lexus, Mercedes, Nissan, Saturn, Toyota, Volvo
17.5mm	weigh sta both lanes
17	Fremont Blvd, Irving Dist, **W**...McDonald's, Courtyard, GoodNite Inn, Homestead Village, La Quinta, Marriott
16.5	Warren St(from sb), **E**...Quality Inn, **W**...Courtyard, Hampton Inn
16	Gateway Blvd(from nb), **W**...Courtyard
15	Mission Blvd, **E**...to I-680, 76, Denny's, Togo's, Quality Inn, MailBoxes Etc, cleaners
13	Dixon Landing Rd, **E**...DENTIST, 7-11, Burger King, Chinese Rest., Residence Inn
10	CA 237, Alviso Rd, Calaveras Rd, Milpitas, **E**...DENTIST, Arco, 76, 7-11, Chili's, Denny's, Marie Callender, RoundTable Pizza, Best Western, Economy Inn, Travelodge, Albertson's, Kragen Parts, SavMart Foods, Walgreen, bank, **W**...Applebee's, Black Angus, HomeTown Buffet, In-n-Out, Macaroni Grill, McDonald's, On the Border, Taco Bell, Candlewood Suites, Hampton Inn, Hilton Garden, Holiday Inn, Best Buy, Borders Books, Chevrolet, Good Guys, MensWearhouse, OfficeMax, PetsMart, Ranch Mkt Foods, Ross, Service Merchandise, SportMart, Wal-Mart/auto
9	Great Mall Parkway, Tasman Dr, **E**...mall
8	Montague Expswy, **E**...MEDICAL CARE, 76, Shell, Carl's Jr, Jack-in-the-Box, U-Haul, auto repair, tires, **W**...Chevron/mart/wash, Dave&Buster's, Sheraton
7	Brokaw Rd, **W**...Ford Trucks, CHP
6	Gish Rd, **W**...auto/diesel repair/transmissions, tires
5	US 101, N to San Francisco, S to LA
4.5	1st St, **E**...Shell/repair, Mexican Rest., **W**...Chevron, 76, Denny's/24hr, Japanese Steaks, McDonald's, Adlon Hotel, Comfort Inn, Day's Inn, Holiday Inn Express, Wyndham Garden, bank
4	Coleman St, **E**...Exxon/mart/atm, George's Rest., **W**...airport
3.5	CA 82, The Alameda, **E**...San Jose Inn/rest., **W**...Shell/repair, Cozy Rest., Bell Motel, Comfort Inn, Co-Z Hotel, Friendship Inn, Valley Inn, Safeway, Santa Clara U, laundry
3	Bascom Ave, to Santa Clara, **E**...HOSPITAL, DENTIST, Shell/repair, **W**...CHIROPRACTOR, Exxon, Rotten Robbie/diesel, Burger King, Japanese Rest., Normandie House Pizza, laundry
2	Stevens Creek Blvd, San Carlos St, **E**...Coast/mart, Korean Rest., **W**...CHIROPRACTOR, Exxon, Arby's, Burger King, Lyon's Rest., Teriyaki Rest., Audi/VW, Chevrolet, Firestone/auto, Goodyear/auto, Isuzu, Longs Drugs, Macy's, Mitsubishi, Nordstrom's, Quality Tuneup, Safeway, Subaru, bank, pizza
1	San Carlos St, **E**...HOSPITAL
0mm	I-880 begins/ends on I-280.

Fremont

San Jose

N

S

Colorado Interstate 25

Exit #	Services

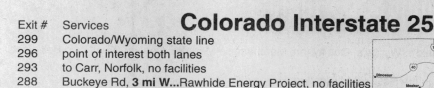

299 Colorado/Wyoming state line

296 point of interest both lanes

293 to Carr, Norfolk, no facilities

288 Buckeye Rd, **3 mi W**...Rawhide Energy Project, no facilities

281 Owl Canyon Rd, **5 mi E**...KOA, phone

278 CO 1 to Wellington, **1/2 mi W**...Conoco/mart, Texaco/Burger King/Taco Bell/mart, Subway, bank

271 Mountain Vista Dr, **W**...Anheiser-Busch Brewery

269b a CO 14, to US 87, Ft Collins, **E**...Mulberry Inn, RV service, **W**...Conoco, Phillips 66/diesel/mart, Denny's, Kettle Rest., Waffle House, Comfort Inn, Day's Inn, Holiday Inn, Motel 6, National 9 Inn, Plaza Inn, Ramada Inn, Super 8, Ace Hardware/gas, to CO St U, stadium

268 Prospect Rd, to Ft Collins, **E**...Powder River RV Ctr, Yamaha/Suzuki, **W**...HOSPITAL, Harley-Davidson

267mm weigh sta both lanes

266mm rest area both lanes, full(handicapped)facilities, info, phone, picnic tables, litter barrels, petwalk

265 CO 68 W, Timnath, **W**...Texaco/mart, **3 mi W**...Carrabba's, McDonald's, Courtyard, Marriott, Residence Inn

262 CO 392 E, to Windsor, **E**...Phillips 66/Subway/diesel/mart, Arby's, McDonald's(3mi), AmericInn, Windsor Inn(3mi), **W**...truck/auto parts

259 Airport Rd, **E**...Wal-Mart Depot, **W**...to airport

257b a US 34, to Loveland, **E**...Total/Taco Bell/diesel/mart/RV camping, Country Inn Suites, **W**...HOSPITAL, Texaco/diesel/mart, Arby's, Chili's, Cracker Barrel, IHOP, LoneStar Steaks, Subway, Wendy's, Best Western, Comfort Inn, Fairfield Inn, Hampton Inn, Holiday Inn Express, Super 8(2mi), Prime Outlets/famous brands, Target, RV camping, racetrack, museum, to Rocky Mtn NP

255 CO 402 W, to Loveland, no facilities

254 to CO 60 W, to Campion, **E**...Johnson's Corner/diesel/café/24hr, Budget Host, RV camping/service

252 CO 60 E, to Johnstown, Milliken, no facilities

250 CO 56 W, to Berthoud, no facilities

245 to Mead, no facilities

243 CO 66, to Longmont, Platteville, **E**...Conoco, Texaco/mart, Blimpie, K&C RV Ctr, Yamaha/Suzuki, **W**...to Rocky Mtn NP, to Estes Park

241mm St Vrain River

240 CO 119, to Longmont, **E**...Phillips 66/Wendy's/TCBY/diesel/mart, Stevinson RV Ctr, **W**...HOSPITAL, Conoco/Subway/diesel/mart/24hr, Texaco/Pizza Hut/diesel/mart, Arby's, Burger King, Dairy Queen, McDonald's, Taco Bell, Waffle House, Comfort Inn, Day's Inn, 1st Interstate Inn, Super 8, Travelodge, Del Camino RV Ctr, museum, to Barbour Ponds SP

235 CO 52, Dacono, **W**...Phillips 66/diesel/LP/mart, to Eldora Ski Area

232 to Erie, no facilities

229 CO 7, to Lafayette, Brighton, **E**...RV camping

223 CO 128, 120th Ave, to Broomfield, **E**...Amoco/diesel/mart, Conoco/mart, Sinclair/mart, Texaco, Applebee's, Burger King, Café Mexico, Damon's, Fuddrucker's, Golden Gate Chinese, LoneStar Steaks, OutBack Steaks, Pizza Hut, Shari's, Day's Inn, Hampton Inn, Holiday Inn, Radisson, Sleep Inn, Albertson's, Checker's Parts, Discount Tire, Econo Lube'n Tune, MailBoxes Etc, PetCo, bank, cleaners, **W**...DENTIST, Conoco/mart/wash, Shamrock/mart, Texaco/Popeye's/mart, Total/diesel/mart/wash, Casa Loma Mexican, CB&Potts Rest., Chili's, Cracker Barrel, Dairy Queen, Jade City Chinese, Perkins, Starbucks, Subway, Village Inn Rest., Wendy's, Comfort Inn, La Quinta, Super 8, Kinko's, bank

221 104th Ave, to Northglenn, **E**...HOSPITAL, CHIROPRACTOR, Conoco/mart/wash, Phillips 66, Total/diesel/mart, Burger King, Chinese Rest.. Denny's, IHOP, McDonald's, Subway, Taco Bell, Taco John's, Winchell's, AutoZone, Big A Parts, Bigg's Foods, Cloth World, Home Depot, King's Sooper's, Midas Muffler, Service Merchandise, Super K-Mart, Target, Wal-Mart, Waxman Camera/video, **W**...OPTOMETRIST, Citgo/7-11, Conoco/mart, Texaco/diesel/mart/wash, Applebee's, Bennigan's, Blackeyed Pea, Cinzinetti's Italian, McDonald's, Taco Bell, La Quinta, Ramada Ltd, Albertson's, Borders Books, Dodge, Eagle Hardware, Ford/FastLube, Firestone/auto, Gart Sports, Goodyear/auto, Mervyn's, Office Depot, Rite Aid, Shepler's Western Store, bank, cinema, mall

220 Thornton Pkwy, **E**...HOSPITAL, Crossland Suites, House Of Fabrics, Thornton Civic Ctr, **W**...MEDICAL CARE, Conoco/mart, Shamrock/mart, Texaco/diesel/mart

219 84th Ave, to Federal Way, **E**...CHIROPRACTOR, Conoco/mart/wash, Shamrock, Arby's, Bonanza, Dos Verdes Mexican, Goodtimes Grill, Taco Bell, Waffle House, Crossland Suites, Radio Shack Outlet, Walgreen, Ward's Outlet World/auto, **W**...HOSPITAL, DENTIST, OPTICIAN, VETERINARIAN, Amoco, Coastal/mart, Total/diesel/mart/wash, Burger King, Dairy Queen, Dunkin Donuts, Mexican Rest., Pizza Hut, Tokyo Bowl Japanese, Village Inn Rest., Motel 6, CarQuest, Discount Tire, Econo Lube'n Tune, Meineke Muffler, Performance Auto Service, SoundTrack Superstore

217 US 36 W, to Boulder, **W**...Subway, Chevrolet/Toyota

Colorado Interstate 25

216 I-76 E, to I-270 E, no facilities

215 58th Ave, **E**...Amoco/mart/wash, McDonald's, Wendy's, Quality Inn, Merchandise Mart, **W**...Conoco, Total/Taco Bell/diesel/ LP/mart, Super 8, Malibu FunPark

214c 48th Ave, **E**...coliseum, airport, **W**...Village Inn Rest., Best Western, Holiday Inn/rest.

b a I-70, E to Limon, W to Grand Junction

213 Park Ave, W 38th Ave, 23rd St, downtown, **E**...Amoco/McDonald's/mart, Citgo/7-11, Burger King, Texaco, Denny's, La Quinta, Goodyear, **W**...Regency Inn, Travelodge

212c 20th St, Denver, downtown, **W**...Paglicci's Italian

b a Speer Blvd, **E**...downtown, convention complex, museum, **W**...Conoco, Texaco, Budget Host, Residence Inn, Super 8, Travel Inn

211 23rd Ave, **E**...funpark

210c CO 33(from nb)

210b US 40 W, Colfax Ave, **W**...Sports Complex, Ramada Inn/rest., Red Lion Inn

a US 40 E, Colfax Ave, **E**...civic center, downtown, U-Haul

209c 8th Ave, **E**...Motel 7, Bob's Auto Parts, Sears Service Ctr

b 6th Ave W, US 6, **W**...Day's Inn, Ryder Trucks

a 6th Ave E, Denver, downtown

208 CO 26, Alameda Ave(from sb), **E**...Amoco, Total/diesel/mart, Burger King, Denny's, same as 207b

207b US 85 S, Santa Fe Dr, **E**...Total/mart, Burger King, Denny's, Home Depot, **W**...Texaco

a Broadway, Lincoln St, **E**...Griff's Burgers, TrueValue, USPO, auto parts

206b Washington St, Emerson St, **E**...WildOats Mkt/café, **W**...HOSPITAL, auto service

a Downing St(from nb), no facilities

205b a University Blvd, **W**...to U of Denver

204 CO 2, Colorado Blvd, **E**...Amoco, Conoco, Total/mart, 7-11, Grisanti's Grill, KFC, Lazy Dog Café, Pizza Hut, Village Inn Rest., Day's Inn, Fairfield Inn, Ramada, Albertson's/drugs, Best Buy, CompUSA, ToysRUs, bank, cinema, **W**...Dave&Buster's, Denny's, Perkins, La Quinta, Ford, cinema

203 Evans Ave, **E**...Texaco, Denny's, KFC, Rockies Inn, Big O Tires, auto parts, **W**...Cameron Motel, Ford

202 Yale Ave, **W**...Total/mart

201 US 285, CO 30, Hampden Ave, to Englewood, Aurora, **E**...MEDICAL CARE, CHIROPRACTOR, DENTIST, VETERINAR-IAN, Amoco, Conoco/mart/LP, Phillips 66/mart/wash, Texaco/mart/wash, Total/mart, Applebee's, Benihana, Blackeyed Pea, Boston Mkt, Chicago Grill, Chili's, Chinese Rest., Domino's, Einstein Bros Bagels, Fresh Fish Co, Houlihan's, Jason's Deli, Le Peep's Café, McDonald's/playplace, Mexican Grill, NY Deli, No Frills Grill, On-the-Border, Piccolo's Café, Skillet's Rest., Starbuck's, Subway, Uno Pizzaria, Embassy Suites, Marriott, Quality Inn/rest., TownePlace Suites, Discount Tire, GreaseMonkey Lube, Just Brakes, King's Sooper Foods, MailBox Express, Walgreen, bank, carwash, cinema, cleaners, **W**...Texaco, Burger King

200 I-225 N, to I-70, no facilities

199 CO 88, Belleview Ave, to Littleton, **E**...OPTICAL CARE, Amoco/wash, Phillips 66, Bagels&Beyond, Chinese Rest., Gandhi's Indian Cuisine, Garcia's Mexican, Harvest Rest., Off Belleview Grill, Sandwiches+, Tosh's Hacienda Rest., Wendy's, Hyatt Regency, Marriott, Wyndham Garden, Kinko's, MailBoxes Etc, bank, cleaners, **W**...Conoco/mart/wash, Total/diesel/mart, European Café, McDonald's, Mountainview Rest., Pappadeaux Café, Pizza Hut, Taco Bell, Day's Inn, Holiday Inn Express, HomeStead Village, Super 8, Wellesley Inn, GreaseMonkey Lube, Midas Muffler, golf

198 Orchard Rd, **E**...Del Frisco's Steaks, Shepler's, **W**...HOSPITAL, Texaco, Bayou Bob's Café, 4Happiness Chinese, Le Peep's Café, Quizno's, Hilton

197 Arapahoe Blvd, **E**...VETERINARIAN, Amoco, Conoco/mart, Applebee's, BBQ, Bennigan's, BlackJack Pizza, Burger King, Carrabba's, Chinese Rest., Country Dinner/playhouse, DeliZone, Denny's, El Parral Mexican, GoldStar Chili, GrandSlam Steaks, Gunther Toody's Rest., Heavenly Ham/turkey, IHOP, Jalapeno Mexican, Landry's Seafood, Outback Steaks, Pizza Hut, Red Lobster, Schlotsky's, Subway, Taco Cabana, TrailDust Steaks, Wendy's, Candlewood Hotel, Courtyard, Embassy Suites, Hampton Inn, Holiday Inn, La Quinta, MainStay Suites, Motel 6, Radisson, Sheraton, Sleep Inn, Summerfield Suites, Woodfield Suites, Big A Parts, Buick/Pontiac, Chrysler/Toyota, Discount Tire, Eagle Hardware, Ford, Gart Sports, Olds/GMC/Cadillac/Subaru, Nissan, Super K-Mart/auto/24hr, Target, USPO, bank, carwash, cinema, cleaners, hardware, **W**...EYECARE, Phillips 66/diesel/mart, 7-11, Arby's, Baker St Gourmet, Blackeyed Pea, Boston Mkt, Brooks Steaks, Burger King, Chevy's Mexican, Dairy Queen, GoodTimes Grill, Grady's Grill, Grisante's Grill, KFC, Lamonica's Rest., Macaroni Grill, McDonald's, Red Robin Café, Ruby Tuesday, Stanford's Grill, Souper Salad, Taco Bell, TCBY, Uno Pizzaria, Residence Inn, Albertson's/drugs, Barnes&Noble, Firestone/auto, Goodyear/auto, GNC, Office Depot, OfficeMax, MailBoxes Etc, 1-Hr Photo, Pier 1, carwash, cinema, cleaners

196 Dry Creek Rd, **E**...Vasil's Rest., Landry's Seafood, Best Western, Bradford Home Suites, Country Inn Suites, Homestead Village, Quality Inn, Ramada Ltd, Studio+, **W**...Drury Inn

195 County Line Rd, **E**...Courtyard, Residence Inn, **W**...OPTICAL CARE, Alexander's Rest., Champp's Rest., Cucina Cucina, Starbucks, AmeriSuites, Barnes&Noble, Bed Bath&Beyond, Best Buy, Borders Books&Café, CompUSA, Costco, Dillard's, Foley's, GolfSmith, Home Depot, HomePlace, JC Penney, Joslin's, Kinko's, Lil Things, Linens'n Things, Michael's, Nordstrom's, OfficeMax, PetCo, PetsMart, Pier 1, Ross, SoundTrack, ToysRUs, cinema

N

S

Denver Area

Cherry Hills

Colorado Interstate 25

194 CO 470 W, CO 470 E(tollway), **1 exit W on Quebec**...Arby's, ClaimJumper, Country Buffet, LoneStar Steaks, McDonald's, AmeriSuites, Comfort Suites, Fairfield Inn, Barnes&Noble, Circuit City, Costco, Firestone, GreaseMonkey, HomeBase, Home Depot, PepBoys, Sam's Club, Wal-Mart/auto/gas

193 Lincoln Ave, to Parker, **E**...Hilton

191 no facilities

190 Surrey Ridge, no facilities

188 Castle Pines Pkwy, **W**...Texaco, Total/mart/phone, Popeye's

187 Happy Canyon Rd, no facilities

184 Founders Pkwy, Meadows Pkwy, to Castle Rock, **E**...DENTIST, Conoco/diesel/mart, Texaco/Carl's Jr/diesel/mart/atm, Chinese Rest., Sonic, Starbucks, GNC, King's Sooper/24hr, Mailboxes Etc, Pearle Vision, Wal-Mart SuperCtr/24hr, bank, cleaners, **W**...Conoco/diesel/mart, Blackeyed Pea, Chili's, Food Court, IHOP, McDonald's, Rockyard Café, Best Western, Comfort Inn, Day's Inn, Holiday Inn Express, Castle Rock Prime Outlet/famous brands

183 US 85 N(from nb), Sedalia, Littleton

182 CO 86, Castle Rock, Franktown, **E**...CHIROPRACTOR, VETERINARIAN, Amoco/repair, Phillips 66/diesel/mart, Western Gas/mart, BagelStop, Chinese Rest., Little Caesar's, Mexicali Café, Nick&Willie's Pizza, Castle Rock Muffler, st patrol, **W**...Shamrock/mart, Texaco/diesel/mart/wash, Burger King, KFC, McDonald's, Mr Manner's Country Rest., Shari's/24hr, Taco Bell, Village Inn Rest., Wendy's, Comfort Inn, Holiday Inn Express, Quality Inn, Super 8, Chrysler/Plymouth/Dodge/Jeep, NAPA/repair, carwash

181 CO 86, Wilcox St, Plum Creek Pkwy, Castle Rock, **E**...MEDICAL CARE, CHIROPRACTOR, EYECARE, Citgo/7-11, Total/diesel/mart, Western/diesel/mart, Dairy Queen, Duke's Rest., Hungry Heifer Steaks, New Rock Café, Nicolo's Pizza, Pino's Café, Pizza Hut, Subway, Castle Rock Motel, Big A Parts, Ford/Lincoln/Mercury, MailBoxes Etc, Meineke Muffler, Midas Muffler, Safeway, USPO, XpressLube, 1-Hr Photo, bank, cleaners, hardware, tires

174 Tomah Rd, **W**...KOA(seasonal)

173 Larkspur(from sb, no return), **1 mi W**...Conoco/diesel/phone

172 South Lake Gulch Rd, Larkspur, **2 mi W**...Conoco/diesel/phone

171mm rest area both lanes, full(handicapped)facilities, phone, picnic tables, litter barrels, petwalk, RV dump

167 Greenland, no facilities

163 County Line Rd, no facilities

16.5mm Monument Hill, elev 7352

162mm weigh sta both lanes

161 CO 105, Woodmoor Dr, **E**...Amoco/repair, Falcon Inn, bank, carwash, **W**...Citgo/7-11, Conoco/diesel/mart, Texaco, Arby's, Boston Mkt, Burger King, Dairy Queen, Lotsa Bagels, McDonald's, Pizza Hut, Subway, Taco Bell, Village Inn Rest., Big O Tires, MailBoxes Etc, Pak'n Ship, Rite Aid, Safeway, RV camping, oil change

158 Baptist Rd, **E**...Texaco/Popeye's/diesel/mart/24hr, Jackson Creek Chinese, Subway, Ace Hardware, King's Sooper/24hr, PakMail, **W**...Total/diesel/mart

156b N Entrance to USAF Academy, **W**...visitors center

a Gleneagle Dr, **E**...mining museum

153 InterQuest Pkwy

151 Briargate Pkwy, **E**...Hilton, Focus on the Family Visitor Ctr, to Black Forest

150b a CO 83, Academy Blvd, **E**...DENTIST, VETERINARIAN, Farmcrest/gas, Texaco/A&W/diesel/mart, Total/Quizno's/diesel/mart, Applebee's, Baskin-Robbins, Blimpie, Boston Mkt, Bruebagger's Bagels, Burger King, Capt D's, Chevy's Mexican, Country Buffet, Cracker Barrel, Denny's, Einstein Bagels, Grady's Grill, IHOP, Joe's Crabshack, KFC, Macaroni Grill, McDonald's, Olive Garden, On-the-Border, Pizza Hut, Red Robin, Sonic, Souper Salad, Starbucks, Subway, Taco Bell, Village Inn Rest., Wendy's, Comfort Inn, Day's Inn, Drury Inn, Radisson, Red Roof Inn, Sleep Inn, Super 8, Advance Parts, Barnes&Noble, Bed Bath&Beyond, Best Buy, Big O Tires, Border's Books, Circuit City, CompUSA, CopyMax, Cub Foods, Dillard's, Econo Lune'nTune, Firestone/auto, Foley's, Gart Sports, HobbyLobby, HomeBase, Home Depot, JiffyLube, King's Soopers, Kinko's/24hr, MailBoxes Etc, Marshall's, MediaPlay, Michael's, Midas Muffler, NAPA, Office Depot, OfficeMax, PepBoys, PetsMart, Rite Aid, Sam's Club, Sears/auto, SteinMart, USPO, Wal-Mart SuperCtr/24hr, Western Auto, bank, cinema, cleaners, mall, to Peterson AFB, **W**...S Entrance to USAF Academy

149 Woodmen Rd, **E**...Carrabba's, **W**...Citgo/7-11, Old Chicago Pizza/pasta, Outback Steaks, TGIFriday, Zio's Italian, Comfort Inn, Embassy Suites, Extended Stay America, Fairfield Inn, Hampton Inn, Holiday Inn Express, Microtel, Radisson Suites, cinema

148b a Corporate Center Dr, Nevada Ave, **E**...RV Ctr, **W**...New South Wales Rest.

147 Rockrimmon Blvd, **W**...Texaco/mart/wash, Bradford Suites, Extended Stay America, Wyndham Garden, bank, to Pro Rodeo Hall of Fame

73

Colorado Interstate 25

146   Garden of the Gods Rd, **E**...Texaco/diesel/mart/wash, Antonio's Pasta/seafood, Carl's Jr, Denny's, McDonald's, Taco John's, Econolodge, La Quinta, Aamco, RV Ctr, **W**...CHIROPRACTOR, DENTIST, Citgo/7-11, Conoco, Phillips 66/mart/wash, Shamrock/mart/wash, Applebee's, Arby's, Blackeyed Pea, Dunkin Donuts, Hungry Farmer Rest., French Café, Italian Market/deli, Quizno's, Subway, Sunbird Rest., Taco Bell, Village Inn Rest., Wendy's, AmeriSuites, Day's Inn, Holiday Inn, Quality Inn, Sumner Suites, Super 8, Action Auto Repair, bank, to Garden of Gods

145   CO 38 E, Fillmore St, **E**...HOSPITAL, Citgo/7-11, Texaco, Total/mart, Dairy Queen, BudgetHost, Ramada Inn, JustBrakes, **W**...Conoco/diesel/mart, Texaco/diesel/mart, Total/diesel/mart, BBQ, CountyLine SmokeHouse, Murray's Rest., Waffle House, Best Western, Motel 6, Super 8, SuperLube, carwash

144   Fontanero St, no facilities

143   Uintah St, **E**...Citgo/7-11, Uintah Fine Arts Ctr

142   Bijou St, Bus Dist, **E**...Firestone/auto, **W**...Total, Denny's, Antlers Doubletree Hotel, Best Western, Le Baron Motel, RainTree Inn, bank

141   US 24 W, Cimarron St, to Manitou Springs, **W**...CHIROPRACTOR, Conoco/mart, Phillips 66/mart/wash, Burger King, Capt D's, Chinese Rest., McDonald's, Papa John's, Popeye's, Sonic, Subway, Taco John's, Waffle House, Western Sizzlin, Holiday Inn Express, Acura, AutoZone, Buick/GMC/Pontiac, Chevrolet, Chrysler/Plymouth, Diahatsu, Ford, GreaseMonkey, HobbyLobby, Hyundai, Infiniti, Isuzu, Lincoln/Mercury, Mazda, Mercedes, NAPA, Nissan, Office Depot, PetSupply, RV Ctr, Saturn, Subaru/Saab/VW, Suzuki, Toyota, Volvo, Wal-Mart SuperCtr/24hr, to Pikes Peak

140b   US 85 S, Tejon St, **W**...Conoco, Circle K, access to same as 141

    a   Nevada Ave, **E**...Chateau Motel, Colorado Springs Motel, Economy Inn, Howard Johnson, Chrysler/Jeep/Eagle, RV camping, **W**...CHIROPRACTOR, DENTIST, Amoco, Citgo/7-11, Texaco/diesel, Total/mart, Burger King, Chinese Rest., Colima Mexican, Dunkin Donuts, KFC, Little Caesar's, McDonald's, Schlotsky's, Subway, Taco Bell, TCBY, Wienerschitzel, Wendy's, American Inn, Chief Motel, Cheyenne Motel, Circle S Motel, Econolodge, Stagecoach Motel, Checker Parts, Midas Muffler, Meineke Muffler, 1-Hr Photo, PakMail, Safeway, USPO, Walgreen, bank, hardware, laundry/cleaners, pet supply, tires, access to auto dealers at 141

139   US 24 E, to Lyman, Peterson AFB

138   CO 29, Circle Dr, **E**...Conoco/mart, Texaco/diesel/mart/wash, Day's Inn/Mexican Rest., Sheraton, RV repair, airport, zoo, **W**...Citgo/7-11, Arby's, Burger King, Carrabba's, Chesapeake Bagel/deli, Carl's Jr, Chili's, Denny's, Fazoli's, Outback Steaks, Papa John's, Village Inn Rest., Subway, Budget Inn, DoubleTree Hotel, Fairfield Inn, Hampton Inn, La Quinta, Quality Inn, Residence Inn, GNC, OfficeMax, Radio Shack, Target, 1-Hr Photo, bank, cinemas

135   CO 83, Academy Blvd, **E**...to airport, same as 132, **W**...Ft Carson

132   CO 16, Wide Field Security, **3 mi E on US 85**...Shamrock/mart, Arby's, Burger King, KFC, McDonald's, Papa John's, Pizza Hut, Subway, Taco Bell, Wendy's, AutoZone, Checker Parts, K-Mart, Wal-Mart SuperCtr/24hr

128   to US 85 N, Fountain, Security, **E**...DENTIST, Citgo/7-11, Conoco/Subway/diesel/mart, USPO, auto/radiator repair, laundry, **W**...Texaco/diesel/rest./24hr, 1st Interstate Inn, Super 8

125   Ray Nixon Rd, no facilities

123   no facilities

122   to Pikes Peak Meadows, **W**...Pikes Peak Intn'l Raceway

119   Midway, no facilities

116   county line rd, no facilities

115mm rest area nb, full(handicapped)facilities, picnic tables, litter barrels, petwalk

114   Young Hollow, no facilities

112mm rest area sb, full(handicapped)facilities, picnic tables, litter barrels, petwalk

110   Pinon, **W**...Sinclair/Subway/diesel/repair, Pinon Inn

108   Bragdon, **E**...racetrack, **W**...KOA

106   Porter Draw, no facilities

104   Eden, **W**...flea market, RV Sales

102   Eagleridge Blvd, **E**...Conoco/diesel/mart/wash, Burger King, Texas Roadhouse, Barnes&Noble, Big O Tire, Home Depot, Sam's Club/gas, **W**...MEDICAL CARE, Texaco/Blimpie/LP/mart, Cracker Barrel, DJ's Steaks, IHOP, Village Inn Rest., Comfort Inn, Day's Inn, Econolodge, Hampton Inn, Holiday Inn, La Quinta, Motel 6, National 9 Inn, Wingate Inn, Fed Ex, Harley-Davidson, frontage rds access 101

101   US 50 W, Pueblo, **E**...Conoco, Denny's, Ruby Tuesday, Sleep Inn, Circuit City, JC Penney, Pier 1, Mervyn's, OfficeMax, PetsMart, Ross, Wal-Mart SuperCtr/24hr, U-Haul, cinema, mall, **W**...EYECARE, CHIROPRACTOR, Conoco/Blimpie/mart/wash/atm, Phillips 66/mart, Shamrock/mart, Total/diesel/mart, Applebee's, Arby's, Blackeyed Pea, Boston Mkt, Burger King, Carl's Jr, Chinese Rest., Country Kitchen, Dairy Queen, Domino's, Fazoli's, Gaetano's Italian, Golden Corral, King's Table Buffet, McDonald's, Pizza Hut, Red Lobster, Stuffy's Rest., Subway, Taco Bell, Wendy's, Holiday Inn, Motel 6, Quality Inn, Super 8, Villager Lodge, Advance Parts/brakes, Albertson's, Batteries+, Checker Parts, Discount Tire, Dodge, Goodyear/auto, K-Mart, Lincoln/Mercury, Lowe's Bldg Supply, National Pride Carwash, Toyota, Western Auto, frontage rds access 102

100b   29th St, Pueblo, **E**...Country Buffet, KFC, King's Sooper Foods, mall, tires, **W**...Amoco, Shamrock/mart, Sinclair/mart, GreaseMonkey, Safeway, Star Muffler/brake

    a   US 50 E, to La Junta, **E**...Conoco/LP, Phillips 66/mart, Texaco/mart, Total/mart/24hr, Dunkin Donuts, Little Caesar's, McDonald's, Pizza Hut, Ramada Inn/rest., Rodeway Inn, AutoZone, Furr's Foods, Goodyear, cinema

Colorado Interstate 25

99b a   Santa Fe Ave, 13th St, downtown, **W...**HOSPITAL, Amoco, Pancake House, Best Western, Brambletree Inn, Travelers Motel, Chevrolet, Chrysler/Jeep/Eagle, Ford/Subaru, Honda, JiffyLube, Mazda, Nissan, Olds, Pontiac, VW

98b   CO 96, 1st St, Union Ave Hist Dist, Pueblo, **W...**Conoco/mart, Carl's Jr, Wendy's, Jubilee Inn, Marriott, bank

a   US 50E bus, to La Junta, **E...**Shamrock/diesel/mart/24hr, **W...**Texaco/diesel/mart, Sonic

97b   Abriendo Ave, **W...**Wendy's, Best Western, auto repair

a   Central Ave, **W...**Conoco/mart, McDonald's, Jim's Automotive, Tire King

96   Indiana Ave, **W...**HOSPITAL, Conoco, Texaco/mart, Cobra Automotive

95   Illinois Ave(from sb), **W...**to dogtrack

94   CO 45 N, Pueblo Blvd, **E...**5J's Auto Parts, **W...**Citgo/7-11, Conoco/mart/wash/atm, Total/Taco Bell/diesel/mart, Hampton Inn, Microtel Suites, Pueblo Greyhound Park, café, RV camping, fairgrounds, to Lake Pueblo SP

91   Stem Beach, **W...**antiques

88   Burnt Mill Rd, no facilities

87   Verde Rd, no facilities

83   no facilities

77   Cedarwood, no facilities

74   CO 165 W, Colo City, **E...**Total/Taco Bell/diesel/mart/24hr, KOA, **W...**Texaco/Subway/TCBY/diesel/mart, Greenhorn Inn/rest., Rocking B Country Store, **rest area both lanes, full(handicapped)facilities, phone, vending, picnic tables, litter barrels, petwalk**

71   Graneros Rd, access to Columbia House

67   to Apache, no facilities

64   Lascar Rd, no facilities

60   Huerfano, no facilities

59   Butte Rd, no facilities

56   Red Rock Rd, no facilities

55   Airport Rd, no facilities

52   CO 69 W, Walsenburg, to Alamosa, **W...**Phillips 66/diesel/mart/24hr, Shamrock/mart, Alpine Rose Café(2mi), Carl's Jr(2mi), Pizza Hut, Subway(2mi), Tex' Mexican(1mi), Best Western, Budget Host, to Great Sand Dunes NM, San Luis Valley, camping

50   CO 10 E, to La Junta, **W...**HOSPITAL, tourist info

49   Lp 25, to US 160 W, Walsenburg, **W...**Lathrop SP, to Cuchara Ski Valley

42   Rouse Rd, to Pryor, no facilities

41   Rugby Rd, no facilities

37mm   **rest area both lanes, full(handicapped)facilities, picnic tables, litter barrels, petwalk**

34   Aguilar, **E...**Amoco/diesel/mart/rest.

30   Aguilar Rd, no facilities

27   Ludlow, **W...**point of interest

23   Hoehne Rd, no facilities

18   El Moro Rd, no facilities

15   US 350 E, Trinidad, Goddard Ave, **E...**Burger King, Super 8, Big R Foods, Family$, **W...**Texaco, B Lee's Café, Frontier Motel/cafe

14b a   CO 12 W, Trinidad, **E...CO Welcome Ctr**, Amoco, Texaco/mart, McDonald's, Pizza Hut, Subway, Taco Bell, **W...**Conoco/mart, Phillips 66/mart, Shamrock/mart, Dairy Queen, Domino's, El Capitan Rest., Prospect Plaza Motel, Parts+, TrueValue, laundry, RV camping, tires, to Trinidad Lake, Monument Lake

13b   Main St, Trinidad, **E...**Conoco/mart/wash, Day's Inn, CarQuest, Safeway

a   Santa Fe Trail, Trinidad, **E...**HOSPITAL, Best Western, RV camping

11   Starkville, **E...**weigh/check sta, Texaco/diesel/mart, Bob&Earl's Rest./24hr, Budget Host/RV park, Budget Summit Inn/RV Park, to Santa Fe Trail, **W...**Country Kitchen, Holiday Inn, Checker Parts/gas, Chevrolet/Buick, Wal-Mart SuperCtr/24hr, camping

8   Springcreek, no facilities

6   Gallinas, no facilities

2   Wootten, no facilities

1mm   scenic area pulloff nb

0mm   Colorado/New Mexico state line, Raton Pass, elev 7834, weigh sta sb

N ↕ S

Pueblo Walsenburg Trinidad

Colorado Interstate 70

Exit #	Services
447mm	Colorado/Kansas state line
438	US 24, Rose Ave, Burlington, **N**...HOSPITAL, Conoco/mart, RedFront/gas/wash, Texaco/mart, Dairy Queen, Comfort Inn, Hi-Lo Motel, Kit Carson Motel, Sloan's Motel, Super 8, Buick/Pontiac/GMC, CarQuest, Chevrolet, Ford, Goodyear, NAPA, **S**...Amoco/diesel/mart/24hr
437.5mm	**Welcome Ctr wb, full(handicapped)facilities, info, phone, picnic tables, litter barrels, petwalk, historical site**
437	US 385, Burlington, **N**...HOSPITAL, Conoco/diesel/mart/rest./24hr, Phillips 66/Taco Bell/diesel/mart/atm/24hr, Arby's, Burger King, Dairy Queen, McDonald's, Pizza Hut, Smoky Hill Tr Rest., Sonic, Subway/TCBY, Burlington Inn, Chaparral Motel, Comfort Inn, Super 8, Western Motel, Alco, Ford/Lincoln/Mercury, Goodyear, Safeway, to Bonny St RA, RV camp
429	Bethune, no facilities
419	CO 57, Stratton, **N**...Ampride/diesel/mart/24hr, Conoco/diesel/mart, Best Western/café, Claremont Inn, Marshall Ash Village Camping, Trails End Camping, auto museum
412	Vona, **1/2 mi N**...gas, phone
405	CO 59, Seibert, **N**...Shady Grove Camping, lodging, **S**...Texaco/A&W/diesel/mart/24hr,
395	Flagler, **N**...Total/mart, Dairy King, Freshway Subs, Central Air Motel, NAPA, **S**...Country Store/diesel/café/atm
383	Arriba, **N**...Phillips 66/diesel/café, lodging, RV camping, **S**...**rest area both lanes full(handicapped)facilities, picnic tables, litter barrels, point of interest, petwalk**
376	Bovina, no facilities
371	Genoa, **N**...point of interest, gas, food, phone
363	US 24, US 40, US 287, to CO 71, Limon, to Hugo, **13 mi S**...HOSPITAL
361	CO 71, Limon, **N**...Chrysler/Dodge/Jeep, airport, **S**...Conoco/diesel/mart, Flying J/diesel/rest./mart/24hr, Texaco/Wendy's/diesel/mart, Dairy Queen, Pizza Hut, Midwest Inn, Silver Spur Motel(1mi), Travel Inn, KOA, st patrol
360.5mm	weigh/check sta both lanes
359	to US 24, CO 71, Limon, **S**...Rip Griffin/Texaco/Subway/diesel/rest./24hr, Total/diesel/mart/24hr, Arby's, Fireside Rest., McDonald's, Best Western Limon, Comfort Inn, Econolodge, Holiday Inn Express, Midwest Inn, Silver Spur Motel, Super 8, camping, tires
354	(from eb), no facilities
352	CO 86 W, to Kiowa, no facilities
348	to Cedar Point, no facilities
340	Agate, **1/4 mi S**...gas/diesel, phone
336	to Lowland, no facilities
332mm	**rest area wb, full(handicapped)facilities, info, phone, picnic tables, litter barrels, vending, petwalk**
328	to Deer Trail, **N**...Texaco/diesel/mart, Dairy King, **S**...Corner Gas/mart, Deer Trail Café
325mm	East Bijou Creek
323.5mm	Middle Bijou Creek
322	to Peoria, **S**...dog race track
316	US 36 E, Byers, **N**...Sinclair/mart, Longhorn Motel/rest., SuperValue Foods, bank, **S**...Amoco/diesel/mart, Country Burger Rest., Lazy 8 Motel(1mi), USPO, carwash
310	Strasburg, **N**...Amoco/mart, Pizza Shop, Corner Mkt, KOA, Western Hardware, USPO, auto parts, bank
306mm	**N**...**rest area both lanes, full(handicapped)facilities, phone, picnic tables, litter barrels, petwalk**
305	Kiowa(from eb), no facilities
304	CO 79 N, Bennett, **N**...Hank's Trkstp/diesel/mart/24hr
299	CO 36, Manila Rd, **S**...Total/Taco Bell/diesel/24hr, phone
295	Lp 70, Watkins, **N**...Texaco/Tomahawk/diesel/rest., motel, diesel repair/wash
292	CO 36, Airpark Rd, **S**...Conoco/diesel/mart, BBQ, BarnStore
289	E-470 Tollway, 120th Ave, CO Springs, no facilities
288	US 287, US 40, Lp 70, Colfax Ave(exits left from wb), no facilities
286	CO 32, Tower Rd, no facilities
285	Airport Blvd, **N**...Denver Int Airport, **S**...Flying J/Conoco/diesel/rest./24hr, Comfort Inn, Crystal Inn
284	I-225 N(from eb)
283	Chambers Rd, **N**...AmeriSuites, Hilton, Marriott, Sleep Inn, **S**...Phillips 66/mart, Total/Taco Bell/mart, Burger King, Hardee's, Subaru
282	I-225 S, to Colorado Springs, no facilities
281	Peoria St, **N**...Amoco/wash, Citgo/7-11, Conoco/Blimpie/diesel/LP/mart/24hr, Phillips 66, Burger King, McDonald's, Village Inn Rest., Best Western/grill, Drury Inn, Executive Hotel, bank, **S**...ARMY MED CTR, VETERINARIAN, Amoco/mart, Total/diesel/mart, Airport Broker Rest., BBQ, Church's Chicken, Denny's, IHOP, KFC, Mexican Grill, Pizza Hut, Subway, Taco Bell, Waffle House, Wendy's, Airport Value Inn, La Quinta, Motel 6, Traveler's Inn, Goodyear/auto, auto/RV repair, carwash
280	Havana St, **N**...Embassy Suites, Office Depot
279	I-270 W, US 36 W(from wb), to Ft Collins, Boulder
278	CO 35, Quebec St, **N**...Sapp Bros/Sinclair/Burger King/diesel, TA/diesel/rest./24hr, Denny's, Best Western, Day's Inn, Hampton Inn, Quality Inn, **S**...Stapleton Airport, Amoco, Phillips 66/mart, Breakfast Shop, Courtyard, Holiday Inn/rest., Ramada/rest., Red Lion Hotel

E

W

Colorado Interstate 70

277	to Dahlia St, Holly St, Monaco St, frontage rd, no facilities
276b	US 6 E, US 85 N, CO 2, Colorado Blvd, **S**...Silver Bullet Rest., U-Haul, Hertz, Penske
a	Vasquez Ave, **N**...Pilot/Wendy's/diesel/24hr, Colonial Motel, Western Inn, Blue Beacon, Ford/Mack Trucks, **S**...Citgo/7-11, Burger King
275c	York St(from eb), **N**...Family Kitchen Rest., Colonial Motel
b	CO 265, Brighton Blvd, Coliseum
a	Washington St, **N**...Citgo/7-11, Pizza Hut, **S**...Conoco, McDonald's, Muneca Mexican, Quizno's, Subway
274b a	I-25, N to Cheyenne, S to Colorado Springs
273	Pecos St, **N**...Safeway, hardware, **S**...Conoco, Phillips 66/diesel, Circle K, Discount Hotel, transmissions
272	US 287, Federal Blvd, **N**...Amoco/service, Sinclair/mart, Burger King, Goodtimes Burgers, Hamburger Stand, Loco Pollo, McCoy's Rest., McDonald's, Pizza Hut, Subway, Taco Bell, Village Inn Rest., Wendy's, Motel 6, K-Mart/auto/drugs, bank, carwash, camping, **S**...Amoco, Conoco/mart/wash, 7-11, Popeye's, Howard Johnson/Las Palmeras Mexican
271b	Lowell Blvd, Tennyson St(from wb), no facilities
a	CO 95, **S**...Wild Chipmunk Amusement Park
270	Sheridan Blvd, **S**...Shamrock, Arby's, Oriental Rest., Sunrise Café, Target, U-Haul, Ward's/auto, amusement park, bank, mall
269b	I-76 E(from eb), to Ft Morgan, Ft Collins
a	CO 121, Wadsworth Blvd, **N**...Citgo/7-11, Conoco, LuckyMart/diesel, Applebee's, Bennigan's, Country Buffet, Fazoli's, Goodberry's Rest., Gunther Toody's Diner, Kokoro Japanese, LoneStar Steaks, McDonald's, Ruby Tuesday, Schlotsky's, Starbucks, Taco Bell, Texas Roadhouse, Advance Parts, Brakes+, Econo Lube'n Tune, Gart Sports, HomeBase, Home Depot, Lowe's Bldg Supply, OfficeMax, Office Depot, PetsMart, Pier 1, Sam's Club, Tires+, Waldenbooks, bank, mall, **S**...Ace Hardware, Discount Tire
267	CO 391, Kipling St, Wheat Ridge, **N**...Amoco/mart, Texaco/Carl's Jr/diesel/mart, 7-11, Burger King, Denny's, Furr's Dining, Subway, American Motel, Motel 6, Chevrolet/Cadillac/Olds, Chrysler/Plymouth/Jeep, **S**...Conoco/mart, Village Inn Rest., Taco Bell, Holiday Inn Express, Motel 6, 17 Inn, Super 8, RV Ctr
266	CO 72, W 44th Ave, Ward Rd, Wheat Ridge, **N**...Texaco/mart, **S**...TA/diesel/rest./24hr, Total/mart, Quality Inn
265	CO 58 W(from wb), to Golden, Central City, no facilities
264	Youngfield St, W 32nd Ave, **N**...Conoco, Country Café, GoodTimes Burgers, La Quinta, **S**...CHIROPRACTOR, Amoco/mart, Conoco/mart, Shamrock/mart, Baskin-Robbins, Chili's, Dairy Queen, Las Carretas Mexican, McDonald's, Pizza Hut/Taco Bell, Starbucks, Subway, King's Sooper/24hr, Camping World RV Supplies/service, Casey's RV Ctr, MailBoxes Etc, PetsMart, Radio Shack, Walgreen, Wal-Mart, bank, cleaners
263	Denver West Blvd, **N**...Marriott/rest., Nat Energy Lab, **S**...Barnes&Noble, cinema
262	US 40 E, W Colfax, Lakewood, **N**...Sinclair, Arby's, Home Depot, Honda, Kohl's, NAPA, U-Haul, transmissions, **S**...Conoco, Texaco/diesel/LP/service, Outback Steaks, Wendy's, Day's Inn/rest., Holiday Inn Express, Pleasant View Motel, Chevrolet, Lexus, Toyota, RV Ctr
261	US 6 E(from eb), W 6th Ave, to Denver, no facilities
260	CO 470, to Colo Springs, no facilities
259	CO 26, Golden, **N**...Conoco/mart, Heritage Sq Funpark, motel, **S**...Music Hall, to Red Rock Park
257mm	runaway truck ramp eb
256	Lookout Mtn, **N**...to Buffalo Bill's Grave
254	Genesee, Lookout Mtn, **N**...to Buffalo Bill's Grave, **S**...DENTIST, VETERINARIAN, MEDICAL CARE, Conoco/LP/mart, Genesee Country Store, Chart House Rest., Chinese Rest., Guido's Pizza, USPO, cleaners
253	Chief Hosa, **S**...Chief Hosa Camping, phone
252	(251 from eb), CO 74, Evergreen Pkwy, **S**...Amoco/mart, Chesapeake Bagel, Burger King, El Rancho Rest., McDonald's, Quality Suites, King's Sooper/deli(2mi), Wal-Mart/auto
248	(247 from eb), Beaver Brook, Floyd Hill, **S**...Gourmet Coffee, to Saddlerack Summit Rest., Floyd Hill Grocery/grill
244	US 6, to CO 119, Central City, to Golden, Eldora Ski Area, **N**...Kermit's Café Food/drink
243	Hidden Valley, no facilities
242mm	tunnel
241b a	rd 314, Idaho Springs West, **N**...VETERINARIAN, Amoco/McDonald's/mart, Phillips 66, Sinclair/mart, 7-11, Texaco, A&W, BBQ, Beaujo's Pizza, Buffalo Rest., Café Expresso, JC Sweet's, King's Derby Rest., Marion's Rest., Subway, Taco Bell, Argo Motel, Club Hotel, Columbine Inn, Hanson Lodge, H&H Motel, Indian Springs Resort, King Henry's Motel, National 9 Motel, Peoriana Motel, 6&40 Motel/rest., The Lodge, CarQuest, NAPA, Radio Shack, Safeway, USPO, antiques, bank, cleaners, hardware, repair
240	CO 103, **N**...Texaco/mart, 2 Bros Deli, Tommy Knockery Brewery, same as 241, **S**...to Mt Evans

Colorado Interstate 70

239	Idaho Springs, **N**...Amoco, Phillips 66, Sandwich Mine Mkt, camping
238	Fall River Rd, to St Mary's Glacier, no facilities
235	Dumont(from wb), no facilities
234	Downeyville, Dumont, weigh sta both lanes, **N**...Conoco/Subway/diesel/mart, Burger King, ski rentals
233	Lawson(from eb), no facilities
232	US 40 W, to Empire, Atacula NP, Rocky Mtn NP, **N**...gas, lodging, food, camping, to Berthoud Pass, Winterpark, Silver Creek ski areas
228	Georgetown, **S**...Conoco/mart, Phillips 66, Total/diesel/mart, Sunmart, Dairy King, Georgetown Lodge, Super 8
226.5mm	scenic overlook eb
226	Georgetown, Silver Plume Hist Dist, **N**...Buckley Bros Mkt, lodging, ski rentals
221	Bakerville, no facilities
220mm	Arapahoe NF eastern boundary
218	no facilities
216	US 6 W, Loveland Valley, Loveland Basin, ski areas
214mm	Eisenhower Tunnel, elev 11013
213mm	parking area eb
205	US 6 E, CO 9 N, Dillon, Silverthorne, **N**...Citgo/7-11, Conoco/wash, Texaco/GoodTimes/diesel/mart, Denny's, Domino's, Old Dillon Inn Mexican, Quizno's, Mtn Lyon Café, Village Inn Rest., Wendy's/24hr, Day's Inn, 1st Interstate Inn, Hampton Inn/rest., Luxury Suites, Sheraton, Silver Inn, CarQuest, Chevrolet/Olds, Chrysler/Plymouth/Dodge, Ford/Mercury, Prime Outlets/famous brands, **S**...CHIROPRACTOR, DENTIST, Coastal/mart, Shamrock/mart, Total/Taco Bell/mart, Arby's, BBQ, Burger King, Chinese Rest., Dairy Queen, McDonald's, Pizza Hut, Ruby Tuesday, SweetPeas, Starbucks, Subway, Best Western, Comfort Suites, Super 8, City Mkt Foods, Dillon Stores/famous brands, Goodyear/auto, MailBoxes Etc, OfficeMax
203.5mm	scenic overlook both lanes, phones
203	CO 9 S, Frisco, to Breckenridge, **S**...MEDICAL CARE, Citgo/7-11, Shamrock/mart, Texaco/24hr, A&W, Back Country Brewery/rest., BBQ, ClaimJumper Rest., Country Kitchen, KFC, Papa John's, Pizza Hut, Starbuck's, Subway, Taco Bell, TCBY, Tex's Café, Alpine Inn, Best Western, Holiday Inn, Luxury Inn, Microtel, Skyview Motel, Summit Inn, Big O Tires, NAPA, Radio Shack, Safeway, Wal-Mart/McDonald's/drugs, bank, carwash, cleaners/laundry, tires, to Breckenridge Ski Area, Tiger Run RV Resort
201	Main St, Frisco, **S**...CHIROPRACTOR, Conoco/mart, A&W, KFC, Pizza Hut, Bighorn Reservations, Pearl Head Lodge, Woodbridge Inn, to Breckenridge Ski Area
198	Officer's Gulch, no facilities, emergency callbox
195	CO 91 S, to Leadville, **1 mi S**...Conoco/diesel/mart, Molly D's Rest., O'Shea's Café, Salsa Mtn Mexican, East West Resort, Fox Pine Inn, Telemark Lodge, to Copper Mtn Ski Resort
190	**S...rest area both lanes, full(handicapped)facilities, phone, picnic tables, litter barrels**
189mm	Vail Pass Summit, elev 10662 ft, parking area both lanes
180	Vail East Entrance, phone, services 3-4 mi S
176	Vail, **S**...MEDICAL CARE, info, Amoco/diesel, Best Western, Vailglo, Holiday Inn, Craig's Mkt, bank, **1 mi S**...Amoco, Chicago Pizza
173	Vail West Entrance, Vail Ski Area, **N**...Phillips 66/mart, Texaco/diesel/mart, 7-11, Dairy Queen, 1/2 Moon Chinese, Jackalope Café, McDonald's, Subway, Taco Bell, Wendy's, West Vail Lodge, Ace Hardware, City Mkt Foods, GNC, MailBoxes Etc, Safeway, Vail Drug, bank, cleaners, **S**...Conoco/LP/mart, Black Bear Inn, Day's Inn, Marriott Streamside Hotel, The Roost Lodge
171	US 6 W, US 24 E, to Minturn, Leadville, NF info, Ski Cooper, **N**...Leadville Café, NAPA, tires, **2 mi S**...gas, food, phone, carwash/parts
167	Avon, **N**...DENTIST, VETERINARIAN, Coastal/mart, Pizza Hut, Goodyear, **S**...VETERINARIAN, Burger King, China Garden, Denny's, Domino's, Quizno's, Starbucks, Subway, Avon Ctr Lodge, Beaver Creek Inn, Beaver Creek Condos, Christy Lodge, Comfort Inn, City Mkt/drugs, GNC, MailBoxes Etc, Wal-Mart, bank, cameras, cleaners, to Beaver Creek/Arrowhead Skiing
163	**Edwards, S...rest area both lanes, full(handicapped)facilities, phone, picnic tables, litter barrels,** Conoco/mart, Texaco/Wendy's/diesel/mart/wash, Canac's Kitchen, Gore Range Brewery, Marko's Pizza, Riverwalk Inn, bank, to Arrowhead Ski Area
162mm	scenic overlook both lanes
159mm	Eagle River
157	CO 131 N, Wolcott, **N**...to gas, phone, to Steamboat Ski Area
147	Eagle, **N**...EYECARE, Texaco/Taco Bell/diesel/mart/wash, Burger King, McDonald's, Mi Pueblo Mexican, AmericInn, Comfort Inn, Holiday Inn Express, City Mkt Foods, bank, cleaners, **S**...Amoco/Subway/diesel/mart, Conoco/mart/subs/pizza, Eagle Diner, Jackie's Rest., Best Western, Prairie Moon Motel, Suburban Lodge, FoodTown Foods, RV camping, TrueValue, airport, auto parts, bank, **rest area both lanes, full(handicapped)facilities, info**
140	Gypsum, **S**...Texaco/diesel/mart, Columbine Mkt Deli, Mexican-American Café, auto/truck repair, to camping, airport

E

W

Colorado Interstate 70

134mm	Colorado River
133	Dotsero, no facilities
129	Bair Ranch, **S...rest area both lanes, full(handicapped)facilities, picnic tables, litter barrels, petwalk**
128.5mm	parking area eb
127mm	tunnel wb
125mm	tunnel
125	to Hanging Lake(no return eb)
123	Shoshone(no return eb), no facilities
122.5mm	exit to river(no return eb)
121	**Grizzly Creek, S...rest area both lanes, full(handicapped)facilities, picnic tables, litter barrels,** to Hanging Lake
119	**No Name, rest area both lanes, full(handicapped)facilities), RV camping,** rafting
118mm	tunnel
116	CO 82 E, Glenwood Springs, to Aspen, **N...**Amoco, Conoco/mart, Texaco, A&W, BBQ, Denny's, Kettle, KFC, Mancinelli's Pizza, Pizza Hut, Rosi Bavarian Kitchen, Village Inn Rest., Best Western, Glenwood Inn, Hampton Inn, Holiday Inn Express, Hotel Colorado, Ramada Inn, Silver Spruce Motel, Starlight Motel, Audi, Mazda, Nissan, Saab, Suzuki, Toyota, VW, PitStop Lube, Hot Springs Bath, bank, **S...**MEDICAL CARE, DENTIST, CHIROPRACTOR, Conoco, Phillips 66, Sinclair, Total/mart, 7-11, Arby's, Baskin-Robbins, Chinese-Japanese Rest., Domino's, Glenwood Sprs Brewpub, 19th St Diner, Quizno's, Taco Bell, Wendy's, Best Western, Frontier Lodge, Hotel Denver, City Mkt Foods, Safeway, Midas Muffler, NAPA, Rite Aid, TrueValue, **3 mi S...**Subway, Goodyear, GreaseMonkey, Honda, Wal-Mart, to Ski Sunlight
115mm	**rest area eb, full(handicapped)facilities, picnic tables, litter barrels**
114	W Glenwood Springs, **N...**DENTIST, Amoco, Citgo/7-11, Texaco/repair, Burger King, CharBurger, Dairy Queen, Dos Hombres Mexican, Fireside Steaks, Los Desperados Mexican, Marshall Dillon Steaks, Ocean Pearl Chinese, Ponderosa, Taco Giant, Affordable Inn, Best Value Inn, Budget Host, Colonial Inn, 1st Choice Inn, Holiday Inn, Motel 5, National 9 Inn, Red Mtn Inn, Super 8, Terra Vista Motel, Super 8, Big O Tires, Chevrolet/Olds/Cadillac, Chrysler/Dodge/Plymouth, Ford, JC Penney, K-Mart, Lincoln/Mercury, Radio Shack, Sears, Taylor's RV Ctr, auto repair, bank, camping, carwash, cleaners/laundry, hardware, mall, tires, **S...**Conoco/diesel/mart, Quality Inn
111	South Canyon, no facilities
109	Canyon Creek, no facilities
108mm	parking area both lanes
105	New Castle, **N...**Conoco/Subway/diesel/mart, Kum&Go/gas, Sinclair/wash(1mi), City Mkt Foods, KOA, bank, **S...**Phillips 66
97	Silt, **N...**Conoco/Blimpie/diesel/mart/24hr, Sinclair, Mexican-American Rest., Red River Inn(1mi), Reed's Auto Service, Viking RV Park, to Harvey Gap SP
94	Garfield County Airport Rd, no facilities
90	CO 13 N, Rifle, **N...rest area both lanes, full(handicapped)facilities, phone, picnic tables, litter barrels, RV dump, NF Info,** HOSPITAL, Amoco/mart/24hr, Conoco, Kum&Go/gas, Phillips 66, Texaco/mart, Homestead Family Rest., KFC, Winchester Motel(1mi), USPO, bank, **S...**Conoco/diesel/mart, Phillips 66/mart, Burger King, McDonald's, Subway, TCBY, Buckskin Motel, Red River Inn/rest., Rusty Cannon Motel
87	West Rifle, **2 mi N...**La Donna Motel, Shaler Motel, Winchester Motel, gas, food, phone
81	Rulison, no facilities
75	Parachute, **N...rest area both lanes, full(handicapped)facilities, info, phone, picnic tables, litter barrels, petwalk,** Kum&Go/gas/24hr, Sinclair/diesel/mart, Texaco/diesel/pizza/subs/mart/24hr, Pat's Rest., Super 8, **S...**Conoco/Taco Bell/mart, Duke's Rest., Family Café, Golf Club Grill, Dave's Trading Post/rest., Parachute Plaza Motel, Good Sam RV Park(4mi)
63mm	Colorado River
62	De Beque, **1 mi N...**Conoco, food, lodging, phone, auto repair
50mm	parking area eastbound, Colorado River, tunnel begins eastbound
49mm	Plateau Creek
49	CO 65 S, to CO 330 E, to Grand Mesa, to Powderhorn Ski Area
47	Island Acres St RA, **N...**CO River SP, RV camping, **S...**Total/diesel/24hr, Rosie's Rest./24hr, Fawn's Gift Shop, motel, RV camping
46	Cameo, no facilities
44	Lp 70 W, to Palisade, **3 mi S...**gas, food, lodging
43.5mm	Colorado River
42	Palisade, US 6, **1 mi S...**Fruitstand/store, gas, lodging, Grand River Winery

E

W

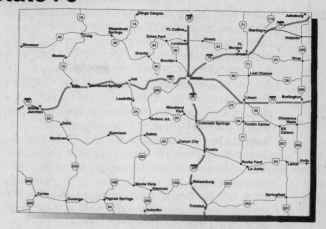

Glenwood Springs

Colorado Interstate 70

Grand Junction

37	to US 6, to US 50 S, Clifton, Grand Jct, **1 mi S on US 6 bus**...MEDICAL CARE, Amoco/mart, Conoco/diesel/mart, Phillips 66/diesel/mart/24hr, Sinclair, Total/diesel/mart, Texaco/mart, Burger King, Dos Hombres Mexican, KFC, Little Caesar's, McDonald's, Papa Murphy's, Pizza Chef, Pizza Hut, Subway, Taco Bell, Texas Roadhouse, Best Western, Albertson's, KOA, Max Foods, Pennzoil, Rite Aid, USPO, carwash, cleaners/laundry, **2-3 mi S**...CHIROPRACTOR, Amoco/mart, Phillips 66, Capt D's, Furr's Dining, Village Inn Rest., Wendy's, Western Sizzlin, Wienerschnitzel, Timbers Motel, AutoZone, Checker Parts, Chrysler/Plymouth/Dodge, Chevrolet, Discount Tire, GreaseMonkey Lube, Hastings Books, K-Mart, RV Ctr, Wal-Mart
31	Horizon Dr, Grand Jct, **N**...Amoco, Texaco/diesel/mart, Diorio's Pizza, Pepper's Rest., Shake Rattle&Roll Diner, Village Inn Rest., Wendy's, Best Western, Comfort Inn, Grand Vista Hotel, Holiday Inn, La Quinta, Motel 6, Ramada Inn, Harley-Davidson, USPO, airport, **S**...HOSPITAL, Conoco/diesel, Phillips 66/mart/atm, Applebee's, Burger King, Denny's, Pizza Hut, Shanghai Garden Chinese, Taco Bell, Adams Mark Hotel, Best Western, Budget Host, Country Inn, Day's Inn/rest., Mesa Inn, Super 8, bank, golf, to Mesa St Coll, CO NM
28	Redlands Pkwy, 24 rd, **N**...camping, **S**...recreational park/sports complex
26	US 6, US 50, Grand Junction, **N**...Mobile City RV Park, **S**...Conoco/A&W/diesel/mart/24hr, Otto's Rest., Westgate Inn/rest., Ford/Lincoln/Mercury, Mazda, Mercedes, Subaru, Toyota, Freightliner, RV Ctr, RV camping, radiators, **3-5 mi S**...BBQ, Golden Corral, McDonald's, Outback Steaks, Red Lobster, JC Penney, Mervyn's, mall
19	US 6, CO 340, Fruita, **N**...HOSPITAL, CHIROPRACTOR, Amoco, Burger King, Balanced Rock Motel, H Motel, City Mkt Foods/24hr, NAPA, bank, **S**...**Welcome Ctr both lanes, full(handicapped)facilities, phone, picnic tables, litter barrels, RV dump, petwalk,** Conoco/Subway/diesel/mart/atm/24hr, Texaco/Wendy's/diesel/mart/wash/24hr, McDonald's, Comfort Inn, Super 8, CO Mon Trading Co, camping, museum, to CO NM
17mm	Colorado River
15	CO 139 N, to Loma, Rangely, **N**...to Highline Lake SP, gas/diesel, phone
14.5mm	weigh/check sta both lanes, phones
11	Mack, **2-3 mi N**...gas/diesel, food
2	Rabbit Valley, to Dinosaur Quarry Trail, no facilities
0mm	Colorado/Utah state line

Colorado Interstate 76

Julesburg Sterling

Exit #	Services
184mm	Colorado/Nebraska state line
180	US 385, Julesburg, **N**...HOSPITAL, Flying J/diesel/rest./mart/24hr, Texaco/mart, ice cream, Holiday Motel(3mi), Platte Valley Inn, **Welcome Ctr, rest area both lanes, full(handicapped)facilities, info, RV dump, S**...Conoco/diesel/mart
172	Ovid, **2 mi N**...gas, food
165	CO 59, Sedgewick, to Haxtun, **N**...Conoco/diesel/mart, Total/Taco Bell/mart
155	Red Lion Rd, no facilities
151mm	**rest area both lanes, full(handicapped)facilities, phone, picnic tables, litter barrels, vending, petwalk**
149	CO 55, Crook, to Fleming, **S**...Sinclair/diesel/café
141	Proctor, no facilities
134	Iliff, no facilities
125	US 6, Sterling, **N**...HOSPITAL, Ampride/diesel/mart, Arby's, Burger King(2mi), McDonald's, Taco Bell, Village Inn Rest., Wendy's, Best Western, 1st Interstate Inn, Lontine's Motel/rest., Sterling Motel, Super 8, N Sterling SP, museum, st patrol, **S**...Phillips 66/Quizno's/diesel/mart, Country Kitchen, Day's Inn, Ramada Inn, Super 8, Jellystone Camping
115	CO 63, Atwood, **N**...HOSPITAL, Sinclair/diesel/mart, **S**...Amoco, Steakhouse
108	**rest area both lanes, full(handicapped)facilities, phone, picnic tables, litter barrels, vending, petwalk, RV dump**
102	Merino, no facilities
95	Hillrose, no facilities
92b a	to US 6 E, to US 34, CO 71 S, no facilities
90b a	CO 71 N, to US 34, Brush, **N**...Texaco/diesel/rest./24hr, Pizza Hut, Wendy's, Best Western, **S**...Conoco/mart, McDonald's, Budget Host, Microtel
89	Hospital Rd, no facilities
83	Dodd Bridge Rd, no facilities
82	Barlow Rd, **N**...Texaco/diesel/mart, Econolodge, **S**...Coastal/diesel/rest., Morgan Manor Motel
80	CO 52, Ft Morgan, **S**...Conoco/diesel/24hr, Sinclair, Texaco/diesel/mart, Total, A&W, Arby's, Carl's Jr, Dairy Queen, KFC, McDonald's, Sonic, Subway/TCBY, Taco John's, Best Western/rest., Central Motel, Day's Inn, Super 8, AutoZone, K-Mart, Rite Aid

E

W

E

W

Colorado Interstate 76

Exit	Services
79	CO 144, to Weldona, **N**...Texaco/diesel/mart
75	US 34 E, to Ft Morgan, **S**...Conoco/A&W/diesel/mart, Quality Inn
74.5mm	weigh sta both lanes
73	Long Bridge Rd, no facilities
66b	US 34 W(from wb), to Greeley
66a	CO 39, CO 52, Wiggins, to Goodrich, **N**...Gas USA/diesel, to Jackson Lake SP, Trophy Room Rest., **S**...Sinclair/diesel/mart
64	Wiggins, no facilities
60	to CO 144 E, to Orchard, no facilities
57	rd 91, no facilities
49	Painter Rd(from wb), no facilities
48	to Roggen, **N**...Texaco/diesel, I-76 Motel
39	Keenesburg, **S**...Phillips 66/diesel/mart, Fine Food Rest., Keene Motel, phone
34	Kersey Rd, no facilities
31	CO 52, Hudson, **S**...Amoco/diesel/mart/wash, Longhorn&Co Rest., Pepper Pod Rest., Piccadilly Piza, auto/truck repair, RV camping
25	CO 7, Lochbuie, **N**...Texaco/diesel/mart
22	Bromley, **N**...HOSPITAL, **S**...Barr Lake SP
21	144th Ave, no facilities
20	136th Ave, **N**...RV camping
16	CO 2 W, Sable Blvd, CommerceCity, **N**...Texaco/Blimpie/Baskin-Robbins/mart/24hr
12	US 85 N, to Brighton, Greeley, no facilities
11	96th Ave, **N**...trailer sales
10	88th Ave, **N**...Amoco/Blimpie/diesel/mart, Conoco, Holiday Inn Express, Super 8, **S**...flea mkt
9	US 6 W, US 85 S, Commerce City, **S**...Texaco/mart, GMC/Freightliner, st patrol
8	CO 224, 74th Ave(no EZ eb return), **1 mi N**...NAPA, **S**...Shamrock/diesel/mart, Crestline Motel
6b a	I-270 E, to Limon, to airport
5	I-25, N to Ft Collins, S to Colo Springs
4	Pecos St, no facilities
3	US 287, Federal Blvd, **N**...Total/diesel/mart/atm, **S**...Taco House
1b	CO 95, Sheridan Blvd, **S**...Homei Chinese, transmissions
a	CO 121, Wadsworth Blvd, **N**...Citgo/7-11, Conoco, LuckyMart/diesel, Applebee's, Bennigan's, Country Buffet, Fazoli's, Goodberry's Rest., Gunther Toody's Diner, Kokoro Japanese, LoneStar Steaks, McDonald's, Ruby Tuesday, Schlotsky's, Starbucks, Taco Bell, Texas Roadhouse, Advance Parts, Brakes+, Econo Lube'n Tune, Gart Sports, HomeBase, Home Depot, Lowe's Bldg Supply, OfficeMax, Office Depot, PetsMart, Pier 1, Sam's Club, Tires+, Waldenbooks, bank, mall, **S**...Ace Hardware, Discount Tire
0mm	I-76 begins/ends on I-70, exit 169b.

E / W

Colorado Interstate 225(Denver)

Exit #	Services
12b a	I-70, W to Denver, E to Limon
10	US 40, US 287, Colfax Ave, **E**...Conoco, Waffle House, **W**...Conoco
9	Co 30, 6th Ave, **E**...Conoco, Pizza Hut, Super 8, **W**...Phillips 66
8	Alameda Ave, **W**...Texaco/mart/repair
7	Mississippi Ave, Alameda Ave, **E**...Denny's, Jason's Deli, Hampton Inn, La Quinta, Best Buy, Circuit City, PetsMart, Sam's Club, Target, Wal-Mart
5	Iliff Ave, **E**...Applebee's, Hop's Rest., Comfort Inn, Motel 6, HomeBase, **W**...DoubleTree Hotel
4	CO 83, Parker Rd, **E**...Holiday Inn, Cherry Creek SP, **W**...Amoco, Denny's
2	DTC Blvd, Tamarac St, no facilities
0mm	I-225 begins/ends on I-25, exit 200.

N / S

Connecticut Interstate 84

Exit #(mm)Services

98mm	Connecticut/Massachusetts state line
74(97)	CT 171, Holland, **S**...BP/diesel, Traveler's Rest./books, Chrysler/Plymouth/Dodge, RV camp, **N**...Goodhall's Rest.
96mm	weigh sta wb
73(95)	CT 190, Stafford Springs, **N**...camping(seasonal), motor speedway, st police
72(93)	CT 89, Westford, **N**...Ashford Motel, camping(seasonal), phone
71(88)	Ruby Rd, **N**...Citgo/diesel/repair, **S**...TA/Shell/Burger King/Dunkin Donuts/diesel/24hr, Sleep Inn
70(86)	CT 32, Willington, **S**...HOSPITAL, Mobil/diesel/mart/24hr, Sunoco/diesel/mart/24hr
85mm	**rest area both lanes, full(handicapped)facilities, info, phone, vending, picnic tables, litter barrels, petwalk, campers, RV dump**
69(83)	CT 74, to US 44, Willington, **S**...gas, food, phone, RV camping, st police
68(81)	CT 195, Tolland, **N**...Gulf/diesel/repair, Mobil/mart, Papa T's Rest., Subway, NAPA, **S**...Texaco/mart, camping(seasonal)
67(77)	CT 31, Rockville, **N**...HOSPITAL, Mobil, Texaco, Burger King, McDonald's, Theo's Rest., **S**...Nathan Hale Mon
66(76)	Tunnel Rd, Vernon, no facilities
65(75)	CT 30, Vernon Ctr, **N**...Mobil/24hr, Shell/DunkinDonuts/service, Bickford's, Brannigan's Rest., Burger King, Denny's/24hr, KFC, Lotus Rest., Olympic Diner/24hr, Rein's Deli, Comfort Inn, Howard Johnson Express, CarQuest, Firestone/auto, Meineke Muffler, Vernon Drug, bank
64(74)	Vernon Ctr, **N**...VETERINARIAN, Mobil/mart/24hr, Sunoco, Angellino's Italian, Anthony's Pizza, Damon's, D'Angelo's, Denny's, Dunkin Donuts/24hr, Friendly's, Kim's Oriental, McDonald's, 99 Rest., Taco Bell, Holiday Inn Express, Adam's IGA, Advance Parts, CVS Drug/24hr, GNC, Goodyear/auto, JiffyLube, K-Mart, Old Navy, Sears Hardware, Staples, Stop'n Shop, TJ Maxx, TownFair Tire, cleaners, **S**...Quality Inn
63(72)	CT 30, CT 83, Manchester, S Windsor, **N**...Texaco, Macaroni Grill, McDonald's, Old Country Buffet, Outback Steaks, TGIFriday, Uno Pizzaria, Circuit City, Computer City, Filene's, HomePlace, JC Penney, Luxury Linens, Marshall's, MediaPlay, Michael's, Office Depot, PetCo, Sears/auto, Wal-Mart, same as 62, **S**...HOSPITAL, Exxon/diesel, Getty Gas, Shell/mart/24hr, Sunoco/diesel/mart, King Buffet, Roy Rogers, Ct Motel, Exit 63 Motel, Big Y Food, Chrysler/Plymouth, Ford/Kia, Lincoln/Mercury/Mazda, Rite Aid
62(71)	Buckland St, **N**...Exxon/diesel/mart/24hr, Boston Mkt, Chili's, Friendly's, Olive Garden, Borders Books&Music, CompUSA, Dick's Sports, Firestone/auto, Home Depot, OfficeMax, Pier 1, Sam's Club, Service Merchandise, ToysRUs, cleaners, mall, same as 63, **S**...CHIROPRACTOR, Citgo/diesel/mart, Mobil/mart, Burger King, ChuckeCheese, Dunkin Donuts, Ground Round, McDonald's, Ames, Buick, Firestone/auto, Honda, JC Penney, Jo-Ann Fabrics, USPO, cinema
61(70)	I-291 W, to Windsor, no facilities
60(69)	US 6, US 44, Burnside Ave(from eb), same as 62
59(68)	I-384 E, Manchester, no facilities
58(67)	Roberts St, Burnside Ave, **N**...Holiday Inn, Wellesley Inn, Freightliner, **S**...Burlington Coats, Super Stop'n Shop, cinema
57(66)	CT 15 S, to I-91 S, NYC, Charter Oak Br, no facilities
56(65)	Governor St, E Hartford, **S**...airport
55(64)	CT 2 E, New London, downtown
54(63)	Old State House, **N**...Ford/Isuzu, Lincoln/Mercury, **S**...Ramada Inn, downtown
53(62)	CT Blvd(from eb), **S**...Olds, Ramada Inn
52(61)	W Main St(from eb), downtown
51(60)	I-91 N, to Springfield
50(59.8)	to I-91 S(from wb)
48(59.5)	Asylum St, downtown, **N**...Crown Plaza Hotel, **S**...HOSPITAL, Hilton, Marriott Residence Inn
47(59)	Sigourney St, downtown, **N**...Hartford Seminary, Mark Twain House
46(58)	Sisson St, downtown, UConn Law School
45(57)	Flatbush Ave(exits left from wb), **N**...Super Stop'n Shop
44(56.5)	Prospect Ave, **N**...Exxon/mart, Motomart, Shell, Texaco/mart/wash, Big Boy, Burger King, Duffy's Donuts, Gold Roc Diner/24hr, HomeTown Buffet, McDonald's, Roy Rogers, carwash
43(56)	Park Rd, W Hartford, **N**...to St Joseph Coll
42(55)	Trout Brk Dr(exits left from wb), to Elmwood, no facilities
41(54)	S Main St, American School for the Deaf
40(53)	CT 71, New Britain Ave, **S**...Shell/mart, Sunoco, American Grill, Chicago Grill, Ruby Tuesday, TGIFriday, Wendy's, Barnes&Noble, Filene's, Kinko's, Old Navy, JC Penney, Radio Shack, Sears/auto, Speedy Mufflers, mall
39a(52)	CT 9 S, Newington, to New Britain, **S**...HOSPITAL
39(51.5)	CT 4, Farmington, **N**...HOSPITAL, food, lodging
38(51)	US 6 W, Bristol, **N**...Texaco
37(50)	Fienemann Rd, to US 6 W, **N**...Shell/service, Mr G's Deli, Marriott
36(49)	Slater Rd, **S**...HOSPITAL, no facilities
35(48)	CT 72, to CT 9(exits left from wb), New Britain, **S**...HOSPITAL
34(47)	CT 372, Crooked St, **N**...Ramada Inn, gas, food, phone

Vernon (vertical label)

Hartford (vertical label)

E ↕ W

Connecticut Interstate 84

33(46)   CT 72 W, to Bristol, no facilities

32(45)   Ct 10, Queen St, Southington, **N**...HOSPITAL, Exxon, Shell/mart, Beijing Chinese, Burger King, Chili's, D'Angelo's, Denny's, Dunkin Donuts, Friendly's, JD's Rest., KFC, McDonald's, Outback Steaks, Randy's Pizza, Ruby Tuesday, Strawberry's Rest., Subway, Taco Bell, TCBY, Motel 6, Ames, CVS Drug, $Tree, FashionBug, GNC, Jo-Ann Fabrics, PetCo, PetSupplies+, Shaw's Foods, Staples, TJ Maxx, cleaners, **S**...Mobil, Sunoco, Bickford's, Brannigan's Ribs, Friendly's, Holiday Inn Express, Susse Chalet, Travelodge, Chevrolet, K-Mart/auto, Speedy Muffler

31(44)   CT 229, West St, **N**...Mobil/Dunkin Donuts/mart, Sunoco/diesel/mart/ 24hr, **S**...DENTIST, Citgo/mart, Gulf/service, Angelo's II Rest., Valentino's Pizzaria, cleaners

30(43)   Marion Ave, W Main, Southington, **N**...ski area, **S**...Mobil/repair

42.5mm rest area eb, full(handicapped)facilities, info, phone, picnic tables, litter barrels, petwalk

29(42)   CT 10, Milldale(exits left from wb), no facilities

28(41)   CT 322, Marion, **S**...MEDICAL CARE, Mobil, TA/Texaco/diesel/rest./24hr, Burger King, Dunkin Donuts, Ramada Inn, cinema, radiators

27(40)   I-691 E, to Meriden, no facilities

26(38)   CT 70, to Cheshire, **N**...Silver Diner, Four Points Hotel/rest., Sheraton/rest., same as 25

25a(37) Austin Rd(from eb), **N**...Sheraton, Costco Whse

25(36)   Scott Rd, E Main St, **N**...Exxon/mart, Gulf/diesel/mart, Chinese Rest., Dunkin Donuts, **S**...Mobil, Burger King, Friendly's, McDonald's, Spago Italian, Ramada Inn, Super 8, BJ's Whse, Bradlee's, Chevrolet, CVS Drug, Super Stop'n Shop, cleaners

24(34)   Harpers Ferry Rd, **S**...Citgo/Subway/DairyMart/24hr, Getty Gas, Texaco/Dunkin Donuts/mart/24hr

23(33.5) CT 69, Hamilton Ave, **N**...HOSPITAL, Bertucci's, Chili's, McDonald's, Ruby Tueday, Barnes&Noble, CopyMax, Filene's, JC Penney, KidsRUs, OfficeMax, Sears/auto, ToysRUs, cinema, mall, **S**...Texaco, same as 24

22(33)   Baldwin St, Waterbury, **N**...HOSPITAL, McDonald's, Courtyard, Quality Inn, same as 23

21(33)   Meadow St, Banks St, **N**...HomeTown Buffet, TGIFriday, Shaw's Foods, **S**...Exxon/diesel/mart/24hr, Home Depot, PetsMart, SportsAuthority

20(32)   CT 8 N, to Torrington, **N**...Texaco, McDonald's

19(32)   CT 8 S(exits left from wb), to Bridgeport, no facilities

18(32)   W Main, Highland Ave, **N**...HOSPITAL, Exxon, A&P, drugs

17(30)   CT 63, CT 64, to Watertown, Naugatuck, **N**...Mobil/diesel, Maples Rest./pizza, funpark, to ski area

16(25)   CT 188, to Middlebury, **N**...Mobil/mart, Hilton/rest., Patty's Deli, airport, funpark

15(22)   US 6 E, CT 67, Southbury, **N**...Mobil, Shell/diesel, Dunkin Donuts, Friendly's, McDonald's, Heritage Inn, Oaktree Inn, K-Mart/Little Caesar's, bank, laundry, to Kettletown SP, Woodbury Ski Area

14(20)   CT 172, to S Britain, **N**...Texaco/mart/24hr, Thatcher's Rest., st police

20mm   motorist callboxes begin eb, end wb

13(19)   River Rd(from eb), to Southbury

11(16)   CT 34, to New Haven, no facilities

10(15)   US 6 W, Newtown, **S**...HOSPITAL, Amoco/diesel/mart/24hr, Shell/mart/24hr, Blue Colony Diner/24hr, Pizza Palace, Drug Ctr, Newtown Hardware, auto repair, bank, cleaners, laundry

9(11)   CT 25, to Hawleyville, **N**...Taunton Rest., gas, **2 mi S**...Best Western

8(8)   Newtown Rd, **N**...Gulf, Mobil/diesel/mart/24hr, Texaco/diesel/mart/24hr, Ramada Inn, **S**...VETERINARIAN, BP, Shell, Bankok Rest., Blimpie, Boston Mkt, Burger King, Chili's, Denny's, Dunkin Donuts, Friendly's, McDonald's, SitDown Diner, Subway, Taco Bell, Best Western, Holiday Inn/rest., Quality Inn, Bradlee's, Buick/Pontiac, Chrysler/Plymouth/ Jeep/Kia, CVS Drug, FashionBug, GMC, Goodyear/auto, Marshall's, Radio Shack, Ritz Camera, Service Merchan- dise, Staples, Stop&Shop, TownFair Tire, bank, cinema, laundry

7(7)   US 7N/202E, to Brookfield, New Milford, **1 exit N on Federal Rd**...Applebee's, Best Inn, Borders Books, Circuit City, Ford, Harley-Davidson, Home Depot, Linens'n Things, cinema, mall

6(6)   CT 37(from wb), New Fairfield, **N**...HOSPITAL, Citgo, Shell, Sunoco, A&P, to Squantz Pond SP, **S**...BP, KFC

5(5)   CT 37, CT 39, CT 53, Danbury, **N**...Exxon/mart, Texaco/diesel/mart/24hr, Exit 5 Motel, **S**...HOSPITAL, Citgo, Mobil/ mart, Deli/Snacks, Pizza Hut, Taco Bell, AvisLube, Midas Muffler, tires, to Putnam SP

4(4)   US 6 W/202 W, Lake Ave, **N**...Amoco/mart, Gulf/diesel, Texaco/diesel/mart/wash, 7-11, Dunkin Donuts, McDonald's, Taj Mahal Rest., Windmill Rest., Radisson, Super 8, Goodyear/auto, Monro Muffler/brake, Super Stop'n Shop/24hr, **S**...Texaco, drugs, to mall

3(3)   US 7 S(exits left from wb), to Norwalk, **S**...Dave&Buster's, JC Penney, Lord&Taylor, Macy's, Sears, ToysRUs, mall, to airport

2b a(1) US 6, US 202, Mill Plain Rd, **N**...Exxon/mart, Dutchess Rest., Hilton Garden, PetLand, Staples, **S**...**Welcome Ctr/ weigh sta, full(handicapped)facilities, info, picnic tables, litter barrels, petwalk,** to Old Ridgebury, Sunoco, Hilton

1(0)   Sawmill Rd, no facilities

0mm   Connecticut/New York state line

E

W

Connecticut Interstate 91

Exit #(mm)Services

58mm Connecticut/Massachusetts state line

49(57) US 5, to Longmeadow, MA, **E**...Citgo/DairyMart, Mobil/mart, Abdow's Rest., Bagel Jct, Friendly's, McDonald's, Rinaldi's Italian, Harley Hotel, Aamco, Meineke Muffler, Weiner's Tires, cleaners, **W**...Cloverleaf Café, LP

48(56) CT 220, Elm St, **E**...Mobil/mart, Burger King, ChiChi's, Denny's, Dunkin Donuts, Figaro's ClamBar, FoodMart Rest., Friendly's, Kenny Rogers, McDonald's, Papa Gino's, Pumpernickel Rest., Wendy's, Caldor, Honda, same as 47

47(55) CT 190, to Hazardville, **E**...HOSPITAL, Citgo, Abdow's, Bickford's, D'Angelo's, Dunkin Donuts, Ground Round, KFC, McDonald's, Old Country Buffet, Olive Garden, Pizza Hut, Red Lobster, Taco Bell, TCBY, Motel 6, Red Roof Inn, Ames, Barnes&Noble, Bradlee's, Edward's Foods, Ford, Goodyear, JC Penney, Radio Shack, RX Drug, Service Merchandise, Stop&Shop Foods, Walgreen, auto repair, bank, carwash, cinema, cleaners, mall, same as 48

46(53) US 5, King St, to Enfield, **E**...Mobil/mart, Astro's Rest., **W**...Carmen's Rest., Super 8

45(51) CT 140, Warehouse Point, **E**...Shell/Dunkin Donuts/mart/wash/atm, Blimpie, Burger King, Friendly's, Kowloon Chinese, Sophia's Pizza, Wal-Mart, bank, cinema, cleaners/laundry, to Trolley Museum, **W**...Sunoco/diesel/mart/ 24hr, Best Western

44(50) US 5 S, to E Windsor, **E**...Citgo/diesel/mart, Dunkin Donuts, E Windsor Rest., Tobacco Valley Grill, Holiday Inn Express

49mm Connecticut River

42(48) CT 159, Windsor Locks, **W**...Gulf, same as 41

41(47) Center St(exits with 39), **W**...Shell/mart/24hr, Ad Pizzaria, Howard Johnson

40(46.5) CT 20, **W**...Old New-Gate Prison, airport

39(46) Kennedy Rd(exits with 41), Community Rd, **W**...Shell/mart/24hr, Edwards Foods/24hr, K-Mart

38(45) CT 75, to Poquonock, Windsor Area, **E**...Mobil/diesel/mart, Beanery Bistro, Chinese Rest., McDonald's, Subway, cleaners/laundry, to Ellsworth Homestead, **W**...Courtyard

37(44) CT 305, Bloomfield Ave, Windsor Ctr, **E**...Mobil/diesel/mart, McDonald's, **W**...Citgo/mart, Residence Inn

36(43) CT 178, Park Ave, to W Hartford, no facilities

35b(41) CT 218, to Bloomfield, to S Windsor, **E**...gas/diesel, food

a I-291 E, to Manchester

34(40) CT 159, Windsor Ave, **E**...Texaco/diesel/mart, **W**...HOSPITAL, Citgo, Flamingo Inn

33(39) Jennings Rd, Weston St, **E**...Buick/GMC, Saturn, repair, **W**...Exxon/diesel/mart, Mobil, Burger King, Dunkin Donuts, McDonald's, Subway, Red Roof Inn, Super 8, Honda, Hyundai, Jaguar, JiffyLube, Mazda, Midas Muffler, Mitsubishi, Nissan, Toyota, carwash

32b(38) Trumbull St, **W**...HOSPITAL, downtown, Holiday Inn, Sheraton, Goodyear

a (exit 30 from sb), I-84 W

29b(37) Hartford, I-84 E

a(36.5) **W**...downtown(exits left from nb), HOSPITAL, capitol, civic ctr

28(36) US 5, CT 15 S(from nb), **W**...Citgo, Burger King, Wendy's

27(35) Brainerd Rd, Airport Rd, **E**...Texaco/diesel/mart/wash, McDonald's, Day's Inn, Susse Chalet, Valle's Steaks, Ford, cinema, to Regional Mkt

26(33.5) Marsh St, **E**...Silas Deane House, Webb House, CT MVD

25(33) CT 3, Glastonbury, Wethersfield, no facilities

24(32) CT 99, Rocky Hill, Wethersfield, **E**...Gulf, Mobil/mart, 7-11, Bickford's, Boston Mkt, McDonald's, On-the-Border, Subway, Susse Chalet, Travelers Motel, Allstar Automotive, Ames, Kawasaki, Midas Muffler, Monro Muffler/ brake, **W**...MEDICAL CARE, DENTIST, Mobil/mart, Texaco/diesel/mart, Angellino's Italian, Bennigan's, China Star Buffet, Denny's, Dunkin Donuts, Ground Round, KFC, Luna Mia Ristorante, McDonald's, Old Country Buffet, Papa Gino's/D'Angelo's, Red Lobster, Motel 6, Ramada Inn, AutoZone, Caldor, CVD Drug/24hr, Goodyear/auto, Lowe's Bldg Supply, MailBoxes Etc, Marshall's, Radio Shack, TJ Maxx, TownFair Tire, True Value, Walgreen, bank

23(29) to CT 3, West St, Rocky Hill, Vet Home, **E**...HOSPITAL, Marriott, to Dinosaur SP, **W**...Citgo/diesel/mart, Mobil, D'Angelo's/Papa Gino's, Italian Rest., McDonald's, Manhattan Bagels, Oscar's Rest., Westside Mkt Foods, bank, cleaners/laundry

22(27) CT 9, to New Britain, Middletown, no facilities

21(26) CT 372, to Berlin, Cromwell, **E**...Sunoco/diesel/repair, Krausnzer's Foods, Friendly's, Comfort Inn, Radisson Inn/ rest., bank, cleaners, **W**...DENTIST, MEDICAL CARE, Citgo/diesel, Mobil/diesel/mart, Blimpie, Burger King, Cromwell Diner/24hr, McDonald's, Yes Buffet, Holiday Inn, Super 8, Firestone/auto, Fabric Place, Wal-Mart

Hartford

Connecticut Interstate 91

20(23) Country Club Rd, Middle St, no facilities
22mm **rest area/weigh sta nb, full(handicapped)facilities, info, phone, picnic tables, litter barrels, vending, RV dumping, petwalk**
19(21) Baldwin Ave(from sb), no facilities
18(20.5) I-691 W, to Marion, access to same as 16 & 17, ski area
17(20) CT 15 N(from sb), to I-691, CT 66 E, Meriden, **W**...Amoco, Getty, Residence Inn
16(19) CT 15, E Main St, **E**...Gulf/repair, Mobil/mart, SmartShop/diesel, Sunoco/Dunkin Donuts, American Steaks, Lido's Pizza, Hampton Inn, Ramada Inn, The Inn, Ames, CVS Drug, Ford, Radio Shack, bank, cinema, **W**...Amoco/mart, Getty, Gulf/repair, Sunoco/diesel, Bess Eaton, Boston Mkt, Burger King, Domino's, Dunkin Donuts, Friendly's, Great Wall Chinese, Hotdogs+, KFC, Manhattan Bagel, McDonald's, Taco Bell, East Inn, Brooks Drug, CarQuest, Hancock Drugs, bank
15(16) CT 68, to Durham, **E**...golf, **W**...Courtyard, Susse Chalet
15mm **rest area sb, full(handicapped)facilities, info, phone, picnic tables, litter barrels, petwalk**
14(12) CT 150(no EZ return), Woodhouse Ave, Wallingford, **3 mi W on US 5**...Citgo/diesel/mart, Dodge
13(10) US 5(exits left from nb), Wallingford, **2 mi W on US 5**...Citgo/diesel/mart, DairyMart/gas, Dunkin Donuts, Audi/Olds/Porche/Saab/Suzuki, Krauszners Foods, to Wharton Brook SP
12(9) US 5, Washington Ave, **E**...CHIROPRACTOR, Shell, Texaco, Boston Mkt, Burger King, China Buffet, D'Angelo's, Dunkin Donuts, Friendly's, McDonald's, Rustic Oak Rest., Subway, Wendy's, Stop&Shop Foods, Walgreen, banking, cleaners/laundry, **W**...Exxon/diesel/mart/24hr, Athena II Diner, Danny's Pizza, Roy Rogers, Holiday Inn
11(7) CT 22(from nb), North Haven, **E**...Super Foodmart, AutoZone, FashionBug, JC Penney Outlet, Pennzoil, Radio Shack, TownFair Tire, same as 12
10(6) CT 40, Hamden, to Cheshire, park n ride
9(5) Montowese Ave, **W**...Berkshire/diesel, Sunoco/diesel/mart, Dynasty Chinese, McDonald's, Sbarro's, Subway, Barnes&Noble, BJ's Wholesale, Circuit City, Home Depot, Lechmere's, OfficeMax, PetCo, Sports Authority, TJ Maxx, Price Club, cinema
8(4) CT 17, CT 80, Middletown Ave, **E**...Amoco/mart/wash, Citgo/diesel, Exxon/mart, Shell/mart/atm, Sunoco/mart/24hr, DairyMart, Burger King, D'Angelo's/Papa Gino's, Dunkin Donuts, Exit 8 Diner, KFC, McDonald's, Taco Bell, Pizza Hut/TCBY, Motel 6, Lowe's Bldg Supply, K-Mart/Little Caesar's/24hr, Oil Express, PartsAmerica, Valvoline, cleaners, **W**...Mobil/diesel, auto parts
7(3) Ferry St(from sb), Fair Haven, no facilities
6(2.5) Willow St(exits left from nb), Blatchley Ave, **W**...Merit/mart
5(2) US 5(from nb), State St, Fair Haven, **E**...Star Diner
4(1.5) State St(from sb), downtown
3(1) Trumbull St, New Haven, downtown, **W**...Peabody Museum
2(.5) Hamilton St, New Haven, downtown
1(.3) CT 34W(from sb), New Haven, **W**...HOSPITAL, downtown
0mm I-91 begins/ends on I-95, exit 48.

Connecticut Interstate 95

Exit #(mm) Services
112mm Connecticut/Rhode Island state line
93(111) CT 216, Clarks Falls, **E**...Shell/repair/24hr, to Burlingame SP, **W**...Citgo/diesel/mart, Republic/Texaco/diesel/mart/rest./atm, Bess Eaton Café, McDonald's, Budget Inn, Stardust Motel
92(107) CT 2, CT 49(no EZ nb return), Pawcatuck, **E**...HOSPITAL, New England Gas/repair, Sunoco, 1-Hr Photo, cinema, funpark, golf, **W**...RV camping
91(103) CT 234, N Main St, to Stonington, **E**...HOSPITAL
90(101) CT 27, Mystic, **E**...Mobil, Bickford's Rest., Friendly's, GoFish Rest., Jam's Rest., McDonald's, Newport Creamery, Steak Loft, AmeriSuites, Old Mystic Motel, Seaport Lodge, Mystic Factory Outlets, aquarium, bank, cinema, **W**...Texaco/Subway/Dunkin Donuts/diesel/mart/atm, Best Western, Comfort Inn, Day's Inn, Residence Inn, TrueValue, Chevrolet/Chrysler/Plymouth/Dodge, laundry
100mm scneic overlook nb
89(99) CT 215, Allyn St, **W**...camping(seasonal)
88(98) CT 117, to Noank, **E**...MEDICAL CARE(7 am-11pm), airport
87(97) Sharp Hwy, Groton, **E**...Griswold SP, airport

Connecticut Interstate 95

New London

86(96)	CT 184(exits left from nb), Groton, **W**...Groton Suites, US Sub Base
85(95)	US 1 N, Groton, downtown, **E**...Citgo/diesel, Norm's Diner, NAPA
94.5mm	Thames River
84(94)	CT 32(from sb), New London, downtown
83(92)	CT 32, New London, **E**...to Long Island Ferry
82a(90.5)	frontage rd, New London, **E**...AutoZone, Adam's Foods, Brooks Drug, TownFair Tire, same as 82, **W**...Fairfield Suites, Holiday Inn, Monro Muffler/brake, same as 82
82(90)	CT 85, New London, **E**...Pizza Hut, Burlington Coats, Staples, tires, **W**...DENTIST, Mobil/mart, ChuckeCheese, Dairy Rest., Red Lobster, Fairfield Suites, Holiday Inn, Boater's World, Filene's, Home Depot, JC Penney, Macy's, Marshall's, Sears, mall
81(89.5)	Cross Road, **W**...VETERINARIAN, Bob's Store, BJ's Wholesale, Odd Jobs, Wal-Mart/McDonald's/drugs, cinema, Waterford Ind Park
80(89.3)	Oil Mill Rd(from sb), **W**...Lamplighter Motel
76(89)	I-395 N, to Norwich
75(88)	US 1, to Waterford, no facilities
74(87)	CT 161, to Flanders, Niantic, **E**...Citgo/mart, Exxon/mart, Mobil, Sunoco/diesel/repair, Bickford's Rest., Burger King, Dunkin Donuts, KFC, Illiano's Grill, Best Western, CT Yankee Inn, Day's Inn, Motel 6, Starlight Motel, Children's Museum, National Tire, **W**...MEDICAL CARE, Shell/mart, Flanders Pizza/foods, Kings Garden Chinese, McDonald's, Brooks Drug, IGA Foods, Ford, True Value, bank, cleaners

Old Lyme

73(86)	Society Rd, no facilities
72(84)	to Rocky Neck SP, **2 mi E**...food, lodging, to Rocky Neck SP
71(83)	4 Mile Rd, River Rd, to Rocky Neck SP, beaches, **1 mi E**...camping(seasonal)
70(80)	US 1, CT 156, Old Lyme, **W**...MEDICAL CARE, Texaco/diesel, Bee&Sissel Inn, Old Lyme Inn/dining, A&P, Griswold Museum, bank
78mm	Connecticut River
69(77)	US 1, CT 9 N, to Hartford, **W**...Saybrook Fish House, Comfort Inn/rest.
68(76.5)	US 1 S, Old Saybrook, **E**...Citgo/wash, Mobil, Texaco/wash, Pat's Country Kitchen, Chrysler/Plymouth/Dodge/Jeep, Isuzu, **W**...Liberty Inn, Chevrolet/Nissan, Lube'n Tune, Pontiac/Olds/GMC, Toyota, antiques
67(76)	CT 154, Elm St(no EZ sb return), Old Saybrook, same as 68
66(75)	to US 1, Spencer Plain Rd, **1 mi E**...Citgo/diesel/24hr, Benny's Pizzeria, CA Pizza, Cucu's Mexican, Gateway Indian Cuisine, Luigi's Italian, Paisan's Pizza, SoleMar Café, Thai Cuisine, TNT Rest., Day's Inn, Heritage Inn, Saybrook Motel, Super 8, ABC Hardware, transmissions, **W**...st police
74mm	**rest area nb, full(handicapped)facilities, st police sb**

Westbrook

65(73)	CT 153, Westbrook, **E**...Exxon/Dunkin Donuts/mart, Denny's, Day's Inn, Honda, Westbrook Factory Stores/famous brands
64(70)	CT 145, Horse Hill Rd, Clinton, **1 mi E**...VETERINARIAN, **3 mi E on US 1**...DENTIST, Shell/mart, Dunkin Donuts, Hungry Lion Rest., Clinton Motel, NAPA, Radio Shack, Stop'n Shop, TJ Maxx, auto/tire repair, bank
63(68)	CT 81, Clinton, **E**...Shell/mart, **1 mi E on US 1**...Citgo/diesel, Texaco/diesel/LP, Friendly's, McDonald's, CVS Drug, **W**...Clinton Crossing Premium Outlets/famous brands
62(67)	**E**...camping, public beach, Hammonasset SP
66mm	**service plaza both lanes, Mobil/diesel, McDonald's**
61(64)	CT 79, Madison, **E**...MEDICAL CARE, Panda House Chinese, **on US 1 E**...Amoco, Citgo/mart, Gulf/mart, CVS Drug, bank
60(63.5)	Mungertown Rd(from sb, no return), **E**...food, lodging
61mm	East River
59(60)	CT 146, Guilford, **E**...MEDICAL CARE, Citgo/DairyMart, Mobil/Donuts/mart/24hr, Texaco/diesel/mart, Friendly's, McDonald's, SeaBreeze Rest., Sheppard's Rest., Shoreline Diner, Wendy's, Tower Motel, Chevrolet/Pontiac, Pepper's Gen Store, Valvoline, transmissions, **W**...Sachem Country Rest.
58(59)	CT 77, Guilford, **on US 1 E**...Getty, Donut Village, Friendly's, Subway, Big Y Foods, CVS Drug, Radio Shack, bank, to Whitfield Museum, **W**...st police
57(58)	US 1, Guilford, **E**...MEDICAL CARE, **W**...CHIROPRACTOR, Land Rover, Saab
56(55)	CT 146, to Stony Creek, **E**...Advanced Motel, **W**...Berkshire/diesel/LP, Mobil/mart/wash, TA/Sunoco/diesel/rest./24hr, Friendly's, McDonald's Motel, Ramada Ltd, Stop'n Shop Foods
55(54)	US 1, **E**...Mobil, Sunoco, Thornton/gas, 7-11/atm/24hr, Dunkin Donuts, McDonald's, My Dad's Rest., Knight's Inn, Motel 6, Ford, Walgreen, cinema, **W**...Citgo/mart/wash, Texaco/mart, Margarita's Mexican, Parthenon Diner/24hr, Su Casa Mexican, Day's Inn

Connecticut Interstate 95

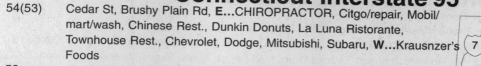

54(53) Cedar St, Brushy Plain Rd, **E**...CHIROPRACTOR, Citgo/repair, Mobil/ mart/wash, Chinese Rest., Dunkin Donuts, La Luna Ristorante, Townhouse Rest., Chevrolet, Dodge, Mitsubishi, Subaru, **W**...Krausnzer's Foods

52mm **service plaza both lanes, Mobil/diesel/24hr, McDonald's, atm**

52(50) CT 100, North High St, **E**...CHIROPRACTOR, DENTIST, bank, to Trolley Museum, **W**...st police

51(49.5) US 1, Easthaven, **E**...Sunoco, Texaco, Chevrolet, **W**...Gulf, Mobil, Sunoco, Champion Parts

50(49) Woodward Ave(from nb), **E**...Shell, US Naval/Marine Reserve, Ft Nathan Hale, airport

49(48.5) Stiles St, no facilities

48(48) I-91 N to Hartford, no facilities

47(47.5) CT 34, New Haven, **E**...Texaco, **W**...HOSPITAL, Mobil, Howard Johnson

46(47) Long Wharf Dr, Sargeant Dr, **W**...Mobil, Brick Oven Pizza, Howard Johnson, bank

45(46.5) (from sb), same as 44, Exxon, Dairy Queen, McDonald's

44(46) CT 10, Kimberly Ave, **W**...Amoco/mart/24hr, Mobil, Dairy Queen, McDonald's, Wendy's, cleaners/laundry, same as 43

43(45) CT 122, 1st Ave(no EZ return), West Haven, **E**...BP, 7-11/24hr, WaWa Mart, McDonald's, ShopRite Foods, InstaLube/wash, Marshall's/repair/tires, **W**...HOSPITAL, Mobil, Chinese Rest., to U of New Haven

42(44) CT 162, Saw Mill Rd, **E**...Mobil, Chinese Rest., Pizza Hut, Yankee Inn, **W**...Shell, Denny's, Friendly's, Day's Inn, cleaners/laundry

41(42) Marsh Hill Rd, to Orange, **W**...cinema

41mm **service plaza both lanes, Mobil/diesel, McDonald's, Denny's(nb)**

40(40) Old Gate Lane, Woodmont Rd, **E**...Getty/diesel, Gulf/diesel/24hr, Mayflower/diesel/rest./motel/24hr, Shell, Sunoco/24hr, Bennigan's, D'Angelo's, Donut Express, Dutchess Rest., Uncle Tony's Pizza, Milford Inn, Comfort Inn, bank

39(39) US 1, to Milford, **E**...Gulf/diesel, Gathering Rest., Pizzaria Uno, CT Tpk Motel, Howard Johnson, Milford Inn, Olds/Mazda/Volvo, Firestone/auto, **W**...CHIROPRACTOR, DENTIST, Mobil, Bagel Maven, Bagel Connection, Baskin-Robbins, Burger King, Chili's, Dunkin Donuts, HoneyBaked Ham, KFC, Little Caesar's, Miami Grill, Mr Sizzzl, McDonald's, Nathan's Famous, Steak&Sword Rest., Subway, Taco Bell, Wendy's, Acura, Chrysler/Jeep, House of Fabrics, JC Penney, Linens'n Things, Pearle Vision, Postal Ctr, Stop&Shop Food, USPO, Waldbaum's Foods, bank, cinema, cleaners, mall

38(38) CT 15, Merritt Pkwy, Cross Pkwy, no facilities

37(37.5) High St(from nb, no EZ return), **E**...Amoco, Citgo, Gulf, Mobil, 7-11, Chinese Rest., Roy Rogers, ShoreLine Motel, Meineke Muffler

36(37) Plains Rd, **E**...HOSPITAL, Exxon/diesel/mart, Sunoco/diesel/mart, Hampton Inn, bank

35(36) Bic Dr, School House Rd, **E**...Citgo, Subway, Wendy's, Susse Chalet, Chevrolet, Chrysler/Jeep, Ford/Lincoln/ Mercury, K-Mart, Pontiac/Nissan

34(34) US 1, Milford, **E**...Gulf, Shell, Denny's, McDonald's, Taco Bell, Devon Motel

33(33.5) US 1(from nb, no EZ return), CT 110, Ferry Blvd, **E**...Exxon/diesel/mart, Sunoco, Danny's Drive-In, Fagan's Rest., Walgreen, Jo-Ann Crafts, Bradlee's, laundry, tires, **W**...BP, Chinese Rest., Ponderosa, PathMark Drug, Marshall's, cinema

32(33) W Broad St, Stratford, **E**...Amoco, Getty, pizza, Ford, cinema, tires, **W**...Gulf/mart, Dunkin Donuts, Pepin's Rest., cleaners

31(32) South Ave, Honeyspot Rd, **E**...Gulf, HoJo Inn, Camelot Motel, antiques

30(31.5) Lordship Blvd, Surf Ave, **E**...Shell/24hr, Ramada/rest, Ryder, bank, airport, **W**...Sunoco

29(31) CT 130, Stratford Ave, HOSPITAL

28(30) CT 113, E Main St, Pembrook St

27(29.5) Lafayette Blvd, downtown, **W**...HOSPITAL, Sears/auto service, Day's Hotel, Barnum Museum

27a(29) CT 25, CT 8, to Waterbury

26(28) Wordin Ave, downtown

25(27) CT 130(from sb, no EZ return), State St, Commerce Dr, Fairfield Ave, **E**...Getty, Mercedes, **W**...Gulf/lube, McDonald's

24(26.5) Black Rock Tpk, **E**...Fairfield Diner, Bridgeport Motor Inn, **W**...Ford

23(26) US 1, Kings Hwy, **E**...Sunoco, Home Depot, auto parts

Connecticut Interstate 95

22(24)	Round Hill Rd(from sb), N Benson Rd, no facilities
23.5mm	**service plaza both lanes, Mobil/diesel, McDonald's/24hr, Denny's**
21(23)	Mill Plain Rd, **E**...CHIROPRACTOR, MEDICAL CARE, Citgo, McDonad's, Dairy Queen, Grotto Rest., P Gordon's Coffee Roasters, Fairfield Inn, Howard Johnson, Brooks Drug, Mail Depot, Midas Muffler, Stop&Shop Food
20(22)	Bronson Rd(from sb), no facilities
19(21)	US 1, Center St, Southport, **W**...Sunoco, Texaco/mart, Athena Diner, Tequot Motor Inn, bank, carwash
18(20)	to Westport, **E**...Sherwood Island SP, public beaches, **W**...CHIROPRACTOR, VETERINARIAN, Gulf, Sunoco, Texaco, Bertucci's Pizza, Burger King, Roy Rogers, Sherwood Diner, Subway, Mazda/Toyota, cleaners, hardware, st police
17(18)	CT 33, CT 136, Westport, **E**...Charles St Arrow Rest., bank, **W**...CHIROPRACTOR, FastStop Mart, Nonna's Pizza, cleaners/laundry
16(17)	E Norwalk, **E**...CHIROPRACTOR, BP, Texaco/diesel/wash, Mike's Deli/foods, Penny's Diner, Rite Aid, **W**...MEDICAL CARE, McDonald's
15(16)	US 7, Norwalk, to Danbury, **W**...HOSPITAL, same as 14
14(15)	US 1, CT Ave, S Norwalk, **E**...st police, **W**...HOSPITAL, CHIROPRACTOR, Amoco, Texaco/Lube, Chinese Rest., Gulf, Firestone, cleaners/laundry
13(13)	US 1(no EZ return), Post Rd, Norwalk, **W**...VETERINARIAN, Amoco, Shell/24hr, Burger King, Dunkin Donuts, IHOP, KFC, McDonald's, PastaFare/pizza, Wendy's, Holiday Inn, Barnes&Noble, Bradlee's, ShopRite Foods, TJ Maxx, Valvoline, Waldbaum's Foods, bank, tires
12.5mm	**service plaza nb, Mobil/diesel, McDonald's**
12(12)	CT 136, Tokeneke Rd(from nb, no return), **W**...deli
11(11)	US 1, Darien, **E**...VETERINARIAN, Amoco, Scot Gas, Chuck's Steaks, Chevrolet, Ford/Lincoln/Mercury/ LandRover/Jaguar, **W**...Citgo, Exxon, Howard Johnson/rest., BMW, antiques, cinema, cleaners/laundry, photo shop
10(10)	Noroton, **E**...DENTIST, MEDICAL CARE, **W**...Getty Gas, Mobil, Texaco/wash
9.5mm	**service plaza sb, Mobil/diesel, McDonald's**
9(9)	US 1, CT 106, Glenbrook, **E**...Stamford Motor Inn, **W**...VETERINARIAN, Gulf, Sergio's Pizza, McDonald's/ playplace, Aamco, Meineke Muffler, cleaners/laundry, drugs
8(8)	Atlantic Ave, Elm St, **W**...HOSPITAL, Exxon, Mobil, Italian Rest., Mexican Rest., Marriott, Ramada Inn, Saturn
7(7)	CT 137, Atlantic Ave, **E**...Sheraton, **W**...mall, JC Penney, same services as 8
6(6)	Harvard Ave, West Ave, **E**...BP, Day's Inn/rest., USPO, **W**...HOSPITAL, DENTIST, VETERINARIAN, BP/wash, Getty Gas/mart, Shell/mart/24hr, Boston Mkt, Corner Deli/pizza, Subway, Taco Bell, Super 8, Stamford Hotel, Firestone, Pennzoil
5(5)	US 1, Riverside, Old Greenwich, **W**...MEDICAL CARE, Mobil/mart, Shell/mart/24hr, Chinese Rest., Italian Ristorante, McDonald's, Hyatt Regency, A&P, Edward's Drug, MailBoxes Etc, Woolworth's
4(4)	Indian Field Rd, no facilities
3	Greenwich, HOSPITAL, no facilities
2	Byram, no facilities
1mm	Connecticut/New York state line

Connecticut Interstate 395

Exit #(mm)	Services
55.5mm	Connecticut/Massachusetts state line
100(54)	to Wilsonville, no facilities
99(50)	CT 200, N Grosvenordale, **E**...W Thompson Lake Camping(seasonal)
98(49)	to CT 12(from nb), Grosvenordale, same as 99
97(47)	US 44, to E Putnam, **E**...Chinese Rest., Dunkin Donuts, KFC, McDonald's, Subway, Wendy's, $Tree, GNC, K-Mart/Little Caesar's, MailBoxes Etc, Meineke Muffler, Radio Shack, Stop'n Shop, TJ Maxx, bank, **W**...BP/mart, Shell/repair/24hr, Wal-Mart/McDonald's, antiques
96(46)	to CT 12, Putnam, **W**...HOSPITAL, Heritage Rd Café, King's Inn
95(45)	Kennedy Dr, to Putnam, **E**...Ford/Mercury, **W**...HOSPITAL
94(43)	Ballouville, **1 mi E on CT 12**...Golden Greek Rest.
93(41)	CT 101, to Dayville, **E**...Shell/repair, Xtra Fuel, Burger King, Carvel Ice Cream/bakery, China Garden, Dunkin Donuts, Fitzgerald's Grill, McDonald's, Subway, A&P, Ames, IGA Foods, Wibberley Tire, **W**...MEDICAL CARE, Mobil/diesel/mart/24hr

Connecticut Interstate 395

92(39)   to S Killingly, st police, **1 mi W**...services, RV camping
91(38)   US 6, to Danielson, to Quinebaug Valley Coll
90(36)   to US 6 E(from nb), to Providence, no facilities
35mm   **rest area both lanes, full(handicapped)facilities, Mobil/diesel/mart**
89(32)   CT 14, to Central Village, **E**...Gulf, bank, RV camping, **W**...Citgo/diesel/
          mart/24hr, Sunoco/diesel/LP/repair, Plainfield Motel/rest., RV camping
88(30)   CT 14A to Plainfield, **1/2 mi W**...Mobil/mart
87(28)   Lathrop Rd, to Plainfield, **E**...Texaco/diesel/mart, Dunkin Donuts, Pizza,
          Plainfield Yankee Motel, Big Y Foods, CVS Drug, Greyhound Park,
          Radio Shack, bank, **W**...MEDICAL CARE, Citgo/mart, Sunoco,
          Bakers Donuts, Gold Eagle Rest., Golden Greek Rest., McDonald's,
          carwash
86(24)   CT 201, Hopeville, **E**...RV camping
85(23)   CT 164, CT 138, to Pachaug, Preston, **E**...Ames, NAPA, bank, RV camping
84(21)   CT 12, Jewett City, **W**...Citgo/7-11/hotdogs, Mobil/Pizza Hut/diesel/mart, Shell/wash/24hr, McDonald's/
          playplace, Better Valu Foods
83a(20)   CT 169(from nb), Lisbon, **E**...RV camping
83(18)   CT 97, Taftville, **E**...Getty/diesel/mart, camping, repair, **W**...Citgo/mart
82(14)   to CT 2 W, CT 32 N, Norwichtown, **E**...CHIROPRACTOR, Friendly's, bank, cleaners, **W**...Mobil/diesel/mart,
          Shell/mart/wash, Texaco/LP, McDonald's, Comfort Suites, Courtyard, Rosemont Suites, Total Vision, True Value
81(14)   CT 2 E, CT 32 S, Norwich, **E**...HOSPITAL, to Mohegan Coll
80(12)   CT 82, Norwich, **E**...VETERINARIAN, Citgo/diesel/mart, Mobil/mart/wash, Shell/repair, Bess Eaton/Subway,
          Burger King, Chinese Buffet, Dominic's Pizza, Friendly's, KFC, McDonald's, Mr Pizza, Papa Gino's, Wendy's,
          Ames, Jo-Ann Crafts, Monro Muffler/brake, Pets+, ShopRite Foods, Staples, TJ Maxx, TownFair Tire, bank,
          cinema, **W**...Ramada Inn, RV camping
79a(10)   CT 2A E, to Ledyard, **E**...to Pequot Res
8.5mm   **nb...st police, phone**
          sb...Mobil/mart, rest area, full facilities
79(6)   CT 163, to Uncasville, Montville, **1 mi E**...Texaco/diesel/LP/repair, McDonald's, Ziggy's Café
78(5)   CT 32(from sb, exits left), to New London, RI Beaches
77(2)   CT 85, to Colchester, **1/2 mi E**...Shell, Oakdale Motel
0mm   I-95. I-395 begins/ends on I-95, exit 76.

N
S

Norwich

Connecticut Interstate 691

Exit #(mm)Services
I-691 begins/ends on I-91, exit 18.
12   Preston Ave, no facilities
11(11)   I-91 N, to Hartford
10(11)   I-91 S, to New Haven, CT 15 S, W Cross Pkwy
8(10)   US 5, Broad St, **N**...Cumberland/gas, Scott/mart, Texaco/diesel/mart, Broad St Pizza, Dairy Queen, cleaners,
          S...DENTIST, bank
7(9)   Meriden(from eb), downtown
6(8)   Lewis Ave(from wb, no EZ return), to CT 71, **N**...HOSPITAL, Ruby Tuesday, Filene's, JC Penney, Sears,
          mall,**S**...Citgo/mart
5(7)   CT 71, to Chamberlain Hill(no return from eb), **N**...HOSPITAL, Super Foodmart, mall, **S**...Getty/diesel/mart,
          Sunoco, McDonald's, JiffyLube, Meineke Muffler, Midas Muffler, VW
4(4)   CT 322, W Main St(no re-entry from eb), **N**...HOSPITAL, Citgo/repair, Sal's Foods
3mm   Quninnipiac River
3(1)   CT 10, to Cheshire, Southington, **N**...Sunoco/mart, Tony's Pizza, Whole Donut, Day's Inn, **S**...cartoon mu-
          seum
2(0)   I-84 E, to Hartford, no facilities
1(0)   I-84 W, to Waterbury, no facilities
I-691 begins/ends on I-84, exit 27.

E
W

Meriden

Delaware Interstate 95

Wilmington Newark

Exit #(mm)Services

23mm — Delaware/Pennsylvania state line, motorist callboxes for 23 miles sb

11(22) — I-495 S to Wilmington, DE 92, Naaman Rd, **W**...Hilton

10(21) — Harvey Rd(from nb, no EZ) **1 mi E on DE 13**...Exxon/diesel, Gulf, Shell, Sunoco/mart, 7-11, McDonald's, Arby's, Burger King, Charlie's Pizza, Eckerd, Precision Tune, bank, laundry

9(19) — DE 3, to Marsh Rd, **E**...to Bellevue SP, st police, **W**...museum

8b a(17) — US 202, Concord Pike, to Wilmington, **W**...HOSPITAL, to gas, food, lodging

7b a(16) — DE 52, Delaware Ave, **E**...HOSPITAL, to gas, food, lodging

6(15) — DE 4, MLK Blvd, **W**...Gulf/mart

5d(12) — I-495 N, to Wilmington, to DE Mem Bridge

5b a(11) — DE 141, to US 202, Newport, to New Castle, no facilities

4b a(8) — DE 1, DE 7, to Christiana, **E**...FoodCourt, JC Penney, Macy's, Sears, cinema, mall, **W**...HOSPITAL, Chi Chi's, Chili's, Michael's Rest., Shoney's Inn/rest., Courtyard, Fairfield Inn, Hilton, Red Roof Inn, Border's Books, OfficeMax, AAA

3b a(6) — DE 273, to Newark, Dover, **E**...Amoco/mart, Exxon/diesel, Bob Evans, Boston Mkt, Chinese Rest., Ruby Tuesday, Wendy's, Best Western, Comfort Inn, Acme Foods, Baby Depot, Burlington Coats, FashionBug, Jo-Ann Fabrics, Luxury Linens, **W**...Getty Gas, Shell/mart, 7-11, Denny's, Donut Connection, Pizza Hut, Hampton Inn, Holiday Inn/Oliver's Rest., McIntosh Motel

5mm — **service area both lanes(exits left from both lanes), info, Exxon/ diesel, Mobil/diesel, Bob's Big Boy, Roy Rogers, Sbarro's, Taco Bell, TCBY**

1b a(3) — DE 896, to Newark, Middletown, to U of DE, **W**...Exxon, Gulf/diesel, Mobil, Shell, Texaco/diesel/24hr, Big Boy, Friendly's, Ground Round, McDonald's, Pizza Pie, Roy Rogers, Comfort Inn, Howard Johnson, White Glove Carwash

1mm — toll booth, st police

0mm — Delaware/Maryland state line, motorist callboxes for 23 miles nb

Delaware Interstate 295(Wilmington)

Wilmington

Exit # — Services

15mm — Delaware River, Delaware Memorial Bridge, Delaware/New Jersey state line

14.5mm — toll plaza

14 — DE 9, New Castle Ave, to Wilmington, **E**...Amoco/mart/24hr, Mobil, Giovanni's Rest., Big A Parts, Family$, Firestone/auto, Harley-Davidson, Rite Aid, SuperFresh Foods, carwash, hardware, laundry, **W**...Gulf, Texaco/ mart, McDonald's, Day's Inn, Motel 6, Travelodge

13 — US 13, US 301, US 40, to New Castle, **E**...Amoco, Exxon/diesel, Sunoco/mart, Texaco, 7-11, Burger King, Denny's, Dunkin Donuts, La Gondola Italian, IHOP, McDonald's, Taco Bell, Wendy's, Quality Inn, Red Rose Inn, Midas Muffler, PepBoys, auto repair, bank, **W**...Shell, Ramada Inn, Nissan

12 — I-495, US 202, N to Wilmington

I-295 begins/ends on I-95, exit 5.

Florida Interstate 4

Exit # (mm)Services

58(134) I-95, S to Miami, N to Jacksonville, FL 400. I-4 begins/ends on I-95, exit 86.

57(129) to US 92(from eb, exits left), **N...**VETERINARIAN

121mm parking area eb, no facilities, no security, litter barrels

56(119) FL 44, to De Land, **N...**HOSPITAL, Texaco/diesel/mart, Quality Inn, **S...**Shell/diesel/fruit

55(116) Lake Helen, **S...**Copper Cove Rest.

54(113) FL 472, Orange City, to De Land, **N...**McDonald's, Monastery Rest., Chimney Corner Motel, City Gate Motel/café, Comfort Inn, auto/transmissions, KOA, Village RV Park, to Blue Sprgs SP

53cba(111) Deltona, **N...**HOSPITAL, DENTIST, drugs, **S...**Chevron/repair, Albertson's

53(108) De Bary, Deltona, **N...**Burger King, Shoney's, Hampton Inn, **S...**Texaco, McDonald's, Best Western/rest., Paradise Lakes Camping(3mi)

52(104) US 17, US 92, Sanford, **S...**HOSPITAL, Citgo/Subway/mart, Holiday Inn(1mi)

51(102) FL 46, Sanford, to Mt Dora, **N...**Amoco/Pizza Hut/diesel/mart, AutoNation USA, **S...**HOSPITAL, Chevron/mart, Mobil/mart, Speedway/mart, Burger King, Cracker Barrel, Denny's, Don Pablo's, Hops Grill, McDonald's, OutBack Steaks, Red Lobster, Steak'n Shake, Waffle House, Day's Inn, Holiday Inn, Super 8, Dillard's, HomePlace, JC Penney, Old Navy, Pier 1, Ross, Sears/auto, Service Merchandise, cinema, mall

E

50(97) Lake Mary Blvd, Heathrow, **N...**DENTIST, Exxon, Courtyard, Barney's Coffee Co, Eckerd, Goodings Food/24hr, AAA, **S...**CHIROPRACTOR, Amoco/mart/wash/24hr, Chevron, Citgo/7-11, Arby's, Bob Evans, Boston Mkt, Burger King, Checkers, Chick-fil-A, Chili's, Dunkin Donuts, Frank&Naomi's, Japanese Steaks, KFC, Krystal, LongHorn Steaks, McDonald's, Papa Joe's, Papa John's, Shoney's, Steak'n Shake, Subway, Taco Bell, Uno Pizzaria, Wendy's, Hilton, HomeWood Suites, Albertson's, Home Depot, K-Mart, Old Time Pottery, Publix, Target, Winn-Dixie, auto repair/tires, cinema, cleaners, mall

95mm **rest areas both lanes, full(handicapped)facilities, phone, vending, picnic tables, litter barrels, petwalk, 24hr security**

49(93) FL 434, Longwood, to Winter Springs, **N...**Chevron, Exxon, Denny's, Kyoto Steaks, Miami Subs, Pizza Hut, Shoney's, Ramada Inn, Quality Inn, **S...**HOSPITAL, Mobil/diesel/mart, Shell/mart, 7-11, Boston Mkt

48(92) FL 436, Altamonte Springs, **N...**Shell, Texaco, LongHorn Steaks, Best Western, Days Inn, Hampton Inn, Holiday Inn, Travelodge, **S...**Amoco, Chevron, Embassy Suites, Hilton, Linens'n Things, TJ Maxx, cinema

47(90) FL 414, Maitland Blvd, **N...**7-11, Sheraton, **S...**art ctr

46(89) FL 423, Lee Rd, **N...**Texaco, Shoney's, Waffle House, Comfort Inn, Econolodge, Holiday Inn, Knight's Inn, Park Inn, Aamco, Infiniti, **S...**Chevron, Mobil/diesel, Denny's, Steak'n Shake, Fairfield Inn, Plaza Inn, BMW

W

45(88) Fairbanks Ave(no eb re-entry), **S...**Amoco/mart

44(87.5) Paar Ave(no eb re-entry), **N...**Citgo, **S...**Texaco/mart

43(87) Princeton St, **S...**HOSPITAL, Exxon, 7-11, Shell/mart, Spur

42(86) Ivanhoe Blvd, downtown, **S...**Radisson, Travelers Hotel, bank

41(85.5) US 17, US 92, FL 50, Amelia St, **N...**Holiday Inn

40(85) FL 526, Robinson St, **N...**Marriott

39(84.5) South St(from wb), downtown

38(84) Anderson St E, Church St Sta Hist Dist, downtown

37(83.5) Gore St(from wb), **S...**HOSPITAL, downtown

36(83) FL 408(toll), to FL 526, no facilities

35(82) Kaley Ave, **S...**HOSPITAL, Mobil, Texaco/mart

34(81) Michigan St(from wb)

33(80) US 17, US 441, US 92, **S...**RaceTrac/mart, Denny's, Wendy's, Day's Inn

32(79) FL 423, 33rd St, John Young Pkwy, **N...**McDonald's, Catalina Inn, **S...**Citgo/7-11, RaceTrac/mart, Texaco/diesel/mart, IHOP, Days Inn/rest.

31(78) FL 527, FL TPK(toll)

30b a(76) FL 435, International Dr(exits left from eb), **N...**Cracker Barrel, Denny's, Day's Inn, Delta Resort, Radisson, Belz Factory Outlet/famous brands, to Universal Studios, **S...**Steak&Ale, Court of Flags Hotel, Econolodge, Hampton Inn, Holiday Inn Express, Howard Johnson, International Gateway Inn, Knight's Inn, Lakefront Inn, Las Palmas Hotel, Motel 6, Rodeway Inn, Super 8, funpark

Lake Mary

Orlando

Florida Interstate 4

Orlando

29(73) FL 482, Sand Lake Rd, **N**...Citgo/7-11, Comfort Suites, K-Mart, Publix, to Universal Studios, **S**...HOSPITAL, Exxon, Chevron, Shell/mart, Texaco/mart, Burger King, Checker's, Denny's, Golden Corral, Italianni's, King Henry's Feast, McDonald's, Morrison's Cafeteria, Perkins, Popeye's, Wendy's, Best Western, Courtyard, Day's Inn, Embassy Suites, Fairfield Inn, Howard Johnson, La Quinta, Marriott, Quality Inn, Peabody Hotel, Radisson Inn, Ramada Inn, Stouffer Resort, RV Park

28b a(72) FL 528 E(toll, no eb re-entry), to Cape Canaveral, **S**...to airport

27a(71) Central FL Pkwy(from eb), **S**...Chevron/mart, Wendy's, to SeaWorld

70mm **rest areas both lanes, full(handicapped)facilities, phone, vending, picnic tables, litter barrels, petwalk, 24hr security**

27(68) FL 535, Lake Buena Vista, **N**...Chevron/mart/24hr, Citgo/7-11, Texaco, AleHouse, Burger King, Chevy's Mexican, Chili's, CrabHouse Seafood, Denny's, Dunkin Donuts, Giordano's, IHOP, Jungle Jim's Rest., Macaroni Grill, McDonald's, Miami Subs, Olive Garden, Pebble's Dining, Perkins, Pizza Hut, Pizzaria Uno, Red Lobster, Shoney's, Sizzler, Steak&Shake, Subway, Taco Bell, TGIFriday, Tony Roma's, Comfort Inn, Day's Inn, Embassy Suites, Hilton, Hotel Orlando, Hyatt Hotel, Radisson, Summerfield Suites, Wyndham Hotel, Crossroads Outlet/FoodCourt, Eckerd's, Gooding's Foods, USPO, Walgreen, **S**...Shell, 7-11, Chick-fil-A, Landry's Seafood, LoneStar Steaks, Holiday Inn

26b a(66) FL 536 E, to FL 417(toll), **N**...DisneyWorld, **1-2 mi S**...Citgo/7-11, Ponderosa, Eckerds, Buena Vista Suites, Caribe Royale Suites, Marriott, VF Factory Outlet, to airport

25b a(64) US 192, FL 536, to FL 417(toll), to Kissimmee, **N**...to DisneyWorld, MGM, **S**...Citgo/7-11, RaceTrac/mart, Boston Mkt, Food Court, IHOP, Kobe Japanese, McDonald's, Shoney's, Steakhouse Foods, Waffle House, Hampton Inn, Homewood Suites, Howard Johnson, Hyatt Hotel, Larsen's Lodge, Radisson, Camping World RV Service/supplies(3mi), factory outlet/famous brands

24cde FL 417(toll, from eb), to airport, Port Canaveral

24(58) FL 532, to Kissimmee, no facilities

23(55) US 27, to Haines City, **N**...Chevron/mart/24hr, Citgo/7-11, Speedway/diesel/mart, Burger King, McDonald's, NY Pizza, Shoney's, Waffle House, Wendy's, Comfort Inn, SouthGate Inn, Motel 6, Super 8, **S**...HOSPITAL, Amoco/diesel/mart, Citgo/7-11, Exxon, RaceTrac/mart/24hr, Shell/mart, Texaco/diesel/service, BBQ, Bob Evans, Denny's, Greenleaf Chinese, Hardee's, KeyWest Seafood, KFC, Perkins, Day's Inn, Holiday Inn Express, to Cypress Gardens

22(48) FL 557, to Winter Haven, Lake Alfred, **S**...BP/diesel/mart

46mm **rest area both lanes, full(handicapped)facilities, phone, vending, picnic tables, litter barrels, petwalk, 24hr security**

21(44) FL 559, Polk City, to Auburndale, **S**...Amoco/mart, Texaco/diesel/rest./mart, lodging

20(38) FL 33, to Lakeland, Polk City, no facilities

19(33) FL 33, Lakeland, **N**...Exxon/mart/wash, Cracker Barrel, Budgetel, Quality Inn, RV Sales/service, **S**...HOSPITAL, Amoco/mart, Speedway/mart, Ryan's, Waffle House, Relax Inn

Lakeland

18(32) US 98, Lakeland, **N**...VETERINARIAN, Chevron/diesel/mart, Spur/mart, Texaco/Blimpie/mart, Bennigan's, Checker's, Chili's, ChuckeCheese, Don Pablo's, Dragon Buffet Chinese, Hooter's, IHOP, Kenny Rogers Roasters, KFC, LoneStar Steaks, McDonald's, Olive Garden, Pizza Hut, Red Lobster, Ruby Tuesday, Shoney's, Stacy's Buffet, Steak'n Shake/24hr, Subway, Taco Bell, TGIFriday, Lakeland Motel, La Quinta, Royalty Inn, Wellesley Inn, Barnes&Noble, Belk, Best Buy, Circuit City, Computer City, Dillard's, $Tree, Food Lion, Goodyear/auto, Kids R Us, K-Mart, Lenscrafters, PepBoys, Pier 1, Sam's Club, Service Merchandise, SportsAuthority, Tire Kingdom, Wal-Mart, RV Ctr, **S**...HOSPITAL, Amoco, Citgo/diesel/mart, Coastal/diesel/mart, RaceTrac/mart, Bob Evans, Burger King, Chinese Rest., Denny's, Long John Silver, McDonald's, Roadhouse Grill, Waffle House, Wendy's, Best Western/rest., Day's Inn, Motel 6, Ramada Inn, AutoZone, Chrysler/Plymouth/Dodge, Family$, Home Depot, Office Depot, Radio Shack, U-Haul, Winn-Dixie, cleaners, transmissions

17(31) FL 539, Lakeland, to Kathleen, **S**...hist dist

16(28) to US 92, Lakeland(from eb re-entry), no facilities

15a(27) new exit, no facilities

15(26) County Line Rd, **N**...Shell/mart, **S**...airport

14(23) FL 553, Plant City, **S**...Shell/Subway/Taco Bell/mart, Texaco/Blimpie/mart, Arby's, Burger King, Denny's, Popeye's, Holiday Inn Express, RV Sales

13(21) FL 39, Plant City, to Zephyrhills, **N**...Texaco/diesel/mart/24hr, **S**...Amoco/diesel/mart, Shell/mart, Day's Inn, Ramada Inn/rest.

E

W

Florida Interstate 4

E

W

11(20) FL 566, Plant City, to Thonotosassa, **N**...Amoco/mart, **S**...HOSPITAL, RaceTrac/mart, BBQ, BuddyFreddy's Dining, McDonald's, Branch Farm

19mm weigh sta both lanes

10(17) Branch Forbes Rd, **N**...Shell/mart, Spur/mart, **S**...farmer's mkt

9(14) McIntosh Rd, **N**...Amoco/mart, **S**...BP/mart, RaceTrac/mart, Burger King, McDonald's, RV Ctr, Green Acres RV Resort

8(10) FL 579, Mango, Thonotosassa, **N**...Amoco/mart, Cracker Barrel, Camping World RV Service/supplies, Lazy Day's RV Ctr, RV camping, **S**...Texaco/diesel/mart, Denny's, Hardee's, Wendy's, Masters Inn

7(8.5) I-75, N to Ocala, S to Naples

6c(7) US 92 W

6ba(6.5) US 92, US 301, Hillsborough Ave, **N**...Citgo/diesel/mart/wash, Circle K/gas, My's Chinese/Vietnamese, Sheraton, to Busch Gardens, **S**...RaceTrac/mart/24hr, Denny's, Red Roof Inn, ForeTravel RV, Holiday Travel RV, FL Expo Fair

5(6) Orient Rd, **N**...same as 6, **S**...Exxon/mart, Shell/mart

4(5) FL 574, MLK Blvd, **N**...McDonald's, **S**...Exxon/mart, Shell/mart, Wendy's, Economy Inn, bank, laundry

3(3) US 41, 50th St, Columbus Dr, **N**...Texaco/Subway/mart, Econolodge, La Quinta, Milner Hotel, **S**...Exxon/diesel/mart, Shell, Speedway/diesel/mart, Burger King, Checker's, Chinese Rest., Church's Chicken, KFC, McDonald's, Pizza Hut, Taco Bell, Wendy's, Eckerd, Family$, Kash&Karry, bank, hardware, laundry

2(2) FL 569, 40th St, **N**...Citgo/diesel/mart, Texaco/diesel, Budget Inn, to Busch Gardens

1(1) FL 585, 22nd, 21st St, Hist District, Port of Tampa, **N**...BP/diesel/mart, **S**...Amoco, Burger King, Hardee's, Ybor City Brewing Co

0mm I-4 begins/ends on I-275, exit 27.

Tampa

Florida Interstate 10

Exit #(mm)Services

370mm I-10 begins/ends on I-95, exit 111.

59(369) Stockton St(from eb), to Riverside, **S**...HOSPITAL, Amoco/mart, Gate Gas

58(368) US 17 S(from wb), downtown

57(367) FL 129, McDuff Ave, **S**...Amoco/mart, Chevron, Popeye's

56(366) Lennox Ave, Edgewood Ave(from wb), no facilities

55(365) FL 111, Cassat Ave, **N**...Amoco/diesel/mart, Hess/mart, Burger King, Firehouse Subs, Popeye's, AutoZone, **S**...RaceTrac/24hr, Dunkin Donuts, Krispy Kreme, Winn-Dixie, transmissions

54(364) FL 103, Lane Ave, **N**...Day's Inn, Quality Inn, Ramada Inn, **S**...Amoco/mart, Chevron/mart, Texaco/diesel, Denny's, Hardee's, Mr Donut, Piccadilly's, Shoney's, Super 8, Battery Depot, Home Depot, Office Depot, Wrangler Outlet

53(359) I-295, N to Savannah, S to St Augustine

52(357) Marietta, **N**...Gate/mart, Texaco/mart, **S**...VETERINARIAN, Amoco/diesel/mart, Domino's, carwash, laundry

51(353) Whitehouse, to Cecil Fields, **N**...Chevron/mart/24hr, Lil Champ/gas, RV Ctr, **S**...Shell/diesel/mart/24hr, Texaco/mart

352mm rest area eb, full(handicapped)facilities, phone, vending, picnic tables, litter barrels, 24hr security

351mm rest area wb, full(handicapped)facilities, phone, vending, picnic tables, litter barrels, 24hr security

50(344) US 301, Baldwin, to Starke, **S**...Amoco/76/diesel/rest./24hr, BP/mart, Exxon/mart, Shell, Speedway/diesel/mart/24hr, Texaco/Pizza Hut/mart, Burger King, McDonald's, Pizza Hut, Waffle House, Best Western

49(337) FL 228, Macclenny, to Maxville, **N**...HOSPITAL

48(335) FL 121, Macclenny, to Lake Butler, **N**...HOSPITAL, CHIROPRACTOR, PODIATRIST, BP/diesel/mart, Exxon/mart, BBQ, Hardee's, KFC, McDonald's, Pier 6 Seafood/steaks, Pizza Hut, Subway, Taco Bell/TCBY, Waffle House, Day's Inn, Discount Parts, Food Lion, JiffyLube, Wal-Mart, Winn-Dixie, laundry, **S**...RaceWay/mart, Exxon/diesel/mart, Burger King, Econolodge, Expressway Motel/rest.

47(333) FL 125, Glen Saint Mary, **N**...Exxon/diesel/mart/24hr

46(327) FL 229, Sanderson, to Raiford, **1 mi N**...gas

45(325) US 90, Sanderson, to Olustee, **S**...Exxon/mart, to Olustee Bfd

318mm rest area both lanes, full(handicapped)facilities, phone, vending, picnic tables, litter barrels, petwalk, 24hr security

Jacksonville

93

Florida Interstate 10

44(305) US 441, Lake City, to Fargo, **N**...BP/repair, Chevron/diesel/mart, KOA(1mi) **S**...HOSPITAL, SuperTest/diesel/mart, Texaco/diesel/mart, Day's Inn

43(302) US 41, to Lake City, **N**...Amoco/mart, Kelly's RV Park(6mi), to Stephen Foster Ctr, **S**...HOSPITAL, **S**...Texaco/ice cream/mart

42b a(298) I-75, N to Valdosta, S to Tampa

294.5mm **rest area both lanes, full(handicapped)facilities, phone, vending, picnic tables, litter barrels, petwalk, 24hr security**

41(293) rd 137, to Wellborn, no facilities

40(284) US 129, to Live Oak, **N**...Penn/diesel/mart, to Boys Ranch, **S**...HOSPITAL, BP/mart, Chevron/mart, Shell/mart/24hr, Texaco/mart, Huddle House, McDonald's, Subway, Taco Bell/TCBY, Waffle House, Wendy's, Best Western, Econolodge, Holiday Inn Express, $Tree, Wal-Mart SuperCtr/24hr

39(276) US 90, Live Oak, **N**...to Suwannee River SP, **S**...HOSPITAL, United 500/diesel/repair

269mm Suwannee River

265mm **rest areas both lanes, full(handicapped)facilities, phone, vending, picnic tables, litter barrels, petwalk, 24hr security**

264mm weigh sta both lanes

38(263) rd 255, Lee, **N**...Exxon/diesel/mart, Kountry Kitchen, to Suwannee River SP, **S**...Citgo/diesel, Texaco, Red Onion Grill

37(258) FL 53, **N**...HOSPITAL, Amoco/Burger King/diesel/mart/24hr, Citgo/diesel/24hr, Texaco/Subway/DQ/Taco Bell/diesel/mart, Latrelle's Rest., Waffle House, Day's Inn, Holiday Inn Express, Super 8, **S**...Deer Wood Inn/resort, Jellystone Camping, Madison Camping

36(252) FL 14, to Madison, **N**...HOSPITAL, Texaco/Arby's/Ice Cream/mart

35(241) US 221, Greenville, **N**...Texaco/Dairy Queen/mart, **S**...BP/diesel/mart

234mm **rest area both lanes, full(handicapped)facilities, phone, picnic tables, litter barrels, petwalk, 24hr security**

34(232) rd 257, Aucilla, **N**...Amoco/diesel, Chevron

33(224) US 19, to Monticello, **N**...Burger King(4mi), Campers World RV Park, **S**...Amoco/mart, Chevron/mart, Exxon/Wendy's/mart/atm, Texaco/Arby's/Ice Cream/diesel/mart, Huddle House, McDonald's, Day's Inn, Super 8, Alligator Lake Camping, KOA, dogtrack

32(216) FL 59, Lloyd, **S**...Amoco/Subway/mart/24hr, BP/diesel/mart/24hr, Citgo/diesel, Villager Lodge

31b a(208) US 90, Tallahassee, **N**...Tallahassee RV Camping, **S**...Citgo/diesel/mart, Best Western, Day's Inn(2mi), golf

30(202) US 319, FL 61, Tallahassee, **N**...Amoco/mart, BP/diesel/mart, Dixie Gas/mart, 76/Circle K, Shell/mart, Texaco/diesel/mart/wash, Applebee's, Baskin-Robbins, Bruegger's Bagels, Dunkin Donuts, Fuddrucker's, KFC, McDonald's, Pizza Hut, Popeye's, Quincy's, Subway, Taco Bell, TCBY, Village Pizza, Wendy's, Motel 6, Albertson's/drugs, Books-A-Million, Discount Tire, Eckerd, GNC, MailBoxes Etc, Publix, Radio Shack, SteinMart, SuperLube, TJ Maxx, Wal-Mart, Winn-Dixie, bank, cinema, cleaners, **S**...HOSPITAL, VETERINARIAN, Citgo/mart, Chick-fil-A, Don Pablo's, Osaka Japanese, Outback Steaks, Miami Subs, Steak'n Shake, TGIFriday, Wings Grill, Cabot Lodge, Hilton Garden, Residence Inn, Studio+ Suites, Home Depot, Office Depot, bank, cleaners

29(199) US 27, Tallahassee, **N**...Citgo, Burger King, Taco Bell, Waffle House, Best Inn, Comfort Inn, Fairfield Inn, Hampton Inn, Holiday Inn, Microtel, Villager Lodge, Boot Country, Big Oak RV Park(2mi), Sam's Club, **S**...VETERINARIAN, EYECARE, Amoco/Pizza Hut/mart/24hr, BP/wash, Chevron/diesel/mart, Shell, 76/Circle K, Texaco, USA/mart, BBQ, Blimpie, Boston Mkt, China Buffet, China Wok, Cracker Barrel, Crystal River Seafood, Dairy Queen, El Maya Mexican, Food Court, Hooters, Julie's Rest., KFC, Krispy Kreme, Lindy's Chicken, Longhorn Steaks, LJ Silver, McDonald's, Melting Pot Rest., Miami Subs, Papa John's, Pizza Hut, Quincy's, Rally's, Red Lobster, Roadhouse Grill, Steak&Ale, Subway, TCBY, Village Inn Rest., Whataburger, Wendy's, Western Sizzlin, Cabot Lodge, Day's Inn, Econolodge, Howard Johnson, La Quinta, Motel 6, Ramada Inn, Red Roof Inn, Shoney's Inn/rest., Super 8, Albertson's, AutoZone, Barnes&Noble, Big 10 Tire, Burlington Coats, CompUSA, Dillard's, Midas Muffler, Oschman's Sports, Parisian, Publix, Service Merchandise, Sun Tire/lube, SuperLube, TuffyAuto, Walgreen, cinema, cleaners, mall

28(195) FL 263, Tallahassee, **S**...Chevron/diesel/mart, Dixie/diesel/deli, Shell/mart, Steak'n Shake, Waffle House, Sleep Inn, Lowe's Bldg Supply, **2-5 mi S**...McDonald's, Subway, Colony Inn, Day's Inn, Lafayette Motel, Skyline Motel, RV camping, civic ctr, museum/zoo, airport

194mm **rest area both lanes, full(handicapped)facilities, phone, vending, picnic tables, litter barrels, petwalk, 24hr security**

27(192) US 90, to Tallahassee, Quincy, **N**...BP/mart, Flying J/Conoco/diesel/LP/mart/24hr, Howard Johnson, **S**...Bennett's RV Park(4mi), Lakeside RV Park(4mi)

26(180) FL 267, Quincy, **N**...HOSPITAL, Texaco/Ice Cream/mart, McDonald's(4mi), **S**...Amoco/mart, BP/diesel/mart, Holiday Inn Express, to Lake Talquin SP

Live Oak

Tallahassee

E

W

Florida Interstate 10

25(174) FL 12, to Greensboro, **N**...Amoco/Burger King/
mart, Texaco/diesel/mart, Beaver Lake Camping

24(165) rd 270A, Chattahoochee, **N**...to Lake Seminole SP,
S...Texaco/diesel/mart, KOA

161mm **rest area both lanes, full(handicapped)facilities, phone,
vending, picnic tables, litter barrels, petwalk, 24hr security**

160mm Apalachicola River, central/eastern time zone

23(158) rd 286, Sneads, **N**...Lake Seminole, to Three Rivers SRA

155mm weigh sta both lanes

22(152) FL 69, to Grand Ridge, **N**...Chevron/mart, GL/mart, Texaco/mart, Golden Lariat Rest./
Western Shop, Durden's Motel/café, antiques

21(142) FL 71, Oakdale, to Marianna, **N**...HOSPITAL, Amoco/diesel/mart/24hr, Chevron/mart,
Pilot/Arby's/diesel/mart/24hr, Shell/mart, Texaco/diesel/mart, Baskin-Robbins, BBQ, Ruby
Tuesday, Shoney's, Waffle House, Comfort Inn, Hampton Inn, Holiday Inn Express, Super
8, to FL Caverns SP(8mi), **S**...BP/TA/diesel/rest./24hr, Chevron, Speedway/diesel/mart,
McDonald's, Best Western, Arrowhead Camping

20(136) FL 276, to Marianna, **N**...HOSPITAL, Chevron/diesel/mart, Day's Inn(3mi), Sandusky Inn,
Villager Lodge(3mi), to FL Caverns SP(8mi)

133mm **rest area both lanes, full(handicapped)facilities, phone, picnic tables, litter barrels,
petwalk, 24hr security**

19(130) US 231, Cottondale, **N**...Amoco/diesel/mart, Chevron/mart, Hardee's, Subway

18(119) FL 77, Chipley, to Panama City, **N**...HOSPITAL, BP/mart(1mi), Exxon/Burger King/TCBY/Pizza Inn/
diesel/mart/atm/24hr, Texaco/Taco Bell/mart, KFC, McDonald's, Waffle House, Wendy's, Budget
Inn(1mi), Chipley Motel(1mi), Day's Inn/rest., Holiday Inn Express, Ramada Inn, Super 8, Wal-Mart SuperCtr/
gas/24hr, Winn-Dixie, **S**...Chevron/Ice Cream/mart, Falling Water SRA

17(111) FL 79, Bonifay, **N**...HOSPITAL, Chevron/mart, Citgo/Simbo's Rest./diesel, Exxon/diesel/mart, Blitch's Rest.,
Hardee's, McDonald's, Pizza Hut, Subway, Waffle House, Best Western, Bonifay Inn, Economy Inn, Tivoli Inn,
AutoParts, Hidden Lakes Camping, **S**...Cypress Springs Camping(8mi)

16(103) rd 279, Caryville, **N**...Dianne's Home Cookin'(1.5mi)

15(96) FL 81, Ponce de Leon, **N**...to Ponce de Leon SRA, camping, **S**...Amoco/diesel/mart/24hr, Exxon/diesel/mart,
Texaco/Subway/Ice Cream/diesel/mart, Ponce de Leon Motel/RV Park, **rest area both lanes,
full(handicapped)facilities, phone, picnic tables, litter barrels, petwalk, 24hr security**

14(84) US 331, De Funiak Springs, **N**...HOSPITAL, EYECARE, PODIATRIST, Amoco/mart, Chevron/mart/24hr,
Exxon/Taco Bell/diesel/mart, Arby's, Burger King, Chinese Rest., Hungry Howie's Pizza/subs, McLain's Steaks,
Old Mill Rest., Pizza Hut, Subway, Waffle House, Comfort Inn, Day's Inn, Ramada Ltd, Sundown Inn, Super 8,
Family$, Food World, Rite Aid, Wal-Mart SuperCtr/gas/24hr, Winn-Dixie, bank, hardware, winery, **S**...Happy
Store/Taco Bell/diesel/mart, Shell/mart, Texaco/mart, Hardee's, KFC, McDonald's, Best Western

13(69) FL 285, Eglin AFB, to Ft Walton Beach, **N**...J&S Plaza/diesel, RaceWay/mart, **S**...Citgo/Subway/diesel/mart/
24hr, Lucky13 Plaza/rest., repair

60mm **rest area wb, full(handicapped)facilities, phone, vending, picnic tables, litter barrels, petwalk, 24hr
security**

58mm **rest area eb, full(handicapped)facilities, phone, vending, picnic tables, litter barrels, petwalk, 24hr
security**

12(56) FL 85, Crestview, Eglin AFB, **N**...HOSPITAL, Amoco/diesel/mart/24hr, Chevron/mart, Exxon/Burger King,
RaceTrac/mart, Shell/diesel/mart, Applebee's, BBQ, Blimpie, Capt D's, Great China, Little Caesar's,
McDonald's, Pizza Hut, Popeye's, Subway, Taco Bell, Western Sizzlin, Budget Host, Econolodge, AutoZone,
Beall's, Big Lots, Discount Parts, Family Vision Ctr, Food World/deli/bakery, PostNet, Scotty's Hardware, Wal-
Mart SuperCtr/24hr, tires, **S**...Amoco/mart, Citgo/diesel/mart, Exxon/Subway/Baskin-Robbins/mart, Arby's,
Cracker Barrel, Hardee's, LaBamba Mexican, Nim's Garden Chinese, Shoney's, Waffle House, Wendy's,
Comfort Inn, Day's Inn, Hampton Inn, Holiday Inn, Jameson Inn, Super 8, AutoValu Parts, Dodge/Plymouth/
Jeep, Ford/Mercury, museum, RV camping

11(46) rd 189, to US 90, Holt, **N**...Chevron(1mi), to Blackwater River SP, Log Lake Rd RV Park, **S**...Amoco/diesel/
mart, River's Edge RV Park

10(32) FL 87, Milton, to Ft Walton Beach, **N**...Rolling Thunder Trkstp/diesel, Waffle House, Holiday Inn Express,
Blackwater River SP, Gulf Pines Camping, **S**...Amoco/mart, RH Express/gas, Comfort Inn, Red Carpet Inn/
diner

31mm **rest area both lanes, full(handicapped)facilities, phone, picnic tables, litter barrels, petwalk, 24hr
security**

E

W

Chipley

Crestview

Milton

Florida Interstate 10

9(28)	rd 89, Milton, **N**...HOSPITAL, Parade Gas/pizza/mart, **S**...Cedar Lakes RV Camping
26mm	Blackwater River
8(25)	rd 191, Bagdad, Milton, **N**...HOSPITAL, Citgo/mart(1mi), **S**...Citgo/Dairy Queen/Stuckey's, TB/diesel/mart, RV camping
7(22)	N FL 281, Avalon Blvd, **N**...Amoco/mart/atm/24hr, Exxon/TasteeFreez/mart, McDonald's, Oval Office Café, golf, 3-4 mi **N**...HOSPITAL, Citgo/mart, Chevrolet, Discount Parts, K-Mart, Nissan, Wal-Mart, **S**...Texaco/diesel/mart, Red Roof Inn, By the Bay RV Park(3mi)
18mm	Escambia Bay
6(16)	US 90, Pensacola, **N**...MEDICAL CARE, BP/mart, **S**...Exxon/mart, Dairy Queen, Ramada Inn/rest., airport
5(14)	FL 291, to US 90, Pensacola, **N**...MEDICAL CARE, Amoco/wash, Chevron/mart/wash, Texaco/diesel/mart, Arby's, Barnhill's Buffet, Burger King, Capt D's, Chinese Rest., Denny's, McDonald's, Montana's Buffet, Subway, Taco Bell, TCBY, Waffle House, Knight's Inn, La Quinta, Motel 6, Shoney's Inn/rest., Villager Inn, AutoZone, Big B Drug/USPO, DelChamps Foods, Eckerd, Food World/24hr, Mailboxes Etc, SuperLube, bank, cleaners/laundry, **S**...HOSPITAL, CHIROPRACTOR, EYECARE, Bennigan's, ChuckeCheese, Cuco's Restaurante, Darryl's, Fazoli's, HoneyBaked Ham, Pizza Hut, Popeye's, Steak&Ale, Waffle House, Wendy's, Whataburger, Fairfield Inn, Hampton Inn, Holiday Inn, Motel 6, Red Roof Inn, Residence Inn, Super 8, Big 10 Tire, Firestone/auto, Goodyear/auto, JC Penney, McRae's, Mr Transmission, Phar-Mor Drug, Sears, SuperLube, TJ Maxx, U-Haul, bank, cinema, mall
4(13)	I-110, to Pensacola, Hist Dist, Islands Nat Seashore
3b a(10)	US 29, Pensacola, **N**...CHIROPRACTOR, Exxon/mart, Williams/diesel/mart, Church's Chicken, Hardee's, Waffle House, **S**...MEDICAL CARE, RaceTrac/mart, Shell/mart, Texaco/diesel/café/mart, BBQ, Burger King, Denny's, IHOP, McDonald's, Ruby Tuesday, Subway, Waffle House, Wendy's, Comfort Inn, Day's Inn, Econolodge, Executive Inn, Holiday Inn Express, Howard Johnson, Hospitality Inn, Landmark Inn, Motel 6, Pensacola Suites, Ramada/rest., Red Roof Inn, Travelodge, Buick, Ford, Hill-Kelly RV Ctr, Jeep, Toyota
2b a(7)	Fl 297, Pine Forest Rd, **N**...Amoco/mart(1mi), Exxon/Krystal/mart, Comfort Inn, Rodeway Inn, Albertson's, Tall Oaks Camping, transmissions, **S**...BP/mart, Citgo/mart, Texaco/diesel/café/mart, BBQ, Burger King, Cracker Barrel, Hardee's, Little Caesar's, McDonald's, Waffle House, Microtel, Ramada Ltd, Sleep Inn, Food World SuperCtr/24hr, Big Lagoon SRA(12mi), museum
1(5)	US 90 A, **N**...Speedway/Subway/diesel/mart, Texaco/mart
4mm	**Welcome Ctr eb, full(handicapped)facilities, info, phone, vending, picnic tables, litter barrels, petwalk, 24hr security**
3mm	weigh sta both lanes
0mm	Florida/Alabama state line

Florida Interstate 75

Exit #(mm)	Services
472mm	Florida/Georgia state line. Motorist callboxes begin sb.
470mm	**Welcome Ctr sb, full(handicapped)facilities, info, phone, vending, picnic tables, litter barrels, petwalk**
87(469)	FL 143, Jennings, **E**...Chevron/mart, Texaco/Ice Cream/mart, Quality Inn/buffet, **W**...Amoco, Exxon/Burger King/diesel/mart, 1990 Jennings House, Jennings Campground
86(462)	FL 6, Jasper, **E**...HOSPITAL, Amoco/Burger King/KrispyKreme/mart/24hr, Exxon/Huddle House/diesel/mart, Raceway/mart, Day's Inn, **W**...Shell/mart, Texaco/mart, Sheffield's Catfish/diesel, Scottish Inn, Suwanee River SP
85(453)	US 129, Jasper, Live Oak, **E**...HOSPITAL, Texaco/DQ/Subway/KrispyKreme/mart/wash, **W**...Shell/Ice Cream/mart, Suwanee Music Park(4mi), to FL Boys Ranch
450mm	weigh sta both lanes
447mm	insp sta both lanes
443mm	Historic Suwanee River
84(441)	to FL 136, White Springs, Live Oak, **E**...BP/mart/24hr, Gate/diesel/mart, Texaco/mart, McDonald's, US Inn/rest., Kelly RV Park, Lee's Camping, to S Foster Ctr, **W**...Amoco/mart, Colonial Inn, Economy Inn
83(436)	I-10, E to Jacksonville, W to Tallahassee
82(429)	US 80, Lake City, to Live Oak, **E**...BP/mart, B&B/Subway/mart, Chevron/diesel/mart/24hr, Exxon, Texaco/diesel/mart, Applebee's, Arby's, BBQ, Blimpie, Burger King, Cracker Barrel, Domino's, Fazoli's, Hardee's, IHOP, KFC, McDonald's, Red Lobster, RJ Gator's Grill, Ryan's, Taco Bell, Texas Roadhouse, Waffle House, Wendy's, A-1 Inn, Day's Inn, Driftwood Motel, Executive Inn, Howard Johnson, Knight's Inn, Rodeway Inn, USA Budget Inn, Villager Lodge, AutoZone, Discount Parts, Eckerd, Food Lion, Ford/Lincoln/Mercury, Goody's, JC Penney, K-Mart, Lowe's Bldg Supply, Publix, Radio Shack, Toyota, Wal-Mart SuperCtr/24hr, In&Out RV Park, cinema, cleaners, mall, tires, **W**...Amoco/LJ Silver/mart, BP, Chevron/mart, Citgo/mart, Shell/mart, Spur/mart, Texaco, Bob Evans, Shoney's, Waffle House, Best Western, Comfort Inn, Econolodge, Hampton Inn, Holiday Inn, Motel 6, Ramada Inn/Pablo's Rest., Red Carpet Inn, Roadmaster Inn, Travelodge, Chrysler/Dodge, FL Sports Hall of Fame

Pensacola

Lake City

E

W

N

S

Florida Interstate 75

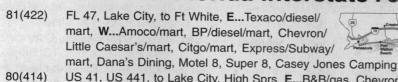

81(422) FL 47, Lake City, to Ft White, **E**...Texaco/diesel/
mart, **W**...Amoco/mart, BP/diesel/mart, Chevron/
Little Caesar's/mart, Citgo/mart, Express/Subway/
mart, Dana's Dining, Motel 8, Super 8, Casey Jones Camping

80(414) US 41, US 441, to Lake City, High Sprs, **E**...B&B/gas, Chevron/diesel/
mart/24hr, Texaco/mart, Travelers Inn, Travelodge, **W**...Amoco/diesel/mart, BP/diesel/
mart, Citgo/repair, Spur/diesel/rest./24hr, Sunshine/gas, Huddle House, Subway,
Diplomat Motel, Econolodge, antiques, to O'Leno SP

413mm **rest areas both lanes, full(handicapped)facilities, phone, vending, picnic tables,
litter barrels, petwalk**

409mm Santa Fe River

79(406) FL 236, to High Sprs, **E**...Chevron/Texaco/fruits/gifts, **W**...High Sprs Camping

78(400) US 441, Alachua, to High Sprs, **E**...BP/mart, Texaco, BBQ, McDonald's, Pizza Hut,
Subway, Waffle House, Comfort Inn, Travelodge, Travelers RV Camping(1mi), **W**...Amoco/
Hardee's/mart, Citgo/diesel/mart, Lil Champ/gas, Mobil/Dairy Queen/Huddle House/mart,
Wayfara Rest., Day's Inn, Ramada Ltd

77(392) FL 222, to Gainesville, **E**...Chevron/mart, Mobil/diesel/LP/mart, BBQ(3mi), Wendy's,
W...VETERINARIAN, Texaco/diesel/mart, Hardee's, Best Western, Buick, Chrysler/Jeep/
Mercedes, Harley-Davidson, auto repair

76(389) FL 26, Gainesville, to Newberry, **E**...HOSPITAL, Chevron/mart, Citgo/mart, Shell/mart/wash, Speed-
way/mart, Texaco/diesel/service, BBQ, Boston Mkt, Burger King, FoodCourt, Heavenly Ham, LJ
Silver, McAlister's Deli, McDonald's, Perkins, Red Lobster, Subway, Wendy's, La Quinta, Belk, Books-
A-Million, Burdine's, Dillard's, JC Penney, Sears/auto, Service Merchandise, bank, cinema, mall, **W**...DENTIST,
VETERINARIAN, BP/mart, Chevron/diesel/mart/repair, Mobil/diesel/LP/mart, Boston Seafood, Cracker Barrel,
Domino's, Hardee's, Pizza Hut, Rocky's Ribs, Shoney's, Taco Bell, Waffle House, Day's Inn, Econolodge,
Fairfield Inn, Holiday Inn, Circuit City, Crafts&Stuff, Discount Parts, Home Depot, JiffyLube, K-Mart, MediaPlay,
PepBoys, Produce Place, Publix, Sports Authority, TJ Maxx, Winn-Dixie, bank, tires/repair

75(385) FL 24, Gainesville, to Archer, **E**...Amoco, Chevron/diesel/mart/24hr, Citgo, Shell, AleHouse, BBQ, Bennigan's,
Bob Evans, Buffalo's Café, Burger King, Capt D's, Checker's, Chili's, Chinese Cuisine, Kenny Rogers, KFC,
LoneStar Steaks, McDonald's, Miami Subs, Olive Garden, OutBack Steaks, Papa John's, Pizza Hut, Steak&Ale,
Subway, Taco Bell, Texas Roadhouse, TGIFriday, Waffle House, Wendy's, Cabot Lodge, Courtyard, Extended
Stay America, Hampton Inn, Motel 6, Ramada Ltd, Red Roof Inn, Super 8, Albertson's, Barnes&Noble,
CarQuest, $Tree, Eckerd, Firestone/auto, GNC, Goody's, JiffyLube, Kinko's, Lowe's Bldg Supply, Michael's,
Midas Muffler, OfficeMax, Old Navy, Publix, PetsMart, Pier 1, Radio Shack, Ross, Santa Fe Parts, Target, Tuffy
Muffler, Wal-Mart, bank, cinema, museum, **W**...Mobil/diesel/mart, Cracker Barrel, Sunshine RV Park, to Bear
Museum

74(384) FL 121, Gainesville, to Williston, **E**...Amoco/repair, Citgo/diesel/mart, Howard Johnson, antiques, **W**...Chevron/
diesel/mart/24hr, Lil Champ/gas, Shell/diesel/mart ChuckWagon Buffet, BrierCliff Inn, Fred Bear Mueum

383mm **rest areas both lanes, full(handiacpped)facilities, phone, vending, picnic tables, litter barrels, petwalk,
24hr security**

73(374) FL 234, Micanopy, **E**...Fina/mart, Shell/ice cream/mart, antiques, fruit, to Paynes Prairie SP, **W**...Amoco/repair,
Citgo/diesel/mart, Texaco/mart, Scottish Inn/rest., Textile Town

72(368) FL 318, Orange Lake, **E**...Petro/Amoco/Wendy's/diesel/mart/24hr, Citgo/BBQ/mart, Mobil, **W**...BP/diesel/repair,
antiques, Calawood RV Camping

71(358) FL 326, **E**...BP/diesel/mart, Speedway/Hardee's/diesel/mart, **W**...Chevron/Gator's/Dairy Queen/diesel/mart

70(354) US 27, Ocala, to Silver Springs, **E**...Amoco/wash/24hr, RaceTrac/mart, SuperTest/diesel, BBQ, Krystal, Quality
Inn, fruits, **W**...BP/mart, Chevron, Shell/diesel/mart, Texaco/diesel/mart, Lorenzo's Pizza, Waffle House, Budget
Host, Day's Inn/café, Howard Johnson, Knight's Inn, Ramada Inn, Red Coach Inn, Skip's Western Outfitter

69(352) FL 40, Ocala, to Silver Springs, **E**...BP/diesel/mart, Chevron/mart, Citgo/Mr Sub/ice cream/mart, RaceTrac/
mart/24hr, McDonald's, Wendy's, Day's Inn/café, Holiday Inn, Scottish Inn, Motor Inn/RV park, fruits,
W...VETERINARIAN, Texaco/mart, Denny's, Waffle House, Comfort Inn, Horne's Motel, Super 8, Holiday TravL
Park, flea mkt

N ↑↓ S

Florida Interstate 75

68(350) FL 200, Ocala, to Hernando, **E...**HOSPITAL, Citgo, RaceTrac/mart, Shell/mart, Texaco/diesel/mart, Arby's, Bella Luna Italian, Blimpie, Baskin-Robbins, Bob Evans, Boston Mkt, Burger King, Checker's, Chik-fil-A, Chili's, Fazoli's, Golden Corral, Hops Grill, Kenny Rogers Roasters, Krystal, LoneStar Steaks, Olive Garden, Outback Steaks, Perkins, Pizza Hut, Quincy's, Red Lobster, Ruby Tuesday, Shell's Rest., Shoney's, Taco Bell, TGIFriday, Wendy's, Hampton Inn, Hilton, La Quinta, Acme Boots, Advance Parts, Barnes&Noble, Belk, Books-A-Million, Burdine's, Chevrolet/Nissan/Mitsubishi, Circuit City, Discount Parts, Eckerd, $General, Goodyear/auto, Home Depot, JC Penney, Kia, K-Mart, Lenscrafters, Lowe's Bldg Supply, Midas Muffler, Pennzoil, PepBoys, PetsMart, Pier 1, Publix, Target, Tire Kingdom, TJ Maxx, ToysRUs, Tuffy Muffler, Walgreen, Wal-Mart, Winn-Dixie, bank, cleaners, fruits, mall, **W...**VETERINARIAN, Amoco/mart, Chevron/Waffle House/mart, Burger King, Cracker Barrel, Dunkin Donuts, Steak'n Shake/24hr, Baymont Inn, Courtyard, Firestone/auto, KOA, Sam's Club, Camper Village RV Park, bank

346mm rest area both lanes, full(handicapped)facilities, phone, vending, picnic tables, litter barrels, petwalk, 24hr security

67(342) FL 484, to Belleview, **E...**Citgo/fruit, Chevron/mart/fruit/24hr, Exxon/mart, BBQ, museums, **W...**Amoco/repair, Pilot/Arby's/Dairy Queen/diesel/mart/24hr, Texaco/diesel/rest./24hr, McDonald's, Waffle House, Ocala Factory Stores, WaterWheel RV Park, fruit

338mm weigh sta both lanes

66(329) FL 44, Wildwood, to Inverness, **E...**Amoco/repair, Gate/Steak'n Shake/diesel/mart, Shell, Texaco/mart/gifts, Burger King, Dairy Queen, McDonald's, Shoney's, Waffle House, Wendy's, KOA, fruits, **W...**BP/76/Pizza Hut/ Subway/diesel/mart/24hr, Citgo/diesel/repair/24hr, Speedway/diesel/mart/24hr, Denny's, KFC, Waffle House, Budget Suites, Day's Inn/rest., Knight's Inn/café, Super 8

65(328) FL TPK(from sb), to Orlando

64(321) FL 470, Lake Panasoffkee, to Sumterville, **E...**Sunshine/diesel/rest./mart/24hr, Coleman Correctional, **W...**Chevron/mart, Lake Panasoffkee Seafood, Motel USA, Turtleback RV Resort

63(315) FL 48, to Bushnell, **E...**BP/mart, Mobil/Dairy Queen, Texaco/mart, KFC/Taco Bell, Best Western, RV Ctr, Red Barn RV Camp, The Oaks Camp(1mi), to Dade Bfd Hist Site, **W...**Citgo/diesel/mart, Shell/mart, McDonald's, Waffle House, flea mkt

62(309) FL 476, to Webster, **E...**Sumter Oaks RV Park(1mi)

307mm rest areas both lanes, full(handicapped)facilities, phone, coffee, vending, picnic tables, litter barrels, petwalk

61(303) US 98, FL 50, to Dade City, **E...**HOSPITAL, CHIROPRACTOR, Amoco, RaceTrac/mart, Shell, Cracker Barrel, Denny's, McDonald's, Toni's Italian, Waffle House, Wendy's, Day's Inn, Winn-Dixie, bank, RV Park, **W...**Mobil/ Subway/LP/mart, Hampton Inn, Holiday Inn, Florida Campland(5mi)

60(294) FL 41, to Dade City, **W...**RV camping

59(287) FL 52, to Dade City, New Port Richey, **E...**HOSPITAL, Flying J/Country Mkt/diesel/LP/24hr, tires, **W...**Shell, Texaco/Blimpie/diesel/mart/24hr, Waffle House

58(280) FL 54, to Land O' Lakes, Zephyrhills, **E...**DENTIST, RaceTrac/mart, Shell/fruit, Texaco/mart/wash, ABC Pizza, BrewMaster Steaks, Burger King, Holiday Café, Subway, Waffle House, Wendy's, Winner's Grill, Discount Parts, MailBoxes Etc, Publix, Walgreen, Winn-Dixie, Saddlebrook RV Resort(1mi), bank, cleaners, **W...**CHIROPRACTOR, VETERINARIAN, Amoco/mart, Citgo/diesel/mart, Circle K/diesel, Cracker Barrel, Denny's, McDonald's, Peacock's Grill, Shoney's, Comfort Inn, Holiday Inn Express, Masters Inn, RV camping, cleaners

278mm rest areas both lanes, full(handicapped)facilities, phone, vending, picnic tables, litter barrels, petwalk, 24hr security

57(275) I-275(from sb), to Tampa, St Petersburg

56(271) FL 581, Bruce B Downs Blvd, no facilities

55(267) FL 582A, Fletcher Ave, Tampa, **W...**HOSPITAL, Shell/Subway/mart/wash, Courtyard, Extended Stay America, Hampton Inn Suites, Residence Inn, Sleep Inn

54(265) FL 582, Fowler Ave, Temple Terrace, **E...**Happy Traveler RV Park(1mi), **W...**Shoney's Inn/rest.(1mi), to Busch Gardens

53(262) I-4, W to Tampa, E to Orlando

52ba(260) FL 574, Tampa, to Mango, **E...**AutoNation USA, **W...**Amoco/mart, Rainbow Gas, Subway, Sumner Suites

Ocala · Wildwood · Tampa

N

S

Florida Interstate 75

51(257)   FL 60, Brandon, **E**...HOSPITAL, DENTIST, CHIROPRACTOR, Amoco/mart, Mobil, Texaco/mart, AleHouse, Bennigan's, BuddyFreddy's Diner, Chesapeake Bagels, Chili's, Domino's, Grady's Grill, Macaroni Grill, Olive Garden, Outback Steaks, Red Lobster, TGIFriday, Waffle House, Holiday Inn Express, HomeStead Village, La Quinta, Aamco, Barnes&Noble, Bed Bath &Beyond, Best Buy, Burdine's, Chrysler/Plymouth/Dodge, Dillard's, JC Penney, Jo-Ann Fabrics, KidsRUs, Kinko's, K-Mart, Lenscrafters, Marshall's, MensWearhouse, Michael's, MobiLube/wash, PakMail, Pearle Vision, PepBoys, Pier 1, Sam's Club, Sears/auto, Staples, TJ Maxx, ToysRUs, Tuffy Muffler, bank, cinema, cleaners, mall, **W**...Citgo, Shell, BBQ, Bob Evans, Hooter's, McDonald's, Sweet Tomato, Wendy's, Baymont Inn, Comfort Inn, Day's Inn, Acme Boots, Chevrolet/Olds, Circuit City, Dodge, Home Depot, Honda, Mitsubishi/Suzuki/Yamaha, Nissan, Office Depot, Pontiac/GMC, Sports Authority, Waccamaw

50(256)   FL 618 W(toll), to Tampa, no facilities

49(254)   US 301, Riverview, no facilities

48(250)   Gibsonton, Riverview, **E**...Amoco(1mi), Burger King, McDonald's(1mi), Alafia River RV Resort, Hidden River RV Resort(4mi), **W**...Amoco(1mi)

47(246)   FL 672, Big Bend Rd, Apollo Beach, **1 mi W**...BP/mart, Ramada Inn/rest.

46(241)   FL 674, Ruskin, Sun City Ctr, **E**...Chevron/mart/24hr, Burger King, Checker's, Shoney's, Comfort Inn, Sun City Ctr Inn/rest., Food Lion, MobiLube, Radio Shack, Walgreen, Wal-Mart, SunLake RV Resort(1mi), bank, hardware, to Little Manatee River SP, **W**...Citgo/Circle K/diesel, Exxon/mart, KFC, Maggie's Rest., McDonald's, Subway, Holiday Inn Express, Eckerd, NAPA, Publix

238mm   **rest area nb, full(handicapped)facilities, phone, vending, picnic tables, litter barrels, petwalk, 24hr security**

45(230)   FL 683, Moccasin Wallow Rd, to Parrish, **E**...Fiesta Grove Camping

44(229)   I-275 N, to St Petersburg

43(225)   US 301, Ellenton, to Bradenton, **E**...DENTIST, EYECARE, Amoco/diesel/LP, Chevron/mart/wash/24hr, RaceTrac/McDonald's, Shell/mart/wash, Checker's, Food Court, Wendy's, K-Mart, Publix, Walgreen, USPO, Ellenton Garden Camping(1mi), Gulf Coast Factory Shops/famous brands, cleaners, **W**...Speedway/Subway/diesel/mart, Crabtrap II Rest., Waffle House, Best Western, Day's Inn(3mi), Shoney's Inn/rest.

42ba(221)   FL 64, Bradenton, to Zolfo Springs, **E**...Lake Manatee SRA, **W**...HOSPITAL, Amoco/diesel, Chevron/mart/wash, Citgo/diesel/mart/24hr, RaceTrac/mart, 76/Circle K/diesel, Shell/mart, Burger King, Cracker Barrel, Denny's, McDonald's, Subway, Waffle House, Comfort Inn, Day's Inn, Econolodge, Holiday Inn, Knight's Inn, Motel 6, Winter Quarters RV Resort, hardware

41ba(217)   FL 70, Bradenton, to Arcadia, **W**...DENTIST, EYECARE, Amoco/diesel/LP, Shell/mart/wash, Circle K, China Village, Denny's, McDonald's, Final Stop Diner, Hungry Howie's, Subway, GNC, PakMail, Publix, HorseShoe RV Park, Pleasant Lake RV Resort, bank, cleaners

40(214)   University Parkway, to Sarasota, **W**...Sarasota Outlet Ctr/famous brands/food court, **3-6 mi W**...Chevron/24hr, Applebee's, Burger King, KFC, McDonald's, Courtyard, Ringling Museum, airport, dogtrack

39(210)   FL 780, Fruitville Rd, Sarasota, **1 mi E**...Texaco/diesel/mart, Sun&Fun RV Park, golf, radiator repair, **W**...Amoco/diesel/LP/mart/wash, BP/diesel/mart, Chevron/mart/wash/24hr, Mobil/diesel, Applebee's, Checker's, Chick-fil-A, KFC, Longhorn Steaks, McDonald's, Subway, Taco Bell, Wings'n Weenies, Hyatt Hotel(6mi), Wellesley Inn(6mi), Eckerd, GNC, 1-Hr Photo, Publix, Radio Shack, Target, cinema

38(208)   FL 758, Sarasota, **W**...MEDICAL CARE, CHIROPRACTOR, VETERINARIAN, DENTIST, Mobil/Subway/diesel/mart/repair, Texaco/Blimpie/mart, Arby's, BBQ, Checker's, Chili's, Chinese Wokery, McDonald's, Pizza Hut, Pizza World, Sarasota Ale House, South Beach Café, Sugar&Spice Rest., Stockyard Steaks, Taco Bell, Wings'n Weenies(2mi), Hampton Inn, Cash'n Carry, Home Depot, K-Mart, PakMail, Publix, Radio Shack, Sports Authority, Waccamaw, Walgreen, Wal-Mart, bank, cleaners

37(206)   FL 72, Sarasota, to Arcadia, **E**...Myakka River SP(9mi), **W**...VETERINARIAN, Citgo/7-11/diesel, Mobil/mart/wash, Burger King, Subway, Waffle House, Wendy's, Comfort Inn, Holiday Inn(5mi), Ramada Ltd, Beach Club RV Resort(6mi), bank

36(201)   FL 681 S(from sb), to Venice, Osprey, same as 35a

35a(196)   Nokomis, Laurel, **E**...Stay'n Play RV Park(1mi), **W**...Royal Coachmen RV Park(2mi), Scherer SP(6mi)

N

S

Florida Interstate 75

35(193) Jaracanda Blvd, Venice, **W**...HOSPITAL, Citgo, Hess/Blimpie/Pizza Hut/diesel/mart, RaceTrac/mart, Cracker Barrel, McDonald's, Best Western, Day's Inn, Veranda Inn

34(191) Englewood Rd, **W**...Ramblers Rest Resort(3mi), Venice Campground(1mi)

33(182) Sumter Blvd, to North Port, **W**...gas, food, lodging

32(179) Toledo Blade Blvd, North Port, no facilities

31(170) FL 765, Port Charlotte, to Arcadia, **E**...Citgo/7-11/diesel/mart, Hampton Inn, Lettuce Lake Camping(7mi), Yogi Bear Camping(4mi), **W**...HOSPITAL, Hess/diesel/mart, Mobil/Blimpie/Dunkin Donuts/diesel/mart, 76/Circle K, Texaco/DQ/diesel/mart, Burger King, Cracker Barrel, McDonald's, Taco Bell, Waffle House, Zorba's Pizza, Eckerd, Publix, Winn-Dixie

30(167) FL 776, Port Charlotte, **2 mi W**...Econolodge, Quality Inn

29(164) US 17, Punta Gorda, Arcadia, **E**...Shell/diesel/mart/24hr, KOA(2mi), **W**...HOSPITAL, VETERINARIAN, EYECARE, 76/Circle K, Fisherman's Village Rest., Karl Ehmer Rest., Best Western(2mi), Howard Johnson(2mi), SeaCove Motel(2mi), auto/tire repair

28(161) FL 768, Punta Gorda, **E...rest area both lanes, full(handicapped)facilities, phone, vending, picnic tables, litter barrels, petwalk, 24hr security,** KOA, **W**...Amoco/Subway/diesel/mart, Hess/Dunkin Donuts/diesel/mart, Shell, Speedway/Blimpie/Church's Chicken/diesel/mart, Burger King, Dairy Queen, Pizza Hut, Shoney's, Taco Bell, McDonald's, Waffle House, Wendy's, Day's Inn, Motel 6, Alligator RV Park(2mi), Gulfview RV Park(2mi)

160mm weigh sta both lanes

27(158) FL 762, **W**...Texaco/diesel/mart, to RV camping

26(143) FL 78, N Ft Myers, to Cape Coral, **E**...Citgo/diesel/mart, Upriver RV Park(1mi), **W**...RaceTrac/diesel/mart/24hr, Cactus Motel, Sable Motel, Pioneer Village RV Resort

25(140) FL 80, Palm Bch Blvd, Ft Myers, **E**...Amoco/mart, Citgo/mart/wash, Cracker Barrel, Waffle House, **W**...MEDICAL CARE, DENTIST, Citgo/7-11, Hess/diesel/mart, RaceTrac/mart, BBQ, Chinese Rest., Dairy Queen, Hardee's, Juicy Lucy's Burgers, KFC(2mi), Perkins, Pizza Hut, Subway, Taco Bell, Holiday Inn(6mi), Sheraton(5mi), Big Lots, $General, Eckerd, Hayloft Western Store, Publix, Radio Shack, RV Ctr, auto/tire repair, cleaners, hardware

24(138) Luckett Rd, Ft Myers, **W**...Pilot/Subway/Grandma's Kitchen/diesel/mart/24hr, Camping World RV Service/supplies, Gulf Coast RV, Lazy J's RV Camping

23(137) FL 82, Ft Myers, to Lehigh Acres, **W**...RaceTrac/diesel, Speedway/diesel

22(135) FL 884, Colonial Blvd, Ft Myers, **1-3 mi W**...HOSPITAL, RaceTrac/mart, Baymont Inn, Courtyard, Residence Inn, Wellesley Inn, flea mkt

21(131) to Cape Coral, **E...rest area both lanes, full(handicapped)facilities, phone, vending, picnic tables, litter barrels, petwalk, 24hr security,** airport, **W**...HOSPITAL, Citgo/7-11/24hr, Hess/mart, RaceTrac/mart/24hr, Shell/mart/wash/24hr, Arby's, Burger King, Denny's, McDonald's, Taco Bell, Waffle House, Wendy's, Comfort Suites, Hampton Inn, Holiday Inn, Homewood Suites, Shoney's Inn/rest., Sleep Inn

20(127) Alico Rd, San Carlos Park, **1-3 mi W**...Hess Gas, Site Gas, Wendy's, Subway

19(123) FL 850, Corkscrew Rd, Estero, **W**...Koreshan St Hist Site, Woodsmoke RV Park(4mi)

18(116) Bonita Bch Rd, Bonita Springs, **1-3 mi W**...Amoco/McDonald's/mart/24hr, Hess/diesel/mart/24hr, Shell/mart/24hr, Waffle House, Wendy's, Perkins, Comfort Inn, Day's Inn, Hampton Inn, RV camping, to gulf beaches

17(111) FL 846, Immokalee Rd, Naples Park, **E**...Mobil, **W**...HOSPITAL, Shell/mart, Circle K, Wendy's(2mi), Fairways Motel(2mi), Vanderbilt Inn(5mi), Publix, to Wiggins SP

16(107) FL 896, Pinebridge Rd, Naples, **E**...MEDICAL CARE, VETERINARIAN, Mobil/McDonald's/diesel/mart/24hr, Bruegger's Bagels, Ben Franklin, Publix, Walgreen, cleaners, **W**...Chevron/Subway/diesel/mart/24hr, Shell/diesel/mart/24hr, Applebee's, Burger King, Waffle House, Knight's Inn, Registry Resort(4mi)

15(101) FL 84, **2 mi E**...BP, Citgo/diesel/lube, Hess/mart, Circle K/76/24hr, BBQ, Chinese Rest., KFC, McDonald's, Pizza Hut, Quality Inn, Discount Parts, Golden Gate Drug, hardware, **W**...Amoco/mart, Mobil/Subway/Dunkin Donuts/diesel/mart/24hr, Shell/diesel/mart/24hr, Circle K/76/24hr, Burger King, Checker's, Cracker Barrel, McDonald's, Waffle House, Baymont Inn, Comfort Inn, Holiday Inn Express, Super 8, Mazda, RV camping

100mm toll plaza both lanes

14a(80) FL 29, to Everglade City, Immokalee, Big Cypress Nat Preserve, no facilities

71mm Big Cypress Nat Preserve, hiking, no security

63mm **W...rest area both lanes, full(handicapped)facilities, phone, vending, picnic tables, litter barrels, petwalk, 24hr security**

14(49) to Indian Reservation, **E**...Shell/diesel/mart

38mm parking area wb

31mm parking area wb

25mm toll plaza both lanes, motorist callboxes begin/end

Ft Myers

Naples

N

S

Florida Interstate 75

13b a(23) US 27, Miami, South Bay, no facilities

12(22) Arvida Pkwy, no facilities

11(19) FL 84 W, Indian Trace, **W**...Mobil/diesel/mart/wash, Antonello's Italian, Café Café, Papa John's Pizza, Mail-it-Here, cleaners

10(17) I-595 E, FL 869(toll), Sawgrass Expsway

8(15) Arvida Pkwy, Weston, Bonaventure, **W**...MEDICAL CARE, Miyoko Sushi, Pampered Chef Café, Pastability Ristorante, Sporting Brew Grill, Wendy's, PillBox Drug, USPO, cleaners, shops

7b a(13) Griffin Rd, **E**...CHIROPRACTOR, DENTIST, MEDICAL CARE, VETERINARIAN, Amoco/mart/wash/24hr, Shell/diesel/mart/wash, Dairy Queen, Subway, Waffle House, Goodyear/auto, Publix

6b a(10) Sheridan St, **E**...HOSPITAL(3mi)

5b a(9) FL 820, Pines Blvd, Hollywood Blvd, **E**...Macaroni Grill, Wendy's, AutoNation USA, BJ's Wholesale Club, JC Penney, Walgreen, CB Smith Park

4b a(7) Miramar Pkwy, no facilities

3b(5) to FL 821(from sb), FL TPK(toll)

3a(4.5) NW 186th, Miami Gardens Dr, **E**...DENTIST, CHIROPRACTOR, PODIATRIST, VETERINARIAN, Amoco/mart/wash/24hr, Chevron/mart/wash/24hr, Subway, Eckerd, GNC, Pak Mail, Publix, cleaners/laundry

2(2) NW 138th, Graham Dairy Rd, **W**...HOSPITAL, Mobil/mart/wash, Dairy Queen, DonTike Chinese, McDonald's, GNC, Publix, cleaners

1b a(0) I-75 begins/ends on FL 826. Multiple services on FL 826.

Florida Interstate 95

Exit #(mm)Services

382mm St Marys River, Florida/Georgia state line, motorist callboxes begin/end.

381mm weigh sta both lanes

130(379) US 17, to Yulee, Kingsland, **E**...Hance's RV Camping, **W**...Amoco/mart/24hr, Shell/mart/24hr, Po Folks, Day's Inn, Holiday Inn Express

378mm Welcome Ctr sb, full(handicapped)facilities, phone, vending, picnic tables, litter barrels, petwalk, 24hr security

129(373) FL 200, FL A1A, to Yulee, Callahan, Fernandina Bch, **E**...Chevron/service, Citgo/Dairy Queen/Stuckey's/mart, Shell/service, McDonald's, Taco Bell, Nassau Holiday Motel, to Ft Clinch SP, **W**...Amoco/diesel/mart, BP/diesel/mart, Citgo/Wayfara Rest./mart, Exxon/diesel/mart, Sunshine Foods/rest., Ten-4/diesel/mart, Waffle House

128(366) Pecan Pk Rd, **E**...BP/diesel/mart/ice cream, Citgo/diesel/mart, **W**...Flea&Farmer's Mkt

127(363) Duval Rd, **E**...Mobil/diesel/mart, **W**...Amoco/Subway/Dunkin Donuts/mart, BP/mart, Chevron/Taco Bell/A&W/mart, Texaco/diesel/mart, Denny's, Waffle House, Admiral Benbow Inn, Airport Hotel, Courtyard, Day's Inn, Hampton Inn, Holiday Inn, Microtel, Red Roof Inn, Valu Lodge, RV Ctr, auto rentals, to airport

126(361) I-295 S, Jacksonville, FL 9A, to Blount Island

125(360) FL 104, Dunn Ave, Busch Dr, **E**...Gate/diesel/mart, Applebee's, Hardee's, Waffle House, Admiral Benbow Inn, NAPA, Sam's Club, **W**...DENTIST, Amoco/mart, BP/wash, Hess/mart, Shell/mart/wash, Texaco, Arby's, BBQ, Burger King, Capt D's, Chinese Rest., Dunkin Donuts, KFC, Krystal, Lee's Chicken, LJ Silver, McDonald's, Pizza Hut, Popeye's, Quincy's, Rally's, Shoney's, Taco Bell, Wendy's, Best Western, La Quinta, Motel 6, Red Carpet Inn, Super 8, Big Lots, Discount Parts, Eckerd, Family$, PepBoys, Publix, Radio Shack, Winn Dixie, bank

124ba(358) FL 105, Broward Rd, Heckscher Dr, **E**...zoo, **W**...Day's Inn

357mm Trout River

123(356) FL 111, Edgewood Ave, **W**...Amoco/repair, Citgo/mart

122(356.5) FL 115, Lem Turner Rd, **E**...Hardee's, auto repair, **W**...Amoco/mart/24hr, Hess/mart, Shell/repair, Texaco/diesel/mart, Burger King, Chinese Rest., Krystal, Popeye's, Rally's, Taco Bell, Discount Parts, JiffyLube, Midas Muffler, 1 Stop Parts, Precision Tune, Sav-A-Lot Foods, Walgreen, flea mkt, tires

121(355) Golfair Blvd, **E**...Shell/mart, Texaco/mart, carwash, flea mkt, tires, **W**...Amoco/repair/24hr, Exxon/mart, RaceTrac/mart, Valu Lodge

120ba(354) US 1, 20th St, to Jacksonville, to AmTrak

119(353) FL 114, to 8th St, **E**...HOSPITAL, Amoco, McDonald's

Florida Interstate 95

Jacksonville

118(352) US 23 N, Kings Rd, downtown

117(351) US 90A, Union St, Sports Complex, downtown

116 Church St, Myrtle Ave, Forsythe St, downtown

115 Monroe St(from nb), downtown

114 Myrtle Ave(from nb), downtown

113 Stockton Rd, HOSPITAL, downtown

111 I-10 W, to Tallahassee

110 College St, HOSPITAL, to downtown, no facilities

351mm St Johns River

107(350) Prudential Dr, Main St, Riverside Ave(from nb), to downtown

106(349) US 90 E(from sb), downtown, to beaches, W...Super 8

105(348) US 1 S(from sb), Philips Hwy, downtown, W...Scottish Inn, Super 8, Cadillac, Chevrolet, VW/Volvo

104(347) US 1A, FL 126, Emerson St, E...Chevron/mart, Shell/mart, Subway, W...Amoco, Gate Gas, McDonald's, Comfort Inn, Howard Johnson, mall

103ba(346) FL 109, University Blvd, E...HOSPITAL, Amoco/mart, Citgo, Hess/diesel/mart, Texaco/mart, Captain D's, Chinese Rest., Dairy Queen, El Potro Mexican, Firehouse Subs, Hungry Howie's Pizza/sub, Krystal, Pizza Hut, Rally's, Firestone/auto, Goodyear/auto, JiffyLube, Meineke Muffler, Midas Muffler, 1 Stop Parts, Tire Kingdom, Winn-Dixie, bank, laundry, W...BP/mart, RaceTrac/mart, BBQ, Buckingham Grill, Burger King, Dunkin Donuts, IHOP, OceanBay Seafood, Ryan's, Shoney's, Taco Bell, Waffle House, Wendy's, Comfort Lodge, Econolodge, Ramada Inn, Red Carpet Inn, Chrysler/Plymouth, auto repair

102(345) FL 109, University Blvd(from nb), same as 103

101(343) FL 202, Butler Blvd, E...HOSPITAL, Doubletree Inn, Economy Inn, Hampton Inn, Marriott, Quality Inn/café, Ramada Inn, W...Shell/mart, Texaco/diesel/mart/wash, Cracker Barrel, Hardee's, Waffle House, Courtyard, Extended Stay America, La Quinta, Microtel, Red Roof Inn, AutoNation USA

100(341) FL 152, E...DENTIST, Amoco/mart, Chevron/mart, Gate/diesel/mart, Shell/mart, Texaco/mart/wash, Applebee's, Arby's, Chili's, Domino's, Dunkin Donuts, Hardee's, India's Rest., Joseph's Italian, Kenny Rogers Roaster, Lee's Chicken, Manhattan Bagels, Quincy's, Subway, TGIFriday, AmeriSuites, Embassy Suites, Fairfield Inn, Holiday Inn, HomeStead Village, Food Lion, Kinko's, Publix, bank, W...Chevron/mart, Exxon/diesel, Shell/mart/wash, Texaco/diesel/mart, Bennigan's, Bombay Bicycle Club, Burger King, Chinese Rest., Denny's, IHOP, KFC, Larry's Subs, McDonald's, Miami Subs, Pagoda Chinese, Pizza Hut, Red Lobster, Renna's Pizza, Rio Bravo Cantina, Shoney's, Steak&Ale, Taco Bell, Wendy's, Best Inn, Comfort Inn, Homewood Suites, Motel 6, Quality Inn, Residence Inn, Travelodge, Goodyear/auto, JiffyLube, Office Depot, bank, cinema

99(339) FL 115, Southside Blvd(from nb), no facilities

98(338) US 1, Philips Hwy, E...Arby's, Baskin-Robbins, Burger King, McDonald's, Olive Garden, Taco Bell, Waffle House, Dillard's, Ford, Gayfer's, Hobby Superstore, JC Penney, Toyota, mall

97(337) I-295 N, to Orange Park

331mm rest area both lanes, full(handicapped)facilities, phone, vending, picnic tables, litter barrels, petwalk, 24hr security

96(329) FL 210, E...Citgo/fruit, Exxon, 76/Shell/diesel/rest, Speedway/Hardee's/diesel/mart/24hr, Hardee's, Waffle House, KOA, W...Amoco/mart, Chevron/diesel, fireworks

95A(323) International Golf Pkwy, E...Shell/mart, W...World Golf Village

95(317) FL 16, St Augustine, Green Cove Sprgs, E...Amoco/mart, BP/Pizza Hut/mart, Citgo/mart, Shell, Speedway/diesel/mart, Burger King, McDonald's, Ocean Palace Chinese, Waffle House, GuestHouse Inn, Holiday Inn Express, W...Exxon, Texaco/diesel/mart, BBQ, Cracker Barrel, Denny's, FoodCourt, KFC, Shoney's, Wendy's, Best Western, Day's Inn, Econolodge, Scottish Inn, Super 8, St Augustine Outlet Ctr/famous brands

Palm Coast

94(310) FL 207, St Augustine, E...HOSPITAL, Chevron/A&W/mart, Indian River Fruit/gas, Indian Forest RV Park(2mi), KOA(7mi), St Johns RV Park, flea mkt, to Anastasia SP, W...BP/mart/repair, Mobil/diesel, Texaco/mart, Comfort Inn

93(305) FL 206, to Hastings, Crescent Beach, E...Texaco/diesel/mart, to Ft Matanzas NM, W...auto/RV/truck repair

302.5mm rest areas both lanes, full(handicapped)facilities, phone, vending, picnic tables, litter barrels, petwalk, 24hr security

92(297) US 1, to St Augustine, E...BP/mart, Citgo/gifts/fruit/mart, Indian River Fuit/gas, Texaco/diesel/mart, to Faver-Dykes SP, W...Charlie T's/diesel/rest./24hr, Shell/Dairy Queen/Stuckey's/mart, Waffle House, RV camping

91C(289) FL A1A(toll br), to Palm Coast, E...VETERINARIAN, CHIROPRACTOR, BP/mart, Exxon/diesel, Shell/Dunkin Donuts/mart, Cracker Barrel, Denny's, Jasper's Café, KFC, McDonald's, Pizza Hut, Shoney's, Wendy's, Hampton Inn, Sleep Inn, Eckerd, 1-Hr Photo, Publix, bank, carwash, hardware, W...BP, Chevron/mart, Texaco, BBQ, Chinese Rest., Hardee's, Perkins, Steak'n Shake/24hr, Subway, Taco Bell, TCBY, Discount Parts, K-Mart, Walgreen, Wal-Mart SuperCtr/24hr, Winn-Dixie, bank

N

S

Florida Interstate 95

286mm weigh sta both lanes, phone

91(285) FL 100, to Bunnell, Flagler Beach,
E...VETERINARIAN, BP/mart, Chevron/diesel/mart/
24hr, Burger King, McDonald's, Oriental Garden,
Subway, Taco Bell, Winn-Dixie/drugs, cleaners, **Oceanshore RV
Wash(100E to A1A, N to 23rd)**, **W**...HOSPITAL, Chrysler

90(278) Old Dixie Hwy, **E**...Publix, to Tomoke SP, RV camping, **W**...Texaco/diesel/mart, Best
Western/rest., Holiday TravL Park

89(273) US 1, **E**...Amoco/mart, BP/mart, Citgo/fruit/mart, Fina/diesel/mart, Mobil/Wendy's/diesel/
mart, Denny's/24hr, McDonald's, Waffle House, Comfort Inn, Day's Inn(2mi), RV Ctr,
W...Exxon/Burger King/mart, Dairy Queen, Budget Host, Day's Inn, Ramada Inn/rest.,
Scottish Inn, Super 8, RV camping

88(267) FL 40, Ormond Beach, **E**...HOSPITAL, DENTIST, BP/Blimpie/mart, Citgo/Taco Bell/mart,
Applebee's, Boston Mkt, Chili's, Chinese Rest., Denny's, Papa John's, Schlotsky's, Steak'n
Shake, Waffle House, Sleep Inn, Beall's, K-Mart, Publix, Wal-Mart SuperCtr, USPO,
XpressLube, cleaners, **W**...MEDICAL CARE, Amoco/Burger King/TCBY/mart, Chevron/mart,
Mobil/diesel/mart, Cracker Barrel

87C(265) LPGA Blvd, Holly Hill, Daytona Beach, no facilities

87(261) US 92, to De Land, Daytona Beach, **E**...Amoco/diesel/mart, Citgo/7-11, Exxon, Hess/Blimpie/
Pizza Hut/diesel/mart, Mobil, RaceTrac/mart, Shell/mart/wash, Bob Evans, Burger King, Carrabba's
Italian, Checker's, Chick-fil-A, Cracker Barrel, Fazoli's, Friendly's, Hooters, Krystal, Olive Garden,
Pizzaria Uno, Red Lobster, Rio Bravo Cantina, Ruby Tuesday, Shoney's, S&S Cafeteria, Subway, Taco
Bell, Waffle House, Holiday Inn, La Quinta, Ramada Inn, Travelodge, Barnes&Noble, Best Buy,
Dillard's, Dodge/Kia, Gayfer's, Home Depot, JC Penney, Marshall's, 1-Hr Photo, PepBoys, PetsMart, Sears/auto,
Target, ToysRUs, Walgreen, cinema, funpark, golf, mall, to Daytona Racetrack, **W**...BP/diesel/mart, Denny's,
IHOP, McDonald's, Day's Inn, Super 8, RV camping, museum

86ba(258) I-4, to Orlando, FL 400, to S Daytona, **E**...BP/diesel/mart

85(256) FL 421, to Port Orange, **E**...Amoco/mart, BBQ, Boston Mkt(2mi), Denny's, Spruce Creek Pizza, Day's Inn, Palm
Plaza Motel(5mi), **W**...CHIROPRACTOR, Amoco, Citgo/7-11, Shell/mart, McDonald's/playplace, Subway, Publix,
XpressLube

255mm **rest area sb, full(handicapped)facilities, phone, vending, picnic tables, litter barrels, petwalk**

253mm **rest area nb, full(handicapped)facilities, phone, vending, picnic tables, litter barrels, petwalk**

84ba(248) FL 44, New Smyrna Beach, to De Land, **E**...HOSPITAL, Shell/diesel/mart/fruit, KOA, **W**...Chevron/diesel/mart

83(244) FL 442, to Edgewater, **E**...Chevron/mart

82(233) FL 5a, Scottsmoor, **E**...BP/Dairy Queen/Stuckey's/diesel/mart, RV camping, **W**...Citgo/Dairy Queen/Stuckey's/
mart

227mm **rest area sb, full(handicapped)facilities, phone, vending, picnic tables, litter barrels, petwalk, 24hr
security**

225mm **rest area nb, full(handicapped)facilities, phone, vending, picnic tables, litter barrels, petwalk, 24hr
security**

81(224) FL 46, Mims, **E**...HOSPITAL, Fina(1mi), McDonald's, **W**...Amoco/mart, Mobil, Shell/deli, KOA/LP, golf/café

80(220) FL 406, Titusville, **E**...HOSPITAL, EYE CLINIC, DENTIST, BP/diesel/mart, Shell/Piccadilly's/subs/mart, KFC,
McDonald's, Subway, Wendy's, Day's Inn, Travelodge/Chinese Rest., Discount Parts, Eckerd, Fabric King,
Publix, Walgreen, auto repair, hardware, **W**...Texaco/diesel/mart

79(215) FL 50, Titusville, to Orlando, **E**...Amoco/diesel/mart, Chevron/mart/fruit, Circle K/gas, Shell/mart, Texaco/diesel/
mart, Denny's, Durango Steaks, McDonald's, Shoney's, Waffle House, Wendy's, Best Western, Ramada Inn,
Lowe's Bldg Supply, Wal-Mart SuperCtr/24hr, to Kennedy Space Ctr, **W**...Cracker Barrel, Day's Inn, Luck'sWay
Inn

78(211) FL 407, to FL 528(toll), no facilities

209mm **rest area both lanes, picnic tables, litter barrels, no restrooms, no security**

77ba(205) FL 528(wb toll), to Cape Canaveral, City Point, no facilities

76(203) FL 524, Cocoa, **E**...Citgo/mart, **W**...Amoco/diesel, Day's Inn, Ramada Inn, Super 8

75(202) FL 520, Cocoa, to Cocoa Beach, **E**...HOSPITAL, BP/diesel/mart, Fina, Speedway/Subway/diesel/mart, IHOP,
Waffle House, Best Western, Budget Inn, truck repair, **W**...Chevron/diesel/mart, McDonald's, Ramada Inn(1mi),
Super 8(1mi)

74(195) Fiske Blvd, **E**...Citgo/7-11, RV camping

N

S

Daytona

Florida Interstate 95

M e l b o u r n e

73(191) FL 509, to Satellite Beach, **E**...Citgo/7-11, Shell/mart, Denny's, McDonald's, Miami Subs, Perkins, Wendy's, Comfort Inn, Pennzoil, auto repair, bank, to Patrick AFB, **W**...Texaco/LJSilver/diesel/mart, Burger King, Cracker Barrel, Baymont Inn

189mm **rest area both lanes, picnic tables, litter barrels, no restrooms, no security**

72(183) FL 518, Melbourne, Indian Harbour Beach, **E**...Amoco, Citgo/7-11, RaceTrac/24hr, museum, **W**...Flea Mkt

71(181) US 192, to Melbourne, **E**...HOSPITAL, VETERINARIAN, BP/diesel/mart, Circle K/gas, Citgo/7-11, Mobil, RaceTrac/mart, Speedway/diesel/mart, Dairy Treats, Denny's, IHOP, Olive Garden, Steak'n Shake/24hr, Waffle House, Courtyard, Hampton Inn, Holiday Inn, Shoney's Inn/rest., Travelodge, York Inn, Sam's Club, Saturn, antiques, **W**...Texaco/diesel/mart, Western Store

70a(176) FL 516a, to Palm Bay, **E**...CHIROPRACTOR, VETERINARIAN, BP/diesel/mart, Fina/diesel/mart, Shell/mart/ wash, Applebee's, Boston Mkt, Denny's, Taco Bell, Ramada Inn, Albertson's, cleaners, **W**...Citgo/7-11, Fina, Mobil Lube, Sav-A-Ton/diesel/mart

70(171) FL 514, to Palm Bay, **E**...MEDICAL CARE, DENTIST, CHIROPRACTOR, Cumberland/gas, Speedway/diesel/ mart, Chinese Rest., Firestone/auto, Ford, Meineke Muffler, MobiLube, bank, **W**...VETERINARIAN, CHIRO- PRACTOR, Amoco/mart, Shell/mart, Speedway/diesel/mart, Arby's, BBQ, IHOP, McDonald's, Subway, Taco Bell, Waffle House, Wendy's, Motel 6, $General, Discount Parts, Eckerd, Goodyear/auto, Publix, Russell Stover, Tire Kingdom, USPO, bank

168.5mm **rest areas both lanes, full(handicapped)facilities, phone, vending, picnic tables, litter barrels, petwalk, 24hr security, motorist aid callboxes begin nb/end sb**

69(156) FL 512, to Sebastian, Fellsmere, **E**...HOSPITAL, Amoco/McDonald's/mart, BP, Citgo/Dairy Queen/Stuckey's/ diesel/mart, Mobil/LP(2mi), Subway(2mi), Sunshine TravL Park, KOA(9mi)

68(147) FL 60, Osceola Blvd, **E**...HOSPITAL, Amoco/76/diesel/rest./24hr, Chevron/mart/repair, Citgo/diesel/mart/24hr, Mobil, Shell/mart, Marotta's Pizza, Waffle House, Wendy's, Best Western, Day's Inn/rest., Howard Johnson, Super 8, Boot Country, NAPA, **W**...Cracker Barrel, McDonald's, Hampton Inn, Holiday Inn Express, Horizon Outlet Ctr/famous brands

F t

P i e r c e

67(138) FL 614, Indrio Rd, **E**...Oceanographic Institute, citrus, **W**...airport, citrus

133mm **rest areas both lanes, full(handicapped)facilities, phone, vending, picnic tables, litter barrels, petwalk, 24hr security**

66ba(131) FL 68, Orange Ave, **E**...Big John's Western, NAPA, auto/truck repair, batteries/tires, hardware, to Ft Pierce SP

65(129) FL 70, to Okeechobee, **E**...HOSPITAL, Citgo/mart, Hess/diesel/mart, Applebee's, BBQ, Golden Corral, Piccadilly's, Discount Parts, Belk, Firestone/auto, Goodyear/auto/truck, Home Depot, Rex Audio/video, Wal-Mart SuperCtr/24hr, Neill's Farm U-Pick, bank, mall, **W**...Amoco/diesel/mart/24hr, Chevron/mart/wash, Exxon/fruit, Mobil/Subway/Dunkin Donuts/mart, Pilot/Arby's/diesel/mart/24hr, Texaco/mart/wash, Burger King, Cracker Barrel, Denny's, KFC, McDonald's, Miami Subs, Old Carolina Buffet, Perkins, Red Lobster, Shoney's, Taco Bell, Waffle House, Wendy's, Day's Inn, Econolodge, Hampton Inn, Holiday Inn Express, Motel 6, Manufacturers Outlet/ famous brands, VisionLand, hardware, to FL TPK

64(126) FL 712, Midway Rd, **E**...Fina/mart, Burger King, Subway

63c(121) St Lucie West Blvd, **E**...Chevron/mart/wash, Mobil(1mi), Shell/Subway/diesel/mart, Little Italy Rest.(1mi), McDonald's, Wendy's, Fairfield Suites

63(117) Gatlin Blvd, to Port St Lucie, **1-2 mi E**...HOSPITAL, Amoco/diesel/LP/mart, Mobil/Subway/diesel/mart, Burger King, Gateway Diner

62(110) FL 714, to Stuart, Palm City, no facilities

106mm **rest areas both lanes, full(handicapped)facilities, phone, vending, picnic tables. litter barrels, petwalk, 24hr security**

61c(102) FL 713, to Stuart, Palm City, no facilities

J u p i t e r

61(101) FL 76, to Stuart, Indiantown, **E**...HOSPITAL, Chevron/mart/24hr, Mobil/diesel/mart, Skipper's/diesel/mart, Texaco, Cracker Barrel, FlashBack Diner(5mi), McDonald's, RV camping, **W**...Speedway/mart(2mi)

60(96) FL 708, to Hobe Sound, **E**...Dickinson SP(11mi), RV/tent camping

59b a(87) FL 706, Jupiter, to Okeechobee, **E**...HOSPITAL, DENTIST, CHIROPRACTOR, VETERINARIAN, PODIATRIST, Chevron/mart/wash, FarmStore/gas/24hr, Hess/mart, Mobil/mart/wash, Texaco/diesel/mart/wash/24hr, Applebee's, Dunkin Donuts, Giuseppe's Italian, Giardino's Rest., IHOP, KFC, Little Caesar's, McDonald's, Nick's Tomato Pie, Pizza Man, Shoney's, Subway, Taco Bell, YumYum Chinese, AutoCare, Discount Parts, Dodge/ Mazda, Eckerd, GNC, Home Depot, MailBoxes Etc, PepBoys, Publix, VisionLand, Walgreen, Wal-Mart, Winn- Dixie, bank, cleaners, hist sites, museum, **W**...info, RV camping, to FL TPK

58(83) Donald Ross Rd, **E**...HOSPITAL

57c(81) FL 809 S(from sb), Military Trail, **W**...to FL TPK, same services as 57b(W)

N

S

Florida Interstate 95

57a b(80) FL 786, PGA Blvd, **E**...HOSPITAL, Mobil/mart, Caper's Cod/Seafood, Chinese Rest., Durango's Steaks, Marriott, Loehmann's Foods, cinema, **W**...Shell/repair, Old Boston Pasta, DoubleTree Hotel, Embassy Suites, bank

56(77) Northlake Blvd, to W Palm Bch, **E**...HOSPITAL, Amoco, Citgo, Hess Gas, Texaco/diesel/mart, Applebee's, Arby's, Burger King, Checker's, KFC, McDonald's, Taco Bell, Wendy's, Chrysler/Jeep/Kia, Costco Whse, Ford, K-Mart, Lincoln/Mercury, Olds/Isuzu, PepBoys, Pennzoil, Pontiac/GMC, Sports Authority, Subaru, Upton's, VW/Mitsubishi, bank, **W**...Chevron/mart, Mobil, Shoney's, Economy Inn, Inns of America, OfficeMax, Publix

55(75) Blue Heron Rd, **E**...Amoco, Shell/mart, **W**...Mobil/mart/wash, Burger King, Denny's, McDonald's, Motel 6, Super 8

54(74) FL 702, 45th St, **E**...MEDICAL CARE, Burger King, Greenhouse Rest., Day's Inn, Knight's Inn, **W**...Cracker Barrel, Courtyard, Red Roof Inn

53(72) Lake Blvd, Palm Beach, **E**...HOSPITAL, Best Western, Best Buy, Burdine's, Firestone/auto, JC Penney, mall, **W**...Comfort Inn, Wellesley Inn, bank

52b a(70) FL 704, Okeechobee Blvd, **E**...Exxon, Sheraton, Kravis Ctr, museum, **1-2 mi E**...beach motels, **W**...Chevron, Goodway/diesel, Hess, Shell, Texaco, Burger King, Mexican Rest., Shell's Rest., Sub Factory, Holiday Inn, Omni Hotel, Radisson, Bed Bath&Beyond, Chevrolet, Circuit City, CompUSA, Dodge, Lenscrafters, Lincoln/Mercury, Mens Wearhouse, Mitsubishi, Pennzoil, cinema

51(69) Belvedere Rd, **W**...Shell, Burger King, Shoney's, Hampton Inn, Holiday Inn/rest., Omni Hotel, U-Haul, to airport

50(68) US 98, Southern Blvd, **W**...Hilton, airport, bank

49(66) Forest Hill Blvd, **E**...golf

48(65) 10th Ave N, **W**...Tri-Rail Sta

47(63) 6th Ave S, no facilities

46(62) FL 812, Lantana Rd, **E**...7-11/24hr, Shell/mart, CrabHouse Rest., KFC, McDonald's, Subway, Motel 6, Eckerd, Publix, bank, **W**...Costco Whse

45(61) Hypoluxo Rd, **E**...Amoco/mart, Mobil, Texaco/mart, Circle K, Pizza Hut, Shoney's, Taco Bell, Inns of America, Super 8, Goodyear, Sam's Club, U-Haul, bank, **W**...Shell/mart, airport

44c(59) Gateway Blvd, **W**...CarMax

44(57) FL 804, Boynton Bch Blvd, **E**...HOSPITAL, CHIROPRACTOR, Texaco/repair, KFC, Holiday Inn Express, USPO, laundry, **W**...Texaco/diesel/mart, 7-11, Waffle House, Wendy's, PetSupplies

43(56) Woolbright Rd, **E**...HOSPITAL, **W**...Cracker Barrel

42b a(53) FL 806, Atlantic Ave, **W**...AmTrak

41(51) Linton Blvd, **E**...HOSPITAL, CHIROPRACTOR, DENTIST, Shell, Abbey Rd Grill, Dairy Queen, McDonald's, OutBack Steaks, Circuit City, Dodge, Ford, LinenStore, Nissan, OfficeMax, Ross, Target, cleaners, **W**...CHIROPRACTOR, DENTIST, Shell/mart, BBQ, NAPA, Winn-Dixie, auto repair, bank, museum

40c(50.5) Congress Ave, no facilities

40(48) FL 794, Yamato Rd, **W**...DoubleTree Guest Suites, Embassy Suites

39(46) FL 808, Glades Rd, **E**...funpark, museum, **W**...Marriott, bank

38(44) Palmetto Park Rd, **E**...museums

37(42) FL 810, Hillsboro Blvd, **E**...MEDICAL CARE, Amoco/mart, Texaco/mart, Clock Rest., McDonald's, Popeye's, Hilton, La Quinta, Discount Parts, bank, **W**...Chevron/mart, Mobil/diesel/mart/wash, Blimpie, Boston Mkt, Checker's, Denny's, Italiano Ristorante, Pizza Hut, Wolfie's Rest., Ramada Inn, Villager Lodge, Wellesley Inn, CompUSA, Eckerd, Home Depot, Lens Express, cleaners

36c(41) FL 869(toll), SW 10th, to I-75, **E**...Mobil/mart/wash, Extended Stay America, **W**...Comfort Suites, Quality Suites, antiques

36(40) FL 834, Sample Rd, **E**...MEDICAL CARE, Hess/mart, Shell/diesel, Texaco/diesel/mart, **W**...Amoco, Chevron, Citgo/mart, Mobil/diesel/mart, Shell/mart, Texaco/wash, 7-11, Arby's, Blimpie, Burger King, Checker's, Chinese Rest., Domino's, McDonald's, Metro Pizzaria, Miami Subs, Subway, Costco, Eckerd, Mr Grocer, Winn-Dixie, bank

35b a(38) Copans Rd, **E**...Amoco/mart, Mobil/7-11, Circuit City, Mercedes/Audi/Porsche, PepBoys, Wal-Mart, **W**...Amoco/diesel/mart, Wendy's, Home Depot, NAPA, NTB

34b a(37) FL 814, Atlantic Blvd, to Pompano Bch, **E**...MEDICAL CARE, KFC, Miami Subs, Taco Bell, **1 mi W**...Amoco/mart, Mobil/diesel/mart, Shell/repair, Chevrolet, Eckerd, Harness RaceTrack, Winn-Dixie, to FL TPK, Power Line Rd has multiple services

N

↑

↓

S

W Palm Beach

Pompano

Florida Interstate 95

Ft Lauderdale	
33(36)	Cypress Creek Rd, **E**...Amoco/mart, Hess/mart, Circle K, Boston Bagels, Domino's, Hampton Inn, Westin Hotel, JiffyLube, Postal Ctr, laundry, **W**...Shell/repair, Arby's, Bennigan's, Burger King, Chili's, Hooters, Longhorn Steaks, McDonald's, Miami Subs, StarLite Diner, Steak&Ale, Sweet Tomatos, Wendy's, La Quinta, Marriott, Office Depot, OfficeMax, bank
32(34)	FL 870, Commercial Blvd, Lauderdale by the Sea, Lauderhill, **W**...Mobil, Shell, McDonald's, Waffle House, Holiday Inn, Red Roof Inn, Wellesley Inn
31(33)	FL 816, Oakland Park Blvd, **E**...Chevron, Phillips 66, Denny's, Wendy's, Firestone, Meineke Muffler, **W**...BP, Day's Inn, Home Depot
30(32)	FL 838, Sunrise Blvd, **E**...Amoco, Texaco/mart, Burger King
29(31)	FL 842, Broward Blvd, Ft Lauderdale, **E**...HOSPITAL
28(30)	FL 736, Davie Blvd, **W**...Miami Subs
27(29)	FL 84, **E**...Chevron, Citgo/7-11, Coastal/diesel, Mobil/diesel, RaceTrac/mart/24hr, Shell, Domino's, Dunkin Donuts, Lil Red's Rest., Little Caesar's, McDonald's, Subway, Wendy's, Best Western, Budget Inn, Motel 6, Sky Motel, Big Lots, U-Haul, Walgreen, Winn-Dixie, XpressLube, **W**...Ramada Inn, Red Carpet Inn, Ford Trucks
26c(28)	I-595, to I-75, **E**...to airport
26b a(27)	FL 818, Griffin Rd, **E**...Hilton, Wyndham Hotel, **W**...Amoco/diesel
25(26)	FL 848, Stirling Rd, Cooper City, **E**...Burger King, Taco Bell, Comfort Inn, Hampton Inn Suites, Hilton, Wyndham Hotel, Barnes&Noble, Discount Parts, GNC, K-Mart/Little Caesar's, Marshall's, OfficeMax, PetsMart, Service Merchandise, bank, funpark, to Lloyd SP, same as 24, **1/2 mi W**...CHIROPRACTOR, MEDICAL CARE, Circle K, Dunkin Donuts, Hunan House, Miami Subs, Subway, Eckerd, Lucy's Auto Clinic, PepBoys, Tire Kingdom/ JiffyLube, Walgreen, hardware
24(25)	FL 822, Sheridan St, **E**...Mobil/diesel/mart, AleHouse Grill, Chevy's Mexican, Pizza/subs, Sweet Tomato, TGIFriday, Cumberland Foods, Pearle Vision, Ross, same as 25, **W**...Shell/mart, Denny's, Day's Inn, Holiday Inn
23(24)	FL 820, Hollywood Blvd, **E**...Howard Johnson, **W**...HOSPITAL, Mobil/mart, AmTrak
22(23)	FL 824, Pembroke Rd, **E**...Shell, Texaco, Howard Johnson, dogtrack
21(22)	FL 858, Hallandale Bch Blvd, **E**...MEDICAL CARE, Amoco/mart, Citgo/7-11, Hess/mart, Shell/Blimpie/mart, Bakehouse Rest., BBQ, Burger King, Carvel Ice Cream, Chinese Rest., Denny's, Dunkin Donuts, IHOP, KFC, Little Caesar's, Long John Silver, McDonald's, Miami Subs, Subway, Holiday Inn Express, Goodyear/auto, Tire Kingdom, Walgreen, Winn-Dixie, laundry, **W**...Amoco/Taco Bell/mart, RaceTrac/mart, Speedway/mart, Discount Parts, bank
20(21)	Ives Dairy Rd, **W**...CHIROPRACTOR, VETERINARIAN, Amoco/mart/24hr, 7-11, Subway
19(20)	FL 860, Miami Gardens Dr, N Miami Beach, **E**...HOSPITAL, Oleta River SRA, **1/2 mi W**...Amoco/mart/24hr, Chevron/mart/wash, Citgo/diesel/mart, Subway, BJ's Club
18(19)	US 441, FL 826, FL 9, **E**...HOSPITAL, HoJo's, **W**...Checker's, Sports Authority
17(18)	US 441(from nb), same as 18
16(17.5)	FL 868(from nb), FL TPK N, no facilities
15(17)	NW 151st, **W**...McDonald's, Discount Parts, Winn-Dixie, services on US 441
14(16)	FL 916, NW 135th, Opa-Locka Blvd, **W**...Amoco/mart, Mobil/mart, QuikStop, motel
13(15)	NW 125th, N Miami, Bal Harbour, **W**...Shell/mart, Wendy's
12(14)	NW 119th(from nb), **W**...Amoco, BBQ, Church's Chicken, Winn-Dixie
11(13)	FL 932, NW 103rd, **E**...Shell/mart/wash, 7-11, **W**...Amoco/mart, Chevron/mart, Mobil/mart, Dunkin Donuts
10(12)	NW 95th, **W**...Mobil/mart, Shell, Burger King, McDonald's, Discount Parts, Ford, Walgreen, bank
9b(11)	FL 934, NW 81st, NW 79th, **E**...BP/mart, Chevron/diesel/mart, **W**...Shell/mart, Checker's, Chinese Rest., Day's Inn, Mercury/Lincoln, Midas Muffler
9a(10)	NW 69th(from sb), no facilities
Miami 8(9)	FL 944, NW 62nd, NW 54th, **W**...McDonald's, Subway, Winn-Dixie
7(8)	I-195 E, Miami Beach, FL 112 W(toll), **E**...downtown, **W**...airport
6(7)	FL 836 W(toll)(exits left from nb), **W**...HOSPITAL, to Orange Bowl, airport
5b(6)	I-395 E(exits left from sb), to Miami Beach
5a(5)	NW 8th, NW 14th(from sb), **E**...Port of Miami, **W**...to Orange Bowl
4(4)	NW 2nd(from nb), downtown Miami
3(3)	US 1(exits left from sb), Biscayne Blvd, downtown Miami
2(2)	US 41, SW 7th, SW 8th, Brickell Ave, downtown, **E**...Hampton Inn, Publix, bank, **W**...Shell/mart
1(1)	SW 25th(from sb), downtown, to Rickenbacker Causeway, Key Biscayne, **E**...Hampton Inn, bank, museum, to Baggs SRA
0mm	I-95 begins/ends on US 1. **1 mi S**...Citgo/mart

N

↑

↓

S

Florida Interstate 275(Tampa)

Exit #	Services
59mm	I-275 begins/ends on I-75, exit 57.
36(52)	Bearss Ave, **E**...Citgo, CarMax, **W**...Amoco/mart/wash, Chevron/diesel/mart, Shell/mart/wash, BBQ, Burger King, McDonald's, Subway, Wendy's, Holiday Inn Express, Albertson's, Big Lots, Eckerd, GNC, Publix, Radio Shack, bank, cleaners, pizzaria
35(51)	Fletcher Ave, **E**...HOSPITAL, Day's Inn, funpark, **W**...BP, Citgo, Americana Inn
34(50)	FL 582, Fowler Ave, **E**...Shell, Burger King, McDonald's, Subway, Howard Johnson, Quality Inn, **W**...Citgo/mart, Discovery Inn, Motel 6, SleepRite Motel, Audi
33(48)	FL 580, Busch Blvd, **E**...Amoco, Exxon, Best Western, Red Roof Inn, Busch Gardens, **W**...KFC, flea mkt, dogtrack
32(47)	Bird Ave(from nb), **W**...Shell/mart, Wendy's, K-Mart
31(46)	Sligh Ave, **E**...Amoco/mart, Coastal/mart, Spur/mart, **W**...BP/mart, PitStop/gas, zoo
30(45)	US 92, to US 41 S, Hillsborough Ave, **E**...Citgo, Discount Parts, **W**...Amoco
29(44)	FL 574, MLK Blvd, **E**...Chevron, Kash&Karry, **W**...Cumberland Farms, McDonald's
28(43)	Floribraska Ave, HOSPITAL
27(42)	I-4 E, to Orlando, I-75
26(41)	Jefferson St, downtown E
25(40.5)	Ashley Dr, Tampa St, downtown W
24(40)	Howard Ave, Armenia Ave, **E**...Citgo, **W**...HOSPITAL, Amoco/mart, Texaco/diesel/mart, Popeye's
23c(39)	Himes Ave(from sb), **W**...Tampa Stadium
23ba(38)	US 92, Dale Mabry Blvd, **E**...MEDICAL CARE, Amoco/mart/24hr, BP, Citgo/Hardee's, Exxon, Shell/mart, Alexander's Rest., Carrabba's Italian, HoneyBaked Ham, Hops Grill, Japanese Steaks, Krystal, Lobster Louie's Crabs, Miami Subs, Pizza Hut, Rio Bravo, Ruby Tuesday, Village Inn Pancakes, Courtyard, Borders Books&Café, Computer City, JiffyLube, Kinko's, Midas Muffler, Office Depot, to MacDill AFB, **W**...Exxon/diesel, Bennigan's, Blimpie, Burger King, Checker's, Chili's, Denny's, Dunkin Donuts, Eastside Mario's Café, LongHorn Steaks, McDonald's, Sweet Tomatos, Taco Bell, Tia's TexMex, Waffle House, Wendy's, Day's Inn, Howard Johnson, Best Buy, Circuit City, Gateway 2000, Home Depot, Wal-Mart, airport
22(37.5)	Lois Ave, **W**...Radiant/diesel/mart, Westin Suites
21(37)	FL 587, Westshore Blvd, **E**...Citgo, Shell, Steak&Ale, Waffle House, Embassy Suites, Ramada Inn, JC Penney, bank, **W**...Shell, Crown Plaza, Doubletree Hotel, Marriott
20ba(36)	FL 60 W, **2 mi W**...Day's Inn, to airport
19(32)	Fl 687 S, 4th St N, to US 92(no sb re-entry), no facilities
18(31)	FL 688 W, Ulmerton Rd, to Largo, info, airport
17(30)	9th St N, MLK St N(exits left from sb), no facilities
16(29)	FL 686, Roosevelt Blvd, no facilities
15(28)	FL 694 W, Gandy Blvd, Indian Shores, no facilities
14(26)	54th Ave N, no facilities
13(25)	38th Ave N, to beaches
12(24)	22nd Ave N, **E**...Sunken Garden, **W**...Home Depot
11(23)	FL 595, 5th Ave N, **E**...HOSPITAL
10(22)	I-375, **E**...The Pier, Waterfront, downtown
9(21)	I-175 E, Tropicana Fields, **E**...HOSPITAL
8(20.5)	46th St, downtown
7(20)	31st Ave(from nb), downtown
6(19.5)	22nd Ave S, Gulfport, **W**...Citgo
5(19)	26th Ave S(from nb)
4(18)	FL 682 W, 54th Ave S, Pinellas Bayway, **W**...Bob Evans, Burger King, McDonald's, Taco Bell, Krystal Inn, Park Inn, to Ft DeSoto Pk, St Pete Beach
3(17)	Pinellas Point Dr, Skyway Lane, to Maximo Park, **W**...marina
16mm	toll plaza sb
2b(13)	N Skyway Fishing Pier, **W**...**rest area both lanes, full(handicapped)facilities, phone, vending, picnic tables, litter barrels, petwalk**
10mm	Tampa Bay

N

S

Tampa

St Petersburg

Florida Interstate 275

2a(7)	S Skyway Fishing Pier, **E…rest area both lanes, full(handicapped)facilities, phone, vending, picnic tables, litter barrels, petwalk**
6mm	toll plaza nb
2(5)	US 19, Palmetto, Bradenton, no facilities
1(2)	US 41, Palmetto, Bradenton, no facilities
0mm	I-275 begins/ends on I-75, exit 44.

Florida Interstate 295(Jacksonville)

Exit #(mm)Services

14b a(35)	I-95, S to Jacksonville, N to Savannah. I-295 begins/ends on I-95, exit 126.
13a(34)	new exit
13(32)	FL 115, Lem Turner Rd, **E…**Wal-Mart, **W…**Flamingo Lake Camping
12(30)	FL 104, Dunn Ave, **E…**Gate Gas, Shell/mart/24hr(1mi), McDonald's, Winn-Dixie/deli, **W…**Pines RV Park
11b a(28)	US 1, US 23, Jackson, to Callahan, **E…**Lil Champ/gas, **W…**RaceTrac/mart
10(25)	Pritchard Rd, **W…**Texaco/Taco Bell/diesel/mart/24hr
9(22)	Commonwealth Ave, **E…**Hardee's, Holiday Inn, Winn-Dixie, dogtrack, **W…**Goodyear
8ba(21)	I-10, W to Tallahassee, E to Jacksonville
7(19)	FL 228, Normandy Blvd, **E…**Amoco/24hr, Texaco/diesel/mart, Burger King, Food Lion, Walgreen, cleaners, st patrol, **W…**RaceTrac/mart, Shell/mart/wash, Hardee's, Pizza Hut, Subway, Discount Parts, Eckerd, Family$, K-Mart, Publix, Radio Shack, Winn-Dixie, bank, cleaners
6(16)	FL 208, Wilson Blvd, **2 mi E…**Hardee's, Ryan's, Winn-Dixie
5(14)	FL 134, 103rd St, **E…**Gate/diesel/mart, Hess, Shell/24hr, Texaco/diesel/mart, Applebee's, Burger King, Popeye's, Shoney's, Hospitality Inn, Discount Parts, XpressLube, **W…**CHIROPRACTOR, VETERINARIAN, Amoco, Chevron/mart, Lil Champ/gas, Shell, Burger King, Chinese Rest., Dunkin Donuts, Gene's Seafood, KFC, McDonald's, Miami Subs, Pizza Hut, Rally's, Subway, Taco Bell, Family$, HobbyWorld, Precision Tune, Publix, Sun Tire, Walgreen, Winn-Dixie
4(12)	FL 21, Blanding Blvd, **E…**Chevron/diesel/mart, RaceTrac, Burger King, Larry's Subs, Pizza Hut, Acura, Burlington Coats, Cadillac, Eckerd, Ford, Luxury Linens, Office Depot, PetsMart, Pier 1, Publix, Saturn, SportsAuthority, Toyota, **W…**HOSPITAL, DENTIST, Amoco, Chevron/mart, Shell/mart/wash, Texaco, Arby's, Bennigan's, Chili's, ChuckeCheese, Cross Creek BBQ/steaks, Denny's, HoneyBaked Ham, Longhorn Steaks, Olive Garden, Outback Steaks, Papa Joyhn's, Red Lobster, Rio Bravo Mexican, Shoney's, Steak&Ale, Taco Bell, Tony Roma's, Economy Motel, Hampton Inn, La Quinta, Motel 6, Red Roof Inn, Circuit City, Discount Tire, Gayfer's, Home Depot, JiffyLube, Kinko's, LensCrafters, Pearle Vision, Pet Supplies+, Sam's Club, Sears/auto, Service Merchandise, Target, VisionWorks, cinema, mall
3(10)	US 17, FL 15, Roosevelt Blvd, Orange Park, **E…**Wilson Inn, **W…**Amoco, Chevron/diesel/mart/24hr, RaceTrac/mart, Shell/mart, Cracker Barrel, Damon's, Krystal, LJ Silver, McDonald's, Pizza Hut, Quincy's, Waffle House, Wendy's, Best Western, Comfort Inn, Day's Inn, Holiday Inn, Villager Lodge, Chrysler, Honda, Nissan, Olds, VW, bank
7mm	St Johns River, Buckman Br
2ba(4)	FL 13, San Jose Blvd, **E…**Exxon/diesel/mart/24hr, Shell, Applebee's, Arby's, Chinese Rest., Famous Amos, Kenny Rogers, TCBY, Budgetel, Ramada Inn, Albertson's, Eckerd, Firestone/auto, Pet SuperMkt, Publix, cinema, **W…**CHIROPRACTOR, DENTIST, Shell/mart/wash, Baskin-Robbins, BBQ, Burger King, Chili's, Golden China, Golden Corral, Krispy Kreme, Mozzarella's Café, Pizza Hut, Renna's Pizza, Schlotsky's, Barnes&Noble, Books-A-Million, Fresh Mkt, MailBoxes Etc, Michael's, NAPA, Pet Supplies+, Radio Shack, Sears, SteinMart, TJ Maxx, Wal-Mart, cleaners
1(2)	Old St Augustine Rd, **E…**Amoco, Shell/mart, Burger King, Chinese Rest., Denny's, McDonald's, Pizza Hut, Taco Bell, Wendy's, $General, Eckerd, JiffyLube, Publix, Sing Food/deli, Waccamaw, Winn-Dixie, cleaners, **W…**MEDICAL CARE, Chevron/mart, Lil Champ/gas, Lowe's Bldg Supply, Walgreen
0mm	I-295 begins/ends on I-95, exit 97.

Jacksonville

Georgia Interstate 16

Exit #	Services
167	W Broad, Montgomery St, Savannah, **N**...Enmark/mart, Speedway/mart, Courtyard, Hampton Inn(1mi), **S**...Burger King, KFC, Popeye's, Wendy's, I-16 begins/ends in Savannah.
166	US 17, Gwinnet St, Savannah, no facilities
165	GA 204, 37th St(from eb), to Ft Pulaski NM, Savannah Coll
164b a	I-516, US 80, US 17, GA 21, **1 mi S**...Budget Inn, Courtyard, Day's Inn, Holiday Inn, La Quinta
162	Chatham Pkwy, **S**...Shell/Taco Maker/diesel/mart, Lexus, Toyota
160	GA 307, Dean Forest Rd, **N**...Shell/diesel/mart, Speedway/diesel/mart, Ronnie's Rest., Waffle House, **S**...st patrol
157b a	I-95, S to Jacksonville, N to Florence, no facilities
155	new exit, no facilities
152	GA 17, to Bloomingdale, no facilities
148	Old River Rd, to US 80, no facilities
144mm	weigh sta both lanes
143	US 280, to US 80, **N**...Citgo(2mi), **S**...BP/mart, Chevron/diesel/café/mart, Black Creek Golf Course(3mi)
137	GA 119, to Pembroke, Ft Stewart, no facilities
132	Ash Branch Church Rd, no facilities
127	GA 67, to Pembroke, Ft Stewart, **N**...BP/diesel(1mi), Texaco, Oasis Camping(1mi), **S**...Chevron/diesel/mart/24hr, Huddle House
116	US 25/301, to Statesboro, **N**...Chevron/diesel/mart/24hr, PoJo's Deli, to GA S U, **S**...Shell/KrispyKreme/diesel/mart, Uncle Gene's Trkstp/diesel, Huddle House, Red Carpet Inn
111	Pulaski-Excelsior Rd, **S**...Citgo/diesel, Grady's Grill, Beaver Run RV Park
104	GA 22, GA 121, Metter, **N**...HOSPITAL, BP/diesel/mart/24hr, Exxon/mart, Phillips 66/diesel/mart, Shell/diesel/mart, Burger King, Dairy Queen, Hardee's, Huddle House, KFC/Taco Bell, McDonald's, Subway, Village Pizza, Waffle House, Western Steer, Comfort Inn, Day's Inn, Holiday Inn Express, Metter Inn, Chevrolet/Olds, Rite Aid, to Smith SP, RV camping, **S**...OakTree Express/gas, Texaco/diesel/mart, Ford
101mm	Canoochee River
98	GA 57, to Stillmore, **S**...Amoco/diesel/mart/24hr, Chevron/diesel/mart
90	US 1, to Swainsboro, **N**...BP/mart, Phillips 66/diesel/rest., Oak Island Crabhouse, repair
87.5mm	Ohoopee River
84	GA 297, to Vidalia, **N**...truck sales
78	US 221, GA 56, to Swainsboro, **N**...BP/diesel(1mi)
71	GA 15, GA 78, to Soperton, **N**...Chevron/diesel/mart/24hr
67	GA 29, to Soperton, **S**...Chevron/diesel/mart, Texaco/mart, Huddle House
58	GA 199, Old River Rd, East Dublin, no facilities
56mm	Oconee River
54	GA 19, to Dublin, **1 mi N**...Chevron
51	US 441, US 319, to Dublin, **N**...HOSPITAL, Amoco/diesel/mart/24hr, BP/Subway/diesel/mart, Exxon/diesel/mart, Speedway/diesel/mart, Texaco/mart/wash/24hr, Arby's, Buffalo's Café, Burger King, Johnnie's Rest., KFC, McDonald's, Pizza Hut, Taco Bell/TCBY, Waffle House/24hr, Wendy's, Best Western, Comfort Inn, Day's Inn, Hampton Inn, Holiday Inn/rest., Jameson Inn, Super 8, **S**...Chevron/mart, Cracker Barrel, antiques, to Little Ocmulgee SP
49	GA 257, to Dublin, Dexter, **N**...Chevron/mart, **3 mi N**...HOSPITAL, Texaco, BBQ, Friendly Gus' Fried Chicken, Huddle House, **S**...257 Trkstp/diesel/mart, truck sales
46mm	**rest area wb, full(handicapped)facilities, phone, picnic tables, litter barrels, vending, petwalk, RV dump**
44mm	**rest area eb, full(handicapped)facilities, phone, picnic tables, litter barrels, vending, petwalk, RV dump**
42	GA 338, to Dudley, no facilities
39	GA 26, Montrose, to Cochran, no facilities
32	GA 112, Allentown, **S**...Chevron/diesel, Me-Ma's Rest.
27	GA 358, to Danville, no facilities
24	GA 96, to Jeffersonville, **N**...Citgo/diesel/mart, **S**...BP/diesel/mart/24hr, Huddle House/24hr, Day's Inn, to Robins AFB, museum
18	Bullard Rd, Bullard, to Jeffersonville, no facilities
12	Sgoda Rd, Huber, **N**...Marathon/diesel/mart
6	US 23, US 129A, East Blvd, Ocmulgee, **N**...Day's Inn(2mi), to airport, GA Forestry Ctr, **S**...Chevron/Stuckey's/diesel/mart, Texaco/diesel/mart, Huddle House, Subway
2	US 80, GA 87, MLK Jr Blvd, **N**...HOSPITAL, Ocmulgee NM, coliseum, **S**...Marathon/diesel/mart, Saf-T-Oil/diesel, to Hist Dist

E ↑

W ↓

Georgia Interstate 16

Macon

1b GA 22, to US 129, GA 49, 2nd St(from wb), **N**...HOSPITAL, coliseum

a US 23, Gray Hwy(from eb), **N**...HOSPITAL, DENTIST, Citgo, Marathon/diesel, Shell/mart, Speedway/mart, Texaco/diesel/mart, Arby's, BBQ, Burger King, Central Park, Dairy Queen, Domino's, HongKong Express, Huddle House, KrispyKreme, Krystal, McDonald's, Papa John's, Popeye's, Subway, Taco Bell, Isuzu, Kroger, Lincoln/Mercury, Medicine Shoppe, Meineke Muffler/brake, Mitsubishi, Radio Shack, U-Haul, XpressLube, repair, transmissions, **S**...Amoco/mart, Conoco/diesel/mart, Exxon/mart, Burger King, Checker's, Hardee's, KFC, Krystal, Pizza Hut, T's Steaks, Waffle House, Wendy's, Crown Plaza Hotel, Pontiac

0mm I-75, S to Valdosta, N to Atlanta. I-16 begins/ends on I-75, exit 53 in Macon.

Georgia Interstate 20

Exit #	Services

201.5mm Savannah River, Georgia/South Carolina state line

201mm **Welcome Ctr wb, full(handicapped)facilities, phone, vending, picnic tables, litter barrels, petwalk**

200 GA 104, Riverwatch Pkwy, Augusta, **N**...Pilot/Wendy's/diesel/mart/24hr, AmeriSuites, Courtyard, Quality Inn, Sleep Inn, **S**...auto repair

Augusta

199 GA 28, Washington Rd, Augusta, **N**...BP/mart/wash, RaceTrac/mart/24hr, Burger King, Capt D's, Chick-fil-A, Chinese Rest., Dairy Queen, Damon's, KFC, McDonald's, Mikoto Rest., Piccadilly's, Pizza Hut, Rhinehart's Seafood, Veracruz Mexican, Waffle House, Day's Inn, Hampton Inn, Holiday Inn, Homewood Suites Hotel, La Quinta, Masters Inn, Scottish Inn, Sunset Inn, Shoney's Inn/rest., BMW/Infiniti, Fabric Whse, Hyundai, Lowe's Bldg Supply, Mazda, Mercedes, XpressLube, bank, cleaners, **S**...MEDICAL CARE, Amoco/mart/24hr, Hess/diesel/mart, 76/diesel, Shell/mart/24hr, Smile Gas/mart, Texaco/diesel/mart, Arby's, Blimpie, Bojangles, Domino's, Fazoli's, Haagen-Daz, Hardee's, Hooters, Krispy Kreme, Kyoto Japanese, Little Caesar's, LoneStar Café, LJ Silver, Mally's Bagels, McDonald's, Michael's Rest., Olive Garden, Outback Steaks, Red Lobster, Subway, Taco Bell, T-Bone's Steaks, TGIFriday, Waffle House, Wendy's, Econolodge, Fairfield Inn, Knight's Inn, Motel 6, Richfield Lodge, Amoco Wash/lube, Books-A-Million, CastroLube, Drug Emporium, Firestone/auto, Goodyear/auto, Kroger/deli/24hr, Midas Muffler, PepBoys, Publix, Thrifty Drug, White's, bank, camera shop, cleaners, hardware

196b GA 232 W, **N**...CHIROPRACTOR, Amoco/mart, Enmark/mart, RaceTrac/mart, 76/mart, Applebee's, Arby's, Athens Pizza, Burger King, Checker's, China Pearl, Gianni's Italian, Golden Corral, Krystal, Po Folks, Ruby Tuesday, Ryan's, Schlotsky's, Shoney's, Waffle House, Wendy's, Comfort Inn, Howard Johnson, Suburban Lodge, Travelers Inn, BiLo Foods, OfficeMax, Sam's Club, Wal-Mart SuperCtr/24hr

a I-520, Bobby Jones Fwy, **S**...HOSPITAL, Amoco, Circuit City, Target, Winn-Dixie, to airport

195 Wheeler Rd, **S**...HOSPITAL, Chevron, 76, Speedway/mart, Blimpie, Day's Inn, Kroger/deli

194 GA 383, Belair Rd, to Evans, **N**...Exxon/Taco Bell/mart, Phillips 66/diesel/mart, 76/diesel/mart, Burger King, Waffle House, Villager Lodge, Funsville Park, bank, carwash, **S**...Amoco/DQ/Stuckey's/diesel, Speedway/diesel/mart/24hr, Cracker Barrel, Huddle House/24hr, Waffle House, Best Western, Econolodge, Ramada Ltd, Red Roof Inn, Wingate Inn

190 GA 388, to Grovetown, **N**...Amoco/diesel, **S**...Chevron/diesel/mart, Smile/gas/mart, Hardee's, McDonald's, TP's Grill

187mm weigh sta both lanes

183 US 221, Appling, to Harlem, **N**...Ford, **S**...Citgo/diesel, 76

182mm **rest area both lanes, full(handicapped)facilities, phone, vending, picnic tables, litter barrels, RV dump, petwalk**

175 GA 150, **N**...Amoco/Samuels/diesel/rest./24hr, Day's Inn, to Mistletoe SP

Thomson

172 US 78, GA 17, Thomson, **N**...BP/diesel/mart, Chevron/diesel/mart, Waffle House, Chrysler/Dodge/Jeep, Ford/Mercury, Olds/Pontiac/GMC, **S**...HOSPITAL, Amoco/DQ/wash/24hr, Fuel City/Blimpie/diesel/mart/24hr, RaceTrac/diesel/mart, 76/diesel, Texaco/diesel/mart, Amigo's Mexican, Arby's, Burger King, Denny's, Domino's, Hardee's, LJ Silver, McDonald's, Pizza Hut, Shoney's, Taco Bell, Waffle House, Wendy's, Western Sizzlin, Best Western, Econolodge, Holiday Inn Express, Ramada Ltd, Advance Parts, AutoZone, BiLo Foods, Buick/Chevrolet/Cadillac, CVS Drug, Family$, Food Lion, K-Mart, Lowe's Bldg Supply, bank, cleaners/laundry, **2-3 mi S**...Sonic, Subway, Wal-Mart, USPO

165 GA 80, Camak, no facilities

160 E Cadley Rd, Norwood, no facilities

154 US 278, GA 12, Barnett, no facilities

148 GA 22, Crawfordville, **N**...Amoco/diesel, to Stephens SP, **S**...Chevron/diesel/rest./motel/repair

138 GA 77, Siloam, **S**...HOSPITAL, Amoco/diesel/mart, BP, Exxon/mart

130 GA 44, Greensboro, **N**...HOSPITAL, Amoco/diesel/mart/24hr, BP/Subway/mart, McDonald's, Pizza Hut, Waffle House, Wendy's, Zaxby's Café, Jameson Inn, Microtel, Thrift Court Motel(3mi), **S**...Chevron/diesel/mart/24hr

121 Buckhead, to Lake Oconee, **S**...Phillips 66

Georgia Interstate 20

114	US 441, US 129, to Madison, **N**...HOSPITAL, VETERINARIAN, Amoco/mart/24hr, Chevron/Subway/mart, Exxon, RaceTrac/mart, Arby's, Burger King, KFC, Krystal, McDonald's, Pizza Hut, Waffle House, Wendy's, Comfort Inn, Day's Inn, Hampton Inn, Chevrolet/Olds/ Pontiac/GMC, Ingles Foods, Wal-Mart/drugs, **S**...TA/BP/Popeye's/Taco Bell/diesel/mart, FuelMart/diesel, Texaco/mart, Waffle House, Budget Inn, Holiday Inn Express, Ramada Inn/Chinese Rest., Super 8, Talisman RV Resort
113	GA 83, Madison, **N**...HOSPITAL, BP/mart, st patrol
108mm	**rest area wb, full(handicapped)facilities, phone, picnic tables, litter barrels, vending, RV dump, petwalk**
105	Rutledge, Newborn, **N**...BP/diesel/mart, Yesterday Café, Hard Labor Creek SP
103mm	**rest area eb, full(handicapped)facilities, phone, picnic tables, litter barrels, vending, RV dump, petwalk**
101	US 278, no facilities
98	GA 11, to Monroe, Monticello, **4 mi N**...Amoco, Blue Willow Inn/rest., Log Cabin Rest., Sycamore Grill, **S**...Chevron/diesel, Citgo
95mm	Alcovy River
93	GA 142, Hazelbrand Rd, **S**...HOSPITAL, Exxon/diesel/mart, Phillips 66/mart, Texaco/mart/atm/24hr, Jameson Inn
92	Alcovy Rd, **N**...Chevron/diesel/mart, Circle K/diesel, Conoco, BBQ, Hardee's, Krystal, McDonald's, Pizza Hut, Waffle House, Wendy's, Best Western/rest., Day's Inn, Econolodge, Holiday Inn Express, **S**...HOSPITAL
90	US 278, GA 81, Covington, **S**...Amoco, Citgo, Exxon/diesel/mart, RaceTrac/mart, Shell/diesel/mart, Texaco, Arby's, Burger King, Capt D's, Checker's, Chick-fil-A, Dairy Queen, Hardee's, Japanese Rest., KFC, Papa John's, Shoney's, Taco Bell/Pizza Hut, Waffle House, Crest Motel, Advance Parts, Cato's, Chevrolet/Olds, $General, Eckerd, GNC, Ingles Foods, K-Mart, Winn-Dixie/deli, tires
88	Almon Rd, to Porterdale, **N**...Chevron/diesel/mart, **S**...BP/mart, Riverside Estates RV Camp
84	GA 162, Salem Rd, to Pace, **N**...K&D Shoes, Saturn, **S**...Amoco/mart, BP, Chevron/mart/24hr, QT/mart, RaceTrac/mart, Burger King, Hardee's, Pike Rest., Subway, Waffle House, Kawasaki/Suzuki, antiques, cleaners
82	GA 138, GA 20, Conyers, **N**...Amoco, BP/diesel/mart/repair, Speedway/diesel/mart, ChuckeCheese, Crabby Bill's Rest., Cracker Barrel, Don Pablo's, Golden Corral, IHOP, O'Charley's, On the Border, Outback Steaks, Piccadilly's, Red Lobster, Day's Inn, Hampton Inn, Jameson Inn, La Quinta Suites, Ramada Ltd, Chevrolet, Ford, Goody's, Home Depot, Kohl's, Michael's, NAPA, Old Navy, PetsMart, Staples, Wal-Mart SuperCtr/24hr, **S**...CHIROPRACTOR, Chevron/mart/wash/24hr, Conoco/mart, Texaco/diesel/mart, Arby's, Applebee's, Athens Pizza, Baskin-Robbins, BBQ, Boston Mkt, Blimpie, Burger King, Capt D's, Checker's, Chianti Italian, Chili's, Chick-fil-A, Daruma Japanese, Folks Rest., Hardee's, Hickory Ham Café, Hooters, Huddle House, KFC, Krystal/24hr, Little Caesar's, LJ Silver, Mandarin Garden, McDonald's, Morrison's Diner, Papa John's, Pizza Hut, Popeye's, Provino's Italian, Ruby Tuesday, Ryan's, 7 Gables Rest., Shoney's, Sonny's BBQ, Subway, Taco Bell, TCBY, Waffle House, Wendy's, Suburban Lodge, Dodge, $General, Eckerd, Firestone/auto, Food Depot, GNC, Goodyear/auto, Honda, JiffyLube, Jo-Ann Fabrics, K-Mart, Kroger/24hr, MailBoxes Etc, Mighty Muffler, NTB, Office Depot, PepBoys, Precision Tune, Publix, Radio Shack, Rite Aid, Target, TJ Maxx, Toyota, TuneUp Clinic, Upton's, USPO, XpressLube, bank, carwash, cinema, cleaners, mall
80	West Ave, Conyers, **N**...Exxon/diesel/mart, Speedway/diesel/mart, Texaco/diesel/mart/wash, Burger King, Dairy Queen, Domino's, Donna Marie's Pizza, Mrs Winner's, Subway, Holiday Inn, Richfield Lodge, BrakeLand, Midas Muffler, Piggly Wiggly, cleaners, tires, **S**...Exxon/diesel/mart/24hr, Longhorn Steaks, McDonald's, Waffle House, Comfort Inn, Chrysler/Plymouth/Jeep, Ford, Hyundai, Mitsubishi, Nissan
79mm	**rest area eb, no facilities, phone, litter barrel**
78	Sigman Rd, **N**...Circle K, Victorian House Rest., Waffle House, **S**...Buick/Pontiac/GMC/Mazda, Crown RV Ctr, st police
75	US 278, GA 124, Turner Hill Rd, **N**...Amoco/diesel/mart, US Discount/diesel/mart
74	GA 124, Lithonia, **N**...Amoco/mart/wash/24hr, Chevron/mart/24hr, Phillips 66, Texaco/mart/wash, Capt D's, Hardee's, KFC, McDonald's, Pizza Hut, Waffle House, Wendy's, bank, hardware, **S**...Speedway/diesel/mart/24hr, Chinese Rest., Dairy Queen, Krystal/24hr, Snuffy's Grille, Waffle House, Howard Johnson, CVS Drug, $General, Piggly Wiggly, Radio Shack, bank, hardware
71	Hillandale Dr, Farrington Rd, Panola Rd, **N**...Conoco, Exxon, QT/diesel/mart, Texaco/diesel/mart, Burger King, Checker's, Cracker Barrel, HotWings/pizza Café, McDonald's, Waffle House, Wendy's, La Quinta, Super 8, bank, **S**...CHIROPRACTOR, Amoco/diesel/mart/24hr, Speedway/mart, Domino's, Sleep Inn, cleaners

E

W

Madison

Conyers

Lithonia

Georgia Interstate 20

68 Wesley Chapel Rd, Snapfinger Rd, **N...**CHIROPRACTOR, DENTIST, MEDICAL CARE, VETERINARIAN, Exxon/diesel/mart, Conoco/mart, Texaco/diesel/mart/wash, Baskin-Robbins, Blimpie, Capt D's, Checker's, Chick-fil-A, Church's Chicken, Hardee's, KFC, LJ Silver, Martinez Pizza, Off Da Hook Grill, Shoney's, Subway, Taco Bell, Waffle House, Wendy's, Day's Inn, Motel 6, Ford, Goodyear/auto, Home Depot, Ingles, JiffyLube, K-Mart/auto, Kroger/drugs/24hr, MailBoxes Etc, NTB, Wal-Mart, bank, cleaners/laundry, **S...**Amoco/mart, Chevron/mart/24hr, Citgo/mart, Speedway/diesel/mart, Burger King, Dairy Queen, McDonald's, Pizza Hut, Dragon Chinese, JiffyLube, PrecisionTune, USPO, bank, cleaners

67b a I-285, S to Macon, N to Greenville

66 Columbia Dr(from eb, no return), **N...**Citgo

65 GA 155, Candler Rd, to Decatur, **N...**DENTIST, CHIROPRACTOR, Amoco/mart/24hr, Citgo/mart, Marathon, Blimpie, Dundee's Café, LJ Silver, Pizza Hut, Red Lobster, Supreme Fish Delight, Wendy's, Discover Inn, Econolodge, Howard Johnson, A-1 Foods, CVS Drug, Service Merchandise, U-Haul, XpressLube/Tune-Up, cleaners, **S...**MEDICAL CARE, DENTIST, Amoco/mart/wash, Chevron/mart/24hr, Conoco/mart, Circle K/gas, Arby's, Burger King, Checker's, China Café, Church's Chicken, Dairy Queen, Dunkin Donuts, KFC, McDonald's, Rally's, Ruby Tuesday, Taco Bell, WK Wings, Best Western, Candler Inn, Ramada Ltd, Sunset Lodge, Advance Parts, Firestone/auto, JC Penney, JiffyLube, Kroger, PepBoys, Rich's, Winn-Dixie, bank, mall, tires

63 Gresham Rd, **S...**Amoco/mart/24hr, Citgo, Shell/mart, Church's Chicken, auto repair

62 Flat Shoals Rd(from eb, no return), **N...**Amoco/mart, Texaco/mart

61b GA 260, Glenwood Ave, **N...**Texaco/mart/24hr, **1 mi N...**Amoco, Chevron, Phillips 66, KFC

a Maynard Terrace(from eb, no return), no facilities

Atlanta Area

60b a US 23, Moreland Ave, **N...**Exxon/mart, Atlanta Motel, **S...**Fina/Chinese Rest./diesel/mart, Shell/mart, Checker's, KFC/Taco Bell, Krystal, LJ Silver, McDonald's, Mrs Winner's, Wendy's, bank, cleaners

59b Memorial Dr, Glenwood Ave(from eb), no facilities

a Blvd, Cyclorama, **N...**Chevron/Blimpie/diesel/mart, Confederate Ave Complex, MLK Site, **S...**Amoco/mart

58b Hill St(from wb, no return), **N...**Fina, Mrs Winner's, **S...**stadium

a Capitol St(from wb, no return), downtown, **N...**to GA Dome

57 I-75/85, no facilities

56b Windsor St(from eb), stadium

a US 19, US 29, McDaniel St, **N...**Chevron/mart

55b Lee St(from wb), Ft McPherson, **S...**BP/mart, Church's Chicken, Popeye's, Taco Bell

a Ashby St, **S...**BP, Exxon/mart, Taco Bell, Eckerd, Sears, auto parts, groceries, mall

54 Langhorn St(from wb), to Cascade Rd

53 MLK Dr, to GA 139, **N...**Texaco/mart, auto parts, laundry, **S...**Amoco/mart/wash, Citgo/mart

52b a GA 280, Holmes Dr, High Tower Rd, **N...**Amoco, Church's Chicken, McDonald's, **S...**Exxon

51b a I-285, S to Montgomery, N to Chattanooga

49 GA 70, Fulton Ind Blvd, **N...**Citgo/diesel/24hr, Conoco/diesel, RaceTrac/mart, Capt D's, Checker's, Dairy Queen, Dunkin Donuts, EJ's Rest., Hardee's, Krystal, Mrs Winners, Shoney's, Subway, Waffle House, Wendy's, Crossroads Inn, Fulton Inn, Masters Inn, Ramada Inn, Suburban Lodge, Summit Inn, airport, bank, **S...**Amoco/mart, Chevron, RaceTrac/mart/24hr, Texaco, Williams/diesel/mart, Arby's, Blimpie, Burger King, Grand Buffet, McDonald's, Waffle House, Wency's Café, Comfort Inn, Executive Inn, Mark Inn, Red Roof Inn, Super 8, Office Depot, Penske Trucks, U-Haul, Wolf Camera, truck parts

48mm Chattahoochee River

47 Six Flags Pkwy(from wb), **N...**Amoco/mart, Waffle House, Mark Inn, Sleep Inn, Arrowhead Camping, **S...**Comfort Inn, Day's Inn, Sam's Club, Six Flags Funpark

46b a Riverside Parkway, **N...**Amoco/Church's Chicken/mart, Conoco, QT/mart/24hr, Denny's, Waffle House, La Quinta, EZ Wash, **S...**Coastal/mart, Wendy's, Day's Inn, Sam's Club, Six Flags Funpark

44 GA 6, Thornton Rd, to Lithia Springs, **N...**HOSPITAL, DENTIST, Amoco/mart/24hr, Conoco/diesel/mart, Exxon, Phillips 66/mart, RaceTrac, Burger King, Chick-fil-A, Chinese Rest., Courtini's Café, Hardee's, IHOP, Krystal, McDonald's, Shoney's Inn/rest., Subway, Taco Bell, Waffle House, Wendy's, Budget Inn, Hampton Inn, Knight's Inn, Suburban Lodge, Chevrolet, CVS Drug, Ford, Kroger, Mazda, Nissan, VW, bank, cleaners, **S...**Country Inn Suites, Courtyard, Fairfield Inn, Motel 6, SpringHill Suites, AutoNation, Buick/Olds/GMC, Chrysler/Plymouth/Jeep, Dodge, Mitsubishi, Toyota, KOA, to Sweetwater Creek SP

42mm weigh sta eb

41 Lee Rd, to Lithia Springs, **N...**Citgo/diesel/mart, Hardee's, Ace Hardware, transmissions, **S...**Speedway/diesel/mart/24hr, Waffle House

37 GA 92, to Douglasville, **N...**HOSPITAL, CHIROPRACTOR, VETERINARIAN, BP/mart/wash/24hr, Chevron, RaceTrac/mart/24hr, Texaco/diesel/mart/wash/atm/24hr, Circle K/gas, Arby's, Burger King, Capt D's, Catfish House, Checker's, Chick-fil-A, Chinese Rest., Church's Chicken, Cracker Barrel, Dairy Queen, Hardee's, KFC, Krystal/24hr, Little Caesar's, Mazzio's Pizza, McDonald's, Monterey Mexican, Mrs Winner's, Papa John's, Pizza Hut, Shoney's,

E ↕ W

Georgia Interstate 20

Subway, Taco Bell, Waffle House, Wendy's, Best Western, Comfort Inn, Day's Inn, Holiday Inn Express, Ramada Inn, Advance Parts, AutoZone, Big Lots, Dodge, $General, Gordy Tires, Kroger/24hr, Midas Muffler/brake, Valvoline, Wal-Mart/24hr, antiques, bank, carwash, laundry, **S**...CHIROPRACTOR, EYECARE, Amoco/repair/ 24hr, Chevron/mart/wash, Conoco/mart, Shell/diesel/mart, Pizza Hut, Aamco, Eckerd, Ingles Food, Mail'n More, Winn-Dixie, cleaners

36   Chapel Hill Rd, **N**...HOSPITAL, **S**...CHIROPRACTOR, Amoco/ McDonald's/diesel/mart/24hr, Citgo/mart, Conoco/mart, QT/mart/24hr, Texaco/mart/atm, Arby's, Blimpie, Hops Café/brewery, Logan's Road-house, NY Pizza, O'Charley's, Olive Garden, Outback Steaks, Rio Bravo, TGIFriday, Waffle House, Hampton Inn, Super 8, Bed Bath&Beyond, Borders, Circuit City, Dillard's, Eckerd, HiFi Buys, Gateway Computers, Goody's, Marshall's, MediaPlay, Michael's, OfficeMax, Old Navy, Parisian, Pier 1, Sears/auto, Target, ToysRUs, bank, cinema, cleaners, mall

34   GA 5, to Douglasville, **N**...MEDICAL CARE, DENTIST, Amoco, Texaco/ mart/atm, Hooters, Huddle House, Waffle House, Holiday Inn Express, Quality Inn, Sleep Inn, $Tree, Honda/Isuzu, Sam's Club, Wal-Mart SuperCtr/24hr, **S**...CHIROPRACTOR, DENTIST, EYECARE, VETERINARIAN, Chevron/mart/wash/24hr, Circle K/gas, Shell, Baskin-Robbins, Burger King, Chili's, Chick-fil-A, Chinese Rest., CiCi's, Dairy Queen, Dunkin Donuts, Folks Rest., Golden Corral, KFC, Krystal, LJ Silver, McDonald's, Mexican Rest., Miano's Pasta, Papa John's, Pizza Hut, Red Lobster, Ruby Tuesday, Ryan's, Subway, Taco Bell, Taco Mac, Tokyo Steaks, Waffle House, Wendy's, Suburban Lodge, Advance Parts, Brake Express, Cub Foods, Eckerd, Food Depot, Garden Ridge, GNC, Goodyear/auto, Goody's, Home Depot, JiffyLube, Jo-Ann Crafts, K-Mart, Kroger/deli, Lowe's Bldg Supply, MailBoxes Etc, PepBoys, Publix, Radio Shack, Rite Aid, USPO, Wolf Camera, XpressLube/ brakes, bank, cleaners

30   Post Rd, **N**...Texaco/diesel

26   Liberty Rd, Villa Rica, **N**...Texaco/diesel, **S**...Wilco/Citgo/diesel/24hr, Leather's Rest., Knight's Inn

24   GA 101, GA 61, Villa Rica, **N**...HOSPITAL, Amoco, Chevron, Exxon/mart, RaceTrac, Shell/diesel, Arby's, El Tio Mexican, Hardee's, KFC/Taco Bell, Krystal, McDonald's, New China Buffet, Pizza Hut, Romero's Italian, Subway, Waffle House, Wendy's, Comfort Inn, Super 8, CVS Drug, Ingles Foods, Mail'n Copy, Winn-Dixie, **S**...Texaco/diesel, Wal-Mart SuperCtr/24hr, to W GA Coll

21mm   Little Tallapoosa River

19   GA 113, Temple, **N**...Flying J/diesel/LP/rest./24hr, Williams/Main St Café/diesel/mart/24hr, BBQ, Hardee's, **S**...Texaco/diesel

15mm   weigh sta wb

11   US 27, Bremen, Bowdon, **N**...HOSPITAL, Texaco/diesel/mart/atm, Arby's, Hardee's(2mi), McDonald's, Pizza Hut, Waffle House, Wendy's, Day's Inn, Hampton Inn, Travelodge, $General, Ingles Foods, Wal-Mart, **S**...Amoco/diesel, BP/diesel/mart/24hr, Waffle House, John Tanner SP

9   Waco Rd, no facilities

5   GA 100, Tallapoosa, **N**...Exxon/diesel/mart/24hr, Shell/diesel, **S**...Citgo/Noble/diesel/mart/24hr, Big O/diesel, Marathon, Mobil/diesel, Pilot/KFC/Taco Bell/diesel/mart/24hr, Dairy Queen, Huddle House, Waffle House, Comfort Inn, to John Tanner SP

1mm   **Welcome Ctr eb, full(handicapped)facilities, phone, vending, picnic tables, litter barrels, petwalk**

0mm   Georgia/Alabama state line, Eastern/Central time zone

Exit #   Services

Georgia Interstate 59

I-59 begins/ends on I-24, exit 127. For I-24, turn to Tennessee Interstate 24.

20mm   I-24, W to Nashville, E to Chattanooga

17   Slygo Rd, to New England, **W**...Citgo/diesel/mart, KOA

11   GA 136, Trenton, **E**...MEDICAL CARE, EYECARE, Exxon/diesel/mart, Chevron/Fastpay, Citgo/mart, Burger King, Hardee's, Larry's Rest., McDonald's, Pizza Hut, Subway, El Rancho Court Motel, BiLo/drugs, CVS Drug, Family$, Ingles Foods, NAPA, Plymouth/Dodge, Smith's Auto Parts, antiques, to Cloudland Canyon SP, **W**...Amoco/TCBY/mart, Citgo/mart, Spur Gas/mart, Huddle House, Little Caesar's, Taco Bell, Food Lion/deli/ drugs, Price Drug, bank, carwash

11mm   scenic area/weigh sta both lanes, litter barrels

4   Rising Fawn, **E**...Citgo/diesel/mart/24hr, **W**...Amoco/Blimpie/TCBY/diesel/mart, Rising Fawn Café, camping

0mm   Georgia/Alabama state line, eastern/central time zone

E ↕ **W**

N ↕ **S**

Douglasville

Trenton

113

Georgia Interstate 75

Exit #	Services
355mm	Georgia/Tennessee state line
354.5mm	Chickamauga Creek
353	GA 146, Rossville, **E**...BP/mart/24hr, Chevron/mart, Knight's Inn, **W**...Cowboy's/mart(1mi), Exxon/diesel/mart/LP/24hr, Texaco/mart, carpet outlet
352mm	**Welcome Ctr sb, full(handicapped)facilities, info, phone, vending, picnic tables, litter barrels, petwalk**
350	GA 2, Bfd Pkwy, to Ft Oglethorpe, **E**...Exxon/diesel/mart, Sav-A-Ton/diesel/mart, **W**...HOSPITAL, Amoco/diesel, RaceTrac/mart/24hr, Texaco/mart, **3 mi W**...Conoco/diesel/rest., BBQ, Fazoli's, Taco Bell, KOA, Wal-Mart SuperCtr/24hr, to Chickamauga NP
348	GA 151, Ringgold, **E**...DENTIST, Citgo/mart/wash, Golden Gallon/diesel/mart, Texaco/diesel/mart, Aunt Effie's Rest., BBQ, Domino's, Hardee's, Krystal/24hr, KFC, McDonald's, Pizza Hut, Subway, Taco Bell, Waffle House, Best Western, Day's Inn, Hampton Inn, Holiday Inn Express, Super 8, Advance Parts, CVS Drug, Chevrolet, Chrysler/Plymouth/Jeep, Family$, Ingles/deli, bank, RV camping, **W**...Amoco, Chevron/mart/wash/24hr, Exxon/mart, Wendy's, Comfort Inn, Ford
345	US 41, US 76, Ringgold, **E**...BP/mart, Cowboys/diesel/mart, **W**...Phillips 66/mart, TA/Citgo/diesel/rest./24hr, Waffle House, Friendship Inn, Bell's Truck Repair/24hr, truckwash
343mm	weigh sta both lanes
341	GA 201, Tunnel Hill, to Varnell, **W**...Chevron/mart, Texaco/mart/wash, carpet outlets
336	US 41, US 76, Dalton, Rocky Face, **E**...HOSPITAL, BP/U-Haul, Chevron/mart, RaceTrac/mart, Texaco/mart, Waffle House, Country Hearth Inn, Stay Lodge, Home Depot, **W**...Amoco/diesel/mart, Phillips 66, Denny's, TaGin Chinese, Best Western/rest., Econolodge, Howard Johnson/rest., Motel 6, carpet outlets
333	GA 52, Dalton, **E**...EYECARE, BP, Chevron/mart/wash/24hr, Exxon/wash, RaceTrac/diesel/mart, Texaco, Applebee's, Burger King, Capt D's, KFC, Chick-fil-A, CiCi's, Cracker Barrel, Dairy Queen, Emperor Garden, Fuddrucker's, IHOP, KFC, LJ Silver, Longhorn Steaks, Los Pablos Mexican, McDonald's, O'Charley's, Outback Steaks, Pizza Hut, Shoney's, Steak'n Shake, Taco Bell, TCBY, Waffle House, Wendy's, Best Inn, Day's Inn, Hampton Inn, Travelodge, Big Lots, Chevrolet, Chrysler/Jeep, Ford, Isuzu, K-Mart, Kroger/deli/24hr, Mr Transmission, Tanger Outlets/famous brands, Walgreen, carpet outlet, bank, hardware, **W**...Texaco/mart, Red Lobster, Comfort Inn, Courtyard, Country Inn Suites, Holiday Inn, Jameson Inn, Wingate Inn, NW GA Trade/Conv Ctr
328	GA 3, to US 41, **E**...BP/Blimpie/KrispyKreme/mart, Pilot/Arby's/diesel/mart/24hr, Wendy's, Super 8, **W**...carpet outlets
326	Carbondale Rd, **E**...Chevron/diesel/mart, **W**...Citgo/diesel/mart, Phillips 66/diesel/mart
320	GA 136, Resaca, to Lafayette, **E**...Flying J/Conoco/diesel/LP/rest./24hr, truckwash
319mm	**Oostanaula River, rest area sb, full(handicapped)facilities, phone, vending, picnic tables, litter barrels, petwalk**
318	US 41, Resaca, **E**...Wilco/DQ/Wendy's/diesel/mart/24hr, Hardee's, Huddle House, Best Western, Knight's Inn, **W**...Exxon/diesel, Right Stuff Gas, Budget Inn, Duffy's Motel, Econolodge, Smith Motel, Super 8
317	GA 225, to Chatsworth, **E**...New Echota HS, Vann House HS, **W**...Express Inn/rest.
315	GA 156, Redbud Rd, to Calhoun, **E**...Citgo/diesel/mart, Exxon/mart, Waffle House, Scottish Inn, KOA(2mi), antiques, **W**...HOSPITAL, BP/diesel, Chevron/mart, Shell/mart, Texaco/diesel, Arby's, Shoney's, Howard Johnson, Ramada Ltd
312	GA 53, to Calhoun, **E**...Chevron/diesel/mart, Texaco/diesel/mart, Cracker Barrel, Denny's, Budget Host/rest., Quality Inn, Red Carpet Inn, Prime Outlets/famous brands, W...CHIROPRACTOR, BP/mart, Citgo/mart, Exxon/mart, Raceway, Shell/mart/wash, Arby's, Atlanta Gate Rest., BBQ, Bojangles/Central Park, Brangus Rest., Capt D's, Checker's, Chinese Rest., Dairy Queen, Golden Corral, Hickory House Rest., Huddle House, IHOP, KFC, Krystal/24hr, LJ Silver, McDonald's, Pizza Hut, Subway, Taco Bell, Waffle House, Wendy's, Day's Inn, Guest Inn, Hampton Inn, Holiday Inn Express, Jameson Inn, Knight's Inn, AutoZone, Chevrolet/Olds, CVS Drug, $General, Goodyear/auto, Ingles/deli, Kroger, MagicLube, Wal-Mart, Winn-Dixie, bank, laundry
308mm	**rest area nb, full(handicapped)facilities, phone, picnic tables, litter barrels, vending, petwalk**
306	GA 140, Adairsville, **E**...Amoco/mart/24hr, Patty's/Texaco/diesel/mart/24hr, QT/diesel/mart, Shell/diesel/mart, Wendy's, Travelodge, **W**...BP/Subway/mart, Citgo/diesel/mart, Exxon/diesel/mart, Burger King, Hardee's, Taco Bell, Waffle House, Comfort Inn, Country Hearth Inn, Ramada Ltd, camping
296	Cassville-White Rd, **E**...Speedway/Hardee's/Subway/diesel/mart, TA/Burger King/Popeyes/Sbarro's/diesel/mart/24hr, Texaco/Blimpie/24hr, **W**...BP/mart, Chevron/mart, Shell/mart, Waffle House/24hr, Budget Host/rest., Howard Johnson, Red Carpet Inn, Travelodge, KOA
293	US 411, to White, **E**...Shell/diesel/mart, Texaco, Scottish Inn, **W**...Chevron/diesel/mart, Coastal/diesel/café, Waffle House, Economy Inn, Holiday Inn, RV camping, museum, st patrol
290	GA 20, to Rome, **E**...Chevron/Subway/diesel/mart, Speedway/diesel/mart, Arby's, BBQ, Huddle House, McDonald's, Morrell's Rest., Wendy's, Best Western, Comfort Inn, Howard Johnson, Econolodge, Motel 6, Ramada Ltd, Super 8, **W**...HOSPITAL, BP/mart, Texaco/mart, Cracker Barrel, Shoney's, Waffle House, Day's Inn, Hampton Inn, Jameson Inn, RV camping(7mi)
288	GA 113, Cartersville, **2 mi W**...Amoco/24hr, BP/diesel, Applebee's, Blimpie, Burger King, Chick-fil-A, Golden Corral, Krystal/24hr, Mrs Winner's, Pizza Hut, Subway, Knight's Inn, Quality Inn, Kroger, to Etowah Indian Mounds(6mi)
286mm	Etowah River
285	Emerson, **E**...Texaco/mart/24hr, to Allatoona Dam, to Red Top Mtn SP
283	Emerson Allatoona Rd, Allatoona Landing Resort, camping
280.5mm	Allatoona Lake

Georgia Interstate 75

278 Glade Rd, to Acworth, **E...**BP/Subway/mart/wash, Shell/mart, to Glade Marina, camping, **W...**DENTIST, CHIROPRACTOR, Chevron/mart/wash, Citgo/mart, Burger King, Country Café, Dunkin Donuts, His&Hers Southern Buffet, Hong Kong Chinese, KFC, Krystal, Pizza Hut, Subway, Taco Bell, Waffle House, Western Sizzlin, Red Roof Inn, AutoZone, Big Lots, CVS Drug, Ingles/deli/24hr, K-Mart, NAPA, Radio Shack, laundry/cleaners, tires/brakes

277 GA 92, **E...**Amoco/mart, Exxon, RaceTrac/mart, Shell/diesel/mart/24hr, Texaco/diesel/mart, Hardee's, Shoney's, Waffle House, Holiday Inn Express, Ramada Ltd, **W...**BP/mart/wash, Shell/DQ/diesel/rest., McDonald's, Sonic, Waffle House, Wendy's, Zaxby's Café, Best Western, Day's Inn, Quality Inn, Super 8, CVS Drug, GNC, PakMail, Publix, cleaners

273 Wade Green Rd, **E...**DENTIST, BP/mart, Citgo/diesel/mart, RaceTrac/mart/wash, Arby's, Burger King, China King, Del Taco/Mrs Winners, Dunkin Donuts, McDonald's, Papa John's, Pizza Hut/Taco Bell, Subway, Waffle House, Rodeway Inn, Big Lots, Eckerd, GNC, Goodyear/auto, Publix/deli, bank, cleaners, **W...**Amoco, Conoco, Texaco/Blimpie/diesel, JiffyLube

271 Chastain Rd, to I-575 N, **E...**Chevron, Cracker Barrel, Best Western, Comfort Inn, Econolodge, Extended Stay America, Fairfield Inn, Residence Inn, Studio+, Suburban Lodge, Circuit City, Goodyear, Malibu Funpark, Outlets Ltd Mall, Wal-Mart/auto, **W...**Amoco/mart, Shell/Blimpie/mart, Texaco/Subway/diesel/mart/wash, Arby's, Del Taco, Mrs Winner's, Waffle House/24hr, Wendy's, Country Inn Suites, SpringHill Suites, Sun Suites, airport, museum

269 to US 41, to Marietta, **E...**CHIROPRACTOR, Chevron/mart/wash/24hr, Texaco/diesel/mart, Applebee's, Atlanta Bread, Baskin-Robbins, Burger King, Fuddrucker's, Grady's Grill, Happy China, HoneyBaked Ham, Itpolotis Italian, Longhorn Steaks, Manhattan Bagel, McDonald's, New China, Olive Garden, Piccadilly's, Pizza Hut, Red Lobster, Rio Bravo Cantina, Shoney's, Sizzler, Starbucks, Subway, $3 Café, Waffle House, Holiday Inn Express, Econolodge, Red Roof Inn, Shoney's Inn, Barnes&Nobel, Big 10 Tires, Drug Emporium, ExpressLube, Firestone/auto, Home Depot, HomePlace, KidsRUs, Kinko's, Macy's, Marshall's, MensWearhouse, Parisian, Publix, Rich's, Ritz Camera, Sears/auto, SteinMart, TJ Maxx, ToysRUs, cleaners, mall, **W...**BP/mart, Exxon/mart/wash, Chick-fil-A, Chili's, Cooker Rest., Golden Corral, Macaroni Grill, On-the-Border, Outback Steaks, Roadhouse Grill, Steak'n Shake, TGIFriday, Comfort Inn, Day's Inn, Hampton Inn, Sleep Inn, Baby SuperStore, Bed Bath&Beyond, Best Buy, Buick/Pontiac/GMC, CarMax, Chevrolet, Goody's, Linens'n Things, MediaPlay, Mitsubishi, NTB, Office Depot, OfficeMax, Old Navy, Olds/Kia/Toyota, PetsMart, Phar-Mor Drug, Sears HomeLife, Service Merchandise, Sports Authority, Target, bank, camping, cinema, mall, to Kennesaw Mtn NP

268 I-575 N, GA 5 N, to Canton

267b a GA 5 N, to US 41, Marietta, **1/2 mi W...**HOSPITAL, BP, Citgo, Burger King, McDonald's, Welcome Inn America

265 GA120, N Marietta Pkwy, **W...**Amoco, Shell/diesel, Arrival Inn, Budget Inn, Crown Inn, Sun Inn, Suburban Lodge, Travelers Motel

263 GA 120, Marietta, to Roswell, **E...**VETERINARIAN, Chevron/diesel/mart/24hr, Shell/mart/24hr, Harry's Farmers Mkt, cleaners, **W...**DENTIST, Amoco, Exxon/diesel, Applebee's, Blackeyed Pea, Chili's, Chinese Rest., Dairy Queen, Hardee's, Piccadilly's, Subway, Best Western, Fairfield Inn, Four Points Hotel, Hampton Inn, Marietta Motel, Mayflower Motel, Ramada Ltd, Sheraton, Super 8, Wyndham Garden, U-Haul, **1-2 mi W...**BBQ, Buffalo's Café, Checker's, Chick-fil-A, Dairy Queen, Krystal, Shoney's, Sonic, Taco Bell, Suburban Lodge, AutoZone, Brake-O, Cub Foods, K-Mart, Sam's Club, XpressLube

261 GA 280, Delk Rd, to Dobbins AFB, **E...**Exxon/mart/wash, Shell/mart/24hr, Texaco/diesel/mart, Chinese Rest., Denny's, Hardee's, KFC, McDonald's, Papa John's, Spaghetti Whse, Waffle House, Courtyard, Drury Inn, Howard Johnson, Knight's Inn, Motel 6, Scottish Inn, Sleep Inn, Super 8, Travelers Inn, JiffyLube, laundry/cleaners, **W...**Amoco/wash, BP/mart/wash, Chevron/mart/wash/24hr, Shell/24hr, Cracker Barrel, D&B Rest., Waffle House, Best Inn, Comfort Inn, Fairfield Inn, Holiday Inn, La Quinta

260 Windy Hill Rd, to Smyrna, **E...**Amoco/mart/wash, BP/mart/wash, Applebee's, Boston Mkt, Bruegger's Bagels, Chinese Rest., Cooker Rest., Fuddrucker's, Houston's Rest., LePeep Rest., Manhattan Bagels, NY Pizza, Pappasito's Cantina, Pappadeaux Seafood, Quizno's Subs, Schlotsky's, Subway, TGIFriday, Clarion Suites, Econolodge, Hawthorn Suites, Hyatt, Marriott, Travelodge, CVS Drug, bank, cleaners, **W...**HOSPITAL, DENTIST, Chevron/wash, Citgo, Texaco/diesel/mart, Arby's, Chick-fil-A, Fuddrucker's, McDonald's, Popeye's, Sub Shop, $3 Café, Waffle House, Wendy's, Best Western, Courtyard, Day's Inn, Hilton, Masters Inn, Red Roof Inn, Target, bank

259b a I-285, W to Birmingham, Montgomery, E to Greenville

258 Cumberland Pkwy

257mm Chattahoochee River

256 to US 41, Northside Pkwy, no facilities

255 US 41, W Paces Ferry Rd, **W...**Amoco/24hr, **E...**HOSPITAL, BP/diesel/wash, Chevron, McDonald's, Dairy Queen, Avanti's Italian, Steak'n Shake, OK Café/24hr, Chinese Rest., Pero's Pizza, Houston's Rest., A&P, Big B Drug, Mailboxes Etc, bank, hardware, laundry

Georgia Interstate 75

254	Moores Mill Rd, no facilities
252b	Howell Mill Rd, **E**...Texaco/mart/wash, Hardee's, McDonald's, Budget Inn, Eckerd, Goodyear, Winn-Dixie, bank, **W**...CHIROPRACTOR, Texaco/diesel/mart, Arby's, Copper Kettle, Hunter's Rest., KFC, LJ Silver, Piccadilly's, Sensational Subs, Taco Bell, Wendy's, Castlegate Inn, Holiday Inn, Firestone, JiffyLube, Kroger/deli/24hr, Precision Tune, Tune-Up Clinic, Valvoline, bank, hardware
a	US 41, Northside Dr, **W**...BP/wash, Waffle House, Day's Inn, Firestone
251	I-85 N, to Greenville
250	Techwood Dr(from sb), 10th St, 14th St, **E**...Travelodge
249d	10th St, Spring St(from nb), **E**...BP, Chevron/mart/24hr, Checker's, Domino's, Pizza Hut, Varsity Drive-In, Fairfield Inn, Regency Suites, Renaissance Hotel, Residence Inn, **W**...HOSPITAL, McDonald's, Courtyard, Comfort Inn, to GA Tech
249c	Williams St(from sb), downtown, to GA Dome
249b	Pine St, Peachtree St(from nb), downtown, **W**...HOSPITAL, Hilton, Marriott
249a	Courtland St(from sb), downtown, **W**...Hilton, Marriott, GA St U
248d	Piedmont Ave, Butler St(from sb), downtown, **W**...HOSPITAL, Courtyard, Fairfield Inn, Radisson, Ford, MLK NHS
248c	GA 10 E, Intn'l Blvd, downtown, **W**...Hilton, Holiday Inn, Marriott Marquis, Radisson
248b	Edgewood Ave(from nb), **W**...HOSPITAL, downtown, hotels
248a	MLK Dr(from sb), **W**...st capitol, to Underground Atlanta
247	I-20, E to Augusta, W to Birmingham
246	Georgia Ave, Fulton St, **E**...Hampton Inn, Holiday Inn Express, stadium, **W**...Amoco/mart, KFC, to Coliseum, GSU
245	Ormond St, Abernathy Blvd, **E**...Hampton Inn, Holiday Inn Express, stadium, **W**...st capitol
244	University Ave, **E**...Exxon, **W**...Amoco, Mrs Winner's, Ford Trucks
243	GA 166, Lakewood Fwy, to East Point, no facilities
242	I-85 S, to airport
241	Cleveland Ave, **E**...Chevron, Checker's, McDonald's, Palace Inn, K-Mart, **W**...Shell/mart, Texaco, Burger King, Church's, Krystal/24hr, American Inn, Day's Inn
239	US 19, US 41, **E**...Chevron, Checker's, Waffle House, **W**...Amoco/mart, IHOP, Krystal, McDonald's, Best Western, USPO, to airport
238b a	I-285 around Atlanta
237a	GA 85 S(from sb), **W**...Denny's, Day's Inn, Day's Lodge, King's Inn
237	GA 331, Forest Parkway, **E**...Chevron/mart/wash, Exxon, Happy Store/gas, Shell, Burger King, McDonald's, Waffle House, Econolodge, Motel 6, Rodeway Inn, Farmer's Mkt, Chevrolet, **W**...Day's Inn, Ramada Ltd
235	US 19, US 41, GA 3, Jonesboro, **E**...Phillips 66/mart, Hardee's, Waffle House, Super 8, K-Mart, RV Ctr, **W**...HOSPITAL, Amoco, Marathon, RaceTrac/diesel, Shell, Texaco, Checker's, Dunkin Donuts, Folks Rest., Johnny's Pizza, KFC, Krystal, McDonald's, Red Lobster, Waffle House, Comfort Inn, Day's Inn, Econolodge, Holiday Inn Express, Shoney's Inn/rest., Cub Foods, Dodge, Office Depot
233	GA 54, Morrow, **E**...MEDICAL CARE, BP/mart/wash, Chevron, Conoco/diesel/mart, Cracker Barrel, Krystal/24hr, Mrs Winner's, Taco Bell, Waffle House, Wendy's, Best Western, Drury Inn, Fairfield Inn, Red Roof Inn, bank, **W**...Exxon/mart/24hr, RaceTrac/mart, Texaco/diesel/mart, Bennigan's, China Café, Japanese Rest., KFC, LJ Silver, McDonald's, Pizza Hut, Shoney's, Waffle House, Hampton Inn, Quality Inn, Acura/Cadillac, Best Buy, Burlington Coats, Chevrolet, Golf Whse, HobbyLobby, JiffyLube, Kroger, Midas Muffler, Nissan, OfficeMax, Olds, Pier 1, Service Merchandise, Sports Authority, Toyota, mall
231	Mt Zion Blvd, **E**...Honda, **W**...EYESMART, FOOTCARE, Citgo, Conoco, Exxon/mart/wash, Arby's, Blimpie, Chick-fil-A, Chili's, Del Taco, Longhorn Steaks, McDonald's, Mrs Winner's, On-the-Border, Papa John's, Philly Subs, Steak'n Shake, TGIFriday, Waffle House, Wendy's, Country Inn Suites, Extended Stay America, Howard Johnson Express, Sleep Inn, Sun Suites, BabysRUs, Baby SuperStore, Circuit City, Goody's, Home Depot, HomePlace, MailBoxes Etc, MediaPlay, Michael's, NTB, Old Navy, PetsMart, Publix, Target, XpressLube, cinema
228	GA 54, GA 138, Jonesboro, **E**...HOSPITAL, DENTIST, CHIROPRACTOR, Exxon/mart/wash, RaceTrac/mart/24hr, Applebee's, Arby's, Burger King, Chick-fil-A, Chinese Rest., CiCi's, Dairy Queen, Damon's, Frontera Mexican, Golden Corral, Gregory's Grill, Hardee's, IHOP, Kenny Rogers, KFC, Krystal, LJ Silver, McDonald's, Piccadilly's, Philly Connection, Shoney's, Subway, Taco Bell, Waffle House, Wendy's, Best Western, Day's Inn, Comfort Inn, Holiday Inn Express, Motel 6, Ramada Ltd, Shoney's Inn, Bruno's Foods, GNC, Goodyear, K-Mart, Kroger, MailBoxes Etc, XpressLube, bank, cleaners, **W**...Amoco/mart/24hr, Chevron/mart, Citgo/diesel, Waffle House, CarMax, Chrysler/Plymouth/Jeep, Garden Ridge
227	I-675 N, to I-285 E(from nb)
224	Hudson Bridge Rd, **E**...HOSPITAL, BP/mart, Citgo/24hr, Phillips 66, Texaco/Chick-fil-A/diesel/mart, Dunkin Donuts, McDonald's, Subway, Waffle House, Wendy's, AmeriHost Inn, Eagle Landing Inn, CVS Drug, **W**...Chinese Rest., Teddy's Diner/24hr, Super 8
222	Jodeco Rd, **E**...Amoco/mart, Citgo/24hr, Hardee's, Waffle House, **W**...BP/wash, Chevron/diesel, KOA
221	Jonesboro Rd, **E**...Williams/Chick-fil-A/diesel/mart/24hr, **W**...flea market, KOA
218	GA 20, GA 81, McDonough, **E**...Amoco/mart, BP, Applebee's, Arby's, Burger King, Dairy Queen, KFC, McDonald's, Mrs Winner's, Pizza Hut, Taco Bell, Waffle House, Wendy's, Best Western, Budget Inn, Hampton Inn, Red Carpet Inn, **W**...Shell/mart/24hr, Speedway/diesel/mart, Subway, Waffle House, Comfort Inn, Econolodge, HoJo's, Master's Inn

Georgia Interstate 75

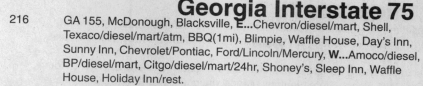

216 GA 155, McDonough, Blacksville, **E...**Chevron/diesel/mart, Shell, Texaco/diesel/mart/atm, BBQ(1mi), Blimpie, Waffle House, Day's Inn, Sunny Inn, Chevrolet/Pontiac, Ford/Lincoln/Mercury, **W...**Amoco/diesel, BP/diesel/mart, Citgo/diesel/mart/24hr, Shoney's, Sleep Inn, Waffle House, Holiday Inn/rest.

212 to US 23, Locust Grove, **E...**BP/Subway/diesel/mart, Citgo, Exxon/diesel/mart, Shell/diesel/mart, Denny's, Hardee's, Huddle House, Waffle House, Executive Inn, Outlet Inn, Red Carpet, Scottish Inn, Travel Inn, Tanger Outlet/famous brands, **W...**Chevron/diesel, Scottish Inn, Super 8

205 GA 16, to Griffin, Jackson, **E...**BP, Citgo/diesel, BBQ, **W...**Amoco, Chevron/diesel, Texaco/diesel/mart

201 GA 36, to Jackson, Barnesville, **E...**Sunshine/diesel/grill/24hr, TA/Citgo/Subway/Taco Bell/diesel/24hr, Wilco/Exxon/DQ/Wendy's/diesel/mart/24hr, Huddle House, Blue Beacon, **W...**BP, Flying J/Conoco/Hardee's/diesel/LP/24hr

198 Highfalls Rd, **E...**Exxon(1mi), High Falls SP, **W...**High Falls RV Park

193 Johnstonville Rd, **E...**BP

190mm weigh sta both lanes

188 GA 42, **E...**Shell/mart/24hr, Best Value Inn, Best Western, Super 8, to Indian Spings SP, RV camping

187 GA 83, Forsyth, **E...**El Tejado Mexican, Econolodge, New Forsyth Inn, Regency Inn, **W...**Amoco/Blimpie/mart, Citgo/diesel/mart, Conoco/diesel/mart, Exxon/mart, Texaco, Burger King, Capt D's, Hardee's, McDonald's, Pizza Hut, Subway, Taco Bell, Waffle House, Wendy's, Day's Inn, Tradewinds Motel, Advance Parts, CVS Drug, Family$, Piggly Wiggly/24hr, Wal-Mart, bank, laundry

186 Tift College Dr, Juliette Rd, Forsyth, **E...**Jarrell Plantation HS, KOA, **W...**HOSPITAL, BP/diesel, Chevron/mart, Shell/mart/24hr, Dairy Queen, Hong Kong Café, Waffle House, Ambassador Inn, Hampton Inn, Holiday Inn/rest., Chrysler/Plymouth/Dodge, Ingles/deli/bakery

185 GA 18, **E...**L&D RV Park, **W...**Amoco/Waffle King/mart/24hr, Texaco/diesel/mart, Shoney's, Comfort Inn, Ford, st patrol

181 Rumble Rd, to Smarr, **E...**BP/Subway/diesel/mart/24hr, Shell/diesel/mart/24hr

179mm **rest area sb, full(handicapped)facilities, phone, vending, picnic tables, litter barrels, petwalk**

177 I-475 S around Macon(from sb)

175 Pate Rd, Bolingbroke(from nb, no re-entry), no facilities

172 Bass Rd, **E...**funpark, **W...**to Museum of Arts&Sciences

171 US 23, to GA 87, Riverside Dr, **E...**BP, Marathon/diesel, Shell, Cracker Barrel, Huddle House/24hr, Jack&Jill's Brewery

169 to US 23, Arkwright Dr, **E...**Shell/mart/24hr, Carrabba's, Logan's Roadhouse, Outback Steaks, Waffle House, Wager's Grill, Courtyard, Fairfield Inn, La Quinta, Red Roof Inn, Residence Inn, Sleep Inn, Super 8, Buick/Cadillac/GMC/Saturn, **W...**HOSPITAL, VETERINARIAN, OPTICIAN, Amoco/diesel/mart, Chevron/mart/24hr, Conoco/diesel/mart, Applebee's, Baskin-Robbins, Burger King, Chick-fil-A, Chili's, Chinese Rest., Cracker Barrel, Dunkin Donuts, Heavenly Ham, Hooters, KFC, Krystal, Longhorn Steaks, McDonald's, Papa John's, Papoulis' Gyros, Popeye's, Rio Bravo Cantina, Ryan's, Steak'n Shake, Subway, Taco Bell, Waffle House, WhataPizza, Hampton Inn, Holiday Inn, Quality Inn, Ramada Ltd, Shoney's Inn/rest., Studio+, Wingate Inn, Acura, Barnes&Noble, BMW/Mitsubishi, Chrysler/Jeep/Dodge, GNC, K-Mart, Kroger, Lexus, Olds/Pontiac, Publix, Radio Shack, Rite Aid, Volvo, bank, carwash, cinema, tires, same as 167

167 GA 247, Pierce Ave, **E...**Budget Inn Suites, **W...**Amoco/Blimpie/mart, BP/diesel, Chevron/mart, Exxon, Marathon/Subway/diesel/mart, Texaco/diesel/mart, Applebee's, Arby's, Bennigan's, Capt D's, Denny's, Pizza Hut, Red Lobster, S&S Cafeteria, Shogun Japanese, SteakOut Rest., Texas Cattle Co, Waffle House, Wendy's, Ambassador Inn, Best Western/rest., Comfort Inn, Holiday Inn Express, Howard Johnson, Eckerd, Goodyear/auto, Piece Goods

165 I-16 E, to Savannah

164 US 41, GA 19, Forsyth Ave, Macon, **E...**Marathon, Sid's Rest., hist dist, **W...**HOSPITAL, CHIROPRACTOR, BP/lube/wash, museum

163 GA 74 W, Mercer U Dr, **E...**to Mercer U, **W...**Citgo, Marathon/diesel

162 US 80, GA 22, Eisenhower Pkwy, **W...**Amoco/mart, Chevron/mart/24hr, Speedway, Capt D's, Checker's, IHOP, Johnnie's Rest., LJ Silver, McDonald's, Mrs Winners, Taco Bell/TCBY, Wendy's, Suburban Lodge, Advance Parts, Burlington Coats, FoodMax, Goodyear, Home Depot, K-Mart, MediaPlay, Meineke Muffler, Midas Muffler, Office Depot, PepBoys, PetsMart, Wal-Mart

160 US 41, GA 247, Pio Nono Ave, **E...**VETERINARIAN, Exxon/diesel/mart, RaceTrac/mart, Waffle House, Masters Inn, camper sales/service, **W...**Citgo/mart, Enmark/diesel, Shell, Arby's, Chinese Rest., Gabby's Diner, KFC, McDonald's, Pizza Hut, Shoney's, Subway, Waffle House, Advance Parts, Discount Parts, Eckerd, JiffyLube, Piggly Wiggly, USPO, bank, same as 162

156 I-475 N around Macon(from nb)

N

S

Forsyth

Macon

Georgia Interstate 75

M a c o n

155 Hartley Br Rd, **E**...BP/KFC/Pizza Hut/diesel/mart/24hr, Marathon/diesel/mart, Phillips 66/diesel/mart, Shell/mart, Golden Wok, Popeye's, Wendy's, Kroger/24hr, bank, cleaners, **W**...Citgo/mart, Exxon/mart, Subway, Waffle House, Ambassador Inn, auto repair

149 GA 49, Byron, **E**...Shell/mart/24hr, Texaco/diesel/mart, Burger King, McDonald's, Pizza Hut, Shoney's, Waffle House, Best Western, Super 8, Coachmen RV Ctr, GNC, Peach Stores/famous brands, antiques, **W**...BP/mart, Citgo/diesel/mart/24hr, Marathon/diesel/mart, RaceTrac/mart/atm, Speedway/diesel/mart, Dairy Queen, Huddle House, Papa's Pizza, Popeye's, Subway, Waffle House, Comfort Inn, Day's Inn, Econolodge, Passport Inn, Chevrolet/Olds, Ford, antiques

146 GA 247, to Centerville, **E**...HOSPITAL, Exxon, Shell, Speedway/diesel/mart, Steak'n Shake(3mi), Subway, Waffle House, GuestHouse Inn, Ramada Inn(4mi), to Robins AFB, museum, **W**...Pilot/Arby's/diesel/mart/24hr, Red Carpet Inn, Robins Travel Park

142 GA 96, Housers Mill Rd, **E**...Shell/repair, Ponderosa RV Park

138 Thompson Rd, **E**...Phillips/ChesterFriedChicken/diesel/mart, **W**...airport

P e r r y

136 US 341, Perry, **E**...HOSPITAL, Amoco/mart/wash, Chevron/Baskin-Robbins/diesel/mart/24hr, Shell, Speedway/diesel/mart, Arby's, BBQ, Burger King, Capt D's, Chick-fil-A, Chinese Rest., Hardee's, Jalisco Grill, KFC, Krystal, McDonald's, Pizza Hut, Quincy's, Red Lobster, Shoney's, Subway, Taco Bell, Waffle House, Wendy's, Zaxby's Café, Fairfield Inn, Hampton Inn, Ramada Ltd, Red Gable Inn, Super 8, Advance Parts, CastroLube, $Tree, GNC, K-Mart, Kroger, Radio Shack, Wal-Mart SuperCtr/24hr, bank, **W**...BP, Chevron, Conoco/diesel/mart, RaceTrac/mart/24hr, Angelina's Café, Applebee's, Comfort Inn, Day's Inn, Econolodge, Holiday Inn/rest., Knight's Inn, Holiday Inn, Jameson Inn, Manor Inn, Quality Inn, Ford, Crossroads Camping

135 US 41, GA 127, Perry, **E**...BP/diesel/mart, Exxon/mart, Shell/mart, Speedway/diesel/mart/24hr, Texaco, Cracker Barrel, Waffle House, Best Western, Day's Inn, Red Carpet Inn, Rodeway Inn, Scottish Inn, Travelodge, Plymouth/Jeep/Dodge, GA Nat Fair, **W**...Chinese Rest., Chevrolet/Buick/Pontiac/Olds/GMC, GA Patrol

134 South Perry Pkwy

127 GA 26, Henderson, **E**...Twin Oaks Camping, **W**...Chevron/Ice Cream/mart, Henderson Lodge(1mi), Judy's RV Park

122 GA 230, Unadilla, **E**...Dixie Gas/mart, Chevrolet/Ford, **W**...Phillips 66/mart, Red Carpet Inn

121 US 41, Unadilla, **E**...BP/mart, Shell/mart, Texaco/Stuckey's/Dairy Queen, Dooly's Kuntry Kitchen, Cotton Patch Rest., Subway, Day's Inn, Economy Inn, Scottish Inn, $General, Food Pride Foods, CarQuest, Lamp Factory Outlet, Southern Trails RV Resort, **W**...Citgo/diesel/rest./atm/24hr, Passport Inn

118mm **rest area sb, full(handicapped)facilities, phone, vending, picnic tables, litter barrels, petwalk**

117 to US 41, Pinehurst, **W**...BP/diesel/mart, New Colony Inn/rest.

112 GA 27, Vienna, **W**...BP/diesel/mart, Marathon

109 GA 215, Vienna, **E**...BP/diesel, **W**...HOSPITAL, Amoco/mart, Chevron/mart, Citgo/diesel/mart, Hardee's, Huddle House, Popeye's, Knight's Inn, antiques, flea mkt

108mm **rest area nb, full(handicapped)facilities, phone, vending, picnic tables, litter barrels, petwalk**

104 Farmers Mkt Rd, Cordele, **E**...Super 8/rest., **W**...Phillips 66/diesel/wash, RV repair

102 GA 257, Cordele, **E**...Shell, **W**...HOSPITAL, Pecan House/gas

C o r d e l e

101 US 280, GA 90, Cordele, **E**...Citgo/mart, Exxon/diesel/mart, Texaco, Williams/Perkins/diesel/mart/24hr, Denny's, Waffle House, Day's Inn, Ramada Inn, Ford/Lincoln/Mercury, st patrol, **W**...Amoco/mart, BP/diesel, Chevron/KrispyKreme/mart/wash/24hr, RaceTrac/24hr, Burger King, Capt D's, Chinese Rest., Cracker Barrel, Dairy Queen, Domino's, Golden Corral, Hardee's, KFC, Krystal/24hr, McDonald's, Pizza Hut, Shoney's, Subway, Taco Bell, Wendy's, Western Steer, Athens Motel, Comfort Inn, Economy Inn, Hampton Inn, Holiday Inn Express, Passport Inn, Quality Inn, Rodeway Inn, Ace Hardware, CVS Drug, $General, Firestone, Flint River Pottery, Goody's, Piggly Wiggly, Wal-Mart SuperCtr/24hr, Winn-Dixie/deli/24hr, bank, cinema, to Veterans Mem SP, J Carter HS

99 GA 300, GA/FL Pkwy, **W**...to Chehaw SP

97 to GA 33, Wenona, **E**...Royal Inn, Cordele RV Park, antiques, flea mkt, **W**...BP/Hardee's/Pizza Hut/TCBY/diesel, KOA

92 Arabi, **E**...Chevron/Plantation House, Phillips 66, Budget Inn, **W**...BP/mart

85mm **rest area nb, full(handicapped)facilities, phone, vending, picnic tables, litter barrels, petwalk**

84 GA 159, Ashburn, **E**...Shell/Ice Cream/diesel, **W**...A-1 Trkstp/Phillips 66/mart, BP/Subway, Waffle King, Knight's Inn/RV Park

82 GA 107, GA 112, Ashburn, **W**...BP/24hr, Chevron/mart, Hardee's, Huddle House/24hr, McDonald's, Pizza Hut, Shoney's, Comfort Inn, Day's Inn, Ramada Ltd, Super 8, Chevrolet, Rite Aid, to Chehaw SP

80 Bussey Rd, Sycamore, **E**...Exxon/Subway/mart, Shell, Budget Lakeview Inn/camping, **W**...Chevron, auto/tire repair

78 GA 32, Sycamore, **E**...to Jefferson Davis Memorial Park(14mi)

76mm **rest area sb, full(handicapped)facilities, phone, vending, picnic tables, litter barrels, petwalk**

75 Inaha Rd, **E**...BP/mart, **W**...Citgo/Stuckey's/Dairy Queen

71 Willis Still Rd, Sunsweet, **W**...BP/diesel/produce, RV camping

69 Chula-Brookfield Rd, **E**...Red Carpet Inn/gas/rest./24hr, antiques

66 Brighton Rd, no facilities

64 US 41, Tifton, **E**...HOSPITAL, Chevron, $General, Food Lion, **W**...Citgo/diesel/mart

63b 8th St, Tifton, **E**...Exxon/mart, Texaco/mart, Chinese Rest., Hardee's, KFC, Krystal, McDonald's, Day's Inn, Belk, JC Penney, Winn-Dixie, **W**...Split Rail Grill, same as 63a

N

S

Georgia Interstate 75

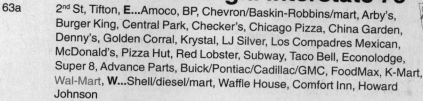

63a   2nd St, Tifton, **E**...Amoco, BP, Chevron/Baskin-Robbins/mart, Arby's, Burger King, Central Park, Checker's, Chicago Pizza, China Garden, Denny's, Golden Corral, Krystal, LJ Silver, Los Compadres Mexican, McDonald's, Pizza Hut, Red Lobster, Subway, Taco Bell, Econolodge, Super 8, Advance Parts, Buick/Pontiac/Cadillac/GMC, FoodMax, K-Mart, Wal-Mart, **W**...Shell/diesel/mart, Waffle House, Comfort Inn, Howard Johnson

62   US 82, to US 319, Tifton, **E**...BP, Citgo, Exxon/diesel/mart, Applebee's, Charles Seafood Rest., Country Buffet, Cracker Barrel, Dairy Queen(1mi), Sonic, Waffle House, Western Sizzlin, Courtyard, Hampton Inn, Masters Inn, Microtel, Ford/Lincoln/Mercury, Pecan Outlet, Staples, auto/tire repair, **W**...Amoco/Subway/mart, Chevron/mart/24hr, RaceTrac/24hr, Shell/Wendy's/mart/wash, Texaco/Burger King/mart, BBQ, Capt D's, Chick-fil-A, Shoney's, Waffle House, Wendy's, Day's Inn(2mi), Holiday Inn, Ramada Ltd, Scottish Inn, Chevrolet/Olds, Chrysler/Jeep, Honda, Mazda, Plymouth/Dodge, Toyota, Volvo

61   Omega Rd, **W**...Citgo/diesel, Waffle King/24hr, Motel 6

60   Central Ave, Tifton, **E**...Amoco/KrispyKreme/mart/wash, **W**...Shell/Steak'n Shake/diesel/mart

59   Southwell Blvd, to US 41, Tifton, no facilities

55   to Eldorado, Omega, **E**...Chevron/Magnolia Plantation Gifts, **W**...Phillips 66/diesel/mart, Shell

49   Kinard Br Rd, Lenox, **E**...Dixie/diesel/mart, Knight's Inn, **W**...BP/diesel/mart/24hr, Texaco/diesel/mart, Blimpie

47.5mm   **rest area both lanes, full(handicapped)facilities, phone, vending, picnic tables, litter barrels, petwalk**

45   Barneyville Rd, **E**...Red Carpet Inn/rest.

41   Rountree Branch Rd, **W**...to Reed Bingham SP

39   GA 37, Adel, Moultrie, **E**...Shell/McDonald's/mart, Texaco, Dairy Queen, Hardee's, Pizza Hut, Subway, Waffle House, Howard Johnson Express, Scottish Inn, Super 8, **W**...BP/IHOP/mart, Citgo/Huddle House/diesel/mart, Burger King, Capt D's/Burger King/Popeye's, Mama's Table Rest., Taco Bell, Western Sizzlin, Day's Inn, Hampton Inn, Adel Factory Stores/famous brands, auto/truck repair, to Reed Bingham SP

37   Adel, **E**...BP/café

32   Old Coffee Rd, Cecil, **W**...Amoco/mart

29   US 41 N, GA 122, Hahira, Sheriff's Boys Ranch, **E**...Subway(.8mi), **W**...BP, Sav-A-Ton/Apple Valley Café/diesel/24hr, Super 8/Citgo/Stuckey's

23mm   weigh sta both lanes

22   US 41 S, to Valdosta, **E**...HOSPITAL, Shell/Baskin-Robbins/diesel/mart/24hr, Chevrolet/Mazda, Chrysler/Jeep/Toyota, Ford/Lincoln/Mercury, golf course, **W**...Citgo/Burger King/Dairy Queen/Stuckey's/Day's Inn/diesel, RV camping

18   GA 133, Valdosta, **E**...Amoco/KrispyKreme/mart/atm, Chevron, Citgo, Shell/mart, Texaco/dieel/mart, Applebee's, Arby's, BBQ, Burger King, Chick-fil-A, Cracker Barrel, Denny's, El Potro Mexican, Fazoli's, Hardee's, KFC, Krystal, Little Caesar's, Longhorn Steaks, McDonald's, Ole Time Country Buffet, Outback Steaks, Quincy's, Red Lobster, Ruby Tuesday, Steak'n Shake, Subway, Taco Bell, Waffle House, Wendy's, Clubhouse Inn, Fairfield Inn, Hampton Inn, Holiday Inn, Jameson Inn, Jolly Inn, Quality Inn, Scottish Inn, Travelodge, Belk, Food Lion, Goody's, Home Depot, InstaLube, JC Penney, K-Mart/24hr, MailBoxes Etc, Michaels, Pier 1, ProWash, Publix, Target, TJ Maxx, ToysRUs, Wal-Mart/auto, cinema, mall, **W**...BP/mart/repair/24hr, Gas'n Go/mart, Best Western/rest.

16   US 84, US 221, GA 94, Valdosta, **E**...Amoco/mart/wash/24hr, BP/diesel/mart/atm, Citgo/mart, Phillips 66, Shell/KrispyKreme/mart/atm/24hr, Williams/diesel/mart, Alligator Oriental Rest., Burger King, IHOP, McDonald's, Pizza Hut, Waffle House, Wendy's, Big 7 Motel, Day's Inn, Motel 6, Quality Inn, Ramada Ltd, Rodeway Inn, Shoney's Inn/rest., Super 8, Sam's Club, to Okefenokee SP, **W**...Texaco/Huddle House/diesel/rest./24hr, Austin's Steaks, Comfort Inn, Knight's Inn

13   Old Clyattville Rd, Valdosta, **W**...Wild Adventures Park

11   GA 31, Valdosta, **E**...Speedway/Subway/diesel/mart/24hr, Wilco/Exxon/Bojangles/diesel/mart/24hr, Waffle House, Villager Lodge, **W**...BP/mart, Texaco/repair, airport

5   GA 376, to Lake Park, **E**...Amoco/mart/wash, Chevron/mart, RaceTrac/mart, Shell/mart, Texaco/mart, Chick-fil-A, Farmhouse Rest., Hardee's, Subway, Waffle House, Holiday Inn Express, Shoney's Inn/rest., MillStore Plaza Outlet, TG&Y, Winn-Dixie/deli/24hr, bank, camping, golf, **W**...Citgo/Baskin-Robbins/Burger King/diesel/mart/24hr, Phillips 66/diesel/mart, Cracker Barrel, McDonald's, Pizza Hut, Taco Bell, Day's Inn, Hampton Inn, Super 8, Travelodge, Factory Stores of America/famous brands, GA Winnebago

3mm   **Welcome Ctr nb, full(handicapped)facilities, phone, vending, picnic tables, litter barrels, petwalk**

2   Lake Park, Bellville, **E**...TA/BP/Taco Bell/yogurt/diesel/rest./24hr, Shell/repair, Texaco/Dairy Queen, **W**...Flying J/Conoco/diesel/LP/rest./24hr, Best Western

0mm   Georgia/Florida state line

N

S

Georgia Interstate 85

Lavonia

Exit #	Services
179mm	Lake Hartwell, Tugaloo River, Georgia/South Carolina state line
177	GA 77 S, to Hartwell, **E**...BP/Dad's Rest./mart/gifts, to Hart SP, Tugaloo SP
176.5mm	**Welcome Ctr sb, full(handicapped)facilities, info, phone, picnic tables, litter barrels, vending, petwalk**
173	GA 17, to Lavonia, **E**...DENTIST, CHIROPRACTOR, BP/mart, RaceTrac/mart/24hr, Texaco/diesel/mart, Fernside Buffet, KFC, McDonald's, Subway, Waffle House, Day's Inn, Sleep Inn, Big A Parts, Comm Cash Food, $General, Rite Aid, bank, tires/lube, **W**...Exxon/mart, Shell/mart, Arby's, Burger King, Hardee's, Mac's Donuts, Pizza Hut, Waffle House, Wendy's, Shoney's Inn/rest., Chrysler/Plymouth, to Tugaloo SP
171mm	weigh sta nb
169mm	weigh sta sb
166	GA 106, to Carnesville, **E**...Shell/Pizza/Subs/mart, **W**...Echo Truckstop/diesel/rest./24hr, repair
164	GA 320, to Carnesville, **E**...Sunshine Travel/Hardee's/diesel/24hr, tires
160.5mm	**rest area nb, full(handicapped)facilities, phone, vending, picnic tables, litter barrels, petwalk**
160	GA 51, to Homer, **E**...Shell/Subway/diesel/mart/24hr, **W**...Petro/Amoco/diesel/rest./24hr, Blue Beacon
154	GA 63, Martin Br Rd, no facilities

Commerce

149	US 441, GA 15, to Commerce, **E**...HOSPITAL, Amoco/24hr, Citgo/diesel/rest./24hr, TA/diesel/rest./24hr, BBQ, Captain D's, Chinese Rest., Hardee's, Shoney's, South Fork Steaks, Stringer's Fishcamp, Taco Bell, Waffle House, Day's Inn, Guest House Inn, Hampton Inn, Holiday Inn Express, Pottery Outlet, Tanger Outlet/famous brands, KOA, bank, to UGA, **W**...BP/diesel/mart, RaceTrac/diesel/mart, Texaco/mart, Arby's, Burger King, Checker's, Dairy Queen, Davis Bros Cafeteria, KFC, La Hacienda Mexican, McDonald's, Pizza Hut, Ryan's, Subway, T-Bone's Steaks, Waffle House, Wendy's, Comfort Inn, $Wise Inn, HoJo Inn, Jameson Inn, Ramada Ltd, Buick/Pontiac/GMC, Tanger Outlet/famous brands, Boot Outlet
147	GA 98, to Commerce, **E**...HOSPITAL, FuelMart/diesel, Speedway/diesel/mart, **W**...Shell/mart
140	GA 82, Dry Pond Rd, **W**...Interstate Truck Sales
137	US 129, GA 11 to Jefferson, **E**...BP/mart, Phillips 66, Shell/mart, Texaco/diesel/mart, Arby's, Hardee's, McDonald's, Waffle House, museum, **W**...Exxon, QT/diesel/mart/24hr, Waffle House, Katherine's Kitchen, flea mkt
129	GA 53, to Braselton, **E**...Shell/mart, Texaco/diesel/mart, Waffle House, Papa's Pizza, Best Western, **W**...BP/Subway/diesel/mart, Fina/diesel, Speedway/diesel/mart, hardware
126	GA 211, to Chestnut Mtn, **W**...BP/mart, Chateau Elan Winery/Hotel/rest., Day's Inn
120	to GA 124, Hamilton Mill Rd, **E**...BP/mart/wash, Circle K/diesel, Buffalo's Café, Dos Copas Mexican, Hamilton Rest., McDonald's, Subway, PakMail, Publix/deli, cleaners, **W**...auto repair
115	GA 20, to Buford Dam, **W**...to Lake Lanier Islands
114mm	**rest area sb, full(handicapped)facilities, phone, vending, picnic tables, litter barrels, petwalk**
113	I-985 N, to Gainesville, no facilities
112mm	**rest area nb, full(handicapped)facilities, phone, vending, picnic tables, litter barrels, petwalk**

Suwanee

111	GA 317, to Suwanee, **E**...CHIROPRACTOR, DENTIST, EYECARE, Amoco/mart/wash, BP/mart/wash, Phillips 66/diesel/mart, Arby's, Baskin-Robbins/DunkinDonuts/Philly Connection, Blimpie, Burger King, Cazadore's Mexican, Checker's, Chick-fil-A, Chinese Rest., Cracker Barrel, Del Taco/Mrs Winner's, Outback Steaks, Pizza Hut, Subway, Taco Bell, Waffle House, Wendy's, Comfort Inn, Courtyard, Holiday Inn, Howard Johnson Express, Sun Suites, CVS Drug, GNC, Ingles Foods, Publix, Wolf Cameras, bank, cleaners, **W**...Chevron/diesel/mart/24hr, Exxon, Denny's, McDonald's, Waffle House, Best Western/rest., Day's Inn, Ramada Ltd, flea mkt
109	Old Peachtree Rd, **E**...QT/diesel/mart/24hr
107	GA 120, to Duluth, **E**...HOSPITAL, **W**...Amoco/diesel/mart, BP/mart/wash, Chevron/mart, McDonald's
106	GA 316 N(from sb, no return), to Athens, **W**...QT/diesel/mart/24hr
104	Pleasant Hill Rd, **E**...VETERINARIAN, Chevron/mart/wash/24hr, Citgo/mart/LP, Exxon/diesel/mart, Phillip 66/mart, Circle K, Burger King, Chick-fil-A, Cooker, Corky's Ribs/BBQ, Grady's Grill, Dunkin Donuts, Hardee's, Krispy Kreme, O'Charley's, Popeye's, Restaurante Mexicano, Ruby Tuesday, TGIFriday, Waffle House, Comfort Suites, Hampton Inn Suites, Holiday Inn Express, Marriott, Baby SuperStore, Best Buy, CVS Drug, Eckerd, Goodyear/auto, Hancock Fabrics, Home Depot, Ingles Foods, K-Mart/auto, MailBoxes, Office Depot, Publix, TuneUp/Faslube, Valvoline, Wal-Mart, Winn-Dixie, cleaners, **W**...Amoco, BP/diesel/mart/wash, Phillips 66/diesel, Texaco/diesel/mart/wash, Applebee's, Arby's, Blackeyed Pea, Burger King, Chili's, Hooters, KFC, McDonald's, Mrs Winner's, Olive Garden, Pizza Hut, Red Lobster, Ryan's, Shoney's, Subway, Taco Bell, Waffle House, Wendy's, AmeriSuites, Courtyard, Day's Inn, Extended Stay America, Fairfield Inn, Ramada Ltd, Sumner Inn, Wingate Inn, Wyndham Garden, BMW, Chevrolet, Dodge, Firestone/auto, Ford/Lincoln/Mercury, Fox Foto, Honda, JiffyLube, Kroger, Macy's, Mazda, Mitsubishi, Nissan, PetsMart, Sears/auto, Service Merchandise, Shoe Expo, bank, cinema, laundry, mall
103	Steve Reynolds Blvd(from nb, no return), **E**...QT/diesel/mart/24hr(1mi), **W**...QT/mart, Texaco/mart, Dave&Buster's, Waffle House, BabysRUs, Circuit City, Costco Whse, CompUSA, Goody's, MediaPlay, Sam's Club, Sports Authority, Waccamaw

N

S

Georgia Interstate 85

102	GA 378, Beaver Ruin Rd, **E**...Conoco/mart/24hr, QT/diesel/mart/24hr(1mi), Shell/mart, Chevrolet, **W**...Amoco/mart/wash/24hr, Texaco/diesel/mart/wash
101	Lilburn Rd, **E**...QT/diesel/mart, Texaco/mart/wash, Burger King, McDonald's, Shoney's Inn/rest., Taco Bell, Suburban Lodge, Super 8, bank, **W**...DENTIST, CHIROPRACTOR, BP/diesel/mart/wash, Chevron/mart/wash/24hr, Citgo/mart, Exxon/mart, Arby's, Blimpie, Chinese Rest., Dairy Queen, Hardee's, Japanese Rest., Little Caeser's, Mexican Rest., Mrs Winner's, Papa John's, Peach's Rest., Waffle House, Wendy's, Red Roof Inn, Villager Lodge, CarMax, CVS Drug, JiffyLube, Outlet Mall, Winn-Dixie
99	GA 140, Jimmy Carter Blvd, **E**...MEDICAL CARE, Amoco/wash/24hr, Chevron, Exxon, Phillips 66/diesel, Texaco/diesel/mart/wash, Bennigan's, Burger King, Chili's, Cracker Barrel, Denny's, Dunkin Donuts, KFC, Krystal, LJ Silver, McDonald's, Morrison's Cafeteria, Pizza Hut, Sizzler, Steak&Ale, Taco Bell, Waffle House, Wendy's, Amberley Suites, Best Western, Comfort Inn, Courtyard, Clubhouse Inn, La Quinta, Motel 6, Quality Inn, Circuit City, Cub Foods, CVS Drug, 1-Hr Photo, auto parts, bank, **1 mi E**...CHIROPRACTOR, DENTIST, Crown, Dairy Queen, ChuckeCheese, Waffle House, IHOP, Subway, Mrs Winner's, Ryan's, El Amigo's Mexican, Japanese Steaks, PoFolks, Firestone/auto, Goodyear, K-Mart/drugs, Midas Muffler, Tune-UpClinic, U-Haul, **W**...Chevron/mart/24hr, Phillips 66/mart, QT/diesel/mart/24hr, Arby's, Barnacle's Grill, Blimpie, Hooters, Pappadeaux Steak/seafood, RW GoodTimes Grill, Shoney's, Waffle House, Wendy's, Drury Inn, Comfort Inn, Country Inn Suites, AutoZone, Big 10 Tire, NTB, PepBoys, Pier 1, bank, cleaners
96	Pleasantdale Rd, Northcrest Rd, **E**...Burger King, Econolodge, E&B Marine, **W**...BP/diesel/mart/wash, QT/diesel/mart, Waffle House, Howard Johnson, bank
95	I-285, no facilities
94	Chamblee-Tucker Rd, **E**...CHIROPRACTOR, Amoco/mart, Fina, Chinese Rest., Masters Inn, Travelodge, **W**...CHIROPRACTOR, QT/diesel/mart, Texaco/diesel/mart, Dairy Queen, Waffle House, Motel 6, Red Roof Inn, Ryder Trucks, to Mercer U
93	Shallowford Rd, to Doraville, **E**...Shell, Waffle House, Pizza 2 U, Global Inn, BrakeLand, U-Haul, laundry, **W**...VETERINARIAN, Citgo/Circle K, Texaco/diesel/mart/wash, Quality Inn
91	US 23, GA 155, Clairmont Rd, **E**...CHIROPRACTOR, Chevron, Speedway/mart, Popeye's, IHOP, Waffle House, Mo's Pizza, CVS Drug, Firestone, Piggly Wiggly, auto repair, bank, **W**...Amoco/diesel/mart/wash/24hr, McDonald's, Waffle House, Williams Seafood Rest., Day's Inn, Marriott, NTB
89	GA 42, N Druid Hills, **E**...Amoco/24hr, QT/diesel/mart/24hr, Crown Gas, Arby's, Burger King, Chick-fil-A, Grady's Grill, McDonald's, Mexican Rest., Miami Subs, Morrison's Cafeteria, Rally's, Taco Bell, Wall St Pizza, Courtyard, CVS Drug, Firestone/auto, Target, OptiWorld, Eckerd, bank, carwash, **W**...VETERINARIAN, BP/diesel/wash, Chevron/wash, Exxon/mart, Hess/diesel/mart, Applebee's, BBQ, Capt D's, Chinese Rest., Denny's, Dunkin Donuts, Folks Rest., Italian Rest., Krystal, Waffle House, Hampton Inn, Radisson, Red Roof Inn, Travelodge, Hi-Speed Lube, laundry
88	Lenox Rd, GA 400 N, Cheshire Br Rd(from sb), **E**...Baymont Suites, Service Merchandise, **W**...Pancho's Mexican
87	GA 400 N(from nb), no facilities
86	GA 13 S, Peachtree St, **E**...Amoco, BP, Denny's, Wendy's, La Quinta, Brake-O, **W**...Holiday Inn, Sleep Inn
85	I-75 N, to Marietta, Chattanooga
84	Techwood Dr, 14th St, **E**...Amoco/mart, BP/diesel/mart, CheeseSteaks, Einstein Bros Bagels, La Bamba Mexican, Thai Cuisine, VVV Ristorante Italiano, Hampton Inn, Marriott, Sheraton, Travelodge, Woodruff Arts Ctr, **W**...AllStar Pizza, Blimpie, Courtyard, CVS Drug, Office Depot, Wolf Camera, cleaners

I-85 and I-75 run together 8 miles. See Georgia Interstate 75, exits 88-103.

77	I-75 S, no facilities
76	Cleveland Ave, **E**...HOSPITAL, CHIROPRACTOR, BP, Conoco, Hess/mart, Phillips 66/Blimpie, Arby's, Burger King, Mrs Winner's/Del Taco, WK Wings Rest., American Inn, Day's Inn, CVS Drug, Kroger/24hr, Radio Shack, **W**...Advance Parts, Chevrolet, Honda
75	Sylvan Rd, **E**...Best Western, **W**...Mark Inn
74	Aviation Commercial Center
73b a	Virginia Ave, **E**...Citgo/diesel/mart, Hardee's, IHOP, KFC, Malone's Grill, McDonald's, Morrison's Cafeteria, Pizza Hut, Schlotsky's, Waffle House, Wendy's, Club Hotel, Courtyard, Drury Inn, Hilton, Renaissance Hotel, Red Roof Inn, Residence Inn, CastroLube, tires, **W**...Chevron/mart/24hr, Texaco, Blimpie, Happy Buddha Chinese, Hardee's, KFC, Steak&Ale, Waffle House, Crowne Plaza, Country Inn Suites, Econolodge, Harvey Hotel, Holiday Inn, Howard Johnson, Ramada Plaza

N

S

Georgia Interstate 85

Atlanta Area

72	Camp Creek Pkwy
71	Riverdale Rd, Atlanta Airport, **E**...Ruby Tuesday, Comfort Suites, Courtyard, GA Conv Ctr, Microtel, Hampton Inn, Sheraton/grill, Sleep Inn, Sumner Suites, Super 8, Wingate Inn, **W**...Bennigan's, Comfort Inn, Day's Inn, Embassy Suites, Marriott, Quality Inn, Ramada, Super 8, Travelodge, Westin Hotel
69	GA 14, GA 279, **E**...Chevron/mart/24hr, Chinese Rest., Denny's, Waffle House, La Quinta, Radisson, CVS Drug, Goodyear/auto, cinema, **W**...CHIROPRACTOR, MEDICAL CARE, BP/wash, Citgo/mart, Conoco/diesel, Exxon/diesel/mart, RaceTrac/mart, Texaco/mart, Arby's, Blimpie, Burger King, Cajun Crabhouse, Checker's, Chinese Rest., Church's Chicken, El Ranchero Mexican, KFC, Krystal, Longhorn Steaks, McDonald's, Pizza Hut, Red Lobster, ShowCase Eatery, Steak&Ale, Subway, Taco Bell, Wendy's, Baymont Inn, Day's Inn, Fairfield Inn, Red Roof Inn, Advance Parts, AutoZone, Cottman Transmissions, Econo Lube'n Tune, Family$, Kroger, Maxway, Meineke Muffler, Midas Muffler, Radio Shack, Target, Tune-Up Clinic, U-Haul, XpressLube, bank, cleaners, tires
68	I-285 Atlanta Perimeter, no facilities
66	Flat Shoals Rd, **W**...Amoco/mart/24hr, Phillips 66(2mi), Speedway/diesel/mart, Texaco/mart, Waffle House, Motel 6
64	GA 138, to Union City, **E**...BP/diesel, Waffle House, Econolodge, Ramada Ltd, Buick/Pontiac/GMC, Chrysler, Dodge, Ford, Honda, Kia/Nissan(1/4mi), **W**...CHIROPRACTOR, Chevron/mart/atm, Conoco/mart, Texaco/Church's/diesel/mart/atm, USA/mart, Arby's, Burger King, Capt D's, China King, Corner Café, Cracker Barrel, Dunkin Donuts, Formosa Chinese, IHOP, KFC, Krystal, La Fiesta Mexican, McDonald's, Papa John's, Shoney's, Subway, Taco Bell, Wendy's, Day's Inn, Holiday Inn Express, Red Roof Inn, Aamco, Chevrolet, Drug Emporium, $Tree, Eckerd, Firestone, Ingles/deli, JiffyLube, Kroger/24hr, NTB, PepBoys, PetsCo, PrecisionTune, Sears/auto, ToysRUs, Wal-Mart, bank, cleaners, cinema, mall
61	GA 74, to Fairburn, **E**...Amoco/mart, Chevron/mart/diesel, RaceTrac/mart, Waffle House, **W**...BP/diesel/mart, Phillips 66/Pit Stop, Blimpie, Efficiency Motel
56	Collinsworth Rd, **W**...Fina/mart, Frank's Rest.
51	GA 154, to Sharpsburg, **E**...BP/diesel, Phillips 66/Blimpie/diesel/mart, Hardee's, Auto Auction, **W**...Chevron/diesel/mart, Shell/Krystal/diesel/mart
47	GA 34, to Newnan, **E**...Chevron/mart, Citgo/Subway, Petro/mart/wash, Shell/mart, Texaco, Applebee's, Arby's, Dunkin Donuts, Hooters, Longhorn Steaks, Red Lobster, Ryan's, Sprayberry's BBQ, Steak'n Shake, Texas Roadhouse, Waffle House, Wendy's, Hampton Inn, Jameson Inn, Springhill Suites, GNC, Goodyear, Goody's, Home Depot, Lowe's, OfficeMax, Peachtree Stores/famous brands, PetsMart, Upton's, Wal-Mart SuperCtr/24hr, cleaners, **W**...HOSPITAL, VETERINARIAN, Amoco/diesel/mart/wash/atm, BP, HotSpot/diesel, Phillips 66, RaceTrac, Banzai, Baskin-Robbins, Burger King, Chick-fil-A, Cracker Barrel, Golden Corral, Hardee's, IHOP, KFC, Krystal, McDonald's(2mi), O'Charley's, Papa John's, Shoney's, Taco Bell, Waffle House, Best Western, Comfort Inn, Day's Inn, Holiday Inn Express, Motel 6, Big Lots, Brake Depot, Chevrolet/Oldsmobile, Ford/Lincoln/Mercury(1/2mi), Hyundai, Mailboxes Etc, Michael's, Office Depot, Plymouth, Pontiac/Buick/GMC, Publix/drugs, Toyota, bank, cleaners

Newnan

41	US 27/29, Newnan, **W**...Amoco/diesel/mart/24hr, HotSpot/gas/24hr, Phillips 66, Speedway/diesel/mart, Denny's, McDonald's, Huddle House, Waffle House, Day's Inn, Comfort Inn, Ramada Ltd, Super 8
35	US 29, to Grantville, **W**...Amoco/diesel, Phillips 66/diesel/mart
28	GA 54, GA 100, to Hogansville, **E**...Shell/diesel/mart, **W**...Amoco/diesel, Chevron/diesel/mart, Citgo, Marathon/diesel/mart, Noble's Trkstp/diesel/rest., Janet's Rest., McDonald's, Subway, Waffle House, Wendy's, Day's Inn, KeyWest Inn, Ingles/deli, Flat Creek RV Park
23mm	Beech Creek
22mm	weigh sta both lanes
21	I-185 S, to Columbus, no facilities
18	GA 109, to Mountville, **E**...Chevron, Best Western, to FDR SP, Little White House HS, **W**...Amoco/diesel, BP, Citgo/diesel/mart, Crown/mart/24hr, RaceTrac/mart, Shell/Church'sChicken/diesel/mart, Banzai, Burger King, Ryan's, Subway, Waffle House, Wendy's, AmeriHost, Comfort Inn, Holiday Inn Express, Jameson Inn, Ramada Inn/rest., Super 8, Chrysler/Plymouth/Dodge/Jeep, Honda, Hooper's RV Park(3mi), JC Penney, cinemas

LaGrange

14	US 27, to La Grange, **E**...Mill Store Carpets, **W**...Amoco/mart/24hr, Shell/mart, Hampton Inn
13	GA 219, to La Grange, **E**...Texaco/diesel, Waffle House, Day's Inn, Motel 6, truckwash, **W**...HOSPITAL, Amoco/diesel, BP, Speedway/diesel/mart, Hardee's, McDonald's
10mm	Long Cane Creek
2	GA 18, to West Point, **E**...Amoco/mart/atm, Shell/Church's/diesel/mart/24hr, Bob's House Rest., Travelodge, **W**...to West Point Lake, camping
.5mm	**Welcome Ctr nb, full(handicapped)facilities, phone, picnic tables, litter barrels, vending, petwalk**
0mm	Chattahoochee River, Georgia/Alabama state line, eastern/central time zone

N

S

Georgia Interstate 95

Exit #	Services

113mm   Savannah River, Georgia/South Carolina state line

111mm   Welcome Ctr/weigh sta sb, full(handicapped)facilities, info, phone, vending, picnic tables, litter barrels, petwalk

109   GA 21, to Savannah, Pt Wentworth, Rincon, **E**...Enmark/diesel/mart, Speedway/diesel/mart, Waffle House, Country Inn Suites, Hampton Inn, **W**...76/Circle K, Smile Gas/diesel/mart, Sea Grill Rest., Holiday Inn Express, Park Inn, Ramada Ltd, Quality Inn, Savannah Inn, Sleep Inn

107mm   Augustine Creek

106   Jimmy DeLoach Pkwy, no facilities

104   Savannah Airport, **E**...Amoco/mart, Shell/Taco Maker/diesel/mart, Fairfield Inn, to airport, **W**...EYECARE, Chevron/mart, Texaco/mart, Arby's, Sonic, Zaxby's Café, Red Roof Inn, Wal-Mart SuperCtr/Radio Grill/gas/24hr, bank

102   US 80, to Garden City, **E**...CHIROPRACTOR, Amoco/mart, Enmark/diesel/mart, Cracker Barrel, Huddle House, KFC, Krystal, McDonald's, Pizza Hut/Taco Bell, Waffle House, Comfort Inn, Jameson Inn, Microtel, Ramada Ltd, Travelodge Suites, Food Lion, Family$, Russell Stover Candy, to Ft Pulaski NM, museum, **W**...Amoco/mart, BP/mart, Gate/diesel/mart, Texaco/mart, BBQ, Burger King, Domino's, El Potro Mexican, Hardee's, Italian Pizza, Subway, Wendy's, Western Sizzlin, Country Hearth Inn, Econolodge, CarCare/lube, auto parts, bank

99b a   I-16, W to Macon, E to Savannah

94   GA 204, to Savannah, to Pembroke, **E**...HOSPITAL, Amoco/diesel/mart, BP/diesel/mart/wash, El Cheapo's/24hr, Exxon/mart, Shell/diesel/mart, Cracker Barrel, Denny's, Hardee's, McDonald's, Best Western, Day's Inn, Hampton Inn, Holiday Inn, La Quinta, Shoney's Inn/rest., Sleep Inn, Travelodge, Wingate Inn, Factory Stores, **W**...Chevron/diesel/mart/24hr, Giovanni's Rest., Huddle House, Shell House Rest., Subway, Waffle House, Econolodge, Red Carpet Inn/pizza, Super 8, Harley-Davidson, RV camping(2mi)

91mm   Ogeechee River

90   GA 144, Richmond Hill, to Ft Stewart, Richmond Hill SP, **E**...Amoco/mart/24hr, BP(1mi), Exxon/mart, El Potro Mexican(1mi), Waterway RV Park, **W**...Texaco/diesel/mart, RV/Marine Sales/service, Volvo Trucks

87   US 17, to Coastal Hwy, **E**...Welcome Ctr, Amoco/Subway/mart, Chevron/diesel/mart/24hr, Citgo/mart, RaceTrac/mart, Shell/mart, Dairy Queen, Denny's/24hr, Huddle House, Sandra D's Steaks/seafood, Waffle House, Day's Inn, Motel 6, Royal Inn, Scottish Inn, Travelodge, **W**...BP/diesel/mart/wash, Speedway/diesel/mart, Texaco/diesel/mart, 76/Amoco/LJSilver/Pizza Hut/diesel/rest./24hr, Arby's, Burger King, KFC/Taco Bell, McDonald's, Waffle House, Wendy's, Econolodge, Holiday Inn, Ramada Inn/rest., KOA

85mm   Elbow Swamp

80mm   Jerico River

76   US 84, GA 38, to Midway, Sunbury, **E**...hist sites, **W**...HOSPITAL, Amoco/mart, El Cheapo/BP/diesel/mart, Shell/mart, Huddle House, RV camping, museum

73mm   N Newport River

67   US 17, to S Newport, **E**...BP/mart, Chevron/diesel/mart, Shell/McDonald's/mart, Texaco/mart, Newport Camping(2mi), **W**...Amoco/mart

58   GA 99, GA 57, Townsend Rd, Eulonia, **E**...MEDICAL CARE, Amoco/diesel/mart, Fina, Eulonia Lodge, **W**...Amoco/diesel/mart, Chevron/diesel/mart/24hr, Shell/diesel/mart, Texaco/Day's Inn/Eulonia Café, Huddle House, Ramada Ltd, Lake Harmony RV Camp

55mm   weigh sta both lanes, phone

49   GA 251, to Darien, **E**...BP/diesel/mart, Chevron/diesel/mart/24hr, Archie's Rest.(1mi), Dairy Queen, McDonald's, Waffle House, Inland Harbor RV Park, Tall Pines RV Park, **W**...Amoco/mart, Mobil/Krispy Kreme/diesel/mart, Shell/diesel/mart, Texaco/diesel/mart, Burger King, FoodCourt, Huddle House, KFC, Sbarro's, TCBY, Comfort Inn, Hampton Inn, Holiday Inn Express, Super 8, Magnolia Bluff Stores/famous brands, auto/truck repair, flea mkt

47mm   Darien River

46.5mm   Butler River

46mm   Champney River

45mm   Altamaha River

42   GA 99, **E**...to Hofwyl Plantation HS

41mm   rest area sb, full(handicapped)facilities, info, phone, vending, picnic tables, litter barrels, petwalk

38   GA 25, to US 17, N Golden Isles Pkwy, Brunswick, **E**...HOSPITAL, Capt D's, Embassy Suites, Fairfield Inn, Jameson Inn, airport, **W**...Amoco/mart, Shell/Krispy Kreme/pizza/mart, Comfort Inn, Econolodge, Guest Cottage Motel, Quality Inn

N

↑
↓

S

Savannah

Georgia Interstate 95

36b a	US 25, US 341, Brunswick, to Jesup, **E**...Amoco/mart, Chevron/mart/24hr, Exxon/diesel/mart, RaceTrac/mart, Shell/diesel/mart, Texaco/diesel/mart, Burger King, Cracker Barrel, IHOP, KFC, Krystal/24hr, McDonald's, Pizza Hut, Shoney's, Taco Bell, Waffle House, Wendy's, Baymont Inn, Day's Inn, Hampton Inn, Howard Johnson, Knight's Inn, Ramada Inn, **W**...Amoco/mart, BP/diesel/mart, Mobil/diesel/mart, Phillips 66/mart, Abe's Buffet, BBQ, Capt Joe's Seafood, Denny's, Huddle House, Matteo's Italian, Quincy's, Subway, Waffle House, Best Western, Comfort Inn, Holiday Inn, Motel 6, Sleep Inn, Super 8, CVS Drug, Discount Parts, $General, Family$, Winn-Dixie, bank, carwash, laundry
33mm	Turtle River
30mm	S Brunswick River
29	US 17, US 82, GA 520, S GA Pkwy, Brunswick, **E**...Exxon/mart, Pilot/Steak'n Shake/Subway/diesel/mart/24hr, Texaco/diesel/mart, BBQ, Huddle House, Blue Beacon, **W**...BP/TA/diesel/rest./mart/24hr, Citgo, Flying J/Conoco/ County Mkt/diesel/LP/mart/24hr, Phillips 66/diesel/mart, Shell/El Cheapo/diesel/mart, Dairy Queen, Waffle House, DayStop, Super 8, QuikLube/tires, RV camping
27.5mm	Little Satilla River
26	Dover Bluff Rd, **E**...Shell/diesel/mart, BBQ, Ocean Breeze Camping(2mi)
21mm	White Oak Creek
19mm	Canoe Swamp
15mm	Satilla River
14	GA 25, to Woodbine, **W**...BP/mart, Sav-a-Ton/diesel/24hr, Sunshine/diesel/mart/rest./24hr, BBQ, Stardust Motel(3mi)
7	Harriett's Bluff Rd, **E**...Shell/mart, Texaco/diesel/mart, Angelo's Rest., Huddle House, Jack's BBQ, Taco Bell
6.5mm	Crooked River
6	Colerain-St Mary's Rd, **E**...BP/Arby's/diesel/mart/24hr, Cone/diesel/mart/24hr, pecans
3	GA 40, Kingsland, to St Mary's, **E**...HOSPITAL, CHIROPRACTOR, Amoco/mart, BP/mart, Chevron/mart, Enmark/mart, Mobil/mart, Shell/Subway/mart, Texaco/mart, Applebee's, BBQ, Burger King, Dairy Queen, Dynasty Chinese, KFC, Larry's Subs, McDonald's, Pizza Hut, Ponderosa, Shoney's, Trolley's Steaks, Waffle House, Wendy's, Best Western, Comfort Inn, Country Inn Suites, Day's Inn, Hampton Inn, PeachTree Inn, Super 8, Chevrolet/Buick, Chrysler/Plymouth/Jeep, CVS Drug, $Tree, Ford/Mercury, K-Mart/Little Caesar's, Publix, Winn-Dixie, bank, to Submarine Base, **W**...Exxon/diesel/mart, RaceTrac/mart, Shell/mart, Texaco, Bennigan's, Cracker Barrel, Country Suites, Econolodge, Holiday Inn, Quality Inn, Mail&More, cleaners, hardware
1	St Marys Rd, **W**...BP/mart, KOA
1mm	**Welcome Ctr nb, full(handicapped)facilities, phone, vending, picnic tables, litter barrels, petwalk**
0mm	St Marys River, Georgia/Florida state line

Georgia Interstate 285(Atlanta)

Exit #	Services
	I-285 and I-85 run together 3 mi. **See GA I-85, exit 69.**
62	GA 14, GA 279, Roosevelt Hwy, Old Nat Hwy
61	I-85, N to Atlanta, S to Montgomery. **Services 1 mi N...see GA I-85, exit 71.**
60	GA 139, Riverdale Rd, **N**...CHIROPRACTOR, MEDICAL CARE, Hess/diesel/mart, McDonald's, NTB, **S**...Amoco/ diesel/mart, BP/mart/wash, QT/Blimpie/mart/24hr, Speedway/diesel/mart, Texaco/diesel/mart/wash, BBQ, Burger King, Checker's, China Café, HoneyBaked Ham, LJ Silver, Waffle House, Wendy's, Comfort Inn, Day's Inn, Ramada Inn, Travelodge, Family$, bank, laundry
59	Clark Howell Hwy, **N**...air cargo
58	I-75, N to Atlanta, S to Macon(from eb), to US 19, US 41, Forest Park, to Hapeville, **S**...Amoco/mart, Chevron/mart/24hr, Waffle House, Home Lodge Motel
55	GA 54, Jonesboro Rd, **S**...Amoco/mart, RaceTrac/mart, Citgo/mart, Hess/mart, Arby's, McDonald's, Shoney's, Waffle House, BrakeO, Home Depot, White/GMC/Volvo, bank, **N**...Comfort Inn
53	US 23, Moreland Ave, to Ft Gillem, **N**...Conoco/diesel/mart, Speedway/diesel/mart, **S**...MEDICAL CARE, Rio Vista Catfish, Edgemont Motel/rest., International Trucks
52	I-675, S to Macon
51	Bouldercrest Rd, **N**...Amoco/mart/24hr, Pilot/Wendy's/diesel/mart/24hr, Domino's, Hardee's, DeKalb Inn, Big B Drug, Wayfield Foods, cleaners, **S**...Chevron/diesel/mart
48	GA 155, Flat Shoals Rd, Candler Rd, **N**...DENTIST, Amoco/mart/wash/24hr, Chevron/mart/24hr, Circle K/gas, Conoco/mart, Shell/diesel/mart, Arby's, Burger King, Checker's, Chinese Rest., Church's Chicken, Dairy Queen, HomeBox Rest., KFC, McDonald's, Rally's, Taco Bell, Waffle King, WK Wings, Econolodge, Gulf American Inn, JiffyLube, Kroger, PepBoys, Rich's, auto parts, bank, laundry, mall, **S**...Citgo, Dynomart/gas, Fina/mart, cleaners
46b a	I-20, E to Augusta, W to Atlanta
44	GA 260, Glenwood Rd, **E**...Rodeway Inn, **W**...Amoco, RaceTrac/mart/24hr, Texaco, Burger King, Mrs Winner's, Taco Bell, laundry

Georgia Interstate 285

43   US 278, Covington Hwy, **E**...DENTIST, Chevron/mart/wash/24hr, Citgo/mart, Crown/diesel/mart/wash, Waffle House, Pontiac\GMC, U-Haul, **W**...DENTIST, Amoco/mart/wash/24hr, BP/mart, Texaco/diesel/mart/wash, Blimpie, Checker's, Hardee's, HoneyBaked Ham, KFC/Taco Bell, Wendy's, Cub Foods, CVS Drug, Firestone/auto, Target

42   Marta Station(from sb)

41   GA 10, Memorial Dr, Avondale Estates, **E**...Texaco/diesel/mart/wash, Applebee's, Arby's, Church's Chicken, Denny's, Hardee's, McDonald's, Steak'n Shake, Waffle House, Wendy's, BMW, Circuit City, Mercedes, Photo Lab, Service Merchandise, Tune-Up Clinic, 10 Minute Lube, auto parts, bank, **W**...Amoco/mart, Fina, Burger King, KFC, Sizzler, Waffle King

40   Church St, to Clarkston, HOSPITAL(no EZ return from nb), **E**...Amoco/mart, Chevron/mart, Texaco/diesel/mart, Waffle House, carwash

39b a   US 78, to Athens, Decatur, **1.5 mi E**...Citgo/mart, Waffle House

38   US 29, Lawrenceville Hwy, **E**...HOSPITAL, BP/mart/wash, Chevron/24hr, Waffle House, Knight's Inn, Super 8, **W**...Amoco/wash, RaceTrac/mart, Waffle House, Economy Inn, Red Roof Inn

37   GA 236, to LaVista, Tucker, **E**...Chevron/wash, Circle K/gas, Checker's, Chili's, IHOP, O'Charley's, Olive Garden, Po Folks, Piccadilly's, Po Folks, Steak&Ale, Waffle House, Best Western, Ramada Inn, Firestone, Pontiac/GMC, **W**...Amoco/mart/wash, BP/wash, Hess, Texaco/wash/mart/Lube, Arby's, Blackeyed Pea, Capt D's, Chinese Rest., Dairy Queen, Dunkin Donuts, Fuddrucker's, McDonald's, Pizza Hut, Taco Bell, Wendy's, Courtyard, Day's Inn, Fairfield Inn, Holiday Inn, A&P, Best Buy, Drug Emporium, Goodyear/auto, JC Penney, Kroger/deli/24hr, Macy's, Mailboxes Etc, Midas Muffler, Pearle Vision, bank, cleaners, mall

33b a   I-85, N to Greenville, S to Atlanta

34   Chamblee-Tucker Rd, **E**...BP/mart/wash, Chevron, Texaco/mart/wash, Arby's/Mrs Winner's, KFC, Eckerd's, bank, **W**...Exxon, Phillips 66, Chinese Rest., LoneStar Steaks, McDonald's, Waffle House, Kroger, bank

32   US 23, Buford Hwy, to Doraville, **E**...CHIROPRACTOR, Amoco/mart/24hr, Burger King, Chick-fil-A, El Azteca Mexican, Grandma's Biscuits, Hardee's, KFC, Korean Rest., Krystal, LJ Silver, Mrs Winners, Steak'n Shake, Wendy's, BigStar Foods, Firestone/auto, Goodyear/auto, K-Mart/drugs, Service Mechandise, Target, TJ Maxx, laundry/cleaners, **W**...Crown/diesel/mart/wash, Fina, Arby's, Capt D's, Sensational Subs, Taco Bell, Waffle House, BrakeO, Midas Muffler, carwash

31b a   GA 141, Peachtree Ind, to Chamblee, **E**...Exxon, RaceTrac

30   Chamblee-Dunwoody Rd, N Shallowford Rd, to N Peachtree Rd, **N**...Amoco/mart, Phillips 66/wash, Waffle House, **S**...CHIROPRACTOR, DENTIST, BP/wash, Citgo/mart, Texaco/wash, Blimpie, Hardee's, KFC, LJ Silver, Malone's Grill, Mad Italian Rest., Steak&Ale, Taco Bell, Wendy's

29   Ashford-Dunwoody Rd, **N**...CHIROPRACTOR, Amoco, BP/mart/wash, Exxon/mart, Applebee's, Calif Pizza, Burger King, Denny's, Food Court, Houlihan's, L&N Seafood, Mick's Rest., Mrs Winner's, Subway, Holiday Inn, Marquis Hotel, Marriott, Ramada Inn, Best Buy, Firestone/auto, Goodyear/auto, Home Depot, Kroger, Rich's, bank, cinema, mall, **S**...Chevron, Conoco/mart, Arby's, Hilton, Holiday Inn, Residence Inn

28   Peachtree-Dunwoody Rd(no EZ return wb), **N**...CHIROPRACTOR, Burger King, Chequer's Grill, Einstein Bagels, Comfort Suites, Concourse Hotel, Courtyard, Hampton Inn, Holiday Inn Express, Marriott, Residence Inn, Westin, Wyndham Garden, Eckerd, Goodyear/auto, Marshall's, OfficeMax, Publix Foods, cleaners, mall, **S**...HOSPITAL

27   US 19 N, GA 400, **2 mi N**...LDS Temple

26   Glenridge Dr(from eb), Johnson Ferry Rd

25   US 19 S, Roswell Rd, Sandy Springs, **N**...HOSPITAL, BP/wash, Chevron, Citgo/diesel, Exxon, American Pie Rest., Athen's Pizza, Billy McHale's Rest., Burger King, Checker's, Chick-fil-A, Church's Chicken, Dunkin Donuts, El Azteca Mexican, El Toro Mexican, IHOP, KFC, Morrison's Cafeteria, Rally's, Ruth's Chris Steaks, Steak&Ale, Taco Mack, Sandy Springs Inn, Firestone/auto, Midas Muffler, K-Mart/auto, Super Shop Auto Repair, Office Depot, Eye Gallery, Service Merchandise, **S**...DENTIST, Chevron/24hr, Texaco/mart/wash, Magic Mart, Mama's Café, El Taco Veloz Mexican, TGIFriday, Day's Inn/rest., Executive Villas, laundry/cleaners

24   Riverside Dr, no facilities

22   New Northside Dr, to Powers Ferry Rd, **N**...Texaco/mart, Bennigan's, bank, **S**...Amoco/mart/wash, BP/mart/wash, Chevron/mart/wash/24hr, Blimpie, Chevy's Mexican, McDonald's, On-the-River- Café, SideLines Grill, Wendy's, Crowne Plaza, CVS Drug, bank

21   (from wb), **N**...Shell/mart, HillTop Café, Homestead Village, BMW

20   I-75, N to Chattanooga, S to Atlanta(from wb), to US 41 N

Georgia Interstate 285

19	US 41, Cobb Pkwy, to Dobbins AFB, **N...**MEDICAL CARE, BP/mart/wash, Amoco/mart/wash/24hr, Chevron/mart/wash/24hr, Citgo/mart, Texaco, Arby's, BBQ, Burger King, Carabba's, Checker's, Chinese Rest., ChuckeCheese, Crabhouse Rest., Denny's, Dunkin Donuts, Hardee's, Indian Cuisine, Jilly's Ribs, KFC, McDonald's, Old Hickory House Rest., Olive Garden, Pizza Hut, Po Folks, Red Lobster, Steak&Ale, Steak'n Shake, Sizzler, Taco Bell, Waffle House, Wendy's, French Quarter Hotel, Hilton, Holiday Inn, Howard Johnson, Red Roof Inn, Sumner Suites, Cadillac, Buick/Pontiac/Subaru, Chevrolet/Olds/Saab, Circuit City, Eckerd, Honda, Pearle Vision, Service Merchandise, Target, cleaners, tires, **S...**Amoco/mart/24hr, Chevron/24hr, Buffalo's Café, El Toro Mexican, Malone's Grill, Ruby Tuesday, Schlotsky's, Courtyard, Hampton Inn, Homewood Suites, Sheraton Suites, Stouffer Waverly Hotel, Sumner Suites, A&P, Barnes&Noble, JC Penney, Lenscrafters, Macy's, Sears/auto, bank, cinema, mall
18	Paces Ferry Rd, to Vinings, **N...**Fairfield Inn, La Quinta, **S...**BP/mart, QT/mart/24hr, Texaco/mart/wash, Atlanta Exchange Café, Blimpie, Mrs Winner's, Subway, Hampton Inn, Studio+, Wyndham, Eckerd, Goodyear/auto, Home Depot, Kinko's, Publix/drug, Wolf Camera
16	S Atlanta Rd, to Smyrna, **N...**Waffle House, Holiday Inn Express, **S...**Exxon/mart, Pilot/KFC/Subway/diesel/mart/24hr, Texaco/diesel/mart/wash, RV Ctr
15	GA 280, S Cobb Dr, **E...**Microtel, U-Haul, **W...**HOSPITAL, Amoco/mart, BP/diesel, Exxon/diesel, RaceTrac/mart, Arby's/Mrs Winners/Taco Bell, Checker's, Church's Chicken, IHOP, Krystal/24hr, McDonald's, Monterrey Mexican, Subway, Waffle House, Wendy's, AmeriHost, Knight's Inn, Sun Suites
14mm	Chattahoochee River
13	Bolton Rd(from nb)
12	US 78, US 278, Bankhead Hwy, **E...**Citgo/diesel, Petro/Chevron/diesel/rest./24hr, McDonald's, Mrs Winner's, Blue Beacon, **W...**Amoco/mart
10b a	I-20, W to Birmingham, E to Atlanta(exits left from nb), **W...**to Six Flags
9	GA 139, MLK Dr, to Adamsville, **E...**Amoco/mart/wash, Church's Chicken, KFC, McDonald's, Mrs Winner's, auto parts, **W...**Chevron
7	Cascade Rd, **E...**Chevron/mart/wash, Coastal/mart, Conoco, bank, **W...**HOSPITAL, Amoco/mart, BP/mart/wash, KFC, Mrs Winner's
5b a	GA 166, Lakewood Fwy, **E...**DENTIST, Texaco/mart/wash, Applebee's, Burger King, Checker's, Dairy Queen, Piccadilly's, Taco Bell, Circuit City, CVS Drug, Goodyear, Kroger/deli, Midas Muffler, Rich's, mall, **W...**VETERINARIAN, Amoco/mart, Citgo/mart, Conoco/mart, RaceTrac/mart, Shell/diesel/mart/24hr, Church's Chicken, KFC, Mrs Winner's, Pizza Hut, Wendy's, Deluxe Inn, Publix, Rite Aid, Home Depot
2	Camp Creek Pkwy, to airport, **E...**Amoco, BP/mart/wash, Citgo/mart, Checker's, McDonald's, Mrs Winner's, Clarion Hotel
1	Washington Rd, **W...**Chevron/mart, Seafood Rest., Motel 6, Mark Inn, carwash

Georgia Interstate 475(Macon)

Exit #	Services
16mm	I-475 begins/ends on I-75, exit 58.
15	US 41, Bolingbroke, **1mi E...**Exxon/diesel/LP/mart, Marathon/diesel/mart
9	Zebulon Rd, **E...**HOSPITAL, VETERINARIAN, BP/Chin's Wok/mart/wash, Chevron/mart/24hr, Conoco/mart, Shell/Pizza Hut/Taco Bell/TCBY/mart/24hr, Texaco/Popeye's/mart24hr, Buffalo's Café, Chick-fil-A, Crabby Bob's Seafood, Fuddrucker's, Hong Kong Chinese, Krystal, McDonald's, Subway, Waffle House, Wendy's, Fairfield Inn, Jameson Inn, Sleep Inn, GNC, Kroger/deli, Lowe's Bldg Supply, MailBoxes Etc, Radio Shack, USPO, bank, cinemas, cleaners, museum, **W...**Citgo, Zebulon Corners Café
8mm	**rest area nb, full(handicapped)facilities, phone, vending, picnic tables, litter barrels, petwalk**
5	GA 74, Macon, **E...**Waffle House, Harley Davidson/Suzuki, auto repair, to Mercer U, **W...**VETERINARIAN, Exxon/Subway/Taco Bell/mart, Marathon/mart, Phillips 66/Church's/diesel/mart, Fazoli's, Wok&Roll Chinese, Family Inn, $General, Food Lion/deli, Olson Tire, Pennzoil, bank, cleaners, to Lake Tobesofkee
3	US 80, Macon, **E...**Citgo/Subway/mart, Marathon/diesel/mart, RaceTrac, BBQ, Cracker Barrel, Johnnie's Rest., Popeye's, Waffle House, Best Western, Comfort Inn, Day's Inn, Discovery Inn, Economy Inn, Hampton Inn, Holiday Inn, Motel 6, Quality Inn, Rodeway Inn, Royal Inn, Super 8, Travelodge, **1 mi E...**Exxon, Applebee's, Chick-fil-A, KFC, Krystal, Dairy Queen, Taco Bell, Books-A-Million, Carolina Pottery, Chrysler/Plymouth, Circuit City, CVS Drug, Daewoo, Dillard's, Firestone/auto, Honda, JC Penney, Jo-Ann Fabrics/crafts, K-Mart, Kroger/deli, Lowe's Bldg Supply, Nissan, Parisian, Pier 1, Rich's, Sears/auto, Toyota, ToysRUs, mall, **W...**Amoco/service, Shell/Blimpie/mart/24hr, Burger King, McDonald's, Econolodge, Knight's Inn, Scottish Inn, Sam's Club
0mm	I-475 begins/ends on I-75, exit 48.

Georgia Interstate 575(Woodstock)

	I-575 begins/ends on GA 5/515.
27	GA 5, Howell Br, to Ball Ground, no facilities
24	Airport Dr, no facilities
20	GA 5, to Canton, **E...**Amoco, Krystal/24hr, Hardee's, McDonald's, Shoney's, Day's Inn, Wal-Mart/McDonald's, **W...**Exxon/diesel, Waffle House
19	GA 20 E, Canton, no facilities

N ↑ / S
N ↑ / S

Georgia Interstate 575

17	GA 140, Canton, to Roswell, no facilities
16	GA 20, GA 140, **W**...Exxon
14	Holly Springs, **E**...Chris' Pinecrest Rest., Pinecrest Motel, **W**...Amoco, BP, Chevron, Exxon, RaceTrac, Kroger
11	Sixes Rd, **E**...Chevron, Citgo, **W**...Conoco/mart
8	Towne Lake Pkwy, to Woodstock, **E**...BP, Chevron/mart, HotRod Café, McDonald's, 1904 House Rest., Waffle House, Waffle King/24hr, Ford
7	GA 92, Woodstock, **E**...Exxon, RaceTrac, Shell, Texaco, Arby's, Burger King, McDonald's, Mrs Winner's, Waffle House, MetroLodge, **W**...Amoco, IHOP, Honda, Home Depot
4	Bells Ferry Rd, **W**...Chevron/mart, QT/mart/24hr, Arby's, Subway, Waffle House, Eckerd
3	Chastain Rd, to I-75 N, **E**...BP, **W**...Amoco, Texaco, Arby's, Cracker Barrel, Subway, Waffle House, Wendy's, Comfort Inn, to Kennesaw St Coll
1	Barrett Pkwy, to I-75 N, US 41, **E**...DENTIST, CHIROPRACTOR, Chevron/mart/wash/24hr, Burger King, Hickory Ham Café, KFC, Waffle House, Barnes&Noble, Drug Emporium, HomePlace, MailBoxes Etc, Pier 1, Publix, TuneUp Clinic/XpressLube, FunPark, cinema, **W**...DENTIST, Chevron, Texaco, McDonald's, Applebee's, Baskin-Robbins, Fuddrucker's, Honeybaked Ham, Olive Garden, Starbuck, Subway, $3 Cafe, Comfort Inn, Day's Inn, Econolodge, Holiday Inn Express, Red Roof Inn, Shoney's Inn, Home Depot, Kinko's, Marshall's, Rich's, TJ Maxx, 1-Hr Photo, mall
0mm	I-575 begins/ends on I-75, exit 115.

Georgia Interstate 675(Stockbridge)

Exit #	Services
10mm	I-285 W, to Atlanta Airport, E to Augusta. I-285 begins/ends on I-285, exit 38.
7	Ft Gillem, Anvil Block Rd, no facilities
5	Ellenwood Rd, to Ellenwood, no facilities
2	US 23, GA 42, **E**...El Puente Mexican, Mo-joe's Café, cleaners, **W**...Speedway/diesel/mart, Chinese Café, Waffle House, Eckerd, Winn-Dixie/drug
1	GA 138, to I-75 N, Stockbridge, **E**...Amoco/mart/wash, Chevron/mart/wash/24hr, Exxon/mart/wash, Hess/mart, RaceTrac, Spur Gas, Arby's, Blimpie, Bruno's Foods, Burger King, Capt D's, Checker's, Chinese Rest., Dairy Queen, Dunkin Donuts, Golden Corral, Hardee's, KFC, McDonald's, Papa John's, Pizza Hut, Ryan's, Shoney's, Taco Bell, Waffle House, Wendy's, Best Western, Comfort Inn, Holiday Inn Express, Motel 6, Ramada Ltd, Shoney's Inn, Advance Parts, Big A Parts, CVS Drug, Eckerd, Goodyear, Ingles Foods, JiffyLube, K-Mart/auto, Radio Shack, Wal-Mart SuperCtr, USPO, bank, cleaners, hardware, **W**...HOSPITAL, RaceTrac/mart, Applebee's, BBQ, Chick-fil-A, Damon's, Frontera Mexican Grill, IHOP, Kenny Rogers, Krystal, LJ Silver, Morrison's Cafeteria, Waffle House, Best Western, Day's Inn, XPressLube

I-675 N begins/ends on I-75, exit 74.

Georgia Interstate 985(Gainesville)

Exit #	Services
	I-985 begins/ends on US 23, 25mm.
24	to US 129 N, GA 369 W, Gainesville, **N**...HOSPITAL, GA Mtn Ctr, **S**...Chevron/diesel, Citgo, café
22	GA 11, Gainesville, **N**...BP, QT/mart/24hr, Burger King, IHOP(2mi), McDonald's, Waffle House, Best Western/rest., Masters Inn, Shoney's Inn/rest., **S**...Citgo/diesel, Exxon/mart, Huddle House, Waffle House
20	GA 60, GA 53, Gainesville, **N**...Hardee's, McDonald's, Mellow Mushroom Pizza, Mrs Winners, Day's Inn, Hampton Inn, Holiday Inn, Shoney's Inn/rest., airport, **S**...Citgo/diesel, Waffle House
16	GA 53, Oakwood, **N**...Amoco/mart, Citgo/mart, Arby's, Baskin-Robbins/Dunkin Donuts, Burger King, Dairy Queen, El Sombrero Mexican, Hardee's, KFC, McDonald's, Pizza Hut, Subway, Taco Bell, Waffle House, Admiral Benbow Inn, Country Inn Suites, Jameson Inn, Shoney's Inn/rest., A&P, Chrysler/Jeep, Food Lion, RV Ctr, Sam's Club, bank, **S**...Amoco, Chevron, Citgo/diesel/mart, Exxon, QT/mart, Checker's, Mrs Winner's, Waffle House, Wendy's, Comfort Inn, Goodyear/auto
12	Spout Springs Rd, Flowery Branch, **N**...Amoco, Taco Bell, **S**...BP/Subway/mart, Chevron/diesel/mart, bank
8	GA 347, Friendship Rd, Lake Lanier, **N**...Chevron, Phillips 66, **S**...RV Ctr
4	US 23 S, GA 20, Buford, **N**...Amoco/24hr, RaceTrac/mart, Texaco, Arby's, Burger King, Hardee's, Krystal, McDonald's, Shoney's, Big B Drug, Ingles Food, **S**...Amoco/mart/wash, Chevron, Speedway/mart, Waffle House, $Tree, Lowe's Bldg Supply, Wal-Mart SuperCtr, Xpert Tire/brake, cleaners
0mm	I-985 begins/ends on I-85.

Idaho Interstate 15

Exit #	Services
196mm	Idaho/Montana state line, Monida Pass, continental divide, elev 6870
190	Humphrey, no facilities
184	Stoddard Creek Area, **E**...RV camping, **W**...Stoddard Creek Camping
180	Spencer, **E**...High Country Opal Store/RV camping, Spencer Opal Mines/gas, **W**...Spencer Grill/gas
172	**E**...US Sheep Experimental Sta, no facilities
167	ID 22, Dubois, **E**...Exxon/diesel/mart/atm, Phillips 66/diesel/mart/atm, Cow Country Kitchen, Crossroads Motel, Scoggins RV Park, **rest area both lanes, full(handicapped)facilities, phone, picnic table, litter barrels, petwalk, hist site, W**...to Craters NM, Nez Pearce Tr
150	Hamer, **E**...Goodyear, gas, food, phone
143	ID 33, ID 28, to Mud Lake, Rexburg, **W**...weigh sta both lanes
142mm	roadside parking
135	ID 48, Roberts, **E**...Amoco/diesel/mart/LP/24hr, Broulim's Foods, TrueValue/auto parts
128	Osgood Area, **E**...Sinclair/diesel/mart
119	US 20 E, Idaho Falls, to Rexburg, **E**...Sinclair/diesel, Texaco/diesel, Applebee's, Chili's, Denny's, Frontier Pies, JB's, Jaker's Steaks&Fish, McDonald's, Outback Steaks, Sandpiper Rest., Smitty's Rest., AmeriTel Inn, Best Western, Comfort Inn, Day's Inn, Quality Inn/rest., Shilo Inn/rest., Super 8, WestCoast Hotel, KOA, LDS Temple, diesel repair, **W**...Texaco/mart
118	US 20, Broadway St, Idaho Falls, **E**...HOSPITAL, LDS Temple, Chevron/service, Phillips 66/diesel/mart, Arctic Circle, Cedric's Rest., Domino's, JB's, Smitty's Pancakes, AmeriTel, Quality Inn, Super 8, Ford, Buick/Olds/GMC, FastLube, Probrake, Subaru, auto parts, hardware, **W**...DENTIST, EYECARE, Amoco/service, Exxon/mart, Flying J/diesel/mart, Maverick/gas, Sinclair/mart, Arby's, Bagelby's, Burger King, Chinese Rest., Dairy Queen, Hong Kong Chinese, Leo's Pizza, Little Caesar's, McDonald's, Papa Murphy's, Pizza Hut, Subway, Comfort Inn, Motel 6, Motel West/rest., Albertson's, American RV Sales, Circle K, Pennzoil, RiteAid, bank, laundry
113	US 26, to Idaho Falls, Jackson, **E**...HOSPITAL, Yellowstone/Exxon/diesel/motel, diesel repair/wash, Sunnyside RV Park
108	Shelley, Firth Area, **3 mi E**...Chevron/wash, Stop'n Go/diesel/mart, RV dump
101mm	**rest area both lanes, full(handicapped)facilities, phone, picnic tables, litter barrels, petwalk, geological site**
98	Rose-Firth Area, no facilities
94.5mm	Snake River
93	US 26, ID 39, Blackfoot, **E**...Chevron/mart/24hr, Flying J/diesel/LP/mart/24hr, Maverick/gas, Arctic Circle, Domino's, Hogy Yogi, Homestead Rest., KFC, Little Caesar's, McDonald's, Papa Murphy's, Pizza Hut, Subway, Taco Bell, Taco Time, Wendy's, Best Western, Riverside Inn/rest., Albertson's, Chrysler/Dodge/Ford/Lincoln/Mercury, Goodyear/auto, Grand Auto Supply, Kesler's Foods, Pennzoil, RiteAid, Schwab Tires, Potato Expo, Wal-Mart SuperCtr/24hr, bank, **1 mi E**...HOSPITAL, Amoco, Texaco, Big A Parts, El Mirador Mexican, Gorman's IGA, Kirkham Auto Parts, Potato Expo, **W**...Cenex, Riverside Boot/saddleshop(4mi)
90.5mm	Blackfoot River
89	US 91, S Blackfoot, **2 mi E**...Y Motel
80	Ft Hall, **W**...Sinclair/diesel/rest./casino, Shoshone Bannock Tribal Museum
72	I-86 W, to Twin Falls, no facilities
71	Pocatello Creek Rd, Pocatello, **E**...HOSPITAL, Chevron/Burger King/mart, Phillips 66/diesel, Circle K/gas, Applebee's, Frontier Pies, Jack-in-the Box, Perkins, Sandpiper Rest., Subway, AmeriTel, Comfort Inn, Cottontree Best Western/rest., Holiday Inn, Quality Inn/rest., Super 8, WestCoast Hotel, KOA(1mi), **W**...CHIROPRACTOR, DENTIST, Exxon/mart, Sinclair, Dairy Queen, Papa Kelsey's Pizza, SF Pizza, Sizzler, RiteAid, WinCo Foods, **1 mi W**...Texaco, Arby's, Eduardo's Mexican, KFC, McDonald's, Papa Murphy's, Shakey's Pizza, Subway, Taco Bell, TCBY, Wendy's, Winger's Diner, Albertson's, Ford, Fred Meyer, Smith's Foods, bank, auto repair
69	Clark St, Pocatello, **W**...HOSPITAL, to ID St U, museum
67	US 30/91, 5th St, Pocatello, **E**...Exxon/mart/24hr, **1-2 mi W**...HOSPITAL, Cowboy/gas, Phillips 66/diesel/mart, Sinclair/diesel, Elmer's Dining, McDonald's, Papa Paul's Rest., Pizza Hut, Taco Bell, Tom's Burgers, Best Western, Econolodge, Rainbow Motel, Sundial Inn, Cowboy RV Park, Del's Foods, Old Fort Hall, Sullivan's RV Tents, museum
63	Portneuf Area, **W**...to Mink Creek RA, RV camping/dump
59mm	**rest area/weigh sta both lanes, full(handicapped)facilities, phone, picnic table, litter barrel, vending, petwalk, hist site**
58	Inkom, **1/2 mi W**...Sinclair/diesel/café, Pebble Creek Ski Area
47	US 30, McCammon, to Lava Hot Springs, **E**...Amoco/mart, Flying J/Conoco/diesel/LP/rest./24hr, Chevron/Taco Time/TCBYdiesel/mart/atm, to Lava Hot Springs RA, camping
44	Lp 15, Jenson Rd, McCammon, **E**...access to food
40	Arimo, **E**...Sinclair/diesel/mart/deli, repairs
36	US 91, Virginia, no facilities

N

S

Idaho Interstate 15

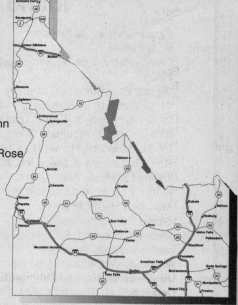

N

↕

S

31	ID 40, to Downey, **E**...Texaco/Flags West/diesel/motel/café/24hr, RV camping
25mm	**rest area sb, full(handicapped)facilities, phone, picnic tables, litter barrels, petwalk**
24.5mm	Malad Summit, elev 5574
22	to Devil Creek Reservoir, RV camping
17	ID 36, to Weston, to Preston, **W**...Deep Creek Rest.
13	ID 38, Malad City, **W**...CHIROPRACTOR, Phillips 66/café/mart, Village Inn Motel, Chevrolet/Olds/Buick, RV camping/repair, carwash, **1 mi W**...HOSPITAL, Chevron/diesel/mart, Sinclair, Texaco/diesel/mart, Red Rose Inn, bank, museum
7mm	**Welcome Ctr nb, full(handicapped)facilities, info, phone, picnic tables, litter barrels, vending, petwalk**
3	Woodruff, to Samaria, no facilities
0mm	Idaho/Utah state line

Idaho Interstate 84

Exit #	Services
275.5mm	Idaho/Utah state line
270mm	**rest area both lanes, full(handicapped)facilities, geological site, phone, picnic tables, litter barrels, petwalk**
263	Juniper Rd, no facilities
257.5mm	Sweetzer Summit, elev 5530
254	Sweetzer Rd, no facilities
245	Sublett Rd, to Malta, **N**...Sinclair/diesel/café/mart, camping
237	Idahome Rd, no facilities
234.5mm	Raft River
229mm	**rest area/weigh sta both lanes, full(handicapped)facilities, phone, picnic tables, litter barrels, petwalk**
228	ID 81, Yale Rd, to Declo, no facilities
222	I-86, US 30 E to Pocatello, no facilities
216	ID 77, ID 25, to Declo, **N**...HOSPITAL, Phillips 66/Blimpie/diesel/mart, RV camping
215mm	Snake River
211	ID 24, Heyburn, Burley, **N**...HOSPITAL, Chevron/A&W/diesel/mart, Sinclair/diesel/café, Tops Motel, RV park
208	ID 27, Burley, **N**...Phillips 66/diesel/café, Super 8, **S**...HOSPITAL, VETERINARIAN, Amoco, Chevron/Subway/diesel/mart/24hr, Texaco/Taco Bell/mart, Arby's, Burger King, Eduardo's Mexican, George K's Chinese, Jack-in-the-Box, JB's, KFC, McDonald's, Perkins, Polo Café, Wendy's, Best Western, Budget Motel, Cal Store, JC Penney, K-Mart/Little Caesar's, Radio Shack, Ray's Mufflers, Wal-Mart, bank, to Snake River RA, **1 mi S**...Amoco, Sinclair/diesel, Burgers Etc, Cancun Mexican, KFC, Starlite Motel, Buick/Pontiac/GMC, CarQuest, Checker Parts, Chrysler/Plymouth/Dodge/Jeep, Kelly Tire, Minico Parts, NAPA, Pennzoil, Stokes Food/deli
201	ID 25, Kasota Rd, to Paul, no facilities
194	ID 25, to Hazelton, **S**...Sinclair/diesel/mart/café, RV camping
188	Valley Rd, to Eden, no facilities
182	ID 50, to Kimberly, Twin Falls, **N**...Sinclair/diesel, Gary's RV Sales/park/dump, **S**...HOSPITAL, Chevron/diesel/mart(7mi), Travelers/Texaco/Blimpie/Taco Bell/mart, Amber Inn, Best Western, to Shoshone Falls scenic attraction
173	US 93, Twin Falls, **N**...Petro 2/Chevron/diesel/rest., Day's Inn, KOA(1mi), Blue Beacon/24hr, to Sun Valley, **5 mi S**...HOSPITAL, Chevron/diesel/mart, Phillips 66/mart, Arby's, Burger King, JB's, McDonald's, Shari's, Best Western, Holiday Inn Express, Motel 6, Super 8, Weston Inn, Home Depot, Coll of S ID
171mm	**rest area/weigh sta eb, full(handicapped)facilities, phone, vending, picnic tables, litter barrels, petwalk**
168	ID 79, to Jerome, **N**...Chevron/diesel/mart/wash, Sinclair/Honker's/diesel/mart, Texaco/Wendy's/diesel/mart, Burger King, Cindy's Rest., McDonald's, Pizza Hut, Best Western, Crest Motel, Brockman RV Ctr, Wal-Mart SuperCtr/gas/24hr, **2 mi N**...HOSPITAL, Dairy Queen, Holiday Motel, Chevrolet, **S**...Subway
165	ID 25, Jerome, **N**...HOSPITAL, Sinclair/diesel/mart, Holiday Motel, RV camping/dump
157	ID 46, Wendell, **1 mi N**...HOSPITAL, Intermountain RV Park, **S**...Texaco/Farmhouse Rest./diesel/mart
155	ID 46, to Wendell, **N**...Intermountain RV Ctr/Park
147	to Tuttle, **S**...to Malad Gorge SP, High Adventure RV Park
146mm	Malad River
141	US 26, to US 30, Gooding, **S**...HOSPITAL, Phillips 66/diesel/mart/café, Sinclair/diesel/mart/24hr, Texaco/diesel/LP, Amber Inn, Hagerman Inn, RV camping, laundry
137	Bliss, to Pioneer Road, **2 mi S**...Sinclair/diesel/mart/24hr, Y Inn Motel, camping
133mm	**rest area both lanes, full(handicapped)facilities, info, picnic tables, litter barrels, petwalk, phone(wb)**
129	King Hill, no facilities

E

↕

W

Burley   Jerome

Idaho Interstate 84

128mm	Snake River
125	Paradise Valley, no facilities
122mm	Snake River
121	Glenns Ferry, **1 mi S**...Amoco/mart/24hr, Sinclair/mart, Redford Motel, Carmela Winery/rest., to 3 Island SP, RV camp
120	Glenns Ferry(from eb), same as 121
114	ID 78(from wb), to Hammett, **1 mi S**...access to gas/diesel, to Bruneau Dunes SP
112	to ID 78, Hammett, **1 mi S**...gas/diesel, food, to Bruneau Dunes SP
99	ID 51, ID 67, to Mountain Home, **2 mi S**...Maple Cove Motel, gas, camping
95	US 20, Mountain Home, **N**...Chevron/diesel/mart/24hr, Pilot/DQ/Subway/diesel/mart/24hr, AJ's Rest., Jack-in-the-Box, Best Western, Sleep Inn, **S**...HOSPITAL, McDonald's, Smoky Mtn Pizza, Taco Maker, Hilander Motel(1mi), Towne Ctr Motel(1mi), K-Mart, Pennzoil, Sears, Wal-Mart SuperCtr/gas/24hr, to Mountain Home AFB, KOA
90	to ID 51, ID 67, W Mountain Home, **S**...Texaco/Burger King/diesel/mart, to Hilander Motel, Towne Ctr Motel, KOA, to Mountain Home AFB
74	Simco Rd, no facilities
71	Orchard, Mayfield, **S**...Sinclair/diesel/StageStop Motel/rest./24hr, phone
66mm	weigh sta both lanes
64	Blacks Creek, Kuna Rd, historical site
62mm	**rest area both lanes, full(handicapped)facilities, OR Trail info, phone, picnic tables, litter barrels, vending, petwalk**
59	S Eisenman Rd, no facilities
57	ID 21, Gowen Rd, to Idaho City, **N**...Jack-in-the-Box, McDonald's, Perkins, Subway, Albertson's/gas, bank, to Micron, **S**...Chevron/mart/24hr, Burger King, FoodCourt, McDonald's, Boise Factory Stores/famous brands, ID Ice World, RV camping
54	US 20/26, Broadway Ave, Boise, **N**...HOSPITAL, VETERINARIAN, Chevron/diesel/mart/24hr, Flying J/Conoco/diesel/LP/mart/24hr, Chili's, Hugo's Deli, Jack-in-the-Box, McDonald's(2mi), Subway, Wendy's, Courtyard, Big O Tires, Dowdie's Automotive, Goodyear/auto, Jo-Ann Fabrics, Meineke Muffler, OfficeMax, Pennzoil, Radio Shack, ShopKO, to Boise St U, **S**...TA/Subway/Taco Bell/diesel/mart/rest./24hr, Shilo Inn, Kenworth, Mtn View RV Park
53	Vista Ave, Boise, **N**...Citgo/7-11, Shell/mart, Sinclair, Texaco/Taco Bell/mart, Extended Stay America, Fairfield Inn, Hampton Inn, Holiday Inn Express, Quality Inn, Super 8, Parts'n Stuff, museums, st capitol, st police, zoo, **S**...Chevron/McDonald's/mart/24hr, Denny's, Kopper Kitchen, Best Western, Comfort Inn, InnAmerica, Motel 6, Sleep Inn, airport, auto rentals, bank
52	Orchard St, Boise, **N**...Texaco/Taco Bell/diesel/mart/atm/24hr, Mazda/Nissan, Olds/GMC, Alamo/Hertz, **1-2 mi N**...Burger King, Jack-in-the-Box, McDonald's, Pizza Hut, Raedean's Rest., Wendy's
50	Cole Rd, Overland Rd, **N**...CHIROPRACTOR, DENTIST, LDS Temple, Chevron/mart/24hr, Circle K/gas, Texaco, Buster's Grill, Cancun Mexican, Eddie's Rest., McDonald's, Outback Steaks, Pizza Hut, Subway, Taco Bell, TCBY, Plaza Suites, Pontiac/Buick, Valvoline, cleaners/laundry, transmissions, **S**...Flying J/Conoco/diesel/rest./24hr, BBQ, Black Angus, Burger King, Carino's, ChuckaRama, Cracker Barrel, Jamba Juice, KFC, Little Mexico, McGrath's FishHouse, On the Border, Pollo Rey Mexican, Yen Ching Chinese, AmeriTel, Best Rest Inn, Commercial Tire, Goodyear, Goody's, Lowe's Bldg Supply, Wal-Mart SuperCtr/24hr, cinema
49	I-184(exits left from eb), to W Boise, **N**...HOSPITAL, Rodeway Inn, National 9 Inn
46	ID 55, Eagle, **N**...MEDICAL CARE, Chevron/McDonald's/diesel/mart/24hr, Texaco/Taco Bell/diesel/mart/atm/24hr, Holiday Inn Express, **S**...Fiesta RV Park
44	ID 69, Meridian, **N**...MEDICAL CARE, CHIROPRACTOR, Chevron/diesel/mart/24hr, Sinclair, Blimpie, Bolo's Eatery, Dairy Queen, KFC, McDonald's, Pizza Hut, Quizno's, RoundTable Pizza, Shari's/24hr, Taco Bell, Taco Time, Under The Onion Steaks, Best Western, Microtel, Bodily RV Ctr, Home Depot, Schwab Tires, WinCo Foods, **S**...Texaco/diesel/mart/24hr, JB's, Knotty Pine Motel, Mr Sandman Motel, Ford, Playground RV Park, funpark
38	Garrity Blvd, Nampa, **1.5 mi N**...Chevron/diesel/mart/24hr, Swiss Village Cheese, golf, **S**...MEDICAL CARE, Chevron/diesel/mart/24hr, Texaco/Taco Ole/diesel/mart/24hr
36	Franklin Blvd, Nampa, **N**...Jack-in-the-Box, Noodles Rest., Shilo Inn/rest., **S**...HOSPITAL, Chevron/diesel/mart/24hr, Texaco/A&W/Taco Bell/diesel/RV dump/24hr, Desert Inn, Sleep Inn, Mason Cr RV Park, Minor RV Ctr
35	ID 55, Nampa, **S**...Shell/diesel/mart, Denny's/24hr, InnAmerica, Shilo Inn, Super 8, golf, **1 mi S**...HOSPITAL, Burger King, McDonald's, Pizza Hut, Taco Time, Desert Inn
29	US 20/26, Franklin Rd, Caldwell, **N**...Flying J/Conoco/diesel/LP/rest./mart/24hr, Guesthaus Rest., RV camping, airport, **S**...Sinclair/diesel/mart/Sage Café, Perkins/24hr, Best Inn, Best Western, **1-2 mi S on Cleveland/Blaine St**...McDonald's, Subway, Taco Time, Desert Inn, Albertson's/gas, Chrysler/Plymouth/Dodge/Jeep, Honda, NAPA, Rite Aid, Wal-Mart SuperCtr/24hr, to Simplot Stadium
28	10th Ave, Caldwell, **N**...Maverick/gas, I-84 Motel, parts/repair, **S**...HOSPITAL, Amoco/mart, Chevron/mart/24hr, 7-11, Carl's Jr, Dairy Queen, Jack-in-the-Box, Mr V's Rest., Pizza Hut, Wendy's, Holiday Motel/café, Sundowner Motel, Paul's Drug, brakes/lube/tires
27	ID 19, to Wilder, **1 mi S**...Amoco/diesel/24hr
26.5mm	Boise River

E

W

B o i s e

N a m p a

Idaho Interstate 84

26   US 20/26, to Notus, **N...**Camp Caldwell RV Park, fireworks, **S...**Chevron/ diesel

25   ID 44, Middleton, **N...**Shell/diesel/mart, Bud's Burgers/shakes, **4 mi N...**Texaco, Taco Bell, **S...**weigh sta eb

17   Sand Hollow, **N...**Sand Hollow Café, Country Corners RV Park

13   Black Canyon Jct, **S...**Sinclair/diesel/motel/rest./24hr, phone

9   US 30, to New Plymouth, no facilities

3   US 95, Fruitland, **N...**Shell/A&W/diesel/mart, Texaco/diesel/mart/atm/ 24hr(3mi), **5 mi N...**Neat Retreat RV Park, to Hell's Cyn RA

1mm   **Welcome Ctr eb, full(handicapped)facilities, info, phone, picnic tables, litter barrels, petwalk**

0mm   Snake River, Idaho/Oregon state line

Idaho Interstate 86

Exit #   Services

63b a   I-15, N to Butte, S to SLC. I-86 begins/ends on I-15, exit 72.

61   US 91, Yellowstone Ave, Pocatello, **N...**CHIROPRACTOR, Amoco/mart, Exxon/mart, Sinclair, Texaco/diesel/mart, Arctic Circle, Burger King, Eddy's Bakery, Johnny B Goode Diner, Pizza Hut, Subway, Vickie's Hitchin Café, Day's Inn, Motel 6, JiffyLube, **S...**Chevron/mart, Flying J/Taco Bell/diesel/ mart/24hr, Phillips 66/diesel, Asian Rest., Denny's, McDonald's, Red Lobster, Taco Bandido, Day's Inn, Nendel's Inn/rest., Pine Ridge Inn, GreaseMonkey Lube, Grocery Outlet, Home Depot, JC Penney, K-Mart, Schwab Tires, ShopKO, Wal-Mart SuperCtr/24hr, ZCMI, Herb's RV, auto parts, bank, cinema, diesel repair, mall

58.5mm   Portneuf River

58   US 30, W Pocatello, **S...**Stinker/diesel, RV dumping

56   **N...**Pocatello Air Terminal, speedway

52   Arbon Valley, **S...**Sinclair/Bannock Peak/diesel/mart, casino

51mm   Bannock Creek

49   Rainbow Rd, no facilities

44   Seagull Bay, no facilities

40   ID 39, American Falls, **N...**HOSPITAL, Cenex/diesel/mart, Pizza Hut, Chevrolet, Schwab Tires, to Am Falls RA, RV camping/dump, **S...**Amoco/diesel/café, Hillview Motel

36   ID 37, American Falls, to Rockland, **N...**HOSPITAL, Falls Motel, **2 mi S...**Indian Springs RV Resort

33   Neeley Area, no facilities

31mm   **rest area wb, full(handicapped)facilities, phone, picnic table, litter barrel, petwalk, hist site**

28   **N...**to Massacre Rock SP, Register Rock Hist Site, RV camping/dump

21   Coldwater Area, no facilities

19mm   **rest area eb, full(handicapped)facilities, phone, picnic table, litter barrel, petwalk, hist site**

15   Raft River Area, **S...**Sinclair/diesel/mart, phone

4mm   weigh sta both lanes

1   I-84 E, to Ogden. I-86 begins/ends on I-84, exit 222.

Idaho Interstate 90

Exit #   Services

74mm   Idaho/Montana state line, Pacific/Mountain time zone, Lookout Pass elev 4680

73mm   scenic area/hist site wb

72mm   scenic area/hist site eb

71mm   runaway truck ramp wb

70mm   runaway truck ramp wb

69   Lp 90, Mullan, **N...**Exxon/diesel/mart/24hr, Maximart/gas, Mullan Café, Lookout Motel(1mi), USPO, museum

68   Lp 90(from eb), Mullan, same as 69

67   Morning District, no facilities

66   Gold Creek(from eb), no facilities

65   Compressor District, no facilities

64   Golconda District, no facilities

62   ID 4, Wallace, **S...**HOSPITAL, Exxon/mart, Pizza Factory, Sweet's Café, Wallace Café, Brooks Hotel, Stardust Motel, Sweet's Hotel, Depot RV Park, Excell Foods, Parts+, TrueValue, USPO, bank, museum, repair

61   Lp 90, Wallace, **S...**DENTIST, Conoco/diesel/mart/rest./24hr, Exxon, Cyndi's Drive-In, Best Western, Brooks Hotel, Molly B-Damm Inn, info ctr, same as 62

60   Lp 90, Silverton, **N...**HOSPITAL, **S...**SilverLeaf Motel, RV camping

57   Lp 90, Osburn, **S...**Texaco/diesel/mart/24hr, Blue Anchor RV Park, auto repair

E

W

W

E

W

Pocatello

Idaho Interstate 90

Kellogg

54	Big Creek, **N**...auto repair
51	Lp 90, Division St, Kellogg, **N**...HOSPITAL, VETERINARIAN, Conoco/diesel/mart, Broken Wheel Rest., Motel 51, Chevrolet/Pontiac/Olds/Buick/Cadillac, Chrysler/Dodge, IGA Food, Radio Shack, Schwab Tires, Sunnyside Drug, cinema, laundry, **S**...Kopper Keg Café/pizza, USPO, bank, mufflers/repair, museum
50	Hill St(from eb), Kellogg, **N**...Jack's Hookshop/gas, Sunshine Café, Trail Motel, Ace Hardware, IGA Foods, NAPA, **S**...Conoco/mart, Pac'n Sav Foods, Silver Mtn Ski/summer resort/rec area, museum
49	Bunker Ave, **N**...HOSPITAL, Chevron/mart, McDonald's, Sam's Drive-In, Subway, Taco John's, Silverhorn Motel/rest., Sunshine Inn, **S**...Silver Mtn/Gondola Café/Timbers Rest., Super 8, museum, RV dump
48	Smelterville, **N**...Silver Valley Car/trkstp/motel/café, airport, RV camping, **S**...motel/café
45	Pinehurst, **S**...Chevron/mart/hardware, Conoco/diesel/mart, Honda/Yamaha, Pinehurst RV, KOA, RV dumping
43	Kingston, **N**...Texaco/diesel/mart/24hr, Snakepit Café, Enaville Resort, RV camping, **S**...Exxon/mart
40	Cataldo, **N**...General Store, 3rd Generation Rest., RV camping
39.5mm	Coeur d' Alene River
39	Cataldo Mission, **S**...Old Mission SP, Nat Hist Landmark
34	ID 3, Rose Lake, to St Maries, **S**...Conoco/diesel/mart, Texaco/diesel/mart, Country Chef Café, Rose Lake Gen Store, White Pines Scenic Rte
32mm	chainup area wb
31.5mm	Idaho Panhandle NF, eastern boundary, 4th of July Creek
28	4th of July Pass, elev 3069, Mullan Tree HS, ski area, snowmobile area
28.5mm	turnout both lanes
24mm	chainup eb, removal wb
22	ID 97, Harrison, to St Maries, L Coeur d' Alene Scenic By-way, Wolf Lodge District, **1 mi N**...Wolf Lodge Campground, **S**...KOA, Squaw Bay Resort(7mi)
20.5mm	Lake Coeur d' Alene
17	Mullan Trail Rd, no facilities

Coeur d' Alene

15	Lp 90, Sherman Ave, Coeur d' Alene, City Ctr, **N**...forest info, Lake Coeur D' Alene RA/HS, **S**...CHIROPRACTOR, Cenex, Exxon/diesel/mart/24hr, Shell/mart, Texaco/mart, Piggy's Gas/mart, BBQ, Down the Street Rest., Henry's Rest., Michael D's Eatery, Roger's Ice Cream, Rustler's Roost Rest., Zip's Rest., Bates Motel, BudgetSaver Motel, Holiday Inn Express, Holiday Motel, Red Rose Motel, Sandman Motel, Star Motel, State Motel, Sundowner Motel, Ace Hardware, Buick/Pontiac/GMC, CompuTune, IGA Foods/24hr, golf, laundromat
14	15th St, Coeur d' Alene, **S**...TAJ Mart, Jordon's Grocery
13	4th St, Coeur d' Alene, **N**...Citgo/7-11, Conoco/diesel/mart, Baskin-Robbins, Bruchi's Café, Dairy Queen, Denny's, Godfather's, IHOP, KFC, Little Caesar's, Taco John's, Wendy's, Comfort Inn, Fairfield Inn, Alton Tires, Big 5 Sports, Hastings Books, Kinko's, MailBoxes Etc, NAPA, Radio Shack, Schuck's Parts, Schwab Tires, same as 12, **S**...Exxon/diesel/mart, BBQ, Hunter's Grill, Subway, Chevrolet, Chrysler/Jeep/Eagle, Nissan/Subaru
12	US 95, to Sandpoint, Moscow, **N**...DENTIST, CHIROPRACTOR, Chevron/mart, Conoco/diesel/mart, Exxon/diesel/mart, Holiday/A&W/mart, Shell/mart, Applebee's, Arby's, Bagelby's/Jungle Juice, BBQ, Boston Mkt, Burger King, Domino's, DragonHouse Chinese, Las Chavelaz Mexican, Log Cabin Rest., McDonald's, Pelican Rest., Pizza Hut, Pizza Pipeline, Red Lobster, Taco Bell, Thai Palace, Tomato St Italian, Village Inn Rest., Comfort Inn, Coeur d'Alene Inn, Holiday Inn, Motel 6, Shilo Inn, Super 8, Travelodge, Al's Auto Supply, Black Sheep Sports, Dodge, Ford/Lincoln/Mercury, Fred Meyer, JiffyLube, K-Mart, Office Depot, Rosauers Food/drugs, Safeway, Super 1 Foods, Tidyman's Food/24hr, Toyota, U-Haul, antiques, cinema, airport, **S**...MEDICAL CARE, CHIROPRACTOR, EYECARE, Texaco, Chopstix Express, Figaro's Italian, Mr Steak, Jack-in-the-Box, Schlotsky's, Shari's, TCBY, Ameritel, Albertson's, GNC, Rite Aid, ShopKO/drugs, Staples, bank, funpark, same as 13
11	Northwest Blvd, **N**...Country Qwikstop, Eagle Hardware, Golf Supply, **S**...HOSPITAL, Exxon/mart, Qwikstop/gas, Outback Steaks, Day's Inn, Garden Motel, Rodeway Inn, FedEx, RV camping, Honda (.5mi)
8.5mm	**Welcome Ctr/weigh sta eb, rest area both lanes, full(handicapped)facilities, info, phone, picnic tables, litter barrels, petwalk**
7	ID 41, to Rathdrum, Spirit Lake, **N**...Exxon/diesel/mart, Carnegie's Rest., Couer d'Alene RV Park, **S**...Chevron/diesel/mart/24hr, Casey's Rest./brewery, Dairy Queen, KFC, Holiday Inn Express, repair
6	Seltice Way, City Ctr, **N**...MEDICAL CARE, Citgo/7-11, Exxon/mart/RV dump, Texaco/mart, La Cabana Mexican, Pizza Hut, Excell Foods, NAPA, Super 1 Foods, bank, **S**...CHIROPRACTOR, EYECARE, Arby's, Godfather's, Hot Rod Café, Little Caesar's, McDonald's, Papa Murphy's, Rancho Viejo Mexican, Subway, Taco Bell, ZukaJuice, Al's Auto Supply, Do-It Hardware, JiffyLube, Tidyman's Food/24hr, Upscale Mail, carwash, laundry
5	Lp 90, Spokane St, Treaty Rock HS, **N**...BP/diesel/mart, Andy's Pantry Rest., BagelWorks, Golden Dragon Chinese, McDuff's Rest., Mazda(1mi), Schwab Tires, same services as 6, **S**...Best Western
2	Pleasant View Rd, **N**...Conoco/diesel/mart, Flying J/Conoco/diesel/rest./mart/LP/24hr, Texaco/mart, Burger King, McDonald's, Subway, Toro Viejo Mexican, Howard Johnson Express, All Lube&Tune, **S**...Exxon/mart/24hr, Chocolate Factory/pizza/pasta, Jack-in-the-Box, Michel's Rest., O'Keely's Rest., Riverbend Inn/rest., Sleep Inn, GNC, Post Falls Stores/famous brands, dogtrack
0mm	Idaho/Washington state line

E

W

Illinois Interstate 24

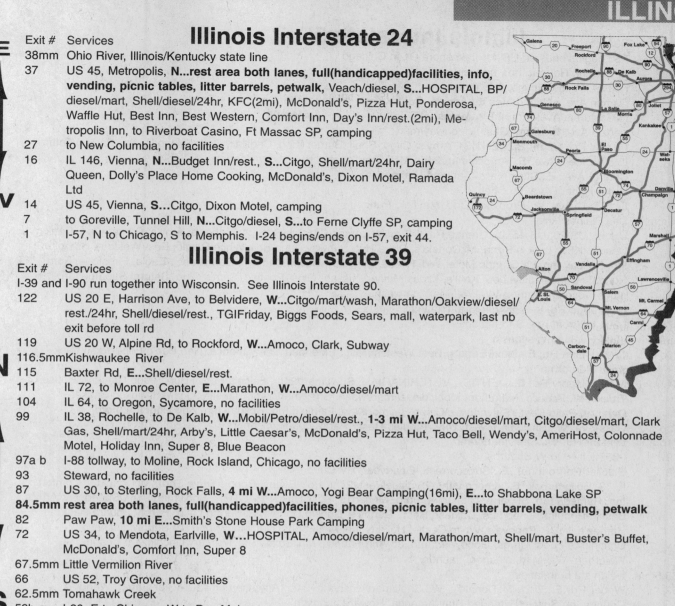

Exit #	Services
38mm	Ohio River, Illinois/Kentucky state line
37	US 45, Metropolis, **N...rest area both lanes, full(handicapped)facilities, info, vending, picnic tables, litter barrels, petwalk,** Veach/diesel, **S...**HOSPITAL, BP/diesel/mart, Shell/diesel/24hr, KFC(2mi), McDonald's, Pizza Hut, Ponderosa, Waffle Hut, Best Inn, Best Western, Comfort Inn, Day's Inn/rest.(2mi), Metropolis Inn, to Riverboat Casino, Ft Massac SP, camping
27	to New Columbia, no facilities
16	IL 146, Vienna, **N...**Budget Inn/rest., **S...**Citgo, Shell/mart/24hr, Dairy Queen, Dolly's Place Home Cooking, McDonald's, Dixon Motel, Ramada Ltd
14	US 45, Vienna, **S...**Citgo, Dixon Motel, camping
7	to Goreville, Tunnel Hill, **N...**Citgo/diesel, **S...**to Ferne Clyffe SP, camping
1	I-57, N to Chicago, S to Memphis. I-24 begins/ends on I-57, exit 44.

Illinois Interstate 39

Exit #	Services
	I-39 and I-90 run together into Wisconsin. See Illinois Interstate 90.
122	US 20 E, Harrison Ave, to Belvidere, **W...**Citgo/mart/wash, Marathon/Oakview/diesel/rest./24hr, Shell/diesel/rest., TGIFriday, Biggs Foods, Sears, mall, waterpark, last nb exit before toll rd
119	US 20 W, Alpine Rd, to Rockford, **W...**Amoco, Clark, Subway
116.5mm	Kishwaukee River
115	Baxter Rd, **E...**Shell/diesel/rest.
111	IL 72, to Monroe Center, **E...**Marathon, **W...**Amoco/diesel/mart
104	IL 64, to Oregon, Sycamore, no facilities
99	IL 38, Rochelle, to De Kalb, **W...**Mobil/Petro/diesel/rest., **1-3 mi W...**Amoco/diesel/mart, Citgo/diesel/mart, Clark Gas, Shell/mart/24hr, Arby's, Little Caesar's, McDonald's, Pizza Hut, Taco Bell, Wendy's, AmeriHost, Colonnade Motel, Holiday Inn, Super 8, Blue Beacon
97a b	I-88 tollway, to Moline, Rock Island, Chicago, no facilities
93	Steward, no facilities
87	US 30, to Sterling, Rock Falls, **4 mi W...**Amoco, Yogi Bear Camping(16mi), **E...**to Shabbona Lake SP
84.5mm	**rest area both lanes, full(handicapped)facilities, phones, picnic tables, litter barrels, vending, petwalk**
82	Paw Paw, **10 mi E...**Smith's Stone House Park Camping
72	US 34, to Mendota, Earlville, **W...**HOSPITAL, Amoco/diesel/mart, Marathon/mart, Shell/mart, Buster's Buffet, McDonald's, Comfort Inn, Super 8
67.5mm	Little Vermilion River
66	US 52, Troy Grove, no facilities
62.5mm	Tomahawk Creek
59b a	I-80, E to Chicago, W to Des Moines
57	US 6, La Salle, to Peru, **1-2 mi W...**Amoco, Casey's Store, Shell/mart/24hr, Hardee's, Daniels Motel
56mm	Illinois River, Abraham Lincoln Mem Bridge
54	Oglesby, **E...**Amoco/mart, Casey's Store, Shell/mart/atm/24hr, Baskin-Robbins/Dunkin Donuts, Delaney's Rest., Burger King, Hardee's, McDonald's, Pizza Grill, Subway, Day's Inn, Holiday Inn Express, Starved Rock SP
52	IL 251, to La Salle, Peru, no facilities
51	IL 71, Oglesby, to Hennepin, no facilities
48	Tonica, **E...**Casey's Store
41	IL 18, to Streator, Henry, no facilities
35	IL 17, to Wenona, Lacon, **2.5 mi E...**Amoco/diesel/24hr, Casey's/gas, Burger King, Buster's Family Rest., Pizza Hut, Super 8/RV/truck parking
27	to Minonk, **E...**Shell/Subway/Woody's Rest./diesel/mart/24hr, **1 mi E...**Casey's
22	IL 116, Benson, to Peoria, no facilities
14	US 24, to El Paso, Peoria, **E...**Casey's, Freedom/mart, Shell/mart/24hr, Dairy Queen, Elm's Buffet, Hardee's/24hr, McDonald's, Subway, Woody's Family Rest., Day's Inn, Ford/Mercury, **W...**Dabney's Rest., Super 8/RV/truck parking, Hickory Hill Camping, antiques/arts/crafts
9mm	Mackinaw River
8	IL 251, Lake Bloomington Rd, **E...**Lake Bloomington, **W...**Evergreen Lake, to Comlara Park, camping
5	Hudson, **1 mi E...**Casey's Store/gas
2	US 51 bus, Bloomington, Normal, no facilities
0mm	I-39 begins/ends on I-55, exit 164. Facilities located N on I-55, exit 165.

E
W

N
S

Rockford

Peru

Normal

Illinois Interstate 55

Exit #	Services
295mm	I-55 begins/ends on US 41, Lakeshore Dr, in Chicago.
293a	to Cermak Rd(from nb), no facilities
292	I-90/94, W to Chicago, E to Indiana
290	Damen Ave, Ashland Ave, **E**...Amoco, Burger King, White Castle, Dominick's Foods
289	to California Ave(from nb), same as 290
288	Kedzie Ave, **E**...Clark Gas, Speedway/mart
287	Pulaski Rd, **1/2 mi E**...DENTIST, Arby's, Citgo, Shell, Burger King, Cadillac, Dodge, Honda, TrakParts, Venture
286	IL 50, Cicero Ave, **E**...Amoco/mart/24hr, McDonald's, airport, laundry/cleaners, **W**...Citgo
285	Central Ave, **E**...Citgo/mart, Clark/mart, airport
283	IL 43, Harlem Ave, **E**...Shell
282	IL 171, 1ˢᵗ Ave, **W**...Brookfield Zoo, Mayfield Park
279mm	Des Plaines River
279b	US 12, US 20, US 45, La Grange Rd, **1-2 mi W**...Amoco, Mobil, Shell, Applebee's, Arby's, Burger King, Dunkin Donuts, KFC, Ledo's Pizza, McDonald's, Pizza Hut, Subway, Taco Bell, White Castle, Best Western, Hampton Inn, Holiday Inn, La Grange Motel, Aldi Foods, Best Buy, Buick, Cadillac, Dodge/Jeep/Eagle, Firestone, Ford, Honda, Hyundai, JiffyLube, Kohl's, Midas Muffler, Nissan, Olds, Pontiac, Sam's Club, Saturn, Subaru, Target/drugs, Toyota, Venture, VW, Wal-Mart, carwash
a	La Grange Rd, to I-294 toll, S to Indiana
277b	I-294 toll(from nb), S to Indiana
a	I-294 toll, N to Wisconsin
276b a	County Line Rd, **E**...Max&Erma's, Best Western/rest., Extended Stay America, Cee Bee's Foods/deli, bank
c	Joliet Rd(from nb)
274	IL 83, Kingery Rd, **E**...DENTIST, VETERINARIAN, Shell/mart/24hr, Falco's Pizza, Burr Ridge Carcare, House of Trucks, **W**...Amoco, Marathon, Mobil/diesel/mart/wash, Shell/mart/24hr, Bakers Square, Chicken Basket, Denny's, Patio Rest., Salvadore's Diner, Baymont Inn, Fairfield Inn, Holiday Inn, Red Roof Inn, Ford
273b a	Cass Ave, **W**...Shell/mart/wash/24hr, Ripples Rest., Parts+, bank, cleaners
271b a	Lemont Rd, **W**...Shell/mart/24hr
269	I-355 toll N, to W Suburbs
268	S Joliet Rd(from nb), **E**...Sunoco/rest., Goodyear, RV Ctr
267	IL 53, Bolingbrook, **E**...Amoco/24hr, 76/diesel/rest./24hr, Bob Evans, Bono's Drive-Thru, McDonald's, Comfort Inn, Ramada Ltd, Super 8, Chevrolet, Ford, Menard's, **W**...CHIROPRACTOR, Shell/mart/24hr, Speedway, Arby's, Baskin-Robbins, Cheddars, Chinese Rest., Denny's, Dunkin Donuts, El Burrito Loco, Family Square Rest., Hardee's, IHOP, Popeye's, White Castle, Holiday Inn, Aldi Foods, Camping World RV Supplies/service, Dominick's Foods/drugs, Goodyear/auto, Jo-Ann Fabrics, MailBoxes Etc, Mercury/Lincoln, Oil Change, Walgreen, Wal-Mart, cleaners, laundry, bank
265.5mm	weigh sta both lanes
263	Weber Rd, **E**...7-11, McDonald's, **W**...Shell/mart/wash/atm/24hr, American Grill, Cracker Barrel, Wendy's, Country Inn Suites, Extended Stay America, Howard Johnson
261	IL 126, to Plainfield, no facilities
257	US 30, to Joliet, Aurora, **E**...Citgo/mart, Shell/mart/24hr, Diamand's Rest., Hardee's, McDonald's, Pizza Hut, Red Lobster, Steak'n Shake/24hr, Subway, Taco Bell, Wendy's, Comfort Inn, Fairfield Inn, Hampton Inn, Motel 6, Ramada Ltd, Super 8, Circuit City, Home Depot, Honda, House of Fabrics, JC Penney, K-Mart, TireAmerica, bank, cinema, mall, **W**...Amoco/diesel/mart/24hr, Clark/Subway/diesel/rest./24hr, Phillips 66/diesel/rest./24hr, Ford
253b a	US 52, Jefferson St, Joliet, **E**...DENTIST, Amoco/repair, Shell/mart/24hr, McDonald's, Stockpot Rest., Wendy's, Day's Inn, Fireside Resort Motel, Wingate Inn, Ford, Rick's RV Ctr, **W**...HOSPITAL, Amoco/diesel/mart/24hr, Baba's Rest., Burger King, Dairy Queen, Subway(1mi), White Hen Pantry, Goodyear
251	IL 59(from nb), to Shorewood, access to same as 253
250b a	I-80, W to Iowa, E to Toledo, no facilities
248	US 6, Joliet, **E**...Speedway/diesel/mart/24hr, Frank's Country Store, Manor Motel, bank, **W**...Amoco/McDonald's/mart, Ivo's Grill, Lone Star Rest.(2mi), to Ill/Mich SP
247	Bluff Rd, no facilities
245mm	Des Plaines River
245	Arsenal Rd, **E**...Joliet Army Ammunition Plant, Mobil Refinery
241	to Wilmington, no facilities
241mm	Kankakee River
240	Lorenzo Rd, **E**...Phillips 66/diesel/24hr, **W**...Citgo/diesel/mart/24hr, Motel 55
238	IL 129, Braidwood, to Wilmington, no facilities

Chicago Area

Joliet

N

S

Illinois Interstate 55

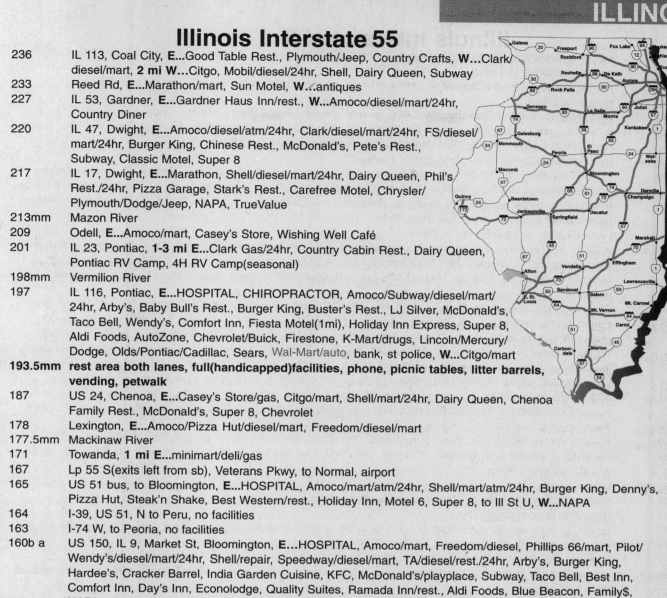

236	IL 113, Coal City, **E**...Good Table Rest., Plymouth/Jeep, Country Crafts, **W**...Clark/diesel/mart, **2 mi W**...Citgo, Mobil/diesel/24hr, Shell, Dairy Queen, Subway
233	Reed Rd, **E**...Marathon/mart, Sun Motel, **W**...antiques
227	IL 53, Gardner, **E**...Gardner Haus Inn/rest., **W**...Amoco/diesel/mart/24hr, Country Diner
220	IL 47, Dwight, **E**...Amoco/diesel/atm/24hr, Clark/diesel/mart/24hr, FS/diesel/mart/24hr, Burger King, Chinese Rest., McDonald's, Pete's Rest., Subway, Classic Motel, Super 8
217	IL 17, Dwight, **E**...Marathon, Shell/diesel/mart/24hr, Dairy Queen, Phil's Rest./24hr, Pizza Garage, Stark's Rest., Carefree Motel, Chrysler/Plymouth/Dodge/Jeep, NAPA, TrueValue
213mm	Mazon River
209	Odell, **E**...Amoco/mart, Casey's Store, Wishing Well Café
201	IL 23, Pontiac, **1-3 mi E**...Clark Gas/24hr, Country Cabin Rest., Dairy Queen, Pontiac RV Camp, 4H RV Camp(seasonal)
198mm	Vermilion River
197	IL 116, Pontiac, **E**...HOSPITAL, CHIROPRACTOR, Amoco/Subway/diesel/mart/24hr, Arby's, Baby Bull's Rest., Burger King, Buster's Rest., LJ Silver, McDonald's, Taco Bell, Wendy's, Comfort Inn, Fiesta Motel(1mi), Holiday Inn Express, Super 8, Aldi Foods, AutoZone, Chevrolet/Buick, Firestone, K-Mart/drugs, Lincoln/Mercury/Dodge, Olds/Pontiac/Cadillac, Sears, Wal-Mart/auto, bank, st police, **W**...Citgo/mart
193.5mm	**rest area both lanes, full(handicapped)facilities, phone, picnic tables, litter barrels, vending, petwalk**
187	US 24, Chenoa, **E**...Casey's Store/gas, Citgo/mart, Shell/mart/24hr, Dairy Queen, Chenoa Family Rest., McDonald's, Super 8, Chevrolet
178	Lexington, **E**...Amoco/Pizza Hut/diesel/mart, Freedom/diesel/mart
177.5mm	Mackinaw River
171	Towanda, **1 mi E**...minimart/deli/gas
167	Lp 55 S(exits left from sb), Veterans Pkwy, to Normal, airport
165	US 51 bus, to Bloomington, **E**...HOSPITAL, Amoco/mart/atm/24hr, Shell/mart/atm/24hr, Burger King, Denny's, Pizza Hut, Steak'n Shake, Best Western/rest., Holiday Inn, Motel 6, Super 8, to Ill St U, **W**...NAPA
164	I-39, US 51, N to Peru, no facilities
163	I-74 W, to Peoria, no facilities
160b a	US 150, IL 9, Market St, Bloomington, **E**...HOSPITAL, Amoco/mart, Freedom/diesel, Phillips 66/mart, Pilot/Wendy's/diesel/mart/24hr, Shell/repair, Speedway/diesel/mart, TA/diesel/rest./24hr, Arby's, Burger King, Hardee's, Cracker Barrel, India Garden Cuisine, KFC, McDonald's/playplace, Subway, Taco Bell, Best Inn, Comfort Inn, Day's Inn, Econolodge, Quality Suites, Ramada Inn/rest., Aldi Foods, Blue Beacon, Family$, NAPA, auto repair, **W**...Shell/mart, Country Kitchen, FoodCourt, Steak'n Shake/24hr, Country Inn Suites, Bloomington-Normal Stores/famous brands
157b	Lp 55 N, Veterans Pkwy, Bloomington, **N**...HOSPITAL, Clark Gas/24hr, Shell/mart/service, CJ's Rest., Jumer's Hotel/rest., Knight's Inn, Parkway Inn/rest./RV/truck parking, UPS, to airport
a	I-74 E, to Indianapolis, US 51 to Decatur
154	Shirley, no facilities
149mm	**rest area both lanes, full(handicapped)facilities, phone, picnic tables, litter barrels, vending, playground, petwalk**
145	US 136, McLean, **W**...Citgo/mart, Dixie/Mobil/diesel/rest./24hr, McDonald's, Super 8
140	Atlanta, **E**...camping, **W**...Phillips 66/diesel/mart, Country-Aire Rest., I-55 Motel
133	Lp 55, Lincoln, **2 mi E**...HOSPITAL, Citgo/mart, Budget Inn, Camp-A-While Camping
127	I-155 N, to Peoria, no facilities
126	IL 10, IL 121 S, Lincoln, **E**...HOSPITAL, Phillips 66/Steak'n Shake/diesel/mart/24hr, Cracker Barrel, Maverick Steaks, Taco Bell, Wendy's, Comfort Inn, Holiday Inn Express, Super 8, camping, **1 mi E**...EYECARE, Baby Bulls Rest., Burger King, Hardee's, LJ Silver, McDonald's, Crossroads Motel, Redwood Motel, Ace Hardware, Aldi Foods, AutoZone, $General, Eagle Foods, FashionBug, Ford/Mercury, JiffyLube, Radio Shack, Stage Store, Staples, Wal-Mart/auto, hardware
123	Lp 55, to Lincoln, no facilities
119	Broadwell, no facilities
115	Elkhart, **W**...Shell/diesel/mart
109	Williamsville, **E**...VETERINARIAN, Shell/mart, IGA Food, bank, **W**...New Salem SP
107mm	weigh sta sb

N
S

Illinois Interstate 55

105	Lp 55, to Sherman, **W**...Amoco/towing, Shell/mart, Cancun Mexican, Subway, to Prairie Capitol Conv Ctr, airport, camping, hist sites
103mm	**rest area sb, full(handicapped)facilities, phone, picnic tables, litter barrels, vending, petwalk**
102.5mm	Sangamon River
102mm	**rest area nb, full(handicapped)facilities, phone, picnic tables, litter barrels, vending, petwalk**
100b	IL 54, Sangamon Ave, Springfield, **W**...Amoco/mart/24hr, Shell/diesel/mart/atm, Speedway/mart, Arby's, Burger King, McDonald's, Parkway Café, Northfield Inn, Ramada Ltd, to Vet Mem
a	**E**...Citgo/diesel/mart, 4Star Café, Ryder Trucks
98b	IL 97, Springfield, **W**...HOSPITAL, Amoco/mart/24hr, Clark/mart, Harper's/diesel, Phillips 66/diesel/mart/atm, Shell/diesel/mart/atm/24hr, Baskin-Robbins, Chesapeake Seafood, Hardee's, Joette's Pizza/ribs, Mario's Italian, McDonald's, Subway, Taco Bell, Wendy's, Best Rest Inn, Dirksen Inn, Parkview Motel, Ford, Goodyear, K-Mart/ Little Caesar's, Nissan, Pennzoil, Tuffy Muffler, Walgreen, bank, to Capitol Complex
a	I-72, US 36 E, to Decatur, no facilities
96b a	IL 29 N, S Grand Ave, Springfield, **W**...Amoco/mart, Citgo/diesel/mart/24hr, Burger King, Godfather's, Nichow's Rest., Knight's Inn, Motel 6, Red Roof Inn, Super 8, AutoZone, Buick, Cub Foods, Dodge/Jeep, Ford/Lincoln/ Mercury, Hyundai, JC Penney, Mazda, Mitsubishi, Pontiac/GMC/Subaru, Shop'n Save Foods, Toyota, VW, museum
94	Stevenson Dr, Springfield, **E**...KOA(7mi), **W**...DENTIST, Amoco/diesel/mart/24hr, Shell/mart/atm, Arby's, Bob Evans, Bombay Bicycle Club, Denny's, Hardee's, HideOut Rest., Little Caesar's, LJ Silver, Maverick Steaks, McDonald's, Mtn Jack's Rest., Outback Steaks, Pizza Hut, Red Lobster, Steak'n Shake, Subway, Taco Bell, Taipan Chinese, Wendy's, Best Western/café, Comfort Suites, Day's Inn/rest., Drury Inn, Hampton Inn, Holiday Inn, Howard Johnson, PearTree Inn, Signature Inn, Big Lots, Chevrolet, CVS Drug, Eagle Food/deli, Honda, Olds/Cadillac, Midas Muffler, Mr Lincoln's Camping/LP, Saturn, USPO, bank, cinema, cleaners, hardware
92b a	I-72 W, US 36 W, 6th St, Springfield, **W**...HOSPITAL, Shell/diesel/mart, Burger King, Dairy Queen, Heritage House Smorgasbord, KFC, McDonald's, Mexican Rest., Subway, Econolodge/rest., Holiday Inn/rest., Super 8, Travelodge, Jewel/Osco Food/drug
90	Toronto Rd, **E**...HOSPITAL, Amoco/mart/24hr, Shell/mart, Antonio's Pizza, Burger King, Cracker Barrel, Hardee's, HenHouse, McDonald's, Subway, Taco Bell, Baymont Inn, Fairfield Inn, Motel 6, Ramada Ltd, cleaners, **W**...Johnny's Pizza
89mm	Lake Springfield
88	E Lake Dr, Chatham, **E**...to Lincoln Mem Garden/Nature Ctr, st police
83	Glenarm, **W**...camping
82	IL 104, to Pawnee, **E**...to Sangchris L SP, **W**...Mobil/Homestead/Subway/diesel/rest./24hr, boots, crafts
80	Divernon, Hist 66, **W**...Citgo/mart, Phillips 66/diesel/mart, antiques/carwash
72	Farmersville, **W**...Mobil/diesel/mart/24hr, Shell/mart/24hr, Subway/TCBY, Art's Motel/rest., antiques
65mm	**rest area both lanes, full(handicapped)facilities, phone, picnic tables, litter barrels, vending, playground, petwalk**
63	IL 48, IL 127, to Raymond, no facilities
60	IL 108, to Carlinville, **W**...Shell/diesel/LP/café, MoonLite Gardens Rest./antiques, Holiday Inn/rest., to Blackburn Coll, **E**...Kamper Kampanion RV camping
56.5mm	weigh sta nb
52	IL 16, Litchfield, Hist 66, **E**...HOSPITAL, DENTIST, EYECARE, Amoco/mart/24hr, Casey's/gas, Clark/mart, Coastal/mart, Phillips 66, Shell/mart/24hr, Ariston Café, Dairy Queen, Hardee's, Jubelt's Rest., KFC, Maverick Steaks, McDonald's, Pizza Hut, Ponderosa, Subway, Taco Bell, Wendy's, Best Western/rest., 66 Motel, Super 8, CarQuest, Dodge, Ford/Mercury, Goodyear/auto, IGA Foods, Kroger, McKay Parts, Medicine Shoppe, Rain- maker Camping(9mi), Rural King Foods, Wal-Mart, bank, cinema, hardware, tires, **W**...st police
44	IL 138, Mt Olive, to White City, **E**...Mobil/diesel/mart, Budget 10 Motel
41	to Staunton, **W**...Casey's/gas, Phillips 66/diesel/rest./24hr, Dairy Queen, Super 8
37	Livingston, **E**...Shady-Oak Camping(4mi), **W**...Amoco/diesel/mart/24hr, Country Inn/café, bank
33	IL 4, Worden, to Staunton, no facilities
30	IL 140, Hamel, **E**...Innkeeper Motel/rest., **W**...Citgo/diesel/mart, Shell/mart/24hr, Earnie's Café, Holiday Inn
23	IL 143, Edwardsville, **E**...Citgo/mart, auto repair
20b	I-270 W, to Kansas City, no facilities
a	I-70 E, to Indianapolis, no facilities
I-55 S and I-70 W run together 18 mi	
18	IL 162, to Troy, **E**...HOSPITAL, DENTIST, Amoco/FastFood/diesel/24hr, Citgo/mart/24hr, Pilot/Arby's/diesel/ mart/24hr, Burger King, Dairy Queen, Imo's Pizza, Jack-in-the-Box, Little Caesar's, McDonald's, Perkins, Pizza Hut, Subway, Relax Inn, Family$, SuperValu Foods, TrueValue, USPO, bank, laundry, truckwash/lube, **W**...Mobil/ Randy's/diesel/mart/24hr, China Garden, Cracker Barrel, Denny's, Taco Bell, Scottish Inn, Super 8

N

S

Illinois Interstate 55

17	US 40 E, to Troy, to St Jacob
15b a	IL 159, Maryville, Collinsville, **E**...Mobil/mart, Phillips 66/diesel/mart/atm/24hr, Sharkey's Rest., **W**...Conoco/mart/repair, Econolodge, bank
14mm	weigh sta sb
11	IL 157, Collinsville, **E**...Amoco/mart/24hr, Shell/mart/24hr, Denny's, Hardee's, LJ Silver, McDonald's, Pizza Hut, Waffle House, Wendy's, Best Western, Day's Inn, Howard Johnson, Motel 6, PearTree Inn, Travelodge, Midas Muffler, cinema, **W**...DENTIST, Motomart/diesel/mart/atm/24hr, Applebee's, Arby's, Bob Evans, Boston Mkt, Burger King, Cancun Mexican, Dairy Queen, Ponderosa, Shoney's, Steak'n Shake, White Castle/24hr, Comfort Inn, Drury Inn, Fairfield Inn, Hampton Inn, Holiday Inn/rest., Microtel, Ramada Ltd, Super 8, Buick/Pontiac/GMC, Chrysler/Plymouth/Dodge/Jeep, bank, funpark, st police
10	I-255, S to Memphis, N to I-270
9	Black Lane(from nb), **E**...Fairmount RaceTrack
6	IL 111, Fairmont City, Great River Rd, **E**...Clark Gas, Rainbo Motel, **W**...Horseshoe SP
5mm	motorist callboxes begin at 1/2 mi intervals nb
4b a	IL 203, Granite City, **W**...Texaco/diesel/mart/rest./24hr, Gateway Int Raceway
3	Exchange Ave, no facilities
2	I-64 E, IL 3 N, St Clair Ave, no facilities
2b	3rd St, no facilities
2a	M L King Bridge, to downtown E St Louis, no facilities
1	IL 3, to Sauget(from sb), no facilities
	I-55 N and I-70 E run together 18 mi
0mm	Mississippi River, Illinois/Missouri state line

N ↕ S

E St Louis Area

Illinois Interstate 57

Exit #	Services
358mm	I-94 E to Indiana, I-57 begins/ends on I-94, exit 63 in Chicago.
357	IL 1, Halsted St, **E**...Amoco/mart/wash, Mobil/mart, laundry, **W**...Citgo/mart, Marathon/mart, Shell/mart/24hr, McDonald's
355	111th St, Monterey Ave, no facilities
354	119th St, no facilities
353	127th St, Burr Oak Ave, **E**...HOSPITAL, Amoco/mart, Shell, Burger King, McDonald's, Mr T's Café, Plaza Inn, Super 8, Meineke Muffler, **W**...Amoco/mart
352mm	Calumet Sag Channel
350	IL 83, 147th St, Sibley Blvd, **E**...Citgo, Mobil/mart, EZ gas/mart, Pancake Rest., mufflers
348	US 6, 159th St, **E**...Citgo/diesel/mart, Mobil/mart, Phillips 66, **W**...King Gas/diesel/mart, Family Rest., Delux Motel, **2 mi W**...facilities on W 159th
346	167th St, Cicero Ave, to IL 50, **E**...Amoco, **W**...Shell
345b a	I-80, W to Iowa, E to Indiana, to I-294 N toll to Wisconsin
342	Vollmer Rd, **E**...HOSPITAL, Shell/mart/wash/24hr
340b a	US 30, Lincoln Hwy, Matteson, **E**...CHIROPRACTOR, Amoco/24hr, Citgo/diesel/mart, Mobil/mart, Shell, American Bagel, Applebee's, Bob Evans, Boston Mkt, Burger King, ChuckeCheese, Cracker Barrel, Denny's, Dock's Seafood, Fazoli's, Fuddrucker's, Iceland Café, JN Michael's Grill, KFC, McDonald's, Nino's Pizza, Old Country Buffet, Olive Garden, Pizza Hut, Red Lobster, Siam Thai/Chinese, Taco Bell, Baymont Inn, Hampton Inn, Holiday Inn/rest., Matteson Motel, Best Buy, Circuit City, CopyMax, Dominick's Foods, 1st Muffler/brake, Goodyear/auto, Home Depot, Isuzu, JC Penney, Jo-Ann Fabrics, Kinko's, K-Mart, OfficeMax, Olds, PetsMart, Pier 1, Sam's Club, Sears/auto, Target, TJ Maxx, ToysRUs, Walgreen, Wal-Mart, bank, mall, **W**...auto mall/multiple dealers
339	Sauk Trail, to Richton Park, **E**...Amoco/mart/wash, Mobil, Shell/mart, Bob's HotDogs, McDonald's, cinema, **1-2 mi E**...CHIROPRACTOR, DENTIST, PODIATRIST, Arby's, Baskin-Robbins, Burger King, Dairy Queen, Dunkin Donuts, Eagle Foods, Hook's Drugs, Howard's AutoCare, cleaners
335	Monee, **E**...Amoco/Subway/mart, Petro/Mobil/diesel/rest./Blue Beacon/24hr, Speedway/Hardee's/diesel/mart, Burger King, Pizza Hut, Country Host Motel, Super 8
332mm	**Prairie View Rest Area both lanes, full(handicapped)facilities, info, phone, vending, picnic tables, litter barrels, RV dump, petwalk**
330mm	weigh sta both lanes
327	to Peotone, **E**...Shell/Taco Bell/mart/24hr, McDonald's

N ↕ S

Chicago

Illinois Interstate 57

Kankakee

322   Manteno, **E**...Amoco/McDonald's/mart/24hr, Phillips 66/Subway/mart/atm, Dairy Queen, Hardee's, Monical's Pizza, Comfort Inn, Harley-Davidson, **W**...GasCity/diesel

315   IL 50, Bradley, **E**...Shell/Burger King/mart/24hr, Cracker Barrel, LoneStar Steaks, Pizza Hut, Red Lobster, TGIFriday, White Castle, Fairfield Inn, Hampton Inn, Holiday Inn Express, Lee's Inn, Barnes&Noble, JC Penney, Marshall's, MC Sports, Michael's, Midas Muffler, Pier 1, Rex Audio/video, Staples, Target, cinema, mall, **W**...Amoco/diesel/mart, Applebee's, Arby's, Bakers Square, Boston Mkt, Coyote Canyon Steaks, Denny's, Hardee's, LJ Silver, McDonald's, Mongolian Buffet, Ponderosa, Steak'n Shake, Sirloin Stockade, Subway, Taco Bell, Wendy's, Howard Johnson, Motel 6, Northgate Motel, Ramada Inn, Super 8, Aldi Foods, Buick/Olds/Nissan, Chevrolet, Hyundai, K-Mart, Lowe's Bldg Supply, Mazda, Menard's, OfficeMax, PetCo, RV Sales, Wal-Mart/24hr, to Kankakee River SP

312   IL 17, Kankakee, **E**...Chrysler/Plymouth/Dodge, **W**...HOSPITAL, Amoco/wash, Clark Gas, Shell/mart/atm/24hr, Burger King, KFC, LJ Silver, McDonald's, PoorBoy Rest., River Oaks Rest., Subway, Uncle Johnni's Rest., Wendy's, Avis Motel, Day's Inn/rest., JustLube, Walgreen, auto repair, laundry, mufflers/brakes

310.5mm   Kankakee River

308   US 45, US 52, to Kankakee, **E**...Redwood Rest., **W**...Phillips 66/diesel/mart, Fairview Motel, Knight's Inn, KOA, airport

302   Chebanse, **W**...BP/mart

297   Clifton, **W**...Phillips 66/diesel/mart, CharGrilled Cheeseburgers, Dairy Queen

293   IL 116, Ashkum, **E**...Amoco/mart, Dairy Queen, **W**...Loft Rest., st police, tires

283   US 24, IL 54, Gilman, **E**...Amoco/diesel/mart/24hr, Apollo/diesel/mart, Citgo/diesel/mart, Shell/mart, Burger King, Dairy Queen, McDonald's, Budget Host, Super 8, Travel Inn, truckwash, **W**...Phillips 66/Subway/diesel/mart

280   IL 54, Onarga, **E**...Phillips 66/mart/wash, camping

272   Buckley, to Roberts, no facilities

268.5mm rest area both lanes, full(handicapped)facilities, vending, phone, picnic tables, litter barrels, petwalk

261   IL 9, Paxton, **E**...Casey's/gas, Citgo/mart/24hr, Hardee's, Monical's Pizza, Pizza Hut, Subway, Chevrolet/Olds/Pontiac/Buick/GMC, Family$, Pennzoil/wash, **W**...Amoco/mart/atm, Marathon/mart, Country Garden Rest., Paxton Inn, Ford/Mercury

Champaign

250   US 136, Rantoul, **E**...Amoco/mart/rest./24hr, Phillips 66/mart, Arby's, Hardee's, LJ Silver, McDonald's, Monical's Pizza, Red Wheel Rest., Taco Bell, Best Western, Day's Inn/rest., Rantoul Motel, Super 8, camping, to Chanute AFB

240   Market St, **E**...Amoco/diesel/wash/rest., camping

238   Olympian Dr, to Champaign, **W**...Mobil/diesel/mart/atm, Dairy Queen, Microtel

237b a   I-74, W to Peoria, E to Urbana

235b   I-72 W, to Decatur

    a   University Ave, to Champaign, **E**...HOSPITAL, U of Ill

229   to Savoy, Monticello, **E**...Speedway/diesel/mart, airport

221.5mm rest area both lanes, full(handicapped)facilities, phones, picnic tables, litter barrels, vending, petwalk

220   US 45, Pesotum, **W**...Citgo/mart, Lakeside Rest./gifts, st police

212   US 36, Tuscola, **E**...FuelMart/diesel, **W**...Amoco/towing/24hr, Citgo/Dixie/diesel/mart/café, Pilot/diesel/mart/24hr, Dairy Queen, Denny's, FoodCourt, Hardee's, McDonald's, Monical's Pizza, Pizza Hut, Subway, AmeriHost, Holiday Inn Express, Super 8, Chevrolet/Pontiac/Olds/Buick/GMC, Firestone/auto, IGA Foods, Pamida, Tuscola Stores/famous brands, auto parts, camping, carwash

203   IL 133, Arcola, **E**...Citgo/mart, CampALot, **W**...VETERINARIAN, Marathon/mart, Phillips 66, Shell/diesel/mart, Dairy Queen, Dutch Kitchen Rest., Hardee's, Hen House, Subway, Arcola Inn, Budget Host, Comfort Inn, Day's Inn, $General, to Rockome Gardens, Arcola Camping

Mattoon

190b a   IL 16, to Mattoon, **E**...HOSPITAL, Citgo/mart, to E IL U, Fox Ridge SP, **W**...Amoco/24hr, Clark Gas, Phillips 66/Subway/mart/atm/24hr, Arby's, Cody's Roadhouse, Country Kitchen, Cracker Barrel, Dairy Queen, McDonald's, Ponderosa, Steak'n Shake/24hr, Taco Bell, Wendy's, Comfort Suites, Fairfield Inn, Hampton Inn, Ramada Inn/rest., Super 8, Aldi Foods, $General, JC Penney, K-Mart/Little Caesar's, Osco Drug, Sears, Staples, Walgreen, Wal-Mart SuperCtr/gas/24hr, mall

184   US 45, IL 121, to Mattoon, **W**...Marathon/diesel/mart, Shell/24hr, Hardee's/24hr(3mi), McDonald's, Subway(3mi), Knight's Inn, US Grant Motel, Villager Lodge, to Lake Shelbyville

177   US 45, Neoga, **E**...BP/diesel/mart, Citgo/mart, **W**...Phillips 66/diesel/mart, Casey's/gas(2mi)

166.5mm rest area both lanes, full(handicapped)facilities, vending, phones, picnic tables, litter barrels, petwalk, RV dump

163   I-70 E, to Indianapolis

I-57 S and I-70 W run together 6 mi

162   US 45, Effingham, **E**...Moto/mart/24hr, Trailways Buffet, Budget Host, **W**...Citgo/mart/atm/24hr, Shell/diesel/mart, Effingham Motel/café

N

S

Illinois Interstate 57

160   IL 33, IL 32, Effingham, **E**...HOSPITAL, Amoco/mart/atm, Shell/repair, KFC, AmeriHost, Hampton Inn, Aldi Foods, K-Mart, Kroger/deli, ProGolf, Radio Shack, carwash, **W**...Bobber/Shell/Blimpie/diesel/café, Marathon, Phillips 66/mart, Shell/Subway/mart, Speedway/diesel/mart, TA/BP/diesel/rest., A&W, Arby's, BBQ, Burger King/24hr, Cracker Barrel, Denny's, FoodCourt, KFC, LJ Silver, McDonald's, Ponderosa, Ryan's, Steak'n Shake/24hr, Taco Bell, TGIFriday, Wendy's, Baymont Inn, Best Inn, Econolodge, Super 8, Ramada Inn/rest., Ford/Lincoln/Mercury, Factory Outlet/famous brands, PartsPro, Wal-Mart, bank, repair, tires, trucklube/wash

159   US 40, Effingham, **E**...Amoco/mart/24hr, Clark, Dixie/Citgo/diesel/rest./24hr, Phillips 66/diesel/mart/24hr, Shell/TCBY/Taco John/mart/24hr, Speedway/diesel/mart, Golden China, Hardee's/24hr, Niemerg's Rest., Papa John's, Subway, Abe Lincoln Motel, Comfort Suites, Day's Inn, Holiday Inn/rest., Howard Johnson, Quality Inn, Firestone, **W**...Petro/Mobil/diesel/rest./24hr, Best Western, Blue Beacon

     I-57 N and I-70 E run together 6 mi

157   I-70 W, to St Louis, no facilities

151   Watson, no facilities

150mm   Little Wabash River

145   Edgewood, **E**...Citgo/mart, **W**...Shell/diesel/rest.

135   IL 185, Farina, **E**...76/diesel/mart/rest., Frontier Village/Gen Store, Chevrolet, Ford

127   to Kinmundy, Patoka, **E**...Marathon/diesel/mart/24hr

116   US 50, Salem, **E**...Amoco/mart/24hr, Clark Gas, Motomart/24hr, Shell, Austin Fried Chicken, Burger King, Denny's, Golden Corral, KFC, LJ Silver, Mazzio's Pizza, McDonald's, Pancake House, Pizza Hut, Pizza Man, Subway, Taco Bell, Wendy's, Western Ribeye, Continental Motel, AutoZone, Chrysler/Dodge, K-Mart, to Forbes SP, **W**...Phillips 66/diesel, Comfort Inn, Day's Inn, Holiday Inn, Super 8, Chevrolet/Buick, Ford/Mercury, tires

114mm   **rest area both lanes, full(handicapped)facilities, phones, picnic tables, litter barrels, vending, petwalk**

109   IL 161, to Centralia, **W**...Phillips 66/mart/24hr, Shell/7-II, camping

103   Dix, **E**...Marathon/diesel/mart, Scottish Inn, camping

96   I-64 W, to St Louis, no facilities

95   IL 15, Mt Vernon, **E**...HOSPITAL, CHIROPRACTOR, DENTIST, Amoco/diesel/mart/24hr, Mobil/mart, Phillips 66/mart, Speedway/mart/24hr, Arby's, Bonanza, Burger King, Denny's, Hardee's, Fazoli's, Hunan Chinese, KFC, LJ Silver, McDonald's, Papa John's, Pizza Hut, Steak'n Shake/24hr, Subway, Taco Bell, Wendy's, Western Sizzlin, Best Inns, Best Western, Drury Inn, Econolodge, Motel 6, Super 8, Thrifty Inn, AutoZone, Chevrolet/Olds, Country Fair Foods, CVS Drug, FastLube, Ford, JC Penney, Jo-Ann Fabrics, K-Mart, Kroger/deli, Midas Muffler, Radio Shack, Walgreen, bank, **W**...Hucks/diesel/rest./24hr, Shell/7-11/24hr, TA/Citgo/Popeye's/diesel/rest./24hr, Applebee's, Arby's, Burger King, Cracker Barrel, LoneStar Steaks, McDonald's, Sonic, Comfort Inn, Day's Inn, Hampton Inn, Holiday Inn, Ramada Hotel, Cadillac/Pontiac/Buick/GMC, FannieMae Candies, Lowe's Bldg Supply, Outlet Mall/famous brands, Quality Times RV Park, Staples, Toyota, Wal-Mart SuperCtr/24hr, White/Volvo/Freightliner, Wheeler's Truckwash, cinema

92   I-64 E, to Louisville, no facilities

83   Ina, **E**...BP/diesel/deli/mart/24hr, camping, **W**...to Rend Lake Coll

79mm   **rest area sb, full(handicapped)facilities, info, vending, phones, picnic tables, litter barrels, petwalk**

77   IL 154, to Whittington, **E**...Shell/mart, Holiday Trav-L Park, **W**...Best Western, Seasons at Rend Lake Lodge/rest., to Rend Lake, golf, Wayne Fitzgerrell SP

74mm   **rest area nb, full(handicapped)facilities, vending, phones, picnic tables, litter barrels, petwalk**

71   IL 14, Benton, HOSPITAL, **E**...CHIROPRACTOR, Amoco/mart/24hr, BP/mart, Citgo/mart, Gas4Less, Phillips 66/wash, HandeeMart, Hardee's, KFC, Pizza Hut, Plaza Rest., Wendy's, Day's Inn/rest., Gray Plaza Motel, Motel Benton, Super 8, Firestone, KOA, Pennzoil, bank, carwash, cinema, laundry, repair, **W**...Shell/diesel/mart/24hr, Bonanza, McDonald's, Subway, Taco John's, Big John Foods, CVS Drug, Wal-Mart/drugs, bank, to Rend Lake

65   IL 149, W Frankfort, **E**...HOSPITAL, Amoco/mart/24hr, BP, Citgo/diesel/mart/24hr, Clark Gas, Phillips 66, Shell/mart/24hr, Hardee's, KFC, LJ Silver, Mike's Drive-In, Pizza Inn, Pancakes/24hr, Gray Plaza Motel, Farm Fresh Foods, SuperX Drugs, Big A Parts, bank, carwash, **W**...Casey's/gas, Triple E+ Steaks, McDonald's, HoJo's, Chevrolet/Olds/Pontiac, CVS Drug, Kroger/deli, K-Mart, Radio Shack, VF Factory Stores

59   Johnston City, to Herrin, **E**...Citgo/mart, Shell/mart/24hr, Dairy Queen, Hardee's, McDonald's, camping(2mi), tires, **W**...HOSPITAL

N

S

Illinois Interstate 57

Marion

54b a	IL 13, Marion, **E**...EYECARE, MEDICAL CARE, Cheker/gas, Citgo, Phillips 66/mart/wash/24hr, Arby's, Fazoli's, Hardee's, KFC, LJ Silver, Papa John's, Pizza Hut, Ponderosa, Wendy's, Western Sizzlin, Day's Inn, Gray Plaza Motel, Red Lion Inn, Chevrolet/Cadillac, Ford/Lincoln/Mercury/Hyundai, Aldi Foods, Kroger, Town&Country Foods, USPO, Valvoline, airport, camping, hardware, tires, **W**...Amoco/mart/24hr, BP/diesel/mart/24hr, Bob Evans, Burger King, Chinese Rest., McDonald's, Ryan's, Shoney's Inn/rest., Sonic, Steak'n Shake, Taco Bell, Best Inn, Drury Inn, Hampton Inn, Holiday Inn Express, Motel 6, Super 8, Buick/GMC/Olds/Pontiac/Mercedes, Sam's Club, Wal-Mart
53	Main St, Marion, **E**...HOSPITAL, CHIROPRACTOR, Shell/diesel/mart, Dairy Queen, TCBY, Comfort Inn, Motel Marion/camping, auto parts, **W**...Motomart/24hr, Cracker Barrel, HideOut Steaks, TCBY, Comfort Inn/Suites
47mm	weigh sta both lanes
45	IL 148, **1 mi E**...BP/diesel/24hr, Toupal's Country Inn/deli, Egyptian Hills Marina(9mi), camping, diesel repair
44	I-24 E to Nashville, no facilities
40	Goreville Rd, **E**...Ferne Clyffe SP, camping, scenic overlook, **W**...Citgo/diesel/mart/24hr
36	Lick Creek Rd, no facilities
32mm	**Trail of Tears Rest Area both lanes, full(handicapped)facilities, info, phones, picnic tables, litter barrels, vending, petwalk, playground**
30	IL 146, Anna, Vienna, **W**...HOSPITAL, Shell/Omelet Hut/mart
24	Dongola Rd, **W**...Shell/mart
18	Ullin Rd, **W**...BP/diesel/24hr, Cheeko's Family Rest./24hr, Best Western, st police
8	Mounds Rd, to Mound City, **E**...K&K AutoTruck/diesel/repair
1	IL 3, to US 51, Cairo, **E**...MEDICAL CARE, Amoco/diesel/mart, BP/diesel/mart, Belvedere Motel, Day's Inn, Garden Inn/camping, Mound City Nat Cem(4mi), **W**...Sinclair/mart, camping
0mm	Mississippi River, Illinois/Missouri state line

Illinois Interstate 64

Exit #	Services
131.5mm	Wabash River, Illinois/Indiana state line
131mm	**Skeeter Mtn Welcome Ctr wb, full(handicapped)facilities, phone, vending, picnic tables, litter barrels, petwalk**
130	IL 1, to Grayville, **N**...Shell/mart/24hr, Gingham House Rest., Best Western/rest., museum, camping, **S**...Phillips 66/diesel/mart
124mm	Little Wabash River
117	Burnt Prairie, **S**...Marathon/diesel/mart, ChuckWagon Charlie's Café, antiques
110	US 45, Mill Shoals, **N**...SS/gas/diesel/mart, Barnhill Camping
100	IL 242, to Wayne City, **N**...Marathon/diesel/mart
89	to Belle Rive, Bluford, no facilities
86mm	**rest area wb, full(handicapped)facilities, phone, vending, picnic tables, litter barrels, petwalk**
82.5mm	**rest area eb, full(handicapped)facilities, phone, vending, picnic tables, litter barrels, petwalk**
80	IL 37, to Mt Vernon, **2 mi N**...Amoco/Burger King/diesel/mart/atm/24hr, Citgo/diesel/mart, Marathon/diesel/mart/24hr, Royal Inn, camping
78	I-57, S to Memphis, N to Chicago
	I-64 and I-57 run together 5 mi. See Illinois Interstate 57, exit 95.
73	I-57, N to Chicago, S to Memphis, no facilities
69	Woodlawn, no facilities
61	US 51, Richview, to Centralia, **S**...access to gas, food
50	IL 127, to Nashville, **N**...to Carlyle Lake, **S**...HOSPITAL, Amoco/24hr, Conoco/diesel/rest./wash, Shell/diesel/mart/24hr, Hardee's(3mi), LN Foodmart/24hr, McDonald's, Subway, Best Western, Little Nashville Inn/rest., Derrick Motel/rest.(3mi)
41	IL 177, Okawville, **S**...Phillips 66/Burger King/diesel/mart/atm/24hr, Dairy Queen, Hen House/24hr, Golfer's Steakhouse, Super 8, Original Springs Motel, Toyo Tires/truck repair
37mm	Kaskaskia River
34	to Albers, **3 mi N**...Casey's Gen Store/gas
27	IL 161, New Baden, **N**...Shell/diesel/mart/24hr, Good Ol' Days Rest., Chevrolet
25mm	**rest area both lanes, full(handicapped)facilities, info, phone, vending, picnic tables, litter barrels, petwalk**
23	IL 4, to Mascoutah, **3 mi N**...Hardee's
19b a	US 50, IL 158, **N**...Motomart/24hr, Hero's Pizza/subs, Comfort Inn, **S**...HOSPITAL, Citgo/diesel/mart, Ivory Chopsticks Chinese, to Scott AFB
18mm	weigh sta eb
15mm	motorist callbox every 1/2 mile wb
14	O'Fallon, **N**...QT/mart, Shell/mart/24hr, Japanese Garden, Steak'n Shake/24hr, Baymont Inn, Extended Stay America, Holiday Inn Express, Sleep Inn, Chevrolet/Olds, CVS Drug, Ford, Nissan/Olds/Cadillac, bank, **S**...Mobil/mart, Chevy's Mexican, Dairy Queen, Emperor's Wok, Hardee's, Jack-in-the-Box, KFC, Lion's Choice, LoneStar

Mt Vernon

N

S

E

W

Illinois Interstate 64

Steaks, McDonald's, O'Charley's, Pepper's Rest., SteakOut, Taco Bell, Western Sizzlin, Econolodge, Ramada Ltd, Garden Ridge, Home Depot, Honda, Hyundai, Kia, Mazda, Mitsubishi, PetsMart, Play It Again Sports, Sam's Club, Subaru, Suzuki, Toyota, VW, Wal-Mart/auto/24hr, cinema

E ↕ W

12   IL 159, to Collinsville, **N**...CFM/diesel, Applebee's, Bob Evans, Carlos O'Kelly's, Damon's, HideOut Steaks, Houlihan's, Joe's Crabshack, Olive Garden, Red Lobster, TGIFriday, Best Western Camelot, Drury Inn, Fairfield Inn, Hampton Inn, Ramada Inn/rest., Super 8, Circuit City, Michael's, 1/2Price Store, Saturn, SportsAuthority, antiques, cinema, **S**...Amoco/mart/24hr, Mobil/mart, Motomart/diesel/24hr, Boston Mkt, Burger King, Capt D's, Casa Gallardo, Chili's, Denny's, Fazoli's, Hardee's, Honeybaked Ham, IHOP, Longhorn Steaks, LJ Silver, Mazzio's Pizza, McDonald's, Old Country Buffet, Pasta House, Ponderosa, Popeye's, Rally's, Ramon's Mexican, Ruby Tuesday, Schlotsky's, Steak'n Shake, Taco Bell, AutoTire, Bed Bath&Beyond, Best Buy, Borders, Burlington Coats, CarX Muffler, Dillard's, Famous Barr, Firestone/auto, Goodyear, JC Penney, JiffyLube, KidsRUs, Kinko's, K-Mart, Kohl's, MensWearhouse, Midas Muffler, OfficeMax, PetCo, Pier 1, Shoe Carnival, Schnuck's Foods, Target, TireAmerica, TJ Maxx, Valvoline, Walgreen, bank, hardware, mall

9   IL 157, to Caseyville, **N**...Amoco/mart/24hr, Phillips 66/Subway/repair, Hardee's, Wendy's, Western Inn, **S**...Amoco/diesel/repair, Cracker Barrel, Domino's, McDonald's, Pizza Hut/Taco Bell, Best Inn, Day's Inn, Motel 6, Quality Inn

7   I-255, S to Memphis, N to Chicago, no facilities

6   IL 111, Kingshighway, **N**...Amoco/mart, Shell/mart, Popeye's, Econo Inn

5.5mm   motorist callbox every 1/2 mi eb

5   25th St, no facilities

4   15th St, Baugh, no facilities

3   I-55 N, I-70 E, IL 3 N, to St Clair Ave, to stockyards

2b a   3rd St, MLK, **S**...Clark Gas

1   IL 3 S, 13th St, E St Louis, **N**...Casino Queen

0mm   Mississippi River, Illinois/Missouri state line

E St Louis

Illinois Interstate 70

Exit #	Services
156mm	Illinois/Indiana state line
154	US 40 W, **S**...Southfork Convenience Mart
151mm	weigh sta wb

E ↕ W

149mm   **rest area wb, full(handicapped)facilities, info, phone, picnic tables, vending, litter barrels, petwalk, camping**

147   IL 1, Marshall, **N**...Ike's Rest., **S**...Casey's/gas, Jiffy/diesel/mart/24hr, Phillips 66/Arby's/diesel, Shell/Subway/mart, Burger King, Dairy Queen, Hardee's, McDonald's, Pizza Hut, Wendy's, Peak's Motel, Lincoln Motel(2mi), Super 8, Lincoln Trail SP, camping

136   to Martinsville, **S**...BP/diesel/mart/24hr

134.5mm   N Fork Embarras River

129   IL 49, Casey, **N**...RV service, KOA(seasonal), **S**...BP/mart/24hr, Citgo/DQ/diesel/mart, Speedway/mart, Hardee's, Joe's Pizza, KFC, McDonald's, Pizza Hut, Comfort Inn, Casey Motel(1mi)

119   IL 130, Greenup, **S**...MEDICAL CARE, Amoco/mart, BP/diesel/mart, Phillips 66/Subway/mart, Dairy Queen, Dutch Pan Rest., Budget Host, Gateway Motel, 5 Star Motel, Chevrolet/Olds, bank, carwash, camping, hist sites

105   Montrose, **S**...Amoco/diesel/mart, Shell/delimart/24hr, Montarosa Motel/café

98   I-57, N to Chicago

I-70 and I-57 run together 6 mi. See Illinois Interstate 57, exits 159-162.

92   I-57, S to Mt Vernon

91mm   Little Wabash River

87mm   **rest area both lanes, full(handicapped)facilities, info, phone, vending, picnic tables, litter barrels, playground, petwalk, RV dump**

82   IL 128, Altamont, **N**...Casey's/gas, Citgo/Stuckey's/Subway/mart/atm/24hr, Marathon, Speedway/diesel/mart, McDonald's, Altamont Motel/rest., Best Western/rest., Fostoria Factory Outlet, **S**...Phillips 66/diesel/mart, Shell/diesel, TJ's Rest., Super 8

Effingham

Illinois Interstate 70

76	US 40, St Elmo, **N**...Marathon/mart, Phillips 66/mart, Waldorf Motel, Timberline Camping(2mi)
71mm	weigh sta eb
68	US 40, Brownstown, **N**...Okaw Valley Camping
63.5mm	Kaskaskia River
63	US 51, Vandalia, **N**...Day's Inn, LJ Silver, Indian crafts, **S**...HOSPITAL, Amoco/24hr, Clark/mart, Marathon/mart/atm, Hardee's/24hr, KFC, McDonald's, Pizza Hut, Wendy's, Jay's Inn, Mabry Motel, Robbins Motel, Travelodge, bank, hist site
61	US 40, Vandalia, **N**...airport, **S**...Fastop/diesel, Ponderosa, Bunyard Café(1mi), Ramada Ltd, Goodyear/auto, TrueValue, Wal-Mart
52	US 40, Mulberry Grove, **N**...Citgo/diesel/mart, Timber Trail Camp-In(2mi), tires/repair, antiques, **S**...Cedar Brook Camping(1mi)
45	IL 127, Greenville, **N**...HOSPITAL, Amoco, Phillips 66/mart, Shell/diesel/mart/24hr, Hardee's(2mi), KFC, Lu-Bob's Rest., McDonald's, Best Western/rest., Budget Host, Super 8, 2 Acre Motel, Ford/Mercury, **S**...to Carlyle Lake, Circle B Steaks, airport
41	US 40 E, to Greenville, **1-3 mi N**...Nuby's Steaks
36	US 40 E, Pocahontas, **S**...Amoco/diesel/mart/24hr, Phillips 66/diesel/24hr, Nuby's Steaks, Powhatan Motel/rest., Wickiup Hotel, Tahoe Motel(3mi), tires/repair
30	US 40, IL 143, to Highland, **S**...Shell/mart/24hr, Blue Springs Café, Tomahawk RV Park(7mi)
26.5mm	**Silver Lake Rest Area both lanes, full(handicapped)facilities, phone, picnic tables, litter barrels, vending, petwalk**
24	IL 143, Marine, **S**...HOSPITAL
23mm	motorist callboxes begin wb every 1/2mile
21	IL 4, Troy, **N**...Mobil/diesel/24hr
15b a	I-55, N to Chicago, S to St Louis, I-270 W to Kansas City
I-70 and I-55 run together 18 mi . See Illinois Interstate 55, exits 1-18.	
0mm	Mississippi River, Illinois/Missouri state line

Illinois Interstate 72

Exit #	Services
182b a	I-57, N to Chicago, S to Memphis, to I-74
176	IL 47, to Mahomet, no facilities
172	IL 10, Lodge, Seymour, no facilities
169	White Heath Rd, no facilities
166	IL 105 W, Market St, **N**...Ford/Mercury, **S**...FastStop/Taco Bell/gas, Best Western
165mm	Sangamon River
164	Bridge St, **1 mi S**...HOSPITAL, Mobil/mart, Dairy Queen, Hardee's, McDonald's, Monical's Pizza, Pizza Hut, Subway
156	IL 48, Cisco, to Weldon, no facilities
153mm	**rest area both lanes, full(handicapped)facilities, phone, picnic tables, litter barrels, vending, petwalk**
152mm	Friends Creek
150	Argenta, **1 mi N**...gas
144	IL 44, Oreana, **N**...fuel, food, **S**...HOSPITAL
141b a	US 51 S, Decatur, **N**...Shell/mart, Applebee's, Cheddar's, Cracker Barrel, Country Kitchen, Hardee's, HomeTown Buffet, McDonald's, Steak'n Shake, Taco Bell, Baymont Inn, Comfort Inn, Country Inn Suites, Fairfield Inn, Hampton Inn, Ramada Ltd, Advance Parts, Buick/GMC, JC Penney, Lowe's Bldg Supply, Kohl's, Menard's, Mitsubishi/Hyundai/Honda, Pennzoil, Sears/auto, Staples, cinema, mall, **S**...Arby's, Burger King, Wal-Mart
138	IL 121, Decatur, **S**...HOSPITAL
133b a	US 36 E, US 51, Decatur, **S**...Phillips 66/mart, Day's Inn, Holiday Inn Select/rest.
128	Niantic, no facilities
122	Illiopolis, to Mt Auburn, **N**...Citgo/mart
114	Buffalo, Mechanicsburg, no facilities
108	Riverton, Dawson, no facilities
107mm	Sangamon River
104	Camp Butler, **N**...Best Rest Inn, Park View Motel, golf(1mi)
103b a	I-55, N to Chicago, S to St Louis, Il 97, to Springfield
I-72 and I-55 run together 6 mi. See IL I-55, exits 92-98.	
97b a	6th St, I-55 S, **N**...Shell, Heritage House Rest., McDonald's, Illini Inn, Ramada Inn, Super 8, Travelodge, Mr Lincoln's Camping/LP

Vandalia

Decatur

E

W

E

W

Illinois Interstate 72

E

↕

W

93	IL 4, Springfield, **N**...Huck's Food/fuel, Applebee's, Arby's, Bakers Square, Best Buffet, Burger King, Chili's, Damon's, Hardee's, Kenny Rogers, LoneStar Steaks, Maverick Rest., McDonald's/playplace, Ned Kelly's Steaks, Olive Garden, Panera Bread Café, Pasta House, Perkins, Taco Bell, TCBY, Comfort Inn, Courtyard, Fairfield Inn, Sleep Inn, Barnes&Noble, Batteries+, Best Buy, Circuit City, Cub Foods/24hr, Jo-Ann Crafts, Kinko's, K-Mart, Kohl's, Lands End, Lowe's Bldg Supply, Menard's, Michael's, Office Depot, PetsMart, Pier 1, Sam's Club, ShopKO, Staples, Target, TJ Maxx, Wal-Mart, cinema, to airport
91	Wabash Ave, to Springfield, **N**...Lincoln/Mercury
82	New Berlin, **S**...Travel Plaza/diesel/mart, bank
76	IL 123, Alexander, to Ashland, no facilities
68	to IL 104, to Jacksonville, **2 mi N**...Hucks Food/gas, Wareco/gas
64	US 67, to Jacksonville, **2 mi N**...HOSPITAL, Amoco, Citgo, Phillips 66, Dairy Queen, Subway, Motel 6, Harpers Foods
60	to US 67 N, to Jacksonville, **6 mi N**...HOSPITAL, gas, food, lodging
52	to IL 106, Winchester, **N**...golf, **2 mi S**...gas, food, lodging
46	IL 100, to Bluffs, no facilities
42mm	Illinois River
35	US 54, IL 107, Griggsville, to Pittsfield, **4 mi N**...gas, food, lodging, **S**...HOSPITAL, st police, camping(6mi)
31	New Salem, to Pittsfield, **5 mi S**...HOSPITAL, gas, food, lodging, camping
20	IL 106, Barry, **N**...AppleBasket Farms, **S**...Amoco, Phillips 66/diesel/mart/24hr, Shell/mart/24hr, Wendy's, bank
10	IL 96, Hull, to Payson, no facilities
4c b a	US 36 W, I-172, to Quincy
0mm	Mississippi River, Illinois Missouri state line

Illinois Interstate 74

E

↕

W

Exit #	Services
221mm	Illinois/Indiana state line
220	Lynch Rd, Danville, **N**...VETERINARIAN, Amoco/diesel/mart, Marathon/pizza/diesel/mart/24hr, Big Boy, Best Western, Comfort Inn, Fairfield Inn, Knight's Inn, Ramada Inn, Super 8
216	Bowman Ave, Danville, **N**...Citgo/subs/mart, Mobil/diesel/mart/24hr, Shell/delimart, Burger King, Godfather's, KFC, airport, **S**...The Diner/Colonial Parkway Rest., camping
215	US 150, IL 1, Gilbert St, Danville, **N**...HOSPITAL, CHIROPRACTOR, Citgo/mart, Speedway/mart/24hr, Arby's, Central Park, Hardee's, LJ Silver, McDonald's, Pizza Hut, Steak'n Shake, Taco Bell, Best Western, Day's Inn, Bass Tires, Care Muffler/brake, Ford/Lincoln/Mercury, QuickLube, bank, **S**...VETERINARIAN, Clark Gas, Speedway/mart/24hr, Burger King, Mike's Grill, Subway, Aldi Foods, Big R Foods, Buick/Pontiac/GMC, $General, Eagle Foods, Harley-Davidson, Pennzoil, Forest Glen Preserve Camping(11mi)
214	Tilton, **S**...Shell/mart, Hooligan's Café
210	US 150, MLK Dr, **2 mi N**...HOSPITAL, Marathon/mart, Little Nugget Steaks, to Kickapoo SP, camping, **S**...PossumTrot Rest.
208mm	**Welcome Ctr wb, full(handicapped)facilities, info, phones, picnic tables, litter barrels, vending, petwalk**
206	Oakwood, **N**...Shell/diesel/mart/rest./24hr, **S**...Casey's Gen Store(1mi), Marathon/diesel/mart/24hr, Phillips 66/mart, McDonald's, camping
200	IL 49 N, to Rankin, **N**...5 Bridges Camping
197	IL 49 S, Ogden, **S**...Citgo/diesel/mart, Colonial Pantry/mart, Godfather's, Rubio's Rest.
192	St Joseph, **S**...Dairy Queen, antiques
185	IL 130, University Ave, no facilities
184	US 45, Cunningham Ave, Urbana, **N**...Park Inn, Farm&Fleet, **S**...VETERINARIAN, Clark Gas, Freedom/diesel/mart/24hr, Shell, Speedway/Taco Bell/mart, Cracker Barrel, Domino's, Longhorn Steaks, Ned Kelly's Steaks, Steak'nShake/24hr, Best Western, Eastland Suites, Manor Motel, Motel 6, $General, Family$, Firestone/auto, NAPA, PDR Auto, Sav-A-Lot Foods, laundry
183	Lincoln Ave, Urbana, **S**...HOSPITAL, Phillips 66/diesel/mart, Speedway/mart, Urbana Garden Café, Holiday Inn/rest., Ramada Ltd, Sleep Inn, Super 8, to U of IL
182	Neil St, Champaign, **N**...Alexander's Steaks, Bob Evans, Chevy's Mexican, ChiChi's, Denny's, FoodCourt, Fortune House Chinese, Grandy's, McDonald's, Olive Garden, Subway, Taco Bell, Baymont Inn, Comfort Inn, Extended Stay America, La Quinta, Red Roof Inn, Super 8, Barnes&Noble, Berdner's, Chevrolet/Olds/Cadillac, Chrysler/Jeep, FashionBug, JC Penney, K's Merchandise, Kohl's, Office Depot, Osco Drug, Pennzoil, Sears/auto, TJ Maxx, mall, same as 181, **S**...Mobil/mart, Perkins, Howard Johnson, Jo-Ann Fabric/crafts, NTB

Illinois Interstate 74

181	Prospect Ave, Champaign, **N**...Meijer/diesel/24hr, Applebee's, Burger King, Cheddar's, Chili's, Damon's, Hardee's, Fazoli's, HomeTown Buffet, LoneStar Steaks, Old Country Buffet, Outback Steaks, Ryan's, Steak'n Shake/24hr, Subway, Wendy's, Courtyard, Drury Inn, Fairfield Inn, Aamco, Advance Parts, Best Buy, Borders Books, Circuit City, $Tree, Hyundai, LandsEnd, Lowe's Bldg Supply, Menard's, Mitsubishi, PetsMart, Pier 1, Sam's Club, Staples, Target, Wal-Mart/auto, same as 182, **S**...Amoco/mart, Freedom/mart, Mobil, KFC, AmeriInn, Day's Inn, Econolodge, K-Mart, Midas Muffler, NAPA, Nissan, Saturn, Subaru
179b a	I-57, N to Chicago, S to Memphis
174	Lake of the Woods Rd, Prairieview Rd, **N**...Amoco/diesel/mart, Casey's/gas, Mobil/diesel/mart, Auggie's Steaks, Subway, Lake of the Woods Pk/museum, RV Ctr, camping, **S**...McDonald's
172	IL 47, Mahomet, **N**...RV Ctr, **S**...Apollo/gas, Citgo, Clark Gas, Marathon, Mobil/mart, Dairy Queen, Hardee's, HenHouse Rest., Monical's Pizza, Subway, Heritage Inn, CVS Drug, IGA Foods, auto parts, bank, laundry
166	Mansfield, **S**...Amoco
159	IL 54, Farmer City, **N**...Farmer Dave's Buffalo Ranch(1mi), **S**...Casey's/gas, Bon-Aire Rest., Dewey's Drive-In, Budget Motel, Day's Inn, to Clinton Lake RA
156mm	**rest area both lanes, full(handicapped)facilities, phone, picnic tables, litter larrels, vending, playground, petwalk**
152	US 136, to Heyworth, no facilities
149	Le Roy, **N**...Amoco/diesel/mart/24hr, Casey's/gas, Hardee's, McDonald's, Pizza Hut, Old Bank Inn, IGA Foods, to Moraine View SP, **S**...Shell/diesel/mart/24hr, Woody's Rest., Super 8/RV/truck parking, Clinton Lake, camping
142	Downs, no facilities
135	US 51, Bloomington, **N**...HOSPITAL, Mobil/diesel/mart, Phillips 66/mart, McDonald's, Steak'n Shake, to IL St U, **S**...Shell/diesel/mart
134b[157]	**N**...Veterans Pkwy, Bloomington, Clark Gas/24hr, Shell/mart/service, CJ's Rest., Jumer's Hotel/rest., Knight's Inn, Parkway Inn/rest./RV/truck parking, UPS, airport
a	I-55, N to Chicago, S to St Louis, I-74 E
I-74 and I-55 run together 6 mi. See IL 55, exit 160b a.	
127[163]	I-55, N to Chicago, S to St Louis, I-74 W to Peoria
125	US 150, Diamond-Star Pkwy, to Bloomington, no facilities
123mm	weigh sta wb
122mm	weigh sta eb
120	Carlock, **N**...Amoco/diesel/repair, Countryside Rest.
114.5mm	**rest area both lanes, full(handicapped)facilities, vending, phone, picnic tables, litter barrels, petwalk**
113.5mm	Mackinaw River
112	IL 117, Goodfield, **N**...Shell/Sub Express/diesel/mart, to Timberline RA, Yogi Bear Camping, Eureka Coll, Reagan Home
102	Morton, **N**...MEDICAL CARE, Amoco/mart/atm, Casey's/gas, Citgo/Blimpie/Dunkin Donuts/pizza/diesel/24hr, Mobil/mart, Burger King, Chief Konrad's Rest., Cracker Barrel, Peppermill Rest., Taco Bell, Comfort Inn, Day's Inn, Holiday Inn Express, Knight's Inn/rest., **S**...CHIROPRACTOR, Clark/24hr, Shell/Subway/diesel/mart/24hr, China Dragon, Golden Corral, McDonald's, Chrysler/Dodge, CVS Drug, Ford, K-Mart/auto, Kroger/deli/24hr, cleaners
101	I-155 S, to Lincoln, no facilities
99	I-474 W, airport
98	Pinecrest Dr, no facilities
96	95c(from eb), US 150, IL 8, E Washington St, E Peoria, **N**...Site Gas/mart, Monical's Pizza, Subway, Wendy's, Super 8
95b	IL 116, to Metamora, **N**...Hampton Inn, Firestone/auto
a	Peoria, N Main St, **S**...Amoco/mart/24hr, Applebee's, Blimpie, Bob Evans, Hardee's, LJ Silver, Best Western, Budget Host, Motel 6, Aldi Foods, CVS Drug, Goodyear/auto, Kroger/deli/24hr, Wal-Mart/drugs, Whitlock Parts, Valvoline, bank
94	IL 40, Industrial Spur, no facilities
93.5mm	Illinois River
93b	US 24, IL 29, Peoria, **N**...Phillips 66/diesel, **S**...HoJo Inn, Sears, civic ctr
a	Jefferson St, Peoria, **S**...Mark Twain Best Western, civic ctr
92	Glendale Ave, Peoria, **S**...HOSPITAL, Hardee's, downtown
92a	IL 40 N, Knoxville Ave, Peoria, HOSPITAL, **N**...Hardee's
91b a	University St, Peoria, **N**...to Expo Gardens, **S**...to Bradley U
90	Gale Ave, Peoria, no facilities
89	US 150, War Memorial Dr, Peoria, **N**...Amoco, Bob Evans, Burger King, Carl's Jr, Cheddar's, Denny's, Dunkin Donuts, Perkins/24hr, Pizza Hut, Red Lobster, Steak'n Shake, Subway, Wendy's, Comfort Suites, Day's Inn, Fairfield Inn, Holiday Inn, Red Roof Inn, Signature Inn, Super 8, Circuit City, Cub Foods, Firestone, JC Penney, Lowe's Bldg Supply, Midas Muffler, PetsMart, Pontiac/Cadillac, Target, Thrifty/Tuffy Muffler, bank, cinema, mall

E

W

Illinois Interstate 74

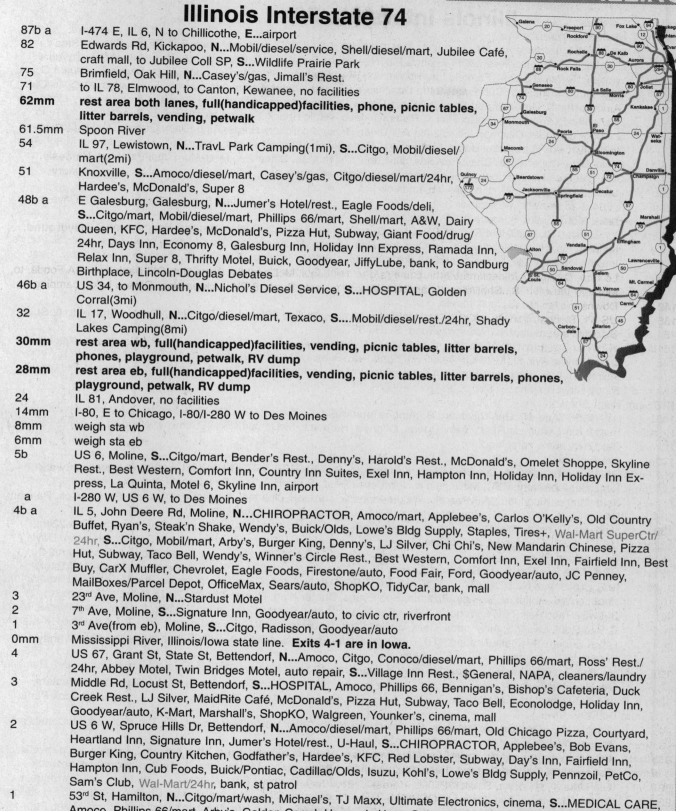

87b a	I-474 E, IL 6, N to Chillicothe, **E**...airport
82	Edwards Rd, Kickapoo, **N**...Mobil/diesel/service, Shell/diesel/mart, Jubilee Café, craft mall, to Jubilee Coll SP, **S**...Wildlife Prairie Park
75	Brimfield, Oak Hill, **N**...Casey's/gas, Jimall's Rest.
71	to IL 78, Elmwood, to Canton, Kewanee, no facilities
62mm	**rest area both lanes, full(handicapped)facilities, phone, picnic tables, litter barrels, vending, petwalk**
61.5mm	Spoon River
54	IL 97, Lewistown, **N**...TravL Park Camping(1mi), **S**...Citgo, Mobil/diesel/mart(2mi)
51	Knoxville, **S**...Amoco/diesel/mart, Casey's/gas, Citgo/diesel/mart/24hr, Hardee's, McDonald's, Super 8
48b a	E Galesburg, Galesburg, **N**...Jumer's Hotel/rest., Eagle Foods/deli, **S**...Citgo/mart, Mobil/diesel/mart, Phillips 66/mart, Shell/mart, A&W, Dairy Queen, KFC, Hardee's, McDonald's, Pizza Hut, Subway, Giant Food/drug/24hr, Days Inn, Economy 8, Galesburg Inn, Holiday Inn Express, Ramada Inn, Relax Inn, Super 8, Thrifty Motel, Buick, Goodyear, JiffyLube, bank, to Sandburg Birthplace, Lincoln-Douglas Debates
46b a	US 34, to Monmouth, **N**...Nichol's Diesel Service, **S**...HOSPITAL, Golden Corral(3mi)
32	IL 17, Woodhull, **N**...Citgo/diesel/mart, Texaco, **S**....Mobil/diesel/rest./24hr, Shady Lakes Camping(8mi)
30mm	**rest area wb, full(handicapped)facilities, vending, picnic tables, litter barrels, phones, playground, petwalk, RV dump**
28mm	**rest area eb, full(handicapped)facilities, vending, picnic tables, litter barrels, phones, playground, petwalk, RV dump**
24	IL 81, Andover, no facilities
14mm	I-80, E to Chicago, I-80/I-280 W to Des Moines
8mm	weigh sta wb
6mm	weigh sta eb
5b	US 6, Moline, **S**...Citgo/mart, Bender's Rest., Denny's, Harold's Rest., McDonald's, Omelet Shoppe, Skyline Rest., Best Western, Comfort Inn, Country Inn Suites, Exel Inn, Hampton Inn, Holiday Inn, Holiday Inn Express, La Quinta, Motel 6, Skyline Inn, airport
a	I-280 W, US 6 W, to Des Moines
4b a	IL 5, John Deere Rd, Moline, **N**...CHIROPRACTOR, Amoco/mart, Applebee's, Carlos O'Kelly's, Old Country Buffet, Ryan's, Steak'n Shake, Wendy's, Buick/Olds, Lowe's Bldg Supply, Staples, Tires+, Wal-Mart SuperCtr/24hr, **S**...Citgo, Mobil/mart, Arby's, Burger King, Denny's, LJ Silver, Chi Chi's, New Mandarin Chinese, Pizza Hut, Subway, Taco Bell, Wendy's, Winner's Circle Rest., Best Western, Comfort Inn, Exel Inn, Fairfield Inn, Best Buy, CarX Muffler, Chevrolet, Eagle Foods, Firestone/auto, Food Fair, Ford, Goodyear/auto, JC Penney, MailBoxes/Parcel Depot, OfficeMax, Sears/auto, ShopKO, TidyCar, bank, mall
3	23rd Ave, Moline, **N**...Stardust Motel
2	7th Ave, Moline, **S**...Signature Inn, Goodyear/auto, to civic ctr, riverfront
1	3rd Ave(from eb), Moline, **S**...Citgo, Radisson, Goodyear/auto
0mm	Mississippi River, Illinois/Iowa state line. **Exits 4-1 are in Iowa.**
4	US 67, Grant St, State St, Bettendorf, **N**...Amoco, Citgo, Conoco/diesel/mart, Phillips 66/mart, Ross' Rest./24hr, Abbey Motel, Twin Bridges Motel, auto repair, **S**...Village Inn Rest., $General, NAPA, cleaners/laundry
3	Middle Rd, Locust St, Bettendorf, **S**...HOSPITAL, Amoco, Phillips 66, Bennigan's, Bishop's Cafeteria, Duck Creek Rest., LJ Silver, MaidRite Café, McDonald's, Pizza Hut, Subway, Taco Bell, Econolodge, Holiday Inn, Goodyear/auto, K-Mart, Marshall's, ShopKO, Walgreen, Younker's, cinema, mall
2	US 6 W, Spruce Hills Dr, Bettendorf, **N**...Amoco/diesel/mart, Phillips 66/mart, Old Chicago Pizza, Courtyard, Heartland Inn, Signature Inn, Jumer's Hotel/rest., U-Haul, **S**...CHIROPRACTOR, Applebee's, Bob Evans, Burger King, Country Kitchen, Godfather's, Hardee's, KFC, Red Lobster, Subway, Day's Inn, Fairfield Inn, Hampton Inn, Cub Foods, Buick/Pontiac, Cadillac/Olds, Isuzu, Kohl's, Lowe's Bldg Supply, Pennzoil, PetCo, Sam's Club, Wal-Mart/24hr, bank, st patrol
1	53rd St, Hamilton, **N**...Citgo/mart/wash, Michael's, TJ Maxx, Ultimate Electronics, cinema, **S**...MEDICAL CARE, Amoco, Phillips 66/mart, Arby's, Golden Corral, Heavenly Ham, Quizno's, Steak'n Shake, Village Inn Rest., Best Buy, Gateway Computers, PetsMart, Staples, Super Target
0mm	I-74 begins/ends on I-80, exit 298. **Exits 1-4 are in Iowa.**

Illinois Interstate 80

Exit #	Services

163mm Illinois/Indiana state line

161 US 6, IL 83, Torrence Ave, **N**...MEDICAL CARE, CHIROPRACTOR, Amoco, Arby's, Bob Evans, Burger King, Checker's, Chili's, Hooters, IHOP, Kenny's Rib, New China Buffet, Old Country Buffet, Olive Garden, Wendy's, Best Western, Comfort Suites, Fairfield Inn, Holiday Inn, Red Roof Inn, Sleep Inn, Super 8, Baby Depot, Burlington Coats, CarX Muffler, Chrysler, Cub Foods/24hr, Dominick's Food/drug/24hr, Fannie May Candy, Firestone/auto, GMC/Olds, Goodyear/auto, Gordon Food Service, Home Depot, Jo-Ann Fabrics, K-Mart/Little Caesar's/drugs, Luxury Linens, Pearle Vision, PepBoys, PetCo, Pier 1, Plymouth/Jeep, Radio Shack, Service Merchandise, SportMart, Trak Auto, **S**...Clark Gas, Gas City/mart, Shell/24hr, Al's Diner, Brown's Chicken, Burger King, Dairy Queen, Dunkin Donuts, Golden Crown Rest., Great Bagel, McDonald's, Papa John's, Pappy's Gyros, Popolano's Italian, Pioneer Motel, Auto Clinic, Baby SuperStore, Chevrolet/Olds, JiffyLube, MufflerMax, OfficeMax, Phar-Mor Drug, PetsMart, Saab, Sam's Club, SunRise HealthFoods, Wolf Camera, bank, cleaners, hardware

160b I-94 W, to Chicago, tollway begins wb, ends eb

a IL 394 S, to Danville, no facilities

159mm Oasis, Mobil/diesel, Burger King, Popeye's

157 IL 1, Halsted St, **N**...CHIROPRACTOR, Clark Gas, Denny's, Mr Philly Pasta, Wendy's, Baymont Inn, Comfort Inn, Hampton Inn, Hilton, Holiday Inn, Junction Inn, Ramada Inn, Red Roof Inn, CarX Muffler, bank, tires/repair, **S**...Arby's, Dunkin Donuts, IHOP, LJ Silver, McDonald's, Shooter's Buffet, Subway, Taco Bell, Washington Square Rest., Best Western, Day's Inn, Motel 6, K-Mart/auto/24hr, Chevrolet, Goldblatt's Foods, Goodyear/auto, Jewel-Osco, Jo-Ann Fabrics, Phar-Mor, Venture, cinema, cleaners, mall

156 Dixie Hwy(from eb), **S**...golf course

155 I-294 N, Tri-State Tollway, toll plaza, no facilities

154 Kedzie Ave(from eb), **N**...Speedway/mart, **S**...HOSPITAL

151b a I-57, N to Chicago, S to Memphis, no facilities

148 IL 43, Harlem Ave, **N**...CHIROPRACTOR, Speedway/Subway/diesel/mart/atm, Burger King, Cracker Barrel, Wendy's, Baymont Inn, Fairfield Inn, Hampton Inn, Holiday Inn, Sleep Inn, Wingate Inn, **S**...CarMax, World Music Theatre, Windy City Camping

147.5mm weigh sta wb

145b a US 45, 96th Ave, **N**...Gas City/mart, **S**...Amoco/mart/wash, Clark/mart, Gas City/diesel/mart, Shell/diesel/mart/24hr, Burger King, Chinese Rest., Dairy Queen, Denny's, Hardee's, McDonald's(3mi), Subway, Wendy's, Super 8, Prime Car Care, bank, camping(2mi)

143mm weigh sta eb

137 US 30, New Lenox, **N**...Les Bros Rest., K-Mart/auto, **S**...DENTIST, VETERINARIAN, Speedway/diesel/mart, Burger King, Hardee's/24hr, KFC, McDonald's, New Lenox Rest., Pizza Hut, Taco Bell, Eagle Food/deli, Jewel-Osco Food/drug, Muffler King/lube, Mail/Parcels, Walgreen, bank, carwash, cleaners, hardware

134 Briggs St, **N**...HOSPITAL, Speedway/mart, **S**...Amoco/repair/24hr, Citgo/diesel, EZLube, bank

133 Richards St, no facilities

132b a US 52, IL 53, Chicago St, no facilities

131.5mm Des Plaines River

131 US 6, Meadow Ave, **S**...to Riverboat Casino

130b a IL 7, Larkin Ave, **N**...HOSPITAL, DENTIST, VETERINARIAN, Citgo, Clark Gas, Marathon/mart/24hr, Shell/wash/24hr, Thornton's/diesel/mart, Bellagio Pizzaria, Bob Evans, Burger King, Dunkin Donuts, Pizza Hut, Steak'n Shake, Subway, Taco Bell, TCBY, Wendy's, White Castle, White Hen Pantry, Comfort Inn, Holiday Inn Express, Microtel, Motel 6, Red Roof, Super 8, Aldi Foods, CarX Muffler, Chevrolet/Mitsubishi, Cub Food/drug, Ford, Goodyear/auto, K-Mart/Little Caesar's, Meineke Muffler, NTB, PepBoys, Radio Shack, Sam's Club, Tuffy Repair, Wal-Mart, to Coll of St Francis, **S**...auto repair

127 Houbolt Rd, to Joliet, **N**...Amoco/deli/mart, Burger King, Cracker Barrel, Wendy's, Fairfield Inn, Hampton Inn, Ramada Ltd, Harley-Davidson

126b a I-55, N to Chicago, S to St Louis

125.5mm Du Page River

122 Minooka, **N**...Citgo/diesel/mart/24hr, **S**...Amoco/mart/24hr, Pilot/Arby's/diesel/mart/24hr, Dairy Queen, McDonald's, Subway, Wendy's, SuperValu Foods, bank, drugs

119mm **rest area wb, full(handicapped)facilities, vending, phones, picnic tables, litter barrels, playground, petwalk**

117mm **rest area eb, full(handicapped)facilities, vending, phone, picnic tables, litter barrels, playground, petwalk**

112 IL 47, Morris, **N**...Amoco/diesel/mart/24hr, Citgo/diesel/mart, Denny's, Best Western, Comfort Inn, Holiday Inn/rest., cinema, **S**...MEDICAL CARE, CHIROPRACTOR, Amoco/mart, Clark Gas, Mobil/mart, Shell/mart/24hr, Boz Hotdogs, Burger King, Hardee's, Hong Kong Chinese, KFC, McDonald's, Pizza Hut, Taco Bell, Wendy's, Morris Motel, Park Motel, Super 8, Aldi Foods, Chevrolet/Buick/Cadillac, Fisher Parts, Morris Drug, Pennzoil, Radio Shack, Walgreen, Wal-Mart SuperCtr/24hr, bank, cleaners, to Stratton SP

105 to Seneca, no facilities

97 to Marseilles, **S**...Prairie Lakes Resort/rest., to Illini SP, RV camping

93 IL 71, Ottawa, **N**...Mobil, Shell/diesel/mart/24hr, J&L/diesel/mart, **S**...HOSPITAL, Hank's Farm Rest.

92.5mm Fox River

Chicago Area

Joliet

E

W

Illinois Interstate 80

90	IL 23, Ottawa, **N**...HOSPITAL, Amoco/Subway/mart, Cracker Barrel, Taco Bell, Holiday Inn Express, Buick/Pontiac/Olds/Cadillac, Farm&Fleet, Ford/Lincoln/Mercury/ Kia, Chrysler/Jeep, Honda, **S**...Amoco/diesel/LP, Shell/mart/24hr, Country Kitchen, Dunkin Donuts, Hardee's, McDonald's, KFC, Ponderosa, Comfort Inn, Ottawa Inn/rest., Sands Motel, Super 8, Surrey Motel, FashionBug, JiffyLube, K-Mart/ drugs, Kroger/deli, USPO, ValueCity, Wal-Mart/drugs, bank
81	IL 178, Utica, **N**...KOA(2mi), **S**...Amoco/mart, Shell/diesel/mart, Duffy's Tavern(2mi), Jimmy Johns Sandwiches, Starved Rock Inn, to Starved Rock SP, camping
79b a	I-39, US 51, N to Rockford, S to Bloomington
77.5mm	Little Vermilion River
77	IL 351, La Salle, **S**...Amoco, Flying J/Country Mkt/diesel/mart/24hr, Daniels Motel, st police
75	IL 251, Peru, **N**...Shell/diesel/rest./24hr, Amoco/mart, Arby's, McDonald's, Shakey's Pizza, Best Inn, Comfort Inn, Econolodge, Motel 6, Super 8, Tiki Inn/ RV Park/rest., antiques, **S**...HOSPITAL, Amoco/mart, Bob Evans, Dairy Queen, Dunkin Donuts, Mi Margarita Rest., Oogie's Rest., Red Lobster, Steak'n Shake, Subway, Wendy's, Fairfield Inn, Ramada Ltd, AutoZone, Big Lots, Buick/Pontiac/GMC, Chevrolet/Mercedes/Nissan, Chrysler/Jeep, EconoFoods, Goodyear/auto, JC Penney, Jo-Ann Fabrics, K-Mart/auto/Little Caesar's, Menard's, Midas Muffler, Pennzoil, Staples, Target, Toyota, Walgreen, Wal-Mart/drugs, bank, cinema, hardware, mall
73	Plank Rd, **N**...Sapp Bros/Texaco/Burger King/diesel, camping, **S**...to airport
70	IL 89, to Ladd, **N**...Amoco/mart, **S**...Casey's/gas, Motel Riviera, **3 mi S**...HOSPITAL, Hardee's, Torie's Café
61	I-180, to Hennepin, no facilities
56	IL 26, Princeton, **N**...Pilot/diesel/mart/24hr, Super 8, **S**...HOSPITAL, Amoco/mart/24hr, Phillips 66/mart, Shell/ Subway/mart, Big Apple Rest., Country Kitchen, McDonald's, Pizza Hut(1mi), Sullivan's Foods, Taco Bell, Wendy's, Comfort Inn, Day's Inn, Motor Lodge/rest., Eagle Foods, Wal-Mart/drugs, bank, hardware
51.5mm	**rest area both lanes, full(handicapped)facilities, phone, picnic tables, litter barrels, vending, playground, petwalk, RV dump**
45	IL 88, **N**...Marathon/diesel/Farmhouse Rest., Day's Inn/rest., antiques, to Ronald Reagan birthplace(21mi), **S**...Hennepin Canal SP, camping
44mm	Hennepin Canal
33	IL 78, Annawan, **S**...Amoco/mart, Phillips 66/diesel/mart, Back Alley Rest., The Loft Rest., to Johnson-Sauk Trail SP
27	to US 6, Atkinson, **N**...Amoco/diesel/mart/24hr, Casey's(1mi), Mobil/diesel/rest./24hr, Wood Shed Rest., Grandma's Motel
19	IL 82, Geneseo, **N**...HOSPITAL, Citgo/mart, Phillips 66/diesel/mart/24hr, Dairy Queen, Hardee's, McDonald's, Pizza Hut, Subway, Deck Motel/diner, Oakwood Motel, Royal Motel, Super 8, Ford, Wal-Mart/drugs, hardware, **S**...Phillips 66, KFC, antiques, camping(4mi)
10	I-74, W to Moline, E to Peoria, no facilities
9	US 6, to Geneseo, no facilities
7	Colona, **N**...Conoco/mart/rest.
5mm	Rock River
4a	IL 5, IL 92, W to Silvis, **S**...to Quad City Downs, Lundeen's Camping
b	I-88, IL 92, E to Rock Falls, no facilities
2mm	weigh sta both lanes
1.5mm	**Welcome Ctr eb, full(handicapped)facilities, info, phone, picnic tables, litter barrels, petwalk, scenic overlook**
1	IL 84, 20th St, E Moline, Great River Rd, **N**...Amoco/mart, Git'n Go/24hr, Brothers Rest., auto parts, camping, carwash, **3 mi S**...Citgo, camping
0mm	Mississippi River, Illinois/Iowa state line

Illinois Interstate 88

Exit #	Services
155.5mm	I-88 begins/ends on I-290.
155	I-294, S to Indiana, N to Milwaukee
154mm	toll plaza
152	Midwest Rd, **N**...Bennigan's, La Quinta
149	Highland Ave, HOSPITAL, **N**...on Butterfield Rd, Burger King, Hooters, Olive Garden, Pizza Hut, Red Lobster, Comfort Inn, Embassy Suites, Red Roof Inn, Best Buy, Circuit City, Firestone, JC Penney, bank, mall, **S**...Highland Grill
148	I-355 N(from wb)
147	I-355 S(from eb)
146	IL 53(from wb), **S**...on Ogden Ave, Amoco, Shell, Hyatt, Hyundai, Honda, Toyota

Illinois Interstate 88

Chicago Area / DeKalb

143	Naperville Rd, **N**...Del Rio Mexican, Radisson, Hilton, **1/2mi S on Ogden Ave**...HOSPITAL, VETERINARIAN, CHIROPRACTOR, Mobil/Dunkin Donuts, Applebee's, Arby's, Bennigan's, Baskin-Robbins, Bob Evans, Boston Mkt, Burger King, Casa Lupita Mexican, Chinese Rest., ChuckeCheese, Connie's Pizza, Edwardo's Pizza, El Centro Mexican, El Torito, Los Sombreros Mexican, Hardee's, Japanese Steaks, McDonald's, Ponderosa, Portillo's Dogs, Subway, Taco Bell, TGIFriday, Wendy's, Exel Inn, Holiday Inn, Travelodge, Wyndham Garden, CarX Muffler, Chrysler/Jeep, Dominick's Foods, Ford, Fresh Fields Foods, JiffyLube, Jo-Ann Fabrics, MailBoxes Etc, Meineke Muffler, Office Depot, Olds, Walgreen/24hr, bank, cinema, cleaners, mall, repair, transmissions
139	IL 59, **N**...HOSPITAL, CHIROPRACTOR, Amoco/mart/wash, 76/mart, Café 59, **S**...DENTIST, Mobil/diesel/mart, Chinese Rest., Subway, Wendy's, White Hen Pantry, Red Roof Inn, cleaners
135	Farnsworth Ave, **N**...Papa Bear Rest., Fox Valley Inn/Best Western, Motel 6, Ford/Mercury/Lincoln/Mitsubishi, **S**...DENTIST, CHIROPRACTOR, Marathon, Shell, McDonald's, Pizza, White Hen Pantry, cleaners/laundry, bank
133mm	toll plaza
131	IL 31, IL 56, to Aurora, Batavia, **N**...Shell, Phillips 66, A&W, Brown's Chicken, Raimondo's Pizza, **S**...HOSPITAL, Amoco, Mobil/mart, Thornton's/mart, Boston Mkt, Hardee's, LJ Silver, McDonald's, Riverdale Family Rest., Ponderosa, Tacos/Burgers, Taco Bell, White Castle, Howard Johnson, Super 8, Aurora Mkt/deli, Chrysler, Firestone, Jewel/Osco, K-Mart/Little Caesar's/drugs, Meineke Muffler, Oil Xchange, Plymouth, Tire Ctr, Trak Parts, U-Haul, Venture, Ziebart TidyCar, bank, hardware
130	Orchard Rd, **S**...Toyota Plant
129	IL 56W(from wb, no EZ return), to Sugar Grove
124	IL 47(from eb, no EZ return), to Black Berry Hist Farm, **5 mi S**...Phillips 66/diesel/mart/wash
108mm	Dekalb Oasis/24hr, both lanes, Mobil/diesel/mart, McDonald's, bank
107mm	toll plaza
107	IL 38, IL 23, Annie Glidden Rd, to DeKalb, **N**...Super 8, **2-3 mi N**...Amoco/wash, 76/ mart, Marathon, Shell, Burger King, Hardee's, Jiordano's Pizza, McDonald's, Pagliai's Pizza, Pizza Hut, Taco Bell, Wendy's, HoJo's, Motel 6, University Inn, cinema, carwash, to N IL U
94	I-39, US 51, S to Bloomington, N to Rockford
92	IL 251, Rochelle, **N**...HOSPITAL, Amoco/Blimpie/diesel/mart, Casey's/gas, Marathon/diesel/mart, Shell/mart, Dodge, Ford/Lincoln/Mercury, Goodyear, GMC, bank, carwash
69mm	toll plaza
69	IL 26, Dixon, **N**...Pizza Hut, Comfort Inn, Super 8, **1-2 mi N**...HOSPITAL, Clark Gas, Golden Corral, Hardee's, Hometown Pantry, McDonald's, to Ronald Reagan Birthplace, antiques, to John Deere HS, to St Parks
44	US 30(last free exit eb), **N**...camping
41	IL 88, Rock Falls, to Sterling, **1-2 mi N**...HOSPITAL, Clark Gas, Mobil/mart/24hr, Shell/deli, Bennigan's, Burger King, Hardee's, KFC, LJ Silver, McDonald's, Perna's Pizza, Red Apple Rest., Subway, All Seasons Motel, Country Inn Suites, Holiday Inn, Super 8, AutoZone, Eagle Foods, Goodyear/auto, Harley-Davidson, Walgreen, Wal-Mart, bank, hardware, laundry, **S**...airport
36	to US 30, Rock Falls, Sterling, no facilities
26	IL 78, to Prophetstown, Morrison, **N**...to Morrison-Rockwood SP, camping, **S**...Conoco/diesel
18	Erie, to Albany, no facilities
10	Hillsdale, to Port Byron, **1 mi S**...Citgo/diesel/mart, Phillips 66/diesel/mart
6	IL 92 E, to Joslin, **1 mi S**...Sunset Lake Camping
2	Former IL 2, no facilities
1b a	I-80, W to Des Moines, E to Chicago
0mm	I-88 begins/ends on I-80, exit 4b. IL 5, IL 92, W to Silvis, to Quad City Downs, Lundeen's Camping

Illinois Interstate 90

Exit #	Services
0mm	Illinois/Indiana state line, Chicago Skyway Tool Rd begins/ends
1mm	US 12, US 20, 106th St, Indianapolis Blvd, **N**...Amoco/diesel/mart, Shell/diesel/mart, **S**...Shell, Burger King, Giappo's Pizza, KFC, McDonald's, Jewel-Osco, Payless Muffler/brake
3mm	Skyway Oasis, McDonald's, toll plaza
5mm	79th St, services along 79th St and Stoney Island Ave, **S**...lube
5.5mm	73rd St(from wb), no facilities
6mm	State St(from wb), no facilities
7mm	I-94 N(mile markers decrease to IN state line)
	I-90 E and I-94 E run together, **see Illinois Interstate 94, exits 43b thru 59a.**
84	I-94 W...Lawrence Ave, **N**...Amoco/mart
83b a	Foster Ave(from wb), **N**...Amoco/mart, Checker's, Dunkin Donuts, Firestone/auto, Goodyear/auto, Walgreen, other facilities on Milwaukee Ave...McDonald's
82c	Austin Ave, to Foster Ave(no EZ return from eb), no facilities
b	Byrn-Mawr(from wb)
a	Nagle Ave, no facilities

Illinois Interstate 90

81b	Sayre Ave(from wb)
a	IL 43, Harlem Ave, **S**...Amoco, Shell/mart
80	Canfield Rd(from wb)
79b a	IL 171 S, Cumberland Ave, **N**...Citgo/7-11, Mobil, Denny's, Hooters, McDonald's, Porter's Steaks, Holiday Inn, Marriott, Dominick's Foods, JiffyLube, bank, **S**...Bennigan's, Wyndham Garden
0mm	River Road Plaza, **N**...Marriott, Westin Hotel, **S**...Hyatt(mile markers increase to Rockford)
1mm	I-294, I-190 W, to O'Hare Airport
2mm	IL 72, Lee St, **N**...Extended Stay America, Quality Inn, Radisson, Studio+, **S**...McDonald's, Best Western, Ramada Plaza, Sheraton Gateway, Travelodge
5mm	Des Plaines Oasis both lanes, Mobil/diesel/24hr, McDonald's/24hr
6mm	Elmhurst Rd(from wb), **N**...Marathon, **S**...Comfort Inn, **2 mi S**...Amoco, McDonald's, Spaghetti Warehouse, Best Western, Day's Inn, Hampton Inn, LaQuinta, Motel 6, JiffyLube, OfficeMax
7.5mm	Arlington Hgts Rd(from wb), **N**...Amoco, Shell, AmeriSuites, Motel 6, Red Roof Inn, **S**...Extended Stay America, Sheraton
11mm	I-290, IL 53, **N**...Holiday Inn, mall, **S**...Extended Stay America, Radisson, Sheraton
13mm	Roselle Rd(from wb), **N**...Medieval Times
15mm	Barrington Rd(from wb), **S**...Baymont Inn, **1-2 mi S**...AmeriSuites, Hampton Inn
19mm	IL 59, **N**...to Poplar Creek Music Theatre
22mm	IL 25, **N**...Day's Inn, K-Mart/auto, Santa's Village, **S**...HOSPITAL, Amoco, Shell, Arby's
24mm	IL 31 N, **N**...Amoco/mart/wash, Alexander's Rest., Cracker Barrel, Wendy's, Baymont Inn, Crown Plaza, Hampton Inn, Super 8, Big Lots, Marshall's, Service Merchandise, **S**...IL 31 S, Elgin Inn/rest., Holiday Inn
25mm	toll plaza, phone
27mm	Randall Rd, **N**...HOSPITAL
32mm	IL 47(from wb), to Woodstock, Prime Outlets/famous brands
37mm	US 20, Marengo, **N**...Mobil/DQ/Subway/Taco Bell/diesel/24hr, Shell/diesel/mart/repair/24hr, TA/Amoco/diesel/rest./24hr, Wendy's, antiques, museums, to Prime Outlets at exit 32(6mi)
41mm	Marengo-Hampshire toll plaza, phone
53mm	Genoa Rd, to Belvidere, **N**...K-Mart(1mi), McDonald's, camping
55mm	Belvidere Oasis both lanes, Mobil/diesel/24hr, McDonald's/24hr, phone
56mm	toll plaza
60.5mm	Kishwaukee River
61mm	I-39 S, US 20, US 51, to Rockford, **S**...amusement park
63mm	US 20, State St, **N**...Phillips 66/Subway/diesel/mart/wash, Cracker Barrel, Toscano's Pizza/pasta, Best Western, Exel Inn, Wingate Inn, Time Museum, cinema, **S**...HOSPITAL, CHIROPRACTOR, Amoco, Citgo/mart, Mobil/Blimpie/diesel/mart, Arby's, Applebee's, BeefARoo, Burger King, Carlos O'Kelly's, Checker's, Cheddar's, ChiChi's, Chili's, Country Kitchen, Dairy Queen, Don Pablo's, Dos Reales Mexican, Gerry's Pizza, Giovanni's Rest., Heavenly Ham, HomeTown Buffet, Kenny Rogers, KFC, Lino's Pizza, LoneStar Steaks, Machine Shed Rest., McDonald's, Old Chicago Grill, Old Country Buffet, Olive Garden, Outback Steaks, Panino's Drive-Thru, Perkins, Red Lobster, Ruby Tuesday, Ryan's, Steak'n Shake, Taco Bell, Thunderbay Grill, Tom&Jerry's Gyros, Tumbleweed Grill, Wendy's, Best Suites, Candlewood Suites, Comfort Inn, Courtyard, Extended Stay America, Fairfield Inn, Hampton Inn, Holiday Inn Express, Ramada Inn/rest., Red Roof Inn, Residence Inn, Sleep Inn, Studio+, Super 8, Advance Parts, Aldi Foods, Barnes&Noble, Bed Bath&Beyond, Best Buy, Borders Books, Buick/Pontiac/GMC, Circuit City, CompUSA, Dick's Sports, Dodge, Elect Ave&More, FashionBug, Gateway Computers, HobbyLobby, Home Depot, Iogli Foods, Jo-Ann Fabrics, Kinko's, K-Mart, Kohl's, Lowe's Bldg Supply, LubePro, Mailboxes Etc, Marshall's, MediaPlay, MensWearhouse, Michael's, Nevada Bob's Golf, Office Depot, OfficeMax, Old Navy, PepBoys, PetCo, PetsMart, PharMor Drug, Pier 1, Sam's Club, Saturn, ShopKO, Target/drugs, TJ Maxx, Toyota/Lexus, Wal-Mart, bank, water funpark
66mm	E Riverside Blvd, Loves Park, **1-2 mi S**...CHIROPRACTOR, DENTIST, PODIATRIST, VETERINARIAN, Amoco/mart/atm/24hr, Mobil/diesel/mart, Phillips 66/diesel/mart, A&W, BeefARoo Diner, Culver's, Dairy Queen, Little Caesar's, McDonald's, Sam's Ristorante, Singapore Grill, Subway, Wendy's, Audi/Honda/Jaguar/Mercedes/Porsche/Subaru, Eagle Foods, Farm&Fleet, LubePro, Mailboxes Etc, Precision Tune, ProSports, bank, cleaners/laundry, to Rock Cut SP
75.5mm	tollbooth, phone(mile markers decrease from W to E to Chicago)
3	Rockton Rd, no facilities
1.5mm	**Welcome Ctr/rest area eb, full(handicapped)facilities, info, picnic tables, litter barrels, phone, petwalk, playground**
1	US 51 N, IL 75 W, S Beloit, **S**...Amoco/diesel, Flying J/Country Mkt/diesel/mart/24hr, Red Apple Rest., Holiday Inn, Knight's Inn, Ford/Lincoln/Mercury, Olds/GMC
0mm	Illinois/Wisconsin state line

E

W

Illinois Interstate 94

Exit #	Services
77mm	Illinois/Indiana state line

I-94 and I-80 run together 3 mi. See Illinois Interstate 80, exit 161.

Exit #	Services
74[160]b	I-80/I-294 W
a	IL 394 S to Danville, no facilities
73 b a	US 6,159th St, **N**...76/diesel, Annie's Rest., Denny's, Chevrolet, truckwash, **S**...Ford
71b a	Sibley Blvd, **N**...Mobil, 7-11, McDonald's, Econolodge, **S**...Amoco, Shell
70b a	Dolton, no facilities
69	Beaubien Woods(from eb)
68b a	130th St, no facilities
66b	115th St, **S**...VETERINARIAN, McDonald's
a	111th Ave, **S**...Amoco, Shell/wash, Firestone
65	103rd Ave, Stony Island Ave, no facilities
63	I-57 S, no facilities
62	US 12, US 20, 95th St, **N**...Citgo, Shell/mart, Subway, Taco Bell, **S**...Amoco
61b	87th St, **N**...Amoco, Shell, Burger King, McDonald's, **S**...GoldBlatt's Foods, Home Depot, Jewel-Osco Food/drug, Zayre Discount, bank
a	83rd St(from eb), **N**...Shell, st police
60c	79th St, **N**...Mobil, Shell, **S**...Amoco, Shell, Church's Chicken, Oldsmobile
b	76th St, **N**...Amoco, Mobil/wash, Shell/mart, Walgreen/24hr, **S**...KFC, Popeye's
a	75th St(from eb), **N**...Amoco/subs, Shell/mart, **S**...KFC, Popeye's
59c	71st St, **N**...Amoco, **S**...McDonald's
a	I-90 E, to Indiana Toll Rd
58b	63rd St(from eb), **N**...Shell, **S**...Amoco
a	I-94 divides into local and express, 59th St, **S**...Amoco
57b	Garfield Blvd, **N**...Checker's, Popeye's, Trak Auto, Walgreen, laundry, **S**...HOSPITAL, Mobil/mart, Shell/mart/24hr, Famous Burritos, Wendy's
a	51st St, **N**...McDonald's
56b	47th St(from eb)
a	43rd St, **S**...Amoco/mart, Citgo/diesel/mart
55b	Pershing Rd, Amoco
a	35th St, **S**...to New Comiskey Park
54	31st St, **S**...HOSPITAL, Shell/mart
53c	I-55, Stevenson Pkwy, N to downtown, Lakeshore Dr
b	I-55, Stevenson Pkwy, S to St Louis
52c	18th St, no facilities
b	Roosevelt Rd, Taylor St(from wb), **N**...Amoco
a	Taylor St, Roosevelt Rd(from eb), **N**...Amoco
51h-i	I-290 W, to W Suburbs
g	E Jackson Blvd, to downtown
f	W Adams St, to downtown
e	Monroe St(from eb), downtown, **S**...Quality Inn
d	Madison St(from eb), downtown, **S**...Quality Inn
c	E Washington Blvd, downtown
b	W Randolph St, downtown
a	Lake St(from wb)
50b	E Ohio St, downtown, **S**...Marathon
a	Ogden Ave, no facilities
49b a	Augusta Blvd, Division St, **S**...Amoco/mart, Shell, Pizza Hut
48b	IL 64, North Ave, **N**...Amoco, Home Depot, **S**...Amoco, Shell, Volvo
a	Armitage Ave, **N**...Phillips 66, **S**...Amoco/mart, Shell, Jaguar, Volvo
47c b	Damon Ave, **N**...Citgo, car/vanwash
a	Western Ave, Fullerton Ave, **N**...Citgo, Burger King, Popeye's, 10MinLube, **S**...Amoco, Marathon
46b a	Diversey Ave, California Ave, **S**...Mobil, IHOP/24hr, Popeye's, Subway
45c	Belmont Ave, **N**...Wendy's
b	Kimball Ave, **N**...Marathon/diesel/mart, **S**...Amoco, Dunkin Donuts, Subway, Wendy's, Dominick's Foods, Radio Shack, Walgreen/24hr
a	Addison St(from eb), no facilities
44b	Pulaski Ave, **N**...Amoco, Mobil, **S**...Midas Muffler
a	IL 19, Keeler Ave, Irving Park Rd, **N**...Amoco, Shell/24hr, to Wrigley Field
43c	Montrose Ave, no facilities
b	I-90 W

E

W

Chicago Area

Illinois Interstate 94

a	Wilson Ave
42	W Foster Ave, **S**...Amoco, 76, Dunkin Donuts, **N**...Venture Store
41mm	Chicago River, N Branch
41c	IL 50 S, to Cicero, to I-90 W
b a	US 14, Peterson Ave, **N**...Amoco, Eden's Motel, bank, carwash
39b a	Touhy Ave, **N**...Amoco/mart, Shell/mart, Dominick's Foods, tires, **S**...Amoco/mart, Mobil/mart, Shell/mart
37 b a	IL 58, Dempster St, no facilities
35	Old Orchard Rd, HOSPITAL, **N**...Amoco, Shell
34c b	E Lake Ave, **S**...Shell
a	US 41 S, Skokie Rd(from eb)
33b a	Willow Rd, **N**...Amoco, **S**...76 gas, Dominick's Foods, Sheraton
31	E Tower Rd, **S**...VETERINARIAN, Mercedes, Toyota
30b	Dundee Rd(from wb), **S**...Amoco, Bennigan's, Olive Garden, Japanese Steaks, Sheraton
29	US 41, to Waukegan, to Tri-state tollway
50mm	IL 43, Waukegan Rd, **N**...Red Roof Inn, to Embassy Suites
53mm	I-294 S
53.5mm	Deerfield Rd toll plaza, phones
54mm	Deerfield Rd, **W**...Marriott Suites
56mm	IL 22, Half Day Rd, no facilities
59mm	IL 60, Town Line Rd, **E**...HOSPITAL
60mm	Lake Forest Oasis both lanes, Mobil/diesel, Baskin-Robbins, Wendy's/24hr, info
62mm	IL 176, Rockland Rd, **E**...to Lamb's Farm
64mm	IL 137, Buckley Rd(from wb), **E**...to VA HOSPITAL, Chicago Med School
67mm	IL 120, Belvidere Rd, **E**...HOSPITAL
68mm	IL 21, Milwaukee Ave(from eb), **E**...HOSPITAL, Six Flags
70mm	IL 132, Grand Ave, **E**...DENTIST, Speedway/diesel/mart, Burger King, Cracker Barrel, IHOP, Little Caesar's, McDonald's/playplace, Ming's Chinese, Outback Steaks, Subway, TCBY, Budgetel, Extended Stay America, Hampton Inn, Piggly Wiggly, Six Flags Park, cleaners, **W**...Mobil, Shell/mart/wash/etd, Applebee's, Baker's Square, Boston Mkt, Chili's, Denny's, Einstein Bros Bagels, LoneStar Steaks, Max&Erma's, McDonald's, Pizza Hut, Pizzaria Uno, Planet Hollywood, RainForest Café, Red Lobster, Schlotsky's, Sizzler, Starbuck's, Steak'n Shake, Taco Bell, TGIFriday, Wendy's, White Castle, Comfort Inn, Fairfield Inn, Holiday Inn, Bed Bath&Beyond, Burlington Coats, Circuit City, Computer City, Dominick's Food/drug, Gurnee Mills Outlet Mall/famous brands, Home Depot, Honda, JC Penney, Jewel-Osco, Kohl's, Lincoln/Mercury, Linens'n Things, LubeMasters, Midas Muffler, OfficeMax, Pearle Vision, PetsMart, Sam's Club, Sears, SportsAuthority, Target/drugs, Waccamaw, Wal-Mart/drugs, Wolf Camera, bank, carwash, cleaners
73.5mm	Waukegan toll plaza, phones
76mm	IL 173(from nb, no return), Rosecrans Ave, **E**...to IL Beach SP
1b	US 41 S, to Waukegan, **E**...Sky Harbor RV Sales/service
a	Russell Rd, **W**...Citgo/diesel/mart/24hr, TA/Mobil/Pizza Hut/diesel/mart/24hr, Peterbilt
0mm	Illinois/Wisconsin state line

Illinois Interstate 255(St Louis)

Exit # Services

I-255 begins/ends on I-270, exit 7.

30	I-270, W to Kansas City, E to Indianapolis
29	IL 162, to Glen Carbon, Granite City, to Pontoon Beach
26	Horseshoe Lake Rd, st police
25b a	I-55/I-70, W to St Louis, E to Chicago, Indianapolis
24	Collinsville Rd, **E**...Amoco/mart/24hr, Hardee's, Jack-in-the-Box, **W**...racetrack
20	I-64, US 50, W to St Louis, E to Louisville, facilities 1 mi E off I-64, exit 9.
19	State St, E St Louis, **E**...Clark Gas
17b a	IL 15, E St Louis, Centreville, to Belleville, no facilities
15	Mousette Lane, HOSPITAL, to Sauget Business Park
13	IL 157, to Cahokia, **W**...CHIROPRACTOR, Amoco/mart/24hr, Burger King, Capt D's, Chinese Rest., Domino's, Hardee's, KFC, Little Caesar's, McDonald's, Pizza Hut, Rally's, Taco Bell, Aldi Foods, AutoZone, Goodyear/auto, Schnuck's/deli/24hr, Valvoline, Wal-Mart/drugs, cleaners, hardware, Cahokia RV Parque(2mi), **E**...Clark/diesel
10	IL 3 N, to Cahokia, E St Louis, no facilities
9	to Dupo, no facilities
6	IL 3 S, to Columbia, **E**...Shell/mart/24hr, Chevrolet

Illinois Interstate 255

4mm	Mississippi River, Missouri/Illinois state line
3	Koch Rd, **N...rest area both lanes, full(handicapped)facilities, picnic tables, litter barrels, phones**
2	MO 231, Telegraph Rd, N...Clark Gas, Hardee's, Jefferson Barracks HS, Nat Cemetary, S...Amoco, Coastal, Dairy Queen, McDonald's
1a	US 50, US 61, US 67, Lindburgh Blvd, Lemay Ferry Rd, accesses same as I-55 exit 197, **N**...Arby's, McDonald's, Subway, Dillard's, K-Mart, Pearle Vision, TireAmerica, Western Auto, cinema, mall, **S**...Citgo/mart, Jack-in-the-Box, Papa John's, White Castle, Firestone, JiffyLube, Sam's Club, Service Merchandise, bank, hardware
1c b	I-55 S to Memphis, N to St Louis. I-255 begins/ends on I-55, exit 196.

Illinois Interstate 294(Chicago)

Exit #   Services

I-294 begins/ends on I-94, exit 74. Numbering descends from west to east.

I-294 & I-80 run together 5 mi. See Illinois Interstate 80, exits 155-160.

5mm	I-80 W, access to I-57
5.5mm	167th St, toll booth, phones
6mm	US 6, 159th St, **W**...Shell/mart/24hr, Baskin-Robbins, Burger King, McDonald's, Holiday Inn Express, Firestone/auto, U-Haul, auto repair, banking, carwash, drugs, hardware, laundry, towing
11mm	Cal Sag Channel
12mm	IL 50, Cicero Ave, **E**...Citgo/mart/24hr, A-1 Subs, Dunkin Donuts, Onion Field Rest., Budget Motel, Tuffy Auto, Valvoline Lube, **W**...Gas City/Subway/diesel/mart/atm/24hr, Dairy Queen, Hardee's, Baymont Inn, Hampton Inn, Radisson, **1 mi W**...Amoco/mart, Phillips 66, Shell, McDonald's, Taco Bell
18mm	US 12/20, 95th St, **E**...HOSPITAL, mall(1mi), **W**...Citgo, Burger King, Denny's, Wendy's, Exel Inn, Cub Foods/24hr, Wal-Mart, auto repair
20mm	toll booth, phones
22mm	75th St, Willow Springs Rd, no facilities
23mm	I-55, Wolf Rd, to Hawthorne Park, no facilities
25mm	Hinsdale Oasis both lanes, Mobil/diesel, Baskin-Robbins, Wendy's/24hr
28mm	US 34, Ogden Ave, **E**...zoo, **W**...HOSPITAL, Amoco/mart, Shell/deli/mart/etd, Dunkin Donuts, McDonald's, Firestone/auto, LandRover, Porche/Audi, Rolls-Royce/Bentley/Ferrari/Lotus, bank
29mm	I-88 tollway, no facilities
30mm	toll booth, phones
31mm	IL 38, Roosevelt Rd(no EZ nb return), **E**...Country Inn Pizza, Galway Café, **1 mi E**...KingMart/diesel, Hillside Manor Motel, Midas Muffler, **1 mi W**...Chevrolet
32mm	I-290 W, to Rockford(from nb), no facilities
34mm	I-290(from sb), to Rockford, no facilities
38mm	O'Hare Oasis both lanes, Mobil/diesel, Burger King, Connie's Pizza, TCBY
39mm	IL 19 W(from sb), Irving Park Rd, **E**...Holiday Inn
40mm	I-190 W, **E**...services from I-90, exit 79...Mobil, McDonald's, Courtyard, Holiday Inn, Hyatt, Marriott, Radisson, Rosemont Suites, Westin
41mm	toll booth, phones
42mm	Touhy Ave, **W**...Mobil/service, Tiffany's Rest., Comfort Inn
43mm	Des Plaines River
44mm	Dempster St(from nb, no return), **E**...HOSPITAL, **W**...Dunkin Donuts, Subway, cleaners
46mm	IL 58, Golf Rd, **E**...shopping ctr
49mm	Willow Rd, **1 mi W**...Amoco, Dunkin Donuts, Johnny's Grill, TGIFriday, Wendy's, Baymont Inn, Courtyard, Doubletree Guest Suites, Fairfield Inn, Motel 6, bank, to Arlington Race Course
53mm	Lake Cook Rd(no nb re-entry), **E**...Hyatt

I-294 begins/ends on I-94.

Illinois Interstate 474(Peoria)

Exit #   Services

15	I-74, E to Bloomington, W to Peoria. I-474 begins/ends on I-74, exit 99.
9	IL 29, E Peoria, to Pekin, **N**...Clark, Pizza Hut, Williams Rest., Ragon Motel, Riverboat Casino, camping, **S**...Citgo, McDonald's, Subway, Toyota
8mm	Illinois River
6b a	US 24, Adams St, Bartonville, **S**...Clark/diesel, Shell/24hr, Carl's Jr, KFC, McDonald's, Tyroni's Café
5	Airport Rd, **S**...Citgo/subs/pizza/mart
3a	IL 116, Farmington, **S**...Wildlife Prairie Park
0b a	I-74, W to Moline, E to Peoria. I-474 begins/ends on I-74, exit 87.

Left margin labels: Chicago Area, Peoria

Indiana Interstate 64

Exit #	Services
124mm	Ohio River, Indiana/Kentucky state line
123	IN 62 E, New Albany, **N**...HOSPITAL, Amoco/mart/24hr, Sunoco, SuperAmerica/mart, Chevrolet, Firestone/auto, Goodyear/auto, Valu Mkt Foods, bank, **S**...BP/mart/24hr, DairyMart/gas/24hr, Marathon/diesel/mart/24hr, Tobacco Road/gas/mart, Waffle Steak, Hampton Inn, Holiday Inn Express
121	I-265 E, to I-65(exits left from eb), no facilities, **N**...access to HOSPITAL
119	US 150 W, to Greenville, **1/2 mi N**...Citgo/diesel, Marathon/mart, Dairy Queen, Sam's Family Rest., Huber Winery, bank, drugs, groceries, hardware, laundry
118	IN 62, IN 64W, to Georgetown, **N**...Jacobi's Gas/mart/wash, Marathon/diesel/24hr, Shell/mart/24hr, Korner Kitchen Rest., McDonald's, Pizza King, Motel 6, Thriftway Foods, bank, drugs, **S**...Marathon/diesel
115mm	**Welcome Ctr wb, full(handicapped)facilities, vending, phone, picnic tables, litter barrels**
113	to Lanesville, no facilities
105	IN 135, to Corydon, **N**...Citgo/diesel/mart, Shell/mart/24hr, Big Boy, Best Western, Holiday Inn Express, **S**...MEDICAL CARE, Amoco/mart, Chevron, 5 Star Gas/mart, Arby's, Burger King, Country Folks Buffet, Cracker Barrel, Hardee's, KFC, Lee's Chicken, LJ Silver, McDonald's, Papa John's, Ponderosa, Subway, Taco Bell, Waffle Steak/24hr, Wendy's, Baymont Inn, Hampton Inn, AutoZone, Big O Tires, Chevrolet/Pontiac/Buick, $General, Ford/Mercury, JayC Food/24hr, Pennzoil, Radio Shack, Wal-Mart SuperCtr/24hr, cinema, laundry, RV camping
100mm	Blue River
97mm	parking area both lanes, no facilities
92	IN 66, Carefree, **S**...BP/mart, Citgo/Taco Bell/diesel/24hr, Shell/KrispyKreme/diesel, Kathy's Kitchen, Noble Roman's, Day's Inn/rest./diesel, to Wyandotte Caves, Harrison Crawford SF
88mm	Hoosier Nat Forest eastern boundary
86	IN 37, to Sulphur, **N**...to Patoka Lake, **S**...gas, food, phone, scenic route
81.5mm	**parking area wb, no facilities**
80mm	**parking area eb, no facilities**
79	IN 37, St Croix, **S**...to Hoosier NF, camping, phone, to OH River Toll Br
76mm	Anderson River
72	IN 145, to Birdseye, **N**...to Patoka Lake, **S**...gas, phone, St Meinrad Coll
63	IN 162, to Ferdinand, **N**...Marathon/diesel/mart, Wendy's, Ferdinand SF, **S**...Holiday World Camping
58mm	**rest areas both lanes, full(handicapped)facilities, info, vending, picnic tables, litter barrels, phone**
57	US 231, to Dale, **N**...231 Trkstp/diesel/rest./atm/24hr, Best Western(8mi), Holiday Inn, Scottish Inn, antiques, **S**...Marathon/Denny's/24hr, Shell/mart/24hr, Baymont Inn, Budget Host(2mi), Motel 6, Lincoln Boyhood Home, Lincoln SP
54	IN 161, to Holland, no facilities
39	IN 61, Lynnville, **N**...Shell/diesel/mart, Best Western, Old Fox Inn/rest., Goodyear, **S**...museum(10mi)
32mm	Wabash & Erie Canal
29b a	I-164 S, IN 57 S, to Evansville, **N**...Sunoco/Subway/TCBY/diesel/mart, **N**...Best Western
25b a	US 41, to Evansville, **N**...Citgo/diesel/mart(2mi), Pilot/Wendy's/diesel/mart/24hr, Williams/MainSt Café/diesel/mart/24hr, Baymont Inn, Log Inn Rest.(1mi), Blue Beacon, **S**...Amoco/mart/motel, Pennzoil/diesel/mart/24hr, Arby's, Burger King, Denny's, McDonald's, Comfort Inn, Holiday Inn Express, Motel 6(10mi), Ramada Ltd, Signature Inn(10mi), Super 8, camping, st police, to U of S IN
18	IN 65, to Cynthiana, **S**...Motomart/diesel/24hr
12	IN 165, Poseyville, **S**...Citgo/diesel/mart, Chevrolet, New Harmonie Hist Area/SP
7mm	**Black River Welcome Ctr eb, full(handicapped)facilities, phone, picnic tables, litter barrels, petwalk**
5mm	Black River
4	IN 69 S, New Harmony, Griffin, **1 mi N**...gas/diesel, food, motel, antiques
2mm	Big Bayou River
0mm	Wabash River, Indiana/Illinois state line

Indiana Interstate 65

Exit #	Services
262	I-90, W to Chicago, E to Ohio, I-65 begins/ends on US 12, US 20.
261	15th Ave, to Gary, **E**...Mack Trucks, **W**....Inland/mart, Marathon
259b a	I-94/80, US 6W, no facilities
258	US 6, Ridge Rd, **E**...Speedway/diesel/mart, Country Lounge Rest., Diner's Choice Rest., transmissions, **W**....Shell/mart/24hr, Lou's Auto Repair, USA Muffler/brake
255	61st Ave, Merrillville, **E**...Amoco/mart/24hr, Marathon/mart/wash, Mobil/Blimpie/diesel/mart/24hr, Speedway/diesel/mart/24hr, Cracker Barrel, McDonald's, Taco Bell, Chevrolet, Chrysler/Jeep, Comfort Inn, $Inn, Lee's Inn, Menard's, **1 mi W**...HOSPITAL, Shell/mart, Burger King, LJ Silver, Subway

Indiana Interstate 65

Merrillville

253b US 30 W, Merrillville, **W**...Amoco/24hr, Gas City/diesel/mart/24hr, Meijer/diesel/mart/24hr, Shell/mart, Speedway/diesel/mart, Arby's, Blimpie, Buster's Drive-Thru, Colorado Steaks, Denny's, Dunkin Donuts, Fannie Mae Candies, Giordano's Italian, Great Wall Chinese, Hooters, Johnny's Diner/24hr, KFC, LoneStar Steaks, Miner Dunn Burger, New Moon Chinese, Pepe's Mexican, Pizza Hut, Rio Bravo, Schlotsky's, Shakey's Pizza, Steak'n Shake, Subway, Texas Corral Steaks, Venezia Mexican, Wendy's, White Castle, Courtyard, $Inn, Fairfield Inn, Hampton Inn, Holiday Inn Express, Radisson, Red Roof Inn, Aldi Foods, Aladdin Lube, All Tune/lube, CarQuest, CarX Muffler, Celebration Sta, Electric Ave, FashionBug, Firestone, Ford/Lincoln/Mercury, Gordon Food Service, Hyundai/Mitsubishi, Jo-Ann Fabrics, Kinko's/24hr, K-Mart, Lentz Mufflers, Midas Muffler, Saturn, Toyota, Walgreen, bank, RV Ctr, transmissions

a US 30 E, **E**...EYECARE, Amoco/diesel/24hr, Clark Gas, Mobil/DQ/mart/24hr, Bakers Square, Bennigan's, Bob Evans, Boston Mkt, Burger King, Casa Gallardo, Chili's, Don Pablo, Great American Bagel, Joe's Crabshack, McDonald's, New China Buffet, Old Country Buffet, Olive Garden, Pizzaria Uno, Popeye's, Red Lobster, Ruby Tuesday, Subway, Taco Bell, TGIFriday, Day's Inn, Extended Stay America, Holiday Inn, Knight's Inn, La Quinta, Motel 6, Super 8, Barnes&Noble, Bed Bath&Beyond, Best Buy, Circuit City, Dodge, Firestone/auto, Gateway Computers, Home Depot, JC Penney, KidsRUs, Kohl's, MensWearhouse, Michael's, Nevada Bob's, Nissan, Office Depot, Old Navy, PetCo, PetsMart, Pier 1, Sam's Club, Sayre's, Sears/auto, Service Merchandise, SportMart, Sports Authority, Target/drugs, Tire Barn, ToysRUs, VW, Wal-Mart, bank, cinema, mall

247 US 231, Crown Point, **W**....HOSPITAL, Mobil/mart/24hr, Vietnam Vet Mem

241mm weigh sta both lanes

240 IN 2, Lowell, **E**...Citgo/diesel/rest., Mobil/Blimpie/Little Caesars/diesel/mart, Shell/Burger King/diesel/mart, Williams/diesel/rest./24hr, Super 8, fireworks, repair, **W**...Marathon/diesel/mart, st police

234mm Kankakee River

231mm **rest area both lanes, full(handicapped)facilities, phones, info, picnic tables, litter barrels, vending, petwalk**

230 IN 10, Roselawn, **E**...Crazy D's/diesel/rest./24hr, **W**...Amoco/mart, Shell/Subway/mart, Marge's Sandwiches, Rascal's Diner, Renfrow's Burgers, CarQuest, CVS Drug, IGA Foods, Jellystone Park Camping, Rose RV Ctr, TrueValue, bank, laundry

215 IN 114, Rensselaer, **E**...HOSPITAL, Amoco/KFC/diesel/24hr, Arby's, McDonald's, Holiday Inn Express, Interstate Motel, **W**...Phillips 66/Grandma's Rest./fireworks, Shell/mart, Burger King, Dairy Queen, Trail Tree Rest., MidContinent Inn, tires/repair/towing/24hr, Catfish Farm/fishing

212mm Iroquois River

205 US 231, Remington, **E**...HOSPITAL, Phillips 66/diesel/mart, Shell/Subs/pizza/diesel/mart, Knight's Inn/rest., to St Joseph's Coll

201 US 24/231, Remington, **W**...Citgo/diesel/mart/24hr, Marathon/Taco Bell/diesel/24hr, Speedway/Subway/diesel/mart/24hr, KFC, McDonald's, Super 8, Sunset Inn, tires

196mm **rest area both lanes, full(handicapped)facilities, vending, phone, info, picnic tables, litter barrels, petwalk**

193 US 231, to Chalmers, **E**...Shell/Wayfara Rest./mart, Dairy Queen

188 IN 18, to Brookston, Fowler, no facilities

178 IN 43, W Lafayette, **E**...GasAmerica/Taco Bell/mart, Phillips 66/Subway/diesel/mart, Shell/Burger King/mart, McDonald's, Holiday Inn, Super 8, to Tippecanoe Bfd, museum, st police, **W**...to Purdue U

176mm Wabash River

175 IN 25, Lafayette, **E**...Citgo/mart/24hr, **W**...HOSPITAL, Marathon(1mi)

172 IN 26, Lafayette, **E**...Meijer/diesel/24hr, Cracker Barrel, Steak'n Shake, White Castle/Church's, Baymont Inn, Budget Inn, Comfort Inn, Hawthorn Suites, Holiday Inn Express, Lee's Inn, Microtel, airport, **W**...HOSPITAL, Amoco/diesel/mart/24hr, Shell/mart, Speedway/diesel/mart/24hr, Arby's, Bob Evans, Burger King, Chili's, China Garden, Country Café, Dairy Queen, Damon's, Denny's, Don Pablo's, KFC, Logan's Roadhouse, McDonald's, Mtn Jack's Rest., Olive Garden, Outback Steaks, Pizza Hut, Taco Bell, $Inn, Fairfield Inn, Hampton Inn, Homewood Suites, Knight's Inn, Radisson, Ramada, Red Roof, Signature Inn, CVS Drug, FashionBug, IGA Foods, Marsh Foods, Buick/Nissan/Cadillac, Lowe's Bldg Supply, Sam's Club, Target, Wal-Mart/auto, bank, to Purdue U

168 IN 38, Dayton, **E**...Amoco/Piccadilly's/mart/atm/24hr, Mobil/diesel/mart

158 IN 28, to Frankfort, **E**...Amoco/diesel/mart/24hr, Marathon/mart, **2 mi W**...HOSPITAL, Lincoln Lodge Motel, camping

150mm **rest area sb, full(handicapped)facilities, info, picnic tables, litter barrels, phone, vending, petwalk**

148mm **rest area nb, full(handicapped)facilities, info, picnic tables, litter barrels, phone, vending, petwalk**

146 IN 47, Thorntown, **W**...camping

141 US 52 W, Lafayette Ave, **E**...HOSPITAL

140 IN 32, Lebanon, **E**...HOSPITAL, Amoco/mart, Big Boy, White Castle, Comfort Inn, AutoZone, Goodyear/auto, Petro Tire, laundry/cleaners, ° mi **E**...Energy+/mart/24hr, Ford/Lincoln/Mercury, **W**...McClure/diesel/mart, Shell/diesel/mart, Arby's, Burger King, KFC, McDonald's, Ponderosa, Steak'n Shake, Subway, Taco Bell, $Inn, Lee's Inn, Super 8, factory outlets

Lafayette

N

S

Indiana Interstate 65

139	IN 39, Lebanon, **E**...HOSPITAL, GasAmerica/atm, Wendy's, **1 mi E**...Hardee's, Chrysler/Plymouth, **W**...Flying J/Country Mkt/diesel/LP/mart/24hr, Old Chicago Rest., Holiday Inn
138	to US 52, Lebanon, **E**...Citgo/diesel/mart, Shell/diesel/mart, Sunshine Café, Ford, antiques, bank
133	IN 267, Whitestown, no facilities
130	IN 334, Zionsville, **E**...Crystal Flash/Subway/diesel/mart, Phillips 66/Stuckey's, **W**...TA/Marathon/diesel/rest./24hr
129	I-465 E, US 52 E(from sb)
126mm	Fishback Creek
124	71st St, **W**...Eagle Creek Park
123	I-465 S, to airport, no facilities
121	Lafayette Rd, **E**...Citgo/7-11/24hr, Lee's Inn, **W**...HOSPITAL, VETERINARIAN, Amoco/mart/wash, Shell/mart/wash/24hr, Applebee's, Chinese Rest., Hardee's, MCL Cafeteria, Subway, Taco Bell, TCBY, $Inn, Batteries+, Cub Foods, Discount Tire, Firestone/auto, Gordon Food Service, IndyLube, JC Penney, Kinko's, Lenscrafters, McKinney Transmissions, Mercury/Lincoln/Mitsubishi, NAPA, Nissan, Olds/Isuzu, PepBoys, Sears/auto, Speedway Parts, Tire Barn, Toyota, Value City, Ziebart TidyCar, carwash, cinemas, cleaners, mall, same as 119
119	38th St, Dodge, **W**...MEDICAL CARE, Meijer's/diesel/24hr, Speedway, Arby's, KFC, McDonald's, Outback Steaks, Pizza Hut, Ryan's, CarX Muffler, Dominick's Foods, Honda, Lazurus, MediaPlay, Midas Muffler, Office Depot, Pearle Vision, Precision Tune, Sears/auto, Service Merchandise, TireAmerica, Toyota, Ziebart's TidyCar, Waccamaw, mall, same as 121
117.5mm	White River
117	MLK St(from sb)
116	29th St, 30th St(from nb), Marian Coll, no facilities
115	21st St, **E**...HOSPITAL, **W**...Shell/mart, museums, zoo
114	MLK St, West St, downtown
113	US 31, IN 37, Meridian St, to downtown, **E**...HOSPITAL, JiffyLube, Midas Muffler
112a	I-70 E, to Columbus, no facilities
111	Market St, Michigan St, Ohio St, **E**...Hardee's, McDonald's, **W**...downtown, City Market, to Market Square Arena, museum
110b	I-70 W, to St Louis, no facilities
a	Prospect St, Morris St, E St(from sb), no facilities
109	Raymond St, **E**...HOSPITAL, **W**...Speedway/mart, White Castle, CVS Drug, Safeway
107	Keystone Ave, **E**...HOSPITAL, $Inn, **W**...Speedway/mart, Big Boy, Burger King, McDonald's, Subway, Holiday Inn Express, Cub Foods/24hr, U of Indianapolis
106	I-465 and I-74, no facilities
103	Southport Rd, **E**...Amoco/McDonald's/mart/24hr, 24hr, Meijer/gas/diesel/24hr, Shell/mart/wash, Dog'n Suds, Noble Roman Pizza, Schlotsky's, Firestone/auto, Harley-Davidson, JiffyLube, Menard's, bank, carwash, **W**...HOSPITAL, Bigfoot/24hr, Citgo/7-11, Speedway/diesel/mart/24hr, Bob Evans, Burger King, Cracker Barrel, Hooters(3mi), KFC, McDonald's, Steak'n Shake, Waffle Steak, Wendy's, Best Western, Country Inn Suites, Courtyard, $Inn, Fairfield Inn, Hampton Inn, Signature Inn, Super 8, bank
101	CountyLine Rd, **W**...HOSPITAL, Arby's
99	Greenwood, **E**...Phillips 66/diesel/repair, camping, **W**...HOSPITAL, Amoco/mart, Citgo/7-11, Marathon/mart, Shell/mart/wash, Arby's, Big Boy, Bob Evans, China Gate, Hardee's, Jonathan Byrd's Cafeteria, KFC, McDonald's, Noble Roman Pizza, Subway, Taco Bell, TCBY, Waffle Steak, White Castle, Comfort Inn, Greenwood Inn, Lee's Inn, LowCost Drug, McQuik's Lube, Sam's Club, Stout's RV Ctr, airport, bank
95	Whiteland, **E**...Marathon/Kathy's Kitchen/diesel, **W**...Pilot/Arby's/diesel/mart/24hr, Speedway/Country Kitchen/diesel/mart, Mayes RV Ctr
90	IN 44, Franklin, **W**...HOSPITAL, Amoco, Shell/mart, Burger King, McDonald's, Subway, Waffle Steak, Carlton Lodge, Day's Inn/rest., Howard Johnson, Quality Inn, Super 8, golf course
85mm	Sugar Creek
82mm	Big Blue River
80	IN 252, Edinburgh, to Flat Rock, **W**...Shell/diesel/mart
76b a	US 31, Taylorsville, **E**...HOSPITAL, Shell/diesel/mart, KFC, Waffle Steak, White Castle(8mi), Comfort Inn, **W**...Amoco, Citgo/Subway/mart, Thornton's/diesel/café/mart, Arby's, Cracker Barrel, FoodCourt, Hardee's, McDonald's, Waffle Steak, Best Western, Hampton Inn, Holiday Inn Express, Edmundson RV Ctr, Goodyear, Prime Outlets/famous brands, RV camping, repair
73mm	**rest area both lanes, full(handicapped)facilities, phone, vending, info, picnic tables, litter barrels, petwalk**
68mm	Driftwood River

Indiana Interstate 65

68 IN 46, Columbus, **E**...HOSPITAL, Amoco/diesel/mart/24hr, Bigfoot/gas, Shell/mart, Speedway/diesel/mart, Burger King, McDonald's, Riviera Diner, Waffle House/24hr, Holiday Inn/rest., Ramada Inn, Super 8, **W**...VETERINARIAN, Marathon/mart, Bob Evans, Taco Bell, Wendy's, Day's Inn, $Inn, Knight's Inn, U-Haul, to IU Bloomington, to Brown Co SP

64 IN 58, Walesboro, **W**...Marathon, Kathy's Express Diner, to RV camping

55 IN 11, Seymour, to Jonesville, **E**...Marathon/mart

54mm White River

51mm weigh sta both lanes

50b a US 50, Seymour, **E**...BP/TA/diesel/rest./24hr, Marathon/diesel/mart, Swifty Gas/mart, Chinese Rest., McDonald's, Waffle House, Allstate Inn, Day's Inn, Econolodge, Super 8, Tanger Outlet/famous brands, **W**...HOSPITAL, Amoco/mart, Shell/diesel/mart, Speedway/diesel/mart, Sunoco/mart, Arby's, Baskin-Robbins, Bob Evans, Burger King, Capt D's, China Buffet, Cracker Barrel, Denny's, KFC(1mi), Little Caesar's, LJ Silver, McDonald's, Papa John's, Pizza Hut, Ponderosa, Rally's, Ryan's, Shoney's, Steak'n Shake, Subway, Taco Bell, Tapatio Mexican, Tumbleweed Grill, Wendy's, Hampton Inn, Holiday Inn/rest., Knight's Inn, Lee's Inn, Aldi Foods, Big Lots, Chevrolet/Buick/Olds/Pontiac/GMC, $General, Family$, Farm&Fleet, FashionBug, Foods+, Ford/Lincoln/Mercury, GNC, Goody's, GreasePit Lube, JC Penney, K-Mart, Radio Shack, Staples, Wal-Mart SuperCtr/24hr, antiques, auto parts, bank, cinema, laundry

41 IN 250, Uniontown, **W**...Marathon/diesel/rest./24hr, auto/truck repair

36 US 31, Crothersville, **E**...Shell/mart, **W**...BP/pizza/subs/mart

34a b IN 256, Austin, **E**...Bigfoot/gas, Dairy Bar, Chevrolet, auto parts, hardware, to Hardy Lake, Clifty Falls SP, **W**...Fuelmart/A&W/diesel/mart, Home Oven Rest.

29b a IN 56, Scottsburg, to Salem, **E**...HOSPITAL, Amoco/diesel, MotoMart, Speedway/diesel/mart, Burger King, KFC, Ponderosa, Sonic, Subway, Taco Bell, Holiday Inn Express, Mariann Motel/rest., Ace Hardware, Grease Pit Lube, **W**...Citgo/diesel/mart, Shell/mart, Arby's, LJ Silver, McDonald's, Pizza Hut, Roadhouse USA, Waffle Steak, Wendy's, Best Western, $Inn, Hampton Inn, Wa-Mart SuperCtr/24hr, bank

22mm **rest area both lanes, full(handicapped)facilities, info, phone, picnic tables, litter barrels, vending, petwalk**

19 IN 160, Henryville, **E**...Citgo/DQ/Stuckey's, Shell/mart/24hr, Schuler's Rest., fireworks

16 Memphis Rd, Memphis, **E**...BP/mart, Citgo/KrispyKreme/diesel/mart, **W**...Pilot/Arby's/diesel/mart/24hr, Customers 1st RV Ctr

9 IN 311, Sellersburg, to New Albany, **E**...Chevron/mart, Shell, Swifty Gas, Cracker Barrel, Dairy Queen, IGA Foods, Ford, Carmerica Tires/service, auto parts, st police, **W**...DM/gas, Arby's, Burger King, McDonald's, Taco Bell, Comfort Inn, city park

7 IN 60, Hamburg, **E**...BP/diesel/mart/24hr, Davis Bros/Ashland/diesel/rest./atm/24hr, **W**...KFC/Pizza Hut, Day's Inn

6b a I-265 W, to I-64 W, New Albany, IN 265 E

4 US 31 N, IN 131 S, Clarksville, New Albany, **E**...Shell/diesel/rest./24hr, Thornton's/Subway/diesel/mart, Crest Motel, RV camping, fireworks, **W**...DENTIST, EYECARE, Speedway/diesel/mart, Applebee's, Arby's, Asian Pearl Chinese, Bailey's Grill, Blimpie, Bob Evans, Boston Mkt, Burger King, Capt D's, ChiChi's, Damon's, Denny's, Don Pablo's, Fazoli's, Hardee's, HomeTown Buffet, Hooters, Jerry's Rest., Little Caesar's, Logan's Roadhouse, LJ Silver, McDonald's, Mr Gatti's, O'Charley's, Old Country Buffet, Papa John's, Pizza Hut, Ponderosa, Rally's, Red Lobster, Shoney's, Steak'n Shake/24hr, Taco Bell, Texas Roadhouse, Wendy's, Best Western, $Inn, Hampton Inn, AutoZone, Bigg's Foods/24hr, Big Lots, Buick/Pontiac/GMC, Circuit City, Dillard's, Firestone/auto, Ford, Goodyear, Hancock Fabrics, Home Depot, Honda, JC Penney, K-Mart, Kroger/deli, Mazda, MensWearhouse, Office Depot, OfficeMax, Olds, PepBoys, Pier 1, Premier RV Ctr, Sears/auto, Service Merchandise, Target, Toyota, USPO, Walgreen, Wal-Mart, bank, cinema, cleaners, mall

2 Eastern Blvd, Clarksville, **E**...Day's Inn, Motel 6, Super 8, U-Haul, **W**...Amoco, Chevron/diesel, Citgo/7-11, Shell/mart, KFC, Omelet Shop, Ponderosa, Bel Aire Motel, Econolodge, Howard Johnson, Knight's Inn, Quality Inn, Rivers Edge Motel/café, Rodeway Inn, Star Motel, Greyhound Supermkt

1 US 31 S, IN 62, Stansifer Ave, **W**...HOSPITAL, Derby Dinner House, Lakeview Hotel, Stansifer Hotel, Stinnett RV Ctr

0 Jeffersonville, **E**...HOSPITAL, Chevron, Thornton's/gas, Hardee's, McDonald's, Waffle Steak, Village Pub, Chrysler/Jeep, Hyundai, Nissan, Walgreen, to Falls of OH SP, **W**...Ramada Inn

0mm Ohio River, Indiana/Kentucky state line

Indiana Interstate 69

Exit #	Services
158mm	Indiana/Michigan state line

157 Lake George Rd, to IN 120, Fremont, Lake James, **E**...Mobil/diesel/mart/LP/24hr, Lake George Inn, **W**...Shell/diesel/mart/24hr, Speedway/Hardee's/diesel/mart/24hr, Holiday Inn Express(1mi), Redwood Lodge, Boot Outlet, Prime Outlets/famous brands(1mi), repairs/tires

156 I-80/90 Toll Rd, E to Toledo, W to Chicago

154 IN 127, to IN 120, IN 727, Fremont, Orland, **E**...E&L Motel, Hampton Inn, Super 8, Golf/rest., U-Haul/repair(1mi), **4 mi E on IN 127**...McDonald's, Santa Fe Grill, Tom's Donuts, **W**...Marathon/Taco Bell/diesel/mart, Budgeteer Motel, Holiday Inn Express, Prime Outlets/famous brands(2mi), to Pokagon SP, Jellystone Camping(7mi)

Indiana Interstate 69

150	rd 200 W, to Lake James, Crooked Lake, **E**...CHIROPRACTOR, DENTIST, PODIATRIST, **W**...Shell/mart, Pennzoil Gas/mart, Marine Ctr
148	US 20, to Angola, Lagrange, **E**...HOSPITAL, Amoco/Subway/mart, Marathon/diesel, Speedway/diesel/mart, McDonald's, Wendy's(1mi), **W**...WestEdge Grill, Best Western
145mm	Pigeon Creek
144mm	**rest area sb, full(handicapped)facilities, info, phone, picnic tables, litter barrels, vending, petwalk**
140	IN 4, Ashley, Hudson, to Hamilton, **1 mi W**...Phillips 66/mart
134	US 6, to Waterloo, Kendallville, **W**...Marathon/diesel/mart/24hr, Western Wear
129	IN 8, Auburn, to Garrett, **E**...DENTIST, Amoco/Dunkin Donuts, BP/diesel/mart/atm, GasAmerica/mart/atm, Marathon/diesel/24hr, Shell/mart/wash, Speedway/mart, Ambrosia Rest., Arby's/24hr, Big Boy, Burger King, Dairy Queen, Fazoli's, KFC, McDonald's, Pizza Hut, Ponderosa, Richard's Rest., Subway, Taco Bell, TCBY, Wendy's, Budget Inn, Country Hearth Inn, Holiday Inn Express, Ramada Suites, Super 8, Ace Hardware, AutoZone, Chevrolet/Olds/Pontiac/Buick, Chrysler/Jeep, Coast Hardware, $General, Ford, Monro Muffler/brake, Pontiac/Olds, Radio Shack, RV Ctr, Sav-A-Lot Foods, Sears Appliances, Staples, Wal-Mart SuperCtr/24hr, bank, museum, tires
126	IN 11-A, to Garrett, Auburn, **E**...Kruse Auction Park, **W**...KOA
124mm	**rest area both lanes, full(handicapped)facilities, phone, picnic tables, litter barrels, vending, petwalk**
116	IN 1 N, Dupont Rd, **E**...Amoco/Burger King/mart, Comfort Suites, **W**...BP/mart, Speedway/mart, Bob Evans, McDonald's(1mi)
115	I-469, US 30 E, no facilities
112b a	Coldwater Rd, **E**...EYECARE, Amoco/diesel/24hr, Marathon/mart, Sunoco, Arby's/24hr, Bill Knapp's, ChiChi's, Cork'n Cleaver, Grand Gourmet Chocolate, Hunan House Chinese, Joe's CrabShack, Krispy Kreme, Lone Star Steaks, Old Country Buffet, Oriental Buffet, Papa John's, Steak'n Shake, Taco Bell(1mi), Taco Cabana, Zesto IceCream, Wendy's, Marriott, Sumner Suites, HobbyLobby, PostNet, ProGolf, Service Merchandise, Walgreen, Wal-Mart, cinema, to coliseum
111b a	US 27 S, IN 3 N, **E**...Shell/mart/wash, Dairy Queen, Denny's, Day's Inn, Residence Inn, Honda, Infiniti, Kia, Nissan, Pontiac/GMC/Isuzu, Toyota, **W**...Amoco/mart, BP/diesel/mart, Marathon/mart, Meijer/diesel/24hr, Applebee's, Cracker Barrel, Dog'n Suds, KFC, McDonald's, Texas Roadhouse, Tumbleweed Grill, Baymont Inn, Courtyard, $Inn, Economy Inn, Fairfield Inn, Guest House Motel/rest., Hampton Inn, Lee's Inn, Signature Inn, Studio+, Sam's Club
109b a	US 33, Ft Wayne, **E**...HOSPITAL, Amoco, Ft Wayne Trk Plaza/Citgo/Subway/mart, Arby's, Big Boy, McDonald's, Best Inn, Comfort Inn, $Inn, Holiday Inn, Knight's Inn, Motel 6, Red Roof Inn, ValuLodge, Blue Beacon, Goodyear, Ryder Trucks, auto parts, to Children's Zoo
105b a	IN 14 W, Ft Wayne, **E**...HOSPITAL, Meijer/diesel/24hr, Shell/wash, Speedway/mart/LP, Steak'n Shake, Red Carpet Inn, BMW, Cadillac, Chevrolet, Chrysler/Plymouth/Jeep, Dodge, Hires Auto Parts, Lexus, Mazda, Olds, Saab, Saturn, Toyota, bank, to St Francis U
102	US 24, to Huntington, Ft Wayne, **E**...HOSPITAL, Subway(1mi), Taco Bell(1mi), Extended Stay America, Hampton Inn, to In Wesleyan U, **W**...Amoco/mart/24hr, Marathon/mart, Applebee's, Arby's, Bob Evans, Capt D's, Carlos O'Kelly's, Coventry Tavern Dining, McDonald's, Pizza Hut, Wendy's, Zesto's Yogurt, Best Western, Comfort Suites, MailBoxes Etc, Scott's Foods, Walgreen, bank, cinema, repair, st police
99	Lower Huntington Rd, no facilities
96b a	I-469, US 24 E, US 33 S, **E**...to airport
93mm	**rest area sb, full(handicapped)facilities, info, phone, vending, picnic tables, litter barrels, pet walk**
89mm	**rest area nb, full(handicapped)facilities, info, phone, vending, picnic tables, litter barrels, pet walk**
86	US 224, Markle, to Huntington, **E**...HOSPITAL, Citgo/diesel(1mi), Marathon/mart/24hr(1mi), Sunoco/Subway/mart, Dairy Queen(1mi), Taco Bell, Sleep Inn, Super 8, antiques, **W**...to Huntington Reservoir
80mm	weigh sta both lanes
78	IN 5, to Warren, Huntington, **E**...Quikstop/gas, Sunoco/diesel/mart, Huggy Bear Motel, **W**...HOSPITAL, Citgo/diesel, Clark Gas/Subway/diesel/mart/atm/24hr, Truck Plaza, Hoosier-Land Rest., McDonald's, Subway, Ramada, Super 8, RV Camping, to Salmonie Reservoir
76mm	Salamonie River
73	IN 218, to Warren, no facilities
64	IN 18, to Marion, Montpelier, **E**...camping, **W**...HOSPITAL, Marathon/mart, Arby's, Harley-Davidson
60mm	Walnut Creek
59	US 35 N, IN 22, to Upland, Gas City, **E**...Citgo/Subway/mart, Burger King, Cracker Barrel, Best Western, Super 8, RV camping, Taylor U, **W**...RV camping, IN Wesleyan
55	IN 26, to Fairmount, **E**...Marathon/mart

N

S

Ft Wayne

Indiana Interstate 69

50mm	**rest area both lanes, full facilities, info, phone, picnic tables, litter barrels, vending, pet walk**
45	US 35 S, IN 28, to Alexandria, Albany, **E**...Amoco/diesel/rest., Shell/Taco Bell/diesel/rest., camping, **W**...camping
41	IN 332, to Muncie, Frankton, **E**...HOSPITAL, Amoco/mart, to Ball St U, **W**...Jack Smith's RV Ctr
34	IN 67, to IN 32, Chesterfield, Daleville, **E**...HOSPITAL, Shell/Burger King, Speedway/Hardee's, Arby's, Denny's, FoodCourt, Taco Bell, Budget Inn, Rise Motel, mall, **W**...Daleville Travel Plaza, GasAmerica/diesel/mart, Cleo's, McDonald's, Wendy's, Subway, Super 8, flea mkt, truckwash
26	IN 9, IN 109, to Anderson, **E**... Major/24hr, Ryan's, Best Western, Courtyard, $Inn, Hampton Inn, info, **W**...HOSPITAL, VETERINARIAN, Amoco/mart/24hr, Citgo/mart, Marathon/diesel, Red Barn/Noble Roman Pizza, Shell/wash, Applebee's, Arby's, Baskin-Robbins, Bob Evans, Burger King, Cracker Barrel, Great Wall Chinese, Grindstone Charley's, Hardee's, La Charreada Mexican, Little Caesar's, Lone Star Steaks, McDonald's, Old Country Buffet, Perkins, Pizza Hut, Red Lobster, Ruby Tuesday, Steak'n Shake, Taco Bell, TCBY, Texas Road-house, Waffle Steak, Wendy's, White Castle, Baymont Inn, Best Inn, Comfort Inn, Holiday Inn, Lee's Inn, Motel 6, Ramada/rest., Super 8, JiffyLube, Kohl's, Lowe's, Monroe Muffler, Office Max, Olds/Cadillac/GMC, Payless Foods/deli/24hr, Pier 1, Radio Shack, Target, Tire Barn, Toyota, Wal-Mart SuperCtr, bank, cinema, to Anderson U, to Mounds SP
22	IN 9, IN 67, to Anderson, **E**...lodging(3mi), **W**...HOSPITAL, Gas America, Anderson Country Inn (1mi), Rick's Mufflers/brakes, st police
19	IN 38, Pendleton, **E**...Marathon/KrispyKreme/mart, Dairy Queen, McDonald's, Subway, **W**...camping
14	IN 13, to Lapel, **W**...Pilot/Subway/diesel/24hr, camping
10	IN 238, to Noblesville, Fortville, **E**...HOSPITAL, furniture, outlet mall
5	IN 37 N, 116th St, Fishers, to Noblesville, **W**...Marathon/mart, Shell/autocare, McDonald's
3	96th St, **E**...DENTIST, Amoco/KrispyKreme/diesel/mart/24hr, Marathon/mart, Meijer/diesel/24hr, Shell/mart/wash, Applebee's, Bennigan's, Blimpie, Cracker Barrel, Einstein Bros, Grindstone Charley's, McDonald's, Muldoon's Grill, Noble Roman Pizza, Ruby Tuesday, Steak'n Shake, Wendy's, Holiday Inn, Holiday Inn Express, Sleep Inn, AutoNation, Cord's Camera, Kinko's, Kohl's, Marsh Foods, PepBoys, Radio Shack, Sam's Club, Staples, Wal-Mart/auto, bank, cinema, cleaners, **W**...Marathon/mart, Arby's, Burger King, Peterson's Steaks/seafood, Schlotsky's, Taco Bell, Residence Inn, JiffyLube, Monro Muffler/brake
1	82nd St, Castleton, **E**...HOSPITAL, DENTIST, $Inn, Howard Johnson, Omni Hotel, Lowe's Bldg Supply, **W**... Amoco, Marathon/Qlube/mart, Shell, Arby's, Burger King, Cancun Mexican, Castleton Grill, ChiChi's, Denny's, Fazoli's, Great China Buffet, Hooters, IHOP, Joe's Grille, KFC, LJ Silver, McDonald's, Olive Garden, Pizza Hut, Priscilla's Café, Skyline Chili, Steak'n Shake, Tony Roma, Wendy's, Day's Inn, Hampton Inn, Aamco, Burlington Coats, Discount Tire, Goodyear/auto, Kinko's, IndyLube, Indy Tires, MensWearhouse, Midas Muffler, Sears/auto, West Marine, mall
0mm	I-465 around Indianapolis. I-69 begins/ends on I-465, exit 37, at Indianapolis.

Indiana Interstate 70

Exit #	Services
156.5mm	Indiana/Ohio state line, weigh sta
156b a	US 40 E, Richmond, **N**...FuelMart/Subway/diesel, Petro/Marathon/diesel/rest./24hr, Swifty Gas, Fairfield Inn, Golden Inn, Blue Beacon, **S**...Amoco/mart/24hr, Shell/mart, Speedway/mart, Applebee's, Bob Evans, Burger King, Chili's, Cracker Barrel, Fazoli's, Garfield's Rest., Jerry's Rest., KFC, LJ Silver, McDonald's, MCL Cafeteria, Pizza Hut, Ponderosa, Red Lobster, Ryan's, Steak'n Shake, Super China Buffet, Texas Roadhouse, TCBY, White Castle, Best Western, Day's Inn, $Inn, Hampton Inn, Holiday Inn, Lee's Inn, Motel 6, Ramada Inn, Big Lots, Chevrolet/Olds/Cadillac, Chrysler/Jeep, $General, Durham Sports, Firestone, Ford/Lincoln/Mercury, Goody's, Goodyear/auto, Hastings Books, HobbyLobby, Honda, IndyLube, JC Penney, Kroger, Lowe's Bldg Supply, Michel Tire, Midas Muffler, Nissan, OfficeMax, Sav-A-Lot Foods, Sears/auto, Target, Tuffy Muffler, U-Haul, Wal-Mart SuperCtr/24hr, bank, cinema
153	IN 227, Richmond, to Whitewater, no facilities
151b a	US 27, Richmond, to Chester, **N**...Fricker's Rest., Best Buy RV Ctr, Dodge, Honda, Olds, KOA, **S**...HOSPITAL, DENTIST, Meijer/diesel/24hr, Speedway, Shell/McDonald's, Sunoco/mart, Big Boy, Burger King, Carver's Rest., Pizza Hut, Ponderosa, Richard's Rest., Subway, Taco Bell, Wendy's, Comfort Inn, Super 8, Cox's Mkt, Harley-Davidson, bank
149	US 35, IN 38, Richmond, to Muncie, **S**...Raper RV Ctr
148mm	weigh sta both lanes
145	Centerville, **N**...Amoco/Stuckey's/Dairy Queen, Super 8, Goodyear, repair, **S**...KOA
145mm	Nolans Fork Creek
144mm	**rest area both lanes, full(handicapped)facilities, info, vending, phone, picnic tables, litter barrels, petwalk**
141mm	Greens Fork River
137	IN 1, to Hagerstown, Connersville, **N**...Amish Cheese, **S**...GasAmerica/mart/24hr, Marathon/diesel/rest., Shell/mart/24hr, Burger King, Flagman Rest., McDonald's
131	Wilbur Wright Rd, New Lisbon, **S**...Marathon/KFC/Taco Bell/diesel24hr, Country Harvest Rest.
126mm	Flatrock River

Anderson (side label)

Richmond (side label)

N
S

E
W

Indiana Interstate 70

123	IN 3, Spiceland, to New Castle, **N**...HOSPITAL, Shell/mart/24hr, Speedway, Denny's/24hr, Ponderosa(4mi), Alford Motel, Best Western(2mi), Day's Inn, Holiday Inn Express(3mi), **S**...BP/Subway/diesel/mart, Marathon/mart, Kathy's Kitchen
117mm	Big Blue River
115	IN 109, to Knightstown, Wilkinson, **N**...GasAmerica/diesel/rest./24hr, Burger King, camping
107mm	**rest area both lanes, full(handicapped)facilities, vending, phone, picnic tables, litter barrels, petwalk, RV dump**
104	IN 9, Greenfield, Maxwell, **N**...GasAmerica/mart/atm, **S**...VETERINARIAN, CHIROPRACTOR, DENTIST, Amoco, GasAmerica/diesel/atm, Shell/mart, Sunoco/mart, Swifty/mart, Applebee's, Arby's, Bob Evans, Burger King, KFC, McDonald's, Pizza Hut, Ponderosa, Shoney's, Taco Bell, Waffle House, Wendy's, White Castle, Comfort Inn, $Inn, Holiday Inn Express, Lee's Inn, Liberty Motel, Super 8, Aldi Foods, Big Lots, CVS Drug, FashionBug, GNC, Kroger/deli, Marsh Foods/deli, Monro Muffler/brake, Radio Shack, TrueValue, Wal-Mart, bank, laundry
96	Mt Comfort Rd, **N**...GasAmerica/mart, Speedway/diesel/mart/24hr, **S**...Pilot/diesel/mart/rest./24hr, Shell/24hr, McDonald's, KOA(seasonal)
91	Post Rd, to Ft Harrison, **N**...Marathon/7-11, Swifty Gas, Big Boy, Cracker Barrel, Joe's Crabshack, McDonald's, Outback Steaks, Steak'n Shake, Wendy's, Baymont Inn, Suburban Lodge, Lowe's Bldg Supply, st police, **S**...Amoco/wash/24hr, Clark Gas, Phillips 66/mart, Shell/mart/wash, Hardee's, KFC, Taco Bell, Waffle Steak, Best Western, $Inn, Quality Inn, Super 8, Travelers Inn, CVS Drug/24hr, Marsh Foods/deli, bank, cleaners
90	I-465(from wb), no facilities
89	Shadeland Ave, I-465(from eb)**N**...DENTIST, Marathon, Phillips 66, Quick Fuel, Waffle House, Country Inn Suites, Hampton Inn, Hawthorn Suites, Motel 6, Quality Inn(1mi), Toyota, U-Haul, bank, mufflers, **S**...VETERINARIAN, BigFoot/gas, Clark Gas, Marathon, Shell/mart/24hr, Speedway, Arby's, Bob Evans, Cattle Co Rest., McDonald's, Omelet Shop, Rally's, Red Lobster, Sizzler, Steak&Ale, Substation/deli, Wendy's, Always Inn, Budget Inn, Day's Inn, Holiday Inn/Damon's, Knight's Inn, La Quinta, Marriott, Super Inn, Aamco, Buick, CarX Muffler, Chevrolet, Dodge, Ford/Lincoln/Mercury, auto parts, bank, hardware
87	Emerson Ave, **N**...Amoco/McDonald's, **S**...HOSPITAL, Speedway/mart
85b a	Rural St, Keystone Ave, **N**...fairgrounds, **S**...Marathon, McDonald's
83b(112)	I-65 N, to Chicago
a(111)	Michigan St, Market St, downtown, **S**...Hardee's
80(110a)	I-65 S, to Louisville
79b	Illinois St, McCarty St, downtown
a	West St, **N**...HOSPITAL, Comfort Inn, to Union Sta, Govt Ctr, RCA Dome, zoo
78	Harding St, to downtown
77	Holt Rd, **S**...Clark Gas, Shell/mart, McDonald's, Mr Dan's Rest., AllPro Parts
75	Airport Expswy, to Raymond St, **N**...Marathon/mart, Speedway/mart, Cracker Barrel, Denny's, JoJo's Rest., Schlotsky's, Waffle Steak, Adam's Mark, Baymont Inn, Fairfield Inn, La Quinta, Motel 6, Red Roof Inn, Residence Inn, access to Day's Inn, Hilton, Holiday Inn, Ramada, to airport
73b a	I-465 N/S, I-74 E/W
66	IN 267, to Plainfield, Mooresville, **N**...Amoco/mart/24hr, Shell/mart/atm, Speedway/diesel/mart, Thornton/diesel/mart, Arby's, Big Boy, Bob Evans, Burger King, Coachman Rest., Cracker Barrel, Dog'n Suds, Golden Corral, McDonald's, Perkins, Ritter's Custard, Royal Line Pizza, Steak'n Shake, Subway, Wendy's, White Castle, AmeriHost, Comfort Inn, Day's Inn, $Inn, Hampton Inn, Holiday Inn Express, Lee's Inn, Super 8, Chateau Thomas Winery
65mm	**rest area both lanes, full(handicapped)facilities, info, vending, phone, picnic tables, litter barrels, petwalk**
59	IN 39, to Belleville, **N**...Marathon/mart, Canary Motel, **S**...TA/76/diesel/rest./24hr, Blue&White/diesel/repair/rest.
51	rd 1100W, **S**...Phillips 66/diesel/rest./24hr, Koger's/diesel/mart, repair/24hr, camping
41	US 231, Cloverdale, to Greencastle, **N**...HOSPITAL, Long Branch Steaks, Midway Motel, **S**...Amoco/diesel/mart, Citgo/Subway/diesel/mart/24hr, Shell/mart/24hr, Arby's, Burger King, Chicago's Pizza, KFC, McDonald's, Taco Bell, Wendy's, Briana Inn, Day's Inn, $Inn, Holiday Inn Express, Ramada Inn, Big A Parts, Goodyear, bank, repair, to Lieber SRA
37	IN 243, to Putnamville, **S**...Marathon/diesel/mart, to Lieber SRA
23	IN 59, to Brazil, **N**...HOSPITAL, Williams/Main St Café/diesel/mart/24hr, **S**...Shell/Brazil 70/diesel/rest./24hr, Speedway/Subway/diesel/mart, Sunoco/Rally's/diesel/mart, Burger King, Howard Johnson Express
15mm	Honey Creek

E

W

Terre Haute

Indiana Interstate 70

11	IN 46, Terre Haute, **N**...Pilot/Arby's/diesel/mart/24hr, Thornton/diesel, Burger King, McDonald's, Goodyear, airport, **S**...KOA
7	US 41, US 150, Terre Haute, **N**...Amoco/mart, BigFoot/gas, Marathon/diesel/mart/atm, Thornton/diesel, Applebee's, Arby's, Bob Evans, Burger King, Cracker Barrel, Fazoli's, Hardee's, Little Caesar's, LoneStar Steaks, Pizza City, Pizza Hut, Shoney's, Steak'n Shake, Taco Casita, Texas Roadhouse, Tumbleweed Mesquite Grill, Comfort Suites, $Inn, Drury Inn, Fairfield Inn, Knight's Inn, PearTree Inn, Signature Inn, Super 8, Travelodge(2mi), AutoZone, Chrysler/Dodge/Jeep, Midas Muffler, S IN Tire, **S**...HOSPITAL, Jiffy/mart/24hr, Kocolene/diesel, Shell/mart, Speedway/mart, Arby's, Baskin-Robbins, ChiChi's, Denny's, Dog'n Suds, FoodCourt, Great China Buffet, Hardee's, KFC, Laughner's Cafeteria, LJ Silver, McDonald's, Olive Garden, Outback Steaks, Papa John's, Ponderosa, Rally's, Red Lobster, Ryan's, Subway, Taco Bell, Wendy's, Best Western, Holiday Inn, Motel 6, Aldi Foods, Books-A-Million, Chevrolet, CountyMkt Foods, Dodge, FashionBug, Goodyear/auto, HobbyLobby, JiffyLube, Jo-Ann Crafts, K-Mart/auto, Lowe's Bldg Supply, OfficeMax, Olds/GMC, Pontiac/Buick/Cadillac, Sam's Club, Sear/auto, Service Merchandise, Toyota, ToyRUs, bank, carwash
5.5mm	Wabash River
3	Darwin Rd, W Terre Haute, **N**...to St Mary of-the-Woods Coll
1.5mm	**Welcome Ctr eb, full(handicapped)facilities, info, picnic tables, litter barrels, phone, vending, petwalk**
1	US 40 E(from eb, exits left), W Terre Haute, to Terre Haute
0mm	Indiana/Illinois state line

Indiana Interstate 74

Exit #	Services
171.5mm	Indiana/Ohio state line
171mm	weigh sta wb
169	US 52 W, to Brookville, no facilities
168.5mm	Whitewater River
164	IN 1, St Leon, **N**...Citgo/rest., Exxon, Shell/diesel, Christina's Rest., Fox's Den Rest., **S**...BP/Blimpie/diesel/mart
156	IN 101, to Sunman, Milan, **N**...Polaris, **S**...Exxon/diesel/mart/atm, 1000Trails Camping
152mm	**rest area both lanes, full(handicapped)facilities, phone, picnic tables, litter barrels, vending, petwalk**
149	IN 229, Batesville, to Oldenburg, **N**...Shell/diesel/mart/24hr, China Wok, McDonald's, Subway, Wendy's, Hampton Inn, Kroger/deli, Pamida, Pennzoil, **S**...HOSPITAL, Amoco/mart, Arby's, Dairy Queen, KFC/Taco Bell, La Rosa's Pizza, Subway, Waffle House, Day's Inn, Salt Creek Inn, Sherman House Inn/rest., CVS Drug
143	to IN 46, New Point, **N**...Marathon/diesel/rest./24hr, camping, golf, **S**...Marathon/mart
134b a	IN 3, Greensburg, to Rushville, **S**...DENTIST, Bigfoot/mart, BP/diesel, Shell/mart/atm/24hr, Speedway, Arby's, Big Boy, Burger King, KFC, McDonald's, Papa John's, Subway, Taco Bell, Waffle House/24hr, Wendy's, Belter Motel, Best Western, Lee's Inn, Advance Parts, Chrysler, CVS Drug, FashionBug, Ford/Mercury, Goody's, JayC Foods, K-Mart, Radio Shack, Staples, Wal-Mart SuperCtr/24hr, bank, carwash
132	US 421(from eb), to Greensburg, no facilities
129.5mm	Clifty Creek
123	Saint Paul, **S**...BP/diesel, camping
119	IN 244 E, to Milroy, **S**...Compton's Cow Palace(1mi)
116	IN 44, to Shelbyville, Rushville, **N**...VETERINARIAN, Bigfoot/diesel/mart, **S**...HOSPITAL, CHIROPRACTOR, Amoco, Marathon/mart, Shell/mart/atm/24hr, Swifty/mart, Applebee's, Arby's, Baskin-Robbins/Dunkin Donuts, Bavarian Haus, Bob Evans, Burger King, Denny's, Domino's, Golden Corral(1mi), LJ Silver, McDonald's, New China, Papa John's, Pizza Hut, Subway, Taco Bell, Wendy's, Lee's Inn, Rasner Motel, Ace Hardware, Aldi Foods, Big Lots, Chevrolet/Olds/Nissan, CVS Drug, $General, Ford/Lincoln/Mercury, GNC, Goody's, IGA Foods, Kroger/deli, Marsh Food/drug, Midas Muffler, NAPA, Osco Drug, Radio Shack, Wal-Mart, carwash
115mm	Little Blue River
113mm	Big Blue River
113	IN 9, to Shelbyville, **N**...CF/diesel/mart, Cracker Barrel, **S**...HOSPITAL, Shell/mart/atm, Cow Palace(1mi), McDonald's, Mr T's Rest., Waffle Steak, Comfort Inn, Hampton Inn, Holiday Inn Express, Ramada Inn, Super 8, Buick/GMC, Chrysler/Plymouth, Ford/Lincoln/Mercury, bank
109	Fairland Rd, **S**...RV camping/funpark, Brownie's Marine
103	London Rd, to Boggstown, no facilities
102.5mm	Big Sugar Creek
101	Pleasant View Rd, **N**...Marathon/mart, Pleasant View 1Stop/gas/groceries/pizza, Golden Royal Boots/saddles, bank, tires
99	Acton Rd, no facilities
96	Post Rd, **N**...Marathon/Subway/diesel/mart/atm, McDonald's, RV/boat rental, **S**...Amoco/mart, Shell/diesel/mart, Chevrolet/GMC
94b a	I-465/I-74 W, I-465 N, US 421 N
	I-74 and I-465 run together 21 miles. **See Indiana Interstate 465, exits 2-16, and 52-53.**
73b	I-465 N, access to same facilities as 16a on I-465

Greensburg

E

W

Indiana Interstate 74

E

W

a	I-465 S, I-74 E, no facilities
71mm	Eagle Creek
66	IN 267, Brownsburg, **N**...Phillips 66/mart/24hr, Shell/mart/24hr, Hardee's, La Charreada Mexican, Pizza King, Holiday Inn Express, Big O Tires, GNC, JiffyLube, bank, cleaners, **S**...DENTIST, MEDICAL CARE, Amoco/diesel, Speedway/diesel/mart, Arby's, Blimpie, Bob Evans, Burger King, McDonald's, Noble Roman's, Papa John's, Taco Bell, Wendy's, White Castle, Comfort Suites, \$Inn, Ace Hardware, CVS Drug, Ford, Kroger/deli/24hr, Mailboxes Etc, Mears Auto, K-Mart, Radio Shack, Wal-Mart SuperCtr/24hr, bank
61	to Pittsboro, **S**...Hap's Place Rest./gas/diesel, Blue&White Service/diesel/24hr, tires
58	IN 39, Lizton, to Lebanon, **N**...Phillips 66/mart, **S**...HOSPITAL
57mm	**rest area both lanes, full(handicapped)facilities, phone, picnic tables, litter barrels, vending, petwalk**
52	IN 75, Jamestown, to Advance, **2 mi S**...gas, food, camping
39	IN 32, to Crawfordsville, **S**...antiques(1mi), to Wabash Coll
34	US 231, Crawfordsville, to Linden, **S**...HOSPITAL, Amoco/mart, Citgo/diesel/ rest./24hr, GasAmerica, Marathon, Shell, Burger King, KFC, McDonald's, Waffle House, Comfort Inn, Day's Inn, \$Inn, Holiday Inn, Super 8, KOA
25	IN 25, Waynetown, to Wingate, no facilities
23mm	**rest area both lanes, full(handicapped)facilities, phone, picnic tables, litter barrels, vending, petwalk**
19mm	weigh sta both lanes
15	US 41, Veedersburg, to Attica, **1/2 mi S**...Phillips 66/Subway/diesel/mart/atm/24hr, Noble Roman's, FastLube, to Turkey Run SP, camping
8	Covington, **N**...Shell/mart, Maple Corner Rest.(1mi), fireworks
7mm	Wabash River
4	IN 63, to Newport, **N**...Pilot/DQ/diesel/mart/24hr, Beefhouse Rest.
1mm	**Welcome Ctr eb, full(handicapped)facilities, info, phone, picnic tables, litter barrels, vending, petwalk**
0mm	Indiana/Illinois state line

Indiana Interstate 80/90

E

W

Exit #	Services
157mm	Indiana/Ohio state line
153mm	toll plaza, litter barrels
146mm	**Booth Tarkington Service Area both lanes, BP/diesel, Dairy Queen, McDonald's, playground**
144	I-69, US 27, Angola, Ft Wayne, **N**...Mobil/diesel/mart/LP/24hr, Shell/diesel/mart/24hr, Speedway/Hardee's/diesel/mart/24hr, Clay's Family Rest., Redwood Lodge, Lake George Inn, Volvo Trucks/repair, boots, tires, **S**...Marathon/Taco Bell/diesel/mart/24hr, Holiday Inn Express, Valley Outlet Ctr/famous brands, **services on IN 120 E**...DENTIST, Herb Garden Rest., E&L Motel, Hampton Inn, Super 8, Golf/rest., U-Haul/repair(1mi), **W**...to Pokagon SP, Jellystone Camping(7mi)
131.5mm	Fawn River
126mm	**Ernie Pyle Travel Plaza eb, Gene S Porter Travel Plaza wb, BP/diesel, Baskin-Robbins, Fazoli's, Hardee's, gifts, RV dump**
121	IN 9, Howe, to Lagrange, **N**...HOSPITAL(4mi), Golden Buddha, Greenbriar Inn, Hampton Inn, Travel Inn, **3 mi N**...Golden Corral, Dairy Queen, Maria's Mexican, Wendy's, Comfort Inn, Knight's Inn, Wood Motel, **S**...HOSPITAL(8mi), Amoco/mart(2mi), Holiday Inn Express, Patriot Inn(2mi), Super 8, Twin Mills Camping Resort(5mi on IN 120W)
120mm	Fawn River
107	US 131, IN 13, to Middlebury, Constantine, **N**...Marathon, Country Table Rest.(1mi), McDonald's(4mi), PatchWork Quilt Inn(1mi), Plaza Motel, Tower Motel, Dick's Auto Parts, **1 mi S**...BP/Blimpie/diesel/mart, Yup's DairyLand, Golf, KOA, **5 mi S**...Dairy Queen, Subway, Coachman RV Factory
101	IN 15, Bristol, to Goshen, **1 mi S**...MEDICAL CARE, Amoco/pizza/subs/mart/atm, River Inn Rest., Revco, Eby's Pines Camping(3mi), bank
96	rd 17, E Elkhart, **2 mi S**...Amoco/diesel/mart, Citgo/7-11, Dairy Queen, McDonald's/playplace
92	IN 19, to Elkhart, **N**...Citgo/7-11, Phillips 66/Lee's Chicken/Subway/pizza/mart, Applebee's, Cracker Barrel, Shoney's, Simonton Lake Drive-In, Steak'n Shake, Best Western, Comfort Inn, Diplomat Motel, Econolodge, Elkhart Inn, Hampton Inn, Knight's Inn, Turnpike Motel, Aldi Foods, CVS Drug, FashionBug, GNC, Goodwill, K-Mart, Martin's Foods, Radio Shack, Revco, Worldwide RV Ctr, bank, carwash, info, RV camping, **S**...HOSPITAL, Clark/mart, Citgo/7-11, Marathon/diesel/mart, Blimpie, Bob Evans, Burger King, Callahan's Rest., Da Vinci's Pizza, King Wha Chinese, McDonald's, Olive Garden, Perkins, Red Lobster, StarDust Café, Budget Inn, Day's

Indiana Interstate 80/90

Inn, Ramada Inn, Red Roof Inn, Signature Inn, Super 8, Westin Plaza, Holiday World RV. Honda Motorcycles, Menard's, Michana RV, Wal-Mart SuperCtr/24hr, carwash, cinema, truck/RV repair, **1 mi S**...Citgo/Lube, Shell, Speedway/mart, Swifty Gas, Dairy Queen, Gyro City, KFC, LJ Silver, Papa John's, Pizza Hut, Matterhorn Rest., Taco Bell, Wendy's, Sleepy Hollow Motel, Alick's Drug, AutoZone, Family$, Gordon Foods, JiffyLube, Osco Drug, TrueValue, Volvo Trucks, Walgreen, cleaners

<table>
<tr><td>91mm</td><td>Christiana Creek</td></tr>
<tr><td>90mm</td><td>Henry Schricker Travel Plaza eb, George Craig Travel Plaza wb, BP/diesel, Arby's, Dunkin Donuts, RV dump, travelstore, USPO</td></tr>
<tr><td>83</td><td>to Mishawaka, 2 mi N on IN 23W...Amoco/diesel/mart/wash, Citgo/7-11, Mobil/Subway/mart, Phillips 66/mart, Arby's, Chinese Rest., Colorado Steaks, Garfield's Rest., Olive Garden, Pizz Hut, Taco Bell, Wendy's, Carlton Lodge, Fairfield Inn, Hampton Inn, Holiday Inn Express, Super 8, Best Buy, Kroger, JC Penney, Marshall Field's, Menard's, Michael's, Office Depot, Pearle Vision, Sears, Service Merchandise, ValueCity, Walgreen, KOA, cinema, mall, 2 mi S on Grape Rd(off IN 23W)...Meijer/diesel/24hr, Burger King, Chili's, LoneStar Steaks, Old Country Buffet, Papa Vino's Italian, Ryan's, Steak'n Shake, TGIFriday, Courtyard, Studio+, Barnes&Noble, Buick/GMC, Chrysler/Plymouth, Circuit City, Discount Tire, Frank's Nursery/crafts, Jo-Ann Fabrics, Lowe's Bldg Supply, MediaPlay, OfficeMax, Sam's Club, ValueCity, VisionCtr, Waccamaw</td></tr>
<tr><td>77</td><td>US 33, US 31bus, South Bend, N...Mobil, Phillips 66, Speedway/diesel/mart, Arby's, Baskin-Robbins/Dunkin Donuts, Bonnie Doon Drive-In, Burger King, Clock Rest./24hr, Dairy Queen, Damon's, Denny's, Fazoli's, KFC, Marco's Pizza, McDonald's, Papa John's, Pizza Hut, Ponderosa, Scharff's Deli, Steak&Ale, Subway, Taco Bell, Day's Inn, Hampton Inn, Motel 6, Ramada Inn, Super 8, BMW/Isuzu/Mazda, Pennzoil, Radio Shack, TrueValue, Walgreen, antiques, bank, 1 mi N on frtge rd...Meijer/diesel/24hr, Phillips 66, Burger King, S...HOSPITAL, VETERINARIAN, Amoco/wash/atm/24hr, Phillips 66/diesel/mart/pizza, Bennitt's Buffet, Bill Knapp's, Bob Evans, Great Wall Chinese, Colonial Pancakes, King Gyro's, Perkins, Pizza King, Raphael's Mexican, Schlotsky's, Wendy's, Best Inn, Holiday Inn/rest., Howard Johnson, Signature Inn, St Marys Inn, CarX Mufflers, JiffyLube, cleaners, to Notre Dame</td></tr>
<tr><td>76mm</td><td>St Joseph River</td></tr>
<tr><td>72</td><td>US 31, to Niles, South Bend, N...Speedway/Hardee's/diesel/mart/24hr, 2 mi S on US 20...4 Seasons Rest., McDonald's, Ponderosa, Taco Bell, Wendy's, Econolodge, RV Ctr, to Potato Creek SP(20mi), airport, st police</td></tr>
<tr><td>62mm</td><td>eastern/central time zone line</td></tr>
<tr><td>56mm</td><td>service area both lanes, BP/diesel, Dairy Queen, McDonald's, cookies/ice cream, phone, RV dump, litter barrel</td></tr>
<tr><td>49</td><td>IN 39, to La Porte, S...Cassidy Motel/RV park, 4mi S...Mobil/diesel, Blue Heron Motel, Holiday Inn Express, Super 8</td></tr>
<tr><td>39</td><td>US 421, to Michigan City, Westville, 5 mi N...Hampton Inn, Holiday Inn, Knight's Inn, S...Purdue U North Cent</td></tr>
<tr><td>38mm</td><td>rest area wb, truckers only, litter barrels, portable toilets</td></tr>
<tr><td>31</td><td>IN 49, to Chesterton, Valparaiso, N...to IN Dunes Nat Lakeshore, S...Yellow Brick Rd Museum/gifts</td></tr>
<tr><td>24mm</td><td>toll plaza</td></tr>
<tr><td>23</td><td>Portage, Port of Indiana, N...Lee's Inn, S...CHIROPRACTOR, VETERINARIAN, Amoco, Marathon/mart, Burger King, Dunkin Donuts, KFC, McDonald's, Subway, Wendy's, Family$, Midas, Walgreen/24hr, Town&Country Mkt/24hr, USPO, bank, carwash, cinema, hardware</td></tr>
<tr><td>22mm</td><td>service area both lanes, info, BP/diesel, Baskin-Robbins, Fazoli's, Hardee's, playground</td></tr>
<tr><td>21mm</td><td>I-90 and I-80 run together eb, separate wb. I-80 runs with I-94 wb. For I-80 exits 1 through 15, see Indiana Interstate 94.</td></tr>
<tr><td>21</td><td>I-94 E to Detroit, I-80/94 W, US 6, IN 51, Lake Station, N...Citgo, Dunes Trkstp/diesel/repair/24hr, E-Z Go/mart, Burger King, McDonald's, Chinese Rest., Ponderosa, Aldi Foods, K-Mart/auto, JiffyLube, Buick, Chrysler/Jeep/Eagle, bank S...Petro/Mobil/diesel/rest./24hr, Phillips 66/Subway/diesel/rest., Shell/diesel/mart, Speedway/diesel/mart, LJ Silver, McDonald's, Red Baron, Blue Beacon, cleaners, hardware</td></tr>
<tr><td>17</td><td>I-65 S, US 12, US 20, Dunes Hwy, to Indianapolis, no facilities</td></tr>
<tr><td>15</td><td>IN 53, Broadway, to Gary, no facilities</td></tr>
<tr><td>13</td><td>Grant St, to Gary, no facilities</td></tr>
<tr><td>10</td><td>IN 912, Cline Ave, to Gary, N...casino</td></tr>
<tr><td>5</td><td>US 41, Calumet Ave, to Hammond, S...Amoco, Arby's, Aurelio's Pizza, KFC, McDonald's, American Inn, Ramada Inn, Super 8, AutoZone, tires</td></tr>
<tr><td>3</td><td>IN 912, Cline Ave, to Hammond, to Gary Reg Airport, S...Shell/mart</td></tr>
<tr><td>1mm</td><td>toll plaza</td></tr>
<tr><td>1mm</td><td>US 12, US 20, 106th St, Indianapolis Blvd, N...Amoco/diesel/mart, Shell/diesel/mart, S...Shell, Burger King, Giappo's Pizza, KFC, McDonald's, Jewel-Osco, Payless Muffler/brake</td></tr>
<tr><td>0mm</td><td>Indiana/Illinois state line</td></tr>
</table>

E

W

South Bend

Indiana Interstate 94

Exit #	Services

46mm Indiana/Michigan state line

43mm **Welcome Ctr wb, full(handicapped)facilities, info, phone, picnic tables, litter barrels, vending, petwalk**

40b a US 20, US 35, to Michigan City, **S**...Amoco/diesel/mart

34b a US 421, to Michigan City, **N**...HOSPITAL, Amoco/WhiteCastle/mart, Clark Gas, Meijer/diesel/mart/24hr, Mobil/NobleRoman's/diesel/mart, Shell/mart, Speedway/diesel/mart/24hr, Applebee's, Arby's, Baskin-Robbins, Bob Evans, Burger King, Chili's, Denny's, Gordon Food Service, KFC, King Gyro's, McDonald's, Pizza Hut, Red Lobster, Schoot's Rest., Steak'n Shake, Subway, Taco Bell, Wendy's, City Manor Motel, Comfort Inn, $Inn, Hampton Inn, Holiday Inn, Knight's Inn, Red Roof Inn, Super 8, Travel Inn(2mi), Aldi Foods, Big Lots, Fannie Mae Candies, JC Penney, Jo-Ann Fabrics, K-Mart, Lowe's Bldg Supply, OfficeMax, Parts+, ToysRUs, Wal-Mart/auto, antiques, cinema, **S**...Harley-Davidson, camping

29mm weigh sta both lanes

26b a IN 49, Chesterton, **N**...to IN Dunes SP, **S**...MEDICAL CARE, CHIROPRACTOR, EYECARE, Amoco/Blimpie/mart, Mobil/mart/wash, Shell/mart/24hr, Arby's, Burger King, Dunkin Donuts, Gelsosomo's Pizza, KFC, Little Caesar's, LJ Silver, McDonald's, Pizza Hut, Subway, Taco Bell, Wendy's, Econolodge, Super 8, Buick/Pontiac, Jewel-Osco, K-Mart/auto, OilXpress, Walgreen, bank, laundry/cleaners, to Valparaiso

22b a US 20, Burns Harbor, **N**...Mobil/Subway/diesel/rest./24hr, Shell/diesel/mart/24hr, Steel City Express/mart, Blue Beacon, **S**...Camp-Land RV Ctr, Chevrolet, Chrysler/Dodge/Jeep, Ford/Mercury, Toyota, fireworks

19 IN 249, Portage, to Port of IN, **S**...Amoco/mart/24hr, Shell/mart/wash24hr, Denny's, McDuffy's Rest., Day's Inn, $Inn, Hampton Inn, Lee's Inn, Motel 6, Ramada Inn, Super 8, **2 mi S**...Burger King, Subway, Wendy's

16 access to I-80/90 toll road E, I-90 toll road W, IN 51N, Ripley St

I-94/I-80 run together wb

15b US 6W, IN 51, **N**...Petro/Mobil/Blimpie/McDonald's/diesel/24hr, Phillips 66/Subway/diesel/24hr, Speedway/diesel/mart/wash, Ponderosa, Red Baron Wash, **N on US 20**...Dunes/diesel repair, Marathon/mart, Burger King, Chinese Rest., Buick/Subaru/Mr Goodwrench, Jeep, JiffyLube

a US 6E, IN 51S, to US 20, **S**...Marathon/diesel/mart, Speedway/diesel/mart, Shell/mart/wash, Burger King/playplace, Dairy Queen, Reuben's Rest., Snak-Time Rest., cleaners, **S on Central Ave**...LJ Silver, Papa John's, Wendy's, transmissions

13 Central Ave(from eb), no facilities

12b I-65 N, to Gary and toll road

a I-65 S(from wb), to Indianapolis

11 I-65 S(from eb)

10b a IN 53, Broadway, **N**...Amoco/mart/24hr, BBQ, carwash, **S**...Amoco/mart/24hr, Shell/mart, Dairy Queen, Rally's, cleaners

9b a Grant St, **N**...Amoco, Pennzoil, Walgreen, laundry, **S**...Shell/mart, Flying J/Conoco/diesel/LP/rest./24hr, Steel City/diesel/rest./24hr, Burger King, Church's Chicken, Dairy Queen, Dunkin Donuts, KFC, Little Caesar's, McDonald's, Subway, Aldi Foods, CarX Muffler, Firestone/auto, Ford, Goodyear/auto, JC Penney, Midas Muffler, cleaners, drugs, parts

6 Burr St, **N**...Amoco/mart/24hr, Pilot/Subway/diesel/mart/24hr, TA/BP/KFC/Taco Bell/diesel/mart, Rico's Pizza, hardware, **S**...Shell/diesel/mart/24hr

5b a IN 912, Cline Ave, **1 mi S**...MEDICAL CARE, Amoco, Clark Gas, Mobil/diesel/mart/wash, Shell/mart/wash, Speedway/mart, Arby's, Bob Evans, Burger King, Dairy Queen, Denny's, Hardee's, McDonald's, Pizza Hut, Popeye's, Subway, Taco Bell, Wendy's, White Castle, Holiday Inn, Motel 6, Super 8, JiffyLube, Jo-Ann Fabrics, Service Merchandise, Venture, cinemas, **1 mi N off 169th St**...HOSPITAL, Amoco, McDonald's, to airport

3b a Kennedy Ave, **N**...HOSPITAL, Citgo, Clark Gas, Martin/mart, Shell, Burger King, McDonald's, **S**...Marathon, Shell, Speedway/mart, Cracker Barrel, Dairy Queen, Fran's Diner, Wendy's, Courtyard, Fairfield Inn, Residence Inn, USPO, cleaners/laundry, mufflers/brakes, tires

2b a US 41S, IN 152N, Indianapolis Blvd, **N**...CHIROPRACTOR, DENTIST, MEDICAL CARE, VETERINARIAN, Amoco, Phillips 66, Shell, Sav-A-Stop/mart, GasCtr/mart/wash, Arby's, Domino's, Dunkin Donuts, The Wheel Rest., Woodmar Rest., CarX Muffler, Midas Muffler, Apex Muffler/brakes, **S**...Martin/diesel/mart, Denny's, Burger King, AmeriHost, Ramada Inn/cafe, K-Mart/LittleCaesar's/auto/drugs, **1/2 mi S**...CHIROPRACTOR, Thornton's/mart, Brown's Chicken/pasta, Taco Bell, Blue Top Drive-In, Purple Steer Rest., Top Notch Rest., Aldi Foods, JiffyLube, Marshall's, Quality Brakes/mufflers, parts

1b a US 41N, Calumet Ave, **N**...Amoco, Clark Gas, GasCtr/Subway/diesel, Gas City/mart, Dunkin Donuts, Rick's Grill, Precision Tune, Dodge, laundry, **S**...CHIROPRACTOR, MEDICAL CARE, VETERINARIAN, Amoco/mart/24hr, Gas City/mart, Marathon, Shell, Arby's, Arnie's Doghouse, Burger King, Casa Del Rio Mexican, Edwardo's Pizza, Mister Donut, Taco Bell, Chinese Rest., Sterk's Foods, bank, cleaners

0mm Indiana/Illinois state line

E

W

Gary Area

Indiana Interstate 465(Indianapolis)

Exit #	Services

I-465 loops around Indianapolis. Exit numbers begin/end on I-65, exit 108.

53b a I-65 N to Indianapolis, S to Louisville

52 Emerson Ave, **N**...HOSPITAL, CHIROPRACTOR, DENTIST, Amoco, Clark Gas, Marathon, Shell/mart/wash/24hr, Speedway/mart, Burger King, KFC, LJ Silver, Noble Roman's, Subway, Taco Bell, Wendy's, Motel 6, CVS Drug, $General, Kroger/deli, cleaners, radiators, tires, **S**...VETERINARIAN, CHIROPRACTOR, Citgo/mart, Shell/repair/mart, Speedway/mart, Arby's, Bamboo House, Dairy Queen, Donato's Pizza, Egg Roll, Fazoli's, Hardee's, Hunan House, McDonald's, Papa John's, Pizza Hut, Ponderosa, Steak'n Shake, Subway, Waffle House, White Castle/24hr, Holiday Inn, Radio Shack, Ramada Inn, Red Roof Inn, Super 8, AutoZone, CarX Muffler, Goodyear/auto, IndyLube, JiffyLube, K-Mart, MailBoxes Etc, Marsh Foods, Osco Drug, Walgreen, bank, laundry

I-74 W and I-465 S run together around S Indianapolis 21 miles

49 I-74 E, US 421 S, no facilities

47 US 52 E, no facilities

46 US 40, Washington St, **E**...VETERINARIAN, Amoco, Crystal Flash/diesel, Gas America/mart, Marathon/mart, Phillips 66, Shell, Arby's, LJ Silver, Shell's Seafood, Steak'n Shake, White Castle, All Tune&Lube, Ford, JiffyLube, Meineke, Osco Drug, Target, **W**...Amoco, Applebee's, Burger King, Bob Evans, ChiChi's, Don Pablo's, Fazoli's, Hardee's, Markpi's Express Chinese, McDonald's, Pizza Hut, Rax, Shoney's, Steak&Waffle, Waffle House, Wendy's, Signature Inn, Cub Foods/24hr, K-mart, PepBoys, Pontiac/GM/Mazda, Suzuki, Thornton's/foodmart, Wholesale Club, car wash, manufacturers outlet

44b I-70 E, to Columbus

a I-70 W, to Indianapolis

42 IN 67, US 36, Pendleton Pike, **E**...Amoco/mart/wash, Crystal Flash/diesel, Burger King, Pancake House, Days Inn, Econolodge, Ramada, Sheraton, **W**...Clark Gas, Speedway, Arby's, Chinese Rest., Denny's, McDonald's, Taco Bell, White Castle, HiTech Tune-up, K-Mart, Meineke Muffler, Nissan, Subaru, fairgrounds

40 56th St, Shadeland Ave, **E**...Marathon/mart, to Ft Harrison

37b a I-69, N to Ft Wayne, IN 37, **W**...HOSPITAL, facilities on frontage rds

35 Allisonville Rd, **N**...Hardee's, McDonald's, MCL Cafeteria, Courtyard, Best Buy, CompUSA, Firestone/auto, JC Penney, Jo-Ann Crafts, Lazarus, Linens'n Things, Osco Drug, Pier 1, Service Merchandise, ToysRUs, Waccamaw, mall, **S**...CHIROPRACTOR, Shell, Speedway/diesel/mart, Bob Evans, ChuckeCheese, Hop's Brewery, Papa John's, Perkins/24hr, White Castle, Signature Inn, Circuit City, JiffyLube, Kroger, Precision Tune, **S on 82nd**...PrimeTime Grill, Marsh Foods, PetsMart

33 IN 431, Keystone Ave, **N**...Amoco/McDonald's, Marathon/diesel/mart, Shell, Arby's, Bob Evans, Burger King, Steak&Ale, Subway, BMW, Chevrolet/Isuzu, Infiniti, Nissan, Toyota **S**...Cooker, Keystone Grill, AmeriSuites, Marriott, Sheraton, Westin Suites, Champ's, GateWay Country, Kohl's, bank, mall, **1 mi S on 82nd**...Applebee's, Boston Mkt, Chili's, Chinese Rest., Don Pablo's, O' Charley's, Logan's Roadhouse, LoneStar Steaks, Pizzeria Uno, Prime Time Amercan Grill, Schlotsky's, Shell's Seafood, Subway, Barnes&Noble, CopyMax, Office Max, cinema

31 US 31, Meridian St, **N**...HOSPITAL, Signature Inn, Wyndham Garden, **S**...Marathon, Shell/diesel/mart, McDonald's

27 US 421 N, Michigan Rd, **N**...DENTIST, Dairy Queen, McDonald's, Red Roof Inn, AutoNation USA, Chevrolet, **S**...Amoco, Marathon/mart, Shell/mart/wash, Sunoco/mart, Arby's, Bob Evans, ChiChi's, Denny's, Fortune House Chinese, IHOP, Max&Erma's, O'Charley's, Ruby Tuesday, Steak'n Shake, White Castle, Wildcat Brewing Co/rest., Yen Ching Rest., Comfort Inn, $Inn, Drury Inn, Embassy Suites, Extended Stay America, Fairfield Inn, Holiday Inn Select, HomeGate Inn, Quality Inn, Residence Inn, Signature Inn, Studio Suites, Aamco, Cub Foods, CVS Drug, Discount Tire, JiffyLube, K-Mart/Little Caesar's, Lowe's Bldg Supply, Meineke Muffler, Midas Muffler, OfficeMax, PepBoys, Radio Shack, Wal-Mart, cleaners

25 I-465 W, no facilities

23 86th St, **E**...HOSPITAL, Shell/mart/wash/24hr, Speedway/mart, Arby's, Burger King, Dog'n Suds, Quizno's, MainStay Suites, Suburban Lodge, bank

21 71st St, **E**...DENTIST, Amoco, Galahad's Café, Hardee's, McDonald's, Steak'n Shake, Subway, Clarion Inn, Courtyard, Hampton Inn, cleaners

20 I-65, N to Chicago, S to Indianapolis

19 56th St(from nb), no facilities

17 38th St, **E**...CHIROPRACTOR, Amoco, Marathon/mart, Shell/mart/24hr, ChiChi's, China Chef, Dairy Queen, Olive Garden, Subway, Day's Inn, Aamco, Circuit City, Kroger, Osco Drug, **W**...Arby's, Burger King, Chili's, Cracker Barrel, Don Pablo's, McDonald's, Mtn Jack's Rest., Ruby Tueday, Taco Bell, TGIFriday, Country Hearth Inn, Signature Inn, Marsh Foods, Target, cinema

I-74 E and I-465 run together around S Indianapolis 21 miles

16b I-74 W, to Peoria

Indianapolis Area

Indiana Interstate 465

a   US 136, to Speedway, **N**...Amoco/mart, Bigfoot/gas, Shell/mart, Thornton's/gas/24hr, Applebee's, Blimpie, Burger King, Denny's, Dunkin Donuts, Grindstone Charlie's, Hardee's, KFC, LJ Silver, McDonald's, Rally's, Subway, Taco Bell, Wendy's, White Castle, $Inn, Motel 6, Red Roof, Super 8, FashionBug, Firestone, Goodyear, JiffyLube, Kohl's, Kroger/drug, Marsh Foods, MensWearhouse, Old Navy, Sears, Speedway Museum, Waxwerks, cinema, **S**...Best Western

14b a   10th St, **N**...HOSPITAL, DENTIST, CHIROPRACTOR, Bigfoot/gas, Marathon, China Wok, Pizza Hut, Wendy's, Cub Food/24hr, Lowe's Bldg Supply, **S**...CHIROPRACTOR, Amoco/mart, GasAmerica/Subway/mart, Shell/mart, Speedway/mart/24hr, Arby's, Fazoli's, Hardee's, Little Caesar's, McDonald's, Noble Roman's, Pizza Hut, Rally's, Taco Bell, CVS Drug, $General, JiffyLube, bank

13b a   US 36, Rockville Rd, **N**...Shell/Burger King/mart, Comfort Inn, Sleep Inn, Sam's Club, **S**...CHIROPRACTOR, DENTIST, Speedway/mart/24hr, Bob Evans, IndyLube, laundry

12b a   US 40 E, Washington St, **N**...Amoco, Burger King, Church's Chicken, Fazoli's, McDonald's, Papa John's, Rally's, Taco Bell, White Castle, Ace Hardware, Family$, Kroger, Osco Drug, Speedway Auto Parts, U-Haul, Walgreen, cleaners, **S**...Bigfoot/gas/24hr, Phillips 66/Noble Roman's/TCBY/diesel/mart, Shell, Thornton/gas/24hr, Arby's, Burger King, Dunkin Donuts, Hardee's, KFC, LJ Silver, McDonald's, Omelet Shoppe, Pizza Hut, Steak'n Shake, Wendy's, $Inn, Aamco, CarX Muffler, Goodyear, JiffyLube, K-Mart, Midas Muffler, Radio Shack, Target, TireBarn, carwash

11b a   Airport Expressway, **N**...Adam's Mark Hotel, Baymont Inn, Day's Inn, Extended Stay America, Motel 6, **S**...Hilton, Holiday Inn, Ramada Inn, airport

9b a   I-70, E to Indianapolis, W to Terre Haute

8   IN 67 S, Kentucky Ave, **N**...HOSPITAL, **S**...Amoco/diesel, Marathon, Speedway/mart, Shell, Big Boy, Hardee's, KFC

7   Mann Rd(from wb), **N**...HOSPITAL

4   IN 37 S, Harding St, **N**...Mr Fuel/diesel/mart, Pilot/Wendy's/DQ/diesel/24hr, Omelet Shoppe, $Inn, Econolodge/rest., Super 8, Blue Beacon, **S**...Flying J/Conoco/diesel/LP/rest./24hr, Marathon/mart, Hardee's, McDonald's, Taco Bell, Waffle&Steak, White Castle, Knight's Inn, Freightliner, Speedco Tires

2b a   US 31, IN 37, **N**...EYECARE, Amoco/mart/24hr, Arby's, Burger King, Dutch Oven, Golden Wok, Hardee's, KFC, King Ribs, J's Burger, Laughner's Cafeteria, LJ Silver, McDonald's, MCL Cafeteria, Old Country Buffet, Papa John's, Pizza Hut, Ponderosa, Rally's, Steak&Ale, Steak'n Shake, Taco Bell, Wendy's, White Castle, Aldi Foods, AutoTire, AutoWorks, AutoZone, CarX Muffler, Chrysler/Jeep, CVS Drug, Dodge, $General, Family$, Firestone, Ford, Goodyear, Hancock Fabrics, JiffyLube, Jo-Ann Fabrics, Kroger/deli/24hr, Lincoln/Mercury, Marsh Foods, Meineke Muffler, Midas Muffler, Office Depot, Osco Drug, Radio Shack, Save-A-Lot Food, Target, U-Haul, bank, laundry, **S**...VETERINARIAN, Bigfoot Gas, Shell/mart/wash, Sunoco/Subway/mart, Bob Evans, Denny's, HH Smorgasbord, McDonald's, Red Lobster, Wendy's, Best Inn, Comfort Inn, Day's Inn, Holiday Inn Express, Quality Inn, Ramada Ltd, Red Roof Inn, IndyLube

53b a   I-65 N to Indianapolis, S to Louisville

I-465 loops around Indianapolis. Exit numbers begin/end on I-65, exit 108.

Indiana Interstate 469(Ft Wayne)

Exit #	Services
31c b a	I-69, US 27 S. I-469 begins/ends.
29.5mm	St Joseph River
29b a	Maplecrest Rd, no facilities
25	IN 37, to Ft Wayne, **W**...Meijer/diesel/24hr, Steak'n Shake
21	US 24 E, no facilities
19b a	US 30 E, to Ft Wayne, **E**...Wendy's, **W**...Amoco, Arby's(2mi), KFC(1mi)
17	Minnich Rd, no facilities
15	Tillman Rd, no facilities
13	Marion Center Rd, no facilities
11	US 27, US 33 S, Ft Wayne, to Decatur, no facilities
10.5mm	St Marys River
9	Winchester Rd, no facilities
6	IN 1, Ft Wayne, to Bluffton, **N**...to airport
2	Indianapolis Rd, **N**...to airport
1	Lafayette Ctr Rd
0b a	I-69, US 24 W, US 33 N, to Ft Wayne, Indianapolis. I-469 begins/ends.

N

S

Iowa Interstate 29

Exit #	Services

Sioux City

152mm Big Sioux River, Iowa/South Dakota state line

151 IA 12 N, Riverside Blvd, **E**...to Stone SP, Pecaut Nature Ctr

149 Hamilton Blvd, **E**...Conoco/mart, Horizon Rest., Hamilton Inn, JiffyLube, to Briar Cliff Coll, **W...Iowa Welcome Ctr sb, full facilities,** Riverboat Museum

148 US 77 S, to S Sioux City, Nebraska, **W**...Conoco/diesel/mart, Phillips 66/mart, McDonald's, Taco Bell, Valentino's Rest., Travelodge, Ford

147b US 20 bus, Sioux City, **E**...HOSPITAL, Amoco, Holiday/mart, Arby's, Burger King, Hardee's, KFC, Perkins/24hr, Best Western, Hilton Inn, Imperial Motel, Riverboat Inn, Chevrolet, JC Penney, Midas Muffler, Speedy Lube, Staples, USPO, Walgreen, bank, **W**...Burger King, Hardee's, Super 8

  a Floyd Blvd, **E**...Total/mart, to Riverboat Casino, stockyards

146.5mm Floyd River

144 b I-129 W, US 20 W, US 75 S, no facilities

  a US 20 E, to Ft Dodge, **1 mi E on Lakeport Rd**...Casey's Store, Texaco/mart, Applebee's, Burger King, Hardee's, Hunan Palace, KFC, LJ Silver, Outback Steaks, Pizza Hut, Red Lobster, Sirloin Stockade, Comfort Inn, Fairfield Inn, Holiday Inn Express, Buick, Cub Foods/24hr, Honda, Hy-Vee Foods/24hr, JiffyLube, Jo-Ann Fabrics/crafts, MailBoxes Etc, Michael's, Pier 1, Sears/auto, Target, TJ Maxx, Toys R Us, Younkers, cinema, golf outlet, mall

143 US 75 N, Industrial Rd, **E**...Cenex/mart, Texaco/diesel/motel/café/24hr, McDonald's, AmericInn, Baymont Inn, Day's Inn, Sam's Club, Sgt Floyd Mon, **W**...Amoco/diesel/motel/café/24hr, Wendy's, Kenworth/Peterbilt, CB Shop, truckwash

141 D38, Sioux Gateway Airport, **E**...CHIROPRACTOR, DENTIST, Conoco, Phillips 66/mart, Texaco/mart, Godfather's, Subway, Rath Inn, bank, carwash, cleaners, mall, **W**...Amoco/diesel/mart, Motel 6, museum

139mm **rest area both lanes, full(handicapped)facilities, phone, info, picnic tables, litter barrels, RV dump**

135 Port Neal Landing, no facilities

134 Salix, **E**...Total/mart

132mm weigh sta both lanes

127 IA 141, Sloan, **E**...Amoco/mart, Texaco/diesel/mart, Homestead Inn, WinnaVegas Inn, **3 mi W**...to Winnebago Indian Res/casino

120 to Whiting, **W**...camping

112 IA 175, Onawa, **E**...HOSPITAL, Conoco/Subway/diesel/mart/atm, Phillips 66/mart, McDonald's, Michael's Rest., Oehler Bros Rest., Super 8, Buick/GMC, Country General, Interchange RV Camp, Pamida, antiques, carwash, **2 mi W**...Lewis&Clark SP, Keelboat Exhibit

110mm **rest area both lanes, full(handicapped)facilities, phone, info, picnic tables, litter barrels, petwalk, RV dump**

105 E60, Blencoe, **1/2 mi E**...TR's Conv/gas

96mm Little Sioux River

95 IA 301, Little Sioux, **E**...G&N OneStop/gas, **W**...Woodland RV Camp

92mm Soldier River

91.5mm **rest area both lanes, litter barrels, no facilities**

89 IA 127, Mondamin, **1 mi E**...Jiffy Mart/gas/diesel

82 IA 300, F50, Modale, **1 mi W**...Cenex

80mm **rest area sb, full(handicapped)facilities, info, phone, picnic tables, litter barrels, RV dump**

78.5mm **rest area nb, full(handicapped)facilities, info, phone, picnic tables, litter barrels, RV dump**

Council Bluffs

75 US 30, Missouri Valley, **E...Iowa Welcome Ctr(5mi)**, HOSPITAL(2mi), Phillips 66/mart, Texaco/FoodCourt/diesel/24hr, McDonald's, Subway, Hillside Motel(2mi), Ford/Mercury, to Steamboat Exibit, **W**...Amoco/diesel/mart/atm, Conoco/Kopper Kettle/mart, Burger King, Oehler Bros Café, Day's Inn, Rath Inn, Super 8, Chevrolet/Pontiac/Olds/Buick, Chrysler/Jeep/Dodge

74mm weigh sta sb

73mm weigh sta nb

72.5mm Boyer River

72 IA 362, Loveland, **E**...Conoco/diesel/mart, **W**...to Wilson Island SP(6mi)

71 I-680 E, to Des Moines, no facilities, **I-29 S & I-680 W run together 10 mi**

66 Honey Creek, **W**...Phillips 66/diesel/mart/LP/Iowa Feed&Grain Co Rest., camping

61b I-680 W, to N Omaha, **I-29 N & I-680 E run together 10 mi, W**...Mormon Trail Ctr

  a IA 988, to Crescent, **E**...Phillips 66, to ski area

56 IA 192 S(sb only, exits left), Council Bluffs, **E**...HOSPITAL, Drog Inn, 1/2 Price Store

55 N 25th, Council Bluffs, **E**...Sinclair/mart/24hr, Travelodge, carwash

54b N 35th St(from nb), Council Bluffs, **E on Broadway**...Amoco, Arby's, Burger King, Wendy's, Best Western, Honda

  a G Ave(from sb), Council Bluffs, **E**...Texaco/mart, **W**...auto repair

53b I-480 W, US 6, to Omaha, no facilities

  a 9th Ave, S 37th Ave, Council Bluffs, **E**...Amoco/mart/24hr, Conoco/mart, Phillips 66/mart, Texaco/atm, Total/mart, Day's Inn, **W**...Harvey's Casino/hotel, RiverBoat Casino, camping

Iowa Interstate 29

52	Nebraska Ave, **E...**Conoco/diesel/mart, Comfort Suites, dogtrack, **W...**AmeriStar Hotel/casino, Holiday Inn, RiverBoat Casino
51	I-80 W, to Omaha, no facilities
	I-29 and I-80 run together 3 miles. See Iowa Interstate 80, exits 1b-3.
48	I-80 E(from nb), to Des Moines, HOSPITAL
47	US 275, IA 92, Lake Manawa, **E...**Phillips 66, Iowa School for the Deaf
42	IA 370, to Bellevue, **W...**K&B Saddlery, to Offutt AFB, camping
38mm	**rest area both lanes, full(handicapped)facilities, phone, info, picnic tables, litter barrels, RV dump, petwalk**
35	US 34 E, to Glenwood, **E...**Western Inn(4mi), **W...**Amoco/diesel/café, Bluff View Motel, Ford
32	US 34 W, Pacific Jct, to Plattsmouth, no facilities
24	L31, Bartlett, to Tabor, **1 mi E...**food
20	IA 145, Thurman, no facilities
15	J26, Percival, **1-2 mi E...**gas/diesel/mart
11.5mm	weigh sta nb
10	IA 2, to Nebraska City, Sidney, **E...**to Waubonsie SP(5mi), **W...**Conoco/diesel/mart, Phillips 66/diesel/rest./24hr, Texaco/diesel/mart/atm/24hr, **6 mi W...**McDonald's, to Arbor Lodge SP, Nebraska Crossing Outlet(10mi), camping(4mi)
1	IA 333, Hamburg, **1 mi E...**HOSPITAL, Casey's/gas, Pizza Hut, Hamburg Motel, FoxLake Food
0mm	Iowa/Missouri state line

Iowa Interstate 35

Exit #	Services
219mm	Iowa/Minnesota state line
214	rd 105, to Northwood, Lake Mills, **W...Welcome Ctr both lanes, full(handicapped)facilities, picnic tables, litter barrels, vending, petwalk, RV dump,** Cenex/Burger King/diesel/mart/24hr
212mm	weigh sta both lanes
208	rd A38, to Joice, Kensett, no facilities
203	IA 9, to Manly, Forest City, **W...**HOSPITAL, Amoco/diesel/mart, to Pilot Knob SP
202mm	Winnebago River
197	rd B20, **8 mi E...**Lime Creek Nature Ctr
194	US 18, Clear Lake, to Mason City, **E...**HOSPITAL(8mi), Coastal/diesel/deli/mart, **W...**Casey's/gas, Conoco/diesel/mart, Kum&Go/gas, Bennigan's, Burger King, Country Kitchen, Dairy Queen, Denny's, KFC, McDonald's, Perkins/24hr, Pizza Hut, Subway, Taco John's, AmericInn, Best Western/rest., Budget Inn, Lake Country Inn, Microtel, airport
193	rd B35, Emery, to Mason City, **E...**Amoco/diesel/mart/24hr, Happy Chef/24hr, Super 8, **W...**Phillips 66/mart, Heartland Inn(3mi), Chevrolet, Ford, antiques, to Clear Lake SP
190	to Mason City, no facilities
188	rd B43, to Burchinal, no facilities
182	rd B60, Swaledale, to Rockwell, no facilities
180	rd B65, to Thornton, **2 mi W...**Cenex, camping
176	rd C13, to Sheffield, Belmond, no facilities
170	rd C25, to Alexander, no facilities
165	IA 3, to Hampton, Clarion, **E...**HOSPITAL(7mi), Texaco/diesel/rest., AmericInn, Hampton Motel
159	rd C47, Dows, **2 mi W...**IA Welcome Ctr
155mm	Iowa River
151	rd R75, to Woolstock, no facilities
147	rd D20, to US 20 E, no facilities
144	US 20, Williams, **E...**Phillips 66/diesel/Boondocks Motel, TH Café, Best Western, RV camping, **W...**Amoco/diesel/mart/24hr
142b a	US 20, to Webster City, Ft Dodge, no facilities
139	rd D41, to Kamrar, no facilities
133	IA 175, Ellsworth, to Jewell, **W...**Cenex/mart, Kum&Go/gas
128	rd D65, Randall, to Stanhope, **5 mi W...**Little Wall Lake Pk

Iowa Interstate 35

Ames

124	rd 115, Story City, **E…**Pella Windows, **W…** Kum&Go/diesel/mart/24hr, Texaco/mart, Dairy Queen, Godfather's, Happy Chef/24hr, McDonald's, Pizza Hut, Subway, Valhalla Rest., Super 8, Viking Motel, Ford, RV Ctr, VF Factory Stores/famous brands, KOA, bank
123	IA 221, rd E18, to Roland, McCallsburg, no facilities
120mm	**rest area nb, full(handicapped)facilities, info, phone, picnic tables, litter barrels, vending, RV dump/scenic prairie area sb, no facilities**
119mm	**rest area sb, full(handicapped)facilities, phone, picnic tables, litter barrels, vending, RV dump**
116	rd E29, to Story, **7 mi W…**Story Co Conservation Ctr
113	213th St, Ames, **W…**HOSPITAL, Amoco/mart, Kum&Go/Burger King/diesel/mart, KFC(2mi), Best Western/rest., Harley-Davidson, bank, to USDA Vet Labs, ISU
111b a	US 30, Ames, to Nevada, **E…**camping(11mi), **W…**Amoco, Texaco/diesel/mart/rest., Happy Chef, Hickory Park Rest., Hombre's Rest., AmericInn, Comfort Inn, Country Inn Suites, Hampton Inn, Heartland Inn, Microtel, Silver Saddle Motel, Super 8, to IA St U
109mm	S Skunk River
106mm	weigh sta both lanes
102	IA 210, to Slater, no facilities
96	to Elkhart, **W…**to Big Creek SP(11mi), Saylorville Lake
94mm	**rest area both lanes, full(handicapped)facilities, info, phone, picnic tables, litter barrels, vending, petwalk**
92	1st St, Ankeny, **W…**Amoco/diesel/mart/24hr, Kum&Go/gas, QT/mart, Texaco, Applebee's, Arby's, Burger King, Cazador Mexican, Fazoli's, Happy Chef, KFC, Stuffy's Rest., Subway, Village Inn Rest., Best Western/rest., Day's Inn, Heartland Inn, Super 8, Goodyear/auto, O'Reilly Parts, Staples, Tires+, auto repair, bank, **1mi W…**HOSPITAL, VETERINARIAN, Casey's/gas, Conoco, Phillips 66/mart, Golden Corral, LJ Silver, MaidRite Café, McDonald's, Pizza Hut, Taco John's, Dillow Foods, Hy-Vee Food/drug, NAPA, Vision Ctr, cleaners

Des Moines

90	IA 160, Ankeny, **E…**Chip's Diner, Country Inn Suites, Holiday Inn Express, airport, **W…**Casey's/gas, Burger King, McDonald's, Chevrolet, Menard's, Wal-Mart SuperCtr/24hr, to Saylorville Lake(5mi)
87b a	I-235, I-35 and I-80, no facilities

I-35 and I-80 run together 14 mi around NW Des Moines. See Iowa Interstate 80 exits 124-136.

72c	University Ave, **N…**HOSPITAL, Amoco/mart/wash, Texaco/mart/wash, Cracker Barrel, Baymont Inn, Country Club Shopping Ctr, **S…**MEDICAL CARE, Phillips 66/mart/wash, Applebee's, Bakers Square, Cheddar's Rest., KFC, McDonald's, Courtyard, Fairfield Inn, Heartland Inn, Holiday Inn/rest., Residence Inn, The Inn, bank
b	I-80 W and I-35 N, no facilities
a	I-235 E, to Des Moines
69b a	Grand Ave, W Des Moines, no facilities
68.5mm	Racoon River
68	IA 5, **7 mi E…**Crystal Inn, Hampton Inn, to airport, to Walnut Woods SP
65	G14, Cumming, to Norwalk, **14 mi W…**John Wayne Birthplace, museum
61mm	North River
56	IA 92, to Indianola, Winterset, **W…**Texaco/diesel/café, Total/pizza/mart, repair
56mm	Middle River
53mm	**rest area nb, litter barrels, no facilities**
52	G50, St Charles, St Marys, **14 mi W…**John Wayne Birthplace, museum
51mm	**rest area sb, litter barrels, no facilities**
47	rd G64, to Truro, no facilities
45.5mm	South River
43	rd 207, New Virginia, **E…**Sinclair/diesel/mart, **W…**Total/diesel/mart
36	rd 152, to US 69, **3 mi E…**lodging
33	US 34, Osceola, **E…**HOSPITAL, Amoco, Ampride/diesel/mart/café/atm/24hr, Conoco, Texaco/diesel/mart/rest., Hardee's, McDonald's, Pizza Hut, Subway, Best Western, Blue Haven Motel(2mi), Holiday Inn Express, Super 8, Goodyear, Hy-Vee/deli/bakery, Ford/Mercury/Chrysler/Plymouth/Dodge/Jeep, Pamida, bank, hardware, st patrol, **W…**KFC/Taco Bell, AmericInn
32mm	**rest area both lanes, full(handicapped)facilities, phone, picnic tables, litter barrels, vending, petwalk, RV dump**
30mm	weigh sta both lanes
29	rd H45, no facilities
22	rd 258, Van Wert, no facilities
18	rd J20, to Grand River, no facilities
12	rd 2, Decatur City, Leon, **E…**Texaco/diesel/mart/rest., **5 mi E…**HOSPITAL, Little River Motel
7.5mm	Grand River
7mm	**Welcome Ctr nb, full(handicapped)facilities, phone, picnic tables, litter barrels, vending, petwalk, RV dump**
4	US 69, Lamoni, to Davis City, **E…**to 9 Eagles SP(10mi), **W…**Amoco/Kum&Go/diesel/pizza/mart, Casey's/gas(2mi), Cenex/diesel(2mi), Subway(2mi), Chief Lamoni Motel, Super 8, antiques, IA Welcome Ctr
0mm	Iowa/Missouri state line

N

S

Iowa Interstate 80

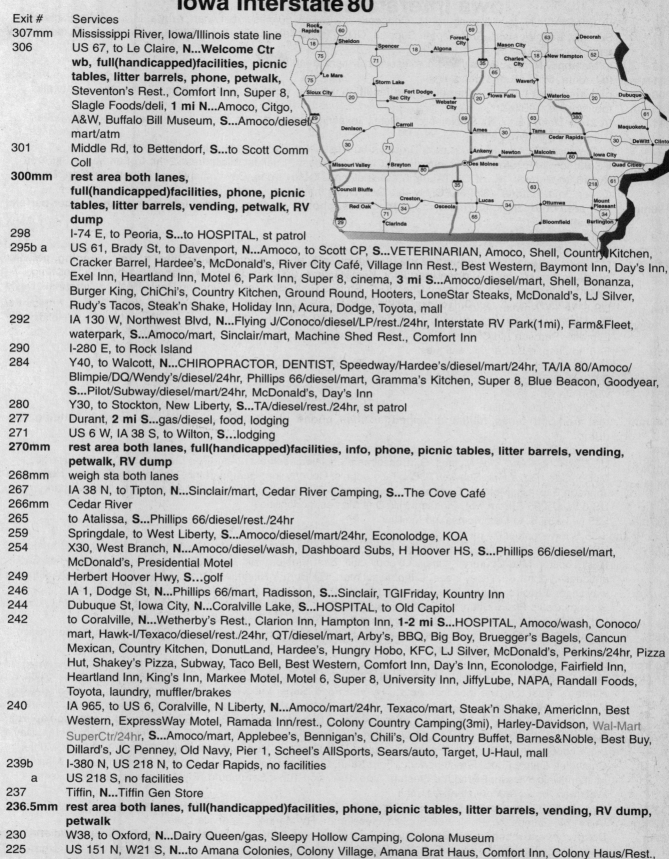

Exit #	Services
307mm	Mississippi River, Iowa/Illinois state line
306	US 67, to Le Claire, **N...Welcome Ctr wb, full(handicapped)facilities, picnic tables, litter barrels, phone, petwalk,** Steventon's Rest., Comfort Inn, Super 8, Slagle Foods/deli, **1 mi N...**Amoco, Citgo, A&W, Buffalo Bill Museum, **S...**Amoco/diesel/mart/atm
301	Middle Rd, to Bettendorf, **S...**to Scott Comm Coll
300mm	**rest area both lanes, full(handicapped)facilities, phone, picnic tables, litter barrels, vending, petwalk, RV dump**
298	I-74 E, to Peoria, **S...**to HOSPITAL, st patrol
295b a	US 61, Brady St, to Davenport, **N...**Amoco, to Scott CP, **S...**VETERINARIAN, Amoco, Shell, Country Kitchen, Cracker Barrel, Hardee's, McDonald's, River City Café, Village Inn Rest., Best Western, Baymont Inn, Day's Inn, Exel Inn, Heartland Inn, Motel 6, Park Inn, Super 8, cinema, **3 mi S...**Amoco/diesel/mart, Shell, Bonanza, Burger King, ChiChi's, Country Kitchen, Ground Round, Hooters, LoneStar Steaks, McDonald's, LJ Silver, Rudy's Tacos, Steak'n Shake, Holiday Inn, Acura, Dodge, Toyota, mall
292	IA 130 W, Northwest Blvd, **N...**Flying J/Conoco/diesel/LP/rest./24hr, Interstate RV Park(1mi), Farm&Fleet, waterpark, **S...**Amoco/mart, Sinclair/mart, Machine Shed Rest., Comfort Inn
290	I-280 E, to Rock Island
284	Y40, to Walcott, **N...**CHIROPRACTOR, DENTIST, Speedway/Hardee's/diesel/mart/24hr, TA/IA 80/Amoco/Blimpie/DQ/Wendy's/diesel/24hr, Phillips 66/diesel/mart, Gramma's Kitchen, Super 8, Blue Beacon, Goodyear, **S...**Pilot/Subway/diesel/mart/24hr, McDonald's, Day's Inn
280	Y30, to Stockton, New Liberty, **S...**TA/diesel/rest./24hr, st patrol
277	Durant, **2 mi S...**gas/diesel, food, lodging
271	US 6 W, IA 38 S, to Wilton, **S...**lodging
270mm	**rest area both lanes, full(handicapped)facilities, info, phone, picnic tables, litter barrels, vending, petwalk, RV dump**
268mm	weigh sta both lanes
267	IA 38 N, to Tipton, **N...**Sinclair/mart, Cedar River Camping, **S...**The Cove Café
266mm	Cedar River
265	to Atalissa, **S...**Phillips 66/diesel/rest./24hr
259	Springdale, to West Liberty, **S...**Amoco/diesel/mart/24hr, Econolodge, KOA
254	X30, West Branch, **N...**Amoco/diesel/wash, Dashboard Subs, H Hoover HS, **S...**Phillips 66/diesel/mart, McDonald's, Presidential Motel
249	Herbert Hoover Hwy, **S...**golf
246	IA 1, Dodge St, **N...**Phillips 66/mart, Radisson, **S...**Sinclair, TGIFriday, Kountry Inn
244	Dubuque St, Iowa City, **N...**Coralville Lake, **S...**HOSPITAL, to Old Capitol
242	to Coralville, **N...**Wetherby's Rest., Clarion Inn, Hampton Inn, **1-2 mi S...**HOSPITAL, Amoco/wash, Conoco/mart, Hawk-I/Texaco/diesel/rest./24hr, QT/diesel/mart, Arby's, BBQ, Big Boy, Bruegger's Bagels, Cancun Mexican, Country Kitchen, DonutLand, Hardee's, Hungry Hobo, KFC, LJ Silver, McDonald's, Perkins/24hr, Pizza Hut, Shakey's Pizza, Subway, Taco Bell, Best Western, Comfort Inn, Day's Inn, Econolodge, Fairfield Inn, Heartland Inn, King's Inn, Markee Motel, Motel 6, Super 8, University Inn, JiffyLube, NAPA, Randall Foods, Toyota, laundry, muffler/brakes
240	IA 965, to US 6, Coralville, N Liberty, **N...**Amoco/mart/24hr, Texaco/mart, Steak'n Shake, AmericInn, Best Western, ExpressWay Motel, Ramada Inn/rest., Colony Country Camping(3mi), Harley-Davidson, Wal-Mart SuperCtr/24hr, **S...**Amoco/mart, Applebee's, Bennigan's, Chili's, Old Country Buffet, Barnes&Noble, Best Buy, Dillard's, JC Penney, Old Navy, Pier 1, Scheel's AllSports, Sears/auto, Target, U-Haul, mall
239b	I-380 N, US 218 N, to Cedar Rapids, no facilities
a	US 218 S, no facilities
237	Tiffin, **N...**Tiffin Gen Store
236.5mm	**rest area both lanes, full(handicapped)facilities, phone, picnic tables, litter barrels, vending, RV dump, petwalk**
230	W38, to Oxford, **N...**Dairy Queen/gas, Sleepy Hollow Camping, Colona Museum
225	US 151 N, W21 S, **N...**to Amana Colonies, Colony Village, Amana Brat Haus, Comfort Inn, Colony Haus/Rest., **S...Welcome Ctr**, Amoco/mart, Phillips 66, Little Amana Rest./Winery, Day's Inn/rest, Holiday Inn, My Little Inn, Super 8

Iowa Interstate 80

220   IA 149 S, V77 N, to Williamsburg, **N**...Conoco/mart, Phillips 66/diesel/rest./mart, Arby's, McDonald's, Pizza Hut, Subway, Best Western, Crest Motel, Super 8, GNC, Tanger/famous brands, VF/famous brands, **S**...Texaco/diesel/rest., Ramada Ltd

216   to Marengo, **N**...HOSPITAL, Texaco/diesel/pizza/mart, Sudbury Court Motel(7mi)

211   to Ladora, Millersburg, no facilities

208mm   **rest area both lanes, full(handicapped)facilities, phone, vending, picnic tables, litter barrels, petwalk**

205   to Victor, no facilities

201   IA 21, to Deep River, **N**...Sinclair/Nick's Rest., Texaco/diesel/mart/24hr, Sleep Inn, RV camping, **S**...KwikStar/Conoco/diesel/rest./24hr

197   to Brooklyn, **N**...Amoco/diesel/mart, Brooklyn-80 Rest./24hr, diesel/tire repair, RV camping

191   US 63, to Montezuma, **N**...Citgo/mart, **S**...Citgo/diesel/rest., FuelMart/diesel/rest./24hr, to Fun Valley Ski Area

182   IA 146, to Grinnell, **N**...HOSPITAL(4mi), Coastal/mart, Casey's/gas, City Limits Rest., Country Kitchen, Dairy Queen(3mi), Hardee's(3mi), KFC, Hy-Vee Deli, Day's Inn, Econolodge, Super 8, **S**...Fun Valley Ski Area

180mm   **rest area both lanes, full(handicapped)facilities, phone, vending, weather info, picnic tables, litter barrels, petwalk**

179   IA 124, to Oakland Acres, Lynnville, no facilities

175.5mm   N Skunk River

173   IA 224, Kellogg, **N**...Citgo/diesel/mart/rest./24hr, Pella Museum, Rock Creek SP(9mi)

168   SE Beltline Dr, to Newton, **N**...Citgo/diesel, Arby's, Mid-Iowa Motel, Rolling Acres Camping, Wal-Mart SuperCtr/24hr

164   US 6, IA 14, Newton, **N**...HOSPITAL, Amoco/diesel/mart/wash/LP, Conoco/diesel/mart, Kum&Go/gas, Phillips 66/Subway/diesel/mart, Country Kitchen, Golden Corral, KFC, Perkins/24hr, Village Inn Rest., Day's Inn, Holiday Inn Express, Ramada Ltd/Chinese Rest., Super 8, museum, **S**...Best Western/rest., to Lake Red Rock

159   F48, to Jasper, Baxter, **N**...antiques

155   IA 117, Colfax, **N**...antiques, **S**...Casey's/gas, Texaco/diesel/mart/24hr

153mm   S Skunk River

151   weigh sta wb

149   Mitchellville, no facilities

148mm   **rest area both lanes, full(handicapped)facilities, phone, picnic tables, litter barrels, petwalk, vending, RV dump**

143   Altoona, Bondurant, **S**...Amoco(2mi), Casey's/gas, Texaco, Settle Inn(2mi), Hy-Vee Foods(2mi)

142b a   US 65, Hubble Ave, Des Moines, **S**...Bosselman/Sinclair/diesel/rest./24hr, Git'n Go/gas, Burger King, Hardee's, McDonald's, Pizza Hut, Subway/TCBY, Taco John's, Country Inn, Heartland Inn, Holiday Inn Express, Howard Johnson Express, Motel 6, Rodeway Inn, Settle Inn, Blue Beacon, Factory Outlet/famous brands, funpark

141   US 6 W, US 65 S(from wb), Pleasant Hill, Des Moines, no facilities

137b a   I-35 N, I-235 S, to Des Moines, no facilities

I-80 W and I-35 S run together 14 mi

136   US 69, E 14th St, Camp Sunnyside, **N**...Amoco/diesel/mart, Phillips 66/mart/wash, Sinclair, Bonanza, Bontel's Rest., Boston Mkt, Country Kitchen, Okoboji Grill, Best Western/rest., Motel 6, K-Mart/drugs, Volvo/White/GMC, antiques, **1-2 mi S**...Casey's/gas, Citgo/diesel/mart, QT/Burger King/diesel/mart, Arby's, Bakers Square, Broadway Diner, Fazoli's, Happy Joe's Pizza, Hardee's, KFC, LJ Silver, McDonald's, Pizza Hut, Schlotsky's, Scornovacca's Pizza, Stuffy's Rest., Subway, Taco Bell, Village Inn Rest., Wendy's, Day's Inn, 14th St Inn, Ramada Ltd, CarQuest, CarX Muffler, Dahl's Food/drug, Ford Trucks, Goodyear, MidState RV, Tires+, bank

135   IA 415, 2nd Ave, Polk City, **N**...diesel repair, **S**...HOSPITAL, Amoco, Coastal/mart, QT/mart, Dairy Queen, Osco Drug, Walgreen, USPO(2mi), st patrol

133mm   Des Moines River

131   IA 28 S, NW 58th St, **N**...CHIROPRACTOR, DENTIST, Casey's/gas, Coastal Gas, QT/mart, Greenbriar Rest., Hardee's, Best Inn, The Inn/diner, Acura, Hy-Vee Food, Super Lube/wash, Goodyear/auto, **S**...VA HOSPITAL, VETERINARIAN, Amoco/diesel/mart/24hr, Phillips 66/mart/wash, QT/mart, Sinclair/diesel, Texaco/diesel, Arby's, BBQ, Bakers Square, Burger King, Country Kitchen, Denny's, Embers Rest., Ground Round, Happy Joe's Pizza, Hardee's, KFC, Marcella's Italian, McDonald's, Perkins, Pizza Hut, Village Inn Rest., Wendy's, Comfort Inn, Day's Inn, Econolodge, Holiday Inn, Howard Johnson, Red Roof Inn, Rodeway Inn, Super 8, Best Buy, CarX Muffler, Chevrolet, Dahl's Food, Firestone/auto, Ford/Saab, Goodyear/auto, Isuzu, JiffyLube, Lincoln/Mercury, Marshall's, Muffler Clinic, Nissan, Plymouth, Western Auto, Precision Tune, Sears/auto, Toyota, XpressLube, Younker's, Walgreen, bank, laundry/cleaners, mall

129   NW 86th St, Camp Dodge, **N**...Phillip 66/mart, **S**...Amoco, Casey's/gas, B-Bop Burgers

127   IA 141 W, Grimes, **N**...Phillips 66/Subway/diesel/mart, RV/Marine, Saylorville Lake

126   Douglas Ave, Urbandale, **N**...Bar-B Truck Plaza/diesel/rest./24hr, Pilot/diesel/rest./24hr, **S**...Phillips 66/diesel/mart/wash, Day's Inn, Econolodge, Extended Stay America

125   US 6, Hickman Rd, **N**...Flying J/Conoco/Hardee's/diesel/LP/24hr, **S**...Phillips 66/mart/wash, IA Machine Shed Rest., Best Western, Comfort Suites, Sheraton/rest., Sleep Inn, GMC, Honda, Hyundai, Olds, Goodyear, hardware, to Living History Farms

E

W

Newton

Des Moines

Iowa Interstate 80

124   (72c from I-35 nb), University Ave, **N...**HOSPITAL, CHIROPRACTOR, Amoco/mart/wash, Kum&Go/Burger King/mart, QT/mart, Cracker Barrel, Mustard's Grill, Baymont Inn, Country Inn Suites, **S...**MEDICAL CARE, Phillips 66/mart/wash, Applebee's, Bakers Square, Cheddar's Café, Chili's, Cuco's Mexican, Don Pablo's, KFC, Macaroni Grill, McDonald's, Outback Steaks, RockBottom Rest./brewery, Courtyard, Fairfield Inn, Heartland Inn, Holiday Inn/Damon's, Residence Inn, The Inn, Wildwood Lodge, Barnes&Noble, Lowe's Bldg Supply, bank

I-80 E and I-35 N run together 14 mi

123b a   I-80/I-35 N, I-35 S to Kansas City, I-235 to Des Moines

121   74th St, W Des Moines, **N...**West End Diner, Hampton Inn, Hawthorn Suites, Hy-Vee Food/drug, **S...**Kum&Go/Blimpie/gas, Arby's, Burger King, McDonald's, Perkins, Taco John's, Best Suites, Candlewood Suites, Marriott, Motel 6, Wingate Inn

119mm   **rest area both lanes, full(handicapped)facilities, info, vending, phone, picnic tables, petwalk, RV dump**

117   R22, Booneville, Waukee, **N...**camping(2mi), **S...**Kum&Go/24hr, Rube's Steaks

115mm   weigh sta eb

113   R16, Van Meter, **2 mi S...**Casey's/gas

112.5mm   N Racoon River

111.5mm   Middle Racoon River

110   US 169, DeSoto, to Adel, **N...**AirCraft Supermkt, Edgetowner Motel, **S...**Amoco/mart, Casey's/gas, DeSoto Motel

106   F90, P58, **N...**KOA

104   P57, Earlham, **S...**Casey's/gas(2mi)

100   US 6, Dexter, to Redfield, no facilities

97   P48, to Dexter, **N...**Casey's/gas(2mi), camping

93   P28, Stuart, **N...**Amoco/diesel/mart/24hr, Conoco/diesel/mart/atm, Cyclone Drive-In, Burger King, McDonald's, AmericInn, Subway, Super 8, Jubilee Foods, NAPA, bank, camping(7mi), laundry, tires, **S...**Phillips 66/diesel/mart/atm, Country Kitchen, New Edgetowner Motel

88   P20, Menlo, camping

86   IA 25, to Greenfield, Guthrie Ctr, **S...**HOSPITAL(13mi), to Preston/Spring Brook SP

85mm   Middle River

83   N77, Casey, **1 mi N...**Kum&Go/gas, camping

80mm   **rest area both lanes, full(handicapped)facilities, phone, picnic tables, litter barrels, vending, petwalk, RV dump**

76   IA 925, N54, Adair, **N...**Amoco/diesel, Casey's/diesel, Kum&Go/diesel, Happy Chef/24hr, Super 8, camping

75   G30, to Adair, **1 mi N...**Casey's/gas, Phillips 66/diesel, same as 76

70   IA 148 S, Anita, **S...**to Lake Anita SP(6mi), camping

64   N28, to Wiota, no facilities

61mm   E Nishnabotna River

60   US 6, US 71, Lorah, to Atlantic, **S...**Phillips 66/diesel/24hr, Texaco, Country Kitchen, Econolodge, Super 8

57   N16, to Atlantic, **S...**HOSPITAL(7mi)

54   IA 173, to Elk Horn, **7 mi N...Welcome Ctr**, AmericInn, Windmill Museum, gas, food

51   M56, to Marne, no facilities

46   M47, Walnut, **N...**Amoco/McDonald's/mart/24hr, Villager Buffet, Super 8, to Prairie Rose SP(8mi), **S...**Kum&Go/pizza/subs/diesel/24hr, Aunt B's Kitchen, Red Carpet Inn/RV camping, repair

44mm   weigh sta both lanes

40   US 59, Avoca, to Harlan, **N...**HOSPITAL(12mi), Conoco/Wings/diesel/rest./24hr, Sinclair/diesel/mart, **S...**Texaco/diesel, Embers Rest., Avoca Motel, Capri Motel, Parkway Camping(2mi), Nishna Museum

39.5mm   W Nishnabotna River

34   M16, Shelby, **N...**Texaco/diesel/mart/atm, Cornstalk Rest., Dairy Queen, Shelby Country Inn, **S...**repair

32mm   rest area both lanes, litter barrels, no facilities

29   L66, to Minden, **S...**Conoco/Kopper Kettle/diesel/mart, Phillips 66/A&W/diesel/mart, Midtown Motel(2mi)

27   I-680 W, to N Omaha, no facilities

23   IA 244, L55, Neola, **S...**Kum&Go/diesel, to Arrowhead Park, camping

20mm   **Welcome Ctr eb/rest area wb, full(handicapped)facilities, phone, picnic tables, litter barrels, vending, petwalk, RV dump**

Iowa Interstate 80

17	G30, Underwood, **N**...Phillips 66/Subway/diesel/mart/24hr, Interstate 80 Inn/rest., Underwood Motel
8	US 6, Council Bluffs, **N**...HOSPITAL, Coastal/mart, Phillip 66/mart, Chalet Motel(5mi)
5	Madison Ave, Council Bluffs, **N**...MEDICAL CARE, Amoco/mart, Burger King, Café Court, Garden Café, Great Wall Chinese, KFC, McDonald's, Pizza Hut, Subway, Heartland Inn, Dillard's, DrugTown, Hy-Vee Food/drug, JC Penney, NTB, Sears/auto, Target, bank, cinema, laundry, mall, **S**...DENTIST, CHIROPRACTOR, Conoco/mart/atm, Texaco/mart, Dairy Queen, Village Inn Rest., Western Inn, No Frills Foods
4	I-29 S, to Kansas City
3	IA 192 N, Council Bluffs, **N**...to Hist Dodge House, **S**...Conoco, Phillips 66/diesel/mart, TA/diesel/rest./24hr, Burger King, Applebee's, Cracker Barrel, Dairy Queen, Fazoli's, Golden Corral, Hardee's, LJ Silver, McDonald's, Perkins/24hr, Red Lobster, Taco Bell, Comfort Inn, Econolodge, Fairfield Inn, Motel 6, Settle Inn, Advance Parts, Aldi Foods, Chevrolet/GMC, Dodge, Ford/Lincoln/Mercury, Mazda, Menard's, Nissan, OfficeMax, Olds, Plymouth/Jeep, Pontiac/Buick, Sam's Club, Subaru, Toyota, TruckOMat, U-Haul, Wal-Mart SuperCtr/24hr, diesel repair
1b	S 24th St, Council Bluffs, **N**...Pilot/Arby's/diesel/mart/atm/24hr, Sinclair/mart, Texaco/Burger King/diesel/rest., Best Western, Interstate Inn, Super 8, Goodyear, RV camping, casino, dog track, truckwash
a	I-29 N, to Sioux City
0mm	Missouri River, Iowa/Nebraska state line

Iowa Interstate 235(Des Moines)

Exit #	Services
15	I-80, E to Davenport
13	US 6, E Euclid Ave, **E**...Perkins, **W**...Denny's, Sutherland's Rest., Midas Muffler, NAPA
12	Guthrie Ave, **N**...Coastal/diesel/mart, Jocko's Auto Parts
11	IA 163, University Ave, Easton Dr, no facilities
10	US 65/69, E 14th, E 15th, **N**...Sinclair, **S**...HOSPITAL, QT/mart, McDonald's, st capitol, zoo
9	E 6th St, Penn Ave(from wb), **S**...HOSPITAL
8.5	3rd St, 5th Ave, **N**...HOSPITAL, Holiday Inn, **S**...Conv Ctr, Best Western, Embassy Suites, Park Inn, Marriott
8	Keo Way, **S**...HOSPITAL, Git'n Go/gas
7.5	Harding Rd(from wb), **S**...airport
7	MLK Blvd, Drake U, **S**...airport
6.5	31st St, Governor's Mansion
6	42nd St, Science & Art Ctr, **N**...Taco John's
5	56th St(from wb), no facilities
4	IA 28, 63rd St, to Windsor Heights, **S**...Hist Valley Jct, zoo
3	8th St, W Des Moines, **N**...Kum&Go/gas, B-Bop's Café, Blimpie, Burger King, Dotty's Donuts, Battery Patrol, NTB, PetCo, Sam's Club/gas, Wal-Mart SuperCtr/24hr, **S**...Amoco, Coastal/gas, Kum&Go, Garcia's Mexican, Ford
2	24th St, W Des Moines, **N**...Sinclair, ChuckeCheese, Hardee's, Hooter's, Loco Joe's Café, McDonald's, Old Country Buffet, Taco Bell, Village Inn Rest., Studio+, Firestone, Goodyear/auto, Hancock Fabrics, Jo-Ann Fabrics, Michael's, Office Depot, 1/2 Price Store, Osco Drug, Pier 1, Rex, ToysRUs, bank, cinema
1	35th St, W Des Moines, **N**...Amoco/diesel/mart, Ramada Inn, Hy-Vee Foods, JC Penney, SteinMart, Target, Younker's, bank, mall, **S**...MEDICAL CARE
0mm	I-235 begins/ends on I-80, exit 123.

Iowa Interstate 280(Davenport)

Exit #	Services
18b a	I-74, US 6, **S**...Denny's, Harold's Rest., McDonald's, Omelet Shoppe, Exel Inn, Comfort Inn, Hampton Inn, Holiday Inn, La Quinta, Motel 6, Ramada Inn,
15	Airport Rd, Milan, **N**...Hardee's, MaidRite Café, **S**...gas
11b a	IL 92, Rock Island, to Andalusia, no facilities
9.5mm	Mississippi River, Iowa/Illinois state line
8	IA 22, Rockingham Rd, to Buffalo
6	US 61, W River Dr, to Muscatine, **W**...food, camping
4	Locust St, rd F65, 160th St, **E**...Texaco(2mi), to Palmer Coll, St Ambrose U
1	US 6 E, IA 927, Kimberly Rd, to Walcott, **E**...lodging
0mm	I-280 begins/ends on I-80, exit 290.

Kansas Interstate 35

Exit #	Services
235mm	Kansas/Missouri state line
235	Cambridge Circle, **E**...QT/mart
234b a	US 169, Rainbow Blvd, **E**...Phillips 66/mart, Applebee's, Arby's, BBQ, Burger King, McDonald's, Wendy's, Best Western, Day's Inn, KU MED CTR
233a	SW Blvd, Mission Rd, no facilities
b	37th Ave(from sb), no facilities
232b	US 69 N, **E**...QT/mart, Phillips 66/mart, McDonald's, Taco Bell
a	Lamar Ave, **E**...QT/mart, Total/mart
231b a	I-635(exits left from nb), no facilities
230	Antioch Rd(from sb), **E**...QT Gas/mart
229	Johnson Dr, **E**...Conoco, Texaco, Arby's, HenHouse Mkt, Home Depot, OfficeMax, Old Navy, PetsMart, Walnut Grove RV Park
228b	US 56, US 69, Shawnee Mission Pkwy, **W**...Burger King, LJ Silver, Perkins, Taco Bell, Wendy's
a	67th St, **E**...QT/mart, Phillips 66/mart, Texaco/mart, Rio Bravo, Shoney's, Comfort Inn, Drury Inn, Econolodge, Fairfield Inn, HomeStead Village, Quality Inn, BMW
227	75th St, **E**...HOSPITAL, DENTIST, Circle K/gas, QT/mart, McDonald's, Mr GoodSense Subs, Perkins, Acura, Georgetown Drug, JC Penney, NW Fabrics, Wal-Mart, **W**...VETERINARIAN, Citgo/7-11, Texaco/mart, Total/diesel/mart, McDonald's, Ryan's, Sonic, Subway, Taco Bell, Wendy's, Hampton Inn, JiffyLube
225b	US 69 S(from sb), Overland Pkwy, no facilities
a	97th St, **E**...Amoco, Buca's Italian, Pippen's Rest., Shoney's, Wendy's, Radisson/rest., Gomer's Foods, cinema, **W**...CHIROPRACTOR, DENTIST, EYECARE, Phillips 66, Texaco/mart, Arby's, Bagel&Bagel, Kahn Chinese, LongBranch Steaks, Mary's Place Gyros, Taco Bell, EconoLube'n Tune, NTB, bank, cleaners, museum
224	95th St, **E**...HOSPITAL, DENTIST, CHIROPRACTOR, Amoco, Phillips 66/mart, Texaco/mart, Applebee's, Burger King, Cielio's Italian, Denny's, Holiday Ham Café, Houlihan's, Italian Oven, JoyLuck Chinese, KFC, McDonald's, Ming Palace, Old Chicago Pizza, Pizza Hut, Ponderosa, Ruby Tuesday, Santa Fe Café, Steak&Ale, Taco Bell, TGIFriday, Day's Inn, GuestHouse Hotel, Holiday Inn, Howard Johnson, La Quinta, Motel 6, Super 8, Best Buy, CircuitCity, Dillard's, HobbyLobby, HyVee Foods, JC Penney, JiffyLube, MailBoxes Etc, MC Sports, MensWearhouse, Nordstrom's, 1/2Price Store, PartsAmerica, PetCo, Pier 1, Sam's Club, SteinMart, bank, cleaners, cinema, mall, **W**...Total/mart
222b a	I-435 W, I-435 E
220	119th St, **E**...CHIROPRACTOR, Conoco/mart, Texaco/mart/wash, Burger King, China Café, Cracker Barrel, Gambucci's Italian, Honeybaked Ham, Joe's Crabshack, LJ Silver, Machine Shed Rest., McDonald's, Rio Bravo Cantina, Schlotsky's, Subway, Tres Hombres Mexican, Wendy's, Zio's Italian, Comfort Suites, Fairfield Inn, Hampton Inn, Aamco, Barnes&Noble, Dodge, Goodyear/auto, Home Depot, Honda, Hyundai, OfficeMax, PetsMart, Target, U-Haul, cinema
218	135th, Santa Fe St, Olathe, **E**...Texaco/mart, Backyard Burgers, McDonald's, Perkins, Ace Hardware, Big Lots, Ford, bank, **W**...VETERINARIAN, Amoco, Phillips 66/mart, QT/mart, Denny's, Ponderosa, Wendy's, Best Western, Chevrolet, Chrysler/Plymouth/Jeep, Mazda, Nissan, Saturn, Suzuki, Subaru, Toyota
217	Old Hwy 56(from sb), same as 215
215	US 169 S, KS 7, Olathe, **E**...Phillips 66/mart/atm/24hr, **W**...HOSPITAL, Conoco/mart, Texaco/Blimpie/mart/24hr, Phillips 66/mart, Applebee's, Burger King, Chili's, Country Kitchen, FoodCourt, Jeeper's Diner, Red Lobster, McDonald's, Wendy's, Econolodge, Fairfield Inn, Holiday Inn, Microtel, Burlington Coats, Dillard's, Linens'n Things, Marshall's, Oschman's Sports, bank, cinema, mall
213mm	weigh sta both lanes
210	US 56 W, Gardner, **W**...Phillips 66/mart, McDonald's, Mr GoodSense Subs, Pizza Hut, Sonic, Taco Bell, Waffle House, Super 8
207	US 56 E, Gardner Rd, **E**...RV Sales, **W**...Conoco/mart
202	Edgerton, no facilities
198	KS 33, to Wellsville, **W**...Total/food mkt, bank
193	LeLoup Rd, Baldwin, no facilities
187	KS 68, Ottawa, **W**...Texaco/diesel/24hr, Blimpie, Dairy Queen, Buick/Olds/GMC/Pontiac
185	15th St, Ottawa, no facilities
183b a	US 59, Ottawa, **W**...Amoco/mart, Citgo/mart, Conoco/diesel/mart, Burger King, Country Kitchen, KFC, LJ Silver, McDonald's, Sirloin Stockade, Wendy's, Best Western, Econolodge, Holiday Inn Express, Chrysler/Plymouth/Dodge, CountryMart Foods, $General, Wal-Mart SuperCtr/24hr, car/truck wash
182b a	US 50, Ottawa, **W**...Village Inn Motel

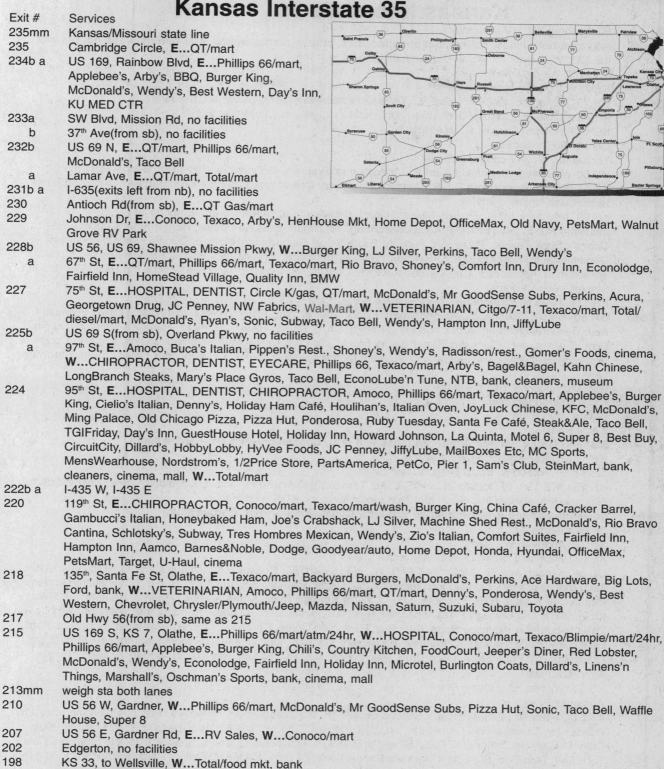

Kansas Interstate 35

176	Homewood, **W**...RV camping
175mm	**rest area both lanes, full(handicapped)facilities, phone, picnic tables, litter barrels, vending, petwalk**
170	KS 273, Williamsburg, **W**...Total/mart/café, phone
162	KS 31 S, Waverly, no facilities
160	KS 31 N, Melvern, no facilities
155	US 75, Burlington, Melvern Lake, **E**...TA/Texaco/diesel/rest./24hr, Total/mart, Beto Inn
148	KS 131, Lebo, **E**...Coastal/diesel/mart, Universal Inn, rest., **W**...to Melvern Lake
141	KS 130, Neosho Rapids, **E**...NWR(6mi)
138	County Rd, no facilities
135	Thorndale, **W**...RV camping/phone
133	US 50 W, 6th Ave, Emporia, **1-3 mi E**...Casey's/gas, Braum's, McDonald's, Pizza Hut, Budget Host Inn
131	KS 57, KS 99, Burlingame Rd, **E**...Amoco/diesel/mart, Coastal/mart, Burger King, Hardee's, Mr GoodSense Subs, bank, sports complex
130	KS 99, Merchant St, **E**...Phillips 66/diesel/mart, Subway(1mi), University Inn, Emporia St U
128	Industrial Rd, **E**...VETERINARIAN, Conoco/mart, Total/mart, Baskin-Robbins, Burger King, Coburn's Rest., Carroll's Book's/music, Pizza Hut, Subway, Econolodge, Ramada Inn, Motel 6, Dillon's Foods, Food4Less, Goodyear/auto, FashionBug, JC Penney, Western Auto/parts, cinema, **W**...Phillips 66/Wendy's/diesel/mart, Texaco/mart, Applebee's, Cracker Barrel, Golden Corral, McDonald's, Taco Bell, Village Inn Rest., Fairfield Inn, Staples, Wal-Mart/auto
127c	KS Tpk, I-335 N, to Topeka
b a	US 50, KS 57, Newton, **E**...Ampride/mart, Conoco/diesel/mart, Flying J/diesel/rest./24hr, Phillips 66/diesel/mart, Texaco/A&W/mart, Arby's, Carlos O'Kelly's, China Buffet, Gambino's Pizza, Hardee's, Western Sizzlin, Best Western/rest., Day's Inn, Super 8, Aldi Foods, Big Lots, Chevrolet/Pontiac/Buick/Olds, Chrysler/Dodge/Plymouth/Jeep/Toyota, $General, Ford/Lincoln/Mercury/Nissan, NAPA, PriceChopper Foods, Sears Hardware, bank
127mm	I-35 and I-335 KS Tpk, toll plaza
I-35 S and KS Tpk S run together	
125mm	Cottonwood River
111	Cattle Pens
97.5mm	**Matfield Green Service Area(both lanes exit left), Coastal/diesel/mart, Hardee's/atm**
92	KS 177, Cassoday, **E**...Fuel'n Service
76	US 77, El Dorado N, **E**...El Dorado Lake SP
71	KS 254, KS 196, El Dorado, **E**...HOSPITAL, Texaco/diesel/mart/rest./showers, Golden Corral, Best Western, Heritage Inn, Sunset Inn, Super 8, Ford/Lincoln/Mercury, KOA, Wal-Mart
65mm	**Towanda Service Area(both lanes exit left), Coastal/diesel/mart, Hardee's/atm**
62mm	Whitewater River
57	21st St, Andover, **W**...golf
53	KS 96, Wichita, **1 mi W**...Acura, Mercedes
50	US 54, Kellogg Ave, **E**...McConnell AFB, **W**...VA HOSPITAL, Coastal/mart, Denny's, McDonald's, Steak&Ale, Taco Bell, Day's Inn, Fairfield Inn, Hampton Inn, Hilton, Marriott, Scotsman Inn, Super 8, Wyndham Garden, Circuit City, K-Mart, NTB, PepBoys, USPO, **W on Kellogg Ave**...Arby's, Carlos O'Kelly's, Hooters, LJ Silver, McDonald's, Old Chicago Pizza, Spangles Rest., LaQuinta, Mark 8 Lodge, Ramada Inn, Sheraton, Wichita Inn, Williamsburg Inn, Advance Parts, Best Buy, Burlington Coats, Chevrolet, Chrysler/Plymouth, CopyMax, Ford, Hancock Fabrics, Honda, JC Penney, Kia/Nissan, KidsRUs, Lincoln/Mercury, MensWearhouse, Office Depot, OfficeMax, 1/2 Price Store, Osco Drug, Pier 1, Pontiac/GMC, Saturn, Sears/auto, Target, TJ Maxx, Toyota, ToysRUs, mall
45	KS 15, Wichita, **E**...Boeing Plant
44.5mm	Arkansas River
42	47TH St, I-135, to I-235, Wichita, **services on 47th St, E**...Coastal/mart, Potbelly's Rest., Comfort Inn, Day's Inn, Holiday Inn Express, Red Carpet Inn, **W**...CHIROPRACTOR, Phillips 66, Applebee's, Braum's, Burger King, Dairy Queen, Godfather's, KFC, LJ Silver, McDonald's, Pizza Hut, Quizno's, Spaghetti Jack's, Spangles Rest., Taco Bell, Taco Tico, Big Lots, Checker's Foods, Dillon's Foods, K-Mart, O'Reilly's Parts, Radio Shack, to airport
39	US 81, Haysville, **W**...Haysville Inn
33	KS 53, Mulvane, **E**...Mulvane Hist Museum, **W**...Wyldewood Winery
26mm	**Belle Plaine Service Area(both lanes exit left), Coastal/diesel/mart, Hardee's/atm**
19	US 160, Wellington, **3 mi W**...OakTree Inn, Old Oxford Mill, RV camping
19mm	toll plaza
I-35 N and KS TPK N run together	
4	US 166, to US 81, South Haven, **E**...Total/diesel/pizza/subs/mart/atm, Economy Motel/rest.
1.5mm	weigh sta nb
0mm	Kansas/Oklahoma state line

Kansas Interstate 70

Exit #	Services
422b a	US 69 N, US 169 N
421b	I-670, no facilities
a	S...railroad yard
420b a	US 69 S, 18th St Expswy, N...HOSPITAL, gas/diesel/mart, Eagle Inn, truck repair
419	38th St, Park Dr, access to 10 motels
418b	I-635 N, N...LJ Silver, McDonald's, Taco Bell, Wendy's, Dillard's, JC Penney, mall
a	I-635 S, no facilities
417	57th St, no facilities
415a	KS 32 E(from eb), no facilities
b	to US 40 W, State Ave, Kansas City, N...Conoco/mart, Sinclair, Arby's, Capt D's, Perkins, McDonald's, Taco Bell
414.5mm	weigh sta wb, parking area both lanes, phone
414	78th St, N...HOSPITAL, QT/mart, Arby's, Burger King, Cracker Barrel, Dairy Queen, Furr's Cafeteria, Hardee's, S...Conoco/mart, Texaco/mart, American Motel, Comfort Inn
411b	I-435 N, access to Woodlands Racetrack, to KCI Airport
a	I-435 S, no facilities, last free exit wb before KS TPK
410	110th St, N...KS Speedway
224.5mm	toll plaza, I-70 W and KS TPK run together
224	KS 7, to US 73, Bonner Springs, Leavenworth, N...Texaco/diesel/mart, KFC/Taco Bell, Waffle House, Wendy's, Holiday Inn Express, Wyandotte Co Museum, Agri Hall of Fame
216.5mm	toll booth
209mm	**Lawrence Service Area(both lanes exit left), full facilities, Conoco/diesel/mart, Hardee's/atm**
204	US 24, US 59, to E Lawrence, S...Citgo/mart, Total/diesel/mart, Burger King, Sonic, Bismarck Inn, JayHawk Motel, Ford/Lincoln/Mercury, Tanger Outlet/famous brands, to KOA/deli/showers/LP
203mm	Kansas River
202	US 59 S, to W Lawrence, **1 mi S**...HOSPITAL, Capt's Galley, Chili's, Hardee's, Kettle, Day's Inn, Holiday Inn, Ramada Inn, Travelodge, Virginia Inn, Super 8, to Clinton Lake SP, to U of KS
197	KS 10, Lecompton, Lawrence, N...Perry Lake SP, S...Clinton Lake SP,
183mm	**Topeka Service Area(both lanes exit left), full facilities, Conoco/diesel/mart, Hardee's/atm**
182	I-70 W(from wb), to Denver
365	toll plaza, I-70 E and KS TPK E run together
365	21st St, Rice Rd, to Lake Shawnee, last free exit eb before KS TPK
364b	US 40 E, Carnahan Ave, no facilities
a	California Ave, S...Amoco/diesel/mart, McDonald's, Pizza Hut, Subway
363	Adams St, downtown
362c	10th Ave(from wb), N...Amoco/mart, Ramada Inn, st capitol
362b a	to 8th Ave, downtown, N...Day's Inn, Ramada Inn, to St Capitol
361b a	to 1st Ave, Topeka Blvd, 4th Ave, N...Ramada Inn, Michelin, Ryder, bank, to KS Expo Ctr
359	MacVicar Ave, no facilities
358b a	Gage Blvd, S...HOSPITAL, Conoco/repair, Judy's Rest., McDonald's, Subway, st patrol
357b a	Fairlawn Rd, 6th Ave, S...Amoco, Phillips 66/mart/wash, Texaco, A&W, Best Western, Holiday Inn/rest., Motel 6, NAPA, zoo-rain forest
356b a	Wanamaker Rd, N...AmeriSuites, KS Museum of History, S...VETERINARIAN, PODIATRIST, Citgo/mart, Conoco/mart, Phillips 66/diesel/mart, Texaco/diesel/mart/rest./24hr, Amarillo Mesquite Grill, Applebee's, Boston Mkt, Burger King, Chili's, ChuckeCheese, Cracker Barrel, Denny's, Golden Corral, GoodCents Subs, Hardee's, IHOP, McDonald's, Old Country Buffet, Olive Garden, Panera Bread, Perkins, Pizza Hut, Red Lobster, Ruby Tuesday, Shoney's, Sirloin Stockade, Steak'n Shake, Taco Bell, Timberline Steaks, Vista Burger, Wendy's, Winstead's Burgers, Candlewood Suites, Clubhouse Inn, Comfort Inn, Courtyard, Day's Inn, Fairfield Inn, Hampton Inn, Motel 6, Holiday Inn Express, Quality Inn, Residence Inn, Sleep Inn, Super 8, Barnes&Noble, Best Buy, Circuit City, Dillard's, Food4Less/24hr, HobbyLobby, Home Depot, JC Penney, JiffyLube, K-Mart/auto, Kohl's, Lowe's Bldg Supply, Office Depot, PetCo, Pier 1, Sam's Club, Sports Outlet, Target, TJ Maxx, ToysRUs, Wal-Mart SuperCtr/24hr, bank, cinema, mall
355	I-470 E, Topeka, to VA MED CTR, air museum, **1 mi S**...same as 356
353	KS 4, to Eskridge, no facilities

E

W

KS City

Topeka

Kansas Interstate 70

351	frontage rd(from eb), Mission Creek, no facilities
350	Valencia Rd, **N**...Carlson's I-70 Auto Auction
347	West Union Rd, no facilities
346	Willard, to Rossville, no facilities
343	frontage rd, no facilities
342	Eskridge Rd, Keene Rd, access to Lake Wabaunsee
341	KS 30, Maple Hill, **S**...Amoco/diesel/mart/atm
338	Vera Rd, **S**...Texaco/Dairy Queen/Stuckey's
336.5mm	**rest area(exits left from both lanes), full(handicapped)facilities, phone, picnic tables, litter barrels, RV camping/dump, petwalk**
335	Snokomo Rd, Paxico, Skyline Mill Creek Scenic Drive
333	KS 138, Paxico, **N**...Phillips 66/diesel/winery/gifts, Mill Creek RV Camping
332	Spring Creek Rd, no facilities
330	KS 185, to McFarland, no facilities
329mm	weigh sta both lanes
328	KS 99, to Alma, **N**...Gas'n Shop/diesel/deli, **S**...Wabaunsee Co Museum
324	Wabaunsee Rd, **N**...Cowboy Café
323	frontage rd, no facilities
318	frontage rd, no facilities
316	Deep Creek Rd, no facilities
313	KS 177, to Manhattan, **N**...Conoco/Hilltop Café/diesel, **8 mi N**...Applebee's, Chili's, McDonald's, Village Inn Rest., Best Western, Fairfield Inn, Hampton Inn, Motel 6, Super 8, Jeep, Nissan, Sears, to KSU
311	Moritz Rd, no facilities
310mm	**rest area both lanes, full(handicapped)facilities, phone, picnic tables, litter barrels, petwalk**
307	McDowell Creek Rd, no facilities
304	Humboldt Creek Rd, no facilities
303	KS 18 E, to Ogden, Manhattan, **8 mi N**...Hampton Inn, Super 8, to KSU
301	Marshall Field, to Ft Riley, **N**...Cavalry Museum, Custer's House, KS Terr Capitol
300	US 40, KS 57, Council Grove, **N**...Dreamland Motel, Sunset Motel, hist church
299	to Jct City, Ft Riley, **N**...Texaco/mart/repair, Total Gas, Subway(2mi), Best Western, Econolodge, Great Western Inn, Red Carpet Inn, Super 8
298	Chestnut St, to Jct City, Ft Riley, **N**...Texaco/Burger King/diesel/mart/atm/24hr, BBQ, Cracker Barrel, Family Buffet, Taco Bell, Holiday Inn Express, Super 8, Cato's, $Tree, Staples, Wal-Mart SuperCtr/24hr, carwash
296	US 40, Washington St, Junction City, **N**...CHIROPRACTOR, DENTIST, Citgo/diesel/mart, Coastal/mart, Texaco/diesel/mart/atm/24hr, Total/diesel/mart, Country Kitchen, Dairy Queen, Denny's/24hr, El Cazador Mexican, KFC, McDonald's, Pizza Hut(1mi), Sirloin Stockade, Sonic, Subway/TCBY, Budget Host/RV Park, Comfort Inn, Day's Inn, Liberty Inn, Ramada Ltd, Food4Less/24hr
295	US 77, KS 18, Marysville, to Milford Lake, **N**...Phillips 66/A&W/diesel/mart/atm/24hr, Motel 6
294mm	rest area both lanes, full(handicapped)facilities, phone, picnic tables, litter barrels, RV dump, petwalk
290	Milford Lake Rd, no facilities
286	KS 206, Chapman, **S**...Citgo/diesel/mart/atm/24hr, **1 mi S**...Casey's/gas
281	KS 43, to Enterprise, **N**...4 Seasons RV Ctr/Park
277	Jeep Rd, no facilities
275	KS 15, Abilene, to Clay Ctr, **N**...Dairy Queen, Holiday Inn Express, **S**...HOSPITAL, Amoco/Blimpie/diesel/mart/atm/24hr, Kwikstop/gas, Phillips 66/Taco Bell/mart/wash, Texaco/repair, Burger King, Evergreen Chinese, McDonald's, Pizza Hut, Sirloin Stockade, Sonic, Subway, Best Western/rest., Super 8, Alco/gas, AutoZone, Chevrolet/Olds/Pontiac/Buick, Country Mart Food/drug, Ford/Lincoln/Mercury, Chrysler, Plymouth/Jeep/Dodge, to Eisenhower Museum
272	Fair Rd, to Talmage, **S**...Econo RV Park, Russell Stover Candy
266	KS 221, Solomon, **S**...Total/diesel/rest./24hr
265mm	**rest area both lanes, full(handicapped)facilities, phone, picnic tables, litter barrels, petwalk**
264.5mm	Solomon River
260	Niles Rd, New Cambria, no facilities
253.5mm	Saline River
253	Ohio St, **S**...HOSPITAL, Flying J/Conoco/Country Mkt/diesel/LP/24hr, Kenworth
252	KS 143, Salina, **N**...Amoco/diesel/rest./atm/24hr, Petro/Mobil/Phillips 66/Baskin-Robbins/Wendy's/Pizza Hut/diesel/mart, Bayard's Café, Dairy Queen, Denny's, McDonald's, Day's Inn, Holiday Inn Express, Motel 6, Salina Inn, Super 8, Blue Beacon, KOA, **S**...Bosselman/Sinclair/Blimpie/diesel/24hr, Best Western/rest., Ramada Inn, museum

Jct City Abilene Salina

E

W

Kansas Interstate 70

250b a   I-135, US 81, N to Concordia, S to Wichita

249   Halstead Rd, to Trenton, no facilities

244   Hedville, **N**...Sundowner West RV Park, **S**...Phillips 66/diesel/mart/RV, OutPost FunCtr/ Rest.

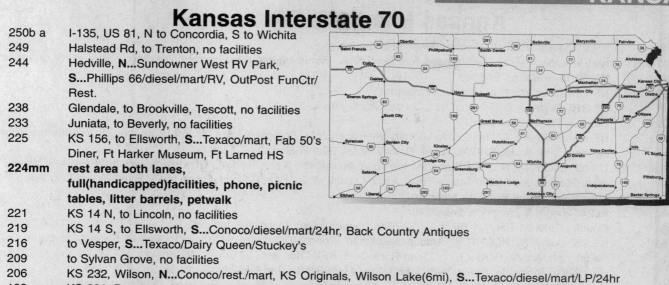

238   Glendale, to Brookville, Tescott, no facilities

233   Juniata, to Beverly, no facilities

225   KS 156, to Ellsworth, **S**...Texaco/mart, Fab 50's Diner, Ft Harker Museum, Ft Larned HS

224mm   **rest area both lanes, full(handicapped)facilities, phone, picnic tables, litter barrels, petwalk**

221   KS 14 N, to Lincoln, no facilities

219   KS 14 S, to Ellsworth, **S**...Conoco/diesel/mart/24hr, Back Country Antiques

216   to Vesper, **S**...Texaco/Dairy Queen/Stuckey's

209   to Sylvan Grove, no facilities

206   KS 232, Wilson, **N**...Conoco/rest./mart, KS Originals, Wilson Lake(6mi), **S**...Texaco/diesel/mart/LP/24hr

199   KS 231, Dorrance, **N**...to Wilson Lake, **S**...Agco/gas

193   Bunker Hill Rd, **N**...Total/diesel/mart/Bearhouse Café/24hr, to Wilson Lake

189   US 40 bus, Pioneer Rd, Russell, **5 mi N**...gas, food, lodging

187mm   **rest area both lanes, full(handicapped)facilities, phone, picnic tables, litter barrels, RV dump, petwalk**

184   US 281, Russell, **N**...HOSPITAL, VETERINARIAN, Amoco/diesel/mart/atm/24hr, Phillips 66/Mesquite Grill/ diesel/mart, Meridy's Rest., McDonald's, Pizza Hut, Sonic, Subway, Day's Inn, Russells Inn, Super 8, Dumler RV Park, JJJ RV Park, Pennzoil, st patrol, **2 mi N**...Conoco, Total, Texaco, A&W, Dairy Queen, Peking Garden, Alco, AllPro Parts, IGA Foods

180   Balta Rd, to Russell, no facilities

175   KS 257, Gorham, **1 mi N**...gas, food, phone

172   Walker Ave, **S**...Walker Shed Café

168   KS 255, to Victoria, **S**...Ampride/diesel/mart, to Cathedral of the Plains

163   Toulon Ave, no facilities

161   Commerce Parkway, no facilities

159   US 183, Hays, **N**...Total/diesel/mart/24hr, Applebee's, Carlos O'Kelly's, Best Western, Comfort Suites, Fairfield Inn, Ford/Lincoln/Mercury, Plymouth/Dodge, Toyota, Wal-Mart SuperCtr/24hr, **S**...HOSPITAL, CHIROPRAC- TOR, DENTIST, Amoco/mart/24hr, Coastal/mart, Conoco/Golden OX/diesel/mart/24hr, Love's/mart, Phillips 66/ mart, Texaco/diesel/mart/atm, A&W, Arby's, Burger King, China Garden, Country Kitchen, El Cazador Mexican, Gutierrez Mexican, KFC, LJ Silver, McDonald's, Montana Mike's Steaks, Pheasant Run Pancakes, Pizza Hut, Pizza Inn, Sonic, Subway, Taco Bell, Taco Tico, Vagabond Rest., Village Inn Rest., Wendy's, Best Western, Comfort Inn, Day's Inn, Econolodge, Hampton Inn, Holiday Inn, Midway Motel, Motel 6, Ramada, Super 8, Advance Parts, Alco, Chevrolet/Mazda, Chrysler, FashionBug, Firestone/auto, Harley-Davidson, JC Penney, NAPA, U-Save Food, Walgreen, bank, cinema, mall, st patrol, **2 mi S**...Burger King, Dairy Queen, Budget Host, I-70 repair

157   US 183 S byp, to Hays, **S**...Gen Hayes Inn, museum, to Ft Hays St U

153   Yocemento Ave, no facilities

145   KS 247 S, Ellis, **S**...Casey's/gas, Texaco/mart, Alloway's Rest./TasteeFreez, Ellis House Inn, to Chrysler Museum, Railroad Museum, antiques

140   Riga Rd, no facilities

135   KS 147, Ogallah, **N**...Schreiner/diesel/café, Goodyear, **S**...to Cedar Bluff SP

132mm   **rest area both lanes, full(handicapped)facilities, picnic tables, litter barrels, petwalk**

128   US 283 N, WaKeeney, **N**...HOSPITAL, Conoco/diesel/mart/24hr, Budget Host, Super 8

127   US 283 S, WaKeeney, **N**...Phillips 66/McDonald's/diesel/mart/24hr, Dairy Queen, Nobody's Rest., Pizza Hut, KS Kountry Inn, Sundowner Motel, radiators, tires, **S**...Amoco/diesel/rest./atm/24hr, Conoco/Subway/diesel/mart, Best Western/rest., KOA, antiques, truck repair

120   Voda Rd, no facilities

115   KS 198 N, Banner Rd, Collyer, **N**...Sinclair/diesel/mart/radiators

107   KS 212, Castle Rock Rd, Quinter, **N**...HOSPITAL, Phillips 66/diesel/mart, Q Rest., Budget Host/rest., Chevrolet, RV camp, **S**...Conoco/diesel/mart/atm/24hr, Dairy Queen

E

W

Hays

Kansas Interstate 70

99	KS 211, Park, **1 mi N**...Sinclair/diesel, phone, camping
97mm	**rest area both lanes, full(handicapped)facilities, picnic tables, litter barrels, vending, petwalk, RV dump**
95	KS 23, to Hoxie, no facilities
93	KS 23, Grainfield, **N**...Conoco/diesel/repair
85	KS 216, Grinnell, **S**...Texaco/Dairy Queen/Stuckey's
79	Campus Rd, no facilities
76	US 40, to Oakley, **S**...Texaco/diesel/mart/café/repair/24hr, Phillips 66/mart, 1st Interstate Inn, KS Kountry Inn, Blue Beacon, **2 mi S**...HOSPITAL, Annie Oakley Motel, Best Western
70	US 83, to Oakley, **N**...Free Breakfast Inn, auto repair/24hr, **S**...HOSPITAL, Conoco/diesel/mart, Phillips 66/diesel/mart, Steakhouse Buffet, Tastee Treat, Annie Oakley Motel, Blue Beacon, Camp In RV Park, Prairie Dog Town, antiques, to Pizza Hut, Fick Museum
62	rd 24, Mingo, **S**...Ampride/diesel/24hr
54	Country Club Dr, Colby, **1 mi N**...HOSPITAL, Country Club Motel, RV camping
53	KS 25, Colby, **N**...HOSPITAL, Amoco/diesel/mart, Conoco/DQ/diesel/café/mart, Phillips 66/Subway/mart/atm, Total/diesel/mart, Arby's, Burger King, Deep Rock Café, KFC/Taco Bell, LJ Silver, McDonald's, MT Mike's Steaks, Pizza Hut, Sirloin Stockade, Sonic, Taco John's, Day's Inn, Holiday Inn Express, Ramada, Super 8, Welk-Um Inn, AutoZone, Dillon Food/drug, $General, Ford/Lincoln/Mercury, Goodyear, Pennzoil, Prairie Art Museum, Radio Shack, Wal-Mart, cinema, diesel repair, **S**...Total/diesel/mart, Village Inn Rest., Best Western, Comfort Inn, outlets
48.5mm	**rest area both lanes, full(handicapped)facilities, phone, picnic tables, litter barrels, RV camp/dump, vending, petwalk**
45	US 24 E, Levant, no facilities
36	KS 184, Brewster, **N**...Citgo/Stuckey's/diesel
35.5mm	Mountain/Central time zone
27	KS 253, Edson, no facilities
19	US 24, Goodland, **N**...Conoco/mart, Pizza Hut, Best Western, Motel 7, KOA, High Plains Museum
17	US 24, KS 27, Goodland, **N**...HOSPITAL, Conoco/mart/atm, Phillips 66/diesel/mart, Sinclair/mart, Dairy Queen, KFC/Taco Bell, McDonald's, Pizza Hut, Subway, Taco Place, Wendy's, Best Western/rest., Comfort Inn, Howard Johnson, Motel 6, Super 8, Pontiac/Olds/GMC, Wal-Mart SuperCtr, tires, **S**...Texaco/diesel/mart/rest./24hr, Total/A&W/diesel/mart/atm/24hr, K Motel, Mid-America Camping
12	rd 14, Caruso, no facilities
9	rd 11, Ruleton, no facilities
7.5mm	**Welcome Ctr eb/rest area wb, full(handicapped)facilities, info, phone, picnic tables, litter barrels, petwalk, vending, RV dump**
1	KS 267, Kanorado, **N**...gas/mart, phone
.5mm	weigh sta eb
0mm	Kansas/Colorado State Line

Colby

Goodland

E

W

Kansas Interstate 135(Wichita)

Exit #	Services
95b a	I-70, E to KS City, W to Denver. I-135 begins/ends on I-70, exit 250. US 81 continues nb.
93	KS 140, Salina, no facilities
92	Crawford St, **E**...Amoco/diesel/mart, Texaco, Braum's, Taco Bell, Best Western, Comfort Inn, Fairfield Inn, Holiday Inn, Super 8, **W**...Phillips 66/diesel/mart, Red Coach Inn
90	Magnolia Rd, **E**...Phillips 66/diesel/mart/atm, Texaco/mart, Carlos O'Kelly's, Chili's, Golden Corral, Hong Kong Buffet, McDonald's, Mesquite Grill, Schlotsky's, 1st Inn, Advance Parts, Buick/Subaru, Dillon's Foods, $General, Food4Less, HobbyLobby, JC Penney, Sears/auto, cinema, mall
89	Schilling Rd, **E**...Applebee's, Burger King, Fazoli's, Pizza Hut, Red Lobster, Sonic, Candlewood Suites, Country Inn Suites, Hampton Inn, Aldi Foods, Chevrolet, Honda, OfficeMax, Sam's Club, Target, Wal-Mart SuperCtr/24hr, bank, **W**...Casey's/gas
86	Mentor, Smolan, no facilities
82	KS 4, Assaria, no facilities
78	KS 4 W, Lindsborg, **E**...Texaco/Dairy Queen/Stuckey's
72	US 81, Lindsborg, **4 mi W**...HOSPITAL, food, phone, gas, lodging, camping, museum
68mm	**rest areas(both lanes exit left), full(handicapped)facilities, phone, picnic tables, litter barrels petwalk**

Salina

N

S

Kansas Interstate 135

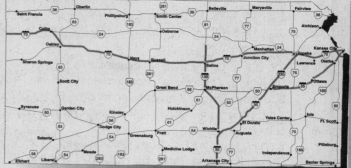

65	Pawnee Rd, no facilities
60	US 56, McPherson, Marion, **W**...HOSPITAL, Conoco/diesel/mart, Phillips 66/mart, Arby's, Braum's, KFC, McDonald's, Perkins, Pizza Hut, Red Coach Rest., Sirloin Stockade, Best Western, Super 8, Wal-Mart/auto
58	to Hutchinson, McPherson, no facilities
54	Elyria, no facilities
48	KS 260 E, Moundridge, **2 mi W**...gas, food, phone
46	KS 260 W, Moundridge, **2 mi W**...gas, food, phone
40	Hesston, **W**...Conoco/diesel/mart, Pizza Hut, Subway, Hesston Heritage Inn
34	KS 15, N Newton, **E**...Phillips 66/diesel/LP/mart, golf, **1/2 W**...LJ Silver, Subway, Kauffman Museum
33	US 50 E, to Peabody, no facilities
32	Broadway Ave, **E**...Applebee's, Chevrolet/Olds/Cadillac, Chrysler/Plymouth/Dodge/Jeep, Ford/Lincoln/Mercury
31	Newton, **E**...Ampride/diesel/mart, Conoco/diesel/mart, Texaco/diesel/motel/mart, KFC, Red Coach Rest., Day's Inn, 1st Inn, Super 8, **W**...Phillips 66/diesel/mart, Braum's, Sirloin Stockade, Best Western
30	US 50 W, KS 15(exits left from nb), Newton, to Hutchinson, **W**...HOSPITAL, Phillips 66/mart, AutoZone, Buick/Pontiac/GMC, Dillon's Foods/24hr, Wal-Mart SuperCtr
28	SE 36th St, **W**...Total/diesel/mart, Burger King, Subway, Taco Bell, Newton Outlets/famous brands
25	KS 196, to Whitewater, El Dorado, no facilities
23mm	**rest areas both lanes, full(handicapped)facilities, phone, picnic tables, litter barrels, vending, petwalk**
22	125th St, no facilities
19	101st St, phone, camping
17	85th St, **E**...Valley Ctr, KS Coliseum
16	77th St, no facilities
14	61st St, **E**...QT/Blimpie/mart, Total/mart, Cracker Barrel, Sonic, Taco Bell, Wendy's, Comfort Inn, **W**...Coastal/mart, KFC, McDonald's, Super 8, Goodyear/auto
13	53rd St, **W**...Phillips 66, Red Coach Rest., Best Western, Day's Inn
11b	I-235 W, KS 96, to Hutchinson, no facilities
a	KS 254, to El Dorado, no facilities
10b	29th St, Hydraulic Ave, no facilities
a	KS 96 E, no facilities
9	21st St, **E**...Amoco, Burger King, Wichita St U
8	13th St, **E**...Total/mart
7b	8th St, 9th St, **E**...School of Medicine
6b	1st St, 2nd St, **E**...AutoZone, **W**...Chevrolet, Chrysler/Jeep
5b	US 54, US 400, Kellogg Ave, **1 mi E**...Amoco, Total/mart, McDonald's, Spangles Rest., Wendy's
a	Lincoln St, **W**...QT/mart
4	Harry St, **1 mi E**...QT/mart, Church's Chicken, Denny's, Dunkin Donuts, McDonald's, Spangles Rest., Wendy's, **W**...Amoco
3	Pawnee Ave, **E**...QT/mart, **W**...Burger King, Boston Mkt, Spangles, Pawnee Inn
2	Hydraulic Ave, **W**...QT/mart, McDonald's
2mm	Arkansas River
1c	I-235 N, to airport
b a	US 81 S, 47th St, **E**...Coastal/mart, Potbelly's Rest., Comfort Inn, Day's Inn, Holiday Inn Express, **W**...CHIROPRACTOR, Phillips 66, Applebee's, Braum's, Burger King, Dairy Queen/Super Wok, Godfather's, KFC, LJ Silver, McDonald's, Pizza Hut, Quizno's, Spaghetti Jack's, Spangles Rest., Taco Bell, Taco Tico, Big Lots, Checker's Foods, Dillon's Foods, K-Mart, O'Reilly's Parts, Radio Shack
0mm	I-135 begins/ends on I-35, exit 42.

N

S

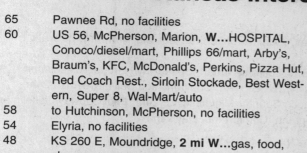

McPherson Newton Wichita

Kentucky Interstate 24

Exit #	Services
93.5mm	Kentucky/Tennessee state line
93mm	**Welcome Ctr wb, full(handicapped)facilities, phones, picnic tables, litter barrels, vending, petwalk**
91.5mm	Big West Fork Red River
89	KY 115, to Oak Grove, **N**...to Jeff Davis Mon St HS, **S**...Citgo/diesel/mart, Speedway/Hardee's/diesel/mart
86	US 41A, to Ft Campbell, Hopkinsville, Pennyrile Pkwy, **N**...HOSPITAL, Chevron/Taco Bell/diesel/mart/24hr, **S**...Amoco/Burger King/diesel/mart/24hr, Flying J/Conoco/Country Mkt/diesel/LP/24hr, Pilot/Subway/diesel/mart/24hr, Williams/diesel/rest./24hr, Waffle House, Baymont Inn, Comfort Inn, Day's Inn
79mm	Little River
73	KY 117, Newstead, to Gracey, no facilities
65	US 68, KY 80, to Cadiz, **S**...HOSPITAL, Amoco, BP/diesel/mart, Phillips 66/diesel/mart/24hr, Shell/diesel/mart/24hr, Cracker Barrel, Hardee's, KFC, Sherlock's Buffet, Holiday Inn Express, Knight's Inn, Super 8, antiques, golf, to NRA's
56	KY 139, to Cadiz, Princeton, **S**...Chevron/diesel/mart, Nat Rec Areas
47mm	Lake Barkley
45	KY 293, Saratoga, to Princeton, **N**...to KY St Penitentiary, **S**...Chevron/diesel/pizza/mart, Old Farmhouse Rest., Regency Inn, Lake Barkley RV Camping
42	to W KY Pkwy eb
40	US 62, US 641, Kuttawa, Eddyville, **N**...Country Hearth Inn(20mi), Relax Inn, Regency Inn(3mi), camping, **S**...BP/Wendy's/diesel/mart/24hr, Chevron, Shell/Burger King/Subway/Taco Bell/24hr, Sunshine Travel Plaza/diesel/mart/24hr, Day's Inn, Hampton Inn, W KY Factory Outlet, to Lake Barkley, KY Lake Rec Areas, camping
36mm	weigh sta both lanes, phones
34mm	Cumberland River
31	KY 453, to Grand Rivers, Smithland, **N**...Amoco/Blimpie/diesel/mart, Microtel, **S**...BP/mart/24hr, Miz Scarlett's Rest., Best Western, Grand Rivers Resort(3mi), NRA's, camping
29mm	Tennessee River
27	US 62, Calvert City, to KY Dam, **N**...BP, Chevron, Shell, Fireside Rest., Dairy Queen, KFC, McDonald's, Willow Pond Rest., Foxfire Motel, KOA, Super 8, Freightliner, **S**...Citgo, Coastal/diesel/rest./24hr, KY Dam Village Resort, BoatMart, Cypress Lakes Camp
25b a	to Calvert City, Purchase Pkwy, services 1 mi N
16	US 68, to Paducah, **S**...BP/diesel, Citgo, Southern Pride/diesel/rest./24hr, truckwash
11	rd 954, Husband Rd, to Paducah, **N**...Citgo, Best Western, **5 mi N**...Broken Spoke Rest., Burger King, Grecian Steaks, Holman House Café, Pizza Hut, Knight's Inn, Quality Inn, Duck Creek RV Park
7	US 45, US 62, to Paducah, **N**...HOSPITAL, Ashland/Subway/diesel/mart, BP/diesel, Citgo, Pet-tro/mart, Shell/diesel/mart/24hr, Burger King, LJ Silver, Taco Bell, BudgetHost(3mi), Quality Inn, **S... Welcome Ctr both lanes, full(handicapped)facilities, phones, vending, picnic tables, litter barrels, petwalk,** DENTIST, BP/diesel/mart, Shell/mart, Scot Gas/mart, Arby's, Chinese Rest., Golden Corral, Hardee's, KFC, Little Caesar's, Mardi Gras Café, McDonald's, Sonic, Taco John's, Denton Motel, Sunset Inn, FoodTown, K-Mart/drugs, Rite Aid, SuperValue Food/24hr, XpressLube, bank, carwash, cleaners/laundry
4	US 60, to Paducah, **N**...EYECARE, Shell/diesel/mart/24hr, Applebee's, Burger King, Chinese Rest., Denny's, McDonald's, O'Charley's, Outback Steaks, Courtyard, Day's Inn/rest., Drury Inn, Holiday Inn Express, Ramada Inn, Red Carpet Inn(1mi), Acme Boot Outlet, Hancock Fabrics, KY Tobacco Outlet, antiques, drugs, laundry, **S**...CHIROPRACTOR, BP/mart, Chevron/Domino's/mart, Pet-tro/mart, Atlanta Bread Co, Capt D's, ChuckeCheese, Cracker Barrel, Damon's, Denny's, El Chico's, Fazoli's, Godfather's, Hardee's/24hr, Logan's Roadhouse, Olive Garden, Ponderosa, Pizza Hut, Red Lobster, Ruby Tuesday, Ryan's, Shoney's, Steak'n Shake, Subway, Taco Bell, TGIFriday, Wendy's, Best Inn, Comfort Suites, Drury Suites, Hampton Inn, Motel 6, PearTree Inn, Thrifty Inn, Advance Parts, Aldi Foods, Books A Million, Circuit City, Dillard's, Goody's, Goodyear, JC Penney, Lowe's Bldg Supply, Office Depot, OfficeMax, PartsAmerica, PetsMart, Sam's Club, Sears/auto, TJ Maxx, ToysRUs, Wal-Mart SuperCtr/24hr, Western Auto, Millsprings Funpark, bank, cinema, mall
3	KY 305, to Paducah, **N**...Citgo/diesel/mart/24hr, Casa Mexicana, Huddle House/24hr, Comfort Inn/rest., Ramada Ltd, Super 8, **S**...BP/diesel/mart/24hr, Pilot/Subway/diesel/mart/24hr, Waffle Hut, Baymont Inn, Fern Lake Camping, antiques
0mm	Ohio River, Kentucky/Illinois state line

E

↑

↓

W

Paducah

Kentucky Interstate 64

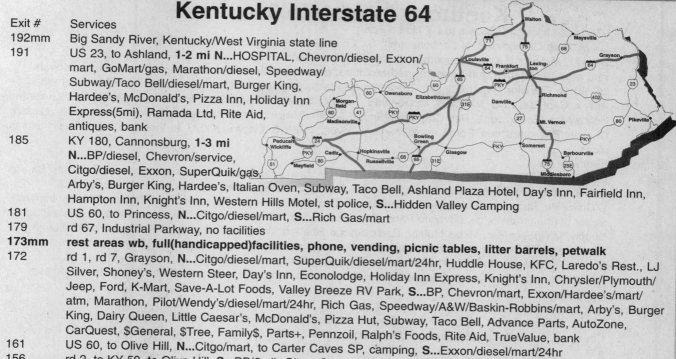

Exit #	Services
192mm	Big Sandy River, Kentucky/West Virginia state line
191	US 23, to Ashland, **1-2 mi N**...HOSPITAL, Chevron/diesel, Exxon/mart, GoMart/gas, Marathon/diesel, Speedway/Subway/Taco Bell/diesel/mart, Burger King, Hardee's, McDonald's, Pizza Inn, Holiday Inn Express(5mi), Ramada Ltd, Rite Aid, antiques, bank
185	KY 180, Cannonsburg, **1-3 mi N**...BP/diesel, Chevron/service, Citgo/diesel, Exxon, SuperQuik/gas, Arby's, Burger King, Hardee's, Italian Oven, Subway, Taco Bell, Ashland Plaza Hotel, Day's Inn, Fairfield Inn, Hampton Inn, Knight's Inn, Western Hills Motel, st police, **S**...Hidden Valley Camping
181	US 60, to Princess, **N**...Citgo/diesel/mart, **S**...Rich Gas/mart
179	rd 67, Industrial Parkway, no facilities
173mm	**rest areas wb, full(handicapped)facilities, phone, vending, picnic tables, litter barrels, petwalk**
172	rd 1, rd 7, Grayson, **N**...Citgo/diesel/mart, SuperQuik/diesel/mart/24hr, Huddle House, KFC, Laredo's Rest., LJ Silver, Shoney's, Western Steer, Day's Inn, Econolodge, Holiday Inn Express, Knight's Inn, Chrysler/Plymouth/Jeep, Ford, K-Mart, Save-A-Lot Foods, Valley Breeze RV Park, **S**...BP, Chevron/mart, Exxon/Hardee's/mart/atm, Marathon, Pilot/Wendy's/diesel/mart/24hr, Rich Gas, Speedway/A&W/Baskin-Robbins/mart, Arby's, Burger King, Dairy Queen, Little Caesar's, McDonald's, Pizza Hut, Subway, Taco Bell, Advance Parts, AutoZone, CarQuest, $General, $Tree, Family$, Parts+, Pennzoil, Ralph's Foods, Rite Aid, TrueValue, bank
161	US 60, to Olive Hill, **N**...Citgo/mart, to Carter Caves SP, camping, **S**...Exxon/diesel/mart/24hr
156	rd 2, to KY 59, to Olive Hill, **S**...BP(3mi), Citgo, Sunoco/diesel, Dairy Queen(3mi), NAPA(3mi)
148mm	weigh sta both lanes
141mm	**rest areas both lanes, full(handicapped)facilities, phone, vending, picnic tables, litter barrels, petwalk**
137	KY 32, to Morehead, **N**...BP/mart, Shell/diesel, China Garden, Dairy Queen, Big Lots, Kroger/deli, bank, laundry, tires, **S**...HOSPITAL, CHIROPRACTOR, Amoco/mart, Chevron/Burger King/diesel/mart/24hr, Exxon/diesel/mart, Sunoco, Fazoli's, Hardee's, KFC, Lee's Chicken, McDonald's, Papa John's, Ponderosa, Subway, Bet Western, Day's Inn, Holiday Inn/rest., Mtn Lodge Motel, Ramada Ltd, Shoney's Inn/rest., Super 8, Ace Hardware, Buick/Olds/Pontiac, Food Lion, Goodyear, Goody's, Radio Shack, Wal-Mart, flea mkt, st police
133	rd 801, Farmers, to Sharkey, **N**...Eagle Trace Golf/rest.(4mi), **S**...BP/diesel/mart, Cave Run Cabins(4mi)
123	US 60, to Salt Lick, Owingsville, **N**...Chevron/diesel/mart
121	KY 36, to Owingsville, **N**...BP/diesel/mart, Citgo/diesel/mart, Sunoco/diesel/mart, Chester Fried Chicken, Dairy Queen, Kountry Kettle, McDonald's, Tom's Pizza, $General, bank, carwash, drugs
113	US 60, to Mt Sterling, **N**...Citgo/diesel/mart
110	US 460, KY 11, Mt Sterling, **N**...Chevron/repair/24hr, Shell/Krystal/diesel/mart, Cracker Barrel, Fairfield Inn, Ramada Ltd, golf, **S**...HOSPITAL, BP/diesel/mart, Exxon/Subway/mart, Marathon, Rich Gas, Shell/mart/wash, SuperAmerica/diesel, Applebee's, Arby's, Burger King, Golden Corral, Jerry's Rest., KFC, Lee's Chicken, LJ Silver, McDonald's, Pizza Hut, Rio's Steaks, Shoney's, Wendy's, Budget Inn, Day's Inn/rest., Scottish Inn, Family$, Ford/Mercury, bank, carwash, **S on KY 686**...Ashland Gas, Hardee's, Little Caesar's, Peking Chinese, Taco Bell, Taco Tico, Advance Parts, Chevrolet/Olds, Chrysler/Plymouth/Dodge/Jeep, Food Lion, JC Penney, Pennzoil, Wal-Mart SuperCtr/24hr
108mm	**rest area wb, full(handicapped)facilities, phone, vending, picnic tables, litter barrels, petwalk**
101	US 60, no facilities
98.5mm	**rest area eb, full(handicapped)facilities, phone, vending, picnic tables, litter barrels, petwalk**
98	KY 402(from eb), no facilities
96b a	KY 627, to Winchester, Paris, **N**...BP/diesel/mart, 96/Sunoco/diesel/rest., **S**...VETERINARIAN, Citgo, Marathon/diesel/mart, Comfort Suites, Day's Inn, Hampton Inn, Chevrolet/Olds
94	KY 1958, Van Meter Rd, Winchester, **N**...Chevron/mart/24hr, Shell/diesel/mart, Holiday Inn Express, Super 8, flea mkt, **S**...HOSPITAL, CHIROPRACTOR, DENTIST, BP/diesel/mart, Citgo/mart, Marathon/mart, Speedway/diesel/mart, SuperAmerica/mart, Applebee's, Arby's, Burger King, Cantuckee Diner, Capt D's, Domino's, Fazoli's, Golden Corral, Great Wall Chinese, Hardee's, KFC, Lee's Chicken, Little Caesar's, LJ Silver, McDonald's, Papa John's, Pizza Hut, Rally's, Shoney's, Sonic, Subway, Tacos Too, Waffle House, Wendy's, Best Western, Travelodge, Advance Parts, AutoZone, Big Lots, Chrysler/Plymouth/Dodge/Jeep, Family$, Ford/Mercury, GNC, Goody's, K-Mart, Kroger/deli, Lowe's Bldg Supply, Peebles, Radio Shack, Valvoline/wash, Wal-Mart, Winn-Dixie/deli, bank, cleaners, tires, to Ft Boonesborough Camping
87	KY 859, Blue Grass Sta, no facilities
81	I-75 S, to Knoxville

I-64 and I-75 run together 7 mi. See Kentucky Interstate 75, exits 113-115.

E

W

Grayson

Lexington

Kentucky Interstate 64

Frankfort

75	I-75 N, to Cincinnati, access to KY Horse Park
69	US 62 E, to Georgetown, **N**...antiques(mi), **S**...Equus Run Vineyards(2mi)
65	US 421, Midway, **S**..Citgo/diesel/mart, Depot Rest., antiques
60mm	**rest area both lanes, full(handicapped)facilities, phone, picnic tables, litter barrels, vending, petwalk**
58	US 60, to Frankfort, **N**...BP/diesel, Chevron/diesel/mart, Citgo/diesel/mart, Shell/mart, SuperAmerica/mart, Arby's, Capt D's, KFC, Mkt Café, McDonald's, Sandy's Steaks, White Castle, Best Western/rest., Blue Grass Inn, Holiday Inn, Econolodge, Red Carpet Inn(3mi), Elkhorn Camping(5mi), Cadillac/Olds/Pontiac, Chevrolet/Nissan, Chrysler/Plymouth/Jeep, Ford/Lincoln/Mercury, Toyota, to KY St Capitol, KYSU, to Viet Vets Mem, tires, transmissions
55mm	Kentucky River
53b a	US 127, Frankfort, **N**...HOSPITAL, Chevron/mart/24hr, Shell/mart/24hr, SuperAmerica(1mi), Applebee's, Baskin-Robbins, Big Boy, Burger King, Chili's, Columbia Steaks, Fazoli's, Hardee's, KFC, LJ Silver, McDonald's, O'Charley's, Pizza Hut, Shoney's, Steak'n Shake, Taco Bell, Wendy's, Day's Inn, Hampton Inn, Holiday Inn(4mi), Super 8(1mi), Travelodge, Advance Parts, Goodyear, Goody's, JC Penney, K-Mart, Kroger/24hr, Lowe's Bldg Supply, MailBoxes Etc, Midas Muffler, Radio Shack, Rite Aid, Wal-Mart SuperCtr/24hr, Winn-Dixie/deli/bakery, Ancient Age Tour, bank, cinema, farmer's mkt, laundry, to KY St Capitol, st police, **S**...Citgo
48	KY 151, to US 127 S, **S**...Chevron/mart/atm/24hr, Shell/Subway/diesel/mart
43	KY 395, Waddy, **N**...Flying J/Conoco/Country Mkt/diesel/LP/mart/RVparking/24hr, **S**...Citgo/Stuckey's/diesel/rest./24hr
38.5mm	weigh sta both lanes
35	KY 53, Shelbyville, **N**...BP/diesel/mart, Chevron/diesel/mart, Cracker Barrel, KFC, McDonald's(1mi), Subway, Chevrolet, Ford/Mercury, Kroger/gas/deli, antiques, **S**...Shell/Noble Roman's/diesel/mart, Holiday Inn Express, golf
32b a	KY 55, Shelbyville, **1-3 mi N**...HOSPITAL, Shell, Arby's, Burger King, Dairy Queen, McDonald's, Pizza Hut, Wendy's, Best Western, Country Hearth Inn, Day's Inn, Shelby Motel, Probus Log Cabin, **S**...Taylorsville Lake SP
29mm	**rest area both lanes, full(handicapped)facilities, info, phone, picnic tables, litter barrels, vending, petwalk**
28	Veechdale Rd, Simpsonville, **N**...Citgo, Pilot/Subway/diesel/mart/24hr, Oasis Mkt/diner, Old Stone Inn, golf, **S**...BP/diesel/mart
19b a	I-265, Gene Snyder Fwy, **N**...to Tom Sawyer SP

Louisville

17	S Blankenbaker, **N**...DM/diesel/mart/atm, Staybridge Suites, Olds, **S**...BP/mart, Chevron/mart/wash, Shell/mart, Thornton's/Subway/diesel/mart, Arby's, Big Boy, Burger King, Cracker Barrel, HomeTown Buffet, King Buffet, Kingfish Rest., McDonald's, Ruby Tuesday, Subway, Waffle House, Wendy's, Best Western Signature, Candlewood Suites, Comfort Suites, Country Inn Suites, Hilton Garden, Holiday Inn Express, MainStay Suites, Microtel, Sleep Inn, Garden Ridge, Sam's Club, Outlets Ltd, bank
15	Hurstbourne Lane, Louisville, **N**...CHIROPRACTOR, Amoco, BP/diesel/wash, Chevron/diesel/wash, Shell/diesel/mart/wash, Arby's, Benihana, Bob Evans, Burger King, Chili's, Don Pablo's, Harper's Rest., McDonald's, Olive Garden, Papa John's, Perkins, Sichuan Garden Chinese, Steakout, Subway, TGIFriday, Tumbleweed Mexican, Waffle House, AmeriSuites, Courtyard, Fairfield Inn, Holiday Inn, Knight's Inn, Red Roof Inn, Travelodge, Barnes&Noble, CompUSA, Kroger, Lowe's Bldg Supply, bank, **S**...BP/diesel/mart, Shell, Meijer/diesel/mart/24hr, Thornton's Gas/mart, Applebee's, Asian Pearl Chinese, Blimpie, China Star, ChuckeCheese, Dairy Queen, Damon's, Dillon's Rest., Hardee's, Macaroni Grill, O'Charley's, Piccadilly's, Shoney's, Shogun Japanese, Steak'n Shake, Wendy's, Day's Inn, Doubletree Hotel, Hampton Inn, Marriott, Radisson, Red Carpet Inn, Cadillac, Chevrolet, EyeMart, FashionBug, Home Depot, Honda, Infiniti, Michael's, Mitsubishi, Pier 1, Pontiac/GMC, Radio Shack, Shoe Carnival, Staples, Target, Town Fair Foods, Wal-Mart/drugs
12b	I-264 E, Watterson Expsway, **1 exit N**...Chevron, Denny's, Hop's Grill, Logan's Roadhouse, McDonald's, Wendy's, Acura, Dillard's, Ford, Goodyear/auto, JC Penney, Kohl's, Lord&Taylor, Sears/auto, Service Merchandise, Suzuki, mall
a	I-264 W, access to HOSPITAL
10	Cannons Lane, no facilities
8	Grinstead Dr, Louisville, **S**...BP, Chevron, Southern Baptist Seminary, Presbyterian Seminary, Jim Porter's Dining
7	US 42, US 62, Mellwood Ave, Story Ave, **N**...Marathon/mart
6	I-71 N(from eb), to Cincinnati, no facilities
5a	I-65, S to Nashville, N to Indianapolis
b	3rd St, Louisville, **N**...Hardee's, Joe's CrabShack, McDonald's, Ramada Inn, **S**...HOSPITAL, Galt House Hotel, Kingfish Rest.
4	9th St, Roy Wilkins Ave, **S**...KY Art Ctr, science museum, downtown
3	US 150 E, to 22nd St, **S**...MEDICAL CARE, Chevron/mart, DairyMart/gas, Shell/mart, Dairy Queen, McDonald's
2	(from eb), **S**...Chevron/mart, Shell/mart, McDonald's
1	I-264 E, to Shively, **S**...airport, zoo
0mm	Ohio River, Kentucky/Indiana state line

E

W

Kentucky Interstate 65

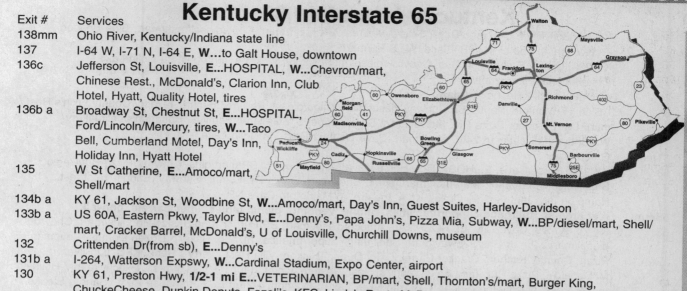

Exit #	Services
138mm	Ohio River, Kentucky/Indiana state line
137	I-64 W, I-71 N, I-64 E, **W**...to Galt House, downtown
136c	Jefferson St, Louisville, **E**...HOSPITAL, **W**...Chevron/mart, Chinese Rest., McDonald's, Clarion Inn, Club Hotel, Hyatt, Quality Hotel, tires
136b a	Broadway St, Chestnut St, **E**...HOSPITAL, Ford/Lincoln/Mercury, tires, **W**...Taco Bell, Cumberland Motel, Day's Inn, Holiday Inn, Hyatt Hotel
135	W St Catherine, **E**...Amoco/mart, Shell/mart
134b a	KY 61, Jackson St, Woodbine St, **W**...Amoco/mart, Day's Inn, Guest Suites, Harley-Davidson
133b a	US 60A, Eastern Pkwy, Taylor Blvd, **E**...Denny's, Papa John's, Pizza Mia, Subway, **W**...BP/diesel/mart, Shell/mart, Cracker Barrel, McDonald's, U of Louisville, Churchill Downs, museum
132	Crittenden Dr(from sb), **E**...Denny's
131b a	I-264, Watterson Expswy, **W**...Cardinal Stadium, Expo Center, airport
130	KY 61, Preston Hwy, **1/2-1 mi E**...VETERINARIAN, BP/mart, Shell, Thornton's/mart, Burger King, ChuckeCheese, Dunkin Donuts, Fazoli's, KFC, Lindy's Rest., McDonald's, Papa John's, Pepper Shaker Chili, PoFolks, Ponderosa, Rally's, Taco Bell, Taco Tico, Waffle House, Wendy's, Red Roof, Super 8, All Pro Parts, Chevrolet/Nissan, Ford, Goodyear, Midas Muffler, PepBoys, Staples, U-Haul, Walgreen, Winn-Dixie
128	KY 1631, Fern Valley Rd, **E**...Amoco/mart/24hr, BP/diesel/mart, Chevron/mart, Shell/mart, Thornton's/Subway/diesel, Arby's, Big Boy, Bojangles, Chinese Rest., Hardee's, McDonald's, Outback Steaks, Shoney's, Substation II, Waffle House, Holiday Inn, Signature Inn, Thrifty Dutchman, Lowe's Bldg Supply, bank, **W**...Ford, UPS Depot
127	KY 1065, outer loop, **W**...McDonald's, to Louisville Motor Speedway
125b a	I-265 E, KY 841, Gene Snyder Fwy
121	KY 1526, Brooks Rd, **E**...Amoco/mart, Chevron/mart, Arby's, Burger King, Cracker Barrel, Shoney's, Baymont Inn, Fairfield Inn, **W**...BP/diesel/mart, Pilot/Subway/Taco Bell/diesel/mart/24hr, Shell/diesel/mart, Waffle House, Comfort Inn, Hampton Inn, Holiday Inn Express, Quality Inn
117	KY 44, Shepherdsville, **E**...BP/mart, Shell/diesel/mart, Hardee's, Pizza Hut, Best Western/rest., Comfort Inn, Day's Inn/rest., Howard Johnson, Travelodge, KOA(2mi), **W**...DENTIST, Amoco/diesel/mart/atm, Chevron/diesel/mart/24hr, SuperAmerica/diesel/mart, Arby's, Burger King, Dairy Queen, Fazoli's, LJ Silver, KFC, McDonald's, Mr Gatti's/Italian, Papa John's, Ponderosa, Shoney's, Sonic, Subway, Taco Bell, Waffle House, Wendy's, White Castle, Hampton Inn, Motel 6, Super 8, Big Lots, $General, Family$, NAPA, Rite Aid, Winn-Dixie/deli, auto repair, cleaners/laundry, hardware
116.5mm	Salt River
116	KY 480, KY 61, **E**...Shell/diesel/mart, **W**...Citgo/mart, antiques
114mm	**rest area sb, full(handicapped)facilities, phone, vending, picnic tables, litter barrels, petwalk**
112	KY 245, Clermont, **E**...Shell/diesel/mart, Jim Beam Outpost, Bernheim Forest, to My Old Kentucky Home SP
105	KY 61, Lebanon Junction, **W**...Shell/Davis Bros. Travel Plaza/diesel/rest./mart/24hr
102	KY 313, to KY 434, to Radcliff, **E**...to Patton Museum
94	US 62, Elizabethtown, HOSPITAL, **E**...Citgo/diesel/mart/wash, Shell/mart, Denny's, Golden Corral, Hardee's, KFC, Waffle House, White Castle, Day's Inn, Red Roof Inn, Super 8(2mi), Travelodge, bingo, KOA(1mi), **W**...HOSPITAL, BP, Chevron/diesel/mart/24hr, SuperAmerica/diesel/mart/atm, Swifty Gas, Blimpie, Burger King, Chinese Rest., Cracker Barrel, McDonald's, Mr Gatti's, Papa John's, Pizza Hut, Ryan's, Shoney's, Subway, Uncle Bud's River Café, Wendy's, Western Steer, Comfort Inn, Hampton Inn, Holiday Inn, Lincoln Trail Motel, Motel 6, Ramada Ltd, Advance Auto Parts, CVS Drug, Kroger, Olymco Muffler/brake, PartsAmerica, Western Wear, bank, tires, st police
93	to Bardstown, to BG Pky, **E**...to My Old KY Home SP, Maker's Mark Distillery
91	US 31 W, KY 61, Elizabethtown, WK Pkwy, **E**...Big T/Chevron/mart/24hr, Citgo/diesel, Shell/mart, LJ Silver, Omelet House, Budget Motel, Commonwealth Lodge, Howard Johnson, Red Carpet, Ryder Trucks, carwash, to A Lincoln B'place, **W**...HOSPITAL, Citgo, Jerry's Rest./24hr, Lee's Chicken, McDonald's(3mi), Best Western, Rodeside Inn, bank
90mm	weigh sta both lanes
86	KY 222, Glendale, **E**...Speedway/diesel/mart, Krispy Kreme, Glendale Camping, Hist Dist, diesel repair, **W**...Texaco/diesel/rest./mart/24hr, Economy Inn
83mm	Nolin River
82mm	**rest area sb, full(handicapped)facilities, phone, vending, picnic tables, litter barrels, petwalk**
81.5mm	**rest area nb, full(handicapped)facilities, phone, vending, picnic tables, litter barrels, petwalk**

N
S

Louisville

Elizabethtown

Kentucky Interstate 65

81	KY 84, Sonora, **E**...Ashland/Davis Bros/diesel/rest./24hr, BP/mart, Citgo/diesel/rest./mart, Blue Beacon Truckwash, to Abe Lincoln B' Place, **W**...Shell/mart
76	KY 224, Upton, **E**...Ashland Gas, Chevron/diesel/mart, Citgo/Stuckey's, motel, **W**...to Nolin Lake
75mm	eastern/central time zone
71	KY 728, Bonnieville, no facilities
65	US 31 W, Munfordville, **E**...BP/Subway/diesel/mart, Citgo/diesel, Phillips 66/mart, Texaco/ County Fixin's Rest./mart, Dairy Queen, Pizza Hut, McDonald's, Sonic, Super 8, $General, Family$, Houchen's/deli/bakery, **W**...Chevron/diesel/mart/24hr, Shell/mart, Cave Country Rest., to Nolin Lake
61mm	Green River
58	KY 218, Horse Cave, **E**...HOSPITAL, Horse Cave Motel, info, **W**...BP/mart/gifts, Chevron/Pizza Hut/Radio Shack/diesel/mart/24hr, Marathon/mart/repair, Shell/mart, Dee's Rest./24hr, Subway, Budget Host/rest., Hampton Inn, Jent Factory Outlet, KOA, theatre, to Mammoth Cave NP
55mm	**rest area sb, full(handicapped)facilities, phone, vending, picnic tables, litter barrels, petwalk**
53	KY 70, KY 90, Cave City, **E**...HOSPITAL, Amoco/diesel/mart, BP/Burger King/Baskin-Robbins/mart, Chevron/diesel/mart/24hr, Jr/Gas/Subway/mart, Shell/mart, SuperAmerica, Texaco, Baker's Dozen Donuts, Caveman Pizza, Country Kitchen, Cracker Barrel, Dairy Queen, Hickory Villa Rest., Jerry's Rest., KFC, LJ Silver, McDonald's, Pizza Hut, Taco Bell, Wendy's, Best Western, Comfort Inn, Day's Inn/rest., Executive Inn, Holiday Inn Express, Quality Inn, Super 8, Barren River Lake SP(24mi), **1/2 mi E**...Farmhouse Rest., Sahara Steaks, Star Motel, Chevrolet/Buick, museum, **W**...Onyx Cave, Mammoth Cave NP, Jellystone Camping
48	KY 255, Park City, **E**...Citgo, Shell/diesel/mart/24hr, Best Western(1mi), Parkland Motel, **W**...to Mammoth Cave NP, RV camping
43	Cumberland Pky(toll), to Barren River Lake SP
39.5mm	**rest area nb, full(handicapped)facilities, phone, vending, picnic tables, litter barrels, petwalk**
38	KY 101, Smiths Grove, **W**...BP/diesel, Chevron/mart, Keystop/gas/mart, Shell/mart, Donita's Diner, McDonald's, Bryce Motel, Victorian House B&B, Seven Springs Park, RV camping, antiques, groceries
36	US 68, KY 80, Oakland, no facilities
30mm	**rest area sb, full(handicapped)facilities, phone, vending, picnic tables, litter barrels, petwalk**
28	to US 31 W, Bowling Green, **W**...HOSPITAL, BP/mart, Shell/Blimpie/mart/wash/24hr, Hardee's, Jerry's Rest., Wendy's, Best Western, Country Hearth Inn, Hatfield Inn, Value Lodge, Corvette Museum, **1-2 mi W**...CHIROPRACTOR, Chevron/diesel/mart, Marathon/Church's Chicken/mart, Swifty Gas/mart, Roxie's Rest., Houchen's Foods, Northgate Drugs, Family$, NAPA Autocare, bank, laundry, **3 mi W**...Camping World RV service/supplies, to WKYU
22	US 231, Bowling Green, **E**...Chevron/mart, Citgo/diesel/mart, Exxon/mart, Shell/mart/wash, Cracker Barrel, Denny's, Domino's, Old KY Home Hams, Ryan's, Tabatha's Café, Waffle House, Best Western, Comfort Inn, Day's Inn, Econolodge, Fairfield Inn, Microtel, Quality Inn, Ramada/rest., Super 8, Mark Mufflers/lube, USPO, RV camping, bank, **W**...HOSPITAL, BP/mart/wash, Chevron/diesel/mart, Exxon/TCBY/mart, Marathon/Bullets/mart, RaceTrac/mart/atm, Shell/Blimpie/diesel/mart, Speedway/mart, SuperAmerica/mart, Applebee's, Arby's, Bob Evans, Beijing Chinese, Burger King, Capt D's, CheeseShop, ChiChi's, CiCi's Pizza, Fazoli's, Garfield's Rest., GD Ritzy's Rest., Hardee's, HomeTown Buffet, Hop's Grill, Italian Oven, KFC, Krystal, Little Caesar's, LoneStar Steaks, Mancino's Grinders, McDonald's, Montana Grille, C'Charley's, Olive Garden, Outback Steaks, Pizza Hut, Ponderosa, Rafferty's, Red Lobster, Santa Fe Steaks, Shoney's, Sonic, Steak'n Shake, Subway, Taco Bell, Toots Diner, Tumbleweed Grill, Waffle House, Wendy's, White Castle, Baymont Inn, Courtyard, Drury Inn, Executive Inn, Hampton Inn, Holiday Inn, Motel 6, News Inn, Plaza Hotel, Scottish Inn, Acme Boots, Advance Parts, Best Buy, Buick/GMC, Castner-Knott, Chevrolet, Chrysler/Jeep, CVS Drug, $General, FashionBug, Firestone, Ford/Lincoln/Mercury, Goodyear, Honda, Houchen's Foods, Hyundai/Isuzu/Subaru, JC Penney, K-Mart, Kroger/24hr, Mazda, Nat's Outdoor Sports, Nissan, Office Depot, Olds/Cadillac, Pets&More, Pontiac, RV camping, Scotty's Parts, Sears, Toyota, ToysRUs, U-Haul, ValuVision, Winn-Dixie, XpressLube, Ziebart TidyCar, airport, antiques, bank, carwash, cinema, cleaners/laundry, mall
20	WH Natcher Toll Rd, to Bowling Green, access to W KY U, KY st police
6	KY 100, Franklin, **E**...BP/diesel/mart/24hr, Citgo/diesel/24hr, Old South Diner, **W**...HOSPITAL, Express/Wendy's/diesel/24hr, Speedway/Subway/diesel/mart/24hr, Loretta Lynn's Kitchen, Miss Penny's Rest./antiques, Day's Inn, Super 8, PetroLube/tires/repair
4mm	weigh sta nb
2	US 31 W, to Franklin, **E**...Flying J/Conoco/diesel/LP/rest./24hr, Texaco/Keystop/Burger King/diesel/24hr, Boot Place, **W**...HOSPITAL, BP/diesel/mart, Cracker Barrel, Huddle House, McDonald's, Shoney's, Comfort Inn, Hampton Inn, Holiday Inn Express, Quality Inn, Kentucky Motel, Ramada Ltd, LottoLand Mkt/grill
1mm	**Welcome Ctr nb, full(handicapped)facilities, phone, vending, picnic tables, litter barrels, petwalk**
0mm	Kentucky/Tennessee state line

N

S

Bowling Green

Franklin

184

Kentucky Interstate 71

Exit #	Services

Ohio River, Kentucky/Ohio state line

I-71 and I-75 run together 19 miles. **See Kentucky Interstate 75, exits 175-192.**

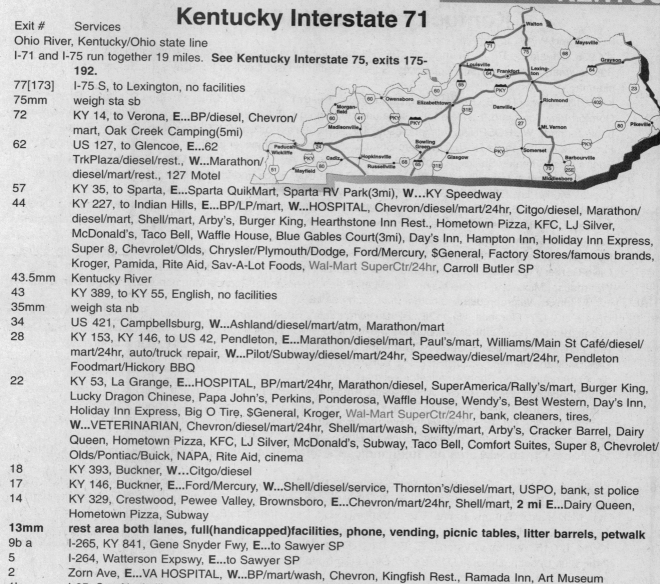

77[173]	I-75 S, to Lexington, no facilities
75mm	weigh sta sb
72	KY 14, to Verona, **E**...BP/diesel, Chevron/mart, Oak Creek Camping(5mi)
62	US 127, to Glencoe, **E**...62 TrkPlaza/diesel/rest., **W**...Marathon/diesel/mart/rest., 127 Motel
57	KY 35, to Sparta, **E**...Sparta QuikMart, Sparta RV Park(3mi), **W**...KY Speedway
44	KY 227, to Indian Hills, **E**...BP/LP/mart, **W**...HOSPITAL, Chevron/diesel/mart/24hr, Citgo/diesel, Marathon/diesel/mart, Shell/mart, Arby's, Burger King, Hearthstone Inn Rest., Hometown Pizza, KFC, LJ Silver, McDonald's, Taco Bell, Waffle House, Blue Gables Court(3mi), Day's Inn, Hampton Inn, Holiday Inn Express, Super 8, Chevrolet/Olds, Chrysler/Plymouth/Dodge, Ford/Mercury, $General, Factory Stores/famous brands, Kroger, Pamida, Rite Aid, Sav-A-Lot Foods, Wal-Mart SuperCtr/24hr, Carroll Butler SP
43.5mm	Kentucky River
43	KY 389, to KY 55, English, no facilities
35mm	weigh sta nb
34	US 421, Campbellsburg, **W**...Ashland/diesel/mart/atm, Marathon/mart
28	KY 153, KY 146, to US 42, Pendleton, **E**...Marathon/diesel/mart, Paul's/mart, Williams/Main St Café/diesel/mart/24hr, auto/truck repair, **W**...Pilot/Subway/diesel/mart/24hr, Speedway/diesel/mart/24hr, Pendleton Foodmart/Hickory BBQ
22	KY 53, La Grange, **E**...HOSPITAL, BP/mart/24hr, Marathon/diesel, SuperAmerica/Rally's/mart, Burger King, Lucky Dragon Chinese, Papa John's, Perkins, Ponderosa, Waffle House, Wendy's, Best Western, Day's Inn, Holiday Inn Express, Big O Tire, $General, Kroger, Wal-Mart SuperCtr/24hr, bank, cleaners, tires, **W**...VETERINARIAN, Chevron/diesel/mart/24hr, Shell/mart/wash, Swifty/mart, Arby's, Cracker Barrel, Dairy Queen, Hometown Pizza, KFC, LJ Silver, McDonald's, Subway, Taco Bell, Comfort Suites, Super 8, Chevrolet/Olds/Pontiac/Buick, NAPA, Rite Aid, cinema
18	KY 393, Buckner, **W**...Citgo/diesel
17	KY 146, Buckner, **E**...Ford/Mercury, **W**...Shell/diesel/service, Thornton's/diesel/mart, USPO, bank, st police
14	KY 329, Crestwood, Pewee Valley, Brownsboro, **E**...Chevron/mart/24hr, Shell/mart, **2 mi E**...Dairy Queen, Hometown Pizza, Subway
13mm	**rest area both lanes, full(handicapped)facilities, phone, vending, picnic tables, litter barrels, petwalk**
9b a	I-265, KY 841, Gene Snyder Fwy, **E**...to Sawyer SP
5	I-264, Watterson Expswy, **E**...to Sawyer SP
2	Zorn Ave, **E**...VA HOSPITAL, **W**...BP/mart/wash, Chevron, Kingfish Rest., Ramada Inn, Art Museum
1b	I-65, S to Nashville, N to Indianapolis
a	I-64 W, to St Louis, Louisville. I-71 begins/ends on I-64, exit 6 in Louisville.

N ↕ **S**

Kentucky Interstate 75

Exit #	Services
193mm	Ohio River, Kentucky/Ohio state line
192	5th St(from nb), Covington, **E**...BP/mart, Chevron, Shell/mart/wash, Speedway/mart, Big Boy, Bonanza, Burger King, McDonald's, Perkins, Waffle House, White Castle, Clarion, Courtyard, Extended Stay America, Holiday Inn, Ford/Lincoln/Mercury, Subaru/VW, **W**...Hampton Inn
191	12th St, Covington, **E**...HOSPITAL, Donut Café, Dodge, beer museum, laundry, same as 192
189	KY 1072(from sb), Kyles Lane, **W**...BP/diesel, Marathon/diesel, Shell/mart/diesel/wash, SuperAmerica, Big Boy, Hardee's, Substation II, Pizza Hut, Day's Inn, Ramada Inn, SteinMart, Thriftway Food/drug, Walgreen, Vision 1, bank, laundry, same as 188
188	US 25, US 42, Dixe Hwy, **W**...DENTIST, Day's Inn, Holiday Inn, Ramada, same as 189
186	KY 371, Buttermilk Pike, Covington, **E**...BP, Citgo/diesel/mart, DrawBridge Rest., Oriental Wok, Best Western, Cross Country Inn, **W**...Ashland/mart, BP/mart, Shell/mart, Sunoco/diesel/mart, Arby's, Bob Evans, Burger King, Domino's, Fazoli's, Hardee's, LJ Silver, Mr Gatti's, McDonald's, Outback Steaks, Pizza Hut, Subway, IGA Foods, Drug Emporium, JiffyLube, bank

N ↕ **S**

Kentucky Interstate 75

185	I-275 E and W, **W**...to airport
184	KY 236, Donaldson Rd, to Erlanger, **E**...BP/deli/mart, Citgo/mart, Double Dragon Chinese, Rally's, **W**...Ashland/ Subway/mart, Marathon/mart, Speedway, Sunoco, Waffle House, Comfort Inn, Day's Inn, Econolodge, HoJo's, cinema
182	KY 1017, Turfway Rd, **E**...BP/mart, Big Boy, Blimpie, Krispy Kreme, Lee's Chicken, Penn Sta Subs, Ryan's, Comfort Inn, Courtyard, Fairfield Inn, Signature Inn, Big Lots, FashionBug, Office Depot, ProGolf, Winn-Dixie, bank, funpark, **W**...HOSPITAL, Meijer/diesel/24hr, Applebee's, Boston Mkt, Burger King, Cracker Barrel, Fuddrucker's, Italianni's Rest., Longhorn Steaks, Ming Garden Chinese, O'Charley's, Rafferty's, Schlotsky's, Shell's Rest., Steak'n Shake, Tumbleweeds Grill, Wendy's, AmeriSuites, Ashley Qtrs Hotel, Extended Stay America, Hampton Inn, Hilton Inn, Studio+ Suites, Best Buy, Biggs Foods, Home Depot, Kohl's, Lowe's Bldg Supply, MediaPlay, OfficeMax, Sam's Club, Turfway Park Racing, Wal-Mart, cinema
181	KY 18, Florence, **E**...CHIROPRACTOR, Speedway/mart, Swifty, TA/Citgo/diesel/rest./24hr, Goodfellow's Dining, Shoney's, Waffle House, Best Western, Cross Country Inn, **W**...BP/diesel/mart/Procare, Citgo/mart, Suburban Lodge, JC Penney, K-Mart, Dodge, Ford, Mazda, Nissan, Toyota
180-A	Mall Rd(from sb), **W**..ChiChi's, ChuckeCheese, GoldStar Chili, Hardee's, Honeybaked Ham, Old Country Buffet, Olive Garden, Taco Bell, Barnes&Noble, Dick's Sports, Home Goods, Jo-Ann Fabrics, Kroger, Lazarus, Mens Wearhouse, Michael's, Pearle Vision, PetsMart, Pier 1, Sears/auto, Service Merchandise, Staples, Sun Electronics, TJ Maxx, Walgreen, bank, cinema, mall, same as 180
180	US 42, US 127, Florence, Union, **E**...BP/diesel/mart/atm, Speedway/mart, Thornton/mart, BBQ, Big Boy, Bob Evans, Burger King, Capt D's, Jalapeno's Mexican, LJ Silver, Main Moon Chinese, McDonald's, Pizza Hut, Rally's, Red Lobster, Substation II, Subway, WarmUps Café, Wendy's, Knight's Inn, Motel 6, Ramada Inn/rest., Super 8, Olds/Cadillac, funpark, **W**...CHIROPRACTOR, DENTIST, VETERINARIAN, BP/mart, Chevron/diesel/mart, Shell/ diesel/mart/atm/24hr, SuperAmerica/diesel/mart, Arby's, Burger King, Dairy Queen, KFC, Little Caesar's, Perkins/ 24hr, Ponderosa, Waffle House, White Castle, Kroger/deli, Budget Host, Envoy Inn, Holiday Inn, Travelodge, AutoZone, CarX Muffler, Circuit City, CVS Drug, Michel Tire, Midas Muffler, NTB, PepBoys, QuikStop, Valvoline
178	KY 536, Mt Zion Rd, **E**...BP/diesel, Shell/diesel/mart/atm, Sunoco/diesel, GoldStar Chili, Hometown Pizza, Jersey Mike's Subs, Rally's, Steak'n Shake, Subway, Goodyear/auto, Winn-Dixie
177mm	**Welcome Ctr sb/rest area nb, full(handicapped)facilities, phone, vending, picnic tables, litter barrels, RV dump**
175	KY 338, Richwood, **E**...TA/BP/Taco Bell/diesel/rest., Pilot/Subway/diesel/mart, Burger King, White Castle/24hr, Holiday Inn Express, Florence RV Park, **W**...BP/Wendy's/mart, Pilot/Subway/diesel/mart/24hr, Shell/diesel/mart/ atm, McDonald's, Snappy Tomato Pizza, Waffle House, Day's Inn/rest., Econolodge, to Big Bone Lick SP
173	I-71 S, to Louisville
171	KY 14, KY 16, Walton, to Verona, **E**...BP, Citgo, Dairy Queen, Waffle House, **W**...Flying J/Conoco/diesel/LP/rest./ 24hr, Blue Beacon, Delightful Dave's RV Ctr, to Big Bone Lick SP
168mm	weigh sta sb
166	KY 491, Crittenden, **E**...BP/mart, Citgo, Marathon/A&W/Taco Bell/diesel/mart, KOA(3mi), **W**...Chevron/mart, Shell/Subway/diesel/mart, Burger King
159	KY 22, Dry Ridge, to Owenton, **E**...HOSPITAL, BP, Marathon/mart, Shell/diesel/mart/wash, Arby's/24hr, Burger King, KFC, McDonald's, Taco Bell, Waffle House, Wendy's, Dry Ridge Inn, Microtel, Super 8, Wal-Mart SuperCtr/ 24hr, **W**...Speedway/mart, Country Grill, Shoney's, Hampton Inn, Holiday Inn Express, Toyota, Dry Ridge Outlets/ famous brands
154	KY 36, Williamstown, **E**...HOSPITAL, Citgo/diesel/LP, Shell/diesel/mart, Alice's Rest., Chester Fried Chicken, Skyway Inn/rest., Red Carpet Inn, to Kincaid Lake SP, **W**...BP/diesel/mart, Marathon/diesel, Day's Inn, HoJo's
144	KY 330, Corinth, to Owenton, **E**...Marathon/Noble's Truck Plaza/diesel/rest., Three Springs Camping, **W**...BP, K&T Motel
136	KY 32, to Sadieville, **W**...Chevron/mart
130.5mm	weigh sta nb
129	KY 620, Delaplain Rd, **E**...Pilot/Subway/diesel/mart/24hr, Waffle House, Day's Inn, Motel 6, Ramada Ltd, **W**...Shell/mart/24hr, Speedway/Hardee's/diesel/mart/24hr
127mm	**rest area both lanes, full(handicapped)facilities, phone, vending, picnic tables, litter barrels, petwalk**
126	US 62, to US 460, Georgetown, **E**...BP/mart, Chevron/mart/24hr, Marathon/mart, Big Boy, McDonald's, Econolodge, Flag Inn, **W**...HOSPITAL, BP/mart/wash, Shell/mart/wash/24hr, Speedway/A&W/diesel/mart/atm, Cracker Barrel, Fazoli's, Golden Corral, KFC, Best Western, Comfort Suites, Country Inn Suites, Hampton Inn, Holiday Inn Express, Microtel, Shoney's Inn/rest., Subway, Waffle House, Wendy's, Super 8, Winner's Circle Motel, Chevrolet/Pontiac/Olds/Buick, FashionBug, Ford, K-Mart/Little Caesar's, Outlets/famous brands, bank, to Georgetown Coll, same as 125

Florence

Kentucky Interstate 75

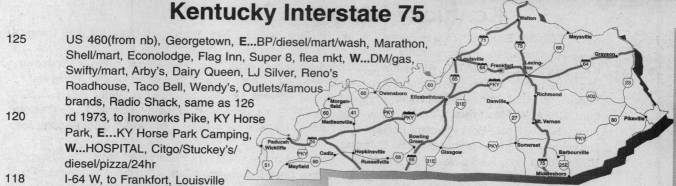

125 US 460(from nb), Georgetown, **E...**BP/diesel/mart/wash, Marathon, Shell/mart, Econolodge, Flag Inn, Super 8, flea mkt, **W...**DM/gas, Swifty/mart, Arby's, Dairy Queen, LJ Silver, Reno's Roadhouse, Taco Bell, Wendy's, Outlets/famous brands, Radio Shack, same as 126

120 rd 1973, to Ironworks Pike, KY Horse Park, **E...**KY Horse Park Camping, **W...**HOSPITAL, Citgo/Stuckey's/ diesel/pizza/24hr

118 I-64 W, to Frankfort, Louisville

115 KY 922, Lexington, **E...**Exxon/diesel/mart, Shell/Subway/mart/wash/24hr, Cracker Barrel, McDonald's, Waffle House, Knight's Inn, La Quinta, Sheraton, SaddleHorse Museum(4mi), **W...**Chevron/diesel, Denny's, Post Rest., Embassy Suites, Holiday Inn, Marriott/rest., museum

113 US 27, US 68, Lexington, to Paris, **E...**BP/diesel(2mi), SuperAmerica, Waffle House, Ramada Inn, **W...**Chevron/ Subway/diesel/mart/24hr, Shell/mart, Fazoli's, LJ Silver, Catalina Motel, Day's Inn/rest., **1 mi W...**Shell/ KrispyKreme/mart, Swifty Gas, Burger King, Capt D's, Hardee's/24hr, Shoney's, Catalina Motel, Chevrolet, Chrysler/Plymouth/Jeep, Hall's RV Ctr, K-Mart, Kroger/deli, Northside RV Ctr, carwash, to UKY, Rupp Arena

111 I-64 E, to Huntington, WV

110 US 60, Lexington, **W...**Shell/wash/deli, Speedway/diesel/mart, SuperAmerica/mart, Thornton/Subway/mart/24hr, Arby's, Bob Evans, Cracker Barrel, Hardee's, McDonald's, Reno's Roadhouse, Shoney's, Waffle House, Wendy's, Baymont Inn, Best Western, BlueGrass Suites, Comfort Inn, Country Inn Suites, Hampton Inn, HoJo's, Holiday Inn Express, Knight's Inn, Microtel, Motel 6, Quality Inn, Ramada Ltd, Signature Inn, Super 8, Wilson Inn

108 Man O War Blvd, **W...**HOSPITAL, BP/diesel, Meijer/diesel/24hr, Shell/Burger King/LJS/Mr Gatti's/mart/24hr, Speedway/mart, Applebee's, Arby's, BackYard Burgers, Burger King, Damon's, Don Pablo's, Fazoli's, Logan's Roadhouse, Max&Erma's, McDonald's, Rafferty's, Ruby Tuesday, Steak'n Shake, Taco Bell, TGIFriday, Waffle House, Courtyard, Hilton Garden, Sleep Inn, BabysRUs, Barnes&Noble, Dick's Sports, Garden Ridge, Goody's, Kohl's, LensCrafters, Linens'n Things, OfficeMax, Old Navy, PetsMart, Pier 1, Radio Shack, Target, cinema

104 KY 418, Lexington, **E...**Exxon/Baskin-Robbins/Wendy's/diesel/mart/24hr, Shell/Krystal/mart, Waffle House, Comfort Suites, Day's Inn, Econolodge, Red Roof Inn, Holiday Inn, **W...**HOSPITAL, BP/diesel/mart/wash, Chevron/mart, Speedway/Taco Bell/mart, Jerry's Rest., **4 mi W...**Bob Evans, Burger King, KFC, Shoney's Inn/ rest.

99 US 25 N, US 421 N, no facilities

98mm Kentucky River

97 US 25 S, US 421 S, **E...**Exxon/Huddle House/diesel/24hr, flea mkt, **W...**RV camping(2mi)

95 KY 627, to Boonesboro, Winchester, **E...**BP/Blimpie/mart, Hall's Diner(6mi), McDonald's, Ft Boonesborough SP, camping, **W...**Shell/Burger King/diesel/mart/atm/24hr

90 US 25, US 421, Richmond, **E...**Shell/mart, Cracker Barrel, Best Western, Knight's Inn, Motel 6, Red Roof Inn, Travelodge, Western Sizzlin, bank, laundry, **W...**BP/diesel, Citgo/mart/24hr, Exxon/diesel/mart, Marathon/mart, Pennzoil Gas, Arby's, Big Boy, Dairy Queen, Hardee's, Pizza Hut, Waffle House, Day's Inn, Super 8, antiques

87 KY 876, Richmond, **E...**HOSPITAL, CHIROPRACTOR, Amoco/diesel/mart/24hr, BP/diesel/mart/wash, Chevron/ mart/24hr, Citgo/diesel/mart, Shell/diesel/mart/atm/24hr, Speedway/diesel/mart, SuperAmerica, Arby's, Bojangles, Burger King, Chinese Rest., Dairy Queen, Denny's, Dunkin Donuts, Fazoli's, Hardee's, KFC, Krystal/ 24hr, Little Caesar's, LJ Silver, McDonald's, Papa John's, Pizza Hut, Rally's, Red Lobster(2mi), Shoney's, Snappy Tomato Pizza, Subway, Taco Bell, Waffle House, Wendy's, Best Western, Econolodge, Holiday Inn, Quality Qtrs Inn, Ace Hardware, Big Lots, $General, Goodyear/auto, MailBoxes Etc, Wholesale Foods, Winn-Dixie, bank, cinema, cleaners, laundry, to EKU, **W...**BP/diesel/mart, DM/gas/24hr, Reno's Rest., Ryan's, Steak'n Shake/24hr, Comfort Suites, Fairfield Inn, Hampton Inn

82.5mm **rest area both lanes, full(handicapped)facilities, phone, vending, picnic tables, litter barrels, petwalk**

N

S

Lexington

Richmond

Kentucky Interstate 75

75	KY 595, Berea, **E**...HOSPITAL, to Berea Coll, **W**...BP/Taco Bell/mart/24hr, Shell/mart, Columbia Grill, Denny's, Day's Inn, Holiday Inn Express
76	KY 21, Berea, **E**...HOSPITAL, BP/mart/wash, Citgo/DQ/Stuckey's, DM/gas, Shell/Burger King/atm, Speedway/ diesel/mart, Arby's, Dinner Bell Rest., KFC, Little Caesar's, LJ Silver, Mario's Pizza, McDonald's, Pizza Hut, Subway, WanTen Chinese/Thai, Wendy's, Holiday Motel, Howard Johnson, Knight's Inn, Super 8, $General, Ford, NAPA, Wal-Mart SuperCtr/24hr, bank, tires, **W**...BP, Chevron/mart/24hr, Marathon/diesel, Lee's Chicken, Pantry Family Rest., Best Western, Econolodge, Fairfield Inn, Mtn View Motel, Chrysler/Plymouth/Dodge/Jeep, O Kentucky Camping, Walnut Meadow Camping
62	US 25, to KY 461, Renfro Valley, **E**...Shell/mart/atm/24hr, Hardee's, Country Hearth Inn, Renfro Valley Inn, KOA(2mi), **W**...HOSPITAL, BP/Blimpie/mart/24hr, Citgo, Shell/Subway/mart/24hr, Marathon/mart, Apollo Pizza, Dairy Queen, Denny's, McDonald's, Rockcastle Steaks, Wendy's, Day's Inn, Econolodge, Mt Villa Motel, to Big South Fork NRA, Lake Cumberland
59	US 25, Mt Vernon, to Livingston, **E**...BP, Shell/diesel/mart, Sunoco/diesel/mart, Jean's Rest., Pizza Hut, Best Western, Kastle Inn/rest., Nicely Camping, **W**...Citgo/mart, Super 8
51mm	Rockcastle River
49	KY 909, to US 25, Livingston, **W**...49er Fuel Ctr/diesel/rest., Shell/diesel/mart
41	KY 80, London, to Somerset, **E**...HOSPITAL, Chevron/mart, Kocolene/mart, Speedway, Arby's, Burger King, Chinese Rest., Dairy Queen, Haymarket Rest., KFC, McDonald's, Pizza Hut, Rax/GoldStar Chili, Sonic, Best Western, Day's Inn, Economy Inn, Holiday Inn Express, Red Roof Inn, Sleep Inn, Super 8, AutoZone, Chrysler/ Jeep, FoodFair, K-Mart, Kroger/deli, Pontiac/GMC, Rite Aid, SuperX Drug, bank, laundry, tires, st police, **W**...DENTIST, Amoco, BP/diesel/mart/24hr, Chevron/McDonald's/mart, Citgo/diesel/mart/24hr, Shell/Subway/ mart/wash/24hr, Bonanza, Jerry's Rest., Kountry Kookin Rest., LJ Silver, Shiloh Roadhouse, Wendy's, Budget Host, Dog Patch Trading Post, Westgate RV Camping
38	KY 192, London, **E**...HOSPITAL, BP/diesel/mart, Shell/mart, Speedway/mart, Arby's, Big Boy, Burger King, Fazoli's, Hardee's, Krystal, McDonald's, Perkins, Ponderosa, RockaBilly Café, Ruby Tuesday, Taco Bell, Hardee's, Comfort Suites, Day's Inn, Hampton Inn, Microtel, Ramada Ltd, Firestone, Wal-Mart, airport, camp- ing, toll rd to Manchester, Hazard, Western Star Trucks, to Levi Jackson SP, **W**...Citgo/diesel, Shell/diesel/mart/ 24hr, to Laurel River Lake RA
34mm	weigh sta both lanes
30.5mm	Laurel River
29	US 25, US 25E, Corbin, **E**...BP, Citgo/diesel/mart/24hr, Exxon, Pilot/Subway/diesel/mart/24hr, Speedway/diesel/ mart, Arby's(2mi), KFC(2mi), Burger King, Shoney's, Western Steer, Western Sizzlin, Economy Inn(3mi), Quality Inn/rest., Super 8, Goodyear, Blue Beacon, to Cumberland Gap NP, **W**...Amoco/Krystal/diesel/mart, Chevron/diesel/mart, Shell/diesel/mart/24hr, BBQ, Cracker Barrel, Taco Bell, Baymont Inn, Fairfield Inn, Hampton Inn, Knight's Inn, KOA, to Laurel River Lake RA
25	US 25W, Corbin, **E**...HOSPITAL, Speedway/diesel/mart, Burger King, Jerry's Rest., McDonald's, Ribeye Rest., Country Inn Suites, Day's Inn, Holiday Inn Express, Landmark Inn, Red Carpet Inn, Suburban Motel, **W**...BP/ diesel, Exxon, Shell/mart/wash/24hr, Arby's, Reno's Roadhouse, Best Western, Ramada Ltd, Regency Inn, to Cumberland Falls SP
15	US 25W, Goldbug, to Williamsburg, **W**...Chevron/diesel, Shell/mart, Cumberland Falls SP
14.5mm	Cumberland River
11	KY 92, Williamsburg, **E**...BP/Subway/diesel/mart, Exxon/diesel/mart, Shell/mart, Arby's, Dairy Queen, Hardee's, KFC, McDonald's, Pizza Hut, TCBY, Adkins Motel, Cumberland Lodge, Holiday Inn Express, Marriott, Super 8, Chevrolet, Chrysler/Plymouth/Dodge/Jeep, $General, Ford, Sav-A-Lot Foods, U-Haul, mufflers/transmissions, **W**...Amoco, Shell/diesel/mart, Williams/Wendy's/diesel/mart, Burger King, Krystal, LJ Silver, Day's Inn, Williamsburg Motel/RV camping, Wal-Mart SuperCtr/24hr, to Big South Fork NRA
1.5mm	**Welcome Ctr nb, full(handicapped)facilities, phone, vending, picnic tables, litter barrels, petwalk**
0mm	Kentucky/Tennesee state line

N

S

Louisiana Interstate 10

Exit #	Services
274mm	Pearl River, Louisiana/Mississippi state line
272mm	West Pearl River
270mm	**Welcome Ctr wb, full(handicapped)facilities, info, phone, picnic tables, litter barrels, petwalk**
267b	I-12 W, to Baton Rouge, no facilities
a	I-59 N, to Meridian, no facilities
266	US 190, Slidell, **N**...HOSPITAL, Exxon/diesel/mart, Shell/mart/ wash, TA/Mobil/diesel/rest./24hr, Texaco/mart/wash, USA/mart, Baskin-Robbins, Burger King, China Wok, Denny's, Harusake Japanese, KFC, LoneStar Steaks, McDonald's, Mongolian Grill, Pizza Hut, Shoney's, Taco Bell, Wendy's, Day's Inn, Motel 6, Super 8, U-Haul, **S**...HOSPITAL, Chevron/diesel/mart/atm/24hr, RaceTrac/ mart, Applebee's, Big Easy Diner, Cracker Barrel, Outback Steaks, Sonic, Waffle House, Econolodge, King's Guest Lodge, La Quinta, Ramada Inn, Travel Value Inn, 5 Minute Lube, Meineke Muffler, Wal-Mart, auto repair/transmissions, casino
265mm	weigh sta both lanes
263	LA 433, Slidell, **N**...Citgo, Exxon/mart/wash, Shell/mart/wash, Shoney's, Waffle House, City Motel(3mi), Comfort Inn, Hampton Inn, Chevrolet/Olds, Mitsubishi, **S**...Speedway/Subway/diesel/mart, Texaco/Domino's/diesel/mart, Food Court, McDonald's, Wendy's, Jeep, Ford/Lincoln/Mercury, Isuzu, Mazda, Nissan, Olds/Pontiac/Buick/GMC, Toyota, Slidell Factory Outlet/famous brands, KOA(1mi), fireworks
261	Oak Harbor Blvd, Eden Isles, **N**...Phil's Marina Café, Sleep Inn, **S**...BP/diesel
255mm	Lake Pontchartrain eb
254	US 11, to Northshore, Irish Bayou, **S**...BP/mart
251	Bayou Sauvage NWR, **S**...swamp tours
248	Michoud Blvd, to NASA, no facilities
246b a	I-510 S, LA 47 N, S to Chalmette, N to Littlewood, **S**...JazzLand Funpark
245	Bullard Ave, **N**...Chevron/mart/24hr, Shell/mart, James Seafood, Kettle, Pizza Hut, Comfort Suites, La Quinta, Daewoo, Honda, Hyundai, Mitsubishi, NAPA, **S**...HOSPITAL, Exxon, Shell/mart/atm, Burger King, KFC, McDonald's, Shoney's, Subway, Taco Bell, Hampton Inn, Motel 6, Big Lots, Buick/GMC, Circuit City, Ford, Home Depot, Nissan, PepBoys, Toyota, ToysRUs
244	Read Blvd, **N**...VETERINARIAN, Shell/mart, Lama's Seafood, McDonald's, DelChamps Foods, Marshall's, Office Depot, Pennzoil, Rite Aid, Sam's Club, Wal-Mart, cinema, **S**...HOSPITAL, Exxon, Popeye's, Wendy's, Best Western, Day's Inn, Holiday Inn Express, Dillard's, Dodge/Jeep, Ford/Lincoln/Mercury, Goodyear/auto, Honda, Kia, Midas Muffler, Sears/auto, Service Merchandise, cinema, mall
242	Crowder Blvd, **S**...Exxon, Denny's, Domino's, Pizza Place, Rally's, Shoney's, Wendy's, La Quinta, Family$, Lincoln/ Mercury, SuperStore Foods, Walgreen
241	Morrison Rd, **N**...Exxon, Shell, Pennzoil, Rite Aid, bank
240b a	US 90 E, Chef Hwy, Downman Rd, **N**...Amoco, Comfort Inn, Family Inn, Ramada Inn, Super 8, Chevrolet, Hyundai, U-Haul, RV camping, **S**...Chevron, Shell
239b a	Louisa St, Almonaster Blvd, **N**...Discount Fuel, Exxon, Burger King, McDonald's, Piccadilly's, Pizza Hut, Howard Johnson, Firestone, K-Mart, Precision Tune, Radio Shack, Toyota, Walgreen, mall, **S**...Fina/mart
238b	I-610 W(from wb)
237	Elysian Fields Ave, **N**...Amoco/diesel
236b	LA 39, N Claiborne Ave, no facilities
a	Esplanade Ave, downtown
235a	Orleans Ave, downtown, to Vieux Carre, French Qtr, **S**...Day's Inn, Marriott, Radisson, Sheraton
234b	Poydras St, **N**...HOSPITAL, **S**...to Superdome, downtown
a	US 90A, Claiborne Ave, to Westbank, Superdome
232	US 61, Airline Hwy, Tulane Ave, **N**...Pennzoil, **S**...Exxon/mart, McDonald's, Piccadilly's, A&P, Chevrolet, Firestone, Radio Shack, Rite Aid, to Xavier U
231b	Florida Blvd, WestEnd
a	Metairie Rd, no facilities
230	I-610 E(from eb), to Slidell
229	Bonnabel Blvd, no facilities
228	Causeway Blvd, **N**...Shell, Best Western, Hampton Inn, Ramada Ltd, **S**...MEDICAL CARE, Amoco/mart, Exxon, Phillips 66, Texaco, Timesaver Gas, Augie's Rest., Denny's, Pasta Etc, Steak'n Egg/24hr, Courtyard, Extended Stay America, Holiday Inn, La Quinta, Quality Hotel, Residence Inn, Rapid Lube, carwash, cinema, laundry
226	Clearview Pkwy, Huey Long Br, **N**...Chevron, A&G Cafeteria, **S**...HOSPITAL, Chevron, Circle K, Burger King, Piccadilly's, Shoney's Inn/rest., Subway, Buick/GMC, Dillard's, Sears, bank, mall
225	Veterans Blvd, **N**...VETERINARIAN, Chevron, EZ Serve/gas, Shell, Texaco, Burger King, Denny's, McDonald's, Rouse's Foods, Econolodge, La Quinta, Aamco, Dodge, Honda, Radio Shack, Rite Aid, **S**...CHIROPRACTOR, Amoco/mart/atm, Shell, Church's Chicken, Godfather's, NO Burger/seafood, Pancake House, Piccadilly's,

E

W

Slidell

New Orleans Area

189

Louisiana Interstate 10

Popeye's, Rally's, Schlotsky's, Thai Cuisine, Wendy's, Evergreen Plaza Inn, Holiday Inn, Acura, Big Lots, Chevrolet, CompUSA, GNC, Home Depot, JiffyLube, Jo-Ann Fabrics, K-Mart, Michael's, Nissan, Office Depot, PepBoys, Service Merchandise, TJ Maxx, VW, Walgreen, Wal-Mart

224	Power Blvd(from wb), N...Texaco
223b a	LA 49, Williams Blvd, N...CHIROPRACTOR, VETERINARIAN, Exxon/mart, Shell, Texaco/diesel/mart/wash, Burger King, Church's Chicken, Fisherman's Cove, Imperial Chinese, Italian Pie Rest., Little Tokyo Japanese, Pizza Hut, Popeye's, Rally's, Shoney's, Subway, Taco Bell, Wendy's, Fairfield Inn, Albertson's, AutoZone, Baby Depot, Burlington Coats, Ford, Hancock Fabrics, Office Depot, OfficeMax, Rite Aid, TrueValue, bank, cleaners, S...Chevron, Exxon/diesel/mart/wash, Shell/mart/wash, Brick Oven, Burger King, Denny's, Jazz Seafood/steaks, McDonald's, Messina's Rest., Pizza Hut, Subway, Best Western, Comfort Inn, Contempra Inn, Day's Inn, Extended Stay America, Holiday Inn, La Quinta, Park Plaza Inn, Radisson, Travelodge, Wingate Inn, Advance Parts, Circuit City, Dillard's, Eckerd, Firestone/auto, Goodyear, Hyundai, Midas Muffler, NAPA, Radio Shack, U-Haul, Wal-Mart, bank, mall
221	Loyola Dr, N...Circle K/gas, Citgo/mart, Exxon/diesel/mart, Shell/repair/wash, Texaco/Blimpie/mart, TimeSaver/gas, Chinese Rest., Church's Chicken, McDonald's, Popeye's, Rally's, Subway, Taco Bell, Sam's Club, S...Amoco/mart, Speedway/mart, Rick's Café, Wendy's, Sleep Inn, airport, info
220	I-310 S, to Houma, no facilities
214mm	Lake Pontchartrain
210	I-55N(from wb)
209	I-55 N, US 51, to Jackson, N...Texaco/diesel/casino, S...Chevron/mart/24hr, Jet/Citgo/mart/24hr, Shell/diesel/mart/atm/wash, Speedway/Blimpie/Church's/diesel/mart, Bully's Seafood, Burger King, KFC, McDonald's, Shoney's, Waffle House, Wendy's, Best Western, Holiday Inn Express, La Place RV Camping
207mm	weigh sta both lanes
206	LA 3188 S, La Place, S...HOSPITAL, Shell, Texaco/diesel, McDonald's, Billy Motel
194	LA 641 S, to Gramercy, 11-15 mi S...Burger King, plantations
187	US 61 S, to Gramercy, no facilities
182	LA 22, Sorrento, N...Chevron/Popeye's/mart, Shell/diesel/mart, S...Texaco/McDonald's/diesel, Lafitte's Café, The CoffeeHouse, Best Western(8mi)
181mm	**rest area both lanes, full(handicapped)facilities, phone, picnic tables, litter barrels, RV dump, petwalk**
179	LA 44, Gonzales, N...Circle K/gas, Exxon/diesel/mart, Mobil/diesel, Redman's Trkstp/diesel/rest., 3 mi S...The Cabin Fine Cajun Food
177	LA 30, Gonzales, N...HOSPITAL, Chevron/USA/diesel/mart/rest./24hr, Exxon/diesel/mart, Shell/deli/mart/wash, Texaco/diesel/mart, Burger King, Church's Chicken, McDonald's, Shoney's, Taco Bell, Waffle House, Best Western, Budget Inn, Day's Inn, Holiday Inn, Quality Inn, White Rose Motel, S...Jet24/Citgo/Blimpie/mart, Texaco/diesel/mart, Cracker Barrel, Popeye's, Subway, Wendy's, Comfort Inn, Cavenders Boots, Tanger/famous brands, Vesta RV Park
173	LA 73, Prairieville, to Geismar, N...Superstop/gas, Texaco/diesel/mart, S...Exxon/KrispyKreme/TCBY/mart, Mobil/Subway/diesel/mart, Burger King, Twin Lakes RV Park(1mi), souveniers
166	LA 42, LA 427, Highland Rd, Perkins Rd, N...Chevron/mart, Church's Chicken, Sonic, Waffle House, Royal Motel(4mi), funpark, S...Shell/BBQ/mart, Texaco/mart, antiques, bank
163	Siegen Lane, N...Chevron/mart/wash/24hr, RaceTrac/mart, Shell/mart/wash, Texaco/Chester's/diesel/mart/wash, Burger King, Fazoli's, McAlister's Deli, McDonald's, Shoney's, Subway, Taco Bell, Waffle House, Baymont Inn, Holiday Inn, Microtel, Motel 6, Advance Parts, $Tree, Honda, K-Mart, Office Depot, Radio Shack, bank, cinema, S...Chili's, Joe's Crabshack, Subway, Wendy's, Courtyard, Residence Inn, Bed Bath&Beyond, Lowe's Bldg Supply, Sam's Club, Wal-Mart SuperCtr/24hr, cinema
162	Bluebonnet Rd, N...Comfort Suites, Quality Suites, S...MEDICAL CARE, Exxon/diesel/mart, RaceTrac, Burger King, Ralph&Kacoo's Rest., AmeriSuites, CompUSA, Dillard's, JC Penney, McRae's, Parisian, Sears/auto, Mall of LA
160	LA 3064, Essen Lane, S...HOSPITAL, Chevron/mart/24hr, Exxon/mart/wash, RaceTrac/mart, Louisiana Pizza Kitchen, McDonald's, Wendy's, NTB
159	I-12 E, to Hammond
158	College Dr, Baton Rouge, N...HOSPITAL, Speedway/mart, Café Louisiana, Damon's, Grady's Grill, Hooters, Jason's Deli, Macaroni Grill, On-the-Border, Ruby Tuesday, Steam Room Grill, Waffle House, Wendy's, Best Western, Hilton, Residence Inn, Corporate Inn, Barnes&Noble, Midas Muffler, cinema, S...DENTIST, OPTICIAN, Chevron/mart/atm/24hr, Exxon/mart/wash, Mobil, Texaco/mart, Burger King, Chili's, Gino's Pizza, Godfather's, Great Wall Chinese, Kyoto Oriental Cuisine, McDonald's, Ninfa's Mexican, On-the-Bayou Café, Ruth's Chris Steaks, Embassy Suites, Hawthorn Suites, Holiday Inn Express, Shoney's, Taco Bell, Hampton Inn, Radisson, Albertson's/drugs, AutoZone, GNC, K-Mart, Mail Express, Office Depot, Radio Shack, Rite Aid, SuperFresh Foods, bank
157b	Acadian Thruway, N...HOSPITAL, Citgo, Shell/mart, Denny's, Rib's Rest., Comfort Inn, La Quinta, Marriott, Rodeway Inn, bank, S...Exxon, Mobil, LoneStar Steaks, Outback Steaks, Courtyard, Books-A-Million, Eckerd, tires
a	Perkins Rd, S...Exxon, Phillips 66, Texaco/diesel/mart, Books-A-Million, Wal-Mart/auto
156b	Dalrymple Dr, S...to LSU
a	Washington St, no facilities
155c	Louise St(from wb), no facilities
b	I-110 N, to Baton Rouge bus dist, airport
a	LA 30, Nicholson Dr, Baton Rouge, downtown, S...to LSU
154mm	Mississippi River
153	LA 1, Port Allen, N...Chevron/mart, Exxon/mart, Texaco/mart, Circle K, Church's Chicken, Popeye's

La Place

Baton Rouge

E

W

Louisiana Interstate 10

151 LA 415, to US 190, **N**...Amoco, Cash's Trk Plaza/diesel/casino, Exxon/diesel/mart, RaceTrac/mart/atm, Texaco/diesel/mart/ 24hr, Williams/diesel/mart/24hr, Burger King, KFC/Taco Bell, McDonald's, Popeye's, Waffle House; Wendy's, Best Western, Day's Inn, Holiday Inn Express, Shoney's Inn/rest., Wal-Mart, Cajun Country Camping(2mi), **S**...Love's/Arby's/diesel/mart/ 24hr, Shell/diesel/mart/24hr, Texaco/diesel/mart/24hr, Motel 6, Ramada Inn, Super 8

139 LA 77, Grosse Tete, **N**...Texaco/diesel/mart, Chevrolet, **S**...Tiger Trkstp/diesel/rest.

137.5mm **rest area both lanes, full(handicapped)facilities, phone, picnic tables, litter barrels, petwalk, RV dump**

135 LA 3000, to Ramah, **N**...Texaco/mart

128mm emergency callboxes begin wb, about 1/2 mi intervals

127 LA 975, to Whiskey Bay, no facilities

126.5mm Pilot Channel of Whiskey Bay

122mm Atchafalaya River

121 Butte La Rose, **rest area both lanes, full(handicapped)facilities, phone, picnic tables, litter barrels, petwalk, RV dump**

117mm emergency callboxes begin eb, about 1/2 mi intervals

115 LA 347, to Cecilia, Henderson, **N**...Exxon/diesel/mart/24hr, Texaco, Boudin's Rest., Crawfish Town Cajun, Landry's Seafood, Holiday Inn Express, **S**...Chevron/mart, Citgo, Exxon/Subway/diesel, Texaco/Dairy Queen/diesel/mart, USA/diesel, Williams/diesel/mart/24hr, Papa's Café, Waffle House, $General, auto/truck repair, casino

109 LA 328, to Breaux Bridge, **N**...Shell/mart, Crawfish Rest., Campers UnLtd RV Ctr, **S**...Chevron/Popeye's/mart, Citgo/mart, Mobil/Burger King/Baskin-Robbins/Domino's/diesel/mart, Texaco, Williams/diesel/mart/24hr, Church's Chicken, McDonald's, Mulate's Cajun Rest., Waffle House, Wendy's, Best Western

108mm weigh sta both lanes

106mm **rest area wb, full(handicapped)facilities, phone, picnic tables, litter barrels, petwalk, RV dump**

104mm **rest area eb, full(handicapped)facilities, phone, picnic tables, litter barrels, petwalk, RV dump**

103b I-49 N, to Opelousas

a US 167 S, to Lafayette, **N**...Texaco/diesel/mart, Church's Chicken, Motel 6, Holiday Inn, **S**...HOSPITAL, Exxon, Chevron/diesel/mart/24hr, RaceTrac/mart, Shell/mart, Texaco/diesel/mart, Checker's, China Palace, Kajun Kitchen, KFC, McDonald's, Mr Gatti's, Pizza Hut, Popeye's, Shoney's, Subway, Taco Bell, Waffle House, Wendy's, Western Sizzlin, Best Western(2mi), Comfort Suites, Fairfield Inn, Hilton, Holiday Inn, La Quinta, Rodeway Inn, Shoney's Inn, Super 8, TravelHost Inn, Albertson's/drugs, $General, Firestone/auto, Ford, K-Mart/auto, Rite Aid, Super 1 Foods, VW, Wal-Mart/auto/gas, transmissions, tires/brakes, camping

101 LA 182, to Lafayette, **N**...Chevron/McDonald's/mart, TA/Mobil/diesel/rest./24hr, Speedway/diesel/mart/24hr, Texaco/ diesel/mart/wash/24hr, Burger King, Waffle House, Red Roof Inn, Honda Motorcycles, **S**...HOSPITAL, Circle K/gas, Exxon/mart/wash, RaceTrac/mart, Shamrock/mart, Shell/mart, Texaco/diesel, Cracker Barrel, Day's Inn/rest., Quality Inn, Isuzu/Kia

100 Ambassador Caffery Pkwy, **N**...Exxon/Subway/diesel/mart/24hr, Sam's Club, **S**...HOSPITAL, Chevron/mart/wash/ atm/24hr, Citgo/mart/24hr, RaceTrac/mart/24hr, Shell/diesel/mart, Texaco/Cracker Barrel/diesel/mart/wash/24hr, Burger King, CiCi's(3mi), McDonald's, Sonic, Taco Bell, Waffle House, Wendy's, Sleep Inn, Lafayette Inn(2mi), Microtel, Goodyear

97 LA 93, to Scott, **N**...Chretien Pt Plantation, **S**...Texaco/Church's/Cracker Barrel/diesel/mart/24hr, KOA

92 LA 95, to Duson, **N**...Texaco/diesel/mart/rest./casino/24hr, **S**...Chevron/diesel/mart, Super 8

87 LA 35, to Rayne, **N**...Chevron/diesel, Exxon/diesel/mart/24hr, Burger King, Chef Roy's Café, McDonald's, RV camping, **S**...HOSPITAL, Citgo/diesel/mart, Mobil/diesel/mart/24hr, Shamrock, Shell/Subway/mart, Texaco/diesel/ mart/towing/24hr, Chinese Rest., Dairy Queen, Gade's Cajun Food, Pizza Hut, Popeye's, Comfort Inn, Champagne's Foods, Eckerd, Family$, NAPA, Winn-Dixie, bank, laundry

82 LA 1111, to E Crowley, **S**...HOSPITAL, Chevron/diesel/mart/atm, Murphy USA/gas, Shamrock/mart, Wendy's, $Tree, Radio Shack, Wal-Mart SuperCtr/24hr

80 LA 13, to Crowley, **N**...Citgo/diesel/mart/24hr, Texaco/diesel/wash, Rice Palace Rest., Waffle House, Best Western, Crowley Inn, **S**...Chevron/diesel/mart, Circle K/gas/24hr, Exxon/mart/wash/24hr, RaceTrac/mart, Shamrock/diesel/ mart/wash, Burger King, Dairy Queen, Domino's, Fezzo's Seafood/steaks, Golden Corral, KFC, Lucky Wok Chinese, McDonald's, Mr Gatti's, Pizza Hut, Popeye's, PJ's Grill, Sonic, Taco Bell, Tootie's Seafood/steaks, AutoZone, Chrysler/Plymouth/Dodge/Jeep, $General, Ford/Mercury/Nissan, Nissan, Radio Shack, Rite Aid, U-Haul, Winn-Dixie, bank, laundry

76 LA 91, to Iota, **S**...Conoco/diesel/mart/café/24hr, phone, camping

72 Egan, **N**...Clark/diesel/mart/RV Park, Cajun Connection Rest., **S**...Cajun Haven Camping

67.5mm **rest areas both lanes full(handicapped)facilities, phone, picnic tables, litter barrels, petwalk**

65 LA 97, to Jennings, **S**...Shell/mart, Spur/diesel(1.5mi), Day's Inn, to SW LA St School

64 LA 29, to Jennings, **N**...Sundown Inn, **S**...HOSPITAL, Citgo/mart, Exxon, Fina/mart, Shamrock/diesel/mart/wash, Shell/diesel/mart/24hr, Texaco/diesel/mart/wash, Burger King, Chinese Rest., Dairy Queen, Denny's, Golden

E

W

Lafayette Crowley Jennings

Louisiana Interstate 10

Corral, McDonald's, Mr Gatti's, Pizza Hut, Shoney's, Sonic, Taco Bell, Waffle House, Wendy's, Comfort Inn, Holiday Inn, Goodyear/auto, HiLo Parts, Wal-Mart SuperCtr/24hr, bank

59 LA 395, to Roanoke, no facilities

54 LA 99, Welsh, **S**...Chevron, Citgo, Conoco/diesel, Exxon/diesel/mart/24hr, Texaco, Cajun Tails Rest., Dairy Queen

48 LA 101, Lacassine, **S**...Citgo/mart

44 US 165, to Alexandria, **N**...Comfort Inn, Quiet Oaks RV Park, Grand Casino, **S**...Wooden Treasures RV Park

43 LA 383, to Iowa, **N**...Exxon/Pitt Grill/diesel/mart/24hr, Loves/Hardee's/diesel/mart/24hr, Burger King, repair, **S**...Chevron/mart/24hr, Conoco/McDonald's/diesel/mart/atm, Shell/mart, Texaco/diesel/mart/atm, Cajun Kwik Deli, Emery's Rest., Fausto's Chicken, Factory Stores of America/famous brands

36 LA 397, to Creole, Cameron, **N**...Jean Lafitte RV Park(2mi), **S**...Chevron/diesel(1mi), Conoco/diesel/RV dump/ 24hr, James Mobile Camping, casino

34 I-210 W, to Lake Charles, no facilities

33 US 171 N, **N**...Chevron/mart/wash, RaceTrac/mart, Shamrock/mart, Texaco/mart/wash, Burger King, Church's Chicken, McDonald's, Taco Bell, Best Western, Day's Inn, Motel 6, Richmond Suites, auto parts, bank, drugs, groceries, to Sam Houston Jones SP, **S**...Exxon/Subway/mart

32 Opelousas St, **N**...Exxon/24hr

31b US 90 E, Shattuck St, to LA 14, **N**...Texaco/diesel/mart/24hr, casino, **S**...Chevron/mart/24hr, McDonald's, Econolodge

a US 90 bus, Enterprise Blvd, **S**...Texaco/mart/wash/24hr, Popeye's

30b a LA 385, N Lakeshore Dr, Ryan St, **N**...Chevron, Travel Inn/rest., **S**...Holiday Inn, Lakeview Motel

29 LA 385, **N**...Chevron/mart, Waffle House, Lakeview Motel, Travel Inn/rest., **S**...Best Suites, Players Island Casino/ hotel

28mm Calcasieu Bayou, Lake Charles

27 LA 378, to Westlake, **N**...Circle K/gas/24hr, Fina, Shell, Burger King, Dairy Queen, Pizza Hut, to Sam Houston Jones SP, **S**...Conoco/mart/24hr, Inn at the Isle, Riverboat Casinos

26 US 90 W, Southern Rd, Columbia, no facilities

25 I-210 E, to Lake Charles, no facilities

23 LA 108, to Sulphur, **N**...DENTIST, CHIROPRACTOR, Citgo/diesel/mart, Conoco/repair, Exxon/Subway/mart/atm, Circle K/gas, Burger King, Cline's Drive-In, McDonald's, Popeye's, Taco Bell, Comfort Suites, Dimmick's Parts, $General, K-Mart, Wal-Mart SuperCtr/24hr, bank, **S**...Citgo/diesel/mart, Texaco/diesel/mart/24hr, Cracker Barrel, Waffle House, Winner's Choice Rest./24hr, Crossland Suites, Holiday Inn Express, Super 8, Goodyear, casino

21 LA 3077, Arizona St, **N**...HOSPITAL, Boling Point Deli, **S**...Chevron/mart, Conoco, Exxon, Dairy Queen, USA/ diesel/Cajun Deli, Hidden Ponds RV Park

20 LA 27, to Sulphur, **N**...Bayou Gas, Circle K/gas, Exxon/diesel/mart/atm/24hr, Shamrock/mart, Shell/mart/atm, Bonanza, Burger King, Cajun Charlie's Rest./gifts, Casa Ole Mexican, Checker's, Dino's Donuts, Hollier's Cajun Kitchen, Hong Kong Chinese, McDonald's, Mr Gatti's, Novrozsky's Burgers, Popeye's, Subway, Taco Bell, Wendy's, Chateau Motel, Hampton Inn, Holiday Inn, Brookshire Foods, Family$, Firestone/auto, Goodyear/auto, InstaLube, ProTire, bank, cleaners, hardware, **S**...HOSPITAL, Conoco, Shamrock, Speedway/diesel/mart/24hr, Texaco/diesel/ mart/wash, Pit Grill Cajun, Pizza Hut, Sunshine Yogurt, Waffle House, Fairfield Inn, La Quinta, Microtel, Wingate Inn, casino, to Creole Nature Trail

15mm **rest area wb, full(handicapped)facilities, phone, picnic tables, litter barrels, petwalk**

8 LA 108, Vinton, **N**...Chevron/diesel/mart, EZ/diesel/mart, KOA

7 LA 3063, Vinton, **N**...Delta/Exxon/diesel/mart/wash, Burger King, Dairy Queen, KOA

4 US 90, LA 109, Toomey, **N**...Exxon/diesel/mart/24hr, Shamrock/diesel/mart/24hr, Shell/diesel/mart, casinos, **S**...Delta Gas, Exxon/diesel/mart, Jackpot/diesel/mart, Cajun Rest., Best Western/rest., bingo, casino

2.5mm weigh sta both lanes

1.5mm **Welcome Ctr eb, full(handicapped)facilities, phone, picnic tables, litter barrels, petwalk**

1 (from wb), Sabine River Turnaround, no facilities

0mm Sabine River, Louisiana/Texas state line

Louisiana Interstate 12

Exit # Services

85c I-10 E, to Biloxi. I-12 begins/ends on I-10, exit 267.

b I-59 N, to Hattiesburg, no facilities

a I-10 W, to New Orleans, no facilities

83 US 11, to Slidell, **N**...Chevron/diesel/mart/24hr, Citgo/mart(1/2mi), Exxon/diesel/mart, Texaco/Cracker Barrel, Burger King, McDonald's, Waffle House, $General, SuperValu Foods, TrueValue, **S**...HOSPITAL, Shell/mart/wash, Speedway/diesel/mart, City Motel(3mi)

80 Airport Dr, North Shore Blvd, **S**...DENTIST, Chevron/mart/atm/24hr, Exxon/diesel/mart, Burger King, McDonald's, Subway/TCBY, Taco Bell, Wendy's, DelChamps Foods, Dillard's, Home Depot, JC Penney, Mervyn's, Pier 1, Sam's Club, Sears, Service Merchandise, Wal-Mart SuperCtr/24hr, cinema, mall

74 LA 434, to Lacombe, **S**...Super D Fuel/diesel(3mi)

65 LA 59, to Mandeville, **N**...Chevron/Subway/mart/atm, Shell/mart/wash, Abbita Brew Pub, **S**...CHIROPRACTOR, Conoco/Burger King/mart, Texaco/Domino's/mart, Winn-Dixie/deli, to Fontainebleau SP, camping

E

W

Louisiana Interstate 12

63b a	US 190, Covington, Mandeville, **N**...MEDICAL CARE, EYECARE, Exxon/mart/wash, RaceTrac/mart, Shell/mart/wash, Texaco/diesel/mart/wash, Applebee's, Burger King, Copeland's Grill, Daquiris&Creams, Denny's, Ground Pati Rest., KFC, Outback Steaks, Pasta Kitchen, Piccadilly's, Sally's Pizza, Shoney's, Subway, Waffle House, Wendy's, Best Western, Courtyard, Hampton Inn, Holiday Inn, Mt Vernon Motel, Super 8, Albertson's, Books-A-Million, GNC, Home Depot, Kirchner's, Midas Muffler, Nissan, Office Depot, Pier 1, QuickLube/wash, Wal-Mart SuperCtr, bank, cinema, **S**...HOSPITAL, Chrysler/Plymouth, st police, to New Orleans via toll causeway
60.5mm	**rest areas both lanes, full(handicapped)facilities, phone, picnic tables, litter barrels, petwalk**
59	LA 21, Madisonville, to Covington, **N**...HOSPITAL, Conoco/Burger King/diesel/mart, Shell/mart/wash, McDonald's, **S**...Texaco/ Domino's/diesel/mart/wash, Fairview Riverside SP
57	LA 1077, to Goodbee, Madisonville, **S**...to Fairview Riverside SP
47	LA 445, to Robert, **1-3 mi N**...Conoco, Hidden Oaks Camping, Jellystone Camping, Sunset Camping, to Global Wildlife Ctr
42	LA 3158, to Airport, **N**...Amoco/mart, Chevron/Stuckey's/diesel/mart/24hr, Lucky$ Rest., Friendly Inn, airport
40	US 51, to Hammond, **N**...MEDICAL CARE, Chevron/mart, RaceTrac/mart, Shell/mart/24hr, Texaco/diesel/mart/ wash, Chinese Rest., Border Café, Burger King, China Garden, Denny's, McDonald's, Pizza Hut, Ryan's, Taco Bell, Wendy's, Best Western, Budget Inn, Comfort Inn, Holiday Inn, Dillard's, Rite Aid, Sears/auto, U-Haul, bank, mall, **S**...Petro/Mobil/diesel/rest./mart/24hr, Pilot/Citgo/Arby's/diesel/mart/24hr, Shell/mart/wash, Waffle House, Ramada Inn, Blue Beacon, Speedco Lube, hardware, **2-3 mi S**...HOSPITAL, Colonial Inn, Day's Inn, KOA
38b a	I-55, N to Jackson, S to New Orleans
36mm	weigh sta both lanes
35	Pumpkin Center, Baptist, **N**...Exxon/mart, Punkin RV Park, **S**...Chevron/mart/24hr
32	LA 43, to Albany, **N**...Chevron, **1 mi S**...Citgo/diesel, to Tickfaw SP
29	LA 441, to Holden, **N**...Coastal/diesel/mart, st police
28mm	**rest areas both lanes, full(handicapped)facilities, phone, picnic tables, litter barrels, petwalk**
22	LA 63, Livingston, to Frost, **N**...Mobil/mart, Texaco/CarQuest/diesel/tires
19	to Satsuma, **1.5 mi S**...Mobil
15	LA 447, to Walker, **N**...Citgo/mart, Shell/mart/wash/24hr, Texaco/mart, Burger King, Church's Chicken, La Fleur's Seafood, McDonald's, Waffle House, Rite Aid, antiques/gifts, **S**...Chevron/diesel/mart, Fina/diesel/mart
10	LA 3002, to Denham Springs, **N**...DENTIST, CHIROPRACTOR, Chevron/mart/atm, Circle K/gas, Exxon/mart/atm, RaceTrac/mart, Texaco/diesel/mart/wash, BBQ, Burger King, Cactus Café, Chinese Inn, Crawford's Cajun, Golden Corral, KFC, Lagniappe TexMex, McDonald's, Pizza Hut, Popeye's, Ron's Seafood, Ryan's, Subway, Waffle House, Wendy's, Best Western, Holiday Inn Express, Advance Parts, DelChamps Foods, Firestone/auto, Radio Shack, Rite Aid, Wal-Mart SuperCtr/24hr, bank, carwash, cleaners, farmers mkt, **S**...EYECARE, Speedway/Church's Chicken/ diesel/mart, Texaco/Cracker Barrel, LaFleurs Seafood, Piccadilly's, Shoney's, Day's Inn, Quality Inn, Dodge/Isuzu, Ford, KOA(1mi), cleaners
8.5mm	Amite River
7	O'Neal Lane, **N**...HOSPITAL, Mobil/mart/wash, Shell/mart/wash, Eckerd, HobbyLobby, Knight's RV Park(2mi), Office Depot, bank, hardware, **S**...MEDICAL CARE, Chevron/mart/wash/24hr, Exxon/diesel, Texaco/diesel/mart, Blimpie, Burger King, China King, Fazoli's, LoneStar Steaks, McDonald's, Popeye's, Sonic, Subway, Taco Bell, Waffle House, Wendy's, $General, MailBoxes Etc, Radio Shack, Super Fresh Foods, Wal-Mart SuperCtr/24hr
6	Millerville Rd, **N**...Chevron/mart/24hr, Super K-Mart/auto, hardware
4	Sherwood Forest Blvd, **N**...Exxon/mart, Jet/24, Texaco/diesel/mart/wash, Burger King, Chinese Rest., ChuckeCheese, Denny's, Godfather's, KFC, McDonald's, Popeye's, Subway, Waffle House, Crossland Suites, Red Roof Inn, Super 8, Cloth World, Eckerd, Firestone/CarQuest, Goodyear, Rite Aid, Sports Ltd, cleaners, **S**...Chevron/ mart/wash/FastPay, Shell/mart/wash/24hr, Baskin-Robbins, BBQ, Daiquiri Café, Ground Patti, Pasta Garden, Pizza Hut, Taco Bell, Quality Inn, Harley-Davidson
2b	US 61 N, **N**...Chevron/Blimpie/mart, Exxon/diesel/mart/wash, Mobil, Shell/mart/24hr, Texaco/wash, Applebee's, Checker's, Cracker Barrel, McDonald's, Rafferty's Rest., Russell's Grill, Shoney's, Subway, Taco Bell, Taste of China, Wendy's, Wienerschnitzel, Hampton Inn, Holiday Inn, Microtel, Motel 6, Shoney's Inn/rest., Sleep Inn, Acura/ Infiniti, Albertson's, Baby Depot, Budget Transmissions, Burlington Coats, Dodge, Ford/Lincoln/Mercury, Kinko's, Luxury Linens, Marshall's, Michael's, Nissan/Suzuki, PepBoys, SteinMart, Toyota, bank, funpark, laundry, tires
a	US 61 S, **S**...Speedway/mart, Texaco/diesel/mart, Circle K/gas, Bonanna's Seafood, China Moon, McDonald's, Waffle House, Day's Inn, MasterHost Inn, Cadillac, Chevrolet, Home Depot, Mitsubishi, Ryder Trucks, Winn-Dixie
1b	LA 1068, to LA 73, Essen Lane, **N**...HOSPITAL, VETERINARIAN, Texaco/diesel/mart/wash, Chinese Rest., Drusilla Seafood Rest., Fast Track Burgers, McDonald's, Eckerd, JiffyLube, Radio Shack, bank, cleaners
a	I-10(from wb). I-12 begins/ends on I-10, exit 159 in Baton Rouge.

E ↑ ↓ **W**

(side labels: Hammond, Baton Rouge)

Louisiana Interstate 20

Exit #	Services
189mm	Mississippi River, Louisiana/Mississippi state line
187mm	weigh sta both lanes
186	US 80, Delta, **N**...Texaco/diesel/mart/24hr, **S**...Chevron/Blimpie/diesel/mart/24hr
184mm	**rest area both lanes, full(handicapped)facilities, phone, picnic tables, litter barrels, petwalk, RV dump**
182	LA 602, Mound, **S**...Winner's Circle Rest./OTB
173	LA 602, Richmond, **S**...Roundaway RV Park
171	US 65, Tallulah, **N**...HOSPITAL, Citgo/Subway/diesel/mart/24hr, Shell/diesel/mart, Johnny's Pizza(1mi), KFC, McDonald's, Popeye's(1mi), Wendy's, Day's Inn, Holiday Capri Motel(2mi), Super 8, **S**...Conoco/diesel/mart, TA/Mobil/diesel/rest./24hr, Texaco/mart
164mm	Tensas River
157	LA 577, Waverly, **N**...Tiger Trkstp/diesel/rest./24hr, **S**...to Tensas River NWR
155mm	Bayou Macon
153	LA 17, Delhi, **N**...HOSPITAL, Chevron/Taco Bell/mart, Exxon/mart, Burger King, Dairy Queen, Moby's Rest., Pizza Hut, Sonic, Subway, **S**...Texaco/mart, Best Western, Day's Inn
150mm	**rest area both lanes, full(handicapped)facilities, phone, picnic tables, litter barrels, petwalk, RV dump**
148	LA 609, Dunn, no facilities
145	LA 183, rd 202, Holly Ridge, no facilities
141	LA 583, Bee Bayou Rd, **N**...Texaco/diesel/mart
138	LA 137, Rayville, **N**...HOSPITAL, Bud's/diesel/mart, EZ Mart/diesel, Pilot/Wendy's/diesel/mart/24hr, Burger King, McDonald's, Day's Inn, Chevrolet/Pontiac/Buick/Olds, Family$, Firestone, $General, Wal-Mart, XpressLube, bank, repair, **S**...Citgo/Subway/TCBY/diesel/mart/24hr, Exxon/mart, Popeye's, Waffle House, Cottonland Inn/RV Park, Goodyear
135mm	Beouf River
132	LA 133, Start, **N**...Exxon/mart
128mm	Lafourche Bayou
124	LA 594, Millhaven, **N**...EZ Mart/diesel, st police, to Arsage Wildlife Area
120	Garrett Rd, Pecanland Mall Dr, **N**...EYECARE, Citgo/pizza/diesel/mart, Shell/mart, Morrison's Cafeteria, Olive Garden, Pizza Hut, Red Lobster, Zipp's Drive-Thru, Comfort Inn, Courtyard, Holiday Inn, Dillard's, Firestone/auto, JC Penney, McRae's, Pier 1, Sears/auto, ToysRUs, airport, mall, **S**...Exxon/diesel/mart, Best Western, Day's Inn, Harley-Davidson, Hope's RV Ctr, Lowe's Bldg Supply, Pecanland RV Park, Sam's Club, Shilo RV Camp
118b a	US 165, **N**...Copeland's Rest., Kettle, Holiday Inn, La Quinta, Home Depot, Plymouth/Dodge, Goodyear, Kenworth Trucks, to NE LA U, **S**...Chevron, Citgo, Exxon, Shell, Texaco, Burger King, McDonald's, Comfort Inn, Hampton Inn, Motel 6, Ramada Ltd, Olds
117b	LA 594, Texas Ave, no facilities
a	Hall St, Monroe, **N**...HOSPITAL, Civic Ctr
116b	US 165 bus, LA 15, Jackson St, **N**...HOSPITAL, **S**...Popeye's, Guesthouse Inn/rest.
a	5th St, Monroe, **N**...Citgo/Circle K
115	LA 34, Mill St, **N**...Citgo/mart, Day's Inn(2mi), **S**...hardware
114	LA 617, Thomas Rd, **N**...HOSPITAL, EYECARE, RaceTrac/mart, Shell, Texaco/mart, BBQ, Bennigan's, Bonanza, Burger King, Capt D's, Chick-fil-A, Chili Verde Mexican, El Chico, KFC, McDonald's, Pizza Hut, Popeye's, Shoney's, Subway, Taco Bell, TCBY, Waffle House, Wendy's, Shoney's Inn, Super 8, Big Lots, HobbyLobby, Office Depot, Rite Aid, Wal-Mart SuperCtr/gas/24hr, XpressLube, bank, cinema, **S**...Citgo/mart/wash, Exxon/diesel/mart, Texaco/diesel/mart, Chili's, Cracker Barrel, Kettle, Logan's Roadhouse, LoneStar Steaks, Outback Steaks, Peking Chinese, Sonic, Waffle House, Western Sizzlin, Baymont Suites, Best Western, Holiday Inn Express, Red Roof Inn, Firestone/auto, Radio Shack, Sears Parts, Super K-Mart/24hr
112	Well Rd, **N**...Shell/diesel/mart/wash/24hr, Texaco/mart/24hr, Dairy Queen, McDonald's, Waffle House, XpressLube, **S**...Williams/Main St Café/diesel/mart/24hr, RV camping
108	LA 546, to US 80, Cheniere, **N**...Exxon/diesel/mart, Texaco/diesel/mart
107	Camp Rd, rd 25, Cheniere, no facilities
103	US 80, Calhoun, **N**...Chevron/diesel/mart/rest., Avant Motel, **S**...Citgo/diesel/mart/rest.
101	LA 151, to Calhoun, **N**...Texaco, **S**...Shell/diesel/mart, Café 101
97mm	**rest area wb, full(handicapped)facilities, phone, picnic tables, litter barrels, petwalk, RV dump**
95mm	**rest area eb, full(handicapped)facilities, phone, picnic tables, litter barrels, petwalk, RV dump**
93	LA 145, Choudrant, **N**...Texaco, **S**...Chevron/mart
86	LA 33, Ruston, **N**...VETERINARIAN, Citgo, Shell/mart/wash, Texaco, Annie's Rest., Log Cabin Rest., Ryan's, Comfort Inn, Lincoln Motel, Ramada Ltd, Chevrolet/Pontiac/Cadillac, Chrysler/Plymouth, Firestone, Ford/Lincoln/Mercury, Olds/Buick/GMC, Toyota, Wal-Mart SuperCtr/24hr, XpressLube, hardware, **S**...Texaco/diesel/mart, Holiday Inn, laundry
85	US 167, Ruston, **N**...Chevron/diesel/mart/wash/24hr, Citgo/Subway/yogurt/mart, Shell/mart/wash, Baskin-Robbins, Burger King, Capt D's, Huddle House, McDonald's, Old Mexico Rest., Peking Chinese, Shoney's, Wendy's, Hampton Inn, Holiday Inn, Relax Inn, Super 1 Foods, JC Penney, Radio Shack, TrueValue, cleaners, **S**...HOSPITAL, Pennzoil/mart, Texaco/diesel/mart, Pizza Hut, Best Western, Snappy Lube, bank

E

W

Louisiana Interstate 20

Exit	Description
84	LA 544, Ruston, **N**...Mobil/mart, **S**...VETERINARIAN, Chevron, Citgo/mart/wash/24hr, Exxon/mart, Shell/mart/24hr, Texaco/diesel/mart, Cuco's Mexican, Dairy Queen, Johnny's Pizza House, Mulligan's Café, Pizza Inn, Rabb's Steaks, Subway, TCBY, Wendy's, Super 8, bank
81	LA 141, Grambling, **S**...Clark/Church's Chicken/diesel/mart/atm/24hr, Exxon, Texaco, to Grambling St U
78	LA 563, Industry, **S**...Texaco/diesel
77	LA 507, Simsboro, **S**...access to gas, phone
69	LA 151, Arcadia, **S**...BP/diesel/mart, Exxon/mart, Texaco/mart, Country Folks Kitchen, McDonald's, Sonic, Subway, Day's Inn, Nob Hill Inn/rest., Arcadia Tire, Brookshire's Foods, Factory Stores/famous brands, Fred's Drugs, NAPA, carwash
67	LA 9, Arcadia, **N**...to Lake Claiborne SP, **S**...Texaco
61	LA 154, Gibsland, **S**...Exxon(1mi)
58mm	**rest area both lanes, full(handicapped)facilities, phone, picnic tables, litter barrels, petwalk, RV dump**
55	US 80, Ada, Taylor, no facilities
52	LA 532, to US 80, Dubberly, **S**...Chevron/mart
49	LA 531, Minden, **N**...Trucker's Paradise/diesel/rest./casino/24hr
47	US 371 S, LA 159 N, Minden, **N**...HOSPITAL, Chevron/diesel/mart, Mobil/diesel/mart, Texaco/mart/24hr, Golden Biscuit, Best Western, Exacta Inn/rest., Southern Inn, Ford/Lincoln/Mercury, bank, **S**...B&B, to Lake Bistineau SP, camping
44	US 371 N, Dixie Inn, Cotton Valley, **N**...Chevron, Crawfish Hole, Hamburger Happiness, Minden Motel(2mi)
38	Goodwill Rd, Ammo Plant, **S**...BP/Rainbow Diner/diesel/mart/24hr, horse trailer sales
36mm	**rest area both lanes, full(handicapped)facilities, phone, picnic tables, litter barrels, petwalk, RV dump**
33	LA 157, Fillmore, **N**...Phillips 66/mart, Johnny's Pizza, Hilltop Camping(2mi), **S**...Exxon(1mi), Lake Bistineau SP
26	I-220 W, Shreveport, **1 mi N**...facilities off of I-220
23	Industrial Dr, **N**...Citgo/Circle K, Exxon/Subway/mart, Phillips 66/diesel, Texaco/Popeye's/mart/24hr, American Steak Buffet, Burger King, Harvest Buffet, McDonald's, Sue's Country Kitchen, Taco Bell, Le Bossier Motel, XpressLube, hardware, st police, **S**...Chevron/diesel/mart, Road Mart/diesel, Quality Inn
22	Airline Dr, **N**...DENTIST, MEDICAL CARE, EYECARE, FOOT CARE, Chevron/McDonald's/diesel/mart, Citgo/Circle K, Exxon/diesel, Mobil/mart/wash, Pennzoil/gas, Applebee's, Arby's, Backyard Burgers, Bennigan's, Burger King, Capt D's, Chili's, ChuckeCheese, Dairy Queen, Grandy's, Jack-in-the-Box, Little Caesar's, Luby's, Mr Gatti's, Mr Jim's Chicken, Pizza Hut, Popeye's, Red Lobster, Sonic, Taco Bell, Waffle House, Best Western, Crossland Suites, Isle of Capri Hotel, Rodeway Inn, Albertson's, Books-A-Million, Dillard's, Eckerd/24hr, Firestone/auto, Goodyear/auto, JC Penney, K-Mart/auto, NAPA, Office Depot, PepBoys, Rite Aid, Sears/auto, Service Merchandise, bank, cinemas, mall, **S**...HOSPITAL, Exxon/diesel/mart, Texaco/diesel/mart/wash/24hr, Circle K/gas, Catfish Rest., Chinese Rest., Church's Chicken, Darryl's Rest., Domino's, Outback Steaks, Popeye's, Baymont Inn, Red Carpet Inn, AutoZone, Fabrics+, Olds, Super1 Food, Time-It Lube, Tune-A-Car, bank, cleaners, to Barksdale AFB
21	LA 72, to US 71 S, Old Minden Rd, **N**...Circle K/gas, Exxon/mart, Burger King, Cowboy's Rest., El Chico, Kettle, McDonald's, Morrison's Cafeteria, Pancho's Mexican, Ralph&Kacoo's, Whataburger, Hampton Inn, Holiday Inn, La Quinta, Residence Inn, Shoney's Inn/rest., Advance Parts, AutoZone, **S**...Dragon House Chinese, Waffle House, Wendy's, Day's Inn, Motel 6
20c	(from wb), to US 71 S, to Barksdale Blvd
20b	LA 3, Benton Rd, same as 21
a	Hamilton Rd, Isle of Capri Blvd, **N**...Circle K/gas, Speedway/mart, Texaco/mart/24hr, BBQ, Comfort Inn, **S**...Chevron, Ramada/rest., casino
19b	Traffic St, Shreveport, downtown **N**...Chevrolet, casino
a	US 71 N, LA 1 N, Spring St, Shreveport, **N**...Don's Seafood, Best Western, Holiday Inn
18b-d	Fairfield Ave(from wb), downtown Shreveport, **N**...radiator repair, **S**...HOSPITAL
18a	Line Ave, Common St(from eb), downtown, **S**...HOSPITAL, Citgo/mart
17b	I-49 S, to Alexandria
a	Lakeshore Dr, Linwood Ave, no facilities
16b	US 79/80, Greenwood Rd, **N**...HOSPITAL, bank, **S**...Citgo/mart, Burger King, El Chico
a	US 171, Hearne Ave, **N**...HOSPITAL, Texaco/mart, Medic Drugs, Tune-A-Car, tires, **S**...Exxon/repair/mart, Fina/mart, Texaco/diesel/mart, KFC
14	Jewella Ave, Shreveport, **N**...Citgo/diesel/mart, Pennzoil/gas, Texaco/diesel/mart/wash/24hr, Burger King, Church's Chicken, John's Seafood Rest., McDonald's, Popeye's, Shortstop Burgers, Subway, Taco Bell, Whataburger, Advance Parts, AutoZone, Brake-O, County Mkt Foods, Eckerd, Family$, Medic Drugs, Rite Aid, Walgreen
13	Monkhouse Dr, Shreveport, **N**...DENTIST, BBQ/Smoked Turkey, Denny's, Best Value Inn, Day's Inn, Econolodge, Holiday Inn Express, Plantation Inn/rest., Shade Tree Parts, bank, **S**...Chevron/mart/wash/24hr, Exxon/Subway/diesel/mart, Texaco/diesel/mart, Dairy Queen, Waffle House, Best Western, Ramada/rest., Super 8, to airport
11	I-220 E, LA 3132 E, to I-49 S

E ↕ **W**

Louisiana Interstate 20

10	Pines Rd, **N**...BP/diesel/mart, Dairy Queen, Pizza Hut, Subway, Western Sizzlin, Brookshire's Foods, KOA, bank, cleaners, **S**...MEDICAL CARE, EYECARE, Chevron/McDonald's/mart, Exxon/Blimpie/TCBY/diesel/mart, Shell/mart, Burger King, Church's Chicken, Cracker Barrel, Domino's, Grandy's, IHOP, KFC, Subway, Taco Bell, Wendy's, Whataburger/24hr, Courtyard, Fairfield Inn, La Quinta, Holiday Inn, Chrysler/Dodge, GNC, Kroger, Pennzoil, Radio Shack, USPO, Wal-Mart SuperCtr/24hr, bank, cleaners
8	US 80, LA 526 E, **N**...Red Roof Inn, Freightliner, RV Ctr, truck repair, **S**...Chevron/diesel/mart, Petro/Mobil/diesel/rest., Speedway/diesel/mart/24hr, Crescent Landing Catfish Rest.(3mi), Blue Beacon, RV camping
5	US 79 N, US 80, to Greenwood, **N**...76/Mobil/diesel/rest./24hr, Texaco/diesel/mart, Country Inn, Mid Continent Motel, auto auction, RV park, **S**...bank
3	US 79 S, LA 169, Mooringsport, **S**...Flying J/Conoco/diesel/LP/rest./24hr
2mm	**Welcome Ctr eb, full(handicapped)facilities, phone, picnic tables, litter barrels, petwalk, RV dump**
1mm	weigh sta both lanes
0mm	Louisiana/Texas state line

Louisiana Interstate 49

Exit #	Services
	I-49 begins/ends in Shreveport on I-20, exit 17.
206	I-20, E to Monroe, W to Dallas
205	King's Hwy, **E**...Piccadilly's, Dillard's, Sears/auto, mall, **W**...HOSPITAL, Shamrock/mart, Burger King, LJ Silver, McDonald's, Subway, Taco Bell
203	Hollywood Ave, Pierremont Rd, **W**...Fina/mart
202	LA 511, E 70th St, **W**...Chevron/mart, Circle K/gas, Sonic, Family$
201	LA 3132, to Dallas, Texarkana, **3 mi E**...multiple facilities at LA 1
199	LA 526, Bert Kouns Loop, **E**...Chevron/Arby's/diesel/mart/24hr, Exxon/mart, Burger King, KFC, Taco Bell, Wendy's, Home Depot, Comfort Inn, bank, **W**...Texaco/diesel/mart/wash, McDonald's, Red Lobster(4mi), Brookshire's/deli/bakery/drugs
196mm	Bayou Pierre
191	LA 16, to Stonewall, no facilities
186	LA 175, to Frierson, Kingston, no facilities
177	LA 509, to Carmel, **E**...Eagle's Truckstp/diesel/rest.
172	US 84, to Grand Bayou, Mansfield, no facilities
169	Asseff Rd, no facilities
162	US 371, LA 177, to Evelyn, Pleasant Hill, no facilities
155	LA 174, to Ajax, Lake End, **W**...Spaulding Gas/diesel/mart, Country Livin' RV
148	LA 485, Powhatan, Allen, no facilities
142	LA 547, Posey Rd, no facilities
138	LA 6, to Natchitoches, **E**...HOSPITAL, Exxon/mart/24hr, RaceTrac/mart, Shell/mart/24hr, Shoney's, Wendy's, Best Western, Day's Inn(5mi), Holiday Inn Express, Super 8(5mi), Albertson's(3mi), Wal-mart SuperCtr/24hr(5mi), **W**...Chevron/diesel/mart, Jet24/BasinCafé/mart, Texaco/diesel/mart, Burger King, McDonald's, Simpatico's Grille, Comfort Inn, Hampton Inn, Microtel, Nakatosh RV Park, to Kisatchie NF
132	LA 478, rd 620, no facilities
127	LA 120, to Cypress, Flora, **E**...to Cane River Plantations, Citgo, Holiday Inn(10mi)
119	LA 119, to Derry, Cloutierville, **E**...to Cane River Plantations
113	LA 490, to Chopin, no facilities
107	to Lena, **E**...USPO
103	LA 8 W, to Flatwoods, **E**...Texaco/mart, to Cotile Lake, RV camping
99	LA 8, LA 1200, to Boyce, Colfax, **6 mi W**...Cotile Lake, RV camping
98	LA 1(from nb), to Boyce, no facilities
94	rd 23, to Rapides Sta Rd, **E**...Shari Sue's/diesel/deli/mart, **W**...I-49 RV Ctr
90	LA 498, Air Base Rd, **W**...Chevron/diesel/mart/24hr, Exxon/diesel/mart, Mobil/Krystal/diesel/mart, Texaco/BBQ/diesel/mart, Burger King, Cracker Barrel, McDonald's, Howard Johnson, La Quinta, Super 8, **4 mi W**...Fina/mart, Pizza Hut, Ryan's, Shoney's, Holiday Inn, Ramada
86	US 71, US 165, MacArthur Dr, **E**...Chevron, Fina, **1-2 mi W**...Citgo, Fina, Kwik Pantry, Burger King, Ryan's, Shoney's, Best Western, Comfort Inn, Hampton Inn, Holiday Inn, MacArthur Inn, Motel 6, Quality Inn, Red Carpet Inn
85b	Monroe St, Medical Ctr Dr(from nb)
a	LA 1, 10th St, Elliot St, downtown
84	US 167 N, LA 28, LA 1, Pineville Expswy(no ez return nb), no facilities
83	Broadway Ave, **E**...gas/mart, **1 mi W**...Texaco/mart, Wendy's, Books-A-Million, Lowe's Bldg Supply, Target, ToysRUs, Wal-Mart SuperCtr/24hr
81	US 71 N, LA 3250, Sugarhouse Rd, MacArthur Dr(from sb), **W**...Phillips 66/diesel, same as 83

Shreveport

Alexandria

Louisiana Interstate 49

80	US 71 S, US 167, MacArthur Dr, Alexandria, **2-4 mi** **W...**Chevron, Coastal/wash, Conoco, Exxon/mart, I-49 Truck Plaza/diesel, Mobil, Phillips 66, Texaco/mart/wash, Burger King, Cuco's Mexican, McDonald's, Pizza Hut, Sonic, Subway, Taco Bell, Western Sizzlin, Best Western, Holiday Inn, Rodeway Inn, Travelodge, Albertson's/drugs, Dillard's, Dodge, Ford, Hyundai, JC Penney, Mazda, Precision Tune, Sam's Club, U-Haul, bank, cinema, mall
73	LA 3265, rd 22, to Woodworth, **W...**Exxon/Blimpie/diesel/mart, LA Conf Ctr, to Indian Creek RA, RV camping
66	LA 112, to Lecompte, **E...**Chevron/Burger King/diesel/mart, Lea's Lunchroom(2mi), **W...**museum
61	US 167, to Meeker, Turkey Creek, **E...**to Loyd Hall Plantation(3mi)
56	LA 181, Cheneyville, no facilities
53	LA 115, to Bunkie, **6 mi E...**HOSPITAL, gas, food, to Avoyelles Grand Casino(24mi)
46	LA 106, to St Landry, **W...**to Chicot SP
40	LA 29, to Ville Platte, **E...**Texaco(1/2mi)
35mm	**E...rest area/rec area both lanes, full(handicapped)facilities, picnic tables, litter barrels, petwalk, RV dump**
27	LA 10, to Lebeau, no facilities
25	LA 103, to Washington, Port Barre, **W...**antique mall/café
23	US 167 N, LA 744, to Ville Platte, **E...**Texaco/diesel/mart, **W...**trkstop/diesel/rest./casino/24hr, to Chicot SP
19b a	US 190, to Opelousas, **W...**HOSPITAL, **1 mi W...**Exxon/mart, Mobil/mart, Checker's, Palace Café, Soileau's Café, bank, RV camping
18	LA 31, to Cresswell Lane, **E...**Chrysler/Plymouth/Dodge/Jeep/Isuzu, antiques, **W...**CHIROPRACTOR, Chevron/diesel/24hr, Mobil/mart, Shamrock/diesel/mart/wash, Shell/wash/24hr, Burger King, Creswell Lane Chinese, Domino's, McDonald's, Pizza Hut, Ryan's, Subway, Taco Bell, Wendy's, Best Western, Day's Inn, Eckerd, Express Lube, Family$, Firestone, Nissan, Pontiac/Buick/Cadillac, Rite Aid, TireWorld, Wal-Mart/auto, bank, cinema
17	Judson Walsh Dr, **E...**Texaco/diesel/rest., **W...**Chevron, Ray's Diner, Ryan's, to Wal-Mart
15	LA 3233, Harry Guilbeau Rd, **W...**HOSPITAL, Chevron/mart, Quality Inn/rest., Courvelles RV Ctr, RV camping
11	LA 93, to Grand Coteau, Sunset, **E...**VETERINARIAN, Chevron/mart, Citgo/diesel/mart/24hr, Exxon/Popeye's/mart, Beau Chere Rest., RV camping, carwash, **W...**Sunset Inn
7	LA 182, no facilities
4	LA 726, Carencro, **E...**Evangeline Downs Racetrack, **W...**Chevron/wash, Jet24/Domino's/mart, Burger King, McDonald's, Kenworth/Mack Trucks, Economy Inn, bank
2	LA 98, Gloria Switch Rd, **E...**VETERINARIAN, Chevron/mart, Jet 24/Citgo/Blimpie/diesel/mart, Prejean's Rest., Edgar RV Ctr, Lowe's Bldg Supply, bank, repair/transmissions, **W...**Texaco/Church's/Cracker Barrel/mart, Picante Mexican
1b	Pont Des Mouton Rd, **E...**Exxon/diesel/mart, Texaco/Subway/diesel/mart, Holiday Inn Express, Motel 6, Plantation Motel, Ramada Inn, Ford Trucks, Gauthier's RV Ctr, st police, **W...**Ford
a	I-10, W to Lake Charles, E to Baton Rouge, **US 167 S...**HOSPITAL, Exxon, Chevron/diesel/mart/24hr, RaceTrac/mart, Shell/mart, Texaco/diesel/mart, Checker's, China Palace, Kajun Kitchen, KFC, McDonald's, Mr Gatti's, Pizza Hut, Popeye's, Shoney's, Subway, Taco Bell, Waffle House, Wendy's, Western Sizzlin, Best Western(2mi), Comfort Suites, Fairfield Inn, Hilton, Holiday Inn, La Quinta, Rodeway Inn, Shoney's Inn, Super 8, TravelHost Inn, Albertson's/drugs, $General, Firestone/auto, Ford, K-Mart/auto, Rite Aid, Super 1 Foods, VW, Wal-Mart/auto/gas, transmissions, tires/brakes, camping

I-49 begins/ends on I-10, exit 103.

Louisiana Interstate 55

Exit #	Services
65.5mm	Louisiana/Mississippi state line
65mm	**Welcome Ctr sb, full(handicapped)facilities, tourist info, phone, picnic tables, litter barrels, petwalk**
64mm	weigh sta nb
61	LA 38, Kentwood, **E...**HOSPITAL, Chevron/diesel/mart/24hr, Citgo, Texaco/repair, Kentwood Plaza Rest., Sonic, Brown Morris Drug, Ford/Mercury, Sunflower Foods, carwash, cleaners, **W...**Texaco/diesel/mart/24hr, Great Discovery Camping
58.5mm	weigh sta sb
57	LA 440, Tangipahoa, **E...**to Camp Moore Confederate Site
54.5mm	**rest area nb, full(handicapped)facilities, phone, picnic tables, litter barrels, petwalk**

Louisiana Interstate 55

N

S

53	LA 10, Fluker, to Greensburg, **W**...casino(3mi)
50	LA 1048, Roseland, **E**...Texaco/diesel/mart/24hr
47	LA 16, Amite, **E**...HOSPITAL, Chevron/Domino's/KrispyKreme/diesel/mart, Citgo/mart, Conoco/mart, RaceTrac/mart, BBQ, Burger King, KFC, Master Chef, Mike's Catfish, McDonald's, Pizza Hut, Popeye's, Sonic, Subway, AutoZone, Winn-Dixie, RV camping, bank, carwash, **W**...Amite Trkstp/OysterGrill/diesel(2mi), Amoco/mart, Citgo/Stuckey's/diesel/mart/24hr, Ardillo's Rest., Colonial Motel
41	LA 40, Independence, **E**...HOSPITAL, Conoco/mart, Kingfish Seafood Rest., **W**...Indian Creek Camping(2mi)
36	LA 442, Tickfaw, **E**...Chevron/diesel/mart, camping, to Global Wildlife Ctr(15mi)
32	LA 3234, Wardline Rd, **E**...Citgo/Blimpie/mart, Texaco/diesel/mart, Burger King, McDonald's, Murphy's Seafood, **W**...Chevron, BBQ(1mi), Chinese Cuisine
31	US 190, Hammond, **E**...HOSPITAL, Chevron/diesel/mart, Citgo/mart, Exxon/Subway/TCBY/mart, Fleet/diesel, Speedway/Blimpie/diesel/mart/atm, Swifty's Gas/mart, Applebee's, Burger King, Cracker Barrel, McDonald's, Pizza Hut, Russell's Grill/24hr, Shoney's, Sicily's Pizza, Sonic, Taco Bell, Waffle House, Wendy's, Country Inn Suites, Econo Motel, Executive Suites, Super 8, Western Inn, Aamco, Advance Parts, AutoZone, BooksAMillion, Chrysler/Plymouth/Buick/Dodge, $General, Eckerd, HobbyLobby, K-Mart, Lowe's Bldg Supply, Office Depot, OfficeMax, Radio Shack, Rite Aid, Speedee Lube, Walgreen/24hr, Wal-Mart SuperCtr/gas/24hr, Winn-Dixie/deli, laundry, muffler/brakes, transmissions/repair
29b a	I-12, W to Baton Rouge, E to Slidell
28	US 51 N, Hammond, **E**...HOSPITAL, RaceTrac/mart, Catfish Charlie's, CiCi's, Don's Seafood Rest., Shoney's(2mi), Econolodge, Holiday Inn/rest., Ramada Inn/rest., KOA, diesel repair
26	LA 22, Ponchatoula, to Springfield, **E**...Amoco, Chevron/diesel/mart, Conoco/mart, Exxon/diesel/mart, Burger King, China King, KFC, McDonald's, Popeye's, Sonic, Subway(1mi), Wendy's, AutoZone, Bohnings Foods, Cato's, $General, Eckerd, Ford, NAPA, USA Parts, Valvoline, Winn-Dixie/deli/24hr, auto repair, bank, cleaners, **W**...Texaco/Domino's/KrispyKreme/diesel/mart/wash/atm
23	US 51, Ponchatoula, no facilities
22	frontage rd(from sb), no facilities
15	Manchac, **E**...Middendorf Café, phone, swamp tours
7	Ruddock, no facilities
1	US 51, to I-10, Baton Rouge, La Place, **S**...HOSPITAL, Chevron/mart/24hr, Jet/Citgo/mart/24hr, Shell/diesel/mart/atm/wash, Speedway/Blimpie/Church's/diesel/mart, Bully's Seafood, Burger King, KFC, McDonald's, Shoney's, Waffle House, Wendy's, Best Western, Holiday Inn Express, La Place RV Camping

I-55 begins/ends on I-10, exit 209.

For Louisiana Interstate 59, see Mississippi Interstate 59.

(side tab: Hammond La Place)

Louisiana Interstate 220(Shreveport)

E

W

Exit #	Services
	I-220 begins/ends on I-20, exit 26.
17b	I-20, W to Shreveport, E to Monroe
a	US 79, US 80, **N**...Exxon(1mi), Citgo/Circle K/atm, RaceTrac/mart, Waffle House, LA Downs Racetrack, Hilltop RV Park(8mi), **S**...Chevron/diesel/mart, **2-3 mi S**...McDonald's, Subway
15	Shed Rd, no facilities
13	Swan Lake Rd, no facilities
12	LA 3105, Airline Dr, **N**...HOSPITAL, **S**...DENTIST, Exxon/diesel/mart, Citgo, Shell/Subway/mart/wash, Total/diesel/mart, Applebee's, Burger King, CiCi's, Grandy's, McDonald's, Quizno's, Sonic, Taco Bell, Wendy's, Best Western, Brake-O, Cavender's Boots, Home Depot, Time-It Lube, Wal-Mart SuperCtr/24hr/gas, cleaners, **1 mi S**...HOSPITAL, DENTIST, CHIROPRACTOR, Mobil, Circle K/gas, Arby's, Chili's, ChuckeCheese, Dairy Queen, Mr Gatti's, Popeye's, Books-A-Million, Dillard's, Eckerd, K-Mart, NAPA, Office Depot, mall
11	LA 3, Bossier City, **N**...HOSPITAL, Buick/Pontiac/GMC, Ford, Toyota/Lexus, Maplewood RV Park(3mi), **S**...Chevron/McDonald's/diesel/mart/24hr, **1 mi S**...DENTIST, PODIATRIST, VETERINARIAN, Texaco, Pizza Hut, Lube Express, auto repair, bank, hardware
7b a	US 71, LA 1, Shreveport, **N**...Citgo/mart, Exxon/diesel/mart, Fina/diesel/mart, Subway, Vince's Rest., Whataburger/24hr, Brookshire's Foods, NAPA, Pennzoil, bank, **S**...CHIROPRACTOR, Chevron/mart/wash/24hr, RaceTrac/mart, Texaco/diesel/mart/24hr, BBQ, Burger King, Church's Chicken, Domino's, KFC, McDonald's, Sammy's Rest., Taco Bell, Howard Johnson/rest., AutoZone, Eckerd, Family$, Kroger/deli, MinitLube, Radio Shack, Rite Aid, auto repair, bank, carwash, cleaners/laundry, transmissions
5	LA 173, Blanchard Rd, **S**...Citgo
2	Lakeshore Dr, no facilities
1a	Jefferson Paige Rd, **S**...Denny's, Grandma's Kitchen, RJ's Rest./24hr, Best Western, Day's Inn, Plantation Inn, Ramada Inn
b	I-20, E to Shreveport, W to Dallas. I-220 begins/ends on I-20, exit 11.

(side tab: Shreveport)

Maine Interstate 95

Exit #(mm)Services

298mm	US/Canada border, Maine state line, US Customs. I-95 begins/ends.
63(297.5)	US 2, to Houlton, **E**...Houlton Airport, Ammex Duty Free Shop
296mm	Meduxnekeag River
62(295)	US 1, Houlton, **E**...HOSPITAL, Irving/diesel/mart/24hr, Burger King, KFC, McDonald's, Pizza Hut, Rite Aid, Sears, VIP Auto, **W...rest area both lanes, full(handicapped)facilities, phone, picnic tables, litter barrels, petwalk,** DENTIST, EYECARE, Citgo/Subway/ diesel/mart, Exxon/diesel/LP/mart, Irving/BigStop Diner/diesel/ mart, Ivey's Motel, Shiretown Motel/rest., Chevrolet/Pontiac/ Olds/Buick, Chrysler/Plymouth/Dodge/Jeep, Ford/Lincoln/ Mercury, Shop'n Save, Wal-Mart
294.5mm	B Stream
61(284)	US 2, to Smyrna, **E**...Smyrna Motel/rest.
60(279)	Oakfield Rd, to ME 11, Eagle Lake, Ashland, **W**...Irving/diesel/ mart, Mobil/diesel/mart, Country Plaza Diner, Oakfield Inn/rest.
270mm	Mattawamkeag River, W Branch
59(269)	ME 159, Island Falls, **E**...Citgo/LP/mart, Mobil/diesel/service/ mart, Bishop's Mkt, **W**...to Baxter SP(N entrance), RV camping
58(257)	Sherman, to ME 11, **E**...Mobil/diesel/mart/rest., **W**...Irving/diesel mart, Katahdin Valley Motel, to Baxter SP(N entrance)
57(252)	Benedicta(from nb, no re-entry), no facilities
245mm	scenic view nb
240mm	Salmon Stream
56(237.5)	ME 157, to Medway, E Millinocket, **W**...MEDICAL CARE, Irving/diesel/mart, Ruthie's Cafe, Gateway Inn, D&M RV, Pine Grove Camping(4mi), USPO, repair, to Baxter SP(S entrance)
237mm	Penobscot River
236.5mm	**rest area both lanes, full(handicapped)facilities, phone, picnic tables, litter barrels, petwalk**
55(220)	to US 2, ME 6, Lincoln, **4 mi E**...HOSPITAL, gas, food, lodging, RV camping
212mm	Piscataquis River
54(211)	ME 6, Howland, **E**...HandyStop, Irving/95/diesel/diner, King's Camping, 95er Towing/repair
194.5mm	Birch Stream
53(193)	ME 16(no nb re-entry), to LaGrange, no facilities
192mm	**rest area/truck check/hunter check both lanes**
52(190)	ME 43, to Old Town, **E**...gas/diesel, food, phone
189mm	Pushaw Stream
51(186)	Stillwater Ave, to Old Town, **E**...DENTIST, EYECARE, Citgo/diesel/mart, Irving/diesel/mart, Texaco/diesel/mart, Burger King, China Garden, KFC, McDonald's, Pizza Dome, Taco Bell, Best Western, Ames, IGA Foods, VIP Auto, cinema, cleaners, mall, **W**...Arctic Cat, Suzuki
50(184)	Kelly Rd, to Orono, **2-3 mi E**...gas, food, lodging, camping
49(181)	Hogan Rd, Bangor Mall Blvd, to Bangor, **E**...HOSPITAL, Citgo/mart, Denny's, Audi/VW, Chevrolet/Cadillac/ Olds, Chrysler/Plymouth/Dodge, Daewoo/Suzuki, Firestone/auto, Ford, GMC/Isuzu, Honda/Nissan/Volvo, Mitsubishi/Hyundai/Mercedes/Saab/Subaru, Saturn, Sam's Club, bank, **W**...Citgo/diesel/mart, Exxon/diesel/LP/ mart/wash, Applebee's, Arby's, Bugaboo Creek Café, Burger King, KFC, McDonald's, Mr Bagel, Olive Garden, Oriental Jade, Papa Gino's, Paul's Rest., Pizza Hut, Red Lobster, Subway, TCBY, Uno Pizzaria, Wendy's, Bangor Motel, Comfort Inn, Country Inn, Hampton Inn, Advance Parts, Bed Bath&Beyond, Best Buy, Borders Books&Café, Filene's, Goodyear/auto, Home Depot, JC Penney, Jo-Ann Fabrics, K-Mart, Lincoln/Mercury/Kia, MailBoxes Etc, Mr Paperback, Mr QuickLube, PetQuarters, ProGolf, Sears/auto, Shaw's Food/deli/24hr, Shop'n Save, Staples, ToysRUs, Turner Sports, VIP Auto, Wal-Mart, bank, mall
48(179)	ME 15, to Broadway, Bangor, **E**...HOSPITAL, Irving/mart, Thrifty Pizza, **W**...Exxon/mart, Mobil/mart, Texaco, 7-11, China Light, Dairy Queen, Friendly's, Governor's Rest., KFC, McDonald's, Pizza Hut, Seafood Rest., Subway, Ames, Broadway Hardware, Firestone, Mazda, Pontiac, Prompto Lube/wash, Rite Aid, Shop'n Save, TJ Maxx, Varney's Parts
47(178)	ME 222, Union St, to Ohio St, Bangor, **E**...Exxon, **W**...Exxon/wash, Mobil/diesel/mart, Texaco, Burger King, Dunkin Donuts, McDonald's, Nicky's Diner, Papa Gino's, Wendy's, AAA, Chrysler/Jeep, Olds/Cadillac, Midas Muffler, Mr Paperback, Radio Shack, Shop'n Save, Staples, airport, bank, cleaners/laundry, RV camping
46(177)	US 2, ME 2, Hammond St, Bangor, **E**...DENTIST, Gulf, Corner Store, Napoli Pizza, Subway, Fairmount Hardware, NAPA, golf, radiators, **W**...airport
45b(176)	US 2, ME 100 W, **W**...Irving/diesel/mart, Mobil/diesel/mart, Dunkin Donuts, Ground Round, Jason's Pizza, Day's Inn, Econolodge, Fairfield Inn, Holiday Inn, Howard Johnson, Motel 6, Ramada Inn, Rodeway Inn, Super 8, auto parts, cinema, RV camping, tires
a	I-395, to US 2, US 1A, Bangor, downtown

Houlton

Bangor

Maine Interstate 95

44(173.5) Cold Brook Rd, to Hampden, **W**...Citgo/diesel/mart, White House Inn Best Western, Mack/Kenworth/Volvo/White/GMC, RV camping

172mm **rest area sb, full(handicapped)facilities, info, phone, vending, picnic tables, litter barrels, petwalk**

170mm Soudabscook Stream

169mm **rest area nb, full(handicapped)facilities, info, phone, vending, picnic tables, litter barrels, petwalk**

43(167) ME 69, to Carmel, **E**...Citgo/mart, **W**...RV camping

42(161) ME 69, ME 143, to Etna, no facilities

41(155) ME 7, to E Newport, Plymouth, **E**...lodging, LP, **W**...RV camping

40(153) Ridge Rd, to Plymouth, Newport, no facilities

39(150) to US 2, ME 7, ME 11, Newport, **E**...HOSPITAL, **W**...CHIROPRACTOR, MEDICAL CARE, Citgo, Exxon/mart, Irving/diesel/mart/atm/24hr, Mobil/diesel/mart, Burger King, China Way, Dunkin Donuts, King's Rest., McDonald's, Sawyers Dairybar, Scotty's Diner, Subway, AAA Motel, Lovley's Motel, Ames, Aubuchon Hardware, Big A Parts, CarQuest, Chrysler/Plymouth/Jeep, Muffler King/tires, NAPA, Radio Shack, Rite Aid, Shop'n Save, Wal-Mart/drugs, bank, cleaners/laundry, RV camping

145mm Sebasticook River

38(144) Somerset Ave, Pittsfield, **E**...HOSPITAL, Mobil, Subway, Ponderosa Motel, Chevrolet, Family$, NAPA, Rite Aid, Shop'n Save, bank, carwash, laundry, RV camping, tires

141mm **rest area both lanes, full(handicapped)facilities, phone, picnic tables, litter barrels, petwalk**

37(131) Hinckley Rd, Clinton, **W**...Citgo/diesel/LP/mart, phone

127.5mm Kennebec River

36(127) US 201, Fairfield, **E**...Jim's Gas/diesel/mart, Purple Cow Pancakes, carwash, **W**...Fairfield Creamery, Old NE CandleStore

35(126) ME 139, Fairfield, **E**...Texaco/deli/mart, Coffee Connection, **W**...Citgo/Subway/diesel/mart/atm

34(124) ME 104, Main St, Waterville, **E**...HOSPITAL, Mobil/mart, Arby's, Fresh Strawberry Pie, Friendly's, Governor's Rest., McDonald's, Wendy's, Atrium Motel, Best Western, Holiday Inn, Howard Johnson, Advance Parts, FashionBug, JC Penney, K-Mart, Mazda/VW, Mr Paperback, Radio Shack, Sears, Shop'n Save, VIP Auto, XpressLube, cleaners/laundry, tires

123mm Messalonskee Stream

33(121) ME 11, ME 137, Waterville, Oakland, **E**...HOSPITAL, Citgo/Burger King/mart, Irving/diesel/mart/atm/24hr, Mobil/mart/wash, Angelo's Diner, Applebee's, Classic Café, McDonald's, Pizza Hut, Subway, Weathervane Seafood, Budget Host, Econolodge, Chevrolet/Olds/Pontiac/Buick/Cadillac/Toyota, Chrysler/Plymouth/Dodge/Jeep, JJ Nissen Bakery, Nissan, Mr Paperback, Rite Aid, Shaw's Food/gas/24hr, Wal-Mart, cinema, RV camping, **W**...Coastal/diesel/mart, Exxon/diesel/mart, China Express, Aubuchon Hardware, CarQuest, Ford/Lincoln/Mercury

32(114) Lyons Rd, Sidney, no facilities

111mm **rest area sb, full(handicapped)facilities, phone, vending, picnic tables, litter barrels, petwalk**

107mm **rest area nb, full(handicapped)facilities, phone, vending, picnic tables, litter barrels, petwalk**

31(106) ME 27, ME 8, ME 11, Augusta, **E**...EYECARE, Citgo/mart, Getty Gas, Capt Core's Seafood, Ground Round, Olive Garden, Holiday Inn, Barnes&Noble, Home Depot, Linens'n Things, Michael's, Old Navy, PetQuarters, Sam's Club, Staples, Wal-Mart SuperCtr/24hr, bank, cinema, **W**...Irving/Taco Bell/diesel/mart/atm/24hr, NorthPark Grille, Comfort Inn, Mitsubishi/Volvo, Vet Cem

105mm Bond Brook

30(103.5) US 202, ME 11, ME 17, ME 100, Augusta, **E**...CHIROPRACTOR, EYECARE, VETERINARIAN, Citgo, Gulf/mart, Irving/diesel/mart, Texaco/mart, Arby's, Burger King, Charlie's Pizza, Damon's, Domino's, Dunkin Donuts, Dairy Queen, Friendly's, Hong Kong Isle, KFC, Little Caesar's, Maine Bagels, McDonald's, Pizza Hut, Subway, Wendy's, Best Western/rest., CarQuest, K-Mart, Midas Muffler, Mr Paperback, PromptoLube, Service Merchandise, Shaw's Foods, SuperLube, U-Haul, USPO, VIP Auto, bank, **W**...Exxon/repair, Getty/mart/atm, Bonanza, TeaHouse Chinese, Motel 6, Super 8, Susse Chalet, Travelodge, Ames, Chrysler/Jeep/Nissan/Subaru, Dodge/Plymouth/Hyundai, Pontiac, Sam Goody, Shop'n Save/24hr, Sears/auto, TJ Maxx, Toyota, antiques, bank

14 ME TPK/I-495. Maine Turnpike and I-95 run together nb, divide sb.

28(100) ME 9, ME 126, to Gardiner, Litchfield, TOLL PLAZA, **W**...gas, last nb exit before toll rd

27(94) US 201, to Gardiner, **E**...gas/diesel, **W**...gas, RV camping

26(88) ME 197, to Richmond, **W**...gas, phone, RV camping

25(81.5) ME 125, Bowdoinham, no facilities

24(75) ME 196, to Lisbon, Topsham, **E**...Arby's, McDonald's, Romeo's Pizza, Meineke Muffler, NAPA, Radio Shack, Rite Aid, Super Shop'n Save, laundry, mall, **W**...Topsham SuperStop/diesel/mart

74.5mm Androscoggin River

22(73) US 1, Bath, **1 mi E on US 1**...HOSPITAL, Citgo/diesel/mart, Irving/diesel/mart, Mobil/diesel/mart, Chinese Rest., Dunkin Donuts, McDonald's, Comfort Inn, Econolodge, Fiesta Motel, Chevrolet/Mazda, Ford, XpressLube, RV camping

71mm parking area southbound

W a t e r v i l l e   A u g u s t a

N
S

Maine Interstate 95

21(68)   to Freeport(from nb), facilities 1 mi E on US 1

20(67)   ME 125, to Pownal, **1/2 mi E on US 1**...DENTIST, MEDICAL CARE, Exxon/
mart, Arby's, Chinese Rest., Friendly's, McDonald's, Taco Bell, LL Bean,
USPO, bank, laundry, outlet mall/famous brands, RV camping, **W**...to
Bradbury Mtn SP

19(65)   Desert Rd, Freeport, **E**...Citgo/mart, Thai Rest., Caroline Inn, Dutch
Village Motel, Super 8, outlet mall/famous brands, RV camping

17(62)   US 1, Yarmouth, **E...rest area both lanes,
full(handicapped)facilities, info**, Texaco/diesel/mart, Bill's Pizza,
Muddy Rudder Rest., Eagle Motel, Freeport Inn/rest., Ford,
Delorme Mapping, **W**...MEDICAL CARE, Texaco/mart, Bill's
Pizza, McDonald's, Pat's Pizza, DownEast Village Motel/rest.,
Royal River Natural Foods, Shop'n Save, XpressLube/wash,
auto/truck repair, hardware, laundry/cleaners, tires

16(59.5)   US 1, to Cumberland, Yarmouth, **W**...CHIROPRACTOR, Exxon,
Mobil/mart, Birchwood Rest., Lox Stock&Bagels, Maine Roaster
Coffee, Romeo's Pizza, Brookside Motel, Rite Aid, tires

15(55)   I-295 S, **US 1 S**...Exxon, Gulf/diesel, Mobil, Dunkin Donuts,
McDonald's, Moose Crossing Rest., Ames, Goodyear, Olds, Rite
Aid, Shaw's Foods, bank

54mm   toll booth

9(49)   I-495 N, to Lewiston, no facilities

8(46)   ME 25, to Portland, **E**...Citgo, Burger King, Subway, Valle's Rest.,
Inn at Portland, Ames, BJ's Whse, Chevrolet/Saab, CVS Drug, Firestone/auto, Jo-Ann Fabrics, Shaw's Foods,
W...CHIROPRACTOR, Exxon, FuelMart/diesel, Mobil/diesel/mart, Texaco/Subway/mart, Buffalo Wings,
Chinese Buffet, Dairy Queen, Denny's, Dunkin Donuts, Governor's Rest., KFC, McDonald's, Pizza Hut,
Verrillo's Rest., Holiday Inn, Howard Johnson, Motel 6, Super 8, Susse Chalet, Bradlee's, Chrysler/Plymouth/
Dodge/Jeep, Crafter's Outlet, DD Auto Parts, Ford/Hyundai, Home Depot, Midas Muffler, NAPA, Shop'n Save,
Tire Whse, VIP Auto

45mm   Stroudwater River

7(43)   to US 1, Maine Mall Rd, S Portland, **E**...CHIROPRACTOR, Citgo/mart, Mobil/mart, Texaco/diesel/mart, Burger
King, Bugaboo Creek Steaks, Chili's, Eastside Mario's Café, Friendly's, Fuji Japanese, Great Wall Chinese,
IHOP, LoneStar Steaks, Macaroni Grill, McDonald's, Old Country Buffet, Olive Garden, On the Border, Pizza
Hut, Sebago Brewing Co, Strawberry's Cafe, Tim Horton's, Uno Pizzaria, Weathervane Seafood, Wendy's,
Coastline Inn, Comfort Inn, Day's Inn, Fairfield Inn, Hampton Inn, Residence Inn, Sheraton Tara, AAA, Ames,
Bed Bath&Beyond, Best Buy, Boaters World, Borders Books, Burlington Coats, Circuit City, Craftmania,
Filene's, Ground Round, Honda, JC Penney, KidsRUs, Lexus/Subaru/Toyota, Luxury Linens, Macy's,
MailBoxes Etc, Marshall's, MVP Sports, Nevada Bob's, OfficeMax, PetCo, ProntoLube, Sam's Club, Sears/
auto, Service Merchandise, Shaw's Foods, Shop'n Save, SportsAuthority, Staples, TJ Maxx, ToysRUs, cinema,
mall, to Jetport, **W**...AmeriSuites, Marriott

6a(42)   I-295 N(from nb), to S Portland, Scarborough, **E**...HOSPITAL, Wal-Mart(1mi)

41mm   Nonesuch River

6(39)   to US 1, **E**...Scarborough Downs Racetrack(seasonal)

5(34)   I-195 E, to Saco, Old Orchard Beach, **2 mi E**...gas, food, phone, lodging, camping

33mm   **E**...Holiday Inn Express/Saco Hotel Conference Ctr

31mm   Saco River

4(29)   ME 111, to Biddeford, **E**...HOSPITAL, Irving/Subway/diesel/mart/atm, Wendy's, Ford, Shaw's Food/drug, Wal-
Mart/auto, **W**...Home Depot

25mm   Kennebunk River

3(24)   ME 35, Kennebunk Beach, **E**...VETERINARIAN, Turnpike Motel

**24mm   service plaza both lanes, sb...Mobil/diesel, Burger King, TCBY, gifts, nb...Mobil/diesel, Burger King,
TCBY, atm, gifts**

23mm   Mousam River

19.5mm   Merriland River

2(18)   ME 9, ME 109, to Wells, Sanford, auto museum, phone, lodging, **W**...to Sanford RA

6mm   Maine Tpk begins/ends, toll booth

4(5.8)   ME 91, to US 1, to The Yorks, **E**...HOSPITAL, DR/gas, Irving/diesel/mart, Mobil/mart, Texaco/mart, Canellie's
Deli, Foodee's Pizza, Norma's Cafe, Ruby's Grill, Family$, Ford, Shop'n Save, TrueValue, cleaners, last exit
before toll rd nb

5.5mm   weigh sta nb

5mm   York River

N

S

Maine Interstate 95

4mm	weigh sta sb	
3mm	**Welcome Ctr nb, full(handicapped)facilities, info, phone, vending, picnic tables, litter barrels, petwalk**	
2(2)	(2 & 3 from nb), US 1, to Kittery, **E**...VETERINARIAN, Citgo/Taco Bell/diesel/mart, Getty/mart, Mobil/diesel/mart, Burger King, Clam Hut Diner, Dairy Queen, McDonald's, Payrin Thai, Quarterdeck Rest., Subway, Sunrise Grill, Weathervane Seafood, Coachman Motel, Super 8, Factory Stores of America/famous brands, Tanger Factory Outlet/famous brands, **W**...Mobil	
1	ME 103(from nb, no re-entry), to Kittery, **E**...Navy Yard, Citgo/service, Blue Roof Motel/diner, Day's Inn, NorthEaster Motel, Rex Motel	
0mm	Piscataqua River, Maine/New Hampshire state line	

N ↕ S

Maine Interstate 295

Exit #(mm)Services

I-295 begins/ends on I-95, exit 15.

10	US 1, to Falmouth, **E**...Exxon, Gulf/diesel, Mobil, Dunkin Donuts, McDonald's, Moose Crossing Rest., Ames, Goodyear, Olds, Rite Aid, Shaw's Foods, bank, carwash
9mm	Presumpscot River
9	US 1 S, ME 26, to Baxter Blvd, no facilities
8	ME 26 S, Washington Ave, **E**...AAA, U-Haul
7	US 1A, Franklin St, **E**...NAPA, West Marine
6b a	US 1, Forest Ave, **E**...HOSPITAL, Mobil, Firestone/auto, carwash, **W**...Arby's, Bleacher's Rest., Burger King, Pizza Hut, Pier 1, Shop'n Save, Sir Speedy, bank
5b a	ME 22, Congress St, **E**...HOSPITAL, Denny's, McDonald's, Susse Chalet, **W**...Gas For Less, Getty Gas, Mobil/diesel/mart, Ananias Italian, DoubleTree Hotel, bank
3mm	Fore River
4	US 1 S, to Main St, to S Portland, facilities on US 1 away from exit
3	ME 9, to Westbrook St, **W**...Irving/diesel/mart, Mobil/diesel/mart, Burger King, LoneStar Steaks, Olive Garden, TGIFriday, Chevrolet, Home Depot, HomeGoods, MVP Sports, to airport
2	to US 1 S, to S Portland, services on US 1, **1/2 mi E on US 1**...Citgo/7-11, Exxon, Irving Gas/mart, Mobil, Dunkin Donuts, Governor's Rest., Tony Roma's, Best Western, Day's Inn, Econolodge, Howard Johnson, Discount Tire
1	to I-95, to US 1, multiple services on US 1, same as 2

I-295 begins/ends on I-95, exit 6a.

N ↕ S

Maine Interstate 495

Exit #(mm)Services

Maine Tpk begins/ends at exit 30 on I-95. I-495 portion of Maine Tpk begins/ends at 100mm.

14b(100)	I-95 S, ME 9, ME 126, to Gardiner
a	I-95 N
97.5mm	toll plaza
95mm	**service plaza nb, Mobil/diesel/mart, Burger King, TCBY, atm, gifts**
82.5mm	Sabattus Creek
81mm	**service plaza sb, Mobil/diesel/mart, Burger King, TCBY, atm, gifts**
13(78)	to ME 196, Lewiston, **W on ME 196**...HOSPITAL, CHIROPRACTOR, Getty Gas, Mobil/diesel, Sunoco/mart, Texaco/mart, Cathay Hut Chinese, Chalet Rest., D'Angelo's, Dunkin Donuts, Fanny's Diner, Governor's Rest., KFC/Taco Bell, McDonald's, Wendy's, Motel 6, Ramada Inn, Super 8, Aubuchon Hardware, Rite Aid, Shaw's Foods, carwash
76.5mm	Androscoggin River
12(73)	US 202, rd 4, rd 100, to Auburn, **E**...HOSPITAL, Irving/Blimpie/Taco Bell/diesel/mart, KnuckleHeads Buffet, Auburn Inn, RV camping, **W**...HOSPITAL
69mm	Royal River
64.5mm	toll plaza
11(61)	US 202, rd 115, rd 4, to ME 26, Gray, **E**...Exxon, Gulf/mart/atm, Mobil/service, Texaco/mart, Dunkin Donuts, Subway, McDonald's, Rite Aid, Thriftway Foods, TrueValue, cleaners/laundry
57mm	**service plaza nb, Mobil/diesel, Burger King, TCBY, atm, gifts**
56mm	**service plaza sb, Mobil/diesel, Burger King, TCBY, atm, gifts**
53mm	Piscataqua River
10(50)	to ME 26, ME 100 W, N Portland, **E**...HOSPITAL, Dunkin Donuts, Patty's Rest., Hannaford's Food/drug, bank, cleaners
49mm	Presumpscot River
9(49)	I-95 N and S, to US 1

I-495 portion of Maine Tpk begins/ends at exit 9, 49mm.

N ↕ S

Maryland Interstate 68

Exit #	Services
82c	I-70 W, to Breezewood. I-68 begins/ends on I-70, exit 1.
b	I-70 E, US 40 E, to Hagerstown
a	US 522, Hancock, **N**...Chevrolet, Chrysler/ Dodge, **S**...Exxon, Sheetz/ mart/24hr, Pizza Hut, Roy Rogers, Park'n Dine, Weaver's Rest., SuperFresh Food, Comfort Inn, Hancock Motel, camping
77	US 40, MD 144, Woodmont Rd, **S**...RV camping, food
75mm	runaway truck ramp eb
74mm	**Sideling Hill rest area/exhibit ctr both lanes, full(handicapped)facilities, info, vending**
74	US 40, Mountain Rd(no return from eb), no facilities
73mm	Sideling Hill Creek
72mm	truck ramp wb
72	US 40, High Germany Rd, Swain Rd, **S**...Amoco/diesel/mart
68	Orleans Rd, **N**...Citgo/diesel/mart
67mm	Town Hill, elevation 940 ft
64	MV Smith Rd, **S**...to Green Ridge SF HQ, scenic overlook, phone
62	US 40, 15 Mile Creek Rd, **N**...Bill Meyer Wildlife Mgt Area, **1/2 mi S**...gas/mart
58.7mm	Polish Mtn, elevation 1246 ft
57mm	Town Creek
56mm	Flintstone Creek
56	MD 144, National Pike, Flintstone, **S**...Sunoco/mart, USPO, groceries
52	MD 144(from wb), Pleasant Valley Rd, National Pike, no facilities
50	Pleasant Valley Rd, **N**...to Rocky Gap SP, info, seasonal camping
47	MD 144, Old National Pike, Dehaven Rd(from wb), same as 46
46	US 220 N, Dehaven Rd, Baltimore Pike, **N**...VETERINARIAN, BP/store, DaVinci's Pizzaria, Lindy's Rest., Cumberland Motel, MD Motel, Cook Bros Parts/lube, Great Value Foods, truck parts, **S**...Exxon/diesel, Mason's Barn Rest., Steak Cellar, Uncle Tucker's Pizza, Exit 46 Motel, Folcks Mill Inn/rest., crafts
45	Hillcrest Dr, **S**...BP/diesel/mart
44	US 40A, Baltimore Ave, Willow Brook Rd, to Allegany Comm Coll
43d	Maryland Ave(from eb), **N**...HOSPITAL, USPO, **S**...Texaco/Subway/diesel/mart
c	(from wb), same as 43b
b	Maryland Ave, **N**...McDonald's, Holiday Inn, **S**...Amoco, Citgo/diesel, Day Light Donuts, Roy Rogers, Taco Bell, Wendy's
a	to WV 28A, Beall St, Industrial Blvd, to Cumberland, **N**...Sheetz/mart
42	US 220 S, Greene St, Ridgedale, phone
41	(from wb)
41mm	Haystack Mtn, elev 1240 ft
40	US 220 S, to US 40A, Vocke Rd, La Vale, **N**...HOSPITAL, Amoco/repair, BP/mart, Citgo/diesel/repair, Exxon, Mobil/ mart, Texaco/Subway, Arby's, Bob Evans, Burger King, Dairy Queen, D'Atri Rest., Denny's, Gehauf's Rest., KFC, Little Caesar's, LJ Silver, McDonald's, Oriental Rest., Pizza Hut, Roy Rogers, Texas Grill, Wendy's, Best Western, Scottish Inn, Super 8, County Mkt Foods/24hr, CVS Drug, Ford, Jo-Ann Fabrics, Lowe's Bldg Supply, Midas Muffler, Revco, Staples, auto parts, bank, st police, **S**...Chi Chi's, Ponderosa, Western Sizzlin, OakTree Inn, Bon-Ton, JC Penney, Kelly Tire, K-Mart, Martin Food/drug, Sears/auto, Wal-Mart/drugs, cinema, mall
39	US 40A(from wb), same as 40
34	MD 36, to Westernport, Frostburg, **N**...HOSPITAL, Amoco, Sheetz/mart, McDonald's, Old Depot Rest., Pizza Hut, Roy Rogers, Comfort Inn, Falinger's Hotel Gunter, Hampton Inn, bank, **S**...to Dans Mtn SP
33	Midlothian Rd, to Frostburg, **N**...HOSPITAL, **S**...to Dans Mt SP
31mm	weigh sta eb
30mm	Big Savage Mtn, elevation 2800 ft
29	MD 546, Finzel, **N**...Mason-Dixon Camping(4mi)
25.8mm	eastern continental divide, elevation 2610 ft
24	Lower New Germany Rd, to US 40A, **S**...to New Germany SP, to Savage River SF
23mm	Meadow Mtn, elevation 2780 ft
22	US 219 N, to Meyersville, **N**...BP/diesel/mart/24hr, Exxon, Burger King, Penn Alps Rest., Subway, Little Meadows Motel/diner, Hilltop Fruit Mkt, **S**...Amoco/mart/wash, Holiday Inn, camping
20mm	Casselman River
19	MD 495, to US 40A, Grantsville, **N**...Exxon/diesel/mart, Mobil/mkt/deli, Penn Alps Rest., Casselman Motel/rest., Elliott House Inn, Beachy's Drug, Big A Parts, Chevrolet, Kelly Tire, USPO, bank, hardware
15mm	Mt Negro, elevation 2980 ft
14mm	Keyser's Ridge, elevation 2880 ft
14b a	US 219, US 40 W, Oakland, **N**...Citgo/diesel/mart, Keyser's Ridge/Mobil/diesel/rest., McDonald's, repair
6mm	**Welcome Ctr eb, full(handicapped)facilities, info, phone, picnic tables, litter barrels, vending, petwalk**

E

W

Maryland Interstate 68

4.5mm	Bear Creek
4mm	Youghiogheny River
4	MD 42, Friendsville, **1 mi N...**Amoco, Citgo/mart, Old Mill Rest., Yough Valley Motel, S&S Mkt, USPO, bank, **S...**to Deep Creek Lake SP, camping
0mm	Maryland/West Virginia state line

Maryland Interstate 70

Exit #	Services
	I-70 begins/ends in Baltimore at Cooks Lane
91b a	I-695, **N off exit 17...**Citgo/7-11, McDonald's, Day's Inn, Motel 6, Ramada, Best Buy, Burlington Coats, Firestone/auto, JC Penney, Old Navy, Sears/auto, Staples, mall
87b a	US 29(exits left from wb)to MD 99, Columbia, **1-2 mi S on US 40...**CHIROPRACTOR, Amoco/diesel, Crown/mart, Mobil, Shell, Burger King, McDonald's, Subway, Wendy's, Advance Parts, Chevrolet, Giant Foods, JiffyLube, Pontiac/Isuzu/Olds/GMC/Volvo, Rite Aid, Valu Food, Wal-Mart
83	US 40, Marriottsville(no EZ wb return), **2 mi S...**Turf Valley Hotel/Country Club/rest.
82	US 40 E(from eb), same as 83
80	MD 32, Sykesville, **N...**phone, fairgrounds, golf, **S...**Citgo
79mm	weigh sta wb, phone
76	MD 97, Olney, **S...**Citgo
73	MD 94, Woodbine, **N...**Texaco, McDonald's, Pizza Hut, Ramblin Pines RV Park(6mi), SuperMkt, bank, cleaners, **1 mi S...**Amoco/diesel, Citgo
68	MD 27, Mt Airy, **N...**Amoco/Blimpie/diesel/mart/24hr, Citgo/7-11, Mobil/diesel, Texaco, Arby's, Big Boy, Burger King, Domino's, KFC, Ledo's Pizza, McDonald's, Pizza Hut, Roy Rogers, TCBY, Ace Hardware, Chrysler/Plymouth/Dodge/Jeep, Food Lion, Rite Aid, Safeway, SuperFresh Foods, **S...**Exxon/diesel/mart/24hr, 4 Seasons Rest.
66mm	truckers parking area eb
64mm	weigh sta eb
62	MD 75, Libertytown, **N...**Mobil/diesel/mart, Texaco, McDonald's, New Market Hist Dist
59	MD 144, no facilities
57mm	Monocacy River
56	MD 144, **N...**Citgo, Sheetz/mart, Burger King, Donut Shoppe, McDonald's, Roy Rogers, Taco Bell, bank, to Hist Dist
55	South St, no facilities
54	Market St, to I-270, **N...**Amoco, Texaco/diesel/rest., Travelodge, **S...**DENTIST, Exxon/diesel/mart, Mobil/diesel, Sheetz/mart/24hr, Shell/mart/24hr, SouStates/diesel, Texaco/diesel/mart/24hr, 7-11, Bob Evans, Burger King, Checker's, Cracker Barrel, Jerry's Subs/pizza, KFC, McDonald's, Pargo's Café, Popeye's, Roy Rogers, Subway, Sunflower Rest., Wendy's, Day's Inn/rest., Econolodge, Fairfield Inn, Hampton Inn, Holiday Inn Express, Aamco, Best Buy, Chrysler/Plymouth, Circuit City, Ford/Lincoln/Mercury, Honda, Hyundai/Buick, Kohl's, OfficeMax, PrecisionTune, Sam's Club, Saturn, Sears/auto, Wal-Mart/auto/drug, bank, cleaners, mall
53	I-270 S(from eb), US 15 N(from wb), to Frederick
52	US 15, US 340, Leesburg, **1 mi N...**Comfort Inn
49	US 40A, Braddock Heights, **N...**Amoco, Crown/mart, Exxon/diesel/mart, Mobil/diesel, Shell/repair, Texaco/mart, 7-11, Big Boy, Bob Evans, Boston Mkt, Burger King, Casa Rico Mexican, Chi Chi's, Denny's, Fritchie's Candystick Rest., Ground Round, Hardee's, Joe's Diner, McDonald's, Pizza Hut, Roy Rogers, Shoney's, Wendy's, Best Western, Masser Motel/rest., Holiday Inn, Ford/Subaru, Goodyear, JC Penney, JiffyLube, Martin's Foods, Merchant Tires, Midas Muffler, PepBoys, cinema, st police, **S...**to Washington Mon SP, camping
48	US 40 E, US 340(from eb, no return), **1 mi N...**same as 49
42	MD 17, Myersville, **N...**Exxon, Sunoco/diesel, Burger King, McDonald's, Village House Rest., to Gambrill SP, **S...**Amoco/diesel, to Greenbrier SP
39mm	**rest area both lanes, full(handicapped)facilities, phone, vending, picnic tables, litter barrels, petwalk**
35	MD 66, to Boonsboro, **S...**Sheetz/mart, to Greenbrier SP, Washington Mon SP, camping
32b a	US 40, Hagerstown, **1-3 mi N...**HOSPITAL, AC&T Gasmart/24hr, Amoco/Blimpie/mart, Citgo/7-11/24hr, Exxon/diesel, Sunoco, BBQ, Bob Evans, Burger King, Denny's, Little Caesar's, McDonald's, Pizza Hut, PoFolks, Taco Bell, Tortuga Dining, Comfort Suites, Day's Inn, Finley Inn, Hampton Inn, Holiday Inn/rest., Quality Inn, Ramada Inn, Sheraton, Super 8, Wellesley Inn, Chevrolet, Chrysler/Plymouth, Goodyear/auto, Martin's Food/drug/24hr, Mercedes, Midas Muffler, Toyota, bank, **S...**Honda, Kia, Mazda, Olds, Pontiac/Buick/GMC, VW
29b a	MD 65, Sharpsburg, **N...**HOSPITAL, Exxon/diesel/mart/24hr, Sheetz/mart, Sunoco/diesel/mart/24hr, FoodCourt, Pizza Hut, Prime Outlets/famous brands, st police, **S...**Shell/Blimpie/diesel/mart/atm/24hr, Burger King, Cracker Barrel, McDonald's, Wendy's, Safari Camping, to Antietam Bfd
28	MD 632, Hagerstown, **N...**Chevron, Pizza Hut, **S...**Yogi Bear Camping
26	I-81, N to Harrisburg, S to Martinsburg, no facilities
24	MD 63, Huyett, **N...**Sheetz/mart(3mi), tires/repair, **S...**KOA, C&O Canal
18	MD 68 E, Clear Spring, **N...**Amoco, BP/mart, McDonald's, bank, **S...**Exxon/diesel/mart, Wendy Hill Café
12	MD 56, Indian Springs, **N...**Indian Springs Camping, **S...**Ft Frederick SP
9	US 40 E(from eb), Indian Springs, no facilities
5	MD 615(no immediate wb return), **N...**Log Cabin Rest.(2mi)

E

W

Baltimore / Frederick / Hagerstown

Maryland Interstate 70

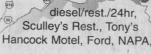

3 MD 144, Hancock(exits left from wb),
S...CHIROPRACTOR, Amoco/diesel/rest./24hr, BP/diesel/mart, Citgo/diesel/mart, Hancock Trkstp/ Shive's Subs, Hardee's, Park'n Dine, Pizza Hut, Italian/Mexican, Weaver's Rest., Econolodge(2mi), Hancock Motel, Ford, NAPA, bank, carwash, repair diesel/rest./24hr, Sculley's Rest., Tony's

1b US 522, Hancock, **N**...Econolodge, Chevrolet, Chrysler/Dodge/Jeep, **S**...Gary's Gas/ mart, Lowest Price Gas/diesel, Sheetz/mart/24hr, CrabDen Seafood, Park'n Dine, Pizza Hut, Weaver's Rest., Hancock Motel, $General, Happy Hills Camp, bank

a I-68 W, US 40, W to Cumberland

0mm Maryland/Pennsylvania state line

Maryland Interstate 81

Exit # Services
12mm Maryland/Pennsylvania state line
10b a Showalter Rd, **E**...airport
9 Maugans Ave, **E**...Amoco, Mobil, Texaco/mart, Antrim House Rest., Hardee's, McDonald's, Pizza Hut, Roy Rogers, Colonial Motel/rest., **W**...Exxon, Family Time Rest., GMC/Kenworth/Volvo
8 Maugansville Rd(from sb, no re-entry), same as 9
7b a MD 58, Hagerstown, no facilities
6b a US 40, Hagerstown, **E**...HOSPITAL, Chevron/diesel, Exxon, Texaco, Dunkin Donuts, Roy Rogers, Best Western, Dagmar Motel, Holiday Inn, Ramada
5 Halfway Blvd, **E**...AC&T/diesel/rest., Exxon/diesel, Sheetz/mart, Shell, Applebee's, Boston Mkt, Chinese Rest., Crazy Horse Steaks, Ground Round, Hardee's, Little Caesar's, McDonald's, Pizza Hut, Raphael's Mexican, Red Lobster, Roy Rogers, Ruby Tuesday, Shoney's, Taco Bell, Wendy's, Western Sizzlin, Howard Johnson, Motel 6, Travelodge, BEST, Bon-Ton, CVS Drug, Firestone/auto, Food Lion, Ford/Lincoln/Mercury, JC Penney, K-Mart/auto, Lowe's Bldg Supply, Martin's Food/drug, Sam's Club, Staples, Value City, Value Vision Ctr, Wal-Mart, Weis Foods, WonderBooks, bank, cinema, mall, **W**...AC&T/mart
4 I-70, E to Frederick, W to Hancock, to I-68
2 US 11, Williamsport, **E**...AC&T/diesel/mart/24hr, **W**...Exxon/service, Shell/mart/atm, Sunoco/diesel/mart/24hr, McDonald's, Waffle House, Red Roof Inn, KOA(4mi), bank
1 MD 63, MD 68, Williamsport, **E**...Bowman/diesel, Safari Camping, to Antietam Bfd
0mm Potomac River, Maryland/West Virginia state line

Maryland Interstate 83

Exit # Services
37.5mm Mason-Dixon Line, Maryland/Pennsylvania state line
37 to Freeland(from sb), no facilities
36 MD 439, Bel Air, **5 mi W**...Morris Meadows Camping
35.5mm weigh sta sb
33 MD 45, Parkton, **E**...Exxon/diesel
31 Middletown Rd, to Parkton, no facilities
27 MD 137, Mt Carmel, Hereford, **E**...MEDICAL CARE, DENTIST, CHIROPRACTOR, VETERINARIAN, Exxon/diesel/mart, High's Dairy, Michael's Pizza, Graul's Foods, Hereford Drug, NAPA, TrueValue, USPO, bank, cleaners
24 Belfast Rd, to Butler, Sparks, no facilities
20 Hunt Valley, **E**...Amoco, Exxon/diesel, Shell, 7-11, Beijing Rest., Burger King, Carrabba's, Cinnamon Tree Rest., Friendly's, McDonald's, Outback Steaks, Wendy's, Chase Suites, Courtyard, Econolodge, Embassy Suites, Hampton Inn, Hunt Valley Marriott, Burlington Coats, Dick's Sports, Giant Food/drug, Goodyear/auto, Metro Food/drug, Sears/ auto, Valley View Farms, Wal-Mart, cinema, mall
18 Warren Rd(from nb, no return), Cockeysville, **E**...Exxon/mart, Southern States Gas/diesel, Residence Inn, JiffyLube, PrecisionTune
17 Padonia Rd, Deereco Rd, **E**...CHIROPRACTOR, Amoco/diesel, Hess/diesel/mart, Shell, Bob Evans, Chili's, Denny's, Macaroni Grill, Day's Hotel, Extended Stay America, Aamco, Chevrolet, Goodyear/auto, Mars Foods, Olds, Rite Aid, USPO, bank, cleaners
16b a Timonium Rd, **E**...Sunoco/diesel, Steak&Ale, Day's Hotel, Holiday Inn, Red Roof Inn, Infiniti, Saab
14 I-695 N, no facilities
13 I-695 S, Falls Rd, HOSPITAL, st police
12 Ruxton Rd(from nb, no return), no facilities
10b a Northern Parkway, **E**...HOSPITAL, Exxon/mart, Texaco/24hr
9b a Cold Spring Lane, no facilities
8 MD 25 N(from nb), Falls Rd
7b a 28th St, Baltimore Zoo, Memorial Stadium, HOSPITAL
6 US 1, US 40T, North Ave, downtown
3 Chase St, downtown
2 Pleasant St(from sb), downtown
1 Fayette St, downtown Baltimore, Bennigan's. I-83 begins/ends.

Maryland Interstate 95

Exit #	Services
110mm	Maryland/Delaware state line
109b a	MD 279, to Elkton, Newark, DE, **E**...Petro/Texaco/diesel/rest./24hr, McDonald's, Taco Bell, Econolodge, Elkton Lodge, Knight's Inn, Motel 6, Blue Beacon, **W**...TA/Mobil/Subway/diesel/rest./24hr, to U of DE
100	MD 272, to North East, Rising Sun, **1 mi E**...Flying J/diesel/rest./LP/24hr, Sunoco, Frank's Pizza, McDonald's, Poor Jimmy's, Roy Rogers, Schroeder's Deli, Crystal Inn, Eckerd, auto parts/repair, carwash, groceries, museum, st police, to Elk Neck SP, **W**...Citgo/mart, Mobil, zoo
96mm	**Chesapeake House service area(exits left from both lanes), Exxon/diesel, Sunoco/diesel, Burger King, Mrs Fields, Pizza Hut, Popeye's, Starbuck's, Taco Bell, TCBY, Farmers Mkt, gifts**
93	MD 275, to Rising Sun, US 222, to Perryville, **E**...HOSPITAL, Exxon/diesel/mart, Pilot/Subway/Dairy Queen/diesel/mart/24hr, Crothers Rest., Denny's, KFC/Taco Bell, Outlet Ctr Café, Comfort Inn, Chesapeake Outlet/famous brands, **W**...Mama's Kitchen, Douglas Motel, st police
92mm	weigh sta/toll booth
91.5mm	Susquehanna River
89	MD 155, to Havre de Grace, last nb exit before toll, **1-5 mi E**...HOSPITAL, KFC, McDonald's, MacGregor's Rest., Red Carpet Inn, Super 8, **W**...to Susquehanna SP
85	MD 22, to Aberdeen, **E**...Amoco/24hr, Citgo/7-11, Crown/diesel/mart, Sunoco/mart, Texaco/diesel/mart/wash/24hr, Arby's, Durango's Steaks, Golden Corral, High's Dairy, KFC, McDonald's, Olive Tree Rest., Sassy's Rest., Taco Bell, Wendy's, Day's Inn, Econolodge/rest., Holiday Inn, Howard Johnson, Quality Inn, Red Roof, Sheraton, Super 8, Farmer's Mkt Food, Mars Foods/deli/bakery, JiffyLube, K-Mart, Olds, Rite Aid, bank, cinema, laundry, museum
81mm	**Maryland House service area(exits left from both lanes), Exxon/diesel, Sunoco/diesel, Big Boy, Cinnamon Gourmet, Hotdog City, Roy Rogers, Sbarro's, TCBY, gifts**
80	MD 543, to Riverside, Churchville, **E**...CHIROPRACTOR, DENTIST, Amoco, Crown/mart, Mobil/diesel, A&W, Burger King, Chinese Rest., Cracker Barrel, McDonald's, Klein's Food/bakery/deli, bank, drugs, laundry, Bar Harbor RV Park
77b a	MD 24, to Edgewood, Bel Air, **E**...Exxon, Shell, Texaco/diesel, Burger King, Denny's/24hr, Giovanni's Rest., McDonald's, Taco Bell, Vitah's Rest., Best Western/rest., Comfort Inn, Day's Inn, Edgewood Motel, Sleep Inn, **W**...Exxon/diesel/mart, cinema
74	MD 152, Joppatowne(no EZ nb return), **E**...HOSPITAL, Citgo/diesel, Exxon/diesel/mart, Shell, High's Dairy, KFC, IHOP, McDonald's, Roy Rogers, Sizzler, Venitian Palace, Wendy's, Super 8, Toyota(1mi)
70mm	Big Gunpowder Falls
67b a	MD 43, to White Marsh Blvd, US 1, US 40, **1 mi E**...Crown/24hr, Shell/mart, Sunoco, White Marsh Diner, Williamsburg Inn, Best Buy, CarMax, Dick's Sports, Service Merchandise, Target, to Gunpowder SP, **3 mi E on US 40**...Burger King, Costco Whse, Home Depot, PepBoys, PetsMart, SportsAuthority, RV Ctr, **W**...Exxon/diesel, Bertoucci's, McDonald's, Olive Garden, Red Brick Station, Red Lobster, Ruby Tuesday, Taco Bell, TGIFriday, Wendy's, Hampton Inn, Barnes&Noble, Hecht's, Macy's, Old Navy, Sears, USPO, bank, mall, to Gunpowder SP
64b a	I-695(exits left), E to Essex, W to Towson
62	to I-895(from sb)
61	US 40, Pulaski Hwy, **E**...Texaco
60	Moravia Rd, **E**...Texaco, Burger King, Denny's, Pontiac
59	Eastern Ave, HOSPITAL, **W**...Amoco/wash/24hr, Exxon
58	Dundalk Ave, no facilities
57	Boston St, **E**...Texaco/diesel, Best Western, Silverado Hotel/rest.
56	Keith Ave, no facilities
56mm	McHenry Tunnel, toll plaza(north side of tunnel)
55	Key Hwy, to Ft McHenry Nat Mon, last nb exit before toll
54	MD 2 S, to Hanover St, access to HOSPITAL, downtown
53	I-395 N, to downtown, to Oriole Park
52	Russell St N, no facilities
51	Washington Blvd, no facilities
50	Caton Ave, **E**...Shell/mart/24hr, Caton House Rest., Pargo's Rest., Shoney's Inn/rest., bank, **W**...HOSPITAL
49b a	I-695, E to Key Bridge, Glen Burnie, W to Towson, to I-70, to I-83
47b a	I-195, to BWI Airport, to Baltimore
46	I-895, to Harbor Tunnel Thruway
43	MD 100, to Glen Burnie, **1 mi E**...Exxon, Best Western
41b a	MD 175, to Columbia, HOSPITAL, **E**...DENTIST, Chevron, Shell, High's Dairy, Jerry's Subs/pizza, McDonald's, Roy Rogers, Holiday Inn, Red Roof Inn, Trucker's Inn, laundry, **W**...Exxon/mart/wash, Blackeyed Pea, Bob Evans, McDonald's, Olive Garden, TGIFriday, Royal Farm Store, to Johns Hopkins U, Loyola U
38b a	MD 32, to Ft Meade, **2 mi E**...Amoco, Exxon, Roy Rogers, Taco Bell, Comfort Inn, to BWI Airport
37mm	**Welcome Ctr both lanes, full(handicapped)facilities, info, phone, vending, picnic tables, litter barrels, petwalk**
35b a	MD 216, to Laurel, **E**...Crown Gas, Weis Food/drug, laundry
34mm	Patuxent River
33b a	MD 198, to Laurel, **E**...HOSPITAL, Exxon, High's Dairy, **2 mi E**...other facilities, **W**...Exxon, Shell, Best Western

A b e r d e e n

B a l t i m o r e

N

S

Maryland Interstate 95

29   MD 212, to Beltsville, **E**...VETERINARIAN, 7-11, gas, pizza, **W**...Exxon/diesel, American Café, Chinese Rest., McDonald's, Plata Grande Grill, Wendy's, Holiday Inn, Ramada Inn, Giant Food/drug, CVS Drug, bank, cinema, RV camping

27   I-495 S around Washington

25   US 1, Baltimore Ave(from nb), to Laurel, College Park, **E**...Amoco, Chevron, Citgo/7-11, Exxon/diesel, Shell, Burger King, Hardee's, Jerry's Subs/pizza, McDonald's, Pizza Hut, Roy Rogers, Wendy's, Holiday Inn, Fresh Value Mkt, Goodyear, JiffyLube, Midas Muffler, Precision Tune, Rite Aid, bank, laundry, US Agri Library, **W**...Amoco/24hr, Shell, Rodeway Inn, Honda, Hyundai, VW, RV Ctr, bank, to U of MD

23   MD 201, Kenilworth Ave, **E**...Marriott/rest., camping, **W**...Shell, Courtyard, Raleigh Inn, Cadillac

22   Baltimore-Washington Pkwy, **E**...to NASA

20   MD 450, Annapolis Rd, Lanham, HOSPITAL, **E**...Mobil/mart, Horn's Smorgasbord, Jerry's Subs, Lanham Inn Pizza, McDonald's, Pizza Hut, Day's Inn/rest., Holiday Inn, Ford, Jeep/Eagle, Midas Muffler, Speedee Muffler, **W**...Chevron/7-11/24hr, Shell, Sunoco, Big Boy, Chesapeake Bay Seafood, Chinese Rest., Dunkin Donuts, KFC, Popeye's, Sheraton, Ames, Dodge, Firestone/auto, Goodyear/auto, Grease'n Oil Express, Meineke Muffler, Pontiac, Safeway

19   US 50, to Annapolis, Washington, no facilities

17   MD 202, Landover Rd, to Upper Marlboro, **E**...Holiday Inn, Capitol Centre, **W**...Chinese Rest., IHOP, Buick, Goodyear/auto, Pace Whse, Sears, mall

15   MD 214, Central Ave, **E**...Hampton Inn/rest., Capitol Centre, **W**...Crown Gas/wash, Exxon/diesel, Shell, Sunoco/24hr, Big Boy, KFC, Jerry's Subs, McDonald's, Pizza Hut, Roy Rogers/Hardee's, Day's Inn/rest., Super 8, Goodyear, Meineke Muffler, Midas Muffler, auto parts, bank, carwash, mall

11   MD 4, Pennsylvania Ave, to Upper Marlboro, st police, **E**...Amoco/mart/24hr, Exxon/mart/wash, Sunoco/diesel, Texaco/mart, Ames, Murry's Whse, SuperFresh Foods, **W**...Shell, 7-11, bank

9   MD 337, to Allentown Rd, **E**...to Andrews AFB, HOSPITAL, Shell/autocare, Roy Rogers/Hardee's, Toddle House, Motel 6, Ramada Inn, Firestone/auto, U-Haul, Woolworth, **W**...Crown Gas, Mobil, Dunkin Donuts, McDonald's, Holiday Inn

7   MD 5, Branch Ave, to Silver Hill, **W**...HOSPITAL, Texaco/diesel, Day's Inn, Ford

4   St Barnabas Rd, **E**...same services as 3

3   MD 210, Indian Head Hwy, to Forest Hts, **W**...DENTIST, MEDICAL CARE, Amoco/24hr, Chevron/diesel/24hrs, 7-11, Mobil/mart, Shell, Aloha Rest., Arby's, Baskin-Robbins, Big Boy, Burger King, Chinese Rest., KFC, Little Caesar's, McDonald's, Pizza Hut, Popeye's, RanchHouse Rest., Roy Rogers, Sizzler, Taco Bell, Wendy's, Ramada Inn, Red Roof, Susse Chalet, Drug Emporium, Giant Foods, Goodyear/auto, K-Mart, Rite Aid, Safeway, XpressLube, auto parts, carwash, laundry

2b a   I-295, to Washingon

0mm   Potomac River, Woodrow Wilson Bridge, Maryland/Virginia state line

Maryland Interstate 270(Rockville)

Exit #   Services

32   I-270 begins/ends on I-70, exit 53.

31b a   MD 85, **N**...Exxon/mart/wash, Sheetz/mart, Applebee's, Bob Evans, Golden Corral, KFC/Taco Bell, LoneStar Steaks, Pargo's, Holiday Inn Express, Borders Books, Goodyear/auto, Home Depot, JiffyLube, ValuCity, **S**...Amoco/Blimpie/Baskin-Robbins/mart/24hr, Cracker Barrel, Honda, Fairfield Inn, Hampton Inn, Holiday Inn

30mm   Monocacy River

28mm   viewpoint wb

26   MD 80, Urbana, **N**...Exxon/mart, **S**...VETERINARIAN

22   MD 109, Hyattstown, to Barnesville, **S**...Comus Inn Rest., gas

21mm   weigh sta both lanes

18   MD 121, Boyds, to Clarksburg, Little Bennett Pk, Blackhill Pk, **S**...gas, camping

16   MD 27, Father Hurley Blvd, to Damascus, **N**...Exxon/mart/wash, Mobil/mart, Applebee's, Bob Evans, Bruegger's Bagels, Japanese Rest., McDonald's, TCBY, Giant Foods, Home Depot, HomePlace, KidsRUs, Kohl's, Michael's, PepBoys, PetsMart, Target, Wal-Mart, bank

15b a   MA 118, to MD 355, **N**...Chevron/diesel/mart, Mobil/mart/wash, Hampton Inn, **S**...Amoco, Exxon/mart/24hr, st police, tire/brake repair

13b a   Middlebrook Rd(from wb), no facilities

11b a   MD 124, Quince Orchard Rd, **N**...Shell, ChuckeCheese, Italian Rest., McDonald's, Starbucks, Hilton, Holiday Inn, Bed Bath&Beyond, Borders Books, Boston Mkt, Roy Rogers, Subway, CompUSA, CVS Drugs, Ford, Sam's Club, Sports Authority, mall, **S**...Seneca Creek SP

10   MD 117, Clopper Rd(from nb), same as 11.

N
S

E
W

Washington DC Area

Frederick

Rockville

Maryland Interstate 270

9b a	I-370, to Gaithersburg, Sam Eig Hwy, **S**...access to Chinese Rest., India Bistro, Pizza Hut, Subway, Festival Foods, Weis Mkt, cleaners/laundry
8	Shady Grove Rd, **S**...HOSPITAL, Marriott, Quality Suites, Sleep Inn
6b a	MD 28, W Montgomery Ave, **N**...Woodfin Suites, **W**...Best Western, bank
5b a	MD 189, Falls Rd
4b a	Montrose Rd, **N**...gas, **S**...st police
2	I-270/I-270 spur diverges eb, converges wb
1b a	Democracy Blvd, **N**...Exxon/diesel, Texaco/diesel, Marriott, Hecht's, JC Penney, Nordstrom's, cinema, mall
0mm	I-270 spur begins/ends on I-495, exit 38.

E
↑
↓
W

Maryland Interstate 695(Baltimore)

Baltimore

Exit #	Services
48mm	Patapsco River, Francis Scott Key Br
44	MD 695(from nb)
43mm	toll plaza
42	MD 151 S, Sparrows Point(last exit before toll sb), **E**...gas/diesel
41	MD 20, Cove Rd, **W**...gas, food
40	MD 150, MD 151, North Point Blvd, **E**...auto repair
39	Merritt Blvd, **W**...Amoco, Ford, Hyundai
38b a	MD 150, Eastern Blvd, to Baltimore, **W**...KFC, JC Penney, Staples
36	MD 702 S(exits left from sb), Essex
35	US 40, **1 mi N**...Dunkin Donuts, McDonald's, Continental, Motel Christlen, Circuit City, NTB, Office Depot, ToysRUs, U-Haul, cinema, **2 mi N**...Citgo, Harley-Davidson, Home Depot, PetsMart, Sports Authority
34	MD 7, Philadelphia Rd, **N**...Susse Chalet
33b a	I-95, N to Philadelphia, S to Baltimore
32b a	US 1, Bel Air, **N**...VETERINARIAN, Exxon/mart/wash, 7-11, Denny's, Dunkin Donuts, IHOP, Taco Bell, Giant Foods, K-Mart, Merchant Tire/auto, Salvo Auto Parts, **S**...DENTIST, Crown Gas, Getty Gas, Chinese Rest., McDonald's, Mr Crab, Subway, Valu Food, Goodyear/auto
31c	MD 43 E(from eb), no facilities
b a	MD 147, Hartford Rd, **N**...DENTIST, CHIROPRACTOR, EYECARE, Amoco/mart/24hr, Citgo/7-11, Mobil, Texaco/diesel, CVS Drugs, Chrysler/Plymouth, Goodyear, **S**...DENTIST, VETERINARIAN
30b a	MD 41, Perring Pkwy, **N**...Burger King, Denny's, Roy Rogers, Burlington Coats, Chevrolet, Ford, Giant Foods, Home Depot, K-Mart, Office Depot, Ross, TJ Maxx, bank, **S**...Dodge, Mazda, Midas Muffler
29b a	MD 542, Loch Raven Blvd, **S**...Holiday Inn
28	Providence Rd, no facilities
27b a	MD 146, Dulaney Valley Rd, **N**...Hampton NHS, **S**...DENTIST, Mobil/diesel/mart, Sheraton, Hecht's, SuperFresh Food, bank, drugs, mall
26b a	MD 45, York Rd, Towson, **S**...Linens'n Things, gas, food
25	MD 139, Charles St, **N**...HOSPITAL, AAA
24	I-83 N, to York
23b a	I-83 S, Baltimore, MD 25 N, **N**...gas, food
22	Green Spring Ave, no facilities
21	Stevenson Rd, no facilities
20	MD 140, Reisterstown Rd, Pikesville, **N**...gas, food, lodging, **S**...gas, food, lodging
19	I-795, NW Expswy, no facilities
18b a	MD 26, Randallstown, Lochearn, **E**...HOSPITAL, Crown/mart
17	MD 122, Security Blvd, **E**...Red Lobster, Comfort Inn, Day's Inn, Motel 6, Mitsubishi, **W**...Citgo/7-11, McDonald's, Holiday Inn, Best Buy, Firestone, JC Penney, Staples, Woolworth's, mall
16b a	I-70, E to Baltimore, W to Frederick
15b a	US 40, Ellicott City, Baltimore, **E**...ChuckeCheese, Day's Inn, Dodge, Value City Foods, **W**...Amoco, Shell
14	Edmondson Ave, no facilities
13	MD 144, Frederick Rd, Catonsville, **W**...gas, food
12c b	MD 372 E, Wilkens, **E**...HOSPITAL
11b a	I-95, N to Baltimore, S to Washington
10	US 1, Washington Blvd, **W**...auto repair
9	Hollins Ferry Rd, Lansdowne, **W**...Goodyear
8	MD 168, Nursery Rd, **N**...Exxon, Shell, KFC, McDonald's, Wendy's, Motel 6
7b a	MD 295, **N**...to Baltimore, **S**...BWI Airport
6b a	Camp Mead Rd(from nb)
5	MD 648, Ferndale, gas, food, lodging
4b a	I-97 S, to Annapolis
3b a	MD 2 N, Brooklyn Park, **S**...gas, food, lodging
2	MD 10, Glen Burnie
1	MD 174, Hawkins Point Rd, **S**...Citgo/mart

Massachusetts Interstate 84

Exit #(mm)Services

E

4(11) I-84 begins/ends on I-90, Exit 9.
3b a(9) US 20, Sturbridge, **W**...HOSPITAL, Citgo/ mart, Mobil/diesel, Burger King, Friendly's, McDonald's, O'Brien's Rest., Piccadilly Rest., Sturbridge Pizza House, Best Western, Sturbridge Coach Motel, Super 8, Crabtree&Evelyn's, antiques, bank, info, RV camping
2(5) MA 131, Sturbridge, to Old Sturbridge Village, **E**...Sturbridge Motor Inn, RV camping
4mm picnic area wb, litter barrels
1(2.5) Mashapaug Rd(from sb), to Southbridge, **E**...Mobil/diesel/24hr, Sturbridge Isle Trkstp/ diesel/rest., Texaco/diesel/24hr, Boston Pizza/deli, Roy Rogers, Sbarro's, RV camp(seasonal)

W

2mm weigh sta wb
0mm Massachusetts/Connecticut state line

Massachusetts Interstate 90

Exit #(mm)Services

E

136mm I-90 begins/ends on I-93, exit 20 in Boston.
22(134) Presidential Ctr, downtown
20(132) MA 28, Alston, Brighton, Cambridge, **N**...HOSPITAL, Guest Quarters
131mm toll plaza
19(130) MA Ave(from eb), **N**...IHOP, McDonald's, Day's Inn
17(128) Centre St, Newton, **N**...Sheraton Tara, Cadillac, Chevrolet, Honda, Nissan
16(125) MA 16, W Newton, **S**...Mobil/repair
15(124) I-95, **N**...Marriott
123mm toll plaza
14(122) MA 30, Weston, no facilities
117mm Natick Travel Plaza eb, Mobil/diesel, Burger King, TCBY, gifts, info
13(116) MA 30, Natick, **S**...HOSPITAL, Shell
114mm Framingham Travel Plaza wb, Mobil/diesel, Burger King, Popeye's, TCBY, gifts
12(111) MA 9, Framington, **N**...HOSPITAL, Motel 6, Sheraton Tara
11a(106) I-495, N to NH, S to Cape Cod
105mm Westborough Travel Plaza wb, Mobil/diesel, Sbarro's, TCBY, gifts
11(96) MA 122, to Millbury, **N**...UMA Med Ctr
10a(95) MA 146, no facilities
94mm Blackstone River
10(90) I-395 S, to Auburn, I-290 N, Worcester, **N**...Arby's, Bickford's Rest., Papa Gino's, Baymont Inn, Day's Inn, Acura, Barnes&Noble, Caldor, Filene's, Midas Muffler, PetCo, Pier 1, Sears/auto, Staples, mall, **S**...HOSPITAL, CVS Drug

W

84mm Charlton Travel Plaza wb, Mobil/diesel, Burger King, Pizza Hut, Taco Bell, TCBY, Starbucks, gifts, info
80mm Charlton Travel Plaza eb, Mobil/diesel, Starbucks, Pizza Hut, TCBY, gifts, info, st police
79mm toll plaza
9(78)mm I-84, to Hartford, NYC, to Sturbridge, access to HOSPITAL
67mm Quaboag River
8(62) MA 32, to Palmer, **S**...HOSPITAL, Mobil, McDonald's, Ames, Goodyear, Midas Muffler, Rite Aid, Super Big Y Food/deli, U-Haul, auto parts, camping, carwash, motel, to UMA
56mm Ludlow Travel Plaza wb, Mobil/diesel, Roy Rogers
55mm Ludlow Travel Plaza eb, Mobil/diesel, Roy Rogers, Mrs Fields, TCBY, gifts
7(54) MA 21, to Ludlow, **S**...HOSPITAL, Gulf/mart, Pride/mart, Burger King, Friendly's, McDonald's, Subway, Big Y Foods, XpressLube, auto parts, bank, carwash, drugs, hardware, tires
6(51) I-291, to Springfield, Hartford CT, **S**...HOSPITAL, Exxon, Motel 6, Ramada Inn, cinema, to Bradley Int Airport, Basketball Hall of Fame
5(49) MA 33, to Chicopee, Westover AFB, **S**...HOSPITAL, Pride/mart, Motomart, Burger King, Chinese Rest., Denny's, Friendly's, IHOP, Pizza Hut, Best Western, Comfort Inn/rest., BJ's Whse, Chevrolet, Ford, Honda, Midas Muffler, Monro Muffler/brakes, Pearle Vision, Pontiac/GMC, bank, mall
46mm Connecticut River
4(46) I-91, US 5, W Springfield, to Holyoke, **1.5mi S**...HOSPITAL, Shell/mart/wash, Huntsman Motel
41mm st police wb
3(40) US 202, to Westfield, **S**...HOSPITAL, Mobil/mart, Texaco/diesel, Bickford's Rest., Dairy Mart, Friendly's, Westfield Motor Inn, bank, drugs, RV camping **209**

Massachusetts Interstate 90

36mm	Westfield River
29mm	**Blandford Travel Plaza wb...Mobil/diesel, Burger King, gifts, info, eb...Mobil/diesel**
20mm	1724 ft, highest point on MA Tpk
20.5mm	litter barrel wb
14.5mm	Appalachian Trail
12mm	parking area both lanes, litter barrels
2(11)	US 20, to Lee, Pittsfield, **N**...Christy's/gas, Shell/mart/24hr, Sunoco/repair, Athena Pizza, Burger King, Cunkin Donuts, Friendly's, McDonald's, Pizza Hut, Best Western, Pilgrim Inn, Sunset Motel, Super 8, Brooks Drug, bank, **S**...Texaco, Lee Plaza Motel/rest., Bershire Outlets/famous brands, NAPA
10.5mm	Hoosatonic River
9mm	**service plaza wb, Mobil/diesel/mart, atm, vending**
8mm	**Lee Travel Plaza eb, Mobil/diesel, Burger King, TCBY, bank, info**
4mm	toll booth, phone
1(2)	MA 41(from wb, no return), to MA 102, W Stockbridge, the Berkshires, **N**...Pleasant Valley Motel, access to Bousquet Ski Area
0mm	Massachusetts/New York state line

Massachusetts Interstate 91

Exit #(mm)Services

55mm	Massachusetts/Vermont state line, callboxes begin/end
54mm	parking area both lanes, picnic tables, litter barrels
28(51)	US 5, MA 10, Bernardston, **E**...RV camping, **W**...Sunoco, RV camping
27(45)	MA 2 E, Greenfield, **E**...HOSPITAL, Citgo, Mobil, Sunoco/mart, Burger King, Denny's Pantry, Friendly's, McDonald's, CVS Drug, Meineke Muffler
26(43)	MA 2 W, MA 2A E, Greenfield, **E**...HOSPITAL, CHIROPRACTOR, Citgo/diesel/mart/atm, Mobil/mart, Palmer/ diesel/mart/wash, China Gourmet, Dunkin Donuts, Howard Johnson/rest., Chevrolet, Ford/Lincoln/Mercury, MailBoxes Etc, carwash, **W**...Mobil/mart/atm/24hr, Texaco/mart/atm/24hr, Abdow's, Bickford's, Bricker's Rest., Friendly's, McDonald's, Turnbull's Rest., CandleLight Inn, Super 8, Ames, Big Y Foods, BJ's Wholesale, Family$, Nissan, Rich's, Staples, Super Foodmart, cinema, to E Berkshire Skiing, Mohawk Tr
39mm	Deerfield River
37mm	weigh sta both lanes
25(36)	MA 116(from sb), S Deerfield, hist dist, camping, same as 24
24(35)	US 5, MA 10, MA 116, Deerfield(no EZ return), **E**...Mobil/mart/atm, Motel 6, Yankee Candle Co, bank, **W**...BP/ diesel/diner/24hr
34.5mm	parking area nb, no facilities
23(34)	US 5(from sb), no facilities
22(30)	US 5, MA 10(from nb), N Hatfield, **W**...Diamond RV Ctr
21(28)	US 5, MA 10, Hatfield, **W**...Sunoco/mart/24hr, Stearn's Motel, Long View RV, st police
20(26)	US 5, MA 9, MA 10(from sb), Northampton, **W**...HOSPITAL, Mobil, Pride/mart, Bickford's, Burger King, D'Angelo's, Friendly's, McDonald's, Big Y Foods, Caldor, Chevrolet/VW, Ford, JiffyLube, Pontiac/GMC/Cadillac
19(25)	MA 9, Northampton, to Amherst, **E**...HOSPITAL, Getty Gas, Webster's Fishook Rest., Hadley BarnShops, to Elwell SP, **W**...Citgo
18(22)	US 5, Northampton, **E**...Mobil/mart, Inn at Northampton, **W**...Shell/24hr, TVI Inn, ProLube, carwash, to Smith Coll
18mm	scenic area both lanes
17b a(16)	MA 141, S Hadley, **E**...Citgo/diesel, Gulf, Mobil/diesel, DairyMart, Bess Eaton, Subway, Super 8, Susse Chalet, Brooks Drug/24hr, Buick, **W**...to Mt Tom Ski Area
16(14)	US 202, Holyoke, **E**...HOSPITAL, Burger King, Denny's, Friendly's, McDonald's, Schermerhorn's Seafood, Yankee Pedlar Inn, to Heritage SP, **W**...Soldier's Home
15(12)	to US 5, Ingleside, **E**...HOSPITAL, CHIROPRACTOR, Shell/mart, Cracker Barrel, Friendly's, Pizzaria Uno, Ruby Tuesday, Holiday Inn, Barnes&Noble, Bed Bath&Beyond, Circuit City, CompUSA, Filene's, JC Penney, JiffyLube, Lord&Taylor, MensWearhouse, PetCo, Pier 1, Sears/auto, Service Merchandise, bank, mall
14(11)	to US 5, to I-90(Mass Tpk), E to Boston, W to Albany, HOSPITAL
13(9)	US 5 N, W Springfield, **E**...Citgo/mart, Bickford's, B'Shara's Rest., Donut Dip, Piccadilly's, Subway, Comfort Inn, Knight's Inn, Red Roof Inn, Super 8, BMW, GolfDay, Home Depot, Honda/Lexus/Toyota, mall, **W**...Citgo/diesel, Mobil/mart, Sunoco/diesel/mart, Abdow's, Arby's, Bertucci's Pizzaria, Boston Mkt, Burger King, ChiChi's, Chili's, D'Angelo's, Friendly's, Ivanhoe Rest., KFC, McDonald's, Pizza Hut, Wendy's, Best Western, Day's Inn, Econolodge, Hampton Inn, Quality Inn/buffet, Red Carpet Inn, Bradlee's, Chrysler/Plymouth, Costco, Dick's Sports, FashionBug, Home Depot, Lincoln/Mercury, Mazda/Mercedes, Michael's, Monro Muffler/brake, Nevada Bob's, Nissan/Subaru, Stop'n Shop Foods, Staples, bank, carwash, cinema, cleaners
12(8.5)	I-391 N, to Chicopee, no facilities

Massachusetts Interstate 91

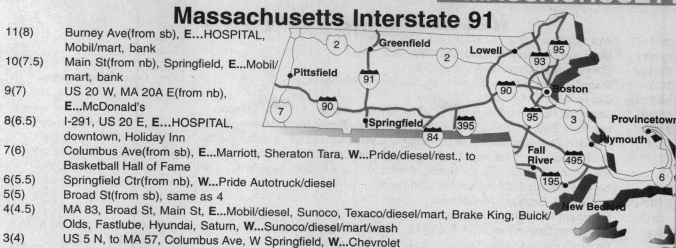

11(8)	Burney Ave(from sb), **E**...HOSPITAL, Mobil/mart, bank
10(7.5)	Main St(from nb), Springfield, **E**...Mobil/mart, bank
9(7)	US 20 W, MA 20A E(from nb), **E**...McDonald's
8(6.5)	I-291, US 20 E, **E**...HOSPITAL, downtown, Holiday Inn
7(6)	Columbus Ave(from sb), **E**...Marriott, Sheraton Tara, **W**...Pride/diesel/rest., to Basketball Hall of Fame
6(5.5)	Springfield Ctr(from nb), **W**...Pride Autotruck/diesel
5(5)	Broad St(from sb), same as 4
4(4.5)	MA 83, Broad St, Main St, **E**...Mobil/diesel, Sunoco, Texaco/diesel/mart, Brake King, Buick/Olds, Fastlube, Hyundai, Saturn, **W**...Sunoco/diesel/mart/wash
3(4)	US 5 N, to MA 57, Columbus Ave, W Springfield, **W**...Chevrolet
2(3.5)	MA 83 S(from nb), to E Longmeadow, **E**...Friendly's
1(3)	US 5 S(from sb), no facilities
0mm	Massachusetts/Connecticut state line, callboxes begin/end

Massachusetts Interstate 93

Exit #(mm)Services

47mm	Massachusetts/New Hampshire state line, callboxes begin/end
48(46)	MA 213 E, to Methuen, **E**...HOSPITAL
47(45)	Pelham St, Methuen, **E**...Shell, Sunoco/mart/24hr, Dunkin Donuts, McDonald's, Outback Steaks, **W**...Getty, Day's Inn/rest., Chrysler/Jeep, Olds/Pontiac/Nissan
46(44)	MA 110, MA 113, to Lawrence, **E**...HOSPITAL, Getty, Mobil/repair, Shell, Burger King, McDonald's/24hr, Papa Gino's, Pizza Hut/D'Angelo, MktBasket Foods, Osco Drug, Rich's, **W**...Coastal Gas, Dunkin Donuts, Jimmy's Rest., Jackson's Rest., Millhouse Rest., 6-11 Deli, Motel 110
45(43)	Andover St, River Rd, to Lawrence, **E**...Courtyard, Marriott, **W**...Mobil/Dunkin Donuts/mart/atm, Grill 93, Tage Inn
44b a(40)	I-495, to Lowell, Lawrence, **E**...HOSPITAL
43(39)	MA 133, N Tewksbury, **E**...Mobil/Dunkin Donuts/mart/atm/24hr, Ramada Inn/rest., **W**...McDonald's(2mi), 99 Rest., Wendy's
42(38)	Dascomb Rd, East St, Tewksbury, no facilities
41(35)	MA 125, Andover, st police
40(34)	MA 62, Wilmington, no facilities
39(33)	Concord St, **E**...Shriners Auditorium
38(31)	MA 129, Reading, **W**...Mobil, Burger King, Dunkin Donuts, Michael's Place Rest., Spelio's Rest.
37b a(29)	I-95, S to Waltham, N to Peabody
36(28)	Montvale Ave, **E**...Mobil/24hr, Radisson, **W**...HOSPITAL, Citgo/diesel, Exxon/24hr, Getty Gas, Bickford's/24hr, Friendly's, McDonald's, Primo's Italian, Spud's Rest., Crown Plaza Hotel, BJ's Wholesale(1mi)
35(27)	Winchester Highlands, Melrose, **E**...HOSPITAL(no EZ return to sb)
34(26)	MA 28 N(from nb, no EZ return), Stoneham, **E**...HOSPITAL, Mobil/24hr, Friendly's
33(25)	MA 28, Fellsway West, Winchester, **E**...HOSPITAL
32(23)	MA 60, Salem Ave, Medford Square, **W**...HOSPITAL, AmeriSuites, to Tufts U
31(22)	MA 16 E, to Revere(no EZ return sb), **E**...Texaco/mart, Bertucci's, Subway, Howard Johnson Rest., Bradlee's, Ford/Lincoln/Mercury, laundry, mall, st police, **W**...Exxon/mart, Chinese Rest., Dunkin Donuts, AutoZone, Dodge
30(21)	MA 28, MA 38, Mystic Ave, Somerville
29(20)	MA 28(from nb), Somerville, **E**...McDonald's, Circuit City, Home Depot, K-Mart, cinema, mall, **W**...Gulf, Merit/gas, Holiday Inn
28(19)	Sullivan's Square, Charles Town, downtown, Gulf, cinema
27	US 1 N(from nb)
26	Storrow Dr, North Station, downtown
25	Haymarket Square, Gov't Center
24(18)	Callahan Tunnel, **E**...airport
23	High St, Congress St, **W**...Marriott
22(17)	Atlantic Ave, Northern Ave, South Station, Boston World Trade Ctr
21(16.5)	Kneeland St, China Town
20(16)	I-90 W, to Mass Tpk

N ↕ **S**

Methuen

Boston Area

Massachusetts Interstate 93

19(15.5) Albany St(from sb), **W**...HOSPITAL, Mobil/diesel
18(15) Mass Ave, to Roxbury, **W**...HOSPITAL
17(14.5) E Berkeley, **E**...New Boston Food Mkt
16(14) S Hampton St, Andrew Square, **W**...Shell/24hr, Bickford's, Holiday Inn Express, Home Depot, K-Mart/Little Caesar's/auto, Marshall's, OfficeMax
15(13) Columbia Rd, Everett Square, **E**...JFK Library, bank, to UMA, **W**...Texaco
14(12.5) Morissey Blvd, **E**...JFK Library, **W**...Shell, Gino's Pizza, Midas Muffler, Pontiac
13(12) Freeport St, to Dorchester, **W**...Bradlee's, CVS Drug, Dodge, Toyota
12(11.5) MA 3A S(from sb), Quincy, **E**...Wendy's, **W**...Merit Gas, Shell, Sunoco, Arby's, Bickford's Rest., Ground Round, Papa Gino's, Phillips Candy House, Wendy's, Susse Chalet, AutoZone, Ford/Lincoln/Mercury, Pontiac/GMC, Staples, Stop'n Shop Foods, Walgreen
11b a(11) to MA 203, Granite Ave, Ashmont
10(10) Squanton Ave(from sb), Milton, **W**...HOSPITAL
9(9) Adams St, Bryant Ave, to N Quincy, **E**...bank, **W**...Shell/repair
8(8) Brook Pkwy, to Quincy, Furnace, **E**...Global Gas, Gulf/mart, Ace Hardware, Saturn
7(7) MA 3 S, Braintree, to Cape Cod, no facilities
6(6) MA 37, Braintree, to Holbrook, **E**...Mobil/mart/24hr, Boardwalk Café, Pizzaria Uno, Sheraton Tara/café, Circuit City, Filene's, Firestone/auto, Lord&Taylor, Macy's, Sears/auto, cinema, mall, **W**...Sunoco, Day's Inn, Barnes&Noble, Ford, Linens'n Things, Nissan, Osco Drug, Pearle Vision
5b a(4) MA 28 S, to Randolph, Milton, **E**...Citgo/mart/wash, Mutual Gas, Shell/repair/24hr, Texaco/wash, D'Angelo's, Domino's, Dunkin Donuts, Friendly's, IHOP, Legends Café, Lombardo's Rest., Sal's Calzone Rest., Wong's Chinese, Holiday Inn/rest., bank
4(3) MA 24 S, to Brockton, no facilities
3(2) MA 138 N, Houghton's Pond, to Ponkapoag Trail, no facilities
2b a(1) MA 138 S, Milton, to Stoughton, **E**...golf, **W**...Gulf/diesel, Mobil/mart, Sunoco/mart, Texaco/diesel/mart, Howard Johnson Rest., zoo
1(0) I-95 N, S to Providence. I-93 begins/ends on I-95, exit 12.

Massachusetts Interstate 95

Exit #(mm)Services
89.5mm Massachusetts/New Hampshire state line, parking area sb
60(89) MA 286(no sb return), to Salisbury, beaches, **E**...Mobil/diesel/mart, Capt Hook's Seafood, Chubby's Diner, East Coast Motel/diner, camping(seasonal)
59(88) I-495 S(from sb)
58b a(87) MA 110, to I-495 S, to Amesbury, Salisbury, **E**...Sunoco/Dunkin Donuts/diesel/mart, Simon's Subs, Sylvan St Grill, Winner's Circle Rest., Dodge, Plymouth/Jeep, ProLube, cinema, radiators, repair, **W**...Mobil/mart, Burger King, Friendly's, McDonald's, Susse Chalet, Chevrolet/VW
86mm Merrimac River
57(85) MA 113, to W Newbury, **E**...HOSPITAL, Mobil/mart/atm/24hr, Shell/repair/24hr, Sunoco/service, Dunkin Donuts, Friendly's, Guiseppe's Rest., McDonald's, Ming Jade Chinese, Papa Gino's, Wendy's, White Hen Pantry, Brooks Drug, K-Mart, MktBasket Foods, Marshall's, Midas Muffler, Radio Shack, Ritz Cameras, Shaw's Foods, Walgreen, bank
56(78) Scotland Rd, to Newbury, **E**...st police
55(77) Central St, to Byfield, **E**...Buddy's Rest., Gen Store Eatery, hardware/LP, **W**...Prime/diesel/repair, antiques
54b a(76) MA 133, E to Rowley, W to Groveland, no facilities
75mm weigh sta both lanes
53b a(74) MA 97, S to Topsfield, N to Georgetown, no facilities
52(73) Topsfield Rd, to Topsfield, Boxford, no facilities
51(72) Endicott Rd, to Topsfield, Middleton, no facilities
50(71) US 1, to MA 62, Topsfield, **E**...Exxon/mart/atm/24hr, Mobil/mart/24hr, Ristorante Italiano, cleaners, **W**...Supino's Rest., Sheraton Ferncroft, CVS Drug, Rich's, Staples, Stop'n Shop, VisionWorks, st police
49(70) MA 62(from nb), Danvers, Middleton, **W**...same as 50
48(69) Hobart St, **W**...CHIROPRACTOR, Italian Rest., Jake's Rest., Comfort Inn, Extended Stay America, Motel 6, Home Depot, Honda
47b a(68) MA 114, Peabody, to Middleton, **E**...Exxon/diesel/mart, Shell/wash, Sunoco/mart, Dunkin Donuts, Friendly's, McDonald's, Papa Gino's, Chevrolet, Toyota, **W**...Chili's, TGIFriday, Motel 6, Residence Inn, Circuit City, Costco, Dodge, Home Depot, LandRover, Mazda, NAPA, Olds, Pontiac/Hyundai, tires
46(67) to US 1, **W**...Best Gas/mart, Gulf/diesel/mart, Texaco/mart, Burger King, Auto Parts+
45(66) MA 128 N, to Peabody, no facilities

Massachusetts Interstate 95

44b a(65) US 1 N, MA 129, **E**...Shell, **W**...Bennigan's, Bickford's, Wendy's

43(63) Walnut St, Lynnfield, **E**...to Saugus Iron Works NHS(3mi), **W**...Sheraton, golf

42(62) Salem St, Montrose, **E**...Sunoco, Montrose Rest., **W**...Sheraton

41(60) Main St, Lynnfield Ctr, **E**...Shell

40(59) MA 129, N Reading, Wakefield Ctr, **W**...Chevrolet, diesel repair

39(58) North Ave, Reading, **E**...HOSPITAL, Best Western, Mazda/Olds/Isuzu, **W**...Exxon/repair, Texaco/diesel/mart, Ford, Volvo

38b a(57) MA 28, to Reading, **E**...DENTIST, Gulf, Texaco/repair, DairyMart, Boston Mkt, Denny's, Dunkin Donuts, Ground Round, KFC, Bed&Bath, Caldor, CVS Drug, GNC, Lenscrafters, **W**...Exxon/Dunkin Donuts/mart, Getty, Mobil/mart, Shell, Friendly's, Harrow's Rest.

37b a(56) I-93, N to Manchester, S to Boston

36(55) Washington St, to Winchester, **E**...Howard Johnson, AutoZone, Buick/Pontiac/GMC, Jaguar, Mitsubishi, Nissan, Peugeot, Toyota, **W**...DENTIST, Texaco, Chinese Rest., Bertucci's, D'Angelo's, Joe's Grill, McDonald's, On-the-Border, Pizza Hut, Pizzaria Uno, TGIFriday, MktBasket Foods, Comfort Inn, Courtyard, Hampton Inn, Red Roof Inn, Susse Chalet, Bradlee's, CVS Drug, GolfDay, NTB, PetCo, Pier 1, TJ Maxx, mall

35(54) MA 38, to Woburn, **E**...Ramada Inn, cinema, **W**...HOSPITAL, Purity Drugs/24hr, Ames, Stop'n Shop Foods, bank

34(53) Winn St, Woburn, no facilities

33b a(52) US 3 S, MA 3A N, to Winchester, **E**...ChuckeCheese, Papa-Razzi's, CVS Drug, Service Merchandise, Marshall's, Lenscrafters, Honda, **W**...HOSPITAL, Citgo, Marriott

32b a(51) US 3 N, MA 2A S, to Lowell, **E**...Shell, Burger King, McDonald's, Wyndham Garden Hotel, Circuit City, Midas Muffler, Tweeter Electronics, **W**...Bagel Café, Bean Café, Boston Mkt, Chili's, Pizzaria Regina, TCBY, Victoria Station Rest., Howard Johnson, Barnes&Noble, Bradlee's, Dodge, Filene's, Lord&Taylor, MailBoxes Etc, Macy's, Sears/auto, bank, cinema, cleaners, mall

31b a(48) MA 4, MA 225, Lexington, **E**...Mobil, Friendly's, Lexington Seafood, **W**...MEDICAL CARE, VETERINARIAN, Denny's, Friendly's, McDonald's, Holiday Inn, Ramada Inn

30b a(47) MA 2A, Lexington, **E**...HOSPITAL, Shell, **W**...Sheraton, to MinuteMan NP, Hanscom AFB, **Lexington Travel Plaza nb, Sunoco/diesel/24hr, Roy Rogers, Dunkin Donuts, TCBY, gifts**

29b a(46) MA 2 W, Cambridge, no facilities

28b a(45) Trapelo Rd, Belmont, **E**...Exxon, **1 mi E**...Mobil, Shell, Burger King, Friendly's, Dunkin Donuts, McDonald's, Papa Gino's, Osco Drugs, GoodCare Brake/muffler

27b a(44) Totten Pond Rd, Waltham, **E**...Shell, Japanese Steaks, Best Western, Home Suites Inn, Susse Chalet, Westin Hotel, **W**...Bertucci's Rest., Doubletree/rest., Costco, Home Depot

26(43) US 20, to MA 117, to Waltham, **E**...HOSPITAL, Gulf, **W**...VETERINARIAN, Mobil

25(42) I-90, MA Tpk

24(41) MA 30, Newton, Wayland, **E**...Mobil, Marriott

23(40) Recreation Rd(from nb), to MA Tpk

22b a(39) Grove St, **E**...Holiday Inn

38.5mm Newton Travel Plaza sb, Gulf/diesel, Roy Rogers, Dunkin Donuts, TCBY, Day's Inn, gifts

21b a(38) MA 16, Newton, Wellesley, **E**...HOSPITAL, **W**...Pillar House Rest.

20b a(36) MA 9, Brookline, Framingham, no facilities

19b a(35) Highland Ave, Newton, Needham, **E**...Charter Gas, Gulf, Chinese Rest., D'Angelo's, Dunkin Donuts, Ground Round, McDonald's, Souper Salad, Sheraton/rest., Midas Muffler, Olds, **W**...Chevrolet, Ford

18(34) Great Plain Ave, W Roxbury, no facilities

33.5mm parking area sb, phone, litter barrels

17(33) MA 135, Needham, Wellesley, no facilities

32mm truck turnout sb

16b a(31) MA 109, High St, Dedham, no facilities

15b a(29) US 1, MA 128, **E**...VETERINARIAN, Italian Rest., TGIFriday, Comfort Inn, Holiday Inn/grill, Ames, Midas Muffler, cinema, **W**...Shell, Sunoco/diesel/24hr, Burger King, Dunkin Donuts, McDonald's/playplace, Ramada Inn, Buick, Dodge, Hyundai, Mr Muffler, LP, multiple services W on US 1

14(28) East St, Canton St, **E**...Hilton

27mm parking area sb, full facilities

13(26.5) University Ave, no facilities

12(26) I-93 N, to Braintree, Boston, motorist callboxes end nb

Massachusetts Interstate 95

11b a(23) Neponset St, to Canton, **E**...Citgo/repair, Sunoco/repair, **W**...HOSPITAL, Gulf/diesel

22.5mm Neponset River

10(20) Coney St(from sb, no EZ return), to US 1, Sharon, Walpole, **1 mi W on US 1**...BP, Exxon/mart, Mobil/mart, Texaco, Bertucci's Pizza, Chinese Rest., Dunkin Donuts, Friendly's, Ground Round, McDonald's, Papa Gino's, Pizza Hut(2mi), Taco Bell, Best Western, Acura, Bradlee's, CVS Drug, Isuzu, Jo-Ann Crafts, Lexus, Mail&Pakit, Stop'n Shop Foods, Waldenbooks, Wal-Mart, cleaners, mall

9(19) US 1, to MA 27, Walpole, **W**...Bickford's, Super 8, Sharon Inn, same as 10

8(16) S Main St, Sharon, **E**...Osco Drug, Shaw's Foods, cinema, cleaners, whaling museum

7b a(13) MA 140, to Mansfield, **E**...99 Rest., Courtyard, Day's Inn, Holiday Inn, Motel 6, Stop'n Shop

6b a(12) I-495, S to Cape Cod, N to NH

10mm **Welcome Ctr/rest area both lanes, full(handicapped)facilities, info, phone, picnic tables, litter barrels, petwalk**

9mm truck parking area sb

5(7) MA 152, Attleboro, **E**...HOSPITAL, **W**...Gulf/diesel/mart, Bill's Pizza, Bliss Rest., Wendy's, Brooks Drug, FashionBug, Gloria's Place, Radio Shack, Shaw's Foods, cinema

4(6) I-295 S, to Woonsocket

3(4) MA 123, to Attleboro, **E**...HOSPITAL, Texaco/diesel/mart, zoo

2.5mm rest area/weigh sta both lanes, no restrooms, litter barrels, motorist callboxes

2b a(1) US 1A, Newport Ave, Attleboro, **E**...Cumberland/gas, Mobil, Shell, Sunoco, McDonald's, Olive Garden, Holiday Inn, Bob's Store, FashionBug, Home Depot, K-Mart, Mazda, OfficeMax, Pearle Vision, Pontiac, Shaw's Foods, Speedy Muffler

1(.5) US 1(from sb), **E**...Brooks Rest., Day's Inn

0mm Massachusetts/Rhode Island state line

Massachusetts Interstate 195

Exit #(mm)Services

22(41) I-495 N, MA 25 S, to Cape Cod. I-195 begins/ends on I-495, exit 1.

21(39) MA 23, to Wareham, **N**...MaxiGas/diesel/24hr, **S**...HOSPITAL, Texaco/Pizza Hut/diesel/mart, NAPA, repair, **1 mi S**...Mobil

37mm **rest area eb, info, phone, picnic tables, litter barrels, petwalk, boat ramp**

36mm Sippican River

20(35) MA 105, to Marion, no facilities

19b a(31) to Mattapoisett, **S**...Mobil

18(26) MA 240 S, to Fairhaven, **1 mi S**...Exxon, Gulf, Shell, Boston Coffee/bagel, Burger King, Dunkin Donuts, Fairhaven Chowder, Great Wall Chinese, McDonald's, Pasta House, Taco Bell, Wendy's, Hampton Inn, AutoZone, Brooks Drug, FashionBug, GMC/Buick/Pontiac, GNC, JiffyLube, K-Mart/Goodyear, Mazda, Midas Muffler, Radio Shack, Sears Hardware, Shaw's Foods, Stop'n Shop Foods, Staples, TownFair Tire, Wal-Mart, bank

25.5mm Acushnet River

17(24) Coggeshall St(from sb)New Bedford, same as 16

16(23) Washburn St, **N**...Shell, Sunoco/mart, McDonald's

15(22) MA 18 S, New Bedford, downtown, **S**...Whaling Museum, hist dist

14(21) New Bedford, Penniman St, downtown

13b a(20) MA 140, **N**...airport, **S**...HOSPITAL

12b a(19) N Dartmouth, **N**...VF Outlet/famous brands, **S**...Gibb's Gas/mart, Mobil/diesel/mart, Burger King, Dunkin Donuts, Friendly's, McDonald's, 99 Rest., Old Country Buffet, Papa Gino's, Peking Garden, Ponderosa, Taco Bell, Wendy's, Comfort Inn, Ann&Hope, Ben Franklin, Chevrolet, Firestone/auto, JC Penney, Midas Muffler, Pier 1, Saturn, Sears/auto, TownFair Tire, Toyota, bank, cinema, golf outlet, mall, st police

11b a(17) Reed Rd, to Dartmouth, **2 mi S**...Shell/mart/24hr, Dartmouth Motel

10(16) MA 88 S, Westport, no facilities

9(15.5) MA 24 N(from nb), Stanford Rd, Westport, **S**...Priscilla Rest., Hampton Inn

8b a(15) MA 24 S, Fall River, Westport

7(14) MA 81 S, Plymouth Ave, Fall River, **N**...Getty, 99 Rest., **S**...Shell, Applebee's, McDonald's, Goodyear/auto, Walgreen

6(13.5) Pleasant St, Fall River, downtown

5(13) MA 79, MA 138, to Taunton, **S**...Citgo/7-11, Merit Gas, Denny's, Day's Inn, Valvoline

12mm Assonet Bay

A t t l e b o r o

F a l l   R i v e r

N

S

N

S

Massachusetts Interstate 195

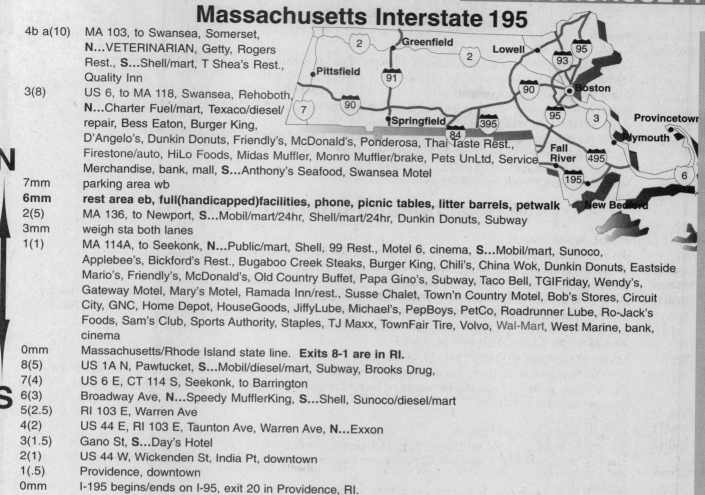

4b a(10)   MA 103, to Swansea, Somerset, **N**...VETERINARIAN, Getty, Rogers Rest., **S**...Shell/mart, T Shea's Rest., Quality Inn

3(8)   US 6, to MA 118, Swansea, Rehoboth, **N**...Charter Fuel/mart, Texaco/diesel/repair, Bess Eaton, Burger King, D'Angelo's, Dunkin Donuts, Friendly's, McDonald's, Ponderosa, Thai Taste Rest., Firestone/auto, HiLo Foods, Midas Muffler, Monro Muffler/brake, Pets UnLtd, Service Merchandise, bank, mall, **S**...Anthony's Seafood, Swansea Motel

7mm   parking area wb

6mm   **rest area eb, full(handicapped)facilities, phone, picnic tables, litter barrels, petwalk**

2(5)   MA 136, to Newport, **S**...Mobil/mart/24hr, Shell/mart/24hr, Dunkin Donuts, Subway

3mm   weigh sta both lanes

1(1)   MA 114A, to Seekonk, **N**...Public/mart, Shell, 99 Rest., Motel 6, cinema, **S**...Mobil/mart, Sunoco, Applebee's, Bickford's Rest., Bugaboo Creek Steaks, Burger King, Chili's, China Wok, Dunkin Donuts, Eastside Mario's, Friendly's, McDonald's, Old Country Buffet, Papa Gino's, Subway, Taco Bell, TGIFriday, Wendy's, Gateway Motel, Mary's Motel, Ramada Inn/rest., Susse Chalet, Town'n Country Motel, Bob's Stores, Circuit City, GNC, Home Depot, HouseGoods, JiffyLube, Michael's, PepBoys, PetCo, Roadrunner Lube, Ro-Jack's Foods, Sam's Club, Sports Authority, Staples, TJ Maxx, TownFair Tire, Volvo, Wal-Mart, West Marine, bank, cinema

0mm   Massachusetts/Rhode Island state line. **Exits 8-1 are in RI.**

8(5)   US 1A N, Pawtucket, **S**...Mobil/diesel/mart, Subway, Brooks Drug,

7(4)   US 6 E, CT 114 S, Seekonk, to Barrington

6(3)   Broadway Ave, **N**...Speedy MufflerKing, **S**...Shell, Sunoco/diesel/mart

5(2.5)   RI 103 E, Warren Ave

4(2)   US 44 E, RI 103 E, Taunton Ave, Warren Ave, **N**...Exxon

3(1.5)   Gano St, **S**...Day's Hotel

2(1)   US 44 W, Wickenden St, India Pt, downtown

1(.5)   Providence, downtown

0mm   I-195 begins/ends on I-95, exit 20 in Providence, RI.

Massachusetts Interstate 290

Exit #(mm) Services

20mm   I-290 begins/ends on I-495, exit 25.

25b a(17)   Solomon Pond Mall Rd, to Berlin, **N**...Super 8, mall

24(15)   Church St, Northborough, **N**...Citgo/mart

23b a(13)   MA 130, Boylston, **N**...Citgo/diesel, Mobil, Dunkin Donuts, Other Place Rest.

22(11)   Main St, Worchester, **N**...HOSPITAL, Exxon/24hr, Bickford's, Friendly's, McDonald's, Wendy's

20(8)   MA 70, Lincoln St, Burncoat St, **N**...EYE DOCTOR, Charter/mart/24hr, Exxon/Subway/mart/24hr, Shell/mart/wash, Sunoco, Bickford's, Denny's, Dunkin Donuts, Friendly's, McDonald's, Papa Gino's, Taco Bell, Wendy's, Day's Inn, Econolodge, Holiday Inn, Auto Palace, CVS Drug, Oil Doctor, Radio Shack, Shaw's Food/24hr, Walgreen, laundry

19(7)   I-190 N, MA 12, no facilities

18   MA 9, Framington, Ware, Worcester Airport, **N**...HOSPITAL

16   Central St, Worcester, **N**...Aku-Aku Islander Rest., Hampton Inn, mall

14   MA 122, Barre, Worcester, downtown

13   MA 122A, Vernon St, Worcester, downtown

12   MA 146 S, to Millbury, no facilities

11   Southbridge St, College Square, **N**...Texaco, Wendy's, **S**...Getty Gas

10   MA 12 N, Hope Ave, **S**...auto repair

9   Auburn St, to Auburn, **E**...Shell/24hr, Arby's, Bickford's, Wendy's, McDonald's, Baymont Inn, Comfort Inn, Ramada Inn, Acura, Firestone, PetCo, Pier 1, Staples

8   MA 12 S, Webster, **W**...Texaco, Ramada Inn

7   I-90, E to Boston, W to Springfield. I-290 begins/ends on I-90.

Massachusetts Interstate 395

Exit #(mm)Services

I-395 begins/ends on I-90, exit 10.

7(12)   to I-90(MA Tpk), MA 12, **E**...Texaco/mart, Chinese Rest., Baymont Inn, bank

6b a(11) US 20, **E**...Ford, transmissions, **W**...Shell

5(8)   Depot Rd, N Oxford, no facilities

4(6)   Sutton Ave, to Oxford, **W**...Cumberland/gas, Mobil/mart/24hr, Dunkin Donuts, McDonald's, Cahill's Tire

3(4)   Cudworth Rd, to N Webster, S Oxford, no facilities

2(3)   MA 16, to Webster, **E**...Suburu, **W**...HOSPITAL, Exxon/Dunkin Donuts/mart/24hr, Shell/24hr, Burger King, Empire Wok, Friendly's, McDonald's, Papa Gino's, Ford

1(1)   MA 193, to Webster, **E**...HOSPITAL, **W**...auto repair

0mm   Massachusetts/Connecticut state line

Massachusetts Interstate 495

Exit #(mm)Services

I-495 begins/ends on I-95, exit 59.

54(118)   MA 150, to Amesbury, **E**...HOSPITAL, Gulf/diesel/diner, Sports Park, **W**...RV camping

53(115)   Broad St, Merrimac, **W**...Texaco/service, Cutting Room Rest.

114mm   parking area sb, picnic tables, litter barrels

52(111)   MA 110, to Haverhill, **E**...HOSPITAL, **W**...Mobil/diesel/mart, Sunoco

110mm   parking area nb, picnic table, litter barrel

51(109)   MA 125, to Haverhill, **E**...HOSPITAL, Gulf, Mobil, bank, **W**...Mobil, Texaco, Burger King, Dunkin Donuts, Friendly's, McDonald's, Taco Bell, Wendy's

50(107)   MA 97, to Haverhill, **E**...HOSPITAL, **W**...Ford, gas

49(106)   MA 110, to Haverhill, **E**...Gulf, Sunoco/mart/24hr, Bickford's, Chinese Rest., Chunky's Diner, Dunkin Donuts, McDonald's, 99 Rest., Papa Gino's, Wendy's, Best Western, Comfort Inn, K-Mart, MktBasket Foods, Oil Change, bank, drugs, laundry

105.8mm Merrimac River

48(105.5) MA 125, to Bradford, no facilities

47(105)   MA 113, to Methuen, **1 mi W**...Burger King, McDonald's, Methuen Mall

46(104)   MA 110, **E**...HOSPITAL, CHIROPRACTOR, Gulf, Sunoco/mart/24hr, Ford/Lincoln/Mercury, bank, laundry, **1 mi W**...Texaco

45(103)   Marston St, to Lawrence, **W**...Chevrolet, Honda, Isuzu

44(102)   Merrimac St, to Lawrence, access to gas

43(101)   Mass Ave, no facilities

42(100)   MA 114, **E**...Exxon/mart, Gulf, Mobil, Denny's, Friendly's, Pizza Hut, Hampton Inn, **W**...HOSPITAL, VETERI-NARIAN, BP, Texaco/diesel/mart/wash, 7-11, Burger King, Dunkin Donuts, Marathon Pizzaria, McDonald's, TCBY, Wendy's, Quality Inn, MktBasket Foods, Speedy Muffler, auto parts, bank, cinema, drugs

41(99)   MA 28, to Andover, **E**...Courtyard Rest., Dunkin Donuts, Chevrolet, bank

40b a(98) I-93, N to Methuen, S to Boston

39(94)   MA 133, to N Tewksbury, **E**...Mobil/diesel/24hr, **W**...Gibb's Gas, Cracker Barrel, McDonald's, Wendy's, Holiday Inn/rest., Ramada/rest., Residence Inn, Susse Chalet

38(93)   MA 38, to Lowell, **E**...Shell/mart, Applebee's, Burger King, Friendly's, IHOP, T-D Waffle, Motel 6, MktBasket Foods, Home Depot, Honda/VW, Mazda, **W**...MEDICAL CARE, Citgo, Mobil/24hr, Sunoco, Texaco/diesel, USA/diesel, Dunkin Donuts, Milan Pizza, McDonald's, Wendy's, Big K-Mart, Chevrolet/Pontiac/Buick/GMC, CVS Drug, Marshall's, Saturn, Service Merchandise, Staples

37(91)   Woburn St, to S Lowell, **W**...Exxon

36(90)   Lowell ConX, to Lowell SP, **1 mi W**...McDonald's, Courtyard, Shop'n Save, Walgreen

35b a(89) US 3, S to Burlington, N to Nashua, NH

34(88)   MA 4, Chelmsford, **E**...EYECARE, Mobil/mart, Sunoco/mart, Chinese Rest., Dunkin Donuts, Gip's Rest., Ground Round, Town Meeting Rest., Howard Johnson, Radisson, **W**...Shell/mart/repair, Best Western

88mm   motorist aid call boxes begin sb

87mm   rest area both lanes, full(handicapped)facilities, phone, picnic tables, litter barrels, vending, petwalk

32(83)   Boston Rd, to MA 225, **E**...DENTIST, CHIROPRACTOR, Exxon/24hr, Gulf/service, Mobil/mart/24hr, Applebee's, B&B Café, Boston Mkt, Burger King, Chili's, D'Angelo's, Dunkin Donuts, McDonald's, CVS Drug, MktBasket Foods, Osco Drug, bank, laundry, to Nashoba Valley Ski Area, **W**...RV parking

31(80)   MA 119, to Groton, **E**...Mobil/diesel/24hr, Dunkin Donuts, Ken's Café, Subway, auto parts

30(78)   MA 110, to Littleton, **E**...HOSPITAL, Shell/24hr, RV parking, **W**...Citgo/diesel, Sunoco/diesel

29b a(77) MA 2, to Leominster, **E**...to Walden Pond St Reserve

Massachusetts Interstate 495

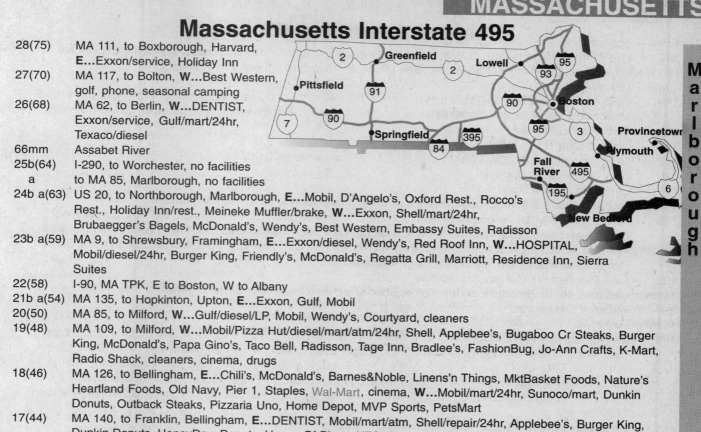

Marlborough

28(75)   MA 111, to Boxborough, Harvard, **E**...Exxon/service, Holiday Inn

27(70)   MA 117, to Bolton, **W**...Best Western, golf, phone, seasonal camping

26(68)   MA 62, to Berlin, **W**...DENTIST, Exxon/service, Gulf/mart/24hr, Texaco/diesel

66mm   Assabet River

25b(64) I-290, to Worcester, no facilities

a   to MA 85, Marlborough, no facilities

24b a(63) US 20, to Northborough, Marlborough, **E**...Mobil, D'Angelo's, Oxford Rest., Rocco's Rest., Holiday Inn/rest., Meineke Muffler/brake, **W**...Exxon, Shell/mart/24hr, Brubaegger's Bagels, McDonald's, Wendy's, Best Western, Embassy Suites, Radisson

23b a(59) MA 9, to Shrewsbury, Framingham, **E**...Exxon/diesel, Wendy's, Red Roof Inn, **W**...HOSPITAL, Mobil/diesel/24hr, Burger King, Friendly's, McDonald's, Regatta Grill, Marriott, Residence Inn, Sierra Suites

22(58)   I-90, MA TPK, E to Boston, W to Albany

21b a(54) MA 135, to Hopkinton, Upton, **E**...Exxon, Gulf, Mobil

20(50)   MA 85, to Milford, **W**...Gulf/diesel/LP, Mobil, Wendy's, Courtyard, cleaners

19(48)   MA 109, to Milford, **W**...Mobil/Pizza Hut/diesel/mart/atm/24hr, Shell, Applebee's, Bugaboo Cr Steaks, Burger King, McDonald's, Papa Gino's, Taco Bell, Radisson, Tage Inn, Bradlee's, FashionBug, Jo-Ann Crafts, K-Mart, Radio Shack, cleaners, cinema, drugs

18(46)   MA 126, to Bellingham, **E**...Chili's, McDonald's, Barnes&Noble, Linens'n Things, MktBasket Foods, Nature's Heartland Foods, Old Navy, Pier 1, Staples, Wal-Mart, cinema, **W**...Mobil/mart/24hr, Sunoco/mart, Dunkin Donuts, Outback Steaks, Pizzaria Uno, Home Depot, MVP Sports, PetsMart

17(44)   MA 140, to Franklin, Bellingham, **E**...DENTIST, Mobil/mart/atm, Shell/repair/24hr, Applebee's, Burger King, Dunkin Donuts, HoneyDew Donuts, House Of Pizza, KFC, Pizza Hut, Taco Bell, AutoZone, Buick/GMC, Chrysler/Jeep, CVS Drug, GNC, JiffyLube, Marshall's, Mens Wearhouse, Stop'n Shop/drug, cinema, hardware, laundry

16(42)   King St, to Franklin, **E**...Sunoco/24hr, Gold Fork Rest.

15(39)   MA 1A, to Plainville, Wrentham, **E**...HOSPITAL, Mobil, **W**...Gulf/mart, Premium Outlets/famous brands, KOA/LP(seasonal)

14b a(37) US 1, to N Attleboro, **E**...RV camping

13(32)   I-95, N to Boston, S to Providence, access to HOSPITAL

12(30)   MA 140, to Mansfield, **3 mi E**...99 Rest., Courtyard, Day's Inn, Holiday Inn, Motel 6, Stop'n Shop

11(29)   MA 140 S(from eb), **1 mi W**...Gulf/mart, McDonald's, drugs, groceries

10(26)   MA 123, to Norton, **E**...Quickstop, Barnside Grill, Mino's Pizza

9(24)   Bay St, to Taunton, **W**...MEDICAL CARE, Dunkin Donuts, Golden Flower Chinese, Pizza Hut, Holiday Inn

8(22)   MA 138, to Raynham, **E**...DENTIST, Mobil/diesel/mart, Christopher's Rest., Honeydew Donuts, dogtrack, **W**...HOSPITAL, VETERINARIAN, Exxon/Dunkin Donuts/diesel/mart, Mobil/mart/wash/24hr, Texaco/repair, Bros Pizza, China Garden, Honeydew Donuts, McDonald's, Pepperoni's Pizza, cleaners

7b a(19) MA 24, to Fall River, Boston, **1/2 mi E**...Burger King

18mm   weigh sta both lanes

17.5mm Taunton River

6(15)   US 44, to Middleboro, **W**...Mobil/diesel/mart, Susse Chalet

5(14)   MA 18, to Lakeville, **E**...MEDICAL CARE, DENTIST, Circle Farm/gas, Citgo, Mobil/mart, Burger King, Dunkin Donuts, Fireside Grill, Friendly's, Lorenzo's Rest., Panda Chinese, Papa Gino's, USA Pizza, Corsini Parts, CVS Drug, Kelly's Tire, Stop'n Shop Foods, **W**...Massasoit SP, RV camping(seasonal), same as 6

4(12)   MA 105, to Middleboro, **E**...Exxon/diesel/mart, Shell/wash/24hr, Sunoco/mart/24hr, Texaco/diesel/repair, Dairy Queen, Dunkin Donuts, McDonald's, Subway, Day's Inn, Brooks Drug, Chevrolet, Osco Drug, bank

11mm   parking area both lanes, picnic tables

3(8)   MA 28, to Rock Village, S Middleboro, **E**...Citgo/diesel/mart, Benson's Cabins, RV camping(seasonal), **W**...Huckleberry's Chickenhouse

2(3)   MA 58, W Wareham, **E**...to RV camping(seasonal), to Myles Standish SF

2mm   Weweantic River

1(0)   I-495 begins/ends on I-195, MA 25 S.

Middleboro

Michigan Interstate 69

Port Huron

Exit #	Services
199	Lp I-69(from eb, no return), to Port Huron, **S**...Shell/diesel/mart, AutoZone, Super K-Mart/24hr, Sam's Club

I-69 E and I-94 E run together into Port Huron. **See Michigan Interstate 94, exit 274-275mm.**

198	I-94, to Detroit and Canada
196	Wadhams Rd, **N**...DENTIST, Bylo/diesel/mart, Marathon/mart, Shell/diesel, Burger King, Hungry Howie's, McDonald's, Ben Franklin, Carter's Foods, Wadham's Drugs, bank, cleaners, car/truckwash, golf, RV camping
194	Barth Rd, **N**...Ft Tridd RV camping
189	Wales Center Rd, to Goodells, no facilities
184	MI 19, to Emmett, **N**...Citgo/diesel/rest./24hr, **S**...Marathon/diesel/mart/24hr
180	Riley Center Rd, no facilities
176	Capac Rd, **2 mi N**...gas, food
174mm	**rest area wb, full(handicapped)facilties, phone, picnic tables, litter barrels, petwalk**
168	MI 53, Imlay City, **N**...Amoco/diesel/mart/atm/24hr, Total/diesel/mart, Speedway/mart, Big Boy, Burger King, Dairy Queen, Hungry Howie's, Jet's Pizza, Little Caesar's, McDonald's, Taco Bell, Tim Horton's, Wendy's, Day's Inn, M53 Motel, Super 8, AutoZone, Chevrolet/Olds/Pontiac, Chrysler/Jeep, Ford, FarmerJack's Foods, Pamida, Pennzoil, bank, carwash, cleaners/laundry, **S**...camping
163	Lake Pleasant Rd, to Attica, no facilities
160mm	**rest area eb, full(handicapped)facilities, phone, picnic tables, litter barrels, petwalk**
159	Wilder Rd, no facilities
158mm	Flint River
155	MI 24, Lapeer, **1mi N**...HOSPITAL, CHIROPRACTOR, DENTIST, EYECARE, VETERINARIAN, Amoco, Clark Gas, Marathon/diesel/mart, Meijer/diesel/24hr, Arby's, Brian's Rest., Burger King, Chinese Rest., Dairy Queen, Hot'n Now, KFC, Little Caesar's, McDonald's, Michael's Dining, Subway, Taco Bell, Tim Horton's/24hr, Tubby's Subs, Wendy's, Best Western, Arbor Drug, FashionBug, K-Mart, JiffyLube, Kroger/drugs, Midas Muffler, Pennzoil, Pet Supplies+, Radio Shack, Sav-A-Lot Foods, bank, cleaners/laundry, st police, **S**...Mobil/diesel/mart, Buick/Pontiac
153	Lake Nepessing Rd, **S**...to Thumb Correctional, camping
149	Elba Rd, **S**...HB's Country Mkt/bait
145	MI 15, Davison, **N**...DENTIST, Shell/diesel/mart, Marathon/mart, Speedway/mart, Arby's, Big Boy, Big John's Rest., Burger King, Country Boy Rest., Dunkin Donuts, Hungry Howie's, Italia Gardens, KFC, Little Caesar's, Lucky's Steaks, McDonald's, Subway, Taco Bell, Comfort Inn, AutoValue Parts, GNC, JiffyLube, Kessel Foods, Pontiac/Buick/GMC, Radio Shack, Rite Aid/24hr, auto repair, carwash, cleaners/laundry, RV/boat storage, **S**...HOSPITAL
143	Irish Rd, **S**...Shell/McDonald's/mart, 7-11, Pizza Chef
141	Belsay Rd, Flint, **N**...MEDICAL CARE, Shell/diesel/mart/24hr, Country Kitchen, McDonald's, Taco Bell, Kessel Foods, K-Mart/Little Caesar's/drugs, Wal-Mart/auto, cinema, hardware, **S**...carwash
139	Center Rd, Flint, **N**...Speedway/mart, Total/mart, Applebee's, Boston Mkt, Chinese Rest., Halo Burger, McDonald's, Old Country Buffet, Ponderosa, Subway, Wendy's, Travelodge, Buick, Goodyear/auto, Home Depot, JC Penney, Jo-Ann Fabrics, Mervyn's, Northwest Tire, Precision Tune, VG Food/drug, cinema, mall, **S**...MEDICAL CARE, VETERINARIAN, Bob Evans, China 1, Walli's Rest., Super 8, FarmerJack's Foods, Pier 1, Staples, Target, TJ Maxx, to IMA Sports Arena

Flint

138	MI 54, Dort Hwy, **N**...HOSPITAL, CHIROPRACTOR, VETERINARIAN, Amoco/24hr, Sunoco, Total/DunkinDonuts/diesel/mart, Big John's Rest., Cruisin 50's Diner, GrapeVine Rest., Little Caesar's, Rally's, Pumpernik's Rest., YaYa's Chicken, Rite Aid, bank, tires, **S**...Speedway/mart, Big Boy/24hr, Bill Knapp's, Empress of China, Hot'n Now, KFC, McDonald's, Subway, Taco Bell, Echo Inn, Travel Inn, Cadillac/Pontiac, Midas Muffler, Tuffy Auto, U-Haul, 10 Min Lube/wash, auto parts, transmissions
137	I-475, UAW Fwy, to Detroit, Saginaw
136	Saginaw St, Flint, **N**...HOSPITAL, Sunoco, U MI at Flint, **S**...ExpertTire, GMC, 10 Min Lube
135	Hammerberg Rd, industrial area, **S**...Sunoco
133b a	I-75, S to Detroit, N to Saginaw, US 23 S to Ann Arbor
131	MI 121, to Bristol Rd, **1/2 mi N on Miller Rd**...Amoco, Sunoco, Total/mart, A&W, Burger King, Chili's, Cony Island Diner, Halo Burger, LJ Silver, Outback Steaks, Ponderosa, Ryan's, Ruby Tuesday, Subway, Taco Bell, BabysRUs, Best Buy, Borders Books, Burlington Coats, Firestone/auto, Goodyear/auto, Hudson's, JC Penney, MailBoxes Etc, Mervyn's, Michael's, Pier 1, Precision Tune, Sears/auto, Service Merchandise, TJ Maxx, bank, cinema, funpark, mall, **S**...airport
129	Miller Rd, **S**...Marathon/mart, Arby's, McDonald's
128	Morrish Rd, **S**...Amoco/mart, Sports Creek Horse Racing
126mm	**rest area eb, full(handicapped)facilities, info, phone, picnic tables, litter barrels, petwalk**
123	MI 13, Lennon, to Saginaw, no facilities

Michigan Interstate 69

118   MI 71, Durand, to Corunna, **N**...Mobil/service, **S**...MEDICAL CARE, DENTIST, Shell/ diesel/mart, Total/mart, Hardee's, McDonald's, Subway, Ace Hardware, Chevrolet/Pontiac/Olds, Durand Muffler/brake, IGA Foods, Radio Shack, Rite Aid, golf, laundry

115mm Shiawassee River

113   Bancroft, **S**...RV camping

105   MI 52, Perry, to Owosso, **S**...Citgo/7-11/wash, Shell/Taco Bell/mart, Total/diesel/LP/mart, Burger King, Café Sports Grill, McDonald's, Westside Deli, Heb's Inn, Ford, IGA Foods, RV camping, truck repair(1mi)

101mm rest area wb, full(handicapped)facilities, phone, picnic tables, litter barrels, petwalk

98.5mm Looking Glass River

98   Woodbury Rd, Shaftsburg, to Laingsburg, **S**...RV camping

94   Lp 69, Marsh Rd, to E Lansing, Okemos, **S**...Admiral/diesel/mart, Speedway/ mart, McDonald's, Your Family Rest., Gillett RV/trailer Ctr, Sav-A-Lot Foods

92   Webster Rd, Bath, no facilities

89   US 127 S, US 27 N, to E Lansing, no facilities

87   Old US 27, Lansing, to Clare, **N**...MEDICAL CARE, DENTIST, VETERINAR- IAN, Amoco/Dunkin Donuts/diesel/mart, Marathon, Total/diesel/mart, Arby's, Bob Evans, Burger King, DraftHouse Rest., FlapJack's Rest., Little Caesar's, McDonald's, Subway, Sleep Inn, Chevrolet, L&L Foods, MailBoxes Etc, Rae RV Ctr, bank, carwash, **S**...Speedway/mart, Lansing Factory Outlet/famous brands

85   DeWitt Rd, to DeWitt, no facilities

84   Airport Rd, no facilities

81   I-96(from sb), W to Grand Rapids, Grand River Ave, Frances Rd

91   I-96, W to Lansing, I-69/US 27 N to Flint

93b a MI 43, Lp 69, Saginaw Hwy, to Grand Ledge, **E**...HOSPITAL, Meijer/diesel/24hr, Shell/mart/wash, Speedway/ diesel/mart, Burger King, Denny's, McDonald's, Roxy's Café, TGIFriday, Best Western, Fairfield Inn, Hampton Inn, Holiday Inn, Motel 6, Quality Suites, Red Roof Inn, bank, **W**...Amoco/mart/24hr, Cracker Barrel, Olds/GMC/ Mazda, hardware

95   I-496, to Lansing, no facilities

72   I-96, E to Detroit, W to Grand Rapids

70   Lansing Rd, **1 mi E**...Citgo/diesel/rest./24hr, st police

66   MI 100, to Grand Ledge, Potterville, **W**...Amoco/mart, BP/mart, McDonald's, to Fox Co Park

61   Lansing Rd, **E**...Mobil/diesel/mart, Total/diesel/mart, Crestview Motel(1mi), AutoZone, Carter's Foods/24hr, Chevrolet/Olds, FashionBug, Plymouth/Dodge, K-Mart, Wal-Mart, **W**...HOSPITAL, CHIROPRACTOR, DENTIST, VETERINARIAN, Speedway/mart, Bay Gas, QD/mart/atm, Arby's, Big Boy, Burger King, El Sombrero Mexican, Hot'n Now, KFC, Little Caesar's, Mancino's Pizza, McDonald's, Pizza Hut, Subway, Taco Bell, Wendy's, CarQuest, Coast to Coast Hardware, Family$, Ford/Mercury, Geldhof Tire/auto, Jo-Ann Fabrics, Pontiac/Buick/GMC, QuikLube/wash, Radio Shack, Sav-A-Lot Foods, bank, cleaners

60   MI 50, Charlotte, **W**...HOSPITAL, Super 8, RV camping

57   Lp 69, Cochran Rd, to Charlotte, **E**...RV camping

51   Ainger Rd, **1 mi E**...access to gas, food, RV camping

48   MI 78, Olivet, to Bellevue, **E**...access to gas, RV camping, to Olivet Coll

42   N Drive N, Turkeyville Rd, **W**...Citgo/diesel/mart

41mm rest area sb, full(handicapped)facilities, phone, picnic tables, litter barrels, petwalk

38   I-94, E to Detroit, W to Chicago

36   Michigan Ave, to Marshall, **E**...HOSPITAL, Mobil/diesel/mart, Clark Gas, Shell/mart/wash, Arby's, Big Boy, Burger King, McDonald's, Pizza Hut, Subway, Taco Bell, Wendy's, YinHai Chinese, Chevrolet/Olds, FarmerJack's Foods, Felpausch Foods, K-Mart/Little Caesar's, Parts+, Pennzoil, Radio Shack, Rite Aid, TrueValue, Tuffy Muffler, auto repair, bank, carwash, cleaners, **W**...Arbor Inn, Imperial Motel, Chrysler/Dodge/Jeep

32   F Drive S, **3/4 mi E**...Shell/mart, RV camping

28mm rest area nb, full(handicapped)facilities, phone, info, picnic tables, litter barrels, petwalk

25   MI 60, to Three Rivers, Jackson, **E**...BP/diesel/mart, Citgo/Subway/Norma's Kitchen/diesel/mart, Sunoco/diesel/ mart/atm, McDonald's, ice cream, RV camping

23   Tekonsha, **E**...access to food, **W**...access to RV camping

N

S

Lansing

Marshall

Michigan Interstate 69

16	Jonesville Rd, **W**...RV camping, Wal-Mart Dist Ctr
13	US 12, Coldwater, to Quincy, **E**...EYECARE, Speedway, Bob Evans, China1, AutoZone, FarmerJack's Food/drug, FashionBug, GNC, JiffyLube, Jo-Ann Fabrics, Radio Shack, Sav-A-Lot Foods, Wal-Mart, bank, cinema, laundry, **W**...Amoco/mart/wash/24hr, Citgo/mart, Speedway/diesel/mart, Sunoco, Arby's, Burger King, Coldwater Garden Rest., Elias Bros Rest., Hot'n Now, Little Caesar's, KFC, McDonald's, Pizza Hut, Ponderosa, Subway, Taco Bell, TCBY, Wendy's, Cadet Motel, Holiday Inn Express, Quality Inn, Super 8, Durham Sports, Ford/Lincoln/Mercury, K-Mart, Midas Muffler, Rite Aid, bank, st police
10	Lp 69, to Coldwater, no facilities
8mm	weigh sta nb
6mm	**Welcome Ctr nb, full(handicapped)facilities, phone, picnic tables, litter barrels, vending, petwalk**
3	Copeland Rd, Kinderhook, **1/2 mi W**...gas, camping
0mm	Michigan/Indiana state line

Michigan Interstate 75

Exit #	Services
395mm	US/Canada Border, state line, I-75 begins/ends at toll bridge to Canada
394	Easterday Ave, **E**...Amoco/mart, McDonald's, Holiday Inn Express, to Lake Superior St U, **W**...**Welcome Ctr/rest area, info**, Holiday/diesel/mart/currency exchange, USA Minimart/diesel
392	3 Mile Rd, Sault Ste Marie, **E**...HOSPITAL, Amoco/diesel/mart, Holiday/diesel/mart, Marathon/mart, Mobil/diesel/mart, Shell/mart, USA/diesel/mart/24hr, Abner's Rest., Ang-Gio's Rest., Arby's, Burger King, Great Wall Chinese, La Senorita Mexican, Mancino's Pizza, McDonald's, Pizza Hut, Robin's Nest Rest., Studebaker's Rest., Subway, Taco Bell, Wendy's, Best Western, Comfort Inn, Day's Inn, Hampton Inn, Kewadin Inn, King's Inn, Plaza Motel, Ramada Inn/rest., Skyline Motel, Super 8, Chevrolet/Pontiac/Olds/Buick/Cadillac, Delta Tires, Family$, FashionBug, Glen's Mkt, Gordon Food Service, JC Penney, JifflyLube/wash, Jo-Ann Fabrics, K-Mart, Mailboxes Etc, NAPA, OfficeMax, Pennzoil, Radio Shack, Wal-Mart, RV camping, bank, cleaners/laundry, st police, Soo Locks Boat Tours, **W**...Sears Hardware, cinema
389mm	**rest area nb, full(handicapped)facilities, info, phone, picnic tables, litter barrels, petwalk**
386	MI 28, **3 mi E**...food, lodging, **W**...to Brimley SP, Clear Creek Camping(5mi)
379	Gaines Hwy, **E**...to Barbeau Area, camping
378	MI 80, Kinross, **E**...Mobil/diesel/mart, Frank&Jim's Dining, to Kinross Correctional, airport
373	MI 48, Rudyard, **2 mi W**...gas/diesel, food, lodging
359	MI 134, to Drummond Island, access to lodging, camping
352	MI 123, to Moran, **5 mi W**...lodging
348	H63, St Ignace, to Sault Reservation, **E**...Shell/mart, Birchwood Motel, Cedars Motel, NorthernAire Motel, Pines Motel, Rockview Motel, KOA, RV camping, st police, to Mackinac Trail, **W**...Castle Rock Gifts
346mm	**rest area/scenic turnout sb, full(handicapped)facilities, picnic tables, litter barrels, petwalk**
345	Portage St(from sb), St Ignace, **E**...Marathon, other services
344b	US 2 W, **W**...Citgo/diesel/rest./atm/24hr, Holiday/diesel/mart, Shell/Subway/diesel/mart/24hr, Big Boy, Burger King, Clyde's Drive-In, Miller's Camp Rest., McDonald's, UpNorth Rest., Howard Johnson/rest., Sunset Motel, Super 8, Ford/Mercury, Goodyear, bank, golf
a	Lp 75, St Ignace, **E**...HOSPITAL, Shell/mart(1mi), Flame Rest., Gandy Dancer Rest., Northern Lights Rest., Aurora Borealis Motel, Collins Motel, Normandy Motel, Quality Inn, Rodeway Inn, Glen's Mkt, USPO, Family$, hardware, to Island Ferrys, Straits SP, KOA, st police
343mm	toll booth to toll bridge, **E**...**Welcome Ctr nb, full(handicapped)facilities, phone, picnic tables, litter barrels**, **W**...museum
341mm	toll bridge
339	US 23, Jamet St, **E**...Amoco/mart, Audie's Rest., Big Boy, KFC, BudgetHost, Econolodge, LaMirage Motel, Motel 6, Parkside Motel, Ramada Inn, Riviera Motel, Super 8, hardware, **W**...Shell, donuts, Chalet Motel, Holiday Inn Express, QuarterDeck Motel
338	US 23(from sb), **E**...**Welcome Ctr/rest area**, Amoco/mart, Marathon, Mobil, Total/diesel/mart, Big Boy, Burger King, Dairy Queen, Ponderosa, Subway, Baymont Inn, Downing's Motel, Ramada Inn, Rodeway Inn, IGA Foods/camping supplies, same services as 337, **W**...Bindel Motel, Chalet Motel, Ft Mackinaw Motel, Trails End Inn
337	MI 108(no EZ return to nb), Nicolet St, Mackinaw City, **E**...Citgo/mart, Embers Rest., Anchor Inn, BeachComber Motel, Best Western, Budget Inn, Capri Motel, Cherokee Shores Inn, Chippewa Inn, Day's Inn, Friendship Inn, Grand Mackinaw Resort, Hamilton Lodge, Hampton Inn, Howard Johnson, King's Inn, Nicolet Inn, North Pointe Inn, Ottawa Motel, Quality Inn, Ramada Ltd, Starlite Inn, Sundown Motel, Surf Motel, Travelodge, Old Mill Creek SP, RV camping, to Island Ferrys, **W**...Wilderness SP
336	US 31 S(from sb), to Petoskey
328mm	**rest area sb, full(handicapped)facilities, info, phone, picnic tables, litter barrels, petwalk**

Michigan Interstate 75

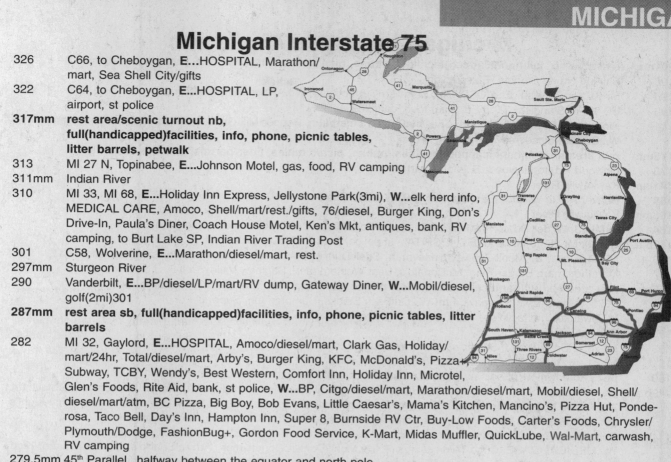

326	C66, to Cheboygan, **E**...HOSPITAL, Marathon/mart, Sea Shell City/gifts
322	C64, to Cheboygan, **E**...HOSPITAL, LP, airport, st police
317mm	**rest area/scenic turnout nb, full(handicapped)facilities, info, phone, picnic tables, litter barrels, petwalk**
313	MI 27 N, Topinabee, **E**...Johnson Motel, gas, food, RV camping
311mm	Indian River
310	MI 33, MI 68, **E**...Holiday Inn Express, Jellystone Park(3mi), **W**...elk herd info, MEDICAL CARE, Amoco, Shell/mart/rest./gifts, 76/diesel, Burger King, Don's Drive-In, Paula's Diner, Coach House Motel, Ken's Mkt, antiques, bank, RV camping, to Burt Lake SP, Indian River Trading Post
301	C58, Wolverine, **E**...Marathon/diesel/mart, rest.
297mm	Sturgeon River
290	Vanderbilt, **E**...BP/diesel/LP/mart/RV dump, Gateway Diner, **W**...Mobil/diesel, golf(2mi)301
287mm	**rest area sb, full(handicapped)facilities, info, phone, picnic tables, litter barrels**
282	MI 32, Gaylord, **E**...HOSPITAL, Amoco/diesel/mart, Clark Gas, Holiday/mart/24hr, Total/diesel/mart, Arby's, Burger King, KFC, McDonald's, Pizza+, Subway, TCBY, Wendy's, Best Western, Comfort Inn, Holiday Inn, Microtel, Glen's Foods, Rite Aid, bank, st police, **W**...BP, Citgo/diesel/mart, Marathon/diesel/mart, Mobil/diesel, Shell/diesel/mart/atm, BC Pizza, Big Boy, Bob Evans, Little Caesar's, Mama's Kitchen, Mancino's, Pizza Hut, Ponderosa, Taco Bell, Day's Inn, Hampton Inn, Super 8, Burnside RV Ctr, Buy-Low Foods, Carter's Foods, Chrysler/Plymouth/Dodge, FashionBug+, Gordon Food Service, K-Mart, Midas Muffler, QuickLube, Wal-Mart, carwash, RV camping
279.5mm	45th Parallel...halfway between the equator and north pole
279	Old US 27, Gaylord, **E**...Marathon/Subway/diesel/mart, Mobil/diesel/mart, Shell/mart, Burger King, Gobblers Rest., Mama Leone's, Willabee's Rest., Alpine Motel, Best Western(2mi), Brentwood Motel, Downtown Motel, Econolodge, Chevrolet/Olds, Chrysler/Jeep/Nissan, Ford/Lincoln/Mercury, Pontiac/Buick/GMC, RV Ctr, st police, **W**...Marsh Ridge Motel(2mi), KOA(3mi)
277mm	**rest area nb, full(handicapped)facilities, info, phone, picnic tables, litter barrels, petwalk**
270	Waters, **E**...Mobil/diesel/rest., **W**...McDonald's, lodging, to Otsego Lake SP, RV service
264	Lewiston, Frederic, **W**...access to food, lodging, camping
262mm	**rest area sb, full(handicapped)facilities, phone, picnic tables, litter barrels, petwalk**
259	MI 93, **E**...Hartwick Pine SP, **2-4 mi W**...Fay's Motel, North Country Lodge, Pointe North Motel, River Country Motel, Woodland Motel, Chevrolet/Olds/Pontiac/Cadillac, Chrysler/Plymouth/Dodge/Jeep, auto repair
256	(from sb), to MI 72, Grayling, access to same as 254
254	MI 72(exits left from nb, no return), Grayling, **1mi W**...HOSPITAL, DENTIST, Amoco/mart/wash, Admiral Gas, BP, Citgo, Mobil/diesel/mart, Phillips 66/mart, 7-11, Shell, Total/mart, A&W, Big Boy, Burger King, Dairy Queen, KFC, Little Caesar's, McDonald's, Pizza Hut, Subway, Taco Bell, Wendy's, Aquarama Motel, Day's Inn, Holiday Inn, Ace Hardware, Carquest, Family$, Ford/Lincoln/Mercury, Glen's Foods/24hr, K-Mart, NAPA, Rite Aid, bank, camping, laundry
251.5mm	**rest area nb, full(handicapped)facilities, info, phone, picnic tables, litter barrels, petwalk, vending**
251	4 Mile Rd, **E**...RV camping, skiing, **W**...Charlie's/Total/Arby's/diesel/24hr, Super 8
249	US 27 S(from sb), to Clare, no facilities
244	MI 18, Roscommon, **W**...Sunoco/diesel/LP/mart, N Higgins Lake SP, camping, museum
239	MI 18, Roscommon, S Higgins Lake SP, **E**...Ford/Mercury, gas, food, lodging, camping
235mm	**rest area sb, full(handicapped)facilities, phone, info, picnic tables, litter barrels, petwalk, vending**
227	MI 55 W, rd F97, to Houghton Lake, **5 mi W**...food, Burnside RV Ctr
222	Old 76, to St Helen, **5 mi E**...food, lodging, camping
215	MI 55 E, West Branch, **E**...HOSPITAL, Total/U-haul/mart
212	MI 55, West Branch, **E**...HOSPITAL, Marathon/diesel/mart, Shell/Subway/mart, Arby's, Big Boy, Burger King, McDonald's, Ponderosa, Taco Bell, Wendy's, Quality Inn/rest., Super 8, Tanger Outlet/famous brands, camping, st police, **W**...BP/diesel, Lk George Camping
210mm	**rest area nb, full(handicapped)facilities, info, phone, picnic tables, litter barrels, petwalk, vending**
202	MI 33, Alger, to Rose City, **E**...BP/Narski's Mkt/gerky(1/2mi), Mobil/diesel/mart, Shell/Subway/Taco Bell/mart/atm, camping

Gaylord

Grayling

N

S

Michigan Interstate 75

201mm **rest area sb, full(handicapped)facilities, phone, picnic tables, litter barrels, petwalk, vending**

195 Sterling Rd, to Sterling, **E**...camping

190 MI 61, to Standish, **E**...HOSPITAL, Standish Correctional, **W**...Amoco/mart, Mobil/mart

188 US 23, to Standish, **2-3 mi E**...gas, food, camping

181 Pinconning Rd, **E**...Mobil/mart, Shell/McDonald's/mart/atm, Pepper Mill Cheese, Pinconning Inn(2mi), camping, **W**...Sunoco/diesel/rest./24hr

175mm **rest area nb, full(handicapped)facilities, phone, picnic tables, litter barrels, petwalk, vending**

173 Linwood Rd, to Linwood, **E**...Mobil/mart

171mm Kawkawlin River

168 Beaver Rd, to Willard, **E**...to Bay City SP, **W**...Mobil/mart

166mm Kawkawlin River

164 to MI 13, Wilder Rd, to Kawkawlin, **E**...Meijer Food/diesel/24hr, McDonald's, AmericInn

162b a US 10, MI 25, to Midland, **E**...HOSPITAL, st police, **3 mi E**...McDonald's

160 MI 84, Delta, **E**...Mobil/Subway/diesel/mart, Shell/Dunkin Donuts/mart, **W**...Amoco/diesel/mart, Citgo/7-11, Marathon/mart, Burger King, McDonald's, Best Western/rest., Saginaw Valley Coll

158mm **rest area sb, full(handicapped)facilities, phone, picnic tables, litter barrels, vending, petwalk**

155 I-675, to downtown Saginaw, **4 mi W**...OutBack Steaks, Hampton Inn, Super 8

154 to Zilwaukee, no facilities

153.5mm Saginaw River

153 MI 13, E Bay City Rd, Saginaw, **2-3 mi W**...lodging

151 MI 81, to Reese, **E**...Sunoco/diesel/mart, Burger King

150 I-675, to downtown Saginaw, **6 mi W**...OutBack Steaks, Hampton Inn, Super 8

149b a MI 46, Holland Ave, to Saginaw, **W**...HOSPITAL, Admiral Gas, Amoco, Speedway/diesel, Sunoco/mart, Total/mart, Arby's, Big Boy, Big John Steaks, Burger King, McDonald's, Taco Bell, Texan Rest., Wendy's, Holiday Inn, Knight's Inn, Red Roof Inn, Rodeway Inn, Celebration Sq, Kessel Foods, K-Mart/drug, bank, hardware

144b a Bridgeport, **E**...Shell/Blimpie/mart, Speedway/diesel/mart/24hr, Total/A&W/diesel/mart, Heidelburg Inn, **W**...MEDICAL CARE, Citgo/TA/diesel/rest./24hr, Mobil/Dunkin Donuts/mart, Arby's/Sbarro's, Big Boy, Burger King, Cracker Barrel, Little Caesar's, McDonald's, Peking City Chinese, Subway, Taco Bell, Wendy's, Baymont Inn, Day's Inn, Motel 6, Fantastic Sam's, IGA Food, Rite Aid, Radio Shack, bank, st police

143mm Cass River

138mm weigh sta both lanes

136 MI 54, MI 83, Birch Run, **E**...Citgo, Marathon/parts, Mobil/diesel/mart/24hr, Shell/Dunkin Donuts, Exit Rest., Halo Burger, Hardee's, KFC, Old Dixie Inn/rest., Subway, Best Western, Comfort Inn, Market St Inn/rest., Hampton Inn, Holiday Inn Express, Super 8, CarQuest, Dixie Speedway, cinema, cleaners/laundry, RV camping, tires, **W**...Amoco/Burger King/mart, Citgo/7-11, Sunoco/diesel/mart, Total/pizza/mart, A&W, Arby's, Applebee's, Bob Evans, Big Boy, Coney Island Rest., Little Caesar's, McDonald's, Schlotsky's, Shoney's, Wendy's, Country Inn Suites, Chevrolet/Buick/Suski, outlet mall/famous brands

131 MI 57, Clio, to Montrose, **E**...DENTIST, MEDICAL CARE, PODIATRIST, Shell/mart, Sunoco/mart, Arby's, Burger King, Chinese Rest., KFC, McDonald's, Subway, Taco Bell, Chrysler/Jeep/Eagle, Chevrolet, Ford, Pamida, bank, **W**...Amoco/diesel, Mobil, Big Boy, Wendy's

129mm **rest area both lanes, full(handicapped)facilities, phone, picnic tables, litter barrels, vending, petwalk**

126 to Mt Morris, **E**...BP/Burger King/diesel/mart, **W**...Amoco/McDonald's/mart, auto auction

125 I-475, UAW Fwy, no facilities

122 Pierson Rd, to Flint, **E**...MEDICAL CARE, Amoco/mart/wash, Citgo, Clark Gas, Marathon, Sunoco, Total/mart, A&W, KFC, McDonald's, Moykong Chinese, Ponderosa, Subway, Walli's Rest., Super 8, Goodyear, 3C Tires, Kessel Foods, K-Mart, Murray's Auto, Muffler Man, Perry Drug, Tuffy Muffler, laundry, **W**...Meijer/diesel/24hr, Shell/mart, Arby's, Bill Knapp's, Bob Evans, Burger King, Cracker Barrel, Denny's, LJ Silver, Pizza Hut, Red Lobster, Taco Bell, Wendy's, YaYa's Chicken, Baymont Inn, Knight's Court, Ramada Inn, Valvoline, bank

118 MI 21, Corunna Rd, **E**...HOSPITAL, Sunoco/wash, 7-11, Big John's Steaks, Burger King, Coney Island Rest., Hardee's, Hungry Howie's, King Chinese Buffet, Little Caesar's, Taco Bell, Wingfong Chinese, AutoValue Parts, Family$, Kessel Foods, Rite Aid, cleaners, hardware, **W**...Citgo/mart, Muffler Man, cinema, st police

117b Miller Rd, to Flint, **E**...Speedway/diesel/mart/atm, Sunoco/diesel/mart, Applebee's, Arby's, Bennigan's, Bill Knapp's, ChiChi's, Chinese Rest., Cottage Inn Pizza, Don Pablo's, Fuddrucker's, KFC, LoneStar Steaks, McDonald's, Subway, Comfort Inn, Motel 6, K-Mart/deli/bakery, Meineke Muffler, NTB, PetSupplies+, Sav-A-Lot Foods, Tuffy Muffler, TireAmerica, Valvoline, cleaners/laundry, **W**...DENTIST, MEDICAL CARE, VETERINARIAN, Amoco/McDonald's/mart, Marathon/Dunkin Donuts/mart, Big Boy, ChuckeCheese, Bob Evans, Burger King, Cajun Joe's Chicken/ribs, Country Grains Bread, HoneyBaked Ham, Mancino's Italian, Old Country Buffet, Olive Garden, Pizza Hut, Ryan's, Salvatori's Ristorante, Taco Bell, TCBY, Wendy's, Howard Johnson/rest., Red Roof

Saginaw

Flint

N

S

Michigan Interstate 75

Inn, Super 8, AllTune/lube, Best Buy, Coat Factory Whse, Fast Eddie's Lube, FastPhoto, Goodyear, Hudson's, JC Penney, Jo-Ann Fabrics, Linens'n More, Midas Muffler, Office Depot, PepBoys, PetsCo, Plymouth/ Dodge, Radio Shack, Service Merchandise, Target, U-Haul, West Marine, bank, carwash, cinema, mall

117a	I-69, E to Lansing, W to Port Huron
116	MI 121, Bristol Rd, **E...**Amoco/24hr, Citgo, Total/diesel/mart/Taco Bell, 7-11, Coney Island Rest., McDonald's, Day's Inn, GM Plant, **W...**airport
115	US 23(from sb), **W...**Mobil, McDonald's, Courtyard, Holiday Inn
111	I-475, UAW Fwy, to Flint
109	MI 54, Dort Hwy(no EZ return to sb), **E...**Food Castle
108	Holly Rd, to Grand Blanc, **E...**Sunoco, AmeriHost, BMW/Mercedes/Toyota, **W...**HOSPITAL, Amoco/McDonald's/mart
106	Dixie Hwy, Saginaw Rd, to Grand Blanc
101	Grange Hall Rd, Ortonville, **E...**Holly RA, Jellystone Park, food, lodging, st police, **W...**to Seven Lakes/Groveland Oaks SP, lodging, RV camping
98	E Holly Rd, **E...**Mobil/diesel/mart/24hr, golf
95.5mm	**rest area both lanes, full(handicapped)facilities, picnic tables, litter barrels, info, phone, vending, petwalk**
93	US 24, Dixie Hwy, Waterford, **E...**Dodge, Saturn, **W...**DENTIST, MEDICAL CARE, Chrysler/Jeep/Eagle, **1-2 mi W...**gas, food, lodging, Pontiac Lake RA
91	MI 15, Davison, Clarkston, **E...**Citgo/mart, **W...**Shell, bank, cleaners
89	Sabashaw Rd, **E...**Shell/diesel/mart/24hr, county park, **W...**VETERINARIAN, Amoco/24hr, Chinese Rest., Roasters Chicken, Subway, Food Town, Arbor Drug, Pine Knob Music Theatre
86mm	weigh sta both lanes
84	Baldwin Ave, **E...**Big Boy, Joe's Crabshack, CompUSA, Kohl's, Linens'n Things, MensWearhouse, Michael's, OfficeMax, World Mkt, **W...**Shell/mart/rest., Mobil/mart, Chili's, McDonald's, Bed Bath&Beyond, Borders Books, Burlington Coats, Great Lakes Crossing Outlet/famous brands, Marshall's, TJ Maxx, cinema
83b a	Joslyn Rd, **W...**Food Town/24hr, K-Mart/Little Caesar's/auto, ProGolf
81	MI 24, Pontiac, no EZ return, **E...**The Palace Arena, **1mi E...**Amoco/mart/wash/24hr
79	University Dr, **E...**CHIROPRACTOR, Amoco/mart/wash, Domino's, Dunkin Donuts, Subway, Thai Rest., cleaners, **W...**HOSPITAL, Speedway/mart, Big Buck Brewery, Mtn Jack's Rest., McDonald's, AmeriSuites, Courtyard, Extended Stay America, Fairfield Inn, Hampton Inn, Holiday Inn, Motel 6, GM, carwash, cinema
78	Chrysler Dr, **E...**Oakland Tech Ctr, Daimler-Chrysler
77b a	MI 59, to Pontiac, **1 mi W...**Shell, Wal-Mart, Silver Dome Stadium
75	Square Lake Rd(exits left from nb), to Pontiac, **W...**HOSPITAL, St Mary's Coll
74	Adams Rd, no facilities
72	Crooks Rd, to Troy, **W...**DoubleTree Guest Suites, Northfield Hilton Inn, bank
69	Big Beaver Rd, **E...**Drury Inn, Marriott, bank, **W...**MEDICAL CARE, Amoco/wash, Shell, 7-11/24hr, Denny's, Papa Romano's Pizza, Ruth's Chris Steaks, bank
67	Rochester Rd, to Stevenson Hwy, **E...**CHIROPRACTOR, DENTIST, Clark Gas, Mobil, Shell, Big Boy, Burger King, Buscemi's Pizza/subs, Chinese Rest., Coney Island Rest., Dunkin Donuts, KFC, Mr Pita Diner, Ram's Horn Rest., Taco Bell, CVS Drug, Discount Tires, Radio Shack, auto repair, bank, cleaners/laundry, hardware, **W...**Mtn Jack's, Holiday Inn, Red Roof Inn
65b a	14 Mile Rd, Madison Heights, **E...**MEDICAL CARE, EYECARE, Mobil, Shell, BeefCarver Rest., Bob Evans, Burger King, ChiChi's, Chili's, Country Oven, Denny's, McDonald's, Panera Bakery, Shoney's, Steak&Ale, Taco Bell, Day's Inn, Motel 6, Red Roof Inn, Belle Tire, Borders Books, Circuit City, CompUSA, Dodge, Fannie Mae Candie's, Ford, Goodyear/auto, Hudson's, JC Penney, JiffyLube, Midas Muffler, NTB, Office Depot, Photo Shop, Sam's Club, Sears, Troy Carcare, bank, cinema, hardware, mall, **W...**Mobil/Blimpie/mart/atm, Applebee's, Bennigan's, Coney Island, McDonald's, Ponderosa, White Castle, Courtyard, Extended Stay America, Fairfield Inn, Day's Inn, Hampton Inn, Knight's Inn, Residence Inn, CVS Drug, FoodLand/deli/bakery, cleaners/laundry
63	12 Mile Rd, **E...**CHIROPRACTOR, DENTIST, VETERINARIAN, Clark/mart, Marathon/mart, Total/mart/wash, Blimpie, Golden Wheel Chinese, Hacienda Azteca Mexican, Marinelli's Pizza, McDonald's, Red Lobster, Sero's Rest., TCBY, AutoLab, K-Mart/drugs, Midas Muffler, Mister Muffler, Radio Shack, Uncle Ed's Lube, Valvoline, transmissions, **W...**Marathon/mart, Total/mart, Denny's, Dunkin Donuts, Chevrolet, Costco, Sparks TuneUp

Michigan Interstate 75

Detroit Area

62	11 Mile Rd, **E**...7-11, **1mi E**...Sunoco, NAPA, U-Haul, **W**...BP, Shell/autocare, Hardee's, Taco Bell, Tire Man, bank
61	I-696 E, to Port Huron, W to Lansing, to Hazel Park Raceway
60	9 Mile Rd, John R St, **E**...BP, Shell, Hardee's, McDonald's, Rally's, Randazo's Rest., Quality Inn, CVS Drug, Farmer Jack's, carwash, transmissions/repair, **W**...VETERINARIAN, Mobil/Dunkin Donuts, Shell, Sunoco, Big Boy, Tubby's Subs, Wendy's, auto parts
59	MI 102, 8 Mile Rd, **3 mi W**...st fairgrounds
58	7 Mile Rd, **W**...DENTIST, Amoco/diesel
57	McNichols Rd, **E**...La Koney Rest./24hr, **W**...BP/mart
56b a	Davison Fwy, no facilities
55	Holbrook Ave, Caniff St, **E**...BP, Mobil/mart, **W**...KFC, Taco Bell
54	E Grand Blvd, Clay Ave, **W**...Shell/autocare, Coney Island Rest.
53 b	I-94, Ford Fwy to Port Huron, Chicago
a	Warren Ave, **E**...Mobil/mart, Shell/mart, **W**...Amoco/mart
52	Mack Ave, HOSPITAL, **E**...Shell/mart, McDonald's
51c	I-375 to civic center, downtown, tunnel to Canada
b	MI 3, Gratiot Ave, downtown
50	Grand River Ave, downtown
49b	MI 10, Lodge Fwy, downtown
a	Rosa Parks Blvd, **E**...Tiger Stadium, **W**...Mobil/mart, Firestone
48	I-96 begins/ends, no facilities
47b	Porter St, **E**...bridge to Canada, DutyFree/24hr
a	MI 3, Clark Ave, **E**...Mobil/mart
46	Livernois Ave, to Hist Ft Wayne, **E**...KFC
45	Fort St, Springwells Ave, **E**...Marathon/diesel/mart, **W**...Mobil/mart
43b a	MI 85, Fort St, to Schaefer Hwy, **E**...Amoco, Sunoco, **W**...Marathon, to River Rouge Ford Plant
42	Outer Dr, **W**...Amoco/diesel/mart, K-Mart
41	MI 39, Southfield Rd, to Lincoln Park, **E**...A&W(seasonal), Bill's Place Rest., Budget Inn, carwash, **W**...Mobil/diesel/mart, Dunkin Donuts, Buick
40	Dix Hwy, **E**...MEDICAL CARE, CHIROPRACTOR, BeeQuik/gas, BP/mart, Citgo, Clark Gas, Total/mart, 7-11, Baffo's Pizza, Chinese Rest., Coney Island Diner, Jelly Donut, Ponderosa, Farmer Jack's, Meineke Mufflers, auto parts, carwash, **W**...Shell/autocare, Total/mart, Big Boy, Burger King, Church's Chicken, Dairy Queen, Dunkin Donuts, Hardee's, LJ Silver, McDonald's, Pizza Hut, Rally's, Taco Bell, Holiday Motel, Firestone, CVS Drug, Foodland, DOC Eye Exam, Midas/Speedy/Top Value Mufflers, Sears/auto, auto repair, tires, bank, hardware
37	Allen Rd, North Line Rd, to Wyandotte, **E**...HOSPITAL, VETERINARIAN, Amoco/mart, Mobil, Total/mart, 7-11/24hr, Anita's Pizza, Chinese Rest., Italian Rest., TCBY, Yum Yum Donuts, Ramada Inn, Sam's Club, XpressLube, bank, cleaners, hardware, transmissions, **W**...Arby's, Burger King, McDonald's, Baymont Inn, Cross Country Inn, Loveday's Lube
36	Eureka Rd, **E**...CHIROPRACTOR, VETERINARIAN, Marathon, Shell/autocare, Total/mart, Bob Evans, Chinese Rest., Denny's, Domino's, Hungry Howie's, Mexican Gardens Rest., Ryan's, Trovano Pizzaria, Holiday Inn/rest., Super 8, Chevrolet, Olds, Pennzoil, bank, **W**...DENTIST, MEDICAL CENTER, Meijer/diesel/24hr, Bakers Square, Chinese Rest., Mtn Jack's Prime Rib, Shoney's, Wendy's, Red Roof Inn, Burlington Coats, Costco, CVS Drug, DOC EyeExam, Hudson's, bank, cinema, mall, tires
34b	Sibley Rd, Riverview, to ski area
a	to US 24, Telegraph Rd, no facilities
32	West Rd, Woodhaven, to Trenton, **E**...DENTIST, VETERINARIAN, Meijer/diesel/mart/24hr, Mobil/diesel/rest./24hr, Total/mart, Burger King, Christoff's Rest., Dunkin Donuts, Japanese Rest., KFC, LJ Silver, Pizza Hut, Subway, Taco Bell, Uncle Harry's Dining, White Castle/24hr, Chevrolet, Firestone/auto, Ford, Guardian Brake/muffler, Kroger/deli, Midas Muffler, Perry Drug, auto parts, bank, cleaners, **W**...Amoco/mart/wash/24hr, Shell/mart/wash, Chinese Rest., Country Skillet, McDonald's, Best Western/rest., Knight's Inn, cleaners

Monroe

29	Gilbralter Rd, to Flat Rock, to Lake Erie Metropark, **E**...bank, **W**...Marathon, Ford, st police
27	N Huron River Dr, to Rockwood, **E**...Total/mart, CVS Drug, Foodtown, bank, **W**...Speedway/diesel/mart, Coney Island Diner
26	S Huron River Dr, to S Rockwood, **E**...Sunoco/mart
21	Newport Rd, to Newport, **W**...Total/diesel/mart/24hr towing
20	I-275 N to Detroit, no facilities
18	Nadeau Rd, **W**...HOSPITAL, Pilot/Arby's/diesel/mart/24hr, RV camping
15	MI 50, Dixie Hwy, to Monroe, **E**...Shell, Bob Evans, Burger King, Cross Country Inn, Day's Inn/rest., Hometown Inn, to Sterling SP, camping, **W**...BP/diesel/rest./24hr, Speedway/diesel/mart/24hr, Big Boy, Denny's, McDonald's, Subway, Wendy's, Holiday Inn, Knight's Inn, hardware, to Viet Vet Mem

Michigan Interstate 75

N

S

14	Elm Ave, to Monroe, info, **W**...Marathon, Sheriff's Dept.
13	Front St, no facilities
11	La Plaisance Rd, to Bolles Harbor, **W**...Amoco/mart/wash, Speedway/mart, Burger King, McDonald's, Outlet Mall/famous brands
10mm	**Welcome Ctr nb, full(handicapped)facilities, phone, info, picnic tables, litter barrels, vending, petwalk**
9	S Otter Creek Rd, to La Salle, **W**...antiques
7mm	weigh sta both lanes
6	Luna Pier, **E**...food, **W**...Julie's/mart/gas, st police
5	to Erie, to Temperance, no facilities
2	Summit St, no facilities
0mm	Michigan/Ohio state line

Michigan Interstate 94

E

W

Exit #	Services
275mm	I-69/I-94 begin/end on MI 25, **Pinegrove Ave in Port Huron**...Amoco/mart/24hr, Shell/mart/24hr, Speedway/mart, Total/mart, Little Caesar's, McDonald's, Tim Horton's, Wendy's, White Castle, Best Western, Day's Inn, Holiday Inn Express, Can-Am DutyFree, Family$, Honda, Pennzoil, Rite Aid, 30 Min Photo, bank, tollbridge to Canada
274.5mm	Black River
274	Water St, Port Huron, **N**...**Welcome Ctr/rest area wb, full facilities,** Cracker Barrel, Ramada Inn, **S**...Bylo/diesel/mart, Total/Taco Bell/diesel/mart, Bob Evans, Comfort Inn, Fairfield Inn, Hampton Inn, Knight's Inn, Lake Port SP, RV camping
271	I-69 E and I-94 E run together eb, **E**...Lp I-69, to Port Huron
269	Dove St, Range Rd, **N**...Total/Subway/diesel/mart/wash/24hr, Burger King, AmeriHost, Horizon Outlet/famous brands
266	Gratiot Rd, Marysville, **S**...Amoco/diesel/mart/24hr, Burger King, **1-2 mi S**...HOSPITAL, Shell/mart/24hr, Total/mart, Big Boy, Chinese Rest., Georgio's Rest., KFC, Little Caesar's, McDonald's, Mancino's Pizza, Oliver's Pizza, Pizza Hut, Budget Host, Days Inn, Goodyear, IGA Foods, Rite Aid, 25 Tire/auto, bank, hardware
262	Wadhams Rd, **N**...camping, **S**...Total/diesel/mart/showers/24hr
257	St Clair, Richmond, **S**...BP/diesel/mart, st police
255mm	**rest area eb, full(handicapped)facilities, phones, info, picnic tables, litter barrels, petwalk**
251mm	**rest area wb, full(handicapped)facilities, phones, info, picnic tables, litter barrels, petwalk**
248	26 Mile Rd, to Marine City, **N**...Macomb Correctional, golf, **S**...airport, diner
247mm	Salt River
247	MI 19(from eb), New Haven, no facilities
245mm	weigh sta both lanes
243	MI 29, MI 3, Utica, New Baltimore, **N**...CHIROPRACTOR, Citgo/diesel/mart, Shell/diesel, Sunoco, Meijer/diesel/mart/24hr, Applebee's, Arby's, Burger King, Horn of Plenty Rest., LeGrand Buffet, Little Caesar's, McDonald's, Papa Romano's Pizza, Ponderosa, Wendy's, White Castle/Church's, Chesterfield Motel, AutoWorks, Discount Tire, FarmerJack's, Foods, Home Depot, JiffyLube, K-Mart, Midas Muffler, NTB, ProGolf, Radio Shack, Sears Hardware, Staples, Target, Walgreen, **S**...VETERINARIAN, Mobil/TCBY/diesel/mart/24hr, Speedway/diesel/mart, Big Boy, Buscemis Pizza, Hot'n Now, Taco Bell, LodgeKeeper, Chevrolet, InstyLube
241	21 Mile Rd, Selfridge, **N**...Marathon/diesel/mart, Shell/mart, Hungry Howie's, Subway, Sero's Dining, Goodyear, Precision Tune, S&K Muffler, RV Ctr, auto parts, bank, hardware, same as 240
240	to MI 59, **N**...Citgo/7-11//24hr, Marathon/mart, Ford, **1/2 mi N on Gratiot**...Shell/mart, Total/mart, Arby's, Burger King, Dairy Queen, KFC, McDonald's, Taco Bell, Pennzoil, Tuffy Muffler
237	N River Rd, Mt Clemens, **N**...Amoco/diesel/mart, Mobil/Subway/diesel/mart, Holiday Inn, General RV Ctr, Gibralter Trade Ctr
236.5mm	Clinton River
236	Metro Parkway, **S**...McDonald's
235	Shook Rd(from wb), no facilities
234b a	Harper Rd, 15 Mile Rd, **N**...VETERINARIAN, Amoco/McDonald's/mart, Speedway/mart, Sunoco/diesel/mart/24hr, JiffyLube, bank, carwash, **S**...Mobil/mart/atm, China Moon, Little Caesar's, Subway, bank
232	Little Mack Ave, **N**...Speedway/mart, Econolodge, Microtel, Red Roof Inn, **S**...Amoco/Dunkin Donuts/24hr, Meijer/diesel/24hr, Total/diesel/mart, Cracker Barrel, IHOP, Baymont Inn, Circuit City, Home Depot, K-Mart, same as 231

Port Huron

Detroit Area

Michigan Interstate 94

Detroit Area

231	(from eb), MI 3, Gratiot Ave, **N**...EYECARE, Shell/mart, Speedway/mart, Sunoco, Big Boy, Bill Knapp's, Boston Mkt, Burger King, BBQ, ChuckeCheese, Denny's, McDonald's, Mtn Jack's Rest., Pizza Hut, Sea Breeze Dining, Day's Inn, Georgian Inn, King Motel, Knight's Inn, Red Roof Inn, Super 8, Brake Shop, Discount Tire, Farmer Jack's, Firestone/auto, Nissan, Sam's Club, Sears, So-Fro Fabrics, Speedy Muffler, Target, Toyota, U-Haul, Uncle Ed's Lube, bank, cinema, mall
230	12 Mile Rd, **N**...DENTIST, 7-11, Mobil/diesel/mart, Burger King, Dunkin Donuts, Outback Steaks, Arbor Drug, Embassy Mkt, Farmer Jack's, Ford/Lincoln/Mercury, Mazda, Service Merchandise, mall **S**...Marathon/mart
229	I-696 W, Reuther Fwy, to 11 Mile Rd, **N**...Amoco, **S**...Total
228	10 Mile Rd, **N**...CHIROPRACTOR, Amoco/24hr, Shell/mart/wash/etd, Mobil/diesel, Baskin-Robbins, Chinese Rest., CVS Drug, auto parts, pizza
227	9 Mile Rd, **N**...Mobil/mart, Speedway/diesel/mart, Sunoco/mart, Dairymart, LJ Silver, McDonald's, Rally's, Taco Bell, Wendy's, Farmer Jack's/bakery/drugs/USPO, JiffyLube, Office Depot, Valvoline, hardware, **S**...Mobil/mart, Ziebart Tidycar, bank
225	MI 102, Vernier Rd, 8 Mile Rd, **S**...CHIROPRACTOR, DENTIST, MEDICAL CARE, 7-11/24hr, Mobil/mart, Speedway/mart, Sunoco/mart, Chinese Rest., KFC, Taco Bell, Wendy's, Kroger, Cadillac
224b	Allard Ave, Eastwood Ave, **S**...Park Crest Inn
a	Moross Rd, **S**...Shell/mart, Farmer Jack's, bank
223	Cadieux Rd, **S**...Amoco, Shell/mart, Boston Mkt, Wendy's
222b	Harper Ave(from eb), no facilities
a	Chalmers Ave, Outer Dr, **N**...Amoco/mart/24hr, BP/Subway/mart, Shell/mart, Ziebart Tidycar
220b	Conner Ave, **N**...Mobil, Regency Inn, city airport
a	French Rd, **S**...Shell/mart
219	MI 3, Gratiot Ave, **N**...KFC, Farmer Jack's
218	MI 53, Van Dyke Ave, **N**...Amoco/mart, **S**...Sunoco/mart
217b	Mt Elliott Ave(from eb), **S**...Mobil/mart
a	E Grand Blvd, Chene St, **S**...BP/diesel/mart
216b	Russell St(from eb), to downtown
a	I-75, Chrysler Fwy, to tunnel to Canada
215c	MI 1, Woodward Ave, John R St, **N**...Marathon
b	MI 10 N, Lodge Fwy
a	MI 10 S, downtown, tunnel to Canada
214b	Trumbull Ave, to Ford Hospital
a	(from wb)Grand River Ave
213b	I-96 W to Lansing, E to Canada, bridge to Canada, to Tiger Stadium
a	W Grand, no facilities
212b	Warren Ave(from eb), no facilities
a	Livernois Ave, **S**...BP/mart
211b	Cecil Ave(from wb), Central Ave, no facilities
a	Lonyo Rd, **S**...Ford
210	US 12, Michigan Ave, Wyoming Ave, **N**...Marathon, Sunoco/mart/wash, Big Boy, Pizza Hut, KFC, Royal Inn, Danny's Auto Service, carwash, **S**...Sunoco/mart, BP/diesel/mart, Mobil, Coney Island Diner, YumYum Donuts, Star Motel, Chevrolet
209	Rotunda Dr(from wb), no facilities
208	Greenfield Rd, Schaefer Rd, **S**...River Rouge Ford Plant
207mm	Rouge River
206	Oakwood Blvd, Melvindale, **N**...Marathon, Oakwood Grill, Ford Plant, UPS, to Greenfield Village, **S**...BP/mart, Mobil, 7-11, Burger King, Domino's, Little Caesar's, O Henry's, Pizza Hut, Subway, YumYum Donuts, Best Western, Frito-Lay, Smirnoff Distillery, bank, drugs, laundry
205mm	Largest Uniroyal Tire in the World
204b a	MI 39, Southfield Fwy, Pelham Rd, **N**...CHIROPRACTOR, Delta/diesel, Mobil/mart, Marathon, 7-11, Manino's Pizza, Ponderosa, to Greenfield Village, **S**...Mobil/mart, BP/mart, pizza, Quality Quick Lube, Precision Tune, bank, **1/2 mi S**...Ecorse Rd parallel, 7-11, Burger King, Dunkin Donuts, McDonald's
202b a	US 24, Telegraph Rd, **N**...Clark, Shell/mart, Sunoco, Andoni's Rest., Benny's Pizza, Chinese Rest., Hardee's, KFC, McDonald's, Rally's, Ram's Horn Rest., Taco Bell, Wendy's, Casa Bianca Motel, bank, auto clinic, **S**...Marathon/diesel/mart, Total/mart, Nu Haven Motel, **1/2 mi S**...VETERINARIAN, Amoco, Mobil, Shell, Burger King, Chinese Rest., Hungry Howie's, Marina's Pizza/subs, Old Country Buffet, Pizza Hut, Red Lobster, YumYum Donuts, Alleo Motel, CVS Drug, JiffyLube, U-Haul, bank, hardware
200	Ecorse Rd, to Taylor, **N**...Granny's Rest., **S**...BP/mart, Speedway/mart, Krazy Jack's Pizzaria, Save-Up Foods, Norm's Mkt, Rainbow Brake/muffler, bank

E

W

Michigan Interstate 94

199	Middle Belt Rd, **S**...Amoco/diesel/mart/24hr, Clark Gas(1mi), McDonald's, Denny's, Rally's, Wendy's, Super 8, Comfort Inn, Day's Inn, rental cars
198	Merriman Rd, **N**...Bob Evans, Baymont Inn, Clarion, Courtyard, Day's Hotel, Fairfield Inn, Hampton Inn, Hilton, Holiday Inn, Howard Johnson, Marriott, Quality Inn, Ramada Inn, Rodeway Inn, Royce Hotel, Signature Inn, **S**...Wayne Co Airport
196	Wayne Rd, Romulus, **N**...Shell/care/wash, McDonald's, bank, **S**...Mobil, Total/mart
194b a	I-275, N to Flint, S to Toledo
192	Haggerty Rd, **N**...Mobil/mart, Roxy Dining, Lower Huron Metro Park
190	Belleville Rd, to Belleville, **N**...CHIROPRACTOR, Amoco/mart/wash/24hr, Marathon, Big Boy, Cracker Barrel, McDonald's, Taco Bell, Tin Lizzie Rest., Wendy's, Red Roof Inn, Camping World RV Service/supplies, CVS Drug, Farmer Jack's Food/24hr, Firestone/auto, Pennzoil, U-Haul, RV camping, **S**...DENTIST, EYECARE, Shell/mart, Burger King, Chinese Rest., Country Kitchen, Domino's, Dos Pesos Mexican, Subway, TCBY, Super 8, USPO, bank, laundry
187	Rawsonville Rd, **S**...Mobil, Burger King, Chinese Rest., Denny's, Hardee's, KFC, McDonald's, Pizza Hut, Woodstone Inn, CVS Drug, K-Mart, So-Fro Fabrics
185	US 12, Michigan Ave, DENTIST, K-Mart/drugs, bank, to frontage rds, to airport
184mm	Ford Lake
183	US 12, Huron St, Ypsilanti, **N**...HOSPITAL, BP/diesel/mart, st police, to E MI U, access to frontage rds, **S**...McDonald's, Radisson
181	US 12 W, Michigan Ave, Ypsilanti, **N**...HOSPITAL, Speedway/mart, Total/diesel/mart, Burger King, Taco Bell, BrakeStop, Busch's Food/24hr, Jo-Ann Fabrics, Perry Drug, Victory Lane Lube, Wal-Mart, bank, carwash, laundry
180	US 23, to Toledo, Flint, **N**...off US 23, Amoco, Citgo, Meijer/diesel/mart/24hr, Mobil, KFC, McDonald's, Ann Arbor Inn/Suites, Home Depot, cinema
177	State St, **N**...Amoco/mart/24hr, Mobil/mart, Shell/mart/wash, Azteca Mexican, Bennigan's, Bicycle Club Rest., Bill Knapp's, Burger King, Graham's Steaks, Macaroni Grill, Max&Erma's, Mediterranean Rest., Olive Garden, Best Western, Crown Plaza Hotel, Courtyard, Fairfield Inn, Hampton Inn, Hilton, Holiday Inn Express, Red Roof Inn, Sheraton, Wolverine Inn, Firestone, Hudson's, JC Penney, Mitsubishi, Sears/auto, VW, World Mkt, bank, mall, to UMI, **S**...Clark/diesel/mart, ChiChi's, Coney Island Rest., McDonald's, Taco Bell, Motel 6, U-Haul
175	Ann Arbor-Saline Rd, **N**...Shell/mart/wash, Applebee's, Old Country Buffet, ABC Whse, Mervyn's, Pier 1, mall, to UMI Stadium, **S**...Meijer/diesel/24hr, Big Boy, Chinese Rest., Joe's Crabshack, McDonald's, Outback Steaks, TGIFriday, Best Buy, CompUSA, Jo-Ann Fabrics, Kohl's, Linens'n Things, OfficeMax, Speedy Auto/Valvoline, Target
172	Jackson Ave, to Ann Arbor, **N**...HOSPITAL, Amoco/mart, BP, Marathon, Shell/mart/wash, Bill Knapp's, KFC, Schlotsky's, Ann Arbor Muffler, K-Mart, Kroger, PhotoShop, Rite Aid, mall, **S**...BP, Weber's Rest., Clarion Hotel, Muffler Man, Chevrolet/Pontiac/Cadillac, Ford, cleaners
171	MI 14(from eb), to Ann Arbor, to Flint by US 23, no facilities
169	Zeeb Rd, **N**...Amoco/24hr, Baxter's Deli, McDonald's, **S**...Mobil, Arby's, PB's Rest., Pizza Hut, Taco Bell, Wendy's, bank
168mm	**rest area eb, full(handicapped)facilities, info, phone, picnic tables, litter barrels, vending, petwalk**
167	Baker Rd, Dexter, **N**...Speedway/PizzaHut/diesel/mart/24hr, **S**...Pilot/Arby's/diesel/mart/24hr, TA/BP/diesel/rest./24hr, Blue Beacon
162	Jackson Rd, Fletcher Rd, **S**...Clark/Subway/diesel/mart/atm/24hr, Stiver's Rest.
159	MI 52, Chelsea, **N**...HOSPITAL, DENTIST, VETERINARIAN, Amoco/diesel/mart/atm/24hr, Mobil/diesel/mart, Speedway/mart, A&W, Big Boy, Chinese Tonite, Little Caesar's, McDonald's, Schumm's Rest., Subway, Taco Bell, TCBY, Wendy's, Comfort Inn, Holiday Inn Express, Arbor Drug, Chevrolet/Buick/Olds, Chrysler/Plymouth/Jeep, Farmer Jack's Foods, Lloyd Bridges RV Ctr, 1-Hr Photo, Pamida, auto parts, bank, carwash, golf, hardware
157	Jackson Rd, Pierce Rd, **N**...Gerald Eddy Geology Ctr, **3 mi S**...Mobil/mart
156	Kalmbach Rd, **N**...to Waterloo RA
153	Clear Lake Rd, **N**...Total/diesel/mart
151.5mm	weigh sta both lanes
150	to Grass Lake, **S**...RV camping, food

E ↑ **W**

Ann Arbor

Jackson

Michigan Interstate 94

150mm	**rest area wb, full(handicapped)facilities, phone, picnic tables, litter barrels, vending, petwalk**
147	Race Rd, **N**...to Waterloo RA, camping, **S**...lodging
145	Sargent Rd, **S**...Amoco/White Castle/diesel/mart, Mobil/diesel/rest./mart/atm/24hr, Jackson Brewing Co, McDonald's, Richard's Rest./24hr, Colonial Inn, RV camping
144	Lp 94(from wb), to Jackson, same as 145
142	US 127 S, to Hudson, **3 mi S**...Meijer/diesel/mart/24hr, Speedway/mart, Domino's, Hot'n Now, McDonald's, Rally's, Wendy's, Advance Parts, K-Mart, Kroger, Parts+, Rite Aid, RV Ctr, Sal Foods, Valvoline
141	Elm Rd, **N**...Travelodge, Dodge, food, **S**...HOSPITAL
139	MI 106, Cooper St, to Jackson, **N**...st police, st prison, **S**...HOSPITAL, Citgo/A&W/mart
138	US 127 N, MI 50, Jackson, to Lansing, **N**...Bill Knapp's, Gilbert's Steaks, Red Lobster, Yen Kang Chinese, Baymont Inn, Fairfield Inn, Hampton Inn, Holiday Inn, Super 8, **S**...Shell/mart/24hr, Bob Evans, Ground Round, Old Country Buffet, Country Hearth Inn, Outback Steaks, Motel 6, Circuit City, Gordon Food Service, Kohl's, Lowe's Bldg Supply, MC Sports, Michael's, Midas Muffler, Muffler Man, OfficeMax, Phar-Mor Drug, Sears, So-Fro Fabrics, Target, Toys R Us
137	Airport Rd, **N**...Meijer/diesel/mart/24hr, Shell/mart/24hr, Burger King, Denny's, Dunkin Donuts, Hot'n Now, McDonald's, Steak'n Shake/24hr, Subway, Wendy's, Vision Ctr, auto parts, bank, carwash, hardware, **S**...Amoco/mart/wash/24hr, Cracker Barrel, Olive Garden, LoneStar Steaks, K-Mart, Sam's Club, Staples
136	Lp 94, MI 60(from eb), to Jackson, no facilities
135mm	**rest area eb, full(handicapped)facilities, phone, picnic tables, litter barrels, vending, petwalk**
133	Dearing Rd, Spring Arbor, to Spring Arbor Coll
130	Parma, **S**...Citgo/Autotruck/diesel/rest./24hr
128	Michigan Ave, **N**...Amoco/Burger King/diesel/mart/24hr, Marathon/diesel/mart/24hr, Cracker Hill Antiques
127	Concord Rd, **N**...St Julian's Winery, **S**...Harley's Antiques
124	MI 99, to Eaton Rapids, **S**...HOSPITAL, food, lodging
121	28 Mile Rd, to Albion, **N**...Mobil/mart, Arby's, Day's Inn, **S**...HOSPITAL, Amoco/diesel/mart/24hr, Shell/mart, Speedway/diesel/mart/24hr, A&W, Big Boy, Burger King, Chinese Rest., Frosty Dan's, KFC, McDonald's, Pizza Hut, Ponderosa, Best Western, AutoZone, Chevrolet/Pontiac, Family$, Ford/Mercury, K-Mart, Radio Shack, Selpausch Foods, bank, RV camping
119	26 Mile Rd, no facilities
115	22.5 Mile Rd, **N**...Total/diesel/rest./mart/24hr
113mm	**rest area wb, full(handicapped)facilities, phone, picnic tables, litter barrels, vending, petwalk**
112	Partello Rd, **S**...Shell/mart/24hr, Schuler's Rest.
110	Old US 27, Marshall, **N**...Shell/Subway/diesel/rest./24hr, Country Kitchen/24hr, Marshall Heights Motel, **S**...HOSPITAL, Howard's Motel(1mi), Sheriff's Dept
108	I-69, US 27, N to Lansing, S to Ft Wayne, no facilities
104	11 Mile Rd, Michigan Ave, **N**...Phillips 66/Te-Khi Trkstp/FoodCourt/24hr, Total/Taco Bell/diesel/mart/24hr, **S**...Mobil/Subway/diesel/mart/24hr, Quality Inn/rest.
103	Lp 94(from wb, no return), to Battle Creek, **N**...HOSPITAL, same as 104
102mm	Kalamazoo River
100	Beadle Lake Rd, **N**...Moonraker Rest., **S**...Citgo/diesel/mart, Whse Mkt, Binder Park Zoo
98b	I-194 N, to Battle Creek, no facilities
a	MI 66, to Sturgis, access to same as 97
97	Capital Ave, to Battle Creek, **N**...Amoco/mart/wash/24hr, Clark/mart, Arby's, Country Kitchen, LoneStar Steaks, McDonald's, Red Lobster, Holiday Inn Express, Knight's Inn, Welcome Inn, bank, **S**...CHIROPRACTOR, Citgo/Blimpie/TCBY/diesel/mart/24hr, Shell/mart/24hr, Bill Knapp's, Bob Evans, Burger King, Cracker Barrel, Denny's, Don Pablo's, Fazoli's, Old Country Buffet, Pizza Hut, Schlotsky's, Steak'n Shake, Subway, Taco Bell, Wendy's, Appletree Inn, Battle Creek Inn, Baymont Inn, Day's Inn, Hampton Inn, Motel 6, Super 8, Borders Books, Firestone/auto, Goodyear/auto, Hudson's, JC Penney, K-Mart, Lowe's Bldg Supply, MailBoxes Etc, Pennzoil, Pier 1, Sam's Club, Sears/auto, Staples, Target, ToysRUs, TJ Maxx, Uncle Ed's Lube, VanHorn's Mkt, Wal-Mart, carwash, mall
96mm	**rest area eb, full(handicapped)facilities, phone, picnic tables, litter barrels, vending, petwalk**
95	Helmer Rd, **N**...Marathon/diesel/mart, Italian Rest., airport, st police
92	Lp 94, to Battle Creek, Springfield, **N**...Citgo/Arlene's Trkstp/diesel/rest./24hr, Shell/mart, RV camping, to Ft Custer RA
88	Climax, **N**...Galesburg Speedway
85	35th St, Galesburg, **N**...Shell/Burger King/diesel/mart/24hr, McDonald's, Galesburg Speedway, to Ft Custer RA, **S**...Winery Tours, RV camping
85mm	**rest area wb, full(handicapped)facilities, phone, picnic tables, litter barrels, vending, petwalk**
81	Lp 94(from wb), to Kalamazoo, no facilities

E

W

Marshall

Michigan Interstate 94

80   Cork St, Sprinkle Rd, to Kalamazoo, **N**...Clark/mart, Double Express/diesel/mart/24hr, Mobil/Blimpie/mart, Wesco/diesel, Burger King, Denny's, Godfather's, HanaEast Sushi, Hot'n Now, Perkins, Sweetwater Donuts, Taco Bell, Best Western, Clarion Hotel, Fairfield Inn, Red Roof, Lentz Muffler/brakes, Pennzoil, PhotoLab, Sears, bank, **S**...Amoco/mart/24hr, Speedway/diesel/mart, Country Kitchen, Holly's Landing Rest., McDonald's, Subway/24hr, Holiday Inn, LaQuinta, Motel 6, Quality Inn

78   Portage Rd, Kilgore Rd, **N**...HOSPITAL, Citgo/mart, Subway, Hampton Inn, Residence Inn, Pennzoil, Uncle Ed's Lube, bank, cleaners, **S**...Mobil, Shell/mart/24hr, Total/mart, Bill Knapp's, Gum Ho Chinese, McDonald's, Olympia Rest., Pizza King, Subs+, Taco Bell, Theo&Stacy's Rest., Economy Inn, Lee's Inn, Sam's Club, auto parts, museum

76b a   Westnedge Ave, **N**...DENTIST, Admiral/Steak'n Shake/mart, Clark/mart/24hr, Meijer/diesel/mart/24hr, Speedway/diesel/mart, Arby's, Big Boy, Golden Palace Chinese, Lee's Chicken, Mancino's Eatery, Outback Steaks, Papa John's, Pappy's Mexican, Peter Piper Pizza, Pizza Hut, RootBeer Stand, Subway, Taco Bell, Quality Inn, Big Lots, Discount Tire, Goodyear/auto, Midas, Office Depot, PetSupplies+, Rite Aid, bank, **S**...Shell/mart/24hr, Applebee's, Bob Evans, Burger King, Chili's, Country Kitchen, Dunkin Donuts, Fannie Mae Candies, Fazoli's, Finley's Rest., Heavenly Ham, Hot'n Now, KFC, Little Caesar's, LJ Silver, Mtn Jack's Rest., McDonald's, Old Country Buffet, Olive Garden, Peking Palace, Pizza Hut, Ponderosa, Red Lobster, Russ' Rest., Ryan's, Schlotsky's, Taco Bell, Taco John's, TCBY, V-Teca Mexican, Wendy's, Holiday Motel, ABC Whse, Barnes&Noble, Best Buy, Cadillac/Pontiac/Nissan, Circuit City, Durham's Sports, Firestone/auto, Gateway Computers, Harding's Foods, Home Depot, Hudson's, JC Penney, JiffyLube, Jo-Ann Fabrics, Kinko's/24hr, K-Mart, Kohl's, Lentz Muffler, MailBoxes Etc, MC Sports, MensWearhouse, Merlin Muffler/brake, Mervyn's, Michael's, OfficeMax, Old Navy, PepBoys, Radio Shack, SuperPetz, Target, TJ Maxx, Tuffy Auto, Uncle Ed's Lube, Walgreen, bank, cinema, cleaners/laundry, hardware, mall

75   Oakland Dr, no facilities

74b a   US 131, to Kalamazoo, **N**...Baymont Inn, to W MI U, Kalamazoo Coll

72.5mm   **rest area eb, full(handicapped)facilities, phone, picnic tables, litter barrels, petwalk**

72   Oshtemo, **N**...Citgo/diesel/mart, Marathon/diesel/mart/rest./24hr, Total/diesel/mart, Burger King, McDonald's, Taco Bell, Saturn, **S**...Cracker Barrel, Fairfield Inn

66   Mattawan, **N**...DENTIST, Mobil/Subway/diesel/mart/24hr, Rossman Auto/towing, drugs, laundry, **S**...Shell/mart, camping

60   MI 40, Paw Paw, **N**...HOSPITAL, Amoco/mart/wash/24hr, Crystal Flash/mart, Speedway/diesel/mart/24hr, Arby's, Big Boy, Burger King, Chicken Coop, Coyote Creek Rest., Hot'n Now, McDonald's, Pizza Hut, Subway/TCBY, Taco Bell, Wendy's, Mroczek Inn, Quality Inn, Chrysler/Dodge/Jeep, Felpausch Foods, Maxi Muffler, St Julian Winery, Warner Winery

56   MI 51, to Decatur, **N**...st police, food, **S**...Citgo/diesel/mart, Total/diesel/mart/rest./24hr

52   Lawrence, **N**...gas(1mi), Waffle House of America, Pizza Place, **S**...RV camping

46   Hartford, **N**...Shell/diesel/mart/24hr, Dowd's Mkt/LP, Panel Room Rest., golf driving range

42mm   **rest area wb, full(handicapped)facilities, phone, picnic tables, litter barrels, vending, petwalk**

41   MI 140, Watervliet, to Niles, **N**...HOSPITAL, DENTIST, Amoco/Burger King/mart/atm/24hr, Citgo/mart, Chicken Coop, Waffle House, RV camping

39   Millburg, Coloma, Deer Forest, **N**...Mobil/mart/wash/24hr, McDonald's, RV Ctr

36mm   **rest area eb, full(handicapped)facilities, phone, picnic tables, litter barrels, vending, petwalk**

34   I-196 N, US 31 N, to Holland, Grand Rapids, no facilities

33   Lp I-94, to Benton Harbor, **2-4 mi N**...airport, Sheriff's Dept

30   Napier Ave, Benton Harbor, **N**...HOSPITAL, Petro II/Mobil/Wendy's/diesel/24hr, Shell/mart/atm/24hr, Super 8, Blue Beacon, **S**...Buick/Olds/Pontiac, Chrysler/Honda

29   Pipestone Rd, Benton Harbor, **N**...MEDICAL CARE, Meijer/diesel/mart/24hr, Speedway/diesel/mart, Applebee's, Big Boy, Burger King, Denny's, El Rodeo Mexican, Hacienda Mexican, Hardee's, Mancino's Pizza, McDonald's/playplace, Pizza Hut, Red Lobster, Steak'n Shake, Subway, Comfort Inn, Courtyard, Motel 6, Red Roof Inn, Aldi Foods, $Tree, FashionBug, Ford/Lincoln/Mercury, Goodyear/auto, JC Penney, Jo-Ann Fabrics, K-Mart, Lowe's Bldg Supply, Nevada Bob's, OfficeMax, PetSupplies+, Quicklube, Radio Shack, Rex Audio/video, Sears, Staples, Walgreen, Wal-Mart, bank, funpark(seasonal), mall, **S**...Citgo/Blimpie/Taco Bell/diesel/mart/24hr, Bob Evans, Holiday Inn Express, Ryder Trucks

E ↑ ↓ **W**

Michigan Interstate 94

28	US 31 S, MI 139 N, to Niles, **N**...HOSPITAL, VETERINARIAN, Citgo/diesel/mart, Shell/repair, Total/diesel/mart, Arby's, Beijing House, Bill Knapp's, Burger King, Capozio's Pizza, Chicken Coop, Country Kitchen, Dairy Queen, Henry's Burgers, Hot'n Now, KFC, Little Caesar's, Pizza Hut, Purple Onion Rest., Subway, Wendy's, Day's Inn/rest., Ramada Inn/rest., AutoZone, Big Lots, Chevrolet, $Tree, DurhamSports, Firestone, Midas Muffler, OfficeDepot, Pennzoil, ProTransmissions, QuikLube/wash, Rite Aid, Target, TJMaxx, bank, camping, cinema, cleaners/laundry, st police, **S**...Marathon, Quality Inn/rest.
27mm	St Joseph River
27	MI 63, Niles Ave, to St Joseph, **N**...DENTIST, PODIATRIST, Amoco/mart/wash/atm/24hr, Nye's Apple Barn, **S**...Total/diesel/mart, Goodyear
23	Red Arrow Hwy, Stevensville, **N**...Amoco/TacoBell/mart, Marathon/diesel/mart, Shell/diesel/mart/24hr, Speedway/mart, Big Boy, Burger King, Cracker Barrel, LJ Silver, McDonald's, Pedro's Rest.(1mi), Popeye's, Subway, Baymont Inn, Park Inn, **S**...Schuler's Rest., Hampton Inn
22	John Beers Rd, Stevensville, **N**...to Grand Mere SP, **S**...CHIROPRACTOR, DENTIST, BP/diesel/mart, Deli Beloved
16	Bridgman, **N**...Amoco/Blimpie/DQ/diesel/mart/wash/24hr, to Warren Dunes SP, **S**...Speedway/diesel/mart/24hr, McDonald's, **1/2 mi S**...VETERINARIAN, Brian's Cove Rest., Chicken Coop Rest., Olympus Rest., Pizza Hut, Roma Pizza, Subway, Chevrolet/Buick, bank, cleaners, drugs, hardware, parts/repair
12	Sawyer, **N**...Citgo/diesel/rest./24hr, camping, **S**...TA/Mobil/BurgerKing/Popeye's/Taco Bell/diesel/24hr, USPO, drugs, hardware, laundry
6	Lakeside, Union Pier, **N**...St Julian's Winery, antiques, **S**...RV camping
4b a	US 12, New Buffalo, to Three Oaks, **N**...J&J Rest., gas/diesel/mart, st police, **S**...Amoco/diesel/mart, auto repair
2.5mm	weigh sta both lanes
1	MI 239, New Buffalo, to Grand Beach, **N**...Marathon/diesel/mart, Edgewood Motel, **S**...New Buffalo/Phillips66/diesel/rest./repair/24hr, Arby's, Wendy's, Comfort Inn
.5mm	**Welcome Ctr eb, full(handicapped)facilities, info, phone, picnic tables, litter barrels, vending, petwalk**
0mm	Michigan/Indiana state line

E

W

Michigan Interstate 96

Exit #	Services
	I-96 begins/ends on I-75, exit 48 in Detroit.
191	I-75, N to Flint, S to Toledo, US 12, to MLK Blvd, to Michigan Ave, **S**...access to Sunoco
190b	Warren Ave, **N**...Citgo/mart, **S**...Marathon/mart
a	I-94 E to Port Huron
189	W Grand Blvd, Tireman Rd, **N**...Amoco
188b	Joy Rd, **N**...Church's Chicken
a	Livernois, **N**...Shell/mart, McDonald's, bank
187	Grand River Ave(from eb), no facilities
186b	Davison Ave, I-96 local and I-96 express divide, no exits from express
a	(from eb), Wyoming Ave
185	Schaefer Hwy, to Grand River Ave, **N**...Amoco/mart/24hr, Mobil/mart, **S**...Citgo/diesel/mart
184	Greenfield Rd, no facilities
183	MI 39, Southfield Fwy, exit from expswy and local, no facilities
182	Evergreen Rd, no facilities
180	Outer Dr, **N**...Mobil/diesel/mart/lube, Michigan AAA
180mm	I-96 local/express unite/divide
179	US 24, Telegraph Rd, **N**...White Castle, Chevrolet, Goodyear, **S**...bank
178	Beech Daly Rd, **N**...Citgo/mart
177	Inkster Rd, **N**...7-11, Subway, Minnesota Fabrics, **S**...U-Haul, bank
176	Middlebelt Rd, **N**...MEDICAL CARE, Bob Evans, ChiChi's, Mitch Housey's Rest., Olive Garden, Comfort Inn, Super 8, **S**...PriceMart/gas, Chinese Rest., Handy Andy Hardware, Oil Dispatch Lube, cleaners/laundry
175	Merriman Rd, **N**...CHIROPRACTOR, Mobil/mart, Total/mart, **S**...Sunoco/mart, Mtn Jack's Rest.
174	Farmington Rd, **N**...Mobil/mart, Total/mart, **S**...Amoco/mart, KFC
173b	Levan Rd, HOSPITAL, **N**...to Madonna U
a	Newburgh Rd, **S**...Top Value Muffler/repair
I-275 and I-96 run together 9 miles	
170	6 Mile Rd, **N**...Big Boy, Best Western, Holiday Inn, Marriott, Quality Inn, Perry Drug, mall, bank, cinema, **S**...Fairfield Inn, Residence Inn
169b a	7 Mile Rd, **N**...Rio Bravo, LoneStar Steaks, Embassy Suites, **S**...Cooker Rest., Del Rio Cantina, Macaroni Grill, Home Depot

E

W

Detroit Area

Michigan Interstate 96

167 8 Mile Rd, to Northville, **S**...Meijer/diesel/24hr, Speedway/mart, Benihana, Big Boy, Café Marie, Chili's, McDonald's, Kyoto Steaks, Taco Bell, Hilton, Hampton Inn, Travelodge, Best Buy, CopyMax, Costco Whse, Dick's Sports, Home Depot, Kohl's, OfficeMax, Target, to Maybury SP

164 I-696, I-275, MI 5, Grand River Ave, **1.5 mi E**...Amoco, Mobil, Shell, McDonald's, Red Roof Inn, Buick/Pontiac, K-Mart, **I-275 and I-96 run together 9 miles**

163 I-696(from eb), no facilities

162 Novi Rd, Novi, to Walled Lake, **N**...HOSPITAL, Amoco/mart/wash, Denny's, McDonald's, Pizza Hut, Subway, DoubleTree Hotel, Hotel Baronette, Sheraton, Circuit City, Hudson's, JC Penney, MailBoxes Etc, Midas Muffler, KidsRUs, K-Mart, Kroger, Marshall's, OfficeMax, Ritz Camera, Sears/auto, Service Merchandise, ToysRUs, bank, cleaners, hardware, mall, **S**...Mobil/ mart, Big Boy, Bob Evans, Hardee's, Grady's Rest., Olive Garden, Red Robin Rest., TGIFriday, Wyndham Garden, Borders Books, Mervyn's, TJ Maxx

161mm rest area eb, full(handicapped)facilities, phone, vending, picnic tables, litter barrels, petwalk

160 Beck Rd, 12 Mile Rd, **N**...HOSPITAL, **S**...to Maybury SP

159 Wixom Rd, Walled Lake, **N**...to Proud Lake RA, **S**...Mobil, Meijer/diesel/ 24hr, Shell/mart/wash, Arby's, Burger King, McDonald's, Taco Bell, Lincoln/ Mercury, RV Ctr

155 New Hudson, to Milford, **N**...to RV camping, **S**...carpool parking

153 Kent Lake Rd, **N**...Kensington Metropark, **S**...Mobil/mart

151 Kensington Rd, **N**...Kensington Metropark, **S**...Island Lake RA, food, lodging

150 Pleasant Valley Rd, **S**...phone

148b a US 23, N to Flint, S to Ann Arbor, facilities 1/2 mi S off of US 23, exit 58...Total/mart

147 Spencer Rd, **N**...st police, **S**...to Brighton St RA

145 Grand River Ave, to Brighton, **N**...Amoco/mart/24hr, Shell/NobleRomans/mart, Arby's, Cracker Barrel, Outback Steaks, Pizza Hut, Courtyard, Ford/Mercury, Olds/Cadillac/GMC, bank, **S**...DENTIST, CHIROPRACTOR, EYECARE, URGENT CARE, Clark/Subway/diesel/mart, Meijer/diesel/24hr, Big Boy, Burger King, Chili's, Dairy Queen, KFC, Lil Chef, Little Caesar's, McDonald's, Ponderosa, Taco Bell, Wendy's, Holiday Inn Express, Farmer Jack's, Home Depot, Honda, Jo-Ann Crafts, K-Mart, Mazda, MC Sports, Midas Muffler, Radio Shack, Staples, Target, USPO, auto parts, bank, cinema, carwash, laundry, mall, to Brighton Ski Area

141 Lp I-96(from wb), to Howell, **N**...Clark/Subway/diesel/mart, Big Boy, Chevrolet

141mm rest area wb, full(handicapped)facilities, phone, vending, picnic tables, litter barrels, petwalk

137 D19, Howell, to Pinckney, **N**...HOSPITAL, Mobil/mart, Shell/diesel/wash, Speedway/mart, Total/mart, Kensington Inn, Ramada Inn, Goodyear, hardware, **S**...Benny's Grill, Best Western/rest., auto parts, auto repair/oil change

135mm rest area eb, full(handicapped)facilities, vending, phone, picnic tables, litter barrels, petwalk

133 MI 59, Highland Rd, **N**...Sunoco/McDonald's/mart, Food Pavilion, Kensington Valley Shops/famous brands

129 Fowlerville Rd, Fowlerville, **N**...Amoco/rest./24hr, Shell/diesel/mart, Big Boy, Fowlerville Rest., McDonald's, Taco Bell, Wendy's, Best Western, **S**...Mobil/Dunkin Donuts/mart, Chysler/Dodge/Jeep

126mm weigh sta both lanes

122 MI 43, MI 52, Webberville, **N**...Mobil/diesel/mart/24hr, McDonald's

117 Williamston, to Dansville, **1 mi N**...food, lodging

111mm rest area wb, full(handicapped)facilities, phone, picnic tables, litter barrels, vending, petwalk

110 Okemos, Mason, **N**...Amoco/Dunkin Donuts/mart/24hr, Marathon/mart/wash, Mobil/mart/wash, 7-11, Applebee's, Arby's, Big Boy, British Isles Rest., Burger King, Café Oriental, Little Caesar's, McDonald's, Comfort Inn, Fairfield Inn, Holiday Inn Express, bank, cleaners, to stadium

106b a I-496, US 127, Lansing, to Jackson, st police, services 1 mi N off of I-496

104 Lp 96, Cedar St, Lansing, to Holt, **N**...HOSPITAL, CHIROPRACTOR, Amoco, Meijer/diesel/mart/24hr, Shell, Speedway/mart, Total/mart, 7-11, Arby's, Blimpie, Bill Knapp's, Bob Evans, Boston Mkt, Denny's, KFC, LJ Silver, Mr Taco, Pizza Hut, Wendy's, Best Western, Day's Inn, Knight's Inn, Regent Inn, Super 8, Aldi Foods, Cadillac, Chevrolet, Chrysler/Jeep, Dodge, Harley-Davidson, Kinko's, Lexus, Lincoln/Mercury, MC Sports, Mitsubishi, NTB, Olds/GMC, PetSupplies+, Saab, Sam's Club, Target, Toyota, bank, cinema, cleaners, **S**...Marathon/mart, Shell, Speedway/mart/24hr, Total/mart, A&W, Big Boy, Burger King, Charlie Kang Chinese, Flapjack's Rest., McDonald's, Ponderosa, Rico's Taco, Subway, Holiday Inn, Ramada Ltd, CarQuest, Fast Eddie's Lube, Kroger/ deli/bakery, Midas Muffler, Muffler Man, NAPA, bank, cleaners/laundry, RV camping

E

W

Lansing

Michigan Interstate 96

101	MI 99, MLK Blvd, to Eaton Rapids, **S**...Total/diesel/mart/24hr, McDonald's, Rich's Country Store/pizza
98b a	Lansing Rd, to Lansing, **S**...Citgo/diesel/rest./24hr, Windmill Rest.
97	I-69, US 27 S, S to Ft Wayne, N to Lansing, no facilities
95	I-496, to Lansing
93b a	MI 43, Lp 69, Saginaw Hwy, to Grand Ledge, **N**...HOSPITAL, Meijer/diesel/mart/24hr, Shell/mart/wash, Speedway/diesel/mart, Burger King, Denny's, McDonald's, TGIFriday, Best Western, Fairfield Inn, Hampton Inn, Holiday Inn, Motel 6, Quality Suites, Red Roof Inn, bank, **S**...Amoco/mart/24hr, Bill Knapp's, Cracker Barrel, Olds/GMC/Mazda, hardware
92mm	Grand River
91	I-69 N, US 27 N(from wb), to Flint
90	Grand River Ave, to airport
89	I-69 N, US 27 N(from eb), to Flint
87mm	**rest area eb, full(handicapped)facilities, phone, picnic tables, litter barrels, vending, petwalk**
86	MI 100, Wright Rd, to Grand Ledge, **S**...Total/Taco Bell/diesel/mart
84	to Eagle, Westphalia, no facilities
79mm	**rest area wb, full(handicapped)facilities, info, phone, picnic tables, litter barrels, vending, petwalk**
77	Lp 96, Grand River Ave, Portland, **N**...CHIROPRACTOR, Amoco/mart/wash/24hr, Marathon/diesel, Shell/mart/atm, Speedway/diesel/mart, Arby's, Burger King, Dina's Subs, McDonald's, Subway, Tommie's Rest., Best Western, Family$, Tom's Food/deli/bakery, QuikLube, Rite Aid, auto parts, bank, hardware, **S**...Ford/Mercury
76	Kent St, Portland, **N**...Marathon/diesel(1mi), food
76mm	Grand River
73	to Lyons-Muir, Grand River Ave, no facilities
69mm	weigh sta both lanes
67	MI 66, to Ionia, Battle Creek, **N**...HOSPITAL, Amoco/mart, Scalehouse Trkstp/diesel/rest./24hr, Total/diesel/mart, Big Boy(3mi), Corner Landing Rest., AmeriHost, Midway Motel, Super 8, antiques, camping, st police, **S**...Wrecker truckwash/tires/repair
64	to Lake Odessa, Saranac, **N**...Ionia St RA, **S**...I-96 Speedway
63mm	rest area eb, full(handicapped)facilities, phone, picnic tables, litter barrels, petwalk, vending
59	Clarksville, **N**...auto repair
52	MI 50, to Lowell, **N**...fairgrounds, **S**...RV camping
46mm	Thornapple River
45mm	**rest area wb, full(handicapped)facilities, info, phone, picnic tables, litter barrels, vending**
43b a	MI 11, 28th St, Cascade, **N**...Marathon/diesel/mart, Meijer/diesel/mart/24hr, Big Boy, Boston Mkt, Brann's Steaks, Burger King, Heavenly Ham, Manhattan Bagel, Shanghai Garden, Sundance Grill, Baymont Inn, Country Inn Suites, Crowne Plaza Hotel, Days Inn, Hawthorne Suites, Lexington Suites, Mailboxes Etc, Wal-Mart, bank, cleaners, **S**...Citgo/mart, Mobil/mart/wash, Shell, Speedway, Arby's, Blimpie, Bob Evans, Burger King, Cheddar's, Chili's, Contino Mexican, Damon's, Denny's, Dunkin Donuts, Faro's Pizza, Fuddrucker's, Grand Rapids Brewery, HoneyBaked Ham, HongKong Garden, Hooters, McDonald's, Mikado Sushi, Olive Garden, Outback Steaks, Perkins, Pizza Hut, Red Lobster, Rio Bravo, Steak'n Shake, Subway, Wendy's, YinChing Chinese, Best Western, Comfort Inn, Courtyard, Econolodge, Exel Inn, Extended Stay America, Hampton Inn, Hilton, Motel 6, Quality Inn, Red Roof Inn, Babys R Us, Barnes&Noble, Bed Bath&Beyond, BoatersWorld, CarQuest, Circuit City, CompUSA, FasTrak Lube, Gander Mtn, Gateway Computers, Gazelle Sports, Golf Galaxy, Home Depot, Kinko's/24hr, K-Mart, Lentz Muffler, Menard's, Nissan/Acura/Audi/Porche/Subaru, NTB, Pennzoil, PetsMart, Sam's Club, Target, U-Haul, cinema
40b a	Cascade Rd, **N**...CHIROPRACTOR, DENTIST, MEDICAL CARE, BP, Marathon/diesel/mart, 7-11/24hr, Forest Hills Food/deli/drug, NAPA Autocare, bank, cleaners, **S**...Shell/mart, Speedway/mart, Travelodge
39	MI 21(from eb), to Flint, no facilities
38	E Beltline Ave, to MI 21, MI 37, MI 44, **N**...RV camping, **S**...HOSPITAL, Duba's Dining
37mm	I-196, Gerald Ford Fwy, to Grand Rapids
36	Leonard St, **N**...LDS Church, Sheriff's Dept
33	Plainfield Ave, MI 44 Connector, **N**...DENTIST, EYECARE, Amoco/mart, 4Star Gas, Meijer/diesel/mart/24hr, Total/mart, 7-11, Arby's, Asiana Rest., Big Boy, Blimpie, Burger King, Cheers Grill, Fred's Pizza, Golden Dragon, KFC, Little Caesar's, LJ Silver, McDonald's, Papa Romano's Italian, Pizza Hut, Russ' Rest., Schlotsky's, Subway, Taco Bell, Taco Boy, Wendy's, Lazy T Motel, President Inn, AutoZone, Chevrolet, Chrysler, Discount Tire, Dodge, Ford/Mercury, Goodyear/auto, Lentz Muffler, Midas Muffler, NAPA, Nissan, NTB, Pennzoil, Plymouth/Jeep, Radio Shack, U-Haul, Valvoline, Walgreen, bank, carwash, cleaners, transmissions, **S**...Amoco/mart/wash/24hr, Bill Knapp's, Denny's
31mm	Grand River
31b a	US 131, N to Cadillac, S to Kalamazoo, **1 mi N**...McDonald's

E

W

Grand Rapids

Michigan Interstate 96

E

W

30b a   Alpine Ave, Grand Rapids, **N**...CHIROPRACTOR, Amoco/mart/wash/24hr, Citgo/7-11, Marathon/mart, Shell/mart, Applebee's, Blimpie, ChuckeCheese, Clock Rest., Cooker Rest., Cracker Barrel, Damon's, McDonald's, Old Country Buffet, Olive Garden, Outback Steaks, Perkins, Russ' Rest., Ryan's, Schlotsky's, Steak'n Shake, Subway, Taco Bell, Village Inn Pizza, Hampton Inn, Belle Tire, Best Buy, Big Lots, CarQuest, Circuit City, Discount Tire, Ford/Kia, GolfUSA, K-Mart/auto, Kohl's, MediaPlay, Menard's, Michael's, NAPA, OfficeMax, PepBoys, PetSupplies+, Radio Shack, Sam's Club, Target, TJ Maxx, ToysRUs, Wal-Mart/auto, bank, cinema, cleaners/laundry, **S**...DENTIST, Admiral/mart, Meijer/diesel/mart/24hr, Total/mart, Arby's, Burger King, Fazoli's, KFC, LJ Silver, McDonald's, Ole Taco, Papa John's, Pizza Hut, Ponderosa, Wendy's, Motel 6, Goodyear/auto, Home Depot, Jo-Ann Fabrics, Midas/Tuffy Muffler, U-Haul, Valvoline, bank, laundry

28   Walker Ave, **S**...Meijer/diesel/mart/24hr, McDonald's, Amerihost

26   Fruit Ridge Ave, **N**...Citgo/diesel/mart, **S**...Amoco/diesel/mart/atm, Riviera Motel

25mm   **rest area eb, full(handicapped)facilities, phone, picnic tables, litter barrels, petwalk**

25   8th Ave, 4Mile Rd(from wb), **S**...Marathon/diesel/mart, Wayside Motel

24   8th Ave, 4Mile Rd(from eb), **S**...Marathon/diesel/mart, Wayside Motel

23   Marne, **N**...Shell/mart, tires, fairgrounds/raceway, **S**...Depot Café, food mkt, golf

19   Lamont, Coopersville, **N**...gas/diesel/mart, **S**...Sam's Joint Rest., LP

16   B-35, Eastmanville, **N**...Amoco/Subway/TCBY/diesel/mart, Shell/diesel/mart, Speedway/Burger King/diesel/mart/24hr, Arby's, Little Caesar's, McDonald's, Taco Bell, AmeriHost, Fun-n-Sun RV, Rite Aid, ShopRite Foods, bank, cleaners/laundry, **S**...Pacific Pride/diesel, RV camping

10   B-31, Nunica, **N**...Turk's Rest., **S**...Mobil/mart, RV camping, golf course/rest.

9   MI 104(from wb, exits left), to Grand Haven, Spring Lake, **S**...Mobil/mart, to Grand Haven SP

8mm   **rest area wb, full(handicapped)facilities, phone, picnic tables, litter barrels, vending, petwalk**

5   Fruitport(from wb, no return), no facilities

4   Airline Rd, **N**...racetrack, **S**...DENTIST, VETERINARIAN, Speedway/diesel/mart, Shell, Wesco/diesel/mart, Burger Crest Diner, Pizza Reaction, Subway, Fruitport Foods/24hr, bank, Pleasure Island Water Park(5mi), to PJ Hoffmaster SP

1c   Hile Rd(from eb), **S**...racetrack

b a   US 31, to Ludington, Grand Haven, **N**...All Seasons RV, Bel-Aire Motel, Motel Haven, **3 mi N on Sherman Blvd**...HOSPITAL, Applebee's, Bill Knapp's, Fazoli's, McDonald's, Old Country Buffet, Comfort Inn/rest., Ramada Inn, Super 8, Circuit City, FashionBug, Lowe's Bldg Supply, MC Sports, OfficeMax, PetsMart, Sam's Club, Staples, Target, Wal-Mart/auto, **S**...Mobil/diesel/mart, Alpine Motel, AmeriHost, Best Western, Quality Inn, airport, hardware, racetrack

I-96 begins/ends on US 31 at Muskegon.

Michigan Interstate 196(Grand Rapids)

N

S

Exit #	Services
81mm	I-196 begins/ends on I-96, 37mm in E Grand Rapids.
79	Fuller Ave, no facilities
78	College Ave, **S**...HOSPITAL, McDonald's, Pizza Hut, Rite Aid, auto repair, museum
77c	Ottawa Ave, downtown, **S**...Gerald R Ford Museum
b a	US 131, S to Kalamazoo, N to Cadillac
76	MI 45 E, Lane Ave, **S**...Amoco, Gerald R Ford Museum, John Ball Park&Zoo
75	MI 45 W, Lake Michigan Dr, **N**...to Grand Valley St U
74mm	Grand River
73	Market Ave, **N**...to Vanandel Arena
72	Lp 196, Chicago Dr E(from eb), no facilities
70	MI 11, Grandville, Walker, **S**...Amoco/diesel/mart/wash, Citgo/7-11, Shell/repair, Muffler Man, NAPA, USPO, bank
69b a	Chicago Dr, **N**...Meijer/diesel/mart/24hr, KFC, McDonald's, Mr Fable's Rest., Sara's Pizza, Subway, Aldi Foods, K-Mart, Pennzoil, Rainbow Brake/lube, Target, Valvoline, **S**...Amoco, Clark Gas, Crystal Flash/mart, Arby's, Burger King, Donato's Pizza, Get-em-n-Go Burger, Little Caesar's, Ole Tacos, Pizza Hut, Rainbow Grill, Russ' Rest., Wendy's, Best Western, Holiday Inn Express, Jerry's Country Inn

Muskegon

Holland

Michigan Interstate 196

67	44th St, **N**...CHIROPRACTOR, Mobil/diesel/mart, Burger King, Cracker Barrel, Steak'n Shake, Comfort Suites, Wal-Mart/auto, carwash, **S**...airport
62	32nd Ave, to Hudsonville, **N**...Amoco/mart/atm/24hr, BP/diesel/mart/wash, Arby's, Burger King, McDonald's, AmeriHost, Chevrolet, FasTrak Lube, bank, camping, **S**...Mobil/Subway/diesel/mart/wash/24hr, Rest-All Inn
58mm	**rest area eb, full(handicapped)facilities, phone, picnic tables, litter barrels, vending, petwalk**
55	Byron Rd, Zeeland, **3-5 mi N**...HOSPITAL, **10 mi N**...facilities on Lp 196, to Holland SP, camping
52	16th St, Adams St, **5 mi N**...food, lodging, **S**...HOSPITAL, Mobil/Subway/diesel/mart, Burger King
49	MI 40, to Allegan, **2 mi N**...lodging, airport, **S**...Tulip City/diesel/mart/wash/24hr
44	US 31 N(from eb), to Holland, **3-5 mi N**...HOSPITAL, Country Inn, gas, food
43mm	**rest area wb, full(handicapped)facilities, info, phone, picnic tables, litter barrels, vending, petwalk**
41	rd A-2, Douglas, Saugatuck, **N**...Marathon/mart/atm, Shell/Subway/diesel/mart, Burger King, Skyline Rest., Timberline Motel(3mi), Twin Oaks Inn, to Saugatuck SP, **S**...Shangrai-la Motel
38mm	Kalamazoo River
36	rd A-2, Ganges, **N**...food, lodging
34	MI 89, to Fennville, **6 mi S**...gas, Winery Tours
30	rd A-2, Glenn, Ganges, to Westside Cnty Park
27.5mm	**rest area eb, full(handicapped)facilities, phone, picnic tables, litter barrels, vending, petwalk**
26	109th Ave, to Pullman, no facilities
22	N Shore Dr, **N**...to Kal-Haven Trail SP, RV camping, **S**...Cousins Camping/rest.
20	rd A-2, Phoenix Rd, **N**...HOSPITAL, Amoco/mart/atm/24hr, Marathon/diesel/mart, Arby's, Hot'n Now, McDonald's, Taco Bell, Old Harbor Inn, AutoZone, camping, cleaners/laundry, st police, **S**...BP/diesel/mart, Big Boy, Sterman's IceCream, GuestHouse Inn, Hampton Inn, Farm&Fleet, Sears, Wal-Mart/auto
18	MI 140, MI 43, to Watervliet, **N**...Shell/diesel/mart/atm/24hr, Ma's Coffeepot Rest./24hr, **1-2 mi N**...DENTIST, Cheker/mart, Speedway/mart, Burger King, McDonald's, Pizza Hut, Platter's Diner, Econolodge, LakeBluff Motel, D&W Foods, Buick/Pontiac/Cadillac/GMC, Chevrolet/Olds, Chrysler/Dodge, auto parts, bank
13	to Covert, **N**...to Van Buren SP, RV camping
7	MI 63, to Benton Harbor, **N**...DiMaggio's Pizza, Vitale's Mkt/subs, camping
4	Riverside, to Coloma, **S**...Marathon/diesel/mart, McDonald's(4mi), KOA
2mm	Paw Paw River
1	Red Arrow Hwy, **N**...Ross Field Airport
0mm	I-94, E to Detroit, W to Chicago

I-196 begins/ends on I-94, exit 34 at Benton Harbor.

Michigan Interstate 275(Livonia)

Exit #	Services
	I-275 and I-96 run together 9 miles. **See Michigan Interstate 96, exits 165-170.**
29	I-96 E, to Detroit, MI 14 W, to Ann Arbor
28	Ann Arbor Rd, Plymouth, **E**...Amoco/mart/24hr, Shell/mart, Denny's, Dunkin Donuts, Red Roof Inn, **W**...Bennigan's, Bill Knapp's, Burger King, Steak&Ale, Quality Inn, K-Mart/drugs, Ford/Lincoln/Mercury, bank
25	MI 153, Ford Rd, Garden City, **W**...MEDICAL CARE, DENTIST, Amoco/mart/24hr, Citgo/diesel/mart, Shell/mart, Speedway/mart, Arby's, Bob Evans, BJ's Bowery/dining, Chinese Rest., ChuckeCheese, Dunkin Donuts, Hardee's, Johnson's Rest., KFC, Little Caesar's, Olive Garden, Outback Steaks, Roman Forum Dining, Wendy's, White Castle, Baymont Inn, Fairfield Inn, Motel 6, Super Drug, Discount Tire, Midas/Speedy Muffler, Pennzoil, K-Mart, Target, Vision Ctr, bank, laundry/cleaners
23	**rest area nb, full(handicapped facilities), phone, info, picnic tables, litter barrels**
22	US 12, Michigan Ave, to Wayne, **E**...Amoco/mart/24hr, Shell/mart/wash/atm, Speedway/diesel/mart, Hardee's, McDonald's, Subway, Fellows Cr Motel, Ramada Ltd, Super 8, Willow-Acres Motel
20	Ecorse Rd, to Romulus, no facilities
17	I-94 E to Detroit, W to Ann Arbor, **E**...airport
15	Eureka Rd, no facilities
13	Sibley Rd, New Boston, **W**...gas, hardware
11	S Huron Rd, **1 mi W**...Sunoco/diesel/mart
8	Will Carleton Rd, to Flat Rock, st police
5	Carleton, South Rockwood, **W**...food
4mm	**rest area sb, full(handicapped)facilties, phone, picnic tables, litter barrels**
2	US 24, to Telegraph Rd, no facilities
0mm	I-275 begins/ends on I-75, exit 20.

E

W

Benton Harbor

N

S

Livonia

Michigan Interstate 475 (Flint)

Exit #   Services

I-475 begins/ends on I-75, exit 125.

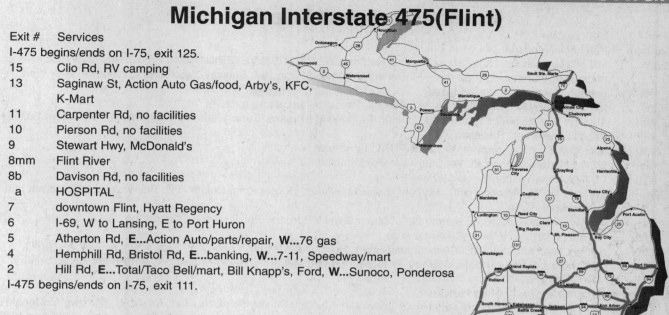

15   Clio Rd, RV camping

13   Saginaw St, Action Auto Gas/food, Arby's, KFC, K-Mart

11   Carpenter Rd, no facilities

10   Pierson Rd, no facilities

9   Stewart Hwy, McDonald's

8mm   Flint River

8b   Davison Rd, no facilities

a   HOSPITAL

7   downtown Flint, Hyatt Regency

6   I-69, W to Lansing, E to Port Huron

5   Atherton Rd, **E**...Action Auto/parts/repair, **W**...76 gas

4   Hemphill Rd, Bristol Rd, **E**...banking, **W**...7-11, Speedway/mart

2   Hill Rd, **E**...Total/Taco Bell/mart, Bill Knapp's, Ford, **W**...Sunoco, Ponderosa

I-475 begins/ends on I-75, exit 111.

N ↕ S

Flint

Michigan Interstate 696 (Detroit)

Exit #   Services

I-696 begins/ends on I-94.

28   I-94 E to Port Huron, W to Detroit, 11 Mile Rd, **E**...DENTIST, Marathon, Total, brakes

27   MI 3, Gratiot Ave, **N**...Amoco/mart, Dimitri's Rest./24hr, **S**...White Castle, Goodyear/auto

26   MI 97, Groesbeck Ave, Roseville, **N**...Marathon, **S**...Shell, Wendy's

24   Hoover Rd, Schoenherr Rd, **N**...Sunoco/mart, Burger King, KFC, **S**...Amoco/Dunkin Donuts, Mobil/mart, ChiChi's, Hardee's, Subway, Taco Bell, Holiday Inn Express, TJ Maxx, bank

23   MI 53, Van Dyke Ave, **N**...MEDICAL CARE, Amoco/mart, Arby's, Coney Island Diner, Dunkin Donuts, Juliano's Italian, McDonald's, Baymont Inn, Cadillac/Olds/Pontiac/GMC, Dodge, Minit-lube, Jo-Ann Fabrics, radiators, **S**...MEDICAL CARE, VETERINARIAN, Amoco/24hr, Burger King, Chevrolet, Ford, Toyota, USPO, auto parts, cleaners, drugs

22   Mound Rd, no facilities

20   Ryan Rd, Dequindre Rd, **N**...Citgo, Shell, Knight's Inn, Red Roof Inn, Ziebart TidyCar, **S**...Bob Evans, McDonald's, Cross Country Inn, transmissions

19   Couzens St, 10 Mile Rd, **S**...Hazel Park Racetrack

18   I-75 N to Flint, S to Detroit

17   Campbell Ave, Hilton Ave, Bermuda, Mohawk, **S**...Citgo/diesel, donuts, pizza, cleaners

16   MI 1, Woodward Ave, Main St, **N**...Big Boy, zoo, **S**...MI AAA

14   Coolidge Rd, 10 Mile Rd, **S**...Mobil, Total/mart, Dunkin Donuts, Hardee's, Little Caesar's, LJ Silver, Pizza Hut, Subway, Taco Bell, Wendy's, Farmer Jack's Foods, Arbor Drugs, Pennzoil, Jo-Ann Fabrics, hardware

13   Greenfield Rd, **S**...Mobil, Sunoco, Ford

12   Southfield Rd, 11 Mile Rd, **N**...Firestone

11   Evergreen Rd, **N**...cleaners, pizzaria, **S**...Residence Inn

10   US 24, Telegraph Rd, **N**...Amoco, Sunoco, Mobil, Hampton Inn, Embassy Suites, Chevrolet, Chrysler/Jeep, Dodge/Plymouth, Ford, Honda, K-Mart, Lincoln/Mercury, Olds, Pontiac/Buick, mall, **S**...Mobil, Sunoco, Courtyard, Hilton, Holiday Inn, Marriott

8   MI 10, Lodge Fwy, no facilities

5   Orchard Lake Rd, Farmington Hills, **N**...Marathon, Mobil/mart, Shell, Sunoco/wash, Arby's, Bill Knapp's, Ruby Tuesday, Silverman's Deli, Steak&Ale, Steamer's Seafood, Subway, Wendy's, Clarion Hotel, Comfort Inn, Discount Tire, Uncle Ed's Lube, cinema, to St Mary's Coll

1   (from wb), I-96 W, I-275 S, to MI 5, Grand River Ave

E ↕ W

Detroit Area

Minnesota Interstate 35

D u l u t h

Exit #	Services
260mm	I-35 begins/ends on MN 61 in Duluth.
259	MN 61, London Rd, North Shore, to Two Harbors, **W**...DENTIST, VETERINARIAN, Amoco/mart/24hr, ICO/diesel/mart, Spur/mart/atm, Burger King, McDonald's, Perkins, Pizza Hut, Subway, Taco John's, Wendy's, Best Western, London Court Motel, Viking Motel
258	21st Ave E, to U of MN at Duluth, **W**...Perkins, Best Western, same as 259
256b	Mesaba Ave, Superior St, **E**...Red Lobster, Timberlodge Steaks, Canal Park Inn, Hampton Inn, Hawthorn Suites, **W**...MH Gas/mart, Radisson
a	Michigan St, **E**...waterfront, **W**...HOSPITAL, downtown
255a	US 53 N, downtown, mall, **W**...Spur Gas, Ford, bank
b	I-535 spur, to Wisconsin, no facilities
254	27th Ave W, **W**...Amoco/mart/wash/24hr, Spur/diesel/mart, Burger King, Duluth Grill, Subway, Motel 6, Select Inn, QuakerState Lube
253b	40th Ave W, **W**...Amoco/diesel/mart/24hr, Mean Gene's Burgers, Perkins/24hr, Comfort Inn, Super 8, bank
a	US 2 E, US 53, to Wisconsin, no facilities
252	Central Ave, W Duluth, **W**...Conoco/diesel/mart, Holiday/diesel/mart/wash/24hr, Jade Fountain Chinese, KFC, McDonald's, Sammy's Café, Subway, Taco Bell, K-Mart, Menard's, Radio Shack, Super 1 Foods, bank, laundry/cleaners
251b	MN 23 S, Grand Ave, no facilities
a	Cody St, zoo, **E**...Allendale Motel
250	US 2 W(from sb), to Grand Rapids, **1/2 mi W**...Holiday/diesel, Mobil/diesel/LP/mart/atm, Blackwoods Grill, Country Kitchen, AmericInn
249	Boundary Ave, Skyline Pkwy, **E**...Holiday/McDonald's/diesel/mart, Country Inn Suites, Sundowner Motel(1mi), to ski area, **W...rest area both lanes, full(handicapped)facilities, info, phone, picnic tables, litter barrels,** Phillips 66/diesel/mart/atm/24hr, Blackwoods Grill, Country Kitchen, Spirit Mtn Lodge
246	MN 13, Midway Rd, Nopeming, **2 mi on frontage rd**...Sundowner Motel, **W**...Dry Dock Rest.
245	MN 61, **E**...Buffalo House Rest./camping
242	rd 1, Esko, Thomson, **E**...Amoco/mart
239.5mm	St Louis River
239	MN 45, Scanlon, to Cloquet, **E**...Jay Cooke SP, KOA(May-Oct), **W**...HOSPITAL, Conoco/Blimpie/mart/wash, Golden Gate Motel, Chevrolet, Pontiac/Buick
237	MN 33, Cloquet, **1 mi W**...HOSPITAL, Amoco/mart/wash/24hr, Conoco/diesel/mart, Spur Gas, Blackwoods Grill, Country Kitchen, Dairy Queen, Hardee's, McDonald's, Perkins/24hr, Pizza Hut, Rudy's Rest., Subway, Taco John's, AmericInn, Driftwood Motel, Super 8, Chrysler/Plymouth/Jeep, Ford/Mercury, NAPA, QuikLube, Super 1 Foods, Wal-Mart/24hr, bank
236mm	weigh sta both lanes
235	MN 210, Carlton, to Cromwell, **E**...Amoco/diesel/mart/rest., Spur/diesel/mart/24hr, AmericInn, Royal Pines Motel, diesel repair, to Jay Cooke SP, **W**...Black Bear Casino/Hotel/rest.
235mm	Big Otter Creek
233mm	Little Otter Creek
227	rd 4, Mahtowa, **E**...camping, **W**...Conoco/mart
226mm	**rest area nb, full(handicapped)facilities, phone, picnic tables, litter barrels, vending, petwalk**
220	rd 6, Barnum, **E**...camping, **W**...Amoco/diesel/mart/café/24hr, Northwoods Motel
219mm	Moose Horn River
218mm	Moose Horn River
216	MN 27(from sb), Moose Lake, **1 mi W**...HOSPITAL, Holiday/mart, Dairy Queen, Moose Lake Motel, Wyndtree Rest., to Munger Trail
214	rd 73, Moose Lake, **E**...Moose Lake SP(2mi), camping, **W**...Conoco/Subway/diesel/mart, AmericInn, **2 mi W**...HOSPITAL, Wyndtree Rest., Moose Lake Motel, Munger Trail, Red Fox Camping
209	rd 46, Sturgeon Lake, **E**...Phillips 66/mart, **W**...Sturgeon Lake Motel, Timberline Camping(3mi)
209mm	**rest area sb, full(handicapped)facilities, phone, picnic tables, litter barrels, vending, petwalk**
206.5mm	Willow River
205	MN 43, Willow River, **E**...camping, **W**...Citgo/diesel/mart, camping(2mi), laundry
198.5mm	Kettle River
198mm	**rest area nb, full(handicapped)facilities, phone, picnic tables, litter barrels, vending, petwalk**
195	MN 18, MN 23 E, to Askov, **E**...Amoco/diesel/café/mart, Super 8, to Banning SP, camping, **W**...camping
191	MN 23, Sandstone, **E**...MEDICAL CARE, Amoco/diesel/LP/mart, Conoco/diesel/mart/wash, Family Drug, **2 mi E**...61 Motel, **W**...hardware
184mm	Grindstone Rover

N

↕

S

Minnesota Interstate 35

183	MN 48, Hinckley, **E**...Conoco/mart, Holiday/gas, Burger King, Dairy Queen, Hardee's/RV parking, Subway, Taco Bell, Tobie's Sta/rest., Day's Inn, Holiday Inn Express, laundry, to St Croix SP(15mi), **W**...Mobil/mart/wash/24hr, Phillips 66/White Castle/mart/atm, Texaco, Cassidy's Rest., Best Western, Gold Pine Motel, Fire Museum
180	MN 23 W, to Mora, no facilities
175	rd 14, Beroun, **E**...Amoco/diesel/mart
171	rd 11, Pine City, **E**...SuperAmerica/gas, McDonald's, **W**...camping
170mm	Snake River
169	MN 324, rd 7, Pine City, **E**...CHIROPRACTOR, DENTIST, VETERINARIAN, Amoco/mart, Conoco/diesel/mart, Holiday/diesel/mart, Sinclair, A&W, Dairy Queen, Grizzly's Grill, Hardee's, KFC, Pizza Hut, Red Shed Rest., Subway, Gail Motel, Schwartzwald Motel, Ford/Mercury, Jubilee Foods, Pamida/drugs, Radio Shack, Wal-Mart, auto parts, bank, hardware, camping, **W**...to NW Co Fur Post HS
165	MN 70, Rock Creek, to Grantsburg, **E**...Tim's/Citgo/diesel/mart, Chalet Motel, camping, **W**...Total/diesel/mart/café/24hr
159	MN 361, rd 1, Rush City, **E**...HOSPITAL, Conoco/Burger King/Baskin-Robbins/diesel/mart, Tank&Tackle/diesel/mart, Jubilee Foods, laundry, **W**...camping(2mi)
154mm	**rest area nb, full(handicapped)facilities, phone, picnic tables, litter barrels, vending, petwalk**
152	rd 10, Harris, **2 mi E**...gas/diesel
147	MN 95, North Branch, to Cambridge, **E**...CHIROPRACTOR, VETERINARIAN, Amoco/mart, Casey's Store, Conoco/diesel/mart, Holiday/diesel/mart, Total/diesel/mart/café, McDonald's, North Branch Café, Pizza Hut, Subway, Oak Inn/rest., Super 8, CarQuest, SuperValu Foods, repair, to Wild River SP(14mi), camping, **W**...Burger King, Denny's, Imperial Lodge(13mi), Chevrolet, Chrysler/Plymouth/Dodge/Jeep, Ford, Tanger Outlet/famous brands, bank, camping(6mi), cinema
139	rd 19, Stacy, **1/2 mi E**...Phillips 66/mart, Rustic Inn Rest., Super 8(6mi), bank, transmissions, **W**...Conoco/pizza/mart
135	US 61 S, rd 22, Wyoming, **E**...Amoco/diesel/mart, Cornerstone Café, Dairy Queen, McDonald's, Subway, CarQuest, IGA Foods, bank, lodging(1mi), camping(2mi), **W**...Citgo/diesel/mart, Village Inn Rest., camping(10mi), golf
132	US 8(from nb), to Taylors Falls
131	rd 2, Forest Lake, **E**...HOSPITAL, CHIROPRACTOR, DENTIST, Amoco/mart/24hr, Holiday/mart, SuperAmerica/diesel/wash, Arby's/Sbarro's, Burger King, Cheung Chinese, Hardee's, KFC, McDonald's, Perkins, Taco Bell, AmericInn, Forest Lake Motel, CarX Muffler/brake, Castrol, Champion Auto, Checker Parts, Rainbow Foods/24hr, Target, Tires+, Wal-Mart, bank, carwash, laundry, **W**...Holiday/mart, Famous Dave's BBQ, Country Inn Suites, Buick/Olds/Pontiac, Chevrolet/Cadillac, Chrysler/Plymouth/Dodge/Jeep, Ford, K-Mart/Little Caesar's, Menard's, Valvoline
131mm	**rest area sb, full(handicapped)facilities, phone, picnic tables, litter barrels, vending, petwalk**
129	MN 97, rd 23, **E**...camping(6mi), **W**...Conoco/pizza/diesel/mart/24hr, camping(1mi)
128mm	weigh sta both lanes
127	I-35W, S to Minneapolis. **See I-35W.**
123	rd 14, Centerville, **E**...Otter Lake RV Ctr, **W**...Amoco/diesel/mart/atm, Conoco/mart, Texaco/mart, Kelly's Café, WiseGuys Pizza, ADL Brake/muffler/tire, carwash
120	rd J(from nb), no facilities
117	MN 96, **E**...SuperAmerica/gas, Total/mart, Burger King, Georgie V's Rest., AmericInn, **W**...MEDICAL CARE, DENTIST, Amoco/mart/wash, PDQ/mart, Applebee's, Arby's/Sbarro's, Boston Mkt, Hardee's, Little Caesar's, McDonald's, Subway, Taipan Chinese, Cub Foods/24hr, NAPA, Tires+, USPO, Valvoline, Walgreen
115	rd E, **E**...Amoco/mart/wash, Citgo/mart, SuperAmerica/gas, Perkins/24hr, **W**...McDonald's, FashionBug, Festival Foods, Target/drug, Wal-Mart/auto, bank
114	I-694 E, no facilities
113	I-694 W, no facilities
112	Little Canada Rd, **E**...Citgo/mart, **W**...Sinclair/mart, Rocco Pizza, cleaners
111b	rd 36 W, to Minneapolis, no facilities
a	rd 36 E, to Stillwater, no facilities
110b	Roselawn Ave, no facilities
a	Wheelock Pkwy, **E**...Amoco/mart, Total/mart, Subway, **W**...Sinclair, Champp's Café

Minnesota Interstate 35

St Paul

109	Maryland Ave, **E**...Citgo, SuperAmerica/gas, **W**...Wendy's, K-Mart/auto/drugs
108	Pennsylvania Ave, downtown
107c	University Ave, downtown, **E**...Amoco, **W**...HOSPITAL, to st capitol
107b a	I-94, W to Minneapolis, E to St Paul. I-35 and I-94 run together.
106c	11th St, Marion St, downtown, Best Western, Sears
106b	Kellogg Blvd, downtown, **W**...HOSPITAL, Denny's, Civic Center Inn, Radisson
a	Grand Ave, **W**...HOSPITAL
105	St Clair Ave, no facilities
104c	Victoria St, Jefferson Ave, no facilities
a	Randolph Ave, no facilities
103b	MN 5, W 7th St, **E**...Burger King, **W**...SuperAmerica/mart, Midas Muffler, radiators, to airport
a	Shepard Rd(from nb)
102mm	Mississippi River
102	MN 13, Sibley Hwy, **W**...Amoco/mart/wash, Holiday/mart, Bogey's Tacos
101b a	MN 110 W, **E**...Ziggy's Deli, **W**...SuperAmerica/gas
99b	I-494 W, no facilities
a	I-494 E, no facilities
98	Lone Oak Rd, **E**...Homestead Village, Microtel, USPO, **W**...DENTIST, Amoco/mart, American Hero Café, Joe Senser's Grill, Magic Thai Café, Hampton Inn
97b	Yankee Doodle Rd, **E**...CHIROPRACTOR, Holiday/mart, Applebee's, Arby's, Burger King, Cattle Co Rest., Chili's, Dairy Queen, Don Pablo's, Houlihan's, KFC, McDonald's, New China Buffet, Perkins, Pizza Hut, Taco Bell, Residence Inn, Springhill Suites, TownePlace Suites, Barnes&Noble, Big Wheel Auto, Byerly's Foods/24hr, GNC, HomePlace, Kinko's, Kohl's, Michael's, Office Depot, OfficeMax, Old Navy, PaperWhse, Pier 1, PrecisionTune, Tires+, TJ Maxx, Valvoline, bank, **W**...Amoco/mart, SuperAmerica/gas, Al Baker's Rest., Dragon Palace Chinese, Italian Pie Shoppe, Z-teca Mexican, Extended Stay America, Yankee Square Inn, bank, cleaners
a	Pilot Knob Rd, same as 97b, **E**...Hardee's, Best Western, Firestone, Goodyear, JiffyLube, mall
94	rd 30, Diffley Rd, to Eagan, **W**...Conoco/mart
93	rd 32, Cliff Rd, **E**...Texaco, Subway, **W**...Total/mart/wash, Burger King, Bakers Square, Boston Mkt, Burger King, Cherokee Sirloin, Dairy Queen, Denny's, Greenmill Rest., KFC, Little Caesar's, McDonald's, Subway, Taco Bell, Hilton, Holiday Inn Express, Staybridge Suites, Meineke Muffler, Target, Valvoline, USPO, Walgreen, bank, carwash, cinema
92	MN 77, Cedar Ave, **E**...Zoo
90	rd 11, **E**...KwikTrip/mart, PDQ/mart
88b	rd 42, Crystal Lake Rd, **E**...Byerly's Rest., Ciatti's Grill, PetsMart, cleaners, **W**...HOSPITAL, DENTIST, CHIRO-PRACTOR, Amoco/mart/wash/atm, PDQ/mart, Arby's/Sbarro's, BBQ, Burger King, Ground Round, McDonald's, Nicollet Grill, Old Country Buffet, Original Roadhouse Grill, Taco Bell, Country Inn Suites, Fairfield Inn, Hampton Inn, Holiday Inn, Home Depot, Midas Muffler, TJ Maxx, USPO, **1/2 mi W**...Holiday/mart, SuperAmerica/gas, Applebee's, Bakers Square, Bruegger's Bagels, Champp's Café, ChiChi's, Chili's, ChuckeCheese, Godfather's, Honeybaked Ham, KFC, Macaroni Grill, Mandarin Chinese, Olive Garden, Outback Steaks, Papa John's, Red Lobster, Schlotsky's, Southern China Café, Suburban Lodge, CarMax, CarX Muffler/brake, Chevrolet, Circuit City, Cub Foods/24hr, Dayton's, Gander Mtn, Goodyear/auto, JC Penney, JiffyLube, KidsRUs, Kinko's, Linens'n Things, MailBoxes Etc, Mervyn's, Michael's, NTB, OfficeMax, Pier 1, Rainbow Foods/24hr, Sears/auto, SportMart, Target, Tires+, Toyota, ToysRUs, Walgreen, bank, cinema, mall
a	I-35W, N to Minneapolis. **See I-35W.**
87	Crystal Lake Rd, **W**...Kwiktrip/gas, Buick, Ford/Lincoln/Mercury, Honda/Nissan, Saturn, Toyota, Beaver Mtn Ski Area
86	rd 46, **E**...Kwiktrip/gas, SuperAmerica/gas, KFC, Harley-Davidson
85	MN 50, Sinclair, **E**...CHIROPRACTOR, DENTIST, MEDICAL CARE, Amoco/mart/24hr, Fleet&Farm/gas, Sinclair, SuperAmerica/mart/wash, Burger King, Chinese Rest., Dairy Queen, Pizza Hut, Subway, Taco Bell, Comfort Inn, Goodyear, 1-Hr Photo, drugs, cleaners, bank, auto parts, **W**...Holiday/diesel/mart, Cracker Barrel, AmeriHost
84	185th St W(from sb), Orchard Trail, no facilities
81	rd 70, Lakeville, **E**...Megastop/diesel/rest., McDonald's, Pizza'n Pasta, Subway, Tacoville, Motel 6, Super 8/rest., bank, laundry, towing/repair, **W**...cinema
76	rd 2, Elko, **E**...Phillips 66/diesel/LP/mart, **W**...Glenno's Eatery/pizza, antiques
76mm	**rest area sb, full(handicapped)facilities, phone, picnic tables, litter barrels, vending, petwalk**
69	MN 19, to Northfield, **E**...Fuel&mart, Taco Bell(6mi), **8 mi E**...Archer House Inn, Country Inn, Super 8, Carleton Coll, St Olaf Coll, **W**...Conoco/A&W/Big Steer/diesel/mart/24hr
68mm	**rest area nb, full(handicapped)facilities, phone, picnic tables, litter barrels, vending, petwalk**
66	rd 1, to Dundas, **1 mi W**...Blue Horse Saloon Rest.

Minnesota Interstate 35

59 — MN 21, Faribault, **E**...Texaco/diesel/rest., Lavender Rest., Microtel, Super 8, **2 mi E**...VETERINARIAN, Kwiktrip/gas, SuperAmerica/gas, A&W, Burger King, Hardee's, KFC, Pizza Hut, Taco John's, AmericInn, Best Western, Galaxie Inn, Knight's Inn, Ford/Lincoln/Mercury, **W**...Go-Kart rides, camping

56 — MN 60, Faribault, **E**...HOSPITAL, Amoco/mart/wash, Mobil, Burger King, Great China Buffet, Hardee's, Perkins, Taco John's, Chevrolet, Chrysler/Plymouth/Jeep, Dodge, Goodyear/auto, Hy-Vee Food/drugs, JC Penney, Jo-Ann Fabrics, Olds/Buick/Pontiac, Tires+, Wal-Mart, bank, cinema, mall, **W**...gas/diesel/mart/atm, Dairy Queen, Happy Chef/24hr, Select Inn, Sakatah Lake SP, camping

55 — (from nb, no return), **1 mi E**...MEDICAL CARE, SuperAmerica/mart, KwikTrip/gas, Mobil/mart, Broaster Rest., Burger King, Dairy Queen, Great China Buffet, Hardee's, KFC, Little Caesar's, Pizza Hut, Subway, Taco John's, AmericInn, Best Western, Ford/Mercury, K-Mart/auto, Valvoline

48 — rd 12, rd 23, Medford, **1/2 mi E**...gas, bank, **W**...McDonald's, Outlet Mall/famous brands

45 — rd 9, Clinton Falls, **W**...Kwiktrip/diesel, Comfort Inn, Cabela's Sporting Goods, museum

43 — rd 34, 26th St, Airport Rd, Owatonna, **W**...Ramada Inn

42b a — US 14 W, rd 45, Owatonna, to Waseca, **E**...Sinclair/diesel/24hr, Texaco, Burger King, Kernel Rest., McDonald's, Budget Host, CashWise Food/drug, Chrysler/Dodge, Ford/Lincoln/Mercury, Wal-Mart, auto parts, **W**...Amoco/diesel/mart/wash, Happy Chef, Perkins, Best Budget Inn, Ramada Inn, Super 8

41 — Bridge St, Owatonna, **E**...HOSPITAL, Citgo/diesel/mart/wash, Applebee's, Arby's, Burger King, Dairy Queen, KFC, Subway, Taco Bell, AmericInn, Country Inn Suites, QuickLube, **W**...Budget Oil/gas, Sears Hardware, Target

40 — US 14 E, US 218, Owatonna, **1 mi E**...HOSPITAL, VETERINARIAN, Texaco, Godfather's, Pizza Hut, Taco John's, Oakdale Motel, Cedar Mall, Walgreen, Hy-Vee Foods/24hr, bank, camping, carwash, hardware, laundry

38mm — Turtle Creek

35mm — **rest area both lanes, full(handicapped)facilities, phone, picnic tables, litter barrels, vending, petwalk**

34.5mm — Straight River

32 — rd 4, Hope, **1/2 mi E**...Hope Oak Knoll Camping, **1 mi W**...gas, food

26 — MN 30, Ellendale, to Blooming Prairie, **E**...Cenex/diesel/mart/rest./24hr, **W**...Amoco/diesel/mart/rest./24hr

22 — rd 35, Geneva, to Hartland, **1 mi E**...gas, food

18 — MN 251, Clarks Grove, to Hollandale, **W**...Phillips 66/diesel/LP/mart

17mm — weigh sta both lanes

13b a — I-90, W to Sioux Falls, E to Austin, **W**...HOSPITAL

12 — US 65 S(from sb), Lp 35, Albert Lea, same as 11

11 — rd 46, Albert Lea, **E**...TA/Texaco/McDonald's/Pizza Hut/diesel/mart/24hr, Trails Rest., KOA(6mi), **W**...HOSPITAL, Amoco/diesel/mart/24hr, Conoco/diesel/mart, Phillips 66, Texaco/mart/atm, Burger King, China Buffet, Country Kitchen, Golden Corral, Perkins, Pizza Hut, Trumble's Rest., Wendy's, Albert Lea Inn, Countryside Inn, Day's Inn, Super 8, Buick/Pontiac/Olds/Cadillac/Honda, Ford Trucks, Mazda, NAPA, Nissan/VW, Plymouth/Jeep, Toyota, diesel repair, to Myre-Big Island SP

9mm — Albert Lea Lake

8 — US 65, Lp 35, Albert Lea, to Glenville, **2 mi W**...Cenex/mart/24hr, Citgo, Holiday/gas, Hardee's

5 — rd 13, Twin Lakes, to Glenville, **3 mi W**...camping

Minnesota Interstate 35W

Exit #	Services
41mm	I-35W begins/ends on I-35, exit 127.
36	rd 23, **E**...Amoco/diesel/mart, bank, **W**...MEDICAL CARE, Phillips 66/diesel/mart
33	rd 17, Lexington Ave, **E**...Amoco/mart, **1 mi E**...Burger King, McDonald's
32	95th Ave NE, to Lexington, Circle Pines, **W**...Nat Sports Ctr
31b	Lake Dr, no facilities
a	rd J, 85th Ave NE(no EZ return to nb), no facilities
30	US 10 W, MN 118, to MN 65, no facilities

Minnesota Interstate 35W

29	rd I, no facilities
28c b	rd 10, rd H, **W**...Amoco/mart/wash, Conoco/mart, Blimpie, Donatelle's Dining, Hardee's, KFC, McDonald's, Mermaid Café, Mr Donut, Perkins, Pizza Hut, RJ Riches Rest., Subway, Taco Bell, MoundsView Inn/steaks, Saturn, carwash
a	MN 96, **W**...Phillips 66/mart
27b a	I-694 E and W
26	rd E2, **W**...Tank'n Tummy/diesel/mart/wash
25b	MN 88, to Roseville(no EZ return to sb), **E**...Spur/mart, Fairfield Inn, Residence Inn, **W**...Kwikmart/gas, Total/mart, Barley John's, Chinese Rest., Godfather's, Jake's Café, Main Event Rest., McDonald's, Perkins/24hr, Subway
a	rd D(from nb), same as 25b
24	rd C, **E**...Burger King, Holiday Inn, **W**...Comfort Inn, Chevrolet/Olds/GMC, Chrysler/Plymouth, Dodge, Volvo
23b	Cleveland Ave, MN 36, **E**...India Palace Rest., Joe Senser's Rest., Day's Inn, Motel 6, Ramada Inn, Super 8, Golf Galaxy
a	MN 280, Industrial Blvd(from sb)
22	MN 280, Industrial Blvd(from nb), **E**...Anchorage Seafood Rest., Sheraton
21b a	rd 88, Broadway St, Stinson Blvd, **W**...Burger King, Country Kitchen, McDonald's, Mr Donut, Taco Bell, Home Depot, OfficeMax, Old Navy, PetsMart, Rainbow Foods/24hr, Target
19	E Hennepin(from nb)
18	US 52, 4th St SE, University Ave, to U of MN, **E**...Amoco/mart, Hardee's
17c	11th St, Washington Ave, **E**...Holiday Inn, **W**...HOSPITAL, Mobil, Goodyear, LubeExpress, to Metrodome
b	I-94 W(from sb)
a	MN 55, Hiawatha
16b a	I-94(from nb), E to St Paul, W to St Cloud, to MN 65
15	31st St(from nb), Lake St, **E**...McDonald's, Taco Bell, **W**...HOSPITAL
14	35th St, 36th St, no facilities
13	46th St, no facilities
13mm	Minnehaha Creek
12b	Diamond Lake Rd, no facilities
a	60th St(from sb), **W**...Total/gas, Perkins, Cub Foods, PrecisionTune
11b	MN 62 E, to airport
a	Lyndale Ave(from sb), **E**...Food'n Fuel, auto parts
10b	MN 62 W, 58th St, no facilities
a	rd 53, 66th St(from sb), no facilities
9c	76th St(from sb), no facilities
b a	I-494, MN 5
8	82nd St, **E**...BMW, Mercury, to airport, **W**...Bennigan's, Dutch Family Rest., Chevrolet, Infiniti, Saturn, Toyota
7b	90th St, no facilities
a	94th St, **E**...Amoco/wash, **W**...Holiday Inn
6	rd 1, 98th St, **E**...EZ Stop/gas, Bakers Square, Dutch Family Rest., Szechuan Chinese, Snyder's Drug, Ford, auto parts, bank, **W**...SuperAmerica/mart, Burger King, Denny's, Subway
5	106th St, no facilities
5mm	Minnesota River
4b	113th St, Black Dog Rd, no facilities
a	Cliff Rd, **E**...Dodge, **W**...VW
3b a	MN 13, Shakopee, Canterbury Downs, **E**...Quality Inn
2	Burnsville Pkwy, **E**...MEDICAL CARE, Citgo/mart, Bob's Café, Denny's, Hardee's, bank, hardware, mall, **W**...Amoco/mart, Chinese Rest., Embers Rest., Perkins, TimberLodge Steaks, Day's Inn, Prime Rate Motel, Red Roof Inn, Super 8, Best Buy, Fabrics-N-Home Outlet
1	rd 42, Crystal Lake Rd, **E**...HOSPITAL, DENTIST, CHIROPRACTOR, Amoco/mart/wash/atm, PDQ/mart, Arby's/ Sbarro's, BBQ, Burger King, Ground Round, McDonald's, Nicollet Grill, Old Country Buffet, Original Roadhouse Grill, Taco Bell, Country Inn Suites, Fairfield Inn, Hampton Inn, Holiday Inn, Home Depot, Midas Muffler, TJ Maxx, USPO, **1/2 mi E**...Byerly's Rest., Ciatti's Grill, PetsMart, cleaners, **W**...Holiday/mart, SuperAmerica/gas, Applebee's, Bakers Square, Bruegger's Bagels, Champp's Café, ChiChi's, Chili's, ChuckeCheese, Godfather's, Honeybaked Ham, KFC, Macaroni Grill, Mandarin Chinese, Olive Garden, Outback Steaks, Papa John's, Red Lobster, Schlotsky's, Southern China Café, TGIFriday, Suburban Lodge, Buick, CarMax, CarX Muffler/brake, Chevrolet, Circuit City, Cub Foods/ 24hr, Dayton's, Gander Mtn, Goodyear/auto, Honda/Nissan, JC Penney, JiffyLube, KidsRUs, Kinko's, Lincoln/ Mercury, Linens'n Things, MailBoxes Etc, Mervyn's, Michael's, NTB, OfficeMax, Pier 1, Rainbow Foods/24hr, Saturn, Sears/auto, SportMart, Target, Tires+, Toyota, ToysRUs, Walgreen, bank, cinema, mall
0mm	I-35W begins/ends on I-35, exit 88a.

N

S

Minnesota Interstate 90

Exit #	Services
277mm	Mississippi River, Minnesota/Wisconsin state line
275	US 14, US 61, to MN 16, La Crescent, **N...Welcome Ctr wb, full(handicapped)facilities, info, phone, picnic tables, litter barrels**, **vending, petwalk**, Pettibone RV Park
272b a	Dresbach, **N**...lodging, **S**...Mobil/service/mart, Gopher Marine, Kubota
270	Dakota, **N**...accesses lodging at 272
269	US 14, US 61, to Winona, **N**...to OL Kipp SP, camping, **S**...camping
266	rd 12, Nodine, **N**...OL Kipp SP, camping, **S**...Amoco/Subway/diesel/24hr, Truckers Inn/rest.
261mm	weigh sta both lanes
257	MN 76, Ridgeway, Witoka, to Houston, **N**...gas, **S**...camping
252	MN 43N, to Winona, **7 mi N**...HOSPITAL, AmeircInn(10mi), Best Western, Quality Inn, Super 8
249	MN 43S, to Rushford, **N**...Peterbilt Trucks/repair, **S**...Ernie Tuff Museum
244mm	**rest area eb, full(handicapped)facilities, phone, picnic tables, litter barrels, vending, petwalk**
242	rd 29, Lewiston, no facilities
233	MN 74, St Charles, to Chatfield, **N**...Kwiktrip/LP/24hr, Whitewater SP, lodging, **S**...Texaco/diesel/mart, Amish Ovens Rest./bakery, RV dumping/LP
229	rd 10, Dover, no facilities
224	rd 7, Eyota, no facilities
222mm	**rest area wb, full(handicapped)facilities, phone, picnic tables, litter barrels, vending, petwalk**
218	US 52, to Rochester, **8 mi N**...Old Country Buffet, Hampton Inn, Motel 6, Brookside RV Park, **S**...Citgo, KOA(Mar-Oct)(1mi)
209b a	US 63, MN 30, to Rochester, Stewartville, **8-10 mi N**...Comfort Inn, Day's Inn, Hampton Inn, Holiday Inn, Super 8, airport, **S**...Amoco/diesel/mart, KwikTrip/gas, Hardee's, AmericInn
205	rd 6, no facilities
202mm	**rest area eb, full(handicapped)facilities, phone, picnic tables, litter barrels, vending, petwalk**
193	MN 16, Dexter, **S**...Amoco/Windmill Rest./diesel/24hr, Mill Inn Motel
189	rd 13, to Elkton, no facilities
187	rd 20, **S**...Beaver Trails Camping
183	MN 56, to Rose Creek, Brownsdale, **S**...Cenex/diesel/LP/mart
181	28th St NE, no facilities
180b a	US 218, 21st St NE, Oakland Place, to Austin, **S**...Texaco/mart, Austin Motel
179	11th Dr NE, to Austin, **N**...Citgo/diesel/rest./24hr
178b	6th St NE, to Austin, downtown
a	4th St NW, **N**...Perkins, AmericInn, Day's Inn, Holiday Inn, **S**...HOSPITAL, Amoco/mart/wash, Conoco/mart, Sinclair/diesel, A&W, Burger King, Goodyear
177	US 218 N, Austin, Mapleview, to Owatonna, **N**...Holiday Gas, Applebee's, KFC, Hy-Vee Foods, IndyLube, JC Penney, K-Mart/auto, Rainbow Foods, ShopKO, Staples, Super Fair Foods, Target, Younkers, cinema, hardware, mall, **S**...Total/diesel/mart, Hardee's, Super 8
175	MN 105, rd 46, to Oakland Rd, **N**...VETERINARIAN, Phillips 66/diesel/mart, Sportts Rest., Countryside Inn, **S**...Amoco/wash, Conoco/diesel/mart, McDonald's(2mi), Downtown Motel(2mi), Sterling Motel(2mi), Ford Trucks/Mercury, camping
171mm	**rest area wb, full(handicapped)facilities, phone, picnic tables, litter barrels, petwalk**
166	rd 46, Oakland Rd, **N**...KOA/LP, golf(par3)
163	rd 26, Hayward, **S**...Cenex Gas/mart, Texaco(4mi), Pizza Hut(4mi), Trails Rest.(4mi), Myre-Big Island SP, antiques, camping
159b a	I-35, N to Twin Cities, S to Des Moines, no facilities
157	rd 22, Albert Lea, **S**...HOSPITAL, Citgo/Subway/diesel/mart, Conoco/mart/wash, Burgers&More, Café Don'l, Dairy Queen, Hardee's, Herberger's, McDonald's, AmericInn, Holiday Inn Express, Chevrolet, Coast Hardware, Harley-Davidson, Hy-Vee Foods/24hr, IndyLube, Jubilee Foods, Pamida, Radio Shack, ShopKO, cinema, hardware, mall
154	MN 13, to US 69, Albert Lea, to Manchester, **N**...SuperAmerica/diesel/mart, Skyline Rest., **S**...Wal-Mart(3mi)
146	MN 109, to Wells, Alden, **S**...Amoco/diesel/mart/rest., Cenex

E

W

Albert Lea

Minnesota Interstate 90

138	MN 22, to Wells, Keister, **S**...camping
134	MN 253, rd 21, to Bricelyn, MN Lake, no facilities
128	MN 254, rd 17, to Frost, no facilities
119	US 169, Blue Earth, to Winnebago, **S**...HOSPITAL, KwikTrip/gas, Sinclair/diesel/mart, Texaco/Dairy Queen/Taco Bell/mart, Country Kitchen, Hardee's, McDonald's, Pizza Hut, Subway, AmericInn, Budget Inn, Super 8, Wal-Mart/drugs, carwash, to Jolly Green Giant
118mm	**rest area both lanes, full(handicapped)facilities, phone, picnic tables, litter barrels, petwalk, playground**
113	rd 1, Guckeen, no facilities •
107	MN 262, rd 53, Granada, to East Chain, **S**...gas/diesel, Flying Goose Camping(May-Oct)(1mi)
102	MN 15, Fairmont, **N**...truckwash/repair, **S**...HOSPITAL, CHIROPRACTOR, Amoco/24hr, Cenex/Taco Bell/mart, Conoco/diesel/mart, Phillips 66/mart, SuperAmerica/diesel/mart/24hr, Burger King, China Rest., Happy Chef/24hr, Hardee's, KFC, McDonald's, Perkins, Pizza Hut, Ranch Family Rest., Subway, Budget Inn, Comfort Inn, Holiday Inn, Super 8, Dodge/Chrysler, K-Mart/drugs, Mazda, Pontiac/Olds/Buick, auto parts, camping
99	rd 39, Fairmont, **S**...access to gas, food, lodging
93	MN 263, rd 27, Welcome, **1/2 mi S**...Cenex/mart, camping
87	MN 4, Sherburn, **1 mi S**...HOSPITAL, Cenex, Ma Faber's Cookin', camping
80	rd 29, Alpha, no facilities
73	US 71, Jackson, **N**...Conoco/diesel/mart, Hardee's, Best Western/rest., Super 8, to Kilen Woods SP, **S**...HOSPITAL, Amoco/Burger King/mart, Pizza Ranch, Subway, Budget Host, Park-Vu Motel, Chevrolet/Buick/Olds/Pontiac, Plymouth/Dodge, to Spirit Lake, KOA(May-Sept)
72.5mm	W Fork Des Moines River
72mm	**rest area wb, full(handicapped)facilities, phone, picnic tables, litter barrels, vending, petwalk**
69mm	**rest area eb, full(handicapped)facilities, phone, picnic tables, litter barrels, vending, petwalk**
64	MN 86, Lakefield, **N**...HOSPITAL, gas/diesel, food, camping, away from exit
57	rd 9, Spafford, to Heron Lake, no facilities
50	MN 264, rd 1, to Brewster, Round Lake, **S**...camping
47	rd 53, no facilities
46mm	weigh sta eb
45	MN 60, Worthington, **N**...Amoco/diesel/mart, **S**...Texaco/diesel/mart/24hr, Hardee's, McDonald's, Best Western, Budget Inn, camping
43	US 59, Worthington, **N**...Ramada Inn, airport, **S**...HOSPITAL, CHIROPRACTOR, VETERINARIAN, Amoco, Casey's/gas, Cenex/Taco Bell/mart, Conoco, Phillips 66/diesel/mart, Texaco/mart, Burger King, Dairy Queen, Godfather's, Happy Chef, Hardee's, KFC, McDonald's, Perkins/24hr, Pizza Hut, Ruttle's Grill, Subway, Taco John's, Wendy's, AmericInn, Budget Inn, Day's Inn, Holiday Inn Express, K-Mart, County Mkt Foods/deli/bakery/24hr, Hy-Vee Foods, Chevrolet/Pontiac/Olds/Buick/Cadillac, Ford, Mazda, Family$, JC Penney, ShopKO, auto parts, bank, drugs, laundry
42	MN 266, rd 25, to Reading, **S**...Super 8
33	rd 13, Rushmore, to Wilmont, no facilities
26	MN 91, Adrian, **S**...Amoco/diesel/mart, Cenex/mart, camping, hardware
25mm	**rest area wb, full(handicapped)facilities, phone, picnic tables, litter barrels, petwalk**
24mm	**rest area eb, full(handicapped)facilities, phone, picnic tables, litter barrels, petwalk**
18	rd 3, Kanaranzi, Magnolia, no facilities
12	US 75, Luverne, **N**...HOSPITAL, Amoco/Subway, Casey's Store, Cenex/diesel, Conoco/diesel, Phillips 66/diesel, Texaco/mart, Country Kitchen, Hardee's, JJ's Tasty Drive-In, McDonald's, Pizza Hut, Taco John's, Comfort Inn, Cozy Rest Motel(1mi), Hillcrest Motel(2mi), Jubilee Foods, Chevrolet/Pontiac/Buick/Olds/GMC, Chrysler/Dodge/Jeep, NAPA, bank, camping, hardware, truckwash, to Blue Mounds SP, Pipestone NM, **S**...Magnolia Steaks, Super 8, Pamida
5	rd 6, Beaver Creek, **N**...Texaco/diesel/mart
1	MN 23, rd 17, to Jasper, **N**...to Pipestone NM, access to gas/diesel
0mm	Minnesota/South Dakota state line, **Welcome Ctr/weigh sta eb, full(handicapped)facilities, info, phone, picnic tables, litter barrels**

E
↕
W

Minnesota Interstate 94

Exit #	Services
259mm	St Croix River, Minnesota/Wisconsin state line
258	MN 95 N, to Stillwater, Hastings, Lakeland, no facilities
257mm	weigh sta wb
256mm	**Welcome Ctr wb, full(handicapped)facilities, phone, picnic tables, litter barrels, vending, petwalk**
253	rd 15, Manning Ave, **S**...KOA, StoneRidge Golf, to Afton Alps SP, ski area
251	rd 19, Keats Ave, Woodbury Dr, **N**...camping, **S**...Amoco/mart, SuperAmerica/mart/wash, Burger King, Dairy Queen, FoodCourt, Subway, Extended Stay America, Holiday Inn Express, Prime Outlets/famous brands, KOA, bank, funpark

Minnesota Interstate 94

250 rd 13, Radio Dr, Inwood Ave, **S**...VETERINARIAN, Holiday/mart, Blimpie, Don Pablo's, Einstein Bagels, Schlotsky's, Sunset Grill, Taco Bell, TGIFriday, Wendy's, Bath&BodyWorks, Borders Books/Café, Champp's, Circuit City, CompUSA, Cub Foods, Galyan's Trading Co, Gateway Computers, GNC, Hepner's Auto Ctr, Home Depot, HomePlace, Jo-Ann Crafts, LandsEnd Inlet, LensCrafters, MailBoxes Etc, Mens Wearhouse, Mervyn's Calif, OfficeMax, Old Navy, PetsMart, Pier 1, Tires+, ToysRUs, auto repair

249 I-694 N & I-494 S, no facilities

247 MN 120, Century Ave, **N**...Denny's, Toby's Rest., AmericInn, Super 8, Harley-Davidson, PrecisionTune, **S**...Sinclair/mart, SuperAmerica/mart, GreenMill Rest., McDonald's, Country Inn/ rest., Chevrolet

246c b McKnight Ave, **N**...3M

a Ruth St(from eb, no return), **N**...Amoco/mart, Sinclair, Domino's, HoHo Chinese, Perkins, Firestone/auto, Michael's

245 White Bear Ave, **N**...SuperAmerica/Subway, Hardee's, Embers Rest., N China Buffet, Exel Inn, Ramada Inn, JiffyLube, **S**...DENTIST, CHIROPRACTOR, Amoco/mart/wash, Arby's, Bakers Square, Burger King, Davanni's Pizza/subs, Ground Round, KFC, McDonald's, Perkins, Taco Bell, Holiday Inn, Chrysler/Plymouth, Firestone/auto, JC Penney, Lenscrafters, Target/drugs, TJ Maxx, bank

244 US 10 E, US 61 S, Mounds/Kellogg, **N**...Peking Buffet, Country Club Mkt, to downtown

243 US 61, Mounds Blvd, **S**...River Centre

242d US 52 S, MN 3, 6th St, **N**...Holiday/mart, XpressLube, Subway, airport

c 7th St, **S**...SuperAmerica

b a I-35E N, US 10 W, I-35E S(from eb)

241c I-35E S(from wb)

b 10th St, 5th St, to downtown

a 12th St, Marion St, Kellogg Blvd, **N**...Best Western Kelly Inn, Sears, **S**...HOSPITAL, Amoco, Holiday Gas, SuperAmerica, Texaco, Savoy Inn, XpressLube, tires, st capitol

240 Dale Ave, **S**...HOSPITAL, Best Western, Holiday Inn, Valvoline Lube

239b a Lexington Pkwy, Hamline Ave, **N**...MEDICAL CARE, Amoco/mart, Hardee's/24hr, Rio Bravo, Sheraton, Cub Foods, K-Mart, PetsMart, Target

238 Snelling Ave, **N**...Hardee's, ChiChi's, Sheraton, Target, **S**...Citgo/mart, Maltshop, same as 239

237 Cretin Ave, Vandalia Ave, to downtown

236 MN 280, University Ave, to downtown

235b Huron Blvd, **N**...HOSPITAL, Citgo, Arnold's Burger Grill, U of MN

235mm Mississippi River

235a Riverside Ave, 25th Ave, **N**...Citgo/mart, Brubaegger's Bagels, Starbucks, **S**...Perkins, Taco Bell

234c Cedar Ave, downtown

234b a MN 55, Hiawatha Ave, 5th St, **N**...Holiday Inn, to downtown

233b I-35W N, I-35W S(exits left from wb)

a 11th St(from wb), **N**...downtown

231b Hennepin Ave, Lyndale Ave, to downtown

a I-394, US 12 W, to downtown

230 US 52, MN 55, 4th St, 7th St, Olson Hwy, **N**...Metrodome, **S**...HOSPITAL, Int Mkt Square

229 W Broadway, Washington Ave, **N**...Holiday/mart, Old Colony/diesel/wash, Burger King, **S**...Taco Bell, Wendy's, Target

228 Dowling Ave N, no facilities

226 53rd Ave N, 49th Ave N, no facilities

225 I-694 E, MN 252 N, to Minneapolis, no facilities

34 to MN 100, Shingle Creek Pkwy, **N**...Amoco/wash, ChiChi's, Cracker Barrel, Denny's, Olive Garden, TGIFriday, AmericInn, Baymont Inn, Comfort Inn, Country Inn Suites, Day's Inn, Extended Stay America, Hilton, Holiday Inn, Park Inn, Super 8, JC Penney, **S**...Chinese Rest., Perkins, Pizza Hut, Subs Etc, Inn on the Farm, Target, TJ Maxx, Tires+, Drug Emporium, cinema

33 rd 152, Brooklyn Blvd, **N**...Holiday/mart, Mobil/mart, Chevrolet, Dodge, Honda, Mazda, Olds, Pontiac, Pitstop Lube/ CarX Muffler, RV Ctr, auto parts, laundry, **S**...CHIROPRACTOR

31 rd 81, Lakeland Ave, **N**...SuperAmerica, Texaco, Dairy Queen, Wagner's Drive-In, Wendy's, Ramada Inn, **S**...Amoco, Budget Host, Super 8(1mi)

30 Boone Ave, **N**...Comfort Inn, Northland Inn/rest., Sleep Inn, **S**...Home Depot, NTB

Minnesota Interstate 94

29b a		US 169, to Hopkins, Osseo, no facilities
28		rd 61, Hemlock Ln, **N**...ChuckeCheese, Olive Garden, Staybridge Suites, Best Buy, Byerly's Foods, Jo-Ann Fabrics, Linens'n Things, Office Depot, **S**...Amoco/mart/wash, Angino's Rest, Arby's/Sbarro's, BBQ, Champp's, Perkins/24hr, Hampton Inn, Red Carpet Inn, bank
216		I-94 W and I-494, no facilities
215		rd 109, Weaver Lake Rd, **N**...CHIROPRACTOR, DENTIST, EYECARE, MEDICAL CARE, VETERINARIAN, Citgo, SuperAmerica/diesel/mart, Angino's Rest., Burger King, Bakers Square, BBQ, Broadway Pizza, Champp's, Dairy Queen, Don Pablo's, Hop's Grill, Joe's Crabshack, KFC, McDonald's, Old Country Buffet, Papa John's, Pizza Hut, Subway, Taco Bell, Wendy's, Hampton Inn, Barnes&Noble, Cub Food/drug, GNC, Goodyear/auto, HomePlace, Kohl's, K-Mart/drugs, MailBoxes Etc, Midas Muffler, Old Navy, PetCo, Tires+, USPO, Walgreen, XpressLube, bank, cinema, laundry, mall, **S**...Applebee's, Fuddrucker's
214mm		**rest area eb, full(handicapped)facilities, phone, picnic tables, litter barrels**
213		rd 30, 95th Ave N, Maple Grove, **N**...Conoco/diesel/mart, SuperAmerica/mart/24hr, Tom Thumb/gas/mart, **S**...Menard's, Rainbow Foods, Target, bank, **2 mi S**...Mama G's Rest., KOA
207		MN 101, Rogers, to Elk River, **N**...Holiday/diesel/mart, SuperAmerica/diesel/mart/atm, TA/Citgo/diesel/rest./24hr, Burger King, DiMagio's Café, McDonald's, Taco Bell, Super 8, Camping World(off Rogers Dr), car museum, **S**...Amoco/mart/wash/24hr, Holiday/mart, Sinclair/diesel, Country Kitchen, Dairy Queen, Ember's Rest., Kernel Rest., Subway, AmercInn, Chevrolet, bank, hardware
205.5mm		Crow River
205		MN 241, rd 36, St Michael, **N**...golf, **S**...SuperAmerica/diesel/mart, **3 mi S**...Mobil, McDonald's, Jubilee Foods
202		rd 37, Albertville, **N**...Conoco/diesel/mart, **S**...Amoco/mart, Phillips 66/mart, Country Pizza, Riverwood Hotel/rest.
201		rd 19(rom eb), Albertville, St Michael, **N**...Albertville Outlets/famous brands, **S**...Amoco/mart
195		rd 75(from wb, no EZ return), Monticello, **N**...HOSPITAL, Conoco/diesel/mart, Mobil/diesel/mart, Hawk's Grill, Jam'n Joe Café
193		MN 25, Monticello, to Buffalo, Big Lake, **N**...Holiday/diesel/mart, Burger King, Country Grill, Dairy Queen, KFC, Perkins, Taco Bell, Wendy's, AmericInn, AutoValue Parts, BMW, Cub Foods, Honda, K-Mart, Snyder Drug, bank, **S**...VETERINARIAN, Amoco/mart/wash, Mobil/diesel/mart, SuperAmerica/diesel/mart, McDonald's, Subway, Best Western, Comfort Inn, Day's Inn, AutoMax, Checker Parts, Chevrolet/Olds, Ford/Mercury, Goodyear/auto, UltraLube, Lake Maria SP
187mm		**rest area eb, full(handicapped)facilities, phone, picnic tables, litter barrels, vending, petwalk**
183		rd 8, Hasty, to Silver Creek, Maple Lake, **N**...Northern RV/boat, **S**...Marathon/diesel/rest./24hr, to Lake Maria SP, camping
178		MN 24, Clearwater, to Annandale, **N**...VETERINARIAN, Citgo/diesel/mart, Holiday/Pizza/mart/24hr, Marathon/Pizza Hut/mart, Dairy Queen, Hardee's, Java Bros Coffee, Ole Kettle, Subway, Budget Inn, Eagle Trace Golf/grill, KOA(1mi), TrueValue, Valvoline, **14 mi N**...Best Western, Day's Inn, **S**...Phillips 66/diesel/mart, Longhorn Steaks, A-J Acres Camping(1mi)
178mm		**rest area wb, full(handicapped)facilities, phone, picnic tables, litter barrels, petwalk, vending**
171		rd 7, rd 75, St Augusta, **5 mi N**...HOSPITAL, Amoco/diesel/mart, Cenex/mart, Holiday/McDonald's/diesel/LP/mart/24hr, FoodCourt, Nathan B's Rest., AmericInn, Comfort Inn, Holiday Inn Express, Travelodge, Toyota, tires, **S**...Conoco/diesel
167b a		MN 15, to St Cloud, to Kimball, **4 mi N**...HOSPITAL, CHIROPRACTOR, SuperAmerica/diesel, Applebee's, Arby's, Bakers Square, Embers Rest., McDonald's, Old Country Buffet, Perkins, Pizza Hut, Taco Bell, TimberLodge Steaks, Baymont Inn, Best Western, Comfort Inn, Country Inn Suites, Fairfield Inn, Holiday Inn, Quality Inn, Ramada Ltd, Radisson, Super 8, Cub Foods, Kohl's, Sam's Club, ShopKO, Valvoline, golf, **2 mi S**...Texaco/mart
164		MN 23, Rockville, to St Cloud, **1-2 mi N**...Conoco, Maid-Rite Café, Taco Bell, Motel 6, Ramada Ltd, antiques, to Waite Park
162.5mm		Sauk River
160		rd 2, St Joseph, to Cold Spring, **1 mi N**...Amoco, SuperAmerica, Super 8, Coll of St Benedict
158		rd 75(from eb), to St Cloud, same as 160, 3 mi N
156		rd 159, St Joseph, **N**...St Johns U, no facilities
153		rd 9, Avon, **N**...Citgo/mart, Texaco/diesel/LP/mart, Burger Treat Rest., Dahlin's Foods, Rascal's Rest., AmericInn, **S**...El Rancho Manana Camping, auto parts
152mm		**rest area both lanes, full(handicapped)facilities, phone, picnic tables, litter barrels, vending, petwalk**
147		MN 238, rd 10, Albany, **N**...HOSPITAL, Holiday/diesel/mart/24hr, Marathon/mart, Texaco/Subway/Piccadilly's/diesel/mart, Dairy Queen, Hillcrest Rest., Country Inn Suites, Chrysler/Plymouth/Dodge/Jeep, IGA Food, NAPA, carwash, golf, **S**...KFC, Ford
140		rd 11, Freeport, **N**...Conoco//Pizza/diesel/mart, Mobil/mart, Charlie's Café, Ackie's Pioneer Inn, USPO, auto repair/tires, **S**...LP
137		MN 237, rd 65, New Munich, no facilities
137mm		Sauk River
135		rd 13, Melrose, **N**...HOSPITAL, Mobil/diesel/mart, Phillips 66/Subway/diesel/mart/atm/24hr, Burger King, Ford, Jublilee Foods, NAPA, bank, camping, **S**...VETERINARIAN, Conoco/diesel/mart/atm, Countryside Rest., Dairy Queen, Super 8, Save Foods

E

↑
↓

W

St Cloud

244

Minnesota Interstate 94

132.5mm	Sauk River
131	MN 4, Meire Grove, to Paynesville, no facilities
128mm	Sauk River
127	US 71, MN 28, Sauk Centre, **N**...HOSPITAL, CHIROPRAC-TOR, Casey's/gas, Cenex/Taco John's/diesel/mart/24hr, Holiday/diesel/mart/atm, SuperAmerica/mart/atm/24hr, Dairy Queen, Hardee's, McDonald's, Pizza Hut, Subway, AmericInn, BestValue Inn, Super 8, EconoMart Foods, Ford/Mercury, antiques, camping, **S**...Texaco/diesel/mart/café/24hr, Chevrolet/Pontiac/Olds/Buick/Chrysler/Jeep, RV Service
124	Sinclair Lewis Ave(from eb), Sauk Centre, no facilities
119	rd 46, West Union, no facilities
114	MN 127, rd 3, Osakis, to Westport, **3 mi N**...gas/diesel, food, lodging, camping
105mm	**rest area wb, full(handicapped)facilities, phone, picnic tables, litter barrels, vending, petwalk**
103	MN 29, Alexandria, to Glenwood, **N**...HOSPITAL, VETERINAR-IAN, Amoco/Subway/mart, Citgo/mart, Texaco/diesel/mart/rest., Burger King, Country Kitchen, Hardee's, KFC, McDonald's, Perkins, Pizza Hut, Sirloin Buffet, Taco Bell, AmercInn, Best Inn, Comfort Inn, Day's Inn, SkyLine Motel, Super 8, Chevrolet/Cadillac/Mazda, OfficeMax, Target/drugs, Wal-Mart/24hr, **S**...B&H/diesel, Conoco/diesel/mart, Country Inn Suites, Holiday Inn, Buick/Pontiac/Olds/GMC, diesel repair
100mm	Lake Latoka
100	MN 27, no facilities
99mm	**rest area eb, full(handicapped)facilities, phone, picnic tables, litter barrels, vending, petwalk**
97	MN 114, rd 40, Garfield, to Lowery, no facilities
90	rd 7, Brandon, **2-3 N**...Texaco/diesel, food, camping, **S**...camping, ski area
82	MN 79, rd 41, Evansville, to Erdahl, **2 mi N**...Amoco/diesel, Kraemer's Rest., **S**...HOSPITAL
77	MN 78, rd 10, Ashby, to Barrett, **N**...gas/diesel, food, camping, **S**...camping
69mm	**rest area wb, full(handicapped)facilities, phone, picnic tables, litter barrels, petwalk, vending**
67	rd 35, Dalton, **N**...Dalton Café(2mi), camping, **S**...camping
61	US 59 S, rd 82, to Elbow Lake, **N**...HOSPITAL, Citgo/diesel/café/LP/24hr, Lakeside Rest., camping(4mi), **S**...camping
60mm	**rest area eb, full(handicapped)facilities, phone, picnic tables, litter barrels, petwalk, vending**
57	MN 210, rd 25, Fergus Falls, **S**...HOSPITAL
55	rd 11, Fergus Falls, to Wendell, no facilities
54	MN 210, Lincoln Ave, Fergus Falls, **N**...HOSPITAL, CHIROPRACTOR, VETERINARIAN, Amoco/mart/24hr, Cenex/diesel/mart, Fleet&Farm/gas, Holiday/mart, Spur/diesel/mart, Blimpie/TCBY, Burger King, Ember's Rest., King Buffet, KFC, McDonald's, Perkins, Pizza Hut, Ponderosa, Sandpiper Rest., Subway, AmericInn, Best Western, Comfort Inn, Day's Inn, Motel 7, Super 8, Chrysler/Jeep/Dodge, Ford/Lincoln/Mercury, Hedahl's Parts, JC Penney, K-Mart, Mazda, More 4 Foods/24hr, Olds/Pontiac/GMC, Radio Shack, RV Ctr, Target, Toyota, Valvoline, bank, carwash, cinema, laundry, mall, tires, **S**...Mabel Murphy's Rest., Wal-Mart/drugs/24hr
50	rd 88, rd 52, to US 59, Elizabeth, to Fergus Falls, **N**...Texaco/diesel/mart/rest./24hr
38	rd 88, Rothsay, **S**...Amoco/diesel/café/24hr, tires
32	MN 108, rd 30, Lawndale, to Pelican Rapids, no facilities
24	MN 24, Barnesville, **N**...Mitzel's Rest., **1 mi S**...Amoco, Cenex/diesel, Dairy Queen
22	MN 9, Barnesville, **1 mi S**...Amoco, Cenex/diesel, Dairy Queen, Wagon Wheel Rest.
15	rd 10, Downer, no facilities
8mm	Buffalo River
6	MN 336, rd 11, to US 10, Dilworth, **N**...Citgo/diesel/café/24hr, camping, **S**...airport
5mm	Red River weigh sta eb
2	rd 52, Moorhead, **1-2 mi N**...HOSPITAL, Holiday/diesel/mart, Perkins, Guesthouse Motel, Travelodge Suites, KOA, Target
2mm	**Welcome Ctr eb, full(handicapped)facilities, info, phone, picnic tables, litter barrels, vending**
1b	20th St, Moorhead(from eb, no return), **1/2 mi S**...CHIROPRACTOR, Stop'n Go, bank, laundry
1a	US 75, Moorhead, **N**...Phillips 66, Spur Gas, Blimpie, Burger King, Domino's, Fry'n Pan Rest., KFC, Little Caesar's, Pizza Hut, Taco Bell(2mi), Regency Inn, Champion Auto Parts, MillEnd Fabrics, antiques, bank, **S**...CHIROPRACTOR, Amoco/mart/atm/24hr, Citgo/mart/atm, Chinese Rest., Hardee's, Village Inn Rest., Best Western, Motel 75, Super 8, SunMart/deli/24hr, Osco Drug, Pontiac/GMC, bank, carwash
0mm	Red River, Minnesota/North Dakota state line

E ↑ W

Alexandria Fergus Falls Moorhead

Minnesota Interstate 494/694(Twin Cities)

Exit #	Services

I-494/I-694 loops around Minneapolis/St Paul.

71   rd 31, Pilot Knob Rd, **N**...Courtyard, Fairfield Inn, **S**...Holiday Inn Select, LoneOak Café, Sidney's Café(1mi)

70   I-35E, N to St Paul, S to Albert Lea

69   MN 149, MN 55, Dodd Rd, **N**...Ziggy's Deli

67   US 52, MN S, Robert Rd, **1 mi N**...CHIROPRACTOR, MEDICAL CARE, Amoco, Holiday/mart, Total/mart, Acre's Rest., Arby's/Sbarro's, Arnold's Burgers, Bakers Square, Bridgeman's Rest., Burger King, Godfather's, KFC, Old Country Buffet, Pizza Hut, Taco Bell, White Castle, Aamco, Best Buy, CarX, Buick/Olds/Pontiac, Chevrolet, Cub Foods, Dodge, Ford/Lincoln/Mercury, Jo-Ann Fabrics, Kia, K-Mart, Mazda, Nissan, PrecisionTune, Rainbow Foods/24hr, Sam's Club, Target, TJ Maxx, Toyota, VW, auto parts, bank, outlet mall, **S**...CHIROPRACTOR, DENTIST, PDQ Gas/mart, cleaners

66   MN 3 N, MN 103 S, no facilities

65   7th Ave, 5th Ave, no facilities

64b a   MN 56, Concord St, **N**...Conoco/diesel/mart, Best Western Drovers, Ford Trucks, Goodyear, Peterbilt, **S**...gas/mart, Golden Steer Motel/rest.

63mm   Mississippi River

63c   Maxwell Ave, no facilities

   b a   US 10, US 61, to St Paul, Hastings, **S**...Amoco/mart/wash, SuperAmerica/mart, Boyd's Motel, Subway, Valvoline

62   Lake Rd, **N**...Country Inn Suites, **S**...SuperAmerica/gas

60   Valley Creek Rd, **N**...Burger King, Cracker Barrel, McDonald's, Pizza Hut, Subway, Hampton Inn, Goodyear, JiffyLube, hardware, **S**...DENTIST, EYECARE, CHIROPRACTOR, Amoco/mart/wash, SuperAmerica/diesel/LP/mart, Applebee's, Ciatti's Rest., Old Mexico Rest., Oriental Rest., Perkins, Pizza, Red Roof Inn, Best Buy, Kohl's, Rainbow Foods/24hr, Target, Walgreen, bank, cinema, cleaners

58b a   I-94, E to Madison, W to St Paul. **I-494 S begins/ends, I-694 N begins/ends**

57   rd 10, 10th St N, **E**...Wingate Inn, cinema, **W**...Tom Thumb Superette, Burger King, Chinese Rest., KFC, K-Mart, PetCo, Rainbow Foods/24hr, Tools&more, bank, cleaners, mall

55   MN 5, **W**...Amoco/diesel/mart/wash/lube, Holiday/gas, Subway, Menard's, st patrol

52b a   MN 36, N St Paul, to Stillwater, **W**...Fleet&Farm/auto/gas, Toyland

51   MN 120, **E**...DENTIST, CHIROPRACTOR, Amoco, Conoco/mart, SuperAmerica/diesel/mart, Best Subs, Taco Bell, Zapata's Café, cleaner

50   White Bear Ave, **E**...Gas4Less/mart/wash, SuperAmerica/mart, K-Mart/drugs, **W**...Amoco, Texaco/mart/Lube, Arby's, Applebee's, Bakers Square, Burger King, ChiChi's, Chinese Rest., Denny's, Dutch Rest., Godfather's, Hardee's/24hr, KFC, Old Country Buffet, Perkins/24hr, Red Lobster, Taco Bell, Wendy's, Best Western, Emerald Inn, Aamco, Best Buy, Goodyear, Jo-Ann Fabrics, Meineke Muffler, Midas Muffler, PrecisionTune, Sears/auto, Snyder's Drug, Tires+, Valvoline, Ziebart TidyCar, bank, mall

48   US 61, **E**...Citgo, KFC, Acura/Honda/Buick, Chrysler/Dodge/Jeep, Ford, Isuzu, Mercury/Lincoln, Saturn, Subaru, Van Sales, auto repair, **W**...HOSPITAL, VETERINARIAN, Chili's, Gulden's Rest., McDonald's, Olive Garden, Best Western, NorthernAire Motel, Toyota, Audi/Porsche, Venburg Tires

47   I-35E, N to Duluth

46   I-35E, US 10, S to St Paul

45   MN 49, Rice St, **N**...Phillips 66/mart/wash, Total/diesel/mart, Taco Bell, auto parts, cleaners, drugs, **S**...Ashland Gas, A&W, Burger King, Hardee's, Hoggsbreath Café, Taco John's

43b   Victoria St, no facilities

   a   Lexington Ave, **N**...Greenmill Rest., Hampton Inn, Hilton Garden, bank, **S**...CHIROPRACTOR, Amoco/wash, Conoco/mart, Sinclair, Total/diesel/mart, Blue Fox Dining, Burger King, Davanni's Pizza, Perkins, Wendy's, Emerald Inn, Ramada Inn, Super 8, Goodyear, Kennedy Transmissions, Target

42b   US 10 W(from wb), to Anoka, no facilities

   a   MN 51, Snelling Ave, **1 mi S**...Amoco, McDonald's, Country Inn Suites, Holiday Inn, Ramada

41b a   I-35W, S to Minneapolis, N to Duluth

40   Long Lake Rd, 10th St NW, no facilities

39   Silver Lake Rd, **N**...Amoco/mart, Sinclair, Los Bandijos Mexican, McDonald's, Ford, Snyder's Drug, U-Haul, hardware

38b a   MN 65, Central Ave, **N**...VETERINARIAN, Amoco, Sinclair, Subway, Shorewood Rest., **S**...CHIROPRACTOR, SuperAmerica/mart, Total/mart, Applebee's, Arby's, Chinese Rest., Cousins Subs, Denny's, Dunkin Donuts, Dutch Rest., Embers Rest., Flameburger Rest., Godfather's, Ground Round, Hardee's, La Casita Mexican, McDonald's, Mr Steak, Ponderosa, Taco Bell, White Castle, Best Western Kelly, Starlite Motel, JiffyLube, Midas Muffler, K-Mart/drugs, Target, Tires+, auto parts, bank, laundry

37   rd 47, University Ave, **N**...SuperAmerica/mart, Hardee's, McDonald's, **S**...Amoco/mart, Coastal/mart

36   E River Rd, no facilities

35mm   I-494 W begins/ends, I-694 E begins/ends

35c   MN 252, **N**...SuperAmerica/mart, motel

   b a   I-94 E to Minneapolis

E

W

Minnesota Interstate 494/694

34	to MN 100, Shingle Creek Pkwy, **N**...Amoco/wash, ChiChi's, Hardee's, Baymont Inn, Day's Inn, Holiday Inn, Park Inn, Super 8, JC Penney, **S**...Chinese Rest., Cracker Barrel, Perkins, Pizza Hut, Subs Etc, The Inn on the Farm, Drug Emporium, Target, Tires+, TJ Maxx, cinema
33	rd 152, Brooklyn Blvd, **N**...Holiday Gas/mart, Total/mart, Chevrolet, Dodge, Honda, Mazda, Olds, Pontiac, Pitstop Lube/CarX Muffler, Photo Shop, RV Ctr, auto parts, laundry, **S**...CHIROPRACTOR
31	rd 81, Lakeland Ave, **N**...SuperAmerica, Dairy Queen, Wagner's Drive-In, Wendy's, Northwest Inn Best Western
30	Boone Ave, **N**...Northland Inn
29b a	US 169, to Hopkins, Osseo, no facilities
28	rd 61, Hemlock Lane, **S**...Amoco/mart/wash, Perkins/24hr, Red Carpet Inn, bank
27	I-94 W to St Cloud, I-94/694 E to Minneapolis
26	rd 10, Bass Lake Rd, **E**...DENTIST, VETERINARIAN, Conoco/mart, Chinese Rest., McDonald's, New Mkt Foods, bank, carwash, cleaners, hardware, mall, **W**...Amoco/mart/wash, Sinclair/LP/U-Haul
23	rd 9, Rockford Rd, **E**...Amoco/mart, Texaco/mart/wash, Bakers Square, Chili's, Drug Emporium, Mercury/Lincoln, Radio Shack, Rainbow Foods/24hr, Target/drugs, TJ Maxx, Walgreen, **W**...Dairy Queen
22	MN 55, **E**...CHIROPRACTOR, Green Mill Rest., Best Western Kelly, Radisson, **W**...Perkins, Day's Inn, Holiday Inn, Goodyear/auto
21	rd 6, no facilities
20	Carlson Pkwy, no facilities
19b a	I-394 E, US 12 W, to Minneapolis, **1/2 mi W**...MEDICAL CARE, ChiChi's, Pizza Hut, Chevrolet, Mitsubishi, Nissan, **1 mi E off of I-394**...Applebee's, Bakers Square, Byerly's Rest., Chinese Rest., Dutch Rest., Godfather's, Uno Pizza, Wendy's, Best Buy, Dayton's, Ford, JC Penney, Jo-Ann Fabrics, Mercedes, Saab, Sears/auto, Target, cleaners, bank, mall
17b a	Minnetonka Blvd, no facilities
16b a	MN 7, no facilities
13	MN 62, rd 62, no facilities
11c	MN 5 W, no facilities
b a	US 169 S, US 212 W, **N**...Courtyard, Hampton Inn, Residence Inn, bank, **S**...Phillips 66, Hardee's, Cub Foods
10	US 169 N, to rd 18, no facilities
8	rd 28(from wb, no return), E Bush Lake Rd, no facilities
7b a	rd 34, Normandale Blvd, MN 100, **N**...Burger King, Chili's, Embers Rest., Radisson, Select Inn, Sofitel Hotel, carwash, **S**...Billabong Aussie Grill, Day's Inn, Holiday Inn, Seville Best Western
6b	France Ave, rd 17, HOSPITAL, **N**...Mobil, Cattle Co Rest., Fuddrucker's, Perkins, Best Western, Bradberry Suites, Hawthorn Suites, Wyndham Garden, **S**...Denny's, Lincoln Dell Rest., Joe Senser's Grill, Olive Garden, Hampton Inn, Ford, Mercedes, Nissan
a	Penn Ave, **N**...Citgo, Buick, Hyundai, Isuzu, **S**...Edward's Café, Starbucks, Subway, Bed Bath&Beyond, Gateway Computers, Chevrolet, Chrysler/Pymouth/Jeep, Dodge, Kohl's, Mailboxes Etc, Rainbow Foods, Target, TJ Maxx, ToysRUs, bank
5b a	I-35W, S to Albert Lea, N to Minneapolis
4b	Lyndale Ave, **N**...Amoco, Conoco/mart, Dairy Queen, Hampton Inn, BabysRUs, Best Buy, Borders, Honda, LandsEnd, Mitsubishi, PetsMart, Radio Shack, SportMart, Tires+, **S**...Phillips 66, Acura, Golf Galaxy, Lincoln/Mercury, Mazda, Subaru
a	MN 52, Nicollet Ave, **N**...SuperAmerica/diesel, Burger King, ChiChi's, Menard's, **S**...Total Gas, Texaco, Kwik Mart, Big Boy, McDonald's, Baymont Inn, Super 8, Home Depot, cleaners
3	Portland Ave, 12th Ave, **N**...Phillips 66/mart/wash, Sinclair/mart, PDQ Mart, Arby's, BBQ, Ground Round, AmericInn, **S**...Amoco/wash/24hr, Denny's, Outback Steaks, Subway, Comfort Inn/rest., Holiday Inn Express, Microtel, SuperValu Foods, Walgreen, Wal-Mart
2c b	MN 77, **N**...Amoco/mart(1mi), Motel 6, **S**...Amoco/mart, SuperAmerica/mart/wash, AmeriSuites, Best Western, Courtyard, Embassy Suites, Exel Inn, Marriott, Registry Hotel, Sheraton, Thunderbird Hotel, Nordstrom's, Sears, Mall of America
a	24th Ave, same as 2c b
1b	34th Ave, Nat Cemetary, **S**...Embassy Suites, Hilton, Holiday Inn
a	MN 5 E, **N**...airport
0mm	Minnesota River. I-494/I-694 loops around Minneapolis/St Paul.

E

W

Minneapolis

Mississippi Interstate 10

Exit #	Services
77mm	Mississippi/Alabama state line, weigh sta wb
76.5mm	weigh sta wb
75	Franklin Creek Rd, no facilities
74.5mm	**Welcome Ctr wb, full(handicapped)facilities, phone, picnic tables, litter barrels, petwalk, RV dump, weigh sta eb**
74mm	Escatawpa River
69	MS 63, to E Moss Point, **N**...Texaco/Domino's/diesel/mart/24hr, Best Inn Suites, Econolodge, Ramada Ltd, bank, **S**...HOSPITAL, Amoco, Chevron/diesel/mart, Cone/diesel/mart, Exxon/Subway/mart/24hr, Burger King, Cracker Barrel, Hardee's, JJ's Rest., McDonald's, Quincy's, Shoney's(1mi), Waffle House, Wendy's, Best Western, Comfort Inn, Day's Inn, Hampton Inn, Holiday Inn Express, La Font Inn, Shular Inn
68	MS 613, to Moss Point, Pascagoula, **N**...Citgo/mart, Conoco/diesel/mart, Texaco/KFC/diesel/mart, Super 8, Riverbend Camping, **S**...HOSPITAL, Chevron/diesel/mart
64mm	Pascagoula River
63.5mm	**rest area both lanes, full(handicapped)facilities, phone, picnic tables, litter barrels, petwalk, RV dump, 24hr security**
61	to Gautier, **N**...MS Nat Golf Course, RV camping, **1-3 mi S**...BP/diesel/mart, Chevron, Citgo, Conoco/mart, Fireside Catfish Café, Food Court, Hardee's, McDonald's, Pizza Hut, Quincy's, Suburban Lodge, RV camping
57	MS 57, to Vancleave, **N**...KOA(1/2mi), Bluff Creek Camping(7mi), **S**...Exxon, Texaco/diesel, Suburban Lodge
50	MS 609 S, Ocean Springs, **N**...Chevron/diesel/mart, Texaco/Domino's/mart, Waffle House, Comfort Inn, Country Inn Suites, Ramada Ltd, Super 8, Martin Lake Camping, **S**...Amoco/McDonald's/mart, BP/mart, Speedway/Blimpie/diesel/mart, Burger King, Denny's, Hardee's, Krystal, Waffle House, Day's Inn, Hampton Inn, Holiday Inn Express, Economy Inn, Sleep Inn, Nat Seashore
46b a	I-110, MS 15 N, to Biloxi, **N**...Chevron/Domino's/mart, Texaco/mart, McDonald's, Waffle House, Wingate Inn, Travelodge, Wal-Mart SuperCtr/24hr, **S**...HOSPITAL, RV camping, to beaches, **S off of I-110, exit 2**...Amoco, Shell, Texaco, Hardee's, McDonald's, Pizza Hut, Subway, Taco Bell, Waffle House, Suburban Lodge
44	Cedar Lake Rd, to Biloxi, **N**...Williams/Main St Café/diesel/mart/2hr, hardware, **S**...**MEDICAL CARE, VETERINARIAN,** Citgo, Exxon/diesel/mart/atm, Shell/diesel/mart, McDonald's, AutoValue Parts, to Jeff Davis Shrine, Biloxi Nat Cem
41	MS 67 N, to Woolmarket, **N**...BP/mart, Chevron/diesel/mart, golf(6mi), **S**...Mazalea RV Park
39.5mm	Biloxi River
38	Lorraine-Cowan Rd, **N**...Exxon/Subway/mart, Texaco/diesel/mart, Domino's, **4-8 mi S**...HOSPITAL, Citgo, Food Court, KFC, Waffle House, Coast Motel, Edgewater Inn, Holiday Inn Express, Oceanview Motel, Ramada Ltd, Baywood Camping, Fox's Camping, to beaches
34b a	US 49, to Gulfport, **N**...Amoco/mart/wash, Speedway/diesel/mart, Texaco/diesel/mart/atm, Chili's, ChuckeCheese, Hardee's, Papa John's, Waffle House, Academy Sports, Albertson's, Barnes&Noble, Circuit City, Eckerd, Goody's, Honda, Office Depot, Old Navy, PetsMart, Pier 1, Service Merchandise, Shoe Carnival, TJ Maxx, cinema, **S**...HOSPITAL, Amoco/mart, BP, Citgo/diesel/mart, FastLane/diesel, RacTrac/mart/atm, Texaco/diesel/mart/wash, Applebee's, Arby's/24hr, Burger King, Fazoli's, Food Court, KFC, Krispy Kreme Donuts, McAlister's, McDonald's, Montana's Seafood/ribs, Perkins, Piccadilly's, Rowdy's Roast Beef/burgers, Schlotsky's, Sonic, Subway, Toucan's Mexican, Waffle House, Wendy's, Best Western/rest., Comfort Inn, Day's Inn, Fairfield Inn, Hampton Inn, Holiday Inn, Holiday Inn Express, Motel 6, Shoney's Inn/rest., Studio Inn, Cottman Transmissions, Ford, Home Depot, Kia, Meineke Muffler, Michael's, Nissan, OfficeMax, Prime Outlets/famous brands, QuickLube, Sam's Club, Wal-Mart SuperCtr/gas/24hr, airport, bank, drugs, RV park
31	Canal Rd, to Gulfport, **N**...Magnolia Plantation Hotel, **S**...BP/diesel, Flying J/Conoco/diesel/LP/rest./24hr, Texaco/diesel/mart/24hr, Crystal Inn, Econolodge
28	to Long Beach, **S**...Chevron/mart, Citgo/mart, Red Creek Colonial B&B, RV camping
27mm	Wolf River
24	Menge Ave, **N**...Amoco/diesel/mart/rest., **S**...Texaco/Stuckey's, golf, to beaches
20	to De Lisle, to Pass Christian, **N**...Shell/mart, Lillie's Burger/Shakes, flea mkt
16	Diamondhead, **N**...EYECARE, VETERINARIAN, BP/Domino's/mart, Chevron/Blimpie/mart/24hr, Texaco, Burger King, Dairy Queen, DragonHouse Chinese, Jungle Rest., Waffle House, Diamondhead Inn, Ramada Inn, DiamondHead Foods, Jerry's/repair/lube, bank, **S**...Texaco/Seafood Rest./diesel/mart, Subway, Comfort Inn, Ace Hardware
15mm	Jourdan River
13	MS 43, MS 603, to Kiln, Bay Saint Louis, **N**...McLeod SP, **S**...Chevron/mart, Conoco/mart/motel, Exxon/Subway/Chester Fried/diesel/mart/atm, **3-7 mi S**...HOSPITAL, McDonald's, BayHouse Inn, Holiday Inn, Key West Inn, Studio Inn, Waveland Resort Inn, Bay Marina RV Park, Casino Magic RV Park, KOA
10mm	parking area eb, litter barrels, no restroom facilities
2	MS 607, to Waveland, NASA Test Site, **S**...**Welcome Ctr both lanes, full(handicapped)facilities, phone, picnic tables, litter barrels, RV dump, petwalk, 24hr security,** Holiday Inn(14mi), Key West Inn(14mi), Studio Inn(14mi), to Buccaneer SP, camping, to beaches
1mm	weigh sta both lanes
0mm	Pearl River, Mississippi/Louisiana state line

Mississippi Interstate 20

Exit #	Services

I-20 W and I-59 S run together to Meridian

172mm Mississippi/Alabama state line

170mm weigh sta both lanes

169 Kewanee, **N**...Dixie Gas/diesel, **S**...Kewanee 1-Stop/diesel/rest., Red Apple Gas/mart

165 Toomsuba, **N**...Shell/mart, Texaco/diesel/Travla Rest./gifts, **S**...FuelMart/Arby's/diesel, KOA(2mi)

164mm **Welcome Ctr wb, full(handicapped)facilities, phone, vending, picnic tables, litter barrels, petwalk, RV dump, 24hr security**

160 to Russell, **N**...TA/BP/diesel/rest./24hr, Nanabe Cr Camping(1mi), **S**...Amoco/diesel/mart/rest.

157b a US 45, to Macon, Quitman, no facilities

154b a MS 19 S, MS 39 N, Meridian, **N**...Amoco/diesel/mart, BP/diesel/mart/wash, Shell/mart/wash, Texaco/Domino's/mart, Applebee's, Backyard Burgers, China King, Cracker Barrel, Waffle House, Best Western, Day's Inn, Econolodge, Hampton Inn, Holiday Inn, Howard Johnson, Relax Inn, Rodeway Inn, Super 8, Western Motel, Dodge/Subaru/Kia, Lincoln/Mercury, Mitsubishi, Olds/Cadillac/Pontiac, U-Haul, antiques, RV camping, **S**...Chevron/mart/wash, Conoco/diesel/mart, Luby's, McDonald's, O'Charley's, Outback Steaks, Popeye's, Red Lobster, Ryan's, Taco Bell, Comfort Inn, Jameson Inn, Pinehaven Motel, Scottish Inn, Books-A-Million, Dillard's, ExpressLube, Goody's, JC Penney, K-Mart, Lenscrafters, McRae's, OfficeMax, Old Navy, Pier 1, Sears/auto, TJ Maxx, Wal-Mart, bank, cinema, mall, same as 153

153 MS 145 S, 22nd Ave, Meridian, **N**...HOSPITAL, Amoco/mart, BP/mart, Arby's, Barnhill's Buffet, Burger King, Capt D's, El Chico, Hardee's, KFC, McDonald's, Pizza Hut, Quincy's, Shoney's, Sub-Wendy's, Western Sizzlin, Super Inn, Firestone, FoodMax, Ford/Nissan, Goodyear/auto, $General, Ford, Sack&Save Foods, bank, cinema, tires **S**...Chevron/diesel/mart, Conoco/mart, Exxon/diesel/mart, Shell/service, Depot Rest., Waffle House, Astro Motel, Baymont Inn, Best Western, Budget 8 Motel, Econolodge, Holiday Inn Express, Motel 6, Sleep Inn, Buick, Chevrolet/Olds/Cadillac, Chrysler/Plymouth/Jeep, Honda, Suzuki

152 29th Ave, 31st Ave, Meridian, **N**...Amoco/diesel/mart, Chevron/ChesterFried/diesel/mart/24hr, Ramada Ltd, **S**...Royal Inn

151 49th Ave, Valley Rd, **N**...tires, **S**...stockyards

150 US 11 S, MS 19 N, Meridian, **N**...BP/diesel, Chevron, McDonald's, Popeye's, Okatibbee Lake, RV camping, **S**...Amoco/diesel/mart, Subway, airport

131[149] I-59 S, to Hattiesburg. **I-20 E and I-59 N run together.**

129 US 80 W, Lost Gap, **S**...Spaceway Gas/Grill King/mart

121 Chunky, no facilities

119mm Chunky River

115 MS 503, Hickory, no facilities

109 MS 15, Newton, **N**...Texaco/Wendy's/diesel/mart/atm/24hr, Bo-Ro Rest., Thrifty Inn, lube, **S**...HOSPITAL, BP/mart, Chevron/diesel/mart/24hr, Conoco/diesel/mart, Hardee's, KFC/Taco Bell, McDonald's, Pizza Hut, Sonic, Subway, Day's Inn, $General, Piggly Wiggly, Wal-Mart

100 US 80, Lake, Lawrence, **N**...BP/diesel/mart/rest.

96 Lake, no facilities

95mm Bienville Nat Forest, eastern boundary

90mm **rest area eb, full(handicapped)facilities, phone, picnic tables, litter barrels, petwalk, RV dump, 24hr security**

88 MS 35, Forest, **N**...HOSPITAL, Amoco/diesel/mart, BP/diesel/mart, Shell/mart, Dairy Queen, KFC, McDonald's, Pizza Hut, Wendy's, Best Western/rest., Comfort Inn, Day's Inn, Scott Motel, Chevrolet/Olds, Honda, Wal-Mart, **S**...Chevron/Stuckey's/diesel/mart/24hr, Santa Fe Steaks

80 MS 481, Morton, no facilities

77 MS 13, Morton, **N**...HOSPITAL, Phillips 66/diesel/mart/rest., to Roosevelt SP, RV camping

76mm Bienville NF, western boundary

75mm **rest area wb, full(handicapped)facilities, phone, picnic tables, litter barrels, petwalk, RV dump, 24hr security**

68 MS 43, Pelahatchie, **N**...Chevron/Subway/diesel/mart/atm/24hr, Conoco/diesel/rest./RV dump/24hr, Little Red Smokehouse, RV camping, **S**...BP/diesel/mart

59 US 80, E Brandon, **2 mi S**...Conoco/mart

56 US 80, Brandon, **N**...Texaco, BBQ, Burger King, CiCi's, Krystal, McDonald's, Taco Bell, AutoZone, cinema, **S**...VETERINARIAN, Amoco, BP/mart/wash, Chevron, Exxon, Shell/diesel, Texaco/Blimpie/diesel/mart/wash, Azteca Mexican, Dairy Queen, Dark Horse Rest., Little Joe's Chicken, Sonic, Day's Inn, Delta Muffler/brake, XpressLube/wash, auto parts/repair, to Ross Barnett Reservoir

54 Crossgates Blvd, W Brandon, **N**...HOSPITAL, Amoco/mart, BP/mart, Exxon, Phillips 66, Burger King, Domino's, KFC, Mazzio's Pizza, Papa John's, Pizza Hut, Popeye's, Subway, Waffle House, Wendy's, Ridgeland Inn, Batteries+, Big Lots, Buick/GMC, Chevrolet, Eckerd, Firestone/auto, Ford, Goodyear/auto, JetLube, Kroger, Wal-Mart

E

W

Meridian

Brandon

Mississippi Interstate 20

52	MS 475, **N**...Chevron, Conoco/diesel, Krystal(3mi), Waffle House, Wendy's(3mi), Ramada Ltd, Super 8, Peterbilt, to Jackson Airport
48	MS 468, Pearl, **N**...CHIROPRACTOR, BP/mart/wash, Conoco/mart, Shell/diesel/mart, Speedway/diesel/mart, Arby's, Bumpers Drive-In, Burger King, Cracker Barrel, Domino's, El Charro Mexican, KFC, McDonald's, O'Charley's, Popeye's, Ryan's, Schlotsky's, Shoney's, Sonic, Waffle House, Best Western, Comfort Inn, Econolodge, Fairfield Inn, Hampton Inn, Holiday Inn Express, Jameson Inn, Motel 6, E-Z Lube, bank, cinema, transmissions, **S**...Chevron/diesel/mart/atm/24hr, Texaco/diesel/mart/24hr, Day's Inn
47b a	US 49 S, Flowood, **N**...BP, Flying J/Conoco/diesel/LP/rest./24hr, Texaco/mart(1mi), Williams/diesel/mart, Western Sizzlin, Travelodge, RV Ctr(1mi), **S**...Pilot/Krystal/Subway/diesel/mart/24hr, Dairy Queen, Waffle House, Best Inn(3mi), Day's Inn(4mi), Ford/Freightliner/Mercedes/Ryder/Kenworth, tires, Magnolia RV Camping
46	I-55 N, to Memphis, no facilities
45b	US 51, State St, to downtown, **N**...Chevron/mart, ExpressLube, **S**...Speedway/Hardee's/diesel/mart, Nissan
a	Gallatin St(from wb), to downtown, **N**...VETERINARIAN, Dixie Gas, Petro/diesel/rest., Texaco, Blue Beacon, **S**...Coachlight Rest., Mazda, Toyota
44	I-55 S, to New Orleans, no facilities
43b a	Terry Rd, **N**...Shell/mart, Kim's Seafood Rest., Krystal, Tarrymore Motel, Big A Parts, RV Ctr, **S**...La Quinta
42b a	Ellis Ave, Belvidere, **N**...BP/mart/wash, Chevron/diesel, Conoco, Shell/mart/wash, Burger King, Capt D's, Denny's, McDonald's, Popeye's, Waffle House, Wendy's, Best Western, Comfort Inn, Day's Inn, Econolodge, Holiday Inn, Ramada, Relax Inn, Scottish Inn, Stonewall Jackson Motel, AutoZone, Chrysler/Plymouth/Dodge, Family$, Firestone, Radio Shack, U-Haul, bank, transmissions, zoo, **S**...MEDICAL CARE, CHIROPRACTOR, Amoco/mart, Conoco/diesel/mart, Exxon/diesel/mart, Pump'n Save, Dairy Queen, Pizza Hut, RapidLube, Sack&Save Foods, Wal-Mart
41	I-220 N, US 49 N, to Jackson
40b a	MS 18 W, Robinson Rd, **N**...Exxon, Phillips 66, Shell/diesel, Arby's, El Chico, Krystal, Mazzio's Pizza, McDonald's, Piccadilly's, Pizza Hut, Popeye's, Shoney's, Wendy's, CarCare, Home Depot, JetLube, Office Depot, Pier 1, Sears/auto, Service Merchandise, **S**...HOSPITAL, Conoco/mart, Mack's Gas/diesel, Phillips 66, Waffle House, Comfort Inn, Wal-Mart SuperCtr/gas/24hr
36	Springridge Rd, Clinton, **N**...MEDICAL CARE, DENTIST, BP/Burger King/mart, Orbit Gas/mart, Shell/mart, Backyard Burger, Bumpers Drive-In, Capt D's, China Garden, Dairy Queen, Mazzio's Pizza, McDonald's, Papa John's, Smoothie King, Subway, Taco Bell, Waffle House, Wendy's, Clinton Inn, Day's Inn, Discount Parts, GNC, Kroger/drugs, Pennzoil, Radio Shack, Walgreen, Wal-Mart/drugs/auto, bank, carwash, **S**...Exxon, Texaco/Blimpie/diesel/mart/wash, Applebee's, Fazoli's, Ned's Diner, Pizza Hut, Popeye's, Shoney's, Comfort Inn, Hampton Inn, Holiday Inn Express, SpringRidge RV Park, cinema
35	US 80 E, Clinton, **N**...Chevron/diesel/mart, Phillips 66/diesel/mart/rest., Texaco/diesel/mart, **S**...Conoco/diesel/rest., Eagle Ridge RV Park
34	Natchez Trace Pkwy, no facilities
27	Bolton, **N**...Chevron/mart, **S**...BP/diesel/mart
19	MS 22, Edwards, Flora, **N**...Askew Landing RV Camping, **S**...Amoco/diesel/mart, Texaco/Dairy Queen/Stuckey's, Relax Inn
17mm	Big Black River
15	Flowers, no facilities
11	Bovina, **N**...Texaco/diesel/mart/24hr, golf(1mi), RV camping
10mm	weigh sta wb
8mm	weigh sta eb
6.5mm	parking area eb
5b a	US 61, MS 27 S, **N**...Conoco/diesel/mart, Shell/diesel/mart, Sonic, **S**...Texaco/Domino's/diesel/mart/wash, Rowdy's Rest., Comfort Inn, Scottish Inn, RV camping, same as 4a
4b a	Clay St, **N**...HOSPITAL, Chevron/mart, Sherman's Rest./museum, Waffle House, Hampton Inn, Holiday Inn, Quality Inn, Super 8, KOA, to Vicksburg NP, **S**...Texaco/Domino's/diesel/mart/wash, China Buffet, Cracker Barrel, Dock Seafood, Pizza Inn, McAlister's, Beechwood Inn/rest., Comfort Inn, Jameson Inn, Scottish Inn, Chrysler/Jeep/Toyota, $General, Lincoln/Mercury, Outlet Mall/famous brands/deli, same as 5
3	Indiana Ave, **N**...Texaco/Blimpie/diesel/mart/atm, Krystal, McDonald's, Waffle House, Best Western, Deluxe Inn, Chevrolet/Olds, Ford/Lincoln/Mercury, Honda, IGA Foods, Mazda, Rite Aid, **S**...Shell/mart/wash, KFC, Ponderosa, Quality Inn, Buick/GMC/Pontica/Subaru, Family$, Sack&Save Foods, Sears Parts, hardware
1c	Halls Ferry Rd, **N**...HOSPITAL, Chevron/mart/24hr, Exxon/mart, Burger King, Guesthouse Inn, **S**...FastLane/diesel/mart/wash, Shell, Baskin-Robbins, Capt D's, Dairy Queen, El Sombrero Mexican, Garfield's Rest., Goldie's BBQ, Hardee's, Little Caesar's, Pizza Hut, Popeye's, Ryan's, Shoney's, Subway, Taco Bell, Taco Casa, TCBY, Wendy's, Day's Inn, Econolodge, Fairfield Inn, Dillard's, Discount Parts, JC Penney, Kroger/drugs/24hr, McRae's, Super K-Mart, USPO, bank, mall
b	US 61 S, **S**...Amoco/mart, Chevron, Mitsubishi, OfficeMax, Wal-Mart SuperCtr/24hr, same as 1c
a	Washington St, Vicksburg, **N**...**Welcome Ctr both lanes, full(handicapped)facilities, phone,** Exxon/mart/atm, Shell/Subway/diesel/mart, BBQ, AmeriStar Hotel, Isle of Capri RV Park, carwash, casino, transmissions, **S**...Waffle House, Ramada Ltd, Ridgeland Inn
0mm	Mississippi River, Mississippi/Louisiana state line

E

↕

W

Mississippi Interstate 55

Exit #	Services
291.5mm	Mississippi/Tennessee state line
291	State Line Rd, Southaven, **E**...EYECARE, 76/Circle K, BBQ, Best Pizza, Burger King, La Hacienda Mexican, McDonald's, Pizza Hut, Subway, Waffle House, Best Western, Comfort Inn, Holiday Inn Express, Quality Inn, Firestone/auto, Goodyear/auto, K-Mart/auto/24hr, Kroger/deli/24hr, MegaMkt Foods, Walgreen, **W**...Amoco/24hr, Exxon/diesel/repair, Capt D's, Dale's Rest., El Porton Mexican, KFC, Mrs Winner's, Rally's, Sonic, Taco Bell, Wendy's, Dalton Carpet Outlet, Fred's Drug, Oliver Drug, Rite Aid, Seessel's Foods, Xpert Tune, bank
289	MS 302, to US 51, Horn Lake, **E**...HOSPITAL, 76/Circle K, Shell/mart, Back-yard Burger, Burger King, Chick-fil-A, Chili's, Danver's Rest., Fazoli's, Heavenly Ham, IHOP, Krystal/24hr, McDonald's, O'Charley's, Outback Steaks, Schlotsky's, Steak'n Shake, Fairfield Inn, Hampton Inn, Chevrolet, Chrysler/Plymouth/Jeep, $Tree, Ford, Hancock Fabrics, HomePlace, Midas Muffler, NAPA/repair, Office Depot, PetCo, Pontiac/GMC, Valvoline, Wal-Mart SuperCtr/24hr, cinema, **W**...Amoco/mart, Phillips 66/diesel/mart, Shell/mart, Applebee's, Arby's, Cracker Barrel, Great Wall Chinese, Hardee's, KFC, McDonald's, Mrs Winner's, Papa John's, Pizza Hut, Popeye's, Roadhouse Grill, Ryan's, Taco Bell, Waffle House, Wendy's, Day's Inn, Drury Inn, Motel 6, Ramada Ltd, Sleep Inn, Super 8, Express Lube, Family$, Gateway Tire/repair, Home Depot, Kroger/deli, OfficeMax, Piggly Wiggly, Super D Drug, Target, Walgreen
287	Church Rd, **E**...Texaco/ChesterFried/diesel/mart, **W**...Shell/DQ/diesel/mart, Waffle House
285mm	weigh sta both lanes
284	to US 51, Nesbit Rd, **E**...Chevrolet/Olds/GMC, **W**...Citgo/mart, Nesbit 1-Stop/gaas, Happy Daze Dairybar, USPO
280	MS 304, US 51, Hernando, **E**...BP, Day's Inn, Hernando Inn, **W**...Chevron/Subway/diesel/mart, Coastal/Taco Bell/mart, Shell/diesel/mart/wash, BBQ, Church's Chicken, McDonald's, Pizza Hut, Sonic, Wendy's, Super 8, Delta Muffler, Kroger, NAPA, Piggly Wiggly, XpressLube, auto repair, bank, tires, Memphis South Camping(2mi), to Arkabutla Lake
279mm	**Welcome Ctr sb, full(handicapped)facilities, phone, picnic tables, litter barrels, petwalk, RV dump, 24hr security**
276mm	**rest area nb, full(handicapped)facilities, phone, picnic tables, litter barrels, petwalk, RV dump, 24hr security**
273mm	Coldwater River
271	MS 306, Coldwater, **W**...Amoco, **1-2 mi W**...BP, Chevron, Citgo, Exxon/mart/24hr, Lake Arkabutla, RV camping
265	MS 4, Senatobia, **W**...HOSPITAL, BP/diesel/mart/atm, Comet/mart/wash, Exxon/mart, FuelMart/diesel, Shell/diesel/mart, Texaco/diesel/mart/atm, BBQ, Country Kitchen, Domino's, Hardee's, KFC, McDonald's, Pizza Hut, Sonic, Subway, Taco Bell, Wendy's, Western Sizzlin, Comfort Inn, Motel 6, CarQuest, $General, Fred's Drug, Goodyear/auto, NAPA, Piggly Wiggly, Pontiac/Buick/GMC, USPO, bank, cinema, laundry
257	MS 310, Como, **E**...N Sardis Lake
252	MS 315, Sardis, **E**...Amoco/diesel/mart/atm, Chevron/diesel/mart/atm, Exxon/mart, BBQ, Lake Inn, Super 8, NAPA, to Kyle SP, Sardis Dam, RV camping, **W**...HOSPITAL, BP/diesel/mart, Texaco/diesel/mart, Sonic, Best Western/rest., $General, Fred's Drug, Piggly Wiggly
246	MS 35, N Batesville, **E**...to Sardis Lake, **W**...Conoco/mart, Texaco/diesel/mart/deli/24hr
243b a	MS 6, to Batesville, **E**...Amoco/diesel/mart/atm, Texaco/diesel/mart/atm, Target, TrueValue, to Sardis Lake, U of MS, **W**...HOSPITAL, DENTIST, EYECARE, Amoco/KrispyKreme/diesel/mart, Chevron/diesel/mart, Exxon/diesel/mart, FastLane/diesel/mart/atm, Phillips 66/diesel/mart, Shell/BBQ/diesel/mart/atm, Burger King, Capt D's, Cracker Barrel, Dairy Queen, Domino's, El Charro Mexican, Hardee's, KFC, McDonald's, Pizza Hut, Popeye's, Sonic, Subway, Taco Bell, Wendy's, Western Sizzlin, AmeriHost, Comfort Inn, Day's Inn, Hampton Inn, Ramada Ltd, Skyline Motel, Advance Parts, AutoZone, $General, Factory Stores/famous brands, Family$, Kroger, Radio Shack, Wal-Mart SuperCtr/24hr, XpressLube, bank, cleaners
240mm	**rest area both lanes, full(handicapped)facilities, phone, picnic tables, litter barrels, petwalk, RV dump, 24hr security**
237	to US 51, Courtland, **E**...Pure/diesel/mart/rest.
233	to Enid Dam, **E**...to Enid Lake, RV camping, **W**...Benson's Gas/groceries
227	MS 32, Oakland, **E**...to Cossar SP, RV camping, **W**...Chevron/diesel/rest./mart
220	MS 330, Tillatoba, **E**...Conoco/diesel, All American Rest.
211	MS 7 N, to Coffeeville, **E**...Oxbow RV camping, **W**...Texaco/ChesterFried/diesel/mart
208	Papermill Rd, **E**...Grenada Airport
206	MS 8, MS 7 S, to Grenada, **E**...HOSPITAL, EYECARE, CHIROPRACTOR, Exxon/mart/24hr, RaceTrac/mart, Shell/pizza/diesel/mart, Texaco/mart, BBQ/Steaks, Burger King, Domino's, Fiori's Rest., Jake&Rip's Café, La Cabana Mexican, McAlister's Deli, McDonald's, Pizza Hut, Pizza Inn, RagTime Grill, Shoney's, Subway, Taco

N

S

Mississippi Interstate 55

Bell, TCBY, Wendy's, Western Sizzlin, Best Western/rest., Comfort Inn, Day's Inn, Hampton Inn, Holiday Inn/rest., Jameson Inn, Ramada Ltd, AutoZone, Cato's, Chrysler/Plymouth/Dodge, $General, Ford/Lincoln/Mercury, GNC, Piggly Wiggly/deli, Toyota, USPO, Wal-Mart SuperCtr/24hr, bank, to Grenada Lake/RV camping, W...Exxon/ HuddleHouse/mart/atm, Texaco, Country Inn Suites, Hilltop Inn

204mm	parking area sb, phone, litter barrels
202mm	parking area nb, phone, litter barrels
199	S Grenada, to Camp McCain
195	MS 404, Duck Hill, **E**...to Camp McCain
185	US 82, Winona, **E**...HOSPITAL, Exxon/mart/24hr, Shell/mart, Texaco/Baskin-Robbins/diesel/mart, KFC, McDonald's, Pizza Hut, Budget Inn, Hitching Post Motel, Magnolia Lodge, Relax Inn, **W**...Amoco/diesel/mart
174	MS 35, MS 430, Vaiden, **E**...BP, Chevron/35/55/diesel/motel/24hr, Comet Gas, Shell/diesel/mart, KOA, NAPA, **W**...Texaco/Stuckey's/24hr
173mm	**rest area sb, full(handicapped)facilities, phone, picnic tables, litter barrels, petwalk, RV dump, 24hr security**
164	to West, **W**...West Trkstp/diesel/mart
163.5mm	**rest area nb, full(handicapped)facilities, phone, picnic tables, litter barrels, petwalk, RV dump, 24hr security**
156	MS 12, Durant, **E**...HOSPITAL, Texaco/diesel/mart/24hr, Durant Motel/rest., Super 8
150	**E**...Holmes Co SP, RV camping
146	MS 14, Goodman, **W**...to Little Red Schoolhouse
144	MS 17, to Pickens, **E**...Texaco/diesel/24hr, HomePlace Rest., J's Deli, **W**...BP/diesel/rest., MGM Motel/rest., to Little Red Schoolhouse
139	MS 432, to Pickens, no facilities
133	Vaughan, **E**...to Casey Jones Museum
128mm	Big Black River
124	MS 16, to N Canton, no facilities
121mm	**rest area sb, picnic tables, litter barrels, no restrooms**
119	MS 22, to MS 16 E, Canton, **E**...HOSPITAL, Amoco/Subway/diesel/mart, BP/mart, Chevron/Nancy'sRest./mart/ 24hr, Exxon/Taco Bell/diesel/mart, Shell/mart, Texaco/Domino's/diesel/mart/atm, McDonald's, Pizza Hut, Popeye's, Two Rivers Steaks, Wendy's, Best Western, Budget Inn(3mi), Comfort Inn, Day's Inn, Econolodge, Holiday Inn Express, Ford/Mercury, to Ross Barnett Reservoir, **W**...Love's/Arby's/diesel/mart/24hr
117mm	**rest area nb, picnic tables, litter barrels, no restrooms**
112	US 51, Gluckstadt, **E**...Amoco/ChesterFried/diesel/mart, Exxon/Krystal/diesel/mart/atm, Super 8
108	MS 463, Madison, **E**...Phillips 66/mart, Shell/diesel/mart, Texaco/Domino's/diesel/mart/wash, Burger King, **2 mi** **E**...McAlister's Deli, McDonald's, Popeye's, Jitney Jungle/deli
105b	Old Agency Rd, **E**...Chevron/diesel/mart, Pontiac/Buick/GMC
a	Natchez Trace Pkwy, no facilities
104	I-220, to W Jackson
103	County Line Rd, **E**...BP/mart/wash, Exxon/diesel/mart, Speedway, Applebee's, Burger King, Chick-fil-A, Cuco's Mexican, Fuddrucker's, Grady's Grill, Hardee's, HoneyBaked Ham, KFC, Krispy Kreme, Mazzio's Pizza, McDonald's, Popeye's, Roadhouse Grill, Ralph&Kacoo's Seafood, Ruby Tuesday, Wendy's, Cabot Lodge, Hilton Garden, Red Roof Inn, Shoney's Inn/rest., Subway, Taco Bell, Day's Inn, Acura, Dillard's, Goodyear, HomeFest, Isuzu, Mazda, McRae's, Midas Muffler, MailBoxes Etc, Sam's Club, Service Merchandise, TJ Maxx, Wal-Mart, mall, to Barnett Reservoir, bank, **W**...EYECARE, Olive Garden, Red Lobster, Subway, Comfort Suites, Motel 6, Drugs For Less, Home Depot, Office Depot, PetsMart, Target, West Marine
102b	Beasley Rd, Adkins Blvd, **E**...Chevron, Cracker Barrel, Highlands Rest., LoneStar Steaks, OutBack Steaks, La Quinta, Ramada Inn, BMW, Chrysler/Plymouth/Jeep, Ford, Lincoln/Mercury, Gateway Computers, Nissan, Olds, Toyota, **W**...McDonald's, Fairfield Inn, Jameson Inn, Suburban Lodge, Travelodge, Super K-Mart/24hr, frontage rds access 102a
a	Briarwood, **E**...Jitney Premier/deli, Steam Room Grill, Extended Stay America, **W**...Capt D's, Chili's, ChuckeCheese, El Chico's, Good Eats Grill, Perkins, Popeye's, Red Lobster, Steak&Ale, Best Suites, Comfort Inn, Hampton Inn, Chevrolet, Dodge, Jaguar/Saab, Mercedes
100	North Side Dr W, **E**...CHIROPRACTOR, VETERINARIAN, BP/mart/wash, Chevron, Sprint Gas, Burger King, Dunkin Donuts, McAlister's Deli, McDonald's, Olde Time Deli/bakery, Papa John's, Piccadilly's, Shoney's, SteakOut, Subway, Western Sizzlin, AutoZone, BooksAMillion, Eckerd, Goodyear/auto, Kroger, Office Depot, Pak&Mail, ProGolf, QuikLube, SteinMart, XpressLube, bank, **W**...EYECARE, Amoco/mart/24hr, Exxon/mart/wash, Shell/mart, Bennigan's, Domino's, IHOP/24hr, Pizza Hut, Waffle House, Holiday Inn, Knight's Inn, Super 8, laundry
99	Meadowbrook Rd, Northside Dr E(from nb)
98c b	MS 25 N, Lakeland Dr, **E**...CHIROPRACTOR, Texaco, Parkside Inn, LaFleur's Bluff SP, **W**...HOSPITAL, museum, airport
a	Woodrow Wilson Dr, downtown
96c	Fortification St, **E**...Residence Inn, **W**...HOSPITAL, Bellhaven College
b	High St, Jackson, downtown, **E**...Chevrolet, Infiniti, Lexus, Saturn, **W**...HOSPITAL, Chevron/mart, Shell/mart, Texaco/Blimpie/diesel/mart, Burger King, Dairy Queen, Dennery's Rest., Dunkin Donuts, Popeye's, Shoney's, Taco Bell, Waffle House, Wendy's, Coliseum Hotel, Hampton Inn, Holiday Inn Express, Red Roof, Wilson Inn, museum, st capitol

N

S

Mississippi Interstate 55

a	Pearl St, Jackson, downtown, **W**...access to same as 96b
94	(46 from nb), I-20 E, to Meridian, US 49 S
45b[I-20]	US 51, State St, to downtown, **N**...Chevron/mart, XpressLube, **S**...Nissan
a	Gallatin St(from sb), to downtown, **S**...Speedway/diesel, Daewoo, Mazda, Toyota
92c	(44 from sb), I-20 W, to Vicksburg, US 49 N
b	US 51 N, State St, Gallatin St
a	McDowell Rd, **E**...Petro/Exxon/diesel, Speedway/diesel/mart, Hardee's, **W**...Dixie Gas/deli, Shell/mart/wash, Texaco/diesel/mart, CiCi's, Waffle House, Wendy's, HoJo Inn, La Quinta, Super 8
90b	Daniel Lake Blvd(from sb), **W**...Shell/atm
a	Savanna St, **E**...Rodeway Inn, **W**...Shell, Spur/mart, BoDon's Catfish, Save Inn RV camping
88	Elton Rd, **W**...Chevron/pizza/mart, Exxon/Subway/ChesterFriedChicken/ diesel/mart, carwash
85	Byram, **E**...BP/HuddleHouse/diesel/mart/24hr, Chevron/Krystal/mart/24hr, Swinging Bridge RV Park, **W**...DENTIST, Bar Gas/mart, Conoco/diesel/ mart/wash, Pump&Save/mart, Spur/diesel/mart/wash, Dragon Garden Chinese, Mazzio's Pizza, McDonald's, Pizza Hut, Popeye's, Sonic, Subway, Taco Bell, Wendy's, Day's Inn, $General, Jitney Jungle Foods, PawPaw's Camper City, Pennzoil, Super D Drug, bank, cleaners
81	Wynndale Rd, **W**...Conoco/diesel/mart
78	Terry, **E**...Conoco/mart, Texaco/Church's/diesel/mart, **W**...Mac's Gas/mart/atm, New Deal Food
72	MS 27, Crystal Springs, **E**...Phillips 66/diesel/mart, McDonald's, Ford/Lincoln/Mercury, RV camp
68	to US 51, S Crystal Springs, no facilities
65	to US 51, Gallman, **E**...Shell/Dairy Queen/Stuckey's
61	MS 28, Hazlehurst, **E**...HOSPITAL, Amoco/mart, Chevron/diesel, Exxon/Subway/mart/wash, Phillips 66/mart/ wash, Pump&Save Gas, Bumpers Drive-In, Burger King, Catfish House, KFC, McDonald's, Pizza Hut, Stark's Family Rest., Wendy's, Western Sizzlin, Ramada Inn/rest., Western Inn Express, CastroLube, Discount Parts, Family$, Jitney Jungle Food/deli, Piggly Wiggly, Wal-Mart, bank
59	to S Hazlehurst, no facilities
56	to Martinsville, no facilities
54mm	**rest area both lanes, full(handicapped)facilities, phone, picnic tables, litter barrels, petwalk, RV dump, vending, 24hr security**
51	to Wesson, **W**...Chevron/Country Jct/diesel/rest./RV camping
48	Mt Zion Rd, to Wesson, no facilities
42	to US 51, N Brookhaven, **E**...HOSPITAL, Exxon, Phillips 66/diesel, Shell/diesel/mart, Church's Chicken, Della's Motel, camping, **W**...Super 8
40	to MS 550, Brookhaven, **E**...HOSPITAL, Amoco/Domino's/mart, BP/mart, Chevron/mart/24hr, CXpress/diesel, Exxon/Subway/mart, Texaco/diesel/mart/24hr, BBQ, Burger King, Cracker Barrel, Dairy Queen, El Sombrero Mexican, Hudgey's Rest., KFC, McDonald's, Pizza Hut, Popeye's, Shoney's, Sonic, Taco Bell, Wendy's, Western Sizzlin, Best Inn Suites, Best Western, Comfort Inn, Day's Inn, Hampton Inn, Spanish Inn, AutoZone, CarQuest, Chevrolet/Olds, Chrysler/Plymouth/Dodge/Jeep/Daewoo, Ford/Lincoln/Mercury, Honda, Nissan, Rite Aid, SaveALot Foods, Super D Drug, Wal-Mart SuperCtr/gas/24hr, bank
38	US 84, S Brookhaven, **1 mi E**...Exxon/diesel, **W**...Chevron/diesel/mart/24hr
30	Bogue Chitto, Norfield, **E**...Papa'sTrkstp/diesel, Shell/diesel, **W**...MEDICAL CARE
26mm	parking area nb, no facilities
24	Johnston Station, **E**...to Lake Dixie Springs
23.5mm	parking area sb, no facilities
20b a	US 98 W, Summit, to Natchez, **E**...BP/Ward'sBurgers/diesel/mart, Shell/diesel, Texaco/diesel/mart, **W**...Exxon/ Subway/Stuckey's/diesel/mart, Phillips 66/diesel/mart
18	MS 570, Smithdale Rd, N McComb, **E**...HOSPITAL, BP/mart, Burger King, FoodCourt, McDonald's, Piccadilly's, JC Penney, McRae's, Sears/auto, Wal-Mart SuperCtr/24hr, mall, **W**...Chevron/diesel/mart/wash, Hawthorn Inn, Ford/Lincoln/Mercury
17	Delaware Ave, McComb, **E**...HOSPITAL, DENTIST, BP/diesel/mart, Chevron/TCBY/diesel/mart, Citgo/mart, Exxon/Krystal/mart, RaceWay/mart, Shell/diesel/mart, SuperSaver/gas, Burger King, Dairy Queen, Golden Corral, Huddle House, McDonald's, Pizza Hut, Pizza Inn, Popeye's, Sonic, Subway, Taco Bell, Wendy's, Ramada Inn, Super 8, AutoZone, Delaware Tire/muffler/brake, $General, Eckerd, Family$, Jitney Jungle Foods, Kroger, Rite Aid, bank, carwash, **W**...Day's Inn, Plymouth/Toyota
15b a	US 98 E, MS 48 W, McComb, **1 mi E**...Amoco, BP/mart, Exxon/Subway/mart, Quikmart/gas, Shell/mart, Church's Chicken, Hardee's, Hudgey's Rest., KFC, Camellian Motel, Economy Inn, Discount Parts, **W**...BP/diesel
13	Fernwood Rd, **W**...Conoco/Fernwood/diesel/rest./24hr, Fernwood Motel, golf, to Percy Quin SP

N

S

Mississippi Interstate 55

10	MS 48, Magnolia, **1 mi E**...BP/mart, Exxon, Pizza Inn Express, Subway, RV camping
8	MS 568, Magnolia, **E**...Pike Co Speedway
4	Chatawa, no facilities
3.5mm	**Welcome Ctr nb, full(handicapped)facilities, phone, picnic tables, litter barrels, petwalk, RV dump, 24hr security**
1	MS 584, Osyka, Gillsburg, **W**...RV camping
0mm	Mississippi/Louisiana state line

Mississippi Interstate 59

Exit #	Services
	I-59 S and I-20 W run together to Meridian
172mm	Mississippi/Alabama state line
	For exits 150 to Mississippi/Alabama state line, see Mississippi Interstate 20.
149mm	I-59 N and I-20 E run together 22 mi
142	to US 11, Savoy, **W**...to Dunn's Falls
137	to N Enterprise, to Stonewall, no facilities
134	MS 513, S Enterprise, **E**...BP/diesel
126	MS 18, Pachuta, to Rose Hill, **E**...Amoco/diesel/mart/24hr, BP/diesel/mart/24hr
118	to Vossburg, Paulding, no facilities
113	MS 528, to Heidelberg, **E**...BP/JR's/diesel/mart/24hr, Exxon/Subway/Pizza Inn/diesel/mart, Stuckey's Express/pizza/diesel/mart, Shell/diesel/mart, Flying Pig BBQ
109mm	**rest area sb, litter barrels, no restrooms**
106mm	**rest area nb, litter barrels, no restrooms**
104	Sandersville, no facilities
99	US 11, **E**...T&B's/diesel/mart, Magnolia Motel, KOA(1mi)
97	US 84 E, **E**...Chevron/Subway/diesel/mart/24hr, Exxon/diesel/mart, Shell, Doc's Rest., Hardee's, **W**...BP/diesel/mart, Texaco/mart/wash/24hr, KFC, Vic's Biscuits/burgers
96b	MS 15 S, Cook Ave, no facilities
a	Masonite Rd, 4th Ave, no facilities
95d	(from nb), no facilities
c	Beacon St, Laurel, **W**...Chevron/diesel/mart, PumpSave Gas, Texaco/mart, BBQ, Burger King, Church's Chicken, McDonald's, Popeye's, Townhouse Motel/rest., Country Mkt Foods, Family$, JC Penney, McRae's, USPO, XpertTire, cinema
b a	US 84 W, MS 15 N, 16th Ave, Laurel, **W**...HOSPITAL, Amoco/mart, BP, Exxon/diesel/mart, Shell/mart, Dairy Queen, KFC, McAlister's Deli, McDonald's, Pasquale's Pizza, Shoney's, Subway, Taco Bell, Waffle House, Wendy's, Comfort Suites, Econolodge, Executive Inn, Holiday Inn Express, Quality Inn, Ramada Inn, Super 8, Discount Parts, $General, Piggly Wiggly, bank
93	US 11, S Laurel, **W**...Citgo/diesel/mart, Exxon/Subway/diesel/mart, Shell/FoodCourt/diesel/mart/wash/24hr, Hardee's, New Laurel Motel, bank
90	US 11, Ellisville Blvd, **W**...Dixie/diesel/mart
88	MS 588, MS 29, Ellisville, **E**...BP/mart, Chevron/diesel/mart, Country Girl Kitchen, KFC, McDonald's, Pizza Hut(1mi), Subway, Ward's Burgers, $General, Food Tiger, Family$, NAPA, TrueValue, bank, **W**...Kwikstop/gas, Shell/mart
85	MS 590, to Ellisville, **W**...Gas/diesel, auto repair
80	to US 11, Moselle, no facilities
78	Sanford Rd, no facilities
76	**W**...Hattiesburg-Laurel Reg Airport
73	to Monroe, no facilities
69	to Glendale, Eatonville Rd, no facilities
67b a	US 49, Hattiesburg, **E**...Amoco/mart, Exxon/mart, Shell/Subway/mart, Texaco/mart, Arby's, Burger King, Cracker Barrel, Dairy Queen, Farmers Mkt Buffet, KFC, Krystal, McDonald's, Pizza Hut, Taco Bell, Waffle House, Budget Inn, Cabot Lodge, Comfort Inn, Day's Inn, Executive Suites, Holiday Inn, Howard Johnson, Motel 6, Ramada Ltd, Scottish Inn, Super 8, $General, cinema, **W**...Amoco/diesel/mart, Chevron/mart/wash, Dandy Dan's/diesel/mart, Pure/diesel/mart, Raceway/mart, Shell/mart, Speedway/diesel/mart, Stuckey's Express/diesel/mart, Waffle House, Ward's Burgers, Best Western, Hawthorn Suites, bank
65b a	US 98 W, Hardy St, Hattiesburg, **E**...MEDICAL CARE, VETERINARIAN, Chevron/diesel/mart, Conoco/mart, Exxon/mart, Shell/diesel/mart, Texaco/mart, Applebee's, Bailey's Chicken, Barnhill's Buffet, Baskin-Robbins, BBQ, Burger King, Checker's, Domino's, IHOP, KFC, Krystal, Lenny's Subs, McDonald's, Mr Gatti's, Panino's Italian, Pizza Hut, Purple Parrot Café, Shoney's, Smoothie King Diner, Steak-Out, Subway, Taco Bell, Ward's Burgers, Xan's Diner, Best Western, Econolodge, Fairfield Inn, Eckerd, Roses, SpeeDee Lube, bank, cleaners, S MS U, **W**...HOSPITAL, DENTIST, Amoco/TCBY/diesel/mart, BP/Subway/diesel/mart, Chevron/mart, Exxon/Domino's/mart, Pump&Save/mart, Shell/KFC/mart/wash, Texaco/Arby's/mart/atm, Burger King, Chili's, Copeland's Rest., Corky's BBQ, Hardee's, Heavenly Ham, La Fiesta Brava, LoneStar Steaks, Luby's, Mack's

Meridian • **Laurel** • **Hattiesburg**

N

S

Mississippi Interstate 59

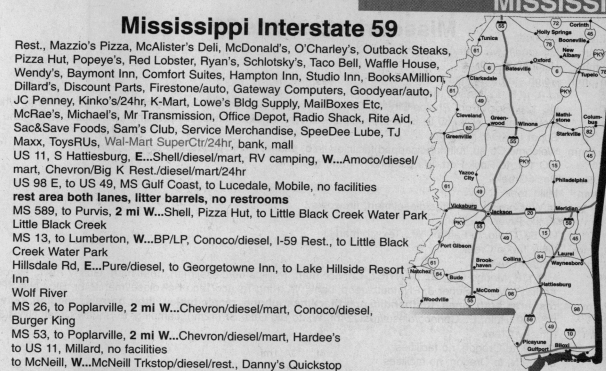

Rest., Mazzio's Pizza, McAlister's Deli, McDonald's, O'Charley's, Outback Steaks, Pizza Hut, Popeye's, Red Lobster, Ryan's, Schlotsky's, Taco Bell, Waffle House, Wendy's, Baymont Inn, Comfort Suites, Hampton Inn, Studio Inn, BooksAMillion, Dillard's, Discount Parts, Firestone/auto, Gateway Computers, Goodyear/auto, JC Penney, Kinko's/24hr, K-Mart, Lowe's Bldg Supply, MailBoxes Etc, McRae's, Michael's, Mr Transmission, Office Depot, Radio Shack, Rite Aid, Sac&Save Foods, Sam's Club, Service Merchandise, SpeeDee Lube, TJ Maxx, ToysRUs, Wal-Mart SuperCtr/24hr, bank, mall

60	US 11, S Hattiesburg, **E...**Shell/diesel/mart, RV camping, **W...**Amoco/diesel/mart, Chevron/Big K Rest./diesel/mart/24hr
59	US 98 E, to US 49, MS Gulf Coast, to Lucedale, Mobile, no facilities
56mm	**rest area both lanes, litter barrels, no restrooms**
51	MS 589, to Purvis, **2 mi W...**Shell, Pizza Hut, to Little Black Creek Water Park
48mm	Little Black Creek
41	MS 13, to Lumberton, **W...**BP/LP, Conoco/diesel, I-59 Rest., to Little Black Creek Water Park
35	Hillsdale Rd, **E...**Pure/diesel, to Georgetowne Inn, to Lake Hillside Resort Inn
32mm	Wolf River
29	MS 26, to Poplarville, **2 mi W...**Chevron/diesel/mart, Conoco/diesel, Burger King
27	MS 53, to Poplarville, **2 mi W...**Chevron/diesel/mart, Hardee's
19	to US 11, Millard, no facilities
15	to McNeill, **W...**McNeill Trkstop/diesel/rest., Danny's Quickstop
13mm	**rest area sb, litter barrels, no restrooms**
10	to US 11, Carriere, **E...**Hilda's/diesel/rest., **W...**tires
8mm	**rest area nb, litter barrels, no restrooms**
6	MS 43 N, N Picayune, **W...**HOSPITAL, VETERINARIAN, Chevron/diesel/mart, Paul's Pastrys, Budget Host, Picayune Motel(1mi), Eckerd, Winn-Dixie/deli, bank, cinema
4	MS 43 S, to Picayune, **E...**Ryan's, Chevrolet/Pontiac/Buick/Olds/Cadillac, Wal-Mart SuperCtr/gas/24hr, **W...**HOSPITAL, EYECARE, BP/diesel/mart, Conoco/mart, Exxon/diesel/mart, Shell/diesel/mart, Spur/mart, Texaco/diesel/mart, Burger King, Domino's, El Mariachi Loco Mexican, Godfather's, Hardee's, KFC, McDonald's, Pizza Hut, Popeye's, Sherral's Rest., Shoney's, Taco Bell, Waffle House, Wendy's, Comfort Inn, Day's Inn, Heritage Inn, AutoZone, Cato's, Chrysler/Plymouth/Dodge/Jeep, Discount Parts, Firestone/auto, Ford/Lincoln/Mercury, Radio Shack, Rite Aid, SpeeDee Lube, Winn-Dixie/deli/24hr, bank
3mm	**Welcome Ctr nb, full(handicapped)facilties, phone, vending, picnic tables, litter barrels, petwalk, RV dump**
1.5mm	weigh sta both lanes
1	US 11, MS 607, NASA, **W...**Chevron/diesel/mart
0mm	**Pearl River, Mississippi/Louisiana state line. Exits 11-1 are in Louisiana.**
11	Pearl River Turnaround. Callboxes begin sb.
5b	Honey Island Swamp, no facilities
a	LA 41, Pearl River, **E...**Chevron/mart/gifts
3	US 11 S, LA 1090, Pearl River, **1 mi W...**Chevron/diesel/mart/rest./24hr
1.5mm	**Welcome Ctr sb, full(handicapped)facilties, info, phone, picnic tables, litter barrels, petwalk**
1c b	I-10, E to Bay St Louis, W to New Orleans
a	I-12 W, to Hammond. I-59 begins/ends on I-10/I-12. **Exits 1-11 are in Louisiana.**

Mississippi Interstate 220(Jackson)

Exit #	Services
11mm	I-220 begins/ends on I-55, exit 104.
9	Hanging Moss Rd, County Line Rd, **E...**BP/mart/wash
8	Watkins Dr, **E...**Shell/mart
5b a	US 49 N, Evers Blvd, to Yazoo City, **E...**Chevron, **W...**Amoco/diesel, Shell/diesel/mart/24hr, Penske Trucks
2b a	Clinton Blvd, Capitol St, **E...**to Jackson Zoo, **W...**Shell, Walco/mart, Burger King, McDonald's, Pizza Hut, Popeye's, Sonic
1b a	US 80, **E...**BP/mart/wash, Citgo/mart, Shell/mart/wash, BBQ, Capt D's, Chinese Rest., Denny's, KFC, McDonald's, Popeye's, Ponderosa, Taco Bell, Wendy's, Western Sizzlin, Best Western, Day's Inn, Econolodge, Holiday Inn, Sleep Inn, Firestone, Pennzoil, transmissions, auto repair, bank, **W...**BP/mart/wash, Exxon/Subway/diesel/mart, Shell, Texaco, Arby's, El Chico's, Krystal, McDonald's, Piccadilly's, Pizza Hut, Popeye's, Ruby Tuesday, Shoney's, Wendy's, Ford, Gayfer's, K-Mart, McRae's, Sears/auto, bank, hardware, mall
0mm	I-220 begins/ends on I-20, exit 41.

Picayune

Jackson

N

S

E

W

Missouri Interstate 29

Exit #	Services
124mm	Missouri/Iowa state line
123mm	Nishnabotna River
121.5mm	weigh sta both lanes
116	rd A, rd B, to Watson, no facilities
110	US 136, Rock Port, Phelps City, **E**...Conoco/Hardee's/diesel/mart, Texaco/diesel/mart, Rockport Inn, White Rock Motel(2mi), to NW MO St U, **W**...Amoco/diesel/24hr, Phillips 66/diesel/mart/24hr, McDonald's, Trails End Rest., Super 8, KOA, fireworks
109.5mm	Welcome Ctr sb, full(handicapped)facilities, info, phone, picnic tables, litter barrels, petwalk
107	MO 111, to Rock Port, **W**...Elk Inn/RV Park, gas/café
106.5mm	Rock Creek
102mm	Mill Creek
99	rd W, Corning, **E**...Citgo/diesel/mart, fireworks
97mm	Tarkio River
92	US 59, Craig, to Fairfax, **W**...Texaco/mart
90.5mm	Little Tarkio Creek
86.5mm	Squaw Creek
84	MO 118, Mound City, **E**...DENTIST, Conoco/mart, Phillips 66/diesel/mart, Total/mart, Galley Rest., Hardee's, Audrey's Motel, Super 8, Plymouth/Jeep, bank, **W**...King/Texaco/Taco Bell/diesel/mart/24hr, Big Lake SP(12mi)
82mm	**rest area both lanes, full(handicapped)facilities, phone, picnic tables, litter barrels, vending, petwalk**
79	US 159, Rulo, **E**...Conoco/diesel/rest./24hr, **W**...to Big Lake SP(12mi), camping, to Squaw Creek NWR(3mi)
78mm	Kimsey Creek
75	US 59, to Oregon, no facilities
67	US 59 N, to Oregon, no facilities
66.5mm	Nodaway River
65	US 59, rd RA, to Fillmore, Savannah, **E**...Conoco/mart
60	rd K, rd CC, Amazonia, **W**...Hunt's Fruit Barn, antiques
58.5mm	Hopkins Creek
56b a	I-229 S, US 71 N, US 59 N, to St Joseph, Maryville, no facilities
55mm	Dillon Creek
53	US 59, US 71 bus, to St Joseph, Savannah, **E**...AOK Camping, **W**...antiques, fireworks
50	US 169, King City, St Joseph, **1-3 mi W**...Conoco/mart, Applebee's, Blackeyed Pea, Hardee's, LJ Silver, McDonald's, Ryan's, Shoney's, Taco Bell, Wendy's, Pony Express Motel
47	MO 6, Frederick Blvd, St Joseph, to Clarksdale, **E**...Conoco/mart, Country Kitchen, Day's Inn, Drury Inn/Golden Grill, **W**...HOSPITAL, DENTIST, VETERINARIAN, Amoco/wash/24hr, Sinclair/mart, Texaco/diesel/mart, Total/diesel/mart, Amigo's Mexican, Applebee's, Boston Mkt, Burger King, Chinese Rest., Church's Chicken, Cracker Barrel, Denny's, Dunkin Donuts, Fazoli's, Godfather's, Ground Round, Hardee's, KFC, McDonald's, Perkins/24hr, Pizza Hut, Red Lobster, Sizzler, Sonic, Speedy's Burgers, Subway, Taco Bell, Winstead's Steakburgers, Wyatt's Cafeteria, Budget Inn, Comfort Suites, Hampton Inn, Motel 6, Pony Express Motel, Ramada Inn, Super 8, Aldi Foods, AutoZone, Chevrolet/Mazda, Cub Food/drug/24hr, Food4Less/24hr, Ford, Firestone, Goodyear/auto, Hasting's Books, Dillard's, JC Penney, Midas Muffler, Osco Drug, Radio Shack, Sears, Wal-Mart, bank, cinema, RV camp
46b a	US 36, St Joseph, to Cameron, **1 mi W**...Phillips 66, Burger King, Wendy's, to MO W St Coll
44	US 169, St Joseph, to Gower, **E**...Phillips 66/Subway/gas, BBQ, Myer's Family Grill, LP, Best Western, tires, **W**...Texaco/diesel/mart/24hr, McDonald's, Taco Bell, Goodyear
43	I-229 N, to St Joseph, no facilities
39.5mm	Pigeon Creek
35	rd DD, Faucett, **W**...Farris/Phillips 66/Subway/diesel/motel/mart/24hr, Oliver's Rest., antiques
33.5mm	Bee Creek
30	rd Z, rd H, Dearborn, New Market, **E**...Conoco/diesel/mart/atm/24hr
29.5mm	Bee Creek
27mm	**rest area both lanes, full(handicapped)facilities, phone, picnic tables, litter barrels, vending, petwalk**
25	rd E, rd U, to Camden Point, **E**...Phillips 66/diesel/mart/rest.
24mm	weigh sta both lanes
20	MO 92, MO 273, to Atchison, Leavenworth, **W**...Phillips 66/mart, antiques, to Weston Bend SP
19.5mm	Platte River
19	rd HH, Platte City, **E**...antiques, **W**...Amoco, Conoco/mart, Dairy Queen, Red Dragon Rest., Mr B's Rest., Best Western/RV camping, Comfort Inn, Super 8, Dodge, Ford, IGA Foods, USPO, auto parts, hardware, same as 18
18	MO 92, Platte City, **E**...Basswood Good Sam RV Park, **W**...Amoco, Conoco/mart, QT/diesel/mart, Burger King, Cracker Barrel, Dairy Queen, Ma&Pa's Kettle, McDonald's, Red Dragon Chinese, Subway, Taco Bell, Waffle House, AmericInn, Comfort Inn, Super 8, Chevrolet, Pontiac/Olds/GMC, Saturn, bank, same as 19
17	I-435 S, to Topeka

St Joseph

N ↑ ↓ S

Missouri Interstate 29

15	Mexico City Ave, no facilities
14	I-435 E(from sb), to St Louis
13	to I-435 E, **E**...Best Western, Comfort Suites, Extended Stay America, Fairfield Inn, Hampton Inn, Holiday Inn, Microtel, Wyndham Garden, **W**...Marriot, KCI Airport
12	NW 112th St, **E**...Amoco/mart, Allie's Rest., Day's Inn, Hampton Inn, Hilton, Holiday Inn Express, Wyndham Garden, bank, **W**...Econolodge
10	Tiffany Springs Pkwy, **E**...EYECARE, CHIROPRACTOR, Texaco/mart, SmokeBox BBQ, Embassy Suites, Homewood Suites, cleaners, **W**...Cracker Barrel, Ruby Tuesday, AmeriSuites, Chase Suites, Courtyard, Drury Inn, MainStay Suites, Ramada Inn, Residence Inn, Sleep Inn
9b a	MO 152, to Liberty, Topeka
8	MO 9, rd T, NW Barry Rd, **E**...HOSPITAL, Total/mart, Applebee's, Bob Evans, Boston Mkt, Chili's, Einstein Bros, Godfather's, Hooters, LoneStar Steaks, On the Border, Pizza Hut/Taco Bell, Starbuck's, Subway, Waid's Rest., Wendy's, Winstead's Rest., Barnes&Noble, Bed Bath&Beyond, Eckerd, HyVee Foods, Lowe's Bldg Supply, OfficeMax, Old Navy, Pier 1, SteinMart, Wal-Mart SuperCtr/24hr, cinema, laundry, **W**...Citgo/mart, Phillips 66, QT/diesel/mart, Hardee's, LJ Silver, McDonald's, Minsky's Pizza, OutBack Steaks, Taco John's, Country Hearth Inn, Motel 6, Super 8, Tires+, bank
7	(from nb), same as 6
6	NW 72nd St, Platte Woods, **E**...Sinclair/diesel, **W**...Phillips 66/mart, Texaco/mart, Jazz Kitchen, Papa John's, Pizza Hut, Pizza Shoppe, Piggly Wiggly, K-Mart/Little Caesar's/drugs
5	MO 45 N, NW 64th St, **W**...DENTIST, CHIROPRACTOR, Texaco/mart, IHOP, O'Quigley's Grill, McDonald's, Paradise Grill, Subway, Super Food Barn, Goodyear, HobbyLobby, TrueValue, bank, carwash
4	NW 56th St(from nb), **W**...Phillips 66/mart, same as 5
3c	rd A(from sb), Riverside, **W**...Phillips 66/mart, Argosy Café, Corner Café
b	I-635 S, no facilities
a	Waukomis Dr, rd AA(from nb), no facilities
2b	US 169 S(from sb), to KC, no facilities
a	US 169 N(from nb), to Smithville, no facilities
1e	US 69, Vivion Rd, **E**...Chrysler/Plymouth, Chevrolet/Cadillac, Ford/Mercury/Lincoln, Home Depot, OfficeMax, PriceChopper Foods/24hr, bank, **W**...Phillips 66, Argosy Café, McDonald's
d	MO 283 S, Oak Trfwy(from sb)
c	Gladstone, Midwestern Seminary(from nb), **E**...Arby's, Luby's, Perkins, Pizza St Buffet, Ryan's, Tippen's Café, K-Mart
b	I-35 N(from sb), to Des Moines
a	Davidson Rd, **E**...Texaco, **W**...Travelodge
8mm	I-35 N. I-29 and I-35 run together 6 mi.

See Missouri Interstate 35, exits 3-8a.

I-29 begins/ends on I-70.

Missouri Interstate 35

Exit #	Services
114mm	Missouri/Iowa state line
114	US 69, to Lamoni, **W**...Conoco/diesel/mart/24hr
113.5mm	Zadie Creek
110	weigh sta both lanes
106	rd N, Blythedale, **E**...Conoco/mart/fireworks, Phillips 66/diesel/café/motel/24hr, diesel repair, Eagle's Landing Motel, **W**...Texaco/diesel/mart/24hr, camping
99	rd A, to Ridgeway, **5 mi W**...Eagle RV Camping
94mm	E Fork Big Creek
93	US 69, Bethany, **W**...RV dumping
92	US 136, Bethany, **E**...Conoco/diesel/mart/24hr, Phillips 66/Taco Bell/diesel/mart/24hr, McDonald's, Best Western, Budget Inn, **W**...HOSPITAL, Amoco/mart,Casey's/gas, Kum&Go/Wendy's/diesel/mart, Country Kitchen, Dairy Queen, Hardee's/24hr, Pizza Hut, Subway, TootToot Rest., Super 8, Chevrolet/Olds/Buick, Wal-Mart, auto parts, **1 mi W**...Ford/Lincoln/Mercury, Hy-Vee Foods
90mm	Pole Cat Creek
88	MO 13, to Bethany, Gallatin, **3 mi W**...Bethany Motel
84	rds AA, H, to Gilman City, no facilities

N ↕ S

Kansas City

Missouri Interstate 35

81mm	**rest area both lanes, full(handicapped)facilities, phone, picnic tables, litter barrels, vending, petwalk**
80	rds B, N, to Coffey, no facilities
78	rd C, Pattonsburg, **W**...Total/diesel/mart, antiques
74.5mm	Grand River
72	rd DD, no facilities
68	US 69, to Pattonsburg, no facilities
64	MO 6, to Maysville, Gallatin, no facilities
61	US 69, Winston, Gallatin, **E**...Texaco/diesel/mart/rest./24hr
54	US 36, Cameron, **E**...Amoco/diesel/mart/24hr, Ampride/diesel/mart/rest./24hr, McDonald's, Best Western, Comfort Inn, Crossroads Inn/RV camping, **W**...HOSPITAL, Phillips 66/mart, Total/mart, Burger King, Dairy Queen, Hardee's, Kettle Diner, KFC/Taco Bell, Pizza Hut, Pringle's Rest., Sonic, Subway, Best Western, Day's Inn, EconoLodge, Holiday Inn Express, Super 8, Advance Parts, Chevrolet, CountryMart Foods, Ford/Mercury, Goodyear, KwikLube, Wal-Mart, auto parts, W MO Corr Ctr
52	rd BB, Lp 35, to Cameron, **2 mi W**...same services as 54
49mm	Brushy Creek
48.5mm	Shoal Creek
48	US 69, Cameron, **E**...to Wallace SP(2mi), **W**...Total/mart, fireworks
40	MO 116, Lathrop, **E**...Phillips 66/Travel Express/diesel/Country Café, **W**...Dinner Time Rest.(4mi)
35mm	**rest area sb, full(handicapped)facilities, phone, picnic tables, litter barrels, vending, petwalk**
34mm	**rest area nb, full(handicapped)facilities, phone, picnic tables, litter barrels, vending, petwalk**
33	rd PP, Holt, **E**...Rock Front Steaks, **W**...Conoco/diesel/mart, Phillips 66/diesel/mart, Holt Steel Inn, flea mkt
30mm	Holt Creek
26	MO 92, Kearney, **E**...CHIROPRACTOR, Casey's/gas, Conoco/mart, Texaco/Taco Bell/diesel/mart/wash, Sonic, Day's Inn, Super 8, Big V Food/drug, KwikLube/wash, repair/towing, bank, hardware, laundry, to Watkins Mill SP, **W**...Conoco/diesel/mart/24hr, Arby's, Hardee's, Hunan Garden Chinese, Subway, Econolodge, Chevrolet, Family$, John's Foods, bank, to Smithville Lake
22mm	weigh sta both lanes
20	US 69, MO 33, to Excelsior Springs, **E**...HOSPITAL, Snackshop/gas
17	MO 291, rd A, **1 mi E**...Boston Mkt, LJ Silver, Ponderosa, Hy-Vee Foods, CountryMart Food, Eckerd, JiffyLube, bank, same as 16, **W**...Phillips 66/mart, KCI Airport
16	MO 152, Liberty, **E**...HOSPITAL, DENTIST, MEDICAL CARE, VETERINARIAN, Conoco/diesel, Texaco/mart, Total/mart, Big Cheese Pizza, Chinese Rest., Godfather's, Golden Corral, Hardee's, Perkins, KFC, Little Caesar's, LJ Silver, Pizza Hut, Ponderosa, Subway, Taco Bell, Village Inn Rest., Wendy's, Winstead's Drive-Thru, Best Western, Super 8, Chevrolet, Firestone/auto, Food Barn/deli/24hr, Ford, K-Mart/drugs, RV Ctr, Treasury Drug, Western Auto, Walton's Lube, auto parts, bank, cinema, cleaners, hardware, **W**...Phillips 66/mart/wash, Applebee's, Bob Evans, Country Kitchen, Cracker Barrel, Golden Corral, McDonald's, Waffle House, Comfort Suites, Fairfield Inn, Hampton Inn, Wal-Mart/auto
14	US 69, Liberty Dr, Pleasant Valley, to Glenaire, **E**...QT/diesel/mart, I-35 RV Ctr
13	US 69(from nb), to Pleasant Valley, **E**...QT/diesel/mart, Sinclair/mart/24hr, Total/diesel/mart, Texaco/A&W/mart, A&W, KFC, McDonald's, auto repair
12b a	I-435 S, to St Louis
11	US 69 N, Vivion Rd, **E**...Amoco, McDonald's, same as 10
10	N Brighton Ave(from nb), **W**...QT/mart, Church's Chicken, Sonic, Stroud's Rest.
9	MO 269 S, Chouteau Trfwy, **E**...Phillips 66, Sinclair, Harrah's Casino/rest., **W**...Amoco, IHOP, Wendy's
8c	MO 1, Antioch Rd, **E**...Citgo/7-11, Conoco, Sinclair, Chinese Rest., Domino's, Antioch Inn, Best Western, Country Inn, InnTowne Inn, **W**...Phillips 66/mart, Waffle House
b	I-29 N, KCI airport
I-35 S and I-29 S run together 6 mi	
8a	Parvin Rd, **E**...Texaco/mart, Holiday Inn, O'Reilly Parts, **W**...Trevalyn Rest., Parks Lodge
6b a	Armour Rd, **E**...HOSPITAL, Phillips 66/mart, Arby's, Big Boy, Burger King, Capt D's, Denny's, McDonald's, Baymont Inn, Day's Inn, bank, to Riverboat Casino, **W**...QT/mart/atm, Texaco/KrispyKreme/mart, Church's Chicken/White Castle, Chinese Rest., Dairy Queen, LJ Silver, Pizza Hut, Taco Bell, Wendy's, American Inn, Country Hearth Suites, bank
5b	16th Ave, industrial district
a	Levee Rd, Bedford St, industrial district
4.5mm	Missouri River
4b	Front St, **E**...Riverboat Casino/rest.
a	US 24 E, access to Royal Inn, Capri Motel, Ford
3	I-70 E, US 71 S, to St Louis, no facilities
2n	I-35 N and I-29 N run together 6 mi
2f	Walnut St, downtown

N

S

Cameron

Liberty

Kansas City

Missouri Interstate 35

d	Wyandotte St, downtown, **E**...Texaco, Midas Muffler
c	Broadway, **S**...Denny's
a	I-70 W, to Topeka
w	12th St, Kemper Arena
u	I-70 E, to St Louis
1d	20th St(from sb), no facilities
c b	SW Blvd, W Pennway(from nb), **E**...HOSPITAL, Conoco/mart
a	SW Trafficway(from sb), no facilities
0mm	Missouri/Kansas state line

Missouri Interstate 44

Exit #	Services
290mm	I-44 begins/ends on I-55, exit 207 in St Louis.
290a	I-55 S, to Memphis
290c	Gravois Ave(from wb), 12th St, **N**...Citgo, **S**...McDonald's
290b	18th St(from eb), downtown
289	Jefferson Ave, St Louis, **N**...Citgo/mart, Vickers/mart, Subway, **S**...McDonald's
288	Grand Blvd, St Louis, **N**...HOSPITAL, bank, **S**...Jack-in-the-Box
287b a	Kingshighway, Vandeventer Ave, St Louis, **N**...Amoco, **S**...Amoco/mart/wash, Chevrolet, Chrysler, Walgreen, to MO Botanical Garden
286	Hampton Ave, St Louis, **N**...DENTIST, Amoco, Clark Gas, Mobil, Checker's, Chinese Rest., Denny's, Jack-in-the-Box, McDonald's, Steak'n Shake, Subway, Taco Bell, bank, hardware, **S**...Shell/mart, Burger King, Chinese Rest., Church's Chicken, Hardee's, Village Inn Rest., Holiday Inn, Red Roof Inn, carwash, museums, zoo
285	SW Ave(from wb, no EZ return), no facilities
284b a	Arsenal St, Jamieson St, no facilities
283	Shrewsbury(from wb), some services same as 282
282	Murdock Ave(from eb), St Louis, **N**...Amoco, Mobil, Chinese Rest., Pantera's Pizza, McDonald's, Subaru, bank, drugs, laundry
280	Elm Ave, St Louis, **N**...Amoco/repair, Citgo/mart, **S**...Shell/mart, Radiator King, laundry
279	(from wb), Berry Rd
278	Big Bend Rd, St Louis, **N**...HOSPITAL, **S**...QT/mart, Sinclair, Denny's
277	US 67, US 61, US 50, Lindbergh Blvd, **N**...HOSPITAL, Shell/mart, Hardee's, PoFolks, Best Western, Howard Johnson, JiffyLube, K-Mart, TireAmerica, Fabric Whse, auto repair, hardware, **S**...CFM/diesel, Citgo/mart, Total/mart, Vickers/mart, Bob Evans, Burger King, Chinese Rest., Denny's, Steak'n Shake, Comfort Inn, Day's Inn, Econolodge, Holiday Inn, Westward Motel, Midas Muffler, cinema
277a	MO 366 E, Watson Rd, no facilities
276b a	I-270, S Memphis, N to Chicago, no facilities
275	N Highway Dr(from wb), Soccer Pk Rd, **N**...Texaco/diesel/rest., Kenworth
274	Bowles Ave, **N**...Texaco/diesel/mart, **S**...Amoco, Citgo/diesel/mart/24hr, QT/mart, Big Boy, Cracker Barrel, Denny's, Krispy Kreme, Quizno's, Tubby's Grill, White Castle, Drury Inn, Fairfield Inn, Holiday Inn Express, Howard Johnson, Motel 6, PearTree Inn, Stratford Inn, Goodyear, bank
272	MO 141, Fenton, Valley Park, **N**...Motomart, Firestone/auto, **S**...Citgo/diesel/mart, QT/mart, Shell/mart/wash, Burger King, McDonald's, Steak'n Shake, Subway, Taco Bell, Pennzoil, Wet Willie's Waterslide
269	Antire Rd, Beaumont, no facilities
266	Lewis Rd, no facilities
266mm	Meramec River
265	Williams Rd(from eb), **S**...Dri-Port Marine
264	MO 109, rd W, Eureka, **N**...CHIROPRACTOR, Citgo/7-11, Phillips 66/mart, Shell/mart, Texaco/mart, BBQ, Burger King, Dairy Queen, KFC, McDonald's, Pizza Hut, Ponderosa, Rich&Charlie's Rest., Subway, Taco Bell, Wendy's, Day's Inn, Byerly RV Ctr, Firestone, NAPA, National Foods/deli, bank, to Babler SP, **S**...VETERINARIAN, QT/diesel/mart, Shell/mart, Texaco
261	Lp 44, to Allenton, **N**...MEDICAL CARE, to Six Flags, Applebee's, Country Kitchen, Denny's, McDonald's, KFC, Lion's Choice Rest., Steak'n Shake, Oak Grove Inn, Ramada, Red Carpet Inn, Super 8, AutoZone, $Tree, FashionBug, Radio Shack, Wal-Mart SuperCtr, same as 264, **S**...Phillips 66/diesel/mart/rest., Shell/diesel/mart/wash, Ford, KOA
257	Lp 44, Pacific, **N**...fireworks, **S**...Citgo/mart, Conoco/mart, Mobil/mart/wash/24hr, Motomart/atm/24hr, Hardee's, McDonald's, Pizza Hut, Taco Bell, Holiday Inn Express, Buick/Chevrolet/Olds, Chrysler/Jeep/Dodge, IGA Foods, NAPA, auto repair, bank

E

W

St Louis

Missouri Interstate 44

253	MO 100 E, to Gray Summit, **N**...antiques, **S**...Phillips 66/mart, Diamond's Rest., Best Western, Gardenway Motel, fireworks, laundry
251	MO 100 W, to Washington, **N**...Citgo/mart, Conoco/diesel/rest./mart/24hr, FuelMart/Blimpie/Burger King/diesel/mart/24hr, Gas Xpress/diesel/mart, Mr Fuel/diesel/mart
247.5mm	Bourbeuse River
247	US 50 W, rd AT, rd O, to Union, **N**...Harley-Davidson, flea mkt, **S**...to Robertsville SP
242	rd AH, no facilities
240	MO 47, St Clair, **N**...Phillips 66/Taco Bell/diesel/mart, Sinclair/mart, Burger King, Reed's RV Ctr, bank, **S**...Mobil/mart, Hardee's, Jay-Bo's BBQ, McDonald's, Subway, Budget Lodge, Super 8
239	MO 30, rds AB, WW, St Clair, **N**...The Arch Motel, repair, **S**...Amoco/mart, Shell
238mm	weigh sta both lanes
235mm	**rest area both lanes(both lanes exit left), full(handicapped)facilities, phone, picnic tables, litter barrels, petwalk**
230	rds W, JJ, Stanton, **N**...Stanton Motel, **S**...Gas Xpress/diesel/mart, Delta Motel, KOA, Meramec Caverns Camping(3mi), Toy Museum
226	MO 185 S, Sullivan, **S**...Flying J/Conoco/Country Mkt/diesel/LP/mart/24hr, **S**...Phillips 66/Burger King/mart, Golden Corral, Hardee's, KFC, Steak'n Shake, Taco Bell, TrueValue, Wal-Mart SuperCtr/24hr, KOA(1mi), to Meramec SP, same as 225
225	MO 185 N, rd D, Sullivan, **N**...Mobil/mart, Shell/diesel/mart, Domino's, Baymont Inn, Best Western, Family Motor Inn, Ramada, Sunrise Motel, Super 8, Chrysler/Plymouth/Dodge/Jeep, Ford/Mercury, **S**...HOSPITAL, Bobber/Texaco/diesel/café, Citgo/mart, Capt D's, Chinese Rest., Jack-in-the-Box, McDonald's, Pizza Hut, Shoney's, Sonic, Subway, Sullivan Motel, Ace Hardware, Aldi Foods, Goodyear, Shop'n Save Foods, bank, cinema, laundry, same as 226,
218	rds N, C, J, Bourbon, **N**...Citgo/diesel/mart, Budget Inn, **S**...Mobil/mart/24hr, HenHouse Rest., Blue Sprgs Camping(6mi), Riverview Ranch Camping(8mi)
214	rd H, Leasburg, **N**...Mobil/diesel/mart, **S**...to Onandaga Cave SP(7mi), camping
210	rd UU, **N**...antiques, **S**...Mt Peasant Winery
208	MO 19, Cuba, **N**...Citgo/mart, Texaco/diesel/mart, Best Western/rest., Super 8, **S**...Mobil/Taco Bell/mart/atm/24hr, Phillips 66, Burger King, Hardee's, Kum&Get It Buffet, McDonald's, Pizza Pro, Subway, Wagon Wheel Motel, Chevrolet/GMC, NAPA, Wal-Mart/drugs, bank, laundry, to Ozark Nat Scenic Riverways
203	rds F, ZZ, **N**...Blue Moon RV Park, **S**...Rosati Winery(2mi)
195	MO 8, MO 68, St James, Maramec Sprg Park, **N**...Mobil/diesel/mart, Conoco/diesel/mart, McDonald's, Pizza Hut, Subway, Taco Bell, Comfort Inn, Economy Inn, Ozark Motel, Nu-Way Foods, St James Winery, tires/repair, laundry, **S**...Burger King
189	rd V, **N**...camping, **S**...Conoco/diesel/mart/24hr, Golden Catfish, Mule Trading Post, antiques
186	US 63, MO 72, Rolla, **N**...Sinclair, Steak'n Shake, Drury Inn, Hampton Inn, Sooter Inn, tires/repair, **S**...HOSPITAL, Amoco/mart/24hr, Phillips 66/mart/atm, Shell/mart/24hr, Denny's, Dunkin Donuts, Lee's Chicken, Pizza Inn, Waffle House, American Motor Inn, Budget Deluxe Motel, Rustic Motel(2mi)
185	rd E, to Rolla, **S**...HOSPITAL, Phillips 66, Dairy Queen, Hardee's/24hr, Subway, Taco Bell, UMO at Rolla, st patrol
184	US 63 S, to Rolla, **S**...HOSPITAL, MotoMart/gas, Phillips 66/mart, Shell/mart, StationBreak/gas, Texaco/mart, Arby's, Burger King, Domino's, Golden Corral, KFC, Kyoto Japanese, Little Caesar's, LJ Silver, Lucky House Chinese, Maid-Rite Rest., McDonald's, Papa John's, Pizza Hut, Pizza Inn, Shoney's, Sirloin Stockade, Subway, Taco Hut, Wendy's, Bestway Motel, Best Western, Chalet Motel, Day's Inn, Econolodge, Holiday Inn Express, Howard Johnson/rest., Interstate Motel, Ramada Inn, Rolla Inn, Super 8, Travelodge, Wayfarer Inn, Western Inn, Zeno's Motel/steaks, Chevrolet, Ford/Lincoln/Mercury, K-Mart, Kroger, Valvoline, Wal-Mart, auto parts, fireworks, photo lab, transmissions
179	rds T, C, to Doolittle, Newburg, **S**...Citgo/mart/24hr, Cookin' From Scratch Rest., bank
178mm	**rest area both lanes, full(handicapped)facilities, phone, picnic tables, litter barrels, vending, petwalk**
176	Sugar Tree Rd, **N**...Vernell's Motel(°mi), **S**...camping
172	rd D, Jerome, **N**...camping
169	rd J, **S**...camping
164mm	Big Piney River
166	to Big Piney
163	MO 28, to Dixon, **N**...Voss Truckport/diesel/mart/café/repair/24hr, Hillbilly Store, DJ's Mkt, Ozark Crafts, fireworks, **S**...Amoco, Conoco/mart, Country Café, Best Western Montis/rest., Day's Inn, Sunset Village Motel, Super 8, RV camping, to The Hi-Way Motel
161b a	rd Y, to Ft Leonard Wood, **N**...Conoco/mart, Aussie Jack's Steaks, Cracker Barrel, Popeye's, Comfort Inn, Fairfield Inn, Red Roof Inn, Toyota, Bill's Farm&Home, **S**...Total/mart, Arby's, Capt D's, Great Wall Chinese, KFC, Korean Rest., McDonald's, Mitch's Café, Papa John's, Pizza Hut, Subway, Taco Bell, Waffle House, Wendy's, Best Budget Inn, Econolodge, Holiday Inn Express, Ramada Inn, AutoZone, Dodge/Plymouth, $General, Firestone/auto, Ford/Lincoln/Mercury/Mazda, IGA Foods, Wal-Mart/auto, bank, carwash

Sullivan

Rolla

E

W

Missouri Interstate 44

159	Lp 44, to Waynesville, St Robert, **N**...Dairy Queen, Star Motel, Covered Wagon RV Park(2mi), Firestone, IGA Foods, TrueValue, auto parts, bank, **S**...Phillips 66/mart, Texaco/repair, DeVille Motel, Ozark Motel, Cadillac, Chevrolet, GMC, Goodyear, Olds, Pontiac, USPO
158.5mm	Roubidoux Creek
156	rd H, Waynesville, **N**...Citgo/mart, McDonald's, Subway, Star Motel(1mi), Covered Wagon RV Park(2mi), Ace Hardware
153	MO 17, to Buckhorn, **N**...Citgo/diesel/mart/rest., Ted Williams Seafood/Steaks, Ft Wood Inn, **S**...Texaco/diesel/mart/atm, Glen Oaks RV Park
150	MO 7, rd P to Richland, **N**...to Lake of the Ozarks, **S**...RV camping
145	MO 133, rd AB, to Richland, **N**...Conoco/diesel/mart/rest./24hr, **S**...Gasconade Hills RV(1°mi)
143mm	Gasconade River
140	rd N, to Stoutland, **S**...Conoco/mart, Phillips 66/diesel/mart, Midway Rest./American/Chinese
139.5mm	Bear Creek
135	rd F, Sleeper, no facilities
130	rd MM, **N**...Conoco/mart/Holiday Motel, Phillips 66/mart, Bell Rest., Best Budget Inn, Best Western, Forest Manor Motel, Munger Moss Motel, RV Park, **S**...Phillips 66/mart
129	MO 5, MO 32, MO 64, Lebanon, to Hartville, **N**...Conoco, Phillips 66/diesel, Baskin-Robbins, Burger King, Country Kitchen, Dairy Queen, Domino's, KFC, LJ Silver, McDonald's, Pizza Hut, Shoney's, Sonic, Subway, Taco Bell, Taco Hut, Wendy's, Western Sizzlin, Advance Parts, AutoZone, Chevrolet, Ford, IGA Foods, K-Mart/auto, NAPA, O'Reilly Parts, Russell Stover's, Walnut Bowl Outlet, bank, cinema, laundry, to Bennett Sprgs SP, **S**...CHIROPRACTOR, HOSPITAL, VETERINARIAN, Amoco, Conoco/diesel/mart, Capt D's, Hardee's, Pizza Hut, Stonegate Sta Rest., Brentwood Motel/rest., $Tree, Goodyear, Wal-Mart SuperCtr/24hr, to Lake of the Ozarks
127	Lp 44, Lebanon, **N**...Citgo, Phillips 66/diesel/mart, Texaco/diesel/rest./mart/24hr, Country Kitchen, Ranch House Rest., Waffle House, Best Way Inn, Econolodge, Hampton Inn, Holiday Inn Express, Quality Inn, Scottish Inns, Super 8, Chrysler/Plymouth, Ford/Lincoln/Mercury, Walnut Bowl Outlet, tires, **S**...Buick/Pontiac/Olds/GMC
123	county rd, **S**...KOA, RV/auto/diesel repair
118	rds C, A, Phillipsburg, **N**...Conoco/mart, **S**...Phillips 66/mart
113	rds J, Y, Conway, **N**...Phillips 66/mart, Shell/diesel/mart, Rockin Chair Café, Budget Inn, to Den of Metal Arts
111mm	**rest area both lanes, full(handicapped)facilities, phone, vending, litter barrels, picnic tables, petwalk**
108mm	Bowen Creek
107	county rd, no facilities
106mm	Niangua River
100	MO 38, rd W, Marshfield, **N**...Amoco/mart, Tiny's BBQ, Fair Oaks Motel, Plaza Motel, Chevrolet/Olds, Ford, Sears Hardware, **S**...CHIROPRACTOR, DENTIST, Conoco/diesel/U-Haul/mart, Phillips 66/mart/24hr, Sinclair, Country Kitchen, Dairy Queen, KFC, McDonald's, Pizza Hut, Sonic, Sunshine House Rest., Taco Bell, Holiday Inn Express, Goodyear/auto, Wal-Mart/auto, auto parts, bank, camping, carwash, laundry, tourist info
96	rd B, Northview, **N**...RV camping, no facilities
89mm	weigh sta both lanes
88	MO 125, Strafford, to Fair Grove, **N**...Phillips 66/diesel/mart/rest./24hr, TA/Amoco/Subway/Taco Bell/diesel/24hr, McDonald's, Goodyear, Paradise RV Park, fireworks, **S**...Citgo/mart, Super 8, Strafford RV Camping, bank
84	MO 744, no facilities
82b a	US 65, to Branson, Fedalia, **S**...to Table Rock Lake, Bull Shoals Lake, Amoco/diesel/mart, Consumer's Mkt Deli, Waffle House, American Inn, Best Western, Village Inn, Truck World, st patrol
80b a	Springfield, rd H to Pleasant Hope, **N**...Conoco/diesel/rest./24hr, Phillips 66/mart, Texaco/mart, Waffle House, Microtel, Motel 6, Quality Inn/rest., Super 8, to SWSU, **S**...to Glenstone Ave, HOSPITAL, Amoco/mart, Citgo/mart, Conoco/diesel, Phillips 66/mart, QT/mart, Texaco/mart, Total/A&W/Subway/diesel/LP/mart, Andy's Café, Applebee's, BBQ, Bob Evans, Burger King, ChinaChina, Cracker Barrel, Country Kitchen, Dairy Queen, Denny's, Frozen Custard, Hardee's, HongKong Chinese, LJ Silver, KFC, McDonald's, Pizza Hut, Pizza Inn, Ruby Tuesday, Shanghai Chinese, Shoney's, Sonic, Taco Bell, Village Inn Rest., Best Inn, Bass Country Inn, Best Western, Comfort Inn, Day's Inn, Drury Inn, Econolodge, Economy Inn, GuestHouse Inn, Holiday Inn, Maple Motel/Rest., PearTree Inn, Ramada Inn, RainTree Inn, Red Roof Inn, Rodeway Inn, Scottish Inn, Sheraton, Skyline Motel, Solar Inn, Goodyear/auto, K-Mart, O'Reilly's Parts, Precision Tune, PriceCutter Foods, Radio Shack, U-Haul, Wal-Mart/auto, cinema, mall

E ↕ W

Lebanon

Springfield

Missouri Interstate 44

| 77 | MO 13, KS Expswy, **N**...Interstate Inn, **S**...DENTIST, Amoco/diesel/mart, Citgo/Git'n Go/bank, Gas+, Phillips 66/mart, QT/mart, Aldi Foods, Arby's, Burger Sta, Dairy Queen, McDonald's, Pizza Inn, Subway, Taco Bell, Tiny's BBQ, Waffle House, Wendy's, Econolodge, Goodyear/auto, Wal-Mart/Super Ctr/24hr, cinema, hardware, zoo |

75 US 160 W byp, to Willard, Stockton Lake, **S**...Conoco/mart(1mi), Courtyard, airport

72 MO 266, to Chesnut Expwy, **N**...camping, tire repair, **S**...Coastal/diesel/trkstp/rest., Hardee's, McDonald's(2mi), Taco Bell(2mi), Best Budget Inn, Best Inn(1mi), Holiday Inn, Redwood Motel, KOA Traveler's Park(˚mi)

70 rds MM, B, **N**...antiques, fireworks, **S**...Phillips 66, Wilson's Creek Nat Bfd, KOA(1mi)

67 rds N, T, Bois D' Arc, **S**...to Republic, Citgo/Ozarks/diesel/mart, Conoco/mart, antiques(1mi)

66mm Pond Creek

64.6mm Dry Branch

64.5mm Pickerel Creek

61 rds K, PP, **N**...Amoco/mart/laundry, Total/diesel/rest./motel, Fostoria Glass outlet

58 MO 96, rds O, Z, Halltown, to Carthage, **S**...Texaco/diesel/mart, Scandinavian Motel, auto repair

57 to rd PP(from wb), no facilities

56.5mm Turnback Creek

56mm Goose Creek

52.5mm rest area both lanes, full(handicapped)facilities, phone, picnic tables, litter barrels, vending, petwalk

49 MO 174E, rd CCW, Chesapeake, **S**...Boondocks BBQ/steaks

46 MO 39, MO 265, Mt Vernon, Aurora, **N**...DENTIST, CHIROPRACTOR, Casey's Store/gas, TA/Conoco/diesel/rest./24hr, Phillips 66/mart, Sinclair/diesel/mart/rest./24hr, Apple Barrel, Chinese Rest., Dalma's Family Rest., Hardee's, KFC, McDonald's, Simple Simon's Pizza, Sonic, Taco Bell, Best Western Bel Aire, Budget Host, Midwest Motel, Family$, Food Faire, O'Reilly's Parts, cleaners/laundry, **S**...Total/diesel/mart/rest., Bamboo Garden, Subway, Best Western, Comfort Inn, Buick/Olds/Pontiac, antiques, flea mkt, to Table Rock Lake, Stockton Lake

44 rd H, Mt Vernon, to Monett, **N**...Citgo(2mi), Subway(1mi), Walnuts Motel, Mid-America Dental/Vision/Hearing

43.5mm Spring River

38 MO 97, to Stotts City, Pierce City, **N**...gas/repair, antiques, U of MO SW Ctr

33 MO 97 S, to Pierce City, **S**...Sinclair/diesel/mart

29 rd U, Sarcoxie, to La Ruffell, **N**...Ozark Village, WAC RV Park, antiques, **S**...Citgo/mart, Rebel's Rest., Sarcoxie Motel, Town&Country Motel/rest.

29mm Center Creek

26 MO 37, Sarcoxie, to Reeds, **N**...truck/trailer repair, **S**...Amoco/mart/gifts, Ford(2mi)

22 rd 100, **N**...RV parts, **S**...Xpress Trkstp/diesel/rest./24hr

21mm Jones Creek

18b a US 71A N, MO 59 S, to Carthage, Neosho, **N**...Phillips 66/mart, BudgetWay Motel, Coachlight RV Park, **S**...Total/mart, George Carver Mon, Ballard's Camping

15 MO 66 W, Lp 44(from wb), Joplin, **N**...Tara Motel, Timber Motel

15mm Grove Creek

14mm Turkey Creek

11 US 71 S, to Neosho, Ft Smith, **S**...Flying J/Conoco/Country Mkt/diesel/LP/rest./24hr

8b a US 71, Joplin, to Neosho, **N**...Amoco, Conoco/mart, Phillips 66/diesel, Sinclair, Texaco/mart, Applebee's, Arby's, BBQ, Benito's Mexican, Bob Evans, Braum's, Capt D's, Carl's Jr, Casa Montez Mexican, Country Kitchen, Denny's, Foretune East Chinese, Gringo's Grill, Jim Bob's Steaks, KFC, Kyoto Steaks, McDonald's, Olive Garden, Pizza Hut, Red Lobster, Ruby Tuesday, Shoney's, Steak'n Shake, Stout's Pizza, Subway, Taco Hut, Waffle House, Wendy's, Western Sizzlin, Best Western, Comfort Inn, Day's Inn, Drury Inn, Fairfield Inn, Hampton Inn, Holiday Inn, Howard Johnson, Motel 6, Ramada Inn, Roadsite Inn, Solar Inn, Super 8, Thunderbird Motel, Tropicana Hotel, Chrysler/Plymouth/Dodge, Food4Less/24hr, Ford/Lincoln/Mercury, Jo-Ann Fabrics, K-Mart, Lowe's Bldg Supply, NAPA, Nissan/Mercedes, Office Depot, Pennzoil, Sam's Club/gas, Toyota, VW, Wal-Mart, airport, bank, **S**...Citgo/diesel/mart, Phillips 66/diesel/mart, Cracker Barrel, Fazoli's, Viva Mexico Buffet, Microtel, RV Ctr

6 MO 86, MO 43 N, Joplin, to Racine, **N**...HOSPITAL, Citgo/mart, McDonald's, Pizza Hut(1.5mi), Capri Motel(1mi), West Wood Motel(2mi), **S**...Texaco, Harley-Davidson

5.5mm Shoal Creek

4 MO 43 to Seneca, **N**... Pilot/DQ/Wendy's/diesel/mart/24hr, Peterbilt, **S**...Amoco/rest., Conoco/Petro/Blimpie/diesel/mart, McDonald's, Comfort Inn, Sleep Inn, Tahoe Motel, KOA, TruckWorld

3mm weigh sta both lanes

2mm Welcome Ctr eb, rest area wb, full(handicapped)facilities, picnic tables, litter barrels, phones, vending

1 US 400, US 166W, to Baxter Springs, KS, no facilities

0mm Missouri/Oklahoma state line

E

W

Joplin

Missouri Interstate 55

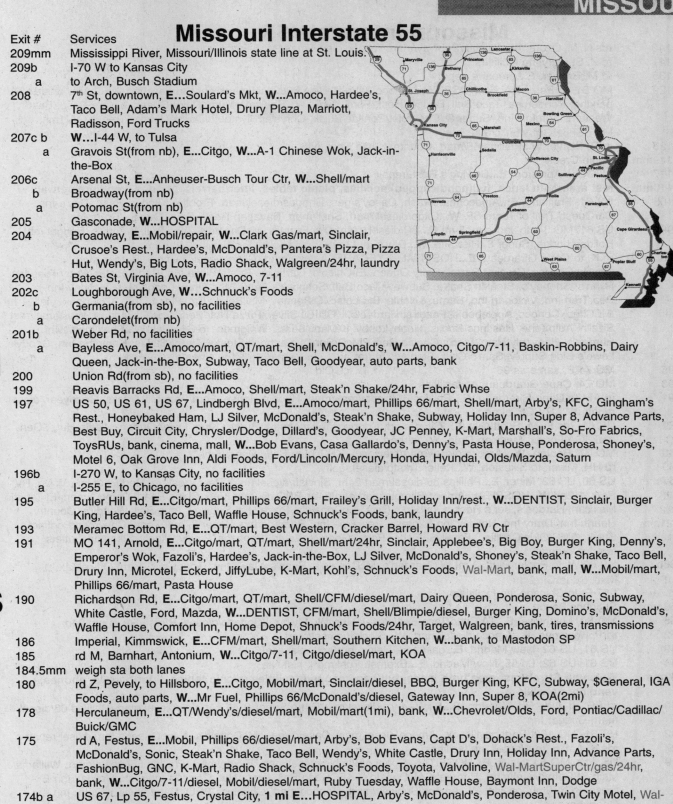

Exit #	Services
209mm	Mississippi River, Missouri/Illinois state line at St. Louis.
209b	I-70 W to Kansas City
a	to Arch, Busch Stadium
208	7th St, downtown, **E**...Soulard's Mkt, **W**...Amoco, Hardee's, Taco Bell, Adam's Mark Hotel, Drury Plaza, Marriott, Radisson, Ford Trucks
207c b	**W**...I-44 W, to Tulsa
a	Gravois St(from nb), **E**...Citgo, **W**...A-1 Chinese Wok, Jack-in-the-Box
206c	Arsenal St, **E**...Anheuser-Busch Tour Ctr, **W**...Shell/mart
b	Broadway(from nb)
a	Potomac St(from nb)
205	Gasconade, **W**...HOSPITAL
204	Broadway, **E**...Mobil/repair, **W**...Clark Gas/mart, Sinclair, Crusoe's Rest., Hardee's, McDonald's, Pantera's Pizza, Pizza Hut, Wendy's, Big Lots, Radio Shack, Walgreen/24hr, laundry
203	Bates St, Virginia Ave, **W**...Amoco, 7-11
202c	Loughborough Ave, **W**...Schnuck's Foods
b	Germania(from sb), no facilities
a	Carondelet(from nb)
201b	Weber Rd, no facilities
a	Bayless Ave, **E**...Amoco/mart, QT/mart, Shell, McDonald's, **W**...Amoco, Citgo/7-11, Baskin-Robbins, Dairy Queen, Jack-in-the-Box, Subway, Taco Bell, Goodyear, auto parts, bank
200	Union Rd(from sb), no facilities
199	Reavis Barracks Rd, **E**...Amoco, Shell/mart, Steak'n Shake/24hr, Fabric Whse
197	US 50, US 61, US 67, Lindbergh Blvd, **E**...Amoco/mart, Phillips 66/mart, Shell/mart, Arby's, KFC, Gingham's Rest., Honeybaked Ham, LJ Silver, McDonald's, Steak'n Shake, Subway, Holiday Inn, Super 8, Advance Parts, Best Buy, Circuit City, Chrysler/Dodge, Dillard's, Goodyear, JC Penney, K-Mart, Marshall's, So-Fro Fabrics, ToysRUs, bank, cinema, mall, **W**...Bob Evans, Casa Gallardo's, Denny's, Pasta House, Ponderosa, Shoney's, Motel 6, Oak Grove Inn, Aldi Foods, Ford/Lincoln/Mercury, Honda, Hyundai, Olds/Mazda, Saturn
196b	I-270 W, to Kansas City, no facilities
a	I-255 E, to Chicago, no facilities
195	Butler Hill Rd, **E**...Citgo/mart, Phillips 66/mart, Frailey's Grill, Holiday Inn/rest., **W**...DENTIST, Sinclair, Burger King, Hardee's, Taco Bell, Waffle House, Schnuck's Foods, bank, laundry
193	Meramec Bottom Rd, **E**...QT/mart, Best Western, Cracker Barrel, Howard RV Ctr
191	MO 141, Arnold, **E**...Citgo/mart, QT/mart, Shell/mart/24hr, Sinclair, Applebee's, Big Boy, Burger King, Denny's, Emperor's Wok, Fazoli's, Hardee's, Jack-in-the-Box, LJ Silver, McDonald's, Shoney's, Steak'n Shake, Taco Bell, Drury Inn, Microtel, Eckerd, JiffyLube, K-Mart, Kohl's, Schnuck's Foods, Wal-Mart, bank, mall, **W**...Mobil/mart, Phillips 66/mart, Pasta House
190	Richardson Rd, **E**...Citgo/mart, QT/mart, Shell/CFM/diesel/mart, Dairy Queen, Ponderosa, Sonic, Subway, White Castle, Ford, Mazda, **W**...DENTIST, CFM/mart, Shell/Blimpie/diesel, Burger King, Domino's, McDonald's, Waffle House, Comfort Inn, Home Depot, Shnuck's Foods/24hr, Target, Walgreen, bank, tires, transmissions
186	Imperial, Kimmswick, **E**...CFM/mart, Shell/mart, Southern Kitchen, **W**...bank, to Mastodon SP
185	rd M, Barnhart, Antonium, **W**...Citgo/7-11, Citgo/diesel/mart, KOA
184.5mm	weigh sta both lanes
180	rd Z, Pevely, to Hillsboro, **E**...Citgo, Mobil/mart, Sinclair/diesel, BBQ, Burger King, KFC, Subway, $General, IGA Foods, auto parts, **W**...Mr Fuel, Phillips 66/McDonald's/diesel, Gateway Inn, Super 8, KOA(2mi)
178	Herculaneum, **E**...QT/Wendy's/diesel/mart, Mobil/mart(1mi), bank, **W**...Chevrolet/Olds, Ford, Pontiac/Cadillac/Buick/GMC
175	rd A, Festus, **E**...Mobil, Phillips 66/diesel/mart, Arby's, Bob Evans, Capt D's, Dohack's Rest., Fazoli's, McDonald's, Sonic, Steak'n Shake, Taco Bell, Wendy's, White Castle, Drury Inn, Holiday Inn, Advance Parts, FashionBug, GNC, K-Mart, Radio Shack, Schnuck's Foods, Toyota, Valvoline, Wal-MartSuperCtr/gas/24hr, bank, **W**...Citgo/7-11/diesel, Mobil/diesel/mart, Ruby Tuesday, Waffle House, Baymont Inn, Dodge
174b a	US 67, Lp 55, Festus, Crystal City, **1 mi E**...HOSPITAL, Arby's, McDonald's, Ponderosa, Twin City Motel, Wal-Mart SuperCtr/gas/24hr, same as 175
170	US 61, **W**...Citgo/diesel/LP/mart
162	rds DD, OO, no facilities
160mm	**rest areas both lanes, full(handicapped)facilities, picnic tables, litter barrels, phones, vending, petwalk**
157	rd Y, Bloomsdale, **E**...Phillips 66/mart, **W**...Texaco/diesel/mart
154	rd O, no facilities
150	MO 32, rds B, A, to St Genevieve, **E**...HOSPITAL, Amoco/mart, Dairy Queen, KFC(4mi), McDonald's(5mi), Microtel(4mi), Hist Site(6mi), **W**...Olds/Pontiac/GMC, Hawn SP(11mi)

N

S

Missouri Interstate 55

P **e** **r** **r** **y** **v** **i** **l** **e**	143	rds N, M, Ozora, **W**...Sinclair/Subway/mart, Texaco/diesel/mart/24hr, Family Inn/rest.
	141	rd Z, St Mary, no facilities
	135	rd M, Brewer, **E**...propane depot
	129	MO 51, to Perryville, **E**...HOSPITAL, Amoco/24hr, Phillips 66/diesel/mart/24hr, Burger King, KFC, McDonald's/RV parking, Ponderosa, Taco Bell, Budget Host, Ford, bank, **W**...Shell/mart/atm, Dairy Queen, Kelly's Rest., Best Western, Comfort Inn, Super 8, Chevrolet/Pontiac/Buick, Chrysler/Plymouth/Dodge/Jeep, $Tree, KOA(1mi), Wal-Mart SuperCtr/24hr, crafts
	123	rd B, Biehle, **W**...Phillips 66/mart, Country Kettle Rest./gifts
	119mm	Apple Creek
	117	rd KK, to Appleton, **E**...Sewing's Rest./repair
	110mm	**rest area both lanes, full(handicapped)facilities, picnic tables, litter barrels, phones, vending, petwalk**
	105	US 61, Fruitland, **E**...Amoco/mart/wash, Casey's/gas, Rhodes/diesel/mart, Frontier Kitchen Rest., Day's Inn, CarQuest, Trail of Tears SP, **W**...Citgo/diesel/mart, Shell/mart, Bavarian Rest., Dairy Queen, Drury Inn
	99	US 61, MO 34, to Jackson, **3 mi E**...BP/diesel, Capt D's, Fazoli's, McDonald's, Taco Bell, Wendy's, Budget Inn, Holiday Lodge, Super 8, Hist Site, **W**...Day's Inn(3mi), Wal-Mart Super Ctr/gas/24hr(2mi)
	96	rd K, to Cape Girardeau, **E**...HOSPITAL, EYECARE, Amoco/mart, Citgo/diesel, Blimpie, Burger King, ChuckeCheese, Cracker Barrel, Dairy Queen, El Chico's, Garfield's Rest., Pizza Inn, Red Lobster, Ruby Tuesday, Ryan's, Shoney's, Steak'n Shake, Subway, Taco Bell, Schnuck's Food/24hr, Drury Lodge/rest., Holiday Inn, PearTree Inn, Victorian Inn, Barnes&Noble, Big Lots, JC Penney, MailBoxes Etc, Pier 1, ToysRUs, bank, mall, **1 mi E**...Citgo, Conoco, Applebee's, Fazoli's, Hardee's, KFC, LJ Silver, Pizza Hut, Ponderosa, Wendy's, Western Sizzlin, AutoZone, Hastings Books, HobbyLobby, K-Mart, Sears, Walgreen, to SEMSU, **W**...Shell/mart/24hr, Hardee's, Heavenly Ham, Lion's Choice Rest., McDonald's, Outback Steaks, Drury Suites, Hampton Inn, Goody's, Lowe's Bldg Supply, Sam's Club, Staples, Target, Wal-Mart SuperCtr/24hr, cinema
	95	MO 74 E, same as 96
	93	MO 74, Cape Girardeau, no facilities
	91	rd AB, to Cape Girardeau, **E**...Phillips 66/diesel/rest./mart, Huddle House, Ford/Peterbilt Trucks, Goodyear, diesel repair, **W**...RV Ctr, airport
	89	US 61, rds K, M, Scott City, **E**...Citgo/mart, Conoco, Texaco/mart, Burger King, Dairy Queen, Waffle Hut, $General, IGA Foods, Medicap Drug, **W**...airport
	80	MO 77, Benton, **W**...Amoco/diesel/mart/fireworks, KFC/Taco Bell, winery(8mi)
	69	rd HH, Miner, to Sikeston, **W**...Keller Trkstp/diesel, golf
M **i** **n** **e** **r**	67	US 60, US 62, Miner, **E**...Phillips 66/diesel/mart/24hr, Sinclair/diesel/mart, Best Western, Red Roof Inn, Volvo, RV Park, **1-2mi W**...HOSPITAL, Amoco/Blimpie/mart/atm/24hr, BBQ, Burger King, China Pearl, Dairy Queen, El Tapatio Mexican, Hardee's, Jer's Rest., Lambert's Café, McDonald's, Pizza Hut, Subway, Taco Bell, Wendy's, Country Hearth Inn, Drury Inn, Holiday Inn Express, PearTree Inn, Ramada Inn, Super 8, AutoZone, Chevrolet/Pontiac/Buick, GMC/Olds/Cadillac, Family$, Fisher Parts, Goodyear/auto, Pennzoil, Piggly Wiggly, Sikeston Outlets/famous brands
	66b	US 60 W, to Poplar Bluff, **3 mi W**...Hardee's, McDonald's, Ryan's, Day's Inn, Goody's, K-Mart, OfficeMax, Wal-Mart SuperCtr/24hr
	a	I-57 E, to Chicago, no facilities
	59mm	St Johns Bayou
	58	MO 80, Matthews, **E**...TA/Citgo/Taco Bell/diesel/24hr, **W**...Flying J/Conoco/diesel/LP/rest./RV dump/24hr
	52	rd P, Kewanee, **E**...Amoco/diesel/mart, **W**...Sinclair/diesel/mart
	49	US 61, US 62, New Madrid, **E**...gas/food, Relax Inn(3mi), hist site(3mi)
	44	US 61, US 62, Lp 55, New Madrid, **E**...Express Fuel/mart, hist site
	42mm	**Welcome Ctr nb/rest area both lanes, full(handicapped)facilities, picnic tables, litter barrels, phones, vending, petwalk**
	40	rd EE, St Jude Rd, Marston, **E**...Pilot/Arby's/diesel/mart/24hr, Super 8, **W**...Amoco/diesel/mart, Phillips 66/diesel/mart, Budget Inn
H **a** **y** **t** **i**	32	US 61, MO 162, Portageville, **W**...Amoco/diesel/mart, Casey's/gas, McDonald's, New Orleans Inn, diesel repair
	27	rds K, A, BB, to Wardell, no facilities
	19	US 412, MO 84, Hayti, **E**...HOSPITAL, Conoco/Blimpie/diesel/mart, Exxon/mart, Phillips 66/diesel/mart, Williams/Main St Café/diesel/mart, Hardee's, McDonald's, KFC/Taco Bell, Pizza Hut, Comfort Inn/rest., Holiday Inn Express, KOA(6mi), casino, **W**...Amoco/Subway/mart, Conoco/mart, Chubby's BBQ, Dairy Queen, Drury Inn, Goodyear
	17a	I-155 E, to TN, US 412, no facilities
	14	rds J, H, U, to Caruthersville, Braggadocio, no facilities
	10mm	weigh sta both lanes
	8	MO 164, Steele, **E**...Duckie's/diesel/mart
	4	rd E, Cooter, to Holland, no facilities
	3mm	**rest area both lanes, full(handicapped)facilities, picnic tables, litter barrels, phones, vending, petwalk**
	1	US 61, rd O, Holland, **E**...Raceway/diesel/mart, **W**...Coastal/diesel/mart
	0mm	Missouri/Arkansas state line

N

S

Missouri Interstate 57

Exit #	Services
22mm	Mississippi River, Missouri/Illinois state line
18.5mm	weigh sta both lanes
12	US 60, US 62, MO 77, Charleston, **E...**Phillips 66/diesel/mart, Sunshine Travel Ctr/diesel/buffet, Economy Motel, **W...**Casey's/gas, Sinclair/diesel/mart, KFC, Charleston Motel/rest.
10	MO 105, Charleston, **E...**Conoco/Boomland/diesel/mart, Pilot/Subway/diesel/mart/24hr, antiques, camping, **W...**Casey's/gas, Dairy Queen, McDonald's, Pizza Hut, Comfort Inn, Town&Country Foods, Wal-Mart, bank
4	rd B, Bertrand, **E...**Citgo/diesel/mart/fireworks
1b a	I-55, N to St Louis, S to Memphis
I-57 begins/ends on I-55, exit 66.	

N ↕ S

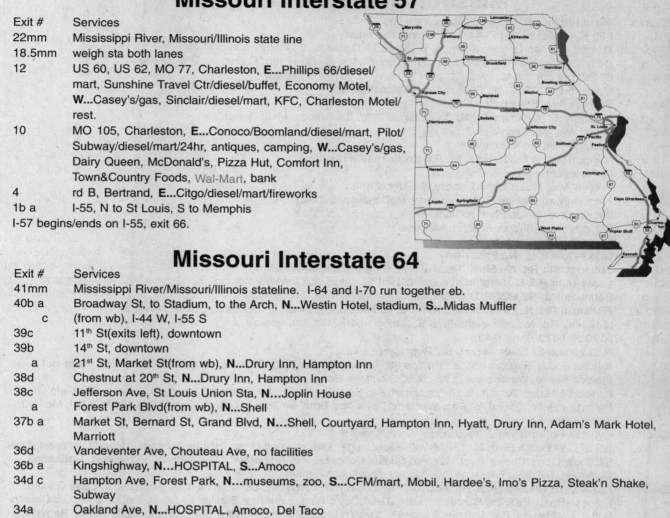

Missouri Interstate 64

Exit #	Services
41mm	Mississippi River/Missouri/Illinois stateline. I-64 and I-70 run together eb.
40b a	Broadway St, to Stadium, to the Arch, **N...**Westin Hotel, stadium, **S...**Midas Muffler
c	(from wb), I-44 W, I-55 S
39c	11th St(exits left), downtown
39b	14th St, downtown
a	21st St, Market St(from wb), **N...**Drury Inn, Hampton Inn
38d	Chestnut at 20th St, **N...**Drury Inn, Hampton Inn
38c	Jefferson Ave, St Louis Union Sta, **N...**Joplin House
a	Forest Park Blvd(from wb), **N...**Shell
37b a	Market St, Bernard St, Grand Blvd, **N...**Shell, Courtyard, Hampton Inn, Hyatt, Drury Inn, Adam's Mark Hotel, Marriott
36d	Vandeventer Ave, Chouteau Ave, no facilities
36b a	Kingshighway, **N...**HOSPITAL, **S...**Amoco
34d c	Hampton Ave, Forest Park, **N...**museums, zoo, **S...**CFM/mart, Mobil, Hardee's, Imo's Pizza, Steak'n Shake, Subway
34a	Oakland Ave, **N...**HOSPITAL, Amoco, Del Taco
33d	McCausland Ave, **N...**Amoco, Del Taco
33c	Bellevue Ave, **N...**HOSPITAL
33b	Big Bend Blvd, no facilities
32b a	Eager Rd, Hanley Rd, **S...**Shell/mart, Bed Bath&Beyond, Home Depot, PetsMart, SportsAuthority, Target
31b a	I-170 N, **N...**Shell/mart, Burger King, Dairy Queen, KFC, Steak'n Shake, Dillard's, Famous Barr, mall, **S...**Amoco, Subway, Goodyear
30	McKnight Rd, no facilities
28c	Clayton Rd(from wb), no facilities
28b a	US 67, US 61, Lindbergh Blvd, **S...**Hilton, Shnuck's Foods, antiques, bank, mall
27	Spoede Rd, no facilities
26	rd JJ, Ballas Rd, **N...**HOSPITAL, **S...**HOSPITAL
25	I-270, N to Chicago, S to Memphis
23	Mason Rd, **N...**Courtyard, Marriott, LDS Temple, hwy patrol
22	MO 141, **N...**HOSPITAL, Courtyard, Marriott, **S...**Pizza Hut, bank
21	Timberlake Manor Pkwy, no facilities
20	Chesterfield Pkwy(from wb), same as 19b a
19b a	MO 340, Chesterfield Pkwy, **N...**Amoco/mart, Shell/mart, Applebee's, DoubleTree Hotel, Hampton Inn, Residence Inn, **S...**Mobil/mart, Casa Gallardo's, Dillard's, bank, cinema, mall
16	Long Rd, Chesterfield Airport Rd, **N...**Amoco, **1 mi S...**Amoco, Phillips 66/mart, Annie Gun's Rest., Longhorn Steaks, McDonald's, Old Country Buffet, Red Robin, SmokeHouse Rest., Subway, Hampton Inn, Hilton Garden, Best Buy, $Tree, Ford, Linens'n Things, Lowe's Bldg Supply, Michael's, OfficeMax, PetsMart, Wal-Mart, WorldMkt
14	Chesterfield Airport Rd(from eb), **N...**QT/mart, Dairy Queen, McDonald's, Target, airport, **S...**Phillip 66/mart
13mm	Missouri River, I-64 begins/ends.

E ↕ W

Charleston

St Louis

Missouri Interstate 70

St Louis

Exit #	Services
252mm	Mississippi River, Missouri/Illinois state line
251a	I-55 S, to Memphis, to I-44, to downtown/no return
250b	Memorial Dr, downtown, Stadium, **S**...Shell, McDonald's, Day's Inn, Midas Muffler
a	Conv Center, Arch, Riverfront, **N**...Embassy Suites, **S**...Drury Inn, Radisson, TWA Dome
249c	6th St(from eb), no facilities
a	Madison St, 10th St, **N**...Phillips 66/diesel/mart/wash
248b	St Louis Ave, Branch St, no facilities
a	Salisbury St, McKinley Br, **N**...truck repair/24hr, **S**... Amoco, Mobil/diesel/mart
247	Grand Ave, **N**...Citgo/diesel
246b	Adelaide Ave, no facilities
a	N Broadway, O'Fallon Park, **N**...Phillips 66/mart, Freightliner
245b	W Florissant, no facilities
a	Shreve Ave, **N**...Pillsbury Factory, **S**...Amoco/mart
244b	Kingshighway, **3/4 mi S**...Burger King, McDonald's, Subway
a	Bircher Blvd, Union Blvd, no facilities
243b	(243c from eb)Bircher Blvd, no facilities
243a	Riverview Blvd, no facilities
243	Goodfellow Blvd, **N**...Shell/mart
242b a	Jennings Sta Rd, **N**...Shell, Texaco/mart, **S**...transmission repair
241b	Lucas-Hunt Rd, **N**...Shell/mart/wash, **3/4 mi S**...Citgo, McDonald's
a	Bermuda Rd, **S**...HOSPITAL, Sinclair
240b a	Florissant Rd, **N**...CHIROPRACTOR, Amoco/McDonald's/mart, Vickers/mart, Sonic, Walgreen
239	N Hanley Rd, **N**...Jack-in-the-Box, **S**...Amoco/mart/24hr, McDonald's
238c b	I-170 N, I-170 S, no return
a	**N**...Lambert-St Louis Airport, **S**...Renaissance Hotel
237	Natural Bridge Rd(from eb), **S**...Phillips 66, Texaco, Burger King, Denny's, Jack-in-the-Box, KFC, Pizza Hut, Steak'n Shake, Waffle House, Wendy's, Stouffer's Hotel
236	Lambert-St Louis Airport, **S**...Amoco/mart, Big Boy, Coco's, Grone Cafeteria, Hardee's, Lombardo's Café Rafferty's Rest., Shoney's, Best Western, Day's Inn, Drury Inn/rest., Hampton Inn, Hilton, Holiday Inn, Marriott, Motel 6, Rent-a-truck
235c	Cypress Rd, rd B W, **N**...to airport
b a	US 67, Lindbergh Blvd, **N**...Shell/mart, Econolodge, Executive Intn'l Inn, Holiday Inn/café, Howard Johnson, bank, **S**...Shell, Lion's Choice Rest., Steak'n Shake, Congress Inn, Embassy Suites, Homestead Village, Radisson, Chevrolet, Dillard's, Firestone, JC Penney, Sears, bank, mall
234	MO 180, St Charles Rock Rd, **N**...HOSPITAL, CHIROPRACTOR, Citgo, Shell/mart, Applebee's, Casa Gallardo's, Chinese Rest., Fazoli's, Hardee's, Hatfield's/McCoy's Rest., LoneStar Steaks, LJ Silver, McDonald's, Old Country Buffet, Ponderosa, Red Lobster, Shoney's, Steak'n Shake, Taco Bell, Tony Bono's Rest., Knight's Inn, BabysRUs, Best Buy, Circuit City, Garden Ridge, Gateway Computers, GrandPa's Food/drug, Honda, JiffyLube, K-Mart/auto, Meineke Muffler, NTB, Office Depot, 1/2 Price Store, PetsMart, Sam's Club, SportsAuthority, Target, ToysRUs, Valvoline, Walgreen/24hr, bank, **S**...CHIROPRACTOR, Isuzu, hardware
232	I-270, N to Chicago, S to Memphis, no facilities, **1 mi S**...Shoney's
231b a	Earth City Expwy, **N**...Candlewood Suites, Courtyard, Sheraton, Studio+, **S**...Dave&Buster's, Doubletree Hotel, Wingate Inn, Harrah's Casino/Hotel, Riverport Ampitheatre
230mm	Missouri River

St Charles

Exit #	Services
229b a	5th St, St Charles, **N**...Amoco/mart/wash, Mobil/diesel/mart, MotoMart, Burger King, Denny's, Hardee's, KFC, McDonald's, Subway, Waffle House, Baymont Inn, St Charles Inn, Meineke Muffler, Walgreen/24hr, bank, **S**...Phillips 66/mart, QT/diesel/mart, Cracker Barrel, Best Western Noah's Ark, Day's Inn, Fairfield Inn, Howard Johnson, Ramada, Suburban Lodge, Townhouse Inn, JC Penney, malls, RV park
228	MO 94, St Charles, to Weldon Springs, **N**...Citgo/mart, Clark Gas, Shell/mart, Arby's, Baskin-Robbins, Chinese Rest., Dairy Queen, Hardee's, Imo's Pizza, LJ Silver, Papa John's, Pizza Hut, Steak'n Shake, Victoria's Creamery, Wendy's, Advance Parts, AutoValue, AutoZone, Chevrolet, Firestone, Midas Muffler, Precision Tune, Radio Shack, Shop'n Save, Walgreen, Valvoline, cleaners, **S**...Mobil/mart, Shell/mart/wash, ChuckeCheese, Gingham's Rest., Fazoli's, Beat Western, K-Mart, access to 227
227	Zumbehl Rd, **N**...Texaco, Burger King, PoFolks, Econolodge, Lowe's Bldg Supply, Sav-A-Lot Foods, cinema, **S**...DENTIST, Amoco/mart/wash, Citgo/mart, Mobil/FoodCourt/mart, Applebee's, BBQ, Blackeyed Pea, Bob Evans, Boston Mkt, Capt D's, Chevy's Mexican, Chinese Rest., Golden Corral, Hardee's, Italian Ristorante, Jack-in-the-Box, LoneStar Steaks, McDonald's, Mr Steak, Old Country Buffet, Popeye's, Subway, Taco Bell, Best Western, Comfort Inn, Red Roof Inn, TownePlace Suites, Deerborg's Foods/24hr, Drug Emporium, FashionBug, JiffyLube, Kia, OfficeMax, PetsMart, Sam's Club, Schnuck's Foods/24hr, Wal-Mart/auto, bank, cinema, cleaners, hardware, access to 228

E

W

Missouri Interstate 70

225 Truman Rd, to Cave Springs, **N**...Amoco/mart, Citgo/mart, Shell/mart, Texaco, Taco Bell, Wendy's, Budget Motel, Hampton Inn, Knight's Inn, Motel 6, Buick/Pontiac, Cadillac/Olds, Dunn's Sports, Mazda, Mercury/Lincoln, RV Ctr, Saturn, Target, Toyota, U-Haul, **S**...HOSPITAL, Conoco, Mobil, QT/mart, Big Boy, Burger King, Dairy Queen, Denny's, El Mezcal Mexican, Fazoli's, Ground Round, IHOP, Jack-in-the-Box, Lion's Choice Rest., LJ Silver, McDonald's, Pasta House, Pizza Hut, Ponderosa, Red Lobster, Steak'n Shake, Subway, White Castle, Holiday Inn Select, Chrysler/Jeep, Dodge, Home Depot, K-Mart, Office Depot, OfficeMax, PetCare, Shop'n Save, TJ Maxx

224 MO 370, no facilities

222 Mall Dr, St Peters, **N**...QT/diesel/mart/24hr, Burger King, Carnection Auto Dealer, Chevrolet, Honda, Mitsubishi, Subaru, **S**...DENTIST, CHIROPRACTOR, VETERINARIAN, MEDICAL CARE, Citgo/wash, Mobil/mart/wash, Arby's, Bob Evans, Chili's, Domino's, Hardee's, Jack-in-the-Box, Joe's Crabshack, McDonald's, Olive Garden, Pizza Hut, Steak'n Shake, Subway, Taco Bell, Wendy's, Drury Inn, Extended Stay America, Aldi Foods, AutoZone, Best Buy, Circuit City, Deerborg's Foods, Dillard's, JC Penney, JiffyLube, KidsRUs, Marshall's, Midas Muffler, Nissan/Hyundai/VW, Schnuck's Foods/24hr, Sears/auto, Service Merchandise, TireAmerica, ToysRUs, Valvoline, Walgreen, bank, cinema, cleaners, hardware

220 MO 79, to Elsberry, **N**...Cherokee Lakes Camping(7mi), **S**...Amoco/mart/wash, Citgo/7-11(1mi), Blimpie(1mi), Hardee's, McDonald's(1mi), Smokehouse Rest., Sonic

217 rd K, rd M, O'Fallon, **N**...OPTOMETRIST, Amoco, QT/mart, Blimpie, Burger King, Hardee's, Jack-in-the-Box, Ponderosa, Rally's, Taco Bell, Waffle House, AutoValue Parts, Firestone, JiffyLube, Radio Shack, Valvoline, bank, **S**...Citgo/diesel/mart, DirtCheap/mart, Shell/mart/wash/atm, Arby's, Big Boy, Fazoli's, IHOP, KFC, Lion's Choice Rest., McDonald's, Pizza Hut, Shoney's, Steak'n Shake/24hr, Stefanina's Pizza, Aldi Foods, AutoZone, K-Mart/drugs, Lowe's Bldg Supply, Meineke Muffler, Midas Muffler, Schnuck's Foods, Shop'n Save Foods, Walgreen, camping

216 Bryan Rd, **N**...VETERINARIAN, Super 8, Ford, **S**...Mobil/mart, QT/mart

214 Lake St Louis, **N**...Freedom RV Ctr(1mi), **S**...HOSPITAL, Phillips 66/mart, Shell/mart, Cutter's Rest., Denny's, Hardee's, Subway, Day's Inn, bank

212 rd A, **N**...Howard Johnson, Ramada Ltd, **S**...Citgo/diesel/mart, Arby's, Burger King, Imo's Pizza, Texas Willie's Steaks, Holiday Inn, Chrysler/Plymouth/Dodge/Jeep, Luxury Linens, Wentzville Outlets/famous brands

210b a US 61, US 40, to I-64 E, **S**...HOSPITAL, VETERINARIAN, Conoco/mart, Phillips 66/diesel, Shell/diesel/mart, Texaco/mart

209 rd Z(from wb), Church St, New Melle, no facilities

208 Pearce Blvd, Wentzville, **N**...HOSPITAL, Citgo/mart, QT/diesel/mart, KFC, McDonald's, Palace Buffet, Pizza Hut, Steak'n Shake/24hr, Taco Bell, Waffle House, Wendy's, Chevrolet/Buick, Family$, Schnuck's Food/deli, Walgreen, **S**...Amoco/mart/wash, Texaco, Capt D's, Hardee's, KwanYin Chinese, Pantera's Pizza, Ruggery's Ristorante, Super 8, $General, GreaseMonkey, Pinewood RV Park, Plymouth/Dodge, Radio Shack, Wal-Mart SuperCtr/24hr, bank

204mm weigh sta both lanes

203 rds W, T, Foristell, **N**...TA/Mobil/diesel/rest./24hr, Mr Fuel/diesel/mart, Best Western, diesel repair, **S**...Texaco/diesel/mart/café

200 rds J, H, F, Wright City, **N**...Citgo/diesel/mart, Shell/mart, Ruiz Castillo's Mexican(1mi), **S**...Phillips 66/mart/Master Mufflers, Big Boy's Rest., Super 7 Inn, Volvo, bank, laundry

199 (from wb), same as 200

198mm **rest area both lanes, full(handicapped)facilities, phone, picnic tables, litter barrels, petwalk**

193 MO 47, Warrenton, **N**...Citgo/mart, Phillips 66/diesel/mart, Sinclair, Burger King, Dairy Queen, Jack-in-the-Box, McDonald's, Pizza Hut, Subway, Waffle House, Day's Inn, Super 8, Ford, Wal-Mart, **S**...Amoco/mart, Conoco/Blimpie/diesel, Texaco/diesel/deli/mart/atm/24hr, Denny's, Hardee's, Imo's Pizza, KFC, Taco Bell, AmeriHost, Budget Host, Chrysler/Plymouth/Dodge/Jeep, bank, laundry, **1/2 mi down frontage rd**...CarQuest, Chevrolet/Pontiac/Olds, Outlet Ctr/famous brands

188 rds A, B, to Truxton, **S**...Flying J/Conoco/diesel/rest./LP/24hr, Budget Inn, Outlet Ctr/famous brands

183 rds E, NN, Y, Jonesburg, **1 mi N**...KOA/LP, **S**...Amoco/mart, Texaco/diesel/mart

179 rd F, High Hill, **S**...Budget Motel, Colonial Inn

175 MO 19, New Florence, **N**...Amoco/Hardee's/diesel/mart, Phillips 66/diesel/mart/24hr, Shell/diesel, Maggie's Café, McDonald's, Royal Inn, Super 8, auto repair, winery tours, **S**...RV parking

170 MO 161, rd J, Danville, **N**...Phillips 66/mart, to Graham Cave SP, Kan-Do RV Park, **S**...Lazy Day RV Park

169.5mm **rest area wb, full(handicapped)facilities, phone, vending, picnic tables, litter barrels, petwalk**

168mm Loutre River

Missouri Interstate 70

167mm **rest area eb, full(handicapped)facilities, phone, vending, picnic tables, litter barrels, petwalk**

161   rds D, YY, Williamsburg, **N**...Crane's Gen Store, **S**...Conoco/diesel/café/mart/24hr

155   rds A, Z, to Calwood, **N**...antiques

148   US 54, Kingdom City, **N**...Amoco, Phillips 66/Taco Bell/diesel, AmeriHost, carpet outlet, to Mark Twain Lake, **S**...Conoco/diesel/mart, Texaco/diesel/rest./atm, Petro/Mobil/diesel/rest./24hr, Phillips 66/McDonald's, Subway, Comfort Inn, Day's Inn, Frontier Motel, Super 8

144   rds M, HH, to Hatton, **S**...Crooked Creek Camping, fireworks

137   rds DD, J, Stephens, to Millersburg, **S**...Walnut Bowls Factory Outlet, to Little Dixie WA

133   rd Z, to Centralia, **N**...Loveall's RV

131   Lake of the Woods Rd, **N**...Amoco/mart, Texaco/Subway/mart, **S**...Shell/diesel/mart/24hr/RV Park

128a   US 63, Columbia, to Jefferson City, **N**...QT/mart, Bob Evans, Burger King, Casa Grande Mexican, Cracker Barrel, Golden Corral, KFC, McDonald's, Pizza Hut, Schlotsky's, Steak'n Shake, Taco Bell, Wendy's, Fairfield Inn, Hampton Inn, Super 8, Aamco, Home Depot, KOA, Cottonwood's RV, same as 128, **S**...HOSPITAL, Breaktime/mart, Texaco/mart, Haymkt Rest., Sonic, Subway, Best Western, Baymont Inn, Comfort Inn, Holiday Inn Express, Hawthorn Suites, Howard Johnson, Wingate Inn, FashionBug, Lowe's Bldg Supply, MegaMkt Foods, Nowell's Foods, Sam's Club, Staples, Wal-Mart SuperCtr/24hr(1mi)

128   Lp 70(from wb), Columbia, **N**...Wendy's, Super 8, **S**...HOSPITAL, Conoco/diesel, Best Western, Baymont Inn, Comfort Inn, Holiday Inn Express, same as 128a

127   MO 763, Columbia, to Moberly, **N**...Breaktime/diesel/mart, Burger King, Mel's Diner, Budget Host, Eastwood Motel, Motel 6, Ramada, Scottish Inn, Travelodge, Acura, Chrysler/Dodge, $General, Pontiac/Cadillac, Saturn, Toyota, same as 126, **S**...Amoco/mart, Dairy Queen, Taco Bell, Econolodge, Super 7 Motel

126   MO 163, Providence Rd, Columbia, **N**...Amoco, Conoco, Big Boy, Country Kitchen, Denny's, Shoney's, Best Value Inn, Motel 6, Red Roof, Buick/GMC, Ford, Honda, groceries, same as 127, **S**...HOSPITAL, Amoco/mart, Breaktime/gas, Burger King, Dairy Queen, McDonald's, Pizza Hut, Sonic, Subway, Taco Bell, Aldi Foods, AutoZone, Chevrolet/Nissan, Chrysler/Jeep, $General, Lincoln/Mercury, Midas Muffler, O'Reilly Parts

125   Lp 70, West Blvd, Columbia, **S**...Breaktime/diesel, Conoco/diesel/mart, Citgo, Phillips 66/diesel/mart, Texaco/mart, Chevy's Mexican, Denny's, El Maguey Mexican, Fazoli's, Hardee's, LJ Silver, Olive Garden, Outback Steaks, Perkins, Pizza Hut, Red Lobster, Ryan's, Wendy's, Western Sizzlin, Yen Ching Chinese, Deluxe Inn, Eastwood Inn, Econolodge, Howard Johnson/rest., Scottish Inn, Advance Parts, Chrysler/Plymouth/Subaru, Firestone/auto, Mitsubishi, Mr Transmissions, Muffler Clinic, PrecisionTune, PriceChopper Foods, U-Haul, laundry, same as 124

124   MO 740, rd E, Stadium Blvd, Columbia, **N**...BreakTime/mart, **S**...Texaco/mart, Alexander's Steaks, Applebee's, Burger King, Bobby Buford's Rest., Denny's, Furr's Cafeteria, Golden Corral, Hardee's, KFC, Little Caesar's, LJ Silver, McDonald's, Old Chicago Pizza, Pasta House, Pizza Hut, Ponderosa, Red Lobster, Ruby Tuesday, Steak'n Shake, Sub Shop, Subway, Taco Bell, Wendy's, Baymont Inn, Day's Inn, Drury Inn, Holiday Inn/rest., Motel 6, Circuit City, Dillard's, Ford, Isuzu, JC Penney, JiffyLube, K-Mart/drugs, MailBoxes Etc, MC Sports, Pier 1, Radio Shack, Sears/auto, Toyota, ToysRUs, Wal-Mart/auto, bank, cinema, laundry, mall, to U of MO, same as 125

122mm   Perche Creek

121   US 40, rd UU, Midway, **N**...Conoco/diesel/rest./mart/atm, Texaco/mart, Budget Inn, Goodyear, Flea Mkt/antiques

117   rds J, O, to Huntsdale, **S**...Furniture World, repair

115   rd BB, Rocheport, **N**...winery, to Katy Tr SP, **S**...Sinclair/mart, antiques

114.5mm   Missouri River

111   MO 98, MO 179, to Wooldridge, Overton, **S**...Phillips 66/diesel/mart

106   MO 87, Bingham Rd, to Boonville, **N**...Conoco/diesel/mart/rest., Texaco/diesel/mart, Atlasta Motel/rest.

104mm **rest area both lanes, full(handicapped)facilities, phone, picnic tables, litter barrels, vending, petwalk**

103   rd B, Main St, Boonville, **N**...HOSPITAL, Phillips 66/TCBY/diesel/mart, Texaco/Subway/mart, KFC, McDonald's, Sonic, Taco Bell, Day's Inn, Super 8, Wal-Mart/café, antiques, camping, **S**...Amoco/diesel/mart, Bobber/Conoco/diesel/rest./mart/24hr, OK Motel

101   US 40, MO 5, to Boonville, **N**...Speedway/Country Kitchen/mart, Burger King, Hardee's, Comfort Inn, Chrysler/Dodge/Jeep, Russell Stovers Candy, **2 mi N**...Homestead Motel, **S**...Texaco/mart, to Lake of the Ozarks

98   MO 41, MO 135, Lamine, **N**...to Arrow Rock HS(13mi), **S**...Conoco/diesel/mart, Texaco/diesel/mart/24hr, Dogwood Rest.

93mm   Lamine River

89   rd K, to Arrow Rock, **N**...to Arrow Rock HS, **S**...Sinclair, motel, NAPA Repair

84   rds J, **N**...Citgo/Dairy Queen/Stuckey's, truck repair

78b a   US 65, to Marshall, **N**...Conoco/mart, Lazy Days RV Park, **S**...Breaktime/diesel/mart

77mm   Blackwater River

74   rds YY, **N**...Betty's Inn/Texaco/rest./garage/24hr, **S**...Amoco/diesel/mart/rest.

71   rds EE, K, to Houstonia, **N**...flea mkt

66   MO 127, Sweet Springs, **S**...HOSPITAL, Breaktime/Phillips 66/diesel/mart, Brownsville Sta Rest., People's Choice Motel, Super 8, auto parts/repair

65.5mm   Davis Creek

62   rds VV, Y, Emma, **N**...Sinclair/diesel/mart/RV dump, Malfunction Jct Motel/rest./RV Park

E

W

Missouri Interstate 70

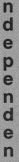

58   MO 23, Concordia, **N**...TA/Phillips 66/Subway/TCBY/ diesel/rest./24hr, KFC/Taco Bell, McDonald's, $General, **S**...Breaktime/diesel/mart, Conoco/diesel/mart, Texaco, BBQ, Hardee's, Palace Rest., Pizza Hut, Taco John's, Best Western, Day's Inn, Golden Award Motel/plaza, Country Mart/deli/bakery, Goodyear, JiffyLube, antiques

57.5mm   **rest area both lanes, full(handicapped)facilities, phone, picnic tables, litter barrels, vending, petwalk**

52   rd T, Aullville, no facilities

49   MO 13, to Higginsville, **N**...Williams/Main St Café/diesel/mart/ 24hr, Best Western/rest., Classic Motel(4mi), to Whiteman AFB, to Confederate Mem, **S**...Super 8/rest., Interstate RV Park(1/ 2mi)

45   rd H, to Mayview, **S**...Interstate RV Park(3mi)

43mm   weigh sta both lanes

41   rds O, M, to Lexington, Mayview, no facilities

38   MO 131(from wb), Odessa, **S**...Amoco, King/Phillips 66/diesel/ mart, Texaco/mart/atm, Hardee's, McDonald's, Sonic, Subway, Taco John's, Wendy's, camping, same as 37

37   MO 131, Odessa, **N**...Countryside Dining, **S**...Amoco, King/Phillips 66/diesel/mart, Texaco/mart/ atm, Hardee's, McDonald's, Morgan's Rest., Pizza Hut, Sonic, Subway, Taco John's, Wendy's, Odessa Inn, Parkside Inn, Chrysler/Plymouth/Dodge/Jeep, Ford, Midas Muffler, O'Reilly Parts, Prime Outlets/famous brands, Thriftway Foods, RV dump, same as 38

31   rds D, Z, to Bates City, Napoleon, **N**...I-70 RV Park, **S**...Amoco/diesel/mart/atm, Sinclair/diesel/mart, fireworks

29.5mm   Horse Shoe Creek

28   rd HF, Oak Grove, **N**...TA/Conoco/Popeye's/Pizza Hut/diesel/24hr, Day's Inn, Blue Beacon, KOA, **S**...QT/diesel/ mart/24hr, Texaco/DQ/Country Kitchen/Wendy's/diesel/mart, Hardee's, KFC/Taco Bell, McDonald's, Subway, Waffle House, Econolodge, Oak Grove Inn, Wal-Mart SuperCtr/24hr

24   US 40, rds AA, BB, to Buckner, **N**...McLeroy/diesel/mart, Phillips 66/diesel/mart/rest., Comfort Inn, Travelodge, Country Campers RV Ctr, **S**...Pilot/Subway/diesel/24hr, Sonic, GoodSam RV Park/LP, Trailside RV Ctr

21   Adams Dairy Pkwy, **N**...VETERINARIAN, Nationwide RV Ctr(1mi), **S**...Conoco/Burger King/diesel/mart, Total/ diesel, Courtyard, **2 mi S**...Arby's, Fazoli's, Ponderosa

20   MO 7, Blue Springs, **N**...CHIROPRACTOR, Phillips 66/diesel/mart, Sinclair, Bob Evans, Country Kitchen, Hardee's, NY Burrito, Motel 6, Ramada Ltd, Super 8, Sleep Inn, Ace Hardware, O'Reilly Parts, Osco Drug, PriceChopper Foods, Valvoline, Walgreen, **S**...HOSPITAL, CHIROPRACTOR, VETERINARIAN, Amoco/diesel/ mart/wash/24hr, Conoco/diesel/mart, QT/mart, Texaco/KrispyKreme/mart, Total/mart, Applebee's, BBQ, Burger King, China Buffet, Denny's, Godfather's, Golden Corral, Hardee's, Jose Pepper's, KFC, LJ Silver, McDonald's, Subway, Texas Tom's Mexican, Village Inn, Wendy's, White Castle, Winstead's Burgers, Day's Inn, Hampton Inn, Holiday Inn Express, Advance Parts, Chevrolet, Firestone/auto, HobbyLobby, Hyundai, JiffyLube, Midas Muffler, Office Depot, PrecisionTune, Radio Shack, Saturn, Wolf Camera, bank, **1 mi S**...Arby's, Fazoli's, AutoZone, Goodyear/auto, Hy-Vee Foods, K-Mart, NAPA, Wal-Mart/auto

18   Woods Chapel Rd, **N**...Amoco, American Inn, Interstate Inn, Microtel, Ford, Harley-Davidson, **S**...Conoco/ diesel/mart, QT/mart, Church's Chicken, McDonald's, Perkins, Pizza Hut, Taco Bell, Waffle House, AllSeasons RV, IGA Foods, Nissan, ProLube/wash, bank, same as 20

17   Little Blue Pkwy, 39ᵗʰ St, **N**...BBQ, Fazoli's, mall entrance

16mm   Little Blue River

15b   MO 291 N, Independence, **1 exit N on 39ᵗʰ St**...HOSPITAL, DENTIST, CHIROPRACTOR, EYECARE, Conoco, QT/mart, Phillips 66/mart, Applebee's, Arby's, BBQ, Bob Evans, Burger King, Chevy's Mexican, Chili's, Denny's, Hops Grill, KFC, LJ Silver, LoneStar Steaks, Longhorn Steaks, Luby's, McDonald's, Mr GoodCents Subs, Souper Salad, Starbucks, TGIFriday, Wendy's, Fairfield Inn, Residence Inn, Albertson's/drugs, Barnes&Noble, Bed Bath&Beyond, Circuit City, Dick's Sports, Dillard's, Gateway Computers, Jo-Ann Fabrics, Kohl's, Marshall's, NTB, PetsMart, Sears/auto, Target, Wal-Mart SuperCtr/24hr, cinema, mall

   a   I-470 S, MO 291 S, to Lee's Summit, no facilities

14   Lee's Summit Rd, **S**...Total/diesel/mart, AppleFarm Rest., Boston Mkt, Cracker Barrel, Olive Garden, Red Mule Grill, Steak&Ale(2mi), Budget Host, Home Depot

12   Noland Rd, Independence, **N**...VETERINARIAN, Amoco, Texaco/diesel/mart, Total/diesel/mart, Big Boy, ChuckeCheese, Denny's, Godfather's, Gold China, Golden Corral, Hardee's, Little Caesar's, Mr GoodCents Pasta, Ramada, Shoney's Inn, Super 8, Advance Parts, Firestone, K-Mart, Osco Drugs, Office Depot, OfficeMax, Walgreen, U-Haul, bank, transmissions, to Truman Library, **S**...CHIROPRACTOR, DENTIST, Phillips 66/mart/wash, Arby's, Baskin-Robbins, Burger King, Country Kitchen, Fuddrucker's, Honeybaked Ham, KFC/ Taco Bell, KrispyKreme, McDonald's, Old Country Buffet, Red Lobster, Wendy's, American Inn, Crossland Inn, Howard Johnson/rest., Red Roof Inn, $Tree, EconoLube'n Tune, HobbyHaven, MailBoxes Etc, Old Navy, Pier 1, PriceChopper Foods, Tires+, TJ Maxx, ToysRUs, U-Haul

Missouri Interstate 70

11	US 40, Blue Ridge Blvd, Independence, **N**...Circle K, Conoco/mart, Big Boy, Burger King, LJ Silver, McDonald's, Sonic, Village House Rest., V's Italiano Ristorante, Day's Inn, Deluxe Inn, Villager Lodge, JiffyLube, Pearle Vision, Radio Shack, tires, **S**...Amoco, Sinclair/7-11, Applebee's, ChiChi's, Hong Kong Buffet, Sports Stadium Motel, JC Penney, Olds/Cadillac, bank, cinema, mall
10	Sterling Ave(from eb), same as 11
9	Blue Ridge Cutoff, **N**...HOSPITAL, Conoco/mart, Denny's/24hr, Wendy's, Adam's Mark Hotel/rest., Day's Inn, Drury Inn, Villager Lodge, **S**...Amoco/mart/24hr, Sinclair/mart, Taco Bell, Holiday Inn, Sports Complex
8b a	I-435, N to Des Moines/Wichita, S to Witchita
7b	Manchester Trafficway, **N**...Day's Inn, Knight's Inn, Villager Lodge, Ryder Trucks, **S**...Hertz
7mm	Blue River
7a	US 40 E, 31st St, **S**...Travelers Inn
6	Van Brunt Blvd, **N**...Citgo/7-11, **S**...VA HOSPITAL, Amoco/mart, Texaco, KFC, McDonald's, Pizza Hut, Taco Bell, Osco Drug, carwash
5c	Jackson Ave(from wb), no facilities
b	31st St(from eb), no facilities
a	27th St(from eb), no facilities
4c	23rd Ave, no facilities
b	18th St, **N**...Total/diesel/mart
a	Benton Blvd(from eb), Truman Rd, **N**...Total/diesel/mart
3c	US 71 S, Prospect Ave, **N**...Service Oil Co/diesel, **S**...BBQ, McDonald's, Church's Chicken
b	Brooklyn Ave(from eb), **S**...same as 3c
3a	Paseo St, **S**...Amoco, tires
2l	I-35, no facilities
2h	US 24 E(from wb), **S**...Total/mart, Super 8
2g	I-29/35(from wb)
2k	Harrison St, Troost Ave, no facilities
2p	13th St, downtown
2q	McGee St, Truman Rd, downtown, **S**...Firestone, Midas Muffler
2s	Broadway St, Oak St
2t	I-35, no facilities
1b	Wyoming St, Genessee St, **N**...Phillips 66/Taco Bell/diesel/mart, Golden Ox Rest., Connie's Genessee Inn, Suitera's Italian, Kemper Arena
1a	Central Ave, downtown
0mm	Kansas River, Missouri/Kansas state line

Vertical label: **Kansas City**

Right margin compass: **E** ↑ ↓ **W**

Missouri Interstate 270(St Louis)

Exit #	Services
	I-270 begins/ends in Illinois on I-55/I-70, exit 20.
15b a	I-55 N to Chicago, S to St Louis
12	IL 159, to Collinsville, **S**...HOSPITAL, **N**...Dairy Queen, Hardee's
9	IL 157, to Collinsville, **S**...CHIROPRACTOR, Phillips 66/mart/24hr, 76/mart, auto repair, **N**...Comfort Inn
7	I-255, to I-55 S to Memphis
6b a	IL 111, **N**...Amoco/mart/24hr, 76, Hen House Rest., Best Western, **S**...McDonald's, Apple Valley Motel, to Pontoon Beach
4	IL 203, Old Alton Rd, to Granite City, no facilities
3b a	IL 3, **N**...RV Ctr, Riverboat Casino, **S**...Amoco/mart, Phillips 66, Jack's Rest./24hr, Hardee's/24hr, Waffle House, Chain of Rocks Motel, KOA
2mm	Chain of Rocks Canal
0mm	Mississippi River, Illinois/Missouri state line, motorist callboxes begin eastbound
34	Riverview Dr, to St Louis, **N**...**Welcome Ctr/rest area both lanes, full(handicapped)facilities, info, picnic tables, litter barrels, phones**
33	Lilac Ave, **S**...Amoco/mart, Phillips 66/diesel, QT/diesel/mart/24hr
32	Bellefontaine Rd, **N**...Mobil/mart, Shell/mart/wash, Burger King, Chinese Rest., Denny's, McDonald's, Pizza Hut, Steak'n Shake, Motel 6, Travelodge, Advance Parts, Tire Station, bank, **S**...Amoco/24hr
31b a	MO 367, **N**...HOSPITAL, Shell, **1 mi S**...Phillips 66/diesel/mart, Denny's, McDonald's, Taco Bell, Schnuck's Foods/drugs, K-Mart, Valvoline Lube
30	Hall's Ferry Rd, rd AC, **N**...Mobil/diesel/mart, QT/mart/24hr, Applebee's, Hardee's, Red Lobster, Shoney's, Taco Bell, White Castle, Wendy's, Super 8, TireAmerica, Target, cinema, **S**...Amoco, Shell/mart, Mobil/mart, Church's Chicken, McDonald's, Pope's Cafeteria, AutoZone, Buick, Hyundai, PriceChopper, bank, hardware
29	W Florissant Rd, **N**...Jack-in-the-Box, Best Buy, Office Depot, TJ Maxx, bank, **S**...Blackeyed Pea, Lion's Choice Rest., McDonald's, Old Country Buffet, Pasta House, Circuit City, PharMor Drug, Saturn, Sam's Club, Wal-Mart

Vertical label: **St Louis**

Right margin compass: **N** ↑ ↓ **S**

Missouri Interstate 270

28	Elizabeth Ave, Washington St, **S**...Amoco/mart/24hr, **N**...Sinclair, Jack-in-the-Box, Mrs O's Rest., Chevrolet, JC Penney, Schnuck's Foods/24hr, Walgreen
27	New Florissant Rd, rd N, **N**...Amoco/mart, Shell/mart
26b	Graham Rd, N Hanley, **N**...HOSPITAL, Shell, 7-11, Denny's, Fazoli's, Hardee's, LJ Silver, Rosemari's Rest., Day's Inn, Hampton Inn, Red Roof Inn, bank
a	I-170 S, no facilities
25b a	US 67, Lindbergh Blvd, **N**...Clark Gas, Shell, Texaco, Arby's, Burger King, Chinese Rest., Hardee's, Imo's Pizza, Jack-in-the-Box, McDonald's, Papa John's, PoFolks, Rally's, Ruiz' Mexican, Taco Bell, Village Inn Rest., Yum-Yum Donuts, Baymont Inn, Fairfield Inn, AutoZone, Cadillac, CarX Muffler, Goodyear, JiffyLube, Midas Muffler, NAPA, Schnuck's Foods/deli/24hr, Toyota, Walgreen, bank, cinema, tires, cleaners/laundry, **S**...Citgo/7-11, Shell, Econolodge, Villa Motel, VW/Volvo
23	McDonnell Blvd, **W**...QT/mart, Hardee's, Pontiac/GMC, **E**...Bicycle Club Rest., Denny's, McDonald's, La Quinta, bank
22	MO 370 W, to MO Bottom Rd, no facilities
20b	MO 180, St Charles Rock Rd, **E**...HOSPITAL, Amoco, Shell/mart, Casa Gallardo's, Ground Round, Hardee's, McDonald's, Fazoli's, LJ Silver, Old Country Buffet, Ponderosa, Red Lobster, Shoney's, Steak'n Shake, Taco Bell, Tony Bono's Rest., Knight's Inn, Best Buy, Circuit City, K-Mart, Office Depot, Target, Walgreen/24hr, bank, **W**...Amoco/24hr, Citgo/mart, Mobil, Bob Evans, Denny's, Olive Garden, Ryan's, Waffle House, Holiday Inn, Motel 6, Red Roof Inn, Super 8, Ford, Ziebart TidyCar
20a	I-70, E to St Louis, W to Kansas City
17	Dorsett Rd, **W**...Amoco/mart, Mobil, QT, Shell/mart, Denny's, Fuddrucker's, McDonald's, Subway, Baymont Inn, **E**...Hardee's, Shoney's, Steak'n Shake, Best Western, Drury Inn, Hampton Inn
16b a	Page Ave, rd D, **E**...Amoco/mart, Citgo/7-11, QT, Chinese Rest., Hooters, Comfort Inn, Courtyard, Holiday Inn, Red Roof Inn, Residence Inn, Sheraton
14	MO 340, Olive Blvd, **E**...HOSPITAL, Mobil/mart, Denny's, McDonald's, Lion's Choice Rest., Chevrolet, **W**...Bristol's Café, Culpepper's Café, TGIFriday, Courtyard
13	rd AB, Ladue Rd, no facilities
12	I-64, US 40, US 61, E to St Louis, W to Wentzville, **E**...HOSPITAL
9	MO 100, Manchester Rd, **E**...Amoco, Houlihan's Rest., IHOP, McDonald's, **W**...Shell/mart, Casa Gallardo's Mexican, Olive Garden, cinema
8	Dougherty Ferry Rd, **S**...HOSPITAL
5b a	I-44, US 50, MO 366, E to St Louis, W to Tulsa
3	MO 30, Gravois Rd, **N**...Amoco/mart, Bob Evans, Best Western, Comfort Inn, Econolodge
2	MO 21, Tesson Ferry Rd, **N**...Amoco/mart, Calico's Rest., Jack-in-the-Box, Olive Garden, TGIFriday, Wendy's, White Castle, bank, **S**...Mobil, Shell
1b a	I-55 N to St Louis, S to Memphis

Missouri Interstate 435(Kansas City)

Exit #	Services
83	I-35, N to KS City, S to Wichita
82	Quivira Rd, Overland Park, **N**...HOSPITAL, Burger King, Old Chicago Pizza, Ponderosa, Taco Bell, **S**...McDonald's, Wendy's, Extended Stay America
81	US 69 S, to Ft Scott, no facilities
79	US 169, Metcalf Rd, **N**...HOSPITAL, Amoco, Texaco, Denny's, Dick Clark's Grill, Hooters, Tippin's Café, Clubhouse Inn, Embassy Suites, Hampton Inn, Wyndham Garden, **S**...Courtyard, Drury Inn, KC Masterpiece, McDonald's
77b a	Nall Ave, Roe Ave, **N**...Amoco, Texaco/mart, Dairy Queen, Winstead's Grill, Fairfield Inn, **S**...Amoco, Cactus Grill, McDonald's, Wendy's, Homestead Suites, Sumner Suites
75b	State Line Rd, **N**...Conoco, McDonald's, Waid's Rest., Buick/Cadillac, Ford, Infiniti, Lexus, **S**...HOSPITAL, Amoco/mart, EBT Rest., city park
a	Wornall Rd, **N**...QT/mart, Texaco/mart, Applebee's, Wendy's, Chevrolet/Olds, VW
74	Holmes Rd, **S**...Amoco, Phillips 66/mart, Burger King, Guacamole Grill, Courtyard, Extended Stay America
71b a	I-470, US 71 S, US 50 E, no facilities
70	Bannister Rd, **E**...Conoco/mart, QT/mart, Bennigan's, McDonald's, **W**...LJ Silver, Old Country Buffet, Home Depot

Missouri Interstate 435

69	87th St, Denny's, **E**...Darryl's, Denny's, IHOP, Luby's, McDonald's, Tippin's Café, Best Western, Motel 6, Wal-Mart HyperMart, mall, **W**...Amoco/diesel/mart, Baymont Inn
67	Gregory Blvd, **W**...Nature Ctr, IMAX Theatre, zoo
66a	63rd St, **E**...Texaco/mart, Applebee's, Waid's Rest., Wendy's
b	MO 350 E, **W**...LC's BBQ, Relax Inn
65	Eastwood Trfcwy, **W**...Conoco/mart, KFC, LC's BBQ, McDonald's, Pizza Hut, Relax Inn, laundry
63c	Raytown Rd, Stadium Dr, **E**...Sports Complex, Day's Inn, Sports Stadium Motel, Villager Lodge
63b a	I-70, W to KC, E to St Louis
61	MO 78, no facilities
60	MO 12 E, Truman Rd, 12th St, **E**...Amoco/repair, Total/mart
59	US 24, Independence Ave, **E**...Queen City Motel(3mi), to Truman Library
57	Front St, **E**...Flying J/Conoco/diesel/rest./24hr, Burger King, Taco Bell, Blue Beacon, **W**...QT/mart, Big Boy, KFC, McDonald's, Pizza Hut, Waffle House, Wendy's, Hampton Inn, Smugglers Inn
55b a	MO 210, **E**...Phillips 66/diesel, Red Roof Inn, Riverboat Casino, Ford/Volvo/GMC/Mercedes Trucks, **W**...Arby's, McDonald's, Baymont Inn(2mi), Country Hearth, Day's Inn
54	48th St, Parvin Rd, **E**...funpark, **W**...QT/mart, Alamo Mexican, Ponderosa, Waffle House, Best Western, Comfort Inn, Country Inn Suites, Crossland Suites, Holiday Inn, Super 8
52a	US 69, **E**...A&W, KFC, **W**...Amoco/repair, Coastal/mart, Texaco/mart, McDonald's, Pizza Hut, Taco Bell
52b	I-35, N to Des Moines, S to KC
51	Shoal Creek Dr, no facilities
49b a	MO 152 E, to Liberty, **E**...Applebee's, Cracker Barrel, Golden Corral, Best Western, Comfort Inn, Hampton Inn, Super 8
47	NE 96th St, no facilities
46	NE 108th St, no facilities
45	MO 291, NE Cookingham Ave, no facilities
42	N Woodland Ave, no facilities
41b a	US 169, Smithville, **N**...McDonald's, Super 8
40	NW Cookingham, no facilities
37	NW Skyview Ave, rd C, **N**...Total
36	to I-29 S, to KCI Airport, **S**...Amoco/mart, Best Western, Comfort Suites, Fairfield Inn, Hilton, Holiday Inn Express, Wyndham Garden
31mm	Prairie Creek
29	rd D, NW 120th St, no facilities
24	MO 154, rd N, NW Berry Rd, no facilities
22	MO 45, Weston, Parkville, no facilities
20mm	Missouri River, Missouri/Kansas state line
18	KS 5 N, Wolcott Dr, **E**...to Wyandotte Co Lake Park
15b a	Leavenworth Rd, **E**...Woodlands Racetrack
14	Parallel Pkwy, **E**...HOSPITAL, QT/mart
13	US 24, US 40, State Ave, **E**...Total, Arby's, Taco Bell
12	I-70, KS Tpk, to Topeka, St Louis
11	Kansas Ave, no facilities
9	KS 32, KS City, Bonner Springs, **W**...Total/mart
8b	Woodend Rd, no facilities
8.8mm	Kansas River
8 a	Holliday Dr, to Lake Quivira, no facilities
6c	Johnson Dr, no facilities
6b a	Shawnee Mission Pkwy, **E**...Amoco/diesel/mart, Sonic, museum
5	Midland Dr, Shawnee Mission Park, **E**...Texaco/mart, Circle K, Arizona's Grill, Barley's Brewhouse, Blimpie, Savannah's Grill, Wendy's, Hampton Inn, bank
3	87th Ave, **E**...Amoco/mart, Phillips 66/mart, Texaco/mart, McDonald's, NY Burrito, museum
2	95th St, no facilities
1b	KS 10, to Lawrence
a	Lackman Rd, **S**...Conoco/mart, QT/mart
0mm	I-435 begins/ends on I-35.

Montana Interstate 15

Exit #	Services
398mm	Montana/US/Canada Border
397	Sweetgrass, **W...rest area both lanes, full(handicapped)facilities, info, picnic tables, litter barrels, petwalk, RV dump,** Gas4Less/mart, Sinclair/mart, Glocca Morra Motel/café, Ammex Duty Free
394	ranch access, no facilities
389	MT 552, Sunburst, **W...**farm mkt, gas/diesel/tires, hardware, laundry
385	Swayze Rd, no facilities
379	MT 215, MT 343, Oilmont, to Kevin, **W...**Four Corners Café
373	Potter Rd, no facilities
369	Bronken Rd, no facilities
366.5mm	weigh sta sb
364	Shelby, **E...**Lewis&Clark RV Park, **W...**airport
363	US 2, Shelby, to Cut Bank, **E...**HOSPITAL, Exxon/Subway/diesel/mart/atm/24hr, Sinclair/mart, Dash Drive-In, Dixie Inn Steaks, Pizza Hut, Comfort Inn, Crossroads Inn, Glacier Motel/RV Park, Albertson's, CarQuest, SpeedyLube, TrueValue, casinos, diesel repair, **1 mi E...**Conoco, Chan's Chinese, Ford/Mercury, GMC, Goodyear/auto, O'Haire Motel, Sherlock Motel, **W...**McDonald's, Pamida/drugs, to Glacier NP
361mm	parking area nb
358	Marias Valley Rd, to Golf Course Rd, **E...**camping
357mm	Marias River
352	Bullhead Rd, no facilities
348	rd 44, to Valier, **W...**Lake Francis RA(15mi)
345	MT 366, Ledger Rd, **E...**to Tiber Dam(42mi)
339	Conrad, **W...**HOSPITAL, VETERINARIAN, Cenex/diesel/mart, Conoco, Exxon/Arby's/diesel, Main Drive-In, Pizza Hut(2mi), Pizza Pro, Conrad Motel, Northgate Motel, Super 8, IGA Foods, Olson's Drug, Parts+, Radio Shack, Chevrolet/Olds/Pontiac/Buick, Ford/Mercury, Pondera RV Park, Village Muffler, auto repair, bank, carwash, hardware, laundry, RV Ctr/LP
335	Midway Rd, Conrad, **4 mi W...**HOSPITAL, gas, food, phone, lodging, RV camping
328	MT 365, Brady, **1 mi W...**Mtn View Co-op/diesel, USPO, antiques, phone
321	Collins Rd, no facilities
319mm	**Teton River, rest area both lanes, full(handicapped)facilities, phones, picnic tables, litter barrels, petwalk**
313	MT 221, MT 379, Dutton, **W...**Mtn View Co-op/diesel, Dutton Café, USPO, phone
302	MT 431, Power, no facilities
297	Gordon, no facilities
290	US 89 N, rd 200 W, to Choteau, **W...**Exxon/mart, Sinclair/diesel/LP/RV dump/mart, USPO
288mm	parking area both lanes
286	Manchester, **W...**livestock auction, same as 290(2mi)
282	US 87 N(from sb), weigh sta, **E...**Yellowstone Trkstp/diesel, **1-3 mi E...**Exxon/A&W/Blimpie/mart, McDonald's, Day's Inn, Wal-Mart
280	US 87 N, Central Ave W, Great Falls, **E...**VETERINARIAN, Conoco/mart, Flying J/diesel/LP/mart/24hr, Sinclair, Arby's/A&W(2mi), Dairy Queen, Hardee's(2mi), Ford's Drive-In, KFC, Taco John's, Adelweiss Motel, Alberta Inn, Day's Inn, Starlit Motel, NAPA, Noble Tires, U-Haul/LP, bank, lube, to Giant Sprgs SP
280mm	Sun River
278	US 89 S, rd 200 E, 10th Ave, Great Falls, **E...**Exxon/A&W/pizza/mart, Sinclair/diesel/mart, China Town, Dairy Queen, Elmer's Rest., McDonald's, Airway Motel, Best Western, Budget Inn, Holiday Inn Express, Kanga Inn, Barnes&Noble, Dick's RV Park, Home Depot, OfficeMax, Pier 1, PostNet, Smith's Food/drug, bank, cinema, **1-3 mi E...**HOSPITAL, CHIROPRACTOR, EYECARE, VETERINARIAN, Cenex/diesel/mart, Conoco/mart/atm, Holiday/diesel/gas, Oasis/mart, Sinclair/diesel/atm, Applebee's, Arby's, Baskin-Robbins, Big Sky Bagel, Burger King, BurgerMaster, Cattin's Rest., Country Kitchen, Dairy Queen, 4B's Rest., Fuddrucker's, Godfather's, Hardee's, Jaker's Rib/fish, JB's Rest., KFC, Little Caesar's, McDonald's, Ming's Chinese, Papa John's, Papa Murphy's, Pizza Hut, PrimeCut Rest., Quizno's, Sting Rest., Subway, Taco Bell, Taco John's, Taco Treats, TCBY, Wendy's, Fairfield Inn, Plaza Inn, Townhouse Inn, Village Motel, Western Motel, Ace Hardware, Albertson's, Checker Parts, Chevrolet/Cadillac/Toyota, County Mkt Food/24hr, Dodge/Hyundai/Suzuki/VW, Ford, Hancock Fabrics, Hastings Books, Herberger's, HiTech Auto, Honda, JC Penney, Jo-Ann Crafts, K-Mart, KOA(6mi), Lincoln/Mercury, MailBoxes Etc, McCollum RVs, Midas Muffler, NAPA, Osco Drug, Nissan, Pennzoil, PetCo, PhotoMax, Plymouth/Jeep, ProLube, Rex Audio/video, Sears/auto, Target, TireRama, USPO, antiques, bank, carwash, casinos, crafts, golf, transmissions, to Malmstrom AFB

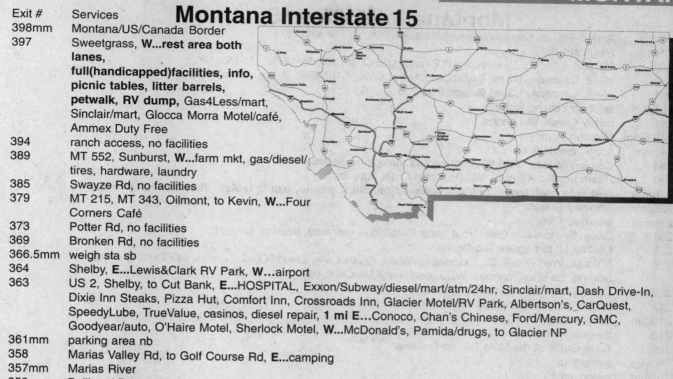

N

S

Montana Interstate 15

277	Airport Rd, **E**...Conoco/diesel/café/casino/24hr, **W**...to airport
275mm	weigh sta nb
270	MT 330, Ulm, **E**...Exxon/diesel/LP/mart, USPO, **W**...Griffin's Rest., to Ulm SP
256	rd 68, Cascade, **1/2 mi E**...Sinclair, Angus Café, A-C Motel, Badger Motel/café, IGA Foods, NAPA, USPO
254	rd 68, Cascade, **1/2 mi E**...same as 256
250	local access, no facilities
247	Hardy Creek, **W**...phone
246.5mm	Missouri River
245mm	scenic overlook sb
244	Canyon Access, **2 mi W**...boat access, camping, rec area
240	Dearborn, **E**...RV park, **W**...Dearborn Café, auto repair
239mm	**rest area both lanes, full(handicapped)facilities, phone, picnic tables, litter barrels, petwalk**
238mm	Stickney Creek
236mm	Missouri River
234	Craig, **E**...Hooker's Grill, Trout Shop Café/lodge, rec area, boating, camping
228	US 287 N, to Augusta, no facilities
226	MT 434, Wolf Creek, **E**...Exxon/diesel/mart, Oasis Café, Lewis&Clark Canoes, MT Outfitters/flyshop/RV/motel, boating, camping, phones, **W**...Frenchman&Me Café, USPO
222mm	parking area both lanes
219	Spring Creek, Recreation Rd(from nb), boating, camping
218mm	Little Prickly Pear Creek
216	Sieben, no facilities
209	**E**...to Gates of the Mtns RA, no facilities
205mm	turnout sb
202mm	weigh sta sb
200	MT 279, MT 453, Lincoln Rd, **W**...Sinclair/Bob's Mkt/diesel, GrubStake Rest., Helena Campground(4mi), Lincoln Rd RV Park, to ski area
193	Cedar St, Helena, **E**...airport, tires, Helena RV Park(5mi), **W**...to MT Ave, VETERINARIAN, Cenex/diesel/mart, Conoco/diesel/mart/atm, Exxon/diesel/mart, Sinclair/mart/24hr, Applebee's, Arby's, Dragon Wall Chinese, Godfather's, Jade Garden Chinese, McDonald's, McKenzie River Pizza, Perkins/24hr, Pizza Hut, Steffano's Pizza/subs, Subway, Taco Bell, Taco John's, Wong Chinese, Ace Hardware, Albertson's/24hr, AutoWerks, Bumper Parts, CarQuest, Checker Parts, Chevrolet, County Mkt Foods, Hastings Books, JiffyLube, Kia, K-Mart/drugs, NAPA, Pennzoil, PostNet, ShopKO, Target, TireRama, USPO, Ward's Sports, bank, carwash, cinema, hardware, radiator/transmissions
192b a	US 12, US 287, Helena, Townsend, **E**...Conoco/diesel/mart/24hr, Burger King, Staples, Wal-Mart/drugs, RV Ctr, car/truckwash, st patrol, **1mi E**...VETERINARIAN, Exxon/diesel/LP/mart, Stageline Pizza, Subway, Buick/Cadillac/GMC, Chrysler/Jeep/Nissan, Ford/Lincoln/Mercury, Honda, Plymouth, Toyota, **W**...HOSPITAL, CHIROPRACTOR, DENTIST, EYECARE, Exxon/diesel/mart, Sinclair/mart/24hr, A&W/KFC, Big Cheese Pizza, Blimpie, Bullseye Rest./casino, Burger King, County Harvest Rest., Dairy Queen, Frontier Pies, JB's, L&D Chinese, Little Caesar's, McDonald's, Overland Express Rest., Papa John's, Taco Bell, Taco Treat, Wendy's, Best Western, Comfort Inn, Country Inn Suites, Day's Inn, Fairfield Inn, Holiday Inn Express, Jorgenson's Inn, Motel 6, Shilo Inn, Super 8, Albertson's/gas, Checker Parts, Dillard's, Goodyear/auto, InstyPrint, JC Penney, J&J Tires, Osco Drug, Pier 1, ProBrake, Safeway/drugs, bank, carwash, laundry
187	MT 518, Montana City, **E**...Papa Ray's Casino/Pizza, to NF, **W**...Conoco/Mtn City Grill/diesel/mart/casino/24hr, Elkhorn Mtn Inn, Parts+, bank, hardware, transmissions
182	Clancy, **E**...RV camping, **W**...Gen Store/gas, Legal Tender Rest., to NF
178mm	**rest area both lanes, full(handicapped)facilities, phone, picnic tables, litter barrels, petwalk**
176	Jefferson City, NF access, no facilities
174.5mm	chain up area both lanes
168mm	chainup area both lanes
164	rd 69, Boulder, **E**...Exxon/diesel/mart/casino/24hr, Dairy Queen, Phil&Tim's Café, Boulder Tire/brake, IGA Foods, Parts+, USPO, camping, carwash
161mm	parking area nb
160	High Ore Rd, no facilities
156	Basin, **E**...Basin Cr Pottery, Merry Widow Health Mine/RV camping, **W**...Silver Saddle Café, camping
154mm	Boulder River
151	Bernice, to Boulder River Rd, **W**...camping, picnic area
148mm	chainup area both lanes
143.5mm	chainup area both lanes
138	Elk Park, **W**...Sheepshead Picnic Area, wildlife viewing
134	Woodville, no facilities
133mm	continental divide, elev 6368

N

S

Helena

Montana Interstate 15

130.5mm scenic overlook sb

129   I-90 E, to Billings, no facilities

I-15 S and I-90 W run together 8 mi

127   Harrison Ave, Butte, **N**...Cenex/pizza/diesel/mart/atm, Conoco/mart/atm, Sinclair/mart, Arctic Circle, Dairy Queen, Denny's, Derby Steaks, Hardee's, John's Rest., Little Caesar's, Taco John's, Day's Inn, Holiday Inn Express, WarBonnett Inn, Ace Hardware, Champion Parts, Honda Motorcycles, Lisac's Tires, NAPA, Nissan/Toyota, Safeway/drugs, Yamaha, auto repair, bank, casinos, cleaners, **S**...EYECARE, Conoco/diesel/mart, Exxon/diesel/mart/24hr, Sinclair/mart, Arby's, Burger King, 4B's Rest., Godfather's, KFC, McDonald's, Perkins, Paul Bunyan's Sandwiches, Pizza Hut, Ray's Rest., Silver Bow Pizza, Subway, Taco Bell, Wendy's, Comfort Inn, Hampton Inn, Ramada Inn/rest., Super 8, CarCare/tires, Checker Parts, Chevrolet/Subaru, Chrysler/Plymouth/Dodge/Jeep, Ford/Honda, GMC/Buick/Olds/Cadillac/Pontiac, Herberger's, Honda, JC Penney, Jo-Ann Crafts, K-Mart, Lincoln/Mercury, MailBoxes Etc, Pennzoil, Smith Food/drug, Staples, Wal-Mart, Whalen Tires/lube, bank, cinema

126   Montana St, Butte, **N**...HOSPITAL, Cenex/diesel/mart, Joker's Wild Rest./casino, Eddy's Motel, KOA, Safeway Food, Schwab Tires, **S**...Conoco/mart, Exxon/diesel/mart

124   I-115(from eb), to Butte, City Ctr

122   Rocker, **N**...Flying J/Arby's/diesel/LP/casino/24hr, Rocker Inn Motel, weigh sta both lanes, RV camping, **S**...Conoco/4B's Rest./diesel/casino/24hr

I-15 N and I-90 E run together 8 mi

121   I-90 W, to Missoula, no facilities

119   Silver Bow, no facilities

116   Buxton, no facilities

112mm Continental Divide, elevation 5879

111   Feely, no facilities

109mm rest area both lanes, full(handicapped)facilities, phone, picnic tables, litter barrels, petwalk

102   rd 43, Divide, to Wisdom, **1 mi W**...Sinclair, to Big Hole Nat Bfd

99   Moose Creek Rd, no facilities

93   Melrose, **W**...Melrose Café/grill/gas/diesel, Sportsman Motel/RV camping, phone

85.5mm Big Hole River

85   Glen, **E**...phone

74   Apex, Birch Creek, no facilities

64mm Beaverhead River

63   Lp 15, rd 41, Dillon, **E**...HOSPITAL, DENTIST, Cenex/diesel/mart, Conoco, Exxon/diesel/mart, Phillips 66/mart, KFC, McDonald's, Pizza Hut, Subway, Best Western/rest., Comfort Inn, GuestHouse Inn, Sundowner Motel, Super 8, KOA, Safeway/deli, Chevrolet, auto parts, camping, carwash, cinema, laundry, museum, tires, W MT Coll

62   Lp 15, Dillon, **3/4 mi E**...HOSPITAL, info, Exxon, Sinclair/mart, Arctic Circle, Dairy Queen, Taco John's, Creston Motel, Crosswinds Motel/rest., KOA, to W MT Coll

60mm Beaverhead River

59   MT 278, to Jackson, **W**...Bannack Ghost Town SP, ski area

56   Barretts, **E**...RV camping, **W**...Big Sky Truckstop/diesel/mart

52   Grasshopper Creek, no facilities

51   Dalys(from sb, no return), no facilities

50mm Beaverhead River

46mm Beaverhead River

45mm Beaverhead River

44   MT 324, **E**...Armstead RV Park, **W**...Clark Canyon Reservoir, rec area, RV camping

38.5mm Red Rock River

37   Red Rock, no facilities

34mm rest areas both lanes, full(handicapped)facilities, phone, picnic tables, litter barrels, petwalk

29   Kidd, no facilities

23   Dell, **E**...Red Rock Inn/rest., auto repair

16.5mm weigh sta both lanes

Montana Interstate 15

15   Lima, **E**...Exxon/diesel/mart, Jan's Café, Lee's Motel/RV camping, Big Sky Service/tires, I&J Mkt, USPO, ambulance, hardware

9   Snowline, no facilities

0   Monida, **E**...phone, to Red Rock Lakes

0mm   Monida Pass, elevation 6870, Montana/Idaho state line

Montana Interstate 90

Exit #	Services
554.5mm	Montana/Wyoming state line
549	Aberdeen, no facilities
544	Wyola, **2 mi S**...gas, food
530	MT 463, Lodge Grass, **1 mi S**...Cenex/diesel/LP, lodging, phone
517.5mm	Little Bighorn River
514	Garryowen, **N**...Conoco/mart, Custer Bfd Museum
511.5mm	Little Bighorn River
510	US 212 E, **N**...HOSPITAL, Exxon/diesel/café/gifts, Crows Nest Café, bank, casino, to Little Bighorn Bfd, **S**...Little Bighorn Camping/motel/laundry
509.5mm	weigh sta both lanes
509.3mm	Little Bighorn River
509	Crow Agency, **N**...Conoco/mart/atm/24hr, **S**...to Bighorn Canyon NRA
503	Dunmore, no facilities
498mm	Bighorn River
497	MT 384, 3rd St, Hardin, **S**...HOSPITAL, visitors ctr, **2 mi S**...Lariat Motel, Western Motel, Casino Rest./lounge
495	MT 47, City Ctr, Hardin, **N**...Texaco/diesel/mart, Purple Cow Rest., KOA, **S**...HOSPITAL, Cenex/diesel/mart/atm, Conoco/Subway/diesel/LP/mart/24hr, Exxon/diesel/mart, Sinclair/Blimpie/diesel/mart, Dairy Queen, KFC, McDonald's, Pizza Hut/Taco Bell, Taco John's, American Inn, Lariat Motel, Super 8, Western Motel, Royal Lube, RV camping
484	Toluca, no facilities
478	Fly Creek Rd, no facilities
477mm	**rest area both lanes, full(handicapped)facilities, phone, picnic tables, litter barrels, petwalk**
469	Arrow Creek Rd, no facilities
462	Pryor Creek Rd, no facilities
456	I-94 E, to Bismarck, ND, no facilities
455	Johnson Lane, **S**...Exxon/A&W/Blimpie/diesel/mart/24hr, Flying J/Conoco/diesel/LP/rest./24hr, Burger King, Fly-In Lube, RV Ct(1mi)
452	US 87 N, City Ctr, Billings, **N**...Conoco/Little Caesar's/diesel/wash/mechanic/LP, Exxon/diesel/mart, Big Dipper Drive-Inn, Bruno's Pizza/pasta, Best Western, Chevrolet, RV Sales, transmissions, **2-4 mi N on US 87**...MEDI-CAL CARE, VETERINARIAN, Cenex/mart, Conoco/mart, Holiday/diesel/mart, Applebee's, Arby's, Burger King, Dairy Queen, Elmer's Pancakes, Godfather's, Golden Grand Bagels, Phoenix Chinese, Guadalajara Mexican, Little Caesar's, McDonald's, Papa John's, Papa Murphy's, Peking Express, Pizza Hut, Pudge Bros Pizza, Subway, Taco Bell, Taco John's, Wendy's, Foothills Inn, Metra Inn, Ace Hardware, Albertson's, CarQuest, Champion Parts, Checkers Parts, County Mkt Foods, K-Mart, MasterLube, Osco Drug, Radio Shack, U-Haul, Wal-Mart SuperCtr/24hr, Western Drug, bank, carwash, casinos, cleaners, diesel repair, tires, **S**...Cenex/diesel/mart, Texaco/mart
451.5mm	Yellowstone River
450	MT 3, 27th St, Billings, **N**...HOSPITAL, Conoco/diesel/mart/24hr, Exxon/Subway/mart, Sinclair/mart/repair, Bungalow Rest., Denny's, Perkins/24hr(1mi), Pizza Hut, Howard Johnson, Radisson(2mi), Rimview Motel(2mi), Sheraton(1mi), War Bonnet Inn/rest., CarQuest, USPO, to airport, **S**...KOA
447	S Billings Blvd, **N**...Conoco/Subway/diesel/mart/24hr, Burger King, 4B's Rest., McDonald's, Day's Inn, Sleep Inn, Super 8, funpark
446	King Ave, Billings, **N**...Conoco/mart/lube, Holiday/diesel/mart/LP/atm/RV dump, Burger King, Country Harvest Buffet, Denny's, Fuddrucker's, Gusick's Rest., Olive Garden, Outback Steaks, Perkins/24hr, Pizza Hut, Red Lobster, Taco John's, C'Mon Inn, Comfort Inn, Day's Inn, Fairfield Inn, Hampton Inn, Quality Inn, Super 8, Best Buy, Bumper Parts, Chrysler/Plymouth/Kia, Costco, Dodge, Ford, Kia, NAPA, Nissan, RV Repair, ShopKO, Subaru/Hyundai, bank, USPO, **N...King Ave to S 24th**...VETERINARIAN, Exxon/Subway/mart, Conoco, Sinclair/mart/atm, Applebee's, Baskin-Robbins, Dos Machos Cantina, 5&Diner, Godfather's, Golden Corral, Hardee's, McDonald's, Pratt's Lunch, Taco Bell, Wendy's, Albertson's, Ace Hardware, Barnes&Noble, Chevrolet/Isuzu, County Mkt Foods, Eagle Hardware, Gart Sports, Hastings Books, Home Depot, K-Mart,

Billings

E

W

Montana Interstate 90

Lincoln/Mercury, Michael's, Midas Muffler, Office Depot, OfficeMax, Old Navy, Olds/Suzuki, Pak'n Mail, Pearle Vision, PetsMart, Pier 1, Target, Wal-Mart/auto, Western Drug, laundry, mall, **S**...Conoco/mart/24hr, Appletree Rest., Cracker Barrel, Fossen's Rest., Japanese Rest., Best Western, Billings Hotel, Fireside Inn, Holiday Inn, Kelly Inn, Motel 6, Ramada Ltd, funpark

443	new exit, **S**...Exxon
439mm	weigh sta both lanes
437	E Laurel, **S**...VETERINARIAN, Sinclair/diesel/rest./casino/motel/24hr, RV camping
434	US 212, US 310, Laurel, to Red Lodge, **N**...Cenex/diesel/mart/wash, Conoco/mart/24hr, Exxon/diesel/mart/24hr, Burger King, Hardee's, Little Big Men Pizza, Pizza Hut, Subway, Taco John's, Best Western/rest., Jan's IGA/24hr, Western Drug, Chevrolet, Ford, Pennzoil, RV Parts, XpertLube/wash, **S**...to Yellowstone NP
433	Lp 90(from eb), no facilities
426	Park City, **S**...Cenex/diesel/café/mart/24hr, motel, auto/tire repair
419mm	**rest area both lanes, full(handicapped)facilities, phone, picnic tables, litter barrels, petwalk**
408	rd 78, Columbus, **N**...Mtn Range RV Park, **S**...HOSPITAL, Conoco/Big Sky Motel, Exxon/KFC/diesel/mart/24hr, Apple Village Café/gifts, McDonald's, Super 8, airport, casino, museum, to Yellowstone
400	Springtime Rd, no facilities
398mm	Yellowstone River
396	ranch access, no facilities
392	Reed Point, **N**...Sinclair/diesel/mart, Hotel Montana, RV camping
384	Bridger Creek Rd, no facilities
381mm	**rest area both lanes, full(handicapped)facilities, phone, picnic tables, litter barrels, petwalk**
377	Greycliff, **S**...Prairie Dog Town SP, KOA, funpark
370	US 191, Big Timber, **1 mi N**...HOSPITAL, Conoco, Exxon/mart, Sinclair, Grand Hotel, Lazy J Motel, Russell's Motel/café, Spring Camp RV Ranch(4mi), auto parts, bank
369mm	Boulder River
367	US 191 N, Big Timber, **N**...Exxon/KFC/Taco Bell/diesel/mart/atm, Conoco/mart/24hr, Crazy Jane's Eatery/casino, Super 8, CarQuest, Chevrolet/Olds, antiques, hist site, laundry, museum, visitor ctr
362	De Hart, no facilities
354	MT 563, Springdale, no facilities
352	ranch acess, no facilities
350	East End access, no facilities
343	Mission Creek Rd, no facilities
340	US 89 N, to White Sulphur Sprgs, **S**...airport
337	Lp 90, to Livingston, **4 mi N**...Exxon/mart, motels
333mm	Yellowstone River
333	US 89 S, Livingston, **N**...HOSPITAL, Holiday/mart, Crazy Coyote Mexican, Dairy Queen, Pizza Hut, Taco John's, Best Western, Budget Host, Econolodge, Livingston Inn, Paradise Inn/rest., Ace Hardware, Chevrolet, County Mkt Food/24hr, Ford/Lincoln/Mercury, Oasis RV Park, Pamida, Radio Shack, Western Drug, laundry, **S**...VETERINARIAN, Cenex/diesel/mart, Conoco/diesel/mart, Exxon/mart, Buffalo Jump Steaks, Hardee's, McDonald's, Subway, Comfort Inn, Super 8, Albertson's, KOA(10mi), Osen's RV Park, Windmill RV Park, antiques, casino, to Yellowstone
330	Lp 90, Livingston, **1 mi N**...Yellowstone Truckstop/diesel/24hr/rest.
326.5mm	chainup area both lanes
324	ranch access, no facilities
323mm	chainup area wb
322mm	Bridger Mountain Range
321mm	turnouts/hist marker both lanes
319	Jackson Creek Rd, no facilities
319mm	chainup area both lanes
316	Trail Creek Rd, no facilities
313	Bear Canyon Rd, **S**...Bear Canyon Camping

E

W

Montana Interstate 90

| 309 | US 191 S, Main St, Bozeman, **N**...Sunrise RV Park, **S**...HOSPITAL, Exxon/mart, Sinclair, MT AleWorks, Alpine Lodge, Best Western, Blue Sky Motel, Continental Motel, Ranch House Motel, Western Heritage Inn, Toyota/Mazda, RV service, Goodyear, auto repair, bank, carwash, laundry, tires, to Yellowstone |

306   US 191, N 7th, Bozeman, **N**...VETERINARIAN, Appletree Rest., Conoco, McDonald's, Rum's Deli, Econolodge, Fairfield Inn, Ramada Ltd, Sleep Inn, Super 8, TLC Inn, RV supplies, airport, ski area, **S**...Conoco/Arby's/diesel/mart, Exxon/mart, Holiday/mart, Sinclair/diesel, Applebee's, Dairy Queen, Ferraro's Italian, Frontier Pies Rest., Hardee's, KFC, McDonald's, Subway, Taco Bell, Taco John's, Village Inn Pizza, Wendy's, Wok Chinese, Best Western, Bozeman/rest., Comfort Inn, Day's Inn, Hampton Inn, Holiday Inn, Rainbow Motel, Royal 7 Motel, CarQuest, County Mkt Foods, Goodyear/auto, K-Mart/drugs, NAPA, SpeedyLube, U-Haul, Wal-mart/auto, bank, carwash, cleaners, casino, Museum of the Rockies

305   MT412, N 19th Ave, **N**...Exxon/mart, **S**...A&W/KFC, Denny's, McKenzie River Pizza, Wingate Inn, Costco, Smith's Foods, Target, UPS, USPO, **2-3 mi S**...Arby's, Burger King, Perkins, Pizza Hut, Spanish Peaks Brewery/Italian Café, Best Western, Bobcat Lodge, museum

298   MT 291, MT 85, Belgrade, **N**...Cenex/diesel/mart, Conoco, Exxon/Subway/diesel/mart, Phillips 66, Burger King, McDonald's, Taco Time, Albertson's/drugs, Goodyear, IGA Foods, Marine Sales/service, NAPA, airport, auto/truck repair, **S**...Flying J/Conoco/Country Kitchen/diesel/LP/mart, Pacific Pride/diesel, Holiday Inn Express, Super 8, Belgrade Tires, ski area, truckwash, to Yellowstone NP

292.5mm   Gallatin River

288   MT 346, MT 288, Manhattan, **N**...Conoco/diesel/mart/atm, RV camping

283   Logan, **S**...Madison Buffalo Jump SP(7mi)

279mm   Madison River

278   MT 205, rd 2, Three Forks, Trident, **N**...Missouri Headwaters SP, **1 mi S**...Broken Spur Motel, Sacajawea Inn, camping, phone

277.5mm   Jefferson River

274   US 287, to Helena, Ennis, **N**...Conoco/diesel, Steer Inn/casino/rest., Prairie Plaza Gen Store/café, Wheat MT Bakery/deli, auto/diesel/repair, to Canyon Ferry SP, Ft 3 Forks, **S**...Exxon/Subway/diesel/casino/atm/24hr, Lewis and Clark Caverns SP, KOA(1mi), to Yellowstone NP

267   Milligan Canyon Rd, no facilities

261.5mm   chain removal area

257mm   Boulder River

256   MT 359, Cardwell, **S**...Conoco/diesel/mart, RV camping, to Yellowstone NP, Lewis&Clark Caverns SP

249   rd 55, to rd 69, Whitehall, **S**...Exxon/Subway/diesel/mart/24hr, A&W, Rice Motel, Chief Motel, Super 8, camping

241   Pipestone, **S**...Pipestone Camping(apr-oct), gas, phone

240.5mm   chainup area both lanes

238.5mm   runaway ramp eb

237.5mm   pulloff eb

235mm   parking area both lanes, litter barrels

233   Homestake, Continental Divide, elev 6393, no facilities

230mm   chain removal area both lanes

228   MT 375, Continental Dr, no facilities

227   I-15 N, to Helena, Great Falls, no facilities

I-90 and I-15 run together 8 mi. **See Montana Interstate 15, exits 122-127.**

219   I-15 S, to Dillon, Idaho Falls, no facilities

216   Ramsay, no facilities

211   MT 441, Gregson, **3-5 mi S**...food, lodging, Fairmont RV Park

210.5mm   parking area wb, Pintlar Scenic route info

208   MT 1, Opportunity, Anaconda, Pintlar Scenic Loop, Georgetown Lake RA, **3-5 mi S**...HOSPITAL, gas, food, lodging, RV camping, ski area

201   Warm Springs, **S**...MT ST HOSPITAL, gas/diesel/mart, casino, laundry

197   MT 273, Galen, **S**...to MT ST HOSPITAL

195   Racetrack, no facilities

187   Deer Lodge(no wb return), **2 mi S**...HOSPITAL, Conoco/repair/LP, R-B Drive-In, Downtowner Motel, Scharf's Motel/rest., IGA Foods, Tow Ford Museum, Grant-Kohrs Ranch NHS, KOA(seasonal), Indian Creek RV Park(seasonal)(4mi)

184   Deer Lodge, **S**...VETERINARIAN, BP/mart/rest., Conoco/diesel/mart/casino, 4B's Rest., McDonald's, Super 8, Indian Cr Camping, **2 mi S**...Exxon/diesel/mart/casino, Pizza Hut, Downtowner Motel, Western Big Sky Inn, Safeway, Schwab Tires

179   Beck Hill Rd, no facilities

175mm   Little Blackfoot River

175   US 12 E(from wb), Garrison, **N**...gas/mart/phone, hist site, RiverFront RV Park

E

W

Montana Interstate 90

174	US 12 E(from eb), same as 175
170	Phosphate, no facilities
168mm	**rest area both lanes, full(handicapped)facilities, phone, picnic tables, litter barrels, petwalk**
166	Gold Creek, **S**...Camp Mak-A-Dream
162	Jens, no facilities
154	MT 1(from wb), Drummond, **S**...Cenex/diesel, Exxon/diesel/mart/atm, Sinclair/diesel/mart/24hr, Wagon Wheel Café/Motel, Sky Motel, Pintlar Scenic Rt, Georgetown Lake RA, Goodtime RV Park(3mi)
153	MT 1(from eb), **N**...Garnet GhostTown, **S**...D-M Café, Drummond Motel, same as 154
150.5mm	weigh sta both lanes
143mm	**rest area both lanes, full(handicapped)facilities, phone, picnic tables, litter barrels, petwalk**
138	Bearmouth Area, **N**...Chalet Bearmouth Camp/rest., to gas, food, lodging
130	Beavertail Rd, **S**...rec area, camping(seasonal), to Beavertail Hill SP
128mm	parking area both lanes, no facilities
126	Rock Creek Rd, **S**...Rock Creek Lodge/gas/casino, Ekstrom camping/rest., rec area
120	Clinton, **1 mi N**...Poor Henry's Café, **S**...Conoco/diesel/mart, USPO
113	Turah, **N**...casino, **1 mi S**...Turah Store/gas/camping, Steaks & BBQ
109.5mm	Clark Fork
109mm	Blackfoot River
109	MT 200 E, Bonner, **N**...Exxon/Arby's/Subway/diesel/casino/mart/LP/24hr, River City Grill, bank, hist site, truck/auto repair
108.5mm	Clark Fork
107	E Missoula, **N**...BP/mart/laundry/atm, Ole's Mkt/Dixie's Diner/Conoco/diesel/mart/24hr, Kolb's Café/casino, Aspen Motel, bank, repair, ski areas, transmissions, **2 mi S**...Holiday Inn Express
105	US 12 W, Missoula, **S**...Cenex/diesel/24hr, Conoco/diesel/pizza/subs/mart/24hr, Pacific Pride/diesel, Sinclair/mart, Burger King, Finnegan's, Little Caesar's, McDonald's, McKay's Rest./casino, Pizza Hut, PressBox Café, Taco Bell, Best Western, Campus Inn, Creekside Inn, DoubleTree Hotel, Family Inn, Holiday Inn, Holiday Inn Express, Ponderosa Motel, Thunderbird Motel, Ace Hardware, Albertson's, Champion Parts, JiffyLube, Mr Muffler, Shipping Depot, U of MT, Vietnam Vet's Mem
104	Orange St, Missoula, **S**...HOSPITAL, Conoco/mart, KFC, Pagoda Chinese, Subway, Budget Motel, Red Lion Inn, Travelodge, bank, laundromat, to City Ctr
101	US 93 S, Reserve St, **N**...Conoco/mart, Best Western, Motel 6, ski area, **S**...Cenex/diesel/LP/mart, Conoco/mart/wash/24hr, Exxon/diesel/mart/24hr, Deano's Rest., Joker's Wild Rest., McDonald's, Rowdy's Rest., Taco Time/TCBY, Comfort Inn, 4B's Inn/rest., Hampton Inn, Ruby's Inn/rest., Super 8, Travelers Inn, JiffyLube, KOA, casinos, diesel repair, **1 mi S**...Conoco, Exxon, Arby's, Burger King, China Bowl, Fuddrucker's, Perkins, Subway, Taco Bell, Wendy's, Albertson's, Barnes&Noble, Chevrolet, Costco, FutureShop, Gart Sports, Mail Depot, Michael's, PetsMart, Pier 1, Staples, Target, bank, cinema, mall
99	Airway Blvd, **S**...airport
96	US 93 N, MT 200W, Kalispell, **N**...Conoco/diesel/mart/wash/rest./24hr, Day's Inn/rest., Jellystone RV Park, Peterbilt/Ford/Freightliner, to Flathead Lake & Glacier NP, **S**...Sinclair/4B's/diesel/24hr, Exxon/diesel/24hr, Redwood Lodge
92.5mm	weigh sta both lanes
89	Frenchtown, **S**...MEDICAL CARE, Conoco/diesel/café/mart/24hr, Sinclair/rest./laundry, bank, USPO, to Frenchtown Pond SP
85	Huson, **S**...Sinclair/café, food, phone
82	Nine Mile Rd, **N**...Hist Ranger Sta/info, food, phone
81.5mm	Clark Fork
80mm	Clark Fork
77	MT 507, Alberton, Petty Creek Rd, **S**...access to gas, food, lodging, phone
75	Alberton, **S**...MT Hotel, River Edge Motel/gas/mart, MT Book Store, RV camp, casino
73mm	parking area wb, litter barrels
72mm	parking area eb, litter barrels
70	Cyr, no facilities

E ↑ ↓ **W**

Missoula

Montana Interstate 90

70mm	Clark Fork
66	Fish Creek Rd, no facilities
66mm	Clark Fork
61	Tarkio, no facilities
58.5mm	Clark Fork
58mm	**rest area both lanes, full(handicapped)facilities, phone, picnic tables, litter barrels, petwalk, NF camping(seasonal)**
55	Lozeau, Quartz, **N**...gas/mart
53.5mm	Clark Fork
49mm	Clark Fork
47	MT 257, Superior, **N**...HOSPITAL, BP/Durango's Rest./mart, Cenex/wash/24hr, Conoco/JG's Rest./mart, Budget Host, Hilltop Motel, Castle's IGA Food, Mineral Drug, NAPA, TrueValue, USPO, auto repair, cinema, **S**...Exxon/ diesel/casino/mechanic/wash/24hr
45mm	Clark Fork
43	Dry Creek Rd, **N**...NP camping(seasonal)
37	Sloway Area, no facilities
34mm	Clark Fork
33	MT 135, St Regis, **N**...Conoco/diesel/rest./mart/gifts, Exxon/pizza/mart, Frosty Drive-In, Jasper's Rest., OK Café/ casino, Little River Motel, St Regis Motel/camping, Super 8, KOA, USPO, antiques
30	Two Mile Rd, **S**...fishing access
29mm	fishing access, wb
26	Ward Creek Rd(from eb), no facilities
25	Drexel, no facilities
22	Camels Hump Rd, Henderson, **N**...Cabin City camping(seasonal)
18	DeBorgia, **N**...Cenex/Beartrap Foods, Pinecrest Rest./casino, Hotel Albert, Riverside Inn/café(1mi), U-Haul, antiques, auto repair, same as 16
16	Haugan, **N**...10,000 Silver $/motel/rest./casino/RV parking, Exxon/diesel/mart
15mm	weigh sta both lanes, exits left from both lanes
10	Saltese, **N**...Elk Glen's Store, Mangold's Motel, USPO, antiques, café/bar, motel
10mm	St Regis River
5	Taft Area, no facilities
4.5mm	**rest area both lanes, full(handicapped)picnic tables, litter barrels, petwalk, chainup/removal**
0	Lookout Pass, Lookout RV Park/deli/rest., access to ski area, info
0mm	Montana/Idaho state line, Mountain/Pacific time zone, Lookout Pass elev 4680

Montana Interstate 94

Exit #	Services
250mm	Montana/North Dakota state line
248	Carlyle Rd, no facilities
242	MT 7(from wb), Wibaux, **S**...rest area both lanes, full(handicapped)facilities, phone, picnic tables, litter barrels, Coop gas/mart, Conoco/mart, Super 8, RV camping
241	MT 261(from eb), to MT 7, Wibaux, **S**...Coop gas/mart, Super 8, same as 242
241mm	weigh sta both lanes
236	ranch access, no facilities
231	Hodges Rd, no facilities
224	Griffith Creek, frontage road, no facilities
222.5mm	Griffith Creek
215	MT 335, Glendive, City Ctr, **N**...Conoco/mart, CC's Café, Comfort Inn, Day's Inn, King's Inn, Super 8, Glendive Camping(apr-oct), TrueValue, museum, **S**...HOSPITAL, Exxon/mart, Sinclair/diesel/repair, Dairy Queen, Hardee's, Best Western, Budget Motel, El Centro Motel, to Makoshika SP
215mm	Yellowstone River
213	MT 16, Glendive, to Sidney, **N**...Exxon/diesel, Green Valley Camping, **S**...Cenex/diesel/mart/atm, Conoco/diesel, Sinclair/diesel, Bachio's Italian, McDonald's, Pizza Hut, Subway, Taco John's, Best Western Downtown, Budget Host, Holiday Lodge, Parkwood Motel, Albertson's, Ford/Lincoln/Mercury, K-Mart, Radio Shack, laundry, repair, st patrol
211	MT 200S(from wb, no EZ return), to Circle, **N**...airport, **1/2 mi S**...Cenex/mart
210	Lp 94, to MT 200S, W Glendive, **S**...McDonald's(2mi), Chevrolet/Pontiac/Olds/Buick, Chrysler/Plymouth/Dodge/ Jeep, Makoshika SP

E

W

E

W

Glendive

Montana Interstate 94

206	Pleasant View Rd, no facilities
204	Whoopup Creek Rd, no facilities
198	Cracker Box Rd, **1 mi N**...Hess Arabians
192	Bad Route Rd, **S...rest area/weigh sta both lanes, full(handicapped)facilities, weather info, phone, picnic tables, litter barrels, camping, petwalk**
187mm	Yellowstone River
185	MT 340, Fallon, **S**...access to café, phone
184mm	O'Fallon Creek
176	MT 253, Terry, **N**...HOSPITAL, Cenex/diesel/mart, Overland Rest., Diamond Motel, Kempton Hotel, Ford, Terry RV Oasis
170mm	Powder River
169	Powder River Rd, no facilities
159	Diamond Ring, no facilities
148	Valley Access, no facilities
141	US 12 E, Miles City, **N**...Flying J/diesel/mart/rest./24hr, Dairy Queen(3mi), Star Motel, Big Sky RV Camping, casino(4mi), to KOA
138	MT 58, Miles City, **N**...HOSPITAL, Cenex/diesel/mart/24hr, Conoco/mart/atm, Exxon/diesel/mart/atm/24hr, A&W/KFC, Dairy Queen, 4B's Rest., Gallagher's Rest., Hardee's, Little Caesar's, McDonald's, Pizza Hut/Taco Bell, Subway, Taco John's, Wendy's, Best Western, Custer's Inn, Day's Inn, Motel 6, Albertson's/drug, AmericanLube, Checker Parts, County Mkt Foods/24hr, Osco Drug, K-Mart, KOA, Wal-Mart/auto, carwash, casinos, **S**...Hunan Chinese, Comfort Inn, Guesthouse Suites, Holiday Inn Express, Super 8
137mm	Tongue River
135	Lp 94, Miles City(no EZ return), **3 mi N**...Olive Dining Room, Friendship Inn, KOA, museum, visitor info
128	local access, no facilities
126	Moon Creek Rd, no facilities
117	Hathaway, **N**...phone
114mm	**rest area eb, full(handicapped)facilities, phone, picnic tables, litter barrels, petwalk**
113mm	**rest area wb, full(handicapped)facilities, phone, picnic tables, litter barrels, petwalk, overlook**
106	Butte Creek Rd, to Rosebud, **N**...food, phone
103	MT 446, MT 447, Rosebud Creek Rd, no facilities
98.5mm	weigh sta both lanes
95	Forsyth, **N**...Conoco/7-11/24hr, Exxon/diesel/mart/24hr, Bloomin Onion Rest., Dairy Queen, Best Western Sundowner, Rails Inn Motel(1mi), MT Inn, WestWind Motel(2mi), Ace Hardware, Ford/Mercury, Yellowstone Drug, carwash, casinos, cinema, to Rosebud RA, **S**...camping
93	US 12 W, Forsyth, **N**...HOSPITAL, Conoco, TownPump/gas/24hr, TopHat Eatery, Rails Inn Motel/café, Westwind Motel, repair/tires
87	MT 39, to Colstrip, no facilities
82	Reservation Creek Rd, no facilities
72	MT 384, Sarpy Creek Rd, no facilities
67	Hysham, **1-2 mi N**...gas, phone, food, lodging, museum
65mm	**rest area both lanes, full(handicapped)facilities, phone, picnic tables, litter barrels, petwalk**
63	ranch access, no facilities
53	Bighorn, access to phone
52mm	Bighorn River
49	MT 47, Custer, to Hardin, **S**...Ft Custer Café/bar, to Little Bighorn Bfd, camping
47	Custer, **S**...Conoco/mart, Custer Food Mkt, D&L Motel/café, Jct City Saloon/café
41.5mm	**rest area wb, full(handicapped)facilities, phone, picnic tables, litter barrels, petwalk**
38mm	**rest area eb, full(handicapped)facilities, phone, picnic tables, litter barrels, petwalk**
36	Waco, frontage rd, no facilities
23	Pompeys Pillar, **N**...Pompeys Pillar Nat Landmark
14	Ballentine, Worden, **1/2 mi N**...Conoco/mart, **S**...Long Branch Café/casino
6	MT 522, Huntley, **N**...gas/mart/phone
0mm	I-90, E to Sheridan, W to Billings, I-94 begins/ends on I-90, exit 456.

E ↕ **W**

Nebraska Interstate 80

Exit #	Services
455mm	Missouri River, Nebraska/Iowa state line
454	13ᵗʰ St, **N**...McDonald's, Comfort Inn, **S**...Phillips 66/diesel/mart, Doorly Zoo, Imax, stadium
453	24ᵗʰ St(from eb), no facilities
452b	I-480 N, US 75 N, to Henry Ford's Birthplace, Eppley Airfield
a	I-480 S, US 75 S, no facilities
451	42ⁿᵈ St, **N**...HOSPITAL, Conoco/diesel/mart, Phillips 66, Texaco/mart, **S**...Phillips 66, Burger King, McDonald's, Taco Bell, Super 8, FastLube, **1 mi S on 50ᵗʰ St**...Boston Mkt, Baker's Foods, Osco Drug, Firestone, NAPA, bank
450	60ᵗʰ St, **N**...Conoco/mart(1mi), Texaco/mart, McDonald's, Honda, NAPA, muffler/brakes, tires, to U of NE Omaha, **S**...Phillips 66/diesel/mart/wash
449	72ⁿᵈ St, to Ralston, **N**...HOSPITAL, Amoco/mart, Burger King, Grover St Grill, LaStrada Italiano, Perkins, Best Western, Cornhusker Inn, Day's Inn, Hampton Inn, Holiday Inn, Homewood Suites, Oak Creek Inn, Ramada Inn, Super 8, **S**...Amoco/repair/wash, Anthony's Rest., Fed Ex
448	84ᵗʰ St, **N**...Amoco/mart, Citgo/7-11, Texaco/mart/wash, Arby's, Denny's, McDonald's, Papa John's, Taco Bell, Econolodge, Ace Hardware, Advance Parts, Baker's Foods, Goodyear/auto, Hancock Fabrics, Kohll's Drug, Mangelson's Crafts, Master Tune-Up, NAPAutocare, Pearle Vision, ShopKO/drugs, bank, **S**...Phillips 66/diesel/mart, QT/diesel/mart, Sinclair/diesel/mart, SuperGas, Arby's, Subway, Wendy's, Acura/Isuzu/Kia, Chevrolet, CarQuest, Discount Tire, JiffyLube, Just Good Meats, U-Haul, bank
446	I-680 N, to Boystown
445	US 275, NE 92, I thru L St, **N**...Total/mart, Austin's Steaks, Residence Inn, Sheraton, Sam's Club, RV Ctr, **S**...Conoco/diesel/mart, Phillips 66, QT/diesel/mart, McDonald's, Arby's, Baskin-Robbins, Burger King, Chinese Rest., Godfather's, Golden Corral, Hardee's, LJ Silver, McDonald's, Nebraska Beef Co, Taco Bell, TCBY, $3 Café, Valentino's Ristorante, Village Inn Rest., Wendy's, Winchell's Donuts, Best Western, Baymont Inn, Clarion Motel, Comfort Inn, Hawthorn Suites/rest., Hampton Inn, Holiday Inn Express, Motel 6, Super 8, Albertson's/drug, Bag'n Sav, MailRoom, GMC, O'Reilly's Parts, bank, laundry
444	Q St, **N**...Cenex/diesel/mart, cinema
442	126ᵗʰ St, Harrison St, **N**...to airport
440	NE 50, to Springfield, **N**...HOSPITAL, Citgo, Conoco/mart, Texaco/Subway/diesel/mart/24hr, Catfish Charlie's Rest., Cracker Barrel, Hardee's, McDonald's, TH Café, Ben Franklin Motel, Best Inn, Budget Inn, Comfort Inn, National 9 Inn, Ramada Ltd, Ford, tires, truckwash, **S**...Amoco/diesel/mart, to Platte River SP
439	NE 370, to Gretna, **N**...Sinclair/diesel/mart, Day's Inn, Suburban Inn, antiques, **S**...HOSPITAL, museum
432	US 6, NE 31, to Gretna, **N**...Bosselman/Texaco/diesel/rest./24hr, McDonald's, Super 8, GNC, Nebraska X-ing/famous brands, **S**...Flying J/Conoco/diesel/LP/rest./24hr, KOA, to Schramm SP
431.5mm	**rest area wb, full(handicapped)facilities, info, phone, picnic tables, litter barrels, petwalk**
427mm	Platte River
426	to Ashland, **N**...gas, food, RV camping, to Mahoney SP
425.5	**rest area eb, full(handicapped)facilities, phone, picnic tables, litter barrels, petwalk, vending**
420	NE 63, Greenwood, **N**...Conoco/diesel/café/mart/atm, **S**...Phillips 66/diesel/mart, Donna's Place Rest., Day's Inn, antiques, wiegh sta, to Platte River SP
409	US 6, Waverly, to E Lincoln, **N**...Gas'n Shop, **S**...HOSPITAL
405	US 77 N, 56ᵗʰ St, Lincoln, **S**...HOSPITAL, Phillips 66/mart, **1 mi S**...Conoco/service, Total/mart, Misty's Rest., Econolodge, Inn at Lincoln/rest., Motel 6, Quality Inn, antiques
404.5mm	**rest area wb, full(handicapped)facilities, info, phone, picnic tables, litter barrels, petwalk**
403	27ᵗʰ St, Lincoln, **S**...Phillips 66/Tubby's/diesel/mart/wash, Cracker Barrel, AmericInn, Country Inn Suites, Red Roof Inn, Settle Inn, Staybridge Suites, Ford, **1-3 mi S**...Amoco/mart, Texaco, Applebee's, Burger King, Carlos O'Kelly's, DaVinci's Italian, Garden Café/bakery, Godfather's, Golden Corral, McDonald's, Papa John's, Runza Rest., Schlotsky's, Sonic, Taco John's, Valentino's Rest., Village Inn, Baymont Inn, Comfort Suites, Fairfield Inn, Microtel, Ramada Ltd, Super 8, Battery Patrol, GNC, Haas Tire, Hy-Vee Foods, Mazda, Michael's, PetsMart, Radio Shack, Sam's Club, ShopKO, ToysRUs, Tuffy Muffler, Wal-Mart/24hr, to U NE, st fairpark
401b	US 34 W, **N**...RV camping
a	I-180, US 34 E, to 9ᵗʰ St, Lincoln
399	Lincoln, **N**...Amoco/diesel/mart/24hr, Baskin-Robbins, Denny's, McDonald's, NY Pizza, Perkins/24hr, Quizno's, TacoMaker, Best Western, Comfort Inn, Day's Inn, Hampton Inn, Holiday Inn Express, Horizon Inn, Motel 6, Quality Inn, Ramada Inn, Sleep Inn, Travelodge, Honda/Kawasaki/Suzuki/Yamaha, to airport, **S**...Sinclair/diesel, Econolodge/rest., Inn4Less
397	US 77 S, to Beatrice, **S**...Congress Inn, Ford/Lincoln/Mercury
396	US 6, West O St(from eb), **S**...Total/mart, Cobbler Inn, Senate Inn, Super 8, Travelodge

Omaha

Lincoln

E

W

Nebraska Interstate 80

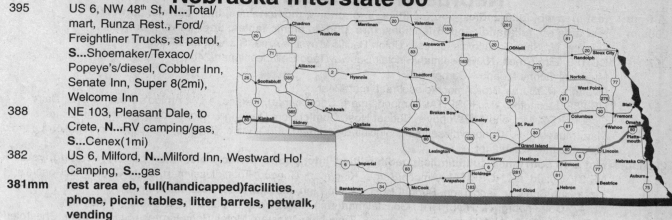

395   US 6, NW 48th St, **N**...Total/mart, Runza Rest., Ford/Freightliner Trucks, st patrol, **S**...Shoemaker/Texaco/Popeye's/diesel, Cobbler Inn, Senate Inn, Super 8(2mi), Welcome Inn

388   NE 103, Pleasant Dale, to Crete, **N**...RV camping/gas, **S**...Cenex(1mi)

382   US 6, Milford, **N**...Milford Inn, Westward Ho! Camping, **S**...gas

381mm   **rest area eb, full(handicapped)facilities, phone, picnic tables, litter barrels, petwalk, vending**

379   NE 15, to Seward, **2-3 mi N**...HOSPITAL, Total/mart, McDonald's, Valentino's Rest., Dandy Lion Inn, Super 8, antiques, **S**...Phillips 66/diesel/mart

375.5mm   **rest area wb, full(handicapped)facilities, info, phone, picnic tables, litter barrels, petwalk, vending**

373   80G, Goehner, **N**...Texaco/service, food

369   80E, Beaver Crossing, **3 mi S**...food, RV camping

366   80F, to Utica, no facilities

360   93B, to Waco, **N**...Burns Bros/Amoco/diesel/rest., **S**...Double Nickel Camping

355mm   **rest area wb, full(handicapped)facilities, info, phone, picnic tables, litter barrels, vending, petwalk**

353   US 81, to York, **N**...HOSPITAL, Amoco/diesel/mart, SappBros/diesel/rest., Sinclair, A&W/Amigo's Mexican, Arby's, Bonanza, Burger King, Chinese Rest., Country Kitchen, Hy-Mark Rest., KFC, McDonald's, Wendy's, Best Western, Comfort Inn, Day's Inn, Econolodge, Quality Inn, Super 8, Buick/GMC/Olds, Chevrolet/Pontiac, Chrysler/Dodge, **S**...Conoco/diesel, Petro/Mobil/Pizza Hut/diesel/rest./24hr, Texaco/diesel/mart/rest./24hr, Holiday Inn, Village Inn Rest., York Inn, USA Inn/rest., Blue Beacon

351mm   **rest area eb, full(handicapped)facilities, info, phone, picnic tables, litter barrels, petwalk, vending**

348   93E, to Bradshaw, no facilities

342   93A, Henderson, **N**...KOA, antiques, **S**...HOSPITAL, FuelMart/diesel, Dell's Rest., Wayfarer Motel/café, Western RV Park

338   41D, to Hampton, **S**...gas

332   NE 14, Aurora, **N**...HOSPITAL, Texaco/diesel/mart, McDonald's(3mi), Pizza Hut(3mi), Budget Host(3mi), Hamilton Motel/rest., Ford/Lincoln/Mercury, to Plainsman Museum, **S**...Citgo/diesel/mart

324   41B, to Giltner, no facilities

318   NE 2, to Grand Island, **S**...Grand Island RV Park, KOA

317mm   **rest area wb, full(handicapped)facilities, info, phone, picnic tables, litter barrels, vending, petwalk**

315mm   **rest area eb, full(handicapped)facilities, info, phone, picnic tables, litter barrels, vending, petwalk**

314.5mm   Platte River

312   US 34/281, to Grand Island, **N**...HOSPITAL, Bosselman/Sinclair/Max's/Subway/Taco Bell/diesel/24hr, Phillips 66, Tommy's Rest., USA Inn/rest., Mormon Island RA, to Stuhr Pioneer Museum, **7 mi N**...McDonald's, Pizza Hut, Wendy's, Comfort Inn, Day's Inn, **S**...Amoco, Holiday Inn/rest., Holiday Inn Express, Ford/Peterbilt, Hastings Musum(15mi), KOA(15mi)

305   40C, to Alda, **N**...TA/Conoco/diesel/rest./24hr, Total/diesel/mart/24hr, Crane Meadows Nature Ctr/rest area

300   NE 11, Wood River, **S**...Bosselman/Sinclair/Subway/diesel/24hr, Wood River Motel

291   10D, Shelton, **N**...War Axe SRA, **S**...gas/diesel/mart

285   10C, Gibbon, **N**...Ampride/mart, Phillips 66, RV camping, **rest area both lanes, full facilities, picnic tables, litter barrels, S**...5 Star Motel, antiques, auto repair

279   NE 10, to Minden, **N**...Texaco/diesel/mart, Cabela's SportsGear, **S**...Pioneer Village Camping

275mm   The Great Platte River Road Archway Monument

272   NE 44, Kearney, **N**...HOSPITAL, Coastal/mart, Gas-Stop/mart, Phillips 66/diesel/mart, Texaco/diesel/mart, Total/mart, Amigo's Mexican, Arby's, Bonanza, Burger King, Carlos O'Kelly's, Chinese Rest., Country Kitchen, Dairy Queen, Golden Corral, LJ Silver, McDonald's, Perkins/24hr, Pizza Hut, Popeye's, Red Lobster, Runza Rest., Taco Bell, Taco John's, Valentino's Rest., Wendy's, AmericInn, Best Western, Budget Motel, Comfort Inn, Country Inn Suites, Day's Inn, Fairfield Inn, Hampton Inn, Holiday Inn, Ramada Inn/rest., Regency Inn, Super 8, Western Motel, Wingate Inn, Boogaart's Foods, Buick/Cadillac, Chevrolet/Mazda, Chrysler/Dodge, Pontiac/Olds, Goodyear, JiffyLube, Wheeler's Tire/repair, Budget$ RV Park, Clyde&Vi's RV Park, RV Ctr, bank, to U NE Kearney, Museum of NE Art, **S**...Amoco/mart, Grandpa's Steaks, Ft Kearney Inn

Nebraska Interstate 80

L **e** **x** **i** **n** **g** **t** **o** **n**	**271mm**	**rest area wb, full(handicapped)facilities, info, phone, picnic tables, litter barrels, petwalk**
	269mm	**rest area eb, full(handicapped)facilities, info, phone, picnic tables, litter barrels, petwalk**
	263	Odessa, **N**...Sapp/Texaco/diesel/rest., Union Pacific Wayside Area, **S**...Aunt Lu's Rest.
	257	US 183, Elm Creek, **N**...Bosselman/Sinclair/Subway/Taco Bell/diesel/mart/24hr, 1st Interstate Inn, Antique Car Museum, Sunny Meadow Camping
	248	Overton, **N**...Burns Bros/Amoco/diesel/rest./mart/24hr
	237	US 283, Lexington, **N**...HOSPITAL, Ampride/mart, Conoco/mart, Texaco, A&W, Amigo's Rest., Arby's, Dairy Queen, KFC, McDonald's, Wendy's, Budget Host, Comfort Inn, Day's Inn, Econolodge, Holiday Inn Express, Wal-Mart, museum, **S**...Conoco/diesel/mart, Sinclair/diesel/mart, Kirk's Café, Super 8, Goodyear, to Johnson Lake RA
	231	Darr, **S**...truckwash/24hr
	228mm	**rest area both lanes, full(handicapped)facilities, info, phone, picnic tables, litter barrels, vending, petwalk**
	222	NE 21, Cozad, **N**...HOSPITAL, Amoco/Burger King/mart, Conoco, Total/diesel/mart, Dairy Queen, McDonald's, Pizza Hut, Subway, Taco Time, Budget Host, Comfort Inn, Motel 6, Alco, Ford, museum
	211	NE 47, Gothenburg, **N**...HOSPITAL, Coastal/mart(1mi), Texaco/diesel/rest./24hr, El Ranchito Mexican, McDonald's, Pizza Hut, Runza Rest., Super 8, Travel Inn, Western Motel, Pony Express Sta Museum, Chevrolet/Pontiac/Olds/Buick, truck permit sta, **S**...Total/mart, KOA
	199	Brady, **N**...Dairy Queen/gas/diesel, RV camping
	194mm	**rest area both lanes, full(handicapped)facilities, phone, picnic tables, litter barrels, vending, petwalk**
N **P** **l** **a** **t** **t** **e**	190	Maxwell, **N**...Sinclair/diesel/mart, **S**...to Ft McPherson Nat Cemetary, RV camping
	181mm	weigh sta both lanes, phones
	179	to US 30, N Platte, **N**...Stanford Motel, airport, **S**...Flying J/Conoco/diesel/LP/rest./24hr
	177	US 83, N Platte, **N**...HOSPITAL, Amoco/Subway/mart, Coastal/mart, Conoco/diesel/mart/atm, Phillips 66/diesel/mart/24hr, Texaco/diesel/mart/atm/24hr, Total/diesel/mart, A&W, Amigo's Rest., Applebee's, Arby's, Baskin-Robbins/Blimpie/Dunkin Donuts, BBQ, Burger King, Dairy Queen, KFC, LJ Silver, McDonald's, Perkins/24hr, Pizza Hut, Roger's Diner, Valentino's Rest., Village Inn Rest., Wendy's, Best Western, Blue Spruce Motel, 1st Interstate Inn, Hampton Inn, Motel 6, Pioneer Motel, Quality Inn, Sands Motel, Stockman Inn, Ace Hardware, Advance Parts, Goodyear/auto, Holiday TravL Park, JC Penney, MailBoxes Etc, Modern Muffler, Oil Exchange, ShopKO, Staples, USave Food, Wal-Mart SuperCtr/24hr, XpressLube, bank, cinema, mall, museum, to Buffalo Bill's Ranch, **S**...info, Phillips 66, Texaco/diesel/rest./24hr, Total/Taco Bell/diesel/mart/atm/24hr, Country Kitchen, Hunan Chinese, Comfort Inn, Day's Inn, Holiday Inn Express, Ramada Ltd, Super 8, Chevrolet, Chrysler/Jeep, Dodge, Ford/Lincoln/Mercury, Honda, Olds/Cadillac, Toyota/Mazda, truck permit sta, to Lake Maloney RA
	164	56C, Hershey, **N**...Sinclair/mart, Texaco/diesel/rest./24hr, Western Wear Outlet
	160mm	**rest area both lanes, full(handicapped)facilities, info, phone, picnic tables, litter barrels, petwalk**
	158	NE 25, Sutherland, **N**...Park Motel, Sutherland Lodge, **S**...Conoco/diesel/pizza/subs/mart/24hr
	149mm	Central/Mountain time zone
	145	51C, Paxton, **N**...Texaco/mart/atm/24hr, Ole's Café, Day's Inn
	133	51B, Roscoe, no facilities
O **g** **a** **l** **l** **a** **l** **a**	**132.5mm**	**rest area wb, full(handicapped)facilities, info, phone, picnic tables, litter barrels, petwalk**
	126	US 26, NE 61, Ogallala, **N**...HOSPITAL, Amoco/mart, Phillips 66/diesel/mart, Sapp/Texaco/diesel/mart/atm/24hr, Texaco/Burger King/mart, Amigo's Rest., Arby's/RV parking, Country Kitchen, McDonald's, Pizza Hut, Taco Bell, Taco John's, Valentino's, Best Western, Day's Inn, Holiday Inn Express, Ramada Ltd, Ace Hardware, Big A Parts, Chevrolet/Buick, Radio Shack, to Lake McConaughy, RV camping, **S**...Conoco/Subway/mart, TA/diesel/rest./24hr, Phillips 66/diesel/mart, Cassel's Pancakes, Dairy Queen, KFC, Wendy's, Comfort Inn, Econolodge, 1st Interstate Inn, Super 8, Open Corral RV Park, Pamida, antiques
	124mm	**rest area eb, full(handicapped)facilities, info, phones, picnic tables, litter barrels, petwalk**
	117	51A, Brule, **N**...Happy Jack's/diesel/mart, RV camping
	107	25B, Big Springs, **N**...Bosselman/Sinclair/Grandma Max's/diesel, Total/diesel/mart, Budget 8, McGreer Camping
	102	I-76 S, to Denver, no facilities
	102mm	S Platte River
	101	US 138, to Julesburg, no facilities
	99mm	scenic turnout eb
	95	NE 27, to Julesburg, no facilities
	88mm	**Lodgepole Creek, rest area wb, full(handicapped)facilities, phone, picnic tables, litter barrels, vending, petwalk**
	85	25A, Chappell, **N**...Ampride/mart, Chevrolet/Buick, Creekside Camping, USPO, wayside park, **S**...Texaco/diesel/mart
	82.5mm	**rest area eb, full(handicapped)facilities, phone, picnic tables, litter barrels, vending, petwalk**
	76	17F, Lodgepole, **1 mi N**...gas/diesel, lodging

E

↕

W

Nebraska Interstate 80

69	17E, to Sunol, no facilities
61mm	**rest area wb, full(handicapped)facilities, phone, picnic tables, litter barrels, vending, petwalk**
59	US 385, 17J, Sidney, **N...**Amoco/mart, Conoco/KFC/Taco Bell/TCBY/diesel/mart, Texaco/Sapp/diesel/mart/24hr, Arby's, High Plains Cache Café, El Ranchito Rest., McDonald's, Runza Rest., Taco John's, AmeriSuites, Comfort Inn, Day's Inn, Motel 6, Cabela's SportsGear, Pamida, RV camping(2mi), golf, truck permit sta, **3 mi N...**HOSPITAL, KFC, Pizza Hut, Generic Motel, Sidney Motel, Super 8, RV camping, **S...**Country Kitchen, Holiday Inn, truckwash
55	NE 19, Sidney, to Sterling, CO, **2 mi N...**Super 8
51.5mm	**rest area/hist marker eb, full(handicapped)facilities, phone, picnic tables, litter barrels, vending, petwalk**
48	to Brownson, no facilities
38	Potter, **N...**Cenex/diesel/LP, Texaco/mart, repair
29	53A, Dix, **1/2 mi N...**gas, food
25mm	**rest area wb, full(handicapped)facilities, phone, picnic tables, litter barrels, vending, petwalk**
22	53E, Kimball, **N...**Day's Inn, KOA, golf
20	NE 71, Kimball, **N...**HOSPITAL, Phillips 66/mart/atm, Total/mart, Beef&Brunch Rest., Pizza Hut, Subway, Day's Inn, 1st Interstate Inn, Super 8, RV Park, truck permit sta, **S...**Burger King
18mm	parking area eb, litter barrel
10mm	**rest area eb, full(handicapped)facilities, info, phone, picnic tables, litter barrels, petwalk**
8	53C, to Bushnell, no facilities
1	53B, Pine Bluffs, **1 mi N...**RV camping
0mm	Nebraska/Wyoming state line

Nebraska Interstate 680(Omaha)

Exit #	Services
29b a	I-80, W to Omaha, E to Des Moines. I-680 begins/ends on I-80, exit 27.
28	IA 191, to Neola, Persia, no facilities
21	L34, Beebeetown, no facilities
19mm	**rest area wb, full(handicapped)facilities, info, phone, picnic tables, litter barrels, petwalk**
16mm	**rest area eb, full(handicapped)facilities, info, phone, picnic tables, litter barrels, petwalk**
71	I-29 N, to Sioux City
66	Honey Creek, **W...**Phillips 66/diesel/rest., Iowa Feed&Grain Co, camping
3b a	I-29, S to Council Bluffs, IA 988, to Crescent, **E...**to ski area
1	County Rd, no facilities
14mm	Missouri River, Nebraska/Iowa state line
13	US 75 S, Florence, **1-2 mi E...**CHIROPRACTOR, MEDICAL CARE, VETERINARIAN, DENTIST, Amoco, Conoco, Texaco/diesel/mart, Burger King, Chinese Rest., Godfather's, KFC, Pizza Hut, Taco Bell, Zesto Diner, Baker's Foods, ShopKO, Walgreen, LDS Temple, Mormon Trail Ctr, auto Parts, hardware, **W...**Phillips 66/mart/wash
12	US 75 N, 48th St, **E...**Phillips 66/mart, Burger King
9	72nd St, **E...**HOSPITAL, **W...**Cunningham Lake Rec Area
6	NE 133, Irvington, **E...**Conoco/mart, McDonald's
5	Fort St, **W...**Sinclair, Texaco, Wal-Mart, USPO
4	NE 64, Maple St, **E...**Amoco, **W...**Perkins, La Quinta, Ramada Ltd
3	US 6, Dodge St, **E...**MEDICAL CARE, Marriott, Ford, Mazda, **W...**Toyota
2	Pacific St, **E...**Amoco, Best Western, shopping
1	NE 38, W Center St, **E...**Don&Millie's Rest., **W...**Total/mart
0mm	I-680 begins/ends on I-80, exit 446.

Nevada Interstate 15

Mesquite

Exit #	Services
123mm	Nevada/Arizona state line, Pacific/Mountain time zone
122	Lp 15, Mesquite, **E...NV Welcome Ctr both lanes, full(handicapped)facilities, petwalk,** Chevron/Subway/diesel/mart/24hr, Shell/Dairy Queen/mart, Texaco/Arby's/mart, Blimpie/Pizza Hut, Burger King, Café Silvestre Mexican, Denny's, Golden West Rest./casino, Jack-in-the-Box, KFC, Panda Garden Chinese, Best Western, Budget Suites, Desert Palms Motel, Hardy's Motel, Mesquite Star Hotel/casino, Up Town Motel, Ace Hardware, El Rancho Mkt, Great Outdoors RV, Mesquite Lube/wash, NAPA, Radio Shack, Rite Aid, Smith's Foods, USPO, bank, diesel repair, laundry, museum, **W...**Virgin River/76/diesel/LP/mart/RV park, Holiday Inn, Mesquite Springs Motel, Rancho Mesquite Hotel/casino, Virgin River Hotel/casino, cinema
120	Lp 15, Mesquite, Bunkerville, **E...**CHIROPRACTOR, Arco/mart/atm/24hr, Chevron/mart, Texaco/diesel/mart, Baskin-Robbins, Carollo's Rest., McDonald's, Casablanca Resort/casino, Oasis Resort/casino, Peppermill Resort/casino, Players Island Resort/casino, Valley Inn, bank, carwash, golf, laundry, RV camping
112	NV 170, Riverside, Bunkerville, no facilities
110mm	truck parking both lanes, litter barrels
100	to Carp, Elgin, no facilities
96mm	truck parking both lanes, litter barrels
93	NV 169, to Logandale, Overton, **E...**Lake Mead NRA, Lost City Museum
91	NV 168, Glendale, **W...**Chevron/diesel/mart/rest., Pat's Rest., auto parts, tires
90.5mm	Muddy River
90	NV 168(from nb), Glendale, Moapa, **W...**Chevron, BJ's Rest., Moapa Indian Reservation
88	Hidden Valley, no facilities
87.5mm	parking area both lanes, litter barrels
84	Byron, no facilities
80	Ute, no facilities
75	NV 169 E, Valley of Fire SP, Lake Mead NRA, **E...**casino/gifts/fireworks
64	US 93 N, Great Basin Hwy, to Ely, Great Basin NP, no facilities
60mm	livestock check sta sb
58	NV 604, Las Vegas Blvd, to Apex, Nellis AFB, no facilities
54	Speedway Blvd, Hollywood Blvd, **E...**Las Vegas Speedway
50	Lamb Ave(from sb), no facilities
48	Craig Rd, **E...**Arco/mart/24hr, Chevron/Subway/mart/24hr, Mobil, Pilot/Dairy Queen/KFC/Pizza Hut/diesel/mart/24hr, Shell/mart, Burger King, Jack-in-the-Box, Speedway Grill, Barcelona Hotel/casino(2mi), Best Western(2mi), Super 8, Nellis AFB
46	Cheyenne Ave, **E...**VETERINARIAN, Arco/mart/24hr, Citgo/7-11, Wendy's(1mi), Ramada Inn/casino, Mario's Mkt, NAPA, carwash, **W...**Citgo/7-11, Flying J/diesel/LP/rest./24hr, McDonald's, Comfort Inn, Blue Beacon, Purcell Tires, RV Ctr, diesel repair
45	Lake Mead Blvd, HOSPITAL, **E...**7-11, McDonald's, tires
44	Washington Ave(from sb), **E...**Best Western, casinos
43	D St(from nb), **E...**Best Western, same as 44, **W...**Office Depot
42b a	I-515 to LV, US 95 N to Reno, US 93 S to Phoenix, **E...**casinos, **W...**Arco/mart/24hr, Texaco
41b a	Charleston Blvd, **E...**Texaco/diesel/mart, 7-11, antiques, **W...**Carl's Jr, Del Taco, McDonald's, Taco Bell, Wendy's
40	Sahara Ave, **E...**The Strip, 76/diesel/rest./24hr, Las Vegas Inn/casino, Travelodge, LV Conv Ctr, Purcell Tires, auto parts, RV Ctr, **W...**Texaco/diesel/LP/mart, 7-11, Denny's, Palace Station Hotel/casino, Lexus, Mazda, Mitsubishi, Mercedes, bank
39	Spring Mtn Rd(from sb), **E...**Budget Suites, Frontier Hotel, Mirage, Treasure Island, **W...**Arco/mart/24hr, Circle K/gas, United/diesel/mart, Schlotsky's, EconoLube/Tune, Discount Tire, Firestone/auto, Goodyear/auto, Pennzoil, auto/transmission repair, bank
38b a	Dunes Flamingo Rd, **E...**The Strip, Caesar's Palace, Flamingo Hilton, Monte Carlo, to UNLV, **W...**MEDICAL CARE, Arco/mart/24hr, Burger King, Outback Steaks, Subway, Gold Coast Hotel, Rio Hotel, Terrible Herbst Wash/lube
37	Tropicana Ave, **E...**Excaliber Hotel, Mandalay Bay, MGM Grand, Motel 6, Tropicana Hotel, GMC/White/Volvo, airport, casinos, funpark, **W...**Arco/mart/24hr, Chevron/diesel/mart, 76, Shell/Mr Subs/mart, WildWest/diesel/LP, Burger King, IHOP, In-n-Out, Jack-in-the-Box, KFC, McDonald's, Taco Bell, Taco Cabana, Wendy's, Best Western, Budget Suites, Hampton Inn, Howard Johnson, Motel 6, Honda Motorcycles, casino
36	Russell Rd, **E...**Texaco, McDonald's, Diamond Inn, Klondike Hotel, Pollyanna Motel, casinos, to airport, **W...**Chevron/Herbst/diesel/mart
34	to I-215 E, Las Vegas Blvd, to The Strip, McCarran Airport
33	NV 160, to Blue Diamond, Death Valley, **E...**Arco/mart/24hr, Chevron/McDonald's/mart, Citgo/7-11/diesel, Mobil/diesel/LP/mart, factory outlet/famous brands, **W...**TA/76/diesel/rest./LP/24hr, Jack-in-the-Box, Firebird Inn, Oasis RV Resort, Silverton Hotel/RV Resort/casino
27	NV 146, to Henderson, Lake Mead, Hoover Dam, **2 mi E...**Wheelers RV/LP, factory outlets
25	NV 161, Sloan, **1 mi E...**Wheelers RV/service/mart/LP
24mm	bus/truck check sta nb
12	NV 161, Jean, to Goodsprings, **E... Welcome Ctr nb, full facilities, info,** Mobil/diesel/mart, Gold Strike Casino/hotel/Burger King, USPO, skydiving, **W...**Nevada Landing Casino/Shell/diesel/mart/24hr, NV Correctional

Las Vegas

Jean

Nevada Interstate 15

1	Primm, **E**...76/diesel, McDonald's, Buffalo Bill's Resort/casino, Primm Valley Resort/casino, Prima Donna RV Park, factory outlets, funpark, **W**...Texaco/diesel, Whiskey Pete's Hotel/casino/steaks
0mm	Nevada/California state line

Nevada Interstate 80

Exit #	Services
411mm	Nevada/Utah state line, Pacific/Mountain time zone
410	US 93A, W Wendover, to Ely, **S...NV Welcome Ctr/info, full(handicapped)facilities, phone, MEDICAL CARE,** Chevron/mart/24hr, Pilot/ Arby's/diesel/mart/24hr, Shell/diesel/mart, Burger King, McDonald's, Pizza Hut, Subway, Taco Burger, Best Western, Day's Inn, Motel 6, Nevada Crossing Hotel, Peppermill Hotel/casino/RV parking, Rainbow Hotel/casino, Red Garter Hotel/casino, Super 8, Coast Hardware, Smith's Food/drug, StateLine RV Park, KOA, bank
407	Ola, no facilities
398	to Pilot Peak, no facilities
387	to Shafter, no facilities
378	NV 233, Oasis, to Montello, **N**...gas/diesel, Oasis Café/motel
376	to Pequop, no facilities
373mm	**rest area both lanes, picnic tables, litter barrels, toilet**
365	to Independence Valley, **N**...prison facilities
360	to Moor, no facilities
354mm	parking area eb
352b a	US 93, Great Basin Hwy, E Wells, **N**...Chevron/repair, Texaco/diesel/mart/café, Burger King, 4Way Rest./casino, Best Western, LoneStar Motel, Motel 6, Rest Inn Motel, Super 8, Mtn Shadows RV Park, Goodyear/auto/diesel/ repair, Schwab Tires, carwash, **1 mi N**...Best Western, Mountainview Motel, Shellcrest Motel, Wagon Wheel Motel, Ford, NAPA, bank, **S**...Flying J/Conoco/diesel/LP/rest./casino/24hr
351	W Wells, **N**...Amoco/mart, Best Western, China Town Motel/RV Park/casino, CarQuest, FoodTown, to Angel Lake RA
348	to Beverly Hills, **N**...RV camping
343	to Welcome, Starr Valley, **N**...Welcome RV Camping, phone, food
333	Deeth, Starr Valley, no facilities
328	to River Ranch, no facilities
321	NV 229, Halleck, Ruby Valley, no facilities
318mm	N Fork Humboldt River
317	to Elburz, no facilities
314	to Ryndon, Devils Gate, **S**...RV camping
312mm	check sta both lanes
310	to Osino, **4 mi S**...Valley View RV Park
303	E Elko, **N**...Chevron/CFN/Blimpie/diesel/mart/24hr, **S**...HOSPITAL, Amoco/mart/24hr, Chevron/mart/24hr, Conoco/ diesel, Texaco/diesel/mart, Baskin-Robbins, Burger King, Dairy Queen, Dinner Sta Rest., JR's Grill, King Buffet, McDonald's, Pizza Barn, Pizza Hut, Subway, Taco Time, Toki Ona Diner, Wendy's, Best Western/casino, Budget Inn, Comfort Inn, Day's Inn, High Desert Inn, Hilton Garden, Holiday Inn Express, Microtel, Motel 6, Parkview Motel, Red Lion Inn/casino, Super 8, Albertson's, American CarCare, Chevrolet/Pontiac/Olds, Buick/Cadillac/GMC, Chrysler/Jeep/Toyota/Honda, Ford/Lincoln/Mercury, Goodyear/auto, JC Penney, NAPA, Pennzoil/wash, Radio Shack, Rite Aid, Western Auto, Double Dice RV Park, Gold Country RV Park, Valleyview RV Park(2mi), bank, cleaners
301	NV 225, Elko, **N**...Maverick/atm, Arby's, Denny's, Hogi Yogi, McDonald's, Papa Murphy's, RoundTable Pizza, OakTree Inn, Shilo Inn Suites, Big 5 Sports, BuildersMart, FashionBug, K-Mart, Raley's Food/drug, Wal-Mart/auto, airport, cinema, laundry, RV camping, **S**...HOSPITAL, EYECARE, Phillips 66/diesel/pizza/subs/mart/wash, Texaco/ diesel/mart/RV park, Dos Amigos Mexican, KFC, Taco Bell, Checker Parts, CopyMax, IGA Foods, OfficeMax, Smith Food/drug/24hr, **1/2 mi S**...Maverick/gas, Arctic Circle, Golden Phoenix Chinese, Los Sanchez Mexican, New China Café, Best Western, Key Motel, Manor Inn, National 9 Inn, Stampede 7 Motel, Star-lite Motel, Stockmen's Motel/ casino, Thunderbird Motel, Towne House Motel, Big A Parts, CarQuest, FoodTown, Sav-On Drug, Schwab Tires, bank, carwash, casinos
298	W Elko, **S**...RV camping
292	to Hunter, no facilities
285.5mm	tunnel
285mm	Humboldt River
282	NV 221, E Carlin, **N**...prison area
280	NV 766, Carlin, **N**...diesel repair, RV camping, **S**...Amoco/diesel/mart, Pilot/Subway/diesel/mart/24hr, Texaco/ Burger King/diesel/mart, Chin's Café, Hungry Miner Café, State Café/casino, Best Inn Suites, USPO
279	NV 278(from eb), to W Carlin, **1 mi S**...Amoco/diesel/mart
271	to Palisade, no facilities

287

Nevada Interstate 80

B	270mm	Emigrant Summit, elev 6114, parking area both lanes, litter barrels
	268	to Emigrant, no facilities
	261	NV 306, to Beowawe, Crescent Valley, no facilities
a	**258.5mm**	**rest area both lanes, full(handicapped)facilities, picnic tables, litter barrels, petwalk**
	257mm	Humboldt River
t	254	to Dunphy, no facilities
	244	to Argenta, no facilities
t	233	NV 304, to Battle Mountain, **1-2 mi N**...HOSPITAL, Conoco/diesel, Shell/wash/24hr, Best Western, Comfort Inn
l	231	NV 305, Battle Mountain, **N**...Big R's/gas/mart, Chevron/mart, Hideaway Steaks, McDonald's, Subway, Super 8, Midway Mkt, NAPA, Radio Shack, USPO, Wagon Wheel Food, bank, drugs, library, **1 mi N**...HOSPITAL, Conoco/diesel/mart, Flying J/Exxon/Blimpie/diesel/casino/mart/24hr, Shell, Donna's Diner, El Aguila Mexican, Limon Mexican Rest., Mama's Pizza, Moon Garden Chinese, Best Western, FoodTown, laundry, **S**...golf
e	229	NV 304, W Battle Mountain, **N**...Flying J/Exxon/Blimpie/diesel/casino/mart/24hr, Texaco/diesel/mart, Best Inn, Best Western, RV camping, same as 231
	222	to Mote, no facilities
M	216	Valmy, **N**...76/USPO/diesel/mart/24hr, Golden Motel/grill, RV camping, **S...rest area both lanes, full(handicapped)facilities, phone, picnic tables, litter barrels, petwalk, RV dump**
t	212	to Stonehouse, no facilities
	205	to Pumpernickel Valley, no facilities
n	203	to Iron Point, no facilities
	200	Golconda Summit, elev 5145, truck parking area both lanes, litter barrels
	194	Golconda, **N**...Waterhole #1 Hotel/gas/groceries, Z Bar/Grill, USPO
W	187	to Button Point, **N...rest area both lanes, full(handicapped)facilities, phone, picnic tables, litter barrels, RV dump, petwalk, RV dump**
	180	E Winnemucca Blvd, **2-4 mi S**...High Desert RV Park
i	178	Winnemucca Blvd, Winnemucca, **S**...Maverick/gas, Pump'n Save/gas, Rte 66 Grill, Budget Inn, Bull Head Motel, Cozy Motel, Frontier Motel, Scott Motel, Valu Inn, CarQuest, carwash, **1 mi S**...HOSPITAL, Exxon/diesel/mart, Shell, Texaco/mart, Burger King, Imachea's Dinnerhouse, Pizza Hut, Subway, La Villa Motel, Pyrenees Motel, Winners Hotel/casino, Winnemucca RV Park, Radio Shack, bank, casinos, to Buckeroo Hall of Fame, same as 176
n	176	US 95 N, Winnemucca, **N**...Pacific Pride/diesel, **S**...HOSPITAL, Chevron/mart/wash/24hr, Flying J/Conoco/diesel/LP/rest./mart/24hr, Exxon/diesel/mart, Shell/mart, Texaco/diesel/mart, A&W, Arby's, Baskin-Robbins, Burger King, Denny's, Flying Pig Rest., Griddle Diner, Jerry's Rest., KFC, Los Sanchez Mexican, McDonald's, Pizza Hut, RoundTable Pizza, Subway, Taco Bell, Taco Time, Best Western, Day's Inn, Holiday Inn Express, La Villa Motel, Model T Motel, Motel 6, Nevada Motel, Park Hotel, Ponderosa Motel, Pyrenees Motel, Quality Inn, Ramada Ltd, Red Lion Inn/casino, Santa Fe Inn, Scottish Inn, Super 8/RV parking, Thunderbird Motel, Townhouse Motel, Winner's Hotel/casino, Ford/Mercury, Ford/Mercury, Goodyear/auto, JC Penney, Kragen Parts, OK Tire, Raley's Food/drug, RV Ctr(1mi), Wal-Mart/auto/gas, Western Auto, bank, casinos, cinema, cleaners, hardware, laundry, RV camping, truck repair
e	173	W Winnemucca, **1 mi S**...airport, industrial area
m	168	to Rose Creek, **S**...prison area
u	158	to Cosgrave, **S...rest area both lanes, full(handicapped)facilities, phone, picnic tables, litter barrels, petwalk, RV dump**
c	151	Mill City, **N**...TA/Arco/Taco Bell/diesel/casino/24hr, Super 8, **S**...Trading Post/RV park
c	149	NV 400, Mill City, **1 mi N**...to TA/Arco/Taco Bell/diesel/24hr, Super 8, **S**...RV camping
a	145	Imlay, no facilities
	138	Humboldt, no facilities
	129	Rye Patch Dam, **N**...to Rye Patch RA, RV camping, **S**...TA/Arco/diesel/café/24hr
	119	Oreana, to Rochester, no facilities
	112	to Coal Canyon, **S**...to correctional ctr
L	107	E Lovelock(from wb), **N**...Exxon/diesel/mart, Sturgeon Rest., Desert Plaza Inn, Ramada Inn, Sage Motel, Super 10 Inn, Lazy K Camping
o	106	Main St, Lovelock, **N**...HOSPITAL, CHIROPRACTOR, Chevron/Taco Bell/diesel/LP/mart/atm, Exxon/diesel, MiniMart/gas, Davin's Dining, McDonald's, Pizza Factory, Ranch House Rest., Covered Wagon Motel, Ace Hardware, NAPA AutoCare, Radio Shack, Safeway, bank, laundry, playground/restrooms, same as 105
v	105	W Lovelock(from eb), **N**...HOSPITAL, Beacon/diesel/repair, Exxon/diesel/mart, La Casita Mexican, Lovelock Inn/rest., Jim's Tires, auto parts, bank, camping, same as 106
e	93	to Toulon, **S**...airport
l	83	US 95 S, to Fallon, **S...rest area both lanes, full(handicapped)facilities, phone, picnic tables, litter barrels**
o	78	to Jessup, no facilities
c	65	to Hot Springs, Nightingale, no facilities
k	48	US 50A, US 95A, E Fernley, to Fallon, **N**...Gas4Less, 76/diesel/rest./casino, Truck Inn, RV camping, **S**...Chevron/diesel/mart/casino, Texaco/Taco Bell/diesel/mart/24hr, 7-11, McDonald's, Pizza Factory, Silverado Rest./gas/casino, Best Western, Super 8, Goodyear/auto, Warehouse Mkt Foods, bank, to Great Basin NP

E

W

Nevada Interstate 80

46	US 95A, W Fernley, **S**...Pilot/DQ/Wendy's/diesel/mart/24hr, Blue Beacon, Fernley RV Park, **1 mi S**...Chevron, Exxon
45mm	Truckee River
43	Wadsworth, to Pyramid Lake, **N**...76/diesel/mart/camping
42mm	**rest area wb, full(handicapped)facilities, phone, picnic tables, litter barrels, petwalk,** check sta eb
40	Painted Rock, no facilities
38	Orchard, no facilities
36	Derby Dam, no facilities
32	Tracy, Clark Station, no facilities
28	NV 655, Patrick, no facilities
27.5mm	scenic view, eb
25mm	check sta wb
23	Mustang, **N**...Mustang Sta Café, RV camping
22	Lockwood, no facilities
21	Vista Blvd, Greg St, Sparks, **N**...HOSPITAL, Chevron/McDonald's/mart, Del Taco, **S**...Chevron/Alamo/diesel/rest./24hr, Super 8
20	Sparks Blvd, Sparks, **N**...Texaco/diesel/mart/wash/24hr, Carl's Jr, Outback Steaks, water funpark, **S**...Chevron/Alamo/diesel/rest./24hr, Super 8
19	E McCarran Blvd, Sparks, **N**...Beacon/diesel/mart, Chevron/diesel/mart, TA/76/diesel/rest., Western Mtn Gas, Applebee's, Black Bear Diner, Burger King, Craig's Rest., El Pollo Loco, Fast Chinese, Jack-in-the-Box, Jerry's Rest., Joe Bob's Chicken, KFC, Sierra Sid's, Wendy's, InnCal, Sunrise Motel, Western Village Motel, Windsor Inn, $Tree, EconoLube'n Tune, Ford/Lincoln/Mercury/Isuzu, GNC, Jo-Ann Fabrics, Mervyn's, Radio Shack, Safeway, Sav-On Drug, Suzuki, Target, Victorian RV Park, bank, books, mall, **1 mi N on Prater Way**...CHIROPRACTOR, 7-11, IHOP, McDonald's, Pizza Hut, Sizzler, Taco Bell, Lariat Motel, Albertson's/drugs, Goodyear, JiffyLube, Longs Drug, Midas Muffler, **S**...Black Forest Rest., Denny's, Best Western
18	NV 445, Pyramid Way(from eb), Sparks, **N**...Citgo/7-11, Courtyard, Silver Club Hotel/casino, **S**...Nugget Hotel/casino
17	Rock Blvd, Nugget Ave, Sparks, **N**...Arco/mart/24hr, Chevron/mart, Exxon/mart, Kragen Parts, casinos, tires, **S**...Nugget Hotel/casino
16	B St, E 4th St, Victorian Ave, **N**...Arco/mart/24hr, 7-11, Motel 6, **S**...Arco/diesel/mart/24hr, Chevron/repair, Gold Coin Motel, Hilton
15	US 395, to Carson City, **1 mi N on McCarran Blvd**...CHIROPRACTOR, Chevron, Shell/mart, Texaco/diesel/mart, Arby's, Burger King, Del Taco, Sezchuan Chinese, Sonic, Subway, Taco Bell, TCBY, Wendy's, Home Depot, Kragen Parts, OfficeMax, Pennzoil, PetCo, Ross, Wal-Mart/auto, WinCo Foods, cleaners, **S**...Bally's, Holiday Inn, 6 Gun Motel, to airport
14	Wells Ave, Reno, **N**...Motel 6, bank, **S**...Chevron, Texaco, Denny's, Day's Inn, Econolodge, Holiday Inn
13	US 395, Virginia St, Reno, **N**...Citgo/7-11, **S**...HOSPITAL, Chevron, Texaco/diesel/mart, Dairy Queen, Giant Burger, Ramada Inn, to downtown hotels/casinos, to UNVReno
12	Keystone Ave, Reno, **N**...EYECARE, Arco/mart/24hr, 7-11, Gateway Inn, Motel 6, Pizza Hut, JiffyLube, Raley's Food/drug, Sav-On Drug, bank, **S**...Chevron/mart/wash, 76/diesel/mart, Baskin-Robbins, Burger King, Chinese Rest., Coffee Grinder Rest., Higgy's Pizza, Jack-in-the-Box, McDonald's, Port of Subs, Shakey's Pizza, Taco Bell, Wendy's, Harrah's, Hilton's, Albertson's, Allied Parts, MailBoxes Etc, Meineke Muffler, Midas Muffler, Olson Tire, Radio Shack, T&S Hardware, 98 Cents Store, bank, casinos, cleaners, RV park
10	McCarran Blvd, Reno, **N**...MEDICAL CARE, CHIROPRACTOR, EYECARE, Arco/mart/24hr, Citgo/7-11, Tesoro/mart, Arby's, Baskin-Robbins, Burger King, Carl's Jr, Chinese Rest., Del Taco, Hacienda Mexican, IHOP, Jack-in-the-Box, KFC, McDonald's, Papa Murphy's, Schlotsky's, Starbucks, StrawHat Pizza, Taco Bell, Albertson's, Kragen Parts, MailBoxes Etc, Mail+, Olson Tire/auto, Pets R Us, QuikLube/gas, Safeway, Sav-On Drug, ShopKO, bank, laundry, **S**...Arco/mart/24hr, Citgo/7-11, Home Depot, Super K-Mart/Little Caesar's
9	Robb Dr, **N**...76/mart
8	W 4th St(from eb), Robb Dr, Reno, **S**...RV camping
7	Mogul, no facilities
6.5mm	truck parking/hist marker/scenic view both lanes
5	to E Verdi(from wb), no facilities
4.5mm	scenic view eb
4	Garson Rd, Boomtown, **N**...Chevron/diesel/mart, Boomtown Hotel/casino, RV park
3.5mm	check sta eb
3	Verdi(from wb), no facilities
2.5mm	Truckee River
2	Lp 80, to Verdi, **N**...Arco/diesel/mart/24hr, Branding Iron Café, Jack-in-the-Box, Taco Bell, Gold Ranch Hotel/casino
1	Gold Ranch Rd(from eb), to Verdi, no facilities
0mm	Nevada/California state line

E

W

New Hampshire Interstate 89

Exit #(mm)Services

L e b a n o n

61mm   Connecticut River, New Hampshire/Vermont state line

20(60)   NH 12A, W Lebanon, **E**...Mobil/mart/24hr, Sunoco/mart/24hr, Board&Basket, Brick Oven Pizza, Einstein Bros Bagels, KFC/Taco Bell, Shorty's Mexican, Subway, Brooks Drug, GNC, JiffyLube, Jo-Ann Crafts, K-Mart, MailBoxes Etc, NE Soap&Herb, Pearle Vision, PowerHouse Mall/shops, Ritz Camera, Shaw's Foods, **W**...Citgo/diesel/repair, Applebee's, Burger King, China Light, D'Angelo's, Denny's, Friendly's, McDonald's, Pizza Hut, TCBY, Weathervane Seafood, Wendy's, Economy Inn, Radisson, Ames, AutoZone, CVS Drug, FashionBug, Golf&Ski Whse, JC Penney, Midas Muffler, Sears, Shaw's Foods, Staples, 7 Barrel Brewery, Wal-Mart, bank, carwash, to airport

19(58)   US 4, NH 10, W Lebanon, **E**...Exxon/diesel/mart/24hr, Mobil/Blimpie/mart, China Sta, Little Caesar's, Grand Union Food/24hr, Ford, Honda, NAPA, Radio Shack, Ryder, Woodworkers Whse, carwash/laundry, cinema, LP, **W**...Sunoco/repair, hardware

57mm   **weigh sta both lanes**

18(56)   NH 120, Lebanon, **E**...Boise/Coastal/diesel/rest./24hr, Holiday Inn Express, Buick/Olds/GMC, Chevrolet/Volvo/VW, Dodge/Mazda, bank, to Dartmouth Coll, **W**...HOSPITAL, Citgo/diesel/mart, Exxon/mart

17(54)   US 4, to NH 4A, Enfield, **E**...Riverside Grill, Shaker Museum, tires

16(52)   Eastman Hill Rd, **E**...Exxon/Dunkin Donuts/diesel/mart/24hr, **W**...Mobil/Burger King/diesel/mart/24hr, Arctic Cat/Kawasaki/Suzuki, Whaleback Ski Area

15(50)   Montcalm, no facilities

14(47)   NH 10(from sb), **E**...RV camping(seasonal)

13(43)   NH 10, Grantham, **E**...Gen Store/gas/diesel, RV camping, **W**...Mobil/mart/repair, bank

40mm   **rest area nb, full(handicapped)facilities, info, phone, picnic table, litter barrels, vending, petwalk**

12A(37)   Georges Mill, **W**...to Sunapee SP, food, phone, lodging, RV camping

12(34)   NH 11 W, New London, **2 mi E**...HOSPITAL, Exxon/diesel/mart, lodging

11(31)   NH 11 E, King Hill Rd, New London, **2 mi E**...Fairway Motel

10(27)   NH 114, Sutton, **E**...to Winslow SP, **1 mi W**...lodging, to Wadleigh SB

26mm   **rest area sb, full(handicapped)facilities, info, phone, picnic table, litter barrels, vending, petwalk**

9(19)   NH 103, Warner, **E**...Citgo/Subway/Pizza Hut/TCBY/mart, Mobil/diesel/mart/24hr, Foothills Rest., McDonald's, MktBasket Food, Rollins SP, **W**...to Sunapee SP, ski area

8(17)   NH 103(no EZ nb return), Warner, **1 mi E**...gas, food, museum, to Rollins SP

15mm   Warner River

7(14)   NH 103, Davisville, **E**...TrueValue, phone, RV camping(seasonal)

12mm   Contoocook River

C o n c o r d

6(10)   NH 127, Contoocook, **1 mi E**...Mobil, Sunoco/mart, **W**...Elm Brook Park, RV camping(seasonal)

5(8)   US 202 W, NH 9(exits left from nb), Hopkinton, **W**...food, RV camping(seasonal)

4(7)   NH 103, Hopkinton(no EZ nb return), **E**...HorseShoe Tavern, gas

3(4)   Stickney Hill Rd(from nb), no facilities

2(2)   NH 13, Clinton St, Concord, **E**...HOSPITAL, food, **W**...NH Audubon Ctr

1(1)   Logging Mill Rd, Bow, **E**...Mobil/mart/24hr, Hampton Inn

0mm   I-93 N to Concord, S to Manchester, I-89 begins/ends on I-93, 36mm.

New Hampshire Interstate 93

Exit#(mm)Services

L i t t l e t o n

2(11)   I-91, N to St Johnsbury, S to White River Jct.   I-93 begins/ends on I-91, exit 19.

1(8)   VT 18, to US 2, to St Johnsbury, **2 mi E**...gas, food, lodging, camping

1mm   **Welcome Ctr nb, full(handicapped)facilities, info, phone, picnic tables, litter barrels, vending, petwalk**

131mm   Connecticut River, Vermont/New Hampshire state line. **Exits 1-2 are in VT.**

44(130)   NH 18, NH 135, **W**...**Welcome Ctr(8am-8pm)/scenic vista both lanes, full(handicapped)facilities, info, phone, picnic tables, litter barrels, petwalk**

43(125)   NH 135(from sb), to NH 18, Littleton, **1-2 mi W**...VETERINARIAN, same as 42

42(124)   US 302 E, NH 10 N, Littleton, **E**...Sunoco/mart/24hr, Burger King, ClamShell Rest., Dunkin Donuts, House of Pizza, Jing Fong Chinese, McDonald's, Pizza Hut, Subway, Big A Parts, Brooks Drug, JiffyLube, Medicine Shoppe, Parts+, Rite Aid, Radio Shack, **W**...Agway/gas, Irving/diesel, Continental Motel/rest., Chevrolet/Buick/Pontiac/Olds, Chrysler/Jeep, KOA, PriceMart Foods, Wal-Mart

41(122)   US 302, NH 18, NH 116, Littleton, **1-2 mi E**...HOSPITAL, DENTIST, Irving/diesel/mart/atm/24hr, Eastgate Motor Inn/rest., **W**...tires

New Hampshire Interstate 93

40(121) US 302, NH 10 E, Bethlehem, **E...**Exxon, Adair Country Inn/rest., RV camping, antiques, to Mt Washington

39(119) NH 116, NH 18(from sb), N Franconia, Sugar Hill, **W...**lodging

38(117) NH 116, NH 117, NH 142, NH 18, Sugar Hill, **E...**Red Coach Inn/rest., **W...**CHIROPRACTOR, Mobil, DairyBar Rest., DutchTreat Rest., HillWinds Rest., Village House Rest., Franconia Inn(2mi), Frost Museum, Kelley's FoodTown, USPO, bank, camping, gifts, hardware, info

37(115) NH 142(from sb), NH 18, Franconia, **W...**services, RV camping, info

36(114) NH 141, to US 3, S Franconia, **W...**golf, food, lodging

35(113) US 3 N(from nb), to Twin Mtn Lake, no facilities

112mm S Franconia, Franconia Notch SP begins sb

3(111) NH 18, Echo Beach Ski Area, view area, info

2(110) Cannon Mtn Tramway, **W...**Boise Rock, Old Man Viewing, Lafayette Place Camping

109.5mm trailhead parking

108mm Lafayette Place Camping(from sb), trailhead parking

107mm The Basin

1(106) US 3, The Flume Gorge, info, camping(seasonal)

104mm Franconia Notch SP begins nb

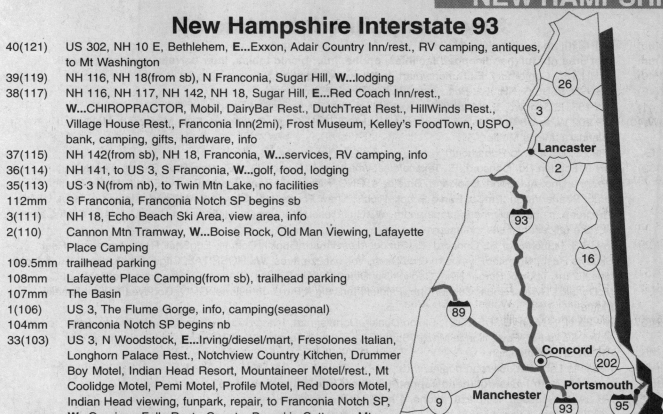

33(103) US 3, N Woodstock, **E...**Irving/diesel/mart, Fresolones Italian, Longhorn Palace Rest., Notchview Country Kitchen, Drummer Boy Motel, Indian Head Resort, Mountaineer Motel/rest., Mt Coolidge Motel, Pemi Motel, Profile Motel, Red Doors Motel, Indian Head viewing, funpark, repair, to Franconia Notch SP, **W...**Gorgiana Falls Rest., Country Bumpkin Cottages, Mt Liberty Motel, Cold Springs Camping, repair

32(101) NH 112, N Woodstock, Loon Mtn Rd, **E...**MEDICAL CARE, Citgo/Dunkin Donuts/diesel/mart, Irving/mart, Mobil/mart, Burger King, Dragon Lite Chinese, Earl of Sandwich Rest., Elvio's Pizza, GH Pizza, House of Pancakes, McDonald's, White Mtn Bagels, Millhouse Inn, Lincoln Motel, River Green Hotel, Grand Union Foods, Rite Aid, USPO, bank, cinema, laundry, to North Country Art Ctr, **W...**Citgo/mart, Mobil, Chalet Rest., Lafayette Dinner Train, Landmark II Rest., Truant's Rest., Carriage Motel, Woodstock Inn/rest., NAPA, camping, candy/fudge/gifts, cleaners/laundry

31(97) to NH 175, Tripoli Rd, **E...**RV camping(seasonal), **W...**camping

30(95) US 3, Woodstock, **E...**Jack-O-Lantern Inn/rest., golf, **W...**flea mkt, RV camping(seasonal)

29(89) US 3, Thornton, **1 mi W...**Citgo/diesel/pizza/mart, Gilcrest Motel, RV camping

28(87) NH 49, Campton, **E...**Citgo/mart, Mobil, Campton Café, TrueValue/LP, USPO, to ski area, RV camping, **W...**Citgo/diesel/mart, Sunset Grill, Scandinavi-Inn, Branch Brook Camping

27(84) Blair Rd, Beebe River, **E...**Covered Bridge B&B, Best Western, Gateway B&B, Red Sleigh Motel/chalets

26(83) US 3, NH 25, NH 3A, Tenny Mtn Hwy, **W on US 3...**HOSPITAL, Calzone's, JC's Roast Beef, McDonald's, Pilgrim Motel, Susse Chalet, camping

25(81) NH 175(from nb), Plymouth, **W...**HOSPITAL, Gulf, Irving/diesel/mart, tires, info, to Holderness School, Plymouth St Coll

24(76) US 3, NH 25, Ashland, **E...**Cumberland/gas, Ecco/diesel, Irving/Subway/diesel/mart, Mobil/Dunkin Donuts/diesel/mart, Burger King, Common Man Dining, John's Breakfast, Comfort Inn, Ashland Parts, Bob's Foods, TrueValue, USPO, bank, RV camp(4mi)

23(71) NH 104, NH 132, New Hampton, to Mt Washington Valley, **E...**Citgo/diesel/mart, Irving/diesel/mart/24hr, Dunkin Donuts, Hart's Turkey Farm Rest.(8mi), Rossi Italian, antiques, RV camping(3mi), **W...**Homestead Rest.(2mi), ski area

22(62) NH 127, Sanbornton, **5 mi W...**HOSPITAL, gas, food, phone

61mm rest area sb, full(handicapped)facilities, info, phone, picnic tables, litter barrels, vending, petwalk

20(57) US 3, NH 11, NH 132, NH 140, Tilton, **E...**Exxon/Subway/diesel/mart/24hr, Irving/diesel/mart/24hr, Mobil/mart, Applebee's, Burger King, FoodCourt, Great American Diner, McDonald's, Oliver's Rest., Tilt'n Rest., UpperCrust Pizza, Super 8, BJ's Wholesale, Lakes Region Factory Outlet/famous brands, RV camping, Shaw's Foods, Staples, VIP Auto Ctr, **W...**Black Swan B&B(2mi), Chrysler/Dodge, Ford/Nissan, USPO, Wal-Mart/auto, bank

New Hampshire Interstate 93

Concord

19(55)	NH 132(from nb), Franklin, **W...**HOSPITAL, NH Vet Home, antiques
51mm	**rest area nb, full(handicapped)facilities, phone, info, picnic tables, litter barrels, vending, petwalk**
18(49)	to NH 132, Canterbury, **E...**Sunoco/mart, to Shaker Village Hist Site
17(46)	US 4 W, to US 3, NH 132, Boscawen, **W...**McDonald's(4mi), gas
16(41)	NH 132, E Concord, **E...**Mobil/diesel, MillStream Mkt, Sunflour's Bakery, police
15W(40)	US 202 W, to US 3, N Main St, Concord, **W...**HOSPITAL, Citgo, Getty/mart, Gulf, Merit/mart, Mobil/mart, Friendly's, Midas Muffler
E	I-393 E, US 4 E, to Portsmouth
14(39)	NH 9, Loudon Rd, Concord, **E...**Texaco/diesel/mart/wash, Boston Mkt, Einstein Bros Bagels, Family Buffet, Pizzaria Uno, AutoZone, Bookland, Bradlee's, GNC, Kinko's, LLBean, MarketBasket Foods, Midas Muffler, Osco Drug, Pearle Vision, PetCo, Radio Shack, Rich's, Shaw Foods/24hr, Shop'n Save Food, Staples, U-Haul, USPO, Walgreen, cinema, cleaners, planetarium, **W...**Gulf/diesel, Merit/diesel, Holiday Inn, Ames, Jo-Ann Crafts, bank, to state offices, hist sites, museum
13(38)	to US 3, Manchester St, Concord, **E...**Sunoco/diesel/mart, Dunkin Donuts, EggShell Rest., Landmark Rest., Skuffy's Rest., Chrysler/Plymouth, Olds/Saab, Volvo/Isuzu, tires, **W...**HOSPITAL, Citgo, Gibb's/diesel, Gulf, Mobil/diesel/24hr, Texaco, Burger King, Capitol City Diner, Dunkin Donuts, D'Angelo's, Hawaiian Isle Rest., KFC, McDonald's, Miami Subs, Comfort Inn, Fairfield Inn, Big A Parts, Firestone, GMC, Goodyear/auto, Meineke Muffler
12N(37)	NH 3A N, S Main, **E...**DENTIST, Exxon/Dunkin Donuts/mart, Irving/Subway/diesel/mart/atm/24hr, Brick Tower Motel, Day's Inn, Ford, Honda/Mazda/Saturn/VW, Suzuki, Toyota, **W...**HOSPITAL, Mobil, Hampton Inn
S	NH 3A S, Bow Junction
36mm	I-89 N to Lebanon, toll road begins/ends
31mm	**rest area both lanes, full(handicapped)facilities, info, phone, vending**
11(30)	NH 3A, to Hooksett, toll plaza, phone, **4 mi E...**truckstop/diesel/rest.
28mm	I-293(from sb), Everett Tpk
10(27)	NH 3A, Hooksett, **E...**Irving/diesel/deli(1mi), cinema, **W...**Exxon/Dunkin Donuts/mart
26mm	Merrimac River

Manchester

9N S(24)	US 3, NH 28, Manchester, **W...**HOSPITAL, Exxon/diesel/mart, Mobil/mart, Sunoco/diesel/24hr, Boston Mkt, Burger King, D'Angelo's, Dunkin Donuts, Happy Garden Chinese, KFC, Little Caesar's, Luigi's Italian, Papa Gino's, Shorty's Mexican, Chrysler/Dodge/Jeep, Ford/Lincoln/Mercury, Mercedes, Photo Shop, Shop'n Save Foods, U-Haul, bank, carwash, cleaners
8(23)	Wellington Rd, **W...**VA HOSPITAL, Currier Gallery
7(22)	NH 101 E, to Portsmouth, no facilities
6(21)	Hanover St, Candia Rd, Manchester, **E...**VETERINARIAN, **W...**HOSPITAL, Citgo/diesel/mart, Mobil/diesel, Angelo's Rest.
19mm	I-293 W, to Manchester, to airport
5(15)	NH 28, to N Londonderry, **E...**Sunoco, Applebee's(3mi), Carboni's Pizza, **W...**MEDICAL CARE, Exxon/diesel/mart, Suzuki Motorcycles

Derry

4(12)	NH 102, Derry, **E...**HOSPITAL, Citgo/diesel/mart, Gulf/diesel, Mobil/mart/24hr, Mutual Gas, Shell/Dunkin Donuts/mart/etd/24hr, Sunoco/diesel, Burger King, C&K Rest., Derry Pizza, Ben Franklin, Ritz Camera, **W...**CHIROPRACTOR, Citgo/7-11, Exxon, Texaco/mart/atm, American Subs, Happy Garden Chinese, McDonald's, Papa Gino's, Wendy's, Chrysler/Plymouth, Ford, JiffyLube, K-Mart/Little Caesar's, Marshall's, MktBasket Foods, Osco Drug, Shaw's Foods, TJ Maxx, Walgreen, carwash, cinema
8mm	weigh sta both lanes
3(6)	NH 111, Windham, **E...**CHIROPRACTOR, DENTIST, Citgo/diesel, Capri Pizza, Dunkin Donuts, House of Pizza, Subway, bakery, bank, cleaners, **W...**Exxon/mart, Sunoco/mart/24hr, Castleton Rest., hardware
2(3)	to NH 38, NH 97, Salem, **E...**Sunoco, Red Roof Inn, **W...**DENTIST, MEDICAL CARE, Citgo/mart, Lucy Country Store/subs, Millstone Manor Rest., Hampshire Hotel, Holiday Inn, Susse Chalet/rest., parts/tires/repair
1(2)	NH 28, Salem, **E...**VETERINARIAN, Citgo, Exxon, Getty, Bickford's, Burger King, Chinese Rest., Denny's, 99 Rest., Park View Inn, Barnes&Noble, Bradlee's, Dodge/Nissan/Toyota, Filene's, Home Depot, JC Penney, Jordan Marsh, K-Mart, MktBasket Foods, Sears/auto, Shaw's Foods, TJ Maxx, Walgreen, racetrack, mall
1mm	**Welcome Ctr nb, full(handicapped)facilities, info, phone, picnic tables, litter barrels, vending, petwalk**
0mm	New Hampshire/Massachusetts

N

S

New Hampshire Interstate 95

Portsmouth

Exit #(mm)Services

17mm	Piscataqua River, New Hampshire/Maine state line
7(16)	Portsmouth, Port Authority, waterfront hist sites, **E...**Sheraton, **W...**Applebee's, Schoolhouse Rest., Wendy's, Courtyard
6(15)	Woodbury Ave(from nb), Portsmouth, **E...**Gulf, Texaco, Anchorage Inn, Econolodge
5(14)	US 1, US 4, NH 16, The Circle, Portsmouth, **E...**HOSPITAL, Gulf, Mobil/diesel/mart, Texaco/diesel/mart, Bickford's, Momma D's Rest., Best Inn, Holiday Inn, Meadowbrook Inn, Port Inn, Wynwood Inn, Nissan/Suzuki, Olds, Pontiac/Cadillac/GMC, U-Haul, **W...**Exxon, Mobil, Dunkin Donuts, McDonald's, Pizza Hut, Hampton Inn, Ford, Home Depot, K-Mart, Mazda/VW, Midas Muffler, MVP Sports, Saturn, ToysRUs, tires
4(13.5)	US 4(from nb), to White Mtns, Spaulding TPK, **E...**HOSPITAL, gas/diesel, food, lodging, RV camping, **W...**to Pease AFB
3a(13)	NH 33, Greenland
3b(12)	NH 101, to Portsmouth, **E...**HOSPITAL, **W...**Mobil/diesel, Sunoco/diesel/mart, TA/diesel/rest./24hr, McDonald's(1mi), camping, golf
2(6)	NH 51, to Hampton, **1 mi E on US 1...**HOSPITAL, McDonald's, gas, lodging
5.5mm	toll plaza
4mm	Taylor River
1(1)	NH 107, to Seabrook, toll rd begins/ends, **E...**CHIROPRACTOR, DENTIST, Getty/mart, Gulf, Mobil/mart, Applebee's, Burger King, D'Angelo's, Dunkin Donuts, KFC, McDonald's, Pizza Hut/Taco Bell, Road Kill Café, Subway, Wendy's, Hampshire Motel, AutoZone, CVS Drug, FashionBug, GNC, JiffyLube, Jo-Ann Crafts, MailBoxes Etc, MarketBasket Foods, Midas Muffler, NAPA, Prompto Lube, Radio Shack, Shaw's Foods, Staples, TJ Maxx, Wal-Mart, West Marine, to Seacoast RA, bank, carwash, **W...**Citgo/mart, Capt K's Seafood, McGrath's Dining, Best Western, GolfDay Outlet, Sam's Club, Seabrook Greyhound Pk, WoodWorkers Whse, tires
.5mm	**Welcome Ctr nb, full(handicapped)facilities, phone, vending, picnic tables, litter barrels, petwalk**
0mm	New Hampshire/Massachusetts state line

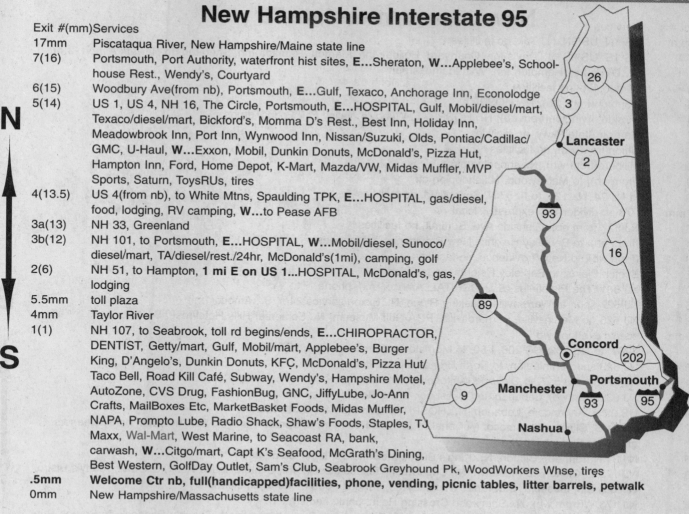

New Hampshire Interstate 293(Manchester)

Manchester

Exit#(mm)Services

8	I-93, N to Concord, S to Derry. I-293 begins/ends on I-93, 28mm.
7	Dunbarton Rd(from nb), no facilities
6	Singer Park, Manchester, **W...**HOSPITAL, Mobil/mart, Texaco/Dunkin Donuts/diesel/mart
5	to Manchester(from nb), **W...**Exxon/mart, Gulf/mart/24hr, Dunkin Donuts
4	US 3, NH 3A, NH 114A, Queen City Br, **W...**Gulf/diesel, Merit/diesel/mart, Mobil, Applebee's, BagelWorks, Burger King, D'Angelo's, Dairy Queen, KFC, McDonald's, Papa John's, Subway, Wendy's, Comfort Inn, Econolodge, Goodyear, Midas Muffler, Osco Drug, Subaru
3	NH 101, **W on US 3...**Outback Steaks, Caldor, Macy's, Marshall's, Radio Shack, Shop'n Save, cinema, mall
2	NH 3A, Brown Ave, **S...**Citgo, Shell/repair, Texaco/mart, Kwikava, McDonald's, Super 8, airport, bank
1	NH 28, S Willow Rd, **N...**Mobil/Pizza Hut/diesel/mart/atm, Sunoco/Dunkin Donuts/diesel/mart/atm, Bickford's, Boston Mkt, Bros Dining, Burger King, Cactus Jack's, Chili's, Friendly's, India Palace, Luisa's Italian, McDonald's, Taco Bell, Wendy's, Yee Dynasty Chinese, Sheraton, Susse Chalet, AutoZone, Bradlee's, Buick/Chevrolet, Circuit City, CopyMax, CVS Drug, GMC, Harley-Davidson, Home Depot, Lenscrafters, Mercedes, Michael's, Nissan/Mitsubishi, OfficeMax, Osco Drug, Pearle Vision, PepBoys, PetCo, Pier 1, Pontiac/Cadillac/Mazda, Service Merchandise, SportsAuthority, Sullivan Tires, TJ Maxx, Tweeter Electronics, U-Haul, VW, cinema, **S...**HOSPITAL, Exxon/diesel/mart, Antics Grill, Bernardo's Bistro, Bickford's, ChuckeCheese, FoodCourt, Ground Round, 99 Rest., Pizzaria Uno, Ruby Tuesday, Courtyard, Barnes&Noble, Best Buy, Chrysler/Plymouth, Filene's, Ford, JC Penney, K-Mart, Lexus, Midas Muffler, NTB, Sears/auto, Toyota, mall
0mm	I-93, N to Concord, S to Derry. I-293 begins/ends on I-93.

New Jersey Interstate 78

Exit #	Services
58b a	US 1N, US 9N, NJ Tpk, no facilities
57	US 1S, US 9S, **S**...Day's Inn, Courtyard, Holiday Inn, Marriott, Ramada Inn, Sheraton, to Newark Airport, **3 mi** **S**...Day's Inn, Hampton Inn, Wyndham Garden
56	Clinton Ave, no facilities
55	Irvington(from wb), **N**...HOSPITAL
54	Hillside, Irvington(from eb), **N**...HOSPITAL, Wendy's, Valley Fair Foods, drugs
52	Garden State Pkwy, no facilities
51	Union(from wb), no facilities
50b a	Millburn(from wb), **N**...shopping ctr
49	(from eb), to Maplewood, **N**...shopping ctr
48	to NJ 24, NJ 124, to I-287 N, Springfield, no facilities
47mm	I-78 eb divides into express & local
45	NJ 527(from eb), Glenside Ave, Summit, no facilities
44	(from eb), to Berkeley Heights, New Providence, no facilities
43	(from wb), to New Providence, no facilities
41	Scotch Plains, to Berkeley Heights, no facilities
40	NJ 531, The Plainfields, **S**...HOSPITAL, Amoco(1mi), phone
36	NJ 527 spur, to Warrenville, Basking Ridge, **N**...Exxon/service/24hr, **S**...Amoco(1mi)
33	NJ 525, to Martinsville, Bernardsville, PGA Golf Museum, **N**...Somerset Hills Hotel/rest., **S**...Amoco
32mm	scenic overlook wb
29	I-287, to US 202, US 206, I-80, to Morristown, Somerville, **S**...HOSPITAL
26	NJ 523 spur, Lamington, to North Branch, no facilities
24	NJ 523, to NJ 517, to Oldwick, Whitehouse, no facilities
20	NJ 639(from wb), Lebanon, to Cokesbury, **S**...to Round Valley RA
18	US 22 E, Annandale, Lebanon, **N**...HOSPITAL, same as 17, **S**...Honda
17	NJ 31 S, Clinton, **N**...Amoco, McDonald's, Clinton Point Inn, to Spruce Run RA, **S**...Chevrolet, bank, cleaners, grill, **3 mi S**...Mobil, pizza
16	NJ 31 N(from eb), Clinton, **N**...King's Buffet, McDonald's, bank, same as 17
15	NJ 173 E, Clinton, to Pittstown, **N**...Citgo/wash, Texaco/diesel/24hr, Clinton House Rest., Subway, Teddy's Bistro, Holiday Inn, museum, **S**...Frank's Pizza, $Tree, Laneco Foods, Wal-Mart, cleaners
13	NJ 173 W(from wb), **N**...Sherwood Crossing Rest., bank, same as 12
12	NJ 173, Norton, to Jutland, **N**...Citgo/Johnny's Rest./diesel, Exxon/diesel/mart, Coach'n Paddock, to Spruce Run RA, **S**...Shell/diesel/mart
11	NJ 173, West Portal, Pattenburg, **N**...Coastal/mart/24hr, Texaco/diesel, Busy B Deli, Ladyslide Rest., Mountainview Chalet, RV camping, to st police
8mm	**rest area both lanes, picnic tables, litter barrels, no restrooms**
7	NJ 173, to Bloomsbury, West Portal, **N**...RV repair(.5 mi), RV/tent camping, **S**...Citgo/diesel/rest./mart/24hr, Pilot/ Subway/diesel/mart/24hr, TA/Mobil/diesel/rest./24hr, bank
6	Warren Glen, Asbury(from eb), no facilities
4	Stewartsville(from wb), no facilities
3.5mm	weigh sta eb
3	US 22, NJ 173, to Phillipsburg, **N**...HOSPITAL, PennJersey/diesel/rest./24hr, USA/diesel, **1-3 mi N**...Burger King, Dunkin Donuts, McDonald's, Ponderosa, Holiday Inn, Phillipsburg Inn, **S**...Chevrolet/Olds/Isuzu
0mm	Delaware River, New Jersey/Pennsylvania state line

New Jersey Interstate 80

Exit #(mm)Services

I-80 begins/ends at G Washington Bridge in Ft Lee, NJ.

125mm	toll plaza eb
72	NJ 4, **S**...Mobil, Shell, Courtesy Inn, Hilton
71	Broad Ave, Leonia, Englewood, **N**...Gulf, Day's Inn, Executive Inn, **S**...Shell
70b a	NJ 93, Leonia, Teaneck, **N**...Marriott, golf
68b a	I-95, N to New York, S to Philadelphia, to US 46
67	to Bogota(from eb), no facilities
66	Hudson St, to Hackensack, **S**...KFC
65	Green St, S Hackensack, **S**...Exxon, Sheraton

E

W

E

W

New Jersey Interstate 80

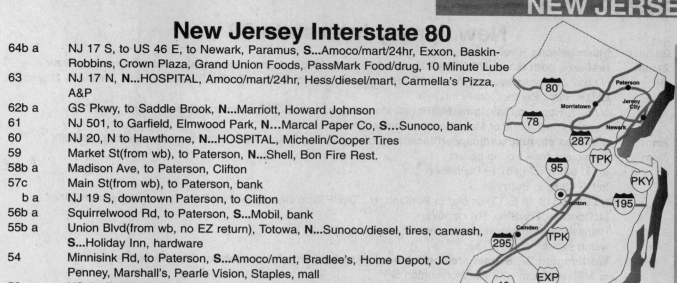

64b a NJ 17 S, to US 46 E, to Newark, Paramus, **S**...Amoco/mart/24hr, Exxon, Baskin-Robbins, Crown Plaza, Grand Union Foods, PassMark Food/drug, 10 Minute Lube

63 NJ 17 N, **N**...HOSPITAL, Amoco/mart/24hr, Hess/diesel/mart, Carmella's Pizza, A&P

62b a GS Pkwy, to Saddle Brook, **N**...Marriott, Howard Johnson

61 NJ 501, to Garfield, Elmwood Park, **N**...Marcal Paper Co, **S**...Sunoco, bank

60 NJ 20, N to Hawthorne, **N**...HOSPITAL, Michelin/Cooper Tires

59 Market St(from wb), to Paterson, **N**...Shell, Bon Fire Rest.

58b a Madison Ave, to Paterson, Clifton

57c Main St(from wb), to Paterson, bank

b a NJ 19 S, downtown Paterson, to Clifton

56b a Squirrelwood Rd, to Paterson, **S**...Mobil, bank

55b a Union Blvd(from wb, no EZ return), Totowa, **N**...Sunoco/diesel, tires, carwash, **S**...Holiday Inn, hardware

54 Minnisink Rd, to Paterson, **S**...Amoco/mart, Bradlee's, Home Depot, JC Penney, Marshall's, Pearle Vision, Staples, mall

53 US 46 E, NJ 23, **S**...BP, Exxon, Gulf, Mobil, Chinese Rest., Deli Express, Di Angelo's El Torito's, Houlihan's Rest., McDonald's, Pizzaria Uno, Red Lobster, TGIFriday, Holiday Inn, King's Inn, Ramada Inn, Sheraton, Firestone/auto, Ford, Macy's, Service Merchandise, cinema, mall

48 to Montville, no facilities

47b US 46 E, to Montclair, **N**...Amoco, Holiday Inn, **S**...Amoco, Gulf, Mobil/diesel, Shell/mart/wash, Dakota Diner/24hr, Dunkin Donuts, Indian Cuisine, Italian Deli, McDonald's, Roy Rogers, Subway, Midas Muffler, **3-5 mi S**...multiple facilities on US 46, Exxon, Hess, Sunoco, Oriental Rest., Best Western, JiffyLube, laundry

a I-280 E, to The Oranges, Newark, no facilities

45 to US 46, Lake Hiawatha, Whippany, **N**...Amoco, Gulf, Bennigan's, Chili's, IHOP, Red Lobster, Taco Bell, Wendy's, Holiday Inn/rest., Howard Johnson, Ramada Ltd, Red Roof Inn, Bradlee's, Drug Fair, K-Mart/auto, ShopRite Foods, RV camping, bank, cinema, laundry, repair, **services on US 46**...Applebee's, Boston Mkt, Burger King, Empire Diner, AC Moore Arts/crafts, CompUSA, Firestone/auto, **S**...Sheraton Tara

43b a I-287, to US 46, Boonton, Morristown

42 US 202, US 46, Parsippany, to Morris Plains, **N**...VETERINARIAN, Gulf/mart, Hess/mart, Texaco/repair, Black Bull Rest., Mtn Lakes Bagels, Paul's Diner, TGIFriday, Hampton Inn, White Deer Motel, Ford, Pontiac, Subaru, bank, cleaners

39 (38 from eb), US 46 E, to NJ 53, Denville, **N**...HOSPITAL, Exxon, Getty, Sunoco, Banzai Steaks, Burger King, Charlie Brown's Rest., Pizza King, Wendy's, BMW, Chevrolet, Discount Tire, Firestone, Grand Union Foods, MotoPhoto, cleaners, tires, **S**...MEDICAL CARE, Texaco, auto parts/repair

37 NJ 513, Rockaway, to Hibernia, **N**...Exxon/diesel, Shell/24hr, Hibernia Diner, Howard Johnson, bank, **S**...HOSPITAL, Mobil, McDonald's(1mi)

35b a Mount Hope, to Dover, **S**...HOSPITAL, Exxon/24hr, Food Works, Maggie's Rest., Peking Garden, Sizzler, 13 Eateries, Acme Foods, Eckerd, JC Penney, Lord&Taylor, Macy's, Pearle Vision, Sears/auto, bank, cinema, mall

34b a NJ 15, Wharton, to Sparta, **N**...HOSPITAL, Exxon, Gulf, Donut Town, McDonald's, Food Town, Goodyear, Rite Aid, **S**...Dunkin Donuts, Townsquare Diner, Costco, Home Depot, OfficeMax, cleaners

32mm truck rest area wb, no facilities

30 Howard Blvd, to Mt Arlington, **N**...Exxon/diesel/24hr, Gulf/diesel/24r, Davy's Hotdogs, McDonald's, Sheraton, bank, **S**...Lucky's Diner(2mi)

28 NJ 10, to Ledgewood, Lake Hopatcong, **S**...G&N/diesel/repair, Cliff's Grill, Day's Inn, Meineke Muffler, Toyota, **1 mi S**...CHIROPRACTOR, Hess/diesel, Texaco, Burger King, Blimpie, Boston Mkt, ChiChi's, Dunkin Donuts, McDonald's, Pizza Hut, Red Lobster, Robury Pizza, Roy Rogers, Subway, Taco Bell, TCBY, Wendy's, Roxbury Circle Motel, Ledgewood Mall, cinema, laundry

27 US 206 S, NJ 182, to Netcong, Somerville, **N**...Perkins, Day's Inn, Ford, **S**...Coastal, Mobil/diesel, Texaco/diesel/mart/24hr, McDonald's

26 US 46 W(from wb, no EZ return), to Budd Lake, **S**...CHIROPRACTOR, MEDICAL CARE, Pennzoil/gas, Texaco/diesel, Dairy Queen, Dunkin Donuts, Perkins, Henry's Motel, Budd Lake Motel, A&P, Goodyear/auto, Service Tires, bank

25 US 206 N, Stanhope, to Newton, **N**...Exxon/mart, Shell, Tasty Family Rest., Black Forest Inn/rest., Wyndham Garden, to Waterloo Village, Int Trade Ctr, **S**...URGENT CARE, PetroStop/gas, Shell/mart/24hr, Budd Lake Diner/rest., Chinese Rest., Domino's, **1-2 mi S**...Amoco/24hr, 7-11

New Jersey Interstate 80

23.5mm	Musconetcong River
21mm	**rest area both lanes, NO TRUCKS, scenic overlook(eb), phone, picnic tables, litter barrels, petwalk**
19	NJ 517, to Hackettstown, Andover, **N**...RV camping, **1-2 mi S**...HOSPITAL, Shell/lube, 7-11, Babalou's Steaks, BLD's Rest., Panther Valley Inn/rest.
12	NJ 521, Hope, to Blairstown, **N**...Honda Motorcycles, **1 mi S**...Texaco/mart, MillRace Pond Inn, RV camping(5mi), Land of Make Believe, Jenny Jump SF
7mm	**rest area eb, full(handicapped)facilities, info, phone, picnic tables, litter barrels, vending, petwalk**
6mm	scenic overlook wb, no trailers
4c	to NJ 94 N(from eb), to Blairstown
b	to US 46 E, to Buttzville
a	NJ 94, to US 46 E, Columbia, to Portland, **N**...TA/BP/Taco Bell/diesel/mart/24hr Texaco/diesel/mart/24hr, McDonald's, DayStop, RV camping
3.5mm	Hainesburg Rd(from wb) accesses services at 4
2mm	weigh sta eb
1mm	**Worthington SF, S...rest area both lanes, restrooms, info, picnic tables, litter barrels, petwalk**
1	to Millbrook(from wb), **N**...Worthington SF
0mm	Delaware River, New Jersey/Pennsylvania state line

E

W

New Jersey Interstate 95

Exit #	Services
35b a	NJ 34, to Brielle, Point Pleasant, Garden State Pkwy. I-195 begins/ends on Garden State Pkwy, exit 99.
31b a	NJ 547, NJ 524, to Farmingdale, **N**...to Allaire SP, golf
28b a	US 9, to Freehold, to Lakewood, **N**...Ivy League Rest., Lino's Pizza, bank, **S**...CHIROPRACTOR, DENTIST, Exxon, Gulf, Mobil/diesel, Boston Mkt, McDonald's, Taco Bell, Grand Union Foods, Radio Shack, Sears Hardware, auto repair
22	to Jackson Mills, Georgia, **N**...to Turkey Swamp Park
21	NJ 526, NJ 527, to Jackson, Siloam, no facilities
16	NJ 537, to Freehold, **N**...HOSPITAL, Amoco/mart/24hr, Sunoco, 6 Flags Outlet/famous brands, **S**...Wawa/diesel/24hr, McDonald's, camping
11	NJ 524, Imlaystown, to Horse Park of NJ
8	NJ 539, Allentown, **S**...Amoco/repair, Mobil(1mi), American Gyro Deli, Crosswicks HP
7	NJ 526, Robbinsville, Allentown, no facilities
6	NJ Tpk, N to NY, S to DE Memorial Br
5b a	US 130, **N**...A Better Pizza, Econolodge, **S**...Gulf/diesel/24hr, Harry's Army Navy, USPO, to state aquarium
3b a	Hamilton Square, Yardville, **N**...HOSPITAL
2	US 206 S, S Broad St, Yardville, **N**...7-11, Pizza Star, Midas Muffler, **1 mi S**...Amoco, Burger King, Ivy Rest., McDonald's
1b	US 206 N, **N**...Anthony's Pizza, Burger King, McDonald's, Radio Shack, Rite Aid
a	I-295. I-195 begins/ends in Trenton.

Trenton

N

S

New Jersey Turnpike

Exit #(mm)	Services
18(117)	US 46 E, Ft Lee, Hackensack, last exit before toll sb
17(116)	Lincoln Tunnel
115mm	**Vince Lombardi Service Plaza nb, Sunoco/diesel, Big Boy, Nathan's, Roy Rogers, TCBY, gifts**
114mm	toll plaza, phone
16W(113)	NJ 3, Secaucus, Rutherford, **E**...Hess, Shell, Hilton, M Plaza Hotel, **W**...Sheraton, Meadowlands
112mm	**Alexander Hamilton Service Area sb, Sunoco, Roy Rogers, gifts**
16E(112)	NJ 3, Secaucus, **E**...Lincoln Tunnel
15W(109)	I-280, Newark, The Oranges, **W**...cinema
15E(107)	US 1, US 9, Newark, Jersey City, **E**...Lincoln Tunnel
14c	Holland Tunnel
14b	Jersey City
14a	Bayonne
14(105)	I-78 W, US 1, US 9, **2 mi W**...Holiday Inn, Marriott, Radisson, airport

N

S

New Jersey Turnpike

102mm   **Halsey Service Area, Sunoco, Roy Rogers, other services in Elizabeth**

13a(102)   Elizabeth, **E**...Ikea Furnishings, **W**...Hilton, Wyndham Garden, airport

13(100)   I-278, to Verrazano Narrows Bridge

12(96)   Carteret, Rahway, **E**...McDonald's, Holiday Inn

93mm   **Cleveland Service Area nb, Sunoco/diesel, Roy Rogers, TCBY, gifts, Thomas Edison Service Area sb, Sunoco/diesel/mart, Big Boy, Dunkin Donuts, Roy Rogers, Starbucks, TCBY, gifts**

11(91)   US 9, Garden State Pkwy, to Woodbridge, **E**...Home Depot

10(88)   I-287, NJ 514, to Perth Amboy, **W**...Hess/diesel, Chinese Rest., Holiday Inn, Ramada/rest., Meineke Muffler

9(83)   US 1, NJ 18, to New Brunswick, E Brunswick, **E**...Gulf, Hess, Merit/diesel, Mobil, Sunoco, Burger King, Chi Chi's, Chinese Rest., Dunkin Donuts, IHOP, KFC, McDonald's, Roy Rogers, Motel 6, Sheraton/rest., Chrysler, FoodTown/drugs, Ford/Lincoln/Mercury, Goodyear/auto, Mazda, Midas Muffler, Sam's Club, Shopper's World, U-Haul, auto repair, cinema, laundry

79mm   **Kilmer Service Area nb, Sunoco/diesel/mart, Big Boy, Roy Rogers, TCBY, gifts**

8a(74)   to Jamesburg, Cranbury, **W**...Holiday Inn/rest.

72mm   **Pitcher Service Area sb, Sunoco/diesel, Big Boy, Roy Rogers, TCBY, Mrs Fields**

8(67)   NJ 33, NJ 571, Highstown, **E**...Exxon, Hess, Mobil, Shell, 8 Diner, Mom's Rest., Silver Eagle Diner, Day's Inn, Ramada Inn, **W**...Coach&Four Rest., Town House Motel, **2 mi W on US 130**...MEDICAL CARE, CHIROPRACTOR, Sunoco, 7-11, American Diner, Burger King, Dunkin Donuts, Golden Coach Rest., McDonald's, Roy Rogers, Colonade Motel, ShopRite Foods, CVS Drug/24hr, Meineke Muffler, Midas Muffler, bank, cinema

7a(60)   I-195 W to Trenton, E to Neptune, no facilities

59mm   **Richard Stockton Service Area sb, Sunoco/diesel, Bob's Big Boy, Roy Rogers, TCBY, gifts, Woodrow Wilson Service Area nb, Sunoco, Roy Rogers, Bob's Big Boy, gifts**

7(54)   US 206, to Bordentown, Trenton, to Ft Dix, McGuire AFB, to I-295, **E**...Mobil/diesel/mart, Petro/diesel/rest./24hr, Sandman Rest, Truck Plaza Motel, Buick/GMC, auto repair, **W**...Citgo, Exxon, Sunoco, Amerigas LP, Burger King, Denny's, Macedonia Rest., Best Western/rest., Day's Inn/rest., Laurel Notch Lodge, Quality Inn, RV Ctr, AG Transmissons, carwash

6(51)   I-276, to Pa Turnpike, no facilities

5(44)   Willingboro, to Mount Holly, **E**...Citgo/diesel, Bob's Big Boy, Charlie Brown's Rest., Howard Johnson, Best Western, **W**...VETERINARIAN, Exxon/diesel/mart, Getty/diesel, Mobil/mart, laundry

39mm   **James Fenimore Cooper Service Area nb, Sunoco/diesel/mart, Bob's Big Boy, Roy Rogers, TCBY, gifts**

4(34)   NJ 73, to Philadelphia, Camden, **E**...Mobil/mart, 7-11, Bennigan's, Chili's, Copperfield's Rest., Denny's, McDonald's, Sage Diner, McIntosh Motor Inn, Red Carpet Inn, Travelodge, Ford, **1-2 mi W**...Exxon, Gulf/diesel, Hess, Mobil, Shell, Bob Evans, Burger King, Colonnade Rest., Dunkin Donuts, Garden Room Rest., KFC, Pizza Hut, Roy Rogers, Bel Air Motel, Budget Motel, Clarion Motel, Clover Motel, Courtyard, Econolodge, Landmark Inn, Motel 6, Ramada Inn/rest., Red Roof Inn, Sharon Motel, Track&Turf Motel, Trav-Lers Motel, Ford/Lincoln/Mercury, Lexus, Mazda, White Glove Carwash, auto repair, transmissions, to st aquarium

30mm   **Walt Whitman Service Area sb, Sunoco, Roy Rogers, Nathan's, TCBY, gifts**

3(26)   NJ 168, Camden, Woodbury, Atlantic City Expwy, Walt Whitman Br, **E**...Citgo, Comfort Inn, Holiday Inn, Chevrolet, **W**...Extra/diesel, Dunkin Donuts, Italia Pizza, Pizza Hut, Sub's/steaks, Wendy's, Bellmawr Motel, Econolodge, Monticello Motel

2(13)   US 322, to Swedesboro, no facilities

5mm   **Barton Service Area sb, Sunoco/mart, Bob's Big Boy, Nathan's, TCBY, gifts, Fenwick Service Area nb, Sunoco/mart, Bob's Big Boy, Taco Bell, TCBY, gifts**

1(1.2)   Deepwater, **E**...HOSPITAL, truckstop/diesel, Del Valley Motel, **W**...Mobil/diesel, Pilot/diesel/mart/24hr/Subway, Texaco/diesel/rest./24hr, Friendship Motor Inn, Holiday Inn, Howard Johnson

1mm   toll road begins/ends

2(I-295)   I-295 N divides from toll road, I-295 S converges with toll road

1(I-295)   NJ 49, to Pennsville, **E**...Sunoco, Burger King, McDonald's, Olds/Cadillac, **W**...Seaview Motel

0mm   Delaware River, Delaware Memorial Bridge, New Jersey/Delaware state line

N

S

Camden

New Jersey Interstate 195

Freehold

Exit #	Services
35b a	NJ 34, to Brielle, Point Pleasant, Garden State Pkwy. I-195 begins/ends on Garden State Pkwy, exit 99.
31b a	NJ 547, NJ 524, to Farmingdale, **N**...to Allaire SP, golf
28b a	US 9, to Freehold, to Lakewood, **N**...Ivy League Rest., Lino's Pizza, bank, **S**...CHIROPRACTOR, DENTIST, Exxon, Gulf, Mobil/diesel, Boston Mkt, McDonald's, Taco Bell, Grand Union Foods, Radio Shack, Sears Hardware, auto repair
22	to Jackson Mills, Georgia, **N**...to Turkey Swamp Park
21	NJ 526, NJ 527, to Jackson, Siloam, no facilities
16	NJ 537, to Freehold, **N**...HOSPITAL, Amoco/mart/24hr, Sunoco, 6 Flags Outlet/famous brands, **S**...Wawa/diesel/24hr, McDonald's, camping
11	NJ 524, Imlaystown, to Horse Park of NJ
8	NJ 539, Allentown, **S**...Amoco/repair, Mobil(1mi), American Gyro Deli, Crosswicks HP
7	NJ 526, Robbinsville, Allentown, no facilities
6	NJ Tpk, N to NY, S to DE Memorial Br
5b a	US 130, **N**...A Better Pizza, Econolodge, **S**...Gulf/diesel/24hr, Harry's Army Navy, USPO, to state aquarium
3b a	Hamilton Square, Yardville, **N**...HOSPITAL
2	US 206 S, S Broad St, Yardville, **N**...7-11, Pizza Star, Midas Muffler, **1 mi S**...Amoco, Burger King, Ivy Rest., McDonald's
1b	US 206 N, **N**...Anthony's Pizza, Burger King, McDonald's, Radio Shack, Rite Aid
a	I-295. I-195 begins/ends in Trenton.

New Jersey Interstate 287

Exit #	Services
68mm	New Jersey/New York state line
66	NJ 17 S, Mahwah, **1-2 mi E**...services
59	NJ 208 S, Franklin Lakes, no facilities
58	US 202, Oakland, **E**...Mobil, **W**...Exxon, auto parts
57	Skyline Dr, Ringwood, no facilities
53	NJ 694, NJ 511, Bloomingdale, Pompton Lakes, no facilities
52b a	NJ 23, Riverdale, Wayne, Butler, no facilities
47	US 202, Montville, Lincoln Park, **E**...DENTIST, CHIROPRACTOR, MEDICAL CARE, bank
45	Myrtle Ave, Boonton, **W**...BP, Exxon, Jack's IGA, Buick/Chevrolet/Olds, cleaners/laundry, drugs
43	Intervale Rd, to Mountain Estates, **W**...Dodge
42	US 46, US 202, **W**...Exxon, Sunoco, Texaco, Fuddrucker's, Roy Rogers, Day's Inn, Embassy Suites, Hampton Inn, Chrysler/Jeep/Pontiac/Subaru, Ford, bank, cleaners/laundry, drugs
41b a	I-80, E to New York, W to Allentown
40	NJ 511, Parsippany Rd, to Whippany, **W**...CHIROPRACTOR, DENTIST, Mobil, Chinese Rest., Pizzaria, motel
39b a	NJ 10, Dover, Whippany, **W**...Marriott
37	NJ 24 E, Springfield, **4 mi E**...mall
22b a	US 202, US 206, Pluckemin, Bedminster, **E**...CHIROPRACTOR, Amoco, McDonald's, Nino's Pizza, TCBY, CVS Drugs, King's Foods, MailBoxes Etc, camera shop, cleaners
21b a	I-78, E to NY, W to PA
17	US 206(from sb), Bridgewater, **W**...Hess, TGIFriday, Super 8, Bradlee's, Borders Books, Linens'n Things, Macy's, OfficeMax, mall
14b a	US 22, to US 202/206, **E**...sports arena
13b a	NJ 28, Bound Brook, **E**...Amoco, Mobil/24hr, **W**...HOSPITAL, UPS
12	Weston Canal Rd, Manville, **E**...Texaco, **W**...Ramada Inn
10	NJ 527, Easton Ave, New Brunswick, **E**...Texaco, Marriott, **W**...Amoco, Coastal/mart, Holiday Inn, Doubletree Garden State Exhibit Ctr
9	NJ 514, River Rd, **W**...Embassy Suites, Wyndham Garden
8.5mm	weigh sta nb

New Jersey Interstate 287

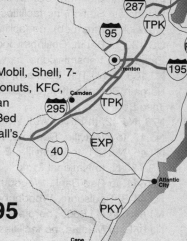

8	Possumtown Rd, Highland Park, no facilities
7	S Randolphville Rd, Piscataway, **E**...Mobil/mart
6	Washington Ave, same as 5
5	NJ 529, Stelton Rd, Dunellen, **E**...Amoco/diesel, BP, Mobil/mart, Shell, Texaco, KFC, Ramada Ltd, AJV Auto Mall, Home Depot, Meineke Muffler, **W**...Exxon, Burger King, Friendly's, Grand Buffet, McDonald's, Piancone's Ristorante, White Castle, Day's Inn, Motel 6, Holiday Inn
4	Durham Ave, S Plainfield, **E**...HOSPITAL, BP/diesel/mart
3	(from sb), **W**...Red Roof Inn
2b a	NJ 27, Metuchen, New Brunswick, no facilities
1b a	US 1, **1-4 mi N on US 1**...Amoco/24hr, Exxon/diesel/mart, Getty Gas, Mobil, Shell, 7-11, Bennigan's, Blimpie, Boston Mkt, ChiChi's, Chinese Rest., Dunkin Donuts, KFC, Macaroni Grill, Red Lobster, Ruby Tuesday, Sizzler, Steak&Ale, Suburban Diner, Uno Pizzaria, Wendy's, White Castle, Day's Inn, Sheraton, A&P, Bed Bath&Beyond, Computer City, Drug Fair, Ford, KidsRUs, Macy's, Marshall's, Midas Muffler, Nordstrom's, Pearle Vision, Sears, Service Merchandise, Staples, cleaners, cinema, mall

New Jersey Interstate 295

Exit #	Services
67b a	US 1. I-295 nb becomes I-95 sb at US 1. **See NJ I-95, exit 9b a.**
65b a	Sloan Ave, **E**...Exxon/mart, Burger King, Dunkin Donuts, Taco Bell, Ames, CVS Drug, Hobby Shop, ShopRite Foods, Sir Speedy, auto parts/repair
64	to NJ 33 E(from sb), same as 63
63b a	NJ 33 W(from nb), rd 535, Mercerville, Trenton, **E**...CHIROPRACTOR, DENTIST, Mobil, Shell/24hr, Applebee's, Boston Mkt, Popeye's, Vincent's Pizza, Lincoln/Mercury, Mufflex, **W**...Exxon/mart/24hr, WawaMart, Burger King, Domenico's Ristorante, Dunkin Donuts, PartsAmerica, cleaners, repair, transmissions
62	Olden Ave N(from sb, no return), **W**...Delta Gas
61b a	Arena Dr, White Horse Ave, **W**...7-11
60b a	I-195, to I-95, W to Trenton, E to Neptune
58mm	scenic overlook both lanes
57b a	US 130, US 206, **E**...Mobil, Shell, Burger King, Denny's, McDonald's, Rosario's Pizza, Acme Foods, Saturn, **2 mi E**...Petro/diesel/rest./24hr, Pilot/Wendy's/diesel/mart/24hr
56	US 206 S(from nb, no return), to NJ Tpk, Ft Dix, McGuire AFB, **E**...Sunoco, Day's Inn, Holiday Inn Express, same as 57
52b a	rd 656, to Columbus, Florence, no facilities
50mm	**rest area both lanes(7am-11pm), full(handicapped)facilities, phone, picnic tables, litter barrels, vending, RV dumping**
47b a	NJ 541, to Mount Holly, Burlington, NJ Tpk, **E**...MEDICAL CARE, VETERINARIAN, CHIROPRACTOR, Exxon/diesel/mart/24hr, Mobil, Burger King, Taco Bell, Howard Johnson, $Express, Home Depot, Kohl's, Sears, Target, cinema, mall, **W**...HOSPITAL, Exxon, Gulf/repair, Hess/diesel/mart, Shell/mart/24hr, Checker's, Shoney's, Subway, Wedgewood Farms Rest., Acme Foods, FashionBug, JiffyLube, K-Mart, Marshall's, Wal-Mart SuperCtr, auto parts, bank, drugs
45b a	to Mt Holly, Willingboro, **E**...HOSPITAL, **W**...Exxon, Mobil, Brake&Go, auto repair
43b a	rd 636, to Rancocas Woods, Delran, **W**...HOSPITAL, Texaco/24hr, Pirates Rest.
40b a	NJ 38, to Mount Holly, **E**...HOSPITAL, Getty, **W**...Texaco/diesel/mart/24hr
36b a	NJ 73, to NJ Tpk, Berlin, Tacony Br, **E**...Exxon, Mobil/diesel/24hr, Bennigan's, Bob Evans, Denny's, McDonald's, Sage Rest., Wendy's, Budget Motel, Courtyard, Day's Inn, Econolodge, Fairfield Inn, McIntosh Inn, Radisson, Red Carpet Inn, Red Roof Inn, Super 8, Travelodge, Wingate Inn, **W**...Amoco/mart, Shell, Texaco/diesel/mart/24hr, Bertucci's Rest., Boston Mkt, Burger King, Don Pablo's, Dunkin Donuts, Perkins, Pizza Hut, Ponderosa, Wendy's, Motel 6, Rodeway Inn, Barnes&Noble, Best Buy, Boaters World, Caldor, Chrysler, CompUSA, Dick's Sports, $Tree, Drug Emporium, Firestone/auto, Filene's, Gateway Computers, Goodyear/auto, Home Depot, JiffyLube, K-Mart/auto, Linens'n Things, Lexus, Loehmann's, MensWearhouse, Mitsubishi, OfficeMax, Old Navy, PetsMart, PharMor Drug, Ross, Sears/auto, ShopRite Foods, bank, cinema

New Jersey Interstate 295

34b a	NJ 70, Cherry Hill, to Camden, **E**...DENTIST, Amoco/mart, Exxon/mart/24hr, Burger King, Dunkin Donuts, Friendly's, Korea Garden Rest., McDonald's, Pizzaria Uno, Extended Stay America, Residence Inn, Bargain Brakes/muffler, GreaseMonkey Lube, Nevada Bob's, STS Tire/auto, bank, **W**...HOSPITAL, Mobil/repair/24hr, Texaco, Boston Mkt, Denny's, Steak&Ale, Holiday Inn(3mi), Howard Johnson(4mi), Sheraton/rest., bank, carwash
32	NJ 561, to Haddonfield, Voorhees, **3 mi E**...HOSPITAL, Mobil/diesel/mart/24hr, Applebee's, Olive Garden, Vito's Pizza, Hampton Inn, USPO, cleaners, drugs, tires, **W**...Citgo, Burger King
31	Woodcrest Station, no facilities
30	Warwick Rd(from sb), no facilities
29b a	US 130, to Berlin, Collingswood, **E**...Amoco, Texaco, Church's Chicken, Pizza Delight, Subway, Super 8, Drug Emporium, transmissions, carwash, **W**...Shell
28	NJ 168, to NJ Tpk, Belmawr, Mt Ephraim, **E**...Shell, Texaco/diesel, Burger King, Club Diner, Dunkin Donuts, Bargain Brake/muffler, JiffyLube, bank, carwash, repair, **W**...Amoco, Exxon/LP/24hr, Carvel Bakery, McDonald's, Taco Bell
26	I-76, NJ 42, to I-676, Walt Whitman Bridge
25b a	NJ 47, to Westville, Deptford, no facilities
24	NJ 45, NJ 551, to Westville, **E**...HOSPITAL, Getty, **W**...Chevrolet
23	US 130 N, to National Park
22	NJ 644, to Red Bank, Woodbury, **E**...Mobil, **W**...repair
21	NJ 44 S, Paulsboro, Woodbury, **W**...WaWaMart, Westwood Motor Lodge
20	NJ 643, to NJ 660, Thorofare, to National Park, **E**...Best Western
19	NJ 656, to Mantua, no facilities
18b a	NJ 667, to NJ 678, Clarksboro, Mt Royal, **E**...Amoco/diesel, TravelPort/diesel/rest., Chinese Rest., KFC/Taco Bell, McDonald's, Wendy's, **W**...Texaco/diesel/mart
17	NJ 680, Gibbstown, to Mickleton, **W**...Mobil, Burger King, Little Caesar's, Ramada Inn, Rite Aid, Thriftway Foods, laundry
16b	NJ 551, to Gibbstown, to Mickleton, no facilities
a	NJ 653, to Paulsboro, Swedesboro, no facilities
15	NJ 607, to Gibbstown, no facilities
14	NJ 684, to Repaupo, no facilities
13	US 130 S, US 322 W, to Bridgeport, no facilities
11	US 322 E, to Mullica Hill, no facilities
10	Ctr Square Rd, to Swedesboro, **E**...Texaco/repair, WaWa/deli, McDonald's, Hampton Inn, Holiday Inn, **W**...Camping World RV Supplies/service
7	to Auburn, Pedricktown, **E**...AutoTruck/diesel/rest./motel/laundry, P Town Diner
4	NJ 48, Woodstown, Penns Grove
3mm	weigh sta nb
2mm	**rest area nb, full(handicapped)facilities, info, phone, picnic table, litter barrels**
2c	to US 130(from sb), Deepwater, **W**...HOSPITAL, Flying J/diesel/LP/rest./mart/24hr TA/Texaco/Blimpie/Popeye's/diesel/rest./24hr
b	US 40 E, to NJ Tpk, **E**...Mobil/diesel/mart, MJS Trkstp/diesel, Pilot/Subway/diesel/mart/24hr, Del Valley Motel, Friendship Motel, Landmark Lodge, Turnpike Inn, Wellesley Inn
a	US 40 W(from nb), to Delaware Bridge
1c	NJ 551 S, Hook Rd, to Salem, **E**...White Oaks Motel, **W**...HOSPITAL
b	US 130 N(from nb), Penns Grove
a	NJ 49 E, to Pennsville, Salem, **E**...Amoco/diesel/LP, Exxon/diesel/repair, Sunoco/24hr, Burger King, Cracker Barrel, KFC, McDonald's, Taco Bell, Hampton Inn, Super 8, hardware, **W**...Seaview Motel
0mm	Delaware River, Delaware Memorial Bridge, New Jersey/Delaware state line

N

S

New Mexico Interstate 10

Exit #	Services
164.5mm	New Mexico/Texas state line
164mm	**Welcome Ctr wb, full(handicapped)facilities, phone, picnic tables, litter barrels, petwalk**
162	NM 404, Anthony, **S**...RV camping
160mm	weigh/insp sta wb
155	NM 227 W, to Vado, **N**...Vado RV Park, **S**...Fina/diesel/pizza/24hr, Shell/Pizza Inn/Vado Rest./diesel, El Camino Real HS
151	Mesquite, no facilities
144	I-25 N, to Las Cruces
142	NM 478, Main St, Las Cruces, **N**...HOSPITAL, Chevron/diesel/mart, Shamrock/mart, Chilito's Mexican, Denny's, Village Inn Rest., Whataburger/24hr, Best Western/rest., Comfort Inn, Day's Inn, Holiday Inn, Holiday Inn Express, Motel 6, Super 8, Chevrolet/Pontiac, Trailer Corral RV Park, NMSU, **S**...Phillips 66/mart, Ace Hardware, USPO
140	NM 28, Las Cruces, to Mesilla, **N**...Exxon/diesel/mart, Blake's Lotaburger, BurgerTime, Choice Rest., Cracker Barrel, McDonald's, Baymont Inn, Best Western/grill, Hampton Inn, La Quinta, SpringHill Suites, **N on Valley Dr**...Shell/diesel/mart, Dairy Queen, Domino's, Old Town Rest., Dodge, Toyota, VW, **S**...Conoco/rest., S&H RV Ctr, Siesta RV Park, Trailer Corral RV Park
139	NM 292, Amador Ave, Motel Blvd, Las Cruces, **N**...Pilot/Subway/diesel/mart/24hr, Shell, TA/diesel/rest./24hr, Dick's Rest., Western Inn, **S**...Economy Inn, Coachlight Inn/RV Park
138mm	Rio Grande River
135.5mm	**rest area eb, full(handicapped)facilities, picnic tables, litter barrels, petwalk, scenic view, RV dump**
135	US 70 E, to W Las Cruces, Alamogordo
132	**N**...to airport, fairgrounds, **S**...Love's/Texaco/Subway/diesel/mart/24hr
127	Corralitos Rd, **N**...Chevron, Bowlin's Trading Post, to fairgrounds
120.5mm	weigh sta wb, parking area eb, litter barrels
116	NM 549, **S**...gas/café/gifts
102	Akela, **N**...Exxon/diesel/mart/gifts
85	East Deming(from wb), **S**...Chevron, Texaco, Motel 6, RV camping, same as 83
83	Deming, **S**...Chevron/service, Fina, Texaco/diesel/mart, Dairy Queen, KFC, Best Western, Day's Inn, Grand Motel, Holiday Inn, Mirador Motel, Motel 6, Chevrolet/Olds/Pontiac/GMC, Chrysler/Dodge/Jeep, Ford, Firestone, Goodyear, K-Mart, RV repair, auto parts, radiators, Little RV Park, Roadrunner RV Park, Wagon Wheel RV Park, to Rock Hound SP, st police
82b a	US 180, NM 26, NM 11, Deming, **N**...Fina/diesel/mart, Texaco, Blake's Lotaburger, **S**...Exxon, Phillips 66/mart, Texaco, Arby's, Burger King, Cactus Café, Chinese Rest., Denny's, K-Bob's, KFC, LJ Silver, McDonald's, Pizza Hut, Subway, Grand Motel, AutoZone, Budget Tire, CarQuest, Furr's Foods, K-Mart, Radio Shack, laundry, museum, to Pancho Villa SP, Rockhound SP
81	NM 11, W Motel Dr, Deming, **S**...Chevron/mart/24hr, Conoco/diesel/mart/atm, DemingTT/diesel, Shamrock/diesel, Texaco/diesel/mart/24hr, Arby's, Burger King, Burger Time, El Camino Real Rest., McDonald's, Sonic, Taco Bell, Balboa Motel, Best Western, Budget Motel, Grand Hotel/rest., Mirador Motel, Super 8, Wagon Wheel Motel, Western Motel/rest., PicQuik Foods, 81 Palms RV Park, to Pancho Villa SP, Rock Hound SP
68	NM 418, **S**...Savoy/diesel/rest., Citgo/Stuckey's/Baskin-Robbins
62	Gage, **S**...Citgo/Dairy Queen/diesel, Butterfield Station RV Park
61mm	**rest area wb, full(handicapped)facilities, picnic tables, litter barrels, vending, petwalk**
55	Quincy, no facilities
53mm	**rest area eb, full(handicapped)facilities, picnic tables, litter barrels, vending, petwalk**
51.5mm	Continental Divide, elev 4585
49	NM 146 S, to Hachita, Antelope Wells, no facilities
42	Separ, **S**...Trading Post/gas, Separ/diesel/mart, Wind Meal Diner, Bowlin's Continental Divide Gifts, tires
34	NM 113 S, Muir, Players, no facilities
29	no facilities
24	US 70, E Motel Dr, Lordsburg, **N**...Chevron/mart, Pilot/Arby's/diesel/mart/24hr, Best Western, RV park
23.5mm	weigh sta both lanes
22	NM 494, Main St, Lordsburg, **N**...Save Gas/wash, Dairy Queen, McDonald's, Saucedo's Foods, auto/diesel/tire repair, carwash/gas, **S**...Chevron/mart, Exxon, Shamrock/mart, SnappyMart/24hr, Texaco/diesel/mart, KFC, Kranberry's Rest., Best Western/rest., Holiday Inn Express, Motel 10, Super 8, KOA, cinema
20b a	W Motel Dr, Lordsburg, **Visitors Ctr/full(handicapped)facilities, info, N**...Love's/Texaco/A&W/Subway/diesel/mart, Denny's, Day's Inn

New Mexico Interstate 10

15	to Gary, no facilities
11	NM 338 S, to Animas, no facilities
5	NM 80 S, to Road Forks, S...Fina/diesel/rest., Desert West Motel, RV park, diesel repair
3	Steins, no facilities
0mm	New Mexico/Arizona state line

New Mexico Interstate 25

Exit #	Services
460.5mm	New Mexico/Colorado state line
460	weigh sta sb, Raton Pass Summit, elev 7834
454	Lp 25, Raton, **2 mi W...**HOSPITAL, Texaco, Budget Host, Capri Motel, El Portal Motel
452	NM 72 E, Raton, **E...**to Sugarite SP, **W...**Conoco/mart, Mesa Vista Motel
451	US 64 E, US 87 E, Raton, **E...**Chevron/diesel/service, Texaco/diesel/mart, Total/diesel/mart/24hr, Subway, to Capulin Volcano NM, **W...**HOSPITAL, CHIROPRACTOR, info, Chevron/diesel/mart, Conoco/mart, Phillips 66/mart, Texaco/mart, Arby's, All Seasons Rest., Chinese Buffet, Dairy Queen, Denny's, K-Bob's, KFC, McDonald's, Sands Rest., Best Western/rest., Comfort Inn, El Kapp Motel, Harmony Manor Motel, Holiday Classic Motel, Motel 6, Super 8, Texan Motel, Travel Motel, K-Mart, NAPA, camping
450	Lp 25, Raton, **W...**HOSPITAL, CHIROPRACTOR, VETERINARIAN, Conoco/diesel/mart, Phillips 66, Shamrock/diesel/mart, Total/mart, K-Bob's, Pappa's Rest., Rainmaker Café, Sonic, Holiday Inn Express, Maverick Motel, Oasis Motel/rest., Robin Hood Motel/café, Village Inn Motel, Family$, K-Mart, KOA, Knight's Auto Parts, Medicine Shoppe, SuperSave Foods, bank
446	US 64 W, to Cimarron, Taos, **4 mi W...**NRA Whittington Ctr, airport
440mm	Canadian River
435	Tinaja, no facilities
434.5mm	**rest area both lanes, full(handicapped)facilities, weather info, picnic tables, litter barrels, petwalk**
426	NM 505, Maxwell, **W...**to Maxwell Lakes, phone
419	NM 58, to Cimarron, **E...**Texaco/diesel/mart/24hr, Heck's Hungry Traveler Rest.
414	US 56, Springer, **E...**Phillips 66/diesel/mart, Shamrock/mart(1mi), Dairy Delite Café, Brown Hotel, RV camping
412	US 56 E, US 412 E, NM 21, NM 468, Springer, **1 mi E...**Shamrock/mart, Del Taco, Brown Hotel
404	NM 569, Colmor, Charette Lakes, no facilities
393	Levy, no facilities
387	NM 120, Wagon Mound, to Roy, **E...**Chevron/diesel/mart, Phillips 66/mart, Santa Clara Café
376mm	**rest area sb, full(handicapped)facilities, phone, picnic tables, litter barrels, petwalk, RV camping/dump**
374mm	**rest area nb, full(handicapped)facilities, phone, picnic tables, litter barrels, petwalk, RV camping/dump**
366	NM 97, NM 161, Watrous, Valmora, **W...**Santa Fe Trail, Ft Union NM, no facilities
364	NM 97, NM 161, Watrous, Valmora, no facilities
361	no facilities
360mm	**rest area both lanes, litter barrels, no facilities**
356	Onava, no facilities
352	**E...**RV camping, **W...**airport
347	to NM 518, Las Vegas, **E...**CHIROPRACTOR, **1-2 mi W...**Fina/diesel/mart, Phillips 66/Burger King/mart, Texaco/rest., Arby's, Dairy Queen, Hillcrest Rest., KFC, McDonald's, Pino's Family Rest., Pizza Hut, Taco Bell, Comfort Inn, Day's Inn, El Camino Motel, Inn of Las Vegas/rest., Super 8, Townhouse Motel, Pennzoil, KOA, Storrie Lake SP, Las Vegas Civic Ctr
345	NM 65, NM 104, University Ave, Las Vegas, **E...**to Conchas Lake SP, camping, **W...**HOSPITAL, Phillips 66/diesel/mart, Fina/diesel/rest., Texaco/mart/24hr, Arby's, Burger King, Dairy Queen, Hillcrest Rest., K-Bob's, KFC, McDonald's, Mexican Kitchen, Pizza Hut, Subway, Taco Bell, Budget Inn, Comfort Inn, El Camino Motel/rest., Palamino Motel, Plaza Hotel/rest., Sante Fe Trail Inn, Sunshine Motel, Townhouse Motel, Firestone, Hist Old Town Plaza, laundry
343	to NM 518 N, Las Vegas, **E...**tires, **1 mi W...**Conoco, Exxon, Phillips 66, Texaco, Burger King, Hillcrest Rest., K-Bob's, McDonald's, Taco Bell, Comfort Inn, El Camino Motel, Plaza Motel, Santa Fe Trail Inn, Thunderbird Motel
339	US 84 S, Romeroville, to Santa Rosa, **E...**Texaco/diesel/mart, KOA, phone
335	Tecolote, no facilities
330	Bernal, no facilities
325mm	**rest area both lanes, picnic tables, litter barrels, no restroom facilities**
323	NM 3 S, Villanueva, **E...**to Villanueva SP, Madison Winery(6mi), RV camping
319	San Juan, San Jose, **3 mi E...**gas/diesel/mart, **W...**Pecos River RV Camp/mart/phone
307	NM 63, Rowe, Pecos, **W...**Pecos NM
299	NM 50, Glorieta, Pecos, **W...**Fina/diesel(4mi), Shell(6mi), Glorieta Conf Ctr, Renate's Rest., Pecos NHP
297	Valencia, no facilities

New Mexico Interstate 25

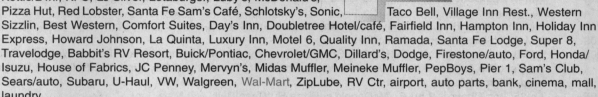

294   Apache Canyon, **W...**KOA, Rancheros Camping(Mar-Nov)

290   US 285 S, to Lamy, S to Clines Corners, **W...**KOA, Rancheros Camping(Mar-Nov)

284   NM 466, Old Pecos Trail, Santa Fe, **W...**HOSPITAL, Chevron/diesel/mart, Sunset Gen Store/gas, to Best Western Inn of Loretta, Garrett's Desert Inn, Pecos Trail Inn, The Sands, museums

282   US 84, US 285, St Francis Dr, Santa Fe, **W...**Conoco/Wendy's/mart, Giant/diesel/mart, **1 mi W...**HOSPITAL, Allsup's/mart, Chevron, Shell/mart, Carrow's Rest., McDonald's, Pizza Hut, Sizzler, El Ray Motel, Residence Inn, Santa Fe Budget Inn, Thunderbird Motel, Travelodge, Cadillac, Nissan, bank

278   NM 14, Cerrillos Rd, Santa Fe, **W...**Sleep Inn, Santa Fe Outlets/famous brands, **1-4 mi W...**CHIROPRACTOR, Allsup's/gas/24hr, Chevron/diesel/mart/24hr, Conoco/mart, Shamrock/diesel, Shell/mart/24hr, Texaco/diesel/café, Applebee's, Arby's, Baskin-Robbins, Blue Corn Café, Burger King, CiCi's, Denny's, Kettle/24hr, KFC, LJ Silver, Lotaburger, Luby's, McDonald's, Pizza Hut, Red Lobster, Santa Fe Sam's Café, Schlotsky's, Sonic, Taco Bell, Village Inn Rest., Western Sizzlin, Best Western, Comfort Suites, Day's Inn, Doubletree Hotel/café, Fairfield Inn, Hampton Inn, Holiday Inn Express, Howard Johnson, La Quinta, Luxury Inn, Motel 6, Quality Inn, Ramada, Santa Fe Lodge, Super 8, Travelodge, Babbit's RV Resort, Buick/Pontiac, Chevrolet/GMC, Dillard's, Dodge, Firestone/auto, Ford, Honda/Isuzu, House of Fabrics, JC Penney, Mervyn's, Midas Muffler, Meineke Muffler, PepBoys, Pier 1, Sam's Club, Sears/auto, Subaru, U-Haul, VW, Walgreen, Wal-Mart, ZipLube, RV Ctr, airport, auto parts, bank, cinema, mall, laundry

276   NM 599, to NM 14, to Madrid, **E...**Phillips 66/diesel/mart/24hr, Delfino's Mexican(1mi), **W...**Shell

271   NM 587, La Cienega, **W...**Pinon RV Park, racetrack, museum

269mm   **rest area nb, full(handicapped)facilities, phone, picnic tables, litter barrels, petwalk**

267   **E...rest area sb, full(handicapped)facilities, phone, picnic tables, litter barrels, petwalk**

264   NM 16, Cochiti Pueblo, **W...**to Cochiti Lake RA

263mm   Galisteo River

259   NM 22, to Santo Domingo Pueblo, **W...**Phillips 66/mart/LP/24hr, to Cochiti Lake RA, RV camping

257   Budaghers, **W...**Mormon Battalion Mon

252   San Felipe Pueblo, **E...**San Felipe Casino

248   Algodones, **E...**phone

242   US 540, NM 44 W, NM 165 E, to Farmington, Aztec, **W...**Chevron/24hr, Circle K/gas/24hr, Giant/Conoco/diesel/mart, Mustang/mart, Texaco/Burger King/diesel/mart, Coronado Rest., KFC, Lotaburger, McDonald's, Pizza Hut, Range Café, Sonic, Subway, Taco Bell, Village Pizza, Day's Inn, Microtel Suites, Super 8, bank, casino, tires, to Coronado SP

240   NM 473, to Bernalillo, **W...**Conoco/diesel/mart, Abuelito's Kitchen, Range Café, to Coronado SP

234   NM 556, Tramway Rd, **E...**Indian Arts/crafts, **W...**Phillips 66/mart, to McDonald's(4mi), casino/bingo

233   Alameda Blvd, **E...**Chevron/Burger King/mart, Comfort Inn, Toyota, foreign auto parts, **W...**Phillips 66/diesel/mart, Ramada Ltd

232   Paseo del Norte, **E...**Motel 6, Gruet Winery, **W...**Courtyard, Crowne Plaza

231   San Antonio Ave, **E...**HOSPITAL, Cracker Barrel, Kettle, Amberley Suites, Comfort Inn, Howard Johnson, La Quinta, Quality Suites, **W...**Baymont Inn, Crossland Suites, Hampton Inn, Mazda, VW

230   San Mateo Blvd, Osuna Rd, Albuquerque, **E...**HOSPITAL, CHIROPRACTOR, Chevron/mart, Circle K/gas, Giant Gas/mart, Pump'n Save, Texaco/diesel/mart, Applebee's, Arby's, Bennigan's, Black-Eyed Pea, Burger King, Chili's, Chinese Rest., Doc&Eddy's. Grady's Grill, Hooters, KFC, LJ Silver, Lotaburger, McDonald's, Olive Garden, Pizza Hut, Schlotsky's, Sizzler, Sonic, Subway, SweetTomato's, Taco Bell, Taco Cabana, Thai Cuisine, Village Inn Rest., Wendy's, Wienerschnitzel, Wild Oats Mkt, Winchell's, Wyndham Garden, Buick/Pontiac/GMC, Diahatsu, Mercedes, Nissan, Olds/Lexus, Subaru/Isuzu, Brake/lube/tune, Firestone/auto, Kinko's, Midas Muffler, NAPA, PepBoy's, Speedee, Valvoline, ZipLube, bank, cinema, **1 mi E...**Chevron, Phillips 66, Texaco, Church's Chicken, Dairy Queen, Grandy's, JB's, Mac's Drive-In, Papa John's, Pancho's Mexican, Peter Piper Pizza, Souper Salad, AutoZone, Batteries+, Big 5 Sports, BrakeMasters, Just Brakes, Lucky SavOn, MailBoxes Etc, U-Haul, Walgreen, **W...**Shamrock/mart, Whataburger, Homestead Village

229   Jefferson St, **E...**HOSPITAL, Carrabba's, Landry's Seafood, Olive Garden, OutBack Steaks, cinema, same as 230, **W...**Blue Corn Café, Fuddrucker's, Texas Land&Cattle Steaks

228   Montgomery Blvd, **E...**HOSPITAL, Chevron/mart, Shamrock/mart, Texaco/Taco Bell/mart/wash, Golden Corral(4mi), Best Western, Discount Tire, **W...**Wendy's, Suburban Lodge, bank, funpark

New Mexico Interstate 25

Albuquerque

227b	Comanche Rd, Griegos Rd, **E**...UPS Depot
a	Candelaria Rd, Albuquerque, **E**...Chevron/rest., Circle K/gas, Fina/Subway/diesel/mart, Shell, TA/diesel/mart/24hr, JB's, IHOP, Village Inn Rest., Comfort Inn, Clubhouse Inn, Fairfield Inn, Hilton, Holiday Inn, LeBaron Inn, Rodeway Inn, Motel 76, Motel 6, Sumner Suites, Super 8, **W**...Waffle House, A-1 Motel, Red Roof Inn, Volvo, Express Truckwash
226b a	I-40, E to Amarillo, W to Flagstaff
225	Lomas Blvd, **E**...HOSPITAL, Chevron/7-11, JB's, Plaza Inn/rest., Chevrolet, Dodge, Ford, Saturn
224b	Grand Ave, Central Ave, **W**...HOSPITALS, Chevron, Shamrock, Econolodge, bank
a	Lead Ave, Coal Ave, **E**...HOSPITAL, Texaco/mart, Burger King, Crossroads Motel
223	Stadium Blvd, **E**...Motel 6, sports facilities
222b a	Gibson Blvd, **E**...HOSPITAL, VETERINARIAN, Phillips 66/mart/wash, Pump'n Save/mart, Applebee's, Burger King, Toastie's Grill, Waffle House, Baymont Inn, Best Western, Courtyard, Crossland Suites, Fairfield Inn, Hampton Inn, Hawthorn Suites, Holiday Inn Express, La Quinta, Quality Suites, Radisson, Ramada Ltd, Sleep Inn, Kirtland AFB, airport, atomic museum, **W**...Fina/7-11/24hr, Church's Chicken, LotaBurger, Thalia's Mexican, auto repair
221	Sunport(from nb), **E**...AmeriSuites, Wyndham Garden, USPO, airport
220	Rio Bravo Blvd, Mountain View, **E**...golf, **2 mi W**...Giant/diesel/mart/24hr, Shamrock, Burger King, Church's Chicken, McDonald's, Pizza Hut, Subway, Taco Bell, Albertson's, Family$, Pet/Vet supply, Walgreen
215	NM 47, **E**...Conoco/diesel/mart, to Isleta Lakes RA/RV Camping, bingo, casino, golf, st police
214mm	Rio Grande
213	NM 314, Isleta Blvd, **W**...Chevron/Subway/diesel/mart/24hr
209	NM 45, to Isleta Pueblo, no facilities
203	NM 6, to Los Lunas, **E**...EYECARE, Chevron/diesel/mart/24hr, Shamrock/mart, Texaco/Wendy's/diesel/mart, Arby's, Little Anita's Café, McDonald's, Village Inn Rest., Comfort Inn, Day's Inn, **W**...Phillips 66/Subway/diesel/mart, Microtel Suites, cinema
195	Lp 25, Los Chavez, **2 mi E**...McDonald's, Pizza Hut/Taco Bell, Wal-Mart SuperCtr/24hr
191	NM 548, Belen, **1 mi E**...HOSPITAL, Chevron, Conoco/mart, Circle K, Golden Corral, KFC, McDonald's, Pizza Hut, Subway, Freeway Inn, Super 8, IGA, Chevrolet/Olds, CarQuest, USPO, bank, laundry, tires, to Chavez SP, **W**...Casa de Mirada Rest./RV Park, Rio Grande Diner, Best Western, OakTree Inn
190	Lp 25, Belen, **1-2 mi E**...Akin's Gas/diesel, Conoco/diesel, Giant Gas/food, Arby's, Golden Corral, KFC, Lesuer's Rest., McDonald's, Pizza Hut, Whiteway Café, Freeway Inn, Mountainview Motel, Resort Motel, Super 8, auto repair
175	US 60, Bernardo, **E**...gas/mart, Salinas NM, phone, **W**...Gen Store/Kiva RV Park
174mm	Rio Puerco
169	**E**...La Joya St Game Refuge, Sevillita NWR
167mm	**rest area both lanes, full(handicapped)facilities, picnic tables, litter barrels, vending, petwalk**
166mm	Rio Salado
165mm	weigh sta nb/parking area both lanes
163	San Acacia, no facilities
156	Lemitar, no facilities
152	Escondida, **W**...to st police
150	US 60 W, Socorro, **W**...Chevron/mart/wash, Circle K/gas, Exxon/diesel, Phillips 66/diesel/mart, Shamrock/diesel, Texaco, Blake's Lotaburger, Burger King, Denny's, Domino's, K-Bob's, KFC, McDonald's/playplace, Pizza Hut, RoadRunner Steaks, Sonic, Subway, Taco Bell, Tina's Rest., Best Inn, Best Western, Economy Inn, Econolodge, El Camino Motel/rest., Holiday Inn Express, Payless Inn, Rio Grande Motel/rest., Sands Motel, San Miguel Motel, Super 8, Ace Hardware, Big Value Drugs/Health Foods, Chevrolet/Pontiac/Olds/Buick, Chrysler/Dodge/Jeep, Ford/Mercury, Furr's Food/deli, NAPA/autocare, Radio Shack, to NM Tech
147	US 60 W, Socorro, **W**...HOSPITAL, Chevron/diesel/mart/RV dump, Conoco/LP, Pump n Save, Shell/diesel/mart, Texaco/diesel/mart, Arby's, Armijo's Mexican, Coffeehouse, Denny's, KFC, Pizza Hut, Holiday Inn Express, Motel 6, Sands Motel, Parts+, TrueValue, antiques, bank, carwash, tires, Socorro RV Park, to airport
139	US 380 E, to San Antonio, **E**...Manny's Burgers, Owl Café, Bosque Del Apache NWR
124	to San Marcial, **E**...to Bosque del Apache NWR, Ft Craig
115	NM 107, **E**...Truck Plaza/diesel/mart/rest./24hr
114mm	**rest areas both lanes, full(handicapped)facilities, picnic tables, litter barrels, petwalk, RV parking, vending**
107mm	Nogal Canyon
100	Red Rock, no facilities
92	Mitchell Point, no facilities
90mm	La Canada Alamosa
89	NM 181, to Cuchillo, to Monticello, **4 mi E**...Chevron/mart, Shamrock/Mean Gene's/mart, Monticello RV Park
83	NM 52, NM 181, to Cuchillo, **3 mi E**...Marina Suites Motel, Elephant Butte Inn/rest., Lakeside RV Park, Elephant Butte SP
82mm	insp sta nb

Socorro

New Mexico Interstate 25

N

S

79   Lp 25, to Truth or Consequences, **E...**HOSPITAL, VETERINAR-IAN, Chevron/diesel/repair, Circle K/gas, Dairy Queen, Denny's, Hilltop Café, K-Bob's, KFC, La Cocina Mexican, Los Apcos Steaks, McDonald's, Pizza Hut, Sonic, Subway, Tin Lizzie's Rest., Ace Lodge, Best Western, Frontier Motel, Holiday Inn, Quality Inn, Red Haven Motel, Sunland Motel, Super 8, Trail Motel, AutoZone, Furr's Foods, Medicine Shoppe, NAPA, USPO, hardware, laundry, (17 gas stations, 15 campgrounds, 26 motels, 2 museums), to Elephant Butte SP

76   (75 from nb)Lp 25, to Williamsburg, **E...**Chevron/mart/24hr, Conoco/diesel/mart, Shamrock/mart, Café Rio, Rio Grande Motel, Shady Corner RV Park, USPO, municipal park, tires, **1 mi E...**Chevrolet/Pontiac/Buick/Olds/GMC, RJ RV Park, (17 gas stations, 15 campgrounds, 26 motels, 2 museums)

71   Las Palomas, no facilities

67mm   **rest area both lanes, picnic tables, litter barrels, no other facilities**

63   NM 152, Caballo, to Hillsboro, **E...**Texaco/KOA/mart

59   Arrey, Derry, **E...**to Caballo-Percha SPs

58mm   Rio Grande

51   Garfield, to Arrey, Derry, no facilities

41   NM 26 W, Hatch, **1 mi W...**MEDICAL CARE, Phillips 66/diesel, Dairy Queen, Village Plaza Motel, Chevrolet, Happy Trails RV Park

35   NM 140 W, Rincon, no facilities

32   Upham, no facilities

27mm   scenic view nb, picnic tables, litter barrels

26mm   insp sta nb

23mm   **rest area both lanes, full(handicapped)facilities, picnic tables, litter barrels, vending, petwalk**

19   Radium Springs, **W...**Leasburg SP, Fort Selden St Mon, RV camping

9   Dona Ana, **W...**Chevron/Pizza Inn/mart, Citgo/mart, Conoco/mart, RV camping

6b a   US 70, Las Cruces, to Alamogordo, **E...**HOSPITAL, Phillips 66/diesel/mart/atm, Texaco/TCBY/mart/wash, IHOP, Pizzaria Uno, Century 21 Motel, Fairfield Inn, Super 8, GNC, K-Mart/drugs, Raley's Foods, **W...**DENTIST, MEDICAL CARE, EYECARE, Chevron/mart/wash, Conoco/mart/wash, Shamrock/mart, Shell/mart, Baskin-Robbins, BurgerTime, Dairy Queen, Domino's, KFC, Lotaburger, McDonald's, Sonic, Subway, Taco Bell, Whataburger/24hr, Albertson's, AutoZone, Better Drugs, Checker Parts, IGA Foods, Lube Express, bank, golf

3   Lohman Ave, Las Cruces, **E...**MEDICAL CARE, DENTIST, EYECARE, Shamrock/mart, Shell/mart, Applebee's, Burger King, Cattle Baron Steak/seafood, Chili's, Farley's Grill, Golden Corral, Hooters, Jack-in-the-Box, KFC, Luby's, Pizza Hut, Popeye's, Red Lobster, Sonic, Village Inn Rest., Hilton, Albertson's/drugs, American RV/Marine, Beall's, Dillard's, Discount Tire, Home Depot, JC Penney, OfficeMax, Saturn(1mi), Sears/auto, Service Merchandise, Target, Walgreen, bank, cinema, mall, **W...**DENTIST, EYECARE, VETERINARIAN, Conoco/diesel/mart, Phillips 66, Shamrock, Arby's, McDonald's, Mesilla Valley Kitchen, Oriental Express, Si Senor Mexican, Subway, TCBY, Wienerschnitzel, Day's End Lodge, Hastings Books, HobbyLobby, Lube'n Go, MailBoxes Etc, Martin Tires, NAPA, Pennzoil, PepBoys, Smith's Food/drug, ToysRUs, Wal-Mart SuperCtr/24hr, bank, cleaners

1   University Ave, Las Cruces, **E...**HOSPITAL, golf course, museum, st police, **W...**Conoco/mart, Bennigan's, Dairy Queen, Lorenzo's Italian, McDonald's, Comfort Suites, Sleep Inn, Furr's Foods, Jo-Ann Fabrics, Kinko's, Postal Annex, NMSU, carwash, cleaners/laundry

0mm   I-25 begins/ends on I-10, exit 144 at Las Cruces.

New Mexico Interstate 40

Exit #	Services
373.5mm	New Mexico/Texas state line, Mountain/Central time zone
373mm	**Welcome Ctr wb, full(handicapped)facilities, phone, picnic tables, litter barrels, petwalk**
369	NM 93 S, NM 392 N, to Endee, no facilities
361	Bard, no facilities
358mm	weigh sta both lanes
356	NM 469, San Jon, **N...**Citgo/KFC/Taco Bell/diesel/mart/24hr, to Ute Lake SP, **S...**Phillips 66/diesel/pizza/mart, Texaco/Burger King/diesel/mart, San Jon Motel
343	ranch access, no facilities
339	NM 278, airport

Las Cruces

E

W

New Mexico Interstate 40

Tucumcari

335 Lp 40, E Tucumcari Blvd, Tucumcari, **N**...Chevron, Conoco/diesel/mart, Shell, Texaco, Denny's, McDonald's(1mi), Best Western, Comfort Inn, Econolodge, Holiday Inn/rest., Howard Johnson, Motel 6, Super 8, to Conchas Lake SP, **S**...KOA

333 US 54 E, Tucumcari, **N**...Love's/Godfather's/diesel/rest./mart, Mtn Rd RV Park, **1 mi N**...Texaco, Circle K, Hardee's, Lotaburger, Pizza Hut, Sonic, Best Western, Comfort Inn, Friendly Motel, Holiday Inn, Palomino Motel, Relax Inn, Travelodge, K-Mart, auto parts

332 NM 209, NM 104, 1ˢᵗ St, Tucumcari, **N**...HOSPITAL, CHIROPRACTOR, VETERINARIAN, Chevron, Exxon/ Subway/diesel/mart/24hr, Phillips 66/mart, Hardee's, KFC, Lotaburger, McDonald's, Best Western/K-Bob's, Comfort Inn, Day's Inn, Friendship Inn, Microtel, Safari Motel, st police, to Conchas Lake SP, **S**...Texaco/mart

331 Camino del Coronado, Tucumcari, **2 mi N**...Texaco, Sonic, Westerner Drive-In, Budget Inn, Friends Inn, Redwood Lodge, Travelers Inn, Furr's Foods

329 US 54, US 66 E, W Tucumcari Ave, **N**...HOSPITAL, Shell/diesel/mart/café/atm/24hr, Budget Inn, Paradise Motel, Tucumcari Inn, golf

321 Palomas, **S**...Texaco/Dairy Queen/Stuckey's/diesel

311 Montoya, no facilities

302mm **rest area both lanes, full(handicapped)facilities, phone, picnic tables, litter barrels, petwalk, RV dump**

300 NM 129, Newkirk, **N**...Phillip 66/diesel/mart, to Conchas Lake SP

291 to Rte 66, Cuervo, **N**...gas/food, auto repair/24hr

284 no facilities

277 US 84 S, to Ft Sumner, **N**...Chevron/diesel/pizza/mart, Phillips 66/mart, Texaco, Dairy Queen, Denny's, Golden Dragon Chinese, Silver Moon Café, Best Western, Budget Inn, Comfort Inn, Holiday Inn Express, Motel 6, Ford, KOA, **S**...Love's/Carl's Jr/diesel/mart/24hr, TA/Shell/Subway/diesel/mart/24hr

Santa Rosa

275 US 54 W, Santa Rosa, **N**...Phillips 66/mart, Texaco/Burger King/mart, McDonald's, Rte 66 Rest., Best Western, Day's Inn, Ramada Ltd, Travelodge, Donnie's RV Park, KOA, **S**...HOSPITAL, Fina, Shell/diesel/mart/24hr, Comet Drive-In, Joseph's Grill, Pizza Hut, American Inn, LaLoma Motel, Sun'n Sand Motel, Sunset Motel, Super 8, Tower Motel, Western Motel, CarQuest, Goodyear/auto, st police, USPO

273.5mm Pecos River

273 US 54 S, Santa Rosa, **N**...Santa Rosa Lake SP, camping, **S**...Chevron/mart, Conoco, Shell, Mateo's Rest., AmericInn, Budget 10 Motel, Economy Inn, Super 8, Tower Motel, Chief Auto Supply, auto repair

267 Colonias, **N**...Texaco/Stuckey's/rest./mart

263 San Ignacio, no facilities

256 US 84 N, NM 219, to Las Vegas, no facilities

252 no facilities

251.5mm **rest area both lanes, full(handicapped)facilities, phone, picnic tables, litter barrels, petwalk, RV dump**

243 Milagro, **N**...Chevron/café/mart

239 no facilities

234 **N**...Citgo/Flying C/Dairy Queen/diesel/mart/gifts, repair/24hr

230 NM 3, to Encino, **N**...to Villanueva SP, towing/repair/24hr

226 no facilities

220mm parking area both lanes, litter barrels

218b a US 285, Clines Corners, **N**...Chevron/diesel/mart/24hr, Shell/diesel/mart/24hr, Clines Corners Rest., **S**...to Carlsbad Caverns NP

208 Wagon Wheel, **N**...Wagon Wheel Gas/repair/motel

207mm **rest area both lanes, full(handicapped)facilities, picnic tables, litter barrels, petwalk**

203 **N**...Zia RV Park, **S**...El Vaquero Motel/rest.

197 to Rte 66, Moriarty, **S**...Lisa's/diesel/rest., **1-2 mi S**...same as 194, 196

196 NM 41, Howard Cavasos Blvd, **S**...Citgo/KFC/pizza/mart, Circle K/diesel, BBQ, Lotaburger, Motel 6, Siesta Motel, USPO, to Salinas NM(35mi)

194 NM 41, Moriarty, **S**...MEDICAL CARE, CHIROPRACTOR, Chevron/mart/24hr, Circle K/gas, Phillips 66, Rip Griffin/Texaco/Pizza Hut/Subway/Taco Bell/diesel/24hr, Arby's, BBQ, El Comedor Rest., Mama Rosa's Rest., McDonald's, Day's Inn, Holiday Inn Express, Howard Johnson, Lariat Motel, Lazy J Motel, Luxury Inn, Ponderosa Motel, Sunset Motel, Super 8, Chevrolet/GMC/Olds, I-40 RV Park, Horizon Drug, IGA Foods, auto parts

187 NM 344, Edgewood, **N**...Citgo, Conoco/mart/LP, Dairy Queen, **S**...MEDICAL CARE, Fina/mart, Phillips 66/ diesel/mart, Shamrock/Stuckey's, BBQ Man, Homestead Rest., Tastee Freez, ExpressLube, Ford, Red Arrow Camping, Smith's Food/drug, USPO

181 NM 217, Sedillo, no facilities

178 Zuzax, **S**...Chevron/diesel/mart/tires, Hidden Valley RV Park

E

W

New Mexico Interstate 40

175 NM 337, NM 14, Tijeras, **N**...Burger Boy(2mi), to Cibola NF, Turquoise Trail RV Park, **1 mi E**...auto/truck repair/24hr

170 Carnuel, no facilities

167 Central Ave, to Tramway Blvd, **S**...Chevron/mart/24hr, Fina/7-11, Phillips 66, Texaco/Burger King/mart, Einstein Bros Bagels, KFC, Lotaburger, McDonald's, Taco Bell, Waffle House, Best Western, Comfort Inn, Day's Inn, Deluxe Inn, Howard Johnson, Motel 6, Travelodge, GNC, Goodyear/U-Haul/auto, KOA, Meyer's RV, Raley's Food/drug, Rocky Mtn RV/marine, Smith's Food/drug, XpressLube, laundry, to Kirtland AFB

166 Juan Tabo Blvd, **N**...VETERINARIAN, Chevron, Phillips 66/mart/wash, Texaco, Blackeyed Pea, Burger King, Carrow's Rest., LJ Silver, McDonald's, Olive Garden, Oriental Buffet, Paul's Rest., Pizza Hut, Taco Bell, Twisters Diner, Village Inn Rest., Whataburger, Wendy's, Best Inn, Super 8, Albertson's, BrakeMasters, Discount Tire, Hastings Books, HobbyLobby, Midas Muffler, PepBoys, Radio Shack, Sav-On Drug, Valvoline, transmissions, **S**...Wienerschnitzel, Furr's Foods, Walgreen, KOA, bank

165 Eubank Blvd, **N**...CHIROPRACTOR, Chevron, Phillips 66, Shamrock/mart, Texaco/mart, Circle K, JB's, Owl Café, Sonic, Day's Inn, Econolodge, Holiday Inn Express, Howard Johnson, Ramada Inn, Rodeway Inn, Best Buy, Target, **S**...Conoco, Burger King, Bob's Burgers, Boston Mkt, Taco Bell, Wendy's, HomeBase, Office Depot, PetsMart, Sam's Club, Wal-Mart SuperCtr/24hr, auto repair

164c Lomas Blvd(from wb), **N**...Texaco, Golden Corral(3mi), Rodeway Inn

  b a Wyoming Blvd, **N**...HOSPITAL, VETERINARIAN, Phillips 66, **S**...Ford, Dodge, Mazda/Toyota, Subaru, Kirtland AFB

162b a Louisiana Blvd, **N**...EYECARE, Amoco/mart, Bennigan's, Macaroni Grill, Japanese Rest., Steak&Ale, TGIFriday, Best Western, Marriott, Bed Bath&Beyond, Dillard's, JC Penney, Marshall's, Walgreen, Ward's/auto, cinema, **S**...atomic museum

161b a San Mateo Blvd, Albuquerque, **N**...VETERINARIAN, Phillips 66, Texaco/A&W/mart, Boston Mkt, Burger King, Denny's, K-Bob's Rest., Kettle, McDonald's, Starbucks, Wendy's, La Quinta, Circuit City, CompUSA, Linens'n Things, Old Navy, **S**...Plateau/gas

160 Carlisle Blvd, Albuquerque, **N**...Shell/mart, Texaco/mart, Conoco, Fina/mart, Giant/mart, 76, Circle K, Baskin-Robbins, JB's, Lotaburger, Pizza Hut, Rudy's BBQ, Sonic, Village Inn Rest., Whataburger, Budget Inn, Candle-wood Suites, Comfort Inn, Courtyard, Econolodge, Fairfield Inn, Hampton Inn, Hilton, Homestead Village, Quality Inn, Radisson, Residence Inn, Rodeway Inn, Sumner Suites, Super 8, Travelodge, BMW, Goodyear/auto, JC Penney, OfficeMax, Smith's/drugs, Walgreen, auto parts, **S**...HOSPITAL, Texaco/Subway/diesel/mart/atm/LP, Circle K, Burger King, Holiday Inn, Motel 6, K-Mart/auto, Wild Oats Mkt, auto supply

159b c I-25, S to Las Cruces, N to Santa Fe

  a 2nd St, 4th St, Albuquerque, **N**...CHIROPRACTOR, Chevron/mart/24hr, Conoco/mart, Love's/Subway/diesel/mart, Furr's Café, General Tire, Hi Mkt, U-Haul, **S**...Tony's Pizza, Village Inn Rest., Howard Johnson, Interstate Inn, Firestone

158 6th St, 8th St, 12th St, Albuquerque, **N**...Love's/Subway/diesel/mart, U-Haul, **S**...Chevron/mart, Interstate Inn

157b 12th St(from eb), no facilities

  a Rio Grande Blvd, Albuquerque, **N**...Shamrock/mart, Texaco/Burger King/mart, JR's Seafood, **S**...Chevron/mart/24hr, Texaco/mart/wash, Best Western Rio Grande, Sheraton

156mm Rio Grande River

155 Coors Rd, Albuquerque, **N**...CHIROPRACTOR, VETERINARIAN, Chevron/mart/wash, Giant/diesel/mart/24hr, Phillips 66, Shamrock/diesel/mart/24hr, Applebee's, Arby's, Burger King, Capt D's, Cuco's Kitchen/24hr, McDonald's, Ristorante Pizzaria, Wendy's, Furr's Foods, Goodyear/auto, JiffyLube, Mail Service, Midas Muffler, Radio Shack, Walgreen, ZipLube, bank, cinema, **S**...Chevron/mart/24hr, Conoco/mart, Giant/mart, Phillips 66/diesel/mart, Texaco/mart, Arby's, Denny's, Furr's Café, Lotaburger, McDonald's, New China, Pizza Hut, Subway, Taco Bell, Village Inn Rest., Comfort Inn, Day's Inn, Holiday Inn Express, La Quinta, Motel 6, Motel 76, Red Roof Inn, Super 8, Checker Parts, Discount Tire, U-Haul, carwash, laundry

154 Unser Blvd, **N**...Shamrock/mart, to Petroglyph NM

153 98th St, **S**...Flying J/Conoco/Country Mkt/diesel/LP/rest./24hr, Tumbleweed Steaks(2mi), Microtel

149 Central Ave, Paseo del Volcan, **N**...Enchanted Trails RV Camping, to Shooting Range SP, **S**...Chevron/diesel/mart/24hr, Shamrock/diesel/mart, Tumbleweed Steaks(2mi), American RV Park

140.5mm Rio Puerco River

140 Rio Puerco, **N**...Exxon/diesel/mart, **S**...Citgo/Dairy Queen/Stuckey's

E ↑

W ↓

Albuquerque

New Mexico Interstate 40

Grants

131	Canoncito, no facilities
126	NM 6, to Los Lunas, no facilities
120mm	Rio San Jose
117	Mesita, no facilities
114	NM 124, Laguna,**1/2 mi N**...gas, food
113.5mm	scenic view both lanes, litter barrels
108	Casa Blanca, Paraje, **S**...Conoco/Casa Blanca/diesel/mart/24hr, Casa Blanca Mkt, casino
104	Cubero, Budville, no facilities
102	Acomita, **N**...HOSPITAL, Acoma/diesel/mart/24hr, Huwak'a Rest., RV Park/laundry, casino, **S...rest area both lanes, full(handicapped)facilities, phone, picnic tables, litter barrels, petwalk**
100	San Fidel, no facilities
96	McCartys, no facilities
89	NM 117, to Quemado, **N**...Citgo/Stuckey's, **S**...El Malpais NM
85	NM 122, NM 547, Grants, **N**...HOSPITAL, Chevron/diesel/mart/24hr, Conoco/diesel/mart, Pump'n Save/gas, Texaco/diesel/mart/24hr, Domino's, 4B's Rest./24hr, House of Pancakes, Subway/TCBY, Best Western, Comfort Inn, Day's Inn, Econolodge/rest., Holiday Inn Express, Motel 6, Sands Motel, Super 8, Travelodge, AutoZone, Checker Parts, Chevrolet/Buick, Chrysler/Plymouth/Dodge, Tire Corral/brakes, Wal-Mart SuperCtr, **S**...Lavaland RV Park
81b a	NM 53 S, Grants, **N**...Chevron/mart/atm24hr, Fina, Bobbie's Kitchen, Burger King, KFC, McDonald's, Pizza Hut, Sands Motel, SW Motel, Ford/Lincoln/Mercury, NAPA, auto/RV repair, USPO, **S**...Blue Spruce RV Park, Cibola Sands RV Park, El Malpais NM
79	NM 122, NM 605, Milan, **N**...Love's/Subway/diesel/mart/24hr, Dairy Queen, Crossroads Motel, Bar-S RV Park, Milan Foods, truckwash, **S**...Petro/Mobil/diesel/rest./24hr, diesel repair, st police
72	Bluewater Village, **N**...Citgo/Dairy Queen/diesel
63	NM 412, Prewitt, **N**...Conoco/mart, **S**...to Bluewater SP, Grants West RV Camp
53	NM 371, NM 612, Thoreau, **N**...Red Mtn Mkt&Deli, St Bonaventure RV camp
47	Continental Divide, 7275 ft, **N**...Chevron/mart, Continental Divide Trdg Post, **S**...USPO
44	Coolidge, no facilities
39	Refinery, **N**...Giant/Conoco/A&W/Pizza Hut/Taco Bell/diesel/24hr, **S...rest area both lanes, full(handicapped)facilities, phone, picnic tables, litter barrels, petwalk,** RV camp
36	Iyanbito, no facilities
33	NM 400, McGaffey, Ft Wingate, **N**...to Red Rock SP, RV camping, museum

Gallup

26	E 66th Ave, E Gallup, **N**...Chevron/diesel/mart/24hr, Denny's/24hr, Sleep Inn, to Red Rock SP, RV parking/museum, st police, **S**...HOSPITAL, Conoco/diesel/mart, Fina/diesel/mart/atm, Mustang/mart, Texaco/mart, Burger King, Dunkin Donuts/24hr, KFC, Lotaburger, McDonald's, Subway, Wendy's, Best Western, Hacienda Motel, Super 8, Roadrunner Motel, Ortega Gifts
22	Montoya Blvd, Gallup, **N...rest area both lanes, full facilities, info, S...**Chevron, Conoco, Giant Gas/mart, Phillips 66, Shell, Texaco, Circle K, 7-11, Baskin-Robbins, Burger King, Carl's Jr, Chinese Rest., Church's Chicken, Earl's Rest., Kristy's CoffeeShop, LJ Silver, McDonald's, Miller's Rest., Pedro's Mexican, Pizza Hut, Subway, Taco Bell, Wendy's, Arrowhead Lodge, Best Western, Blue Spruce Motel, El Capitan Motel, El Rancho Motel/rest., Lariat Lodge, Redwood Lodge, Albertson's/drugs, Master Tune, TG&Y, Walgreen, tires
20	US 666 N, Gallup, to Shiprock, **N**...VETERINARIAN, Chevron, Giant/diesel/mart/24hr, Malco/gas, Shell/mart, Texaco/mart, Arby's, Big Cheese Pizza, Burger King, Church's Chicken, Cracker Barrel, Dairy Queen, Furr's Café, KFC, LotaBurger, McDonald's, Pizza Hut, Sizzler, Sonic, Taco Bell, Wendy's, Ramada Ltd, Holiday Inn Express, AutoZone, CarQuest, Checker Parts, Chrysler/Plymouth/Dodge/Jeep, Family$, Firestone, JC Penney, JiffyLube, K-Mart, Lube'n Tune, MailBoxes Etc, Midas Muffler, NAPA, Nissan, PepBoys, PriceRite Foods, Safeway, Sears, Wal-Mart SuperCtr/24hr, carwash, cinema, laundry, mall, radiators, **S**...HOSPITAL, Allsup's/mart, Conoco/diesel, Texaco, El Dorado Rest., Little Caesar's, Lotaburger, Sonic, Best Western, Day's Inn, Economy Inn, Royal Holiday Inn, Shalimar Inn, Super 8, Thunderbird Motel, Travelodge, RV camping
16	NM 118, W Gallup, Mentmore, **N**...Love's/Subway/diesel/mart/24hr, Shell/TA/Blimpie/diesel/24hr, Texaco/Pizza Inn/diesel/mart/24hr, A&W, Howard Johnson, Blue Beacon, diesel repair, **S**...Chevron/mart, Conoco, Exxon, Fina/Allsup's, GasMan/mart, Phillips 66, Texaco/mart, Ranch Kitchen Rest., Taco Bell, Best Western, Budget Inn, Comfort Inn, Day's Inn, Econolodge, Holiday Inn, Microtel, Motel 6, Red Roof Inn, Travelers Inn, Travelodge, KOA
12mm	inspection sta eb
8	to Manuelito, no facilities
2mm	**Welcome Ctr eb, full(handicapped)facilities, phone, picnic tables, litter barrels, petwalk**
0mm	New Mexico/Arizona state line

New York Interstate 81

Exit #(mm)Services

184mm	US/Canada border, New York state line. I-81 begins/ends.
183.5mm	US Customs(sb), Ammex Duty Free(sb)
52(183)	Island Rd, to De Wolf Point, last US exit nb, no facilities
51(180)	Island Rd, to Fineview, Islands Parks, **2-3 mi**...Citgo/diesel/mart, Sunoco, Thousand Islands Club, Seaway Island Resort, Torchlite Motel, Nature Ctr, camping, golf
179mm	St Lawrence River
178.5mm	Thousand Islands Toll Bridge Booth, **rest area sb, full(handicapped)facilities, phone, picnic tables, litter barrels, petwalk**
50NS(178)	**N**...NY 12, to Alexandria Bay, HOSPITAL, Mobil/diesel/mart, Kountry Kottage Rest., Subway, Bonnie Castle/Rest., Green Acres River Motel, Pinehurst Cottages, funpark(seasonal), to Thousand Island Region **S**...NY 12, to Clayton, VETERINARIAN, Citgo/diesel/mart, Mobil/mart/atm, Yazell's Rest., Bridgeview Motel, PJ's Motel, NAPA, to RV camping
174mm	**rest area nb, full(handicapped)facilities, phone, vending, picnic tables, litter barrels, petwalk, st police**
49(171)	NY 411, to Theresa, Indian River Lake, **E**...Sunoco/diesel/mart, **W**...6T's Diner
168mm	parking area sb, picnic table
161mm	parking area nb
48(158)	US 11, NY 37, **1-4 mi E**...Citgo/24hr, Mobil/mart, Sugar Creek/diesel/diner, Arby's, McDonald's, Longway's Diner, Allen's Budget Motel, Hotis Motel, Microtel, Royal Inn, st police
156.5mm	parking area both lanes
47(155)	NY 12, Bradley St, Watertown, **E**...HOSPITAL, Mobil/mart, Frosty Dairy Bar, Lighthouse Motel, The Maples Motel, **W**...Rainbow Motel, antiques
154.5mm	Black River
46(154)	NY 12F, Coffeen St, Watertown, **E**...Mobil/mart, Chappy's Diner, Cracker Barrel, Subway, Ryder Trucks, airport, cleaners, **W**...Citgo/mart
45(152)	NY 3, to Arsenal St, Watertown, **E**...MEDICAL CARE, DENTIST, Citgo/mart, Mobil/mart, Sunoco/mart, Applebee's, Arby's, Benny's Steaks, Burger King, China Café, Denny's, Dunkin Donuts, Friendly's, Jreck Subs, KFC, LJ Silver, McDonald's, Panda Buffet, Pizza Hut, Ponderosa, Smiley's Subs, Taco Bell, Wendy's, Best Western(2mi), Budget Inn, Econolodge, Day's Inn, The Inn, Advance Parts, Aldi Foods, AutoZone, Big Lots, Cole Muffler, $Tree, Eckerd, FashionBug, Jo-Ann Fabrics, Kost Tire/muffler, Midas Muffler, Monro Muffler/brake, Pearle Vision, Pennzoil, PriceChopper Foods/24hr, Radio Shack, Staples, TJ Maxx, USPO, cinema, **W**...Ann's Rest., Bob Evans, Red Lobster, Ramada Inn, Ames, Bon-Ton, Ford, Hannaford's Foods, JC Penney, K-Mart/Little Caesar's, Lowe's Bldg Supply, Old Navy, Pier 1, Sam's Club, Sears/auto, Wal-Mart, West Marine, bank, cinema, mall, to Sackets Harbor
149mm	parking area nb, phone
44(148)	NY 232, to Watertown Ctr, **3 mi E**...HOSPITAL, Hess/diesel, Mobil, Cityline Motel, Hi-Hat Motel, Hillside Motel, Holiday Inn, New Parrott Motel/rest., K-Mart/rest.
147mm	**rest area sb, full(handicapped)facilities, phone, picnic tables, litter barrels, vending, petwalk**
43(146)	US 11, to Kellogg Hill, no facilities
42(144)	NY 177, Adams Center, **E**...Mobil/mart, Harley-Davidson, Polaris
41(140)	NY 178, Adams, **E**...Citgo/diesel/mart/atm, McDonald's, **1 mi E**...Tomacy's Rest., Chevrolet/Buick, Pontiac, Chrysler/Jeep/Dodge, Ford, NAPA, laundry, **W**...Stony Creek Marina
138mm	South Sandy Creek
40(135)	NY 193, Pierrepont Manor, to Ellisburg, no facilities
134mm	parking area both lanes
39(133)	Mannsville, no facilities
38(131)	US 11, **E**...81-11 Motel
37(128)	Lacona, Sandy Creek, **E**...Lacona Gas/mart, J&R Diner, Harris Lodge, Lake Effect Inn, **W**...Citgo, Sunoco/diesel/mart, Angler's Roost, Snackery Rest., Subway, Salmon River Motel, KOA, Pennzoil, antiques, **4 mi W**...Colonial Camping
36(121)	NY 13(no immediate return sb or nb), Pulaski, **E**...Agway/diesel, Citgo/mart, C&M Diner, Ponderosa, Redwood Motel, Whitaker's Motel, **W**...DENTIST, Citgo, Mobil/diesel/mart, Sunoco/diesel/mart/atm, Arby's, Burger King, McDonald's, WaffleWorks, Country Pizza Motel, Salmon Acres Lodge, Super 8, Ames, Chevrolet/Olds/Buick, Kinney Drug, NAPA, Radio Shack, to Selkirk Shores SP, bank, camping, fish hatchery, **1 mi W on US 11**...Kwikfill, Dunkin Donuts, Log Cabin Inn, CarQuest, Rite Aid, Cole Muffler
35(118)	to US 11, Tinker Tavern Rd, no facilities
34(115)	NY 104, to Mexico, **E**...Sunoco/EZ/diesel/mart/24hr, **2-12 mi W**...Citgo, La Siesta Motel, Eis House, KOA, Salmon Country/Dowiedale/Yogi Camping

New York Interstate 81

Syracuse

33(111) NY 69, Parish, **E**...MEDICAL CARE, Sunoco/diesel/24hr, Grist Mill Rest./24hr, Wayne Drugs, **W**...Citgo/mart, Mobil/diesel/mart, Talk of the Town Rest., Montclair Motel/camping, East Coast Resort, Wooden Acres Camping(8mi)

32(103) NY 49, to Central Square, **E**...Mobil/diesel/mart, Sunoco/diesel/mart, Wilborn's Rest., NAPA AutoCare/lube, Pennzoil, Yamaha, **W**...BP/diesel/mart, Coastal/gas, Mobil, Arby's, Burger King, Quinto's NY Pizza, Pine Grove Motel, Ford, IGA Foods, NAPA, Rite Aid, bank

101mm **rest area sb, full(handicapped)facilities, phone, picnic tables, litter barrels, vending, petwalk**

31(99) to US 11, Brewerton, **E**...Oneida Shores Camping, **W**...CHIROPRACTOR, VETERINARIAN, Citgo/mart/atm, Mobil/diesel, Sunoco, Burger King, Castaways Rest., Little Caesar's, McDonald's, Sam's Lakeside Rest., Subway, BelAir Motel, Brewerton Motel, Holiday Inn Express, Kinney Drugs, bank, cleaners

30(96) NY 31, to Cicero, **E**...Hess/diesel/mart, Mobil, Sunoco/mart, Arby's, Cicero Country Pizza, Cracker Barrel, Denny's, Dunkin Donuts, McDonald's, Lamplighter Motel, **W**...MEDICAL CARE, Kwikfill/mart, BelAir Motel(2mi), Plainville Farms Rest., Dodge, RV Ctr, auto parts

29(93) I-481 S, Syracuse, NY 481, to Oswego, **1 mi W on US 11**...VETERINARIAN, Hess/mart, Mobil/Blimpie/mart, Burger King, Carmella's Café, Chinese Buffet, Cicero Diner, Dunkin Donuts, Friendly's, KFC, McDonald's, Papa John's, Perkins, Pizza Hut, Red Lobster, Tully's Rest., Rodeway Inn, Ames, Advance Parts, AutoZone, Burlington Coats, Chrysler/Plymouth, Dodge, Eckerd, Firestone/auto, Ford/Lincoln/Mercury, Goodyear/auto, Home Depot, NAPA, P&C Food/drug, PepBoys, PriceChopper Foods, Revco, Service Merchandise, Valvoline, VW/Porsche/Audi, bank, cinema, cleaners/laundry, mall

28(91) N Syracuse, Taft Rd, **E**...Sunoco, Eckerd, rental cars, **W**...Mobil/mart

27(90) N Syracuse, **E**...airport

26(89) US 11, Mattydale, **E**...DENTIST, Mobil/mart, Asian 98 Buffet, Doug's Fishfry, Friendly's, Paladino's Pizza, Zebb's Grill, Red Carpet Inn, Big Lots, Eckerd, Ford, Goodyear/auto, Kinko's, K-Mart/auto, MediaPlay, Michael's, PetCo, Staples, TJ Maxx, **W**...Hess/diesel/mart, Kwikfill, Sunoco, Applebee's, Burger King, Carvel Ice Cream, Clam Bar Seafood, Denny's, McDonald's, Ponderosa, Subway, Taco Bell, Wendy's, Airflite Motel, Aamco, Aldi Foods, Cole Muffler, Empire Vision Ctr, Lexus, Mazda, Midas Muffler, Olds/Toyota/VW, P&C Food/drug, USA Baby, bank, tires

25a(88) I-90, NY Thruway

25(87.5) 7th North St, **E**...Pilot/KFC/Subway/diesel/mart/atm/24hr, auto repair, **W**...Mobil/mart, Sunoco/diesel/service, Bob Evans, Burger King, Colorado Steaks, Denny's, Friendly's, Ground Round, Jreck Subs, Day's Inn, Econolodge, Hampton Inn, Holiday Inn, Quality Inn, Ramada, Super 8, bank

24(86) NY 370, to Liverpool, **W**...Hess/mart, Bon-Ton, JC Penney, Kaufmann's, mall

23(86) Hiawatha Blvd, industrial area, **W**...Hess/mart, Bon-Ton, JC Penney, mall, same as 25

22(85) NY 298, Court St, no facilities

21(84.5) Spencer St, Catawba St(from sb), industrial area

20(84) I-690 W(from sb), Franklin St, West St

19(84) I-690 E, Clinton St, Salina St, to E Syracuse

18(84) Harrison St, Adams St, **E**...Best Western, Holiday Inn, Sheraton, to Syracuse U, Civic Ctr, **W**...HOSPITAL

17(82) Brighton Ave, S Salina St, **E**...HOSPITAL, KwikFill, Carrier Dome, Syracuse U, **W**...Mobil

16a(81) I-481 N, to DeWitt

16(78) US 11, Onondaga Nation, to Nedrow, **1-2 mi W**...Citgo, Hess, Mobil, McDonald's, Pizza Hut, Smoke Signals Diner

15(73) US 20, La Fayette, **E**...DENTIST, Mobil/diesel/pizza/subs/mart, Nice'n Easy Deli, NY Pizza, IGA Foods, NAPA, USPO, bank, cleaners, st police, **W**...McDonald's

70mm truck insp area both lanes, phones

14(67) NY 80, Tully, **E**...Mobil/diesel/mart, Best Western/rest., Nice'n Easy Deli, bank, **W**...Burger King

13(63) NY 281, Preble, **E**...Sunoco/diesel/mart, to Song Mtn Ski Resort

60mm **rest area nb, full(handicapped)facilities, phone, picnic tables, litter barrels, vending, petwalk**

Cortland

12(53) US 11, NY 281, to Homer, **W**...HOSPITAL, BP/diesel, Citgo, Mobil/diesel/24hr, Burger King, Coachlite Family Rest., Little Italy Pizzaria, Rusty Nail Rest., Budget Motel, to Fillmore Glen SP

11(52) NY 13, Cortland, **E**...Denny's, Comfort Inn, Super 8, **W**...Mobil/mart, Arby's, Bob Evans, China Moon, Friendly's, Little Caesar's, McDonald's, Subway, Taco Bell, Wendy's, Holiday Inn, Eckerd, Family$, JiffyLube, Jo-Ann Fabrics, Kost Brake/muffler, PartsAmerica, P&C Foods/24hr, Yellow Lantern Camping, museum

10(50) US 11, NY 41, to Cortland, McGraw, **W**...Agway/diesel, Citgo/Burger King/Pizza Hut/diesel/mart, Mobil/Subway/diesel/mart/24hr, Sunoco/diesel/mart, Cracker Barrel, Skyliner Diner, Econolodge, Evergreen Motel

45mm parking area nb, picnic tables

9(38) US 11, NY 221, **W**...Citgo/mart, Sunoco/diesel/24hr, NY Pizzaria, Taco Express, 3 Bear Inn/rest., Greek Peak Lodge, Country Hills Camping

33mm **rest area sb, full(handicapped)facilities, phone, picnic tables, litter barrels, vending, petwalk**

8(30) NY 79, to US 11, NY 26, NY 206(no EZ return), Whitney Pt, **E**...Hess/mart, Kwikfill/gas, Mobil/diesel/mart/24hr, Subway, Aiello's Pizzaria, Arby's, Country Kitchen/24hr, Point Motel, Chevrolet, NAPA, Parts+, Radio Shack, Strawberry Valley Farms(3mi), bank, to Dorchester Park

N

S

New York Interstate 81

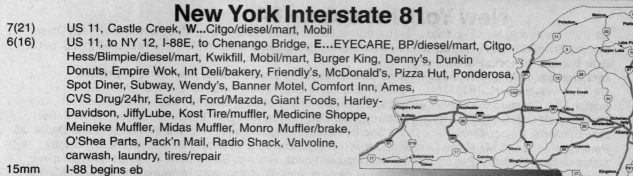

7(21)   US 11, Castle Creek, **W**...Citgo/diesel/mart, Mobil

6(16)   US 11, to NY 12, I-88E, to Chenango Bridge, **E**...EYECARE, BP/diesel/mart, Citgo, Hess/Blimpie/diesel/mart, Kwikfill, Mobil/mart, Burger King, Denny's, Dunkin Donuts, Empire Wok, Int Deli/bakery, Friendly's, McDonald's, Pizza Hut, Ponderosa, Spot Diner, Subway, Wendy's, Banner Motel, Comfort Inn, Ames, CVS Drug/24hr, Eckerd, Ford/Mazda, Giant Foods, Harley-Davidson, JiffyLube, Kost Tire/muffler, Medicine Shoppe, Meineke Muffler, Midas Muffler, Monro Muffler/brake, O'Shea Parts, Pack'n Mail, Radio Shack, Valvoline, carwash, laundry, tires/repair

15mm   I-88 begins eb

14mm   **rest area nb, full facilities, picnic tables, phones, free coffee**

5(14)   US 11, Front St, **1 mi W**...Mobil/diesel, Day's Inn, Howard Johnson, Motel 6, Super 8

4(13)   NY 17, Binghamton, **W**...BP, Econolodge

3(12)   Broad Ave, Binghamton, **W**...BP/mart, KFC, Papa John's, CVS Drug, Giant Foods

2(8)   US 11, NY 17, **2-3 mi W**...BP/diesel/mart, Hess/diesel, Lechner's/gas, Sunoco/mart, TravelPort/diesel/rest.; Arby's, Burger King, Far Out Pizza, McDonald's, Subway, Econolodge, Foothills Motel, Super 8, Thruway Motel, Midas Muffler, auto parts

1(4)   US 11, NY 7, Kirkwood, **1-2 mi W**...Citgo, Mobil/diesel/24hr, Texaco/diesel, Kirkwood Motel, golf

2mm   **rest area nb, full(handicapped)facilities, phone, picnic tables, litter barrels, vending, petwalk**

1mm   weigh sta nb

0mm   New York/Pennsylvania state line

New York Interstate 84

Exit #(mm)Services

71.5mm   New York/Connecticut state line

21(68)   US 6, US 202, NY 121(from wb), N Salem, same as 20

20(67.5)   I-684, US 6, US 202, NY 22, **N**...ATI/diesel, Mobil/mart/24hr, Shell/mart, Bob's Diner/24hr, Burger King, Cracker Barrel, Keltie's Steak/seafood, McDonald's, BelAire Motel, Fox Ridge Motel, Heidi's Motel, Henry Van Motel, Ford

19(65)   NY 312, Carmel, **3-4 mi S**...HOSPITAL, Shell/24hr, Court Rest./24hr, Friendly's, KFC, McDonald's, Roy Rogers, Wendy's, st police

18(62)   NY 311, Lake Carmel, **2-4 mi S**...Bob's Towing/gas/diesel, Mobil, Texaco, Burger King, Villa Carmel Rest., Hollywood Motel

17(59)   Ludingtonville Rd, **S**...Hess/diesel/mart/24hr, Sunoco/24hr, Lou's Deli, Tina's Rest.

56mm   elevation 970 ft

55mm   **rest area both lanes, full(handicapped)facilities, phone, vending, picnic tables, litter barrels, petwalk**

16(53)   Taconic Parkway, N to Albany, S to New York, no facilities

15(51)   Lime Kiln NY, 3 mi **N**...Mobil/mart/24hr, Dunkin Donuts, Olympic Diner, Royal Inn

13(46)   US 9, to Ploughkeepsie, **N**...Mobil/diesel/mart, Boston Mkt, Burger King, Charley Brown's Steaks, Cracker Barrel, Denny's, Pizza Hut, Taco Bell, Wendy's, Courtyard, Hampton Inn, Holiday Inn/Stanley's Eatery, MainStay Suites, Residence Inn, Wellesley Inn, Drug World, Sam's Club, Wal-Mart SuperCtr/24hr, cinema, **S**...Hess/diesel/mart/24hr, Blue Dolphin Rest.(1mi), Bridle Ridge Rest., McDonald's, Countryside Motel(2mi), bank

12(45)   NY 52 E, Fishkill, **N**...Coastal/diesel/mart, Gulf, Friendly's, **S**...Shell/mart/24hr, Sunoco/diesel, Dunkin Donuts, Hometown Deli, I-84 Diner/24hr, KFC

11(42)   NY 9D, to Wappingers Falls, **1 mi N**...Citgo/mart/24hr, Mobil/diesel, Sunoco/mart

41mm   toll booth wb

40mm   Hudson River

10(39)   US 9W, NY 32, to Newburgh, **N**...Citgo/diesel/24hr, Mobil, Texaco, Blimpie, Burger King, Lexus Diner/24hr, McDonald's, Perkins, Knight's Inn, Balmville Motel, **S**...HOSPITAL, Exxon/diesel/mart/24hr, Sunoco/diesel/mart/24hr, DairyMart, Subway, Green Valley Motel, Havarest Motel, Imperial 400 Motel, Newburgh Motel, Jo-Ann Fabrics

8(37)   NY 52, to Walden, **N**...Citgo/diesel, Sunoco/24hr, **S**...Gulf/mart

7(36)   NY 300, Newburgh, to I-87(NY Thruway), **N**...MEDICAL CARE, Exxon/24hr, Mobil/mart/24hr, CB Driscoll's Rest., Dunkin Donuts, King Buffet, McDonald's, Perkins, Taco Bell, Wendy's, AutoZone, Bon-Ton, KOA, Mavis Tire, Midas Muffler, PetCo, Sears, Stop'n Shop Foods, Valvoline, Weis Foods, bank, cinema, mall, **S**...CHIROPRACTOR, Citgo, Getty Gas, Sunoco/mart, Applebee's, Burger King, Cosimos Ristorante, Denny's, Hibachi Steaks, Neptune Diner, Yobo Asian, Day's Inn, Hampton Inn, Holiday Inn, Ramada Inn/rest., Super 8, Adams Farms Foods, Auto Mall/dealers, Home Depot, Meineke Muffler, Nissan, Wal-Mart SuperCtr/24hr, bank

6(34)   NY 17K, to Newburgh, **N**...Mobil/mart/24, Comfort Inn, Hampton Inn, **S**...Exxon/diesel/mart, Bob Evans, Courtyard, **3 mi S**...Getty Gas, Burger King, Johnny D's Diner/24hr, McDonald's, Neptune Diner, Day's Inn, Holiday Inn, Howard Johnson/rest.

Binghamton

Newburgh

Middletown Pt Jervis

New York Interstate 84

5(29) NY 208, Maybrook, **N**...Citgo/mart/24hr, Exxon/mart/24hr, Mobil, Burger King, McDonald's, Maybrook Diner, Eckerd, ShopRite Food/Café, bank, **S**...Stewarts Shop/gas, Sunoco/diesel/rest., Pizza Hut, DayStop, Super 8, Blue Beacon, Winding Hills Camping

24mm **rest area wb, full(handicapped)facilities, phone, vending, picnic tables, litter barrels, petwalk**

4(19) NY 17, Middletown, **N**...HOSPITAL, PODIATRIST, Getty Gas, Mobil/mart/24hr, Sunoco, Americana Diner, Boston Mkt, Cosimo's Brick Oven, Denny's, Dunkin Donuts, Friendly's, Galleria Rest., KFC, McDonald's, Olive Garden, Red Lobster, Ruby Tuesday, Subway, Taco Bell, Howard Johnson, Middletown Motel, Super 8, Best Buy, Bradlee's, Caldor, Circuit City, Dick's Sports, Filene's, Firestone/auto, Hannaford's Foods, Home Depot, JC Penney, JiffyLube, K-Mart/Little Caesar's, Linens'n Things, Lowe's Bldg Supply, MailBoxes Etc, MediaPlay, Midas Muffler, Old Navy, PetCo, Pier 1, PriceChopper Foods, Rite Aid, Sam's Club, ShopRite Food, Stop'n Shop Foods, TJ Maxx, U-Haul, Wal-Mart, bank, cinema, mall, **S**...Citgo/mart, Mobil/mart, El Bandido Mexican, Rusty Nail Rest., Holiday Inn, st police

17mm **rest area eb, full(handicapped)facilities, phone, vending, picnic tables, litter barrels, petwalk**

3(15) US 6, to Middletown, **N**...HOSPITAL, Citgo/diesel/24hr, Mobil/mart, Texaco/mart/24hr, Burger King, Cancun Mexican, Chipper's Steak/seafood, McDonald's, Perkins, Pizza Hut, Ponderosa, 6-17 Diner, Taco Bell, Wendy's, Acura/Honda, Chevrolet/Isuzu, Mazda, Subaru, Toyota, bank, cinema, **S**...84 Kwikstop/gas, Sunoco/diesel, Day's Inn, Global Budget Inn, Chrysler/Jeep, Nissan/BMW/Suzuki, auto repair, transmissions

2(5) Mountain Rd, no facilities

4mm elevation 1254 ft wb, 1272 ft eb

3mm parking area both lanes

1(1) US 6, NY 23, Port Jervis, **N**...Sunoco, Marie&Tom's Diner, Dunkin Donuts, Homer's Rest., KFC, Deerdale Motel, Painted Aprons Motel, Ford/Lincoln/Mercury, **S**...HOSPITAL, Citgo/diesel/mart, Gulf/diesel, Xtra/diesel/ mart, Dairy Queen, McDonald's, Ponderosa, Comfort Inn, Country Inn(6mi), PetStore, ShopRite Foods, cleaners, mall

0mm Delaware River, New York/Pennsylvania state line

New York Interstate 87

Plattsburgh

Exit #(mm)Services

176mm US/Canada Border, NY state line, I-87 begins/ends.

43(175) US 9, Champlain, **E**...Ammex Duty Free, **W**...GMC, Peterbilt, Fruehoff, diesel repair

42(174) US 11 S, Champlain, to Rouse's Point, **E**...Citgo/mart, Mobil/Pizza+/mart, Art's Family Rest., Chinese Rest., Peppercorn Rest., AmCan Motel, Ames, Chevrolet(2mi), Grand Union Foods, Kinney Drug, Rite Aid, bank, **W**...Mobil/diesel/mart/atm, Neverett Bros/diesel, PetroCanada/Dunkin Donuts/diesel/mart, Sunoco/diesel/mart/ 24hr, Burger King, McDonald's, Ammex, Miromar Outlet/famous brands

41(167) NY 191, Chazy, **E**...st police, **W**...Miner Institute

162mm **rest area both lanes, full(handicapped)facilities, info, phone, picnic tables, litter barrels, petwalk**

40(160) NY 456, Beekmantown, **E**...Mobil/mart, Stonehelm Motel/café, **W**...Twin Elks Camping

39(156) NY 314, Moffitt Rd, Plattsburgh Bay, **E**...Atlantic/24hr, Stewart's/gas, Domenic's Rest., Gus' Rest., McDonald's, Chateau Motel, Rip van Winkle Motel, Sundance Motel, Super 8, A&P, Plattsburgh RV Park, hardware, **W**...Shady Oaks Camping, to Adirondacks

38(154) NY 22, NY 374, to Plattsburgh, **E**...Atlantic/24hr, Mobil, **W**...Buck's/diesel/rest.

37(153) NY 3, Plattsburgh, **E**...HOSPITAL, Mobil/mart, ShortShop/gas, Sunoco/mart/24hr, Bootleggers Rest., Burger King, Dairy Queen, Domino's, Dunkin Donuts, IHOP, Jade Buffet, KFC, Little Caesar's, Mangia Pizza, McDonald's, Pizza Hut, Silver Star Café, Subway, Wendy's, Comfort Inn, Holiday Inn, Ames, Big Lots, Buick, Family$, Firestone/auto, Ford, Grand Union Foods, Honda, Isuzu, Jo-Ann Crafts, Kinney Drug, Olds/Cadillac/ GMC, P&C Foods, Pearle Vision, Pontiac, Radio Shack, Sam's Club, Staples, TJ Maxx, TrueValue, Vision Ctr, Wal-Mart, **W**...DENTIST, MEDICAL CARE, Exxon/mart, PetroCanada/mart, Sunoco/diesel/mart, Anthony's Bistro, Barkin' Dog Rest., Friendly's, Ground Round, Lum's, Ponderosa, Red Lobster, Baymont Inn, Day's Inn, Econolodge, Howard Johnson/rest., Travelers Inn, Advance Parts, AutoZone, Eckerd, Henry's Lube, JC Penney, JiffyLube, K-Mart, Midas Muffler, Monro Muffler/brake, PriceChopper Food/24hr, Sears/auto, USPO, bank, cinema, to airport

151mm Saranac River

36(150) NY 22, Plattsburgh AFB, **E**...Mobil/diesel/mart/24hr, Burger King(4mi), **W**...st police

35(144) NY 442, Peru, to Port Kent, **2-8 mi E**...Valcour Lodge, Iroquois/Ausable Pines Camping, **W**...Sunoco/diesel/ mart, **1 mi W**...Citgo/SugarCreek Diner, Mobil, Leaning Pine Rest., McDonald's, Grand Union Foods, Buirchwood Camping(2mi)

143mm emergency phones at 2 mi intervals begin sb/end nb

34(137) NY 9 N, Ausable Chasm, **E**...VETERINARIAN, Sunoco/diesel, Pleasant Corner Dining, Tastee-Freez, **W**...Family Produce, Ausable River RV Camping

136mm Ausable River

33(135) US 9, NY 22, to Willsboro, **E**...Mobil/diesel, Chesterfield Motel, RV camping

125mm N Boquet River

New York Interstate 87

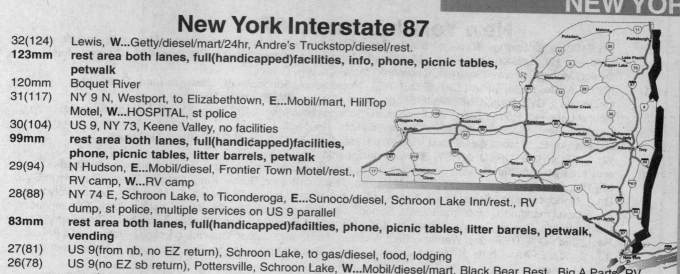

32(124)   Lewis, **W**...Getty/diesel/mart/24hr, Andre's Truckstop/diesel/rest.

123mm   **rest area both lanes, full(handicapped)facilities, info, phone, picnic tables, petwalk**

120mm   Boquet River

31(117)   NY 9 N, Westport, to Elizabethtown, **E**...Mobil/mart, HillTop Motel, **W**...HOSPITAL, st police

30(104)   US 9, NY 73, Keene Valley, no facilities

99mm   **rest area both lanes, full(handicapped)facilities, phone, picnic tables, litter barrels, petwalk**

29(94)   N Hudson, **E**...Mobil/diesel, Frontier Town Motel/rest., RV camp, **W**...RV camp

28(88)   NY 74 E, Schroon Lake, to Ticonderoga, **E**...Sunoco/diesel, Schroon Lake Inn/rest., RV dump, st police, multiple services on US 9 parallel

83mm   **rest area both lanes, full(handicapped)facilties, phone, picnic tables, litter barrels, petwalk, vending**

27(81)   US 9(from nb, no EZ return), Schroon Lake, to gas/diesel, food, lodging

26(78)   US 9(no EZ sb return), Pottersville, Schroon Lake, **W**...Mobil/diesel/mart, Black Bear Rest., Big A Parts, RV camping

25(73)   NY 8, Chestertown, **W**...Mobil, Sunoco/diesel, Adirondack Rest., RV camping

24(67)   Bolton Landing, **E**...RV camping

66mm   Schroon River

65mm   parking area sb, picnic tables, no facilities

63mm   parking area nb, picnic tables, no facilities

23(58)   to US 9, Diamond Point, Warrensburg, **W**...Citgo/diesel/mart, Mobil/mart, McDonald's, Super 8, Ford/Mercury, hardware, repair, RV camping, Central Adirondack Tr, ski area

22(54)   US 9, NY 9 N, Lake George, to Diamond Pt, **E**...Citgo, Mobil, China Wok, Dunkin Donuts, Guiseppe's Rest., Jasper's Steaks, Lobster Pot, Luigi's Italian, Mario's Italian, McDonald's, Mr B's Subs, Pizza Hut, Subway, Taco Bell, Trattoria Siciliano, Trolley Steaks, Balmoral Motel, Balsam Motel, Blue Moon Motel, Brookside Motel, Cedarhurst Motel, Econolodge, Ft Henry Resort, Georgian Lodge, Heritage Motel, Knight's Inn, Lakecrest Motel, Lakehaven Motel, Mohawk Cottages, Motel Montreal, Nordick's Motel/rest., Oasis Motel, O'Sullivan's Motel, 7 Dwarfs Motel, Sundowner Motel, Surfside Motel, Windsor Lodge, Rexall Drug, **W**...parking area both lanes

21(53)   NY 9 N, Lake Geo, Ft Wm Henry, **E on US 9**...Sunoco, A&W, Barnsider Smokehouse, Dairy Queen, Mama Riso's Italian, Mountaineer Rest., Prospect Mt Diner, Best Western, Colonial Manor, Comfort Inn, Holiday Inn, Holly Tree Inn, Howard Johnson/rest., Nomad Motel, Northland Motel, Ramada Inn, funpark, **W**...Mobil/diesel/mart/LP, Kathy's Motel

51mm   Adirondack Park

20(49)   NY 149, to Ft Ann, **E**...Mobil/Subway/mart/atm, Shell/diesel/mart, BrickOven Pizza, Frank's Italian, Logjam Rest., Meeting Place Rest., Montcalm Rest., Day's Inn, French Mtn Motel, Mohican Motel, Rodeway Inn, Ledgeview RV Park(3mi), Whipporwill/King Phillip/Lake George Camping(2mi), Factory Outlets/famous brands, funpark, st police

19(47)   NY 254, Glens Falls, **E**...MEDICAL CARE, Citgo, Hess/Blimpie/mart, Mobil/mart, Sunoco/mart, Bruebagger's Bagels, Burger King, Dunkin Donuts, Fazoli's, FoodCourt, Friendly's, Ground Round, KFC, McDonald's, Old China Buffet, Olive Garden, Pizza Hut, Red Lobster, 7 Steers Grill, Silo Rest., Taco Bell, Wendy's, Econolodge, Welcome Inn, Advance Parts, AutoZone, CVS Drug, Eckerd, Firestone/auto, JC Penney, Jo-Ann Crafts, Monro Muffler/brake, Pearle Vision, Rex, Rite Aid, Sears, Staples, TJ Maxx, TrueValue, Valvoline, cinema, mall, **W**...Mobil/mart, Ramada/rest., st police

18(45)   Glens Falls, **E**...Citgo/mart, Gulf/Subway/mart/24hr, Hess/diesel/mart/24hr, Mobil/24hr, Carl R's Café, Pizza Hut, Steve's Place Rest., Susse Chalet, Queensbury Hotel, U-Haul, **W**...McDonald's, Super 8

43mm   **rest area both lanes, full(handicapped)facilities, picnic tables, litter barrels, phone, vending, petwalk**

42mm   Hudson River

17(40)   US 9, S Glen Falls, **E**...Citgo, Mobil/diesel/24hr, Agway/diesel/LP, Beverly's Diner, Moreau Diner, Taco Maker, Wishing Well Rest., Landmark Motel, Sara-Glen Motel, Sunhaven Motel, Town&Country Motel, **W**...Moreau Lake SP, American RV Camping

16(36)   Ballard Rd, Wilton, **E**...Coldbrook Campsites, **W**...Mobil/mart, Stewart's/gas, Sunoco/Scotty's Rest./diesel/rest./24hr, MtnView Acres Motel, Ernie's Grocery

15(30)   NY 50, NY 29, Saratoga Springs, **E**...Hess/diesel/mart/24hr, Mobil/Blimpie/mart, Burger King, FoodCourt, McDonald's, Ponderosa, Ruby Tuesday, BJ's Whse, Bon-Ton, Dick's Sports, Dodge, Eckerd, Hannaford's Foods, JC Penney, K-Mart, PriceChopper Foods, Sears/auto, Service Merchandise, Staples, Subaru, Toyota, Wal-Mart, cinema, mall, **W**...HOSPITAL, Exxon/diesel, Birches Motel, Gateway Motel, Care Away Travel Park

14(28)   NY 9 T, Schuylerville, **2 mi E**...Mobil/Lakeside Mkt/deli, Bayshore Rest./marina, Waterfront Rest., Anchor Inn/rest., Longfellow Inn/rest., Saratoga Lake Inn, Saratoga Springs Motel, Lee's RV Park, museum, racetrack, **2 mi W**...Citgo/repair/LP

N

S

New York Interstate 87

13(25) US 9, Saratoga Springs, **E...**Chez Souffe Rest., ElmTree Rest., Locust Grove Motel, Unique Posthouse Motel, Northway RV, Ballston Spa SP, **W...**Mobil/mart, Sunoco/diesel/mart/24hr, Joe Collins Rest., Coronet Motel, Design Motel, Roosevelt Inn/rest., Thorobred Motel

12(21) NY 67, Malta, **E...**MEDICAL CARE, DENTIST, Mobil/diesel, Sunoco/mart, Stewart's Mart, Malta Diner, McDonald's, CVS Drug, GNC, Malta Shops, bank, st police, **W...**Saratoga NHP

11(18) Round Lake Rd, Round Lake, **W...**Mobil/mart, Sunoco/diesel/mart/24hr, Stewart's Mart, Adirondack Rest., CiderMill Rest., Good Times Rest., Grand Prix Motor Inn/rest.

10(16) Ushers Rd, **E...**Sunoco/diesel/mart, Ferretti's Rest., **3 mi E...**Parkwood Rest., Rusty Nail Rest., **W...**CHIROPRACTOR, DENTIST, StewartsMart, bank

14mm **rest area nb, full(handicapped)facilities, info, phone, picnic tables, litter barrels, vending, petwalk**

9(13) NY 146, Clifton Park, **E...**VETERINARIAN, Hess/diesel/mart/24hr, Burger King, Cracker Barrel, Pizza Hut, CVS Drug, Goodyear/auto, Grand Union Foods, JiffyLube, Midas Muffler, OfficeMax, TJ Maxx, TrueValue, Valvoline, **W...**Exxon/diesel/mart, Mobil/mart/wash, Sunoco/diesel, Applebee's, Boston Mkt, Chinese Rest., Conservatory Grill, Denny's, Dunkin Donuts, Friendly's, KFC, Manhattan Bagel, McDonald's, Snyder's Rest., Strawberrys Rest., Taco Bell, TGIFriday, Wendy's, Best Western, AutoZone, Caldor, Chevrolet, Eckerd, Hannaford's Foods, JC Penney, Jo-Ann Crafts, K-Mart, Marshall's, Pier 1, PriceChopper Foods, Rex, Steinback's, bank, cinema, mall, st police

8a(12) Grooms Rd, to Waterford, **2 mi E on US 9...**Getty/mart, Gulf/mart, °Moon Diner, Pusateres Mkt, Subway, cleaners

8(10) Crescent, Vischer Ferry, **E...**McDonald's, **3 mi E...**Krause's Deli, **W...**Mobil/Mr Subb/diesel/mart/24hr, Sunoco/mart/24hr, StewartsMart, Romano's Deli, CVS Drug, bank, carwash, cleaners

8mm Mohawk River

7(7) NY 7, Troy, **E on US 9...**HOSPITAL, Citgo, Mobil, McDonald's, Subway, Hampton Inn, Holiday Inn Express, Acura, Bradlee's, CompUSA, Dick's Sports, Ford, Hannaford's Foods, JC Penney, Latham Stores/famous brands, Lincoln/Mercury, Marshall's, Midas Muffler, NAPA, Nissan, Parts+, PetsMart, PriceChopper Foods, Sam's Club, Staples, Wal-Mart, bank, cinema, mall, transmissions

6(6) NY 2, to US 9, Schenectady, **E...**Mobil/mart, Circle Diner, Dakota Steaks, Ginza Chinese, Ground Round, Burlington Coats, Caldor, CVS Drug, Goodyear/auto, SteinMart, VW, same as 7, **E on US 9...**Getty, Boston Mkt, Burger King, Friendly's, Old Country Buffet, 76 Diner, Ponderosa, Taco Bell, Wendy's, Cocca's Inn, Audi, Firestone, Monro Muffler/brake, OfficeMax, TJ Maxx, Toyota, **W...**Mobil/mart/24hr, Bennigan's, McDonald's, Clarion, Microtel, Super 8

5(5) NY 155 E, Latham, **E...**Quality Inn/rest., st police, **W...**MEDICAL CARE

4(4) NY 155 W, Wolf Rd, **E on Wolf Rd...**Hess/diesel/mart, Mobil/mart, Sunoco/mart, Arby's, Ben&Jerry's, Burger King, ChiChi's, Denny's, Lexington Grill, Macaroni Grill, Maxie's Grill, McDonald's, Olive Garden, Outback Steaks, Pizza Hut, Ponderosa, Real Seafood Co, Subway, Weathervane Seafoods, Wolf Rd Diner, Best Western, Courtyard, Hampton Inn, Holiday Inn, Marriott, Red Roof Inn, Chevrolet, Ford/Lincoln/Mercury, Hannaford's Foods, Pearle Vision, Pier 1, Service Merchandise, bank, cleaners, **W...**Gateway Grill, Peony Oriental Rest., Desmond Hotel, Wingate Inn, to Heritage Park, airport

2(2) NY 5, Central Ave, **E on Wolf Rd...**Mobil/mart, Applebee's, Bangkok Thai, Cranberry Bog Rest., Dunkin Donuts, Friendly's, Ground Round, Honeybaked Ham, IHOP, LoneStar Steaks, Papa Gino's, Platt's Place Rest., Starbuck's, Cocca's Suites, Day's Inn, Susse Chalet, Barnes&Noble, Bed Bath&Beyond, BJ's Whse, Borders Books, Firestone/auto, Goodyear/auto, HobbyTown, Macy's, MensWearhouse, OfficeMax, Sears/auto, cinema, mall

1W(1) NY State Thruway(from sb), I-87 S to NYC, I-90 W to Buffalo

1E(1) I-90(from sb), E to Albany, Boston

1S(1) to US 20(from nb), Western Ave, **on Western Ave...**Tom Sawyer Inn, JC Penney, mall

1N(1) I-87(from nb), N to Plattsburgh

NY State Thruway goes west to Buffalo(I-90), S to NYC(I-87), I-87 N to Montreal

24(148) I-90 and I-87 N

23(142) I-787 to Albany, US 9 W, **E...**Big M Truckstop, Exxon, Mobil, Day's Inn, Howard Johnson, to Knickerbocker Arena, **W...**Stone Inn/diner

139mm parking area sb, phone, picnic tables, litter barrel

22(135) NY 396, to Selkirk, no facilities

21a(134) I-90 E, to MA Tpk, Boston

127mm **New Baltimore Travel Plaza sb, Mobil/diesel, Big Boy, Gourmet Bean, Mrs Fields, Roy Rogers, TCBY, atm, gifts, info, UPS**

21b(125) US 9 W, NY 81, to Coxsackie, **W...**Citgo/FoxRun/diesel/motel/rest./24hr, Sunoco/diesel/mart, McDonald's(5mi), repair

21(114) NY 23, Catskill, **E...**Mobil/mart, Sunoco/diesel/mart/24hr, Astoria Motel, Catskill Motel/rest., Day's Inn/rest., to Rip van Winkle Br, **W...**Anthony's Italian, Indian Ridge Camping, to Hunter Mtn/Windham Ski Areas

103mm **Malden Service Area nb, Mobil/diesel, Carvel Ice Cream/bakery, McDonald's, Uncle Rick's Store, parking area sb, phone**

A
l
b
a
n
y

N

S

New York Interstate 87

20(102) NY 32, to Saugerties, **E...**McDonald's, **W...**Hess, Comfort Inn, Howard Johnson/rest., to Catskills, KOA

99mm parking area nb, phone, picnic tables

96mm **Ulster Travel Plaza sb, Mobil/diesel, Big Apple Bagels, Nathan's, Roy Rogers, Mrs Fields, TCBY, gifts, phone**

19(91) NY 28, Kingston, **E...**BP/mart, Mobil/mart/atm, Blimpie, Friendly's, Gateway Diner, Grand Buffet Rest., Plaza Pizza, Holiday Inn, Super 8, Advance Parts, Ames, FashionBug, Grand Union Foods, Radio Shack, Walgreen, auto repair, **W...**Howard Johnson/rest., Knight's Inn, Ramada Inn/rest., Skytop Motel/steaks, Super Lodge, Buick, Camper's Barn RV, Ford, Nissan, bank, access to I-587, US 209

18(76) NY 299, New Paltz, to Ploughkeepsie, **E...**Citgo/repair, Mobil/mart, Cumberland Farms, Austrian Rest., China Buffet, College Diner/24hr, Day's Inn(1mi), Econolodge, 87 Motel, to Mid-Hudson Br, carwash, **W...**MEDICAL CARE, DENTIST, EYECARE, Sunoco/mart/24hr, Blue Jeans Steaks, Burger King, Chinese Rest., Dunkin Donuts, Friendly's, McDonald's, Pasquale's Pizza, Plaza Diner, Pizza Hut, TCBY, Mohonk Lodge(6mi), Super 8, Advance Parts, Ames, Big A Parts, Eckerd, Midas Muffler, Radio Shack, Rite Aid, ShopRite Foods, laundry, KOA(10mi), Lazy River Camping(9mi)

66mm **Modena service area sb, Mobil/diesel, Arby's, Carvel's Ice Cream/bakery, Mama Ilardo's Pizza, McDonald's, atm, fax, gifts, UPS**

65mm **Plattekill Travel Plaza nb, Mobil/diesel, Big Boy, Nathan's, Roy Rogers, atm, gifts, info**

17(60) I-84, NY 17K, to Newburgh, **E...**Getty/diesel/mart, Sunoco/diesel/mart/24hr, Burger King, Neptune Diner, Pizza Hut, Subway, Day's Inn, Holiday Inn, Howard Johnson/rest., Adams Foods, Ames, Caldor, Buick/Pontiac, Chevrolet/Olds/Cadillac, Dodge/Plymouth, Ford, Harley-Davidson, Lincoln/Mercury, NAPA, Nissan, Radio Shack, Rite Aid, ShopRite Foods, bank, laundry, **W...**Citgo/diesel, **E on NY 300...**CHIROPRACTOR, Exxon, Mobil/mart, Applebee's, Cosimo's Rest., Denny's, Dunkin Donuts, King Buffet, McDonald's, Taco Bell, Wendy's, Yobo Oriental, Hampton Inn, Ramada Inn, Super 8, AutoZone, Discount Tire, Home Depot, Meineke Muffler, Midas Muffler, Pitstop Lube, Sears/auto, Wal-Mart/auto, Weis Foods, cinema, mall

16(45) US 6, NY 17, Harriman, to West Point, **W...**Exxon/Subway/diesel/mart, American Budget Inn, Chevrolet/Buick, laundry, Woodbury Outlet/famous brands, st police

34mm **Ramapo Service Area sb, Sunoco/diesel/24hr, McDonald's, atm**

33mm **Sloatsburg Travel Plaza nb, Sunoco/diesel/24hr, Burger King, Dunkin Donuts, Sbarro's, TCBY, atm, gifts, info**

15a(31) NY 17 N, NY 59, Sloatsburg

15(30) I-287 S, NY 17 S, to NJ. I-87 S & I-287 E run together.

14b(27) Airmont Rd, Montebello, **E...**Holiday Inn, **W...**MEDICAL CARE, CHIROPRACTOR, VETERINARIAN, Exxon, Friendly's, Hong Kong Chinese, Pasta Cucina, Sutter's Mill Rest., Wellesley Inn, DrugMart, Grand Union Foods, cleaners

14a(23) Garden State Pkwy, Chestnut Ridge, to NJ

14(22) NY 59, Spring Valley, Nanuet, **E...**Amoco/mart, Gulf, Mobil, Burger King, Chinese Rest., Denny's, McDonald's, Subway, O'Toole's, Susse Chalet, BMW, Pergament, Pier 1, ShopRite Foods, Staples, TJ Maxx, bank, cinema, mall, **W...**Bagel Bin, Carvel Bakery, Dunkin Donuts, Franco's Pizza, IHOP, New King Buffet, Red Lobster, Taco Bell, Thruway Deli, Nanuet Inn, Expressway Lube, VIP Lodge, Midas Muffler, cleaners

13(20) Palisades Pkwy, N to Bear Mtn, S to NJ

12(19) NY 303, Palisades Ctr Dr, W Nyack, **E...**Saturn, **W...**Shell, Rainforest Café, The Eatery, Nyack Motel, Barnes&Noble, Bed Bath&Beyond, BJ's Whse, CompUSA, Dave&Buster's, Filene's, Home Depot, JC Penney, Jeeper's, Jo-Ann's Etc, Lord&Taylor, Sports Authority, Staples, Target, cinema, mall

11(18) US 9W, to Nyack, **E...**HOSPITAL, Mobil/mart, Best Western, **W...**Exxon, Texaco/mart, KFC, McDonald's, Super 8, JiffyLube, Midas Muffler, Nissan, cleaners, farmers mkt, **on 9 W...**Gulf, Shell/mart/atm, Mitsubishi

10(17) Nyack(from nb), same as 11

14mm Hudson River, Tappan Zee Br

13mm toll plaza

9(12) to US 9, to Tarrytown, **E...**Shell/repair, Stop'n Shop, Marriott(1mi), bank, **W...**Mobil/mart, Tarrytown Diner, Hilton, Honda

8(11) I-287 E, Saw Mill Parkway, to White Plains

7a(10) Saw Mill River Pkwy S, to Saw Mill River SP, Taconic SP

7(8) NY 9A(from nb), Ardsley, **W...**HOSPITAL, Ardsley Acres Motel

6mm **Ardsley Travel Plaza nb, Sunoco/diesel, Burger King, Popeye's, TCBY, vending**

5.5mm toll plaza, phone

6b(5.3) Corporate Dr(from nb), no facilities

6a(5) Corporate Dr(from sb), to Ridge Hill, no facilities

New York Interstate 87

6(4.5) Tuckahoe Dr, **E**...Getty/repair, SuperValue/gas, McDonald's, Pizza Pasta Café, Tuckahoe Motel, Buick/Pontiac/GMC, ShopRite Foods, **W**...Gulf, Mobil/mart, Chinese Rest., Domino's, Dunkin Donuts, Treetops Rest., Holiday Inn, Regency Hotel

5 NY 100 N, Central Park Ave(from nb), White Plains, **E**...Getty, Sunoco, Texaco/wash, Ground Round, Ford/Subaru, XpressLube

4(4) Cross Country Pkwy, **W**...Texaco/diesel/mart/24hr

3(3) Mile Square Rd, **E**...Stop'n Shop, Circuit City, FashionBug, GNC, LensCrafters, Sports Authority, Stern's, TJ Maxx, Thriftway Drug, mall, **W**...Shell/mart/24hr

2(2) Yonkers Ave(from nb), Westchester Fair, **E**...Mobil, Yonkers Speedway

1(1) Hall Place, McLean Ave, **E**...Texaco, Dunkin Donuts

0mm New York St Thruway and I-87 N run together to Albany

14(11) McLean Ave, **E**...Texaco, Dunkin Donuts

13(10) E 233rd , NE Tollway, service plaza both lanes/Mobil

12(9.5) Hudson Pkwy(from nb), Sawmill Pkwy

11(9) Van Cortland, no facilities

10(8.5) W 230th St(from sb), W 240th(from nb), **E**...Getty/mart/24hr, bank, laundry

9(8) Fordham Rd, **E**...HOSPITAL, Gaseteria, Jimmy's Bronx Café, Toyota

8(7) W 179th (from nb), **W**...Roberto Clemente SP

7(6) I-95, US 1, to NJ

6(5) E 161st, Stadium Rd

I-87 begins/ends in NYC

(vertical label: NYC Area)

New York Interstate 88

Exit #(mm)Services

25a(118) I-90/NY Thruway. I-88 begins/ends on I-90, exit 25a.

117mm toll booth (to enter or exit NY Thruway)

25(116) NY 7, to Rotterdam, Schenectady, **3 mi S**...Burger King, Dunkin Donuts, 5 Corners Pizzaria, McDonald's, Top's Diner, Best Western, L&M Motel

24(112) US 20, NY 7, to Duanesburg, **N**... Mobil/mart, Dunkin Donuts, Perillo's Pizza/Pasta, st police, **S**...Stewart's/gas, Duanesburg Diner, Frosty Acres Camping

23(101) NY 30, to Schoharie, Central Bridge, **N**...Mobil, **2 mi S**...Mobil, Dunkin Donuts, McDonald's, Hyland House B&B, Parrott House 1870 Inn, Wedgewood B&B, Hideaway Camping, Locust Park Camping

22(95) NY 7, NY 145, to Cobleskill, Middleburgh, **2-4 mi N**...HOSPITAL, Hess/diesel, Kwikfill/gas, Mobil/mart, Apollo Diner/24hr, Boreali's Diner, Burger King, McDonald's, Holiday Motel, Howe Caverns Motel/rest., Twin Oaks Camping, to Howe Caverns, **S**...st police

21(90) NY 7, NY 10, to Cobleskill, Warnerville, **N**...HOSPITAL, Hess/mart, Mobil/diesel, PriceChopper Foods, **3 mi N**...Delaney's Rest., Best Western, Gables B&B

20(87) NY 7, NY 10, to Richmondville, **S**...Mobil/diesel/mart/24hr, Blue Spruce B&B, Motel 88, **1 mi S**...Sunoco/diesel, Hi-View Camping

79mm **rest area wb, full(handicapped)facilities, phone, vending, picnic tables, litter barrels, petwalk**

19(76) to NY 7, Worcester, **N**...Stewart's/gas, Sunoco

73mm **rest area eb, full(handicapped)facilities, phones, vending, picnic tables, litter barrels, petwalk**

18(71) to Schenevus, **N**...Citgo

17(61) NY 7, to NY 28 N, Colliersville, Cooperstown, **2 mi N**...Mobil/diesel, Taylor's/gas, Homestead Rest., Knott's Motel, Lorenzo's Motel, Redwood Motel, to Baseball Hall of Fame

16(59) NY 7, to Emmons, **N**...Farmhouse Rest., Perrucci's Pizza, Masterhost Inn, Eckerd, PriceChopper Foods, **1 mi N**...Arby's, Burger King, Pizza Hut, bank, **S**...Brooks BBQ(4mi)

15(56) NY 28, NY 23, Oneonta, **N**...to Soccer Hall of Fame, **S**...Citgo, Getty/mart, Hess/mart/wash, Mobil, Red Barrel/Taco Bell/gas/wash, Burger King, McDonald's, Neptune Diner/24hr, Perkins, Sabatini's Italian, Wendy's, Christopher's Lodge/rest., Riverview Motel, Super 8, Aldi Foods, BJ's Whse, Empire Vision Ctr, Hannaford's Food/drug, JC Penney, K-Mart/Little Caesar's, Midas Muffler, NAPA, OfficeMax, Wal-Mart SuperCtr/24hr, cinema

14(55) Main St(from eb), Oneonta, **N**...Citgo, Kwikfill/gas, Stewart's/gas, Sunoco/mart, Alfresco's Italian, Golden Guernsey Ice Cream, Pepper Joe's Deli, CVS Drug, cleaners, **S**...McDonald's, Perkins, Taco Bell, Kountry Livin B&B, Subaru

13(53) NY 205, Citgo/mart, **1-2 mi N**...Citgo/mart/atm, Hess/mart, Mobil, Sunoco/diesel, Burger King, Duke Diner, Dunkin Donuts, McDonald's, Ponderosa, Celtic Motel, Maple Terrace Motel, Oasis Motor Inn, Cathedral Country Motel/rest., Ames, Chevrolet, Chrysler/Jeep, Honda/Mitsubishi, Nissan, Pontiac/Olds/Buick/GMC/Cadillac, Parts+, Rite Aid, to Susquehanna Tr, Gilbert Lake SP(11mi), camping

12(47) NY 7, to Otego, **S**...Sunoco/diesel/mart, Country Store Kitchen

43mm **rest area wb, full(handicappe)facilities, phone, picnic tables, litter barrels, vending, petwalk**

11(41) NY 357, to Unadilla, KOA

39mm **rest area eb, full(handicapped)facilities, phone, picnic tables, litter barrels, vending, petwalk**

(vertical label: Cobleskill Oneonta)

(compass directions: N, S, E, W)

New York Interstate 88

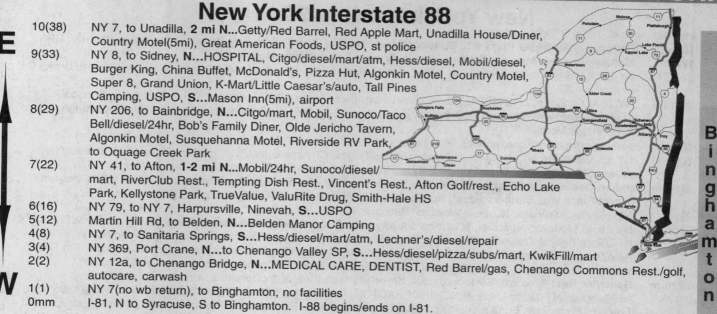

E
W

10(38) NY 7, to Unadilla, **2 mi N**...Getty/Red Barrel, Red Apple Mart, Unadilla House/Diner, Country Motel(5mi), Great American Foods, USPO, st police

9(33) NY 8, to Sidney, **N**...HOSPITAL, Citgo/diesel/mart/atm, Hess/diesel, Mobil/diesel, Burger King, China Buffet, McDonald's, Pizza Hut, Algonkin Motel, Country Motel, Super 8, Grand Union, K-Mart/Little Caesar's/auto, Tall Pines Camping, USPO, **S**...Mason Inn(5mi), airport

8(29) NY 206, to Bainbridge, **N**...Citgo/mart, Mobil, Sunoco/Taco Bell/diesel/24hr, Bob's Family Diner, Olde Jericho Tavern, Algonkin Motel, Susquehanna Motel, Riverside RV Park, to Oquage Creek Park

7(22) NY 41, to Afton, **1-2 mi N**...Mobil/24hr, Sunoco/diesel/mart, RiverClub Rest., Tempting Dish Rest., Vincent's Rest., Afton Golf/rest., Echo Lake Park, Kellystone Park, TrueValue, ValuRite Drug, Smith-Hale HS

6(16) NY 79, to NY 7, Harpursville, Ninevah, **S**...USPO

5(12) Martin Hill Rd, to Belden, **N**...Belden Manor Camping

4(8) NY 7, to Sanitaria Springs, **S**...Hess/diesel/mart/atm, Lechner's/diesel/repair

3(4) NY 369, Port Crane, **N**...to Chenango Valley SP, **S**...Hess/diesel/pizza/subs/mart, KwikFill/mart

2(2) NY 12a, to Chenango Bridge, **N**...MEDICAL CARE, DENTIST, Red Barrel/gas, Chenango Commons Rest./golf, autocare, carwash

1(1) NY 7(no wb return), to Binghamton, no facilities

0mm I-81, N to Syracuse, S to Binghamton. I-88 begins/ends on I-81.

New York Interstate 90

Exit #(mm)Services

B24.5mm New York/Massachusetts state line

B3(B23) NY 22, to Austerlitz, New Lebanon, W Stockbridge, **N**...Citgo/diesel/mart, Mobil/mart, Depot 22 Rest., **S**...Sunoco/diesel/mart, Berkshire Spur Motel, Woodland Hills Camp

B18mm toll plaza, phone

B2(B15) NY 295, Taconic Pkwy, **1-2 mi S**...gas/mart

B1(B7) US 9, NY Thruway W, to I-87, toll booth, phone

12(20) US 9, to Hudson, **1-3 mi S**...Mobil/diesel/24hr, Sunoco/diesel/mart, McDonald's, Bel Air Motel, Blue Spruce Motel, to Van Buren HS

11(15) US 9, US 20, E Greenbush, Nassau, **N**...Citgo/diesel/mart, Hess/diesel/mart, Countryside Diner, st police, **S**...DENTIST, VETERINARIAN, Burger King, Jimmie's Sandwiches, Econolodge, Grand Union, Rite Aid, USPO, laundry

10(10) Miller Rd, to E Greenbush, **1-3 mi S**...Mobil/mart, Stewarts/gas, Sunoco/mart, Dunkin Donuts, Pizza Hut, Ponderosa, Weathervane Seafood, Dewitt Motel

E
W

9(9) US 4, to Rensselaer, Troy, **N**...Bruebagger's Bagels, Grand Union Foods, Radio Shack, Wal-Mart, bank, cinema, **1-2 mi S**...DENTIST, Citgo/diesel/mart, Mobil/Mr Subb/mart, Stewart's/gas, Cracker Barrel, Denny's, Ground Round, McDonald's, Wendy's, Econolodge, 4 Seasons Motel, Mt Vernon Motel, Susse Chalet

8(8) NY 43(from eb), Defreestville

7(7) Washington Ave(from eb), Rensselaer

6.5mm Hudson River

6mm I-787, to Albany, no facilities

6(4.5) US 9, Northern Blvd, to Loudonville, **N**...HOSPITAL, access to food, lodging

5a(4) Corporate Woods Blvd, no facilities

5(3.5) Everett Rd, to NY 5, **S**...HOSPITAL, Bob&Ron's Fishfry, Bruebagger's Bagels, Denny's, Friendly's, Gateway Diner, Ground Round, McDonald's, Mr Subb, Pizza Hut, Taco Bell, Motel 6, Quality Inn, Advance Parts, AutoZone, Chevrolet, Chrysler/Jeep, Cole Muffler, CVS Drug, Ford, Hannaford's Foods/24hr, Honda/Nissan, JiffyLube, Midas Muffler, Monro Muffler/brake, PepBoys, Plymouth/Dodge, Pontiac, PriceChopper Foods, Radio Shack, Suzuki, Valvoline, multiple facilities on NY 5

4(3) NY 85 S, to Slingerlands, no facilities

3(2.5) State Offices, no facilities

2(2) Fuller Rd, Washington Ave, **S**...CrossGates Rest., Sunoco, Courtyard, same as 1S

1N(1) I-87 N, to Montreal, to Albany Airport

1S(1) US 20, Western Ave, **S**...CrossGates Mall Rd, FoodCourt, Pizzaria Uno, Caldor, Dick's Sports, Filene's, Home Depot, JC Penney, Lord&Taylor, Macy's, Old Navy, Sam's Club, Wal-Mart, cinema, mall

24(149) I-87 N, I-90 E

153mm Guilderland Service Area eb, Sunoco/diesel, Ben&Jerry's, McDonald's, Mr Subb

25(154) I-890, NY 7, NY 146, to Schenectady, **1/2 mi N**...exit 9 off I-890, Stardust Motel/rest., **1 mi N**...exit 8 off I-890, Mobil

25a(159) I-88 S, NY 7, to Binghamton

New York Interstate 90

26(162) I-890, NY 5 S, Schenectady

168mm **Pattersonville Travel Plaza wb, Sunoco/diesel, Big Boy, Mrs Fields, Roy Rogers, atm, fax, gifts, info, UPS**

172mm **Mohawk Service Area eb, Sunoco/diesel, Bryer's Sandwiches, McDonald's**

27(174) NY 30, Amsterdam, **N...**Getty/mart, Mobil, Super 8/diner/24hr, Valleyview Motel, **1 mi N...**Best Western, bank, cinema

28(182) NY 30A, Fonda, **N...**HOSPITAL, Getty/diesel/mart, Glen Trkstp/Citgo/diesel/rest./24hr, Gulf/repair, Sunoco/Travelport/diesel/rest./motel, McDonald's, SugarShack Grill, Cloverleaf Motel, Holiday Inn(7mi), Poplars Inn, Super 8, Riverside Motel

184mm parking area both lanes, phones, litter barrels

29(194) NY 10, Canajoharie, **1/2 mi N...**Gulf/mart, Stewart's/gas, Chinese Buffet, McDonald's, Pizza Hut, Rodeway Inn, Ames, Chevrolet/Olds, Grand Union Foods, Radio Shack, Rite Aid, **S...**Mobil/mart, Sunoco/mart

210mm **Indian Castle Travel Plaza eb...Sunoco/diesel, Big Boy, Mrs Fields, Roy Rogers, atm, gifts, UPS. Iroquois Travel Plaza wb...Sunoco/diesel, Burger King, Dunkin Donuts, TCBY, atm, gifts, UPS**

29a(211) NY 169, to Little Falls, **N...**Best Western, Herkimer Home

30(220) NY 28, to Mohawk, Herkimer, **N...**Citgo/Subway/mart, Mobil/diesel, Stewart's/gas, Burger King, Denny's, Friendly's, KFC/Taco Bell, McDonald's, Pizza Hut, Tony's Pizzaria, Yetty's Pizza, Budget Motel, Herkimer Motel, Inn Towne, AutoZone, FastLube, Ford/Lincoln/Mercury, Goodyear, K-Mart, Rite Aid, Sears Hardware, **S...**Mohawk Sta Rest., WhiffleTree Inn, Big M Foods, Factory Depot/apparel, antiques, to Cooperstown(Baseball Hall of Fame)

227mm **Schuyler Rest Area wb, Sunoco/diesel, Bryer's, Fresh Fudge, McDonald's, atm, fax, st police**

31(233) I-790, NY 8, NY 12, to Utica, **N...**CHIROPRACTOR, Citgo/diesel/mart, Mobil/mart, Burger King, Byrne Dairymart, Franco's Pizza, Lota'Burger, Lupino's Pizza, Paesano's Pizza, Big Lots, BJ's Whse, Eckerd, Empire Vision Ctr, Lowe's Bldg Supply, PriceChopper Foods, Rite Aid, Wal-Mart, bank, cinema, **S...**GasCard/diesel, Hess/diesel/mart, Mobil, Sunoco/diesel, Babe's Grill, Friendly's, Jreck Subs, Pizza Hut, Taco Bell, Wendy's, A-1 Motel, Best Western, Happy Journey Motel, Motel 6, Red Roof Inn, Super 8, Harley-Davidson, Monro Muffler/brake, Pennzoil

237.5mm Erie Canal

238mm Mohawk River

32(243) NY 232, Westmoreland, **S...**Carriage House Motel, **1 mi S...**gas

244mm **Oneida Travel Plaza eb, Sunoco/diesel, Burger King, Sbarro's, TCBY**

250mm parking area eb, phones, picnic tables, litter barrel

33(253) NY 365, Verona, to Vernon Downs, **N...**HOSPITAL, Mobil/mart/24hr, Comfort Suites(6mi), Super 8, KOA, to Griffiss AFB, **S...**SavOn Gas/mart/atm, Super 8, Turning Stone Resort/casino, repair

256mm parking area wb, phones, picnic tables, litter barrel

34(262) NY 13, to Canastota, **S...**Mobil/diesel/mart, SavOn/mart/24hr, Arby's, Canastota Rest., McDonald's, Days Inn, Graziano Motel/rest., to Sylvan Beach, to Boxing Hall of Fame

266mm **Chittenango Travel Plaza wb, Sunoco/diesel, Dunkin Donuts, Sbarro's, TCBY, atm, gifts**

34a(277) I-481, to Syracuse, Chittenango, no facilities

35(279) NY 298, The Circle, Syracuse, **S...**Kwikfill/mart, Mobil/repair/mart, Sunoco/mart/atm, Burger King, Denny's, Dunkin Donuts, EastWok Chinese, Franco's Pizzaria, Hub Diner, Joey's Italian, Jreck Subs, McDonald's, Pronto Pizza, Seniora Pizza, Candlewood Suites, Comfort Inn, Courtyard, Day's Inn, Embassy Suites, Extended Stay America, Fairfield Inn, Hampton Inn, Holiday Inn, Howard Johnson Lodge, John Milton Inn, Marriott, Microtel, Motel 6, Ramada Ltd, Red Roof Inn, Residence Inn, Super 8, Midas Muffler, Speedy Transmissions, Valvoline, carwash

280mm **Dewitt Service Area eb, Sunoco/diesel, McDonald's, ice cream**

36(283) I-81, N to Watertown, S to Binghamton

37(284) 7th St, Electronics Pkwy, to Liverpool, **S...**Hess/Blimpie/Godfather's/diesel/mart/24hr, Holiday Inn, Homewood Suites, Knight's Inn, **1 mi S on 7th St...**Mobil/mart, Sunoco/diesel, Bob Evans, Burger King, Colorado Steaks, Denny's, Friendly's, Ground Round, Jreck Subs, Day's Inn, Econolodge, Hampton Inn, Quality Inn, Ramada Inn, Super 8

38(286) NY 57, Syracuse, to Liverpool, **1/2 mi N...**DENTIST, Hess/mart, KwikFill/mart, Sunoco, Ground Round, Hooligan's Grill, Kirby's American Rest., Pier 57 Diner, Pizza Hut, Super 8, Eckerd, Midas Muffler, NAPA, Sears Hardware, Valvoline, **1 mi S...**Mobil/mart, Sunoco/mart/24hr, Burger King, Cole Muffler, Onondaga Lake Park, Salt Museum

39(290) I-690, NY 690, Syracuse, **N...**Comfort Inn/rest.

292mm **Warners Service Area wb, Mobil/diesel/rest., Ben&Jerry's, McDonald's, Mama Iraldo's Café**

40(304) NY 34, Weedsport, to Owasco Lake, **N...**Riverforest RV Park, **S...**KwikFill/mart, Mobil/diesel, Sunoco/diesel/mart/atm, Arby's, Arnold's Rest., DB's Drive-In, Mais Oriental Food, The Diner, Best Western, Day's Inn, Holiday Inn(9mi), Ace Hardware, Big M Foods, NAPA, bank, repair

310mm **Port Byron Service Area eb, Mobil/diesel/rest., Mama Iraldo's Café, McDonald's, ice cream**

318mm parking area wb, picnic tables, phones

41(320) NY 414, Waterloo, to Cayuga Lake, **S...**Mobil/diesel/mart, McGee Country Diner, Holiday Inn(8mi), Microtel, Prime Outlets/famous brands(3mi), NWR

324mm **Junius Ponds Travel Plaza wb, Sunoco/diesel, Dunkin Donuts, Mrs Fields, Roy Rogers, TCBY**

42(327) NY 14, to Geneva, Lyons, **N...**RV camping, **S...**Mobil/diesel/mart, Goody's Rest., Belhurst Castle Inn, Chanticleer Motel, Motel 6, Ramada Inn(6mi), Relax Inn, Prime Outlets/famous brands(3mi), tires

U t i c a

S y r a c u s e

E

W

New York Interstate 90

337mm **Clifton Springs Travel Plaza eb, Sunoco/diesel, Roy Rogers, Sbarro's, TCBY, atm, gifts**

43(340) NY 21, Manchester, to Palmyra, **N...**Hill Cumorah LDS HS(6mi), **S...**Mobil/diesel/ mart/24hr, McDonald's, Steak-Out, Roadside Inn

44(347) NY 332, Victor, **S...**MEDICAL CARE, Coastal/Subway/diesel/ mart, Mobil/mart, Sunoco/diesel/mart, Cassidy's Rest., Di Pacific's Rest., Happy Days Rest., Park Place Diner, KFC, King's Wok Chinese, McDonald's, Pizza Hut, Wendy's(2mi), Best Western, Budget Inn, Economy Inn, CVS Drug, KOA, laundry, st police

350mm **Seneca Travel Plaza wb, Mobil/diesel, Burger King, Mrs Fields, Sbarro's, TCBY, atm, info, phone**

45(351) I-490, NY 96, to Rochester, **N...**Hampton Inn, **S...**KwikFill, Burger King, Chili's, Dede's Rest., Denny's, India House Rest., Exit 45 Motel, Microtel, Chevrolet, Monro Muffler/brake, RV Ctr, bank, cinema

353mm parking area eb, phone, litter barrels

46(362) I-390, to Rochester, **1 exit N off I-390, W...**Citgo/mart, Hess/mart, Sunoco/mart, Denny's, McDonald's, Peppermints Rest., Wendy's, Country Inn Suites, Day's Inn, Fairfield Inn, Marriott, Microtel, Red Carpet Inn, Red Roof Inn, Super 8, UPS

366mm **Scottsville Travel Plaza eb, Mobil/diesel, Burger King, Dunkin Donuts, TCBY, atm, info**

376mm **Ontario Service Area wb, Mobil/diesel/deli, McDonald's**

47(379) I-490, NY 19, to Rochester, no facilities

48(390) NY 98, to Batavia, **N...**Comfort Inn, **S...**Bob Evans, Best Western, Crown Inn, Day's Inn, Holiday Inn, Microtel, Park-Oak Motel, Sheraton, Super 8, OfficeMax, Wal-Mart/auto, **1 mi S on NY 63...**EYECARE, Hess/diesel/mart, Mobil/mart, Sunoco, Arby's, Burger King, Denny's, Dunkin Donuts, McDonald's, Taco Bell, Perkins, Ponderosa, Red Dragon Chinese, Wendy's, Advance Parts, AutoZone, BJ's Whse, CarQuest, Chevrolet/Cadillac/Olds, Cole Muffler, CVS Drug, $Tree, GNC, Jo-Ann Fabrics, K-Mart, NAPA, Radio Shack, Top's Foods/deli

397mm **Pembroke Travel Plaza eb, full(handicapped)facilities, Sunoco/diesel, Burger King, Mrs Fields, Popeye's, Starbucks, TCBY, atm, gifts, phone, UPS**

48a(402) NY 77, Pembroke, **S...**Flying J/diesel/LP/rest./24hr, TA/Citgo/diesel/rest./24hr, Roy's Diner, Econolodge, NAPA

412mm **Clarence Travel Plaza wb, full(handicapped)facilities, Sunoco/diesel, Burger King, Cinnabon, Nathan's, Pizza Hut, TCBY, info, phone**

49(417) NY 78, Depew, **1-2 mi N...**VETERINARIAN, Mobil/mart, Sunoco, Applebee's, Arby's, Bennigan's, Boston Mkt, Burger King, ChiChi's, Cracker Barrel, Denny's, Don Pablo's, Eatery Rest., Fazoli's, Golden Corral, KFC, McDonald's, Mighty Taco, Old Country Buffet, Perkins, Picasso's Pizza, Pizza Hut, Ponderosa, Protocol Rest., Red Lobster, Roadhouse Grill, Ruby Tuesday, Shogun Japanese, Spilio's Rest., Starbucks, Subway, Ted's Hotdogs, TGIFriday, Wendy's, Best Value Inn, Econolodge, Fairfield Inn, Holiday Inn Express, Microtel, Ramada Ltd, Acura/Jaguar, Ames, Barnes&Noble, Bon-Ton, Buick, Burlington Coats, Cole's Mufflers, Dick's Sports, Dodge, Eckerd/24hr, Firestone/auto, Ford, Home Depot, HomePlace, JC Penney, Jo-Ann Fabrics, Kinko's, K-Mart, MediaPlay, Midas Muffler, Mitsubishi, Monro Muffler/brake, NTB, OfficMax, Pearle Vision, Saturn, Sears, Target, TireMax, ToysRUs, TJ Maxx, Top's Food/deli, Valvoline, Wal-Mart/auto, Wegman's Food/deli, auto parts, bank, cinema, mall, **S...**Kwikfill, Mobil, Bob Evans, John&Mary's Cafe, Salvatore's Italian, Garden Place Hotel, Hospitality Inn, Howard Johnson, Red Roof Inn, Sleep Inn, Aamco, Hutchin's Parts, Top's Food/drug

419mm toll booth

50(420) I-290 to Niagara Falls, no facilities

50a(421) Cleveland Dr(from eb)

51(422) NY 33 E, Buffalo, **S...**airport, st police

52(423) Walden Ave, to Buffalo, **N...**Arby's, Bob Evans, TGIFriday, Wendy's, Hampton Inn, Buick/Olds, Ford, Linens'n Things, Office Depot, Old Navy, PetsMart, Target, **S...**Coastal/wash, Sunoco/diesel/mart/rest./24hr, Alton's Rest., Denny's, McDonald's, Mighty Taco, Olive Garden, Pizza Hut, Sheraton, Bed Bath&Beyond, Borders Books, Burlington Coats, Cole Muffler, Dodge, Hutchin's Automotive, JC Penney, Kaufmann's, KidsRUs, K-Mart/auto, Lord&Taylor, MediaPlay, Niagara Hobby, PetCo, Pier 1, ToysRUs, Wegman's Food/drug, cinema, mall

52a(424) William St(from eb)

53(425) I-190, to Buffalo, Niagara Falls, **N...**McDonald's, Holiday Inn Express

54(428) NY 400, NY 16, to W Seneca, E Aurora, no facilities

55(430) US 219, Ridge Rd, Orchard Park, to Rich Stadium, **N...**Midas Muffler, **S...**Coastal/wash, Arby's, Denny's, Ponderosa, Wendy's, Aldi, Big Lots, FashionBug, Home Depot, K-Mart, Monro Muffler/brake, OfficeMax, Tops Foods, bank

431mm toll booth

56(432) NY 179, Mile Strip Rd, **N...**Sunoco, Burger King, Cracker Barrel, Odyssey Rest., Econolodge, Ames, CVS Drug, Eckerd, Jubilee Foods, repair, **S...**EYECARE, Citgo/mart, Applebee's, Boston Mkt, ChiChi's, ChuckeCheese, Lin Chinese, McDonald's, Olive Garden, Outback Steaks, Pizza Hut, Red Lobster, Roadhouse Grill, Ruby Tuesday,

New York Interstate 90

	Starbucks, TGIFriday, Wendy's, McKinley Park Inn(2mi), BJ's Whse, Bon-Ton, Circuit City, DressBarn, Firestone/auto, Home Depot, JC Penney, Jo-Ann Etc, Kaufmann's, KidsRUs, MailBoxes Etc, MediaPlay, OfficeMax, PepBoys, PetsMart, Pier 1, Sears/auto, SK Menswear, TJ Maxx, ToysRUs, Wegman's Food/deli, cinema, mall
57(436)	NY 75, to Hamburg, **N**...Mobil/diesel/mart, Bob Evans, Denny's, McDonald's, Wendy's, Comfort Suites, Day's Inn, PennyWise Inn, Red Roof Inn, Chrysler/Plymouth/Jeep/Mitsubishi, Dodge, Ford, GMC/Pontiac, Meineke Muffler, **S**...VETERINARIAN, Kwikfill, Mobil, Stop&Gas/mart, Arby's, Burger King, Pizza Hut, Subway, Holiday Inn, Camp Rd Drugs, Chevrolet, Goodyear/auto, Midas Muffler, Monro Muffler/brake, USPO, Valvoline, bank, cleaners
442mm	parking area both lanes, phone, litter barrels
57a(445)	to Eden, Angola, no facilities
447mm	**Angola Service Area both lanes, full(handicapped)facilities, Mobil/diesel, Denny's, McDonald's, phone/fax/atm, gifts**
58(456)	US 20, NY 5, Irving, to Silver Creek, **N**...HOSPITAL, Citgo, Coastal/mart/wash, Kwikfill/mart, Burger King, Aunt Millie's Rest., Tom's Rest., auto repair, to Evangola SP
59(468)	NY 60, Fredonia, Dunkirk, **N**...Dunkirk Motel(4mi), Sheraton(2mi), Lake Erie SP(7mi), camping, **S**...Citgo/mart, Kwikfill/diesel/mart, Mobil/diesel, Sunoco, Applebee's, Arby's, Bob Evans, Burger King, China King, Cracker Barrel, Denny's, KFC, Little Caesar's, McDonald's, Perkins, Pizza Hut, Ponderosa, Taco Bell, Wendy's, Best Western, Comfort Inn, Day's Inn, Advance Parts, Aldi, Ames, AutoZone, Big Lots, Eckerd, FashionBug, Ford/Lincoln/Mercury, GNC, JC Penney, Jo-Ann Fabrics, K-Mart, Lightning Lube, Midas Muffler, NAPA, Rite Aid, TJ Maxx, Top's Foods, Wal-Mart, Woodbury Winery, Arkwright Hills Camping(6mi), bank, cinema
60(485)	NY 394, Westfield, **N**...to Lake Erie SP, camping, **S**...HOSPITAL, Holiday Motel, gas
494mm	toll booth
61(495)	Shortman Rd, to Ripley, **N**...Shell/diesel/rest./24hr, **S**...Colonial Squire Motel/rest.
496mm	New York/Pennsylvania state line

New York Interstate 95

Exit #(mm)	Services
32mm	New York/Connecticut state line
22(30)	Midland Ave, Port Chester, Rye, no facilities
21(29)	I-287 W, US 1 N, to White Plains, Port Chester
20(28)	US 1 S(from nb), Port Chester
19(27)	Playland Pkwy, Rye, Harrison, no facilities
18b(25)	Mamaroneck Ave, to White Plains, **E**...Shell, cleaners
18a(24)	Fenimore Rd(from nb), Mamaroneck, **E**...Sunoco
17(20)	Chatsworth Ave(from nb, no return), Larchmont, no facilities
19.5mm	toll plaza
16(19)	North Ave, Cedar St, New Rochelle, HOSPITAL, **E**...Ramada Inn
15(16)	US 1, New Rochelle, The Pelhams, **E**...Pitstop/gas/24hr, Italian Rest., Thru-Way Diner
14(15)	Hutchinson Pkwy(from sb), no facilities
13(16)	Conner St, to Mt Vernon, **E**...Mack Trucks, auto repair, **W**...HOSPITAL, Amoco/mart, McDonald's, Andrea Motel, Holiday Motel
12(15.5)	Baychester(from nb)
11(15)	Bartow Ave, Co-op City Blvd, **E**...Mobil, **W**...Shell, facilities on frontage rds
10(14.5)	Gun Hill Rd(from nb)
9(14)	Hutchinson Pkwy, **W**...Hail Hotel/diner
8c(13.5)	Pelham Pkwy W, no facilities
b(13)	Orchard Beach, City Island, no facilities
a(12.5)	Westchester Ave(from sb), no facilities
7c(12)	Pelham Bay Park(from nb), Country Club Rd
b(11.5)	E Tremont(from sb), **E**...Citgo/diesel/mart/24hr, Mobil, diner
a(11)	I-695(from sb), to I-295 S, no facilities
6b(10.5)	I-278 W(from sb), I-295 S(from nb), no facilities
a(10)	I-678 S, Whitestone Bridge, no facilities
5b(9)	Castle Hill Ave, **W**...Sunoco, McDonald's
a(8.5)	Westchester Ave, White Plains Rd, no facilities
4b(8)	Bronx River Pkwy, Rosedale Ave, **W**...Mobil
a(7)	I-895 S, Sheridan Expsy, no facilities
3(6)	3rd Ave, HOSPITAL
2b(5)	Webster Ave, **E**...HOSPITAL
a(4)	Jerome Ave, to I-87, no facilities
1c(3)	I-87, Deegan Expswy, to Upstate, no facilities
b(2)	Harlem River Dr(from sb)
a(1)	US 9, NY 9A, H Hudson Pkwy, 178th St, HOSPITAL
0mm	Hudson River, Geo Washington Br, New York/New Jersey state line

New York Interstate 190(Buffalo)

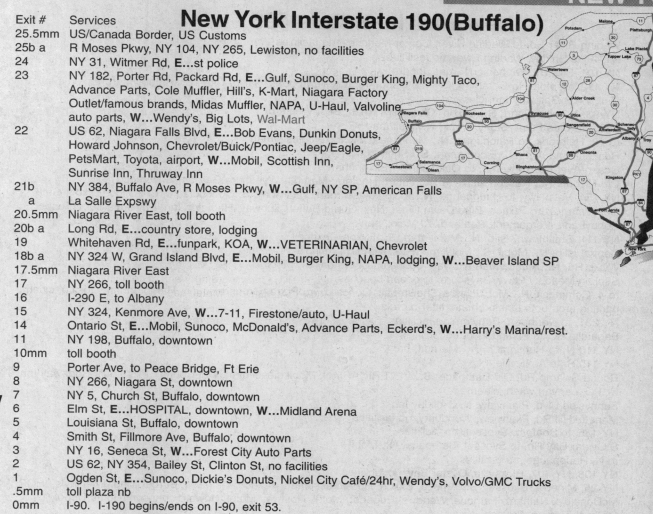

Exit #	Services
25.5mm	US/Canada Border, US Customs
25b a	R Moses Pkwy, NY 104, NY 265, Lewiston, no facilities
24	NY 31, Witmer Rd, **E**...st police
23	NY 182, Porter Rd, Packard Rd, **E**...Gulf, Sunoco, Burger King, Mighty Taco, Advance Parts, Cole Muffler, Hill's, K-Mart, Niagara Factory Outlet/famous brands, Midas Muffler, NAPA, U-Haul, Valvoline auto parts, **W**...Wendy's, Big Lots, Wal-Mart
22	US 62, Niagara Falls Blvd, **E**...Bob Evans, Dunkin Donuts, Howard Johnson, Chevrolet/Buick/Pontiac, Jeep/Eagle, PetsMart, Toyota, airport, **W**...Mobil, Scottish Inn, Sunrise Inn, Thruway Inn
21b	NY 384, Buffalo Ave, R Moses Pkwy, **W**...Gulf, NY SP, American Falls
a	La Salle Expswy
20.5mm	Niagara River East, toll booth
20b a	Long Rd, **E**...country store, lodging
19	Whitehaven Rd, **E**...funpark, KOA, **W**...VETERINARIAN, Chevrolet
18b a	NY 324 W, Grand Island Blvd, **E**...Mobil, Burger King, NAPA, lodging, **W**...Beaver Island SP
17.5mm	Niagara River East
17	NY 266, toll booth
16	I-290 E, to Albany
15	NY 324, Kenmore Ave, **W**...7-11, Firestone/auto, U-Haul
14	Ontario St, **E**...Mobil, Sunoco, McDonald's, Advance Parts, Eckerd's, **W**...Harry's Marina/rest.
11	NY 198, Buffalo, downtown
10mm	toll booth
9	Porter Ave, to Peace Bridge, Ft Erie
8	NY 266, Niagara St, downtown
7	NY 5, Church St, Buffalo, downtown
6	Elm St, **E**...HOSPITAL, downtown, **W**...Midland Arena
5	Louisiana St, Buffalo, downtown
4	Smith St, Fillmore Ave, Buffalo, downtown
3	NY 16, Seneca St, **W**...Forest City Auto Parts
2	US 62, NY 354, Bailey St, Clinton St, no facilities
1	Ogden St, **E**...Sunoco, Dickie's Donuts, Nickel City Café/24hr, Wendy's, Volvo/GMC Trucks
.5mm	toll plaza nb
0mm	I-90. I-190 begins/ends on I-90, exit 53.

E ↑↓ **W**

New York Interstate 290(Buffalo)

Exit #	Services
8	I-90, NY Thruway, I-290 begins/ends on I-90, exit 50.
7	NY 5, **N**...Citgo, Kwikfill, Arby's, Pizza Plant, **S**...Coastal/mart
6	NY 324, NY 240, **N**...Cadillac, **S**...MEDICAL CARE, Chrysler/Dodge, cleaners
5	NY 263, to Millersport, **N**...Hampton Inn, Marriott, Red Roof Inn, Super 8, **S**...Mobil, Boston Mkt, Dakota Grill, Fuddruckers, Hooters, Taco Bell, Motel 6, cinema
4	I-990, to St U, no facilities
3b a	US 62, to Niagara Falls Blvd, **N**...Bob Evans, Extended Stay America, Holiday Inn, RV camping, **S**...Mobil, K-Mart, Valvoline
2	NY 425, Colvin Blvd, HOSPITAL, **N**...KFC, McDonald's, Wendy's, Microtel, Ames, BJ's Wholesale, Goodyear/auto, Top's Foods, **S**...Kwikfill/mart, Big Lots
1	Elmwood Ave, NY 384, NY 265, **N**...Kwikfill, Rite Aid, **S**...HOSPITAL, Mobil/diesel/mart, Urbank Transmissons
0mm	I-190. I-290 begins/ends on I-190 in Buffalo.

E ↑↓ **W**

New York Interstate 495(Long Island)

Exit #	Services
	I-495 begins/ends on NY 25.
73	rd 58, Old Country Road, to Greenport, Orient, **S**...Coastal/diesel/mart, Tanger/famous brands, **1 mi S**...Exxon/mart, Gulf, Mobil/mart, Taco Bell, Wendy's, CVS Drug, Ford/Lincoln/Mercury, Nissan/Hyundai/Suzuki, Pergament, Tire Country, Waldbaum's
72	NY 25, Riverhead, Calverton(no EZ eb return), **S**...Ramada Inn/grill, Tanger/famous brands/foodcourt
71	NY 24, Calverton, to Hampton Bays, no facilities
70	NY 111, Manorville, to Eastport, **S**...Mobil/mart

N ↑↓ **S**

New York Interstate 495

Exit	Description
69	Wading River Rd, to Wading River, Center Moriches, no facilities
68	NY 46, to Shirley, Wading River, no facilities
66	NY 101, Sills Rd, Yaphank, **N**...gas, food
64	NY 112, Medford, to Coram, **N**...Home Depot, King Kullen Foods, Sam's Club, **S**...Amoco/mart, Exxon/mart, Gulf, 7-11, Gateway Inn, cinema
63	NY 83, **N**...Hess/mart, **S**...Exxon, Churchill's Rest., Best Western
62	rd 97, to Blue Point, Stony Brook, **N**...Gulf/mart
61	rd 19, Holbrook, to Patchogue, **N**...Mobil/mart, **S**...Exxon/diesel/mart
60	Ronkonkoma Ave, **N**...Exxon/mart, **S**...Shell
59	Ocean Ave, Ronkonkoma, to Oakdale, **S**...Exxon/mart
58	Old Nichols Rd, Nesconset, **N**...Exxon/mart, Hooters, Marriott, Wyndham Garden, **S**...Amoco
57	NY 454, Vets Hwy, to Hauppauge, **N**...Exxon/diesel/mart, TGIFriday, **S**...Amoco/24hr, BP, Citgo, Exxon, Gulf/diesel/mart/24hr, Texaco, Blue Dawn Diner, New Grand Buffet, Subway, Hampton Inn, Edward's SuperFoods, HobbyLand, Pergament, PetLand, TJ Maxx, bank
56	NY 111, Smithtown, Islip, **N**...Mobil/mart, **S**...Exxon, Texaco
55	Central Islip, **N**...Mobil, **S**...Mobil/mart, Texaco/mart, Kinko's
54	Wicks Rd, **N**...Amoco/mart, Howard Johnson, Sheraton, **S**...Mobil
53	Sunken Meadow Pkwy, Bayshore, to ocean beaches
52	rd 4, Commack, **N**...Mobil/diesel, Shell/repair, Conca d'Oro Pizza, Ground Round, Hampton Inn, Costco, cinema
51.5mm	parking area both lanes, phone, litter barrels
51	NY 231, to Northport, Babylon, no facilities
50	Bagatelle Rd, to Wyandanch, no facilities
49N	NY 110 N, to Huntington, **N**...Marriott, bank
S	NY 110 S, to Amityville
48	Round Swamp Rd, Old Bethpage, **S**...VETERINARIAN, Mobil/diesel/mart, Old Country Pizza/deli, Rodeway Inn, USPO, car/van wash, cinema
46	Sunnyside Blvd, Plainview, **N**...Holiday Inn
45	Manetto Hill Rd, Plainview, Woodbury, no facilities
44	NY 135, to Seaford, Syosset, no facilities
43	S Oyster Bay Rd, to Syosset, Bethpage, **N**...Mobil
42	rd N, Hauppauge, no facilities
41	NY 106, NY 107, Hicksville, Oyster Bay, no facilities
40	NY 25, Mineola, Syosset, **S**...Amoco, Exxon, Hess/diesel/mart, Texaco, 7-11/atm, Burger King, Friendly's, IHOP, McDonald's, Nathan's Famous, Wendy's, Edgerton Motel, Hostway Motel, Howard Johnson, Home Depot, OffTrack Betting, Staples
39	Glen Cove Rd, no facilities
38	Northern Pkwy E, Meadowbrook Pkwy, to Jones Beach
37	Willis Ave, to Roslyn, Mineola, **N**...Exxon, Shell/mart, repair, **S**...Mobil/mart,
36	Searingtown Rd, to Port Washington, **S**...HOSPITAL
35	Shelter Rock Rd, Manhasset, no facilities
34	New Hyde Park Rd, no facilities
33	Lakeville Rd, to Great Neck, **N**...HOSPITAL
32	Little Neck Pkwy, **N**...pizza
31	Douglaston Pkwy, **S**...Amoco/mart, Stern's, USPO, mall
30	E Hampton Blvd, Cross Island Pkwy
29	Springfield Blvd, **S**...Exxon/mart, Texaco/repair, Chinese Rest., McDonald's
27	I-295, Clearview Expswy, Throgs Neck, **N**...Gulf/mart, 7-11, Blue Bay Diner, Rockbottom Drug
26	Francis Lewis Blvd, S
25	Utopia Pkwy, 188th St, **N**...Citgo, Mobil, Sunoco, **S**...Amoco/mart, Gaseteria, Mobil/mart, Shell/repair, Sunoco, Chinese Rest., Filene's Basement, Fresh Meadows, Goodyear, Radio Shack, cinema
24	Kissena Blvd, **N**...Exxon/diesel/mart, Dunkin Donuts, **S**...Mobil
23	Main St, **N**...Palace Diner
22	Grand Central Pkwy, to I-678, College Pt Blvd, **N**...Citgo, Mobil, Paris Suites, **S**...Mobil, Best Western
21	108th St, **N**...Gaseteria, cleaners
19	NY 25, Queens Blvd, Woodhaven Blvd, to Rockaways, **N**...JC Penney, Macy's, MailCtr, Stern's, mall, **S**...Amoco/mart, Bed Bath&Beyond, Circuit City, Marshall's, Old Navy, Sears
18.5	69th Ave, Grand Ave(from wb), no facilities
18	Maurice St, **N**...Amoco/mart, **S**...Gaseteria/Dunkin Donuts, McDonald's, diesel repair
17	48th St, to I-278, no facilities
16	I-495 begins/ends in NYC.

Long Island

N

S

North Carolina Interstate 26

Exit #	Services
40mm	North Carolina/South Carolina state line
38mm	N Pacolet River
36.5mm	**Welcome Ctr wb, full(handicapped)facilities, phone, picnic tables, litter barrels**
36	US 74 E, to NC 108, Columbus, Tryon, **N**...HOSPITAL, Amoco, BP/Burger King/diesel, Exxon/diesel, Texaco, Hardee's, KFC, McDonald's, Subway, Waffle House, Western Steer, Day's Inn, Food Lion
28	Saluda, **N**...Heaven's View Motel, camping, **S**...Amoco/diesel/mart, Texaco/mart, Apple Mill Rest./gifts, repair
25mm	Green River
23	US 25, to Greenville, E Flat Rock, access to Carl Sandburg Home
22.5mm	Eastern Continental Divide, 2130 ft
22	Upward Rd, Hendersonville, **N**...Texaco/diesel/mart, Dish Barn Giftshop, **S**...Exxon/McDonald's/diesel/mart, Shell/Pizza Inn/mart, Cracker Barrel, Holiday Inn Express, RV camping, to Carl Sandburg Home, **3 mi S**...Kelsey's Rest., Quincy's, Toyota, RV Ctr
18b a	US 64, Hendersonville, **N**...HOSPITAL, Chevron/mart, Shell/Blimpie/diesel/mart/24hr, Texaco/diesel/mart, Waffle House, Best Western, Hampton Inn, Quality Inn, Ramada Ltd, CarQuest, Ingles/deli, World of Clothing, camping, **S**...HOSPITAL, EYECARE, Chevron, Exxon/KrispyKreme/diesel/mart, Shell/mart, Applebee's, Arby's, Baskin-Robbins, Binion's Roadhouse, Bojangles, Burger King, Checker's, Denny's, Fatz Café, Hardee's, Honeybaked Ham, KFC, LJ Silver, McDonald's, McGuffey's Café, Outback Steaks, Pizza Hut, Ryan's, Savannah Beach Grill, Schlotsky's, Shoney's, Subway/TCBY, Taco Bell, Wendy's, Comfort Inn, Day's Inn, Belk, BiLo Foods, CVS Drug, Eckerd, Goodyear/auto, Home Depot, JC Penney, JiffyLube, K-Mart/auto, Lowe's Bldg Supply, Parts+, Radio Shack, Wal-Mart, Winn-Dixie, bank, cinema, laundry, mall
15mm	weigh sta both lanes, phones
13	US 25, Fletcher, **N**...HOSPITAL, Citgo/diesel/mart, Hardee's, Subway, **S**...Exxon/mart, Phillips 66/diesel, Shell/diesel/rest./repair, Burger King, Huddle House, Skyway Diner/24hr, Ledford's RV/marine, to Uncle John's
10mm	**rest area both lanes, full(handicapped)facilities, phone, picnic tables, litter barrels, vending**
9	NC 280, Arden, **N**...Citgo/mart, Exxon/diesel/mart/wash, Texaco/Arby's/diesel/mart, McDonald's, Pizza Hut, Yesterday's Diner, Waffle House, Budget Motel, Comfort Inn, Day's Inn, Econolodge, Hampton Inn, Holiday Inn/rest., Sun Valley Motel, Honda/Acura/Kia, multiple facilities on US 25(4mi), **S**...Amoco/diesel/mart, J&S Cafeteria, Fairfield Inn, Asheville Airport
6	NC 146, Skyland, **N**...VETERINARIAN, Exxon/KrispyKreme/diesel/mart, Texaco/McDonald's/diesel/mart/24hr, Arby's, Shoney's, Waffle House, Quality Inn, **2 mi N on US 25**...CHIROPRACTOR, Citgo/diesel/mart, BBQ, Bojangles, Sagebrush Steaks, Taco Bell, Advance Parts, Firestone/auto, NAPA, **S**...Cider Mill Rest., **2 mi S**...Shell/mart/wash
3mm	French Broad River
2	NC 191, Brevard Rd, **N**...Saturn/Toyota, Bear Creek RV Camp, **2 mi N**...Phillips 66/mart, Asheville Farmers Mkt, **S**...HotSpot/gas, Phillips 66/diesel/mart, China Garden, Harbor House Seafood, LJ Silver, McDonald's, Mtn View Rest., Subway, Taco Bell, Waffle House, Comfort Suites, Country Inn Suites, Hampton Inn, Holiday Inn Express, Super 8, Wingate Inn, Belk, Ben Franklin, Dillard's, K-Mart, Ingles Foods, Profitt's, SuperPetz, cinema, mall, to Blue Ridge Pkwy
1b a	I-40, E to Statesville, W to Knoxville, no facilities. I-26 begins/ends on I-40, exit 46.

E ↑↓ **W**

North Carolina Interstate 40

Exit #	Services
420.1mm	I-40 begins/ends at Wilmington. Facilities on US 17, **N**...Bob Evans, Chrysler/Jeep, Lincoln/Mercury/Isuzu, Home Depot, Kia, Nissan, Subaru/Saab/Audi, Toyota, VW, **S**...Crown Gas/diesel/mart/24hr, Dodge's Store/gas, Exxon/diesel/mart/24hr, Shell, Burger King, Carrabba's, Chick-fil-A, Cracker Barrel, Hardee's, Hooters, Hieronymus' Seafood, KFC, McDonald's, Quincy's, Sticky Fingers Rest., Taco Bell, Waffle House, Wendy's, Best Western, Day's Inn, Fairfield Inn, GreenTree Inn, Hampton Inn, Holiday Inn, Howard Johnson, Innkeeper, Motel 6, Ramada Inn, Rodeway Inn, Sheraton, Sleep Inn, Super 8, Wingate Inn, Advance Parts, AutoZone, Batteries+, Buffalo Tire/auto, Buick, Circuit City, Food Lion, Marshall's, OfficeMax, PetsMart, Saturn/Suzuki, Target, Wal-Mart SuperCtr/24hr, bank. **Facilities 2-4 mi S on NC 132**...Amoco, Citgo, Texaco/mart, Applebee's, Bennigan's, Bojangles, Burger King, Checker's, Chili's, Domino's, Golden Corral, Golden Phoenix Chinese, Hardee's, Honeybaked Ham, Jeter's HotDogs, KFC, McAlister's Deli, McDonald's, Outback Steaks, Perkins/24hr, Rockola Café, Subway, Swensen's Café, Taco Bell, TCBY, Wendy's, Comfort Inn, Courtyard, Fairfield Inn, Holiday Inn Express, Barnes&Noble, Chevrolet, Chrysler/Plymouth, CVS Drug, Dick's Sports, $Tree, Ford, Harris-Teeter Foods, Honda/Acura, Hyundai, K-Mart, Lowe's Bldg Supply, OfficeAmerica, Phar-Mor Drug, Pontiac/GMC/Mercedes, Sam's Club, TJ Maxx, XpressLube, bank, cinema.

E ↑↓ **W**

North Carolina Interstate 40

420b a	Gordon Rd, NC 132 N, **2 mi N**...Citgo, Exxon/diesel, McDonald's, Perkins, Smithfield's Chicken/BBQ, Camelot RV Camping(4mi), **S**...Amoco/diesel, Shell, Subway
414	Castle Hayne, to Brunswick Co beaches, **1/2mi S**...Shell/mart, Sprint Gas, Hardee's
412	NE Cape Fear River
408	NC 210, **S**...Exxon/diesel/mart, Paul's Place Café, Country Café(7mi), to Moore's Creek Nat Bfd
398	NC 53, Burgaw, **2 mi S**...HOSPITAL, Hardee's, Skat's Café, Subway, Burgaw Motel, camping
390	to US 117, Wallace, no facilities
385	NC 41, Wallace, **N**...Exxon/mart, Mad Boar Rest., Holiday Inn Express, River Landing Cottages, **2-3 mi S**...to US 117, Hardee's, McDonald's, Liberty Inn
384	NC 11, Wallace, **S**...Bed&Breakfast
380	Rose Hill, **S**...Amoco, Citgo/mart/wash, lodging(1mi)
373	NC 903, Magnolia, **5 mi N**...HOSPITAL, BP/diesel, Bed&Breakfast, Cowan Museum, **2 mi S**...Bed&Breakfast
369	US 117, Warsaw, no facilities
364	NC 24, to NC 50, Clinton, **rest area both lanes, full(handicapped)facilities, phone, picnic tables, litter barrels, vending, petwalk, 3 mi N**...Hardee's, Country Squire Inn(7mi), Warsaw Inn, **S**...Amoco/Bojangles/diesel/mart, BP/diesel/mart, Crown/mart/24hr, Phillips 66/diesel/mart/24hr, Texaco/diesel/mart, KFC, McDonald's, Smithfield's Chicken&BBQ, Subway, Taste of Country Buffet, Waffle House, Wendy's, Day's Inn, Holiday Inn Express
355	NC 403, Faison, to US 117, to Goldsboro, **3 mi N**...Exxon, Faison Bed&Breakfast
348	Suttontown Rd, no facilities
343	US 701, Newton Grove, **1 mi N**...Two Dogs Pizza, to Bentonville Bfd
341	NC 50, NC 55, to US 13, Newton Grove, **1.5 mi N**...Exxon/diesel, Hardee's, William&Mary Motel, **S**...BP/McDonald's/mart, Texaco, Smithfield's Chicken&BBQ, Subway
334	NC 96, Meadow, **S**...BP(1mi)
328b a	I-95, N to Smithfield, S to Benson, no facilities
325	NC 242, to US 301, to Benson, **S**...Citgo/diesel
324mm	**rest area both lanes, full(handicapped)facilities, phone, picnic tables, litter barrels, vending, petwalk, no overnight parking**
319	NC 210, McGee's Crossroads, **N**...I-40 SuperGas/mart, Texaco/diesel/mart/24hr, Papa's Pizza/subs
312	NC 42, to Clayton, Fuquay-Varina, **N**...Exxon/Wendy's/diesel/mart/24hr, Phillips 66/diesel/mart, China Buffet, Papa's Subs/pizza, Smithfield's Chicken/BBQ, Best Western, Holiday Inn Express, Jameson Inn, Goodyear/auto, Lowe's Bldg Supply, **S**...VETERINARIAN, Amoco/Subway/diesel/mart/wash/24hr, BP/Burger King/mart/24hr, Citgo/diesel/mart, Texaco/diesel/mart, Black Angus Steaks, Bojangles, Dairy Queen, KFC/Taco Bell, McDonald's, Waffle House, Hampton Inn, Sleep Inn, CVS Drug, Food Lion, bank, cleaners, laundry
306	US 70, Garner, to Smithfield, Goldsboro, **1 mi N**...Texaco/diesel/mart
303	Jones Sausage Rd, Rd, **N**...Shell/diesel/mart, Burger King, **S**...Texaco/diesel/mart, Sandiford Sausage Co
301	I-440 E, US 64/70 E, to Wilson, no facilities
300b a	Rock Quarry Rd, **S**...Texaco, Burger King, Hardee's, Subway
299	Person St, Hammond Rd, Raleigh(no EZ return eb), **1 mi N**...Amoco, Texaco/diesel, King's Motel, to Shaw U
298b a	US 401 S, US 70 E, NC 50, **N**...Parts+, **S**...Amoco/mart/wash, Citgo/diesel/mart/atm, Crown/mart, Servco/mart/wash, Bojangles, Burger King, Domino's, KFC, Claremont Inn, Comfort Inn, Day's Inn, Innkeeper, Sam's Club, CarQuest
297	Lake Wheeler Rd, **N**...HOSPITAL, Exxon, Burger King, Farmer's Mkt, **S**...Citgo
295	Gorman St, **1 mi N**...BP, Exxon/diesel, Texaco, Hardee's, Subway, to NCSU, Reynolds Coliseum
293	US 1, US 64 W, Raleigh, to inner 440 Lp, **S**...Exxon, Motel 6
291	Cary Towne Blvd, Cary, **1 mi S**...Burger King, Dairy Queen, Hardee's, McDonald's, Olive Garden, Ragazzi's, Taco Bell
290	NC 54, Cary, **1 mi N**...Citgo, **2 mi S**...BP, Citgo, Shell, Hampton Inn, state fairgrounds
289	US 1, Wade Ave, to Raleigh, **N**...HOSPITAL, museum, **S**...to fairgrounds
287	Harrison Ave, Cary, **N**...to Wm B Umstead SP, **S**...Phillips 66, Shell, Texaco, Arby's, Burger King, McDonald's, Newton's SW Rest., Subway, Wendy's, Embassy Suites, Studio+
285	Aviation Pkwy, to Morrisville, Raleigh/Durham Airport, **N**...Hilton Garden
284	Airport Blvd, **N**...ChopHouse Rest., to RDU Airport, **S**...Amoco/diesel/mart, Citgo/diesel, Exxon, Shell/diesel/mart, FoodCourt, ChopHouse, Jersey Mike's Subs, Quizno's, Texas Steaks, Waffle House, Wendy's, Baymont Inn, Courtyard, Day's Inn, Extended Stay America, Fairfield Inn, Hampton Inn, Holiday Inn Express, Microtel, La Quinta, Prime Outlets/famous brands
283	I-540, to US 70, Aviation Pkwy, no facilities
282	Page Rd, **S**...Comfort Suites, Holiday Inn/rest., Sheraton/rest., Wingate Inn, World Trade Ctr
281	Miami Blvd, **N**...Wendy's, Best Western Crown Park, Marriott, Wyndham Garden, **S**...BP, Shell/mart, HomeGate Inn, Homewood Suites, Wellesley Inn
280	Davis Dr, **N**...to Research Triangle, **S**...Radisson Governor's Inn, bank
279b a	NC 147, Durham Fwy, to Durham, HOSPITAL, no facilities

E

↑
↓

W

North Carolina Interstate 40

278 NC 55, to NC 54, Apex, Foreign Trade Zone 93, World Trade Ctr, **N**...Citgo/mart, China One Rest., Waffle House, DoubleTree Suites, Fairfield Inn, Innkeeper, La Quinta, Red Roof Inn, **S**...CHIROPRACTOR, MEDI-CAL CARE, BP, Citgo/mart, Crown Gas/mart, Exxon/die-   sel/mart, Arby's, Bojangles, Burger King, Chick-fil-A, Fazoli's, Golden Corral, Hardee's, KFC, McDonald's, Miami Grill, Mike's Subs, O!Brian's, Papa John's, Pizza Hut, Schlotsky's, Subway, Taco Bell, Wendy's, Candlewood Suites, Court-yard, Crossland Suites, Homestead Village, Residence Inn, CVS Drug, Econo Lube/tune, Eckerd, Food Lion/deli, JiffyLube, PrecisionTune, Winn-Dixie/deli, bank, cleaners/laundry, transmissions/tires

276 Fayetteville Rd, **N**...Exxon/diesel/mart, Phillips 66/diesel/mart, McDonald's, Quizno's, Ruby Tuesday, Rudino's Pizza, Souper Salad, Subs Etc, TCBY, Waffle House, Wendy's, Eckerd, GNC, Harris-Teeter, Kroger/deli, to NC Central U

274 NC 751, to Jordan Lake, **1-2 mi N**...Burger King, McDonald's, Waffle House, Wendy's

273 NC 54, to Durham, UNC-Chapel Hill, **N**...Shell/diesel, **S**...VETERINARIAN, Amoco/diesel/mart, BP/diesel/mart/wash, Citgo(2mi), Texaco/diesel/mart, Hardee's, Pizza Chef, Best Western(2mi), University Inn, Hampton Inn, Howard Johnson, bank, cleaners

270 US 15, US 501, Chapel Hill, Durham, **N**...HOSPITALS, Speedway(1mi), Bob Evans, Bojangles, Outback Steaks, Philly Steaks, Tripp's Rest., Comfort Inn, Homewood Suites, Barnes&Noble, Best Buy, Dick's Sports, $Tree, Home Depot, MensWearhouse, OfficeMax, Old Navy, Linens'n Things, Marshall's, Michael's, Saab, Wal-Mart, to Duke U, **S**...BP, Exxon, Applebee's, Golden Corral, Hardee's, McDonald's, Subway, Wendy's, Hampton Inn, Holiday Inn, Omni Europa, Red Roof Inn, Siena Hotel, BMW, Chevrolet, CVS Drug, JiffyLube, Lowe's Bldg Supply, Saturn, bank, mall

266 NC 86, to Chapel Hill, **2 mi S**...BP, Citgo, Exxon, Wilco/diesel, Subway, Harris-Teeter/FoodCourt

263 New Hope Church Rd, no facilities

261 Hillsborough, **1.5 mi N**...Amoco, Citgo/diesel, Shell, Hardee's, KFC, McDonald's, Pizza Hut, Waffle House, Wendy's

259 I-85 N, to Durham, no facilities

I-40 and I-85 run together 38 mi. See North Carolina Interstate 85, exits 124-161.

219 1-85 S, to Charlotte, US 29, US 70, US 220, no facilities

218b a US 220 S, to I-85 S, Freeman Mill Rd, no facilities

217b a High Point Rd, Greensboro, **N**...Exxon/diesel/mart, Texaco/diesel/mart, Arby's, Astor's Grill, Bennigan's, Best Bagels, Biscuitville, Blue Marlin Seafood, Burger King, Chili's, Dunkin Donuts, Ghassan's Steak/sub, Grady's American Grill, Hooters, Jake's Diner/24hr, La Bamba Mexican, KFC, LoneStar Steaks, Mrs Winner's, McDonald's, Olive Garden, Perkins/24hr, PoFolks, Stephen's Seafood, Subway, Taco Bell, TCBY, Coliseum Motel, Howard Johnson/rest., Park Lane Hotel, Red Roof, Super 8, Travelodge, Better Vision, Burlington Coats, $General, Office Depot, PetSupplies+, bank, cleaners, **S**...CHIROPRACTOR, DENTIST, VETERINARIAN, Amoco/diesel/repair, Citgo, Crown/mart, Phillips 66/mart, Texaco/diesel/mart/wash, Bojangles, Burger King, Chen's Chinese, Darryl's, Hoolihan's Rest., Krispy Kreme, La Bamba Mexican, Libby Hill Seafood, McDonald's, Miami Grill, Oh! Brian's, Pizza Inn, Quincy's, Sally's Hotdogs, Shoney's, Sonic, Subway, Taco Bell, Tides Seafood, Waffle House, Wendy's, Best Western, Comfort Inn, Day's Inn, Drury Inn Suites, Econolodge, Fairfield Inn, Hampton Inn, Holiday Inn, Residence Inn, Aamco, Advance Parts, AutoZone, Belk, Baby Superstore, Borders Books&Café, CompUSA, Dillard's, Family$, Firestone/auto, Goodyear/auto, GreaseMonkey, Hancock Fabrics, JC Penney, Kerr Drug, Lowe's Foods, Merchants Tire/auto, Midas Muffler, NAPA, NTB, OfficeMax, PartsAmerica, PepBoys, PharMor Drug, Pier 1, Radio Shack, S&K Menswear, Sears/auto, TJ Maxx, TireAmerica, VisionWorks, bank, cinema, cleaners, mall

216 NC 6(from eb, exits left), to Greensboro Coliseum

214b a Wendover Ave, **N**...Citgo/diesel, Exxon/diesel/mart, Texaco/diesel/mart, BBQ, Blimpie, Burger King, Chinese Rest., Herbie's Diner/24hr, K&W Cafeteria, Ruby Tuesday, TCBY, Waffle House, Extended Stay America, Innkeeper, Microtel, BMW, Chevrolet, Chrysler, Dodge, Ford, Goodyear, Honda/Acura/Nissan, Isuzu/Subaru, Mercedes, Mitsubishi, Saab/Jaguar, Saturn, Staples, Volvo, bank, funpark, **S**...CHIROPRACTOR, Applebee's, Arby's, Bojangles, Chick-fil-A, Cracker Barrel, Fuddrucker's, Golden Corral, IHOP/24hr, McDonald's, Red Lobster, Steak'n Shake, Subway, TGIFriday, Taco Bell, Tripp's Rest., Wendy's, AmeriSuites, Courtyard, Shoney's Inn/rest., Alfa Romeo, BabysRUs, Best Buy, Circuit City, DriversMart, EconoLube'n Tune, Ferrari, Home Depot, HomePlace, Kohl's, Kroger, Lowe's, Meineke Muffler, PetsMart, Porsche, Sam's Club, Service Merchandise, Sports Authority, Super K-Mart/Little Caesar's/auto/24hr, Target, Waccamaw, Wal-Mart SuperCtr/24hr

213 Jamestown, **N**...BP/diesel, Damon's Ribs, Radisson, to Guilford Coll, **S**...Exxon/diesel/mart, DairyMart/gas, laundry, accesses same as 214

212 Chimney Rock Rd, no facilities

E

W

North Carolina Interstate 40

210	NC 68, to High Point, Piedmont Triad, **N**...Phillips 66/diesel, Shell(1mi), Arby's, Hardee's, Mr Omelet, Wendy's, Day's Inn, Embassy Suites, Holiday Inn, Homewood Suites, Ford Trucks, Freightliner, to airport, **S**...Exxon/diesel, Mobil, Bojangles, McDonald's, Shoney's, Subway, Best Inns, Candlewood Suites, Comfort Suites, Hampton Inn, Motel 6, Ramada, Red Roof Inn, Peterbilt, bank
208	Sandy Ridge Rd, **N**...Exxon/diesel/mart, RV Country, **S**...Citgo/diesel/mart, Out Of Doors Mart/Airstream, Farmer's Mkt
206	Lp 40(from wb), to Kernersville, Winston-Salem, downtown
203	NC 66, to Kernersville, **N**...Amoco/mart, BP/mart, Citgo/McDonald's/diesel/mart, BBQ, Capt Tom's Seafood, OutWest Steaks, Suzie's Diner, Waffle House, Sleep Inn, Ford, Merchant's Tire/repair, Southern Battery, **2 mi N**...Bojangles, Pizza Hut, King's Inn, Quality Inn, **S**...Shell/diesel/mart
201	Union Cross Rd, **N**...QM/mart(1mi), Burger King, China Café, Jerry's Pizza/Subs, CVS Drug, Food Lion
196	US 311 S, to High Point, no facilities
195	US 311 N, NC 109, to Thomasville, **S**...Amoco/mart, Wilco/Citgo/diesel/mart
193c	Silas Creek Pkwy(from wb), same as 192
193b a	US 52, NC 8, to Lexington, **N**...Wilco/mart, Hardee's
192	NC 150, to Peters Creek Pkwy, Winston-Salem, **N**...MEDICAL CARE, CHIROPRACTOR, Shell, Texaco/mart, Wilco/mart, Arby's, Bojangles, Boston Mkt, Burger King, Char's Hotdogs, Checker's, Chinese Rest., IHOP, KFC, Little Caesar's, Monterrey Mexican, Mr BBQ, Old Country Buffet, Perkins, Pizza Hut, Quincy's, Red Lobster, Shoney's, Sonic, Taco Bell, Wendy's, Comfort Inn(1mi), Innkeeper, Knight's Inn, Acura, Audi, AutoZone, Cloth World, Eckerd, Ford, Hyundai, Kroger, Lowe's Foods, Mazda, NAPA, Office Depot, PharMor Drug, Shoppers Paradise, Suzuki, Woolworth's, bank, cinema, laundry, **S**...EYECARE, BP/mart/wash, Libby Hill Seafood, McDonald's, K&W Cafeteria, Wendy's, Advance Parts, BiLo, Buick/BMW/Saturn, CVS Drug, Food Lion, Goodyear, Harris Teeter, Honda, K-Mart, bank, cleaners
190	Hanes Mall Blvd(from wb, no re-entry), **N**...HOSPITAL, Don Pablo's, McDonald's, Tripp's Rest., Day's Inn, Belk, Dillard's, Drug Emporium, Firestone, JC Penney, Marshall's, MensWearhouse, Sears/auto, mall, same as 189, **S**...Bojangles, Outback Steaks, Comfort Suites, Microtel, Sleep Inn
189	US 158, Stratford Rd, Hanes Mall Blvd, **N**...HOSPITAL, CHIROPRACTOR, DENTIST, BP, Exxon, Bojangles, Chili's, Cozumel Mexican, Honeybaked Ham, McDonald's, O'Charley's, Olive Garden, Red Lobster, Sagebrush Steaks, Taco Bell, Best Western, Courtyard, Chevrolet, Hecht's, Jo-Ann Fabrics, OfficeMax, Marshall's, Michael's, Natural Health Foods, NTB, TireAmerica, VisionWorks, mall, **S**...VETERINARIAN/PetsMart, BP/mart, Shell/mart, Applebee's, Burger King, Chick-fil-A, Chinese Rest., Little Caesar's, LoneStar Steaks, Macaroni Grill, Outback Steaks, Oyster Bay Seafood, Subway, Tripp's Rest., Village Tavern, Extended Stay America, Hampton Inn, La Quinta, Sleep Inn, Barnes&Noble, Best Buy, Circuit City, CVS Drug, Home Depot, Kohl's, Lowe's Bldg, PetsMart, Sam's Club, Sports Authority, Food Lion, bank, laundry
188	US 421, to Yadkinville, Winston-Salem, to WFU(no EZ wb return), **1/2mi N off US 421**...Amoco/mart, BP/mart, Shell/mart, Texaco/mart, Boston Mkt, Burger King, McDonald's, Waffle House, Wendy's, CVS Drug, Kroger, Mercedes, Pennzoil/wash, Wal-Mart SuperCtr/24hr, cinema, cleaners/laundry
184	to US 421, Clemmons, **N**...VETERINARIAN, Shell/mart, Texaco/mart, Capt's Galley Seafood, KFC, Quincy's, Holiday Inn Express, carwash, **S**...DENTIST, Amoco/mart, BP/diesel/mart/wash, Chevron/mart, Citgo/diesel/mart, Etna/Dunkin Donuts/diesel/mart, Exxon/diesel/mart, Texaco/mart, Arby's, Baskin-Robbins, BBQ, Biscuitville, Burger King, Chinese Rest., Cozumel Mexican, Cracker Barrel, Dockside Seafood, Domino's, House of Pizza, Japanese Rest., Little Caesar's, McDonald's, Mi Pueblo Mexican, Mountain Fried Chicken, Pizza Hut, Sagebrush Steaks, Schlotsky's, Sonic, Subway, Taco Bell, Waffle House, Wendy's, Holiday Inn, Ramada Ltd, Super 8, Advance Parts, CarQuest, CVS Drug, Firestone, Food Fair, Food Lion, JiffyLube, K-Mart, Lowe's Foods, MailBoxes Etc, XpressLube, USPO, bank, carwash, hardware, laundry/cleaners
182	Bermuda Run(from wb, no re-entry), Tanglewood, no facilities
181.5mm	Yadkin River
180	NC 801, Tanglewood, **S**...DENTIST, CHIROPRACTOR, MEDICAL CARE, VETERINARIAN, Chevron/diesel/mart, Citgo/Subway/diesel, Shell/mart, Venezie's Italian, CVS Drug, Eckerd, Food Lion, Radio Shack, bank, hardware, laundry
177mm	**rest areas both lanes, full(handicapped)facilities, phone, vending, picnic tables, litter barrels, petwalk**
174	Farmington Rd, **N**...Exxon/diesel/mart, antiques, **S**...Furniture Gallery
170	US 601, Mocksville, **N**...Citgo/diesel/LP, Shell/diesel/rest./24hr, **S**...HOSPITAL, VETERINARIAN, Amoco/mart, BP/diesel/mart, Texaco/mart, Burger King, Cap'n Stevens Seafood, Dynasty Chinese, Ketchie Creek Café, KFC, McDonald's, Pizza Hut, Little Caesar's, Subway, Wendy's, Western Steer, Comfort Inn, Hi-Way Inn, Scottish Inn, Advance Parts, Ben Franklin Crafts, CVS Drug, Food Lion, Ford/Mercury, Goodyear/auto, USPO, Wal-Mart, bank, cinema, cleaners
168	US 64, to Mocksville, **N**...Mobil, Lake Myers RV Resort(3mi), **S**...Amoco/mart
162	US 64, Cool Springs, **N**...Lake Myers RV Resort(5mi), **S**...Texaco/mart, TravL Park
161mm	S Yadkin River
154	to US 64, Old Mocksville Rd, **N**...HOSPITAL, **S**...VETERINARIAN, Citgo, Jaybee's Hotdogs, Hallmark Inn

North Carolina Interstate 40

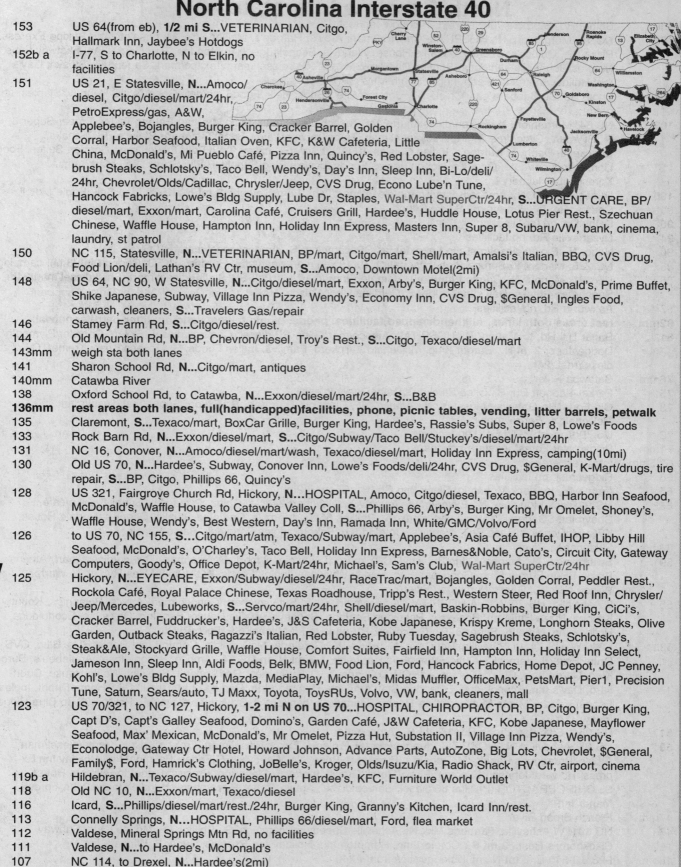

153 US 64(from eb), **1/2 mi S...**VETERINARIAN, Citgo, Hallmark Inn, Jaybee's Hotdogs

152b a I-77, S to Charlotte, N to Elkin, no facilities

151 US 21, E Statesville, **N...**Amoco/ diesel, Citgo/diesel/mart/24hr, PetroExpress/gas, A&W, Applebee's, Bojangles, Burger King, Cracker Barrel, Golden Corral, Harbor Seafood, Italian Oven, KFC, K&W Cafeteria, Little China, McDonald's, Mi Pueblo Café, Pizza Inn, Quincy's, Red Lobster, Sagebrush Steaks, Schlotsky's, Taco Bell, Wendy's, Day's Inn, Sleep Inn, Bi-Lo/deli/24hr, Chevrolet/Olds/Cadillac, Chrysler/Jeep, CVS Drug, Econo Lube'n Tune, Hancock Fabricks, Lowe's Bldg Supply, Lube Dr, Staples, Wal-Mart SuperCtr/24hr, **S...**URGENT CARE, BP/diesel/mart, Exxon/mart, Carolina Café, Cruisers Grill, Hardee's, Huddle House, Lotus Pier Rest., Szechuan Chinese, Waffle House, Hampton Inn, Holiday Inn Express, Masters Inn, Super 8, Subaru/VW, bank, cinema, laundry, st patrol

150 NC 115, Statesville, **N...**VETERINARIAN, BP/mart, Citgo/mart, Shell/mart, Amalsi's Italian, BBQ, CVS Drug, Food Lion/deli, Lathan's RV Ctr, museum, **S...**Amoco, Downtown Motel(2mi)

148 US 64, NC 90, W Statesville, **N...**Citgo/diesel/mart, Exxon, Arby's, Burger King, KFC, McDonald's, Prime Buffet, Shike Japanese, Subway, Village Inn Pizza, Wendy's, Economy Inn, CVS Drug, $General, Ingles Food, carwash, cleaners, **S...**Travelers Gas/repair

146 Stamey Farm Rd, **S...**Citgo/diesel/rest.

144 Old Mountain Rd, **N...**BP, Chevron/diesel, Troy's Rest., **S...**Citgo, Texaco/diesel/mart

143mm weigh sta both lanes

141 Sharon School Rd, **N...**Citgo/mart, antiques

140mm Catawba River

138 Oxford School Rd, to Catawba, **N...**Exxon/diesel/mart/24hr, **S...**B&B

136mm **rest areas both lanes, full(handicapped)facilities, phone, picnic tables, vending, litter barrels, petwalk**

135 Claremont, **S...**Texaco/mart, BoxCar Grille, Burger King, Hardee's, Rassie's Subs, Super 8, Lowe's Foods

133 Rock Barn Rd, **N...**Exxon/diesel/mart, **S...**Citgo/Subway/Taco Bell/Stuckey's/diesel/mart/24hr

131 NC 16, Conover, **N...**Amoco/diesel/mart/wash, Texaco/diesel/mart, Holiday Inn Express, camping(10mi)

130 Old US 70, **N...**Hardee's, Subway, Conover Inn, Lowe's Foods/deli/24hr, CVS Drug, $General, K-Mart/drugs, tire repair, **S...**BP, Citgo, Phillips 66, Quincy's

128 US 321, Fairgrove Church Rd, Hickory, **N...**HOSPITAL, Amoco, Citgo/diesel, Texaco, BBQ, Harbor Inn Seafood, McDonald's, Waffle House, to Catawba Valley Coll, **S...**Phillips 66, Arby's, Burger King, Mr Omelet, Shoney's, Waffle House, Wendy's, Best Western, Day's Inn, Ramada Inn, White/GMC/Volvo/Ford

126 to US 70, NC 155, **S...**Citgo/mart/atm, Texaco/Subway/mart, Applebee's, Asia Café Buffet, IHOP, Libby Hill Seafood, McDonald's, O'Charley's, Taco Bell, Holiday Inn Express, Barnes&Noble, Cato's, Circuit City, Gateway Computers, Goody's, Office Depot, K-Mart/24hr, Michael's, Sam's Club, Wal-Mart SuperCtr/24hr

125 Hickory, **N...**EYECARE, Exxon/Subway/diesel/24hr, RaceTrac/mart, Bojangles, Golden Corral, Peddler Rest., Rockola Café, Royal Palace Chinese, Texas Roadhouse, Tripp's Rest., Western Steer, Red Roof Inn, Chrysler/Jeep/Mercedes, Lubeworks, **S...**Servco/mart/24hr, Shell/diesel/mart, Baskin-Robbins, Burger King, CiCi's, Cracker Barrel, Fuddrucker's, Hardee's, J&S Cafeteria, Kobe Japanese, Krispy Kreme, Longhorn Steaks, Olive Garden, Outback Steaks, Ragazzi's Italian, Red Lobster, Ruby Tuesday, Sagebrush Steaks, Schlotsky's, Steak&Ale, Stockyard Grille, Waffle House, Comfort Suites, Fairfield Inn, Hampton Inn, Holiday Inn Select, Jameson Inn, Sleep Inn, Aldi Foods, Belk, BMW, Food Lion, Ford, Hancock Fabrics, Home Depot, JC Penney, Kohl's, Lowe's Bldg Supply, Mazda, MediaPlay, Michael's, Midas Muffler, OfficeMax, PetsMart, Pier1, Precision Tune, Saturn, Sears/auto, TJ Maxx, Toyota, ToysRUs, Volvo, VW, bank, cleaners, mall

123 US 70/321, to NC 127, Hickory, **1-2 mi N on US 70...**HOSPITAL, CHIROPRACTOR, BP, Citgo, Burger King, Capt D's, Capt's Galley Seafood, Domino's, Garden Café, J&W Cafeteria, KFC, Kobe Japanese, Mayflower Seafood, Max' Mexican, McDonald's, Mr Omelet, Pizza Hut, Substation II, Village Inn Pizza, Wendy's, Econolodge, Gateway Ctr Hotel, Howard Johnson, Advance Parts, AutoZone, Big Lots, Chevrolet, $General, Family$, Ford, Hamrick's Clothing, JoBelle's, Kroger, Olds/Isuzu/Kia, Radio Shack, RV Ctr, airport, cinema

119b a Hildebran, **N...**Texaco/Subway/diesel/mart, Hardee's, KFC, Furniture World Outlet

118 Old NC 10, **N...**Exxon/mart, Texaco/diesel

116 Icard, **S...**Phillips/diesel/mart/rest./24hr, Burger King, Granny's Kitchen, Icard Inn/rest.

113 Connelly Springs, **N...**HOSPITAL, Phillips 66/diesel/mart, Ford, flea market

112 Valdese, Mineral Springs Mtn Rd, no facilities

111 Valdese, **N...**to Hardee's, McDonald's

107 NC 114, to Drexel, **N...**Hardee's(2mi)

North Carolina Interstate 40

Morganton

106	Bethel Rd, **S**...Exxon/diesel/mart, Rainbow Inn/rest.
105	NC 18, Morganton, **N**...HOSPITAL, Chevron/mart, Arby's, Hardee's, McDonald's, Mr Omelet/24hr, Peking Express, Pizza Inn, Quincy's, Shoney's, Waffle House, Wendy's, Econolodge, Hampton Inn, Red Carpet Inn, Fastway Lube, Natures Bounty Healthfoods, Pontiac/Cadillac/GMC, **S**...QM/diesel/mart, Texaco/mart, Sagebrush Steaks, Day's Inn, Holiday Inn/rest., Sleep Inn, to South Mtns SP
104	Enola Rd, **S**...Cubbard/gas, Chick-fil-A, Jersey Mike's Subs, Belk, Food Lion/deli, Goody's, JC Penney, Staples, st patrol
103	US 64, Morganton, **N**...Exxon/diesel/mart/24hr, Texaco/diesel/mart, Max' Mexican Eatery, Tastee Freez, Super 8, **S**...BP/mart, Citgo/tires, Phillips 66/diesel/mart, RaceTrac/mart/atm, SuperTest/mart, Blake's BBQ/steaks, Checker's, Denny's, Hardee's, Italian Cuisine, KFC, LJ Silver, Subway, Taco Bell, Tokyo Diner, Comfort Suites, Food Lion, Goodyear/auto, Goody's, Honda, Ingles Foods, Lowe's Bldg Supply, Radio Shack, Wal-Mart/drugs, XpressLube, laundry
100	Jamestown Rd, **N**...Amoco/KrispyKreme/diesel/mart/24hr, Ford/Mercury, **2 mi N**...KFC, Taco Bell, Eagle Motel
98	Causby Rd, to Glen Alpine, **S**...B&B/food
96	Kathy Rd, no facilities
94	Dysartsville Rd, no facilities
90	Nebo, **N**...Exxon/diesel/rest., to Lake James SP, camping, **S**...Amoco/diesel/mart, RV Ctr
86	NC 226, Marion, to Spruce Pine, hwy patrol, **N**...Exxon/mart/wash/atm, Hardee's, KFC, Comfort Inn(3mi), camping
85	US 221, Marion, **N**...Comfort Inn(5mi), Hampton Inn, to Mt Mitchell SP, **S**...BP/diesel/mart, Shell/diesel/mart/24hr, Sagebrush Steaks, Day's Inn, Super 8, Carolina Chocolate
83	Ashworth Rd, no facilities
82mm	**rest areas both lanes, full(handicapped)facilities, phone, picnic tables, vending, litter barrels, petwalk**
81	Sugar Hill Rd, to Marion, **N**...HOSPITAL, Amoco/mart/24hr, BP/diesel, Chevron/diesel, Exxon, Chrysler/Plymouth/ Dodge/Jeep, **2 mi N**...Burger King, McDonald's, Eckerd, Family$, Ingles Foods, **S**...Texaco/KrispyKreme/TCBY/ diesel/rest./24hr
76mm	Catawba River
75	Parker Padgett Rd, **S**...Citgo/Stuckey's/Dairy Queen/diesel
73	Old Fort, Mtn Gateway Museum, **N**...Amoco, BP, Chevron, Hardee's, **S**...Exxon, Super Test/diesel/mart, McDonald's, Parts+/repair
72	US 70(from eb), Old Fort, no facilities
71mm	Pisgah Nat Forest, eastern boundary
67.5mm	truck rest area eb
66	Ridgecrest, no facilities
65	(from wb), to Black Mountain, Black Mtn Ctr, **N**...Super 8(2mi)
64	NC 9, Black Mountain, **N**...Exxon/mart, Texaco/Subway/mart/24hr, #1China, Pizza Hut, Super 8, BiLo/café, Chevrolet, $General, XpressLube, bank, **S**...Phillips 66/mart, Arby's, Campfire Steaks, Denny's, Huddle House, KFC, McDonald's, Taco Bell, Wendy's, Comfort Inn, Eckerd, Ingles Food/deli, Radio Shack
63mm	Swannanoa River
59	Swannanoa, **N**...Amoco, Citgo/Subway/TCBY/mart, Exxon/diesel/mart, Servco/mart, Texaco/diesel/mart, Athens Pizza, BBQ, Burger King, Goodyear/auto, Harley-Davidson, Ingles Foods, KOA(2mi), Miles RV Ctr, to Warren Wilson Coll, **S**...Gant's Furniture Outlet
55	E Asheville, US 70, **N**...VA HOSPITAL, Amoco/mart, Citgo/Subway/mart, Conoco/mart, Arby's, Bojangles, Kountry Kitchen, Pizza Hut(1mi), Popeye's, Poseidon Seafood, Shoney's, Waffle House, Best Inn, Day's Inn, Econolodge, Holiday Inn, Motel 6, Super 8, Go Groceries, RV park, to Mt Mitchell SP, Folk Art Ctr
53b a	I-240 W, US 74, to Asheville, Bat Cave, **N**...KFC, J&S Cafeteria, McDonald's, Comfort Inn, BabiesRUs, BiLo, CVS Drug, Eckerd, Hamrick's Outlet, PartsAmerica, bank, **1-2 mi N on US 74**...Amoco, Exxon/diesel, Applebee's, Burger King, Carrabba's, Chili's, Damon's, IHOP, O'Charley's, Olive Garden, Red Lobster, Subway, Waffle House, Court-yard, Day's Inn, Econolodge, Extended Stay America, Hampton Inn, Ramada Ltd, Circuit City, Home Depot, Ingles Foods, K-Mart, Midas Muffler, Office Depot, Sears/auto, bank, mall, **S**...Amoco/diesel/LP, Phillips 66, to Blue Ridge Pkwy
51	US 25A, Sweeten Creek Rd, **1/2 mi S**...Amoco/mart
50	US 25, Asheville, **N**...HOSPITAL, Amoco/mart, BP/diesel, Exxon/diesel/mart, Texaco/Krispy Kreme/diesel/mart, Arby's, Bruegger's Bagels, McDonald's, Pizza Hut, Popeye's, Wendy's, TGIFriday, Baymont Inn, Holiday Inn Express, Howard Johnson Express, Plaza Motel, Quality Inn/rest., Sleep Inn, PrecisionTune, to Biltmore House, **S**...CHIROPRACTOR, Phillips 66/diesel, Servco/diesel, Apollo Flame Pizza/pasta, Huddle House/24hr, Forest Manor Inn
47mm	French Broad River
47	NC 191, W Asheville, Farmer's Mkt, **N**...Asheville Speedway, camping, **S**...Phillips 66, Moose Café, Subway, Blackberry's Rest., **2 mi S**...Comfort Inn, Hampton Inn, Holiday Inn Express, Super 8, Wingate Inn
46b a	I-26 & I-240 E, **2 mi N**...multiple facilities from I-240
44	W Asheville, US 74, **N**...CHIROPRACTOR, Amoco/KrispyKreme/mart/wash/atm, Chevron/diesel, Conoco, Exxon/ mart, Servco/mart, Shell/Substation II/mart/24hr, Burger King, Cracker Barrel, El Chapala Mexican, Fatz Café,

Asheville

E

↑

↓

W

North Carolina Interstate 40

Pizza Hut, Popeye's, Szechwan Chinese, Waffle House, Wendy's, Best Western, Comfort Inn, Red Roof Inn, Sleep Inn, Super 8, Whispering Pines Motel, CarQuest, Chevrolet, Chrysler/Plymouth/Jeep, CVS Drug, Family$, Ingles Foods, Lowe's Bldg Supply, Mercedes/Mazda, bank, **S...**McDonald's, Shoney's, Subway, Budget Motel, Ramada Inn/rest., Home Depot, **1 mi S...**VETERINARIAN, Shell/mart, BBQ, J&S Cafeteria, Little Caesar's, Subway, TCBY, Eckerd, Food Lion

41mm	weigh sta both lanes
37	E Canton, **N...**Amoco/mart, Citgo/diesel/rest./24hr, Goodyear, **S...**Exxon/diesel/mart, Texaco, Day's Inn, Plantation Motel, KOA, Big Cove Camping(2mi)
33	Newfound Rd, to US 74, no facilities
31	Canton, **N...**Sagebrush Steaks, Econolodge, **S...**Amoco/diesel/mart, Exxon/McDonald's/mart/24hr, Texaco/Arby's/diesel/mart, Burger King, McDonald's, Pizza Hut(2mi), Subway(1mi), Taco Bell, Waffle House, Comfort Inn, Ingles Foods, Chevrolet/Pontiac/Olds/Buick, Ford, RV/truck repair
27	US 19/23, to Waynesville, Great Smokey Mtn Expswy, **3 mi S...**HOSPITAL, Shell/Burger King/mart, Papa's Pizza, Shoney's, Subway, Taco Bell, Food Lion, GNC, Lowe's Bldg Supply, Wal-Mart/drugs, to WCU(25mi)
24	NC 209, to Lake Junaluska, **N...**Pilot/diesel/rest./24hr, motel, **S...**Citgo/diesel/mart/24hr
20	US 276, to Maggie Valley, Lake Junaluska, **S...**Amoco/diesel/mart, Exxon/diesel/mart/café, Phillips 66/diesel/mart, Quality Inn(5mi), Creekwood Farm RV Park
16mm	Pigeon River
15	Fines Creek, no facilities
12mm	Pisgah NF eastern boundary
10mm	**rest area both lanes, full(handicapped)facilities, phone, vending, picnic tables, litter barrels, petwalk**
7	Harmon Den, no facilities
4mm	tunnel both lanes
0mm	North Carolina/Tennessee state line

North Carolina Interstate 77

Exit #	Services
105.5mm	North Carolina/Virginia state line
105mm	**Welcome Ctr sb, full(handicapped)facilities, info, phone, vending, picnic tables, litter barrels, petwalk**
103mm	weigh sta both lanes
101	NC 752, to US 52, to Mt Airy, Greensboro, **E...**HOSPITAL(12mi), Hampton Inn
100	NC 89, to Mt Airy, **E...**HOSPITAL(12mi), Brintle's/Citgo/diesel/rest./mart/24hr, Exxon/diesel/mart, Marathon/Subway/diesel/mart/atm, Texaco/diesel/mart/atm, Wagon Wheel Rest., Best Western, Comfort Inn(2mi), clothing outlet
93	to Dobson, **E...**Citgo/diesel/diner, Texaco/DQ/Stuckey's/diesel/mart, Surry Inn
85	NC 1138, CC Camp Rd, to Elkin, **W...**Exxon/diesel/mart, **3 mi W...**Burger King, Hardee's, KFC, McDonald's, Elk Inn
83	US 21 byp, to Sparta, no facilities
82.5mm	Yadkin River
82	NC 67, Elkin, **E...**Amoco/diesel/mart/24hr, BP/Case Knife Outlet, Chevron/diesel/mart/atm, Citgo/Subway/mart, Arby's, Cracker Barrel, Jordan's Rest., Holiday Inn Express, Holly Ridge Camping(8mi), **W...**Exxon/Baskin-Robbins/TCBY/diesel/mart, Bojangles, McDonald's, New China, Shoney's, Waffle House, Wendy's, Comfort Inn, Day's Inn, Hampton Inn, Roses Motel, AutoValue Parts, Buick/Olds/Pontiac/GMC, D-Rex Drug, Family$, Food Lion/deli, tires
79	US 21 S, to Arlington, **E...**Shell/mart, Super 8, **W...**Texaco/diesel/mart, Sally Jo's Kitchen, Country Inn, auto repair
73b a	US 421, to Winston-Salem(20mi), **E...**Amoco/diesel/24hr, Exxon/diesel/mart, Sleep Inn(8mi), Welborn Motel, Yadkin Inn
72mm	**rest area nb, full(handicapped)facilities, phone, vending, picnic tables, litter barrels, petwalk**
65	NC 901, to Union Grove, Harmony, **E...**Van Hoy Farms Camping, **W...**Amoco/diesel/mart, BP/diesel/mart, Citgo, Texaco/Subway/TCBY/diesel/mart/atm/24hr, Burger Barn, Honey's Café, Ace Hardware, Fiddler's Grove Camping, bank
63mm	**rest area sb, full(handicapped)facilities, phone, vending, picnic tables, litter barrels, petwalk**
59	Tomlin Mill Rd, **W...**Citgo/mart
56.5mm	S Yadkin River
54	US 21, to Turnersburg, **E...**Citgo/mart, auto auction, **W...**Texaco/diesel/mart, flea mkt

Elkin

North Carolina Interstate 77

S t a t e s v i l l e

51b a	I-40, E to Winston-Salem, W to Hickory, no facilities
50	E Broad St, Statesville, **E**...MEDICAL CARE, EYECARE, Amoco, BP/mart, Crown/diesel/mart/wash, Etna/diesel, Arby's, Bojangles, Burger King, Domino's, Hardee's, IHOP, McDonald's, Pizza Hut, Shoney's, TCBY, Village Inn Pizza, Wendy's, Fairfield Inn, Red Roof Inn, Belk, Eckerd, Harris-Teeter, JC Penney, JR Outlet, K-Mart, Midas Muffler, Peebles, Radio Shack, Sears, USPO, Winn-Dixie, bank, cinema, laundry, mall, **W**...Amoco(1mi)
49b a	US 70, G Bagnal Blvd, to Statesville, **E**...BP/mart/wash, CircleK/gas, Etna/mart, PDI/gas, Texaco/diesel/mart, KFC, Waffle House, Best Western, Comfort Inn, Holiday Inn, Motel 6, Super 8, Buick/GMC/Olds/Cadillac, Dodge, Ford, Honda, Lincoln/Mercury, Nissan, Subaru, Toyota, mufflers/brakes, **W**...Amoco/diesel/mart, Chevron, Citgo/mart, BestStay Inn, Microtel
45	to Troutman, Barium Springs, **E**...KOA, **W**...Amoco/diesel(2mi), Chevron/BBQ/diesel, antiques
42	US 21, NC 115, Oswalt, to Troutman, **E**...Citgo/Subway/Taco Bell/TCBY/diesel/mart/24hr, **W**...to Duke Power SP, camping
39mm	**rest area/weigh sta both lanes, full(handicapped)facilities, phone, picnic tables, litter barrels, petwalk, vending**
36	NC 150, Mooresville, **E**...EYECARE, Accel/diesel/mart, Exxon, Shell/mart/24hr, Bob Evans, Burger King, Denny's, FatBoy's Cafe, McDonald's, Peking Palace, Taco Bell, Texas Steaks, Waffle House, Wendy's, Day's Inn, Ramada Ltd, Belk, FastLube, Goody's, Griffin Tire/repair, Harris-Teeter, K-Mart, Staples, Suzuki, Wal-Mart SuperCtr/24hr, cinema, **W**...VETERINARIAN, BJ's Whse/gas, BP/diesel/mart/wash, Citgo/KrispyKreme/BBQ/mart/atm, Servco/mart, Texaco/diesel/mart/24hr, Arby's, Chick-fil-A, Cracker Barrel, Golden Corral, Hardee's, Kyoto Japanese, McDonald's, Monterrey Mexican, Subway, Hampton Inn, Super 8, Wingate Inn, CVS Drug, Food Lion/deli, Lowe's Bldg Supply, bank, laundry
33	US 21 N, **E**...HOSPITAL, Phillips 66/diesel, **W**...BP/mart, Citgo/diesel/mart/24hr
30	Davidson, **E**...Exxon/diesel, to Davidson College
28	US 21 S, NC 73, Cornelius, Lake Norman, **E**...Amoco, Cashion's/diesel/mart, Citgo/mart/24hr, Bojangles, Denny's, Hardee's, Mom's Country Store/Rest., Prime Buffet Rest., Quincy's, Shoney's, Subway, Hampton Inn, Holiday Inn/rest., bank, **W**...CHIROPRACTOR, Burger King, El Cancun Mexican, Honeybaked Ham, Jersey Mike's Subs, KFC, Kobe Japanese, Lake Norman Brewing, Little Caesar's, LoneStar Steaks, Lotus 28 Chinese, McDonald's, Papa John's, Pizza Hut, Taco Bell, Wendy's, Best Western, Comfort Inn, Microtel, Quality Inn, BiLo/24hr, Chrysler/Jeep, Dodge, Eckerd, Food Lion, Harris-Teeter, Radio Shack, SteinMart, West Marine, bank, cinema
25	NC 73, Concord, Lake Norman, **E**...MEDICAL CARE, EYECARE, Shell/diesel/mart, Texaco/mart, Atlanta Bread Factory, Burger King, Chili's, Fuddrucker's, Hops Rest., Longhorn Steaks, McDonald's, O'Charley's, US Subs, Wendy's, Country Suites, Hawthorn Suites, Ramada Ltd, Home Depot, Kohl's, Lowe's Bldg Supply, MailBoxes Etc, Old Navy, Target/drugs, Winn-Dixie/deli, bank, cinema, **W**...76/Circle K/Blimpie, Arby's, Bob Evans, Bojangles, Carrabba's, Dairy Queen, Outback Steaks, Subway, Taipei House, Candlewood Suites, Courtyard, Food Lion/deli, bank, cleaners, to Energy Explorium
23	NC 73, Huntersville, **E**...MEDICAL CARE, EYECARE, Amoco/mart, Texaco/mart/24hr, Hardee's, Mama Mia Mexican, Palace of China, Waffle House, Wendy's, Holiday Inn Express, Red Roof Inn, Eckerd, Food Lion, Ford, Hancock Fabrics, Radio Shack, USPO, Lake Lube, bank, laundry, **W**...Sam's Mart/diesel, BiLo/deli, CVS Drug, Harris-Teeter/deli
18	Harris Blvd, Reames Rd, **E**...HOSPITAL, Phillips 66/Arby's/diesel/mart, Texaco/diesel/mart, Bob Evans, Harris Grill, Jack-in-the-Box, Waffle House, Fairfield Inn, Hampton Inn, Hilton Garden, Suburban Lodge, to UNCC, Univ Research Park
16b a	US 21, Sunset Rd, **E**...Amoco/Circle K/atm, Jakes Trkstp/diesel/rest., Shell/diesel/mart/24hr, Capt D's, Hardee's, KFC, McDonald's, Subway, Taco Bell, Waffle House, Wendy's, Best Stay Inn, Day's Inn, Super 8, AllPro Parts, AutoZone, Kerr Drug, Winn-Dixie, bank, **W**...Citgo/mart/wash, Texaco/diesel/mart, BBQ, Bojangles, Burger King, Denny's, Domino's, Waffle House, Sleep Inn, Sunset Motel, RV Ctr
13b a	I-85, S to Spartanburg, N to Greensboro
12	La Salle St, **W**...Texaco/mart
11b a	I-277, Brookshire Fwy, no facilities

C h a r l o t t e

10	Trade St, 5th St(from nb), **E**...downtown, to Discovery Place, **W**...Citgo/mart, Bojangles
10a	US 21(from sb), Moorhead St, downtown
9	I-277, John Belk Fwy, downtown, **E**...HOSPITAL, stadium
8	Remount Rd(from nb, no re-entry), no facilities
7	Clanton Rd, **E**...Citgo/diesel/mart, Krystal/24hr, McDonald's, Waffle House, Wendy's, Day's Inn, Econolodge, Holiday Inn Express, Motel 6, Ramada, Super 8, carwash, **W**...Amoco/mart
6b a	US 521, Billy Graham Pkwy, **E**...Citgo/mart, Exxon/diesel/mart, 76/diesel/mart/wash, Speedway/diesel/mart, Texaco/diesel/mart/wash, Bojangles, Burger King, Capt D's, Checker's, Chinese Rest., IHOP, KFC, Krispy Kreme, McDonald's, Rockola Café, Shoney's, Steak&Ale, Wendy's, Comfort Inn, Day's Inn, Holiday Inn, Howard Johnson, Registry Hotel, to airport, **1 mi E on US 521**...Sunoco/mart, Arby's, Waffle House, Tripp's Rest., CVS Drug, to Queens Coll, **W**...Phillips 66, McDonald's, Best Western, Clarion, Embassy Suites, Homestead Suites, Summerfield Suites

N

↑

↓

S

North Carolina Interstate 77

5 Tyvola Rd, **E...**Citgo/diesel/mart, BBQ, Blackeyed Pea, Chili's, Japanese Rest., LoneStar Steaks, McDonald's, Comfort Inn, Econolodge, Hampton Inn, Hilton, Marriott, Orchard Inn, Radisson, Residence Inn, Studio+, Wingate Inn, BiLo/24hr, Buick/GM, Jaguar, Target, bank, **1 mi E on US 521...**Texaco, El Cancun Mexican, Flamingo Rest., Hooters, McDonald's, Ryan's, Taco Bell, Pizza Hut, Wendy's, Pontiac, Office Depot, OfficeMax, Pearle Vision, Meineke Muffler, TireAmerica, Western Auto, **W...**to Coliseum

4 Nations Ford Rd, **E...**Amoco, Citgo/Circle K/24hr, Caravel Seafood, Hardee's, Shoney's, Best Inn, Innkeeper Inn, La Quinta, Red Roof, Villager Lodge, Sam's Club, **W...**Shell/Burger King/mart

2 Arrowood Rd, to I-485(from sb), **E...**Shell, Bob Evans, Jack-in-the-Box, LJ Silver, McDonald's, Wendy's, AmeriSuites, Courtyard, Fairfield Inn, TownePlace Suites, **W...**Bojangles, Hampton Inn, MainStay Suites, SpringHill Suites

1.5mm **Welcome Ctr nb, full(handicapped)facilities, info, phone, vending, picnic tables, litter barrels, petwalk**

1 Westinghouse Blvd, to I-485(from nb), **E...**Amoco/diesel/mart, Citgo/diesel/mart, Jack-in-the-Box, Subway, Waffle House, Super 8, **W...**Texaco/diesel/mart/wash, Burger King, McDonald's, Rodeway Inn, PDQ Lube, Sears, bank

0mm North Carolina/South Carolina state line

North Carolina Interstate 85

Exit #	Services

234mm North Carolina/Virginia state line

233 US 1, to Wise, **E...**Wise Trkstp/Shell/diesel/rest.

231mm **Welcome Ctr sb, full(handicapped)facilities, phone, picnic tables, litter barrels, petwalk**

229 Oine Rd, to Norlina, no facilities

226 Ridgeway Rd, **W...**to Kerr Lake, to St RA, no facilities

223 Manson Rd, **E...**BP/diesel/mart, camping, **W...**to Kerr Dam

220 US 1, US 158, Fleming Rd, to Middleburg, **E...**Amoco/diesel/mart, **W...**Citgo/diesel/mart, Chex Motel/rest.

218 US 1 S(from sb), to Raleigh

217 Nutbush Bridge, **E...**same as 215 on US 158, **W...**Exxon/diesel/mart, Kerr Lake RA

215 US 158 BYP E, Henderson(no EZ return from nb), **E...**Citgo/diesel/mart, Shell/mart, Winoco/diesel/mart, BBQ, Burger King, Mystic Grill, PD Quix Burgers, Tastee Freez, 220 Seafood Rest., Subway, Budget Host, Comfort Inn, Howard Johnson, Quality Inn, Scottish Inn, Food Lion, Goodyear/auto, Roses, bank, laundry, services on US 158

214 NC 39, Henderson, **E...**BP/mart, Texaco/mart, TrueValue, **W...**Amoco/diesel/mart, Shell/HotStuff Pizza/mart

213 US 158, Dabney Dr, to Henderson, **E...**EYECARE, Amoco, Shell/mart, Bamboo Garden, Bojangles, Dairy Queen, Denny's, KFC, McDonald's, Papa John's, Pizza Inn, Subway, Wendy's, CVS Drug, Eckerd, Food Lion, Goodyear/auto, Lowe's Foods, Radio Shack, Roses, Winn-Dixie, bank, cinema, **W...**Exxon/mart, Shell/mart, Golden Corral, Hurricanes Grill, Chevrolet/Buick, Chrysler/Plymouth/Dodge, $Tree, Ford/Lincoln/Mercury, K-Mart, Lowe's Bldg Supply, Pontiac/Olds, Tires+, same as 212

212 Ruin Creek Rd, **E...**Shell/diesel/mart/wash, Cracker Barrel, Mazatlan Mexican, SiLo Rest., Day's Inn, **W...**HOSPITAL, Amoco/Burger King/mart, BBQ, Western Sizzlin, Hampton Inn, Holiday Inn Express, Jameson Inn, Sleep Inn, Belk, Goody's, JC Penney, Wal-Mart/auto, mall

209 Poplar Creek Rd, **W...**Vance-Granville Comm Coll

206 US 158, Oxford, **E...**Citgo/mart, **W...**Texaco/diesel/mart, airport

204 NC 96, Oxford, **E...**Amoco/diesel/mart, King's Inn, Honda, Olds/Pontiac/Buick/GMC, **W...**HOSPITAL, Exxon/DQ/mart, Shell/mart/atm/24hr, Texaco, Trade/diesel/mart, Burger King, Chinese Rest., KFC, Little Caesar's, McDonald's, Pizza Hut, Subway, Taco Bell, Wendy's, Zeko's Italian, Ramada Inn, Byrd's Foods, GNC, Radio Shack, Wal-Mart, bank, laundry

202 US 15, Oxford, **2 mi W...**Crown Motel

199mm **rest area both lanes, full(handicapped)facilities, phone, picnic tables, litter barrels, petwalk**

198mm Tar River

191 NC 56, Butner, **E...**DENTIST, EYECARE, Amoco/diesel/mart, BP/mart, TradeMart/diesel, BBQ, Bojangles, Burger King, China Taste Rest., Domino's, KFC/Taco Bell, McDonald's, Pizza Hut, Subway, Wendy's, Comfort Inn, AutoValue Parts, Eckerd, Food Lion, Herbs4U, M&H Tires, Pennzoil, bank, carwash, to Falls Lake RA, **W...**Exxon/diesel/mart/24hr, Shell/diesel/mart, Hardee's, Econolodge, Holiday Inn Express, Ramada Ltd, Sunset Inn, Goodyear/auto

189 Butner, no facilities

North Carolina Interstate 85

186b a	US 15, to Creedmoor, no facilities
185mm	Falls Lake
183	Redwood Rd, **E**...Day's Inn/gas/24hr, Redwood Café
182	Red Mill Rd, **E**...Exxon/repair, Kenworth/Volvo Trucks
180	Glenn School Rd, **W**...Heritage Gas/mart
179	E Club Blvd, **E**...Exxon/mart, **W**...BP/mart
178	US 70 E, to Raleigh, Research Triangle, RDU Airport, Falls Lake RA
177b c	Avondale Dr, NC 55, same as 177a, **E**...Super 8, **W**...Golden Corral
a	Durham, downtown, **W**...Amoco/mart, BP/mart, Joy Gas, Texaco/mart, American Hero Subs, Arby's, Dunkin Donuts, Hardee's, KFC, McDonald's, Pizza Hut, Shoney's, Sizzler, Chesterfield Motel, Advance Parts, Family$, K-Mart/auto, Quick10 Lube, cleaners, laundry
176b a	Gregson St, US 501 N, **E**...HOSPITAL, Crown/mart, Shell, Biscuitville, Burger King, Subway, Tripp's Rest., Belk, Museum of Life&Science, Pearle Vision, Sears/auto, bank, mall
175	Guess Rd, **E**...Texaco/diesel/mart/wash, Pizza Hut, Carolina Duke Motor Inn, Holiday Inn Express, Super 8, **W**...CHIROPRACTOR, DENTIST, VETERINARIAN, Amoco/diesel, BP/diesel/mart, Etna/mart, Exxon/nmart, Bojangles, Honey's Diner/24hr, Tokyo Express, Zero's Subs, Red Roof Inn, CVS Drug, Home Depot, Kroger/deli, PetsMart, antiques
174a	Hillandale Rd, **W**...MEDICAL CARE, BP/diesel/mart, PanPan Diner/24hr, Courtyard, Hampton Inn, Howard Johnson, Shoney's Inn, Kerr Drug, Winn-Dixie/deli
b	US 15 S, US 501 S(from sb), **E**...Forest Inn
173	US 15, US 501, US 70, W Durham, **E**...HOSPITAL, CHIROPRACTOR, DENTIST, VETERINARIAN, BP, Exxon/diesel, Texaco/mart, Arby's, BBQ, Bojangles, Burger King, Checker's, Cracker Barrel, DogHouse Rest., Domino's, Galley Seafood, Italian Garden Rest., McDonald's, Miami Subs, Subway, Taco Bell, Waffle House, Wendy's, Econolodge, Fairfield Inn, Hilton, Holiday Inn, Byrd's Foods, Dodge/Jeep, Eckerd, Kroger/deli/bakery/24hr, Rite Aid, Western Auto/service, bank, tires
172	NC 147 S, to US 15 S, US 501 S(from nb), Durham
170	to NC 751, to Duke U(no EZ return from nb), **E**...Harbor Bay Seafood, Scottish Inn, Skyland Inn Best Western/rest., **W**...to Eno River SP
165	NC 86, to Chapel Hill, **E**...Amoco/mart, ExpressAmerica Trkstp/diesel/rest., **W**...BP/diesel
164	Hillsborough, **E**...Amoco, Citgo, McDonald's, Holiday Inn Express, **W**...CHIROPRACTOR, DENTIST, Exxon/diesel/mart, Shell/mart, Bojangles, Burger King, Canton House Chinese, Casa Ibarra Mexican, Domino's, Hardee's, KFC, Mayflower Seafood, Occoneechee Steaks, Pizza Hut, Subway, Waffle House, Wendy's, Microtel, Southern Country Inn, Chevrolet/Buick, Food Lion, Ford, Firestone/auto, GNC, Goodyear/auto, Kerr Drug, Lowe's Foods, Merchant Tire, Wal-Mart, auto parts, bank, laundry
163	I-40 E, to Raleigh. **I-85 S and I-40 W run together 38 mi.**
161	to US 70 N, NC 86 N, no facilities
160	Efland, **W**...BP/diesel, Texaco/diesel, Mary's Grill
158mm	weigh sta both lanes
157	Buckhorn Rd, **E**...Amoco/diesel/mart, Citgo, Exxon, Petro/Mobil/diesel/rest./24hr
154	Mebane-Oaks Rd, **E**...CHIROPRACTOR, Shell/Blimpie/diesel/mart/24hr, Biscuitville, **W**...DENTIST, Amoco/tires/24hr, Citgo/mart, Texaco/diesel/mart/24hr, Bojangles, McDonald's, Budget Inn/rest., KidsWear Outlet, Mast Drugs, Winn-Dixie, bank
153	NC 119, Mebane, **E**...BP/KFC/Taco Bell/Pizza Hut/mart, Hampton Inn, **W**...VETERINARIAN, Exxon/Burger King/mart/atm, Jersey Mike's Subs, Subway, CVS Drug, Food Lion/deli
152	Trollingwood Rd, **E**...Speedway/Country Kitchen/diesel/mart/24hr, **W**...Amoco/diesel/mart, Fuel City/diesel/mart, Cactus Café, funpark(1mi)
150	Haw River, to Roxboro, **E**...Dockside Dolls Rest., **W**...Citgo/Wilco/DQ/Wendy's/diesel/mart/24hr, Flying J/Conoco/diesel/LP/rest./24hr
148	NC 54, Graham, **E**...Amoco/diesel/mart/24hr, BP/diesel/mart/24hr, QP/mart, Waffle House, to Harbor House Seafood, **W**...Econolodge/rest., truckwash/repair
147	NC 87, Graham, to Pittsboro, **E**...BP/mart/24hr, Citgo/diesel, Phillips 66/diesel, QP/mart, Servco/diesel/mart/wash, Arby's, Bojangles, Burger King, Domino's, Harbor House Seafood, Sagebrush Steaks, Subway, Wendy's, Chevrolet, Chrysler/Jeep, CVS Drug, Food Lion, Ford, Goodyear, RV Ctr, Winn-Dixie/deli, bank, laundry, **W**...HOSPITAL, Citgo/diesel/mart, Exxon/diesel/mart, Texaco/diesel, Hardee's, McDonald's, Taco Bell, bank
145	NC 49, Burlington, **E**...Speedway/mart, Texaco/diesel/mart/lube, Capt D's, Day's Inn, Microtel, Motel 6, Red Roof Inn, **W**...BP/mart/wash, Exxon/diesel, Biscuitville, Bojangles, Burger King, Chinese Rest., Hardee's, KFC, Perkins/24hr, Quincy's, Speedo's Burgers, Ship Ahoy Seafood, Subway, Waffle House, Western Sizzlin, Comfort Inn, Holiday Inn, Scottish Inn, Eckerd, Food Lion/deli, K-Mart, Pontiac/Dodge/Chrysler, factory oulet/famous brands, bank, carwash, laundry
143	NC 62, Burlington, **E**...Archdale/diesel/mart, Citgo/Wendy's/mart, Bob Evans, Hardee's, Waffle House, JR Discount Ctr, airport, to Alamance Battleground, **W**...Exxon, 76/Circle K, Cutting Board Rest., La Fiesta Mexican, Libby Hill Seafood, Nick's Cuisine, Ramada, Chevrolet, Food Lion, Ford, Kerr Drug, Mitsubishi, Olds/Cadillac, bank

Durham

Burlington

N

S

North Carolina Interstate 85

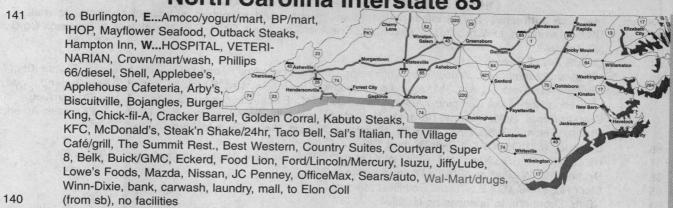

141   to Burlington, **E**...Amoco/yogurt/mart, BP/mart, IHOP, Mayflower Seafood, Outback Steaks, Hampton Inn, **W**...HOSPITAL, VETERINARIAN, Crown/mart/wash, Phillips 66/diesel, Shell, Applebee's, Applehouse Cafeteria, Arby's, Biscuitville, Bojangles, Burger King, Chick-fil-A, Cracker Barrel, Golden Corral, Kabuto Steaks, KFC, McDonald's, Steak'n Shake/24hr, Taco Bell, Sal's Italian, The Village Café/grill, The Summit Rest., Best Western, Country Suites, Courtyard, Super 8, Belk, Buick/GMC, Eckerd, Food Lion, Ford/Lincoln/Mercury, Isuzu, JiffyLube, Lowe's Foods, Mazda, Nissan, JC Penney, OfficeMax, Sears/auto, Wal-Mart/drugs, Winn-Dixie, bank, carwash, laundry, mall, to Elon Coll

140   (from sb), no facilities

139mm   **rest area both lanes, full(handicapped)facilities, phone, picnic tables, litter barrels, vending**

138   NC 61, Gibsonville, **W**...BP/TA/diesel/rest., Shell, Texaco/diesel/mart(1mi), Burger King, Popeye's, Days Inn

135   Rock Creek Dairy Rd, **W**...Citgo, Exxon, Jersey Mike's Subs, McDonald's

132   Mt Hope Church Rd, **W**... Exxon/Wilco/Wendy's/diesel/mart/atm/24hr, Shell/diesel/mart, Texaco/mart, Hampton Inn

131   new exit

130   McConnell Rd, **E**...VETERINARIAN, Texaco/mart/repair, **W**...Replacements LTD Outlet

128   NC 6, E Lee St, **W**...BP/Blimpie/diesel/wash, Phillips 66, Holiday Inn Express, I-85 Family Rest., access to Coliseum, KOA

127   US 29 N, US 70, US 220 N(from nb), to Reidsville, **W**...KOA

126   US 421 S, to Sanford, **E**...Exxon/mart, Arby's, Biscuitville, Burger King, Domino's, Golden Pizza, McDonald's, Szechuan Chinese, Subway, Wendy's, Advance Parts, BiLo, CVS Drug, Food Lion, Goodyear/auto, bank, cleaners, tires, **W**...auto repair

125   S Elm St, Eugene St, **E**...Amoco/mart/wash, Texaco/diesel/mart/24hr, DairyMart/gas, Cricket Inn, Day's Inn, Howard Johnson, Super 8, CarQuest, **W**...Citgo/diesel/mart, Chevron/mart, Crown/mart, Bojangles, Sonic, Homestead Lodge, Ramada Ltd, AutoZone, Family$, Food Lion, Kerr Drug, cleaners/laundry, tires

124   Randleman Rd, **E**...Exxon, Texaco/mart/24hr, Cookout Drive-Thru, Mayflower Seafood Rest., Quincy's, Waffle House, Wendy's, Cavalier Inn, K-Mart, Optical Place, bank, **W**...Amoco/mart, Citgo, Crown/mart, Shell/mart, Arby's, BBQ, Biscuitville, Burger King, Chinese Rest., Dairy Queen, KFC, McDonald's, Pizza Hut, Substation II, Taco Bell, SouthGate Motel, Advance Parts, GreaseMonkey, Harley-Davidson, Meineke Muffler, Precision Tune, bank, carwash, laundry, tires

123   I-40 W(from sb), to Winston-Salem. **I-85 N and I-40 E run together 38 mi.**

122c   US 220, Rehobeth Church Rd, to Asheboro, **E**...Motel 6, **W**...Citgo/diesel/mart/rest., auto repair

122b a   (b only from nb), **N**...to I-40 W, **S**...US 220 S, to Asheboro

121   Holden Rd, **E**...Texaco/mart/wash, Arby's, Burger King, Dairy Queen, Denny's, K&W Cafeteria, FashionBug, GNC, K-Mart, Winn-Dixie, waterpark, **W**...Americana Motel, Howard Johnson

120   Groometown Rd, **W**...Phillips 66/diesel/mart/24hr

118   US 29 S, US 70 W, Jamestown, to High Point, **W**...HOSPITAL, Grandover Resort/golf

115mm   Deep River

113   NC 62, Archdale, **W**...BP/diesel/mart, Best Western

111   US 311, Archdale, to High Point, **E**...Bamboo Garden, Big Guy Subs, Hardee's, Little Caesar's, Wendy's, Innkeeper, CVS Drug, $General, Food Lion/deli, Lowe's Foods/24hr, bank, fabric outlet, **W**...HOSPITAL, Amoco/diesel/mart/atm/24hr, Citgo/mart, Exxon/McDonald's/mart/atm, Phillips 66/diesel/mart, Texaco/diesel/mart/wash, Waffle House, Comfort Inn, Hampton Inn, USPO, **2 mi W**...Bojangles, Burger King, Hardee's, KFC, Papa John's, Pizza Hut, SaltMarsh Annie's Seafood, Taco Bell, Advance Parts, Eckerd, Firestone, NAPA

108   Hopewell Church Rd, no facilities

106   Finch Farm Rd, **E**...Exxon/diesel/mart

103   NC 109, to Thomasville, **E**...DENTIST, Texaco/mart, Arby's, Taco Bell, CVS Drug, Ingles Foods/deli, K-Mart, Radio Shack, **W**...CHIROPRACTOR, EYECARE, Amoco/mart, BP/diesel/mart, Coastal/mart, Crown/mart/24hr, Mobil, Phillips 66, Texaco, Wilco/mart, Biscuitville, Burger King, Capt Tom's Seafood, China Garden, Flashburger, Golden Corral, Hardee's, KFC, Loflin's Rest., McDonald's, Mr Gatti's, Papa John's, PieZone Pizza, Pizza Hut, Sonic, Sunrise Grille, Waffle House, Wendy's, Howard Johnson, Ramada Ltd, Thomasville Inn, Advance Parts, AutoZone, Eckerd, Family$, Food Lion, Guy's Drug, Merchant's Tire/auto, Winn-Dixie, XpressLube, bank, hardware

102   Lake Rd, **W**...HOSPITAL, Phillips 66/diesel/mart, Texaco/KrispyKreme/diesel/mart, Day's Inn/rest., Bapt Children Home

N   **S**

North Carolina Interstate 85

100mm	**rest area both lanes, full(handicapped)facilities, phone, vending, picnic tables, litter barrels, petwalk**
96	US 64, Lexington, to Asheboro, E...Exxon, W...Chevron/diesel/mart, Citgo, Gant/diesel(1mi), to Davidson Co Coll, NC Zoo
94	Old US 64, E...Texaco
91	NC 8, to Southmont, E...Amoco/diesel/mart/atm, Citgo/mart, Phillips 66/mart, Texaco/diesel/mart, Biscuit King, Capt Stevens Seafood, KFC, McDonald's, Pizza Oven, Sonic, Subway, Wendy's, Comfort Suites, Super 8, Food Lion, High Rock Lake Camping, carwash, W...HOSPITAL, Exxon/diesel/mart, QM/mart, Arby's, Burger King, Cracker Barrel, Golden Corral, Hardee's, Hunan Chinese, Little Caesar's, Taco Bell, Holiday Inn Express, American Childrens Home, Belks, GNC, Goody's, Hamrick's Clothing, Ingles Foods/deli, Wal-Mart
88	Linwood, W...HOSPITAL, BP/diesel/mart
87	US 29, US 70, US 52(from nb), High Point, W...HOSPITAL, to Best Western, airport, mall
86	Belmont Rd, W...Phillips 66/Bill's Trkstp/diesel/rest./24hr
85	Clark Rd, to NC 150, E...Tracks End Rest., flea mkt
83	NC 150(from nb), to Spencer, no facilities
82	US 29, US 70(from sb), to Spencer, no facilities
81.5mm	Yadkin River
81	Spencer, E...Amoco/diesel/mart, W...Texaco/mart
79	Spencer, E Spencer, Spencer Shops St HS, E...Chanticleer Motel, auto repair
76b a	US 52, Salisbury, to Albemarle, E...VETERINARIAN, DENTIST, OPTOMETRIST, Amoco/mart, Citgo, RaceTrac/mart/24hr, Speedway/mart, Applebee's, Athena Diner, China Rainbow, Golden Bee Rest., IHOP, Italy Café, Lighthouse Seafood, Little Caesar's, LoneStar Steaks, Schlotsky's, Shoney's, Sleep Inn, Aldi Foods, Circuit City, CVS Drug, Dodge, Eckerd, Food Lion/deli, GNC, Lowe's Bldg Supply, Meineke Muffler, Staples, Super 8, auto repair, bank, cleaners, cinema, W...HOSPITAL, Exxon/diesel/mart, Servco/mart, 76/diesel/mart/wash, Shell/mart, Bojangles, Burger King, Capt D's, Chick-fil-A, China Garden, Christo's Rest., Dunkin Donuts, El Cancun Mexican, Ham's Rest., Hardee's, KFC, McDonald's, Outback Steaks, Pizza Hut, Taco Bell, Village Inn Pizza, Waffle House/24hr, Wendy's, Comfort Suites, Howard Johnson, Advance Parts, AutoZone, Family$, Firestone/auto, Goodyear/auto, K-Mart, Office Depot, USPO, XpressLube/wash
75	US 601, Jake Alexander Blvd, E...Arby's, Farmhouse Rest., Ramada Ltd, Dan Nicholas Park, W...Amoco/mart/24hr, BP/mart, Citgo/mart, Exxon/diesel/mart, Texaco/diesel, Ichiban Japanese, Ryan's, Sagebrush Steaks, Subway, Waffle House, Wendy's, Best Western, Day's Inn, Hampton Inn, Holiday Inn, Chevrolet/Olds/Cadillac, Chrysler/Plymouth/Jeep, Ford/Toyota, Honda/Kia, Nissan/Pontiac/GMC, Wal-Mart, hardware
74	Julian Rd, no facilities
72	Peach Orchard Rd, W...airport
71	Peeler Rd, E...Texaco/diesel/rest./24hr, W...Citgo/Wilco/Bojangles/Taco Bell/diesel/24hr
70	Webb Rd, E...flea mkt, W...st patrol
68	US 29, US 601, China Grove, to Rockwell, no facilities
63	Kannapolis, E...Speedway/Subway/diesel/mart, Waffle House, Best Western, 3 mi W...Exxon/diesel/mart, China Buffet, Hardee's, KFC, repair
60	Earnhardt Rd, E...HOSPITAL, Exxon/mart(1mi), Shell/Burger King/mart, Cracker Barrel, Kramer&Eugene's Café(1mi), Schlotsky's, Texas Roadhouse, Hampton Inn, Sleep Inn, MailBoxes Etc(1mi), W...BP/diesel, visitor info
59mm	**rest area both lanes, full(handicapped)facilities, phone, vending, picnic tables, litter barrels, petwalk**
58	US 29, US 601, Concord, E...HOSPITAL, Crown/mart/24hr, Exxon/mart, Texaco/diesel/mart, Wilco/diesel/mart, Applebee's, BearRock Café, Burger King, Capt D's, Chick-fil-A, China Orchid, El Cancun Mexican, El Vallarta Mexican, Golden Corral, Italian Oven, KFC, Little Caesar's, Longhorn Steaks, Mayflower Seafood, McDonald's, Mr C's Rest., Pizza Hut, Schlotsky's, Shoney's, Subway, Taco Bell, Texas Steaks, Villa Maria Italian, Waffle House, Wendy's, Colonial Inn, Holiday Inn Express, Mayfair Motel, Rodeway Inn, Belk, BrakeXperts, Cadillac/Olds, Chrysler/Plymouth/Jeep, Eckerd, Food Lion, JC Penney, Sears/auto, U-Haul, bank, cinema, cleaners, mall, st patrol, W...VETERINARIAN, Phillips 66, CiCi's, Bojangles(2mi), Domino's, Hardee's(2mi), IHOP, Ryan's, Cabarrus Inn, Comfort Inn, Fairfield Inn, Microtel, Park Inn, Studio 1 Suites, $General, Drug Emporium, Eddie's Pizza/funpark, Harris-Teeter, Home Depot, JC Penney, OfficeMax, Target
55	NC 73, Concord, to Davidson, E...Exxon/diesel/mart/atm, Waffle House, W...Phillips 66/diesel/mart, 76/diesel/mart/atm, Huddle House, Day's Inn
54	new exit
52	Poplar Tent Rd, E...BP/diesel, Texaco/diesel/mart, to Lowe's Speedway, museum, W...Exxon/24hr
49	Speedway Blvd, Concord Mills Blvd, E...Shell/mart, Texaco/mart, BBQ, Bob Evans, Bojangles, Texas Roadhouse, Zaxby's, Hampton Inn, Hawthorn Suites, Holiday Inn Express, Sleep Inn, SpringHill Suites, Wingate Inn, XpressLube, to Lowes Motor Speedway, W...Citgo/mart, On-the-Border, Roadhouse Grill, Steak'n Shake, Bed Bath&Beyond, Books-A-Million, Discount Tire, TJ Maxx, bank, cinema, Concord Mills Mall
48	I-485, to US 29, no facilities
46	Mallard Creek Church Rd, E...Wilco/Citgo/diesel/mart, Research Park, W...Exxon/24hr, Subway(1mi)

Salisbury

Concord

N

S

North Carolina Interstate 85

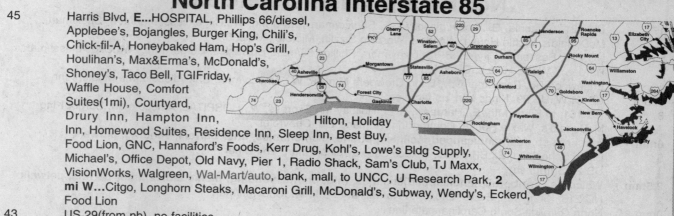

45 Harris Blvd, **E**...HOSPITAL, Phillips 66/diesel, Applebee's, Bojangles, Burger King, Chili's, Chick-fil-A, Honeybaked Ham, Hop's Grill, Houlihan's, Max&Erma's, McDonald's, Shoney's, Taco Bell, TGIFriday, Waffle House, Comfort Suites(1mi), Courtyard, Drury Inn, Hampton Inn, Hilton, Holiday Inn, Homewood Suites, Residence Inn, Sleep Inn, Best Buy, Food Lion, GNC, Hannaford's Foods, Kerr Drug, Kohl's, Lowe's Bldg Supply, Michael's, Office Depot, Old Navy, Pier 1, Radio Shack, Sam's Club, TJ Maxx, VisionWorks, Walgreen, Wal-Mart/auto, bank, mall, to UNCC, U Research Park, **2 mi W**...Citgo, Longhorn Steaks, Macaroni Grill, McDonald's, Subway, Wendy's, Eckerd, Food Lion

43 US 29(from nb), no facilities

41 Sugar Creek Rd, **E**...RaceTrac/mart, Texaco/diesel/mart/wash, Bojangles, McDonald's, Taco Bell, Wendy's, Best Western, Brookwood Inn, Continental Inn, Econolodge, Microtel, Quality Inn, Red Roof Inn, **W**...BP/diesel/mart/wash, Exxon/diesel/mart, Shoney's, Texas Ranch Steaks, Waffle House, Comfort Inn, Day's Inn, Fairfield Inn, Holiday Inn Express, Ramada Inn, Rodeway Inn, Super 8

40 Graham St, **E**...Hereford Barn Steaks, Travelodge, Ford Trucks, UPS, **W**...Citgo/diesel/mart/wash, Goodyear

39 Statesville Ave, **E**...Pilot/Subway/diesel/mart/24hr, CarQuest, carwash, **W**...Citgo/mart/wash, Bojangles, Knight's Inn, auto repair

38 I-77, US 21, N to Statesville, S to Columbia

37 Beatties Ford Rd, **E**...CHIROPRACTOR, Citgo, Petro Express/mart, Phillips 66/diesel/mart, Texaco/diesel/mart/wash, Burger King, McDonald's, Food Lion, **W**...Amoco/mart, McDonald's Cafeteria, Travelodge, Sentry Hardware, funpark, to JC Smith U

36 NC 16, Brookshire Blvd, **E**...HOSPITAL, Amoco/diesel/mart, Burger King, China City, Hornet's Rest Inn, repair, **W**...Exxon/mart/wash, RaceTrac/mart, Speedway/mart, Bojangles, Burger King, BrakeXperts

35 Glenwood Dr, **E**...Citgo/Circle K, Innkeeper, **W**...Old Europe Rest., White/GMC

34 NC 27, Freedom Dr, **E**...MEDICAL CARE, VETERINARIAN, Amoco/diesel, BP/mart/wash, Bojangles, Burger King, Capt D's, IHOP, McDonald's, Pizza Hut, Ruby Palace Chinese, Shoney's, Subway, Taco Bell, Tung Hoy Chinese, Wendy's, Advance Parts, Big Lots, BiLo, Firestone, Goodyear, Meineke Muffler, Midas Muffler, NAPA, Peebles, Pizza Hut, Vision Ctr, Walgreen, **W**...IHOP, Chinese Rest., Howard Johnson, Quality Suites, Ramada Ltd, JiffyLube, repair

33 US 521, Billy Graham Pkwy, **E**...BP/Blimpie/diesel/mart, Bojangles, Krystal, Wendy's, Day's Inn, Economy Inn, Sheraton, to Coliseum, airport, **W**...Exxon/diesel/mart, Cracker Barrel, Prime Buffet, Waffle House/24hr, Fairfield Inn, Hampton Inn, La Quinta, Microtel, Red Roof, Save Inn

32 Little Rock Rd, **E**...Waffle House, Best Western, Courtyard, Econolodge, Holiday Inn, Toyota, tires, **W**...Amoco/Circle K, Exxon, Arby's, Hardee's, Quincy's, Subway, Comfort Inn, Country Inn Suites, Motel 6, Shoney's Inn/rest., Family$, Food Lion

29 Sam Wilson Rd, **E**...Phillips 66, **W**...Texaco/diesel/mart, Stinger's Steaks

28mm weigh sta both lanes

27.5mm Catawba River

27 NC 273, Mt Holly, **E**...Citgo, Exxon, Arby's, Burger King, Capt's Seafood, Subway, Waffle House, Wendy's, Western Sizzlin, College Park Drug, **W**...BP/diesel/mart, Texaco/diesel

26 Belmont Abbey Coll, **E**...McDonald's, Quincy's, Ford, **W**...Amoco, Bojangles, Hardee's

24mm South Fork River

23 NC 7, McAdenville, **W**...Chevron/diesel/mart, Shell, Hardee's, Hillbilly's BBQ/Steaks

22 Cramerton, Lowell, **E**...Applebee's, Burger King, LJ Silver

21 Cox Rd, **E**...DENTIST, Citgo/diesel/mart, Exxon, Boston Mkt, Chick-fil-A, Chinese Rest., Chili's, Houlihan's, Krispy Kreme, Longhorn Steaks, Max' Mexican Eatery, McDonald's, Pizza Inn, Ryan's, Subway, Best Buy, Books-A-Million, BiLo/24hr, Dodge, $Tree, Hannaford's Food/drug, Harris-Teeter/24hr, Home Depot, Honda, Kohl's, K-Mart, Lowe's Bldg Supply, MediaPlay, Michael's, Nissan, OfficeMax, Old Navy, PepBoys, PetsMart, Pier 1, Sam's Club, SpeeDee Lube, SportsAuthority, Tire Kingdom, Upton's, Wal-Mart, antiques, bank, mall, **W**...HOSPITAL, Exxon, Texaco/mart/wash, IHOP, Super 8, Eckerd, Harley-Davidson, Med Ctr Drug, Suzuki

20 NC 279, New Hope Rd, **E**...FOOT CLINIC, Amoco, Shell, Texaco/wash, Arby's, Burger King, Capt D's, ChiChi's, Checker's, Hong Kong Buffet, Italian Oven, Mario's Pizza, McDonald's, Morrison's Cafeteria, Pizza Hut, Red Lobster, Sake Japanese, Shoney's, Taco Bell, Holiday Inn Express, Belk, BrakeXperts, Dillard's, Firestone/auto, JC Penney, K-Mart, NAPA, Office Depot, PartsAmerica, Precision Tune, Service Merchandise, Target, TJ Maxx, Winn-Dixie, bank, cinema, mall, **W**...HOSPITAL, Bojangles, Cracker Barrel, Hickory Ham Café, KFC, Outback Steaks, Waffle House, Fairfield Inn, Hampton Inn, Innkeeper, Circuit City, cinema, hardware

19 NC 7, E Gastonia, **E**... Shell, **W**...Servco/mart, Hardee's

N

S

Charlotte

Gastonia

North Carolina Interstate 85

17	US 321, Gastonia, **E**...Exxon/diesel/wash, Servco/mart, Day's Inn, **W**...Citgo/diesel/mart/atm, Shell/mart, McDonald's, Waffle House, Wendy's, Western Sizzlin, Microtel, Motel 6
14	NC 274, Bessemer City, **E**...BP/Subway/mart, Phillips 66/mart, Bojangles, Burger King, **W**...Citgo/diesel/mart, Hardee's, Waffle House, Econolodge
13	Edgewood Rd, Bessemer City, **E**...to Crowders Mtn SP, **W**...Amoco/mart, Shell/diesel/24hr, Masters Inn
10b a	US 74 W, US 29, Kings Mtn, no facilities
8	NC 161, to Kings Mtn, **E**...Hardee's, KFC, Holiday Inn Express, **W**...HOSPITAL, Amoco/mart, Burger King, McDonald's, Waffle House, Comfort Inn, Ramada Ltd, Chevrolet
6mm	**rest area sb, full(handicapped)facilities, phone, picnic tables, litter barrels, vending, petwalk**
5	Dixon School Rd, **E**...BP/diesel, Texaco/Subway/diesel/mart, The Diner
4	US 29 S(from sb)
2.5mm	**Welcome Ctr nb, full(handicapped)facilities, info, phone, vending, picnic tables, litter barrels, petwalk**
2	NC 216, Kings Mtn, **E**...to Kings Mtn Nat Military Park, **W**...Chevron/mart
0mm	North Carolina/South Carolina state line

North Carolina Interstate 95

Exit #	Services
181mm	North Carolina/Virginia state line, **Welcome Ctr sb, full(handicapped)facilities, phone, picnic tables, litter barrels, vending, petwalk**
180	NC 48, to Gaston, Pleasant Hill, to Lake Gaston, **W**...Speedway/Subway/diesel/mart/24hr
176	NC 46, to Garysburg, **E**...Texaco/Stuckey's, repair, **W**...Shell/mart, Burger King, Comfort Inn/Aunt Sarah's
174mm	Roanoke River
173	US 158, Roanoke Rapids, Weldon, **E**...HOSPITAL, Shell/Blimpie/mart, Speedway/diesel/mart, Texaco/diesel/mart/wash, BBQ, Chinese Rest., Trigger's Steaks, Waffle House, Day's Inn, Interstate Inn, Orchard Inn, antiques, **W**...CHIROPRACTOR, Amoco/diesel/mart, BP/mart/wash, Exxon/mart, RaceTrac/mart, Burger King, Cracker Barrel, Fisherman's Paradise, Hardee's, KFC, Little Caesar's, McDonald's, New China, Pizza Hut, Pizza Inn, Ryan's, Shoney's, Subway, Taco Bell, Texas Roadhouse, Waffle House, Wendy's, Comfort Inn, Fairfax Motel, Hampton Inn, Jameson Inn, Motel 6, Sleep Inn, The Inn, Advance Parts, AutoZone, Big Lots, $General, Eckerd, Firestone/auto, Food Lion, Ford/Lincoln/Mercury/Honda, Goody's, Piggly Wiggly, Roses, Wal-Mart, bank, cleaners
171	NC 125, Roanoke Rapids, **W**...Texaco/diesel/mart/atm, Holiday Inn Express, st patrol
168	NC 903, to Halifax, **E**...Exxon/Hardee's/diesel/mart, Shell/Burger King/diesel/mart, Speedway/diesel/mart, **W**...Citgo/diesel/rest./24hr
160	NC 561, to Brinkleyville, **E**...Exxon/mart, **W**...Citgo/diesel/rest.
154	NC 481, to Enfield, **E**...Mobil, **W**...KOA(1mi)
152mm	weigh sta both lanes
150	NC 33, to Whitakers, **E**...golf, **W**...Texaco/Dairy Queen/Stuckey's
145	NC 4, to US 301, Battleboro, **E**...Amoco/diesel/mart, Exxon/mart, Shell/diesel/mart, Texaco/mart, BBQ, Dairy Queen, Denny's, Hardee's, Waffle House, Wendy's, Best Western/rest., Comfort Inn, Day's Inn, Deluxe Inn, Howard Johnson/rest., Masters Inn, Quality Inn, Red Carpet Inn, Scottish Inn, Shoney's Inn/rest., Super 8, Travelodge
142mm	**rest area both lanes, full(handicapped)facilities, phone, picnic tables, litter barrels, vending, petwalk**
141	NC 43, Red Oak, **E**...Exxon/diesel/mart/LP, Texaco/diesel/mart/wash, **W**...BP/diesel, Holiday Inn/café
138	US 64, Rocky Mount, **5 mi E**...HOSPITAL, Comfort Inn, Hampton Inn, Holiday Inn, Residence Inn, to Cape Hatteras Nat Seashore
132	to NC 58, **E**...Citgo/diesel, **1 mi W**...Amoco/diesel
128mm	Tar River
127	NC 97, to Stanhope, **E**...BP/diesel, airport, **2 mi W**...gas/diesel/mart
121	US 264, Wilson, **E**...HOSPITAL, Amoco/mart, EastCoast/diesel/LP/mart, Exxon/Subway/mart, Shell/mart, Aunt Sarah's, Hardee's, Waffle House, Comfort Inn, **4 mi E**...Shell/diesel/mart, Applebee's, Arby's, Boston Mkt, Burger King, Chick-fil-A, Denny's, Golden Corral, Sonic, Subway, Hampton Inn, Books-A-Million, Eckerd, Goody's, JiffyLube, Jo-Ann Crafts, K-Mart/Little Caesar's, Harris Teeter, Lowe's Bldg Supply, Monro Muffler/brake, Staples, Wal-Mart SuperCtr/24hr, laundry, **W**...BP/Blimpie/diesel/mart, Bojangles, Burger King, Cracker Barrel, McDonald's, Taco Bell, Fairfield Inn, Holiday Inn Express, Jameson Inn, Microtel, Sleep Inn, to Country Doctor Museum
119b a	to US 264 W byp, no facilities
116	NC 42, to Clayton, Wilson, **E**...HOSPITAL, Shell/diesel/mart, **W**...Texaco/diesel/mart, Rex Gas/diesel/repair, Rock Ridge Camping
107	US 301, Kenly, **E**...Amoco/repair, BP/mart, Coastal/Subway/mart, Exxon/McDonald's/diesel/mart, Phillips 66/mart, Shell/diesel/mart/24hr, Texaco, BBQ, Burger King, Golden China, Nik's Pizza, Patrick's Rest., Waffle

North Carolina Interstate 95

	House, Willoughby's Seafood, Budget Inn, Deluxe Inn, Econolodge, Food Lion, Ford, Family$, IGA Food, Tobacco Museum, bank, repair/tires
106	Kenly, **W**...TA/Texaco/Wendy's/diesel/mart/24hr, Wilco/Exxon/diesel/mart/24hr, Kenly Kitchen, Waffle House, Best Western, Day's Inn, Blue Beacon, Speedco
105.5mm	Little River
105	Bagley Rd, Kenly, **E**...Citgo/Stormin Norman/diesel/rest./24hr
102	Micro, **E**...BP/repair, **W**...Phillips 66/mart, USPO
101	Pittman Rd, no facilities
99mm	**rest areas both lanes, full(handicapped)facilities, vending, phone, picnic tables, litter barrels, petwalk, hist marker**
98	to Selma, **E**...KOA
97	US 70 A, Selma, to Pine Level, **E**...Citgo/diesel/mart/24hr, Denny's, Subway, Holiday Inn Express, J&R Outlet, antiques, **W**...HOSPITAL, Amoco/diesel/mart/24hr, BP/diesel/mart, Exxon/diesel/mart/24hr, Texaco/diesel/mart, Bojangles, China Buffet, Golden Corral, Hardee's, KFC, McDonald's, Oliver's Rest., Pizza Hut, Shoney's, Waffle House, Comfort Inn, Day's Inn, Hampton Inn, Luxury Inn, Masters Inn, Regency Inn, Royal Inn
95	US 70, Smithfield, **E**...Log Cabin Motel/rest., Howard Johnson Express, Village Motel, **W**...Speedway/diesel/mart, Burger King, Cracker Barrel, Smithfield BBQ(2mi), Jameson Inn, Super 8, Trot Motel(2mi), Harley-Davidson, Valvoline, Factory Outlets/famous brands
93	Brogden Rd, Smithfield, **W**...Amoco/diesel/mart, Shell/mart
91.5mm	Neuse River
90	US 301, US 701, to Newton Grove, **E**...Citgo/diesel/mart, Roz's Rest., Travelers Inn, Holiday TravLPark, to Bentonville Bfd, **W**...Phillips 66/diesel, Four Oaks Motel
87	NC 96, **W**...BP/diesel/repair/24hr, Texaco/mart
81b a	I-40, E to Wilmington, W to Raleigh
79	NC 50, to NC 27, Benson, Newton Grove, **E**...BP/mart, Citgo/mart, Golden Corral, Olde South BBQ, Waffle House, Dutch Inn, Food Lion, **W**...Amoco/diesel/wash, Coastal/mart, Exxon/mart, Phillips 66, Burger King, McDonald's, KFC, Pizza Hut, Subway, Day's Inn, Byrd's Foods, Family$, Rite Aid, Valvoline, laundry, tires
77	Hodges Chapel Rd, **E**...BP/diesel/mart/rest./24hr, Speedway/diesel/mart/24hr, RV/truckwash
75	Jonesboro Rd, **W**...Citgo/mart/diner, Sadler/Shell/diesel/rest./mart/24hr
73	US 421, NC 55, to Dunn, Clinton, **E**...Wendy's, **W**...Exxon/diesel/mart, Chevron/Servco/mart, Texaco/KrispyKreme/mart, Bojangles, Burger King, Dairy Freeze, McDonald's(2mi), Sagebrush Steaks, Taco Bell, Triangle Waffle, Comfort Inn, Econolodge, Express Inn, Holiday Inn Express, Jameson Inn, Ramada Inn/rest., IGA Foods, bank, museum, tires
72	Pope Rd, **E**...BP, Best Western, Comfort Inn, **W**...Amoco/mart, Chevron/diesel/mart, Brass Lantern Steaks, Budget Inn, Super 8, Olds/Cadillac/GMC, auto repair
71	Longbranch Rd, **E**...Shell/Hardee's/Travel World, **W**...to Averasboro Bfd
70	SR 1811, **E**...Relax Inn
65	NC 82, Godwin, **W**...Sam's/diesel/mart, Children's Home
61	to Wade, **E**...Citgo/diesel/mart, KOA, **W**...BP/Subway/diesel/mart, Queen Anne Motel
58	US 13, to Spivey's Corner, **E**...Texaco/diesel/mart, Day's Inn/rest.
56	Lp 95, to US 301(from sb), Fayetteville, **W**...HOSPITAL, Exxon(1mi), Budget Lodge(1mi), to Ft Bragg, Pope AFB
55	NC 1832, **W**...Exxon/diesel, Texaco, Budget Inn
52	NC 24, Fayetteville, **W**...to Ft Bragg, Pope AFB, botanical gardens, museum
49	NC 53, NC 210, Fayetteville, **E**...Amoco/diesel/mart, Citgo/mart, Exxon/mart, Texaco/diesel/mart, Burger King, Denny's, McDonald's, Pizza Hut, Taco Bell, Waffle House, Day's Inn/rest, Deluxe Inn, Motel 6, Quality Inn, **W**...Amoco/Subway/diesel/mart, Exxon/diesel/mart, Shell/diesel/mart, Beaver Dam Seafood, Cracker Barrel, Shoney's, Best Western, Comfort Inn, Econolodge, Fairfield Inn, Hampton Inn, Holiday Inn, Innkeeper, Red Roof Inn, Sleep Inn, Sheraton, Super 8
48mm	**rest areas both lanes, full(handicapped)facilities, phone, picnic tables, litter barrels, vending, petwalk**
47mm	Cape Fear River
46b a	NC 87, to Fayetteville, Elizabethtown, **W**...HOSPITAL, Civic Ctr, to Agr Expo Ctr
44	Snow Hill Rd, **W**...Lazy Acres Camping, to airport
41	NC 59, to Hope Mills, Parkton, **E**...Texaco, **W**...BP, Spring Valley RV Park
40	Lp 95, to US 301(from nb), to Fayetteville, facilities on US 301(5-7mi)
33	US 301, St Pauls, **E**...Amoco/diesel/mart/repair/24hr, tires

N

S

Selma

Dunn

Fayetteville

North Carolina Interstate 95

L u m b e r t o n

31	NC 20, to St Pauls, Raeford, **E**...Amoco/mart, BP/mart, Shell/Blimpie/TCBY/diesel/mart/atm/24hr, Texaco/mart, Burger King, Hardee's, McDonald's, Day's Inn, **W**...Citgo/mart, Exxon/diesel/mart
25	US 301, no facilities
24mm	weigh sta both lanes
22	US 301, **E**...CHIROPRACTOR, Citgo, Exxon/diesel/mart, Denny's, Hardee's, Huddle House, John's Rest., Ruby Tuesday, Ryan's, Texas Steaks, Waffle House, Best Western, Comfort Suites, Hampton Inn, Holiday Inn, Redwood Lodge, Super 8, Goodyear/auto, Lowe's Bldg Supply, Lowe's Foods, OfficeMax, Wal-Mart SuperCtr/24hr, st patrol, **W**...BP/diesel/mart, Circle B/gas, Texaco/Subway/diesel/mart, Uncle Geo's Pizza
20	NC 211, to NC 41, Lumberton, **E**...HOSPITAL, Amoco/mart, Citgo/mart, Exxon/diesel/mart, Carolina Steaks, Golden City Chinese, Hardee's, Little Caesar's, McDonald's, New China, Subway, Village Sta Rest., Waffle House, Western Sizzlin, Howard Johnson, Quality Inn, Ramada Ltd, Food Lion/deli, K-Mart/drugs, **1 mi E**...BP/Blimpie/diesel/mart, Bojangles, Burger King, KFC, OutWest Steaks, Papa John's, Shoney's, Sonic, Taco Bell, Wendy's, Advance Parts, Belk, JC Penney, PostNet, ProTire, Winn-Dixie, cinema, **W**...Texaco/diesel/mart, BBQ, Cracker Barrel, LungWah Chinese, Comfort Inn, Country Inn Suites, Day's Inn/rest., Econolodge, Fairfield Inn, National 9 Inn, flea mkt, outlets
19	Carthage Rd, Lumberton, **E**...BP/diesel, San Jose Mexican, Travelers Inn, auto repair, **W**...Exxon/diesel/mart, Texaco/mart, Knight's Inn, Motel 6
18mm	Lumber River
17	NC 72, **E**...BP/diesel/mart, Citgo, Exxon/mart, Mobil/diesel/mart, Texaco/mart, Burger King, Hardee's, McDonald's, Old Foundry Rest., Subway, Waffle House, Budget Inn, Economy Inn, Southern Inn, **W**...auto repair
14	US 74, Maxton, to Laurinburg, **W**...BP/diesel/mart, Exit Inn, Sleepy Bear's RV Camp
10	US 301, to Fairmont, no facilities
7	to McDonald, Raynham, no facilities
5mm	**Welcome Ctr nb, full(handicapped)facilities, phone, picnic tables, litter barrels, vending, petwalk**
2	NC 130, to NC 904, Rowland, no facilities
1b a	US 301, US 501, Dillon, **E**...BP, Citgo, Exxon, Shell/mart, Hot Tamale Rest., Pedro's Diner, Pedro's Ice Cream Fiesta, Sombrero Rest., Golden Triangle Motel, South-of-the-Border Motel, Pedro's Campground, Golf of Mexico Golf Shop, El Drug Store, bank, fireworks, gifts, outlets, **W**...Amoco/mart/24hr, BP, Texaco/diesel/mart, Denny's, Hardee's, Waffle House, Budget Motel, Day's Inn, Family Inn/rest., Holiday Inn Express
0mm	North Carolina/South Carolina state line

North Carolina Interstate 240(Asheville)

A s h e v i l l e

Exit #	Services
9mm	I-240 begins/ends on I-40, exit 53b a.
8	Fairview Rd, **N**...Shell/Blimpie/mart/wash, Burger King, J&S Cafeteria, KFC, Little Caesar's, Mandarin Chinese, McDonald's, Subway, Comfort Inn, Advance Parts, Bilo Foods, CVS Drug, Eckerd, Hamrick's Clothing, **S**...Pizza Now, Home Depot, Oakley Foods, laundry
7.5mm	Swannanoa River
7	US 70, **N**...U-Haul, **S**...Amoco/mart, Exxon/diesel, Applebee's, Burger King, Chili's, Cornerstone Rest., Damon's, Greenery Rest., IHOP, Joe's Crabshack, Kell's Grill, McGuffy's Grill, O'Charley's, Olive Garden, Philly Connection, Rio Bravo, Subway, Taco Bell, Waffle House, Courtyard, Day's Inn, Econolodge, Hampton Inn, Holiday Inn, Ramada Ltd, Belk, Circuit City, Firestone, Food Lion, Goody's, K-Mart, Lowe's Bldg Supply, MensWearhouse, Midas Muffler, Michael's, Office Depot, OfficeMax, Sears/auto, VisionWorks, Wash'n Lube, cinema
6	Tunnel Rd(from eb), same as 7
5b	US 70, US 74A, Charlotte St, **N**...Amoco, Exxon, **S**...Civic Ctr, bank
a	**N**...Exxon/diesel, Texaco/mart, Bojangles, 3 Pigs BBQ, **S**...Interstate Motel
4c	Haywood St, Montford, downtown
b	Patton Ave(from eb), downtown
a	US 19 N, US 23 N, US 70 W, to Weaverville
3b	Westgate, **N**...Servco/gas, EarthFare Foods, Holiday Inn, CVS Drug, NTB
a	US 19 S, US 23 S, W Asheville, **N**...CHIROPRACTOR, Amoco/mart, Exxon, Arby's, Bojangles, Denny's, KFC, Little Caesar's, LJ Silver, Mandarin Chinese, McDonald's, Pizza Hut, Ryan's, Taco Bell, Wendy's, Advance Parts, AutoZone, Best Foods, Bilo, Cadillac/Pontiac/GMC, Checker Parts, Goodyear/auto, Ingles, K-Mart, Radio Shack, Sam's Club/gas, cleaners
2	US 19, US 23, W Asheville, **N**...Shell/mart/wash, **S**...Eblen Gas, B&B Drug
1c	Amboy Rd(from eb), no facilities
b	NC 191, to I-40 E, Brevard Rd, **S**...farmers mkt, speedway, camping
a	I-40 W, to Knoxville
0mm	I-240 begins/ends on I-40, exit 46b a.

North Dakota Interstate 29

Exit #	Services
218mm	North Dakota/Canada border. I-29 begins/ends at US/Canada border.
217.5mm	US Customs sb
216.5mm	historical site nb, tourist info sb
215	ND 59, rd 55, Pembina, **E**...Citgo/Subway/DutyFree/diesel/mart, Gastrak/pizza/diesel/mart/24hr, Depot Café, Gifthouse, MapleLeaf DutyFree, Red Roost Motel
212	no facilities
208	rd 1, to Bathgate, no facilities
203	US 81, ND 5, to Hamilton, Cavalier, **W**...Jubilee Express/diesel/café/24hr, weigh sta both lanes
200	no facilities
196	rd 3, Bowesmont, no facilities
193	no facilities
191	ND 11, to St Thomas, no facilities
187	ND 66, to Drayton, **E**...Amoco/diesel, Cenex/diesel/mart, SpurGas/diesel/mart, Curnal's Café, Dairy Queen, Motel 66, V&R Motel, Ford, auto parts, car/truckwash/laundry, USPO, camping
184	to Drayton, **2 mi E**...gas/diesel/mart/hardware, USPO, camping
180	rd 9, no facilities
179mm	**rest area both lanes, full(handicapped)facilities, phone, picnic tables, litter barrels, vending, petwalk**
176	ND 17, to Grafton, **10 mi W**...HOSPITAL, gas, food, lodging
172	no facilities
168	rd 15, Warsaw, to Minto, no facilities
164	no facilities
161	ND 54, rd 19, to Ardoch, Oslo, no facilities
157	no facilities
152	US 81, Manvel, to Gilby, **W**...Coop/diesel/mart
145	US 81 bus, N Washington St, to Grand Forks, no facilities
141	US 2, Gateway Dr, Grand Forks, **E**...Amoco/diesel/mart/24hr, Conoco/diesel/mart, Buffet House, Burger King, Chuckhouse Rest., Fire Island Grill, McDonald's, Comfort Inn(6mi), Dakota Inn, Econolodge, Holiday Inn, Ramada Inn, Roadking Inn, Rodeway Inn, Select Inn, Super 8, Westward Ho Motel, AutoValue Parts, Checker Parts, Ford/Lincoln/Mercury, to U of ND, **1 mi E**...HOSPITAL, Fat Albert's Subs, Figaro's Pizza, Hardee's, Subway, Taco John's, Plainsman Motel, Exhaust Pros, Goodyear/auto, Hugo's Foods, Muffler Master, Mailboxes Etc, U-Haul, auto repair, **W**...VETERINARIAN, Simonson/diesel/café/24hr, Stamart/diesel/RV dump, Emerald Grill, Perkins, Prairie Inn, Budget RV Ctr, Chrysler, GMC, Honda/Nissan/Chevrolet, NAPA, Plymouth/Jeep, Toyota/Saab, port of entry/weigh sta, airport, to AFB
140	DeMers Ave, **E**...HOSPITAL, Amoco/mart/wash/bank/LP, Aurora Ctr, to U of ND
138	32nd Ave S, **E**...Amoco/mart/wash, Conoco/mart, Holiday/Subway/diesel/mart/wash, SuperPumper/diesel/mart, Applebee's, Burger King, ChiChi's, China Garden, China King, Domino's, Ground Round, Happy Joe's Pizza, McDonald's, Pizza Hut, Ponderosa, Red Lobster, Shakey's Pizza, Taco Bell, C'mon Inn, Comfort Inn, Country Inn Suites, Day's Inn, Fairfield Inn, Roadking Inn, Chevrolet, Cadillac, Dayton's, FashionBug, GreaseMonkey, Honda/Nissan, Hugo's Food/deli, JC Penney, Jo-Ann Fabrics, K-Mart, LumberMart, Mazda, Menard's, Midas Muffler, National Muffler, OfficeMax, Osco Drug, PetCo, Pier 1, Rex, Sam's Club, Sears/auto, Super1 Foods/deli, Target, TJ Maxx, Target, Tires+, Wal-Mart/photo, cinema, funpark, hardware, mall, **W**...VETERINARIAN, Conoco/Subway/diesel/mart, Grand Forks Camping
130	ND 15, rd 81, Thompson, **1 mi W**...gas, food
123	CountyLine rd, to Reynolds, **E**...to Central Valley School
120mm	weigh sta both lanes
118	to Buxton, no facilities
111	ND 200 W, to Cummings, Mayville, **W**...Big Top Fireworks, to Mayville St U
104	Hillsboro, **E**...HOSPITAL, Amoco/pizza/diesel/mart/24hr, Cenex/diesel/LP/atm/24hr, Burger King, Country Hearth Rest., Dairy Bar, Sunset View Motel, Hillsboro Inn, Ford/Mercury, Goodyear, Rick's Foods, bank, drugs
100	ND 200 E, ND 200A, to Blanchard, Halstad, no facilities
99mm	**rest area both lanes, full(handicapped)facilities, phone, picnic tables, litter barrels, vending, petwalk**
92	rd 11, Grandin, **E**...Coop/diesel, **W**...Citgo/diesel/mart/hardware
86	Gardner, **W**...Gardner Service/Sinclair/diesel/mart/café

Grand Forks

North Dakota Interstate 29

79	Argusville, no facilities
74.5mm	Sheyenne River
74mm	**rest area both lanes, full(handicapped)facilities, phone, picnic tables, litter barrels, vending, petwalk**
73	rd 17, rd 22, Harwood, **E**...Cenex/pizza/diesel/mart/café/24hr
69	rd 20, no facilities
67	US 81 bus, 19th Ave N, **E**...VA HOSPITAL, Hector Int Airport
66	12th Ave N, **E**...HOSPITAL, to ND St U, **W**...Ampride/diesel/mart, Arby's, Microtel, OK Tire
65	US 10, Main Ave, W Fargo, **E**...VETERINARIAN, Amoco/mart/wash/24hr, Stamart/gas, Kroll's Kitchen, Bernie's RV Ctr, Big Wheel Auto, NAPA, Nissan, transmissions, to Heritage Ctr, **W**...Cenex/Subway/diesel/mart, Mobil/mart, Hardee's, Outback Steaks, Best Western Kelly/rest., Aamco, Buick, CarQuest, Chrysler/Dodge/Jeep, Ford/Subaru/VW, Hyundai, Mac's Hardware, Mercury/Lincoln, Saturn, Toyota, White/Volvo/GMC, Kenworth, Cummins Diesel
64	13th Ave, Fargo, **E**...MEDICAL CARE, DENTIST, Amoco/mart/wash24hr, Cenex/mart, Conoco/mart, Kum&Go/gas, Phillips 66/mart/wash/atm, Stamart/diesel/mart, Applebee's, Arby's, Burger King, Buck's Roadhouse, ChuckeCheese, Dairy Queen, Giant Panda Chinese, GreenMill Rest., Ground Round, Guadalajara Mexican, Hardee's, Mr Steak, Perkins/24hr, Ponderosa, Quizno's, Subway, Taco John's, TCBY, Wendy's, AmericInn, Best Western, Comfort Inn, Country Suites, Econolodge, Hampton Inn, Motel 6, Motel 75, Super 8, CashWise Foods/24hr, Goodyear, JiffyLube, Kinko's, Osco Drug, Scheel's Sports, SunMart Foods, Tires+/transmissions, Vision Express, White Drug, bank, laundry, **W**...Amoco/mart/wash/24hr, Stop'n Go/gas, Blimpie, ChiChi's, Chili's, Denny's, McDonald's, Fuddrucker's, Grandma's Saloon/deli, Godfather's, Hooters, Kroll's Diner, LoneStar Steaks, Olive Garden, Paradiso Mexican, Pizza Hut, Randy's Rest., Red Lobster, Royal Fork Buffet, Schlotsky's, Taco Bell, TGIFriday, Valentino's Ristorante, Comfort Inn, Day's Inn, Fairfield Inn, Holiday Inn Express, Kelly Inn, Ramada Suites, Red Roof Inn, Select Inn, Barnes&Noble, Best Buy, Chevrolet, CopyMax, Herberger's, Hornbacher's Foods, JC Penney, Jo-Ann Crafts, K-Mart/auto, Kohl's, MailBoxes Etc, MediaPlay, Menard's, Michael's, Midas Muffler, Office Depot, OfficeMax, PetCo, Pier 1, Sam's Club, Savers Dept Store, Sears/auto, SunMart Foods, Target, TJ Maxx, ToysRUs, Valvoline, Wal-Mart/photo, Walgreen, bank, carwash, cinema, mall
63b a	I-94, W to Bismarck, E to Minneapolis
62	32nd Ave S, Fargo, **E**...Amoco/mart/wash/24hr, Fleet&Farm/diesel/mart/wash, Country Kitchen, Papa John's, **W**...Flying J/Conoco/diesel/LP/mart/motel/24hr, Goodyear/auto
60	52nd Ave S, to Fargo, **W**...Starr Fireworks
56	to Wild Rice, Horace, no facilities
54	rd 16, to Oxbow, Davenport, no facilities
50	rd 18, Hickson, no facilities
48	ND 46, to Kindred, no facilities
44	to Christine, **1 mi E**...gas
42	rd 2, to Walcott, no facilities
40.5mm	**rest area both lanes, full(handicapped)facilities, phone, picnic tables, litter barrels, vending, petwalk**
37	rd 4, Colfax, to Abercrombie, **E**...to Ft Abercrombie HS, **3 mi W**...gas
31	rd 8, Galchutt, **1 mi E**...Cenex
26	to Dwight, no facilities
24mm	weigh sta both lanes
23b a	ND 13, Mooreton, to Wahpeton, **10 mi E**...HOSPITAL, ND St Coll of Science
15	rd 16, Great Bend, to Mantador, no facilities
8	ND 11, to Hankinson, Fairmount, **E**...Mobil/diesel/mart, **3 mi W**...camping
3mm	**Welcome Ctr nb, full(handicapped)facilities, phone, picnic tables, litter barrels, petwalk**
2	rd 22, no facilities
1	rd 1E, **E**...Dakota Magic Casino/Hotel/rest.
0mm	North Dakota/South Dakota state line

North Dakota Interstate 94

Exit #	Services
352.5mm	Red River, North Dakota/Minnesota state line
351	US 81, Fargo, **N**...HOSPITAL, CHIROPRACTOR, VETERINARIAN, Conoco, Mobil/diesel/service, Sinclair/mart, Stop'n Go/gas, Godfather's, Great Wall Chinese, Pizza Hut, Taco Shop, Radisson(2mi), Brake Shop, Hornbacher's/deli/24hr, MailBoxes Etc, Medicine Shoppe, PetCtr, Scheel's Sports, Thrifty Drug, bank, camping, **S**...CHIROPRACTOR, DENTIST, Amoco/mart/wash, Kum&Go/gas, Phillips 66/diesel/wash, Stop'n Go/diesel, Burger King, Domino's, Embers Rest., Happy Joe's Pizza, KFC, MayDay's Steaks/pasta, McDonald's, Papa Murphy's, Pepper's Café, Quality Bakery, Randy's Rest., Subway, Taco Bell, Expressway Inn/casino, Rodeway Inn, Hornbacher's Foods, K-Mart, USPO, **S on 32nd Ave**...Kroll's Diner

N

S

E

W

Fargo

Fargo

North Dakota Interstate 94

350	25th St, Fargo, **N**...HOSPITAL, Stop'n Go/gas, **S**...HOSPITAL, EYECARE, Conoco/diesel/mart, Holiday/mart, Texaco, Blimpie, Pazzo's Pasta/pizza, Subway, Taco John's, Village Inn Rest., MailBoxes Etc, SunMart/deli/24hr, XpressLube, cleaners, health foods
349b a	I-29, N to Grand Forks, S to Sioux Falls, facilities 1 mi N, exit 64
348	**45th St, N...Visitor Ctr/full facilities,** Amoco/pizza/mart, Petro/Mobil/diesel/ LP/rest./mart/24hr, Alien's Grill, Bennigan's, McDonald's, C'mon Inn, Ramada Inn, Sleep Inn, Wingate Inn, **1-2 mi N**...Cenex/mart, Stop'n Go, KFC, Olive Garden, Pizza Hut, Schlotsky's, Big K-Mart, JC Penney, Sam's Club, Target, TJ Maxx, Valvoline, Wal-Mart/photo
347mm	Sheyenne River
346b a	W Fargo, to Horace, **S**...Conoco/diesel/mart
343	US 10, Lp 94, W Fargo, **N**...Smoky's Steaks, Pioneer Village, fairgrounds, **3 mi N**...Cenex, Kum&Go, Mobil/ diesel/mart/atm, Hardee's, McDonald's, Taco John's, Harley-Davidson
342	no facilities
341.5mm	weigh sta both lanes
340	to Kindred, no facilities
338	Mapleton, **N**...Ampride/diesel/mart, golf
337mm	truck parking wb, litter barrels
331	ND 18, Casselton, to Leonard, **N**...Phillips 66/diesel/mart/café/motel, NAPA
328	to Lynchburg, no facilities
327mm	truck parking eb, litter barrels
324	Wheatland, to Chaffee, no facilities
322	Absaraka, no facilities
320	to Embden, no facilities
317	to Ayr, no facilities
314	ND 38, Buffalo, to Alice, **3 mi N**...gas, food
310	no facilities
307	to Tower City, **N**...Mobil/diesel/café/repair/24hr, motel
304.5mm	**rest area both lanes(both lanes exit left), full(handicapped)facilities, info, phone, picnic tables, litter barrels, vending, petwalk**
302	ND 32, Oriska, to Fingal, **1 mi N**...gas, food
298	no facilities
296	no facilities
294	Lp 94, Valley City, to Kathryn, **N**...HOSPITAL, camping
292	Valley City, **N**...HOSPITAL, Amoco/diesel/café/atm/24hr, Sabir's Rest., AmericInn, Super 8, Wagon Wheel Inn/ rest., to Bald Hill Dam, camping, **S**...auto/diesel repair, Ft Ransom SP(35mi)
291	Sheyenne River
290	Lp 94, Valley City, **N**...HOSPITAL, Amoco(1mi), Kenny's Rest., Pizza Hut, Bel Air Motel, Valley City Motel, Chrysler/Plymouth/Dodge, Pamida, camping, **S**...Flickertail Inn
288	ND 1 S, to Oakes, no facilities
283	ND 1 N, to Rogers, no facilities
281	Sanborn, to Litchville, **1-2 mi N**...gas, food, lodging
276	Eckelson, **S**...Prairie Haven Camping/gas
275mm	continental divide, elev 1490
272	to Urbana, no facilities
269	Spiritwood, no facilities
262	Bloom, **N**...airport
260	US 52, Jamestown, **N**...to St HOSPITAL, Amoco/diesel/café, Stop'n Go/gas, Starlite Motel, Chevrolet/Buick/ Dodge, Harley-Davidson, camping

E ↑

W ↓

Fargo

North Dakota Interstate 94

Jamestown	259.5mm	James River
	258	US 281, Jamestown, **N**...HOSPITAL, Amoco/mart/wash/24hr, Cenex/24hr, Holiday/mart, Phillips 66/diesel, Sinclair/diesel/mart, Arby's, Dairy Queen, McDonald's, Taco Bell, Wagonmaster Rest., Comfort Inn, Day's Inn, Holiday Inn Express, Ranch House Motel, Sundownere Motel, Buffalo Herd/museum, Firestone/auto, Frontier Village, Cadillac/Olds/GMC, CarQuest, Chrysler/Plymouth/Toyota/Jeep, Goodyear/auto, Mazda, **S**...VETERINARIAN, Conoco/Subway/diesel/mart/24hr, Burger King, Embers Rest., Paradiso Mexican, Perkins, Ponderosa, Super Buffet, Best Western Dakota, Super 8, Ford, JC Penney, K-Mart/Little Caesar's, Sears, Snow's RV Ctr, Valvoline, Wal-Mart/drugs, cinema, mall
	257	Lp 94(from eb), to Jamestown, **N**...diesel repair
	256	Lp 94, **S**...Wiest truck/trailer repair, **1 mi S**...KOA
	254mm	**rest area both lanes, full(handicapped)facilities, phone, picnic tables, litter barrels, petwalk, vending**
	251	Eldridge, no facilities
	248	no facilities
	245	no facilities
	242	Windsor, **1/4 mi N**...gas
	238	Cleveland, to Gackle, **N**...Prairie Oasis/gas/café
	233	no facilities
	230	Medina, **1 mi N**...MEDICAL CARE, Cenex/diesel/LP/mart, Home Cooking Café, Ozzie's Café, Chase Lake Country Inn, Cozy Corners Motel, North Country Lodge, RV camping, USPO
	228	ND 30 S, to Streeter, no facilities
	224mm	**rest area wb, full(handicapped)facilities, phone, picnic tables, litter barrels, vending, petwalk, RV dump**
	221	Crystal Springs, no facilities
	221mm	**rest area eb, full(handicapped)facilities, phone, picnic tables, litter barrels, vending, petwalk, RV dump**
	217	Pettibone, no facilities
	214	Tappen, **S**...Amoco/mart, Roadhouse Café
	208	ND 3 S, Dawson, **N**...RV camping, **1/2 mi S**...gas, food, to Camp Grassick, RV camping
	205	Robinson, no facilities
	200	ND 3 N, Steele, to Tuttle, **S**...Cenex/mart, Conoco/diesel/mart/24hr, Lone Steer Motel/café/24hr
	195	no facilities
	190	Driscoll, **1 mi S**...food
	182	US 83 S, ND 14, Sterling, to Wing, **S**...Cenex/diesel/mart/24hr, Top's Café, Top's Motel(1mi)
	176	McKenzie, no facilities
	170	Menoken, **S**...to McDowell Dam, no facilities
	168.5mm	**rest area both lanes, full(handicapped)facilities, phone, picnic tables, litter barrels, vending, petwalk**
Bismarck	161	Lp 94, Bismarck Expswy, Bismarck, **N**...Cenex/diesel/mart/24hr, KOA(1mi), **S**...Amoco/diesel/mart/rest./24hr, Ramada Ltd, Capitol RV Ctr, Dakota Zoo, Goodyear, airport, diesel repair
	159	US 83, Bismarck, **N**...CHIROPRACTOR, Amoco/mart, Minimart/gas, Sinclair/diesel/mart, Arby's, Burger King, HongKong Chinese, KFC, Kroll's Kitchen, McDonald's, Paradiso Mexican, Perkins, Red Lobster, Royal Fork Buffet, Schlotsky's, Space Alien Grill, Taco Bell, AmericInn, Comfort Inn, Country Inn Suites, Fairfield Inn, Motel 6, Chevrolet, Dan's Food/drug, K-Mart, Menard's, Mr Lube, Mr Muffler, Osco Drug, Sears, U-Haul, bank, mall, tires, **S**...HOSPITAL, DENTIST, Amoco/mart/wash/24hr, Conoco/diesel/mart, StaMart/diesel, Dairy Queen, Donut Hole, 83 Diner, Hardee's, Intn'l Rest., Panda House Chinese, Pizza Hut, Steak Buffet, Subway, Taco John's, Wendy's, Woodhouse Rest., Best Western, Day's Inn, Hillside Motel, Kelly Inn/rest, Select Inn, Super 8, bank
	157	Divide Ave, Bismarck, **N**...Conoco/diesel/mart/LP/24hr, Cracker Barrel, McDonald's, **S**...Cenex/A&W/mart, Econo Food/drug/deli/24hr
	156.5mm	Missouri River
	156	Bismarck Expswy(from eb), Bismarck, City Ctr, **1/2 mi S**...Colonial Motel, RiverTree Inn
	155	to Lp 94(from wb), Mandan, City Ctr, same as 153
	153	ND 6, Mandan Dr, Mandan, **1/2 mi S**...Amoco/mart/wash, Cenex/mart/24hr, StaMart/diesel, Burger King, Bonanza, Dairy Queen, Dakota Farms Rest., Domino's, Godfather's, Hardee's, McDonald's, Pizza Hut, Subway, Taco John's, North Country Inn, CarQuest, Chevrolet, Ford/Mercury, Goodyear/auto, NAPA, Dacotah Centennial Park, Ft Lincoln SP(5mi)
	152	Sunset Dr, Mandan, **N**...Conoco/mart, Best Western, Ridge Motel, **S**...HOSPITAL, Amoco/mart/wash/RV dump, Pride Rest., Sunset Rest., Best Western 7 Seas, Day's Inn, Hospitality Inn, camping
	152mm	scenic view eb

E

W

North Dakota Interstate 94

147	ND 25, to ND 6, Mandan, **S**...Sinclair/ diesel/café/atm/24hr
143mm	central/mountain time zone
140	to Crown Butte
135mm	scenic view wb, litter barrel
134	to Judson, Sweet Briar Lake, no facilities
127	ND 31, to New Salem, **N**...Knife River Indian Village(35mi), **S**...VETERINARIAN, Amoco/diesel/mart, Cenex/diesel/mart, Sunset Inn/café, Golden West Shopping Ctr, Randy's Food/drug, Biggest Cow in the World
123	to Almont, no facilities
120	no facilities
119mm	**rest area both lanes, full(handicapped)facilities, phone, picnic tables, litter barrels, petwalk**
117	no facilities
113	no facilities
110	ND 49, to Glen Ullin, no facilities
108	to Glen Ullin, Lake Tschida, **3 mi S**...gas, food, lodging, camping
102	Hebron, to Glen Ullin, to Lake Tschida, **3 mi S**...gas, food, lodging, camping
97	Hebron, **2 mi N**...gas, food, lodging
90	no facilities
84	ND 8, Richardton, **N**...HOSPITAL, Cenex/diesel/mart, to Assumption Abbey, Schnell RA
78	to Taylor, no facilities
72	Gladstone, no facilities
71mm	**rest area both lanes, full(handicapped)facilities, phone, picnic tables, litter barrels, petwalk, RV dump**
64	Dickinson, **S**...Amoco/diesel/mart/rest./24hr, Firestone/auto, Ford/Lincoln/Mercury, Honda/Toyota, diesel repair
61	ND 22, Dickinson, **N**...CHIROPRACTOR, Cenex/diesel/mart/24hr, Sinclair/diesel/mart, Applebee's, Arby's, Bonanza, Burger King, Dairy Queen, Happy Joe's Pizza, Taco Bell, Taco John's, TCBY, Wendy's, AmericInn, Comfort Inn, Travelodge, Albertson's/24hr, Dan's Food/drug/24hr, Goodyear/auto, Herberger's, JC Penney, K-Mart, Midas Muffler, NAPA, Pennzoil, Sears, Wal-Mart/24hr, Yamaha, bank, laundry, **S**...HOSPITAL, CHIRO-PRACTOR, Amoco/mart/wash/atm, Cenex/diesel/mart, Conoco/mart/repair, Holiday/diesel/mart/atm/24hr, China Doll, Country Kitchen, Domino's, KFC, King Buffet, McDonald's, Perkins, Pizza Hut, Subway, Best Western, Budget Inn, Select Inn, Super 8, Ace Hardware, bank, info
59	Lp 94, to Dickinson, **S**...to Patterson Lake RA, camping, **3 mi S**...facilities in Dickinson
51	South Heart, no facilities
42	US 85, Belfield, to Grassy Butte, Williston, **N**...T Roosevelt NP(52mi), **S**...info, Amoco/Food Court/diesel/mart/ 24hr, Conoco/mart, Dairy Queen, Trapper's Inn Motel/rest., NAPA, camping
36	Fryburg, no facilities
32	**T Roosevelt NP, Painted Canyon Visitors Ctr, N...rest area both lanes, full(handicapped)facilities, phone, picnic tables, litter barrels, petwalk**
27	Historic Medora(from wb), T Roosevelt NP
24.5mm	Little Missouri Scenic River
24	Medora, Historic Medora, Chateau de Mores HS, T Roosevelt NP, visitors ctr, **S**...services away from exit
23	West River Rd, no facilities
22mm	scenic view eb
18	Buffalo Gap, **N**...Buffalo Gap Camping(seasonal)
12mm	**rest area eb, full(handicapped)facilities, phone, picnic tables, litter barrels, petwalk**
10	Sentinel Butte, Camel Hump Lake, **S**...gas
7	Home on the Range, no facilities
1	ND 16, Beach, **N**...Outpost Motel, camping, **S**...HOSPITAL, Cenex/diesel/LP/mart/24hr, Flying J/Conoco/diesel/ rest./mart/24hr, Dairy Queen, Buckboard Inn, Westgate Motel
1mm	weigh sta wb, weigh sta/tourist info eb, litter barrel
0mm	North Dakota/Montana state line

E

W

Dickinson

Ohio Interstate 70

Exit #	Services
225.5mm	Ohio River, Ohio/West Virginia state line
225	US 250 W, OH 7, Bridgeport, **N**...Citgo, StarFire Express/gas, Sunoco/24hr, Pizza Hut, **S**...Domino's
220	US 40, rd 214, **N**...Citgo, Exxon/diesel/mart, Texaco/mart, Big Boy, Hardee's, Shoney's, Hillside Motel(2mi), Plaza Motel, Aldi Foods, **S**...Marathon, Day's Inn
218	Mall Rd, to US 40, to Blaine, **N**...BP/diesel/mart, Citgo, Marathon, Applebee's, Big Boy, DeEselice's Pizza, Denny's, Hardee's, Pizza Hut, Shoney's, Hampton Inn, Knight's Inn, Red Roof, Super 8, Lowe's Bldg Supply, Staples, Wal-Mart/McDonald's, Chevrolet/Pontiac/Buick, Midas Muffler, RV Ctr, **S**...USA/mart, Bob Evans, Bonanza, LJ Silver, McDonald's, Rax, CVS Drug, BEST, Kroger/24hr, Jo-Ann Fabrics, JC Penney, Kaufmann's, K-Mart, OfficeMax, Sears/auto, Sun Electronics, AmeriLube, Firestone, TireAmerica, cinema, mall
216	OH 9, St Clairsville, **N**...BP/service
215	National Rd, **1/2mi N**...BP/diesel, Burger King, Domino's, Subway, WenWu Chinese, Riesbeck's Foods, Parcel Place
213	OH 331, Flushing, **S**...Exxon/mart, Marathon/diesel/mart, Twin Pines Motel, carwash
211mm	**rest area both lanes, full(handicapped)facilities, phone, picnic tables, litter barrels, petwalk**
208	OH 149, Morristown, **N**...BP/diesel/mart, 208/diesel/rest., Arrowhead Motel(1mi), Schlepp's Rest., **S**...Marathon/diesel/mart, Barkcamp SP, RV camping
204	US 40 E(from eb), National Rd, no facilities
202	OH 800, to Barnesville, **S**...HOSPITAL, Citgo/mart, flea mkt
198	rd 114, Fairview, no facilities
193	OH 513, Middlebourne, **N**...BP/mart, FuelMart/diesel, Shell/mart, fireworks
189mm	**rest area eb, full(handicapped)facilities, phone, picnic tables, litter barrels, petwalk**
186	US 40, OH 285, to Old Washington, **N**...Marathon, Shell/mart, RV camping, **S**...Citgo/diesel, GoMart/diesel, Shenandoah Inn/rest.
180b a	I-77 N, to Cleveland, to Salt Fork SP, I-77 S, to Charleston
178	OH 209, Cambridge, **N**...HOSPITAL, BP/diesel/mart, Shell/mart, Big Boy, Bob Evans, Cracker Barrel, Deer Creek Steaks, KFC, McDonald's, Rax, Best Western, BudgetHost, Comfort Inn, Day's Inn, Executive Inn, Holiday Inn, Travelodge, Kroger/deli, **1 mi N**...Papa John's, Pizza Hut, Subway, Taco Bell, Wendy's, AutoZone, Buick/Pontiac/Olds/GMC/Cadillac, CVS Drug, Family$, Monro Muffler/brake, Odd Lots, Valvoline/wash, cinema, **S**...Speedway/Blimpie/diesel/mart, Burger King, AmeriHost, K-Mart, Spring Valley RV Park(1mi)
176	US 22, US 40, to Cambridge, **1/2mi N**...Budget Inn, Cambridge Deluxe Inn, Western Shop, RV camping, st patrol
173mm	weigh sta both lanes
169	OH 83, New Concord, to Cumberland, **N**...BP/mart, RV camping, to Muskingum Coll
164	US 22, US 40, Norwich, **N**...Shell/mart, Hickory Creek Rest., Baker's Motel, Nat'l Rd/Zane Grey Museum, **2 mi S on US 22**...antiques/gifts
163mm	**rest area wb, full(handicapped)facilities, phone, picnic tables, litter barrels, petwalk**
160	OH 797, Airport, **S**...BP/mart, Exxon, Shell/diesel/mart/24hr, Jake's Rest., McDonald's, Wendy's(RV parking), Day's Inn/rest., Holiday Inn, Red Roof Inn, st patrol, access to antiques
157	OH 93, Zanesville, **N**...Duke/mart, **S**...BP, Marathon, Shell/Blimpie/mart, Mickey's Country Cooking/Grandma's Store, KOA(2mi)
155	OH 60, OH 146, Underwood St, Zanesville, **N**...HOSPITAL, info, Bob Evans, Olive Garden, Red Lobster, Shoney's, Steak'n Shake, Tumbleweed Grill, Comfort Inn, Fairfield Inn, Hampton Inn, Pick'n Save Foods, camping, **S**...Exxon/Wendy's/mart, Adornetto Café, Cracker Barrel, Subway, AmeriHost, Best Western, Old Mkt House Inn, ThriftLodge, Family$, Rite Aid
154	5th St(from eb)
153b	Maple Ave(no EZ return from wb), **N**...HOSPITAL, BP, Big Boy, Dairy Queen, Wendy's
a	State St, **N**...Speedway/diesel/mart, Harley-Davidson, **S**...BP/mart
153mm	Licking River
152	US 40, National Rd, **N**...Shell/mart, Big Boy, McDonald's, Super 8, White/GMC/Volvo, RV camp
142	US 40(from wb, no EZ return), Gratiot, **N**...RV camping
141	OH 668, US 40(from eb), to Gratiot, same as 142
132	OH 13, to Thornville, Newark, **N**...Dawes Arboretum(3mi), **S**...BP/mart, Shell/mart, RV camping
131mm	**rest area both lanes, full(handicapped)facilities, phone, picnic tables, litter barrels, petwalk**
129b a	OH 79, Hebron, to Buckeye Lake, **N**...AmeriHost, **S**...Amoco/Duke/diesel/rest./mart, BP/mart, Shell, Burger King, McDonald's, Subway, Taco Bell, Wendy's, Super 8, CarQuest, KOA(2mi), Blue Goose Marina(2mi)
126	OH 37, to Granville, Lancaster, **N**...Citgo/diesel/mart, Pilot/DQ/diesel/mart/24hr, Starfire Express/diesel, RV camping(1mi), **S**...Sunoco/mart, TA/BP/Sbarro's/Popeye's/diesel/24hr, Deluxe Inn, truckwash, KOA(4mi)
122	OH 158, Kirkersville, to Baltimore, **N**...Regal Inn, **S**...Flying J/CountryMkt/diesel/LP/rest./24hr
118	OH 310, to Pataskala, **N**...Sunoco/mart, **S**...Duke's/diesel/mart, antiques/gifts
112	OH 256, to Pickerington, Reynoldsburg, **N**...Shell/McDonald's/mart, Sunoco/diesel, McDonald's, O'Charley's, TGIFriday, Country Inn Suites, Lenox Inn/rest., FashionBug, PetMart, Wal-Mart SuperCtr/24hr, carwash, laundry/cleaners, **S**...Arby's, Bob Evans, Cracker Barrel, Damon's, JinXing Chinese, KFC, Montana Steaks, Schlotsky's, Steak'n Shake, Wendy's, Hampton Inn, Hawthorn Suites, CVS Drug, FasLube, Hix-Pix, bank

Blaine

Cambridge

Zanesville

E

W

Ohio Interstate 70

110 Brice Rd, to Reynoldsburg, **N**...Shell/repair, Speedway/mart, Sunoco/mart, SuperAmerica/mart, Abner's Country Rest., Arby's, Baskin-Robbins, Burger King, Bob Evans, ChiChi's, Gengi Japanese, Golden China, McDonald's, Max&Erma's, Pizza Hut, Popeye's, Roadhouse Grill, Ryan's, Subway, TeeJaye's Rest., Tim Horton, Waffle House, White Castle, Best Western, Comfort Inn, Cross Country Inn, Extended Stay America, La Quinta, Ramada Inn, Red Roof, Super 8, BP/Procare, Goodyear/auto, Home Depot, Midas Muffler, Odd Lots, Pennzoil, Tuffy Auto, bank, **S**...BP/mart, Meijer/diesel/24hr, SuperAmerica/diesel/mart, Sunoco/mart, Applebee's, Arby's, Big Boy, Boston Mkt, Burger King, China Paradise, KFC, McDonald's, Ponderosa, Ruby Tuesday, Subway, Steak'n Shake, Taco Bell, Waffle House, Wendy's, Comfort Suites, Day's Inn, Motel 6, Travelodge, Acura, Aldi, All Tune/lube, Batteries+, Burlington Coats, Circuit City, CVS Drug, Drug Emporium, FashionBug, Firestone/auto, GFS, GNC, Honda, Hyundai, JC Penney, Jo-Ann Fabrics, Kinko's, Kroger/deli, Lowe's Bldg Supply, Magic Mtn Funpark, Meineke Muffler, Michael's, NTB, OfficeMax, Old Navy, Olds, PetsMart, Pier 1, PrecisionTune, Sam's Club, Sears/auto, SprintLube, Target, Toyota, Valvoline, bank, carwash, cinema, mall

108b a I-270 N to Cleveland, access to HOSPITAL, I-270 S to Cincinnati

107 OH 317, Hamilton Rd, to Whitehall, **S**...BP/mart/wash, Shell/wash, Sunoco/diesel/mart/wash, Arby's, Bob Evans, Burger King, Denny's, McDonald's, MCL Cafeteria, Olive Garden, Ponderosa, Red Lobster, Subway, Taco Bell, Holiday Inn/rest., Knight's Inn, Park Inn, Residence Inn, Sheraton/rest., Phar-Mor Drugs, Service Merchandise, BP/Procare, Goodyear/auto, TireAmerica, Pearle Vision, auto parts, bank, cinema

105b a US 33, James Rd, to Lancaster, Bexley, no facilities

103 Livingston Ave, to Capital University, **N**...BP/mart/wash, Citgo, Speedway/mart, Thornton's/mart, Chinese Rest., Domino's, LJ Silver, Mr Hero Subs, Subway, Taco Bell, Wendy's, Nationwise Auto Parts, Speedy Mufflers, Muffler King, carwash, **S**...Rich Gas/mart, Shell/wash, McDonald's, Rally's, White Castle

102 Kelton Ave, Miller Ave, HOSPITAL

101a I-71 N, to Cleveland, no facilities

100b US 23, to 4th St, downtown

99c Rich St, Town St, **N**...Sunoco, Ford

b OH 315 N, downtown

a I-71 S, to Cincinnati, no facilities

98b Mound St(from wb, no EZ return), Cooper Stadium, **S**...DENTIST, Bonded Gas, Omega/mart, Chinese Rest., LJ Silver, McDonald's, Rally's, Sister's Chicken/biscuits, GMC, Goodyear, bank, hardware

a US 62, OH 3, Central Ave, to Sullivan, access to same as 98b

97 US 40, W Broad St, **N**...Omega Gas/mart, Burger King, KFC, McDonald's, Taco Bell, White Castle, Knight's Inn

96 Grandview Ave(from eb), no facilities

95 Hague Ave(from wb), **S**...BP/service/mart

94 Wilson Rd, **N**...United/gas, **S**...VETERINARIAN, BP/mart/wash, Shell/mart/wash/24hr, Speedway/mart/24hr, Waffle House, Econolodge, Buick, Sam's Club, tires

93b a I-270, N to Cleveland, S to Cincinnati, access to HOSPITAL

91b a New Rome, to Hilliard, **N**...Meijer/diesel/24hr, Shell/mart, Speedway/mart, Applebee's, Arby's, Big Boy, Burger King, Cracker Barrel, Fazoli's, KFC, McDonald's, Outback Steaks, Perkins, Salvy's Bistro, Taco Bell, White Castle, Wendy's, Comfort Suites, Cross Country Inn, Fairfield Inn, Hampton Inn, Hawthorn Inn, Knight's Inn, Motel 6, Red Roof Inn, Ford, PetsMart, Sam's Club, Valvoline, Wal-Mart SuperCtr/24hr, bank, cinema, **S**...BP/diesel, Marathon/diesel, Bob Evans, Steak'n Shake, Best Western, Country Inn Suites

85 OH 142, W Jefferson, to Plain City, no facilities

80 OH 29, to Mechanicsburg, no facilities

79 US 42, to London, Plain City, **N**...Discount Fuel/diesel, MPG/diesel, Williams/Main St Café/diesel/mart/atm/24hr, Waffle House, **S**...HOSPITAL, Sunoco/mart/24hr, Speedway/diesel/mart/24hr, TA/BP/Popeye's/Pizza Hut/diesel/24hr, McDonald's, Subway, Taco Bell, Wendy's, Holiday Inn Express, Knight's Inn, truckwash

72 OH 56, Summerford, to London, **N**...Shell/mart/24hr, **4 mi S**...HOSPITAL, lodging

71mm **rest area both lanes, full(handicapped)facilities, phone, picnic tables, litter barrels, vending, petwalk**

66 OH 54, South Vienna, to Catawba, **N**...FuelMart/diesel, **S**...Speedway/mart, RV camping(1mi),

62 US 40, Springfield, **N**...Speedway/mart, Harmony Motel, antiques, auto repair, to Buck Creek SP, **S**...RV camp

59 OH 41, to S Charleston, **N**...HOSPITAL, Harley-Davidson, fairgrounds, st patrol, **S**...BP/diesel/mart, Prime Fuel/diesel/mart, Dan's Towing/24hr

Ohio Interstate 70

Springfield

| 54 | OH 72, Springfield, to Cedarville, **N**...HOSPITAL, BP/diesel/mart/wash, Shell/mart/wash, Speedway, Sunoco/diesel/24hr, Arby's, Bob Evans, Cassano's Pizza, Cracker Barrel, Denny's, Domino's, El Toro Mexican, Hardee's/24hr, KFC, Lee's Chicken, Little Caesar's, LJ Silver, McDonald's, Panda Chinese, Perkins/24hr, Pizza Hut/Taco Bell, Ponderosa, Rally's, Shoney's, Subway, Wendy's, Comfort Suites, Day's Inn, Hampton Inn, Holiday Inn, Red Roof Inn, Super 8, Aldi Foods, Drug Castle, Family$, Kroger/deli, Midas Muffler, Odd Lots, Valvoline, auto parts, bank, carwash, **S**...Swifty Gas, tires |

52b a	US 68, to Urbana, Xenia, no facilities
48	OH 4(from wb), to Enon, Donnelsville, **N**...RV camping, **S**...Speedway
47	OH 4(from eb), to Springfield, **5 mi N**...gas, food, lodging
44	I-675 S, to Cincinnati, no facilities
43mm	Mad River
41	OH 4, OH 235, to Dayton, New Carlisle, **N**...BP/diesel, Sunoco/diesel, McDonald's, Wendy's
38	OH 201, Brandt Pike, **N**...Shell/mart, **S**...DENTIST, Amoco, Shell/mart, Denny's/24hr, Waffle House, Wendy's(1mi), Comfort Inn, Travelodge, antiques, auto parts/mufflers
36	OH 202, Huber Heights, **N**...VETERINARIAN, Speedway/diesel/mart/24hr, Applebee's, Big Boy, Fazoli's, Pizza Hut, Ruby Tuesday, Steak'n Shake, Subway, Taco Bell, Uno Pizzaria, Waffle House, Wendy's, Super 8, Cub Foods/24hr, FashionBug, GNC, Kohl's, Lowe's Bldg Supply, Marshall's, MC Sports, Radio Shack, Saturn, Target, Wal-Mart/auto/24hr, **S**...BP/diesel/mart, Sunoco/diesel/mart, Arby's, Bob Evans, Burger King, Cadillac Jack's Grill, LJ Silver, McDonald's, Old Country Buffet, SkyLine Chili, TCBY, White Castle, Days Inn, Hampton Inn, Holiday Inn Express, CVS Drug, $Plus, K-Mart, Kroger/deli/24hr, bank, cinema
33 b a	I-75, N to Toledo, S to Dayton
32	Vandalia, Dayton Intn'l Airport
29	OH 48, Englewood, to Dayton, **N**...VETERINARIAN, BP/mart, Shell/repair, Speedway, Sunoco/mart/24hr, Arby's, Big Boy, Bob Evans, Perkins, Pizza Hut, Ponderosa, Steak'n Shake, Taco Bell, Wendy's, $Inn, Hampton Inn, Holiday Inn, Motel 6, Aldi Foods, McNulty RV Ctr(1mi), NAPA, Rite Aid, bank, **S**...Meijer/diesel/24hr, McDonald's, Waffle House, Cross Country Inn
26	OH 49 S, **S**...Shell/mart
24	OH 49 N, Clayton, to Greenview, **S**...KOA(seasonal)
21	rd 533, Brookville, **S**...BP/diesel, Speedway/Subway/diesel/mart, Arby's, Hardee's, KFC/Taco Bell, McDonald's, Rob's Family Dining, Waffle House, Wendy's, Day's Inn, Chevrolet, $General, auto parts, bank, carwash
14	OH 503, Lewisburg, to West Alexandria, **N**...Sunoco/Subway/mart, Barb's Home Cookin', Super Inn, SuperValu Foods, bank, **S**...Marathon/diesel/mart
10	US 127, to Eaton, Greenville, **N**...Marathon/mart, TA/BP/Burger King/diesel/24hr, Sandman Motel, st patrol, **S**...Pilot/DQ/diesel/mart/24hr, Omega/diesel/mart, Country Folks Rest., Subway, Econolodge
3mm	**Welcome Ctr eb/rest area both lanes, full(handicapped)facilities, phone, vending, picnic tables, litter barrels, petwalk**
1	US 35 E(from eb), to Eaton, New Hope
0mm	Welcome Arch, Ohio/Indiana state line, weigh sta eb

Ohio Interstate 71

Cleveland

Exit #	Services
247b	I-90 W, I-490 E. I-71 begins/ends on I-90, exit 170 in Cleveland.
a	W 14th, Clark Ave, no facilities
246	Denison Ave, Jennings Rd(from sb)
245	US 42, Pearl Rd, **E**...HOSPITAL, BP, Sunoco, Zoo
244	W 65th, Denison Ave, no facilities
242b a	W 130th, to Bellaire Rd, **W**...Citgo/mart, Sunoco, Burger King(1/2mi)
240	W 150th, **E**...Shell, Speedway/mart, Denny's, Jack's Steaks, Marriott, LubeStop, **W**...BP/mart/wash, Country Kitchen/24hr, McDonald's, Taco Bell, Baymont Inn, Holiday Inn
239	OH 237 S(from sb), **W**...to airport
238	I-480, Toledo, Youngstown, **W**...airport
237	Snow Rd, Brook Park, **E**...Shell, McDonald's, Best Western, Fairfield Inn, Holiday Inn Express, **W**...to airport
235	Bagley Rd, **E**...Bob Evans, Clarion, BP ProCare, **W**...HOSPITAL, BP/diesel/mart, Shell, Speedway/mart, BackHome Buffet, Bruegger's Bagels, Burger King, Brown Derby Roadhouse, China's Best Buffet, Cooker, Damon's, Denny's, Dunkin Donuts, Friendly's, McDonald's, Olive Garden, Perkins, Pizza Hut, Taco Bell, Comfort Inn, Courtyard, Cross Country Inn, Motel 6, Radisson, Red Roof Inn, Residence Inn, Studio+, TownePlace Suites, K-Mart/auto, Olds/GMC, LubeStop, cinema
234	US 42, Pearl Rd, **E**...DENTIST, Shell/repair, Sunoco, Hunan Chinese, Islander Grill, Katherine's Rest., Mr Hero, Santo's Pizza, PetSupply+, Audi/Porsche, Honda, Hyundai, Saturn, laundry, **W**...Marathon/diesel, Jennifer's Rest., Day's Inn, Extended Stay America, La Siesta Motel, Villager Lodge, Home Depot, Wal-Mart, auto repair

E

W

N

S

Ohio Interstate 71

233   I-80 and Ohio Tpk, to Toledo, Youngstown

231   OH 82, Strongsville, **E**...Shell/24hr, Speedway, Holiday Inn, Red Roof Inn, **W**...MEDICAL CARE, BP/diesel/mart, Marathon/diesel/mart, Sunoco/diesel, Applebee's, Country Kitchen/24hr, Demetrio's Rest., Longhorn Steaks, Macaroni Grill, Rio Bravo, TGIFriday, Borders Books, Dillard's, Giant Eagle Foods, JC Penney, Jo-Ann Fabrics, Kaufmann's, Kohl's, LubeStop, Medic Drug, Midas Muffler, OfficeMax, Sears/auto, Target, bank, cleaners, mall, tires

226   OH 303, Brunswick, **E**...Shell/diesel/mart/24hr, Camping World RV Supplies/service, Chrysler/Plymouth/Jeep/Toyota, **W**...BP/mart, Marathon/diesel/mart, Sunoco/diesel/mart, Arby's, Big Boy, Bob Evans, Burger King, Denny's, McDonald's, Subway, Taco Bell, Wendy's, Howard Johnson, Sleep Inn, Ford, Giant Eagle Food, K-Mart, Radio Shack, bank, carwash

225mm   **rest area nb, full(handicapped)facilities, phone, picnic tables, litter barrels, petwalk**

224mm   **rest area sb, full(handicapped)facilities, phone, picnic tables, litter barrels, petwalk**

222   OH 3, **W**...st patrol

220   I-271 N, to Erie, Pa

218   OH 18, to Akron, Medina, **E**...BP/diesel/mart/24hr, Citgo/diesel/mart, Shell/mart/wash/atm, Speedway/diesel/mart, Sunoco/diesel/mart, Big Boy, Blimpie, Burger King, Cracker Barrel, Dairy Queen, McDonald's, Medina Rest., Best Western, Day's Inn, Holiday Inn Express, Suburbanite Motel, Super 8, $General, Medina Hardware, Olds/GMC, cleaners, **W**...HOSPITAL, Arby's, Bob Evans, Denny's, Pizza Hut, Wallaby's Grill, Wendy's, Cross Country Inn, Motel 6, Dodge, Honda, Mueller Tire/brake, Pontiac/Buick/Cadillac

209   I-76 E, to Akron, US 224 **W**...TA/diesel/rest./24hr, Speedway/diesel/rest./24hr, Blimpie, Country Kitchen, McDonald's, Taco Bell, HoJo's, Super 8, Blue Beacon, Chippewa Valley Camping(1mi)

204   OH 83, Burbank, **E**...BP/diesel/mart, Duke/diesel/mart, Shell/mart, Taco Bell, Plaza Motel, **W**...HOSPITAL, Pilot/Wendy's/diesel/mart/24hr, Bob Evans, Burger King, FoodCourt, McDonald's, Prime Outlets/famous brands

198   OH 539, W Salem, **W**...Town&Country RV Park(2mi)

196.5mm   **rest area both lanes, full(handicapped)facilities, phone, vending, picnic tables, litter barrels, petwalk**

196   OH 301(from nb, no re-entry), W Salem, no facilities

191mm   weigh sta sb

186   US 250, Ashland, **E**...Citgo/mart, Perkins, Goschinski Outfitters, Grandpa's Village/cheese/gifts, Hickory Lakes Camping(7mi), **W**...HOSPITAL, TA/BP/diesel/rest./24hr, Bob Evans, Denny's, McDonald's, Subway(3mi), Wendy's, AmeriHost, Day's Inn, Holiday Inn Express, Super 8, st patrol, to Ashland U

180mm   **rest area both lanes, full(handicapped)facilities, phone, vending, picnic tables, litter barrels, petwalk**

176   US 30, to Mansfield, **E**...Marathon, Best Western, Econolodge, Old Town Motel, antiques, fireworks

173   OH 39, to Mansfield, no facilities

169   OH 13, Mansfield, **E**...BP/diesel/rest., Cracker Barrel, Steak'n Shake, AmeriHost, Baymont Inn, **W**...HOSPITAL, Citgo/mart, Arby's, Big Boy, Bob Evans, Burger King, Denny's, McDonald's(truck parking), Taco Bell, Super 8, 42 Motel(2mi), Travelodge, st patrol

165   OH 97, to Bellville, **E**...BP/mart/wash, Shell/diesel/mart/atm/24hr, Speedway/diesel/mart, Burger King, Der Dutchman Rest., McDonald's, Comfort Inn, $Motel, Ramada Ltd, Star Inn, to Mohican SP, **W**...Dinner Bell Rest., Hardee's(2mi), Subway(2mi), Wendy's

151   OH 95, to Mt Gilead, **E**...Duke/diesel/mart/rest./24hr, Marathon, Gathering Inn Rest., McDonald's, Wendy's, Best Western, st patrol, **W**...HOSPITAL, BP/diesel/mart, Sunoco/diesel/mart, Leaf Rest., Knight's Inn, KOA, Mt Gilead SP(6mi)

140   OH 61, Mt Gilead, **W**...Amoco/mart, Sunoco/Subway/TCBY/mart, Ole Farmstead Rest., Taco Bell

131   US 36, OH 37, to Delaware, **E**...Flying J/Conoco/diesel/LP/rest./24hr, Speedway/Blimpie/diesel/mart/24hr, Burger King, RV camping, **W**...HOSPITAL, BP/diesel/mart, Shell/mart/24hr, Sunoco/diesel, A&W, Arby's, Bob Evans, Cracker Barrel, KFC, McDonald's, Subway, Taco Bell, Waffle House, Wendy's, Hampton Inn, Holiday Inn Express, Microtel, Alum Cr SP, RV camping, to OH Wesleyan U, truck repair

129mm   weigh sta nb

128mm   **rest area both lanes, full(handicapped)facilities, phone, vending, picnic tables, litter barrels, petwalk**

Ohio Interstate 71

121	Polaris Pkwy, **E**...BP/Blimpie/mart/wash, Shell/mart/wash/atm/24hr, McDonald's, Polaris Grill, Skyline Chili, Steak'n Shake, Wingate Inn, Polaris Ampitheatre, **W**...Marathon/A&W/mart/atm, Meijer/diesel/24hr, Applebee's, Arby's, Bob Evans, Hoggy's Grille, Hop's Grill, Magic Mtn Grill, Martini Italian, Max&Erma's, O'Charley's, Panera's Bread, Quaker Steak/lube, Starbuck's, Wendy's, Hilton Garden, Wellesley Inn, Barnes&Noble, Best Buy, GNC, JC Penney, Jo-Ann Fabrics, Kaufmann's, Kroger/deli, Lazurus, Linens'n Things, Lowe's Bldg Supply, OfficeMax, Old Navy, Target, World Mkt, Wal-Mart SuperCtr/24hr(3mi), mall
119	I-270, to Indianapolis, Wheeling
117	OH 161, to Worthington, **E**...BP/diesel/mart, Shell/autocare, Sunoco, Burger King, LJ Silver, Max&Erma's, Ponderosa, Rally's, Red Lobster, Subway, TeeJaye's Rest., White Castle, Comfort Inn, Day's Inn, Holiday Inn Express, Knight's Inn, Motel 6, Big Bear/deli, Firestone, PrecisionTune, Staples, bank, cleaners, **W**...Shell/mart/wash/atm, Speedway/diesel/mart, Bob Evans, Casa Fiesta Mexican, Chinese Express, Elephant Rest., McDonald's, Otani Japanese, Pizza Hut, Waffle House, Wendy's, Best Western, Country Inn Suites, Cross Country Inn, Econolodge, Extended Stay America, Hampton Inn, Harley Hotel, Marriott, Ramada Ltd, Residence Inn, Super 8, Travelodge, Chevrolet, NTB, Grand Prix Amusement, cinema
116	Morse Rd, Sinclair Rd, **E**...MEDICAL CARE, BP/mart/wash, Speedway/mart, Shell/mart/atm, Max&Erma's, McDonald's, Fairfield Inn, Buick, Office Depot, PepBoys, PrecisionTune, Sun Electronics, cleaners, **W**...BP, TeeJaye's Rest., Cross Country Inn, Ramada Inn, Red Roof Inn, TireAmerica
115	Cooke Rd, **W**...Clark Gas
114	N Broadway, **W**...BP/mart, Sunoco/mart
113	Weber Rd, **W**...Speedway/mart, CarQuest, NAPA
112	Hudson St, **E**...BP/mart/wash, Shell/mart/wash/atm, Wendy's, **W**...fairgrounds
111	17th Ave, **W**...McDonald's, Day's Inn
110b	11th Ave, **E**...HOSPITAL
a	5th Ave, **W**...BP/mart, Shell/wash, Auddino's Bread/Pastries, McDonald's, Snapp's Burgers, Wendy's, White Castle, airport
109a	I-670 W, no facilities
108b	US 40, Broad St, downtown, **W**...Honda
101a[70]	I-70 E, to Wheeling
100b a[70]	US 23 S, Front St, High St, downtown
106a	I-70 W, to Indianapolis
b	OH 315 N, Dublin St, Town St, HOSPITAL
105	Greenlawn, HOSPITAL
104	OH 104, Frank Rd, **W**...tires
101b a	I-270, Wheeling, Indianapolis
100	Stringtown Rd, **E**...BP/mart, Bob Evans, Church's Chicken/White Castle, Best Western, Cross Country Inn, Hampton Inn, Holiday Inn, Hilton Garden, Microtel, Ramada Inn, Super 8, auto repair, **W**...MEDICAL CARE, DENTIST, EYECARE, BP/mart, Certified/diesel, Speedway/diesel/mart, Sunoco/24hr, Arby's, Burger King, Capt D's, China Bell, Cracker Barrel, Damon's, Fazoli's, KFC, McDonald's, Perkins/24hr, Pizza Hut, Ponderosa, Rally's, Roadhouse Grill, Subway, Taco Bell, TeeJaye's Rest., Tim Horton, Waffle House, Wendy's, Comfort Inn, Heritage Inn, Red Roof Inn, Saver Motel, Value Inn, Aldi Foods, All Tune/lube, Big Bear Food/drug, CVS Drug, GNC, Goodyear/auto, K-Mart, Kroger/24hr, Midas Muffler, NAPA, PetMart, PostNet, Radio Shack, Sears/hardware, Tuffy Auto Ctr, Valvoline, bank, camera/video shop, cleaners/laundry
97	OH 665, London-Groveport Rd, **E**...Citgo/DM/Taco Bell/mart, Sunoco/Subway/mart, Arby's, McDonald's, Tim Horton/Wendy's, to Scioto Downs, **W**...Eddie's Auto/lube
94	US 62, OH 3, Orient, **E**...BP/diesel/mart, **W**...Sunoco/mart
84	OH 56, Mt Sterling, **E**...BP/Subway/mart, Sunoco/mart, Royal Inn, to Deer Creek SP(9mi)
75	OH 38, Bloomingburg, **E**...fireworks, **W**...Sunoco/mart, Fayette Village
69	OH 41, OH 734, Jeffersonville, **W**...HOSPITAL, BP, Shell/Subway/diesel/mart, Arby's, TCBY, Wendy's, White Castle, AmeriHost, Prime Outlets II/famous brands, RV camping
68mm	**rest area both lanes, full(handicapped)facilities, phone, vending, picnic tables, litter barrels, petwalk**
65	US 35, Washington CH, **E**...HOSPITAL, Amoco/diesel/mart/24hr, MPG/diesel, Shell/mart/atm, TA/BP/Pizza Hut/Popeye's/diesel/mart/24hr, Bob Evans, Burger King, KFC, McDonald's, Taco Bell, Waffle House, Wendy's, AmeriHost, Hampton Inn, Prime Outlets/famous brands, **W**...TA/BP/diesel/rest./24hr, $Motel
58	OH 72, to Sabina, no facilities
54mm	weigh sta sb
50	US 68, to Wilmington, **E**...HOSPITAL, repair, **W**...BP/diesel/mart, Shell/diesel/mart/etd/24hr, Speedway/Subway/diesel/mart/24hr, Dairy Queen, McDonald's, Knight's Inn, Goodyear, RV Ctr, Tack Outfitters
49mm	weigh sta nb
45	OH 73, to Waynesville, **E**...HOSPITAL, BP/mart, Marathon/diesel/mart, Thousand Trails RV camping, **W**...Citgo/diesel/mart, flea mkt, Caesar Creek SP(5mi)
36	Wilmington Rd, **E**...to Ft Ancient St Memorial, RV camping
35mm	Little Miami River

N

S

Ohio Interstate 71

34mm rest area both lanes, full(handicapped)facilities, scenic view, phone, vending, picnic tables, litter barrels, petwalk

32 OH 123, to Lebanon, Morrow, **E**...BP/mart, Clark/mart, Country Kitchen, **3 mi W**...Bob Evans, Skyline Chili, Taco Bell, Knight's Inn, RV camping

28 OH 48, S Lebanon, **E**...Countryside Inn(1mi), **W**...Lebanon Raceway(6mi), hwy patrol

25 OH 741 N, Kings Mills Rd, **E**...Shell, Speedway/diesel/mart, Bill Knapp's, McDonald's, Taco Bell, Outback Steaks, Comfort Suites, Kings Island Inn, Quality Inn, Harley-Davidson, camping, **W**...VETERINARIAN, EYECARE, CHIROPRACTOR, AmeriStop/Subway/gas, BP, Big Boy, Bob Evans, Burger King, Dairy Queen, GoldStar Chili, Perkins, Pizza Hut, Skyline Chili, Tabby's Grill, Waffle House, Wendy's, Best Western, Hampton Inn, Holiday Inn Express, Microtel, GNC, Kroger/deli, MailBoxes Etc, bank, cinema, waterpark

24 Western Row, King's Island Dr(from nb), **E**...Sunoco/mart, King's Island, Jellystone Resort Camping

19 US 22, Mason Montgomery Rd, **E**...VETERINARIAN, Meijer/diesel/24hr, Shell/mart, SuperAmerica/diesel/mart, Sunoco/mart/24hr, United Dairy Farmers, Arby's, Big Boy, Bennigan's, Bob Evans, Boston Mkt, Cooker, Cracker Barrel, Fazoli's, KFC, McDonald's, Olive Garden, Pizza Hut, Pizza Tower, Ponderosa, Taco Bell, TGIFriday, Wendy's, White Castle, Comfort Inn, Signature Inn, SpringHill Suites, TownePlace Suites, AutoZone, Barnes&Noble, CarX Muffler, Circuit City, Chevrolet, Firestone/auto, GMC/Buick/Pontiac, Goodyear, HomePlace, JiffyLube, Kohl's, Kroger/deli, Lincoln/Mercury, Meineke Muffler, Michael's, Michel's Tires, OfficeMax, Old Navy, PetsMart, Radio Shack, Target, Toyota, Tuffy AutoCtr, Wal-Mart/auto, Walgreen, cleaners **W**...BP/diesel/bakery/deli/mart, Marathon/diesel/mart/24hr, Shell/mart/wash/atm, Applebee's, Burger King, Carrabba's, Chipotle Mexican, Copeland's Rest., Fuddrucker's, LoneStar Steaks, O'Charley's, RiverCity Grill, Steak'n Shake, Skyline Chili, Subway, Tumbleweed Grill, Waffle House, Wendy's, AmeriSuites, Baymont Inn, Best Western, Country Hearth Inn, Day's Inn, Marriott, Quality Inn, Red Roof Inn, Bigg's HyperMart, Drug Emporium, Home Depot, Lowe's Bldg Supply, MailBoxes Etc, Staples

17b a I-275, to I-75, OH 32, no facilities

15 Pfeiffer Rd, **E**...HOSPITAL, **W**...BP/mart/wash, Shell/diesel/mart, Sunoco/diesel/mart/24hr, Applebee's, Bennigan's, Bob Evans, Subway, Best Western, Courtyard, Embassy Suites, Hampton Inn, MainStay Suites, Ramada Inn, Red Roof Inn, Office Depot, airport, bank

14 OH 126, Reagan Hwy, Blue Ash, no facilities

12 US 22, OH 3, Montgomery Rd, **E**...Sheraton, **W**...HOSPITAL, BP/diesel/mart, Chevron/mart, Shell/mart/24hr, Sunoco/mart, Arby's, Bob Evans, Bandstand Grill, Jalapeno's Café, KFC, LoneStar Steaks, LJ Silver, Max&Erma's, McDonald's, OutBack Steaks, Red Lobster, Ruby Tuesday, Subway, Taco Bell, TGIFriday, Wendy's, Hannaford Suites, Harvey Hotel, Barnes&Noble, Computer City, Cost+ World Mkt, Dillard's, Dodge/Plymouth, Firestone, Goodyear, Hyundai, JiffyLube, Lazurus, Linens'n Things, OfficeMax, Parisian, PepBoys, Pier 1, Ritz Camera, Staples, ToysRUs, bank, cinema, cleaners, mall

11 Kenwood Rd, **W**...HOSPITAL, same as 12

10 Stewart Rd(from nb), to Silverton, **W**...Marathon/diesel

9 Redbank Rd, to Fairfax, **1 mi W**...Marathon/diesel

8 Kennedy Ave, Ridge Ave W, **E**...Red Roof Inn, **W**...BP, Marathon/diesel/mart, Speedway/diesel/mart, Denny's, KFC, LJ Silver, McDonald's, Old Country Buffet, Pizza Hut, Rally's, Subway, Taco Bell, Wendy's, Bigg's Food/drug, Big Lots, Circuit City, Ford, Goodyear, Home Depot, K-Mart/Little Caesar's, Office Depot, Sam's Club, Tire Discounter, ValueCity, Wal-Mart, transmissions

7b a OH 562, Ridge Ave E, Norwood, **1/2mi E**...VETERINARIAN, BP/mart, Ponderosa, AutoZone

6 Edwards Rd, **E**...BP/diesel, Shell, Speedway/mart, Boston Mkt, Chang's China Bistro, Don Pablo's, Fuddrucker's, GoldStar Chili, J Alexander's Rest., Longhorn Steaks, Starbuck's, Drug Emporium, HomeGoods, JiffyLube, SteinMart, TJ Maxx, **W**...Shell/mart, CarX Muffler, Midas Muffler

5 Dana Ave, Montgomery Rd, **W**...Xavier Univ, Zoo

3 Taft Rd(from sb), U of Cincinnati

2 US 42 N, Reading Rd, **W**...HOSPITAL, downtown, art museum

1k j I-471 S

1d Main St, downtown

1c b Pete Rose Way, Fine St, stadium, downtown

1a I-75 N, US 50, to Dayton, no facilities

I-71 S and I-75 S run together

0mm Ohio River, Ohio/Kentucky State line

Cincinnati

Ohio Interstate 74

Exit #	Services
20	I-75(from eb), N to Dayton, S to Cincinnati, I-74 begins/ends on I-75.
19	Gilmore St, Spring Grove Ave, no facilities
18	US 27 N, Colerain Ave, no facilities
17	Montana Ave(from wb), **N**...BP/mart/wash, Dairymart
14	North Bend Rd, Chevoit, HOSPITAL, **N**...Speedway, SuperAmerica, Shoney's, Sam's Club, Tuffy Auto, Valvoline, **S**...BP/mart, Shell, Bob Evans, McDonald's, Wendy's, TriStar Motel
11	Rybolt Rd, Harrison Pike, **N**...BP, **S**...Sunoco, Imperial House Hotel
9	I-275 N, to I-75, N to Dayton
8mm	Great Miami River
7	OH 128, Cleves, to Hamilton, **N**...BP/diesel/mart, Shell/diesel/mart, Wendy's
5	I-275 S, to Kentucky
3	Dry Fork Rd, **N**...Citgo, **S**...BP, Chevron/diesel, Shell/Burger King/diesel/mart/atm, Motel Deluxe, Chevrolet
2mm	weigh sta eb
1	New Haven Rd, to Harrison, **N**...BP/diesel/mart/wash, Buffalo Wings, Cracker Barrel, Biggs Foods, Ford, Michel Tire, **S**...PODIATRIST, VETERINARIAN, Marathon, Shell/mart/24hr, Speedway/mart, Sunoco, Super America, Arby's, Big Boy, Burger King, Domino's, Hardee's, KFC, Little Caesar's, McDonald's, Perkins, Pizza Hut, Ponderosa, Shoney's, Skyline Chili, Taco Bell, Waffle House, Wendy's, Holiday Inn Express, Quality Inn, FashionBug, Firestone, GNC, Goodyear/auto, Kroger/bakery/deli/24hr, Lobill's Foods, K-Mart, NAPA, Walgreen, bank
0mm	Ohio/Indiana state line

Ohio Interstate 75

Exit #	Services
211.5mm	Ohio/Michigan state line
210	OH 184, Alexis Rd, to Raceway Park, **E**...access to BP, Foodtown, **W**...BP/mart/24hr, Meijer/diesel/24hr, Speedway/diesel/mart/24hr, Burger King, McDonald's, Taco Bell, Ford Trucks
210mm	Ottawa River
209	Ottawa River Rd(from nb), **E**...BP, Foodtown
208	I-280, to I-80/90, access to Sunoco 1 mi E off I-280, exit 12.
207	Stickney Ave, Lagrange St, **E**...FoodTown, Kroger, Rite Aid, AutoWorks
206	to US 24, Phillips Ave, **W**...transmissions
205a	to Willys Pkwy, to Jeep Pkwy
204	I-475 W, to US 23, to Maumee, Ann Arbor
203b	US 24, to Detroit Ave, **W**...BP/mart/wash/24hr, Shell/24hr, KFC, McDonald's, Rite Aid, auto parts, bank
a	Bancroft St, downtown
202	Washington St, Collingwood Ave(from sb, no EZ return), **E**...HOSPITAL, BP, Art Museum, **W**...McDonald's
201b a	OH 25 S, **W**...Toledo Zoo
200	South Ave, Kuhlman Dr, no facilties
200mm	Maumee River
199	OH 65, Miami St, to Rossford, **E**...Econolodge, **W**...Subway(1mi)
198	Wales Rd, Oregon Rd, to Northwood, **E**...Shell/diesel/mart/24hr, Speedway, Comfort Inn, bank
197	Buck Rd, to Rossford, **E**...Fuel&Deli/diesel, Wendy's/Tim Horton's, **W**...BP/mart/24hr, Sunoco, Denny's, Ike's Rest., McDonald's, Knight's Inn, Super 8
195	to I-80/90, Ohio Tpk(toll rd)
194	OH 795, Perrysburg, to Millbury, **E**...BP/diesel/mart/wash, Courtyard
193	US 20, US 23 S, Perrysburg, **E**...BP/diesel/mart/24hr, Sunoco/mart, Big Boy, Bob Evans, Burger King, Chinese Rest., Cracker Barrel, McDonald's, Ralphie's Burgers, Ranch Steaks/seafood, Subway, Taco Bell, Wendy's, Best Western, Day's Inn, French Qtr Motel, Holiday Inn Express, Farm Mkt, GNC, Kroger/deli/24hr, K-Mart/drugs, KOA(7mi), **W**...Marathon/diesel/mart, Baymont Inn, AutoZone, bank
192	I-475, US 23 N, to Maumee, 1 mi W off I-475, Amoco, BP, Speedway, Ponderosa, Subway, motel
187	OH 582, to Luckey, Haskins, no facilities
181	OH 64, OH 105, Bowling Green, to Pemberville, **E**...Citgo/mart, **W**...HOSPITAL, BP/mart/wash, Citgo/Baskin-Robbins/Subway/diesel/mart, Speedway/diesel/mart, Sunoco/diesel/mart, SuperAmerica/Blimpie/diesel/mart, Big Boy, Bob Evans, Burger King, Chi Chi's, DiVanetti's Italian, Domino's, Fricker's Rest., Hunan Palace, Kaufman's Rest., Little Caesar's, McDonald's, Ranch Steak/seafood, Wendy's, Best Western, Buckeye Motel, Day's Inn, Quality Inn, bank, to Bowling Green U
179	US 6, to Fremont, Napoleon, **W**...FireLake Camper Park
179mm	**rest area both lanes, full(handicapped)facilities, phone, picnic tables, litter barrels, vending, petwalk**
175.5mm	weigh sta nb
171	OH 25, Cygnet, no facilities
168	Eagleville Rd, Quarry Rd, **E**...Fuelmart/diesel

Ohio Interstate 75

167   OH 18, North Baltimore, to Fostoria, **E**...Petro/Mobil/diesel/ rest./24hr, McDonald's, Blue Beacon, truck repair, **W**...Citgo/ diesel/mart, Denny's, Crown Inn

165mm   Rocky Ford River

164   OH 613, to McComb, Fostoria, **E**...Shadylake Camping, to Van Buren SP, **W**...Pilot/DQ/Subway/Taco Bell/diesel/mart/ 24hr, Shell/Burger King/Grandma's Kitchen/diesel

162mm   weigh sta sb, phones

161   OH 99, **W**...antiques, grill, hwy patrol

159   US 224, OH 15, Findlay, **E**...HOSPITAL, BP/diesel/mart, Marathon/mart, Shell/mart, Big Boy, Bob Evans, Burger King, Domino's, McDonald's, Pizza Hut, Ponderosa, Rally's, Spaghetti Shop, Taco Bell, Wendy's, Great Scott Foods, Cross Country Inn, Knight's Inn, Ramada Inn, Rodeway Inn, Super 8, carwash, **W**...Cracker Barrel, Country Hearth Inn, Hampton Inn, Holiday Inn

158   Blanchard River

157   OH 12, Findlay, **E**...Citgo/TCBY, Marathon/Blimpie/diesel/mart/wash, Bill Knapp's, Day's Inn, **W**...VETERINARIAN, BP, OH W Truck Plaza/diesel, Pacific Pride/diesel, Fricker's Rest., Pilgrim's Rest., Econolodge, Findlay Inn

156   US 68, OH 15, to Carey, **E**...HOSPITAL

153mm   **rest area both lanes, full(handicapped)facilities, phone, picnic tables, litter barrels, vending, petwalk**

145   OH 235, Mount Cory, to Ada, **E**...TwinLakes Camping

142   OH 103, Bluffton, to Arlington, **E**...Citgo/mart, Sunoco/diesel/mart, Denny's, HoJo's, **W**...Marathon/diesel/mart, Shell/diesel, Arby's, KFC, McDonald's, Subway, Taco Bell, to Bluffton Coll

140   Bentley Rd, to Bluffton, no facilities

135   to US 30, Beaverdam, to Delphos, **E**...Citgo/Subway/diesel/mart, Speedway/Blimpie/diesel/mart, Grandma's Pantry Rest., Yesterday's Family Rest., **W**...Flying J/diesel/rest./LP/24hr, tires, truck repair

134   Napolean Rd, to Beaverdam, no facilities

130   Blue Lick Rd, **E**...Citgo/diesel/mart, Subway, Ramada Inn, marine sales

127b a   OH 81, Lima, to Ada, **W**...BP, Clark/diesel, Chinese Rest., Dari Hut, Waffle House, Comfort Inn, Day's Inn/rest., Econolodge, auto parts, tires, truck repair

126mm   Ottawa River

125   OH 309, OH 117, Lima, **E**...BP/diesel/mart/wash, Speedway/mart, Arby's, Big Boy, Bob Evans, Burger King, Capt D's, Cracker Barrel, Hunan Garden Chinese, McDonald's, Olive Garden, Pizza Hut, Ponderosa, Rally's, Red Lobster, Ryan's, Subway, Taco Bell, TCBY, Wendy's, Hampton Inn, Holiday Inn, Motel 6, Jo-Ann Fabrics, K-Mart/Little Caesar's/drugs, PartsAmerica, Pharm Drug, Ray's MktPlace, Sam's Club, Wal-Mart, bank, carwash, cinemas, **W**...HOSPITAL, Citgo, Lima Truck Plaza/diesel, Shell, Shoney's, Crossroads Motel, Economy Inn, Knight's Court, JiffyLube, SavALot Foods

124   4ᵗʰ St, **E**...hwy patrol

122   OH 65, Lima, **E**...Speedway/diesel/mart, **W**...Shell, Volvo

120   Breese Rd, Ft Shawnee, **W**...Citgo/diesel/mart, auto repair, truckwash

118   to Cridersville, **W**...Fuelmart/diesel, Speedway/diesel/mart, Subway, U-Haul, antiques

114mm   **rest area both lanes, full(handicapped)facilities, phones, picnic tables, litter barrels, vending, petwalk**

113   OH 67, Wapakeneta, to Uniopolis, no facilities

111   Bellefontaine St, Wahpakeneta, **E**...L&G Truckstop/diesel/mart, Day's Inn, Dodge City Rest., TwinLakes KOA, **W**...BP/mart/wash, Citgo, Shell/mart/wash, Arby's, Burger King, Capt D's, Chalet Rest., Dairy Queen, KFC, Lucky Steer Rest., McDonald's, Pizza Hut, Ponderosa, Taco Bell, Waffle House, Wendy's, $Inn, Holiday Inn, Super 8, Advance Auto Parts, Big Bear Foods, Family$, Pennzoil, Radio Shack, bank, carwash, st patrol

110   US 33, to St Marys, Bellefontaine, **E**...hwy patrol

104   OH 219, Botkins, **W**...Sunoco, Budget Host/Western Ohio Inn/rest.

102   OH 274, to Jackson Ctr, New Breman, no facilities

99   OH 119, Anna, to Minster, **E**...Citgo/diesel/mart/grill, 99 Truckstop/diesel/rest., SavATon/gas, **W**...5Star Truckstop/diesel, Sunoco/mart, Subway, Wendy's, truckwash

94   rd 25A, Sidney, no facilities

93   OH 29, Sidney, to St Marys, no facilities

N

S

Findley

Lima

Ohio Interstate 75

Sidney

92	OH 47, Sidney, to Versailles, **E**...HOSPITAL, CHIROPRACTOR, Shell/mart/wash/etd, Speedway/mart, Arby's, Subway, Taco Bell, Wendy's, NAPA, Odd Lots, Pharm Drugs, **W**...Amoco/diesel, BP/mart, Sunoco, Bob Evans, Burger King, Hardee's, Highmarks Café, KFC, McDonald's, Perkins, Pizza Hut, Ponderosa, Rally's, Super Subbies, Waffle House, Comfort Inn, Day's Inn, Econolodge, Holiday Inn/rest., Aldi Foods, Buick/Pontiac/GMC, Campbell's Meats/deli, Chevrolet, CVS Drug, Dodge/Plymouth, FashionBug, Ford/Lincoln/Mercury, Kroger/deli, Radio Shack, Sidney AutoLube, SuperWash, Wal-Mart, bank, cleaners
90	Fair Rd, to Sidney, **E**...Sunoco/diesel, Fairington Rest.
88mm	Great Miami River
83	rd 25A, Piqua, **W**...Knight's Inn, Red Carpet Inn, Chevrolet/Plymouth, mufflers, to Piqua Hist Area
82	US 36, Piqua, to Urbana, **E**...MEDICAL CARE, BP/mart, Arby's, KFC, LJ Silver, Ponderosa, Taco Bell, Waffle House, Wendy's, Big Lots, Goodyear, Pennzoil, Sears/auto, mall, **W**...Speedway/mart, Bob Evans, Cracker Barrel, McDonald's, Red Lobster, Terry's Cafeteria, Comfort Inn, Howard Johnson, Ramada Ltd, Aldi Foods, JC Penney, cinema
81mm	**rest area both lanes, full(handicapped)facilities, phones, picnic tables, litter barrels, vending**
78	rd 25A, **E**...HOSPITAL

Troy

74	OH 41, Troy, to Covington, **E**...HOSPITAL, CHIROPRACTOR, BP/diesel/mart, Bagel Boomers Deli, El Sombrero Mexican, Little Caesar's, LJ Silver, McDonald's, Perkins, Pizza Hut, Sally's Café, Subway, Day's Inn, Goodyear/auto, Pharm Drugs, Radio Shack, SuperPetz, bank, cleaners, to Hobart Arena, **W**...Meijer/diesel/24hr, Shell/mart/wash/etd, Speedway/diesel/mart, Applebee's, Big Boy, Bob Evans, Burger King, Chinese/Japanese Rest., Dairy Queen, Fazoli's, Friendly's, Golden Corral, Highmarks Café, KFC, Steak'n Shake, Fairfield Inn, Hampton Inn, Holiday Inn Express, Knight's Inn, Residence Inn, AutoZone, County Mkt Foods, FashionBug, 1-Hr Photo, CVS Drug, Sears Hardware, Staples, Wal-Mart, bank, cleaners
73	OH 55, Troy, to Ludlow Falls, **E**...HOSPITAL, Amoco/mart, BP/mart, Waffle House, Holiday Inn, Microtel, Motel 6, Quality Suites, Super 8, Country Folks Gen Store
69	rd 25A, **E**...DairyMart/gas, Marathon/mart, Sunoco, Buick/Pontiac/GMC
68	OH 571, Tipp City, to West Milton, **E**..BP/dieselmart, Shell/mart, Speedway, Burger King, Lee's Chicken, McDonald's, Ponderosa, Subway, Taco Bell, Honda, Goodyear, **W**...Citgo, SuperAmerica/Blimpie/mart, Arby's, Big Boy, Tipp' O the Town Rest., Wendy's, Heritage Motel, Holiday Inn Express, Sears Parts/service
64	Northwoods Blvd, no facilities
63	US 40, Vandalia, to Donnelsville, **E**...SuperAmerica/diesel/mart, Big Boy, Chinese Rest., Crossroads Motel, Kroger, auto repair, **W**...BP/diesel/mart, Shell/etd, Speedway/mart, Arby's, Cassano's Pizza/subs, KFC, McDonald's, Pizza Hut, Taco Bell, Wendy's, Cross Country Inn, airport, drugs
61b a	I-70, E to Columbus, W to Indianapolis, to Dayton Int Airport
60	Little York Rd, Stop 8 Rd, **E**...Cooker Rest., Damon's, Olive Garden, Red Lobster, Ryan's, Subway, Residence Inn, Rodeway Inn, Kinko's/24hr, NTB, SpeedyPrint, Volvo/BMW/VW, hwy patrol, **W**...Sunoco, Arby's, Bennigan's, Bob Evans, Chinese Rest., Cracker Barrel, Don Pablo's Mexican, Joe's Crabshack, LoneStar Steaks, Max&Erma's, O'Charley's, Outback Steaks, Perkins, Skyline Chili, Wendy's, Comfort Inn, Day's Inn, Fairfield Inn, Knight's Inn, Motel 6, Ramada Inn, Red Roof Inn, Sam's Club, auto/mufflers/shocks
58	Needmore Rd, to Dayton, **E**...BP/diesel/mart/atm, Shell/McDonald's/wash, Big Boy, Hardee's, Quality Suites, Goodyear/auto, Meineke Muffler, to AF Museum, hwy patrol, **W**...Speedway/Subway/diesel/mart, SuperAmerica/Taco Bell/mart, Friendly's, Waffle House
57b a	Neva Rd, Wagner Ford Rd, to Dayton, **E**...Marathon, Sunoco/mart, Holiday Inn, **W**...Denny's, Best Western, Econolodge, Economy Inn, Holiday Motel
56	Stanley Ave, Dayton, **E**...Amoco, **W**...BP, McDonald's, Taco Bell, Plaza Motel
55b a	Keowee St, Dayton, **E**...Sunoco, Arrow Batteries, **W**...BP/mart, Plaza Motel
54c	OH 4 N, Webster St, to Springfield, no facilities
54mm	Great Miami River

Dayton

54b	OH 48, Main St, Dayton, **E**...Chevrolet/Cadillac, **W**...HOSPITAL, BP/mart, auto parts
a	Grand Ave(from sb), Dayton, downtown
53b	OH 49, 1st St, Salem Ave, Dayton, downtown, **E**...Day's Inn, Radisson, Country Inn
a	OH 49, 3rd St, downtown
52b a	US 35, E to Dayton, W to Eaton
51	Edwin C Moses Blvd, Nicholas Rd, **E**...Marriott, to U of Dayton, **W**...BP/diesel/mart, Shell, McDonald's, Perkins, Econolodge, SunWatch Indian Village
50b a	OH 741, Kettering St, Dryden Rd, **W**...Sunoco, Holiday Inn/rest., Super 8, U-Haul
47	Central Ave, W Carrollton, Morraine, **E**...Citgo/diesel/mart, **W**...McDonald's
44	OH 725, to Centerville, Miamisburg, **E**...HOSPITAL, Amoco, BP/Procare/mart, Shell/mart, Speedway/mart, Applebee's, Arby's, Big Boy, Bill Knapp's, Blimpie, Brubaegger's Bagels, Burger King, Captain D's, Chi Chi's, Chinese Rest., Denny's, Dunkin Donuts, Friendly's, Hardee's, KFC, La Pinata Mexican, McDonald's, O'Charley's, Olive Garden, Pizza Hut, Ponderosa, Rally's, Red Lobster, Taco Bell, TGIFriday, Wendy's, Courtyard, Doubletree

Ohio Interstate 75

Guest Suites, Hampton Inn, Holiday Inn, Motel 6, Quality Inn, Residence Inn, Barnes&Noble, Burlington Coats, CompUSA, Cub Foods, Best Buy, Dick's Sports, Firestone/auto, Goodyear/auto, Honda/Nissan/Mazda, JC Penney, Jo-Ann Fabrics, Kia, K-Mart/drugs, Midas Muffler, Office Depot, OfficeMax, Olds, Pier 1, Rex, Sears/JiffyLube/auto, SuperPetz, TireAmerica, TJ Maxx, Wal-Mart, bank, mall, **W...**HOSPITAL, BP/mart, Marathon, Shell/autocare, Bob Evans, China Hut, Perkins, Day's Inn, Knight's Inn, Red Roof Inn, Signature Inn, Aamco Transmissions, Main Auto Parts, McQuik Lube, Precision Tune, bank

43   I-675 N, to Columbus

38   OH 73, Springboro, Franklin, **E...**BP, Sunoco/diesel, Arby's, McDonald's, Perkins, Wendy's, Holiday Inn Express, Ramada Ltd, **W...**Amoco/mart, Speedway, Big Boy, KFC, Louise's Place Rest., Taco Bell, Econolodge, Knight's Inn

36   OH 123, Franklin, to Lebanon, **E...**Citgo/diesel/24hr, Speedway/ Hardee's/diesel/mart, McDonald's, Waffle House, Royal Inn, Super 8, **W...**BP/mart/ wash

32   OH 122, Middletown, **E...**CHIROPRACTOR, BP/mart, Duke gas/mart, McDonald's, Waffle House, Best Western, Comfort Inn, Ramada Inn, Super 8, Jeep/Eagle, Ford/Mercury/Lincoln, **W...**HOSPITAL, Amoco, Citgo/ diesel, Meijer/diesel/24hr, Speedway/mart, Applebee's, Big Boy, Bill Knapp's, Bob Evans, Cracker Barrel, Hardee's, KFC, LoneStar Steaks, McDonald's, Old Country Buffet, Olive Garden, Ponderosa, Schlotsky's, Shell's Rest., Shoney's, Sonic, Steak'n Shake, Wendy's, Garden Inn Suites, Fairfield Inn, Holiday Inn Express, Buick/GMC/Olds, Kohl's, Kroger/deli/24hr, Lowe's Bldg Supply, Pearle Vision, Sears/auto, Target, Tire Discounters, Wal-Mart, Goodyear/auto, bank, cinema, cleaners, mall

29   OH 63, Monroe, to Hamilton, **E...**Chevron/diesel/mart, Shell/diesel, Stony Ridge Inn/truck plaza/diesel/rest., Burger King/playplace, GoldStar Chili, KFC, Waffle House, Wendy's/Tim Horton's, Day's Inn, Trader's World Mkt, **W...**BP/diesel/mart, Sunoco/diesel/mart, SuperAmerica/Subway/mart, GoldStar Chili, McDonald's, Perkins, Sara Jane's Rest., Econolodge, Hampton Inn, bank, carwash, flea mkt

27.5mm   **rest area both lanes, full(handicapped)facilities, info, phone, picnic tables, litter barrels, vending, petwalk**

23   new exit

22   Tylersville Rd, to Mason, Hamilton, **E...**Amoco, BP/mart/atm, Sunoco/mart, Arby's, Bob Evans, Boston Mkt, Burger King, KFC, LJ Silver, McDonald's, Perkins, Pizza Hut, Waffle House, Wendy's, Rodeway Inn, Big Lots, Goodyear, JiffyLube, Radio Shack, **W...**Shell/mart/wash, Speedway/diesel/mart/24hr, Meijer/diesel/24hr, Brubaegger's Bagels, O'Charley's, Steak'n Shake, CarX Muffler, Complete PetMart, OilXpress, Sears Hardware, Tire Discounters, Wal-Mart/auto, bank

21   Cin-Day Rd, **E...**Big Boy, Holiday Inn Express, **W...**BP, Shell, Sunoco/diesel/mart, Skyline Chili, Econolodge, Knight's Inn

18   Union Centre Blvd, no facilities

16   I-275 to I-71, to I-74

15   Sharon Rd, to Sharonville, Glendale, **E...**BP/mart, Chevron, Marathon/diesel/mart, Shell, Sunoco/mart, BBQ, Big Boy, Bob Evans, Pizza Hut, Shoney's, Skyline Chili, Waffle House, Fairfield Inn, Hampton Inn, Holiday Inn/ rest., Motel 6, Ramada Inn, Red Roof Inn, Woodfield Suites, auto parts, UPS, **W...**Bill Knapp's, Captain D's, Friendly's, McDonald's, Shogun Japanese, Subway, Taco Bell, Wendy's, Comfort Inn, Econolodge, Extended Stay America, Howard Johnson, Marriott, Radisson, Red Roof Inn, Residence Inn, Signature Inn, Super 8, Nissan, Top Value Mufflers

14   OH 126, to Woodlawn, Evendale, **E...**GE Plant, **W...**Quality Inn, bank

13   Shepherd Lane, to Lincoln Heights, **E...**GE Plant, **W...**Taco Bell, Wendy's

12   Wyoming Ave, Cooper Ave, to Lockland, no facilities

10a   OH 126, Ronald Reagan Hwy, **E...**Chevrolet, Ramada Inn

  b   Galbraith Rd, Arlington Heights

9   OH 4, OH 561, Paddock Rd, Seymour Ave, to Cincinnati Gardens

8   Towne St, Elmwood Pl(from nb), **E...**United Mail, cinema

7   OH 562, to I-71, Norwood, Cincinnati Gardens, no facilities

6   Mitchell Ave, St Bernard, **E...**Marathon, Speedway/mart, Wendy's, to Cincinnati Zoo, to Xavier U, **W...**BP/diesel/ mart, McDonald's, Ford, Honda, Kroger, tires

4   I-74 W, US 52, US 27 N, to Indianapolis, no facilities

N

S

Middletown

Cincinnati

Ohio Interstate 75

3	to US 27 S, US 127 S, Hopple St, U of Cincinnati, **E**...Marathon, White Castle, access to Day's Inn, **W**...HOSPITAL
2b	Harrison Ave, industrial district, **E**...foodmart, Ford Trucks, **W**...BP, McDonald's
a	Western Ave, Liberty St(from sb)
1h	Ezzard Charles Dr, **W**...Holiday Inn
g	US 50W, Freeman Ave, **E**...Regal Hotel, bank, **W**...Site gas/mart, Big Boy, Wendy's, Holiday Inn, Ford
f	7th St(from sb), downtown, **W**...Holiday Inn
e	5th St(from sb), downtown, **E**...Clarion, bank
1a	I-71 N, to Cincinnati, downtown, to stadium
0mm	Ohio River, Ohio/Kentucky state line

Ohio Interstate 76

Exit #	Services
	Ohio/Pennsylvania state line
	For exits 15 through 16A, see Ohio Turnpike.
60mm	I-76 eb joins Ohio TPK(toll)
57	OH 45, Bailey Rd, to Warren, **S**...access to gas, food, lodging
54	OH 534, Lake Milton, to Newton Falls, **N**...RV camping, **S**...BP/mart, Citgo/mart, to Berlin Lake, camping
52mm	Lake Milton
48	OH 225, to Alliance, **N**...to W Branch SP, camping, **S**...to Berlin Lake
45.5mm	**rest area both lanes, full(handicapped)facilities, phone, picnic tables, litter barrels, petwalk**
43	OH 14, to Alliance, Ravenna, **N**...truck repair, to W Branch SP, **S**...Citgo/diesel/mart, fireworks
38b a	OH 5, OH 44, to Ravenna, HOSPITAL, **N**...BP/mart, SuperAmerica, McDonald's, Wendy's, **S**...DM/Subway/mart, Edie's Rest., Giant Eagle Foods, RV camping
33	OH 43, to Hartville, to Kent, **N**...BP/diesel, Duke/mart, Bob Evans(2mi), Burger King(2mi), Country Manor Rest., Country Kitchen, Alden Inn, Day's Inn, Holiday Inn, Knight's Inn, Super 8, to Kent St U, **S**...DENTIST, Marathon/mart, Speedway/mart, McDonald's, Wendy's, Brimfield's Steaks
31	OH 18, Tallmadge, **S**...Brimfield's Steaks(1mi)
29	OH 532, Tallmadge, Mogadore, **N**...Marathon
27	OH 91, Canton Rd, Gilchrist Rd, **S**...Marathon/diesel, Shell, Hardee's, Wendy's, Best Western
26	OH 18, E Market St, Mogadore Rd, **N**...Sunoco/diesel, **S**...McDonald's
25b a	Martha Ave, General St, industrial
24	Arlington St, Kelly Ave, **N**...Clark Gas
23b	OH 8, Buchtell Ave, to Cuyahoga, to U of Akron
a	I-77 S, to Canton
22b	Wolf Ledges, Akron, downtown, **S**...BP/mart, McDonald's
a	Main St, Broadway, downtown
21c	OH 59 E, Dart Ave, **N**...HOSPITAL
b	Lakeshore St, Bowery St(from eb), no facilities
a	East Ave(from wb), no facilities
20	I-77 N(from nb), to Cleveland
19	Battles Ave, Kenmore Blvd, no facilities
18	I-277, US 224 E, to Canton, Barberton, no facilities
17b a	OH 619, Wooster Rd, State St, to Barberton, HOSPITAL
16	Barber Rd, **S**...Sunoco/diesel, Shamrock Motel, Chrysler/Jeep, Nissan, Olds
14	Cleve-Mass Rd, to Norton, **S**...BP/mart, Clark Gas, Marathon, Berlin's Motel(2mi)
13b a	OH 21, N to Cleveland, S to Massillon, **N**...Duke/mart
11	OH 261, Wadsworth, **N**...Speedway/mart, Marathon
9	OH 94, Wadsworth, to N Royalton, **N**...Arby's, Bob Evans, McDonald's, Pizza Hut, Ponderosa, Taco Bell, Wendy's, Ramada Ltd, Giant Eagle Foods, Goodyear/auto, K-Mart, **S**...BP/mart, Marathon, Shell/mart, Sunoco/mart, 76, Denny's, Friendly's, Rizzi's Ristorante, Knight's Inn, bank
7	OH 57, to Rittman, to Medina, **N**...HOSPITAL, Marathon/diesel/mart, airport
6mm	weigh sta both lanes
2	OH 3, Seville, to Medina, **N**...Citgo, Dairy Queen, Hardee's, Subway, Comfort Inn, Maple Lakes Camping(seasonal), **S**...76/diesel/mart, RV camping
1	I-76 E, to Akron, US 224 **W**...on US 224...TA/diesel/rest./24hr, Speedway/diesel/rest./24hr, Blimpie, Country Kitchen, McDonald's, Taco Bell, HoJo's, Super 8, Blue Beacon, Chippewa Valley Camping(1mi)
0mm	I-76 begins/ends on I-71, exit 209.

N
S

E

W

Akron

Ohio Interstate 77

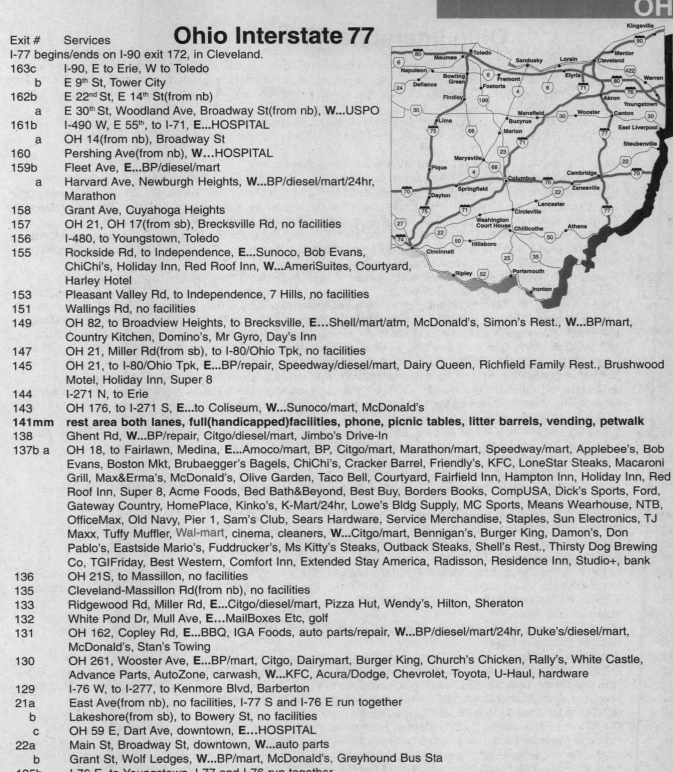

Exit #   Services

I-77 begins/ends on I-90 exit 172, in Cleveland.

163c I-90, E to Erie, W to Toledo

 b E 9th St, Tower City

162b E 22nd St, E 14th St(from nb)

 a E 30th St, Woodland Ave, Broadway St(from nb), **W...**USPO

161b I-490 W, E 55th, to I-71, **E...**HOSPITAL

 a OH 14(from nb), Broadway St

160 Pershing Ave(from nb), **W...**HOSPITAL

159b Fleet Ave, **E...**BP/diesel/mart

 a Harvard Ave, Newburgh Heights, **W...**BP/diesel/mart/24hr, Marathon

158 Grant Ave, Cuyahoga Heights

157 OH 21, OH 17(from sb), Brecksville Rd, no facilities

156 I-480, to Youngstown, Toledo

155 Rockside Rd, to Independence, **E...**Sunoco, Bob Evans, ChiChi's, Holiday Inn, Red Roof Inn, **W...**AmeriSuites, Courtyard, Harley Hotel

153 Pleasant Valley Rd, to Independence, 7 Hills, no facilities

151 Wallings Rd, no facilities

149 OH 82, to Broadview Heights, to Brecksville, **E...**Shell/mart/atm, McDonald's, Simon's Rest., **W...**BP/mart, Country Kitchen, Domino's, Mr Gyro, Day's Inn

147 OH 21, Miller Rd(from sb), to I-80/Ohio Tpk, no facilities

145 OH 21, to I-80/Ohio Tpk, **E...**BP/repair, Speedway/diesel/mart, Dairy Queen, Richfield Family Rest., Brushwood Motel, Holiday Inn, Super 8

144 I-271 N, to Erie

143 OH 176, to I-271 S, **E...**to Coliseum, **W...**Sunoco/mart, McDonald's

141mm rest area both lanes, full(handicapped)facilities, phone, picnic tables, litter barrels, vending, petwalk

138 Ghent Rd, **W...**BP/repair, Citgo/diesel/mart, Jimbo's Drive-In

137b a OH 18, to Fairlawn, Medina, **E...**Amoco/mart, BP, Citgo/mart, Marathon/mart, Speedway/mart, Applebee's, Bob Evans, Boston Mkt, Brubaegger's Bagels, ChiChi's, Cracker Barrel, Friendly's, KFC, LoneStar Steaks, Macaroni Grill, Max&Erma's, McDonald's, Olive Garden, Taco Bell, Courtyard, Fairfield Inn, Hampton Inn, Holiday Inn, Red Roof Inn, Super 8, Acme Foods, Bed Bath&Beyond, Best Buy, Borders Books, CompUSA, Dick's Sports, Ford, Gateway Country, HomePlace, Kinko's, K-Mart/24hr, Lowe's Bldg Supply, MC Sports, Means Wearhouse, NTB, OfficeMax, Old Navy, Pier 1, Sam's Club, Sears Hardware, Service Merchandise, Staples, Sun Electronics, TJ Maxx, Tuffy Muffler, Wal-mart, cinema, cleaners, **W...**Citgo/mart, Bennigan's, Burger King, Damon's, Don Pablo's, Eastside Mario's, Fuddrucker's, Ms Kitty's Steaks, Outback Steaks, Shell's Rest., Thirsty Dog Brewing Co, TGIFriday, Best Western, Comfort Inn, Extended Stay America, Radisson, Residence Inn, Studio+, bank

136 OH 21S, to Massillon, no facilities

135 Cleveland-Massillon Rd(from nb), no facilities

133 Ridgewood Rd, Miller Rd, **E...**Citgo/diesel/mart, Pizza Hut, Wendy's, Hilton, Sheraton

132 White Pond Dr, Mull Ave, **E...**MailBoxes Etc, golf

131 OH 162, Copley Rd, **E...**BBQ, IGA Foods, auto parts/repair, **W...**BP/diesel/mart/24hr, Duke's/diesel/mart, McDonald's, Stan's Towing

130 OH 261, Wooster Ave, **E...**BP/mart, Citgo, Dairymart, Burger King, Church's Chicken, Rally's, White Castle, Advance Parts, AutoZone, carwash, **W...**KFC, Acura/Dodge, Chevrolet, Toyota, U-Haul, hardware

129 I-76 W, to I-277, to Kenmore Blvd, Barberton

21a East Ave(from nb), no facilities, I-77 S and I-76 E run together

 b Lakeshore(from sb), to Bowery St, no facilities

 c OH 59 E, Dart Ave, downtown, **E...**HOSPITAL

22a Main St, Broadway St, downtown, **W...**auto parts

 b Grant St, Wolf Ledges, **W...**BP/mart, McDonald's, Greyhound Bus Sta

125b I-76 E, to Youngstown, I-77 and I-76 run together

 a OH 8 N, to Cuyahoga Falls, U of Akron

124b Lover's Lane, Cole Ave, no facilities

 a Archwood Ave, Firestone Blvd, no facilities

123b OH 764, Wilveth Rd, **E...**to airport

 a Waterloo Rd, **W...**BP, Acme Foods

122b a I-277, US 224 E, to Barberton, Mogadore

N

S

Ohio Interstate 77

120	Arlington Rd, to Green, **E...**CHIROPRACTOR, DENTIST, MEDICAL CARE, Pennzoil/gas, Speedway/mart/atm, Big Boy, Denny's, Friendly's, Kenny Rogers, Pizza Hut, Ryan's, White Castle, Comfort Inn, Red Roof Inn, Holiday Inn Express, JiffyLube, K-Mart/auto, Rite Aid, Wal-Mart/auto, **W...**BP/mart, Bob Evans, Burger King, McDonald's, Taco Bell, TGIFriday, Wendy's, Day's Inn, Econolodge, Ames, Buick/Pontiac/GMC, Chevrolet, Honda, Meineke Muffler, Sirpilla RV Ctr, bank, cinema
118	OH 241, to OH 619, Massillon, **E...**Speedway/diesel/mart, Belgrade Garden Dining, **W...**MEDICAL CARE, Citgo/ DM, Duke Gas/mart, Arby's, McDonald's, Meneche Bros Rest., Pancho's Mexican(1mi), Super 8, Honda Motrocycles
113	Akron-Canton Airport, **2 mi W...**Clay's RV, London Chocolate Factory
111	Canal Fulton, N Canton, **E...**TA/Marathon/diesel/rest./24hr, DairyMart/gas, Shell/mart/atm, Sunoco/diesel/mart, BrewHouse Grill, Burger King, Golden Corral, KFC, Midas Muffler, TrueValue, Winn's RV Ctr, **W...**BP/diesel/mart/ atm, ChuckeCheese, Cracker Barrel, Don Pablo, Einstein Bagels, Joe's Crabshack, Longhorn Steaks, McDonald's, Pizza Hut, Quizno's, Red Robin, Taco Bell, Wendy's, Best Western, Microtel, Motel 6, BabysRUs, Bed Bath&Beyond, BJ's Whse, Borders Books, DrugMart, Gander Mtn, Giant Eagle Foods, Goodyear/auto, Harley-Davidson, Home Depot, Lowe's Bldg Supply, Marshall's, MensWearhouse, OfficeMax, Old Navy, PetSupplies+, SteinMart, Sun Electronics, Wal-Mart/auto, cinema
109b a	Everhard Rd, Whipple Ave, **E...**Shell/mart/wash/24hr, Speedway/diesel/mart, Blimpie, Burger King, Denny's, Fazoli's, McDonald's, Taco Bell, Comfort Inn, Fairfield Inn, Hampton Inn, Residence Inn, Ames, $General, Ford, Saturn, cinema, **W...**Marathon/mart, Speedway/mart, Applebee's, Arby's, Bob Evans, Boston Mkt, Brubaegger's Bagels, Chang's Chinese, ChiChi's, Country Kitchen, Damon's, Donato's Pizza, Dragon Buffet, Eastside Mario's, Eat'n Park, Friendly's, Fuddrucker's, Great Wall Chinese, HomeTown Buffet, Italian Oven, KFC, LoneStar Steaks, Manchu Café, Olive Garden, Papa Bear's Italian, Perkins, Pizza Hut, Ponderosa, Red Lobster, TGIFriday, Thirsty Dog Grill, Wendy's, Best Suites, Day's Inn, Holiday Inn, Red Roof Inn, Sheraton, Super 8, B Dalton's, Best Buy, Burlington Coats, Circuit City, Dick's Sports, Dillard's, FashionBug, Firestone/auto, Goodyear/auto, HomePlace, Jo-Ann Crafts, Kaufmann's, KidsRUs, Kinko's, Kohl's, London Fog, Marc's Foods, Mens Wearhouse, Michael's, NTB, Pearle Vision, PetsMart, Pier 1, Radio Shack, Sears/auto, Service Merchandise, Staples, Target, TJ Maxx, Valvoline, WorldMkt, bank, cinema, mall
107b a	US 62, OH 687, Fulton Rd, to Alliance, **E...**All Tune&Lube, City Park, **1/2mi E on Cleveland Ave...**Burger King, KFC, Taco Bell, Midas/Top Value Muffler, tires, **W...**VETERINARIAN, Pro Football Hall of Fame
106	13th St NW, **E...**HOSPITAL
105b	OH 172, Tuscarawas St, downtown
a	6th St SW(no EZ return from sb), **E...**Sunoco, auto repair, **W...**HOSPITAL, VETERINARIAN, Shell, McDonald's, Subway, carwash
104	US 30, US 62, to E Liverpool, Massillon, no facilities
103	OH 800 S, **E...**BP/mart, Shell, Speedway, Arby's, Subway, Taco Bell, **W...**Pennzoil/diesel/24hr, Firestone/auto
101	OH 627, to Faircrest St, **E...**Citgo/diesel, Gulliver's 77 Plaza/diesel/rest.
99	Fohl Rd, to Navarre, **W...**Shell/Subway/mart, KOA(4mi)
93	OH 212, Bolivar, to Zoar, **E...**VETERINARIAN, DM/gas, Shell/mart/24hr, Der Dutchman Rest., McDonald's, Pizza Hut, Sisters Inn Rest., Wendy's, Wilkshire Family Rest., Zoar Tavern(3mi), Sleep Inn, IGA/drugs, NAPA, cleaners, to Lake Atwood Region, **W...**MEDICAL CARE, BP/diesel, Citgo/Dairy Queen, antiques
92mm	weigh sta both lanes
87	US 250, to Strasburg, **E...**Arby's(2mi), **W...**Citgo/mart, Grinders, Hardee's/24hr, Lugnut Café, McDonald's, Ramada Ltd, Twins Motel
85mm	**rest areas both lanes, full(handicapped)facilities, info, phone, picnic tables, litter barrels, petwalk**
83	OH 39, OH 211, Dover, to Sugarcreek, **E...**HOSPITAL, BP/diesel/mart, Shell/Blimpie/mart/24hr, Speedway, Big Boy, Bob Evans, Denny's, KFC, McDonald's, Shoney's, Wendy's, Knight's Inn, Ford/Chrysler/Plymouth/Dodge/ Jeep, Lincoln/Mercury/Nissan, RV American, **W...**Marathon/DQ/mart, Comfort Inn
81	US 250, to Uhrichsville, **1 mi E...**BP, Lee's Chicken, mall
	OH 39, New Philadelphia, **E...**BP/autocare, Shell/mart/24hr, Speedway, Big Boy, Burger King, Denny's, Don Pancho's Mexican, Hong Kong Buffet, LJ Silver, McDonald's, Pizza Hut, Spaghetti Shop, Taco Bell, Texas Road-house, Water's Edge Rest., Hampton Inn, Holiday Inn, Day's Inn, Motel 6, Schoenbrunn Inn, Super 8, Travelodge, Advance Parts, Aldi Foods, $General, Farm&Fleet, K-Mart/Little Caesar's/auto/24hr, Odd Lots, Wal-Mart SuperCtr/24hr, antiques, **W...**Eagle/diesel/rest., Marathon/repair, Ford/Peterbilt
73	OH 751, to rd 21, Stone Creek, **W...**Marathon/mart
65	US 36, Port Washington, Newcomerstown, **W...**BP, Duke/diesel/rest./atm, McDonald's, Super 8
64mm	Tuscarawas River
54	OH 541, rd 831, Kimbolton, to Plainfield, **E...**RV camp, **W...**Shell/mart, Rocky's Rest.
47	US 22, to Cadiz, Cambridge, **E...**to Salt Fork SP(6mi), lodging, RV camping, **W...**HOSPITAL, BP/mart/repair, to Glass Museum, info

Canton

N

S

Ohio Interstate 77

46b a   US 40, Cambridge, to Old Washington, **E**...Stoney's RV(2mi), **W**...VETERINARIAN, BP/mart/atm/24hr, Exxon/Wendy's/ diesel/mart, FuelMart/diesel, Speedway/mart/24hr, Burger King, Hardee's, Hunan Chinese, Lee's Rest., LJ Silver, McDonald's, Frisbee Motel(2mi), Long's Motel, Ames, Riesbeck's Food/drug/deli, Save-A-Lot Food, auto parts/repair, bank, cleaners/laundry

44b a   I-70, E to Wheeling, W to Columbus

41   OH 209, OH 821, Byesville, **W**...BP/diesel/mart, McDonald's, IGA Foods, Byesville Drugs, Family$, bank

39mm   **rest area nb, full(handicapped)facilities, phone, picnic tables, litter barrels, petwalk**

37   OH 313, Buffalo, **E**...Duke/diesel/mart, StarFire/mart, Buffalo Grill, Subway, truck repair, to Senecaville Lake

36.5mm   **rest area sb, full(handicapped)facilities, phone, picnic tables, litter barrels, petwalk**

28   OH 821, Belle Valley, **E**...Sunoco/diesel/mart, Marianne's Rest., USPO, to Wolf Run SP, RV camping

25   OH 78, Caldwell, **E**...BP/mart, Certified Gas(1mi), Sunoco/diesel, Dairy Queen, Lori's Rest., McDonald's, Best Western, Chevrolet/Pontiac/Olds/Buick, **W**...Pilot/Arby's/diesel/mart/24hr

16   OH 821, Macksburg, **E**...MiniMart Rest., antiques

6   OH 821, to Devola, **E**...Exxon/mart, **W**...HOSPITAL, BP/diesel/mart/LP, RV camping

3mm   **rest area nb, full(handicapped)facilities, info, phone, vending, picnic tables, litter barrels, petwalk**

1   OH 7, to OH 26, Marietta, **E**...CHIROPRACTOR, Ashland Gas, GoMart/diesel/atm/24hr, Rich Gas, Dairy Queen, Damon's, Ryan's, Comfort Inn, Econolodge, Holiday Inn, Aldi Foods, Chevrolet/Olds/Cadillac, Chrysler/Plymouth, Dodge, Ford, IGA Foods, Harley-Davidson, Pontiac/Buick, River City Tire/repair, Toyota, Wal-Mart/auto, carwash, **W**...MEDICAL CARE, EYECARE, BP/diesel/mart, Duke/diesel/mart, Exxon/repair, Marathon/diesel/mart, SuperAmerica/mart/atm, Applebee's, Big Boy, Bob Evans, Bonanza, Burger King, Capt D's, Chicago Pizza, LJ Silver, McDonald's, Napoli's Pizza, Papa John's, Pizza Hut, Rax, Shoney's, Subway, Taco Bell, Wendy's, Knight's Inn, Super 8, Ames, AutoZone, Big Lots, CVS Drug, Food4Less, Jo-Ann Crafts, K-Mart, Kroger/deli/24hr, Medicine Shoppe, Midas Muffler, MotoCare Lube, True Value, bank, cinema, museum, st patrol

0mm   Ohio River, Ohio/West Virginia state line

Ohio Interstate 80

Exit #   Services

237.5mm   Ohio/Pennsylvania state line

237mm   **Welcome Ctr wb, full(handicapped)facilities, info, phone, picnic tables, litter barrels, vending, petwalk**

234b a   US 62, OH 7, Hubbard, to Sharon, PA, **N**...Shell/diesel/mart/rest./motel/atm/24hr, Arby's, McDonald's, Blue Beacon, RV camping(2mi), tire repair, **S**...Marathon, motel

232mm   weigh sta wb

229   OH 193, Belmont Ave, to Youngstown, **N**...DENTIST, MEDICAL CARE, Pennzoil/gas, Speedway/mart, Handel's Ice Cream, Day's Inn, Holiday Inn, Knight's Inn, Ramada Inn, Super 8, TallyHo-tel, **S**...CHIROPRACTOR, Amoco/mart/24hr, Citgo, Rich Gas, Speedway/mart, Antone's Italian, Arby's, Armando's Rest., Bob Evans, Burger King, Cancun Mexican, Denny's, Golden Hunan, KFC, LJ Silver, McDonald's, Papa John's, Perkins/24hr, Pizza Hut, Subway, Taco Bell, Treacher's Fish'n Chips, Comfort Inn, Econolodge, Advance Parts, Big Lots, Firestone/auto, Giant Eagle Foods, Goodyear/auto, Hill's, Monro Muffler/brake, Phar-Mor Drug, bank, cleaners

228   OH 11, to Warren, Ashtabula, no facilities

227   US 422, Girard, Youngstown, **N**...Amoco/mart, Clark, Shell/diesel/mart/wash/24hr, Big Boy, Burger King, Subway

226   Salt Springs Rd, to I-680, **N**...BP/diesel/mart/atm, Sheetz/diesel/mart/24hr, McDonald's, **S**...Mr Fuel/diesel/ mart/24hr, Petro/Mobil/diesel/rest./24hr, Pilot/Arby's/diesel/mart/24hr, Blue Beacon, diesel repair

224b   I-680(from eb), to Youngstown

a   OH 11 S, to Canfield

223   OH 46, to Niles, **N**...Citgo/diesel/mart/atm, Bob Evans, Burger King, Country Kitchen, Economy Inn, **S**...BP/ diesel/mart, FuelMart/diesel, Speedway/diesel/mart, Sunoco/Subway/mart, TA/diesel/mart/rest./24hr, Arby's, Cracker Barrel, McDonald's, Perkins, Taco Bell, Wendy's, Best Western, Econolodge, Hampton Inn, Motel 6, Rebel Inn, Sleep Inn, Super 8, Blue Beacon, Sears Parts, Freightliner/24hr

221mm   Meander Reservoir

219mm   I-80 wb joins Ohio Tpk(toll)

For I-80 exits 1 through 14, see Ohio Turnpike.

Ohio Interstate 90

Exit #	Services
243mm	Ohio/Pennsylvania state line
242mm	**Welcome Ctr/weigh sta wb, full(handicapped)facilities, info, phone, picnic tables, litter barrels, petwalk**
241	OH 7, Conneaut, to Andover, **N**...HOSPITAL, Burger King, McDonald's(2mi), Day's Inn, Giant Eagle/deli/bakery, CVS Drug, K-Mart, RV camping, **S**...café
235	OH 84, OH 193, N Kingsville, to Youngstown, **N**...Citgo/diesel/mart, Dav-Ed Motel, **S**...Amoco, 76/diesel/rest./24hr, Shell/diesel/24hr, Speedway/Subway/diesel/mart, Jonathan's Family Dining, Still's Family Rest., Kingsville Motel
228	OH 11, to Ashtabula, Youngstown, **N**...HOSPITAL(4mi)
223	OH 45, to Ashtabula, **N**...Mr C's Family Rest., Holiday Inn, Travelodge, camping, **S**...BP/mart/rest., Speedway/Subway/diesel/mart, Austinburg Family Rest., Burger King, McDonald's, Hampton Inn
218	OH 534, Geneva, **N**...HOSPITAL, BP/mart, Sunoco, KFC(3mi), Granny Smith Rest., McDonald's, Wendy's, Geneva Inn/rest., Howard Johnson, Goodyear, truck service/diesel, KOA(8mi), Indian Chief Camping, to Geneva SP, **S**...KwikFill/diesel/rest./24hr, Shell/Applewood Rest.
212	OH 528, Madison, to Thompson, **N**...McDonald's, PotBellie's Rest., **S**...auto repair, radiator/towing, camping(4mi)
205	Vrooman Rd, **S**...BP/diesel/mart, Capp's Eatery(1mi)
200	OH 44, to Painesville, Chardon, **N**...HOSPITAL, **S**...BP/diesel/mart, McDonald's, Red Hawk Grill, Quail Hollow Resort, hwy patrol
198mm	**rest area both lanes, full(handicapped)facilities, phone, picnic tables, litter barrels, vending, petwalk**
193	OH 306, Kirtland, to Mentor, **N**...BP/mart, Shell, McDonald's, Knight's Inn, **S**...HOSPITAL, Marathon/mart, Speedway, Bob Evans, Brown Derby Rest., Burger King, Long John Silver, Ponderosa, Red Lobster, Roadhouse Steaks, TGIFriday, Arborgate Inn, Day's Inn/café, Red Roof Inn, Super 8, Travelodge
190.5mm	weigh sta eb
190	Express Lane to I-271
189	OH 91, to Willoughby, Willoughby Hills, **N**...HOSPITAL, CHIROPRACTOR, DENTIST, BP/diesel/mart/wash, Shell/mart/wash/24hr, Bob Evans, Brubaegger's Bagels, Café Europa, Chinese Rest., Fairfield Inn, Harley Hotel, Travelodge, CVS Drug, MailBoxes Etc, 1-Hr Photo, Walgreen, cleaners/laundry, **S**...Fazio's
188	I-271 S, to Akron
187	OH 84, Bishop Rd, to Wickliffe, to Willoughby, **S**...HOSPITAL, VETERINARIAN, CHIROPRACTOR, DENTIST, EYECARE, BP/diesel/mart/wash, Gulf, 76, Shell/mart/wash, Dairymart, Bakers Square, Burger King, Chinese Rest., Friendly's, McDonald's, Pizza, Sbarro's, Subway, TCBY, Quality Inn, CVS Drug, Chevrolet, Hill's, Marc's Foods, Mazda, Rainbow Muffler, Rini's Foods, Sam's Club, Shop'n Save, TireAmerica, VW, bank, cinema, laundry
186	US 20, Euclid Ave, **N**...DENTIST, Shell/wash, Speedway/mart, Sunoco, Dairymart, Denny's, Dock Seafood, Donuts, McDonald's, Papa Joe's Subs, Hampton Inn, Holiday Inn, Plaza Motel, Dodge/Saturn/Subaru, Ford, Olds, auto/radiator repair, carwash, **S**...Shell/mart/atm, Conv/mart, Arby's, BBQ, Jenny's Chicken, KFC, Long John Silver, Pizza Hut, SideWalk Café, Taco Bell, Wendy's, Innvoy Motel, CVS Drug, Firestone/auto, K-Mart, LubeStop, Midas, Sparks TuneUp, Speedee Lube, bank,
185	OH 2 E, to Painesville
184b	OH 175, E 260th St, **N**...Shell/autocare, **S**...Ryder Trucks, tires, transmissions
a	Babbitt Rd, same as 183
183	222nd St, **S**...Sunoco/mart/24hr, auto parts
182b a	E 200th St, E 185th St, **N**...BP/mart/24hr, Clark Gas, Honda, **S**...HOSPITAL, BP/mart/wash, 76/mart, Shell/mart/24hr, Speedway/mart, auto repair
181b a	E 156th St, **N**...Clark Gas, **S**...BP/mart/24hr
180b a	E 140th St, E 152nd St, no facilities
179	OH 283 E, to Lake Shore Blvd, no facilities
178	Eddy Rd, to Bratenahl, no facilities
177	University Circle, MLK Dr, **S**...HOSPITAL
176	E 72nd St, no facilities
175	E 55th St, Marginal Rds, **S**...AAA
174b	OH 2 W, downtown Cleveland, to Lakewood
a	Lakeside Ave
173c	Superior Ave, St Clair Ave, downtown, **N**...BP/mart/wash
b	Chester Ave, **S**...BP
a	Prospect Ave(from wb), downtown
172d	Carnegie Ave, downtown, **S**...Burger King, Cadillac
c b	E 9th St, **S**...HOSPITAL, to Cleveland St U
a	I-77 S, to Akron
171b a	US 422, OH 14, Broadway St, Ontario St, **N**...downtown, to Stadium
170c b	I-71 S, to I-490
a	US 42, W 25th St, **N**...Sunoco
169	W 44th St, W 41st St, **N**...HOSPITAL
167b a	OH 10, West Blvd, 98th St, to Lorain Ave, **N**...HOSPITAL, **S**...BP/mart

Ohio Interstate 90

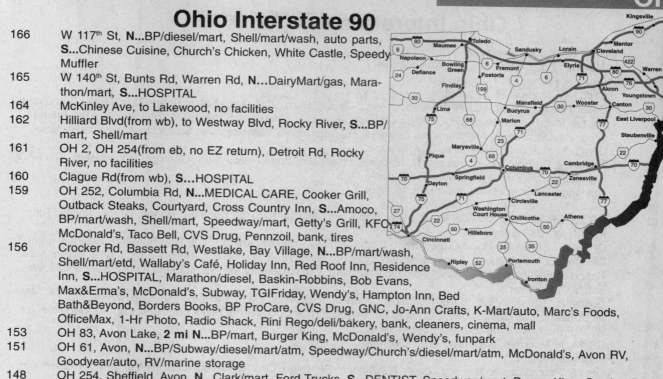

166   W 117th St, **N**...BP/diesel/mart, Shell/mart/wash, auto parts, **S**...Chinese Cuisine, Church's Chicken, White Castle, Speedy Muffler

165   W 140th St, Bunts Rd, Warren Rd, **N**...DairyMart/gas, Marathon/mart, **S**...HOSPITAL

164   McKinley Ave, to Lakewood, no facilities

162   Hilliard Blvd(from wb), to Westway Blvd, Rocky River, **S**...BP/mart, Shell/mart

161   OH 2, OH 254(from eb, no EZ return), Detroit Rd, Rocky River, no facilities

E

160   Clague Rd(from wb), **S**...HOSPITAL

159   OH 252, Columbia Rd, **N**...MEDICAL CARE, Cooker Grill, Outback Steaks, Courtyard, Cross Country Inn, **S**...Amoco, BP/mart/wash, Shell/mart, Speedway/mart, Getty's Grill, KFC, McDonald's, Taco Bell, CVS Drug, Pennzoil, bank, tires

156   Crocker Rd, Bassett Rd, Westlake, Bay Village, **N**...BP/mart/wash, Shell/mart/etd, Wallaby's Café, Holiday Inn, Red Roof Inn, Residence Inn, **S**...HOSPITAL, Marathon/diesel, Baskin-Robbins, Bob Evans, Max&Erma's, McDonald's, Subway, TGIFriday, Wendy's, Hampton Inn, Bed Bath&Beyond, Borders Books, BP ProCare, CVS Drug, GNC, Jo-Ann Crafts, K-Mart/auto, Marc's Foods, OfficeMax, 1-Hr Photo, Radio Shack, Rini Rego/deli/bakery, bank, cleaners, cinema, mall

153   OH 83, Avon Lake, **2 mi N**...BP/mart, Burger King, McDonald's, Wendy's, funpark

151   OH 61, Avon, **N**...BP/Subway/diesel/mart/atm, Speedway/Church's/diesel/mart/atm, McDonald's, Avon RV, Goodyear/auto, RV/marine storage

W

148   OH 254, Sheffield, Avon, **N**...Clark/mart, Ford Trucks, **S**...DENTIST, Speedway/mart, Burger King, Cracker Barrel, KFC, Marco's Pizza, McDonald's, Pizza Hut, Subway, Sugarcreek Rest., Taco Bell, Aldi Foods, CarCare USA, CVS Drug, Giant/Eagle/Rini Rego/deli/bakery, Mueller Tire/brake, Sears Hardware, auto parts, bank

147mm   Black River

145   OH 57, to Lorain, Elyria, I-80/90 E, Ohio Tpk E, **N**...HOSPITAL, Burger King, Cracker Barrel(2mi), Goodyear, U-Haul, **S**...BP/diesel, Shell, Speedway/diesel, Arby's, McDonald's, Pizza Hut, Wendy's, Best Western, Comfort Inn, Day's Inn, Econolodge, Holiday Inn, Howard Johnson, Ramada Inn, **2 mi S**...mall, same as Ohio Tpk, exit 8.

144   OH 2 W(from wb, no return), to Sandusky, **1 mi N**...Broadway Ave, HOSPITAL, Amoco/mart/24hr, Marathon, McDonald's

143   I-90 wb joins Ohio Tpk

For wb exits to Ohio/Indiana state line, see Ohio Turnpike.

Ohio Interstate 271(Cleveland)

Exit #	Services
38mm	I-271 begins/ends on I-90, exit 188.
36	Wilson Mills Rd, Mayfield, Highland Hts, **E**...Holiday Inn, **W**...BP/mart, Sunoco/diesel/mart, Home Depot
34	US 322, Mayfield Rd, **E**...BP, Giant Eagle Food, Mueller Tire/brake, Walgreen, **W**...HOSPITAL, Bob Evans, Longhorn Steaks, McDonald's, Ponderosa, Baymont Inn, Best Buy, HomePlace
32	Brainerd Rd, Cedar Rd, **E**...Alexander's Rest., Champ's Grill

N

29   US 422 W, OH 87, Chagrin Blvd, Harvard Rd, **E**...Gulf, Uno Café, Holiday Inn, Homestead Suites, Travelodge, **W**...Shell, Embassy Suites, Marriott, Radisson

28   OH 175, Richmond Rd, Emery Rd, **E**...HOSPITAL, BP, Country Kitchen, Lowe's Bldg Supply, **W**...BJ's Whse

27b   I-480 W, no facilities

26   Rockside Rd, **E**...Sunoco, Perkins, Red Roof Inn, Lowe's Bldg Supply

23   OH 14 W, Forbes Rd, Broadway Ave, **E**...Sunoco/mart, McDonald's, Pizza Hut, Subway, Taco Bell, Wendy's, Holiday Inn Express, OfficeMax, PetsMart, Sam's Club, **W**...HOSPITAL, BP, LubeStop

21   I-480 E, OH 14 E(from sb), to Youngstown, no facilities

19   OH 82, Macedonia, **E**...Speedway, **W**...DENTIST, Shell, Sunoco, Arby's, Burger King, Dairy Queen, Little Caesar's, Outback Steaks, Pizza Hut, Taco Bell, Wendy's, Cord Camera, CVS Drug, Giant Eagle Food, Home Depot, K-Mart/auto, Top's Foods, Valvoline, bank, cinema, cleaners

18   OH 8, Boston Hts, to Akron, **E**...BP/Subway/mart, Shell/diesel/mart, Speedway/mart, Bob Evans, Denny's, Baymont Inn, Country Inn Suites, Knight's Inn, Motel 6, Travelodge, **W**...Applebee's, Ground Round, KFC, McDonald's, Kohl's, Wal-Mart/auto

12   OH 303, Richfield, Peninsula, no facilities

10   I-77, to I-80, OH Tpk(from nb), to Akron, Cleveland

S   9   I-77 S, OH 176(from nb), to Richfield, **E**...Sunoco

8mm   **rest area both lanes, full(handicapped)facilities, phone, picnic tables, litter barrels, petwalk**

3   OH 94, to I-71 N, Wadsworth, N Royalton, **W**...Marathon/mart

0mm   I-271 begins/ends on I-71, exit 220.

Ohio Interstate 475(Toledo)

Exit #	Services
20	I-75. I-475 begins/ends on I-75, exit 204.
19	Jackman Rd, Central Ave, **S**...Amoco/mart
18b	Douglas Rd(from wb), no facilities
a	OH 51 W, Monroe St, no facilities
17	Secor Rd, **N**...MEDICAL CARE, Amoco/mart/24hr, BP/mart, Shell, Applebee's, Betsy Ross' Rest., Bob Evans, Boston Mkt, Burger King, KFC, Netty's ChiliDog, Best Buy, EyeGlass SuperStore, JiffyLube, Kroger, Pharm Drug, SavOn, Service Merchandise, cleaners, RV/auto repair, transmissions, **S**...McDonald's, Clarion Hotel, Comfort Inn, Valvoline, cinema, U of Toledo
16	Talmadge Rd(from wb, no return), **N**...BP/diesel/mart/atm, Speedway, Arby's, Charcoal House Rest., Chi Chi's, Jacobson's, JC Penney, Marshall Field, bank, mall
15	Talmadge Rd(from eb, no return), no facilities
14	US 23 N, to Ann Arbor, no facilities
13	US 20, OH 120, Central Ave, **E**...Big Boy, Bob Evans, Burger King, McDonald's, Honda, Lexus, Toyota, Tuffy Muffler, **W**...BP, Speedway
8b a	OH 2, **E**...BP/diesel, HoneyBaked Ham, Extended Stay America, Knight's Inn, Red Roof Inn, Residence Inn, to OH Med Coll **W**...Amoco, Speedway/mart, Sunoco, Bob Evans, Boston Mkt, Burger King, Chili's, McDonald's, Ranch Steak/seafood, Subway, Wendy's, Courtyard, Cross Country Inn, Fairfield Inn, Dick's Sports, OfficeMax, Sam's Club, cinema
6	Dussel Dr, Salisbury Rd, to I-80-90/tpk, **E**...Amoco, Speedway, Bill Knapp's, Burger King, Max&Erma's, McDonald's, Wendy's, Subway, Country Inn Suites, Cross Country Inn, Day's Inn, Hampton Inn, Red Roof Inn, **W**...BP/mart, Airborne Express, Sears Parts
4	US 24, to Maumee, Napolean, **N**...HOSPITAL
3mm	Maumee River
2	OH 25, Perrysburg, to Bowling Green, **N**...Amoco, BP, Citgo, Charlie's Rest., McDonald's, Subway, Goodyear/auto, **S**...Speedway/mart, Red Carpet Inn
0mm	I-475 begins/ends on I-75, exit 192.

Ohio Turnpike

N → **S**

Exit #(mm)	Services
242mm	Ohio/Pennsylvania state line
239mm	toll plaza, phone
237mm	**Mahoning Valley Travel Plaza eb, Glacier Hills Travel Plaza wb, Sunoco/diesel/24hr, McDonald's, gifts, phone**
16a(235)	I-680(from wb), to Youngstown
16(233)	OH 7, to Boardman, to Youngstown, **N**...Kocolene/gas/mart, Palmer House Rest., Smaldino's Family Rest., Budget Inn, Comfort Inn, Day's Inn, Economy Inn/camping, Knight's Inn, Super 8, **S**...Amoco/diesel/rest., Speedway/diesel/mart, Road House Diner, Liberty Inn, North Lima Motel, truck repair
15(218)	I-80 E, to Youngstown, Niles, OH Tpk runs with I-76 eb, I-80 wb.
14b(216)	Lordstown(from wb), **N**...GM Plant
a(215)	Lordstown(from eb), **N**...GM Plant, **S**...BP/diesel
210mm	Mahoning River
14(209)	OH 5, to Warren, **N**...ShortStop/diesel/repair(3mi), Budget Lodge, **S**...Marathon/diesel/mart/rest./repair, Rodeway Inn, RV camping
197mm	**Portage Travel Plaza wb, Brady's Leap Travel Plaza eb, Sunoco/diesel/24hr, Dunkin Donuts, Popeye's, Taco Bell, TCBY, gifts, phone**
13a(193)	OH 44, to Ravenna, no facilities
191.5mm	Cuyahoga River
13(187)	OH 14 S, I-480, to Streetsboro, **S**...Comfort Inn, Fairfield Inn, Hampton Inn, Palms Motel, Firestone/auto, **1 mi S**...CHIROPRACTOR, Amoco/diesel/mart, BP/mart, Clark, DairyMart/gas, Arby's, Big Boy, Bob Evans, Burger King, Burger Central, Dairy Queen, Domino's, Golden Flame Steaks, KFC, LJ Silver, McDonald's, Mr Hero, Perkins, Pizza Hut, Rally's, Subway, Taco Bell, Best Western, Holiday Inn Express, Microtel, Super 8, Wendy's, Farm&Fleets, FashionBug, Midas Muffler, Save-A-Lot Foods, Staples, Van's Tire, Wal-Mart, to Kent St U
12(180)	OH 8, to I-90, **N**...Marathon/mart, MoonBeam's Burgers, Comfort Inn, Holiday Inn, golf, **S**...BP/Subway/diesel/mart, Starfire Express/mart, to Cuyahoga Valley NRA
177mm	Cuyahoga River
11(173)	OH 21, to I-77, **N**...Clark Gas, My Place Rest., Valley Forge Rest., Howard Johnson, Lake Motel/rest., Scottish Inn, tires, **S**...BP/repair, Speedway/diesel/mart, Dairy Queen, Richfield Family Rest., Brushwood Motel, Holiday Inn, Super 8
170mm	**Towpath Travel Plaza eb, Great Lakes Travel Plaza wb, Sunoco/diesel/24hr, McDonald's, gifts, phone**
10(161)	US 42, to I-71, Strongsville, **N**...Marathon/diesel, Shell, DairyMart, Jennifer's Rest., Mad Cactus Mexican, Colony Motel, Day's Inn, Extended Stay America, La Siesta Motel, Murphy's Motel, Villager Lodge, Wal-Mart, **S**...DENTIST, CHIROPRACTOR, Dairy Queen, Friendship Inn, Chrysler/Plymouth, Staples, bank

E → **W**

Youngstown

Ohio Turnpike

9(152) OH 10, to Oberlin, Cleveland, I-480, **N**...Citgo, Marathon, Shell/McDonald's, Speedway/diesel/mart/atm, Sunoco/diesel, Kartel's Rest, Super 8, Travelers Inn, Giant-Eagle Foods, U-Haul, **S**...BP, drive-in theatre(seasonal)

9a(151) I-480 E(from eb), to Cleveland, airport

146mm Black River

8(145) OH 57, to Loraine, Elyria, to I-90, **N**...DENTIST, BP/diesel/mart/wash, Speedway/mart, Arby's, Bob Evans, Country Kitchen, McDonald's, Mtn Jack's Rest., Pizza Hut, Red Lobster, Subway, Wendy's, White Castle, Best Western, Camelot Inn, Comfort Inn, Day's Inn, Econolodge, Holiday Inn, Knight's Inn, Dillard's, Firestone/auto, Goodyear/auto, GreaseMonkey, JC Penney, Jo-Ann Crafts, Kaufman's, K-Mart, Lowe's Bldg Supply, NTB, RiniRego Foods, Sam's Club, Sears/auto, Staples, TJ Maxx, Tuffy Brake/muffler, Wal-Mart/auto, TireAmerica, bank, carwash, cinema, mall, **S**...HOSPITAL, Shell/mart/24hr, Howard Johnson, Ramada Inn/rest., laundry, st patrol

8a(142) I-90(from eb), OH 2, to W Cleveland

139.5mm **Middle Ridge Travel Plaza wb, Vermilion Travel Plaza eb, Sunoco/diesel/24hr, Bob's Big Boy, Burger King, TCBY, gifts, phone, RV parking**

132mm Vermilion River

119.5mm Huron River

7a(135) rd 51, Baumhart Rd, to Vermilion, no facilities

7(118) US 250, to Norwalk, to Sandusky, **N**...Gastown/mart, Marathon/mart, Subway, Comfort Inn, Day's Inn, Hampton Inn, Super 8, Outlet Mall, RV camping, **5 mi N**...Brown Derby, Perkin's, Econolodge, Fairfield Inn, Ramada Inn, **S**...Crown Motel, HoJo's Rest., Homestead Inn/rest., Chevrolet/RV sales, to T Edison's Birthplace

100mm **service plaza both lanes, Sunoco/diesel/24hr, phone**

93mm Sandusky River

6(92) OH 53, to Fremont, Port Clinton, **N**...Best Western, Comfort Inn(12mi), **S**...BP/service, Shell/diesel/mart/24hr, AA Motel/rest.(2mi), Freemont Motel, Holiday Inn

80.5mm Portage River

77mm **service plaza both lanes, Sunoco/diesel/24hr, Hardee's/chicken, phone**

5(72) I-280, OH 420, to Stony Ridge, to Toledo, **N**...Petro/Amoco/diesel/rest./24hr, Flying J/Conoco/diesel/rest./24hr/LP, Charter House Rest., Country Diner, Howard Johnson, Fairlane Motel, Knight's Inn, Metro Inn, Stony Ridge Inn/truck plaza/diesel, Blue Beacon, **S**...TA/diesel/rest./wash/24hr, Speedway/diesel/mart, McDonald's

4a(65) I-75 N, to Toledo, Perrysburg

63mm Maumee River

4(59) US 20, to I-475, Maumee, Toledo, **N**...CHIROPRACTOR, Amoco/mart, BP/diesel/mart, Speedway/mart, SuperAmerica/diesel/mart, Arby's, Bob Evans, Bombay Bicycle Club, China Gate Rest., Church's, Dominic's Italian, Indian Cuisine, Max's Diner, McDonald's, Pizza Hut, Budget Inn, Motel 6, Ramada Inn, Advance Parts, Frank's Nursery/crafts, Goodyear/auto, Jo-Ann Fabrics, Kinko's/24hr, K-Mart/Little Caesar's/drugs, MC Sports, NAPA, Radio Shack, Savers Foods, Sears/auto, cinema, to Toledo Stadium, **S**...Meijer/diesel/24hr, Speedway/mart, Big Boy, ChiChi's, Chinese Rest., Fazoli's, Fricker's, Friendly's, Ralphie's Burgers, Red Lobster, Schlotsky's, Taco Bell, Best Western, Comfort Inn, Cross Country Inn, Day's Inn, Hampton Inn, Red Roof Inn, Chevrolet, Ford/Lincoln/Mercury, Hill's, Honda/Olds, Toyota, bank, carwash, cleaners

3a(52) OH 2, to Toledo, **S**...airport, RV/truck repair

49mm **service plaza both lanes, Sunoco/diesel/24hr, Charlie's Rest./gifts/ice cream**

3b(39) OH 109, no facilities

3(34) OH 108, to Wauseon, **S**...HOSPITAL, Shell/diesel/mart/24hr, Smith's Rest., Arrowhead Motel, Best Western, Super 8, Woods Trucking/repair(1mi), RV sales, **2 mi S on US 20A**...DairyMart/gas, Mobil/mart, Burger King, McDonald's, Pizza Hut, Subway, Wendy's, Wauseon Motel, Wal-Mart/auto

2a(25) OH 66, Burlington, **3 mi S**...Sauder Village Museum

24.5mm Tiffen River

21mm **Indian Meadow Service Plaza both lanes, info, Sunoco/diesel/24hr, Hardee's, travel trailer park**

2(13) OH 15, to Bryan, Montpelier, **S**...Marathon/Subway/diesel/mart, Pennzoil/diesel/mart, Country Fair Rest., Econolodge, Holiday Inn, Rainbow Motel, Hutch's Diesel Repair

11.5mm St Joseph River

3mm toll plaza, phone

1(2) OH 49, to US 20, **N**...Mobil/Subway/diesel/mart, Burger King, info

0mm Ohio/Indiana state line

E

W

Loraine

Toledo

Oklahoma Interstate 35

Exit #	Services
236mm	Oklahoma/Kansas state line
231	US 177, Braman, **E**...Conoco/Kanza Café/diesel/mart/motel
230	Braman Rd, **W**...Texaco/diesel/mart/tire repair
229.5mm	Chikaskia River
225mm	**Welcome Ctr sb, rest area nb, full(handicapped)facilities, phones, picnic tables, litter barrels, vending, petwalk**
222	OK 11, to Blackwell, Medford, Alva, Newkirk, **E**...HOSPITAL, Conoco/diesel/mart/rest., Phillips 66/diesel/mart, Texaco, Braum's, McDonald's, Pizza Hut(2mi), Sonic(3mi), Subway, Taco Bell, Comfort Inn, Day's Inn, Super 8(2mi), **W**...Texaco/mart
218	Hubbard Rd, no facilities
216.5mm	weigh sta both lanes
214	US 60, to Tonkawa, Lamont, Ponka City, N OK Coll, **E**...Holiday Inn, **W**...Phillips 66/diesel/rest., New Western Inn, Woodland RV Park
213.5mm	Salt Fork of Arkansas River
211	Fountain Rd, **E**...Love's/Pizza Hut/diesel/mart/24hr
209mm	parking area both lanes, litter barrels
203	OK 15, Billings, to Marland, **E**...Conoco/Dairy Queen/diesel/mart/24hr
199.5mm	Red Rock Creek
195.5mm	parking area both lanes, no facilities, litter barrels
194b a	US 412, US 64 W, **E**...Cimarron Tpk(eb), to Tulsa, **W**...to Enid, Phillips U
193	to US 412, US 64 W(from nb), no facilities
191mm	Black Bear Creek
186	US 64 E, to Fir St, Perry, **E**...HOSPITAL, Shamrock/Subway/diesel/mart, Braum's, KFC(2mi), McDonald's, Pizza Hut, Taco Mayo, museum, **W**...Conoco/diesel/mart/24hr, Day's Inn
185	US 17, Perry, to Covington, **E**...Phillips 66/mart, Best Western/rest., **W**...Texaco/diesel/motel/rest./atm/24hr, RV camping
180	Orlando Rd, no facilities
174	OK 51, to Stillwater, Hennessee, **E**...Phillips 66/diesel/mart, Texaco/diesel/mart, Fairfield Inn(12mi), Holiday Inn(13mi), Motel 6(12mi), Lake Carl Blackwell RV Park
173mm	parking area sb, no facilities, litter barrels
171mm	parking area nb, no facilities, litter barrels
170	Mulhall Rd, **E**...Sinclair/diesel/auto/tire repair
165.5mm	Cimarron River
157	OK 33, Guthrie, to Cushing, **E**...Sinclair/diesel/repair, **W**...HOSPITAL, Conoco/mart, Love's/Subway/diesel/mart/24hr, Phillips 66/diesel/mart, Texaco/diesel/mart, Total/mart, Burger King, Carl's Jr, Dairy Queen, El Rodeo Mexican, KFC, Best Western, Interstate Motel, Sleep Inn, OK Terr Museum, Langston U, RV camping
153	US 77 N, Guthrie, **E**...McDonald's, Taco Mayo, Dodge Trucks, Ford/Mercury
151	Seward Rd., **E**...Conoco/diesel/mart, Lazy E Arena(4mi), RV park
149mm	weigh sta sb
146	Waterloo Rd, **E**...Conoco/diesel/mart/24hr, café, ° mi **W**...Citgo/diesel/mart
143	Cobell Rd, Danforth, no facilities
141b	US 77 S, OK 66E, to 2nd St, Edmond, Tulsa, **W**...Phillips 66/diesel, Texaco/mart, Carl's Jr, Coyote Café, Denny's, McDonald's, Western Sizzlin, Hampton Inn, Ramada Inn, Stafford Inn
140	SE 15th St, Spring Creek, Arcadia Lake, Edmond Park, **E**...Braum's, Central SP
139	SE 33rd St, **W**...Applebee's, Sleep Inn
138d	Memorial Rd, Enterprise Square USA
c	Sooner Rd, no facilities
b	Kilpatrick Tpk, no facilities
a	I-44 Tpk E to Tulsa
I-35 S and I-44 W run together 8 mi	
137	NE 122nd St, to OK City, **E**...Texaco/diesel/mart, Total/diesel, Kettle, Motel 6, **W**...Flying J/Conoco/diesel/LP/mart/24hr, Love's/diesel/mart/24hr, Texaco/diesel/mart/24hr, Carl's Jr, Cracker Barrel, McDonald's, Waffle House, Comfort Inn, Day's Inn, Motel 6, Quality Inn, Red Carpet, RV camp, Frontier City Fun Pk, Morgan Boots
136	Hefner Rd, **W**...Conoco/Little Italy Rest./diesel/mart, Dairy Queen, funpark, same as 137
135	Britton Rd, **E**...Conoco/diesel/mart/24hr
134	Wilshire Blvd, **E**...Conoco/diesel/mart/24hr, Braum's, Gordon Inland Marine, **W**...Bruce's/diesel/rest./24hr, Econolodge, Blue Beacon, Goodyear
I-35 N and I-44 E run together 8 mi	
133	I-44 W, to Amarillo, **W**...st capitol, Cowboy Hall of Fame
132b	NE 63rd St(from nb), **1/2 mi E**...Conoco/diesel/mart, Braum's, Remington Inn
a	NE 50th St, Remington Pk, **W**...funpark, info, museum, zoo
131	NE 36th St, **E**...VETERINARIAN, golf, **W**...Phillips 66/diesel/mart/24hr, 45th Inf Division Museum

The vertical side labels read: **Blackwell**, **Guthrie**, **Okla City**

N
S

Oklahoma Interstate 35

130   US 62 E, NE 23rd St, **W**...to st capitol

129   NE 10th St, **E**...Conoco/ McDonald's/diesel/mart, Total/ mart, **W**...Tom's BBQ

128   I-40 E, to Ft Smith, no facilities

127   Eastern Ave, OK City, **W**...Petro/diesel/rest./Blue Beacon/ 24hr, Pilot/Wendy's/diesel/mart/24hr, Texaco/diesel/mart, BBQ, Waffle House, Best Western, Central Plaza Hotel, Blue Beacon, Lewis RV Ctr

126a   I-40, W to Amarillo, I-235 N, to st capitol

  b   I-35 S to Dallas

125d   SE 15th St, **E**...Conoco/diesel/mart, Total/repair, Choice Rest., Howard Johnson

  b   SE 22nd St(from nb), no facilities

  a   SE 25th, **E**...Denny's, McDonald's, Taco Bell, Waffle House, Day's Inn, Plaza Inn, Ramada Inn, Royal Inn, Super 8, **W**...Phillips 66

124b   SE 29th St, **W**...Mama Lou's Rest., McClain's RV Ctr, same as 125a

  a   Grand Blvd, **E**...Texaco, Courtesy Inn, **W**...Sinclair, Cattle Rustler Steaks, Taco Mayo, Drover's Inn, Executive Inn

123b   SE 44th St, **E**...MEDICAL CARE, Conoco/mart, Sonic, BestValue Inn, **W**...Dairy Queen, Pizza 44, Taco Mayo, Sav A Lot Foods, USPO, bank

  a   SE 51st St, **E**...Conoco/diesel/mart, **W**...Southgate Inn, tires

122b   SE 59th St, **E**...Texaco/diesel/LP/mart, **W**...Phillips 66, Total/diesel/mart, EZMart/gas, U-Haul

  a   SE 66th St, **E**...Burger King, Luby's, McDonald's, Taco Bell, Village Inn Rest., Ramada, BabysRUs, Best Buy, Dillard's, EyeMart, Firestone, JC Penney, Jumbo Sports, KidsRUs, MensWearhouse, ToysRUs, mall, **W**...Citgo/7-11, Arby's

121b   US 62 W, I-240 E, **W**...to Will Rogers World Airport

  a   SE 82nd St, **W**...Denny's, Green Carpet Inn, La Quinta, Classic Auto Parts

120   SE 89th St, **E**...Total/diesel/mart, **W**...Love's/Subway/diesel/mart/24hr, Carl's Jr, Tom's BBQ

119b   N 27th St, **E**...Conoco, Texaco/diesel, Circle K, Best Western, Luxury Inn/rest., RVs4Less, laundry, **W**...Phillips 66/mart, Pickles American Grill, Harley-Davidson

118   N 12th St, **E**...Total/mart, Super 8, **W**...Phillips 66/diesel/mart, Texaco/mart, Braum's, Dairy Queen, Fajita's Mexican, Grandy's, KFC, LJ Silver, Mama Lou's Rest., Mazzio's Pizza, McDonald's, Taco Bell, Wendy's, Western Sizzlin, Comfort Inn, Day's Inn, Motel 6, Ace Hardware, Family$

117   OK 37, S 4th St, **E**...HOSPITAL, Del Rancho Steaks, **W**...Total/mart

116   S 19th St, **E**...Conoco, BBQ, Braum's, Capt D's, Carl's Jr, Dairy Queen, McDonald's, Taco Bell, Microtel, Firestone/auto, Goodyear/auto, Ross, **W**...Burger King, Wal-Mart

114   Indian Hill Rd, **E**...Barry's Chicken Ranch, Indian Hill Steaks, Kerr RV Ctr

113   US 77 S(from sb), Norman, no facilities

112   Tecumseh Rd, no facilities

110b a   Robinson St, **E**...HOSPITAL, CHIROPRACTOR, DENTIST, Petrostop, Hardee's, Sonic, Taco Bell, Western Sizzlin, Day's Inn, Albertson's/drugs, Chrysler/Jeep, Ford, GMC/Pontiac, Honda, Hyundai, Lincoln/Mercury/Mazda, Subaru, Tires+, Toyota, **W**...Conoco/Subway/mart, 7-11/gas, Arby's, Blackeyed Pea, Braum's, Cracker Barrel, Domino's, Little Mexico Rest., Outback Steaks, Ryan's, Santa Fe Grill, Waffle House, Holiday Inn, Kia/Isuzu, cinema

109   Main St, **E**...CHIROPRACTOR, Texaco/diesel/mart, Arby's, Braum's, CiCi's, Denny's, Golden Corral, LJ Silver, Prairie Kitchen, Subway, Waffle House, Wendy's, Econolodge, Guest Inn, Ramada Inn, Super 8, Thunderbird Lodge, Travelodge, AutoZone, AvisLube, Buy4Less, Chevrolet, Dodge, Nissan, Target, Wal-Mart SuperCtr/24hr, cleaners, **W**...EYECARE, Conoco/diesel/mart/wash, Applebee's, Burger King, Don Pablo's, Charleston Rest., Chili's, Cracker Barrel, IHOP, Marie Callender's, McDonald's, Olive Garden, On the Border, Piccadilly's, Red Lobster, Red River Steaks, Souper Salad, Village Inn, Rest., Fairfield Inn, Hampton Inn, La Quinta, Barnes&Noble, Bed Bath&Beyond, Dillard's, Home Depot, IGA Foods, Mailboxes Etc, Michael's, OfficeMax, Old Navy, PetsMart, Pier 1, Ross, Saab, Sears/auto, ToysRUs, mall

108b a   OK 9 E, Norman, **E**...CHIROPRACTOR, Citgo, Texaco, Total/mart, Arby's, Braum's, Del Rancho Burgers, Residence Inn, Thunderbird Lodge, Midas Muffler, to U of OK

107mm   Canadian River

106   OK 9 W, to Chickasha, **W**...Conoco/Burger King/mart, Love's/Taco Bell/diesel/mart/24hr, Chevrolet

104   OK 74 S, Goldsby, **E**...Floyd's RV Ctr/LP, **W**...Texaco/diesel/mart, Total/diesel/mart

101   Ladd Rd, no facilities

98   Johnson Rd, **E**...Texaco/diesel/mart, RV camping

95   US 77(exits left from sb), Purcell, **1-3 mi E**...HOSPITAL, to Braum's, Carl's Jr, KFC, Mazzio's Pizza, Pizza Hut, Ford

Oklahoma Interstate 35

91	OK 74, to OK 39, Maysville, **E**...Phillips 66/diesel/mart, Total/mart, Casa Rosa Mexican, McDonald's, Ruby's Inn/rest., Summit 2 Rest., Econolodge, flea mkt, **W**...Texaco/Blimpie/diesel/mart/atm/24hr
86	OK 59, Wayne, Payne, no facilities
79	OK 145 E, Paoli, no facilities
76mm	Washita River
74	OK 19, Kimberlin Rd, no facilities
72	OK 19, Paul's Valley, **E**...DENTIST, Citgo/mart, Total/diesel/mart, Braum's, Carl's Jr, Denny's, KFC/Taco Bell, McDonald's, Pizza/pasta/steaks, Sonic, Subway, Taco Mayo, Tio's Mexican, Amish Inn, Best Western, Day's Inn, Garden Inn, Sands Motel, Buick/Pontiac/Cadillac/GMC, Chrysler/Dodge/Jeep, $General, Ford/Lincoln/Mercury, Wal-Mart/auto(1mi), **W**...Love's/A&W/diesel/mart/24hr, Phillips 66/diesel/mart/24hr, Chaparral Steaks
70	Airport Rd, **E**...HOSPITAL
66	OK 29, Wynnewood, **E**...Kent's/Motel/diesel/rest., **W**...Texaco/diesel
64	OK 17A E, to Wynnewood, no facilities
60	Ruppe Rd, no facilities
59mm	**rest area both lanes, full(handicapped)facilities, phone, picnic table, litter barrels, petwalk, RV dump**
55	OK 7, Davis, **E**...Phillips 66/A&W/diesel/mart/24hr, Total/mart, to Chickasaw NRA, **W**...Oak Hill RV Park, to Arbuckle Ski Area
54.5mm	Honey Creek Pass
54mm	weigh sta both lanes
51	US 77, Turner Falls, **E**...Arbuckle Mtn Motel, Mountainview Inn(3mi), **W**...Sinclair/grill, Canyon Breeze Motel/RV camping, to Arbuckle Wilderness, Botanic Gardens
49mm	scenic turnout both lanes
47	US 77, Turner Falls Area, no facilities
46mm	scenic turnout both lanes
42	OK 53 W, Springer, Comanche, **W**...Shamrock/diesel/mart/24hr
40	OK 53 E, Gene Autry, **E**...Total/diesel/mart/café/24hr, Ardmore Airpark
33	OK 142, Ardmore, **E**...Phillips 66/diesel/mart/24hr, Ponder's Rest., Ryan's, Day's Inn, Guest Inn, Super 8
32	12th St, Ardmore, **1 mi E**...HOSPITAL, Conoco/mart, Arby's, Braum's, Burger King, Carl's Jr, Grandy's, LJ Silver, Taco Bell, Taco Bueno, Chevrolet, Hastings Books, HobbyLobby, MailBoxes Etc, O'Reilly Parts, PitPro Lube, Staples, Toyota, **W**...Love's/Subway/diesel/mart/24hr
31b a	US 70 W, OK 199 E, Ardmore, **E**...CHIROPRACTOR, Conoco/diesel/mart/24hr, Sinclair/diesel/mart, Texaco/diesel/mart, Total/mart/24hr, Applebee's, Burger King, Denny's, El Chico's, KFC, Mazzio's Pizza, McDonald's, Pecos Red's Roadhouse, Pizza Hut, Prairie Kitchen, 2Frogs Grill, Best Western, Comfort Inn, Dorchester Inn, Hampton Inn, Holiday Inn, Motel 6, Buy4Less/24hr, cinema, **W**...Conoco/diesel/mart, Total/diesel/mart, Ford/Lincoln/Mercury, Honda, Mazda, Nissan, Olds/Cadillac
29	US 70 E, Ardmore, **E**...to Lake Texoma SP
24	OK 77 S, **E**...Total/diesel/mart, Peabody Gen Store, Red River Livestock Mkt, to Lake Murry SP
22.5mm	Hickory Creek
21	Oswalt Rd, **W**...KOA
15	OK 32, Marietta, **E**...HOSPITAL, Sinclair, Total/diesel/mart, Jiffy Jim's/gas, Carl's Jr, Denim's Rest., Pizza Hut, Robertson's Ham Sandwiches, Sonic, $General, Ford, Winn-Dixie Foods, bank, laundry, to Lake Texoma SP, **W**...Phillips 66/Subway/mart/24hr, Texaco/mart, BBQ
5	OK 153, Thackerville, **W**...Total/diesel/mart, Indian Nation RV Park
3.5mm	**Welcome Ctr nb/rest area sb, full(handicapped)facilities, phone, picnic tables, litter barrels, vending, petwalk**
1	US 77 N, **W**...Shamrock
0mm	Red River, Oklahoma/Texas state line

Oklahoma Interstate 40

Exit #	Services
331mm	Oklahoma/Arkansas state line
330	OK 64D S(from eb), Ft Smith, no facilities
325	US 64, Roland, Ft Smith, **N**...Total/diesel/mart/rest./24hr, Day's Inn, Roland Inn, Cherokee Casino, AllPro Parts, **S**...CHIROPRACTOR, DENTIST, Citgo/diesel/mart, Conoco/McDonald's/diesel/mart, Phillips 66/mart, Pilot/Wendy's/diesel/mart/24hr, AJ's Pizza, Hardee's, KFC, Mazzio's, Sonic, Taco Bell, Interstate Inn, Quality Inn, $General, IGA Foods
321	OK 64b N, Muldrow, **N**...Texaco/diesel/mart/24hr, Sonic(1mi), **S**...Curt's/diesel/mart/Economy Motel, Carolyn's Rest., antiques, truckwash
316mm	**rest area eb, full(handicapped)facilities, info, phone, picnic tables, litter barrels, vending, petwalk, RV dump**
314mm	**Welcome Ctr wb, full(handicapped)facilities, info, phone, picnic tables, litter barrels, vending, petwalk, RV dump**

Oklahoma Interstate 40

311	US 64, Sallisaw, **N...**HOSPITAL, Conoco, Phillips 66/mart, Texaco/mart, Hardee's, KFC, Pizza Hut, Simon's Pizza, Sonic, Subway, Taco Bell, Taco Mayo, Econolodge, Motel 6, AutoZone, Bill's Tire/repair, Pontiac, Sallisaw Drug, Super H Foods, bank, **1 mi N...**Lessley's Café, Day's Inn, to Sallisaw RA, Brushy Lake SP(10mi), Sequoyah's Home(12mi), **S...**golf
308	US 59, Sallisaw, **N...**HOSPITAL, Citgo/diesel/mart/rest., Phillips 66/mart, Braum's, JR's Dinner Bell, Mazzio's, McDonald's, Skip's Rest., Western Sizzlin, Best Western, Day's Inn, Magnolia Inn, McKnight Motel, Microtel, Super 8/RV, Cherokee Indian Gifts, $General, Wal-Mart SuperCtr/24hr, antiques, auto repair, bank, to Blue Ribbon Downs, **S...**Texaco/diesel/mart, Ole South Rest., Chevrolet, Chrysler/Plymouth/Dodge/Jeep, Ford, KOA, to Kerr Lake, airport, repair
303	Dwight Mission Rd, **3 mi N...**Blue Ribbon Downs
297	OK 82 N, Vian, **N...**DENTIST, Phillips 66/mart/atm, Texaco/diesel/mart(1mi), Total/diesel/mart, Eddie's Drive-In, Siesta Motel, NAPA, Pennzoil, to Tenkiller Lake RA
291	OK 10 N, to Gore, **N...**Greenleaf SP(10mi), Tenkiller SP(21mi)
290.5mm	Arkansas River
287	OK 100 N, to Webbers Falls, **N...**Love's/Subway/Taco Bell/diesel/mart/24hr, Charlie's Chicken, Knight's Inn, OK Trading Post, tires/repair
286	Muskogee Tpk, to Muskogee, no facilities
284	Ross Rd, no facilities
283mm	scenic turnout both lanes, litter barrels
278	US 266, OK 2, Warner, **N...**Texaco/diesel/mart, Big Country Rest., Western Sands Motel
270	Texanna Rd, to Porum Landing, **S...**gas/mart
265	US 69 bs, Checotah, **N...**Total/mart, Dairy Queen, Pizza Hut, Sonic, **S...**Citgo/Taco Shack/Subs/diesel/mart/24hr, Phillip 66/mart, Budget Host, Chevrolet/Olds, Chrysler/Plymout/Dodge/Jeep
264b a	US 69, to Eufaula, **1 mi N...**Citgo/diesel/mart/24hr, Flying J/Conoco/Country Mkt/diesel/LP/24hr, Midway Inn, $General, Wal-Mart/McDonald's, auto repair
262	to US 266, Lotawatah Rd, **N...**Conoco/diesel/café/24hr, Phillips 66/mart
261mm	Lake Eufaula
259	OK 150, to Fountainhead Rd, **S...**Citgo/mart/LP, Texaco/mart/café, Fountainhead Motel, Lake Eufaula Inn, RV camping, to Fountainhead SP
255	Pierce Rd, **N...**KOA
251mm	parking area both lanes, no facilities
247	Tiger Mtn Rd, **S...**Quilt Barn/antiques
240b a	US 62 E, US 75 N, Henryetta, **N...**Citgo/diesel/mart, Conoco/mart, Love's/diesel/mart, Phillips 66, Texaco, Total/mart, Arby's, Braum's, Dairy Queen, Mazzio's Pizza, McDonald's, Sisters Two Rest., Subway, Taco Mayo, Colonial Motel, Gateway Inn, GuestHouse Inn/rest., Henryetta Inn/dome, LeBaron Motel, Relax Inn, Chrysler/Jeep, Ford, G&W Tire, O'Reilly's Parts, Wal-Mart, **Henryetta RV Park, S...**Indian Nation Tpk
237	US 62, US 75, Henryetta, **N...**HOSPITAL, Citgo/diesel/mart/24hr, Texaco/diesel/rest./mart/atm/24hr, Cow Creek Rest., Sleepy Traveler Motel, Trail Motel, WhiteHorse Trading Post, **Henryetta RV Park(rt 1.5 mi to Main, rt on 10th 5blks, left on Corp, 1blk), S...**Hungry Traveler Rest., Super 8
231	US 75 S, to Weleetka, **N...**Phillips 66/diesel/mart/café
227	Clearview Rd, no facilities
221	US 62, OK 27, Okemah, **N...**HOSPITAL, Conoco/mart, Total/Subway/diesel/mart/24hr, Mazzio's Pizza, Sonic, OK Motor Lodge, Goodyear, U-Haul, auto parts, tires, **S...**Love's/A&W/Taco Bell/diesel/mart/24hr, Texaco/diesel/mart/rest., BBQ, truck repair
217	OK 48, Bearden, to Bristow, **N...**Total/mart, Old West Museum, **S...**Total/mart
216mm	N Canadian River
212	OK 56, to Cromwell, Wewoka, **N...**Conoco/mart, **S...**Texaco/diesel/mart/atm, to museum
208mm	Gar Creek
202mm	Turkey Creek
200	US 377, OK 99, to Little, Prague, **N...**Phillips 66/diesel/rest./mart/24hr, **S...**Citgo/diesel/mart/24hr, Conoco/diesel/mart/24hr, Love's/Subway/diesel/mart/24hr, Catfish Round-Up, Robertson's Ham Sandwiches, Village Inn Motel/rest.
197.5mm	**rest area both lanes, full(handicapped)facilities, phone, picnic tables, litter barrels, petwalk**
192	OK 9A , Earlsboro, **S...**Texaco/diesel/mart, Biscuit Hill Rest., Roadside Motel, CB Shop
189mm	N Canadian River

E ↑ ↓ W

Oklahoma Interstate 40

186 OK 18, to Shawnee, **N**...Citgo/diesel/mart, Texaco/mart, Denny's, American Inn, Day's Inn, Motel 6, Ramada Inn, Super 8, **S**...Colonial Inn, FastLube, Honda Motorcycles, antique auto museum, bank, **1-2 mi S**...Conoco/diesel/mart, Sinclair, Total/mart, Burger King, Carl's Jr, Chinese Rest., Golden Corral, Mazzio's, PJ's Charcoal Rest., Subway, Western Sizzlin, Best Western, Big A Parts, Chevrolet/Olds/Cadillac/GMC, Chrysler/Jeep, Ford, HomeLand Foods, K-Mart, Pratt's Foods/24hr, tires

185 OK 3E, Shawnee Mall Dr, to Shawnee, **N**...Garcia's Mexican, Red Lobster, Taco Bueno, Dillard's, JC Penney, Sears, Wal-Mart/auto, cinema, mall, **S**...Phillips 66/mart/atm/24hr, Applebee's, Braum's, Charlie's Chicken, CiCi's, Cracker Barrel, Delta Café, Garfield's Rest., Mazzio's, McDonald's, Schlotsky's(2mi), Sonic, Best Western(3mi), Hampton Inn, Lowe's Bldg Supply, Staples, airport

181 US 177, US 270, to Tecumseh, **1-2 mi S**...Citgo/mart, Best Western, Budget Host, antiques

180mm N Canadian River

178 OK 102 S, Dale, **N**...trailer sales

176 OK 102 N, McLoud Rd, **S**...Love's/Subway/diesel/mart/24hr, Curtis Watson Rest.

172 Newalla Rd, to Harrah, **S**...Texaco/mart

169 Peebly Rd, no facilities

166 Choctaw Rd, to Woods, **N**...Love's/Taco Bell/diesel/mart/24hr, KOA, **S**...Texaco/diesel/mart/atm/24hr, Subway, Diamond B Rest., to Little River SP(11mi)

165 I-240 W(from wb), to Dallas, no facilities

162 Anderson Rd, **N**...gas/mart/LP

159b Douglas Blvd, **N**...Phillips 66/mart, Denny's, McDonald's, Sonic, Taco Bell, RV camping, **S**...Tinker AFB, HOSPITAL

a Hruskocy Gate, **N**...Sinclair/mart, Texaco, China Grill, Executive Inn, Chrysler/Dodge, Family$, Firestone/auto, Nissan, U-Haul, same as 157, **S**...Gate 7, Tinker AFB

157c Eaker Gate, Tinker AFB, same as 159

b Air Depot Blvd, **N**...Texaco/mart, Golden Corral(1mi), Golden Griddle Rest., Pizza Inn, JC Penney, Chevrolet, Firestone, S&S Mufflers, bank, same as 159, **S**...Gate 1, Tinker AFB

a SE 29th St, Midwest City, **N**...Conoco/diesel/mart, Texaco, Planet Inn, Super 8, NAPA, O'Reilly Parts, **S**...Sam's Club, Ford

156b a Sooner Rd, **N**...Conoco/mart, Texaco, Cracker Barrel, Ray's Rest., Comfort Inn, Hampton Inn, La Quinta, Motel 6, **S**...Motel 6, Chevrolet/Pontiac/GMC, Toyota

155b SE 15th St, Del City, **N**...Conoco, Texaco/mart, Holiday Inn, La Quinta, Beachler's Food/drug, **S**...Ashley's Rest.

155a Sunny Lane Rd, Del City, **N**...Conoco/diesel/mart, U-Haul, **S**...Circle K/gas, Citgo, Texaco XpressLube, Braum's, Dunkin Donuts, Pizza Hut, Taco Bell

154 Reno Ave, Scott St, **N**...Conoco/mart, Sinclair/dieselmart, RV Ctr, **S**...Phillips 66/mart/24hr, 7-11/gas

152 I-35 N, to Wichita

127 Eastern Ave(from eb), Okla City, **N**...Petro/diesel/rest, Pilot/Wendy's/diesel/mart/24hr, Texaco, Gary Dale's BBQ, Hammett House Rest., Waffle House, Best Western, Central Plaza Hotel, Sunshine Inn, Blue Beacon, Lewis RV Ctr

151b c I-35, S to Dallas, I-235 N, to downtown, st capitol

a Lincoln Blvd, **N**...Bricktown Stadium

150c Robinson Ave(from wb), OK City, **N**...Spaghetti Warehouse, Westin Hotel, U-Haul, Ford

b Harvey Ave(from eb), downtown, **N**...Rennaisance Hotel, Westin Hotel, Hertz

a Walker Ave(from eb), **N**...Goodyear, **S**...Phillips 66, transmissions

149b Classen Blvd(from wb), to downtown

a Western Ave, Reno Ave, **N**...Total/Subway/TCBY/mart, McDonald's, Taco Bell, **S**...Conoco/diesel/mart, Burger King, TuneUp

148c Virginia Ave(from wb), to downtown

b Penn Ave(from eb), **N**...Shamrock, Total, **S**...Isuzu/Ryder Trucks

a Agnew Ave, Villa Ave, **N**...Phillips 66/diesel/mart, Ford Trucks, **S**...Conoco/diesel/mart/wash, Braum's, BBQ, Freddie's Tires

147c May Ave, **N**...Dairy Queen

b a I-44, E to Tulsa, W to Lawton

146 Portland Ave(from eb, no return), **N**...Texaco/Subway/diesel/mart/atm, water funpark

145 Meridian Ave, OK City, **N**...Conoco/mart, Texaco/mart, Boomerang Grill, Denny's, McDonald's, Outback Steaks, Steak&Ale, Two Pesos Mexican, Best Western, Biltmore Hotel, Extended Stay America, Howard Johnson, Motel 6, Red Roof Inn, Residence Inn, Super 8, Travelers Inn, Chevrolet, some same as 144, **S**...Conoco, Phillips 66, Texaco/mart, Bennigan's, Burger King, Cracker Barrel, IHOP, Kettle, Waffle House, Western Sizzlin, Wendy's, Clarion Hotel, Comfort Suites, Courtyard, Day's Inn, Hilton, Holiday Inn Express, La Quinta, Motel 6, Ramada Inn, Sleep Inn, Celebration Sta

144 MacArthur Blvd, **N**...Texaco/mart, Applebee's, McDonald's, Taco Cabana, Cummins Diesel, Fruehauf/Kenworth/Ryder Trucks, HobbyLobby, Office Depot, Wal-Mart Super Ctr/24hr, laundry, **S**...Conoco/diesel, Texaco/mart, Apple Jack's Rest., Microtel, Quality Inn, Super 10, Travelodge, Garden Ridge, Freightliner/Peterbilt, HomeBase, Sam's Club, some same as 145

143 Rockwell Ave, **N**...Rockwell Inn, Home Depot, McClain's RV, **S**...Sands Motel/RV Park/LP, Rockwell RV Park

E

W

Oklahoma Interstate 40

142   Council Rd, **N**...Sinclair/mart, Texaco/mart, BBQ, Braum's, McDonald's, Taco Bell, Waffle House, Wendy's, Best Budget Inn, Ford Trucks, Isuzu, **S**...Conoco/mart, TA/diesel/rest./24hr, Burger King, Econolodge, truckwash, Council Rd RV Park

140   Morgan Rd, **N**...TA/Phillips 66/Sbarro's/Popeye's/diesel/24hr, truckwash/lube, **S**...Flying J/Conoco/diesel/LP/rest./24hr, Love's/Carl's Jr/diesel/mart/24hr, Sonic

138   OK 4, to Yukon, Mustang, **N**...Denny's, Comfort Inn, Green Carpet Inn, Super 8, **S**...Conoco/Blimpie/diesel/mart, Burger King, Classic Grill, Sonic, Homeland Foods, Best Western

137   Cornwell Dr, Czech All Rd, **N**...Albertson's, bank

136   OK 92, to Yukon, **N**...Conoco, Texaco/mart, Braum's, Carl's Jr, KFC, LJ Silver, McDonald's, Wendy's, Green Carpet Inn, Hampton Inn, AutoZone, Chevrolet/Olds, Radio Shack, Rite Aid, Wal-Mart SuperCtr/gas/24hr, cinema, **S**...HOSPITAL

132   Cimarron Rd, **N**...Cimarron Pottery/rest., **S**...airport

130   Banner Rd, **N**...Texaco/diesel/mart/rest.

129mm   weigh st both lanes

125   US 81, to El Reno, **N**...Conoco/mart, Love's/diesel/mart, Brass Apple Rest., Pit BBQ, Taco Mayo, Best Budget Inn, Economy Express Inn, El Reno Inn, Merit Inn, Ramada Ltd, Sands Motel, Super 8, Trail Motel, Chevrolet, Ford/Lincoln/Mercury, Plymouth, Pontiac/GMC/Buick

123   Country Club Rd, to El Reno, **N**...HOSPITAL, Conoco/mart, Texaco/mart, Braum's, Carl's Jr, KFC, Little Caesar's, LJ Silver, McDonald's, Pizza Hut, Subway, Taco Bell, Radio Shack, Shop'nSave Food, Wal-Mart SuperCtr/24hr, **S**...Phillips 66/diesel/mart, Texaco/diesel, Denny's, Sirloin Stockade, Best Western, Comfort Inn, Day's Inn, Red Carpet Inn, Don's Mufflers/repair, Family RV Park

119   Lp 40, to El Reno, no facilities

115   US 270, to Calumet, no facilities

111mm   picnic area eb, picnic tables, litter barrels

108   US 281, to Geary, **N**...Texaco/Subway/diesel/mart, Cherokee Motel/rest., KOA/Indian Trading Post, motel, to Roman Nose SP, **S**...Love's/diesel/mart, RV camping

105mm   S Canadian River

104   Methodist Rd, no facilities

101   US 281, OK 8, to Hinton, **N**...to Roman Nose SP, **S**...Phillips 66/mart, Texaco/DQ/diesel/mart, Microtel, picnic area, to Red Rock Canyon SP

95   Bethel Rd, no facilities

94.5mm   picnic area wb, picnic tables, litter barrels

88   OK 58, to Hydro, Carnegie, **N**...Texaco/diesel/mart/deli

84   Airport Rd, **N**...Conoco/diesel/mart, Texaco/diesel/mart, Holiday Inn Express, Travel Inn, Buick/Pontiac/GMC, Chevrolet/Olds/Cadillac, Chrysler/Plymouth/Dodge/Jeep

82   E Main St, Weatherford, **N**...HOSPITAL, DENTIST, Citgo, Conoco/diesel/mart, Phillips 66/mart, Texaco/mart, Total/diesel/mart, Braum's, Alfredo's Café, Arby's, Carl's Jr, City Diner, K-Bob's, Ken's Pizza, KFC, McDonald's, Pizza Hut, Subway, Taco Bell, Taco Mayo, T-Bone Rest., Best Western, Day's Inn, AutoZone, $General, United Foods, Wal-Mart, auto repair, to SW OSU

80a   W Main St(from eb), **N**...Conoco, Taco Mayo, Best Western, Scottish Inn, Ford/Lincoln/Mercury, Plymouth

80   OK 54, Weatherford, **N**...Conoco/diesel/mart, Texaco/LP, Little Mexico Rest./gifts, Econolodge, Ford, Plymouth

71   Custer City Rd, **N**...Conoco/diesel/Trading Post/rest./24hr, Love's/A&W/Taco Bell/diesel/rest./mart

69   Lp 40(from wb), to Clinton, **2 mi N**...Dairy Queen, Glancey's Motel/rest.

67.5mm   Washita River

66   US 183, Clinton, **N**...Pop Hick's Rest., Glancey Motel, Western Motel, Chevrolet/Olds, **S**...Texaco/diesel, Ford/Lincoln/Mercury, Chrysler/Plymouth/Dodge

65a   10th St, Neptune Dr, Clinton, **N**...Braum's, KFC, Pizza Hut, Subway, Ramada Inn, Relax Inn, Super 8, **S**...Phillips 66/diesel/mart, Budget Inn, RV camping

65   Gary Blvd, Clinton, **N**...HOSPITAL, Conoco, Texaco, Carl's Jr, Dairy Queen, LJ Silver, McDonald's, Pancake Inn, Taco Mayo, Western Sizzlin, Best Western, Comfort Inn, MidTown Inn, Red Roof Inn, Travelodge, K-Mart/drugs, RV camping

62   Parkersburg Rd, **N**...Conoco/diesel/mart, **S**...Hargus RV Ctr

61   Haggard Rd, no facilities

57   Stafford Rd, no facilities

53   OK 44, Foss, **N**...to Foss RA, **S**...Texaco/diesel/mart

50   Clinton Lake Rd, **N**...KOA/gas/LP/mart/phone

47   Canute, **S**...Texaco/mart, Sunset Inn

E

W

El Reno

Clinton

Oklahoma Interstate 40

E l k C i t y

41	OK 34(exits left from eb), Elk City, **N**...HOSPITAL, Conoco/diesel/mart, Love's/Subway/diesel/mart, Texaco, Kettle Rest., Steak Rest./BBQ, Best Western, Budget Inn/rest., Day's Inn, Executive Inn, 1st Interstate Inn, Holiday Inn, Howard Johnson, Motel 6, Travelodge, Super 8, Elk Run RV Park
40	E 7th St, Elk City, same as 41
38	OK 6, Elk City, **N**...Citgo/mart, Conoco/diesel/mart, Texaco/diesel/mart, Arby's, Denny's, K-Bob's, LJ Silver, McDonald's, Western Sizzlin, Bedford Inn, Ramada Inn, Anthony's Food/drugs, RV camping, **S**...Phillips 66/diesel/mart, Total/mart, Econolodge, Holiday Inn, Luxury Inn, Quality Inn, RV camping, U-Haul, to Quartz Mtn SP
34	Merritt Rd, no facilities
32	OK 34 S(exits left from eb), Elk City, **3 mi N**...HOSPITAL, Conoco, KFC, Subway, Taco Bell
26	Cemetery Rd, **N**...Citgo/diesel/café/24hr, **S**...TA/Texaco/diesel/rest./24hr
25	Lp 40, Sayre, **N**...HOSPITAL, Conoco, Shamrock/diesel/mart, Sands Motel, Ford
23	OK 152, Sayre, **N**...Phillips 66/diesel/mart, **S**...Texaco/diesel/mart/Big Sky Rest.
22.5mm	N Fork Red River
20	US 283, Sayre, **N**...Flying J/Conoco/CountryMkt/diesel/mart/24hr, to Washita Battle Site(25mi)
14	Hext Rd, no facilities
13.5mm	check sta both lanes, litter barrels
11	Lp 40, to Erick, Hext, no facilities
10mm	**Welcome Ctr/rest area both lanes, full(handicapped)facilities, phone, picnic tables, litter barrels, petwalk, RV dump**
7	OK 30, Erick, **N**...Texaco/diesel/café, Cal's Café, Comfort Inn, **S**...Love's/A&W/Taco Bell/diesel/mart, Cowboy's Rest./Trading Post, Day's Inn
5	Lp 40, Honeyfarm Rd, no facilities
1	Texola, **S**...gas/diesel/mart/rest., BBQ, RV camping
0mm	Oklahoma/Texas state line

Oklahoma Interstate 44

M i a m i

Exit #	Services
329.5mm	Oklahoma/Missouri state line
321mm	Spring River
314mm	**Oklahoma Welcome Ctr, service plaza wb, Phillips 66/diesel/mart**
313	OK 10, Miami, **N**...HOSPITAL, CHIROPRACTOR, VETERINARIAN, Citgo/diesel/mart, Conoco/diesel/mart, Love's/diesel/mart/Taco Bell/24hr, Arby's, Chinese Rest., Golden Corral, McDonald's, Papazano's Pizza, Stables Rest., Tastee Freez, Best Western/rest., Super 8, Thunderbird Motel/rest./24hr, Townsman Motel, Pennzoil, Miami RV Park, carwash, cinema, repair, tires, to NE OK A&M Coll, **S**...Dick's BBQ, Chrysler/Jeep, Goodyear, RVCtr, hardware
312mm	Neosho River
312mm	picnic area eb, phone, picnic table, litter barrel
310mm	picnic area wb, phone, picnic table, litter barrel
302	US 59, US 69, Afton, **3 mi S**...gas, Best Western, Grand Lake Country Inn, Shangri-La Inn, Bears Den Resort Camping
299.5mm	picnic area eb, picnic table, litter barrel
289	US 60, Vinita, **N**...HOSPITAL, Citgo/mart, Phillips 66/Subway, Braum's, Carl's Jr, Clanton's Rest., Golden Spike Rest., KFC, McDonald's, Pizza Hut, Vinita Motel, Ford/Mercury, Super H Foods, Wal-Mart, tires, st patrol, **S**...Sinclair/diesel/mart
288mm	**service plaza both lanes, Phillips 66/diesel/mart/24hr, McDonald's, phone**
286mm	toll plaza
283	US 69, Big Cabin, **N**...Texaco/diesel/mart/rest./24hr, Pizza Hut, Super 8
271mm	picnic area eb, picnic table, litter barrel
269	OK 28(from eb), to Adair, Chelsea, **S**...Citgo/mart, golf
269mm	picnic area eb, picnic table, litter barrel
256mm	rest area wb, picnic tables, litter barrel, phone
255	OK 20, Claremore, to Pryor, **N**...HOSPITAL, Citgo, Arby's, Carl's Jr, KFC, McDonald's, Best Western, Day's Inn, to Rogers St Coll, Will Rogers Memorial, museum
244mm	Kerr-McClellan Navigation System
241mm	Will Rogers Tpk begins eb, ends wb, phones
241	OK 66 E, to Catoosa, **N**...Phillips 66/mart, Texaco, Phil's Rest., U-Haul
240b	OK 33, to Chouteau, **S**...Citgo/mart, Conoco/mart, Phillips 66/mart, Texaco/A&W/diesel/mart/wash, Harden's Chicken/BBQ, Lot-a-Burger, Mazzio's Pizza, Sonic, Subway, $General, Homeland Foods, NAPA, RV Ctr, cleaners, tires/repair
a	OK 167 N, 193rd E Ave, **N**...Phillips 66/diesel/rest./24hr, Shamrock/diesel/mart/24hr, McDonald's, Pauline's Buffet, Pizza Hut, Waffle House, Wendy's, Super 8, Traveler's Inn, KOA
238	161st E Ave, **N**...Texaco/diesel/rest./mart/24hr, **S**...QT/diesel/mart/atm/24hr, Arby's, Burger King, Microtel, I-44 Auto Auction

Oklahoma Interstate 44

236b	I-244 W, to downtown Tulsa, airport
a	129th E Ave, **N**...RV Ctr, **S**...Flying J/Conoco/diesel/LP/rest./24hr
235	E 11th St, Tulsa, **N**...Phillips 66/mart, QT/mart, Carl's Jr, Fajita Rita's, Lot-a-Burger, Mazzio's Pizza, RedBud Foods, Sonic, Waffle House, Garnett Inn, Motel 6, Stratford House Inn, Super 8, $General, JiffyLube, May's Drug, O'Reilly's Parts, repair, laundry, **S**...QT/diesel/mart, Braum's, Denny's, Taco Bueno, Econolodge, National Inn, Whse Foods
234b a	US 169, N to Owasso, to airport, S to Broken Arrow, no facilities
233	E 21st St(from wb), **N**...hardware, **S**...MEDICAL CARE, Citgo, El Chico's, Comfort Suites, K-Mart, Midas Muffler, Dean's RV Ctr
232	E 31st St, Memorial Dr, **N**...Phillips 66, Whataburger, Day's Inn, Georgetown Plaza Motel, Ramada Inn, Travelodge, Homeland Foods, **S**...Texaco/A&W/mart/wash, Cracker Barrel, IHOP, McDonald's, Village Inn Rest., Courtyard, Embassy Suites, Extended Stay America, Hampton Inn, Holiday Inn Express, Super 8, Super Inn Suites, Chevrolet, Chrysler, Nissan
231	US 64, OK 51, to Muskogee, no facilities
230	E 41st St, Sheridan Rd, **N**...DENTIST, Texaco/mart/wash, Oriental Rest., Ricardo's Mexican, Whataburger/24hr, Goodyear/auto, bank, cinema, **S**...Phillips 66/diesel, Quality Inn, auto parts
229	Yale Ave, Tulsa, **N**...Texaco/mart, **S**...HOSPITAL, Conoco/rest., Phillips 66/lube/wash, QT/mart/24hr, Applebee's, Arby's, Braum's, Burger King, Delta Café, Denny's, Don Pablo's, Joseph's Steaks/Seafood, Outback Steaks, Red Lobster, Subway, Taco Bell, Taco Cabana, Baymont Inn, Comfort Inn, Holiday Inn Select, Salar Inn, Celebration Sta, Mazda, Olds
228	Harvard Ave, Tulsa, **N**...CHIROPRACTOR, Texaco/diesel/mart/atm, Subway, Best Western, JiffyLube, bank, **S**...CHIROPRACTOR, Citgo/mart, Phillips 66, Blimpie, Chili's, Chimi's Mexican, LoneStar Steaks, LJ Silver, Marie Callender's, McDonald's, Papa John's, Perry's Rest., Piccadilly's, Pizza Hut, Sol's Grill, Thai Rest., Holiday Inn Express, Ramada Inn, Albertson's, HobbyLobby, K-Mart/drugs, auto repair, bank, camera shop, cinema, 1-Hr photo, tires
227	Lewis Ave, Tulsa, **S**...Citgo/Circle K, Conoco/mart, Gasoline/mart, QT/mart, AZ Mexican Rest., El Chico, Goldie's Patio Grill, SteakSuffers USA, Wendy's, bank, laundry/cleaners
226b	Peoria Ave, Tulsa, **N**...CiCi's, Pizza Hut, Waffle House, Parkside Hotel, Super 8, Hancock Fabrics, **S**...QT, Texaco, Braum's, Burger King, Stratford House Inn, AutoZone, Buy4Less/24hr, $General, Food City, Meineke Muffler
226a	Riverside Dr(from eb), **S**...Villager Lodge
225	Elwood Ave(from wb), **N**...Ford, **S**...Oil Capitol Motel
225	Arkansas River
224	US 75, to Okmulgee, Bartlesville, **N**...Citgo/mart, QT/diesel/mart, KFC, Mazzio's Pizza, Chevrolet, $General, Whse Mkt, **S**...Rio Motel/rest.
223c	33rd W Ave, Tulsa, **N**...Braum's, Little Caesar's, **S**...Phillips 66/mart, Rib Crib BBQ
223b	51st St(from wb)
223a	I-244 E, to Tulsa, downtown
222c	(from wb), **S**...Value Inn
222b	55th Place, **N**...Crystal Motel, Super 9 Inn, **S**...Day's Inn, Economy Inn
222a	49th W Ave, Tulsa, MEDICAL CARE, **N**...Carl's Jr, Mama Lou's Rest., Monterey Café, Gateway Motel, Motel 6, Tulsa Inn, Big Lots, $General, May Drug, Radio Shack, auto parts, tires, **S**...QT/Wendy's/diesel/mart/24hr, Texaco/Subway/diesel/mart/atm, Arby's, McDonald's, Village Inn Rest., Waffle House, Super 8
221b	OK 66 W, to Sapulpa, to Bristow
221a	57th Ave W, **S**...Windgate Motor Inn, Buick/Pontiac
221.5mm	Turner Tkp begins wb, ends eb
215	OK 97, Sapulpa, to Sand Sprgs, **S**...HOSPITAL, Phillips 66, Arby's, Super 8, Chrysler/Jeep/Dodge, RV Ctr
211	OK 33, to Kellyville, Drumright, Heyburn SP, **S**...Phillip 66/diesel/mart
207mm	service plaza wb, Phillips 66/diesel/mart
205.5mm	picnic area wb, litter barrels, picnic tables
204mm	picnic area eb, litter barrels, picnic tables
197mm	**service plaza eb,** Phillips 66/diesel/mart/24hr, McDonald's
196	OK 48, Bristow, **S**...HOSPITAL, Carolyn Inn
191.5mm	picnic area wb, litter barrels, picnic tables
189.5mm	picnic area eb, litter barrels, picnic tables
182.5mm	toll plaza

E

W

Tulsa

Tulsa

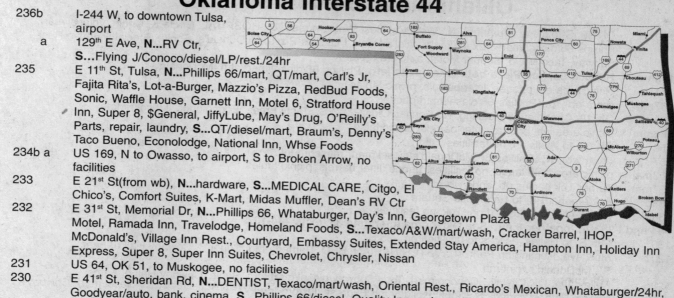

Oklahoma Interstate 44

179	OK 99, Stroud, to Drumright, **N**...Wendy's, Best Western/rest., **S**...HOSPITAL, Circle K, Conoco/diesel/mart, Phillips 66/diesel/mart, Texaco/mart, Jalal's Rest., Mazzio's Pizza, McDonald's, Sonic, Specialty House Rest., Subway, Taco Mayo, Sooner Motel, auto/tire repair
178mm	**Hoback Service Plaza both lanes(exits left),** Phillips 66/diesel/mart/atm, McDonald's
171mm	picnic area eb, phones, litter barrels, picnic tables
167mm	(from eb), **S**...Phillips 66/diesel/mart/phone
166	OK 18, Chandler, to Cushing, **N**...Pontiac/Olds/GMC, **S**...Tasty Freez, Econolodge, Lincoln Motel
166mm	picnic area wb, phones, litter barrels, picnic tables
158	OK 66, to Wellston, no facilities
157	(from wb), **N**...Phillips 66/McDonald's/diesel/mart, museum info
153mm	picnic area both lanes, no facilities, phone, litter barrels, picnic tables
137mm	Turner Tpk begins eb, ends wb
138d	to Memorial Rd, to Enterprise Square
138a	I-35, I-44 E to Tulsa, Turner Tpk, no facilities
I-44 and I-35 run together 8 mi. See Oklahoma Interstate 35, exits 134-137.	
130	I-35 S, to Dallas, access to facilities on I-35 S
129	MLK Ave, Remington Park, **N**...VETERINARIAN, BBQ, Ramada Ltd, Super 8, Cowboy Hall of Fame, **S**...McDonald's, cinema
128b	Kelley Ave, OK City, **N**...Conoco/mart, Total/Subway/diesel/mart/atm, Sonic, Cowboy Hall of Fame
a	Lincoln Blvd, st capitol, **S**...Total, Holiday Inn Express, Oxford Inn
127	I-235 S, US 77, City Ctr, Broadway St, **N**...Conoco, Best Western, Travel Master Motel
126	Western Ave, **E**...Guest House Motel, **W**...Deep Fork Grill
125b	Classen Blvd, OK City, **E**...McDonald's, Courtyard, Richmond Suites, **W**...Acura, Dillard's, Foley's, Full Circle Books, JC Penney, Tuneup Masters, mall
a	OK 3A, Penn Ave, to NW Expswy, **E**...Braum's, Habana Inn, tires, **W**...Conoco/Burger King/diesel/mart24hr
124	N May, **E**...Aamco, Ford, **W**...Texaco/Subway/mart, Decker's Rest., Day's Inn, Super 8, Dodge
123b	OK 66 W, NW 39th, to Warr Acres, **W**...Conoco/McDonald's/diesel/mart/atm, Burger King, Carl's Jr, Chinese Buffet, Denny's, Kettle, Village Inn Rest., Comfort Inn, Ramada Inn, Travelodge, Villager Lodge, Pratt's Foods
a	NW 36th St, no facilities
122	NW 23rd St, **E**...Conoco, Arby's, Big O Tires, **W**...Conoco, 7-11, Church's Chicken, Taco Mayo, Hibdon Tire
121	NW 10th St, **E**...7-11/gas, Texaco/mart, $General, Whitaker's Foods/24hr, antiques, fairgrounds, **W**...Stax/gas/atm
120b a	I-40, W to Amarillo, E to Ft Smith
119	SW 15th St, no facilities
118	OK 152 W, SW 29th St, OK City, **E**...VETERINARIAN, EYECARE, Conoco/mart, Total, Burger King, Capt D's, Grandy's, KFC, McDonald's, Sonic, Taco Bueno, AutoZone, $General, Grider's Foods, Walgreen, bank, **W**...Texaco/mart/atm, Michelle's Café, Bel Air Motel, U-Haul, transmissions
117	SW 44th St, **E**...Sinclair, **W**...gifts
116b	Airport Rd, **W**...airport
a	SW 59th St, **W**...Will Rogers Airport
115	I-240 E, US 62 E, to Ft Smith
114	SW 74th St, OK City, **E**...Sinclair/repair, Stax/gas/atm, Texaco, Total/diesel/mart/atm, Braum's, Burger King, Perry's Rest., Taco Bell, Ramada Ltd, Villager Lodge, same as I-240, exit 1b
113	SW 89th St, **E**...Conoco, Love's/diesel/mart/24hr, Total/diesel/mart, 7-11/gas, McDonald's, Sonic, Taco Mayo
112	SW 104th St, **W**...Texaco/rest./mart
111	SW 119th St, **E**...Mid-America Bible Coll
110	OK 37 E, to Moore, no facilities
109	SW 149th St, **E**...JR's Grill
108.5mm	S Canadian River
108	OK 37 W, to Tuttle, **W**...CHIROPRACTOR, MEDICAL CARE, Conoco/Subway/diesel/mart, Phillips 66/mart/wash/atm, Braum's, Carl's Jr, City Diner, Mazzio's Pizza, Ford, Pennzoil, Wal-Mart, bank, cleaners
107	US 62 S(no wb return), to Newcastle, **E**...Conoco/diesel/mart/24hr, Texaco, Total/mart, Sonic, Taco Mayo, Newcastle Motel, Newcastle RV, Walker RV
100mm	picnic area eb, picnic tables, litter barrels
97mm	toll booth, phone
95.5mm	picnic area wb, picnic tables, litter barrels
85.5mm	**service plaza, both lanes exit left, Texaco/diesel/mart, McDonald's**
83	Chickasha, US 62, **W**...Citgo/diesel/mart, Total/diesel/mart, museum

E

W

Oklahoma Interstate 44

E

↕

W

80	US 81, Chickasha, **E...**HOSPITAL, Conoco/diesel/mart, Texaco, Burger King, Eduardo's Café, Western Sizzlin, Day's Inn, Deluxe Inn, Royal American Inn, Buick/Chevrolet/Pontiac/Cadillac, Chrysler/Plymouth/Dodge/Jeep, Olds/GMC, Wal-Mart/auto, **W...**CHIROPRACTOR, Coastal/mart, Conoco, Love's/mart, Texaco/diesel/mart, Arby's, Braum's, Carl's Jr, Domino's, El Rancho Mexican, KFC, LJ Silver, McDonald's, Peking Dragon, Pizza Hut, Sonic, Subway, Taco Bell, Taco Mayo, Best Western, RanchHouse Motel, Ace Hardware, AutoZone, CrossRoads Books, Eckerd, Family$, Firestone/auto, Ford/Lincoln/Mercury, Lightner's Foods/24hr, O'Reilly Parts, bank, carwash
78mm	toll plaza, phone
63mm	**rest area wb, picnic tables, litter barrels**
62	to Cyril(from wb)
60.5mm	**rest area eb, picnic tables, litter barrels**
53	US 277, Elgin, Lake Ellsworth, **E...**Fina/mart/repair, Total/pizza/mart, Super H Foods
46	US 62 E, US 277, US 281, to Elgin, Apache, Comanche Tribe, last free exit nb
45	OK 49, to Medicine Park, **W...**Love's/A&W/Subway/diesel/mart/24hr, Burger King
41	Key Gate, to Ft Sill, Key Gate, **W...**museum
40c	Gate 2, to Ft Sill
40a	US 62 W, to Cache, **E...**Fina/mart
39a	US 281, Cache Rd, Lawton, **W...**Citgo/Circle K, Conoco/diesel/mart, Phillips 66, Total/A&W/Subway/diesel/mart, Applebee's(6mi), KFC, Ryan's(2mi), Day's Inn, Economy Inn, Holiday Inn, Super 8, Super 9 Motel, Midas Muffler, U-Haul, carwash
37	Gore Blvd, Lawton, **E...**VETERINARIAN, Monterey Jack's Café, Sonic, Taco Mayo, Howard Johnson, **W...**Arby's(3mi), Cracker Barrel, Mike's Grill, Fairfield Inn, Ramada Inn(2mi), Lincoln/Mercury
36	OK 7, Lee Blvd, Lawton, **W...**HOSPITAL, VETERINARIAN, Citgo/mart, Fina/diesel/mart/repair, Leo&Ken's/diesel/rest., Phillips 66(1mi), Texaco/mart, Big Chef Rest., Sonic, KFC, Motel 6
33	US 281, 11ᵗʰ St, Lawton, **W...**HOSPITAL, gas, food, lodging, airport, to Ft Sill
30	OK 36, Geronimo, no facilities
20.5mm	**Elmer Graham Plaza, both lanes exit left, Texaco/diesel/mart, McDonald's, info**
20	OK 5, to Walters, no facilities
19.5	toll plaza
5	US 277 N, US 281, Randlett, no facilities, last free exit nb
1	OK 36, to Grandfield, no facilities
0mm	Red River, Oklahoma/Texas state line

Oklahoma Interstate 240(Okla City)

E

↕

W

Exit #	Services
16mm	I-240 begins/ends on I-40.
14	Anderson Rd, **S...**Conoco/mart
11b a	Douglas Blvd, **N...**Tinker AFB
9	Air Depot Blvd, no facilities
8	OK 77, Sooner Rd, **N...**Love's/diesel/mart/24hr
7	Sunnylane Ave, no facilities
6	Bryant Ave, no facilities
5	S Eastern Ave, no facilities
4c	Pole Rd, **N...**Luby's, Best Buy, Ramada Inn, Dillard's, JC Penney, mall
4b a	I-35, N to OK City, S to Dallas, US 77 S, US 62/77 N
3b a	S Santa Fe, **N...**Circuit City, Home Depot, **S...**KM, Texaco, Buick, Nissan
2b	S Walker Ave, **N...**Texaco/mart, 7-11, Johnny's Charcoal Broiler, **S...**ChuckeCheese, PepBoys
2a	S Western Ave, **N...**Conoco/mart, KM Gas, Burger King, Nino's Mexican, **S...**7-11, Country Cookin Rest., Grandy's, HomeTown Buffet, KFC, LJ Silver, McDonald's, Red Lobster, Steak&Ale, Comfort Inn, Honda, Office Depot
1c	S Penn Ave, **N...**Denny's, Golden Corral, Hardee's, Harrigan's Rest., Olive Garden, Ryan's, Service Merchandise, Wal-Mart/auto, bank, **S...**Texaco/mart, 7-11, Blackeyed Pea, Pancho's Mexican, Taco Bueno, Wendy's, PetsMart
b	S May Ave(from wb)
a	I-44, I-240 begins ends on I-44.

Oregon Interstate 5

Exit #	Services

Portland

308.5mm Columbia River, Oregon/Washington state line

308 Jansen Beach Dr, **E**...Chevron/repair/24hr, Burger King, Taco Bell, Waddles Rest., Oxford Suites, DoubleTree Suites, Safeway, bank, cleaners, **W**...Arco/mart, 76/diesel, BJ's Brewery, Damon's, Denny's, McDonald's, Newport Bay Rest., Stanford's Rest., Subway, Tom's Pizza, DoubleTree Hotel, Barnes&Noble, Boaters World, CarToys, Circuit City, CompUSA, Copeland's Sports, Firestone, Home Depot, Kit's Cameras, K-Mart, Linens'n Things, Old Navy, Pier 1, Staples, bank, mall

307 OR 99E S, MLK Blvd, Union Ave, Marine Dr, Expo Ctr, **E**...Burger King, El Burrito, Elmer's Rest., Best Western, to stockyards

306b Interstate Ave, Delta Park, **E**...76, Burger King, Burrito House, Chinese Cuisine, Elmer's Rest., Shari's Rest., Best Western, Delta Inn, Baxter Parts, Eric's Oilery, Portland Meadows, West Marine

  a Columbia(from nb), **W**...76, Texaco/diesel/mart, 7-11, Wendy's, Winchell's, Fred Meyer

305b a US 30(from nb)

304 Portland Blvd, U of Portland, **W**...Arco

303 Alberta St, Swan Island, **E**...HOSPITAL

302b I-405, US 30 W, **W**...to ocean beaches, zoo

  a Rose Qtr, City Ctr, **E**...HOSPITAL, Texaco, 7-11, McDonald's, Holiday Inn, Ramada Inn, Travelodge, Ford, **W**...Cucina Cucina, Best Western, Red Lion Inn, coliseum

301 I-84 E, US 30 E, facilities E off I-84 exits

300 US 26 E(from sb), Milwaukie Ave, **W**...Hilton, Marriott

299b I-405, no facilities

  a US 26 E, OR 43(from nb), City Ctr, to Lake Oswego

297 Terwilliger Blvd, **W**...HOSPITAL, Shell, Burger King, KFC, Fred Meyer, bank, cleaners

296b Multnomah Blvd, no facilities

  a Barbur, **W**...VETERINARIAN, Chevron/mart/24hr, Texaco, 7-11, Szechuan Chinese, Subway, Taco Time, Wendy's, Aladdin Motor Inn, Capitol Hill Motel, Portland Rose Motel, JiffyLube, MailBoxes Etc, bank

295 Capitol Hwy, **E**...Shell, Texaco, Pacific Pride/diesel, Dunkin Donuts, McDonald's, RoundTable Pizza, Hospitality Motel, Ranch Inn, Oil Change, 1-Hr Photo, tire repair

294 OR 99W, to Tigard, **W**...Arco/mart/24hr, BP, Shell, Texaco, Arby's, Banning's Rest., Burger King, Carrow's Rest., Chinese Rest., Crab Bowl Rest., Italian Rest., KFC, King's Table, Mexican Rest., Newport Bay Rest., Day's Inn, Tigard Inn, Value Inn, Wayside Inn, Fred Meyer, U-Haul, auto parts, tires

293 Haines St, **W**...Ford/Diahatsu

292 OR 217, Lake Oswego, **E**...LDS Temple, Texaco/mart, Applebee's, Chili's, Olive Garden, Sizzler, Holiday Inn Crown, Phoenix Inn, Residence Inn, Deseret Books

291 Carman Dr, **W**...Chevron, Shell, Burgerville, Houlihan's, Best Western, Courtyard, Sherwood Inn, Home Depot, Office Depot

290 Lake Oswego, **E**...76, Burger King, Denny's, Skipper's, Taco Bell, Motel 6, **W**...Arco, Chevron, 76/diesel, Texaco, Chinese Rest., Fuddrucker's, JB's Roadhouse, Village Inn Rest., Best Suites, Best Western, Quality Inn, Shilo Inn, Bed Bath&Beyond, Borders Books, CarQuest

289 Tualatin, **E**...BP, Texaco/mart, Sweetbrier Inn, **W**...HOSPITAL, Arco/mart/24hr, Chevron, Shell, Lee's Kitchen, McDonald's, Wendy's, Century Hotel, Fred Meyer, K-Mart, Safeway, bank, camping

288 I-205, to Oregon City, no facilities

Wilsonville

286 Stafford, **E**...Chevron/Burns Bros/diesel/24hr, 76/diesel/mart/24hr, Cheyenne Deli, IHOP, Best Inn, Super 8, Mercedes, Pleasant Ridge RV Park, cinema, **W**...Holiday Inn/rest., Camping World RV Service/supplies, Chevrolet, Dodge

283 Wilsonville, **E**...MEDICAL CARE, EYECARE, VETERINARIAN, DENTIST, 76/diesel, Applebee's, Arby's, Bullwinkle's Rest., Dairy Queen, Denny's, Domino's, Isleta Bonita Mexican, Izzy's Pizza, Jamba Juice, McDonald's, Papa Murphy's, Quizno's, Red Robin, Shari's/24hr, Subway, Taco Bell, TCBY, Wanker's Café, Wendy's, Wok Chinese, Best Western, Comfort Inn, SnoozInn, Ace Hardware, Fry's Electronics, GNC, Lamb's Foods, MailBoxes Etc, NAPA, Rite Aid, Scwab Tire, USPO, bank, cinema, funpark, **W**...CHIROPRACTOR, Chevron/mart, Shell/24hr, 7-11, Burger King/24hr, Chili's, Marvel's Pizza, New Century Chinese, Phoenix Inn, Animal Crossing PetStore, IGA Food, bank, cleaners

282.5mm Willamette River

282 Charbonneau District, **E**...Charbonneau-on-the-Green Rest., Langdon Farms Golf

281.5mm rest area both lanes, full(handicapped)facilities, phone, info, picnic tables, litter barrels, petwalk, vending, coffee

278 Donald, **E**...PepStop/diesel/mart/RV Park, **W**...TA/Shell/diesel/rest./24hr, Texaco/CFN/diesel, to Champoeg SP

N

S

Oregon Interstate 5

275mm	weigh sta sb
274mm	weigh sta nb
271	OR 214, Woodburn, **E...**VETERINARIAN, Arco/mart/24hr, Chevron/mart/24hr, Exxon/diesel/mart, 76/repair, Shell, Burger King, Dairy Queen, Denny's, Don Pedro's Express, KFC, McDonald's, Patterson's Rest., Shari's, Taco Bell, Wendy's, Best Western, Fairway Inn/RV Park, Super 8, Fairway Drug, Lind's Mkt, Trailer World, Wal-Mart, bank, **W...**Texaco/diesel/mart, Arby's, Jack-in-the-Box, Hawthorn Suites, Chevrolet, Chrysler/Plymouth/Dodge/Jeep, Ford, Outlet/famous brands, Woodburn RV Park
263	Brooks, Gervais, **E...**Astro Gas/LP, Brooks Mkt/deli, **W...**Pilot/Subway/Taco Bell/diesel/mart/24hr, Willamette Mission SP(4mi)
260b a	OR 99E, Keizer, Salem Pkwy, **2 mi E...**camping, **2 mi W...**Arby's, Bob's Burger Express, Dairy Queen, McDonald's, Pietro's Pizza, Subway, Wittenburg Inn, JiffyLube
259mm	45th parallel, halfway between the equator and N Pole
258	N Salem, **E...**76/Circle K, Figaro's Italian, Guesthouse Rest., McDonald's, Original Pancakes, Best Western, Sleep Inn, Cottman Transmissions, Hwy RV Ctr, Maximum Brake/muffler, Roth's Foods, flea mkt, **W...**Chevron/mart, 76/Circle K, Pacific Pride/diesel, Texaco/diesel/mart, Big Shots Steaks, Jack-in-the-Box, LunYuen Chinese, Rock'n Rogers Diner, Rodeway Inn, Travelers Inn, JiffyLube, Staples, Stuart's Parts, to st capitol
256	to OR 213, Market St, Salem, **E...**Shell, Alberto's Mexican, Carl's Jr, Chalet Rest., China Faith, Denny's, Elmer's, El Mirador Mexican, Figaro's Italian, Jack-in-the-Box, Kyoto Japanese, Olive Garden, Outback Steaks, Skipper's, Taco Bell, Best Western, Cozzzy Inn, Crossland Suites, Tiki Lodge, Albertson's, American Tire, Best Buy, Bon-Ton, Borders Books, Fred Meyer, Goodyear/auto, Midas Muffler, NAPA, Pic'n Save Foods, Ross, Safeway, Schwab Tires, Target, Walgreen, mall, **W...**CHIROPRACTOR, Arco/mart/24hr, Chevron, Pacific Pride/diesel, Texaco/diesel/mart/24hr, Baskin-Robbins, Canton Garden, McDonald's, Newport Bay Seafood, O'Callahan's Rest., Outback Steaks, Pietro's Pizza, Roger's 50's Diner, Subway, Tony Roma's, Village Inn, Holiday Lodge, Motel 6, Phoenix Inn, Quality Inn, Red Lion Hotel, Salem Inn, Shilo Inn, Super 8, Buick/GMC, Heliotrope Natural Foods, InStock Fabrics, Jack's IGA, Kia/Mazda/Isuzu, Nissan
253	OR 22, Salem, Stayton, **E...**Chevron/repair, Texaco/diesel/mart, Burger King, Carl's Jr, Las Polomas Mexican, McDonald's, Shari's, Subway, Home Depot, HomeBase, ShopKO, WinCo Foods, Salem Camping, to Detroit RA, **W...**HOSPITAL, Texaco/diesel, Dairy Queen, Denny's, Sybil's Omelet, Best Western, Comfort Suites, Economy Inn, Holiday Inn Express, Motel 6, Travelodge, AAA, Chevrolet/Cadillac/Subaru, Chrysler/Plymouth, Costco, Jeep/Pontiac, K-Mart/auto, Roberson RV Ctr, Schwab Tire, Toyota, airport, st police
252	Kuebler Blvd, **1-3 mi W...**HOSPITAL, Arco/24hr, 76/24hr, Burger King, McDonald's, Neufeldt's Rest., Shari's/24hr, Phoenix Inn
249	to Salem(from nb), no facilities
248	Sunnyside, **E...**Arco/mart/24hr, Burger King, Thrillville Funpark, Willamette Valley Vineyards, RV camping, **W...**Pacific Pride/diesel, Phoenix Inn, towing
244	to Jefferson, no facilities
243	Ankeny Hill, no facilities
242	Talbot Rd, no facilities
241mm	**rest area both lanes, full(handicapped)facilities, info, phone, picnic tables, litter barrels, RV dump, petwalk**
240.5mm	Santiam River
240	Hoefer Rd, no facilities
239	Dever-Conner, no facilities
238	Scio, no facilities
237	Viewcrest(from sb), **W...**RV/truck repair
235	Millersburg, no facilities

Oregon Interstate 5

234   OR 99E, Albany, **E**...Comfort Suites, Harley-Davidson, airport, **W**...HOSPITAL, Arco/mart/24hr, Chevron/mart/ 24hr, 76/mart, Texaco/diesel, Arby's, Burger King, China Buffet, Dairy Queen, Hereford Steer Rest., McDonald's, Pizza Hut, Subway, Taco Bell, TomTom Rest., Wendy's, Bamboo Terrace Motel, Best Western/rest., Budget Inn, Hawthorn Inn, Motel 6, K-Mart, Mervyn's, Chrysler/Jeep, Nissan, Lincoln/Mercury, LubeExpress, RV camping, RV repair, to Albany Hist Dist, tires

233   US 20, Albany, **E**...Chevron/diesel/24hr, 76/diesel/mart, Burgundy's Rest., LumYuen Chinese, Best Inn Suites, Motel Orleans, Phoenix Inn, Chevrolet/Toyota, Home Depot, Honda, Lassen RV Ctr, Mazda, RV camping, mufflers, st police, **W**...HOSPITAL, Texaco, Abby's Pizza, AppleTree Rest., Baskin-Robbins, Burger King, Burgerville, Cameron's Rest., Carl's Jr, Chalet Rest., Denny's, Elmer's, Los Tequilos Mexican, McDonald's, Sizzler, Skipper's, Taco Time, Valu Inn, Albertson's, Battery Xchange, BiMart, Chrysler/Dodge/Jeep/Subaru/ Hyundai, CraftWorld, $Tree, Fred Meyer, Goodyear, Jo-Ann Fabrics, Knecht's Parts, Rite Aid, Schwab Tires, Schuck's Parts, Staples, bank, cinema

228   OR 34, to Lebanon, Corvallis, **E**...Texaco/diesel/mart/24hr, Pine Cone Rest., **W**...Arco/mart/24hr, Chevron/A&W/ diesel, Shell/CFN/diesel, 76/mart, to OSU, KOA

222mm   Butte Creek

216   OR 228, Halsey, Brownsville, **E**...76/Blimpie/diesel/24hr, McDonald's, Best Western/rest., parts/repair/towing, **W**...Texaco/Subway/Taco Bell/diesel/mart

209   Harrisburg, to Jct City, **W**...Hungry Farmer Café, GoodSam RV Park

206mm   **rest area both lanes, full(handicapped)facilities, info, phone, picnic tables, litter barrels, petwalk**

199   Coburg, **E**...Fuel'n Go/diesel, Country Squire Inn, Coburg Hills RV Park, **W**...TA/Shell/diesel/rest./24hr, Texaco/ LP/mart, Destinations RV Ctr, Ford/Freightliner/GMC, Guaranty RV Outlet, Marathon RV Ctr, diesel repair, hist dist

197mm   McKenzie River

195b a   N Springfield, **E**...Arco/mart/24hr, Chevron/mart/24hr, 76/diesel/mart/24hr, DariMart, A&W, Denny's, Elmer's Rest., FarMan Rest., Gateway Chinese, HomeTown Buffet, IHOP, Jack-in-the-Box, KFC, McDonald's, Outback Steaks, Roadhouse Grill, Schlotsky's, Shari's/24hr, Sizzler, SweetRiver Grill, Taco Bell, Best Western, Comfort Suites, Courtyard, Doubletree Hotel/rest., Gateway Inn, Holiday Inn Express, Motel Orleans, Motel 6, Pacific 9 Motel, Rodeway Inn, Shilo Inn/rest., Big 5 Sports, Circuit City, Kinko's, Sears/auto, Target, USPO, Western Outfitters, bank, cinema, laundry, mall, st police, **W**...Texaco, Taco Bell, Costco, ShopKO, to airport

194b a   **E**...OR 126 E, Springfield, **W**...I-105 W, Eugene, HOSPITAL, **2 mi W**...76/repair, Trader Joe's, Red Lion Inn, Albertson's, Honda, TJ Maxx, mall

193mm   Willamette River

192   OR 99(from nb), to Eugene, **W**...Burger King, Wendy's, Best Western, Quality Inn

191   Glenwood, **W**...76/diesel/mart, Texaco/diesel/mart/24hr, Denny's, Lyon's Rest., Motel 6, Phoenix Inn, RV camping

189   30th Ave S, **E**...Texaco/Subway/Taco Bell/diesel/LP/mart, Harley-Davidson, marine ctr, **W**...Exxon/LP/mart/24hr, Rainbow Mtn Rest., RV Ctr, Shamrock RV Park

188b a   OR 58, to Oakridge, OR 99S, **W**...Goshen Trkstp/diesel/rest./24hr

186   to Goshen(from nb), no facilities

182   Creswell, **E**...Emerald Valley Rest.(1mi), airport, auto repair, **W**...DENTIST, Arco/mart/24hr, 76/Taco Bell/diesel/ mart, Texaco/diesel/mart, Country Mouse Deli, Dairy Queen, Mr Macho's Pizza, Pizza Sta, TJ's Rest., Creswell Inn, Motel Orleans, Century Foods, Creswell Drugs, KOA, Kragen Parts, Taylor's Travel Park, bank

180mm   Coast Fork of Willamette River

178mm   **rest area both lanes, full(handicapped)facilities, phone, picnic tables, litter barrels, petwalk, coffee**

176   Saginaw, **W**...Taylor's Travel Park, camping

175mm   Row River

174   Cottage Grove, **E**...Chevron/diesel/repair/24hr, China Garden, Subway, Taco Bell, Best Western/rest., Chevrolet/ Pontiac/Buick/GMC, Chrysler/Plymouth/Dodge/Jeep, Wal-Mart/auto, golf, RV Ctr/camp, **W**...HOSPITAL, Chevron/ mart, Texaco/diesel, Arby's, Burger King, Carl's Jr, Dairy Queen, KFC, McDonald's, Vintage Inn Rest./24hr, City Ctr Motel, Comfort Inn, Holiday Inn Express, Relax Inn, Safeway, Mazda, Village Green Motel/RV Park, airport

172   6th St, Cottage Grove Lake(from sb), **2 mi W**...RV camping

170   to OR 99, London Rd, Cottage Grove Lake, **6 mi E**...RV camping

163   Curtin, **E**...Curtin Café, Stardust Motel, Pass Creek RV Park, USPO, antiques, **W**...Coach House Rest.

162   OR 38, OR 99 to Drain, Elkton, **E**...Stardust Motel

161   Anlauf(from nb), no facilities

160   Salt Springs Rd, no facilities

159   Elk Creek, Cox Rd, no facilities

N
↑
↓
S

Oregon Interstate 5

154	Yoncalla, Elkhead, no facilities
150	OR 99, to OR 38, Yoncalla, Red Hill, **W**...Trees of Oregon RV Park
148	Rice Hill, **E**...Chevron/LP/mart/atm/24hr, Pacific Pride/diesel, Pilot/Subway/diesel/mart/24hr, Homestead Rest./24hr, Peggy's Rest., Best Western, Ranch Motel/rest., **W**...K-R Drive-In
146	Rice Valley, no facilities
144mm	**rest area sb, full(handicapped)facilities, phone, picnic tables, litter barrels, petwalk**
143mm	**rest area nb, full(handicapped)facilities, phone, picnic tables, litter barrels, petwalk**
142	Metz Hill, no facilities
140	OR 99(from sb), Oakland, **E**...Medley Mkt Deli, Tolly's Rest., Oakland Hist Dist
138	OR 99(from nb), Oakland, **E**...Medley Mkt Deli, Tolly's Rest., Oakland Hist Dist
136	OR 138W, Sutherlin, **E**...Chevron/A&W/mart/24hr, 76/mart, Texaco/Subway/diesel/mart, Burger King, Dory's Italian, McDonald's, Papa Murphy's, Red Apple Rest., Microtel, Town&Country Motel, Umpqua Regency Inn, Schwab Tires, **W**...Astro Gas/LP, Dairy Queen, Taco Bell, West Wild Rest./24hr, Best Budget Inn, Hi-Way Haven RV Camp, diesel repair/towing
135	Wilbur, Sutherlin, **E**...muffler repair
130mm	weigh sta sb, phone
129	OR 99, Winchester, **E**...gas, food, camping, diesel repair
128mm	N Umpqua River
127	Stewart Pkwy, N Roseburg, **E**...Chevron/diesel/mart/24hr, Motel 6, Super 8, **W**...VETERINARIAN, Texaco/Taco Maker/diesel/mart, Applebee's, Carl's Jr, IHOP, McDonald's, Subway, Taco Bell, Sleep Inn, Big 5 Sports, Big O Tire, K-Mart, Oilcan Henry's Lube, OfficeMax, Parkway Drug, Sears/auto, Sherm's Foods, Staples, Wal-Mart/auto, Mt Nebo RV Park
125	Garden Valley Blvd, Roseburg, **E**...VETERINARIAN, Chevron, 76/diesel/mart, Texaco/diesel/mart/24hr, Jack-in-the-Box, KFC, McDonald's, Sandpiper Rest., Taco Bell, WagonWheel Rest., Best Inn Suites, Comfort Inn, Windmill Inn/rest., Albertson's, CarQuest, Ford, NAPA, Olds/GMC, Pennzoil, Rite Aid, laundry, transmissions, **W**...MEDICAL CARE, DENTIST, EYECARE, Chevron/mart, 76, Shell/LP/repair, Arby's, Burger King, Carl's Jr, IHOP, Izzy's Pizza, La Hacienda Mexican, RoundTable Pizza, Sizzler, Tequila's Mexican, Wendy's, Best Western, BiMart, Fred Meyer, House of Fabrics, JC Penney, Sears/auto, Schuck's Parts, bank, mall
124	OR 138, Roseburg, City Ctr, **E**...MEDICAL CARE, 76/diesel/mart/24hr, Tesoro/mart, Denny's, HiHo Rest., Best Western, Dunes Motel, Holiday Inn Express, Holiday Motel, Travelodge, Chevrolet/Pontiac/Buick, Honda, Mazda, Rite Aid, Safeway, **W**...HOSPITAL, CHIROPRACTOR, VETERINARIAN, Chevron, Texaco/Subway, Gay 90's Deli, KFC, Pete's Drive-In, Taco Time, Grocery Outlet, SpeedyLube, cleaners
123	Roseburg, **E**...to Umpqua Park, camping, fairgrounds, museum
121	McLain Ave, no facilities
120.5mm	S Umpqua River
120	OR 99 N(no EZ nb return), Green District, Roseburg, **E**...Shady Oaks Motel, **3 mi E**...HiHo Rest, Best Western, **W**...RV parts
119	OR 99 S, OR 42 W, Winston, **W**...Chevron/A&W/diesel/mart/24hr, Texaco/diesel/mart/wash, Callahan's Cove Rest., McDonald's, Ocampo's Mexican, Papa Murphy's, Subway, PriceLess Foods, UPS, RV Park, **3 mi W**...Greentree Inn/rest., Hillside Court Motel, Wildlife Safari Rest., Sweet Breeze Inn
113	Clarks Branch Rd, Round Prairie, **W**...QuickStop Motel/mkt, Cubby Hole Mkt, On the River RV Park, diesel repair
112.5mm	S Umpqua River
112	OR 99, OR 42, Dillard, **E**...Rivers West RV park, Southfork Lodge Rest./store, **rest area nb, W...rest area sb, full(handicapped)facilities, phone, picnic tables, litter barrels, petwalk**
111mm	weigh sta nb
110	Boomer Hill Rd, no facilities
108	Myrtle Creek, **E**...Dairy Queen, The Hub Rest., Myrtle Creek RV Park, gas
106	Weaver Rd, **E**...airport

Oregon Interstate 5

103	Tri City, Myrtle Creek, **E...**Broaster House Diner, Kelly's Steaks(1mi), **W...**Chevron/A&W/diesel/mart/24hr, McDonald's	
102	Gazley Rd, **W...**Surprise Valley RV Park	
101.5mm	S Umpqua River	
101	Riddle, Stanton Park, **W...**camping	
99	Canyonville, **E...**Burger King, Cow Creek Café/RV Park/casino, Riverside Motel, 7 Feathers Hotel/RV Park/casino, Valley View Motel, diesel/RV repair, **W...**Fat Harvey's/diesel/LP/café, Shell/diesel/mart/24hr, Best Western	
98	OR 99, Canyonville, Days Creek, **E...**HOSPITAL, 76/diesel/mart, Texaco/diesel/mart/24hr, Bob's Country Jct Rest., Feedlot Rest., Heaven-on-Earth Café, Natural Foods, Master Host Inn, Leisure Inn, antiques, bank, **W...**Tastee Burger, museum, auto repair	
95	Canyon Creek, no facilities	
90mm	Canyon Creek Pass, elev 2020	
88	Azalea, **W...**Azalea Gen Store/gas	
86	Barton Rd, Quine's Creek, **E...**Texaco/diesel/mart, Heaven on Earth Rest./RV camp	
83	Barton Rd(from nb), **E...**Meadow Wood RV Park(3mi)	
82mm	**rest area both lanes, full(handicapped)facilities, phone, picnic tables, litter barrels, petwalk**	
80	Glendale, **W...**Sawyer Sta/gas/LP, Village Inn Rest., Glendale Inn, camping	
79.5mm	Stage Road Pass, elev 1830	
78	Speaker Rd, no facilities	
76	Wolf Creek, **E...**Stage Coach Pass Motel, **W...**Exxon/diesel/24hr, Pacific Pride/diesel, 76, Bite of Wyo Rest., Wolf Creek Inn Rest., RV park	
74mm	Smith Hill Summit, elev 1730	
71	Sunny Valley, **E...**76/diesel, Sunny Valley Motel, **W...**Aunt Mary's Tavern, KOA	
69mm	Sexton Mtn Pass, elev 1960	
66	Hugo, **W...**Joe Creek Waterfalls RV Camping	
63mm	**rest area both lanes, full(handicapped)facilities, phone, info, picnic tables, litter barrels, vending, petwalk**	
61	Merlin, **E...**Twin Pines RV Park, **W...**VETERINARIAN, Texaco/diesel/mart, Earl's RV, airport	
58	OR 99, to US 199, Grants Pass, **W...**HOSPITAL, CHIROPRACTOR, DENTIST, Arco/mart, Chevron, Gas4Less/U-Haul, 76/diesel/mart, Shell/repair, Texaco/diesel/mart, TownePump Gas, Angela's Mexican, Baskin-Robbins, BeeGee's Rest., Burger King, Carl's Jr, Dairy Queen, Della's Rest., Denny's, Marco's Pizza, McDonald's, Papa Murphy's, Pizza Hut, Senor Sam's Mexican, Sizzler, Skipper's, Stanbrook's Rest., Subway, Wendy's, Budget Inn, Comfort Inn, Hawk's Inn, Hawthorn Suites, Motel 6, Parkway Lodge, Regal Lodge, Riverside Inn, Royal Vue Motel, Shilo Inn, Super 8, SweetBreeze Inn, TownHouse Motel, Travelodge, AutoZone, Chevrolet/Nissan/Honda, Chrysler/Plymouth/Dodge/Jeep, PriceChopper Foods, Radio Shack, Schwab Tire, ToolBox Repair, bank, carwash/lube, laundry, st police, towing	
55	US 199, Redwood Hwy, E Grants Pass, **W...**HOSPITAL, Arco/mart/24hr, CFN/diesel, Exxon/diesel/LP/24hr, Abby's Pizza, Arby's, Carl's Jr, Elmer's, JJ North's Buffet, McDonald's, Shari's/24hr, Si Casa Flores Mexican, Taco Bell, Wild River Brewing Co, Best Western, Discovery Inn, Holiday Inn Express, Knight's Inn, Albertson's, Big 5 Sports, Big O Tire, FashionBug, Fred Meyer, Gottchaulk's, Grocery Outlet, JC Penney, Rite Aid, RiverPark RV Park, Shuck's Parts, Siskiyou RV Ctr, Staples, Wal-Mart	
48	Rogue River, Rogue River RA, **E...**Exxon/LP, Fireball Gas, Texaco/diesel/mart, Homestead Rest., **W...**Abby's Pizza, Betty's Kitchen, Karen's Kitchen, Tarasco Mexican, Best Western, Rogue River Inn, Rod&Reel Motel, Weasku Inn, Circle RV Park, MktBasket Foods	
45b	**W...Valley of the Rogue SP/rest area both lanes, full(handicapped) facilities, phone, picnic tables, litter barrels, petwalk, camping**	
45mm	Rogue River	
45a	OR 99, Savage Rapids Dam, **E...**Cypress Grove RV Park, gas, food, lodging	
43	OR 99, OR 234, Gold Hill, to Crater Lake, **E...**Rock Point Bistro, Lazy Acres Motel/RV Park	
40	OR 99, OR 234, Gold Hill, **E...**Sammy's Rest., Gold Hill Mkt/deli, to Shady Cove Trail, KOA, **W...**76/mart, Gato Gordo Rest., Jacksonville Nat Hist Landmark	
35	OR 99, Blackwell Rd, Central Point, **2-4 mi W...**gas, food, lodging, st police	
33	Central Point, **E...**Chevron/mart, Pilot/Subway/Taco Bell/diesel/mart/24hr, Burger King, RV Ctr, RV camping, funpark, **W...**DENTIST, EYECARE, 76/diesel/mart, Texaco/diesel/mart/24hr, Abbey's Pizza, BeeGee's Rest., BigTown Hero Sandwiches, Little Caesar's, McDonald's, Pappy's Pizza, Albertson's, airport, bank	

Grants Pass (vertical label, left margin)

N / S (compass, right margin)

Oregon Interstate 5

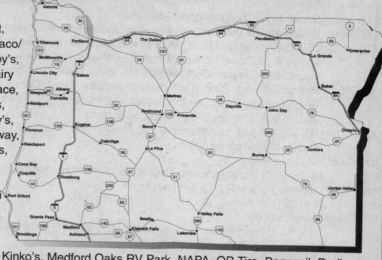

30   OR 62, Medford, to Crater Lake, **E...**VETERINARIAN, Arco/diesel, 76/mart, Chevron/diesel/mart/24hr, Gas4Less, Texaco/diesel/mart, Abby's Pizza, Applebee's, Arby's, Asian Grill, Carl's Jr, Chevy's Mexican, Dairy Queen, Denny's, Elmer's, IHOP, India Palace, Marie Callender, McDonald's, Papa John's, Pizza Hut, Quizno's, Red Robin, Schlotsky's, Si Casa Mexican, Sizzler, Starbucks, Subway, Taco Bell, Taco Delite, Thai Bistro, Wendy's, Williams Bakery, Best Western, Cedar Lodge, Comfort Inn, Motel 6, Reston Inn, Rogue Regency Hotel, Shilo Inn, ValleyHai Inn, Windmill Inn, AAA RV, Albertson's, Aamco, Ace Hardware, Bach's Cameras, Barnes&Noble, BiMart Foods, Chevrolet, Circuit City, Ford, Fred Meyer, Food4Less, Kinko's, Medford Oaks RV Park, NAPA, OR Tire, Pennzoil, Radio Shack, River City RV Ctr, Safeway, Sears, Sir Speedy, TJ Maxx, ToysRUs, USPO, airport, bank, cinema, laundry, st police, **W...**HOSPITAL, Chevron/diesel/mart, Texaco/diesel/mart, Burger King, Domino's, Jack-in-the-Box, KFC, King Wah Chinese, Red Lobster, Skipper's, Wendy's, JC Penney, JiffyLube, Krecht's Parts, Mervyn's, Midas Muffler, Target, U-Haul, mall

27   Barnett Rd, Medford, **E...**Exxon/diesel/mart, Shell/mart/24hr, Kopper Kitchen, Best Western, Day's Inn/rest., Economy Inn, Motel 6, golf, **W...**Chevron/mart/24hr, Exxon/diesel/mart, Gas4Less, 76, Texaco/diesel/mart/24hr, Abby's Pizza, Apple Annie's, Burger King, Carl's Jr, HomeTown Buffet, Jack-in-the-Box, KFC, Kim's Chinese, McDonald's, McGrath's FishHouse, Pizza Hut, Rooster's Rest., Senor Sam's Mexican, Shari's, Starbucks, Subway, Taco Bell, Wendy's, Zach's Deli, Best Inn, Comfort Inn, Red Lion Inn, Royal Crest Motel, Big 5 Sports, Ford/Mercury/Lincoln, Food4Less, Fred Meyer, Harry&David's, Hyundai, K-Mart, OfficeMax, Radio Shack, Saturn, Schwab Tires, Staples, Toyota, WinCo Foods, auto parts, bank, cinema

24   Phoenix, **E...**Petro/diesel/rest./mart/RV dump/24hr, Texaco/mart, PearTree Motel/rest./RV park, Peterbilt/GMC, **W...**DENTIST, VETERINARIAN, Exxon/mart, Angelo's Pizza, Courtyard Café, Luigi's Café, McDonald's, Randy's Café, Bavarian Inn, Phoenix Motel, Ray's Foods, CarQuest, Holiday RV Park, Factory Stores/famous brands, laundry

22mm   **rest area sb, full(handicapped)facilities, phone, picnic tables, litter barrels, vending, petwalk**

21   Talent, **W...**VETERINARIAN, Arco/mart/24hr, Gas4Less, Talent's/diesel/rest./repair, Arbor House Rest., Expresso Café, Figaro's Italian, Senor Sam's Cantina, GoodNight Inn, OR RV Roundup Camping, Wal-Mart/auto, bank, hardware

19   Valley View Rd, Ashland, **W...**HOSPITAL, Exxon/diesel/LP, Pacific Pride/diesel, Texaco/diesel/mart, Burger King, El Tapatio Mexican, Best Western, Hawthorn Inn, Regency Inn/rest./RV Park, Stratford Inn, U-Haul

18mm   weigh sta both lanes

14   OR 66, Ashland, to Klamath Falls, **E...**Chevron/mart, 76/diesel/LP/mart, Texaco/diesel/mart, Denny's, OakTree Rest., Ashland Hills Inn/rest., Best Western, Rodeway Inn, Vista Motel, Windmill Inn, Emigrant Lake Camping(3mi), KOA(3mi), Nat Hist Museum, **W...**Arco/mart/24hr, Astro/mart, 76/LP/24hr, Azteca Mexican, Dairy Queen, McDonald's, Omar's Rest(1mi), Pizza Hut, Taco Bell, Thai Rest., Knight's Inn/rest., Super 8, Albertson's, BiMart Foods, Rite Aid, Schwab Tire, bank

11.5mm   chainup area

11   OR 99, Siskiyou Blvd(no nb return), **2-4 mi W...**VETERINARIAN, CHIROPRACTOR, Exxon, Shell, BBQ, Dairy Queen, Figaro's Italian, Goodtimes Café, House of Thai, Little Caesar's, Omar's Rest., Senor Sam's Mexican, Subway, Wendy's, Best Western, Cedarwood Inn, Hillside Inn, Rodeway Inn, Stratford Inn, NAPA, Radio Shack, Weisinger Brewery, tires

6   to Mt Ashland, **E...**Callahan's Siskiyou Lodge/rest., phone, ski area

4mm   Siskiyou Summit, elev 4310, brake check both lanes

1   to Siskiyou Summit(from nb)

0mm   Oregon/California state line

N

S

Oregon Interstate 84

Exit #	Services
378mm	Snake River, Oregon/Idaho state line
377.5mm	**Welcome Ctr wb, full(handicapped)facilities, info, phone, picnic tables, litter barrels, vending, petwalk**
376	US 30, Ontario, to US 20/26, Payette, **N**...Chevron/mart/24hr, Burger King, Country Kitchen, Dairy Queen, Denny's, McDonald's, Taco Time, Best Western, Colonial Inn, Holiday Inn, Motel 6, Sleep Inn, Super 8, Akins Food/24hr, Chrysler/Dodge/Plymouth/Jeep, K-Mart, Staples, Wal-Mart SuperCtr/24hr, st police, **S**...HOSPITAL, Phillips 66/mart, Pilot/Arby's/diesel/mart/24hr, Shell, Domino's, Klondike Pizza, Rusty's Steaks, Sizzler, Taco Bell, Wendy's, Economy Inn, Holiday Motel/rest., OR Trail Motel, Stockmen's Motel, Ace Hardware, Commercial Tire, NAPA, Pennzoil, Radio Shack, Schwab Tires, U-Haul, RV repair, 4 Wheeler Museum, laundry
374	US 30, OR 201, to Ontario, **N**...to Ontario SP, **S**...HOSPITAL, Shell/diesel/mart, North Star Café, Budget Motel, Carlile Motel, parts/repair
373.5mm	Malheur River
371	Stanton Blvd, no facilities, **2 mi S**...to correctional institution
362	Moores Hollow Rd, no facilities
356	OR 201, to Weiser, ID, **3 mi N**...Oasis RV Park
354.5mm	weigh sta eb
353	US 30, to Huntington, **N**...weigh sta wb, Farewell Bend Travel Plaza/diesel/rest./motel, to Farewell Bend SP, truck repair, RV/tent camping, info
351mm	Pacific/Mountain time zone
345	US 30, Lime, Huntington, **1 mi N**...gas, food, lodging
342	Lime(from eb), no facilities
340	Rye Valley, no facilities
338	Lookout Mountain, **N**...lodging
337mm	Burnt River
335	**to Weatherby, N...rest area both lanes, full(handicapped)facilities, Oregon Trail Info, picnic tables, litter barrels, vending, petwalk**
330	Plano Rd, to Cement Plant Rd, **S**...cement plant
329mm	pulloff eb
327	Durkee, **N**...Oregon Trail Travel Ctr/diesel/café
325mm	Pritchard Creek
321mm	Alder Creek
319	to Pleasant Valley, no facilities
317	to Pleasant Valley(from wb)
315	to Pleasant Valley, no facilities
306	US 30, Baker, **2 mi S**...HOSPITAL, Chevron/diesel, Dairy Queen, Baker City Motel, Budget Inn, Friendship Inn, Lariat Motel/RV, OR Tr Motel/rest., Royal Motel, to st police, same as 304
304	OR 7, Baker, **N**...Chevron/Burger King/TCBY/diesel/mart, Super 8, Welcome Inn, **S**...HOSPITAL, Sinclair/diesel/mart/rest./24hr, Texaco/Blimpie/diesel/mart/wash/24hr, Dairy Queen, Joltin Java, Klondike Pizza, McDonald's, Papa Murphy's, Pizza Hut, Subway, Sumpter Jct Rest., Taco Time, Best Western, Eldorado Inn, Quality Inn, Royal Motel, Western Motel, Albertson's, Ford/Lincoln/Mercury, Mel's Auto Parts, Mtn View RV Park(3mi), Paul's Transmissions, Rite Aid, Safeway, laundry, museum, tires, to hist dist
302	OR 86 E to Richland, **S**...HOSPITAL, OR Tr RV Park/gas
298	OR 203, to Medical Springs, no facilities
297mm	Baldock Slough
295mm	**rest area both lanes, full(handicapped)facilities, info, phone, picnic tables, litter barrels, vending, petwalk**
289mm	Powder River
287.5mm	45th parallel...halfway between the equator and north pole
286mm	N Powder River
285	US 30, OR 237, North Powder, **N**...Cenex/diesel/café, Powder River Motel, **S**...to Anthony Lakes, ski area
284mm	Wolf Creek
283	Wolf Creek Lane, no facilities
278	Clover Creek, no facilities
273	Ladd Canyon, no facilities
270	Ladd Creek Rd(from eb, no return), no facilities
269mm	**rest area both lanes, full(handicapped)facilities, info, phone, picnic tables, litter barrels, vending, petwalk**
268	Foothill Rd, no facilities
265	OR 203, LaGrande, **N**...airport, **S**...Flying J/Exxon/TCBY/diesel/rest./mart, SmokeHouse Rest., to Broken Arrow Lodge, Quail Run Motel, Travelodge, Radio Shack, RV camping, to E OR U

E

W

Ontario

Baker

Oregon Interstate 84

261   OR 82, LaGrande, **N...**Chevron/diesel/mart, Shell/diesel/mart, Texaco/diesel/mart, Denny's, Pizza Hut, Quizno's, Subway, Taco Bell, Howard Johnson, Chrysler/Plymouth/Dodge/ Jeep, Ford/Lincoln/Mercury, Grocery Outlet, Island Lube/wash, Shop'n Kart, Wal-Mart/ drugs, LaGrande Rendezvous RV Resort, bank, st police, **S...**HOSPITAL, CHIROPRAC-TOR, DENTIST, EYECARE, Exxon/diesel/ mart/wash, 76/Subway/diesel/mart, Dairy Queen, Klondike Pizza, Little Caesar's, McDonald's, Skipper's, Taco Time, Wendy's, Best Western, Mr Sandman Motel, Super 8, Albertson's, Cottman Transmissions, Rite Aid, Safeway, Shuck's Parts, E OR U, Wallowa Lake

260mm   Grande Ronde River

259   US 30 E(from eb), to La Grande, **1-2 mi S...**Chevron, Exxon/ Burger King/diesel/mart, Smokehouse Rest., Greenwell Rest., Stardust Lodge, same as 261

257   Perry(from wb), no facilities

258mm   weigh sta eb

256   Perry(from eb), no facilities

255mm   Grande Ronde River

254mm   scenic wayside

252   OR 244, to Starkey, Lehman Springs, **S...**Hilgard SP, camping, chainup area

251mm   Wallowa-Whitman NF, eastern boundary

248   Spring Creek Rd, to Kamela, no facilities

246mm   Wallowa-Whitman NF, western boundary

243   Summit Rd, Mt Emily Rd, to Kamela, Oregon Trail info, **2 mi N...**Emily Summit SP

241mm   Summit of the Blue Mtns, elev 4193

238   Meacham, **1 mi N...**Blue Mountain Lodge/mart, Oregon Trail info

234   Meacham, **N...**Emigrant Sprs SP, RV camping

231.5mm   Umatilla Indian Reservation, eastern boundary

228mm   Deadman Pass, Oregon Trail info, **N...rest area both lanes, full(handicapped)facilities, phone, picnic table, litter barrel, petwalk, vending, RV dump**

227mm   weigh sta wb, brake check area, phone

224   Poverty Flats Rd, Old Emigrant Hill Rd, to Emigrant Springs SP, no facilities

223mm   wb viewpoint, no restrooms

220mm   wb runaway truck ramp

216   Mission, McKay Creek, **N...**Texaco/Arrowhead/diesel, Cody's Rest., Wildhorse Casino/RV Park

213   US 30(from wb), Pendleton, **3-5 mi N...**HOSPITAL, Chevron/diesel, Exxon/mart/deli, Budget Inn, Economy Inn, Pioneer Motel, Tapadera Budget Motel, Travelers Inn, Pendleton NHD

212mm   Umatilla Indian Reservation western boundary

210   OR 11, Pendleton, **N...**HOSPITAL, museum, **S...**76/diesel/mart, Texaco/diesel/LP/mart, Kopper Kitchen, Shari's/24hr, Best Western, Holiday Inn Express, Motel 6, Red Lion Inn/rest., Super 8, Mtn View RV Park, st police

209   US 395, Pendleton, **N...**Arco/mart/24hr, Dean's Mkt/deli, KFC, Taco Bell, Oxford Suites, Travelodge, Rite Aid, Safeway/24hr, Shuck's Parts, Wal-Mart SuperCtr/24hr, bank, laundry, **S...**MEDICAL CARE, Exxon/diesel/24hr, Texaco/diesel/mart, Arby's, Burger King, Denny's/24hr, Klondike Pizza, McDonald's, Rooster's Rest., Subway, Wendy's, Chaparral Motel, Chevrolet/Olds/Pontiac/Buick, Goodyear, Honda, K-Mart, KubeLube/wash, Schwab Tire, Thompson RV Ctr, to Pilot Rock

208mm   Umatilla River

207   US 30, W Pendleton, **N...**Chevron/diesel/mart, Texaco/diesel/LP/mart, Dairy Queen, Godfather's, Longhorn Motel, Tapadera Budget Motel, Vagabond Inn, Brooke RV West, airport

202   Barnhart Rd, to Stage Gulch, **N...**Woodpecker Truck Repair, **S...**Floyd's Truck Ranch/diesel/motel/café, 7 Inn, Oregon Trail info

199   Stage Coach Rd, Yoakum Rd, no facilities

198   Lorenzen Rd, McClintock Rd, **N...**trailer/reefer repair

193   Echo, Oregon Trail Site

Oregon Interstate 84

188	US 395 N, Hermiston, **N**...Chevron/diesel(1mi), Pilot/Subway/diesel/mart/24hr/RV park, **5 mi N**...HOSPITAL, Denny's/24hr, Jack-in-the-Box, McDonald's, Shari's/24hr, Best Western, Economy Inn, Oxford Suites, **S**...Henrietta RV Park
187mm	**rest area both lanes, full(handicapped)facilities, phone, info, picnic tables, litter barrels, petwalk**
182	OR 207, to Hermiston, **N**...Buffalo Jct Gas/mart, **5 mi N**...Butter Creek/gas/diesel/RV park, McDonald's, Shari's
180	Westland Rd, to Hermiston, McNary Dam, **N**...trailer repair, **S**...Shell/diesel/mart
179	I-82 W, to Umatilla, Kennewick, WA
177	Umatilla Army Depot, no facilities
171	Paterson Ferry Rd, to Paterson, no facilities
168	US 730, to Irrigon, **8 mi N**...Oregon Trail info, RV Park
167mm	Coyote Springs Wildlife Area
165	Port of Morrow, **S**...Pacific Pride/diesel
164	Boardman, **N**...Chevron/diesel/mart, Texaco/diesel/mart, C&D Drive-In, Lynard's Cellar Café, Dodge City Motel, Riverview Motel, USPO, Boardman RV/Marina Park, bank, **S**...76/Taco Bell/diesel/mart/24hr, Nomad Deli/24hr, Econolodge, NAPA Repair/parts, Sentry Foods
161mm	**rest area both lanes, full(handicapped)facilities, phone, picnic tables, litter barrels, vending, petwalk**
159	Tower Rd, no facilities
151	Threemile Canyon, no facilities
147	OR 74, to Ione, Heppner, Oregon Trail Site, Blue Mtn Scenic Byway, no facilities
137	OR 19, Arlington, **S**...Chevron/mart, 76/diesel/mart, Happy Canyon Chicken/pizza, Pheasant Grill, Arlington Motel, Village Inn/rest., Ace Hardware, Thrifty Foods, bank, city park, Columbia River RV Park(1mi)
136.5mm	view point wb, picnic tables, litter barrels
131	Woelpern Rd(from eb, no return), no facilities
129	Blalock Canyon, Lewis&Clark Trail, no facilities
123	Philippi Canyon, Lewis&Clark Trail, no facilities
114.5mm	John Day River
114	**S**...to John Day River RA
112	**N**...John Day Dam
109	Rufus, **N**...John Day Visitor Ctr, **S**...Pacific Pride/diesel, Texaco/diesel/mart, BBQ, Bob's T-Bone, Frosty's Café, Larry's Rest., Arrowhead Motel, Tyee Motel, Rufus Inn, RV camping
104	US 97, Biggs, **N**...Des Chutes Park Bridge, **S**...Astro/diesel, Chevron, 76/diesel/mart/atm, Texaco/diesel/mart, Biggs Motel/café, Jack's Rest., Linda's Rest., Subway, Dinty's Motel, Riviera Motel, Maryhill Museum
100mm	Deschutes River, Columbia River Gorge Scenic Area
97	OR 206, Celilo, **N**...Celilo SP, restrooms, **S**...Deschutes SP, Indian Village
88	**N**...to The Dalles Dam
87	US 30, US 197, to Dufur, **N**...76/Taco Bell/mart/24hr, Texaco/diesel/mart, Big Jim's Drive-In, McDonald's, Inn at the Dalles, Lone Pine Motel/rest., Shilo Inn/rest., airport, st police, RV camping
85	The Dalles, **N**...**Riverfront Park, restrooms, phone, picnic tables, litter barrels, playground, marina,** **S**...HOSPITAL, 76, BurgervilleUSA, Holstein's Coffee, Best Western, Oregon Motel, camping, tires, to Nat Hist Dist
83	(84 from wb)W The Dalles, **N**...Orient Café, Windseeker Rest., Tire Factory, **S**...HOSPITAL, CHIROPRAC-TOR, Chevron, Exxon/mart, Texaco, Burger King, Cousins Rest., Denny's, McDonald's, Papa Murphy's, Skipper's, Subway, Taco Bell, Taco Time, Best Western, Day's Inn, Quality Inn, Shilo Inn, Tillicum Inn, Super 8, Albertson's, Chrysler/Plymouth/Dodge, $Store, Fred Meyer, Jo-Ann Fabrics, K-Mart, Rite Aid, Safeway, Shuck's Parts, Staples, bank, carwash
82	Chenowith Area, **S**...Texaco/diesel/mart, **1-2 mi S**...Astro/diesel/24hr, Exxon/mart, Arby's, Burger King, Cousins Rest., KFC, Pietro's Pizza, Stooky's Café, Subway, Wendy's, Day's Inn, BiMart, Chevrolet/Pontiac/Olds/Buick/Jeep/Subaru, Ford/Isuzu/Nissan, Columbia River Discovery Ctr, museum, same as 84
76	Rowena, **N**...Mayer SP, Lewis & Clark info, windsurfing
73mm	**Memaloose SP, rest area both lanes, full(handicapped)facilities, picnic tables, litter barrels, phone, petwalk, RV dump, camping**
69	US 30, Mosier, **S**...gas, lodging, auto service, groceries
66mm	**N**...**Koberg Beach SP, rest area wb, full facilities, picnic table, litter barrels**
64	US 30, OR 35, Hood River, to White Salmon, Gov't Camp, **N**...Chevron, Texaco/mart/pizza/subs/24hr, McDonald's, Riverside Grill, Taco Time, Best Western, marina, museum, st police, **S**...Exxon, China Gorge Café
63	Hood River, City Ctr, **N**...Texaco/mart, Visitor Ctr, **S**...HOSPITAL, Exxon/diesel/24hr, 76/mart, Burger King, China Gorge, Pietro's Pizza, Hood River Hotel, USPO, hardware, same as 64

E

↑

↓

W

The Dalles

Oregon Interstate 84

62 US 30, W Hood River, W Cliff Dr, **N**...Charburger, Columbia Gorge Hotel, Meredith Motel, Vagabond Lodge, FruitTree Gifts, **S**...HOSPITAL, Chevron/diesel/LP/mart, Texaco/diesel/LP/mart, Carolyn's Rest., DQ, McDonald's, Red Carpet Rest., Shari's/24hr, Subway, Taco Bell, Comfort Suites, Econolodge, Prater's Motel, Riverview Lodge, Stonehedge Inn, Chevrolet/Pontiac/Olds/Buick, Pets'n More, Rite Aid, Safeway, Schwab Tire, Wal-Mart

61mm pulloff wb

60 service rd wb(no return), no facilities

58 Mitchell Point Overlook(from eb)

56 **N**...Viento SP, RV/tent camping, phone

55 Starvation Peak Tr Head(from eb), restrooms

54mm weigh sta wb

51 Wyeth, **S**...camping

49mm pulloff eb

47 Forest Lane, Hermon Creek(from wb), camping

45mm weigh sta eb

44 US 30, to Cascade Locks, **1-4 mi N**...Chevron/LP/mart, Shell/diesel, Texaco/mart, Charburger, Best Western, Cascade Motel, Econolodge, Scanamia Lodge, to Bridge of the Gods, KOA

41 Eagle Creek RA(from eb), to fish hatchery

40 **N**...Bonneville Dam NHS, info, to fish hatchery

37 Warrendale(from wb), no facilities

35 **S**...Ainsworth SP, scenic loop highway, waterfall area, RV/tent camping

31 Multnomah Falls(exits left from eb), **S**...Multnomah Falls Lodge(hist site), camping

30 **S**...Benson SP(from eb)

29 Dalton Point(from wb)

28 to Bridal Veil(7 mi return from eb)

25 **N**...Rooster Rock SP

23mm viewpoint wb, hist marker

22 Corbett, **S**...Corbett Mkt/gas, Chinook Inn/rest., Crown Point RV Camping

19mm Columbia River Gorge scenic area

18 Lewis&Clark SP, to Oxbow SP, lodging, food, RV camping

17.5mm Sandy River

17 Marine Dr, Troutdale, **N**...Wendy's, Holiday Inn Express, to airport, **S**...Chevron/mart/24hr, Flying J/Conoco/Burger King/diesel/LP/mart/24hr, TA/Arco/Subway/diesel/24hr, Arby's, McDonald's, Shari's/24hr, Taco Bell, Burns West Motel, Motel 6, Phoenix Inn, Columbia Outlets/famous brands

16 238th Dr, Fairview, **N**...Arco/diesel/mart/24hr, Yazzi's Rest., Travelodge, Wal-Mart/auto, **S**...Chevron, Best Western, Edgefield Motel, Gresham RV Ctr(3mi)

14 207th Ave, Fairview, **N**...Texaco/A&W/Taco Bell/diesel/mart, GinSun Chinese, Rolling Hills RV Park

13 181st Ave, Gresham, **N**...Chevron/mart, Hampton Inn, **S**...VETERINARIAN, Arco/mart/24hr, 76/Circle K, 7-11, Blimpie, Burger King, Jung's Chinese, McDonald's, PlumTree Rest., Shari's/24hr, Wendy's, Comfort Suites, Extended Stay America, Hawthorn Suites, Sheraton/rest., Sleep Inn, Cand yBasket Chocolates, FashionBug, Safeway, auto repair

10 122nd Ave(from eb), no facilities

9 I-205, S to Salem, N to Seattle, to airport, (to 102nd Ave from eb)

8 I-205 N(from eb), to airport

7 Halsey St(from eb), Gateway Dist

6 I-205 S(from eb)

5 OR 213, to 82nd Ave, **N**...Capri Motel, Motel Cabana, JiffyLube, **1-3 mi N**...76, Texaco, 7-11, Arby's, Domino's, Pizza Hut, Quality Inn, Shilo Inn

4 68th Ave(from eb), to Halsey Ave

3 58th Ave(from eb), **S**...76, HOSPITAL

2 43rd Ave, 39th Ave, Halsey St, **N**...HOSPITAL, 76/mart, Texaco/mart/wash, McDonald's, Winchell's, Banfield Motel, HoJo's, Rodeway Inn, Rite Aid, **S**...Buick/Pontiac, Jeep, bank, laundry, same as 1

1 33rd Ave, Lloyd Blvd, **N**...DENTIST, 76, Pacific Pride/diesel, Texaco/mart/wash, Burger King, Banfield Motel, Holiday Inn, Red Lion Inn, Residence Inn, JC Penney, JiffyLube, Rite Aid, **S**...Buick/Olds/GMC

0mm I-84 begins/ends on I-5, exit 301.

Oregon Interstate 205(Portland)

Exit #	Services
37mm	I-405 begins/ends on I-5. **Exits 36-27 are in Washington.**
36	NE 134th St(from nb), **W**...Burger King, Burgerville, McDonald's, RoundTable, Taco Bell, Comfort Inn, Salmon Creek Motel, Shilo Inn
32	NE 83rd St, Andreson Rd, **W**...Meadows Kitchen
30	Orchards, Vancouver, **E**...CHIROPRACTOR, Texaco/24hr, KFC, Burger King, Burgerville, Cisco's Rest., KFC, McDonald's, Subway, Albertson's, HomeBase, Midas Muffler, Rite Aid, Toyota, RV park, **W**...Chevron/mart/24hr, Chevy's Mexican, Grand Chinese Buffet, Olive Garden, Red Lobster, Red Robin Rest., RoundTable Pizza, Skipper's, Shari's/24hr, Subway, Taco Bell, TGIFriday, Tony Roma, Comfort Suites, Holiday Inn Express, Residence Inn, Sleep Inn, America's Tire, JC Penney, Mailboxes Etc, Mervyn's, Pic'n Save Foods, Pier 1, Ross, Sears, Target, TJ Maxx, mall, RV park
28	Mill Plain Rd, **E**...CHIROPRACTOR, DENTIST, Chevron, 76/Circle K, Texaco/mart, 7-11, Applebee's, Baskin-Robbins, Burgerville, Dairy Queen, McDonald's, Ming Chinese, Mongolian BBQ, Pizza Hut, RJ's Rest., Shari's, Sweet Tomato's, Taco Bell, Extended Stay America, Phoenix Inn, Rodeway Inn, Travelodge, Fred Meyer, Schwab Tires, Schuck's Parts, **W**...HOSPITAL, Arco/mart/24hr, Texaco/mart, Jack-in-the-Box, BrakeShop, Chevrolet, Hyundai, JiffyLube, Tire Factory, Wal-Mart/auto
27	WA 14, Vancouver, Camas, no facilities
25mm	Columbia River, Oregon/Washington state line. **Exits 27-36 are in Washington.**
24	122nd Ave, **E**...Courtyard, Shilo Inn/rest., Super 8, Home Depot, **W**...Burger King, McDonald's, Shari's, Hampton Inn, Holiday Inn Express, Sheraton/rest., to airport
23b a	US 30 byp, Columbia Blvd, **E**...HOSPITAL, Shell/diesel/mart/atm, Best Western, Econolodge, Travelodge, **W**...Texaco/mart/wash, Bill's Steaks, Elmer's Rest., Best Western, Clarion, Day's Inn, Holiday Inn, Quality Inn, Ramada Inn, bank, camping
22	I-84 E, US 30 E, to The Dalles
21b	I-84 W, US 30 W, to Portland
a	Glisan St, **E**...76, Applebee's, Carl's Jr, Izzy's Pizza, Quizno's, Szechuan Chinese, Chesnut Tree Inn, Circuit City, Fred Meyer, JiffyLube, Mervyn's, Office Depot, Ross, Tower Books, WinCo Foods, mall, **W**...76, Burger King, Elmer's Rest., McDonald's, Best Value Inn, Travelodge, Nissan, drugs
20	**Stark St, Washington St, E**...**Elmer's Rest., McMenamin's Café, Newport Bay Café, Tony Roma's, Holiday Inn Express, Cottman Transmissions, Ward's/auto, mall, W**...**7-11, ChuckeCheese, Izzy's Pizza, McDonald's, Stark St Pizza, Taco Bell, Village Inn, Rodeway Inn**
19	US 26, Division St, **E**...HOSPITAL, Exxon, Burger King, Radio Shack, **W**...CHIROPRACTOR, Chevron/24hr, Texaco/diesel/mart, 7-11, Izzy's Pizza, McDonald's, Sizzler, Wendy's, Best Value Inn, Travelodge
17	Foster Rd, **E**...7-11, **W**...Chevron, Texaco, Arby's, Burger King, IHOP, McDonald's, New Copper Penny Grill, Wendy's, Econolodge, HomeBase, Home Depot
16	Johnson Creek Blvd, **W**...Applebee's, Burger King, Carl's Jr, Denny's, Fuddrucker's, Old Chicago Pizza, Old Country Buffet, Outback Steaks, Papa Joe's, Ron's Café, Starbucks, Subway, Taco Bell, BabysRUs, Best Buy, Fred Meyer, HomeBase, Home Depot, Linens'n Things, OfficeMax, PetsMart, Postal Annex, RV Ctrs, Shuck's Parts
14	Sunnyside Rd, **E**...HOSPITAL, 76, Baja FreshMex, Burger King, Domino's, Izzy's Pizza, KFC, McDonald's, McMenamin's Rest., Subway, Taco Time, Thai House, Wendy's, Best Western, Day's Inn, Kinko's, Office Depot, bank, **W**...76/LP/24hr, Chevy's Mexican, Gustav's Grill, Olive Garden, Red Robin Rest., Courtyard, Monarch Hotel/rest., Sunnyside Inn/rest., Barnes&Noble, FabricLand, Gart Sports, JC Penney, Mervyn's, Nordstrom's, Old Navy, Sears/auto, SportMart, Target, ToysRUs, World Mkt, mall
13	OR 224, to Milwaukie, **W**...K-Mart
12	OR 213, to Milwaukie, **E**...MEDICAL CARE, Chevron/24hr, Texaco, 7-11, Denny's, Elmer's, KFC, McDonald's, Skipper's, Taco Bell/24hr, Wendy's, Clackamas Inn, Cypress Inn, Hampton Inn, Fred Meyer, bank, **W**...Comfort Suites
11	82nd Dr, Gladstone, **W**...Arco/mart/24hr, High Rock Steakery, Safeway, Oxford Suites
10	OR 213, Park Place, **E**...HOSPITAL, to Oregon Trail Ctr
9	OR 99E, Oregon City, **E**...HOSPITAL, Arco/repair, 76/mart, KFC, Clackamas Parts, Olds/Pontiac/GMC, Mazda, Subaru, Yamaha, **W**...76, HongKong Express, La Hacienda Mexican, McDonald's, Shari's, Subway, Budget Inn, Rivershore Hotel, Firestone/auto, Fishermans Marine/outdoor, Jo-Ann Fabrics, Michael's, PostNet, Rite Aid, bank
8.5mm	Willamette River
8	OR 43, W Linn, Lake Oswego, **E**...76/mart, museum, **W**...Texaco/diesel
7mm	viewpoint nb, hist marker
6	10th St, W Linn St, **E**...VETERINARIAN, Chevron/LP/mart, 76, Ixtapa Mexican, McDonald's, McMenamin's Café, Shari's/24hr, Thriftway/deli, Willamette Coffeehouse, Oilcan Henry's Lube, Postal Annex, **W**...Bugatti's Pizza, Jack-in-the-Box, Subway, Albertson's, bank, cleaners
4mm	Tualatin River
3	Stafford Rd, Lake Oswego, **W**...HOSPITAL, Wanker's Country Store/phone
0mm	I-205 begins/ends on I-5, exit 288.

N

S

Portland Area

Pennsylvania Interstate 70

Exit #(mm)Services

173mm   Pennsylvania/Maryland state line. **Welcome Ctr wb, full(handicapped)facilities, info, phone, vending, picnic tables, litter barrels, petwalk**

33(171)   US 522 N, Warfordsburg, **N**...Exxon/mart, Tastee Freez, **S**...fireworks

32(166)   PA 731 S, Amaranth, no facilities

31(158)   PA 643, Town Hill, **N**...Texaco/Subway/diesel/ mart/24hr, 4 Seasons Rest., Day's Inn

156mm   **rest area eb, full(handicapped)facilities, phone, picnic tables, litter barrels, vending, petwalk**

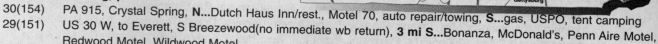

30(154)   PA 915, Crystal Spring, **N**...Dutch Haus Inn/rest., Motel 70, auto repair/towing, **S**...gas, USPO, tent camping

29(151)   US 30 W, to Everett, S Breezewood(no immediate wb return), **3 mi S**...Bonanza, McDonald's, Penn Aire Motel, Redwood Motel, Wildwood Motel

28(150)   US 30, Breezewood, **Services on US 30**...Amoco/TA/diesel/rest./24hr, BP/diesel, Citgo/Dunkin Donuts/mart, Exxon/Stuckey's/diesel/mart, 76/diesel/rest./24hr, Sheetz/mart/24hr, Shell/Blimpie/Perkins/Tacomaker/diesel/ mart, Sunoco/diesel/café, Texaco/mart, Arby's, Bob Evans, Bonanza, Burger King, Dairy Queen, Family House Rest., Fish'n Chips, Hardee's, KFC, Little Caesar's, McDonald's, Pizza Hut, Subway, Taco Bell, Wendy's, Best Western, Breezewood Motel, Comfort Inn/rest., Econolodge, Penn Aire Motel, Quality Inn, Ramada Inn, Village Motel, Wiltshire Motel, camping, museum, truck/tire repair

I-70 W and I-76/PA Turnpike W run together.

For I-70, I-76/PA Turnpike exits 9-11, see PENNSYLVANIA INTERSTATE 76/PA Tpk

27(61)   PA Tpk. I-70 E and I-76/PA Turnpike E run together.

26b a(60)   I-70 W, US 119, PA 66(toll), New Stanton, **N**...Exxon/mart, Sunoco, Texaco/mart, Bob Evans, Chinese Rest., Eat'n Park, KFC, McDonald's/24hr, Donut Chef, Northern Italian Rest., Pagano's Motel/Rest., Pizza Hut, Subway, Wendy's, Budget Inn, Conley Inn, Day's Inn, Howard Johnson/rest., Plaza Hotel, Super 8, bank, **S**...BP/diesel/mart, Citgo/diesel/mart, Sunoco/24hr, cleaners

25A(54)   Madison, **N**...camping, **S**...Citgo/diesel/rest., truckstop/diesel

25(53)   Yukon, no facilities

24b a(51)   PA 31, West Newton, no facilities

23(49)   Smithton, **N**...Citgo/diesel, 76/diesel/rest./24hr, Brass Saddle Rest., **S**...truck repair

22b a(46)   PA 51, Pittsburgh, **N**...CHIROPRACTOR, Exxon/24hr, Budget Host, Howard Johnson, **S**...Texaco/diesel/mart/ 24hr, Holiday Inn, Knotty Pine Motel, 5M's Motel

21(44)   Arnold City, no facilities

20b a(43)   PA 201, Fayette City, **N**...gas/mart, **S**...PODIATRIST, Exxon/mart/24hr, Sunoco/mart/24hr, Texaco/mart/24hr, Burger King, Chinese Rest., Denny's, Hoss' Rest., KFC, LJ Silver, McDonald's, Pizza Hut, Subway, Best Val-U Motel, Advance Parts, Foodland, Giant Auto Parts, Jo-Ann Fabrics, Kelly Parts, K-Mart/auto, Lowe's Bldg Supply, Pearle Vision, Radio Shack, Thrift Drug, Wal-Mart SuperCtr/24hr, cleaners/laundry, bank, hardware

19A(42)   Monessen, no facilities

19(41.5)   N Belle Vernon, **S**...BP/mart, Sunoco, Hardee's

18(41)   PA 906, Belle Vernon, no facilities

40mm   Monongahela River

17(39)   PA 88, Charleroi, HOSPITAL

16(38)   Speers, **N**...Loraine's Rest., **S**...Exxon, camping

15b a(37)   PA 43, no facilities

14(36)   Lover(from wb, no re-entry), no facilities

13(35)   PA 481, Centerville, no facilities

12b a(32)   PA 917, Bentleyville, **N**...camping, **S**...MEDICAL CARE, DENTIST, Amoco/diesel, Pilot/Subway/diesel/mart/ 24hr, Sheetz/mart/24hr, Hardee's, McDonald's, Advance Parts, Ford, Giant Eagle Foods, Rite Aid, 1-Hr Photo

11(30)   to PA 136, Kammerer, **N**...Carlton Motel/rest.

10(27)   Dunningsville, **S**...Avalon Motel/rest.

9(25)   PA 519, to Eighty Four, **S**...Amoco/mart/24hr, Citgo/diesel/mart

21mm   I-79 S, to Waynesburg. **I-70 W and I-79 N run together 3.5 mi.**

8(19)   PA 136, Beau St, **N**...KOA, **S**...to Washington&Jefferson Coll

E

W

Pennsylvania Interstate 70

E

Washington

7b a(18) US 19, Murtland Ave, **N**...Applebee's, Cracker Barrel, McDonald's, Texas Roadhouse, TGIFriday, SpringHill Suites, Dick's Sports, $Tree, Giant Eagle Foods, GNC, Lowe's Bldg Supply, Michael's, OfficeMax, PetsMart, Radio Shack, Sam's Club, Target, Toyota/Honda, Wal-Mart SuperCtr/24hr, bank, **S**...HOSPITAL, Amoco/mart, BP/diesel/mart/wash, Crossroads/diesel, Exxon, Sunoco/mart, Arby's, Bob Evans, Big Boy, Burger King, Donut Connection, Eat'n Park/24hr, Evergreen Chinese, Jade Garden Chinese, KFC, LJ Silver, Old Mexico, Papa John's, Pizza Hut, Shoney's, Taco Bell, Hampton Inn, Motel 6, Buick/Olds/Cadillac/GMC, $General, Firestone/auto, Home Depot, JC Penney, Jo-Ann Fabrics, K-Mart, Midas Muffler, Monro Muffler, Staples, Subaru, ToysRUs, cinema, cleaners, mall, st police

17.5mm I-79 N, to Pittsburgh. **I-70 E and I-79 S run together 3.5 mi.**

6(17) PA 18, Jefferson Ave, Washington, **S**...Citgo, Texaco/diesel, CoGo's/mart, Burger King, FourStar Pizza, Dairy Queen, McDonald's, Foodland, Rite Aid, groceries

5(16) Jessop Place, no facilities

4(15) US 40, Chesnut St, Washington, **N**...DENTIST, Bonanza, StoneCrab Inn Rest., carwash, **S**...Amoco, BP/mart, Exxon/mart/24hr, Sunoco/24hr, Texaco, Big Boy, Dairy Queen, Denny's, Donut Connection, Eat'n Park/24hr, Hardee's/24hr, LJ Silver, McDonald's, Pizza Hut, Pizza Inn, Taco Bell, Wendy's, Best Western, Century Plaza Motel, Day's Inn, Interstate Motel, Ramada Inn, Red Roof Inn, Sears/auto, bank, cleaners, hardware, mall

3(11) PA 221, Taylorstown, **N**...Amoco/diesel, truck repairs

2(7) to US 40, to PA 231, Claysville, **N**...Exxon/mart/24hr, hubcaps **S**...TA/Citgo/Sbarro's/diesel/rest./24hr, King's Rest., yogurt, camping

5mm **Welcome Ctr eb, full(handicapped)facilities, phone, vending, picnic tables, litter barrels, petwalk**

1(1) W Alexander, **S**...Texaco/mart

0mm Pennsylvania/West Virginia state line

W

Pennsylvania Interstate 76

E

Philadelphia

Exit #(mm)Service

354mm Delaware River, Walt Whitman Bridge, state line

47(353) I-95, N to Trenton, S to Chester

46(352) Packer Ave, **S**...Holiday Inn, to sports complex

45(351) to I-95, PA 611, Broad St

44(350) PA 291, W to Chester, Ramada Inn

43a(349) to I-95 S

b Passyunk Ave, Oregon Ave, sports complex

42(348) 28th St

41(347) Gray's Ferry Ave(from eb), University Ave, civic center, **S**...Amoco

40(346) South St, no facilities

39(345) 30th St, Station Market St,

38(345) I-676, to Philadelphia(no return from eb)

37(344) Spring Garden St, Haverford

36(343) US 13, US 30 W, Girard Ave, Philadelphia Zoo, E Fairmount Park

35(342) Montgomery Dr, W River Dr, W Fairmount Park

34(341) US 1 N, Roosevelt Blvd, to Philadelphia

33(340) US 1 S, **S**...Holiday Inn

32(339) Lincoln Dr, Kelly Dr, to Germantown

31(339) Belmont Ave, Green Lane, **S**...Sunoco/mart

30(338) Hollow Rd

29(331) PA 23, Conshohocken(from wb)

28(331) I-476, Conshohocken, to Chester

27(330) PA 320, Gulph Mills, no facilities, access to Villanova U

26(327) US 202 N, to King of Prussia, **N**...Exxon/diesel/mart, Mobil/mart, Sizzler, WaWa Mart, Charlie's Place Rest., Chinese Rest., Dunkin Donuts, Friendly's, Pizzaria Uno, TGIFriday, Econolodge, Holiday Inn, Howard Johnson, McIntosh Inn, JC Penney, Macy's, Sears/auto, Thrift Drug, bank, cinema, mall, **S**...George Washington Lodge

25(327) US 202 S, Goddard Blvd, to Valley Forge, Valley Forge Park

24(326) I-76 wb becomes I-76/PA Tpk to Ohio

For I-76 westbound to Ohio, see I-76/PA Turnpike.

W

Pennsylvania Interstate 76/Turnpike

Exit #(mm)Services

PA Tpk runs wb as I-276

30	Delaware River Bridge, Pennsylvania/New Jersey state line
29	US 13, Delaware Valley
352mm	**North Neshaminy Travel Plaza wb...Welcome Ctr, Sunoco/diesel/mart/24hr, Burger King, Nathan's, TCBY, gifts**
28	US 1, to Philadelphia
27	PA 611, Willow Grove
26	PA 309, Ft Washington
25A	PA Tpk NE Extension, I-476 S
25	Germantown Pike, to Norristown
328mm	**service plaza wb...Sunoco/diesel/mart/24hr, Hardee's**

PA Tpk runs eb as I-276, wb as I-76

27(330) PA 327, Gulph Mills

26(327) US 202, to King of Prussia, N...Exxon/diesel/mart, Mobil/mart, WaWa Mart, Charlie's Place Rest., Chinese Rest., Dunkin Donuts, Friendly's, Pizzaria Uno, Sizzler, TGIFriday, Econolodge, Holiday Inn, Howard Johnson, McIntosh Inn, JC Penney, Macy's, Sears/auto, Thrift Drug, bank, cinema, mall, S...George Washington Lodge

25(327) US 202, Goddard Blvd, to Valley Forge, Valley Forge Park

24(326) I-76 E, to US 202, to I-476, to Valley Forge, Hoss' Steak/seafood

325mm Valley Forge Travel Plaza eb...Sunoco/diesel/mart/24hr, Burger King, Mrs Fields, TCBY, gifts

23(312) PA 100, to Downingtown, Pottstown, N...Mobil/mart, S...HOSPITAL, Gulf, Sunoco/diesel/mart/24hr, WaWa Mart, Hoss' Steak/seafood, Lion's Share Rest., Comfort Inn, Hampton Inn, Holiday Inn, bank, carwash

305mm Brandywine Travel Paza wb...Sunoco/diesel/mart/24hr, Roy Rogers, Mrs Fields, Nathan's, Sbarro's, TCBY, gifts

22(298) I-176, PA 10, Morgantown, to Reading, HOSPITAL, N...Economy Lodge, Heritage Motel/rest., S...McDonald's, Holiday Inn, Manufacturers Outlet Mall

290mm Bowmansville Travel Plaza eb...Sunoco/diesel/mart/24hr, Mrs Fields, Taco Bell, TCBY, gifts

21(286) US 322, PA 272, to Reading, Ephrata, HOSPITAL, info, N...VETERINARIAN, Citgo/mart/Penn Amish Motel, Exxon/diesel, Texaco/diesel/mart, Zia Maria Italian, Zinn's Diner, Black Horse Lodge/rest., Dutchland Motel, Penn Dutch Motel, camping, S...Country Pride Drive-In, Econolodge, Holiday Inn, antiques

20(267) PA 72, to Lebanon, to Lancaster, HOSPITAL, N...Texaco/diesel/mart, Little Corner of Germany Cafe, S...Amoco/mart, Penn's Pantry Rest., Friendship Inn, L&L Motel, Mt Hope Motel

259mm Lawn Travel Plaza wb...Sunoco/diesel/mart/24hr, Burger King, Mrs Fields, TCBY, gifts

250mm Highspire Travel Plaza eb...Sunoco/diesel/mart/24hr, Sbarro's, gifts

19(247) I-283, PA 283, Harrisburg East, to Harrisburg, Hershey, N...Best Trkstp/diesel/mart, Getty Gas, Eat'n Park, Wendy's, Congress Inn/Angie's Rest., Day's Inn, Rodeway Inn, facilities 3-5 mi N

246mm Susquehannah River

18(242) I-83, Harrisburg West, HOSPITAL, N...VETERINARIAN, BP/mart, Exxon/diesel/mart, Bob Evans, Eat'n Park, Pizza Hut, Best Western, Day's Inn/rest., Fairfield Inn, Holiday Inn, Knight's Inn, McIntosh Inn, Motel 6

17(236) US 15, Gettysburg Pike, to Gettysburg, Harrisburg, HOSPITAL, N...VETERINARIAN, Texaco, Subway, Amber Inn, Econolodge, Hampton Inn/rest., Homewood Suites, U-Haul, cleaners, S...Sheetz/mart/24hr, Hoss' Steaks(1mi), McDonald's(1mi), Best Western, Wingate Inn, Giant Food/24hr(1mi)

16(226) US 11, to I-81, Carlisle, to Harrisburg, N...HOSPITAL, Amoco/diesel/mart/24hr, BP/mart, Citgo/diesel/24hr, Flying J/diesel/rest/24hr, Pilot/Wendy's/diesel/mart/24hr, Shell/diesel/mart, Texaco/diesel/mart, All American Rest., Arby's, BBQ, Big Boy, Bob Evans, Carelli's Subs/pizza, Eat'n Park, Hardee's/24hr, Dunkin Donuts, McDonald's, Middlesex Diner/24hr, Subway, Western Sizzlin, Appalachian Trail Inn, Best Western, Budget Host, Comfort Inn, Econolodge, Ember's Inn/rest., Hampton Inn, Holiday Inn, Quality Inn, Rodeway Inn, Super 8, ThriftLodge, Blue Beacon, S...HOSPITAL, VETERINARIAN, Exxon/Taco Bell/mart/24hr, Sheetz/mart/24hr, Hoss' Steak/seafood, Motel 6, Pike Motel, Chrysler/Jeep, U-Haul

219mm Plainfield Travel Plaza eb...Sunoco/diesel/mart/24hr, Roy Rogers, TCBY, gifts

203mm Blue Mountain Travel Plaza wb...Sunoco/diesel/mart/24hr, Mrs Fields, Roy Rogers, TCBY, gifts

15(202) PA 997, Blue Mountain, to Shippensburg, S...Johnnie's Motel/rest., Kenmar Motel

199mm Blue Mountain Tunnel

197mm Kittatinny Tunnel

E ↕ W

Pennsylvania Interstate 76/Turnpike

B r e e z e w o o d

14(189) PA 75, Willow Hill, **S**...Willow Hill Motel/rest.

187mm Tuscarora Tunnel

13(180) US 522, Ft Littleton, **N**...HOSPITAL, Amoco, Texaco/mart, The Fort Rest., Downs Motel

172mm Sideling Hill Travel Plaza both lanes...Sunoco/diesel/mart/24hr, Big Boy, Burger King, Mrs Fields, Pretzel Time, TCBY, gifts

12(161) US 30, Breezewood, **Services on US 30**...Amoco/TA/diesel/rest./24hr, BP/diesel, Citgo/Dunkin Donuts/mart, Exxon/Stuckey's/diesel/mart, 76/diesel/rest./24hr, Sheetz/mart/24hr, Shell/Blimpie/Perkins/Tacomaker/diesel/mart, Sunoco/diesel/café, Texaco/mart, Arby's, Bob Evans, Bonanza, Burger King, Dairy Queen, Family House Rest., Fish'n Chips, Hardee's, KFC, Little Caesar's, McDonald's, Pizza Hut, Subway, Taco Bell, Wendy's, Best Western, Breezewood Motel, Comfort Inn/rest., Econolodge, Penn Aire Motel, Quality Inn, Ramada Inn, Village Motel, Wiltshire Motel, camping, museum, truck/tire repair

161mm I-70 W and I-76/PA Turnpike W run together

150mm roadside park wb

148mm Midway Travel Plaza both lanes, eb...Sunoco/diesel/mart/24hr, Mrs Fields, Sbarro's, TCBY, gifts, wb...Sunoco/diesel/mart/24hr, KFC, Mrs Fields, Sbarro's, TCBY, gifts

11(147) US 220, Bedford, HOSPITAL, Amoco, Exxon/diesel, Sunoco, Burger King, Mr Donut, Sbarro's, Pizza Hut, Best Western, Quality Inn, Terrace Motel, camping

142mm parking area wb

123mm Allegheny Tunnel

112mm Somerset Travel Plaza both lanes, eb...Sunoco/diesel/mart/24hr, Roy Rogers, TCBY, gifts, wb...Sunoco/diesel/mart/24hr, Burger King, TCBY, gifts

10(110) PA 601, Somerset, Amoco/diesel/rest., BP, Burger King, Dunkin Donuts, KFC, McDonald's, Pizza Hut, Summit Diner, Highlands Motel, Holiday Inn, Knight's Inn, Quality Inn, outlet shopping, tires

9(91) PA 711, Donegal, HOSPITAL, Exxon, Sunoco, Candlelight Family Rest., Dairy Queen, pizza, Sandman Motor Lodge, Hidden Valley Golf, camping

78mm service plaza wb...Sunoco/diesel/mart/24hr, McDonald's, Kings Family Rest.

I-70 E runs with I-76/PA Turnpike eb

8(75) I-70 W, US 119, PA 66(toll), New Stanton, **S**...BP/diesel/mart, Citgo/diesel/mart, Exxon/mart, Sunoco/mart/24hr, Texaco/mart, Chinese Rest., Donut Chef, Eat'n Park, KFC, McDonald's/24hr, Northern Italian Rest., Pet&Ernie's Steaks, Pizza Hut, Wendy's, Subway, Wendy's, Budget Inn, Conley Inn, Day's Inn, Howard Johnson/rest., Knight's Court, Pagano's Motel/rest., Plaza Hotel, Super 8, bank

74.6mm service plaza eb...Sunoco/diesel/mart/24hr, Breyer's, McDonald's, ATM

7(67) US 30, Irwin, to Greensburg, HOSPITAL, **N**...BP/mart/service, Knight's Court, auto parts, laundry, **S**...Gulf, SuperAmerica, Angelo's Rest., Arby's, Bob Evans, Burger King, KFC, LJ Silver, McDonald's/24hr, Taco Bell, Teddy's Rest., Vincent's Pizza, Wendy's, Conley Inn, Holiday Inn Express, Ames, Ford, Giant Eagle Foods/24hr, JiffyLube, Shop'n Save, Speedy Mufflers, bank, cinema

61mm parking area eb, picnic table, litter barrel

6(56) I-376, US 22, Monroeville, to Pittsburgh, **1-2 mi N**...Gulf, Sunoco/mart, Birdie's Rest., Jaden's Family Dining, East Exit Motel, Acura, CVS Drug, Giant Eagle Foods, Hyundai, Mazda, tires, **S**...HOSPITAL, BP, Exxon, Gulf/diesel, Big Boy, Burger King, Chinese Rest., Dunkin Donuts, McDonald's, Pizza Hut, Taco Bell, Day's Inn, Holiday Inn, Palace Inn, Red Roof Inn, William Penn Motel, Buick/Jeep/Eagle, Infiniti, Saturn, bank, to Three Rivers Stadium

49mm service plaza eb...Sunoco/diesel/mart/24hr, Arby's/Sub Shop, picnic tables, litter barrels, phone

48.5mm Allegheny River

P i t t s b u r g h

5(48) PA 28, Allegheny Valley, to Pittsburgh, New Kensington, **N**...Amoco/wash/24hr, Italian Rest., **S**...Exxon/diesel/mart, Bob Evans, Burger King, Denny's, Eat'n Park, Fuddrucker's, KFC, King's Family Rest., McDonald's, Ponderosa, Taco Bell, Wendy's, Comfort Inn, Day's Inn, Holiday Inn, Super 8, Valley Motel, Ames, Ford, Giant Eagle/drugs/24hr, bank, cinema

41mm parking area/call box eb

4(39) PA 8, Butler Valley, to Pittsburgh, HOSPITAL, **N**...CHIROPRACTOR, DENTIST, BP, Exxon/diesel, Sheetz/mart/24hr, Chinese Rest., Eat'n Park, Ponderosa, Pittsburgh North Motor Lodge/rest., Ames, CVS Drug, Jo-Ann Fabrics, JiffyLube, Goodyear/auto, Shop'n Save Foods, auto parts, cinema, mall, **S**...DENTIST, VETERINARIAN, Sunoco, Arby's, Baskin-Robbins, Burger King, Denny's, Dunkin Donuts, LJ Silver, McDonald's, Pizza Hut, Subway, Taco Bell, Buskey's Motel, Conley Inn, Day's Inn, BP ProCare, Goodyear/auto, Midas Muffler, Thrift Drug, Valvoline, Woolworth's, bank, carwash, laundry

E

W

Pennsylvania Interstate 76/Turnpike

E

W

31mm   **Butler Travel Plaza wb...Sunoco/diesel/mart/24hr, Burger King, Mrs Fields, Popeye's, TCBY, gifts**

3(28)   to I-79, to Cranberry, Pittsburgh, **N...HOSPITAL, CHIROPRACTOR,** Amoco/mart/wash/24hr, Exxon/diesel/mart, Sheetz/mart, Boston Mkt, Burger King, Climo's Pizza/chicken, Denny's, Dunkin Donuts, Hardee's, Italian Rest., LJ Silver, Shoney's, Wendy's, Conley Inn, Fairfield Inn, Hampton Inn, Giant Eagle Food/drug, Goodyear/atuo, GreaseMonkey Lube, Jo-Ann Fabrics, K-Mart, MailBox Etc, Meineke Muffler, Phar-Mor Drug, Shop'n Save/24hr, Wal-Mart, bank, camping, cinema, mall, **S...**Atlantic/mart, BP/diesel/mart, Gulf/mart, Sunoco/24hr, Arby's, Bob Evans, Eat'n Park, Hot Dog Shop, McDonald's, Perkins/24hr, Day's Inn, Holiday Inn Express, Motel 6, Red Roof Inn, Sheraton, Toyota

23mm   picnic area eb, tables, litter barrels

22mm   **Zelienople Travel Plaza eb...Welcome Ctr, Sunoco/diesel/mart/24hr, Roy Rogers, Mrs. Fields, TCBY, crafts, gifts**

17mm   parking area eb

13.4mm   parking area eb

13mm   Beaver River

2(12)   PA 8, Beaver Valley, to Ellwood City, Beaver Falls, HOSPITAL, **N...**Atlantic Gas(1mi), Beaver Valley Motel, Danny's Motel, Hill Top Motel, Holiday Inn, Lark Motel, **S...**Giuseppe's Italian, Conley Motel/rest., auto repair, **2 mi S...**Big Beaver Rest., Wendy's(3mi), Comet Foods

1A(9)   PA 60(toll), to New Castle, Pittsburgh, **S...**services(6mi), to airport

6mm   picnic area eb

2mm   truck pulloff eb

1.5mm   toll plaza, phone, call boxes located at 1 mi intervals

0mm   Pennsylvania/Ohio state line

Pennsylvania Interstate 78

Exit #(mm)Services

77   Delaware River, Pennsylvania/New Jersey state line

76mm   **Welcome Ctr wb, full(handicapped)facilities, phone, vending, picnic tables, litter barrels, petwalk,** toll booth wb

E

22(75)   to PA 611, Easton, **N...**Citgo/mart/24hr, McDonald's, Perkins, Best Western, Laneco Foods, **S...**Shell/mart

21(66)   PA 412, Hellertown, **N...**HOSPITAL, Citgo/mart/24hr, Comfort Suites(3mi), Wendy's, **S...**Exxon/Subway/mart, Burger King, McDonald's, Vassi's Drive-In, Chevrolet

20b a(61)   PA 309 S, Quakertown, **S...**The Motor Lodge

19(59)   to PA 145, Summit Lawn, no facilities

18(57)   Emmaus Ave, Lehigh St, **N...**Hess/diesel/mart, Arby's, Chinese Rest., Denny's, Queen City Diner, Willy Joe's Rest., Day's Inn, Aldi Foods, Laneco Foods, Dodge/Kia/Isuzu, Ford/Lincoln/Mercury, STS Tire/auto, **S...**Getty Gas/repair, Mobil, Texaco/repair, Brass Rail Rest., Burger King, Dunkin Donuts, Friendly's, McDonald's, Perkins, Pancho&Sonny's Mexican, Subway,Taco Bell, Wendy's, AA Auto Parts, Buick/Jeep, Chevrolet, Chrysler, Food4Less, Honda/Hyundai, JiffyLube, MailBoxes Etc, Mazda/Volvo, Midas Muffler, PetCo, PharMor, Pontiac/Audi/Mercedes, Saturn, Staples, Weis Foods

17(56)   PA 29, Cedar Crest Blvd, **S...**HOSPITAL

16(55)   US 222, Hamilton Blvd, **N...**DENTIST, Ambassador Rest., Boston Mkt, Burger King, Ice Cream World, Perkins, TGIFriday, Wendy's, Comfort Suites/rest., Holiday Inn Express, Dorney Funpark, King's Food/drug, Shipping+, cleaners, **S...**Sunoco/wash, Charcoal Drive-In, Pizza Hut, Tom Sawyer Diner/24hr, Subaru

15(52)   to I-476, US 22 E, PA 33 N, Whitehall

W

14(50)   PA 100, Fogelsville, **N...**Arby's, Chinese Rest., Cracker Barrel, Gyro's King, LJ Silver, Pizza Hut, Comfort Inn, Hotel-Suites, Eckerd, STS Tire/repair, bank, **S...**Exxon/Subway/mart, Sunoco/mart/24hr, Burger King, Starlite Buffet, Taco Bell, Yocco's Hotdogs, Cloverleaf Inn, Hampton Inn, Holiday Inn, Holiday Inn Express(4mi), Toyota, st police

Pennsylvania Interstate 78

13(45) PA 863, to Lynnport, **N**...Texaco/diesel/mart, **S**...Super 8

12(40) PA 737, Krumsville, **N**...Top Motel, **Pine Hill Campground**, **S**...Texaco/diesel/mart, Skyview Rest., RV Camping(12mi), to Kutztown U

11(36) PA 143, Lenhartsville, **3 mi S**...RV Camping

10(30) Hamburg, **S**...gas

9(29) PA 61, to Reading, Pottsville, **N**...Cracker Barrel, Wendy's, Harley-Davidson(8mi), st police, **S**...Pot O' Gold Diner

8(23) Shartlesville, **N**...Citgo/Stuckey's/Dairy Queen, Texaco/diesel, K&M Burger Ranch, Dutch Motel, Appalachian Campsites, **S**...Blue Mtn Family Rest., Haag's Motel/rest., Fort Motel/rest., antiques

7(19) PA 183, Strausstown, **N**...Mobil/Dutch Kitchen Rest.

6(17) PA 419, Rehrersburg, **N**...gas

5(16) Midway, **N**...Exxon/diesel/mart, Midway Diner, Comfort Inn, **S**...auto/truck repair

4(15) Grimes, **S**...Midway Motel

3(13) PA 501, Bethel, **N**...Texaco/diesel/mart, **S**...Amoco, Exxon/diesel, Western Star/repair

2(11) PA 645, Frystown, **N**...Fairview Motel, **S**...Shell/diesel/mart/rest./24hr, Texaco/diesel/mart/24hr, Frystown Motel

1(6) PA 343, Fredricksburg, **S**...Esther's Rest.(1mi), Whse Mkt/fuel(1mi), KOA(3mi)

0mm I-78 begins/ends on I-81, 89mm.

Pennsylvania Interstate 79

Exit #(mm)Services

I-79 begins/ends in Erie, exit 44.

44(183) PA 5, 12th St, Erie, **E**...HOSPITAL, **W**...BP/mart/wash, Sunoco, Blimpie, Bob Evans, Eat'n Park/24hr, Applebee's, KFC, Hooters, McDonald's, Serafini's Italian, Taco Bell, Hampton Inn(2mi), Advance Parts, Big Lots, CVS Drug, Dunn Tire, Eckerd, Erie Tire, Giant Eagle/deli, GNC, Monro Muffler/brake, PharMor, PostNet, ValueCity, to Presque Isle SP, bank

43(182) US 20, 26th St, **E**...HOSPITAL, CHIROPRACTOR, Citgo/mart, KwikFill/mart, Dee's Deli, Haggerty's Diner, Marco's Pizza, Advance Parts, CVS Drug, Family$, Pennzoil, Tops Food/24hr, antiques, repair, **W**...BP/mart, Citgo/mart, Arby's, Burger King, Domino's, Hoss' Rest., Little Caesar's, McDonald's, Pizza Hut, Subway, Taco Bell, Aamco, CVS Drug, Eckerd, Ford, Giant Eagle/deli, Goodyear/auto, K-Mart/auto, Radio Shack, TrueValue, bank, laundry

41(180) US 19, to Kearsarge, **E**...HOSPITAL, Arby's, ChiChi's, Don Pablo's, Eat'n Park/24hr, KFC, LoneStar Steaks, Max&Erma's, McDonald's, Olive Garden, Outback Steaks, Ponderosa, Red Lobster, Red River Roadhouse, Wendy's, Ames, Baby Depot, Barnes&Noble, Borders Books, Burlington Coats, Cadillac/Olds, Chrysler/Plymouth/Jeep, HomePlace, Isuzu/Audi, JC Penney, Kaufmann's, KidsRUs, Kinko's, Lazarus, LensCrafters, Luxury Linens, MensWearhouse, Michael's, Sears/auto, SteinMart, TJ Maxx, ToysRUS, cinema, mall, **W**...Citgo/diesel/mart

40(177) I-90, E to Buffalo, W to Cleveland, no facilities

39(174) to McKean, **E**...access to gas/diesel, phone, **W**...camping

38(166) US 6N, to Edinboro, **2-3 mi E**...Burger King, McDonald's, Perkins, Taco Bell, Edinboro Inn, Ramada Inn, **W**...RV camping

163mm rest area both lanes, full(handicapped)facilities, phone, vending, picnic tables, litter barrels, petwalk

37(154) PA 198, to Saegertown, **W**...Citgo/repair

36(147) US 6, US 322, to Meadville, **E**...HOSPITAL, VETERINARIAN, Citgo/diesel/mart, Marathon/mart/wash, Sheetz/diesel/mart/LP/24hr, Sunoco, Applebee's, Big Boy, Cracker Barrel, Dairy Queen, Hoss' Rest., Perkins/24hr, Pizza Hut, Sandalini's Buffet, Day's Inn/rest., Holiday Inn Express, Motel 6, Advance Parts, Bean's Parts, Buick/Pontiac, Giant Eagle/24hr, Peebles, Value City, carwash, repair, **W**...Burger King, King's Rest./24hr, McDonald's, Ponderosa, Red Lobster, Super 8, AutoZone, Chevrolet, $Tree, FashionBug, Goodyear/auto, K-Mart/auto, Nissan, Olds/Cadillac/GMC, SavALot, Staples, Toyota, Wal-Mart SuperCtr/24hr, bank, repair, to Pymatuning SP

35(141) PA 285, to Geneva, **E**...to Erie NWR(20mi), **W**...Gulf/mart, Aunt Bee's Rest., repair

135mm rest area/weigh sta both lanes, full(handicapped)facilities, phone, vending, picnic tables, litter barrels, petwalk

34(130) PA 358, to Sandy Lake, **E**...Lakeway/diesel/rest./atm, Flag Inn, Lakeview Inn, House of Cheese, to Goddard SP, camping, **W**...HOSPITAL, gas/diesel/mart, to Pymatuning SP

33(121) US 62, to Mercer, **E**...Citgo/mart, **W**...Sunoco/diesel/mart, st police

32(117) I-80, E to Clarion, W to Sharon, no facilities

31(113) PA 208, PA 258, to Grove City, **E**...HOSPITAL, BP/Blimpie/mart/atm, Citgo/diesel/mart/24hr, **W**...Pennzoil/mart/24hr, Sunoco/Subway/diesel/mart/24hr, Eat'n Park/24hr, Elephant&Castle Rest., FoodCourt, Hoss' Rest., King's Rest., McDonald's, Wendy's, Amerihost, Comfort Inn, Super 8, KOA(3mi), Prime Outlets/famous brands

E
↑
↓
W

Erie

N
↑
↓
S

Meadville

388

Pennsylvania Interstate 79

110mm **rest area sb, full(handicapped)facilities, phone, vending, picnic tables, litter barrels, petwalk**

107mm **rest area nb, full(handicapped)facilities, phone, vending, picnic tables, litter barrels, petwalk**

30(105) PA 108, to Slippery Rock, **E**...Pennzoil/mart, Dairy Queen, Apple Butter Inn(3mi), Evening Star Motel/rest., Slippery Rock Camping, to Slippery Rock U

29(99) US 422, to New Castle, **E**...to Moraine SP, camping, **W**...Coopers Lake Camping, to Rose Point Camping

28(96) PA 488, Portersville, **E**...Bear Run Camping, **US 19**...gas/diesel, food, lodging

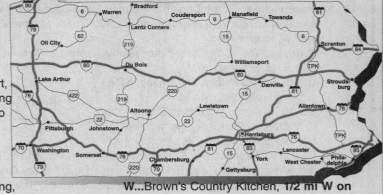

W...Brown's Country Kitchen, **1/2 mi W on**

27(88) to US 19, PA 68, Zelienople, **1-2 mi W**...CHIROPRACTOR, Exxon/diesel/24hr, Burger King, Dairy Queen, Pizza Hut, Subway, Zelienople Motel, Bilo Foods/deli, RV Ctr, Goodyear, Indian Brave RV Camping, carwash, drugs, laundry

26(85) PA 528(no quick return), to Evans City, **E**...GoCarts, auto auction, **W**...Ford, Olds/Pontiac/Buick

81mm picnic area both lanes, phone, picnic table, litter barrel

25(78) US 19, PA 228, to Mars, access to I-76, PA TPK(exits left from nb), **E**...Staples, Pittsburgh North Camping, **W**...CHIROPRACTOR, DENTIST, MEDICAL CARE, VETERINARIAN, Amoco/mart/24hr, BP/diesel/mart, Exxon/diesel/24hr, Gulf, 76/wash, Sheetz/mart/24hr, Sunoco/mart/24hr, Arby's, Big Boy, Bob Evans, Boston Mkt, Burger King, Chen Chinese, Denny's, Dunkin Donuts, Eat'n Park, Einstein Bagels, Italian Oven, Hardee's, Hartners Rest., Hotdog Shoppe, King's Rest., LoneStar Steaks, LJ Silver, Max&Erma's, McDonald's, Montecello's Italian, Ogle Sta Cafe, Perkins, Roma Pizza, Shoney's, Subway, TCBY, Wendy's, Comfort Inn, Conley Inn, Fairfield Inn, Hampton Inn, Holiday Inn Express, Motel 6, Oak Leaf Motel, Red Roof Inn, Sheraton, Ames, Barnes&Noble, CarQuest, Dick's Sports, FashionBug, GNC, Giant Eagle, Goodyear/auto, Home Depot, HomePlace, Jo-Ann Fabrics, Kinko's, K-Mart/drugs, MailBoxes Etc, MailStop, Meineke Muffler, Michael's, OfficeMax, Pennzoil, PepBoys, Phar-Mor, Pier 1, Shop'n Save/24hr, Toyota, ToysRUs, Wal-Mart/drugs, bank, cinema, cleaners, laundry, mall

23(75) US 19 S(from nb), to Warrendale, services same as 25 along US 19

22(73) PA 910, to Wexford, **E**...MEDICAL CARE, DENTIST, BP/mart/atm/24hr, King's Family Rest./24hr, Best Inn, **W**...Exxon/diesel/mart/24hr, Carmody's Rest.

21(71) I-279 S(from sb), to Pittsburgh

20(68) Mt Nebo Rd, **W**...HOSPITAL, gas, phone

19(67) to PA 65, Emsworth, no facilities

18(65) to PA 51, Coraopolis, Neville Island, **E**...Gulf/mart, Penske/Hertz Trucks

64.5mm Ohio River

17(64) PA 51(from nb), to Coraopolis, McKees Rocks

16(60) PA 60, Crafton, **E**...HOSPITAL, Exxon/diesel/mart/24hr, King's Rest./24hr, Day's Inn, Econolodge, Motel 6, **W**...CHIROPRACTOR, Sunoco/24hr, Primati Bros Rest., Pittsburgh Motel, Red Roof Inn, airport, auto parts

15(59) US 22 W, US 30(from nb), **W**...airport

14(58) I-279 N, to Pittsburgh

13(56) to Carnegie, **1-3 mi E**...BP, Exxon, LJ Silver, McDonald's, TCBY, Shop'n Save, Ford, mall

12(54) to Heidelberg, **1-2 mi E**...HOSPITAL, CHIROPRACTOR, DENTIST, Sunoco, Arby's, ChuckeCheese, Dairy Queen, Eat'n Park/24hr, Grand China, KFC, Little Caesar's, Old Country Buffet, Pizza Hut, Ponderosa, Subway, Taco Bell, Wendy's, Ames, Eckerd, Firestone/auto, Giant Eagle/drugs/24hr, Goodyear/auto, JC Penney, Jo-Ann Fabrics, K-Mart, Meineke Muffler, PharMor, Procare/auto, Radio Shack, Service Merchandise, Speedy Muffler, TJ Maxx, Valvoline, carwash, laundry, mall, **W**...LJ Silver, McDonald's

11(53) PA 50, to Bridgeville, **E**...HOSPITAL, CHIROPRACTOR, DENTIST, BP/diesel/mart, Burger King, King's Rest./24hr, McDonald's, Wendy's, Chevrolet, Dodge, Midas Muffler, bank, laundry, parts/repair, **W**...Exxon/diesel/mart/24hr, Texaco, Knights Inn, Firestone/auto

50mm **rest area/weigh sta both lanes, full(handicapped)facilities, phone, vending, picnic tables, litter barrels, petwalk**

10a(48) South Pointe, no facilities

10(45) to PA 980, Canonsburg, **E**...Sheetz/mart, **W**...Amoco, BP/mart/24hr, Hoss' Rest., KFC/Taco Bell, LJ Silver, McDonald's, Pizza Hut, Wendy's, Super 8, Advance Parts, U-Haul, auto repair

N

↕

S

Pennsylvania Interstate 79

Washington

9(43)   PA 519, Houston, **E**...Amoco/diesel, **W**...HOSPITAL, Sunoco/24hr

8b(42.5)   Racetrack Rd, **E**...McDonald's, Comfort Inn, Holiday Inn, **W**...BP/diesel/mart, Trolley Museum

8a(42)   Meadow Lands(from nb, no re-entry), **W**...Trolley Museum(3mi), golf, racetrack

38mm   I-70 W, to Wheeling

I-79 and I-70 run together 4 mi. See Pennsylvania Interstate 70, exits 7-8.

34mm   I-70 E, to Greensburg

7(33)   US 40, to Laboratory, **W**...Amoco, KOA

31mm   parking area/weigh sta sb

6(30)   US 19, to Amity, **W**...Exxon/Subway/mart

5(23)   to Marianna, Prosperity, no facilities

4(19)   US 19, to Ruff Creek, **W**...Amoco/mart

3(14)   PA 21, to Waynesburg, **E**...Amoco/mart/24hr, Comfort Inn, **W**...HOSPITAL, DENTIST, BP/diesel/mart/24hr, Citgo/diesel/mart, Exxon/Taco Bell/diesel, Sheetz/mart, Dairy Queen, Golden Corral, Hardee's, KFC, LJ Silver, McDonald's, Wendy's, Econolodge, Super 8, Ames, Chevrolet, CVS Drug, FashionBug, Shop'n Sav/deli, bank, st police

2(7)   to Kirby, no facilities

6mm   **Welcome Ctr/weigh sta nb, full(handicapped)facilities, phone, picnic tables, litter barrels, vending, petwalk**

1(1)   Mount Morris, **E**...Citgo/diesel/mart/rest., Honda/Mazda, **W**...BP/diesel/mart, Marathon/mart, Mt Morris Campground

0mm   Pennsylvania/West Virginia state line

Pennsylvania Interstate 80

Stroudsburg

Exit #(mm)Services

311mm   Delaware River, Pennsylvania/New Jersey state line

310.5   toll booth wb, phone

53(310)   PA 611, Delaware Water Gap, **S...Welcome Ctr/rest area, full facilities, info**, Amoco/mart/atm, Gulf/repair, Skillet Rest., Water Gap Diner, Glenwood Hotel/rest., Ramada Inn/rest., Christmas Factory Gifts

52(309)   US 209, PA 447, to Marshalls Creek, **N**...HOSPITAL, Exxon, Gulf, Dairy Queen, Landmark Rest., Wendy's, Shannon Motel

51(308)   East Stroudsburg, **N**...HOSPITAL, Citgo, Texaco/mart, **1 mi N**...Arby's, Bruno's Pizza, Burger King, China King, McDonald's, Goodyear/auto, K-Mart, NAPA, ShopRite Foods, Wal-Mart SuperCtr, Weis Foods, **S**...Budget Motel, Super 8

50(307)   PA 191, Broad St, **N**...HOSPITAL, Gulf, KFC, McDonald's

49(306)   Dreher Ave(from wb, no EZ return), **N**...Wawa Mart, Buick

48(305)   US 209, Main St, **1 mi N**...Exxon/mart, Gulf, Mobil/mart, Perkins/24hr, Sheraton, **S**...CHIROPRACTOR, Texaco/mart/pizza/24hr, Colony Motel, auto repair

46a(305)   (47 from eb), US 209, 9th St(no EZ return from wb), no facilities

b(303)   PA 611, to Bartonsville, **N**...Citgo/diesel, Mobil/Subway/Pizza Hut/Taco Bell/diesel/24hr, Texaco/mart, McDonald's, Ponderosa, Comfort Inn, Holiday Inn, Goodyear

45(298)   PA 715, Tannersville, **N**...BP/mart, Mobil/mart, Pocono Diner, Romano's Pizzaria, Pocono Lodge, The Crossing Factory Outlet/famous brands, **1-3 mi N**...Brookdale Inn, Chateau Inn, camping, **S**...Amoco, Exxon/mart, Tannersville Diner, Hill Motel, Summit Resort, to Camelback Ski Area, carwash, to Big Pocono SP

44(297)   PA 611(from wb), to Scotrun, **N**...CHIROPRACTOR, MEDICAL CARE, Sunoco/diesel, Texaco/lube, Scotrun Diner/motel, cleaners, to Mt Pocono

295mm   **rest area eb, full(handicapped)facilities, phone, picnic tables, litter barrels, vending, petwalk**

293mm   I-380 W, to Scranton

43(284)   PA 115, Blakeslee, to Wilkes-Barre, **N**...Fern Ridge Camping, st police, **S**...Exxon/mart, Pizza Nut/Subs, to Pocono Raceway

42(277)   PA 940, to PA Tpk(NE Ext), Lakek Harmony, to Pocono, Allentown, **N**...Amoco/mart/atm/24hr, Exxon, Texaco/mart, Arby's, Burger King, McDonald's, Day's Inn, Pocono Motel/HoJo's Rest., Ramada Inn

41(274)   PA 534, **N**...Hickory Run/Amoco/diesel/rest./24hr, Texaco/diesel/mart/24hr, camping, **S**...to Hickory Run SP(6mi), camping

273mm   Lehigh River

40(272)   PA 940, PA 437, White Haven, to Freeland, **N**...DENTIST, Amoco/mart, Mobil, laundry, **S**...Powerhouse Eatery

270mm   **rest area eb, full(handicapped)facilities, info, phone, picnic tables, litter barrels, vending, petwalk**

39(263)   PA 309, to Hazleton, Mountain Top, **N**...Texaco/mart, 39Trkstp/diesel/24hr, Econolodge, MtnView Motel/rest., **S**...Amoco/mart, Citgo

260mm   I-81, N to Wilkes-Barre, S to Harrisburg

Pennsylvania Interstate 80

38(256) PA 93, Conyngham, to Nescopeck, **N**...Pilot/
Subway/diesel/mart/24hr, Sunoco/repair,
Lookout Motel(1mi), **S**...HOSPITAL, Texaco/
diesel/mart, Day's Inn, Hampton Inn(4mi)

251mm Nescopeck River

246mm **rest area/weigh sta both lanes,
full(handicapped)facilities, weather info,
phone, picnic tables, litter barrels,
vending, petwalk**

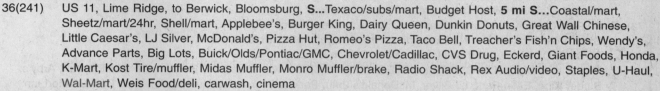

37(242) PA 339, Mifflinville, to Mainville, **N**...Gulf,
Shell/diesel/mart/24hr, McDonald's, Spur's
Steak/ribs, Super 8/rest., **S**...Citgo/diesel/
mart, repair

241.5mm Susquehanna River

36(241) US 11, Lime Ridge, to Berwick, Bloomsburg, **S**...Texaco/subs/mart, Budget Host, **5 mi S**...Coastal/mart,
Sheetz/mart/24hr, Shell/mart, Applebee's, Burger King, Dairy Queen, Dunkin Donuts, Great Wall Chinese,
Little Caesar's, LJ Silver, McDonald's, Pizza Hut, Romeo's Pizza, Taco Bell, Treacher's Fish'n Chips, Wendy's,
Advance Parts, Big Lots, Buick/Olds/Pontiac/GMC, Chevrolet/Cadillac, CVS Drug, Eckerd, Giant Foods, Honda,
K-Mart, Kost Tire/muffler, Midas Muffler, Monro Muffler/brake, Radio Shack, Rex Audio/video, Staples, U-Haul,
Wal-Mart, Weis Food/deli, carwash, cinema

35(235) PA 487, Lightstreet, to Bloomsburg, **N**...camping, **S**...HOSPITAL, Coastal/mart, Denny's/24hr, Ridgeway's
Rest., Tennytown Motel(2mi), to Bloomsburg U

34(232) PA 42, Buckhorn, **N**...Amoco/Travelport/Subway/diesel/rest./24hr, Texaco/TCBY/mart, Burger King, KFC,
Perkins, Wendy's, Western Sizzlin, Buckhorn Inn, Econolodge, Quality Inn/rest., Hill's, JC Penney, Sears, bank,
mall, repair, **S**...Indian Head camping(3mi), to Bloomsburg Fairground

33(224) PA 54, to Danville, **N**...Amoco/Subway/diesel/mart, Mobil/mart, Howard Johnson, **S**...HOSPITAL, Citgo/mart,
Texaco/diesel, Dutch Pantry, Friendly's, McDonald's, Day's Inn, Hampton Inn, Red Roof Inn, Travelodge

219.5mm **rest area eb, full(handicapped)facilities, info, phone, picnic tables, litter barrels, vending, petwalk**

32(215) PA 254, Limestoneville, **N**...All American/diesel/rest./24hr, **S**...Petro/Mobil/diesel/rest./24hr, Shell/diesel/24hr, All
American Rest.

31(212) I-180 W, PA 147 S, to Muncy, Williamsport, **N**...Texaco/diesel, **S**...Coastal/24hr(1mi), st police

210.5mm Susquehanna River, W Branch

30(210) US 15, to Williamsport, Lewisburg, **1 mi N**...Sunoco/repair, **S**...HOSPITAL, Citgo/mart, Bonanza, Comfort Inn,
Holiday Inn Express

29(199) Mile Run Rd, no facilities

194mm **rest area/weigh sta both lanes, full(handicapped)facilities, phone, picnic tables, litter barrels, vending,
petwalk**

28(192) PA 880, to Jersey Shore, **N**...Citgo/diesel/mart/24hr, Pit-Stop Rest., **S**...BP/mart

27(185) PA 477, Loganton, **N**...Gulf, camping, **S**...Citgo/Watt's Rest.

26(178) US 220, Lock Haven, **N**...HOSPITAL(7mi), Garden Rest., auto auction, **S**...Citgo/diesel/mart, Exxon

25(173) PA 64, Lamar, **N**...Gulf/repair, Texaco/diesel/mart, Cottage Family Rest., McDonald's, Comfort Inn/rest.,
Travelers Delite Motel, **S**...Citgo, TA/BP/Subway/diesel/rest./24hr

24(161) PA 26, to Bellafonte, **S**...Exxon/mart, Texaco/diesel/repair, KOA, to PSU

23(158) US 220 S, PA 150, Milesburg, to Altoona, **N**...Amoco/diesel/rest./24hr, Citgo/diesel/rest./24hr, Mobil,
McDonald's, Bestway Inn(1mi), Holiday Inn, **S**...st police

22(147) PA 144, to Snow Shoe, **N**...Citgo/diesel/24hr, Exxon/SnowShoe/diesel/repair, MountainTop Rest., Snow Shoe
Sandwich Shop, IGA Food/deli, repair

146mm **rest area both lanes, full(handicapped)facilities, phone, picnic tables, litter barrels, vending, petwalk**

138mm Moshanna River

21(133) PA 53, Kylertown, to Philipsburg, **N**...KwikFill/motel/rest., Sunoco/mart/LP, Roadhouse Rest., MktPlace Foods,
USPO, bank, diesel repair

20(123) PA 970, Woodland, to Shawville, **N**...st police, camping, **S**...Amoco(1mi), Pacific Pride/diesel

120mm Susquehanna River, W Branch

19(119) PA 879, Clearfield, **N**...Sapp Bros/Texaco/diesel/24hr, Aunt Lu's Café, Best Western/rest., **S**...HOSPITAL,
Amoco/mart, BP/Subway/mart, Sheetz/mart/24hr, Arby's, Auntie Lu's Café, Burger King, Dutch Pantry,
McDonald's, Comfort Inn, Day's Inn, Rodeway Inn(1mi), Super 8, Buick/Olds/Cadillac/GMC, Wal-Mart SuperCtr

111mm highest point on I-80 east of Mississippi River, 2250 ft

Pennsylvania Interstate 80

Du Bois

18(110) PA 153, to Penfield, **N**...to Parker Dam, to SB Elliot SP, **S**...HOSPITAL

17(101) PA 255, Du Bois, **N**...Amoco/mart/24hr, camping, **S**...HOSPITAL, Ramada Inn, st police, **1mi S**...Citgo, KwikFill/mart, Arby's, Bailey's Drive-In, BBQ, Eat'n Park/24hr, Goodyear/auto, Italian Oven, LJ Silver, McDonald's, Perkins/24hr, Ponderosa, Red Lobster, Subway, Taco House, TCBY, Wendy's, Best Western, Hampton Inn, Ames, BiLo Foods, Eckerd, JC Penney, K-Mart, Midas Muffler, Monro Muffler/brake, Sears, Wal-Mart, bank, mall, **2 mi S**...Hoss' Steaks, DuBois Manor Motel, Ramada Inn

16(96) US 219, Du Bois, to Brockway, **S**...HOSPITAL, BP/Dutch Pantry/mart, Pilot/Arby's/diesel/mart/24hr, Sheetz/diesel/mart/24hr, Hoss' Steaks(2mi), Du Bois Manor Motel(2mi), Holiday Inn, Pennzoil

87.5mm **rest area both lanes, full(handicapped)facilities, phone, picnic tables, litter barrels, vending, petwalk**

15(87) PA 830, to Reynoldsville, **S**...J's/diesel/rest./motel/24hr, auto repair

14(81) PA 28, to Brookville, to Hazen, **S**...to Brookville, hist dist

13(79) PA 36, Brookville, to Sigel, **N**...TA/BP/Blimpie/Taco Bell/diesel/rest./24hr, Dairy Queen, KFC, McDonald's, Pizza Hut, Howard Johnson, Super 8, Buick/Olds/Pontiac/Cadillac, NAPA, to Cook Forest SP, **S**...MEDICAL CARE, BP/mart, Citgo/mart, Exxon/mart, Sunoco/mart/24hr, Arby's, Burger King, Subway, American Hotel/rest., Budget Host, Day's Inn, Holiday Inn Express, Pennzoil, bank

12(73) PA 949, Corsica, **N**...Corsica/diesel/rest./24hr, to Clear Creek SP, **S**...USPO

11(71) US 322, to Strattanville, **N**...Exxon/diesel/mart, Keystone/Shell/diesel/rest./24hr

10(65) PA 66 S, Clarion, to New Bethlehem, **N**...to Clarion U

Clarion

9(62) PA 68, to Clarion, **N**...HOSPITAL, VETERINARIAN, BP/Subway/mart/24hr, Exxon/mart, KwikFill/mart, Arby's, Burger King, Eat'n Park, LJ Silver, McDonald's(2mi), Perkins/24hr, Pizza Hut, Sizzler, Taco Bell, Comfort Inn, Day's Inn/rest., Holiday Inn/rest., Super 8, County Mkt/deli/24hr, JC Penney, K-Mart/drugs, Staples, Wal-Mart SuperCtr/24hr, Western Auto/Pennzoil, bank, mall

61mm Clarion River

8(60) PA 66, to Shippenville, **N**...Citgo/mart, st police, to Cook Forest SP, camping

56mm parking area/weigh sta both lanes

7(54) to PA 338, to Knox, **N**...Gulf/diesel/mart, Wolf's Den Rest., Wolf's Camping Resort, golf driving range, **S**...Good Tire Service

6(46) PA 478, Emlenton, to St Petersburg, **4 mi S**...American Golf Hall of Fame

44.5mm Allegheny River

5(42) PA 38, to Emlenton, **N**...Citgo/diesel/rest./24hr, Exxon/Subway/diesel/mart/24hr, Texaco/diesel/mart/24hr, Emlenton Motel, Gaslight RV Park

4(35) PA 308, to Clintonville, **N**...Gulf/diesel/mart/atm/LP/24hr

30.5mm **rest area both lanes, full(handicapped)facilities, phone, picnic tables, litter barrels, vending, petwalk**

3(29) PA 8, Barkeyville, to Franklin, **N**...Gulf/Country Kettle/diesel/24hr, Burger King, King's Rest., Day's Inn, **S**...TA/BP/Subway/diesel/rest./24hr, Citgo/pizza/diesel/mart/24hr, KwikFill/diesel/rest./24hr, to Slippery Rock U

3a(24) PA 173, to Grove City, Sandy Lake, **S**...HOSPITAL

19mm I-79, N to Erie, S to Pittsburgh

2(15) US 19, to Mercer, **N**...Marathon/subs/mart, Burger King, McDonald's, Howard Johnson/rest., st police, camping, **2 mi S**...Iron Bridge Inn/rest.

1(4) PA 18, PA 60, to Sharon-Hermitage, New Castle, **1mi N**...Sheetz/diesel/mart/24hr, Sunoco/diesel/24hr, Comfort Inn, Holiday Inn, Radisson, lube, **S**...DairyMart, Dairy Queen, MiddleSex Diner/24hr, AllPro Parts

2.5mm Shenango River

1mm **Welcome Ctr eb, full(handicapped)facilities, phone, picnic tables, litter barrels, vending, petwalk**

0mm Pennsylvania/Ohio state line

Pennsylvania Interstate 81

Exit #(mm)Services

233mm Pennsylvania/New York state line

68(231) PA 171, Great Bend, **E**...Mobil/mart/24hr, **W**...Exxon/Arby's/diesel/mart/24hr, Sunoco/diesel, Beaver's Rest., Dobb's Country Kitchen, McDonald's, Subway, Colonial Brick Motel/rest., BiLo Foods, Rexall Drug, Rob's Mkt, bank, RV camping

67(223) PA 492, New Milford, **W**...Amoco/diesel/mart/atm, Mobil/24hr, Green Gables Rest., All Season Motel/camping

66(220) PA 848, to Gibson, **E**...Amoco/diesel/mart/rest./24hr, camping, repair, st police

65(217) PA 547, Harford, **E**...Mobil/diesel/mart/rest./24hr, Texaco/diesel/mart/24hr

64(211) PA 92, Lenox, **W**...Mobil/diesel/mart/24hr, Texaco/mart/rest., Bingham's Rest., Lenox Drug, bank

209mm **Welcome Ctr sb, full(handicapped)facilities, phone, picnic tables, litter barrel, vending, petwalk**

63(206) PA 374, Lenoxville, to Glenwood, **E**...to Elk Mountain Ski Resort

E

W

N

S

Pennsylvania Interstate 81

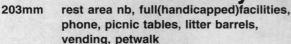

203mm **rest area nb, full(handicapped)facilities, phone, picnic tables, litter barrels, vending, petwalk**

62(202) PA 107, to Fleetville, Tompkinsville, no facilities

61(201) PA 438, East Benton, **W...**Mobil/mart, North 40 Rest.

60(199) PA 524, Scott, **E...**Amoco/diesel/mart, Texaco/diesel/mart/rest./24hr, **W...**Citgo/deli/ mart, Motel 81, camping, to Lackawanna SP

59(198) PA 632, Waverly, **W...**Sunoco/24hr

58(194) US 6, US 11, to PA Tpk, Clark's Summit, **1/2 mi W...**Griffin's/diesel, Shell, Sunoco/diesel/ mart, Texaco, Big Boy, Burger King, Dino's&Francesco's Pizza, Donut Connection, Friendly's, Manhattan Bagel, McDonald's, Pizza Hut, So-Journer Dining, Subway, Taco Bell, Wendy's, Day's Inn, Nichol's Village Inn, Ramada Inn, Summit Inn, Advance Parts, Eckerd, JiffyLube, Kost Tire, MailBoxes Etc, Monro Muffler/brake, Radio Shack, Weis Foods, bank, carwash

57(191) US 6, US 11, to Carbondale, **E...**MEDICAL CARE, Shell/mart, Arby's, Burger King, Denny's, Fresno's SW Rest., Golden Corral, Heavenly Ham, LJ Silver, McDonald's, Old Country Buffet, Pavilion Steaks, Perkins, Pizza Hut, Ponderosa, Ruby Tuesday, TCBY, Wendy's, Quality Hotel, Aldi Foods, Circuit City, CopyMax, Dick's Sports, Firestone/auto, Home Depot, Honda/Suzuki/Kawasaki, JC Penney, Jo-Ann Crafts, K-Mart, Lowe's Bldg Supply, OfficeMax, Pepboys, PetCo, PetsMart, PetSupplies+, Pier 1, Radio Shack, Rex, Sam's Club, Sears/auto, Service Merchandise, Staples, TJ Maxx, Wal-Mart SuperCtr/24hr, Wegman's Food/drug, cinema, mall, **W...**to Anthracite Museum

56(190) Main Ave, Dickson City, **E...**same as 57, **W...**Manhattan Bagel, Schiff's Mkt

55(189) PA 347, Throop, **E...**Sunoco/diesel/mart, Big Boy, Boston Mkt, China World Buffet, Donut Connection, McDonald's, Wendy's, Day's Inn, Dunmore Inn, Econolodge, Advance Parts, Big Lots, PriceChopper Foods, Radio Shack, bank, st police, **W...**Mobil/mart, Burger King, Friendly's, Mazda

54(187) PA 435, Drinker St, to I-84, I-380(no return from nb)

53(185) Central Scranton Expwy, **W...**HOSPITAL, Clarion

52(184) to PA 307, to River St, **W...**HOSPITAL, Exxon, Mobil, Shell, Texaco, Dunkin Donuts, Clarion, CVS Drug, Gerrity Foods, JiffyLube

51(182) Davis St, Montage Mtn Rd, **E...**Mugg's Rest., Stadium Club Rest., Comfort Suites, Courtyard, Hampton Inn, cinema, to Montage Ski Area, **W...**Mobil/mart/24hr, Sunoco/diesel/24hr, Texaco/24hr, LJ Silver, McDonald's, Econolodge/rest., Rodeway Inn, USPO

50(179) to US 11, PA 502, to Moosic, **1-2 mi W...**Amoco/24hr, food, lodging, phone

49(178) to US 11, Avoca, **E...**Damon's, Holiday Inn Express, airport, **W...**VETERINARIAN, Petro/Sunoco/diesel/rest./ 24hr

48b a(175) PA 315 S, to I-476, Dupont, **E...**Mobil, Sunoco/24hr, Perkins, Day's Inn, Howard Johnson's, Knight's Inn, Chevrolet, **W...**Citgo/diesel/mart, Pilot/Wendy's/diesel/mart/24hr, Best Hotel, Goodyear, Wal-Mart

47(170) PA 115, PA 309, Wilkes-Barre, **E...**Exxon/diesel, Best Western, Carriage Stop Inn, Melody Motel, to Pocono Downs, **W...**HOSPITAL, Sunoco/24hr, Big Boy, Friendly's, LJ Silver, McDonald's, Perkins, Pizza Hut, TGIFriday, Day's Inn, Hampton Inn, Holiday Inn, Ramada Inn, Red Roof Inn, Kia, BMW, Subaru, VW, mall

46(167) Highland Park Blvd, Wilkes-Barre, **E...**McDonald's, K-Mart, **W...**Barnes&Noble, Bon-Ton, Dick's Sports, Lowe's Bldg Supply, Michael's, Pier 1, TJ Maxx, Wegman's Foods

45(165) PA 309 S, Wilkes-Barre, **W...**Amoco/diesel/mart/wash, Texaco, Dunkin Donuts, Mark II Rest., McDonald's, Perkins, Taco Bell, Comfort Inn, Econolodge, Advance Parts, Eckerd, K-Mart/auto

44(164) PA 29, Ashley, to Nanticoke, no facilities

43(159) Nuangola, **W...**Amoco/24hr, Godfather's, camping

157mm **rest area/weigh sta sb, full(handicapped)facilities, vending, phone, picnic tables, litter barrels, petwalk**

156mm **rest area/weigh sta nb, full(handicapped)facilities, vending, phone, picnic tables, litter barrels, petwalk**

42(155) to Dorrance, **E...**Sunoco/diesel/24hr, Econolodge(2mi), camping

151mm I-80, E to Mountaintop, W to Bloomsburg

41(146) PA 93, W Hazleton, **E...**Gulf, Sunoco/diesel/mart/24hr, TurkeyHill/gas/mart, Friendly's, LJ Silver, McDonald's, Perkins/24hr, Pizza Hut, Rossi's Rest., Taco Bell, Wendy's, Best Value Inn, Best Western(2mi), Comfort Inn, Day's Inn, Econolodge, Fairfield Inn, Forest Hill Inn, Ramada Inn, Aamco, Ames, Olds/Cadillac, KwikLube, Mazda, Plymouth/Jeep, Sears, bank, carwash, laundry, **W...**Texaco, Tom's Kitchen, Top of the 80's Rest., Hampton Inn, st police

Pennsylvania Interstate 81

40(143) PA 924, to Hazleton, **E**...Carmen's Family Rest., Red Rooster Motel, **W**...Texaco

39a(141) PA 424, S Hazleton Beltway, no facilities

39(138) PA 309, to McAdoo, **2 mi E**...Pines Motel

135mm scenic area nb

38(134) to Delano, no facilities

132mm parking area both lanes

37(131) PA 54, Mahanoy City, **E**...to Tuscarora/Locust Lake SP, **W**...Exxon, Texaco/diesel

36(123) PA 61, to Frackville, **E**...Cracker Barrel, McDonald's, Holiday Inn Express, K-Mart, Phar-Mor Drugs, Sears/auto, auto parts, cinema, mall, **W**...HOSPITAL, Gulf, Hess/mart, Mobil/Taco Bell/mart/24hr, Dutch Kitchen, Subway, Econolodge, Granny's Motel, Motel 6, Goodyear/auto, Plymouth/Dodge, Rite Aid, bank, hardware, st police

35a(118) to Gordon, no facilities

35(116) PA 901, to Minersville, **E**...901 Rest.

34(112) PA 25, to Hegins, **W**...camping

33(107) US 209, to Tremont, no facilities

32(104) PA 125, Ravine, **E**...Exxon/mart/24hr, Mobil/diesel/rest./atm, camping

31(100) PA 443, to Pine Grove, **E**...Exxon/mart/atm, Arby's, McDonald's, Ulsh's Rest., Colony Lodge, Comfort Inn, Econolodge, **W**...Shell/Subway/diesel/mart/atm/24hr, Texaco/diesel/mart, All American Rest./24hr

30(91) PA 72, to Lebanon, **E**...Exxon, Hess/mart, Day's Inn, KOA, **Lickdale Camping**, **W**...Super 8

89mm I-78 E, to Allentown

29(85) PA 934, to Annville, **2 mi W**...Texaco/rest./mart, to Indian Town Gap Nat Cem

28(80) PA 743, Grantville, **E**...Mobil/diesel, Econolodge, Hampton Inn, antiques, **W**...Exxon/mart/atm, Texaco/mart/24hr, Comfort Suites, Holiday Inn, racetrack

79mm **rest area/weigh sta both lanes, full(handicapped)facilities, phone, vending, picnic tables, litter barrels, petwalk**

27(77) PA 39, to Hershey, **E**...BP/mart, Pilot/diesel/mart/24hr, Texaco, Ramada Ltd, Sleep Inn, to Hershey Attractions, st police, **W**...Shell/Subway/diesel/mart/24hr, TA/diesel/24hr, Wilco/Perkins/diesel/mart/24hr, McDonald's, Comfort Inn, Daystop, Goodyear

26(72) to US 22, Linglestown, **E**...CHIROPRACTOR, Citgo/diesel/mart, Hess Gas, Sunoco/diesel, Texaco/diesel/mart/24hr, Applebee's, Burger King, LJSilver, McDonald's, Malley's Rest., Old Country Buffet, Baymont Inn, Harrisburg Inn, Holiday Inn Express, Advance Parts, C&P Repair, Chrysler/Jeep, CVS Drug, Festival Food/drug, K-Mart, Monro Muffler/brake, Toyota, U-Haul, bank, cleaners, **W**...Country Oven Rest., Best Western

25(70) I-83 S, to York, airport

24(69) Progress Ave, **E**...st police, **W**...DENTIST, Western Sizzlin, Best Western, Red Roof

23(67) US 22, US 322 W, PA 230, Cameron St, to Lewistown, no facilities

22(66) Front St, **E**...HOSPITAL, **W**...McDonald's, Wendy's, Day's Inn

21(65) US 11/15, to Enola, **1 mi E**...Mobil, Sunoco/diesel/24hr, Big Boy, Chinese Rest., Dunkin Donuts, Eat'n Park, Hardee's, KFC, McDonald's, Roy Rogers, Wendy's, Friendship Inn, Quality Inn, K-Mart, Rite Aid, SuperFresh Foods, bank, tires

20(61) PA 944, to Wertzville, no facilities

19(60) PA 581, to US 11(from sb), to Harrisburg, **3 mi E**...Bob Evans, Burger King, Friendly's, McDonald's, Wendy's, Comfort Inn, Hampton Inn, Holiday Inn

18(57) PA 114, to Mechanicsburg, **1-2 mi E**...McDonald's, Pizza Hut, Shoney's, Taco Bell, **W**...Visaggio's Motel/rest.

17(52) US 11, to I-76, Middlesex, **E**...Citgo/diesel, Flying J/diesel/LP/rest./24hr, Bob Evans, Hardee's/24hr, Middlesex Diner/24hr, Appalachian Trail Inn, Comfort Inn, Econolodge, Embers Inn/rest., Holiday Inn, Super 8, **W**...Amoco/diesel/rest., BP/mart, Pilot/Wendy's/diesel/mart/24hr, Shell/diesel/24hr, Texaco/diesel/mart/24hr, Arby's, BBQ, Big Boy/24hr, Dunkin Donuts, Eat'n Park, McDonald's, Subway, Western Sizzlin, Best Western, Budget Host/rest., Hampton Inn, Quality Inn, Rodeway Inn, ThriftLodge, Carlisle Camping

16(49) PA 74(no EZ return from sb), **W**...Getty/mart, Burger King, LJ Silver, McDonald's, same as 15

15(48) PA 74, York St(no EZ return from nb), **E**...CHIROPRACTOR, **W**...CHIROPRACTOR, Gulf, KwikFill, Little John's Rest., Rillo's Rest., Farmer's Mkt(fri-sat), **1 mi W**...Hess/mart, Burger King, LJ Silver, McDonald's, Pizza Hut, Giant Food/drugs, JC Penney, K-Mart, Midas Muffler, People's Drug, Sears/auto, mall

14EW(47) PA 34, Hanover St, **E**...Gulf, **W**...Ward's/auto, laundry, mall

13(46) College St, **E**...Mobil, Texaco/mart, Bonanza, Dunkin Donuts, Jerry's Subs/pizza, McDonald's, Shoney's, Day's Inn, Super 8

12(44) PA 465, to Plainfield, **E**...Day's Inn, Super 8, to st police

38mm **rest area both lanes, full(handicapped)facilities, phone, picnic tables, litter barrels, petwalk**

11(37) PA 233, to Newville, **E**...Pine Grove Furnace SP, **W**...Col Denning SP

10(29) PA 174, King St, **E**...Sunoco/24hr, Texaco/diesel, Budget Host, **W**...Ford, Burger King(1mi), Dairy Queen, McDonald's(2mi)

Pennsylvania Interstate 81

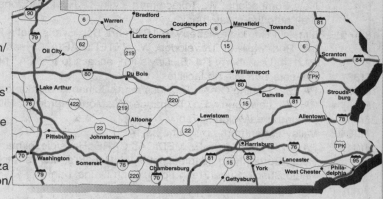

9(24)   PA 696, Fayette St, **W**...Mobil/mart

8(20)   PA 997, Scotland, **E**...Shell/diesel, McDonald's, Comfort Inn, antiques(4mi), **W**...Amoco/diesel/mart, Sunoco/mart/wash/24hr, bank

6(16)   US 30, to Chambersburg, **E**...info, Exxon/mart, Sheetz/mart, KFC, Roy Rogers, Hoss' Steak&Seahouse, Five Points Diner, Shoney's, Day's Inn, Rodeway Inn, Meineke Muffler, 1-Hr Photo, **W**...HOSPITAL, Hess/diesel/mart, Burger King, Chinese Rest., Dunkin Donuts, LJ Silver, McDonald's, Pizza Hut, Ponderosa, Taco Bell, Howard Johnson/rest., Travelodge

N

5(14)   PA 316, Wayne Ave, **E**...Bob Evans, Cracker Barrel, Fairfield Inn, Hampton Inn, **W**...CHIROPRACTOR, Exxon/mart/24hr, KwikFill, Sheetz Gas/mart, Sunoco, Texaco/diesel/mart/24hr, Aplpebee's, Chinese Rest., Denny's, Dino's Kitchen, Dunkin Donuts, Little Caesar's, Pizza Hut, Red Lobster, Roy Rogers, Subway, Wendy's, Econolodge, Holiday Inn, Giant Food/drugs, JetLube, K-Mart, Rite Aid, bank

12mm   weigh sta sb

4(10)   PA 914, Marion, no facilities

7mm   rest area nb, no facilities

3(5)   PA 16, Greencastle, **E**...Mobil/diesel/mart, Texaco/diesel/mart, Antrim House Rest., Arby's, Buckhorn Rest., Hardee's, McDonald's, Roy Rogers, Castle Green Motel, Econolodge, Rodeway Inn, Whitetail Ski Resort, **W**...Exxon/diesel/mart

S

2(3)   US 11, **E**...CHIROPRACTOR, Exxon/mart, Comfort Inn/rest., **W**...Sunoco

2mm   **Welcome Ctr nb, full(handicapped)facilities, phone, picnic tables, litter barrels, petwalk**

1(1)   PA 163, Mason-Dixon Rd, **1 mi E**...gas/diesel, Lantern House Rest., **W**...Black Steer Rest., Mason-Dixon Rest., Best Western, Econolodge, State Line Motel

0mm   Mason-Dixon Line, Pennsylvania/Maryland state line

Pennsylvania Interstate 83

Exit #(mm)Services

50mm   I-83 begins/ends on I-81, exit 25.

30EW(49) US 22, Jonestown Rd, Harrisburg, **E**...EYECARE, Mobil, Sunoco/mart, Sheetz/mart, Texaco/wash, Applebee's, Arby's, Boston Mkt, Chef Wong's, LJ Silver, McDonald's, Pizza Hut, Red Lobster, Subway, Taco Bell, Aamco, Ames, Bon-Ton, Borders Books, CVS Drug, Dick's Sports, Gateway Computers, Goodyear/auto, Home Depot, JiffyLube, Kinko's, K-Mart/auto, Kohl's, MediaPlay, Meineke Muffler, MensWearhouse, Michael's, NTB, PepBoys, Pier 1, Sears/auto, SuperPetz, Target, TireAmerica, Weis Foods, bank, cinema, cleaners/laundry, mall, **W**...Citgo/7-11, Gulf, Dunkin Donuts, Friendly's, KFC, Outback Steaks, JiffyLube, cleaners/laundry

N

29(48.5) Union Deposit Rd, **E**...HOSPITAL, CHIROPRACTOR, Sunoco/mart, Arby's, Bonanza, Burger King, Chinese Rest., Denny's, LoneStar Steaks, The Point Rest./steaks, Wendy's, Hampton Inn, Staples, bank, mall, **W**...Texaco/mart, ChiChi's, Finley's Rest., Hardee's, Little Caesar's, McDonald's, OutBack Steaks, Rita's Italian, Subway, TGIFriday, Your Place Rest., Comfort Inn, Food Festival, Lowe's Bldg Supply, Phar-Mor Drugs, Radio Shack, Weis Foods

28(48)   US 322 E, to Hershey, Derry St, **E**...Hess Gas, Home Depot, PetsMart

27(47.7) I-283 S, to PA Tpk/I-76, **facilities E off I-283, exit 1**...Amoco, Citgo, Exxon/diesel, Sunoco, Bob Evans, Dempsey's/24hr, Eat'n Park, Leed's Ltd Grill, McDonald's, Best Western, Baymont Inn, Crown Park Plaza, Econolodge, Howard Johnson, Marriott, Red Roof Inn, Super 8, Buick, Speedy Muffler King, VW

26(47)   Paxtang St, **E**...Texaco/mart, Burger King, Dunkin Donuts, Pizza Hut, Ponderosa, Santo's Pizza, S Philly Hoagies, Vietnamese Foods/rest., Wendy's, Chrysler/Plymouth, Hecht's, JC Penney, Lord&Taylor, Mazda/Subaru/Toyota, Meineke Muffler, Monro Muffler/brake, So-Fro Fabrics, cinema, mall

25(46)   17th St, 19th St, **E**...Pacific Pride/diesel, Sunoco/mart, A+ Minimart, Benihana Japanese, Chinese Rest., Hardee's, Massimo's Pizza/subs, Royal Thai Rest., Big A Parts, Cadillac/Olds/GMC, Dodge, Firestone, Honda, Hyundai/Suzuki/Isuzu, Midas Muffler, Nissan, Pontiac, cleaners/laundry

S

24(45)   PA 230, 13th St, Harrisburg, downtown, **W**...Chevrolet, Saturn

23(44)   2nd St, Harrisburg, downtown, st capitol, **W**...HOSPITAL, Ramada Inn

43mm   Susquehanna River

22(42.5) Lemoyne, **E**...Hess/mart, Mobil, Turkey Hill Gas/food, Burger King, KFC, **W**...Getty Gas

21(42)   Highland Park, **E**...Hess/mart, Mobil, Bubba's Sandwiches, KFC, Rascal's Pizza, Royal Subs, Weis Foods, auto parts, drugs

20(41.5) US 15, PA 581 W, to Gettysburg, no facilities

Pennsylvania Interstate 83

19(41) New Cumberland, **W**...Texaco/mart, Chinese Rest., McDonald's, Tasty Donuts, Fox's Foods, CVS Drug, $General, auto/tire repair, bank

18a(40) Limekiln Rd, to Lewisberry, **E**...BP/mart, Bob Evans, Eat'n Park, Pizza Hut, Fairfield Inn, Holiday Inn, McIntosh Inn, Rodeway Inn, Travelodge, **W**...VETERINARIAN, Hess/diesel, Knight's Inn, Motel 6

18(39) PA 114, Lewisberry Rd, **E**...Day's Inn, access to I-76/PA Tpk

17(38) Reesers Summit, no facilities

16(37) PA 262, Fishing Creek, **E**...Hess/mart, Culhanes Steaks, bank, **W**...Texaco

15(36) PA 177, Lewisberry, **E**...OilChange, **W**...Mobil, Hardee's, Hillside Café, camping

35mm parking area/weigh sta sb

14a(34.5) Valley Green(from nb), same as 14

14(34) PA 392, Yocumtown, **E**...MEDICAL CARE, Hess/diesel/mart, Henry's Trkstop/diesel, Rutter's Gas/mart, Alice's Rest., Burger King, Maple Donuts, McDonald's, Mom's Kitchen, Robin Hood Diner, Super 8, MailBoxes Etc, Radio Shack, SuperFresh Food, Thrift Drugs, bank

33mm parking area/weigh sta nb

13(32.5) PA 382, Newberrytown, **E**...Sunoco/diesel, Rutter's/diesel/mart, bank, **W**...Exxon/diesel/mart

12(28) PA 295, Strinestown, **W**...Exxon/Rutter's/atm, Wendy's

11(23.5) PA 238, Emigsville, **W**...VETERINARIAN, Sibold's Steaks, Dixie Motel(2mi)

10(21.5) PA 181, N George St, **E**...Exxon/mart/24hr, Getty Gas, Comfort Inn, **W**...Sunoco/24hr, Hardee's, LJ Silver, McDonald's, Wendy's, Day's Inn, Holiday Inn, Motel 6, Super 8, same as 9W

9E W(21) US 30, Arsenal Rd, to York, **E**...Big Bopper's Ice Cream, Bob Evans, McDonald's(2mi), Shoney's(2mi), Day's Inn, Hampton Inn(2mi), Ramada Inn, Red Roof Inn, Sheraton, Pontiac, **W**...CHIROPRACTOR, Crown Gas, Exxon/mart/atm, Mobil, Texaco/diesel/mart/wash, Anthony's Italian, Boston Mkt, Burger King, Chinese Rest., Domino's, Dunkin Donuts, El Rodeo Mexican, Friendly's, Hardee's/24hr, Italian Cousin Rest., KFC, LoneStar Steaks, LJ Silver, McDonald's, Old Country Buffet, Pizza Hut, Ponderosa, Subway, Taco Bell, Wendy's, Holiday Inn, Motel 6, Super 8, Chevrolet/Subaru, Chrysler/Mitsubishi, County Mkt Foods/24hr, CVS Drug, Econo Lube'n Tune, Fisher Auto Parts, Harley-Davidson, JiffyLube, Jo-Ann Fabrics, Monro Muffle/brakes, OfficeMax, PepBoys, Radio Shack, Staples, Thrift Drug, TireAmerica, Tuffy Service Ctr, Weis Foods, cleaners

8(19) PA 462, Market St, **E**...HOSPITAL, Sunoco/mart, Perkins, Quality Inn Suites, Nissan, Saturn, **W**...DENTIST, Sunoco/mart, carwash

7(18) PA 124, Mt Rose Ave, Prospect St, **E**...Exxon/service/mart, Mobil, Sunoco/Pizza Hut/mart, Burger King, Denny's, Little Caesar's, Penrod's Rest., Viet Garden Rest., Budget Host, Spirit of 76 Motel, Big Lots, CVS Drug, K-Mart, Weis Foods, bank, cleaners, **W**...Mobil(1mi)

6E W(16) PA 74, Queen St, **E**...Mobil/24hr, Cracker Barrel, Hardee's/24hr, Maple Donuts, Pizza Hut, Ruby Tuesday, Subs Ltd, Country Inn Suites, Ford/Lincoln/Mercury, Giant Foods, auto repair, bank, cinema, **W**...CHIROPRACTOR, Exxon, Sunoco, Chap's Rest., McDonald's, S York Diner/24hr, Subs2Go, Subway, Taco Bell, CVS Drug, Cadillac/BMW/Hyundai, CVS Drug, $General, Honda, Jo-Ann Fabrics, Monro Muffler/brakes, Olds, Phar-Mor, Weis Foods, bank

5(15) S George St, I-83 spur into York, **W**...HOSPITAL

4(14) PA 182, Leader Heights, **E**...Mobil, Burger King, ChopHouse, Duet's Café, camping, **W**...Exxon/Pizza Hut/mart, Sunoco/diesel/mart/24hr, McDonald's, Comfort Inn, bank

3(11) PA 214, Loganville, **W**...Lee's Drive-In, Mamma's Pizza, Midway Motel, hardware, st police

2(7.5) PA 216, Glen Rock, **E**...Rocky Ridge Motel

1(3) PA 851, Shrewsbury, **E**...Mobil/diesel/mart/24hr, RapidLube, TrueValue, **W**...DENTIST, VETERINARIAN, Exxon/diesel/mart/24hr, Arby's, Coachlight Rest., Dairy Queen, KFC/Taco Bell, McDonald's, Subway, Wendy's, CVS Drug, $Tree, FashionBug, Giant Foods/24hr, GNC, K-Mart, MailBoxes Etc, Radio Shack, Wal-Mart SuperCtr/24hr, bank

2mm **Welcome Ctr nb, full(handicapped)facilities, phone, vending, picnic tables, litter barrels, petwalk**

0mm Pennsylvania/Maryland state line

Pennsylvania Interstate 84

Exit #(mm)Services

54mm Delaware River, Pennsylvania/New York state line

11(53) US 6, PA 209, Matamoras, **N**...**Welcome Ctr/both lanes, full(handicapped)facilities, phone, vending, picnic tables, litter barrels, petwalk,** Amoco/mart, Citgo/mart, Go24/TacoBell/gas, Texaco, Polar Bear Rest., WestFalls Motel, AutoZone, NAPA, Roe's Produce, Tri-State Canoe Camping, bank, **S**...Mobil/mart, McDonald's, Perkins, Wendy's, Best Western, Blue Swan Motel, Tourist Village Motel, Eckerd, FashionBug, Grand Union Foods, K-Mart, Wal-Mart, Riverbeach Camping

10(46) US 6, to Milford, **N**...Mobil/24hr, **2 mi S**...Citgo/diesel/24hr, Sunoco, Apple Valley Rest., Milford Diner, Black Walnut Bed/Breakfast, Hilltop Inn/rest., Sherelyn Motel, Tom Quick Inn/rest.

Pennsylvania Interstate 84

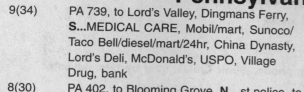

E

9(34) PA 739, to Lord's Valley, Dingmans Ferry, **S...**MEDICAL CARE, Mobil/mart, Sunoco/Taco Bell/diesel/mart/24hr, China Dynasty, Lord's Deli, McDonald's, USPO, Village Drug, bank

8(30) PA 402, to Blooming Grove, **N...**st police, to Lake Wallenpaupack

7(26) PA 390, to Tafton, **N...**to Lake Wallenpaupack, **S...**Exxon/diesel/mart/24hr, Promised Land TrkStop/diesel, **2-3 mi S...**Old Rangers Inn, Promised Land Inn/rest., to Promised Land SP

26mm **rest area/weigh sta both lanes, full(handicapped)facilities, phone, vending, picnic tables, litter barrels, petwalk**

6(20) PA 507, Greentown, **N...**CHIROPRACTOR, MEDICAL CARE, Exxon/diesel/mart/24hr, Mobil/mart, John's Rest., Claws'n Paws Animal Park, camping, **S...**Lakewood 84/diesel/24hr

5(17) PA 191, to Newfoundland, Hamlin, **N...**Exxon/24hr, Howe's 84/diesel/24hr, Twin Rocks Rest., Comfort Inn/rest.

4(8) PA 247, PA 348, Mt Cobb, **N...**Gulf, bank, camping, **S...**Mobil/diesel/mart/24hr

3(4) I-380 E, to Mount Pocono, no facilities

2(3) PA 435 S, to Elmhurst, no facilities

W

1(1) Tigue St, **N...**Holiday Inn, **S...**Mobil

0mm I-84 begins/ends on I-81, exit 54.

Pennsylvania Interstate 90

Exit #(mm)Services

46mm Pennsylvania/New York state line, Welcome Ctr/weigh sta wb, full(handicapped)facilities, phone, vending, picnic tables, litter barrels, petwalk

12(45) US 20, to State Line, **N...**KwikFill/diesel/mart, TA/Shell/Taco Bell/diesel/rest., McDonald's, Pizza Hut, Heritage Wine Cellars, **S...**BP/Subway/diesel/mart/atm, Red Carpet Inn, Niagara Falls Info

11(41) PA 89, North East, **N...**Shell/mart, Super 8, camping, **S...**winery

10a(36) I-86 E, to Jamestown, no facilities

10(35) PA 531, to Harborcreek, **N...**TA/Sunoco/diesel/rest./24hr, Pizza Hut, Rodeway Inn, Blue Beacon

9(32) PA 430, to Wesleyville, **S...**Sunoco/diesel/mart, Freightliner, camping, st police

E

8(29) PA 8, to Hammett, **N...**HOSPITAL, BP/mart, Citgo/mart, Wendy's, **S...**Ramada/rest., Ford/Peterbilt, Cummins Repair

7(27) PA 97, State St, Waterford, **N...**HOSPITAL, BP/diesel/mart, Citgo/diesel/mart/24hr, Kwikfill/mart, Arby's, Big Boy, McDonald's, Best Western, Day's Inn, Red Roof Inn, TalleyHo Inn, **S...**Pilot/Subway/diesel/mart/24hr, Shell/diesel/mart, Quality Inn, Super 8

6(24) US 19, Peach St, to Waterford, **N...**HOSPITAL, Citgo/diesel/mart, KwikFill/mart, Applebee's, Burger King, China Garden, Cracker Barrel, Damon's, Eat'n Park, Fazoli's, Longhorn Steaks, McDonald's, Old Country Buffet, Panera Bread, Ponderosa, QuakerSteak/lube, Taco Bell, TGIFriday, Wendy's, Courtyard, Motel 6, Advance Parts, Best Buy, Circuit City, Country Fair/deli, $Tree, Gateway Computers, Giant Eagle/deli, Home Depot, Jo-Ann Fabrics, K-Mart, Kohl's, Lowe's Bldg Supply, MediaPlay, Pennzoil, PetsMart, Pontiac, Sam's Club, Staples, Target, VW, Wal-Mart/auto, Wegman's Foods, bank, cinema, **S...**BP/diesel/mart, Citgo/mart, Shell/diesel/mart, Bob Evans, Comfort Inn, Econolodge, Hampton Inn, Microtel, Residence Inn

22mm I-79 N to Erie, **3-5 mi N...**facilities in Erie

5(18) PA 832, Sterrettania, **N...**Citgo/diesel/mart/24hr, Shell/diesel/mart/24hr, Burger King, Hill's Family Camping, to Presque Isle SP, **S...**Shell/diesel/mart/24hr, AM Best/diesel/rest./repair/24hr, Best Western, KOA, bank, tires/service

W

4(16) PA 98, Fairview, to Franklin Center, **S..**Follys Inn Camping(2mi), Mar-Jo-Di Camping(5mi)

3(9) PA 18, Platea, to Girard, **N...**diesel repair, st police

2(6) PA 215, East Springfield, to Albion, **S...**Sunoco/mart, Miracle Motel

1(3) US 6N, West Springfield, to Cherry Hill, **N...**lodging on US 20, **S...**BP/diesel/rest./repair/24hr

2.5mm **Welcome Ctr/weigh sta eb, full(handicapped)facilities, info, phone, picnic tables, litter barrels, vending, petwalk**

0mm Pennsylvania/Ohio state line

Erie

Pennsylvania Interstate 95

Levittown

Philadelphia Area

Exit #(mm)Services

51mm   Delaware River, Pennsylvania/New Jersey state line

31b a(50)   PA 32, to New Hope, no facilities

49mm   Welcome Ctr sb, full(handicapped)facilities, vending, phone, picnic tables, litter barrels, petwalk

30(47)   PA 332, to Yardley, Newtown, **W**...to Tyler SP

29b a(45)   US 1, to I-276, Langhorne, Oxford Valley, no facilities

28(43)   US 1, PA 413, to Penndel, Levittown, **E**...HOSPITAL, Mobil/diesel/24hr, Texaco/diesel/mart/wash/24hr, ChuckeCheese, Dunkin Donuts, Friendly's, Great American Diner, Acura, Chevrolet/Buick/Olds, Chrysler/ Dodge, Drug Emporium, Ford/Lincoln/Mercury, Goodyear/auto, Honda, Jeep/Hyundai, K-Mart, Marshall's, Midas Muffler, NTB, Radio Shack, Sam's Club, Subaru, VW/Volvo, **W**...Amoco/mart(1mi), Mobil/diesel/mart, Denny's, McDonald's, Toyota, U-Haul

26(39)   to PA 413, I-276(from nb), to Bristol Bridge, Burlington, **E**...HOSPITAL, McDonald's, Taco Bell

25(36)   PA 132, to Street Rd, **1 mi W**...Amoco, Chef's Place/24hr, Denny's, Friendly's, McDonald's, Old Country Buffet, Pizza Hut, Taco Bell

24(34)   PA 63, to US 13, Woodhaven Rd, to Bristol Park, **W**...Exxon/mart, Mobil/mart/wash, Perkins, Sizzler, Wendy's, White Castle, Day's Inn, Hampton Inn, JC Penney, Marshall's, drugs, groceries, mall

23(32)   Academy Rd, **W**...HOSPITAL, K-Mart

22(30)   PA 73, Cottman Ave, **W**...Sunoco

21(27)   Bridge St, **W**...HOSPITAL, Getty Gas, Eckerd

20(26)   to Betsy Ross Brdg, Pennsauken, NJ, **W**...Sunoco/mart, Burger King, McDonald's, Hampton Inn

19(25)   Allegheny Ave, **W**...HOSPITAL, Sunoco, WawaMart, pizza

18(23)   Lehigh Ave, Girard Ave, **W**...HOSPITAL, Exxon, Dunkin Donuts, Pizza Hut, Ruby Tuesday, Radio Shack

17(21)   I-676, US 30, to Central Philadelphia, Independence Hall, **W**...Comfort Inn

16(20)   Columbus Blvd, Penns Landing, **E**...Amoco, Mobil, Boston Mkt, ChuckeCheese, McDonald's, Home Depot, OfficeMax, PepBoy's, SavALot Foods, ShopRite Foods, Wal-Mart

15(18)   I-76 E, to Walt Whitman Bridge, **W**...Sunoco/diesel/24hr, Burger King, Church's Chicken, McDonald's, Subway, Wendy's, Holiday Inn, K-Mart, auto parts, to stadium

14(16)   PA 611, to Broad St, Pattison Ave, **W**...HOSPITAL, to Naval Shipyard, to stadium

15mm   Schuykill River

13(14)   Enterprise Ave, Island Ave(from sb), no facilities

12(13)   Bartram Ave, Essington Ave(from sb), no facilities

11(11)   PA 291, to I-76 W(from nb), to Central Philadelphia, **E**...Exxon/diesel/mart, Day's Inn, Guest Quarter's, Hilton, Marriott, Residence Inn, Westin Suites

10(10)   PA 291 E, to Philly Airport, **E**...Marriott, **W**...Courtyard, Embassy Suites

9b a(9)   PA 420, to Essington, Prospect Park, **E**...Coastal/diesel/mart, Denny's, Shoney's, Comfort Inn, Econolodge, Holiday Inn, Motel 6, Radisson, Ramada Inn, Red Roof Inn

8(8)   Ridley Park, to Chester Waterfront, no facilities

7(7)   I-476 N, to Plymouth, Meeting, no facilities

6(6)   PA 352, PA 320, to Edgmont Ave, **W**...Howard Johnson, to Widener U

5(5)   Kerlin St(from nb), **E**...Amoco

4(4)   US 322 E, **E**...Highland Motel

3b a(3)   US 322 W, Highland Ave(no EZ nb return), **E**...Sunoco/diesel, Ford, Goodyear

2(2)   PA 452, to US 22, Market St, **E**...Getty Gas, **W**...Citgo/mart, Exxon/mart, Marchianti's

1(1)   Chichester Ave(from sb), **E**...Sunoco, **W**...Amoco/mart

0mm   Pennsylvania/Delaware state line, Welcome Ctr nb, full(handicapped)facilities, phone, picnic tables, litter barrels, petwalk

N

S

Rhode Island Interstate 95

Exit #(mm)Services

43mm	Rhode Island/Massachusetts state line
30(42)	East St, to Central Falls, **E**...Dunkin Donuts
29(41)	US 1, Cottage St, **W**...D'Angelo's, Firestone, Olds
28(40)	RI 114, School St, **E**...HOSPITAL, Sunoco, to hist dist
27(39)	Pawtucket, **W**...Shell/autocare, Sunoco/diesel/mart/24hr, Burger King, Dunkin Donuts, Ground Round, Comfort Inn, laundry
26(38)	RI 122, Lonsdale Ave, no facilities
25(37)	US 1, RI 126, N Main St, Providence, **E**...HOSPITAL, Chili's, **W**...Celo's Italian
24(36.5)	Branch Ave, Providence, downtown, **W**...Sears Parts
23(36)	RI 146, RI 7, Providence, **E**...HOSPITAL, Shell/mart/wash, Marriott, **W**...USPO
22(35.5)	US 6, RI 10, Providence, downtown
21(35)	Broadway St, Providence, **W**...Engle Tires, Goodyear/auto, Speedy Mufflers
20(34.5)	I-195, to E Providence, Cape Cod
19(34)	to US 1(from sb), **W**...HOSPITAL
18(33.5)	US 1A, Thurbers Ave, **W**...HOSPITAL, Cadillac
17(33)	US 1(from sb), Elmwood Ave, no facilities
16(32.5)	RI 10, Cranston, **W**...zoo
15(32)	Jefferson Blvd, **E**...Getty/diesel, Global/diesel, Motel 6, **W**...Ford Trucks, Ryder
14(31)	RI 37, Post Rd, to US 1, **W**...HOSPITAL, Exxon
13(30)	**1 mi E**...TF Green Airport, MainStay Suites, Quality Suites, Residence Inn
12b(29)	RI 2, I-295 N(from sb)
a	RI 113 E, to Warwick, **E**...Sunoco, Texaco/diesel, Holiday Inn/rest., **W**...mall
11(29)	I-295 N(exits left from nb), to Woonsocket
10b a(28)	RI 117, to Warwick, **E**...HOSPITAL
9(25)	RI 4, E Greenwich, no facilities
8b a(24)	RI 2, E Greenwich, **E**...MEDICAL CARE, Texaco/diesel/mart, Jason's Grill, McDonald's, Outback Steaks, Ruby Tuesday, Ro-Jack's Foods, Sears HomeLife, Walgreen, West Marine, cinema, **W**...CHIROPRACTOR, VETERINARIAN, Gulf, Sunoco/diesel/mart, Applebee's, Little Chopsticks, Papa Gino's, Wendy's, Fairfield Inn, Open Gate Motel, Arlington RV, Daewoo, Goodyear/auto, Honda/Volvo/VW, Mazda/Hyundai, Mitsubishi, Nissan, PepBoys, Pontiac/GMC, PostNet, Stop'n Shop Foods, bank, mall
7(21)	to Coventry, **E**...Mobil/diesel, **W**...Wendy's/rv parking
6a(20)	Hopkins Hill Rd, **W**...park'n ride
6(18)	RI 3, to Coventry, **W**...Sunoco/diesel/repair, Texaco/diesel/mart/24hr, Bess Eaton, Dunkin Donuts, Mark's Grill, Best Western, Congress Inn/rest.
5b a(15)	RI 102, **1-2 mi E**...Mobil/diesel, Trkstp/diesel/mart/rest./24hr, **W**...Classic Motel
10mm	**weigh sta both lanes**
4(9)	RI 3, to RI 165, Arcadia, **W**...Arcadia SP, camping
3b a(7)	RI 138 E, Wyoming, to Kingston, **E**...Bess Eaton, Dunkin Donuts, McDonald's, Wendy's, Job Lots, NAPA, Stop'n Shop Foods, **W**...EYECARE, Exxon/repair, Gulf/mart, Mobil, Sunoco/mart, Bess Eaton, Bickford's Rest., Pizza Kingdom, Village Pizza, Sun Valley Inn/rest., Big A Parts, Ocean Drug
6mm	**Welcome Ctr nb, full(handicapped)facilities, phone, picnic tables, litter barrels, vending, petwalk**
2(4)	to RI 3, Hope Valley, **W**...gas, food
1(1)	RI 3, to Hopkinton, **E**...HOSPITAL, Frontier Camping, to Misquamicut SP, beaches
0mm	Rhode Island/Connecticut state line, motorist callboxes begin nb/end sb

N

S

Rhode Island Interstate 295(Providence)

Exit #	Services
2b a(4)	I-95, N to Boston, S to Providence. I-295 begins/ends on I-95, exit 4 in MA.
1b a(2)	US 1, **E**...Charlie's Rest., Borders, Circuit City, Filene's, JC Penney, Lenscrafters, Linens'n Things, Marshall's, Michael's, Pier 1, Sears, Wal-Mart/McDonald's, mall, **W**...Exxon/mart, Shell/Subway/mart/wash/atm, Applebee's, Holiday Inn Express, Pineapple Inn, Super 8, Ford, Dodge/Kia, Nissan, Toyota, cinema
0mm	Rhode Island/Massachusetts state line. **Exits 1-2 are in MA.**
11(24)	RI 114, to Cumberland, **E**...Sunoco/diesel/mart, Texaco/diesel/mart, Diamond Hill Pizza, Honeydew Donuts, Chapel Drug, IGA Foods, bank, **W**...DENTIST, CHIROPRACTOR, VETERINARIAN, Sakis Pizza/subs, Diamond Hill SP
10(21)	RI 122, **E**...Burger King, Dunkin Donuts/bagels, Jim's Deli, Ronzio's Pizza, SpeeDee Lube'n Tune, **W**...DENTIST, Pamfilio's Italian, Pastry Gourmet, Brooks Drug, CVS Drug, Newport Creamery, NHD Hardware, Ro-Jack's Foods, bank, cleaners
20mm	Blackstone River
19.5mm	parking area both lanes, phone
9b a(19)	RI 146, Woonsocket, Lincoln, **E**...mall, **2 mi W**...Mobil, Chrysler/Plymouth, Honda, Isuzu
8b a(16)	RI 7, N Smithfield, **W**...Citgo/diesel/mart, House of Pizza, Smith-Appleby House
7b a(13)	US 44, Centerdale, **E**...HOSPITAL, Sunoco/diesel/repair, Texaco/mart/atm, Anthony's Rest., Club 44 Rest., Parts+, cleaners, **W**...MEDICAL CARE, Exxon, Mobil/mart, Texaco, Applebee's, Burger King, Chilo's Rest., D'Angelo's, Dunkin Donuts, Ho-Ho Chinese, KFC, McDonald's, Papa Gino's, Pizza Hut, Subway, Wendy's, Brooks Drug, CVS Drug, Radio Shack, Stop'n Shop Foods, TJ Maxx, bank, cinema, to Powder Mill Ledges WR
6(10)	US 6, to Providence, **E**...Mobil, Shell/Dunkin Donuts/wash/atm, Burger King, KFC, Luigi's Italian, McDonald's, Quikava, Wendy's, Hi-Way Motel, AutoZone, BJ's Wholesale, Chevrolet/Buick, CVS Drug, Honda, JiffyLube, Pontiac/GMC, Midas, Shaw's Foods, USPO, bank, **W**...Bel-Air Motel
5(9)	US 6 E Expswy
4(7)	RI 14, Plainfield Pk, **W**...Mobil/diesel
3b a(4)	rd 37, Phenix Ave, **E**...TF Green Airport
2(2)	RI 2 S, to Warwick, **E**...Pizzaria Uno, Ames, Filene's, JC Penney, Macy's, Marshall's, Radio Shack, cinema, mall, **W**...VETERINARIAN, Mobil/mart, Sunoco/mart, Texaco/repair, Bickford's Rest., Chili's, Dunkin Donuts, HomeTown Buffet, Kwikava, LoneStar Steaks, McDonald's, Papa Gino's, Subway, Taco Bell, Walt's Roast Beef, Brooks Drug, CompUSA, Dodge/Subaru, Home Depot, JiffyLube, Kinko's, MensWearhouse, Midas Muffler, NAPA, OfficeMax, PetsMart, Pier 1, Sam's Club, Saturn, Sears/auto, Sports Authority, Staples, TJ Maxx, TownFair Tire, mall
1(1)	RI 113 W, to W Warwick, same as 2
0mm	I-295 begins/ends on I-95, exit 11.

N

S

South Carolina Interstate 20

Exit #	Services
141b a	I-95, N to Fayetteville, S to Savannah. I-20 begins/ends on I-95, exit 160. **See Interstate 95, exit 160a for services.**
137	SC 340, to Timmonsville, Darlington, **S**...BP/diesel/mart
131	US 401, SC 403, to Hartsville, Lamar, **N**...Exxon/diesel/mart, Phillips 66/diesel/repair/rest./24hr, to Darlington Int Raceway
129mm	**rest area both lanes, no facilities, litter barrels**
123	SC 22, Lamar, **N**...Amoco/mart, Lee SP
121mm	Lynches River
120	SC 341, Bishopville, Elliot, **N**...HOSPITAL, Bishopville Motel(3mi), **S**...Citgo/diesel, Taste of Country Rest., Howard Johnson Express
116	US 15, Bishopville, to Sumter, Shaw AFB, **N**...Shell/KFC/diesel/mart/24hr, McDonald's, Pizza Hut, Subway(1mi), Waffle House, Econolodge, **2 mi N**...Citgo, Hardee's, Southland Motel, SC Cotton Museum, **S**...Wilco/BP/DQ/Wendy's/Stuckey's/diesel/mart/24hr
108	SC 34, to SC 31, Manville, **N**...Shell/diesel/mart, **S**...Citgo/diesel/mart
101	SC 91, SC 189, no facilities
98	US 521, to Camden, **N**...HOSPITAL, Exxon/mart, Shell/mart, Fairfield Inn, **3-5 mi N**...Golden Corral, Huddle House, McDonald's, Colony Inn, Greenleaf Inn, Knight's Inn, Revolutionary War Pk
96mm	Wateree River
93mm	**rest area both lanes, full(handicapped)facilities, phones, vending, picnic tables, litter barrels, petwalk**
92	US 601, to Lugoff, **N**...Pilot/DQ/Subway/diesel/mart/24hr, Shell/diesel/mart, Hardee's, Waffle House, Day's Inn, Ramada Ltd, Camden RV Park(1mi), **2-3 mi N**...KFC, McDonald's, Shoney's Inn/rest., Holiday Inn, Ford/Lincoln/Mercury
87	SC 47, to Elgin, **N**...BP/diesel/mart, Texaco/diesel
82	SC 53, to Pontiac, **S**...Clothing World Outlet
80	Clemson Rd, **N**...Exxon/diesel/mart/atm, Hess/diesel/mart, 76/Circle K, Shell/Bojangles/mart, McDonald's, Waffle House, Zaxby's Rest., Holiday Inn Express, CVS Drug
76b	Alpine Rd, to Ft Jackson, **N**...Sesquicentennial SP, **S**...gas(1mi)
a	(76 from eb), I-77, N to Charlotte, S to Charleston
74	US 1, Two Notch Rd, to Ft Jackson, **N**...MEDICAL CARE, Amoco/mart/wash/24hr, Exxon/mart, 76/mart/wash, Arby's, Burger King, Chili's, Denny's, Fazoli's, Hops Grill, IHOP, Lizard's Thicket, Outback Steaks, Ryan's, Waffle House, AmeriSuites, Best Western, Baymont Inn, Comfort Inn, Fairfield Inn, Hampton Inn, Holiday Inn, Microtel, Motel 6, Ramada Plaza, Red Roof Inn, Travelodge, Home Depot, TrueValue, USPO, U-Haul, bank, cleaners, to Sesquicentennial SP, **S**...EYECARE, Hess/diesel/mart, Texaco/diesel/mart/atm, Applebee's, BBQ, Bojangles, Capt D's, Church's Chicken, Godfather's, Golden Phoenix Rest., Hardee's, Honeybaked Ham, McDonald's, O'Charley's, Pizza Hut, Quincy's, Schlotsky's, Splendid China Buffet, Substation II, Wendy's, Day's Inn, Advance Parts, Best Buy, Dillard's, Family$, Firestone/auto, JC Penney, JiffyLube, Kinko's, K-Mart/auto, Lowe's Bldg Supply, MailBox, Marshall's, Midas Muffler, NAPA, Old Navy, Pier 1, PrecisionTune, Sears/auto, Staples, Subaru/Hyundai/Porsche/VW, Toyota, ToysRUs, Volvo, Winn-Dixie, bank, cinema, mall
73b	SC 277 N, to I-77 N, no facilities
a	SC 277 S, to Columbia, **S**...HOSPITAL
72	SC 555, Farrow Rd, **S**...Courtyard
71	US 21, N Main, Columbia, to Blythewood, **N**...Amoco/Subway/Pizza Hut/diesel, BP/mart, Exxon/mart, United Gas, Denny's, McDonald's, Day's Inn, truckwash, **S**...Shell/mart, bank
70	US 321, Fairfield Rd, **S**...Exxon/mart, Flying J/Conoco/Hardee's/diesel/LP/atm/24hr, Super 8
68	SC 215, Monticello Rd, to Jenkinsville, **N**...Exxon/diesel/mart, Texaco/diesel/mart/atm, **S**...Shell/mart, Sunrise Café
66mm	Broad River
65	US 176, Broad River Rd, to Columbia, **N**...Exxon/mart/atm, 76/Circle K/wash, Applebee's, Bojangles, Monterrey Mexican, Rush's Rest., Subway, Waffle House, Best Inn, All Tune/lube/wash, Aamco, Champion Brake/muffler, **S**...DENTIST, Hess Gas, RaceTrac/mart, Arby's, Capt Tom's Seafood, Chicago Smokehouse, Church's Chicken, Golden Corral, Hooters, Julie's Place, KFC, Lizard's Thicket, McDonald's, Pizza Hut, Sandy's HotDogs, Taco Bell, Taco Cid, Wendy's, American Inn, Econolodge, La Quinta, Royal Inn, Suburban Lodge, Advance Parts, A&P, Belk, CVS Drug, Eckerd, Family$, Food Lion, JB White, JiffyLube, Links Golf Ctr, MailBoxes Etc, Mr Transmission, PepBoys, PrecisionTune, Service Merchandise, Tapp's, bank, cleaners, mall

E

W

South Carolina Interstate 20

Columbia

64b a	I-26, US 76, E to Columbia, airport, W to Greenville, Spartanburg
63	Bush River Rd, **N**...DENTIST, 76/Circle K/diesel/wash, Burger King, Cracker Barrel, Steak-Out, Subway, Travelodge, CVS Drug, Hamrick's, Waccamaw Outlet, **S**...Exxon/mart/wash, El Chico, Fuddrucker's, Key West Grill, Villa Rest., Waffle House, Best Western, Courtyard, Knight's Inn, Sheraton, Sleep Inn
61	US 378, W Cola, **5 mi N**...Ryan's, Wendy's, **S**...HOSPITAL, Amoco/diesel/mart/24hr, 76/mart(2mi), Waffle House, antiques
58	US 1, W Columbia, **N**...Exxon/mart/wash, Texaco/Subway/diesel/mart/atm, Waffle House, **2 mi S**...Hess/mart, Burger King, KFC, McDonald's, Barnyard RV Park, to airport
55	SC 6, to Lexington, **N**...Texaco/mart, Hardee's, Comfort Inn(2mi), Hampton Inn(2mi), Holiday Inn Express(3mi), CarQuest, bank, **S**...BP/diesel/mart, Citgo/mart, Depot/DQ/mart, Bojangles, Fox's Pizza Den, Golden Town Chinese, McDonald's, Substation II, Subway, Waffle House, Ramada Ltd, CVS Drug, Piggly Wiggly, bank, cleaners
52.5mm	weigh sta wb
51	SC 204, to Gilbert, **N**...Exxon/diesel/mart, **S**...Shell/Gertie's/KrispyKreme/diesel/mart/atm/24hr, Texaco/Subway/diesel/mart/atm/24hr
48mm	parking area both lanes, picnic tables, litter barrels
44	SC 34, to Gilbert, **N**...Citgo/44Trkstp/diesel/rest.
39	US 178, to Batesburg, **N**...Exxon/diesel/mart, **S**...Citgo/diesel/mart/24hr
35.5mm	weigh sta eb
33	SC 39, to Wagener, **N**...BP/diesel, Shell/diesel/mart, Huddle House
29	SC 49, no facilities
22	US 1, to Aiken, **S**...BP/diesel/mart, RaceTrac/mart, Shell/Blimpie/diesel/mart, Smile/diesel/mart, Hardee's, McDonald's, Waffle House, Day's Inn, Holiday Inn Express, Pine Acres RV Camp(5mi)
20mm	**parking areas both lanes, litter barrels**
18	SC 19, to Aiken, **S**...HOSPITAL, Exxon/Subway/diesel/mart, Shell/Blimpie/diesel/mart, Waffle House, Deluxe Inn, Inn of Aiken
11	SC 144, Graniteville, no facilities
5	US 25, SC 121, **N**...Bryants/diesel/mart, Fuel City/Blimpie/pizza/diesel/mart/24hr, Shell/Bojangles/diesel/mart/24hr, Burger King, Hardee's, Huddle House, Sonic, Winn-Dixie/24hr, cleaners, **S**...Exxon/Taco Bell/diesel/mart, Speedway/diesel/mart/24hr, Waffle House
1	SC 230, Martintown Rd, N Augusta, **S**...Smile/Phillips 66/Blimpie/diesel/mart/24hr, Tastee Freez, Waffle House, to Garn's Place
.5mm	**Welcome Ctr eb, full(handicapped)facilities, phone, picnic tables, litter barrels, vending, petwalk**
0mm	Savannah River, South Carolina/Georgia state line

South Carolina Interstate 26

Charleston

Exit #	Services
	I-26 begins/ends on US 17 in Charleston, SC.
221	Meeting St, Charleston, **2 mi E**...Visitors Ctr, Best Western, Hampton Inn
221b	US 17 N, to Georgetown
a	US 17 S, to Kings St, to Savannah, HOSPITAL
220	Romney St(from wb), no facilities
219b	Morrison Dr, East Bay St, **N**...Shell/mart, Sam's Gas/mart, Huddle House
a	Rutledge Ave(from eb, no EZ return), to The Citadel, College of Charleston
218	Spruill Ave(from wb), N Charleston, no facilities
217	N Meeting St(from eb), no facilities
216b a	SC 7, Cosgrove Ave, **N**...HOSPITAL, DENTIST, Murray's Deli, StayOver Lodge, **S**...to Charles Towne Landing
215	SC 642, Dorchester Rd, N Charleston, **N**...Hess/diesel, Hardee's, Howard Johnson, Bus Depot, laundry, **S**...Amoco/diesel/mart/wash, Alex's Rest./24hr, Villager Lodge, NAPA AutoCare/tires
213b a	Montague Ave, Mall Dr, **N**...CHIROPRACTOR, Burger King, Piccadilly's, Red Lobster, Hilton, Charles Towne Square, Firestone, Ward's/auto, bank, **1 mi N on US 52/78**...Exxon/mart, Hess/mart, Catalina Inn, tires, **S**...DENTIST, Ashley Gas, Citgo/mart, Sports Rock Café, Waffle House, Comfort Inn, Day's Inn, Extended Stay America, Hampton Inn, Orchard Inn, Quality Suites, Ramada
212c b	I-526, E to Mt Pleasant, W to Savannah, airport
a	Remount Rd(from wb, no EZ return), Hanahan, **N on US 52/78**...DENTIST, Exxon, Hess/diesel, Shell/mart, BBQ, Burger King, Charleston Cattle Co, Chinese Rest., KFC, McDonald's, Old Country Buffet, Taco Bell, Tomtortaro Italian, Baby Superstore, Ford, Dodge, GreaseMonkey, Office Depot, Piece Goods Whse, bank, cleaners/laundry
211b a	Aviation Pkwy, **N on US 52/78**...Amoco, Exxon/mart/wash, Shell/mart, Arby's, Burger King, Capt D's, Chinese Rest., Church's Chicken, Grandy's, Huddle House, La Hacienda Mexican, McDonald's Express, Pizza Hut, Pizza Inn, Seafare Seafood, Shoney's, Substation II, Subway, Tokyo Japanese, Wendy's, Aviation House Hotel, Masters Inn, Radisson, Aamco, BabysRUs, Batteries+, Circuit City, Goodyear, PepBoys, Photo Shop, Sam's Club, U-Haul, bank, **S**...Speedway/diesel/mart, Waffle House, Budget Inn, Holiday Inn, bank, cinema

South Carolina Interstate 26

E

W

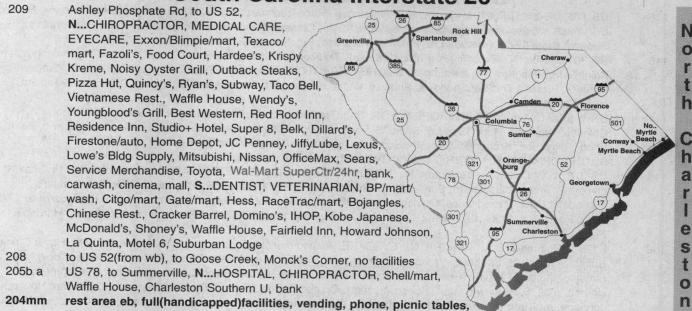

Exit	Description
209	Ashley Phosphate Rd, to US 52, **N**...CHIROPRACTOR, MEDICAL CARE, EYECARE, Exxon/Blimpie/mart, Texaco/mart, Fazoli's, Food Court, Hardee's, Krispy Kreme, Noisy Oyster Grill, Outback Steaks, Pizza Hut, Quincy's, Ryan's, Subway, Taco Bell, Vietnamese Rest., Waffle House, Wendy's, Youngblood's Grill, Best Western, Red Roof Inn, Residence Inn, Studio+ Hotel, Super 8, Belk, Dillard's, Firestone/auto, Home Depot, JC Penney, JiffyLube, Lexus, Lowe's Bldg Supply, Mitsubishi, Nissan, OfficeMax, Sears, Service Merchandise, Toyota, Wal-Mart SuperCtr/24hr, bank, carwash, cinema, mall, **S**...DENTIST, VETERINARIAN, BP/mart/wash, Citgo/mart, Gate/mart, Hess, RaceTrac/mart, Bojangles, Chinese Rest., Cracker Barrel, Domino's, IHOP, Kobe Japanese, McDonald's, Shoney's, Waffle House, Fairfield Inn, Howard Johnson, La Quinta, Motel 6, Suburban Lodge
208	to US 52(from wb), to Goose Creek, Monck's Corner, no facilities
205b a	US 78, to Summerville, **N**...HOSPITAL, CHIROPRACTOR, Shell/mart, Waffle House, Charleston Southern U, bank
204mm	**rest area eb, full(handicapped)facilities, vending, phone, picnic tables, litter barrels, petwalk**
203	College Park Rd, Ladson, **N**...BP/mart/wash, Speedway/diesel/mart, McDonald's, Waffle House, Day's Inn, **S**...VETERINARIAN, Citgo/mart, Hess, Burger King, Taco Bell, CVS Drug, Piggly Wiggly, KOA, auto repair, flea mkt, laundry
202mm	**rest area wb, full(handicapped)facilities, vending, phone, picnic tables, litter barrels, petwalk**
199b a	US 17 A, to Moncks Corner, Summerville, **N**...CHIROPRACTOR, DENTIST, VETERINARIAN, Citgo/diesel/mart/wash, Hess/diesel/mart, Shell/mart, Speedway/diesel/mart, China Chef, KFC, Pizza Hut, Ruthy's Diner, Subway, Tasty Express Burgers, Advance Parts, BiLo, Buick/Pontiac/GMC, CVS Drug, Eckerd, Food Lion, Ford/Mercury, Keith's Auto Parts, auto repair, bank, cleaners/laundry, **S**...CHIROPRACTOR, Amoco/TCBY/diesel/mart, BP/diesel/mart/wash, Enmark/diesel/mart, Blimpie, Bojangles, Burger King, Fazoli's, Hardee's, Huddle House, La Hacienda Mexican, Papa John's, Perkins/24hr, Quincy's, Ryan's, Shoney's, Sticky Fingers Rest., Waffle House, Comfort Inn, Econolodge, Hampton Inn, Holiday Inn Express, Ramada, Advance Parts, Belk, Chevrolet/Olds, GNC, JiffyLube, Lowe's Bldg Supply, MailBoxes Etc, Meineke Muffler, Park Auto Parts, Radio Shack, Staples, SuperPetz, Wal-Mart/drugs, Winn-Dixie/deli, bank, tires
194	SC 16, to Jedburg, access to Foreign Trade Zone 21, no facilities
187	SC 27, to Ridgeville, St George, **S**...BP/diesel/mart, **10 mi S**...Francis Beidler Forest
177	SC 453, Harleyville, to Holly Hill, **S**...BP/diesel/mart/wash/LP
175mm	weigh sta wb
173mm	weigh sta eb
172b a	US 15, to Santee, St George, **N**...OneStop/diesel/café, **S**...Amoco/diesel/rest./24hr
169b a	I-95, N to Florence, S to Savannah
165	SC 210, to Bowman, **N**...Exxon/diesel/mart, **S**...BP/diesel/rest./mart, Texaco/Blimpie/diesel/mart/repair
159	SC 36, to Bowman, **N**...Speedway/diesel/mart/24hr/repair, Blimpie, Church's Chicken
154b a	US 301, to Santee, Orangeburg, **N**...Days Inn, **S**...BP/Blimpie/diesel/mart, Texaco/diesel/mart/24hr, Best Western(7mi), Holiday Inn(8mi), auto repair
152mm	**rest area wb, full(handicapped)facilities, vending, phone, picnic tables, litter barrels, petwalk**
150mm	**rest area eb, full(handicapped)facilities, vending, phone, picnic tables, litter barrels, petwalk**
149	SC 33, to Cameron, Orangeburg, to SC State Coll, Claflin Coll
145b a	US 601, to Orangeburg, St Matthews, **S**...HOSPITAL, Amoco/mart/24hr, Citgo/diesel/mart, Exxon/mart, Speedway/diesel/mart, Texaco/Burger King/mart, Fatz Café, Hardee's, KFC, McDonald's, Rita's Café, Waffle House, Best Western, Carolina Lodge, Comfort Inn, Day's Inn, Hampton Inn, Howard Johnson, Southern Lodge, Cadillac/Olds/Nissan, Chevrolet, bank
139	SC 22, to St Matthews, **S**...Shell/mart(1mi), Sweetwater Lake Camping(2.5mi)
136	SC 6, to North, Swansea, **N**...BP/mart, Exxon/diesel/LP/rest./24hr, auto/tire repair
129	US 21, **N**...Shell/diesel/mart
125	SC 31, to Gaston, **N**...Wolfe's Truck/trailer repair
123mm	**rest area both lanes, full(handicapped)facilities, vending, phone, picnic tables, litter barrels, petwalk**
119	US 176, US 21, to Dixiana, **N**...BP/diesel/mart, **S**...Cheapo's/diesel/rest./mart/showers
116	I-77 N, to Charlotte, US 76, US 378, to Ft Jackson

South Carolina Interstate 26

115 US 176, US 21, US 321, to Cayce, **N**...Amoco/mart, RaceTrac/mart/24hr, Waffle House, auto repair, bank, **S**...Pilot/ DQ/Wendy's/diesel/mart/24hr, United/diesel/mart/24hr, Chinese Rest., Hardee's, Huddle House, McDonald's, Subway, Steak&Egg, Ramada Ltd, PakMail, Firestone, Piggly Wiggly, cleaners

113 SC 302, Cayce, **N**...Exxon/mart, Phillips 66/mart, Texaco/diesel/mart, Airport Inn, Best Western/RV parking/rest., Knight's Inn, Masters Inn, tires/repair, **S**...CHIROPRACTOR, Exxon/mart, RaceTrac/mart, 76/Circle K, Burger King, Denny's, Lizard's Thicket, Shoney's, Subway, Waffle House/24hr, Comfort Inn, Courtyard, Day's Inn, Sleep Inn, NAPA, airport

111b a US 1, to W Columbia, **N**...CHIROPRACTOR, RaceTrac/diesel/mart, 76/Circle K, BBQ, Domino's, Dragon City Chinese, Hardee's, Ruby Tuesday, Sonic, Subway, TCBY, Waffle House, Delta Motel, Holiday Inn, Super 8, BiLo Foods, JiffyLube, Kroger/deli/24hr, PetSupplies+, Wal-Mart SuperCtr/gas/24hr, carwash, to USC, **S**...CHIROPRACTOR, VETERINARIAN, CVS Drug, Lowe's Bldg Supply, U-Haul

110 US 378, to W Columbia, Lexington, **N**...DENTIST, MEDICAL CARE, VETERINARIAN, Conoco/diesel, Phillips 66/ diesel/LP, Texaco/diesel/mart, Burger King, Chinese Rest., Greek Rest., Lizard's Thicket Rest., McDonald's, Millender's BBQ, Pizza House, Quincy's, Rush's Rest., Waffle House, Day's Inn, Hampton Inn, Ramada/rest., CVS Drug, Food Lion, Eckerd's, U-Haul, bank, cleaners/laundry, **S**...HOSPITAL, BP/mart/wash, Bojangles, Hardee's, Pizza Hut

108 I-126 to Columbia, Bush River Rd, **N**...EYECARE, Texaco/diesel/mart/atm, Capt D's, Hardee's, KFC, Royal China Buffet, Schlotsky's, Shoney's, Wendy's, Baymont Inn, Villager Lodge, Baby Depot, Burlington Coats, CopyMax, Firestone/auto, JB White, K-Mart/auto, Luxury Linens, Midas Muffler, Office Depot, OfficeMax, PartsAmerica, cinema, mall, **S**...Raceway/mart, Speedway/mart, Cracker Barrel, Lizard's Thicket, Courtyard, Day's Inn, Howard Johnson, **multiple facilities 1-3 mi N off I-126, Greystone Blvd, N**...Amoco, Texaco/mart, Waffle House, Embassy Suites, Extended Stay America, Residence Inn, Studio+, Chrysler/Jeep, Dodge, Ford, Lincoln/Mercury, Honda, Mitsubishi, Olds/GMC, **S**...Riverbanks Zoo

107b a I-20, E to Florence, W to Augusta

106b a St Andrews Rd, **N**...Exxon/Blimpie/mart, Armadillo Oil Co Rest., ChuckeCheese, IHOP, Papa John's, Economy Inn, Motel 6, Buick/Pontiac, Infiniti, Jaguar, Kroger/deli, Nissan, RV Ctr, **S**...DENTIST, VETERINARIAN, BP/diesel/mart/ wash/24hr, Hess/mart, 76/Circle K, Shell/mart, Bojangles, Burger King, Chinese Rest., Domino's, E Bourbon St Café, Famous HotDogs, Maurice's BBQ, McDonald's, Old Country Buffet, Pizza Hut, Ryan's, Steak&Ale, Substation II, TCBY, Waffle House, Wendy's, Western Steer, Holiday Inn Express, Red Roof Inn, Relax Inn, Eckerd, Food Lion, Piece Goods, Piggly Wiggly, Rite Aid, TireAmerica, TJ Maxx, cinema, cleaners/laundry

104 Piney Grove Rd, **N**...BP/diesel/mart/wash, Speedway/diesel/mart, Texaco/diesel/rest., Hardee's, Lizard's Thicket, Waffle House, Comfort Inn, Econolodge, Knight's Inn, $General, to Subaru, RV/Marine Ctr, **S**...Shell/mart, auto repair

103 Harbison Blvd, **N**...Applebee's, Hops Grill, Hampton Inn, Wingate Inn, Chevrolet, Home Depot, Lowe's Bldg Supply, funpark, **S**...MEDICAL CARE, EYECARE, Amoco/mart/wash/24hr, Exxon, Hess/diesel/mart, Shell, Blimpie, Carrabas, Chili's, Chick-fil-A, Denny's, Fazoli's, Kenny Rogers, McAlister's Grill, McDonald's, Olive Garden, Outback Steaks, Roadhouse Grill, Ruby Tuesday, Rush's Rest., Shoney's, Sonic, Subway, Texas Roadhouse, Country Inn Suites, Fairfield Inn, Suite 1, TownePlace Suites, BiLo, Belk, Best Buy, Books-A-Million, Dillard's, Gateway Computers, JB White, KidsRUs, Marshall's, MensWearhouse, OfficeMax, Parisian, PharMor Drug, Pier 1, Rite Aid, Sam's Club, Saturn, Sears/auto, SpeeDee Lube, Staples, Wal-Mart SuperCtr/24hr, bank, cameras, mall

102 SC 60, Ballentine, Irmo, **N**...AmeriSuites, HomeGate Suites, Wellesley Inn, **S**...DENTIST, Exxon, Texaco, BBQ, Cookies By Design, Papa John's, TCBY, CVS Drug, Piggly Wiggly, same as 103

101 US 76, US 176, to N Columbia, **1/2 mi N**...DENTIST, Exxon/Subway/diesel/mart/wash, China House, Food Lion/ deli, Package Express, carwash, **S**...Amoco/mart, 76/repair/mart, Baskin-Robbins/Burger King, Waffle House

97 US 176, to Ballentine, **N**...Kathryn's Seafood(1mi), Wood Smoke Camping, **S**...Amoco/diesel/mart

94mm weigh sta wb, phones

91 SC 48, to Chapin, **S**...Amoco/diesel/mart, Pitt Stop/fuel/mart, Texaco/diesel, Capital City Subs, Hardee's(3mi), McDonald's, Subway(2mi), Waffle House, to Dreher Island SP

88mm parking area wb, no facilities, litter barrels

85 SC 202, Little Mountain, Pomaria, **S**...to Dreher Island SP

84mm parking area eb, no facilities, litter barrels

82 SC 773, to Prosperity, Pomaria, **N**...Amoco/Subway/diesel/mart, Farmer's/Flea Mkt

81mm weigh sta eb

76 SC 219, to Pomaria, Newberry, **S**...to Newberry Opera House

74 SC 34, to Newberry, **N**...BP/mart, Shell/diesel/mart/24hr, Best Western/rest., **S**...HOSPITAL, Texaco/diesel/mart, Bill&Fran's Café, Capt D's, Hardee's(2mi), McDonald's(2mi), Waffle House, Comfort Inn, Day's Inn, Economy Inn(2mi), Holiday Inn Express(2mi), to NinetySix HS

72 SC 121, to Newberry, **N**...Exxon/mart, **S**...HOSPITAL, Shell/Blimpie/diesel/mart, to Newberry Coll

66 SC 32, to Jalapa, no facilities

South Carolina Interstate 26

63.5mm	**rest area both lanes, full(handicapped)facilities, phone, vending, picnic tables, litter barrels, petwalk**
60	SC 66, to Joanna, **S**...Exxon/diesel/mart, Chevron/mart, KOA(1mi)
54	SC 72, to Clinton, **N**...Citgo/diesel/mart/24hr, **S**...HOSPITAL, Texaco/diesel/mart, to Presbyterian Coll, Thornwell Home
52	SC 56, to Clinton, **N**...Speedway/diesel/mart/24hr, Blue Ocean Rest., McDonald's, Waffle House, Comfort Inn, truck repair, **S**...HOSPITAL, Phillips 66/diesel, Texaco/diesel/mart, Hardee's, Waffle House, Wendy's, Day's Inn, Ramada Inn, Travelers Inn(2mi)
51	I-385, to Greenville(from wb), no facilities
45.5mm	Enoree River
44	SC 49, to Cross Anchor, Union, no facilities
43mm	parking area both lanes, picnic tables, litter barrels
41	SC 92, to Enoree, **S**...Phillips 66/mart
38	SC 146, to Woodruff, **N**...Hot Spot/Shell/Hardee's/diesel/mart/24hr, Big Country Rest./lounge
35	SC 50, to Woodruff, **S**...HOSPITAL, Citgo/diesel/mart/rest./24hr
33mm	S Tyger River
32mm	N Tyger River
28	US 221, to Spartanburg, **N**...HOSPITAL, Citgo/diesel/mart/wash/atm/24hr, Crown/mart(2mi), Shell/Subway/diesel/mart/atm/24hr, Burger King, Pine Ridge Camping(3mi), Walnut Grove Seafood Rest., to Walnut Grove Plantation
22	SC 296, Reidville Rd, to Spartanburg, **N**...CHIROPRACTOR, MEDICAL CARE, Amoco/diesel/mart, Exxon/mart, Arby's, BBQ, Capri's Italian, Chinese Rest., Fuddrucker's(1mi), Little Caesar's, McDonald's, Outback Steaks, Ryan's, Waffle House, Advance Parts, Midas Muffler, Nick's Wash/lube, Precision Tune, Rite Aid, bank, cinema, to Croft SP, **S**...DENTIST, BP/mart/wash, Crown/mart, Hess/diesel/mart, Phillips 66/mart, Baskin-Robbins, Burger King, Denny's, Domino's, Hardee's, Subway, TCBY, Waffle House/24hr, Sleep Inn, Southern Suites, Super 8, Abbot Farms/fruit/fireworks, BiLo, BMW, CVS Drug, Eckerd, Harris-Teeter/deli, Rick's Muffler's, Toyota, XpressLube, auto repair, bank
21b a	US 29, to Spartanburg, **N**...OPTICIAN, BP/mart/wash, Crown/mart/24hr, Exxon/Subway/mart, Phillips 66/wash, Burger King, Calzone's Rest., Checker's, Chick-fil-A, Chinese Rest., Dunkin Donuts, Hardee's, Hop's Grill, Japanese Rest., KFC, LoneStar Steaks, LJ Silver, Mr Gatti's, O'Charley's, Old Country Buffet, Papa John's, Pizza Hut, Pizza Inn, Quincy's, Red Lobster, Rock-Ola Café, Substation II, Wendy's, Barnes&Noble, Belk, Circuit City, Dillard's, Fabric Whse, Firestone/auto, Home Depot, JB White's, JC Penney, JiffyLube, K-Mart/24hr, LensCrafters, Office Depot, OfficeMax, PetsMart, Proffitt's, Rick's Wash/lube, Sears/auto, TJ Maxx, Upton'sbank, cinema, cleaners, mall, **S**...Citgo/diesel/mart, Texaco/diesel, Applebee's, Aunt M's Café, IHOP, McDonald's, Piccadilly's, Prime Sirloin, Taco Bell, Advance Parts, Best Buy, Ingles/deli, Lowe's Bldg Supply, Rite Aid, Sam's Club, Wal-Mart, Western Auto, bank, cleaners
19b a	Lp I-85, Spartanburg, **N**...Amoco/mart, Phillips 66/diesel/mart, Cracker Barrel, Budget Inn, Ramada Inn, Residence Inn, Quality Hotel, Wilson World Hotel, **S**...Hampton Inn, Tower Motel
18b a	I-85, N to Charlotte, S to Greenville
17	New Cut Rd, **S**...Chevron/diesel/mart, Speedway/diesel/mart, Burger King, Fatz Café, Hardee's, McDonald's, Waffle House, Comfort Inn, Day's Inn, Howard Johnson Express, Relax Express Inn, Foothills Factory Outlet/famous brands, Waccamaw Pottery
16	John Dodd Rd, to Wellford, **N**...Citgo/diesel/mart, Phillips 66/mart, Aunt M's
15	US 176, to Inman, **N**...HOSPITAL, FastStop/mart, 76/Circle K/diesel, Phillips 66/mart, Waffle House, **S**...Exxon/Blimpie/diesel/mart, Taco Bell(2mi)
10	SC 292, to Inman, **N**...HotSpot/Shell/Subway/KrispyKreme/diesel/mart/24hr
9.5mm	parking area both lanes, litter barrels
7.5mm	Lake William C. Bowman
5	SC 11, Foothills Scenic Dr, to Chesnee, Campobello, **N**...Citgo/Aunt M's Café/diesel/mart, **S**...Phillips 66/mart
3mm	**Welcome Ctr eb, full(handicapped)facilities, info, phones, picnic tables, litter barrels, vending, petwalk**
1	SC 14, to Landrum, **S**...BP/Burger King/diesel/mart, HotSpot/gas(2mi), Denny's, Pizza Hut(1mi), Ingles/deli/café/24hr
0mm	South Carolina/North Carolina state line

E

W

Spartanburg

South Carolina Interstate 77

Exit #	Services
91mm	South Carolina/North Carolina state line
90	US 21, Carowinds Blvd, E...HOSPITAL, Shell/mart/wash, World Gas, Denny's, Day's Inn, RV Ctr, RV camping(4mi), W...Citgo/mart, Exxon/mart/café, 76/Circle K/Wendy's/diesel, El Cancun Mexican, KFC, Mom's Rest., Shoney's, Comfort Inn, Holiday Inn Express, Motel 6, Ramada Inn, Sleep Inn, Carowinds Funpark, Carolina Pottery/outlet mall/famous brands, bank, camping(1mi), fireworks
89.5mm	**Welcome Ctr sb, full(handicapped)facilities, info, phone, vending, picnic tables, litter barrels, petwalk/weigh sta nb**
88	Gold Hill Rd, to Pineville, E...Texaco/dieselmart, W...Exxon/diesel, McDonald's(2mi), Ford, KOA
85	SC 160, Ft Mill, Tega Cay, E...Exxon, Subway, Winn-Dixie/café, New Peach Stand, W...BP/diesel/mart, Crown Gas, Burger King
84.5mm	weigh sta sb
83	SC 49, Sutton Rd, no facilities
82.5mm	Catawba River
82b a	US 21, SC 161, Rock Hill, Ft Mill, E...Exxon/mart/lube, Phillips 66, BBQ, IHOP, Holiday Inn/rest., Home Depot, PetsMart, XpressLube, flea mkt, museum, W...HOSPITAL, EYECARE, Amoco, Citgo/mart, Exxon/mart, Phillips 66/mart, RaceTrac/diesel/mart, Speedway/mart, Texaco/mart, Arby's, BBQ, Bojangles, Burger King, Capt's Galley Seafood, Chick-fil-A, CiCi's, Denny's, El Cancun Mexican, Godfather's, Little Caesar's, LJ Silver, McDonald's, Outback Steaks, Pizza Hut, Pizza Inn, Sagebrush Steaks, Sakura Chinese, Shoney's, Subway, Taco Bell, TCBY, Waffle House, Wendy's, Bestway Inn, Best Western, Comfort Inn, Country Inn Suites, Courtyard, Day's Inn, Econolodge, Howard Johnson, Microtel, Ramada/rest., Regency Inn, Advance Parts, Aldi, AutoZone, BiLo, Brake Depot, Brake Xperts, Chevrolet, Chrysler/Jeep, Eckerd's, Family$, Firestone/auto, Ford, Honda, K-Mart/drugs, Midas Muffler, Muffler Masters/lube, Nissan, Office Depot, Peebles, PepBoys, PrecisionTune, SpeeDee Lube, Toyota, U-Haul, Winn-Dixie, XpertTire, bank, cinema, hardware, laundry
79	SC 122, Dave Lyle Blvd, to Rock Hill, E...BP/mart/wash, Applebee's, Cracker Barrel, Dairy Queen, Hardee's, J&K Cafeteria, Longhorn Steaks, O'Charley's, Ryan's, Hampton Inn, Wingate Inn, Belk, Food Lion, Honda, JC Penney, Lowe's Bldg Supply, Meineke Muffler, Sears/auto, Staples, Wal-Mart/drugs/24hr, cinema, cleaners, mall, W...Citgo/diesel/mart, Bob Evans, Roadside Grill, Wendy's, Hillside Inn, Hilton Garden, cinema, visitor ctr
77	US 21, SC 5, to Rock Hill, E...BP/Subway/diesel/mart, Citgo/diesel/mart/wash/24hr, to Andrew Jackson SP(12mi), W...Exxon/diesel/mart/wash, Phillips 66/diesel/mart, Bojangles, KFC, McDonald's, Waffle House, to Winthrop Coll
75	Porter Rd, E...Citgo/mart, Shell/mart
73	SC 901, to Rock Hill, York, E...Exxon/diesel/mart, W...HOSPITAL, Exxon(1mi)
66mm	**rest area both lanes, full(handicapped)facilities, phone, picnic tables, litter barrels, vending, petwalk**
65	SC 9, to Chester, Lancaster, E...Amoco/Subway/mart/24hr, BP/mart, Texaco/diesel/mart, Waffle House, Day's Inn, Econolodge, Relax Inn, W...HOSPITAL, Exxon/diesel/mart, Burger King, Country Omelet/24hr, KFC, McDonald's, Comfort Inn, Microtel, Super 8
62	SC 56, Richburg, to Fort Lawn, no facilities
55	SC 97, Great Falls, to Chester, E...BP/diesel/mart, Home Place Rest., W...HOSPITAL
48	SC 200, to Great Falls, E...Shell/diesel/mart, Grand Central Rest.
46	SC 20, to White Oak, no facilities
41	SC 41, to Winnsboro, E...to Lake Wateree SP
34	SC 34, Ridgeway, to Winnsboro, E...BP/diesel/24hr, Ridgeway Motel, camping(1mi), W...Citgo/diesel/store, Exxon/diesel, Waffle House, Ramada Ltd
27	Blythewood Rd, E...DENTIST, Citgo/diesel/mart/wash, Exxon/Bojangles/diesel/mart/24hr, Blythewood Pizza/subs, McDonald's, Oaks Farm/rest., Waffle House, Wendy's, WG's Wings, Comfort Inn, Day's Inn, Holiday Inn Express, IGA Foods, cleaners, tires
24	US 21, to Blythewood, E...BP/mart, Texaco/Subway/diesel/mart/atm, Taco Bell
22	Killian Rd, no facilities
19	SC 555, Farrow Rd, E...Texaco/mart/wash, Cracker Barrel, Courtyard, Longs Drug, bank, W...Waffle House, Carolina Research Pk, SC Archives
18	to SC 277, to I-20 W(from sb), Columbia, W...76/Circle K/diesel/mart/wash
17	US 1, Two Notch Rd, E...DENTIST, CHIROPRACTOR, Amoco/mart/wash/24hr, Exxon/mart, 76/Circle K/diesel, Arby's, Burger King, Denny's, Waffle House, Fairfield Inn, InTown Suites, Ramada Plaza, Wingate Inn, U-Haul, USPO, bank, to Sesquicentennial SP, W...MEDICAL CARE, Chili's, Fazoli's, Hop's Grill, IHOP, Lizard's Thicket, Outback Steaks, Waffle House, AmeriSuites, Comfort Inn, Econolodge, Hampton Inn, Holiday Inn, Microtel, Quality Inn, Red Roof Inn, Home Depot, Luxury Fabrics, TrueValue
16b a	I-20, W to Augusta, E to Florence, Alpine Rd
15b a	SC 12, to Percival Rd, W...Shell/mart, **1/2 mi** W...Exxon/mart, Phillips 66, Texaco/mart, auto repair
13	Decker Blvd(from nb), W...Exxon/mart, Phillips 66, Spinx Gas, Texaco/mart
12	Forest Blvd, Thurmond Blvd, E...to Ft Jackson, W...HOSPITAL, DENTIST, VETERINARIAN, BP/diesel, 76/diesel, Shell/diesel/mart/24hr, Bojangles, Chick-fil-A, Fatz Café, Golden Corral, Hardee's, McDonald's, RedBone Rest., Steak&Ale, Subway, Wendy's, Extended Stay America, Super 8, Cato's, $Tree, FastLube, Sam's Club, Uncle TV's, Wal-Mart SuperCtr/24hr, cinema, museum
10	SC 760, Jackson Blvd, E...to Ft Jackson, **2 mi** W...multiple services

Rock Hill

Columbia

South Carolina Interstate 77

9b a	US 76, US 378, Columbia, to Sumter, **E**...CHIROPRACTOR, EYECARE, MEDICAL CARE, VETERINARIAN, Amoco/mart/wash, Hess/mart, Speedway/diesel/mart, Capt D's, China Chef, Domino's, KFC, McDonald's, Pizza Hut, Rush's Rest., Subway, Waffle House, Best Western, Comfort Inn, Country Inn Suites, Day's Inn, Hampton Inn, Holiday Inn Express, Radisson, Shoney's Inn/rest., Sleep Inn, Advance Parts, BiLo, Family$, Firestone/auto, Food Lion, Goodyear/auto, Harris-Teeter/24hr, Lowe's Bldg Supply, MufflerWorks, NAPA, Piggly Wiggly, Pontiac/Olds/GMC, Radio Shack, U-Haul, USPO, Wal-Mart/auto, bank, cinema, cleaners/laundry, **W**...HOSPITAL, 76/Circle K(1mi), Hardee's, Sonic, Wendy's, Economy Inn, Big Lots, Eckerd, PrecisionTune
6b a	Shop Rd, **W**...to USC Coliseum, fairgrounds
5	SC 48, Bluff Rd, **1 mi E**...Exxon, **W**...Texaco/Burger King/Popeye's/diesel/mart, Bojangles
3mm	Congaree River
2	Saxe Gotha Rd, no facilities
1	US 21, US 176, US 321(from sb), Cayce, **W**...accesses same as SC I-26, exit 115.
0mm	I-77 begins/ends on I-26, exit 116.

South Carolina Interstate 85

Exit #	Services
106.5mm	South Carolina/North Carolina state line
106	US 29, to Grover, **E**...casino, **W**...Crown/mart/atm, Exxon/diesel/mart, Wilco/BP/DQ/Wendy's/diesel/mart
104	SC 99, **E**...Citgo/diesel/rest./24hr
103mm	**Welcome Ctr sb, full(handicapped)facilities, info, phone, picnic tables, litter barrels, vending, petwalk**
102	SC 198, to Earl, **E**...Exxon/diesel/mart, Phillips 66/diesel/mart, Hardee's, **W**...Flying J/Conoco/diesel/LP/rest./24hr, Texaco/diesel/mart, McDonald's, Waffle House
100mm	Buffalo Creek
100	SC 5, to Blacksburg, Shelby, **W**...Speedway/Subway/diesel/mart/24hr, Texaco/diesel/mart
98	Frontage Rd(from nb), **E**...Broad River Café
97mm	Broad River
96	SC 18(from sb), **E**...antiques, **W**...Exxon/diesel/mart
95	SC 18, to Gaffney, **E**...HOSPITAL, Citgo/mart/24hr, Exxon, Mr Petro/mart/Mr Waffle, Aunt M's Rest., Fatz Café, Pontiac/Buick/GMC, to Limestone Coll, **W**...Norma's I-85/diesel/rest./24hr
92	SC 11, to Gaffney, **E**...DENTIST, MEDICAL CARE, Amoco, Citgo/mart, Phillips 66/mart, Spinx/mart, Applebee's, Baker St Grill, Bojangles, Burger King, Chinese Rest., Domino's, McDonald's, Pizza Hut, Quincy's, Sagebrush Steaks, Shoney's, Sonic, Subway, Taco Bell, Wendy's, Holiday Inn Express, Jameson Inn, Advance Parts, Belk, BiLo, Eckerd, Ford/Mercury, Ingles Foods, Radio Shack, Wal-Mart, bank, to Limestone Coll, **W**...Chevron, Texaco/diesel/mart, Waffle House, Comfort Inn, Day's Inn, to The Peach, Foothills Scenic Hwy
90	SC 105, SC 42, to Gaffney, **E**...BP/DQ/diesel/mart/atm, Sleep Inn, **W**...Citgo/Burger King/mart, QuickStop/gas, Cracker Barrel, FoodCourt, Hampton Inn, Prime Outlets/famous brands
89mm	**rest area both lanes, full(handicapped)facilities, phone, picnic tables, litter barrels, vending, petwalk**
87	SC 39, **E**...PineCone Camping(1mi), **W**...World of Clothing, fruit stand/peaches/fireworks
83	SC 110, **1 mi E**...Hot Spot/gas, Subway, **W**...Citgo/Mr Waffle/diesel/24hr, to Cowpens Bfd
82	Frontage Rd(from nb), no facilities
81	Frontage Rd(from sb), no facilities
80.5mm	Pacolet River
80	SC 57, to Gossett, **E**...Citgo/Bullets/diesel/mart
78	US 221, Chesnee, **E**...Shell/mart, Hardee's, Waffle House, Motel 6, Sun'n Sand Motel, fruit stand, **W**...BP/mart/wash, Exxon/Burger King/diesel/mart, RaceTrac/diesel, BBQ, McDonald's, Wendy's, Best Western, Advance Parts, $General, Ingles Foods
77	Lp 85, Spartanburg, facilities along Lp 85 exits E
75	SC 9, Spartanburg, **E**...Red Roof Inn, **1 mi E**...Best Western, Comfort Inn, Day's Inn, **W**...BP/Burger King/LJSilver/diesel/mart, Phillips 66, RaceTrac/mart, Capri's Italian, Fatz Café, McDonald's, Pizza Hut, Waffle House, Jameson Inn, Advance Parts, CVS Drug, Ingles Food/café, USPO, Valvoline, carwash

South Carolina Interstate 85

Spartanburg

72	US 176, to I-585, **E**...Ramada Inn, Super 8, to USC-S, Wofford Coll, Converse Coll, **W**...Shell, RaceTrac
70b a	I-26, E to Columbia, W to Asheville
69	Lp 85, SC 41(from nb), to Fairforest, **E**...Hampton Inn, Tower Motel, camping
68	SC 129(from sb), to Greer, no facilities
67mm	N Tyger River
66	US 29, to Lyman, Wellford, no facilities
63	SC 290, to Duncan, **E**...Citgo, Denny's, Pizza Inn, Jameson Inn, Microtel, **W**...Amoco/mart, Pilot/Wendy's/TCBY/ diesel/mart/24hr, TA/BP/DQ/diesel/rest., Arby's, BBQ, Bojangles, Hardee's, McDonald's, Waffle House, Comfort Inn, Day's Inn, Super 8, Travelodge, RV Ctr, bank
62.5mm	S Tyger River
60	SC 101, to Greer, **E**...Amoco/mart, Citgo/diesel/mart, **W**...Coastal/gas, Exxon/Burger King/mart, Waffle House, Super 8
57	**W**...Greenville-Spartanburg Airport
56	SC 14, to Greer, **E**...Citgo/Stuckey's/DQ, Texaco/diesel/mart, **W**...HOSPITAL, Amoco/diesel/mart, Exxon/mart, Speedway/diesel/mart, Huddle House, Waffle House, Goodyear/auto, RV Ctr
55mm	Enoree River
54	Pelham Rd, **E**...Conoco/mart, BP/diesel/mart/wash, Citgo/diesel/mart/24hr, Brady's Rest., Burger King, Holiday Inn Express, **W**...Amoco/diesel/mart, Exxon/mart/wash, Applebee's, Atlanta Bread Co, Bluff's Rest., Boston Mkt, California Dreaming Rest., Chinese Rest., Hardee's, Joe's Crabshack, J&S Cafeteria, Logan's Roadhouse, Macaroni Grill, Max&Erma's, Mayflower Seafood, McDonald's, On the Border, Ruby Tuesday, Schlotsky's, Tony Roma, Comfort Inn, Extended Stay America, Fairfield Inn, Hampton Inn, MainStay Suites, Marriott, Microtel, Residence Inn, Wingate Inn, BiLo/24hr, CVS Drug, Ingles, MailExpress, bank, cinema, cleaners/laundry
52mm	weigh sta nb
51	I-385, SC 146, Woodruff Rd, **E**...VETERINARIAN, Hess/mart, Fatz Café, IHOP, Garden Ridge, Goodyear/auto, **W**...Amoco, BP/mart/wash, RaceTrac/mart/atm/24hr, Shell, Burger King, Capri's Italian, Cracker Barrel, McDonald's, Morrison's Cafeteria, Prime Sirloin, Rio Bravo, TGIFriday, Waffle House, Day's Inn, Embassy Suites, Fairfield Inn, Holiday Inn, La Quinta, Marriott(3mi), Microtel, Dillard's, Home Depot, JB White, Lowe's Bldg Supply, Proffitt's, Target, bank, cinema, mall
48b a	US 276, Greenville, **E**...Speedway/diesel/mart, BBQ, Waffle House, Red Roof, CarMax, **W**...BP, Exxon, Arby's, Bojangles, Boston Mkt, Burger King, Chinese Rest., Hooters, McDonald's, Old Country Buffet, Olive Garden, Ryan's, Shoney's, Taco Bell, T-Bone's Rest., Ted's Kitchen, Embassy Suites, Howard Johnson, Valu-Lodge, Villager Lodge, Acura, Audi/VW, Best Buy, BiLo, BMW, Books-A-Million, Buick, Chevrolet, Chrysler/Jeep, Circuit City, Dodge, Ford, GreaseMonkey, Honda, Hyundai, Infiniti, Jaguar, Lenscrafters, Lexus, Marshall's, Mazda, Mercedes, NAPA, Office Depot, Office Max, PepBoys, PetsMart, Sam's Club, Service Merchandise, Subaru, Toyota, Volvo, Wal-Mart, Western Auto, bank

Greenville

46	Mauldin Rd, **W**...Amoco/Arby's/mart, Crown Gas, Texaco/diesel/mart, Steak-Out, Subway, Comfort Inn, Ramada Ltd, Suburban Lodge, Travelodge, Aamco, BiLo/24hr, CVS Drug, Harris-Teeter, JiffyLube, Precision Tune, bank
45b a	US 25 bus, Augusta Rd, **E**...Chevron, Hess/mart, Phillips 66/mart, Bojangles, Burger King, Hardee's, Waffle House, Camelot Inn, Cross Hill Inn, Holiday Inn, Motel 6, **W**...Amoco, Denny's, Dixie Family Rest., Day's Inn, Comfort Inn, Travelodge
44	US 25, White Horse Rd, **E**...Exxon/Subway, Economy Inn, **W**...HOSPITAL, Amoco/mart, Texaco/McDonald's, Capt D's, Quincy's, Shoney's, Waffle House
43	SC 20(from sb), to Piedmont, no facilities
42	I-185, US 29, to Greenville, **W**...HOSPITAL
40	SC 153, to Easley, **E**...Exxon/diesel/mart, Waffle House, **W**...RaceTrac/mart, Texaco/diesel/mart/24hr, Arby's, Burger King, Hardee's, KFC, Executive Inn, Super 8, **1 mi W**...Shell/mart, Little Caesar's, McDonald's, BiLo Foods, Eckerd, GNC, Ingles Foods
39	SC 143, to Piedmont, **E**...BP/diesel/mart, **W**...Texaco/diesel/mart, antiques, hardware
35	SC 86, to Easley, Piedmont, **E**...Speedway/Country Kitchen/diesel/mart, Hardee's(1mi)
34	US 29(from sb), to Williamston, no facilities
32	SC 8, to Pelzer, Easley, **E**...Citgo/mart
27	SC 81, to Anderson, **E**...HOSPITAL, Citgo, Exxon, Phillips 66/diesel, Arby's, Carolina Country Café, McDonald's, Waffle House, Holiday Inn Express, **W**...Shell/diesel/mart/24hr
23mm	**rest area sb, full(handicapped)facilities, phone, vending, picnic tables, litter barrels, petwalk**
21	US 178, to Anderson, **E**...Texaco/diesel/mart, **2 mi E**...Applebee's, La Quinta, Quality Inn, Super 8
19b a	US 76, SC 28, to Anderson, **E**...HOSPITAL, Exxon/diesel/mart, Hardee's, Katherine's Kitchen, Day's Inn, Park Inn, Royal American Inn, **3 mi E**...Applebee's, La Quinta, **W**...RaceTrac/diesel/mart/atm/24hr, Buffalo's Café, Cracker Barrel, Outback Steaks, Waffle House, Wendy's, Comfort Suites, Country Inn Suites, Hampton Inn, Jameson Inn, to Clemson U

N

S

South Carolina Interstate 85

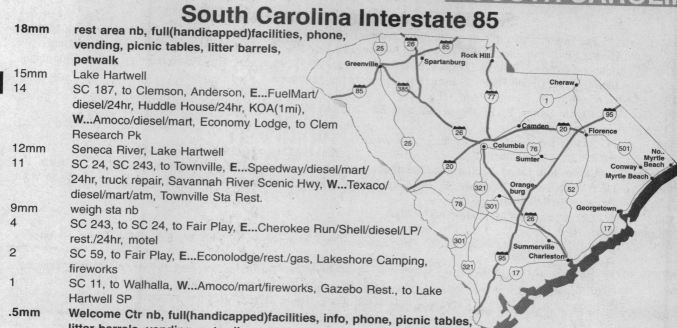

18mm   **rest area nb, full(handicapped)facilities, phone, vending, picnic tables, litter barrels, petwalk**

15mm   Lake Hartwell

14   SC 187, to Clemson, Anderson, **E...**FuelMart/diesel/24hr, Huddle House/24hr, KOA(1mi), **W...**Amoco/diesel/mart, Economy Lodge, to Clem Research Pk

12mm   Seneca River, Lake Hartwell

11   SC 24, SC 243, to Townville, **E...**Speedway/diesel/mart/24hr, truck repair, Savannah River Scenic Hwy, **W...**Texaco/diesel/mart/atm, Townville Sta Rest.

9mm   weigh sta nb

4   SC 243, to SC 24, to Fair Play, **E...**Cherokee Run/Shell/diesel/LP/rest./24hr, motel

2   SC 59, to Fair Play, **E...**Econolodge/rest./gas, Lakeshore Camping, fireworks

1   SC 11, to Walhalla, **W...**Amoco/mart/fireworks, Gazebo Rest., to Lake Hartwell SP

.5mm   **Welcome Ctr nb, full(handicapped)facilities, info, phone, picnic tables, litter barrels, vending, petwalk**

0mm   Lake Hartwell, South Carolina/Georgia state line

N ↕ S

South Carolina Interstate 95

Exit #   Services

198mm   South Carolina/North Carolina state line

196mm   **Welcome Ctr sb, full(handicapped)facilities, info, phone, vending, picnic tables, litter barrels, petwalk**

195mm   Little Pee Dee River

193   SC 9, SC 57, Dillon, to N Myrtle Beach, **E...**HOSPITAL, Amoco/mart/24hr, Exxon/mart, Speedway/diesel/mart, Bojangles, Burger King, Golden Corral, Shoney's, Waffle House, Wendy's, Comfort Inn, Day's Inn/rest., Hampton Inn, Holiday Inn Express, Howard Johnson, Ramada Ltd, **W...**BP/diesel, Hubbard House Rest., Waffle House, Econolodge, Super 8

190   SC 34, to Dillon, **E...**Citgo/Stuckey's/Dairy Queen, repair

181   SC 38, Oak Grove, **E...**BP/mart, Flying J/Conoco/diesel/LP/rest./24hr, Texaco/Subway/mart, **W...**Exxon/mart, Wilco/Citgo/DQ/Wendy's/diesel/mart, auto/truck repair

175mm   Pee Dee River

172mm   **rest area both lanes, full(handicapped)facilities, phone, picnic tables, vending, litter barrels, petwalk**

170   SC 327, **E...**Amoco/mart, Speedway/Hardee's/diesel/mart/24hr, to Myrtle Beach, Missile Museum, **W...**Quick Fill/mart/rest., Shell/diesel/mart

169   TV Rd, to Florence, **E...**KOA(1mi), Peterbilt, diesel repair, **W...**Amoco/mart, BP/mart, Petro/Texaco/diesel/rest./24hr, Rodeway Inn, Blue Beacon

164   US 52, Florence, to Darlington, **E...**HOSPITAL, Amoco, Exxon/Pizza Hut/Baskin-Robbins/diesel/mart, RaceTrac/mart/24hr, Shell/diesel/mart, Texaco/diesel/mart/atm/24hr, Cracker Barrel, Denny's, Hardee's, Kobe Japanese, McDonald's, Quincy's, Waffle House, Wendy's, Best Western, Comfort Inn, Econolodge, Hampton Inn, Holiday Inn, Knight's Inn, Motel 6, Super 8, Travelers Inn, Chrysler/Jeep, Pontiac/Buick, hardware, **W...**Amoco/mart, BP, Hess/diesel/mart, Pilot/Subway/Taco Bell/diesel/mart, Williams/Main St Café/diesel/mart/24hr, Arby's, Bojangles, Burger King, KFC, Country Inn Suites, Day's Inn, Microtel, Ramada Inn, Shoney's Inn/rest., Sleep Inn, Thunderbird Inn, Wingate Inn, to Darlington Raceway

160b   I-20 W, to Columbia

a   Lp 20, to Florence, **E...**Shell/diesel/mart, Burger King, Chick-fil-A, Huddle House, Morrison's Cafeteria, Outback Steaks, Pizza Hut, Red Lobster, Ruby Tuesday, Shoney's, Waffle House, Western Sizzlin, Courtyard, Fairfield Inn, Hampton Inn Suites, Holiday Inn Express, Red Roof Inn, Belk, Chevrolet, $Tree, Goody's, Hamrick's, JC Penney, PartsAmerica, Phar-Mor, Roses, Sears/auto, Staples, ToysRUs, Wal-Mart/drugs, cinema, mall, **1-2 mi E...**Exxon, Shell, Arby's, Hardee's, McDonald's, Percy&Willie's Rest., Subway, Taco Bell, Wendy's, Books-A-Million, Chevrolet/BMW/Isuzu, Circuit City, Harris-Teeter, Lowe's Bldg Supply, Piggly Wiggly, CVS Drug, K-Mart, Peebles, Firestone/auto, Nissan, JiffyLube, Meineke, Midas Muffler, NAPA, Nissan, Sam's Club, bank, carwash, laundry, mall

N ↕ S

South Carolina Interstate 95

157	US 76, Timmonsville, Florence, **E**...Amoco/diesel, BP, Exxon/McDonald's/diesel/mart, Texaco/diesel/mart/24hr, Burger King(1mi), BB's Deli, Red's&Ed's Seafood, Swamp Fox Diner, Waffle House, Day's Inn/gas, Howard Johnson Express, Villager Lodge, fireworks, **W**...Citgo, Shell/diesel/mart, Tree Room Inn/rest., Young's Plantation Inn/Magnolia Dining Room, Swamp Fox Camping(1mi), antiques
153	Honda Way, **W**...Honda Plant
150	SC 403, to Sardis, **E**...BP/diesel/mart/atm, Ramada Ltd, **W**...Exxon/diesel/mart, Honeydew RV Camp
147mm	Lynches River
146	SC 341, to Lynchburg, Olanta, **E**...Exxon, Belinda's BBQ, Relax Inn
141	SC 53, SC 58, to Shiloh, **E**...Exxon/diesel/mart, DonMar RV Ctr, to Woods Bay SP, **W**...Texaco/mart
139mm	**rest area both lanes, full(handicapped)facilities, phone, vending, picnic tables, litter barrels, petwalk**
135	US 378, to Sumter, Turbeville, **E**...BP, Citgo/mart, Compass Rest., Day's Inn, Knight's Inn, **W**...Exxon/diesel/mart, Pineland Golf Course
132	SC 527, to Sardinia, Kingstree, no facilities
130mm	Black River
122	US 521, to Alcolu, Manning, **E**...MD Fried Chicken(2mi), Manning Motel(3mi), **W**...Exxon/diesel/mart, Shrimper Rest.
119	SC 261, Manning, to Paxville, **E**...HOSPITAL, Shell/mart/24hr, TA/Amoco/Blimpie/LJS/Sbarro's/Taco Bell/diesel/24hr, Texaco/mart, Huddle House, Lodge Rest., Shoney's, Waffle House, Wendy's, Western Steer, Comfort Inn, Economy Inn, Holiday Inn Express, Ramada Ltd, Ford, **1 mi E**...EYECARE, Exxon/mart/wash, Burger King, KFC, Subway, CVS Drug, $General, Food Lion, Goodyear/auto, Wal-Mart/drugs, to McDonald's(2mi), Campers Paradise RV Park, **W**...BP/mart, Exxon/mart/24hr, Super 8, auto repair
115	US 301, to Summerton, Manning, **E**...Exxon/diesel, Carolina Inn, Travelers Inn/rest., **W**...Texaco/diesel/mart/24hr, Georgio's Greek, Colonial Inn, Day's Inn/rest./gas/24hr, Sunset Inn, repair
108	SC 102, Summerton, **E**...Citgo/DQ/Stuckey's, Shell/mart/24hr, RV camping, **W**...BP, Econolodge, Knight's Inn, Travelodge
102	US 15, US 301 N, to Santee, **E**...Lake Marion Trkstp/diesel/rest., Texaco/diesel/mart, Santee Resort/rest., RV camping, **W**...Shell/KrispyKreme/diesel/mart, to Santee NWR
100mm	Lake Marion
99mm	**rest area both lanes, full(handicapped)facilities, phone, info, vending, picnic tables, litter barrels, petwalk**
98	SC 6, Santee, to Eutawville, **E**...Chevron/mart/LP, Citgo/mart, Texaco/Subway/mart/wash, Antonio's Italian, Huddle House, KFC, Pizza Hut, Shoney's, Veranda Seafood, Ashley Inn, Best Western/rest., Day's Inn/rest., Hampton Inn, Ramada Inn/rest., Super 8, $General, Piggly Wiggly, Santee Factory Store/famous brands, RV camping(3mi), **W**...MEDICAL CARE, BP/diesel/mart/wash, BYLO/mart, Citgo/mart, Exxon/mart, Horizon/Pizza Hut/TCBY/diesel/mart/24hr, Shell/mart, Burger King, Cracker Barrel, Hardee's, Denny's, McDonald's, Peking Chinese, Waffle House, Budget Motel, Clark Inn/rest., Comfort Inn, Country Inn Suites, Economy Inn, Holiday Inn, Lake Marion Inn, Mansion Park Motel, BiLo, CarQuest, CVS Drug, Family$, Food Lion, NAPA, Radio Shack, Santee Gen Store/gas, USPO, bank, hardware, laundry, to Santee SP(3mi)
97	US 301 S(from sb, no return), to Orangeburg, no facilities
93	US 15, to Santee, Holly Hill, no facilities
90	US 176, to Cameron, Holly Hill, **W**...Exxon/diesel/mart
86b a	I-26, W to Columbia, E to Charleston, no facilities
82	US 178, to Bowman, Harleyville, **E**...Amoco/mart, Wilco/Exxon/Wendy's/DQ/diesel/mart/24hr, **W**...BP/diesel/mart/rest., Texaco/diesel/mart, Jerry's Rest.
77	US 78, St George, to Bamberg, **E**...Exxon/KFC/mart, Shell/Subway/KrispyKreme/TCBY/mart, Texaco/diesel/mart, Georgio's Rest., Hardee's, McDonald's, Pizza Hut, Waffle House, Western Steer, Comfort Inn/RV park, Economy Inn, Holiday Inn, Chevrolet/Pontiac/Olds/GMC, CVS Drug, Family$, Food Lion, Ford, USPO, bank, **W**...BP/mart, Texaco/diesel/mart/wash, Huddle House, Taco Bell, Best Western, Day's Inn, Southern Inn, Super 8
74mm	parking area both lanes, litter barrels
68	SC 61, Canadys, **E**...Amoco/mart, Citgo/mart, Texaco/Subway/diesel/mart, to Colleton SP(3mi)
62	SC 34, **W**...Lakeside RV camping
57	SC 64, Walterboro, **E**...HOSPITAL, BP/Subway/Pizza Hut/KrispyKreme/TCBY/diesel/mart/24hr, Chevron/mart/24hr, Citgo/mart, Exxon/diesel, Shell/DQ/mart, Speedway, Texaco/Blimpie/diesel/mart/wash, Burger King, Capt D's, Huddle House, Little Caesar's, Longhorn Steaks, McDonald's, Olde House Café, Taco Bell, Waffle House, Wendy's, Carolina Lodge, Howard Johnson, Southern Inn, Advance Parts, AutoZone, BiLo, CVS Drug, Eckerd, Food Lion, GNC, K-Mart, Piggly Wiggly, Wal-Mart, 1-Hr Photo, cleaners, **W**...Amoco/diesel/mart, Howard Johnson Express, Super 8

Santee

South Carolina Interstate 95

53 SC 63, Walterboro, to Varnville, Hampton, **E...**Amoco, Citgo/diesel/mart, Texaco/mart, BBQ, Burger King, Dairy Queen, Huddle House, KFC, Longhorn Steaks, McDonald's, Seafood/steak Rest., Shoney's, Waffle House, Best Western, Comfort Inn/rest., Econolodge, Holiday Inn, Rice Planter's Inn, Thunderbird Inn, Town&Country Inn/rest., **W...**BP/Krispy Kreme/mart, Cracker Barrel, Day's Inn, Deluxe Inn, Hampton Inn, Green Acres Camping

47mm **rest area both lanes, full(handicapped)facilities, phone, vending, picnic tables, litter barrels, petwalk**

42 US 21, to Yemassee, Beaufort, no facilities

40mm Combahee River

38 SC 68, to Hampton, Yemassee, **E...**Chevron/mart, **W...**BP/Subway/ TCBY/mart, Shell/mart, Texaco/diesel, Village Rest., Holiday Inn, Palmetto Lodge/rest., Super 8

33 US 17 N, to Beaufort, **E...**BP/diesel/mart, Citgo/mart, Exxon/mart, Shell/ mart, Texaco/mart/24hr, Denny's, McDonald's, Subway, Waffle House, Best Western, Day's Inn/rest., Holiday Inn Express, Knight's Inn, Jerry's Garage(6mi), KOA

30.5mm Tullifinny River

29mm Coosawhatchie River

28 SC 462, to Coosawhatchie, Hilton Head, Bluffton, **W...**Amoco/mart, Chevron/diesel/mart, Citgo/Stuckey's/Dairy Queen, Texaco/diesel/mart/rest.

22 US 17, Ridgeland, **E...**El Ranchito Mexican, **W...**HOSPITAL, Shell/mart, Plantation Inn

21 SC 336, to Hilton Head, Ridgeland, **W...**HOSPITAL, Chevron/Blimpie/diesel/mart/24hr, Exxon/mart, Texaco/ diesel/mart, Burger King, Hardee's, Huddle House, KFC, Pizza Station II, Scoops&Slices Café, Subway, Waffle House, Comfort Inn/rest., Econolodge, Ramada Ltd, $General, Eckerd, Food Lion, tires

18 SC 13, to US 17, US 278, to Switzerland, Granville, Ridgeland, no facilities

17mm parking area both lanes, litter barrels

8 US 278, Hardeeville, to Bluffton, **E...**BP/Joker Joe's/diesel/rest./24hr, Exxon/mart, Huddle House, Wendy's, **W...**Chevron/mart, Sprint/mart/24hr, Hardee's, KrispyKreme, Pizza Hut, Subway, TCBY, Ramada Ltd, Travelworld Lodge/rest.

5 US 17, US 321, Hardeeville, to Savannah, **E...**Chevron/mart, Exxon/Blimpie/mart, Shell/mart/24hr, Denny's, KFC, Waffle House, Day's Inn, Economy Inn, to Savannah NWR, **W...**Amoco/mart, BP/diesel/mart/wash, Exxon/ mart, Speedway/diesel/mart, Burger King, McDonald's, New China Rest., Shoney's, Subway, Wendy's, Budget Inn, Comfort Inn, Deluxe Inn, Holiday Inn Express, Howard Johnson Express, Knight's Inn, Scottish Inn, Super 8, NAPA, antiques

4.5mm **Welcome Ctr nb, full(handicapped)facilities, info, phone, picnic tables, litter barrels, vending, petwalk**

3.5mm weigh sta both lanes

0mm Savannah River, South Carolina/Georgia state line

South Carolina Interstate 385(Greenville)

Exit # Services

42 US 276, Stone Ave, to Travelers Rest., to Greenville Zoo, **W...**Amoco, **1-2 mi W...**multiple facilities on US 276, I-385 begins/ends on US 276.

40b a SC 291, Pleasantburg Dr, **E...**Exxon, Speedway, Bojangles, Hardee's, Jalisco Mexican, Pizza Hut, S&S Cafeteria, Steak&Ale, Subway, Taco Casa, Wendy's, CVS Drug, to BJU, Furman U, **W...**CHIROPRACTOR, Shell/mart, Chinese Rest., Comfort Inn

39 Haywood Rd, **E...**BP/mart/wash, Exxon/diesel/mart, Domino's, LongHorn Steaks, Outback Steaks, Shoney's, Amerisuites, Courtyard, Hilton, Quality Inn, Residence Inn, JiffyLube, TJ Maxx, bank, **W...**Crown/mart, Arby's, Applebee's, Bennigan's, Blackeyed Pea, Burger King, Chili's, ChuckeCheese, Hardee's, O'Charley's, Quincy's, Wendy's, Hampton Inn, Belk, Circuit City, Dillard's, JC Penney, NTB, Phar-Mor Drug, Pier 1, Rich's, TireAmerica, TJ Maxx, VisionWorks, XpressLube, bank, mall

37 Roper Mtn Rd, **E...**Harris-Teeter, **W...**Amoco, BP/diesel/mart/wash, Exxon/Pantry, RaceTrac/mart/atm, Backyard Burgers, Burger King, Capri's Italian, Cracker Barrel, Fazoli's, LoneStar Steaks, McDonald's, Morrison's Cafeteria, Miyabi Japanese, Rafferty's, Rio Bravo, Waffle House, Day's Inn, Crowne Plaza, Embassy Suites, Fairfield Inn, Holiday Inn Select, La Quinta, Travelodge, Home Depot, Lowe's Bldg Supply, Oschman's Sports, Target, Ward's, bank, cleaners, mall

South Carolina Interstate 385

36b a	I-85, N to Charlotte, S to Atlanta
35	SC 146, Woodruff Rd, **E**...Amoco/mart, BP, Applebee's, Bojangles, Chick-fil-A, Dairy Queen, Hardee's, KFC, King Buffet, Little Caesar's, Perkins, Schlotsky's, Sonic, Subway, Taco Bell, Topper's Rest., Waffle House, Wendy's, Aldi Foods, BiLo Foods, $Tree, Legends Sports, LubeKing, 1Price&More, PrecisionTune, Publix, Staples, USPO, Wal-Mart SuperCtr/24hr, bank, carwash, cleaners, **W**...CHIROPRACTOR, VETERINARIAN, Hess/diesel/mart, Burger King, Chinese Rest., Fatz Café, Fuddrucker's, IHOP, Midori's Rest., Monterrey Mexican, TCBY, Hampton Inn, Garden Ridge, Goodyear/auto, Hamrick's Outlet, bank, cleaners, funpark
34	Butler Rd, Mauldin, **E**...bank, **1/2mi E**...CVS Drug
33	Bridges Rd, Mauldin, no facilities
31	SC 417, to Laurens Rd, to Mauldin, **2 mi W**...services
30	US 276, Standing Springs Rd, **W**...Chevrolet, Ford
29	Georgia Rd, to Simpsonville, **W**...Amoco/diesel/mart/24hr, BP/mart, Clancy's Grill
27	Fairview Rd, to Simpsonville, **E**...HOSPITAL, Amoco, Texaco/mart, Corral Rest., Hardee's, McDonald's, Palmetto Inn, Advance Parts, BiLo Food, CVS Drug, $General, **W**...CHIROPRACTOR, Amoco/Arby's/mart/wash, BP/mart/wash, Applebee's, Burger King, Dragon Den Chinese, KFC, Ryan's, Pizza Hut, Taco Bell, Waffle House, Wendy's, Comfort Inn, Holiday Inn Express, Jameson Inn, Microtel, Ingles/deli, K-Mart/drugs, Lowe's Bldg Supply, Radio Shack, Wal-Mart SuperCtr/24hr, bank
26	Harrison Bridge Rd(from sb), no facilities
24	Fairview St, Fountain Inn, **E**...76/mart, Exxon, Phillips 66/mart, Hardee's, Waffle House, BiLo Foods, CVS Drug, carwash
23	SC 418, to Fountain Inn, Fork Shoals, **E**...Exxon/Subway/diesel/mart/atm/24hr
22	SC 14 W, Old Laurens Rd, to Fountain Inn, **E**...to Flowers Bakery
19	SC 14 E, to Gray Court, Owings, no facilities
16	SC 101, to Woodruff, Gray Court, no facilities
10	SC 23, to Ora, no facilities

South Carolina Interstate 526(Charleston)

Exit #	Services
33mm	I-526 begins/ends. **S**...Exxon/mart
32	US 17, **S**...MEDICAL CARE, Arby's, Baldino's Subs, Chick-fil-A, Domino's, KC, McDonald's, Niko's Café, Papa John's, T-Bonz, $General, JiffyLube, K-Mart, Parks Auto Parts, USPO, bank, cinema, cleaners, **1 mi S**...VETERINARIAN, BP/mart, Shell/mart, Capt D's, Hardee's, Huddle House, Shoney's, Wendy's, Comfort Inn, Day's Inn, Extended Stay America, Holiday Inn, Masters Inn, Red Roof Inn, Acura, BiLo, CVS Drug, Firestone/auto, Jiffylube, Radio Shack, Staples
30	Long Point Rd, **N**...Waffle House, Charles Pinckney NHS, **S**...Wando River Terminal
26mm	Wando River
23b a	SC 33, no facilities
21mm	Cooper River
20	Virginia Ave(from eb), **S**...Marathon Refinery
19	N Rhett Ave, **S**...gas/mart
18b a	US 52, US 78, Rivers Ave, **N**...Amoco/mart, Phillips 66, McDonald's, Advance Parts, Family$, Meineke Muffler, Piggly Wiggly, **S**...Catalina Inn
17b a	I-26, E to Charleston, W to Columbia
16	Montague Ave, Airport Rd, **N**...Alamo Rentals, to airport
15	SC 642, Dorchester Rd, Paramount Dr, **S**...Exxon/mart, Burger King, Huddle House, Super 8, Buick/Pontiac/GMC, U-Haul
14	Leeds Ave, **S**...HOSPITAL, boat marina
13mm	Ashley River
12	SC 61, Ashley River Rd, **N**...HOSPITAL
11	Sam Rittenburg Blvd
10	US 17, SC 7, **services from US 17 S**...Speedway/mart, Checker's, CiCi's, McDonald's, Quincy's, Red Lobster, Taco Bell, Holiday Inn Express, Motel 6, Dillard's, Food Lion, JC Penney, Piggly Wiggly, Sears/auto, Toyota, bank, mall

I-526 begins/ends on US 17.

South Dakota Interstate 29

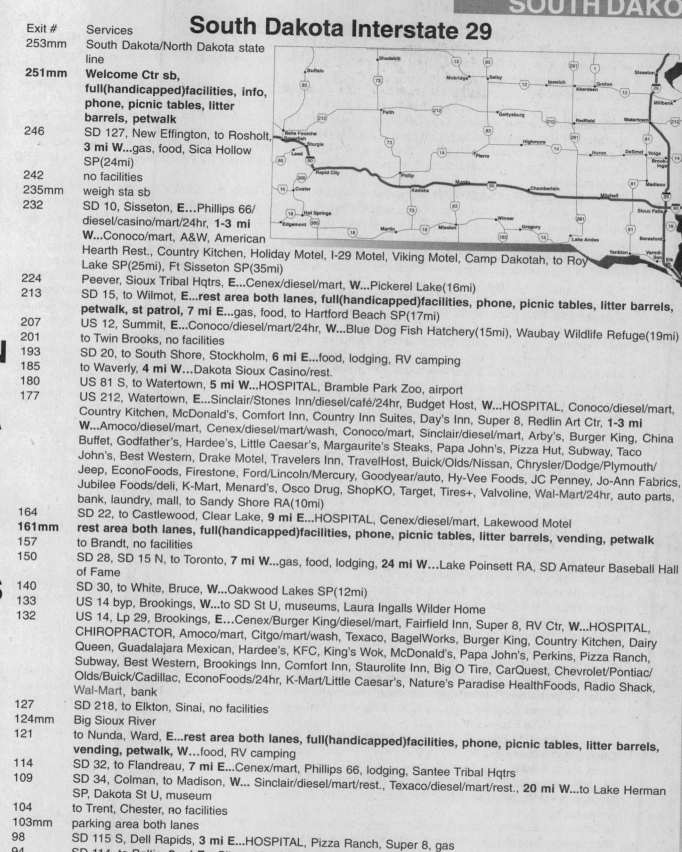

Exit #	Services
253mm	South Dakota/North Dakota state line
251mm	**Welcome Ctr sb, full(handicapped)facilities, info, phone, picnic tables, litter barrels, petwalk**
246	SD 127, New Effington, to Rosholt, **3 mi W**...gas, food, Sica Hollow SP(24mi)
242	no facilities
235mm	weigh sta sb
232	SD 10, Sisseton, **E**...Phillips 66/diesel/casino/mart/24hr, **1-3 mi W**...Conoco/mart, A&W, American Hearth Rest., Country Kitchen, Holiday Motel, I-29 Motel, Viking Motel, Camp Dakotah, to Roy Lake SP(25mi), Ft Sisseton SP(35mi)
224	Peever, Sioux Tribal Hqtrs, **E**...Cenex/diesel/mart, **W**...Pickerel Lake(16mi)
213	SD 15, to Wilmot, **E**...**rest area both lanes, full(handicapped)facilities, phone, picnic tables, litter barrels, petwalk, st patrol, 7 mi E**...gas, food, to Hartford Beach SP(17mi)
207	US 12, Summit, **E**...Conoco/diesel/mart/24hr, **W**...Blue Dog Fish Hatchery(15mi), Waubay Wildlife Refuge(19mi)
201	to Twin Brooks, no facilities
193	SD 20, to South Shore, Stockholm, **6 mi E**...food, lodging, RV camping
185	to Waverly, **4 mi W**...Dakota Sioux Casino/rest.
180	US 81 S, to Watertown, **5 mi W**...HOSPITAL, Bramble Park Zoo, airport
177	US 212, Watertown, **E**...Sinclair/Stones Inn/diesel/café/24hr, Budget Host, **W**...HOSPITAL, Conoco/diesel/mart, Country Kitchen, McDonald's, Comfort Inn, Country Inn Suites, Day's Inn, Super 8, Redlin Art Ctr, **1-3 mi W**...Amoco/diesel/mart, Cenex/diesel/mart/wash, Conoco/mart, Sinclair/diesel/mart, Arby's, Burger King, China Buffet, Godfather's, Hardee's, Little Caesar's, Margaurite's Steaks, Papa John's, Pizza Hut, Subway, Taco John's, Best Western, Drake Motel, Travelers Inn, TravelHost, Buick/Olds/Nissan, Chrysler/Dodge/Plymouth/Jeep, EconoFoods, Firestone, Ford/Lincoln/Mercury, Goodyear/auto, Hy-Vee Foods, JC Penney, Jo-Ann Fabrics, Jubilee Foods/deli, K-Mart, Menard's, Osco Drug, ShopKO, Target, Tires+, Valvoline, Wal-Mart/24hr, auto parts, bank, laundry, mall, to Sandy Shore RA(10mi)
164	SD 22, to Castlewood, Clear Lake, **9 mi E**...HOSPITAL, Cenex/diesel/mart, Lakewood Motel
161mm	**rest area both lanes, full(handicapped)facilities, phone, picnic tables, litter barrels, vending, petwalk**
157	to Brandt, no facilities
150	SD 28, SD 15 N, to Toronto, **7 mi W**...gas, food, lodging, **24 mi W**...Lake Poinsett RA, SD Amateur Baseball Hall of Fame
140	SD 30, to White, Bruce, **W**...Oakwood Lakes SP(12mi)
133	US 14 byp, Brookings, **W**...to SD St U, museums, Laura Ingalls Wilder Home
132	US 14, Lp 29, Brookings, **E**...Cenex/Burger King/diesel/mart, Fairfield Inn, Super 8, RV Ctr, **W**...HOSPITAL, CHIROPRACTOR, Amoco/mart, Citgo/mart/wash, Texaco, BagelWorks, Burger King, Country Kitchen, Dairy Queen, Guadalajara Mexican, Hardee's, KFC, King's Wok, McDonald's, Papa John's, Perkins, Pizza Ranch, Subway, Best Western, Brookings Inn, Comfort Inn, Staurolite Inn, Big O Tire, CarQuest, Chevrolet/Pontiac/Olds/Buick/Cadillac, EconoFoods/24hr, K-Mart/Little Caesar's, Nature's Paradise HealthFoods, Radio Shack, Wal-Mart, bank
127	SD 218, to Elkton, Sinai, no facilities
124mm	Big Sioux River
121	to Nunda, Ward, **E**...**rest area both lanes, full(handicapped)facilities, phone, picnic tables, litter barrels, vending, petwalk, W**...food, RV camping
114	SD 32, to Flandreau, **7 mi E**...Cenex/mart, Phillips 66, lodging, Santee Tribal Hqtrs
109	SD 34, Colman, to Madison, **W**... Sinclair/diesel/mart/rest., Texaco/diesel/mart/rest., **20 mi W**...to Lake Herman SP, Dakota St U, museum
104	to Trent, Chester, no facilities
103mm	parking area both lanes
98	SD 115 S, Dell Rapids, **3 mi E**...HOSPITAL, Pizza Ranch, Super 8, gas
94	SD 114, to Baltic, **2 mi E**...Citgo, **10 mi E**...to EROS Data Ctr
86	to Renner, Crooks, no facilities
84b a	I-90, W to Rapid City, E to Albert Lea
83	SD 38 W, 60th St, **E**...Freightliner Trucks, Stan's RV Ctr, repair, **W**...fireworks

N

S

Watertown Brookings

South Dakota Interstate 29

S i o u x   F a l l s

81   SD 38 E, Russell St, Sioux Falls, **E**...Amoco/mart/wash/24hr, Burger King, Country Kitchen, Roll'n Pin Rest., Arena Motel, Best Western Ramkota/rest., Brimark Inn, Exel Inn, Kelly Inn, Motel 6, Oaks Inn/rest., Sheraton, Sleep Inn, Schaap's RV Ctr, st patrol

79   SD 42, 12ᵗʰ St, **E**...HOSPITAL, Amoco/mart/24hr, Cenex/mart, Citgo/mart, Phillips 66/diesel/mart, Sinclair, Burger King, BurgerTime, Chris' Rest., Fry'n Pan Rest., Godfather's, Golden Harvest Chinese, KFC, Lucky Kruser's Café, McDonald's, Pizza Hut, Subway, Taco Bell, Taco John's, Tomacelli's Pizza, Wendy's, Nite's Inn, Ramada Ltd, Ace Hardware, BMW/Lincoln/Mercury, CarQuest, Chevrolet/Mitsubishi, JiffyLube, K-Mart/auto, Lewis Drug, Saturn, Sunshine Foods, TrueValue, Walgreen, bank, to Great Plains Zoo/museum, **W**...DENTIST, Cenex/diesel/mart, Citgo/mart, Phillips 66/mart, Hardee's, Sunset Motel, Westwick Motel, Tower RV Park, WheelCo Brake, to USS SD Battleship Mem

78   26ᵗʰ St, Empire St, **E**...Phillips 66/diesel/mart/wash, Cracker Barrel, Culver's Custard, Domino's, KFC, Mkt Diner, Outback Steaks, Rio Bravo, Runza Rest., Hampton Inn, Holiday Inn Express, EconoFoods, Michael's, Sam's Club, USPO, cinema, **W**...CHIROPRACTOR, Dairy Queen, Dynasty Chinese/Vietnamese, Westwinds Rest, MainStay Suites, Hy-Vee Foods, bank

77   41ˢᵗ St, Sioux Falls, **E**...CHIROPRACTOR, VETERINARIAN, EYECARE, DENTIST, Amoco/mart/wash/24hr, Sinclair/diesel/mart/wash, SuperAmerica/diesel, Texaco, Andolini's Italian, Arby's, BBQ, Burger King, Carlos O'Kelly's, ChiChi's, Chili's, China Express, Fry'n Pan Rest., Fuddrucker's, Ground Round, KFC, LoneStar Steaks, McDonald's, Mulligan's Grill, Nick's Gyros, Olive Garden, Perkins, Pizza Hut, Pizza Inn, Red Lobster, Schlotsky's, Subway, Taco Bell, Taco John's, TimberLodge Steaks, TGIFriday, Valentino's Rest., Wendy's, Yesterday's Café, Z-teca Mexican, Best Western Empire, Comfort Suites, Fairfield Inn, Microtel, Radisson, Residence Inn, Super 8, Ace Hardware, Advance Parts, Barnes&Noble, Batteries+, Best Buy, Big O Tire, Checker Parts, Chrysler/Ply-mouth/Nissan/Hyundai, Ford, Gateway Computers, Goodyear/auto, HobbyLobby, JC Penney, JiffyLube, Kinko's, MailBoxes Etc, Menard's, MensWearhouse, OfficeMax, 1/2 Price Store, PetCo, Radio Shack, Randall's Foods, Sears/auto, ShopKO, SuperWash, Target, Tires+, Toyota, ToysRUs, Village Automotive, Wal-Mart/24hr, VW, Younkers, bank, cinema, mall, **W**...CHIROPRACTOR, DENTIST, Citgo/mart, Texaco/diesel/mart, Burger King, Denny's, Godfather's, Little Caesar, Perkins/24hr, AmericInn, Baymont Inn, Day's Inn, Select Inn, Lewis Drug, bank

75   I-229 E, to I-90

73   Tea, **E**...Larry's Texaco/diesel/café/24hr, Pat's Steaks, Midwest Tire/muffler, **1.5mi W**...Red Barn Camping

71   to Harrisburg, Lennox, **W**...RV camping

68   to Lennox, Parker, no facilities

64   SD 44, Worthing, **1 mi E**...Amoco, **W**...Chevrolet/Pontiac/Buick

62   US 18 E, to Canton, **E**...Conoco/pizza/diesel/mart, Charlie's Motel, antiques(9mi), **W**...repair

59   US 18 W, to Davis, Hurley, no facilities

56   to Fairview, **E**...to Newton Hills SP(12mi)

53   to Viborg, no facilities

50   to Centerville, Hudson, no facilities

47   SD 46, Beresford, to Irene, **E**...Ampride/diesel/mart, Casey's/gas, Cenex/Burger King/mart, Sinclair/diesel/mart/atm, Emily's Café, Crossroads Motel, Starlite Motel, Super 8, Chevrolet, Fiesta Foods, carwash, **W**...Cenex/diesel/mart/24hr, camping

42   to Alcester, Wakonda, no facilities

38   to Volin, **E**...to Union Co SP(7mi)

31   SD 48, Spink, to Akron, no facilities

26   SD 50, to Vermillion, **E...Welcome Ctr/rest area both lanes, full(handicapped)facilities, info, phone, picnic tables, litter barrels, petwalk, W**...Conoco/diesel/mart/atm/24hr, **6-7 mi W**...Casey's/gas, Phillips 66/mart/wash, Amigo's Mexican, Burger King, Cherry St Grille, Pizza Hut, Subway, Taco John's, Comfort Inn, Super 8, Travelodge, Lube Express, to U of SD

18   Lp 29, Elk Point, to Burbank, **E**...Amoco/diesel/mart/wash, Phillips 66/pizza/subs/mart, Cody's Rest., Dairy Queen, HomeTowne Inn

15   to Elk Point, **1 mi E**...gas, food, lodging

9   SD 105, Jefferson, **E**...Amoco/Choice Cut Rest./mart

4   McCook, **1 mi W**...KOA(seasonal)

3mm   weigh sta nb

2   N Sioux City, **E**...Amoco/diesel/mart/24hr, Ampride/diesel/mart/rest./24hr, Glass Palace Rest.(1mi), McDonald's, Taco John's, Gateway Computers, USPO, **W**...Casey's/gas, Citgo/Taco Bell/mart, Comfort Inn, Hampton Inn, Super 8, to Sodrac Dogtrack

1   **E**...Dakota Dunes Golf Resort, **W**...Citgo/diesel/mart, Country Inn Suites, Milligan's Rest.

0mm   Big Sioux River, South Dakota/Iowa state line

South Dakota Interstate 90

Exit #	Services
412.5mm	South Dakota/Minnesota state line
412mm	**Welcome Ctr wb/rest area eb, full(handicapped)facilities, info, phone, picnic tables, litter barrels, petwalk, RV dumping(wb), weigh sta both lanes**
410	Valley Springs, **N**...Palisades SP, **S**...Beaver Creek Nature Area, gas, food, fireworks
406	SD 11, Brandon, Corson, **S**...Amoco, Holiday Inn Express, **1 mi S**...Ampride/mart, Phillips 66/mart, Texaco/mart, Dairy Queen, Hardee's, to Big Sioux RA
402	EROS Data Ctr, **N**...Jellystone Park RV Camping, fireworks
400	I-229 S, no facilities
399	SD 115, Cliff Ave, Sioux Falls, **N**...Phillips 66/diesel/mart, Sinclair/diesel/mart, Cody's Grill, Frontier Village, Spader RV Ctr, KOA, airport, **S**...HOSPITAL, Cenex/diesel/mart, Holiday/gas, Pilot/Subway/diesel/mart/24hr, Arby's, Burger King, Great American Rest., McDonald's, Perkins, Best Western Townhouse, Cloud Nine Motel, Comfort Inn, Country Inn Suites, Day's Inn, Econolodge, Super 8, Blue Beacon, Goodyear, Kenworth, Peterbilt, Volvo, bank, fireworks
396b a	I-29, N to Brookings, S to Sioux City, no facilities
390	SD 38, Hartford, **N**...Camp Dakota RV Camping, Goos RV Ctr, **S**...Phillips 66/diesel/mart
387	rd 17, Hartford, **N**...Phillips 66/mart
379	SD 19, Humboldt, **1 mi N**... Phillips 66, Home Café/Grill, Town&Country Store
375mm	E Vermillion River
374	to SD 38, Montrose, **2 mi N**...gas, food, **5 mi S**...Lake Vermillion RA, RV camping
368	Canistota, **4 mi S**...Best Western
364	US 81, Salem, to Yankton, **N**...Phillips 66/mart, Edith's Café, Home Motel, RV camp
363.5mm	W Vermillion River
363mm	**rest area both lanes, full(handicapped)facilities, phone, picnic tables, litter barrels, vending, petwalk, RV dumping, st patrol**
357	to Bridgewater, Canova, no facilities
353	Spencer, Emery, **S**...TA/Amoco/Subway/diesel/mart/24hr
352mm	Wolf Creek
350	SD 25, Emery, Farmer, **N**...to DeSmet, Home of Laura Ingalls Wilder
344	SD 262, Alexandria, to Fulton, **N**...Texaco/diesel/mart/wash, Alex's Drive-In(˚mi)
337mm	parking area both lanes
335	Riverside Rd, **N**...KOA
334.5mm	James River
332	SD 37 S, Mitchell, to Parkston, **N**...HOSPITAL, Amoco/Burger King/mart/wash/24hr, Cenex/Dairy Queen/diesel/mart, Phillips 66/Blimpie/mart, Texaco/Subway/diesel/mart/24hr, Arby's, Bonanza, Chinese Rest., Country Kitchen, Happy Chef/24hr, Hardee's/playland, KFC, McDonald's, Perkins, Pizza Hut, AmericInn, Best Western, Chief Motel, Comfort Inn, Corn Palace Motel, Day's Inn, Super 8/RV parking, Thunderbird Motel, Chrysler/Dodge, K-Mart, Menard's, car/truckwash, diesel repair, transmissions, **1-2 mi N**...Conoco, Burger King, McDonald's, to Corn Palace, Museum of Pioneer Life, RV camping
330	SD 37 N, Mitchell, weigh sta, **1/2mi N**...HOSPITAL, CHIROPRACTOR, Amoco/mart/24hr, Cenex/mart/24hr, Texaco/mart, Dairy Stop, Happy Chef/24hr, Anthony Motel, Econolodge, Holiday Inn/rest., Motel 6, Siesta Motel, County Fair Foods, RV Ctr, to Corn Palace, museum, **S**...Dakota Camping
325	Betts Rd, RV camping
319	Mt Vernon, **1 mi N**...Sinclair/mart
310	US 281, to Stickney, **S**...Conoco/A&W/diesel/mart/24hr, to Ft Randall Dam
308	Lp 90, to Plankinton, **N**...Phillips 66/Al's Café/diesel/mart, I-90 Motel, Super 8, RV camping
302mm	**rest area wb, full(handicapped)facilities, phones, picnic tables, litter barrels, RV dump**
301mm	**rest area eb, full(handicapped)facilities, phones, picnic tables, litter barrels, RV dump**

E

W

Sioux Falls

Mitchell

South Dakota Interstate 90

C h a m b e r l a i n

296	White Lake, **1 mi N**...Texaco/mart, I-90 Motel, White Lake Motel, **S**...Warrell Motel, RV Park
294mm	Platte Creek
293mm	parking area both lanes
289	SD 45 S, to Platte, **S**...to Snake Cr/Platte Cr RA(25mi)
284	SD 45 N, Kimball, **N**...Amoco/mart/tires, Phillips 66/Chief Rudy's Buffet/diesel/24hr, Frosty King Drive-In, Ponderosa Café, Super 8, Travlers Motel, RV camping/mart, **S**...Kimball Motel
272	SD 50, Pukwana, **2 mi N**...gas, food, lodging, **S**...Snake/Platte Creek Rec Areas(25mi)
265	SD 50, Chamberlain, **N**...HOSPITAL, VETERINARIAN, Amoco/mart, Dairy Queen, Firestone, **3 mi N**...McDonald's, Subway, Bel Aire Motel, Best Western, Lakeshore Motel, Honda, St Joseph Akta Lakota Museum, KOA, **S**...Conoco/diesel/mart, Roadrunner/mart
264mm	**rest area both lanes, full(handicapped)facilities, scenic view, info, phones, picnic tables, litter barrels, RV dumping**
263	Chamberlain, Crow Creek Sioux Tribal Hqtrs, **N**...Amoco, Sinclair/diesel/mart, A&W, Casey's Café, McDonald's, Pizza Hut, Rainbow Café, Taco John's, The Plaza Motel/groceries/gas, Best Western(1mi), Bel Aire Motel, Hillside Motel, Lakeshore Motel, Super 8, Buche's IGA/deli
262mm	Missouri River
260	SD 50, Oacoma, **N**...Amoco/Burger King/diesel/mart, Conoco/diesel/mart, Taco John's, Al's Oasis Motel, Comfort Inn, Day's Inn, Econolodge, Chevrolet/Buick/Pontiac, Familyland Camping, Cedar Shore Resort, Old West Museum
251	SD 47, to Winner, Gregory, no facilities
248	SD 47, Reliance, **N**...Amoco/diesel/mart, Sioux Tribal Hqtrs, to Big Bend RA
241	to Lyman, no facilities
235	SD 273, Kennebec, **N**...Conoco/mart/repair, Pony Express Café, Budget Host, Kings Inn, KOA
226	US 183 S, Presho, **N**...Amoco/diesel, Conoco/diesel/mart, Norma's Café/Drive-In, Coachlight Inn, Hutch's Motel/café
225	Presho, same as 226
221mm	**rest area wb, full(handicapped)facilities, info, phone, picnic tables, litter barrels, RV dump, petwalk**
220	no facilities, sunflowers in August
218mm	**rest area eb, full(handicapped)facilities, info, phone, picnic tables, litter barrels, RV dump, petwalk**
214	Vivian, no facilities
212	US 83 N, SD 53, to Pierre, **N**...HOSPITAL(34mi), Phillips 66/diesel/mart, Vivian Jct Rest.
208	no facilities
201	Draper, **N**...gas/diesel/mart, Hilltop Café
194mm	parking area both lanes
192	US 83 S, Murdo, **N**...Amoco/diesel/mart, HHH Autotruck/diesel, Phillips 66, Sinclair/diesel/Goodyear, Texaco/diesel, Total/mart, Buffalo Rest., Dairy Bar, Food Court, KFC, Murdo Drive-In, Star Rest., Substation, Anchor Inn, Best Western, Chuck's Motel, Day's Inn, Hospitality Inn, Lee Motel, Super 8, TeePee Motel/RV Camp, Murdo Super Value Foods, camping, **S**...Landmark Country Inn, to Rosebud
191	Murdo, **N**...access to same as 192
190mm	Central/Mountain time zone
188mm	parking area both lanes, litter barrels
183	Okaton, **S**...gas/mart, Ghost Town, antiques
177	no facilities
172	to Cedar Butte, no facilities
170	SD 63 N, to Midland, **N**...Texaco/mart, 1880's Town, KOA, tent camping
167mm	**rest area wb, full(handicapped)facilities, phones, picnic tables, litter barrels, RV dump, petwalk**
165mm	**rest area eb, full(handicapped)facilities, phones, picnic tables, litter barrels, RV dump, petwalk**
163	SD 63, Belvidere, **S**...Amoco/diesel/café, Sinclair/diesel/mart, motel
152	Lp 90, Kadoka, **N**...Badlands/diesel/rest./24hr, **S**...Best Western, Budget Host, Leewood Motel, Badlands Petrified Gardens, camping
150	SD 73 S, Kadoka, **N**...Sinclair/diesel/mart, Dakota Inn/rest., **S**...Amoco/mart, Conoco/diesel, Texaco/mart/café, Happy Chef, Sidekick Rest., Best Western, Budget Host, Downtowner Motor Inn, Ponderosa Motel/RV parking, Super 8, Kadoka Camping, to Buffalo Nat Grasslands
143	SD 73 N, to Philip, **15 mi N**...HOSPITAL
138mm	scenic overlook wb, restrooms

M u r d o

E

W

South Dakota Interstate 90

131	SD 240, **S**...to Badlands NP, Amoco(seasonal), Conoco(11mi), Circle 10 Camping, Prairie Home NHS, Badlands Inn(9mi), Cedar Pass Lodge/rest.(9mi), KOA(15mi), Good Sam Camping
129.5mm	scenic overlook eb
127	no facilities
121	no facilities
116	no facilities
112	US 14 E, to Philip, no facilities
110	SD 240, Wall, **N**...Amoco/diesel/mart, Conoco/mart, Exxon/mart, Texaco, Dairy Queen, Subway, Best Western, Day's Inn, Econolodge, Elk Motel, Elkton House/rest., Fountain Hotel, Hillcrest Motel, Hitching Post Motel, Homestead Motel, Kings Inn, Knight's Inn, Sands Motor Inn, Super 8, The Wall Motel, Welsh Motel, Ace Hardware, Wall Drug/gifts/rest., laundry, **S**...to Badlands NP, antiques, RV camp, fireworks
109	Wall, **1-2 mi N**...access to same as 110
107	Cedar Butte Rd, no facilities
101	rd T-504, Jensen Rd, to Schell Ranch, no facilities
100mm	**rest area both lanes, full(handicapped)facilities, info, phone, picnic tables, litter barrels, RV dump, vending, petwalk**
99.5mm	Cheyenne River
99	Wasta, **N**...gas/mart/café, The Redwood Motel, Bryce's RV Camping
90	to Owanka, no facilities
88	rd 473(from eb, no re-entry), no facilities
84	rd 497, **N**...Olde Glory Fireworks
78	New Underwood, °mi **S**...gas/mart, Jake's Motel, Steve's General Store
69mm	parking area both lanes
66	Box Elder, **N**...Ellsworth AFB, SD Space & Air Museum, DENTIST, Flying J/Conoco/KFC/diesel/mart, Texaco/mart/wash(1mi), McDonald's, Pizza Hut, Taco John's, Box Elder Trading Post, fireworks
61	Elk Vale Rd, **N**...Flying J/Conoco/diesel/LP/rest./mart/RV dump/24hr, Langland RV Ctr, fireworks, **S**...airport, KOA(2mi)(seasonal), transmissions
60	Lp 90(exits left from wb), Rapid City, to Mt Rushmore, **1-3 mi S**...HOSPITAL, Bonanza, JB's, KFC, LJ Silver, Holiday Inn Express, Stardust Motel, K-Mart, Nat Coll of Mines/Geology
59	La Crosse St, Rapid City, **N**...Amoco/wash/24hr, Phillips 66/A&W/mart, Texaco, Blimpie, Denny's, Happy Chef, Minerva's Rest., Royal Fork Buffet, Econolodge, Holiday Inn Express, Rodeway Inn, Super 8, BEST, Sears/auto, Pearle Vision, cinema, mall, st patrol, **S**...Cenex/diesel/mart, Exxon/mart/24hr, Golden Corral, McDonald's, MillStone Family Dining, Perkins/24hr, AmericInn, Comfort Inn, Day's Inn, Fair Value Inn, Foothills Inn, Holiday Inn, Motel 6, Quality Inn, Ramada Inn, Rushmore Motel, Thrifty Motel, Travelodge, Sam's Club, Wal-Mart/auto, **2 mi S on North St**...Amoco/mart, Cenex/diesel/mart, Exxon, 7-11, Bonanza, Burger King, Chinese Rest., Hardee's, JB's, KFC, Little Caesar's, McDonald's, Pizza Hut, Subway, Taco Bell, Wendy's, Budget Inn, Colonial Motel, 4 Seasons Motel, Goldstar Motel, Stardust Motel, K-Mart, Ford/Lincoln/Mercury/Subaru
58	Haines Ave, Rapid City, **N**...Conoco/mart, Texaco/mart/wash, Applebee's, Fuddrucker's, Hardee's, Red Lobster, Su Casa Mexican, Travelodge, JC Penney, NW Fabrics, Target, Tires+, to Rushmore Mall, **S**...HOSPITAL, DENTIST, Minimart/gas, Taco John's, OfficeMax, ShopKO, bank, cinema, accesses same as 59
57	I-190, US 16, to Rapid City, to Mt Rushmore, **1-2 mi S**...Conoco, Exxon, Texaco, Shakey's Pizza, Best Western, Day's Inn, Hilton, Quality Inn, Radisson, Rodeway Inn, Sands of the Black Hills Motel, Albertson's, mufflers
55	Deadwood Ave, **S**...Sinclair/Subway/diesel/mart/rest., A&W, Crack'd Pot, Best Western Sun Inn(3mi), Black Hills Dogtrack, diesel repair
51	Black Hawk Rd, **N**...Fort Welikit Camping(1mi)
48	Stagebarn Canyon Rd, **N**...Cattleman's Steaks, RV camping, Marine Sales/service, **S**...MEDICAL CARE, DENTIST, Sinclair/Stagebarn Plaza/groceries, Classick's Rest., Mike's Pizza, carwash, laundromat
46	Piedmont Rd, Elk Creek Rd, **N**...Elk Creek Steaks, Covered Wagon Resort, auto parts, to Petrified Forest, camping, **1/2 mi S**...Conoco/Subway/diesel/mart, camping

E ↑ ↓ W

Wall

Rapid City

South Dakota Interstate 90

44	Bethlehem Rd, no facilities
42mm	**rest area both lanes, full(handicapped)facilities, info, phones, picnic tables, litter barrels, RV dump, petwalk, vending**
40	Tilford, **N**...antiques
39mm	weigh sta eb
37	Pleasant Valley Rd, **S**...Bulldog Camping, Rush-No-More Camping
34	**S**...Black Hills Nat Cemetary, no facilities
32	SD 79, Sturgis, **N**...HOSPITAL, Amoco, Conoco/Subway/mart, Exxon/Phil-town Gen Store/diesel, Country Kitchen, Lynn's Country Mkt/café, Lantern Motel, National 9 Inn, Phil-town Best Western/rest., StarLite Motel, Sturgis Livestock Exchange, Ford/Lincoln/Mercury, Motorcycle SuperStore, to Bear Butte SP, RV camping
30	US 14A W, SD 34E, Sturgis, to Deadwood, **N**...Cenex, Conoco, Sinclair, Pizza Hut, McDonald's, Pizza Hut, Lynn's Country Mkt/café, Subway, Taco John's, Big A Parts, Dakota Farms Cheese, Pamida/drugs, SpeeDee Lube, Day's End Camping, carwash, laundry, **S**...Phillips 66/Molly B's Motel/diesel/rest., Texaco/mart, Burger King, Trailside Gen Store/gas, Boulder Canyon Rest., Day's Inn, Super 8/rest., Chevrolet
23	SD 34 W, Whitewood, to Belle Fourche, **N**...Northern Hills RV Ctr, **S**...Amoco, Conoco, Pronto Diner, Tony's Motel
17	US 85 S, to Deadwood, **9-12 mi S in Deadwood**...Depot Rest., Motherload Café, Silverado Café, Best Western Hickock, Best Western Phil-town, Day's Inn, Palace Hotel, Super 8, White House Inn, Deadwood Gulch Resort, KOA, Winner's Casino
14	US 14A, Spearfish Canyon, **N**...Happy Chef, Comfort Inn, Fairfield Inn, Holiday Inn/rest., **S**...Amoco/mart, Conoco, KFC, Perkins, The Sluice Rest., All American Inn, Best Western/rest., Country Hearth Inn/rest., Super 8, Ford/Lincoln/Mercury, K-Mart/drugs, Randall's Foods, auto museum, cinema, hardware, Mt View Camping, Chris' Camping
12	Spearfish, **S**...HOSPITAL, Amoco/mart/wash, Exxon, Sinclair/diesel, Texaco, Arby's, Country Kitchen, Burger King, Dairy Queen, McDonald's, Millstone Rest., Pizza Ranch, Subway, Wendy's, Allstar Inn, Best Western, Cottonwood Lodge, Day's Inn, Kelly Inn, L Rancho Lodge, Sundown Best Western/rest., Chevrolet/Buick/Olds/Pontiac, Midas Muffler, Safeway, U-Haul, Black Hills St U, fish hatchery, same as 10
10	US 85 N, to Belle Fourche, **S**...EYECARE, Burger King, Cedar House Rest., Chinese Rest., Dairy Queen, Domino's, McDonald's, Subway, Wendy's, Best Western, Day's Inn, Chevrolet/Buick/Olds/Cadillac, Jo's Camping, KOA, Pamida/drugs, Pennzoil, Safeway, Wal-Mart, same as 12
2	**1 mi N**...McNenny St Fish Hatchery, no facilities
1mm	**Welcome Ctr eb, full(handicapped)facilities, info, phone, picnic tables, litter barrels, RV dumping, petwalk**
0mm	South Dakota/Wyoming state line

E

W

Sturgis · Spearfish

South Dakota Interstate 229(Sioux Falls)

Exit #	Services
10b a	I-90 E and W. I-229 begins/ends on I-90, exit 400.
9	Benson Rd, no facilities
7.5mm	Big Sioux River
7	Rice St, **E**...winter sports, **W**...to stockyards
6	SD 38, 10th St, **E**...Citgo, Phillips 66, Arby's, Baxter Café, Dairy Queen, Domino's, Fry'n Pan Rest., KFC, Pizza Hut, Taco Bell, Tomacelli's Italian, Budget Host, Suburban Motel, Big O Tire, Checker Parts, K-Mart, PrecisionTune, Pronto Parts, ShopKO, Sunshine Foods/24hr, Valvoline, bank, **W**...Amoco, Phillips 66/diesel, Texaco/mart, BurgerTime, Godfather's, Hardee's, Pizza Inn, Puerto Vallarta, Taco John's, 10th St Sirloin/Buffet, Wendy's, Harvey's Motel, Rushmore Motel, JiffyLube, Lewis Drug, Meineke Muffler, Radio Shack, bank
5.5mm	Big Sioux River
5	26th St, **E**...DENTIST, HOSPITAL, Texaco/diesel/mart, McDonald's, bank
4	Cliff Ave, **E**...Sinclair/mart, bank
3	SD 115, Lp 229, Minnesota Ave, **1 mi W**...DENTIST, VETERINARIAN, Burger King, Hardee's, Little Caesar's, McDonald's, Subway, Holiday Inn, GMC, K-Mart, Lewis Drug, Scheel's Allsports, Sunshine Foods, tires
2	Western Ave, **E**...Texaco/mart, **W**...MEDICAL CARE, CHIROPRACTOR, DENTIST, EYECARE, Amoco, Cenex/diesel/mart, Burger King, Champ's Café, China Buffet, Pizza Hut, Valentino's Italian, Advance Parts, Campbell's Home/auto/hardware, Goodyear/auto, 1/2 Price Store, PrecisionTune, cleaners
1.5mm	Big Sioux River
1c	Louise Ave, **W**...Amoco, Phillips 66/diesel, A&W, Burger King, McDonald's, Royal Fork Buffet, Wendy's, Radisson, Honda/Mercedes, mall
1b a	I-29 N and S. I-229 begins/ends on I-29, exit 75.

N

S

Sioux Falls

Tennessee Interstate 24

Exit #	Services
185b a	I-75, N to Knoxville, S to Atlanta. I-24 begins/ends on I-75, exit 2 in Chattanooga.
184	Moore Rd, **N**...Sears, USPO, mall, **S**...China Garden, K-Mart/drugs, Radio Shack, Winn-Dixie/deli, cinema
183	(183a from wb), Belvoir Ave, Germantown Rd, no facilities
181a	US 41 S, to East Ridge(from eb), **S**...Westside Grill, King's Lodge, Ford Trucks
181	Fourth Ave, Chattanooga, to TN Temple U, **N**...VETERINARIAN, Amoco, Citgo/mart, Exxon/diesel, Bojangles, Burger King, Capt D's, Central Park, Hardee's, Krystal/24hr, Nick's Diner/24hr, Subway, Waffle House, Quality Inn, BiLo, $General, Gates Diesel Repair, Goodyear, Mr Transmission, **S**...Citgo/mart, McDonald's, Wendy's
180b a	US 27 S, TN 8, Rossville Blvd, **N**...tires, to Chickamauga, UT Chatt, **S**...Phillips 66/mart, RaceTrac/diesel/mart/24hr, Speedway/diesel/mart/24hr, LJ Silver, 1-Hr Optical, carwash
178	US 27 N, Market St, Chattanooga, to Lookout Mtn, **N**...Amoco/diesel/mart, BP/mart, Day's Inn, Guest House Inn, Marriott(2mi), Ramada Inn, Ford, Midas Muffler, Nissan, U-Haul, to Chattanooga ChooChoo, **S**...KFC, Comfort Inn, Hampton Inn, Microtel
175	Browns Ferry Rd, to Lookout Mtn, **N**...Citgo/mart/atm, Exxon/diesel, Mike's Pizza, Knight's Inn, CVS Drug, $General, Food Lion/deli, **S**...CHIROPRACTOR, Conoco/diesel/mart, Texaco/diesel/mart, Hardee's, Subway, Comfort Inn, La Plaza Motel, AutoValue Parts, BiLo/drugs, bank
174	US 11, US 41, US 64, Lookout Valley, **N**...Waffle House, Day's Inn, Lookout Mtn Inn, Racoon Mtn Camping(2mi), **S**...Amoco/diesel/mart/24hr, BP/mart/rest./24hr, Exxon/mart, BBQ, Cracker Barrel, Taco Bell, Waffle House, Baymont Inn, Best Western, Comfort Inn, Country Inn Suites, Hampton Inn, Econolodge, Erick's Motel, Ramada Ltd, Super 8, Lookout Valley Camping, hardware, st patrol
172mm	**rest area eb, full(handicapped)facilities, phone, picnic tables, litter barrels, vending, petwalk**
171mm	Tennessee/Georgia state line
169	GA 299, to US 11, **N**...Sav-a-Ton/diesel/mart/24hr, **S**...Amoco/Subway/TCBY/diesel/mart/24hr, Cone/diesel/mart/24hr, RaceTrac/mart/24hr
167	I-59 S, to Birmingham
167.5mm	Tennessee/Georgia state line, Central/Eastern time zone
161	TN 156, to Haletown, New Hope, **N**...Phillips 66, On The Lake Camping, **S**...Chevron/fireworks
160mm	Tennessee River/Nickajack Lake
159.5mm	**Welcome Ctr wb/rest area eb, full(handicapped)facilities, phone, vending, picnic tables, litter barrels, petwalk**
158	US 41, TN 27, Nickajack Dam, **N**...Texaco, **S**...Phillips 66/diesel/mart/fireworks
155	TN 28, Jasper, **N**...Amoco/diesel, Citgo(1mi), Dairy Queen(1mi), Hardee's, Western Sizzlin, Acuff Country Inn, **S**...HOSPITAL
152	US 41, US 64, US 72, Kimball, S Pittsburg, **N**...HOSPITAL, Amoco/fireworks, Phillips 66/fireworks, RaceTrac/mart, Arby's, Domino's, Hardee's, KFC, Krystal, LJ Silver, McDonald's, Pizza Hut, Shoney's, Subway, Taco Bell, Waffle House, Wendy's, Budget Host, Day's Inn, Holiday Inn Express, BiLo, Chevrolet/Pontiac/Buick, $Tree, Goody's, Radio Shack, Wal-Mart SuperCtr/24hr, to Russell Cave NM, **3 mi S**...Lodge Cast Iron
143	Martin Springs Rd, **N**...Chevron/diesel/mart/fireworks
135	US 41 N, Monteagle, **N**...Citgo/mart, Phillips 66/diesel/rest./24hr, HappyLand Rest., AutoSure Parts, **S**...Day's Inn
134	US 64, US 41A, Monteagle, to Sewanee, **N**...Amoco, CVS Drug, to S Cumberland SP, **S**...Citgo/mart, Chevron/diesel/mart, Texaco/diesel/mart, Hardee's, Tubby's Diner, Waffle House, SmokeHouse Lodge/Rest., Budget Host, Country Inn of America, Firestone/U-Haul/auto, Monteagle Winery, Piggly Wiggly, auto parts, to U of The South
133.5mm	**rest area both lanes, full(handicapped)facilities, phone, picnic tables, litter barrels, vending, petwalk**
128.5mm	Elk River
127	US 64, TN 50, Pelham, to Winchester, **N**...Amoco/mart, Phillips 66/mart, Texaco/Stuckey's, **S**...Exxon/diesel/mart, to Tims Ford SP
119mm	parking area both lanes
117	to Tullahoma, USAF Arnold Ctr, UT Space Institute

E

W

Tennessee Interstate 24

116mm	weigh sta both lanes
114	US 41, Manchester, **N**...BP, Jiffy/diesel/café/24hr, Phillips 66/mart, Texaco/diesel/mart/24hr, Best Western, Comfort Inn/rest., Country Inn Rest., Ramada Inn, Super 8, Chrysler/Dodge/Toyota, KOA, Nissan, museum, **S**...EYECARE, VETERINARIAN, BP/mart, Cone/diesel/mart, Exxon/TCBY/mart/24hr, RaceWay/mart, Arby's, Burger King, Capt D's, Hunan Chinese, KFC, McDonald's, Shoney's, Taco Bell, Waffle House, Wendy's, Budget Motel, Econolodge, Executive Inn Suites, Holiday Inn, Villager Lodge, Advance Parts, AutoZone, BiLo, Chevrolet, Family$, Ford/Lincoln/Mercury, Goodyear, Wal-Mart/drugs, carwash, laundry
111	TN 55, Manchester, **N**...Amoco/mart, Citgo/diesel/mart/24hr, Chevron/mart, Jameson Inn, to Rock Island SP, **S**...HOSPITAL, BP, to Jack Daniels Dist HS, Old Stone Fort
110	TN 53, Manchester, **N**...BP/mart, Exxon/TCBY/mart/24hr, Shell/mart, Cracker Barrel, Davy Crockett's Roadhouse, Oak Rest., Shoney's, Ambassador Inn, Day's Inn, Hampton Inn, **S**...HOSPITAL, Texaco/diesel/mart/24hr, Ponderosa, Waffle House, Manchester Motel
109.5mm	Duck River
105	US 41, **N**...Busy Corner/BP/diesel/24hr, Texaco, RanchHouse Rest., **S**...Amoco, to Normandy Dam
97	TN 64, Beechgrove, to Shelbyville, **N**...auto parts/repair, **S**...Phillips 66/diesel
89	Buchanan Rd, **N**...Texaco/diesel/mart, **S**...Citgo/diesel/rest./24hr, Huddle House
81	US 231, Murfreesboro, **N**...HOSPITAL, Amoco/mart/wash, BP/diesel/mart/24hr, RaceTrac/mart/24hr, Arby's, Burger King, Cracker Barrel, King's Table Rest., Krystal, Parthenon Steaks, Ponderosa, Shanghai Chinese, Waffle House/24hr, Wendy's, Baymont Inn, Day's Inn, Econolodge, Ramada Ltd, Scottish Inn, Shoney's Inn/Rest., Travelodge, Dodge, Honda, Mazda, fireworks, laundry, **S**...Citgo/mart/24hr, Express/diesel/mart/24hr, Phillips 66/diesel/mart/wash/24hr, Texaco/mart/24hr, BBQ, McDonald's, Subway, Taco Bell, Waffle House, Howard Johnson, Quality Inn, Safari Inn, Toyota, outlets
78	TN 96, Murfreesboro, to Franklin, **N**...BP, Phillips 66/diesel/mart, Shell/mart, Texaco/Burger King/mart/24hr, Applebee's, Baskin-Robbins, Cracker Barrel, Dairy Queen, El Chico's, Ezra's Rest., IHOP, KFC, Luby's, McDonald's, Mrs Winner's, OutBack Steaks, Red Lobster, Ryan's, Subway, Waffle House, Wendy's, Best Western, Comfort Inn, Country Inn Suites, Hampton Inn, Holiday Inn, Microtel, Motel 6, Red Roof Inn, Super 8, Wingate Inn, Goody's, Home Depot, OfficeMax, Sears, Staples, Target, Wal-Mart SuperCtr/24hr, bank, cinema, mall, to Stones River Bfd, **S**...Amoco/diesel/mart/wash/24hr, BP/mart/wash, Chevron/mart/24hr, Citgo/diesel/mart/24hr, Express/mart, Exxon/mart/24hr, Texaco/mart, Corky's Ribs/BBQ, Hardee's, Waffle House, River Rock Outlet/famous brands, Sam's Club
74	TN 840, no facilities
70	TN 102, Lee Victory Pkwy, Almaville Rd, to Smyrna, **S**...Amoco/mart, Citgo/mart/24hr, Express/mart, Exxon, Phillips 66, Dad's Rest., McDonald's
66	TN 266, Sam Ridley Pkwy, to Smyrna, **N**...MEDICAL CARE, Cone/diesel/mart/24hr, Texaco/Baskin-Robbins/diesel/mart, Catfish House Rest., Nashville I-24 Camping(3mi), cleaners, **S**...Cracker Barrel, I-24 Expo
64	Waldron Rd, to La Vergne, **N**...Exxon/mart/24hr, KwikSak/24hr, Speedway/Subway/diesel/mart/24hr, Arby's, Hardee's, Krystal/24hr, McDonald's, Waffle House, Comfort Inn, Holiday Inn Express, RV Ctr, **S**...Express/diesel/mart/24hr, Pizza Neatza, RiceBowl Chinese, Driftwood Inn, cleaners
62	TN 171, Old Hickory Blvd, **N**...Amoco/mart, BP/TA/diesel/rest./24hr, Speedway/diesel/mart, Texaco/diesel/mart/24hr, Waffle House, Best Western
60	Hickory Hollow Pkwy, **N**...DENTIST, CHIROPRACTOR, MEDICAL CARE, OPTOMETRIST, BP, Citgo/mart/24hr, Express/mart, Shell/mart, Applebee's, Arby's, Burger King, Chinese Rest., Courtyard Café, Cracker Barrel, KFC, Logan's Roadhouse, McDonald's, O'Charley's, Outback Steaks, Pizza Hut, Rax, Red Lobster, Subway, Texana Grill, TGIFriday, Wendy's, Day's Inn, Fairfield Inn, Hampton Inn, Holiday Inn, Ramada Inn, Burlington Coats, Chevrolet, Cloth World, Dillard's, Eckerd, EconoPet, Firestone/auto, JiffyLube, Kastner-Knott, Kinko's, Kroger/drug/24hr, Mazda, MediaPlay, MegaMkt Food, Piece Goods, Pier 1, Postal Annex, Service Merchandise, TJ Maxx, bank, cinema, cleaners, mall, transmissions, **S**...VETERINARIAN, Amoco/mart/24hr, BP/diesel/mart/wash, Texaco/mart/wash/XpressLube, Evergreen Chinese, Olive Garden, Shoney's, Waffle House/24hr, Knight's Inn, Quality Inn, Scottish Inn, The Quarters Motor Inn, Acura, Home Depot, Target, Valvoline, cleaners
59	TN 254, Bell Rd, same as 60
57	Haywood Lane, **N**...Amoco/24hr, KwikSak/gas/24hr, BBQ, Hardee's, Waffle House/24hr, Food Lion, Peterbilt, auto parts, carwash, laundry, **S**...DENTIST, Exxon/Taco Bell/TCBY/diesel/24hr, Phillips 66/diesel/mart/24hr, Chanello's Pizza
56	TN 255, Harding Place, **N**...CHIROPRACTOR, Amoco/mart/wash/24hr, Express/mart, Exxon/mart/wash, Texaco/diesel/mart/24hr, Applebee's, Arby's, Chinese Rest., Denny's, KFC, LJ Silver, McDonald's, Subway, Taco Bell, Waffle House, Wendy's, White Castle, Drury Inn, HoJo's, Holiday Inn, Motel 6, PearTree Inn, Super 8, Sam's Club, to airport, **S**...HOSPITAL, CHIROPRACTOR, Citgo/mart, KwikSak, Shell/mart, Burger King, Hooters, Ponderosa, Waffle House, Comfort Inn, Econolodge, Motel 6

E

W

Tennessee Interstate 24

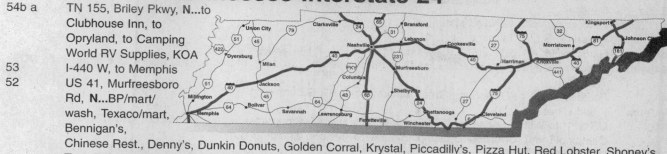

54b a	TN 155, Briley Pkwy, **N**...to Clubhouse Inn, to Opryland, to Camping World RV Supplies, KOA
53	I-440 W, to Memphis
52	US 41, Murfreesboro Rd, **N**...BP/mart/wash, Texaco/mart, Bennigan's, Chinese Rest., Denny's, Dunkin Donuts, Golden Corral, Krystal, Piccadilly's, Pizza Hut, Red Lobster, Shoney's, Taco Bell, Waffle House, Day's Inn, Economy Inn, Holiday Inn Express, Howard Johnson, Quality Inn, Ramada Inn, Scottish Inn, CarQuest, Office Depot, 1-Hr Photo, bank, cleaners, **S**...Chevrolet, Dodge
52b a	I-40, E to Knoxville, W to Memphis
211a	I-65 S, to Birmingham, I-40 W to Memphis
b	I-65 N, to Louisville, I-24 W to Clarksville
212	Hermitage Ave, Sessiers Lane, Citgo, Exxon, Texaco/mart, Burger King, McDonald's, Shoney's, Wendy's, Drake Inn, Freightliner/Mack Trucks
83mm	Cumberland River
84	Shelby Ave, Nashville, **S**...BP/diesel, Exxon/diesel/mart/24hr, Shell/mart, TA/diesel/mart, BBQ, Shoney's, Best Host, Inn, Day's Inn, Econolodge, Ramada Ltd, Scottish Inn
85	James Robertson Pkwy, downtown, st capitol, same as 84
85a	US 31E N, Ellington Pkwy, **N**...Citgo, **S**...Best Western, Day's Inn, U-Haul, Budget Rental Car
86	I-265, to Memphis
87b a	US 431, Trinity Lane, **N**...Citgo/Circle K, Pilot/Arby's/diesel/mart/24hr, Krystal, White Castle, Cumberland Inn, Red Carpet Inn, Scottish Inn, Trinity Inn, carwash, to American Bapt Coll, **S**...DENTIST, Amoco/mart/wash, BP/diesel/mart/wash, Exxon/mart, Texaco/diesel/mart, BBQ, Burger King, Capt D's, Chugger's Rest., Denny's, McDonald's, Ponderosa, Shoney's, Taco Bell, Waffle House/24hr, Comfort Inn, Day's Inn, Drury Inn, Econolodge, Global Hotel, Hallmark Inn, Hampton Inn, Holiday Inn Express, Howard Johnson/rest., Knight's Inn, Liberty Inn, Motel 6, Ramada Inn, RV Ctr
88b	I-24 W and I-65 N, no facilities
44a	I-24 E and I-65 S, no facilities
43	TN 155, Briley Pkwy, Brick Church Pike, **N**...Citgo
40	TN 45, Old Hickory Blvd, **N**...Citgo/Subway/diesel/mart, Phillips 66/diesel/mart/24hr, Super 8
35	US 431, to Joelton, Springfield, **S**...Amoco/diesel/mart/24hr, BP/diesel/mart/24hr, McDonald's, Subway, Day's Inn, OK Camping
31	TN 249, New Hope Rd, **N**...Shell/mart, **S**...BP/diesel/mart/24hr, Buddy's ChopHouse, to Nashville Zoo
24	TN 49, to Springfield, Ashland City, **N**...HOSPITAL, Amoco/mart, Express/diesel/mart/24hr, Texaco, **S**...Shell, Piggly Wiggly
19	TN 256, Maxey Rd, to Adams, **N**...BP/diesel/mart, Phillips 66/diesel/rest./24hr, **S**...Shell/diesel/mart/pizza/24hr
11	TN 76, to Adams, Clarksville, **N**...Texaco/diesel/mart/24hr, **S**...HOSPITAL, Amoco/diesel/mart/24hr, Citgo/Subway/mart/24hr, Phillips 66/Home Place Rest., McDonald's, Waffle House, Comfort Inn, Day's Inn, Holiday Inn Express
9mm	Red River
8	TN 237, Rossview Rd, no facilities
4	US 79, to Clarksville, Ft Campbell, **N**...BP/diesel/mart/24hr, Cracker Barrel, Spring Creek Camping(2mi), RV Ctr, **S**...Amoco/diesel/mart/24hr, Exxon/Blimpie/diesel/mart/wash/24hr, Speedway/diesel/mart/24hr, Texaco/diesel/mart, Applebee's, Arby's, Burger King, ChiChi's, Cracker Barrel, KFC, Krystal, Logan's Roadhouse, LJ Silver, McDonald's, O'Charley's, Olive Garden, Outback Steaks, Ponderosa, Rafferty's Rest., Red Lobster, Ruby Tuesday, Ryan's, Sub's Grill, Suzie's Drive-Thru, Taco Bell, Waffle House, Wendy's, Wilson's Catfish, Best Western, Comfort Inn, Day's Inn, Econolodge, Fairfield Inn, Hampton Inn, Holiday Inn, Microtel, Motel 6(6mi), Ramada Ltd, Red Roof Inn, Royal Inn, Shoney's Inn/rest., Super 8, Travelodge, Wingate Inn, BootCountry, Goodyear/auto, Hess, JiffyLube, K-Mart/drugs, Lowe's Bldg Supply, OfficeMax, PharMor Drug, Piece Goods, Target, Wal-Mart SuperCtr/24hr, Western Auto/repair, bank, cinema, mall, winery, to Austin Peay St U, to Land Between the Lakes
1	TN 48, to Clarksville, Trenton, **N**...Shell/diesel/mart, Clarksville Camping, antiques, **S**...Amoco, Exxon/mart
.5mm	**Welcome Ctr eb, full(handicapped)facilities, phones, vending, picnic tables, litter barrels, petwalk**
0mm	Tennessee/Kentucky state line

Tennessee Interstate 40

Exit #	Services
451.5mm	Tennessee/North Carolina state line
451	Waterville Rd, no facilities
447	Hartford Rd, **N**...giftshop, **S**...Exxon/mart/towing, River Co Deli, whitewater rafting
446mm	**Welcome Ctr wb, full(handicapped)facilities, phones, vending, picnic tables, litter barrels, petwalk, NO TRUCKS**
443	Foothills Pkwy, to Gatlinburg, Great Smokey Mtns NP, **S**...camping
443mm	Pigeon River
440	US 321, to Wilton Spgs Rd, Gatlinburg, **N**...Coastal/diesel/mart/rest./repair, **S**...Amoco, Arrow Creek Camping(14mi), CrazyHorse Camping(12mi), Jellystone Camping(12mi)
439.5mm	Pigeon River
435	US 321, Newport, to Gatlinburg, **N**...HOSPITAL, Exxon/diesel/mart/24hr, Marathon/mart/24hr, Texaco/mart/wash, Arby's, Burger King, Capt D's, Hardee's/24hr, KFC, La Carreta Mexican, McDonald's, Pizza Hut, Pizza+, SageBrush Steaks, Shoney's, Subway, Taco Bell, Bryant Town Motel, Motel 6/RV parking, Goodyear/auto, bank, cinema, **S**...Amoco/mart, BP/TCBY/mart, Cracker Barrel, Ryan's, Shiner's BBQ, Waffle House/24hr, Wendy's, Best Western, Family Inn, Holiday Inn, Cato's, CVS Drug, Goody's, Sav-A-Lot, Wal-Mart SuperCtr/gas/24hr, Pennzoil
432b a	US 70, US 411, US 25W, to Newport, **N**...BP/mart/wash, Exxon/diesel/mart/24hr, Lois' Country Kitchen, Kathy's Rest., Krystal, Sonic, Comfort Inn, Relax Inn, Chevrolet/Pontiac/Buick, Chrysler/Jeep/Dodge, Ford/Mercury, KOA, TMC Camping, **S**...Amoco/mart/atm, Citgo/mart/24hr, Shell/mart/wash, Texaco/diesel/mart/atm, Family Inn/rest.
426mm	**rest area wb, full(handicapped)facilities, phone, vending, picnic tables, litter barrels, petwalk**
425mm	French Broad River
424	TN 113, Dandridge, **N**...Marathon/diesel/mart
421	I-81 N, to Bristol, no facilities
420mm	**rest area eb, full(handicapped)facilities, phone, vending, picnic tables, litter barrels, petwalk**
417	TN 92, Dandridge, **N**...Amoco/diesel/mart, Exxon/Subway/Taco Bell/diesel/mart, Marathon/Baskin-Robbins/24hr, Hardee's, McDonald's, Perkins, Tennessee Mtn Inn, **S**...Citgo/KFC/diesel/mart, Shell/Wendy's/diesel/mart, Shoney's, Waffle House, Best Western, Comfort Inn, Holiday Inn Express, Super 8
415	US 25W, US 70, to Dandridge, **S**...Texaco/diesel/mart
412	Deep Sprgs Rd, to Douglas Dam, **N**...Sunshine/diesel, Apple Valley Café, **S**...Chevron/diesel
407	TN 66, to Sevierville, Pigeon Forge, Gatlinburg, **N**...Citgo/diesel/mart, McDonald's, Country Inn, KOA, **S**...Amoco/mart, BP/DQ/Pizza Inn/mart, Exxon/diesel, Shell/diesel, Texaco/Baskin-Robbins/diesel/mart, Chambers Rest., Ole Southern Rest., Subway, Wendy's, Best Western, Comfort Inn, Day's Inn, Holiday Inn Express, Quality Inn, Foretravel RV Ctr, Honda, RV camping, fireworks, flea mkt, **3-10 mi S**...Citgo/mart, Clarion Inn, Ramada Ltd, Wingate Inn, Lee Greenwood Theatre, USPO, bank, multiple services/outlets
402	Midway Rd, no facilities
398	Strawberry Plains Pk, **N**...Amoco/Baskin-Robbins/diesel/mart/wash, BP/diesel/mart, Phillips 66/Pizza Hut/diesel/mart, Shell/mart, McDonald's, Waffle House, Wendy's, Comfort Inn, Country Inn Suites, Hampton Inn, Holiday Inn Express, Ramada Ltd, Super 8, Ryder, **S**...Citgo/mart, Pilot/DQ/Subway/diesel/mart/24hr, Speedway/diesel/mart/24hr, Arby's, Burger King, Cracker Barrel, HomeFolks Rest., KFC, Krystal, Perkins/24hr, StarLite Diner, Baymont Inn, Fairfield Inn
395mm	Holston River
394	US 70, US 11E, US 25W, Asheville Hwy, **N**...BP/mart/wash, Dixie Motel, Sunbeam Motel(2mi), K-Mart/auto, city park, **1 mi N**...RaceTrac/mart, Subway, Wendy's, Gateway Inn, Advance Parts, Food Lion, flea mkt, **S**...Phillips 66/diesel/mart, Texaco/diesel/mart/atm, Waffle House/24hr, Day's Inn, Economy Inn, **1 mi S**...VETERINARIAN, DENTIST, DairyMart/gas, Family Inn/rest., CVS Drug, Kroger, Walgreen, bank
393	I-640 W, to I-75 N, no facilities
392	US 11 W, Rutledge Pike, **N**...Shell/diesel/mart, CookHouse Rest., $General, U-Haul, repair, tires, **S**...BP/mart/wash, BBQ, Hardee's, Shoney's, Family Inn, AutoValue Parts, Quick-O Mufflers, bank, transmissions, to Knoxville Zoo
390	Cherry St, Knoxville, **N**...Citgo/Taco Bell/mart, Texaco/diesel/mart/wash, Pilot/diesel/mart, Country Table Rest., Hardee's, Red Carpet Inn, Goodyear, **1 mi S**...Amoco, Exxon/mart/wash, Arby's, Krystal, KFC, LJ Silver, Mrs Winner's, Subway, Wendy's, Budget Motel, Regency Inn, Advance Parts, AutoZone, Walgreen, auto repair
389	US 441 N, Broadway, 5th Ave, **1/2 mi N**...DENTIST, Amoco, BP/mart/wash, Conoco, Spur/mart, Texaco, Burger King, Capt D's, KFC, Krystal, Oriental Garden Chinese, Sonic, Subway, Taco Bell, Wendy's, CVS Drug, $General, Family$, Firestone, Kroger/deli/24hr, Radio Shack, USPO, Walgreen/24hr, tires, transmissions
388	US 441 S(exits left from wb), downtown, **S**...BP, Radisson, to Smokey Mtns, to U of TN
387b	TN 62, 17th St, **1/2 mi S**...Hilton, Holiday Inn, Goodyear, bank
a	I-275 N, to Lexington
386b a	US 129, University Ave, to UT, airport
385	I-75 N, I-640 E, no facilities

E

W

Tennessee Interstate 40

I-40 W and I-75 S run together 17 mi

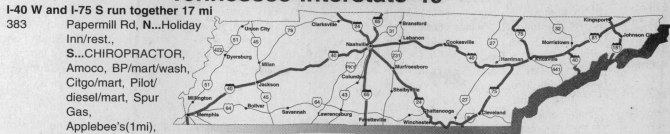

383	Papermill Rd, **N**...Holiday Inn/rest., **S**...CHIROPRACTOR, Amoco, BP/mart/wash, Citgo/mart, Pilot/diesel/mart, Spur Gas, Applebee's(1mi), Big Boy, Bicycle Club Rest., Burger King, Darryl's, Dunkin Donuts, IHOP, Longhorn Steaks, McDonald's, Morrison's Cafeteria, Pizza Hut, Red Lobster, Ruby Tuesday, Showbiz Pizza, Taco Bell, Taco Rancho Mexican, Waffle House, Wendy's, Western Sizzlin, Econolodge, Howard Johnson, Super 8, Buick/GMC, Firestone, Pennzoil, same as 380
380	US 11, US 70, West Hills, **S**...VETERINARIAN, BP/mart/wash, Conoco/mart, Hess/mart, Pilot/mart, Texaco/diesel/mart, Arby's, Applebee's, Backyard Burger, Blackeyed Pea, Cancun Mexican, Checker's, Chili's, Chick-fil-A, Cozymel's Grill, Honeybaked Ham, KFC, Macaroni Grill, Michael's Prime Rib, Mr Gatti's, O'Charley's, Olive Garden, Papa John's, PlumTree Chinese, Schlotsky's, Stefano's Pizza, Subway, Taco Bell, Texas Steaks, Comfort Hotel, Nations Best Motel, Quality Inn, Bed Bath&Beyond, Borders Books, Cloth World, Dillard's, Food Lion, Goody's, JC Penney, K-Mart, Kohl's, Office Depot, Old Navy, PetsMart, TireAmerica, U-Haul, Walgreen, cinema, mall, st patrol, **1 mi S on Kingston Pike**...Citgo/7-11, Morrison's Cafeteria, Central Park Burgers, Little Caesar's, Family Inn, Office Depot, Pearle Vision, Radio Shack
379	Gallaher View Rd, Walker Sprgs Rd, **N**...Exxon/Pizza Hut/Taco Bell/mart, Texaco/diesel/mart, Red Carpet Inn, Sam's Club, Wal-Mart SuperCtr/24hr, **S**...BP/diesel/mart, Pilot/diesel/mart, Texaco/diesel/mart, Burger King, ChuckeCheese, Don Pablo's, Logan's Roadhouse, Mrs Winners, Old Country Buffet, Omelet House, Ryan's, Shoney's, Tony Roma, Wendy's, Family Inn, Holiday Inn, Scottish Inn, AutoZone, Books-A-Million, Buy4Less Foods, Cadillac, Firestone/auto, Ford, Dodge, Goodyear, Mitsubishi, Nissan, Olds/GMC, Pier 1, SuperX Drug, cleaners
378	Cedar Bluff Rd, **N**...HOSPITAL, Amoco/mart/24hr, Pilot/Taco Bell/mart, Texaco, Arby's, Burger King, Cracker Barrel, KFC, LJ Silver, McDonald's, Papa John's, Pizza Hut, Prince Deli, Subway, Waffle House, Wendy's, Econolodge, Hampton Inn, Holiday Inn, Ramada Inn, Sleep Inn, Food Lion, Walgreen, bank, **S**...DENTIST, CHIROPRACTOR, Applebee's, Bob Evans, Carrabba's, Corky's Ribs/BBQ, Denny's, Fazoli's, Friendly's, Grady's Grill, Hops Brewery, IHOP, Outback Steaks, Clubhouse Inn, Courtyard, Extended Stay America, La Quinta, Luxbury Best Western, Microtel, Red Roof Inn, Residence Inn, Signature Inn, Wyndham Garden, Best Buy, Celebration Sta, Chevrolet, Chrysler, Circuit City, Drug Emporium, HomePlace, Lowe's Bldg Supply, Michael's, Waccamaw Pottery, Walgreen, bank, cleaners
376	TN 162 N, I-140 E, to Maryville, **N**...to Oak Ridge Museum, **1 exit S**...Calhoun's Rest., BabysRUs, HomePlace, Michael's
374	TN 131, Lovell Rd, **N**...Amoco/mart, BP/mart/wash, Texaco/diesel/mart, TA/diesel/rest./24hr, McDonald's, Taco Bell, Waffle House, Best Western, Knight's Inn, La Quinta, Travelodge, Acme Boots, Passport RV Ctr, **S**...Citgo/mart, Pilot/Wendy's/diesel/mart/24hr, Speedway/mart, Arby's, Chili's, Krystal, Shoney's, Candlewood Suites, Day's Inn/rest., Motel 6, Land Rover, Mercedes, Toyota
373	Campbell Sta Rd, **N**...Amoco/mart, Texaco/diesel/mart, Comfort Suites, Ramada Ltd, Super 8, Buddy Gregg RV Ctr, **S**...BP/mart, Pilot/diesel/mart, Speedway/diesel/mart, Cracker Barrel, Hardee's, Baymont Inn, Holiday Inn Express
372mm	weigh sta both lanes
369	Watt Rd, **N**...Flying J/Conoco/diesel/LP/rest./24hr, Texaco, Speedco Lube, **S**...Exxon/mart, Petro/Mobil/diesel/rest./24hr, TA/BP/Burger King/Perkins/Pizza Hut/diesel/24hr, Blue Beacon

I-40 E and I-75 N run together 17 mi

368	I-75 and I-40, no facilities
364	US 321, TN 95, Lenoir City, Oak Ridge, **N**...Shell, Crosseyed Cricket Camping(2mi)
363mm	parking area wb, phone, litter barrel
362mm	parking area eb, phone, litter barrel
360	Buttermilk Rd, **N**...Soaring Eagle RV Camping
356	TN 58 N, Gallaher Rd, to Oak Ridge, **N**...Citgo/diesel, BP/diesel/mart, Texaco, Huddle House, Simply Subs, King's Inn, Day's Inn, Family Inn/rest., 4 Seasons Camping
355	Lawnville Rd, **N**...Texaco/diesel/mart/souveniers/fireworks
352	TN 58 S, Kingston, **N**...Knight's Inn, **S**...Exxon/diesel, Shell, Raceway, Dairy Queen, Hardee's, McDonald's, Pizza Hut, Sonic, Subway, Taco Bell, Comfort Inn, to Watts Bar Lake
351mm	Clinch River

Tennessee Interstate 40

350   US 70, Midtown, **S**...Jiggers Rest., Mid-Town Motel, Patterson RV Supplies

347   US 27, Harriman, **N**...Phillips 66/diesel/mart, Cancun Mexican, Cracker Barrel, Hardee's, KFC, LJ Silver, McDonald's, Pizza Hut, Taco Bell, Wendy's, Best Western, Acme Boots, tires, to Frozen Head SP, Big South Fork NRA, **S**...BP/mart, Exxon/TCBY/mart, Shell/DQ/diesel/mart/atm/24hr, Cracker Barrel, Shoney's, Holiday Inn Express, Super 8, **2-3 mi S**...HOSPITAL, Capt D's, Domino's, Little Caesar's, Subway, Advance Parts, Goody's, Ingles Foods, Kroger/deli/24hr, K-Mart, Wal-Mart/auto

340   TN 299 N, Airport Rd, no facilities

339.5mm   eastern/central time zone line

338   TN 299 S, Westel Rd, **N**...BP/diesel/mart, **S**...Shell/mart, Trkstp/diesel/mart/rest./24hr, Top Mtn Camping

336mm   parking area/weigh sta eb, litter barrel

329   US 70, Crab Orchard, **N**...BP/diesel/mart, Exxon/mechanic, groceries

327mm   **rest area wb, full(handicapped)facilities, phone, picnic tables, litter barrels, petwalk, vending**

324mm   **rest area eb, full(handicapped)facilities, phone, picnic tables, litter barrels, petwalk, vending**

322   TN 101, Peavine Rd, Crossville, **N**...BP/Bean Pot Rest., Exxon/Taco Bell/TCBY/diesel/mart/atm, Phillips 66/mart, Hardee's, McDonald's, Holiday Inn Express, Super 8, RV camping, to Fairfield Glade Resort, **S**...HOSPITAL, Comfort Inn, Chesnut Hill Winery/rest., camping(7mi), Cumberland Mtn SP

320   TN 298, Crossville, **N**...antiques, winery, **S**...HOSPITAL, BP/DQ/Pizza Hut/diesel/mart, Shell/mart/24hr, Baskin-Robbins, Catfish Cove Rest., Krystal, Factory Outlet/famous brands, antiques, auto repair, tires

318mm   Obed River

317   US 127, Crossville, **N**...BP/diesel, Citgo/diesel/mart, Exxon/Baskin-Robbins/Subway/diesel/mart/ATM/24hr, Shell/diesel/mart, Huddle House/24hr, Best Western, Hampton Inn, Ramada Inn, **S**...Amoco/mart, Phillips 66/diesel/mart, Texaco/mart, Cracker Barrel, Ponderosa(2mi), Ruby Tuesday, Ryan's, Schlotsky's, Shoney's, Taco Bell(2mi), Vegas Steaks, Waffle House, Wendy's(2mi), Days Inn, Heritage Inn, Scottish Inn, Villager Lodge(3mi), Chevrolet/Olds, Chrysler/Plymouth/Dodge, $ Tree, GNC, Goodyear, Goody's, Lowe's, RV sales/service, True-Value, Wal-Mart SuperCtr/gas/24hr, XpressLube, antiques, tires, truck parts, to Cumberland SP

311   Plateau Rd, **N**...Chevron/diesel/mart/fireworks, **S**...BP/mart, Exxon/mart

306.5mm   parking area/litter barrels/weigh sta wb

305mm   parking area/litter barrels/weigh sta eb

301   US 70 N, TN 84, Monterey, **N**...Amoco/mart, Phillips 66, Dairy Queen, Subway, groceries

300   US 70, Monterey, **N**...Citgo/diesel, Phillips 66/diesel/mart, Hardee's, RV camping

291mm   Falling Water River

290   US 70, Cookeville, **N**...BP, **S**...Amoco, Alpine Inn

288   TN 111, Cookeville, to Livingston, Sparta, **S**... Mid Tenn. Truck Stop, Phillips 66/diesel/rest., Shell, Huddle House, Subway, Knight's Inn

287   TN 136, Cookeville, **N**...Amoco/TCBY/mart, Chevron/diesel/mart, Citgo/mart, Shell/mart/wash/24hr, Texaco/diesel/mart, Applebee's, Arby's, BarbWire's Steaks, Baskin-Robbins, Burger King, Capt D's, Chili's, China Star, Cracker Barrel, Dairy Queen, Fazoli's, Golden Corral, Huddle House, Island Rest., KFC, Krystal, Logan's Roadhouse, LJ Silver, McDonald's, Mexican Rest., Mr Gatti's, O'Charley's, Outback Steaks, Pizza Hut, Po Folks, Ponderosa, Quincy's, Red Lobster, Ryan's, Schlotsky's, Shoney's, Sonic, Subway, Taco Bell, Uncle Bud's Catfish, Waffle House, Wendy's, Best Western/El Tapatio, Comfort Suites, Day's Inn, Executive Inn, Hampton Inn, Holiday Inn/rest., Ramada Ltd, Red Carpet, Scottish Inn, Super 8, Thunderbird Inn, Acme Boot Outlet, Auto Value Parts, Big Lots, CVS Drug, $Tree, Fashion Bug, Firestone, Goody's, JC Penney, K-Mart, Kroger/24hr, Lowe's Bldg Supply, Olds/Cadillac/Honda, 1-Hr Photo, Peebles, Pennzoil, Radio Shack, Service Merchandise, Toyota, Wal-Mart SuperCtr/24hr, bank, cinema, mall, st patrol, tires, transmissions, **S**...Exxon/diesel/mart, Pilot/Blimpie/diesel/mart, Gondola Pizza, KFC, Waffle House, Country Inn Suites, Econolodge, Dodge/Dodge Trucks

286   TN 135, Burgess Falls Rd, **N**...HOSPITAL, BP/mart, Exxon/mart/wash, RaceTrac/diesel/mart, Shell/diesel/mart/wash, Arby's, Hardee's, Thai Cuisine, Waffle House, Chrysler/Plymouth/Jeep, Goodyear, Kawasaki/Yamaha, Kia, Mitsubishi, Mazda, Midas Muffler, Nissan, Toyota, U-Haul, bank, to TTU, **S**...Amoco/diesel, Texaco/diesel/mart, Star Motor Inn/rest.

280   TN 56 N, Baxter, **N**...Shell/diesel/mart, Camp Discovery

276   Old Baxter Rd, **N**...auto parts, **S**...Texaco/diesel/mart, fireworks

273   TN 56 S, to Smithville, **S**...BP/diesel, Phillips 66, crafts

268   TN 96, Buffalo Valley Rd, **N**...motel, **S**...to Edgar Evins SP

267.5mm   Caney Fork River

267mm   **rest area both lanes, full(handicapped)facilities, info, phone, picnic tables, litter barrels, petwalk, vending**

266mm   Caney Fork River

263mm   Caney Fork River

258   TN 53, Gordonsville, **N**... Amoco/mart/atm, Exxon/KFC/Taco Bell, Shell(1/2mi), Connie's BBQ, McDonald's, Timberloft, Comfort Inn, to Cordell Hull Dam, **S**... BP/mart, Citgo/mart/atm, Key Stop/diesel, Courtyard County Buffet

E

W

Cookeville

Tennessee Interstate 40

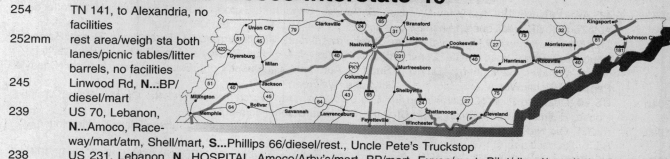

254   TN 141, to Alexandria, no facilities

252mm   rest area/weigh sta both lanes/picnic tables/litter barrels, no facilities

245   Linwood Rd, **N**...BP/diesel/mart

239   US 70, Lebanon, **N**...Amoco, Raceway/mart/atm, Shell/mart, **S**...Phillips 66/diesel/rest., Uncle Pete's Truckstop

238   US 231, Lebanon, **N**...HOSPITAL, Amoco/Arby's/mart, BP/mart, Exxon/mart, Pilot/diesel/mart/24hr, Shell/mart/wash, BBQ, Cracker Barrel, Gondola Rest., Hardee's, McDonald's, Mrs Winner's, Pizza Hut, Ponderosa, Subway, Sunset Family Rest., Taco Bell, Waffle House, Wendy's, Best Western, Comfort Inn, Hampton Inn, Holiday Inn Express, Scottish Inn, Shoney's Inn/rest., Goody's, Wal-Mart SuperCtr/24hr, Western Wear, bank, mall, **S**...Cone/diesel/mart, Express/Perkins/diesel/mart, Pure/diesel, Texaco/Stuckey's, Day's Inn, Knight's Inn, O'Charley's, Super 8, flea mkt, outlet mall/food court, to Cedars of Lebanon SP, RV camping

235   TN 840 W, to Murfreesboro, no facilities

232   TN 109, to Gallatin, **N**...Amoco/McDonalds, Citgo/mart, Express/Main Street Café/diesel/mart/24hr, **2 mi S**...Pilot Gas, camping

228mm   parking area/weigh st wb

226mm   parking area/weigh st eb

226   TN 171, Mt Juliet Rd, **N**...Amoco/diesel/mart/wash, BP/McDonald's/mart, Citgo/mart, Express/diesel/mart, Exxon/diesel/mart, Shell/mart, Texaco/diesel/mart/24hr, Arby's, Capt D's, **S**...Express/diesel, Cracker Barrel, Waffle House, to Long Hunter SP

221   TN 45 N, Old Hickory Blvd, to The Hermitage, **N**...HOSPITAL, CHIROPRACTOR, DENTIST, BP/mart/wash, Citgo, Exxon/mart/wash, Express/diesel, RaceTrac/mart, Shell/mart/24hr, Applebee's, Dairy Queen, Hardee's, Music City Café, O'Charley's, Waffle House, Comfort Inn, Holiday Inn Express, Ramada Ltd, Suburban Lodge, CVS Drug, Kroger/deli, cleaners, **S**...Speedway/mart

219   Stewart's Ferry Pike, **N**...Express/mart, **S**...VETERINARIAN, Express/Subway/mart, Texaco/diesel/mart/wash, Cracker Barrel, Uncle Bud's Catfish, Waffle House, Best Western, Day's Inn, Family Inn, Howard Johnson, Sleep Inn, $General, cleaners

216   (216 c from eb)TN 255, Donaldson Pk, **N**...CHIROPRACTOR, Amoco/mart/wash/24hr, BP/diesel/mart/wash/atm, Citgo/mart, Express/mart, RaceTrac/mart, Texaco/Subway/diesel/mart/24hr, Arby's, Backyard Burger, BBQ, Barsons Rest., Blimpie, Burger King, Domino's, Jalisco's Cantina, KFC, Little Caesar's, McDonald's, New China, Papa John's, Ruby Tuesday, Shoney's, Sonic, Subway, Taco Bell, Waffle House, Wendy's, Baymont Suites, Econolodge, Hampton Inn, Holiday Inn Express, Howard Johnson, Red Roof, Super 8, Wingate Inn, Wyndham Garden, Advance Parts, Custom Cycle, Kinko's, Big K-Mart, Piggly Wiggly, Walgreen/24hr, cleaners, **S**...airport

b a   (from eb)**S**...Nashville Intn'l Airport

215b a   TN 155, Briley Pkwy, to Opreyland, **N**...Texaco, Denny's, Day's Inn, Econolodge, Embassy Suites, Holiday Inn, La Quinta, Marriott, Park Suite Hotel, Quality Inn, Rodeway Inn, Sheraton, bank, **2 mi N on Lebanon Pike**...to Camping World RV Service, KOA, Capt D's, Waffle House, Hampton Inn, **S**...Shell/mart/wash, Clarion, Ramada Inn, Royal Inn, Villager Lodge

213b   I-24 W

a   I-24 E/I-440, E to Chattanooga

213   US 41(from wb), to Spence Lane, **N**...Texaco/diesel/mart/wash, **S**...Denny's, McDonald's, Red Lobster, Shoney's, Waffle House, Western Sizzlin, Day's Inn, Econolodge, Quality Inn, Ramada Inn, Scottish Inn, Tudor Inn, same as 212

212   Fessler's Lane(from eb, no return), **N**...Harley-Davidson, RV service, **S**...Citgo, Texaco/diesel/mart/atm, Burger King, Krystal, McDonald's, Mrs Winners, Sonic, Wendy's, same as 213

211mm   Cumberland River

211b   I-65 N, I-24 W, no facilities

a   I-65 S, I-40 W

210c   US 31 S, US 41A, 2nd Ave, 4th Ave, **N**...Stouffer Hotel, **S**...museum

210b a   I-65(from eb), N to Louisville, S to Birmingham, no facilities

209b a   US 70, Charlotte Ave, Nashville, **N**...Holiday Inn, Chevrolet, Lincoln/Mercury, Nissan, Olds, Pontiac/GMC, Suzuki, Toyota, Country Music Hall of Fame, Conv Ctr, transmissions, **S**...Amoco/diesel/mart, Exxon/mart, Shoney's Inn/rest., Subway, White Castle, Chrysler/Plymouth, JiffyLube

208   I-265, to Louisville, no facilities

207   28th Ave, Nashville, TN St U, **N**...Texaco, Lee's Chicken, pizza

206   I-440 E, to Knoxville, no facilities

Tennessee Interstate 40

N a s h v i l l e

D i c k s o n

205	46th Ave, W Nashville, **N**...Harley-Davidson, **S**...Amoco, Citgo/mart, Express/mart, McDonald's, Mrs Winner's, AutoZone, USPO, antiques, transmissions
204	TN 155, Briley Pkwy, **N**...Exxon/mart/atm, Texaco/diesel/mart, **S**...Citgo/mart, 76, Texaco/mart, Burger King, Church's Chicken/White Castle, Domino's, KFC, Krystal, New China, Shoney's, Uncle Bud's Catfish, Waffle House, Wendy's, Baymont Inn, Daystop, Super 8, AutoZone, CarQuest, Ford, Goodyear/auto, Kroger, NTB, PepBoys, Piggly Wiggly, Precision Tune, JiffyLube, Walgreen
201b a	US 70, Charlotte Pike, **N**...Exxon/mart, Phillips 66/diesel/mart, Texaco/diesel/mart/24hr, Cracker Barrel, Waffle House, Hallmark Inn, Wal-Mart SuperCtr/24hr, **S**...Howard Johnson/rest.
199	rd 251, Old Hickory Blvd, **N**...Texaco/mart/atm, **S**...CHIROPRACTOR, Express/Subway/mart, Waffle House, Sam's Club, cleaners
196	US 70, Newsom Sta, to Bellevue, **N**...Express/mart, Shoney's, cinema, **S**...MEDICAL CARE, DENTIST, OPTOM-ETRIST, BP/mart, Express/diesel/mart, Shell/mart/24hr, Texaco/diesel/mart/atm, Applebee's, Baskin-Robbins, McGillicudy's Grill, Pizza Hut, Sir Pizza, Subway, Taco Bell, Waffle House, Wendy's, Hampton Inn, Circuit City, Dillard's, Firestone/auto, Home Depot, Kastner-Knott, Piggly-Wiggly, Sears, USPO, Walgreen/24hr, bank, cleaners, mall
195mm	Harpeth River
192	McCrory Lane, to Pegram, **S**...Natchez Trace Pkwy
190mm	Harpeth River
188.5mm	Harpeth River
188	rd 249, Kingston Springs, **N**...BP, Express/Blimpie/diesel/mart, Shell/Arby's/mart/wash, Pizza Pro, Sonic, Best Western, Econolodge, Scottish Inn, USPO, bank, **S**...VETERINARIAN, Petro/Chevron/Pizza Hut/diesel/showers/24hr
182	TN 96, to Dickson, Fairview, **N**...BP/mart/atm, Citgo/diesel/mart, Dickson Motel, M Bell SP(16mi), **S**...BP/diesel/mart/atm, Flying J/Conoco/Country Mkt/diesel/LP/mart/24hr
172	TN 46, to Dickson, **N**...HOSPITAL, Amoco/diesel/mart/24hr, BP/Subway/mart, Express/diesel/mart/24hr, Shell/mart, Texaco/Burger King/Baskin-Robbins/mart/atm, Arby's, Cracker Barrel, I-40 Rest., McDonald's, Waffle House/24hr, Wang's China, Baymont Inn, Comfort Inn, Econolodge, Eye40 Motel, Hampton Inn, Knight's Inn, Quality Inn, Ramada Ltd, Super 8, Acme Boots, KOA, to M Bell SP, **S**...Chevron/mart/24hr, Exxon/mart, Day's Inn, Holiday Inn
170mm	**rest area both lanes, full(handicapped)facilities, phone, picnic tables, litter barrels, vending, petwalk**
166.5mm	Piney River
163	rd 48, to Dickson, **N**...Phillips 66/Reed's/diesel, Citgo/mart, **S**...Shell/mart, Rolling Hills Rest., Tanbark Camping
152	rd 230, Bucksnort, **N**...BP/mart, Rudy's Rest., Bucksnort Motel, Travelodge
149mm	Duck River
148	rd 50, Barren Hollow Rd, to Turney Center, **S**...Texaco/mart
143	TN 13, to Linden, Waverly, **N**...BP/diesel/repair, Phillips 66/diesel/mart/atm, Pilot/Subway/diesel/mart/24hr, Shell/mart, BBQ, Buffalo River Grill, Country Kitchen, JJ's Rest., Loretta Lynn's Kitchen, McDonald's, Tyree's Rest./24hr, Day's Inn, Best Western/rest., Buffalo Inn, Holiday Inn Express, Super 8, KOA, **S**...Exxon/repair, Texaco/mart, Southside Motel/rest.
141mm	Buffalo River
137	Cuba Landing, **S**...Citgo/diesel/mart/24hr
133.5mm	Tennessee River
133	rd 191, Birdsong Rd, **N**...Birdsong RV Resort/marina(9mi)
131mm	**rest area both lanes, full(handicapped)facilities, phone, vending, picnic tables, litter barrels, petwalk**
126	US 641, TN 69, to Camden, **N**...Amoco/mart, Exxon/Subway/TCBY/diesel/mart/atm, Phillips 66/North 40/diesel/rest., Texaco/diesel/mart, Burger Barn, tire/truck repair, to NB Forrest SP, **S**...HOSPITAL, BP/diesel/mart, Citgo/Stuckey's/Dairy Queen, Shell/diesel/mart, Day's Inn/rest.
116	rd 114, **S**...to Natchez Trace SP, RV camping
110.5mm	Big Sandy River
108	TN 22, Parkers Crossroads, to Lexington, **N**...Amoco/diesel/mart, BP, Citgo/diesel/mart/24hr, Phillips 66/diesel/mart, Bailey's Rest., Little Venice Pizza, Knight's Inn, USPO, bank, city park, **S**...HOSPITAL, Duckie's/Subway/diesel, Exxon/mart, Texaco/mart, Best Western/rest., RV camping, to Shiloh Nat Bfd(51mi)
103mm	parking area/weigh sta eb, litter barrels, phone
102mm	parking area/weigh sta wb, litter barrels
101	rd 104, **N**...Exxon/mart
93	rd 152, Law Rd, **N**...Phillips 66/diesel/deli/24hr, **S**...Texaco/diesel/mart
87	US 70, US 412, Jackson, **N**...Citgo/diesel/mart, antique, **S**...Exxon, Texaco/diesel/mart
85	Christmasville Rd, to Jackson, **N**...Amoco/Subway/diesel/mart, Exxon/diesel/mart/24hr, Howard Johnson Express

E

W

Tennessee Interstate 40

E

W

82b a US 45, Jackson, **N**...Texaco, Cracker Barrel, Knight's Inn, Microtel, **S**...DENTIST, EYECARE, CHIROPRACTOR, VETERINARIAN, Citgo/mart, Exxon/mart, RaceTrac/mart, Backyard Burger, Baskin-Robbins, Barley's Brewhouse, Burger King, Capt D's, China Palace, Dairy Queen, KFC, Krystal, LJ Silver, Los Portales Mexican, McDonald's, Pizza Hut, Pizza Inn, PoFolks, Shoney's, Sonic, Subway, Taco Bell, Waffle House, Wendy's, Baymont Inn, Quality Inn, Ramada Ltd, Sheraton/rest., Super 8, Travelers Motel, Advance Parts, AutoZone, Belk, Big Lots, CarQuest, Firestone/auto, Goldsmith's, Goodyear/auto, Hobby Lobby, JC Penney, Kinko's, Kroger/24hr, Office Depot, Pier 1, Precision Tune, Radio Shack, Sears/auto, Service Merchandise, TJ Maxx, Upton's, WaldenBooks, Wolf Camera, mall

80b a US 45 Byp, Jackson, **N**...Exxon, Chili's, Chick-fil-A, Domino's, Fazoli's, 5&Diner, IHOP, KFC, LoneStar Steaks, Peking Chinese, Schlotsky's, AmeriHost, Country Inn Suites, Red Roof Inn, Circuit City, Home Depot, Kinko's, Lowe's Bldg Supply, PetsMart, Sam's Club, Wal-Mart SuperCtr, **S**...HOSPITAL, Amoco/mart/wash, BP/mart/wash, Citgo/diesel/mart, Phillips 66/diesel/mart, Applebee's, Arby's, Barnhill's Buffet, Baudo's Rest., BBQ, Burger King, Casey Jones' Village, Dunkin Donuts, El Chico's, Logan's Roadhouse, Madison's Rest., McDonald's, Mrs Winner's, O'Charley's, Old Town Spaghetti, Pizza Hut, Sonic, Subway, Taco Bell, Village Pizza, Waffle House, Best Western, Budget Inn, Comfort Inn, Day's Inn, Econolodge, Fairfield Inn, Garden Plaza Hotel, Hampton Inn, Holiday Inn, Hancock Fabrics, King Tires, K-Mart, Midas Muffler, Olds/Cadillac/Pontiac/Buick/Toyota, bank, to Pinson Mounds SP, Chickasaw SP

79 US 412, Jackson, **S**...BP/repair, Citgo/diesel/mart, Exxon/mart, GG's Café, Day's Inn, Jackson RV Park

78mm Forked Deer River

76 rd 223 S, **S**...McKellar-Sites Airport

74 Lower Brownsville Rd, no facilities

73mm **rest area both lanes, full(handicapped)facilities, info, phone, picnic tables, litter barrels, vending, petwalk**

68 rd 138, Providence Rd, **N**...Amoco/diesel/mart, Econolodge, **S**...BP/TA/diesel/rest./atm/24hr, Citgo/mart, Joy-O RV Park, truck repair

66 US 70, to Brownsville, **S**...Citgo/diesel, FuelMart/Blimpie/diesel/24hr, Motel 6

60 rd 19, Mercer Rd, no facilities

56 TN 76, to Brownsville, **N**...Amoco/mart, Citgo/diesel/mart/24hr, Baskin-Robbins/Taco Bell, Dairy Queen, Fiesta Garden Mexican, KFC, McDonald's, Best Western, Comfort Inn, Day's Inn, Holiday Inn Express, **S**...BP/diesel, Exxon/Huddle House/diesel/mart/24hr

55mm Hatchie River

52 TN 76, rd 179, Koko Rd, to Whiteville, no facilities

50mm weigh sta both lanes, phone

47 TN 179, to Stanton, Dancyville, **S**...truckstop/diesel/rest./24hr

42 TN 222, to Stanton, **N**...Countryside Inn, **S**...Exxon/diesel, Phillips 66/mart

35 TN 59, to Somerville, **S**...BP/diesel/mart, Longtown Rest.

29.5mm Loosahatchie River

25 TN 205, Airline Rd, to Arlington, **N**...Exxon/Subway/diesel/mart

24 new exit

20 Canada Rd, Lakeland, **N**...Amoco/McDonald's/mart, BP/diesel/mart/24hr, Waffle House, Day's Inn/rest., Relax Inn, Super 8, **S**...Exxon/Subway/TCBY/diesel, Factory Outlet/famous brands, KOA

18 US 64, to Bartlett, **N**...Exxon/Burger King/mart/atm, Backyard Burger, Don Pablo's, Luby's, Schlotsky's, Country Inn Suites, Fairfield Inn, Holiday Inn Express, Goodyear/auto, Wal-Mart SuperCtr/24hr, same as 16, **S**...Amoco/mart, Citgo/mart, Circle K/STP Gas, Kroger/deli/bakery/24hr

16b a TN 177, to Germantown, **N**...BP/mart, Shell/mart/atm, Alexander's Rest., Bahama Breeze, Burger King, Chili's, Chick-fil-A, Danver's, Heavenly Ham, IHOP, Joe's Crabshack, Logan's Roadhouse, McDonald's, On-the-Border, Red Lobster, Sessel's Café, Taco Bell, Wendy's, AmeriSuites, Hampton Suites, Barnes&Noble, Bed Bath&Beyond, Best Buy, Chevrolet, Circuit City, Dillard's, Ford, Home Depot, HomePlace, JC Penney, Just For Feet, Linens'n Things, MensWearhouse, Michael's, OfficeMax, Old Navy, Pier 1, Sears, Service Merchandise, Target, TJ Maxx, Walgreen, bank, cinema, **1 mi S**...Phillips 66, Shell/mart, Arby's, Bronx Deli, Burger King, El Porton Mexican, Great Wall Chinese, Krystal, Shogun Japanese, Best Suites, Wingate Inn, AutoZone, Rite Aid

15b a Appling Rd, **N**...Phillips 66, Red Lobster

Tennessee Interstate 40

14	Whitten Rd, **N**...BP/diesel/mart/atm, Express/Blimpie/mart, Shelby Motel, waterpark, **S**...Citgo/Stuckey's/diesel/24hr, Burger King, Travelers Inn
12	Sycamore View Rd, **N**...Shell/mart, Texaco/diesel/mart/atm, Total/diesel/mart, Capt D's, Cracker Barrel, Dunkin Donuts, Gridley's BBQ, Geribaldi's Pizza, Hungry Fisherman, KFC, McDonald's, Mrs Winner's, Perkins, Pizza Inn, Raphael's Mexican, Shoney's, Sonic, Subway, Taco Bell, Waffle House, Baymont Inn, Drury Inn, Hampton Inn, Holiday Inn Express, Howard Johnson, Red Roof, JiffyLube, **S**...BP/diesel/mart, Exxon/mart, Burger King, China Cuisine, Denny's, Old Country Buffet, Wendy's, Best Western, Day's Inn, Fairfield Inn, La Quinta, Memphis Inn, Motel 6, Shoney's Inn/rest., Super 8, Celebration Sta, Midas Muffler, Wal-Mart
10.5mm	Wolf River
10b a	(from wb)I-240 W around Memphis, I-40 E to Nashville
12c	(from eb)I-240 W, to Jackson, I-40 E to Nashville
12b	Sam Cooper Blvd(from eb)
12a	US 64/70/79, Summer Ave, **N**...Express/diesel/mart, Texaco/diesel/mart/atm, Kathy&Jimmie's Rest., Luby's, Waffle House, Holiday Inn Express, Ford, U-Haul, mall, **S**...Amoco, Exxon/mart/atm, Arby's, NamKing Chinese, McDonald's, Pizza Hut, Western Sizzlin, Concord Inn, Brake-O, Firestone/auto, Goodyear/atuo, K-Mart, Kroger/drugs, Sam's Club, Sears
10	TN 204, Covington Pike, **N**...Pontiac/Olds,Cadillac
8b a	TN 14, Jackson Ave, **N**...Amoco, Texaco/mart/wash, McDonald's, Red Lobster, **S**...RaceTrac, Shell, Morrison's Dining, Comfort Inn, Camper Sales
6	Warford Rd, no facilities
5	Hollywood St, **N**...BP, Exxon, Express/mart, Shell/mart, BC Rest., Hardee's, Walgreen, laundry, access to Memphis Zoo, Rhodes Coll
3	Watkins St, **N**...Amoco, Citgo/mart, Total/mart, Donuts, U-Haul, flea mkt
2a	US 51 N, to Millington, **N**...Meeman-Shelby SP
2	Smith Ave, Chelsea Ave, **S**...Mr Complete Auto Parts
1g f	TN 14, Jackson Ave, **N**...Amoco, **S**...Exxon/mart, Express/mart, Rainbow Inn, auto repair, transmissions
1e	I-240 E, no facilities
1d c b	US 51, Danny Thomas Blvd, **N**...St Jude Research Ctr
1a	2nd St(from wb), downtown, **S**...Brownestone Hotel, Crown Plaza, Holiday Inn, Sheraton, Conv Ctr, banks, auto transmission repair
1	Riverside Dr, Front St(from eb), Memphis, **S**...Comfort Inn, Conv Ctr, Riverfront
0mm	Mississippi River, Tennessee/Arkansas state line

Tennessee Interstate 55

Exit #	Services
13mm	Mississippi River, Tennessee/Arkansas state line
12c	Delaware St, Memphis, **E**...Riverside Rest., **W**...Day's Inn
b	Riverside Dr, downtown Memphis, **E**...Texaco
a	E Crump Blvd(from nb), **E**...museum
11	McLemore Ave, Presidents Island, industrial area
10	S Parkway, **1/2mi E**...Conoco/diesel/mart
9	Mallory Ave, industrial area
8	Horn Lake Rd(from sb), no facilities
7	US 61, 3rd St, **1 mi E**...Amoco, Exxon/mart, Capt D's, Church's Chicken, McDonald's, Taco Bell, AutoZone, Family$, Kroger, Radio Shack, SuperValu Foods, Walgreen, **W**...Conoco/diesel, Phillips 66/mart, Williams/diesel/mart, KFC, McDonald's, Subway, Regency Inn, Rest Inn, CarQuest, TO Fuller SP, Indian Museum
6b a	I-240, no facilities
5b	US 51 S, Elvis Presley Blvd, to Graceland, **W**...HOSPITAL, DENTIST, Amoco, Citgo, Dodge Store/gas, Exxon/mart, Phillips 66/diesel/mart, Texaco/mart/wash, Best Pizza, Capt D's, KFC, Peking Chinese, Rally's, Taco Bell, Tony's Grill, American Inn, Day's Inn, Graceland Inn, Heartbreak Hotel, Advance Parts, Davis RV Ctr, Dodge, Honda, KOA, Walgreen, bank, transmissions, to Graceland
a	Brooks Rd, **E**...Amoco/mart/wash, Exxon/mart/wash, Williams/diesel/mart, Kettle, Popeye's, Airport Inn, Comfort Inn, Day's Inn, Ramada Inn/rest., Travelodge, Toyota
3mm	**Welcome Ctr nb, full(handicapped)facilities, phone, vending, picnic tables, litter barrels, petwalk**
2b a	TN 175, Shelby Dr, Whitehaven, **E**...Amoco/mart/24hr, Conoco/mart/atm, Exxon/mart/atm, Shelby Inn, Super 8, cleaners/laundry, **W**...76/Circle K, Texaco, Burger King, CK's CoffeeShop, KeKe's Steaks, McDonald's, Mrs Winner's, Taco Bell, Ace Hardware, Family$, Goodyear/auto, SaveALot Food, Seessel's Food/drug/gas, U-Haul, bank
0mm	Tennessee/Mississippi state line

E ↑↓ W

N ↑↓ S

M e m p h i s

Tennessee Interstate 65

Exit #	Services
121.5mm	Tennessee/Kentucky state line
121mm	**Welcome Ctr sb, full(handicapped)facilities, phone, picnic tables, litter barrels, vending, petwalk**

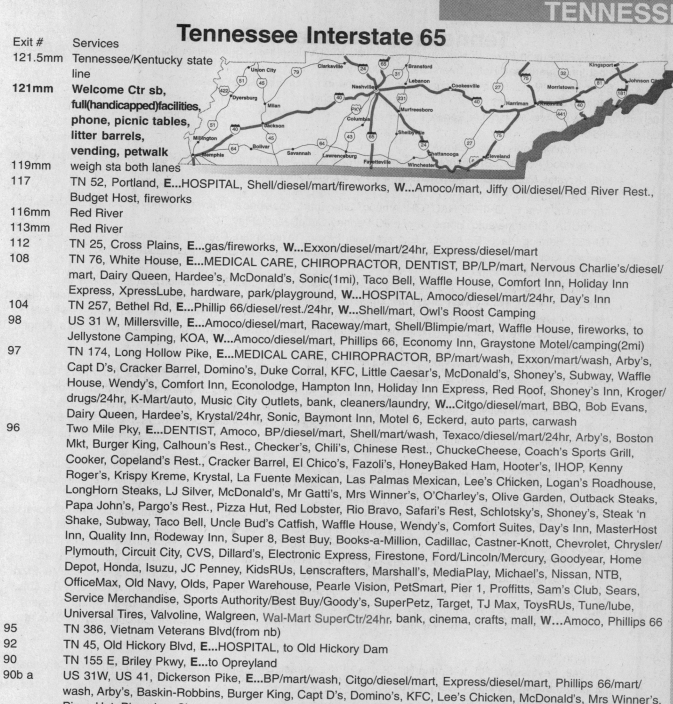

Exit #	Services
119mm	weigh sta both lanes
117	TN 52, Portland, **E**...HOSPITAL, Shell/diesel/mart/fireworks, **W**...Amoco/mart, Jiffy Oil/diesel/Red River Rest., Budget Host, fireworks
116mm	Red River
113mm	Red River
112	TN 25, Cross Plains, **E**...gas/fireworks, **W**...Exxon/diesel/mart/24hr, Express/diesel/mart
108	TN 76, White House, **E**...MEDICAL CARE, CHIROPRACTOR, DENTIST, BP/LP/mart, Nervous Charlie's/diesel/mart, Dairy Queen, Hardee's, McDonald's, Sonic(1mi), Taco Bell, Waffle House, Comfort Inn, Holiday Inn Express, XpressLube, hardware, park/playground, **W**...HOSPITAL, Amoco/diesel/mart/24hr, Day's Inn
104	TN 257, Bethel Rd, **E**...Phillip 66/diesel/rest./24hr, **W**...Shell/mart, Owl's Roost Camping
98	US 31 W, Millersville, **E**...Amoco/diesel/mart, Raceway/mart, Shell/Blimpie/mart, Waffle House, fireworks, to Jellystone Camping, KOA, **W**...Amoco/diesel/mart, Phillips 66, Economy Inn, Graystone Motel/camping(2mi)
97	TN 174, Long Hollow Pike, **E**...MEDICAL CARE, CHIROPRACTOR, BP/mart/wash, Exxon/mart/wash, Arby's, Capt D's, Cracker Barrel, Domino's, Duke Corral, KFC, Little Caesar's, McDonald's, Shoney's, Subway, Waffle House, Wendy's, Comfort Inn, Econolodge, Hampton Inn, Holiday Inn Express, Red Roof, Shoney's Inn, Kroger/drugs/24hr, K-Mart/auto, Music City Outlets, bank, cleaners/laundry, **W**...Citgo/diesel/mart, BBQ, Bob Evans, Dairy Queen, Hardee's, Krystal/24hr, Sonic, Baymont Inn, Motel 6, Eckerd, auto parts, carwash
96	Two Mile Pky, **E**...DENTIST, Amoco, BP/diesel/mart, Shell/mart/wash, Texaco/diesel/mart/24hr, Arby's, Boston Mkt, Burger King, Calhoun's Rest., Checker's, Chili's, Chinese Rest., ChuckeCheese, Coach's Sports Grill, Cooker, Copeland's Rest., Cracker Barrel, El Chico's, Fazoli's, HoneyBaked Ham, Hooter's, IHOP, Kenny Roger's, Krispy Kreme, Krystal, La Fuente Mexican, Las Palmas Mexican, Lee's Chicken, Logan's Roadhouse, LongHorn Steaks, LJ Silver, McDonald's, Mr Gatti's, Mrs Winner's, O'Charley's, Olive Garden, Outback Steaks, Papa John's, Pargo's Rest., Pizza Hut, Red Lobster, Rio Bravo, Safari's Rest, Schlotsky's, Shoney's, Steak 'n Shake, Subway, Taco Bell, Uncle Bud's Catfish, Waffle House, Wendy's, Comfort Suites, Day's Inn, MasterHost Inn, Quality Inn, Rodeway Inn, Super 8, Best Buy, Books-a-Million, Cadillac, Castner-Knott, Chevrolet, Chrysler/Plymouth, Circuit City, CVS, Dillard's, Electronic Express, Firestone, Ford/Lincoln/Mercury, Goodyear, Home Depot, Honda, Isuzu, JC Penney, KidsRUs, Lenscrafters, Marshall's, MediaPlay, Michael's, Nissan, NTB, OfficeMax, Old Navy, Olds, Paper Warehouse, Pearle Vision, PetSmart, Pier 1, Proffitts, Sam's Club, Sears, Service Merchandise, Sports Authority/Best Buy/Goody's, SuperPetz, Target, TJ Max, ToysRUs, Tune/lube, Universal Tires, Valvoline, Walgreen, Wal-Mart SuperCtr/24hr, bank, cinema, crafts, mall, **W**...Amoco, Phillips 66
95	TN 386, Vietnam Veterans Blvd(from nb)
92	TN 45, Old Hickory Blvd, **E**...HOSPITAL, to Old Hickory Dam
90	TN 155 E, Briley Pkwy, **E**...to Opreyland
90b a	US 31W, US 41, Dickerson Pike, **E**...BP/mart/wash, Citgo/diesel/mart, Express/diesel/mart, Phillips 66/mart/wash, Arby's, Baskin-Robbins, Burger King, Capt D's, Domino's, KFC, Lee's Chicken, McDonald's, Mrs Winner's, Pizza Hut, Pizza Inn, Shoney's, Subway, Taco Bell, Waffle House, Wendy's, Colony Motel, Congress Inn, Day's Inn, Econolodge, Sleep Inn, Super 8, Advance Parts, AutoZone, Camping World RV Service, Crossman Muffler, CVS Drug, Family$, KOA, Kroger, Sam's Club, Walgreen, bank, laundry
88b a	I-24, W to Clarksville, E to Nashville, no facilities
87b a	US 431, Trinity Lane, **E**...Citgo/Circle K, Pilot/Arby's/diesel/mart, Krystal, White Castle, Cumberland Inn, Trinity Inn, carwash, **W**...DENTIST, Amoco/mart/wash, BP/mart/wash, Exxon, Shell/24hr, Texaco/diesel/mart, BBQ, Burger King, Capt D's, Denny's, McDonald's, Pizza Etc, Ponderosa, Shoney's, Taco Bell, Waffle House, Baymont, Belvedere Inn, Day's Inn, Drury Inn, Econolodge, Hampton Inn, Holiday Inn Express, Howard Johnson/rest, Knight's Inn, Liberty Inn, Motel 6, Quality Inn, RV Ctr
86	I-265 W to Memphis, no facilities
85b	N 1st St, Jefferson St, **E**...Citgo/mart, Mrs Winner's, diesel repair, **W**...Conoco/mart, Express/mart, Best Western
85a	US 31, Spring St, **W**...Spur Gas, Best Western, Day's Inn, U-haul

N

S

Tennessee Interstate 65

N a s h v i l l e

85   James Robertson Pkwy, downtown, st capitol, **W...**BP/TA/diesel/rest., Shell/wash, Burger King, Shoney's, Best Host Inn, Heartbreak Hotel, Interstate Inn, Quality Inn

84   Shelby Ave, **W...**Exxon/mart, Econolodge, Hampton Inn, Ramada Ltd, to stadium

83a   I-40/I-24 E, no facilities

210c[I-40] 2nd Ave, 4th Ave, no facilities

210[I-40] I-65 S, I-240

82b a   I-40, W to Memphis, E to Nashville

81   Wedgewood Ave, **W...**BP/mart/wash, Exxon/mart, U-Haul, **services on US 31...**Citgo, Cone/diesel//mart, Burger King, Krystal, McDonald's, Mrs Winner's, CVS Drug, $General, Kroger, tires

80   I-440, to Memphis, Knoxville

79   Armory Dr, **1 mi E...**CHIROPRACTOR, Amoco, Citgo/mart, Applebee's, Subway, Wendy's, Burlington Coats, CompUSA, Firestone/auto, Home Depot, JC Penney Outlet, MediaPlay, TJ Maxx, mall

78b a   TN 255, Harding Place, **E...**Amoco, Express/mart, Exxon, Texaco/mart, Cracker Barrel, Darryl's Rest., Garcia's Mexican, Golden Griddle Pancakes, Santa Fe Grill, Waffle House, La Quinta, Ramada/rest., Red Roof Inn, Traveler's Rest Hist Home

74   TN 254, Old Hickory Blvd, to Brentwood, **E...**Capt D's, Shoney's, AmeriSuites, Hilton Suites, Holiday Inn, Steeplechase Inn, **W...**CHIROPRACTOR, DENTIST, Amoco, BP, 76, Shell/mart/wash, Texaco/diesel/mart, August Moon Chinese, BBQ, Mrs Winner's, O'Charley's, Pargo's Rest., Wendy's, Hampton Inn, Courtyard, Traveler's Rest Inn, Boot Town, USPO, Valvoline, bank, laundry, **facilities on US 31...**Burger King, Little Caesar's, Kroger, OfficeMax, cinema

F r a n k l i n

71   TN 253, Concord Rd, to Brentwood, no facilities

69   Moore's Lane, Galleria Blvd, **E...**Mapco/Blimpie/diesel/mart, Applebee's, Backyard Burgers, Cozymel's, Eat at Joe's, Houlihan's, Outback Steaks, Shogun Japanese, AmeriSuites, Bruno's Food/drug, Circuit City, Home Depot, PetsMart, PostNet, Rooms to Go, Vision Ctr, cleaners, **W...**DENTIST, Amoco/mart/wash, BP/mart, Kwiksak/gas, J Alexander's Rest., Macaroni Grill, McDonald's, Red Lobster, Subway, Taco Bell, Sleep Inn, Barnes&Noble, Castner-Knott, Computer City, Dillard's, Infiniti, JC Penney, Sears/auto, Service Merchandise, Target, Upton's, bank, mall

68   Cool Springs Blvd, **W...**MEDICAL CARE, Chili's, Luby's, Panda Chinese, Ruby Tuesday, Country Inn Suites, Hampton Inn Suites, HomePlace, Jo-Ann Fabrics, Kroger, Lowe's, MailBoxes Etc, Marshall's, Parisian, Saturn, Staples, bank, cleaners, funpark, to Galleria Mall

65   TN 96, Franklin, to Murfreesboro, **E...**MEDICAL CARE, CHIROPRACTOR, Exxon/mart, Texaco/mart, Cracker Barrel, Sonic, Steak'n Shake, Baymont, Comfort Inn, Day's Inn, Howard Johnson, Quality Inn, Ramada Ltd, Buick/Pontiac/GMC/Kia, Chevrolet, Food Lion, Honda/Volvo, Pennzoil, cleaners, **W...**CHIROPRACTOR, DENTIST, VETERINARIAN, OPTOMETRIST, Amoco/mart/wash, BP/mart/wash, Mapco/diesel/mart, Shell/mart, Arby's, Backyard Burgers, ChopHouse Chinese, Hardee's, KFC, Little Caesar's, McDonald's, O' Charley's, Pizza Hut, Shoney's, Subway, Taco Bell, TCBY, Waffle House, Wendy's, Best Western, Holiday Inn Express, Big Lots, BiLo Food, Chrysler/Plymouth/Dodge, CVS Drug, $General, Eckerd, Ford/Mercury, Goody's, K-Mart, Kroger, Nissan, Piece Goods, Toyota, Wal-Mart, antiques, bank, cinema, cleaners, hardware, to Confederate Cem at Franklin

64mm   Harpeth River

61   TN 248, Peytonsville Rd, to Spring Hill, **E...**TA/BP/diesel/rest./24hr, Speedway/diesel/mart/24hr, **W...**Cone/mart, Mapco/mart/24hr, Goose Creek Inn/rest.

58mm   W Harpeth River

53   TN 396, Saturn Pkwy, to Spring Hill, Columbia, TN Scenic Pkwy

48mm   parking area/weigh sta nb

46   US 412, TN 99, to Columbia, Chapel Hill, **E...**Chevron/mart, Bear Creek Rest., **W...**HOSPITAL, BP/mart, Exxon/Burger King/mart, Texaco/diesel/rest., Econolodge, Holiday Inn Express, Relax Inn, antiques

40.5mm   Duck River

37   TN 50, to Columbia, Lewisburg, **E...**HOSPITAL, TN Walking Horse HQ, **W...**Texaco/diesel/mart, Richland Inn, to Polk Home

32   TN 373, Mooresville, to Lewisburg, **E...**Exxon/diesel/mart

27   TN 129, Cornersville, to Lynnville, **E...**Texas T Camping

25mm   parking area sb, litter barrels

24mm   parking area nb, litter barrels

22   US 31A, to Pulaski, **E...**Amoco/diesel/mart/rest., Econolodge, tire/diesel repair, truckwash, **W...**BP/diesel/mart/24hr, Exxon, Express/diesel/mart, flea mkt

Tennessee Interstate 65

14   US 64, to Pulaski, **E**...BP/ diesel/mart, Texaco/diesel/ mart, Sands Rest., Super 8, KOA, bank, to Jack Daniels Dist, **W**...to David Crockett SP

6   TN 273, Bryson, **E**...Phillips 66/diesel/ rest., Economy Inn/rest.

4mm   Elk River

3mm   **Welcome Ctr nb, full(handicapped)facilities, info, phone, picnic tables, litter barrels, petwalk**

1   US 31, TN 7, Ardmore, **E**...Chevron/diesel/mart/24hr, HP/diesel/mart, antiques, **1-2 mi E**...Conoco/repair, Dairy Queen, Hardee's, McDonald's, Subway

0mm   Tennessee/Alabama state line

Tennessee Interstate 75

Exit #   Services

161.5mm Tennessee/Kentucky state line

161mm   **Welcome Ctr sb, full(handicapped)facilities, phone, vending, picnic tables, litter barrels, petwalk**

160   US 25W, Jellico, **E**...Amoco, Citgo, Exxon/Subway/diesel, Texaco/Stuckey's, KFC, Jellico Motel/rest., auto parts, **W**...HOSPITAL, Shell/Arby's/diesel/mart, Hardee's, Wendy's, Best Western, Day's Inn/rest., Flowers Bakery, camping, fireworks, to Indian Mtn SP

147mm   parking area sb

144   Stinking Creek Rd, no facilities

141   TN 63, to Royal Blue, **E**...BP/Stuckey's/diesel/mart, Perkins/24hr, **W**...Exxon/mart, Texaco/diesel/mart/24hr, Dairy Queen, Comfort Inn, fireworks, to Big South Fork NRA

134   US 25W, TN 63, Caryville, **E**...HOSPITAL, Exxon/diesel, Shell/mart/24hr, Dinner Bell Rest., LaFollete House Rest., Louie's Rest., Pizza Hut, Waffle House/24hr, Western Sizzlin, Hampton Inn, Holiday Inn, Family Inn/rest., Lakeview Inn/rest., Super 8, Thacker Christmas Inn, to Cove Lake SP, **W**...Amoco, BP, Shoney's, Scotty's Hamburgers, Budget Host

130mm   weigh sta both lanes, phones

129   US 25W S, Lake City, **E**...antiques, LP, **W**...BP, Citgo/mart, Exxon/Subway/diesel/24hr, Shell/mart, Texaco/ Blimpie/diesel/mart, Burger King, Cracker Barrel, KFC, McDonald's, Pizza King&Steaks, Day's Inn, Blue Haven Motel, Lamb's Inn/rest., fireworks, same as 128

128   US 441, to Lake City, **E**...BP, Coastal, Mountain Lake Marina(4mi), **W**...Exxon/diesel/mart/24hr, Shell/mart, Cottage Rest., Subway, Blue Haven Motel, Lake City Motel, Lamb's Inn, $General, Hillbilly Mkt/deli, antique cars, to Norris Dam SP, same as 129

126mm   Clinch River

122   TN 61, Bethel, Norris, **E**...Phillips 66/rest., Shell/mart, antiques, camping, museum, **W**...Citgo/diesel/mart, Exxon/ Subway/diesel/24hr, Git'n Go/diesel/mart, Marathon/diesel/mart/24hr, Texaco, Burger King, Hardee's, Krystal/ 24hr, McDonald's, Waffle House, Wendy's, Comfort Inn, Holiday Inn Express, Jameson Inn, Super 8, Big Pine Ridge SP

117   TN 170, Racoon Valley Rd, **E**...BP, Phillips 66/rest./24hr, Williams/Burger King/Subway/diesel/mart/24hr, Shell/ mart, **W**...Valley Inn, Jellystone Park, KOA

112   TN 131, Emory Rd, to Powell, **E**...BP/mart/wash, Chevron/Pizza Inn/mart, Pilot/DQ/Taco Bell/diesel/mart/24hr, Aubrey's Rest., BBQ, McDonald's, KFC, Krystal, Steak'n Shake, Wendy's, Baymont Inn, Holiday Inn Express, CVS Drug, $General, Family$, Ingles/deli, Toyota, bank, **W**...Phillips 66/mart/24hr, Shell/diesel/mart/24hr, Hardee's, Shoney's, Waffle House/24hr, Comfort Inn

110   Callahan Dr, **E**...Weigel's/gas, Day's Inn(2mi), Knight's Inn, Quality Inn/rest., **W**...Amoco/mart, Phillips 66, Scottish Inn, Burger King, Chick-fil-A, McDonald's, Wendy's, GMC/Volvo, camping

108   Merchant Dr, **E**...BP/diesel/mart/wash, Citgo/diesel/mart, Pilot/diesel/mart, Texaco/diesel/mart, Applebee's, Cracker Barrel, Denny's, El Chico's, Logan's Roadhouse, O'Charley's, Olive Garden, Pizza Hut, Sagebrush Steaks, Ryan's, Waffle House, Best Inn, Comfort Inn, Day's Inn, Hampton Inn, Holiday Inn/rest., Ramada Ltd, Sleep Inn, CVS Drug, Ingles/deli, Valvoline, **W**...CHIROPRACTOR, Citgo/mart, Conoco/mart, Exxon/mart, Pilot/ mart, Shell/mart, Arby's, Baskin-Robbins, Bob Evans, Burger King, Capt D's, Chinese Rest., Darryl's Rest.,

Knoxville

Tennessee Interstate 75

Knoxville

Godfather's, IHOP, McDonald's, Nixon's Deli, Outback Steaks, Pizza Inn, Red Lobster, Stacy's Buffet, Subway, Waffle House/24hr, Econolodge, Family Inn, Friendship Inn, La Quinta, Red Roof Inn, Super 8, Food Lion, Hancock Fabrics, Walgreen, XpressLube, laundry, access to same as US 25W

107	I-640 & I-75
3b[I-640]	US 25W, **N**...Amoco/mart, BP, Texaco/diesel/mart/wash, Hardee's, KFC, Krystal, LJ Silver, Mr Gatti's, Pizza Hut, Quincy's, Taco Bell, Wendy's, Chevrolet, CVS Drug, Firestone, Ford, K-Mart, Kroger, Midas Muffler, Pearle Vision, SuperX Drug, hardware, transmissions, same as 108
1[I-640]	TN 62, Western Ave, **N**...CHIROPRACTOR, Cargo/diesel/mart, RaceTrac/mart, Texaco/diesel/mart, Baskin-Robbins, Central Park, Golden Corral, KFC, LJ Silver, McDonald's, Mixon's Deli, Ruby Tuesday, Shoney's, Subway, Taco Bell, Wendy's, CVS Drug, Kroger, Walgreen, **S**...DENTIST, BP/mart, Dad's Donuts, Domino's, Hardee's, Krystal, Advance Parts, Mighty Muffler, USPO, auto repair, bank

I-75 and I-40 run together 17 mi. See Tennessee Interstate 40, exits 369 through 385.

84[368]	I-40, W to Nashville
81	US 321, TN 95, to Lenoir City, **E**...BP/diesel/mart/wash, Exxon/Subway/diesel/mart, Phillips 66/mart, Shell, Texaco/diesel/mart, BelAir Grill, Burger King(1mi), Dinner Bell Rest., KFC, McDonald's, Shoney's, Waffle House, Wendy's, Crossroads Inn, Inn of Lenoir, King's Inn/rest., to Great Smokies NP, Ft Loudon Dam, **W**...Citgo/mart/fireworks, Shell/diesel/mart/fireworks, Krystal, Comfort Inn, Econolodge, Ramada Ltd, Chevrolet, Pontiac/Buick/GMC/Olds
76	rd 324, Sugar Limb Rd, **W**...to TN Valley Winery
74mm	Tennessee River
72	TN 72, to Loudon, **E**...BP/atm, Shell/Wendy's/mart, BBQ, McDonald's, Holiday Inn Express, Super 8, to Ft Loudon SP, **2 mi E**...Hardee's, Pizza Hut, Advance Parts, $General, Food City/deli, Rite Aid, bank, museum, **W**...Citgo, Phillips 66/mart, Marie's Kitchen, Knight's Inn, Express Camping
68	TN 323, to Philadelphia, **E**...BP/diesel/mart/fireworks
62	TN 322, Oakland Rd, to Sweetwater, **E**...Dinner Bell Rest., **W**...KOA
60	TN 68, Sweetwater, **E**...HOSPITAL, Marathon, Raceway/mart, Texaco/diesel/mart/atm, Burger King, Huddle House, KFC, McDonald's, Pizza Hut(1mi), Subway(2mi), Wendy's, Budget Host, Comfort Inn, Day's Inn, Super 8(2mi), Sweetwater Hotel, to Lost Sea Underground Lake, **W**...BP/diesel/mart, Exxon/diesel/mart, Phillips 66/mart/wash, BBQ, Cracker Barrel, Denny's, Best Western, flea mkt, camping, to Watts Bar Dam
56	TN 309, Niota, **E**...Trkstop/diesel/rest./fireworks/atm, Country Music Camping, **W**...tires
52	TN 305, Mt Verd Rd, to Athens, **E**...BP(1mi), Conoco(2mi), Phillips 66/mart, Overniter RV Park, fireworks, **W**...Exxon/mart, Ramada Ltd
49	TN 30, to Athens, **E**...HOSPITAL, BP/mart/wash, Exxon/TCBY/mart, Raceway/mart, Shell/DQ/diesel/mart/24hr, Texaco/diesel/mart/atm, Applebee's, Burger King, Hardee's, KFC, Krystal, McDonald's, Monterrey Mexican, Shoney's, Subway, Taco Bell, Waffle House, Wendy's, Western Sizzlin(1mi), Day's Inn, Hampton Inn, Homestead Inn, Knight's Inn, Motel 6, Super 8, KOA, Russell Stover Candy, to TN Wesleyan Coll, **W**...Shell/diesel/mart, Homestead Inn, Jeep, Pontiac/Buick/Cadillac/Olds/GMC/Daewoo
45.5mm	**rest area both lanes, full(handicapped)facilities, phone, vending, picnic tables, litter barrels, petwalk**
42	TN 39, to Riceville, **E**...39 Express/diesel/mart, Relax Inn, Rice Inn(2mi)
36	TN 163, to Calhoun, no facilities
35mm	Hiwassee River
33	TN 308, to Charleston, **E**...Citgo/mart, 33 Camping, antiques, **W**...Texaco/diesel/rest./24hr, Ponderosa
27	Paul Huff Pkwy, **1 mi E**...EYECARE, Phillips 66, Applebee's, BelAir Grill, Central Park Burgers, CiCi's, Fazoli's, McDonald's, O'Charley's, Panera Bread, Ryan's, Steak'n Shake/24hr, Taco Bell, Jameson Inn, AutoZone, BooksAMillion/Joe Muggs, Buick/Pontiac, CVS Drug, FashionBug, Food Lion, Goodyear/auto, Goody's, HobbyLobby, JC Penney, K-Mart, Lowe's Bldg Supply, OfficeMax, Pennzoil, PetCo, Proffit's, Sears, Staples, TJ Maxx, Wal-Mart SuperCtr/gas/24hr, bank, cinema, mall, **W**...BP/diesel/mart, Conoco, Shell/mart, Denny's/24hr, Hardee's, Waffle House, Wendy's, Comfort Inn, EQ Motel, Hampton Inn, Ramada Ltd, Royal Inn, Super 8
25	TN 60, Cleveland, **E**...HOSPITAL, BP/mart/wash, Chevron/diesel/mart, Exxon, RaceTrac/mart, Shell/diesel/mart, Texaco/diesel/mart, Burger King, Checker's, Cracker Barrel, Dragon 168 Buffet, El Toro Mexican, Hardee's, McDonald's, Roblyn's Steaks, Schlotsky's, Shoney's, Waffle House, Wendy's, Zaxby's Diner, Colonial Inn, Day's Inn, Econolodge, Economy Inn, Heritage Inn, Knight's Inn, Quality Inn, Travel Inn, auto parts, bank, to Lee Coll, **W**...DENTIST, Amoco/mart, Tennessee Jack's Rest., Baymont Inn, Holiday Inn

Athens Cleveland

N
↑
↓
S

Tennessee Interstate 75

23mm	parking area/weigh sta nb
20	US 64 byp, to Cleveland, **1-3 mi E**...Exxon/mart, Shell/mart, Golden Corral, Hardee's, McDonald's, Subway, Taco Bell, Chrysler/ Jeep/Plymouth, Food Lion, Honda, Toyota, **W**...Citgo/mart/fireworks, Exxon/KrispyKreme/diesel/mart/24hr, KOA(1mi)
16mm	scenic view, sb
13mm	parking area/weigh sta sb, phone
11	US 11 N, US 64 E, Ooltewah, **E**...MEDICAL CARE, Amoco/mart, Chevron, Citgo, RaceTrac/mart, Arby's, Burger King, Hardee's, McDonald's, Taco Bell, Wendy's, BiLo Foods, auto parts, bank, cleaners, **W**...Exxon/TCBY/diesel/ mart/atm, Krystal, Waffle House, Super 8, to Harrison Bay SP
7b a	US 11, US 64, Lee Hwy, **W**...Chevron, Conoco, Texaco/mart, Denny's, Waffle House, Best Inn, Best Western/ rest., Comfort Inn, Country Hearth Inn, Day's Inn, Econolodge, Motel 6, Suburban Lodge, Land Rover
5	Shallowford Rd, **E**...MEDICAL CARE, EYECARE, Alexander's Rest., Arby's, Backyard Burgers, Blimpie, Capt D's, Central Park, CiCi's, Cooker, Country Place Rest., Hardee's, Honeybaked Ham, Hop's Grill, Krystal, Logan's Roadhouse, McDonald's, Old Country Buffet, Perkins, Pizza Hut, Schlotsky's, Steak'n Shake/24hr, Taco Bell, TGIFriday, Tia's Mexican, Comfort Suites/rest., Wingate Inn, AAA, Albertson's/24hr, Belk, Books-A-Million, Firestone/auto, FoodMaxx/24hr, Ford, FreshMkt Foods, K-Mart/drugs, Lowe's Bldg Suppiy, MediaPlay, OfficeMax, Precision Tune, TJ Maxx, Waccamaw Pottery, Walgreen, Wal-Mart SuperCtr/24hr, bank, cleaners, **W**...HOSPITAL, CHIROPRACTOR, EYE CARE, DENTIST, Amoco/mart, Citgo/mart, Exxon/mart, Texaco/mart, Applebee's, Blimpie, Buckhead Roadhouse, Burger King, Cancun Mexican, Capt's Seafood, Cracker Barrel, Domino's, El Chico's, Fazoli's, Glen Gene Deli, Godfather's, Japanese Rest., KFC, McDonald's, Ocean Ave Seafood, O'Charley's, Papa John's, Pizza Hut, Rio Bravo, S&S Cafeteria, SteakOut Rest., Sonic, Shoney's, Subway, Texas Steaks, Waffle House, Wendy's, Country Suites, Day's Inn, Fairfield Inn, Hampton Inn, Holiday Inn, Holiday Inn Express, La Quinta, MainStay Suites, Microtel, Ramada Ltd, Red Roof Inn, Sleep Inn, BiLo Foods, CVS Drug, Goodyear, bank, U of TN/Chatt
4	TN 153, Chickamauga Dam Rd, **E**...Dairy Queen, Kanpai Japanese, El Meson Mexican, FoodCourt, Olive Garden, Outback Steaks, Piccadilly's, Red Lobster, Ruby Tuesday, Courtyard, Barnes&Noble, Circuit City, Dick's Sports, Dillard's, Goody's, HomePlace, JC Penney, Michael's, Old Navy, Parisian, PetsMart, Pier 1, Proffitt's, Sears/auto, Service Merchandise, Staples, Valvoline, cinema, mall, **W**...Microtel, to airport
3b a	TN 320, Brainerd Rd, **E**...CHIROPRACTOR, Amoco/mart/LP, Exxon/mart, Shell/mart, Baskin-Robbins, Subway
2	I-24 W, to I-59, to Chattanooga, Lookout Mtn, no facilities
1.5mm	**Welcome Ctr nb, full(handicapped)facilities, phone, vending, picnic tables, litter barrels, petwalk**
1b a	US 41, Ringgold Rd, to Chattanooga, **E**...BP, Exxon/mart, Trip's Seafood, Best Western, Econolodge, Hawthorn Inn, Howard Johnson, Ramada Inn, BiLo Foods, CVS Drug, Family$, Shipp's RV Ctr/park, **W**...CHIROPRACTOR, BP, Conoco/diesel/mart, Texaco/diesel/mart/wash, A&W, Arby's, Burger King, Central Park Burger, Cracker Barrel, Hardee's, Kay's Ice Cream, Krystal, McDonald's, Pizza Hut, Porto Fino Italian, Shoney's, Subway, Taco Bell, Uncle Bud's Catfish, Waffle House, Wally's Rest., Britcan Inn, Red Roof Inn, Super 8, antiques, cleaners
0mm	Tennessee/Georgia state line

Tennessee Interstate 81

Exit #	Services
75mm	**Tennessee/Virginia state line, Welcome Ctr sb, full(handicapped)facilities, info, phone, vending, picnic tables, litter barrels, petwalk**
74b a	US 11 W, to Bristol, Kingsport, **E**...HOSPITAL, Day's Inn, Hampton Inn, **1-2 mi E**...Texaco, Burger King, KFC, LJ Silver, McDonald's, Pizza Hut, Shoney's, Taco Bell, Wendy's, **W**...Conoco/diesel, Exxon/mart, Uno Pizzaria, Best Western, to King Coll
69	TN 394, to Blountville, **E**...BP/Subway/diesel/mart, Chevron, Arby's, Burger King(2mi), Bristol Int Raceway
66	TN 126, to Kingsport, Blountville, **E**...Amoco/diesel/mart/24hr, Exxon/diesel/mart, **W**...Chevron/mart/24hr, McDonald's, Carolina Pottery, Factory Stores/famous brands, GA Carpets

N

S

Chattanooga

Bristol

Tennessee Interstate 81

<table>
<tr><td>63</td><td>TN 357, Tri-City Airport, E...Exxon/Taco Bell/Krystal/diesel/mart/atm, Shell/Subway/KrispyKreme/diesel/mart/ 24hr, Cracker Barrel, Wendy's, La Quinta, Sleep Inn, Hamrick's Clothing Outlet, W...Amoco/mart/24hr, Phillips 66/diesel/mart, Red Carpet Inn, $General, KOA, Sam's Club, Rocky Top Camping, antiques</td></tr>
<tr><td>60mm</td><td>Holston River</td></tr>
<tr><td>59</td><td>TN 36, to Johnson City, Kingsport, E...Amoco/diesel/mart, Super 8, Travelers Inn, W...MEDICAL CARE, DENTIST, EYECARE, CHIROPRACTOR, VETERINARIAN, Amoco/mart/LP, Coastal/diesel/mart, Exxon, Texaco/diesel/mart, Arby's, Burger King, Chinese Rest., Domino's, Fazoli's, Hardee's, HotDog Hut, Huddle House, La Carreta Mexican, Little Caesar's, McDonald's, Motz's Italian, Pal's HotDogs, Perkins/24hr, Piccadilly's, Pizza Hut, Raffaele's Pizza, Shoney's, Subway, TCBY, Wendy's, Comfort Inn, Holiday Inn Express, Advance Parts, CVS Drug, $Store, Firestone/auto, Food Lion, Ingles/deli, 1-Hr Photo, USPO, Wrap-It Mail, XpressLube, bank, carwash, hardware, laundry, to Warrior's Path SP</td></tr>
<tr><td>57b a</td><td>I-181, US 23, to Johnson City, Kingsport, to ETSU, 6 mi E...Cracker Barrel, Jameson Inn</td></tr>
<tr><td>50</td><td>TN 93, Fall Branch, W...auto auction, st patrol</td></tr>
<tr><td>44</td><td>Jearoldstown Rd, E...Amoco/mart, W...Exit 44 Mkt/gas</td></tr>
<tr><td>41mm</td><td>rest area sb, full(handicapped)facilities, phone, vending, picnic tables, litter barrels, petwalk</td></tr>
<tr><td>38mm</td><td>rest area nb, full(handicapped)facilities, phone, vending, picnic tables, litter barrels, petwalk</td></tr>
<tr><td>36</td><td>TN 172, to Baileyton, E...Speedway/Blimpie/diesel/mart, repair, W...BP/diesel/mart/24hr, Marathon/diesel/rest./ 24hr, Texaco/Subway/diesel/mart/24hr, 36 Motel, Baileyton Camp(2mi)</td></tr>
<tr><td>30</td><td>TN 70, to Greeneville, E...Exxon/Stuckey's/DQ/diesel/mart</td></tr>
<tr><td>23</td><td>US 11E, to Greeneville, E...Amoco/Wendy's/TCBY/mart, BP/mart, to Andrew Johnson HS, W...Citgo/Taco Bell/ diesel/mart, Phillips 66/diesel/mart/rest., McDonald's, NY Pizza/Italian, Tony's Rest., Comfort Inn, Super 8</td></tr>
<tr><td>21mm</td><td>parking area/weigh sta sb</td></tr>
<tr><td>20mm</td><td>parking area/weigh sta nb</td></tr>
<tr><td>15</td><td>TN 340, Fish Hatchery Rd, E...Amoco/diesel/24hr, W...truck/trailer repair</td></tr>
<tr><td>12</td><td>TN 160, to Morristown, E...Citgo/mart, W...Shell/mart, to Crockett Tavern HS</td></tr>
<tr><td>8</td><td>US 25E, to Morristown, E...Shell/repair, Catawba Rest., Sonic(2mi), Twin Pines Motel(3mi), W...BP/diesel/mart/ wash, Phillips 66/diesel/mart, Cracker Barrel, Hardee's, Holiday Inn, Parkway Inn, Super 8, to Cumberland Gap NHP</td></tr>
<tr><td>4</td><td>TN 341, White Pine, E...Exxon/diesel/mart, Williams/Main St Café/diesel/mart/24hr, Crown Inn, W...Citgo/mart, Phillips 66/diesel/rest./repair, Texaco/diesel/mart, Huddle House/24hr, Day's Inn, Hillcrest Inn, to Panther Cr SP</td></tr>
<tr><td>2.5mm</td><td>rest area sb, full(handicapped)facilities, phone, picnic tables, litter barrels, vending, petwalk</td></tr>
<tr><td>1b a</td><td>I-40 E to Asheville, W to Knoxville. I-81 begins/ends on I-40, exit 421.</td></tr>
</table>

Tennessee Interstate 640(Knoxville)

<table>
<tr><td>Exit #</td><td>Services</td></tr>
<tr><td>9mm</td><td>I-640 begins/ends on I-40, exit 393.</td></tr>
<tr><td>8</td><td>Millertown Pike, Mall Rd N, N...Conoco/Backyard Burgers/mart, Exxon/DQ/mart/atm/24hr, Applebee's, Burger King/Church's, Don Pablo's, KFC, McDonald's, Morrison's Cafeteria, Pizza Inn, Ruby Tuesday, Taco Bell, Texas Roadhouse, Wendy's, Circuit City, Dillard's, Hess, JC Penney, Kohl's, OfficeMax, Pier 1, Sam's Club, Sears/auto, Service Merchandise, Wal-Mart/auto, cleaners, farmer mkt, mall, S...Conoco/mart, Subway, Food Lion, Home Depot, PepBoys, bank, cinema</td></tr>
<tr><td>6</td><td>US 441, to Broadway, N...DENTIST, BP, Conoco, Phillips 66/mart, Pilot/diesel/mart, Texaco/diesel/mart, Arby's, Cancun Mexican, Checker's, CiCi's Pizza, Domino's, Fazoli's, Hardee's, Krispy Kreme, LJ Silver, McDonald's, Quincy's, Subway, Taco Bell, Pizza, Best Western, AutoZone, CVS Drug, $General, Firestone, Goodyear, Kroger/ 24hr, NAPA, 9 MinuteLube, Pennzoil, PetSupplies+, Piece Goods, Precision Tune, Target, Walgreen, S...BBQ, CVS Drug, Eddie's Parts, Food City, K-Mart, Winn-Dixie, XPert Tune, transmissions</td></tr>
<tr><td>3a</td><td>I-75 N to Lexington, I-275 S to Knoxville</td></tr>
<tr><td>b</td><td>US 25W, Clinton Hwy, to Hardee's, Quincy's, Chevrolet, Ford, facilities on frontage rds</td></tr>
<tr><td>1</td><td>TN 62, Western Ave, N...CHIROPRACTOR, Cargo/diesel/mart, RaceTrac/mart, Texaco/diesel/mart, Baskin-Robbins, Central Park, Golden Corral, KFC, LJ Silver, McDonald's, Mixon's Deli, Ruby Tuesday, Shoney's, Subway, Taco Bell, Wendy's, CVS Drug, Kroger, Walgreen, S...DENTIST, BP/mart, Dad's Donuts, Domino's, Hardee's, Krystal, Advance Parts, Mighty Muffler, USPO, auto repair, bank</td></tr>
<tr><td>0mm</td><td>I-640 begins/ends on I-40, exit 385.</td></tr>
</table>

Texas Interstate 10

Exit #	Services

880.5mm Sabine River, Texas/Louisiana state line

880 Sabine River Turnaround, RV camping

879mm Welcome Ctr wb, full(handicapped)facilities, phone, picnic tables, litter barrels, vending, petwalk

878 US 90, Orange, **N**...Mobil/diesel/mart, **S**...Western Store

877 TX 87, 16th St, Orange, **N**...Chevron/mart, RaceTrac/mart, Shamrock/diesel/mart, Cajun Cookery, Gary's Café, Pizza Hut, Waffle House, Best Western, Day's Inn, Holiday Inn Express, King's Inn, Motel 6, Ramada Inn, Super 8, Buick/Olds/Pontiac, Toyota, $General, Eckerd, MktBasket/deli, Radio Shack, **S**...CHIROPRACTOR, DENTIST, Chevron/diesel/mart, Exxon/24hr, Shell/diesel, Texaco/mart, Burger King, Cody's Rest., CrabTrap Rest., Church's Chicken, Dairy Queen, Mexican Rest., Little Caesar's, McDonald's, Popeye's, Subway, Taco Bell, Cadillac/GMC, HEB Foods, Olds/Pontiac/Buick, Modica Bros Tires, RV Ctr, Toyota, cleaners/laundry

876 Adams Bayou, frontage rd, **N**...Chevron/mart, RaceTrac/mart/24hr, Waffle House, Kings Inn, Ramada/rest., access to Motel 6, same as 877

875 FM 3247, MLK Dr, **N**...Luby's, **S**...HOSPITAL, Chrysler/Dodge, RV Ctr

874 US 90, to Orange, **S**...HOSPITAL, Oakleaf Park RV Park

873 TX 62, TX 73, to Bridge City, **N**...Chevron/Subway/pizza/diesel/mart/24hr, Flying J/Conoco/diesel/LP/rest./24hr, **S**...Shamrock/mart, Texaco/diesel/mart, Williams/diesel/mart/24hr, Jack-in-the-Box, Burger King, McDonald's, Waffle House, camping

870 FM 1136, no facilities

869 FM 1442(from wb), to Bridge City, **S**...Lloyd RV Ctr

867 frontage rd(from eb), no facilities

866.5mm rest areas both lanes, full(handicapped)facilities, vending, picnic tables, litter barrels

865 Doty Rd(from wb), frontage rd, no facilities

864 FM 1132, FM 1135, **N**...Burr's BBQ, Casey's Motel

862c Timberlane Dr(from eb), **N**...RV Ctr

b Old Hwy(from wb), **N**...RV Ctr

a Railroad Ave, no facilities

861d TX 12, Deweyville, no facilities

c Denver St, **N**...Conoco/diesel

b Lamar St(from wb), **N**...Conoco/diesel/mart

a FM 105, Vidor, **N**...Fina, Shell/mart, Dairy Queen, Domino's, McDonald's, Popeye's, Waffle House, Greenway Motel, Eckerd, Family$, Radio Shack, **S**...CHIROPRACTOR, MEDICAL CARE, Chevron/mart/wash, Citgo/mart, Conoco/mart, Exxon/Blimpie/diesel/mart, Mobil, Phillips 66/mart, Burger King, Church's Chicken, Golden Corral, KFC, Pizza Hut, Sonic, Subway, Taco Bell, Whataburger, Wal-Mart/drugs/auto, antiques, auto parts, groceries, tires

860b a Dewitt Rd, frontage rd, no facilities

859 Bonner Turnaround, no facilities

858b Asher Turnaround, **N**...funpark, RV camping, **S**...Chevron/diesel/rest., Denny's

a Rose City East, **S**...EZ/diesel/mart

857 Rose City, no facilities

856 Old Hwy, no facilities

855a US 90 bus, to downtown, Port of Beaumont, no facilities

854 ML King Pkwy, Beaumont, **N**...Burger King, **S**...Exxon, Texaco, McDonald's, Chrysler/Jeep, Olds

853c 7th St, **N**...Fina, Burger King, Denny's, McDonald's, Ninfa's Mexican, Red Lobster, Waffle House, Holiday Inn, Scottish Inn, Super 8, Travel Inn, bank, same as 853b

b 11th St, **S**...HOSPITAL, Chevron/mart, EZ/diesel/mart, Phillips 66, Dunkin Donuts, Jack-in-the-Box, Japanese Rest., Luby's, Best Western, HoJo's, Motel 6, Quality Inn, Ramada Inn, Rodeway Inn, Chevrolet/Cadillac/GMC

a US 69 N, to Lufkin

852 Harrison Ave, Calder Ave, Beaumont, **N**...Chevron/diesel/mart/24hr, Texaco/diesel/mart, BBQ, Bennigan's, Cajun Cookery, Casa Ole Mexican, Chili's, Chinese Rest., Olive Garden, Steak&Ale, bank, **S**...EZ Mart/gas, Chinese Rest., Church's Chicken, McDonald's, Day's Inn, La Quinta, Chevrolet, RV Ctr, drugs, transmissions

851 US 90, College St, **N**...DENTIST, Exxon/diesel/mart/wash, RaceTrac/mart, China Border, Fujiyama Japanese, Golden Corral, Hardee's, Outback Steaks, Post Oak Grill, Waffle House, Best Western, Roadrunner Motel, AutoZone, O'Reilly Parts, Pennzoil, Subaru/Suzuki, **S**...HOSPITAL, DENTIST, Chevron/Burger King/mart/wash/24hr, Mobil, Shell/mart, Texaco/mart, China Hut, Dairy Queen, IHOP, KFC, Pancho's Mexican, Peter Piper Pizza, Pizza Hut, Shoney's, Taco Bell, Wendy's, Whataburger, Courtyard, Econolodge, Fairfield Inn, Hilton/rest., Motel 6, Premier Inn, Dodge, Eckerd, Ford/Lincoln/Mercury, Honda, Jeep, MktBasket Foods, Mazda, Mercedes, Mitsubishi, Nissan, NTB, Office Depot, Olds/GMC/Cadillac, Sam's Club/gas, U-Haul, Volvo, Walgreen, cinema

E

W

Orange

Beaumont

Texas Interstate 10

850 Washington Blvd(from wb), same as 851
849 US 69 S, Washington Blvd, to Port Arthur, airport, no facilities
848 Walden Rd, **N...**Texaco/mart, Pappadeaux Seafood, Holiday Inn/rest., **S...**Chevron/Subway/diesel/mart/24hr, Petro/Mobil/diesel/rest./motel/24hr, Cheddar's, Cracker Barrel, Jack-in-the-Box, Joe's Crabshack, Waffle House, Hampton Inn, PetroLube, cinema
846 Brooks Rd(from eb), no facilities
845 FM 364, Major Dr, no facilities
843 Smith Rd, **N...**New World RV Ctr/park
838 FM 365, Fannett, **N...**Shell, RV camping
837.5mm picnic areas both lanes, tables, litter barrels, handicapped accessible
833 Hamshire Rd, **N...**Chevron/diesel, Bergeron's Rest., Little TX BBQ
829 FM 1663, Winnie, **N...**Exxon/diesel/mart, Mobil/Burger King/diesel/mart, Texaco/Blimpie/diesel/mart/24hr, McDonald's, Taco Bell, Holiday Inn Express, **S...**HOSPITAL, Bingo/Subway/mart, Chevron/diesel/mart, Exxon/diesel, Mobil, Cajun Cuisine, Hunan Chinese, Jack-in-the-Box, Pizza Inn, Waffle House, Best Western/rest., Winnie Inn, Chevrolet, Ford, Chrysler/Dodge/Plymouth, bank, repair
828 TX 73, TX 124(from eb), to Winnie, **S...**HOSPITAL, same as 829
827 FM 1406, **S...**HOSPITAL, Exxon/diesel/mart, BBQ
822 FM 1410, no facilities
819 Jenkins Rd, **S...**Texaco/Stuckey's
817 FM 1724, no facilities
815mm weigh sta both lanes
813 TX 61(from wb), Hankamer, **S...**Exxon/diesel/mart
812 TX 61, Hankamer, **S...**Exxon/diesel/mart, DJ's Diner
811 Turtle Bayou Turnaround, **N...**Terry's Rest., Turtle Bayou RV Park, **S...**Shell/diesel/mart, Grandma's Diner, Ford
810 FM 563, to Anahuac, Liberty, **N...**auto parts, **S...**Texaco/Blimpie/diesel/mart, Ford
807 to Wallisville, **S...**Heritage Park
806 frontage rd(from eb), turnaround
805.5mm Trinity River
804mm Old, Lost Rivers
803 FM 565, Cove, no facilities
800 FM 3180, no facilities
798 (797 from eb)TX 146, to Baytown, **N...**Conoco/diesel/rest., Texaco/diesel/mart, Dairy Queen, IHOP, McDonald's, Pizza Inn, Waffle House, Motel 6, **S...**HOSPITAL, Chevron/diesel/mart, Exxon/mart, Mobil/Popeye's/diesel/mart, RaceTrac/diesel/mart, Chili Bean's Rest., Jack-in-the-Box, Subway, Texas Burger, Scanadian Inn, KOA, carwash, laundry
796 frontage rd, **N...**Amoco/Chevron Chemical Refineries, no facilities
795 Sjolander Rd, no facilities
793 N Main St, **N...**Shell/mart, **S...**Pilot/Subway/KFC/diesel/mart/24hr
792 Garth Rd, **N...**HOSPITAL, Chevron/mart/wash/24hr, Burger King, Cracker Barrel, Denny's, Jack-in-the-Box, Red Lobster, Waffle House, Whataburger/24hr, Best Western, Baymont Inn, Hampton Inn, La Quinta, Chrysler/Jeep, Lincoln/Mercury, Nissan, Western Wear Outlet, bank, **S...**EYECARE, RaceTrac/mart, Shell/mart/wash, Chinese Rest., Cozumel Mexican, McDonald's, Outback Steaks, Pancho's Mexican, Piccadilly's, Popeye's, Taco Bell, Tortuga Mexican, Wendy's, Holiday Inn Express, Ford, Marshall's, Mervyn's, OfficeMax, Sears/auto, Service Merchandise, Vaughn RV Ctr, Ward's/auto, cinema, mall
791 John Martin Rd, no facilities
790 Wade Rd, no facilities
789 Thompson Rd, **N...**Williams/diesel/mart, Speedway, Buick/Pontiac/GMC/Toyota, **S...**TA/diesel/rest./24hr, truck repair
788.5mm rest areas both lanes, full(handicapped)facilities, phone, picnic tables, litter barrels, petwalk
788 sp 330(from eb), to Baytown, no facilities
787 sp 330, Crosby-Lynchburg Rd, to Highlands, **N...**Mobil/diesel/mart, **S...**Coastal/mart, Lynchburg Grill, to San Jacinto SP, camping
786.5mm San Jacinto River
786 Monmouth Dr, no facilities
785 Magnolia Ave, to Channelview, **S...**Key Trkstp/diesel, same as 784
784 Cedar Lane, Bayou Dr, **N...**Citgo/mart, Shell/diesel/rest., Econolodge/rest., Budget Lodge, Ramada Ltd
783 Sheldon Rd, **N...**Coastal/diesel/mart, Shell, Burger King, Jack-in-the-Box, KFC, Pizza Hut, Pizza Inn, Popeye's, Sonic, Subway, Taco Bell, Whataburger, Best Western/rest., Day's Inn, I-10 Motel, Leisure Inn, Super 8, Travelodge, AutoZone, DiscountTire, Eckerd, Family$, Gerland's Foods/24hr, Goodyear/auto, Radio Shack, Western Auto, **S...**Chevron/mart/24hr, Shell/mart/wash/etd, Capt D's, McDonald's, Wendy's
782 Dell-Dale Ave, **N...**HOSPITAL, Chevron/diesel/mart, Shamrock/diesel/mart, Holiday Inn/rest., Howard Johnson, transmissions, tires, **S...**Mobil/diesel/mart
781b a TX 8, Sam Houston Pkwy

Baytown

E ↕ W

Texas Interstate 10

E

W

780	Uvalde Rd, Freeport St, **N...**HOSPITAL, Chevron/diesel/mart, Exxon, Shamrock, Cracker Barrel, IHOP, Jack-in-the-Box, KFC, Pancho's Mexican, Subway, Taco Bell, Interstate Motel/rest., Ace Hardware, Kinko's/24hr, Office Depot, bank, cinema, **S...**Mobil/diesel/mart, BBQ, Blackeyed Pea, Marco's Mexican, Whataburger, Circuit City, Home Depot, NTB, Sam's Club, U-Haul, Wal-Mart SuperCtr/24hr
778b	Normandy St, **N...**Conoco/diesel/mart, Shell/mart/wash, Golden Corral, **S...**CHIROPRACTOR, DENTIST, VETERINARIAN, EYE CLINIC, Texaco/diesel/mart, Mexican Rest.
a	FM 526, Federal Rd, Pasadena, **N...**CHIROPRACTOR, Chevron/mart, Shell/mart, BBQ, Blimpie, Denny's, Golden Corral, Jack-in-the-Box, LJ Silver, Luby's, Mexican Cantina, Pizza Hut, Popeye's, Shoney's, Wendy's, La Quinta, Champ's Foods, DiscountTire, Eckerd, Fiesta Foods, Radio Shack, Target, **S...**CHIROPRACTOR, MEDICAL CARE, Citgo, Shell/mart, Bennigan's, Bradley's Steaks, Chili's, Church's Chicken, Jack-in-the-Box, James Coney Island, Joe's Crabshack, McDonald's, Ninfa's Mexican, Pappa's Seafood, Peking Chinese, Scottish Inn, AutoZone, BootTown, Family$, Midas Muffler, O'Reilly's Parts, cinema, same as 776
776b	John Ralston Rd, Holland Ave, **N...**Citgo/mart, Mobil/mart, Chinese Rest., Raintree Rest., Best Western, Arandasc Foods, Kroger/deli/24hr, NTB, Walgreen, **S...**Meineke Muffler/brakes, same as 778
a	Mercury Dr, **N...**Conoco, Shamrock/diesel, Burger King, E China Rest., Island Gourmet Seafood, Kettle, McDonald's, Pizza Inn, Day's Inn, Fairfield Inn, Hampton Inn, Premier Inn, **S...**Chevron/mart, Shamrock, Texaco/diesel/mart/wash, Dairy Queen, Luby's, Steak&Ale, bank, drugs
775b a	I-610, no facilities
774	Gellhorn(from eb) Blvd, Anheuser-Busch Brewery
773b	McCarty St(from eb), **N...**Exxon, Shell/mart, bank, **S...**Don Chile Mexican, truck parts
a	US 90A, N Wayside Dr, **N...**Chevron, Exxon, Jack-in-the-Box, Whataburger/24hr, **S...**Coastal/diesel, Shell/mart/wash, Texaco, BBQ, Church's Chicken, bank, diesel repair
772	Kress St, Lathrop St, **N...**Conoco/mart, Exxon/mart, **S...**Burger King, 7 Mares Seafood, cleaners/laundry
771b	Lockwood Dr, **N...**Chevron/diesel/mart/atm, McDonald's, Fiesta Foods, Eckerd, **S...**King/diesel/rest., Texaco, Palace Inn
a	Waco St, **N...**Frenchey's Chicken
770c	US 59 N, no facilities
b	Jenson St, Meadow St, Gregg St, no facilities
a	US 59 S, to Victoria, no facilities
769c	McKee St, Hardy St, Nance St, downtown, no facilities
a	Smith St(from wb), to downtown
768b a	I-45, N to Dallas, S to Galveston, no facilities
767b	Taylor St, no facilities
a	Yale St, Heights Blvd, **N...**Chevron, **S...**Exxon/diesel/mart
766	(from wb), Heights Blvd, Yale St, same as 767a
765b	Patterson St, **N...**VETERINARIAN, Exxon, Shamrock/gas, Texaco, Wendy's, Howard Johnson, Midas Muffler, bank, **S...**Cadillac Bar Mexican
a	TC Jester Blvd, **S...**Exxon/diesel/mart, Texaco/diesel
764	Westcott St, Washington Ave, Katy Rd, **N...**Denny's, Day's Inn, **S...**Chevron/mart, IHOP
763	I-610, no facilities
762	Silber Rd, Post Oak Rd, **N...**Aubrey's Ribs, cinemas, **S...**Shell/mart/wash, Jack-in-the-Box, Holiday Inn, Ramada Inn, Chevrolet, carwash
761b	Antoine Rd, **N...**Country Harvest Buffet, Hunan Chinese, IKEA, **S...**Shamrock/mart, Blue Oyster Grill, Denny's, Papa John's, Whataburger/24hr, La Quinta, Ramada Inn, Wellesley Inn, MailBoxes Etc, NTB, REI Sports
a	Wirt Rd, Chimney Rock Rd, **N...**Capt Penny's Seafood, **S...**Chevron/mart, Exxon/TCBY/mart/wash, Shell/mart/wash/atm, Shamrock, Dixie's Roadhouse, 59 Diner, McDonald's, Steak&Ale, Planter's Seafood, bank
760	Bingle Rd, Voss Rd, **S...**HOSPITAL, DENTIST, Exxon/diesel/mart, Texaco/diesel/mart, Chinese Rest., Goode Co Café, Las Alamedas Mexican, Marie Callender's, Mason Jar Rest., Pappy's Café, SaltGrass Steaks, Southwell's Burgers, Sweet Tomatos, TX BBQ, Ugo's Italian, bank
759	Campbell Rd(from wb), same as 758b
758b	Blalock Rd, Campbell Rd, **N...**Chinese Rest., Ciro's Italian, Fiesta Foods, Sonic, Adam's Automotive, LubeStop, Mail It, cleaners, **S...**HOSPITAL, Chevron/McDonald's/diesel/mart, Exxon, Texaco, Cathy's Café, Kroger, MensWearhouse, Walgreen, bank

Texas Interstate 10

Houston

a   Bunker Hill Rd, **N**...PepBoys, **S**...Texaco/diesel, Circle K, Charlie's Burgers, Chinese Rest., Prince's Burgers, Quizno's, Subway, Day's Inn, Howard Johnson, Super 8, Ford, Goodyear/auto, Marshall's, Nissan, 1-Hr Photo, Target, ToysRUs, cinema

757   Gessner Rd, **N**...DENTIST, MEDICAL CARE, Exxon/mart, Stop'n Go, BBQ, Bennigan's, Chili's, Chinese Rest., CiCi's, Dairy Queen, McDonald's, Schlotsky's, ThunderCloud Subs, Wendy's, Whataburger/24hr, Babys R Us, BrakeCheck, Eckerd, HobbyLobby, Home Depot, Honda, Kroger/drugs, NAPA, OfficeMax, PetsMart, Radio Shack, U-Haul, Wal-Mart, **S**...HOSPITAL, Exxon, Shell/mart, Texaco/diesel/wash, Fuddrucker's, Jack-in-the-Box, Jaon's Deli, Macaroni Grill, Olive Garden, Pappasito's, Papadeaux Seafood, Taste of TX Rest., Radisson, Ramada Ltd, Sheraton, Bed Bath&Beyond, Best Buy, BridesMart, Circuit City, Cost+ World Mkt, Ford, Goodyear, Mervyn's, Office Depot, Oschman Sports, PetCo, Sears/auto, Service Merchandise, TJ Maxx, Ward's/auto, bank, carwash, mall

756   TX 8, Sam Houston Tollway, no facilities

755   Willcrest Rd, **N**...Discount Tire, Mazda, Mike's Muffler/brake, NTB, **S**...CHIROPRACTOR, Citgo, Exxon/McDonald's/mart/atm/24hr, Texaco/mart, China Buffet, IHOP, Steak&Ale, Extended Stay America, La Quinta

754   Kirkwood Rd, **N**...Lincoln/Mercury, Saturn, same as 753b auto dealers, **S**...Chevron/diesel/mart, Shell/mart/24hr, Carrabba's, IHOP/24hr, Mesa Grill, Original Pasta Co, Subway, Taco Cabana, Hampton Inn, Chevrolet, Kirkwood Drugs, Randall's Foods, Subaru

753b   Dairy-Ashford Rd, **N**...Buick, Chrysler/Dodge, Infiniti, K-Mart, Lexus, Nissan, Olds, Toyota, Volvo, **S**...Exxon/TCBY/diesel/mart/wash, Shamrock/mart, Beck's Prime Rest., Subway, TX Cattle Steaks, Whataburger, Houston West Inn, Ramada Inn, Shoney's Inn, Audi/Porsche, Cadillac, Cavender's Boots, Kinko's, Mitsubishi, Pontiac/GMC/Suzuki, bank

a   Eldridge Rd, **N**...Conoco/diesel/mart, Marriott, **S**...Shamrock/mart/24hr, KwikKar Lube/clean

751   TX 6, to Addicks, **N**...Texaco/mart, Cattlegard Rest., Waffle House, Drury Inn, Holiday Inn, Red Roof Inn, AutoCheck, Sam's Club, **S**...Exxon, Chevron/mart, Conoco/diesel/mart, Texaco/diesel/mart, Blimpie, Chinese Rest., Dairy Queen, Denny's/24hr, El Yucatan Mexican, Jack-in-the-Box, Wendy's, Bradford HomeSuites, Fairfield Inn, Hearthside Extended Stay, La Quinta, Motel 6, Super 8, TownePlace Suites, JiffyLube, USPO, bank, carwash, cleaners

748   Barker-Cypress Rd, **S**...Cracker Barrel, Larry's BBQ, Chrysler/Plymouth/Jeep, Hoover RV Ctr

747   Fry Rd, **N**...DENTIST, MEDICAL CARE, VETERINARIAN, Chevron, Phillips 66/mart, Shamrock/diesel/mart, Shell/mart/wash/atm/24hr, Burger King, Church's Chicken, Dairy Queen, Denny's, Food Court, Godfather's, McDonald's, Pizza Hut, Sonic, Subway, Taco Bell, Victor's Mexican, Whataburger, Garden Ridge, Goodyear, HobbyLobby, Home Depot, Kroger/deli/24hr, MailBoxes Etc, Midas Lube, Pennzoil, Randall's Foods, Ross, Walgreen, Wal-Mart/auto, bank, **S**...DENTIST, Chevron/mart/wash, Citgo, Shell/mart/wash, Baskin-Robbins, Boston Mkt, Capt Tom's Seafood, El Chico, Fazoli's, IHOP, Jack-in-the-Box, Ninfa's Café, Omar's Mexican, Outback Steaks, Quizno's, Wendy's, Albertson's/drugs, Ford, Gerland's Foods, Lowe's Bldg Supply, Mike's Muffler/brakes, Pearle Vision, PetsMart, Radio Shack, Target, TJ Maxx, bank

745   Mason Rd, **S**...CHIROPRACTOR, VETERINARIAN, Chevron/mart/24hr, Exxon/mart/24hr, Shamrock/mart/wash, Shell/mart/etd/24hr, Stop'n Go/gas, BBQ, Blackeyed Pea, Burger King, Carino's Italian, Chick-fil-A, Chili's, Chinese Rest., CiCi's, Dairy Queen, Hartz Chicken, Jack-in-the-Box, Jason's Deli, KFC, Kettle, Landry's Seafood, Little Caesar's, LJ Silver, Luby's, Marco's Mexican, McDonald's, Monterey Mexican, Papa John's, Pizza Hut, SaltGrass Steaks, Schlotsky's, SpagEddie's, Subway, Taco Bell, Taco Cabana, TCBY, Whataburger/24hr, Sleep Inn, CarQuest, Discount Tire, Eckerd, Firestone/auto, Gerland's Foods, Goodyear/auto, JiffyLube, Kroger, LubeExpress, Mail&Copies, Pennzoil, Randall's Food/drug, Walgreen, bank, cinema, cleaners/laundry, hardware, transmissions

743   TX 99, Grand Pkwy, Peek Rd, **S**...Texaco/diesel/mart, A&W, Kettle, Popeye's, Best Western, Holiday Inn Express, Ramada Ltd, Super 8, Chevrolet

742   Katy-Fort Bend County Rd, **N**...Citgo, **S**...Mobil/diesel/mart/24hr

740   FM 1463, to Pin Oak Rd, **N**...McDonald's, Sonic, Yamaha, **S**...HOSPITAL, Chevron, RainForest Café, Books-A-Million, Burlington Coats, Katy Mills Outlet/famous brands, OutDoor World, Pyle RV Ctr, cinema, tires

737   Pederson Rd, no facilities

732   FM 359, to Brookshire, **N**...Exxon/diesel/mart, Phillips 66/diesel/rest./24hr, Texaco, Executive Inn, KOA, **S**...Chevron/diesel/mart, Coastal/Burger King/diesel/mart/atm/24hr, Conoco/diesel/mart, Jack-in-the-Box, Ford Trucks, truckwash

731   FM 1489, to Koomey Rd, **N**...Exxon/diesel/mart, Shell/mart, Villa Fuentes Mexican, Brookshire Motel, Travelers Inn, KOA

730mm   picnic areas(both lanes exit left), tables, litter barrels

729   Peach Ridge Rd, Donigan Rd, no facilities

726   Chew Rd(from eb), no facilities

725   Micak Rd(from wb), no facilities

724mm   weigh sta wb

Sealy

723   FM 1458, to San Felipe, **N**...Knox/Subway/diesel/mart/24hr, Ford, to Stephen F Austin SP, **S**...Outlet Ctr/famous brands, Food Court, Goodyear/auto

721   (from wb), no facilities

720   TX 36, to Sealy, **N**...Exxon, Texaco/diesel/mart, Dairy Queen, Hartz Chicken, McDonald's, Sonic, Tony's Rest., Chevrolet/Pontiac/Buick/GMC, **S**...CHIROPRACTOR, Chevron/diesel/mart/wash/24hr, Mobil/diesel/mart, Shell/diesel/mart/wash/24hr, BBQ, Chinese Rest., KFC/Taco Bell, Omar's Mexican, Pizza Hut, Subway, Whataburger/24hr, Best Western, Holiday Inn Express, Rodeway Inn, Chrysler/Plymouth/Dodge/Jeep, Wal-Mart, bank

E

W

Texas Interstate 10

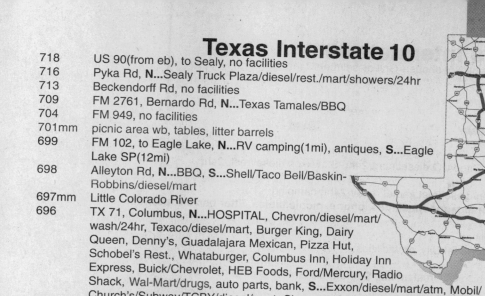

E

W

718	US 90(from eb), to Sealy, no facilities
716	Pyka Rd, **N**...Sealy Truck Plaza/diesel/rest./mart/showers/24hr
713	Beckendorff Rd, no facilities
709	FM 2761, Bernardo Rd, **N**...Texas Tamales/BBQ
704	FM 949, no facilities
701mm	picnic area wb, tables, litter barrels
699	FM 102, to Eagle Lake, **N**...RV camping(1mi), antiques, **S**...Eagle Lake SP(12mi)
698	Alleyton Rd, **N**...BBQ, **S**...Shell/Taco Bell/Baskin-Robbins/diesel/mart
697mm	Little Colorado River
696	TX 71, Columbus, **N**...HOSPITAL, Chevron/diesel/mart/wash/24hr, Texaco/diesel/mart, Burger King, Dairy Queen, Denny's, Guadalajara Mexican, Pizza Hut, Schobel's Rest., Whataburger, Columbus Inn, Holiday Inn Express, Buick/Chevrolet, HEB Foods, Ford/Mercury, Radio Shack, Wal-Mart/drugs, auto parts, bank, **S**...Exxon/diesel/mart/atm, Mobil/Church's/Subway/TCBY/diesel/mart, Shamrock/diesel/mart, McDonald's, Sonic, Country Hearth Inn, RV Park
695	TX 71(from wb), to La Grange
693	FM 2434, to Glidden, no facilities
692mm	**rest areas both lanes, full(handicapped)facilities, vending, phone, picnic tables, litter barrels, RV dump, petwalk**
689	US 90, to Hattermann Lane, **N**...Motorcoach RV Park
682	FM 155, to Wiemar, **N**...HOSPITAL, Exxon/diesel/mart, Texaco/mart, BBQ, Dairy Queen, Ford, **S**...Buick/Chevrolet
678mm	E Navidad River
677	US 90, **N**...Nannie's Café, Fostoria Glass, OutPost Art Gallery
674	US 77, Schulenburg, **N**...Chevron/diesel/mart/wash, Exxon/diesel, BBQ, McDonald's, Oakridge Motel/rest., **S**...Phillip 66/diesel, Shamrock/Subway/mart, Texaco, Burger King, Dairy Queen, Diamond S Rest., Frank's Rest., Schulenberg RV Park
672mm	W Navidad River
668	FM 2238, to Engle, no facilities
661	TX 95, FM 609, to Flatonia, **N**...Mobil/diesel, BBQ, **S**...Exxon/mart/24hr, Shell/Grumpy's Rest./mart/motel, Dairy Queen, Sav-Inn
657.5mm	picnic areas both lanes
653	US 90, **N**...Texaco/diesel/mart
649	TX 97, to Waelder, no facilities
642	TX 304, to Gonzales, **N**...Noah's Land RV Park(6mi)
637	FM 794, to Harwood, no facilities
632	US 90/183, to Gonzales, **N**...Love's/Subway/diesel/mart/24hr, Coachway Inn(2mi), **S**...to Palmetto SP, camping
630mm	San Marcos River
628	TX 80, to Luling, **N**...HOSPITAL, Shamrock/mart, Dairy Queen, Riverbend RV Park, **S**...Exxon/mart, Texaco
625	Darst Field Rd, **S**...Trucks of Texas
624.5mm	Smith Creek
623mm	Allen Creek
621mm	**rest areas both lanes, full(handicapped)facilities, phone, picnic tables, litter barrels, vending, petwalk**
620	FM 1104, no facilities
617	FM 2438, to Kingsbury, no facilities
616mm	truck weigh sta both lanes
614.5mm	Mill Creek
612	US 90, no facilities
611mm	Geronimo Creek
610	TX 123, to San Marcos, **N**...Exxon/mart, Chevron/diesel/mart, Adobe Mexican Café, K&G Steaks, Luby's, Schlotsky's, Subway, Econolodge, Holiday Inn, Tire Whse, **S**...HOSPITAL, Shamrock/diesel/mart, Taco Cabana, River Shade RV
609	TX 123, Austin St, **S**...Mobil/diesel/mart
607	TX 46, FM 78, to New Braunfels, **N**...Texaco/Jack-in-the-Box/diesel/mart, **S**...Exxon/DQ/diesel/mart, Chevron/mart, Fina/mart/24hr, BBQ, Kettle, McDonald's, Soilitas Mexican, Best Western, Super 8, Chevrolet, Chrysler/Plymouth/Dodge, antiques
605	FM 464, **N**...Twin Pines RV Park, **S**...Fina/mart/24hr
605mm	Guadalupe River
604	FM 725, to Lake McQueeney, no facilities
603	US 90 E, US 90A, to Seguin, **S**...S Texas Truck/RV

Texas Interstate 10

601 FM 775, to New Berlin, **N**...Mobil/Roadhouse Grill/diesel/mart/24hr

600 Schwab Rd, no facilities

599 FM 465, to Marion, no facilities

599mm Santa Clara Creek

597 Santa Clara Rd, no facilities

595 Zuehl Rd, no facilities

594mm Cibolo Creek

593 FM 2538, Trainer Hale Rd, **N**...Citgo/DQ/diesel/mart/24hr, **S**...Exxon/diesel/rest./24hr

593mm Woman Hollering Creek

591 FM 1518, to Schertz, **N**...Exxon/diesel/rest./motel/repair/24hr, camping

590.5mm rest areas both lanes, full(handicapped)facilities, phone, picnic tables, litter barrels, petwalk

589mm Salitrillo Creek

589 Pfeil Rd, Graytown Rd, no facilities

587 LP 1604, Randolph AFB, to Universal City, no facilities

585.5mm Escondido Creek

585 FM 1516, to Converse, **N**...Mobil/diesel/rest./motel, **S**...Peterbilt/GMC/Freightliner

585mm Martinez Creek

583 Foster Rd, **N**...Flying J/Conoco/diesel/LP/mart/rest./24hr, Shamrock/mart/wash/24hr, Jack-in-the-Box/24hr, camping, **S**...TA/FoodCourt/diesel/mart/24hr

582.5mm Rosillo Creek

582 Ackerman Rd, Kirby, **N**...Pilot/Arby's/diesel/mart/24hr, **S**...Petro/Mobil/diesel, KFC, Wendy's, Relay Station Motel, Blue Beacon, Petrolube

581 I-410, no facilities

580 LP 13, WW White Rd, **N**...Chevron/mart, Fina, Hailey's Rest./24hr, Wendy's, Comfort Inn/rest., Motel 6, Scotsman Inn, RV camping, **S**...Exxon/diesel/mart/atm, Texaco, BBQ, McDonald's, Mexican Rest., Pizza Hut, Econolodge, Quality Inn, Super 8, Ford/Freightliner, Volvo, bank, tires

579 Houston St, **N**...Shamrock/mart, Ramada Ltd, **S**...Chevron/mart, Day's Inn, Passport Inn

578 Pecan Valley Dr, ML King Dr, **N**...Phillips 66/mart, **S**...Shell/diesel/mart

577 US 87 S, to Roland Ave, **S**...Whataburger/24hr, transmissions

576 New Braunfels Ave, Gevers St, **S**...Shamrock, McDonald's, Taco Snack

575 Pine St, Hackberry St, **S**...Little Red Barn Steaks

574 I-37, US 281, no facilities

573 Probandt St, **N**...Miller's BBQ, **S**...Conoco/mart, Shamrock, to SA Missions HS

572 I-10 and I-35 run together 3 miles

See TX I-35, exits 153-155a b.

570 I-10 and I-35 run together 3 miles

569c Santa Rosa St, downtown, to Our Lady of the Lake U

568 spur 421, Culebra Ave, Bandera Ave, **S**...to St Marys U

567 Lp 345, Fredericksburg Rd(from eb upper level accesses I-35 S, I-10 E, US 87 S, lower level accesses I-35 N)

566b Hildebrand Ave, Fulton Ave, **S**...Exxon/mart, 7-11, Galaxy Motel, brakes/tune-up, bank

   a Fresno Dr, **N**...Exxon/mart, Shamrock/mart, McDonald's

565c (from wb), access to same as 565 a b

   b Vance Jackson Rd, **N**...Miller's BBQ, Quality Inn, auto repair, **S**...DENTIST, Exxon, Texaco/mart/wash, Kettle, La Quinta

   a Crossroads Blvd, Balcone's Heights, **N**...Exxon, Texaco/diesel/mart, Denny's, Whataburger, Comfort Suites, Rodeway Inn, **S**...McDonald's, Rodeway Inn, Sumner Suites, Taco Loco, BrakeCheck, Burlington Coats, Firestone/auto, Mazda, RV Ctr, Toyota, mall, transmissions

564b a I-410, facilities off of I-410 W, Fredericksburg Rd, Chevron, Wendy's, K-Mart

563 Callaghan Rd, **N**...CHIROPRACTOR, Mobil, Chinese Rest., Las Palapas Mexican, Subway, TGIFriday, Embassy Suites, Ford, Lexus, Malibu Grand Prix, Mazda, Ritz Camera, Sun Harvest Foods, Toyota, bank, cinema, **S**...Exxon/mart, Mama's Rest., CarMax

561 Wurzbach Rd, **N**...MEDICAL CARE, Mobil, Phillips 66, CountyLine Steaks, Golden Corral, Jason's Deli, Orchid Rest., Pappasito's Cantina, Popeye's, Sea Island Seafood, Texas Land&Cattle, AmeriSuites, Homewood Suites, Ramada Ltd, Wyndham Hotel, Albertson's/drugs, HEB Foods, Eckerd/24hr, 1-Hr Photo, carwash, laundry, **S**...Texaco/diesel/mart/wash, Arby's, Boiling Pot Café, Church's Chicken, Denny's, IHOP, Jack-in-the-Box, Luby's, McDonald's, Pizza Hut, Sombrero Rosa Café, Taco Bell, Taco Cabana, Tony Roma, Wendy's, Candlewood Suites, Hawthorn Suites, Holiday Inn Express, La Quinta, Motel 6, Sleep Inn, Thrifty Inn, JiffyLube, bank

560b frontage rd(from eb), **N**...Water St Seafood, Chrysler/Plymouth, Nissan

   a Huebner Rd, **N**...BBQ, On the Border, SaltGrass Steaks, Bed Bath&Beyond, Borders, Land Rover, Old Navy, Ross, cinema, **S**...Exxon, Shell/Jack-in-the-Box/mart, Burger King, Cracker Barrel, Jim's Rest., AmeriSuites, Day's Inn, Hampton Inn, Homestead Village, Cadillac, Midas Muffler

559 Lp 335, US 87, Fredericksburg Rd, **N**...Carrabba's, Chili's, Outback Steaks, Joe's Crabshack, Logan's Roadhouse, Buick/GMC, **S**...Texaco/mart, Acura

San Antonio

E

W

Texas Interstate 10

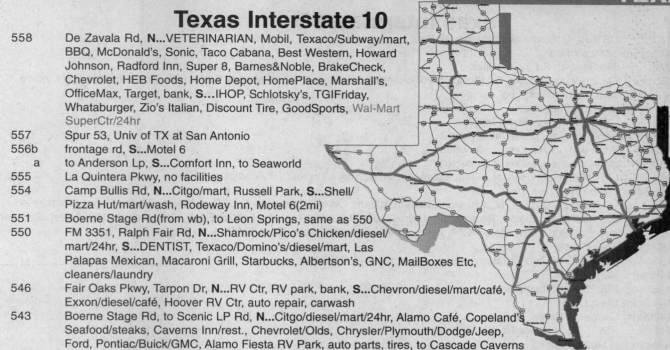

558	De Zavala Rd, **N**...VETERINARIAN, Mobil, Texaco/Subway/mart, BBQ, McDonald's, Sonic, Taco Cabana, Best Western, Howard Johnson, Radford Inn, Super 8, Barnes&Noble, BrakeCheck, Chevrolet, HEB Foods, Home Depot, HomePlace, Marshall's, OfficeMax, Target, bank, **S**...IHOP, Schlotsky's, TGIFriday, Whataburger, Zio's Italian, Discount Tire, GoodSports, Wal-Mart SuperCtr/24hr
557	Spur 53, Univ of TX at San Antonio
556b	frontage rd, **S**...Motel 6
a	to Anderson Lp, **S**...Comfort Inn, to Seaworld
555	La Quintera Pkwy, no facilities
554	Camp Bullis Rd, **N**...Citgo/mart, Russell Park, **S**...Shell/ Pizza Hut/mart/wash, Rodeway Inn, Motel 6(2mi)
551	Boerne Stage Rd(from wb), to Leon Springs, same as 550
550	FM 3351, Ralph Fair Rd, **N**...Shamrock/Pico's Chicken/diesel/ mart/24hr, **S**...DENTIST, Texaco/Domino's/diesel/mart, Las Palapas Mexican, Macaroni Grill, Starbucks, Albertson's, GNC, MailBoxes Etc, cleaners/laundry
546	Fair Oaks Pkwy, Tarpon Dr, **N**...RV Ctr, RV park, bank, **S**...Chevron/diesel/mart/café, Exxon/diesel/café, Hoover RV Ctr, auto repair, carwash
543	Boerne Stage Rd, to Scenic LP Rd, **N**...Citgo/diesel/mart/24hr, Alamo Café, Copeland's Seafood/steaks, Caverns Inn/rest., Chevrolet/Olds, Chrysler/Plymouth/Dodge/Jeep, Ford, Pontiac/Buick/GMC, Alamo Fiesta RV Park, auto parts, tires, to Cascade Caverns
540	TX 46, to New Braunfels, **N**...MEDICAL CARE, Exxon/Taco Bell/diesel/mart/yogurt, Shamrock/mart, Texaco/mart/ wash/24hr, Burger King, Church's Chicken, Dairy Queen, Denny's, Family Korner Buffet, Little Caesar's, Margarita's Café, Pizza Hut, Shanghai Chinese, Sonic, Taco Cabana, Wendy's, Best Western, Key to the Hills Motel, HEB Food/gas, Walgreen, Wal-Mart/auto, cleaners
539	Johns Rd, **S**...Shamrock/diesel/mart/LP, Pico's Chicken, Tapatio Springs Motel(4mi)
538mm	Cibolo Creek
537	US 87, to Boerne, **2 mi N**...Chevron, Shamrock, La Hacienda Rest., Maque's Mexican, Pete's Place, Pico's Chicken, Subway, Key to the Hills Motel
533	FM 289, Welfare, **N**...Po-Po Family Rest.
532mm	Little Joshua Creek
531mm	picnic area wb, tables, litter barrels
530mm	Big Joshua Creek
529.5mm	picnic area eb, tables, litter barrels
527	FM 1621(from wb), to Waring
526.5mm	Holiday Creek
524	TX 27, FM 1621, to Waring, **N**...VETERINARIAN, **S**...Texaco, DD Family Rest., antiques
523.5mm	Guadalupe River
523	US 87 N, to Comfort, **S**...Exxon/diesel/mart/24hr, Texaco/diesel/mart, Dairy Queen, Inn at Comfort, USA RV Park
521.5mm	Comfort Creek
520	FM 1341, to Cypress Creek Rd, no facilities
515mm	Cypress Creek
514mm	**rest areas both lanes, full(handicapped)facilities, vending, phone, picnic tables, litter barrels, petwalk, playground**
508	TX 16, Kerrville, **N**...Exxon/diesel, Chevrolet/Pontiac/Buick/Olds/Cadillac, RV camping, **S**...HOSPITAL, Chamber of Commerce, Chevron/Subway/diesel/mart, Shamrock/diesel/mart, TC/diesel/mart, Texaco/McDonald's/diesel/mart/ 24hr, Acapulco Mexican, BBQ, Cracker Barrel, Dairy Queen, Luby's, Schlotsky's, Best Western, Budget Inn, Comfort Inn, Day's Inn, Econolodge, Hampton Inn, Holiday Inn, Motel 6, Super 8, YoYo Ranch Hotel, Kerrville RV Ctr, Tom's Tires, **1-2 mi S**...DENTIST, CHIROPRACTOR, VETERINARIAN, Conoco/Circle K, Burger King, Jack-in-the-Box, KFC, McDonald's, Pizza Hut, Taco Bell, AutoZone, $General, Eckerd's, Firestone, Hastings Books, HEB Foods, Pennzoil, Walgreen, cleaners, to Kerrville-Schreiner SP
505	FM 783, to Kerrville, **S**...Exxon/diesel/mart, **3 mi S on TX 27**...Exxon/diesel, Inn of the Hills, Chili's, Dairy Queen, Pizza Hut, Sonic, Taco Casa, Rexall Drug, Superette Foods, Wal-Mart SuperCtr/24hr, Classic Car Museum
503.5mm	scenic views both lanes, litter barrels
501	FM 1338, **S**...KOA(2mi)
497mm	picnic area both lanes, tables, litter barrels
492	FM 479, **S**...Lone Oak Store/diesel, RV camping
490	TX 41, no facilities
488	TX 27, Mountain Home, to Ingram, no facilities
484	Midway Rd, no facilities

Texas Interstate 10

477	US 290, to Fredericksburg, no facilities
476.5mm	service rd eb
472	Old Segovia Rd, no facilities
465	FM 2169, to Segovia, **S**...Phillips 66/diesel/mart, Shell/diesel/mart/RV park, Cedar Hills Motel/laundry, River Valley Best Western/rest., RV camping
464.5mm	Johnson Fork Creek
462	US 83 S, to Uvalde, no facilities
461mm	picnic area eb, tables, litter barrels
460	(from wb), to Junction, no facilities
459mm	picnic area wb, tables, litter barrels
457	FM 2169, to Junction, **N**...Texaco/diesel/mart, **S**...Day's Inn/rest., RV camping
456.5mm	Llano River
456	US 83/377, Junction, **N**...Chevron/diesel/mart/24hr, Conoco/diesel/mart, BBQ, Jct Rest., Tastee Freez, Comfort Inn, **S**...HOSPITAL, Exxon, Texaco/mart, Dairy Queen, Isaack's Rest., La Familia Rest., Carousel Inn, Hills Motel, Slumber Inn/rest., Sun Valley Motel, KOA, Lakeview RV Park(3mi), Lazy Daze RV Park, to S Llano River SP
452.5mm	Bear Creek
451	RM 2291, to Cleo Rd, **S**...camping
448mm	North Creek
445	RM 1674, **S**...camping
444.5mm	Stark Creek
442.5mm	Copperas Creek
442	RM 1674, to Ft McKavett, no facilities
439mm	N Llano River
438	Lp 291(from wb), to Roosevelt, same as 437
437	Lp 291(from eb, no EZ return), to Roosevelt, **1 mi N**...Simon Bros Mercantile/diesel, USPO
429	RM 3130, to Harrell, no facilities
423mm	parking area both lanes, litter barrels
420	RM 3130, to Baker Rd, no facilities
412	Allison Rd, RM 3130, no facilities
404	RM 3130, RM 864, **N**...to Ft McKavett St HS, **S**...HOSPITAL
400	US 277, Sonora, **N**...Texaco/diesel/mart, Sutton Co Steaks, Day's Inn, Chevrolet/Buick/Pontiac/Olds, **S**...Chevron/diesel/mart, Conoco/diesel, Exxon/diesel/mart/24hr, Fina/diesel/mart, Phillips 66, Shell/diesel/mart, Country Kitchen, Dairy Queen, Just Home Cookin, La Mexicana Rest., Pizza Hut, Sonic, Best Western, Twin Oaks Motel, Alco Discount, AutoValue Parts, Buster's RV Park, bank, museum
399	LP 467, Sonora, **S**...HOSPITAL, VETERINARIAN, Chevron/diesel, Fina/mart/24hr, Phillips 66, Exxon/mart/diesel/24hr, Dairy Queen, Subway, Day's Inn, Devil's River Inn, Holiday Host Motel, Twin Oaks Motel, Zola's Motel, RV camping
394mm	rest area both lanes, full(handicapped)facilities, phone, picnic tables, litter barrel, petwalk, RV dump
392	RM 1989, Caverns of Sonora Rd, **8 mi S**...Caverns of Sonora Camping, phone
388	FM1312(from eb), no facilities
381	RM 1312, no facilities
372	Taylor Box Rd, **N**...Chevron/diesel/rest./24hr, Auto Museum, Circle Bar Motel/RV Park
368	LP 466, **N**...same as 365 & 363
365	TX 163, Ozona, **N**...HOSPITAL, Chevron/diesel/mart, Texaco/diesel/mart/café, Burger King, T&C/Country Kitchen/ Taco Bell/gas, Café NextDoor, Dairy Queen, El Chateau Rest., Comfort Inn, Daystop, Economy Inn/RV Park, Hillcrest Motel, Travelodge, NAPA, TrueValue, tires, to David Crockett Mon, **S**...Chevron/mart, Frank's Fuels, GC/gas, Phillips 66
363	Lp 466, to Ozona, **2-3 mi N**...Chevron/diesel/mart, Hitchin' Post Rest., Day's Inn, Flying W Lodge, Hillcrest Motel, Travelodge
361	RM 2083, Pandale Rd, no facilities
357mm	Eureka Draw
351mm	Howard Draw
350	FM 2398, to Howard Draw, no facilities
349mm	parking area wb, litter barrels
346mm	parking area eb, litter barrels
343	US 290 W, no facilities
337	Live Oak Rd, no facilities
336.5mm	Live Oak Creek
328	River Rd, Sheffield, **S**...GC/gas, Phillips 66/diesel, groceries/deli/RV camping
327.5mm	Pecos River
325	US 290, TX 349, to Iraan, Sheffield, **N**...HOSPITAL, gas, food, lodging
320	frontage rd, no facilities
314	frontage rd, no facilities
308mm	**rest area both lanes, full(handicapped)facilities, phone, picnic tables, litter barrels, petwalk**

Junction

Sonora

Ozona

Texas Interstate 10

307	US 190, FM 305, to Iraan, **N**...HOSPITAL
298	RM 2886, no facilities
294	FM 11, Bakersfield, **N**...Exxon/mart, **S**...Chevron/diesel/café, phone
288	Ligon Rd, no facilities
285	McKenzie Rd, **S**...Domaine Cordier Ste Genevieve Winery
279mm	picnic area eb, tables, litter barrels
277	FM 2023, no facilities
273	US 67/385, to McCamey, **N**...picnic area wb, tables, litter barrels
272	University Rd, no facilities
264	Warnock Rd, **N**...KOA
261	US 290 W, US 385 S, **N**...Exxon/TCBY/diesel/mart, RV camping, **S**...HOSPITAL
259b a	(259 from eb)TX 18, FM 1053, Ft Stockton, **N**...Citgo/diesel/mart, Shell/Burger King/mart, Texaco/diesel/mart, RV camping
257	US 285, Ft Stockton, to Pecos, **N**...RV camping, **1/2-1 mi S**...Chevron/diesel/mart, Exxon/diesel/rest./mart, Phillips 66/diesel, Shamrock/diesel/mart, Texaco, BBQ, KFC/Taco Bell, McDonald's, Pizza Hut, Sonic, Steakhouse Rest., Subway, Atrium West Motel, Comfort Inn, Day's Inn, Holiday Inn Express, La Quinta, Motel 6, Sands Motel, Chevrolet/Pontiac/Buick, Ford/Lincoln/Mercury, Furr's Foods, Goodyear/auto, Wal-Mart, auto/truck/RV repair, truckwash
256	to US 385 S, Ft Stockton, **1-2 mi S**...HOSPITAL, VETERINARIAN, Shamrock/diesel, Shell/diesel/mart, Texaco/diesel/mart/RV, Brazen Bean Rest., China Inn, Dairy Queen, K-Bob's Steaks, KFC, Pizza Express, Sonic, Subway, Alpine Lodge/rest., Best Western, Comfort Inn, Econolodge, Motel 6, Sands Motor Inn, Super 8, to Ft Stockton Hist Dist, Big Bend NP
253	FM 2037, to Belding, no facilities
248	US 67, FM 1776, to Alpine, **S**...to Big Bend NP, no facilities
246	Firestone, no facilities
241	Kennedy Rd, no facilities
235	Mendel Rd, no facilities
233mm	**rest area both lanes, full(handicapped)facilities, phone, picnic tables, litter barrels, petwalk**
229	Hovey Rd, no facilities
222	Hoefs Rd, no facilities
214	(from wb), FM 2448, no facilities
212	TX 17, FM 2448, to Pecos, **S**...Fina/diesel/mart/rest./service
209	TX 17, **2 mi S**...Citgo, Greasewood RV Park, to Davis Mtn SP, Ft Davis NHS
206	FM 2903, to Balmorhea, **1 mi S**...gas, food, lodging, camping, to Balmorhea SP
192	FM 3078, to Toyahvale, **S**...to Balmorhea SP
188	Giffin Rd, no facilities
186	I-20, to Ft Worth, Dallas
185mm	picnic area both lanes, tables, litter barrels
184	Springhills, no facilities
181	Cherry Creek Rd, **S**...Chevron/mart/24hr
176	TX 118, RM 2424, to Kent, **N**...Chevron/diesel/mart, **S**...to McDonald Observatory, Davis Mtn SP, Ft Davis
173	Hurd's Draw Rd, no facilities
166	Boracho Sta, no facilities
159	Plateau, **N**...Citgo/diesel/rest./24hr
153	Michigan Flat, no facilities
146	Wild Horse Rd, no facilities
146mm	weigh sta wb
144.5mm	**rest area both lanes, full(handicapped)facilities, picnic tables, litter barrels, petwalk**
140b	Ross Dr, Van Horn, **N**...Chevron/mart, Exxon/diesel/mart, Love's/Subway/diesel/mart/24hr, Bell's Motel, Comfort Inn, Day's Inn/rest., Royal Motel, Sands Motel/rest., El Campo RV Park, Van Horn Drug, **S**...MtnView Trkstp/diesel/mart/24hr/repair, Mountain RV Park/dump
a	US 90, TX 54, Van Horn Dr, **N**...HOSPITAL, Shamrock, Fina/diesel/mart, Texaco/Burger King, Best Western, Comfort Inn, Sun Valley Motel, Goodyear, **S**...Pilot/Wendy's/diesel/mart/24hr
138	Lp 10, to Van Horn, **1-2 mi N**...Shell/diesel, Texaco, Chili's Rest., Dairy Queen, Pizza Hut, Best Western, Budget Inn, Comfort Inn, Economy Inn, King's Inn, Motel 6, Ramada Ltd, Ford/Mercury, IGA Foods, RV camping, diesel repair, **S**...Chevron/diesel/mart/24hr, McDonald's, Holiday Inn Express, Super 8
137mm	truck weigh sta eb
136mm	scenic overlook wb, picnic tables, litter barrels
135mm	Mountain/Central time zone line

E

W

Ft Stockton

Van Horn

Texas Interstate 10

129	Allamore, to Hot Wells, no facilities
108	to Sierra Blanca(from wb), same as 107
107	RM 1111, Sierra Blanca Ave, **N**...Exxon/diesel/mart/24hr, Texaco/diesel/mart, Michael's Rest., El Camino Motel, Sierra Motel, truck/tire repair, to Hueco Tanks SP, **S**...Chevron/mart, Lena's Rest.
105	Lp 10, Sierra Blanca, **1-2 mi N**...Exxon/diesel, Texaco/diesel, Best Café, Chester Fried Chicken, El Camino Motel, Sierra Motel, R&G Foods, High Country RV park, USPO
102.5mm	inspection sta eb
99	Lasca Rd, **N**...rest area both lanes, picnic tables, litter barrels, no restrooms
98mm	picnic area eb, tables, litter barrels, no restrooms
95	frontage rd(from eb), **S**...gas/motel/café
87	FM 34, **S**...Tiger Trkstp/diesel/rest., motel
85	Esperanza Rd, no facilities
81	FM 2217, no facilities
78	TX 20 W, to McNary, no facilities
77mm	truck parking area wb
72	spur 148, to Ft Hancock, **S**...Shell/diesel/mart, Texaco/mart, Angie's Rest., Budget Host, RV park
68	Acala Rd, no facilities
55	Tornillo, no facilities
51mm	**rest area both lanes, full(handicapped)facilities, picnic tables, litter tables, petwalk**
49	FM 793, Fabens, **S**...Texaco/diesel, Church's Chicken, Cowboy Café, McDonald's, Fabens Motel, Texas Red's Motel/grill, Barton Bros Parts, Big 8 Foods, bank, camping
42	FM 1110, to Clint, **S**...Exxon/diesel/mart, Cotton Valley Motel/RV Park/rest./dump
37	FM 1281, Horizon Blvd, **N**...Exxon, Love's/Pizza Hut/Subway/Taco Bell/diesel/mart/24hr, Americana Motel, **S**...Petro/Mobil/Blimpie/diesel/mart/rest./24hr, McDonald's, Deluxe Inn, Blue Beacon, PetroLube
34	TX 375, Americas Ave, **N**...Chevron/diesel/mart/24hr, Shamrock/diesel/mart, Texaco/Taco Bell/diesel/mart/24hr, **S**...RV camping, El Paso Mus of Hist
32	FM 659, Zaragosa Rd, **N**...VETERINARIAN, Chevron/diesel/mart, Citgo/7-11, Phillips 66, Arby's, Cheddar's, Logan's Roadhouse, McDonald's, Outback Steaks, Village Inn Rest., Whataburger/24hr, Red Roof Inn, Hertz-Penske Trucks, Office Depot, cinema, **S**...Chevron/mart, Conoco, Shamrock/diesel/mart, Howah Chinese, casino
30	Lee Trevino Dr, **N**...Golden Corral(1mi), Baby Depot, Burlington Coats, Cowtown Boots, Discount Tire, Firestone, Ford, Isuzu, Home Depot, Luxury Linens, NTB, Toyota, bank, mall, **S**...Shamrock/mart/24hr, Chevrolet, Harley-Davidson, Nissan, cinema
29	Lomaland Dr, **N**...Exxon, Chinese Rest., Denny's, Edelweiss Rest., KwikWok, Whataburger, AmeriSuites, La Quinta, Motel 6, Lexus, Ryder Trucks, Toyota, **S**...Fina/7-11, Dot's BBQ, Julio's Café, Tony Roma, HomeGate Studios, Westar Suites, Harley-Davidson, RV camping
28b	Yarbrough Dr, El Paso, **N**...Chevron/24hr, Baskin-Robbins, Beijing Lili, Bennigan's, Bobby Q's Ribs, Burger King, Dunkin Donuts, Grandy's, LJ Silver, Marie Callender's, McDonald's, Peter Piper Pizza, Pistol Pete's Pizza, Sonic, Subway, Wendy's, Whataburger, Wienerschnitzel, Wyatt's Cafeteria, Day's Inn, EyeMart, MailBoxes Etc, Marshall's, Mervyn's, Midas Muffler, PepBoys, PetsMart, Radio Shack, Ross, Wal-Mart/24hr, cinema **S**...Shamrock/mart/24hr, Applebee's, LaMalenche Café, Pizza Hut, Baymont Inn, Comfort Inn, Suburban Lodge
a	FM 2316, McRae Blvd, **N**...HOSPITAL, Chevron/mart/24hr, Chico's Taco, Jack-in-the-Box, KFC, Marie Callender's, Pancho's Mexican, Pepe's Tomales, Pizza Hut, Best Buy, DiscoveryZone, Firestone/auto, Goodyear/auto, K-Mart/auto/drugs, Michael's, OfficeMax, ToysRUs, TuneUp Masters, Walgreen, bank, cleaners, **S**...VETERINARIAN, Chevron/mart/wash, Fina/7-11, Phillips 66/diesel/mart, Gabriel's Mexican, Hyundai/Saturn
27	Hunter Dr, Viscount Blvd, **N**...Shamrock/mart, Fina/7-11, Carrow's Rest., K-Bob's, Red Lobster, Taco Bell, La Quinta, Barnes&Noble, CompUSA, Firestone, Home Express, laundry, **S**...Exxon/Subway/diesel, Shell, Whataburger/24hr, Family$, Food City
26	Hawkins Blvd, El Paso, **N**...Chevron, Shamrock/mart, Texaco/mart, Arby's, Burger King, Country Kitchen, Dairy Queen, Golden Corral, IHOP, Landry's Seafood, Luby's, Taco Cabana, Wyatt's Cafeteria, Howard Johnson, Circuit City, Dillard's, JC Penney, Office Depot, Pennzoil, Pier 1, Sam's Club, Sears/auto, Wal-Mart, diesel service, **S**...Shamrock/diesel/mart, China King, McDonald's, Village Inn Rest., Best Western
25	Airway Blvd, El Paso Airport, **N**...Shell/diesel/mart/24hr, Landry Seafood, Hampton Inn, Holiday Inn, Radisson, GMC/Volvo/Mercedes Trucks, **S**...Chevron/diesel/rest./atm/24hr, Goodyear, Subaru, VW
24b	Geronimo Dr, **N**...MEDICAL CARE, Chevron/mart/24hr, Pancake House, Steak&Ale, Wyatt's Cafeteria, Daystop, Quality Inn/rest., Rodeway Inn, Dillard's, Mervyn's, Office Depot, Service Merchandise, Target, bank, mall, mufflers/brakes, **S**...Fina/7-11, Phillips 66/mart, Bombay Bicycle Club, Denny's, Kettle, Embassy Suites, La Quinta, Sumner Suites
a	Trowbridge Dr, **N**...CHIROPRACTOR, Fina, Thunderbird Gas, Alexandrio's Mexican, Luby's, McDonald's, Steak&Ale, Whataburger, Budget Motel, Ford, Nissan, Toyota
23b	US 62/180, to Paisano Dr, **N**...Jack-in-the-Box, McDonald's, Ford, Tune-Up Ctr, U-Haul, auto parts, to Carlsbad, **S**...McDonald's, Furr's Foods
a	Raynolds St, **S**...HOSPITAL, Arby's, Motel 6, Super 8, Central Auto Parts
22b	US 54, Patriot Fwy, no facilities

El Paso

E

W

444

Texas Interstate 10

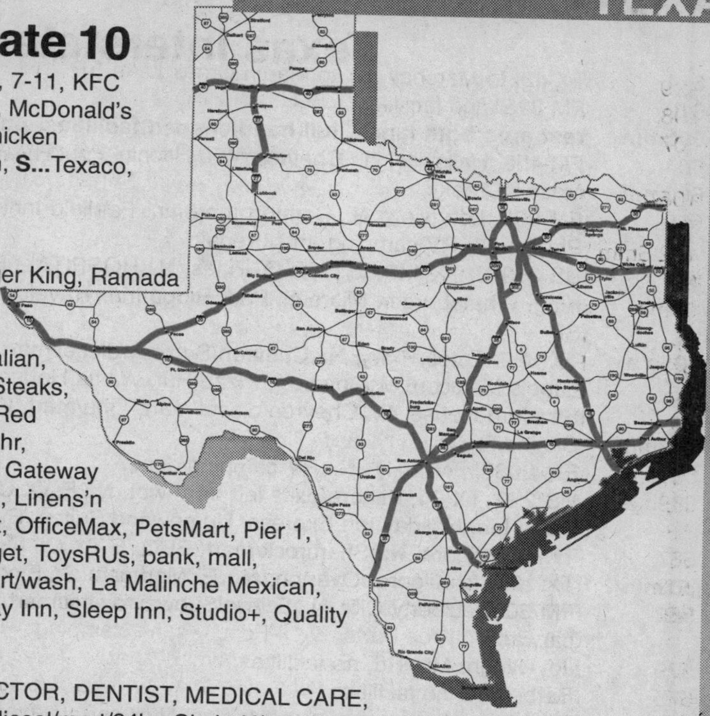

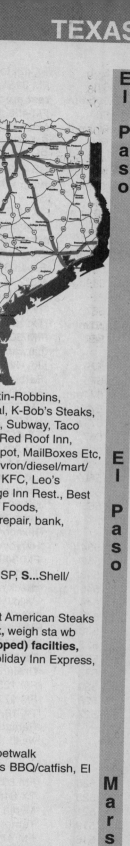

a	Copia St, El Paso, **N**...Fina/mart, Shamrock/mart, 7-11, KFC
21	Piedras St, El Paso, **N**...Exxon/mart, Burger King, McDonald's
20	Dallas St, Cotton St, **N**...Exxon/mart, Church's Chicken
19	TX 20, El Paso, downtown, **N**...Chevron/mart/atm, **S**...Texaco, Holiday Inn, International Hotel, Travelodge, bank
18b	Franklin Ave, Porfirio Diaz St, no facilities
a	Schuster Ave, **N**...Sun Bowl, **S**...to UTEP
16	Executive Ctr Blvd, **N**...Shamrock/mart/24hr, Burger King, Ramada Inn, Ford, ProntoLube
13b a	US 85, Paisano Dr, to Sunland Park Dr, **N**...Shamrock/mart/24hr, Texaco/mart, Carino's Italian, Chinese Rest., ChuckeCheese, Great American Steaks, HomeTown Buffet, Kenny Rogers, Olive Garden, Red Lobster, Sunland Grill, TGIFriday, Whataburger/24hr, Barnes&Noble, Big 5 Sports, Circuit City, Dillard's, Gateway Computers, HomeBase, JC Penney, K-Mart/drugs, Linens'n Things, Marshall's, MensWearhouse, Office Depot, OfficeMax, PetsMart, Pier 1, Sears, Service Merchandise, SportsAuthority, Target, ToysRUs, bank, mall, **S**...Shamrock/diesel/mart/24hr, Shell/Taco Bell/mart/wash, La Malinche Mexican, McDonald's, Best Western, Comfort Suites, Holiday Inn, Sleep Inn, Studio+, Quality Foods, Sunland Park RaceTrack, cleaners
12	Resler Dr(from wb), no facilities
11	TX 20, to Mesa St, Sunland Park, **N**...CHIROPRACTOR, DENTIST, MEDICAL CARE, EYECARE, Chevron/diesel/mart/24hr, Shamrock/diesel/mart/24hr, Circle K/gas, AJ's Diner, Baskin-Robbins, Carrow's/24hr, Chili's, Chinese Buffet, CiCi's, Cracker Barrel, Denny's, El Taco Tote, Golden Corral, K-Bob's Steaks, LJ Silver, Manhattan Bagel, OutBack Steaks, Papa John's, Popeye's, Red Lobster, Souper Salad, Subway, Taco Bell, Tony Roma, Wienerschnitzel, Wendy's, Whataburger, Baymont Inn, Comfort Inn, La Quinta, Red Roof Inn, Albertson's, Checker Parts, Fabric Whse, Firestone/auto, Furr's Foods, Goodyear/auto, Home Depot, MailBoxes Etc, PepBoys, Wal-Mart SuperCtr/24hr, bank, cleaners/laundry, **S**...CHIROPRACTOR, DENTIST, Chevron/diesel/mart/24hr, Shamrock/diesel/mart/24hr, Burger King, Church's Chicken, Dunkin Donuts, Golden China, KFC, Leo's Mexican, Luby's, McDonald's, NJ Deli, Peter Piper Pizza, Pizza Hut, Subway, Taco Cabana, Village Inn Rest., Best Value Inn, Day's Inn, Extended Stay America, Motel 6, Travelodge, AutoZone, Big 8 Foods, Furr's Foods, HobbyLobby, Martin Tires, Postal Annex, ProntoLube, Radio Shack, Sam's Club, Walgreen, auto repair, bank, cleaners
9	Redd Rd, **S**...Phillips 66/diesel/mart/atm, auto auction
8	Artcraft Rd, no facilities
6	Lp 375, to Canutillo, **N**...Texaco/Taco Bell/TCBY/diesel/mart, to Trans Mountain Rd, Franklin Mtns SP, **S**...Shell/Subway/diesel/mart, RV camping
5mm	truck check sta eb
2	Westway, Vinton, **N**...Petro/Mobil/Blimpie/diesel/rest./Petrolube/tires/24hr, **S**...gas/mart/24hr, Great American Steaks
1mm	**Welcome Ctr eb, full(handicapped)facilities, info, phone, picnic tables, litter barrels, petwalk,** weigh sta wb
0	FM 1905, Anthony, **N**...Flying J/Conoco/diesel/LP/rest./24hr, Super 8, **rest area wb, full(handicapped) facilties, picnic tables, litter barrels, S**...Chevron/deli/diesel/mart, Exxon/Burger King/diesel/mart/24hr, Holiday Inn Express, Big 8 Foods, Fun Country RV/Marine(1mi), truckwash
0mm	Texas/New Mexico state line

The **E** / **W** direction arrow appears at left.

Texas Interstate 20

Exit #	Services
636mm	Texas/Louisiana state line
635.5mm	Welcome Ctr wb/parking area eb, full(handicapped)facilities, phone, picnic tables, litter barrels, petwalk
635	TX 9, TX 156, to Waskom, **N**...EYECARE, Chevron/diesel/mart, Texaco/mart, Dairy Queen, Jim's BBQ/catfish, El Nortenio's Mexican, bank
633	US 80, FM 9, FM 134, to Waskom, **N**...Texaco/diesel/mart/24hr, **S**...RV camping
628	to US 80(from wb), to frontage rd(from eb), no facilities
624	FM 2199, to Scottsville, no facilities
620	FM 31, to Elysian Fields, no facilities
617	US 59, Marshall, **N**...Exxon/mart/24hr, Texaco/mart/24hr, Applebee's, Burger King(2mi), Catfish Express, Domino's, Golden Corral, LJ Silver(2mi), McDonald's, Scholotsky's, Subway, Waffle House, Whataburger(2mi), Best Western, Economy Inn Express, Hampton Inn, Holiday Inn Express, Ramada Ltd, Texas Inn(2mi), Chevrolet/Olds/Cadillac, Chrysler/Plymouth/Dodge/Jeep, Ford/Lincoln/Mercury, Country Pines RV Park(8mi), **S**...Chevron/diesel/mart, Shamrock/diesel/mart, Total/mart, Hungri Maverick Rest., Comfort Inn, Day's Inn, Guest Inn, Motel 6, Super 8

The **E** / **W** direction arrow appears at left.

Texas Interstate 20

614	TX 43, to Marshall, **S**...to Martin Creek Lake SP
610	FM 3251, no facilities
608mm	**rest area both lanes, full(handicapped)facilities, phone, vending, picnic tables, litter barrels**
604	FM 450, Hallsville, **N**...Chevron, 450 Hitchin' Post RV Park, to Lake O' the Pines
600mm	Mason Creek
599	FM 968, Longview, **N**...Comfort Inn(8mi), Fairfield Inn(8mi), Kenworth Trucks, **S**...Exxon/mart/24hr, Fina/diesel/rest., Phillips 66, RV camping, truck repair
596	US 259 N, TX 149, to Lake O' Pines, **N**...HOSPITAL, Exxon/Grandy's/mart, Texaco/Taco Bell/diesel/mart, Burger King, Whataburger, Microtel, Park Ridge Inn, Travelodge, **S**...Total/Arby's/diesel/mart, Holiday Inn, to Martin Lake SP
595b a	TX 322, Estes Pkwy, **N**...Chevron/Subway/diesel/mart, Texaco/diesel/mart/24hr, Dairy Queen, Jack-in-the-Box, Lupe's Mexican, McDonald's, Pizza Hut, Waffle House, Best Western, Day's Inn, Econolodge, Longview Inn, Stratford House, **S**...Chevron/diesel/mart, Fina/mart, Kettle, KFC/Taco Bell, Hampton Inn, La Quinta, Motel 6
593mm	Sabine River
591	FM 2087, FM 2011, **N**...RV camping/dump
589b a	US 259, TX 31, Kilgore(exits left from wb), **N**...E TX Oil Museum, **1-3 mi S**...Kilgore Café, Budget Inn, Comfort Inn, Day's Inn, Holiday Inn Express, Homewood Suites, Ramada Inn
587	TX 42, Kilgore, **N**...Shamrock/mart, BBQ, E TX Oil Museum, **S**...Chevron/pizza/subs
583	TX 135, to Kilgore, Overton, **N**...EZMart/gas, **S**...Exxon/diesel/mart(1mi)
582	FM 3053, Liberty City, **N**...Exxon/Subway/diesel/mart, FasTrak/gas, Texaco/mart/24hr, Dairy Queen, auto parts, carwash
579	Joy-Wright Mtn Rd, no facilities
575	Barber Rd, no facilities
574mm	picnic area both lanes, picnic tables, litter barrels, handicapped accessible
571b	FM 757, Omen Rd, to Starrville, no facilities
571a	US 271, to Gladewater, Tyler, **S**...Chevron/diesel
567	TX 155, **N**...Citgo/diesel/pizza/subs/mart/24hr, **S**...Day's Inn
565	FM 2015, to Driskill-Lake Rd, no facilities
562	FM 14, **N**...Bodacious BBQ, Northgate RV Park, to Tyler SP, **S**...CB Shop
560	Lavender Rd, no facilities
557	Jim Hogg Rd, **N**...Texaco/diesel/mart, Tio's Rest.
556	US 69, to Tyler, **N**...RaceTrac/mart, Total/diesel/mart, Burger King, Juanita's Mexican, McDonald's, Paco's Mexican, Pizza Inn, Subway, Taco Bell, Comfort Inn, Day's Inn, Medder's Tire/lube, bank, hardware, **S**...Chevron/mart, Texaco/DQ/diesel/mart/24hr, Cracker Barrel, Denny's, Tio's Rest., Best Western, Econo Inn, antiques, LP
554	Harvey Rd, **S**...Yellow Rose RV Park
552	FM 849, **N**...MEDICAL CARE, Chevron/mart/24hr, auto repair, bank, **S**...BBQ
548	TX 110, to Grand Saline, **N**...Exxon/diesel/mart, **S**...Chevron/diesel/mart, Margie's Rest.
546mm	check sta both lanes
544	Willow Branch Rd, **N**...Conoco/diesel/rest., Running W Rest., Willow Branch RV Park
540	FM 314, to Van, **N**...Love's/Carl's Jr/diesel/mart/24hr, Dairy Queen, Van Inn
538mm	**rest area both lanes, full(handicapped)facilities, phone, vending, picnic tables, litter barrels, petwalk**
537	FM 773, FM 16, no facilities
536	Tank Farm Rd, no facilities
533	Oakland Rd, to Colfax, **N**...Citgo/Blimpie/diesel/mart, Hot Biscuit Rest.
530	FM 1255, no facilities
528	FM 17, to Grand Saline, no facilities
527	TX 19, **N**...Citgo/Burger King/mart, Texaco/diesel/mart/24hr, Jewel's Rest., Whataburger/24hr, Holiday Inn Express, Ramada Ltd, Super 8, **S**...Chevron/diesel/mart/wash/24hr, Fina/diesel/mart, Phillips 66/diesel/mart/24hr, Shell/mart/wash, Dairy Queen, Dairy Palace, Juanita's Mexican, KFC/Taco Bell, McDonald's, Ranchero Rest., Senorita's Mexican, Best Western, Day's Inn, Chrysler/Dodge/Plymouth/Jeep, Ford/Mercury, to First Monday SP, RV camping, LP
526	FM 859, to Edgewood, **N**...Edgewood Heritage Park, **S**...to First Monday SP
523	TX 64, to Canton, no facilities
521	Myrtle Springs Rd, **N**...trailer sales, **S**...Marshall's RV Ctr, RV camp/dump, U-Haul
519	Turner-Hayden Rd, **3 mi N**...gas, food, phone
516	FM 47, to Wills Point, **N**...to Lake Tawakoni, **S**...Shamrock/mart/24hr, Robertson's Café/gas, Interstate Motel/café
512	FM 2965, Hiram-Wills Point Rd, no facilities
511.5mm	weigh sta both lanes
511mm	**rest area both lanes, full(handicapped)facilities, phone, vending, picnic tables, litter barrels, petwalk**
509	Hiram Rd, **S**...Phillip 66/diesel/mart/BBQ café
506	FM 429, FM 2728, College Mound Rd, no facilities
503	Wilson Rd, **S**...Rip Griffin/Texaco/Subway/Pizza Hut/diesel/mart/LP/24hr

E

↑

W

↓

Canton

Texas Interstate 20

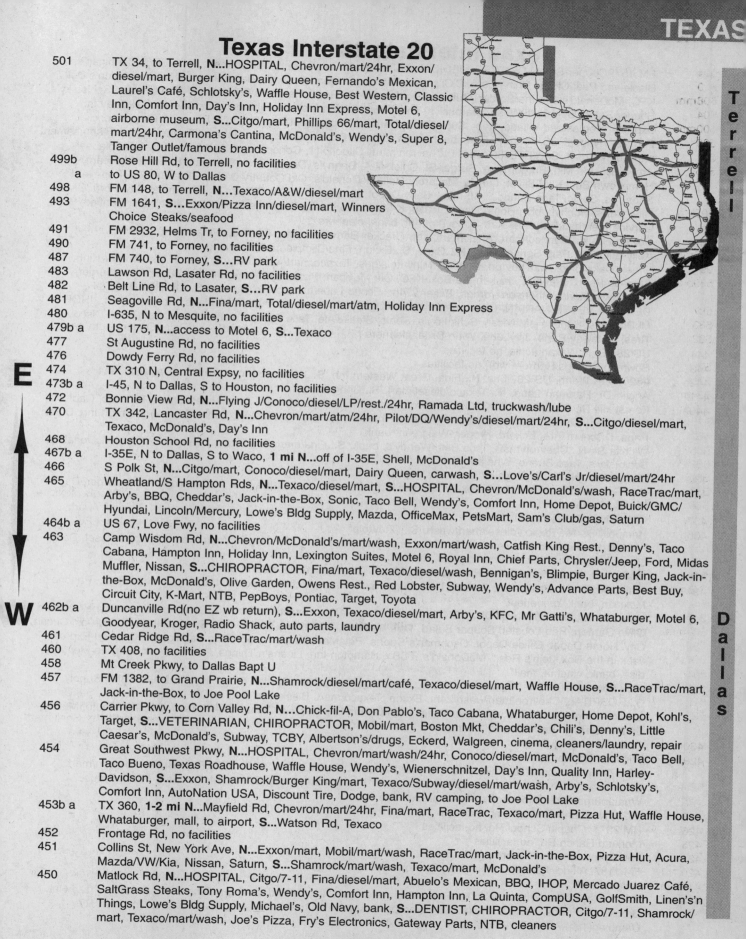

501 TX 34, to Terrell, **N**...HOSPITAL, Chevron/mart/24hr, Exxon/diesel/mart, Burger King, Dairy Queen, Fernando's Mexican, Laurel's Café, Schlotsky's, Waffle House, Best Western, Classic Inn, Comfort Inn, Day's Inn, Holiday Inn Express, Motel 6, airborne museum, **S**...Citgo/mart, Phillips 66/mart, Total/diesel/mart/24hr, Carmona's Cantina, McDonald's, Wendy's, Super 8, Tanger Outlet/famous brands

499b Rose Hill Rd, to Terrell, no facilities

a to US 80, W to Dallas

498 FM 148, to Terrell, **N**...Texaco/A&W/diesel/mart

493 FM 1641, **S**...Exxon/Pizza Inn/diesel/mart, Winners Choice Steaks/seafood

491 FM 2932, Helms Tr, to Forney, no facilities

490 FM 741, to Forney, no facilities

487 FM 740, to Forney, **S**...RV park

483 Lawson Rd, Lasater Rd, no facilities

482 Belt Line Rd, to Lasater, **S**...RV park

481 Seagoville Rd, **N**...Fina/mart, Total/diesel/mart/atm, Holiday Inn Express

480 I-635, N to Mesquite, no facilities

479b a US 175, **N**...access to Motel 6, **S**...Texaco

477 St Augustine Rd, no facilities

476 Dowdy Ferry Rd, no facilities

474 TX 310 N, Central Expsy, no facilities

473b a I-45, N to Dallas, S to Houston, no facilities

472 Bonnie View Rd, **N**...Flying J/Conoco/diesel/LP/rest./24hr, Ramada Ltd, truckwash/lube

470 TX 342, Lancaster Rd, **N**...Chevron/mart/atm/24hr, Pilot/DQ/Wendy's/diesel/mart/24hr, **S**...Citgo/diesel/mart, Texaco, McDonald's, Day's Inn

468 Houston School Rd, no facilities

467b a I-35E, N to Dallas, S to Waco, **1 mi N**...off of I-35E, Shell, McDonald's

466 S Polk St, **N**...Citgo/mart, Conoco/diesel/mart, Dairy Queen, carwash, **S**...Love's/Carl's Jr/diesel/mart/24hr

465 Wheatland/S Hampton Rds, **N**...Texaco/diesel/mart, **S**...HOSPITAL, Chevron/McDonald's/wash, RaceTrac/mart, Arby's, BBQ, Cheddar's, Jack-in-the-Box, Sonic, Taco Bell, Wendy's, Comfort Inn, Home Depot, Buick/GMC/Hyundai, Lincoln/Mercury, Lowe's Bldg Supply, Mazda, OfficeMax, PetsMart, Sam's Club/gas, Saturn

464b a US 67, Love Fwy, no facilities

463 Camp Wisdom Rd, **N**...Chevron/McDonald's/mart/wash, Exxon/mart/wash, Catfish King Rest., Denny's, Taco Cabana, Hampton Inn, Holiday Inn, Lexington Suites, Motel 6, Royal Inn, Chief Parts, Chrysler/Jeep, Ford, Midas Muffler, Nissan, **S**...CHIROPRACTOR, Fina/mart, Texaco/diesel/wash, Bennigan's, Blimpie, Burger King, Jack-in-the-Box, McDonald's, Olive Garden, Owens Rest., Red Lobster, Subway, Wendy's, Advance Parts, Best Buy, Circuit City, K-Mart, NTB, PepBoys, Pontiac, Target, Toyota

462b a Duncanville Rd(no EZ wb return), **S**...Exxon, Texaco/diesel/mart, Arby's, KFC, Mr Gatti's, Whataburger, Motel 6, Goodyear, Kroger, Radio Shack, auto parts, laundry

461 Cedar Ridge Rd, **S**...RaceTrac/mart/wash

460 TX 408, no facilities

458 Mt Creek Pkwy, to Dallas Bapt U

457 FM 1382, to Grand Prairie, **N**...Shamrock/diesel/mart/café, Texaco/diesel/mart, Waffle House, **S**...RaceTrac/mart, Jack-in-the-Box, to Joe Pool Lake

456 Carrier Pkwy, to Corn Valley Rd, **N**...Chick-fil-A, Don Pablo's, Taco Cabana, Whataburger, Home Depot, Kohl's, Target, **S**...VETERINARIAN, CHIROPRACTOR, Mobil/mart, Boston Mkt, Cheddar's, Chili's, Denny's, Little Caesar's, McDonald's, Subway, TCBY, Albertson's/drugs, Eckerd, Walgreen, cinema, cleaners/laundry, repair

454 Great Southwest Pkwy, **N**...HOSPITAL, Chevron/mart/wash/24hr, Conoco/diesel/mart, McDonald's, Taco Bell, Taco Bueno, Texas Roadhouse, Waffle House, Wendy's, Wienerschnitzel, Day's Inn, Quality Inn, Harley-Davidson, **S**...Exxon, Shamrock/Burger King/mart, Texaco/Subway/diesel/mart/wash, Arby's, Schlotsky's, Comfort Inn, AutoNation USA, Discount Tire, Dodge, bank, RV camping, to Joe Pool Lake

453b a TX 360, **1-2 mi N**...Mayfield Rd, Chevron/mart/24hr, Fina/mart, RaceTrac, Texaco/mart, Pizza Hut, Waffle House, Whataburger, mall, to airport, **S**...Watson Rd, Texaco

452 Frontage Rd, no facilities

451 Collins St, New York Ave, **N**...Exxon/mart, Mobil/mart/wash, RaceTrac/mart, Jack-in-the-Box, Pizza Hut, Acura, Mazda/VW/Kia, Nissan, Saturn, **S**...Shamrock/mart/wash, Texaco/mart, McDonald's

450 Matlock Rd, **N**...HOSPITAL, Citgo/7-11, Fina/diesel/mart, Abuelo's Mexican, BBQ, IHOP, Mercado Juarez Café, SaltGrass Steaks, Tony Roma's, Wendy's, Comfort Inn, Hampton Inn, La Quinta, CompUSA, GolfSmith, Linen's'n Things, Lowe's Bldg Supply, Michael's, Old Navy, bank, **S**...DENTIST, CHIROPRACTOR, Citgo/7-11, Shamrock/mart, Texaco/mart/wash, Joe's Pizza, Fry's Electronics, Gateway Parts, NTB, cleaners

E ↑↓ W

Texas Interstate 20

449   FM 157, Cooper St, **N**...DENTIST, MEDICAL CARE, Fina, Mobil, Shell, Texaco/mart, 7-11, BBQ, Bennigan's, Blackeyed Pea, Chili's, China Café, CiCi's, Don Pablo's, Golden Corral, Grandy's, Jack-in-the-Box, Jason's Deli, KFC, McDonald's, On the Border, Outback Steaks, Owens Rest., Razzoo's Cajun Café, Red Lobster, Schlotsky's, Souper Salad, Spaghetti Whse, Starbuck's, Thai Cuisine, Wendy's, Whataburger, Best Western, Holiday Inn Express, Homestead Village, Baby SuperStore, Barnes&Noble, Best Buy, Dillard's, Discount Tire, Hancock Fabrics, Hyundai, JC Penney, JiffyLube, Mervyn's, Office Depot, Sears/auto, Subaru, Target, TJ Maxx, amusement park, bank, mall, **S**...VETERINARIAN, Chevron/mart, Citgo/7-11, Conoco/mart, Mobil/mart/wash, Applebee's, Arby's, Boston Mkt, Burger King, Burger St, Chick-fil-A, Denny's, Dos Gringo's Mexican, El Fenix Mexican, HomeTown Buffet, LJ Silver, Luby's, Macaroni Grill, McDonald's, Old Country Buffet, Olive Garden, Popeye's, Shoney's, Sonic, Taco Bueno, TGIFriday, Baby Depot, Burlington Coats, Circuit City, Computer City, Ford, GMC, HiLo Parts, HobbyLobby, Home Depot, Isuzu, K-Mart/Little Caesar's, Kroger/deli/24hr, Office Max, Pontiac, Ross, Service Merchandise, Wal-Mart SuperCtr/24hr, bank, cleaners

448   Bowen Rd, **N**...Conoco/mart, Shell/mart/24hr, Cracker Barrel, cinema, **S**...Texaco/mart/wash

447   Kelly-Elliot Rd, Park Springs Blvd, **N**...Citgo, **S**...Exxon, Fina/Blimpie/mart, camping

445   Green Oaks Blvd, **N**...Chevron, Conoco, Minyard's/gas, Texaco/mart/wash, Arby's, Boston Mkt, Burger King, Chinese Rest., Grandy's, Jack-in-the-Box, KFC, Ole Mexican, Pizza Hut, Pizza Inn, Taco Bell, Whataburger, Wyatt's Cafeteria, Albertson's/drugs, Eckerd/24hr, Econo Lube/tune, bank, cinema, **S**...CHIROPRACTOR, DENTIST, VETERINARIAN, Chevron/mart/24hr, Citgo/7-11, Shamrock/mart, Cheddar's, Chinese Rest., IHOP, McDonald's, Pancho's Mexican, Schlotsky's, Sonic, Steak&Ale, Taco Bueno, Waffle House, AutoZone, Discount Tires, Gateway Parts, JiffyLube, Winn-Dixie, cleaners

444   US 287 S, to Waxahatchie, no facilities

443   Bowman Springs Rd(from wb), no facilities

442b a   I-820 to Ft Worth, US 287 bus, **N**...Fina, Great Western Inn, **S**...Phillips 66/mart, Texaco, Dairy Queen

441   Anglin Dr, Hartman Lane, **N**...Texaco/diesel/mart, **S**...Conoco/mart

440b   Forest Hill Dr, **S**...CHIROPRACTOR, Chevron/mart/24hr, Phillips 66/mart, Texaco/diesel/mart, Braum's, Capt D's, CiCi's, Dairy Queen, Denny's, Domino's, Jack-in-the-Box, Luby's, McDonald's, Sonic, Subway, Comfort Inn, Chief Parts, Discount Tire, Eckerd, Kroger, Walgreen, bank

     a   Wichita St, **N**...Chevron/mart, Taco Bell, Wendy's, bank, **S**...Fina/mart, Total/mart, Chicken Express, McDonald's, Schlotsky's, Taco Bueno, Whataburger

439   Campus Dr, **N**...Ford, Chrysler/Plymouth/Jeep, Olds, **S**...Sam's Club

438   Oak Grove Rd, **1 mi N**...Mobil, Texaco, Burger King, Denny's, Jack-in-the Box, McDonald's, Whataburger, Day's Inn, **S**...Conoco

437   I-35W, N to Ft Worth, S to Waco

436b   Hemphill St, **N**...Texaco/diesel/mart/wash, **S**...Chevrolet

     a   FM 731(from eb), to Crowley Ave, **N**...$General, Sav-A-Lot Food, **S**...Chevron/mart/wash, Conoco/diesel, Dairy Queen, Pizza Hut/Taco Bell, auto parts, repair

435   McCart St, **N**...Conoco/diesel, Fina/mart, **S**...Mobil, Texaco/mart/wash/24hr, bank

434   Trail Lake Dr, **N**...Phillips 66, **S**...DENTIST, Citgo, Exxon/mart, Mobil, Texaco, 7/11/24hr, Chinese Rest., Pancho's Mexican, bank, groceries

433   Hulen St, **N**...CHIROPRACTOR, DENTIST, MEDICAL CARE, Texaco/diesel/mart, Grady's Grill, Honeybaked Ham, Olive Garden, Red Lobster, Souper Salad, Tia's Mexican, TGIFriday, TownePlace Suites, Albertson's/drugs, Circuit City, Home Depot, Office Depot, Oschman's Sports, PetsMart, TJ Maxx, **S**...BBQ, Bennigan's, Chicken Express, Jack-in-the-Box, Jojo's Rest., McDonald's, TCBY, Hampton Inn, Linens'n Things, Pearle Vision, Service Merchandise, bank, cinema, mall

431   (432 from wb), TX 183, Bryant-Irvin Rd, **N**...Best Buy, CompUSA, Garden Ridge, Kohl's, Lowe's Bldg Supply, **S**...HOSPITAL, Chevron/mart/wash/24hr, Exxon, Texaco/mart, Blackeyed Pea, IHOP, Outback Steaks, SaltGrass Steaks, Sam's Rest., Subway, AmeriSuites, Holiday Inn Express, La Quinta, Buick, Ford, Goodyear/auto, Mazda, Saturn, Suzuki, Target, bank, cinema

430mm   Clear Fork Trinity River

429b   Winscott Rd, no facilities

     a   US 377, to Granbury, **N**...Phillips 66, Cracker Barrel, **S**...CHIROPRACTOR, VETERINARIAN, Chevron/mart, Exxon/mart/wash, RaceTrac/mart/24hr, Shamrock, Texaco/mart, Dairy Queen, McDonald's, Waffle House, Whataburger/24hr, Albertson's, KwikKar Oil/Lube

428   I-820, N around Ft Worth, no facilities

426   RM 2871, Chapin School Rd, no facilities

425   Markum Ranch Rd, no facilities

421   I-30 E(from eb), to Ft Worth, no facilities

420   FM 1187, Aledo, Farmer, parking, **S**...Cowtown RV Camping

419mm   weigh sta eb

418   Ranch House Rd, Willow Park, **N**...Exxon/Popeye's/mart/24hr, Shell/diesel/mart/24hr, BBQ, Subway, Pizza Hut, Brookshire Foods, bank, **S**...Texaco/ChickenExpress/diesel/mart, McDonald's, Ramada Ltd, Cowtown RV Camping(1mi), Winn-Dixie, cleaners

E

W

Dallas

Ft Worth

Ft Worth

Texas Interstate 20

417mm	weigh sta wb
415	FM 5, Mikus Rd, Annetta, **S**...DENTIST, Citgo/diesel/rest./mart, Texaco/mart, laundry
413	(414 from wb), US 180 W, Lake Shore Dr, **N**...Phillips 66/mart, RaceTrac/mart, Texaco/diesel/mart, Dairy Queen, Sonic, Chevrolet/Buick, Ford, Jeep, Lincoln/Mercury, Nissan, Olds/ Pontiac/GMC, Toyota, **S**...Chevron/diesel
410	Bankhead Hwy, **S**...Love's/Baskin-Robbins/Subway/diesel/mart/ 24hr
409	FM 2552 N, Clear Lake Rd, **N**...HOSPITAL, Petro/ Mobil/diesel/mart/rest./24hr, Catfish O'harlie's, Jack- in-the-Box, Best Western, Bed Bath Motel, Blue Beacon, to Weatherford Coll, **S**...Chevron/diesel/mart
408	TX 171, FM 1884, FM 51, Tin Top Rd, Weatherford, **N**...Exxon/Popeye's/mart/24hr, Applebee's, BBQ, Braum's, Denny's, Grandy's, Golden Corral, McDonald's, Montana Rest., Schlotsky's, Subway, Taco Bell, Taco Bueno, Whataburger/24hr, AutoZone, Home Depot, OfficeMax, Pennzoil, Wal-Mart SuperCtr/gas/24hr, bank, **S**...Shell/ diesel/mart/24hr, Texaco/Burger King/diesel/mart/24hr, Waffle House, Hampton Inn, Holiday Inn Express, Super 8
407	Tin Top Rd(from eb), same as 408
406	Old Dennis Rd, **N**...Conoco/diesel/24hr, Day's Inn, repair, **S**...Pilot/Wendy's/diesel/mart/ 24hr
402	(403 from wb), TX 312, to Weatherford, **S**...RV camping
397	FM 1189, to Brock, **N**...Shamrock/diesel/deli/mart, **S**...tires
394	FM 113, to Millsap, **N**...Fina/diesel/mart/RV camping, phone
393mm	Brazos River
391	Gilbert Pit Rd, no facilities
390mm	**rest area both lanes, full(handicapped)facilities, phone, vending, picnic tables, litter barrels, petwalk**
386	US 281, to Mineral Wells, **N**...Chevron/diesel/rest./24hr, Fina/mart
380	FM 4, Santo, **S**...Windmill Acres RV Park/Roadrunner Rest./gas
376	Blue Flat Rd, Panama Rd, no facilities
373	TX 193, Gordon, **N**...Chevron/diesel/mart/BBQ
370	TX 108 S, FM 919, Gordon, **N**...Citgo/Bar-B/diesel/rest., Texaco/mart, repair, **S**...Citgo/diesel/mart, Longhorn Inn/ RV Park
367	TX 108 N, Mingus, **N**...Thurber Sta/gas, Smoke Stack Café, antiques, **S**...NY Hill Rest., gunshop, museum
364mm	Palo Pinto Creek
363	Tudor Rd, picnic area, tables, litter barrels
362mm	Bear Creek, picnic area both lanes, tables, litter barrels
361	TX 16, to Strawn, no facilities
358	(from wb), frontage rd, no facilities
356mm	Russell Creek
354	Lp 254, Ranger, **2 mi N**...Lillie's Café, Relax Inn, antiques
351	(352 from wb), College Blvd, no facilities
349	FM 2461, Ranger, **N**...Conoco(2mi), Love's/Subway/Pizza Hut/diesel/mart/24hr, Dairy Queen, Day's Inn, Relax Inn(2mi), **S**...Chevron/diesel/mart, repair
347	FM 3363(from wb), Olden, **S**...TX Cattle Exchange Steaks
345	FM 3363(from eb), Olden, no facilities
343	TX 112, FM 570, Eastland, Lake Leon, **N**...DENTIST, Conoco/diesel, Fina/Subway/mart, Texaco, Asia Rest., BBQ, Dairy Queen, Jordan's Steaks, Ken's Diner, McDonald's, Pizza Inn(1mi), Rip's Diner, Sonic, Taco Bell/TCBY, Super 8/RV park, AutoZone, $General, Ford/Mercury, Olds/Buick/Chevrolet/Cadillac/GMC/Pontiac, Wal-Mart/ drugs, auto repair, hardware, **S**...Exxon/Burger King/diesel/mart/24hr, Kettle, Pulido's Mexican, Econolodge, Ramada Inn, Chrysler/Plymouth/Dodge/Jeep, antiques
340	TX 6, Eastland, **N**...HOSPITAL, Chevron/diesel/service, **S**...Texaco/diesel/mart
337	spur 490, no facilities
332	US 183, Cisco, **N**...Citgo/diesel/mart, BBQ, Cisco Café, Dairy Queen, Pizza Heaven, Sonic, Oak Motel, carwash, RV camping, truck repair, **S**...Fina/diesel/mart, Ford, LP
330	TX 206, Cisco, **N**...HOSPITAL, Chevron/diesel/mart/White Elephant Rest./24hr, Best Western, RV camping
329mm	picnic area wb, tables, litter barrels, handicapped accessible
327mm	picnic area eb, tables, litter barrels, handicapped accessible
324	Scranton Rd, no facilities

Texas Interstate 20

322	Cooper Creek Rd, no facilities
320	FM 880 N, FM 2945 N, to Moran, no facilities
319	FM 880 S, Putnam, **N...**gas/diesel/café, USPO
316	Brushy Creek Rd, no facilities
313	FM 2228, no facilities
310	Finley Rd, no facilities
308	Lp 20, Baird, **1 mi S...**antiques
307	US 283, Clyde, **N...**Dairy Queen, **S...**Conoco/diesel/mart, Fina/Allsup's/mart, Robertson's Café, Baird Motel/RV park
306	FM 2047, Baird, **N...**Chevrolet/Pontiac/GMC, Chrysler/Plymouth/Dodge/Jeep
303	Union Hill Rd, **S...**Union Hill Steaks, golf
301	FM 604, Cherry Lane, **N...**Chevron/diesel/mart/24hr, Whataburger/24hr, **S...**Fina/diesel/café, Texaco/diesel/mart/24hr, Dairy Queen/24hr, Franklin RV Ctr, IGA Food/drug, bank, carwash
300	FM 604 N, Clyde, **N...**U-Haul, auto parts/repair, **S...**White's RV Park
299	FM 1707, Hays Rd, no facilities
297	FM 603, Eula Rd, no facilities
296.5mm	**rest area both lanes, full(handicapped)facilities, phone, picnic tables, litter barrels, petwalk**
294	Buck Creek Rd, no facilities
292b	Elmdale Rd, **S...**Gym Diner, flea mkt, RV camping
a	Lp 20(exits left from wb), **S...**Travelodge
290	TX 36, Lp 322, **S...**airport, zoo
288	TX 351, **N...**Chevron/mart, Fina/Subway/diesel/mart, Shamrock/Allsup's/diesel/mart, Dairy Queen, Skillet's Café, Best Inn, Comfort Inn, Day's Inn, Executive Inn, **S...**HOSPITAL, Super 8
286c	FM 600, Abilene, **N...**Exxon/diesel/mart, Fina/diesel/mart, Denny's/24hr, La Quinta, **S...**Chevron/diesel/mart
286	US 83, Pine St, Abilene, **N...**Firestone, **S...**HOSPITAL, Fina, Shamrock/mart, Budget Host
285	Old Anson Rd, **N...**Texaco/mart, Travel Inn, carwash, **S...**Texaco/diesel/mart, Econolodge, Quality Inn(2mi)
283	US 277 S, US 83(exits left from wb), **2 mi S...**IHOP, Clarion, Courtyard, Quality Inn, Royal Inn
282	FM 3438, Shirley Rd, **S...**Motel 6, KOA
280	Fulwiler Rd, no facilities
279	US 84 E, to Abilene, **1-3 mi S...**access to facilities
278	Lp 20, **N...**Conoco/diesel/rest./24hr, GMC, diesel repair, **S...**Conoco/diesel
277	FM 707, Tye, **N...**Flying J/pizza/diesel/LP/mart/24hr, RV camping, **S...**Texaco/diesel, Fina/mart
274	Wells Lane, no facilities
272	Wimberly Rd, **1 mi N...**truck/trailer repair
270	FM 1235, Merkel, **N...**Shamrock/diesel/mart, Texaco/diesel/café/24hr
269	FM 126, **N...**Mexican Rest., Subway, Scottish Inn, Chevrolet/GMC, **S...**Fina/diesel/rest./24hr, Texaco/repair, Dairy Queen, Merkel Motel/Rest., Badger Lube
267	Lp 20, Merkel, **1mi S...**access to gas, food, lodging
266	Derstine Rd, no facilities
264	Noodle Dome Rd, no facilities
263	Lp 20, Trent, no facilities
262	FM 1085, **S...**Fina/diesel/mart/24hr
261	Lp 20, Trent, no facilities
259	Sylvester Rd, no facilities
258	White Flat Rd, oil wells
257mm	**rest area both lanes, full(handicapped)facilities, phone, vending, picnic tables, litter barrels, petwalk**
256	Stink Creek Rd, no facilities
255	Adrian Rd, no facilities
251	Eskota Rd, no facilities
249	FM 1856, no facilities
247	TX 70 N, Sweetwater, **2 mi N...**Chinese Rest., Dairy Queen, Pizza Hut, Sonic, Subway, Whataburger, Long Horn Motel
246	Alabama Ave, Sweetwater, no facilities
245	Arizona Ave(from wb), same as 244
244	TX 70 S, Sweetwater, **N...**HOSPITAL, Fina/diesel/mart/24hr, Shamrock/diesel, Texaco/diesel/mart, Dairy Queen, Kettle, McDonald's, Holiday Inn/rest., Motel 6, AutoZone, Ford/Mercury, **S...**Chevron/mart, Fina/mart, Texaco/mart, Gill's Fried Chicken, Golden Chick, Jack's Steaks/BBQ, Schlotsky's, Taco Bell/TCBY, Best Western/rest., Comfort Inn, Ramada Inn, Ranch House Motel, Chevrolet/Pontiac/Buick/Cadillac/Olds, IGA Foods, K-Mart/drugs
243	Hillsdale Rd, Robert Lee St, **N...**Bewley's RV parts/service
242	Hopkins Rd, **S...**TA/Conoco/Pizza Hut/diesel/24hr, Goodyear, Rolling Plains RV Park, lube
241	Lp 20, Sweetwater, **N...**gas, food, lodging, **S...**RV camping
240	Lp 170, **N...**airport, camping, **1 mi S...**auto parts/repair
239	May Rd, no facilities

Abilene

E ↕ W

Texas Interstate 20

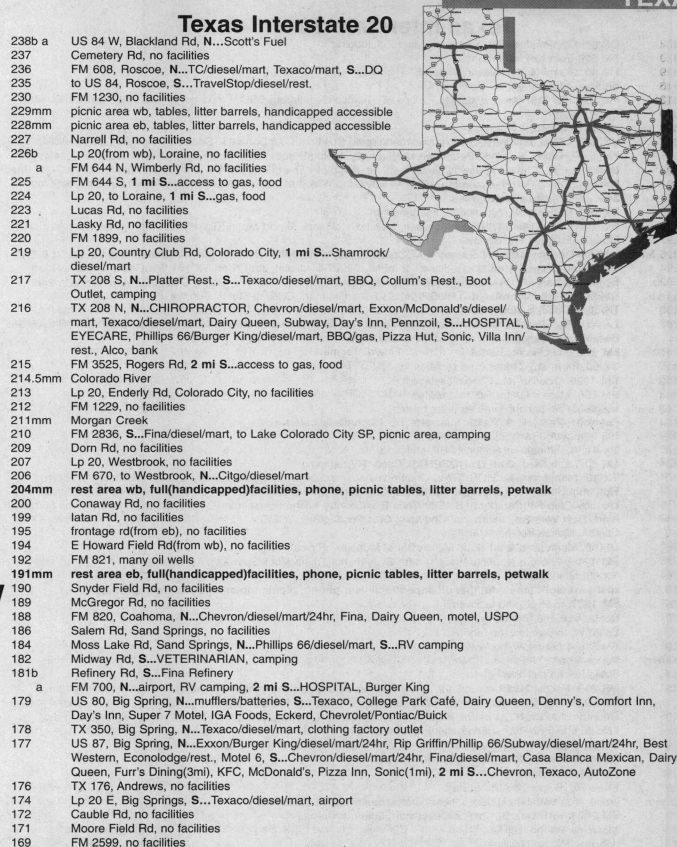

238b a	US 84 W, Blackland Rd, **N**...Scott's Fuel
237	Cemetery Rd, no facilities
236	FM 608, Roscoe, **N**...TC/diesel/mart, Texaco/mart, **S**...DQ
235	to US 84, Roscoe, **S**...TravelStop/diesel/rest.
230	FM 1230, no facilities
229mm	picnic area wb, tables, litter barrels, handicapped accessible
228mm	picnic area eb, tables, litter barrels, handicapped accessible
227	Narrell Rd, no facilities
226b	Lp 20(from wb), Loraine, no facilities
a	FM 644 N, Wimberly Rd, no facilities
225	FM 644 S, **1 mi S**...access to gas, food
224	Lp 20, to Loraine, **1 mi S**...gas, food
223	Lucas Rd, no facilities
221	Lasky Rd, no facilities
220	FM 1899, no facilities
219	Lp 20, Country Club Rd, Colorado City, **1 mi S**...Shamrock/diesel/mart
217	TX 208 S, **N**...Platter Rest., **S**...Texaco/diesel/mart, BBQ, Collum's Rest., Boot Outlet, camping
216	TX 208 N, **N**...CHIROPRACTOR, Chevron/diesel/mart, Exxon/McDonald's/diesel/mart, Texaco/diesel/mart, Dairy Queen, Subway, Day's Inn, Pennzoil, **S**...HOSPITAL, EYECARE, Phillips 66/Burger King/diesel/mart, BBQ/gas, Pizza Hut, Sonic, Villa Inn/rest., Alco, bank
215	FM 3525, Rogers Rd, **2 mi S**...access to gas, food
214.5mm	Colorado River
213	Lp 20, Enderly Rd, Colorado City, no facilities
212	FM 1229, no facilities
211mm	Morgan Creek
210	FM 2836, **S**...Fina/diesel/mart, to Lake Colorado City SP, picnic area, camping
209	Dorn Rd, no facilities
207	Lp 20, Westbrook, no facilities
206	FM 670, to Westbrook, **N**...Citgo/diesel/mart
204mm	**rest area wb, full(handicapped)facilities, phone, picnic tables, litter barrels, petwalk**
200	Conaway Rd, no facilities
199	Iatan Rd, no facilities
195	frontage rd(from eb), no facilities
194	E Howard Field Rd(from wb), no facilities
192	FM 821, many oil wells
191mm	**rest area eb, full(handicapped)facilities, phone, picnic tables, litter barrels, petwalk**
190	Snyder Field Rd, no facilities
189	McGregor Rd, no facilities
188	FM 820, Coahoma, **N**...Chevron/diesel/mart/24hr, Fina, Dairy Queen, motel, USPO
186	Salem Rd, Sand Springs, no facilities
184	Moss Lake Rd, Sand Springs, **N**...Phillips 66/diesel/mart, **S**...RV camping
182	Midway Rd, **S**...VETERINARIAN, camping
181b	Refinery Rd, **S**...Fina Refinery
a	FM 700, **N**...airport, RV camping, **2 mi S**...HOSPITAL, Burger King
179	US 80, Big Spring, **N**...mufflers/batteries, **S**...Texaco, College Park Café, Dairy Queen, Denny's, Comfort Inn, Day's Inn, Super 7 Motel, IGA Foods, Eckerd, Chevrolet/Pontiac/Buick
178	TX 350, Big Spring, **N**...Texaco/diesel/mart, clothing factory outlet
177	US 87, Big Spring, **N**...Exxon/Burger King/diesel/mart/24hr, Rip Griffin/Phillip 66/Subway/diesel/mart/24hr, Best Western, Econolodge/rest., Motel 6, **S**...Chevron/diesel/mart/24hr, Fina/diesel/mart, Casa Blanca Mexican, Dairy Queen, Furr's Dining(3mi), KFC, McDonald's, Pizza Inn, Sonic(1mi), **2 mi S**...Chevron, Texaco, AutoZone
176	TX 176, Andrews, no facilities
174	Lp 20 E, Big Springs, **S**...Texaco/diesel/mart, airport
172	Cauble Rd, no facilities
171	Moore Field Rd, no facilities
169	FM 2599, no facilities
168mm	picnic area both lanes, tables, littter barrels
165	FM 818, no facilities
158	Lp 214 W, to Stanton, RV camping
156	TX 137, Lamesa, **S**...Fina/diesel, Shamrock/Subway/diesel/mart/24hr, Dairy Queen

E

W

Texas Interstate 20

M i d l a n d   O d e s s a

154	US 80, Stanton, **2 mi S**...access to gas, food, lodging
151	FM 829(from wb), no facilities
144	US 80, **2-3 mi N**...facilities in Midland
143mm	frontage rd(from eb), no facilities
138	TX 158, FM 715, Greenwood, **N**...Phillips 66, Texaco/diesel/mart, BBQ, Whataburger, **S**...Fina/Taco Bell/diesel/mart, tires
137	Old Lamesa Rd, no facilities
136	TX 349, Midland, **N**...Phillips 66/diesel/mart/repair, Shell/mart, McDonald's, Sonic, Comfort Inn, Super 8/café, Fiesta Food/fuel, Family$, Petroleum Museum, **S**...Exxon/Burger King/mart, Fina/mart/24hr, Texaco/diesel/repair
134	Midkiff Rd, **N**...Chevron/Subway/diesel/mart, Texaco/diesel/mart, Bo's RV Ctr, **1 mi N on Wall St**...Fina/7-11, Shell, Denny's, Day's Inn, Hampton Inn, Motel 6, Royal Inn, Sleep Inn, Chevrolet, Chrysler/Plymouth/Jeep, Kia, Lincoln/Mercury, Mercedes/Volvo, Nissan
131	TX 158, Midland, **N**...Travelodge, RV camping
126	FM 1788, **N**...Chevron/diesel/mart/repair, Mobil/diesel/mart, Airport Motel, Blakely RV Ctr, Confederate Air Force Museum, airport, RV camping
121	Lp 338, Odessa, **1/2mi N**...Denny's, McDonald's, Day's Inn, Hilton, Holiday Inn, La Quinta, Motel 6, Super 8, U of TX Permian Basin, **1 mi N**...Radisson, **2 mi N**...Fina/7-11/diesel, **3 mi N on TX 191 W**...Chili's, Logan's Roadhouse, McDonald's, Whataburger, Albertson's, Circuit City, Home Depot, Sam's Club, Wal-Mart SuperCtr/24hr
118	FM 3503, Grandview Ave, **N**...HOSPITAL, Shell/diesel/mart, Odessa Motel, Goodyear, Peterbilt, RV Ctr
116	US 385, Odessa, **N**...HOSPITAL, Shell/mart, TC/mart, Dairy Queen, Best Western, Delux Inn, Villa West Inn, Odessa Coll, Presidential Museum, truck repair, **1/2 mi N**...services between Odessa and Midland, **S**...Phillips 66/diesel/mart, Texaco/diesel/mart, Motel 6, city park
115	FM 1882, **N**...Fina/mart/deli
113	TX 302(from eb), Odessa, no facilities
112	FM 1936, Odessa, **N**...Citgo/diesel/mart
104	FM 866, Meteor Crater Rd, no facilities
103.5mm	weigh sta wb/parking area eb, litter barrels
101	FM 1601, Penwell, **S**...Texas Interstate Truckstop/diesel/café/service
92	(93 from wb), FM 1053, no facilities

M o n a h a n s

86	TX 41, **N**...Monahans Sandhills SP, camping
83	US 80, Monahans, **2 mi N**...HOSPITAL, Citgo, RV camping
80	TX 18, Monahans, **N**...HOSPITAL, Chevron/diesel/mart/24hr, Exxon/mart, Dairy Queen, McDonald's, Pizza Hut(1mi), Sonic, Alco Discount, CarQuest, Chrysler/Plymouth/Dodge/Jeep, Furr's Foods, Lowe's Food/drug, Country Club RV Park(2mi), **S**...Fina/Taco Bell/Country Kitchen/diesel/mart/24hr, Kent Gas/mart, Phillips 66/diesel/mart, Best Western, Texan Inn, Chevrolet/Olds/Pontiac/Buick/GMC
79	Lp 464, Monahans, no facilities
76	US 80, Monahans, **2 mi N**...to Million Barrel Museum, RV camping
73	FM 1219, Wickett, **N**...Fina/Allsup's/mart, **S**...National Trkstp/Mobil/Subway/diesel/mart/24hr
70	TX 65, no facilities
69.5mm	**rest area both lanes, full(handicapped)facilities, phone, picnic tables, litter barrels, petwalk**
66	FM 1927, to Pyote, no facilities
58	frontage rd, multiple oil wells
52	Lp 20 W, to Barstow, no facilities
49	FM 516, to Barstow, no facilities
48mm	Pecos River
44	Collie Rd, no facilities

P e c o s

42	US 285, Pecos, **N**...Chevron/24hr, Exxon, Flying J/Conoco/diesel/mart/24hr, GC/gas, Shamrock/diesel(2mi), Texaco/diesel/mart, McDonald's, Pizza Hut, Super 8, Quality Inn, Wal-Mart/drugs, civic ctr, repair, laundry, museum
40	Country Club Dr, **N**...st patrol, **S**...Chevron/Subway/mart, Dairy Queen, Best Western/rest., Chevrolet, Ford/Lincoln/Mercury, RV camping, municipal park
39	TX 17, Pecos, **N**...Goodyear, st patrol, **S**...Dairy Queen, Best Western, Chevrolet/Buick
37	Lp 20 E, **2 mi N**...KFC, Town&Country Motel
33	FM 869, no facilities
29	Shaw Rd, **S**...to TX AM Ag Sta
25mm	picnic area both lanes, tables, litter barrels, handicapped accessible
22	FM 2903, to Toyah, **S**...Texaco/diesel/mart, auto/truck repair
13	McAlpine Rd, no facilities
7	Johnson Rd, no facilities
3	Stocks Rd, no facilities
0mm	I-20 begins/ends on I-10, 187mm.

E

W

Texas Interstate 27

Exit #   Services

I-27 begins/ends on I-40, exit 70 in Amarillo.

123b   I-40, W to Albuquerque, E to OK City

a   26th Ave, **E**...Conoco/mart

122b   34th Ave, Tyler St, no facilities

a   FM 1541, Washington St, Parker St, Moss Lane, **E**...Shamrock/
mart, Texaco/mart/atm, Sonic, bank, **W**...Texaco/mart, Taco Bell

121b   Hawthorne Dr, Austin St, **E**...Amarillo Motel, Honda Motorcycles,
W...Texaco/diesel/mart/wash, Scottie's Transmis-
sions

a   Georgia St, **E**...Phillips 66/diesel/mart, BBQ, Honda,
Mazda, Nissan, Pontiac/GMC, Sizemore RV, Toyota,
W...DENTIST, Texaco, Traveler Motel, Family$, Ford/
Kenworth

120b   45th Ave, **E**...Fina/mart, Waffle House, repair,
W...VETERINARIAN, Shamrock/mart, Hardee's, Luby's, Mr
Burger, BMW, Buick/Olds, High Plains Transmissions, Isuzu,
Lincoln/Mercury, Land Rover, Subaru, carwash, donuts

a   Republic Ave, no facilities

119b   Western St, 58th Ave, **E**...$General, HomeLand Foods, transmissions,
W...CHIROPRACTOR, Texaco/diesel/mart/U-Haul, Pizza Hut, Western Bowl Rest., RV
Ctr, transmissions

a   W Hillside, no facilities

117   Bell St, Arden Rd, **E**...Camper Roundup RV, transmissions, **W**...Fina/mart, Phillips 66, Texaco/mart/24hr, LJ Silver,
Sonic

116   Lp 335, Hollywood Rd, **E**...Love's/diesel/mart/24hr, McDonald's, Waffle House, Day's Inn, **W**...HOSPITAL(8mi),
cinema, fireworks

115   Sundown Lane, no facilities

113   McCormick Rd, **E**...Ford/Suzuki, **W**...Texaco, RV Ctr, RV camping

112   FM 2219, no facilities

111   Rockwell Rd, **E**...Chevrolet, farmers mkt, **W**...Buick/GMC/Olds/Pontiac

110   US 87 S, US 60 W, Canyon, **W**...McDonald's(2mi), Buick/GMC/Olds/Pontiac, WTSU

109   Buffalo Stadium Rd, **W**...stadium

108   FM 3331, Hunsley Rd, no facilities

106   TX 217, Canyon, to Palo Duro Cyn SP, **E**...Palo Duro RV Park, RV camping, **W**...Holiday Inn Express(2mi), Plains
Museum

103   FM 1541 N, Cemetery Rd, no facilities

99   Hungate Rd, no facilities

97.5mm   parking area both lanes, litter barrels

96   Dowlen Rd, no facilities

94   FM 285, to Wayside, no facilities

92   Haley Rd, no facilities

90   FM 1075, Happy, **W**...Phillips 66/diesel/mart, Texaco, Imperial Café, auto repair

88b a   US 87 N, FM 1881, Happy, same as 90

83   FM 2698, no facilities

82   FM 214, no facilities

77   US 87, Tulia, **1 mi E**...HOSPITAL, Pizza Hut, gas, lodging, airport

75   NW 6th St, Tulia, **1 mi E**...Dairy Queen, Best Western, **W**...same as 74

74   TX 86, Tulia, **E**...HOSPITAL, Best Western(1mi), Lasso Motel, **W**...Rip Griffin/Texaco/Grandy's/Subway/diesel/mart/
atm/24hr, Select Inn

70mm   parking area both lanes, litter barrels

68   FM 928, no facilities

63   FM 145, Kress, **1 mi E**...gas/diesel, food, phone

61   US 87, County Rd, no facilities

56   FM 788, no facilities

54   FM 3183, to Plainview, **W**...truck service

53   Lp 27, Plainview, **E**...HOSPITAL, access to gas, food, lodging

51   Quincy St, no facilities

50   TX 194, Plainview, **E**...HOSPITAL, Phillips 66/mart, Subway, Wal-Mart Dist, to Wayland Bapt U

N

S

Texas Interstate 27

<table>
<tr><td>P
l
a
i
n
v
i
e
w</td><td>

49 US 70, Plainview, **E**...CHIROPRACTOR, EYECARE, VETERINARIAN, Chevron, Conoco/diesel, Fina/mart, Phillips 66/mart, Shamrock/diesel/mart, Texaco/Subway/diesel/mart, Carlito's Mexican, Domino's, FarEast Chinese, Furr's Café, Kettle, LJ Silver, Pizza Hut, Sonic, Taco Bell, Best Western, Day's Inn, Beall's, Buick/Pontiac/Olds/Cadillac/GMC, Checker Parts, Chrysler/Jeep, Ford/Lincoln/Mercury/Toyota, GNC, Hastings Books, NAPA, Pinnell's Drug, Radio Shack, Revco, bank, **W**...VETERINARIAN, Chevron/mart, Conoco, Shell/diesel/mart, Burger King, Chicken Express, Little Mexico, McDonald's, Mr Gatti's, Sonic, Subway, Taco Villa, Plainview Motel, Cato's, JC Penney, K-Mart/Little Caesar's, QState/lube/wash, Staples, Wal-Mart

45 Lp 27, to Plainview, no facilities

43 FM 2337, no facilities

41 County Rd, no facilities

38 Main St, no facilities

37 FM 1914, Cleveland St, **E**...Dairy Queen, **W**...HOSPITAL

36 FM 1424, Hale Center, no facilities

32 FM 37 W, no facilities

31 FM 37 E, no facilities

29mm **rest area both lanes, full(handicapped)facilities, phone, picnic tables, litter barrels, petwalk**

27 County Rd, no facilities

24 FM 54, **W**...RV park/dump

22 Lp 369, Abernathy, **E**...Phillips 66, **W**...Phillips 66/mart, Dairy Queen

21 FM 597, Main St, Abernathy, **W**...Conoco/mart, Coop/diesel, Cyclone Rest., Dairy Queen, repair

20 Abernathy(from nb), no facilities

17 County Rd 53, no facilities

15 Lp 461, to New Deal, **1/2mi E**...Texaco/Subway/diesel/mart/24hr, phone

14 FM 1729, same as 15

13 Lp 461, to New Deal, no facilities

12 access rd(from nb), no facilities

11 FM 1294, Shallowater, no facilities

10 Keuka St, **E**...Fed Ex

9 Airport Rd, **E**...airport, **W**...camping

8 Regis St, **W**...Lubbock RV Park

7 Yucca Lane, **W**...Country Creek Motel, Texas Star Inn/rest.

6b a Lp 289, Ave Q, Lubbock, **E**...Pharr RV, **W**...Texas Motel/rest.

5 Ave H, Municipal Dr, **E**...Mackenzie SP

4 US 82, US 87, 4th St, to Crosbyton, **E**...fairgrounds, **W**...to TTU

3 US 62, TX 114, 19th St, **W**...Stubb's BBQ, Cadillac

2 34th St, **E**...Fina, Phillips 66, Budget Motel, **W**...MEDICAL CARE, Burgers/fries

1c 50th St, **E**...El Jalapeno Café, Jose's Mexican, auto/truck repair, **W**...MEDICAL CARE, Fina/7-11, Shamrock, Texaco/XpressLube/diesel, Bryan's Steaks, Burger King, ChinaStar Buffet, Church's Chicken, Dairy Queen, Fins&Hens Rest., KFC, LJ Silver, McDonald's, Pizza Hut/Taco Bell, Subway, Taco Villa, Whataburger, Wienerschnitzel, Chrysler/Plymouth, Howard Johnson, Saver's Foods, ThriftyLube/repair, United Food/drug/gas, USPO, Walgreen, Woody Tires, carwash

</td></tr>
</table>

b US 84, **E**...Circus Inn, Day's Inn, Super 8, **W**...Skillet's Burgers, Motel 6, Ramada Inn, USA Truckwash

a Lp 289, no facilities

0mm 82nd St, **W**...Conoco/diesel/mart, Phillips 66/mart

I-27 begins/ends on 82nd St in S Lubbock.

(left margin: Lubbock)

Texas Interstate 30

Exit # Services

223mm Texas/Arkansas state line

223b a US 59, US 71, State Line Ave, Texarkana, **N**...Shell/mart, Texaco/repair, Denny's, IHOP, Red Lobster, Waffle House, Best Western, Baymont Inn, Holiday Inn, Hampton Inn, Holiday Inn Express, Sheraton, Shoney's Inn/rest., Super 8, Pennzoil, **S**...CHIROPRACTOR, Conoco/diesel/mart, Exxon/mart/wash, Mobil, Texaco, Total/mart, Arby's, Backyard Burger, Baskin-Robbins, Bennigan's, Burger King, Cattleman's Steaks, CiCi's, El Chico, Kettle/24hr, KFC, La Carreta Mexican, Little Caesar's/TCBY, LJ Silver, Mandarin House, McDonald's, Mr Gatti's, Pizza Hut, Pizza Inn, Schlotsky's, Subway, Taco Bell, Taco Tico, Wendy's, Western Sizzlin, Whataburger/24hr, Best Western, Comfort Inn, Day's Inn, Econolodge, Economy Inn, La Quinta, Motel 6, Villager Lodge, Albertson's/drugs, AutoZone, Chevrolet/Buick, Chrysler/Jeep, Harvest Foods, K-Mart, Midas Muffler, 1-Hr Photo, Wal-Mart/auto, bank, cleaners/laundry

223mm **Welcome Ctr wb, full(handicapped)facilities, info, phone, picnic tables, litter barrels, vending, petwalk**

222 TX 93, FM 1397, Summerhill Rd, **N**...MEDICAL CARE, Shell, Total/Subway/mart, Applebee's, CJ's Rest., McDonald's, Waffle House, Motel 6, Ford/Freightliner, Goodyear/auto, Honda, RV Ctr, bank, **S**...DENTIST, Chevron, Exxon, Texaco/mart/wash/24hr, Bryce's Rest., Ramada, Mazda/Isuzu/Olds, bank

Texas Interstate 30

E

W

220b   FM 559, Richmond Rd, **N...**EYECARE, Exxon/mart/wash, Texaco/ Popeye's/diesel, BBQ, Burger King, Carino's Italian, Cracker Barrel, Dairy Queen, Pizza Hut, Taco Bell, Tamolly's Mexican, Texas Roadhouse, Sam's Club, Super 1/drugs, cinema, cleaners, **S...**DENTIST, Chevron, Total/diesel/mart, Arby's, Baskin-Robbins, Chili's, El Chico's, Grandy's, Luby's McDonald's, Outback Steaks, Subway, AutoZone, Albertson's/drugs, Books A Million, Dillard's, JC Penney, Office Depot, Pier 1, Sears/auto, Target, ToysRUs, bank, mall

220a   US 59 S, Texarkana, **S...**Exxon/Wendy's/mart, Texaco, Arby's, Dairy Queen, Lowe's Bldg Supply, Sam's Club, Sears, Wal-Mart SuperCtr/gas/24hr, bank, mall

218   FM 989, Nash, **N...**Citgo/diesel/mart, cinema, **S...**Exxon/ Burger King/diesel/mart, White/GMC/Volvo, to Lake Patman

213   FM 2253, Leary, no facilities

212   spur 74, Lone Star Army Ammo Plant, **S...**Texaco/mart

208   FM 560, Hooks, **N...**Conoco/diesel/mart/24hr, Texaco/diesel/mart/24hr, Main St Deli, Hooks Camper Ctr, **S...**Total/diesel/mart/24hr, Dairy Queen, Family$, auto repair, laundry, tires

206   TX 86, **S...**Red River Army Depot

201   TX 8, New Boston, **N...**Texaco/diesel/mart, Total/diesel/mart, Pitt Grill, Tex Inn, Chevrolet, Chrysler/Dodge, **S...**HOSPITAL, Chevron/diesel/mart, Exxon/Burger King/mart, EZ/mart/24hr, Phillips 66/diesel/ mart, Brookshire's Foods, Catfish King, Church's Chicken, Dairy Queen, Maxi's Drive-In, McDonald's, Pizza Hut, Taco Delite, TCBY, Best Western, Bostonian Inn, Eckerd, Ford/Mercury, Wal-Mart/drugs, bank, RV Park

199   US 82, New Boston, **N...**Texaco/mart

198   TX 98, **N...**Texaco

193mm   Anderson Creek

192   FM 990, **N...**FuelStop/Pitt Grill/diesel

191mm   rest area both lanes, full(handicapped)facilities, phone, picnic tables, litter barrels, vending, petwalk

186   FM 561, no facilities

181mm   Sulphur River

178   US 259, to DeKalb, Omaha, no facilities

174mm   White Oak Creek

170   FM 1993, **N...**Mobil/mart

165   FM 1001, **S...**Chevron/diesel/mart

162   US 271, FM 1402, FM 2152, Mt Pleasant, **N...**Chevron/mart/24hr, BBQ, Super 8, **S...**HOSPITAL, Exxon, Shell/ diesel/mart, Texaco/diesel/mart, Total/Subway/diesel/mart, Burger King, Dairy Queen, McDonald's, Best Western/ rest., Chevrolet/Olds/Cadillac

160   US 67, US 271, TX 49, Mt Pleasant, **N...**Senorita's Mexican, Buick/Pontiac/GMC(1mi), **S...**Exxon/diesel/mart, Texaco/mart, El Chico, Elmwood Café, Sally's Café, Schlotsky's, Western Sizzlin, Comfort Inn, Day's Inn, Holiday Inn/rest., Sands Motel, hardware

157.5mm   weigh sta both lanes

156   frontage rd, no facilities

153   spur 185, to Winfield, Miller's Cove, **N...**Chevron/mart, Citgo/diesel/café, diesel repair, laundry, RV camping, **S...**Citgo/mart

150   Ripley Rd, **N...**Lowe's Distribution

147   spur 423, **N...**Love's/Subway/Taco Bell/diesel/mart/24hr, Super 8, **S...**LP

146   TX 37, Mt Vernon, **N...**HOSPITAL, Exxon/mart, Super 8, **S...**Chevron/diesel/mart/24hr, Fina/diesel/mart/24hr, BarnStormer's Café, BBQ, Burger King, Dairy Queen, Cypress Glen Rest./Motel, Grandma's Fried Chicken, Peppertree Mexican, Sonic, Pennzoil, bank, hardware, tires, to Lake Bob Sandlin SP

144mm   rest area both lanes, full(handicapped)facilities, phone, picnic tables, litter barrels, vending, petwalk

142   County Line Rd, no facilities

141   FM 900, Saltillo Rd, **N...**Exxon/diesel/mart

136   FM 269, Weaver Rd, **N...**Phillips 66/mart

135   US 67 N, no facilities

131   FM 69, no facilities

127   US 67, Lp 301, **N...**Kettle, Budget Inn, Comfort Inn, Holiday Inn, **S...**Chevron/mart/lube, Texaco/mart, Burton's Rest./24hr, Best Western

126   FM 1870, College St, **S...**CHIROPRACTOR, Chevron/diesel/rest., Exxon, Texaco/mart/24hr, Burton's Rest./24hr, Best Western, Firestone/auto

Texas Interstate 30

**G
r
e
e
n
v
i
l
l
e**

125   Frontage Rd, same as 124

124   TX 11, TX 154, Sulphur Springs, **N**...HOSPITAL, Exxon, Phillips 66/deli/24hr, Shamrock, Total/mart, Catfish King Rest., Chinese Rest., Domino's, KFC, Pitt Grill, Pizza Hut, Sonic, Subway, Tamolly's Mexican, Western Sizzlin, Royal Inn, AutoZone, Chevrolet/Buick/GMC, Chrysler/Jeep, Ford, Kroger/24hr, Nissan, USPO, VF Outlet/famous brands, bank, **S**...CHIROPRACTOR, DENTIST, Exxon/diesel/mart, Texaco/diesel/mart, BBQ, Braum's, Burger King, China House, Furr's Rest., Grandy's/Baskin-Robbins, Jack-in-the-Box, K-Bob's Steaks, Ken's Pizza, McDonald's, Pizza Inn, San Remo's Italian, Taco Bell, Wendy's, Holiday King Motel, Eckerd, Pennzoil, Wal-Mart/auto, Winn-Dixie, tires, factory outlet

123   FM 2297, League St, **N**...Chevron/diesel/mart, Texaco/mart

122   TX 19, to Emory, **N**...HOSPITAL, Chevron/mart, **S**...VETERINARIAN, Phillips 66/diesel/rest./24hr, Blimpie(2mi), Shady Lakes RV Park, to Cooper Lake

120   US 67 bus, no facilities

116   US 67, FM 2653, Brashear Rd, no facilities

110   FM 275, Cumby, **N**...Phillips 66/mart, **S**...Texaco/mart/24hr

104   FM 513, FM 2649, Campbell, **S**...to Lake Tawakoni

101   TX 24, TX 50, FM 1737, to Commerce, **N**...Phillips 66/diesel, to E TX St U

97   Lamar St, **N**...Exxon/diesel/mart

96   US 67, Lp 302, **S**...HOSPITAL

95   Division St, **S**...HOSPITAL, Puddin Hill Fruit Cakes

94b a US 69, US 380, Greenville, **N**...Texaco/mart, Total/diesel/mart, Kettle, American Inn, Best Western, Day's Inn, Royal Inn, bank, **S**...Chevron/mart/wash/24hr, Exxon, Fina, Arby's, Burger King, Catfish King Rest., McDonald's, Comfort Inn, Economy Inn, Gold Key Inn, Motel 6, Super 8, Chrysler/Eagle/Dodge

93b a US 67, TX 34 N, Greenville, **N**...HOSPITAL, CHIROPRACTOR, EYECARE, Mobil, Phillips 66, Texaco, Applebee's, Capt D's, CiCi's, Grandy's, IHOP, Jack-in-the-Box, KFC, LJ Silver, Royal Drive-In, Ryan's, Schlotsky's, Sonic, Subway, Taco Bell, Taco Bueno, Tamolly's Mexican, TCBY, Tony's Italian,.Wendy's, Whataburger/24hr, Ace Hardware, Belk, Big Lots, Brookshire's Foods, Buick/Pontiac, JC Penney, Olds/Cadillac, bank, cinema, cleaners, mall, 1-Hr Photo, **S**...Chevron/diesel/mart, Cracker Barrel, Denny's, Luby's, Red Lobster, Spare Rib Café, Ford/Lincoln/Mercury, Honda/Yamaha/Polaris, Radio Shack, RV Ctr, Wal-Mart SuperCtr/24hr

90mm Farber Creek

89   FM 1570, **S**...Meadowview Motel, Lake Country RV Park

89mm E Caddo Creek

87   FM 1903, **S**...Fina/Pancake House/diesel/mart, truck/tire repair

87mm Elm Creek

85   FM 36, Caddo Mills, **N**...RV camping, **S**...auto repair

85mm W Caddo Creek

83   FM 1565 N, no facilities

79   FM 2642, no facilities

77b   FM 35, Royse City, **N**...Exxon/diesel/mart, Knox/Subway/diesel/mart/24hr, BBQ

a   TX 548, **N**...Texaco/mart/wash, Jack-in-the-Box, **S**...Sun Royse Inn

73   FM 551, Fate, no facilities

70   FM 549, no facilities

69   (from wb), frontage rd

68   TX 205, to Rock Wall, **N**...RaceTrac, Shell, Texaco, Braum's, Dairy Queen, KFC, Pizza Hut, Pizza Inn, Whataburger, Holiday Inn Express, Super 8, Chevrolet, Ford/Mercury, HobbyLobby, **S**...VETERINARIAN, TA/diesel/rest./24hr, Total/mart

67b   FM 740, Ridge Rd, **N**...Chevron/mart/wash, Mobil/mart, Arby's, Burger King, Cajun Catfish, Grandy's, IHOP, McDonald's, Schlotsky's, Waffle House, Goodyear/auto, Kwik Kar Lube, Wal-Mart SuperCtr/24hr, cleaners, **S**...CHIROPRACTOR, DENTIST, VETERINARIAN, Chevron, Exxon/Pizza Inn/mart, Texaco/mart, Total/mart, Applebee's, Blackeyed Pea, Boston Mkt, Carino's Italian, Chili's, El Chico, Jack-in-the-Box, McDonald's, On-the-Border, Starbucks, Subway, Taco Bell, TCBY, Country Inn Suites, Albertson's, Beall's, Discount Tire, Eckerd, GNC, Home Depot, Kohl's, Kroger/drugs, Linens'n Things, Lowe's Bldg Supply, Michael's, OfficeMax, Old Navy, PetCo, Radio Shack, Ross, SteinMart, Target, Wolf Camera, bank, to Lake Tawakoni

a   Horizon Rd, Village Dr, no facilities

66mm Ray Hubbard Reservoir

64   Dalrock Rd, Rowlett, **N**...HOSPITAL, Shamrock/diesel/mart/wash

63mm Ray Hubbard Reservoir

62   Chaha Rd, **N**...Best Western, cleaners, to Hubbard RA, **S**...Shamrock/mart, Texaco/mart

61   Zion Rd, **N**...Conoco/diesel/mart, **S**...Chinese Rest., cleaners

60b   Bobtown Rd, **N**...Mobil/mart, Jack-in-the-Box, **S**...Chevron, Texaco/mart, Catfish King Rest.

E

W

Texas Interstate 30

a Rose Hill Dr, no facilities

59 Beltline Rd, Garland, **N**...MEDICAL CARE, Conoco, Mobil, Total/
mart, Chinese Rest., Church's Chicken, Denny's, LJ Silver,
McDonald's, Whataburger, Taco Bell, Albertson's/drugs, Chief
Parts, Eckerd, Goodyear, K-Mart/drugs, hardware,
S...CHIROPRACTOR, DENTIST, VETERINARIAN, Exxon/diesel,
Shell/mart/wash, Burger King, Chinese Rest., Dairy Queen,
Grandy's, Waffle House, Kroger/deli/drugs, Day's
Inn, Motel 6, bank

58 Northwest Dr, **N**...DENTIST, Texaco/mart, Total/
diesel/mart/atm, Nissan, **S**...Fina/mart, Jack-in-the-
Box, Food Lion, Lowe's Bldg Supply

56c b I-635 S, I 635 N

a Galloway Ave, Gus Thomasson Dr, **N**...Shamrock, KFC,
S...MEDICAL CARE, Citgo/7-11, Checker's, Chinese Rest.,
Grandy's, Kenny Rogers, Luby's, On-the-Border, Subway, Wendy's,
Kroger, Courtyard, Crossland Studios, Fairfield Inn, Mesquite Camper Ctr, Midas
Muffler, Western Auto, other multiple facilities, Bigtown Mall...Olive Garden,
Outback Steaks, Red Lobster, Steak&Ale, Celebration Sta, Circuit City, NTB, Sears,
cinema, cleaners

55 Motley Dr, **N**...Texaco/diesel/mart, Astro Inn, Ramada Ltd, **S**...HOSPITAL, Chevron/
mart, Fina, Golden Ox Café, Mesquite Inn, Microtel, Dodge, auto repair, to Eastfield Coll

54 Bigtown Blvd, **N**...Shell/wash, Texaco, Best Western, **S**...TA/diesel/mart/rest./24hr, Ward's/auto, Texas Motel,
Holiday World RV Ctr, mall

52c US 80 E, to Terrell, **N**...Toyota

a Lp 12, **N**...Super 8, Chevrolet, Toyota, **S**...CiCi's, Whataburger, Holiday Inn Express, Ford, K-Mart, Sam's Club,
Wal-Mart SuperCtr/gas/24hr

51 Highland Rd, **N**...Exxon, Denny's, Kettle, Luby's, McDonald's, Owen's Rest., La Quinta, **S**...RaceTrac/mart,
Texaco/diesel/mart/wash, Burger King, Capt D's, Furr's Dining, Grandy's, KFC, Pizza Hut, Wendy's, Exel Inn, Red
Roof, Eckerd, bank

50 Ferguson Rd, **N**...Shamrock/mart, Phillips 66/mart, Texaco/mart, FairPark Inn, **S**...Brake-O, U-Haul

49b a Dolphin Rd, Lawnview Ave, Samuell Ave, **N**...Welcome Inn, Ken's Muffler/shocks

48b Winslow St, **N**...Fina/mart, Phillips 66/mart, Texaco/diesel/mart/repair, Texas FoodLand, bank, **S**...Phillips 66,
Shell/mart/24hr

a TX 78, E Grand, **S**...fairpark, arboreteum

47b Carroll Ave, Central Ave, Peak St, Haskell Ave, **N**...HOSPITAL, Shamrock, **S**...Fina/mart, Joe's Burgers, tires

a 2nd Ave, **S**...Cotton Bowl, fairpark, McDonald's

46b I-45, US 75, to Houston

a Central Expswy, downtown

45 I-35 E, N to Denton, to Commerce St, Lamar St, Griffin St, **S**...McDonald's, Ambassador Plaza Hotel, Ramada
Inn

44b I-35E S, Industrial Blvd, no facilities

a I-35E N, Beckley Ave(from eb)

43b a Sylvan Ave(from wb), **N**...HOSPITAL, Texaco/mart, USPO/24hr

42 Hampton Rd N, **1 mi N**...Burger King, KFC

41 Hampton Rd S, **N**...Mobil, **1 mi S**...Exxon/mart, Texaco/Luby's/mart/wash, Checker's, Dairy Queen, Wendy's,
Miramar Hotel, Ranch Motel, AutoZone, Goodyear/auto

38 Lp 12, **N**...Exxon, **3 mi N**...Chevron/diesel, RaceTrac, Whataburger, Motel 6

36 McArthur Blvd, **S**...U-Haul

34 Belt Line Rd, **N**...Chevron/mart, Day's Inn, Motel 6, Ramada Ltd, Ripley's Museum, **S**...Fina/Blimpie/TCBYmart,
RaceTrac/mart, Shamrock/mart, Dairy Queen, McDonald's

32 NW 19th, **N**...Exxon/mart, **S**...VETERINARIAN, Mobil, Chinese Rest., Denny's, LJ Silver, McDonald's, Pizza Hut,
Taco Bell, Whataburger, La Quinta, bank

30 Six Flags Drive, TX 360, Arlington Stadium, **N**...Shell, Golden Corral, TrailDust Steaks, Day's Inn, Fairfield Inn,
Flagship Inn, Hilton, Radisson Suites, Studio+ Suites, Super 8, Wingate Inn, **1-2 mi N**...Texaco/wash, Mobil,
McDonald's, Wendy's, Hampton Inn, Holiday Inn, Quality Inn, auto parts, to DFW, **S**...Texaco/diesel/mart, Total/
mart, Bennigan's, Chinese Rest., Denny's, Jack-in-the-Box, Luby's, McDonald's, Owens Rest., Steak&Ale,
Baymont Inn, La Quinta, Dillard's, JC Penney, K-Mart, Plymouth, Sears, Six Flags Funpark, mall

29 Ball Park Way, **N**...China Coast, Grady's Grill, Macaroni Grill, TrailDust Steaks, Fairfield Inn, Wet n' Wild, Toyota,
S...Fina/diesel/mart, On the Border, Texas Steaks, Day's Inn, Marriott, Stadium Inn, Six Flags Funpark, bank

E

W

Dallas

Dallas

Dallas

Texas Interstate 30

28b TX 157, Nolan Ryan Expswy, **N...**Mobil/mart, Waffle House, Country Inn Suites, Olds, **S...**Chili's, Cozymel's, Harrigan's Grill, Joe's Crabshack, Landry's Seafood, Olive Garden, Pappadeaux Seafood, Courtyard, Day's Inn, Wyndham Garden, Barnes&Noble, Kinko's, TX Stadium
a FM 157, Collins St, **N...**HOSPITAL, **S...**CHIROPRACTOR, Pappasito's Cantina, Home Depot, bank, cinema
27 Lamar Blvd, Cooper St, **N...**VETERINARIAN, Mobil/mart, Jack-in-the-Box, Eckerd, Kroger, Pennzoil, 1-Hr Photo, bank, **S...**MEDICAL CARE, Citgo/7-11, Texaco/mart/wash, Burger King, Denny's, bank
26 Fielder Rd, **S...**to Six Flags
25mm Village Creek
24 Eastchase Pkwy, **N...**CarMax, Wal-Mart SuperCtr/24hr, golf, **S...**Chevron, Texaco/diesel/mart/wash, IHOP, McDonald's, Whataburger, Office Depot, Old Navy, PetsMart, Ross, Target, ToysRUs, cleaners
23 Cooks Lane, **S...**Mobil/mart
21c Bridgewood Dr, **N...**Mobil, Texaco, Bennigan's, Luby's, **S...**CHIROPRACTOR, DENTIST, MEDICAL CARE, EYECARE, Fina, Shamrock, Texaco/diesel/24hr, Burger King, McDonald's, Pizza Hut, Taco Bueno, Whataburger/24hr
b a I-820, no facilities
19 Brentwood Stair Rd, **N...**Chevron/mart, Texaco/diesel, Kroger, Steak&Ale, frontage rds access facilities from several exits, **S...**Citgo/diesel/mart, Texaco
18 Oakland Blvd, **N...**Phillips 66/mart, Texaco/diesel/mart/wash, Burger King, Taco Bell, Waffle House, Motel 6, cleaners
16c Brentwood Stair Rd, Beach St, **S...**Cattle Drive Rest., Comfort Inn, Holiday Inn, bank
16b a Riverside Dr, no facilities
16mm US 287, **S...**Citgo, Conoco/diesel, no facilities
15b I-35W N to Denton, S to Waco
14.5mm Jones St, Commerce St, Ft Worth, **N...**Ramada
14mm TX 199, Henderson St, Ft Worth, **N...**Chevron, Goodyear/auto
13.5mm Ballenger St, Summit Ave, downtown, no facilities
13 15th Ave, downtown
12d Forest Park Blvd, **N...**HOSPITAL, bank
12b Rosedale St., no facilities
a University Dr, City Parks, **S...**Ramada
11 Montgomery St, **S...**Texaco/diesel/mart/wash, Whataburger, visitor info
10 Hulen St, Ft Worth, no facilities
9b US 377, Camp Bowie Blvd, Horne St, **S...**Citgo/7-11, Exxon/wash, Fina, Mobil, Texaco, Burger King, Dunkin Donuts, Jack-in-the-Box, Subway, Taco Bell, Taco Bueno, Tejano Café, Wendy's, Whataburger, Brake-O, bank, to Carswell AFB
a Bryant-Irvin Rd, **S...**Texaco/mart
8b Ridgmar, Ridglea, **N...**Texaco/mart
a TX 183, Green Oaks Rd, **N...**MEDICAL DOCTOR, Jack-in-the-Box, Olive Garden, Taco Bueno, TGIFriday, Albertson's, Dillard's, JC Penney, Mervyn's, Sears, Service Merchandise, cinema, laundry, bank
7b a Cherry Lane, TX 183, spur 341, to Green Oaks Rd, Carswell AFB, **N...**CHIROPRACTOR, DENTIST, Conoco/diesel/mart, Texaco/repair, Beefer's Burgers, ChuckeCheese, CiCi's, IHOP/24hr, Lenny's Rest., Luby's, Papa John's, Popeye's, Ryan's, Subway, Taco Bell, La Quinta, Stratford House, Ford, Home Depot, K-Mart/drugs, Sam's Club, Wal-Mart/auto, **S...**DENTIST, CHIROPRACTOR, MEDICAL CARE, EYECARE, Best Western, Green Oaks Inn, Hampton Inn, Nissan, Target, U-Haul, cleaners, mail service, auto dealers
6 Las Vegas Trail, **N...**VETERINARIAN, Chevron/mart/wash/24hr, Waffle House, Day's Inn, cinema, **S...**Exxon/repair, Fina, Shamrock, Texaco/diesel/mart/wash, Kettle, Pancake House, Motel 6, Royal Western Suites, auto parts, laundry
5b c I-820 S and N
5a Alemeda St, (no EZ return eb, no return wb), no facilities
3 RM 2871, Chapel Creek Blvd, no facilities
2 spur 580 E, no facilities
1b Linkcrest Dr, **S...**Chevron/mart, Fina/diesel/mart/atm, Mobil/diesel/mart, Shamrock/mart
0mm I-20 W. I-30 begins/ends on I-20, exit 421.

Ft Worth

E ↕ W

Texas Interstate 35

Exit # Services
504mm Red River, Texas/Oklahoma state line
504 frontage rd, no facilities
503mm parking area both lanes, no facilities
502 Welcome Ctr sb, full(handicapped)facilities, TX Tourist Bureau/info, phone, picnic tables, litter barrels
501 FM 1202, **E...**Ford/Mercury, **W...**Hilltop/Conoco/diesel/mart/café, Prime Outlets/famous brands, Western Outfitter, Harper's Steaks, RV camping
500 FM 372, Gainesville, **E...**Hitchin' Post/Texaco/diesel/mart/rest./24hr, Shamrock/diesel/mart/24hr, XpressLube, tires

Texas Interstate 35

498b a   US 82, Gainesville, to Wichita Falls, Sherman, **E**...HOSPITAL, Exxon, Little Red's/diesel, Shamrock/mart, Denny's, Golden Corral(2mi), Whatburger/24hr, Comfort Inn, Day's Inn, Ramada Inn, 12 Oaks Inn, **W**...Exxon/diesel/mart, Ranchito Mexican, Bed&Bath Inn, Best Western, Budget Host, Texas Motel, antiques

497   frontage rd, no facilities

496b   TX 51, FM 51, California St, Gainesville, **E**...Conoco/mart, Shamrock/mart, Braum's, Burger King, McDonald's, Taco Bell, Taco Mayo, Wendy's, Holiday Inn, Dodge/Plymouth/Eagle, Goodyear/auto, Olds/Cadillac, N Central TX Coll, **W**...Texaco

496a   to Weaver St, no facilities

496mm   Elm Fork of the Trinity River

495   frontage rd, no facilities

494   FM 1306, no facilities

492mm   picnic area sb, tables, litter barrels

491   Spring Creek Rd, no facilities

490mm   picnic area nb, picnic tables, litter barrels

489   FM 1307, to Hockley Creek Rd, no facilities

487   rd 922, Valley View, **W**...Chevron/diesel/mart/24hr, Dairy Queen, USPO

486   frontage rd, no facilities

485   frontage rd(from sb), no facilities

483   FM 3002, Lone Oak Rd, **E**...RV Ranch

482   Chisam Rd, no facilities

481   View Rd, no facilities

480   Lois Rd(from sb), no facilities

479   Belz Rd, Sanger, **E**...Phillips 66, Texaco, BBQ, Dairy Queen, USPO, cleaners/laundry, **W**...Chevron/Subway/mart/24hr, Jack-in-the-Box, IGA Foods, KwikLube, bank, RV camping

478   FM 455, to Pilot Pt, Bolivar, services at 479, **E**...Sanger Inn **W**...Chevrolet, RV Ctr, Ray Roberts Lake and SP

477   Keaton Rd, **E**...Phillips 66/mart, Cattle Rustler's Steaks, **W**...Shamrock/mart

475b   Rector Rd, no facilities

a   FM 156, to Krum, no facilities

474   Cowling(from nb), no facilities

473   FM 3163, Milam Rd, **E**...Love's/Subway/diesel/mart/24hr

472   Ganzer Rd, no facilities

471   US 77, FM 1173, Lp 282, to Denton, Krum, **E**...HOSPITAL, TA/Shell/diesel/rest./24hr, Good Eats, Denton Stores/ famous brands, **W**...Citgo/diesel/café/24hr, Fina, Foster's Western Shop, to Camping World

470   Lp 288, same services as 469 from sb

469   US 380, University Dr, to Decatur, McKinney, **E**...Phillips 66/diesel/mart, Braum's, Cracker Barrel, McDonald's, Albertson's/drugs, Ford, K-Mart/drugs, **W**...HOSPITAL, VETERINARIAN, Conoco/diesel/mart, Shamrock/mart/ service, Shell/mart, Texaco/diesel/mart, Dairy Queen, Denny's, Waffle House, Best Western, Day's Inn, Exel Inn, Howard Johnson, Motel 6, Camping World RV Supplies, to TX Woman's U

468   FM 1515, Airport Rd, W Oak St, no facilities

467   I-35W, S to Ft Worth, no facilities

I-35 divides into E and W sb, converges into I-35 nb. **See Texas I-35 W.**

466b   Ave D, **E**...Citgo, Exxon/mart, Burger King, IHOP, McDonald's, NY Subs, Pancho's Mexican, Royal Hotel Sak'n Save Foods, to NTSU, **W**...Radisson

a   McCormick St, **E**...Phillips 66, Texaco/diesel, La Quinta, **W**...Citgo, Fina/diesel/mart/U-Haul

465b   US 377, Ft Worth Dr, **E**...Citgo, Shamrock/mart, Kettle, Taco Bueno, Whataburger/24hr, La Quinta, Home Depot, U-Haul, **W**...Citgo, Conoco, Phillips 66, Total/diesel/mart, Outback Steaks, Day's Inn

a   FM 2181, Teasley Ln, **E**...Citgo/7-11, Applebee's, Braum's, KFC, Holiday Inn, Winn-Dixie, **W**...Shell, Little Caesar's, Pizza Hut, Ramada Inn, Super 8, Honda/Polaris

464   US 77, Pennsylvania Dr, Denton, **E**...Burger King, Wendy's, Dillard's, Kroger, Mervyn's, Office Depot, PetCo, Ross, Sears/auto, Wal-Mart, mall, same as 463

463   Lp 288, to Denton, **E**...El Chico, Discount Tire, Hancock Fabrics, HobbyLobby, Kroger, Midas Muffler, Old Navy, **W**...VETERINARIAN, Mobil/mart, Blackeyed Pea, Chili's, Jack-in-the-Box, Luby's, Red Lobster, Red Pepper's Rest., Albertson's, same as 464

462   State School Rd, Mayhill Rd, **E**...HOSPITAL, Texaco, Hyundai, **W**...Exxon/café/mart, Chevrolet, Olds/Cadillac, Pontiac/Buick/GMC

461   Sandy Shores Rd, Post Oak Dr, **E**...Ford, McClain RV Ctr, **W**...Cattle Co Steaks, Chrysler/Plymouth/Jeep/Kia, Lincoln/Mercury, Nissan, bank

Texas Interstate 35

Grapevine

460	Corinth, **E**...Chevron/diesel/repair, McClain RV Ctr, camping
459	frontage rd, **W**...RV camping
458	FM 2181, Swisher Rd, **E**...Phillips 66/mart, Texaco/Subway/mart/atm, **W**...Chevron/McDonald's/mart, Exxon/Wendy's/mart, Burger King, Chick-fil-A, Jack-in-the-Box, Taco Bell, Albertson's, KwikLube, Radio Shack, cleaners
457	Hundley Dr, Lake Dallas, **E**...Chevron/mart
456	Highland Village, no facilities
456mm	Lewisville Lake
455	McGee Lane, no facilities
454b	Garden Ridge Blvd, no facilities
a	FM 407, Justin, **E**...Citgo/mart, Phillips 66/repair, Shamrock/mart, **W**...Fina/diesel/mart, Texaco/mart, Domino's, McDonald's
453	Valley Ridge Blvd, **E**...Ford, May RV Ctr
452	FM 1171, to Flower Mound, **E**...HOSPITAL, Mobil, Taco Bueno, Holiday Inn Express, JustBrakes, bank, **W**...Chevron/mart/wash, Exxon/diesel/mart, Texaco, Golden Corral, Whataburger/24hr, Home Depot, Kohl's, Lowe's Bldg Supply, U-Haul, same as 451
451	Fox Ave, **E**...Texaco/diesel/mart/wash, Braum's, **W**...Chevron/mart/wash/24hr, Conoco/Quizno's/diesel/mart, Shamrock/mart, Blackeyed Pea, Cracker Barrel, El Chico, Hampton Inn, Microtel, Atlas Transmissions, Midas Muffler
450	TX 121, Grapevine, **E**...Citgo/7-11, RaceTrac/mart, BBQ, Owens Rest., Pancho's Mexican, Pines Motel, Ramada Inn, Chevrolet/Subaru, Dodge, **W**...Chevron/mart/wash, Citgo/7-11, Fina/mart, Mobil/mart, Shamrock/mart, Texaco/diesel/mart/wash, Alfredo's Pizza, Burger King, Chili's, Church's Chicken, Don Pedro's, Furr's Dining, Ming Garden Chinese, KFC, LJ Silver, McDonald's, Rodarte's Cantina, Subway, Taco Bell, Tia's Mexican, Waffle House, Whataburger/24hr, Budget Inn, Day's Inn, J&J Motel, Spanish Trails Inn, Super 8, Chief Parts, Eckerd, Firestone/auto, Food Lion, Kroger/deli/bakery, Pennzoil, bank, carwash, cleaners/laundry, transmissions
449	Corporate Drive, **E**...China Dragon, Hooters, On the Border, Water St Seafood, Extended Stay America, Hearthside Inn, Motel 6, AutoNation USA, Cavender's Boots, cinema, **W**...Exxon/Quizno's/Pizza Inn/TCBY/mart, Fina, Chili's, Jack-in-the-Box, Kettle, Best Western, La Quinta, Sun Suites, Honda, NTB
448b a	FM 3040, Round Grove Rd, **E**...Bennigan's, ChuckeCheese, Olive Garden, SaltGrass Steaks, Subway, Homewood Suites, Best Buy, Garden Ridge, PetsMart, Pier 1, Ross, Service Merchandise, Toys R Us, Ward's/auto, **W**...Chevron/mart, Exxon/mart, RaceTrac/mart/atm, Texaco, Abuelo's Mexican, Applebee's, BBQ, Carino's Italian, Chick-fil-A, Christina's Mexican, Cotton Patch Café, Don Pablo's, Good Eats Grill, IHOP, Logan's Roadhouse, Luby's, Macaroni Grill, McDonald's, OutBack Steaks, Red Lobster, Schlotsky's, Sonic, SpagEddie's Italian, Taco Cabana, TGIFriday, Water Garden Chinese, Wendy's, Comfort Inn, Country Inn Suites, Babys R Us, Barnes&Noble, Borders Books, Circuit City, CompUSA, Dillard's, Discount Tire, Homeplace, JC Penney, Kinko's, Linens'n Things, Marshall's, Michael's, Office Depot, OfficeMax, Oschman's Sports, Old Navy, PartyCity, Pearle Vision, PetLand, Sears/auto, Target, bank, mall
445	Trinity Mills Rd, **E**...HOSPITAL, Buick/GMC/Pontiac, Home Depot, PepBoys, RV Ctr
444	Whitlock Lane, Sandy Lake Rd, **E**...Texaco, Taco Bell, RV camping, **W**...Chevron/mart/wash, McDonald's, Delux Inn
443	Belt Line Rd, Crosby Rd, **E**...Fina/mart, RaceTrac/mart, Shell/mart, NTB
442	Valwood Pkwy, **E**...Texaco, El Chico, Dairy Queen, Denny's, Grandy's, Jack-in-the-Box, Rosita's Mexican, Taco Bueno, Waffle House, Comfort Inn, Guest Inn, Red Roof, Royal Inn, **W**...Fina/diesel, Day's Inn, Chevrolet, Ford, Pennzoil, U-Haul, transmissions
441	Valley View Lane, **W**...Exxon, Mobil/mart, Best Western, Day's Inn, Econolodge, La Quinta
440b	I-635 E, no facilities
c	I-635 W, to DFW Airport, no facilities
439	Royal Lane, **E**...Texaco, McDonald's, Whataburger/24hr, **W**...Chevron/mart, Jack-in-the-Box, Japanese Rest., Wendy's, Ken's Mufflers/Shocks
438	Walnut Hill Lane, **E**...Mobil/mart/wash, Bennigan's, Burger King, Chili's, Denny's, Old San Francisco Steaks, Red Lobster, Steak&Ale, Taco Bell, TGIFriday, Tony Roma's, Trail Dust Steaks, Country Inn Suites, Drury Inn, Hampton Inn, Midas Muffler, **W**...Citgo, Texaco/diesel/mart/wash, cinema, funpark, RV Ctr
437	Manana Rd(from nb), same as 438
436	TX 348, Irving, to DFW, **E**...Texaco/diesel/mart, IHOP, Luby's, Schlotsky's, Waffle House, Best Western, Clarion Suites, Courtyard, Hearthside Suites, Homestead Suites, Radisson, **W**...Exxon, Blackeyed Peas, Chili's, Don Pablo's, Joe's Crabshack, McDonald's, Olive Garden, Outback Steaks, Papadeaux Seafood, Red Lobster, Shoney's, Taco Bell, Tony Roma's, Wendy's, Red Roof Inn, bank
435	Harry Hines Blvd(from nb), **E**...RaceTrack/mart, Arby's, U-Haul, same as 436
434b	Regal Row, **E**...Chevron/mart, Denny's, Whataburger/24hr, La Quinta, **W**...Fairfield Inn
a	Empire, Central, **E**...Shell/mart/wash, McDonald's, Wendy's, Budget Suites, Candlewood Suites, Suburban Lodge, Wingate Inn, Office Depot, **W**...Exxon, Schlotsky's, Taco Bell/Pizza Hut, Kinko's
433b	Mockingbird Lane, Love Field Airport, **E**...Mobil/mart, Jack-in-the-Box, Clarion Hotel, Crowne Plaza, Harvey Hotel, Residence Inn, Sheraton, **W**...Chevron, McDonald's, Goodyear, bank
a	(432b from sb)TX 356, Commonwealth Dr, **E**...Texaco/diesel/mart, Marriott, OfficeMax, **W**...Delux Inn, Brake-O, Optical Mart
432a	Inwood Rd, **E**...HOSPITAL, Chevrolet, **W**...Exxon, Fina/mart, Texaco/mart, Whataburger, Homewood Suites, carwash

Dallas

Texas Interstate 35

431   Motor St, **E**...Mobil/mart, Stouffer Hotel, **W**...Shell/mart/wash, Wok Chinese, Zuma Mexican Grill, Embassy Suites, Mariott Suites

430c   Wycliff Ave, **E**...JoJo's Rest., Renaissance Hotel, Intn'l Apparel Mart, bank

    b   Mkt Ctr Blvd, **E**...World Trade Ctr, **W**...Denny's, Best Western, Courtyard, Fairfield Inn, Sheraton, Travelodge, Wilson World Hotel, Wyndham Garden

    a   Oak Lawn Ave, **W**...Texaco/diesel, Denny's, Medieval Times Rest., Best Western, Holiday Inn, to Merchandise Mart, bank

429c   HiLine Ave(from nb)

    b   Continental Ave, downtown, **W**...Exxon, McDonald's

    a   to I-45, US 75, to Houston, no facilities

428e   Commerce St E, Reunion Blvd, Dallas, downtown

    d   I-30 W, to Ft Worth

    a   I-30 E, to I-45 S

    b   Industrial Blvd

427b   I-30 E, no facilities

    a   Colorado Blvd, **E**...HOSPITAL

426c   Jefferson Ave, carwash

    b   TX 180 W, 8th St, **E**...Texaco/mart, **W**...Shell/wash, Mustang Inn

    a   Ewing Ave, **W**...Circle Inn, Chrysler/Jeep, Pontiac/GMC

425c   Marsalis Ave, **E**...Chevron/24hr, Dallas Inn, **W**...Mobil, Shamrock

    b   Beckley Ave, 12th St, **W**...Exxon/mart, Wendy's, bank

    a   Zang Blvd, **W**...Shamrock, Texaco/diesel/mart, mufflers

424   Illinois Ave, **E**...HOSPITAL, Chevron/mart, BBQ, Dunkin Donuts, Bud&Ben Mufflers, **W**...Exxon, Church's Chicken, IHOP, Jack-in-the-Box, Little Caesar's, Sonic, Taco Bell, OakTree Inn, Brake Ctr, Eckerd, Midas Muffler, bank

423b   Zaner Ave, **W**...Big Boy, Spin's Rest., Westerner Motel

    a   (422b from nb)US 67 S, Kiest Blvd, **W**...Shell/repair, Texaco/diesel/mart, McDonald's, Wendy's, Quality Inn, Subaru/VW, bank

421b   Lp 12W, Ann Arbor St, **W**...Luby's, Red Lobster, Big Couple Motel, Sunbelt Motel, Sam's Club

    a   Lp 12E, **E**...RaceTrak/mart, Texaco/diesel/mart, Budget Inn, Howard Johnson, Motel 6, K-Mart, Western Auto

420   Laureland, **E**...VETERINARIAN, Master Suite Motel, **W**...Fina/mart, Mobil/Blimpie, Embassy Motel

419   Camp Wisdom Rd, **E**...Exxon/mart, **W**...Chevron/mart/wash/24hr, Shell/mart/wash/24hr, McDonald's, Oak Cliff Inn, Suncrest Inn, U-Haul

418c   Danieldale Rd(from sb)

418b   I-635/I-20 E, to Shreveport, no facilities

    a   I-20 W, to Ft Worth, no facilities

417   Wheatland Rd(from nb), no facilities

416   Wintergreen Rd, **E**...CHIROPRACTOR, repair, **W**...Citgo/7-11, Cracker Barrel, Golden Corral, Waffle House, Holiday Inn, Red Roof Inn

415   Pleasant Run Rd, **E**...CHIROPRACTOR, Chevron/mart/wash/24hr, RaceTrac/mart, Texaco/Blimpie/diesel/mart, Benavida's Rest., Chow Line Buffet, Evergreen Buffet, Grandy's, Subway, Yogurt, Waffle House, Wayne's Diner, Great Western Inn, Royal Inn, Spanish Trails Motel, Super 8, Chrysler/Plymouth/Jeep, Midas Muffler, PepBoys, cinema, transmissions, **W**...HOSPITAL, Exxon/diesel/mart, Mobil/mart, Burger King, El Chico, Golden Corral, KFC, LJ Silver, Luby's, McDonald's, On the Border, Outback Steaks, Pizza Inn, Taco Bueno, Wendy's, Best Western, Chevrolet, Firestone, Ford, Kroger, K-Mart, Office Depot

414   FM 1382, Desoto Rd, Belt Line Rd, **E**...Whataburger, Wal-Mart SuperCtr/24hr, **W**...Texaco

413   Parkerville Rd, **W**...Total/mart, Honda, U-Haul

412   Bear Creek Rd, **W**...Texaco/diesel/mart, Jack-in-the-Box, camping, transmissions

411   FM 664, Ovilla Rd, **E**...Exxon/Church's/Taco Bell/TCBY/mart, RaceTrac/mart, McDonald's, Whataburger, Brookshire's Foods, Eckerd, bank, **W**...VETERINARIAN, Exxon/mart, Texaco/mart

410   Red Oak Rd, **E**...Fina/Subway/diesel/mart/24hr, Texaco/diesel/mart, Denny's, Day's Inn, **W**...RV Park

408   US 77, no facilities

406   Sterrett Rd, fireworks

405   FM 387, **E**...Chevron/diesel/mart

404   Lofland Rd, industrial area, no facilities

403   US 287, to Ft Worth, **E**...Texaco, McDonald's, Chevrolet/Olds/Cadillac, Jeep, **W**...Buick/Pontiac/GMC, Ford/Mercury

401b   US 287 bus, Waxahatchie, **E**...Best Western, Super 8, Chrysler/Plymouth/Dodge

    a   Brookside Rd, **E**...Lazy Fisherman Rest., Denham Ranch Inn/rest., Ramada Ltd

399b   FM 1446, **E**...Texas Inn

Texas Interstate 35

a	FM 66, FM 876, Maypearl, **E...**Shamrock, Texas Inn, **W...**Chevron/diesel/mart, Texaco/diesel/mart/24hr
397	to US 77, no facilities
393mm	**rest area both lanes, full(handicapped)facilities, phone, picnic tables, litter barrels, vending, petwalk**
391	FM 329, Forreston Rd, no facilities
386	TX 34, Italy, **E...**Texaco/Blimpie/diesel/mart/24hr, Dairy Queen
384	County Rd, no facilities
381	FM 566, Milford Rd, no facilities
377	FM 934, no facilities
374	FM 2959, Carl's Corner, **W...**Carl's/diesel/mart/rest.

I-35E**********

371mm	I-35 divides into E and W nb, converges sb. **See Texas I-35 W.**
370	US 77 N, FM 579, Hillsboro, no facilities
368b	FM 286, **E...**LoneStar Café, Taco Bell, Wendy's, Prime Outlets/famous brands, **W...**HOSPITAL, Chevron/TCBY/diesel/mart, Shamrock/diesel/mart/atm, Texaco/diesel/wash, Braum's, Dairy Queen, El Conquistador Mexican, McDonald's, Pizza Hut, Best Western, Comfort Inn
a	TX 22, TX 171, to Whitney, **E...**Citgo/7-11, Love's/Subway/Pizza Hut/diesel/24hr, Arby's, Blackeyed Pea, Burger King, Golden Corral, Grandy's, IHOP, McDonald's, Holiday Inn Express, Ramada Inn, **W...**Citgo/diesel, Exxon, Mobil/mart, Shell/mart/wash, KFC, Schlotsky's, Whataburger/24hr, Thunderbird Motel/rest., Buick/Pontiac/GMC, Chevrolet/Olds/Cadillac, Chrysler/Plymouth/Dodge/Jeep, Wal-Mart SuperCtr/24hr, bank
367	Old Bynum Rd(from nb), no facilities
364b	TX 81 N, to Hillsboro, no facilities
a	FM 310, **E...**Conoco/diesel/mart/24hr, Texaco/mart, Walnut Bowls, **W...**Fina/diesel/mart/24hr, BBQ
362	Chatt Rd, no facilities
359	FM 1304, **W...**Mobil/diesel/mart/rest./24hr, antiques
358	FM 1242 E, Abbott, **E...**Exxon/mart, Turkey Shop Café, **W...**truckwash
356	Abest Rd, no facilities
355	County Line Rd, **W...**KOA
354	Marable St, **W...**KOA
353	FM 2114, West, **E...**Circle K/gas, Citgo/diesel, Shell/mart, Skinny's/gas, Dairy Queen, Little Czech Bakery, Ford, **W...**Citgo/diesel/mart/repair, Exxon, Chevrolet, Goodyear
351	FM 1858, **E...**antiques, **W...**West Auction
349	Wiggins Rd, no facilities
347	FM 3149, Tours Rd, no facilities
346	Ross Rd, **E...**Shell/diesel/mart/café, **W...**Conoco/diesel/mart/24hr, **I-35 RV Park**, antiques
345.5mm	picnic area/litter barrels both lanes
345	Old Dallas Rd, **E...**antiques, **W...I-35 RV Park**
343	FM 308, Elm Mott, **E...**Fina/diesel/mart, Texaco/diesel/mart/café/24hr, Dairy Queen, dentures, **W...**Chevron/diesel/mart/24hr, Heitmiller Steaks
342b	US 77 bus, no facilities
a	FM 2417, Northcrest, **W...**Shamrock/diesel/mart, Dairy Queen, Everyday Inn, Chuck's Mufflers
341	Craven Ave, Lacy Lakeview, **E...**Chevron/mart, **W...**Mobil, Shell//pizza/mart, Interstate North Motel, Chief Parts
340	Myers Lane(from nb), no facilities
339	to TX 6 S, FM 3051, Lake Waco, **E...**CHIROPRACTOR, MEDICAL CARE, Shamrock/mart, Domino's, El Conquistador Rest., Jack-in-the-Box, Luby's, Pizza Hut, Popeye's, Sonic, Wendy's, Whataburger/24hr, Country Inn Suites, Albertson's, KwikKar Lube, Wal-Mart SuperCtr/24hr, **W...**Burger King, Cracker Barrel, KFC, McDonald's, Hampton Inn, Winn-Dixie/drugs, cinema, to airport
338b	Behrens Circle, **E...**Texaco/mart, same as 339, **W...**Texaco/diesel/mart/LP, Day's Inn, Delta Inn, Knight's Inn, Motel 6, Waco Inn, Eckerd
337	(338a from nb)US 84, to TX 31, Waco Dr, **E...**Fina/mart, AutoZone, Family$, HEB Food/drugs, Sam's Club, **W...**HOSPITAL, Texaco/mart
336mm	Brazos River
335c	Lake Brazos Dr, **E...**Texaco, Summer Palace Chinese Buffet, Holiday Inn, RiverPlace Inn/café, **W...**HOSPITAL, Econolodge, Victorian Inn
b	FM 434, **E...**Café China Buffet, IHOP, Best Western, Baylor U, **W...**Arby's, Jack-in-the-Box, Tanglewood Rest., Hilton, Lexington Inn, bank
a	4th St, 5th St, **E...**Exxon/diesel/mart, Texaco/diesel/wash, Taco Bueno, Best Western, Baylor U, **W...**Shamrock, Shell/mart/wash, Fazoli's, Golden Fried Chicken, LJ Silver, McDonald's, Pizza Inn, Taco Bell, Taco Cabana, Wendy's, Whataburger/24hr, Clarion, to Baylor U
334b	US 77 S, 17th St, 18th St, **E...**Chevron, Mobil, Texaco/diesel, Burger King, Denny's, Pizza Hut, Schlotsky's, Budget Inn, Comfort Inn, Friendship Inn, La Quinta, Super 8, Harley-Davidson, **W...**HOSPITAL, Chevron/mart, Phillips 66/diesel/mart, Texaco, Dairy Queen, Mexico Lindo

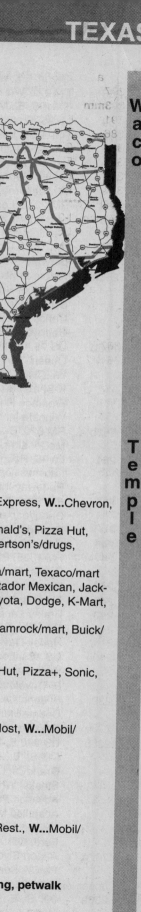

Texas Interstate 35

333a   Lp 396, Valley Mills Dr, **E**...Chevron, El Chico, Elite Café, Texas Roadhouse, Astro Motel, Motel 6, Isuzu/Mazda, **W**...RaceTrac/mart, Shamrock, Texaco, Jack-in-the-Box, Comfort Inn, Volvo/Freightliner

330   Lp 340, TX 6, New Rd, **E**...Chevron, Citgo/mart, Lazy Fisherman Rest., New Road Inn, Rodeway Inn, Thunderbird Motel, **2 mi W**...Texaco/24hr, Chick-fil-A, Luby's, McDonald's, Outback Steaks, TGIFriday, Wienerschnitzel, Fairfield Inn, CompUSA, Ford, Honda, K-Mart, Lowe's Bldg Supply, PetsMart, Pier 1, Wal-Mart SuperCtr/24hr

328   FM 2063, FM 2113, Moody, **E**...Williams/Dairy Queen/diesel/mart/24hr, **W**...Shamrock/mart, Texaco/mart/wash/24hr

325   FM 3148, Robinson Rd, **W**...Conoco, Shell/Subway/TCBY/Pizza Inn/diesel/mart/24hr

323   FM 2837, Lorena, **W**...Brookshire Bros/Conoco/mart, Chicken Express, $General

322   Lorena, **E**...Phillips 66/mart, antiques, bank, **W**...Chevron/mart, Chicken Express

319   Woodlawn Rd, no facilities

318.5mm picnic area both lanes, no facilities

318b   Bruceville, **W**...camping

   a   Frontage Rd, **E**...Chevron/diesel/mart

315   TX 7, FM 107, Eddy, **1 mi E**...Orky's RV Park, **W**...Mobil/mart, Texaco/Pizza Hut/diesel/mart/24hr, to Mother Neff SP

314   Old Blevins Rd, no facilities

311   Big Elm Rd, **W**...picnic area(handicapped), no restrooms

308   FM 935, Troy, **E**...Texaco, **W**...Exxon/mart/pizza, bank

306   FM 1237, Pendleton, **W**...Love's/Citgo/Subway/A&W/diesel/mart/24hr, Temple RV Park

305   Berger Rd, **W**...Conoco/diesel/mart/24hr, Temple RV Park, repair

304   Lp 363, **W**...Shamrock/diesel/mart

303   spur 290, N 3rd St, Temple, **E**...Shamrock/mart, TX Sands Motel, **W**...Continental Motel

302   Nugent Ave, **E**...Exxon/diesel, Texaco/diesel, Kettle, Comfort Suites, Econolodge, Holiday Inn Express, **W**...Chevron, Mobil, Denny's, Day's Inn, Motel 6, Ramada Inn, Stratford House Inn

301   TX 53, FM 2305, Adams Ave, **E**...Shamrock/mart, Grandy's, Jim's Rest., KFC, LJ Silver, McDonald's, Pizza Hut, Taco Bell, Taco Cabana, Wendy's, Whataburger, La Quinta, Ford, bank, **W**...Catfish Shack, Albertson's/drugs, Lincoln/Mercury, Harley-Davidson, Jeep

300   Ave H, 49th –57th Sts, **E**...Shell/mart, Texaco, Las Casas Mexican, Oasis Motel, Dodge, **W**...Fina/mart, Texaco/mart

299   US 190 E, TX 36, **E**...HOSPITAL, Mobil, Shamrock Gas, Shell, Texaco/diesel/mart, El Conquistador Mexican, Jack-in-the-Box, Luby's, Phillip's Steaks, Budget Inn, Skylark Motel, Cavender's Boots, Chevrolet/Toyota, Dodge, K-Mart, RV Ctr, Saturn, **W**...Exxon/diesel/mart/24hr, Texaco, Burger King

297   FM 817, Midway Dr, **E**...Citgo/mart, Day's Inn, Super 8/rest., BMW, Nissan, Saab, Volvo, **W**...Shamrock/mart, Buick/Pontiac/GMC, Fed Ex, VW

294b   FM 93, 6th Ave, **E**...Texaco, McDonald's, **W**...Best Western/rest., U of Mary Hardin Baylor

   a   Central Ave, **W**...Shell, Texaco/diesel, Bobby's Burgers, Charlie's Rest., Great SW Rest., Pizza Hut, Pizza+, Sonic, Texas Café, Whataburger, Best Western, Ramada Ltd, AutoZone, Goodyear

293b   TX 317, FM 436, Main St, no facilities

   a   US 190 W, to Killeen, Ft Hood, no facilities

292   (same as 293a), **E**...Citgo, Fina/mart, Shamrock/diesel/mart/rest./24hr, Bloom's Motel, Budget Host, **W**...Mobil/Blimpie/mart, Oxbow Steaks, Ford, KOA, Sunbelt RV Ctr, auto/tire repair

290   Shanklin Rd, **W**...KOA(1mi)

289   Tahuaya Rd, no facilities

287   Amity Rd, no facilities

286   FM 2484, **W**...to Stillhouse Hollow Lake, no facilities

285   FM 2268, Salado, **E**...Brookshire Foods, bank, **W**...BBQ, Robertson's Ham Sandwiches

284   Robertson Rd, **E**...Exxon/Burger King/mart, Texaco, Dairy Queen, Stagecoach Inn, Peppermill Rest., **W**...Mobil/diesel/mart, Pfalzerhof German Rest., Super 8

283   FM 2268, FM 2843, Salado, to Holland, no facilities

282   FM 2115, **E**...Texaco/diesel/mart/lube/repairs, RV camping

281.5mm rest area both lanes, full(handicapped)facilities, phone, picnic tables, litter barrels, vending, petwalk

280   Prairie Dell, no facilities

279   Frontage Rd, **W**...Emerald Lake RV Park

277   Yankee Rd, no facilities

275   FM 487, Jarrell, to Florence, **E**...Exxon, Shamrock/diesel/mart, **W**...Texaco/mart, USPO

Texas Interstate 35

271   Theon Rd, **W**...Texaco/Subway/diesel/mart/24hr

268   FM 972, Walburg, no facilities

266   TX 195, **E**...Mobil/Arby's/diesel/mart/24hr, **W**...Exxon/diesel/mart

264   Lp 35, Georgetown, **E**...San Gabriel Motel, RV camping

262   RM 2338, Lake Georgetown, **E**...KFC, Luby's, McDonald's, Albertson's, Eckerd, **W**...Phillips 66/mart, Texaco/mart, Chuck Wagon Café, Dairy Queen, Little River Food&Drink, Popeye's, Taco Bueno, Wendy's, Whataburger/24hr, Day's Inn, Georgetown Inn/rest., La Quinta, Ramada, bank

261   TX 69, Georgetown, **E**...Texaco/diesel/mart, Applebee's, Burger King, Chili's, Schlotsky's, HEB Foods, Wal-Mart/auto, **W**...Chevron, antiques

260   RM 2243, Leander, **E**...HOSPITAL, **W**...Chevron/mart, Circle K/gas, Jack-in-the-Box, Comfort Inn

259   Lp 35, **E**...Candle Factory, **W**...RV Outlet Mall, to Interspace Caverns

257   Westinghouse Rd, **E**...Chevrolet, Chrysler/Plymouth/Dodge/Jeep

256   RM 1431, Chandler Rd, **rest area sb, full(handicapped)facilities, phone, picnic tables, litter barrels, vending, petwalk**

255mm   rest area nb, full(handicapped)facilities, vending, phone, picnic tables, litter barrels, vending, petwalk

254   FM 3406, Round Rock, **E**...Chevron/mart, Citgo/7-11, Texaco/mart, Arby's, Giovanni's Italian, McDonald's, Mr Gatti's, Schlotsky's, Sonic, Honda, Olds/Pontiac/GMC/Olds, Pennzoil/wash, Toyota, cinema, **W**...Phillips 66/mart/24hr, Carino's Italian, Cracker Barrel, Denny's, Golden Corral, Kona Ranch Steaks, SaltGrass Steaks, Best Western, Courtyard, Hilton Garden, Ramada Inn, Nissan, bank

253b   US 79, to Taylor, **E**...HOSPITAL, VETERINARIAN, Phillips 66/mart, Shamrock/diesel/mart, Baskin-Robbins, Dairy Queen, Famous Sam's Café, KFC, LoneStar Café, LJ Silver, AutoZone, HEB/deli, MailBoxes Etc, bank, **W**...CHIROPRACTOR, OPTOMETRIST, Exxon/diesel/mart/atm, Texaco/diesel/mart/24hr, BBQ, Chinese Rest., IHOP, K-Bob's Rest., Taco Bell, Thundercloud Subs, La Quinta, Sleep Inn, Eckerd, USPO, laundry

  a   Frontage Rd, **E**...Shell, Arby's, Castaway's Seafood/steaks, Dairy Queen, Damon's, Sirloin Stockade, Best Western, Wingate Inn, **W**...La Marguarita, Popeye's, Ramada Ltd, Eckerd, HobbyLobby

252b a   RM 620, **E**...Texaco/diesel/mart/wash/24hr, Crossland Suites, bank, **W**...Citgo/7-11, Grandy's, Little Caesar's, McDonald's, Comfort Suites, Albertson's, Mail Ctr USA, cleaners, mall

251   Lp 35, Round Rock, **E**...Macaroni Grill, Whataburger, Residence Inn, Aamco, BrakeCheck, Diahatsu, JiffyLube, Olds, transmissions, **W**...Exxon/mart, Texaco/mart, Bagelry IV, Burger King, Jack-in-the-Box, Luby's, Taco Cabana, Rodeway Inn, Albertson's, Hastings Books, NTB, PrecisionTune, Walgreen, bank

250   FM 1325, **E**...Mobil/diesel/mart, Applebee's, BBQ, Chick-fil-A, Chili's, El Chico, McDonald's, Subway, Discount Tire, Goodyear/auto, OfficeMax, PetsMart, Ross, RV Ctr, Target, Wal-Mart SuperCtr/24hr, bank, **W**...Texaco, Hooters, Baymont Suites, Garden Ridge, XpressLube

248   Grand Ave Pkwy, **E**...Circle K/diesel/mart

247   FM 1825, Pflugerville, **E**...RaceTrac/mart, Jack-in-the-Box, Firestone/auto, HEB Foods, **W**...Exxon/mart, KFC, Quality Suites, Goodyear/auto

246   Dessau Rd, Howard Lane, **E**...Shell/diesel/mart, Texaco, **W**...Shamrock/mart/atm, IHOP, Whataburger

245   FM 734, Parmer Lane, to Yeager Lane, **W**...Citgo/7-11, Exxon/mart

243   Braker Lane, **E**...Shamrock, Jack-in-the-Box, **W**...Texaco/diesel/mart, Whataburger/24hr

241   Rundberg Lane, **E**...CHIROPRACTOR, VETERINARIAN, Chevron, Phillips 66, Golden Corral, Mr Gatti's, Albertson's, Harley-Davidson, U-Haul, **W**...Citgo/7-11, Motel 6, bank

240a   US 183, Lockhart, **E**...Exxon, Dairy Queen, Jack-in-the-Box, Old San Francisco Steaks, Day's Inn, Ramada Inn, Wellesley Inn, Chevrolet, **W**...Chevron/diesel/mart/24hr, Texaco/diesel/mart, Austin Inn, Motel 6, Red Roof Inn, Super 8, Travelodge Suites, Wingate Inn, XpressLube

239   St John's Ave, **E**...Texaco/mart, Bennigan's, Chili's, Jim's Rest., Steak&Egg, Budget Host, Day's Inn, Hampton Inn, Dodge, Home Depot, Volvo, Wal-Mart SuperCtr/24hr, bank, **W**...Conoco/diesel/mart, Exxon, Bennigan's, Chariot Inn Rest., Denny's, Comfort Inn, Holiday Inn Express, La Quinta, Sheraton, Sumner Suites

238b   US 290 E, RM 222, same as 238a, frontage rds connect several exits

  a   51st St, **E**...Exxon, Chevron/mart, Phillips 66/mart, Shell, Burger King, Dixie's Roadhouse, El Torito, Fuddrucker's, Grandy's, La Pala Rest., LJ Silver, McDonald's, Owens Rest., Palmera's Rest., Pappasito's, Shoney's, Sonic, Texas Steaks, Whataburger, Doubletree Hotel, Drury Inn, Econolodge, Embassy Suites, Homestead Village, Red Lion Inn, Advance Parts, BrakeCheck, Firestone, FoodLand, Jo-Ann Crafts, OfficeMax, Volvo, bank, **W**...Texaco/mart, Baby Acapulco Mexican, Bombay Bicycle Club, Capt's Seafood, Carrabba's, IHOP, India Cuisine, Outback Steaks, Quizno's, Courtyard, Drury Inn, Fairfield Inn, Hilton, Motel 6, Quality Inn, Ramada Ltd, Rodeway Inn, Super 8, Dillard's, Ford, Office Depot, cinema

237a   Airport Blvd, **E**...BBQ, Motel Rio, True Value Inn, U-Haul, **W**...Jack-in-the-Box, GNC, Goodyear, HEB Foods, Old Navy, PetCo, Sears/auto

upper level is I-35 thru, lower level accesses downtown

236b   39th St, **E**...Chevron/mart, Fiesta Foods, HiLo Parts, **W**...Texaco/diesel/mart/wash, Chinese Rest., Tune&Lube, tires, to U of TX

N

S

Texas Interstate 35

a   26th-32nd Sts, **E**...Mi Madre's Mexican, Day's Inn, **W**...HOSPITAL, Rodeway Inn

235b   Manor Rd, same as 236a, **W**...VETERINARIAN, Rodeway Inn, U of TX, st capitol

a   MLK, 15th St, **W**...HOSPITAL

upper level is I-35 thru, lower level accesses downtown

234c   11th St, 12th St, downtown, **E**...Club Hotel, DoubleTree Hotel, Super 8, Eckerd, **W**...HOSPITAL, Exxon/mart, Shell, Texaco/diesel/mart, Marriott, Radisson, Sheraton, museum, st capitol

b   7th St, **E**...Chevron, Exxon, Texaco, **W**...Mobil, Waller Creek Plaza Hotel

a   1st St, Holly St, downtown, **W**...IHOP, tires

233   Riverside Dr, Town Lake, **E**...Shell/mart, TX Camper Corral, **W**...Chevron/diesel/mart/atm, Texaco/mart, Holiday Inn, Riverside Quarters

232.5mm Little Colorado River

232b   Woodland Ave, **E**...Wellesley Inn

a   Oltorf St, Live Oak, **E**...Texaco/diesel/mart, Carrow's, Luby's, La Quinta, **W**...Chevron, Exxon, Shell/mart, Texaco/repair, Denny's, Marco Polo Rest., Clarion, Quality Inn, Town Lake Inn

231   Woodward St, **E**...Shell/diesel/mart, Country Kitchen, Kettle, Exel Inn, Holiday Inn, Motel 6, Ramada Ltd, Super 8, same as 232, **W**...MEDICAL CARE

230b a   US 290 W, TX 71, Ben White Blvd, St Elmo Rd, **E**...Texaco/mart, Chinese Rest., Domino's, Jim's Rest., McDonald's, Western Choice Steaks, Best Western, Comfort Suites, Courtyard, Fairfield Inn, Red Roof Inn, Residence Inn, Celebration Sta, bank, **W**...HOSPITAL, Burger King, Furr's Cafeteria, IHOP, Pizza Hut, Candlewood Suites, Day's Inn, Hawthorne Suites, La Quinta, BMW, Chrysler/Jeep, Dodge, Ford, GMC, Honda, Hyundai, Lincoln/Mercury, NTB, Nissan, Pontiac/Cadillac/Mazda, Suzuki/Kia, Toyota

229   Stassney Lane, **E**...Sam's Club, Wal-Mart, **W**...same as 230b

228   Wm Cannon Drive, **E**...DENTIST, OPTICAL CARE, Exxon/mart/wash, Applebee's, McDonald's, Taco Bell, Chief Parts, Chrysler/Plymouth, Discount Tire, HEB Foods, Hyundai/Subaru, K-Mart, Mitsubishi, Target, RV camping, **W**...DENTIST, VETERINARIAN, Texaco/diesel/mart, Burger King, Chinese/Thai Rest., KFC, LJ Silver, Peter Piper Pizza, Taco Cabana, Wendy's, Whataburger/24hr, Chevrolet, Eckerd, Firestone

227   Slaughter Lane, Lp 275, S Congress, **E**...LoneStar RV Park, U-Haul, **W**...Shamrock/mart, Sonic

226   Slaughter Creek Overpass, **E**...LoneStar RV Park

225   FM 1626, Onion Creek Pkwy, **E**...Shamrock/mart, Texaco/mart, cleaners, **W**...Marshall's RV Ctr

224   frontage rd(from nb), no facilities

223   FM 1327, no facilities

221   Lp 4, Buda, **E**...Chevron/McDonald's/mart, Phillips 66/diesel, **W**...Chevron/mart/24hr, Texaco/diesel/mart/wash, RV Outlet Mall, Town&Country Mkt, bank

220   FM 2001, Niederwald, **E**...Texaco/mart, **W**...Crestview RV Ctr/Park

217   Lp 4, Buda, **E**...Conoco/diesel/mart/24hr, **W**...Exxon/mart, Shamrock/Pizza Hut/mart, Burger King, RV Ctr

215   Bunton Overpass, no facilities

213   FM 150, Kyle, **E**...Shamrock/mart, Dairy Queen, laundromat, **W**...Conoco/mart, Texaco/diesel/mart, Carquest/repair, tires

211.5mm **rest area both lanes, full(handicapped)facilities, phone, picnic tables, litter barrels, vending, petwalk**

211mm   truck check sta nb

210   Yarrington Rd, no facilities

209mm   truck check sta sb

208.5mm Blanco River

208   Frontage Rd, Blanco River Rd, no facilities

206   Lp 82, Aquarena Springs Rd, **E**...Conoco/mart/RV parking, Shamrock, **W**...Conoco/mart, Exxon/diesel/mart, Mobil/diesel/mart, Phillips 66/diesel/mart, Texaco/mart, Jim's Rest., Popeye's, Sonic, Comfort Inn, Executive Hotel, Howard Johnson Express, La Quinta, Motel 6, Ramada Ltd, Stratford House, Super 8, University Inn, to SW TX U, RV camping

205   TX 80, TX 142, Bastrop, **E**...EYECARE, Conoco/mart, Mobil/mart, Shamrock/mart, Shell/diesel/mart, Arby's, BBQ, Dairy Queen, Jason's Deli, Pantera's Pizza, Schlotsky's, Subway, Weiner's, Zapata's Café, Eckerd, GreaseMonkey, Hastings Books, Mail&More, Wal-Mart/24hr, bank, cinema, cleaners, **W**...VETERINARIAN, Chevron, Shamrock, Circle K/gas/24hr, Burger King, Church's Chicken, Furr's Cafeteria, KFC, LJ Silver, McDonald's, Montana Mike's Steaks, Pizza Hut, Sirloin Stockade, Taco Cabana, Wendy's, Best Western, Budget Inn, Day's Inn, Microtel, Rodeway Inn, Shoney's Inn, BrakeCheck, HEB Foods, JC Penney, OfficeMax, Target, Walgreen, carwash

204.5mm San Marcos River

N

S

Texas Interstate 35

204b CM Allen Pkwy, **W**...Chevron, Conoco/diesel, Texaco/diesel/mart, Chinese Buffet, Dairy Queen, Sonic, Best Western, Econolodge, Mustang Motel

   a Lp 82, TX 123, to Seguin, **E**...HOSPITAL, Conoco/mart, Phillips 66/mart, Shamrock, Texaco/diesel/mart, Burger King, Chili's, FasTaco, Golden Corral, Hardee's, Luby's, McDonald's, Red Lobster, Whataburger/24hr, Holiday Inn Express, Chevrolet/Buick/Olds, Ford/Mercury, Jeep/Chrysler, Pennzoil, repair, transmissions

202 FM 3407, Wonder World Dr, **E**...HOSPITAL, Jack-in-the-Box, Lowe's Bldg Supply, cinema, **W**...Shamrock/mart, BBQ, Plymouth/Dodge, repair

201 McCarty Lane, **E**...Auto Ranch/Pontiac/GMC, Subaru, **W**...Nissan

200 Center Point Rd, **E**...Center Point Sta Rest., Food Court, LoneStar Café, Outback Steaks, Subway, Taco Bell, Wendy's, Prime Outlets/famous brands, Tanger Outlet/famous brands, **W**...Shamrock/diesel/mart, Schlotsky's, Whataburger/24hr, AmeriHost

199 Posey Rd, no facilities

196 FM 1106, York Creek Rd, **W**...Acapulco Motel/rest.

195 Watson Lane, Old Bastrop Rd, **E**...antiques

193 Conrads Rd, Kohlenberg Rd, **W**...Rip Griffin/Shell/Subway/diesel/mart/24hr, Country Fair Rest.

191 FM 306, FM 483, Canyon Lake, **E**...Conoco/Quizno's/diesel/mart, Fina, RV camping, Wal-Mart Dist Ctr, **W**...Exxon/mart, Mobil/wash/24hr, Burger King, Camping World RV Supplies

190c Post Rd, no facilities

   b Lp 35 S, New Braunfels, **W**...Conoco/diesel/mart, Capparelli's Italian

   a frontage rd, **E**...Shell, Discount Tire, Home Depot, **W**...Best Western, Comfort Suites

189 TX 46, Seguin, **E**...Conoco/diesel/mart, Exxon, Texaco/diesel/mart/wash, Luby's, Oma's Haus Rest., Oakwood Inn, Stratford House, Super 8, K-Mart, Office Depot, **W**...HOSPITAL, Chevron/mart/24hr, Shamrock, Applebee's, IHOP, McDonald's, New Braunfels Smokehouse, Skillet's Rest., Taco Bell, Taco Cabana, Wendy's, Adelweiss Inn, Day's Inn, Fountain Motel, Hampton Inn, Holiday Inn, Motel 6, Rodeway Inn, factory stores

188 Frontage Rd, **W**...Mamacita's, Ryan's, Hastings Books, MillStore/famous brands

188mm Guadalupe River

187 FM 725, Lake McQueeny Rd, **E**...Chevron/mart/24hr, Arby's, Burger King, CiCi's, Guadalajara Mexican, LJ Silver, Subway, Whataburger/24hr, Big Lots, JiffyLube, Radio Shack, RV camping, hardware, **W**...HOSPITAL, CHIROPRACTOR, Exxon/diesel, Adobe Café, Dairy Queen, Jack-in-the-Box, Rally's, Steaks to Go, Budget Host, New Braunfels Motel, Shafer Inn/café, NAPA AutoCare, Precision Tune, bank, transmissions

186 Walnut Ave, **E**...Exxon/Subway/mart, Shamrock/mart, Chick-fil-A, Marina's Mexican, McDonald's, Popeye's, Sizzler, Econolodge, Chevrolet, Wal-Mart/auto, bank, **W**...CHIROPRACTOR, DENTIST, Conoco/mart, Shamrock/mart, Texaco/diesel/mart, KFC, Mr Gatti's, Papa John's, Schlotsky's, Shanghai Chinese, AutoZone, HEB/deli/gas, Mail-It+, Ross, Target, U-Haul, XpressLube, cinema, cleaners, hardware

185 FM 1044, **W**...Ford/Mercury/Lincoln

184 FM 482, Lp 337, Ruekle Rd, **E**...Texaco/Blimpie/diesel/mart, Hill Country RV Park

183 Solms Rd, **W**...Exxon/mart, Mazda/Kia

182 Engel Rd, **E**...Pam's Country Kitchen, Stahmann's RV Ctr, **W**...Mazda/Kia

180 Schwab Rd, no facilities

179.5mm   **rest area both lanes, full(handicapped)facilities, phone, picnic tables, litter barrels, vending, petwalk, RV dump**

178 FM 1103, Cibolo Rd, Hubertus Rd, **E**...Shell/diesel/mart, RV camping

177 FM 482, FM 2252, **W**...Stone Creek RV Park

176 Weiderstein Rd, no facilities

175 FM 3009, Natural Bridge, **E**...Shamrock/mart, BBQ, McDonald's, Schlotsky's, Taco Cabana, HEB Food/gas, Radio Shack, **W**...Shamrock/Subway/mart, Texaco/diesel, Arby's, Denny's, Jack-in-the-Box, Wendy's, Atrium Inn, Factory Shoestore, RV camping

174b Schertz Pkwy, **E**...La Pasadita Mexican

   a FM 1518, Selma, **E**...Diahatsu, Honda, Mitsubishi, Saturn, **W**...Texall RV Ctr

173 Old Austin Rd, Olympia Pkwy, **W**...Retama Park RaceTrack

172 TX 218, Anderson Lp, P Booker Rd, **E**...Comfort Inn, to Randolph AFB, **W**...to SeaWorld

171 Topperwein Rd, same as 170

170 Judson Rd, to Converse, **E**...HOSPITAL, CHIROPRACTOR, Texaco/diesel/mart/wash, India Cuisine, Kettle, Subway, Whataburger, La Quinta, Chevrolet, Chrysler/Plymouth, Ford, Gunn Auto Parts, Nissan, Pontiac/GMC, Toyota, cinema, **W**...Exxon/mart/wash, Holiday Inn Express, Mazda, Sam's Club, Subaru

169 O'Conner Rd, Wurzbach Pkwy, **E**...Conoco/diesel/mart, **W**...Shamrock/mart/wash, Texaco/diesel/mart, Jack-in-the-Box, Jim's Rest., Econolodge, K-Mart SuperCtr/Little Caesar's, RV camping

168 Weidner Rd, **E**...Exxon/mart, Shamrock/mart, Day's Inn, **W**...Chevron/mart/wash, Quality Inn, Ramada Inn, Super 8, Harley-Davidson

167b Thousand Oaks Dr, Starlight Terrace, **E**...Stop'n Go/diesel, Best Western

   a Randolph Blvd, **E**...TravelTown TX RV Ctr, **W**...Best Western, Classic Inn, Comfort Inn, Day's Inn, Motel 6, Ruby Inn, Suzuki

New Braunfels

N

S

Texas Interstate 35

166   I-410 W, Lp 368 S, **W**...to Sea World

165   FM 1976, Walzem Rd, **E**...CHIROPRACTOR, OPTOMETRIST, Chevron, Shamrock/mart, Texaco/mart/wash, Applebee's, Baskin-Robbins, Blackeyed Pea, Burger King, Chinese Rest., Church's Chicken, CiCi's, IHOP, Jack-in-the-Box, Jim's Rest., KFC, LJ Silver, Luby's, Marie Callender's, McDonald's, Olive Garden, Pancho's Mexican, Pizza Hut, Red Lobster, Schlotsky's, Shoney's, Subway, Taco Bell, Taco Cabana, Taco Kitchen, Wendy's, Whataburger, Drury Inn, Hampton Inn, Albertson's, AutoZone, CarQuest, Cavender's Boots, Circuit City, Dillard's, Discount Tire, Eckerd, Firestone/auto, HEB Foods, Home Depot, JC Penney, MailBoxes Etc, Mervyn's, Michael's, Office Depot, OfficeMax, Pearle Vision, PepBoys, PetsMart, Radio Shack, Service Mechandise, Target, Texas Mail, ToysRUs, Wal-Mart SuperCtr/24hr, bank, cleaners/laundry, mall, **W**...Mobil, Sonic, NTB

164   Eisenhauer Rd, **E**...Exxon/diesel/mart, $General, Hancock Fabrics

163   Rittiman Rd, **E**...VETERINARIAN, Conoco/diesel/mart, Exxon/TacoMaker/mart, RaceTrac/mart, Shamrock, Texaco/diesel/mart/wash, Circle K/24hr, Burger King, Chinese Rest., Church's Chicken, Cracker Barrel, Denny's, Kettle, McDonald's, Taco Cabana, Whataburger/24hr, Comfort Suites, Knight's Inn, La Quinta, Motel 6, Ramada Inn, Scotsman Inn, Stratford House, Super 8, HEB Foods, **W**...Chevron/diesel/mart, Mobil, Shamrock/mart, BBQ, Sonic, Wendy's

162   Petroleum Dr(from nb), frontage rd, no facilities

162mm   I-410 S(exits left from sb), no facilities

161.5mm   Holbrook Rd, same as 160

160   Coliseum Rd, **E**...Shamrock/mart/24hr, Delux Inn, **W**...Conoco/mart, BBQ, Day's Inn, Economy Inn, Executive Inn, Holiday Inn/rest., Quality Inn, Stratford House, Super 8, Spashtown Funpark, RV camping

159b   Walters St, **E**...McDonald's, Flores Drive-In, transmissions, **W**...Howard Johnson, to Ft Sam Houston

   a   New Braunfels Ave, **E**...Exxon/mart, Texaco/diesel/mart/wash, **W**...Chevron, Shamrock/diesel/24hr, BBQ, Sonic, Ramada Suites, to Ft Sam Houston

158c   N Alamo St, Broadway, **W**...Ace Hardware, Chas Supermkt

   b   I-37 S, US 281 S, to Corpus Christi

   a   US 281 N(from sb), to Johnson City

157b a   Brooklyn Ave, Lexington Ave, N Flores, downtown, **E**...HOSPITAL, **W**...Luby's, Rodeway Inn

156   I-10 W, US 87, to El Paso

155b   Durango Blvd, downtown, **E**...HOSPITAL, Courtyard, Fairfield Inn, Holiday Inn, La Quinta, Residence Inn, Woodfield Suites, **W**...Motel 6, Radisson

   a   South Alamo St, **E**...Exxon, Texaco, Church's Chicken, McDonald's, Pizza Hut, Wendy's, Comfort Inn, Day's Inn, Ramada Ltd, **W**...Conoco, Microtel

154b   S Laredo St, Ceballos St, same as 155b

   a   Nogalitos St, no facilities

153   I-10 E, US 90 W, US 87, to Kelly AFB, Lackland AFB

152b   Malone Ave, Theo Ave, **E**...Taco Cabana/24hr, **W**...Shamrock/mart/24hr, Texaco/diesel/mart

   a   Division Ave, **E**...Chevron/mart, BBQ, Whataburger/24hr, Holiday Inn Express, **W**...Exxon, Phillips 66, transmissions

151   Southcross Blvd, **E**...Citgo/mart, Exxon/mart, Shop'n Save, camping, **W**...Circle K, Texaco/diesel/mart/wash

150b   Lp 13, Military Dr, **E**...CHIROPRACTOR, EYECARE, Chevron/mart/wash, Conoco, Texaco/mart, Chinese Rest., Denny's, KFC, Pizza Hut, Taco Cabana, La Quinta, AutoZone, BrakeCheck, Discount Tire, U-Haul, cinema, **W**...Exxon/mart, Mobil/mart, Shamrock, Golden Corral, Hungry Farmer Steaks, Jack-in-the-Box, KFC, LJ Silver, Luby's, McDonald's, Pizza Hut, Shoney's, Wendy's, Whataburger, Albertson's, Eckerd, Goodyear, HEB Foods, JC Penney, Mervyn's, Office Depot, Sears/auto, Target, ToysRUs, bank, mall,

   a   Zarzamora St(149 fom sb), same as 150b

149   Hutchins Blvd(from sb), **E**...HOSPITAL, Shamrock/mart, Motel 6

148b   Palo Alto Rd, **W**...Shamrock/mart

   a   TX 16 S, spur 422(exits left from sb), Poteet, **E**...Chevron, **W**...Phillips 66/mart, $General

147   Somerset Rd, **E**...Texaco/diesel/mart

146   Cassin Rd(from nb), no facilities

145b   Lp 353 N, no facilities

   a   I-410, TX 16, no facilities

144   Fischer Rd, **E**...Shamrock/Subway/diesel/mart/24hr, D&D Motel, RV camping

142   Medina River Turnaround(from nb), no facilities

141   Benton City Rd, Von Ormy, **W**...Texaco/diesel/café/mart

N

S

Texas Interstate 35

140	Anderson Lp, 1604, **E**...Exxon/Burger King/diesel/mart/24hr, Fina, **W**...Texaco/Pizza Hut/Church's/diesel/mart/24hr, to Sea World
139	Kinney Rd, **E**...RV camping
137	Shepherd Rd, **W**...JC's Gas/mart
135	Luckey Rd, no facilities
133	TX 132 S, Lytle, access to same as 131
131	FM 3175, FM 2790, Benton City Rd, **E**...KwikLube, bank, **W**...Conoco/diesel/mart, Dairy Queen, PigStand Rest., Day's Inn, HEB Foods, bank
129mm	**rest area both lanes, full(handicapped)facilities, phone, picnic tables, litter barrels, vending, petwalk**
127	FM 471, Natalia, no facilities
124	FM 463, Bigfoot Rd, **E**...Ford
122	TX 173, Divine, **E**...Exxon/Church's/diesel/mart, Steak&Burger, Chevrolet, Plymouth/Jeep/Dodge, **W**...Chevron/Subway/diesel/mart, CCC Steaks, Country Corner Motel
121	TX 132 N, Devine, **E**...RV camping
118.5mm	weigh sta both lanes
114	FM 462, Bigfoot, **E**...Shamrock/mart, **W**...gas/mart/lube
111	US 57, to Eagle Pass, no facilities
104	Lp 35, **2 mi E**...HOSPITAL, Executive Inn
101	FM 140, Pearsall, **E**...Chevron/mart, BBQ, Royal Inn, **W**...Exxon/Blimpie/diesel/mart, Porterhouse Inn/rest., Budget Inn, bank
99	FM 1581, Pearsall, to Divot, **W**...LP
93mm	parking/picnic area both lanes, litter barrels, handicapped accessible
91	FM 1583, Derby, no facilities
90mm	Frio River
86	Lp 35, Dilley, no facilities
85	FM 117, **E**...Garcia's Café/motel, **W**...Exxon/diesel/mart, Dairy Queen, Safari Motel
84	TX 85, Dilley, **E**...HOSPITAL, Super S Foods, Chevrolet/Pontiac, NAPA, bank, **W**...Shamrock/diesel/mart/24hr, Texaco/Church's/diesel/mart/24hr, Executive Inn
82	County Line Rd, Dilley, no facilities
77	FM 469, Millett, no facilities
74	Gardendale, no facilities
68	Lp 35(from sb), Cotulla, **1 mi E**...Mobil, Chuck Wagon Steaks, Cotulla Motel
67	FM 468, Big Wells, **E**...Exxon/Wendy's/diesel/mart/24hr, Conoco/Country Store/rest., Shamrock/Church's/Pizza Hut/diesel/mart/24hr, Dairy Queen, Executive Inn, Village Inn, RV camping, tire repair, **W**...Chevron/McDonald's/diesel/mart
65	Lp 35, Cotulla, **1-2 mi E**...Mobil, Chuck Wagon Steaks, Cotulla Motel
63	Elm Creek Interchange, no facilities
59mm	parking/picnic area both lanes, litter barrels, handicapped accessible
56	FM 133, Artesia Wells, **E**...gas/mart
48	Caiman Creek Interchange, no facilities
39	TX 44(from sb), Encinal, no facilities
38	TX 44(from nb), Encinal, no facilities
32	San Roman Interchange, no facilities
27	Callaghan Interchange, no facilities
22	Webb Interchange, no facilities
18	US 83 N, to Carrizo Springs, **Welcome Ctr/rest area(both lanes exit left), full(handicapped)facilities, phone, picnic tables, litter barrels, vending, petwalk**
15mm	parking/picnic area sb, insp area nb
13	Uniroyal Interchange, **E**...Pilot/Subway/diesel/mart/24hr, Blue Beacon
8	FM 1472, Milo, no facilities
7	Shilo Dr, Las Cruces Dr, **E**...Motel 9
6mm	Texas Tourist Bureau, info
4	FM 1472, Del Mar Blvd, **E**...Chevron/repair, Exxon/diesel, Applebee's, Carino's Italian, Dairy Queen, Jack-in-the-Box, Las Asadas Mexican, McDonald's, Shoney's, Hampton Inn, Albertson's/drugs, Marshall's, Target, cinema, **W**...Fina/mart, Texaco/diesel/mart, Danny's Rest., Golden Corral, Executive House Motel, Motel 6
3b	Mann Rd, **E**...Sirloin Stockade, Family Garden Inn, Buick/Olds, Dillard's, Foley's, Ford/Lincoln/Mercury, Honda, Mazda, Mervyn's, Pontiac/Cadillac, Sears/auto, Toyota, Ward's, bank, cinema, mall, **W**...Chili's, Outback Steaks, Subway, Taco Palenque, Motel 6, Best Buy, EyeMart, Michael's, Office Depot, OfficeMax, Ross, Service Merchandise, ToysRUs, Wal-Mart/auto/24hr, bank
a	San Bernardo Ave, **E**...CHIROPRACTOR, Texaco/diesel/mart, Chick-fil-A, Chinese Rest., Fuddrucker's, LJ Silver, Luby's, Peter Piper Pizza, HEB Foods, K-Mart, NAPA, PepBoys, SteinMart, cinema, hardware, mall, **W**...Citgo/Circle K, Chevron, Phillips 66/diesel, Acapulco Steak/shrimp, Baskin-Robbins, Burger King, Coyote Creek Rest., Dairy Queen,

Pearsall

N

S

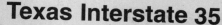

Texas Interstate 35

N

S

Dunkin Donuts, El Pollo Loco, Kettle, McDonald's, Ming Chinese, Pizza Hut, Popeye's, Red Line Burgers, Taco Bell, Taco Palenque, Wendy's, Whataburger, Best Western, Gateway Inn, La Hacienda Motel, Monterey Inn, Red Roof Inn, Burlington Coats, Gonzales Auto Parts, Goodyear/auto, Radio Shack, Sam's Club, bank, carwash

2   US 59, Saunders Rd, **E...**HOSPITAL, Conoco/mart, Shamrock/mart, Texaco/mart, Jack-in-the-Box, Holiday Inn, **2 mi E...**Conoco/diesel/ mart, Exxon/Circle K, Texaco/Subway/mart, Burger King, Church's Chicken, McDonald's, Pizza Hut, Taco Bell, Whataburger, AutoZone, Chevrolet, HEB Foods, USPO, **W...**Chevron/diesel/mart/24hr, Exxon, Phillips 66, Church's Chicken, Denny's, Courtyard, Econolodge, El Cortez Motel, Holiday Inn, La Quinta, Mayan Inn, AutoZone, Mexico Insurance

1b   Park St, to Sanchez St, **W...**CHIROPRACTOR, Conoco/diesel/mart, KFC, La Mexicana Rest., Pizza Hut, Popeye's

1a   Victoria St, Scott St, Washington St, **E...**Exxon, Citgo/Circle K, Texaco/mart, **W...**Exxon/24hr, Texaco/mart, Circle K, KFC, McDonald's, Wendy's, Dutyfree, Firestone/auto, Goodyear/auto

I-35 begins/ends in Laredo at Victoria St...access to multiple facilities

Texas Interstate 35W

Exit #	Services
I-35W begins/ends on I-35, exit 467.	
85a	I-35E S, no facilities
84	FM 1515, Bonnie Brae St, **E...**HOSPITAL
82	FM 2449, to Ponder, no facilities
79	Crawford Rd, no facilities
76	FM 407, Argyle, to Justin, no facilities
76mm	picnic area both lanes, litter barrels, picnic tables
74	FM 1171, to Lewisville, no facilities
70	TX 114, to Dallas, Bridgeport, **E...**Texaco/Church'sChicken/diesel/mart, to DFW Airport, **W...**TX Motor Sreedway, golf
68	Eagle Pkwy, **W...**airport
67	Alliance Blvd, **W...**to Alliance Airport, FedEx
66	to Westport Pkwy, Keller-Haslet Rd, **E...**Hampton Inn, Hawthorn Suites, **W...**DENTIST, Mobil/Wendy's/diesel/mart, BBQ, Bryan's Smokehouse, Cactus Flower Café, Oriental Garden, Schlotsky's, Subway, Taco Bueno, USPO, bank
65	TX 170 E, no facilities
64	Golden Triangle Blvd, to Keller-Hicks Blvd, no facilities
63	Park Glen, no facilities
60	US 287 N, US 81 N, to Decatur, no facilities
59	I-35 W to Denton(from sb), turnaround
58	Western Ctr Blvd, **E...**EYECARE, MEDICAL CARE, Citgo/7-11, Texaco/Church'sChicken/mart, BBQ, Blackeyed Pea, Braum's, Chili's, Domino's, Donut Palace, Macaroni Grill, On-the-Border, SaltGrass Steaks, Wendy's, Best Western, Residence Inn, cinema, cleaners
57b a	I-820 E&W, no facilities
56	Meacham Blvd, **E...**Texaco/mart/wash, Comfort Inn, Hilton Garden, La Quinta, Dillard's, bank, **W...**Mobil/mart/ wash, Cracker Barrel, McDonald's, Hampton Inn, Holiday Inn, USPO, cleaners
55	Pleasantdale Ave(from nb), no facilities
54c	33rd St, Long Ave(from nb), no facilities
b a	TX 183 W, Papurt St, **W...**Citgo/diesel/mart, Shamrock/mart, McDonald's, Classic Inn, Motel 6, bank, truckwash
53	North Side Dr, Yucca Dr, **E...**Texaco/mart, **W...**Mercado Juarez Café, Country Inn Suites
53mm	Trinity River
52d	Carver St(from nb), no facilities
c	Pharr St, no facilities
b	Belknap(from sb), no facilities
a	US 377 N, TX 121, to DFW, no facilities
51	spur 280, downtown Ft Worth
50c a	I-30 W, E to Dallas
b	TX 180 E(from nb), no facilities

N

S

Texas Interstate 35W

49b   Rosedale St, **W**...HOSPITAL

a   Allen Ave, **E**...Chevron, **W**...HOSPITAL

48b   Morningside Ave(from sb), same as 48a

a   Berry St, **E**...Chevron/McDonald's, Citgo, bank, **W**...RaceTrac/mart, Sack'n Save Foods, U-Haul, cleaners, tires

47   Ripy St, **E**...Meineke Mufflers, transmissions

46b   Seminary Dr, **E**...Grandy's, Golden Corral, Jack-in-the-Box, Taco Cabana, Wyatt's Cafeteria, Day's Inn, Delux Inn, Regency Inn, Super 7 Inn, Pearle Vision, **W**...EYECARE, Exxon, Texaco/mart, Denny's, El Chico, Peony Chinese, Dillard's, Firestone/auto, JC Penney, Sears, bank, cinema

a   Felix St(from nb), **E**...Mobil, Pulido's Mexican, S Oaks Inn, **W**...Burger King, McDonald's, Dodge

45b a   I-20, E to Dallas, W to Abilene

44   Altamesa, **E**...Holiday Inn, **W**...Citgo/mart, Rig Steaks, Waffle House, Best Western, Comfort Suites, Motel 6, South Lp Inn, RV Ctr, bank

43   Sycamore School Rd, **W**...Exxon/mart/wash, Beefer's Rest./24hr, Jack-in-the-Box, Subway, Taco Bell, Whataburger/24hr, Aaron's Foods, $General, Home Depot, Radio Shack

42   Everman Pkwy, **E**...Exxon/mart/wash, bank, **W**...RV Ctr, hardware

41   Risinger Rd, **W**...auto repair

40   Garden Acres Dr, **E**...HOSPITAL, Love's/diesel/mart/24hr, Microtel, **W**...RV Lite Ctr

39   FM 1187, McAlister Rd, **E**...HOSPITAL, **W**...Citgo/diesel/mart, Shamrock/diesel/mart/24hr, Taco Bell, Waffle House, Howard Johnson, cinema

38   Alsbury Blvd, **E**...Chevron/mart, Mobil/diesel/mart, Chili's, Country Steaks, Cracker Barrel, McDonald's, Taipan Chinese, Super 8, Ford, Lowe's Bldg Supply, **W**...VETERINARIAN, Citgo, Texaco/24hr, Arby's, Burger King, Denny's, Grandy's, Pancho's Mexican, Albertson's/drugs, Chevrolet, K-Mart, bank

37   TX 174, Wilshire Blvd, to Cleburne, **W**...Exxon, Shell, Eckerd

36   FM 3391, Burleson, **E**...Citgo, Mobil/mart, Luby's, Steak Exchange, Waffle House, Comfort Suites, Day's Inn, bank, **W**...Chevron, Fina/diesel, Shamrock, auto repair, golf

34   Briaroaks Rd(from sb), **W**...RV camping

33mm   **rest area sb, full(handicapped)facilities, phone, picnic tables, litter barrels**

32   Bethesda Rd, **E**...Citgo/mart, 5 Star Inn/rest., Bill&Martin's Seafood Rest., **W**...RV Ranch Park

31mm   **rest area nb, full(handicapped)facilities, phone, picnic tables, litter barrels**

30   FM 917, Mansfield, **W**...Fina/mart, Texaco/Snappy Jack's/diesel

27   County Rds 604 & 707, **E**...Travel Villa RV Ctr

26b a   US 67, Cleburne, **E**...Citgo/Pop'sChicken/diesel/mart, Exxon/Subway/mart/24hr, Texaco, bank

24   FM 3136, FM 1706, Alvarado, **E**...Chevron, Texaco/diesel/mart/24hr, Alvarado House Rest., **W**...McClain's RV Ctr

21   Barnesville Rd, to Greenfield, no facilities

17   FM 2258, no facilities

16   TX 81 S, County Rd 201, Grandview, **E**...RV camping, **W**...RV camping

15   FM 916, Maypearl, **W**...Phillips 66(1mi), Shamrock/diesel/mart/wash, BBQ

12   FM 67, no facilities

8   FM 66, Itasca, **E**...Chevron/diesel/café/24hr, **W**...Citgo/diesel/mart, Dairy Queen, Ford, picnic tables, litter barrels

7   FM 934, **E**...picnic tables, litter barrels

3   FM 2959, **E**...to Hillsboro Airport

I-35W begins/ends on I-35, 371mm.

Texas Interstate 37

Exit #	Services
143mm	I-37 begins/ends on I-35 in San Antonio.
141c	Brooklyn Ave, Nolan St(from sb), downtown
b	Houston St, downtown, **W**...Denny's, Crockett Hotel, Day's Inn, Hampton Inn, Holiday Inn, Marriott, to The Alamo
a	Commerce St, downtown, **W**...La Quinta, Marriott
140b	Durango Blvd, downtown, **E**...Bill Miller BBQ, to Aladome
a	Carolina St, Florida St, no facilities
139	I-10 W, US 87, US 90, to Houston, **W**...to Sea World
138c	Fair Ave, Hackberry St, **E**...Chevron/mart, Dairy Queen, KFC, Peter Piper Pizza, Rally's, RedLine Burgers, Taco Bell, BrakeCheck, Family$, Hancock Fabrics, Home Depot, **W**...Exxon, Texaco/mart
b	E New Braunfels Ave, **E**...Chick-fil-A, Jim's Rest., Luby's, McDonald's, Taco Cabana, Wendy's, EyeMart Express, bank, mall
a	Southcross Blvd, W New Braunfels Ave, **W**...Exxon, Mobil, Burger King, Pizza Hut, Sonic
137	Hot Wells Blvd, **E**...gas/mart, bank, **W**...IHOP, Motel 6
136	Pecan Valley Dr, HOSPITAL, **E**...Coastal/mart, KFC, hardware
135	Military Dr, Lp 13, **E**...Shamrock/mart, **W**...Shamrock/mart, Burger King, Chinese Rest., Coachman Inn, HEB Food/drugs, K-Mart/drugs, 1-Hr Photo, to Brooks AFB
133	I-410, US 281 S, no facilities

Texas Interstate 37

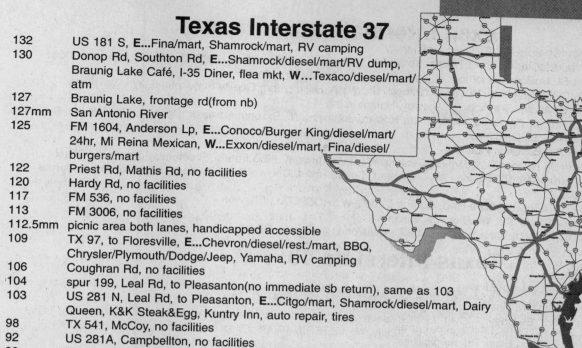

132	US 181 S, **E**...Fina/mart, Shamrock/mart, RV camping
130	Donop Rd, Southton Rd, **E**...Shamrock/diesel/mart/RV dump, Braunig Lake Café, I-35 Diner, flea mkt, **W**...Texaco/diesel/mart/atm
127	Braunig Lake, frontage rd(from nb)
127mm	San Antonio River
125	FM 1604, Anderson Lp, **E**...Conoco/Burger King/diesel/mart/24hr, Mi Reina Mexican, **W**...Exxon/diesel/mart, Fina/diesel/burgers/mart
122	Priest Rd, Mathis Rd, no facilities
120	Hardy Rd, no facilities
117	FM 536, no facilities
113	FM 3006, no facilities
112.5mm	picnic area both lanes, handicapped accessible
109	TX 97, to Floresville, **E**...Chevron/diesel/rest./mart, BBQ, Chrysler/Plymouth/Dodge/Jeep, Yamaha, RV camping
106	Coughran Rd, no facilities
104	spur 199, Leal Rd, to Pleasanton(no immediate sb return), same as 103
103	US 281 N, Leal Rd, to Pleasanton, **E**...Citgo/mart, Shamrock/diesel/mart, Dairy Queen, K&K Steak&Egg, Kuntry Inn, auto repair, tires
98	TX 541, McCoy, no facilities
92	US 281A, Campbellton, no facilities
88	FM 1099, to FM 791, Campbellton, no facilities
83	FM 99, Whitsett, Peggy, **E**...Texaco/diesel/mart, **W**...Chevron/mart, Exxon/diesel/deli/RV Park/dump
82mm	**rest area sb, full(handicapped)facilities, phone, picnic tables, litter barrels**
78mm	**rest area nb, full(handicapped)facilities, phone, picnic tables, litter barrels**
76	US 281Alt, FM 2049, Whitsett, no facilities
75mm	truck weigh sta sb
74mm	truck weigh sta nb
72	US 281 S, Three Rivers, **3 mi W**...Staghorn Rest., Best Western, to Rio Grande Valley
69	TX 72, Three Rivers, **W**...Shamrock/diesel/mart/café/24hr, to Choke Canyon SP
65	FM 1358, Oakville, **E**...Texaco, BBQ, **W**...Chevron/mart, RedLine Café
59	FM 799, no facilities
56mm	parking area sb
56	US 59, George West, **E**...Texaco/Circle K/diesel/mart/24hr, **W**...Chevron/Subway/TCBY/mart/24hr, Exxon/Burger King/diesel/mart/24hr, Rockin'N RV Park
51	Hailey Ranch Rd, no facilities
47	FM 3024, FM 534, Swinney Switch Rd, **W**...to KOA
44mm	parking area sb
42mm	parking area nb
40	FM 888, no facilities
36	TX 359, Mathis, to Skidmore, **W**...Exxon/diesel/mart, Pizza Hut, Mathis Motel, Ranch Motel/rest.
34	TX 359 W, **W**...Shamrock/diesel/mart, BBQ, Ford, to Lake Corpus Christi SP, RV camping
31	TX 188, Rockport, to Sinton
22	TX 234, FM 796, Edroy, to Odem, no facilities
20b	Cooper Rd, no facilities
19.5mm	**rest area both lanes, full(handicapped)facilities, phone, picnic tables, litter barrels, vending, petwalk**
17	US 77 N, to Victoria, no facilities
16	Nueces River Park, info, picnic tables
15	Sharpsburg Rd, Redbird Lane, no facilities
14	US 77 S, Robstown, to Kingsville, **W**...HOSPITAL, Chevron/Burger King/mart, Exxon/Circle K, RaceTrac/mart, Good'n Crisp Chicken, K-Bob's, Pizza Hut, Popeye's, Subway, Wienerschnitzel, Whataburger/24hr, Beall's, Eckerd, Firestone, HobbyLobby, Radio Shack, Wal-Mart Super Ctr/24hr, bank, cinema, cleaners
13b a	FM 1694, Callicoatte Rd, Leopard St, no facilities
11b	FM 24, Violet Rd, Hart Rd, **E**...Conoco/diesel/mart/24hr, **W**...Citgo/mart, Exxon/Circle K, BBQ, Dairy Queen, Domino's, KFC, McDonald's, Schlotsky's, Sonic, Subway, Taco Bell, Whataburger/24hr, Best Western, AutoZone, HEB Food/gas
a	McKinzie Rd, no facilities
10	Carbon Plant Rd, no facilities
9	FM 2292, Up River Rd, Rand Morgan Rd, **W**...Texaco/mart, Whataburger/24hr
7	Suntide Rd, Tuloso Rd, no facilities

N

S

Texas Interstate 37

6   Southern Minerals Rd, **E**...Citgo Refinery

5   Corn Products Rd, **E**...GMC Trucks, Kenworth, Mack, **W**...PetroFleet Gas, Kettle, Red Roof Inn, Travelodge

4b   Lantana St, McBride Lane(from sb), **W**...Drury Inn, Holiday Inn, Motel 6

a   TX 358, to Padre Island, **W**...Texaco/diesel, Drury Inn, Holiday Inn, Quality Inn, Motel 6, Wal-Mart(6mi)

3b   McBride Lane, **W**...Greyhound Raceway, Hampton Inn

a   Navigation Blvd, **E**...Texaco/diesel/mart, Howard Johnson, **W**...Exxon/mart/wash, Denny's, Comfort Inn, Day's Inn, Hampton Inn, La Quinta

2   Up River Rd, **E**...Citgo Refinery

1e   Lawrence Dr, Nueces Bay Blvd, **E**...Texaco, **W**...Citgo/Circle K, HEB Foods, Firestone, Ford, Radio Shack

1d   Port Ave(from sb), **E**...Texaco, **W**...Citgo/mart, Coastal/mart, Chevrolet, HEB Foods, bank, to Port of Corpus Christi

1c   US 181, TX 286, Shoreline Blvd, Corpus Christi, **W**...HOSPITAL, Chevron

1b   Brownlee St(from nb), **W**...Chevron, Circle K

1a   Buffalo St(from sb), Best Western. I-37 begins/ends in Corpus Christi.

Texas Interstate 40

Exit #	Services
177mm	Texas/Oklahoma state line
176	spur 30(from eb), to Texola, **S**...Shamrock/mart, BBQ, RV camping
175mm	picnic area wb, picnic tables, litter barrels
173mm	picnic area eb, picnic tables, litter barrels
169	FM 1802, Carbon Black Rd, no facilities
167	FM 2168, Daberry Rd, **N**...Texaco/diesel/mart/Butch's BBQ/24hr
164	Lp 40(from wb), to Shamrock, **1 mi S**...HOSPITAL, Irish City Trkstp/diesel, Budget Inn, Economy Inn, Econolodge, camping, museum
163	US 83, Shamrock, to Wheeler, **N**...Chevron/diesel, Conoco, Texaco/diesel/mart/24hr, Hasty's Burgers, Mitchell's Rest., Pizza Hut, Best Western Irish/rest., Western Motel, Ace Hardware, mall, **S**...Phillips 66/Subway/mart/24hr, Dairy Queen, McDonald's, Budget Inn, Econolodge
161	Lp 40, Rte 66(from eb), to Shamrock, **S**...Chevron
157	FM 1547, Lela, **2 mi S**...W 40 RV Park
152	FM 453, Pakan Rd, **S**...camping
150mm	**rest area wb, full(handicapped)facilities, picnic tables, litter barrels, petwalk**
149mm	**rest area eb, full(handicapped)facilities, picnic tables, litter barrels, petwalk**
148	FM 1443, Kellerville Rd, no facilities
146	County Line Rd, **S**...camping
143	Lp 40(from wb), to McLean, to gas, phone, food, lodging, museum
142	TX 273, FM 3143, to McLean, **N**...Conoco, Texaco/diesel/mart, Texas Motel/rest., CarQuest, Country Corner RV Park, auto parts
141	Rte 66(from eb), McLean, **N**...Red River Steaks, Cactus Inn, same as 142
135	FM 291, Old US 66, Alanreed, **S**...Conoco/motel/café/RV hookups, USPO, auto repair
132	Johnson Ranch Rd, ranch access, no facilities
131mm	picnic area wb, picnic tables, litter barrels
129mm	picnic area eb, picnic tables, litter barrels
128	FM 2477, to Lake McClellan, **S**...access to camping
124	TX 70 S, to Clarendon, **N**...gas/diesel, café, RV camping
121	TX 70 N, to Pampa, no facilities
114	Lp 40, Groom, **N**...auto/diesel/service, **S**...to gas/diesel, food, lodging
113	FM 2300, Groom, **S**...Texaco/diesel/mart, Dairy Queen, Chalet Inn
112	FM 295, Groom, **S**...Biggest Cross, **1 mi S**...gas/mart
110	Lp 40, Rte 66, no facilities
109	FM 294, no facilities
108mm	parking area wb, litter barrels
106mm	parking area eb, picnic tables, litter barrels
105	FM 2880, grain silo, no facilities
98	TX 207 S(from wb), to Claude, no facilities
96	TX 207 N, to Panhandle, **N**...Love's/Subway/diesel/mart/atm/24hr, **S**...Shell/repair/mart, A&W, Budget Host, Conway Inn
89	FM 2161, to Old US 66, no facilities
87	FM 2373, no facilities
86.5mm	**rest areas both lanes, full(handicapped)facilities, phone, picnic tables, litter barrels, weather info, petwalk**
85	Amarillo Blvd, Durrett Rd, access to camping
81	FM 1912, **N**...Texaco/diesel/mart, **S**...TA/Subway/diesel/rest./mart/24hr

N

S

E

W

Texas Interstate 40

80	FM 228, **N**...Texaco/diesel, AOK RV Park, to Texas Tech Institute
78	US 287 S(from eb), FM 1258, Pullman Rd, same as 77
77	FM 1258, Pullman Rd, **N**...Pilot/Arby's/diesel/mart/24hr, **S**...Texaco/diesel/mart/24hr, Custom RV Ctr
76	spur 468, **N**...Flying J/Conoco/diesel/LP/rest./24hr, Texaco/ diesel/mart, to airport, tourist info
75	Lp 335, Lakeside Rd, **N**...Phillips 66/mart, Shell/mart/24hr, Williams/diesel/mart/rest./24hr, Country Barn Rest., Waffle House, Quality Inn, Super 8, Trailer Inn, UPS, Peterbilt Trucks, KOA(2mi), **S**...Petro/Mobil/ diesel/rest., Blue Beacon
74	Whitaker Rd, **N**...Big Texan Inn/café, RV camping, **S**...Love's/diesel/mart, TA/FoodCourt/diesel/mart/24hr, Coachlight Inn, Blue Beacon
73	Eastern St, Bolton Ave, Amarillo, **N**...Shamrock/diesel/mart, Polly's Rest., Fiesta Motel, Motel 6, Honda/Suzuki/Yamaha, **S**...Chevron/diesel/mart, Citgo/diesel/mart, Pacific Pride/diesel, Cally's Corner Rest., Best Western, Ford Trucks
72b	Grand St, Amarillo, **N**...Conoco, LoneStar Fuel, Texaco/mart, Henk's BBQ, KFC, Motel 6, Nissan, U-Haul, **S**...Phillips 66/mart, Braum's, McDonald's, Pizza Hut, Sonic, Subway, Taco Cabana, Taco Villa, Motel 6, Advance Parts, Big Lots, Meineke Muffler, United Drugs, Wal-Mart SuperCtr/gas/24hr, bank, same as 73
a	Nelson St, **N**...Cracker Barrel, KFC, Budget Host, Econolodge, Ramada Inn, Sleep Inn, Super 8, Travelodge, Qtrhorse Museum, **S**...Chevron, Shamrock/diesel, Shell, Camelot Inn
71	Ross St, Osage St, Amarillo, **N**...Conoco/diesel/mart, Shamrock/mart/wash, Shell/mart, Texaco/diesel/mart, Burger King, Grandy's, Kettle, Longhorn Diner, LJ Silver, McDonald's, Popeye's, Schlotsky's, Coachlight Inn, Comfort Inn, Day's Inn, Holiday Inn, **S**...Shamrock/mart, Texaco, Arby's, Denny's, La Fiesta Mexican, Taco Bell, Wendy's, Hampton Inn, La Quinta, Red Roof Inn, Chevrolet, Ford, Hyundai, Sam's Club, USPO
70	I-27 S, US 60 W, US 87, US 287, to Canyon, Lubbock, **N**...Goodyear/auto, bank
69b	Washington St, Amarillo, **N**...Albertson's, **S**...Phillips 66/mart, Texaco, Dairy Queen, Eckerd
a	Crockett St, access to same as 68b
68b	Georgia St, **N**...Shell/Subway/mart/wash, Texaco/mart, Ambassador Hotel, **S**...HOSPITAL, CHIROPRACTOR, Texaco/wash, Burger King, BBQ, Church's Chicken, Denny's, Furr's Café, Taco Villa, Whataburger, Econolodge, Quality Inn, K-Mart, Chrysler/Dodge, Hastings Books, Mitsubishi, Office Depot, Saturn, bank
a	Julian Blvd, Paramount Blvd, **N**...Chevron, Shell, Texaco, Arby's, Nick's Rest., Pancho's Mexican, Pizza Hut, Schlotzsky's, Wendy's, Harley Hotel, mufflers, same as 67, **S**...Shamrock/mart, Bennigan's, Caboose Diner, Cactus Grill, Cajun Magic, Calico Country Café, El Chico, Godfather's, Hattie's Rest., Kabuki's Japanese Steaks, Kettle, LJ Silver, Orient Express, Peking Rest., Pizza Planet, Red Lobster, Ruby Tequila's Mexican, Steak&Ale, Western Sizzlin, Comfort Suites, Holiday Inn Express, Motel 6, Travelodge, Cadillac, Discount Tire, Mr Muffler, Pennzoil
67	Western St, Amarillo, **N**...Citgo/mart, Phillips 66/diesel/mart, Texaco/Blimpie/mart, Beef Rigger Rest., Blackeyed Pea, Burger King, Chili's, Llano Grill, Marie Callender's, McDonald's, Pancho's Mexican, Taco Bell, **S**...CHIROPRACTOR, Phillips 66/mart, Shamrock/mart, Bagel Place, Catfish Shack, Dairy Queen, Furr's Café, Gary's BBQ, IHOP, Olive Garden, Taco Cabana, Vince's Pizza, Waffle House, Wienerschnitzel, Holiday Inn Express, Firestone/auto, Kinko's, Michael's, NAPA, Radio Shack, Service Merchandise, XpressLube, cinema, same 68
66	Bell St, Amarillo, **N**...Citgo/diesel/mart, Shamrock/mart, AmeriSuites, Fairfield Inn, HomeGate Suites, Motel 6, Residence Inn, Big A Parts, Harley-Davidson, **S**...Chevron, Texaco/mart, King & I Chinese, Regal 8, Albertson's/ drugs, bank, laundry
65	Coulter Dr, Amarillo, **N**...HOSPITAL, CHIROPRACTOR, Phillips 66/diesel/mart/wash, Arby's, Golden Corral, Luby's, My Thai Café, Taco Bell, Waffle House, Best Western, Comfort Inn, Day's Inn, Executive Inn, La Quinta, Chevrolet, Firestone/auto, Nissan, bank, cleaners, **S**...Chevron/KFC/pizza/mart, Citgo/mart, Texaco/mart, Braum's, ChinaStar Buffet, ChuckeCheese, CiCi's, Hoffbrau Steaks, Home Cookin', McDonald's, Outback Steaks, Pizza Hut, Santa Fe Rest., Sonic, Subway, Taco Villa, TCBY, Wendy's, Whataburger, Wienerschnitzel, Ramada Inn, Best Buy, Goodyear/auto, IGA Foods, Lowe's Bldg Supply, MailBoxes Etc, Pennzoil, Sears/auto, TJ Maxx, XpressLube, cinema, cleaners
64	Soncy Rd, to Pal Duro Cyn, **N**...Joe's Crabshack, Logan's Roadhouse, Extended Stay America, Big 5 Sports, Cavender's Boots, Discount Tire, auto repair, cinema, **S**...MEDICAL CARE, Shamrock/diesel/mart/wash/atm/ 24hr, Applebee's, Dairy Queen, Hooters, Legends Steaks, McDonald's, On-the-Border, Ruby Tuesday, Barnes&Noble, Circuit City, Dillard's, Ford, K-Mart, Mervyn's, OfficeMax, Old Navy, PetsMart, Pier 1, Target, ToysRUs, auto repair, bank, mall

E ↑ a

W ↓

Texas Interstate 40

Amarillo

62b	Lp 40, Amarillo Blvd, RV camping, antiques
a	Hope Rd, Helium Rd(from eb), **S**...RV camping, antiques
60	Arnot Rd, **S**...Love's/A&W/Subway/diesel/mart, Texas Trading Co/gifts, camping
57	RM 2381, to Bushland, **N**...grain silo, **S**...USPO, RV camping/dump(1mi)
55mm	parking area wb, litter barrels
54	Adkisson Rd, **S**...RV park
53.5mm	parking area eb, litter barrels
49	FM 809, to Wildorado, **S**...Phillips 66, Shamrock/diesel/mart, Cookie's Café, Royal Inn, repair
42	Everett Rd, no facilities
37	Lp 40, to Vega, **1 mi N**...same as 36, Walnut RV Park, **S**...airport
36	US 385, Vega, **N**...Conoco/diesel/mart/24hr, Phillips 66/mart, Shamrock/mart, Dairy Queen, Best Western/café, Comfort Inn, CarQuest, **S**...Texaco/diesel/mart/atm, RV camping
35	to Rte 66, to Vega, **N**...Best Western/café, same as 36
32mm	picnic area both lanes, litter barrels
28	to Rte 66, Landergin, no facilities
22	TX 214, to Adrian, **N**...Fabulous 40 Motel, **S**...Phillips 66/diesel/café, Tommy's Foods
18	FM 2858, Gruhlkey Rd, **S**...Texaco/Stuckey's/rest.
13mm	picnic area both lanes, picnic tables, litter barrels
5.5mm	turnout
0	Lp 40, to Glenrio, no facilities
0mm	Texas/New Mexico state line, Central/Mountain time zone

Texas Interstate 44

Wichita Falls

Exit #	Services
15mm	Red River, Texas/Oklahoma state line
14	Lp 267, E 3rd St, **W**...KOA
13	Glendale St, **W**...Family$, Sav-A-Lot Foods, Wal-Mart
12	Burkburnett, **E**...Conoco/mart, Ed's Tires, **W**...Fina/7-11, Shamrock/mart, Braum's, Carl's Jr, Mazzio's Pizza, McDonald's, Subway, Whataburger, Chevrolet/Olds/Pontiac, Ford
11	FM 3429, Daniels Rd, no facilities
9mm	**rest area both lanes, full(handicapped)facilities, phone, picnic table, litter barrels, petwalk**
7	East Rd, no facilities
6	Bacon Switch Rd, no facilities
5a	FM 3492, Missile Rd, **E**...Tee Pee Rest./jewelry, st patrol
5	Access Rd, no facilities
4	City Loop St, no facilities
3c	FM 890(from sb), **E**...VETERINARIAN, **W**...Shamrock/mart, airport
3b	sp 325, Sheppard AFB, no facilities
a	US 287 N, to Amarillo
2	Maureen St, **E**...Fina/7-11, Texaco/diesel, Comfort Inn, Day's Inn, Motel 6, Chevrolet, Mazda, Nissan/Volvo/VW, **W**...China Star, Dairy Queen, Denny's, El Chico's Mexican, LJ Silver, Whataburger/24hr, Hampton Inn, La Quinta, Red Roof Inn, Super 8
1d	US 287 bus, Lp 370, **E**...Casa Manana, **W**...Conoco/mart, Rodeway Inn
1c	Texas Tourist Bureau, **E**...Citgo/mart
1b	Scotland Park(from sb), **E**...Scotland Park Motel, **W**...RV park
1a	US 277 S, to Abilene
0mm	I-44 begins/ends in Witchita Falls. **1-2 mi S in Wichita Falls**...HOSPITAL, Circle K/Citgo, Conoco/repair, Fina/7-11, Arby's, Carl's Jr, IHOP, Mama Josie's Mexican, McDonald's, Popeye's, Subway, Whataburger/24hr, Wendy's, Econolodge, Holiday Inn, TradeWinds Motel/rest.

Texas Interstate 45

Dallas

Exit #	Services
286	to I-35 E, to Denton. I-45 begins/ends in Dallas.
285	Bryan St E, US 75 N, no facilities
284b a	I-30, W to Ft Worth, E to Texarkana, access to HOSPITAL
283b	Pennsylvania Ave, to MLK Blvd, **E**...to Shamrock, Shell/wash, BBQ, auto parts
a	Lamar St, no facilities
281	Oberton St(from sb), no facilities
280	Illinois Ave, Linfield St, **E**...Linfield Motel, Star Motel, **W**...Texaco/mart
279b a	Lp 12, no facilities
277	Simpson Stuart Rd, **W**...Chevron/mart, to Paul Quinn Coll
276b a	I-20, W to Ft Worth, E to Shreveport
275	TX 310 N(from nb, no re-entry), no facilities

Texas Interstate 45

N

S

274   Dowdy Ferry Rd, Hutchins, **E**...Texaco/diesel/mart, Gold Inn, **W**...AFCO/diesel, Dairy Queen, Jack-in-the-Box, bank

273   Wintergreen Rd, no facilities

272   Fulghum Rd, **E**...TX Rose Rest.

271   Pleasant Run Rd, no facilities

270   Belt Line Rd, to Wilmer, **E**...Texaco/mart, Total/diesel/mart, **W**...Citgo/Church's/diesel/mart, Exxon/Subway/mart, Dairy Queen

269   Mars Rd, no facilities

268   Malloy Bridge Rd, no facilities

267   Frontage Rd, no facilities

266   FM 660, **E**...Fina/diesel/mart, Get'n Go Mexican Café, bank, **W**...Shamrock/mart, Dairy Queen

265   Lp 45, Ferris, no facilities

263   TX 561, no facilities

262   Newton Rd, to Trumbull, no facilities

261   County Rd(from nb), no facilities

260   Hampel Rd, **W**...Goodyear, bank

259   FM 813, FM 878, **W**...bank

258   Lp 45, Palmer, **E**...Knox/Subway/diesel/mart/24hr, golf, **W**...Mobil/diesel/mart/24hr, Palmer Motel

255   FM 879, Garrett, **E**...Exxon/diesel/mart, **W**...Chevron/diesel/mart

253   Lp 45, **W**...Texaco/diesel/mart, Papa Tino's Mexican, Jeff's RV Park

251   TX 34, Ennis, **E**...Fina/mart, Texaco/diesel/mart, Bubba's BBQ, McDonald's, Best Western, Holiday Inn Express, **W**...HOSPITAL, Chevron/diesel/mart, Exxon/diesel/mart/24hr, Arby's, Braum's, Burger King, Dairy Queen, Golden Corral, Jack-in-the-Box, KFC, Mr Gatti's, Sonic, Taco Bell/Pizza Hut, Waffle House, Wall Chinese Café, WagonWheel Steaks, Whataburger/24hr, Ennis Inn, Quality Inn, AutoZone, Chevrolet, Chrysler/Plymouth/Dodge, Ford, Pennzoil, Pontiac/Buick, Wal-Mart/auto, RV camping

249   FM 85, Ennis, **W**...Fina/diesel/mart, Total/diesel/mart, Ennis Rest., Ennis Inn, Blue Beacon

247   US 287 N, to Waxahatchie, **W**...to Texas MotorPlex

246   FM 1183, Alma, **W**...Chevron/mart, Phillips 66, auto parts

244   FM 1182, no facilities

243   Frontage Rd, no facilities

242   Rice, **W**...bank

239   FM 1126, **W**...Phillips 66/mart

238   FM 1603, **E**...Fina/diesel/mart, Casita RV Trailers, tires

237   Frontage Rd, no facilities

235b   Lp I-45(from sb), to Corsicana

   a   Frontage Rd, no facilities

232   Roane Rd, no facilities

231   TX 31, Corsicana, **E**...Mobil, Texaco/diesel/mart, Colonial Inn, Chevrolet, **W**...HOSPITAL, Chevron/mart, Exxon/diesel/mart, Shell/diesel, Bill's Cafeteria, Cisco's Mexican, Dairy Queen, McDonald's, Comfort Suites, Chrysler/Plymouth/Dodge/Jeep, Ford, to Navarro Coll

229   US 287, **E**...Exxon/diesel/mart, Shell/diesel/mart/24hr, Russell Stover Candies, VF Outlet/famous brands, **W**...Chevron, Catfish King Rest., Waffle House, Day's Inn, Ramada Inn, Royal Inn, Travelers Inn

228b   Lp 45, Corsicana, **2 mi W**...facilities in Corsicana

   a   15th St, Corsicana, **E**...Phillips 66/diesel/mart

224   FM 739, Angus, **E**...Chevron/mart, Exxon/diesel/mart, to Chambers Reservoir, **W**...Citgo/diesel/mart, Shell/mart, Camper Depot

221   Frontage Rd, no facilities

220   Frontage Rd, no facilities

219b   Frontage Rd, no facilities

   a   TX 14(from sb), Richland, to Mexia

218   FM 1394, Richland, **W**...Shell

217mm   **rest area both lanes, full(handicapped)facilities, phone, picnic tables, litter barrels, vending, petwalk**

213   FM 246, to Wortham, **W**...Chevron/mart, Texaco/diesel/mart

211   FM 80, to Streetman, no facilities

206   FM 833, no facilities

198   FM 27, to Wortham, **E**...HOSPITAL, Shell/BBQ/mart, PJ's Café, **W**...Exxon/mart, Budget Inn, I-45 RV Park

197   US 84, Fairfield, **E**...CHIROPRACTOR, Conoco, Exxon/Jack-in-the-Box/mart, Fina/Sam's/mart, Texaco/mart, Dairy Queen, Ponte's Diner, Subway/Texas Burger, Holiday Inn Express, Chevrolet, Chrysler/Plymouth/Dodge, carwash, **W**...Shell/diesel, Texaco/diesel/mart, I-45 Rest., KFC/Taco Bell, Pizza Hut, Sammy's Rest., Regency Inn, Sam's Motel, Ford/Mercury

189   TX 179, to Teague, **E**...Exxon/diesel/mart, Fina/diesel/mart, **W**...Dinner Bell Rest., Tatum's BBQ

187mm   picnic area both lanes, tables, litter barrels

Texas Interstate 45

180	TX 164, to Groesbeck, no facilities
178	US 79, Buffalo, **E**...Conoco, Shell/diesel/mart, Texaco/diesel/mart, Pizza Inn, Weathervane Rest., Brookshire Foods, bank, **W**...Chevron/mart, Mobil/Arby's/diesel/mart, Shamrock/mart/24hr, Dairy Queen, Pitt Grill/24hr, Subway/Texas Burger, Best Western, Economy Inn
175.5mm	Bliss Creek
166mm	weigh sta sb
164	TX 7, Centerville, **E**...Mobil/mart, Texaco/diesel, BBQ, Doc's Café, Texas Burger, Day's Inn, **W**...Exxon/mart/24hr, Shell/diesel/mart, Jack-in-the-Box, Mama Mike's/BBQ, Dairy Queen
160mm	picnic area sb, tables, litter barrels, hist marker, handicapped accessible
159mm	Boggy Creek
156	FM 977, to Leona, **W**...Exxon/diesel/mart
155mm	picnic area nb, handicapped accessible
152	TX OSR, to Normangee, **W**...Chevron/mart, Yellow Rose RV Park/café
146	TX 75, no facilities
142	US 190, TX 21, Madisonville, **E**...Chevron/diesel, Exxon, Shell/mart/24hr, Texaco/diesel/24hr, Church's Chicken, Corral Café, Madisonville Motel, Ramada Inn, **W**...HOSPITAL, Citgo/mart, Exxon, Shamrock/diesel/mart, Shell/Subway/mart/24hr, Alma's Mexican, Lakeside Rest./24hr, McDonald's, Pizza Hut, Sonic, Texas Burger, Budget Motel, Tejas Inn, Western Lodge, CarQuest, Ford/Mercury
136	spur 67, no facilities
132	FM 2989, no facilities
126mm	**rest area sb, full(handicapped)facilities, phone, picnic tables, litter barrels, vending, petwalk, RV dump**
124mm	**rest area nb, full(handicapped)facilities, phone, picnic tables, litter barrels, vending, petwalk, RV dump**
123	FM 1696, no facilities
121mm	weigh sta nb, phone
118	TX 75, **E**...Texaco/diesel/mart/24hr, truckwash, **W**...Pilot/Wendy's/diesel/mart/24hr
116	US 190, TX 30, **E**...HOSPITAL, Citgo, Phillips 66, Shamrock/diesel/mart, Chinese Rest., Church's Chicken, El Chico, Golden Corral, Jct Steaks/seafood, Kettle, McDonald's, TCBY, Tejas Café, Texas Burger, Whataburger/24hr, Comfort Inn, Econolodge, La Quinta, Motel 6, Plymouth/Jeep/Dodge, Pontiac/Olds/GMC, bank, **W**...Exxon/mart, Chevron/mart/24hr, Shell/mart/24hr, Texaco/diesel, Burger King, Chili's, Denny's, Floyd's Rest., IHOP, KFC, Subway, Taco Bell, Huntsville Inn, Discount Tire, Hastings Books, Kroger, Office Depot, Radio Shack, Walgreen, Wal-Mart SuperCtr/gas/24hr
114	FM 1374, **E**...Fina/diesel/mart, Texaco/mart, Corner Pantry/gas, Casa Tomas Mexican, Dairy Queen, Gateway Inn, Ford/Lincoln/Mercury, **W**...HOSPITAL, Chevron, Shamrock/diesel/mart, Country Inn Steaks, Best Western, Sam Houston Inn
113	TX 19(from nb), Huntsville, **E**...Catfish Palace Rest.(2mi)
112	TX 75, **E**...Citgo, Baker Motel, Houston Statue, to Sam Houston St U, museum
109	Park 40, **W**...to Huntsville SP
105mm	picnic area both lanes, tables, litter barrels
103	FM 1374/1375(from sb), to New Waverly, **W**...Citgo/diesel/mart, Texaco/mart
102	FM 1374/1375, TX 150(from nb), to New Waverly, **1 mi W**...Citgo/diesel/mart, Texaco/mart
101mm	rest area nb, full(handicapped)facilities, phone, picnic tables, litter barrels, vending, petwalk
98	TX 75, Danville Rd, Shepard Hill Rd, **E**...Fishpond RV Park/dump, repair
95	Longstreet Rd, no facilities
94	FM 1097, to Willis, **E**...Jack-in-the-Box, **W**...Chevron/mart, Exxon/Burger King/mart, Texaco/Taco Bell/diesel/mart/24hr, McDonald's, Popeye's, Subway, Best Western
92	FM 830, Seven Coves Dr, **W**...RV Park(3mi)
91	League Line Rd, **E**...Chevron/McDonald's/mart, LoneStar Café, Wendy's, Comfort Inn, MasterLube, Prime Outlets/famous brands, **W**...Citgo/mart, Cracker Barrel
90	FM 3083, Teas Nursery Rd, Montgomery Co Park
88	Lp 336, to Cleveland, Navasota, **E**...VETERINARIAN, MEDICAL CARE, CHIROPRACTOR, Exxon/mart, Shamrock/diesel/mart, Shell/mart/wash/24hr, Arby's, Blackeyed Pea, Domino's, Grandy's, Hofbrau Steaks, Hunan Chinese, KFC, Little Caesar's, LJ Silver, Margarita's Mexican, McDonald's, Papa John's, Sonic, Subway, Whataburger, Buick/Pontiac, Discount Tire, Goodyear/auto, Hastings Books, HobbyLobby, Kroger/drugs/24hr, Michael's, Pennzoil, Toyota, Walgreen, bank, laundry, **W**...Chevron/mart/wash/24hr, Eckerd, Goody's, GNC, JC Penney, Lowe's Bldg Supply, Randall's Foods, Radio Shack, PetCo, Sam's Club, Wal-Mart SuperCtr/24hr
87	TX 105, Conroe, **E**...HOSPITAL, Coastal/mart, Shamrock, BBQ, Burger King, CiCi's, Imperial Garden Chinese, Jack-in-the-Box, McDonald's, Outback Steaks, Steak&Ale, Taco Bell, Wyatt's Cafeteria, Big Lots, Eckerd/24hr, Firestone/auto, NTB, bank, **W**...Exxon/TCBY/mart/wash, Texaco/diesel/mart, Golden Corral, Luby's, Buick/GMC/Isuzu, Home Depot, Midas Muffler, Office Depot, Target, Toyota/Mazda
85	FM 2854, Gladstell St, **E**...HOSPITAL, CHIROPRACTOR, Ford/Mercury, Honda, Mitsubishi, Nissan, **W**...Shamrock/mart, Texaco/diesel/mart, Day's Inn, Motel 6, Cadillac/Olds, Chrysler/Plymouth/Jeep, Dodge
84	TX 75 N, Frazier St, **E**...Conoco/diesel, Exxon/diesel/mart, Texaco/diesel, Holiday Inn, Ramada Ltd, **W**...Shell/mart, Pizza Hut, Taco Cabana, Baymont Inn, Albertson's/drugs, Discount Tire, K-Mart, MailBoxes Etc, cleaners

476

Texas Interstate 45

83	Crighton Rd, Camp Strake Rd, **W**...VETERINARIAN
82mm	San Jacinto River
81	FM 1488, to Hempstead, Magnolia, **E**...Citgo/mart, **W**...Shamrock/mart, CamperLand RV Ctr
80	Needham Rd(from sb), no facilities
79	TX 242, Needham, **E**...Texaco/McDonald's/mart, RV Resort(1mi)
78	Needham Rd(from sb), Tamina Rd, access to same as 77
77	Woodlands Pkwy, Robinson, Chateau Woods, **E**...CHIROPRACTOR, VETERINARIAN, Conoco/diesel/mart, Texaco/mart, BBQ, Pancho's Mexican, Pizza Hut, Red Lobster, Howard Johnson, Home Depot, NTB, Office Depot, Pier 1, **W**...HOSPITAL, Shamrock/mart, Texaco/mart, Landry's Seafood, Luby's, Macaroni Grill, Pancho's Mexican, Comfort Inn, Drury Inn, Hampton Inn, La Quinta, Roadrunner Motel, Best Buy, Dillard's, Linens'n Things, Marshall's, Mervyn's, OfficeMax, Service Merchandise, Target, ToysRUs, World Mkt, cinema, mall
76	Research Forest Dr, Tamina Rd, **E**...MEDICAL CARE, Chevron/mart, Coastal/lube, Shamrock/mart, Stop'n Go, Texaco, Casa Elena Mexican, LJ Silver, Food Basket, American Muffler/brakes, Firestone, cleaners, **W**...VETERINARIAN, IHOP, Carrabba's, El Chico, Kyoto Japanese, Olive Garden, Tortuga Mexican, Crossland Suites, Goodyear/auto, bank
73	Rayford Rd, Sawdust Rd(to Hardy Toll Rd from sb), **E**...Conoco, Shell, BBQ, Jack-in-the-Box, LJ Silver, McDonald's, Popeye's, Sonic, AutoZone, Meineke Muffler, O'Reilly Parts, **W**...CHIROPRACTOR, Mobil, Shell/mart/wash, Texaco, Pantry Rest., Grandy's, Sam's Rest., Woodland Chinese, Red Roof Inn, Discount Tire, Eckerd, HEB Foods, Kroger, carwash
72	Frontage Rd(from nb), no facilities
70b	Spring-Stuebner Rd, Texaco/mart
a	FM 2920, to Tomball, **E**...Exxon, El Palenque Mexican, McDonald's, Pizza Hut, Wendy's, Burditt RV Ctr, Radio Shack, Walgreen, **W**...Chevron, Shell/wash, Burger King, Taco Bell, Whataburger/24hr, Travelodge, Ford, U-Haul
68	Holzwarth Rd, Cypress Wood Dr, **E**...Exxon/mart, Burger King, Hartz Chicken, Taco Bell, Wendy's, Albertson's, $Tree, Kroger, Toyota, Wal-Mart SuperCtr/24hr, bank, **W**...VETERINARIAN, CHIROPRACTOR, Chevron/mart/24hr, Jack-in-the-Box, Mexican Rest., Day's Inn, Motel 6, Home Depot, Lowe's Bldg Supply, cinema
66	FM 1960, to Addicks, **E**...Chevron/mart, RaceTrac/mart, Texaco, Jack-in-the-Box, Day's Inn, AutoNation USA, HobbyShop, PetsMart, Radio Shack, **W**...Exxon/mart, Shell/diesel/wash/24hr, Texaco/diesel/mart, Bennigan's, Grandy's, Jack-in-the-Box, Jojo's Rest., Luby's, McDonald's, Outback Steaks, Pizza Hut, Popeye's, Red Lobster, Steak&Ale, Subway, Taco Bell, Comfort Suites, Fairfield Inn, Hampton Inn, La Quinta, GolfSmith, Lexus, Lube Ctr, MailBoxes Etc, NTB, bank, carwash, mall
64	Richey Rd, **E**...Texaco/Church's/mart, Atchafalaya River Café, Pappasito's Cantina, Holiday Inn, Lexington Suites, CarMax, Discount Tire, Sam's Club, **W**...Cracker Barrel, Huey's Brewery/Rest., Jack-in-the-Box, Joe's Crabshack, SaltGrass Steaks, Shoney's, Whataburger, La Quinta, U-Haul, cinema
63	Airtex Dr, **E**...Acura, Cadillac, LoneStar RV Ctr, Nissan, Howard Johnson, Scottish Inn, **W**...Chevron, RaceTrac/mart, McDonald's
62	Rankin Rd, Kuykendahl, **W**...Shell/wash, Baymont Inn, Rodeway Inn, Sun Suites, BMW, Buick/Subaru/Olds, Kia, Mercedes, Mitsubishi, Saab, Suzuki, Volvo, VW, Celebration Sta
61	Greens Rd, **E**...Whataburger, Day's Inn, Marriott, Wyndham Hotel, Circuit City, Dillard's, Foley's, JC Penney, Mervyn's, Ward's/auto, bank, mall, **W**...Burger King, Luby's, Kroger, Computer City, Cloth World, MediaPlay, Office Depot, OfficeMax, Target
60c	Beltway E, no facilities
b	TX 8, **E**...La Quinta, Marriott, Dillard's, Sears/auto, cinema, mall, **W**...Mexican Rest., Dodge
a	TX 525, **E**...Exxon/mart, Shamrock/mart, Shell/mart, Texaco, Burger King, Champ's Rest., China Border, Denny's/24hr, Furr's Cafeteria, LJ Silver, Mucho Mexico, Pizza Hut, Steak&Ale, Firestone/auto, Honda, Pontiac/GMC, Randall's Food/drugs/24hr, Walgreen, **W**...Pappa's Rest., Best Western, AutoNation, Best Buy, Wal-Mart SuperCtr/24hr
59	FM 525, West Rd, **E**...Exxon, Shell, McDonald's, Pizza Hut, Wendy's, Fiesta Foods, Midas Muffler, auto parts, **W**...Exxon, Taco Bell, Taco Cabana, GreenChase Inn, Home Depot, K-Mart/drugs, Lincoln/Mercury, PepBoys, Wal-Mart Super Ctr/24hr
57	TX 249, Gulf Bank Rd, Tomball, **E**...Conoco/mart, Exxon/diesel, Mobil/diesel/mart, Ricardo's Mexican, Discount Tire, Kroger, **W**...Shell, Day's Inn, KFC, Pizza Inn, Sonic, Chevrolet, Chrysler/Plymouth/Jeep, Daewoo, Eckerd, Ford, Mazda, Saturn, bank
56	Canino Rd, **E**...RV Ctr, **W**...Shell/wash, Denny's, Luby's, JiffyLube, Nissan, Toyota

N ↑↓ S

Houston

Texas Interstate 45

Houston

55b a Little York Rd, Parker Rd, **E...**HOSPITAL, Chevron/mart, Exxon/mart, Mobil/diesel/mart, Shamrock/mart, Shell/mart, Schlotsky's, Whataburger, Olympic Motel, FoodTown, Goodyear, auto parts, **W...**Chevron/mart/24hr, Shell/mart/wash, Capt D's, Hartz Chicken, Jack-in-the-Box, KFC, LJ Silver, Popeye's, Rally's, Econolodge, Guest Motel, Walgreen

54 Tidwell Rd, **E...**Exxon, Chinese Rest, Subway, **W...**BBQ, Scottish Inn, Eckerd

53 Airline Dr, same as 52

52 Crosstimbers Rd, **E...**Exxon, Shell/mart/wash, Texaco/diesel/mart, BBQ, Chinese Rest., CiCi's, Denny's, James Coney Island, McDonald's, Pancho's Mexican, Taco Bell, Taco Cabana, Wendy's, Discount Tire, Fiesta Foods, Firestone, Jo-Ann Fabrics, cinema, mall, **W...**Chevron/diesel/mart, Citgo, Exxon, BBQ, Little Mexico Rest., Whataburger/24hr, Ambassador Hotel, Howard Johnson, Luxury Inn, Super 8, U-Haul

51 I-610, no facilities

50b Calvacade St, Link Rd, **E...**Exxon/Taco Bell, Texaco/mart/wash, **W...**Astro Inn

a Patton St, **E...**Chevron/mart/24hr, Pilot/Wendy's/diesel/mart/24hr, Best Value Inn, **W...**Texaco/diesel/mart/24hr

49b N Main St, Houston Ave, **E...**Casa Grande Mexican, **W...**Exxon/diesel/mart, KFC, McDonald's, Whataburger/24hr, Auto Supply

48b a I-10, E to Beaumont, W to San Antonio

47b Houston Ave, Memorial Dr, downtown

a Allen Pkwy

46b a US 59, N to Cleveland, S to Victoria, **E...**BBQ, BrakeCheck, **W...**Exxon, McDonald's, Taco Bell, BMW, bank

45 South St, Scott St, Houston, **E...**Conoco, Texaco/Subway/mart

44 Cullen Blvd, Houston, **E...**Shamrock/gas, Burger King, McDonald's, Taco Bell, **W...**bank, to U of Houston

43b Telephone Rd, Houston, no facilities

a Tellepsen St, **E...**Luby's, **W...**U of Houston

41b US 90A, Broad St, Wayside, **E...**Shell/mart/wash, Stop'n Go, Day's Inn, Red Carpet Inn/rest., **W...**Mobil/mart, Jack-in-the-Box, McDonald's, Monterrey Mexican, K-Mart

a Woodridge Dr, **E...**CHIROPRACTOR, Shell/mart, Bennigan's, Bonnie's Beef/seafood, Grandy's, James Coney Island, McDonald's, Schlotsky's, cleaners, **W...**BBQ, Dot's Coffee Shop, IHOP, Sonic, Whataburger, Wendy's, Chevrolet/Buick, Dillard's, Office Depot, mall

40c I-610 W

b I-610 E, to Pasadena

a Frontage Rd(from nb), no facilities

39 Park Place Blvd, Broadway Blvd, **E...**Texaco, Dante Italian, **W...**DENTIST, Shell/mart, Phillips 66/mart, Texaco/Blimpie/mart, BBQ, Kelly's Rest., May Moon Café, BrakeCheck, airport

38b Howard Dr, Bellfort Dr(from sb), **E...**Texaco, **W...**Citgo, Chilo's Seafood, Palace Inn, HEB Foods, PepBoys

38 TX 3, Monroe Rd, **E...**Chevron, Exxon, Shell, Stop'n Go/gas, Dairy Queen, Fat Maria's Mexican, Jack-in-the-Box, Lonesome Steer Steaks, Wendy's, NTB, U-Haul, auto parts, **W...**DENTIST, Chevron/diesel/mart, Texaco/diesel/mart, Get'n Go, Chinese Rest., Luby's, Best Western, Quality Inn, Smile Inn, Firestone, Radio Shack, Suzuki, U-Haul

36 College Ave, Airport Blvd, **E...**Shamrock, Shell, Dairy Queen, Waffle House, Fairfield Inn, Rodeway Inn, RV Ctr, **W...**Exxon/mart/wash, Stop'n Go/gas, Texaco, Church's Chicken, Denny's, Lucky Dragon Chinese, Mexican Rest., Taco Cabana, Courtyard, Drury Inn, Radisson/Damon's, Red Roof Inn, Sumner Suites, Super 8, Travel Inn, Discount Tire, RV Ctr

35 Edgebrook Dr, **E...**Chevron, Exxon, RaceTrac/mart, Burger King, Grandy's, Jack-in-the-Box, Popeye's, Subway, Taco Bell, Waffle House, Airport Inn, Eckerd, Fiesta Foods, Firestone, Office Depot, Walgreen, **W...**Exxon, Shell, Circle K, Kettle, KFC, LJ Silver, McDonald's, Pizza Hut, Whataburger, La Quinta, Cavender's Boots, NTB, RV camping

34 S Shaver Rd, **E...**Conoco, McDonald's, Honda, Kia, Pontiac/GMC, Saturn, Toyota, Vaughn's RV Ctr, tires, **W...**Chevron/mart/24hr, Exxon/mart, Mobil, Burger King, Pancho's Mexican, Wendy's, Best Buy, Circuit City, Holiday World RV Ctr, Jo-Ann Fabrics, K-Mart, Marshall's, Nissan/Olds, OfficeMax, Target, ToysRUs

33 Fuqua St, **E...**Shamrock, Chili's, Fuddrucker's, Luby's, Mexico Lindo, Olive Garden, TGIFriday, Sun Suites, Acura, Chrysler/Plymouth/Jeep, Dodge, Ford, Honda, Hyundai, Isuzu, Lincoln/Mercury, cinema, **W...**Barco's Mexican, Blackeyed Pea, Boston Mkt, CiCi's, Golden Corral, IHOP, Joe's Crabshack, Mi Amigo's Mexican, Steak&Ale, Subway, Taco Cabana, Taco Bell, Tejas Café, Whataburger, AutoNation USA, Big Lots, CarMax, Foley's, HobbyShop, Home Depot, JC Penney, Kroger, MediaPlay, Radio Shack, Ross, Sam's Club, Subaru, Tire Station, Wal-Mart, cinema, mall

32 Sam Houston Tollway

31 FM 2553, Scarsdale Blvd, **W...**Exxon/mart, Shell/mart, Dairy Queen, McDonald's, Chevrolet, Mitsubishi

30 FM 1959, Ellington Field, Dixie Farm Rd, **E...**HOSPITAL, Conoco/mart, Dodge, **W...**Texaco/diesel/mart, RV Ctr

29 FM 2351, Clear Lake City Blvd, Friendswood, to Clear Lake RA

27 El Dorado Blvd, **E...**Dairy Queen, mall, **W...**Lexus

26 Bay Area Blvd, **E...**HOSPITAL, Kettle, Red Lobster, Barnes&Noble, CompUSA, Michael's, cinema, to Houston Space Ctr, **W...**Shell/mart, Baskin-Robbins, Bennigan's, Chinese Rest., Denny's/24hr, McDonald's, Olive Garden, On-the-Border, Steak&Ale, Circuit City, Dillard's, Gateway Computers, Linens'n Things, Macy's, Mervyn's, Sears/auto, Service Merchandise, Target, Outlet Mall/famous brands, cinema, mall, U of Houston

25 FM 528, NASA rd 1, **E...**HOSPITAL, DENTIST, Conoco, Shamrock/mart, Texaco, Chili's, CiCi's, IHOP, Macaroni Grill, Mason Jar Grill, Pappasito's Cantina, Saltgrass Steaks, Tony Roma's, Waffle House, Motel 6, Audi, Bed Bath&Beyond, Best Buy, Home Depot, Honda, K-Mart, Mazda, Office Depot, Oschman's Sports, Pier 1, SteinMart,

Texas Interstate 45

League City   Galveston

N ↑↓ **S**

Volvo, cinema, **W**...BBQ, Capt's Rest., China Square, Dairy Queen, Hooters, Hot Wok Chinese, Burlington Coats, Computer City, Garden Ridge, Radio Shack, Wal-Mart/auto

23   FM 518, League City, **E**...CHIROPRACTOR, RaceTrac/mart, Texaco, Burger King, Jack-in-the-Box, KFC, Little Caesar's, Pancho's Mexican, Sonic, Subway, Eckerd, JiffyLube, Kroger, Walgreen, cleaners, **W**...Exxon/24hr, Mobil/mart, Cracker Barrel, Hartz Chicken, McDonald's, Taco Bell, Waffle House, Wendy's, Super 8, Discount Tire, U-Haul

22   Calder Dr, Brittany Bay Blvd, **E**...EYECARE, VETERINARIAN, Toyota, camping

20   FM 646, Santa Fe, Bacliff, **W**...RV Ctr

19   FM 517, Dickinson Rd, Hughes Rd, **E**...Shell, Shamrock, Jack-in-the-Box, Monterey Mexican, Eckerd, Family$, Food King, Pontiac/GMC/Subaru, Radio Shack, **W**...Exxon/mart/wash, Texaco/diesel/mart/wash, Circle K/gas, Burger King, Kettle/24hr, KFC, McDonald's, Pizza Hut, Pizza Inn, Subway, Taco Bell, Wendy's, Whataburger/24hr, Ford, Kroger, Walgreen

17   Holland Rd, **W**...to Gulf Greyhound Park

16   FM 1764 E(from sb), Texas City, **E**...HOSPITAL, Demontrond RV Ctr, mall, same as 15

15   FM 2004, FM 1764, Hitchcock, **E**...HOSPITAL, Shell/mart, Carino's Italian, Jack-in-the-Box, Fairfield Inn, Hampton Inn, Chevrolet/Olds/Toyota, Dillard's, Foley's, JC Penney, mall, **W**...Mobil/mart, Texaco, Waffle House, Whataburger, Gulf Greyhound Park

13   Johnny Palmer Rd, **E**...Shamrock/mart, **W**...Ramada Inn, Super 8, Outlet Mall/famous brands

12   FM 1765, La Marque, **E**...Chevron/mart/wash/24hr, Domino's, Jack-in-the-Box, Kelley's Rest., Sonic, camping, **W**...Mobil/diesel

11   Vauthier Rd, no facilities

10   FM 519, Main St, **E**...Shamrock/mart, McDonald's, **W**...Texaco/diesel/mart

9   Frontage Rd(from sb), no facilities

8   Frontage Rd(from nb), no facilities

7c   Frontage Rd, no facilities

b   TX 146, TX 6, Texas City, **W**...Conoco/gas, Texaco/mart

a   TX 146, TX 3, no facilities

6   Frontage Rd(from sb), no facilities

5   Frontage Rd, no facilities

4   Village of Tiki Island, Frontage Rd, **W**...Conoco/Tiki Store/diesel, public boat ramp

4mm   West Galveston Bay

1c   TX 275, FM 188(from nb), Port of Galveston, Port Ind Blvd, Teichman Rd, **E**...Chevron/mart, Conoco/mart, Texaco/diesel/mart, Motel 6, Chevrolet, Chrysler/Dodge/Plymouth/Jeep, Ford, Mazda, Nissan, Toyota

1b   71st St(from sb), same as 1c

1a   TX 342, 61st St, to W Beach, **E**...EZMart/gas, Denny's, **W**...Exxon/diesel/mart, RaceTrac/mart, Burger King, Church's Chicken, Kettle, McDonald's, Taco Bell, Day's Inn, Chevrolet/Pontiac/Buick/Cadillac, Chrysler/Jeep, Honda, U-Haul, **1-2 mi W**...CHIROPRACTOR, VETERINARIAN, Chevron/mart, Exxon/diesel/mart, Shell/mart/wash, Stop'n Go/gas, Texaco/mart, Arby's, Burger King, Domino's, Golden Wok Chinese, Jack-in-the-Box, KFC, LJ Silver, Luby's, Marco's Mexican, Mario's Italian, McDonald's, Papa John's, Pizza Hut, Popeye's, Subway, Taco Cabana, TCBY, Whataburger, Best Western, Econolodge, Super 8, Eckerd, Firestone/auto, Hastings Books, K-Mart, Kwik Kar Lube/tune, Office Depot, Randall's Food/drug, Walgreen, Wal-Mart, auto parts, bank, camping, carwash, laundry

I-45 begins/ends on TX 87 in Galveston.

Texas Interstate 410

Exit #	Services
53	I-35, S to Laredo, N to San Antonio
51	FM 2790, Somerset Rd, no facilities
49	TX 16 S, spur 422, **N**...HOSPITAL, Chevron/mart, Church's Chicken, to Palo Alto Coll
48	Zarzamora St, no facilities
46	Moursund Blvd, no facilities
44	US 281 S, spur 536, Roosevelt Ave, **N**...Texaco/diesel/mart, RV camping
42	spur 122, S Presa Rd, to San Antonio Missions Hist Park, **S**...Whiteside/diesel, RV camping
41	I-37, US 281 N, no facilities
39	spur 117, WW White Rd, no facilities
37	Southcross Blvd, Sinclair Rd, Sulphur Sprs Rd, **N**...HOSPITAL

Texas Interstate 410(San Antonio)

35   US 87, Rigsby Ave, **E...**Exxon/TacoMaker/mart, Shamrock/diesel/mart, BorderTown, Denny's, Jack-in-the-Box, McDonald's, Taco Bell, **W...**Chevron, BBQ, Luby's, Subway, Taco Cabana, Whataburger, Day's Inn, Aamco, XpressLube

34   FM 1346, E Houston St, **W...**Shamrock/mart, McDonald's, Wendy's, Comfort Inn, Econolodge, Motel 6

33   I-10 E, US 90 E, to Houston, I-10 W, US 90 W, to San Antonio

32   Dietrich Rd(from sb), FM 78(from nb), to Kirby, no facilities

31b   Lp 13, WW White Rd, no facilities

a   FM 78, Kirby, no facilities

30   Space Center Dr(from nb), **E...**Coastal/mart, Shamrock/mart

I-410 and I-35 S run together 7 mi, **See Texas I-35, exits 162 thru 165.**

27   I-35, N to Austin, S to San Antonio

26   Lp 368 S, Alamo Heights, **S...**Mobil, Red Lobster, Shoney's, Drury Inn, NTB

25b   FM 2252, Perrin-Beitel Rd, **N...**Chevron/mart, Mobil, Denny's, KFC/Taco Bell, Schlotsky's, Taco Cabana, Wendy's, Wienerschnitzel, Comfort Inn, BrakeCheck, **S...**Jim's Rest., bank, cinema

25a   Starcrest Dr, **N...**DENTIST, Texaco/mart, Los Patios Mexican, Hawthorn Suites, Toyota, **S...**Conoco/mart

24   Harry Wurzbach Hwy, **N...**Taco Cabana, bank, **S...**MEDICAL CARE, DENTIST, Texaco, BBQ

23   Nacogdoches Rd, **N...**BBQ, Church's Chicken, El Caribe Mexican, Jack-in-the-Box, Mama's Café, Pizza Hut, Schlotsky's, Wendy's, Club Hotel, **S...**Exxon, Texaco/diesel, Audi/VW, bank

22   Broadway St, **N...**VETERINARIAN, Texaco/diesel/mart, Crystal Steaks, Luby's, McDonald's, Pizza Hut, Taco Cabana, Wendy's, Courtyard, Ramada Inn, antiques, **S...**Exxon, Shamrock/mart, Burger King, Jim's Rest., Best Western, Residence Inn, BrakeCheck, Infiniti, JiffyLube, Olds, bank

21   US 281 S, Airport Rd, Jones Maltsberger Rd, **N...**Applebee's, Drury Suites, Hampton Inn, Holiday Inn Select, La Quinta, PearTree Inn, to airport, **S...**Texaco/diesel/mart/wash, Pappadeaux, Red Lobster, Texas Land&Cattle, Courtyard, Day's Inn, Fairfield Inn, BabysRUs, Hyundai/Kia, Mitsubishi, Subaru, Wal-Mart/24hr, bank

20   TX 537, **N...**Jason's Deli, TGIFriday, DoubleTree Hotel, Hilton, Barnes&Noble, Best Buy, Cavender's Boots, Chevrolet, Circuit City, Computer City, Honda, Lincoln/Mercury, Linens'n Things, Mazda, Oschman's Sports, Pearle Vision, **S...**Arby's, Luby's, McDonald's, Taco Cabana, Dillard's, Dodge, Macy's, Mervyn's, Saks 5th, Sears, mall

19b   FM 1535, FM 2696, Military Hwy, **N...**Chinese Rest., Gyro/Sub Shop, Souper Salad, Hilton, **S...**CHIROPRACTOR, Exxon, Denny's, Jim's Rest., bank

19a   Honeysuckle Lane, Castle Hills

17b   (from wb), **S...**Dairy Queen, Dunkin Donuts, HEB Foods

17   Vance Jackson Rd, **N...**Mobil, Phillips 66, Shamrock, Burger King, Jack-in-the-Box, KFC, McDonald's, Steak&Ale, Taco Cabana, Tom's Ribs, Embassy Suites, CarQuest, Discount Tire, Target, Tire Sta, carwash, transmissions, **S...**Exxon, Subway, U-Haul

16b a   I-10 E, US 87 S, to San Antonio, I-10 W, to El Paso, US 87 N

15   Lp 345, Fredericksburg Rd, **E...**Texaco, Church's Chicken, Dave&Buster's, Jack-in-the-Box, Jim's Rest., Kettle, Luby's, N China Rest., Sumner Suites, BMW, Kinko's, **W...**Mobil, Holiday Inn, K-Mart, bank

14   Callaghan Rd, Babcock Ln, **E...**Chevron, Exxon/mart/wash, Shamrock/mart, Marie Callender's, Comfort Inn, Hampton Inn, Travelodge Suites, GMC/Pontiac, bank, **W...**HOSPITAL, Conoco, Burger King, Chili's, ChopSticks Chinese, DingHow Chinese, El Chico, Golden Corral, Joe's Crabshack, Landry's Seafood, Pizza Hut, Quizno's, Red Lobster, Wendy's, Whataburger, Cavender's Boots, Home Depot, Sam's Club, Wal-Mart SuperCtr/24hr, cinema

13b   Rolling Ridge Dr, **W...**Jack-in-the-Box, KFC, Las Palapas Mexican, same as 14

13a   TX 16 N, Bandera Rd, Evers Rd, Leon Valley, **E...**Outback Steaks, Albertson's, Office Depot, OfficeMax, Old Navy, PearApple Funpark, PetCo, JiffyLube, Olds, Saturn, Toyota, U-Haul, **W...**Conoco/diesel/mart, Exxon, Shamrock, Applebee's, BBQ, Blackeyed Pea, IHOP, Jack-in-the-Box, Jim's Rest., KFC, Luby's, McDonald's, Olive Garden, Schlotsky's, Taco Cabana, Super 8, Aamco, Best Buy, Big Lots, Chevrolet, Circuit City, Dillard's, Eckerd, Honda, Kinko's, Marshall's, NTB, Service Merchandise, bank

11   Ingram Rd, **E...**Texaco/diesel/mart, Day's Inn, Econolodge, BrakeCheck, Olds, **W...**Burger King, Denny's, Jason's Deli, Olive Garden, Whataburger, Best Western, Barnes&Noble, Best Buy, Dillard's, Foley's, JC Penney, MensWearhouse, Mervyn's, Michael's, Shepler's, Sears/auto, bank, mall

10   FM 3487, Culebra Rd, **E...**Barnacle Bill's Seafood, BBQ, Denny's, McDonald's, Wendy's, Holiday Inn Express, La Quinta, Red Roof Inn, Chrysler/Plymouth/Dodge/Jeep, Garden Ridge, Mazda/Nissan/Daewoo, to St Mary's U, **W...**Exxon/mart/wash, Phillips 66/mart, RaceTrac/mart, Fuddrucker's, BabysRUs, EyeMart, Firestone, Ford, Mitsubishi, ToysRUs

9b a   TX 151, **W...**Home Depot, Wal-Mart SuperCtr/gas/24hr, cinema, to Sea World

7   (8 from sb)Lakeside Dr, **E...**Exxon/mart/wash, McDonald's, PepBoys, **W...**Chevron/diesel/mart, Texaco, Acadiena Café, Burger King, Jack-in-the-Box, Jim's Rest., KFC, LJ Silver, Luby's, McDonald's, Mr Gatti's, Pancho's Mexican, Peter Piper Pizza, Red Lobster, Sirloin Stockade, Sonic, Taco Bell, Taco Cabana, Whataburger/24hr, Motel 6, Super 8, Advance Parts, Discount Tire, Eckerd, EconoLube'nTune, Firestone, HEB Food, HobbyLobby, K-Mart/24hr, Midas Muffler, Target, bank, cinema

6   US 90, to Lackland AFB, **E...**U-Haul, West Point Inn, **W...**Shamrock/mart, Texaco/diesel/mart/atm, El Pescador Mexican

4   Valley Hi Dr, San Antonio, to Del Rio, **E...**Mobil, Phillips 66/diesel/mart, Shamrock/mart, Church's Chicken, McDonald's, Pizza Hut, Sonic, AutoZone, HEB Food/drugs, Radio Shack, bank, **W...**Texaco/diesel/mart/wash, to Lackland AFB

3b a   Ray Ellison Dr, Medina Base, **W...**Shamrock/mart

2   FM 2536, Old Pearsall Rd, **E...**BBQ, McDonald's, Sonic

1   Frontage Rd, no facilities

Texas Interstate 610(Houston)

Exit #	Services
38c a	TX 288 N, downtown, access to zoo
37	Scott St, **N**...Texaco/mart
36	FM 865, Cullen Blvd, **S**...Chevron/McDonald's/mart, Mobil, Shamrock/mart, Chinese Rest., Crown Inn, Cullen Inn
35	Calais Rd, Crestmont St, MLK Blvd, **N**...Exxon/mart, Burger King
34	S Wayside Dr, Long Dr, **N**...Exxon/mart, Texaco, Wendy's, Fiesta Foods, **S**...Coastal/diesel/mart, Conoco/mart, bank
33	Woodridge Dr, Telephone Rd, **N**...Texaco/diesel/mart, IHOP, McDonald's, Papa John's, Wendy's, Service Merchandise, **S**...Shell/mart/wash, Burger King, KFC, Piccadilly's Cafeteria, Spanky's Pizza, Whataburger, BrakeCheck, Dodge, Ford, auto parts, cinema, drugs, mall
32b a	I-45, S to Galveston, N to Houston, to airport
31	Broadway Blvd, **N**...Conoco, **S**...Texaco
30c b	TX 225, to Pasadena, San Jacinto Mon, no facilities
29	Port of Houston Main Entrance
28	Clinton Dr, no facilities
27	Turning Basin Dr, industrial area
26b	Market St, no facilities
a	I-10 E, to Beaumont, I-10 W, to downtown
24b	Wallisville Rd, **E**...Mobil/diesel, Texaco/diesel/mart/wash, McDonald's
24	US 90 E, **E**...Chevron, Phillips 66, Luby's, Wendy's, **W**...Texaco/mart
23b	N Wayside, **E**...Exxon
23a	Kirkpatrick Blvd, no facilities
22	Homestead Rd, Kelley St, **N**...Texaco/diesel/mart, Whataburger
21	Lockwood Dr, **N**...Chevron/McDonald's, Texaco, Burger King, Church's Chicken, Popeye's, Family$, **S**...HOSPITAL, Conoco, CJ's Rest.
20	US 59, to downtown
19	Hardy Toll Rd, no facilities
18	Irvington Blvd, Fulton St, **N**...Chevron, **S**...Texaco/diesel/mart, Jack-in-the-Box
17	I-45, N to Dallas, S to Houston
16	Yale St, N Main St, Shamrock, **N**...Exxon/mart, **S**...Mobil/mart, Shell/wash, Burger King, Church's Chicken, Western Inn
15	TX 261, N Shepherd Dr, **N**...Sonic, Taco Cabana, **S**...Chevron/mart, Shell/mart, Texaco/diesel/mart, Wendy's, Whataburger, Home Depot, PepBoys
14	Ella Blvd, **N**...Exxon, Shell/mart/wash/24hr, Texaco/wash, Blimpie, Burger King, Jack-in-the-Box, KFC, LJ Silver, McDonald's, Popeye's, Taco Bell, auto parts, **S**...HOSPITAL, Shell, BBQ, BrakeCheck, Lowe's Bldg Supply
13c	TC Jester Blvd, **N**...Chevron, Texaco, Atchafalaya Kitchen, Denny's, Golden Gate Chinese, Courtyard, SpringHill Suites, **S**...Phillips 66/mart
13b a	US 290, no facilities
12	W 18th St, **N**...CiCi's, Sheraton, Foley's, JC Penney, OfficeMax, mall, **S**...Whataburger
11	I-10, W to San Antonio, E to downtown Houston
10	Woodway Dr, Memorial Dr, **E**...Shamrock/mart, Steak'n Egg, **W**...Exxon, Shell/mart, Shug's Rest., bank
9b	Post Oak Blvd, **E**...Drury Inn, LaQuinta, **W**...Champ's Rest., McCormick&Schmick's Café
9a	San Felipe Rd, Westheimer Rd, FM 1093, **E**...DENTIST, Exxon/diesel, Texaco, Courtyard, Hampton Inn, Circuit City, NTB, Target, **W**...Shell, Luke's Burgers, Crowne Plaza Hotel, HomeStead Village Studios, Marriott, Sheraton, Best Buy, Dillard's, Nieman-Marcus, bank
8	US 59, Richmond Ave, no facilities
7	Bissonet St, Fournace Place, **E**...Home Depot, **W**...Texaco/diesel/mart/repair
6	Bellaire Blvd, no facilities
5b	Evergreen St, no facilities
5a	Beechnut St, **E**...Chevron, Boston Mkt, IHOP, McDonald's, Outback Steaks, bank, **W**...Shell/mart, Escalante Mexican Grill, James Coney Island, Saltgrass Steaks, Smoothie King, Borders Books, Marshall's, Service Merchandise, cinema, mall
4a	Brasswood, S Post Oak Rd, **E**...Exxon, Day's Inn
3	Stella Link Rd, **N**...Chevron, Jack-in-the-Box, AllLube'nTune, DiscountTire, Food City, Radio Shack, **S**...Exxon/mart, Phillips 66, Texaco, BrakeCheck, tune-up
2	US 90A, **N**...Chevron, Conoco, Shamrock, Arby's, Bennigan's, Burger King, Church's Chicken, Denny's/24hr, KFC, McDonald's, Taco Bell, Wendy's, Howard Johnson, Villa Motel, Ford, Honda, **S**...Chevron, Whataburger/24hr, La Quinta, Motel 6, Super 8, Firestone, Nissan, Toyota, amusement park, bank, to Buffalo Speedway
1c	Kirby Dr(from eb), **N**...Radisson, Shoney's Inn/rest., **S**...Joe's Crabshack, Papadeaux Seafood, Pappasito's Cantina, Cavender's Boots, Chevrolet, NTB, Pontiac/GMC, Toyota
1b a	FM 521, Almeda St, Fannin St, **N**...Chevron, Conoco, Astro Arena, **S**...Texaco/mart, McDonald's, Aamco, Sam's Club, to Six Flags

Houston

Utah Interstate 15

Exit #	Services

403mm Utah/Idaho state line

402 Portage, **W**...gas, phone

394 UT 13, Plymouth, **E**...Chevron/Subway/diesel/mart, rec area, RV camping

387 UT 30 E, to Riverside, Fielding, **3-5 mi E**...gas, phone, lodging, camping

383 Tremonton, Garland, **2 mi E**...HOSPITAL, Burger King, Denny's/24hr, McDonald's, Taco Time, Western Inn

382 I-84 W, to Boise

379 UT 13, to Tremonton, **2-3 mi E**...Chevron, Conoco/diesel, Arctic Circle, Crossroads Rest., JC's Country Diner, Subway, Taco Time, Marble Motel, Sandman Motel

375 UT 240, to UT 13, Honeyville, to rec area, **E**...Crystal Hot Springs Camping

370mm rest area sb, full(handicapped)facilities, info, phone, picnic tables, litter barrels, vending, petwalk

368 UT 13, Brigham City, **W**...to Golden Spike NHS

366 Forest St, Brigham City, **W**...Bear River Bird Refuge

364 US 91, Brigham City, to US 89, Logan, **E**...HOSPITAL, Amoco, Chevron/Blimpie/mart/24hr, Citgo/7-11, Flying J/diesel/rest./24hr, Arby's, Beto's Mexican, Burger King, Hunan Chinese, J&D's Rest., KFC/Taco Bell, Little Caesar's, McDonald's, Pizza Hut, Pizza Press, Subway, Taco Time, Wendy's, Bushnell Motel, Crystal Inn, Galaxie Motel, Howard Johnson Express, AutoZone, Checker Parts, Chevrolet/Pontiac/Olds/Buick/Cadillac, Chrysler/Dodge/Jeep, JiffyLube, K-Mart/drugs, Radio Shack, ShopKO, RV camping, to Yellowstone NP via US 89

363mm rest area nb, full(handicapped)facilities, phone, picnic tables, litter barrels, vending, petwalk

361 Port of Entry both lanes

360 UT 315, to Willard, Perry, **E**...Flying J/Country Mkt/diesel/LP/mart, KOA(2mi)

354 UT 126, to US 89, Willard Bay, to Utah's Fruit Way, no facilities

352 UT 134, Farr West, N Ogden, **E**...DENTIST, Citgo/7-11, Maverick/gas/atm, Phillips 66/Wendy's/diesel/mart, Arby's, McDonald's, Subway, bank, **W**...Conoco/diesel/mart

349 to Harrisville, Defense Depot, **W**...Texaco/diesel/mart/wash, diesel repair

347 UT 39, 12ᵗʰ St, Ogden, **E**...Flying J/diesel/mart, Phillips 66/mart, Best Western/rest., Comfort Inn, **1-2 mi E**...Chevron, Denny's, Country Kitchen, KFC, McDonald's, Sizzler, Motel 6, to Ogden Canyon RA, **W**...Pilot/DQ/Subway/Taco Bell/diesel/24hr, Country Kitchen, Sleep Inn

346 UT 104, 21ˢᵗ St, Ogden, **E**...Chevron/Arby's/diesel, Conoco/mart, Flying J/Conoco/diesel/LP/mart/24hr, Phillips 66/diesel, Cactus Red's Rest., Tamarack Rest., Big Z Motel/rest., Comfort Suites, Holiday Inn Express, Travelodge(2mi), RV Repair, funpark, **W**...Texaco/diesel/café, Super 8/café/diesel, Cookery Rest., Century RV Park

345 (from nb), UT 53, Ogden, 24ᵗʰ St, **E**...Sinclair/diesel/mart, Texaco/mart

344b a UT 79 W, 31ˢᵗ St, Ogden, **1-2 mi E**...HOSPITAL, Citgo/7-11, Arby's, Golden Corral, JJ North's Buffet, Mexican Rest., Sizzler, Holiday Inn, Radisson, Dillard's, K-Mart, Sam's Club, Chevrolet, Ford/Lincoln/Mercury, RV Ctr, mall, to Weber St U, **W**...airport

343 I-84 E(from sb), to Cheyenne, Wyo, no facilities

342 UT 26(from nb), to I-84 E, Riverdale Rd, **E**...Conoco/diesel/mart, Sinclair, Applebee's, Boston Mkt, Carl's Jr, Chili's, La Salsa Mexican, McDonald's, Motel 6, Chrysler/Plymouth/Jeep, Circuit City, Gart Sports, Harley-Davidson, Home Depot, Honda/Nissan, Isuzu, Lincoln/Mercury/Toyota/Kia, Mazda, MediaPlay, Mitsubishi, OfficeMax, PetsMart, Pearle Vision, Pontiac/Buick/GMC, Saturn, SuperCuts, Target, Toyota, Wal-Mart, Wilderness RV, bank

341 UT 97, Roy, Sunset, **E**...Air Force Museum, **W**...EYECARE, Citgo/7-11, Phillips 66, Sinclair, Texaco, Arby's, Arctic Circle, Blimpie, Burger King, Central Park, Chinese Rest., Dairy Queen, Denny's, JB's, KFC, McDonald's, Pizza Hut, Ponderosa, Subway, Taco Bell, Taco Time, Wendy's, Quality Inn, Motel 6, Albertson's/drugs, AutoZone, BrakeWorks, Checker Parts, Early Tires, Goodyear, Discount Tires, JiffyLube, Masterlube, Midas Muffler, Pennzoil, Precision Tune, Radio Shack, RiteAid, Smith's/drugs, cinema, hardware, transmissions, same as 342

338 UT 103, Clearfield, **E**...Hill AFB, **W**...Chevron/mart/wash, Conoco/mart/wash, Citgo/7-11, PetroMart, Texaco/mart/wash, Circle K, Arby's, Carl's Jr, Central Park, KFC, McDonald's, Skipper's, Subway, Taco Bell, Alana Motel, Crystal Cottage Inn, Super 8, Big O Tires, Sierra RV, Smith's Food/drugs, bank, tires, same as 341

336 UT 193, Clearfield, to Hill AFB, **E**...Maverick/gas, Texaco/diesel/mart/wash, **3/4 mi W**...Chevron, Flying J, Coffee Ranch, carwash

335 UT 108, Syracuse, **E**...Chevron/mart/wash, Phillips 66/diesel/mart/wash, Circle K, Applebee's, Cracker Barrel, Golden Corral, JB's, Marie Callender's, Outback Steaks, Quizno's, Red Robin's, SF Pizza, TimberLodge Steaks, Tony Roma's, Fairfield Inn, Hampton Inn, Holiday Inn Express, La Quinta, Barnes&Noble, JiffyLube, Lowe's Bldg Supply, MensWearhouse, Michael's, Office Depot, Target, cinema, **W**...HOSPITAL, Citgo/7-11, Arby's, McDonald's, RV Ctr

334 UT 232, UT 126, Layton, **E**...Phillips 66, Texaco, Denny's, Garcia's, McDonald's, Olive Garden, Red Lobster, Sizzler, Wendy's, Comfort Inn, JC Penney, Mervyn's, ZCMI, bank, cinema, mall, to Hill AFB South Gate, **W**...Flying J, Blimpie, Burger King, Chinese Rest., Fuddrucker's, KFC, LoneStar Steaks, Taco Bell, Taco Time, Chevrolet, Dodge, NTB, OfficeMax, PetsMart, Pontiac/Cadillac/GMC, RV Ctr, Sam's Club, ShopKO, Ultimate Electronics, Wal-Mart/drugs, auto repair

332 to UT 126(from nb), Layton, **E**...Chinese Rest., **W**...Texaco/mart, Blaine Jensen's RV

331 UT 273, Kaysville, **E**...Chevron/McDonald's/mart, Citgo/7-11, Phillips 66/diesel/mart, Sinclair, Dairy Queen, David's Pizza, KFC, Subway, Taco Time, Wendy's, Far West Motel, Albertson's, Checker Parts, Valvoline, XpressLube, **W**...1ˢᵗ Choice Cars, Blaine Jensen's RV

329mm parking area both lanes

Utah Interstate 15

327   UT 225, Lagoon Dr, Farmington, **E...**Subway, amusement park, camping

326   US 89 N, UT 225(from nb), **1 mi E...**VETERINARIAN, Maverick/gas, Arby's, Aunt Pam's, Burger King, Little Caesar's, K-Mart/drugs, Smith's Food/drug/ 24hr, Goodyear/auto, laundry, RV camping, to I-84

325   UT 227(from nb), Lagoon Dr, to Farmington, **E...**Lagoon RV Camping

322   Centerville, **E...**Chevron, Phillips 66/diesel/mart, Arby's, Arctic Circle, Burger King, Carl's Jr, Dairy Queen, Del Taco, Jake's Shakes, LoneStar Steaks, McDonald's, Subway, Taco Bell, TacoMaker, Wendy's, Albertson's/drugs, Home Depot, LandRover, Radio Shack, Target/foods, **W...**RV Ctr

321   US 89 S(exits left from sb), UT 131, 500W, S Bountiful, **E...**CHIROPRACTOR, Amoco/mart, Chevron/mart/wash/24hr, Exxon/diesel, Phillips 66/diesel/mart/ atm, Texaco/mart, Alicia's Rest., Country Inn Suites, Goodyear, JiffyLube, tires

320   UT 68, 500 S, W Bountiful, Woods Cross, **E...**HOSPITAL, Exxon/mart, Texaco/ mart, Circle K, Blimpie, Boston Mkt, Burger King, Carl's Jr, Christopher's Steaks, ChuckaRama, Dee's Rest., HogiYogi, KFC, McDonald's, Papa Murphy's, Pizza Hut, SF Pizza, Sizzler, Subway, SuCasa Mexican, Taco Bell, Winger's, Aamco, Albertson's, AllA$, AutoZone, Barnes&Noble, CarMerica Tires, Checker Parts, Early's Tires, FashionBug, Firestone/auto, Fred Meyer, Gart Sports, GNC, Kinko's, MailBoxes Etc, Meineke Muffler, Midas Muffler, OfficeMax, Radio Shack, Ross, Rite Aid, ShopKO, TJMaxx, Walgreen, cinema, **W...**Chevron/diesel/mart, Phillips 66/ diesel

318   26th S, N Salt Lake, **E...**CHIROPRACTOR, VETERINARIAN, Amoco, Chevron/mart, Texaco/mart, Apollo Burger, Arby's, Burger King, La Frontera Mexican, KFC, McDonald's, Peter Piper Pizza, Pizza Hut, Skipper's, Village Inn, Wendy's, Best Western, Fairfield Inn, Cloth World, Chevrolet/Pontiac/Olds/Buick/Kia, Diahatsu, Ford/Lincoln/Mercury, Honda, K-Mart, Mazda, Nissan, Smith's/drugs/24hr, Toyota, bank, **W...**Conoco, Denny's, Hampton Inn, Motel 6, Goodyear

317   Center St, Cudahy Lane(from sb), N Salt Lake, **E...**gas/mart

316   I-215 W(from sb), to airport, no facilities

315   US 89 S, to Beck St, N Salt Lake, no facilities

314   Redwood Rd, Warm Springs Rd, no facilities

313   9th St W, 7-11, Self's conv/rest., Tia Maria's Mexican, Regal Inn

312   6th St N, **E...**HOSPITAL, LDS Temple, downtown, to UT State FairPark

311   I-80 W, to Reno, airport, no facilities

310   6th S, SLC City Center, info, **1 mi E...**Chevron, Phillips 66/diesel, Circle K, Burger King, Dairy Queen, Denny's, McDonald's, Wendy's, Embassy Suites, Hampton Inn, Little America, Motel 6, Quality Inn, Ramada Inn, Residence Inn, Super 8, Ford, Toyota, to Temple Square, LDS Church Offices

309   13th S, SLC, downtown, **E...**Texaco, **W...**to Flying J/Conoco/diesel/LP/rest./24hr, Wendy's, Chevrolet, Mitsubishi, Goodyear, Blue Beacon

308   UT 201, 21st St S, 13th St S, 9th St S(from nb), **E...**Chevron/mart, Phillips 66, Atlantis Burgers, Carl's Jr, McDonald's, CompUSA, Costco, Home Depot, Office Depot, PetsMart, U-Haul, same as 309

307   I-80 E, to Denver, Cheyenne, no facilities

306   UT 171, 33rd S, S Salt Lake, **E...**Citgo/7-11, Phillips 66/mart, Burger King, McDonald's, Taco Bell, Bonneville Inn, Day's Inn, Roadrunner Motel, tires, **W...**Coastal Gas, GMC

304   UT 266, 45th S, Murray, Kearns, **E...**McDonald's, UpTown Tires, Zim's Crafts, **W...**Chevron, Shell/diesel, Sinclair/ Burger King, Texaco/mart, Denny's, Fairfield Inn, Hampton Inn, Quality Inn, InterMtn RV Ctr, Lowe's Bldg Supply

303   UT 173, 53rd S, Murray, Kearns, **E...**HOSPITAL, **W...**Chevron, Conoco, Sinclair, 7-11, Schlotsky's, Reston Hotel, Smith's Food/drug, bank, Blaine Jenson's RV Ctr, FunDome

302b a   I-215 E and W, no facilities

301   UT 48, 72nd S, Midvale, **E...**Chevron/mart, Conoco, Phillips 66/mart, Sinclair/mart, Texaco/mart/LP, Denny's/24hr, John's Place, KFC, McDonald's, Midvale Mining Café, Sizzler, South Seas Café, Taco Bell, Village Inn Pancakes, Best Western, Day's Inn, Discovery Inn/café, Executive Inn, La Quinta, Motel 6, Rodeway Inn, Sandman Inn, ZCMI, Cadillac/Olds/Buick, bank, carwash, to Brighton, Solitude Ski Areas

298   UT 209, 90th S, Sandy, **E...**Sinclair/mart, Holiday/gas, Arby's, Burger King, Hardee's, Johanna's Kitchen, Sconecutter's Rest., Sizzler, Comfort Inn, Majestic Rockies Motel, Early Tires, Ford, Lowe's Bldg Supply, bank, to Snowbird, Alta Ski Areas, **W...**HOSPITAL, Chevron(2mi), Texaco, Chinese Rest., KFC, NTB, RV Ctr

297   106th S, Sandy, S Jordan, **E...**DENTIST, Amoco, Conoco, Phillips 66/mart, BBQ, Bennigan's, Carver's Prime Rib, Chili's, Eat a Burger, Fuddrucker's, HomeTown Buffet, Johanna's Rest., Ponderosa, Shoney's, Subway, TGIFriday, Village Inn, Best Western, Courtyard, Extended Stay America, Hampton Inn, Marriott, Residence Inn, TownePlace Suites, Chevrolet/Olds, Chrysler/Jeep, Costco, Dillard's, Gart Sports, Goodyear, Honda, JC Penney, Mervyn's, Target, Toyota, ZCMI, bank, cinema, mall, **W...**Denny's, Country Inn Suites, Sleep Inn, Super 8

295   US 89 N, to Sandy(from nb, no return), same as 297

Utah Interstate 15

Draper

294   UT 71, Draper, Riverton, **E**...Flying J/diesel/mart, Holiday/diesel, Texaco, Arby's, Arctic Circle, Carl's Jr, Guadalahonky's Mexican, Jamba Juice, KFC, McDonald's, Neil's Broiler, Pizza Hut, Ruby Tuesday, Quizno's, Teriyaki Express, Wingers Diner, Holiday Inn Express, Ramada Ltd, Travelodge, Camping World RV Supplies(1mi), Goodyear/auto, Greenbax, Mountain Shadows Camping, Smith's Foods, TJ Maxx, FSA Outlets/famous brands, cleaners

291   UT 140, Bluffdale, **E**...Isuzu/Suzuki, Camping World RV Supplies(2mi), **W**...st prison

287   UT 92, to Alpine, Highland, **E**...to Timpanogas Cave, **W**...Amoco/Burger King/diesel/mart/wash/24hr, Lone Peak RV Ctr, Thanksgiving Point/café

285   12th W, to UT 73, Lehi, no facilities

282   UT 73, to Lehi, **E**...1 Man Band Diner, Motel 6, **W**...Chevron/Pizza Hut/diesel/mart/wash/24hr, Phillips 66/Wendy's/mart/ 24hr, Arctic Circle, Country Kitchen, KFC, McDonald's, Papa Murphy's, Subway, Best Western, Comfort Inn, Day's Inn, Super 8, Albertson's/drugs, Big O Tires, Checker Parts, GNC, JiffyLube, Uncle Dave's, USPO, bank, museum

281   Main St, American Fork, **E**...HOSPITAL, Phillips 66/diesel/mart/24hr, Texaco/mart, Chevrolet, Chrysler/Plymouth/ Dodge/Jeep, K-Mart, Smith's Food/drugs

279   5th E, Pleasant Grove, **E**...Conoco/Blimpie/mart, Carl's Jr, Denny's, Taco Bell, Quality Inn, OfficeMax, Stewart's RV Ctr, Wal-Mart, **1-2 mi E**...HOSPITAL, Circle K, Phillips 66/mart, Texaco/mart, Arby's, Del Taco, Golden Corral, Hardee's, KFC, McDonald's, Subway, Wendy's, American Camping, Chevrolet, JiffyLube, **W**...Buick/GMC/Pontiac, Ford

Orem

276   Orem, Lindon, **E**...Phillips 66/mart, Del Taco(1mi), Home Depot, **W**...Conoco/diesel/café/mart/showers/24hr, Texaco/ diesel/mart

275   UT 52, to US 189, 8th N, Orem, **E**...Phillips 66/diesel/mart/wash, Best Inn Suites, **1 mi E**...Arby's, Dairy Queen, to Sundance RA

274   Orem, Center St, **E**...HOSPITAL, Citgo/7-11, Conoco/mart, water park, **1-2 mi E**...Burger King, Hardee's, Kenny Rogers, KFC, Taco Bell, **W**...Amoco/mart, Econolodge, LP

272   UT 265, 12th St S, University Pkwy, **E**...Phillips 66/Wendy's/diesel/mart/24hr, Sinclair, Texaco/mart/wash, Chevy's Mexican, IHOP, Krispy Kreme, McDonald's, Subway, Fairfield Inn, Hampton Inn, Lakeview Manor Motel, La Quinta, Ford, HomeBase, JiffyLube, Saturn, Wal-Mart SuperCtr/24hr, **1-3 mi E**...Chevron, Applebee's, Arby's, Black Angus, Chili's, Fuddrucker's, Golden Corral, Honeybaked Ham, Ponderosa, Red Lobster, Sizzler, Village Inn, Barnes&Noble, Bed Bath&Beyond, Circuit City, Gateway Computers, JC Penney, KidsRUs, Linens'n Things, Lowe's Bldg Supply, MediaPlay, Mervyn's, Mitsubishi, Office Depot, OfficeMax, Old Navy, PetsMart, Subaru, TJ Maxx, ToysRUs, ZCMI, bank, mall, to BYU, many services on US 89

Provo

268b a   UT 114, Center St, Provo, **E**...HOSPITAL, Conoco/mart, Phillips 66/wash, Shell/mart, Sinclair, Texaco, 7-11, auto/brake/ muffler repair, **1 mi E**...Albertson's, Checker Parts, Park Motel, Travelodge, **W**...Chevron/mart, Conoco/mart, Econolodge, KOA, Lakeside RV, Utah Lake SP, airport

266   US 189 N, University Ave, Provo, **E**...Chevron/mart/wash/24hr, Conoco/diesel/mart, Maverick/gas, Phillips 66/mart/ wash, Texaco/diesel/mart, Arby's, Blimpie, Burger King, ChuckaRama, Hogi Yogi, KFC, McDonald's, Papa Murphy's, Ruby River Steaks, Shoney's, Sizzler, Taco Bell, Taco Time, Village Inn Rest., Wendy's, Best Western, Colony Inn, Fairfield Inn, Hampton Inn, Holiday Inn, Howard Johnson, Motel 6, National 9 Inn, Sleep Inn, Super 8, Dillard's, GoodEarth Foods, JC Penney, JiffyLube, K-Mart/drugs, MasterMuffler, NAPA, OfficeMax, Sam's Club, Sears/auto, SportsMans Whse, Staples, ZCMI, Silver Fox RV Camping, RV Ctr, bank, hardware, mall, to BYU, **1 mi E**...Los Three Amigos Mexican, Best Western, Hotel Roberts, Safari Motel, Western Inn, CarQuest, VW/Audi, auto repair

265   UT 75, Springville, **E**...Flying J/diesel/rest./24hr, Maverick/gas, Best Western, E Bay RV Park

263   UT 77, Springville, Mapleton, **E**...Phillips 66/MeanGene Burger/diesel/mart, Wal-Mart SuperCtr/24hr, **W**...VETERINARIAN, Chevron/Arby's/diesel/mart/wash, Cracker Barrel, Day's Inn, RV Ctr, truck repair/tires

261   US 89 S, US 6 E(from sb), to Price, **E**...Chevron/pizza/mart, Phillips 66, Texaco, Arby's, Burger King, JB's, KFC, McDonald's, Papa Murphy's, Taco Bell, Taco Time, Wendy's, Winger's, Holiday Inn Express, Western Inn, Albertson's/ gas, Checker Parts, Fakler's Tire, Food4Less, K-Mart, Pennzoil, Radio Shack, RV Ctr, cinema

260   US 6 E, UT 156, Spanish Fork, **E**...Amoco, Chevron/mart, Conoco/Bob's Pizza/mart, Phillips 66, Texaco/diesel/LP, Arby's, Hogi Yogi, JB's, KFC, Little Caesar's, Macy's Food/deli/drugs, McDonald's, North's Buffet, Subway, Taco Bell, Taco Time, Tela's Tacos, Escalante Bed/Breakfast, Checker's AutoWorks, Macey's Foods, Ken's Service/repair, K-Mart, ShopKO, bank, hardware, **W**...Conoco/diesel/mart, Chevrolet, Chrysler/Plymouth/Jeep, Ford, RV Ctr, cleaners/laundry

256   UT 164, to Spanish Fork, no facilities

254   UT 115, Payson, **E**...HOSPITAL, Flying J/diesel/LP/rest./mart/24hr, Sinclair, Texaco/diesel/mart, McDonald's, Subway/ Fruizle, Comfort Inn, Payson Foods, RiteAid, diesel repair, Mt Nebo Loop

252   Payson, Salem, **E**...Chevron/mart, Texaco/mart, Hogi Yogi, Taco Time(1mi)

248   US 6 W, Santaquin, **E**...TrueValue, **W**...Amoco/mart/atm, Conoco/diesel/mart, Texaco/diesel/mart, Main St Pizza, Main St Mkt, SantaQueen Burgers, Sorenson's Apple Farm, Nat Hist Area, Tintic Mining Dist, auto/tire care

245   to S Santaquin, no facilities

236   UT 54, Mona, no facilities

Nephi

228   UT 28, to Nephi, **2-4 mi W**...HOSPITAL, CashSaver/diesel, Citgo/7-11, Conoco/mart, Reed's Drive-In, Safari Motel, Chevrolet/Pontiac/Olds/Buick, Goodyear, Thriftway Foods, to Great Basin NP

225   UT 132, Nephi, **E**...Amoco/mart/LP, Taco Time, KOA(5mi), auto repair, **W**...HOSPITAL, Chevron/Arby's/diesel/mart, Phillips 66/Wendy's/diesel/mart, Econolodge, Hi-Country RV Park

N

S

Utah Interstate 15

222   UT 28, Nephi, to I-70, **E**...Chevron/diesel/mart/atm, Sinclair/Hogi Yogi/diesel/mart/24hr, Texaco/diesel/mart/atm, Burger King, Denny's, Mickelson's Rest., Subway/Fruizle, Motel 6, Roberta's Cove Motel, Super 8, **W**...HOSPITAL, Flying J/Pepperoni's/diesel/LP/rest./24hr, Tiara Café, Wayne's Rest., Best Western, Safari Motel, Ford/Mercury, Big A Parts, diesel repair

207   to US 89, Mills, no facilities

202   Yuba Lake, phone, access to boating, camping, rec facilities

188   US 50 E, Scipio, to I-70, **E**...Amoco/diesel/mart/24hr, Chevron/mart, Conoco/diesel/repair

184   ranch exit, no facilities

178   US 50, Delta, Holden, **W**...gas, phone, to Great Basin NP

174   to US 50, Holden, **W**...gas, phone, to Great Basin NP

167   Lp 15, Fillmore, **1-3 mi E**...HOSPITAL, Amoco/LP, Chevron/diesel/mart, Shell/diesel/LP, Best Inn/suites, Best Western/rest., Fillmore Motel, WagonsWest RV Park, **W**...Amoco/mart, Texaco/Subway/diesel/mart/24hr

163   Lp 15, to UT 100, Fillmore, **E**...HOSPITAL, Chevron/Arby's/diesel/mart/24hr, Maverick/gas, Best Inn/suites, Best Western, Fillmore Motel, **W**...Phillips 66/Burger King/mart/24hr

158   UT 133, Meadow, **E**...Chevron/mart, Texaco/diesel/mart

153mm   view area sb

151mm   view area nb

146   Kanosh, **2 mi E**...Chevron, chainup area

138   ranch exit, no facilities

137mm rest area sb, full(handicapped)facilities, picnic tables, litter barrels, vending, petwalk

135   Cove Fort, Hist Site, **E**...Chevron/mart

132   I-70 E, to Denver, Capitol Reef NP, Fremont Indian SP

129   Sulphurdale, chainup area, no facilities

126mm rest area nb, full(handicapped)facilities, picnic tables, litter barrels, vending, petwalk

125   ranch exit, no facilities

120   Manderfield, chainup area nb, no facilities

112   to UT 21, Beaver, Manderfield, **E**...HOSPITAL, Arco/mart/24hr, Chevron/diesel/mart, Conoco/diesel, Sinclair/diesel, Arby's, El Bambi Café, McDonald's, Subway, Best Western/rest., Country Inn, Day's Inn, Stag Motel, KOA(1mi), **W**...Amoco/diesel/mart/atm, Wendy's, Super 8, to Great Basin NP

109   to UT 21, Beaver, **E**...HOSPITAL, Phillips 66/mart, Sinclair/mart, Texaco/Burger King/Firestone/diesel/mart/24hr, Arshel's Café, Aspen Lodge, Best Western, Comfort Inn, Delano Motel/RV Park, Granada Inn, Sleepy Lagoon Motel, Mike's Foodtown, Cache Valley Cheese, NAPA, RV Parts/repair/dump, laundry, **W**...Chevron/diesel/mart/tires/24hr, KanKun Mexican, Quality Inn Timberline/RV park, truckwash, to Great Basin NP

100   ranch exit, no facilities

95   UT 20, to US 89, to Panguitch, Bryce Canyon NP, no facilities

88mm rest area both lanes, full(handicapped)facilities, phone, picnic table, litter barrel, petwalk

82   UT 271, Paragonah, no facilities

78   UT 141, Parowan, **E**...Best Western/rest., Crimson Hills Motel, Day's Inn, ski areas, **W**...TA/Arco/Subway/Taco Bell/diesel/mart/24hr

75   UT 143, Parowan, **2 mi E**...Jedadiah's Inn/rest., Best Western, to Brian Head/Cedar Breaks Ski Resorts

71   Summit, **W**...Sunshine TravelPlaza/diesel/rest., lube/tires

62   UT 130, Cedar City, **E**...Phillips 66/diesel/mart, Best Western(3mi), Holiday Inn(1mi), KOA(2mi), st patrol, **W**...Sinclair/mart, Texaco/diesel/mart/24hr, Steak&Stuff Rest., Travelodge

59   UT 56, Cedar City, **E**...MEDICAL CARE, CHIROPRACTOR, Amoco/mart, Chevron/mart/atm, Conoco/diesel/mart, Maverick/gas, Phillips 66/diesel/mart/LP, Texaco/diesel/mart, Arby's, Burger King, Denny's, KFC, McDonald's, Shoney's, Taco Bell, Wendy's, Abbey Inn, Comfort Inn, Econolodge, AutoValue, Chevrolet/Buick, L&S Repair, Rolling Rubber Tires/repair, **1 mi E**...Texaco/service, China Garden, Escobar's Mexican, Godfather's, Pizza Factory, Sizzler, Sullivan's Café, Best Western, Rodeway Inn, Valu Inn, Zion Motel, Ace Hardware, Bradshaw's Parts, Bullock's Drug, Dodge/Chrysler/Jeep, GMC/Olds, Lin's Mkt, USPO, **W**...Texaco/diesel/mart/atm, Subway, Holiday Inn/rest., Motel 6, Super 8, Goodyear

57   Lp 15, to UT 14, Cedar City, **1-2 mi E**...HOSPITAL, CHIROPRACTOR, Chevron/service/24hr, Phillips 66/diesel/mart, Shell/mart, Sinclair/diesel, Texaco/diesel, Dairy Queen, Domino's, Hogi Hogi, Hunan Chinese, JB's, Pizza Hut, Subway, Taco Time, CedarRest Motel, Days Inn, Rodeway Inn, Super 7 Motel, Zion Motel, Albertson's, Allied Tire/brake, Checker Parts, Gart Sports, K-Mart/drugs, LubeAmerica, Radio Shack, Smith's Foods/24hr, Wal-Mart/drugs, bank, carwash, laundry, museum, to Cedar Breaks, Navajo Lake, Bryce Cyn, Duck Crk, **W**...Chevron/A&W/diesel/mart, cinema

51   Kanarraville, Hamilton Ft, no facilities

44mm rest area both lanes, full(handicapped)facilities, phone, picnic tables, litter barrels, petwalk, hist site

Utah Interstate 15

42	New Harmony, Kanarraville, **W**...Texaco/diesel/mart
40	to Kolob Canyon, Zion's NP, **E**...tourist info/phone, scenic drive
36	ranch exit, no facilities
33	ranch exit, no facilities
31	Pintura, no facilities
30	Browse, no facilities
27	UT 17, Toquerville, **9 mi E**...Subway, Super 8, to Zion NP, Grand Cyn, Lake Powell
23	Leeds, Silver Reef(from sb), **E**...Catfish Charlie's, **3 mi E**...Harrisburg RV Resort/mart/gas, hist site, museum
22	Leeds, Silver Reef(from nb), same as 23
16	UT 9, to Hurricane, **E**...Wal-Mart Dist Ctr, **10 mi E**...Chevron, Shell/mart, Texaco/mart, Comfort Inn, Super 8, to Zion NP, Grand Canyon, Lake Powell, RV Camping
10	Middleton Dr, Washington, **E**...Amoco/diesel/mart, Chevron, Phillips 66/diesel/mart, Arby's, Arctic Circle, Burger King, Godfather's, Little Caesar's, Tom's Deli, Red Cliffs Inn/rest., Ace Hardware, Albertson's/drugs, Costco, PostNet, JC Penney, QwikLube/wash, Wal-Mart/McDonald's, Redlands RV Park, ZCMI, bank, cleaners, mall, same as 8, **W**...Arco/mart/24hr, Texaco/diesel/LP/lube, Dixie Radiator, auto repair
8	St George Blvd, St George, **E**...Shell/mart, Texaco/Subway/mart, Café Rio Mexican, Carl's Jr, Chili's, ChuckaRama, Red Lobster, Village Inn Rest., Hampton Inn, Ramada Inn, Shoney's Inn/rest., Big 5 Sports, DownEast Outfitters, Old Navy, Rebel Sports, SunRise Tires, Settler's RV Park, Staples, Zion Factory Stores/famous brands, bank, same as 10, **W**...DENTIST, Amoco/mart, Chevron/mart, Maverick/gas, Shell, Sinclair/LP, Texaco/diesel, Burger King, Denny's, Dick's Rest., Fairway Grill, Frontier Rest., Kenny Rogers, KFC, Luigi's Italian, McDonald's, Panda Garden Chinese, Pizza Hut, Sizzler, Taco Bell, Taco Time, Wendy's, Best Western, Chalet Motel, Comfort Inn, Day's Inn, Desert Edge Inn, Econolodge, Howard Johnson, Motel 6, Sands Motel, Singletree Inn, SunTime Inn, Travelodge, Checker Parts, Econo Lube'n Tune, Honda Motorcycles, NAPA, Old Home Bakery, Rite Aid, St Geo RV, bank, cinema, to LDS Temple
6	UT 18, Bluff St, St George, **E**...Arco/mart/24hr, Chevron/diesel/mart/repair/24hr, Texaco/A&W/diesel/mart, Fairfield Inn, Sleep Inn, Buick/Pontiac/Olds/GMC, Saturn, U-Haul, funpark, tires, **W**...HOSPITAL, VETERINARIAN, DENTIST, C-mart/gas, Texaco/mart, Burger King, Carl's Jr, Dairy Queen, Denny's, McDonald's, NorthStar Buffet, Pizza Hut, Tony Roma's, Best Western, Budget 8 Inn, Claridge Inn, Comfort Suites, Holiday Inn, Quality Inn, Ranch Inn, Sheraton, Ford/Lincoln/Mercury, Honda, Mazda, Nissan, Painters RV Ctr, Plymouth/Chrysler/Dodge, TempleView RV Park, Toyota, cinema
4	Bloomington, **E**...Shell/Burger King/diesel/mart/wash/atm/24hr, **W**...Chevron/Taco Bell/mart, Courtyard Cafeteria, bagels
2	**Welcome Ctr nb, full(handicapped)facilities, picnic tables, litter barrels, phone, petwalk**
1mm	Port of Entry/weigh sta both lanes
0mm	Utah/Arizona state line

Utah Interstate 70

Exit #	Services
230mm	Utah/Colorado state line
226mm	**view area wb, restrooms(handicapped), litter barrels**
225	Westwater, no facilities
220	ranch exit, no facilities
212	to Cisco, no facilities
202	UT 128, to Cisco, no facilities
190	ranch exit, no facilities
188mm	**Welcome Ctr wb, full(handicapped)facilities, info, picnic tables, litter barrels**
185	Thompson, **N**...Texaco/diesel/mart, café, camping, lodging
180	US 191 S, Crescent Jct, to Moab, **N**...Amoco/diesel/café/mart, **S**...lodging
178.5mm	**rest area eb, full(handicapped)facilities, scenic view, picnic tables, litter barrels**
173	ranch exit, no facilities
162	UT 19, Green River, info, **1-2 mi N**...Amoco/Blimpie/diesel/mart, Shell/mart, Sinclair/diesel, Phillips 66/Burger King/diesel/mart/24hr, Texaco/diesel/mart, Ben's Café, Chow Hound Drive-In, Lemieux Café, McDonald's, Westwinds Rest., Best Western/rest., Book Cliff Lodge/rest., Budget Host, Comfort Inn, Luxury Inn, Motel 6, National 9 Inn, Rodeway Inn, Super 8, OK Anderson City Park, RV camping, museum, same as 158
158	UT 19, Green River, **N**...MEDICAL CARE, Amoco/diesel/mart/wash, Chevron/diesel/mart/24hr, Conoco/Arby's/diesel/mart/atm/24hr, Phillips 66/diesel/mart/24hr, Texaco/diesel/mart, Shell/mart, Sinclair/diesel/mart, Ben's Café, Cathy's Rest., ChowHound Drive-In, Budget Inn, Mancose Rose Hotel, Oasis Motel/café, Big A Parts, Goodyear/auto, NAPA Repair, USPO, KOA, Shady Acres RV Park, bank, Green River SP, same as 162, NO SERVICES WB FOR 100 MILES
156	US 6 W, US 191 N, to Price, Salt Lake, no facilities
147	UT 24 W, to Hanksville, to Capitol Reef, Lake Powell, no facilities
144mm	**rest area wb, restrooms(handicapped), litter barrels**
141.5mm	runaway truck ramp eb
140.5mm	**rest area both lanes, view area, restrooms(handicapped), litter barrels**

486

Utah Interstate 70

139mm	runaway truck ramp eb
136mm	brake test area eb
129	ranch exit, no facilities
120mm	**rest area both lanes, restrooms(handicapped), litter barrels**
114	to Moore, **N...rest area both lanes, restrooms(handicapped), litter barrels**
105	ranch exit, no facilities
102.5mm	**rest area both lanes, restrooms(handicapped), litter barrels**
91	ranch exit, no facilities
89	UT 10 N, UT 72, to Emery, Price, **12 mi N...**gas, **S...**to Capitol Reef NP
84mm	**S...rest area both lanes, full(handicapped)facilities, litter barrels, petwalk**
72	ranch exit, no facilities
61	Gooseberry Rd, no facilities
54	US 89 N, to Salina, US 50 W, to Delta, **N...**Amoco/diesel/deli/mart/24hr, Chevron/Burger King/diesel/mart, Shell, Texaco/diesel/mart/repair, Denny's, Subway, Best Western/rest., Budget Host, Safari Motel/café, Scenic Hills Motel, Chevrolet/Olds, Butch Cassidy RV Camp, Salina Creek RV Camp, **1 mi N...**Conoco/mart, Mom's Café, Henry's Motel, Ranch Motel, NAPA, hardware, tires, NEXT SERVICES 108 MI EB
48	UT 24, to US 50, Sigurd, Aurora, **1-2 mi S...**gas, food, to Fishlake NF, Capitol Reef NP
40	Lp 70, Richfield, **S...**HOSPITAL, DENTIST, Chevron/repair/mart, Flying J/diesel/LP/rest./24hr, Arby's, Frontier Café/store, Subway, Super 8, Pennzoil, RV/truck repair, **1-2 mi S...**KFC, McDonald's, Taco Bell, Best Western, Day's Inn/rest., Budget Host, Quality Inn, Buick/Olds/Pontiac/Cadillac/GMC, Lin's Mkt, Thriftway Foods
37	Lp 70, Richfield, **1-2 mi S...**Amoco/pizza/diesel/mart, Chevron/mart, Conoco/Munchie's RV Park/mart, Maverick/gas, Premium/diesel/mart, Ideal Dairy/burgers, JB's, McDonald's, Pizza Hut, HogyYogi, Best Western, Budget Host, Day's Inn, New West Motel, Quality Inn, Romanico Motel, Weston Inn/rest., Ace Hardware, Albertson's, Big A Parts, Checker's Parts, Chevrolet, Ford/Mercury, K-Mart, Richfield Drug, KOA, RV repair, mufflers, st patrol, tires, to Fish Lake, Capitol Reef Parks
32	Elsinore, Monroe, **1/2 mi S...**Chevron/mart/wash
26	UT 118, Joseph, Monroe, **S...**Shell/mart/RV Park, foodmart
23	US 89 S, to Panguitch, Bryce Canyon, no facilities
17	**N...**Fremont Indian Museum, info, phone, camping
13mm	brake test area eb
8	Ranch Exit, no facilities
3mm	Western Boundary Fishlake NF
1	Historic Cove Fort, **N...**Chevron/mart(2mi)
0mm	I-15, N to SLC, S to St George. I-70 begins/ends on I-15, exit 132.

E

W

Salina   Richfield

Utah Interstate 80

Exit #	Services
197.5mm	Utah/Wyoming state line, no facilities
193	Wahsatch, no facilities
189	ranch exit, no facilities
185	Castle Rock, no facilities
182mm	Port of Entry/weigh sta wb
180	Emery(from wb), no facilities
170	**Welcome Ctr wb/rest area eb, full(handicapped)facilities, phone, vending, picnic tables, litter barrels, petwalk, RV dump**
169	Echo, **1 mi N...**gas/diesel, Kozy Rest./lodging
168	I-84 W, to Ogden, I-80 E, to Cheyenne, no facilities
166	view area both lanes, litter barrels
164	Coalville, **N...**Amoco/diesel/mart, Holiday Hills RV Camp/LP, CamperWorld RV Park, **S...**Chevron/diesel/mart, Sinclair, Texaco, Holiday Hills Rest., to Echo Res RA
156	UT 32 S, Wanship, **N...**Spring Chicken Café, **S...**Sinclair/diesel/mart, to Rockport SP
152	ranch exit, no facilities
148b a	US 40 E, to Heber, Provo, **N...**Sinclair/Blimpie/diesel/mart, phone
147mm	**rest area wb, full(handicapped)facilities, phone, vending, picnic tables, litter barrels, petwalk**
145	UT 224, Kimball Jct, to Park City, **N...**VETERINARIAN, DENTIST, Ford/Mercury, auto repair, RV camping, **S...**DENTIST, Chevron/mart/wash, Texaco/diesel/mart/wash, Arby's, Chinese Rest., Denny's, Kenny Roger's Roasters, Loco Lizard Cantina, McDonald's, Subway, Taco Bell, Wendy's, Best Western, Hampton Inn, Holiday Inn Express, GNC, K-Mart, MailBoxes Etc, Smith's Food/drug, USPO, Wal-Mart, Outlet Mall/famous brands, bank, RV camping, **4 mi S...**Citgo/7-11, Olympia Park Hotel, Radisson, ski areas
144mm	view area eb
143	ranch exit, **N...**Amoco/Blimpie/mart, Pizza Hut, to Jeremy Ranch, **S...**camping, ski area
140	Parley's Summit, **S...**Sinclair/diesel/repair/mart, phone
137	Lamb's Canyon, no facilities

E

W

Kimball Jct

Utah Interstate 80

134	UT 65, Emigration Canyon, East Canyon, Mountaindale RA
133	utility exit(from eb), no facilities
132	ranch exit, no facilities
131	(from eb) Quarry, no facilities
130	I-215 S(from wb), no facilities
129	UT 186 W, Foothill Dr, Parley's Way, **N**...HOSPITAL
128	I-215 S(from eb), no facilities
127	UT 195, 23rd E St, to Holladay, no facilities
126	UT 181, 13th E St, to Sugar House, **N**...Chevron, Texaco, Olive Garden, Red Lobster, Sizzler, Training Table Rest., Wendy's, ShopKO, Photo Shop, cinema
125	UT 71, 7th E St, Texaco, Firestone, **N**...McDonald's
124	US 89, S State St, **N**...VETERINARIAN, Citgo/7-11, Chevron, Texaco/diesel/mart, Burger King, Skipper's, Taco Bell, Uncle Sid's Rest., Wendy's, Woody's Drive-In, Buick, Chrysler/Jeep, Discount Tire, Dodge, Honda, Jeep, Suzuki, bank, transmissions, **S**...KFC, Pizza Hut, Ramada Inn
123mm	I-15, N to Ogden, S to Provo, no facilities

I-80 and I-15 run together approx 4 mi. See Utah Interstate 15, exits 308-310.

121	600 S, to downtown
120	I-15 N, to Ogden
118	UT 68, Redwood Rd, to N Temple, **2-3 mi N on N Temple**...Amoco/mart, Chevron/Subway/diesel/mart, Circle K/gas, Burger King, Denny's, KFC, Taco Bell, Airport Inn, Candlewood Suites, Comfor tSuites, Day's Inn, Holiday Inn Express, Motel 6, Radisson, Thrifty Car Rental, Utah St Fairpark, **S**...HOSPITAL
117	I-215, N to Ogden, S to Provo, no facilities
115b a	40th W, W Valley Fwy, **N**...to Salt Lake Airport
114	Wright Bros Dr(from wb), **N**...La Quinta, Microtel, Residence Inn, same as 113
113	5600 W(from eb), **N**...Phillips 66/mart, Perkins, Pizza Hut, Comfort Inn, Courtyard, Fairfield Inn, Hilton, Holiday Inn, La Quinta, Microtel, Quality Inn, Residence Inn, Sheraton, Super 8
111	7200 W, no facilities
104	UT 202, Saltair Dr, to Magna, **N**...Great Salt Lake SP, beaches
102	UT 201(from eb), to Magna, no facilities
101mm	view area wb
99	UT 36, to Tooele, **S**...HOSPITAL, Amoco/Subway/diesel/mart, Flying J/Conoco/Country Mkt/diesel/LP/mart/24hr, TA/Sinclair/Burger King/Taco Bell/diesel/rest./24hr, Texaco/diesel/mart, KFC(10mi), McDonald's, Oquirrh Motel/RV Park
88	to Grantsville, no facilities
84	UT 138, to Grantsville, Tooele, no facilities
77	UT 196, to Rowley, Dugway, no facilities
70	to Delle, **S**...Delle/Sinclair/diesel/café/motel/24hr, phone
62	to Lakeside, Eagle Range, military area, no facilities
56	to Aragonite, no facilities
54mm	**rest area both lanes, full(handicapped)facilities, phone, picnic tables, litter barrels, petwalk, vending**
49	to Clive, no facilities
41	Knolls, no facilities
26mm	architectural point of interest
10mm	**rest area both lanes, full(handicapped)facilities, phone, picnic tables, litter barrels, vending, petwalk, observation area**
4	Bonneville Speedway, **N**...Amoco/diesel/mart/café/24hr
3mm	Port of Entry, weigh sta both lanes
2	UT 58(no EZ wb return), Wendover, **S**...Amoco, Sinclair/diesel, Shell/diesel/mart, Texaco/diesel/mart, McDonald's, Subway, Taco Burger, Best Western, Bonneville Motel, Day's Inn, Heritage Motel, Motel 6, Super 8, Western Ridge Motel, Fred's Foods, Bonneville Speedway Museum, Silversmith Casino, Stateline Inn/casino, RV park, USPO
0mm	Utah/Nevada state line, Mountain/Pacific time zone

Utah Interstate 84

Exit #	Services
120	I-84 begins/ends on I-80, exit 168 near Echo, Utah.
115	to Henefer, Echo, **1/2 mi S**...USPO, to E Canyon RA
112	Henefer, **S**...gas, food, lodging
111	Croydon, no facilities
111mm	Devil's Slide Scenic View
108	Taggart, **N**...food, phone
106	ranch exit, no facilities
103	UT 66, Morgan, E Canyon RA, **S**...MEDICAL CARE, Chevron/mart, Phillips 66/mart, Texaco, Chicken Hut, Spring Chicken Café, Steph's Drive-In, Ford, IGA Foods, USPO, bank, city park, golf, hardware

Utah Interstate 84

96   Peterson, **N**...Sinclair/mart(3mi), to Snow Basin, Powder Mtn, Nordic Valley Ski Areas, **S**...Phillips 66/diesel/mart/lube/wash, phone

94mm   **rest area wb, full(handicapped)facilities, picnic tables, litter barrels, petwalk**

92   UT 167(from eb), to Huntsville, **N**...Sinclair/diesel/mart/atm(2mi), trout farm(1mi), to ski areas

91mm   **rest area eb, full(handicapped)facilities, picnic tables, litter barrels, petwalk**

87b a   US 89, Ogden, to Layton, Hill AFB, **N**...McDonald's(2mi), **1/2 mi S**...Texaco/mart

85   S Weber, Uintah, no facilities

81   to I-15 S, UT 26, Riverdale Rd, **N**...Conoco/diesel/mart, Sinclair, Applebee's, Boston Mkt, Carl's Jr, Chili's, La Salsa Mexican, McDonald's, Chrysler/Plymouth/Jeep, Circuit City, Gart Sports, Harley-Davidson, Home Depot, Honda/Nissan, Isuzu, Lincoln/Mercury/Toyota/Kia, Lowe's Bldg Supply, Mazda, MediaPlay, Mitsubishi, OfficeMax, PepBoys, PetsMart, Pearle Vision, Pontiac/Buick/GMC, Target, Wal-Mart, Wilderness RV, bank, **S**...McDonald's, Motel 6

I-84 and I-15 run together. See Utah Interstate 15, exits 344 through 379.

41mm   I-15 N to Pocatello

40   UT 102, Tremonton, Bothwell, **N**...HOSPITAL(4mi), Chevron/Burger King/diesel/mart/24hr, Texaco/Quizno's/diesel/mart/wash/24hr, Denny's/24hr, McDonald's, Western Inn, Jack's RV, **1 mi N**...Amoco/diesel, Sinclair, to Marble Motel, Sandman Motel, Alco, **S**...to Golden Spike NM

39   to Garland, Bothwell, no facilities

32   ranch exit, no services

26   UT 83 S, to Howell, no services

24   to Valley, no services

20   to Blue Creek, no facilities

17   ranch exit, no facilities

16   to Hansel Valley, no facilities

12   ranch exit, no facilities

7   Snowville, **N**...Chevron/diesel/mart, Flying J/Pepperoni's/diesel/LP/mart/24hr, Mollie's Café, Ranch House Diner, Outsiders Inn, RV camping

5   UT 30, to Park Valley, no facilities

0mm   Utah/Idaho state line

Exit #   Services

Utah Interstate 215(Salt Lake)

29   I-215 begins/ends on I-15.

28   UT 68, Redwood Rd, **W**...Flying J/diesel/rest./mart/24hr

25   22nd N, no facilities

23   7th N, **E**...Amoco/mart, KFC, McDonald's, Taco Bell, Day's Inn, Holiday Inn, Motel 6, **W**...Holiday Inn Express, Radisson

22b a   I-80, W to Wendover, E to Cheyenne

21   California Ave, **E**...Chevron/diesel, Sapp Bros/Sinclair/Burger King/diesel, Great American Diner, Subway

20b a   UT 201, W to Magna, 21st S, no facilities

18   UT 171, 3500 S, W Valley, **E**...HOSPITAL, Denny's, Country Kitchen, Country Inn Suites, Extended Stay America, Sleep Inn, **W**...Mervyn's, Michael's, ZCMI, mall

15   UT 266, 47th S, **E**...Conoco, Arby's, KFC, Taco Bell, Village Inn, Wendy's, **W**...Amoco/mart, Chevron/mart

13   UT 68, Redwood Rd, **E**...Applebee's, Amoco/mart, Arby's, Frontier Pies, Fuddrucker's, Hometown Buffet, Taco Bell, Homewood Suites, Gart Sports

12   I-15 S(from eb)

11   I-15(from wb), N to SLC, S to Provo

10   UT 280 E, **E**...Red Lobster, Sam's Club, carwash, **W**...HOSPITAL, Applebee's, Arby's, La Salsa Mexican, Olive Garden, Taco Bell

9   Union Park Ave, **E**...Amoco/mart, Chili's, LaSalsa Mexican, Outback Steaks, Tony Roma's, Best Western, Crystal Inn, HomeWood Suites

8   UT 152, 2000 E, **E**...Chevron, KFC, Taco Bell, **W**...Wendy's

6   6200 S, **E**...Alta, Brighton, Snowbird, Solitude/ski areas

5   UT 266, 45th S, Holladay, no facilities

4   39th S, **E**...Chevron/mart, Sinclair/mart, Rocky Mtn Pizza, Dan's Food/drug, Ace Hardware, bank, cinema, **W**...HOSPITAL

3   33rd S, Wasatch, **W**...Amoco, Burger King, McDonald's

2   I-80 W, no facilities

I-215 begins/ends on I-80, exit 130.

489

Vermont Interstate 89

Exit #(mm) Services

St Albans

130mm   US/Canada Border, Vermont state line, I-89 begins/ends

129.5mm   rest area sb, full facilities, phone, picnic tables, litter barrels, petwalk

22(129)   US 7 S, Highgate Springs, **E...**Ammex DutyFree, **3 mi E...**Mobil/diesel/mart

129mm   Latitude 45 N, midway between N Pole and Equator

128mm   Rock River

21(123)   US 7, VT 78, Swanton, **E...**Exxon/diesel/mart/24hr, **W...**Mobil/diesel/mart/atm, Sunoco/diesel, Texaco/mart, Chinese Rest., Dunkin Donuts, McDonald's, Grand Union Foods

20(118)   US 7, VT 207, St Albans, **W...**Mobil/mart, Shell/diesel/mart, Burger King, Dunkin Donuts, Jim's Grille, KFC, MailBoxes Etc, McDonald's, Pizza Hut, Ames, Chevrolet/Olds, Ford, Hannaford's Foods, Jo-Ann Crafts, Kinney Drug, PriceChopper Foods, Radio Shack, Sears, bank, cleaners/laundry, golf

19(114)   US 7, VT 36, VT 104, St Albans, **W...**HOSPITAL, Gulf/diesel/mart, Comfort Suites, USPO, st police

111mm   rest area both lanes, full(handicapped)facilities, info, phone, picnic tables, litter barrels, vending, petwalk

18(107)   US 7, VT 104A, Georgia Ctr, **E...**DENTIST, EYECARE, Citgo/mart, Mobil/diesel/mart, Texaco/mart, Michaeleo's Rest., GA Auto Parts, Homestead Camping, bank

17(98)   US 2, US 7, Lake Champlain Islands, **E...**Mobil/service, Texaco/diesel/mart, camping, **W...**to NY Ferry

96mm   weigh sta both lanes

16(92)   US 7, US 2, Winooski, **E...**DENTIST, Shell/mart, Lighthouse Rest., Shoney's, Hampton Inn, Ace Hardware, Shaw's Foods, Target, cleaners, **W...**Texaco/diesel/mart/atm, Burger King, Liby's Diner, McDonald's, Rathskellar Rest., Fairfield Inn, Motel 6

15(91)   VT 15(no EZ nb return), Winooski, **E...**Day's Inn, **W...**Exxon, Mobil/mart, **1 mi W...**Little Caesar's, hardware, USPO

Burlington

90mm   Winooski River

14(89)   US 2, Burlington, **E...**DENTIST, Citgo/repair, Coastal/mart, Mobil/mart, Shell/diesel/mart, Sunoco, Texaco/diesel, Al's Frys, Burger King, CheeseTraders, Friendly's, McDonald's, Orchid Chinese Buffet, TCBY, Anchorage Inn, Best Western, Holiday Inn, Howard Johnson/rest., Ramada Inn, Swiss Hotel, Ames, Barnes&Noble, City Drug, Grand Union Foods, JC Penney, Jo-Ann Crafts, Midas Muffler, Sears/auto, laundry, **W...**Exxon/mart, Mobil, Sheraton, PartsAmerica, PriceChopper Foods, Ritz Camera, Staples, West Marine

13(87)   I-189, to US 7, Burlington, **2-3 mi W on US 7...**Citgo, Getty/mart, Gulf/diesel/mart, Mobil/diesel/mart, Sunoco/mart, Texaco/XpressLube, Bagel Bakery, Bonanza, Burger King, Chinese Rest., Denny's, Friendly's, KFC, Little Caesar's, McDonald's, Pepper's BBQ, Pizza Hut, Perry's Fish House, Red Lobster, Wendy's, Colonial Motel, Del Ray Motel, Ho-Hum Motel, Maple Leaf Motel, Super 8, Town&Country Motel, Ben Franklin, Chrysler/Saturn, Grand Union Foods, K-Mart, Kwik-Photo, Olds/Pontiac/Cadillac, Price Chopper Food/drug, Radio Shack, Sears/auto, Service Merchandise, Shop'n Save Foods, Tire Whse, TJ Maxx, VW/Audi, bank, camping, cinema

12(84)   VT 2A, to US 2, Williston, to Essex Jct, **E...**CHIROPRACTOR, DENTIST, Citgo/mart, Mobil/mart/24hr, Sunoco/diesel/mart/subs/24hr, Café Expresso, Friendly's, Mexicali Rest., Ponderosa, Whole Donut, Wilderness Grill, Susse Chalet, Bed Bath&Beyond, Circuit City, City Drug, GolfWorld, Hannaford's Foods, Home Depot, PetsMart, Wal-Mart, UPS, st police, **W...**Residence Inn

82mm   rest area both lanes(7am-11pm), full(handicapped)facilities, phone, picnic tables, litter barrels, vending, petwalk

Montpelier

11(79)   US 2, to VT 117, Richmond, **E...**Chequered House Motel, **W...**Mobil/diesel/mart/24hr

67mm   rest area/weigh sta sb, no facilities

66mm   rest area/weigh sta nb, no facilities

10(64)   VT 100, to US 2, Waterbury, **E...**Exxon/diesel/mart/24hr, Mobil/diesel/mart/24hr, Caforia Mkt, Blush Hill Inn, Holiday Inn, Thatcher Brook Inn/rest., TrueValue, carwash, **W...**DENTIST, Citgo/mart, Zachary's Pizza

9(59)   US 2, to VT 100B, Middlesex, **W...**Citgo/repair, Camp Meade Motel/rest., museum, st police

8(53)   US 2, Montpelier, **2 mi E...**BP/mart, Cumberland/gas, Exxon/diesel/mart, Gulf/diesel, Mobil, Sunoco/repair, Texaco/mart, Sarducci's Rest., Capitol Plaza Hotel, Montpelier Inn, Grand Union Foods, auto parts, camping(6mi), to VT Coll

7(50)   VT 62, to US 302, Barre, **E...**Mobil/diesel/mart/24hr, Shoney's, Comfort Suites, Honda, Shaw's Foods, Staples, **2 mi E...**HOSPITAL, Lague Inn, Cadillac/GMC/Toyota, JC Penney, airport, bank, camping(7mi)

6(47)   VT 63, to VT 14, S Barre, **4 mi E...**gas, food, lodging, camping, info

5(43)   VT 64, to VT 12, VT 14, Williamstown, **6 mi E...**gas/diesel, food, lodging, camping, **W...**to Norwich U

N
↑
|
↓
S

Vermont Interstate 89

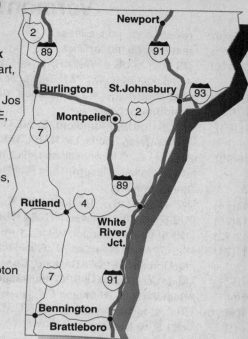

	41mm	highest elevation on I-89, 1752 ft
	34.5mm	**rest area/weigh sta both lanes(7am-11pm),** **full(handicapped)facilities, info, phone, litter barrels, petwalk**
	4(31)	VT 66, Randolph, **E**...RV camping(1mi), **W**...HOSPITAL, Mobil/mart, McDonald's, lodging(3mi), RV camping(5mi)
	3(22)	VT 107, Bethel, **E**...Texaco/diesel/mart, VT Sugarhouse Rest., to Jos Smith Mon(8mi), **1 mi W**...DENTIST, EYECARE, MEDICAL CARE, Citgo/diesel/mart/LP, USPO, bank, st police
	14mm	White River
	2(13)	VT 14, VT 132, Sharon, **E**...° Acre Motel, **W**...Citgo/diesel, Gulf/diesel/mart, Columns Motel, Sharon Trading Post, USPO, antiques, Jos Smith Mon(6mi)
	9mm	**rest area/weigh sta both lanes(7am-11pm), full facilities, phone, info, picnic tables, litter barrels, vending, NO HANDI-CAPPED FACILITIES**
	7mm	White River
	1(4)	US 4, to Woodstock, Quechee, **3 mi W**...Exxon/diesel/24hr, Hampton Inn, Super 8
	1mm	I-91, N to St Johnsbury, S to Brattleboro
	0mm	Connecticut River, Vermont/New Hampshire state line

Vermont Interstate 91

Exit #(mm)Services

	178mm	US/Canada Border, Vermont state line, US Customs, I-91 begins/ends
	29(177)	US 5, Derby Line, **E**...Ammex Dutyfree, **1 mi W**...Border Minimart/gas
	176.5mm	**Welcome Ctr sb, full(handicapped)facilities, info, picnic tables, litter barrels, phone, petwalk, Midpoint between the Equator and N Pole**
	28(172)	US 5, VT 105, Derby Ctr, **E**...Exxon/diesel/mart/24hr, Petrol King/diesel/mart, camping, st police, **W**...HOSPITAL, VETERINARIAN, Gulf/Dunkin Donuts/diesel/mart, Texaco/diesel/mart/atm/24hr, McDonald's, Village Pizza Substation, Pepin's Motel, Super 8, Ames, Chevrolet/Olds/GMC/Pontiac/Cadillac, Chrysler/Jeep, Plymouth/Dodge, Rite Aid, Shop'n Save Foods, auto repair, bank, RV camping
	27(170)	VT 191, to US 5, VT 105, Newport, **3 mi W**...HOSPITAL, Border Patrol, camping, info
	167mm	**nb**...rest area/weigh sta, **sb**...rest area/weigh sta
	26(161)	US 5, VT 58, Orleans, **E**...CHIROPRACTOR, BP, Sunoco, Orleans MinitMart/gas/repair, Cook's Mkt, Rexall Drugs, bank, camping(8mi)
	156.5mm	Barton River
	25(156)	VT 16, Barton, **1 mi E**...Gulf/mart, Irving/diesel/mart, Orlean Minimart/gas, Sunoco/mart/atm, Orlean Pizza, Austin's Drugs, CC Foods, Cole's Mkt, Ford, Orlean Gen Store, USPO, bank, drugs, hardware, camping(2mi)
	154mm	rest area nb, no facilities
	150.5mm	highest elevation on I-91, 1856 ft
	143mm	scenic overlook nb
	141mm	**rest area sb, full(handicapped)facilities, info, picnic tables, litter barrels, phone**
	24(140)	VT 122, Wheelock, **2 mi E**...gas, food, lodging
	23(137)	US 5, to VT 114, Lyndonville, **E**...Gulf/diesel/mart, Mobil/Dunkin Donuts/mart, McDonald's, Day's Inn, Agway Foods, Rite Aid, auto repair, bank, **W**...Lyndon Motel
	22(132)	to US 5, St Johnsbury, **1-2 mi E**...HOSPITAL, Citgo/diesel/mart/LP, Cindy's Pasta, Pizza Hut, St Jay Diner, Buick/Olds/Pontiac/GMC, PriceChopper Foods, Rite Aid, repair, **4 mi E on US 5**...Irving/mart, Ames, FashionBug, Firestone, JC Penney, Sears
	21(131)	US 2, to VT 15, St Johnsbury, **1-2 mi E**...HOSPITAL, services
	20(129)	US 5, to US 2, St Johnsbury, **E**...Irving/diesel/mart, Mobil/repair, Texaco/diesel/mart, Chun Bo Chinese, Dunkin Donuts, McDonald's, Subway, Grand Union Foods, Brooks Drug, bank, truck repair, **W**...st police
	19(128)	I-93 S to Littleton NH, no facilities
	122mm	scenic view nb
	18(121)	to US 5, Barnett, **E**...camping(5mi), **W**...camping(5mi)

Derby Ctr

St Johnsbury

Vermont Interstate 91

White River Jct

115mm	rest area sb, no facilities
114mm	rest area nb, no facilities
17(110)	US 302, to US 5, Wells River, NH, **E**...HOSPITAL(5mi), Mobil/diesel/LP, Gallery Rest., flea mkt, camping(10mi), **W**...P&H Trkstp/diesel/rest./24hr, camping(9mi)
100mm	**rest area/weigh sta both lanes, full(handicapped)facilities, phone, info, picnic tables, litter barrels, vending, petwalk**
16(98)	VT 25, to US 5, Bradford, **E**...VETERINARIAN, Mobil/diesel/mart/café, Hungry Bear Rest., Bradford Motel, Grand Union Foods, LP, NAPA, **W**...st police
15(92)	Fairlee, **E**...Citgo/diesel/mart, Gulf/mart, Mobil/deli/atm, Texaco/diesel/LP/mart, Fairlee Gen Store/gas, Fairlee Diner, Potlatch Tavern, Third Rail Rest., Your Place Rest., Silver Maple Lodge, TrueValue, USPO, antiques, bank, camping
14(84)	VT 113, to US 5, Thetford, **1 mi E**...food, camping, **W**...camping
13(75)	US 5, VT 10, Hanover, NH, **E**...HOSPITAL, camping, to Dartmouth
12(72)	US 5, White River Jct, Wilder, **1 mi E**...gas/diesel/24hr, Shady Lawn Motel
11(71)	US 5, White River Jct, **E**...Gulf/diesel/mart, Mobil/lube, Sunoco/repair, Crossroads Café, Julie's Diner, McDonald's, Steak&Seafood, Coach an' Four Motel, Comfort Inn, Ramada Inn, Ford/Lincoln/Mercury, Jct Mktplace, Toyota, USPO, bank, carwash, gifts, **W**...HOSPITAL, Citgo/mart, Exxon/mart, Texaco/diesel/mart, Best Western, Hampton Inn, Howard Johnson/rest., Super 8
10N(70)	I-89 N, to Montpelier
S	I-89 S, to NH, airport
68mm	**rest area/weigh sta both lanes, full(handicapped)facilities, picnic tables, litter barrels, phone, vending, petwalk**
9(60)	US 5, VT 12, Hartland, **W**...HOSPITAL, info, Mobil/mart(1mi)
8(51)	US 5, VT 12, VT 131, Ascutney, **E**...HOSPITAL, Citgo/diesel, Mobil, Sunoco/diesel, Texaco/diesel/mart, Ascutney House Rest., Country Village Store, Mr G's Rest.
7(42)	US 5, VT 106, VT 11, Springfield, **W**...HOSPITAL, Mobil, Texaco/diesel/LP/mart, Holiday Inn Express, camping
39mm	rest area both lanes, picnic tables, litter barrels
6(34)	US 5, VT 103, Rockingham, to Bellows Falls, **E**...Citgo/mart, Leslie's Rest., Rockingham Motel/rest., **W**...Sunoco/diesel/mart/24hr, st police(6mi)
5(29)	VT 121, to US 5, Westminster, to Bellows Falls, **3 mi E**...U-Haul, repair
24mm	rest area both lanes, litter barrels
22mm	weigh sta sb
20mm	weigh sta nb
4(18)	US 5, Putney, **E**...Putney Inn/rest., **W**...Sunoco/diesel/mart/atm/LP/24hr, camping(3mi)
3(11)	US 5, VT 9 E, Brattleboro, **E**...Agway/gas, Citgo/diesel/mart/24hr, Mobil/mart, Sunoco, Bickford's, Brattleboro Bowl Rest., Dunkin Donuts, Friendly's, KFC, McDonald's, Pizza Hut, Village Pizza, Day's Inn, Super 8, Ames, Ford, GNC, Hannaford's Foods, Herbs&Spices Etc, MailBoxes Etc, Monro Muffler/brake, NAPA, Pontiac/Buick/GMC, Rite Aid, TrueValue, U-Haul, USPO, bank
2(9)	VT 9 W, Brattleboro, **E**...CHIROPRACTOR, Sunoco, **W**...Mobil, Texaco, Country Deli, to Marlboro Coll, st police
1(7)	US 5, Brattleboro, **E**...HOSPITAL, Coastal/diesel/mart/wash, Mobil/Dunkin Donuts/mart, Sunoco, Texaco/Subway/diesel/mart, Burger King, Econolodge, CarQuest, Chevrolet/Olds/Cadillac, Chrysler/Jeep, PriceChopper Foods, Brooks Drug, FashionBug, camping, factory outlet, **1 mi E**...EYECARE, Getty, Village Pizza
3mm	**Welcome Ctr nb, full(handicapped)facilities, info, phone, picnic tables, litter barrels, vending, petwalk**
0mm	Vermont/Massachusetts state line

Brattleboro

N

S

Vermont Interstate 93

See New Hampshire Interstate 93

Virginia Interstate 64

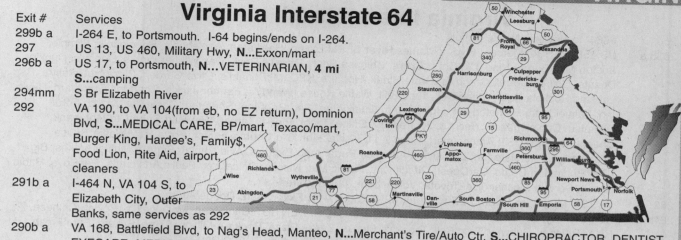

Exit #	Services
299b a	I-264 E, to Portsmouth. I-64 begins/ends on I-264.
297	US 13, US 460, Military Hwy, **N**...Exxon/mart
296b a	US 17, to Portsmouth, **N**...VETERINARIAN, **4 mi** **S**...camping
294mm	S Br Elizabeth River
292	VA 190, to VA 104(from eb, no EZ return), Dominion Blvd, **S**...MEDICAL CARE, BP/mart, Texaco/mart, Burger King, Hardee's, Family$, Food Lion, Rite Aid, airport, cleaners
291b a	I-464 N, VA 104 S, to Elizabeth City, Outer Banks, same services as 292
290b a	VA 168, Battlefield Blvd, to Nag's Head, Manteo, **N**...Merchant's Tire/Auto Ctr, **S**...CHIROPRACTOR, DENTIST, EYECARE, MEDICAL CARE, Amoco/diesel/mart/24hr, Shell/repair, Texaco/Blimpie/mart, Applebee's, Chik-fil-A, ChuckeCheese, Dunkin Donuts, Golden Corral, Grand China Buffet, Hardee's, Ryan's, Taco Bell, Waffle House, Wendy's, Day's Inn, Super 8, $Tree, Lowe's Bldg Supply, Postal Express, Sam's Club, USPO, Wal-Mart SuperCtr/24hr, bank
289b a	Greenbrier Pkwy, **N**...Citgo/mart, Burger King, Nathan's Deli, Subway, Taco Bell, Wendy's, Hampton Inn, Holiday Inn, Motel 6, Red Roof Inn, Wellesley Inn, Acura, BMW, Chevrolet, Cloth World, Dodge, Food Lion, Ford, Honda, Hyundai, Isuzu, Mitsubishi, Pontiac, Subaru, Volvo, U-Haul, bank, transmissions, **S**...DENTIST, Blimpie, Boston Mkt, Cheers Grill, Cooker Rest., Don Pablo's, Fazoli's, LoneStar Café, McDonald's, Old Country Buffet, Olive Garden, Pargo's, Ruby Tuesday, Starbucks, TCBY, Winston's Café, Food Lion, Harris-Teeter/24hr, Comfort Suites, Courtyard, Extended Stay America, Fairfield Inn, Sun Suites, Barnes&Noble, Bed Bath&Beyond, BEST, Circuit City, Dillard's, $General, Drug Emporium, Hecht's, Marshall's, Michael's, Nevada Bob's, Office Depot, OfficeMax, Old Navy, PakMail, PetsMart, Sears/auto, Target, bank, cinema, cleaners, mall
286b a	Indian River Rd, **N**...DENTIST, VETERINARIAN, Amoco/mart/24hr, Citgo/Wilco/diesel/mart, Crown Gas, Exxon/mart, Shell/mart/wash, Texaco/repair, 7-11, Hardee's, bank, cleaners, **S**...DENTIST, Exxon/diesel, 7-11, Capt D's, Deli/pizzaria, Golden Corral, Shoney's, Waffle House, Founder's Inn, laundry
285	E Branch Elizabeth River
284b a	I-264, to Norfolk, VA 44, to VA Beach(exits left from eb), **1/2 mi E off 1 exit N, Newtown Rd**...Amoco, Chevron/wash, Exxon, Texaco, 7-11, Adam's Rest., Denny's/24hr, Golden Corral, Texas Steaks, Wendy's, Hampton Inn, Holiday Inn, La Quinta, Ramada Inn, JiffyLube, bank
282	US 13, Northampton Blvd, **N**...Quality Inn, to Chesapeake Bay Br Tunnel
281	VA 165, Military Hwy(no EZ eb return), **N**...CHIROPRACTOR, Amoco/diesel/mart, Robo Gas/wash, Shell/repair, Texaco/diesel/mart/wash/24hr, Andy's Pizza, Pizza Hut, Wendy's, Chevrolet/Kia, Chrysler/Nissan, Mazda/Suzuki, Econolodge, Aamco, Alamo, FabricHut, Food Lion, **S**...Exxon/mart/wash, BBQ, Dairy Queen, Hampton Inn, Hilton, Firestone, Target
279	Norview Ave, **N**...7-11, Golden Corral, K-Mart, to airport, botanical garden
278	VA 194 S, Chesapeake Blvd(no EZ wb return), **1/2 mi N**...Crown/mart, Eckerd, NAPA, **1/2 mi S**...Burger King
277b a	VA 168, to Tidewater Dr, **N**...Citgo/7-11, Chinese Rest., Hardee's, **S**...HOSPITAL
276c	to US 460 W, VA 165, Little Creek Rd, **N**...Texaco/mart, Big Al's Muffler/brake, **S**...Amoco, Exxon, McDonald's, Mr Jim's Submarines, Taco Bell, Wendy's, AutoZone, Eckerd, Farm Fresh, Hannaford's Foods, Pearle Vision
276b a	I-564 to Naval Base
274	Bay Ave(from wb), to Naval Air Sta
273	US 60, 4th View St, Oceanview, **N**...Exxon/diesel/24hr, Chesabay Motel, Econolodge, Harrison Boathouse/Pier, **S**...to Norfolk Visitors Ctr, info
272	W Oceanview Ave, **N**...7-11, Willoughby's Seafood, **S**...Fisherman's Wharf Rest., Day's Inn
270mm	Chesapeake Bay Tunnel
269mm	weigh sta eb
268	VA 169 E, to Buckroe Beach, Ft Monroe, VA Ctr, **N**...Hardee's, McDonald's, **S**...Strawberry Banks Inn/rest.
267	US 60, to VA 143, County St, **N**...gas/diesel, **S**...HOSPITAL, Burger King, Subway Sta Sandwiches, Radisson, Kinko's, laundry, to Hampton U
265c	(from eb)**N**...Armistead Ave, to Langley AFB
265b a	VA 134, VA 167, to La Salle Ave, **N**...Citgo, Super 8, Hampton Coliseum, Home Depot, **S**...HOSPITAL, McDonald's, Seafood Rest.

E

W

N
o
r
f
o
l
k

Virginia Interstate 64

Hampton

264	I-664, to Newport News, Suffolk
263b a	US 258, VA 134, Mercury Blvd, to James River Br, **N**...Exxon, Shell/mart, Applebee's, Bennigan's, Blimpie, Boston Mkt, Burger King, Chi Chi's, Chili's, Chinese Rest., Darryl's Rest., Das Wiener Works, Denny's, Dunkin Donuts, Golden Corral, Grandy's, Grate Steak, Hooters, KFC, McDonald's, Olive Garden, Pizza Hut, Rally's, Red Lobster, Rockola Café, Steak&Ale, Taco Bell, Waffle House, Wendy's, Farmer Jack's, PharMor, Courtyard, Day's Inn, Fairfield Inn, Hampton Inn, Holiday Inn, Quality Inn, Red Roof Inn, Sheraton, Chevrolet/Mazda, Ford/Diahatsu, Chrysler/Jeep, Hecht's, JC Penney, Office Depot, Pier 1, Target, Wal-Mart SuperCtr/24hr, bank, cinema, **S**...DENTIST, Amoco/mart, Citgo/7-11, Kettle/24hr, Old Country Buffet, Pizza Hut, Waffle House, Econolodge, Interstate Inn, La Quinta, Advance Parts, Auto Express, Avis Rental, Big Al's Muffler/brakes, Big Lots, Burlington Coats, CarQuest, Circuit City, Fabric Mkt, Nissan, OfficeMax, Paul's Arts/crafts, PepBoys, Radio Shack, Service Mechandise, Toyota, Volvo, bank, laundry, tires
262	VA 134, Magruder Blvd(no EZ wb return), **N**...Citgo/7-11, Exxon/mart, Dodge/Acura
261b a	Center Pkwy, to Hampton Roads, **S**...CHIROPRACTOR, DENTIST, Citgo/mart, McDonald's, Peking Chinese, Ruby Tuesday, Subway, Taco Bell, Topeka's Steaks, Zero's Pizza, Books-A-Million, $Tree, Eckerd, FarmFresh Foods, MailBoxes Etc, TJ Maxx, Winn-Dixie, cinema
258	US 17, J Clyde Morris Blvd, **N**...Amoco/mart, Crown/mart, Exxon/mart, 7-11/24hr, Domino's, Hardee's, New China, Waffle House, Wendy's, BudgetLodge, Host Inn, Ramada Inn/rest., Super 8, Advance Parts, Food Lion, Rite Aid, Pack'n Mail, bank, laundry, **S**...HOSPITAL, DENTIST, Citgo/diesel/mart/24hr, Exxon/mart/wash, 7-11/24hr, BBQ, Burger King, Fuddrucker's, Hong Kong Chinese, Papa John's, Pizza Hut, Sammy&Angelo's Steaks, Subway, Taco Bell, Motel 6, Omni Hotel/rest., BMW/Honda, $Tree, Eckerd, JiffyLube, PetWorld, Radio Shack, ValueCity, bank, cleaners, museum
256b a	Victory Blvd, Oyster Point Rd, **N**...Amoco, Citgo/diesel, Texaco, BBQ, Bagel Bakery, Burger King, Fuddrucker's, Hardee's, Starbucks, **S**...Coll of Wm&Mary
255b a	VA 143, to Jefferson Ave, **N**...HOSPITAL, Amoco, Shell/diesel/mart, Golden Corral, Shoney's, Capri Motel, Regency Inn, Travelodge, Aamco, Acura, Cadillac/Olds/GMC, Home Depot, Lowe's Bldg Supply, Sam's Club, Wal-Mart SuperCtr/24hr, Saturn, airport, **S**...Citgo, Exxon/diesel, 7-11, Applebee's, Burger King, Cheddar's Café, Chick-fil-A, Cracker Barrel, Don Pablo's, KFC, Krispy Kreme, McDonald's, OutBack Steaks, Ruby Tuesday, Shoney's, Subway, Waffle House, Wendy's, Comfort Inn, Day's Inn, Econolodge, Hampton Inn, Studio+, Barnes&Noble, Belk, Circuit City, Dillard's, Hecht's, NTB, OfficeMax, PetsMart, Target, Upton's, bank, cinema, laundry, mall
250b a	to US Army Trans Museum, **N**...Amoco/diesel, Citgo/7-11, Shell/mart, **2-4 mi N**...Exxon, Burger King, Capri Motel, Tudor Inn, Newport News Camp, to Yorktown Victory Ctr, **S**...RaceTrac/mart, McDonald's, Wendy's, Ft Eustis Inn, Holiday Inn Express, Mulberry Inn, TDY Inn
247	VA 143, to VA 238(no EZ return wb), **N**...BP/diesel, Citgo/7-11, to Yorktown, **S**...to Jamestown Settlement
243	VA 143, to Williamsburg, **S**...same as 242a

Williamsburg

242b a	VA 199, to US 60, to Williamsburg, **N**...BP(2mi), Day's Inn/rest., water funpark, to Yorktown NHS, **1 mi S**...Frank's Trkstp/diesel/rest., Citgo/7-11, Crown Gas, Texaco, Burger King, LJ Silver, McDonald's, Wendy's, Best Western, Courtyard, Howard Johnson, Marriott/rest., Quality Inn, Rodeway Inn, 1st Settlers Camping, to William&Mary Coll, Busch Gardens, auto parts, to Jamestown NHS
238	VA 143, to Colonial Williamsburg, Camp Peary, **2-3 mi S**...HOSPITAL, Citgo, EastCoast/diesel/mart, Texaco, Cracker Barrel, Golden Corral, Hardee's, Ledo's Pizza, McDonald's, Williamsburg Woodlands Grill, Best Western/rest., Comfort Inn, Day's Inn, Econolodge, Holiday Inn/rest., Homewood Suites, Howard Johnson/rest., King William Inn, Quality Inn/rest., Anvil Camping
234	VA 646, to Lightfoot, **1-2 mi N**...KOA, Colonial Camping, Pottery Camping, **2-3 mi S**...Exxon/diesel, Mobil/diesel, Shell/diesel, BBQ, Burger King, Hardee's, KFC, Lightfoot Rest., McDonald's, Subway, Colonial Hotel, Day's Inn, Quality Inn, Super 8, Kincaid Camping
231b a	VA 607, to Norge, Croaker, **N**...Citgo/7-11, to York River SP, **1-2 mi S**...Texaco, KFC, Candle Factory Rest., Wendy's, Econolodge, Williamsburg Camping(5mi)
227	VA 30, to US 60, Toano, to West Point, **S**...Exxon/diesel(2mi), Shell/Stuckey's/diesel/mart, Texaco/diesel/mart, McDonald's, Welcome South Rest., Anderson's Corner Motel(2mi), Ed Allen's Camping(10mi)
220	VA 33 E, to West Point, **N**...Exxon/diesel/mart
214	VA 155, to New Kent, Providence Forge, **S**...Allen's Camping(10mi), Colonial Downs Racetrack
213.5mm	**rest area both lanes, full(handicapped)facilities, phone, vending, picnic tables, litter barrels, petwalk**
211	VA 106, to Talleysville, to James River Plantations, no facilities
205	VA 33, VA 249, to US 60, Bottoms Bridge, Quinton, **N**...Amoco/diesel/mart, Exxon/diesel/mart, New Peking Chinese, Subway, Eckerd, Food Lion, USPO, bank, cleaners, **S**...Shell/diesel/mart, Texaco/McDonald's/mart, MarketPlace Foods/deli, golf
204mm	Chickahominy River

E

↑

↓

W

Virginia Interstate 64

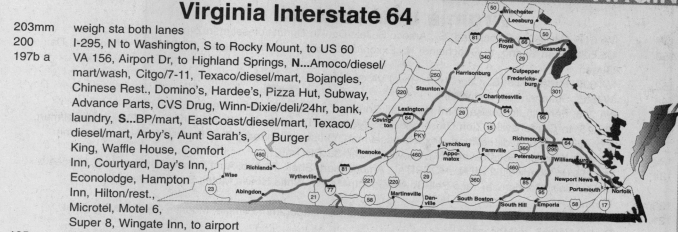

203mm   weigh sta both lanes

200   I-295, N to Washington, S to Rocky Mount, to US 60

197b a   VA 156, Airport Dr, to Highland Springs, **N**...Amoco/diesel/ mart/wash, Citgo/7-11, Texaco/diesel/mart, Bojangles, Chinese Rest., Domino's, Hardee's, Pizza Hut, Subway, Advance Parts, CVS Drug, Winn-Dixie/deli/24hr, bank, laundry, **S**...BP/mart, EastCoast/diesel/mart, Texaco/ diesel/mart, Arby's, Aunt Sarah's, Burger King, Waffle House, Comfort Inn, Courtyard, Day's Inn, Econolodge, Hampton Inn, Hilton/rest., Microtel, Motel 6, Super 8, Wingate Inn, to airport

195   Laburnum Ave, **N**...Chevron/repair, Shell/repair, **1 mi N**...Amoco, Exxon/diesel/mart/wash, Texaco/diesel/mart, 7-11, Shoney's, Chevrolet, CVS Drug, K-Mart, Midas Muffler, Sears, bank, tires, **S**...Applebee's, Subway, Taco Bell, Ukrop's Food/drug, bank, **3/4 mi S**...MEDICAL CARE, OPTOMETRIST, Amoco/24hr, Exxon/mart, 7-11, Aunt Sarah's, Bojangles, Burger King, Capt D's, China King, Dairy Queen, Hardee's, KFC, McDonald's, Papa John's, Wendy's, Western Sizzlin, Airport Inn, Hampton Inn, Holiday Inn, Sheraton/rest., Super 8, Wyndham Garden, CVS Drug, $Tree, Firestone/auto, Ford, Hannaford Food/drug, Radio Shack, Rite Aid, Trak Auto, hardware, laundry

193b a   VA 33, Nine Mile Rd, **N**...Exxon/Subway/mart, GMC, transmissions, **2 mi N**...Amoco, Texaco, Burger King, McDonald's, Shoney's, Taco Bell, **S**...HOSPITAL

192   US 360, to Mechanicsville, **N**...Chevron/wash, Mobil/diesel/mart, McDonald's, Tuffy AutoCtr, transmissions, **S**...Amoco/mart/wash, Citgo, Church's Chicken, KFC

190   I-95 S, to Petersburg, 5th St, **N**...Richmond Nat Bfd Park, **S**...Marriott, st capitol, coliseum. **I-64 W and I-95 N run together. See Virginia Interstate 95, exits 76-78.**

187   I-95 N(exits left from eb), to Washington. **I-64 E and I-95 S run together.**

186   I-195, Richmond, to Powhite Pkwy

185b a   US 33, Staples Mill Rd, Dickens Rd, no facilities

183   US 250, Broad St, Glenside Dr, **N**...Chevron/DQ/mart, Bob Evans, Day's Inn, Embassy Suites, Shoney's Inn/ rest., K-Mart, bank, same as 181, **S**...HOSPITAL, to U of Richmond

181   Parham Rd, **1 mi N on Broad**...HOSPITAL, VETERINARIAN, Amoco/24hr, Citgo, Crown/diesel, Exxon/mart, Shell, Texaco, Aunt Sarah's, Bennigan's, Blue Marlin Seafood, Bojangles, Burger King, Casa Grande Mexican, Friendly's, Fuddruckers, Hardee's, Hooters, KFC, LoneStar Steaks, LJ Silver, Morrison's Cafeteria, Olive Garden, OutBack Steaks, Piccadilly's, Red Lobster, Shoney's, Steak&Ale, Taco Bell, TGIFriday, Waffle House, Wendy's, Wood Grill, Fairfield Inn, Quality Inn, Suburban Lodge, Super 8, Books-A-Million, CVS Drug, Cloth World, Fabric Whse, Ford/Lincoln/Mercury, Honda/Mitsubishi, Hyundai/Subaru, Infiniti, Mazda/BMW, Mercedes, Midas Muffler, Nissan, Olds, PepBoys, PetsMart, Staples, Toyota, Tuffy Muffler, Ukrop's Foods, 1-Hr Photo, bank, transmissions, **1 mi S**...Amoco, Exxon, Texaco, McDonald's, Topeka Steaks, Ramada Inn

180   Gaskins Rd, **N**...BP/mart/24hr, Exxon/mart, Chevron/mart, Texaco/diesel/mart, Courtyard, Residence Inn, Studio+, **1/2 mi N on Broad**...Citgo, East Coast/Blimpie/diesel/mart/24hr, Applebee's, Arby's, Blackeyed Pea, Boston Mkt, Burger King, Golden Corral, IHOP, KFC, McDonald's, Peking Chinese, Rockola Café, Ruby Tuesday's, Ryan's, Subway, Taco Bell, Tripp's Rest., Wendy's, Zorba's Greek/Italian, Bed Bath&Beyond, Borders Books, Cadillac, Chrysler/Jeep, Circuit City, Computer City, Costco, Drug Emporium, Evans Foods, Food Lion, Fresh Fields Foods, Goodyear/auto, Hannaford's Foods, Jumbo Sports, Linens'n Things, Lowe's Bldg Supply, PharMor, Rack&Sack, Sam's Club, Service Merchandise, TJ Maxx, VW, XpressLube, bank, laundry, mall, **S**...Citgo

178b a   US 250, Short Pump, **N**...Shell, Rennie's/diesel, Texaco/diesel, Chesapeake Bagels, Leonardo's Pizza, Starbucks, Thai Garden Rest., Zack's Yogurt, AmeriSuites, Comfort Suites, Hampton Inn, Homestead Village, Homewood Suites, CarMax, CompUSA, Firestone/auto, Kinko's, OfficeMax, **1 mi S**...DENTIST, Amoco/mart, Citgo/7-11, Crown/mart, Burger King, Capt D's, Domino's, McDonald's, Taco Bell, TGIFriday, Wendy's, Candle- wood Suites, Barnes&Noble, Best Buy, CarQuest, Goodyear/auto, Home Depot, HomePlace, JiffyLube, Kohl's, Lowe's Bldg Supply, Midas Muffler, Pier 1, Target, Wal-Mart SuperCtr/24hr, cinemas

177   I-295, to I-95 N to Washington, to Norfolk, VA Beach, Williamsburg

173   VA 623, to Rockville, Manakin, **N**...Citgo/mart(2mi), **S**...Amoco/DQ/diesel/mart, Exxon/diesel(1mi), Shell, Texaco/diesel/mart, BBQ, Alley's Motel(1mi), $General, Food Lion

169mm   **rest area both lanes, full(handicapped)facilities, vending, phone, picnic tables, litter barrels, petwalk**

Virginia Interstate 64

167	VA 617, Oilville, to Goochland, **S**...Amoco/Bullets/Dunkin Donuts/diesel/mart/24hr
159	US 522, Gum Spring, to Goochland, **N**...Exxon/dieselmart, auto/diesel repair, **S**...BP(2mi), Citgo/mart, Dairy Queen
152	VA 629, Hadensville, **S**...Citgo(1mi), Royal VA Golf/rest.
148	VA 605, Shannon Hill, no facilities
143	VA 208, Ferncliff, to Louisa, **7 mi N**...Small Country Camping, **S**...Citgo/diesel, Exxon/diesel
136	US 15, to Gordonsville, Zion Crossroads, **S**...Amoco/McDonald's/diesel/mart/24hr, Citgo/Blimpie/diesel/mart, Exxon/Burger King/diesel/mart, Texaco/diesel/mart, Crescent Inn/rest., Zion Roads Motel, camping(1mi)
129	VA 616, Keswick, Boyd Tavern, **N**...Keswick Hotel, **S**...BP(1mi)
124	US 250, to Shadwell, **2 mi N**...HOSPITAL, Amoco/mart, Shell, Texaco, Aunt Sarah's, Burger King, Hardee's, McDonald's, Ponderosa, Taco Bell, Town&Country Motel, White House Motel, **S**...Ramada Inn
123mm	Rivanna River
121	VA 20, to Charlottesville, Scottsville, **N**...Amoco/diesel, Citgo(2mi), Blimpie, Moore's Creek Family Rest., **S**...KOA(10mi), to Monticello
120	VA 631, 5th St, to Charlottesville, **N**...VETERINARIAN, Exxon/diesel/wash, Texaco/mart, Burger King, Hardee's, Domino's, McDonald's, Pizza Hut/Taco Bell, Waffle House, Wendy's, Holiday Inn, Howard Johnson(2mi), Omni Hotel(2mi), CVS Drug, Family$, Food Lion, laundry
118b a	US 29, Charlottesville, to Lynchburg, **1-4 mi N**...HOSPITAL, Amoco/mart, Citgo/diesel, Exxon/mart, Shell, Blimpie, Hardee's, Kealie's Town Mkt, Shoney's, Subway, Best Western, Boar's Head Inn, Budget Inn, Comfort Inn, Econolodge, UVA, **S**...Exxon/diesel
114	VA 637, to Ivy, no facilities
113mm	**rest area wb, full(handicapped)facilities, phone, vending, picnic tables, litter barrels, petwalk**
111mm	Mechum River
108mm	Stockton Creek
107	US 250, Crozet, **1 mi N**...Citgo, Exxon, Texaco, **1 mi S**...Misty Mtn Camping
105mm	**rest area eb, full(handicapped)facilities, phone, vending, picnic tables, litter barrels, petwalk**
104mm	scenic area eb, litter barrels, no truck or buses
100mm	scenic area eb, litter barrels, hist marker, no trucks or buses
99	US 250, Afton, to Waynesboro, **N**...Colony Motel, to Shenandoah NP, Skyline Drive, ski area, to Blue Ridge Pkwy, **S**...Chevron/mart, Afton Inn
96	VA 622, Waynesboro, to Lyndhurst, **3 mi N**...Shell, Texaco, Tastee Freez, Comfort Inn
95mm	South River
94	US 340, Waynesboro, to Stuarts Draft, **N**...HOSPITAL, VETERINARIAN, Citgo/7-11/diesel, Exxon/diesel, Arby's, Burger King, KFC, Shoney's, Taco Bell(1mi), Wendy's, Western Sizzlin, Best Western, Day's Inn, Deluxe Budget Motel(2mi), Holiday Inn Express, Skyline Motel(2mi), Super 8, Waynesboro N 340 Camping, **S**...Amoco/mart(2mi), Pennzoil/gas, Waynesboro Outlet Village, museum
91	Va 608, Fishersville, to Stuarts Draft, **N**...HOSPITAL, Exxon, Subway, **S**...Exxon/McDonald's/mart, Walnut Hills Camping, tires
89mm	Christians Creek
87	I-81, N to Harrisonburg, S to Roanoke
I-64 and I-81 run together 20 miles. See Virginia Interstate 81, exits 195-220.	
56	I-81, S to Roanoke, N to Harrisonburg
55	US 11, to VA 39, Lexington, info, **N**...HOSPITAL(3mi), Exxon/diesel/mart, Burger King, Crystal Chinese, Naples Pizza, Waffle House, Best Western, Super 8, Wingate Inn, $Tree, Radio Shack, Wal-Mart SuperCtr/24hr, Long's Camping(3mi), VA Horse Ctr, **S**...Texaco/mart, Golden Corral, Shoney's, Comfort Inn, Econolodge, Holiday Inn Express, to Wash&Lee U, VMI
50	US 60, VA 623, to Lexington, **3 mi S**...Day's Inn
43	VA 780, to Goshen, no facilities
35	VA 269, VA 850, Longdale Furnace, **S**...Longdale Inn, **2 mi S**...North Mtn Motel
29	VA 269, VA 850, **S**...Exxon/diesel/rest., Circle H Camping
27	US 60 W, US 220 S, VA 629, Clifton Forge, **N**...to Douthat SP, Alleghany Highlands Arts/crafts, Buckhorne Camping(2mi), **S**...Citgo/diesel/mart, Exxon, Texaco/mart, The Park Motel
24	US 60, US 220, Clifton Forge, **1 mi S**...Shell/mart/24hr, Texaco/diesel/mart, Dairy Queen, Hardee's, Taco Bell
21	to VA 696, Low Moor, **S**...HOSPITAL
16	US 60 W, US 220 N, Covington, to Hot Springs, **N**...Amoco, Etna/mart, Exxon/Burger King/24hr, Shell, Texaco/KrispyKreme/mart/24hr, Western Sizzlin, Best Western, Holiday Inn Express, Knight's Court Inn, Pinehurst Motel, Buick/Pontiac/GMC, hardware, to ski area, **S**...LJ Silver, McDonald's, Comfort Inn, $General, K-Mart/auto, Radio Shack, bank

E

W

C
h
a
r
l
o
t
t
e
s
v
i
l
l
e

L
e
x
i
n
g
t
o
n

Virginia Interstate 64

E ↑↓ **W**

14	VA 154, Covington, to Hot Springs, **N**...Coastal/mart/24hr, Exxon/ Arby's/KrispyKreme/mart, Great Wall Chinese, Hardee's, KFC, Little Caesar's, Subway, Wendy's, Budget Motel, Highland Motel, Townhouse Motel, Advance Parts, CVS Drug, Family$, Food Lion, Kroger, **S**...Casa Pizza, China House, $Tree, GNC, Wal-Mart SuperCtr/24hr
10	US 60 E, VA 159 S, Callaghan, **S**...Marathon/diesel/LP
7	VA 661, no facilities
2.5mm	**Welcome Ctr eb, full(handicapped)facilities, phone, picnic tables, litter barrels, petwalk**
1	Jerry's Run Trail, **N**...to Allegheny Trail
0mm	Virginia/West Virginia state line

Virginia Interstate 66

Exit #	Services
77mm	Independence Ave, to Lincoln Mem. I-66 begins/ends in Washington, DC.
76mm	Potomac River, T Roosevelt Memorial Bridge
75	US 50 W(from eb), to Arlington Blvd, G Wash Pkwy, I-395, US 1, **S**...Iwo Jima Mon
73	US 29, Lee Hwy, Key Bridge, to Rosslyn, **N**...Marriott, **S**...Best Western
72	to US 29, Lee Hwy, Spout Run Pkwy(from eb), **N**...Texaco, Econolodge, **S**...CVS Drug, Eckerd/24hr, Starbucks, Giant Foods, bank
71	VA 120, Glebe Rd(no EZ return from wb), **N**...HOSPITAL, **S**...Mobil/mart, Blackeyed Pea, Comfort Inn, Holiday Inn
69	US 29, Sycamore St, Falls Church, **N**...Exxon/mart, bank, **S**...VETERINARIAN, 7-11, Econolodge
68	Westmoreland St(from eb), same as 69
67	to I-495 N(from wb), to Baltimore, Dulles Airport
66	VA 7, Leesburg Pike, to Tyson's Corner, Falls Church, **N**...Exxon/mart, Mobil, Chinese Cuisine, HoneyBaked Ham, Jerry's Subs/Pizza, Ledo's Pizza, TCBY, Fresh Fields Foods, SuperCrown Books, Linens'n Things, laundry
64b a	I-495 S, to Richmond
62	VA 243, Nutley St, to Vienna, **S**...Exxon, **1 mi S along Lee Hwy**...VETERINARIAN, Citgo, Shell, Sunoco, 7-11, Applebee's, Dunkin Donuts, Esposito's Italian, IHOP/24hr, McDonald's, NY Pizzaria, Subway, TCBY, Treacher's Café, Econolodge, Jeep/Eagle, Olds, Saturn, Precision Tune, Radio Shack, Trak Parts, RV Ctr, cleaners
60	VA 123, to Fairfax, **S**...MEDICAL CARE, Exxon/mart, Shell/mart/24hr, Sunoco, Chinese Rest., Denny's, Fuddrucker's, Hooters, Indian Rest., Japanese Rest., Roy Rogers, Holiday Inn/rest., Wellesley Inn, Chevrolet, Chrysler/Dodge, CVS Drug, Ford/Lincoln/Mercury, Honda, Hyundai, Mazda, Toyota, JiffyLube, bank, to George Mason U
57	US 50, to Dulles Airport, **N**...MEDICAL CARE, VETERINARIAN, Bennigan's, Holiday Inn, Hecht, JC Penney, Sears/auto, cinema, mall, **1 mi N**...Grady's Grill, Silver Diner, Linens'n Things, **S**...EYECARE, Amoco, Shell, 7-11, Italian Rest., Wendy's, Comfort Inn/rest., Courtyard, Ford, Ward's/auto, bank, laundry
56mm	weigh sta both lanes
55	Fairfax Co Pkwy, to US 29, **N**...HOSPITAL, 4 Lakes Mall, Exxon, Mobil/mart/wash, Applebee's, Blue Iguana Café, Burger King, Cooker Rest., Don Pablo's Mexican, Eatery Rest., Food Mall, Fresh Fare Rest., Olive Garden, Pizza Hut, Red Robin Rest., TCBY, Tony's Pizza, Wendy's, Hyatt Hotel, Best Buy, BJ's Whse, Food Lion, PetsMart, Radio Shack, Waccamaw Outlet, Wal-Mart, bank, cleaners
53	VA 28, to Centreville, Dulles Airport, Manassas Museum, **S**...Exxon, Mobil/diesel, 7-11, Chesapeake Bay Seafood, Jake's Rest., LoneStar Steaks, McDonald's, Jo-Ann Fabrics, Peoples Drug, Drug Emporium, bank, hardware, shopping ctr, same as 52
52	US 29, to Bull Run Park, **N**...Texaco/ExpressLube, Goodyear/auto, camping, **S**...DENTIST, Mobil/diesel/mart, Exxon/mart, 7-11, BBQ, Copeland's Café, Domino's, Hunter Mill Deli, LoneStar Steaks, McDonald's, Ruby Tuesday, Wendy's, CVS Drug, Jo-Ann Fabrics, Mail Services Etc, Radio Shack, Shoppers Foods, Tire Express, bank, same as 53
49mm	**Welcome Ctr both lanes, full(facilities)facilities, phone, picnic tables, litter barrels, petwalk**

E ↑↓ **W**

Virginia Interstate 66

47   VA 234, to Manassas, **N...**Texaco/diesel, Cracker Barrel, Courtyard, Holiday Inn, Borders, Kohl's, RoomStore, Manassas Nat Bfd, **S...**MEDICAL CARE, DENTIST, VETERINARIAN, Amoco, Citgo/7-11, Exxon/mart, RaceTrac/diesel/mart/24hr, Shell/repair, Sunoco/24hr, American Steak/buffet, Bertucci's Pizza, Bob Evans, Boston Mkt, Burger King, Casa Chimayo Mexican, Chesapeake Bay Seafood, Checker's, Chili's, Chinese Rest., Denny's, Domino's, Don Pablo Mexican, Hooters, KFC, Little Caesar's, La Tolteca Mexican, Logan's Roadhouse, McDonald's, Old Country Buffet, Olive Garden, Pizza Hut, Red Lobster, Ruby Tuesday, Schlotsky's, Shoney's, Starbucks, Subway, Taco Bell, TGIFriday, Wendy's, Zipani Bread Café, Best Western, Day's Inn, Hampton Inn, Red Roof Inn, Super 8, Aamco, Barnes&Noble, Burlington Coats, Camping World RV Service/supples, Circuit City, CVS Drug, Chevrolet, Economy Parts, Food Lion, Giant Foods, Hecht's, Home Depot, Honda, JC Penney, K-Mart, Linens'n Things, Lowe's Bldg Supply, LV Golf Discount, Marshall's, Merchant Auto Ctr, Midas Muffler, NTB, Office Depot, Olds/GMC/Pontiac, PepBoys, PetCo, PetsMart, Pier 1, Saturn, Sears, Service Merchandise, Shopper's Club, Speedy MufflerKing, Staples, Super Fresh Foods, Trak Parts, U-Haul, Upton's, Wal-Mart, bank, cinema, cleaners/laundry

43   US 29, Gainesville, to Warrenton, **S...**Citgo/7-11, Exxon/diesel/mart, Mobil, RaceTrac/mart, Texaco/diesel/mart/24hr, Blue Ridge Seafood Too, Joe's Italian/pizza, McDonald's, Wendy's, Hillwood Camping

40   US 15, Haymarket, **N...**Yogi Bear's Camping(5mi), **S...**Citgo/diesel/mart/24hr, Sheetz/diesel/mart/24hr, Matthew's Family Rest.(1mi), Lee-Hi Motel(5mi)

31   VA 245, to Old Tavern, **1 mi N...**BP, **S...**Texaco/diesel, The Railstop Rest.

28   US 17, Marshall, **N...**HOSPITAL, Amoco/diesel, Citgo, McDonald's

27   VA 55 E, VA 647, Marshall, **1-2 mi N...**Chevron/diesel, Citgo, Exxon/diesel, Old Salem Rest.

23   US 17, VA 55, Delaplane(no re-entry from eb), no facilities

20mm   Goose Creek

18   VA 688, Markham, no facilities

13   VA 79, to VA 55, Linden, Front Royal, **S...**Chevron/diesel/mart/24hr, Mobil/mart/24hr, Applehouse Rest., BBQ, **5-6 mi S...**Burger King, Donut Express, KFC, Quality Inn, Pioneer Motel, Super 8, KOA, to Shenandoah NP, Skyline Drive, ski area

11mm   Manassas Run

7mm   Shenandoah River

6   US 340, US 522, Front Royal, to Winchester, **S...**HOSPITAL, Exxon/Baskin-Robbins/mart, Mobil/mart/24hr, McDonald's, **1-2 mi S...**East Coast/mart, Arby's, Hardee's, Pizza Hut, Tastee Freez Rest., Wendy's, Bluemont Motel, Blue Ridge Motel, Front Royal Motel, Shenandoah Motel, Twin Rivers Motel, Gooney Creek Camping(7mi), KOA(6mi), Poe's Southfork Camping(2mi)

1b a   I-81, N to Winchester, S to Roanoke

0mm   I-66 begins/ends on I-81, exit 300.

Virginia Interstate 77

Exit #	Services
67mm	East River Mtn, Virginia/West Virginia state line
66	VA 598, to East River Mtn, no facilities
64	US 52, VA 61, to Rocky Gap, no facilities
62	VA 606, to South Gap, no facilities

61.5mm   **Welcome Ctr sb, full(handicapped)facilities, info, phone, vending, picnic tables, litter barrels, petwalk**

60mm   **rest area nb, full(handicapped)facilities, phone, vending, picnic tables, litter barrels, petwalk**

58   US 52, to Bastian, **E...**BP/diesel/mart, **W...**Citgo/diesel/mart, Exxon/diesel/mart

56mm   runaway ramp nb

52   US 52, VA 42, Bland, **E...**Citgo, Bland Square Rest./mart, **W...**Shell/DQ/diesel/mart/atm, Log Cabin Rest., Big Walker Motel, antiques, tires

51.5mm   weigh sta both lanes

48mm   Big Walker Mtn

47   VA 717, **6 mi W...**to Deer Trail Park/NF Camping

41   VA 610, Peppers Ferry, Wytheville, **E...**Sagebrush Steaks, Rodeway Inn, Sleep Inn, Super 8, **W...**Shell/diesel/mart/atm/24hr, TA/BP/Subway/Taco Bell/diesel/24hr, Country Kitchen, Wendy's, Hampton Inn, Ramada/rest.

40   I-81 S, to Bristol, US 52 N

I-77 and I-81 run together 9 mi. See Virginia Interstate 81, exits 73-80.

32   I-81 N, to Roanoke

26mm   New River

24   VA 69, to Poplar Camp, **E...**to Shot Tower SP, New River Trail Info Ctr, **W...**Citgo/diesel/mart

19   VA 620, airport

Virginia Interstate 77

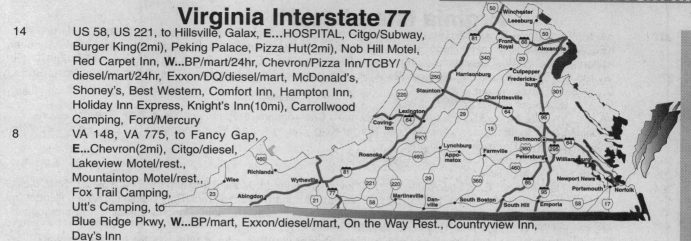

14	US 58, US 221, to Hillsville, Galax, **E...**HOSPITAL, Citgo/Subway, Burger King(2mi), Peking Palace, Pizza Hut(2mi), Nob Hill Motel, Red Carpet Inn, **W...**BP/mart/24hr, Chevron/Pizza Inn/TCBY/diesel/mart/24hr, Exxon/DQ/diesel/mart, McDonald's, Shoney's, Best Western, Comfort Inn, Hampton Inn, Holiday Inn Express, Knight's Inn(10mi), Carrollwood Camping, Ford/Mercury
8	VA 148, VA 775, to Fancy Gap, **E...**Chevron(2mi), Citgo/diesel, Lakeview Motel/rest., Mountaintop Motel/rest., Fox Trail Camping, Utt's Camping, to Blue Ridge Pkwy, **W...**BP/mart, Exxon/diesel/mart, On the Way Rest., Countryview Inn, Day's Inn
6.5mm	runaway truck ramp sb
4.5mm	runaway truck ramp sb
3mm	runaway truck ramp sb
1	VA 620, no facilities
.5mm	**Welcome Ctr nb, full(handicapped)facilities, info, phone, picnic tables, litter barrels, petwalk**
0mm	Virginia/North Carolina state line

Virginia Interstate 81

Exit #	Services
324.5mm	Virginia/West Virginia state line
323	VA 669, to US 11, Whitehall, **E...**Exxon/mart, auto auction, **W...**Flying J/Country Mkt/Pepperoni's/diesel/LP/mart/24hr
321	VA 672, Clearbrook, **E...**VETERINARIAN, Mobil/diesel/mart, Olde Stone Rest.
320mm	**Welcome Ctr sb, full(handicapped)facilities, phone, vending, picnic tables, litter barrels, petwalk**
317	US 11, Stephenson, **W...**HOSPITAL, Amoco/diesel, Exxon/diesel, Mobil/mart, Shell/Blimpie/diesel/mart, Burger King, Denny's, McDonald's, Pizza Hut/Taco Bell, Tastee Freez, Econolodge, Ramada Inn, Candy Hill Camping(4mi), bank
315	VA 7, Winchester, **E...**Exxon, Sheetz/mart/24hr, Dodge/Volvo, mall, **W...**VETERINARIAN, Amoco/mart, Chevron/diesel, Shell/diesel/mart, Texaco/diesel/mart, Arby's, Capt D's, Hardee's, KFC, Little Caesar's, LJ Silver, McDonald's, Pizza Hut, Seafood Rest., Wendy's, Hampton Inn, Shoney's Inn/rest., Wendy's, Phar-Mor Drug, auto repair, bank
314mm	Abrams Creek
313	US 17/50/522, Winchester, **E...**Amoco/diesel/mart, Exxon/Baskin-Robbins/Dunkin Donuts/mart, Shell/diesel/mart, Sheetz/mart, Texaco/diesel/mart, Chason's Buffet, China Town, Cracker Barrel, Hardee's, Hoss' Steaks, Texas Roadhouse, Waffle House, Comfort Inn, Holiday Inn/rest., Super 8, Travelodge/rest., Big Lots, Costco, Food Lion, Ford/Lincoln/Mercury/Nissan, Harley-Davidson, Jo-Ann Fabrics, Target, bank, cleaners, **W...**CHIROPRACTOR, EYECARE, Sheetz/mart/24hr, Bob Evans, Castiglia's Italian, Chili's, China Jade, Joe Muggs Coffee, KFC, McDonald's, Pargo's, Rally's, Subway, Taco Bell, TCBY, Wendy's, Baymont Inn, Best Western/rest., Hampton Inn, Quality Inn, Wingate Inn, BooksAMillion, Circuit City, $Tree, Home Depot, JC Penney, JiffyLube, K-Mart, Kohl's, Kroger, Lowe's Bldg Supply, Martin Food/drug, Monro Muffler/brake, OfficeMax, PepBoys, PetsMart, Precision Tune, Sears/auto, Staples, TJ Maxx, Trak Parts, Target, Wal-Mart SuperCtr/24hr, bank, cinema, mall, to Shenandoah U
310	VA 37, to US 11, **1 mi W...**HOSPITAL, Bonds Motel, Day's Inn, Echo Village Motel, Howard Johnson, Quality Inn, Relax Inn, Candy Hill Camping
307	VA 277, Stephens City, **E...**MEDICAL CARE, Citgo/7-11, Shell/mart, Texaco/diesel/mart, Burger King, KFC/Taco Bell, McDonald's, Roma Pizza, Subway, Tastee Freez, Waffle House, Wendy's, Western Steer, Comfort Inn, Holiday Inn Express, White Oak Camping, bank, hardware, **W...**Exxon/Dunkin Donuts/mart, Sheetz/mart/24hr, Redwood Budget Motel
304mm	weigh sta both lanes
302	VA 627, Middletown, **E...**Exxon/diesel, **W...**Amoco/Blimpie/diesel, Citgo/Stuckey's/diesel, Godfather's, Super 8, Wayside Inn/rest., to Wayside Theatre
300	I-66 E, to Washington, to Shenandoah NP, Skyline Dr
298	US 11, Strausburg, **E...**Amoco/mart, Mobil/diesel/LP, Burger King, McDonald's, Hotel Strausburg/rest., Battle of Cedar Creek Camping, **W...**to Belle Grove Plantation
296	VA 55, Strausburg, **E...**to Hupp's Hill Bfd Museum
291	VA 651, Toms Brook, **W...**Virginian/Texaco/diesel/rest., Wilco/Citgo/DQ/diesel/mart/24hr, Budget Inn, truckwash

Virginia Interstate 81

283	VA 42, Woodstock, **E...**HOSPITAL, Chevron/7-11/wash, Sheetz/mart, Texaco/DunkinDonuts, Arby's, Hardee's, KFC, McDonald's, Pizza Hut, Taco Bell, Wendy's, Budget Host, Comfort Inn, Ramada/rest., bank, to Massanutten Military Academy, **W...**Coastal/mart/24hr, Exxon/diesel/mart/24hr, Subway, Cato's, $Tree, Wal-Mart SuperCtr/24hr
279	VA 185, VA 675, Edinburg, **E...**Amoco/diesel, Mobil/diesel, Shell/diesel/mart/atm/24hr, Edinburg Mill Rest., Subway, Creek Side Camping(2mi)
277	VA 614, Bowmans Crossing, no facilities
273	VA 292, VA 703, Mt Jackson, **E...**Amoco/diesel/mart, Chevron/diesel/repair/24hr, Citgo/7-11, Sheetz/Baskin-Robbins/Wendy's/diesel/mart/24hr, Blimpie, Burger King, Denny's, Best Western/rest., $General, Food Lion/24hr, USPO, **W...**Davis Tire/repair, to ski area
269	VA 730, to Shenandoah Caverns, **E...**Chevron/diesel/mart
269mm	N Fork Shenandoah River
264	US 211, New Market, **E...**Amoco/diesel/mart, Chevron/diesel/mart, Mobil/Blimpie/mart, Exxon/diesel, Shell/diesel/mart, Burger King, Godfather's, McDonald's, Pizza Hut, Southern Kitchen, Taco Bell, Battlefield Motel, Budget Inn, Quality Inn/Johnny Appleseed Rest., Shenvalee Motel/rest., Pottery Museum, Rancho Camping, to Shenandoah NP, Skyline Dr, **W...**Citgo/7-11, Day's Inn, to New Market Bfd
262mm	**rest area both lanes, full(handicapped)facilities, phone, vending, picnic tables, litter barrels, petwalk**
257	US 11, VA 259, to Broadway, **3-5 mi E...**KOA, Endless Caverns Camping
251	US 11, Harrisonburg, **W...**Exxon/diesel/mart, Economy Inn
247	US 33, Harrisonburg, **E...**Amoco/Blimpie/diesel, Royal/mart, Texaco, Applebee's, Boston Beanery, Bravo Italian, Capt D's, Chili's, El Charro Mexican, Little Caesar's, LJ Silver, Pargo's Steaks, Red Lobster, Ruby Tuesday, Shoney's, Taco Bell, TCBY, Texas Steaks, Waffle House, Wendy's, WoodFired Italian, Comfort Inn, Courtyard, Econolodge, Hampton Inn, Jameson Inn, Motel 6, Sheraton/rest., Shoney's Inn, Belk, BooksAMillion, Circuit City, CVS Drug, FashionBug, Food Lion, JiffyLube, Kroger, K-Mart, Lowe's Bldg Supply, MailBoxes Etc, Nissan, OfficeMax, PepBoys, Pier 1, Staples, TJ Maxx, ToysRUs, Wal-Mart, to Shenandoah NP, Skyline Dr, bank, cinema, mall, ski area, **W...**HOSPITAL, Amoco, Chevron/diesel, Exxon/DunkinDonuts/diesel, Royal/mart, Texaco, Arby's, Dragon Palace, Golden China, Hardee's, KFC, McDonald's, Mr Gatti's, Pizza Hut, Sam's Hotdogs, Subway, Marvilla Motel, Advance Parts, Big Lots, CVS Drug, Farmer Jack's Foods, Radio Shack, bank, cleaners
245	VA 659, Port Republic Rd, **E...**Citgo/diesel/mart, Exxon/Subway/diesel, Mobil/diesel, Texaco/mart, Dairy Queen, J Willoby's Rest., RoadHouse Rest., Day's Inn, Howard Johnson, **W...**HOSPITAL, to James Madison U
243	US 11, to Harrisonburg, **W...**Amoco/diesel, Citgo/7-11, Exxon/DunkinDonuts/diesel/mart/atm, Travel Ctr/diesel/rest., Burger King, Chinese Rest., Cracker Barrel, Golden Corral, McDonald's, Taco Bell, Waffle House, Belle Meade Motel/rest., Holiday Inn Express, Ramada Inn, Rockingham Motel, Super 8, BMW, Chrysler, Ford/Lincoln/Mercury, Hyundai, Kia, Subaru, Toyota, Volvo/GMC/Kenworth
240	VA 257, VA 682, Mount Crawford, **1-3 mi W...**Exxon/Arby's/diesel/mart, Mobil, D'n D Mart, Burger King, Ever's Family Rest., Village Inn
235	VA 256, Weyers Cave, **E...**Mobil/diesel/mart, **W...**Amoco/Subway/diesel/mart, Exxon/diesel, Augusta Motel, to Grand Caverns, airport
232mm	**rest area both lanes, full(handicapped)facilities, phone, vending, picnic tables, litter barrels, petwalk**
227	VA 612, Verona, **E...**Amoco/diesel/mart, Waffle Inn/24hr, North 340 Camping(11mi), auto repair, **W...**Citgo/diesel/mart/wash, Exxon/mart, Arby's, Brooks Rest., Burger King, Chinese Rest., Ciro's Pizza, Hardee's, KFC, McDonald's, Subway, Wendy's, Ramada Ltd, Scottish Inn, CVS Drug, Food Lion, KOA(3mi), laundry
225	VA 275, Woodrow Wilson Pkwy, **E...**Best Inn/rest., **W...**Denny's/24hr, VA Steakhouse, Day's Inn, Holiday Inn/rest., Ingleside Resort, golf
222	US 250, Staunton, **E...**Exxon/McDonald's/mart/24hr, Texaco/diesel/mart, Cracker Barrel, Rowe's Rest., Texas Steaks, Best Western, Shoney's Inn/rest., Sleep Inn, Wendy's, **W...**Raceway/mart, Wilco Fuel/diesel, Burger King, Shorty's Diner, Waffle House, Comfort Inn, Econolodge, Microtel, Super 8, AutoZone, Lowe's Bldg Supply, TireMart, Toyota, Wal-Mart SuperCtr/24hr, bank, American Frontier Culture Museum
221	I-64 E, to Charlottesville, to Skyline Dr, Shenandoah NP
220	VA 262, to US 11, Staunton, **1 mi W...**Citgo, Etna/mart, Exxon, Applebee's, Arby's, Burger King, Country Cookin', Hardee's, KFC, McDonald's, Red Lobster, Taco Bell, Hampton Inn, Staunton Budget Motel, Advance Parts, Honda, JC Penney, Pontiac/Buick, Rule RV Ctr
217	VA 654, to Mint Spring, Stuarts Draft, **E...**Amoco/Subway/diesel/mart, Exxon/diesel/mart, Day's Inn, Shenandoah Acres Resort(8mi), **W...**Chevron/mart/LP, Citgo/diesel/mart/24hr, Crown/diesel/mart, Armstrong Motel/rest., Walnut Hill Camping(4mi)
213b a	US 11, US 340, Greenville, **1 mi E...**Amoco/Subway/mart, Shell/KrispyKreme/mart, Econolodge, **3 mi E...**$General, Greenville Grocery, hardware
205	VA 606, Raphine, **E...**Coastal/diesel/mart, Exxon/Burger King/mart, Fuel City/diesel, Whites/diesel/rest./24hr/motel, **W...**Texaco/diesel/mart, Wilco/Wendy's/diesel/mart/24hr, Day's Inn/rest., Willow Lake Travelpark, camping
200	VA 710, Fairfield, **E...**Amoco/diesel/mart/24hr, Texaco, McDonald's, **W...**Citgo/Sunshine/diesel/rest., Exxon/diesel, Shell/diesel, Fairfield Coffee Shop
199mm	**rest area sb, full(handicapped)facilities, phone, vending, picnic tables, litter barrels, petwalk**
195	US 11, Lee Highway, **E...**Maple Hall Lodging/dining, **W...**Citgo/diesel/mart/24hr, Shell/diesel, Lee Hi Trkstp/rest., Aunt Sarah's, Best Western, Comfort Inn, Howard Johnson/rest., Ramada Inn, Travelodge, Long's Camping

Harrisonburg

N

S

Virginia Interstate 81

191   I-64 W(exits left from nb), US 60, to Charleston

188   US 60, to Lexington, Buena Vista, **3-5 mi E...**HOSPITAL, Citgo, Exxon, Country Cookin', Hardee's, KFC, LJ Silver, McDonald's, Pizza Hut, Taco Bell, Wendy's, Budget Inn, Buena Vista Inn, Comfort Inn, Day's Inn, Thrifty Inn, to Glen Maury Park, camping, to Stonewall Jackson Home, George C Marshall Museum, to Blue Ridge Pkwy, **W...**Citgo, Exxon/diesel/mart/24hr, Econolodge, Hampton Inn, to Washington&Lee U, VMI

180   US 11, Natural Bridge, **E...**Pure/diesel, Shell, Econo Inn, **W...**Texaco/diesel/mart, Pink Cadillac Diner, Budget Inn, KOA

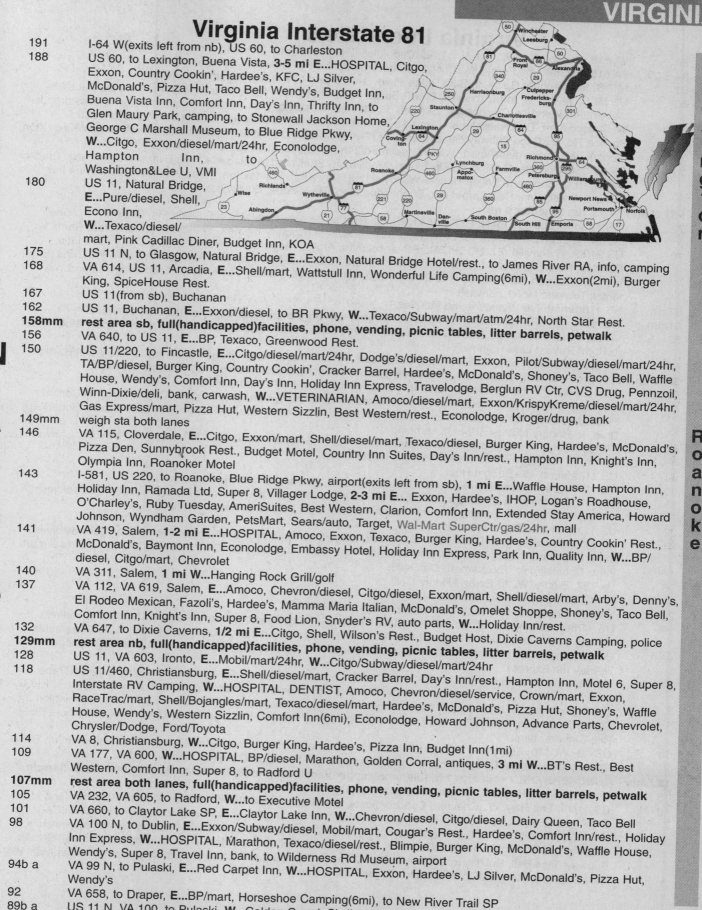

175   US 11 N, to Glasgow, Natural Bridge, **E...**Exxon, Natural Bridge Hotel/rest., to James River RA, info, camping

168   VA 614, US 11, Arcadia, **E...**Shell/mart, Wattstull Inn, Wonderful Life Camping(6mi), **W...**Exxon(2mi), Burger King, SpiceHouse Rest.

167   US 11(from sb), Buchanan

162   US 11, Buchanan, **E...**Exxon/diesel, to BR Pkwy, **W...**Texaco/Subway/mart/atm/24hr, North Star Rest.

158mm   **rest area sb, full(handicapped)facilities, phone, vending, picnic tables, litter barrels, petwalk**

156   VA 640, to US 11, **E...**BP, Texaco, Greenwood Rest.

150   US 11/220, to Fincastle, **E...**Citgo/diesel/mart/24hr, Dodge's/diesel/mart, Exxon, Pilot/Subway/diesel/mart/24hr, TA/BP/diesel, Burger King, Country Cookin', Cracker Barrel, Hardee's, McDonald's, Shoney's, Taco Bell, Waffle House, Wendy's, Comfort Inn, Day's Inn, Holiday Inn Express, Travelodge, Berglun RV Ctr, CVS Drug, Pennzoil, Winn-Dixie/deli, bank, carwash, **W...**VETERINARIAN, Amoco/diesel/mart, Exxon/KrispyKreme/diesel/mart/24hr, Gas Express/mart, Pizza Hut, Western Sizzlin, Best Western/rest., Econolodge, Kroger/drug, bank

149mm   weigh sta both lanes

146   VA 115, Cloverdale, **E...**Citgo, Exxon/mart, Shell/diesel/mart, Texaco/diesel, Burger King, Hardee's, McDonald's, Pizza Den, Sunnybrook Rest., Budget Motel, Country Inn Suites, Day's Inn/rest., Hampton Inn, Knight's Inn, Olympia Inn, Roanoker Motel

143   I-581, US 220, to Roanoke, Blue Ridge Pkwy, airport(exits left from sb), **1 mi E...**Waffle House, Hampton Inn, Holiday Inn, Ramada Ltd, Super 8, Villager Lodge, **2-3 mi E...** Exxon, Hardee's, IHOP, Logan's Roadhouse, O'Charley's, Ruby Tuesday, AmeriSuites, Best Western, Clarion, Comfort Inn, Extended Stay America, Howard Johnson, Wyndham Garden, PetsMart, Sears/auto, Target, Wal-Mart SuperCtr/gas/24hr, mall

141   VA 419, Salem, **1-2 mi E...**HOSPITAL, Amoco, Exxon, Texaco, Burger King, Hardee's, Country Cookin' Rest., McDonald's, Baymont Inn, Econolodge, Embassy Hotel, Holiday Inn Express, Park Inn, Quality Inn, **W...**BP/diesel, Citgo/mart, Chevrolet

140   VA 311, Salem, **1 mi W...**Hanging Rock Grill/golf

137   VA 112, VA 619, Salem, **E...**Amoco, Chevron/diesel, Citgo/diesel, Exxon/mart, Shell/diesel/mart, Arby's, Denny's, El Rodeo Mexican, Fazoli's, Hardee's, Mamma Maria Italian, McDonald's, Omelet Shoppe, Shoney's, Taco Bell, Comfort Inn, Knight's Inn, Super 8, Food Lion, Snyder's RV, auto parts, **W...**Holiday Inn/rest.

132   VA 647, to Dixie Caverns, **1/2 mi E...**Citgo, Shell, Wilson's Rest., Budget Host, Dixie Caverns Camping, police

129mm   **rest area nb, full(handicapped)facilities, phone, vending, picnic tables, litter barrels, petwalk**

128   US 11, VA 603, Ironto, **E...**Mobil/mart/24hr, **W...**Citgo/Subway/diesel/mart/24hr

118   US 11/460, Christiansburg, **E...**Shell/diesel/mart, Cracker Barrel, Day's Inn/rest., Hampton Inn, Motel 6, Super 8, Interstate RV Camping, **W...**HOSPITAL, DENTIST, Amoco, Chevron/diesel/service, Crown/mart, Exxon, RaceTrac/mart, Shell/Bojangles/mart, Texaco/diesel/mart, Hardee's, McDonald's, Pizza Hut, Shoney's, Waffle House, Wendy's, Western Sizzlin, Comfort Inn(6mi), Econolodge, Howard Johnson, Advance Parts, Chevrolet, Chrysler/Dodge, Ford/Toyota

114   VA 8, Christiansburg, **W...**Citgo, Burger King, Hardee's, Pizza Inn, Budget Inn(1mi)

109   VA 177, VA 600, **W...**HOSPITAL, BP/diesel, Marathon, Golden Corral, antiques, **3 mi W...**BT's Rest., Best Western, Comfort Inn, Super 8, to Radford U

107mm   **rest area both lanes, full(handicapped)facilities, phone, vending, picnic tables, litter barrels, petwalk**

105   VA 232, VA 605, to Radford, **W...**to Executive Motel

101   VA 660, to Claytor Lake SP, **E...**Claytor Lake Inn, **W...**Chevron/diesel, Citgo/diesel, Dairy Queen, Taco Bell

98   VA 100 N, to Dublin, **E...**Exxon/Subway/diesel, Mobil/mart, Cougar's Rest., Hardee's, Comfort Inn/rest., Holiday Inn Express, **W...**HOSPITAL, Marathon, Texaco/diesel/rest., Blimpie, Burger King, McDonald's, Waffle House, Wendy's, Super 8, Travel Inn, bank, to Wilderness Rd Museum, airport

94b a   VA 99 N, to Pulaski, **E...**Red Carpet Inn, **W...**HOSPITAL, Exxon, Hardee's, LJ Silver, McDonald's, Pizza Hut, Wendy's

92   VA 658, to Draper, **E...**BP/mart, Horseshoe Camping(6mi), to New River Trail SP

89b a   US 11 N, VA 100, to Pulaski, **W...**Golden Corral, Skyline Motel

Virginia Interstate 81

86	VA 618, Service Rd, **W**...I-81 AutoTruck/Citgo/diesel/rest., Gateway Motel II
84	VA 619, to Grahams Forge, **W**...Shell/DQ/diesel/mart/24hr, Fox Mtn Inn/rest., Trail Motel
81	I-77 S, to Charlotte, Galax, to Blue Ridge Pkwy. **I-81 S and I-77 N run together 9 mi.**
80	US 52 S, VA 121 N, to Ft Chiswell, **E**...BP/Burger King/diesel/mart, Petro/Exxon/KrispyKreme/Little Caesar's/diesel/24hr, Wendy's, Comfort Inn, Super 8, Ft Chiswell Camping, Blue Beacon, Factory Stores/famous brands, **W**...Amoco/mart, Citgo/diesel/mart, Arnold's Diner, McDonald's, Country Inn Suites, Little Valley RV Ctr
77	Service Rd, **E**...Citgo/Subway/diesel/mart/24hr, Flying J/diesel/LP/rest./24hr, Shell/diesel/mart, Burger King, KOA, funpark, **W**...Exxon/diesel/mart, Wilco/diesel/mart/24hr, antiques, st police, wicker furniture
73	US 11 S, Wytheville, **E**...HOSPITAL, Exxon, Mobil/mart, Shell/mart, Applebee's, Bob Evans, Burger King, Cracker Barrel, KFC, Shoney's, Waffle House, Wendy's(1mi), Day's Inn, Holiday Inn, Interstate Motel, MacWythe Inn, Motel 6, Quality Inn, Red Carpet Inn, Shenandoah Inn, Travelodge, Buick/Olds

I-81 N and I-77 S run together 9 mi

72	I-77 N, to Bluefield, **1 mi N, I-77 exit 41, E**...Sagebrush Steaks, Rodeway Inn, Sleep Inn, Super 8, **W**... Shell/diesel/mart/atm/24hr, TA/BP/Subway/Taco Bell/diesel/24hr, Country Kitchen, Wendy's, Hampton Inn, Ramada
70	US 21/52, Wytheville, **E**...HOSPITAL, BP/diesel/mart, Arby's, McDonald's, Wendy's, CVS Drug, Food Lion, Peebles, Pennzoil, bank, **1 mi E**...DENTIST, Exxon, Pizza Hut, Hampton Inn, Ramada Inn, IGA, **W**...Citgo, Etna/mart, Exxon/mart, Dairy Queen, LJ Silver, Pizza Hut, Scrooge's Rest., Comfort Inn, Econolodge, Johnson's Motel, Travelite Motel, carwash
67	US 11(from nb), to Wytheville, no facilities
61mm	**rest area nb, full(handicapped)facilities, phone, vending, picnic tables, litter barrels, petwalk**
60	VA 90, Rural Retreat, **E**...Chevron/diesel/mart, Citgo/Yesterdaze Diner/diesel/mart, Country Kitchen, McDonald's, to Rural Retreat Lake, camping
54	VA 683, to Groseclose, **E**...Exxon/DQ/diesel/mart, Texaco/diesel/mart, Village Motel/rest., Pottery Outlet, airport
53.5mm	**rest area sb, full(handicapped)facilities, phone, vending, picnic tables, litter barrels, petwalk**
50	US 11, Atkins, **W**...Citgo/diesel/mart/24hr, Subs on the Way, Day's Inn/trk parking
47	US 11, to Marion, **W**...Amoco, Chevron/diesel/mart/wash/24hr, Exxon/diesel, Arby's, Burger King, McDonald's, Pizza Hut, Best Western/rest., Econolodge, VA House Inn, K-Mart, Quality Farm/fleet, Radio Shack, SuperX Drug, to Hungry Mother SP(4mi), camping
45	VA 16, Marion, **E**...Shell/diesel/mart, AppleTree Rest., to Grayson Highlands SP, Mt Rogers NRA, **W**...Chevron, Exxon, Hardee's, KFC, LJ Silver, Wendy's, Lincoln Motel
44	US 11, Marion, **W**...Budget Host
39	US 11, VA 645, Seven Mile Ford, **E**...Budget Inn/rest., **W**...Interstate Camping/groceries
35	VA 107, Chilhowie, **E**...GasHaus, Texaco/mart, Knight's Inn/rest., **W**...Chevron, Exxon/mart/atm, Rouse/diesel, McDonald's, Subway, Budget Inn(1mi), Rainbow Autel/rest., $General, Food City, bank
32	US 11, to Chilhowie, no facilities
29	VA 91, Glade Spring, to Damascus, **E**...Amoco, Citgo/Cardinal/diesel/rest./showers/24hr, Shell/Subway/diesel/mart, Giardino's Italian, Wendy's, Economy Inn/rest., Swiss Inn/rest., **W**...OPTOMETRIST, Chevron/diesel/mart/24hr, Coastal/mart, Exxon/mart, CarQuest
26	VA 737, Emory, **W**...to Emory&Henry Coll
24	VA 80, Meadowview Rd, **W**...auto repair
22	VA 704, Enterprise Rd, **W**...racing museum
19	US 11/58, to Abingdon, **E**...VETERINARIAN, Texaco/Subway/diesel/mart/atm, Pizza+, Cherokee Motel/rest., Holiday Lodge, bank, to Mt Rogers NRA, **W**...Chevron/diesel/mart/24hr, Exxon/diesel/mart, Bella Pizza/subs, Burger King, Cracker Barrel, Harbor House Seafood, Omelet Shoppe, Alpine Motel, Empire Motor Lodge/rest., Holiday Inn Express
17	US 58A, VA 75, Abingdon, **E**...CHIROPRACTOR, Amoco/mart, Domino's, LJ Silver, Hampton Inn, Mr Transmission, XpressLube, **W**...HOSPITAL, EYECARE, Chevron, Citgo/diesel/mart, Exxon/mart, Arby's, Hardee's, KFC, Los Arcos Mexican, McDonald's, Papa John's, Pizza Hut, Shoney's, Taco Bell, Wendy's, Martha Washington Inn, Super 8, Advance Parts, CVS Drug, $General, Food City, Food Lion, Kroger/deli/24hr, K-Mart, MailBoxes Etc, Ritz Camera, bank
14	US 19, VA 140, Abingdon, **E**...VETERINARIAN, **W**...Chevron/diesel/mart/24hr, Exxon, Texaco/diesel/mart, Dairy Queen, McDonald's, Budget Inn(3mi), Comfort Inn, Ford, Riverside Camping(10mi)
13.5mm	**TRUCKERS ONLY rest area nb, full(handicapped)facilities, phone, vending, picnic tables, litter barrels**
13	VA 611, to Lee Hwy, **W**...Texaco/diesel/mart, airport
10	US 11/19, Lee Hwy, **W**...Chevron, Conoco/diesel, Evergreen Motel, Red Carpet Inn, Scottish Inn, Skyland Motel
7	Old Airport Rd, **E**...Amoco/mart, Texaco/diesel/mart/atm, Bojangles, Sonic, La Quinta, bank, **W**...MEDICAL CARE, CHIROPRACTOR, BP/diesel/mart/atm, Citgo/Wendy's/mart, Conoco/diesel/mart, Damon's, Domino's, Fazoli's, IHOP, Ming Chinese, O'Charley's, Perkins, Pizza Hut, Prime Sirloin, Ruby Tuesday, Sagebrush Steaks, Subway, Taco Bell, Holiday Inn, Advance Parts, Cato's, Food Country USA, Lowe's Bldg Supply, Muffler Man, Wal-Mart SuperCtr/24hr, GMC/Volvo/White, Sugar Hollow Camping, cinema

N

↑

↓

S

Virginia Interstate 81

N ↕ **S**

5 US 11/19, Lee Hwy, **E**...Amoco/DQ/mart, Shell/mart/atm, Arby's, Burger King, Hardee's, KFC, LJ Silver, McDonald's, Shoney's, Wendy's, Budget Inn, Crest Motel, Siesta Motel, Super 8, CVS Drug, Family$, Food Lion/deli, Parts+, USPO, bank, flea mkt, **W**...Chevron, Exxon/mart, Comfort Inn, Blevins Tire, Buick/ Pontiac, Honda, Lee Hwy Camping

3 I-381 S, to Bristol, **1 mi E**...Citgo/diesel, Chevron, Conoco, Duff's Smorgasbord, Omelet Shoppe/24hr, Pizza Hut, Econolodge, Regency Inn

1 US 58/421, Bristol, **1 mi E**...HOSPITAL, CHIROPRACTOR, VETERINARIAN, Citgo/diesel, Exxon, Burger King, Chick-fil-A, Hong Kong Buffet, KFC, Krispy Kreme, LJ Silver, McDonald's, Shoney's, Taco Bell, Vinyard Italian, Wendy's, Briscoe Motel, Ramada Inn, CarQuest, CVS Drug, Dodge, Family$, Food Lion, Goody's, Kroger/drugs, Lincoln/Mercury, Muffler Repair, Proffit's, Sears/auto, Wal-Mart/grill, bank, cinema, laundry, mall, **W**...Citgo(1mi)

0mm **Virginia/Tennessee state line, Welcome Ctr nb, full(handicapped)facilities, info, phone, vending, picnic tables, litter barrels, petwalk**

Virginia Interstate 85

Exit #	Services
69mm	I-85 begins/ends on I-95.
69	US 301, Petersburg, Wythe St, Washington St, **W**...Regency Inn
68	I-95 S, US 460 E, to Norfolk, Crater Rd, **E**...HOSPITAL, **3/4 mi E**...Exxon/diesel/mart/wash, Texaco, RaceTrac/ diesel/mart, Hardee's, California Inn, to Petersburg Nat Bfd
65	Squirrel Level Rd, **E**...to Richard Bland Coll, **W**...BP/diesel/mart
63	US 1, to Petersburg, **E**...Chevron/diesel, Exxon/diesel/mart//24hr, Shell/mart, Texaco/diesel/mart/24hr, Thriftmart Plaza/diesel, Burger King, Hardee's, Waffle House, **W**...Amoco, McDonald's
61	US 460, to Blackstone, **E**...Amoco, EastCoast/Subway/TCBY/diesel/LP/mart, Day's Inn(3mi), **W**...Texaco/ Dunkin Donuts/diesel/mart, Picture Lake Camping(2mi), airport
55mm	**rest area both lanes, full(handicapped)facilities, phone, picnic tables, litter barrels, vending, petwalk**
53	VA 703, Dinwiddie, **W**...Exxon/diesel, to 5 Forks Nat Bfd
52mm	Stony Creek
48	VA 650, DeWitt, no facilities
42	VA 40, McKenney, **W**...Citgo/mart, Exxon/mart, auto repair, **1.5 mi W**...Economy Inn
40mm	Nottoway River
39	VA 712, to Rawlings, **E**...VA Battlerama, **W**...Chevron/diesel/mart/24hr, Citgo/diesel, Nottoway Motel/rest.
34	VA 630, Warfield, **W**...Exxon/diesel
32mm	**rest area both lanes, full(handicapped)facilities, phone, picnic tables, litter barrels, vending, petwalk**
28	US 1, Alberta, **W**...Exxon/mart
27	VA 46, to Lawrenceville, **E**...to St Paul's Coll
24	VA 644, to Meredithville, no facilities
22mm	weigh sta both lanes
20mm	Meherrin River
15	US 1, to South Hill, **E**...Citgo/deli, **W**...Shell, Kahill's Diner
12	US 58, VA 47, to South Hill, **E**...RaceTrac/mart/atm, Shell/SlipIn/mart, Bojangles, Domino's, Hampton Inn, Holiday Inn Express, Super 8, $Tree, Valvoline, Wal-Mart, **W**...HOSPITAL, Amoco/diesel/wash, Chevron/diesel/ repair, Exxon/diesel/mart, Petrol/mart, Texaco/diesel/mart/wash, Brian's Steaks, Burger King, Denny's, Golden Corral, Hardee's, KFC, McDonald's, New China, Pizza Hut, Subway, Taco Bell, Wendy's, Best Western, Comfort Inn, Econolodge, Super 8, CVS Drug, $General, Farmers Foods, Goodyear/auto, Peeble's, Roses, Winn-Dixie/ deli, bank, cinema, tires
4	VA 903, to Bracey, Lake Gaston, **E**...Amoco/DQ/Subway/mart, Citgo/diesel/mart, Exxon/diesel/rest./mart/24hr, Americamps Camping(5mi), **W**...Texaco/Pizza Hut/mart, Day's Inn/rest.
3mm	Lake Gaston
1mm	**Welcome Ctr nb, full(handicapped)facilities, phone, picnic tables, phones, litter barrels, vending, petwalk**
0mm	Virginia/North Carolina state line

Virginia Interstate 95

Exit #	Services
178mm	Potomac River, W Wilson Mem Br, Virginia/Maryland state line
177(1)	US 1, to Alexandria, Ft Belvoir, HOSPITAL, **E**...Amoco/mart, Mobil/mart, Sunoco, Texaco/diesel/mart, Western Sizzlin, Howard Johnson, Traveler's Motel, Chevrolet, Dodge/Chrysler/Jeep, **W**...Chevron/24hr, Merit Gas/diesel
176(2)	VA 241, Telegraph Rd, **E**...Amoco, Citgo/diesel, Exxon, Hess/diesel, **1 mi W on VA 236(Duke St)**...CHIROPRACTOR, VETERINARIAN, Mobil, Shell, 7-11, Dunkin Donuts, Generous George's Pizza, McDonald's, Wendy's, Courtyard, Holiday Inn, Midas Muffler, JiffyLube, 1-Hr Photo, auto repair, bank, drugs, groceries, transmissions
172(3)	rd 613, Van Dorn St, to Franconia, **E**...Comfort Inn
170a	I-495 N to Rockville
b	I-395 N to Washington
169b a	VA 644, Springfield, Franconia, **E**...HOSPITAL, Mobil/mart, 7-11, Bennigan's, Japanese Rest., Best Western, Day's Inn, Hampton Inn, Hilton, JC Penney, Macy's, Firestone/auto, Ford, bank, cinema, laundry, mall, **W**...EYECARE, Citgo, Mobil, Shell, Bob Evans, Chesapeake Bay Seafood, Chi Chi's, Chili's, Houlihan's Rest., LJ Silver, McDonald's, Shoney's, Econolodge, Holiday Inn Express, Motel 6, K-Mart, Diahatsu, Dodge, drugs, bank, laundry
167	VA 617, Back Lick Rd(from sb), no facilities
166b a	to Ft Belvoir, Newington, **E**...Exxon/diesel/mart, Hunter Motel/rest., auto parts/tires, **W**...McDonald's, Kenworth Trucks, bank
163	VA 642, Lorton, **E**...Citgo/mart, Shell/24hr, camping, **W**...Texaco, Best Western
161	US 1 S(no nb re-entry, exits left from sb), to Ft Belvoir, Mt Vernon, Woodlawn Plantation, Gunston Hall, **1 mi E**...access to same as 160, along US 1
160.5mm	Occoquan River
160b a	VA 123 N, Woodbridge, Occoquan, **E**...Amoco/diesel/mart, Exxon, Shell, Texaco, 7-11, Astoria Pizza, Lum's, Roy Rogers, Shoney's, Taco Bell, Comfort Inn, Econolodge, Friendship Inn, Scottish Inn, Econolodge, Ames, Ford, JiffyLube, auto repair, **services 1 mi E on US 1**...Chevron/diesel, Dunkin Donuts, Roy Rogers, laundry, **W**...Exxon/diesel/mart, Mobil/mart(1mi), Shell/mart, KFC, McDonald's, Toby's Café, VA Grill, Wendy's
158b a	VA 3000, Prince William Pkwy, Woodbridge, **W**...Exxon, Mobil, Shell, Boston Mkt, Taco Bell, Wendy's, Fairfield Inn,
156	VA 784, Potomac Mills, **E**...to Leesylvania SP, **W**...HOSPITAL, VETERINARIAN, Mobil, Exxon, Shell, Texaco, Blackeyed Pea, Burger King, Chinese Rest., Denny's, Domino's, El Charro's Mexican, McDonald's, Popeye's, Silver Diner, Wendy's, Day's Inn, K-Mart/auto, Marshall's, NTB, ToysRUs, U-Haul, auto parts, bank
155mm	**rest area both lanes, full(handicapped)facilities, phone, vending, picnic tables, litter barrels, petwalk, no trucks**
154mm	truck rest area/weigh sta both lanes
152	VA 234, Dumfries, to Manassas, **E**...Amoco/diesel/mart/24hr, Texaco/diesel, Cracker Barrel, Golden Corral, KFC, McDonald's, Taco Bell, Super 8, Meineke Muffler, Precision Tune, GreaseMonkey Lube, auto parts/repair, **W**...Citgo, Exxon, Shell, 7-11, Cervantes Rest., MontClair Rest., Econolodge, Holiday Inn Express, Quality Inn/rest., Weems-Botts Museum, camping
150	VA 619, to Triangle, Quantico, **E**...Amoco/diesel, Exxon, Shell/diesel/mart, Citgo/7-11, Burger King, Dunkin Donuts, Chinese Rest., Jim's Subs/pizza, McDonald's, True Grit Family Rest., Wendy's, Ramada Inn, US Inn, auto repair, to Marine Corps Base, **W**...Locust Shade Park, Prince William Forest Park, camping
148	to Quantico, **2 mi E**...Mobil/diesel, Spring Lake Motel, to Marine Corps Base
143b a	to US 1, VA 610, Aquia, **E**...PODIATRIST, Exxon/mart, Shell/Jerry's/diesel/mart/24hr, Texaco, Carlos O'Kelly's, Chinese Rest., DaVanzo's Italian, KFC, Little Caesar's, McDonald's, Pizza Hut, Ruby Tuesday, Shoney's, Subway, Wendy's, Day's Inn, Hampton Inn, Weyant's Motel, Shopper's Foods, Radio Shack, auto parts, bank, **W**...CHIROPRACTOR, Amoco/diesel/mart/wash, BP/Hardee's, Citgo/7-11/diesel, Exxon/diesel/mart, Texaco/diesel/mart, Baskin-Robbins, Burger King, Chinese Rest., Dunkin Donuts, Golden Corral, Kobe Japanese, McDonald's/playplace, Popeye's, Taco Bell, Wendy's, Comfort Inn, Country Inn, Super 8, Ames, Giant Food/drug, JiffyLube, Peebles, Wal-Mart/auto, Aquia Pines Camping, bank, carwash, laundry, tires
140	VA 630, Stafford, **E**...Amoco/mart, Mobil/mart, Texaco/diesel, Donut Shoppe, Family Café, Jerry's Rest., McDonald's, Ribhouse Rest., Bradshaw Motel(3mi), **W**...BP/diesel, Shell/diesel/mart
137mm	Potomac Creek
133b a	US 17 N, to Warrenton, **E**...Mobil/diesel/mart, RaceTrac/mart, Texaco/diesel, 7-11, Arby's, Howard Johnson, Motel 6, **W**...Amoco/diesel/mart, Citgo/mart, EastCoast/Subway/diesel/mart, Shell/diesel/mart, Texaco/diesel/mart, Burger King, Hardee's, McDonald's, Pizza Hut, Ponderosa, Taco Bell, Waffle House, Wendy's, ServiceTown Rest., Best Western/rest., Comfort Inn, Day's Inn, Holiday Inn, Ramada Ltd, Sleep Inn, Travelodge, Blue Beacon, Food Lion, XpressLube, auto repair
132.5mm	Rappahannock River
132mm	**rest area sb, full(handicapped)facilities, phone, picnic tables, litter barrels, petwalk, vending**

N

S

Virginia Interstate 95

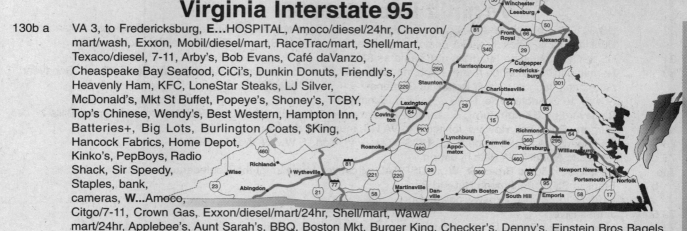

130b a   VA 3, to Fredericksburg, **E...**HOSPITAL, Amoco/diesel/24hr, Chevron/mart/wash, Exxon, Mobil/diesel/mart, RaceTrac/mart, Shell/mart, Texaco/diesel, 7-11, Arby's, Bob Evans, Café daVanzo, Cheaspeake Bay Seafood, CiCi's, Dunkin Donuts, Friendly's, Heavenly Ham, KFC, LoneStar Steaks, LJ Silver, McDonald's, Mkt St Buffet, Popeye's, Shoney's, TCBY, Top's Chinese, Wendy's, Best Western, Hampton Inn, Batteries+, Big Lots, Burlington Coats, $King, Hancock Fabrics, Home Depot, Kinko's, PepBoys, Radio Shack, Sir Speedy, Staples, bank, cameras, **W...**Amoco, Citgo/7-11, Crown Gas, Exxon/diesel/mart/24hr, Shell/mart, Wawa/mart/24hr, Applebee's, Aunt Sarah's, BBQ, Boston Mkt, Burger King, Checker's, Denny's, Einstein Bros Bagels, Fuddrucker's, IHOP, Italian Oven, Little Caesar's, McDonald's, Morrison's Cafeteria, NY Diner, Old Country Buffet, Outback Steaks, Pizza Hut, Prime Rib, Red Lobster, Ruby Tuesday, Subway, Taco Bell, Tia's TexMex, Waffle House, Best Western, Econolodge, Ramada Inn, Sheraton, Super 8, Belk, Best Buy, BJ's Whse, Borders Books, Circuit City, FashionBug, Giant Foods, GNC, Hecht's, JiffyLube, Kohl's, K-Mart, Lowe's Bldg Supply, Meineke Muffler, MensWearhouse, Merchants Tire, NTB, Office Depot, Old Navy, PetsMart, Pier 1, Precision Tune, Sears/auto, Service Merchandise, Shoppers Club, SportsAuthority, Target, Wal-Mart, bank, cinema, funpark, laundry, mall, tires

126   US 1, US 17 S, to Fredericksburg, **E...**Amoco, BP, Citgo/7-11, Exxon/Subway/mart, Mobil, Shell/mart, Texaco/diesel, Arby's, Aunt Sarah's, Denny's, El Charro Mexican, McDonald's, Pizza Hut, Ruby Tuesday, Shoney's, Taco Bell, Waffle House, Wendy's, Western Sizzlin, Day's Inn/rest., Econolodge, Fairfield Inn, Holiday Inn, Howard Johnson, Super 8, Buick/Pontiac/GMC, CarQuest, CVS Drug, Goodyear/auto, GreaseMonkey, Honda/Nissan/Mazda/VW, Olds/Cadillac, Merchants Tires, Midas Muffler, Rite Aid, RV Ctr, **W...**Exxon/Blimpie/mart, RaceTrac/diesel/mart/24hr, Aunt Sarah's, Burger King, Cracker Barrel, Damon's, Golden China, KFC, McDonald's, Pasta Fantastic, Comfort Inn, WyteStone Suites, Massaponax Outlet Ctr/famous brands, KOA(6mi), bank

118   VA 606, to Thornburg, **E...**Shell/diesel, to Stonewall Jackson Shrine, **W...**Amoco/mart, Citgo/diesel, Exxon/mart, Q-Stop/diesel/24hr, Burger King, McDonald's, Holiday Inn Express, Lamplighter Motel, KOA(7mi), to Lake Anna SP

110   VA 639, to Ladysmith, **E...**Shell/diesel, **W...**Citgo, Exxon/KrispyKreme/pizza/mart, Stone Hearth Rest., Speedy's Repair/lube

108mm   **rest area both lanes, full(handicapped)facilities, vending, phone, picnic tables, litter barrels, petwalk**

104   VA 207, to US 301, Bowling Green, **E...**Amoco/diesel/mart, Chevron/mart/atm, Exxon/Krispy Kreme/diesel/mart, Mr Fuel/diesel/mart, Petro/diesel/rest./24hr, Pilot/Subway/DQ/diesel/mart/24hr, Texaco/diesel/mart, McDonald's, Wendy's, Holiday Inn Express, Howard Johnson, Blue Beacon, Russell Stover Candies, to Ft AP Hill, **W...**Exxon/diesel/mart, Flying J/Country Mkt/diesel/LP/mart/24hr, Aunt Sarah's, Waffle House, Day's Inn/rest., Ramada Inn

98   VA 30, Doswell, **E...**to King's Dominion Funpark, Citgo/7-11, Texaco/diesel/mart/atm, Burger King, Denny's, Best Western, Econolodge, All American Camping, King's Dominion Camping, truckwash

92   VA 54, Ashland, **E...** Mobil, **W...**Amoco/mart, BP/TA/diesel/rest., Chevron, EastCoast/Blimpie/diesel/mart, Exxon/Subway/mart, Pilot/diesel/mart/24hr, Shell, Texaco/diesel/mart, 7-11, Arby's, Burger King, Chinese Rest., Country Pride Rest., Cracker Barrel, KFC, Little Caesar's, McDonald's, Omelet Shoppe, Pizza Hut, Ponderosa, Popeye's, Shoney's, Taco Bell, Wendy's, Budget Inn, Comfort Inn, Econolodge, HoJo's, Holiday Inn/rest., PalmLeaf Motel, Ramada Inn, Shady Grove Motel/rest., Super 8, Travelodge, Twin Oaks Motel, CarQuest, Family$, Food Lion, Radio Shack, Rite Aid, Trak Auto, Tuffy Auto, Ukrop's Foods, bank, cleaners

89   VA 802, to Lewistown Rd, **E...**TA/Pizza Hut/Taco Bell/diesel/café/24hr, Shell/mart, Americamps RV Camp, **W...**Cadillac Motel, Kosmo Village Camping

86   VA 656, to Atlee, Elmont, **E...**Citgo/diesel/mart, McDonald's, **W...**Amoco/diesel/mart, Mobil, Texaco/Subway/diesel/mart/atm, Applebee's, Burger King, Country Kitchen, Dairy Queen, FoodCourt, McDonald's, Ruby Tuesday, Sbarro's, Shoney's, Spaghetti Whse, Wendy's, Best Western/rest., Dillard's, Firestone/auto, Hecht's, JCPenney, RV Ctr, Sears/auto, Target, TireAmerica, airport, camping, cinema, mall

84b a   I-295 W, to I-64, E to Norfolk

83b a   VA 73, Parham Rd, **W...**CHIROPRACTOR, Amoco/24hr, EastCoast/diesel/mart, Exxon, Texaco/diesel/mart, Aunt Sarah's, Burger King, Denny's, El Paso Mexican, Hardee's, Little Caesar's, McDonald's, Papa Lou's Pizza, Subway, Waffle House, Wendy's, Alpine Motel, Broadway Motel, Econolodge, HoJo's, Holiday Inn Express, Northbrook Inn, Quality Inn, Shoney's Inn/suites, Sleep Inn, Big Lots, CVS Drug, Food Lion, Hannaford's Foods, Lowe's Bldg Supply, Uncle Cal's, Wal-Mart SuperCtr/24hr

Virginia Interstate 95

82 US 301, Chamberlayne Ave, **E**...Amoco, BP, Chevron, Exxon/diesel, Mobil/diesel, Texaco, Blimpie, Bojangles, Dunkin Donuts, Friendly's, McDonald's, Wendy's, Ramada Ltd, Super 8, Town Motel, Va Motel

81 US 1, Chamberlayne Ave(from nb), same as 82

80 Hermitage Rd, Lakeside Ave(from nb, no return), **W**...Amoco, EastCoast/Subway/mart, Lube'n Tune, Goodyear/auto, bank, laundry, Ginter Botanical Gardens

79 I-64 W, to Charlottesville, I-195 S, to U of Richmond

78 Boulevard(no EZ nb return), **E**...Texaco/diesel/mart, Holiday Inn, **W**...HOSPITAL, Amoco/diesel/mart, Bill's Burgers, Day's Inn, Howard Johnson, to VA HS, museum, stadium

76 Chamberlayne Ave, Belvidere, **E**...HOSPITAL, VA Union U, **W**...VA War Memorial

75 I-64 E, to Norfolk, VA Beach, airport

74c US 33, US 250, to Broad St, **W**...HOSPITAL, st capitol, Museum of the Confederacy

  b Franklin St, **E**...Richmond Nat Bfd Park

  a I-195 N, to Powhite Expswy, downtown

73.5mm James River

73 Maury St, to US 60, US 360, industrial area

69 VA 161, Bells Rd, **E**...Port of Richmond, **W**...Exxon/diesel/mart/24hr, Texaco/diesel/mart, Hardee's, McDonald's, Holiday Inn, Red Roof Inn

67 VA 150, to Chippenham Pkwy, Falling Creek, **W**...Amoco, BP, Chevron/diesel, RaceTrac/mart, Texaco/diesel, Blimpie, Burger King, Hardee's, Wendy's

64 VA 613, to Willis Rd, **E**...Exxon, Shell/wash, Arby's, Aunt Sarah's, Waffle House, Whataburger, Econolodge, Ramada, Drewy's Bluff Bfd, **W**...Citgo/7-11, Crown/mart, Mobil/diesel, Texaco/diesel/mart/rest., Dairy Queen, McDonald's, Economy House Motel, Sleep Inn, VIP Inn, flea mkt

62 VA 288 N, to Chesterfield, Powhite Pkwy, to airport

61 VA 10, Chester, **E**...HOSPITAL, RaceTrac/mart/24hr, Hardee's, Imperial Seafood, Comfort Inn, Hampton Inn, Holiday Inn Express, to James River Plantations, City Point NHS, **W**...Amoco, Citgo/7-11/diesel/mart, Crown/diesel/mart, EastCoast/diesel/mart, Exxon/diesel/mart, Texaco/mart/wash, Applebee's, Burger King, Capt D's, Chinese Rest., Cracker Barrel, Denny's, Friendly's, Hardee's, KFC, McDonald's, Pizza Hut, Shoney's, Subway, Taco Bell, Waffle House, Wendy's, Western Sizzlin, Day's Inn, Fairfield Inn, Howard Johnson, Super 8, Chevrolet, CVS Drug, Rite Aid, Winn-Dixie/deli, K-Mart/drugs, Lowe's Bldg Supply, Target, Ukrops Foods, Roadrunner Camping(2mi), Aamco, Trak Auto, drugs, laundry, to Pocahontas SP

58 VA 746, **E**...Texaco/diesel, Travelodge/rest., **W**...Amoco/Bullets/mart, Chevron/diesel, Interstate Inn/rest., Day's Inn, **1 mi W**...AGIP Gas/diesel/mart, Citgo/7-11/diesel, Little Caesar's, Pizza/subs, Patrick Henry Motel, hardware, laundry

54 VA 144, Temple Ave, Colonial Heights, to Ft Lee, Hopewell, **E**...MEDICAL CARE, DENTIST, Amoco, Chevron/KrispyKreme/diesel/mart, Citgo/mart/wash, Crown/mart, Exxon/mart, Applebee's, Arby's, Burger King, Golden Corral, La Carreta Mexican, LoneStar Steaks, McDonald's, Morrison's Cafeteria, Old Country Buffet, Red Lobster, Ruby Tuesday, Sagebrush Steaks, Subway, Taco Bell, Wendy's, Comfort Inn, Hampton Inn, Innkeeper, Belk, Books-A-Million, Chevrolet/Cadillac/Buick/Nissan, Circuit City, Cloth World, Dillard's, FashionBug, Hecht's, JC Penney, Jo-Ann Crafts, K-Mart, Marshall's, PetsMart, Pier 1, Sam's Club, Sears/auto, Staples, Target, TireAmerica, Wal-Mart SuperCtr, mall, **W**...Texaco/mart/wash, Hardee's, to VSU

53 S Park Blvd(from nb), **E**...same as 54

52.5mm Appomattox River

52 Washington St, Wythe St, **E**...Amoco, BP, Crown/mart, Texaco/diesel/mart, Dairy Queen, Steak&Ale, Econolodge, Holiday Inn, Howard Johnson, King Motel, Knight's Inn, Ramada, Royal Inn, Star Motel, Super 8, Precision Tune, Petersburg Nat Bfd, **W**...HOSPITAL, Shell/mart, Aunt Sarah's, Best Western, Travelodge

51 I-85 S, to South Hill, US 460 W

50a b c d US 301, US 460 E, to Crater Rd, County Dr, **E**...HOSPITAL, Citgo/7-11, Hardee's, McDonald's, Wendy's, Country Side Inn, Flagship Inn

48b a Wagner Rd, **E**...Exxon, McDonald's, **W**...Amoco, Chevron/diesel, Citgo, Exxon, Texaco, Arby's, BBQ, Burger King, Dairy Queen, Hardee's, Honey B's Rest., KFC, McDonald's, Pizza Hut, Ponderosa, Shoney's, Subway, Taco Bell, Crater Inn, Dodge, Food Lion, Ford/Lincoln/Mercury, Pennzoil, PepBoys, auto parts

47 VA 629, to Rives Rd, **1-2 mi E**...McDonald's, **W**...Citgo, Texaco, Heritage Motel, **1-2 mi W**...Crater Inn, LaSalle Motel, Softball Hall of Fame Museum, same as 48 on US 301

46 I-295 N(exits left from sb), to Washington

45 US 301, **E**...Texaco/diesel, **W**...Exxon, Steven-Kent Rest., Comfort Inn, Continental Motel, Day's Inn/rest., Hampton Inn, Holiday Inn Express, Quality Inn

41 US 301, VA 35, VA 156, **E**...Chevron/diesel/mart, Econolodge, KOA(1mi), **W**...Rose Garden Inn, Super 8

40mm weigh sta both lanes

37 US 301, Carson, **W**...Amoco/diesel, Shell/mart/RV parking

36mm **rest area nb, full(handicapped)facilities, phone, vending, picnic tables, litter barrel, petwalk**

N

S

Virginia Interstate 95

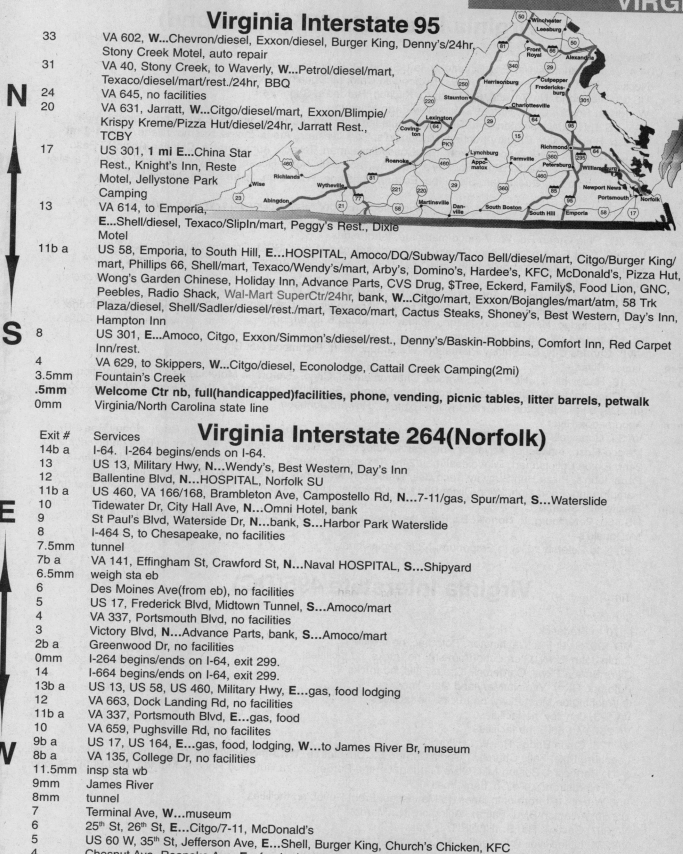

33	VA 602, **W**...Chevron/diesel, Exxon/diesel, Burger King, Denny's/24hr, Stony Creek Motel, auto repair
31	VA 40, Stony Creek, to Waverly, **W**...Petrol/diesel/mart, Texaco/diesel/mart/rest./24hr, BBQ
24	VA 645, no facilities
20	VA 631, Jarratt, **W**...Citgo/diesel/mart, Exxon/Blimpie/Krispy Kreme/Pizza Hut/diesel/24hr, Jarratt Rest., TCBY
17	US 301, **1 mi E**...China Star Rest., Knight's Inn, Reste Motel, Jellystone Park Camping
13	VA 614, to Emporia, **E**...Shell/diesel, Texaco/SlipIn/mart, Peggy's Rest., Dixie Motel
11b a	US 58, Emporia, to South Hill, **E**...HOSPITAL, Amoco/DQ/Subway/Taco Bell/diesel/mart, Citgo/Burger King/mart, Phillips 66, Shell/mart, Texaco/Wendy's/mart, Arby's, Domino's, Hardee's, KFC, McDonald's, Pizza Hut, Wong's Garden Chinese, Holiday Inn, Advance Parts, CVS Drug, $Tree, Eckerd, Family$, Food Lion, GNC, Peebles, Radio Shack, Wal-Mart SuperCtr/24hr, bank, **W**...Citgo/mart, Exxon/Bojangles/mart/atm, 58 Trk Plaza/diesel, Shell/Sadler/diesel/rest./mart, Texaco/mart, Cactus Steaks, Shoney's, Best Western, Day's Inn, Hampton Inn
8	US 301, **E**...Amoco, Citgo, Exxon/Simmon's/diesel/rest., Denny's/Baskin-Robbins, Comfort Inn, Red Carpet Inn/rest.
4	VA 629, to Skippers, **W**...Citgo/diesel, Econolodge, Cattail Creek Camping(2mi)
3.5mm	Fountain's Creek
.5mm	**Welcome Ctr nb, full(handicapped)facilities, phone, vending, picnic tables, litter barrels, petwalk**
0mm	Virginia/North Carolina state line

Virginia Interstate 264(Norfolk)

Exit #	Services
14b a	I-64. I-264 begins/ends on I-64.
13	US 13, Military Hwy, **N**...Wendy's, Best Western, Day's Inn
12	Ballentine Blvd, **N**...HOSPITAL, Norfolk SU
11b a	US 460, VA 166/168, Brambleton Ave, Campostello Rd, **N**...7-11/gas, Spur/mart, **S**...Waterslide
10	Tidewater Dr, City Hall Ave, **N**...Omni Hotel, bank
9	St Paul's Blvd, Waterside Dr, **N**...bank, **S**...Harbor Park Waterslide
8	I-464 S, to Chesapeake, no facilities
7.5mm	tunnel
7b a	VA 141, Effingham St, Crawford St, **N**...Naval HOSPITAL, **S**...Shipyard
6.5mm	weigh sta eb
6	Des Moines Ave(from eb), no facilities
5	US 17, Frederick Blvd, Midtown Tunnel, **S**...Amoco/mart
4	VA 337, Portsmouth Blvd, no facilities
3	Victory Blvd, **N**...Advance Parts, bank, **S**...Amoco/mart
2b a	Greenwood Dr, no facilities
0mm	I-264 begins/ends on I-64, exit 299.
14	I-664 begins/ends on I-64, exit 299.
13b a	US 13, US 58, US 460, Military Hwy, **E**...gas, food lodging
12	VA 663, Dock Landing Rd, no facilities
11b a	VA 337, Portsmouth Blvd, **E**...gas, food
10	VA 659, Pughsville Rd, no facilites
9b a	US 17, US 164, **E**...gas, food, lodging, **W**...to James River Br, museum
8b a	VA 135, College Dr, no facilities
11.5mm	insp sta wb
9mm	James River
8mm	tunnel
7	Terminal Ave, **W**...museum
6	25th St, 26th St, **E**...Citgo/7-11, McDonald's
5	US 60 W, 35th St, Jefferson Ave, **E**...Shell, Burger King, Church's Chicken, KFC
4	Chesnut Ave, Roanoke Ave, **E**...food mkt
3	Aberdeen Rd, **E**...Citgo/diesel/mart, **W**...Hardee's, McDonald's, Wendy's, museum
2	Powhatan Pkwy, no facilities
1b a	I-64, W to Richmond, E to Norfolk. I-664 begins/ends on I-64.

N

S

E

W

Emporia

Portsmouth

507

Virginia Interstate 295(Richmond)

N

Mechanicsville

Exit #	Services
53b a	I-64, W to Charlottesville, E to Richmond, to US 250, I-295 begins/ends.
51b a	Nuckols Rd, **1 mi N**...Amoco/mart, **S**...Texaco/Mkt Café, McDonald's, Wendy's
49b a	US 33, Richmond, **N**...Texaco, **S**...Debbie's Kitchen, Hardee's, Shoney's
45b a	Woodman Rd, **S**...Citgo, Hardee's, Meadow Farm Museum
43	I-95, US 1, N to Washington, S to Richmond(exits left from nb), **services on US 1, N**...Amoco/diesel, Mobil, Texaco, Burger King, Country Kitchen, McDonald's, Best Western, AmeriCamps, Kosmo Village, mall, **1-2 mi S**...CHIROPRACTOR, Amoco, Exxon, EastCoast/diesel/mart, Texaco, Aunt Sarah's, Chinese Rest., El Paso Mexican, Hardee's, McDonald's, Shoney's, Subway, Taco Bell, Waffle House, Wendy's, Broadway Motel, Cavalier Motel, Econolodge, Quality Inn, Sleep Inn, Eckerd, Firestone, Food Lion, Lowe's Bldg Supply, Wal-Mart SuperCtr/24hr
41b a	US 301, VA 2, **E**...Amoco/diesel, Citgo, Texaco/diesel, Burger King, McDonald's, **W**...Exxon/diesel, Friendly's, McDonald's, HoJo Inn, Ramada Ltd, Super 8, Virginia Inn
38b a	VA 627, Pole Green Rd, **W**...Amoco/mart/café, fairgrounds
37b a	US 360, **1 mi E**...Amoco, Citgo, Crown, Texaco/diesel/mart, Arby's, BBQ, Cracker Barrel, Mexico Rest., Shoney's, Taco Bell, Wendy's, InnKeeper, Wal-Mart SuperCtr, **W**...Amoco, Citgo/7-11, Mobil/diesel, NAPA, bank, drugs, to Mechanicsville
34b a	VA 615, Creighton Rd, **E**...Amoco/diesel/mart, Citgo/7-11, **5 mi W**...Citgo, McDonald's
31b a	VA 156, **E**...Citgo/diesel, to Cold Harbor, **4 mi W**...Amoco/diesel, Chevron, Texaco, Bojangles, Hardee's, Day's Inn, Econolodge, Hampton Inn, Hilton, Holiday Inn, Motel 6, to airport
28	US 60, to I-64, **W**...airport, museum
22b a	VA 5, Charles City, **E**...Shirley Plantation, **W**...Amoco/mart, Richmond Nat Bfd
18mm	James River
15b a	VA 10, Hopewell, **E**...HOSPITAL, Amoco, Chevron/diesel, Citgo, Evergreen Motel, James River Plantations, **W**...Chevron/diesel, Citgo/diesel, Burger King, Denny's, Friendly's, McDonald's, Waffle House, Wendy's, Comfort Inn, Day's Inn, Hampton Inn, Holiday Inn Express, Howard Johnson
13mm	Appomattox River
9b a	VA 36, Hopewell, **E**...DENTIST, VETERINARIAN, Chevron/mart, Petrol/mart, Chinese Rest., Honey Bee's Rest., Mexico Rest., Innkeeper, Advance Auto Parts, AutoZone, $ General, carwash, **W**...Amoco/BreezIn/diesel/mart/24hr, Exxon/Bullets/mart, Pilot/diesel/mart/24hr, Texaco, Checker's, Denny's, Japanese Steaks, McDonald's, Papa John's, Pizza Hut, Subway, Taco Bell, Waffle House, Wendy's, Western Sizzlin, Comfort Inn, Day's Inn, Hampton Inn, Food Lion, Winn-Dixie, Rite Aid, Chevrolet, U-Haul, US Army Museum, Valvoline, bank, laundry
5.5mm	Blackwater Swamp
3b a	US 460, Petersburg, to Norfolk, **E**...EastCoast/diesel/mart, Dunkin Donuts, Subway, **W**...Exxon, Hardee's, McDonald's
1	I-95, N to Petersburg, S to Emporium, I-295 begins/ends.

Hopewell

S

Virginia Interstate 495(DC)

N

DC Area

Exit #	Services
38	I-270 to Frederick
39	MD 190, River Rd, Washington, Potomac, no facilities
40	Cabin John Pkwy, Glen Echo(from sb), no trucks, no facilities
41	Clara Barton Pkwy, Carderock, Great Falls, no trucks
15mm	Potomac River, Virginia/Maryland state line
14	G Washington Mem Pkwy, no trucks, no facilities
13	VA 193, Langley, no facilities
12b a	VA 267 W, I-66 E, no facilities
11b a	VA 123, Chain Bridge Rd, **W**...Hilton
10	Leesburg Pike, Falls Church, Tyson's Corner, **E**...Doubletree, **W**...Amoco/diesel, Crown Gas, Exxon, Mobil, Shell, &-11, Bertucci's, Boston Mkt, Olive Garden, On-the-Border, Pizza Hut, Roy Rogers, Marriott, Bed Bath&Beyond, Ford, Nordstrom's, PetCo, bank, mall
9	I-66 W(exits left from both lanes, to Manassas, Front Royal, no facilities
8	US 50, Arlington Blvd, Fairfax, Arlington, **N**...Marriott
7	VA 657, Gallows Rd, **S**...HOSPITAL, gas
6	VA 236, Little River Tpk, Fairfax, **N**...gas/diesel, food
5	VA 620, Braddock Rd, Ctr for the Arts, Patriot Ctr, Geo Mason U, **S**...Safeway
4c b a	I-95 S, I-395 N, I-95 N. I-495 & I-95 run together.

S

Washington Interstate 5

Exit #	Services

277mm USA/Canada Border, Washington state line, customs

276 WA 548 S, Blaine, **E...**Blaine Gas/mart, Border Gas, Exxon/diesel/mart, MP Gas/mart/atm, 76/mart/24hr, USA/diesel/mart/24hr, Denny's, Pasa del Norte Mexican, Northwoods Motel, Ammex Duty Free, MailBoxes+, to Peace Arch SP, **W...**info, Chevron/repair, Exxon/mart, Shell/mart, Contos Pizza/pasta, Harbor Café, Pizza Factory, Subway, WheelHouse Grill, Anchor Inn, Motel Internacional, $Store, antiques, bank, hardware

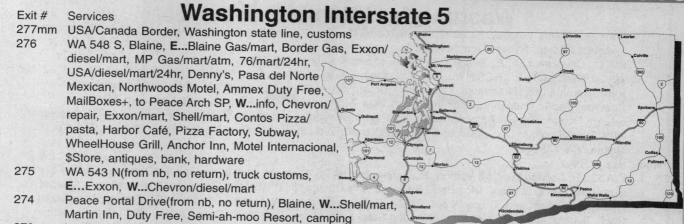

275 WA 543 N(from nb, no return), truck customs, **E...**Exxon, **W...**Chevron/diesel/mart

274 Peace Portal Drive(from nb, no return), Blaine, **W...**Shell/mart, Martin Inn, Duty Free, Semi-ah-moo Resort, camping

270 Lynden, Birch Bay, **W...**Texaco/Subway/Taco Bell/diesel/mart/atm, Semi-ah-moo Resort, Peace Arch Outlet/famous brands, SeaBreeze RV Park(5mi)

269mm Welcome Ctr sb, full(handicapped)facilities, info, phone, picnic tables, litter barrels, petwalk, vending

267mm rest area nb, full(handicapped)facilities, info, phone, picnic tables, litter barrels, petwalk, vending

266 WA 548 N, Custer, Grandview Rd, **W...**Arco/mart/24hr, Birch Bay SP

263 Portal Way, **E...**DENTIST, Pacific Pride/diesel, Texaco/diesel/mart, Cedars RV Park

263mm Nooksack River

262 Main St, Ferndale, **E...**VETERINARIAN, BP/diesel/mart, Chevron/diesel/rest./24hr, Denny's, McDonald's, Best Western, Super 8, RV Park, U-Haul, Whatcum Tires, bingo, **W...**Cenex/mart, Chevron/mart, Exxon/diesel/mart/LP, Texaco/A&W/diesel/mart, Bob's Burgers, Costcutter FoodCourt, Dairy Queen, Grant's Drive-In, Scottish Lodge, Hagan's Food/drug, NAPA, golf

260 Slater Rd, Lummi Island, **E...**Arco/mart/24hr, El Monte RV, antiques, **5 mi W...**BP, Texaco/diesel, Eagle Haven RV Park, Lummi Casino/café(10mi), Lummi Ind Res

258 Bakerview Rd, **W...**Arco/mart/24hr, BP/mart/24hr, Chevron, Exxon/diesel/mart/24hr, Mikono's Greek Rest., Hampton Inn, Shamrock Motel, airport, st patrol

257 Northwest Ave, **E...**Chevrolet/Olds/Cadillac, **W...**Texaco/mart/wash

256b Bellis Fair Mall Pkwy, **E...**Sears, Target, mall

a WA 539 N, Meridian St, **E...**DENTIST, CHIROPRACTOR, BP, Chevron/mart/wash, Exxon/diesel, Shell, Texaco/diesel/mart/atm, Burger King, China Palace, Dairy Queen, Denny's, Godfather's, Izzy's Pizza, Kowloon Garden, McDonald's, McHale's Rest., Mi Mexico Rest., Mitzel's Kitchen, Old Country Buffet, Olive Garden, Red Robin Rest., Shari's/24hr, Sizzler, Taco Time, Wendy's, Best Western, Comfort Inn, Day's Inn, Holiday Inn Express, Quality Inn, Al's Parts, Barnes&Noble, Bon-Marche, Circuit City, Costco Whse, Costcutter Foods, FutureShop, HomeBase, Home Depot, House of Fabrics, JC Penney, JiffyLube, MailBoxes Etc, Mervyn's, Midas Muffler, Nordstrom's, Office Depot, PetsMart, Pier 1, Rite Aid, Ross, RiteWay Auto, Safeway, Schuck's Parts, Schwab Tires, Sears, Target, TJ Maxx, Wal-Mart/auto, bank, cinema, mall, st patrol, to Nooksack Ind Res, **W...**Elmer's Rest., Rodeway Inn, Travelers Inn

255 WA 542 E, Sunset Dr, Bellingham, **E...**Chevron/diesel/24hr, Exxon, 7-11, Jack-in-the-Box, RoundTable Pizza, Scotty B's Bagels, Taco Bell, White Spot Rest., Jo-Ann Crafts, K-Mart/auto, cinema, to Mt Baker, **W...**HOSPITAL

254 Iowa St, State St, Bellingham, **E...**BP/mart/24hr, Exxon/diesel, Texaco, Mitsubishi, Nissan, RV Sales/service, tires, **W...**CHIROPRACTOR, Chevron/mart, 76, Dairy Queen, McDonald's, Skipper's, CarQuest, Chrysler/Dodge, Ford/Lincoln/Mercury, Isuzu, NAPA, Schuck's Parts, Schwab Tires, Subaru, hardware

253 Lakeway Dr, Bellingham, **E...**CHIROPRACTOR, DENTIST, USA/gas, 7-11, Horseshoe Café, Little Caesar's, Mexican Rest., Pizza Hut, Ricky's Rest., Best Western, Shangrila Motel, Valu Inn, Discount Tire, Fred Meyer, bank, **W...**Taco Time, same as 252

252 Samish Way, Bellingham, **E...**Evergreen Motel, cinema, same as 253, **W...**Arco/mart/24hr, BP/mart, Chevron, 76, Texaco/diesel/24hr, A&W, Arby's, Bamboo Rest., Black Angus, Burger King, Christo's Rest., Denny's, Fryday's Rest., Godfather's, IHOP, McDonald's, Rib'n Reef, RoundTable Pizza, Subway, Aloha Motel, Bell Northwoods Motel, Coachman Motel, Key Motel, Motel 6, Park Motel, Ramada Inn, Travelodge, Audi, Chrysler/Jeep, MasterLube, Pennzoil, Rite Aid, VW, bank

250 WA 11 S, Chuckanut Dr, Bellingham, Fairhaven Hist Dist, **W...**Arco/mart/24hr, Chevron, Dos Padres Mexican, Tony's Coffee House, Win's Drive-In, Bullie's Rest., Larrabee SP, to Alaska Ferry, camping

246 N Lake Samish, **W...**Texaco/diesel/mart, Lake Padden RA, RV camping

242 Nulle Rd, S Lake Samish, no facilities

240 Alger, **E...**Texaco/diesel/LP/RV dump, **W...**Alger Grille, Sudden Valley Motel/RV Park

238mm rest area both lanes, full(handicapped)facilities, phone, picnic tables, litter barrels, vending, petwalk

236 Bow Hill Rd, **E...**Harrah's Casino/rest., **W...**fish mkt

235mm weigh sta sb

234mm Samish River

Washington Interstate 5

Mt Vernon

232	Cook Rd, Sedro, Woolley, **E**...Shell/Blimpie/mart/24hr, Texaco/diesel/mart, Burger King(4mi), Dairy Queen, Iron Skillet Rest.(4mi), 3 Rivers Inn(5mi), KOA
231	WA 11, Chuckanut Dr, **E**...Skagit RV/marine, st patrol, to Larrabee SP
230	WA 20, Burlington, **E**...HOSPITAL, BP, Chevron/mart, Texaco/mart, Beary Patch Rest., Red Robin Rest., Teriyaki Rest., Bon-Marche, Sears/auto, OfficeMax, Target, VW, mall, to N Cascades NP, **W**...Arco/diesel/mart/24hr, Chevron, Pacific Pride/mart, Jack-in-the-Box/24hr, McDonald's, Harley-Davidson, to San Juan Ferry
229	George Hopper Rd, **E**...Arco/mart/24hr, Burlington Store/famous brands, Costcutter Foods, **W**...USA/diesel/mart/24hr, Buick/Pontiac, Chrysler/Jeep/Plymouth/Dodge, Ford/Lincoln/Mercury, Mazda/Honda, Nissan, Subaru/Toyota, VW, RV Ctr
228mm	Skagit River
227	WA 538 E, College Way, Mt Vernon, **E**...Cenex, Exxon, 76, Denny's, Jack-in-the Box, McDonald's, Shakey's Pizza, Skipper's, Subway, Taco Bell/24hr, Winchell's Donuts, Best Western, Day's Inn, Ace Hardware, Albertson's, Goodyear, FabricLand, Office Depot, PetCo, Rite Aid, Safeway, Wal-Mart/auto, crafts, **W**...Chevron/mart/wash, Shell, Texaco, Arby's, Burger King, Buzz Inn Steaks, Cranberry Tree Rest., Dairy Queen, KFC, Mitzel's Kitchen, RoundTable Pizza, Royal Fork Buffet, Taco Time, Best Western, Comfort Inn, Travelodge, Tulip Inn, Chevrolet, Firestone, Honda Motorcycles, JiffyLube, Lowe's Bldg Supply, Valley RV, RV camp
226	WA 536 W, Kincaid St, City Ctr, **E**...HOSPITAL, **W**...Pacific Pride/diesel, CarQuest, NAPA, RV camping, Skagit River Brewing Co, bank
225	Anderson Rd, **E**...BP/mart, CFN/diesel, **W**...Chevron/mart
224	WA 99 S(from nb, no return), S Mt Vernon, **E**...gas/diesel, food
221	WA 534 E, Conway, Lake McMurray, **E**...Texaco/diesel/mart/24hr, **W**...Channel Lodge Rest., Ridgeway B&B, Valentine House B&B, Wild Iris B&B, Blake's RV Park/marina
218	Starbird Rd, **W**...Hillside Motel
215	300th NW, **W**...gas/mart
212	WA 532 W, Stanwood, Bryant, **W**...Shell/mart/24hr, Texaco/diesel/mart, **4 mi W**...Burger King, Dairy Queen, McDonald's, Hagan Food/drug/24hr, Camano Island SP(19mi)
210	236th NE, no facilities
209mm	Stillaguamish River
208	WA 530, Silvana, Arlington, **E**...HOSPITAL, Arco/mart/24hr, BP/mart, Chevron/mart/24hr, Shell, Texaco/diesel/mart/24hr, Denny's, O'Brien Manor Rest., Arlington Motel, Weller's Chalet Inn, to N Cascades Hwy, **W**...Exxon/diesel/LP/24hr
208mm	**rest area both lanes, full(handicapped)facilities, phone, picnic tables, litter barrels, coffee, vending, RV dump, petwalk**
206	WA 531, Lakewood, **E**...CHIROPRACTOR, VETERINARIAN, Arco/mart/24hr, 76/diesel/LP/RV dump, Texaco/mart/24hr, 7-11/Citgo, Alfy's Pizza, Buzz Inn Steaks, Jack-in-the-Box, McDonald's, Taco Time, Take'n Bake Pizza, AllCraft Store, Chrysler/Jeep, Lowe's Bldg Supply, Max Foods, Pontiac/Olds/Buick/GMC, Rite Aid, Safeway/24hr, bank, cleaners, **W**...Chevron/mart/24hr, Petosa's Rest., Smokey Point Motor Inn, Cedar Grove Shores RV(6mi), to Wenburg SP
202	116th NE, **E**...Texaco/diesel/mart/24hr, **W**...Chevron/diesel/mart/24hr, st patrol
200	88th St NE, Quilcida Way, no facilities
199	WA 528 E, Marysville, Tulalip, **E**...Arco/mart/24hr, BP/mart, Chevron/mart/24hr, Texaco/Subway/diesel/mart, Burger King, Dairy Queen, Jack-in-the-Box, Las Margaritas Mexican, Royal Fork Buffet, Village Motel/rest., Albertson's, FastLube, JC Penney, Lamont's, Rite Aid, Tires4Less, **W**...76, Arby's, Golden Corral, McDonald's, Taco Time, Wendy's, Best Western/rest., Fairfield Inn, Holiday Inn Express, Chevrolet/Subaru, RV SuperMall, to Tulalip Indian Res
198	Port of Everett(from sb), Steamboat Slough, st patrol
195mm	Snohomish River
195	Port of Everett(from nb), Marine View Dr, no facilities
194	US 2 E, Everett Ave, City Ctr, **W**...Texaco/diesel/mart/atm, Schwab Tires
193	WA 529, Pacific Ave, **W**...HOSPITAL, BP/mart, Chevron/mart, Denny's, Roaster Rest., Best Western, Everett Pacific Hotel, Howard Johnson, Marina Village Inn(3mi)
192	Broadway(exits left from nb), Evergreen Way, City Ctr, **W**...Arco/mart/24hr, BP, Chevron, Texaco/mart, Alfy's Pizza, Jack-in-the-Box, King's Table Rest., McDonald's, O'Donell's Café, Day's Inn, Royal Motor Inn, Travelodge, Welcome Motel, Ford

Everett

189	WA 526 W, WA 527, Everett Mall Way, Everett, **E**...CHIROPRACTOR, Arco/mart/24hr, Texaco/diesel, Burger King/24hr, Buzz Inn Steaks, McDonald's, Orchid Thai Cuisine, Travelodge, Costco Whse, HomeBase, mall, **1 mi W on Evergreen Way**...Chevron, 7-11, Shell, Texaco/diesel/mart, Chinese Rest., Denny's, Village Inn Pancakes, Taco Bell, Taco Time, Vietnamese Rest., Cherry Motel, Comfort Inn, Extended Stay America, Motel 6, Sunrise Inn, Pontiac/Olds/Buick, Dodge/Jeep, Discount Tires, Firestone/auto, K-Mart, Fred Meyer, House of Fabrics, hardware, mall
188mm	**rest area/weigh sta sb, full(handicapped)facilities, info, phone, picnic tables, litter barrels, coffee, RV dump, weigh sta nb**

N

S

Washington Interstate 5

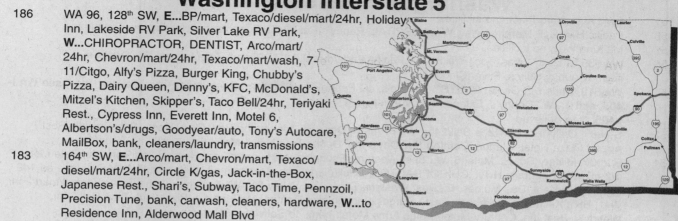

186 WA 96, 128th SW, **E...**BP/mart, Texaco/diesel/mart/24hr, Holiday Inn, Lakeside RV Park, Silver Lake RV Park, **W...**CHIROPRACTOR, DENTIST, Arco/mart/24hr, Chevron/mart/24hr, Texaco/mart/wash, 7-11/Citgo, Alfy's Pizza, Burger King, Chubby's Pizza, Dairy Queen, Denny's, KFC, McDonald's, Mitzel's Kitchen, Skipper's, Taco Bell/24hr, Teriyaki Rest., Cypress Inn, Everett Inn, Motel 6, Albertson's/drugs, Goodyear/auto, Tony's Autocare, MailBox, bank, cleaners/laundry, transmissions

183 164th SW, **E...**Arco/mart, Chevron/mart, Texaco/diesel/mart/24hr, Circle K/gas, Jack-in-the-Box, Japanese Rest., Shari's, Subway, Taco Time, Pennzoil, Precision Tune, bank, carwash, cleaners, hardware, **W...**to Residence Inn, Alderwood Mall Blvd

182 **E...**I-405 S to Bellevue, **W...**WA 525, Alderwood Mall Blvd, to Alderwood Mall, Arco/mart/24hr, Japanese Rest., The Keg Steaks/seafood, TCBY, Residence Inn, Nordstrom's, Sears/auto, Target, TuneTown Repair, Evan's Tires, Lenscrafters, MailCtr/UPS, cinema

181 44th Ave W, to WA 524, Lynnwood, **E...**BP, Shell/mart, Embassy Suites, Hampton Inn, Circuit City, Linens'n Things, Old Navy, **W...**MEDICAL CARE, BP/diesel/wash, Chevron/mart/24hr, 7-11, Texaco/service, Applebee's, Black Angus, Buca Italian, Burger King, Chevy's FreshMex, Country Harvest Rest., Denny's, El Torito's, Evergreen Donuts, IHOP, Jack-in-the-Box, Japanese Rest., Kenny Rogers, KFC, McDonald's, Old Country Buffet, Olive Garden, Red Lobster, Starbucks, Subway, Tony Roma, Wendy's, Best Western/rest., Courtyard, Holiday Inn Express, FabricLand, Firestone/auto, Fred Meyer, Gart Sports, Goodyear/auto, Kinko's/24hr, MailMart, Precision Tune, Radio Shack, bank, cleaners, mall

179 220th SW, Mountlake Terrace, **W...**CHIROPRACTOR, VETERINARIAN, Arco/mart, Shell, Texaco/diesel/mart, 7-11, Azteca Mexican, Chinese Rest., Countryside Donuts, Japanese Rest., Port of Subs, Subway, bookstore, cleaners/laundry, drugs

178 236th St SW(from nb), Mountlake Terrace, no facilities

177 WA 104, Edmonds, **E...**CHIROPRACTOR, Chevron/24hr, Texaco/Taco Bell/diesel/mart, Buzz Inn Steaks, Canyons Rest., McDonald's, Subway, Homestead Village Suites, CompUSA, Qlube, Schuck's Parts, Thriftway Foods, bank, cinema, cleaners, **1-2 mi W on WA 99...**BP, Shell/mart, Texaco/mart, China Clipper Rest., Denny's, Godfather's, Big 5 Sports, Burlington Coats, Computer City, Costco Whse, GNC, GreaseMonkey, Home Depot, Kinko's, OfficeMax, QFC Foods, Radio Shack, Schwab Tires, VW

176 NE 175th St, Aurora Ave N, to Shoreline

175 WA 523, NE 145th, 5th Ave NE, **1 mi W on WA 99...**Las Margaritas Mexican, Shari's, Taco Time, Wendy's, $Store, QLube, Schuck's Parts

174 NE 130th, Roosevelt Way, **1 mi W on WA 99...**Burger King, KFC, Best Western, Albertson's, Buick/GMC, Firestone, K-Mart, Midas Muffler, Office Depot, PetsMart, Rite Aid, Sam's Club

173 1st Ave NE, Northgate Way, **E...**CHIROPRACTOR, EYECARE, Arco/mart/24hr, BP, Azteca Mexican, Baskin-Robbins, KrazyBird Grill, Marie Callender's, Pancake Haus, Red Robin Rest., Seattle Crab Co, Subway, Taco Time, Tony Roma, B&B Parts, Bon-Marche, CamerasWest, Discount Tire, JC Penney, JiffyLube, Kinko's, Longs Drug, MailBoxes Etc, MensWearhouse, Nordstrom's, Northgate MailBox, Pacific Fabrics, QFC Foods, TJ Maxx, bank, cinema, mall, **W...**BP/mart, Chevron/mart, Texaco/diesel, 7-11, Arby's, Barnaby's Rest., Denny's, McDonald's, Ramada Inn, Mr Clutch, cleaners

172 N 85th, Aurora Ave, **1 mi W on Aurora...**Arco, BP/diesel/mart, Jack-in-the-Box, Meineke Muffler, Precision Tune

171 WA 522, Lake City Way, Bothell, no facilities

170 Ravenna Blvd, **E...**MEDICAL CARE, Arco, Texaco/diesel/mart, QFC Foods

169 NE 45th, NE 50th, **E...**HOSPITAL, 76, University Inn, bank, cinema, U of WA, **W...**University Plaza Hotel, to Seattle Pacific U, zoo

168b WA 520, to Bellevue, no facilities

a Lakeview Blvd, downtown

167 Mercer St(exits left from nb), Fairview Ave, Seattle Ctr, **W...**Shell, bank

166 Olive Way, Stewart St, **E...**HOSPITAL, **W...**Honda

165a Seneca St(exits left from nb), James St, **E...**HOSPITAL, **W...**Renaissance Hotel

b Union St, downtown

164b 4th Ave S, **1 mi W...**same as 163

a I-90 E, to Spokane, no facilities

163 6th Ave, S Spokane St, W Seattle Br, Columbian Way, **1 mi W on 4th Ave S...**CHIROPRACTOR, Arco, BP/mart, Texaco/diesel/mart, Arby's, Burger King/24hr, Denny's, McDonald's, Subway, Taco Bell, OfficeMax, Sears, auto repair, transmissions

162 Courson Ave, Michigan St(exits left from nb), no facilities

N

S

Seattle

Washington Interstate 5

S e a t t l e

161   Swift Ave, Albro Place, no facilities

158   Pacific Hwy S, E Marginal Way, **W...**HOSPITAL, Randy's Rest/24hr, Econolodge, Holiday Inn, Travelodge

157   ML King Way, no facilities

156   WA 539 N, Long Acres, Tuck Willow, **E...**Pacific Pride/diesel, **W...**BP/mart, Texaco/mart, 76, Denny's, Town&Country Suites, Silver Cloud Inn

154b   WA 518, Burien, S Center Blvd, **E...**Domino's, JC Penney, Nordstrom's, Sears/auto, Target, airport, mall, **see WA I-405, exit 1. W...**Denny's, Extended Stay America

   a   I-405, N to Bellevue

152   S 188th, Orillia Rd, **E...**Taco Bell/24hr, **W...**BP/mart, Denny's, Jack-in-the-Box, Omni Rest., Schaumski's Rest., Airport Plaza Hotel, Clarion, Comfort Inn, Day's Inn, DoubleTree Hotel, Motel 6, Super 8, KOA, bank

151   S 200th, Military Rd, **E...**Motel 6, **W...**Best Western, Econolodge, Hampton Inn, Howard Johnson, MiniRate Motel, **1/2 mi W on Military Hwy...**CHIROPRACTOR, Shell/mart/wash, Texaco/mart, Midas Muffler, bank, same as 149

149   WA 516, to Kent, Des Moines, **E...**Best Western, Day's Inn, Valley RV, **W...**Arco/mart/24hr, Texaco/diesel/mart/24hr, 7-11, BK's Kitchen, Blockhouse Rest., Burger King, Dunkin Donuts, McDonald's, Pizza Hut, Taco Bell/24hr, Wendy's, Best Inn, Century Motel, Kings Arms Motel, bank, to Saltwater SP, same as 151

147   S 272nd, **1 mi W on Pacific Hwy...**Arco/mart/24hr, Citgo/7-11, Texaco/diesel/mart, Chinese Rest., Dairy Queen, KFC, Little Caesar's, Skipper's, Subway, Taco Bell, Taco Time, Viva Mexican, Wendy's, Travel Inn, Albertson's, Bartel Drug, Firestone/auto, Rite Aid, Safeway/24hr, RV Sales, Shuck's Parts, cleaners

143   Federal Way, S 320th, **W...**MEDICAL CARE, CHIROPRACTOR, VISION Ctr, Arco/mart/24hr, BP/Circle K, Texaco/diesel/mart/24hr, Applebee's, Arby's, Azteca Mexican, Black Angus, Burger King, Country Harvest Rest., Cucina Cucina, Denny's, Dunkin Donuts, Godfather's, KFC, McDonald's, McHale's Rest., Old Country Buffet, Outback Steaks, Pietro's Pizza, Pizza Hut, Red Lobster, Red Robin Rest., RoundTable Pizza, Subway, Taco Time, TCBY, Tony Roma, Torero's Mexican, Wendy's, Best Western, Comfort Inn, Al's Parts, Firestone/auto, Goodyear/auto, Kinko's, K-Mart/auto, Lamont's, Longs Drug, Mervyn's, Midas Muffler, Rite Aid, Ross, Safeway/24hr, Sears/auto, Target, Top Foods, bank, cinema, mall, tires

142b a   WA 18 E, S 348th, Enchanted Pkwy, **E...**funpark, **W...**HOSPITAL, CHIROPRACTOR, Arco/diesel, Chevron/mart/24hr, Ernie's/diesel, Flying J/diesel//LP/mart/rest./24hr Texaco/diesel/mart, Burger King, Chinese Rest., Dairy Queen, Denny's, Jack-in-the-Box, McDonald's, Shari's, Subway, Teriyaki Rest., Eastwind Motel, Holiday Inn Express, Roadrunner Motel, Super 8, Chevrolet, Ford, Home Depot, NAPA, RV Ctr, cleaners

141mm   weigh sta sb

137   WA 99, Fife, Milton, **E...**Arco/mart/24hr, BP/diesel/mart/atm/RV dump, Chevron/diesel/mart/24hr, Texaco/subs/mart/24hr, Dairy Queen, Johnnie's Rest., Best Western, King's Motor Inn, Motel 6, Acura, Nissan, Volvo, **W...**Arco/mart, BP/Circle K, Pacific Xpress/diesel, Arby's, Baskin-Robbins, Burger King, Castle Fife Rest., Denny's, KFC, King's Palace Rest., McDonald's, Mitzel's Kitchen, Skipper's, Taco Bell, Wendy's, Comfort Inn, Ace Hardware, Bucky's Muffler/brake, Camping World/RV Service/supplies, RV Sales, Schwab Tires, ShopRite Drug, auto parts, bank, cleaners

T a c o m a

136b a   Port of Tacoma, **E...**Sam's Club, BMW, Honda, Mercedes, RV Sales, **W...**Chevron, Texaco/diesel/mart/wash/24hr, Flying J/diesel/LP/rest./mart/24hr, Shell, Fife City Grill, Jack-in-the-Box, Day's Inn, Econolodge, Extended Stay America, HomeTel, Port Tacoma Inn, Ramada Ltd, Travelers Inn, Goodyear/auto, World Trade Ctr

135   Bay St, Puyallup, **E...**bingo, **W...**Arco/mart/24hr, La Quinta, repair, to Tacoma Dome

133   WA 7, I-705, City Ctr, **W...**Tacoma Dome, Ramada Inn, Sheraton, Travel Inn, musum

132   WA 16 W, S 38th, to Bremerton, Gig Harbor, **W...**HOSPITAL, Azteca Mexican, ChuckeCheese, Cucina Cucina, TGIFriday, Extended Stay America, CarToys, Circuit City, Diahatsu, Ford, Hyundai, JiffyLube, ProGolf, Sears/auto, Sports Authority, Tire Sta, Toyota, Pearle Vision, mall, to Pt Defiance Pk/Zoo

130   S 56th, Tacoma Mall Blvd, **W...**Texaco/mart, El Torito's, Subway, Tony Romas, Firestone, House of Fabrics, JC Penney, Sears/auto, bank, cinema

129   S 72nd, S 84th, **E...**Arco/mart/24hr, BP/mart, Chevron/mart/24hr, Exxon, Texaco/diesel/mart, Alligator Grill, Applebee's, Burger King, Copperfield's Rest., Dairy Queen, Denny's, Elmer's Rest., IHOP, Jack-in-the-Box, Mitzel's Kitchen, Olive Garden, Red Lobster, Shari's, Subway, Taco Bell, Best Western, Crossland Studios, Holiday Inn Express, Howard Johnson, King Oscar Motel, Motel 6, Rothem Inn, Shilo Inn, Travelodge, HomeBase, Longs Drug, Mega Foods, Precision Tune, cinema, **W...**Arco/mart/24hr, Calzone's Italian, Jack-in-the-Box, Yankee Diner, Day's Inn, Discount Tire, Ernst, Gart Sports, Home Depot, Nevada Bob's, to Steilacoom Lake

128   (from nb), same as 129

127   WA 512, S Tacoma Way, Puyallup, Mt Ranier, **W...**Arco/mart/24hr, BP/mart, Chevron, Texaco/diesel/mart, Chinese Buffet, Fred's Deli, Frisko Freeze, Dairy Queen, Dahari Buffet, Denny's, IHOP, Ivar's Seafood, Jim Moore's Rest., McDonald's, McHale's Rest., Sizzler, Subway, Taco Time, Wendy's, Best Western, Budget Inn, Econolodge, Western Inn, Quality Inn, bank, RV Ctr, Schuck's Parts, transmissions

125   Lakewood, to McChord AFB, **W...**HOSPITAL, Chevron/LP/mart, Exxon/diesel, 76/diesel, Texaco/diesel/mart, Black Angus, Denny's, El Toro Mexican, KFC, Morey's Rest., Pizza Casa Italian, Pizza Hut, Wendy's, Best Western, Lakewood Lodge, Rose Garden Motel, Aamco, CarQuest, Goodyear/auto, Walt's Muffler/brake, U-Haul, mall

124   Gravelly Lake Dr, **W...**Arco/repair, BP/mart/24hr, same as 125

123   Thorne Lane, Tillicum Lane, no facilities

N

S

Washington Interstate 5

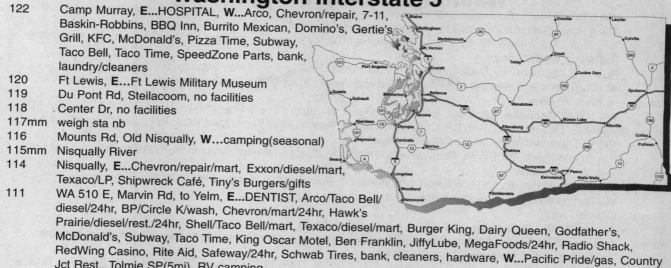

122 Camp Murray, **E**...HOSPITAL, **W**...Arco, Chevron/repair, 7-11, Baskin-Robbins, BBQ Inn, Burrito Mexican, Domino's, Gertie's Grill, KFC, McDonald's, Pizza Time, Subway, Taco Bell, Taco Time, SpeedZone Parts, bank, laundry/cleaners

120 Ft Lewis, **E**...Ft Lewis Military Museum

119 Du Pont Rd, Steilacoom, no facilities

118 Center Dr, no facilities

117mm weigh sta nb

116 Mounts Rd, Old Nisqually, **W**...camping(seasonal)

115mm Nisqually River

114 Nisqually, **E**...Chevron/repair/mart, Exxon/diesel/mart, Texaco/LP, Shipwreck Café, Tiny's Burgers/gifts

111 WA 510 E, Marvin Rd, to Yelm, **E**...DENTIST, Arco/Taco Bell/diesel/24hr, BP/Circle K/wash, Chevron/mart/24hr, Hawk's Prairie/diesel/rest./24hr, Shell/Taco Bell/mart, Texaco/diesel/mart, Burger King, Dairy Queen, Godfather's, McDonald's, Subway, Taco Time, King Oscar Motel, Ben Franklin, JiffyLube, MegaFoods/24hr, Radio Shack, RedWing Casino, Rite Aid, Safeway/24hr, Schwab Tires, bank, cleaners, hardware, **W**...Pacific Pride/gas, Country Jct Rest., Tolmie SP(5mi), RV camping

109 Martin Way, **E**...Mitzel's Kitchen, Taco Bell, Smith's Food, ShopKO/drug, Top Food/24hr, Discount Tire, HomeBase, **W**...HOSPITAL, CHIROPRACTOR, NATUROPATH, Arco/mart/24hr, BP/Circle K, 76, Shell/mart/24hr, Texaco/diesel/mart/atm/24hr, Burger King, Casa Mia Rest., Denny's, El Serabe Mexican, IHOP, Jack-in-the-Box, Mandarin House, Red Lobster, Shari's Rest./24hr, Subway, Bailey Motel, Comfort Inn, Day's Inn, Holiday Inn Express, Holly Motel, Super 8, Firestone/auto, K-Mart/drugs, NAPA, cinema

108 Sleator-Kenny Rd, **W**...HOSPITAL, same as 109, **E**...Texaco/diesel/mart, Arby's, Baskin-Robbins, Godfather's, McDonald's, Red Coral Rest., Starbuck's, Wendy's, Winchell's/24hr, Firestone/auto, Fred Meyer, FutureShop, GNC, Jo-Ann Fabrics, MailBoxes Etc, Mervyn's, Michael's, Office Depot, PetsMart, Radio Shack, Rite Aid, Sears/auto, Sports Whse, bank, cleaners

107 Pacific Ave, **E**...Texaco/diesel/mart, Sizzler, CarToys, mall, **W**...Ford

106 St Capitol, same as 105

105 St Capitol, City Ctr, **E**...CHIROPRACTOR, Figaro's Pizza, Izzy's Pizza, Pacific Terrace Rest., Rainier Rest., Ribeye Rest., Sizzler, Taco Time, Food Pavilion, Foreign Auto Works, RV Sales/service, laundry, mall, **W**...Chevron/mart/24hr, Texaco/diesel, Casa Mia Rest., Crystal Palace Rest., Dairy Queen, Jack-in-the-Box, McDonald's, Plum St Deli, Saigon Rest., Best Western, Carriage Inn, Golden Gavel Motel, Ramada, Pontiac/Cadillac/Saturn, Ford

104 US 101 N, W Olympia, to Aberdeen, **W**...HOSPITAL, Holiday Inn, to Capitol Mall

103 2nd Ave, no facilities

102 Trosper Rd, Black Lake, **E**...CHIROPRACTOR, Arco/mart/24hr, Citgo/7-11, Texaco/diesel/mart/24hr, Arby's, Burger King, Cattin's Rest./24hr, Jack-in-the-Box, KFC, McDonald's, Pizza Hut, Subway, Taco Bell, Teriyaki Rest., Best Western, Motel 6, Al's Parts, Goodyear, Rexall Drug, ShopRite Foods, auto repair/towing, bank, hardware, laundry/cleaners, **W**...BP, Chevron/mart/24hr, Shell/mart/24hr, Iron Skillet Rest., Nickleby's Rest., Tyee Hotel/rest., Costco Whse

101 Airdustrial Way, **E**...Texaco, Olympia Camping, airport

99 WA 121 S, 93rd Ave, Scott Lake, **E**...American Heritage Camping, **W**...Exxon(2mi), Shell/diesel/LP/mart, Hannah's Pantry Rest., Restover Motel

95 WA 121, Littlerock, **3 mi E**...Millersylvania SP, RV camping, **W**...Farmboy Drive-In

93.5mm **rest area sb, full(handicapped)facilities, info, phone, picnic tables, litter barrels, vending, coffee, petwalk**

91mm **rest area nb, full(handicapped)facilities, info, phone, picnic tables, litter barrels, vending, coffee, petwalk**

88 US 12, Rochester, **W**...Arco/mart/24hr, Shell/diesel/LP/repair/mart, Texaco/diesel/mart/24hr, Dairy Queen, Maharaja Indian Cuisine, Little Red Barn Rest./24hr, Outback RV Park(2mi), Harrison RV Park(3mi), bank

82 Centralia, **E**...HOSPITAL, Arco/mart/24hr, Shell/diesel/mart, Burger King, Burgerville, Casa Ramos Rest., Cattin's Rest./24hr, Dairy Queen, Godfather's, Pizza Hut, Shari's/24hr, TCBY, Wendy's, YumYum Chinese, Ferryman's Inn, King Oscar Motel, Riverside Motel, JiffyLube, Rite Aid, VF/famous brands, **W**...BP/mart/24hr, Chevron/mart/atm, Rocky's/gas, Texaco/mart, Arby's, Bill&Bea's Café, Country Cousin Rest., Denny's, Jack-in-the-Box, McDonald's, Papa Murphy's, Subway, Taco Bell, Motel 6, Park Motel, Al's Parts, Fuller's Foods, GNC, Outlet Mall/famous brands, RV park, Safeway/drug, Scwab Tires, bank, cleaners

82mm Skookumchuck River

81 WA 507, Mellen St, **E**...Shell/mart, Texaco/mart/24hr, King Solomon Rest., Peppermill Rest., Holiday Inn Express, Knight's Inn, Pepper Tree Motel/RV Park/rest.

79 Chamber Way, **E**...Shell/mart, Texaco/diesel/mart/24hr, Burger King, McDonald's, Plaza Jalisco Mexican, Subway, Yardbird's Rest., Ford/Lincoln/Mercury/Toyota, Goodyear/auto, RV Ctr, museum, **W**...BP/Burger King/Taco Bell/diesel/LP/mart, K-Mart/Little Caesar's, Wal-Mart/auto, st patrol

Washington Interstate 5

77	WA 6 W, Chehalis, **E...**Arco/mart/24hr, Exxon/mart, Jackpot Foodmart, Dairy Bar, St Helens Inn/rest., **W...**Alva's Place Rest., Rainbow Falls SP(16mi), RV camping
76	13ᵗʰ St, **E...**Arco/mart/24hr, BP/mart, Chevron/mart, Denny's, Jack-in-the-Box, Kit Carson Rest., Howard Johnson, Relax Inn, **W...**RV camping
72	Rush Rd, Napavine, **E...**Shell/diesel/mart, McDonald's, RibEye Rest./24hr, Subway, RV Sales, RV park, **W...**Chevron/diesel/mart/24hr, Pacific Pride/diesel, Texaco/Taco Bell/diesel/mart/atm
72mm	Newaukum River
71	WA 508 E, Onalaska, Napavine, **E...**76/diesel/mart, **W...**Suzy's Place Rest.(1mi)
68	US 12 E, **E...**Arco/mart/24hr, BP, Chevron/diesel/RV Park, Spiffy's Rest./24hr, KOA, to Lewis&Clark SP, Mt Ranier NP, **W...**Texaco/diesel/mart, The Mustard Seed Rest.
63	WA 505, Winlock, **W...**Exxon, Shell/diesel/LP/mart/atm, Texaco/LP/mart, Sunrise Motel/RV Park
60	Toledo, no facilities
59	WA 506 W, Vader, **E...**BP/diesel/mart, Mrs Beesley's Burgers, Cowlitz Motel/RV Park, River Oaks Camp, **W...**Texaco/mart/24hr, Country House Rest., Grandma's Café
59mm	Cowlitz River
57	Barnes Dr, **W...**Texaco/diesel/café/24hr, repair, RV camping
55mm	**rest area both lanes, full(handicapped)facilities, phone, picnic tables, litter barrels, vending, petwalk**
52	Toutle Park Rd, **E...**Paradise Cove RV Park/gas
50mm	Toutle River
49	WA 504 E, Castle Rock, **E...**CHIROPRACTOR, Arco/mart/24hr, Chevron/mart, Shell/mart, Texaco/diesel/mart/24hr, Burger King, El Compadre Mexican, 49er Diner, Pizza Parlor, RoseTree Rest., Mt St Helens Motel, Motel 7, Silver Lake Motel/resort, Timberline Inn, Seaquest SP(5mi), cinema, RV camping
48	Huntington Ave, no facilities
46	Pleasant Hill Rd, **W...**Cedars RV Camping
44mm	weigh sta sb, phone
42	Ostrander Rd, no facilities
40	to WA 4, Kelso, **1 mi W...**HOSPITAL, BP/mart, Best Western, Budget Inn, Townhouse Motel, transmissions
39	WA 4, Kelso, to Longview, **E...**Arco/mart/24hr, Shell/mart, Denny's, Hilander Rest., Jitter's Drive-Thru, Little Caesar's, McDonald's, Shari's/24hr, Subway, DoubleTree Hotel, Motel 6, Super 8, Rite Aid, Safeway, **W...**CHIROPRACTOR, EYECARE, Chevron/repair, Texaco/diesel/mart, Azteca Mexican, Burger King, Dairy Queen, Izzy's Pizza, JJ North's Buffet, Red Lobster, Taco Bell, Comfort Inn, JC Penney, JiffyLube, Sears/auto, Target, Top Foods, cinema, mall, museum
36	WA 432 W, to WA 433, US 30, Kelso, **E...**RV Sales, **3 mi W...**HOSPITAL, BP/mart, Texaco/mart/24hr, Holiday Inn Express, Oaks RV Park, st patrol
32	Kalama River Rd, **E...**Camp Kalama RV Park/gifts
31mm	Kalama River
30	Kalama, **E...**BP, Burger Bar, Kalama Café, Columbia Inn/rest., Big A Parts, bank, **W...**Texaco/diesel, RV camping
27	Todd Rd, Port of Kalama, **E...**Texaco/diesel/café/24hr
22	Dike Access Rd, **W...**Columbia Riverfront/Lewis River/Woodland Shores RV Parks
21	WA 503 E, Woodland, **E...**CHIROPRACTOR, Arco/mart/24hr, Chevron/mart, Shell/diesel/mart/24hr, Texaco/diesel/LP/mart, Pacific Pride/diesel, Burgerville, Casa Maria's, Dairy Queen, Expresso, Grandma's House(8mi), OakTree Rest., Rosie's Rest., South China Rest., Subway, Lewis River Inn(2mi), Woodlander Inn, Ace Hardware, Hi-School Drug, LetterBox, Sav-On Foods, **W...**Exxon/mart/24hr, ParkPlace Pizza, Whimpy's Rest., Hansen's Motel, Lakeside Motel, Scandia Motel, Chevrolet/Olds, NAPA, RV Park, bank
20mm	N Fork Lewis River
18mm	E Fork Lewis River
16	NW 319ᵗʰ St, La Center, **E...**Texaco/diesel/mart/24hr, The Inn B&B(2mi), Paradise Point SP
15mm	weigh sta nb
14	WA 501 W, NW 269ᵗʰ St, **E...**Arco/mart/24hr, BP/Circle K, Country Jct Rest., to Battleground Lake SP(14mi), Big Fir RV Park(4mi), **W...**Chevron/diesel/mart
13mm	**rest area sb, full(handicapped)facilities, info, phone, picnic tables, litter barrels, vending, petwalk, RV dumping**
11mm	**rest area nb, full(handicapped)facilities, info, phone, picnic tables, litter barrels, vending, petwalk, RV dumping**
9	NE 179ᵗʰ St, **E...**Jollie's Rest./24hr, Adventure RV, **W...**Chevron/diesel/mart, bank
7	I-205 S, to I-84, WA 14, NE 134ᵗʰ St, Portland Airport, **E...**HOSPITAL, DENTIST, VETERINARIAN, BP/mart, Citgo/7-11, Burger King, Burgerville, JB's Roadhouse, McDonald's, Round Table Pizza, Taco Bell, Comfort Inn, Olympia Motel, Salmon Creek Inn, Shilo Inn, StageCoach Inn/rest., Albertson's/drugs, Hi-School Drugs, Zupan's Mkt
5	NE 99ᵗʰ St, **E...**DENTIST, CHIROPRACTOR, Arco/mart/24hr, Citgo/7-11, Texaco/mart, Burgerville, Fat Dave's Rest., Super Taco, Nissan/Kia, Winco Foods/24hr, cinema, **W...**Arco/mart/24hr, Chevron/mart/24hr, Bortolami's Pizzaria, Clancy's Seafood, Albertson's, Hi-School Drug

K e l s o   W o o d l a n d

N

S

Washington Interstate 5

N
↕
S

4   NE 78th St, Hazel Dell, **E...**CHIROPRACTOR, NATUROPATH, Chevron/mart/24hr, Citgo/7-11, 76, Texaco/mart, Baskin-Robbins, BBQ, Burger King, Burgerville, Dragon King Chinese, KFC, McDonald's, PeachTree Rest., Pizza Hut, Skipper's, Smokey's Pizza, Steakburger, Subway, Taco Bell, Taco Time, Totem Pole Rest., Edelweiss Inn, Quality Inn, Value Motel, Americas Tires, Cub Foods, Firestone, Fred Meyer, Dodge, Ford, Goodyear, Mazda, Midas Muffler, Nissan, OilCan Henry's, PetCo, Precision Tune, Radio Shack, Schuck's Parts, U-Haul, bank, cleaners/laundry, mufflers/transmissions, **W...**VETERINARIAN, Arco/mart/24hr, Shell/diesel/LP/mart, Texaco/diesel/mart, City Grill, Denny's, RoundTable Pizza, Wendy's, Best Western, Rite Aid, Safeway, laundry

3   NE Hwy 99, Hazel Dell, Main St, **E...**HOSPITAL, **W...**transmissions/tuneups, **1/2 mi W...**Arco, Safeway/

2   WA 500 E, 39th St, to Orchards, no facilities

1d   E 4th, Plain Blvd W, to WA 501, Port of Vancouver, no facilities

1c   Mill Plain Blvd, City Ctr, **W...**Chevron, Texaco, Cattle Co Rest., Burgerville, Denny's, DoubleTree Hotel, Fort Motel, Shilo Inn, Vancouver Inn, Ford, Lincoln/Mercury, Pontiac/Cadillac/GMC, Jeep/Eagle, Mitsubishi, Suzuki, Clark Coll, st patrol

1a   WA 14 E, to Camus

0mm   Columbia River, Washington/Oregon state line

Washington Interstate 82

Exit #	Services

11mm   I-82 Oregon begins/ends on I-84, exit 179.

10   Westland Rd, **E...**HOSPITAL, to Umatilla Army Depot, to Best Western

5   Power Line Rd, no facilities

1.5mm   Umatilla River

1   US 395/730, Umatilla, **E...**Best Western, Desert Inn, Hatrock Camping(8mi), to McNary Dam, **W...**Amoco/diesel/LP/mart, Arco/diesel/mart/24hr, Texaco/diesel/rest., Dojack's Rest., G&J Burgers, Heather Inn, McNary Hotel, Rest-a-Bit Motel, Tillicum Motel, NAPA, Red Apple Mkt, USPO, st police, Welcome Ctr, weight sta

132mm   Columbia River, Washington/Oregon state line

131   WA 14 W, Plymouth, **N...**RV camping, to McNary Dam

130mm   weigh sta wb

122   Coffin Rd, no facilities

114   Locust Grove Rd, no facilities

113   US 395 N, to I-182, Kennewick, Pasco, st patrol, **2-5 mi N...**Arco/mart/24hr, Exxon/mart, Billigan's Roadhouse, Bruchi's Café, Carl's Jr, China Café, Denny's, Godfather's, Jack-in-the-Box, KFC, McDonald's, Starbucks, Subway, Taco Bell/24hr, Best Western, Hawthorn Suites, Holiday Inn Express, Nendel's Inn, Tapadera Budget Motel, Travelodge, GNC, Harley-Davidson, Hastings Books, JiffyLube, ProLube, Radio Shack, Rite Aid, Safeway

109   Badger Rd, W Kennewick, **N...**Shell/BYOBurger/diesel/mart/24hr, **3 mi N...**Chico's Tacos, McDonald's, Subway, Clearwater Inn, Silver Cloud Inn, Super 8, West Coast Inn

104   Dallas Rd, **3 mi N...**Conoco/diesel/mart/24hr, Hotstuff Pizza/subs

102   I-182, US 12 E, to US 395, to Spokane, HOSPITAL, facilities in Richland, Pasco

96   WA 224, Benton City, **N...**Conoco/diesel/mart/24hr, BearHut Rest., Beach RV Park

93   Yakitat Rd, no facilities

88   Gibbon Rd, **N...**gas/diesel/mart, phone

82   WA 22, WA 221, Mabton, **2 mi S...**HOSPITAL, Blue Goose Rest., Prosser Motel, to WA St U Research, to Wine Tasting Facilities, museum

82mm   Yakima River

80   Gap Rd, **S...**HOSPITAL, Texaco/diesel/mart, Blue Goose Rest., Burger King, KFC, McDonald's, NorthWoods Rest., Barn Motel/RV Park/rest., Best Western, Ford/Mercury, **rest area both lanes, full(handicapped)facilities, phone, picnic tables, litter barrels, 1 mi S...**Exxon/mart, Prosser Motel, Chevrolet/Buick, Chukar Cherries, Schwab Tires, auto repair

75   County Line Rd, Grandview, **1 mi S...**Conoco/diesel/24hr, Apple Valley Motel, Grandview Motel, same as 73

73   Stover Rd, Wine Country Rd, Grandview, **S...**Conoco/diesel/mart/24hr, Shell/Quizno's/TCBY/diesel/mart/atm, Eli&Kathy's Breakfast, Morrie's Pizza, New Hong Kong, 10-4 Café, Apple Valley Motel, Grandview Motel, Plymouth/Jeep/Dodge, Red Apple Foods, RV Ctr, RV Park/dump, auto repair

E
↕
W

Washington Interstate 82

69 WA 241, to Sunnyside, **N**...Arco/diesel/mart/24hr, Shell/diesel/mart(1mi), Texaco/TacoMaker/diesel/mart/24hr, Arby's, Burger King, Dairy Queen, McDonald's, Pizza Hut, Skipper's, Subway, Taco Bell, Rodeway Inn, Townhouse Motel, BatteryStore, Buick/Chevrolet/Nissan, GNC, JC Penney, K-Mart/Little Caesars, Shuck's Parts, Wal-Mart/auto

67 Sunnyside, Port of Sunnyside, **N**...HOSPITAL, Cenex/diesel/lube/wash(1mi), **S**...DariGold Cheese

63 Outlook, Sunnyside, **3 mi N**...Sunnyside B&B, Townhouse Motel, Travelodge, RV camping

58 WA 223 S, to Granger, **S**...Conoco/diesel/mart, OK Tires, RV camping

54 Division Rd, Yakima Valley Hwy, to Zillah, **N**...El Ranchito Rest., **S**...Teapot Dome NHS, diesel/LP, RV dump

52 Zillah, Toppenish, **N**...Arco/mart/24hr, BP/diesel, Chevron/mart, McDonald's, Subway, Comfort Inn, RV dump, Wine Tasting

50 WA 22 E, to US 97 S, Toppenish, **N**...Gold Nugget Café/casino, **3 mi S**...HOSPITAL, Legends Buffet/casino, McDonald's, Toppenish Inn, Murals Museum, RV Park, to Yakima Nation Cultural Ctr

44 Wapato, **N**...Texaco/diesel/mart, **2 mi S**...Dairy Queen

40 Thorp Rd, Parker Rd, Yakima Valley Hwy, no facilities

39mm Yakima River

38 Union Gap, **1 mi S**...gas, food, lodging, museum

37 US 97(from eb), **1 mi S**...Exxon/mart, Texaco, Alpenlite RV, NAPA

36 Valley Mall Blvd, Yakima, **N**...st patrol, **S**...MEDICAL CARE, Arco/diesel/mart/24hr, Cenex, Texaco/Gearjammer/ diesel/rest./24hr, Burger King, Country Harvest Rest., Denny's/24hr, Godfather's, IHOP, Jack-in-the-Box, KFC, Miner's Drive-In, Oriental Rest., SeaGalley Rest., Skipper's, Taco Bell, Day's Inn, Quality Inn, Settlers Inn/rest., Super 8, Big A Parts, FashionBug, FutureShop, Goodyear, Lowe's Bldg Supply, Office Depot, Rite Aid, Sears/ auto, ShopKO, XpressLube, cinema, diesel/repair, mall

34 WA 24 E, Nob Hill Blvd, Yakima, **N**...K-Mart/auto, Sportsman SP, KOA, diesel/repair, **S**...HOSPITAL, Arco/mart/ 24hr, Chevron/Subway/TCBY/diesel/mart/atm/24hr, Citgo/7-11, Exxon/diesel/mart, Shell/diesel/mart, Arby's, McDonald's, Al's Auto Supply, Circle H Ranch RV Park, Family Foods, museum

33 Yakima Ave, Yakima, **N**...Burger King, Marti's Café, McDonald's, Oxford Suites, Chevrolet/Olds/GMC/Isuzu, Honda, Mazda, Wal-Mart/auto, **1 mi N**...Chevron/Quizno's/TCBY/mart/24hr, Shell/Chester's/diesel/mart, **S**...Arco/mart/24hr, Citgo/7-11, Dairy Queen, Taco Bell, Budget Suites, Holiday Inn Express, West Coast Inn, Food Pavilion, JC Penney, OfficeMax, Pier 1, Red Apple Mkt, Schwab Tires, Target, mall

31 US 12 W, N 1st St, to Naches, **S**...Arco/diesel/mart/24hr, Conoco/diesel/mart, Exxon/diesel/mart, Texaco/repair, Arctic Circle, Black Angus, Boca Del Rio Seafood, Denny's, Espinosa's Mexican, Golden Phoenix Buffet, Mel's Diner/24hr, Peking Palace, Pizza Hut, Red Apple Rest., Red Lobster, Settler's Inn Rest., Sizzler, Subway, Waffles'n More, Wendy's, AllStar Motel, Bali Hai Motel, Best Western, Big Valley Motel, DoubleTree Hotel, Motel 6, Nendel's, Niska Inn, Red Lion Inn, Sun Country Inn, Tourist Motel, Twin Bridges Inn, Vagabond Inn, RV Park

30 WA 823 W, Resthaven Rd, to Selah, no facilities

29 E Selah Rd, **N**...fruits/antiques

26 WA 821, **N**...Shell/Quizno's/TCBY/mart/atm/24hr

25mm Selah Creek

24mm **rest area eb, full(handicapped)facilities, phone, picnic tables, litter barrels**

22mm **rest area wb, full(handicapped)facilities, phone, picnic tables, litter barrels**

21mm S Umptanum Ridge, 2265 elev

19mm Burbank Creek

17mm N Umptanum Ridge, 2315 elev

15mm Lmuma Creek

11 Military Area, no facilities

8mm view point both lanes, Manastash Ridge, 2672 elev

3 WA 821, Thrall Rd, no facilities

0mm I-90, E to Spokane, W to Seattle. I-82 begins/ends on I-90, exit 110.

Washington Interstate 90

Exit #	Services
300mm	Spokane River, Washington/Idaho state line

299 State Line, Port of Entry, **N**...**Welcome Ctr/rest area both lanes, weigh sta, full(handicapped)facilities, phone, picnic tables, litter barrels, petwalk**

296 Otis Orchards, **N**...Shell/Subway/pizza/diesel/mart, HomePlate Grill, **S**...BP/mart, Chevron/mart/wash, Burger King, Fitzbillie's Bagels, McDonald's, Papa Murphy's, Taco Bell, Albertson's/drugs, GNC, Land Rover, bank

294 Sprague Ave(no EZ wb return), **S**...North Country RV/marine, auto repair

293 Barker Rd, Greenacres, **N**...GTX Trkstp/diesel/café, Alpine Motel/RV Park, KOA, **S**...Conoco/mart/24hr, Exxon/ Subway/24hr, Tattrie's Saddle&Tack

516

Yakima

E

W

Washington Interstate 90

291 Sullivan Rd, Veradale, **N**...Barnes&Noble, Circuit City, Gart Sports, JC Penney, Sears/auto, mall, **1/2 mi N**...Chevron/Blimpie/TCBY/mart, McDonald's, Country Inn, **S**...Chevron/mart/wash, Conoco, Texaco/diesel/mart, Bruchi's Rest., Dairy Queen, Godfather's, Jack-in-the Box, KFC, McDonald's, Mongolian BBQ, Noodle Express, Pizza Hut, Sak's Rest., Schlotsky's, Shari's/24hr, Subway, Taco Bell, Comfort Inn, DoubleTree Hotel, Red Lion Inn, CarToys, Fred Meyer, FutureShop, Goodyear/auto, Hastings Books, Kinko's, K-Mart, MailBox Ctr, Michael's, PetCo, Rite Aid, Wal-Mart/auto, cleaners

289 WA 27 S, Pines Rd, Opportunity, **N**...Citgo/7-11, TakeABreak Rest., **S**...HOSPITAL, CHIROPRACTOR, BP/mart/wash, Cenex, Shell/diesel/mart, Applebee's, Dairy Queen, Denny's, Jack-in-the-Box, Best Western, MailCtr, **1/2 mi S**...Meineke Muffler

287 Argonne Rd, Millwood, **N**...CHIROPRACTOR, Holiday/diesel/mart, Burger King, Denny's, Godfather's, Hungry Farmer Rest., Jack-in-the-Box, Longhorn BBQ, Marie Callender's, McDonald's, Papa Murphy's, Subway, Taco Bell, Wolffy's Burgers, Day's Inn, Motel 6, Super 8, Ace Hardware, Albertson's/drugs, Schuck's Parts, Tidyman's Foods, Walgreen, bank, **S**...BP/Circle K, Chevron, Casa de Oro Mexican, Godfather's, GoodTymes Grill, Perkins, Holiday Inn Express, Quality Inn, Rite Aid, Safeway

286 Broadway Ave, **N**...Conoco, Flying J/Exxon/Sak's Rest./diesel/LP/mart/24hr, Texaco, Zip's Drive-In, Broadway Motel, Comfort Inn, Goodyear, Schwab Tires, White/Volvo/GMC

285 Sprague Ave, **N**...Texaco, Denny's, Dragon Garden Chinese, Jack-in-the-Box, McDonald's, MapleTree Motel/RV Park, Big 5 Sports, Buy-Rite RV, Eagle Hardware, Home Depot, K-Mart, Radio Shack, Tidyman's Foods, West Marine, **S**...VETERINARIAN, Exxon/mart, Little German Rest., Mandarin House, Puerta Vallarta Mexican, Skipperville's Rest., Subway, Taco Time, Zip's Drive-In, Alton's Tires, Chrysler/Plymouth, Dodge, Ford, Hyundai, L&L RV Ctr, Milestone RV, Nissan, carwash, RV camping, transmissions

284 Havana St(from eb, no EZ return), **N**...Jack-in-the-Box, McDonald's

283b Freya St, Thor St, **N**...Conoco/diesel/mart, Park Lane Motel RV camp, **S**...Texaco/mart

a Altamont St, no facilities

282b 2nd Ave, **N**...Shilo Inn

a Trent Ave, Hamilton St, downtown, **N**...Shell, Shilo Inn, Costco, Office Depot

281 US 2, US 395, to Colville, **N**...Conoco, Shell/repair, Texaco/diesel, Arby's, McDonald's, Dick's Burgers, Doodle's Rest., Just Like Home Buffet, McDonald's, Perkins, Pizza Hut, Rancho Chico's Rest., Subway, Waffles'n More, Best Western/rest., Howard Johnson, Firestone, Ford, Schwab Tires, U-Haul, **S**...HOSPITAL, Cavanaugh's Motel

280b Lincoln St, **N**...Chevron, Conoco/diesel, Exxon/mart, Burger King, Coyote Café, Godfather's, IHOP, Jack-in-the-Box, Taco Bell, Wendy's, Zip's Burgers, Best Western, Ramada Inn, Ford/Lincoln/Mercury, Honda, Mazda, Olds/Cadillac, Saturn, Mercedes, Toyota, **S**...HOSPITAL

a downtown, **N**...Conoco/diesel/mart, Shell, Texaco, Arctic Circle, McDonald's, Perkins, Subway, Select Inn, Chevrolet, Ford/Lincoln/Mercury, NAPA, Safeway, cleaners

279 US 195 S, to Colfax, Pullman, **N**...West Wynn Hotel, Shangri-la Motel

277b a US 2 W, Fairchild AFB, to Grand Coulee Dam, **N**...Day's Inn, Hampton Inn, Motel 6, Quality Inn, Ramada, Rodeway Inn, Shangri-la Motel, airport

276 Geiger Blvd, **N**...BP, Flying J/Exxon/diesel/LP/rest./24hr, Denny's, Subway, Best Western, Starlite Motel, st patrol, **S**...Shell/diesel/LP/mart/RV parking, Honda/Yamaha, Sunset RV Park/dump

272 WA 902, Medical Lake, **N**...Texaco/diesel/mart, Overland Sta/RV camping, **S**...Super 8, Yogi Bear Camping

270 WA 904, Cheney, Four Lakes, **S**...Exxon/mart, Rosebrook Inn(6mi), Willow Springs Motel(6mi), Peaceful Pines RV Park, E WA U

264 WA 902, Salnave Rd, to Cheney, Medical Lake, **2 mi N**...camping

257 WA 904, Tyler, to Cheney, no facilities

254 Fishtrap, **S**...Fishtrap RV camping/tents

245 WA 23, Sprague, **S**...Chevron/diesel/mart, Stan's Drive-In, Viking Drive-In, Last Roundup Motel, Purple Sage Motel, Sprague Lake Resort/RV camping

242mm **rest area both lanes, full(handicapped)facilities, phone, picnic tables, litter barrels, tourist/weather info, petwalk, free coffee**

231 Tokio, **S**...weigh sta both lanes, Exxon/Templin's Café/diesel, fresh produce mkt

226 Coker Rd, no facilities

221 WA 261 S, Ritzville, City Ctr, **N**...HOSPITAL, Chevron/McDonald's/mart, Shell/Taco Bell/mart, Texaco/diesel/pizza/mart, Perkins, Zip's Drive-In, Best Western, Circle T Inn, Empire Motel, Portico Inn, Top Hat Motel, RV camping, hist dist

Washington Interstate 90

220	US 395 S, Ritzville, **N**...Cenex, Chevron, Exxon/diesel/café, Jake's Café, Texas John's Café, Caldwell Inn, Westside Motel, st patrol
215	Paha, Packard, no facilities
206	WA 21, to Lind, Odessa, no facilities
199mm	**rest area both lanes, full(handicapped)facilities, phone, picnic tables, litter barrels, vending, RV dumping, petwalk**
196	Deal Rd, to Schrag, no facilities
188	to Warden, Ruff, **10 mi S**...food
184	Q Rd, no facilities
182	O Rd, to Wheeler, no facilities
179	WA 17, Moses Lake, **N**...HOSPITAL, Arco/mart/24hr, Conoco/diesel/mart, Exxon/mart, Shell/mart/24hr, Ernie's/Chevron/diesel/café/atm/24hr, Arby's, Bob's Café/24hr, Burger King, Denny's, McDonald's, Shari's Rest./24hr, Holiday Inn Express, Moses Lake Inn, Shilo Inn/rest./24hr, Travelodge, **1 mi N**...MEDICAL CARE, CHIROPRACTOR, VETERINARIAN, Dairy Queen, Subway, El Rancho Motel, Sage'n Sand Motel, Honda, Nissan, Tri-State Outfitters, USPO, **2 mi N**...Godfather's, KFC, Pizza Hut, Buick/Chevrolet/Pontiac, CarQuest, Chrysler/Plymouth, Ford/Lincoln/Mercury, MailBoxes Etc, Radio Shack, Rite Aid, Safeway, Scwab Tires, **S**...I-90 RV, Sun Country RV, Willows RV Park(2mi), Potholes SP(15mi)
177mm	Moses Lake
176	Moses Lake, **1 mi N**...HOSPITAL, VETERINARIAN, BP, Cenex/24hr, Chevron/diesel/mart, Exxon/diesel/mart, Shell/mart, Texaco/diesel/mart, El Abuelo Mexican, Godfather's, Kiyoki's Rest., Perkins/24hr, Best Western/rest., Interstate Inn, Motel 6, Oasis Budget Inn, Super 8, Motel 6, OK Tires, Big Sun Resort/RV Hookups, **S**...Lakeshore Motel
175	Westshore Dr(from wb), **N**...Moses Lake SP, to Mae Valley, **S**...st patrol
174	Mae Valley, **N**...Suncrest Resort/RV, funpark, **S**...Conoco/diesel/mart, Barbara Ann's Café, st patrol
169	Hiawatha Rd, no facilities
164	Dodson Rd, no facilities
162mm	**rest area wb, full(handicapped)facilities, phone, picnic tables, litter barrels, petwalk, RV dumping**
161mm	**rest area eb, full(handicapped)facilities, phone, picnic tables, litter barrels, petwalk, RV dumping**
154	Adams Rd, no facilities
151	WA 281 N, to Quincy, **N**...HOSPITAL(12mi), Texaco/diesel/pizza/subs/mart, Shady Tree/RV camp
149	WA 281 S, George, **N**...HOSPITAL(12mi), **S**...BP/Subway/diesel/mart, Exxon/diesel/mart/24hr, Martha Inn Café, RV camp
143	Silica Rd, to The Gorge Ampitheatre
139mm	Wild Horses Mon, scenic view both lanes
137	WA 26 E, to WA 243, Othello, Richland, no facilities
137mm	Columbia River
136	Huntzinger Rd, Vantage, **N**...Conoco/mart, Texaco/diesel/repair, Blustery's Café, Golden Harvest Café, Wanapum Rest., Motel Vantage, KOA, to Ginkgo/Wanapum SP, **4 mi S**...Getty's Cove RV Camping
126mm	**Ryegrass, elev 2535, rest area both lanes, full(handicapped)facilities, phone, picnic tables, litter barrels, petwalk**
115	Kittitas, **N**...BP/mart/LP/24hr, Texaco/diesel/mart, RJ's Café, Olmstead Place SP
110	I-82 E, US 97 S, to Yakima, no facilities
109	Canyon Rd, Ellensburg, **N**...HOSPITAL, VETERINARIAN, BP/mart/atm, Chevron/mart/24hr, Exxon/diesel/mart, Pacific Pride/diesel/24hr, Shell/mart/atm/24hr, Texaco/mart, Appleseed Rest., Arby's, Baskin-Robbins, Bruchi's Café, Burger King, Casa de Blanca Mexican, Godfather's, KfC, McDonald's, Pizza Hut, Red Robin Rest., Skipper's, Subway, Taco Bell, RanchHouse Rest., Best Western, Comfort Inn, Goose Creek Inn, Nites Inn, Super 8, Al's Parts, CarQuest, Chevrolet, NAPA, Pennzoil, Radio Shack, Rite Aid, Scwab Tires, Super 1 Food/24hr, TrueValue, auto repair, fruit mkt, **S**...Flying J/Exxon/Sak's/diesel/LP/rest./24hr, Leaton's Rest./mart, R&R Inn/RV park
106	US 97 N, to Wenatchie, **N**...BP/mart, Chevron/mart, Conoco/mart/RV dump, Pilot/Subway/diesel/mart/24hr, Blue Grouse Rest., Copper Kettle(2mi), Dairy Queen, Perkins, I-90 Inn, Regal Lodge(2mi), Thunderbird Inn(2mi), **S**...KOA, st patrol
101	Thorp Hwy, **N**...antiques/fruits/vegetables
93	Elk Heights Rd, Taneum Creek, no facilities
92.5mm	Elk Heights, elev 2359
89mm	**Indian John Hill, elev 2141, rest area both lanes, full(handicapped)facilities, phone, picnic tables, litter barrels, RV dumping, petwalk, vending**
85	WA 970, WA 903, to Wenatchie, **N**...BP/mart, Conoco/repair, Shell/diesel/24hr, Cottage Café/24hr, Dairy Queen, Homestead Rest., Sunset Café, Aster Inn, Cascade Mtn Inn, Cedars Motel, Chalet Motel, Wind Blew Inn, Ace Hardware, CarQuest, PriceChopper Foods, Safeway, Twin Pines RV, auto repair
84	Cle Elum, **N**...HOSPITAL, VETERINARIAN, Chevron/diesel/mart, Conoco, Shell, Texaco/diesel/mart, Buttercup Rest., Dairy Queen, Golden Ocean Chinese, Sunset Café, Jack-in-the-BeanShop, Mexican Rest., Cle Elum Motel, Timber Lodge Motel, Stewarts Lodge, Ace Hardware, bank, museum
81mm	Cle Elum River

Washington Interstate 90

80	Roslyn, Salmon la Sac, no facilities
80mm	weigh sta both lanes, phone
78	Golf Course Rd, **S**...Sun Country Golf/RV Park
74	W Nelson Siding Rd, no facilities
71	Easton, **N**...Mtn High Burgers, **S**...Exxon/diesel/ rest., Silver Ridge Ranch B&B, USPO, John Wayne Tr, Iron Horse SP, camping
71mm	Yakima River
70	Lake Easton SP, **N**...Mtn High Burger, Parkside Café, **S**...Silver Ridge Ranch RV Park
63	Cabin Creek Rd, **S**...Cabin RV/Ufish
62	Stampede Pass, elev 3750, to Lake Kachess, access to towing
54	Hyak, Gold Creek, no facilities
53	Snoqualmie Pass, elev 3022, info, **S**...Chevron/mart, Family Pancake House, Webb's Grill, Best Western/Summit Inn, to rec areas
52	W Summit(from eb), same as 53
47	Denny Creek, Tinkham Rd, **N**...chain area, **S**...RV camping/dump
45	USFS Rd 9030, **N**...to Lookout Point Rd, no facilities
42	Tinkham Rd, no facilities
38	fire training ctr, no facilities
35mm	S Fork Snoqualmie River
34	468th Ave SE, Edgewick Rd, **N**...BP/Pizza Hut/Taco Bell/diesel/mart, Pacific Pride/diesel, Seattle E/diesel/Ken's Rest./ 24hr, Shell/diesel/rrepair, Texaco/Subway/diesel/mart/24hr, Edgewick Inn, NorWest Motel/RV Park, st patrol
32	436th Ave SE, North Bend Ranger Sta, **1 mi N**...gas, food, lodging
31	WA 202 W, North Bend, Snoqualmie, **N**...HOSPITAL, Chevron/diesel/mart, Shell, Texaco/diesel/mart, Dairy Queen, McDonald's, Mitzel's Kitchen, Taco Time, North Bend Motel, Old Honey Farm Inn, Sallish Lodge, Sunset Motel, NorthBend Stores/famous brands, Safeway/24hr, museum, st patrol
27	North Bend, Snoqualmie(from eb), no facilities
25	WA 18 W, to Auburn, Tacoma, **N**...weigh sta
22	Preston, **N**...Chevron/diesel/mart/atm/24hr, Martinelli's Kitchen, Savannah's Burgers, USPO, cleaners, Snoqualmie River RV Park(4mi), **S**...Blue Sky RV Park
20	High Point Way, **1 mi N**...Country Store
18	E Sunset Way, Issaquah(from wb, no EZ return), **1 mi S**...Cat Hospital, Texaco/mart, Jake's Grill, Ben Franklin, Radio Shack, Red Apple Foods, auto parts, bank, cleaners, same as 17
17	E Sammamish Rd, Front St, Issaquah, **N**...76/repair, Bake's Marine, RV park, **S**...CHIROPRACTOR, Arco/repair, Chevron/mart/24hr, Shell/mart, Texaco/diesel/mart/24hr, Domino's, Skipper's, Issaquah Village RV Park, TrueValue, U-Haul, bank, cinema
15	WA 900, Issaquah, Renton, **N**...Arco/mart/24hr, Georgio's Subs, IHOP, Red Robin Rest., Tully's Coffee, Holiday Inn, Motel 6, Barnes&Noble, Big 5 Sports, FutureShop, Harley-Davidson, Linens'n Things, Lowe's Bldg Supply, Office Depot, PetsMart, Pier 1, Trader Joe's, Unique Foods, cinema, to Lk Sammamish SP, **S**...DENTIST, Cenex, Texaco/ mart/wash/atm, Baskin-Robbins, Boston Mkt, Burger King, Denny's, Georgio's Subs, Godfather's, Honey B Ham, Jack-in-the-Box, JayBerry's Rest., KFC/Taco Bell, La Costa Mexican, Lombardi's Rest., McDonald's, RoundTable Pizza, Schlotsky's, Subway, Taco Del Mar, Taco Time, Teriyaki Bowl, 12th Ave Café, Al's Parts, Big O Tires, Firestone/ auto, Ford, GNC, Kinko's, Lamont's, Longs Drug, MailBoxes Etc, Mail Clinic, PetCo, Precision Tune, ProGolf, QFC Foods, Rite Aid, Ross, Safeway Foods, Schuck's Parts, Target, USPO, antiques, cleaners
13	SE Newport Way, W Lake Sammamish, no facilities
11	SE 150th, 156th, 161st, Bellevue, **N**...CHIROPRACTOR, VETERINARIAN, LDS Temple, Texaco/mart, 7-11, Dairy Queen, Little Jon's Rest., Mandarin Chinese, McDonald's, Day's Inn, Embassy Suites, Kane's Motel, Eastgate Motel, Ford, Jo-Ann Fabrics/crafts, Mazda/Subaru, Safeway, RV camping, **S**...Chevron/mart/24hr, Texaco/diesel/mart/24hr, Baskin-Robbins, Denny's, Domino's, Mondo's Coffee, Outback Steaks, Pizza Hut, Candlewood Suites, Albertson's/ drugs, Mail+, Rite Aid, Schuck's Parts, cleaners
10	I-405, N to Bellevue, S to Renton, facilities located off I-405 S, exit 10
9	Bellevue Way, no facilities
8	E Mercer Way, Mercer Island, no facilities
7b a	SE 76th Ave, 77th Ave, Island Crest Way, Mercer Island, **S**...BP/mart, 76, Texaco, Denny's, McDonald's, Subway, Travelodge, MailBoxes Etc, cleaners/laundry
6	W Mercer Way(from eb), same as 7
5mm	Lake Washington
3b a	Ranier Ave, Seattle, downtown, **N**...Texaco/mart
2c b	I-5, N to Vancouver, S to Tacoma
a	4th Ave S, to King Dome

I-90 begins/ends on I-5 at exit 164.

E

W

Issaquah Bellevue

Washington Interstate 182(Richland)

Exit #	Services
14	US 395 N, WA 397 S, OR Ave, **N...**Flying J/diesel/mart/24hr, Texaco, Burger King, Subway, NW Marine/sports, **S...**Budget Inn. **I-182 begins/ends on US 395 N.**
13	N 4th Ave, Cty Ctr, **N...**CFN/diesel, Airport Motel, Starlite Motel, **S...**RV park
12b	N 20th Ave, **N...**DoubleTree Hotel
a	US 395 S, Court St, **S...**Arco/mart/24hr, Conoco, Exxon, Shell, Arby's, Burger King, McDonald's, Pizza Hut, Subway, Taco Bell, Wendy's, Chief RV Ctr, Chevrolet, Food Pavilion/café, Ford, K-Mart, Toyota, U-Haul
9	WA 68, Trac, **N...**Arco/mart, Shell/mart, Heidi Haus Rest.
7	Broadmoor Blvd, **N...**Sleep Inn, Broadmoor Outlet Mall/famous brands, **S...**Broadmoor RV Ctr, Sandy Heights RV Park
6.5mm	Columbia River
5b a	WA 240 E, Geo Washington Way, to Kennewick, no facilities
4	WA 240 W, **N...**CFN/diesel, Conoco/diesel, 7-11, McDonald's, Fred Meyer
3.5mm	Yakima River
3	Queensgate, **N...**Arco/mart, Conoco/diesel/mart/24hr(2mi), **S...**Texaco/diesel/mart, RV Park(3mi)
0mm	I-182 begins/ends on I-82, exit 102.

Washington Interstate 405(Seattle)

Exit #	Services
30	I-5, N to Canada, S to Seattle, I-405 begins/ends on I-5, exit 182.
26	WA 527, Bothell, Mill Creek, **E...**MEDICAL CARE, Canyon's Rest., Denny's, McDonald's, QFC Foods, Lake Pleasant RV Park, bank, **W...**BP, Texaco/diesel/mart/atm, 7-11, Burger King, Dairy Queen, Godfather's, Grand Café, Mitzel's Kitchen, SubShop, Taco Time, Teriyaki Rest., Thai Rest., Comfort Inn, Albertson's, Bartel Drugs, MailRoom, Rite Aid
24	NE 195th St, Beardslee Blvd, **E...**Residence Inn, Wyndham Garden
23b	WA 522 W, Bothell
a	WA 522 E, to WA 202, Woodinville, Monroe, no facilities
22	NE 160th St, **E...**Chevron/mart, Texaco/diesel/mart/24hr, carwash
20	NE 124th St, **E...**HOSPITAL, CHIROPRACTOR, Arco/mart/24hr, BP, Chevron/mart/24hr, Texaco/diesel/mart/24hr, 7-11, Burger King, Denny's, Pietro's Pizza, Pizza Hut, Shari's, Skipper's, Best Western, Clarion Inn, Motel 6, Silver Cloud Inn, Larry's Mkt, Discount Tire, OfficeMax, Larry's Mkt, Rite Aid, hardware, books, mall, **W...**Arco/mart/24hr, BP, McDonald's, Taco Time, Wendy's, All Tune&Lube, Buick/Pontiac/GMC, Fred Meyer
18	WA 908, Kirkland, Redmond, **E...**Arco/mart/24hr, BP, Chevron/24hr, Texaco/diesel/mart, Burger King, McDonald's, Subway, TGIFriday, Chevrolet, Costco
17	NE 70th Pl, no facilities
14b a	WA 520, Seattle, Redmond, no facilities
13b	NE 8th St, **E...**HOSPITAL, Arco/mart/24hr, Chevron, Denny's, Extended Stay America, Ford, Home Depot, Larry's Mkt, **W...**Dairy Queen
a	NE 4th St, **W...**Azteca Mexican, Jonah's Rest., Bellevue Inn, Best Western, Doubletree Hotel, Hilton, Hyatt, **E...**Ford
12	SE 8th St, no facilities
11	I-90, E to Spokane, W to Seattle
10	Cold Creek Pkwy, Factoria, **1-2 mi E...**CHIROPRACTOR, DENTIST, BP/mart, 76/Circle K, 7-11, Godfather's, KFC, McDonald's, McHale's Rest., Old Country Buffet, Peking Wok, Presto Café, Subway, Taco Bell, Taco Time, Tokyo Rest., Winchell's, Homestead Suites, Al's Parts, Big 5 Sports, Ernst, FastLube, Honda, Mervyn's, Midas Muffler, Nordstrom's, Oriental Foods, QFC Foods, Rite Aid, Safeway, Target, cinema, cleaners, mall
9	112th Ave SE, Newcastle, phone
7	NE 44th St, **E...**Denny's/24hr, McDonald's, Travelers Inn, cleaners
6	NE 30th St, **E...**Arco/mart/24hr, **W...**Chevron/mart/24hr, Texaco/mart, 7-11
5	WA 900 E, Park Ave N, Sunset Blvd NE, no facilities
4	WA 169 S, Wa 900 W, Renton, **E...**Shari's, Silver Cloud Inn, Aqua Barn Ranch Camping, **W...**Texaco/diesel/mart/atm, Burger King, 7-11, Golden Palace Chinese, IGA Foods, Precision Tune
2	WA 167, Rainier Ave, to Auburn, **E...**HOSPITAL, **W...**DENTIST, Arco/mart/24hr, BP/mart, Chevron, Texaco/mart, USA Minimart/tires, Arby's, Baskin-Robbins, Boston Mkt, Burger King, China Garden, Diamond Lil's Rest., Dunkin Donuts, IHOP, Jack-in-the-Box, KFC, Mazatlan Mexican, McDonald's, Pizza Hut, Wendy's, Yankee Grill, Holiday Inn, Al's Parts, Chevrolet/Olds, Chrysler/Plymouth/Jeep, Dodge/Toyota, Ford, Fred Meyer, Firestone, Honda/Mazda/Suzuki/Kia, K-Mart/Little Caesar's, Lincoln/Mercury, Midas Muffler, Mitsubishi, Pontiac/Buick/GMC, Radio Shack, Schwab Tires, Subaru/Peugeot, Thriftway Foods, Wal-Mart/auto, bank
1	WA 181 S, Tukwila, **E...**CHIROPRACTOR, DENTIST, BP/diesel/mart/wash, Chevron/diesel, Texaco/diesel, 7-11, Barnaby's Rest., Burger King, Chili's, Jack-in-the-Box, McDonald's, Quizno's, Sizzler, Starbuck's, Taco Bell, Teriyaki Wok, Wendy's, Best Western, Courtyard, DoubleTree Suites, Embassy Suites, Hampton Inn, Homestead Village, Residence Inn, Acura, Barnes&Noble, Bon-Marche, CamerasWest, Circuit City, Computer City, Eagle Hardware, Firestone, FutureShop, Goodyear, JC Penney, Kinko's, Mervyn's, Nordstrom's, Office Depot, Pearle Vision, Puetz Golf, Sears, Target, cinema, **W...**Homewood Suites, Studio+
0mm	I-5, N to Seattle, S to Tacoma, WA 518 W. I-405 begins/ends on I-5, exit 154.

West Virginia Interstate 64

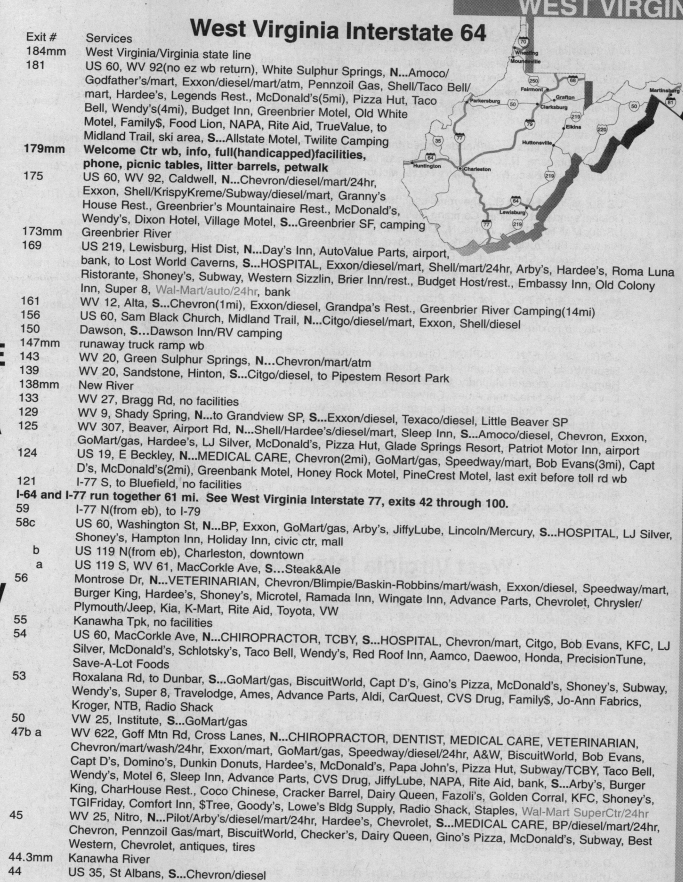

Exit #	Services
184mm	West Virginia/Virginia state line
181	US 60, WV 92(no ez wb return), White Sulphur Springs, **N**...Amoco/Godfather's/mart, Exxon/diesel/mart/atm, Pennzoil Gas, Shell/Taco Bell/mart, Hardee's, Legends Rest., McDonald's(5mi), Pizza Hut, Taco Bell, Wendy's(4mi), Budget Inn, Greenbrier Motel, Old White Motel, Family$, Food Lion, NAPA, Rite Aid, TrueValue, to Midland Trail, ski area, **S**...Allstate Motel, Twilite Camping
179mm	**Welcome Ctr wb, info, full(handicapped)facilities, phone, picnic tables, litter barrels, petwalk**
175	US 60, WV 92, Caldwell, **N**...Chevron/diesel/mart/24hr, Exxon, Shell/KrispyKreme/Subway/diesel/mart, Granny's House Rest., Greenbrier's Mountainaire Rest., McDonald's, Wendy's, Dixon Hotel, Village Motel, **S**...Greenbrier SF, camping
173mm	Greenbrier River
169	US 219, Lewisburg, Hist Dist, **N**...Day's Inn, AutoValue Parts, airport, bank, to Lost World Caverns, **S**...HOSPITAL, Exxon/diesel/mart, Shell/mart/24hr, Arby's, Hardee's, Roma Luna Ristorante, Shoney's, Subway, Western Sizzlin, Brier Inn/rest., Budget Host/rest., Embassy Inn, Old Colony Inn, Super 8, Wal-Mart/auto/24hr, bank
161	WV 12, Alta, **S**...Chevron(1mi), Exxon/diesel, Grandpa's Rest., Greenbrier River Camping(14mi)
156	US 60, Sam Black Church, Midland Trail, **N**...Citgo/diesel/mart, Exxon, Shell/diesel
150	Dawson, **S**...Dawson Inn/RV camping
147mm	runaway truck ramp wb
143	WV 20, Green Sulphur Springs, **N**...Chevron/mart/atm
139	WV 20, Sandstone, Hinton, **S**...Citgo/diesel, to Pipestem Resort Park
138mm	New River
133	WV 27, Bragg Rd, no facilities
129	WV 9, Shady Spring, **N**...to Grandview SP, **S**...Exxon/diesel, Texaco/diesel, Little Beaver SP
125	WV 307, Beaver, Airport Rd, **N**...Shell/Hardee's/diesel/mart, Sleep Inn, **S**...Amoco/diesel, Chevron, Exxon, GoMart/gas, Hardee's, LJ Silver, McDonald's, Pizza Hut, Glade Springs Resort, Patriot Motor Inn, airport
124	US 19, E Beckley, **N**...MEDICAL CARE, Chevron(2mi), GoMart/gas, Speedway/mart, Bob Evans(3mi), Capt D's, McDonald's(2mi), Greenbank Motel, Honey Rock Motel, PineCrest Motel, last exit before toll rd wb
121	I-77 S, to Bluefield, no facilities

I-64 and I-77 run together 61 mi. See West Virginia Interstate 77, exits 42 through 100.

59	I-77 N(from eb), to I-79
58c	US 60, Washington St, **N**...BP, Exxon, GoMart/gas, Arby's, JiffyLube, Lincoln/Mercury, **S**...HOSPITAL, LJ Silver, Shoney's, Hampton Inn, Holiday Inn, civic ctr, mall
b	US 119 N(from eb), Charleston, downtown
a	US 119 S, WV 61, MacCorkle Ave, **S**...Steak&Ale
56	Montrose Dr, **N**...VETERINARIAN, Chevron/Blimpie/Baskin-Robbins/mart/wash, Exxon/diesel, Speedway/mart, Burger King, Hardee's, Shoney's, Microtel, Ramada Inn, Wingate Inn, Advance Parts, Chevrolet, Chrysler/Plymouth/Jeep, Kia, K-Mart, Rite Aid, Toyota, VW
55	Kanawha Tpk, no facilities
54	US 60, MacCorkle Ave, **N**...CHIROPRACTOR, TCBY, **S**...HOSPITAL, Chevron/mart, Citgo, Bob Evans, KFC, LJ Silver, McDonald's, Schlotsky's, Taco Bell, Wendy's, Red Roof Inn, Aamco, Daewoo, Honda, PrecisionTune, Save-A-Lot Foods
53	Roxalana Rd, to Dunbar, **S**...GoMart/gas, BiscuitWorld, Capt D's, Gino's Pizza, McDonald's, Shoney's, Subway, Wendy's, Super 8, Travelodge, Ames, Advance Parts, Aldi, CarQuest, CVS Drug, Family$, Jo-Ann Fabrics, Kroger, NTB, Radio Shack
50	VW 25, Institute, **S**...GoMart/gas
47b a	WV 622, Goff Mtn Rd, Cross Lanes, **N**...CHIROPRACTOR, DENTIST, MEDICAL CARE, VETERINARIAN, Chevron/mart/wash/24hr, Exxon/mart, GoMart/gas, Speedway/diesel/24hr, A&W, BiscuitWorld, Bob Evans, Capt D's, Domino's, Dunkin Donuts, Hardee's, McDonald's, Papa John's, Pizza Hut, Subway/TCBY, Taco Bell, Wendy's, Motel 6, Sleep Inn, Advance Parts, CVS Drug, JiffyLube, NAPA, Rite Aid, bank, **S**...Arby's, Burger King, CharHouse Rest., Coco Chinese, Cracker Barrel, Dairy Queen, Fazoli's, Golden Corral, KFC, Shoney's, TGIFriday, Comfort Inn, $Tree, Goody's, Lowe's Bldg Supply, Radio Shack, Staples, Wal-Mart SuperCtr/24hr
45	WV 25, Nitro, **N**...Pilot/Arby's/diesel/mart/24hr, Hardee's, Chevrolet, **S**...MEDICAL CARE, BP/diesel/mart/24hr, Chevron, Pennzoil Gas/mart, BiscuitWorld, Checker's, Dairy Queen, Gino's Pizza, McDonald's, Subway, Best Western, Chevrolet, antiques, tires
44.3mm	Kanawha River
44	US 35, St Albans, **S**...Chevron/diesel

E

W

Lewisburg

Charleston

West Virginia Interstate 64

39	WV 34, Winfield, **N**...BP/Arby's/Baskin-Robbins/mart, GoMart/diesel, Applebee's, Bob Evans, HalfTime Café, Hardee's, Rio Grande Mexican, Day's Inn, Holiday Inn Express, Red Roof Inn, Advance Parts, Ames, Big Bear Food/drug, Big Lots, $General, USPO, cinema, **S**...MEDICAL CARE, CHIROPRACTOR, Exxon/diesel/mart, GoMart/gas, TA/diesel/rest./24hr, Burger King, Capt D's, China Chef, Fazoli's, Gino's Pizza, KFC, Kobe Japanese, McDonald's, Papa John's, Schlotsky's, Shoney's, Subway, Taco Bell, TCBY, Wendy's, Hampton Inn, K-Mart, Kroger, Rite Aid, bank
38mm	weigh sta both lanes
35mm	**rest area both lanes, full(handicapped)facilities, phone, vending, picnic tables, litter barrels, petwalk**
34	WV 19, Hurricane, **N**...Chevrolet, Ford/Lincoln/Mercury, Saturn, **S**...VETERINARIAN, Chevron/diesel, GoMart/gas, Sunoco/diesel, BiscuitWorld/Gino's Pizza, McDonald's, Nancy's Rest., Pizza Hut, Subway, Ramada Ltd, Pennzoil, bank
28	US 60, Milton, **1-3 mi S**...Chevron/diesel, Exxon, GoMart/diesel, Marathon, Granny's Rest., McDonald's, The Colonel's Inn/rest., Foxfire Camping Resort
20	US 60, Mall Rd, Barboursville, **N**...MEDICAL CARE, Chevron, Exxon/diesel/mart, Applebee's, Bob Evans, Hardee's, Fuddrucker's, Logan's Roadhouse, McDonald's, Olive Garden, Wendy's, Comfort Inn, Holiday Inn, Borders Books, Chrysler/Jeep, Circuit City, Drug Emporium, Firestone/auto, JC Penney, Jo-Ann Fabrics, Kohl's, Lazarus, Linens'n Things, Lowe's Bldg Supply, Michael's, OfficeMax, Pier 1, Sears/auto, ToysRUs, cinema, mall, **S**...DENTIST, CHIROPRACTOR, BP/KrispyKreme/mart, Exxon, RaceTrac/mart, Cracker Barrel, Fiesta Bravo Mexican, Gino's Pizza, Johnny's Pizza, LoneStar Steaks, Outback Steaks, Ponderosa, Sam's Hotdogs, Sonic, Subway, TCBY, Thai Am Rest., Hampton Inn, Ramada Ltd, FasLube, Lincoln/Mercury, Mazda, Olds/Cadillac, laundry, to Foxfire Camping Resort
18	new exit
15	US 60, 29th St E, **N**...HOSPITAL, Chevron, GoMart/diesel, Speedway/mart, Sunoco, Arby's, Burger King(2mi), BiscuitWorld, Nanna's Country Rest., Omelet Shoppe, Pizza Hut, Wendy's, Colonial Inn, Econolodge, Holiday Inn, Ramada Inn, cleaners/laundry, **S**...Exxon/mart, Fazoli's, Golden Corral, KFC, McDonald's, Shoney's, Taco Bell, Day's Inn, Red Roof Inn, Ames, Chrysler/Dodge/Jeep, CVS Drug, Office Depot, Nissan/Daewoo/Mitsubishi/VW, Office Depot, Pontiac/GMC/Buick/Isuzu, Subaru, Wal-Mart/auto
11	WV 10, Hal Greer Blvd, **1-3 mi N**...HOSPITAL, GoMart/gas, Marathon, Hardee's, McDonald's, Holiday Inn, Radisson, **S**...Beech Fork SP(8mi), Sleepy Hollow Camping
10mm	**Welcome Ctr eb, full(handicapped)facilities, phone, vending, picnic tables, litter barrels, petwalk**
8	WV 152 S, WV 527 N, **N**...MEDICAL CARE, **S**...Speedway/mart, Dairy Queen, Laredo Steaks, Ames
6	US 52 N, W Huntington, Chesapeake, **N**...VA HOSPITAL, Chevron/diesel, GoMart/gas, Marathon, Speedway/Blimpie/mart/24hr, Hardee's, Pizza Hut, Shoney's, Coaches Inn, Family$
1	US 52 S, Kenova, **1-3 mi N**...Exxon/mart, Burger King, McDonald's, Pizza Hut, Hollywood Motel, Camden Park Camping, airport
0mm	Big Sandy River, West Virginia/Kentucky state line

West Virginia Interstate 68

Exit #	Services
32mm	West Virginia/Maryland state line
29	WV 5, Hazelton Rd, **N**...Mobil/diesel/mart, **S**...Casteel's Dairy King, RV camping
23	WV 26, Bruceton Mills, **N**...CHIROPRACTOR, Ashland/diesel/rest./atm/24hr, Sandy's/Mobil/diesel/rest./atm/24hr, Country Fixins Rest., MillPlace Rest., Tastee Queen, MapleLeaf Motel, Better Health Foods, Murphy's Foods, USPO, antiques, auto repair, bank, **15 mi S**...Heldreth Motel
18mm	Laurel Run
17mm	runaway truck ramp eb
15	WV 73, WV 12, Cooper's Rock, **N**...Chestnut Ridge SF, **S**...Cooper's Rock Camping
12mm	runaway truck ramp wb
10	WV 857, Fairchance Rd, Cheat Lake, **N**...DENTIST, VETERINARIAN, BP/diesel/mart/24hr, Exxon/diesel, Café Delight, La Casita Mexican, Mario's Pizza, Ruby&Ketchy's Rest., Subway, Lakeview Resort, bank, hardware, laundry, **S**...MEDICAL CARE, AJ's Rest., Burger King, Stone Crab Inn
9mm	Cheat Lake
7	WV 857, Pierpont Rd, **N**...HOSPITAL, CHIROPRACTOR, Exxon/Taco Bell/diesel/mart/atm, Bob Evans, EuroSuites Hotel(4mi), Wal-Mart SuperCtr, to WVU Stadium, airport, bank, **S**...Getty/mart, Tiberio's Rest.
4	WV 7, to Sabraton, **1 mi N**...DENTIST, Amoco/mart, BP, Exxon/Blimpie/mart/24hr, Archie's Rest., Hardee's, Hero's Hut, KFC, LJ Silver, McDonald's, Pizza Hut, Ponderosa, Subway, Wendy's, Advance Parts, Chrysler/Dodge, CVS Drug, Food Lion, Ford, Kroger/deli/24hr, NAPA, bank, carwash, laundry, **S**...BP/diesel/repair, Express/gas/atm, Sunoco/mart
3mm	Decker's Creek
1	US 119, Morgantown, **N**...Exxon/diesel(2mi), Fabian's Rest., Subway, Comfort Inn, Morgantown Motel(2mi), Ramada Inn/rest., **S**...to Tygart L SP
0mm	I-79, N to Pittsburgh, S to Clarksburg. I-68 begins/ends on I-79, exit 148.

West Virginia Interstate 70

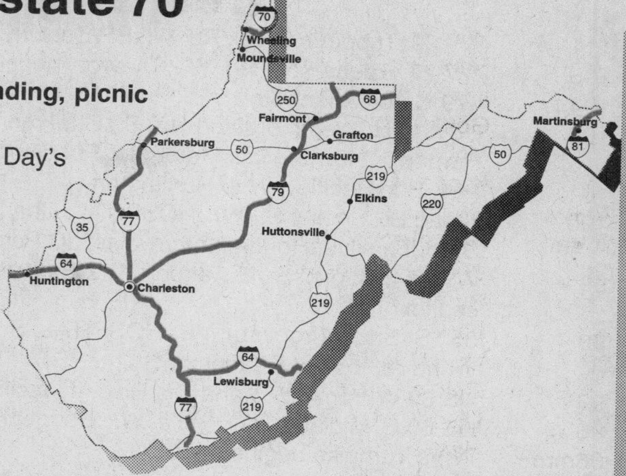

E — W

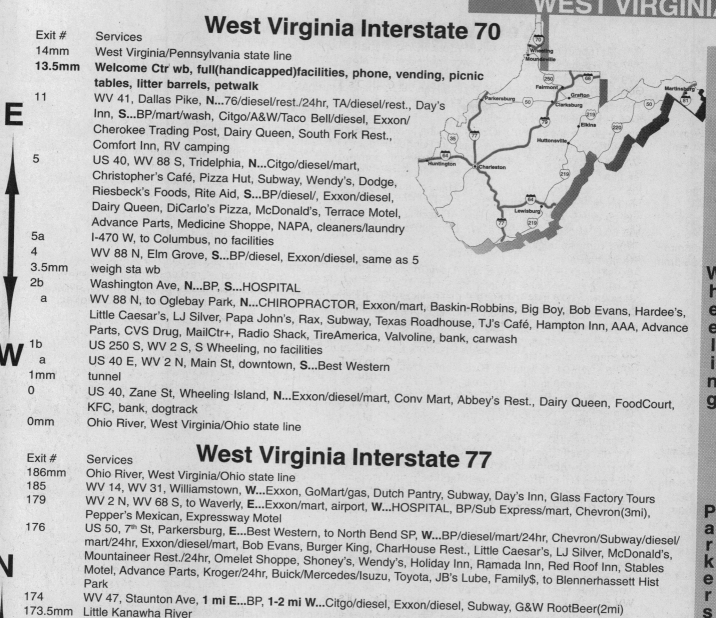

Exit #	Services
14mm	West Virginia/Pennsylvania state line
13.5mm	**Welcome Ctr wb, full(handicapped)facilities, phone, vending, picnic tables, litter barrels, petwalk**
11	WV 41, Dallas Pike, **N...**76/diesel/rest./24hr, TA/diesel/rest., Day's Inn, **S...**BP/mart/wash, Citgo/A&W/Taco Bell/diesel, Exxon/Cherokee Trading Post, Dairy Queen, South Fork Rest., Comfort Inn, RV camping
5	US 40, WV 88 S, Tridelphia, **N...**Citgo/diesel/mart, Christopher's Café, Pizza Hut, Subway, Wendy's, Dodge, Riesbeck's Foods, Rite Aid, **S...**BP/diesel/, Exxon/diesel, Dairy Queen, DiCarlo's Pizza, McDonald's, Terrace Motel, Advance Parts, Medicine Shoppe, NAPA, cleaners/laundry
5a	I-470 W, to Columbus, no facilities
4	WV 88 N, Elm Grove, **S...**BP/diesel, Exxon/diesel, same as 5
3.5mm	weigh sta wb
2b	Washington Ave, **N...**BP, **S...**HOSPITAL
a	WV 88 N, to Oglebay Park, **N...**CHIROPRACTOR, Exxon/mart, Baskin-Robbins, Big Boy, Bob Evans, Hardee's, Little Caesar's, LJ Silver, Papa John's, Rax, Subway, Texas Roadhouse, TJ's Café, Hampton Inn, AAA, Advance Parts, CVS Drug, MailCtr+, Radio Shack, TireAmerica, Valvoline, bank, carwash
1b	US 250 S, WV 2 S, S Wheeling, no facilities
a	US 40 E, WV 2 N, Main St, downtown, **S...**Best Western
1mm	tunnel
0	US 40, Zane St, Wheeling Island, **N...**Exxon/diesel/mart, Conv Mart, Abbey's Rest., Dairy Queen, FoodCourt, KFC, bank, dogtrack
0mm	Ohio River, West Virginia/Ohio state line

West Virginia Interstate 77

N — S

Exit #	Services
186mm	Ohio River, West Virginia/Ohio state line
185	WV 14, WV 31, Williamstown, **W...**Exxon, GoMart/gas, Dutch Pantry, Subway, Day's Inn, Glass Factory Tours
179	WV 2 N, WV 68 S, to Waverly, **E...**Exxon/mart, airport, **W...**HOSPITAL, BP/Sub Express/mart, Chevron(3mi), Pepper's Mexican, Expressway Motel
176	US 50, 7th St, Parkersburg, **E...**Best Western, to North Bend SP, **W...**BP/diesel/mart/24hr, Chevron/Subway/diesel/mart/24hr, Exxon/diesel/mart, Bob Evans, Burger King, CharHouse Rest., Little Caesar's, LJ Silver, McDonald's, Mountaineer Rest./24hr, Omelet Shoppe, Shoney's, Wendy's, Holiday Inn, Ramada Inn, Red Roof Inn, Stables Motel, Advance Parts, Kroger/24hr, Buick/Mercedes/Isuzu, Toyota, JB's Lube, Family$, to Blennerhassett Hist Park
174	WV 47, Staunton Ave, **1 mi E...**BP, **1-2 mi W...**Citgo/diesel, Exxon/diesel, Subway, G&W RootBeer(2mi)
173.5mm	Little Kanawha River
173	WV 95, Camden Ave, **E...**RV camping, **1-4 mi W...**HOSPITAL, BP, Marathon/diesel, Bonanza, Hardee's, Blennerhassett Hotel
170	WV 14, Mineral Wells, **E...**BP/diesel/mart/repair/24hr, Chevron/mart, Citgo/subs/diesel/mart, Exxon/mart/24hr, Liberty Trkstp/diesel/24hr, McDonald's, Subway, Wendy's, Western Steer, Comfort Suites, Hampton Inn, USPO, bank, **W...**Cracker Barrel, Napoli's Pizza, AmeriHost, Microtel
169mm	weigh sta both lanes, phone
166mm	**rest area both lanes, full(handicapped)facilities, phone, picnic tables, litter barrels, vending, petwalk**
161	WV 21, Rockport, **1/2 mi W...**Marathon
154	WV 1, Medina Rd, no facilities
146	WV 2 S, Silverton, Ravenswood, **W...**BP/diesel, Marathon/diesel/mart/24hr, Bob's Café, Scottish Inn
138	US 33, Ripley, **E...**VETERINARIAN, Duke/diesel/mart/24hr, Exxon/mart/atm/24hr, Marathon/diesel/mart, Cozumel Mexican, KFC, LJ Silver, McDonald's, Pizza Hut, Twister's Grill, Wendy's, Best Western/rest., Super 8, Kroger/drug/24hr, NAPA, Rite Aid, Sav-A-Lot, Wal-Mart, Ruby Lake Camping, bank, **W...**HOSPITAL, DENTIST, Exxon/diesel/mart/24hr, Ponderosa, Shoney's, Subway, Holiday Inn Express
132	WV 21, Fairplain, **E...**BP/Burger King/mart, GoMart/diesel, Marathon/diesel, Ford/Mercury, 4 Seasons RV, **W...**77 Motor Inn, Country Treasures Gifts, camping
124	WV 34, Kenna, **E...**Exxon, Pattie's Country Cookin', Your Family Rest., Craft Shop
119	WV 21, Goldtown, no facilities
116	WV 21, Haines Branch Rd, Sissonville, **4 mi E...**Rippling Waters Campground
114	WV 622, Pocatalico Rd, **E...**B&D Market/gas, camping

West Virginia Interstate 77

C **h** **a** **r** **l** **e** **s** **t** **o** **n**	111	WV 29, Tupper's Creek Rd, **W**...BP/Subway/diesel/mart, Tudor's Biscuit World
	106	WV 27, Edens Fork Rd, **W**...Chevron/diesel/country store
	104	I-79 N, to Clarksburg
	102	US 119 N, Westmoreland Rd, **E**...Chevron/7-11/24hr, GoMart/gas/24hr, Pennzoil/mart/24hr, China Garden, Hardee's/24hr, Cutlip's Motor Inn, Foodland, Rite Aid
	101	I-64, E to Beckley, W to Huntington
	100	Broad St, Capitol St, **W**...HOSPITAL, Charleston House Rest., Fairfield Inn, Holiday Inn, Marriott, Super 8, CVS Drug, Family$, Honda Motorcycles, JC Penney, Kroger, Olds/GMC, Ryder Trucks
	99	WV 114, Capitol St, **E**...airport, **W**...Exxon/mart, Domino's, KFC/Taco Bell, McDonald's, Wendy's, to museum, st capitol
	98	35th St Bridge(from sb), **W**...HOSPITAL, Sunoco, KFC/Taco Bell, McDonald's, Steak Escape, Shoney's, Subway, Wendy's, to U of Charleston
	97	US 60 W(from nb), Kanawha Blvd, no facilities
	96	US 60 E, Midland Trail, Belle, **W**...Budget Host
	96mm	W Va Turnpike begins/ends
	95.5mm	Kanawha River
	95	WV 61, to MacCorkle Ave, **E**...BP/DairyCheer/diesel/mart, GoMart/diesel/24hr, Bob Evans, LoneStar Steaks, McDonald's, Wendy's, Country Inn Suites, Day's Inn, Knight's Inn, Motel 6, Red Roof, Advance Parts, K-Mart/auto, **W**...MEDICAL CARE, DENTIST, VETERINARIAN, Ashland/Blimpie/TCBY/mart/24hr, Chevron/7-11, Exxon/diesel/mart/24hr, Marathon, GoMart/24hr, Applebee's, Azteca Mexican, Burger King, Capt D's, ChiChi's, China Buffet, Hooters, Pizza Hut/Taco Bell, Ponderosa, Shoney's, Southern Kitchen/24hr, Shoney's, Ames, BrakeShop, Drug Emporium, Foodland/24hr, Kroger/drugs/24hr, Lowe's Bldg Supply, NAPA, Subaru, Suzuki, bank, cinema, cleaners, mall, repair
	89	WV 61, WV 94, to Marmet, **E**...Exxon/Subway/diesel/mart/atm/24hr, GoMart/gas/24hr, Sunoco, Biscuit World, Gino's Pizza/spaghetti, Hardee's, KFC, LJ Silver, Wendy's, Family$, Kroger/deli, Rite Aid, auto parts, **W**...Ford(1mi)
	85	US 60, WV 61, East Bank, **E**...Citgo/mart, GoMart/diesel/24hr, Gino's Pizza, McDonald's, Shoney's, Chevrolet, Gilmer Foods, Kroger, tire repair
	82.5mm	toll booth
	79	Cabin Creek Rd, Sharon, no facilities
	74	WV 83, Paint Creek Rd, no facilities
	72mm	**Morton Service Area nb, Exxon/diesel, Hotdog City, Roy Rogers, TCBY, atm**
	69mm	**rest area sb, full(handicapped)facilities, phone, picnic tables, litter barrels**
	66	WV 15, to Mahan, **1/2 mi W**...Sunoco/diesel/mart/24hr
	60	WV 612, to Mossy, Oak Hill, **1/2 mi E**...Exxon/diesel/repair/24hr, RV camping
	56.5mm	toll plaza, phone
	54	rd 2, rd 23, Pax, **E**...BP/rest., **W**...Long Branch Rest.
B **e** **c** **k** **l** **e** **y**	48	US 19, N Beckley, **1 mi E**...Amoco/Subway/diesel/mart, LeFayette's Rest., Comfort Inn, Ramada Inn, Belk, JC Penney, Sears/auto, **4 mi E on US 19/VA 16 S**...Chevron, Shell, Bob Evans, Burger King, Checker's, Fazoli's, Hardee's, LoneStar Steaks, LJ Silver, McDonald's, Rio Grande Mexican, Ryan's, Wendy's, Acme Foods, Acura, AutoZone, Big Lots, Buick, Chevrolet/Olds/Cadillac, Chrysler/Plymouth, CVS Drug, Dodge, Goody's, Honda, Hyundai, K-Mart, Staples, Subaru/Kia, U-Haul, Wal-Mart SuperCtr/24hr, bank
	45mm	**Tamarack Service Area both lanes, W...Exxon/diesel, BiscuitWorld, Sbarro's, Starbucks, Taco Bell, TCBY, gifts**
	44	WV 3, Beckley, **E**...MEDICAL CARE, EYECARE, Chevron/diesel/mart, Exxon/mart/24hr, Shell/Hardee's/TCBY/mart/24hr, Applebee's, BBQ, Burger King, Dairy Queen, Japanese Steaks, McDonald's, Outback Steaks, Pancake House, Pizza Hut, Pizza Inn, Best Western, Comfort Inn, Courtyard, Fairfield Inn, Holiday Inn, Howard Johnson, Super 8, Big Lots, CVS Drug, Kroger/deli/24hr, bank, museum(3mi), **W**...BP/Subway/Baskin-Robbins/mart/24hr, GoMart/diesel/24hr, Bob Evans, Cracker Barrel, Fox's Pizza, Pasquale's Italian, Texas Roadhouse, Wendy's, Country Inn Suites, Day's Inn, Hampton Inn, Microtel, Shoney's Inn/rest.
	42	WV 16, WV 97, to Mabscott, **2 mi E**...HOSPITAL, Exxon, Tudor's Biscuits, Budget Inn, **W**...Amoco/diesel, Godfather's, Subway
	40	I-64 E, to Lewisburg, no facilities
	30mm	toll booth, phone
	28	WV 48, to Ghent, **E**...BP, Exxon/diesel, Marathon/diesel, Glade Springs Resort(1mi), Appalachian Resort Inn(12mi), to ski area, **W**...Econolodge
	26.5mm	Flat Top Mtn, elevation 3252
	20	US 19, to Camp Creek, **E**...Exxon/diesel/mart, Tolliver's T-stop, **W**...Camp Creek SP
	18.5mm	scenic overlook/parking area/weigh sta sb, no facilities
	18mm	Bluestone River
	17mm	**Bluestone Service Area/weigh sta nb, full(handicapped)facilities, scenic view, picnic area, Exxon/diesel, Hotdog City, Roy Rogers, TCBY, atm/fax**
	14	WV 20, Athens Rd, **E**...Citgo, Green Acres RV Park, Pipestem Resort SP
	9mm	WV Turnpike begins/ends

N

S

West Virginia Interstate 77

N ↑↓ S

9	US 460, Princeton, **E...Welcome Ctr**, I-77 Trkstp/diesel, Cato's, FashionBug, Goody's, Wal-Mart SuperCtr/24hr, **W...HOSPITAL**, Amoco/diesel/mart, Chevron/diesel/mart/24hr, Exxon/diesel/mart/24hr, Marathon/Subway/diesel/mart/atm, Applebee's, Bob Evans, Burger King, Capt D's, Cracker Barrel, Dairy Queen, Hardee's, McDonald's, Omelet Shoppe, RomaX Italian, Shoney's, Taco Bell, Texas Steaks, Wendy's, Budget Inn, Comfort Inn, Day's Inn, Hampton Inn, Johnston's Inn/rest., Ramada Ltd, Sleep Inn, Super 8, Town&Country Motel, Turnpike Motel, K-Mart/auto, camping
7	WV 27, Ingleside Rd(from nb, no re-entry), no facilities
5	WV 112(from sb, no re-entry), to Ingleside, no facilities
3mm	East River
1	US 52 N, to Bluefield, **4 mi W...HOSPITAL**, LJ Silver, Wendy's, Brier Motel, Econolodge, Highlander Motel, Holiday Inn/rest., Knight's Inn, Ramada Inn
.5mm	East River Mtn
0mm	West Virginia/Virginia state line

West Virginia Interstate 79

Exit #	Services
160.5mm	West Virginia/Pennsylvania state line
159	**Welcome Ctr sb, full(handicapped)facilities, info, picnic tables, litter barrels, phone, vending, petwalk**
155	US 19, WV 7, **E...**Sheetz/mart, **3 mi E...HOSPITAL**, GoMart/gas, Burger King, Eat'n Park, Hardee's, LJ Silver, Shoney's, Wendy's, Day's Inn, Econolodge, Euro Suites, Friends Inn, Hampton Inn, Holiday Inn, to WVU, **W...**Exxon/diesel
152	US 19, to Morgantown, **E...MEDICAL CARE**, BP/mart, Exxon/mart, Getty Gas, McDonald's, Pizza Hut, Subway, Taco Bell, Western Sizzlin, Clarion, Econolodge, Big Lots, Midas Muffler, **W...**Bob Evans, Burger King, Capt D's, K-Mart, mall
150mm	Monongahela River
148	I-68 E, to Cumberland, MD, **1 mi E...**Exxon/diesel/mart/24hr, Comfort Inn, Morgantown Motel, Ramada Inn/rest.
146	WV 77, to Goshen Rd, no facilities
141mm	weigh sta both lanes
139	WV 33, E Fairmont, **E...**Chevron, **W...**Exxon/24hr, K&T/BP/diesel, RV camping
137	WV 310, to Fairmont, **E...**Exxon/diesel/24hr, Simmering Pot Rest., Subway, Holiday Inn, antiques, to Valley Falls SP, **W...HOSPITAL**, Chevron/mart/wash, KFC, McDonald's, Wendy's, **3 mi W...**Belmont Motel/rest.
135	WV 64, Pleasant Valley Rd, no facilities
133	Kingmont Rd, **E...**BP/diesel/mart, Exxon/diesel/mart, Cracker Barrel, Super 8, **W...**Chevron/diesel/mart/24hr, Citgo, DJ's Diner, Comfort Inn Suites, Country Club Motel, Day's Inn, Fairmont Motel, Red Roof Inn
132	US 250, S Fairmont, **E...**Bob Evans, Hardee's, McDonald's, Mi Pueblo Mexican, Shoney's, Subway, TeeJaye's Rest., Taco Bell, Day's Inn, Red Roof Inn, Ames, Chrysler/Dodge, $General, FashionBug, FoodLand, GreaseMonkey, Harley-Davidson, Hess, JC Penney, Jo-Ann Fabrics, Sam's Club, Shop'n Save, Wal-Mart/drugs, RV Ctr, bank, cinema, mall, to Tygart Lake SP, **W...HOSPITAL**, Citgo, Exxon/diesel/mart/24hr, GoMart/diesel/24hr, Sunoco, USA Steaks(4mi), Country Club Motel(4mi), Fairmont Motel(4mi), Buick/Olds/Pontiac/GMC, Ford/Lincoln/Mercury, Toyota, cleaners, tires
125	Saltwell Rd, to Shinnston, airport, **E...**Oliverio's Rest., **W...**Exxon/Subway/diesel/mart/24hr
124	WV 131, FBI Ctr Rd, no facilities
123mm	**rest area both lanes, full(handicapped)facilities, info, phone, picnic tables, litter barrels, vending, petwalk**
121	WV 24, Meadowbrook Rd, **E...MEDICAL CARE**, GoMart/Blimpie/24hr, Sheetz/mart, Bob Evans, Hampton Inn, bank, **W...VETERINARIAN**, Exxon/Subway/diesel/mart, Arby's, Burger King, Garfield's Rest., Jo-Ann Fabrics, Outback Steaks, Ponderosa, Super 8, Honda, JC Penney, NTB, OfficeMax, Phar-Mor Drugs, ToysRUs, Ward's/auto, cinema, mall
119	US 50, to Clarksburg, **E...HOSPITAL**, Chevron/diesel/mart/24hr, Exxon/diesel/mart, GoMart/gas, Speedway/mart, Big Boy, China Buffet, Damon's, Eat'n Park, Hardee's, KFC, Little Caesar's, LJ Silver, Maxey's Rest., McDonald's, Shoney's, Taco Bell, Texas Roadhouse, USA Steaks, Wendy's, Western Steer, Comfort Inn, Day's Inn, Holiday Inn, Knight's Inn, Ramada Ltd, Sleep Inn, Towne House Motel, Ames, Big Lots, Chevrolet, K-Mart/drugs, Kroger/gas/24hr, Lowe's Bldg Supply, Midas Muffler, Nissan, Radio Shack
117	WV 58, to Anmoore, **E...**Applebee's, Arby's, Burger King, Heavenly Ham, Ryan's, Subway, Circuit City, $Tree, FashionBug, Goody's, Staples, Wal-Mart SuperCtr/24hr, **W...**BP/diesel
115	WV 20, to Stonewood, Nutter Fort, **E...**Chevron/mart, Exxon/diesel, Texaco/diesel/mart, Marty's Italian Bread, Sandy Dandy's Rest., **W...HOSPITAL**, Speedway/mart, Mountaineer Family Rest., Greenbrier Motel
110	Lost Creek, **1 mi E...**BP/diesel, Citgo, bank

West Virginia Interstate 79

105	WV 7, to Jane Lew, **E**...Chevron/diesel/rest., I-79 Trkstp/diesel, Wilderness Plantation Inn/rest., **W**...Exxon, glass factory tours
99	US 33, US 119, to Weston, **E**...BP/mart, Exxon/diesel/mart, Sheetz/mart/24hr, McDonald's, Subway, Western Sizzlin, Bicentennial Motel, Budget Host, Comfort Inn/rest., Super 8, Advance Parts, Cato's, Family$, GNC, Kroger, Radio Shack, Wal-Mart/drugs, bank, cinema, **2 mi W**...HOSPITAL, BP, Rich/mart, Donut Connection, Hardee's, KFC, Pizza Hut, Wendy's, $General, Ford/Mercury, NAPA, Twin Lakes Camper Sales, to Canaan Valley Resort, Blackwater Falls, camping
96	WV 30, to S Weston, **E**...to S Jackson Lake SP, Broken Wheel Camping, **W**...Weston Motel
91	US 19, to Roanoke, **E**...Whisper Mtn Camping, to S Jackson Lake SP
85mm	**rest area both lanes, full(handicapped)facilities, info, picnic tables, litter barrels, phone, vending, petwalk, RV dump**
79.5mm	Little Kanawha River
79	WV 5, Burnsville, **E**...Exxon/24hr, 79 Motel/rest., NAPA, **W**...GoMart/repair, bank, Cedar Cr SP
78mm	Saltlick Creek
67	WV 4, to Flatwoods, **E**...Chevron/diesel/mart, Exxon/diesel, GoMart/diesel/rest., Dairy Queen, KFC/Taco Bell, McDonald's, Subway, Waffle Hut/24hr, Western Steer, Day's Inn, Sutton Lake Motel, Ace Hardware, Chevrolet/ Olds/Buick, to Sutton Lake RA, camping, **W**...Ashland/diesel, Citgo/mart, Shoney's, Wendy's, Flatwood Stores/ famous brands
62	WV 4, to Sutton, Gassaway, **1-2 mi E**...GoMart/gas/24hr, Century Inn/rest., Elk Motel, Dodge, **W**...HOSPITAL, GoMart/diesel, LJ Silver, Pizza Hut, CVS Drug, Ford/Mercury, Kroger/deli, Super$
57	US 19 S, to Beckley, **E**...Exxon, Holiday Inn Express
52mm	Elk River
51	WV 4, to Frametown, no facilities
49mm	**rest area both lanes, full(handicapped)facilities, picnic tables, phone, litter barrels, vending, petwalk, RV dump**
46	WV 11, Servia Rd, no facilities
40	WV 16, to Big Otter, **E**...GoMart/diesel/24hr, **W**...Exxon/diesel/mart, Country Inn
34	WV 36, to Wallback, **E**...Gino's Diner, BiscuitWorld
25	WV 29, to Amma, **E**...Exxon/diesel
19	US 119, VW 53, to Clendenin, **E**...BP/diesel/24hr, BiscuitWorld, Gino's Diner, Shafer's Superstop, Kopper Kettle(3mi), Jackson Motel(3mi)
9	WV 43, to Elkview, **W**...Exxon/Arby's/diesel/mart, Speedway/diesel/mart/24hr, McDonald's, Pizza Hut, Ponde-rosa, Subway, Holiday Inn Express, Advance Parts, CVS Drug, GNC, K-Mart/drugs, Kroger/deli/24hr, Radio Shack
5	WV 114, to Big Chimney, **1 mi E**...Exxon, Hardee's/24hr, Rite Aid, Smith's Foods, hardware
1	US 119, Mink Shoals, **E**...Harding's Family Rest., Sleep Inn
0	I-77, S to Charleston, N to Parkersburg. I-79 begins/ends on I-77, exit 104.

Charleston

West Virginia Interstate 81

Exit #	Services
26mm	Potomac River, West Virginia/Maryland state line
25mm	**Welcome Ctr sb, full(handicapped)facilities, info, phone, picnic tables, litter barrels, petwalk**
23	US 11, Marlowe, Falling Waters, **1 mi E**...Falling Waters Camping, **W**...Texaco/diesel/mart, Outdoor Express RV Ctr
20	WV 901, Spring Mills Rd, **E**...TA/diesel/rest./24hr, Barney's Rest., Econolodge, I-81 Flea Mkt
16	WV 9, N Queen St, Berkeley Springs, **E**...Citgo/mart, Exxon/diesel/mart, Sheetz/mart/24hr, Denny's, Dos Amigos Mexican, Dunkin Donuts, Hardee's, Hoss's, KFC, LJ Silver, McDonald's, Ollie's Rest., Anchor Inn, Bavarian Inn, Comfort Inn/rest., Knight's Inn, Stone Crab Inn, Super 8, Travelodge, bank, **1 mi E**...Texaco, 7-11, Cruiser's Rest., Dairy Queen, McDonald's, Pizza Hut, Popeye's, Wendy's, CVS Drug, Food4Less, JiffyLube, MailBoxes Etc, carwash, hardware, **W**...Shell/diesel, Holiday Resort
13	WV 15, Kings St, Martinsburg, **E**...HOSPITAL, DENTIST, Sheetz/mart/24hr, Texaco/Subway/diesel/mart, Applebee's, Asia Garden, Burger King, Cracker Barrel, El Ranchero Mexican, Fazoli's, Outback Steaks, Pizza Hut, Shoney's, Wendy's, Day's Inn, Holiday Inn/rest., Sports Depot, Tanger Outlet/famous brands, same as 12
12	WV 45, Winchester Ave, VA MED CTR, **E**...Citgo/diesel, Sheetz/mart/24hr, Southern States/diesel, Texaco/ diesel/mart, Arby's, Bob Evans, Fazoli's, Hardee's, McDonald's, Ponderosa, Taco Bell, Towers Rest., Wendy's, Comfort Inn, Economy Inn, Hampton Inn, Heritage Motel, Relax Inn, Scottish Inn, Bon-Ton, County Mkt Foods, Hess's, JC Penney, Lowe's Bldg Supply, Martin Foods, Sears/auto, Wal-Mart, mall, Nahkeeta RV Camping(3mi), same as 13
8	WV 32, Tablers Station Rd, **2 mi E**...Pikeside Motel, to airport
5	WV 51, Inwood, to Charles Town, **E**...Exxon/mart, Sheetz/mart, Texaco/diesel, Hardee's, Luwegee's Rest., McDonald's, Viva Mexican, Food Lion, carwash, **W**...Lazy-A Camping(9mi)
2mm	**Welcome Ctr nb, full(handicapped)facilities, info, phone, vending, picnic tables, litter barrels, petwalk**
0mm	West Virginia/Virginia state line

Martinsburg

Wisconsin Interstate 39

Exit #	Services

I-39 N begins/ends, nb continues as US 51.

211 US 51, rd K, Merrill, **E**...antiques

210.5mm Prairie River

208 WI 64, WI 70, Merrill, **W**...Amoco/diesel/24hr, KwikTrip/diesel/mart, Mobil/diesel/mart, Burger King, Diamond Dave's Tacos, Hardee's, McDonald's, Pine Ridge Rest., Taco Bell, Three's Company Dining, Best Western, Super 8, Piggly Wiggly, QuikLube, Radio Shack, Wal-Mart, bank, to Council Grounds SP

206mm Wisconsin River

205 US 51, rd Q, Merrill, **E**...Citgo/diesel/rest./24hr, antiques

197 rd WW, to Brokaw, **W**...Citgo/diesel

194 US 51, rd U, rd K, Wausau, **E**...Fleet&Farm/gas, KwikTrip/gas, Mobil, A&W, McDonald's, Taco Bell, Chrysler/Jeep, Toyota/Isuzu/Mercedes

193 Bridge St, **W**...HOSPITAL

192 WI 29 W, WI 52 E, Wausau, to Chippewa Falls, **E**...Annie's Rest., King Buffet, Little Caesar's, Baymont Inn, Exel Inn, Ramada, Subway, Country Inn Suites, County Mkt Foods, same as 191

191 Sherman St, **E**...HOSPITAL, DENTIST, Applebee's, Big Apple Bagels, Cousins Subs, McDonald's, Courtyard, Hampton Inn, Super 8, **W**...CHIROPRACTOR, Burger King, Grecian Garden Rest., Hardee's, Schlotsky's, 25-10 Rest., Audi/Nissan/VW, K-Mart, Menard's, bank

190.5mm Rib River

190 rd NN, **E**...Emma Krumbee's Bakery, Park Inn, cleaners, **W**...MEDICAL CARE, Citgo/diesel/mart/24hr, Mobil/diesel/mart, Hoffman House Rest., Shakey's Pizza, Subway/TCBY, Best Western, Rib Mtn Ski Area, st patrol

188 rd N, **E**...Amoco/mart/24hr, Phillips 66/mart, Country Kitchen/24hr, Fazoli's, HongKong Buffet, McDonald's, Wendy's, Country Inn Suites(1mi), Day's Inn, Aldi, Best Buy, Chevrolet, King's RV Ctr, Mazda, OfficeMax, Sam's Club, Saturn, TJ Maxx, Volvo/Peterbilt, Wal-Mart

187 WI 29 E, to Green Bay, no facilities

186mm Wisconsin River

185 US 51, **E**...Mobil, Culver's, Denny's, Green Mill Rest., Hardee's, Ponderosa, Tony Roma's, Best Western, Comfort Inn, Country inn Suites, Holiday Inn, Rodeway Inn, Stoney Creek Inn, Cedar Creek Factory Stores/famous Brands, Gander Mtn, bank, cinema, mall

183mm wayside sb, restroom facilities, drinking water, picnic tables

181 Maple Ridge Rd, no facilities

179 WI 153, Mosinee, **E**...airport, **W**...Amoco, Shell/diesel/mart/24hr, Hardee's, McDonald's, Subway, AmeriHost

178mm wayside nb, restroom facilities, drinking water, picnic tables

175 WI 34, Knowlton, to WI Rapids, **1 mi W**...Mullins Cheese Factory

171 rd DB, Knowlton, **W**...to gas, food, lodging, Mullins Cheese Factory

165 rd X, rd N, **E**...food, camping

164mm weigh sta both lanes

161 US 51, Stevens Point, **W**...EYECARE, Amoco, Citgo, KwikTrip/24hr, SuperAmerica/diesel/mart, Burger King, China Garden, Country Kitchen, Domino's, Hardee's, KFC, McDonald's, Mesquite Grill, Olympic Rest., Papa John's, Perkins, Ponderosa, Rococo's Pizza, Subway, Taco Bell, Comfort Suites, Country Inn Suites, Holiday Inn, Point Motel, Roadstar Motel, Super 8, County Mkt Foods, Jeep, K-Mart, Pontiac/Buick, Radio Shack, laundry

159 WI 66, Stevens Point, **W**...Ford/Lincoln/Mercury, Honda, Plymouth

158 US 10, Stevens Point, **E**...VETERINARIAN, Fleet&Farm/gas, Mobil/diesel/mart/24hr, Applebee's, Arby's, Culver's, Fazoli's, McDonald's, Shoney's, Taco Bell, Wendy's, Fairfield Inn, Aldi, Frank's Hardware, Mazda, Nissan, Staples, Target, Tires+, Wal-Mart, bank, **W**...Amoco/mart/24hr, Hilltop Grill, Baymont Inn, Best Western

156 rd HH, Whiting, no facilities

153 rd B, Plover, **W**...Amoco/Burger King/mart, Mobil/diesel/wash, McDonald's, Subway, AmericInn, IGA Foods, Manufacturers Mall/famous brands, Menard's, ShopKO, waterpark

151 WI 54, to Waupaca, **E**...Shell/Arby's/diesel/mart/wash/24hr, 4Star Family Rest., Shooter's Rest., Day's Inn, Elizabeth Inn Conv Ctr, farm mkt, **W**...McDonald's

143 rd W, Bancroft, to WI Rapids, **E**...Citgo/diesel, Cedarwood Rest.

139 rd D, Almond, no facilities

136 WI 73, Plainfield, to WI Rapids, **E**...Amoco/mart, Mobil/diesel/atm, R&R Motel, NAPA

131 rd V, Hancock, **E**...Citgo, 51 Motel, camping

127mm weigh sta both lanes, exits left

124 WI 21, Coloma, **E**...Mobil/diesel/mart, camping

120mm **rest area sb, full(handicapped)facilities, phone, picnic tables, vending, litter barrels, petwalk, RV dump**

118mm **rest area nb, full(handicapped)facilities, phone, picnic tables, vending, litter barrels, petwalk, RV dump**

113 rd E, rd J, Westfield, **E**...Sandman Motel, **W**...Amoco/Burger King/mart, Marathon/mart/wash, Mobil/diesel, McDonald's, Subway, Wild Goose Grill, Pioneer Motel/rest., bank

N ↑
S ↓

Wausau

Wisconsin Interstate 39

106	WI 82 W, WI 23 E, Oxford, **E**...Crossroads Motel
104	(from nb, no EZ return)rd D, Packwaukee, no facilities
100	WI 23 W, rd P, Endeavor, **E**...gas, food
92	US 51 S, Portage, **E**...CHIROPRACTOR, KwikTrip/diesel/atm, Mobil, Culver's, Dino's Rest., KFC, McDonald's, Subway, Taco Bell, Wendy's, Best Western, Ridge Motel, Super 8, Ford/Lincoln/Mercury, $Tree, FashionBug, GNC, K-Mart, Pick'n Save Foods, Piggly Wiggly, Radio Shack, Staples, Wal-Mart, bank
89b a	WI 16, to WI 127, Portage, **E**...Amoco/Burger King/diesel/mart, Hitching Post Rest., SharpShooters Sports Grill, Chrysler/Plymouth/Dodge/Jeep
88.5mm	Wisconsin River
87	WI 33, Portage, **5 mi E**...HOSPITAL
86mm	Baraboo River
85	Cascade Mt Rd, no facilities
84	I-39, I-90 & I-94 run together sb.

Wisconsin Interstate 43

Exit #	Services
192mm	I-43 begins/ends at Green Bay on US 41.
192b	US 41 S, US 141 S, to Appleton, services on Velp Ave, **1 exit S**...Citgo/diesel/24hr, Express/diesel/24hr, Mobil, Shell/diesel/24hr, Burger King, Taco Bell, Wendy's
a	US 41 N, US 141 N
189	Atkinson Dr, to Velp Ave, Port of Green Bay, **1 mi W**...Glen's Gas, Jubilee Foods, auto repair
188	Fox River
187	East Shore Dr, Webster Ave, **W**...Shell/diesel/mart/24hr, McDonald's, Wendy's, Day's Inn, Holiday Inn, RV dump
185	WI 54, WI 57, University Ave, to Algoma, **W**...CHIROPRACTOR, OPTOMETRIST, Citgo/mart, Marathon/mart, Mobil/mart/wash, Japanese Steaks, Lee's Cantonese Chinese, Pizza Hut, Ponderosa, Subway, Taco Bell, Tower Motel, Lindy's Foods, Pennzoil, Walgreen, bank, cleaners/laundry, U of WI GB
183	Mason St, rd V, **E**...AmeriHost, **1 mi W**...HOSPITAL, CHIROPRACTOR, DENTIST, Amoco, Citgo, Applebee's, Burger King, Country Kitchen, Fazoli's, Godfather's, Julie's Café, McDonald's, Perkins, Pizza Hut, Schlotsky's, Taco Bell, Champion Parts, Cub Food/drug/24hr, Dodge, Goodyear/auto, HobbyLobby, K-Mart, Kohl's, Midas Muffler, Nissan, OfficeMax, Osco Drug, ShopKO, Tires+, Wal-Mart, XpressLube, bank, cinema, mall
181	Eaton Rd, rd JJ, **E**...Citgo/Blimpie/diesel/mart, Hardee's, Ford, **W**...Farm&Fleet, Festival Foods, Menard's
180	WI 172 W, to US 41, **W**...HOSPITAL, Amoco/mart/24hr, Citgo/diesel/mart/24hr, Burger King, Country Express Rest./24hr, McDonald's, Best Western, Hampton Inn, Radisson Inn, Ramada, airport, to stadium
178	US 141, to WI 29, rd MM, Bellevue, **E**...Shell/Arby's/diesel/mart/wash, **W**...Amoco/diesel/mart/repair/atm/24hr
171	WI 96, rd KB, Denmark, **E**...Citgo/diesel/mart, Phillips 66(1mi), McDonald's, Steve's Cheese, Uncle Sub's Pizza, camping
168mm	**rest area both lanes, full(handicapped)facilities, phone, vending, picnic tables, litter barrels, RV dump, petwalk**
166mm	Devils River
164	WI 147, rd Z, Maribel, **W**...Citgo/diesel/mart, Cedar Ridge Rest., Shady Acres Camping
160	rd K, Kellnersville, **2-3 mi W**...food
157	rd V, Hillcrest Rd, Francis Creek, **E**...Mobil/Rolo's Diner/diesel/mart
154	US 10 W, WI 310, Two Rivers, to Appleton, **E**...Mobil
153mm	Manitowoc River
152	US 10 E, WI 42 N, rd JJ, Manitowoc, **E**...HOSPITAL, Inn on Maritime Bay(3mi), airport, antiques, maritime museum
149	US 151, WI 42 S, Manitowoc, **E**...HOSPITAL, Mobil/diesel/mart/24hr, Shell/diesel/mart/24hr, Applebee's, Burger King, Country Kitchen, Culver's, Fazoli's, 4Seasons Rest., McDonald's, Perkins, Ponderosa, Wendy's, Birch Beach Inn, Comfort Inn, Holiday Inn, Super 8, AllCar Parts/repair, Buick/Pontiac/Olds/GMC/Cadillac, OfficeMax, Tires+, Wal-Mart, museum
144	rd C, Newton, **E**...Mobil/diesel/mart, antiques
142mm	weigh sta sb, phone
137	rd XX, Cleveland, **E**...Citgo/Burger King/diesel/mart, Cleveland Family Rest.
128	WI 42, Howards Grove, **E**...HOSPITAL, Amoco/diesel/24hr, Mobil/diesel/mart, Country Kitchen, Hardee's, Comfort Inn, Jo-Ann Fabrics, Menard's, tires, **W**...Citgo/Cousins Subs/diesel/mart
126	WI 23, Sheboygan, **E**...HOSPITAL, DENTIST, EYECARE, Applebee's, Culver's, Hardee's, IHOP, Luigi's Italian, McDonald's, Ponderosa, Rococo's Pizza, Taco Bell, Baymont Inn, Ramada Inn, Select Inn, Super 8, Woodlake Inn, Aldi Foods, Sentry Foods, Chrysler/Jeep/Subaru, Firestone/auto, Ford/Lincoln/Mercury, Hyundai, JC Penney, Kohl's, OfficeMax, Olds, Piggly Wiggly, Sears/auto, ShopKO, Toyota/Mazda, Walgreen, Wal-Mart/24hr, bank, cinema, mall, **W**...Pinehurst Inn
123	WI 28, rd A, Sheboygan, **E**...Citgo/diesel/mart/24hr, Mobil/McDonald's/diesel/mart, Cruisers Burgers, Shoney's, Taco Bell, Wendy's, AmericInn, Holiday Inn Express, CarX, Harley-Davidson, Pennzoil, **1 mi E**...Pizza Hut, Chevrolet, K-Mart, Piggly Wiggly, hardware

Wisconsin Interstate 43

120 rds OK, V, Sheboygan, **E...**Phillips 66/diesel/mart, Hill Farm Rest., Parkway Motel, to JM Kohler-Terry Andrae SP, camping, **1 mi W...**Horn RV Ctr

116 rd AA, Foster Rd, Oostburg, **W...**food

113 WI 32 N, rd LL, Cedar Grove, **W...**Citgo/diesel/mart/repair, Dr Zwann Rest., Lakeshore Motel

107 rd D, Belgium, **E...**Harrington Beach SP, **W...**Amoco/diesel/24hr, Mobil/diesel/rest./24hr, Dairy Queen, Hobo's Korner Kitchen, AmericInn

100 WI 32 S, WI 84 W, Port Washington, **E...**DENTIST, EYECARE, Citgo/diesel/mart, Arby's, Burger King, McDonald's, Subway, Best Western, Allen-Edmonds Shoes, Goodyear/auto, PetSupply, SuperSaver Foods, Smith Bros Fish, cleaners, hardware, **W...**Nisleit's Country Inn

97 (from nb), WI 57, Fredonia, no facilities

96 WI 33, to Saukville, **E...**OSTEOPATH, Mobil/diesel/mart/24hr, Subway, Best Western, Buick/Pontiac/Cadillac, Ford/Lincoln/Mercury, Piggly Wiggly, Wal-Mart, carwash, **W...**DENTIST, Amoco/mart/24hr, Bublitz' Rest., Hardee's, Super 8, Tate's Foods

93 WI 32 N, WI 57 S, Grafton, **1-2 mi E...**Dairy House Rest., Pied Piper Rest., Smith Bros Fish, Best Western

92 WI 60, rd Q, Grafton, **E...**GhostTown Rest., **W...**Citgo/DQ/diesel/mart, Home Depot, Target

89 rd C, Cedarburg, **W...**HOSPITAL, Mobil/mart, Cedar Crk Settlement Café(6mi), StageCoach Inn

85 WI 167, Mequon Rd, **W...**DENTIST, Amoco/mart/24hr, Citgo/diesel/mart/wash, Mobil/mart/24hr, Shell/diesel/mart, Chancery Rest., Chinese Rest., Damon's, Dairy Queen, Freddy's Diner, Leonardo's Pizza, McDonald's, Subway, Waltzin' Matilda Steaks, Wendy's, Best Western, Ace Hardware, Kohl's, MailBoxes Etc, Mail+, Vision Ctr, Walgreen, bank, cameras, cinema, laundry

82b a WI 32 S, WI 100, Brown Deer Rd, **E...**MEDICAL CARE, Amoco/mart, Cousins Subs, Heineman's Rest., McDonald's, SpeakEasy Café, Lands End, Walgreen, bank, cinema, cleaners/laundry, **W...**Sheraton/rest.

80 Good Hope Rd, **E...**Amoco, Chinese Rest., Manchester East Hotel, Residence Inn, bank, to Cardinal Stritch U

78 Silver Spring Dr, **E...**DENTIST, EYECARE, Amoco, Mobil, Applebee's, Boston Mkt, Burger King, Denny's, Ground Round, Little Caesar's, LK Chinese, McDonald's, Papa John's, Pizza Hut, Subway, Taco Bell, Tumbleweed Grill, Wendy's, Baymont Inn, Exel Inn, NorthShore Motel, Woodfield Suites, Barnes&Noble, Cadillac, CarX Muffler, Firestone/auto, Goodyear/auto, Kohl's, Midas Muffler, PetSupplies+, Pier 1, QuickLube, Radio Shack, Sears/auto, Walgreen, bank, carwash, cleaners/laundry, mall, **W...**HOSPITAL

77b a (from nb), **E...**Anchorage Rest., Hilton

76b a WI 57, WI 190, Green Bay Ave, **E...**Anchorage Rest., Hilton, **W...**Citgo/mart, Burger King, Buick, Jaguar/Volvo, tires

75 Atkinson Ave, Keefe Ave, **E...**Mobil/mart, **W...**Amoco, Citgo

74 Locust St, no facilities

73c North Ave(rom sb), **E...**Wendy's

a WI 145 E, 4th St(exits left from sb), Broadway, downtown

72c Wells St, **E...**HOSPITAL, Civic Ctr, museum

b (from nb), I-94 W, to Madison

a (310c from nb, exits left from sb), I-794 E, to Lakefront, downtown, I-94 W to Madison

311 WI 59, National Ave, 6th St, downtown

312a Lapham Blvd, Mitchell St, Old Towne Serbian Gourmet House

b (from nb), Becher St, Lincoln Ave

314a Holt Ave, **E...**Citgo/diesel/mart, FoodCourt, Builders Square, Pick'n Save Food, **W...**HOSPITAL, to Alverno Coll

b Howard Ave, no facilities

10b (316 from sb), I-94 S to Chicago, airport, no facilities

9b a US 41, 27th St, **E...**MEDICAL CARE, Amoco, Citgo, Mobil, Arby's, Burger King, Chancery Rest., Chinese Rest., Cousins Subs, Pizza Hut, Village Inn, Suburban Motel, Dodge, K-Mart, Sentry Foods, Target, Valvoline, auto repair/parts, bank, hardware, **W...**HOSPITAL, Denny's, Hardee's, McDonald's, Mexican Rest., Rich's Cakes, Taco Bell, Hospitality Suites, AAA, CarX Muffler, Chevrolet, Chrysler/Jeep, Goodyear, Olds/GMC/Hyundai, JiffyLube, Midas Muffler, Pennzoil/wash, Ziebart TidyCar, bank, laundry

8a WI 36, Loomis Rd, **E...**Amoco, Mobil/mart, **W...**to Alverno Coll

7 60th St, **E...**SuperAmerica/mart

5b 76th St(from sb, no EZ return), **E...**Amoco/wash, Burger King, Ground Round, KFC, McDonald's, Olive Garden, Pizza Hut, Red Lobster, Thomas' Rest., Wendy's, Best Buy, Circuit City, Cub Food/drugs, Isuzu, JiffyLube, Midas Muffler, Office Depot, bank

a WI 24 W, Forest Home Ave, **E...**Citgo/mart, Boerner Botanical Gardens

61 (4 from sb), I-894/US 45 N, I-43/US 45 S

60 US 45 S, WI 100, 108th St(exits left from sb), **E...**Amoco, A&W, motel, **W...**Phillips 66, McDonald's, Walgreen, Wal-Mart

Wisconsin Interstate 43

59	WI 100, Layton Ave(from nb), Hales Corner, same as 60
57	Moorland Rd, **E**...cinema, **2-3 mi W**...Mobil/24hr, SuperAmerica/mart/24hr, Fazoli's, Hardee's, McDonald's, Rococo's Pizza, Taco Bell, Tumbleweed Grill, Best Western, Embassy Suites/rest., Residence Inn
54	rd Y, Racine Ave, **1-2 mi E**...Citgo/diesel, KwikTrip/mart, Culver's, Cousins Subs, McDonald's
50	WI 164, Big Bend, **W**...Arm's/diesel/mart, Antoine's Bakery, McDonald's
44mm	Fox River
43	WI 83, Mukwonago, **E**...Mobil/diesel/mart/atm, **W**...Citgo/FoodCourt/mart, Shell, Burger King, Culver's, Dairy Queen, Subway, Taco Bell, Sleep Inn, Pennzoil, Wal-Mart, bank
38	WI 20, East Troy, **W**...Amoco/mart/wash, Citgo/Subway/diesel/mart/repair/24hr, Burger King, Family Kitchen Rest., Roma Ristorante, Chrysler/Plymouth/Dodge, NAPA, carwash
36	WI 120, East Troy, **W**...Country Inn Suites, gas/diesel
33	Bowers Rd, **E**...to Alpine Valley Music Theatre
32mm	**rest area both lanes, full(handicapped)facilities, phone, picnic tables, litter barrels, vending, petwalk**
29	WI 11, Elkhorn, fairgrounds, no facilities
27b a	US 12, to Lake Geneva, no facilities
25	WI 67, Elkhorn, **E**...MEDICAL CARE, VETERINARIAN, Mobil/diesel/mart/wash, AmericInn, Chevrolet, Chrysler/Dodge **W**...SuperAmerica/diesel/24hr, Burger King, Jeep, Delavan RV Ctr
21	WI 50, Delavan, **E**...Greyhound DogTrack, **W**...MEDICAL CARE, Mobil/24hr, Shell, SuperAmerica/diesel/mart/24hr, Burger King, Cousins Subs, KFC, McDonald's, Pizza Hut, Shoney's, Subway, Taco Bell, Wendy's, Holiday Inn, Super 8, Chevrolet/Olds/Cadillac, FashionBug, Ford/Lincoln/Mercury, K-Mart, Pick'n Save, Piggly Wiggly, Schmaling's Food, Sentry Food/24hr, ShopKO, Walgreen, bank, hardware
17	rd X, Delavan, Darien, **W**...Amoco/A&W/mart
15	US 14, Darien, **E**...Citgo/diesel/mart
6	WI 140, Clinton, **E**...Amoco, Citgo, Subway, Ford
2	rd X, Hart Rd, **E**...Butterfly Supper Club
1b a	I-90, E to Chicago, W to Madison, **W**...Amoco/mart/24hr, BP/Subway/diesel/mart, Citgo/diesel, Phillips 66, Pilot/Shell/DQ/Taco Bell/diesel/24hr, SuperAmerica/diesel/mart, Arby's, Burger King, Country Kitchen, Culver's, Fazoli's, McDonald's, Perkins, Shirley's Rest./24hr, Wendy's, Comfort Inn, Econolodge, Fairfield Inn, Holiday Inn Express, Super 8, Aldi Foods, Buick/Pontiac, Chevrolet/Cadillac/Nissan, Staples, Wal-Mart SuperCtr/24hr, bank, cinema

I-43 begins/ends on I-90, exit 185 in Beloit.

Wisconsin Interstate 90

Exit #	Services
187.5mm	Wisconsin/Illinois state line. I-90 & I-39 run together nb.
187mm	**Welcome Ctr wb, full(handicapped)facilities, info, phones, picnic tables, litter barrels, vending, petwalk**
185b	I-43 N, to Milwaukee
a	WI 81, Beloit, **S**...Amoco/mart, Citgo/diesel/mart/24hr, Phillips 66/mart, Pilot/DQ/Taco Bell/diesel/24hr, SuperAmerica/diesel/mart, BP/Subway/diesel/mart, Applebee's, Arby's, Burger King, Country Kitchen, Culver's, Dairy Queen, Fazoli's, McDonald's, Perkins, Shirley's Rest./24hr, The Manor Dining, Wendy's, Comfort Inn, Econolodge, Fairfield Inn, Holiday Inn Express, Super 8, Buick/Pontiac, Chevrolet/Cadillac/Nissan, Staples, Wal-Mart SuperCtr/24hr, bank, cinema, hardware
183	Shopiere Rd, rd S, to Shopiere, **S**...HOSPITAL, Citgo/diesel/24hr, camping
177	WI 351, Janesville, **S**...to Blackhawk Tec Coll, Rock Co Airport
175b a	**WI 11, Janesville, to Delavan, N**...**Shell/Subway/diesel/24hr, Denny's, Baymont Inn,** S...**Amoco/mart, Dairy Queen, Hardee's, Lannon Stone Motel**
171c b	US 14, WI 26, Janesville, **N**...Mobil/Wendy's/diesel/mart, Damon's, Old Country Buffet, Shoney's, Steak'n Shake, Subway, Microtel, Batteries+, Home Depot, Kamp Dakota, Michael's, Old Navy, PetCo, Staples, Tires+, TJ Maxx, cinema, **S**...HOSPITAL, Amoco/mart, Citgo/diesel/mart/24hr, KwikTrip/diesel/mart/24hr, Phillips 66/diesel/mart, Applebee's, Arby's, BBQ, Big Boy, Burger King, ChiChi's, Chinese Rest., Country Kitchen, Cousins Subs, Fazoli's, Ground Round, Gyro's Greek, Hardee's, KFC, McDonald's, McRaven's Rest., Olive Garden, Perkins, Pizza Hut, Prime Qtr Steaks, Shakey's Pizza, Taco Bell, Ramada Inn/rest., Select Inn, Super 8, Aldi Foods, Farm&Fleet, Firestone/auto, Ford/Lincoln/Mercury, Harley-Davidson, JC Penney, JiffyLube, Jo-Ann Fabrics, Mazda, Mercedes, Menard's, Midas Muffler, K-Mart/Little Caesar's, ShopKO, Target, VW, Wal-Mart/drugs, bank, cinema, cleaners, mall, tires
171a	**S**...same as 171c b, **N**...Phillips 66/mart, Alfresco Café, Cracker Barrel, Best Western/rest., Motel 6, Hampton Inn, Holiday Inn Express, Foxy's Motorcycle, crafts
168mm	**rest area eb, full(handicapped)facilities, phone, picnic tables, litter barrels, vending, petwalk**
163.5mm	Rock River
163	WI 59, Edgerton, to Milton, **N**...Mobil, Shell/Burger King/mart, Cousins Subs, McDonald's, Red Apple Rest., Comfort Inn, Hickory Hill/Lakeland Camping, marina, **S**...Amoco/mart
160	US 51S, WI 73, WI 106, to Deerfield, Oaklawn Academy, **N**...Hickory Hill Camping, **S**...HOSPITAL, KwikTrip/diesel/café, Towne Edge Motel
156	US 51N, to Stoughton, **S**...HOSPITAL, Coachman's Inn
147.5mm	weigh sta wb

Wisconsin Interstate 90

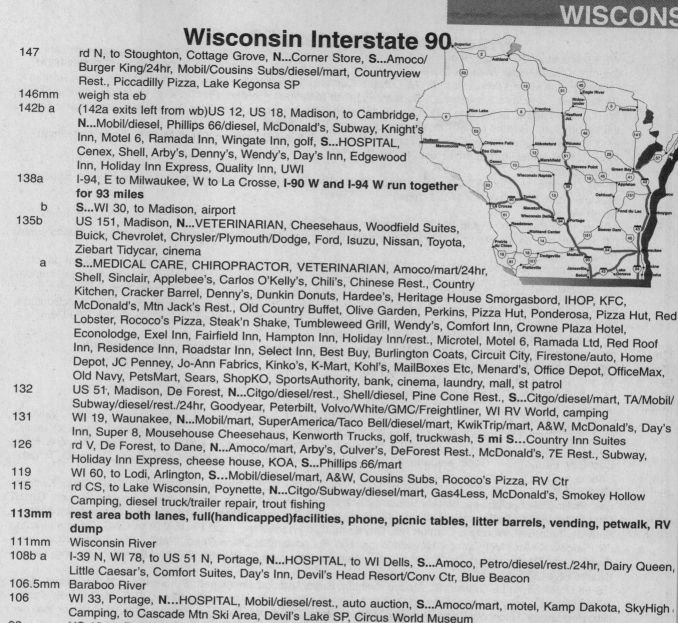

E

W

147	rd N, to Stoughton, Cottage Grove, **N**...Corner Store, **S**...Amoco/ Burger King/24hr, Mobil/Cousins Subs/diesel/mart, Countryview Rest., Piccadilly Pizza, Lake Kegonsa SP
146mm	weigh sta eb
142b a	(142a exits left from wb)US 12, US 18, Madison, to Cambridge, **N**...Mobil/diesel, Phillips 66/diesel, McDonald's, Subway, Knight's Inn, Motel 6, Ramada Inn, Wingate Inn, golf, **S**...HOSPITAL, Cenex, Shell, Arby's, Denny's, Wendy's, Day's Inn, Edgewood Inn, Holiday Inn Express, Quality Inn, UWI
138a	I-94, E to Milwaukee, W to La Crosse, **I-90 W and I-94 W run together for 93 miles**
b	**S**...WI 30, to Madison, airport
135b	US 151, Madison, **N**...VETERINARIAN, Cheesehaus, Woodfield Suites, Buick, Chevrolet, Chrysler/Plymouth/Dodge, Ford, Isuzu, Nissan, Toyota, Ziebart Tidycar, cinema
a	**S**...MEDICAL CARE, CHIROPRACTOR, VETERINARIAN, Amoco/mart/24hr, Shell, Sinclair, Applebee's, Carlos O'Kelly's, Chili's, Chinese Rest., Country Kitchen, Cracker Barrel, Denny's, Dunkin Donuts, Hardee's, Heritage House Smorgasbord, IHOP, KFC, McDonald's, Mtn Jack's Rest., Old Country Buffet, Olive Garden, Perkins, Pizza Hut, Ponderosa, Pizza Hut, Red Lobster, Rococo's Pizza, Steak'n Shake, Tumbleweed Grill, Wendy's, Comfort Inn, Crowne Plaza Hotel, Econolodge, Exel Inn, Fairfield Inn, Hampton Inn, Holiday Inn/rest., Microtel, Motel 6, Ramada Ltd, Red Roof Inn, Residence Inn, Roadstar Inn, Select Inn, Best Buy, Burlington Coats, Circuit City, Firestone/auto, Home Depot, JC Penney, Jo-Ann Fabrics, Kinko's, K-Mart, Kohl's, MailBoxes Etc, Menard's, Office Depot, OfficeMax, Old Navy, PetsMart, Sears, ShopKO, SportsAuthority, bank, cinema, laundry, mall, st patrol
132	US 51, Madison, De Forest, **N**...Citgo/diesel/rest., Shell/diesel, Pine Cone Rest., **S**...Citgo/diesel/mart, TA/Mobil/ Subway/diesel/rest./24hr, Goodyear, Peterbilt, Volvo/White/GMC/Freightliner, WI RV World, camping
131	WI 19, Waunakee, **N**...Mobil/mart, SuperAmerica/Taco Bell/diesel/mart, KwikTrip/mart, A&W, McDonald's, Day's Inn, Super 8, Mousehouse Cheesehaus, Kenworth Trucks, golf, truckwash, **5 mi S**...Country Inn Suites
126	rd V, De Forest, to Dane, **N**...Amoco/mart, Arby's, Culver's, DeForest Rest., McDonald's, 7E Rest., Subway, Holiday Inn Express, cheese house, KOA, **S**...Phillips 66/mart
119	WI 60, to Lodi, Arlington, **S**...Mobil/diesel/mart, A&W, Cousins Subs, Rococo's Pizza, RV Ctr
115	rd CS, to Lake Wisconsin, Poynette, **N**...Citgo/Subway/diesel/mart, Gas4Less, McDonald's, Smokey Hollow Camping, diesel truck/trailer repair, trout fishing
113mm	**rest area both lanes, full(handicapped)facilities, phone, picnic tables, litter barrels, vending, petwalk, RV dump**
111mm	Wisconsin River
108b a	I-39 N, WI 78, to US 51 N, Portage, **N**...HOSPITAL, to WI Dells, **S**...Amoco, Petro/diesel/rest./24hr, Dairy Queen, Little Caesar's, Comfort Suites, Day's Inn, Devil's Head Resort/Conv Ctr, Blue Beacon
106.5mm	Baraboo River
106	WI 33, Portage, **N**...HOSPITAL, Mobil/diesel/rest., auto auction, **S**...Amoco/mart, motel, Kamp Dakota, SkyHigh Camping, to Cascade Mtn Ski Area, Devil's Lake SP, Circus World Museum
92	US 12, to Baraboo, **N**...Amoco/Ponderosa, BP/diesel, Cenex/diesel/24hr, Citgo/24hr, Mobil/diesel/24hr, Burger King, Culver's Custard, Danny's Diner, Houlihan's, KFC, McDonald's, Subway, Alakiai Hotel, Best Western(4mi), Black Wolf Lodge, Camelot Inn, Country Squire Motel, Dell Rancho Motel, Grand Marquis Inn, Kalahari Conv Ctr, Mayflower Motel, Ramada Ltd, Wintergreen Motel, **1 mi N**...Cheese Factory Rest., Fischer's Rest., Dell Creek Motel, Hochunk Lodge, Holiday Motel, museum, **S**...HOSPITAL, Motel 6, Vagabond Motel, Scenic Traveler RV Ctr, Dell Boo Camping, Pioneer Park Camping, Red Oak Camping, antiques/flea mall, Mirror Lake SP
89	WI 23, Lake Delton, **N**...Marathon/mart, Mobil/diesel/mart, Grandma Jakes Rest., Houlihan's, KFC, Copacabana Motel, Malibu Inn, Olympia Motel, RainTree Motel, Rodeway Inn, Sahara Motel, Crystal Grand Music Theatre, Springbrook Camping, Jellystone Camping, antiques, bank
87	WI 13, Wisconsin Dells, **N**...Amoco/diesel/mart, Citgo/diesel, Mobil/diesel/mart, Shell, Burger King, Country Kitchen, Denny's, Perkins, Rococo's Pizza, Taco Bell, Wendy's, Best Western, Christmas Mtn Village, Comfort Inn, Day's Inn, Dells Eagle Motel, Holiday Inn/rest., Polynesian Hotel, Super 8, KOA, Sherwood Forest Camping, Tepee Park Camping, golf, info
85	US 12, WI 16, Wisconsin Dells, **N**...to Rocky Arbor SP, Stand Rock Camping, Sherwood Forest Camping, **S**...flea mkt
79	rd HH, Lyndon Sta, **N**...Greenfield's RV Sales/service, **S**...Shell/diesel/mart, The Wright Place/rest., camping
76mm	**rest area wb, full(handicapped)facilities, phone, picnic tables, litter barrels, vending, petwalk**
74mm	**rest area eb, full(handicapped)facilities, phone, picnic tables, litter barrels, vending, petwalk**

Wisconsin Interstate 90

69	WI 82, Mauston, **N**...Cenex/Taco Bell/diesel/mart/24hr, Phillips 66/mart/24hr, Pilot/Wendy's/diesel/mart/24hr, Country Kitchen, Best Western Oasis, Country Inn, Super 8, to Buckhorn SP, **S**...HOSPITAL, VETERINARIAN, Citgo/diesel, KwikTrip/diesel/mart, A&W, Culver's Custard, Garden Valley Rest., Hardee's, McDonald's, Roman Castle Rest., Sunday's Cheesemart, Subway, Alaskan Motel/rest., K-Mart, Pick'n Save, Walgreen, antiques, repair
61	WI 80, New Lisbon, to Necedah, **N**...Citgo/Subway/Bunkhouse Rest./diesel/24hr, Mobil/A&W/diesel/mart, Edge O' the Wood Motel, Rafters Motel, Travelodge, Buckhorn SP
55	rd C, Camp Douglas, **N**...wayside, to Camp Williams, Volk Field, **S**...Amoco/diesel, Mobil/Subway/diesel/mart, Walsh's K&K Motel, Target Bluff Rest., to Mill Bluff SP
51mm	weigh sta eb
48	rd PP, Oakdale, **N**...Citgo/Old Rig Diner/diesel/mart/24hr, Granger's Camping, Kamp Dakota/showers, antiques, **S**...Mobil/mart/24hr, Oakdale Motel, repair
47.5mm	weigh sta wb
45	I-94 W, to St Paul, I-90 E and I-94 E run together for 93 miles
43	US 12, WI 16, Tomah, **N**...HOSPITAL, Citgo/mart, Burnstadt's Café, Dairy Queen, European Café, Filippo's Rest., Pizza Pit, Budget Host
41	WI 131, Tomah, to Wilton, **N**...VA MED CTR, A&W, European Café, Brentwood Inn, Budget Host, **S**...st patrol
28	WI 16, Sparta, Ft McCoy, **N**...HOSPITAL, Amoco/mart, Kwiktrip/diesel/mart, Hardee's, Downtown Motel, camping
25	WI 27, Sparta, to Melvina, **N**...Casey's Store/gas, Cenex/diesel/mart, Citgo/mart, Conoco/Taco Bell/mart, KwikTrip/diesel/mart/24hr, Burger King, Country Kitchen, Dairy Queen, Happy Chef, Hardee's, KFC, McDonald's, Pizza Hut, Subway, Country Inn, Super 8, Chrysler/Dodge, Family$, Ford/Mercury, Jubilee Foods, Our Own Hardware, Wal-Mart SuperCtr, Zzip Lube, **S**...camping
22mm	**rest area wb, full(handicapped)facilities, phone, picnic tables, litter barrels, vending, petwalk**
20mm	**rest area eb, full(handicapped)facilities, phone, picnic tables, litter barrels, vending, petwalk**
15	WI 162, Bangor, to Coon Valley, **N**...gas, **S**...Chevrolet
12	rd C, W Salem, **N**...Cenex/diesel, Coulee Region RV Ctr, NAPA, Neshonoc Camping, **S**...AmericInn
10mm	weigh sta eb
5	WI 16, La Crosse, **N**...Baymont Inn, Hampton Inn, Microtel, Woodman's Food/gas/lube/wash, Pennzoil, Valvoline, bank, **S**...HOSPITAL, KwikTrip/diesel/24hr, Phillips 66, A&W, Big Apple Bagels, ChuckeCheese, Fazoli's, Garcia's, Hardee's, McDonald's, Old Country Buffet, Olive Garden, Perkins, Wendy's, Holiday Inn Express, Barnes&Noble, Best Buy, Farm&Fleet, Goodyear/auto, Kohl's, Michael's, OfficeMax, Paper Whse, Sears/auto, ShopKO, Target, cinema, mall
4	US 53 N, WI 16, to WI 157, La Crosse, **N**...KwiTrip/gas, bank, **S**...HOSPITAL, CHIROPRACTOR, Citgo, Kwiktrip/gas, Phillips 66, Applebee's, Bakers Square, BBQ, Burger King, ChiChi's, Ciatti's Italian, Country Kitchen, Grizzly's Café, Hardee's, Little Caesar's, McDonald's, Red Lobster, Rococo's Pizza, Shakey's Pizza, Shoney's Inn/rest., Subway, Taco Bell, Wendy's, Comfort Inn, Day's Inn, Cub Foods/24hr, Food Festival/24hr, GNC, Goodyear/auto, JC Penney, K-Mart, MN Fabrics, Nevada Bob's, NW Fabrics/crafts, Office Depot, PetCo, Pier 1, ShopKO/drugs, Sam's Club, Tires+, Vision Ctr, Wal-Mart/24hr, WonderLube/wash, Zzip Lube, Bluebird Springs Camping, La Crosse River St Trail, mall
3	US 53 S, WI 35, to La Crosse, **S**...Amoco/Kwiktrip, SuperAmerica/diesel/mart, Burger King, Chinese Rest., Coney Island Café, Country Kitchen, Edwardo's Pizza, Ember's Rest./24hr, Happy Chef/24hr, KFC, McDonald's, Pizza Hut, Ponderosa, Rococo's Pizza, Subway, Best Western, Courtyard, Exel Inn, Hampton Inn, Nightsaver Inn, Roadstar Inn, Super 8, Sir Speedy, U-Haul, bank, carwash, to Great River St Trail, Viterbo Coll
2.5mm	Black River
2	rd B, French Island, **N**...Mobil/diesel, airport, **S**...Citgo/diesel/IGA Food, Day's Inn/rest.
1mm	**Welcome Ctr eb, full(handicapped)facilities, info, phone, picnic tables, litter barrels, vending, petwalk**
0mm	Mississippi River, Wisconsin/Minnesota state line

Wisconsin Interstate 94

Exit #	Services
349.5mm	Wisconsin/Illinois state line, rd ML(from nb), weight sta nb
347	WI 165, rd Q, Lakeview Pkwy, **E**...Amoco/diesel/mart, McDonald's, Radisson/rest., Lakeside MktPlace/famous brands, **Welcome Ctr, full facilities, W**...GolfShack Whse
345	rd C, **E**...Phillips 66/diesel/mart, **4 mi W**...BP/mart/wash
345mm	Des Plaines River
344	WI 50, to Kenosha, Lake Geneva, **E**...HOSPITAL, Shell/Arby's/mart/wash, Annie's Café, Schnitzel Haus, Shoney's, Baymont Inn, Super 8, Woodman's Mkt/gas, **W**...Amoco/mart, Speedway/Subway/diesel/mart, Burger King, CheeseShop, Chef's Table, Cracker Barrel, Denny's, KFC, LJ Silver, McDonald's, Perkins, Savannah's Rest., Taco Bell, Wendy's, Best Western, Country Inn Suites, Day's Inn, Knight's Inn, Quality Suites, **1/2mi W**...auto mall, BMW, Cadillac, Chevrolet, Ford, GM, Jeep/Eagle, Mitsubishi, Subaru, Toyota
342	WI 158, to Kenosha, **E**...Holiday Inn Express(7mi), airport, dog track
340	WI 142, rd S, to Kenosha, **E**...HOSPITAL, Mobil/diesel/mart, Citgo/diesel/mart, **W**...Mars Cheese Castle Rest., StarBar Rest., Easterday Motel, to Bong RA
339	rd E, **E**...Speedway/diesel/mart

Wisconsin Interstate 94

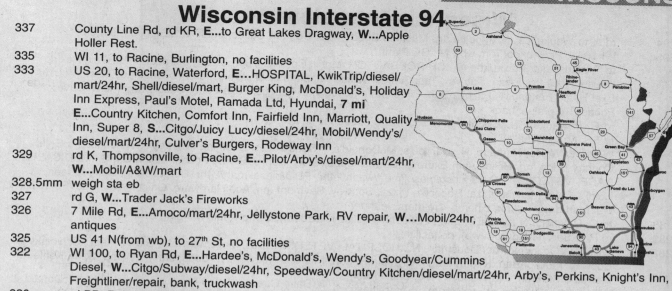

337 County Line Rd, rd KR, **E...**to Great Lakes Dragway, **W...**Apple Holler Rest.

335 WI 11, to Racine, Burlington, no facilities

333 US 20, to Racine, Waterford, **E...**HOSPITAL, KwikTrip/diesel/mart/24hr, Shell/diesel/mart, Burger King, McDonald's, Holiday Inn Express, Paul's Motel, Ramada Ltd, Hyundai, **7 mi E...**Country Kitchen, Comfort Inn, Fairfield Inn, Marriott, Quality Inn, Super 8, **S...**Citgo/Juicy Lucy/diesel/24hr, Mobil/Wendy's/diesel/mart/24hr, Culver's Burgers, Rodeway Inn

329 rd K, Thompsonville, to Racine, **E...**Pilot/Arby's/diesel/mart/24hr, **W...**Mobil/A&W/mart

328.5mm weigh sta eb

327 rd G, **W...**Trader Jack's Fireworks

326 7 Mile Rd, **E...**Amoco/mart/24hr, Jellystone Park, RV repair, **W...**Mobil/24hr, antiques

325 US 41 N(from wb), to 27th St, no facilities

322 WI 100, to Ryan Rd, **E...**Hardee's, McDonald's, Wendy's, Goodyear/Cummins Diesel, **W...**Citgo/Subway/diesel/24hr, Speedway/Country Kitchen/diesel/mart/24hr, Arby's, Perkins, Knight's Inn, Freightliner/repair, bank, truckwash

320 rd BB, Ralston Ave, **E...**Amoco, Mobil, Burger King, Baymont Inn, Honda, cinema

319 rd ZZ, College Ave, **E...**Shell, SuperAmerica/mart, McDonald's, Shoney's, Exel Inn, Hampton Inn, Ramada/rest., Red Roof, Burlington Coats, Luxury Linens, **W...**Amoco/mart/24hr, Citgo/mart, VIP Foodmart, Boy Blue DairyMart, laundry

E

318 WI 119, **E...**airport

317 Layton Ave, **W...**Howard Johnson

316 I-43 S, I-894 W, to Beloit

314b Howard Ave, to Milwaukee, Alverno Coll

a Holt Ave, **E...**Citgo/diesel/mart, ChuckeCheese, K-Mart/auto, Pik'n Save Foods, Sears Hardware, Target, **W...**HOSPITAL

312b a Becher St, Mitchell St, Lapham Blvd, **W...**BP/mart, Citgo/mart, Old Town Gourmet House

311 WI 59, National Ave, 6th St, downtown

310a 13th St(from eb), **E...**HOSPITAL

b I-43 N, to Green Bay

c I-794 E, **E...**Holiday Inn, Ramada Inn, to downtown and Lake Michigan Port of Entry

309b 26th St, 22nd St, Clybourn St, St Paul Ave, **N...**HOSPITAL, to Marquette U

a 35th St, **N...**Amoco/mart/wash

308c b US 41, no facilities

a VA Ctr, **N...**Miller Brewing, **S...**County Stadium

307b 68th-70th St, Hawley Rd, **S...**to County Stadium

W

a **N...**68th St, Pennzoil Lube, carwash

306 WI 181, to 84th St, **N...**HOSPITAL, **S...**Olympic Training Facility, carwash

305b US 45 N, to Fond du Lac, **N...**HOSPITAL

a I-894 S, to Chicago, to airport

304b a WI 100, **N...**Citgo, Clark Gas, Shell, Bagel Tracks, Cactus Jack's, Cousins Subs, Edwardo's Natural Pizza, Ground Round, Giuseppi's Rest., Honeybaked Ham, McDonald's, Mongolian BBQ, Pizza Hut, Taco Bell, Best Western/rest., Camelot Inn, Day's Inn, Exel Inn, Ramada Inn, Kinko's Copies, zoo, **S...**Phillips 66, SuperAmerica/diesel/mart/atm, Dunkin Donuts, PJ's Subs, Batteries+, CarX Muffler, JiffyLube, U-Haul, Walgreen

301b a Moorland Rd, **N...**DENTIST, VETERINARIAN, Amoco/diesel/24hr, Mobil/mart, Bakers Square, Fuddrucker's, Maxwell's Rest., McDonald's, Pizzaria Uno, Schlotsky's, Whitney's Dining, Courtyard, Marriott, Sheraton, Wyndham Garden, Audi, Barnes&Noble, BMW, Boston Store, Firestone/auto, Goodyear/auto, JC Penney, Linens'n Things, Office Depot, Optical Express, Saab, Sears/auto, SportMart, Walgreen, bank, golf, **S...**Mobil/mart, Zarder's Rest., Best Western, Country Inn Suites, Embassy Suites/rest., Residence Inn, Walgreen, golf

297 US 18, rd JJ, Blue Mound Rd, Barker Rd, **1-2 mi N...**MEDICAL CARE, Amoco/mart/wash/24hr, Mobil, Aliota's Steaks, Annie's Café, Boston Mkt, Brennan's Fruit Mkt, Burger King, ChiChi's, ChuckeCheese, Einstein Bros Bagels, KFC, Kopp's Custard, Marie Callender's, McDonald's, Olive Garden, Perkins, Starbucks, Tony Roma's, Wendy's, Zorba's Grill, Baymont Inn, Comfort Inn, Courtyard, Fairfield Inn, Hampton Inn, Motel 6, Wyndham Garden, Acura, Advance Parts, Best Buy, Circuit City, CompUSA, Computer City, Drug Emporium, Kinko's, K-Mart/Little Caesar's/drugs, Lenscrafters, Mazda/Lexus, NTB, Pick'n Save Foods, cinema, **S...**PDQ Gas/mart, Arby's, Dunkin Donuts, McDonald's, Schlotsky's, Taco Bell, Extended Stay America, Holiday Inn, Select Inn, Super 8, Chrysler, Farm&Fleet, JiffyLube, PetWorld, Target, Tires+, Walgreen, bank, cinema, st patrol

295 WI 164, Waukesha, **N...**HOSPITAL, Citgo/diesel/mart, **S...**hardware, to Carroll Coll

Milwaukee

Wisconsin Interstate 94

294	rd J, to Waukesha, **N**...Mobil/mart, Machine Shed Rest., Comfort Suites, bank, **S**...Expo Ctr, airport
293c	WI 16 W, Pewaukee(from wb), **N**...GE Plant
b a	rd T, Wausheka, Pewaukee, **S**...CHIROPRACTOR, Mobil, Cousins Subs, Denny's, Fazoli's, Mr Wok, Peking House, Rococo's Pizza, Subway, Taco Amigo, Waldo Pepper's Rest., Weissgerber's Gasthaus Rest., Wendy's, Exel Hotel, Batteries+, FashionBug, Firestone/auto, Jo-Ann Fabrics, LazerLube, Osco Drug, Pick'n Save Foods, TrueValue, bank
291	rd G, Pewaukee, **N**...Country Inn, **S**...Amoco/mart/24hr
290	rd SS, no facilities
287	WI 83, to Wales, Hartland, **N**...Hardee's, McDonald's, 7 Seas Lakeside Dining, Shoney's, Winchester's Grill, Country Pride Inn, Holiday Inn Express, Barn Shoppe, GNC, Kohl's, Sentry Foods, Vision Ctr, **S**...MEDICAL CARE, VETERINARIAN, Amoco/diesel/mart/wash/24hr, PDQ/diesel/mart/24hr, Burger King, Dairy Queen, Heidi's Café, Marty's Pizza, Rococo's Pizza, Subway, Baymont Inn, Ace Hardware, OfficeMax, Target/Pennzoil, Wal-Mart/drugs, bank, cinema, cleaners, laundry
285	rd C, Delafield, **N**...Mobil(1/2mi), to St John's Military Academy, **S**...to Kettle Moraine SF
283	rd P, to Sawyer Rd(from wb), no facilities
282	WI 67, to Oconomowoc, Dousman, **N**...HOSPITAL, VETERINARIAN, Mobil/wash, Mr Slow's Rest., Schlotsky's, Subway, FashionBug, K-Mart/drugs, Olympia Resort, Pick'n Save, Sentry Foods, bank, cinema, **S**...to Kettle Morraine SF(8mi), Old World WI Hist Sites(13mi)
277	Willow Glen Rd(from eb, no return), no facilities
275	rd F, to Sullivan, Ixonia, **S**...camping
267	WI 26, Johnson Creek, to Watertown, **N**...Shell/diesel/mart/rest./24hr, Arby's, Day's Inn, Super 8(6mi), Goodyear/auto, Johnson Creek Outlet Ctr/famous brands, **S**...HOSPITAL, Amoco, Citgo/Hardee's/diesel/mart/ 24hr, Gobbler Rest., King Arthur Inn, cinema, to Aztalan SP
265.5mm	Rock River
264mm	**rest area wb, full(handicapped)facilities, phone, picnic tables, litter barrels, vending, petwalk**
263mm	Crawfish River
261mm	**rest area eb, full(handicapped)facilities, phone, picnic tables, litter barrels, vending, petwalk**
259	WI 89, Lake Mills, to Waterloo, **N**...Phillips 66/diesel/rest./mart/24hr, Lake Country Inn, **S**...Amoco/diesel/mart/ 24hr, KwikTrip/diesel/mart, Lakeside Rest., McDonald's, Pizza Pit, Subway, Pyramid Motel/RV parking, Pennzoil, camping, laundry, to Aztalan SP
250	WI 73, to Marshall, Deerfield, no facilities
246mm	weigh sta eb
244	rd n, Sun Prairie, Cottage Grove, **N**...Amoco/diesel/mart, Citgo/diesel/mart
240	I-90 E. I-94 and I-90 run together 93 miles. **See Wisconsin I- 90, exits 48-138.**
147	I-90 W, to La Crosse
143	US 12, WI 21, Tomah, **N**...Mobil/diesel/mart, Country Kitchen, Perkins, AmericInn, Holiday Inn, Super 8, Humbird Cheese/gifts, antiques, fireworks, **S**...HOSPITAL, Amoco/diesel/rest./24hr, Citgo/mart, Shell/diesel/ mart/cheese, Burnstad's Amish Café, Culver's, Hardee's, KFC, McDonald's, Subway, Taco Bell, Comfort Inn, Cranberry Suites, Econolodge, Ford/Lincoln/Mercury, Plymouth/Dodge, U-Haul, Wal-Mart SuperCtr/24hr, to Ft McCoy(9mi), **S on US 12**...Burger King, Pizza Hut, Chevrolet/Olds/Buick, Firestone/auto, NAPA
135	rd EW, Warrens, **N**...Jellystone Park Camping, **S**...Texas Longhorn Grill
128	rd O, Millston, **N**...Black River SF, camping, **S**...Hartland/diesel/mart, Millston Motel
122mm	**rest area/scenic view both lanes, full(handicapped)facilities, phone, picnic tables, litter barrels, vending, petwalk**
116	WI 54, **N**...Cenex/Subway/Taco Bell/diesel/mart/LP, Perkins, Best Western Arrowhead/rest., Holiday Inn Express, Super 8, Black River RA, Parkland Camp, **S**...DENTIST, Amoco/diesel/mart/24hr, Citgo/diesel/rest./ mart, Burger King, Dairy Queen, Hardee's, McDonald's, Oriental Kitchen, Pizza Hut, Day's Inn, $Tree, Wal-Mart, bank
115.5mm	Black River
115	US 12, Black River Falls, to Merrillan, **N**...Pines Motel, **S**...HOSPITAL, CHIROPRACTOR, Amoco/diesel/mart, Conoco/mart, Holiday/mart, Phillips 66/DQ/mart, Hardee's/24hr, KFC, Pizza Hut, Subway, Sunrise Rest., Ace Hardware, Harley-Davidson, NAPA, Pontiac/Buick/GMC, carwash, cleaners
105	to WI 95, Hixton, to Alma Center, **N**...Motel 95/camping, **S**...Cenex/diesel/mart, Phillips 66/Jeffrey's Rest./ diesel/24hr, KOA
98	WI 121, Northfield, to Alma Center, Pigeon Falls, **S**...Cenex/diesel/mart, Jackie's Café/Inn
88	US 10, Osseo, to Fairchild, **N**...Holiday/diesel/mart/rest., Mobil/mart, Dairy Queen, Embers Rest., Hardee's, Heckel's Rest., Budget Host, Super 8, Chevrolet, Osseo Camping, antiques, carwash, museum, **S**...HOSPITAL, Amoco/Subway/mart, Speedway/diesel/mart, McDonald's, Taco Bell, Rodeway Inn/rest.
81	rd HH, rd KK, Foster, **S**...Cenex/diesel/LP/mart/atm, Foster Cheesehaus

E

W

M a d i s o n

T o m a h

Wisconsin Interstate 94

E ↕ **W**

70	US 53, Eau Claire, **N off Golf Rd**...Conoco/diesel/mart, A&W, Applebee's, Bakers Square, Oakwood Café Court, Fazoli's, Garfield's Café, McDonald's, Northwoods Grill, Olive Garden, Shoney's, Country Inn Suites, Heartland Inn, Aldi Foods, AllSports, Best Buy, Borders Books, Dayton's, Electric Ave, Gander Mtn Outfitter, Kohl's, Menard's, Michael's, OfficeMax, PakMail, PetCo, Pier 1, Sam's Club, Target, TJ Maxx, Wal-Mart SuperCtr/24hr, cinema, mall, **S**...st police
68	WI 93, to Eleva, **N**...DENTIST, Amoco/mart, KwikTrip/gas, SuperAmerica/mart/24hr, Burger King, Dairy Queen, Hardee's, Econolodge, Chrysler, Firestone/auto, Goodyear/auto, Jeep, Lincoln/Mercury, Nissan, Saturn, Subaru, VW, bank, cinema, same as 70, **S**...Citgo/mart, Buick/Olds/Pontiac/Cadillac/GMC/Hyundai, Ford
65	WI 37, WI 85, Eau Claire, to Mondovi, **N**...HOSPITAL, Amoco, Citgo/diesel/mart, Phillips 66, Arby's, ChiChi's, China Buffet, Fireside Rest., Godfather's, GreenMill Rest., Hardee's, Heckel's Rest., Jericho Grill, McDonald's, Randy's Rest., Red Lobster, Subway, Taco Bell, TCBY, Wendy's, YenKing Chinese, Best Western, Comfort Inn, Day's Inn/rest., Exel Inn, Hampton Inn, Highlander Inn, Holiday Inn/rest., Park Inn, Quality Inn, Ramada Inn, Roadstar Inn, Castle Foods, Mazda/Toyota, Radio Shack, ShopKO, bank
64mm	Chippewa River
59	to US 12, rd EE, to Eau Claire, **N**...HOSPITAL, Citgo/Burger King/diesel/mart/24hr, Holiday/Subway/diesel/mart/24hr, Embers Rest., McDonald's, Pizzaria, AmericInn, Day's Inn, Super 8, Fox RV Ctr(10mi), auto repair/towing, **S**...US RV Ctr, trailer repair
52	US 12, WI 29, WI 40, Elk Mound, to Chippewa Falls, no facilities
49mm	weigh sta wb
45	rd B, Menomonie, **N**...Phillips 66/diesel/mart, **S**...HOSPITAL, Amoco/Subway/diesel/mart/24hr, Heckel's Rest., Wal-Mart Dist Ctr, diesel repair
44mm	Red Cedar River
43mm	**rest areas both lanes, full(handicapped)facilities, phone, picnic tables, litter barrels, vending, petwalk, weather info**
41	WI 25, Menomonie, **N**...HOSPITAL, Cenex/Burger King/mart, Wal-Mart/drugs, Edgewater Camping, KOA, Twin Sprgs Camping, antiques, bank, **S**...Amoco/mart/atm/24hr, Citgo/mart, Fleet&Farm/gas, SuperAmerica/diesel/mart/24hr, Arby's/Sbarro's, Bolo Rest., Country Kitchen, Dairy Queen, Hardee's, Kernel Rest., KFC, Little Caesar's, McDonald's, Perkins/24hr, Pizza Hut, Taco Bell, Taco John's, Wendy's, AmericInn, Best Western, Motel 6, Super 8, Chevrolet, Chrysler, K-Mart, NAPA, Pennzoil, PetCtr, Radio Shack, Sears, laundry, mufflers, museum, to Red Cedar St Tr
32	rd Q, to Knapp, no facilities
28	WI 128, Wilson, to Glenwood City, Elmwood, **N**...Amoco/diesel/mart/24hr, Hearty Platter Rest., **S**...Eau Galle RA, camping
24	rd B, Woodville, to Baldwin, **N**...Mobil/diesel/mart, Woodville Motel, **S**...Eau Galle RA, camping
19	US 63, Baldwin, to Ellsworth, **N**...HOSPITAL, Freedom/diesel/mart, KwikTrip/Subway/gas, A&W, Dairy Queen, Hardee's, McDonald's, AmericInn, Baldwin Motel(2mi), **S**...Citgo/diesel/LP/mart/24hr, Coachman Rest., Super 8, truck repair
16	rd T, Hammond, **N**...Arctic Glass Windows, Sheepskin Outlet
10	WI 65, Roberts, to New Richmond, **S**...trailer sales
8mm	weigh sta eb
4	US 12, rd Q, Somerset, **N**...Citgo/diesel/mart/24hr, TA/Mobil/diesel/rest./24hr, JR Ranch Motel/steaks, to Willow River SP
3	WI 35, to River Falls, U of WI River Falls, no facilities
2	rd F, Carmichael Rd, Hudson, **N**...HOSPITAL, CHIROPRACTOR, Amoco, Conoco/diesel/mart, Freedom/mart, Domino's, KFC, Taco John's, Royal Inn, Muffler Shop, Olds/Pontiac/GMC, Target, TrueValue, bank, golf, **S**...**Welcome Ctr both lanes, full(handicapped)facilities, phone, picnic table, litter barrel, info, petwalk,** CHIROPRACTOR, Amoco/diesel/mart/wash, Conoco/Subway/mart, Arby's, Burger King, Country Kitchen, McDonald's, Perkins/24hr, Pizza Hut, Shoney's, Taco Bell, Wendy's, Best Western, Comfort Inn, Fairfield Inn, Holiday Inn Express, Super 8, Chevrolet/Winnebago, County Mkt Food/drug, FashionBug, Fleet&Farm, Ford/Mercury, K-Mart, Menard's, NAPA, Pack'n Mail, Tires+, Valvoline, Wal-Mart, XpressLube, Yamaha, bank, cinema, dogtrack, to Kinnickinnic SP
1	WI 35 N, Hudson, **1 mi N**...CHIROPRACTOR, EYE CLINIC, Conoco/mart, Freedom/mart, Dairy Queen, McCarthy's Steaks, Pizza Man, City Mkt Foods, Yacht Club, bank
0mm	St Croix River, Wisconsin/Minnesota state line

Eau Claire

Menomonie

Hudson

Wyoming Interstate 25

Exit #	Services

B u f f a l o

300 I-90, E to Gillette, W to Billings. I-25 begins/ends on I-90, exit 56.

299 US 16, Buffalo, **E**...VETERINARIAN, Cenex/diesel, Conoco/diesel/mart/24hr, Exxon/diesel/mart, Sinclair/diesel, Texaco/diesel/mart/24hr, Sundance Grill, Winchester Steaks, Bunkhouse Motel, Comfort Inn, Motel 6, KOA, NAPA, **W**...HOSPITAL, Bighorn/diesel/mart/24hr, Cenex/diesel/rest./24hr, BBQ, Bozeman's Rest., Breadboard Rest., Dash Inn Rest., Hardee's, McDonald's, Pizza Hut, Subway, Taco John's, Crossroads Inn/rest., Econolodge, Horsemen's Inn, Motel Wyoming, Super 8, Indian RV Camping, antiques, auto parts, bank

298 US 87, Buffalo, **W**...Sinclair/Domino's/diesel/mart, Nat Hist Dist Info

291 Trabing Rd, no facilities

280 Middle Fork Rd, no facilities

274mm parking area both lanes, litter barrels

265 Reno Rd, no facilities

254 Kaycee, **E**...Exxon/mart, Texaco/diesel/mart, Country Inn Diner, Cassidy Inn, Siesta Motel, Kaycee Foods, NAPA Repair, Powder River RV Park, USPO, **W**...**rest area both lanes, full(handicapped)facilities, phone, picnic tables, litter barrels, petwalk,** Sinclair/diesel/LP/mart/motel/atm, KC RV Park

249 TTT Rd, no facilities

246 Powder River Rd, no facilities

235 Tisdale Mtn Rd, no facilities

227 WY 387 N, Midwest, Edgerton, no facilities

223 no facilities

219mm parking area both lanes, litter barrels

216 Ranch Rd, no facilities

210 Horse Ranch Creek Rd, Midwest, Edgerton, no facilities

197 Ormsby Rd, no facilities

191 Wardwell Rd, to Bar Nunn, **W**...Conoco/diesel/mart, TM camping

189 US 20, US 26 W, to Shoshone, Port of Entry, airport

C a s p e r

188b WY 220, Poplar St, **E**...Sinclair, El Jarro Mexican, JB's, La Costa Mexican, Econolodge, Hampton Inn, Holiday Inn, Motel 6, Radisson, Ryder Trucks, **W**...Exxon/mart, Texaco/Burger King/diesel/mart/atm, Casper's Rest./24hr, Dairy Queen, Harley-Davidson, to Ft Casper HS

188a Center St, Casper, **E**...Conoco/diesel/mart, Texaco/diesel/mart, JB's, Taco John's, Holiday Inn, National 9 Inn, **W**...Cenex/mart, Denny's, Day's Inn, Parkway Plaza Motel

187 McKinley St, Casper, **E**...Minimart/gas/diesel, Ranch House Motel, repair/transmissons

186 US 20, US 26, US 87, Yellowstone St, Casper, **E**...VW/Audi, **W**...HOSPITAL, Texaco, Best Western, Chevrolet/ Subaru, Olds/Pontiac, VW, diesel repair

185 WY 258, Wyoming Blvd, E Casper, **E**...Conoco/diesel/mart/atm, Kum&Go/gas, Texaco/diesel/mart/24hr, Applebee's, Outback Steaks, Comfort Inn, Shilo Inn, KOA(1mi), bank, **W**...VETERINARIAN, Cenex/mart, Exxon/mart, Flying J/ Conoco/diesel/LP/rest./24hr, Texaco/Burger King/mart/atm, Arby's, Hamburger Stand, Hardee's, KFC, LJ Silver, McDonald's/Playland, Perkins/24hr, Pizza Hut, Red Lobster, Taco Bell, Taco John's, Village Inn Rest., Wendy's, 1st Interstate Inn, American CarCare/tires, AutoZone, Chevrolet, Ford/Mercury, Hyundai, JC Penney, K-Mart, Lube Express, Mailboxes Etc, Nissan, OfficeMax, Olds/GMC, Pennzoil, Safeway/drugs, Sam's Club, Target, Volvo, Wal-Mart SuperCtr, to Oregon Trail, Casper Coll, bank, carwash, cinema, mall

182 WY 253, Brooks Rd, Hat Six Rd, **E**...Exxon/mart/24hr, to Wilkins SP

175mm parking area sb, litter barrels

171mm parking area nb, litter barrels

165 Glenrock, **2 mi E**...Phillips 66, food, lodging, Deer Creek Village RV Camping(apr-nov)

160 US 87, US 20, US 26, E Glenrock, **E**...Phillips 66, Miller's Four Acres Rest., Paisley Café, All American Motel, Glenrock Motel, Hotel Higgins B&B, Deer Creek RV Camping, to Dave Johnston Power Plant

156 Bixby Rd, no facilities

154 Barber Rd, no facilities

153mm parking area both lanes, litter barrels

151 Natural Bridge, no facilities

150 Inez Rd, no facilities

146 La Prele Rd, no facilities

D o u g l a s

140 WY 59, Douglas, **E**...HOSPITAL, VETERINARIAN, Conoco/Subway/mart, Gas4Less/diesel, Arby's, McDonald's, Four Winds Motel, Holiday Inn, Plains Motel, Super 8, Vagabond Motel, Buick/Chevrolet/GMC/Pontiac/Olds, CarQuest, Chevrolet, Jackalope Camping, WY St Fair, Pioneer Museum, tires, **W**...KOA

135 US 20, US 26, US 87, Douglas, **E**...HOSPITAL, Broken Wheel/Sinclair/diesel/rest., Country Inn Rest., Alpine Inn, 1st Interstate Inn, auto repair, **1-2 mi E**...CHIROPRACTOR, Sinclair/mart, Clementine's Pizza, KFC/Taco Bell, Pizza Hut, Taco John's, Village Inn Rest., Chieftain Motel, Dodge/Plymouth, Ford/Mercury, Frontier Drug, Pamida, Safeway, hardware, laundry

N

S

Wyoming Interstate 25

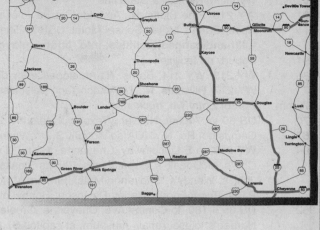

129mm	parking area both lanes
126	US 18, US 20 E, Orin, **E...Orin Jct Rest Area both lanes, full(handicapped)facilities, phones, picnic tables, litter barrels, petwalk, RV dump,** Orin Jct Trkstp/diesel/café
125mm	N Platte River
111	Glendo, **E...**Sinclair/diesel/mart, Glendo Marina Café, Howard's Motel/deli, to Glendo SP, RV camping
104	to Middle Bear, no facilties
100	Cassa Rd, no facilities
94	El Rancho Rd, **W...**gas, food, lodging, RV camping
92	US 26 E, Dwyer, **E...**to Guernsey SP, Ft Laramie NHS
91.5mm	**rest area both lanes, full(handicapped)facilities, picnic tables, litter barrel, petwalk, RV dump**
87	Johnson Rd, no facilities
84	Laramie River Rd, no facilities
83.5mm	Laramie River
80	US 87, Wheatland, Laramie Power Sta, **E...**Pizza Hut, Timberland Rest., Best Western, Chevrolet/Pontiac/Olds/Buick/Cadillac, Chrysler/Dodge/Plymouth/Jeep, Pamida, Safeway, bank, laundry, same as 78
78	US 87, Wheatland, **E...**HOSPITAL, Conoco/diesel, Co-op Gas/diesel, Texaco/diesel/mart/atm, Arby's, Burger King, Subway, Taco John's, Best Western, Motel 6, Plains Motel, Vimbo's Motel/rest., Westwinds Motel, Wyo Motel, Safeway, diesel/auto repair/24hr towing, **W...**Exxon/diesel/mart, Conoco/diesel/mart, Ford/Mercury, RV park
73	WY 34 W, to Laramie, Sybille Wildlife Ctr(seasonal)
70	Bordeaux Rd, no facilities
68	Antelope Rd, no facilities
67.5mm	parking area nb, litter barrels
66	Hunton Rd, no facilities
65.5mm	parking area both lanes, litter barrels
65	Slater Rd, no facilities
64mm	Richeau Creek
57	TY Basin Rd, Chugwater, no facilities
54	Lp 25, Chugwater, **E...**Sinclair/diesel/mart/atm, Super 8, RV camping, **rest area both lanes, full(handicapped)facilities, phone, picnic tables, litter barrels, petwalk, RV dump**
47	Bear Creek Rd, no facilities
39	Little Bear Community, no facilities
36mm	Little Bear Creek
34	Nimmo Rd, no facilities
33mm	Horse Creek
29	Whitaker Rd, no facilities
25	ranch exit, no facilities
21	Ridley Rd, no facilities
17	US 85 N, to Torrington, no facilities
16	WY 211, Horse Creek Rd, **W...**Little Bear Rest.(2mi)
13	Vandehei Ave, **E...**Conoco/mart/atm, bank, **W...**Total/diesel/mart
12	Central Ave, Cheyenne, **E...**HOSPITAL, Conoco/mart/atm, Applebee's, Fairfield Inn, Quality Inn, Frontier Days Park, airport, fairgrounds, golf, museum
11b	Warren AFB, Gate 1, Randall Ave, **E...**to WY St Capitol, museum
10b d	Warren AFB, Gate 2, Missile Dr, WY 210, HappyJack Rd, **E...**RV camping, **W...**to Curt Gowdy SP
9	US 30, W Lincolnway, Cheyenne, **E...**Crossroads/Conoco/diesel/rest., Country Kitchen, Denny's, Best Western, Day's Inn, La Quinta, Luxury Inn, Motel 6, Ramada Ltd, Super 8, Chevrolet, Ford, Honda, Olds/Cadillac, Saturn, Toyota, **W...**Little America/Sinclair/diesel/rest./motel
8d b	I-80, E to Omaha, W to Laramie
7	WY 212, College Dr, **E...**Total/Subway/diesel/mart/24hr, Bailey Tire, RV camping(2mi), **W...WY Info Ctr/rest area both lanes, full(handicapped)facilities, phone, picnic tables, litter barrels, petwalk,** Flying J/Conoco/diesel/LP/mart/rest./24hr, McDonald's, Comfort Inn
6.5mm	Port of Entry, nb
2	Terry Ranch Rd, **2 mi E...**Terry Bison Ranch/rest./hotel, RV camping
0mm	Wyoming/Colorado state line

Wyoming Interstate 80

Exit #	Services
402mm	Wyoming/Nebraska State line
401	WY 215, Pine Bluffs, **N...**Ampride/Subway/diesel/mart/atm/24hr, Conoco/diesel/mart, Sinclair/repair, Uncle Fred's Rest., Gator's Motel, Sunset Motel, USPO, Pine Bluff RV Park, **S...Welcome Ctr/rest area both lanes, full(handicapped)facilities, info, phone, playground, nature trail, picnic tables, litter barrels, petwalk**
391	Egbert, no facilities
386	WY 213, WY 214, Burns, **N...**Antelope/Cenex/diesel/rest./24hr
377	WY 217, Hillsdale, **N...**TA/Amoco/Taco Bell/diesel/motel/24hr, Wyo RV Camping
372mm	Port of Entry wb, truck insp
370	US 30 W, Archer, **N...**Sapp/Texaco/pizza/subs/diesel/rest./24hr, repair, RV park
367	Campstool Rd, **N...**KOA(seasonal), **S...**to Wyoming Hereford Ranch
364	WY 212, to E Lincolnway, Cheyenne, **1-3 mi N on Lincolnway...**HOSPITAL, Conoco/mart/atm, Burger King, Chili's, McDonald's, Papa John's, Parry's Steaks, Perkins, Pizza Hut, Shari's Rest., Subway, Taco Bell, Taco John's, Wendy's, Cheyenne Motel, Firebird Motel, Fleetwood Motel, Home Ranch Motel, AutoZone, Checker Parts, EconoFoods, Harley-Davidson, HobbyLobby, JiffyLube, Midas Muffler, **S...**AB Camping(4mi)
362	US 85, I-180, to Central Ave, Cheyenne, Greeley, **1-2 mi N...**HOSPITAL, Applebee's, Arby's, Golden Corral, Hardee's, diesel repair, museum, st capitol, **S...**Conoco/diesel/mart, Total/diesel/mart, Burger King, Subway, Taco John's, Holiday Inn, Roundup Motel, Safeway, RV camping, transmissions
359c a	I-25, US 87, N to Casper, S to Denver
358	US 30, W Lincolnway, Cheyenne, **N...**HOSPITAL, Conoco/diesel/mart/24hr, Little America/Sinclair/diesel/motel, Denny's, Best Western, Day's Inn, Econolodge, La Quinta, Luxury Inn, Motel 6, Super 8, Chevrolet
348	Otto Rd, no facilities
345	Warren Rd, no facilities
344mm	parking area wb
343.5mm	parking area eb
342	Harriman Rd, no facilities
341mm	parking area both lanes
339	Remount Rd, no facilities
335	Buford, **S...**Sinclair/Lone Tree Jct/diesel/mart/24hr
333mm	point of interest, parking area both lanes
329	Vedeauwoo Rd, **S...**to Ames Monument, Nat Forest RA
323	WY 210, Happy Jack Rd, **N...rest area both lanes, full(handicapped)facilities, phone, picnic tables, litter barrels, petwalk, Lincoln Monument, elev 8640,** to Curt Gowdy SP
322mm	chain up area both lanes
316	US 30 W, Grand Ave, Laramie, **1-2 mi N...**HOSPITAL, Conoco/mart, Applebee's, Arby's, JB's, McDonald's, Sizzler, Taco John's, Winger's Rest., Circle S Motel, Comfort Inn, Sleep Inn, University Inn, Wyoming Motel, Wal-Mart SuperCtr/24hr
313	US 287, to 3rd St, Laramie, Port of Entry, **N...**HOSPITAL, Conoco, Exxon, Phillips 66, Sinclair, Texaco/diesel/mart, Arby's, Chuck Wagon Rest., Country Kitchen, Denny's, Great Wall Chinese, Hardee's/24hr, JB's, Mexican Rest., Sizzler, Best Western, 1st Inn Gold, Laramie Inn/rest., Motel 8, Sunset Inn, Travelodge(1mi), Honda/Isuzu, Laramie Plains Museum, **S...**Holiday Inn, Motel 6, Gibson's, flea mkt
312mm	Laramie River
311	WY 130, WY 230, Snowy Range Rd, Laramie, **S...**Conoco/diesel, Phillips 66/diesel/mart, Sinclair/diesel, Foster's Rest., McDonald's, Best Western/rest., Camelot Motel, Motel 6, Travel Inn, tires, to Snowy Range Ski Area, WY Terr Park
310	Curtis St, Laramie, **N...**HOSPITAL, Pilot/Wendy's/diesel/mart/24hr, Sinclair/diesel, Total/diesel/Outrider Café, Econolodge, Super 8, Goodyear/service/towing, KOA, **1-2 mi N...**Pizza Hut, Shari's Rest., Buttrey Foods, K-Mart, **S...**Petro/diesel/mart/24hr
307mm	parking area both lanes, litter barrels
297	WY 12, Herrick Lane, no facilities
290	Quealy Dome Rd, **S...**Total/diesel/mart/24hr
279	Cooper Cove Rd, no facilities
272mm	Rock Creek
272	WY 13, to Arlington, **N...**Exxon/mart, RV camping
267	**Wagonhound Rd, S...rest area both lanes, full(handicapped)facilities, phone, picnic tables, litter barrels, petwalk**
262mm	parking area both lanes
260	WY 402, **S...**towing

Wyoming Interstate 80

259mm	Medicine Bow River, E Fork
257mm	Medicine Bow River
255	WY 72, Elk Mtn, to Hanna, **N**...Conoco/diesel/mart, to lodging
238	Peterson Rd, no facilities
235	US 30/87, WY 130, Walcott, **N**...Texaco/diesel/mart/café, camping, lodging
229mm	N Platte River
228	**Ft Steele HS, N...rest area both lanes, full(handicapped)facilities, phone, picnic tables, litter barrels, petwalk**
221	E Sinclair, **N**...Burns/Amoco/Mrs B's/diesel/24hr, to Seminoe SP, camping
219	W Sinclair, **N**...to Seminoe SP, camping
215	Cedar St, Rawlins, **N**...EYECARE, Coastal/mart, Conoco/diesel/mart, Kum&Go/gas, Phillips 66/mart, Texaco/KFC/Taco Bell/diesel/mart, McDonald's/playplace, Pizza Hut, Subway, Taco John's, Wendy's, Bridger Inn/rest., Day's Inn, Key Motel, Weston Inn, Alco, Buick/Chevrolet/Pontiac/Olds/GMC, CarQuest, Checker Parts, Chrysler/Plymouth/Dodge/Jeep, City Mkt Food, FastLube, Pamida, bank, camping, diesel repair, drugs, museum, Frontier Prison NHS, to Yellowstone/Teton NP
214	Higley Blvd, Rawlins, **N**...KOA, **S**...Rip Griffin/Texaco/Subway/diesel/24hr, Sleep Inn
211	WY 789, to US 287 N, Spruce St, Rawlins, **N**...HOSPITAL, Conoco/diesel/mart, Exxon/diesel/mart, Phillips 66/mart, Sinclair/diesel, Cappy's Rest., JB's, Best Western/rest., Bucking Horse Lodge, Cliff Motel, Ideal Motel, National 9 Inn, Sunset Motel, Super 8, RV World Camping, auto parts, **1 mi N**...Economy Inn, Rawlins Motel, Ford/Lincoln/Mercury
210mm	weigh sta wb
209	Johnson Rd, **N**...Flying J/diesel/LP/rest./mart/24hr
206	Hadsell Rd, no facilities
205.5mm	continental divide, elev 7000
204	Knobs Rd, no facilities
201	Daley Rd, no facilities
196	Riner Rd, **13 mi N**...Strong Ranch
190mm	parking area wb, litter barrels
189mm	parking area eb, picnic tables, litter barrels
187	WY 789, Creston, Baggs Rd, **S**...fireworks, **50 mi S**...Baggs Jct Gas/diesel/mart
184	Continental Divide Rd, no facilities
173	Wamsutter, **N**...Texaco/diesel/mart/LP, **S**...Conoco/diesel/repair/café/24hr, Phillips 66/mart, Broadway Café, Wamsutter Motel
170	Rasmussen Rd, no facilities
168	Frewen Rd, no facilities
166	Booster Rd, no facilities
165	Red Desert, **S**...SaveWay/gas/mart, lodging, 24hr towing
158	Tipton Rd, continental divide, elev 6930
156	GL Rd, no facilities
154	BLM Rd, no facilities
152	Bar X Rd, no facilities
150	Table Rock Rd, **S**...Sinclair/Major Gas/diesel/mart, camping
146	Patrick Draw Rd, no facilities
144mm	**rest area both lanes, full(handicapped)facilities, phone, picnic tables, litter barrels, petwalk**
143mm	parking area both lanes, litter barrels
142	Bitter Creek Rd, no facilities
139	Red Hill Rd, no facilities
136	Black Butte Rd, no facilities
135mm	parking area both lanes, litter barrels
130	Point of Rocks, **N**...Conoco/diesel/mart/motel, laundry, phone, fireworks
122	WY 371, to Superior, no facilities
111	Baxter Rd, **S**...airport

E ↕ W

Rawlins

Wyoming Interstate 80

Rock Springs

Green River

107	Pilot Butte Ave, Rock Springs, **S**...HOSPITAL, Conoco/diesel, Exxon/diesel/lodging, Gas4Less, Pacific Pride/diesel, Phillips 66, Texaco/diesel/mart, 7-11/gas/24hr, Lew's Rest., El Rancho Motel, Sands Café, Knotty Pine Motel, Rocky Mt Lodge, Springs Motel, Thunderbird Motel, RV camping, repair
104	US 191 N, Elk St, Rock Springs, **N**...Exxon/mart, Flying J/Conoco/diesel/LP/rest./mart/24hr, Kum&Go/gas, Phillips 66/diesel/24hr, Sinclair, Texaco/Burger King/diesel/mart, McDonald's, Renegade Rest., Taco Time, Best Western Outlaw Inn/rest., Econolodge/rest., Buick/Pontiac/GMC, Chrysler/Jeep, to Teton, Yellowstone Nat Parks via US 191, **S**...Exxon, Day's Inn, Fine Arts Ctr, museum
103	Rock Springs, College Dr, **S**...HOSPITAL, Conoco/diesel/mart/24hr, W WY Coll
102	WY 430, Dewar Dr, Rock Springs, **N**...Conoco, Exxon/mart, Sinclair/gas, China Buffet, Taco Time, Motel 6, The Inn, Ramada Ltd, Smith's Food/drug, Chevrolet/Olds/Cadillac, Chrysler/Dodge/Subaru, NAPA, JC Penney, K-Mart, RV Ctr, **S**...HOSPITAL, DENTIST, VETERINARIAN, Amoco/A&W, Chevron/diesel, Kum&Go/gas, Texaco/mart, Arby's, Baskin-Robbins, Burger King, Golden Corral, JB's, KFC, McDonald's, Pizza Hut, Sizzler, Subway, Village Inn Rest., Wendy's, Wonderful House Chinese, Comfort Inn, Holiday Inn, Park Inn, Rodeway Inn, Super 8, Albertson's, AutoZone, Big O Tires, Checker Parts, Midas Muffler, NAPA, Pennzoil, Wal-Mart SuperCtr/24hr, bank
99	US 191 S, E Flaming Gorge Rd, **N**...KOA(1mi), **S**...Conoco/diesel/café/24hr, Big Wheeler Rest., Log In Steaks, fireworks, transmissions
94mm	Kissing Rock
91	US 30, to WY 530, Green River, **S**...McDonald's, Pizza Hut, Subway, Taco John's, Taco Time, Super 8, Western Motel, RV park, Expedition NHS, to Flaming Gorge NRA, same as 89
89	US 30, Green River, **S**...Conoco/mart, Exxon/mart, Texaco/diesel/mart, Arctic Circle, Chinese Rest., Cowboy Café, McDonald's, Penny's Diner, Pizza Hut, Taco Time, Tex's Travel Camp, Trudel Rest., Coachman Inn, Desmond Motel, OakTree Inn, Super 8, Western Motel, Adams RV Service, NAPA, bank, hardware, to Flaming Gorge NRA
87.5mm	Green River
85	Covered Wagon Rd, **S**...Tex's Travel Camp, Adam's RV parts/service
83	WY 372, La Barge Rd, **N**...to Fontenelle Dam
78	(from wb), no facilities
77mm	Blacks Fork River
72	Westvaco Rd, no facilities
71mm	parking area both lanes
68	Little America, **N**...Sinclair/diesel/Little America Hotel/rest./24hr, RV camping
66	US 30 W, Kemmerer, to Teton, Yellowstone, Fossil Butte NM, no facilities
61	Cedar Mt Rd, to Granger, no facilities
60mm	parking area both lanes, litter barrels
54mm	parking area eb, litter barrels
53	Church Butte Rd, no facilities
49mm	parking area wb, litter barrels
48	Lp 80, Lyman, Ft Bridger, Hist Ft Bridger, **S**...Valley West Motel, gas, food
45mm	Blacks Fork River
41	WY 413, Lyman, **N**...Terry's Diesel Repair/café, Gas'n Go/diesel/mart/café, **S...rest area both lanes, full(handicapped)facilities, phone, picnic tables, litter barrels, petwalk,** Valley West Motel(2mi), KOA(1mi)
39	WY 412, WY 414, to Carter, Mountain View, no facilities
34	Lp 80, to Ft Bridger, **S**...Phillips 66/diesel/LP/mart, Wagon Wheel Motel, Ft Bridger NHS, to Flaming Gorge NRA
33.5mm	parking area eb, litter barrels
33	Union Rd, no facilities
30	Bigelow Rd, **N**...TA/Burns/Amoco/Mrs B's/diesel/mart, **S**...fireworks
28	French Rd, no facilities
27.5mm	parking area both lanes
24	Leroy Rd, no facilities
23	Bar Hat Rd, no facilities
21	Coal Rd, no facilities
18	US 189 N, to Kemmerer, to Nat Parks
15	Guild Rd(from eb), no facilities
14mm	parking area both lanes
13	Divide Rd, no facilities
10	Painter Rd, to Eagle Rock Ski Area

Wyoming Interstate 80

E

W

6 US 189, Bear River Dr, Evanston, **N...**Amoco/diesel/ mart, Pilot/diesel/mart/24hr, Texaco/diesel/mart, Back Forty Rest., Blimpie, Last Outpost Café, Bear River Inn, Evanston Inn, Motel 6, Super 8, Prairie Inn, Vagabond Motel, Ford/Mercury, Phillips RV Park, Sunset RV Park, tires, Wyo Downs Racetrack, camp- ing, **S...Welcome Ctr both lanes, full(handicapped)facilities, phone, picnic tables, litter barrels, petwalk, RV dump(seasonal), play- ground**

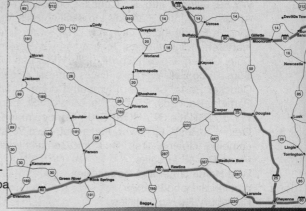

5 WY 89, Evanston, **N...**HOSPITAL, Amoco/mart, Chevron/MeanGeneBurgers/diesel/mart/wash, Maver- ick/gas, Arby's, DragonWall Chinese, McDonald's, Papa Murphy's, Shakey's Pizza, Subway, Taco John's, Wendy's, Super 8, AutoZone, Chevrolet/Olds/Buick, JiffyLube, Jubilee Foods, Radio Shack, Wal-Mart/drugs, auto repair, bank, hardware, **S...**WY St Hospital

3 US 189, Harrison Dr, Evanston, **N...**HOSPITAL, Amoco, Chevron/diesel/mart, Flying J/diesel/mart/rest./24hr, Phillips 66/mart, Sinclair/mart, Texaco/diesel/mart, Burger King, JB's, Taco Time, Best Western/rest., Day's Inn, Economy Inn, National 9 Inn, Super Budget Inn, Weston Inn, Chrysler/Plymouth/Jeep, Corral Ranch Wear, GMC/Pontiac/Cadillac, Goodyear/auto, airport, bank, laundry, **S...**Phillips 66/A&W/diesel/mart, KFC, fireworks

.5mm Port of Entry eb, weigh sta wb

0mm Wyoming/Utah state line

Wyoming Interstate 90

Exit #	Services
207mm	Wyoming/South Dakota state line
205	Beulah, **1/2 mi N...**Conoco/Golden Buffalo Café/diesel/LP, Camp-Fish Camping
204.5mm	Sand Creek
199	WY 111, to Aladdin, no facilities
191	Moskee Rd, no facilities
189	US 14 W, Sundance, **N...**HOSPITAL, Amoco/diesel/mart/24hr, Aro Rest., Subway(2mi), Log Cabin Café, Bear Lodge Motel, Best Western, Budget Host Arrowhead, Mountain View Camping, to Devil's Tower NM, museum, **S...rest area both lanes, full(handicapped)facilities, info, phone, picnic tables, litter barrels, playground, petwalk,** RV dump, port of entry/weigh sta
187	WY 585, Sundance, **1/2 mi N...**Amoco/diesel/mart, Conoco/mart/24hr, Flo's Place Café, Subway, Best West- ern, Decker's Motel, Sundance Mtn Inn, Pennzoil Lube
185	to WY 116, to Sundance, **S...**Texaco/diesel/mart/service
178	Coal Divide Rd, no facilities
177mm	parking area both lanes, litter barrels
172	Inyan Kara Rd, no facilities
171mm	parking area both lanes, litter barrels
165	Pine Ridge Rd, to Pine Haven, **N...**to Keyhole SP, no facilities
163mm	parking area both lanes, litter barrels
160	Wind Creek Rd, no facilities
154	US 14, Moorcroft, **S...**Conoco/diesel, Texaco, Hub Café, Subway, Cozy Motel, Moorcourt Motel
153	US 16 E, US 14, W Moorcroft, **N...rest area both lanes, full(handicapped)facilities, phone, picnic tables, litter barrels, petwalk**
152mm	Belle Fourche River
141	Rozet, **2 mi S...**gas, food, phone
138mm	parking area both lanes
132	Wyodak Rd, no facilities
129	Garner Lake Rd, **S...**High Plains Camping, WY Marine RV Ctr
128	US 14, US 16, Gillette, Port of Entry, **N...**Conoco/mart, Kum&Go/gas, Phillips 66/LP, Texaco, KFC, Mona's Café American/Mexican, Taco John's, Village Inn Rest., Econolodge, Motel 6, National 9 Inn, Quality Inn, Rodeway Inn/rest., Rolling Hills Motel, Thrifty Inn Motel, Don&Mo's Exhaust, Crazy Woman Camping(2mi)

E

W

Evanston

Sundance Gillette

Wyoming Interstate 90

126	WY 59, Gillette, info, **N**...Cenex, Conoco, Co-op Gas/diesel, Hardee's, McDonald's, Prime Rib Rest., Subway, Arrowhead Motel, Ramada Ltd, Rolling Hills Motel, Buttrey Food/drug, 1-Hr Photo, Optical Shop, Plains Tire, Smith's Foods, bank, city park, tires, waterslide, **S**...Conoco/mart, Exxon/mart, Phillips 66/Flying J/diesel/rest./mart/24hr, Minimart/gas, Texaco, Arby's, Baskin-Robbins, Blimpie, Burger King, Café Trinidad, Chinese Rest., Dairy Queen, Golden Corral, JB's, KFC, Las Margaritas Mexican, Papa Murphy's, Perkins/24hr, Pizza Hut, Simply Subs, Taco Bell, Wendy's, Day's Inn/rest., Holiday Inn, Ace Hardware, Advance Parts, Albertson's, Big O Tires, Checker Parts, Chrysler/Plymouth/Dodge, Drug Fair, GNC, Goodyear/auto, IGA Foods, K-Mart, Midas Muffler, Pamida, Pennzoil, RV Ctr, Wal-Mart/drugs, bank, carwash, laundry
124	WY 50, Gillette, **N**...HOSPITAL, Conoco/mart/24hr, Texaco/Burger King/diesel/mart, Chinese Rest., Fireside Café, Granny's Rest., LJ Silver, Pizza Hut, Best Western/rest., Motel 6, Rodeway Inn, Super 8, Dan's Mkt, airport, camping, cinema, tires, **S**...Kum&Go/diesel/mart
116	Force Rd, no facilities
113	Wild Horse Creek Rd, no facilities
110mm	parking area both lanes
106	Kingsbury Rd, no facilities
102	Barber Creek Rd, no facilities
91	Dead Horse Creek Rd, no facilities
89mm	Powder River
88	**Powder River Rd, N...rest area both lanes, full(handicapped)facilities, phone, picnic tables, litter barrels, petwalk**
82	Indian Creek Rd, no facilities
77	Schoonover Rd, no facilities
73.5mm	Crazy Woman Creek
73	Crazy Woman Creek Rd, no facilities
69	Dry Creek Rd, no facilities
68.5mm	parking area both lanes, litter barrels
65	Red Hills Rd, Tipperary Rd, no facilities
59mm	parking area both lanes, litter barrels
58	US 16, Buffalo, to Ucross, **1-2 mi S**...HOSPITAL, VETERINARIAN, Nat Hist Dist, Cenex/diesel, Conoco/diesel/mart/24hr, Exxon/diesel/mart, Sinclair/diesel, Texaco/diesel/mart, Bozeman Rest., McDonald's, Pizza Hut, Subway, Winchester Steaks, Bunkhouse Motel, Comfort Inn, Crossroads Inn, Econolodge, Motel 6, Mountainview Motel, Super 8, Wyoming Motel, Deer Park Camping, Indian Campground, NAPA, KOA
56b	I-25 S, US 87 S, to Buffalo
a	25 Bus, 90 Bus, to Buffalo
53	Rock Creek Rd, no facilities
51	Lake DeSmet, **1 mi N**...LakeStop Truckstp/motel, Lake DeSmet RV Park
47	Shell Creek Rd, no facilities
44	US 87 N, Piney Creek Rd, to Story, Banner, **N**...Ft Phil Kearney, museum, **5 mi S**...Wagon Box Cabins/rest.
39mm	scenic turnout wb, litter barrel
37	Prairie Dog Creek Rd, to Story, no facilities
33	Meade Creek Rd, to Big Horn, **S**...Powderhorn Golf Club
31mm	parking area eb, litter barrel
25	US 14 E, Sheridan, **N**...Comfort Inn, **S**...CHIROPRACTOR, VETERINARIAN, Conoco/diesel/mart, Holiday/A&W/diesel/mart, Sinclair/diesel, Texaco/diesel/mart, Arby's, Burger King, Chinese Rest., Subway, Taco Bell, Taco John's, Appletree Inn, Day's Inn, Holiday Inn, Lariat Motel, Mill Inn, Parkway Motel, Albertson's, Firestone, Goodyear, IGA Foods, NAPA, Pamida, Pontiac/Cadillac, Sheridan RV Park, Toyota, Wal-Mart SuperCtr/McDonald's, airport, carwash, laundry, mall, to Hist Dist, Sheridan Coll
23	WY 336, 5th St, Sheridan, **N...rest area both lanes, full(handicapped)facilities, info, phone, picnic tables, litter barrels, petwalk,** RockShop/Subway/diesel/mart, **S**...Holiday/mart, Texaco/mart, BBQ, Quizno's, Best Western, Evergreen Inn, Sheridan Inn, Honda, **1-2 mi S**...HOSPITAL, Texaco, Arctic Circle, Blimpie, Dairy Queen, JB's, KFC, Las Magaritas, LBM Rest., Papa Murphy's, Perkins, Subway, Wendy's, Energy Inn, Trails End Motel, Checker Parts, JiffyLube, Plains Tires, museum
20	Main St, Sheridan, **S**...Cenex/mart/atm, Conoco/Subway/mart, Exxon/diesel/mart/24hr, Maverick/gas, Texaco/Mean Gene's Burgers/diesel/mart, Cattleman's Cut Rest., Country Kitchen, Domino's, KFC, McDonald's, Pizza Hut, Subway, Taco John's, Holiday Lodge, Super 8, SuperSaver Inn, Trails End Motel, Triangle Motel, K-Mart/Little Caesar's, KOA, Pennzoil, Safeway, tires
16	to Decker, Montana, no facilities
15mm	parking area wb, litter barrels
14.5mm	Tongue River
14	Acme Rd, no facilities
9	US 14 W, Ranchester, **S**...to Yellowstone, Teton NPs, Conner Bfd NHS, Ski Area
1	Parkman, no facilities
0mm	Wyoming/Montana state line

E

W

Help a friend with

Updated and published annually, we will continue to provide the best highway information you can get. Please pass along this order form to a friend, or order *the Next EXIT*® as a present for someone special. Thank you!

Please send_____copies of *the Next EXIT* to the address(es) below. I enclose my check or money order for $12.95 per copy + $4.00 shipping.

name..

street...

city..

state...zip...

name..

street...

city..

state...zip...

Please add $3.00 for priority(2nd day) delivery

**mail to: the Next EXIT, inc.   or: call: 1-800-NEX-EXIT(639-3948),
                PO Box 888   and use your visa/mastercard.
                Garden City, UT 84028   or: fax: 1-435-946-8808,
                                                             and use your visa/mastercard.**

nextexit@cut.net